BIOCHEMISTRY

BIOCHEMISTRY

DONALD VOET

University of Pennsylvania

JUDITH G. VOET

Swarthmore College

Illustrators:
IRVING GEIS
JOHN AND BETTE WOOLSEY
PATRICK LANE

WILEY

JOHN WILEY & SONS

New York • Chichester • Brisbane • Toronto • Singapore

To:

Our parents, who encouraged us,
Our teachers, who enabled us, and
Our children, who put up with us.

Cover Art: One of a series of color studies of horse heart cytochrome *c* designed to show the influence of amino acid side chains on the protein's three-dimensional folding pattern. We have selected this study to symbolize the discipline of biochemistry: Both are beautiful but still in process and hence have numerous "rough edges." Drawing by Irving Geis in collaboration with Richard E. Dickerson.

Cover and part opening illustrations copyrighted by Irving Geis.

Cover Designer: Madelyn Lesure

Photo Research: John Schultz, Eloise Marion

Photo Research Manager: Stella Kupferberg

Illustration Coordinator: Edward Starr

Copy Editor: Jeannette Stiefel

Production Manager: Lucille Buonocore

Senior Production Supervisor: Linda Muriello

Library of Congress Cataloging in Publication Data:

Voet, Donald.
    Biochemistry / by Donald Voet and Judith G. Voet.
        p.   cm.
    Includes bibliographical references.

    1. Biochemistry. I. Voet, Judith G. II. Title.
QP514.2.V64   1990   89-16727
574.19′2—dc20   CIP

Printed in the United States of America

10 9 8 7 6 5 4 3 2

FOREWORD

Contrary to what I once thought, scientific progress did not consist simply in observing, in accumulating experimental facts and drawing up a theory from them. It began with the invention of a possible world, or a fragment thereof, which was then compared by experimentation with the real world. And it was this constant dialogue between imagination and experiment that allowed one to form an increasingly fine-grained conception of what is called reality.
François Jacob, The Statue Within, Basic Books, 1988.

This text is a grand synthesis of biochemical science for the 1990s. It incorporates traditional intermediary metabolism along with the most recent advances in molecular genetics and macromolecular structure.

There have been other inspirational grand syntheses of science in the past. In 1543, Andreas Vesalius produced his great work, *De Humani Corporis Fabrica* (Fabric of the Human Body), an elaborately illustrated compendium of anatomy that laid the foundation for the science of modern medicine. Closer to our time, in the 1970s, Albert Lehninger produced a classic biochemistry text that eloquently summarized metabolism and bioenergetics. Although he also described macromolecular structure and function along with aspects of molecular biology, these descriptions could only point to the great advances yet to come.

The flood of new ideas and experiments in the late 1970s and 1980s—recombinant DNA techniques, site-directed mutagenesis, highly mechanized DNA sequencing, spectacular advances in genetic engineering and vivid new methods for visualizing complex molecular structures—all of these advances demanded a fresh, comprehensive and coherent presentation of current biochemical science. The new complexities revealed by recent experimental data have, paradoxically, made the science easier to comprehend. The new data have uncovered new organizing principles. New methods of presentation have made it possible to dramatically visualize the three-dimensional structure of complex macromolecules by artistic rendition and advanced computer graphics. These advances are incorporated into the communications system of this book. The authors begin each chapter with succinct statements of relevant principles, followed by the ideas and experimental evidence that generated these principles. Presentations of structural chemical formulas, renditions of macromolecular structures and schematic portrayals of processes are all woven into the fabric of the text, helping to make evident the connection between structure and function. The two thousand illustrations along with the clarity of the writing make this book a contemporary classic.

Myoglobin, the first protein structure to be determined, provides an illuminating case history demonstrating how a knowledge of function can be a key to understanding structure. The complete visualization of myoglobins's 153 amino acids seems at first bewilderingly complex. The structure, even at low resolution, proved to be daunting. Max Perutz, who pioneered protein structure determination, remarked on first seeing myoglobin:

Could the search for ultimate truth really have revealed so hideous and visceral-looking an object? Was the nugget of gold a lump of lead? Fortunately, like many other things in nature, myoglobin gains in beauty the more you look at it.
(Scientific American, November, 1964.)

Not only in beauty, but also in logical structure. It can now be seen to assume a simple folding pattern, whose very clear function is to provide a watertight pocket for the oxygen-binding iron atom.

Nature cherishes successful designs and tends to preserve them over aeons of evolutionary time. Cytochrome *c* shows remarkably similar three-dimensional folding patterns for species as different as mammals and bacteria, although their amino acid sequences may differ widely. Families of enzymes, such as the serine proteases, share recognizably similar folding patterns, even though there are differences in their active sites. Connectivity is rampant in Nature, as is reflected in this book. The separate disciplines of cell biology, enzyme kinetics electron microscopy, x-ray crystallography, NMR spectroscopy, DNA sequencing, and gene splicing—all share a common goal, to understand the structure and function of the machinery of life on the molecular level. Living systems maintain their steady state (homeostasis), through the control of metabolic pathways. These, in turn, are monitored by regulatory enzymes. Elucidation of complex regulatory enzyme structure is now beginning to explain how their specific motions exercise feedback control at the molecular level. The Germans have a picturesque word for this kind of coherence, *zusammenhängen*, (literally, "hanging together"). This classic text is comprehensive, clear and concise, but above all, it has *coherence.*

Concise summaries and clear illustrations are of great help in teaching—but biochemistry is, in essence, an experimental science. Ideas generated in the mind must be tested in the laboratory. Marshall McLuhan, a popular sociologist of the 1960s, explained the wide influence of television by coining the phrase, "The Medium is the Message." In a sense, the laboratory is the *message center* where questions are asked in terms of laboratory procedure and answered in coded machine language: that is, in numbers, graphs, spots of differing intensity on film, recordings of spectra, and tracings of polypeptide chains on computer screens. Students must learn to read and interpret these coded signs. Furthermore, they must be able to read the special languages of the research papers in which experiments are recorded. The authors are well aware of the necessity of mastering the skills and codes of the laboratory and the literature. They have provided concise overviews of laboratory practices by describing, for example, the variety of methods for purifying proteins and nucleic acids and determining their sequences. Each chapter has copious references, and there is a valuable guide on how to read a research paper. The student learns to read codes (among others, the three-letter and one-letter codes for amino acids), numerical recordings, spots and tracings on film, arcane jargon for antibodies and newly discovered genes—all with the purpose of seeing life on the molecular level.

Biochemistry in the 1990s may be a mature science, but it is far from a finished one. Imaginative concepts, tested by experiment, and organized into working principles, lead to more concepts. The upward spiral of research continues, but much remains mysterious. Science marches on. This book gives the student and the professional the equipment to understand state of the art biochemistry. It is also preparation for comprehending investigations yet to come. To this end, the authors and the publisher plan to issue periodic updates so that the seeds that have been planted will take root, grow, blossom, and bear fruit.

IRVING GEIS

New York City
August 1989

PREFACE

Biochemistry is a field of enormous fascination and utility, arising, no doubt, from our own self-interest. Human welfare, particularly its medical and nutritional aspects, has been vastly improved by our rapidly growing understanding of biochemistry. Indeed, scarcely a day passes without the report of a biomedical discovery that benefits a significant portion of humanity. Further advances in this expanding field of knowledge will no doubt lead to even more spectacular gains in our ability to understand nature and to control our destinies. It is therefore of utmost importance that individuals embarking on a career in biomedical sciences be well versed in biochemistry.

This textbook is a distillation of our experiences in teaching undergraduate and graduate students at the University of Pennsylvania and Swarthmore College and is intended to provide such students with a thorough grounding in biochemistry. In writing this text we have emphasized several themes. First, biochemistry is a body of knowledge compiled by people through experimentation. In presenting what is known, we therefore stress how we have come to know it. The extra effort the student must make in following such a treatment, we believe, is handsomely repaid, since it engenders the critical attitudes required for success in any scientific endeavor. Although science is widely portrayed as an impersonal subject, it is, in fact, a discipline shaped through the often idiosyncratic efforts of individual scientists. We therefore identify some of the major contributors to biochemistry (the majority of whom are still professionally active) and, in many instances, consider the approaches they have taken to solve particular biochemical puzzles. The student should realize, however, that most of the work described could not have been done without the dedicated and often indispensable efforts of numerous co-workers.

The unity of life and its variation through evolution is a second dominant theme running through the book. Certainly one of the most striking characteristics of life on earth is its enormous variety and adaptability. Yet, biochemical research has amply demonstrated that all living things are closely related at the molecular level. As a consequence, the molecular differences among the various species have provided intriguing insights into how organisms have evolved from one another and have helped to delineate the functionally significant portions of their molecular machinery.

A third major theme is that biological processes are organized into elaborate and interdependent control networks. Such systems permit organisms to maintain relatively constant internal environments, to respond rapidly to external stimuli, and to grow and to differentiate. A fourth theme is that biochemistry has important medical consequences. We therefore frequently illustrate biochemical principles by examples of normal and abnormal human physiology.

We assume that students who use this text have had the equivalent of one year of college chemistry and at least one semester of organic chemistry (perhaps being taken concurrently) so that they are familiar with both general chemistry and the basic principles and nomenclature of organic chemistry. We also assume that students have taken a one-year college course in general biology in which elementary biochemical concepts were discussed. Students who lack these prerequisites are advised to consult the appropriate introductory textbooks in these subjects.

This book is organized into five parts:

I. Introduction and Background: An introductory chapter followed by chapters that review the properties of aqueous solutions and the elements of thermodynamics.

II. Biomolecules: A description of the structures and functions of proteins, carbohydrates, and lipids.

III. Mechanisms of Enzyme Action: An introduction to the properties, reaction kinetics, and catalytic mechanisms of enzymes.

IV. **Metabolism:** A discussion of how living things synthesize and degrade carbohydrates, lipids, amino acids, and nucleotides with emphasis on energy generation and consumption.

V. **The Expression and Transmission of Genetic Information:** An exposition of nucleic acid structures and both prokaryotic and eukaryotic molecular biology.

This organization permits us to cover the major areas of biochemistry in a logical and coherent fashion. Yet, modern biochemistry is a subject of such enormous scope that to maintain a relatively even depth of coverage throughout the text, we include more material than most one-year biochemistry courses will cover in detail. This depth of coverage, we believe, is one of the strengths of this book; it permits the instructor to teach a course of his/her own design and still provide the student with a resource on biochemical subjects not emphasized in the course.

The order in which the subject matter of the book is presented more or less parallels that of most biochemistry courses. However, several aspects of the book's organization deserve comment:

1. We present nucleic acid structures (Chapter 28) as part of molecular biology (Part V) rather than in our discussions of structural biochemistry (Part II) because nucleic acids are not mentioned in any substantive way until Part V. Instructors who, nevertheless, prefer to consider nucleic acid structures in a sequence different from that in the text can easily do so because Chapter 28 requires no familiarity with enzymology or metabolism.

2. We have split our presentation of thermodynamics between two chapters. Basic thermodynamic principles—enthalpy, entropy, free energy, and equilibrium—are discussed in Chapter 3 because these subjects are prerequisite for understanding structural biochemistry, enzyme mechanisms, and kinetics. Metabolic aspects of thermodynamics—the thermodynamics of phosphate compounds and oxidation–reduction reactions—are presented in Chapter 15, since knowledge of these subjects is not required until the chapters that follow.

3. Techniques of protein purification are described in a separate chapter (Chapter 5) that precedes the discussion of protein structure and function. We have chosen this order so that students will not feel that proteins are somehow "pulled out of a hat." Nevertheless, Chapter 5 has been written as a resource chapter to be consulted repeatedly as the need arises.

4. Chapter 9 describes the properties of hemoglobin in detail so as to illustrate concretely the preceding discussions of protein structure and function. This chapter introduces allosteric theory to explain the cooperative nature of hemoglobin oxygen binding. The subsequent extension of allosteric theory to enzymology (Chapter 12) is a relatively simple matter.

5. Concepts of metabolic control are presented in the chapters on glycolysis (Chapter 16) and glycogen metabolism (Chapter 17) through discussions of flux generation, allosteric regulation, substrate cycles, covalent enzyme modification, and cyclic cascades. We believe that these concepts are best understood when they are studied in metabolic context rather than as independent topics.

6. There is no separate chapter on coenzymes. These substances, we feel, are more logically studied in the context of the enzymatic reactions in which they participate.

7. Glycolysis (Chapter 16), glycogen metabolism (Chapter 17), the citric acid cycle (Chapter 19), and oxidative phosphorylation (Chapter 20) are detailed as models of general metabolic pathways with the emphasis placed on many of the catalytic and control mechanisms of the enzymes involved. The principles illustrated in these chapters are reiterated in somewhat less detail in the other chapters of Part IV.

8. Consideration of membrane transport (Chapter 18) precedes that of mitochondrially based metabolic pathways, including the citric acid cycle and oxidative phosphorylation. In this manner, the idea of the compartmentalization of biological processes can be easily assimilated.

9. Discussions of both the synthesis and the degradation of lipids have been placed in a single chapter (Chapter 23) as have the analogous discussions of amino acids (Chapter 24) and nucleotides (Chapter 26).

10. Energy metabolism is summarized and integrated in terms of organ specialization in Chapter 25, following the descriptions of carbohydrate, lipid, and amino acid metabolism.

11. The basic principles of both prokaryotic and eukaryotic molecular biology are introduced in sequential chapters on transcription (Chapter 29), translation (Chapter 30), and DNA replication, repair and recombination (Chapter 31). Viruses (Chapter 32) are the considered as paradigms of more complex cellular functions followed by discussions of newly emerging concepts of eukaryotic gene expression (Chapter 33).

12. Chapter 34, the final chapter, is a series of minichapters that describe the biochemistry of a variety

of well-characterized human physiological processes: blood clotting, the immune response, muscle contraction, hormonal communication, and neurotransmission.

The old adage that you learn a subject best by teaching it simply indicates that learning is an active rather than a passive process. The problems we provide at the end of each chapter are therefore designed to make students think rather than to merely regurgitate poorly assimilated and rapidly forgotten information. Few of the problems are trivial and some of them (particularly those marked with an asterisk) are quite difficult. Yet, successfully working out such problems can be one of the most rewarding aspects of the learning process. Only by thinking long and hard for themselves can students make a body of knowledge truly their own. The answers to the problems are worked out in detail in the Solutions Manual that accompanies this text. This manual, however, can only be an effective learning tool if the student makes a serious effort to solve a problem before looking up its answer.

We have included lists of references at the end of every chapter to provide students with starting points for independent biochemical explorations. The enormity of the biochemical research literature precludes us from giving all but a few of the most important research reports. Hence, we list what we have found to be the most useful reviews and monographs on the various subjects covered in each chapter.

Biomedical research is advancing at such an astonishing pace that a seminal discovery often leads to the development of a mature subdiscipline within the period of a year or so. Consequently, a textbook on biochemistry can never be truly up to date. To alleviate this problem, we shall periodically bring out supplements to this textbook that review the recent biochemical literature and list some of its most important reviews and research reports. Nevertheless, students should be encouraged to peruse the current biochemical literature for only then will they acquire a feeling for the scope and excitement of modern biochemistry.

Finally, although we have made every effort to make this text error free, we are under no illusions that we have done so. We therefore request that readers provide us with their comments and criticisms.

DONALD VOET
JUDITH G. VOET

ACKNOWLEDGMENTS

This textbook represents the dedicated efforts of numerous individuals, many of whom deserve special mention:

Dennis Sawicki, our editor, has shepherded the manuscript through its long and difficult gestation and has given us tremendous support and encouragement at every step of the way.

John and Bette Woolsey and Patrick Lane have created an extraordinary collection of illustrations, many of which, we believe, will become classics.

Irving Geis has generously written the foreword to this text, given us invaluable advice, and provided us with his justly celebrated molecular art, some of which is published here for the first time.

Michael Carson, Richard Feldmann, Arthur Lesk, Arthur Olson, and Michael Pique supplied us with their spectacular state of the art molecular graphics drawings, and Jane Richardson contributed several of her marvelous protein ribbon drawings.

Stella Kupferberg and John Schultz have done an extraordinary job of locating and acquiring every one of the hundreds of photographs we had requested.

Martin Silberberg of Manuscript Associates, the development editor for the first half of the manuscript, taught us how to organize a chapter and to write coherent sentences.

Jeanette Steifel, our copy editor, displayed dedication to producing a perfect text and introduced us to the wonders of "formal writing."

Linda Muriello supervised the production of the textbook.

The following collegues have put in many long hours in reviewing the manuscript. We are greatly indebted to them for their expert advice and encouragement.

Joseph Babitch, Texas Christian University
Kenneth Brown, University of Texas at Arlington
Larry G. Butler, Purdue University
Carol Caparelli, Fox Chase Cancer Center
Glenn Cunningham, University of Central Florida
Don Dennis, University of Delaware
Walter A. Deutsch, Louisiana State University
William A. Eaton, National Institutes of Health
David Eisenberg, University of California at Los Angeles
David Fahrney, Colorado State University
Robert Fletterick, University of California at San Francisco
Scott Gilbert, Swarthmore College
James H. Hammons, Swarthmore College
Edward Harris, Texas A & M University
James H. Hageman, New Mexico State University
Ralph A. Jacobson, California Polytechnic State University
Eileen Jaffe, University of Pennsylvania
Jan G. Jaworski, Miami University
William P. Jencks, Brandeis University
Mary Ellen Jones University of North Carolina
Tokuji Kimura, Wayne State University
Daniel J. Kosman, State University of New York at Buffalo
Albert Light, Purdue University
Robert D. Lynch, University of Lowell
Ronald Montelaro, Louisiana State University
Harry F. Noller, University of California at Santa Cruz
Alan R. Price, University of Michigan
Thomas I. Pynadath, Kent State University
Ivan Rayment, University of Wisconsin
Raghupathy Sarma, State University of New York at Stony Brook
Paul R. Schimmel, Massachusetts Institute of Technology
Thomas Schleich, University of California at Santa Cruz
Allen Scism, Central Missouri State University

Thomas Sneider, Colorado State University
Jochanan Stenish, Western Michigan University
JoAnne Stubbe, Massachusetts Institute of Technolgy
William Sweeney, Hunter College
John Tooze, European Molecular Biology Organization

Francis Vella, University of Saskatchewan
Harold White, University of Delaware
William Widger, University of Houston
Jeffrey T. Wong, University of Toronto
James Zimmerman, Clemson University

D. V.
J. G. V.

BRIEF CONTENTS

CONTENTS

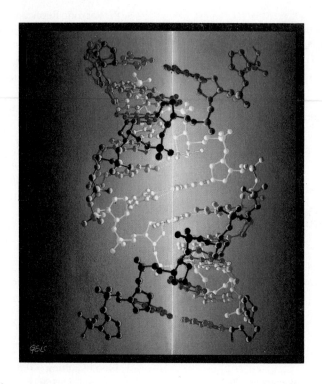

"Hot wire" A-DNA illuminated by its vertical axis.

Introduction
and
Background

Chapter 1
LIFE

It is usually easy to decide whether or not something is alive. This is because living things share many common attributes such as the capacity to extract energy from nutrients to drive their various functions, the power to actively respond to changes in their environment, and the ability to grow, differentiate and perhaps most telling of all, to reproduce. Of course, a given organism may not have all of these traits. For example, mules, which are obviously alive, rarely reproduce. Conversely, inanimate matter may exhibit some lifelike properties. For instance, crystals may grow larger when immersed in a supersaturated solution of the crystalline material. Therefore, life, as are many other complex phenomena, is perhaps impossible to define in a precise fashion. Norman Horowitz, however, has proposed a useful set of criteria for living systems: *Life possesses the properties of replication, catalysis, and mutability.* Much of this text is concerned with the manner in which living organisms exhibit these properties.

Biochemistry is the study of life on the molecular level. The significance of such studies are greatly enhanced if they are related to the biology of the corresponding organisms or even communities of such organisms. This introductory chapter therefore begins with a synopsis of the biological realm. This is followed by an outline of biochemistry, a discussion of the origin of life, and finally, an introduction to the biochemical literature.

1. PROKARYOTES

It has long been recognized that life is based on morphological units known as **cells.** The formulation of this concept is generally attributed to an 1838 paper of Matthias Schleiden and Theodor Schwann, but its origins may be traced to the seventeenth century observations of early microscopists such as Robert Hooke. There are two major classifications of cells: the **eukaryotes** (Greek: *eu,* good or true + *karyon,* kernel or nut), which have a membrane-enclosed **nucleus** encapsulating their **DNA (deoxyribonucleic acid);** and the **prokaryotes** (Greek: *pro,* before), which lack this organelle. Prokaryotes, which comprise the various types of bacteria, have relatively simple structures and are invariably unicellular (although they may form filaments or colonies of independent cells). Eukaryotes, which may be multicellular as well as unicellular, are vastly more complex than prokaryotes. (Viruses, which are much simpler entities than cells, are not classified as living because they lack the metabolic apparatus to reproduce outside their host cells. They are essentially large molecular aggregates.) This section is a discussion of prokaryotes. Eukaryotes are considered in the following section.

A. Form and Function

Prokaryotes are the most numerous and widespread organisms on earth. This is because their varied and often highly adaptable metabolisms suit them to an enormous variety of habitats. Besides inhabiting our familiar temperate and aerobic environment, certain types of bacteria may thrive in or even require conditions that are hostile to eukaryotes such as lack of oxygen, high temperatures, and unusual chemical environments. Moreover, the rapid reproductive rate of prokaryotes (optimally < 20 min per cell division for many species) permits them to take advantage of transiently favorable conditions, and conversely, the ability of many bacteria to form resistant **spores** allows them to survive adverse conditions.

Prokaryotes Have Relatively Simple Anatomies

Prokaryotes, which were first observed in 1683 by the inventor of the microscope, Antoni van Leeuwenhoek, have sizes that are mostly in the range 1 to 10 μm. They have one of three basic shapes (Fig. 1-1): spheroidal **(cocci),** rodlike **(bacilli),** and helically coiled **(spirilla),** but all have the same general design (Fig. 1-2). They are bounded, as are all cells, by an ~ 70-Å thick **cell membrane (plasma membrane)** that consists of a lipid bilayer containing embedded proteins that control the passage of molecules in and out of the cell and catalyze a variety of reactions. The cells of most prokaryotic species are surrounded by a rigid, 30 to 250-Å thick polysaccharide **cell wall** that mainly functions to protect the

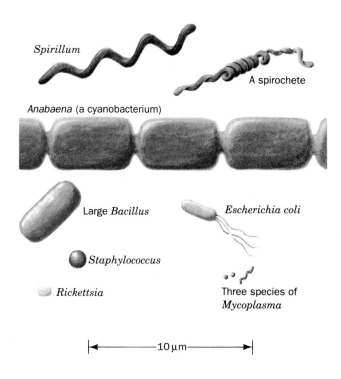

Figure 1-1
Scale drawings of some prokaryotic cells.

cell from mechanical injury and to prevent it from bursting in media more osmotically dilute than its contents. Some bacteria further encase themselves in a gelatinous polysaccharide **capsule** that protects them from the defenses of higher organisms. Although prokaryotes lack the membranous subcellular organelles characteristic of eukaryotes (Section 1-2), their plasma membranes may be infolded to form multilayered structures known as **mesosomes.** The mesosomes are thought to serve as the site of DNA replication and other specialized enzymatic reactions.

The prokaryotic **cytoplasm** (cell contents) is by no means a homogeneous soup. Its single **chromosome** (DNA molecule, several copies of which may be present in a rapidly growing cell) is condensed to form a body known as a **nucleoid.** The cytoplasm also contains numerous species of **RNA (ribonucleic acid),** a variety of soluble **enzymes** (proteins that catalyze specific reactions), and many thousands of 250 Å in diameter particles known as **ribosomes,** which are the sites of protein synthesis.

Many bacterial cells bear one or more whiplike appendages known as **flagella,** which are used for locomotion. Certain bacteria also have filamentous projections of mostly unknown function that are named **pili,** some types of which function as conduits for DNA during sexual conjugation (a process in which DNA is

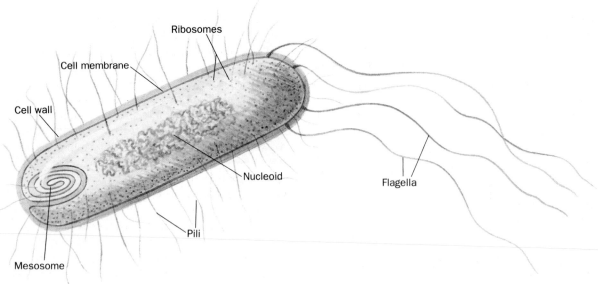

Ribosomes

Cell membrane

Cell wall

Nucleoid

Flagella

Pili

Mesosome

Figure 1-2
A schematic diagram of a prokaryotic cell.

transferred from one cell to another, Section 27-1D; prokaryotes usually reproduce by binary fission).

The bacterium *Escherichia coli* (abbreviated *E. coli*) is the biologically most well-characterized organism as a result of its intensive biochemical and genetic study over the past 40 years. Indeed, much of the subject matter of this text deals with the biochemistry of *E. coli*. Cells of this normal inhabitant of the higher mammalian colon (Fig. 1-3) are typically 2-μm long rods that are 1 μm in diameter and weigh $\sim 2 \times 10^{-12}$ g. Its DNA, which has a molecular mass of 2.5×10^9 **daltons (D)***, is thought to encode ~ 3000 proteins (of which only ~ 1000 have been identified) although not all of them are simultaneously present in a given cell. Altogether an *E. coli* contains some 3 to 6 thousand different types of molecules including proteins, nucleic acids, polysaccharides, lipids, and various small molecules and ions (Table 1-1).

Prokaryotes Employ a Wide Variety of Metabolic Energy Sources

The nutritional requirements of the prokaryotes are enormously varied. **Autotrophs** (Greek: *autos,* self + *trophikos,* to feed) can synthesize all their cellular constituents from simple molecules such as H_2O, CO_2, NH_3, and H_2S. Of course they need an energy source to

---

* The **molecular mass** of a particle may be expressed in units of daltons, which are defined as $\frac{1}{12}$th the mass of a ^{12}C atom [atomic mass units (amu)]. Alternatively, this quantity may be expressed in terms of **molecular weight**, a dimensionless quantity defined as the ratio of the particle mass to $\frac{1}{12}$th the mass of a ^{12}C atom and symbolized M_r (for relative molecular mass). In this text, we shall refer to the molecular mass of a particle rather than its molecular weight.

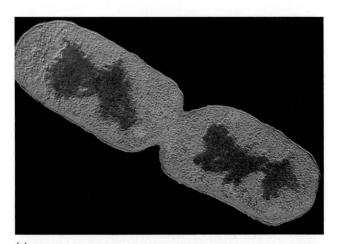

(a)

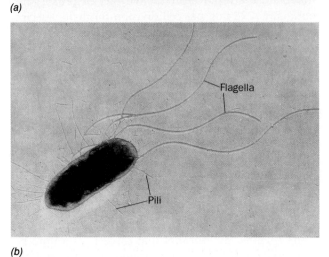

Flagella

Pili

(b)

Figure 1-3
Electron micrographs of *E. coli* cells: *(a)* Stained to show its internal structure. [CNRI.] *(b)* Stained to reveal its flagella and pili (rodlike organelles of mostly unknown function). [Courtesy of Howard Berg, Harvard University.]

Table 1-1
Molecular Composition of *E. coli*

Component	Percentage by Weight
H_2O	70
Protein	15
Nucleic acids:	
DNA	1
RNA	6
Polysaccharides and precursors	3
Lipids and precursors	2
Other small organic molecules	1
Inorganic ions	1

Source: Watson, J. D., *Molecular Biology of the Gene* (3rd ed.), p. 69, Benjamin (1976).

do so as well as to power their other functions. **Chemolithotrophs** (Greek: *lithos*, stone) obtain their energy through the oxidation of inorganic compounds such as NH_3, H_2S, or even Fe^{2+}:

$$2NH_3 + 4O_2 \longrightarrow 2HNO_3 + 2H_2O$$
$$H_2S + 2O_2 \longrightarrow H_2SO_4$$
$$4FeCO_3 + O_2 + 6H_2O \longrightarrow 4Fe(OH)_3 + 4CO_2$$

Photoautotrophs do so via **photosynthesis** (Chapter 22), a process in which light energy powers the transfer of electrons from inorganic donors to CO_2 yielding **carbohydrates** $[(CH_2O)_n]$.

In the most widespread form of photosynthesis, the electron donor in the light-driven reaction sequence is H_2O.

$$nCO_2 + nH_2O \longrightarrow (CH_2O)_n + nO_2$$

This process is carried out by **cyanobacteria** (e.g., the green slimy organisms that grow on the walls of aquariums; cyanobacteria were formerly known as **blue-green algae**), as well as by plants. This form of photosynthesis is thought to have generated the O_2 in the earth's atmosphere. Some species of cyanobacteria have the ability to convert N_2 from the atmosphere to organic nitrogen compounds. This **nitrogen fixation** capacity gives them the simplest nutritional requirements of all organisms: With the exception of their need for small amounts of minerals, they can literally live on sunlight and air.

In a more primitive form of photosynthesis, substances such as H_2, H_2S, thiosulfate, or organic compounds are the electron donors in light-driven reactions such as:

$$nCO_2 + 2nH_2S \longrightarrow (CH_2O)_n + nH_2O + 2nS$$

The **purple** and the **green photosynthetic bacteria** that carry out these processes occupy such oxygen-free habitats as shallow muddy ponds in which H_2S is generated by rotting organic matter.

Heterotrophs (Greek: *hetero*, other) obtain energy through the oxidation of organic compounds and hence are ultimately dependent on autotrophs for these substances. **Obligate aerobes** (which include animals) must utilize O_2 while **anaerobes** employ oxidizing agents such as sulfate (**sulfate-reducing bacteria**) or nitrate (**denitrifying bacteria**). Many organisms can partially metabolize various organic compounds in intramolecular oxidation–reduction processes known as **fermentation**. **Facultative anaerobes** such as *E. coli*, can grow either in the presence or absence of O_2. **Obligate anaerobes**, in contrast, are poisoned by the presence of O_2. Their metabolisms are thought to resemble those of the earliest life forms (which arose some 3.5 billion years ago when the earth's atmosphere lacked O_2; see Section 1-4B). At any rate, there are few organic compounds that cannot be metabolized by some prokaryotic species.

B. Prokaryotic Classification

The traditional methods of **taxonomy** (the science of biological classification), which are based largely on the anatomical comparisons of both contemporary and fossil organisms, are essentially inapplicable to prokaryotes. This is because the relatively simple cell structures of prokaryotes, including those of ancient bacteria as revealed by their microfossil remnants, provide little indication of their phylogenetic relationships (**phylogenesis:** evolutionary development). Compounding this problem is the observation that prokaryotes exhibit little correlation between form and metabolic function. Moreover, the eukaryotic definition of a species as a population that can interbreed is meaningless for the asexually reproducing prokaryotes. Consequently, the conventional prokaryotic classification schemes are rather arbitrary and lack the implied evolutionary relationships of the eukaryotic classification scheme (Section 1-2B).

In the most widely used prokaryotic classification scheme, the **prokaryotae** (also known as **monera**) have two divisions: the cyanobacteria and the **bacteria**. The latter are further subdivided into 19 parts based on their various distinguishing characteristics, most notably cell structure, metabolic behavior, and staining properties.

A simpler classification scheme, which is based on cell wall properties, distinguishes three major types of prokaryotes: the **mycoplasma**, the **gram-positive bacteria**, and the **gram-negative bacteria**. Mycoplasma lack the rigid cell wall of other prokaryotes. They are the smallest of all living cells (as small as 0.12 μm in diameter, Fig. 1-1) and possess ~20% of the DNA of an *E. coli*. Presumably this quantity of genetic information approaches the minimum amount necessary to specify the essential metabolic machinery required for cellular life. Gram-positive and gram-negative bacteria are distinguished according to whether or not they take up **gram stain** (a procedure developed in 1884 by Christian Gram in which heat-fixed cells are successively treated with the dye crystal violet and iodine and then destained with either ethanol or acetone). Gram-positive bacteria possess a monolayered cell wall, whereas those of gram-negative bacteria, which include cyanobacteria, possess at least two structurally distinct layers (Section 10-3B).

The development, in recent years, of techniques for the determination of amino acid sequences in proteins (Section 6-1) and base sequences in nucleic acids (Section 28-6) has provided abundant indications as to the geneological relationships between organisms. Indeed, these techniques make it possible to place these relationships on a quantitative basis, and thus to construct a phylogenetically based classification system for prokaryotes.

By the analysis of ribosomal RNA sequences, Carl Woese demonstrated that a group of prokaryotes he named the **archaebacteria** seem as distantly related to the other prokaryotes, the **eubacteria** ("true" bacteria), as both these groups are to eukaryotes. The archaebacteria constitute three different kinds of unusual organisms: the **methanogens,** obligate anaerobes that produce methane (marsh gas) by the reduction of CO_2 with H_2; the **halobacteria,** which can only live in concentrated brine solutions ($> 2M$ NaCl); and certain **thermoacidophiles,** organisms that inhabit acidic hot springs ($\sim 90°C$ and pH < 2). On the basis of a number of fundamental biochemical traits that differ among the archaebacteria, the eubacteria and the eukaryotes, but which are common within each group, Woese proposed that these groups of organisms constitute the three primary kingdoms of life (rather than the traditional division into prokaryotes and eukaryotes). Moreover, the observation that these three kingdoms are genealogically equidistant suggests that all of them arose independently from some simple primordial life form (Fig. 1-4).

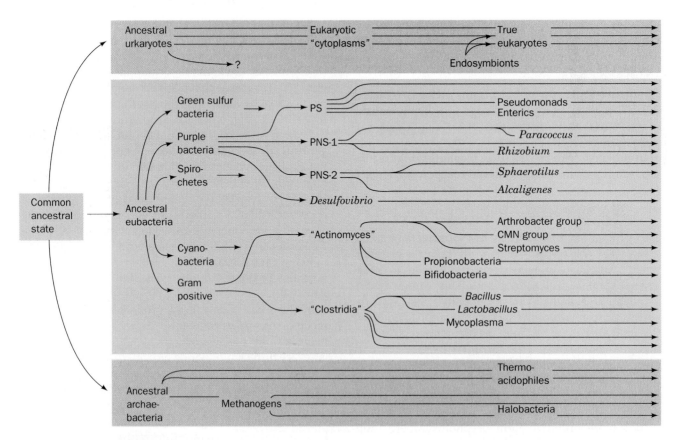

Figure 1-4
The major lines of prokaryotic descent. [After Fox, G. E., Stackebrandt, E., Hespell, R. B., Gibson, J., Maniloff, J., Dyer, T. A., Wolfe, R. S., Balch, W. E., Tanner, R. S., Magrum, L. J., Zablen, L. B., Blakemore, R., Gupta, R., Bonen, L., Lewis, B. J., Stahl, D. A., Leuhrsen, K. R., Chen, K. N., and Woese, C. R., *Science* **209,** 459 (1980).]

2. EUKARYOTES

Eukaryotic cells are generally 10 to 100 μm in diameter and thus have a thousand to a million times the volume of typical prokaryotes. It is not size, however, but a profusion of membrane-enclosed organelles, each with a specialized function, that best characterizes eukaryotic cells (Fig. 1-5). In fact, *eukaryotic organization and function is more complex than that of prokaryotes at all levels of organization,* from the molecular level on up.

Eukaryotes and prokaryotes have developed according to fundamentally different evolutionary strategies. Prokaryotes have exploited the advantages of simplicity and miniaturization: Their rapid growth rates permit them to occupy ecological niches in which there may be drastic fluctuations of the available nutrients. In contrast, the complexity of eukaryotes, which renders them larger and more slowly growing than prokaryotes, gives them the competitive advantage in stable environments with limited resources (Fig. 1-6). It is therefore erroneous to consider prokaryotes as evolutionarily primitive with respect to eukaryotes. Both types of organisms are well adapted to their respective life styles.

The earliest known microfossils of eukaryotes date from ~ 1.4 billion years ago, some 2.1 billion years after life arose. This supports the classical notion that eukaryotes are descended from a highly developed prokaryote, possibly a mycoplasma. The differences between eukaryotes and modern prokaryotes, however, are so profound as to render this hypothesis improbable. Perhaps the early eukaryotes, which according to Woese's evidence evolved from a primordial life form, were relatively unsuccessful and hence rare. Only after they had developed some of the complex organelles described in the following section did they become common enough to generate significant fossil remains.

A. Cellular Architecture

Eukaryotic cells, like prokaryotes, are bounded by a plasma membrane. The large size of eukaryotic cells results in their surface-to-volume ratios being much

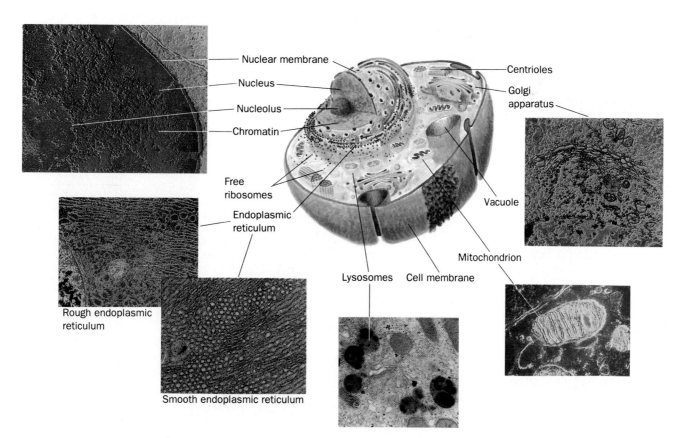

Figure 1-5
A schematic diagram of an animal cell accompanied by electron micrographs of its organelles. [Nucleus: Tektoff-Rhone-Merieux, CNRI; Rough endoplasmic reticulum and Golgi: Secchi-Lecaque/Roussel-UCLAF/CNRI; Smooth endoplasmic reticulum, David M. Phillips/Visuals Unlimited; Mitochondria: CNRI; Lysosome: Biophoto Associates/Photo Researchers, Inc.]

Figure 1-6
[Drawing by T. A. Bramley, *in* Carlile, M., *Trends Biochem. Sci.* **7,** 128 (1982). Copyright ©
Elsevier Biomedical Press, 1982. Used by permission.]

smaller than those of prokaryotes (the surface area of an object increases as the square of its radius, whereas volume does so as the cube). This geometrical constraint, coupled with the fact that many essential enzymes are membrane associated, partially rationalizes the large amounts of intracellular membranes in eukaryotes (the plasma membrane typically constitutes <10% of the membrane in a eukaryotic cell). Since all the matter that enters or leaves a cell must somehow pass through its plasma membrane, the surface areas of many eukaryotic cells are increased by numerous projections and/or invaginations (Fig. 1-7). Moreover, portions of the plasma membrane often bud inward, in a process known as **endocytosis,** such that the cell surrounds portions of the external medium. Thus eukaryotic cells can engulf and digest food particles such as bacteria, whereas prokaryotes are limited to the absorption of individual nutrient molecules. The reverse of endocytosis, a process termed **exocytosis,** is a common eukaryotic secretory mechanism.

Figure 1-7
A scanning electron micrograph of a fibroblast. [Courtesy of Guenther Albrecht-Buehler, Northwestern University.]

The Nucleus Contains the Cell's DNA

The nucleus, the eukaryotic cell's most conspicuous organelle, is the repository of its genetic information. This information is encoded in the base sequences of DNA molecules that form the discrete number of chromosomes characteristic of each species. The chromosomes consist of **chromatin,** a complex of DNA and protein. The amount of genetic information carried by eukaryotes is enormous; for example, a human cell has over 700 times the DNA of *E. coli* (in the terms commonly associated with computer memories, the genetic complement in each human cell specifies over 700 megabytes of information—about 200 times the information content of this text). Within the nucleus, the genetic information encoded by the DNA is transcribed into molecules of RNA (Chapter 29) which, after extensive processing, are transported to the cytoplasm (in eukaroytes, the cell contents exclusive of the nucleus) where they direct the ribosomal synthesis of proteins (Chapter 30). The nuclear envelope consists of a double membrane, which is perforated by numerous ~90-Å wide pores that regulate the flow of matter between the nucleus and the cytoplasm. These pores are of sufficient size to permit the passage of all but large molecular assemblies such as chromosomes and mature ribosomes.

The nucleus of most eukaryotic cells contains at least one dark-staining body known as the **nucleolus,** which is the site of ribosomal assembly. It contains chromosomal segments bearing multiple copies of genes specifying ribosomal RNA. These genes are transcribed in the nucleolus and the resulting RNA is combined with ribosomal proteins that have been imported from their site of synthesis in the **cytosol** (the cytoplasm exclusive of its membrane-bound organelles). The resulting immature ribosomes are then exported to the cytosol where their assembly is completed. Thus protein synthesis can only occur in the cytosol.

The Endoplasmic Reticulum and the Golgi Apparatus Function to Modify Membrane-Bound and Secretory Proteins

The most extensive membrane in the cell, which was discovered in 1945 by Keith Porter, forms a labyrinthine compartment named the **endoplasmic reticulum.** A large portion of this organelle, which is called the **rough endoplasmic reticulum,** is studded with ribosomes that are engaged in the synthesis of proteins that are either membrane bound or destined for secretion. The **smooth endoplasmic reticulum,** which is devoid of ribosomes, is the site of lipid synthesis. Many of the products synthesized in the endoplasmic reticulum are eventually transported to the **Golgi apparatus** (named after Camillo Golgi who first described it in 1898), a stack of flattened membranous sacs, in which these products are further processed (Section 21-3B).

Mitochondria Are the Site of Oxidative Metabolism

The **mitochondria** (Greek: *mitos,* thread + *chondros,* granule) are the site of cellular **respiration** (aerobic metabolism) in almost all eukaryotes. These cytoplasmic organelles, which are large enough to have been discovered by nineteenth century cytologists, vary in their size and shape but are often ellipsoidal with dimensions of around $1.0 \times 2.0 \ \mu m$—much like a bacterium. A eukaryotic cell typically contains on the order of 2000 mitochondria, which occupy roughly one fifth of its total cell volume.

The mitochondrion, as the electron microscopic studies of George Palade and Fritjof Sjöstrand first revealed, has two membranes: a smooth outer membrane and a highly folded inner membrane whose invaginations are termed **cristae** (Latin: crests). Thus the mitochondrion contains two compartments, the **intermembrane space** and the internal **matrix space.** The enzymes that catalyze the reactions of respiration are components of either the gellike **matrix** or the inner mitochondrial membrane. *These enzymes couple the energy-producing oxidation of nutrients to the energy-requiring synthesis of adenosine triphosphate* (**ATP;** Section 1-3B and Chapter 20). Adenosine triphosphate, after export to the rest of the cell, fuels its various energy-consuming processes.

Mitochondria are bacterialike in more than size and shape. Their matrix space contains mitochondrion-specific DNA, RNA, and ribosomes that participate in the synthesis of several mitochondrial components. Moreover, they reproduce by binary fission, and the respiratory processes that they mediate bear a remarkable resemblance to those of modern aerobic bacteria. These observations led to the now widely accepted hypothesis championed by Lynn Margulis that mitochondria evolved from originally free-living aerobic bacteria, which formed a symbiotic relationship with a primordial anaerobic eukaryote. The eukaryote-supplied nutrients consumed by the bacteria were presumably repaid severalfold by the highly efficient oxidative metabolism that the bacteria conferred on the eukaryote. This hypothesis is corroborated by the observation that the amoeba *Pelomyxa pelustris,* one of the few eukaryotes that lack mitochondria, permanently harbors aerobic bacteria in such a symbiotic relationship.

Lysosomes and Peroxysomes Are Containers of Degradative Enzymes

Lysosomes, which were discovered in 1949 by Christian de Duve, are single membrane-bounded organelles of variable size and morphology although most have diameters in the range 0.1 to 0.8 μm. Lysosomes, which are essentially membranous bags containing a large variety of hydrolytic enzymes, function to digest materials ingested by endocytosis and to recycle cellular components (Section 30-6). Cytological investigations have revealed that lysosomes form by budding from the Golgi apparatus.

Peroxisomes (also known as **microbodies**) are membrane-enclosed organelles, typically 0.5 μm in diameter, that contain oxidative enzymes. Some peroxisomal reactions generate **hydrogen peroxide** (H_2O_2), a reactive substance that is either utilized in the enzymatic oxidation of other substances or degraded through a disproportionation reaction catalyzed by the enzyme **catalase:**

$$2H_2O_2 \longrightarrow 2H_2O + O_2$$

It is thought that peroxisomes function to protect sensitive cell components from oxidative attack by H_2O_2. The peroxisome forms by budding from the endoplasmic reticulum. Certain plants contain a specialized type of peroxisome, the **glyoxisome,** so-named because it is the site of a series of reactions that are collectively termed the **glyoxylate pathway** (Section 21-2).

The Cytoskeleton Organizes the Cytosol

The cytosol, far from being a homogeneous solution, is a highly organized gel that can vary significantly in its composition throughout the cell. Much of its internal variability arises from the action of the **cytoskeleton,** an extensive array of filaments that gives the cell its shape, the ability to move, and is responsible for the arrangement and internal motions of its organelles (Fig. 1-8).

The most conspicuous cytoskeletal components, the **microtubules,** are $\sim$250-Å diameter tubes that are composed of the protein **tubulin** (Section 34-3F). They form the supportive framework that guides the movements of organelles within a cell. For example, the **mitotic spindle** is an assembly of microtubules and associated proteins that participates in the separation of replicated chromosomes during cell division. Microtubules are also major constitutents of **cilia,** the hairlike appendages extending from many cells, whose whiplike motions move the surrounding fluid past the cell or propel single cells through solution. Very long cilia, such as sperm tails, are termed **flagella** (prokaryotic flagella, which are composed of the protein **flagellin,** are quite different from those of eukaryotes).

The **microfilaments** are $\sim$90-Å in diameter fibers that consist of the protein **actin.** Microfilaments, as do microtubules, have a mechanically supportive function. Furthermore, through their interactions with the protein **myosin,** microfilaments form contractile assemblies that are responsible for many types of intracellular movements such as cytoplasmic streaming and the formation of cellular protuberances or invaginations. More conspicuously, however, actin and myosin are the major protein components of muscle (Section 34-3A).

The third major cytoskeletal component, the **intermediate filaments,** are protein fibers 100 to 150 Å in

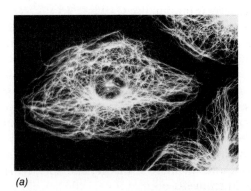

(a)

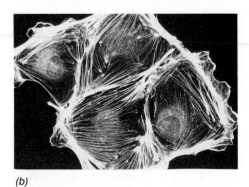

(b)

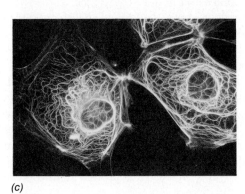

(c)

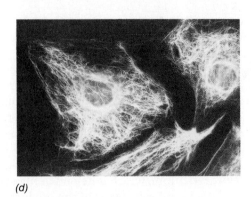

(d)

Figure 1-8
Immunofluorescence micrographs of rat kangaroo cells. The cells have been stained with fluorescently labeled antibodies raised against *(a)* tubulin, *(b)* actin, *(c)* keratin, and *(d)* **vimentin** (a protein constituent of a type of intermediate filament). [Courtesy of Mary Osborn, Max-Planck Institut für Molecular Biologie.]

diameter. Their prominence in parts of the cell that are subject to mechanical stress suggests that they have a load-bearing function. For example, skin in higher animals contains an extensive network of intermediate filaments made of the protein **keratin** (Section 7-2A), which is largely responsible for the toughness of this protective outer covering. In contrast to the case with microtubules and microfilaments, the proteins forming intermediate filaments vary greatly in size and composition, both among the different cell types within a given organism and among the corresponding cell types in different organisms.

Plant Cells Are Enclosed by Rigid Cell Walls

Plant cells (Fig. 1-9) contain all of the previously described organelles. They also have several additional features, the most conspicuous of which is a rigid cell wall that lies outside the plasma membrane. These cell walls, whose major component is the fibrous polysaccharide **cellulose** (Section 10-2C), account for the structural strength of plants.

A **vacuole** is a membrane-enclosed space filled with fluid. Although vacuoles occur in animal cells, they are most prominent in plant cells where they typically occupy 90% of the volume of a mature cell. Vacuoles function as storage depots for nutrients, wastes, and specialized materials such as pigments. The relatively high concentration of solutes inside a plant vacuole causes it to osmotically take up water thereby raising its internal pressure. This effect, combined with its cell walls' resistance to bursting, is largely responsible for the turgid rigidity of nonwoody plants.

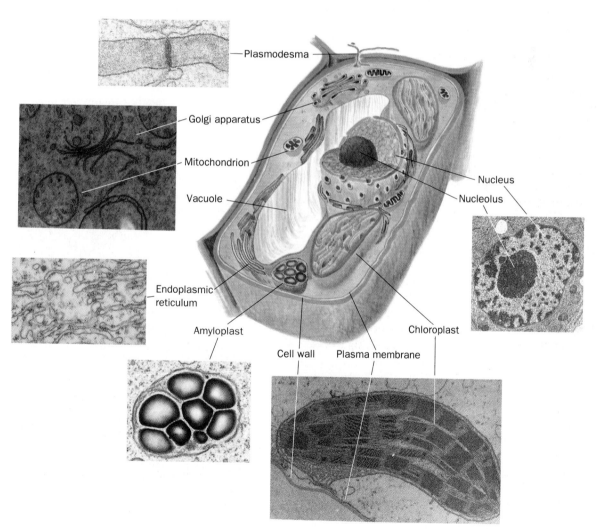

Figure 1-9
A drawing of a plant cell accompanied by electron micrographs of its organelles. [Plasmodesma: Courtesy of Hilton Mollenhauer, USDA; nucleus: Courtesy of Myron Ledbetter, Brookhaven National Laboratory; Golgi: Courtesy of W. Gordon Whaley, University of Texas; chloroplast: Courtesy of Lewis Shumway, College of Eastern Utah; leucoplast: Biophoto Associates; endoplasmic reticulum: Biophoto Associates/Photo Researchers, Inc.]

Chloroplasts Are the Site of Photosynthesis in Plants

One of the definitive characteristics of plants is their ability to carry out photosynthesis. The site of photosynthesis is an organelle known as the **chloroplast** which, although generally several times larger than a mitochondrion, resembles it in that both organelles have an inner and an outer membrane. Furthermore, the chloroplast's inner membrane space, the **stroma,** is similar to the mitochondrial matrix in that it contains many soluble enzymes. However, the inner chloroplast membrane is not folded into cristae. Rather, the stroma encloses a third membrane system that forms interconnected stacks of disklike sacs called **thylakoids,** which contain the photosynthetic pigment **chlorophyll.** The thylakoid uses chlorophyll-trapped light energy to generate ATP, which is used in the stroma to drive biosynthetic reactions forming carbohydrates and other products (Chapter 22).

Chloroplasts, as do mitochondria, contain their own DNA, RNA and ribosomes, and reproduce by fission. Apparently chloroplasts, much like mitochondria, evolved from an ancient cyanobacterium that took up symbiotic residence in an ancestral nonphotosynthetic eukaryote. In fact, several modern nonphotosynthetic eukaryotes have just such a symbiotic relationship with authentic cyanobacteria. Hence *most modern eukaryotes are genetic "mongrels" in that they simultaneously have nuclear, mitochondrial, and in the case of plants, chloroplast lines of descent.*

B. Phylogeny and Differentiation

One of the most remarkable characteristics of eukaryotes is their enormous morphological diversity, both on the cellular and organismal levels. Compare, for example, the architecture of the various human cells drawn in Fig. 1-10. Similarly, recall the great anatomical differences among say, an amoeba, an oak tree, and a human being.

Taxonometric schemes based on gross morphology as well as on protein and nucleic acid sequences (Sections 6-1 and 28-6) indicate that eukaryotes may be classified into three kingdoms: **fungi, plantae** (plants), and **animalia** (animals). The relative structural simplicity of many unicellular eukaryotes, however, makes their classification under this scheme rather arbitrary. Consequently, these organisms are usually assigned a fourth eukaryotic kingdom, the **protista.** (Note that biological classification schemes are for the convenience of biologists; nature is rarely neatly categorized.) Fig. 1-11 is a phylogenetic tree for eukaryotes.

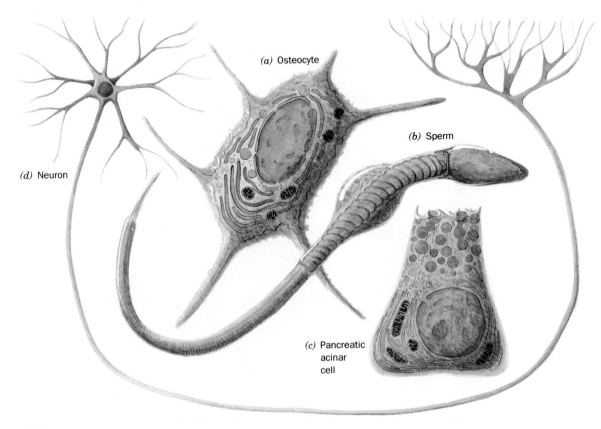

Figure 1-10
Drawings of some human cells: *(a)* an osteocyte (bone cell), *(b)* a sperm, *(c)* a pancreatic acinar cell (which secretes digestive enzymes), and *(d)* a neuron (nerve cell).

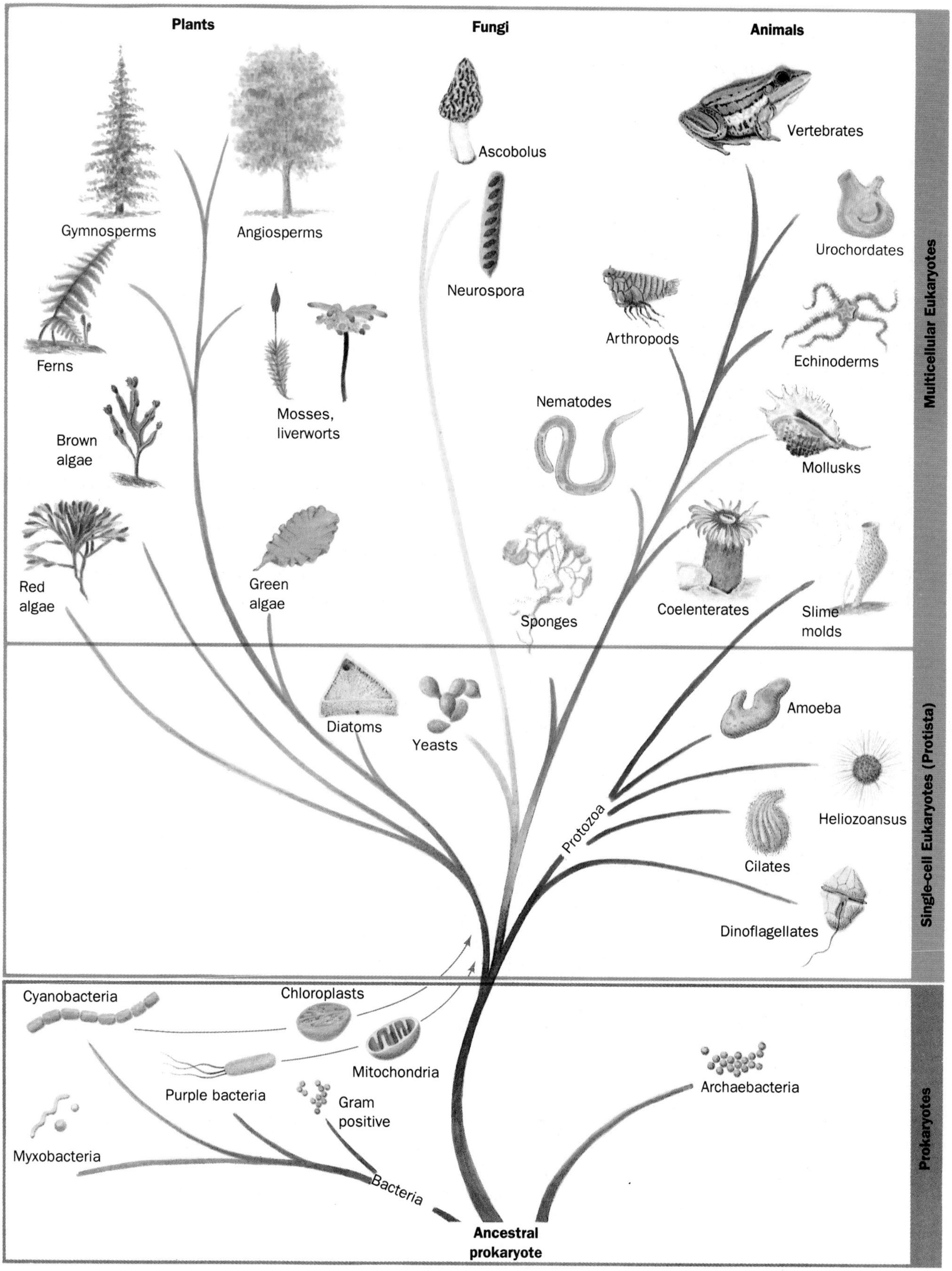

Plants

Gymnosperms

Angiosperms

Ferns

Mosses,
liverworts

Brown
algae

Green
algae

Red
algae

Fungi

Ascobolus

Neurospora

Animals

Vertebrates

Urochordates

Arthropods

Echinoderms

Nematodes

Mollusks

Sponges

Coelenterates

Slime
molds

Diatoms

Yeasts

Protozoa

Amoeba

Heliozoansus

Cilates

Dinoflagellates

Multicellular Eukaryotes

Single-cell Eukaryotes (Protista)

Cyanobacteria

Chloroplasts

Mitochondria

Purple bacteria

Gram
positive

Archaebacteria

Myxobacteria

Bacteria

Ancestral
prokaryote

Prokaryotes

Figure 1-11
An evolutionary tree indicating the lines of descent of cellular life on earth.

Anatomical comparisons among living and fossil organisms indicate that the various kingdoms of multicellular organisms independently evolved from protista (Fig. 1-11). The programs of growth, differentiation, and development followed by multicellular eukaryotes in their transformation from fertilized ova to adult organisms provide a remarkable indication of this evolutionary history. For example, all vertebrates exhibit gill-like pouches in their early embryonic stages, which presumably reflect their common fish ancestry (Fig. 1-12). Indeed, these early embryos are closely similar in size and anatomy even though their respective adult forms are vastly different in these characteristics. Such observations led Ernst Haeckel to formulate his famous although overstated dictum: *ontogeny recapitulates phylogeny* (ontogeny: biological development). The elucidation of the mechanism of cellular differentiation in eukaryotes is one of the major long-range goals of modern biochemistry.

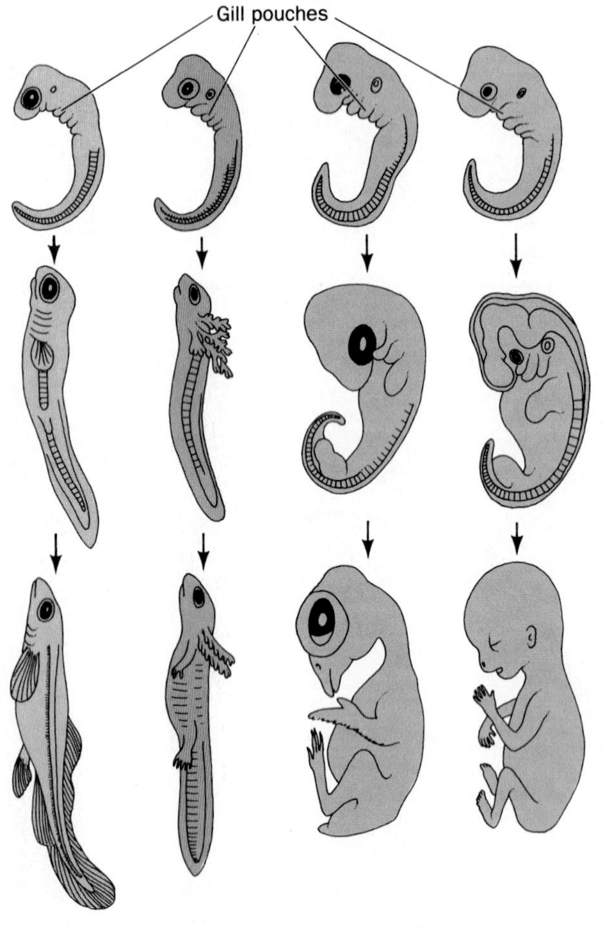

3. BIOCHEMISTRY: A PROLOGUE

Biochemistry, as the name implies, is the chemistry of life. It therefore bridges the gap between chemistry, the study of the structures and interactions of atoms and molecules, and biology, the study of the structures and interactions of cells and organisms. Since living things are composed of inanimate molecules, *life, at its most basic level, is a biochemical phenomenon.*

Although living organisms, as we have seen, are enormously diverse in their macroscopic properties, there is a remarkable similarity in their biochemistry, which provides a unifying theme with which to study them. For example, hereditary information is encoded and expressed in an almost identical manner in all cellular life. Moreover, the series of biochemical reactions, which are termed **metabolic pathways,** as well as the structures of the enzymes that catalyze them, are, for many basic processes, nearly identical from organism to organism. This strongly suggests that all known life forms are descended from a single primordial ancestor in which these biochemical features first developed.

Although biochemistry is a highly diverse field, it is largely concerned with a limited number of interrelated issues. These are

1. What are the chemical and three-dimensional structures of biological molecules and assemblies, and how do their properties vary with these structures?

2. How do proteins work; that is, what are the molecular mechanisms of enzymatic catalysis, how do receptors recognize and bind specific molecules, and what are the intramolecular and intermolecular mechanisms by which receptors transmit information concerning their binding states?

3. How is genetic information expressed and how is it transmitted to future cell generations?

4. How are biological molecules and assemblies synthesized?

5. What are the control mechanisms that coordinate the myriads of biochemical reactions that take place in cells and in organisms?

6. How do cells and organisms grow, differentiate, and reproduce?

Gill pouches

Fish   Salamander   Chick   Human

Figure 1-12
The embryonic development of a fish, an amphibian (salamander), a bird (chick), and a mammal (human). At early stages they are closely similar, both in size and anatomy (the top drawings have about the same scale). Later they diverge in both these properties. [After Haeckel, E., *Anthropogenie oder Entwickelungsgeschichte des Menschen*, Engelmann (1874).]

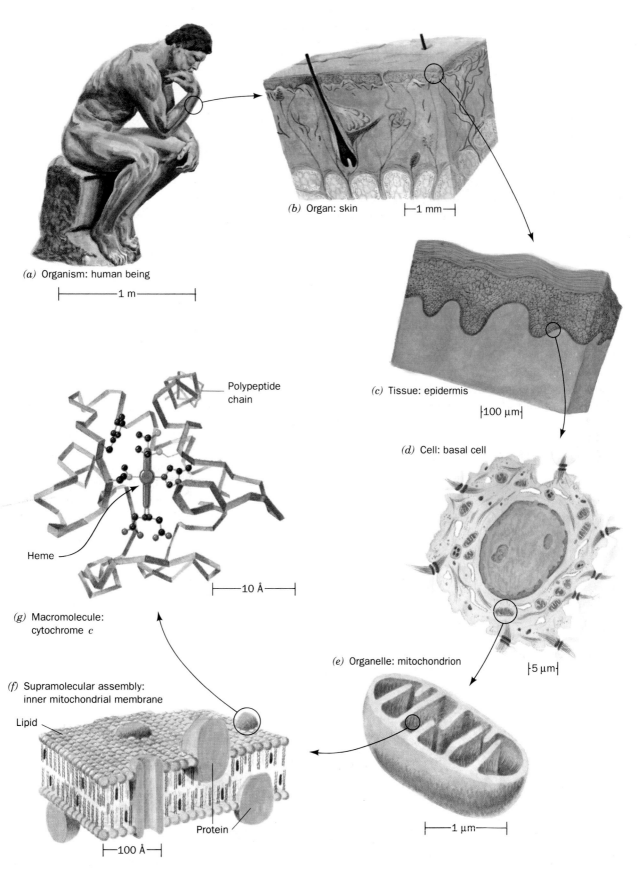

(a) Organism: human being
├─────1 m─────┤

(b) Organ: skin ├─1 mm─┤

(c) Tissue: epidermis ├100 μm┤

(d) Cell: basal cell

(e) Organelle: mitochondrion ├5 μm┤

├───1 μm───┤

(f) Supramolecular assembly: inner mitochondrial membrane

Lipid

Protein

├──100 Å──┤

(g) Macromolecule: cytochrome *c*

Polypeptide chain

Heme

├───10 Å───┤

Figure 1-13
An example of the hierachical organization of biological structures.

These issues are previewed in this section and further illuminated in later chapters. However, as will become obvious as you read further, in all cases, our knowledge, extensive as it is, is dwarfed by our ignorance.

A. Biological Structures

Living things are enormously complex. As indicated in Section 1-1A, even the relatively simple *E. coli* cell contains some 3 to 6 thousand different compounds, most of which are unique to *E. coli*. Higher organisms have a correspondingly greater complexity. *Homo sapiens* (human beings), for example, may contain 100,000 different types of molecules, although only a minor fraction of them have been characterized. One might therefore conclude that to obtain a coherent biochemical understanding of any organism would be a hopelessly difficult task. This, however, is not the case. *Living things have an underlying regularity that derives from their being constructed in a hierarchical manner.* Anatomical and cytological studies have shown that multicellular organisms are organizations of organs, which are made of tissues consisting of cells, composed of subcellular organelles (e.g., Fig. 1-13). At this point in our hierarchical descent, we enter the biochemical realm since organelles consist of **supramolecular assemblies,** such as membranes or fibers, which are organized clusters of **macromolecules** (polymeric molecules with molecular masses from thousands of daltons on up).

As Table 1-1 indicates, *E. coli,* and living things in general, contain only a few different types of macromolecules: **proteins** (Greek: *proteios,* of first importance), **nucleic acids,** and **polysaccharides** (Greek: *sakcharon,* sugar). *All of these substances have a modular construction; they consist of linked monomeric units that occupy the lowest level of our structural hierarchy.* Thus, as Fig. 1-14 indicates, proteins are polymers of amino acids (Section 4-1B), nucleic acids are polymers of nucleotides (Section 28-1), and polysaccharides are polymers of sugars (Section 10-2). **Lipids** (Greek: *lipos,* fat) the fourth major class of biological molecules, are too small to be classified as macromolecules but also have a modular construction (Section 11-1).

The task of the biochemist has been vastly simplified by the finding that *there are relatively few species of monomeric units that occur in each class of biological macromolecule.* Proteins are all synthesized from the same 20 species of **amino acids,** nucleic acids are made from 8 types of **nucleotides** (4 each in DNA and RNA), and there are ~8 commonly occurring types of **sugars** in polysaccharides. The great variation in properties observed among macromolecules of each type largely arises from the enormous number of ways its monomeric units can be arranged and, in many cases, derivatized.

One of the central questions in biochemistry is how are biological structures formed. As is explained in later chapters, the monomeric units of macromolecules are either directly acquired by the cell as nutrients or enzymatically synthesized from simpler substances. Macromolecules are synthesized from their monomeric precursors in complex enzymatically mediated processes.

Newly synthesized proteins spontaneously fold to assume their native conformations (Section 8-1A); that is, they undergo **self-assembly.** Apparently their amino acid sequences specify their three-dimensional structures. Thus proteins, as well as other macromolecules, require no external influence to direct their proper folding. The principle of self-assembly extends at least to the level of supramolecular assemblies. However, the way in which higher levels of biological structures are generated is largely unknown. The elucidation of the mechanisms of cellular and organismal growth and differentiation is a major area of biological research.

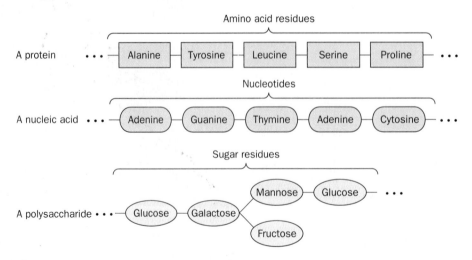

Figure 1-14
The polymeric organization of proteins, nucleic acids, and polysaccharides.

B. Metabolic Processes

There is a bewildering array of chemical reactions that simultaneously occur in any living cell. Yet, these reactions follow a pattern that organizes them into the coherent process we refer to as life. For instance, most biological reactions are members of a metabolic pathway; that is, they function as one of a sequence of reactions that produce one or more specific products. Moreover, one of the hallmarks of life is that the rates of its reactions are so tightly regulated that there is rarely an unsatisfied need for a reactant in a metabolic pathway or an unnecessary buildup of some product.

Metabolism has been traditionally (although not necessarily logically) divided into two major categories:

1. **Catabolism** or degradation, in which nutrients and cell constituents are broken down so as to salvage their components and/or to generate energy.

2. **Anabolism** or biosynthesis, in which biomolecules are synthesized from simpler components.

The energy required by anabolic processes is provided by catabolic processes largely in the form of **adenosine triphosphate (ATP)**. For instance, such energy-generating processes as photosynthesis and the biological oxidation of nutrients produce ATP from **adenosine diphosphate (ADP)** and a phosphate ion.

Adenosine diphosphate (ADP)

Adenosine triphosphate (ATP)

Conversely, such energy-consuming processes as biosynthesis, the transport of molecules against a concentration gradient, and muscle contraction, are driven by the reverse of this reaction, the hydrolysis of ATP:

$$ATP + H_2O \rightleftharpoons ADP + HPO_4^{2-}$$

Thus *anabolic and catabolic processes are coupled together through the mediation of the universal biological energy "currency," ATP.*

C. Expression and Transmission of Genetic Information

Deoxyribonucleic acid (DNA) is the cell's master repository of genetic information. This macromolecule, as is diagrammed in Fig. 1-15, consists of two strands of

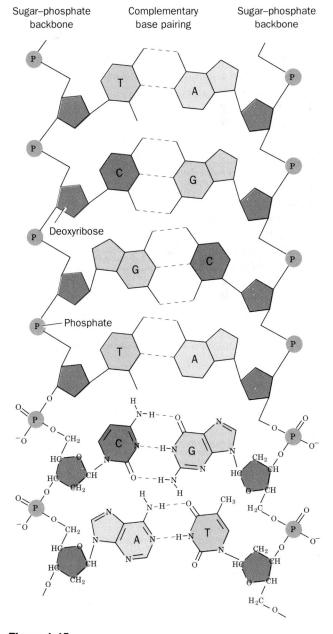

Figure 1-15
Double-stranded DNA. The two polynucleotide chains associate through complementary hydrogen bonding.

linked **nucleotides,** each of which is composed of a **deoxyribose** sugar residue, a phosphoryl group, and one of four bases: **adenine (A), thymine (T), guanine (G), or cytosine (C).** Genetic information is encoded in the sequence of these bases. Each DNA base is hydrogen bonded to a base on the opposite strand. However, A can only hydrogen bond with T, and G with C, so that the two strands are **complementary;** that is, the sequence of one strand implies the sequence of the other.

The division of a cell must be accompanied by the replication of its DNA. In this enzymatically mediated process, each DNA strand acts as a template for the formation of its complementary strand (Fig. 1-16; Section 31-1). Consequently, every progeny cell contains a complete DNA molecule, each of which consists of one parental strand and one daughter strand. Mutations arise when, through rare copying errors or damage to a parental strand, one or more wrong bases are incorporated into a daughter strand. Most mutations are either innocuous or deleterious. Occasionally, however, one results in a new characteristic that confers some sort of selective advantage on its recipient. Individuals with such mutations, according to the tenets of the Darwinian theory of evolution, have an increased probability of reproducing. New species arise through a progression of such mutations.

The expression of genetic information is a two-stage process. In the first stage, which is termed **transcription,** a DNA strand serves as a template for the synthesis of a complementary strand of ribonucleic acid (RNA; Section 29-2). This nucleic acid, which is generally single stranded, differs chemically from DNA only in that it has **ribose** sugar residues in place of DNAs deoxyribose and **uracil (U)** replacing DNAs thymine base.

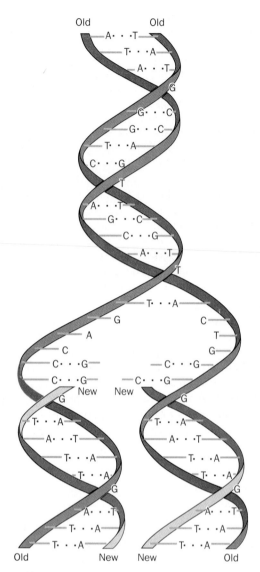

Figure 1-16
Schematic diagram of DNA replication.

Ribose

Uracil

In the second stage of genetic expression, which is known as **translation,** ribosomes enzymatically link together amino acids to form proteins (Section 30-3). The order in which the amino acids are linked together is prescribed by the RNA's sequence of bases. Consequently, since proteins are self-assembling, the genetic information encoded by DNA serves, through the intermediacy of RNA, to specify protein structure and function. Just which genes are expressed in a given cell under a particular set of circumstances is controlled by complex regulatory systems whose workings are just beginning to come to light.

4. THE ORIGIN OF LIFE

Man has always pondered the riddle of his existence. Indeed, all known cultures, past and present, primitive and sophisticated, have some sort of a creation myth that rationalizes how life arose. Only in the modern era, however, has it been possible to consider the origin of life in terms of a scientific framework, that is, in a manner subject to experimental verification. One of the first to do so was Charles Darwin, the originator of the theory of evolution. In 1871, he wrote in a letter to a colleague:

> *It is often said that all the conditions for the first production of a living organism are now present, which could ever have been present. But if (and oh what a big if) we could conceive in some warm little*

pond, with all sorts of ammonia and phosphoric salts, light, heat, electricity, etc., present, that a protein compound was chemically formed ready to undergo still more complex changes, at the present day such matter would be instantly devoured, or absorbed, which would not have been the case before living creatures were formed.

Radioactive dating studies indicate that the earth formed some 4.6 billion years ago. Yet the earliest fossil evidence of life, which was generated by organisms resembling modern bacteria, is ~3.5 billion years old. Apparently the preceding "prebiotic era" left no direct record. *Clearly, then, we cannot hope to determine exactly how life arose. Through laboratory experimentation, however, we can at least demonstrate what sorts of abiotic chemical reactions might have led to the formation of a living system.* Moreover, we are not entirely without traces of prebiotic development. The underlying biochemical and genetic unity of modern organisms suggests that life as we know it arose but once (if life arose more than once, the other forms must have rapidly died out, possibly because they were "eaten" by the present form). Thus, by comparing the corresponding genetic messages of a wide variety of modern organisms it may be possible to derive reasonable models of the primordial messages from which they were descended.

It is generally accepted that the development of life occupied three stages (Fig. 1-17):

1. Chemical evolution, in which simple geologically occurring molecules reacted to form complex organic polymers.

2. The self-organization of collections of these polymers to form replicating entities. At some point in this process, the transition from a lifeless collection of reacting molecules to a living system occurred.

3. Biological evolution to ultimately form the complex web of modern life.

In this section, we outline what has been surmised about these processes. We precede this discussion by a consideration of why only carbon, of all the elements, is suitable as the basis of the complex chemistry required for life.

A. The Unique Properties of Carbon

Living matter, as Table 1-2 indicates, consists of a relatively small number of elements. C, H, O, N, P, and S, all of which readily form covalent bonds, comprise some 92% of the dry weight of living things (most organisms are ~70% water). The balance consists of elements that are mainly present as ions and for the most part occur only in trace quantities (they usually carry out their functions at the active sites of enzymes). Note, however, that there is no known biological requirement

Table 1-2
Elemental Composition of the Human Body

Element	Dry Weight[a] (%)
C	61.7
N	11.0
O	9.3
H	5.7
Ca	5.0
P	3.3
K	1.3
S	1.0
Cl	0.7
Na	0.7
Mg	0.3
B	Trace
F	Trace
Si	Trace
V	Trace
Cr	Trace
Mn	Trace
Fe	Trace
Co	Trace
Cu	Trace
Zn	Trace
Se	Trace
Mo	Trace
Sn	Trace
I	Trace

[a] Calculated from Frieden, E., *Sci. Am.* **227**(1): 54–55 (1972).

for 65 of the 90 naturally occurring elements. Conversely, with the exceptions of oxygen and calcium, the biologically most abundant elements are but minor components of the earth's crust (the most abundant components of which are O, 47%; Si, 28%; Al, 7.9%; Fe, 4.5%; and Ca, 3.5%).

The predominance of carbon in living matter is no doubt a result of to its tremendous chemical versatility compared with all the other elements. Carbon has the unique ability to form a virtually infinite number of compounds as a result of its capacity to make as many as four highly stable covalent bonds (including single, double, and triple bonds) combined with its ability to form covalently linked C—C chains of unlimited extent. Thus, of the over 7 million chemical compounds that are presently known, nearly 90% are organic (carbon containing) substances. Let us examine the other elements in the periodic table to ascertain why they lack these combined properties.

Only five elements, B, C, N, Si, and P, have the capacity to make three or more bonds each and thus form chains of covalently linked atoms that can also have

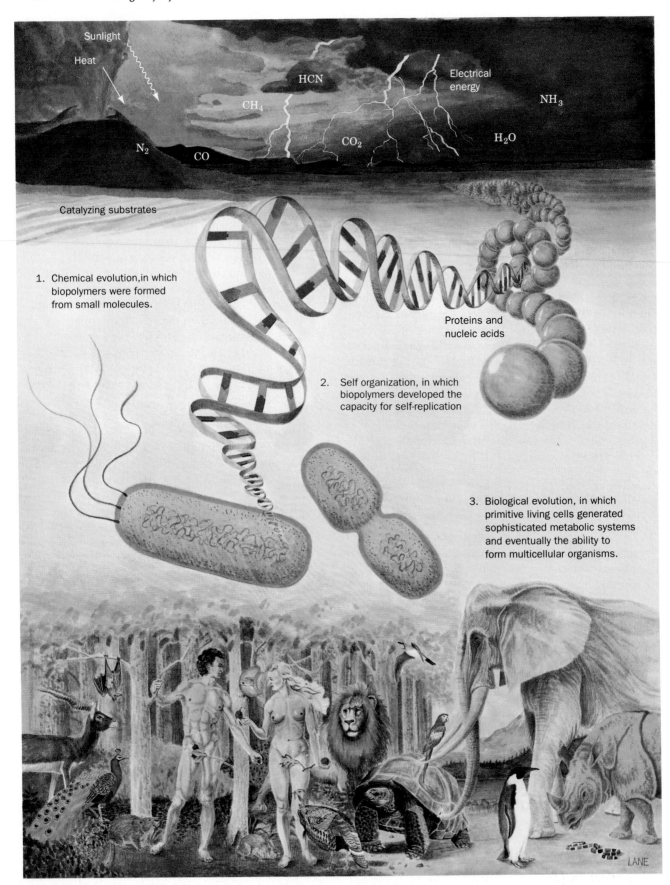

Figure 1-17
The three stages in the evolution of life.

pendant side chains. The other elements are either metals, which tend to form ions rather than covalent bonds; noble gases, which are essentially chemically inert; or atoms such as H or O that can each make only one or two covalent bonds. However, although B, N, Si, and P can each participate in at least three covalent bonds, they are, for reasons indicated below, unsuitable as the basis of a complex chemistry.

Boron, having fewer valence electrons (three) than valence orbitals (four), is electron deficient. This severely limits the types and stabilities of compounds that B can form. Nitrogen has the opposite problem; its five valence electrons make it electron rich. The repulsions between the lone pairs of electrons on bonded N atoms serve to greatly reduce the bond energy of an $N\!-\!N$ bond (171 kJ·mol^{-1}) relative to that of a $C\!-\!C$ bond (348 kJ·mol^{-1}) so that extended chains of bonded N atoms are highly unstable. Silicon and carbon, being in the same column of the periodic table, might be expected to have similar chemistries. Silicon's large atomic radius, however, prevents two Si atoms from approaching each other closely enough to gain much orbital overlap. Consequently $Si\!-\!Si$ bonds are weak (177 kJ·mol^{-1}) and the corresponding multiple bonds are rarely stable. The $Si\!-\!O$ bonds, in contrast, are so stable (369 kJ·mol^{-1}) that chains of alternating Si and O atoms are essentially inert (silicate minerals, whose frameworks consist of such bonds, form the earth's crust). Indeed, science fiction writers have speculated that **silicones,** which are oily or rubbery organosilicon compounds with backbones of linked $Si\!-\!O$ units; for example, **methyl silicones,**

$$\cdots\!-\!\underset{\underset{\text{CH}_3}{|}}{\overset{\overset{\text{CH}_3}{|}}{\text{Si}}}\!-\!\text{O}\!-\!\underset{\underset{\text{CH}_3}{|}}{\overset{\overset{\text{CH}_3}{|}}{\text{Si}}}\!-\!\text{O}\!-\!\underset{\underset{\text{CH}_3}{|}}{\overset{\overset{\text{CH}_3}{|}}{\text{Si}}}\!-\!\text{O}\!-\!\underset{\underset{\text{CH}_3}{|}}{\overset{\overset{\text{CH}_3}{|}}{\text{Si}}}\!-\!\text{O}\!-\!\cdots$$

Methyl silicones

could form the chemical basis of extraterrestrial life forms. Yet, the very inertness of the $Si\!-\!O$ bond makes this seem unlikely. Phosphorus, being below N in the periodic table, forms even less stable chains of covalently bonded atoms.

The foregoing does not imply that heteronuclear bonds are unstable. On the contrary, proteins contain $C\!-\!N\!-\!C$ linkages, carbohydrates have $C\!-\!O\!-\!C$ linkages, and nucleic acids possess $C\!-\!O\!-\!P\!-\!O\!-\!C$ linkages. However, *these heteronuclear linkages are less stable than are $C\!-\!C$ bonds. Indeed, they usually form the sites of chemical cleavage in the degradation of macromolecules and, conversely, are the bonds formed when monomer units are linked together to form macromolecules.* In the same vein, homonuclear linkages other than $C\!-\!C$ bonds are so reactive that they are, with the exception of $S\!-\!S$ bonds in proteins, extremely rare in biological systems.

B. Chemical Evolution

In the remainder of this section, we describe the most commonly accepted scenario for the origin of life. *Note, however, that there are valid scientific objections to this scenario as well as to the several others that have been seriously entertained so that we are far from certain as to how life arose.*

The solar system is thought to have formed by the gravitationally induced collapse of a large interstellar cloud of dust and gas. The major portion of this cloud, which was composed mostly of hydrogen and helium, condensed to form the sun. The rising temperature and pressure at the center of the proto-sun eventually ignited the self-sustaining thermonuclear reaction that has since served as the sun's energy source. The planets, which formed from smaller clumps of dust, were not massive enough to support such a process. In fact the smaller planets, including Earth, consist of mostly heavier elements because their masses are too small to gravitationally retain much H_2 and He.

The primordial earth's atmosphere was quite different from what it is today. It could not have contained significant quantities of O_2, a highly reactive substance. Rather, in addition to the H_2O, N_2, and CO_2 that it presently has, the atmosphere probably contained smaller amounts of CO, CH_4, NH_3, SO_2, and possibly H_2, all molecules that have been spectroscopically detected in interstellar space. The chemical properties of such a gas mixture make it a **reducing atmosphere** in contrast to the earth's present atmosphere, which is an **oxidizing atmosphere** (although recent contradictory evidence suggests that the primoridial earth had an oxidizing atmosphere).

In the 1920s, Alexander Oparin and J. B. S. Haldane independently suggested that *ultraviolet (UV) radiation from the sun [which is presently largely absorbed by an ozone (O_3) layer high in the atmosphere] or lightning discharges caused the molecules of the primordial reducing atmosphere to react to form simple organic compounds such as amino acids, nucleic acid bases, and sugars.* That this process is possible was first experimentally demonstrated in 1953 by Stanley Miller and Harold Urey who, in the apparatus diagrammed in Fig. 1-18, simulated effects of lightning storms in the primordial atmosphere by subjecting a refluxing mixture of H_2O, CH_4, NH_3, and H_2 to an electric discharge for about a week. The resulting solution contained significant amounts of water-soluble organic compounds, the most abundant of which are listed in Table 1-3, together with a substantial quantity of insoluble tar (polymerized material). Several of the soluble compounds are amino acid components of proteins and many of the others, as we shall see, are also of biochemical significance. Similar experiments in which the reaction conditions, the gas mixture, and/or the energy source are varied have resulted in the

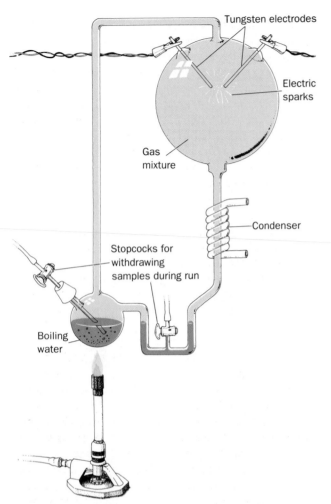

Table 1-3

Yields from Sparking a Mixture of CH$_4$, NH$_3$, H$_2$O, and H$_2$

Compound	Yield (%)
Glycine[a]	2.1
Glycolic acid	1.9
Sarcosine	0.25
Alanine[a]	1.7
Lactic acid	1.6
N-Methylalanine	0.07
α-Amino-*n*-butyric acid	0.34
α-Aminoisobutyric acid	0.007
α-Hydroxybutyric acid	0.34
β-Alanine	0.76
Succinic acid	0.27
Aspartic acid[a]	0.024
Glutamic acid[a]	0.051
Iminodiacetic acid	0.37
Iminoaceticpropionic acid	0.13
Formic acid	4.0
Acetic acid	0.51
Propionic acid	0.66
Urea	0.034
N-Methyl urea	0.051

[a] Amino acid constituent of proteins.

Source: Miller, S. J. and Orgel, L. E., *The Origins of Life on Earth, p.* 85, Prentice–Hall (1974).

Figure 1-18
Apparatus for emulating the synthesis of organic compounds on the prebiotic earth. A mixture of gases thought to resemble the primitive earth's reducing atmosphere is subjected to an electric discharge to simulate the effects of lightning while the water in the flask is refluxed so that the newly formed compounds dissolve in the water and accumulate in the flask. [After Miller, S. L. and Orgel, L. E., *The Origins of Life on Earth, p.* 84, Prentice–Hall (1974).]

synthesis of many other amino acids. This, together with the observation that carbonaceous meteorites contain many of the same amino acids, strongly suggests that these substances were present in significant quantities on the primordial earth.

Nucleic acid bases have also been synthesized under supposed prebiotic conditions. In particular, adenine is formed by the condensation of HCN, a plentiful component of the prebiotic atmosphere, in a reaction catalyzed by NH$_3$ [note that the chemical formula of adenine is (HCN)$_5$]. The other bases have been synthesized in similar reactions involving HCN and H$_2$O. Sugars have been synthesized by the polymerization of formaldehyde (CH$_2$O) in reactions catalyzed by divalent cations,

alumina, or clays. It is probably no accident that these compounds are the basic components of biological molecules. *They were apparently the most common organic substances in prebiotic times.*

The above described prebiotic reactions probably occurred over a period of hundreds of millions of years. Ultimately, it has been estimated, the oceans attained the organic consistency of a thin boullion soup. Of course there must have been numerous places, such as tidal pools and shallow lakes, where the prebiotic soup became much more concentrated. In such environments its component organic molecules could have condensed to form, for example, polypeptides and polynucleotides (nucleic acids). Quite possibly these reactions were catalyzed by the adsorption of the reactants on minerals such as clays. If, however, life were to have formed, the rates of synthesis of these complex polymers would have had to be greater than their rates of hydrolysis. Therefore, the "pond" in which life arose may have been cold rather than warm, possibly even below 0°C (seawater freezes solidly only below −21°C), since hydrolysis reactions are greatly retarded at such low temperatures.

C. The Rise of Living Systems

Living systems have the ability to replicate themselves. The inherent complexity of such a process is such that no man-made device has even approached having this capacity. Clearly there is but an infinitesimal probability that a collection of molecules can simply gather at random to form a living entity. How then did life arise? The answer, most probably, is that it was guided according to the Darwinian principle of the survival of the fittest as it applies at the molecular level.

Life Probably Arose through the Development of Self-Replicating RNA Molecules

The primordial self-replicating system is widely believed to have been a collection of nucleic acid molecules because such molecules, as we have seen in Section 1-3C, can direct the synthesis of molecules complementary to themselves. RNA, as does DNA, can direct the synthesis of a complementary strand. In fact, RNA serves as the hereditary material of many viruses (Chapter 32). The polymerization of the progeny molecules would, at first, have been a simple chemical process and hence could hardly be expected to be accurate. The early progeny molecules would therefore have been only approximately complementary to their parents. Nevertheless, repeated cycles of nucleic acid synthesis would eventually exhaust the supply of free nucleotides so that the synthesis rate of new nucleic acid molecules would be ultimately limited by the hydrolytic degradation rate of old ones. Suppose, in this process, a nucleic acid molecule randomly arose, which through folding was more resistant to degradation than its cousins. The progeny of this molecule, or at least its more faithful copies, could then propagate at the expense of the nonresistant molecules; that is, the resistant molecules would have a Darwinian advantage over their fellows. Theoretical studies suggest that such a system of molecules would evolve so as to optimize its replication efficiency under its inherent physical and chemical limitations.

In the next stage of the evolution of life, it is thought the dominant nucleic acids evolved the capacity to influence the efficiency and accuracy of their own replication. This process occurs in living systems through the nucleic acid-directed ribosomal synthesis of enzymes that catalyze nucleic acid synthesis. How nucleic acid-directed protein synthesis could have occurred before ribosomes arose is unknown because nucleic acids are not known to selectively interact with particular amino acids. This difficulty exemplifies the major problem in tracing the pathway of prebiotic evolution. Suppose some sort of rudimentary nucleic acid-influenced system arose, which increased the efficiency of nucleic acid replication. This system must have eventually been replaced, presumably with almost no trace of its existence, by the much more efficient ribosomal system. Our hypothetical nucleic acid synthesis system is therefore analogous to the scaffolding used in the construction of a building. After the building has been erected, the scaffolding is removed leaving no physical evidence that it ever was there. *Most of the statements in this section must therefore be taken as educated guesses.* Without having witnessed the event, it seems unlikely that we shall ever be certain of how life arose.

A plausible hypothesis for the evolution of self-replicating systems is that they initially consisted entirely of RNA. This idea is based, in part, on the recent observation that certain species of RNA exhibit enzymelike catalytic properties (Section 29-4B). Moreover, since ribosomes are $\sim \frac{2}{3}$ RNA and only $\frac{1}{3}$ protein, it is plausible that the primordial ribosomes were entirely RNA. A cooperative relationship between RNA and protein might have arisen when these self-replicating proto-ribosomes evolved the ability to influence the synthesis of proteins that increased the efficiency and/or the accuracy of RNA synthesis. *From this point of view, RNA is the primary substance of life; the participation of DNA and proteins were later refinements that increased the Darwinian fitness of an already established self-replicating system.*

The types of systems that we have so far described were bounded only by the primordial "pond." A self-replicating system that developed a more efficient component would therefore have to share its benefits with all the "inhabitants" of the "pond," a situation that minimizes the improvement's selective advantage. Only through compartmentalization; that is, the generation of cells, could developing biological systems reap the benefits of any improvements that they might have acquired. Of course, cell formation would also hold together and protect any self-replicating system and therefore help it spread beyond its "pond" of origin. Indeed, the importance of compartmentalization is such that it may have preceded the development of self-replicating systems. The erection of cell boundaries is not without its price, however. Cells, as we shall see in later chapters, must expend much of their metabolic effort in selectively transporting substances across their cell membranes. How cell boundaries first arose, or even what they were made from, is presently unknown.

Competition for Energy Resources Led to the Development of Metabolic Pathways, Photosynthesis, and Respiration

At this stage in their development, the entities we have been describing already fit Horowitz's criteria for life (exhibiting replication, catalysis, and mutability). The polymerization reactions through which these primitive organisms replicated were entirely dependent on the environment to supply the necessary monomeric units and the energy-rich compounds such as ATP or, more likely, just polyphosphates, that powered these reactions. As some of the essential components in the prebiotic soup became scarce, organisms developed the enzymatic systems that could synthesize these substances from simpler but more abundant precursors. As a consequence, energy-producing metabolic pathways arose. This latter development only postponed an "energy crisis," however, because these pathways consumed other preexisting energy-rich substances. The increasing scarcity of all such substances ultimately stimulated the development of photosynthesis to take advantage of a practically inexhaustible energy supply, the sun. Yet, this process, as we saw in Section 1-1A, consumes reducing agents such as H_2S. The eventual exhaustion of these substances led to the refinement of the photosynthetic process so that it used the ubiquitous H_2O as its reducing agent. This refinement led to yet another problem since this more advanced form of photosynthesis generates the highly reactive O_2 as a byproduct. The accumulation of O_2, which over the eons converted the reducing atmosphere of the prebiotic earth to the modern oxidizing atmosphere ($21\% \ O_2$), eventually interfered with the existing metabolic apparatus, which had evolved to operate under reducing conditions. The O_2 accumulation therefore stimulated the development of metabolic refinements, which protected organisms from oxidative damage. More importantly, it led to the evolution of a much more efficient form of energy metabolism than had previously been possible, **respiration** (oxidative metabolism), which used the newly available O_2 as an oxidizing agent.

As previously outlined, the basic replicative and metabolic apparatus of modern organisms evolved quite early in the history of life on earth. Indeed, many modern prokaryotes appear to resemble their very ancient ancestors. The rise of eukaryotes, as Section 1-2 indicates, occurred perhaps 2 billion years after prokaryotes had become firmly established. Multicellular organisms are a relatively recent evolutionary innovation, having not appeared, according to the fossil record, until ~ 700 million years ago.

5. THE BIOCHEMICAL LITERATURE

The biochemical literature contains the results of the work of tens of thousands of scientists extending over more than a century. Consequently a biochemistry text can only report selected highlights of this vast amount of information. Moreover, the tremendous rate at which biochemical knowledge is presently being acquired, which is perhaps greater than that of any other intellectual endeavor, guarantees that there will have been significant biochemical advances even in the year or so that it took to produce this text from its final draft. A serious student of biochemistry must therefore regularly read the biochemical literature to flesh out the details of subjects covered in (or omitted from) this text, as well as to keep abreast of new developments. This section provides a few suggestions on how to do so.

A. Conducting a Literature Search

The primary literature of biochemistry, those publications that report the results of biochemical research, is presently being generated at a rate of tens of thousands of papers per year appearing in around 200 periodicals. An individual can therefore only read this voluminous literature in a highly selective fashion. Indeed, most biochemists tend to "read" only those publications that are likely to contain reports pertaining to their interests. By "read" it is meant that they scan the tables of contents of these journals for the titles of articles that seem of sufficient interest to warrant further perusal (a convenient way of doing so is by using *Current Contents,* which is simply the collected tables of contents of recently published journals).

It is difficult to learn about a new subject by beginning with its primary literature. Rather, in order to obtain a general overview of a particular biochemical subject it is best to first peruse appropriate reviews and monographs (the update supplements to this textbook that we plan to publish from time to time will, hopefully, also be useful in this respect). These usually present a synopsis of recent (at the time of their writing) developments in the area, often from the authors' particular point of view. There are more or less two types of reviews: those that are essentially a compilation of facts and those that critically evaluate the data and attempt to place it in some larger context. The latter type of review is of course more valuable, particularly for a novice in the field. Most reviews are published in specialized books or journals although many journals that publish research reports also occasionally print reviews. Table 1-4 provides a list of many of the important biochemical review publications.

Reviews and monographs relevant to a subject of interest are usually easy to find through the use of a library card catalog and the subject indexes of the major review publications (the chapter-end references of this text may also be helpful in this respect). An important part of any review is its reference list. It usually lists previous reviews in the same or allied fields as well as indicating the most significant research reports in the area. Note the authors of these articles and the journals in which they tend to publish. When the most current reviews and research articles you have found tend to refer to the same group of earlier articles, you can be reasonably confident that your search for these earlier articles is largely complete. Finally, to familiarize yourself with the latest developments in the field, search the recent

Table 1-4
Some Important Biochemical Review Publications

Accounts of Chemical Research
Advances in Carbohydrate Chemistry and Biochemistry
Advances in Enzymology
Advances in Protein Chemistry
Advances in Lipid Research
Angewandte Chemie, International Edition in English[a]
Annals of the New York Academy of Sciences
Annual Review of Biochemistry
Annual Review of Biophysics and Biophysical Chemistry
Annual Review of Cell Biology
Annual Review of Genetics
Annual Review of Immunology
Annual Review of Medicine
Annual Review of Microbiology
Annual Review of Physiology
Annual Review of Plant Physiology and Plant Molecular Biology
Biochemistry[a]
BioEssays
Cell[a]
CRC Critical Reviews in Biochemistry
Current Topics in Bioenergetics
Current Topics in Cell Regulation
Essays in Biochemistry
FASEB Journal[a]
Harvey Lectures
Methods in Enzymology
Methods of Biochemical Analysis
Nature[a]
Progress in Biophysics and Molecular Biology
Progress in Nucleic Acid Research and Molecular Biology
Quarterly Reviews of Biophysics
Science[a]
Scientific American
Trends in Biochemical Sciences
Trends in Genetics
Vitamins and Hormones

[a] Periodicals that mainly publish research reports.

primary literature for the work of its most active research groups.

Biological Abstracts, Chemical Abstracts, and *Science Citation Index* are useful aids for locating references. These compendia list the articles in the many journals they cover by both author and subject (permuted title index). *Biological Abstracts* and *Chemical Abstracts* contain short English language abstracts of the articles listed (including many of foreign language articles). *Science Citation Index* lists all articles in a given year that cite a particular earlier paper so that it can be used to follow the developments in a field that build on a particular body of work.

Most academic libraries subscribe to computerized reference search services. If used properly, they can be highly efficient tools for locating specific information. Their disadvantage is their expense.

B. Reading a Research Article

Research reports more or less all have the same five-part format. They usually have a short abstract or summary located before (or, in some journals, after) the main body of the paper. The paper then continues (or begins) with an introduction, which often contains a short synopsis of the field, the motivation for the research reported, and a preview of its conclusions. The next section contains a description of the methods used to obtain the experimental data. This is followed by a presentation of the results of the investigation. Finally, there is a discussion section wherein the conclusions of the investigation are set forth and placed in the context of other work in the field. Most articles are so-called "full papers," which may be tens of pages long. However, many journals also contain "communications," which are usually only a page or two in length and are often published more quickly than are full papers.

It is by no means obvious how to read a scientific paper. Perhaps the worst way to do so is to read it from beginning to end as if it were some kind of a short story. In fact, most practicing scientists only occasionally read a research article in its entirety. It simply takes too long and is rarely productive. Rather, they scan selected parts of a paper and only dig deeper if it appears that to do so will be profitable. The following paragraph describes a reasonably efficient scheme for reading scientific papers. *This should be an active process in which the reader is constantly evaluating what is being read and relating it to his/her previous knowledge.* Moreover, the reader should maintain a healthy skepticism since there is a reasonable probability that any paper, particularly in its interpretation of experimental data and in its speculations, may be erroneous.

If the title of a paper indicates that it may be of interest then this should be confirmed by a reading of its abstract. For many papers, even those containing useful information, it is unnecessary to read further. If you choose to continue, it is probably best to do so by scanning the introduction so as to obtain an overview of the work reported. At this point most experienced scientists scan the conclusions section of the paper to gain a better understanding of what was found. If further effort seems warranted, they scan the results section to ascertain whether the experimental data supports the conclusions. The methods section is usually not read in detail because it is often written in a condensed form that is only fully interpretable by an expert in the field. However, for such experts, the methods section may be the most valuable part of the paper. At this point, what to read next, if anything, is largely dictated by the remaining points of confusion. In many cases this confusion can only be eliminated by reading some of the references given in the paper. At any rate, unless you plan to repeat or extend some of the work described, it is rarely necessary to read an article in detail. To do so in a critical manner, you will find, takes several hours for a paper of even moderate size.

Chapter Summary

Prokaryotes are single-celled organisms that lack a membrane-enclosed nucleus. Most prokaryotes have similar anatomies: a rigid cell wall surrounding a cell membrane that encloses the cytoplasm. The cell's single chromosome is condensed to form a nucleoid. *Escherichia coli*, the biochemically most well-characterized organism, is a typical prokaryote. Prokaryotes have quite varied nutritional requirements. The chemolithotrophs metabolize inorganic substances. Photolithotrophs, such as cyanobacteria, carry out photosynthesis. Heterotrophs, which live by oxidizing organic substances, are classified as aerobes if they use oxygen in this process and as anaerobes if some other oxidizing agent serves as their terminal electron acceptor. Traditional prokaryotic classification schemes are rather arbitrary because of poor correlation between bacterial form and metabolism. Sequence comparisons of nucleic acids and proteins, however, have established that a class of bacteria, the archaebacteria, are as different from the other prokaryotes, the eubacteria, as these two groups are from eukaryotes.

Eukaryotic cells, which are far more complex than those of prokaryotes, are characterized by having numerous membrane-enclosed organelles. The most conspicuous of these is the nucleus, which contains the cell's chromosomes and the nucleolus where ribosomes are assembled. The endoplasmic reticulum is the site of synthesis of lipids and of proteins that are destined for secretion. Further processing of these products occurs in the Golgi apparatus. The mitochondria, wherein oxidative metabolism occurs, are thought to have evolved from a symbiotic relationship between an aerobic bacterium and a primitive eukaryote. The chloroplast, the site of photosynthesis in plants, similarly evolved from a cyanobacterium. Other eukaryotic organelles include the lysosome, which functions as an intracellular digestive chamber, and the peroxisome, which contains a variety of oxidative enzymes including some that generate H_2O_2. The eukaryotic cytoplasm is pervaded by a cytoskeleton whose components include microtubules, which consist of tubulin; microfilaments, which are composed of actin; and intermediate filaments, which are made of different proteins in different types of cells. Eukaryotes have enormous morphological diversity on the cellular as well as the organismal levels. They have been classified into four kingdoms: the protista, plantae, fungi, and animalia. The pattern of embryonic development in multicellular organisms partially mirrors their evolutionary history.

Organisms have a hierarchical structure that extends down to the submolecular level. They contain but three basic types of macromolecules; proteins, nucleic acids, and polysaccharides, as well as lipids, each of which are constructed from only a few different species of monomeric units. Macromolecules and supramolecular assemblies form their native biological structures through a process of self-assembly. The assembly mechanisms of higher biological structures is largely unknown. Metabolic processes are organized into a series of tightly regulated pathways. These are classified as catabolic or anabolic depending on whether they participate in degradative or biosynthetic processes. The common energy "currency" in all these processes is ATP whose synthesis is the product of many catabolic pathways and whose hydrolysis drives most anabolic pathways. DNA, the cell's hereditary molecule, encodes genetic information in its sequence of bases. The complementary base sequences of its two strands permits them to act as templates for their own replication and for the synthesis of complementary strands of RNA. Ribosomes synthesize proteins by linking amino acids together in the order specified by the base sequences of RNAs.

Life is carbon-based because only carbon, among all the elements in the periodic table, has a sufficiently complex chemistry together with the ability to form virtually infinite stable chains of covalently bonded atoms. Reactions among the molecules in the reducing atmosphere of the prebiotic earth formed the simple organic precursors from which biological molecules developed. Eventually, in reactions that may have been catalyzed by minerals such as clays, polypeptides and polynucleotides formed. These evolved under the pressure of competition for the available monomeric units. Ultimately, a nucleic acid (most probably RNA) developed the capability of influencing its own replication by directing the synthesis of proteins that catalyze polynucleotide synthesis. This was followed by the development of cell membranes so as to form living entities. Subsequently, metabolic processes evolved to synthesize necessary intermediates from available precursors as well as the high-energy compounds required to power these reactions. Likewise, photosynthesis and respiration arose in response to environmental pressures brought about by the action of living organisms.

The sheer size and rate of increase of the biochemical literature requires that it be read to attain a thorough understanding of any aspect of biochemistry. The review literature provides an *entrée* into a given subspeciality. To remain current in any field, however, requires a regular perusal of its primary literature. This should be read in a critical but highly selective fashion.

References

Prokaryotes and Eukaryotes

Attenborough, D., *Life on Earth*, Little, Brown (1980). [A beautifully illustrated exposition of evolutionary development.]

Becker, W. M., *The World of the Cell*, Benjamin/Cummings (1986). [A highly readable cell biology text.]

Campbell, N. A., *Biology*, Benjamin/Cummings (1987). [A comprehensive general biology text. There are several others available of similar content.]

de Duve, C., *A Guided Tour of the Living Cell*, Vols. 1 and 2, Scientific American Books (1984). [A fanciful but highly enlightening examination of the cell's inner workings.]

Dulbecco, R., *The Design of Life*, Yale University Press (1987). [An incisive introduction to modern concepts of biology and biochemistry.]

Fox, G. E., Stackebrandt, E., Hespell, R. B., Gibson, J., Maniloff, J., Dyer, T. A., Wolfe, R. S., Balch, W. E., Tanner, R. S., Magrum, L. J., Zablen, L. B., Blakemore, R., Gupta, R., Bonen, L., Lewis, B. J., Stahl, D. A., Leuhrsen, K. R., Chen, K. N., and Woese, C. R., The phylogeny of prokaryotes, *Science* **209**, 457–463 (1980). [A presentation of the evidence that archaebacteria, eubacteria, and eukaryotes form separate lines of descent.]

Fawcett, D. W., *The Cell*, Saunders (1981). [A collection of electron micrographs of cells and organelles.]

Frieden, E., The chemical elements of life, *Sci. Am.* **227**(1): 52–60 (1972).

Holtzman, E. and Novikoff, A. B., *Cells and Organelles* (3rd ed.), Holt, Rinehart, & Winston (1984).

Lewin, R., *The Thread of Life*, Random House (1982). [A lavishly illustrated presentation of evolutionary biology.]

Margulis, L., Symbiosis and evolution, *Sci. Am.* **225**(2): 48–57 (1972). [A discussion of the evolution of mitochondria and chloroplasts.]

Margulis, L. and Schwartz, K. V., *Five Kingdoms. An Illustrated Guide to the Phyla of Life on Earth* (2nd ed.), Freeman (1987).

Sagan, D. and Margulis, L., *Garden of Microbial Delights*, Harcourt Brace Jovanovich (1988).

Stanier, R. Y., Ingrahan, J. L., Wheelis, M. L., and Painter, P. R., *The Microbial World* (5th ed.), Prentice–Hall (1986).

Origin of Life

Dickerson, R. E., Chemical evolution and the origin of life, *Sci. Am.* **239**(3): 70–86 (1978).

Dyson, F., *Origins of Life*, Cambridge University Press (1985). [A fascinating philosophical discourse on theories of life's origins by a respected theoretical physicist.]

Folsome, C. E., *The Origin of Life*, Freeman (1979).

Miller, S. L. and Orgel, L. E., *The Origins of Life on Earth*, Prentice–Hall (1974). [An authoritative account of prebiotic evolution.]

Schopf, J. W., The evolution of the earliest cells, *Sci. Am.* **239**(3): 110–138 (1978).

Schuster, P., Prebiotic evolution, *in* Gutfreund, H. (Ed.), *Biochemical Evolution*, pp. 15–87, Cambridge University Press (1981).

Shapiro, R., *Origins. A Skeptics Guide to the Creation of Life on Earth*, Summit Books (1986). [An incisive and entertaining critique of the reigning theories of the origin of life.]

Valentine, J. W., The evolution of multicellular plants and animals, *Sci. Am.* **239**(3): 140–158 (1978).

Watson, J. D., Hopkins, N. H., Roberts, J. W., Steitz, J. A., and Weiner, A. M., *Molecular Biology of the Gene* (4th ed.), Chapter 26, Benjamin/Cummings (1987).

Problems

It is very difficult to learn something well without somehow participating in it. The chapter-end problems are therefore an important part of this book. They contain few problems of the regurgitory type. Rather they are designed to make you think and to offer insights not discussed in the text. Their difficulties range from those that require only a few moments reflection to those that might take an hour or more of concentrated effort to work out. The more difficult problems are indicated by a leading asterisk (*). The answers to the problems are worked out in detail in the Solutions Manual by Donald Voet and Judith G. Voet that accompanies this text. You should, of course, make every effort to work out a problem before consulting the Solutions Manual.

1. Under optimal conditions for growth, an *E. coli* cell will divide around every 20 min. If no cells died, how long would it take a single *E. coli* cell, under optimal conditions in a 10-L culture flask, to reach its maximum cell density of 10^{10} cells $\cdot$ mL^{-1} (a "saturated" culture)? Assuming that optimum conditions could be maintained, how long would it take for the total volume of the cells alone to reach 1 km^3? (Assume an *E. coli* cell to be a cylinder 2 μm long and 1 μm in diameter.)

2. Without looking them up, draw schematic diagrams of a bacterial cell and an animal cell. What are the functions of their various organelles?

3. (a) Compare the surface-to-volume ratios of a typical *E. coli* cell (its dimensions are given in Problem 1) and a spherical eukaryotic cell that is 20 μm in diameter. How does this difference affect the life styles of these two cell types? (b) In order to improve their ability to absorb nutrients, the **brush border cells** of the intestinal epithelium have velvetlike patches of **microvilli** facing into the intestine. How does the surface-to-volume ratio of this eukaryotic cell change if 20% of its surface area is covered with cylindrical microvilli that are 0.1 μm in diameter, 1 μm in length, and occur on a square grid with 0.2-μm center-to-center spacing?

4. (a) Many proteins in *E. coli* are normally present at concentrations of two molecules per cell. What is the molar concentration of such a protein (the dimensions of *E. coli* are given in Problem 1)? (b) Conversely, how many glucose molecules does an *E. coli* cell contain if it has an internal glucose concentration of 1.0 m*M*?

5. (a) The DNA of an *E. coli* chromosome measures 1.4 mm in length, when extended, and 20 Å in diameter. What fraction of an *E. coli* cell is occupied by its DNA (the dimensions of an *E. coli* are given in Problem 1)? (b) A human cell has some 700 times the DNA of an *E. coli* cell and is typicaly spherical with a diameter of 20 μm. What fraction of such a human cell is occupied by its DNA?

*6. A new planet has been discovered that has approximately the same orbit about the sun as the earth but is invisible from earth because it is always on the opposite side of the sun. Interplanetary probes have already established that this planet has a significant atmosphere. The National Aeronautics and Space Administration is preparing to launch a new unmanned probe that will land on the surface of the planet. Outline a simple experiment for this lander that will test for the presence of life on the surface of this planet (assume that the life forms, if any, on the planet are likely to be microorganisms and therefore unable to walk up to the lander's video cameras and say "Hello")?

7. It has been suggested that an all out nuclear war will so enshroud the earth with clouds of dust and smoke that the entire surface of the planet will be quite dark and therefore intensely cold (well below 0°C) for several years (the so-called "Nuclear Winter"). In that case, it is thought, eukaryotic life would die out and bacteria would inherit the earth. Why?

8. Green and purple photosynthetic bacteria are thought to resemble the first organisms that could carry out photosynthesis. Speculate on the composition of the earth's atmosphere when these organisms first arose.

9. Explore your local biochemistry library (it may be disguised as a biology, chemistry, or medical library). Locate where the current periodicals, the bound periodicals, and the books are kept. Browse through the contents of a current major biochemistry journal, such as *Biochemistry, Cell,* or *Proceedings of the National Academy of Sciences,* and pick a title that interests you. Scan the corresponding paper and note its organization. Likewise, peruse one of the articles in the latest volume of *Annual Review of Biochemistry.*

Chapter 2

AQUEOUS SOLUTIONS

Life, as we know it, occurs, in aqueous solution. Indeed, terrestial life apparently arose in some primordial sea (Section 1-4B) and, as the fossil record indicates, did not venture onto dry land until comparatively recent times. Yet, even those organisms that did develop the capacity to live out of water still carry the ocean with them: The compositions of their intracellular and extracellular fluids are remarkably similar to that of seawater. This is true even of organisms that live in such unusual environments as saturated brine, acidic hot sulfur springs, and petroleum.

Water is so familiar, we generally consider it to be a rather bland fluid of simple character. It is, however, a chemically reactive liquid with such extraordinary physical properties that, if chemists had discovered it in recent times, it would undoubtedly have been classified as an exotic substance.

The properties of water are of profound biological significance. *The structures of the molecules on which life is based, proteins, nucleic acids, lipid membranes, and complex carbohydrates, result directly from their interactions with their aqueous environment. The combination of solvent properties responsible for the intramolecular and intermolecular associations of these substances is peculiar to water; no other solvent even closely resembles water in this respect.* Although the hypothesis that life could be based on organic polymers other than proteins and nucleic acids seems plausible, it is all but inconceivable that the complex structural organization and chemistry of living systems could exist in other than an aqueous medium. Indeed, direct observations on the surface of Mars, the only other planet in the solar system with temperatures compatible with life, indicate that it is both devoid of water and of life.

Biological structures and processes can only be understood in terms of the physical and chemical properties of water. We therefore begin this chapter with a discussion of the molecular and solvent properties of water. In the following section we review its chemical behavior, that is, the nature of aqueous acids and bases.

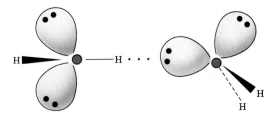

Figure 2-2
A hydrogen bond between two water molecules. The strength of this interaction is maximal when the O—H covalent bond points directly along a lone pair electron cloud of the oxygen atom to which it is hydrogen bonded.

1. PROPERTIES OF WATER

Water's peculiar physical and solvent properties stem largely from its extraordinary internal cohesiveness compared to that of almost any other liquid. In this section, we explore the physical basis of this phenomenon.

A. Structure and Interactions

The H_2O molecule has a bent geometry with an O—H bond distance of 0.958 Å and an H—O—H bond angle of 104.5° (Fig. 2-1). The large electronega-

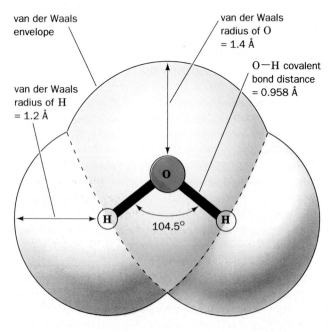

Figure 2-1
The structure of the water molecule. The outline represents the van der Waals envelope of the molecule (where the attractive components of the van der Waals interactions balance the repulsive components). The skeletal model of the molecule indicates its covalent bonds.

tivity difference between H and O confers a 33% ionic character on the O—H bond as is indicated by water's dipole moment of 1.85-debye units. Water is clearly a highly polar molecule, a phenomenon with enormous implications for living systems.

Water Molecules Associate through Hydrogen Bonds

The electrostatic attractions between the dipoles of two water molecules tends to orient them such that the O—H bond on one water molecule points towards a lone pair electron cloud on the oxygen atom of the other water molecule. This results in a directional intermolecular association known as a **hydrogen bond** (Fig. 2-2), an interaction that is crucial to both the properties of water itself and to its role as a biochemical solvent. In general, *a hydrogen bond may be represented as D—H · · · A, where D—H is a weakly acidic "donor group" such as N—H or O—H, and A is a lone pair bearing and thus weakly basic "acceptor atom" such as N or O.* The peculiar requirement of a hydrogen atom in the D—H · · · A interaction stems from the hydrogen atom's small size: Only a hydrogen nucleus can approach the lone pair electron cloud of an acceptor atom closely enough to permit an electrostatic association of significant magnitude.

Hydrogen bonds are structurally characterized by an H · · · A distance that is at least 0.5 Å shorter than the calculated van der Waals distance (distance of closest approach between two nonbonded atoms) between these atoms. In water, for example, the O · · · H hydrogen bond distance is ~1.8 versus 2.6 Å for the corresponding van der Waals distance. The energy of a hydrogen bond (~ 20 kJ·mol^{-1} in H_2O) is small compared to covalent bond energies (for instance, 460 kJ·mol^{-1} for an O—H covalent bond). Nevertheless, most biological molecules have so many hydrogen bonding groups that hydrogen bonding is of paramount importance in determining their three-dimensional structures and their intermolecular associations. Hydrogen bonding is further discussed in Section 7-4B.

The Physical Properties of Ice and Liquid Water Largely Result from Intermolecular Hydrogen Bonding

The structure of ice provides a striking example of the cumulative strength of many hydrogen bonds. X-ray and neutron diffraction studies have established that water molecules in ice are arranged in an unusually open structure. Each water molecule is tetrahedrally surrounded by four nearest neighbors to which it is hydrogen bonded (Fig. 2-3). In two of these hydrogen bonds, the central H_2O molecule is the "donor" and in the other two, it is the "acceptor." As a consequence of its open structure, water is one of the very few substances that expands on freezing (at 0°C, liquid water has a density of $1.00 \; g \cdot mL^{-1}$, whereas ice has a density of $0.92 \; g \cdot mL^{-1}$).

The expansion of water on freezing has overwhelming consequences for life on earth. Suppose that water contracted upon freezing, that is, became more dense rather than less dense. Ice would then sink to the bottoms of lakes and oceans rather than float. This ice would be insulated from the sun so that oceans, with the exception of a thin surface layer of liquid in warm weather, would be permanently frozen solid (the water at great depths in even tropical oceans is close to 4°C, its temperature of maximum density). The reflection of sunlight by these frozen oceans and their cooling effect on the atmosphere would ensure that land temperatures would also be much colder than at present, that is, the earth would have a permanent ice age. Furthermore, since life apparently evolved in the ocean, it seems unlikely that life could have developed at all if ice contracted upon freezing.

Although the melting of ice is indicative of the cooperative collapse of its hydrogen bonded structure, hydrogen bonds between water molecules persist in the liquid state. The heat of sublimation of ice at 0°C is $46.9 \; kJ \cdot mol^{-1}$. Yet, only $\sim 6 \; kJ \cdot mol^{-1}$ of this quantity can be attributed to the kinetic energy of gaseous water molecules. The remaining $41 \; kJ \cdot mol^{-1}$ must therefore represent the energy required to disrupt the hydrogen bonding interactions holding an ice crystal together. The heat of fusion of ice ($6.0 \; kJ \cdot mol^{-1}$) is $\sim 15\%$ of the energy required to disrupt the ice structure. *Liquid water is therefore only $\sim 15\%$ less hydrogen bonded than ice at 0°C.* Indeed, the boiling point of water is 264°C higher than that of methane (CH_4), a substance with nearly the same molecular mass as H_2O but which is incapable of hydrogen bonding (in the absence of intermolecular associations, substances with equal molecular masses should have similar boiling points). This reflects the extraordinary internal cohesiveness of liquid water resulting from its intermolecular hydrogen bonding.

Liquid Water Has a Rapidly Fluctuating Structure

X-ray scattering measurements of liquid water reveal

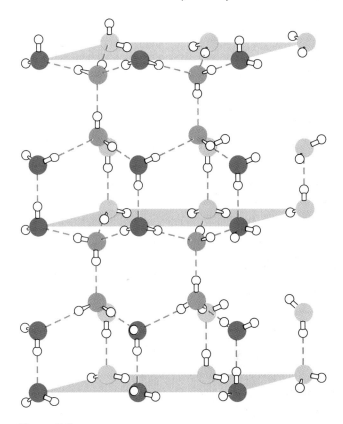

Figure 2-3
The structure of ice. The tetrahedral arrangement of the water molecules is a consequence of the roughly tetrahedral disposition of each oxygen atom's sp^3-hybridized bonding and lone pair orbitals (Fig. 2-2). Oxygen and hydrogen atoms are represented, respectively, by red and white spheres, and hydrogen bonds are indicated by dashed lines. Note the open structure that gives ice its low density relative to liquid water. [After Pauling, L., *The Nature of the Chemical Bond* (3rd ed.), p. 465, Cornell University Press (1960).]

a complex structure. Near 0°C, water exhibits an average nearest-neighbor O···O distance of 2.82 Å, which is slightly greater than the corresponding 2.76-Å distance in ice despite the greater density of the liquid. The X-ray data further indicate that each water molecule is surrounded by an average of 4.4 nearest neighbors, which strongly suggests that the short range structure of liquid water is predominantly tetrahedral in character. This picture is corroborated by the additional intermolecular distances in liquid water of around 4.5 and 7.0 Å, which are near the expected second and third nearest-neighbor distances in an icelike tetrahedral structure. Liquid water, however, also exhibits a 3.5-Å intermolecular distance, which cannot be rationalized in terms of an icelike structure. These average distances, moreover, become less sharply defined as the temperature increases into the physiologically significant range, thereby signaling the thermal breakdown of the short-range water structure.

The structure of liquid water is not simply described. This is because each water molecule reorients about once every 10^{-12} s, which makes the determination of water's instantaneous structure an experimentally and theoretically difficult problem (very few experimental techniques can make measurements over such short time spans). Indeed, only with the recent advent of molecular dynamics simulations, which calculate the trajectories of the individual atoms in a large collection of molecules over a period of time, have theoreticians felt that they had a reasonable understanding of liquid water on the molecular level.

For the most part, molecules in liquid water are each hydrogen bonded to four nearest neighbors as they are in ice. These hydrogen bonds are distorted, however, so that the networks of linked molecules are irregular and varied. For example, 4- to 7-membered rings of hydrogen bonded molecules commonly occur in liquid water, in contrast to the cyclohexanelike 6-membered rings characteristic of ice (Fig. 2-3). Moreover, these networks are continually breaking up and reforming over time periods on the order of 2×10^{-11} s. *Liquid water therefore consists of a rapidly fluctuating, space-filling network of hydrogen bonded H_2O molecules that, over short distances, resembles that of ice.*

B. Water as a Solvent

Water is said to be the "universal solvent." Although this statement cannot literally be true, water certainly dissolves more types of substances and in greater amounts than any other solvent. In particular, the polar character of water makes it an excellent solvent for polar and ionic materials, which are therefore said to be **hydrophilic** (Greek: *hydor*, water + *philos*, loving). On the other hand, nonpolar substances are virtually insoluble in water ("oil and water don't mix") and are consequently described as being **hydrophobic** (Greek: *phobos*, fear). Nonpolar substances, however, are soluble in nonpolar solvents such as CCl_4 or hexane. This information is summarized by another maxim, "like dissolves like."

Why do salts dissolve in water? Salts, such as NaCl or K_2HPO_4, are held together by ionic forces. The ions of a salt, as do any electrical charges, interact according to **Coulomb's law:**

$$F = \frac{k q_1 q_2}{D r^2} \qquad [2.1]$$

where F is the force between two electrical charges, q_1 and q_2, that are separated by the distance r, D is the **dielectric constant** of the medium between them, and k is a proportionality constant (8.99×10^9 J·m·C^{-2}). In a vacuum, D is unity and in air, it is only negligibly larger. The dielectric constants of several common solvents, together with their permanent molecular dipole mo-

Table 2-1

Dielectric Constants and Permanent Molecular Dipole Moments of Some Common Solvents

Substance	Dielectric Constant	Dipole Moment (debye)
Formamide	110.0	3.37
Water	78.5	1.85
Dimethyl sulfoxide	48.9	3.96
Methanol	32.6	1.66
Ethanol	24.3	1.68
Acetone	20.7	2.72
Ammonia	16.9	1.47
Chloroform	4.8	1.15
Diethyl ether	4.3	1.15
Benzene	2.3	0.00
Carbon tetrachloride	2.2	0.00
Hexane	1.9	0.00

Source: Brey, W. S., *Physical Chemistry and Its Biological Applications*, p. 26, Academic Press (1978).

ments, are listed in Table 2-1. Note that these quantities tend to increase together, although not in any regular way.

The dielectric constant of water is among the highest of any pure liquid, whereas those of nonpolar substances, such as hydrocarbons, are relatively small. The force between two ions separated by a given distance in nonpolar liquids such as hexane or benzene is therefore 30 to 40 times greater than that in water. Consequently, in nonpolar solvents (low D), ions of opposite charge attract each other so strongly that they coalesce to form a salt, whereas the much weaker forces between ions in water solution (high D) permit significant quantities of the ions to remain separated.

An ion immersed in a polar solvent attracts the oppositely charged ends of the solvent dipoles as is diagrammed in Fig. 2-4 for water. The ion is thereby surrounded by several concentric shells of oriented solvent molecules. Such ions are said to be **solvated** or, if water is the solvent, to be **hydrated.** The electric field pro-

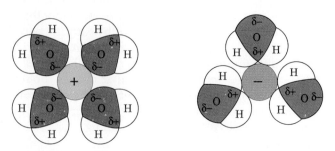

Figure 2-4
Solvation of ions by oriented water molecules.

duced by the solvent dipoles opposes that of the ion so that, in effect, the ionic charge is spread over the volume of the solvated complex. This arrangement greatly attenuates the coulombic forces between ions, which is why polar solvents have such high dielectric constants.

The orienting effect of ionic charges on dipolar molecules is opposed by thermal motions, which continually tend to randomly reorient all molecules. The dipoles in a solvated complex are therefore only partially oriented. The reason why the dielectric constant of water is so much greater than that of other liquids with comparable dipole moments is that liquid water's hydrogen bonded structure permits it to form oriented structures that resist thermal randomization, thereby more effectively distributing ionic charges.

The bond dipoles of uncharged polar molecules makes them soluble in aqueous solutions for the same reasons that ionic substances are water soluble. The solubilities of polar and ionic substances are enhanced if they carry functional groups, such as hydroxyl ($-$OH), keto ($>$C$=$O), carboxyl ($-$CO$_2$H or COOH), or amino groups ($-$NH$_2$), that can form hydrogen bonds with water as is illustrated in Fig. 2-5. Indeed, water soluble biomolecules such as proteins, nucleic acids, and carbohydrates, bristle with just such groups. Nonpolar substances, in contrast, lack both hydrogen bonding donor and acceptor groups.

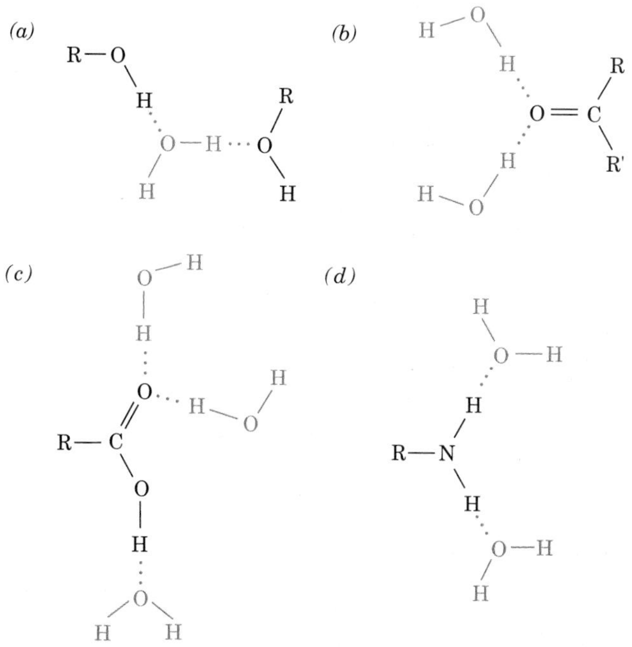

Amphiphiles Form Micelles and Bilayers

Most biological molecules have both polar (or ionically charged) and nonpolar segments and are therefore simultaneously hydrophilic and hydrophobic. Such molecules, for example, fatty acid ions (soap ions; Fig. 2-6), are said to be **amphiphilic** or, synonomously, **amphipathic** (Greek: *amphi*, both + *pathos*, passion). How do amphiphiles interact with an aqueous solvent? Water, of course, tends to hydrate the hydrophilic portion of an amphiphile, but it also tends to exclude its hydrophobic portion. Amphiphiles consequently tend to form water-dispersed structurally ordered aggregates. Such aggre-

Figure 2-5
Hydrogen bonding between water and *(a)* hydroxyl groups; *(b)* keto groups, *(c)* carboxyl groups, and *(d)* amino groups.

gates may take the form of **micelles,** which are globules of up to several thousand amphiphiles arranged with their hydrophilic groups at the globule surface so that they can interact with the aqueous solvent while the hydrophobic groups associate at the center so as to exclude solvent (Fig. 2-7*a*). Alternatively, the amphiphiles may arrange themselves to form bilayered sheets or vesicles (Fig. 2-7*b*) in which the polar groups face the aqueous phase.

The interactions stabilizing a micelle or bilayer are collectively described as **hydrophobic forces** or **interactions** to indicate that they result from the tendency of water to exclude hydrophobic groups. Hydrophobic interactions are relatively weak compared to hydrogen

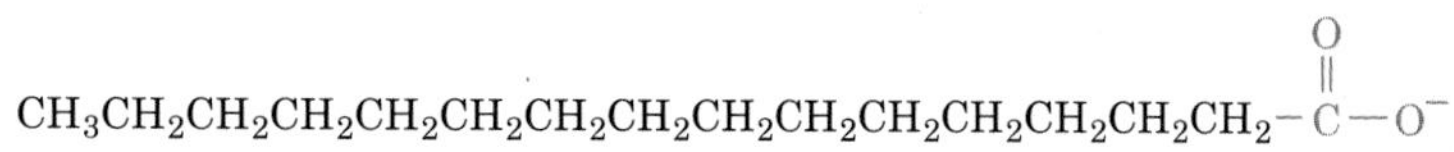

$$CH_3CH_2CH_2CH_2CH_2CH_2CH_2CH_2CH_2CH_2CH_2CH_2CH_2CH_2 - \overset{\overset{\textstyle O}{\|}}{C} - O^-$$

Palmitate $(C_{15}H_{31}COO^-)$

$$CH_3CH_2CH_2CH_2CH_2CH_2CH_2CH_2 - \overset{\overset{\textstyle H}{|}}{C} = \overset{\overset{\textstyle H}{|}}{C} - CH_2CH_2CH_2CH_2CH_2CH_2CH_2 - \overset{\overset{\textstyle O}{\|}}{C} - O^-$$

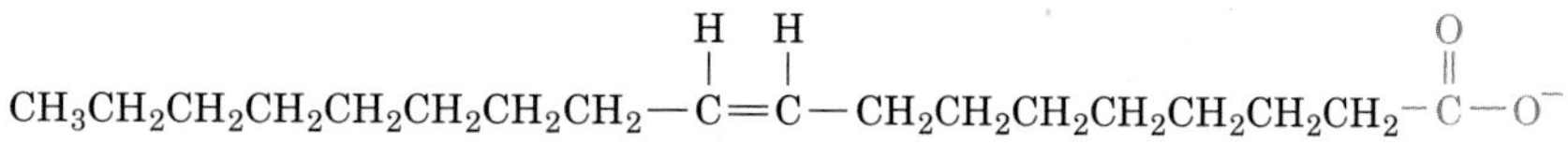

Oleate $(C_{17}H_{33}COO^-)$

Figure 2-6
Examples of fatty acid anions. They consist of a polar carboxylate group coupled to a long nonpolar hydrocarbon chain.

(a) Micelle   *(b)* Bilayer

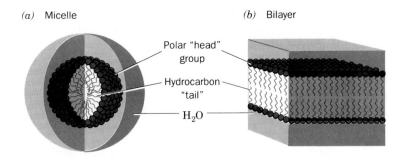

Polar "head" group

Hydrocarbon "tail"

H_2O

Figure 2-7
The associations of amphipathic molecules in aqueous solutions. The polar "head" groups are hydrated, whereas the nonpolar "tails" aggregate so as to exclude the aqueous solution: (*a*) A spheroidal aggregate of amphipathic molecules known as a micelle. (*b*) An extended planar aggregate of amphipathic molecules called a bilayer. The bilayer may form a closed spheroidal shell, known as a vesicle, that encloses a small amount of aqueous solution.

bonds and lack directionality. Nevertheless, hydrophobic interactions are of pivotal biological importance because, as we shall see in later chapters, they are largely responsible for the structural integrity of biological macromolecules, as well as that of supramolecular aggregates such as membranes. Note that hydrophobic interactions are peculiar to an aqueous environment. Other polar solvents do not promote such associations.

C. Proton Mobility

When an electrical current is passed through an ionic solution, the ions migrate towards the electrode of opposite polarity at a rate proportional to the electrical field and inversely proportional to the frictional drag experienced by the ion as it moves through the solution. This latter quantity, as Table 2-2 indicates, varies with the size of the ion. Note, however, that the ionic mobilities of both H_3O^+ and OH^- are anomalously large compared

to those of other ions. For H_3O^+ (the **hydronium ion,** which is abbreviated H^+; a bare proton has no stable existence in aqueous solution), this high migration rate results from the ability of protons to rapidly jump from one water molecule to another as is diagrammed in Fig. 2-8. Although a given hydronium ion can physically migrate through solution in the manner of say, a Na^+ ion, the rapidity of the proton-jump mechanism makes the H_3O^+ ion's effective ionic mobility much greater than it otherwise would be (the mean lifetime of a given H_3O^+ ion is 10^{-12} s at 25°C). The anomalously high ionic mobility of the OH^- ion is likewise accounted for by the proton-jump mechanism but, in this case, the apparent direction of ionic migration is opposite to the direction of proton jumping. Proton jumping is also responsible for the observation that *acid–base reactions are among the fastest reactions that take place in aqueous solutions* and it is probably of importance in biological proton-transfer reactions.

Table 2-2
Ionic Mobilities[a] in H_2O at 25°C

Ion	Mobility × 10^{-5} (cm²·V⁻¹s⁻¹)
H_3O^+	362.4
Li^+	40.1
Na^+	51.9
K^+	76.1
NH_4^+	76.0
Mg^{2+}	55.0
Ca^{2+}	61.6
OH^-	197.6
Cl^-	76.3
Br^-	78.3
CH_3COO^-	40.9
SO_4^{2-}	79.8

[a] Ionic mobility is the distance an ion moves in one second under the influence of an electric field of one volt per cm.

Source: Brey, W. S., *Physical Chemistry and Its Biological Applications,* p. 172, Academic Press (1978).

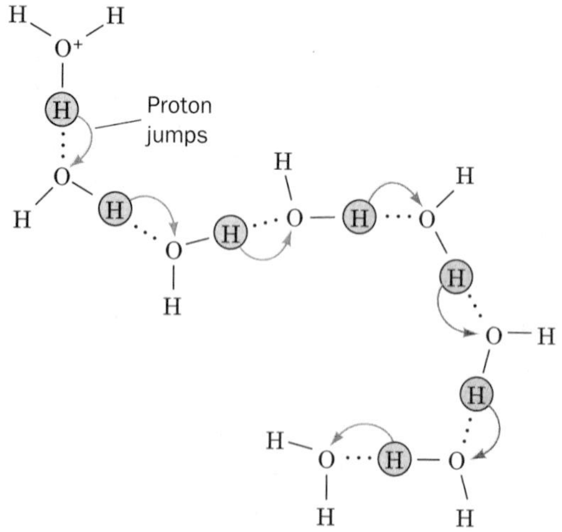

Figure 2-8
The mechanism of hydronium ion migration in aqueous solution via proton jumps. Proton jumps, which mostly occur at random, take place rapidly compared with direct molecular migration thereby accounting for the observed high ionic mobilities of hydronium and hydroxyl ions in aqueous solutions.

2. ACIDS, BASES, AND BUFFERS

Biological molecules, such as proteins and nucleic acids, bear numerous functional groups, such as carboxyl and amino groups, that can undergo acid–base reactions. Many properties of these molecules therefore vary with the acidities of the solutions in which they are immersed. In this section we discuss the nature of acid–base reactions and how acidities are controlled, both physiologically and in the laboratory.

A. Acid–Base Reactions

Acids and **bases,** in a definition coined in the 1880s by Svante Arrhenius, are, respectively, substances capable of donating protons and hydroxide ions. This definition is rather limited, because, for example, it does not account for the observation that NH_3, which lacks an OH^- group, exhibits basic properties. In a more general definition, which was formulated in 1923 by Johannes Brønsted and Thomas Lowry, *an acid is a substance that can donate protons (as in the Arrhenius definition) and a base is a substance that can accept protons.* Under this definition, in every acid–base reaction,

$$HA + H_2O \rightleftharpoons H_3O^+ + A^-$$

an acid (HA) reacts with a base (H_2O) to form the **conjugate base** of the acid (A^-) and the **conjugate acid** of the base (H_3O^+) (this reaction is usually abbreviated $HA \rightleftharpoons H^+ + A^-$ so that the participation of H_2O is implied). Accordingly, the acetate ion (CH_3COO^-) is the conjugate base of acetic acid (CH_3COOH) and the ammonium ion (NH_4^+) is the conjugate acid of ammonia (NH_3).

> In a yet more general definition of acids and bases, Gilbert Lewis described an acid as a substance that can accept an electron pair and a base as a substance that can donate an electron pair. This definition, which is applicable to both aqueous and nonaqueous systems, is unnecessarily broad for describing most biochemical phenomena.

The Strength of an Acid Is Specified by Its Dissociation Constant

The above acid dissociation reaction is characterized by its **equilibrium constant** which, for acid–base reactions, is known as a **dissociation constant,**

$$K = \frac{[H_3O^+][A^-]}{[HA][H_2O]} \qquad [2.2]$$

a quantity that is a measure of the relative proton affinities of the HA/A^- and H_3O^+/H_2O conjugate acid–base pairs. Here, as throughout the text, quantities in square brackets symbolize the molar concentrations of the indicated substances. Since in dilute aqueous solutions the water concentration is essentially constant with $[H_2O] = 1000 \ g \cdot L^{-1}/18.015 \ g \cdot mol^{-1} = 55.5M$, this term is customarily combined with the dissociation constant, which then takes the form

$$K_a = K[H_2O] = \frac{[H^+][A^-]}{[HA]} \qquad [2.3]$$

For brevity, however, we shall henceforth omit the subscript "*a*." The dissociation constants for acids useful in preparing biochemical solutions are listed in Table 2-3.

Acids may be classified according to their relative strengths, that is, according to their abilities to transfer a proton to water. Acids with dissociation constants smaller than that of H_3O^+ (which, by definition, is unity in aqueous solutions) are only partially ionized in aqueous solutions and are known are **weak acids** ($K < 1$). Conversely, **strong acids** have dissociation constants larger than that of H_3O^+ so that they are almost completely ionized in aqueous solutions ($K > 1$). The acids listed in Table 2-3 are all weak acids. However, many of the so-called "mineral acids," such as $HClO_4$, HNO_3, HCl, and H_2SO_4 (for the first ionization) are strong acids. Since strong acids rapidly transfer all their protons to H_2O, the strongest acid that can stably exist in aqueous solutions is H_3O^+. Likewise, there can be no stronger base in aqueous solutions than OH^-.

Water, being an acid, has a dissociation constant:

$$K = \frac{[H^+][OH^-]}{[H_2O]}$$

As above, the constant $[H_2O] = 55.5M$ can be incorporated into the dissociation constant to yield the expression for the ionization constant of water,

$$K_w = [H^+][OH^-] \qquad [2.4]$$

The value of K_w at 25°C is $10^{-14}M$. Pure water must contain equimolar amounts of H^+ and OH^- so that $[H^+] = [OH^-] = (K_w)^{1/2} = 10^{-7}M$. Since $[H^+]$ and $[OH^-]$ are reciprocally related by Eq. [2.4], if $[H^+]$ is greater than this value, $[OH^-]$ must be correspondingly less and *vice versa.* Solutions with $[H^+] = 10^{-7}M$ are said to be **neutral,** those with $[H^+] > 10^{-7}M$ are said to be **acidic,** and those with $[H^+] < 10^{-7}M$ are said to be **basic.** Most physiological solutions have hydrogen ion concentrations near neutrality. For example, human blood is normally slightly basic with $[H^+] = 4.0 \times 10^{-8}M$.

The values of $[H^+]$ for most solutions are inconveniently small and difficult to compare. A more practical quantity, which was devised in 1909 by Søren Sørensen, is known as the **pH:**

$$pH = -\log[H^+] \qquad [2.5]$$

The pH of pure water is 7.0, whereas acidic solutions have pH < 7.0 and basic solutions have pH > 7.0. For a

Table 2-3
**Dissociation Constants and pK's at 25°C of Some Acids in Common
Laboratory Use as Biochemical Buffers**

Acid	K	pK
Oxalic acid	5.37×10^{-2}	1.27 (pK_1)
H_3PO_4	7.08×10^{-3}	2.15 (pK_1)
Citric acid	7.41×10^{-4}	3.13 (pK_1)
Formic acid	1.78×10^{-4}	3.75
Succinic acid	6.17×10^{-5}	4.21 (pK_1)
Oxalate⁻	5.37×10^{-5}	4.27 (pK_2)
Acetic acid	1.74×10^{-5}	4.76
Citrate⁻	1.74×10^{-5}	4.76 (pK_2)
Succinate⁻	2.29×10^{-6}	5.64 (pK_2)
2-(N-Morpholino)-ethane sulfonic acid (MES)	7.94×10^{-7}	6.10
Cacodylic acid	5.37×10^{-7}	6.27
H_2CO_3	4.47×10^{-7}	6.35 (pK_1)
Citrate²⁻	3.98×10^{-7}	6.40 (pK_3)
N-(2-Acetamido)-iminodiacetic acid (ADA)	2.82×10^{-7}	6.55
Piperazine-N, N'-bis(2-ethanesulfonic acid) (PIPES)	1.78×10^{-7}	6.75
N-(2-Acetamido)-2-aminoethanesulfonic acid (ACES)	1.58×10^{-7}	6.80
3-(N-Morpholino)propanesulfonic acid (MOPS)	6.76×10^{-8}	7.17
$H_2PO_4^-$	6.31×10^{-8}	7.20 (pK_2)
N-2-Hydroxyethylpiperazine-N'-2-ethanesulfonic acid (HEPES)	3.31×10^{-8}	7.48
N-2-Hydroxyethylpiperazine-N'-3-propanesulfonic acid (HEPPS)	1.10×10^{-8}	7.96
Tris(hydroxymethyl)aminomethane (TRIS)	8.32×10^{-9}	8.08
N-[Tris(hydroxymethyl)methyl]glycine (Tricine)	8.91×10^{-9}	8.05
N, N-Bis(2-hydroxyethyl)glycine (Bicine)	5.50×10^{-9}	8.26
Glycylglycine	5.50×10^{-9}	8.26
Boric acid	5.75×10^{-10}	9.24
NH_4^+	5.62×10^{-10}	9.25
Glycine	1.66×10^{-10}	9.78
HCO_3^-	4.68×10^{-11}	10.33 (pK_2)
Piperidine	7.58×10^{-12}	11.12
HPO_4^{2-}	4.17×10^{-13}	12.38 (pK_3)

Source: Dawson, R. M. C., Elliott, D. C., Elliott, W. H., and Jones, K. M., *Data
for Biochemical Research* (2nd ed.), *pp.* 481–482, Oxford University Press (1969)
and Good, N. E., Winget, G. D., Winter, W., Connolly, T. N., Izawa, S., and
Singh, R. M. M., *Biochemistry* **5**, 467 (1966).

$1M$ solution of a strong acid, pH = 0 and for a $1M$ solution of a strong base, pH = 14. Note that if two solutions differ in pH by one unit, they differ in [H⁺] by a factor of 10. The pH of a solution may be accurately and easily determined through electrochemical measurements with a device known as a **pH meter**.

easily derived by rearranging Eq. [2.3]

$$[H^+] = K \left(\frac{[HA]}{[A^-]} \right)$$

and substituting it into Eq. [2.5]

$$pH = -\log K + \log \left(\frac{[A^-]}{[HA]} \right)$$

Defining $pK = -\log K$ in analogy with Eq. [2.5], we obtain the **Henderson–Hasselbalch equation**:

$$pH = pK + \log \left(\frac{[A^-]}{[HA]} \right) \qquad [2.6]$$

The pH of a Solution Is Determined by the Relative Concentrations of Acids and Bases

The relationship between the pH of a solution and the concentrations of an acid and its conjugate base can be

This equation indicates that the pK of an acid is numerically equal to the pH of the solution when the molar concentrations of the acid and its conjugate base are equal. Table 2-3 lists the pK values of several acids.

B. Buffers

A 0.01-mL droplet of 1M HCl added to 1 L of pure water will change the water's pH from 7 to 5, which represents a 100-fold increase in [H$^+$]. Yet, since the properties of biological substances vary significantly with small changes in pH, they require environments in which the pH is insensitive to additions of acids or bases. To understand how this is possible, let us consider the titration of a weak acid with a strong base.

Fig. 2-9 shows how the pH values of 1-L solutions of 1M acetic acid, H$_2$PO$_4^-$, and ammonium ion (NH$_4^+$), vary with the quantity of OH$^-$ added. Several details should be noted:

1. The curves have a similar shape but are shifted vertically along the pH axis.

2. The pH at the **equivalence point** of each titration (where the equivalents of OH$^-$ added equal the equivalents of HA initially present) is > 7 because of the reaction of A$^-$ with H$_2$O to form HA + OH$^-$; similarly, each initial pH is < 7.

3. The pH at the midpoint of each titration is numerically equal to the pK of its corresponding acid; here, according to the Henderson–Hasselbalch equation, [HA] = [A$^-$].

4. The slope of each titration curve is much less near its midpoint than it is near its wings. This indicates that *when [HA] ≈ [A$^-$], the pH of the solution is relatively insensitive to the addition of strong base or strong acid. Such a solution, which is known as an acid–base buffer, is resistant to pH changes because small amounts of added H$^+$ or OH$^-$, respectively, react with the A$^-$ or HA present without greatly changing the value of log([A$^-$]/[HA]).*

Titration curves such as those in Fig. 2-9, as well as distribution curves such as those in Fig. 2-10, may be calculated using the Henderson–Hasselbalch equation. Near the beginning of the titration, a significant fraction of the A$^-$ present arises from the dissociation of HA. Similarly, near the endpoint, much of the HA derives from the reaction of A$^-$ with H$_2$O. Throughout most of the titration, however, the OH$^-$ added reacts essentially completely with the HA to form A$^-$ so that

$$[A^-] = \frac{x}{V} \qquad [2.7]$$

where x represents the equivalents of OH$^-$ added and V is the volume of the solution. Then, letting c_0 represent

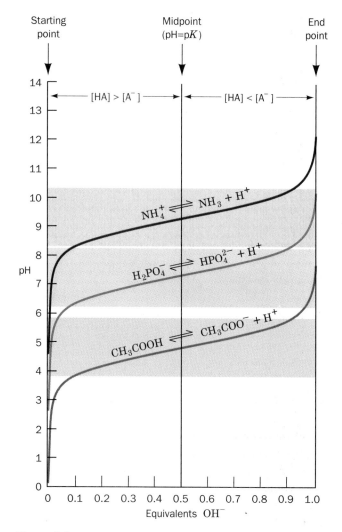

Figure 2-9
The acid–base titration curves of 1-L solutions of 1M acetic acid, H$_2$PO$_4^-$, and NH$_4^+$ by strong base. At the starting point of each titration, the acid form of the conjugate acid–base pair overwhelmingly predominates. At the midpoint of the titration, where pH = pK, the concentration of the acid is equal to that of its conjugate base. Finally, at the endpoint of the titration, where the equivalents of strong base added equal the equivalents of acid at the starting point, the conjugate base is in great excess. The shaded bands indicate the pH ranges over which the corresponding solution can function effectively as a buffer.

the equivalents of HA initially present,

$$[HA] = \frac{c_0 - x}{V} \qquad [2.8]$$

Incorporating these relationships into Eq. [2.6] yields

$$pH = pK + \log\left(\frac{x}{c_0 - x}\right) \qquad [2.9]$$

which accurately describes a titration curve except near its wings (these regions require more exact treatments that take into account the ionizations of water).

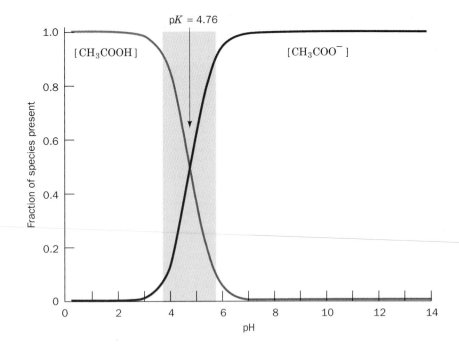

Figure 2-10
The distribution curves for acetic acid and acetate ion. The fraction of species present is given as the ratio of the concentration of CH_3COOH or CH_3COO^- to the total concentrations of these two species. The customarily accepted useful buffer range of p$K \pm 1$ is indicated by the shaded region.

Buffers Stabilize a Solution's pH

The ability of a buffer to resist pH changes with added acid or base is directly proportional to the concentration of the conjugate acid–base pair, $[HA] + [A^-]$. It is maximal when pH = pK and decreases rapidly with a change in pH from that point. A good rule of thumb is that a weak acid is in its useful buffer range within 1 pH unit of its pK (the shaded regions of Figs. 2-9 and 2-10). Above this range, where the ratio $[A^-]/[HA] > 10$, the pH of the solution changes rapidly with added strong base. A buffer is similarly impotent with addition of strong acid when its pK exceeds the pH by more than a unit.

Biological fluids, both those found intracellularly and extracellularly, are heavily buffered. For example, the pH of the blood in healthy individuals is closely controlled at pH 7.4. The phosphate and carbonate ions that are components of most biological fluids are important in this respect. Moreover, many biological molecules, such as proteins, nucleic acids, and lipids, as well as numerous small organic molecules, bear multiple acid–base groups that are effective as buffer components in the physiological pH range.

The concept that the properties of biological molecules vary with the acidity of the solution in which they are dissolved was not fully appreciated before the beginning of the twentieth century so that the acidities of biochemical preparations made before that time were rarely controlled. Consequently these early biochemical experiments yielded poorly reproducible results. More recently, biochemical preparations have been routinely buffered to simulate the properties of naturally occurring biological fluids. Many of the weak acids listed in Table 2-3 are commonly used as buffers in biochemical

preparations. In practice, the chosen weak acid and one of its soluble salts are dissolved in the (nearly equal) mole ratio necessary to provide the desired pH and, with the aid of a pH meter, the resulting solution is "fine tuned" by titration with strong acid or base.

C. Polyprotic Acids

Substances that bear more than one acid–base group, such as H_3PO_4 or H_2CO_3, as well as most biomolecules, are known as **polyprotic acids**. The titration curves of such substances, as is illustrated in Fig. 2-11 for H_3PO_4, are characterized by multiple pK's, one for each ionization step. Exact calculations of the concentrations of the various ionic species present at a given pH is clearly a more complex task than for a **monoprotic acid**.

The pK's of two closely associated acid–base groups are not independent. The ionic charge resulting from a proton dissociation electrostatically inhibits further proton dissociation from the same molecule, thereby increasing the values of the corresponding pK's. This effect, according to Coulomb's law, decreases as the distance between the ionizing groups increases. For example, the pK's of **oxalic acid's** two adjacent carboxyl groups differ by 3 pH units (Table 2-3), whereas those of **succinic acid,** in which the carboxyl groups are separated by two methylene groups, differ by 1.4 units.

$$H-O-\overset{\overset{\displaystyle O}{\|}}{C}-\overset{\overset{\displaystyle O}{\|}}{C}-O-H \qquad H-O-\overset{\overset{\displaystyle O}{\|}}{C}-CH_2CH_2-\overset{\overset{\displaystyle O}{\|}}{C}-O-H$$

Oxalic acid   **Succinic acid**

Likewise, successive ionizations from the same center,

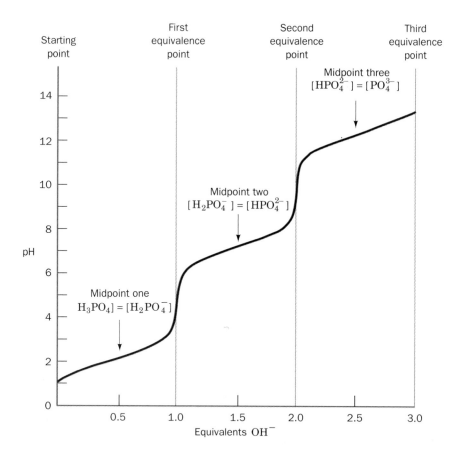

Figure 2-11
The titration curve of H_3PO_4. The two intermediate equivalence points occur at the steepest parts of the curve. Note the flatness of the curve near its starting point and endpoint in comparison with the curved ends of the titration curves in Fig. 2-9. This indicates that H_3PO_4 ($pK_1 = 2.15$) is verging on being a strong acid and PO_4^{3-} ($pK_3 = 12.38$) is verging on being a strong base.

such as in H_3PO_4 or H_2CO_3, have pK's that differ by 4 to 5 pH units. If the pK's for successive ionizations of a polyprotic acid differ by at least 3 pH units, it can be accurately assumed that, at a given pH, only the members of the conjugate acid–base pair characterized by the nearest pK are present in significant concentrations. This, of course, greatly simplifies the calculations for determining the concentrations of the various ionic species present.

Polyprotic Acids with Closely Spaced pK's Have Molecular Ionization Constants

If the pK's of a polyprotic acid differ by less than ~ 2 pH units, as is true in perhaps the majority of biomolecules, the ionization constants measured by titration are not true group ionization constants but rather, reflect the average ionization of the groups involved. The resulting ionization constants are therefore known as **molecular ionization constants.**

Consider the acid–base equilibria shown in Fig. 2-12 in which there are two nonequivalent protonation sites. Here, the quantities K_A, K_B, K_C, and K_D, the ionization constants for each group, are alternatively called **microscopic ionization constants.** The molecular ionization constant for the removal of the first proton from HAH is

$$K_1 = \frac{[H^+]([AH^-] + [HA^-])}{[HAH]} = K_A + K_B \quad [2.10]$$

Similarly, the molecular ionization constant K_2 for the removal of the second proton is

$$K_2 = \frac{[H^+][A^{2-}]}{[AH^-] + [HA^-]} = \frac{1}{(1/K_C) + (1/K_D)}$$
$$= \frac{K_C K_D}{(K_C + K_D)}$$

If $K_A \gg K_B$, then $K_1 \approx K_A$; that is, the first molecular ionization constant is equal to the microscopic ionization constant of the more acidic group. Likewise, if $K_D \gg K_C$, then $K_2 \approx K_C$, so that the second molecular ionization constant is the microscopic ionization constant of the less acidic group. If the ionization steps differ sufficiently in their pK's the molecular ionization constants, as expected, become identical to the microscopic ionization constants.

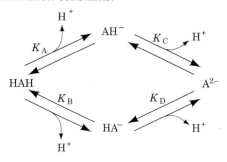

Figure 2-12
The ionization of an acid that has two nonequivalent protonation sites.

Chapter Summary

Water is an extraordinary substance, the properties of which are of great biological importance. A water molecule can simultaneously participate in as many as four hydrogen bonds: two as a donor and two as an acceptor. These hydrogen bonds are responsible for the open, low density structure of ice. Much of this hydrogen bonded structure exists in the liquid phase as is evidenced by the high boiling point of water compared to those of other liquids. Physical and theoretical evidence indicates that liquid water maintains a rapidly fluctuating, hydrogen bonded molecular structure that, over short ranges, resembles that of ice. The unique solvent properties of water derive from its polarity as well as its hydrogen bonding properties. In aqueous solutions, ionic and polar substances are surrounded by multiple concentric hydration shells of oriented water dipoles that act to attenuate the electrostatic interactions between the charges in the solution. The thermal randomization of the oriented water molecules is resisted by their hydrogen bonding associations, thereby accounting for the high dielectric constant of water. Nonpolar substances are essentially insoluble in water. However, amphipathic substances aggregate in aqueous solutions to form micelles and bilayers due to the combination of hydrophobic interactions among the nonpolar portions of these molecules and the hydrophilic interactions of their polar groups with the aqueous solvent. The H_3O^+ and OH^- ions have anomalously large ionic mobilities in aqueous solutions because the migration of these ions through solution occurs largely via proton jumping from one H_2O molecule to another.

A Brønsted acid is a substance that can donate protons, whereas a Brønsted base can accept protons. Upon losing a proton, a Brønsted acid becomes its conjugate base. In an acid–base reaction, an acid donates its proton to a base. Water can react as an acid to form hydroxide ion, OH^-, or as a base to form hydronium ion, H_3O^+. The strength of an acid is indicated by the magnitude of its dissociation constant, K. Weak acids, which have a dissociation constant less than that of H_3O^+, are only partially dissociated in aqueous solution. Water has the dissociation constant $10^{-14} M$ at 25°C. A practical quantity for expressing the acidity of a solution is the $pH = -\log [H^+]$. The relationship between pH, pK, and the concentrations of the members of its conjugate acid–base pair is expressed by the Henderson–Hasselbalch equation. An acid–base buffer is a mixture of a weak acid with its conjugate base in a solution that has a pH near the pK of the acid. The ratio $[A^-]/[HA]$ in a buffer is not very sensitive to the addition of strong acids or bases so that the pH of a buffer is not greatly affected by these substances. Buffers are operationally effective only in the pH range of $pK \pm 1$. Outside of this range, the pH of the solution changes rapidly with the addition of strong acid or base. Buffer capacity also depends on the total concentration of the conjugate acid–base pair. Biological fluids are generally buffered near neutrality. Many acids are polyprotic. However, unless the pK's of their various ionizations differ by less than 2 or 3 pH units, pH calculations can effectively treat them as if they were a mixture of separate weak acids. For polyprotic acids with pK's that differ by less than this amount, the observed molecular ionization constants are simply related to the microsopic ionization constants of the individual dissociating groups.

References

Edsall, J. T. and Wyman, J., *Biophysical Chemistry*, Vol. 1, Chapters 2, 8, and 9, Academic Press (1958). [Contains detailed treatments on the structure of water and on acid–base equilibria.]

Eisenberg, D. and Kauzman, W., *The Structure and Properties of Water*, Oxford University Press (1969). [A comprehensive monograph with a wealth of information.]

Cooke, R. and Kuntz, I. D., The properties of water in biological systems, *Ann. Rev. Biophys. Bioeng.* **3**, 95–126 (1974).

Montgomery, R. and Swenson, C. A., *Quantitative Problems in Biochemical Sciences* (2nd ed.), Chapters 7 and 8, Freeman (1978). [Contains additional problems.]

Segal, B. G., *Chemistry* (2nd ed.), Chapters 7, 9, and 10, Wiley (1989). [Discusses the properties of water and acid–base chemistry. Most other general chemistry textbooks have similar discussions.]

Segel, I. H., *Biochemical Calculations* (2nd ed.), Chapter 1, Wiley (1976). [An intermediate level discussion of acid–base equilibria with worked out problems.]

Stillinger, F. H., Water revisited, *Science* **209**, 451–457 (1980). [An outline of water structure on an elementary level.]

Tanford, C., *The Hydrophobic Effect: Formation of Micelles and Biological Membranes* (2nd ed.), Chapters 5 and 6, Wiley–Interscience (1980). [Discussion of the structure of water and of micelles.]

Problems

1. Draw the hydrogen bonding pattern that water forms with acetamide (CH_3CONH_2) and pyridine (benzene with a CH group replaced by N).

2. Explain why the dielectric constants of the following pairs of liquids have the order given in Table 2-1. (a) carbon tetrachloride and chloroform; (b) ethanol and methanol; and (c) acetone and formamide.

3. "Inverted" micelles are made by dispersing amphipathic molecules in a nonpolar solvent, such as benzene, together with a small amount of water (counterions are also provided if the head groups are ionic). Draw the structure of an inverted micelle and describe the forces that stabilize it.

*4. Amphipathic molecules in aqueous solutions tend to concentrate at surfaces such as liquid–solid or liquid–gas interfaces. They are therefore said to be **surface-active molecules** or **surfactants**. Rationalize this behavior in terms of the properties of the amphiphiles and indicate the effect that surface-active molecules have on the surface tension of water (surface tension is a measure of the internal cohesion of a liquid as manifested by the force necessary to increase its surface area). Explain why surfactants such as soaps and detergents are effective in dispersing oily substances and oily dirt in aqueous solutions.

5. Indicate how hydrogen bonding forces and hydrophobic forces vary with the dielectric constant of the medium.

6. Using the data in Table 2-2, indicate the times it would take a K^+ and an H^+ ion to each move 1 cm in an electric field of $100 \ V \cdot cm^{-1}$.

7. Explain why the mobility of H^+ in ice is only about an order of magnitude less than that in liquid water, whereas the mobility of Na^+ in solid NaCl is zero.

8. Calculate the pH of: (a) $0.1M$ HCl; (b) $0.1M$ NaOH; (c) $3 \times 10^{-5}M$ HNO_3; (d) $5 \times 10^{-10}M$ $HClO_4$; and (e) $2 \times 10^{-8}M$ KOH.

9. The volume of a typical bacterial cell is on the order of $1.0 \ \mu m^3$. At pH 7, how many hydrogen ions are contained inside a bacterial cell? A bacterial cell contains thousands of macromolecules, such as proteins and nucleic acids, that each bear multiple ionizable groups. What does your result indicate about the common notion that ionizable groups are continuously bathed with H^+ and OH^- ions?

10. Using the data in Table 2-3, calculate the concentrations of all molecular and ionic species and the pH in aqueous solutions that have the following formal compositions: (a) $0.01M$ acetic acid; (b) $0.25M$ ammonium chloride; (c) $0.05M$ acetic acid $+ 0.10M$ sodium acetate; and (d) $0.20M$ boric acid [$B(OH)_3$] $+ 0.05M$ sodium borate [Na $B(OH)_4$].

11. **Acid–base indicators** are weak acids that change color upon changing ionization states. When a small amount of an appropriately chosen indicator is added to a solution of an acid or base being titrated, the color change "indicates" the **endpoint** of the titration. **Phenolphthalein** is a commonly used acid–base indicator that, in aqueous solutions, changes from colorless to red-violet in the pH range between 8.2 and 10.0. Referring to Figs. 2-9 and 2-11, indicate the effectiveness of phenolphthalein for accurately detecting the endpoint of a titration with strong base of: (a) Acetic acid; (b) NH_4Cl; and (c) H_3PO_4 (at each of its three equivalence points).

*12. The formal composition of an aqueous solution is $0.12M$ $K_2HPO_4 + 0.08M$ KH_2PO_4. Using the data in Table 2-3, calculate the concentrations of all ionic and molecular species in the solution and the pH of the solution.

13. Distilled water in equilibrium with air contains dissolved carbon dioxide at a concentration of $1.0 \times 10^{-5}M$. Using the data in Table 2-3, calculate the pH of such a solution.

14. Calculate the formal concentrations of acetic acid and sodium acetate necessary to prepare a buffer solution of pH 5 that is $0.20M$ in total acetate. The pK of acetic acid is given in Table 2-3.

15. An enzymatic reaction takes place in a 10-mL solution that has a total citrate concentration of 120 mM and an initial pH of 7.00. During the reaction, 0.2 milliequivalents of acid are produced. Using the data in Table 2-3, calculate the final pH of the solution. What would the final pH of the solution be in the absence of the citrate buffer assuming that the other components of the solution have no significant buffering capacity and that the solution is initially at pH 7?

*16. A solution's **buffer capacity**, β, is defined as the ratio of an incremental amount of base added, in equivalents, to the corresponding pH change. This is the reciprocal of the slope of the titration curve, Eq. [2.9]. Derive the equation for β and show that it is maximal at pH $=$ pK.

17. Using the data in Table 2-3, calculate the microscopic ionization constants for oxalic acid and for succinic acid. How do these values compare with their corresponding molecular ionization constants?

Chapter 3
THERMODYNAMIC PRINCIPLES: A REVIEW

You can't win.
First law of thermodynamics

You can't even break even.
Second law of thermodynamics

You can't stay out of the game.
Third law of thermodynamics

Living things require a continuous throughput of energy. Thus through photosynthesis, plants convert radiant energy from the sun, the primary energy source for life on earth, to the chemical energy of carbohydrates and other organic substances. The plants, or the animals that eat them, then metabolize these substances to power such functions as the synthesis of biomolecules, the maintenance of concentration gradients, and the movement of muscles. These processes ultimately transform the energy to heat, which is dissipated to the environment. A considerable portion of the cellular biochemical apparatus must therefore be devoted to the acquisition and utilization of energy.

Thermodynamics (Greek: *therme*, heat + *dynamis*, power) is a marvelously elegant description of the relationships among the various forms of energy and how energy affects matter on the macroscopic as opposed to the molecular level; that is, it deals with amounts of matter large enough for their average properties, such as temperature and pressure, to be well defined. Indeed, the basic principles of thermodynamics were developed in the nineteenth century before the atomic theory of matter had been generally accepted.

With a knowledge of thermodynamics we can determine whether a physical process is possible. Thermodynamics is therefore essential for understanding why macromolecules fold to their native conformations, how metabolic pathways are designed, why molecules cross biological membranes, how muscles generate mechanical force, and so on. The list is endless. Yet, the reader should be cautioned that thermodynamics does not indicate the rates at which possible processes actually occur. For instance, although thermodynamics tells us that glucose and oxygen react with the release of copious amounts of energy, it does not indicate that this mixture is indefinitely stable at room temperature in the absence of the appropriate enzymes. The prediction of reaction rates requires, as we shall see in Section 13-1C, a mechanistic description of molecular processes. Yet, thermodynamics is also an indispensible guide in formulating such mechanistic models because such models must conform to thermodynamic principles.

Thermodynamics, as it applies to biochemistry, is most frequently concerned with describing the conditions under which processes occur *spontaneously* (by themselves). We shall consequently review the elements of thermodynamics that enable us to predict chemical and biochemical spontaneity: the first and second laws of thermodynamics, the concept of free energy, and the nature of processes at equilibrium. Familiarity with these principles is indispensible for understanding many of the succeeding discussions in this text. We shall, however, postpone consideration of the thermodynamic aspects of metabolism until Sections 15-4 through 15-6.

1. FIRST LAW OF THERMODYNAMICS: ENERGY IS CONSERVED

In thermodynamics, a **system** is defined as that part of the universe that is of interest, such as a reaction vessel or an organism; the rest of the universe is known as the **surroundings**. A system is said to be **open** or **closed** according to whether or not it can exchange matter and energy with its surroundings. Living organisms, which take up nutrients, release waste products, and generate work and heat, are examples of open systems; if an organism were sealed inside a perfectly insulated box, it would, together with the box, constitute a closed system.

A. Energy

The **first law of thermodynamics** is a mathematical statement of the law of conservation of energy: *Energy can neither be created nor destroyed.*

$$\Delta U = U_{final} - U_{initial} = q - w \qquad [3.1]$$

Here U is energy, q represents the **heat** absorbed *by* the system *from* the surroundings, and w is the **work** done *by* the system *on* the surroundings. Heat is a reflection of random molecular motion whereas work, which is defined as force times the distance moved under its influence, is associated with organized motion. Force may assume many different forms including the gravitational force exerted by one mass on another, the expansional force exerted by a gas, the tensional force exerted by a spring or muscle fiber, the electrical force of one charge on another or the dissipative forces of friction and viscosity. Processes in which the system releases heat, which by convention are assigned a negative q, are known as **exothermic processes** (Greek: *exo*, out of); those in which the system gains heat (positive q) are known as **endothermic processes** (Greek: *endon*, within). Under this convention, work done by the system against an external force is defined as a positive quantity.

The SI unit of energy, the **joule (J),** is steadily replacing the **calorie (cal)** in modern scientific usage. The **large calorie (Cal,** with a capital C) is a unit favored by nutritionists. The relationships among these quantities and other units, as well as the values of constants that will be useful throughout this chapter, are collected in Table 3-1.

Table 3-1
Thermodynamic Units and Constants

Joule (J)

   $1\ J = 1\ kg \cdot m^2 \cdot s^{-2}$ $1\ J = 1\ C \cdot V$ (coulomb volt)

   $1\ J = 1\ N \cdot m$ (newton meter)

Calorie (cal)

   1 cal heats 1 g of H_2O from 14.5 to 15.5°C

   $1\ cal = 4.184\ J$

Large calorie (Cal)

   $1\ Cal = 1\ kcal$ $1\ Cal = 4184\ J$

Avogadro's number (N)

   $N = 6.0221 \times 10^{23}$ molecules $\cdot mol^{-1}$

Coulomb (C)

   $1\ C = 6.241 \times 10^{18}$ electron charges

Faraday ($\mathscr{F}$)

   $1\ \mathscr{F} = N$ electron charges

   $1\ \mathscr{F} = 96,494\ C \cdot mol^{-1} = 96,494\ J \cdot V^{-1}\ mol^{-1}$

Kelvin temperature scale (K)

   $0\ K =$ absolute zero $273.15\ K = 0°C$

Boltzmann constant (k_B)

   $k_B = 1.3807 \times 10^{-23}\ J \cdot K^{-1}$

Gas constant (R)

   $R = Nk_B$ $R = 1.9872\ cal \cdot K^{-1}\ mol^{-1}$

   $R = 8.3145\ J \cdot K^{-1}\ mol^{-1}$ $R = 0.08206\ L \cdot atm \cdot K^{-1}\ mol^{-1}$

State Functions Are Independent of the Path a System Follows

Experiments have invariably demonstrated that the energy of a system depends only on its current properties or state, not on how it reached that state. For example, the state of a system composed of a particular gas sample is completely described by its pressure and temperature. The energy of this gas sample is a function only of these so-called **state functions** (quantities that depend only on the state of the system) and is therefore a state function itself. Consequently, there is no net change in energy ($\Delta U = 0$) for any process in which the system returns to its initial state (a **cyclic process**).

Neither heat nor work are separately state functions because they are each dependent on the **path** followed by a system in changing from one state to another. For example, in the process of changing from an initial to a final state, a gas may do work by expanding against an external force, or do no work by following a path in which it encounters no external resistance. If Eq. [3.1] is to be obeyed, heat must also be path dependent. It is therefore meaningless to refer to the heat or work content of a system. To indicate this property, the heat or work produced during a change of state is never referred to as Δq or Δw but rather as just q or w.

B. Enthalpy

Any combination of only state functions must also be a state function. One such combination, which is known as the **enthalpy** (Greek: *enthalpein*, to warm in), is defined

$$H = U + PV \qquad [3.2]$$

where V is the volume of the system and P is its pressure. Enthalpy is a particularly convenient quantity with which to describe biological systems because *under constant pressure, a condition typical of most biochemical processes, the enthalpy change between the initial and final states of a process, ΔH, is the easily measured heat that it generates or absorbs.* To show this, let us divide work into two categories: pressure–volume (P–V) work, which is work performed by expansion against an external pressure ($P\Delta V$), and all other work (w'):

$$w = P\Delta V + w' \qquad [3.3]$$

Then, by combining Eqs. [3.2] and [3.3], we see that

$$\Delta H = \Delta U + P\Delta V = q_P - w + P\Delta V = q_P - w' \qquad [3.4]$$

where q_P is the heat transferred at constant pressure. Thus if $w' = 0$, as is often true of chemical reactions, $\Delta H = q_P$. Moreover, the volume changes in most biochemical processes are negligible so that the differences between their ΔU and ΔH values are usually insignificant.

We are now in the position to understand the utility of state functions. For instance, suppose we wished to determine the enthalpy change resulting from the complete oxidation of 1 g of glucose to CO_2 and H_2O by muscle tissue. To directly make such a measurement would present enormous experimental difficulties. For one thing, the enthalpy changes resulting from the numerous metabolic reactions not involving glucose oxidation that normally occur in living muscle tissue would greatly interfere with our enthalpy measurement. Since enthalpy is a state function, however, we can measure glucose's enthalpy of combustion in any apparatus of our choosing, say a constant pressure calorimeter rather than a muscle, and still obtain the same value. This, of course, is true whether or not we know the mechanism through which muscle converts glucose to CO_2 and H_2O, so long as we can establish that these substances actually are the final metabolic products. *In general, the change of enthalpy in any hypothetical reaction pathway can be determined from the enthalpy change in any other reaction pathway between the same reactants and products.*

We stated earlier in the chapter that thermodynamics serves to indicate whether a particular process occurs spontaneously. Yet, the first law of thermodynamics cannot, by itself, provide the basis for such an indication as the following example demonstrates. If two objects at different temperatures are brought into contact, we know that heat spontaneously flows from the hotter object to the colder one, never *vice versa*. Yet, either process is consistent with the first law of thermodynamics since the aggregate energy of the two objects is independent of their temperature distribution. Consequently we must seek a criterion of spontaneity other than only conformity to the first law of thermodynamics.

2. SECOND LAW OF THERMODYNAMICS: THE UNIVERSE TENDS TOWARDS MAXIMUM DISORDER

When a swimmer falls into the water (a spontaneous process), the energy of the coherent motion of his body is converted to that of the chaotic thermal motion of the surrounding water molecules. The reverse process, the swimmer being ejected from still water by the sudden coherent motion of the surrounding water molecules, has never been witnessed even though such a phenomenon violates neither the first law of thermodynamics nor Newton's laws of motion. This is because *spontaneous processes are characterized by the conversion of order (in this case the coherent motion of the swimmers body) to chaos (here the random thermal motion of the water molecules).* The **second law of thermodynamics,** which expresses this phenomenon, therefore provides a criterion for determining whether a process is spontaneous. Note that thermodynamics says nothing about the rate of a process; that is the purview of **chemical kinetics** (Chapter 13). Thus a spontaneous process might only proceed at an infinitesimal rate.

A. Spontaneity and Disorder

The second law of thermodynamics states, in accordance with all experience, that *spontaneous processes occur in directions that increase the overall disorder of the universe,* that is, of the system and its surroundings. Disorder, in this context, is defined as the number of equivalent ways, W, of arranging the components of the universe. To illustrate this point, let us consider an isolated system consisting of two bulbs of equal volume containing a total of N identical molecules of ideal gas (Fig. 3-1). When the stopcock connecting the bulbs is open, there is an equal probability that a given molecule will occupy either bulb so that there are a total of 2^N equally probable ways that the N molecules may be distributed among the two bulbs. Since the gas molecules are indistinguishable from one another, there are only $(N + 1)$ different states of the system: those with 0, 1, 2, . . . , $(N − 1)$, or N molecules in the left bulb. Probability theory indicates that the number of (indistinguishable) ways, W_L, of placing L of the N molecules in the left bulb is

$$W_L = \frac{N!}{L!(N - L)!}$$

The probability of such a state occurring is its fraction of the total number of possible states: $W_L/2^N$.

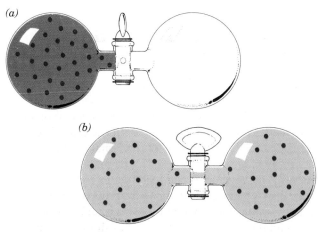

Figure 3-1
Two bulbs of equal volumes connected by a stopcock. In (a), a gas occupies the left bulb, the right bulb is evacuated, and the stopcock is closed. When the stopcock is opened (b), the gas molecules diffuse back and forth between the bulbs and eventually become distributed such that one half of them occupy each bulb.

For any value of N, the state that is most probable, that is, the one with the highest value of W_L, is the one with one half of the molecules in one bulb ($L = N/2$ for N even). As N becomes large, the probability that L is nearly equal to $N/2$ approaches unity: For instance, when $N = 10$ the probability that L is within 20% of $N/2$ (that is, 4, 5, or 6) is 0.66, whereas for $N = 50$ this probability (that L is in the range 20–30) is 0.88. For a chemically significant number of molecules, say $N = 10^{23}$, the probability that the number of molecules in the left bulb differs from those in the right by as insignificant a ratio as 1 molecule in every 10 billion is 10^{-434}, which, for all intents and purposes, is zero. Therefore, the reason the number of molecules in each bulb of the system in Fig. 3-1b is always observed to be equal is not because of any law of motion; the energy of the system is the same for any arrangement of the molecules. *It is because the aggregate probability of all other states is so utterly insignificant (Fig. 3-2).* By the same token, the reason that our swimmer is never thrown out of the water or even noticeably disturbed by the chance coherent motion of the surrounding water molecules is that the probability of such an event is nil.

B. Entropy

In chemical systems, W, the number of equivalent ways of arranging a system in a particular state, is usually inconveniently immense. For example, when the above twin bulb system contains N gas molecules, $W_{N/2} \approx 10^{N \ln 2}$ so that for $N = 10^{23}$, $W_{5 \times 10^{22}} \approx 10^{7 \times 10^{22}}$. In order to be able to deal with W more easily, we define, as did Ludwig Boltzmann in 1877, a quantity known as **entropy** (Greek: *en,* in + *trope,* turning):

$$S = k_B \ln W \qquad [3.5]$$

that increases with W but in a more manageable way. Here k_B is the **Boltzmann constant** (Table 3-1). Thus for our twin-bulb system, $S = k_B N \ln 2$ so that the entropy of the system in its most probable state is proportional to the number of gas molecules it contains. Note that *entropy is a state function because it depends only on the parameters that describe a state.*

The laws of random chance cause any system of reasonable size to spontaneously adopt its most probable arrangement, the one in which entropy is a maximum, simply because this state is so overwhelmingly probable. For example, assume that all N molecules of our twin-bulb system are initially placed in the left bulb (Fig. 3-1a; $W_N = 1$ and $S = 0$ since there is only one way of doing this). After the stopcock is opened, the molecules will randomly diffuse in and out of the right bulb until eventually they achieve their most probable (maximum entropy) state, that with one half of the molecules in each bulb. The gas molecules will subsequently continue to diffuse back and forth between the bulbs but there will be no further macroscopic (net) change in the system.

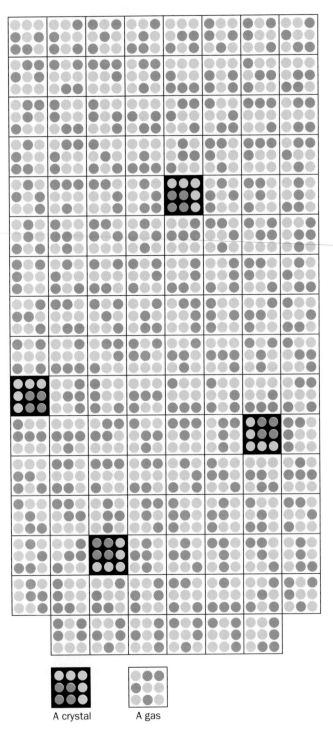

A crystal   A gas

Figure 3-2
The improbability of even a small amount of order is illustrated by a simple "universe" consisting of a square array of 9 positions that collectively contain 4 identical "molecules" *(red dots)*. If the 4 molecules are arranged in a square, we shall call the arrangement a "crystal"; otherwise we shall call it a "gas." The total number of distinguishable arrangements of our 4 molecules in 9 positions is given by:

$$W = \frac{9 \cdot 8 \cdot 7 \cdot 6}{4 \cdot 3 \cdot 2 \cdot 1} = 126$$

Here, the numerator indicates that the first molecule may occupy any of the universe's 9 positions, the second molecule may occupy any of the 8 remaining unoccupied positions, etc.,whereas the denominator corrects for the number of indistinguishable arrangements of the 4 identical molecules. Of the 126 arrangements this universe can have, only 4 are crystals *(black squares)*. Thus, even in this simple universe, there is a more than 30-fold greater probability that it will contain a disordered gas, when arranged at random, than an ordered crystal. [Figure copyrighted © by Irving Geis.]

different forms but can neither be created nor destroyed), *any spontaneous process must cause the entropy of the universe to increase:*

$$\Delta S_{system} + \Delta S_{surroundings} = \Delta S_{universe} > 0 \qquad [3.6]$$

Equation [3.6] is the usual expression for the second law of thermodynamics. It is a statement of the general tendency of all spontaneous processes to disorder the universe; that is, *the entropy of the universe tends towards a maximum.*

The conclusions based on our twin-bulb apparatus may be applied to explain, for instance, why blood transports O_2 and CO_2 between the lungs and the tissues. Solutes in solution behave analogously to gases in that they tend to maintain a uniform concentration throughout their occupied volume because this is their most probable arrangement. In the lungs, where the concentration of O_2 is higher than that in the venous blood passing through them, more O_2 enters the blood than leaves it. On the other hand, in the tissues, where the O_2 concentration is lower than that in the arterial blood, there is net diffusion of O_2 from the blood to the tissues. The reverse situation holds for CO_2 transport since the CO_2 concentration is low in the lungs but high in the tissues. Keep in mind, however, that thermodynamics says nothing about the rates that O_2 and CO_2 are transported to and from the tissues. The rates of these processes depend on the physicochemical properties of the blood, the lungs, and the cardiovascular system.

The system is therefore said to have reached **equilibrium.**

According to Eq. [3.5], the foregoing spontaneous expansion process causes the system's entropy to increase. In general, *for any constant energy process* ($\Delta U = 0$), *a spontaneous process is characterized by* $\Delta S > 0$. Since the energy of the universe is constant (energy can assume

Equation [3.6] does not imply that a particular system cannot increase its degree of order. As is explained in Section 3-3, however, *a system can only be ordered at the expense of disordering its surroundings to an even greater extent by the application of energy to the system.* For example, living organisms, which are organized from the molecular level upwards and are therefore particularly well ordered, achieve this order at the expense of disordering the nutrients they consume. Thus, *eating is as much a way of acquiring order as it is of gaining energy.*

A state of a system may constitute a distribution of more complicated quantities than those of gas molecules in a bulb or simple solute molecules in a solvent. For example, if our system consists of a protein molecule in aqueous solution, its various states differ, as we shall see, in the conformations of the protein's amino acid residues and in the distributions and orientations of its associated water molecules. The second law of thermodynamics applies here because a protein molecule in aqueous solution assumes its native conformation largely in response to the tendency of its surrounding water structure to be maximally disordered (Section 7-4C).

C. Measurement of Entropy

In chemical and biological systems, it is impractical, if not impossible, to determine the entropy of a system by counting the number of ways, W, it can assume its most probable state. An equivalent and more practical definition of entropy was proposed in 1864 by Rudolf Clausius: For spontaneous processes

$$\Delta S \geq \int_{initial}^{final} \frac{dq}{T} \qquad [3.7]$$

where T is the absolute temperature at which the change in heat occurs. The proof of the equivalence of our two definitions of entropy, which requires an elementary knowledge of statistical mechanics, can be found in many physical chemistry texts. It is evident, however, that any system becomes progressively disordered (its entropy increases) as its temperature rises (e.g., Fig. 3-3). The equality in Eq. [3.7] holds only for processes in which the system remains in equilibrium throughout the change; these are known as **reversible processes.**

For the constant temperature conditions typical of biological processes, Eq. [3.7] reduces to

$$\Delta S \geq \frac{q}{T} \qquad [3.8]$$

Thus the entropy change of a reversible process at constant temperature can be straightforwardly determined from measurements of the heat transferred and the temperature at which this occurs. However, since a process at equilibrium can only change at an infinitesimal rate (equilibrium processes are, by definition, unchanging), real processes can approach, but can never quite attain, reversibility. Consequently, *the universe's entropy change in any real process is always greater than its ideal (reversible) value.* This means that when a system departs from and then returns to its initial state via a real process, the entropy of the universe must increase even though the entropy of the system (a state function) does not change.

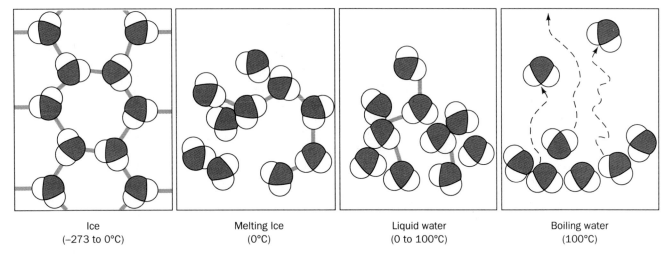

| Ice (−273 to 0°C) | Melting Ice (0°C) | Liquid water (0 to 100°C) | Boiling water (100°C) |

Figure 3-3
The structure of water, or any other substance, becomes increasingly disordered, that is, its entropy increases, as its temperature rises.

3. FREE ENERGY: THE INDICATOR OF SPONTANEITY

The disordering of the universe by spontaneous processes is an impractical criterion for spontaneity because it is rarely possible to monitor the entropy of the entire universe. Yet, the spontaneity of a process cannot be predicted from a knowledge of the system's entropy change alone. This is because exothermic processes ($\Delta H_{system} < 0$) may be spontaneous even though they are characterized by $\Delta S_{system} < 0$. For example, 2 mol of H_2 and 1 mol of O_2, when sparked, react in a decidedly exothermic reaction to form 2 mol of H_2O. Yet two water molecules, each of whose three atoms are constrained to stay together, are more ordered than are the three diatomic molecules from which they formed. Similarly, under appropriate conditions, many **denatured** (unfolded) proteins will spontaneously fold to assume their highly ordered **native** (normally folded) conformations (Section 8-1A). What we really want therefore is a state function that predicts whether or not a given process is spontaneous. In this section, we consider such a function.

A. Gibbs Free Energy

The **Gibbs free energy**:

$$G = H - TS \qquad [3.9]$$

which was formulated by J. Willard Gibbs in 1878, is the required indicator of spontaneity for constant temperature and pressure processes. For systems that can only do pressure–volume work ($w' = 0$), combining Eqs. [3.4] and [3.9] while holding T and P constant yields

$$\Delta G = \Delta H - T\Delta S = q_P - T\Delta S \qquad [3.10]$$

But Eq. [3.8] indicates that $T\Delta S \geq q$ for spontaneous processes at constant T. Consequently, $\Delta G \leq 0$ *is the criterion of spontaneity we seek* for the constant T and P conditions that are typical of biochemical processes.

Spontaneous processes, that is, those with negative ΔG values, are said to be **exergonic** (Greek: *ergon*, work); they can be utilized to do work. Processes that are not spontaneous, those with positive ΔG values, are termed **endergonic**; they must be driven by the input of free energy (through mechanisms discussed in Section 3-4C). Processes at equilibrium, those in which the forward and backward reactions are exactly balanced, are characterized by $\Delta G = 0$. Note that the value of ΔG varies directly with temperature. This is why, for instance, the native structure of a protein, whose formation from its denatured form has both $\Delta H < 0$ and $\Delta S < 0$, predominates below the temperature at which $\Delta H = T\Delta S$ (the **denaturation temperature**), whereas the denatured protein predominates above this temper-

Table 3-2

The Variation of Reaction Spontaneity (Sign of ΔG) with the Signs of ΔH and ΔS

ΔH	ΔS	$\Delta G = \Delta H - T\,\Delta S$
−	+	The reaction is both enthalpically favored (exothermic) and entropically favored. It is spontaneous (exergonic) at all temperatures.
−	−	The reaction is enthalpically favored but entropically opposed. It is spontaneous only at temperatures *below* $T = \Delta H/\Delta S$.
+	+	The reaction is enthalpically opposed (endothermic) but entropically favored. It is spontaneous only at temperatures *above* $T = \Delta H/\Delta S$.
+	−	The reaction is both enthalpically and entropically opposed. It is *unspontaneous* (endergonic) at all temperatures.

ature. The variation of the spontaneity of a process with the signs of ΔH and ΔS is summarized in Table 3-2.

B. Free Energy and Work

When a system at constant temperature and pressure does non-P–V work, Eq. [3.10] must be expanded to

$$\Delta G = q_P - T\Delta S - w' \qquad [3.11]$$

or, because $T\Delta S \geq q_P$, Eq. [3.8],

$$\Delta G \leq -w'$$

so that

$$\Delta G \geq w' \qquad [3.12]$$

Since P–V work is unimportant in biological systems, *ΔG for a biological process must represent its maximum recoverable work*. The ΔG of a process is therefore indicative of the maximum charge separation it can establish, the maximum concentration gradient it can generate (Section 3-4A), the maximum muscular activity it can produce, and so on. In fact, for real processes, which can only approach reversibility, the inequality in Eq. [3.12] holds, so that *the work put into any system can never be fully recovered*. This is indicative of the inherent dissipative character of nature. Indeed, as we have seen, it is precisely this dissipative character that provides the overall driving force for any change.

It is important to reiterate that a large negative value of ΔG does not ensure a chemical reaction will proceed at a measurable rate. This depends on the detailed mechanism of the reaction, which is independent of the ΔG. For instance, most biological molecules, including proteins, nucleic acids, carbohydrates, and lipids, are thermodynamically unstable to hydrolysis but, nevertheless, spontaneously hydrolyze at biologically insignificant rates. Only with the introduction of the proper enzymes will the hydrolysis of these molecules proceed at a reasonable pace. Yet, a catalyst, which by definition is unchanged by a reaction, cannot affect the ΔG of a reaction. Consequently, *an enzyme can only accelerate the attainment of thermodynamic equilibrium; it cannot, for example, promote a reaction that has a positive ΔG.*

4. CHEMICAL EQUILIBRIA

The entropy (disorder) of a substance increases with its volume. For example, as we have seen for our twin-bulb apparatus (Fig. 3-1), a collection of gas molecules, in occupying all of the volume available to it, maximizes its entropy. Similarly, dissolved molecules become uniformly distributed throughout their solution volume. Entropy is therefore a function of concentration.

If entropy varies with concentration, so must free energy. Thus, as is shown in this section, the free energy change of a chemical reaction depends on the concentrations of both its reactants and its products. This phenomenon is of great biochemical significance because enzymatic reactions can operate in either direction depending on the relative concentrations of their reactants and products. Indeed, the directions of many enzymatically catalyzed reactions depend on the availability of their substrates and on the metabolic demand for their products (although most metabolic pathways operate unidirectionally; Section 15-6C).

A. Equilibrium Constants

The relationship between the concentration and the free energy of a substance A, which is derived in the appendix to this chapter, is approximately

$$\overline{G}_A - \overline{G}_A^\circ = RT \ln[A] \qquad [3.13]$$

where $\overline{G}_A$ is known equivalently as the **partial molar free energy** or the **chemical potential** of A (the bar indicates the quantity per mole), $\overline{G}_A^\circ$ is the partial molar free energy of A in its **standard state** (see Section 3-4B), R is the gas constant (Table 3-1), and [A] is the molar concentration of A. Thus for the general nonequilibrium reaction,

$$a\text{A} + b\text{B} \rightleftharpoons c\text{C} + d\text{D}$$

since free energies are additive and the free energy change of a reaction is the sum of the free energies of the products less those of the reactants, the free energy change for this reaction is

$$\Delta G = c\overline{G}_C + d\overline{G}_D - a\overline{G}_A - b\overline{G}_B \qquad [3.14]$$

Substituting this relationship into Eq. [3.13] yields

$$\Delta G = \Delta G^\circ + RT \ln\left(\frac{[C]^c [D]^d}{[A]^a [B]^b}\right) \qquad [3.15]$$

where ΔG° is the free energy change of the reaction when all of its reactants and products are in their standard states. Thus the expression for the free energy change of a reaction consists of two parts: (1) a constant term whose value depends only on the reaction taking place, and (2) a variable term that depends on the concentrations of the reactants and the products, the stoichiometry of the reaction, and the temperature.

For a reaction at equilibrium, there is no *net* change because the free energy of the forward reaction exactly balances that of the backward reaction. Consequently, $\Delta G = 0$ so that Eq. [3.15] becomes

$$\Delta G^\circ = -RT \ln K_{eq} \qquad [3.16]$$

where K_{eq} is the familiar **equilibrium constant** of the reaction:

$$K_{eq} = \frac{[C]^c_{eq} [D]^d_{eq}}{[A]^a_{eq} [B]^b_{eq}} = e^{-\Delta G^\circ/RT} \qquad [3.17]$$

and the subscript "eq" in the concentration terms indicates their equilibrium values. (The equilibrium condition is usually clear from the context of the situation so that equilibrium concentrations are often expressed without this subscript.) *The equilibrium constant of a reaction may therefore be calculated from standard free energy data and vice versa.* Table 3-3 indicates the numerical relationship between ΔG° and K_{eq}. Note that a 10-fold variation of K_{eq} at 25°C corresponds to a 5.7-kJ·mol^{-1} change in ΔG°, which is less than one half of the free energy of even a weak hydrogen bond.

Equations [3.15] to [3.17] indicate that when the reactants in a process are in excess of their equilibrium concentrations, the net reaction will proceed in the forward direction until the excess reactants have been converted to products and equilibrium is attained. Conversely,

Table 3-3

The Variation of K_{eq} with ΔG° at 25°C

K_{eq}	ΔG° (kJ·mol^{-1})
10^6	34.3
10^4	22.8
10^2	11.4
10^1	5.7
10^0	0.0
10^{-1}	−5.7
10^{-2}	−11.4
10^{-4}	−22.8
10^{-6}	−34.3

when products are in excess, the net reaction proceeds in the reverse reaction as so to convert products to reactants until the equilibrium concentration ratio is likewise achieved. Thus, as **Le Chatelier's principle** states, *any deviation from equilibrium stimulates a process that tends to restore the system to equilibrium. All closed systems must therefore inevitably reach equilibrium.* Living systems escape this thermodynamic *cul-de-sac* by being open systems (Section 15-6A).

The manner in which the equilibrium constant varies with temperature is seen by substituting Eq. [3.10] into Eq. [3.16] and rearranging:

$$\ln K_{eq} = \frac{-\Delta H^\circ}{R}\left(\frac{1}{T}\right) + \frac{\Delta S^\circ}{R} \qquad [3.18]$$

If ΔH° and ΔS° are independent of temperature, as they often are to a reasonable approximation, a plot of $\ln K_{eq}$ versus $1/T$, known as a **van't Hoff plot,** yields a straight line of slope $-\Delta H^\circ/R$ and intercept $\Delta S^\circ/R$. This relationship permits the values of ΔH° and ΔS° to be determined from measurements of K_{eq} at two (or more) different temperatures. Calorimetric data, which have been quite difficult to measure for biochemical processes, are therefore not required to obtain the values of ΔH° and ΔS°. Indeed, most biochemical thermodynamic data have been obtained through the application of Eq. [3.18]. However, the recent development of the **scanning microcalorimeter** has made the direct measurement of ΔH for biochemical processes a practical alternative.

B. Standard Free Energy Changes

Since only free energy differences, ΔG, can be measured, not free energies themselves, it is necessary to refer these differences to some standard state in order to compare the free energies of different substances (likewise, we refer the elevations of geographic locations to sea level, which is arbitrarily assigned the height of zero). By convention, the free energy of all pure elements in their standard state: 25°C, 1 atm, and in their most stable form (e.g., O_2 not O_3), is defined to be zero. The **free energy of formation** of any nonelemental substance, ΔG_f°, is then defined as the change in free energy accompanying the formation of one mole of that substance, in its standard state, from its component elements in their standard states. The standard free energy change for any reaction can be calculated according to

$$\Delta G^\circ = \sum \Delta G_f^\circ(\text{products}) - \sum \Delta G_f^\circ(\text{reactants}) \qquad [3.19]$$

Table 3-4 provides a list of standard free energies of formation, ΔG_f°, for a selection of substances of biochemical significance.

Standard State Conventions in Biochemistry

The standard state convention commonly used in physical chemistry defines the standard state of a solute

Table 3-4

Free Energies of Formation of Some Compounds of Biochemical Interest

Compound	$-\Delta G_f^\circ$ (kJ·mol^{-1})
Acetaldehyde	139.7
Acetate$^-$	369.2
Acetyl-CoA	374.1[a]
cis-Aconitate^{3-}	920.9
CO_2 (*g*)	394.4
CO_2 (*aq*)	386.2
HCO_3^-	587.1
Citrate^{3-}	1166.6
Dihydroxyacetone phosphate^{2-}	1293.2
Ethanol	181.5
Fructose	915.4
Fructose-6-phosphate^{2-}	1758.3
Fructose-1,6-bisphosphate^{4-}	2600.8
Fumarate^{2-}	604.2
α-D-Glucose	917.2
Glucose-6-phosphate^{2-}	1760.3
Glyceraldehyde-3-phosphate^{2-}	1285.6
H$^+$	0.0
H$_2$ (*g*)	0.0
H$_2$O (*l*)	237.2
Isocitrate^{3-}	1160.0
α-Ketoglutarate^{2-}	798.0
Lactate$^-$	516.6
L-Malate^{2-}	845.1
OH$^-$	157.3
Oxaloacetate^{2-}	797.2
Phosphoenolpyruvate^{3-}	1269.5
2-Phosphoglycerate^{3-}	1285.6
3-Phosphoglycerate^{3-}	1515.7
Pyruvate$^-$	474.5
Succinate^{2-}	690.2
Succinyl-CoA	686.7[a]

[a] For formation from free elements + free CoA (Coenzyme A).

Source: Metzler, D. E., *Biochemistry, The Chemical Reactions of Living Cells*, pp. 162–164, Academic Press (1977).

as that with unit **activity** at 25°C and 1 atm (activity is concentration corrected for nonideal behavior as is explained in the appendix to this chapter; for the dilute solutions typical of biochemical reactions, such corrections are small so that activities can be replaced by concentrations). However, because biochemical reactions usually occur in dilute aqueous solutions near neutral pH, a somewhat different standard state convention for biological systems has been adopted:

- Water's standard state is defined as that of the pure liquid, so that the activity of pure water is taken to be unity despite the fact that its concentration is $55.5M$. In essence, the $[H_2O]$ term is incorporated into the value of the equilibrium constant. This procedure simplifies the free energy expressions for reactions in dilute aqueous solutions involving water as a reactant or product because the $[H_2O]$ term can then be ignored.

- The hydrogen ion activity is defined as unity at the physiologically relevant pH of 7 rather than at the physical chemical standard state of pH 0 where many biological substances are unstable.

- The standard state of a substance that can undergo an acid–base reaction is defined in terms of the total concentration of its naturally occurring ion mixture at pH 7. In contrast, the physical chemistry convention refers to a pure species whether or not it actually exists at pH 0. The advantage of the biochemistry convention is that the total concentration of a substance with multiple ionization states, such as most biological molecules, is usually easier to measure than the concentration of one of its ionic species. Since the ionic composition of an acid or base varies with pH, however, the standard free energies calculated according to the biochemistry convention are only valid at pH 7.

Under the biochemistry convention, the standard free energy changes of substances are customarily symbolized by $\Delta G^{\circ}{}'$ in order to distinguish them from physical chemistry standard free energy changes, ΔG° (note that the value of ΔG for any process, being experimentally measurable, is independent of the chosen standard state; i.e., $\Delta G = \Delta G'$). Likewise, the biochemical equilibrium constant, which is defined by using $\Delta G^{\circ}{}'$ in place of ΔG° in Eq. [3.17], is represented by K'_{eq}.

The relationship between $\Delta G^{\circ}{}'$ and ΔG° is often a simple one. There are three general situations:

1. If the reacting species include neither H_2O nor H^+, the expressions for $\Delta G^{\circ}{}'$ and ΔG° coincide.

2. For a reaction in dilute aqueous solution that yields nH_2O molecules:

$$A + B \rightleftharpoons C + D + nH_2O$$

Equations [3.16] and [3.17] indicate that

$$\Delta G^{\circ} = -RT \ln K_{eq} = -RT \ln \left(\frac{[C][D][H_2O]^n}{[A][B]} \right)$$

Under the biochemistry convention, which defines the activity of pure water as unity,

$$\Delta G^{\circ}{}' = -RT \ln K'_{eq} = -RT \ln \left(\frac{[C][D]}{[A][B]} \right)$$

Therefore

$$\Delta G^{\circ}{}' = \Delta G^{\circ} + nRT \ln[H_2O] \qquad [3.20]$$

where $[H_2O] = 55.5M$ (the concentration of water in aqueous solution), so that for a reaction at 25°C, which yields 1 mol of H_2O, $\Delta G^{\circ}{}' = \Delta G^{\circ} + 9.96 \text{ kJ} \cdot \text{mol}^{-1}$.

3. For a reaction involving hydrogen ions, such as

$$A + B \rightleftharpoons C + HD$$
$$\Updownarrow K$$
$$D^- + H^+$$

where

$$K = \frac{[H^+][D^-]}{[HD]}$$

manipulations similar to those above lead to the relationship

$$\Delta G^{\circ}{}' = \Delta G^{\circ} - RT \ln(1 + [H^+]_0/K) + RT \ln[H^+]_0 \qquad [3.21]$$

where $[H^+]_0 = 10^{-7}M$, the only value of $[H^+]$ for which this equation is valid. Of course, if more than one ionizable species participates in the reaction and/or if any of them are polyprotic, Eq. [3.21] is correspondingly more complicated.

C. Coupled Reactions

The additivity of free energy changes allows an endergonic reaction to be driven by an exergonic reaction under the proper conditions. This phenomenon is the thermodynamic basis for the operation of metabolic pathways since most of these reaction sequences are comprised of endergonic as well as exergonic reactions. Consider the following two-step reaction process:

(1) $\qquad A + B \rightleftharpoons C + D \qquad \Delta G_1$

(2) $\qquad D + E \rightleftharpoons F + G \qquad \Delta G_2$

If $\Delta G_1 \geq 0$, Reaction (1) will not occur spontaneously. However, if ΔG_2 is sufficiently exergonic so that $\Delta G_1 + \Delta G_2 < 0$, then although the equilibrium concentration of D in Reaction (1) will be relatively small, it will be larger than that in Reaction (2). As Reaction (2) converts D to products, Reaction (1) will operate in the forward direction in order to replenish the equilibrium concentration of D. The highly exergonic Reaction (2) therefore *drives* the endergonic Reaction (1) and the two reactions are said to be **coupled** through their common intermediate, D. That these coupled reactions proceed spontaneously (although not necessarily at a finite rate) can also be seen by summing Reactions (1) and (2) to yield the overall reaction

(1 + 2) $\quad A + B + E \rightleftharpoons C + F + G \qquad \Delta G_3$

where $\Delta G_3 = \Delta G_1 + \Delta G_2 < 0$. *So long as the overall pathway (reaction sequence) is exergonic, it will operate in the forward direction.* Thus, the free energy of ATP hydrolysis, a highly exergonic process, is harnessed to drive many otherwise endergonic biological processes to completion (Section 15-4C).

Appendix

Concentration Dependence of Free Energy

To establish that the free energy of a substance is a function of its concentration, consider the free energy change of an ideal gas during a reversible pressure change at constant temperature ($w' = 0$, since an ideal gas is incapable of doing non-P–V work). Substituting Eqs. [3.1] and [3.2] into Eq. [3.9] and differentiating the result yields

$$dG = dq - dw + PdV + VdP - TdS \qquad [3.A1]$$

Upon substitution of the differentiated forms of Eqs. [3.3] and [3.8] in this expression, it reduces to

$$dG = VdP \qquad [3.A2]$$

The ideal gas equation is $PV = nRT$, where n is the number of moles of gas. Therefore

$$dG = nRT\frac{dP}{P} = nRTd\ln P \qquad [3.A3]$$

This gas phase result can be extended to the more biochemically relevant area of solution chemistry by application of **Henry's law** for a solution containing the volatile solute A in equilibrium with the gas phase:

$$P_A = K_A X_A \qquad [3.A4]$$

Here P_A is the partial pressure of A when its mole fraction in the solution is X_A and K_A is the **Henry's law constant** of A in the solvent being used. It is generally more convenient, however, to express the concentrations of the relatively dilute solutions of chemical and biological systems in terms of molarity rather than mole fractions. For a dilute solution

$$X_A \approx \frac{n_A}{n_{\text{solvent}}} = \frac{[A]}{[\text{solvent}]} \qquad [3.A5]$$

where the solvent concentration [solvent] is approximately constant. Thus

$$P_A \approx K'_A[A] \qquad [3.A6]$$

where $K'_A = K_A/[\text{solvent}]$. Substituting this expression into Eq. [3.A3] yields

$$dG_A = n_A RTd(\ln K'_A + \ln[A]) = n_A RTd\ln[A] \qquad [3.A7]$$

Free energy, as are energy and enthalpy, is a relative quantity that can only be defined with respect to some arbitrary standard state. The standard state is customar-

ily taken to be 25 °C, 1-atm pressure and, for the sake of mathematical simplicity, [A] = 1. The integration of Eq. [3.A7] from the standard state, [A] = 1, to the final state, [A] = [A], results in

$$G_A - G_A^\circ = n_A RT\ln[A] \qquad [3.A8]$$

where G_A° is the free energy of A in the standard state and [A] really represents the concentration ratio [A]/1. Since Henry's law is valid for real solutions only in the limit of infinite dilution, however, the standard state is defined as the entirely hypothetical state of $1M$ solute with the properties that it has at infinite dilution.

The free energy terms in Eq. [3.A8] may be converted from **extensive quantities** (those dependent on the amount of material) to **intensive quantities** (those independent of the amount of material) by dividing both sides of the equation by n_A. This yields

$$\overline{G}_A - \overline{G}_A^\circ = RT\ln[A] \qquad [3.A9]$$

Equation [3.A9] has the limitation that it refers to solutions that exactly follow Henry's law although real solutions only do so in the limit of infinite dilution if the solute is, in fact, volatile. These difficulties can all be eliminated by replacing [A] in Eq. [3.A9] by a quantity, a_A, known as the **activity** of A. This is defined

$$a_A = \gamma_A[A] \qquad [3.A10]$$

where γ_A is the **activity coefficient** of A. Equation [3.A9] thereby takes the form

$$\overline{G}_A - \overline{G}_A^\circ = RT\ln a_A \qquad [3.A11]$$

in which all departures from ideal behavior, including the provision that the system may perform non-P–V work, are incorporated in the activity coefficient, which is an experimentally measurable quantity. Ideal behavior is only approached at infinite dilution; that is, $\gamma_A \to 1$ as [A] $\to 0$. The standard state in Eq. [3.A11] is redefined as that of unit activity.

The concentrations of reactants and products in most biochemical reactions are usually so low (on the order of millimolar or less) that the activity coefficients of these various species are nearly unity. Consequently, the activities of most biochemical species under physiological conditions can be satisfactorily approximated by their molar concentrations:

$$\overline{G}_A - \overline{G}_A^\circ = RT\ln[A] \qquad [3.13]$$

Chapter Summary

The first law of thermodynamics,

$$U = q - w \qquad [3.1]$$

where q is heat and w is work, is a statement of the law of conservation of energy. Energy is a state function because the energy of a system depends only on the state of the system. Enthalpy,

$$H = U + PV \qquad [3.2]$$

where P is pressure and V is volume, is a closely related state function that represents the heat at constant pressure under conditions where only pressure–volume work is possible. Entropy, which is also a state function, is defined

$$S = k_B \ln W \qquad [3.5]$$

where W, the disorder, is the number of equivalent ways the system can be arranged under the conditions governing it and k_B is the Boltzmann constant. The second law of thermodynamics states that the universe tends towards maximum disorder and hence $\Delta S > 0$ for any real process. The Gibbs free energy of a system

$$G = H - TS \qquad [3.9]$$

decreases in a spontaneous, constant pressure process. In a process at equilibrium, the system suffers no net change so that $\Delta G = 0$. An ideal process, in which the system is always at equilibrium, is said to be reversible. All real processes are irreversible since processes at equilibrium can only occur at an infinitesimal rate. For a chemical reaction

$$a\mathrm{A} + b\mathrm{B} \rightleftharpoons c\mathrm{C} + d\mathrm{D}$$

the change in the Gibbs free energy is expressed

$$\Delta G = \Delta G^\circ + RT \ln\left(\frac{[\mathrm{C}]^c[\mathrm{D}]^d}{[\mathrm{A}]^a[\mathrm{B}]^b}\right) \qquad [3.15]$$

where ΔG°, the standard free energy change, is the free energy change at 25°C, 1-atm pressure, and unit activities of reactants and products. The biochemical standard state, $\Delta G^{\circ\prime}$, is similarly defined but in dilute aqueous solution at pH 7 in which the activities of water and H^+ are both defined as unity. At equilibrium

$$\Delta G^{\circ\prime} = -RT \ln K'_{eq} = -RT \ln\left(\frac{[\mathrm{C}]^c_{eq}[\mathrm{D}]^d_{eq}}{[\mathrm{A}]^a_{eq}[\mathrm{B}]^b_{eq}}\right) \qquad [3.17a]$$

where K'_{eq} is the equilibrium constant under the biochemical convention. An endergonic reaction ($\Delta G > 0$) may be driven by an exergonic reaction ($\Delta G < 0$) if they are coupled and if the overall reaction is exergonic.

References

Atkins, P. W., *Physical Chemistry* (3rd ed.), Chapters 1–10, Freeman (1986). [Most physical chemistry texts treat thermodynamics in some detail.]

Atkins, P. W., *The Second Law*, Scientific American Books (1984). [An insightful but nonmathematical exposition of the second law of thermodynamics.]

Brey, W. S., *Physical Chemistry and Its Biological Applications*, Chapters 3 and 4, Academic Press (1978).

Dickerson, R. E., *Molecular Thermodynamics*, Benjamin (1969).

Edsall, J. T. and Gutfreund, H., *Biothermodynamics*, Wiley (1983).

Eisenberg, D. and Crothers, D., *Physical Chemistry with Applications to Life Sciences*, Chapters 1–5, Benjamin (1979).

Nash, L. K., *CHEMTHERMO: A Statistical Approach to Classical Chemical Thermodynamics*, Addison–Wesley (1971). [A delightfully written text on an elementary level.]

Segel, I. H., *Biochemical Calculations* (2nd ed.), Chapter 3, Wiley (1976). [Contains instructive problems accompanied by detailed solutions.]

Tinoco, I., Jr., Sauer, K., and Wang, J. C., *Physical Chemistry. Principles and Applications in Biological Sciences* (2nd ed.), Chapters 2–5, Prentice–Hall (1985).

van Holde, K. E., *Physical Biochemistry* (2nd ed.), Chapters 1–3, Prentice–Hall (1985). [The equivalence of the Boltzmann and Clausius formulations of the second law of thermodynamics is demonstrated on pp. 13–14.]

Wood, W. B., Wilson, J. H., Benbow, R. M., and Hood, L. E., *Biochemistry, A Problems Approach* (2nd ed.), Chapter 9, Benjamin/Cummings (1981). [A question and answer book.]

Problems

1. A common funeral litany is the Biblical verse: "Ashes to ashes, dust to dust." Why might a bereaved family of thermodynamicists be equally comforted by a recitation of the second law of thermodynamics?

2. How many flights of 4-m high stairs must an overweight person weighing 75 kg climb to atone for the indiscretion of eating a 500-Cal hamburger? Assume that there is a 20% efficiency in converting nutritional energy to mechanical energy. The gravitational force on an object of mass m kg is $F = mg$ where the gravitational constant, $g = 9.8$ m·s^{-2}.

3. In terms of thermodynamic concepts, why is it more difficult to park a car in a small space than it is to drive it out from such a space?

4. It has been said that an army of dedicated monkeys, typing at random, would eventually produce all of Shakespeare's works. How long, on average, would it take 1 million monkeys, each typing on a 46-key typewriter (space included but no shift key) at the rate of 1 key stroke per second, to type "to be or not to be"? What does this result indicate about the probability of order randomly arising from disorder?

5. Show that the transfer of heat from an object of higher temperature to one of lower temperature, but not the reverse process, obeys the second law of thermodynamics.

6. Carbon monoxide crystallizes with its CO molecules arranged in parallel rows. Since CO is a very nearly ellipsoidal molecule, in the absence of polarity effects, adjacent CO molecules could equally well line up in a head-to-tail or a head-to-head fashion. In a crystal consisting of 10^{23} CO molecules, what is the entropy of all the CO molecules being aligned head to tail?

7. The U.S. Patent Office has received, and continues to receive, numerous applications for perpetual motion machines. Perpetual motion machines have been classified as those of the first kind, which violate the first law of thermodynamics, and those of the second kind, which violate the second law of thermodynamics. The fallacy in a perpetual motion machine of the first kind is generally easy to detect. An example would be a motor-driven electrical generator that produces energy in excess of that input by the motor. The fallacy in a perpetual motion machine of the second type, however, is usually more subtle. Take, for example, a ship that uses heat energy extracted from the sea by a heat pump to boil water so as to power a steam engine that drives the ship as well as the heat pump. Show, in general terms, that such a propulsion system would violate the second law of thermodynamics.

8. Using the data in Table 3-4, calculate the values of $\Delta G°$ at 25°C for the following metabolic reactions:

(a) $\underset{\text{Glucose}}{C_6H_{12}O_6} + 6O_2 \rightleftharpoons 6CO_2(aq) + 6H_2O\ (l)$

(b) $\underset{\text{Glucose}}{C_6H_{12}O_6} \rightleftharpoons 2\underset{\text{Ethanol}}{CH_3CH_2OH} + 2CO_2\ (aq)$

(c) $\underset{\text{Glucose}}{C_6H_{12}O_6} \rightleftharpoons 2\underset{\text{Lactate}^-}{CH_3CHOHCOO^-} + 2H^+$

[These reactions, respectively, constitute oxidative metabolism, alcoholic fermentation in yeast deprived of oxygen, and homolactic fermentation in skeletal muscle requiring more energy than oxidative metabolism can supply (Section 16-1B).]

*9. The native and denatured forms of a protein are generally in equilibrium as follows:

$$\text{Protein } (denatured) \rightleftharpoons \text{protein } (native)$$

For a certain solution of the protein **ribonuclease A,** in which the total protein concentration is $2.0 \times 10^{-3}M$, the concentrations of the denatured and native proteins at both 50 and 100°C are given in the following table:

Temperature (°C)	[Ribonuclease A *(denatured)*] (M)	[Ribonuclease A *(native)*] (M)
50	5.1×10^{-6}	2.0×10^{-3}
100	2.8×10^{-4}	1.7×10^{-3}

(a) Determine $\Delta H°$ and $\Delta S°$ for the folding reaction assuming that these quantities are independent of temperature. (b) Calculate $\Delta G°$ for ribonuclease A folding at 25°C. Is this process spontaneous under standard state conditions at this temperature? (c) What is the denaturation temperature of ribonuclease A under standard state conditions?

*10. Using the data in Table 3-4, calculate $\Delta G_f°'$ for the following compounds at 25°C:
(a) $H_2O(l)$; (b) sucrose (sucrose + $H_2O \rightleftharpoons$ glucose + fructose: $\Delta G°' = -29.3$ kJ·mol^{-1}); and (c) ethyl acetate (ethyl acetate + $H_2O \rightleftharpoons$ ethanol + acetate$^-$ + H$^+$: $\Delta G°' = -19.7$ kJ·mol^{-1}; the pK of acetic acid is 4.76).

11. Calculate the equilibrium constants for the hydrolysis of the following compounds at pH 7 and 25°C: (a) Phosphoenolpyruvate ($\Delta G°' = -61.9$ kJ·mol^{-1}); (b) pyrophosphate ($\Delta G°' = -33.5$ kJ·mol^{-1}); and (c) glucose-1-phosphate ($\Delta G°' = -20.9$ kJ·mol^{-1}).

The β pleated sheet of silk.

II

Biomolecules

Chapter 4
AMINO ACIDS

It is hardly surprising that much of the early biochemical research was concerned with the study of proteins. Proteins form the class of biological macromolecules that have the most well-defined physicochemical properties and consequently are generally easier to isolate and characterize than nucleic acids, polysaccharides, or lipids. Furthermore, proteins, particularly in the form of enzymes, have an obvious biochemical function. The central role that proteins play in biological processes has therefore been recognized since the earliest days of biochemistry. In contrast, the task of nucleic acids in the transmission and expression of genetic information was not realized until the late 1940s, the role of lipids in biological membranes was not appreciated until the 1960s, and the biological functions of polysaccharides are still somewhat mysterious.

In this chapter we study the properties of the monomeric units of proteins, the **amino acids.** It is from these substances that proteins are synthesized through processes that are discussed in Chapter 30. Amino acids are also energy metabolites and many of them are essential nutrients (Chapter 24). In addition, as we shall see, many amino acids and their derivatives are of biochemical importance in their own right (Section 4-3B).

Table 4-1

Covalent Structures of the "Standard" Amino Acids of Proteins

Name	Structural Formula [a]	Residue Mass (D)	Average Occurrence in Proteins (%) [b]
Amino acids with nonpolar side chains			
Glycine	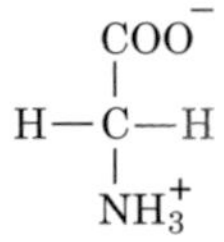	57.0	7.5
Alanine	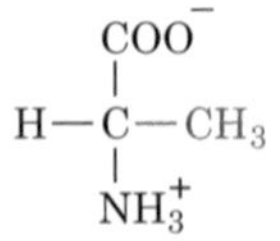	71.0	9.0
Valine	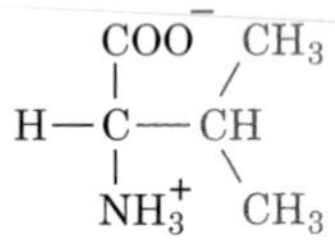	99.1	6.9
Leucine	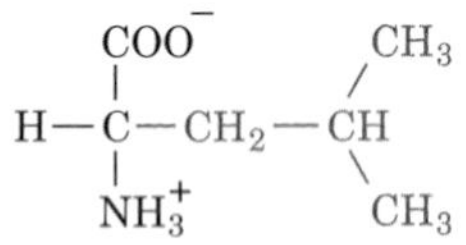	113.1	7.5
Isoleucine	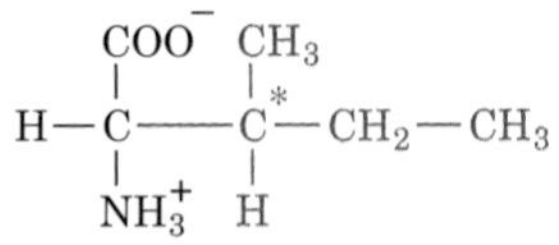	113.1	4.6
Methionine	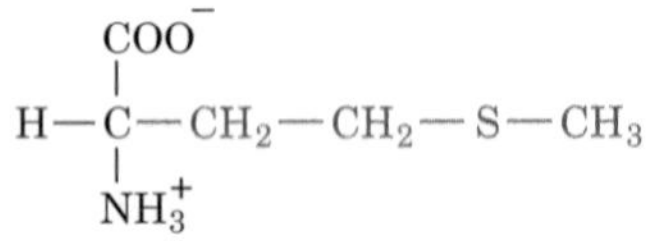	131.1	1.7
Proline	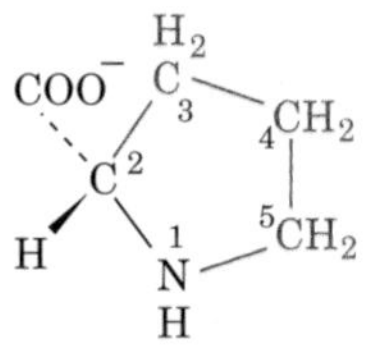	97.1	4.6
Phenylalanine	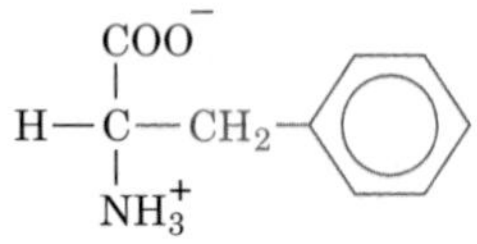	147.1	3.5
Tryptophan	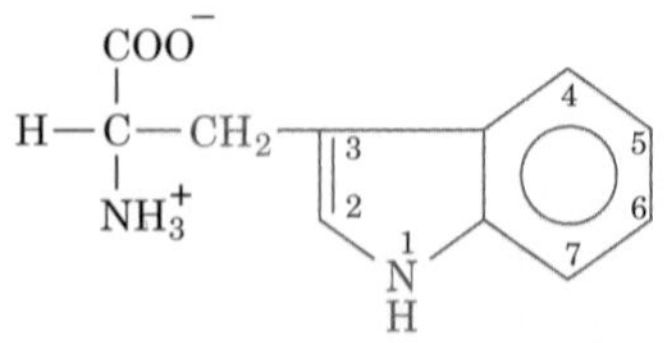	186.2	1.1

[a] The ionic forms shown are those predominating at pH 7.0. The C_α atoms, as well as those atoms marked with an asterisk, are chiral centers with configurations as indicated according to Fischer projection formulas. The standard organic numbering system is provided for heterocycles. For the molecular masses of the parent amino acids, add 18.0 D, the molecular mass of H_2O, to the residue masses. For side chain masses, subtract 56.0 D, the formula mass of a peptide group, from the residue masses.
[b] Estimated from the compositions of 207 unrelated proteins as compiled by Klapper, M. H., *Biochem. Biophys. Res. Commun.* **78**, 1020 (1977).

Name	Structural Formula [a]	Residue Mass (D)	Average Occurrence in Proteins (%) [b]
Amino acids with uncharged polar side chains			
Serine	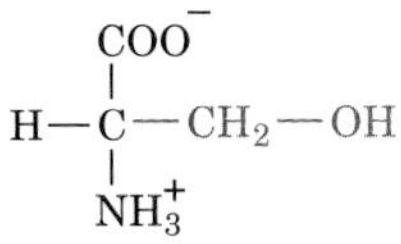 $H-C(COO^-)(NH_3^+)-CH_2-OH$	87.0	7.1
Threonine	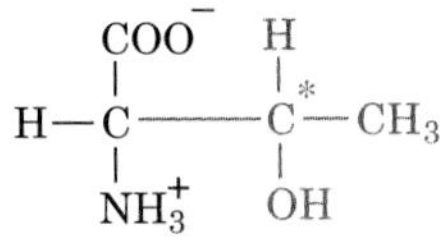 $H-C(COO^-)(NH_3^+)-C^*(H)(OH)-CH_3$	101.1	6.0
Asparagine	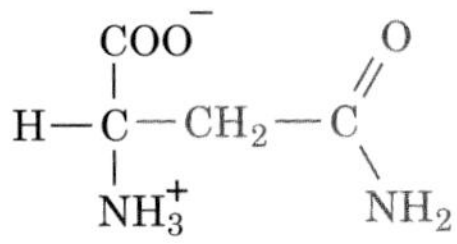 $H-C(COO^-)(NH_3^+)-CH_2-C(=O)-NH_2$	114.1	4.4
Glutamine	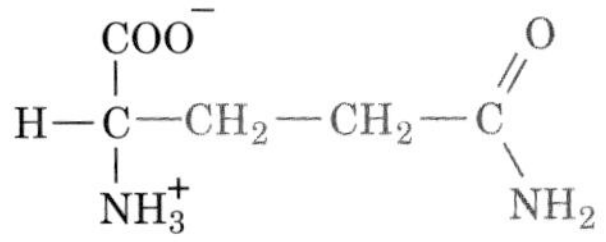 $H-C(COO^-)(NH_3^+)-CH_2-CH_2-C(=O)-NH_2$	128.1	3.9
Tyrosine	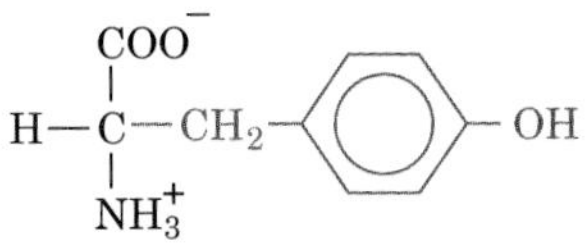 $H-C(COO^-)(NH_3^+)-CH_2-C_6H_4-OH$	163.1	3.5
Cysteine	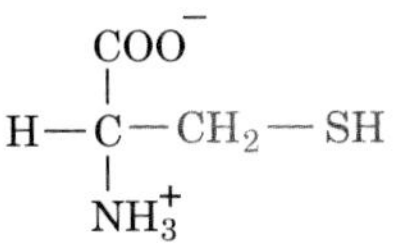 $H-C(COO^-)(NH_3^+)-CH_2-SH$	103.1	2.8
Amino acids with charged polar side chains			
Lysine	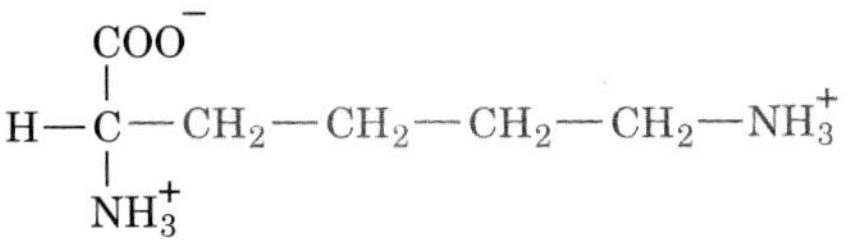 $H-C(COO^-)(NH_3^+)-CH_2-CH_2-CH_2-CH_2-NH_3^+$	129.1	7.0
Arginine	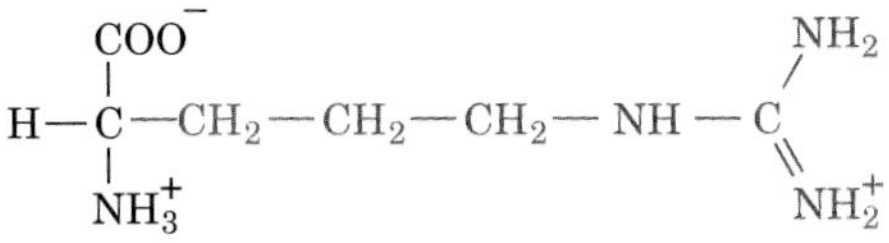 $H-C(COO^-)(NH_3^+)-CH_2-CH_2-CH_2-NH-C(NH_2)(=NH_2^+)$	157.2	4.7
Histidine	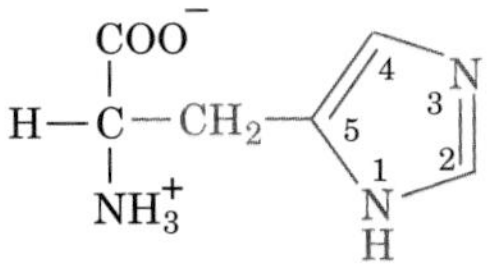 $H-C(COO^-)(NH_3^+)-CH_2-$ (imidazole ring)	137.1	2.1
Aspartic acid	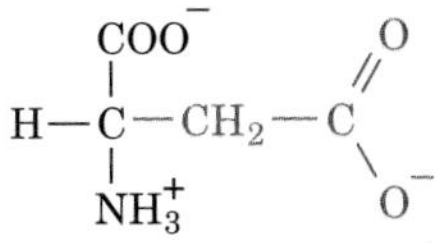 $H-C(COO^-)(NH_3^+)-CH_2-C(=O)-O^-$	114.0	5.5
Glutamic acid	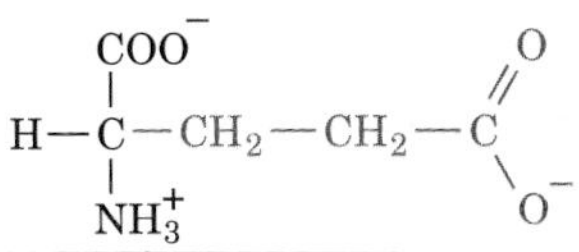 $H-C(COO^-)(NH_3^+)-CH_2-CH_2-C(=O)-O^-$	128.1	6.2

1. THE AMINO ACIDS OF PROTEINS

The analyses of a vast number of proteins from almost every conceivable source have shown that *all proteins are composed of the 20 "standard" amino acids listed in Table 4-1.* These substances are known as **α-amino acids** because, with the exception of **proline**, they have a primary amino group and a carboxylic acid group substituent on the same carbon atom (Fig. 4-1). (Proline, the one exception to this general structure, has a secondary amino group and therefore is actually an **α-imino acid,** but is nevertheless commonly referred to as an amino acid.)

A. General Properties

The pK values of the 20 "standard" α-amino acids of proteins are tabulated in Table 4-2. Here pK_1 and pK_2, respectively, refer to the α-carboxylic acid and α-amino groups, and pK_R refers to the side groups with acid–base properties. Table 4-2 indicates that the pK values of the carboxylic acid groups lie in a small range around 2.2 so that above pH 3.5 these groups are almost entirely in their carboxylate forms. The α-amino groups all have pK values near 9.4 and are therefore almost entirely in the ammonium ion form below pH 8.0. This leads to an important structural point: *In the physiological pH range, both the carboxylic acid and the amino groups of α-amino acids are completely ionized (Fig. 4-2).* An amino acid can therefore act either as an acid or a base. Substances with this property are said to be **amphoteric** and are referred to as **ampholytes** (*amphoteric electrolytes*). In Section 4-1D, we shall delve a bit deeper into the acid–base properties of the amino acids.

Molecules that bear charged groups of opposite polarity are known as **zwitterions** or **dipolar ions**. The zwitterionic character of the α-amino acids has been established by several methods including spectroscopic measurements and X-ray crystal structure determinations (in the solid state the α-amino acids are zwitterionic because the basic amine group abstracts a proton from the nearby acidic carboxylic acid group). Because amino acids are zwitterions, their physical properties are characteristic of ionic compounds. For instance, most

$$H_2N-\underset{\underset{H}{|}}{\overset{\overset{R}{|}}{C_\alpha}}-COOH$$

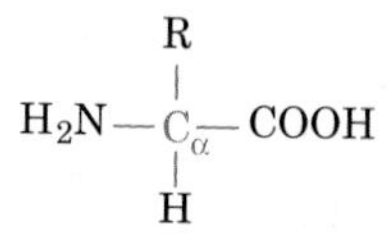

Figure 4-1
The general structural formula for α-amino acids. There are 20 different R groups in the commonly occurring amino acids (Table 4-1).

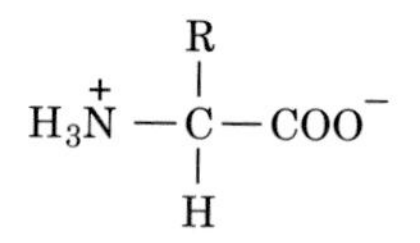

$$H_3\overset{+}{N}-\underset{\underset{H}{|}}{\overset{\overset{R}{|}}{C}}-COO^-$$

Figure 4-2
The zwitterionic form of the α-amino acids that occurs at physiological pH values.

Table 4-2

The pK Values for the Ionizing Groups of the "Standard" α-Amino Acids at 25°C

α-Amino Acid	pK_1 α-COOH	pK_2 α-NH$_3^+$	pK_R Side Chain
Alanine	2.35	9.87	
Arginine	1.82	8.99	12.48 (guanidino)
Asparagine	2.1	8.84	
Aspartic acid	1.99	9.90	3.90 (β-COOH)
Cysteine	1.92	10.78	8.33 (sulfhydryl)
Glutamic acid	2.10	9.47	4.07 (γ-COOH)
Glutamine	2.17	9.13	
Glycine	2.35	9.78	
Histidine	1.80	9.33	6.04 (imidazole)
Isoleucine	2.32	9.76	
Leucine	2.33	9.74	
Lysine	2.16	9.18	10.79 (ε-NH$_3^+$)
Methionine	2.13	9.28	
Phenylalanine	2.16	9.18	
Proline	2.95	10.65	
Serine	2.19	9.21	
Threonine	2.09	9.10	
Tryptophan	2.43	9.44	
Tyrosine	2.20	9.11	10.13 (phenol)
Valine	2.29	9.74	

Source: Dawson, R. M. C., Elliott, D. C., Elliott, W. H., and Jones, K. M., *Data for Biochemical Research* (2nd ed.), *pp. 1–63*, Oxford University Press (1969).

α-amino acids have melting points near 300°C, whereas their nonionic derivatives usually melt around 100°C. Furthermore, amino acids, like other ionic compounds, are more soluble in polar solvents than in nonpolar solvents. Indeed, most α-amino acids are very soluble in water but are largely insoluble in most organic solvents.

B. Peptide Bonds

The α-amino acids polymerize, at least conceptually, through the elimination of a water molecule as is indicated in Fig. 4-3. The resulting CO—NH linkage is known as a **peptide bond.** Polymers composed of two, three, a few (3–10), and many **amino acid residues**

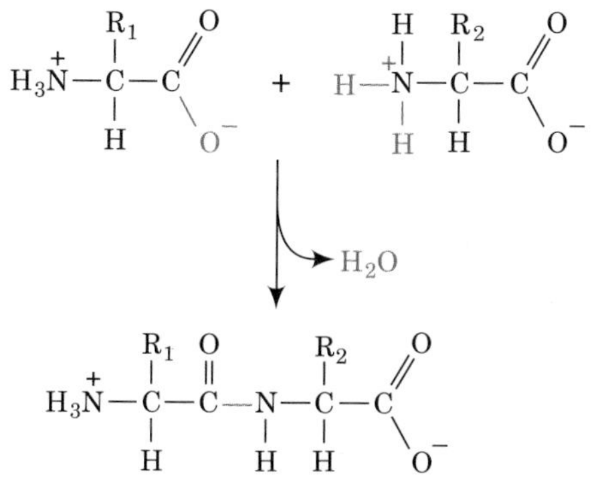

Figure 4-3
The condensation of two α-amino acids to form a dipeptide. The peptide bond is shown in red.

(alternatively called **peptide units**) are known, respectively, as **dipeptides, tripeptides, oligopeptides,** and **polypeptides.** These substances, however, are often referred to simply as "peptides." *Proteins are molecules that consist of one or more polypeptide chains.* These polypeptides range in length from ~40 to over 4000 amino acid residues and, since the average mass of an amino acid residue is ~110 D, have molecular masses that range from ~4 to over 440 kD.

Polypeptides are linear polymers; that is, each amino acid residue is linked to its neighbors in a head-to-tail fashion rather than forming branched chains. This observation reflects the underlying elegant simplicity of the way living systems construct these macromolecules for, as we shall see, the nucleic acids that encode the amino acid sequences of polypeptides are also linear polymers. This permits the direct correspondence between the monomer (nucleotide) sequence of a nucleic acid and the monomer (amino acid) sequence of the corresponding polypeptide without the added complication of specifying the positions and sequences of any branching chains.

With 20 different choices available for each amino acid residue in a polypeptide chain, it is easy to see that a huge number of different protein molecules can exist. For example, for dipeptides, each of the 20 different choices for the first amino acid residue can have 20 different choices for the second amino acid residue, for a total of $20^2 = 400$ distinct dipeptides. Similarly, for tripeptides, there are 20 possibilities for each of the 400 choices of dipeptides to yield a total of $20^3 = 8000$ different tripeptides. A more or less typical protein molecule consists of a single polypeptide chain of 100 residues. There are $20^{100} = 1.27 \times 10^{130}$ possible unique polypeptide chains of this length, a quantity vastly greater than the estimated number of atoms in the universe (9×10^{78}). Clearly, nature could have made only a tiny fraction of the possible different protein molecules. Nevertheless, *the various organisms on earth collectively synthesize an enormous number of different protein molecules whose great range of physicochemical characteristics stem largely from the varied properties of the 20 "standard" amino acids.*

C. Classification and Characteristics

The most common and perhaps the most useful way of classifying the 20 "standard" amino acids is according to the polarities of their side chains (**R groups**). This is because proteins fold to their native conformations largely in response to the tendency to remove their hydrophobic side chains from contact with water and to solvate their hydrophilic side chains (Chapters 7 and 8). According to this classification scheme, there are three major types of amino acids: (1) those with nonpolar R groups, (2) those with uncharged polar R groups, and (3) those with charged polar R groups.

The Nonpolar Amino Acid Side Chains Have a Variety of Shapes and Sizes

Nine amino acids are classified as having nonpolar side chains. **Glycine** (which, when it was found to be a component of gelatin in 1820, was the first amino acid to be identified in protein hydrolysates) has the smallest possible side chain, an H atom. **Alanine, valine, leucine,** and **isoleucine** have aliphatic hydrocarbon side chains ranging in size from a methyl group for alanine to isomeric butyl groups for leucine and isoleucine. **Methionine** has a thiol ether side chain, which resembles an *n*-butyl group in many of its physical properties (C and S have nearly equal electronegativities and S is about the size of a methylene group). **Proline,** a cyclic α-imino acid, has conformational constraints imposed by the cyclic nature of its pyrrolidine side group which is unique among the "standard" 20 amino acids. **Phenylalanine,** with its phenyl moiety, and **tryptophan** with its indole group, contain aromatic side groups, which are characterized by bulk as well as nonpolarity.

Uncharged Polar Side Chains Have Hydroxyl, Amide, or Thiol Groups

Six amino acids are commonly classified as having uncharged polar side chains. **Serine** and **threonine** bear hydroxylic R groups of different sizes. **Asparagine** and **glutamine** have amide-bearing side chains of different sizes. **Tyrosine** has a phenolic group. **Cysteine** has a thiol group that is unique among the 20 amino acids in that it often forms a disulfide bond to another cysteine residue (Fig. 4-4) through the oxidation of their thiol groups. This dimeric compound is referred to in the older biochemical literature as the amino acid **cystine.** The disulfide bridge has great importance in protein structure: *It can join separate polypeptide chains or cross-link two cysteines in the same chain.* The confusing similarity between the names cysteine and cystine has led to the former occasionally being referred to as a **half-cys-**

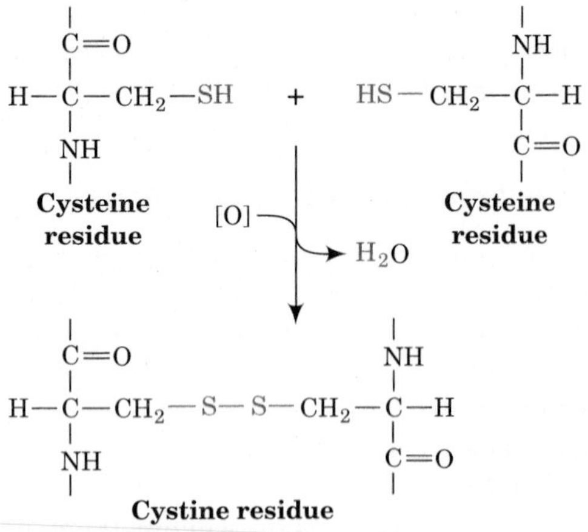

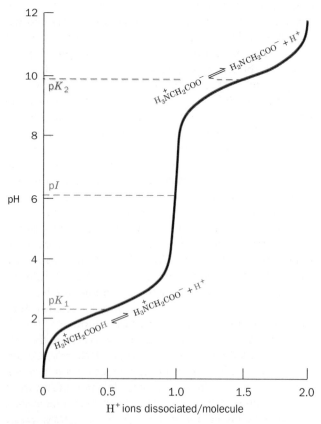

Figure 4-4
The cystine residue consists of two disulfide-linked cysteine residues.

tine residue. However, the realization that cystine arises through the cross-linking of two cysteine residues after polypeptide biosynthesis has occurred, caused the name cystine to become less commonly used.

Charged Polar Side Chains May Be Positively or Negatively Charged

Five amino acids have charged side chains. The basic amino acids are positively charged at physiological pH values and comprise **lysine,** which has a butylammonium side chain, **arginine,** which bears a guanidino group, and **histidine,** which carries an imidazolium moiety. Of the 20 α-amino acids, only histidine, with $pK_R = 6.0$, ionizes within the physiological pH range. At pH 6.0, its imidazole side group is only 50% charged so that at the basic end of the physiological pH range, histidine is neutral. As a consequence, histidine side chains often participate in the catalytic reactions of enzymes. The acidic amino acids, **aspartic acid** and **glutamic acid,** are negatively charged above pH 3; in their ionized state, they are often referred to as **aspartate** and **glutamate.** Asparagine and glutamine are, respectively, the amides of aspartic acid and glutamic acid.

The allocation of the 20 amino acids among the three different groups is, of course, rather arbitrary. For example, glycine and alanine, the smallest of the amino acids, and tryptophan, with its heterocyclic ring, might just as well be classified as uncharged polar amino acids. Similarly, tyrosine and cysteine, with their ionizable side chains, might also be thought of as charged polar amino acids, particularly at higher pH values, while asparagine and glutamine are nearly as polar as their corresponding carboxylates, aspartate and glutamate.

The 20 amino acids vary considerably in their physicochemical properties such as polarity, acidity, basicity, aromaticity, bulk, conformational flexibility, ability to cross-link, ability to hydrogen bond, and chemical reactivity.

These several characteristics, many of which are interrelated, are largely responsible for proteins' great range of properties.

D. Acid–Base Properties

Amino acids and proteins have conspicuous acid–base properties. The α-amino acids have two or, for those with ionizable side groups, three acid–base groups. The titration curve of glycine, the simplest amino acid, is shown in Fig. 4-5. At low pH values, both acid–base groups of glycine are fully protonated so that it assumes the cationic form $^+H_3NCH_2COOH$. In the course of the titration with a strong base, such as NaOH, glycine loses two protons in the stepwise fashion characteristic of a polyprotic acid.

The pK values of glycine's two ionizable groups are sufficiently different so that the Henderson–Hasselbalch equation:

$$pH = pK + \log \frac{[A^-]}{[HA]} \qquad [2.6]$$

closely approximates each leg of its titration curve. Consequently, the pK for each ionization step is that of the

Figure 4-5
The titration curve of glycine. Other monoamino, monocarboxylic acids ionize in a similar fashion. [After Meister, A., *Biochemistry of Amino Acids* (2nd ed.), Vol. 1, p. 30, Academic Press (1965).]

midpoint of its corresponding leg of the titration curve (Sections 2-2A and C): at pH 2.35 the concentrations of the cationic form, $^+H_3NCH_2COOH$, and the zwitterionic form, $^+H_3NCH_2COO^-$, are equal and similarly, at pH 9.78 the concentrations of this zwitterionic form and the anionic form, $H_2NCH_2COO^-$, are equal. Note that *amino acids never assume the neutral form in aqueous solution.*

The pH at which a molecule carries no net electric charge is known as its **isoelectric point, p*I*.** For the α-amino acids, the application of the Henderson–Hasselbalch equation indicates that, to a high degree of precision,

$$pI = \tfrac{1}{2}(pK_i + pK_j) \qquad [4.1]$$

where K_i and K_j are the dissociation constants of the two ionizations involving the neutral species. For monoamino, monocarboxylic acids such as glycine, K_i and K_j represent K_1 and K_2. However, for aspartic and glutamic acids, K_i and K_j are K_1 and K_R whereas, for arginine, histidine and lysine, these quantities are K_R and K_2.

Acetic acid's pK (4.76), which is typical of aliphatic monocarboxylic acids, is ~ 2.4 pH units higher than the pK_1 of its α-amino derivative glycine. This large difference in pK values of the same functional group is caused, as is discussed in Section 2-2C, by the electrostatic influence of glycine's positively charged ammonium group; that is, its NH_3^+ group helps repel the proton from its COOH group. Conversely, glycine's carboxylate group increases the basicity of its amino group (p$K_2 = 9.78$) with respect to that of glycine methyl ester (p$K = 7.75$). However, the NH_3^+ groups of glycine and its esters are significantly more acidic than are aliphatic amines (p$K \approx 10.7$) because of the electron-withdrawing character of the carboxylate group.

The electronic influence of one functional group upon another is rapidly attenuated as the distance between the groups increases. Hence, the pK values of the α-carboxylate groups of amino acids and the side chain carboxylates of aspartic and glutamic acids form a series that is progressively closer in value to the pK of an aliphatic monocarboxylic acid. Likewise, the ionization constant of lysine's side chain amino group is indistinguishable from that of an aliphatic amine.

Proteins Have Complex Titration Curves

The titration curves of the α-amino acids with ionizable side chains, such as that of glutamic acid, exhibit the expected three pK values. However, the titration curves of polypeptides and proteins, an example of which is shown in Fig. 4-6, rarely provide any indication of individual pK values because of the large numbers of ionizable groups they represent (typically 25% of a protein's amino acid side chains are ionizable; Table 4-1). Furthermore, the covalent and three-dimensional structure of a protein may cause the pK of each ionizable

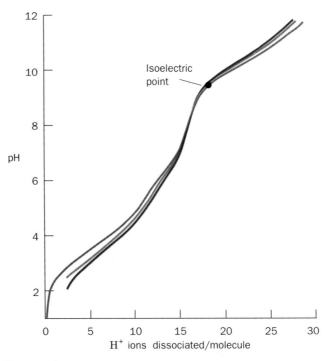

Figure 4-6
Titration curves of the enzyme ribonuclease A at 25°C. The concentration of KCl is 0.01*M* for the blue curve, 0.03*M* for the red curve, and 0.15*M* for the green curve. [After Tanford, C. and Hauenstein, J. D., *J. Am. Chem. Soc.* **78**, 5287 (1956).]

group to shift by as much as several pH units from its value in the free α-amino acid as a result of the electrostatic influence of nearby charged groups, medium effects arising from the proximity of groups of low dielectric constant, and the effects of hydrogen bonding associations. The titration curve of a protein is also a function of the salt concentration, as is shown in Fig. 4-6, because the salt ions act electrostatically to shield the side chain charges from one another, thereby attenuating these charge–charge interactions.

E. A Few Words on Nomenclature

The three letter abbreviations for the 20 amino acid residues are given in Table 4-3. It is worthwhile memorizing these symbols because they are widely used throughout the biochemical literature, including this text. These abbreviations are, in most cases, taken from the first three letters of the corresponding amino acid's name; they are conversationally pronounced as read.

The symbol **Glx** means Glu or Gln and, similarly, **Asx** means Asp or Asn. These ambiguous symbols stem from laboratory experience: Asn and Gln are easily hydrolyzed to aspartic acid and glutamic acid, respectively, under the acidic or basic conditions that are usually used to excise them from proteins (Section 6-1D). Therefore, without special precautions, we cannot determine whether a detected Glu was originally Glu or Gln and likewise for Asp.

Table 4-3
The Three and One Letter Symbols for the "Standard" α-Amino Acid Residues

Amino Acid	Three Letter Symbol	One Letter Symbol[a]
Alanine	Ala	A
Arginine	Arg	R
Asparagine	Asn	N
Aspartic acid	Asp	D
Asparagine *or* aspartic acid	Asx	B
Cysteine	Cys	C
Glutamic acid	Glu	E
Glutamine	Gln	Q
Glutamine *or* glutamic acid	Glx	Z
Glycine	Gly	G
Histidine	His	H
Isoleucine	Ile	I
Leucine	Leu	L
Lysine	Lys	K
Methionine	Met	M
Phenylalanine	Phe	F
Proline	Pro	P
Serine	Ser	S
Threonine	Thr	T
Tryptophan	Trp	W
Tyrosine	Tyr	Y
Valine	Val	V

[a] The one letter symbol for an undetermined or "nonstandard" amino acid is X.

The one letter symbols for the amino acids are also given in Table 4-3. This more compact code is particularly useful when comparing the amino acid sequences of several similar proteins.

Amino acid residues in polypeptides are named by dropping the suffix **-ine** in the name of the amino acid and replacing it by **-yl.** Polypeptide chains are described by starting at the amino terminus (known as the **N-terminus**) and sequentially naming each residue until the carboxyl terminus (the **C-terminus**) is reached. The amino acid at the C-terminus is given the name of its parent amino acid. Thus alanyltyrosylaspartylglycine is the compound shown in Fig. 4-7. Of course such names for polypeptide chains of more than a few residues are extremely cumbersome. The use of abbreviations for amino acid residues partially relieves this problem. Thus the foregoing tetrapeptide is Ala-Tyr-Asp-Gly using the three letter abbreviations and AYDG using the one letter symbols. Note that these abbreviations are always written so that the N-terminus of the polypeptide chain is to the left and the C-terminus is to the right.

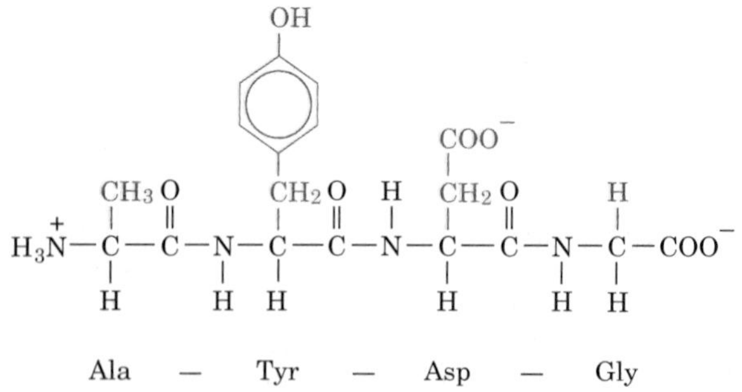

Figure 4-7
The tetrapeptide Ala-Tyr-Asp-Gly.

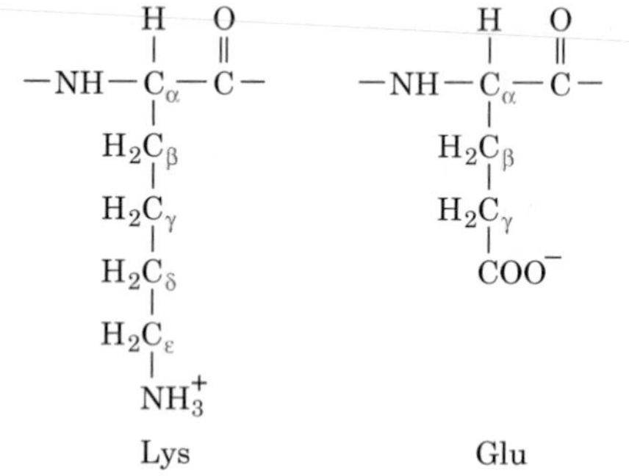

Figure 4-8
The Greek lettering scheme used to identify atoms in lysyl and glutamyl R groups.

The various atoms of the amino acid side chains are often named in sequence with the Greek alphabet starting at the carbon atom adjacent to the peptide carbonyl group. Therefore, as Fig. 4-8 indicates, the Lys residue is said to have an ε-amino group and Glu has a γ-carboxyl group. Unfortunately, this labeling system is ambiguous for several amino acids. Consequently, standard numbering schemes for organic molecules are also employed. These are indicated in Table 4-1 for the heterocyclic side chains.

2. OPTICAL ACTIVITY

The amino acids as isolated by the mild hydrolysis of proteins are, with the exception of glycine, all **optically active;** that is, they rotate the plane of plane-polarized light (see below).

Optically active molecules have an asymmetry such that they are not superimposable on their mirror image in the same way that a left hand is not superimposable on its mirror image, a right hand. This situation is characteristic of substances that contain tetrahedral carbon atoms that have four different substituents. The two molecules depicted in Fig. 4-9 are not superimposable since they are

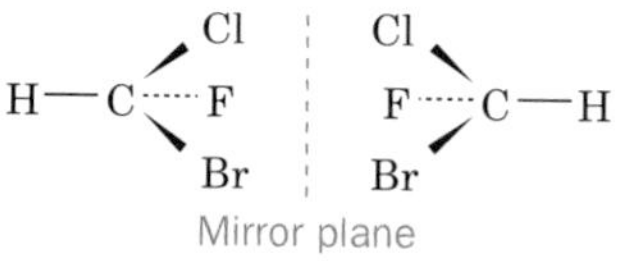

Cl | Cl
H—C····F F····C—H
Br | Br
Mirror plane

Figure 4-9
The two enantiomers of fluorochlorobromomethane. The four substituents are tetrahedrally arranged about the central atom with the dotted lines indicating that a substituent lies behind the plane of the paper, a triangular line indicating that it lies above the plane of the paper, and a thin line indicating that it lies in the plane of the paper. The mirror plane relating the enantiomers is represented by a vertical dashed line.

mirror images. The central atoms in such atomic constellations are known as **asymmetric centers,** or **chiral centers,** and are said to have the property of **chirality** (Greek: *cheir,* hand). The C_α atoms of all the amino acids, with the exception of glycine, are asymmetric centers. Glycine, which has two H atoms substituent to its C_α atom, is superimposable on its mirror image and is therefore not optically active.

Molecules that are nonsuperimposable mirror images are known as **enantiomers** of one another. Enantiomeric molecules are physically and chemically indistinguishable by most techniques. *Only when probed asymmetrically, for example, by plane-polarized light or by reactants that also contain chiral centers, can they be distinguished and/or differentially manipulated.*

There are three commonly used systems of nomenclature whereby a particular stereoisomer of an optically active molecule can be classified. These are explained as follows:

A. An Operational Classification

Molecules are classified as **dextrorotatory** (Greek: *dextro,* right) or **levorotatory** (Greek: *levo,* left) depending on whether they rotate the plane of plane-polarized light clockwise or counterclockwise from the point of view of the observer. This can be determined by an instrument known as a **polarimeter** (Fig. 4-10). A quantitative measure of the optical activity of the molecule is known as its **specific rotation:**

$$[\alpha]_D^{25} = \frac{\text{observed rotation (degree)}}{\text{optical path} \times \text{concentration}} \quad [4.2]$$
$$\text{length (dm)} \quad (g \cdot cm^{-3})$$

where the superscript 25 refers to the temperature at which polarimeter measurements are customarily made (25°C) and the subscript D indicates the monochromatic light that is traditionally employed in polarimetry, the so-called D-line in the spectrum of sodium (589.3 nm). Dextrorotatory and levorotatory molecules are assigned positive and negative values of $[\alpha]_D^{25}$. Dextrorotatory molecules are therefore designated by the prefix (+) and their levorotatory enantiomers have the prefix (−). In an equivalent but archaic nomenclature, the lower case letters *d (dextro)* and *ℓ (levo)* are used.

The sign and magnitude of a molecule's specific rotation depend on the structure of the molecule in a complicated and poorly understood manner. It is not yet possible to predict reliably the magnitude or even the sign of a given molecule's specific rotation. For example, proline, leucine, and arginine, which are isolated from proteins, have specific rotations in pure aqueous solu-

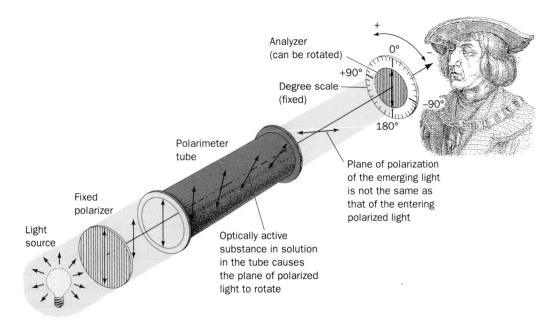

Figure 4-10
Schematic diagram of a polarimeter, a device used to measure optical rotation.

tions of -86.2, -10.4, and $+12.5°C$, respectively. Their enantiomers exhibit values of $[\alpha]_D^{25}$ of the same magnitude but of opposite sign. As might be expected from the acid–base nature of the amino acids, these quantities vary with the solution pH.

A problem with this operational classification system for optical isomers is that it provides no presently interpretable indication of the **absolute configuration** (spatial arrangement) of the chemical groups about a chiral center. Furthermore, a molecule with more than one asymmetric center may have an optical rotation that is not obviously related to the rotatory powers of the individual asymmetric centers. For this reason, the following relative classification scheme is more useful.

B. The Fischer Convention

In this system, the configuration of the groups about an asymmetric center is related to that of **glyceraldehyde,** a molecule with one asymmetric center. By a convention introduced by Emil Fischer in 1891, the $(+)$ and $(-)$ stereoisomers of glyceraldehyde are designated **D-glyceraldehyde** and **L-glyceraldehyde,** respectively (note the use of small upper case letters). With the realization that there was only a 50% chance that he was correct, Fischer assumed that the configurations of these molecules were those shown in Fig. 4-11. Fischer also proposed a convenient shorthand notation for these molecules, known as **Fischer projections,** which are also given in Fig. 4-11. In the Fischer convention, horizontal bonds extend above the plane of the paper and vertical bonds extend below the plane of the paper as is explicitly indicated by the accompanying geometrical formulas.

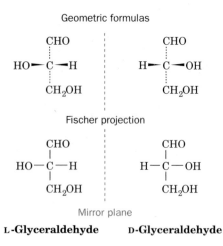

Geometric formulas

CHO   CHO

HO—C—H   H—C—OH

CH₂OH   CH₂OH

Fischer projection

CHO   CHO

HO—C—H   H—C—OH

CH₂OH   CH₂OH

Mirror plane

L-Glyceraldehyde   **D**-Glyceraldehyde

Figure 4-11
The Fischer convention configurations for naming the enantiomers of glyceraldehyde as represented by geometric formulas (*top*) and their corresponding Fischer projection formulas (*bottom*). Note that in Fischer projection, all horizontal bonds point above the page and all vertical bonds point below the page. The mirror planes relating the enantiomers are represented by a vertical dashed line.

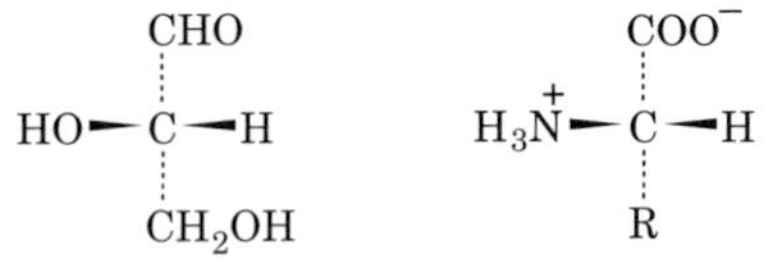

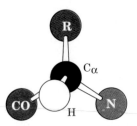

L-Glyceraldehyde   **L-α-Amino Acid**

Figure 4-12
The configurations of L-glyceraldehyde and L-α-amino acids.

Figure 4-13
The "corncrib" mnemonic for the hand of L-amino acids. Looking at the C_α atom from its H atom substituent, its other substituents should read **CO—R—N** in the clockwise direction as shown. Here CO, R, and N, respectively, represent the carbonyl group, side chain, and main chain nitrogen atom. [After Richardson, J. S., *Adv. Protein Chem.* **34,** 171 (1981).]

The configuration of groups about a chiral center can be related to that of glyceraldehyde by chemically converting these groups to those of glyceraldehyde using reactions of known stereochemistry. For α-amino acids, the arrangement of the amino, carboxyl, R, and H groups about the C_α atom is related to that of the hydroxyl, aldehyde, CH_2OH, and H groups, respectively, of glyceraldehyde. In this way, L-glyceraldehyde and L-α-amino acids are said to have the same relative configurations (Fig. 4-12). Through the use of this method, the configurations of the α-amino acids can be described without reference to their specific rotations.

All α-amino acids derived from proteins have the L-stereochemical configuration; that is, they all have the same relative configuration about their C_α atoms. In 1949, it was demonstrated by a then new technique in X-ray crystallography that Fischer's arbitrary choice was correct: The designation of the relative configuration of chiral centers is the same as their absolute configuration. The absolute configuration of L-α-amino acid residues may be easily remembered through the use of the "corncrib" mnemonic that is diagrammed in Fig. 4-13.

Diastereomers Are Chemically and Physically Distinguishable

A molecule may have multiple asymmetric centers. For such molecules, the terms **stereoisomers** and **optical isomers** refer to molecules with different configurations about at least one of their chiral centers, but which are otherwise identical. The term enantiomer still refers

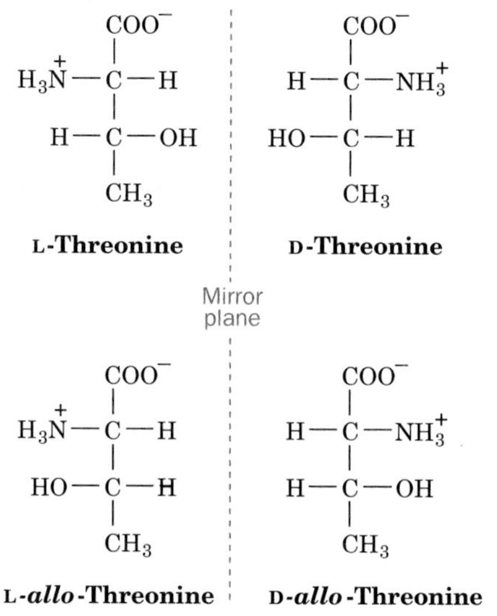

```
      COO⁻                    COO⁻
       |                       |
H₃N⁺ — C — H           H — C — NH₃⁺
       |                       |
 H — C — OH           HO — C — H
       |                       |
      CH₃                     CH₃
   L-Threonine            D-Threonine
```

Mirror
plane

```
      COO⁻                    COO⁻
       |                       |
H₃N⁺ — C — H           H — C — NH₃⁺
       |                       |
HO — C — H             H — C — OH
       |                       |
      CH₃                     CH₃
 L-allo-Threonine      D-allo-Threonine
```

Figure 4-14
Fischer projections of threonine's four stereoisomers. The D and L forms are mirror images as are the D-*allo* and L-*allo* forms. D- and L-threonine are each diastereomers of both D-*allo*- and L-*allo*-threonine.

to a molecule that is the mirror image of the one under consideration, that is, different in all its chiral centers. Since each asymmetric center in a chiral molecule can have two possible configurations, a molecule with n chiral centers has 2^n different possible stereoisomers and 2^{n-1} enantiomeric pairs. Threonine and isoleucine each have two chiral centers and hence $2^2 = 4$ possible stereoisomers. The forms of threonine and isoleucine that are isolated from proteins, which are by convention called the L forms, are indicated in Table 4-1. The mirror images of the L forms are the D forms. Their other two optical isomers are said to be **diastereomers** (or allo forms) of the enantiomeric D and L forms. The relative configurations of all four stereoisomers of threonine are given in Fig. 4-14. Note the following points:

1. The D-*allo* and L-*allo* forms are mirror images of each other as are the D and L forms. Neither allo form is symmetrically related to either of the D or L forms.

2. In contrast to the case for enantiomeric pairs, diaste-

reomers are physically and chemically distinguishable from one another by ordinary means such as melting points, spectra, and chemical reactivity; that is, they are really different compounds in the usual sense.

A special case of diastereoisomerism occurs when the two asymmetric centers are chemically identical. Two of the four Fischer projections of the sort shown in Fig. 4-14 then represent the same molecule. This is because the two asymmetric centers in this molecule are mirror images of each other. Such a molecule is superimposible on its mirror image and is therefore optically inactive. This so-called meso form is said to be **internally compensated**. The three optical isomers of cystine are shown in Fig. 4-15 where it can be seen that the D and L isomers are mirror images of each other as before. Only L-cystine occurs in proteins.

C. The Cahn – Ingold – Prelog System

Despite its usefulness, the Fischer scheme is awkward and sometimes ambiguous for molecules with more than one asymmetric center. For this reason, the following absolute nomenclature scheme was formulated in 1956 by Robert Cahn, Christopher Ingold, and Vladimir Prelog. In this system, the four groups surrounding a chiral center are ranked according to a specific priority scheme: *Atoms of higher atomic number bonded to a chiral center are ranked above those of lower atomic number.* For example, the oxygen atom of an OH group takes precedence over the carbon atom of a CH_3 group that is bonded to the same chiral C atom. If any of the first substituent atoms are of the same element, the priority of these groups is established from the atomic numbers of the second, third, *etc.*, atoms outward from the asymmetric center. Hence a CH_2OH group takes precedence over a CH_3 group. There are other rules (given in the references and in many organic chemistry texts) for assigning priority ratings to substituents with multiple bonds or differing isotopes. The order of priority of some common functional groups is

$$SH > OH > NH_2 > COOH > CHO$$
$$> CH_2OH > C_6H_5 > CH_3 > {}^2H > {}^1H$$

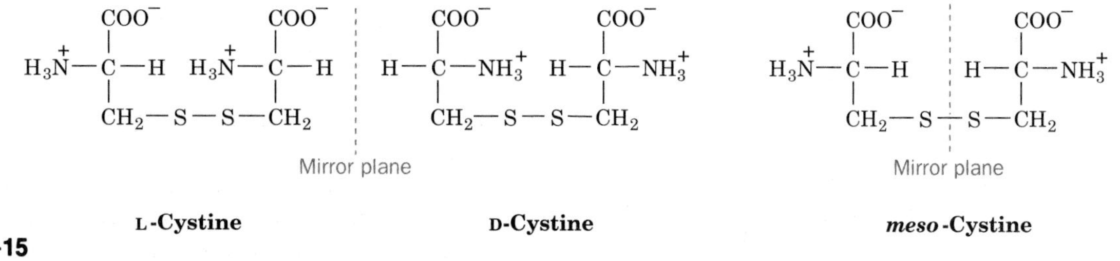

```
    COO⁻        COO⁻            COO⁻       COO⁻            COO⁻        COO⁻
     |           |               |          |               |           |
H₃N⁺—C—H   H₃N⁺—C—H    H—C—NH₃⁺  H—C—NH₃⁺   H₃N⁺—C—H   H—C—NH₃⁺
     |           |               |          |               |           |
    CH₂—S — S—CH₂        CH₂— S—S—CH₂          CH₂—S ⁞ S—CH₂
          Mirror plane                                      Mirror plane
      L-Cystine              D-Cystine                  meso-Cystine
```

Figure 4-15
The three stereoisomers of cystine. The D and L forms are related by mirror symmetry, whereas the meso form has internal mirror symmetry and therefore lacks optical activity.

Note that each of the groups substituent to a chiral center must have a different priority rating; otherwise the center could not be asymmetric.

The prioritized groups are assigned the letters *W, X, Y, Z* such that their order of priority rating is $W > X > Y > Z$. To establish the configuration of the chiral center, it is viewed from the asymmetric center towards the *Z* group (lowest priority). *If the order of the groups* $W \to X \to Y$ *as seen from this direction is clockwise, then the configuration of the asymmetric center is designated (R)* (Latin: *rectus*, right). *If the order of* $W \to X \to Y$ *is counterclockwise, the asymmetric center is designated (S)* (Latin: *sinistrus*, left). L-Glyceraldehyde is therefore designated (*S*)-glyceraldehyde (Fig. 4-16) and similarly, L-alanine is (*S*)-alanine (Fig. 4-17). In fact, all the L-amino acids from proteins are (*S*)-amino acids, with the exception of L-cysteine, which is (*R*)-cysteine.

A major advantage of this so-called **Cahn–Ingold–Prelog** or **(RS) system** is that the chiralities of compounds with multiple asymmetric centers can be unambiguously described. Thus, in the (*RS*) system, L-threonine is (2*S*,3*R*)-threonine, whereas L-isoleucine is (2*S*,3*S*)-isoleucine (Fig. 4-18).

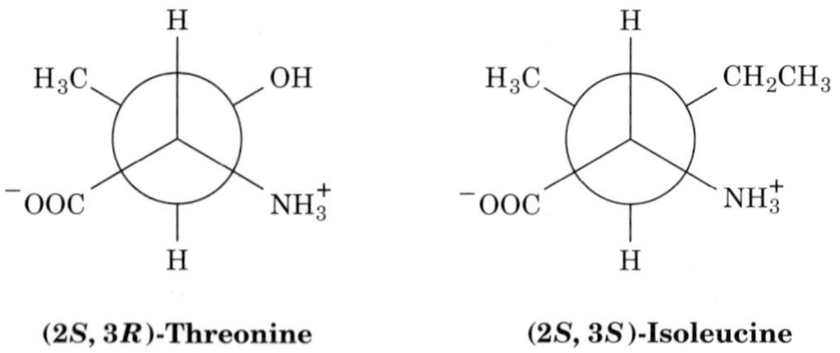

Prochiral Centers Have Distinguishable Substituents

Two chemically identical substituents to an otherwise chiral tetrahedral center are geometrically distinct; that is, the center has no rotational symmetry so that it can be

(2*S*, 3*R*)-Threonine   (2*S*, 3*S*)-Isoleucine

Figure 4-18
Newman projection diagrams of the stereoisomers of threonine and isoleucine derived from proteins. Here the C_α—C_β bond is viewed end on. The nearer atom, C_α, is represented by the confluence of the three bonds to its substituents, whereas the more distant atom, C_β, is represented by a circle from which its three substituents project.

unambiguously assigned left and right sides. Consider, for example, the substituents to the C(1) atom of ethanol (the CH_2 group). If one of the H atoms was converted to another group (not CH_3 or OH), C(1) would be a chiral center. The two H atoms are therefore said to be **prochiral**. If we arbitrarily assign the H atoms the subscripts *a* and *b* (Fig. 4-19), then H_b is said to be *pro-R* because in sighting from C(1) towards H_a (as if it was the Z group of a chiral center), the order of priority of the other substituents decreases in a clockwise direction. Similarly, H_a is said to be *pro-S*.

Planar objects with no rotational symmetry also have the property of prochirality. For example, in many enzymatic reactions, stereospecific addition to a trigonal carbon atom occurs from a particular side of that carbon atom to yield a chiral center (Section 12-2A). If a trigonal carbon is facing the viewer such that the order of priority of its substituents decreases in a clockwise manner (Fig. 4-20*a*), that face is designated as the *re* **face** (after *rectus*). The opposite face is designated as the *si* **face** (after *sinistrus*) since the priorities of its substituents decrease in the counterclockwise direction (Fig. 4-20*b*). Comparison of Figs. 4-19*a* and 4-20*a* indicates that an H atom adding to the *re* side of acetaldehyde atom C(1) occupies the *pro-R* position of the resulting tetrahedral center. Conversely, a *pro-S* H atom is generated by *si* side addition to this trigonal center (Figs. 4-19*b* and 4-20*b*).

The (*RS*) system is presently not widely used in biochemistry. One reason for this is that closely related compounds, which have the same configurational representation under the Fischer DL convention, may have different representations under the (*RS*) system. Consequently, we shall use the Fischer convention in most cases. The (*RS*) system, however, is indispensable for describing prochirality and stereospecific reactions so that we shall find it invaluable for describing enzymatic reactions.

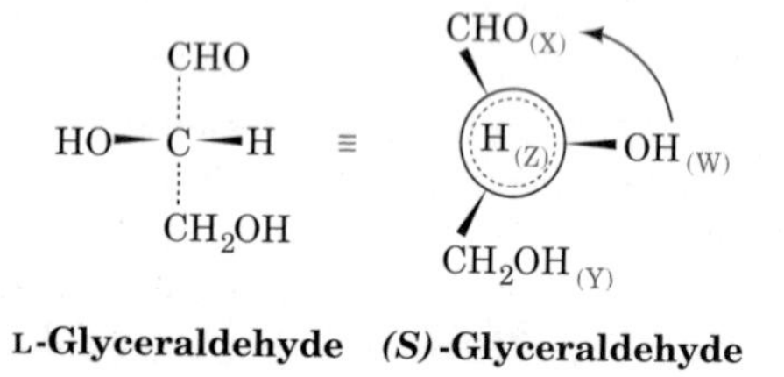

L-Glyceraldehyde   (*S*)-Glyceraldehyde

Figure 4-16
The structural formula of L-glyceraldehyde and its equivalent (*RS*)-system representation indicating that it is (*S*)-glyceraldehyde. In the latter drawing, the chiral C atom is represented by the large circle, and the H atom, which is located behind the plane of the paper, is represented by the smaller concentric dashed circle.

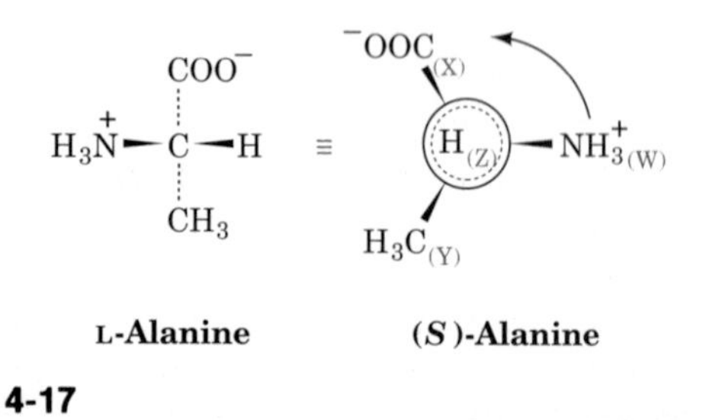

L-Alanine   (*S*)-Alanine

Figure 4-17
The structural formula of L-alanine and its equivalent (*RS*)-system representation indicating that it is (*S*)-alanine.

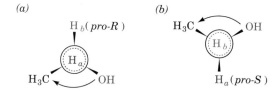

Figure 4-19
Views of ethanol: (*a*) Looking from C(1) to H$_a$, the *pro-S* H atom (the dotted circle). (*b*) Looking from C(1) to H$_b$, the *pro-R* H atom.

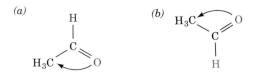

Figure 4-20
Views of acetaldehyde onto (*a*) its *re* face, and (*b*) its *si* face.

D. Chirality and Biochemistry

The ordinary chemical synthesis of chiral molecules produces **racemic** mixtures of these molecules (equal amounts of each member of an enantiomeric pair) because ordinary chemical and physical processes have no stereochemical bias. Consequently, there are equal probabilities for an asymmetric center of either hand to be produced in any such process. In order to produce a product with net optical activity, a chiral process must be employed. This usually takes the form of using chiral reagents although, at least in principle, the use of any asymmetric influence such as light that is plane polarized in one direction can produce a net asymmetry in a reaction product.

One of the most striking characteristics of life is its production of optically active molecules. *The biosynthesis of a substance possessing asymmetric centers almost invariably produces a pure stereoisomer.* The fact that the amino acid residues of proteins all have the L configuration is just one example of this phenomenon. This observation has prompted the suggestion that a simple diagnostic test for the past or present existence of extraterrestrial life, be it on moon rocks or in meteorites that have fallen to earth, would be the detection of net optical activity in these materials. Any such finding would suggest that the asymmetric molecules thereby detected had been biosynthetically produced. Thus, even though α-amino acids have been extracted from carbonaceous meteorites, the observation that they come in racemic mixtures suggests that they are of chemical rather than biological origin.

One of the enigmas of the origin of life is why terrestrial life is based on certain chiral molecules rather than their enantiomers; that is, on L-amino acids, for example, rather than D-amino acids. Arguments that physical effects such as polarized light might have promoted significant net asymmetry in prebiotically synthesized molecules (Section 1-4B) have not been convincing. Perhaps L-amino acid-based life forms arose at random and simply "ate" any D-amino acid-based life forms.

3. "NONSTANDARD" AMINO ACIDS

The 20 common amino acids are by no means the only amino acids that occur in biological systems. "Nonstandard" amino acids are often important constituents of proteins and biologically active polypeptides. Many amino acids, however, are not constituents of proteins. Together with their derivatives, they play a variety of biologically important roles.

A. Amino Acid Derivatives in Proteins

The "universal" genetic code, which is nearly identical in all known life forms (Section 30-1), specifies only the 20 "standard" amino acids of Table 4-1. Nevertheless, many other amino acids, a selection of which is given in Fig. 4-21, are components of certain proteins. *In all known cases but one (Section 30-2D), however, these unusual amino acids result from the specific modification of an amino acid residue after the polypeptide chain has been synthesized.* Among the most prominent of these modified amino acid residues are **4-hydroxyproline** and **5-hydroxylysine.** Both of these amino acid residues are important structural constituents of the fibrous protein **collagen,** the most abundant protein in mammals (Section 7-2C). Amino acids of proteins that form complexes with nucleic acids are often modified. For example, ribosomal proteins (Section 30-3A) and the chromosomal proteins known as **histones** (Section 33-1A) may be specifically methylated, acetylated, and/or phosphorylated. A selection of these derivatized amino acid residues is presented in Fig. 4-21. *N*-**Formylmethionine** is initially the N-terminal residue of all prokaryotic proteins, but is usually removed as part of the protein maturation process (Section 30-3C). **γ-Carboxyglutamic acid** is a constituent of several proteins involved in blood clotting (Section 34-1B).

B. Specialized Roles of Amino Acids

Besides their role in proteins, amino acids and their derivatives have many biologically important functions. A few examples of these substances are shown in Fig. 4-22. This alternative use of amino acids is an example of the biological opportunism that we shall repeatedly encounter: *Nature tends to adapt materials and processes that are already present to new functions.*

Amino acids and their derivatives often function as

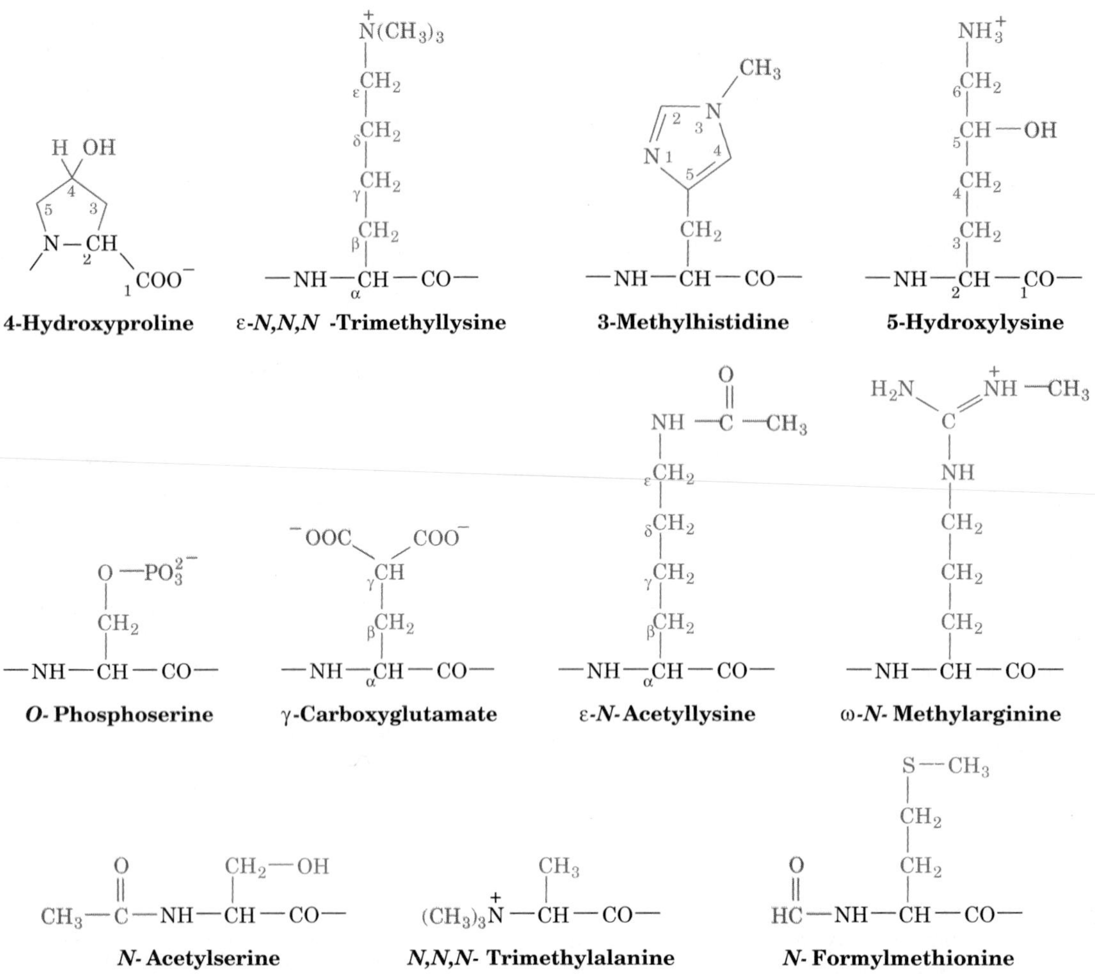

Figure 4-21
Some uncommon amino acid residues that are components of certain proteins. All of these residues are modified from one of the 20 "standard" amino acids after polypeptide chain biosynthesis. Those amino acid residues that are derivatized at their N_α position occur at the N-termini of proteins.

chemical messengers in the communications between cells. For example, glycine, **γ-aminobutyric acid (GABA;** a glutamate decarboxylation product), and **dopamine** (a tyrosine product) are neurotransmitters (substances released by nerve cells to alter the behavior of their neighbors; Section 34-4C); **histamine** (the decarboxylation product of histidine) is a potent local mediator of allergic reactions; and **thyroxine** (a tyrosine product) is an iodine-containing thyroid hormone that generally stimulates vertebrate metabolism (Section 34-4A).

Certain amino acids are important intermediates in various metabolic processes. Among them are **citrulline** and **ornithine,** intermediates in urea biosynthesis (Section 24-2B); **homocysteine,** an intermediate in amino acid metabolism (Section 24-3E); and *S*-**adenosylmethionine,** a biological methylating reagent (Section 24-3E).

Nature's diversity is remarkable. About 250 different amino acids have been found in various plants and fungi. For the most part, their biological roles are ob-

scure although the fact that many are toxic suggests that they have a protective function. Indeed, some of them, such as **azaserine,** are medically useful antibiotics. Many of these amino acids are simple derivatives of the 20 "standard" amino acids although some of them, including azaserine and **β-cyanoalanine** (Fig. 4-22), have unusual structures.

Although only L-amino acids occur in proteins, **D-amino acids** are present in many organisms. These substances, which are directly synthesized, are perhaps most widely distributed as constituents of bacterial cell walls (Section 10-3B). Perhaps their function in this capacity is a defensive one as D-amino acids render the bacterial cell walls less susceptible to attack by the **peptidases** (enzymes that hydrolyze peptide bonds) that many organisms employ to digest bacterial cell walls. D-Amino acids also occur as components of many antibiotics including **valinomycin** (Section 18-2C), **actinomycin D** (Section 29-2D), and **gramicidin S** (Section 30-7).

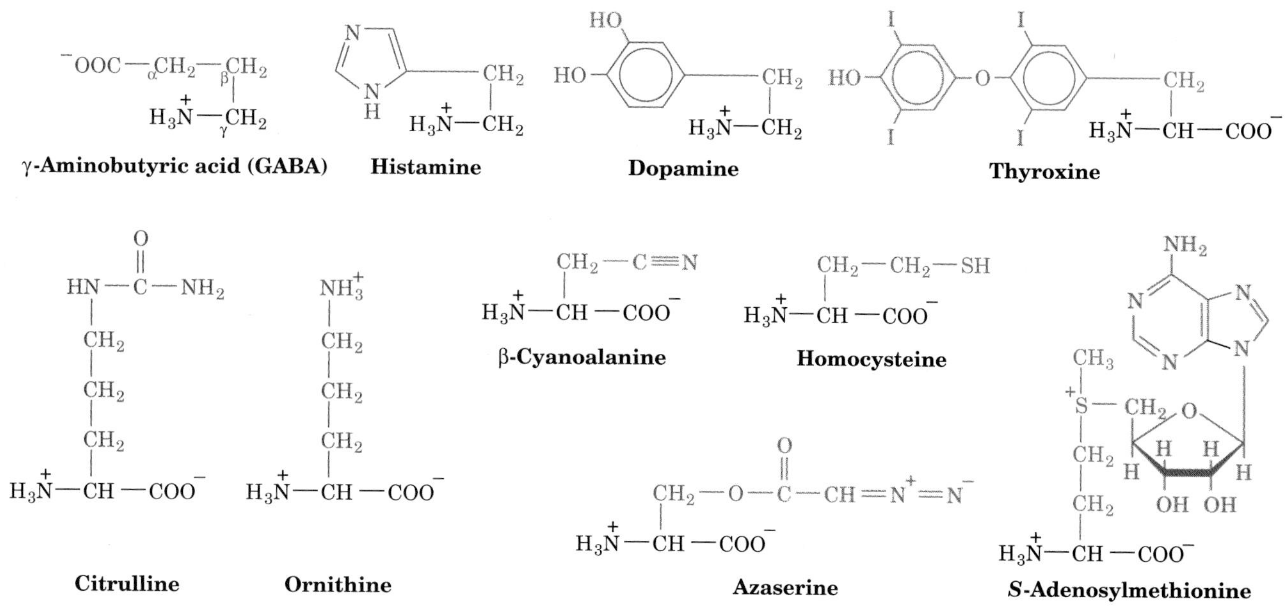

Figure 4-22
Some biologically produced derivatives of "standard" amino acids and amino acids that are not components of proteins.

Chapter Summary

Proteins are linear polymers that are synthesized from the same 20 "standard" α-amino acids through their condensation to form peptide bonds. These amino acids all have a carboxyl group with a pK near 2.2 and an amino substituent with a pK near 9.4 attached to the same carbon atom, the C_α atom. The α-amino acids are zwitterionic compounds, ^+H_3N—CHR—COO$^-$, in the physiological pH range. The various amino acids are usually classified according to the polarities of their side chains, R, which are also substituent to the C_α atom. Glycine, alanine, valine, leucine, isoleucine, methionine, proline (which is really an α-imino acid), phenylalanine, and tryptophan are nonpolar amino acids; serine, threonine, asparagine, glutamine, tyrosine, and cysteine are uncharged and polar; and lysine, arginine, histidine, aspartic acid, and glutamic acid are charged and polar. The side chains of many of these amino acids bear acid–base groups and hence the properties of the proteins containing them are pH dependent.

The C_α atoms of all α-amino acids except glycine each bear four different substituents and are therefore chiral centers. According to the Fischer convention, which relates the configuration of D- or L-glyceraldehyde to that of the asymmetric center of interest, all the amino acids of proteins have the L configuration; that is, they all have the same absolute configuration about their C_α atom. According to the Cahn–Ingold–Prelog (*RS*) system of chirality nomenclature, they are, with the exception of cysteine, all (*S*)-amino acids. The side chains of threonine and isoleucine also contain chiral centers. A prochiral center has no rotational symmetry and hence its substituents, in the case of a central atom, or its faces, in the case of a planar ring, are distinguishable.

Amino acid residues other than the 20 from which proteins are synthesized also have important biological functions. These "nonstandard" residues result from the specific chemical modifications of amino acid residues in preexisting proteins. Amino acids and their derivatives also have independent biological roles such as neurotransmitters, metabolic intermediates, and poisons.

References

History

Vickery, H. B. and Schmidt, C. L. A., The history of the discovery of amino acids, *Chem. Rev.* **9**, 169–318 (1931).

Vickery, H. B., The history of the discovery of the amino acids. A review of amino acids discovered since 1931 as components of native proteins, *Adv. Protein Chem.* **26**, 81–171 (1972).

Properties of Amino Acids

Barrett, G. C. (Ed.), *Chemistry and Biochemistry of Amino Acids,* Chapman & Hall (1985).

Cohn, E. J. and Edsall, J. T., *Proteins, Amino Acids and Peptides as Ions and Dipolar Ions,* Academic Press (1943). [A classic work in its field.]

Davies, J. S. (Ed.), *Amino Acids and Peptides,* Chapman & Hall (1985). [A "sourcebook" on amino acids.]

Edsall, J. T. and Wyman, J., *Biophysical Chemistry,* Vol. 1, Academic Press (1958). [A detailed treatment of the physical chemistry of amino acids.]

Jakubke, H.-D. and Jeschkeit, H., *Amino Acids, Peptides and Proteins,* translated into English by Cotterrell, G. P., Wiley (1977).

Meister, A., *Biochemistry of the Amino Acids* (2nd ed.), Vol. 1, Academic Press (1965). [A compendium of information on amino acid properties.]

Optical Activity

Cahn, R. S., An introduction to the sequence rule, *J. Chem. Educ.* **41**, 116–125 (1964). [A presentation of the Cahn–Ingold–Prelog system of nomenclature.]

Mislow, K., *Introduction to Stereochemistry,* Benjamin (1966).

"Nonstandard" Amino Acids

Amino Acids and Peptides, The Royal Society of Chemistry. [An annual series containing literature reviews on amino acids.]

Fowden, L., Lea, P. J., and Bell, E. A., The non-protein amino acids of plants, *Adv. Enzymol.* **50**, 117–175 (1979).

Fowden, L., Lewis, D., and Tristram, H., Toxic amino acids: their action as antimetabolites, *Adv. Enzymol.* **29**, 89–163 (1968).

Problems

1. Name the 20 standard amino acids without looking them up. Give their three letter and one letter abbreviations.

2. Draw the following oligopeptides in their predominant ionic forms at pH 7. (a) Phe-Met-Arg, (b) tryptophanyllysylaspartic acid, and (c) Gln-Ile-His-Thr.

3. How many different pentapeptides are there that contain one residue each of Gly, Asp, Tyr, Cys, and Leu?

4. Draw the structures of the following two oligopeptides with their cysteine residues cross-linked by a disulfide bond: Val-Cys; Ser-Cys-Pro.

*5. What are the concentrations of the various ionic species in a 0.1M solution of lysine at pH 4, 7, and 10?

6. Derive Eq. [4-1] for a monoamino, monocarboxylic acid (use the Henderson–Hasselbalch equation).

*7. The **isoionic point** of a compound is defined as the pH of a pure water solution of the compound. What is the isoionic point of a 0.1M solution of glycine?

8. Normal human hemoglobin has an isoelectric point of 6.87. A mutant variety of hemoglobin, known as **sickle-cell hemoglobin,** has an isoelectric point of 7.09. The titration curve of hemoglobin indicates that, in this pH range, ~13 groups change ionization states per unit change in pH. Calculate the difference in ionic charge between molecules of normal and sickle-cell hemoglobin.

9. Indicate whether the following familiar objects are chiral, prochiral, or nonchiral.

(a) A glove	(g) A snowflake
(b) A tennis ball	(h) A spiral staircase
(c) A good pair of scissors	(i) A normal staircase
(d) A screw	(j) A paper clip
(e) This page	(k) A shoe
(f) A toilet paper roll	(l) A pair of glasses

10. Draw all four equivalent Fischer projection formulas for L-alanine (see Figs. 4-11 and 4-12).

*11. (a) Draw the structural formula and the Fischer projection formula of (S)-3-methylhexane. (b) Draw all the stereoisomers of 2,3-dichlorobutane. Name them according to the (RS) system and indicate which of them has the meso form.

*12. Identify and name the prochiral centers or faces of the following molecules.

(a) Acetone	(c) Glycine	(e) Lysine
(b) Propene	(d) Alanine	(f) 3-Methylpyridine

Chapter 5
TECHNIQUES OF PROTEIN PURIFICATION

A major portion of most biochemical investigations involves the purification of the materials under consideration because these substances must be relatively free of contaminants if they are to be properly characterized. This is often a formidable task because a typical cell contains thousands of different substances, many of which closely resemble other cellular constituents in their physical and chemical properties. Furthermore, the material of interest may be unstable and exist in vanishingly small amounts. Typically, a substance that comprises $<0.1\%$ of a tissue's dry weight is brought to $\sim98\%$ purity. Purification problems of this magnitude would be considered unreasonably difficult by most synthetic chemists. It is therefore hardly surprising that our understanding of biochemical processes has by and large paralleled our ability to purify biological materials.

This chapter presents an overview of the most commonly used techniques for the isolation, purification and, to some extent, the characterization of proteins as well as other types of biological molecules. These methods are the basic tools of biochemistry whose operation dominates the day-to-day efforts of the practicing biochemist. Furthermore, many of these techniques are routinely used in clinical applications. Indeed, *a basic comprehension of the methods described here is necessary for an appreciation of the significance and the limitations of much of the information presented in this text.* This chapter should therefore be taken as reference material to be repeatedly consulted as the need arises while reading other chapters. Techniques that are specific for the purification of biological molecules other than proteins are described in the appropriate chapters.

1. PROTEIN ISOLATION

Proteins constitute a major fraction of the mass of all organisms. A particular protein, such as **hemoglobin** in red blood cells, may be the dominant substance present in a tissue. Alternatively, a protein such as the *lac repressor* of *E. coli* (Section 29-3B) may normally have a population of only a few molecules per cell. Similar techniques are used for the isolation and purification of both proteins although, in general, the lower the initial concentration of a substance, the more effort is required to isolate it in pure form.

In this section we discuss the care and handling of proteins and outline the general strategy for their purification. For many proteins, the isolation and purification procedure is an exacting task requiring days of effort to obtain only a few milligrams or less of the desired product. However, as we shall see, modern analytical techniques have achieved such a high degree of sensitivity that this small amount of material is often sufficient to characterize a protein extensively. You should note that the techniques described in this chapter are applicable to the separations of most types of biological molecules.

A. Selection of a Protein Source

Proteins with identical functions generally occur in a variety of organisms. For example, most of the enzymes that mediate basic metabolic processes or that are involved in the expression and transmission of genetic information are common to all cellular life. Of course, there is usually considerable variation in the properties of a particular protein from various sources. In fact, different variants of a given protein may occur in different tissues from the same organism or even in different compartments in the same cell. Therefore, if flexibility of choice is possible, the isolation of a protein may be greatly simplified by a judicious choice of the protein source. This choice should be based on such criteria as the ease of obtaining sufficient quantities of the tissue from which the protein is to be isolated, the amount of the chosen protein in that tissue, and any properties peculiar to the specific protein chosen that would aid in its stabilization and isolation. Tissues from domesticated animals such as chickens, cows, pigs, or rats are often chosen. Alternative sources might be easily obtainable microorganisms such as *E. coli* or baker's yeast (*Saccharomyces cerevisiae*). We shall see, however, that proteins from a vast variety of organisms have been studied.

The recent development of **molecular cloning techniques** (Section 28-8) has generated an entirely new protein production method. Almost any protein-encoding gene can be isolated from its parent organism, specifically altered (genetically engineered) if desired, and expressed at high levels (overproduced) in a conveniently grown organism such as *E. coli* or yeast. Indeed, the cloned protein may constitute up to 40% of the overproducer's total cell protein. This high level of protein production generally renders the cloned protein far easier to isolate than it would be from its parent organism (in which it may normally occur in vanishingly small amounts).

B. Methods of Solubilization

The first step in the isolation of a protein, or any other biological molecule, is to get it into solution. In some cases, such as with blood serum proteins, nature has already done so. However, a protein must usually be liberated from the cells that contain it. The method of choice for this procedure depends on the mechanical characteristics of the source tissue as well as on the location of the required protein in the cell.

If the protein of interest is located in the cytosol of the cell, its liberation requires only the breaking open **(lysis)** of the cell. In the simplest and gentlest method of doing so, which is known as **osmotic lysis,** the cells are suspended in a **hypotonic solution;** that is, the cells are in a solution in which the total molar concentration of the solutes is less than that of the cell in its normal physiological state. Under the influence of osmotic forces, water diffuses into the more concentrated intracellular solution, thereby causing the cells to swell and burst. This method works well with animal cells, but with cells that have a cell wall, such as bacteria or plant cells, it is usually ineffective. The use of an enzyme, such as **lysozyme,** which chemically degrades bacterial cell walls (Section 14-2), is sometimes effective with such cells. Detergents or organic solvents such as acetone or toluene are also useful in lysing cells but care must be exercised in their use as they may denature the protein of interest (Section 7-4E).

Many cells require some sort of mechanical disruption process to break them open. This may include grinding with sand or alumina, the use of a high speed blender (similar to the familiar kitchen appliance), a homogenizer (an implement for crushing tissue between a closely fitting piston and sleeve), a French press (a device that shears open cells by squirting them at high pressure through a small orifice), or sonication (breaking open cells through the use of ultrasonic vibrations). Once the cells have been broken open, the crude **lysate** may be filtered or centrifuged to remove the particulate cell debris, thereby leaving the protein of interest in the supernatant solution.

If the required protein is a component of subcellular assemblies such as membranes or mitochondria, a considerable purification of the protein can be effected by first separating the subcellular assembly from the rest of the cellular material. This is usually accomplished by **differential centrifugation,** a process in which the cell lysate is centrifuged at a speed that removes only the cell components denser than the desired organelle followed by centrifugation at a speed that spins down the component of interest. The required protein is then usually separated from the purified subcellular component by extraction with concentrated salt solutions or, in the case of proteins tightly bound to membranes, with the use of detergent solutions or organic solvents, such as butanol or glycerol, that solubilize lipids.

C. Stabilization of Proteins

Once a protein has been removed from its natural environment, it becomes exposed to many agents that can irreversibly damage it. These influences must be carefully controlled at all stages of a purification process or the yield of the desired protein may be greatly reduced or even eliminated.

The structural integrity of many proteins is a sensitive function of pH as a consequence of their numerous acid–base groups. To prevent damage to biological materials due to variations in pH, they are routinely dissolved in buffer solutions effective in the pH range over which the material is stable.

Proteins are easily destroyed by high temperatures. Although the thermal stabilities of proteins vary widely, many of them slowly denature above 25°C. Therefore, the purification of proteins is normally carried out at temperatures near 0°C. However, there are numerous proteins that require lower temperatures, some even lower than −100°C, for stability. Conversely, some **cold-labile** proteins become unstable below characteristic temperatures.

The thermal stability characteristics of a protein can sometimes be used to advantage in its purification. A heat stable protein in a crude mixture can be greatly purified by briefly heating the mixture so as to denature and precipitate most of the contaminating proteins without affecting the desired protein.

Cells contain **proteases** (enzymes that cleave the peptide bonds of proteins) and other degradative enzymes which, upon lysis, are liberated into solution along with the protein of interest. Care must be taken that the protein is not damaged by these enzymes. Degradative enzymes may often be rendered inactive at pH's and temperatures that are not harmful to the protein of interest. Alternatively, these enzymes can often be specifically inhibited by chemical agents without affecting the desired protein. Of course, as the purification of a protein progresses, more and more of these degradative enzymes are eliminated.

Some proteins are more resistant than others to proteolytic degradation. The purification of a protein that is particularly resistant to proteases may be effected by maintaining conditions in a crude protein mixture under which the proteolytic enzymes present are active. This so-called **autolysis** technique simplifies the purification of the resistant protein because it is generally far easier to remove selectively the degradation products of contaminating proteins than it is the intact proteins.

Proteins are often denatured at low protein concentrations and by contact with surfaces so that a protein solution should be kept relatively concentrated and handled so as to minimize frothing. There are, of course, other factors to which a protein may be sensitive, including the oxidation of cysteine residues to form disulfide bonds; heavy metal contaminants, which may irreversibly bind to the protein; and the salt concentration and polarity of the solution, which must be kept within the stability range of the protein. Finally, many microorganisms consider proteins to be delicious so that proteins should be stored under conditions that inhibit the growth of microorganisms.

D. Assay of Proteins

In order to purify any substance, some means must be found for quantitatively detecting its presence. A protein rarely comprises more than a few percent by weight of its tissue of origin and is usually present in much smaller amounts. Yet, much of the material from which it is being extricated closely resembles the protein of interest. Accordingly, an assay must be specific for the protein being purified and highly sensitive to its presence. Furthermore, the assay must be convenient to use because it is done repeatedly at every stage of the purification process.

Among the most straightforward of protein assays are those for enzymes that catalyze reactions with readily detectable products. Perhaps such a product has a characteristic spectroscopic absorption or fluorescence that can be monitored. Alternatively, the enzymatic reaction may consume or generate acid so that the enzyme can be assayed by acid–base titrations. If an enzymatic reaction product is not easily quantitated, its presence may still be revealed by further chemical treatment to yield a more readily observable product. Often, this takes the form of a **coupled enzymatic reaction,** in which the product of the enzyme being assayed is converted, by an added enzyme, to an observable substance.

Proteins that are not enzymes may be assayed through the observation of their biological effects. For example, the presence of a hormone may be revealed by its effect on some standard tissue sample or on a whole organism. Such assays are often rather lengthy procedures because the response elicited by the assay may take days to develop. In addition, their reproducibility is often less than satisfactory because of the complex behavior of living systems. Such assays are therefore used only when no alternative procedure is available.

Immunochemical Techniques Can Readily Detect Small Quantities of Specific Proteins

Immunochemical procedures provide protein assay techniques of high sensitivity and discrimination. These methods employ **antibodies,** proteins produced by an

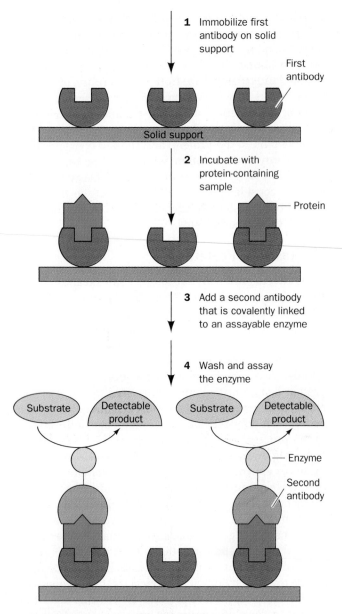

1 Immobilize first antibody on solid support

First antibody

Solid support

2 Incubate with protein-containing sample

Protein

3 Add a second antibody that is covalently linked to an assayable enzyme

4 Wash and assay the enzyme

Substrate — Detectable product — Substrate — Detectable product

Enzyme

Second antibody

Figure 5-1
An enzyme-linked immunosorbant assay (ELISA).

animal's immune system in response to the introduction of a foreign protein and which specifically bind to this foreign protein (antibodies and the immune system are discussed in Section 34-2).

Antibodies extracted from the blood serum of an animal that has been immunized against a particular protein are the products of many different antibody-producing cells. They therefore form a heterogeneous mixture of molecules, which vary in their exact specificities and binding affinities for their target protein. Antibody-producing cells normally die after a few cell divisions so that one of them cannot be cultured (cloned) to produce a single species of antibody in useful quantities. Such **monoclonal antibodies** may be obtained, however, by fusing a cell producing the desired antibody

with a cell of an immune system cancer known as a **myeloma** (Section 34-2B). The resulting **hybridoma** cell has an unlimited capacity to divide and, when raised in cell culture, produces large quantities of the monoclonal antibody.

A protein can be directly detected through its precipitation by its corresponding antibodies or, in a so-called **radioimmunoassay,** it can be indirectly detected by determining the degree with which it competes with a radioactively labeled standard for binding to the antibody (Section 34-4A). In an **enzyme-linked immunosorbant assay (ELISA;** Fig. 5-1):

1. Antibody against the protein of interest is immobilized on an inert solid such as polystyrene.

2. The solution being assayed for the protein is applied to the antibody-coated surface under conditions which the antibody binds the protein.

3. The resulting protein-antibody complex is further reacted with a second protein-specific antibody to which an easily assayed enzyme has been covalently linked.

4. After washing away any unbound antibody-linked enzyme, the enzyme in the immobilized antibody–protein–antibody–enzyme complex is assayed thereby indicating the amount of the protein present.

Both radioimmunoassays and ELISAs are widely used to detect small amounts of specific proteins and other biological substances in both laboratory and clinical applications. For example, a commonly available pregnancy test, which is reliably positive within a few days post-conception, uses an ELISA to detect the placental hormone **chorionic gonadotropin** (Section 34-4A) in the mother's urine.

E. General Strategy of Protein Purification

The fact that proteins are well-defined substances was not widely accepted until after 1926 when James Sumner first crystallized an enzyme, jack bean **urease.** Before that, it was thought that the high molecular masses of proteins resulted from a colloidal aggregation of rather ill defined and mysterious substances of lower molecular masses. Once it was realized that it was possible, in principle, to purify proteins, work to do so began in earnest.

In the first half of the twentieth century, the protein purification methods available were extremely crude by today's standards. Protein purification was an arduous task that was as much an art as a science. Usually, the development of a satisfactory purification procedure for a given protein was a matter of years of labor ultimately involving huge quantities of starting material. Nevertheless, by 1940, ~20 enzymes had been obtained in pure form.

Since then, several thousand proteins have been purified and characterized to varying extents. Modern techniques of separation have such a high degree of discrimination that one can now obtain, in quantity, a series of proteins with such similar properties that only a few years ago their mixture was thought to be a pure substance. Nevertheless, the development of an efficient procedure for the purification of a given protein may still be an intellectually challenging and time-consuming task.

Proteins are purified by fractionation procedures. In a series of independent steps, the various physicochemical properties of the protein of interest are utilized to separate it progressively from other substances. The idea here is not necessarily to minimize the loss of the desired protein but rather, to eliminate selectively the other components of the mixture such that only the required substance remains.

It may not be philosophically possible to prove that a substance is pure. However, *the operational criterion for establishing purity takes the form of the method of exhaustion: the demonstration, by all available methods, that the sample of interest consists of only one component.* Therefore, as new separation techniques are devised, standards of purity may have to be revised. Experience has shown that when a sample of material previously thought to be pure is subjected to a new separation technique, it is occasionally found to be a mixture of several components.

The characteristics of proteins and other biomolecules that are utilized in the various separation procedures are solubility, ionic charge, molecular size, adsorption properties, and binding affinity to other biological molecules. In the remainder of this chapter, we discuss these separation procedures.

2. SOLUBILITIES OF PROTEINS

A protein's multiple acid–base groups make its solubility properties dependent on the concentration of dissolved salts, the polarity of the solvent, the pH, and the temperature. Different proteins vary greatly in their solubilities under a given set of conditions: Certain proteins precipitate from solution under conditions in which others remain quite soluble. This effect is routinely used as the basis for protein purification.

A. Effects of Salt Concentrations

The solubility of a protein in aqueous solution is a sensitive function of the concentrations of dissolved salts (Figs. 5-2–5-4). The salt concentration in Figs. 5-2 and 5-3 is expressed in terms of the **ionic strength**, *I*, which is defined

$$I = \tfrac{1}{2} \sum c_i Z_i^2 \qquad [5.1]$$

where c_i is the molar concentration of the *i*th ionic species and Z_i is its ionic charge. The use of this parameter to account for the effects of ionic charges results from theoretical considerations of ionic solutions. However, as

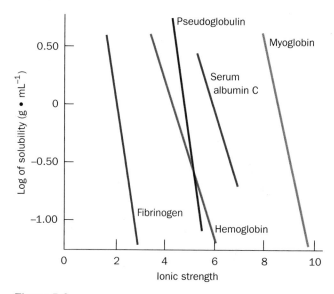

Figure 5-2
Solubilities of several proteins in ammonium sulfate solutions. [After Cohn, E. J. and Edsall, J. T., *Proteins, Amino Acids and Peptides, p.* 602, Academic Press (1943).]

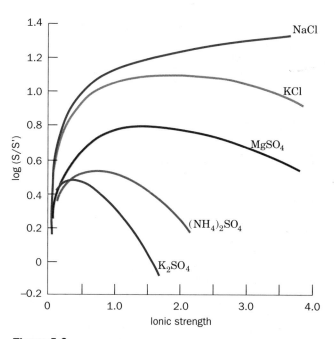

Figure 5-3
Solubility of carboxy-hemoglobin at its isoelectric point as a function of ionic strength and ion type. Here S and S' are, respectively, the solubilities of the protein in the salt solution and in pure water. The logarithm of their ratios is plotted so that the solubility curves can be placed on a common scale. [After Green, A. A., *J. Biol. Chem.* **95,** 47 (1932).]

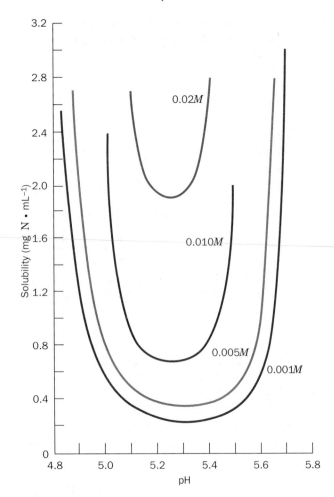

Figure 5-4
Solubility of β-lactoglobulin as a function of pH at several NaCl concentrations. [After Fox, S. and Foster, J. S., *Introduction to Protein Chemistry*, p. 242, Wiley (1957).]

sufficient to dissolve other solutes. In thermodynamic terms, the solvent's activity (effective concentration; Appendix to Chapter 3) is decreased. Hence, solute–solute interactions become stronger than solute–solvent interactions and the solute precipitates.

Salting out is the basis of one of the most commonly used protein purification procedures. Figure 5-2 shows that the solubilities of different proteins vary widely as a function of salt concentration. *By adjusting the salt concentration in a solution containing a mixture of proteins to just below the precipitation point of the protein to be purified, many unwanted proteins can be eliminated from the solution. Then, after the precipitate is removed by filtration or centrifugation, the salt concentration of the remaining solution is increased so as to precipitate the desired protein.* In this manner, a significant purification and concentration of large quantities of protein can be conveniently effected. Consequently, salting out is most often the initial step in protein purification procedures. Ammonium sulfate is the most commonly used reagent for salting out proteins because its high solubility ($3.9M$ in water at $0°C$) permits the achievement of solutions with high ionic strengths (up to 23.4 in water at $0°C$).

Certain ions, notably I^-, ClO_4^-, SCN^-, Li^+, Mg^{2+}, Ca^{2+}, and Ba^{2+} increase the solubilities of proteins rather than salting them out. These ions also tend to denature proteins (Section 7-4E). Conversely, ions that decrease the solubilities of proteins stabilize their native structures so that proteins that have been salted out are not denatured.

B. Effects of Organic Solvents

Water-miscible organic solvents, such as acetone and ethanol, are generally good protein precipitants because their low dielectric constants lower the solvating power of their aqueous solutions for dissolved ions such as proteins. The different solubilities of proteins in these mixed solvents form the basis of a useful fractionation technique. This procedure is normally used near $0°C$ or less because, at higher temperatures, organic solvents tend to denature proteins. The lowering of the dielectric constant by organic solvents also magnifies the differences in the salting out behavior of proteins so that these two techniques can be effectively combined. Some water miscible organic solvents, however, such as dimethyl sulfoxide (DMSO) or *N,N*-dimethylformamide (DMF), are rather good protein solvents because of their relatively high dielectric constants.

C. Effects of pH

Proteins generally bear numerous ionizable groups, which have a variety of pK's. At a pH characteristic for each protein, the positive charges on the molecule exactly balance its negative charges. At this pH, the protein's **isoelectric point,** p*I* (Section 4-1D), the protein

Fig. 5-3 indicates, a protein's solubility at a given ionic strength varies with the type of ions in solution. The order of effectiveness of these various ions in influencing protein solubility is quite similar for different proteins and is apparently mainly due to the ions' size and hydration.

The solubility of a protein at low ionic strength generally increases with the salt concentration (left side of Fig. 5-3 and the different curves of Fig. 5-4). The explanation of this **salting in** phenomenon is that as the salt concentration of the protein solution increases, the additional counterions more effectively shield the protein molecules' multiple ionic charges and thereby increase the protein's solubility.

At high ionic strengths, the solubilities of proteins, as well as those of most other substances, decrease. This effect, known as **salting out,** is primarily a result of the competition between the added salt ions and the other dissolved solutes for molecules of solvation. At high salt concentrations, so many of the added ions are solvated that the amount of bulk solvent available becomes in-

Table 5-1
The Isoelectric Points of Several Common Proteins

Protein	Isoelectric pH
Pepsin	<1.0
Ovalbumin (hen)	4.6
Serum albumin (human)	4.9
Tropomyosin	5.1
Insulin (bovine)	5.4
Fibrinogen (human)	5.8
γ-Globulin (human)	6.6
Collagen	6.6
Myoglobin (horse)	7.0
Hemoglobin (human)	7.1
Ribonuclease A (bovine)	7.8
Cytochrome *c* (horse)	10.6
Histone (bovine)	10.8
Lysozyme (hen)	11.0
Salmine (salmon)	12.1

molecule carries no net charge and is therefore immobile in an electric field.

Figure 5-4 indicates that the solubility of the protein **β-lactoglobulin** is a minimum near its p*I* of 5.2 in dilute NaCl solutions and increases more or less symmetrically about the p*I* with changes in pH. This solubility behavior, which is shared by most proteins, is easily explained. Physicochemical considerations suggest that the solubility properties of uncharged molecules are insensitive to the salt concentration. To a first approximation, therefore, a protein at its isoelectric point should not be subject to salting in. Conversely, as the pH is varied from a protein's p*I*, that is, as the protein's net charge increases, it should be increasingly subject to salting in because the electrostatic interactions between neighboring molecules that promote aggregation and precipitation should likewise increase. Hence, *in solutions of moderate salt concentrations, the solubility of a protein as a function of pH is expected to be at a minimum at the protein's pI and to increase about this point with respect to pH.*

Proteins vary in their amino acid compositions and therefore, as Table 5-1 indicates, in their p*I*'s. This phenomenon is the basis of a protein purification procedure known as **isoelectric precipitation** in which the pH of a protein mixture is adjusted to the p*I* of the protein to be isolated so as to minimize selectively its solubility. In practice, this technique is combined with salting out so that the protein being purified is usually salted out near its p*I*.

D. Crystallization

Once a protein has been brought to a reasonable state of purity, it may be possible to crystallize it. This is usually done by bringing the protein solution just past its saturation point with the types of precipitating agents discussed above. Upon standing for a time (as little as a few minutes, as much as several months), often while the concentration of the precipitating agent is being slowly increased, the protein may precipitate from the solution in crystalline form. It may be necessary to attempt the crystallization under different solution conditions and with various precipitating agents before crystals are obtained. The crystals may range in size from the microscopic to 1 mm or more across. Crystals of the latter size, which generally require great care to grow, may be suitable for X-ray crystallographic analysis (Section 7-3A). Several such crystals are shown in Fig. 5-5.

Repeated recrystallizations of a protein can serve to purify it in much the same manner that smaller molecules are purified by this process. Nevertheless, the ability to crystallize a protein does not reliably indicate that it is pure. Conversely, many highly purified proteins have resisted all efforts to crystallize them.

3. CHROMATOGRAPHIC SEPARATIONS

In 1903, the Russian botanist Mikhail Tswett described the separations of plant leaf pigments in solution

(a)   (b)   (c)

(d)   (e)   (f)

Figure 5-5
Protein crystals: (*a*) azurin from *Pseudomonas aeruginosa*, (*b*) flavodoxin from *Desulfovibrio vulgaris*, (*c*) rubredoxin from *Clostridium pasteurianum*, (*d*) azidomet myohemerythrin from the marine worm *Siphonosoma funafuti*, (*e*) lamprey hemoglobin, and (*f*) bacteriochlorophyll *a* protein from *Prosthecochloris aestuarii*. These proteins are colored because of their associated chromophores (light-absorbing groups); proteins are colorless in the absence of such bound groups. [Parts (*a*)–(*c*) courtesy of Larry Sieker, University of Washington; Parts (*d*) and (*e*) courtesy of Wayne Hendrikson, Columbia University; and (*f*) courtesy of John Olsen, Brookhaven National Laboratories and Brian Matthews, University of Oregon.]

through the use of solid adsorbents. He named this process **chromatography** (Greek: *chroma*, color + *graphein*, to write), presumably resulting from the colored bands that formed in the adsorbents as the components of the pigment mixtures separated from one another (and possibly because Tswett means color in Russian).

Modern separation methods rely heavily on chromatographic procedures. In all of them, a mixture of substances to be fractionated is dissolved in a liquid or gaseous fluid known as the **mobile phase.** The resultant solution is percolated through a column consisting of a porous solid matrix, which in certain types of chromatography may be associated with a bound liquid, and which is known as the **stationary phase.** The interactions of the individual solutes with the stationary phase acts to retard their progress through the matrix in a manner that varies with the properties of each solute. If the mixture being fractionated starts its journey through the column in a narrow band, the different retarding forces on each component that cause them to migrate at different rates will eventually cause the mixture to separate into bands of pure substances.

The power of chromatography derives from the continuous nature of the separation processes. A single purification step (or "plate" as it is often termed in analogy with distillation processes) may have very little tendency to separate a mixture into its components. However, since this process is applied in a continuous fashion so that it is, in effect, repeated hundreds or even hundreds of thousands of times, the segregation of the mixture into its components ultimately occurs. The separated components can then be collected into separate fractions for analysis and/or further fractionation.

The various chromatographic methods are classified according to their mobile and stationary phases. For example, in gas–liquid chromatography the mobile and stationary phases are gaseous and liquid, respectively, whereas in liquid–liquid chromatography they are immiscible liquids, one of which is bound to an inert solid support. Chromatographic methods may be further classified according to the nature of the dominant interaction between the stationary phase and the substances being separated. For example, if the retarding force is ionic in character, the separation technique is referred to as ion exchange chromatography whereas, if it is a result of the adsorption of the solutes onto a solid stationary phase, it is known as adsorption chromatography.

As has been previously mentioned, a cell contains huge numbers of different components, many of which closely resemble one another in their various properties. Therefore, the isolation procedures for most biological substances incorporate a number of independent chromatographic steps in order to purify the substance of interest according to several criteria. In this section, the most commonly used of these chromatographic procedures are described.

A. Ion Exchange Chromatography

*In the process of **ion exchange**, ions that are electrostatically bound to an insoluble and chemically inert matrix are reversibly replaced by ions in solution:*

$$R^+A^- + B^- \rightleftharpoons R^+B^- + A^-$$

Here, R^+A^- is an **anion exchanger** in the A^- form and B^- represents anions in solution. **Cation exchangers** similarly bear negatively charged groups that reversibly bind cations. Polyanions and polycations therefore bind to anion and cation exchangers, respectively. However, proteins and other **polyelectrolytes** (polyionic polymers) that bear both positive and negative charges can bind to both cation and anion exchangers depending on their net charge. *The affinity with which a particular polyelectrolyte binds to a given ion exchanger depends on the identities and concentrations of the other ions in solution because of the competition among these various ions for the binding sites on the ion exchanger. The binding affinities of polyelectrolytes bearing acid–base groups are also highly pH dependent because of the variation of their net charges with pH.* These principles are used to great advantage in isolating biological molecules by ion exchange chromatography.

In purifying a given protein (or some other polyelectrolyte) the pH, the salt concentration, and the ion exchanger are chosen such that the protein to be isolated is immobilized on the ion exchanger. Under a given set of conditions, many proteins are likely to bind to an ion exchanger. Proteins that do not stick to the ion exchanger under these conditions can be removed by suspending the ion exchanger in the buffer in which the protein mixture was dissolved. The protein of interest can later be washed off the ion exchanger with a buffer of a pH and salt concentration that reduces the affinity of this protein for the ion exchanger. Of course, other proteins are also likely to be washed off the ion exchanger under these conditions.

The effectiveness of this **batch** method of purification can be greatly enhanced by the simple expedient of packing the ion exchanger in a column (Fig. 5-6). Various proteins bind to the ion exchanger with different affinities. As the column is washed, a process known as **elution,** *those proteins with relatively low affinities for the ion exchanger will move through the column faster than the proteins that bind to the ion exchanger with higher affinities.* This occurs because the progress of a given protein through the column is retarded relative to that of the solvent resulting from interactions between the protein molecules and the ion exchanger. The greater the binding affinity of a protein for the ion exchanger, the more it will be retarded. With the use of a column and fraction collector, further purification of a protein can be effected by selecting only that fraction of the column effluent that contains the desired protein. This technique,

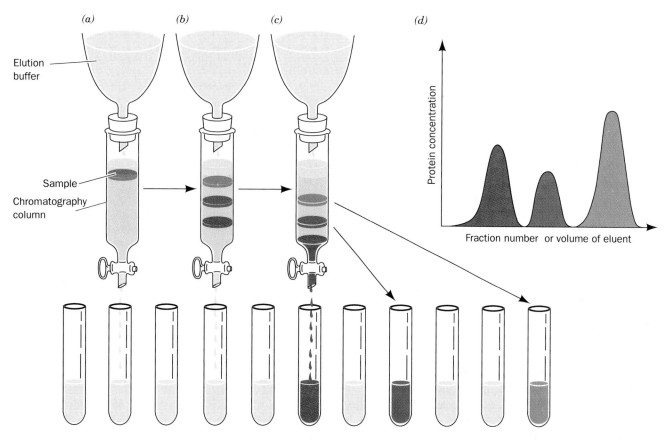

Figure 5-6
A schematic diagram illustrating the separation of several proteins by ion exchange chromatography. Here the shaded area represents the ion exchanger and the colored bands represent the various proteins. (*a*) The protein mixture is bound to the topmost portion of ion exchanger in the chromatography column. (*b*) As the elution progresses, the various proteins separate into discrete bands as a consequence of their different mobilities on the ion exchanger under the prevailing solution conditions. (*c*) The first band of protein has been isolated as a separate fraction of the column effluent. The remaining bands will also be isolated as the elution progresses. (*d*) The elution diagram of the protein mixture from the column.

which is diagrammed in Fig. 5-6, is known as **ion exchange chromatography.**

Gradient Elution Improves Chromatographic Separations

The purification process can be further improved by washing the protein-loaded column using the method of **gradient elution.** Here the salt concentration and/or pH is continuously varied as the column is eluted so as to release sequentially the various proteins that are bound to the ion exchanger. This procedure usually leads to a better separation of proteins than the elution of the column by a single solution or by using a series of different solutions in a stepwise manner.

Many different types of elution gradients have been successfully employed in purifying biological molecules. The most widely used of these is the **linear gra-**dient in which the concentration of the eluant solution varies linearly with the volume of solution passed. A simple device for generating such a gradient is illustrated in Fig. 5-7. Here the solute concentration, c, in the solution being withdrawn from the mixing chamber, is expressed by

$$c = c_2 - (c_2 - c_1)f \qquad [5.2]$$

where c_1 is the solution's initial concentration in the mixing chamber, c_2 is its concentration in the reservoir chamber, and f is the remaining fraction of the combined volumes of the solutions initially present in both reservoirs. Linear gradients of increasing salt concentration are probably more commonly used than all other means of column elution. However, gradients of different shapes can be generated by using two or more chambers of different cross-sectional areas or programmed mixing devices.

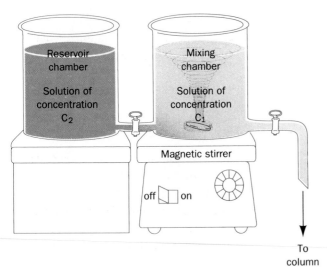

Figure 5-7
A device for generating a linear concentration gradient. Two connected open chambers, which have identical cross-sectional areas, are initially filled with equal volumes of solutions of different concentrations. As the solution of concentration c_1 drains out of the mixing chamber, it is partially replaced by a solution of concentration c_2 from the reservoir chamber. The concentration of the solution in the mixing chamber varies linearly from its initial concentration, c_1, to the final concentration, c_2, as is expressed by Eq. [5.2].

Several Types of Ion Exchangers Are Available

Ion exchangers consist of charged groups covalently attached to a support matrix. The chemical nature of the charged groups determines the types of ions that bind to the ion exchanger and the strength with which they bind. The chemical and mechanical properties of the support matrix govern the flow characteristics, ion accessibility, and stability of the ion exchanger.

Three classes of materials are in general use as support matrices for ion exchangers: (1) **resins** such as **polystyrene** cross-linked with **divinylbenzene** (Fig. 5-8a); (2) cellulose; and (3) cross-linked polyacrylamide or polydextran gels (Section 5-3C; the latter two classes are also colloquially referred to as resins). Table 5-2 contains descriptions of some commercially available ion exchangers in common use.

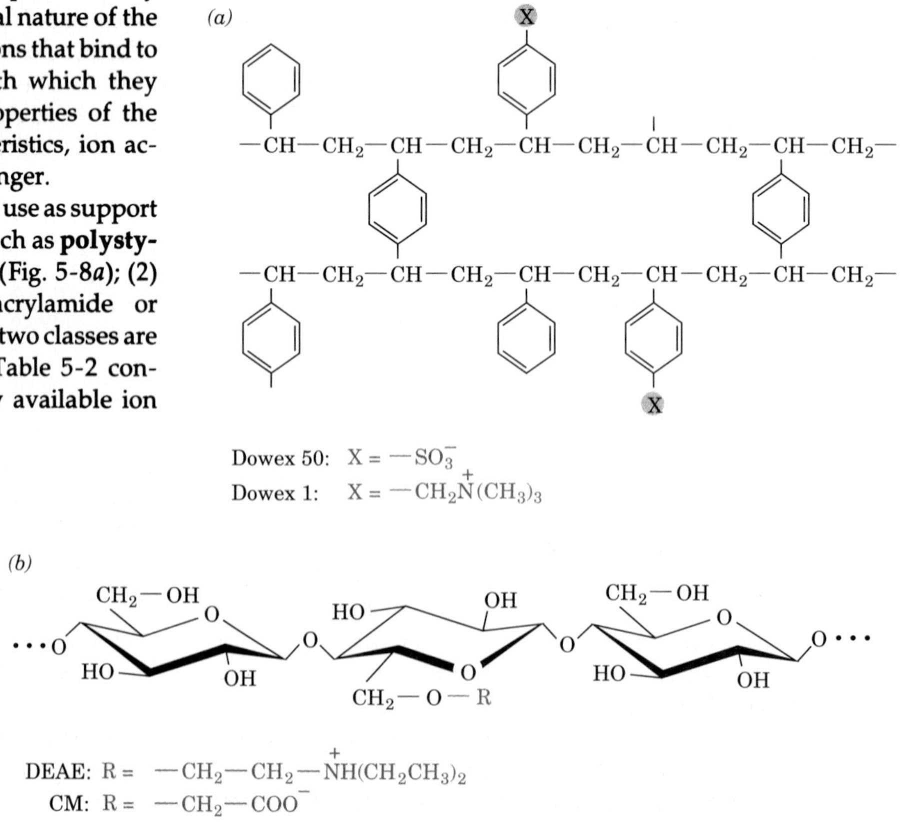

Ion exchange resins are available in granular form. The polymers of cation exchange resins are heavily substituted with anionic groups such as the strongly acidic sulfonic acid residue or weakly acidic carboxyl groups. Similarly, anion exchange resins bear cationic groups such as quaternary or tertiary amines, which are, respectively, strongly and weakly basic. Ion exchange resins are routinely used in the separation of small molecules but have proven to be less effective in the separation of proteins and other macromolecules than are the other types of ion exchangers.

Cellulosic ion exchangers are among the materials most commonly employed to separate biological molecules. The cellulose, which is derived from wood or cotton, is lightly derivatized with ionic groups to form the ion exchanger. The most often used cellulosic anion exchanger is **diethylaminoethyl (DEAE)-cellulose** (Fig. 5-8b), whereas **carboxymethyl (CM)-cellulose** is the most popular cellulosic cation exchanger. An advantage of cellulosic over resin-type ion exchangers for use in separating biological molecules stems from the greater permeability of the former to macromolecular electrolytes. This permits the macromolecules access to the charged groups without the need of using extremely fine particles that would drastically impede the flow of eluant. A further advantage of cellulosic ion exchangers is that their density of charged groups is relatively low, which permits the elution of large polyelectrolytes

(a)

$$-CH-CH_2-CH-CH_2-CH-CH_2-CH-CH_2-CH-CH_2-$$

$$-CH-CH_2-CH-CH_2-CH-CH_2-CH-CH_2-CH-CH_2-$$

Dowex 50: $X = -SO_3^-$
Dowex 1: $X = -CH_2\overset{+}{N}(CH_3)_3$

Figure 5-8
Schematic diagrams of ion exchangers:
(a) Divinylbenzene cross-linked polystyrene ion exchange resin.
(b) Cellulose-based ion exchangers.

(b)

DEAE: $R = -CH_2-CH_2-\overset{+}{N}H(CH_2CH_3)_2$
CM: $R = -CH_2-COO^-$

Table 5-2
Some Commonly Used Ion Exchangers

Name[a]	Type	Ionizable Group	Remarks
Dowex 1	Strongly basic polystyrene resin	$\phi\text{-CH}_2\overset{+}{\text{N}}(\text{CH}_3)_3$	Anion exchange
Dowex 50	Strongly acidic polystyrene resin	$\phi\text{-SO}_3\text{H}$	Cation exchange
DEAE-cellulose	Basic	Diethylaminoethyl $-\text{CH}_2\text{CH}_2\text{N}(\text{C}_2\text{H}_5)_2$	Used in fractionation of acidic and neutral proteins
ECTEOLA-cellulose	Basic	Mixed amines	Used for chromatography of nucleic acids
CM-cellulose	Acidic	Carboxymethyl $-\text{CH}_2\text{COOH}$	Used for fractionation of basic and neutral proteins
P-cellulose	Strongly and weakly acidic	Phosphate $-\text{OPO}_3\text{H}_2$	Dibasic; binds basic proteins strongly
DEAE-Sephadex	Basic cross-linked dextran gel	Diethylaminoethyl $-\text{CH}_2\text{CH}_2\text{N}(\text{C}_2\text{H}_5)_2$	Combined chromatography and gel filtration of acidic and neutral proteins
CM-Sephadex	Acidic cross-linked dextran gel	Carboxymethyl $-\text{CH}_2\text{COOH}$	Combined chromatography and gel filtration of basic and neutral proteins
Bio-Gel CM 100	Acidic cross-linked polyacrylamide gel	$-\text{CH}_2\text{COOH}$	Combined chromatography and gel filtration of basic and neutral proteins

[a] Dowex resins are manufactured by the Dow Chemical Co.; Sephadex gels are manufactured by Pharmacia Fine Chemicals AB; and Bio-Gel gels are manufactured by BioRad Laboratories.

under mild conditions. Of course, this low density of charged groups also limits the loading capacity of cellulosic ion exchangers.

Gel-type ion exchangers can have the same sorts of charged groups as do cellulosic ion exchangers. The advantage of gel-type ion exchangers is that they combine the separation properties of gel filtration (Section 5-3C) with those of ion exchange. Because of their high degree of substitution, which results from their porous structures, these gels have a higher loading capacity than cellulosic ion exchangers.

B. Paper Chromatography

Paper chromatography, developed in 1941 by Archer Martin and Richard Synge, has had an indispensable role in biochemical analyses. It is among the most conveniently used of analytical techniques and requires only the simplest of equipment. It is normally used for the separation of small molecules such as amino acids and oligopeptides.

In paper chromatography, a few drops of solution containing a mixture of the components to be separated is applied (spotted) ~2 cm above one end of a strip of filter paper. After drying, that end of the paper is dipped into a solvent mixture consisting of aqueous and organic components; for example, water/butanol/acetic acid

in 4:5:1 ratio, 77% aqueous ethanol, 6:7:7 water/t-amyl alcohol/pyridine, or 1:6:3 water/n-propanol/concentrated NH_4OH. The paper should also be in contact with the equilibrium vapors of the solvent. The solvent soaks into the paper by capillary action because of the fibrous nature of the paper. The solvent migration direction can be arranged to be upwards **(ascending chromatography)** or downwards **(descending chromatography)** as indicated in Fig. 5-9. The aqueous component of the solvent binds to the cellulose of the paper and thereby forms a stationary gel-like phase with it. The organic component of the solvent continues migrating, thus forming the mobile phase.

The rates of migration of the various substances being separated are governed by their relative solubilities in the polar stationary phase and the nonpolar mobile phase. In a single step of the separation process, a given solute is distributed between the mobile and stationary phases according to its **partition coefficient,** an equilibrium constant defined as

$$K_p = \frac{\text{concentration in stationary phase}}{\text{concentration in mobile phase}} \qquad [5.3]$$

The molecules are therefore separated according to their polarities, with nonpolar molecules moving faster than polar ones.

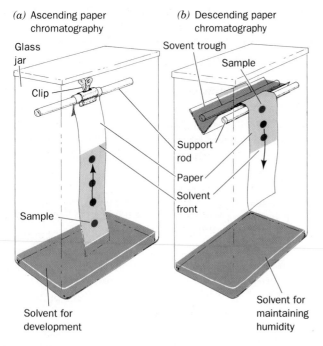

(a) Ascending paper chromatography

(b) Descending paper chromatography

Figure 5-9
The experimental arrangements for (a) ascending paper chromatography; and (b) descending paper chromatography.

After the solvent front has migrated an appropriate distance, the **chromatogram** is removed from the solvent and dried. The separated materials, if not colored, may be detected by several means:

1. Radioactively labeled materials may be located by a variety of radiation detection methods.

2. Materials that are fluorescent or quench the normal fluorescence of the paper can be seen under ultraviolet (UV) light.

3. Materials may be visualized by spraying the chromatogram with a reagent solution that forms a colored product upon reaction with the substance under investigation. For example, α-amino acids and other primary amines react with **ninhydrin** to form an intensely purple compound (Fig. 5-10). Secondary amines such as proline also react with ninhydrin but form a yellow compound.

The migration rate of a substance may be expressed according to the ratio

$$R_f = \frac{\text{distance traveled by substance}}{\text{distance traveled by solvent front}} \quad [5.4]$$

For a given solvent system and paper type, each substance has a characteristic R_f value.

Paper chromatography can be used as a preparative technique for purifying small amounts of materials. A solution containing the substance of interest is applied in a line (streaked) across the bottom of a sheet of filter paper and the entire sheet is chromatographed as has

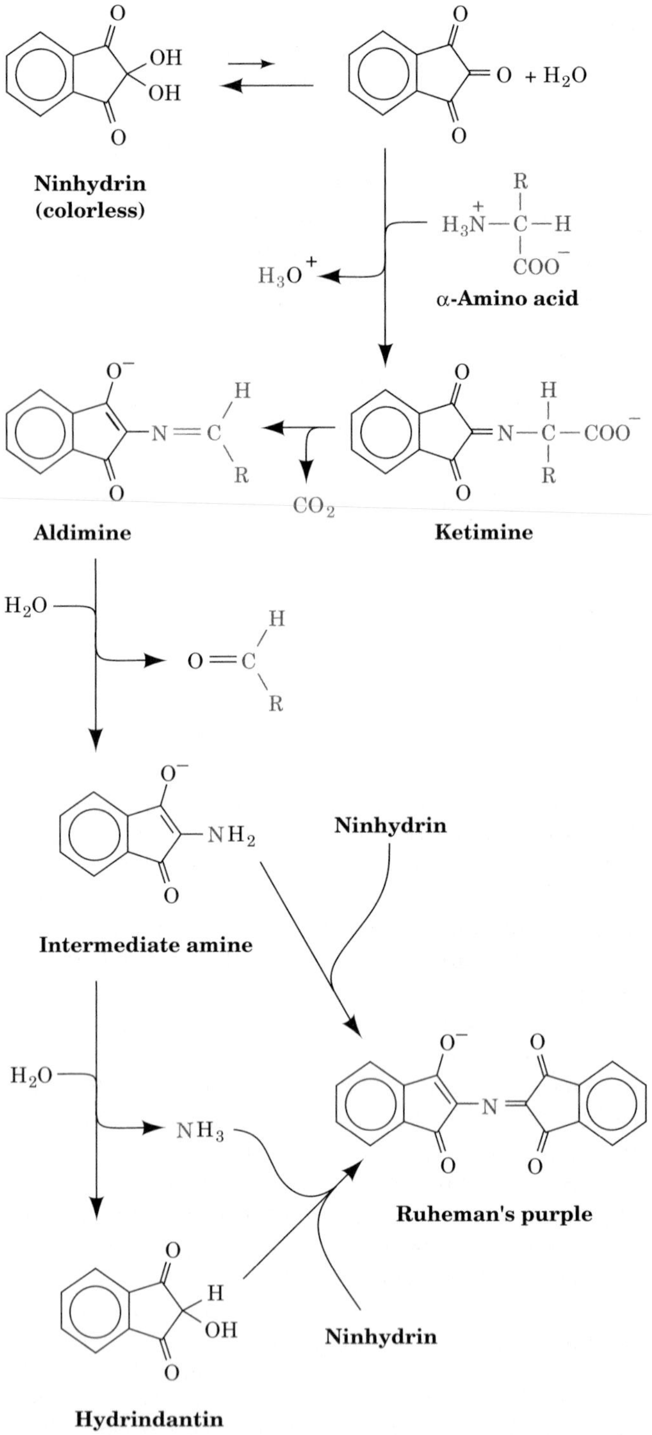

Ninhydrin (colorless)

α-**Amino acid**

Aldimine

Ketimine

Intermediate amine

Ninhydrin

Hydrindantin

Ninhydrin

Ruheman's purple

Figure 5-10
The reaction of ninhydrin with α-amino acids. In the initial reaction, the α-amino acid forms a Schiff base (a ketimine) with the ninhydrin, which is subsequently oxidatively decarboxylated to form an aldimine. This hydrolyzes to form an intermediate amine which, in turn, can react directly or indirectly with a second molecule of ninhydrin to form the intensely colored **Ruheman's purple**. Note that only the nitrogen atom of this pigment arises from the α-amino acid. Primary amines and peptides also react with ninhydrin to form Ruheman's purple, but in these cases, a proton rather than CO_2 is lost to form the aldimine. Ninhydrin preparations used in quantitative analysis usually contain **hydrindantin** to assure maximum color development.

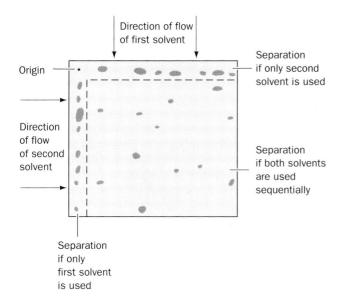

Direction of flow
of first solvent

Origin

Separation
if only second
solvent is used

Direction
of flow
of second
solvent

Separation
if both solvents
are used
sequentially

Separation
if only
first solvent
is used

Figure 5-11
Two-dimensional paper chromatography.

been described. The substance of interest is located on the chromatogram by applying an appropriate detection technique to a small strip that has been cut out from the chromatogram along the direction of solvent migration or according to its known R_f value. Finally, the band of purified substance is cut out and the purified substance recovered by eluting it from the paper with a suitable solvent.

A complex mixture that is incompletely separated in a single paper chromatogram can often be fully resolved by **two-dimensional paper chromatography** (Fig. 5-11). In this technique, a chromatogram is made as previously described except that the sample is spotted onto one corner of a sheet of filter paper and the chromatogram is run parallel to an edge of the paper. After the chromatography has been completed and the paper dried, the chromatogram is rotated 90° and is chromatographed parallel to the second edge using another solvent system. Since each compound migrates at a characteristic rate in a given solvent system, the second chromatographic step should greatly enhance the separation of the mixture into its components.

C. Gel Filtration Chromatography

In gel filtration chromatography, which is also called size exclusion or molecular sieve chromatography, molecules are separated according to their size and shape. The stationary phase in this technique consists of beads of a hydrated, spongelike material containing pores that span a relatively narrow size range of molecular dimensions. If an aqueous solution containing molecules of various sizes is passed through a column containing such "molecular sieves," the molecules that are too large

to pass through the pores will be excluded from the solvent volume inside the gel beads. These larger molecules therefore traverse the column more rapidly, that is, in a smaller eluent volume, than the molecules that pass through the pores (Fig. 5-12).

The molecular mass of the smallest molecule unable to penetrate the pores of a given gel is said to be the gel's **exclusion limit.** This quantity is to some extent a function of molecular shape because elongated molecules, as a consequence of their higher radius of hydration, are less likely to penetrate a given gel pore than spherical molecules of the same molecular volume.

The behavior of a molecule on a particular gel column can be quantitatively characterized. If V_x is the volume occupied by the gel beads and V_0, the **void volume,** is the volume of the solvent space surrounding the beads, then V_t, the total **bed volume** of the column, is simply their sum:

$$V_t = V_x + V_0 \qquad [5.5]$$

V_0 is typically ~35% of V_t.

The **elution volume** of a given solute, V_e, is the volume of solvent required to elute the solute from the column after it has first contacted the gel. The void volume of a column is easily measured as the elution volume of a solute whose molecular mass is larger than the exclusion limit of the gel. The behavior of a particular solute on a given gel is therefore characterized by the ratio V_e/V_0, the **relative elution volume,** a quantity that is independent of the particular column used.

Molecules with molecular masses ranging below the exclusion limit of a gel will elute from the gel in the order of their molecular masses with the largest eluting first. This is because the pore sizes in any gel vary over a limited range so that larger molecules have less of the gel's interior volume available to them than smaller molecules. This effect is the basis of gel filtration chromatography.

Gel Filtration Chromatography Can Be Used to Estimate Molecular Masses

There is a linear relationship between the relative elution volume of a substance and the logarithm of its molecular mass over a considerable molecular mass range (Fig. 5-13). If a plot such as Fig. 5-13 is made for a particular gel filtration column using macromolecules of known molecular masses, *the molecular mass of an unknown substance can be estimated from its position on the plot. The precision of this technique is limited by the accuracy of the underlying assumption that the known and unknown macromolecules have identical shapes.* Nevertheless, gel filtration chromatography is often used to estimate molecular masses because it can be applied to quite impure samples (providing that the molecule of interest can be identified) and because it can be rapidly carried out using simple equipment.

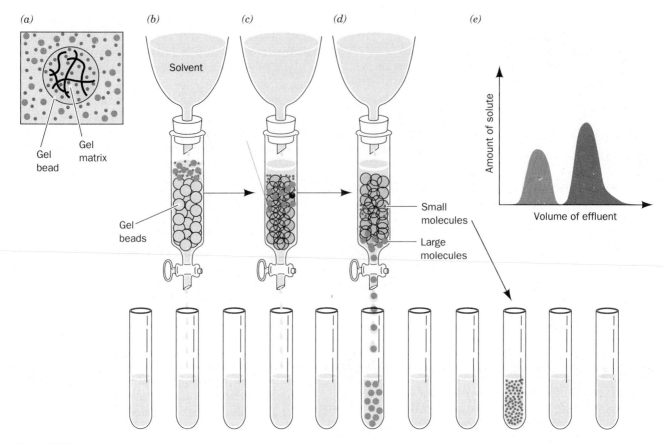

Figure 5-12
A schematic illustration of gel filtration chromatography. (*a*) A gel bead, whose periphery is represented by a dashed line, consists of a gel matrix (*wavy solid lines*) that encloses an internal solvent space. Smaller molecules (*small red dots*) can freely enter the internal solvent space of the gel bead from the external solvent space. However, larger molecules (*large blue dots*) are too large to penetrate the gel pores. (*b*) The sample solution begins to enter the gel column (in which the gel beads are now represented by circles). (*c*) The smaller molecules can penetrate the gel and consequently migrate through the column more slowly than the larger molecules that are excluded from the gel. (*d*) The larger molecules emerge from the column to be collected separately from the smaller molecules, which require additional solvent for elution from the column. (*e*) The elution diagram of the chromatogram indicating the complete separation of the two components with the larger component eluting first.

Most Gels Are Made from Dextran, Agarose, or Polyacrylamide

The most commonly used materials for making chromatographic gels are **dextran** (a high molecular mass polymer of glucose produced by the bacterium *Leuconostoc mesenteroides*), **agarose** (a linear polymer of alternating D-galactose and 3,6-anhydro-L-galactose from red algae), and **polyacrylamide** (see Section 5-4B). The properties of several gels that are commonly employed in separating biological molecules are listed in Table 5-3. The porosity of dextran-based gels, sold under the trade

Table 5-3
Some Commonly Used Gel Filtration Materials

Name[a]	Type	Fractionation Range (kD)
Sephadex G-10	Dextran	0.05–0.7
Sephadex G-25	Dextran	1–5
Sephadex G-50	Dextran	1–30
Sephadex G-100	Dextran	4–150
Sephadex G-200	Dextran	5–600
Bio-Gel P-2	Polyacrylamide	0.1–1.8
Bio-Gel P-6	Polyacrylamide	1–6
Bio-Gel P-10	Polyacrylamide	1.5–20
Bio-Gel P-30	Polyacrylamide	2.4–40
Bio-Gel P-100	Polyacrylamide	5–100
Bio-Gel P-300	Polyacrylamide	60–400
Sepharose 6B	Agarose	10–4,000
Sepharose 4B	Agarose	60–20,000
Sepharose 2B	Agarose	70–40,000

[a] Sephadex and Sepharose gels are products of Pharmacia Fine Chemicals AB.; Bio-Gel gels are manufactured by BioRad Laboratories.

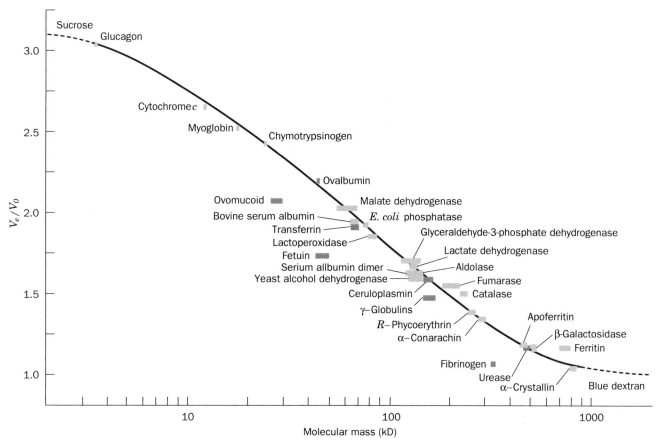

Figure 5-13
A plot of relative elution volume versus the logarithm of molecular mass for a variety of proteins on a cross-linked dextran column (Sephadex G-200) at pH 7.5. Orange bars represent glycoproteins (proteins with attached carbohydrate groups). [After Andrews, P., *Biochem. J.*, **96**, 597 (1965).]

name Sephadex, is controlled by the molecular mass of the dextran used and the introduction of glyceryl ether units that cross-link the hydroxyl groups of the polyglucose chains. The several classes of Sephadex that are available have exclusion limits between 0.7 and 800 kD. The pore size in polyacrylamide gels is similarly controlled by the extent of cross-linking of neighboring polyacrylamide gels (Section 5-4B). They are commercially available under the trade name of Bio-Gel and have exclusion limits between 0.2 and 400 kD. Very large molecules and supramolecular assemblies can be segregated using agarose gels, which have exclusion limits ranging up to 40,000 kD.

Gel filtration is often used to "desalt" a protein solution. For example, an ammonium sulfate precipitated protein can be easily freed of ammonium sulfate by dissolving the protein precipitate in a minimum volume of suitable buffer and applying this solution to a column of gel with an exclusion limit less than the molecular mass of the protein. Upon elution of the column with

buffer, the protein will precede the ammonium sulfate through the column.

Dextran and polyacrylamide gels can be derivatized with ionizable groups such as DEAE and CM to form ion exchange gels (Section 5-3A). Substances that are chromatographed on these gels are therefore subject to separation according to their ionic charges as well as their sizes and shapes.

Dialysis Is a Form of Molecular Filtration

Dialysis is a process that separates molecules according to size through the use of semipermeable membranes containing pores of less than macromolecular dimensions. These pores allow small molecules, such as those of solvents, salts, and small metabolites, to diffuse across the membrane but block the passage of larger molecules. **Cellophane** (cellulose acetate) is the most commonly used dialysis material although many other substances such as **nitrocellulose** and **collodion** are similarly employed.

Dialysis (which is not considered to be a form of chromatography) is routinely used to change the solvent in which macromolecules are dissolved. A macromolecular solution is sealed inside a dialysis bag (usually made by knotting dialysis membrane tubing at both ends), which is immersed in a relatively large volume of the

(a)   At start of
        dialysis

(b)   At equilibrium

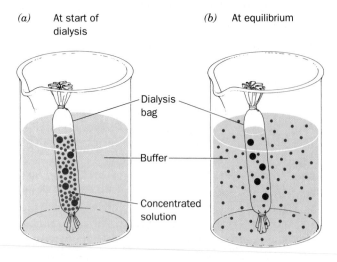

Dialysis
bag

Buffer

Concentrated
solution

Figure 5-14
The separation of small and large molecules by dialysis. (*a*)
Only small molecules can diffuse through the pores in the
bag. (*b*) At equilibrium the concentrations of small molecules
are nearly the same inside and outside the bag, whereas the
macromolecules remain in the bag.

new solvent (Fig. 5-14*a*). After several hours of stirring,
the solutions will have equilibrated but with the macromolecules remaining inside the dialysis bag (Fig. 5-14*b*).
The process can be repeated several times in order to
replace completely one solvent system by another.

Dialysis can be used to concentrate a macromolecular solution by packing a filled dialysis bag in a polymeric desiccant, such as **polyethylene glycol**
[$HOCH_2(CH_2—O—CH_2)_nCH_2OH$], which cannot
penetrate the membrane. Concentration is effected as
water diffuses across the membrane to be absorbed by
the polymer. A related technique that is used to concentrate macromolecular solutions is known as **ultrafiltration**. Here a macromolecular solution is forced, under
pressure, through a semipermeable membranous disk
or bag. Solvent and small solutes pass through the
membrane leaving behind a more concentrated macromolecular solution. Ultrafiltration can also be used to
separate different sized macromolecules.

D. Affinity Chromatography

A striking characteristic of many proteins is their ability to bind specific molecules tightly but noncovalently.
This property can be used to purify such proteins by
affinity chromatography (Fig. 5-15). In this technique,
a molecule known as a **ligand,** which is specifically
bound by the protein of interest, is covalently attached
to an inert and porous matrix. *When an impure protein
solution is passed through this chromatographic material,
the desired protein binds to the immobilized ligand,
whereas other substances are washed through the column
with the buffer. The desired protein can then be recovered in*

*highly purified form by changing the elution conditions
such that the protein is released from the chromatographic
matrix.* The great advantage of affinity chromatography
is its ability to exploit the desired protein's unique biochemical properties rather than the small differences in
physicochemical properties between proteins that other
chromatographic methods must utilize.

The chromatographic matrix in affinity chromatography must be chemically inert, have high porosity, and
large numbers of functional groups capable of forming
covalent linkages to ligands. Of the few materials available that meet these criteria, agarose, which has numerous free hydroxyl groups is, by far, the most widely
used. If the ligand has a primary amino group that is not
essential for its binding to the protein of interest, the
ligand can be covalently linked to the agarose in a two-
step process (Fig. 5-16):

1. Agarose is reacted with **cyanogen bromide** to form
   an "activated" but stable intermediate (which is
   commercially available).

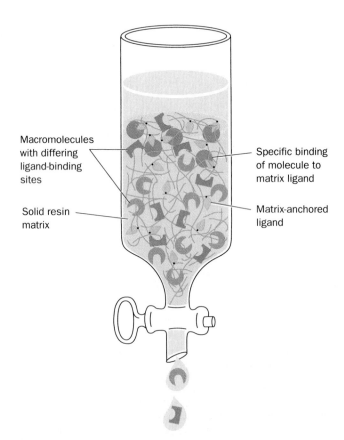

Macromolecules
with differing
ligand-binding
sites

Specific binding
of molecule to
matrix ligand

Solid resin
matrix

Matrix-anchored
ligand

Figure 5-15
The separation of macromolecules by affinity
chromatography. The cutout squares, semicircles, and
triangles are schematic representations of ligand-binding
sites on macromolecules. Only those ligand-binding sites
represented by triangle cutouts specifically bind the
chromatographic matrix-anchored ligands *(yellow)*.

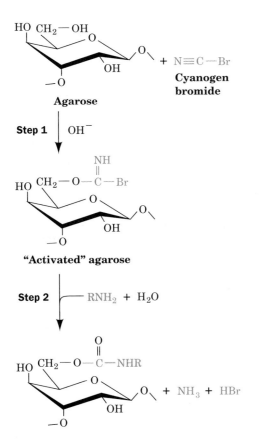

Figure 5-16
The formation of cyanogen bromide-activated agarose (*top*) and its reaction with a primary amine to form a covalently attached ligand for affinity chromatography (*bottom*).

2. Ligand reacts with the activated agarose to form covalently bound product.

Many proteins are unable to bind their cyanogen bromide-coupled ligands due to steric interference with the agarose matrix. This problem is alleviated by attaching the ligand to the agarose by a flexible "spacer" group. A convenient way of doing so is diagrammed in Fig. 5-17. An appropriate diamine, such as **hexamethylene diamine,** is cyanogen bromide coupled to the agarose by ω-aminoalkylation. The resulting product is coupled to the ligand in a reaction that depends on the ligand's functional groups. Figure 5-17 illustrates the overall reaction coupling a ligand possessing a carboxyl group to the end of the spacer through the use of a water soluble carbodiimide such as **1-ethyl-3-(3-dimethylaminopropyl)carbodiimide (EDAC)** to activate the carboxyl group. Many different coupling reactions have been developed so that almost any sort of ligand can be attached to agarose. In fact, many such liganded agarose derivatives are commercially available.

The ligand used in the affinity chromatography isolation of a particular protein must have an affinity high enough to immobilize the protein on the agarose gel but not so high as to prevent its subsequent release. If the ligand is a substrate for an enzyme being isolated, the chromatography conditions must be such that the enzyme does not function catalytically or the ligand will be destroyed.

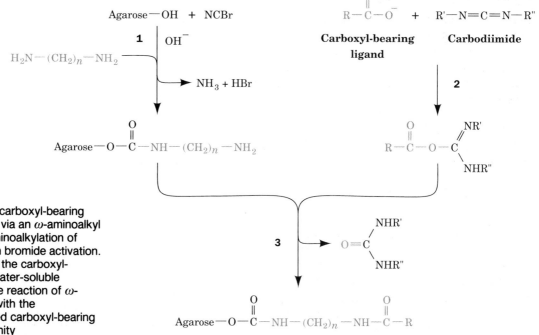

Figure 5-17
The attachment of a carboxyl-bearing ligand to agarose gel via an ω-aminoalkyl group. **(1)** The ω-aminoalkylation of agarose by cyanogen bromide activation. **(2)** The activation of the carboxyl-bearing ligand by a water-soluble carbodiimide. **(3)** The reaction of ω-aminoalkyl-agarose with the carbodiimide-activated carboxyl-bearing ligand to form an affinity chromatography material capable of binding a protein that specifically binds the ligand R.

After a protein has been bound to an affinity chromatography column and washed free of impurities, it must be released from the column. One method of doing so is to elute the column with a solution of a compound that has higher affinity for the protein-binding site than the bound ligand. Another is to alter the solution conditions such that the protein–ligand complex is no longer stable, for example, by changes in pH, ionic strength, and/or temperature. However, care must be taken that the solution conditions are not so inhospitable to the protein being isolated that it is irreversibly damaged.

Affinity chromatography has been used to isolate substances such as enzymes, antibodies, transport proteins, hormone receptors, membranes, and even whole cells. For instance, the protein hormone **insulin** (Section 25-2) has been covalently attached to agarose and used to isolate **insulin receptor protein,** a cell-surface protein whose other properties were unknown and which is only present in very small amounts. The separation power of affinity chromatography for a specific protein is often far greater than that of other chromatographic techniques (e.g., Fig. 5-18 and Table 5-4). Indeed, the replacement of many chromatographic steps in a tried-and-true protein isolation protocol by a single affinity chromatographic step often results in purer protein in higher yield.

Immunoaffinity Chromatography Employs the Binding Specificity of Monoclonal Antibodies

A melding of immunochemistry with affinity chromatography has generated a powerful method for purifying biological molecules. Cross-linking monoclonal antibodies (Section 5-1D) to a suitable column material yields a substance that will bind only the protein against which the antibody was raised. Such **immunoaffinity chromatography** can achieve a 10,000-fold purification in a single step. Disadvantages of immunoaffinity chromatography include the technical difficulty of producing monoclonal antibodies and the harsh conditions that are often required to elute the bound protein.

E. Other Chromatographic Techniques

A number of other chromatographic techniques are of biochemical value. These are briefly discussed below.

Adsorption Chromatography Separates Nonpolar Substances

In **adsorption chromatography** (the original chromatographic method), molecules are physically adsorbed on the surface of insoluble substances such as **alumina** (Al_2O_3), charcoal, **diatomaceous earth** (also called **kieselguhr,** the silicaceous fossils of unicellular organisms known as diatoms), finely powdered sucrose, or **silica gel** (silicic acid), through van der Waals and hydrogen bonding associations. The molecules are then

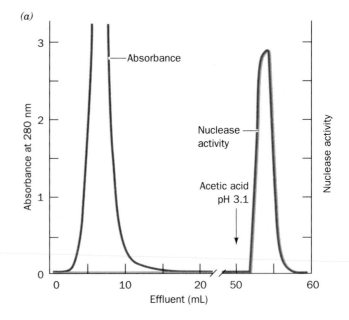

(a)

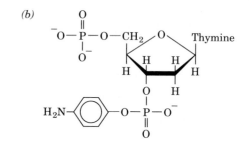

(b)

Figure 5-18
(a) The purification of ***Staphylococcal* nuclease** (a DNA-hydrolyzing enzyme) by affinity chromatography on a nuclease-specific agarose column. The compound shown in Part (b) whose diphosphothymidine moiety specifically binds to the enzyme, was covalently linked to cyanogen bromide-activated agarose. An 0.8 × 5-cm column was equilibrated with 0.05M borate buffer, pH 8.0, containing 0.01M CaCl$_2$. Approximately 40 mg of partially purified material was applied to the column in 3.2 mL of the same buffer. After 50 mL of buffer had been passed through the column at a flow rate of 70 mL · h⁻¹, 0.1M acetic acid was added to elute the enzyme. All of the original enzymatic activity comprising 8.2 mg of pure nuclease was recovered. [After Cuatrecasas, P., Wilchek, M., and Anfinsen, C. B., *Proc. Natl. Acad. Sci.* **61,** 636 (1968).]

eluted from the column by a pure solvent such as chloroform, hexane, or ethyl ether or by a mixture of such solvents. The separation process is based on the partition of the various substances between the polar column material and the nonpolar solvent. This procedure is most often used to separate nonpolar molecules rather than proteins.

Hydroxyapatite Chromatography Separates Proteins

Proteins are adsorbed by gels of crystalline **hydroxyapatite,** an insoluble form of calcium phosphate with

Table 5-4
Purification of Rat Liver Glucokinase

Stage	Specific Activity (nkat · g^{-1})[a]	Yield (%)	Fold[b] Purification
Scheme A: A "traditional" chromatographic procedure			
1. Liver supernatant	0.17	100	1
2. (NH$_4$)$_2$SO$_4$ precipitate	*c*	*c*	*c*
3. DEAE-Sephadex chromatography by stepwise elution with KCl	4.9	52	29
4. DEAE-Sephadex chromatography by linear gradient elution with KCl	23	45	140
5. DEAE-cellulose chromatography by linear gradient elution with KCl	44	33	260
6. Concentration by stepwise KCl elution from DEAE-Sephadex	80	15	480
7. Bio-Gel P-225 chromatography	130	15	780
Scheme B: An affinity chromatography procedure			
1. Liver supernatant	0.092	100	1
2. DEAE-cellulose chromatography by stepwise elution with KCl	20.1	104	220
3. Affinity chromatography[d]	**420**	**83**	**4,500**

[a] A **katal** (abbreviation **kat**) is the amount of enzyme that catalyzes the transformation of one mole of substrate per second under standard conditions. One nanokatal (nkat) is 10^{-9} kat.
[b] Calculated from specific activity; the first step is arbitrarily assigned unity.
[c] The activity could not be accurately measured at this stage because of uncertainty in correcting for contamination by other enzymes.
[d] The affinity chromatography material was made by linking glucosamine (an inhibitor of glucokinase) through a 6-aminohexanoyl spacer arm to NCBr activated agarose.

Source: Cornish-Bowden, A., *Fundamentals of Enzyme Kinetics, p.* 48, Butterworths (1979) as adapted from Parry, M. J. and Walker, D. G., *Biochem. J.* **99**, 266 (1966) for Scheme A and from Holroyde, M. J., Allen, B. M., Storer, A. C., Warsey, A. S., Chesher, J. M. E., Trayer, I. P., Cornish-Bowden, A., and Walker, D. G., *Biochem. J.* **153**, 363 (1976) for Scheme B.

empirical formula Ca$_5$(PO$_4$)$_3$OH. The separation of the proteins occurs upon gradient elution of the column with phosphate buffer (the presence of other anions is unimportant). The physicochemical basis of this fractionation procedure is not fully understood but apparently involves the adsorption of anions to the Ca^{2+} sites and cations to the PO$_4^{3-}$ sites of the hydroxyapatite crystalline lattice.

Thin Layer Chromatography Is Used to Separate Organic Molecules

In **thin layer chromatography (TLC),** a thin (~ 0.25 mm) coating of a solid material spread on a glass or plastic plate is utilized in a manner similar to that of the paper in ascending paper chromatography. In the case of TLC, however, the chromatographic material can be a variety of substances such as ion exchangers, gel filtration agents, and physical adsorbents. According to the choice of solvent for the mobile phase, the separation may be based on adsorption, partition, gel filtration or ion exchange processes, or some combination of these. The advantages of thin layer chromatography in convenience, rapidity, and high resolution have led to its routine use in the analysis of organic molecules.

Reverse-Phase Chromatography Separates Nonpolar Substances

Reverse-phase chromatography (RPC) is a form of liquid–liquid partition chromatography in which the polar character of the phases is reversed relative to that of paper chromatography: The stationary phase consists of a nonpolar liquid immobilized on a relatively inert solid and the mobile phase is a more polar liquid. Reverse-phase chromatography was first developed to separate mixtures of nonpolar substances such as lipids but has also been found to be effective in separating polar substances such as oligonucleotides.

HPLC Has Permitted Greatly Improved Separations

In **high-performance liquid chromatography (HPLC),** a separation may be based on adsorption, ion exchange, size exclusion, or RPC as was previously described. In comparison with these methods, however, the separations are greatly improved through the use of high resolution columns while the column retention times are much reduced. The narrow and relatively long columns are packed with fine glass or plastic beads coated with a thin layer of the stationary phase. The mobile phase is one of the solvent systems previously discussed including gradient elutions with binary or even ternary mixtures. In the case of HPLC, however, the mobile phase is forced through the tightly packed column at pressures of up to 10,000 psi (pounds per square inch) leading to greatly reduced analysis times. The elutants are detected as they leave the column by such methods as UV absorption, refractive index, and fluorescence measurements.

The advantages of HPLC are

1. Its high resolution, which permits the routine purification of mixtures that have defied separation by other techniques.

2. Its speed, which permits most separations to be accomplished in significantly <1 h.

3. Its high sensitivity, which, in favorable cases, permits the quantitative estimation of less than picomole quantities of materials.

4. Its capacity for automation.

Its disadvantages are its low capacity, which limits its use in large-scale purifications (although preparative HPLC columns are becoming available), and the high expense of HPLC instrumentation. The advantages far outweigh the disadvantages so that few biochemistry laboratories could now function without access to an HPLC system.

Gas–Liquid Chromatography Separates Volatile Substances

In **gas–liquid chromatography (GLC)**, the stationary phase is an inert solid, such as diatomaceous earth or powdered fire brick, that has been impregnated with a liquid of low volatility, such as silicone oil or polyethylene glycol. The solid is packed into a long and relatively narrow column, which is maintained at an elevated temperature (usually near 200°C) and which may be systematically varied during an analysis. The mobile phase is a stream of inert gas such as Ar, He, or N_2 into which the sample to be analyzed is flash evaporated. The rate of migration of a compound is determined by its differential solubility between the carrier gas and the immobilized liquid. The sample components in the effluent gas stream are detected by monitoring such physical properties as thermal conductivity or ionization phenomena. They may also be analyzed by directly feeding them into a mass spectrometer.

Substances subject to analyses by GLC must be volatile and stable at high temperatures. This limits the biochemical applications of GLC to relatively low molecular mass compounds. The suitability of compounds for GLC analysis may, however, be greatly enhanced by forming their more volatile derivatives. For example, amino acids derivatized by trimethylsilane

$$RCH(NH_2)COOH + (C_2H_5)_2NSi(CH_3)_3 \longrightarrow$$

$$RCH[NHSi(CH_3)_3]COOH + (C_2H_5)_2NH$$

and similarly treated carbohydrates, are sufficiently volatile to be analyzed by GLC, whereas their parent compounds decompose at the higher temperatures required to volatilize them.

Gas–liquid chromatography and HPLC are being increasingly utilized in the clinical analyses of body fluids. This is because both techniques can rapidly, routinely, and automatically yield reliable quantitative estimates of nanogram quantities of biological materials such as vitamins, steroids, lipids, and drug metabolites.

4. ELECTROPHORESIS

Electrophoresis, the migration of ions in an electric field, is widely used for the analytical separation of biological molecules. The laws of electrostatics state that the electrical force, $F_{electric}$, on an ion with charge q in an electric field of strength E is expressed by

$$F_{electric} = qE \qquad [5.6]$$

The resulting electrophoretic migration of the ion through the solution is opposed by a frictional force

$$F_{friction} = vf \qquad [5.7]$$

where v is the rate of migration (velocity) of the ion and f is its **frictional coefficient.** *The frictional coefficient is a measure of the drag that the solution exerts on the moving ion and is dependent on the size, shape, and state of solvation of the ion as well as on the viscosity of the solution.* In a constant electric field, the forces on the ion balance each other:

$$qE = vf \qquad [5.8]$$

so that each ion moves with a constant characteristic velocity. An ion's **electrophoretic mobility,** μ, is defined

$$\mu = \frac{v}{E} = \frac{q}{f} \qquad [5.9]$$

The electrophoretic (ionic) mobilities of several common small ions in H_2O at 25°C are listed in Table 2-2.

Equation [5.9] really applies only to ions at infinite dilution in a nonconducting solvent. In aqueous solutions, polyelectrolytes such as proteins are surrounded by a cloud of counterions, which impose an additional electric field of such magnitude that Eq. [5.9] is, at best, a poor approximation of reality. Unfortunately, the complexities of ionic solutions have, so far, precluded the development of a theory that can accurately predict the mobilities of polyelectrolytes. Equation [5.9], however, properly indicates that molecules at their isoelectric points, pI, have zero electrophoretic mobility. Furthermore, for proteins and other polyelectrolytes that have acid–base properties, the ionic charge, and hence the electrophoretic mobility, is a function of pH.

The use of electrophoresis to separate proteins was first reported in 1937 by the Swedish biochemist Arne

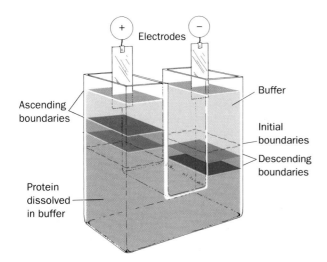

Figure 5-19
A schematic diagram of a Tiselius cell for measuring electrophoretic mobility by the moving boundary method. In the example shown, two different species of negatively charged proteins with different electrophoretic mobilities are dissolved in the buffer. As the proteins migrate towards the anode, the leading and trailing edges of the faster moving protein move past those of the slower protein thereby forming two ascending and two descending moving boundaries of protein.

Tiselius. In the technique he introduced, **moving boundary electrophoresis,** a protein solution is placed in a U-shaped tube and a protein-free buffer solution is carefully layered over both ends of the protein solution. An electric field between electrodes immersed in the buffer on either side of the protein solution causes the charged protein molecules to migrate towards the electrodes of opposite polarity (Fig. 5-19). Different proteins move at different rates as a result of their diverse charges and frictional coefficients so that the leading and trailing edges of the migrating protein columns of each species form separate moving boundaries (fronts) in the buffer solution.

Moving boundary electrophoresis was one of the few powerful analytical techniques available in the early years of protein chemistry. However, preventing the convective mixing of the migrating proteins necessitates a cumbersome apparatus that requires very large samples. Moving boundary electrophoresis has therefore been supplanted by **zone electrophoresis,** a technique in which the sample is constrained to move in a solid support such as filter paper, cellulose, or a gel. This largely eliminates the convective mixing of the sample, which limits the resolution achievable by moving boundary electrophoresis. Moreover, the small quantity of material that is used in zone electrophoresis permits the various sample components to migrate as discrete bands (zones).

A. Paper Electrophoresis

In **paper electrophoresis,** the sample is applied to a point on a strip of filter paper or cellulose acetate mois-

tened with buffer solution. The ends of the strip are immersed in separate reservoirs of buffer in which the electrodes are placed (Fig. 5-20). Upon application of a direct current, the ions of the sample migrate towards the electrodes of opposite polarity at characteristic rates to eventually form discrete bands. An ion's migration rate is influenced, to some extent, by its interaction with the support matrix, but is largely a function of its charge. Upon completion of the electrophoretogram (which usually takes several hours), the strip is dried and the sample components are located using the same detection methods employed in paper chromatography (Section 5-3B).

An electric field of ~ 20 V $\cdot$ cm^{-1} is adequate for many separations. The high diffusion rates of small molecules such as amino acids and small peptides, however, limits their resolution in complex mixtures. This difficulty can be reduced and the separation greatly speeded up by using electric fields of ~ 200 V $\cdot$ cm^{-1}. In this **high voltage paper electrophoresis,** the paper is clamped between two cooled plates to carry off the heat generated by the high voltage current flow.

Paper electrophoresis and paper chromatography are superficially similar. However, *paper electrophoresis separates ions largely on the basis of their charge, whereas paper chromatography separates molecules on the basis of their polarity.* The two methods are often combined in a

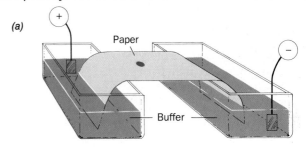

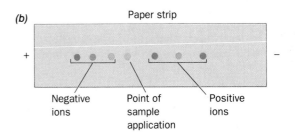

Figure 5-20
Paper electrophoresis. (*a*) A schematic diagram of the apparatus used. The sample is applied to a point on the buffer-moistened paper. The ends of the paper are dipped into reservoirs of buffer in which the electrodes are immersed and an electric field is applied. (*b*) A schematic representation of a completed paper electrophoretogram. Note that positive ions (cations) have migrated towards the cathode and negative ions (anions) have migrated towards the anode. Uncharged molecules remain at the point of sample application.

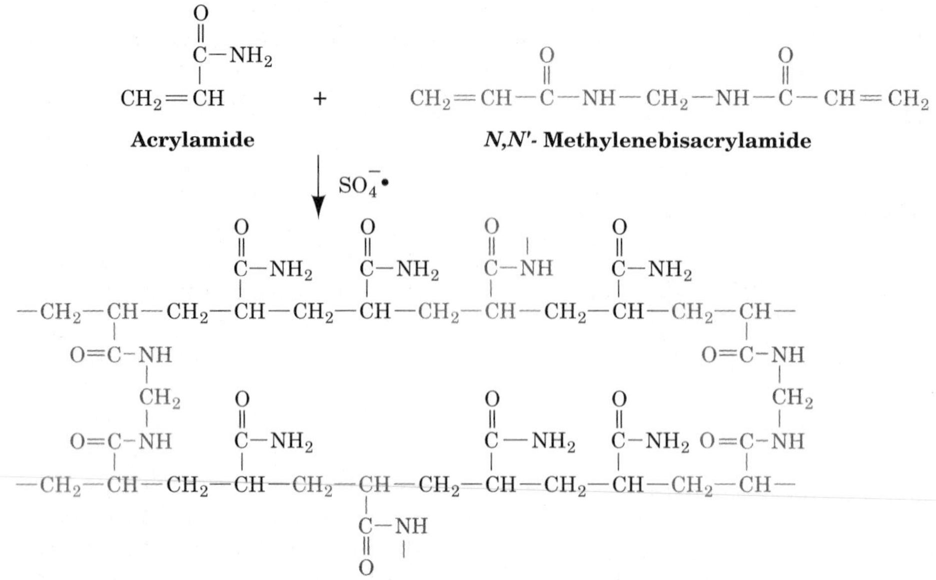

$$
\begin{array}{ccc}
& \overset{\displaystyle O}{\underset{\displaystyle \overset{\|}{C}-NH_2}{|}} & \\
CH_2\!=\!CH & + & CH_2\!=\!CH-\overset{O}{\overset{\|}{C}}-NH-CH_2-NH-\overset{O}{\overset{\|}{C}}-CH\!=\!CH_2
\end{array}
$$

Acrylamide   ***N,N*'- Methylenebisacrylamide**

$$\downarrow\ SO_4^-\!\bullet$$

(chemical structure of cross-linked polyacrylamide gel network)

Figure 5-21

The polymerization of acrylamide and *N,N*'-methylenebisacrylamide to form a cross-linked polyacrylamide gel. The polymerization is induced by free radicals resulting from the chemical decomposition of **ammonium persulfate** ($S_2O_8^{2-} \rightarrow 2SO_4^- \cdot$) or the photodecomposition of riboflavin in the presence of traces of O_2. In either case, ***N,N,N*'*,N*'-tetramethylethylenediamine** (TEMED), a free radical stabilizer, is usually added to the gel mixture. The physical properties of the gel and its pore size are controlled by the proportion of polyacrylamide in the gel and its degree of cross-linking. The most commonly used polyacrylamide concentrations are in the range 3 to 15% with the amount of *N,N*'-methylenebisacrylamide usually fixed at 5% of the total acrylamide present.

two-dimensional technique known as **fingerprinting** in which a sample is first treated as in two-dimensional paper chromatography (Section 5-3B) but is subjected to electrophoresis in place of the second chromatographic step. Molecules are thereby separated according to both their charge and their polarity. Molecules may also be separated by two-dimensional electrophoresis using a different pH buffer for each dimension so that they are separated according to their ionic charges at two pH's.

B. Gel Electrophoresis

Gel electrophoresis is among the most powerful and yet conveniently used methods of macromolecular separation. The gels in common use, polyacrylamide and agarose, have pores of molecular dimensions whose sizes can be specified. *The molecular separations are therefore based on gel filtration as well as the electrophoretic mobilities of the molecules being separated.* The gels in gel electrophoresis, however, retard large molecules relative to smaller ones, the reverse of what occurs in gel filtration chromatography, because there is no solvent space in gel electrophoresis analogous to that between the gel beads in gel filtration chromatography (electrophoretic gels are directly cast in the electrophoresis device). The movement of the larger molecules in gel electrophoresis is therefore impeded relative to that of the smaller molecules as the molecules migrate through the gel.

In **polyacrylamide gel electrophoresis (PAGE)**, gels are made by the free radical-induced polymerization of **acrylamide** and ***N,N*'-methylenebisacrylamide** in the buffer of choice (Fig. 5-21). In the simplest form of PAGE, a tube 3 to 10 cm long, in which the gel has been polymerized, is suspended vertically between an upper and a lower buffer reservoir (Fig. 5-22). The gel may also be cast as a thin rectangular slab in which several samples can be simultaneously analyzed in parallel lanes (Fig. 5-23; a good way of comparing similar samples). The buffer, which is the same in both reservoirs and the gel, has a pH (usually ~9 for proteins) such that the macromolecules have net negative charges and hence migrate to the anode in the lower reservoir. Each sample, which can contain as little as 10 μg of macromolecular material, is dissolved in a minimal amount of a relatively dense glycerol or sucrose solution to prevent it from mixing with the buffer in the upper reservoir, and is applied to the top of the gel. For slab gels, the samples are applied in preformed slots on top of the gel (Fig. 5-23). Alternatively, the sample may be contained in a short length of gel, the "sample gel," whose pores are too large to impede macromolecular migration. A direct current of ~300 V is passed through the gel for a time sufficient to separate the macromolecular components into a series of discrete bands (30–90 min), the gel is removed from its holder, and the bands visualized by an appropriate method (see below). Using this technique, a protein mixture of 0.1 to 0.2 mg can be resolved into as many as 20 discrete bands.

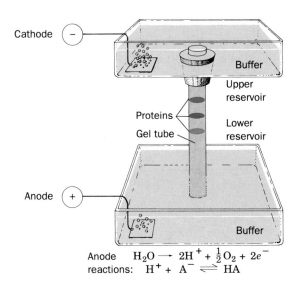

Cathode reactions:
$$2e^- + 2H_2O \rightarrow 2OH^- + H_2$$
$$HA + OH^- \rightleftharpoons A^- + H_2O$$

Anode reactions:
$$H_2O \rightarrow 2H^+ + \tfrac{1}{2}O_2 + 2e^-$$
$$H^+ + A^- \rightleftharpoons HA$$

Figure 5-22
A diagram of a gel electrophoresis apparatus indicating the electrode reactions. The gel is polymerized in the tube which is then placed in the buffer reservoirs. The sample, in a minimal volume of solution, is layered on top of the gel and the current is switched on. After the electrophoresis is complete, the gel is extruded from the tube for staining or other treatment.

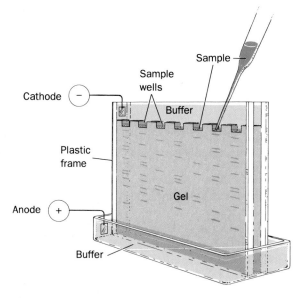

Figure 5-23
Apparatus for slab gel electrophoresis. Samples are applied in slots that have been cast in the top of the gel and electrophoresed in parallel lanes.

Disc Electrophoresis Has Improved Resolution

The narrowness of the bands in the foregoing method, and therefore the resolution of the separations, is limited by the length of the sample column as it enters the gel. The bands are greatly sharpened by an ingenious technique known as **discontinuous pH** or **disc electrophoresis**, which requires a two-gel system and several different buffers (Fig. 5-24). The "running gel," in which the separation takes place, is prepared as described previously and then overlayered by a short (1 cm), large-pored "stacking" or "spacer gel." The buffer in the lower reservoir and in the running gel is as described before, while that in the sample solution (or gel) and in the stacking gel has a pH about two units less than that of the lower reservoir. The pH of the buffer in the upper reservoir, which must contain a weak acid (usually glycine, $pK_2 = 9.6$), is adjusted to a pH near that of the lower reservoir.

When the current is switched on, the buffer ions from the upper reservoir migrate into the stacking gel as the stacking gel buffer ions migrate ahead of them. As this occurs, the upper reservoir buffer ions encounter a pH that is much lower than their pK. They therefore assume their uncharged (or, in the case of glycine, zwitterionic) form and become electrophoretically immobile. This causes a deficiency of charge carriers; that is, a high electrical resistance R, in this region which, because of the requirement of a constant current I throughout the electrical circuit, results, according to Ohm's law ($E = IR$), in a highly localized increase in the electric field, E. In response to this increased field, the macromolecular anions migrate rapidly until they reach the region containing the stacking gel buffer ions, whereupon they slow down because at that point there is no ion deficiency. *This effect causes the macromolecular ions to approach the running gel as stacks of very narrow (~ 0.01 mm thick) bands or discs that are ordered according to their mobilities* and lie between the migrating ions of the upper reservoir and those of the stacking gel. As the macromolecular ions enter the running gel, they slow down as a result of gel filtration effects. This permits the upper reservoir buffer ions to overtake the macromolecular bands and, because of the running gel's higher pH, assume their fully charged form as they too enter the gel.

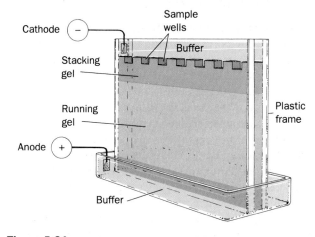

Figure 5-24
A diagram of a disc electrophoresis apparatus.

denaturing it and complexes the dye to the protein. Excess dye is removed by extensively washing the gel with an acidic solution or by electrophoretic destaining. Proteins in fractional microgram quantities can thereby be detected. Gel bands containing less than this amount of protein can be visualized with **silver stain,** which is ~50 times more sensitive but somewhat more difficult to apply. **Fluorescamine,** an alternative type of protein stain, is a nonfluorescent molecule that reacts with primary amines, such as lysine, to yield an addition product that is highly fluorescent under UV irradiation.

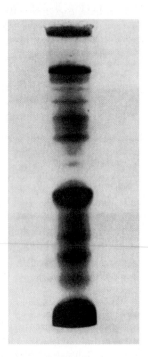

Figure 5-25
A photograph of the bands appearing in a 0.5 × 4.0-cm column of polyacrylamide gel after disc electrophoresis of human serum. The protein was stained with **amido black.** [Courtesy of Robert W. Hartley, NIH]

Fluorescamine (nonfluorescent)   **Fluorescamine adduct** (highly fluorescent)

Proteins, as well as other substances, can be detected through the UV absorption of a gel along its length. If the sample is radioactive, the gel may be dried under vacuum to form a cellophanelike material or, alternatively, covered with plastic wrap, and then clamped over a sheet of X-ray film. After a time (from a few minutes to many weeks depending on the radiation intensity), the film is developed and the resulting **autoradiograph** shows the positions of the radioactive components by a blackening of the film [alternatively, a position-sensitive radiation detector (electronic film) can be used to reveal the locations of the radioactive components within even a few seconds]. A gel may also be sectioned widthwise into many slices and the level of radioactivity in each slice determined using a **scintillation counter.** This latter method yields quantitatively more accurate results than autoradiography. Sample materials can also be eluted from gel slices for identification and/or further treatment.

The charge carrier deficiency therefore disappears and from this point on the electrophoretic separation proceeds normally. However, *the compactness of the macromolecular bands entering the running gel greatly increases the resolution of the macromolecular separations* (e.g., Fig. 5-25).

Agarose Gels Are Used to Electrophoretically Separate Large Molecules

The very large pores needed for the PAGE of large molecular mass compounds (>200 kD) requires gels with such low polyacrylamide concentrations (<2.5%) that they are too soft to be usable. This difficulty is circumvented by using agarose (Section 5-3C), alone or mixed with polyacrylamide. For example, a 0.8% agarose gel is used for the electrophoretic separation of nucleic acids with molecular masses of up to 50,000 kD.

Gel Bands May Be Detected by Staining or Radioactive Counting

Bands resulting from a gel electrophoretic separation can be located by a variety of techniques. Proteins are often visualized by staining. **Coomassie brilliant blue,** which is the most widely used dye for this purpose, is applied by soaking the gel in an acidic, alcoholic solution containing the dye. This fixes the protein in the gel by

C. SDS–PAGE

Soaps and detergents are amphipathic molecules (Section 2-1B) that are strong protein denaturing agents for reasons explained in Section 7-4E. **Sodium dodecyl sulfate (SDS),**

$$[CH_3-(CH_2)_{10}-CH_2-O-SO_3^-]\,Na^+$$

Sodium dodecyl sulfate (SDS)

a detergent that is often used in biochemical preparations, binds quite tenaciously to proteins causing them

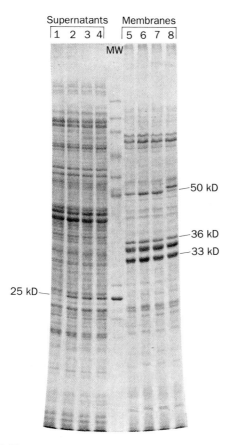

Figure 5-26
The SDS–polyacrylamide disc electrophoresis pattern of the
supernatant (*left*) and membrane fractions (*right*) of various
strains of the bacterium *Salmonella typhimurium*. Samples of
200-μg of protein each were run in parallel lanes on a 35-cm
long × 0.8-mm thick slab gel containing 10% polyacrylamide. The sample marked MW contains molecular weight
standards. [Courtesy of Giovanna F. Ames, University of
California, Berkeley.]

to assume a rodlike shape. Most proteins bind SDS in
the same ratio, 1.4 g of SDS/g of protein (about one SDS
molecule for every two amino acid residues). The large
negative charge that the SDS imparts masks the protein's intrinsic charge so that SDS-treated proteins tend
to have identical charge-to-mass ratios and similar
shapes. Consequently, *the electrophoresis of proteins in
an SDS-containing gel separates them in order of their molecular masses because of gel filtration effects.* Figure 5-26
provides an example of the resolving power and the
reproducibility of SDS–PAGE.

The molecular masses of "normal" proteins are routinely determined to an accuracy of 5 to 10% through
SDS–PAGE. The relative mobilities of proteins on such
gels vary linearly with the logarithm of their molecular
masses (Fig. 5-27). In practice, a protein's molecular
mass is determined by electrophoresing it together with
several "marker" proteins of known molecular masses
that bracket that of the protein of interest.

Many proteins contain more than one polypeptide

chain (Section 7-5A). SDS treatment disrupts the noncovalent interactions between these subunits. Therefore, SDS–PAGE yields the molecular masses of the
subunits rather than that of the intact protein unless the
subunits are disulfide linked. However, mercaptoeth-
anol is usually added to SDS-PAGE gels so as to reduce
these disulfide bonds (Section 6-1B).

D. Isoelectric Focusing

A protein has charged groups of both polarities and
therefore has an isoelectric point, the pH at which it is
immobile in an electric field (Section 4-1D). *If a mixture
of proteins is electrophoresed through a solution having a
stable **pH gradient** in which the pH smoothly increases
from anode to cathode, each protein will migrate to the
position in the pH gradient corresponding to its isoelectric
point.* If a protein molecule diffuses away from this position, its net charge will change as it moves into a region
of different pH and the resulting electrophoretic forces
will move it back to its isoelectric position. Each species
of protein is thereby "focused" into a narrow band
about its isoelectric point that may be as thin as 0.01 pH
unit.

A pH gradient produced by mixing two different
buffers together in continuously varying ratios is unstable in an electric field because the buffer ions migrate to
the electrode of opposite polarity. Rather, the pH gra-

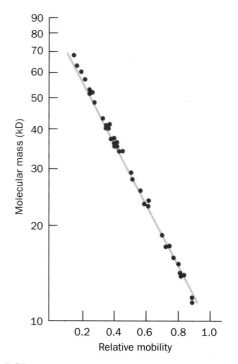

Figure 5-27
A logarithmic plot of the molecular masses of 37 different
polypeptide chains ranging from 11 to 70 kD versus their
relative electrophoretic mobilities on an SDS–polyacrylamide
gel. [After Weber, K. and Osborn, M., *J. Biol. Chem.* **244**,
4406 (1969).]

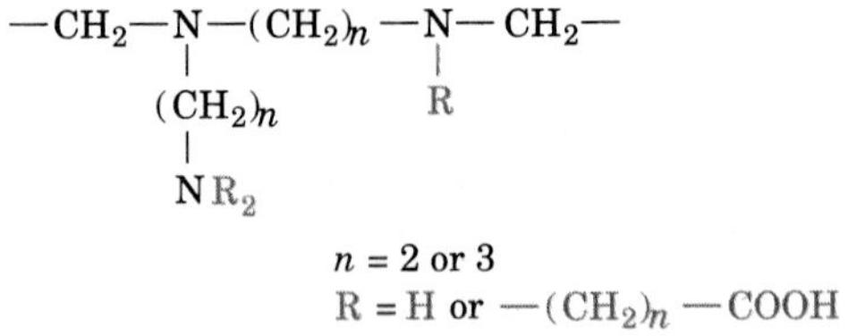

$$-CH_2-N-(CH_2)_n-N-CH_2-$$

$$(CH_2)_n \qquad R$$

$$NR_2$$

$$n = 2 \text{ or } 3$$
$$R = H \text{ or } -(CH_2)_n-COOH$$

Figure 5-28
The general formula of the carrier ampholytes used in isoelectric focusing.

dient in isoelectric focusing is formed by a mixture of low molecular mass (300–600 D) oligomers bearing aliphatic amino and carboxylic acid groups (Fig. 5-28) that have a range of isoelectric points. Under the influence of an electric field, a solution of these **polyampholytes** will segregate according to their isoelectric points such that the most acidic gather at the anode with the progressively more basic positioning themselves ever closer to the cathode. The pH gradient, which is maintained by the electric field, arises from the buffering action of these polyampholytes.

The pH gradient can be stabilized against convection by preparing it in a sucrose concentration gradient in which the sucrose concentration and therefore the density of the viscous solution increases from top to bottom. Convection may instead be eliminated through use of a lightly cross-linked polyacrylamide gel in the form of a tube or a thin slab. The gel often contains **urea**

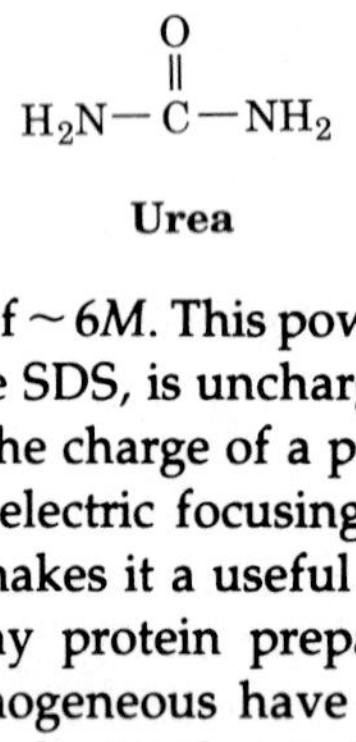

$$\overset{\displaystyle O}{\underset{\displaystyle \|}{H_2N-C-NH_2}}$$

Urea

at a concentration of $\sim 6M$. This powerful protein denaturing agent, unlike SDS, is uncharged and hence cannot directly affect the charge of a protein.

The fact that isoelectric focusing separates proteins into sharp bands makes it a useful analytical and preparative tool. Many protein preparations previously thought to be homogeneous have been resolved into several components by isoelectric focusing. Isoelectric focusing can be combined with electrophoresis in an extremely powerful two-dimensional separation technique (Fig. 5-29).

5. ULTRACENTRIFUGATION

If a container of sand and water is shaken and then allowed to stand quietly, the sand will rapidly sediment to the bottom of the container due to the influence of the earth's gravity (an acceleration $g = 9.81$ m · s^{-2}). Yet macromolecules in solution, which experience the same gravitational field, do not exhibit any perceptible sedimentation because their random thermal (Brownian)

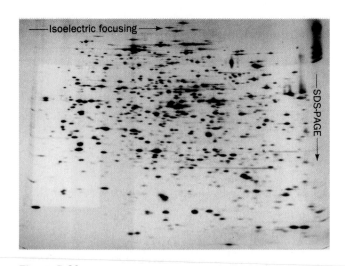

Figure 5-29
An autoradiogram showing the separation of *E. coli* proteins using isoelectric focusing in one dimension and SDS–PAGE in the second dimension. A 10-μg sample of proteins from *E. coli* that had been labeled with [^{14}C]amino acids were subjected to isoelectric focusing in a 2.5 × 130-mm tube of urea containing polyacrylamide gel. The gel was then extruded from its tube, placed at one edge of an SDS–polyacrylamide slab gel, and subjected to electrophoresis. Over 1000 spots were counted on the original autoradiogram, which resulted from an 825-h exposure. [Courtesy of Patrick O'Farrell, University of California at San Francisco.]

motion keeps them uniformly distributed throughout the solution. *Only when they are subjected to enormous accelerations will the sedimentation behavior of macromolecules begin to resemble that of sand grains.*

The ultracentrifuge, which was developed around 1923 by the Swedish biochemist The Svedberg, can attain rotational speeds as high as 80,000 rpm (revolutions per minute) so as to generate centrifugal fields in excess of 600,000g. Using this instrument, Svedberg first demonstrated that proteins are macromolecules with homogeneous compositions and that many proteins are composed of subunits. More recently, ultracentrifugation has become an indispensable tool for the isolation of proteins, nucleic acids, and subcellular particles. In this section we outline the theory and practice of ultracentrifugation.

A. Sedimentation

The rate at which a particle sediments in the ultracentrifuge is related to its mass. The force, $F_{sedimentation}$, acting to sediment a particle of mass m that is located a distance r from a point about which it is revolving with angular velocity ω (in rad · s^{-1}), is the centrifugal force ($m\omega^2 r$) on the particle less the buoyant force ($V_p \rho \omega^2 r$) exerted by the solution:

$$F_{sedimentation} = m\omega^2 r - V_p \rho \omega^2 r \qquad [5.10]$$

Here V_p is the particle volume and ρ is the density of the

Table 5-5
Physical Constants of Some Proteins

Protein	Molecular Mass (kD)	Partial Specific Volume, $\overline{V}_{20,w}$ ($cm^3 \cdot g^{-1}$)	Sedimentation Coefficient, $s_{20,w}$ (S)	Frictional Ratio f/f_0
Lipase (milk)	6.7	0.714	1.14	1.190
Ribonuclease A (bovine pancreas)	12.6	0.707	2.00	1.066
Cytochrome c (bovine heart)	13.4	0.728	1.71	1.190
Myoglobin (horse heart)	16.9	0.741	2.04	1.105
α-Chymotrypsin (bovine pancreas)	21.6	0.736	2.40	1.130
Crotoxin (rattlesnake)	29.9	0.704	3.14	1.221
Concanavalin B (jack bean)	42.5	0.730	3.50	1.247
Diphtheria toxin	70.4	0.736	4.60	1.296
Cytochrome oxidase (*P. aeruginosa*)	89.8	0.730	5.80	1.240
Lactate dehydrogenase H (chicken)	150	0.740	7.31	1.330
Catalase (horse liver)	222	0.715	11.20	1.246
Fibrinogen (human)	340	0.725	7.63	2.336
Hemocyanin (squid)	612	0.724	19.50	1.358
Glutamate dehydrogenase (bovine liver)	1015	0.750	26.60	1.250
Turnip yellow mosaic virus protein	3013	0.740	48.80	1.470

Source: Smith, M. H., *in* Sober, H. A. (Ed.), *Handbook of Biochemistry and Molecular Biology (2nd ed.), p. C-10, CRC Press (1970).*

solution. However, the motion of the particle through the solution, as we have seen in our study of electrophoresis, is opposed by the frictional force:

$$F_{friction} = vf \qquad [5.7]$$

where $v = dr/dt$ is the rate of migration of the sedimenting particle and f is its frictional coefficient. The particle's frictional coefficient can be determined from measurements of its rate of diffusion.

Under the influence of gravitational (centrifugal) force, the particle accelerates until the forces on it exactly balance:

$$m\omega^2 r - V_p \rho \omega^2 r = vf \qquad [5.11]$$

The particle mass, M, is defined according to

$$m = \frac{M}{N} \qquad [5.12]$$

where N is Avogadro's number (6.022×10^{23}). Thus

$$V_p = \overline{V}m = \frac{\overline{V}M}{N} \qquad [5.13]$$

where $\overline{V}$, the particle's **partial specific volume,** is the volume change when 1 g (dry weight) of particles is dissolved in an infinite volume of the solute. For most proteins dissolved in pure water at 20°C, $\overline{V}$ is near $0.73 \ cm^3 \cdot g^{-1}$ (Table 5-5). Indeed, for proteins of known amino acid composition, $\overline{V}$ is closely approximated by the sum of the partial specific volumes of its

component amino acid residues thereby indicating that the atoms in proteins are closely packed (Section 7-3B).

A particle may be characterized by its sedimentation rate. Substituting Eqs. [5.12] and [5.13] into Eq. [5.11] yields

$$vf = \frac{M(1 - \overline{V}\rho)\omega^2 r}{N} \qquad [5.14]$$

Now define the **sedimentation coefficient,** s, as

$$s = \frac{v}{\omega^2 r} = \frac{1}{\omega^2}\left(\frac{d \ln r}{dt}\right) = \frac{M(1 - \overline{V}\rho)}{Nf} \qquad [5.15]$$

The sedimentation coefficient, a quantity that is analogous to the electrophoretic mobility (Eq. [5.9]) in that it is a velocity per unit force, is usually expressed in units of 10^{-13} s, which are known as **Svedbergs (S).** For the sake of uniformity, the sedimentation coefficient is customarily corrected to the value that would be obtained at 20°C in a solvent with the density and viscosity of pure water. This is symbolized $s_{20,w}$. Table 5-5 and Fig. 5-30 indicate the values of $s_{20,w}$ in Svedbergs for a variety of biological materials.

Equation [5.15] indicates that a particle's mass, M, can be determined from the measurement of its sedimentation coefficient, s, if its frictional coefficient, f, is known. Indeed, before about 1970, most macromolecular mass determinations were made using the **analytical ultracentrifuge,** a device in which the sedimentation rates of molecules under centrifugation can be optically measured (the masses of macromolecules are too high to be

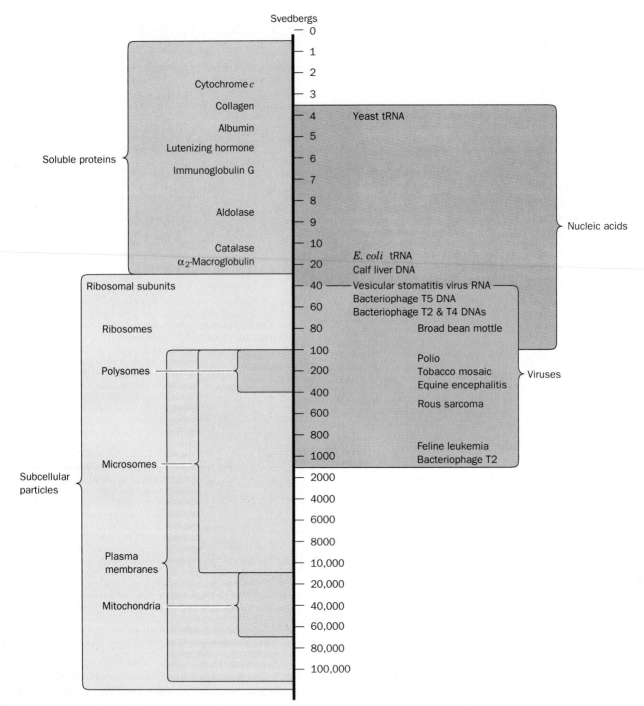

Figure 5-30
The sedimentation coefficients in Svedbergs for some biological materials. [After a diagram supplied by Beckman Instruments, Inc.]

accurately determined by such classical physical techniques as melting point depression or osmotic pressure measurements). With the advent of much simpler molecular mass determination methods, however, such as gel filtration chromatography (Section 5-3C) and SDS–PAGE (Section 5-4C), analytical ultracentrifugation has largely faded from use. Nevertheless, preparative ultracentrifuges remain important laboratory tools (see below).

The Frictional Ratio Is Indicative of Molecular Solvation and Shape

For an unsolvated spherical particle of radius r_p, the frictional coefficient is determined according to the **Stokes equation:**

$$f = 6\pi\eta r_p \qquad [5.16]$$

where η is the **viscosity** of the solution. Solvation increases the frictional coefficient of a particle by increasing its effective or **hydrodynamic volume.** Furthermore, f is minimal when the particle is a sphere. This is because a nonspherical particle has a larger surface area than a sphere of equal volume and therefore must, on the average, present a greater surface area towards the direction of movement than a sphere.

The effective or **Stokes radius** of a particle in solution can be calculated by solving Eq. [5.16] for r_p, given the experimentally determined values of f and η. Conversely, the minimal frictional coefficient of a particle, f_0, can be calculated from the mass and the partial specific volume of the particle by assuming it to be spherical ($V_p = \frac{4}{3}\pi r_p^3$) and unsolvated:

$$f_0 = 6\pi\eta \left(\frac{3M\overline{V}}{4\pi N} \right)^{1/3} \qquad [5.17]$$

If the **frictional ratio,** f/f_0, of a particle is much greater than unity, it must be concluded that the particle is highly solvated and/or significantly elongated. The frictional ratios of a selection of proteins are presented in Table 5-5. The "globular" proteins, which are known from structural studies to be relatively compact and spheroidal (Section 7-3B), have frictional ratios ranging up to ~1.5. Fibrous molecules such as DNA and the blood clotting protein **fibrinogen** (Section 34-1A) have larger frictional ratios. Upon denaturation, the frictional coefficients of globular proteins increase by as much as twofold because denatured proteins assume flexible and fluctuating **random coil** conformations in which all parts of the molecule are in contact with solvent (Section 7-1D).

B. Preparative Ultracentrifugation

Preparative ultracentrifuges, which as their name implies, are designed for sample preparation, differ from analytical ultracentrifuges in that they lack sample observation facilities. Preparative rotors contain cylindrical sample tubes whose axes may be parallel, at an angle, or perpendicular to the rotor's axis of rotation, depending on the particular application (Fig. 5-31).

In the derivation of Eq. [5.15], it was assumed that sedimentation occurs through a homogeneous medium. Sedimentation may be carried out in a solution of an inert substance, however, such as sucrose or CsCl, in which the concentration, and therefore the density of the solution, increases from the top to the bottom of the centrifuge tube. The use of such **density gradients** greatly enhances the resolving power of the ultracentrifuge. Two applications of density gradients are widely employed: (1) **zonal ultracentrifugation** and (2) **equilibrium density gradient ultracentrifugation.**

Zonal Ultracentrifugation Separates Particles According to Their Sedimentation Coefficients

In zonal ultracentrifugation, a macromolecular solution is carefully layered on top of a density gradient

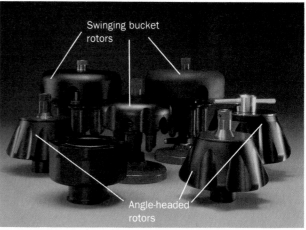

Figure 5-31
A selection of preparative ultracentrifuge rotors. The sample tubes of the swinging bucket rotors (*rear*) are hinged so that they swing from the vertical to the horizontal position as the rotor starts spinning, whereas the sample tubes of the other rotors have a fixed angle relative to the rotation axis. [Courtesy of Beckman Instruments, Inc.]

prepared by use of a device resembling that diagrammed in Fig. 5-7. The purpose of the density gradient is to allow smooth passage of the various macromolecular zones by damping out convectional mixing of the solution. Sucrose, which forms a syrupy and biochemically benign solution, is commonly used to form a density gradient for zonal ultracentrifugation. The density gradient is normally rather shallow because the maximum density of the solution must be less than that of the least dense macromolecule of interest. Nevertheless, consideration of Eq. [5.15] indicates that the sedimentation rate of a macromolecule is a more sensitive function of molecular size than density. Consequently, *zonal ultracentrifugation separates macromolecules largely on the basis of their molecular masses.*

During centrifugation, each species of macromolecule moves through the gradient at a rate largely determined by its sedimentation coefficient and therefore travels as a zone that can be separated from other such zones as is diagrammed in Fig. 5-32. After centrifugation, fractionation is commonly effected by puncturing the bottom of the celluloid centrifuge tube with a needle, allowing its contents to drip out, and collecting the individual zones for subsequent analysis.

Equilibrium Density Gradient Ultracentrifugation Separates Particles According to Their Densities

In equilibrium density gradient ultracentrifugation [alternatively, **isopycnic ultracentrifugation;** (Greek: *isos*, equal + *pyknos*, dense)], the sample is dissolved in a relatively concentrated solution of a dense, fast-diffusing (and therefore low molecular mass) substance such as CsCl or Cs_2SO_4, and is spun at high speed until the solution achieves equilibrium. *The high centrifugal field causes the low molecular mass solute to form a steep density*

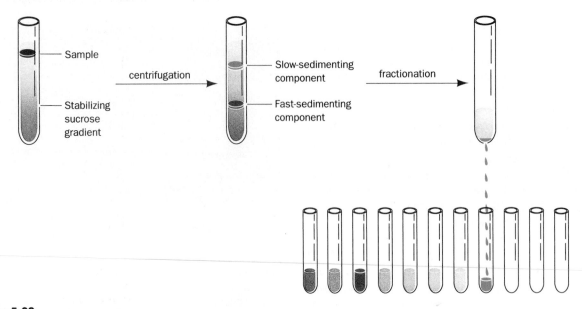

Figure 5-32
A diagrammatic representation of zonal ultracentrifugation. The sample is layered onto a sucrose gradient (*left*). Under centrifugation (*middle*), each particle sediments at a rate that depends largely on its mass. After the end of the run, the centrifugation tube is punctured and the separated particles (zones) are collected (*right*).

gradient (Fig. 5-33) in which the sample components band at positions where their densities are equal to that of the solution; that is, where $(1 - \overline{V}\rho)$ *in Eq. [5.15] is zero (Fig. 5-34). These bands are collected as separate fractions when the sample tube is drained as described above. The salt concentration in the fractions and hence the solution density is easily determined with an* **Abbé refractometer,** *an optical instrument that measures the refractive index of a solution. The equilibrium density gradient technique is often the method of choice for separating mixtures whose components have a range of densities. These substances include nucleic acids, viruses, and certain subcellular organelles such as ribosomes. However, isopycnic ultracentrifugation is rather ineffective for the fractionation of protein mixtures because most proteins have similar densities (high salt concentrations also salt out or possibly denature proteins).*

Figure 5-33
The equilibrium density distribution of a CsCl solution in an ultracentrifuge spinning at 39,460 rpm. The initial density of the solution was 1.7 g · mL⁻¹. [After Ifft, J. B., Voet, D. H., and Vinograd, J., *J. Phys. Chem.* **65,** 1138 (1961).]

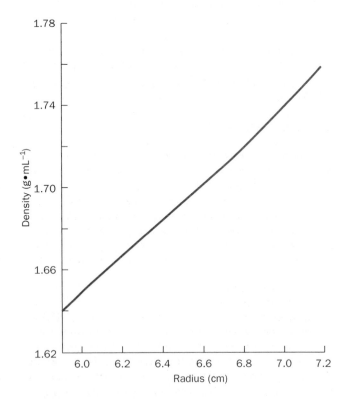

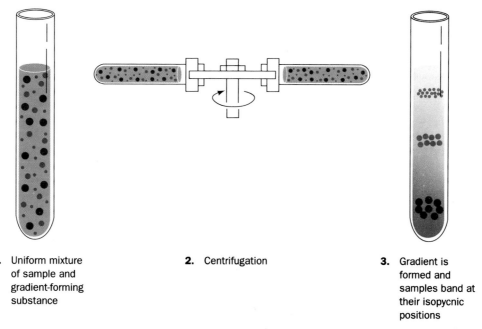

1. Uniform mixture of sample and gradient-forming substance

2. Centrifugation

3. Gradient is formed and samples band at their isopycnic positions

Figure 5-34
Isopycnic ultracentrifugation. The centrifugation of a uniform mixture of a macromolecular sample dissolved in a solution of a dense, fast diffusing solute such as CsCl (*left*). At equilibrium in a centrifugational field, the solute forms a density gradient in which the macromolecules migrate to their positions of buoyant density (*right*).

Chapter Summary

Macromolecules in cells are solubilized by disrupting the cells by various chemical or mechanical means such as detergents or blenders. Partial purification by differential centrifugation is used after cell lysis to remove cell debris or to isolate a desired subcellular component. When out of the protective environment of the cell, proteins and other macromolecules must be treated so as to prevent their destruction by such influences as extremes of pH and temperature, enzymatic and chemical degradation, and rough mechanical handling. The state of purity of a substance being isolated must be monitored throughout the purification procedure by a specific assay.

Proteins are conveniently purified on a large scale by fractional precipitation in which protein solubilities are varied by changing the salt concentration or pH.

Ion exchange chromatography employs support materials such as polystyrene resins or cellulose. Separations are based on differential electrostatic interactions between charged groups on the ion exchange materials and those on the substances being separated. In paper chromatography, compounds are separated by partition between a moving nonpolar solvent phase and a stationary aqueous phase that is bound to the paper fibers. Separation may be enhanced by repeating the chromatographic separation in a second dimension using a different solvent system. Molecules may be located by the use of specific stains, such as ninhydrin for amino acids and polypeptides, or through radioactive labeling. In gel filtration chromatography, molecules are separated according to their size and shape through the use of cross-linked dextran, polyacrylamide, or agarose beads that have pores of molecular dimen-

sions. A suitable calibrated gel filtration column can be used to estimate the molecular masses of macromolecules. Affinity chromatography separates biomolecules according to their unique biochemical abilities to bind other molecules specifically. High-performance liquid chromatography (HPLC) utilizes any of the foregoing separation techniques but uses high resolution chromatographic materials, high solvent pressures, and automatic solvent mixing and monitoring systems so as to obtain much greater degrees of separation than are achieved with the more conventional chromatographic procedures. Adsorption chromatography, thin layer chromatography (TLC), and gas–liquid chromatography (GLC) also have valuable biochemical applications.

In electrophoresis, charged molecules are separated according to their rates of migration in an electric field on a solid support such as paper, cellulose acetate, cross-linked polyacrylamide gel, or agarose. Paper electrophoresis may be combined with paper chromatography in the two-dimensional technique of fingerprinting. Gel electrophoresis employs a cross-linked polyacrylamide or agarose gel support, molecules being separated according to size by gel filtration and according to charge. The anionic detergent sodium dodecyl sulfate (SDS) denatures proteins and uniformly coats them so as to give most proteins a similar charge density and shape. SDS–PAGE may be used to estimate macromolecular masses. In isoelectric focusing, macromolecules are immersed in a stable pH gradient and subjected to an electric field that causes them to migrate to their isoelectric positions.

In ultracentrifugation, molecules are separated by subjecting them to gravitational fields large enough to counteract diffusional forces. Molecules may be separated and their molecular masses estimated from their rates of sedimentation through a solvent or a preformed gradient of an inert low molecular mass material such as sucrose. Alternately, molecules may be separated according to their buoyant densities in a solution with a density gradient of a dense, fast-diffusing substance such as CsCl. The deviation of a molecule's frictional ratio from unity is indicative of its degrees of solvation and elongation.

References

General

Burgess, R. R., Protein purification, *in* Oxender, D. L. and Fox, C. F. (Eds.), *Protein Engineering, pp.* 71–82, Liss (1987).

Cooper, T. G., *The Tools of Biochemistry,* Wiley–Interscience (1977). [A how-to-do-it manual of biophysical techniques.]

Freifelder, D., *Physical Biochemistry* (2nd ed.), Freeman (1983). [A textbook on the techniques used in biophysical analysis.]

Jakoby, W. B. (Ed.), Enzyme Purification and Related Techniques, Part C, *Methods Enzymol.* **104,** (1984).

Robyt, J. F. and White, B. J., *Biochemical Techniques,* Brooks/Cole (1987).

Scopes, R., *Protein Purification: Principles and Practice* (2nd ed.), Springer–Verlag (1987).

Tinoco, I., Jr., Sauer, K., and Wang, J. C., *Physical Chemistry. Principles and Applications in Biological Sciences* (2nd ed.), Chapter 6, Prentice–Hall (1985).

van Holde, K. E., *Physical Biochemistry* (2nd ed.), Prentice–Hall (1985). [An introductory theoretical treatment.]

Walker, J. M., *Methods in Molecular Biology, Vol. 1, Proteins,* Humana Press (1984). [A collection of practical procedures.]

Solubility and Crystallization

Arakawa, T. and Timasheff, S. N., Theory of protein solubility, *Methods Enzymol.* **114,** 49–77 (1985).

Edsall, J. T. and Wyman, J., *Biophysical Chemistry,* Vol. 1, Academic Press (1958). [A detailed treatise on the acid–base and electrostatic properties of amino acids and proteins.]

McPherson, A., Crystallization of macromolecules: general principles, *Methods Enzymol.* **114,** 112–120 (1985).

Chromatography

Ackers, G. K., Molecular sieve methods of analysis, *in* Neurath, H. and Hill, R. L. (Eds.), *The Proteins* (3rd ed.), Vol. 1, *pp.* 1–94, Academic Press (1975).

Dean, P. D. G., Johnson, W. S., and Middle, F. A. (Eds.), *Affinity Chromatography. A Practical Approach,* IRL Press (1985).

Fallon, H., Booth, R. F. G., and Bell, L. D., Applications of HPLC in biochemistry, *in* Burdon, R. H. and van Knippenberg, P. H. (Eds.), *Laboratory Techniques In Biochemistry and Molecular Biology,* Vol. 17, Elsevier (1987).

Fischer, L., Gel filtration chromatography (2nd ed.), *in* Work, T. S. and Burdon, R. H. (Eds.), *Laboratory Techniques In Biochemistry and Molecular Biology,* Vol. 1, Part II, North–Holland Biomedical Press (1980).

Hughes, G. J. and Wilson, K. J., High-performance liquid chromatography: analytic and preparative applications in protein-structure determination, *Methods Biochem. Anal.* **29,** 59–135 (1983).

Lowe, C. R., An introduction to affinity chromatography, *in* Work, T. S. and Work, E. (Eds.), *Laboratory Techniques In Biochemistry and Molecular Biology,* Vol. 7, Part II, North–Holland (1979).

Petersen, E. A., Cellulosic ion exchangers, *in* Work, T. S. and Work, E. (Eds.), *Laboratory Techniques In Biochemistry and Molecular Biology,* Vol. 2, Part II, North–Holland (1970).

Porath, J. and Kristiansen, T., Biospecific affinity chromatography and related methods, *in* Neurath, H. and Hill, R. L. (Eds.), *The Proteins* (3rd ed.), Vol. 1, *pp.* 95–178, Academic Press (1975).

Schott, H., *Affinity Chromatography,* Dekker (1984).

Electrophoresis

Cantor, C. R. and Schimmel, P. R., *Biophysical Chemistry,* Chapter 12, Freeman (1980).

Celis, J. E. and Bravo, R. (Eds.), *Two-Dimensional Gel Electrophoresis of Proteins,* Academic Press (1984).

Gordon, A. H., Electrophoresis of proteins in polyacrylamide and starch gels, *in* Work, T. S. and Work, E. (Eds.), *Laboratory Techniques In Biochemistry and Molecular Biology,* Vol. 1, Part I, revised edition, North–Holland (1975).

Hames, B. D. and Richwood, D. (Eds.), *Gel Electrophoresis of Proteins. A Practical Approach,* IRL Press (1984).

Righetti, R. G., Isoelectric focusing: theory, methodology and applications, *in* Work, T. S. and Burdon, R. H. (Eds.), *Laboratory Techniques In Biochemistry and Molecular Biology,* Vol. 11, Elsevier (1983).

Weber, K. and Osborn, M., Proteins and sodium dodecyl sulfate: Molecular weight on polyacrylamide and related procedures, *in* Neurath, H. and Hill, R. L. (Eds.), *The Proteins* (3rd ed.), Vol. 1, *pp.* 179–223, Academic Press (1975).

Ultracentrifugation

Cantor, C. R. and Schimmel, P. R., *Biophysical Chemistry,* Chapters 10 and 11, Freeman (1980).

Hinton, R. and Dobrata, M., Density gradient ultracentrifugation, *in* Work, T. S. and Work, E. (Eds.), *Laboratory Techniques In Biochemistry and Molecular Biology,* Vol. 6, Part I, North–Holland (1978).

Schachman, H. K., *Ultracentrifugation in Biochemistry,* Academic Press (1959). [A classic treatise on ultracentrifugation.]

van Holde, K. E., Sedimentation analyses of proteins, *in* Neurath, H. and Hill, R. L. (Eds.), *The Proteins* (3rd ed.), Vol. 1, *pp.* 225–291, Academic Press (1975).

Problems

1. What are the ionic strengths of $1.0M$ solutions of NaCl, $(NH_4)_2SO_4$, and K_3PO_4? In which of these solutions would a protein be expected to be most soluble; least soluble?

2. An **isotonic saline solution** (one that has the same salt concentration as blood) is 0.9% NaCl. What is its ionic strength?

3. In what order will the following amino acids be eluted from a column of Dowex 50 ion exchange resin by a buffer at pH 6: arginine, aspartic acid, histidine, and leucine?

4. In what order will the following proteins be eluted from a CM-cellulose ion exchange column by an increasing salt gradient at pH 7: fibrinogen, hemoglobin, lysozyme, pepsin, and ribonuclease A (see Table 5-1)?

5. What is the order of the R_f values of the following amino acids in their paper chromatography with a water/butanol/acetic acid solvent system in which the pH of the aqueous phase is 4.5: alanine, aspartic acid, lysine, glutamic acid, phenylalanine, and valine?

6. What is the order of elution of the following proteins from a Sephadex G-50 column: catalase, α-chymotrypsin, concanavalin B, lipase, and myoglobin (see Table 5-5)?

7. Estimate the molecular mass of an unknown protein that elutes from a Sephadex G-50 column between cytochrome c and ribonuclease A (see Table 5-5).

8. A gel-chromatography column of Bio-Gel P-30 with a bed volume of 100 mL is poured. The elution volume of the protein hexokinase (96 kD) on this column is 34 mL. That of an unknown protein is 50 mL. What is the void volume of the column, the volume occupied by the gel, and the relative elution volume of the unknown protein?

9. What chromatographic method would be suitable for separating the following pairs of substances? (a) Ala-Phe-Lys, Ala-Ala-Lys; (b) lysozyme, ribonuclease A (see Table 5-1); and (c) hemoglobin, myoglobin (see Table 5-1)

10. The neurotransmitter γ-aminobutyric acid is thought to bind to specific receptor proteins in nerve tissue. Design a procedure for the partial purification of such a receptor protein.

11. A mixture of amino acids consisting of arginine, cysteine, glutamic acid, histidine, leucine, and serine is applied to a strip of paper and subjected to electrophoresis using a buffer at pH 7.5. What are the directions of migration of these amino acids and what are their relative mobilities?

*12. Sketch the appearance of a fingerprint of the following tripeptides: Asn-Arg-Lys, Asn-Leu-Phe, Asn-His-Phe, Asp-Leu-Phe, and Val-Leu-Phe. Assume the paper chromatographic step is carried out using a water/butanol/acetic acid solvent system (pH 4.5) and the electrophoretic step takes place in a buffer at pH 6.5.

13. What is the molecular mass of a protein that has a relative electrophoretic mobility of 0.5 in an SDS-polyacrylamide gel such as that of Fig. 5-27.

14. Explain why the molecular mass of fibrinogen is significantly underestimated when measured using a calibrated gel filtration column (Fig. 5-13) but can be determined with reasonable accuracy from its electrophoretic mobility on SDS-polyacrylamide gel (see Table 5-5).

15. What would be the relative arrangement of the following proteins after they had been subjected to isoelectric focusing: insulin, cytochrome c, histone, myoglobin, and ribonuclease A (see Table 5-1)?

16. Calculate the centrifugal acceleration, in gravities (g's), on a particle located 6.5 cm from the axis of rotation of an ultracentrifuge rotating at 60,000 rpm ($1\,g = 9.81$ m $\cdot$ s^{-2}).

17. In a dilute buffer solution at 20°C, rabbit muscle aldolase has a frictional coefficient of 8.74×10^{-8} g $\cdot$ s^{-1}, a sedimentation coefficient of 7.35 S, and a partial specific volume of 0.742 cm^3 $\cdot$ g^{-1}. Calculate the molecular mass of aldolase assuming the density of the solution to be 0.998 g $\cdot$ cm^{-3}.

*18. The sedimentation coefficient of a protein was measured by observing its sedimentation at 20°C in an ultracentrifuge spinning at 35,000 rpm.

Time, t (min)	Distance of Boundary from Center of Rotation, r (cm)
4	5.944
6	5.966
8	5.987
10	6.009
12	6.032

The density of the solution is 1.030 g $\cdot$ cm^{-3}, the partial specific volume of the protein is 0.725 cm^3 $\cdot$ g^{-1}, and its frictional coefficient is 3.72×10^{-8} g $\cdot$ s^{-1}. Calculate the protein's sedimentation coefficient, in Svedbergs, and its molecular mass.

Chapter 6
COVALENT STRUCTURES OF PROTEINS

Proteins are at the center of the action in biological processes. They function as enzymes, which catalyze the complex set of chemical reactions that are collectively referred to as life. Proteins serve as regulators of these reactions, both directly as components of enzymes and indirectly in the form of chemical messengers known as hormones as well as receptors for those hormones. They act to transport and store biologically important substances such as metal ions, O_2, glucose, lipids, and many other molecules. In the form of muscle fibers and other contractile assemblies, proteins generate the coordinated mechanical motion of numerous biological processes including the separation of chromosomes during cell division and the movement of your eyes as you read this page. Proteins, such as **rhodopsin** in the retina of your eye, acquire sensory information that is processed through the action of nerve cell proteins. The proteins of the immune system, such as the **immunoglobulins,** form an essential biological defense system in higher animals. Proteins are the active elements in, as well as the products of, the expression of genetic information; the nucleic acids are, for the most part, information banks upon which proteins act. However, proteins also have important passive roles such as that of **collagen,** which provides bones, tendons, and ligaments with their characteristic tensile strength. Clearly, there is considerable validity to the old cliché that proteins are the ''building blocks'' of life.

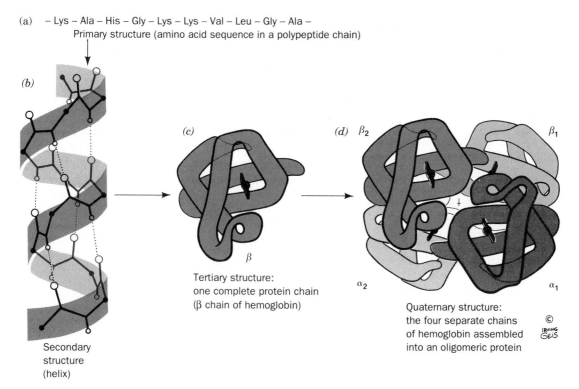

Figure 6-1
The structural hierarchy in proteins: (*a*) primary structure, (*b*) secondary structure, (*c*) tertiary structure, and (*d*) quaternary structure. [Figure copyrighted © by Irving Geis.]

Protein function can only be understood in terms of protein structure, that is, the three-dimensional relationships between a protein's component atoms. The structural descriptions of proteins, as well as those of other polymeric materials, have been traditionally described in terms of four levels of organization (Fig. 6-1):

1. A protein's **primary structure (1° structure)** is the amino acid sequence of its polypeptide chain(s).

2. **Secondary (2°) structure** is the local spatial arrangement of a polypeptide's backbone atoms without regard to the conformations of its side chains.

3. **Tertiary (3°) structure** refers to the three-dimensional structure of an entire polypeptide. The distinction between secondary and tertiary structures is, of necessity, somewhat vague; in practice, the term secondary structure alludes to easily characterized structural entities such as helices.

4. Many proteins are composed of two or more polypeptide chains, loosely referred to as **subunits,** which associate through noncovalent interactions and, in some cases, disulfide bonds. A protein's **quaternary (4°) structure** refers to the spatial arrangement of its subunits.

In this, the first of four chapters on protein structure, we discuss the 1° structures of proteins: How they are elucidated and their biological and evolutionary significance. We also survey methods of chemically synthesizing polypeptide chains. The 2°, 3°, and 4° structures of proteins which, as we shall see, are a consequence of their 1° structures, are treated in Chapter 7. In Chapter 8 we take up protein folding, dynamics, and structural evolution, and in Chapter 9 we analyze hemoglobin as a paradigm of protein structure and function.

1. PRIMARY STRUCTURE DETERMINATION

The first determination of the complete amino acid sequence of a protein, that of the bovine polypeptide hormone **insulin** by Frederick Sanger in 1953, was of enormous biochemical significance in that it definitively established that proteins have unique covalent structures. Since that time, the amino acid sequences of several thousand proteins have been elucidated. This extensive information has been of central importance in the formulation of modern concepts of biochemistry for several reasons:

1. The knowledge of a protein's amino acid sequence is essential for an understanding of its molecular mechanism of action as well as being prerequisite for the elucidation of its X-ray structure (Section 7-3A).

2. Sequence comparisons among analogous proteins from the same individual, from members of the same species, and from members of related species, have yielded important insights into how proteins function and have indicated the evolutionary relationships among the proteins and the organisms that produce them. These analyses, as we shall see in Section 6-3, complement and extend analogous taxonometric studies based on anatomical comparisons.

3. Amino acid sequence analyses have important clinical applications because many inherited diseases are caused by mutations leading to an amino acid change in a protein. Recognition of this fact has led to the development of valuable diagnostic tests for many such diseases and, in several cases, to symptom-relieving therapy.

The elucidation of the 51-residue primary structure of insulin (Fig. 6-2) was the labor of many scientists over the period of a decade that altogether utilized ~ 100 g of protein. Procedures for primary structure determination have since been so refined and automated that proteins of similar size can be sequenced by an experienced technician in a few days using only a few micrograms of protein. The sequencing of the 1021 residue enzyme **β-galactosidase** in 1978 signaled that the sequence analysis of almost any protein could be reasonably attempted. Despite these technical advances, the basic procedure for primary structure determination using the techniques of protein chemistry is that developed by Sanger. The procedure consists of three conceptual parts, each of which can be broken down into several laboratory steps:

1. **Prepare the protein for sequencing:**
   (a) Determine the number of chemically different polypeptide chains (subunits) in the protein.
   (b) Cleave the protein's disulfide bonds.
   (c) Separate and purify the unique subunits.
   (d) Determine the subunits' amino acid compositions.

2. **Sequence the polypeptide chains:**
   (a) Fragment the individual subunits at specific points to yield peptides small enough to be sequenced directly.
   (b) Separate and purify the fragments.
   (c) Determine the amino acid sequence of each peptide fragment.
   (d) Repeat Step 2(a) with a fragmentation process of different specificity so that the subunit is cleaved at different peptide bonds than before. Separate these peptide fragments as in Step 2(b) and determine their amino acid sequences as in Step 2(c).

3. **Organize the completed structure:**
   (a) Span the cleavage points between one set of peptide fragments by the other. By comparison, the sequences of these sets of polypeptides can be arranged in the order that they occur in the subunit, thereby establishing its amino acid sequence.
   (b) Elucidate the positions of the disulfide bonds, if any, between and within the subunits.

We discuss these various steps in the following sections.

A. End Group Analysis: How Many Different Subunits?

Each polypeptide chain (if it is not chemically blocked or circular) has an N-terminal residue and a C-terminal residue. By identifying these **end groups,** we can establish the number of chemically distinct polypeptides in a protein. For example, insulin has equal amounts of the N-terminal residues Phe and Gly, which indicates that it has equal numbers of two nonidentical polypeptide chains.

N-Terminal Identification

There are several effective methods by which a polypeptide's end groups may be identified. **1-Dimethyl-aminonaphthalene-5-sulfonyl chloride (dansyl chloride)** reacts with primary amines (including the ε-amino group of Lys) to yield dansylated polypeptides (Fig. 6-3). Acid hydrolysis (Section 6-1D) liberates the N-terminal residue as a **dansylamino acid,** which exhibits

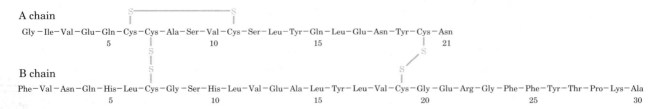

Figure 6-2
The primary structure of bovine insulin. Note the intrachain and interchain disulfide bond linkages.

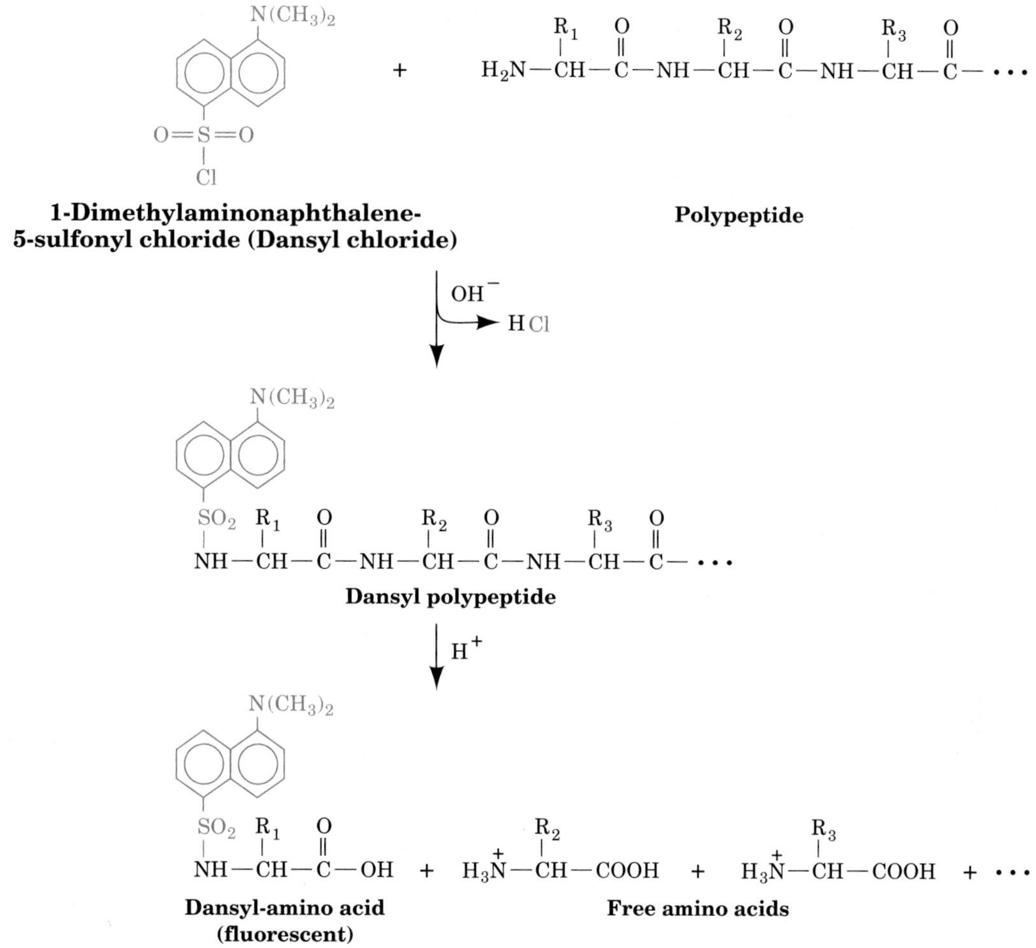

1-Dimethylaminonaphthalene-5-sulfonyl chloride (Dansyl chloride)

Polypeptide

Dansyl polypeptide

Dansyl-amino acid (fluorescent)

Free amino acids

Figure 6-3
The reaction of dansyl chloride in end group analysis.

such intense yellow fluorescence that it can be chromatographically identified from as little as 100 pmol of material (1 pmol = 10^{-12} mol).

In the most useful method of N-terminal residue identification, the **Edman degradation** (named after its inventor, Pehr Edman), **phenyl isothiocyanate (PITC, Edman's reagent)** reacts with the N-terminal amino groups of proteins under mildly alkaline conditions to form their **phenylthiocarbamyl (PTC) adduct** (Fig. 6-4). This product is treated with anhydrous hydrofluoric acid, which cleaves the N-terminal residue as its **thiazolinone** derivative but does not hydrolyze other peptide bonds. *The Edman degradation therefore releases the N-terminal amino acid residue but leaves intact the rest of the polypeptide chain.* The thiazolinone-amino acid is selectively extracted into an organic solvent and is converted to the more stable **phenylthiohydantoin (PTH)** derivative by treatment with aqueous acid. This PTH amino acid may be identified by comparing it with known standards using TLC, electrophoresis, high-performance liquid chromatography (HPLC), or gas-

liquid chromatography (GLC) possibly combined with mass spectroscopy (Sections 5-3 and 5-4).

The most important difference between the Edman degradation and other methods of N-terminal residue identification is that *we can determine the amino acid sequence of a polypeptide chain from the N-terminus inwards by subjecting the polypeptide to repeated cycles of the Edman degradation and, after every cycle, identifying the newly liberated PTH-amino acid.* This technique has been automated resulting in great savings of time and materials (Section 6-1G).

C-Terminus Identification

There is no reliable chemical procedure comparable to the Edman degradation for the sequential end group analysis from the C-terminus of a polypeptide. This can be done enzymatically, however, using **exopeptidases** (enzymes that cleave a terminal residue from a polypeptide). One class of exopeptidases, the **carboxypeptidases,** catalyzes the hydrolysis of the C-terminal residues of polypeptides:

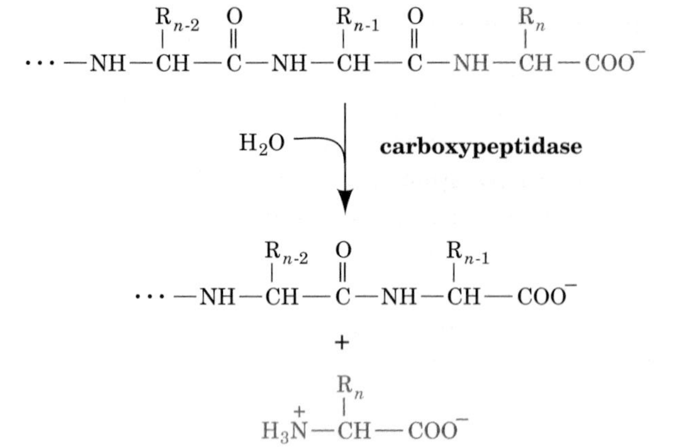

$$\cdots -NH-\underset{\underset{R_{n-2}}{|}}{CH}-\underset{\underset{O}{\|}}{C}-NH-\underset{\underset{R_{n-1}}{|}}{CH}-\underset{\underset{O}{\|}}{C}-NH-\underset{\underset{R_n}{|}}{CH}-COO^-$$

$$H_2O \searrow \Bigg\downarrow \quad \textbf{carboxypeptidase}$$

$$\cdots -NH-\underset{\underset{R_{n-2}}{|}}{CH}-\underset{\underset{O}{\|}}{C}-NH-\underset{\underset{R_{n-1}}{|}}{CH}-COO^-$$

+

$$\overset{+}{H_3N}-\underset{\underset{R_n}{|}}{CH}-COO^-$$

Carboxypeptidases, like all enzymes, are highly specific (selective) for the chemical identities of the substances whose reactions they catalyze (Section 12-2). The side

chain specificities of the various carboxypeptidases in common use are listed in Table 6-1. The second type of exopeptidases listed in Table 6-1, the **aminopeptidases,** sequentially cleave amino acids from the N-terminus of a polypeptide and have been similarly used to determine N-terminal sequences.

Why can't carboxypeptidases be used to determine amino acid sequences? If a carboxypeptidase cleaved all C-terminal residues at the same rate, irrespective of their identities, then by following the course of appearance of the various free amino acids in the reaction mixture (Fig. 6-5a), the sequence of several amino acids at the C-terminus could be determined. If, however, the second amino acid residue, for example, was cleaved at a much faster rate than the first, both amino acids would appear to be released simultaneously (Fig. 6-5b). Carboxypeptidases, in fact, exhibit selectivity towards side chains so that their use, either singly or in mixtures, rarely reveals

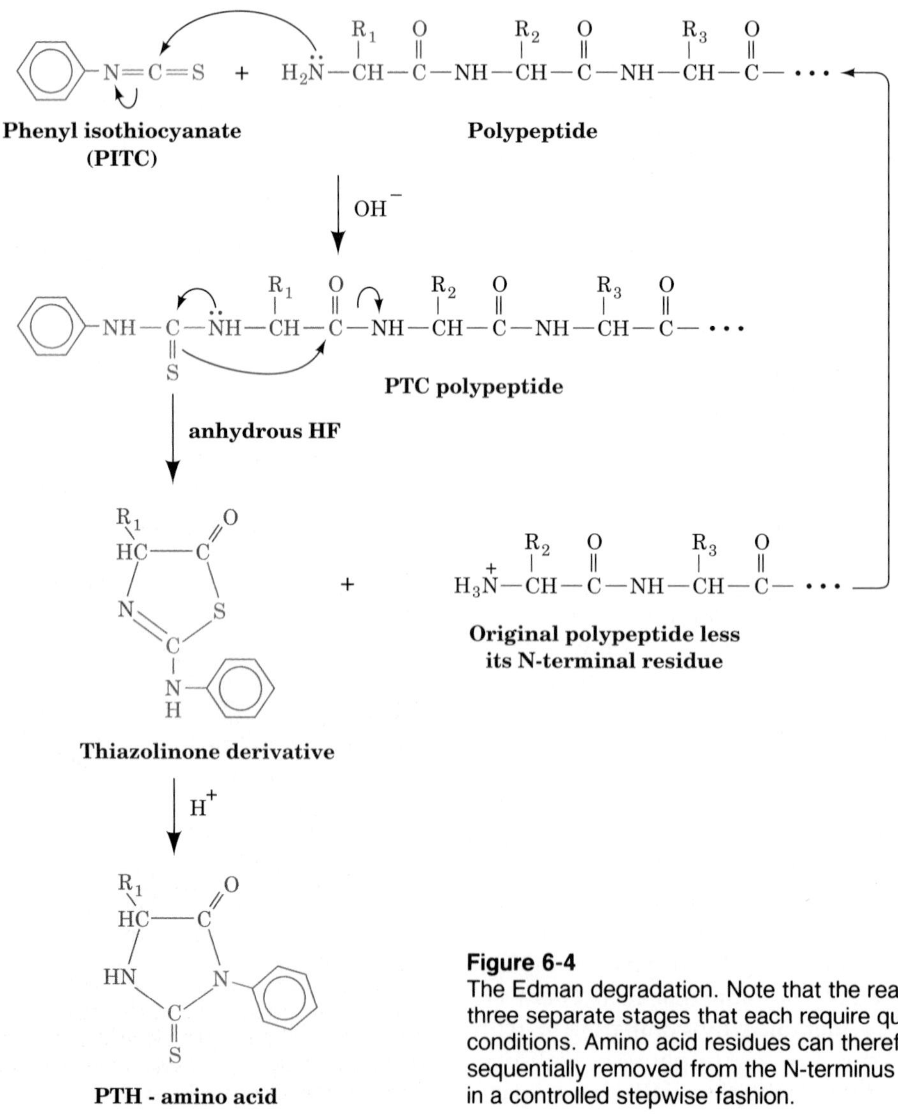

Figure 6-4
The Edman degradation. Note that the reaction occurs in three separate stages that each require quite different conditions. Amino acid residues can therefore be sequentially removed from the N-terminus of a polypeptide in a controlled stepwise fashion.

Table 6-1
Specificities of Various Exopeptidases

Enzyme	Source	Specificity[a]
Carboxypeptidase A	Bovine pancreas	$R_n \neq$ Arg, Lys, Pro; $R_{n-1} \neq$ Pro
Carboxypeptidase B	Bovine pancreas	$R_n =$ Arg, Lys; $R_{n-1} \neq$ Pro
Carboxypeptidase C	Citrus leaves	All free C-terminal residues; pH optimum = 3.5
Carboxypeptidase Y	Yeast	All free C-terminal residues, but slowly with $R_n =$ Gly
Leucine aminopeptidase	Porcine kidney	$R_1 \neq$ Pro
Aminopeptidase M	Porcine kidney	All free N-terminal residues

[a] $R_1 =$ the N-terminal residue; $R_n =$ the C-terminal residue.

the order of more than the first few C-terminal residues of a polypeptide.

C-Terminal residues with a preceding Pro residue are not subject to cleavage by carboxypeptidases A and B (Table 6-1). Chemical methods are therefore usually employed to identify their C-terminal residue. In the most reliable such chemical method, **hydrazinolysis,** a polypeptide is treated with anhydrous **hydrazine** at 90°C for 20 to 100 h in the presence of a mildly acidic ion exchange resin (which acts as a catalyst). All the peptide bonds are thereby cleaved yielding the aminoacyl hydrazides of all the amino acid residues except that of the C-terminal residue, which is released as the free amino acid and therefore can be identified chromatographically. Unfortunately, hydrazinolysis is subject to a great many side reactions, which have largely limited its application to carboxypeptidase-resistant polypeptides.

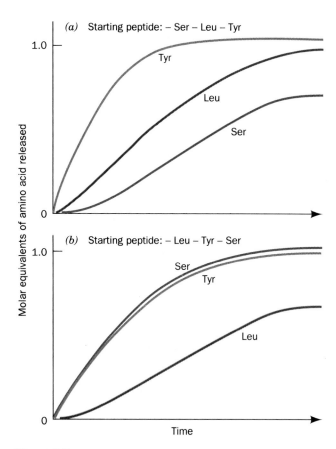

(a) Starting peptide: – Ser – Leu – Tyr

(b) Starting peptide: – Leu – Tyr – Ser

Figure 6-5
The rate of carboxypeptidase A-catalyzed release of amino acids from peptides having the indicated C-terminal sequences. (a) All bonds cleaved at the same rate. (b) Ser removed slowly, Tyr cleaved rapidly, and Leu cleaved at an intermediate rate.

Polypeptide

+

$NH_2—NH_2$

Hydrazine

↓ **acidic ion exchange resin catalyst**

Aminoacyl hydrazides

+

Free amino acid

B. Cleavage of the Disulfide Bonds

The next step in the sequence analysis is to cleave the disulfide bonds between Cys residues. This is done for two reasons:

1. To permit the separation of polypeptide chains (if they are disulfide linked).

2. To prevent the native protein conformation, which is stabilized by disulfide bonds, from obstructing the action of the proteolytic (protein-cleaving) agents used in primary structure determinations (Section 6-1E).

Disulfide bond locations are established in the final step of the sequence analysis (Section 6-1I).

Disulfide bonds may be cleaved oxidatively by **performic acid** or reductively by mercaptans (compounds containing SH groups). Performic acid oxidation, which was pioneered by Sanger, converts all Cys residues, whether linked by S—S bridges or not, to **cysteic acid** residues:

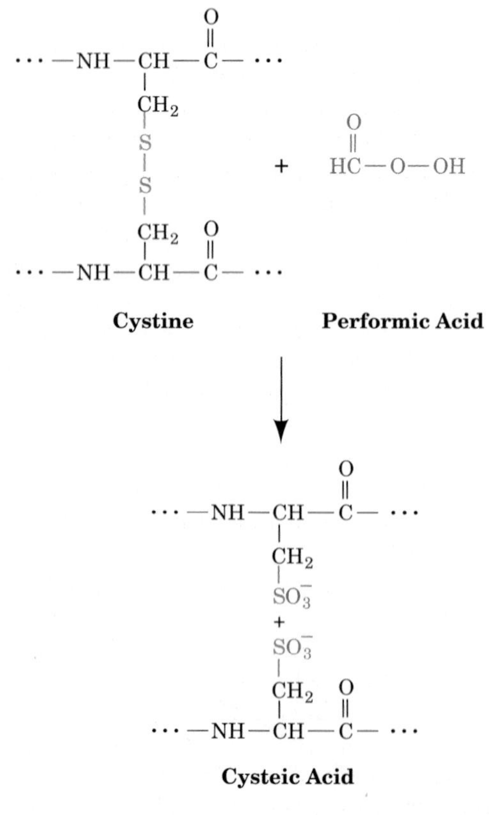

$$
\begin{array}{c}
\text{O} \\
\parallel \\
\cdots \text{—NH—CH—C—} \cdots \\
\mid \\
\text{CH}_2 \\
\mid \\
\text{S} \\
\mid \\
\text{S} \\
\mid \\
\text{CH}_2 \quad \text{O} \\
\mid \quad \parallel \\
\cdots \text{—NH—CH—C—} \cdots
\end{array}
\quad + \quad
\begin{array}{c}
\text{O} \\
\parallel \\
\text{HC—O—OH}
\end{array}
$$

Cystine **Performic Acid**

↓

$$
\begin{array}{c}
\text{O} \\
\parallel \\
\cdots \text{—NH—CH—C—} \cdots \\
\mid \\
\text{CH}_2 \\
\mid \\
\text{SO}_3^- \\
+ \\
\text{SO}_3^- \\
\mid \\
\text{CH}_2 \quad \text{O} \\
\mid \quad \parallel \\
\cdots \text{—NH—CH—C—} \cdots
\end{array}
$$

Cysteic Acid

Cysteic acid is stable in both acidic and basic solutions so that the total Cys content of a protein may be determined as cysteic acid. A major disadvantage of performic acid treatment, however, is that it also oxidizes Met residues to **methionine sulfone** residues,

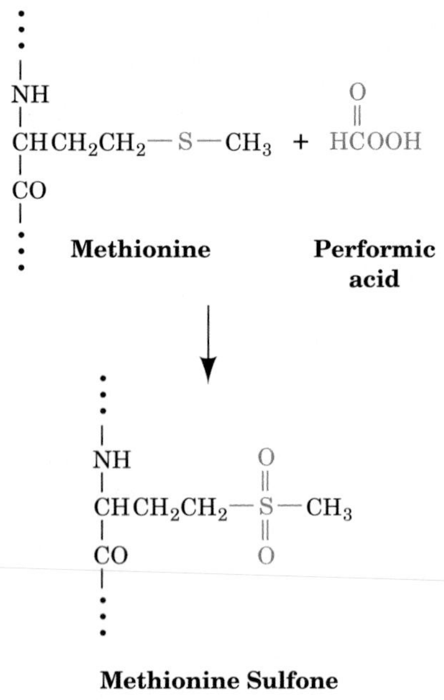

$$
\begin{array}{c}
\vdots \\
\mid \\
\text{NH} \\
\mid \\
\text{CHCH}_2\text{CH}_2\text{—S—CH}_3 \quad + \quad \text{HCOOH} \\
\mid \\
\text{CO} \\
\mid \\
\vdots
\end{array}
$$

Methionine **Performic acid**

↓

$$
\begin{array}{c}
\vdots \\
\mid \\
\text{NH} \quad\quad\quad \text{O} \\
\mid \quad\quad\quad\quad \parallel \\
\text{CHCH}_2\text{CH}_2\text{—S—CH}_3 \\
\mid \quad\quad\quad\quad \parallel \\
\text{CO} \quad\quad\quad \text{O} \\
\mid \\
\vdots
\end{array}
$$

Methionine Sulfone

thereby eliminating the option of using one of the mainstays of primary structure determination—specific peptide bond cleavage at Met residues by reaction with cyanogen bromide (Section 6-1E). Performic acid also partially destroys the indole side chain of Trp.

Reductive cleavage of disulfide bonds by mercaptans is usually more compatible with the overall strategy of primary structure determination. Reductive cleavage is most often achieved by treatment with **2-mercaptoethanol**

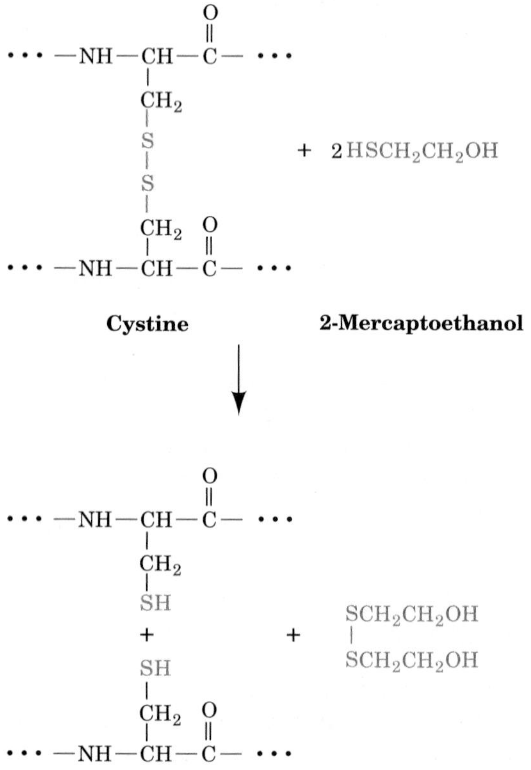

$$
\begin{array}{c}
\text{O} \\
\parallel \\
\cdots \text{—NH—CH—C—} \cdots \\
\mid \\
\text{CH}_2 \\
\mid \\
\text{S} \\
\mid \\
\text{S} \\
\mid \\
\text{CH}_2 \quad \text{O} \\
\mid \quad \parallel \\
\cdots \text{—NH—CH—C—} \cdots
\end{array}
\quad + \quad 2\,\text{HSCH}_2\text{CH}_2\text{OH}
$$

Cystine **2-Mercaptoethanol**

↓

$$
\begin{array}{c}
\text{O} \\
\parallel \\
\cdots \text{—NH—CH—C—} \cdots \\
\mid \\
\text{CH}_2 \\
\mid \\
\text{SH} \\
+ \\
\text{SH} \\
\mid \\
\text{CH}_2 \quad \text{O} \\
\mid \quad \parallel \\
\cdots \text{—NH—CH—C—} \cdots
\end{array}
\quad + \quad
\begin{array}{c}
\text{SCH}_2\text{CH}_2\text{OH} \\
\mid \\
\text{SCH}_2\text{CH}_2\text{OH}
\end{array}
$$

or either of the diastereomers **dithiothreitol** or **dithio-erythritol (Cleland's reagent).**

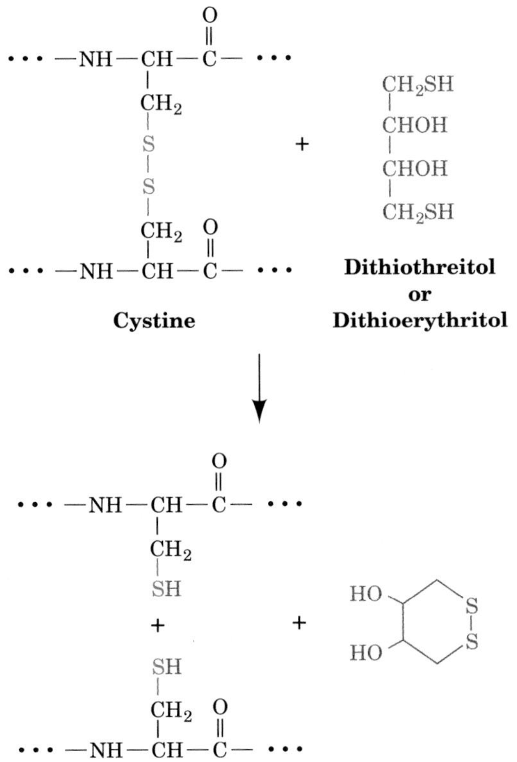

Cystine

Dithiothreitol or Dithioerythritol

S-**Carboxymethylcysteine**

In order to expose all disulfide groups to the reducing agent, the reaction is usually carried out under conditions that denature the protein (disrupt its native conformation; see the next section). The resulting free sulfhydryl groups are alkylated, usually by treatment with **iodoacetic acid,**

$$Cys-CH_2-SH \ + \ ICH_2COO^-$$

Cysteine　　　　**Iodoacetate**

$$Cys-CH_2-S-CH_2COO^- \ + \ HI$$

S-**Carboxymethylcysteine**

to prevent the reformation of disulfide bonds through oxidation by O_2. *S*-Alkyl derivatives are stable in air and under the conditions used for the subsequent cleavage of peptide bonds.

C. Separation and Purification of the Polypeptide Chains

A protein's nonidentical polypeptides must be separated and purified in preparation for their amino acid sequence determination. Subunit dissociation, as well as

denaturation, occurs under acidic or basic conditions, at low salt concentrations, at elevated temperatures, or through the use of denaturing agents such as urea, its iminium analog **guanidinium ion,**

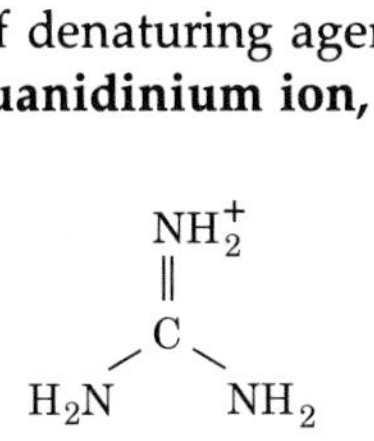

Guanidinium ion

or detergents such as sodium dodecyl sulfate (SDS; Section 5-4C). The dissociated subunits can then be separated by methods described in Chapter 5 that capitalize on small differences in polypeptide size and polarity. Ion exchange and gel filtration chromatography are most often used.

D. Amino Acid Composition

Before we begin the actual sequencing of a polypeptide chain, it is desirable to know its amino acid composition; that is, the number of each type of amino acid residue present. *The amino acid composition of a subunit is determined by its complete hydrolysis followed by the quantitative analysis of the liberated amino acids.* Polypeptide hydrolysis can be accomplished by either chemical (acid or base) or enzymatic means although none of these methods alone is fully satisfactory. For acid catalyzed hydrolysis, the polypeptide is dissolved in 6*N* HCl, sealed in an evacuated tube to prevent the air oxidation of the sulfur-containing amino acids, and heated at 100 to 120°C for 10 to 100 h. The long hydrolysis times are required for the complete liberation of the aliphatic amino acids Val, Leu, and Ile. Unfortunately, not all side chains are impervious to these harsh conditions. Ser, Thr, and Tyr are partially degraded although by following their disappearance as a function of hydrolysis time, correction factors for these losses can be established. A more serious problem is that acid hydrolysis largely destroys the Trp residues. Moreover, Gln and Asn are converted to Glu and Asp plus NH_4^+ so that only the amounts of Asx (= Asp + Asn), Glx (= Glu + Gln), and NH_4^+ (= Asn + Gln) can be independently measured after acid hydrolysis.

Base-catalyzed hydrolysis of polypeptides is carried out in 2 to 4*N* NaOH at 100°C for 4 to 8 h. This treatment is even more problematic because it causes the decomposition of Cys, Ser, Thr, and Arg and partially deaminates and racemizes the other amino acids. Hence alkaline hydrolysis is principally used to measure Trp content.

The complete enzymatic digestion of a polypeptide requires mixtures of peptidases because individual peptidases do not cleave all peptide bonds. Tables 6-1 and

Table 6-2
Specificities of Various Endopeptidases

Enzyme	Source	Specificity	Comments

Enzyme	Source	Specificity	Comments
Trypsin	Bovine pancreas	R_{n-1} = positively charged residues: Arg, Lys; $R_n \neq$ Pro	Highly specific
Chymotrypsin	Bovine pancreas	R_{n-1} = bulky hydrophobic residues: Phe, Trp, Tyr; $R_n \neq$ Pro	Cleaves more slowly for R_{n-1} = Asn, His, Met, Leu
Elastase	Bovine pancreas	R_{n-1} = small neutral residues: Ala, Gly, Ser, Val; $R_n \neq$ Pro	
Thermolysin	*Bacillus thermoproteolyticus*	R_n = Ile, Met, Phe, Trp, Tyr, Val; $R_{n-1} \neq$ Pro	Occasionally cleaves at R_n = Ala, Asp, His, Thr; heat stable
Pepsin	Bovine gastric mucosa	R_n = Leu, Phe, Trp, Tyr; $R_{n-1} \neq$ Pro	Also others; quite nonspecific; pH optimum = 2
Endopeptidase V8	*Staphylococcus aureus*	R_{n-1} = Glu	

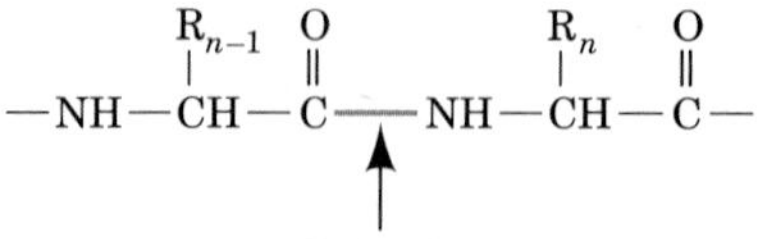

6-2 indicate the specificities of the exopeptidases and **endopeptidases** (enzymes that catalyze the hydrolysis of internal peptide bonds) commonly used for this purpose. **Pronase,** a mixture of relatively nonspecific proteases from *Streptomyces griseus* is also often used to effect complete proteolysis. The amount of enzyme used is limited to ~ 1% by weight of the polypeptide to be hydrolyzed because proteolytic enzymes, being proteins themselves, are self-degrading so that they will, if used too generously, significantly contaminate the final digest. Enzymatic digestion is most often used for determining the amounts of Trp, Asn, and Gln in a polypeptide, which are destroyed by the harsher chemical methods.

Amino Acid Analysis Has Been Automated

The amino acid content of a polypeptide hydrolysate can be quantitatively determined through the use of an automated **amino acid analyzer.** Such an instrument separates amino acids by ion exchange chromatography, a technique pioneered by William Stein and Stanford Moore, or by reverse-phase chromatography using HPLC (Section 5-3E). The amino acids are pre- or post-column derivatized by treatment with either dansyl chloride, Edman's reagent, or **o-phthalaldehyde (OPA)** + 2-mercaptoethanol. The latter reagents react with amino acids to form highly fluorescent adducts:

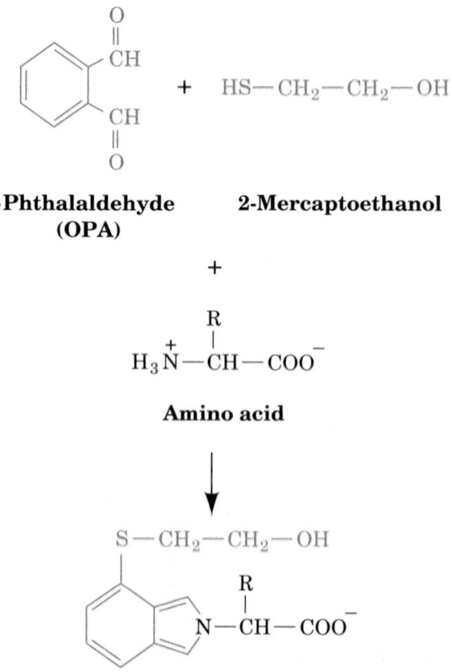

The amino acids are then identified according to their characteristic elution volumes (retention times on HPLC; Fig. 6-6) and quantitatively estimated from their fluorescence intensities (UV absorbances for PTC-amino acids). With modern amino acid analyzers, the

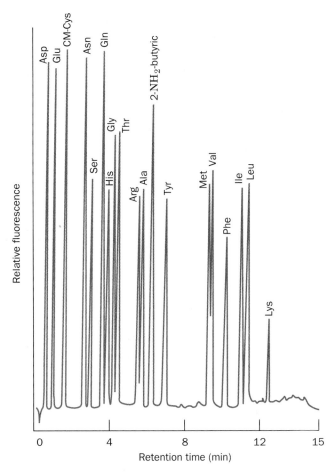

Figure 6-6
The reverse-phase HPLC separation of precolumn OPA-derivatized amino acids. [After Hunkapiller, M. W., Strickler, J. E., and Wilson, K. J., *Science* **226**, 309 (1984).]

complete analysis of a protein digest can be performed in <1h with a sensitivity that can detect as little as 1 pmol of each amino acid.

The Amino Acid Compositions of Proteins Are Indicative of Their Structures

The amino acid analyses of a vast number of proteins indicates that they have considerable variation with respect to their amino acid compositions. Ala, Gly, Leu, Ser, Lys, and Val are the most common amino acid residues, whereas His, Met, and Trp occur least frequently (Table 4-1). Indeed, many proteins lack one or more amino acids. The ratio of polar to nonpolar residues is generally >1 for globular proteins and tends to decrease with increasing protein size. This is because, as we shall see in Chapters 7 and 8, globular proteins have a hydrophobic core and a hydrophilic exterior; that is, they have a micellelike structure. Nonpolar residues predominate in membrane-bound proteins, however, because these proteins, being immersed in a nonpolar environment (Section 11-3A), must also have a hydrophobic exterior.

E. Specific Peptide Cleavage Reactions

Polypeptides that are longer than 40 to 80 residues cannot be directly sequenced (Section 6-1G). Polypeptides of greater length must therefore be cleaved, either enzymatically or chemically, to fragments small enough to be sequenced. In either case, the cleavage process must be complete and highly specific so that the aggregate sequence of a subunit's peptide fragments, when correctly ordered, is that of the intact subunit.

Trypsin Specifically Cleaves Peptide Bonds after Positively Charged Residues

Endopeptidases, like exopeptidases, have side chain requirements for the residues flanking the **scissile** (to be cleaved) peptide bond. The side chain specificities of the endopeptidases most commonly used to fragment polypeptides are listed in Table 6-2. The digestive enzyme **trypsin** has the greatest specificity and is therefore the most valuable member of the arsenal of peptidases used to fragment polypeptides. It cleaves peptide bonds on the C-side (toward the carboxyl terminus) of the positively charged residues Arg and Lys if the next residue is not Pro.

Since trypsin cleaves peptide bonds that follow positively charged residues, trypsin cleavage sites may be added or deleted to a polypeptide by chemically adding or deleting positive charges to or from its side chains. For example, the positive charge on Lys is eliminated by treatment with a dicarboxylic anhydride such as **citraconic anhydride.** The reagent forms a negatively charged derivative of the Lys ε-amino group that trypsin does not recognize.

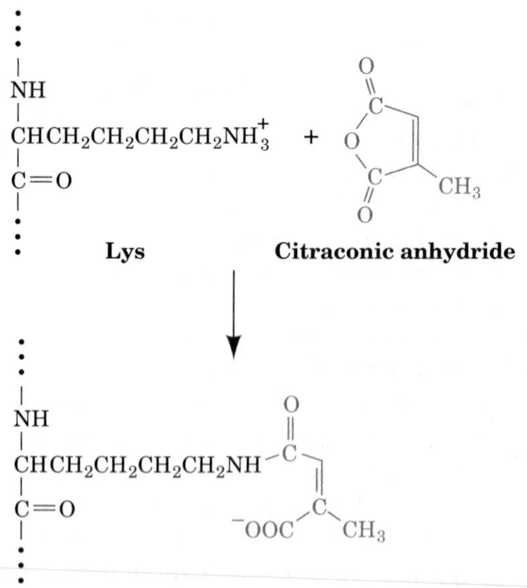

Lys + Citraconic anhydride

(reaction arrow down)

After trypsin hydrolysis of the polypeptide, the Lys residue can be deblocked for identification by mild acid (pH 2–3) hydrolysis. Conversely, Cys may be **aminoalkylated** by a **β-haloamine** to yield a positively charged residue that is subject to tryptic cleavage.

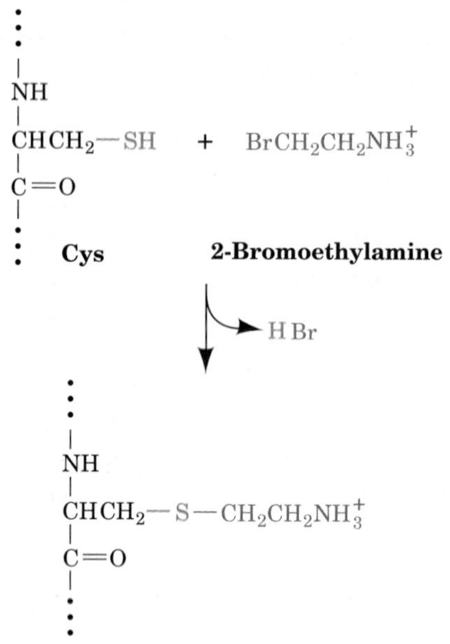

Cys + 2-Bromoethylamine

Such reactions extend the use of trypsin to take further advantage of its great specificity.

The other endopeptidases listed in Table 6-2 exhibit broader side chain specificities than trypsin and often yield a series of peptide fragments with overlapping sequences. However, through **limited proteolysis,** that is, by adjusting reaction conditions and limiting reaction times, these less specific endopeptidases can yield useful peptide fragments. This is because the complex native structure of a protein (subunit) buries many otherwise enzymatically susceptible peptide bonds beneath the surface of the protein molecule. With proper conditions and reaction times, only those peptide bonds in the native protein that are initially accessible to the peptidase will be hydrolyzed. Limited proteolysis is particularly useful in generating peptide fragments of useful size from subunits that have too many or too few Arg and Lys residues to do so with trypsin (although if too many are present, limited proteolysis with trypsin might also yield valuable results).

Cyanogen Bromide Specifically Cleaves Peptide Bonds after Met Residues

Several chemical reagents promote peptide bond cleavage at specific residues. The most useful of these, **cyanogen bromide** (NCBr), causes specific and quantitative cleavage on the C-side of Met residues to form a **peptidyl homoserine lactone.**

Cyanogen bromide

Methyl thiocyanate

Peptidyl homoserine lactone

Aminoacyl peptide

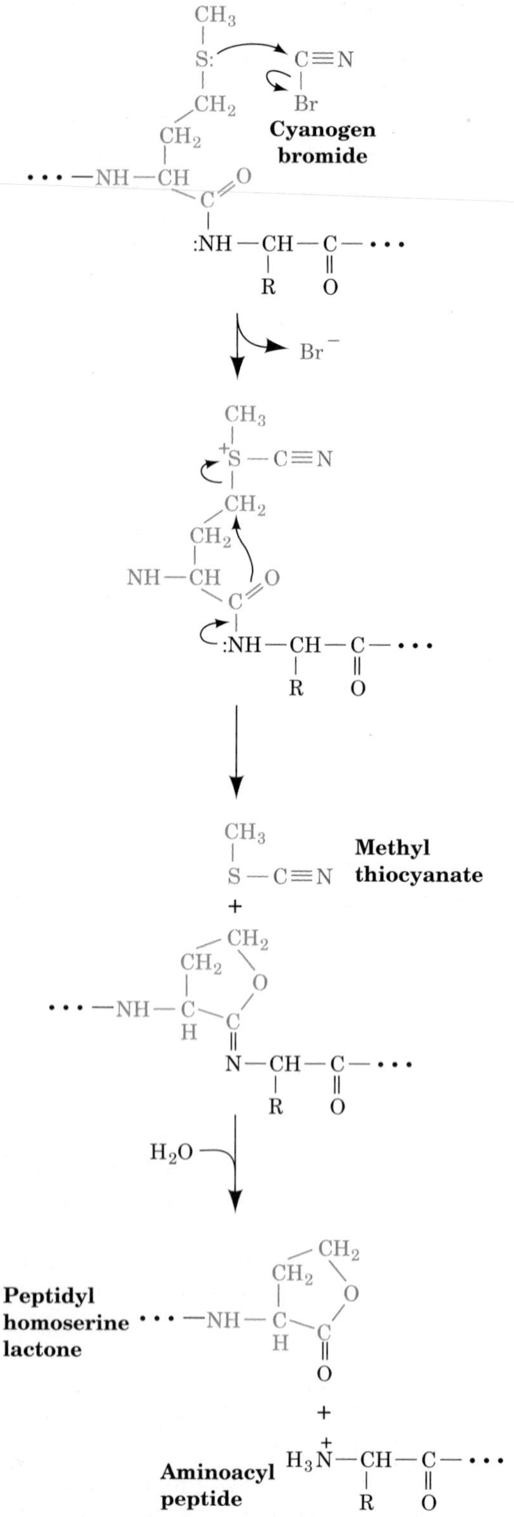

The reaction is performed in an acidic solvent (0.1*N* HCl or 70% formic acid) that denatures most proteins so that cleavage normally occurs at all Met residues.

A peptide fragment generated by a specific cleavage process may still be too large to sequence. In that case, after its purification, it can be subjected to a second round of fragmentation using a different cleavage process.

F. Separation and Purification of the Peptide Fragments

Once again we must employ separation techniques, this time to isolate the peptide fragments of specific cleavage operations for subsequent sequence determinations. The nonpolar residues of peptide fragments are not excluded from the aqueous environment as they are in native proteins (Chapters 7 and 8). Consequently, many peptide fragments aggregate, precipitate, and/or strongly adsorb to chromatographic materials, which often results in unacceptable peptide losses. Until recently, the trial-and-error development of methods that could satisfactorily separate a mixture of peptide fragments constituted the major technical challenge of a protein sequence determination, as well as its most time consuming step. Such methods involved the use of denaturants, such as urea and SDS, to solubilize the peptide fragments, and the selection of chromatographic materials and conditions that would reduce their adsorptive losses. The advent of reverse-phase chromatography by HPLC (Section 5-3E), however, has largely reduced the separation of peptide fragments to a routine procedure.

G. Sequence Determination

Once the manageably sized peptide fragments that were formed through specific cleavage reactions have been isolated, their amino acid sequences can be determined. *This is best accomplished through repeated cycles of the Edman degradation* (Section 6-1A). An automated device for doing so, a **spinning cup sequenator,** was developed by Edman and Geoffrey Begg. It contains a glass cup, spinning at 1000 to 4000 rpm, over whose inner surface the peptide is immobilized in a thin film. Alternatively, in a **solid phase sequencer,** the peptide is covalently coupled to an inert insoluble support such as a polystyrene resin so that only filtration is required to isolate the thiazolinone-amino acid product of each Edman degradation cycle. In the most advanced instruments, the peptide is embedded in a thin film of a polymeric quaternary ammonium salt, **polybrene,** that immobilizes the peptide but is readily penetrated by Edman reagents (delivered as vapors in a stream of argon). In all these instruments, accurately measured quantities of reagents are added to the reaction cell at programmed intervals. The thiazolinone-amino acids are automatically removed, converted to the corresponding PTH-amino acids (Fig. 6-4) and identified chromatographically. Such instruments are capable of processing up to one residue per hour.

Usually, a peptide's 40 to 60 N-terminal residues can be identified (80 or more with the most advanced systems) before the effects of incomplete reactions, side reactions, and peptide loss accumulate to the extent that amino acid identification is no longer reliable. Since less than a pmol of a PTH amino acid can be detected and identified by a reverse-phase HPLC system equipped with a UV detector, sequence analysis can be carried out on as little as 5 to 10 pmol (<0.1 *μg*—an invisibly small amount) of a peptide.

H. Ordering the Peptide Fragments

With the peptide fragments individually sequenced, what remains is to elucidate the order in which they are connected in the original polypeptide. *We do so by comparing the amino acid sequences of one set of peptide fragments with those of a second set whose specific cleavage sites overlap those of the first set (Fig. 6-7).* The overlap-

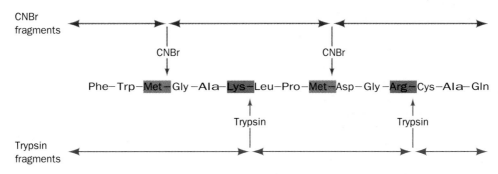

Figure 6-7
The amino acid sequence of a polypeptide chain is determined by comparing the sequences of two sets of mutually overlapping peptide fragments. In this example, the two sets of peptide fragments are generated by cleaving the polypeptide after all its Arg and Lys residues with trypsin and, in a separate reaction, after all its Met residues by treatment with cyanogen bromide. The order of the first two tryptic peptides is established, for example, by the observation that the Gly-Ala-Lys-Leu-Pro-Met cyanogen bromide peptide has its N- and C-terminal sequences in common with the C- and N-termini, respectively, of the two tryptic peptides. In this manner the order of the peptide fragments in their parent polypeptide chain is established.

ping peptide segments must be of sufficient length to identify each cleavage site uniquely, but as there are 20 possibilities for each amino acid residue, an overlap of only a few residues is usually enough.

I. Assignment of Disulfide Bond Positions

The final step in an amino acid sequence analysis is to determine the positions (if any) of the disulfide bonds. This is done by cleaving a sample of the native protein, with its disulfide bonds intact, so as to yield pairs of peptide fragments, each containing a single Cys, that are linked by a disulfide bond.

Such peptide fragments may be identified through **diagonal electrophoresis** (Fig. 6-8). In this technique, the partial peptide digest is electrophoretically separated in two dimensions by identical procedures. After the first separation, the electrophoretogram is exposed to performic acid vapor, which oxidizes the disulfide linkages to cysteic acid (Section 6-1B). After the second separation, those peptide fragments that were not modified by the performic acid treatment will be located along the diagonal of the electrophoretogram because their rates of migration are the same in both directions. Those polypeptides that were originally joined by a disulfide bond will each be decomposed to two peptides that have different migration rates from before and therefore lie off the diagonal of the electrophoretogram.

After the isolation of a disulfide-linked polypeptide fragment, the disulfide bond is cleaved and alkylated (Section 6-1B) and the sequences of the two polypeptides are determined (Fig. 6-9). The various pairs of such polypeptide fragments are identified by the comparison of their sequences with that of the protein, thereby establishing the locations of the disulfide bonds.

J. Peptide Mapping

The sequence determination of a protein is an exacting and time-consuming process. Once the primary structure of a protein has been elucidated, however, that of a nearly identical protein, such as one arising from a closely related species, a mutation, or a chemical modification, can be more easily determined. This is done through combined chromatography and electrophoresis (Section 5-4A) of partial protein digests, a technique synonomously known as **fingerprinting** or **peptide mapping.** The peptide fragments incorporating the amino acid variations migrate to different positions on their fingerprint (peptide map) than do the corresponding peptides of the original protein (Fig. 6-10). The variant peptides can then be eluted from the fingerprint and sequenced to establish the differences between the primary structures of the original and variant proteins.

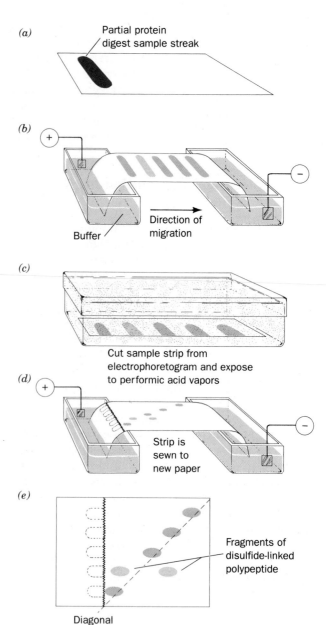

Figure 6-8
Diagonal electrophoresis: (*a*) A partial protein digest with its disulfide bonds intact is streaked onto a sheet of paper, and (*b*) subjected to electrophoresis. (*c*) A guide strip is cut from the edge of the electrophoretogram and exposed to performic acid vapor that oxidizes each —S—S— linkage to two —SO₃H groups. (*d*) The guide strip is sewn to a second sheet of paper and subjected to electrophoresis in the perpendicular direction under the same conditions as before, followed by staining. (*e*) Peptides not on the diagonal of the second electrophoretogram, such as the two derived from the yellow fragment, contain cysteic acid residues and hence were originally linked by at least one disulfide bond. The parent peptide fragment can be located on the first electrophoretogram and eluted for further analysis.

K. Nucleic Acid Sequencing

The amino acid sequences of proteins are specified by the base sequences of nucleic acids (Section 30-1) so

$$-S-S-$$

Polypeptide fragment containing disulfide bond

Reduce disulfide and block with iodoacetate

$$-S-CH_2CO_2^- \quad + \quad {}^-O_2CCH_2-S-$$

Separate and sequence the polypeptides

Figure 6-9
A protein's disulfide bond positions are determined by identifying pairs of disulfide-linked peptide fragments.

that, with a knowledge of the genetic code, a protein's primary structure can be inferred from that of a corresponding nucleic acid. Techniques for sequencing nucleic acids initially lagged far behind those for proteins, but by the late 1970s, DNA sequencing methods had advanced to the point that it became easier to sequence a DNA segment than the protein it specifies (nucleic acid sequencing is the subject of Section 28-6). Although protein primary structures are now routinely inferred from DNA sequences, direct protein sequencing remains an indispensable biochemical tool for several important reasons:

1. Disulfide bonds can only be located by protein sequencing.

2. Many proteins are modified after their biosynthesis by the excision of certain residues and by the specific derivatization of others (Section 30-5). The identities of these modifications, which are often essential for the protein's biological function, can only be determined by directly sequencing the protein.

3. It is often difficult to identify and isolate a nucleic acid that encodes the protein of interest.

4. A common error in DNA sequencing is the inadvertent insertion or deletion of a single nucleotide. The genetic code consists of consecutive triplets of nucleotides, each specifying a single amino acid residue (Section 30-1B). The erroneous addition or deletion of a nucleotide from a gene sequence therefore changes the gene's apparent reading frame and thus

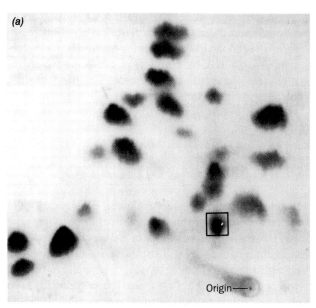

(a)

Origin——

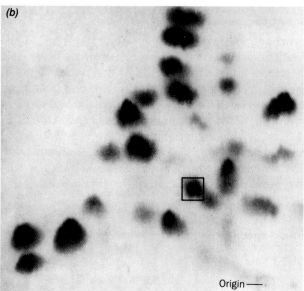

(b)

Origin——

Figure 6-10
A comparison of the ninhydrin-stained fingerprints of trypsin-digested (*a*) hemoglobin A (HbA) and (*b*) hemoglobin S (HbS). The peptides that differ in these two forms of hemoglobin are boxed. These peptides constitute the eight N-terminal residues of the β-subunit of hemoglobin. Their amino acid sequences are

Hemoglobin A	Val-His-Leu-Thr-Pro-**Glu**-Glu-Lys
Hemoglobin S	Val-His-Leu-Thr-Pro-**Val**-Glu-Lys
	β1　2　3　4　5　**6**　7　8

[Courtesy of Corrado Baglioni, State University of New York at Albany.]

changes the predictions for all the amino acid residues past the point of error. Double checking the predicted amino acid sequence by directly sequencing a series of oligopeptides scattered throughout the protein readily detects such errors.

5. The "standard" genetic code is not universal: Those of mitochondria and certain protozoa are slightly different (Section 30-1E). These genetic code variants were elucidated by comparing the amino acid sequences of proteins and the base sequences of their corresponding genes. If there are other genetic code variants, they will, no doubt, be discovered in a like manner.

2. PROTEIN MODIFICATION

A common strategy for identifying the residues of a protein essential for its biological function is to treat the protein with reagents that react with specific types of side chains. A protein will probably be inactivated by such a **group-specific reagent** if it chemically alters one of the protein's essential residues. If a protein has been sequenced, it is a usually simple matter to identify the modified protein's altered residues through fingerprinting. Even without the knowledge of a protein's primary structure, the comparison of the fingerprints of the modified and unmodified proteins can yield valuable information concerning the altered residues. Some of the most useful group specific reagents and their products are listed in Table 6-3.

A catalytically active group may be rendered unusually reactive by its environment in the protein such that it will react with certain reagents in an abnormal manner. For example, the family of enzymes known as **serine proteases,** of which trypsin, chymotrypsin, and

elastase are members, is characterized by an unusually reactive active site Ser [Section 14-3; an enzyme's **active site** constitutes the groups involved in the binding and catalysis of its **substrates** (reactants)]. This Ser reacts with **diisopropylphosphofluoridate (DIPF)** to form the following derivative.

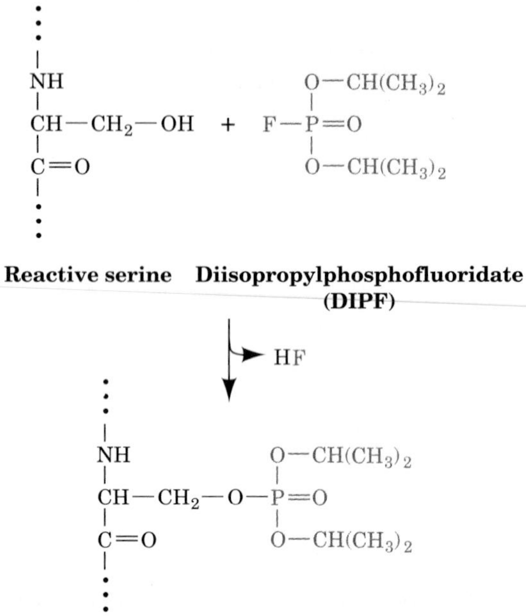

Reactive serine   Diisopropylphosphofluoridate
(DIPF)

Other Ser residues, even in the same protein, do not react with DIPF so that this reaction constitutes a method of uniquely labeling the active site Ser of serine proteases and thus is a diagnostic test for serine proteases.

Table 6-3
Group-Specific Reagents for Amino Acid Residue Modifications

Side Chain	Reagent	Product	Other Reactive Groups
Lys	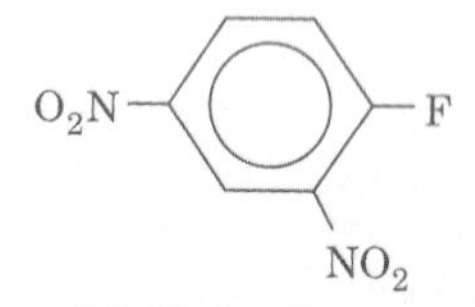 **1-Fluoro-2,4-dinitrobenzene (FDNB)**	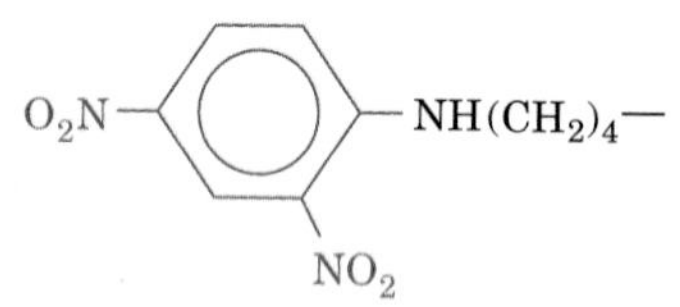 **Dinitrophenylated (DNP)-Lys**	Cys, His, N-terminal amine
	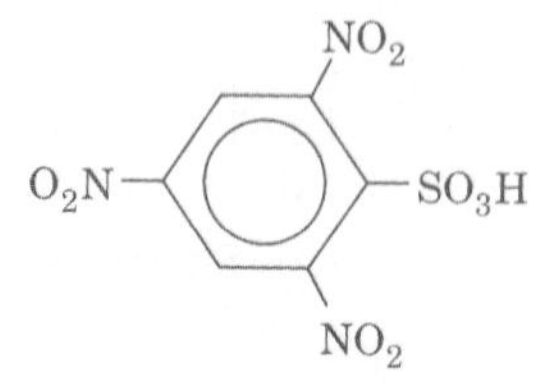 **Trinitrobenzene sulfonic acid**	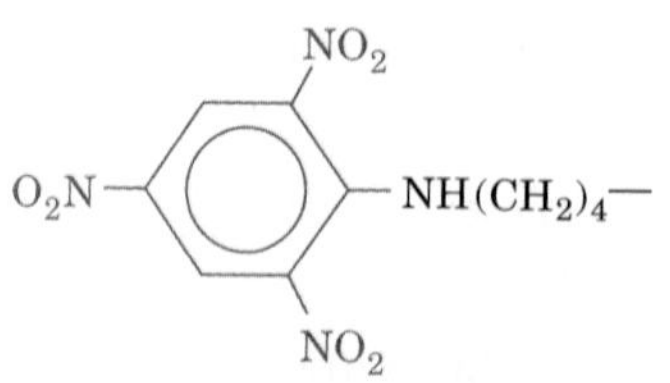 **Trinitrophenyl-Lys**	N-terminal amine
	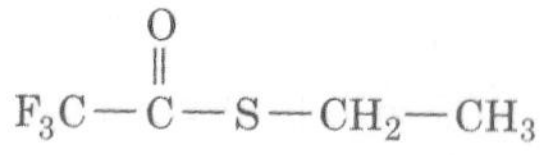 **Ethylthiotrifluoroacetate**	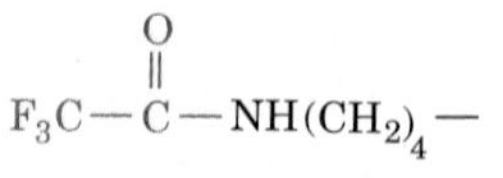 **Trifluoroacetyl-Lys**	N-terminal amine

Side Chain	Reagent	Product	Other Reactive Groups
Lys *(continued)*	**Succinic anyhdride**	$^-OOC-CH_2-CH_2-\overset{\overset{\displaystyle O}{\|}}{C}-NH(CH_2)_4-$ **Succinyl-Lys**	N-terminal amine
Arg	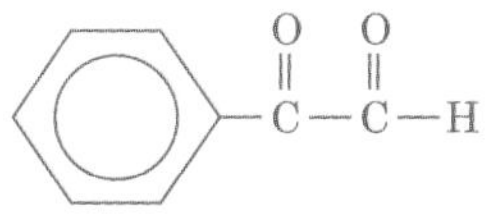 **Phenylglyoxal**	Arg reacts with two phenyl-gloxal molecules to form a product of unknown structure	
	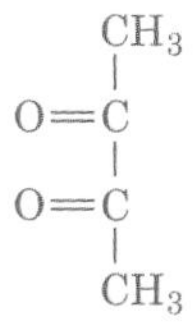 **2,3-Butanedione**	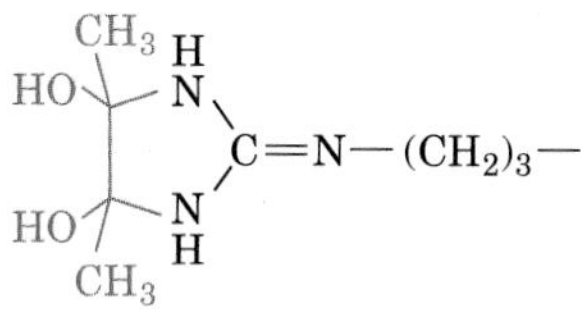 $\underset{CH_3}{\overset{CH_3}{\rangle}}C=N-(CH_2)_3-$ (with HO, N–H groups)	
Cys	ICH_2-COO^- **Iodoacetate**	$^-OOC-CH_2-S-CH_2-$ **S- Carboxymethyl-Cys**	Asp, Glu, His, Lys, Met
	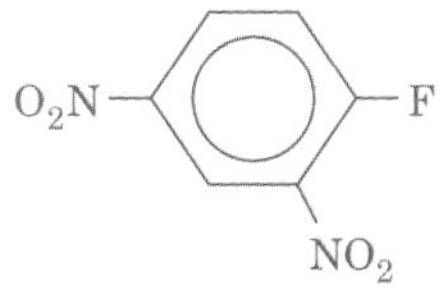 **1-Fluoro-2,4-dinitrobenzene (FDNB)**	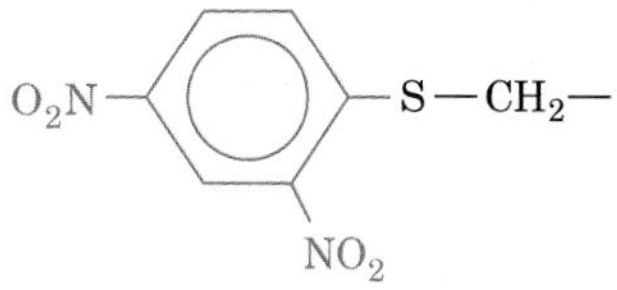 **Dinitrophenylated (DNP)-Cys** $O_2N-\bigcirc-S-CH_2-$	Lys, His, Tyr
	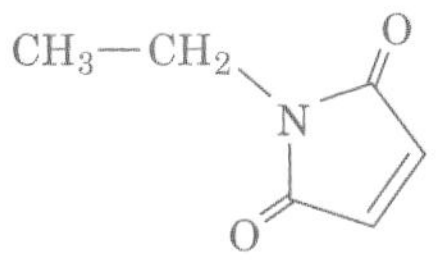 **N-Ethylmaleimide**	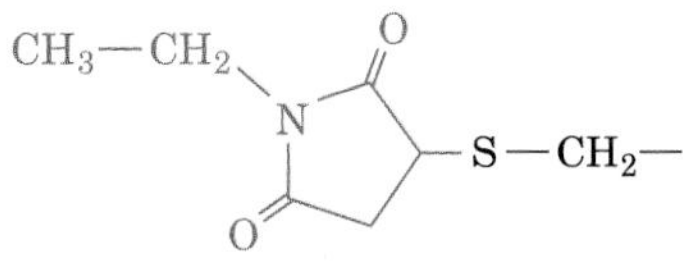 $CH_3-CH_2-N\cdots-S-CH_2-$	Specific at slightly acidic pH's
	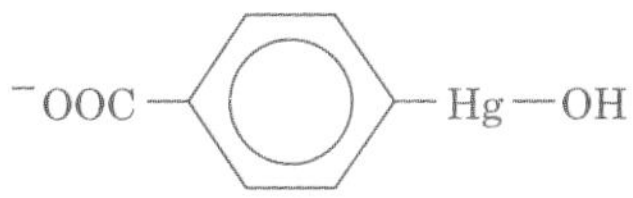 **p-Hydroxymercuribenzoate** $^-OOC-\bigcirc-Hg-OH$	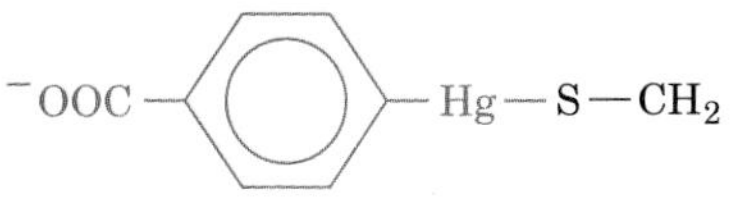 $^-OOC-\bigcirc-Hg-S-CH_2$	
	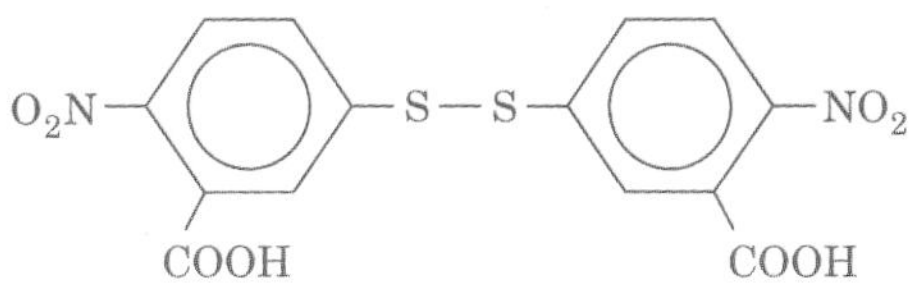 **5,5'-Dithiobis(2-nitrobenzoic acid) (DTNB)**	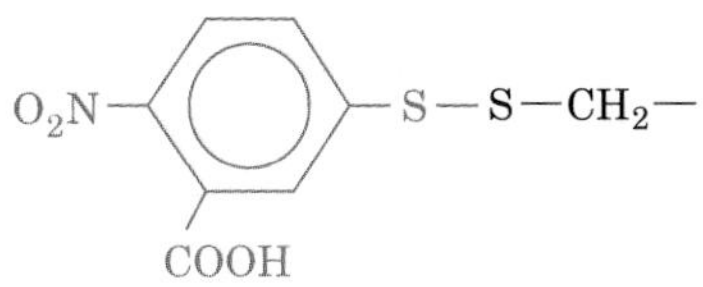 $O_2N-\bigcirc-S-S-CH_2-$ (COOH)	

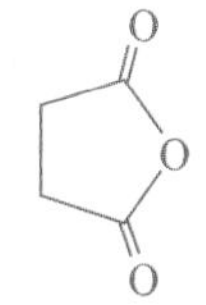

Table 6-3   *(continued)*

Side Chain	Reagent	Product	Other Reactive Groups
Cys *(continued)*	$H-\overset{\overset{O}{\|}}{C}-O-O-H$ **Performic acid**	$^-O_3S-CH_2-$ **Cysteic acid**	Met, Cys-S-S-Cys
Cys-S-S-Cys	$HS-CH_2-CH_2-OH$ **2-Mercaptoethanol** $\begin{array}{c} CH_2-SH \\ \| \\ HO-C-H \\ \| \\ H-C-OH \\ \| \\ CH_2-SH \end{array}$ **Dithiothreitol** $\begin{array}{c} CH_2-SH \\ \| \\ H-C-OH \\ \| \\ H-C-OH \\ \| \\ CH_2-SH \end{array}$ **Dithioerythritol**	$HS-CH_2-$ **Cys-SH**	
	$H-\overset{\overset{O}{\|}}{C}-O-O-H$ **Performic acid**	$^-O_3S-CH_2-$ **Cysteic acid**	Cys, Met
Met	$N\equiv C-Br$ **Cyanogen bromide**	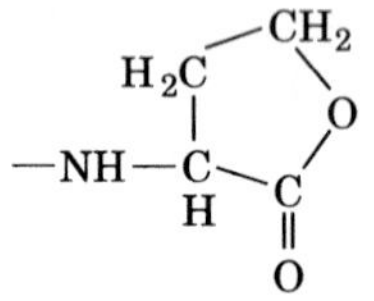 **Peptidyl homoserine lactone**	
	ICH_2-COO^- **Iodoacetate**	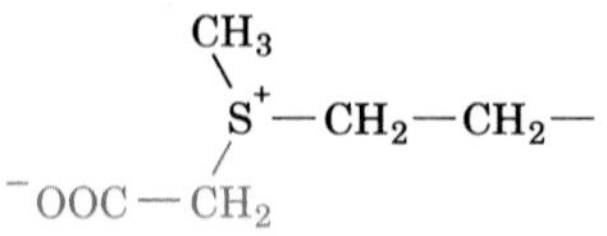 $\overset{CH_3}{\underset{^-OOC-CH_2}{\overset{\|}{S^+}}}-CH_2-CH_2-$ **S-Carboxymethyl-Met**	Asp, Cys, Glu, His, Lys
	$H-\overset{\overset{O}{\|}}{C}-O-O-H$ **Performic acid**	$H_3C-\overset{\overset{O}{\|}}{\underset{\underset{O}{\|}}{S}}-CH_2-CH_2-$ **Methionine sulfone**	Cys, Cys-S-S-Cys
Asp, Glu	$CH_2{=}\overset{+}{N}{=}N^-$ **Diazomethane**	$H_3C-O-\overset{\overset{O}{\|}}{C}-CH_2-$ **Methyl ester**	

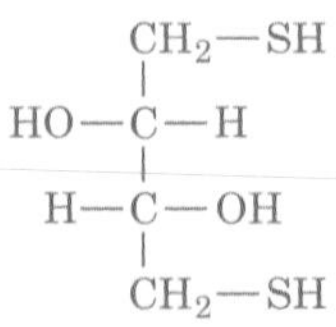

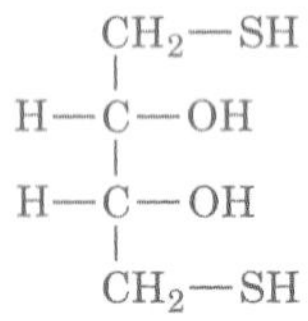

Side Chain	Reagent	Product	Other Reactive Groups
Asp, Glu *(continued)*	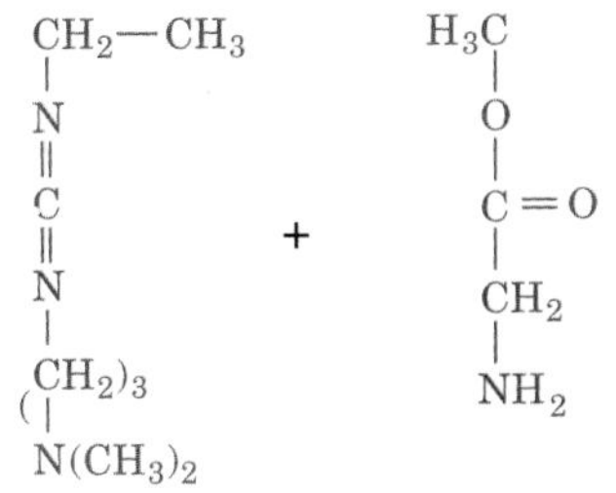 EDAC[a] + Glycine methyl ester	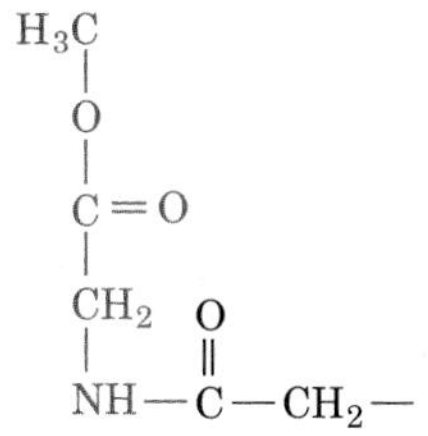	
His	Iodoacetate	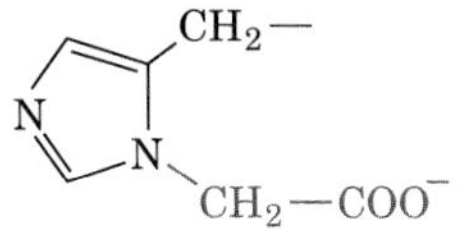	Asp, Cys, Glu, Lys, Met
	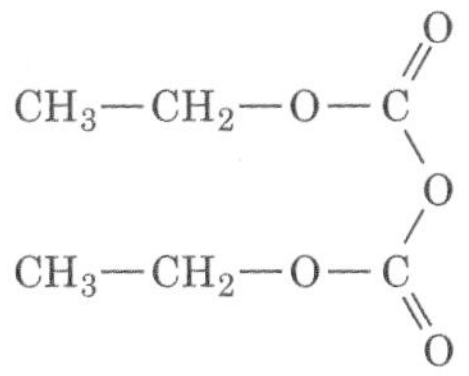 Diethylpyrocarbonate	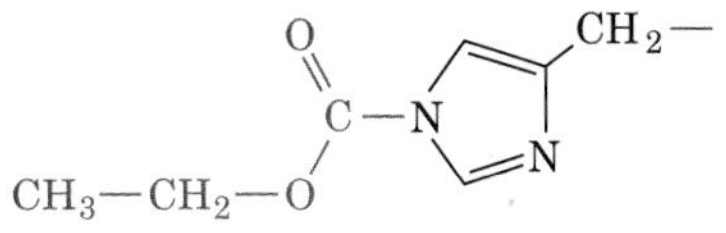 Ethylcarboxamido-His	
Trp	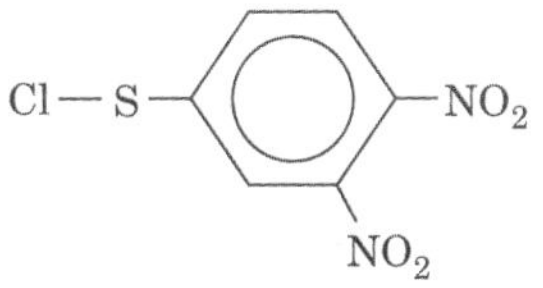 2,4-Dinitrophenylsulfenyl chloride	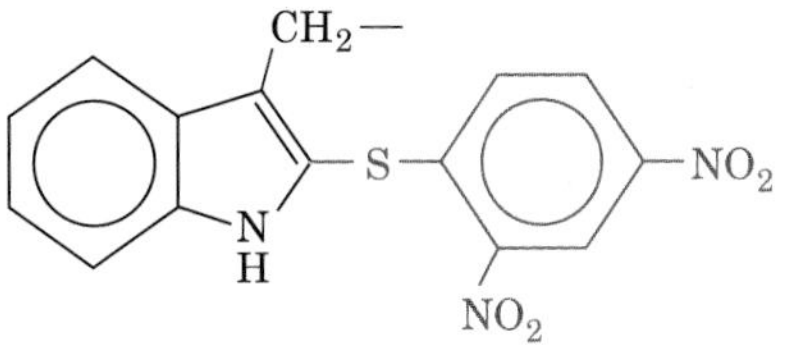	Cys (can be decomposed by thiols)
	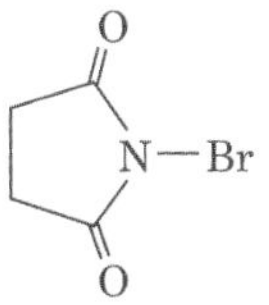 *N*-Bromosuccinimide	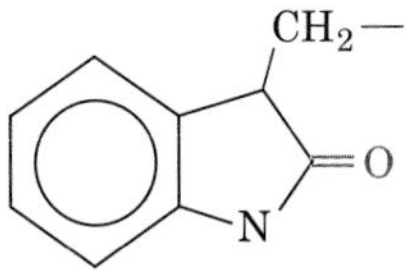	His, Cys, Arg, Lys, Tyr
Tyr	Tetranitromethane	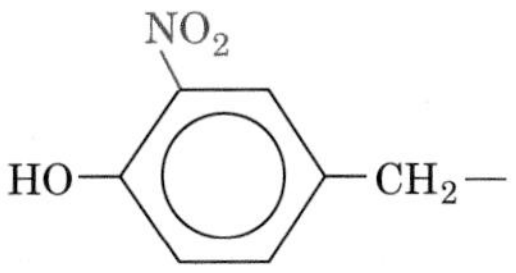 3-Nitrotyrosine	

[a] 1-Ethyl-3-(3-dimethyl)aminopropylcarbodiimide (or other water soluble carbodiimide).

3. CHEMICAL EVOLUTION

Individuals, as well as whole species, are characterized by their inherited genetic compositions. An organism's genetic complement, as we shall see in Part V, specifies the amino acid sequences of all of its proteins together with their quantity and schedule of appearance in each cell. An organism's protein composition is therefore the direct expression of its genetic composition.

In this section, we concentrate on the evolutionary aspects of amino acid sequences, the study of the **chemical evolution** of proteins. Evolutionary changes, which stem from random mutational events, often alter a protein's primary structure. A mutational change in a protein, if it is to be propagated, must somehow increase, or at least not decrease, the probability that its owner will survive to reproduce. Many mutations are deleterious and often lethal in their effects and therefore rapidly die out. On rare occasions, however, a mutation arises that, as we shall see below, improves the fitness of its host in its natural environment.

A. Sickle-Cell Anemia: The Influence of Natural Selection

Hemoglobin, the red blood pigment, is a protein whose major function is to transport oxygen throughout the body. A molecule of hemoglobin is an $\alpha_2\beta_2$ tetramer; that is, it consists of two identical α chains and two identical β chains (Fig. 6-1d). Hemoglobin is contained in the **erythrocytes** (red blood cells; Greek: *erythrose,* red + *kytos,* a hollow vessel) of which it forms $\sim 33\%$ by weight in normal individuals. In every cycle of their voyage through the circulatory system, the erythrocytes, which are normally flexible biconcave disks (Fig.6-11a), must squeeze through capillary blood vessels smaller in diameter than they are.

In individuals with the inherited disease **sickle-cell anemia,** many erythrocytes assume an irregular crescentlike shape under conditions of low oxygen concentration typical of the capillaries (Fig. 6-11b). This "sickling" increases the erythrocytes' rigidity, which hinders their free passage through the capillaries. The sickled cells therefore impede the flow of blood in the capillaries such that, in a sickle-cell "crisis," the blood flow in some areas may be completely blocked, thereby giving rise to extensive tissue damage and excruciating pain. Moreover, individuals with sickle-cell anemia suffer from severe **hemolytic anemia** (a condition characterized by red cell destruction) because the increased mechanical fragility of their erythrocytes halves the normal 120-day lifetime of these cells. The debilitating effects of this disease are such that, before the latter half of this century, individuals with sickle-cell anemia rarely survived to maturity (although modern treatments by no means constitute a cure).

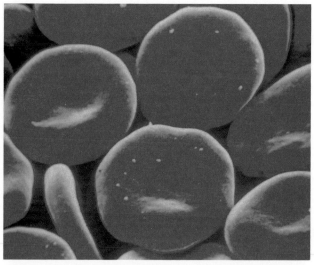

(a)

(b)

Figure 6-11
Scanning electron micrographs of: (*a*) Normal human erythrocytes revealing their biconcave disklike shape. [David M. Phillips/Visuals Unlimited.] (*b*) Sickled erythrocytes from a patient with sickle-cell anemia. [Bill Longcore/Photo Researchers, Inc.]

Sickle-Cell Anemia Is a Molecular Disease

In 1945, Linus Pauling correctly hypothesized that *sickle-cell anemia, which he termed a **molecular disease**, is a result of the presence of a mutant hemoglobin.* Pauling and his coworkers subsequently demonstrated, through electrophoretic studies, that normal human hemoglobin (**HbA**) has an anionic charge that is around two units more negative than that of sickle-cell hemoglobin (**HbS;** Fig. 6-12).

In 1956, Vernon Ingram developed the technique of peptide mapping in order to pinpoint the difference between HbA and HbS. Ingram's fingerprints of tryptic digests of HbA and HbS revealed that their α subunits are identical but that their β subunits differ by a variation in one tryptic peptide (Fig. 6-10). Sequencing studies indicated that this difference arises from the replacement of the Glu $\beta6$ of HbA (the Glu in the sixth position of each β chain) with Val in HbS (Glu $\beta6 \rightarrow$ Val), thus

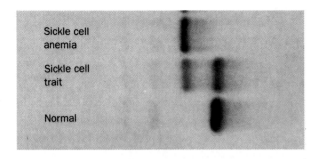

Figure 6-12
The electrophoretic pattern of hemoglobins from normal
individuals and those with the sickle-cell trait and sickle-cell
anemia. [From Montgomery, R., Dryer, R. L., Conway,
T. W., and Spector, A. A., *Biochemistry, A Case Oriented
Approach* (4th ed.), *p. 87.* Copyright © 1983 C. V. Mosby
Company, Inc.]

accounting for the charge difference observed by Pauling. This was the first time an inherited disease was
shown to arise from a specific amino acid change in a
protein. *This mutation causes HbS to aggregate into filaments of sufficient size and stiffness to deform erythrocytes*
—a remarkable example of the influence of primary
structure on quaternary structure. The structure of these
filaments is further discussed in Section 9-3B.

The Sickle-Cell Trait Confers Resistance to Malaria

Sickle-cell anemia is inherited according to the laws of
Mendelian genetics (Section 27-1B). The cells of all
higher organisms but germ cells have two homologous
copies of each chromosome with the exception of sex
chromosomes. An organism carrying a particular gene is
classified as **heterozygous** or **homozygous** for that gene
if its cells, respectively, bear one or two copies of that
gene. The hemoglobin of individuals who are homozygous for sickle-cell anemia is almost entirely HbS. In
contrast, individuals heterozygous for sickle-cell anemia have hemoglobin that is ~40% HbS (Fig. 6-12).
Such persons, who are said to have the **sickle-cell trait,**
lead a normal life even though their erythrocytes have a
shorter lifetime than those of normal individuals.

The sickle-cell trait and disease occur mainly in persons of equatorial African descent. The regions of equatorial Africa where **malaria** is a major cause of death
(contributing to childhood mortality rates as high as
50%), as Fig. 6-13 indicates, coincide closely with those
areas where the sickle-cell gene is prevalent (possessed
by as much as 40% of the population in some places).
This observation led Anthony Allison to the discovery
that *individuals heterozygous for HbS are resistant to malaria.*

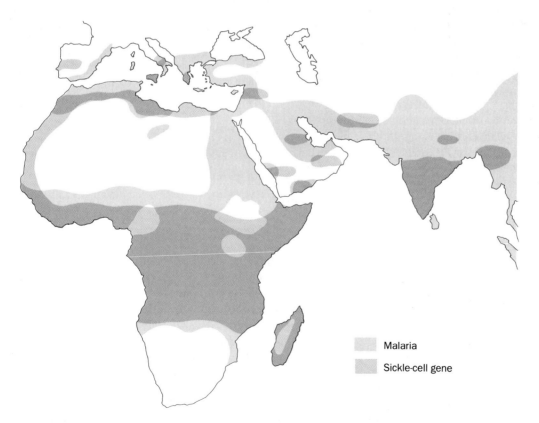

    Malaria

    Sickle-cell gene

Figure 6-13
A map indicating the regions of the world where malaria caused by *P.
falciparum* was prevalent before 1930, together with the distribution of the
sickle-cell gene.

Table 6-4
Amino Acid Sequences of Cytochromes *c* from 38 Species[a]

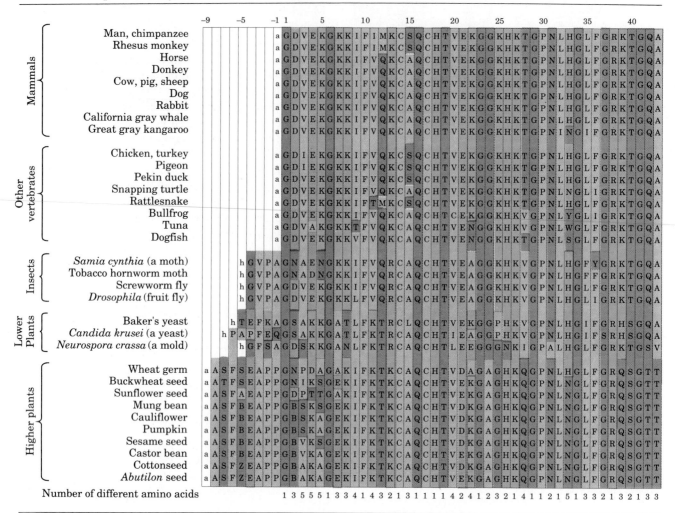

[a] The amino acid side chains have been shaded according to their polarity characteristics so that an invariant or conservatively substituted residue is identified by a vertical band of a single color. The letter a at the beginning of the chain indicates that the N-terminal amino group is acetylated; an h indicates the acetyl group is absent.

Source: After Dickerson, R. E., *Sci. Am.* **226**(4): 58–72 (1972) with corrections from Dickerson, R. E., and Timkovich, R., *in* Boyer, P. D. (Ed.), *The Enzymes* (3rd ed.), Vol. 11, *pp.* 421–422, Academic Press (1975). Table copyrighted © by Irving Geis.

Malaria is a parasitic disease. In Africa it is caused by the mosquito-borne protozoan *Plasmodium falciparum,* which resides within an erythrocyte during much of its 48-h life cycle. *Plasmodia* increase the acidity of the erythrocytes they infect by ~0.4 pH units and cause them to adhere to a specific protein lining blood vessel walls by protein knobs that develop on the erythrocyte surfaces (the spleen would otherwise remove the infected erythrocytes from the circulation thereby killing the parasites). Death often results when so many erythrocytes are lodged in a vital organ (such as the brain in cerebral malaria) that its blood flow is significantly impeded.

How does the sickle-cell trait confer malarial resistance? Normally, ~2% of the erythrocytes of individuals with the sickle-cell trait are observed to sickle under the low oxygen concentration conditions found in the capillaries. However, the lowered pH of infected erythrocytes increases their proportion of sickling in the capillaries to ~40%. A normal erythrocyte maintains a high internal concentration of K^+ relative to that of the blood serum through processes discussed in Section 18-3A. When an erythrocyte sickles, the permeability of its cell membrane to K^+ increases so that the K^+ concentration in sickled cells is lower than in normal erythrocytes. The malarial parasite requires a high K^+ concentration and

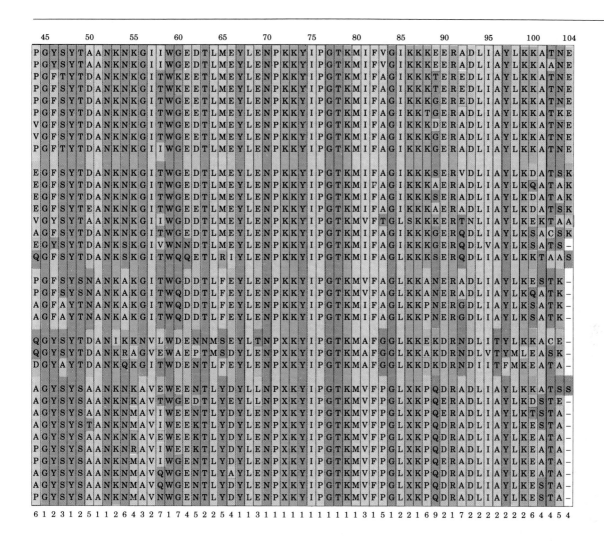

Hydrophobic, acidic: 　D Asp　　E Glu

Hydrophilic, basic: 　H His　　K Lys　　R Arg　　X Methylated Arg

Polar, uncharged: 　B Asn or Asp　　G Gly　　N Asn　　Q Gln
　　　　　　　　　　S Ser　　T Thr　　W Trp　　Y Tyr　　Z Gln or Glu

Hydrophobic: 　A Ala　　C Cys　　F Phe　　I Ile　　L Leu
　　　　　　　M Met　　P Pro　　V Val

therefore cannot long survive in a sickled cell. Consequently, bearers of the sickle-cell trait in a malarial region have an adaptive advantage: The fractional population of heterozygotes (sickle-cell trait carriers) in such areas increases until their reproductive advantage becomes balanced by the inviability of the correspondingly increasing proportion of homozygotes (those with sickle-cell disease). Thus *sickle-cell anemia provides a classical Darwinian example of a single mutation's adap-* tive consequences in the ongoing biological competition among organisms for the same resources.

B. Species Variations in Homologous Proteins: The Effects of Neutral Drift

The primary structures of a given protein from related species closely resemble one another. If one assumes, according to evolutionary theory, that related species evolved from a common ancestor, then it follows that each of their proteins must have likewise evolved from the corresponding protein in that ancestor.

A protein that is well adapted to its function, that is, one that is not subject to significant physiological improvement, nevertheless continues evolving. The random nature of mutational processes will, in time, change such a protein in ways that do not significantly affect its function, a process called **neutral drift** (deleterious mutations are, of course, rapidly rejected through natural selection). *Comparisons of the primary structures of homologous proteins* (evolutionarily related proteins) therefore indicate which of the proteins' residues are essential to its function, which are of lesser significance, and which have little specific function. If, for example, we find the same side chain at a particular position in the amino acid sequence of a series of related proteins, we can reasonably conclude that the chemical and/or structural properties of that so-called **invariant residue** uniquely suit it to some essential function of the protein. Other amino acid positions may have less stringent side chain requirements so that only residues with similar characteristics (e.g, those with acidic properties: Asp and Glu) are required; such positions are said to be **conservatively substituted.** On the other hand, many different amino acid residues may be tolerated at a particular amino acid position, which indicates that the functional requirements of that position are rather nonspecific. Such a position is called **hypervariable.**

Cytochrome *c* Is a Well-Adapted Protein

To illustrate these points, let us consider the primary structure of a nearly universal eukaryotic protein, **cytochrome *c*.** Cytochrome *c* has a single polypeptide chain that, in vertebrates, consists of 103 or 104 residues, but in other phyla has up to 8 additional residues at its N-terminus. It occurs in the mitochondrion as part of the **electron-transport chain,** a complex metabolic system that functions in the terminal oxidation of nutrients to produce ATP (Section 20-2). The role of cytochrome *c* is to transfer electrons between an enzyme complex known as **cytochrome *c* reductase** and one called **cytochrome *c* oxidase.**

It is believed that the electron-transport chain took its present form between 1.5 and 2 billion years ago as organisms evolved the ability to respire (Section 1-4C). Since that time, the components of this multienzyme system have changed very little as is evidenced by the observation that the cytochrome *c* from any eukaryotic organism, say a pigeon, will react *in vitro* (in the test tube) with the cytochrome oxidase from any other eukaryote, for instance, wheat. Indeed, hybrid cytochromes *c* consisting of covalently linked fragments from such distantly related species as horse and yeast (prepared by methods discussed in Section 6-4B) exhibit biological activity.

Protein Sequence Comparisons Yield Taxonometric Insights

Emanuel Margoliash, Emil Smith, and others have elucidated the amino acid sequences of the cytochromes *c* from nearly 100 widely diverse eukaryotic species ranging in complexity from yeast to humans. The sequences from 38 of these organisms are arranged in Table 6-4 so as to maximize the similarities between vertically aligned residues. The various residues in this table have been shaded according to their physical properties in order to illuminate the conservative character of the amino acid substitutions. Inspection of Table 6-4 indicates that cytochrome *c* is an evolutionary conservative protein. A total of 38 of its 105 residues (23 in all that have been sequenced) are invariant and most of the remaining residues are conservatively substituted. In contrast, there are eight positions that each accommodate six or more different residues and, accordingly, are described as being hypervariable.

The clear biochemical role of certain residues makes it easy to surmise why they are invariant. For instance, His 18 and Met 80 form ligands to the redox-active Fe atom of cytochrome *c*; the substitution of any other

Table 6-5

Amino Acid Difference Matrix for 26 Species of Cytochrome *c*[a]

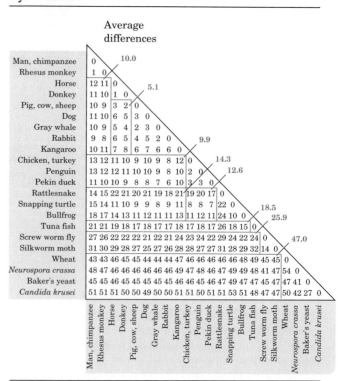

[a] Each table entry indicates the number of amino acid differences between the cytochromes *c* of the species noted to the left of and above that entry.

[Table copyrighted © by Irving Geis.]

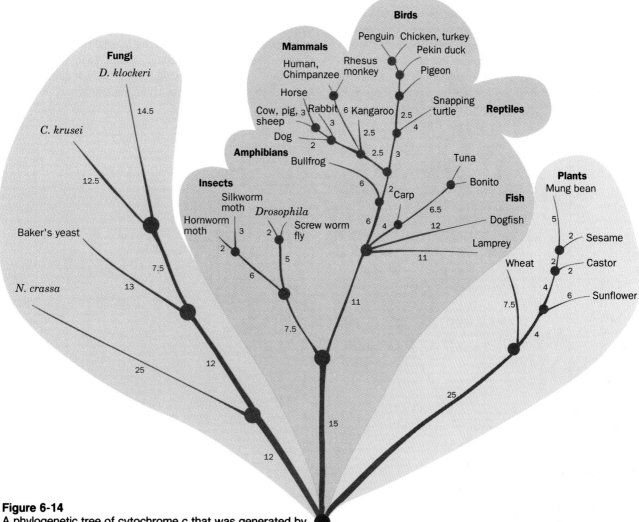

Figure 6-14
A phylogenetic tree of cytochrome *c* that was generated by the computer-aided analysis of difference data such as that in Table 6-5. Each branch point indicates the existence of an organism deduced to be ancestral to the species connected above it. The numbers beside each branch indicate the inferred differences, in PAM units, between the cytochromes *c* of its flanking branch points or species. [After Dayhoff, M. O., Park, C. M., and McLaughlin, P. J., *in* Dayhoff, M. O. (Ed.), *Atlas of Protein Sequence and Structure, p.* 8, National Biomedical Research Foundation (1972).]

residues in these positions inactivates this protein. However, the biochemical significance of most of the invariant and conservatively substituted residues of cytochrome *c* can only be profitably assessed in terms of the protein's three-dimensional structure and is therefore deferred until Section 8-3A. In what follows, we consider what insights can be gleaned solely from the comparisons of the amino acid sequences of related proteins. The conclusions we draw are surprisingly far-reaching.

The easiest way to compare the evolutionary differences between two homologous proteins is simply to count the amino acid differences between them (more

realistically, we should infer the minimum number of DNA base changes to convert one protein to the other but, because of the infrequency with which mutations are accepted, counting amino acid differences yields similar information). Table 6-5 is a tabulation of the amino acid sequence differences among 22 of the cytochromes *c* listed in Table 6-4. It has been boxed off to emphasize the relationships among groups of similar species. The order of these differences largely parallels that expected from classical taxonomy. Thus primate cytochromes *c* more nearly resemble those of other mammals than they do, for example, those of insects (8–12 differences for mammals vs 26–31 for insects). Similarly, the cytochromes *c* of fungi differ as much from those of mammals (45–51 differences) as they do from those of insects (41–47) or higher plants (47–54).

By computer analysis of data such as those in Table 6-5, *a kind of family tree, known as a **phylogenetic tree**, can be constructed, which indicates the ancestral relationships among the organisms that produced the proteins.* That for cytochrome *c* is sketched in Fig. 6-14. Similar trees have been derived for other proteins. Each branch point of a tree indicates the probable existence of a common ancestor for all the organisms above it. The relative evolutionary distances between neighboring branch points are expressed as the number of amino acid differences per 100 residues of the protein (Percentage of Accepted point Mutations, or **PAM units**). This fur-

nishes a quantitative measure of the degree of relatedness of the various species that macroscopic taxonomy cannot provide. Note that the evolutionary distances of modern cytochromes *c* from the lowest branch point on their tree are all approximately equal. Evidently, *the cytochromes c of the so-called lower forms of life have evolved to the same extent as those of the higher forms.*

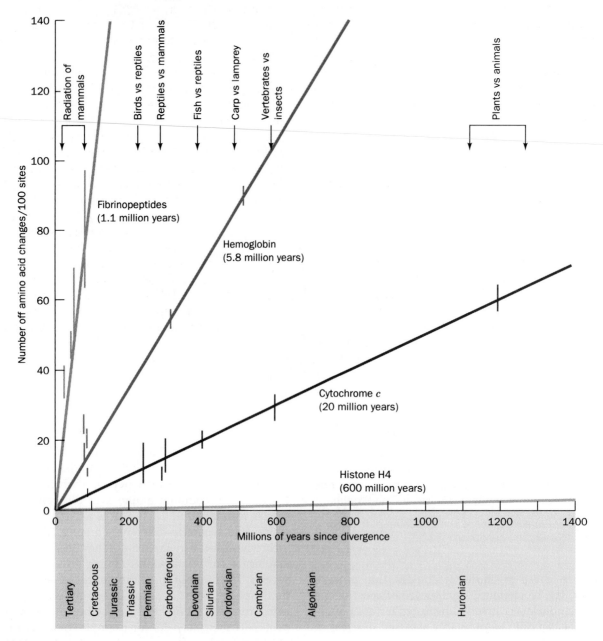

Figure 6-15
The rates of evolution of four unrelated proteins. The graph was constructed by plotting the average differences, in PAM units, of the amino acid sequences on two sides of a branch point of a phylogenetic tree (corrected to allow for more than one mutation at a given site) versus the time, according to the fossil record, since the corresponding species diverged from their common ancestor. The error bars indicate the experimental scatter of the sequence data. Each protein's rate of evolution is proportional to the slope of its curve. The unit evolutionary period, which is inversely proportional to the rate of evolution, is indicated beside the curve. [Figure copyrighted © by Irving Geis.]

Proteins Evolve at Characteristic Rates

The evolutionary distances between various species can be plotted against the time when, according to radiodated fossil records, the species diverged. For cytochrome *c*, this plot is essentially linear, thereby indicating that cytochrome *c* has accumulated mutations at a constant rate over the geological time scale (Fig. 6-15). This is also true for the other three proteins whose rates of evolution are plotted in Fig. 6-15. Each has its characteristic rate of change, known as a **unit evolutionary period,** which is defined as the time required for the amino acid sequence of a protein to change by 1% after two species have diverged. For cytochrome *c*, the unit evolutionary period is 20.0 million years. Compare this with the much less variant **histone H4** (600 million years) and the more variant hemoglobin (5.8 million years) and **fibrinopeptides** (1.1 million years).

The foregoing information does not imply that the rates of mutation of the DNAs specifying these proteins differ, but rather that *the rate that mutations are accepted into a protein depends on the extent that amino acid changes affect its function.* Cytochrome *c*, for example, is a rather small protein that, in carrying out its biological function, must interact with large protein complexes over much of its surface area. Any mutational change to cytochrome *c* will, most likely, affect these interactions unless, of course, the complexes simultaneously mutate to accommodate the change, a very unlikely occurrence. This accounts for the evolutionary stability of cytochrome *c*. Histone H4 is a protein that binds to DNA in eukaryotic chromosomes (Section 33-1A). Its central role in packaging the genetic archives evidently makes it extremely intolerant of any mutational changes. Indeed, histone H4 is so well adapted to its function that the histones H4 from peas and cows, species that diverged 1.2 billion years ago, differ by only two conservative changes in their 102 amino acids. Hemoglobin, like cytochrome *c*, is an intricate molecular machine (Section 9-2). It functions as a free floating molecule, however, so that its surface groups are usually more tolerant of change than are those of cytochrome *c* (although not in the case of HbS; Section 9-3B). This accounts for hemoglobin's greater rate of evolution. The fibrinopeptides are polypeptides of ~ 20 residues that are proteolytically cleaved from the vertebrate protein **fibrinogen** when it is converted to **fibrin** in the blood clotting process (Section 34-1A). Once they have been excised, the fibrinopeptides are discarded so that there is relatively little selective pressure on them to maintain their amino acid sequence and thus their rate of variation is high. If it is assumed that the fibrinopeptides are evolving at random then the foregoing unit evolutionary periods indicate that in hemoglobin only $\frac{1.1}{5.8} = \frac{1}{5}$ of the random amino acid changes are acceptable, that is, innocuous, whereas this quantity is $\frac{1}{18}$ for cytochrome *c* and $\frac{1}{550}$ for histone H4.

Mutational Rates Are Constant in Time

Amino acid substitutions in a protein mostly result from single base changes in the gene specifying the protein (Section 30-1). If such **point mutations** mainly occur as a consequence of errors in the DNA replication process, then the rate at which a given protein accumulates mutations would be constant with respect to numbers of generations. If, however, the mutational process results from the random chemical degradation of DNA, then the mutation rate would be constant with absolute time. To choose between these alternative hypotheses, let us compare the rate of cytochrome *c* divergence in insects with that in mammals.

Insects have shorter generation times than mammals so that if DNA replication were the major source of mutational error, then from the time the insect and mammalian lines diverged, insects would have evolved further from plants than have mammals. However, a simple phylogenetic tree (Fig. 6-16) indicates that the average number of amino acid differences between the cytochromes *c* of insects and plants (45.2) is essentially the same as that between mammals and plants (45.0). We must therefore conclude that cytochrome *c* accumulates mutations at a uniform rate with respect to time rather than number of generations. This, in turn, implies that *point mutations in DNA accumulate at a constant rate with time, that is, through random chemical change, rather than being due mainly to errors in the replication process.*

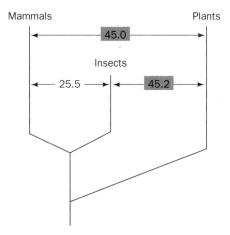

Figure 6-16
A phylogenetic tree showing the average number of amino acid differences between cytochromes *c* from mammals, insects, and plants. Mammals and insects have diverged equally far from plants since their common branch point. [Adapted from Dickerson, R. E. and Timkovitch, R., *in* Boyer, P. D. (Ed.), *The Enzymes* (3rd ed.), Vol. 11, *p.* 447, Academic Press (1975).]

Protein Evolution Is Not the Basis of Organismal Evolution

Despite the close agreement between phylogenetic trees derived from sequence similarities and classical taxonometric analyses, evidence is accumulating which suggests that protein sequence evolution is not the only or even the most important basis of organismal evolution. There is, for example, more than a 99% sequence homology between the corresponding proteins of man and his closest relative, the chimpanzee (e.g., their cytochromes *c* are identical). This is the level of homology observed among sibling species of fruit flies and mammals. Yet, the anatomical and behavioral differences between man and chimpanzee are so great that they have been classified in separate families. *This suggests that the rapid divergence of man and chimpanzee stems from relatively few mutational changes in the segments of DNA that control gene expression; that is, how much of each protein will be made, where, and when.* Such mutations do not change protein sequences but can result in major organismal alterations.

C. Evolution through Gene Duplication

Most proteins have extensive sequence similarities with other proteins from the same organism. Such proteins arose through **gene duplication,** a result, it is thought, of an aberrant genetic recombination event in which a single chromosome acquired both copies of the primordial gene in question (the mechanism of genetic recombination is discussed in Section 31-6A). *Gene duplication is a particularly efficient mode of evolution because one of the duplicated genes can evolve a new functionality through natural selection while its counterpart continues to direct the synthesis of the presumably essential ancestral protein.*

The **globin** family of proteins, which includes hemoglobin and **myoglobin,** provides an excellent example of evolution through gene duplication. Hemoglobin transports oxygen from the lungs (or gills or skin) to the tissues. Myoglobin, which occurs in muscles, facilitates rapid oxygen diffusion through these tissues and also functions as an oxygen storage protein. *The sequences of hemoglobin's α and β subunits (recall that hemoglobin is an $\alpha_2\beta_2$ tetramer) and myoglobin (a monomer) are quite similar.*

The globin family's phylogenetic tree indicates that its members, in humans, arose through the following chain of events (Fig. 6-17):

1. The primordial globin probably functioned simply as an oxygen-storage protein. Indeed, the globins in certain modern invertebrates still have this function. For example, treating a *Planorbis* snail with CO (the binding of which prevents globins from binding O_2; Section 9-1A) does not affect its behavior in well-aerated water but if the oxygen concentration is re-duced, a poisoned *Planorbis* becomes even more sluggish than a normal one.

2. Duplication of a primordial globin gene, ~ 1.1 billion years ago, permitted the resulting two genes to evolve separately so that, largely by a series of point mutations, a monomeric hemoglobin arose that had the lower oxygen-binding affinity required for it to transfer oxygen to the developing myoglobin. Such a monomeric hemoglobin can still be found in the blood of the **lamprey,** a primitive vertebrate that, according to the fossil record, has maintained its eel-like morphology for over 425 million years.

3. Hemoglobin's tetrameric character is a structural feature that greatly increases its ability to transport oxygen efficiently (Section 9-2C). This provided the adaptive advantage that gave rise to the evolution of the β chain from a duplicated α chain.

4. In fetal mammals, oxygen is obtained from the maternal circulation. **Fetal hemoglobin,** an $\alpha_2\gamma_2$ tetramer in which the γ **chain** is a gene-duplicated β chain variant, evolved to have an oxygen-binding affinity between that of normal adult hemoglobin and myoglobin.

5. Human embryos, in their first 8 weeks post-conception, make a $\zeta_2\varepsilon_2$ hemoglobin in which the ζ and ε **chains** are, respectively, gene-duplicated α and β variants.

6. In primates, the β chain has undergone a relatively recent duplication to form a δ **chain.** The $\alpha_2\delta_2$ hemoglobin, which occurs as a minor hemoglobin component in normal adults ($\sim 1\%$), has no known unique function. Perhaps it may eventually evolve one (although the human genome contains the relics of globin genes that are no longer expressed; Section 33-2F).

Our discussion of the globin family implies that protein evolution through gene duplication leads to homologous proteins of similar structural and functional properties. Another well-documented example of this phenomenon has resulted in the formation of a family of endopeptidases, which include trypsin, chymotrypsin, and elastase. These homologous digestive enzymes, which are secreted by the pancreas into the small intestine, are quite similar in their properties, differing mainly in their side chain specificities (Table 6-2). We examine how these functional variations are structurally rationalized in Section 14-3B. Individually, these three enzymes are limited in their abilities to degrade a protein, but in concert, they form a potent digestive system.

Occasionally, proteins with apparently unrelated functions are found to be evolutionarily related. **Lysozyme,** for example, is an enzyme that catalyzes the hy-

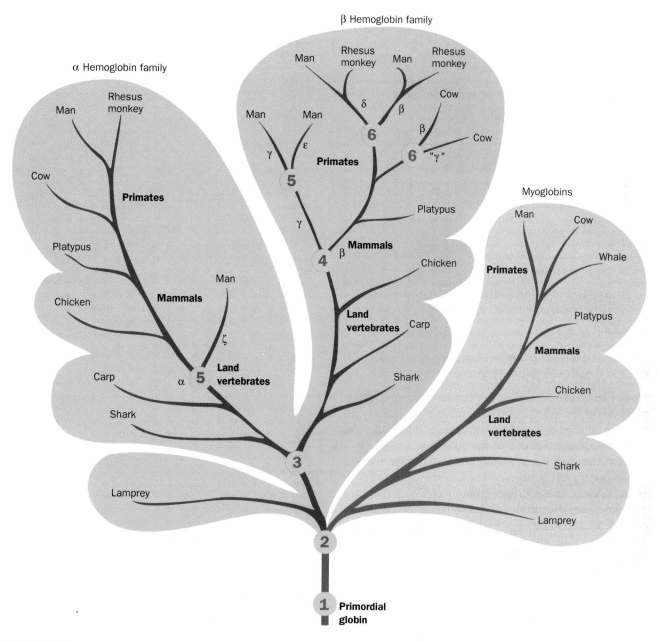

Figure 6-17
The phylogenetic tree of the globin family. The circled branch points represent gene duplications and unmarked branches are species divergences. [After Dickerson, R. E. and Geis, I., *Hemoglobin*, p. 82, Benjamin/Cummings (1983).]

drolysis of the polysaccharide components of bacterial cell walls (Section 14-2). The protein **α-lactalbumin,** which comprises 15% of the protein in milk, is a subunit of the enzyme **lactose synthase,** which forms the disaccharide **lactose** (milk sugar) from its component monosaccharides (Section 21-3A). Despite their widely different physiological functions, lysozyme and α-lactalbumin are 40% homologous; it is estimated that they

diverged, after gene duplication, some 300 million years ago, around the time that mammals diverged from reptiles. The only functional indication of the common origin of these two proteins is that both participate in reactions involving carbohydrates.

As we have stated previously and will explore in detail in Section 8-1, *the three-dimensional structure of a protein, and hence its function, is dictated by its amino acid sequence.* Most proteins that have been sequenced are more or less similar to several other known proteins. Therefore, it seems likely that most of the myriad of proteins in any given organism arose through gene duplications. This suggests that the appearance of a protein with a novel sequence and function is an extremely rare event in biology — one that may not have occurred since early in the history of life.

4. POLYPEPTIDE SYNTHESIS

In this section we describe the chemical synthesis of polypeptides from amino acids. The ability to manufacture polypeptides not available in nature has enormous biomedical potential:

1. To investigate the properties of polypeptides by systematically varying their side chains.

2. To obtain polypeptides with unique properties.

3. To manufacture pharmacologically active polypeptides that are biologically scarce or unavailable.

One of the most promising applications of polypeptide synthesis is the production of synthetic vaccines. Vaccines, which have consisted of viruses that had been "killed" (inactivated) or attenuated ("live" but mutated so as not to proliferate in humans), stimulate the immune system to synthesize antibodies specifically directed against these viruses thereby conferring immunity to them (the immune response is discussed in Section 34-2A). The use of such vaccines, however, is not without risk; attenuated viruses, for example, may mutate to a virulent form and "killed" virus vaccines have, on several occasions, caused disease because they contained "live" viruses. Moreover, it is difficult to culture many viruses and therefore to obtain sufficient material for vaccine production. Such problems would be eliminated by preparing vaccines from synthetic polypeptides that have the amino acid sequences of viral antigenic determinants (molecular groupings that stimulate the immune system to manufacture antibodies against them). Indeed, several such synthetic vaccines are already in general use.

The first polypeptides to be chemically synthesized were composed of only one type of amino acid and are therefore known as **homopolypeptides.** Such compounds as **polyglycine, polyserine,** and **polylysine** are easily synthesized according to classical methods of polymer chemistry. They have served as valuable model compounds in studying the physicochemical properties of polypeptides such as conformational behavior and interactions with the aqueous environment.

The first chemical synthesis of a biologically active polypeptide was that of the nonapeptide (9-amino acid

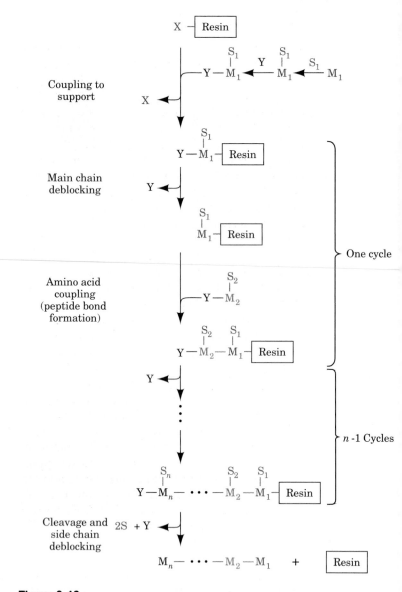

Figure 6-18
Flow diagram for polypeptide synthesis by the solid phase method. The symbol M_i represents the ith amino acid residue to be added to the polypeptide, S_i is its side chain protecting group, and Y represents the main chain protecting group. The specific reactions are discussed in the text. [After Erikson, B. W. and Merrifield, R. B., *in* Neurath, H. and Hill, R. L. (Eds.), *The Proteins* (3rd ed.), Vol. 2, p. 259, Academic Press (1979).]

residue) hormone **oxytocin** by Vincent du Vigneaud in 1953.

$$\text{Gly} - \text{Leu} - \text{Pro} - \text{Cys} - \text{Asn} - \text{Gln} - \text{Ile} - \text{Tyr} - \text{Cys}$$
(with an S—S bond connecting the two Cys residues)

Oxytocin

Improvements in polypeptide synthesis methodology since then have led to the synthesis of numerous biologically active polypeptides and several proteins.

A. Synthetic Procedures

Polypeptides are chemically synthesized by covalently linking (coupling) amino acids, one at a time, to the terminus of a growing polypeptide chain. Imagine that a polypeptide is being synthesized from its C-terminus towards its N-terminus, that is, the growing chain ends with a free amino group. Then each amino acid being added to the chain must already have its own α-amino group chemically protected (blocked) or it would react with other like molecules instead of with the N-terminal amino group of the chain. Once the new amino acid is coupled, its now N-terminal amino group must be deprotected (deblocked) so that the next peptide bond can be formed. *Every cycle of amino acid addition therefore requires a coupling step and a deblocking step.* Furthermore, reactive side chains must be blocked to prevent their participation in the coupling reactions, and then deblocked in the final step of the synthesis.

The reactions that were originally developed for synthesizing polypeptides such as oxytocin take place entirely in solution. The losses incurred upon isolation and purification of the reaction product in each of the many steps, however, contribute significantly to the low yields of final polypeptide. This difficulty was ingeniously circumvented in 1962 by Bruce Merrifield, through his development of **solid phase synthesis.** In this method, a growing polypeptide chain is covalently anchored, usually by its C-terminus, to an insoluble solid support such as beads of polystyrene resin (Fig. 5-8a), and the appropriately blocked amino acids and reagents are added in the proper sequence (Fig. 6-18). This permits the quantitative recovery and purification of intermediate products by simply filtering and washing the beads.

When polypeptide chains are synthesized by amino acid addition to their N-terminus (the opposite direction to that in protein biosynthesis; Section 30-3B), the α-amino group of each sequentially added amino acid must be chemically protected during the coupling reaction. The *tert*-**butyloxycarbonyl (Boc)** group is frequently used for this purpose.

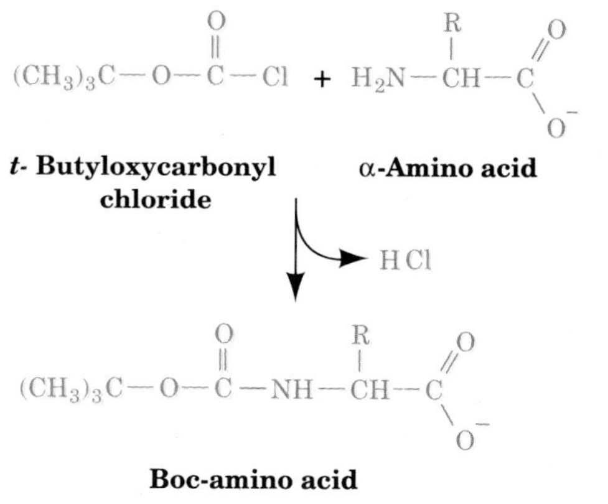

Peptide synthesis then proceeds in a stepwise fashion.

Anchoring the Chain to the Inert Support

The first step of a solid phase polypeptide synthesis is coupling the C-terminal amino acid to a solid support. The most commonly used support is a cross-linked polystyrene resin with pendant chloromethyl groups. Resin coupling occurs through the following reaction:

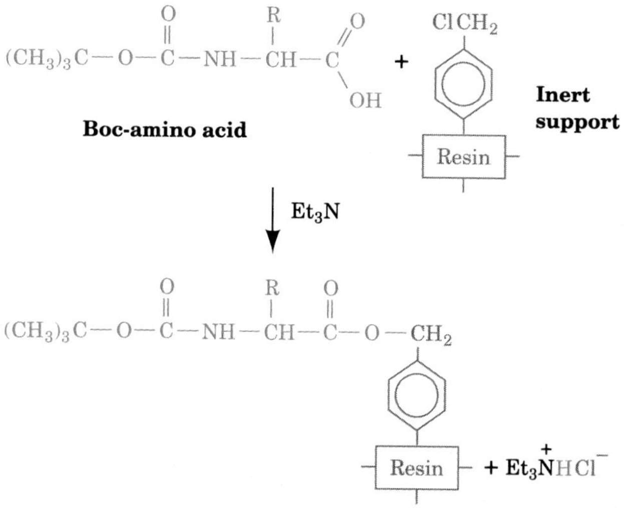

and the resulting α-amino acid-derivatized resin is filtered and washed. The amino group is then deblocked by treatment with an anhydrous acid such as 50% trifluoroacetic acid in dichloromethane, which leaves intact the alkylbenzyl ester bond to the support resin.

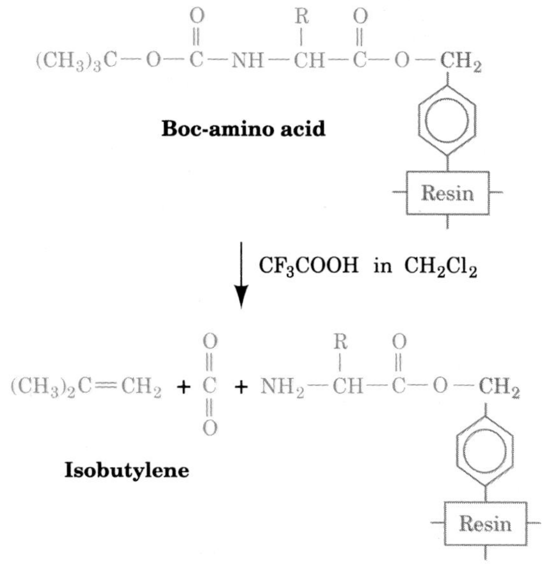

Coupling the Amino Acids

The reaction coupling two amino acids through a peptide bond is endergonic and therefore must be activated to obtain significant yields. The activating agent most often used is **dicyclohexylcarbodiimide (DCC):**

Boc-amino acid

Dicyclohexyl-carbodiimide (DCC)

O-Acylurea intermediate

Resin-bound amino acid

Dipeptidyl-resin

+

N,N′- Dicyclohexylurea

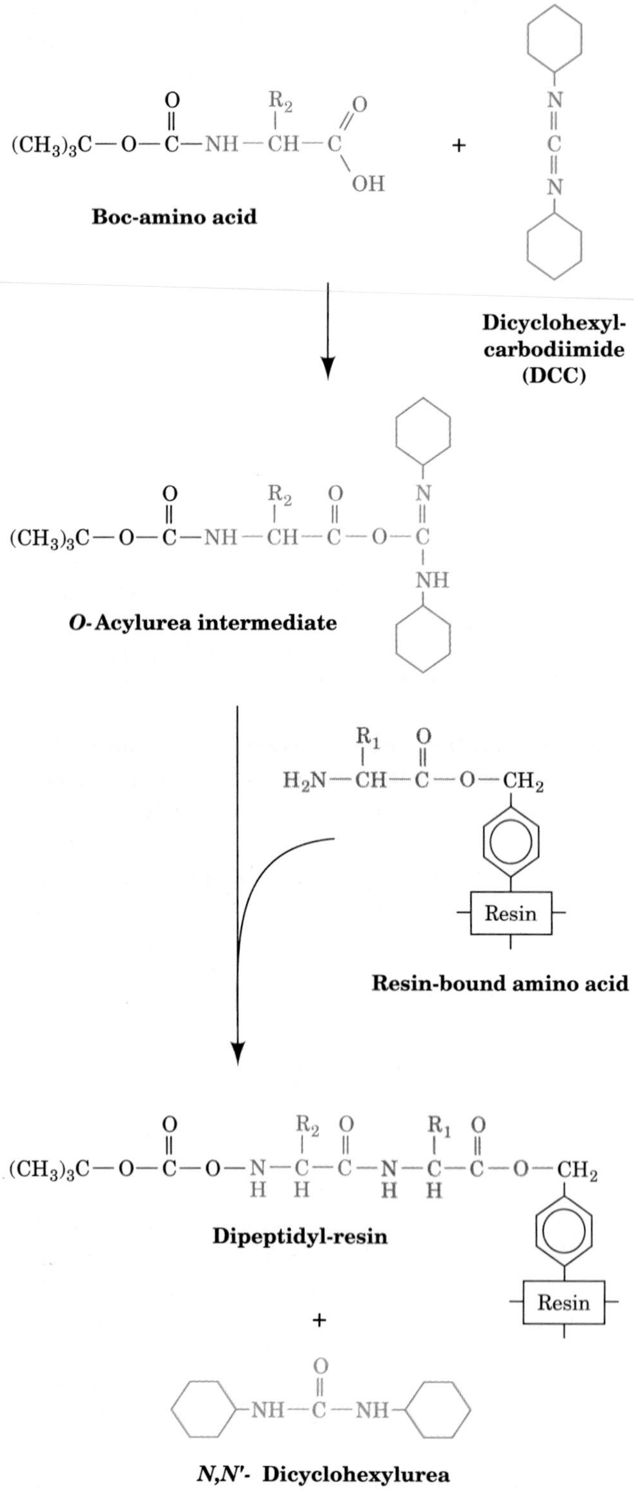

The *O*-acylurea intermediate that results from the reaction of DCC with the carboxyl group of a Boc-protected

α-amino acid, readily reacts with the resin-bound α-amino acid to form the desired peptide bond in high yield. By subsequently alternating the deblocking and coupling reactions, a polypeptide with the desired amino acid sequence can be synthesized. *The repetitive nature of these operations allows the solid phase synthesis method to be easily automated.*

During the course of peptide synthesis, many of the side chains also require protection to prevent their reaction with the coupling agent. Although there are many different blocking groups, the benzyl group is the most widely used (Fig. 6-19).

Releasing the Polypeptide from the Resin

The final step in solid phase polypeptide synthesis is the cleavage of the polypeptide from the solid support. The benzyl ester link from the polypeptide's C-terminus to the support resin may be cleaved by treatment with liquid HF:

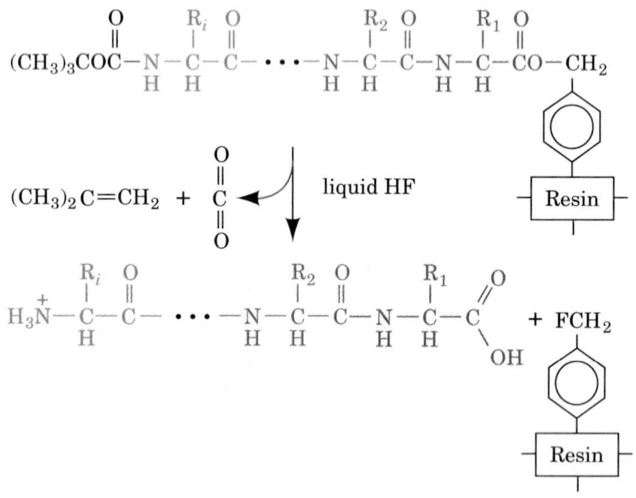

The Boc group linked to the polypeptide's N-terminus, as well as the benzyl groups protecting its side chains, are also removed by this treatment.

B. Problems and Prospects

The steps just outlined seem simple enough although they are not as straightforward as we implied. A major difficulty with the entire procedure is its low cumulative yield. Let us examine the reasons for this. To synthesize a polypeptide chain with n peptide bonds requires at least $2n$ reaction steps—one for coupling and one for deblocking each residue. If a protein-sized polypeptide is to be synthesized in reasonable yield, then each reaction step must be essentially quantitative; anything less greatly reduces the yield of final product. For example, in the synthesis of a 101-residue polypeptide chain, in which each reaction step occurs with an admirable 98% yield through 200 reaction steps, the overall yield is only $0.98^{200} \times 100 = 2\%$. Therefore, although oligopeptides

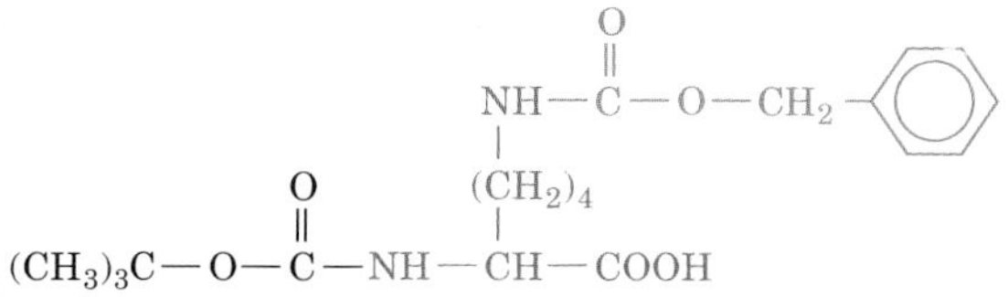

***N*-Boc, N^{ε}-benzyloxycarbonyl-Lys**

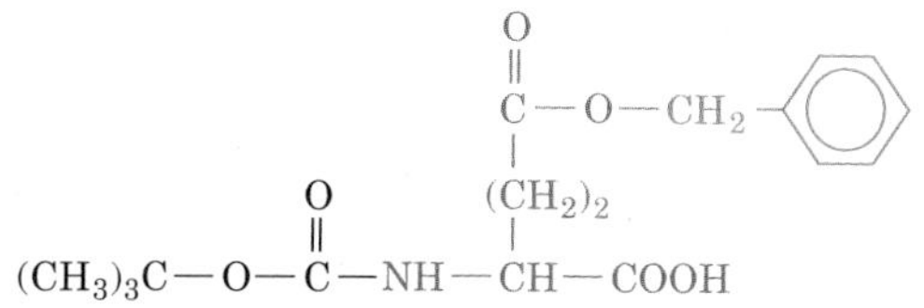

***N*-Boc-Glu, γ-Benzyl ester**

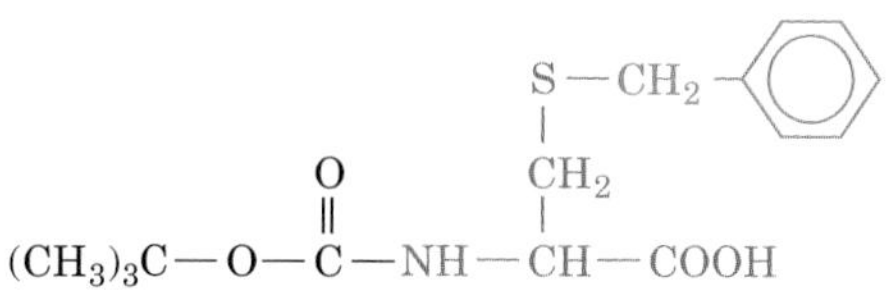

Boc, *S*-benzyl -Cys

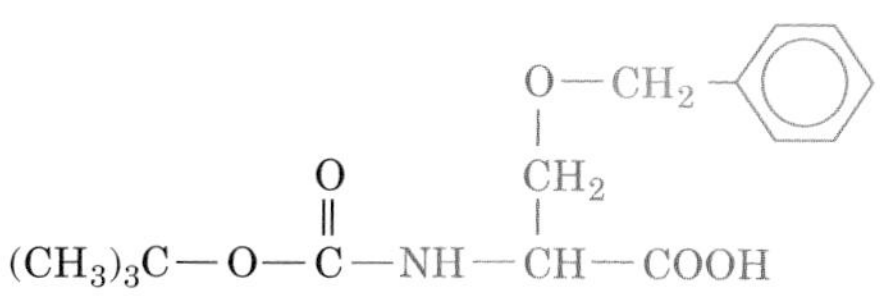

Boc, *O*-benzyl -Ser

Figure 6-19
A selection of amino acids with benzyl-protected side chains and a Boc-protected α-amino group. These substances can be used directly in the coupling reactions forming peptide bonds.

can now be easily made, the synthesis of large polypeptides requires almost fanatical attention to chemical detail.

A greater problem is that the newly liberated synthetic polypeptide must be purified. This is generally a difficult task because incomplete reactions, as well as numerous low yield side reactions, at every stage of the solid phase synthesis result in almost a continuum of closely related products for large polypeptides. The recent development of reverse-phase HPLC techniques (Section 5-3E), however, has greatly facilitated this process.

Using the automated solid phase technique, Merrifield and his coworkers synthesized the nonapeptide hormone **bradykinin** in 85% yield in 27 h.

Arg—Pro—Pro—Gly—Phe—Ser—Pro—Phe—Arg

Bradykinin

They also synthesized the 124-residue bovine pancreatic enzyme **ribonuclease A (RNase A),** which, after purification and renaturation (being folded to its native conformation; Section 8-1A), had 78% of the biologically produced enzyme's specific activity. These results are unusually good ones. Perhaps more typical is the solid phase synthesis of the 129-residue hen egg white enzyme lysozyme, which had only 4 to 6% of the biosynthetically produced enzyme's specific activity. Although steady progress is being made in improving yields and purity, synthetic methods are presently of practical importance only for the synthesis of polypeptides with less than ~50 residues. However, a recently developed hybrid strategy, in which several solid phase-synthesized peptide segments are joined in solution, has yielded encouraging results in the synthesis of 50 to 150-residue polypeptides.

Modified Proteins May Be Obtained by Protein Semisynthesis or Genetic Engineering Techniques

The difficulties of producing specifically modified proteins are partially circumvented by **protein semisynthesis,** in which segments of naturally occurring proteins are used to construct novel proteins. The native protein is fragmented by specific peptide cleavage (Section 6-1E) and the resulting polypeptides are separated and purified. Specific residues are modified through the use of group-specific reagents or a polypeptide segment might be entirely synthesized by the methods described above. The modified protein is then constructed by linking together its component polypeptides using solution techniques.

Genetic manipulations have, since the early 1980s, enabled the biosynthesis of proteins with one or more amino acid residues that differ in a specified manner from those of the natural protein. These "genetic engineering" methods are already in routine use for the production of large quantities of specifically altered proteins for medical, industrial, and agricultural purposes. We shall have much more to say about these methods in Section 28-8. However, polypeptides containing "nonstandard" amino acid residues or even nonamino acid linkages must be synthetically or semisynthetically produced since biologically generated polypeptides are made from only the 20 "standard" amino acids.

Chapter Summary

The initial step in the amino acid sequence determination of a protein is to ascertain its content of chemically different polypeptides by end group analysis. The protein's disulfide bonds are then chemically cleaved, the different polypeptides are separated and purified, and their amino acid compositions are determined. Next, the purified polypeptides are specifically cleaved, by enzymatic or chemical means, to smaller peptides that are separated, purified, and then sequenced by (automated) Edman degradation. Repetition of this process, using a cleavage method of different specificity, generates overlapping peptides whose amino acid sequences, when compared to those of the first group of peptide fragments, indicate their order in the parent polypeptide. The primary structure determination is completed by establishing the positions of the disulfide bonds. This requires the degradation of the protein with its disulfide bonds intact. Then, by sequencing the pairs of disulfide-linked peptide fragments, their positions in the intact protein can be deduced. Once a primary structure is known, its minor variants, which may arise from mutations or chemical modifications, can be easily analyzed by peptide mapping.

Inactivation of a protein by treatment with group-specific reagents may serve to identify its essential residues. An enzyme's active site residues may be unusually reactive.

Sickle-cell anemia is a molecular disease of individuals who are homozygous for a gene specifying an altered β chain of hemoglobin. Fingerprinting and sequencing studies have identified this alteration as arising from a point mutation that changes Glu $\beta6 \rightarrow$ Val. In the heterozygous state, the sickle-cell trait confers resistance to malaria without causing deleterious effects. This accounts for its high incidence in populations living in malarial regions. The cytochromes c from many eukaryotic species contain many amino acid residues that are invariant or conservatively substituted. Hence, this protein is well adapted to its function. The amino acid differences between the various cytochromes c have permitted the generation of their phylogenetic tree, which closely parallels that determined by classical taxonometry. The number of sequence differences between homologous proteins from related species plotted against the time when, according to the fossil record, these species diverged from a common ancestor, reveals that acceptable point mutations in proteins occur at a constant rate. Proteins whose functions are relatively intolerant to sequence changes evolve more slowly than those that are more tolerant to such changes. Phylogenetic analysis of the globin family, myoglobin and the α- and β-chains of hemoglobin, reveals that these proteins arose through gene duplication. In this process, the original function of the protein is maintained while the duplicated copy evolves a new function. Many, if not most, proteins have evolved through gene duplication.

The strategy of polypeptide chemical synthesis involves coupling amino acids, one at a time, to the N-terminus of a growing polypeptide chain. The α-amino group of each amino acid must be chemically protected during the coupling reaction and then unblocked before the next coupling step. Reactive side chains must also be chemically protected but then unblocked at the conclusion of the synthesis. The difficulty in recovering the intermediate product of each of the many steps of such a synthesis has been eliminated by the development of solid phase synthesis techniques. These methods have led to the synthesis of numerous biologically active polypeptides. Semisynthetic techniques also have been useful in producing proteins of novel sequence.

References

General

Creighton, T. E., *Proteins,* Chapters 1 and 3, Freeman (1983).

Wood, W. B., Wilson, J. H., Benbow, R. M., and Hood, L. E., *Biochemistry, A Problems Approach* (2nd ed.), Chapter 3, Benjamin/Cummings (1981). [A question-and-answer approach to biochemical instruction.]

Primary Structure Determination

Allen, G., Sequencing of proteins and peptides, *in* Burdon, R. M. and van Knippenberg, P. H. (Eds.), *Laboratory Techniques in Biochemistry and Molecular Biology,* Vol. 9 (2nd revised ed.), Elsevier (1989).

Bhown, A. S. (Ed.), *Protein/Peptide Sequence Analysis: Current Methodologies,* CRC Press (1988).

Croft, L. R., *Handbook of Protein Sequence Analysis* (2nd ed.), Wiley (1980). [A review of methods together with a compilation of all protein sequences published through 1978.]

Hirs, C. H. W. and Timasheff, S. N. (Eds.), Enzyme structure, Parts B, E, and I, *Methods Enzymol.* **25, 47,** and **91** (1972, 1977, and 1983). [Collections of articles and recipes on sequence analysis and other aspects of protein chemistry.]

Hunkapiller, M. W., Strickler, J. E., and Wilson, K. J., Contemporary methodology for protein structure determination, *Science* **226,** 304–311 (1984).

James, G. T., Peptide mapping of proteins, *Methods Biochem. Anal.* **26,** 165–200 (1980).

Konigsberg, W. H. and Steinman, H. M., Strategy and methods of sequence analysis, *in* Neurath, H. and Hill, R. L. (Eds.), *The Proteins* (3rd ed.), Vol. 3, *pp.* 1–178, Academic Press (1977).

Sanger, F., Sequences, sequences, and sequences, *Annu. Rev. Biochem.* **57,** 1–28 (1988). [A scientific autobiography that provides a glimpse of the early difficulties in sequencing proteins.]

Walsh, K. A., Ericsson, L. H., Parmalee, D. C., and Titani, K., Advances in protein sequencing, *Annu. Rev. Biochem.* **50,** 261–284 (1981).

Protein Modification

Glazer, A. N., The chemical modification of proteins by group-specific and site-specific reagents, *in* Neurath, H. and Hill, R. L. (Eds.), *The Proteins* (3rd ed.), Vol. 2, *pp.* 1–103, Academic Press (1976).

Glazer, A. N., DeLange, A. J., and Sigman, D. S., Chemical modifications of proteins, *in* Work, T. S. and Work, E. (Eds.), *Laboratory Techniques in Biochemistry and Molecular Biology,* Vol. 4, Part 1, North–Holland (1975).

Chemical Evolution

Dayhoff, M. O., (Ed.), *Atlas of Protein Sequence and Structure,* Vol. 5 and Supplements No. 1–3, Biomedical Research Foundation (1972, 1973, 1976 and 1978). [A compendium of published amino acid sequences together with their analyses.]

Dickerson, R. E., The structure and history of an ancient protein, *Sci. Am.* **226**(4): 58–72 (1972).

Dickerson, R. E. and Geis, I. *Hemoglobin,* Chapter 3, Benjamin/Cummings (1983). [A detailed discussion of globin evolution.]

Dickerson, R. E. and Timkovich, R., Cytochromes *c, in* Boyer, P. D. (Ed.), *The Enzymes* (3rd ed.), Vol. 11, *pp.* 397–547, Academic Press (1975). [Contains an authoritative analysis of cytochrome *c* sequence studies.]

Doolittle, R. F., *Of Urfs and Orfs. A Primer of How to Analyze Derived Amino Acid Sequences,* University Science Books (1986).

Doolittle, R. F., The genealogy of some recently evolved vertebrate proteins, *Trends Biochem. Sci.* **10**, 233–237 (1985).

Doolitle, R. F., Similar amino acid sequences: Chance or common ancestry? *Science* **214**, 149–159 (1981). [A discussion of the criteria for establishing the relationship between two amino acid sequences.]

Doolittle, R. F., Protein evolution, *in* Neurath, H. and Hill, R. L. (Eds.), *The Proteins* (3rd ed.), Vol. 4, *pp.* 1–118, Academic Press (1979).

Friedman, M. J. and Trager, W., The biochemistry of resistance to malaria, *Sci. Am.* **244**(3): 154–164 (1981).

Kimura, M., The neutral theory of molecular evolution, *Sci. Am.* **241**(5): 98–126 (1979).

King, M.-C. and Wilson, A. C., Evolution at two levels in humans and chimpanzees, *Science* **188**, 107–116 (1975).

Schultz, G. E. and Schirmer, R. H., *Principles of Protein Structure,* Chapter 9, Springer–Verlag (1979). [Contains an analysis of protein evolution.]

Wilson, A. C., The molecular basis of evolution, *Sci. Am.* **253**(4): 164–173 (1985).

Polypeptide Synthesis

Chaiken, I. M., Semisynthetic peptides and proteins, *CRC Crit. Rev. Biochem.* **11**, 255–301 (1981).

Erickson, B. W. and Merrifield, R. B., Solid phase peptide synthesis, *in* Neurath, H. and Hill, R. L. (Eds.), *The Proteins* (3rd ed.), Vol. 2, *pp.* 255–527, Academic Press (1977).

Finn, F. M. and Hofmann, K., The synthesis of peptides by solution methods with emphasis on peptide hormones, *in* Neurath, H. and Hill, R. L. (Eds.), *The Proteins* (3rd ed.), Vol. 2, *pp.* 105–253, Academic Press (1977).

Kaiser, E. T., Mihara, H., Laforet, G. A., Kelly, J. W., Walters, L., Findeis, M. A., and Sasaki, T., Peptide and protein synthesis by segment synthesis-condensation, *Science,* **243**, 187–192 (1989).

Kent, S. B. H., Chemical synthesis of peptides and proteins, *Annu. Rev. Biochem.* **57**, 957–989 (1988).

Lerner, R. A., Synthetic vaccines, *Sci. Am.* **248**(2): 48–56 (1983).

Merrifield, B., Solid phase synthesis, *Science* **232**, 342–347 (1986).

Offord, R. E., *Semisynthetic Proteins,* Wiley (1980).

Udenfriend, S. and Meienhofer, J. (Eds.), *The Peptides. Analysis, Synthesis, Biology,* Academic Press (1979–). [An ongoing series containing articles on state-of-the-art synthetic methods of peptide synthesis.]

Problems

Note: Amino acid compositions of polypeptides with unknown sequences are written in parentheses with commas separating amino acid abbreviations such as in (Gly, Tyr, Val). Known amino acid sequences are written with residue names in order and separated by dashes; for example, Tyr-Val-Gly.

1. State the cleavage pattern of the following polypeptides by the indicated agents.

   (a) Ser-Ala-Phe-Lys-Pro   by chymotrypsin

   (b) Thr-Cys-Gly-Met-Asn   by NCBr

   (c) Leu-Arg-Gly-Asp   by carboxypeptidase A

   (d) Gly-Phe-Trp-Pro-Phe-Arg   by thermolysin

   (e) Val-Trp-Lys-Pro-Arg-Glu   by trypsin

2. A protein is subjected to end group analysis by dansyl chloride. The liberated dansyl amino acids are found to be present with a molar ratio of two parts Ser to one part Ala. What conclusions can be drawn about the nature of the protein?

3. A protein is subjected to degradation by carboxypeptidase B. Within a short time, Arg and Lys are liberated, following which no further change is observed. What does this information indicate concerning the primary structure of the protein?

4. Consider the following polypeptide:

   Asp-Trp-Val-Arg-Asn-Ser-Phe-Cys-Gln-Gly-Pro-Tyr-Met

   (a) What amino acids would be liberated upon its complete acid hydrolysis? (b) What amino acids would be liberated upon its complete alkaline hydrolysis?

5. Before the advent of the Edman degradation, the primary structures of proteins were elucidated through the use of partial acid hydrolysis. The resulting oligopeptides were separated and their amino acid compositions were determined. Consider a polypeptide with amino acid composition (Ala$_2$, Asp, Cys, Leu, Lys, Phe, Pro, Ser$_2$, Trp$_2$). Treatment with carboxypeptidase A released only Leu.

Oligopeptides with the following compositions were obtained by partial acid hydrolysis:

(Ala, Lys)   (Ala, Ser₂)   (Cys, Leu)

(Ala, Lys, Trp) (Ala, Trp)   (Cys, Leu, Pro)

(Ala, Pro)   (Asp, Lys, Phe) (Phe, Ser, Trp)

(Ala, Pro, Ser) (Asp, Phe)   (Ser, Trp)

(Ser₂, Trp)

Determine the amino acid sequence of the polypeptide.

*6. A polypeptide is subjected to the following degradative techniques resulting in polypeptide fragments with the indicated amino acid sequences. What is the amino acid sequence of the entire polypeptide?

  I. Cyanogen bromide treatment:

    (1) Asp-Ile-Lys-Gln-Met

    (2) Lys

    (3) Lys-Phe-Ala-Met

    (4) Tyr-Arg-Gly-Met

  II. Trypsin hydrolysis:

    (5) Gln-Met-Lys

    (6) Gly-Met-Asp-Ile-Lys

    (7) Phe-Ala-Met-Lys

    (8) Tyr-Arg

7. Treatment of a polypeptide by dithiothreitol yields two polypeptides that have the following amino acid sequences:

    (1) Ala-Phe-Cys-Met-Tyr-Cys-Leu-Trp-Cys-Asn

    (2) Val-Cys-Trp-Val-Ile-Phe-Gly-Cys-Lys

Chymotrypsin catalyzed hydrolysis of the intact polypeptide yields polypeptide fragments with the following amino acid compositions:

    (3) (Ala, Phe)

    (4) (Asn, Cys₂, Met, Tyr)

    (5) (Cys, Gly, Lys)

    (6) (Cys₂, Leu, Trp₂, Val)

    (7) (Ile, Phe, Val)

Indicate the positions of the disulfide bonds in the original polypeptide.

*8. The following treatments of a polypeptide yielded the indicated results. What is the primary structure of the polypeptide?

  I. Acid hydrolysis:

    (1) (Arg, Asx, Cys₂, Gly, Ile, Leu, Lys, Met, Phe, Pro, Ser)

  II. Edman degradation (one cycle):

    (2) (Leu)

    (3) (Ser)

  III. Carboxypeptidase A (sufficient time to remove one residue per chain):

    (4) (Asp)

  IV. Dithioerythritol + iodoacetic acid followed by trypsin hydrolysis:

    (5) (Arg, Ser)

    (6) (Asp, Met)

    (7) (Cys, Gly, Ile, Leu, Phe, Pro)

    (8) (Cys, Lys)

  V. Dithioerythritol + 2-bromoethylamine followed by trypsin hydrolysis:

    (9) (Arg, Ser)

    (10) (Asp, Met)

    (11) (Cys)

    (12) (Cys, Gly, Leu)

    (13) (Ile, Phe, Pro)

    (14) (Lys)

  VI. Chymotrypsin:

    (15) No fragments.

  VII. Pepsin:

    (16) (Arg, Asp, Cys₂, Gly, Leu, Lys, Met, Ser)

    (17) (Ile, Phe, Pro)

9. A polypeptide was subjected to the following treatments with the indicated results. What is its primary structure?

  I. Acid hydrolysis:

    (1) (Ala, Arg, Cys, Glx, Gly, Lys, Leu, Met, Phe, Thr)

  II. Aminopeptidase M:

    (2) No fragments.

  III. Carboxypeptidase A + carboxypeptidase B:

    (3) No fragments.

  IV. Trypsin followed by Edman degradation of the separated products:

    (4) Cys-Gly-Leu-Phe-Arg

    (5) Thr-Ala-Met-Gln-Lys

*10. While on an expedition to the Amazon jungle, you isolate a polypeptide you suspect of being the growth hormone of a newly discovered species of giant spider. Unfortunately, your portable sequenator was so roughly treated by the airport baggage handlers that it refuses to provide the sequence of more than four consecutive amino acid residues. Nevertheless, you persevere and obtain the following data:

  I. Hydrazinolysis:

    (1) (Val)

  II. Dansyl chloride treatment followed by acid hydrolysis:

    (2) (Dansyl-Pro)

  III. Trypsin followed by Edman degradation of the separated fragments:

    (3) Gly-Lys

(4) Phe-Ile-Val

(5) Pro-Gly-Ala-Arg

(6) Ser-Arg

(a) Provide as much information as you can concerning the amino acid sequence of the polypeptide. (b) Considering the poor condition of your sequenator, what additional analytical technique would most conveniently permit you to complete the sequence determination of the polypeptide?

11. A desiccated dodo bird in a reasonable state of preservation has been found in a cave on Mauritius Island. You have been given one of its legs in order to perform biochemical analyses and have managed to sequence its cytochrome *c*. The amino acid difference matrix for a number of birds including the dodo is shown here.

Chicken, turkey	0
Penguin	2 0
Pigeon	4 4 0
Pekin duck	3 3 3 0
Dodo	4 4 2 3 0

(a) To which of the other birds does the dodo appear most closely related? (b) Sketch the phylogenetic tree for birds based on cytochromes *c* (Fig. 6-14) and include the dodo.

12. Using Table 6-5, compare the relatedness of fungi to higher plants and to animals. Fungi are often said to be nongreen plants. In light of your analysis, is this a reasonable classification?

13. The inherited hemoglobin disease *β*-thalassemia is common to people from around the Mediterranean Sea and areas of Asia where malaria is prevalent (Fig. 6-13). The disease is characterized by a reduction in the rate of synthesis of the *β* chain of hemoglobin. Heterozygotes for the *β*-thalassemia gene, who are said to have **thalassemia minor,** are only mildly affected with adverse symptoms. Homozygotes for this gene, however, suffer from **Cooley's anemia** or **thalassemia major;** they are so severely afflicted that they do not survive their childhood. About 1% of the children born in the malarial regions around the Mediterranean Sea have Cooley's anemia. Why do you suppose the *β*-thalassemia gene is so prevalent in this area? Justify your answer.

14. Leguminous plants synthesize a monomeric oxygen-binding globin known as **leghemoglobin.** From your knowledge of biology, sketch the evolutionary tree of the globins (Fig. 6-17) with leghemoglobin included in its most likely position.

15. In a tragic accident, the desiccated dodo discussed in Problem 11 has been eaten by a deranged cat. The sequence of the dodo cytochrome *c* that you had determined before the accident led you to suspect that this cytochrome *c* has some unique biochemical properties. To test this hypothesis, you are forced to synthesize dodo cytochrome *c* by chemical means. As with other known avian cytochromes *c*, that of the dodo consists of 104 amino acid residues. In planning a solid phase synthesis, you expect a 98% yield for each coupling step and a 97% yield for each deblocking step. The cleavage of the completed polypeptide from the resin and the side chain deblocking step should yield an 80% recovery of product. What percentage of the original resin-bound C-terminal amino acid can you expect to form unmodified dodo cytochrome *c*?

Chapter 7

THREE-DIMENSIONAL STRUCTURES OF PROTEINS

The properties of a protein are largely determined by its three-dimensional structure. One might naively suppose that since proteins are composed of the same 20 types of amino acid residues, they would be more or less alike in their properties. Indeed, **denatured** (unfolded) proteins have rather similar characteristics, a kind of homogeneous "average" of their randomly dangling side chains. However, the three-dimensional structure of a **native** (physiologically folded) protein is specified by its primary structure so that it has a unique set of characteristics.

In this chapter, we shall discuss the structural features of proteins, the forces that hold them together, and their hierarchical organization to form complex structures. This will form the basis for understanding the structure–function relationships necessary to comprehend the biochemical roles of proteins. Detailed consideration of the dynamic behavior of proteins and how they fold to their native structures is deferred until Chapter 8.

1. SECONDARY STRUCTURE

A polymer's **secondary structure (2° structure)** is defined as the local conformation of its backbone. For proteins, this has come to mean the specification of regular polypeptide backbone folding patterns: helices, pleated sheets, and turns. However, before we begin our discussion of these basic structural motifs let us consider the geometrical properties of the peptide group because its understanding is prerequisite to that of any structure containing it.

A. The Peptide Group

In the 1930s and 1940s, Linus Pauling and Robert Corey determined the X-ray structures of several amino acids and dipeptides in an effort to elucidate the structural constraints on the conformations of a polypeptide chain. These studies indicated that *the peptide group has a rigid, planar structure (Fig. 7-1) which, Pauling pointed out, is a consequence of resonance interactions that give the peptide bond an ∼ 40% double-bond character:*

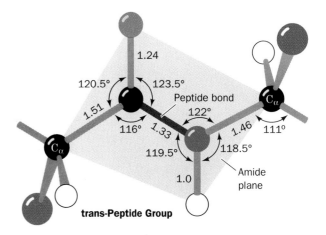

Figure 7-1
The standard dimensions (in angstroms, Å, and degrees, °) of the planar trans-peptide group derived by averaging the results of X-ray crystal structure determinations of amino acids and peptides. [After Marsh, R. E. and Donohue, J., *Adv. Protein Chem.* **22**, 249 (1967).]

This explanation is supported by the observations that a peptide's C—N bond is 0.14 Å shorter than its N—C_α single bond and that its C=O bond is 0.02 Å longer than that of aldehydes and ketones. The peptide bond's resonance energy has its maximum value, ∼ 85 kJ·mol^{-1}, when the peptide group is planar because its π-bonding overlap is maximized in this conformation. This overlap, and thus the resonance energy, falls to zero as the peptide bond is twisted to 90° out of planarity thereby accounting for the planar peptide group's rigidity.

Peptide groups, with few exceptions, assume the trans conformation: that in which successive C_α atoms are on opposite sides of the peptide bond joining them (Fig. 7-1). This is partly a result of steric interference, which causes the cis conformation (Fig. 7-2) to be ∼ 8 kJ·mol^{-1} less stable than the trans conformation (this energy difference is somewhat less in peptide bonds followed by a Pro residue; indeed, Pro residues precede most of the few cis peptides that occur in proteins).

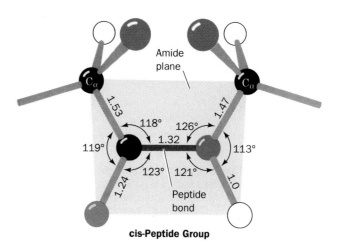

Figure 7-2
The cis-peptide group.

Figure 7-3
A polypeptide chain in its fully extended conformation showing the planarity of each of its peptide groups. [Figure copyrighted © by Irving Geis.]

Polypeptide Backbone Conformations May Be Described by Their Torsion Angles

The above considerations are important because they indicate that *the backbone of a protein is a linked sequence of rigid planar peptide groups (Fig. 7-3).* We can therefore specify a polypeptide's backbone conformation by the **torsion angles** (rotation angles) about the C_α—N bond

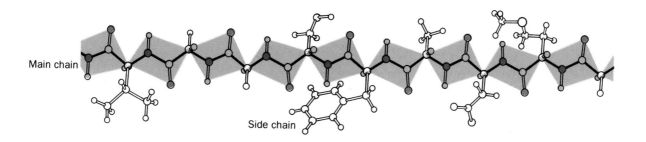

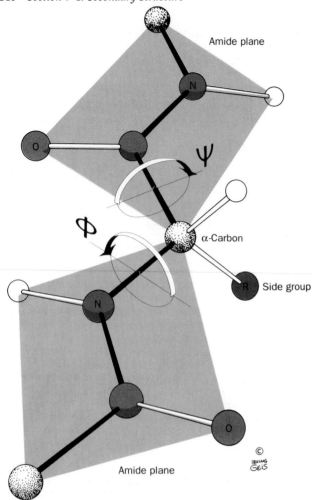

Figure 7-4
A portion of a polypeptide chain indicating the torsional degrees of freedom of each peptide unit. The only reasonably free movements are rotations about the C_α — N bond (ϕ) and the C_α — C bond (ψ). The torsion angles are both 180° in the conformation shown and increase, as is indicated, in a clockwise manner when viewed from C_α. [Figure copyrighted © by Irving Geis.]

(ϕ) and the C_α—C bond (ψ) of each of its amino acid residues. These angles, ϕ and ψ, are both defined as 180° when the polypeptide chain is in its planar, fully extended (all-trans) conformation and increase for a clockwise rotation when viewed from C_α (Fig. 7-4). (In earlier literature, this fully extended conformation was defined as having $\phi = \psi = 0°$. These values can be made to correspond to the present convention by subtracting 180° from both angles. Old habits are often difficult to put aside so that this older convention still appears in contemporary publications.)

There are several steric constraints on the torsion angles, ϕ and ψ, of a polypeptide backbone that limit its conformational range. The electronic structure of a single (σ) bond, such as a C—C bond, is cylindrically symmetrical about its bond axis so that we might expect such a bond to exhibit free rotation. If this were the case, then in ethane, for example, all torsion angles about the

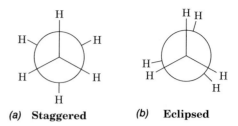

(a) **Staggered**   (b) **Eclipsed**

Figure 7-5
Newman projections indicating the (a) staggered conformation and (b) eclipsed conformation of ethane.

C—C bond would be equally likely. Yet, certain conformations in ethane are favored because of repulsions between electrons in the C—H bonds. The **staggered conformation** (Fig. 7-5a), which occurs at a 180° torsion angle, is ethane's most stable arrangement because opposing hydrogen atoms are maximally separated in this position. In contrast, the **eclipsed conformation** (Fig. 7-5b), which is characterized by a torsion angle of 0°, is least stable because in this arrangement the hydrogen atoms are closest to each other. The energy difference between the staggered and eclipsed conformations in ethane is ~12 kJ·mol⁻¹, a quantity that represents an **energy barrier** to free rotation about the C—C single bond. Substituents other than hydrogen exhibit greater steric interference, that is, they increase the size of this energy barrier due to their greater bulk. Indeed, with large substituents, some conformations may be sterically forbidden.

Allowed Conformations of Polypeptides Are Indicated by the Ramachandran Diagram

The sterically allowed values of ϕ and ψ can be determined by calculating the distances between the atoms of

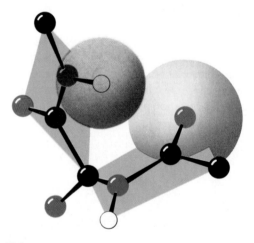

Figure 7-6
Steric interference between the carbonyl oxygen and the amide hydrogen on adjacent residues prevents the occurrence of the conformation $\phi = -60°$, $\psi = 30°$. [Figure copyrighted © by Irving Geis.]

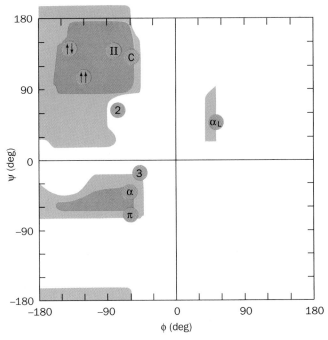

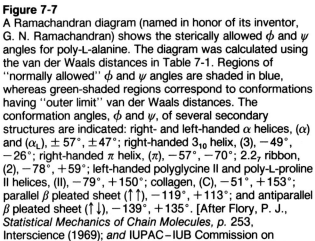

Figure 7-7
A Ramachandran diagram (named in honor of its inventor, G. N. Ramachandran) shows the sterically allowed ϕ and ψ angles for poly-L-alanine. The diagram was calculated using the van der Waals distances in Table 7-1. Regions of "normally allowed" ϕ and ψ angles are shaded in blue, whereas green-shaded regions correspond to conformations having "outer limit" van der Waals distances. The conformation angles, ϕ and ψ, of several secondary structures are indicated: right- and left-handed α helices, (α) and (α_L), $\pm 57°$, $\pm 47°$; right-handed 3_{10} helix, (3), $-49°$, $-26°$; right-handed π helix, (π), $-57°$, $-70°$; 2.2_7 ribbon, (2), $-78°$, $+59°$; left-handed polyglycine II and poly-L-proline II helices, (II), $-79°$, $+150°$; collagen, (C), $-51°$, $+153°$; parallel β pleated sheet ($\uparrow\uparrow$), $-119°$, $+113°$; and antiparallel β pleated sheet ($\uparrow\downarrow$), $-139°$, $+135°$. [After Flory, P. J., *Statistical Mechanics of Chain Molecules, p. 253,* Interscience (1969); *and* IUPAC–IUB Commission on Biochemical Nomenclature, *Biochemistry* **9**, 3475 (1970).]

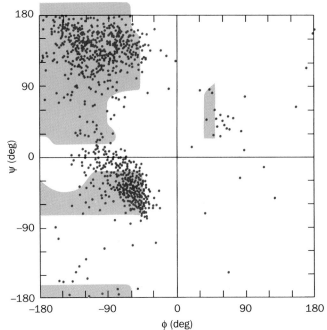

Figure 7-8
The conformation angle distribution of $\sim$1000 non-Gly residues in eight proteins as determined by X-ray crystal structure analyses with a superimposed Ramachandran diagram. [After Richardson, J. S., *Adv. Protein Chem.* **34**, 174 (1981).]

Table 7-1

Limiting Distances for Interatomic Contacts

Contact Type	Normally Allowed (Å)	Outer Limit (Å)
H $\cdots$ H	2.0	1.9
H $\cdots$ O	2.4	2.2
H $\cdots$ N	2.4	2.2
H $\cdots$ C	2.4	2.2
O $\cdots$ O	2.7	2.6
O $\cdots$ N	2.7	2.6
O $\cdots$ C	2.8	2.7
N $\cdots$ N	2.7	2.6
N $\cdots$ C	2.9	2.8
C $\cdots$ C	3.0	2.9
C $\cdots$ CH$_2$	3.2	3.0
CH$_2$ $\cdots$ CH$_2$	3.2	3.0

Source: Ramachandran, G. N. and Sasisekharan, V., *Adv. Protein Chem.* **23**, 326 (1968).

a tripeptide at all values of ϕ and ψ for the central peptide unit. Sterically forbidden conformations, such as the one shown in Fig. 7-6, are those in which any nonbonding interatomic distance is less than its corresponding van der Waals distance. Such information is summarized in a **conformation map** or **Ramachandran diagram** (Fig. 7-7).

Figure 7-7 indicates that most areas of the Ramachandran diagram (most combinations of ϕ and ψ) are conformationally inaccessible to a polypeptide chain. The particular regions of the Ramachandran diagram that represent allowed conformations depend on the van der Waals radii chosen to calculate it. But with any realistic set of values, such as that in Table 7-1, *only three small regions of the conformational map are physically accessible*

to a polypeptide chain. Nevertheless, as we shall see, all of the common types of regular secondary structures found in proteins fall within allowed regions of the Ramachandran diagram. Indeed, the observed conformational angles of most non-Gly residues in proteins whose X-ray structures have been determined lie in these allowed regions (Fig. 7-8).

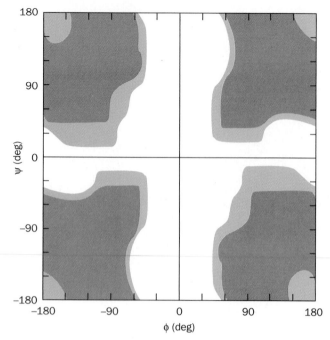

Figure 7-9
The Ramachandran diagram of Gly residues in a polypeptide chain. "Normally allowed" regions are shaded in blue, whereas green-shaded regions correspond to "outer limit" atomic distances. Gly residues have far greater conformational freedom than do other (bulkier) amino acid residues as the comparison of this figure with Fig. 7-7 indicates. [After Ramachandran, G. N. and Sasisekharan, V., *Adv. Protein Chem.* **23**, 332 (1968).]

Most points that fall in forbidden regions of Fig. 7-8 lie between its two fully allowed areas. However, these "forbidden" conformations are allowed if twists of only a few degrees about the peptide bond are permitted. This is not unreasonable since the peptide bond offers little resistance to small deformations from planarity.

Gly, the only residue without a C_β atom, is much less sterically hindered than the other amino acid residues. This is clearly apparent in comparing the Ramachandran diagram for Gly in a polypeptide chain (Fig. 7-9) with that of other residues (Fig. 7-7). In fact, Gly often occupies positions where a polypeptide backbone makes a sharp turn where any other residue would be subject to steric interference.

Figure 7-7 was calculated for three consecutive Ala residues. Similar plots for larger residues that are unbranched at C_β, such as Phe, are nearly identical. In Ramachandran diagrams of residues that are branched at C_β, such as Thr, the allowed regions are somewhat smaller than for Ala. The cyclic side chain of Pro limits its ϕ to the range $-60° \pm 25°$, making it, not surprisingly, the most conformationally restricted amino acid residue. The conformations of residues in chains longer than tripeptides are even more restricted than the Ramachandran diagram indicates because, for instance, a polypeptide chain cannot assume a conformation in which it passes through itself. We shall see, however, that despite the great restrictions that peptide bond planarity and side chain bulk place on the conformations of a polypeptide chain, every unique primary

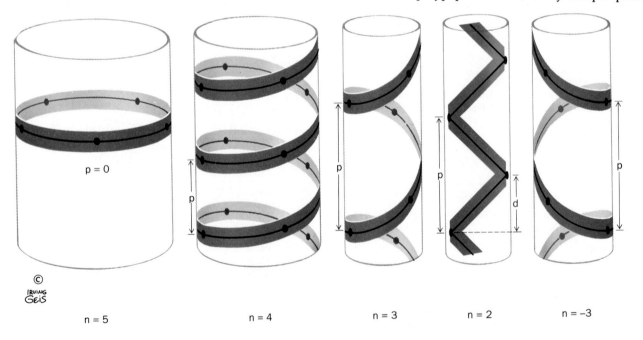

Figure 7-10
Examples of helices indicating the definitions of the helical pitch, *p*, the number of repeating units per turn, *n*, and the helical rise per repeating unit, $d = p/n$. Right- and left-handed helices are defined, respectively, as having positive and negative values of *n*. For $n = 2$, the helix degenerates to a nonchiral ribbon. For $p = 0$, the helix degenerates to a closed ring. [Figure copyrighted © by Irving Geis.]

structure has a correspondingly unique three-dimensional structure.

B. Helical Structures

Helices are the most striking elements of protein 2° structure. If a polypeptide chain is twisted by the same amount about each of its C_α atoms, it assumes a helical conformation. As an alternative to specifying its ϕ and ψ angles, a helix may be characterized by the number, n, of peptide units per helical turn, and its **pitch,** p, the distance the helix rises along its axis per turn. Several examples of helices are diagrammed in Fig. 7-10. Note that a helix has chirality; that is, it may be either right handed or left handed (a right-handed helix turns in the direction that the fingers of a right-hand curl when its thumb points along the helix axis in the direction that the helix rises). In proteins, moreover, n need not be an integer and, in fact, rarely is.

A polypeptide helix must, of course, have conformation angles that fall within the allowed regions of the Ramachandran diagram. As we have seen, this greatly limits the possibilities. Furthermore, if a particular conformation is to have more than a transient existence, it must be more than just allowed — it must be stabilized. The "glue" that holds polypeptide helices and other 2° structures together is, in part, hydrogen bonds.

The α Helix

Only one helical polypeptide conformation has simultaneously allowed conformation angles and a favorable hydrogen bonding pattern: The α helix (Fig. 7-11), a particularly rigid arrangement of the polypeptide chain. Its discovery through model building, by Pauling in 1951, ranks as one of the landmarks of structural biochemistry.

For a polypeptide made from L-α-amino acid residues, the α helix is right handed with torsion angles $\phi = -57°$ and $\psi = -47°$, $n = 3.6$ residues per turn and a pitch of 5.4 Å. (An α helix of D-α-amino acid residues is the mirror image of that made from L-amino acid residues: It is left handed with conformation angles $\phi = +57°$ and $\psi = +47°$ but with the same values of n and p.)

Figure 7-11 indicates that the hydrogen bonds of an α helix are arranged such that the peptide C=O bond of the nth residue points along the helix towards the peptide N—H group of the $(n + 4)$th residue. This results in a strong hydrogen bond that has the nearly optimum $N \cdot \cdot \cdot O$ distance of 2.8 Å. In addition, the core of the α helix is tightly packed, that is, its atoms are in van der Waals contact across the helix thereby maximizing their association energies (Section 7-4A). The R groups, whose positions, as we saw, are not fully dealt with by the Ramachandran diagram, all project backward (downward in Fig. 7-11) and outward from the helix so as to avoid steric interference with the polypeptide

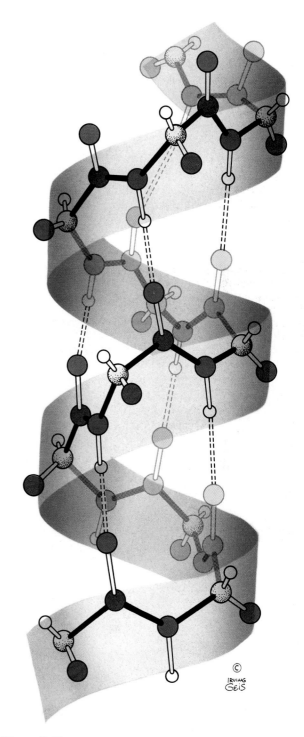

Figure 7-11
The right-handed α helix. Hydrogen bonds between the N — H groups and the C = O groups that are four residues back along the polypeptide chain are indicated by dashed lines. [Figure copyrighted © by Irving Geis.]

backbone and with each other. Such an arrangement can also be seen in Fig. 7-12. Indeed, a major reason why the left-handed α helix has never been observed (its helical parameters are but mildly forbidden; Fig. 7-7) is that its side chains too closely contact its polypeptide backbone. Note, however, that 1 to 2% of the individual non-Gly residues in proteins assume this conformation (Fig. 7-8).

The α helix is a common secondary structural element of both fibrous and globular proteins. In globular proteins, α helices have an average span of $\sim$11 residues, which corresponds to over three helical turns and a length of 17 Å. However, α helices with as many as 53 residues have been found.

Other Polypeptide Helices

Figure 7-13 indicates how hydrogen bonded polypeptide helices may be constructed. The first two, the **2.2₇ ribbon** and the **3₁₀ helix,** are described by the notation n_m where n, as before, is the number of residues per

helical turn, and m is the number of atoms, including H, in the ring that is closed by the hydrogen bond. With this notation, an α helix is a 3.6_{13} helix.

The right-handed 3_{10} helix (Fig. 7-14a), which has a pitch of 6.0 Å, is thinner and rises more steeply than does the α helix (Fig. 7-14b). Its torsion angles place it in a mildly forbidden zone of the Ramachandran diagram that is rather near the position of the α helix (Fig. 7-7) while its R groups experience some steric interference. This explains why the 3_{10} helix is only occasionally observed in proteins, and then only in short segments that are frequently distorted from the ideal 3_{10} conformation. The 3_{10} helix most often occurs as a single turn transition between the end of an α helix and the next portion of a polypeptide chain.

The $\boldsymbol{\pi}$ **helix** (4.4_{16} helix), which also has a mildly forbidden conformation (Fig. 7-7), has only been observed at the ends of a few α helices. This is probably because its comparatively wide and flat conformation (Fig. 7-14c) results in an axial hole that is too small to admit

Figure 7-12
A stereo, space-filling representation of an α helical segment of sperm whale myoglobin (its E helix) as determined by X-ray crystal structure analysis. Nitrogen atoms are blue, oxygen atoms are red, hydrogen atoms are white, and carbon atoms are gray. R groups are represented by large gray balls. Instructions for viewing stereo diagrams are given in the appendix to this chapter. [Courtesy of Richard J. Feldmann, NIH.]

Figure 7-13
The hydrogen bonding pattern of several polypeptide helices. In the cases shown, the polypeptide chain curls around such that the C═O group on residue n forms a hydrogen bond with the N—H group on residue $(n + 2)$, $(n + 3)$, $(n + 4)$, or $(n + 5)$. [Figure copyrighted © by Irving Geis.]

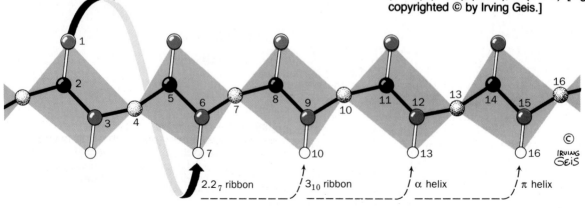

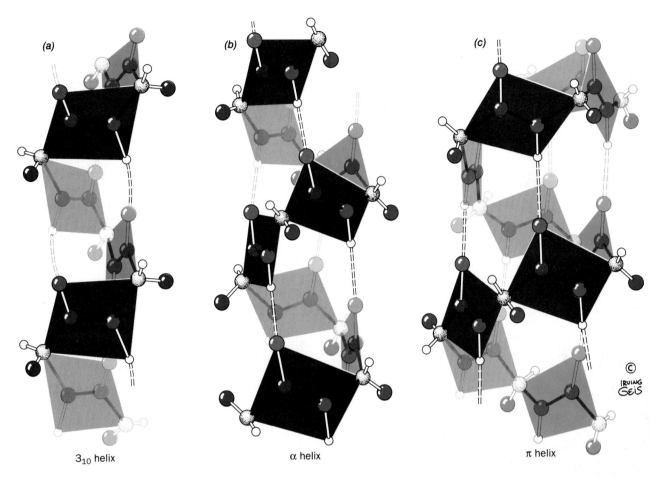

(a)

3_{10} helix

(b)

α helix

(c)

©
IRVING
GEIS

π helix

Figure 7-14
Two polypeptide helices that occasionally occur in proteins compared with the commonly occurring α helix. (*a*) The 3_{10} helix is characterized by 3.0 peptide units per turn and a pitch of 6.0 Å, which makes it thinner and more elongated than the α helix. (*b*) The α helix has 3.6 peptide units per turn and a pitch of 5.4 Å (also see Fig. 7-11). (*c*) The π helix, with 4.4 peptide units per turn and a pitch of 5.2 Å, is wider and shorter than the α helix. The peptide planes are indicated. [Figure copyrighted © by Irving Geis.]

water molecules but yet too wide to allow van der Waals associations across the helix axis; this greatly reduces its stability relative to more closely packed conformations. The 2.2_7 ribbon, which as Fig. 7-7 indicates, has strongly forbidden conformation angles, has not been observed in proteins.

Certain synthetic homopolypeptides assume conformations that are models for helices in particular proteins. **Polyproline** is unable to assume any common secondary structure due to the conformational constraints imposed by its cyclic pyrrolidine side chains. Furthermore, the lack of a hydrogen substituent on its imide nitrogen precludes any polyproline conformation from being stabilized by hydrogen bonding. Nevertheless, under the proper conditions, polyproline precipitates from solution as a left-handed helix of all trans peptides that has 3.0 residues per helical turn and a pitch of 9.4 Å (Fig. 7-15). This rather extended conformation permits the Pro side chains to avoid each other. Curiously, **polyglycine,** the least conformationally constrained polypeptide, precipitates from solution as a helix whose parameters are essentially identical to those

of polyproline, the most conformationally constrained polypeptide (although the polyglycine helix may be either right or left handed because Gly is nonchiral). The structures of the polyglycine and polyproline helices are of biological significance because they form the basic structural motif of collagen, a structural protein that contains a remarkably high proportion of both Gly and Pro (Section 7-2C).

C. Beta Structures

In 1951, the year that they proposed the α helix, Pauling and Corey also postulated the existence of a different polypeptide secondary structure, the **β pleated sheet.** As with the α helix, the β pleated sheet's conformation falls in the allowed region of the Ramachandran diagram (Fig. 7-7) and utilizes the full hydrogen bonding capacity of the polypeptide backbone. *In β pleated sheets, however, hydrogen bonding occurs between neighboring polypeptide chains* rather than within one as in α helices.

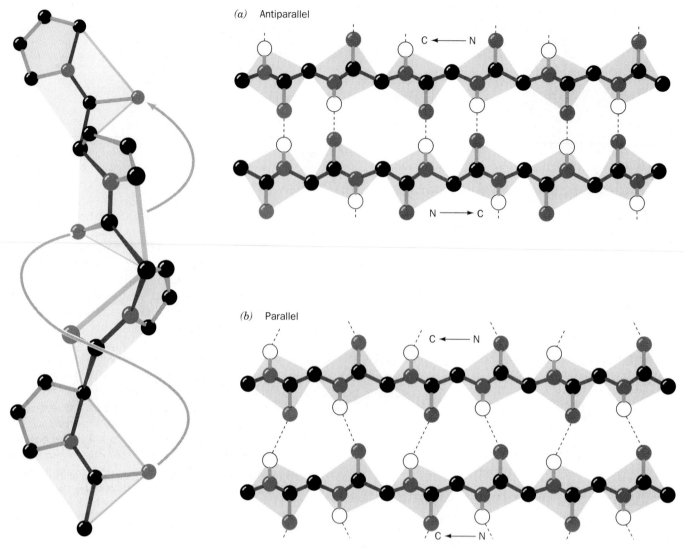

(a) Antiparallel

C ← N

N → C

(b) Parallel

C ← N

C ← N

Figure 7-15
The **polyproline II** helix. Polyglycine forms a nearly identical helix **(polyglycine II)**. [Figure copyrighted © by Irving Geis.]

Figure 7-16
The hydrogen bonding associations in β pleated sheets. Side chains are omitted for clarity. (*a*) The antiparallel β pleated sheet. (*b*) The parallel β pleated sheet. [Figure copyrighted © by Irving Geis.]

β Pleated sheets come in two varieties.

1. The **antiparallel β pleated sheet,** in which neighboring hydrogen bonded polypeptide chains run in opposite directions (Fig. 7-16*a*).

2. The **parallel β pleated sheet,** in which the hydrogen bonded chains extend in the same direction (Fig. 7-16*b*).

The conformations in which these **β structures** are optimally hydrogen bonded vary somewhat from that of a fully extended polypeptide ($\phi = \psi = \pm 180°$) as indicated in Fig. 7-7. They therefore have a rippled or pleated edge-on appearance (Fig. 7-17), which accounts for the appellation "pleated sheet." In this conformation, successive side chains of a polypeptide chain ex-

tend to opposite sides of the pleated sheet with a two-residue repeat distance of 7.0 Å.

β Sheets are common structural motifs in proteins. In globular proteins, they consist of from 2 to as many as 15 polypeptide strands, the average being 6 strands, which have an aggregate width of ~25 Å. The polypeptide chains in a β sheet are known to be up to 15 residues long, with the average being 6 residues that have a length of ~21 Å. A 6-stranded antiparallel β sheet, for example, occurs in the jack bean protein **concanavalin A** (Fig. 7-18).

Parallel β sheets of less than five strands are rare. This observation suggests that parallel β sheets are less stable than antiparallel β sheets, possibly because the hydrogen bonds of parallel sheets are distorted in comparison to those of the antiparallel sheets (Fig. 7-16). Mixed

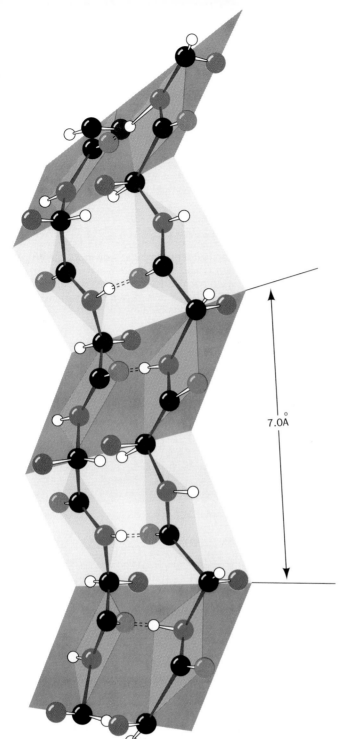

Figure 7-17
A two-stranded β antiparallel pleated sheet drawn so as to emphasize its pleated appearance. Dashed lines indicate hydrogen bonds. Note how the R groups of the polypeptide chains alternately extend to opposite sides of the sheet. [Figure copyrighted © by Irving Geis.]

parallel–antiparallel β sheets are common but occur with far less frequency than is expected for the random mixing of strand directions.

The β pleated sheets in globular proteins invariably exhibit a pronounced right-handed twist or curl when viewed along their polypeptide strands (e.g., Fig. 7-19). Such twisted β sheets are important architectural features of globular proteins since β sheets often form their central cores (Fig. 7-19). Conformational energy calculations indicate that a β sheet's right-handed curl is a consequence of nonbonded interactions between the chiral L-amino acid residues in the sheet's extended polypeptide chains. These interactions tend to give the polypeptide chains a slight right-handed helical twist (Fig. 7-19), which distorts and hence weakens the β sheet's interchain hydrogen bonds. A particular β sheet's geometry is thus the result of a compromise between optimizing the conformational energies of its polypeptide chains and preserving its hydrogen bonds.

The **topology** (connectivity) of the polypeptide strands in a β sheet can be quite complex; the connecting links of these assemblies often consist of long runs of

Figure 7-18
A stereo, space-filling representation of the six-stranded antiparallel β pleated sheet in jack bean concanavalin A as determined by X-ray crystal structure analysis. Nitrogen atoms are blue, oxygen atoms are red, hydrogen atoms are white, and carbon atoms are gray. R groups are represented by large gray balls. Instructions for viewing stereo drawings are given in the appendix to this chapter. [Courtesy of Richard J. Feldmann, NIH.]

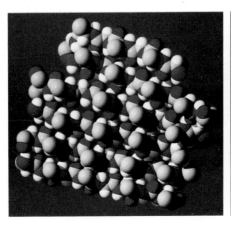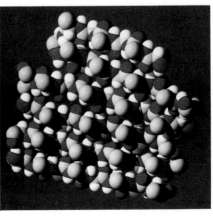

(a)

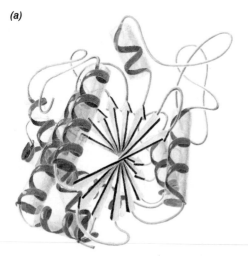

Figure 7-19
Polypeptide chain folding in proteins illustrating the right-handed twist of β sheets. Here the polypeptide backbones are represented by ribbons with α helices shown as coils and the strands of β sheets indicated by arrows pointing towards the C-terminus. Side chains are not shown. (*a*) Bovine carboxypeptidase A, a 307-residue protein, contains an eight-stranded mixed β sheet that forms a saddle-shaped curved surface with a right-handed twist. (*b*) **Triose phosphate isomerase,** a 247-residue chicken muscle enzyme, contains an eight-stranded parallel β sheet that forms a cylindrical structure known as a β **barrel** [here viewed from the top (*left*) and from the side (*right*)]. Note that the cross-over connections between successive strands of the β barrel, which each consist predominantly of an α helix, are outside the β barrel and have a right-handed helical sense. [After drawings by Jane Richardson, Duke University.]

(b)

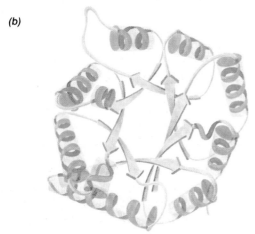 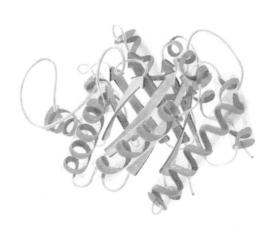

(a)   *(b)*

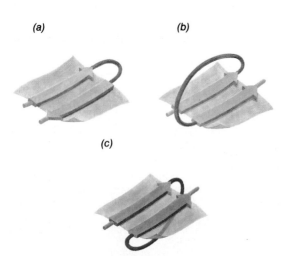

(c)

Figure 7-20
The connections between adjacent polypeptide strands in β pleated sheets: (*a*) The hairpin connection between antiparallel strands is topologically in the plane of the sheet. (*b*) A right-handed cross-over connection between successive strands of a parallel β sheet. Nearly all such cross-over connections in proteins have this chirality (see, e.g., Fig. 7-19*b*). (*c*) A left-handed cross-over connection between parallel β sheet strands. Connections with this chirality are rare. [After Richardson, J. S., *Adv. Protein Chem.* **34**, 290, 295 (1981).]

polypeptide chain, which frequently contain helices (e.g., Fig. 7-19). The link connecting two consecutive antiparallel strands is topologically equivalent to a simple hairpin turn (Fig. 7-20*a*). However, tandem parallel strands must be linked by a cross-over connection that is out of the plane of the β sheet. Such cross-over connections almost always have a right-handed helical sense (Fig. 7-20*b*), which is thought to better fit the β sheets' inherent right-handed twist (Fig. 7-21).

D. Nonrepetitive Structures

Regular secondary structures—helices and β sheets—comprise slightly less than one half of the average globular protein. The protein's remaining polypeptide segments are said to have a **coil** or **loop conformation.** That is not to say, however, that these nonrepetitive secondary structures are any less ordered than are helices or β sheets; they are just more difficult to describe. You should therefore not confuse the term coil conformation with the term **random coil,** which refers to the totally disordered and rapidly fluctuating set of conformations assumed by denatured proteins and other polymers in solution.

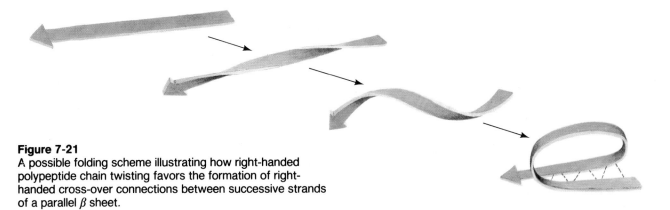

Figure 7-21
A possible folding scheme illustrating how right-handed polypeptide chain twisting favors the formation of right-handed cross-over connections between successive strands of a parallel β sheet.

Globular proteins largely consist of approximately straight runs of secondary structure joined by stretches of polypeptide that abruptly change direction. Such **reverse turns** or **β bends** (so-named because they often connect successive strands of antiparallel β sheets) almost always occur at protein surfaces; indeed, they partially define these surfaces. Most reverse turns involve four successive amino acid residues more or less arranged in one of two ways, Type I and Type II, that differ by a 180° flip of the peptide unit linking residues 2 and 3 (Fig. 7-22). Both types of β bends are stabilized by a hydrogen bond although deviations from these ideal conformations often disrupt this hydrogen bond. Type I β bends may be considered to be distorted sections of 3_{10} helix. In Type II β bends, the oxygen atom of residue 2 crowds the C_β atom of residue 3, which is therefore usually Gly. Residue 2 of either type of β bend is often Pro since it can facilely assume the required conformation.

Almost all proteins of >60 residues contain one or more loops of 6 to 16 residues that are not components of helices or β sheets and whose end-to-end distances are <10 Å. Such **Ω loops** (so-named because they have the necked-in shape of the Greek letter omega; Fig. 7-23), which may contain reverse turns, are compact globular entities because their side chains tend to fill in their internal cavities. Since Ω loops are almost invariably located on the protein surface, they may have an important role in biological recognition processes.

Many proteins have regions that are truly disordered. Extended, charged surface groups such as Lys side chains or the N- or C-termini of polypeptide chains are good examples: They often wave around in solution because there are few forces to hold them in place (Section 7-4). Sometimes entire polypeptide chain segments are disordered. Such segments may have functional roles, such as the binding of a specific molecule, so that they may be disordered in one state of the protein (mole-

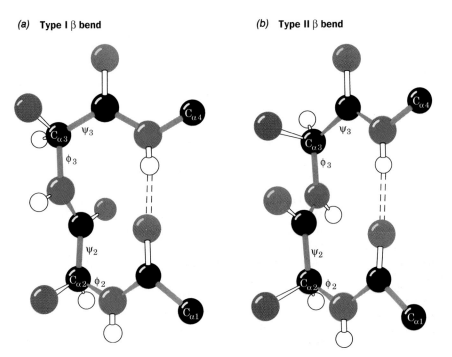

(a) **Type I β bend**

(b) **Type II β bend**

Figure 7-22
Reverse turns in polypeptide chains: *(a)* A Type I β bend, which has the torsion angles $\phi_2 = -60°$, $\psi_2 = -30°$, $\phi_3 = -90°$, and $\psi_3 = 0°$. *(b)* A Type II β bend, which has the approximate torsion angles $\phi_2 = -60°$, $\psi_2 = 120°$, $\phi_3 = 90°$, and $\psi_3 = 0°$. Variations from these ideal conformation angles by as much as 30° are common. Hydrogen bonds are represented by dashed lines. [Figure copyrighted © by Irving Geis.]

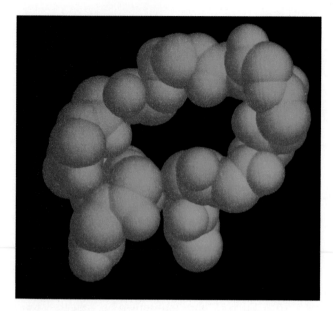

Figure 7-23
Space-filling representation of an Ω loop comprising
residues 40 to 54 of cytochrome *c*. Only backbone atoms
are shown; the addition of side chains would fill in the loop.
[Courtesy of George Rose, Pennsylvania State University
College of Medicine.]

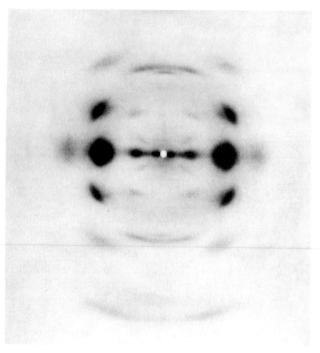

Figure 7-24
The X-ray diffraction photograph of a fiber of *Bombyx mori*
silk obtained by shining a collimated beam of monochromatic
X-rays through the silk fiber and recording the diffracted
X-rays on a sheet of photographic film placed behind the
fiber. The photograph has only a few spots and thus
contains little structural information. [From March, R. E.,
Corey, R. B., and Pauling, L., *Biochim. Biophys. Acta* **16,** 5
(1955).]

cule absent) and ordered in another (molecule bound).
This is one mechanism whereby a protein can interact
flexibly with another molecule in the performance of its
biological function.

2. FIBROUS PROTEINS

*Fibrous proteins are highly elongated molecules whose
secondary structures are their dominant structural motifs.*
Many fibrous proteins, such as those of skin, tendon,
and bone, function as structural materials that have a
protective, connective, or supportive role in living organisms. Others, such as muscle and ciliary proteins, have
motive functions. In this section, we shall discuss
structure–function relationships in four common and
well-characterized fibrous proteins: keratin, silk fibroin,
collagen, and elastin (muscle and ciliary proteins are
considered in Section 34-3). The structural simplicity of
these proteins relative to those of globular proteins
(Section 7-3) makes them particularly amenable to understanding how their structures suit them to their biological roles.

Fibrous molecules rarely crystallize and hence are
usually not subject to structural determination by single-crystal X-ray structure analysis (Section 7-3A).
Rather than crystallize, they associate as fibers in which
their long molecular axes are more or less parallel to the
fiber axis but in which they lack specific orientation in
other directions. The X-ray diffraction pattern of such a

fiber, Fig. 7-24, for example, contains little information,
far less than would be obtained if the fibrous protein
could be made to crystallize. Consequently, the structures of fibrous proteins are not known in great detail.
Nevertheless, the original X-ray studies of proteins were
carried out in the early 1930s by William Astbury on
such easily available protein fibers as wool and tendon.
Since the first X-ray crystal structure of a protein was not
determined until the late 1950s, these studies constituted the first tentative steps in the elucidation of the
structural principles governing proteins and formed
much of the experimental basis for Pauling's formulation of the α helix and β pleated sheet.

A. α Keratin — A Helix of Helices

Keratin is a mechanically durable and chemically
unreactive protein occurring in all higher vertebrates. It
is the principal component of their horny outer epidermal layer and its related appendages such as hair, horn,
nails, and feathers. Keratins have been classified as either **α keratins,** which occur in mammals or **β keratins,**
which occur in birds and reptiles.

Electron microscopic studies indicate that hair, which
is composed mainly of α keratin, consists of a hierarchy

(a) Macroscopic organization

Microfibril

Macrofibril

Cell

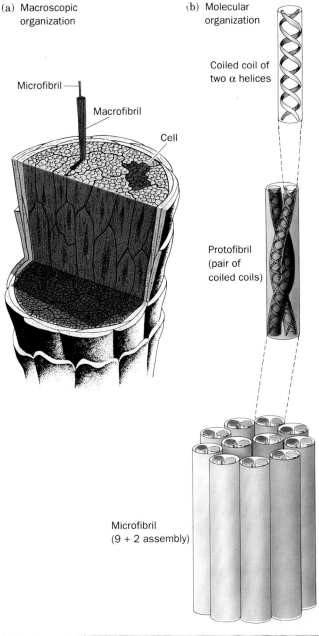

Microfibril (9 + 2 assembly)

(b) Molecular organization

Coiled coil of two α helices

Protofibril (pair of coiled coils)

Figure 7-25
The (a) macroscopic organization, and (b) molecular organization of hair. The manner in which two coiled coils associate to form a protofibril is poorly understood. [Figure copyrighted © by Irving Geis.]

of structures (Fig. 7-25). A typical hair is ~20 μm in diameter and is constructed from dead cells, each of which contains packed **macrofibrils** (2000 Å in diameter) that are oriented parallel to the hair fiber. The macrofibrils are constructed from **microfibrils** (80 Å in diameter) that are cemented together by an amorphous protein matrix of high sulfur content. Microfibrils, in turn, consist of **protofibrils** (20 Å in diameter) that appear to be organized in a ring of nine protofibrils surrounding a central core of two protofibrils (Fig. 7-25b and 7-26). This "9 + 2" motif also occurs in cilia, the whiplike organelles that eukaryotic cells use to swim or to move their surrounding media (Section 34-3F).

X-ray diffraction photographs of α keratin exhibit a pattern resembling that expected for an α helix (hence Pauling's α helix). Yet, α keratin exhibits a 5.1-Å spacing rather than the 5.4-Å distance corresponding to the pitch of the α helix. This observation, together with a variety of physical and chemical data, suggests that *α keratin's protofibrils each consist of two closely associated pairs of α helices in which each pair is twisted into a left-handed coil* (Fig. 7-25b). The normal 5.4-Å repeat distance of each α helix is thereby tilted with respect to the protofibril axis, yielding the observed 5.1-Å spacing. This assembly is said to have a **coiled coil** structure because each α helix axis itself follows a helical path.

α Keratin's coiled coil conformation is a consequence of its primary structure: It has a 7-residue pseudorepeat, *a-b-c-d-e-f-g,* with nonpolar residues predominating at positions *a* and *d.* Since an α helix has 3.6 residues per turn, α keratin's *a* and *d* residues line up on one side of the α helix to form a hydrophobic strip that promotes its lengthwise association with a similar strip on another

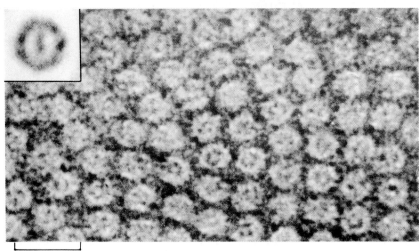

200 Å

Figure 7-26
An electron micrograph of a Merino wool fiber in cross section showing its microfibrillar structure. The inset is the superimposed images of several microfibrils in the same orientation, an averaging process that tends to enhance real detail and to suppress background fluctuations. A microfibril apparently consists of a group of nine outer and two inner protofibrils. [Courtesy of George E. Rogers, University of Adelaide.]

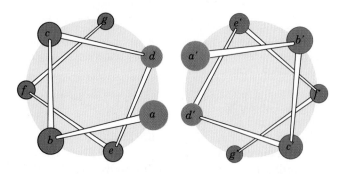

Figure 7-27
The interactions between the nonpolar edges of α helices in a coiled coil as viewed down the coil axis. The α helices have the pseudorepeating heptameric sequence *a-b-c-d-e-f-g* in which residues *a* and *d* are predominantly nonpolar. [After McLachlan, A. D. and Stewart, M., *J. Mol. Biol.* **98**, 295 (1975).]

such α helix (Fig. 7-27; hydrophobic residues, as we shall see in Section 7-4C, have a strong tendency to associate). Indeed, the slight discrepancy between the 3.6 residues per turn of a normal α helix and the ~3.5 residue repeat of α keratin's hydrophobic strip is responsible for the coiled coil's coil. The resulting 18° inclination of the α helices relative to one another permits their contacting side chains to interdigitate efficiently thereby greatly increasing their favorable interactions. The way in which two coiled coils associate to form a protofibril is poorly understood.

α Keratin is rich in Cys residues, which cross-link adjacent polypeptide chains. This accounts for its insolubility and resistance to stretching, two of α keratin's most important biological properties. The α keratins are classified as "hard" or "soft" according to whether they have a high or low sulfur content. Hard keratins, such as those of hair, horn, and nail, are less pliable than soft keratins, such as those of skin and callus, because the disulfide bonds resist any forces tending to deform them. The disulfide bonds can be reductively cleaved with mercaptans (Section 6-1B). Hair so treated can be curled and set in a "permanent wave" by application of an oxidizing agent, which reestablishes the disulfide bonds in the new "curled" conformation. Although the insolubility of α keratin prevents most animals from digesting it, the clothes moth larva, which has a high concentration of mercaptans in its digestive tract, can do so to the chagrin of owners of woolen clothing.

The springiness of hair and wool fibers is a consequence of the coiled coil's tendency to untwist when stretched and to recover its original conformation when the external force is relaxed. After some of its disulfide bonds have been cleaved, however, an α keratin fiber can be stretched to over twice its original length by the application of moist heat. In this process, as X-ray analysis indicates, the α helical structure extends with concomitant rearrangement of its hydrogen bonds to form a β pleated sheet. β Keratin, such as that of feathers, exhibits a similar X-ray pattern in its native state (hence the β sheet).

B. Silk Fibroin — A β Pleated Sheet

Insects and arachnids (spiders) produce **silk** to fabricate structures such as cocoons, webs, nests, and egg stalks. The silk is stored as a fluid in the gland that produces it but, during spinning, is converted to a water-insoluble form. Most silks consist of the fibrous protein **fibroin** and a gummy amorphous protein named **sericin** that cements the fibroin fibers together. An adult moth emerging from its sealed cocoon secretes a protease (**cocoonase**) that digests sericin, thereby enabling the moth to push the fibroin filaments aside and escape from the cocoon. In the preparation of silk cloth, which consists only of fibroin, the sericin is removed by treatment with boiling soap solution.

Silk fibroin from the cultivated larvae (silkworms) of the moth *Bombyx mori* (Fig. 7-28) exhibits an X-ray diffraction pattern (Fig. 7-24) indicating that *its polypeptide chains form antiparallel β pleated sheets in which the chains extend parallel to the fiber axis.* Sequence studies have shown that long stretches of the chain are comprised of the six-residue repeat

$$(-\text{Gly-Ser-Gly-Ala-Gly-Ala-})_n$$

This sequence forms β sheets with its Gly side chains extending from one surface and its Ser and Ala side chains extending from the other surface (as in Fig. 7-17). The β sheets thereby stack to form a microcrystalline array such that layers of contacting Gly side chains from neighboring sheets alternate with layers of contacting

Figure 7-28
The domestic silkworm *Bombyx mori* in the process of constructing its cocoon. [Hans Pfletschinger/Peter Arnold, Inc.]

(a)

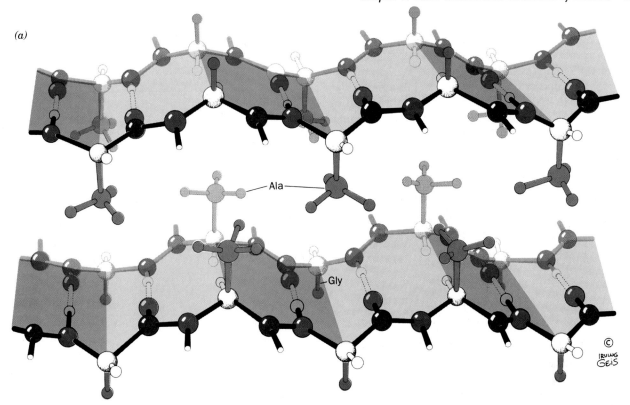

(b)

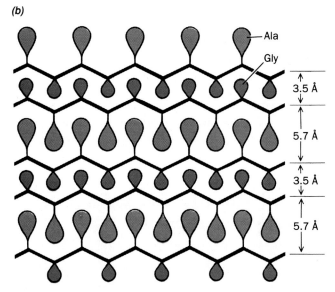

Figure 7-29
The three-dimensional architecture of silk fibroin. (a) Silk's alternating Gly and Ala (or Ser) residues extend to opposite sides of a given β sheet so that the Ala side chains extending from one β sheet efficiently nestle between those of the neighboring sheet and likewise for the Gly side chains. (b) Gly side chains from neighboring β sheets are in contact as are those of Ala and Ser. The intersheet spacings consequently have the alternating values 3.5 and 5.7 Å. [Figure copyrighted © by Irving Geis.]

Ser and Ala side chains (Fig. 7-29). This structure, in part, accounts for silk's mechanical properties. *Silk fibers are strong but only slightly extensible because appreciable stretching would require breaking the covalent bonds of its nearly fully extended polypeptide chains.* Yet, the fibers are flexible because neighboring β sheets associate only through relatively weak van der Waals forces.

Although large segments of silk fibroin have the repeating hexameric amino acid sequence, it also has regions in which bulky residues, such as Tyr, Val, Arg, and Asp occur. These residues distort and therefore disorder the microcrystalline array drawn in Fig. 7-29. Silk fibers are consequently composed of alternating crystalline and amorphous regions. The amorphous regions are largely responsible for the extensibility of the silk fibers. Silks from different species have different proportions of bulky side groups and therefore different mechanical properties. The greater a fibroin's proportion of bulky groups, the less its elasticity (the ability to resist deformation and to recover its original shape when the deforming forces are removed) and the greater its extensibility. Hence, *the mechanical properties of silk fibroin can be understood in terms of its structure which, in turn, depends on its amino acid sequence.*

C. Collagen — A Triple Helical Cable

Collagen occurs in all multicellular animals and is the most abundant protein of vertebrates. It is an extracellular protein that is organized into insoluble fibers of great tensile strength. This suits collagen to its role as the major stress bearing component of **connective tissues** such as bone, teeth, cartilage, tendon, ligament, and the fibrous matrices of skin and blood vessels. Collagen occurs in virtually every tissue.

A single molecule of Type I collagen has a molecular mass of ~285 kD, a width of ~14 Å, and a length of ~3000 Å. It is composed of three polypeptide chains. Mammals have at least 17 genetically distinct polypeptide chains comprising 10 collagen variants that occur in different tissues of the same individual. The most prominant of these are listed in Table 7-2.

Collagen has a distinctive amino acid composition: Nearly one third of its residues are Gly; another 15 to 30% of its residues are Pro and 4-hydroxyproline (Hyp).

Table 7-2

The Most Abundant Types of Collagen

Type	Chain Composition	Distribution
I	$[\alpha 1(I)]_2 \alpha 2(I)$	Skin, bone, tendon, blood vessels, cornea
II	$[\alpha 1(II)]_3$	Cartilage, intervertebral disk
III	$[\alpha 1(III)]_3$	Blood vessels, fetal skin

Source: Eyre, D. R., *Science* **207**, 1316 (1980).

tures at 39°C (denatured collagen is known as **gelatin**). Prolyl hydroxylase requires **ascorbic acid (vitamin C)**

Ascorbic acid (Vitamin C)

to maintain its enzymatic activity. In the vitamin C deficiency disease **scurvy**, the collagen synthesized cannot form fibers properly. This results in the skin lesions, blood vessel fragility, and poor wound healing that are symptomatic of scurvy.

The amino acid sequence of bovine collagen α1(I), which is similar to that of other collagens, consists of monotonously repeating triplets of sequence Gly-X-Y over a continuous 1011-residue stretch of its 1042-residue polypeptide chain (Fig. 7-30). Here X is often Pro and Y is often

4-Hydroxyprolyl residue (Hyp)

3-Hydroxyprolyl residue

5-Hydroxylysyl residue (Hyl)

3-Hydroxyproline and **5-hydroxylysine (Hyl)** also occur in collagen but in smaller amounts. Radioactive labeling experiments have established that these nonstandard hydroxylated amino acids are not incorporated into collagen during polypeptide synthesis: If ¹⁴C-labeled 4-hydroxyproline is administered to a rat, the collagen synthesized is not radioactive, whereas radioactive collagen is produced if the rat is fed ¹⁴C-labeled proline. The hydroxylated residues appear after the collagen polypeptides are synthesized, when certain Pro residues are converted to Hyp in a reaction catalyzed by the enzyme **prolyl hydroxylase.**

Hyp confers stability upon collagen, probably through intramolecular hydrogen bonds that may involve bridging water molecules. If, for example, collagen is synthesized under conditions that inactivate prolyl hydroxylase, it loses its native conformation (denatures) at 24°C, whereas normal collagen dena-

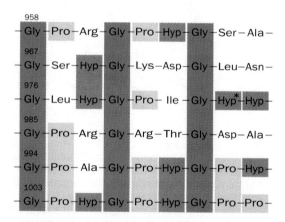

Figure 7-30

The amino acid sequence at the C-terminal end of the triple helical region of the bovine α1(I) collagen chain. Note the repeating triplets Gly-X-Y, where X is often Pro and Y is often Hyp. Here, Hyp* represents 3-hydroxy Pro. [From Bornstein, P. and Traub, W., *in* Neurath, H. and Hill, R. L. (Eds.), *The Proteins* (3rd ed.), Vol. 4 *p.* 483, Academic Press (1979).]

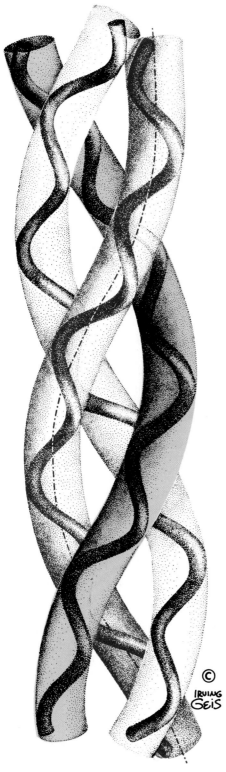

Hyp. The restriction of Hyp to the Y position stems from the specificity of prolyl hydroxylase. 5-Hydroxylysine is similarly restricted to the Y position.

The high Gly, Pro, and Hyp content of collagen suggests that its polypeptide backbone conformation resembles those of the polyglycine II and polyproline II helices (Fig. 7-15). X-ray and model building studies indicate that *collagen's three polypeptide chains, which individually resemble polyproline II helices, are parallel and wind around each other with a gentle, right-handed, ropelike twist to form a triple-helical structure (Fig. 7-31).* Every third residue of each polypeptide chain passes through the center of the triple helix, which is so crowded that only a Gly side chain can fit there. This crowding explains the absolute requirement for a Gly at every third position of a collagen polypeptide chain (Fig. 7-30). It also requires that the three polypeptide chains be staggered so that the Gly, X, and Y residues from the three chains occur at similar levels (Fig. 7-32). The staggered peptide groups are oriented such that the N—H

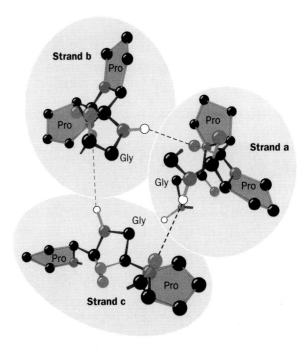

Figure 7-32

A projection down the triple helix axis of the collagenlike polymer (Gly-Pro-Pro)$_n$ as viewed from its carboxyl end. The residues in each chain, Gly-X-Y, are vertically staggered such that a Gly, an X, and a Y residue from different chains are on the same level along the helix axis. The dashed lines represent hydrogen bonds between each Gly N — H group and the oxygen of the succeeding X residue on a neighboring chain. Every third residue on each chain must be Gly because there is no room near the helix axis for the side chain of any other residue. The bulky pyrrolidine side chains (*brown*) of the Pro residues are situated on the periphery of the triple helix where they are sterically unhindered. [After Yonath, A. and Traub, W., *J. Mol. Biol.* **43**, 461 (1969).]

Figure 7-31

The triple helix of collagen indicating how the left-handed polypeptide helices are twisted together to form a right-handed superhelical structure. Ropes and cables are similarly constructed from hierarchies of fiber bundles that are alternately twisted in opposite directions. An individual polypeptide helix has 3.3 residues per turn and a pitch of 10.0 Å (in contrast to polyproline II's 3.0 residues per turn and pitch of 9.4 Å; Fig. 7-15). The collagen triple helix has 10 Gly-X-Y units per turn and a pitch of 86.1 Å. [Figure copyrighted © by Irving Geis.]

of each Gly makes a strong hydrogen bond with the carbonyl oxygen of an X residue on a neighboring chain. The bulky and relatively inflexible Pro and Hyp residues confer rigidity on the entire assembly.

Collagen's well-packed, rigid, triple helical structure is responsible for its characteristic tensile strength. As with the twisted fibers of a rope, the extended and twisted polypeptide chains of collagen convert a longitudinal tensional force to a more easily supported lateral compressional force on the almost incompressible triple helix. This occurs because the oppositely twisted directions of collagen's polypeptide chains and triple helix (Fig. 7-31) prevent the twists from being pulled out under tension (note that successive levels of fiber bundles in ropes and cables are likewise oppositely twisted). The successive helical hierarchies in other fibrous proteins exhibit similar alternations of twist directions, for example, keratin (Section 7-2A) and muscle (Section 34-3A).

Collagen Is Organized into Fibrils

Types I, II, and III collagens form distinctive banded fibrils (Fig. 7-33) that have a periodicity of 680 Å and a diameter of 100 to 2000 Å depending on the collagen type and its tissue of origin (the other collagen types form different sorts of aggregates such as networks; we will not discuss them further). Computerized model building studies indicate that collagen molecules are laterally organized in a precisely staggered array (Fig.

Figure 7-33
An electron micrograph of collagen fibrils from skin. [Courtesy of Jerome Gross, Massachusetts General Hospital.]

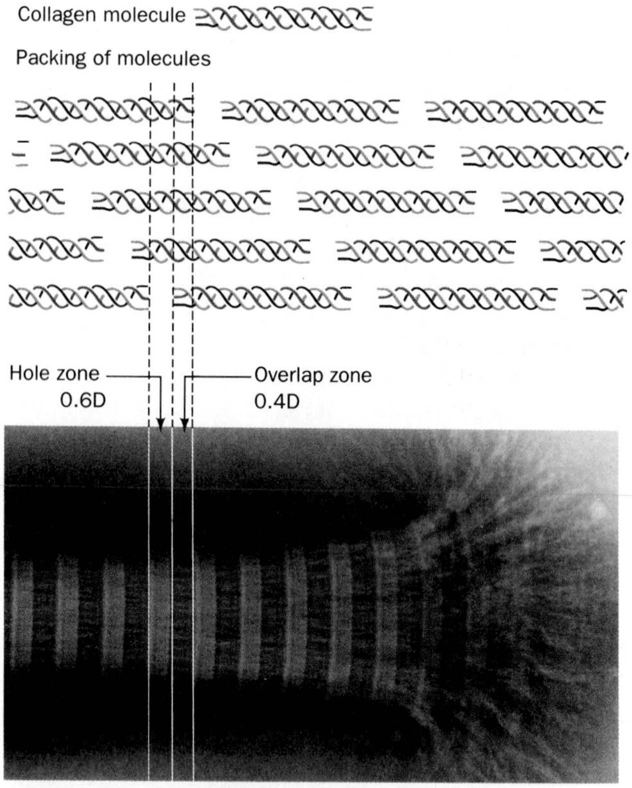

Collagen molecule

Packing of molecules

Hole zone —— 0.6D

Overlap zone 0.4D

Figure 7-34
The banded appearance of collagen fibrils in the electron microscope arises from the schematically represented staggered arrangement of collagen molecules (*above*) that results in a periodically indented surface. *D*, the distance between cross striations is ~680 Å so that the length of a 3000-Å long collagen molecule is 4.4*D*. [Courtesy of Karl A. Piez, Collagen Corporation.]

7-34). The darker portions of the banded structures correspond to the 400-Å "holes" on the surface of the fibril between head-to-tail aligned collagen molecules.

Collagen contains covalently attached carbohydrates in amounts that range from ~0.4 to 12% by weight, depending on the collagen's tissue of origin. The carbohydrates, which consist mostly of glucose, galactose,

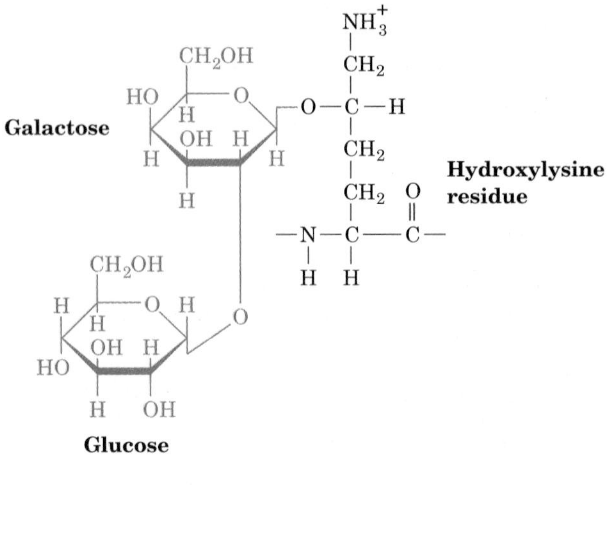

Galactose

CH_2OH

Hydroxylysine residue

Glucose

and their disaccharides, are covalently attached to collagen at its 5-hydroxylysyl residues by specific enzymes. Although the function of carbohydrates in collagen is unknown, the observation that they are located in the "hole" regions of the collagen fibrils suggests that they are involved in directing fibril assembly.

The structure of collagen fibrils plays an important role in the formation of bone. Bone consists of an organic phase, mainly collagen, and an inorganic phase, mainly **hydroxyapatite,** $Ca_5(PO_4)_3OH$. During bone formation, the initial crystals of hydroxyapatite form at intervals of 680 Å, the periodicity of the collagen fibrils. This observation suggests that the fibrils' "hole" regions act as nucleation sites in bone mineralization.

Collagen Fibrils Are Covalently Cross-Linked

Collagen's insolubility in solvents that disrupt hydrogen bonding and ionic interactions is explained by the observation that it is both intramolecularly and intermolecularly covalently cross-linked. The cross-links cannot be disulfide bonds, as in keratin, because collagen is almost devoid of Cys residues. Rather, they are derived from Lys and His side chains in reactions such as those in Fig. 7-35. **Lysyl oxidase,** a Cu-containing enzyme that converts Lys residues to those of the aldehyde **allysine,** is the only enzyme implicated in this cross-linking process. Up to four side chains can be covalently bonded to each other. The cross-links do not form at random but, instead, tend to occur near the N- and C-termini of the collagen molecules.

The importance of cross-linking to the normal functioning of collagen is demonstrated by the disease **lathyrism,** which occurs in humans and other animals as a result of the regular ingestion of seeds from the sweet pea *Lathyrus odoratus*. The symptoms of this condition are serious abnormalities of the bones, joints, and large blood vessels, which are caused by an increased fragility of the collagen fibers. The causative agent of lathyrism, **β-aminopropionitrile,**

$$N \equiv C - CH_2 - CH_2 - NH_3^+$$

β-Aminopropionitrile

inactivates lysyl oxidase by covalently binding to its active site. This results in markedly reduced cross-linking in the collagen of lathrytic animals.

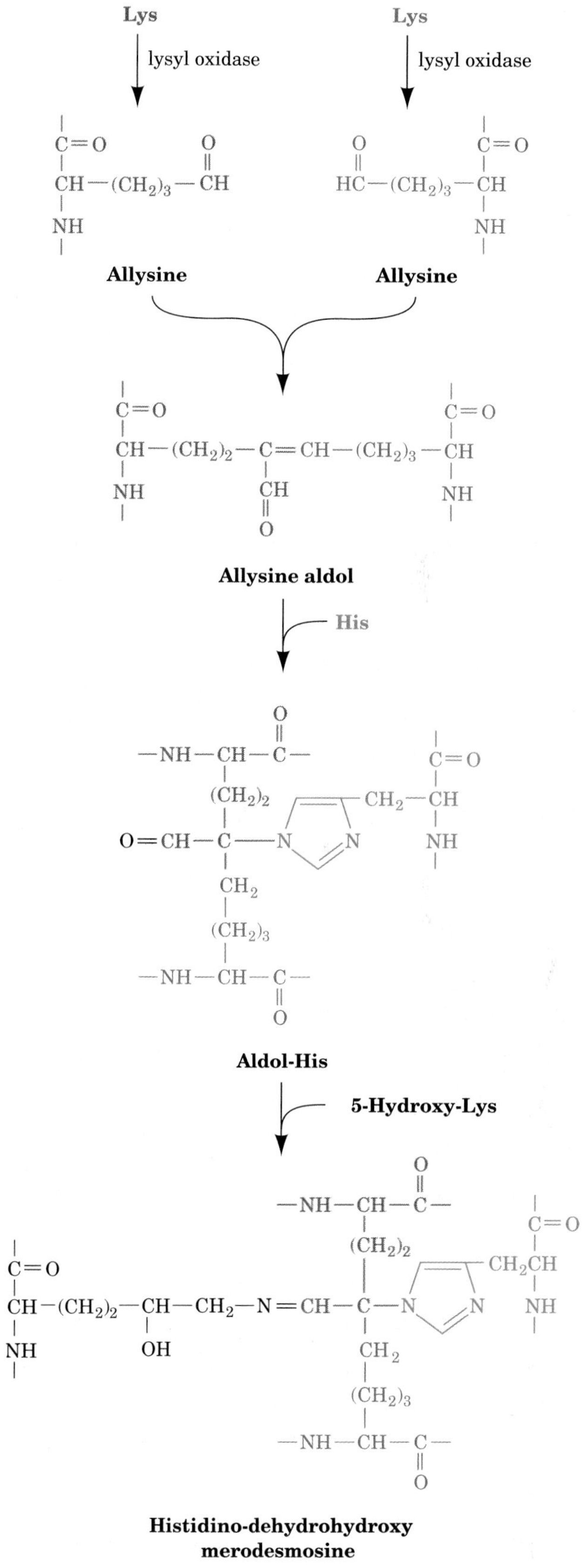

Figure 7-35
A biosynthetic pathway for cross-linking Lys, 5-hydroxylysyl, and His side chains in collagen. The first step in the reaction is the lysyl oxidase-catalyzed oxidative deamination of Lys to form the aldehyde allysine. Two such aldehydes then undergo an aldol condensation to form **allysine aldol.** This product can react with His to form **aldol histidine.** This, in turn, can react with 5-hydroxylysine to form a Schiff base (an imine bond), thereby cross-linking four side chains.

Table 7-3

The Arrangement of Collagen Fibrils in Various Tissues

Tissue	Arrangement
Tendon	Parallel bundles
Skin	Sheets of fibrils layered at many angles
Cartilage	No distinct arrangement
Cornea	Planar sheets stacked crossways so as to minimize light scatter

The degree of cross-linking of the collagen from a particular tissue increases with the age of the animal. This is why meat from older animals is tougher than that from younger animals. In fact, individual molecules of collagen (called **tropocollagen**) can only be extracted from the tissues of very young animals. Collagen cross-linking is not the central cause of aging, however, as is demonstrated by the observation that lathyrogenic agents do not slow the aging process.

The collagen fibrils in various tissues are organized in ways that largely reflect the functions of the tissues (Table 7-3). Thus tendons (the "cables" connecting muscles to bones), skin (a tear-resistant outer fabric), and cartilage (which has a load-bearing function) must support stress in predominantly one, two, and three dimensions, respectively, and their component collagen fibrils are arrayed accordingly. How collagen fibrils are laid down in these arrangements is unknown. However, some of the factors guiding collagen molecule assembly are discussed in Sections 30-5A and B.

Collagen Defects Are Responsible for a Variety of Human Diseases

Several rare heritable disorders of collagen are known. A few result from an amino acid substitution in a collagen polypeptide that presumably disrupts the collagen triple helix or from the deletion of a polypeptide segment. Most collagen disorders, however, are characterized by deficiencies in the amount of a particular collagen type synthesized, or by abnormal activities of collagen-processing enzymes such as lysyl hydroxylase or lysyl oxidase. One group of at least 10 different collagen deficiency diseases, the **Ehlers–Danlos syndromes,** are all characterized by hyperextensibility of the joints and skin. The "India-rubber man" of circus fame had an Ehlers–Danlos syndrome. **Marfan's syndrome** has similar symptoms. The nineteenth century violin virtuoso, Niccolo Paganini (Fig. 7-36), may have suffered (and profited) from this disorder. Defects that reduce the stability or amount of Type I collagen are associated with fragile bones as in **osteogenesis imperfecta.** Many degenerative diseases exhibit collagen abnormalities in the affected tissues. Examples of such tissues are the cartilage in **osteoarthritis** and the fibrous **atherosclerotic plaques** in human arteries.

Figure 7-36
The great violinist Niccolo Paganini (1782–1840) was of such outstanding virtuosity that he was rumored to be in league with the Devil. Portraits and sketches of him show he had an angular physique and features, and while playing the violin, he assumed a distinctive posture that minimized the weight he would have to support with his muscles. This has prompted the retrospective diagnosis that he was suffering from the collagen disorder known as Marfan's syndrome. This would explain his long, thin, hyperextensible fingers that he used so effectively in his legendary violin playing. [New York Public Library, Performing Arts Collection/PAR Archive.]

D. Elastin — A Nonrepetitive Coil

Elastin is a protein with rubberlike elastic properties whose fibers can stretch to several times their normal length. It is the principle protein component of the elastic yellow connective tissue that occurs in the lungs, the walls of large blood vessels such as the aorta, and elastic ligaments such as those in the neck. The inelastic white connective tissue of tendons contains only a small amount of elastin. The hyperextensibility of the joints and skin characteristic of certain collagen deficiency diseases results from the loss of rigidity ordinarily conferred by collagen coupled with the normal presence of elastin.

Elastin, like silk fibroin and collagen, has a distinctive amino acid composition. It consists predominantly of small, nonpolar residues: It is one-third Gly, over one-third Ala + Val, and is rich in Pro. However, it contains little hydroxyproline, no hydroxylysine, and few polar residues.

Elastin forms a three-dimensional network of fibers

that exhibit no recognizable periodicity in the electron microscope. Furthermore, according to X-ray analyses, the fibers are devoid of regular secondary structure.

The covalent cross-links in elastin are formed by allysine aldol, which also occurs in collagen (Fig. 7-35), and the compounds **lysinonorleucine, desmosine,** and **isodesmosine.**

Desmosine

Lysinonorleucine

Isodesmosine

Lysinonorleucine, which likewise occurs in collagen, results from the reduction of the Schiff base (imine bond) formed by the condensation of a Lys side chain with that of allysine (Fig. 7-35). Desmosine and isodesmosine are unique to elastin and are responsible for its yellow color; they result from the condensation of three allysine and one lysine side chains. These cross-links are apparently responsible for elastin's elastic properties and its insolubility.

3. GLOBULAR PROTEINS

Globular proteins comprise a highly diverse group of substances that, in their native state, exist as compact spheroidal molecules. Enzymes are globular proteins as

are transport and receptor proteins. In this section we consider the tertiary structures of globular proteins. However, since most of our detailed structural knowledge of proteins, and thus to a large extent their function, has resulted from X-ray crystal structure determinations of globular proteins [although recently developed nuclear magnetic resonance (NMR) techniques are beginning to have an impact], we begin this section with a discussion of the capabilities and limitations of this powerful technique.

A. Interpretation of Protein X-Ray Structures

X-ray crystallography is a technique that directly images molecules. X-rays must be used to do so because, according to optical principles, the uncertainty in locating an object is approximately equal to the wavelength of the radiation used to observe it (both X-ray wavelengths and covalent bond distances are ~1.5 Å; individual molecules cannot be seen in a light microscope because visible light has a minimum wavelength of 4000 Å). There is, however, no such thing as an X-ray microscope because there are no X-ray lenses. Rather, a crystal of the molecule to be visualized is exposed to a collimated beam of monochromatic X-rays and the consequent diffraction pattern is recorded on photographic film (Fig. 7-37) or by a radiation counter. The intensities of the

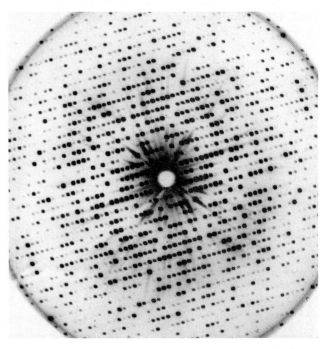

Figure 7-37
An X-ray diffraction photograph of a single crystal of sperm whale myoglobin. The intensity of each diffraction maximum (the darkness of each spot) is a function of the myoglobin crystal's electron density. The photograph contains only a small fraction of the total diffraction information available from a myoglobin crystal. [Courtesy of John Kendrew, Cambridge University.]

(a)

(b)

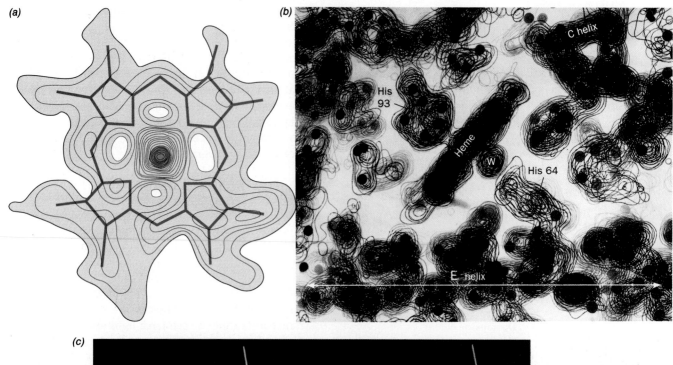

(c)

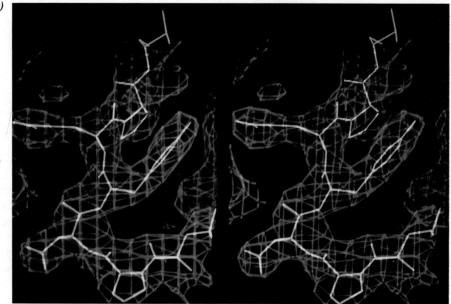

Figure 7-38
Electron density maps of proteins. (a) A section through the
2.0-Å resolution electron density map of sperm whale
myoglobin, which contains the heme group (*red*). The large
peak at the center of the map represents the electron dense
Fe atom. [After Kendrew, J. C., Dickerson, R. E.,
Strandberg, B. E., Hart, R. G., Davies, D. R., Phillips, D. C.,
and Shore, V. C., *Nature* **185,** 434 (1960).] (b) A portion of
the 1.4-Å resolution electron density map of myoglobin
constructed from a stack of contoured transparencies. Dots
have been placed at the positions deduced for the
nonhydrogen atoms. The heme group is seen edge-on
together with its two associated His residues and a water
molecule, W. An α helix, the so-called E helix (Fig. 7-12),

extends across the bottom of the map. Another α helix, the
C helix, extends into the plane of the paper on the upper
right. Note the hole along its axis. [Courtesy of John
Kendrew, Cambridge University.] (c) A portion of the 3.0-Å
resolution electron density map of a human rhinovirus (the
cause of the common cold, Section 32-2C) contoured in
three dimensions on a graphics computer and shown in
stereo. Only a single contour level (*orange*) is shown
together with an atomic model of the corresponding
polypeptide segment (*white*). Instructions for viewing stereo
diagrams are given in the appendix to this chapter.
[Courtesy of Michael Rossmann, Edward Arnold, and Gerrit
Vriend, Purdue University.]

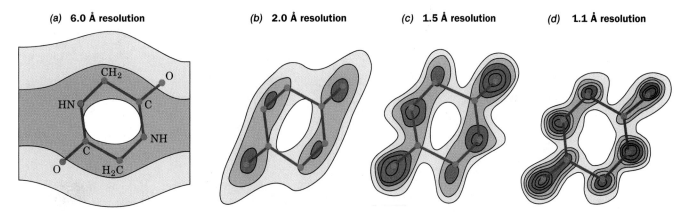

Figure 7-39
A section through the electron density map of diketopiperazine calculated at the indicated resolution levels. Hydrogen atoms are not apparent in this map because of their low electron density. [After Hodgkin, D. C., *Nature* **188**, 445 (1960).]

diffraction maxima (darkness of the spots on the film) are then used to construct mathematically the three-dimensional image of the crystal structure through methods that are outside the scope of this text. In what follows, we discuss some of the special problems associated with interpreting the X-ray crystal structures of proteins.

X-rays interact almost exclusively with the electrons in matter; not the nuclei. An X-ray structure is therefore an image of the electron density of the object under study. Such **electron density maps** may be presented as a series of parallel sections through the object. On each section, the electron density is represented by contours (Fig. 7-38a) in the same way that altitude is represented by the contours on a topographic map. A stack of such sections, drawn on transparencies, yields a three-dimensional electron density map (Fig. 7-38b). Modern structural analysis, however, is often carried out with the aid of graphics computers, on which electron density maps are contoured in three dimensions (Fig. 7-38c).

Protein Crystal Structures Exhibit Less Than Atomic Resolution

The molecules in protein crystals, as in other crystalline substances, are arranged in a regularly repeating three-dimensional lattices. Protein crystals, however, differ from those of most small organic and inorganic molecules in being highly hydrated; they are typically 40 to 60% water by volume. The aqueous solvent of crystallization is necessary for the structural integrity of the protein crystals as John D. Bernal and Dorothy Crowfoot Hodgkin first noted in 1934 when they carried out the original X-ray studies of protein crystals. This is because water is required for the structural integrity of native proteins themselves (Section 7-4).

The large solvent content of protein crystals gives them a soft, jellylike consistency so that their molecules lack the rigid order characteristic of crystals of small molecules such as NaCl or glycine. The molecules in a protein crystal are typically disordered by a few angstroms so that the corresponding electron density map lacks information concerning structural details of smaller size. The crystal is therefore said to have a resolution limit of that size. Protein crystals typically have resolution limits in the range 2 to 3.5 Å although a few are better ordered (have higher resolution) and many are less ordered (have lower resolution).

Since an electron density map of a protein must be interpreted in terms of its atomic positions, the accuracy and even the feasibility of a crystal structure analysis depends on the crystal's resolution limit. Figure 7-39 indicates how the quality (degree of focus) of an electron density map varies with its resolution limit. At 6-Å resolution, the presence of a molecule the size of diketopiperazine is difficult to discern. At 2.0-Å resolution, its individual atoms cannot yet be distinguished although its molecular shape has become reasonably evident. At 1.5-Å resolution, which roughly corresponds to a bond distance, individual atoms become partially resolved. At 1.1-Å resolution, atoms are clearly visible.

Most protein crystal structures are too poorly resolved for their electron density maps to reveal clearly the positions of individual atoms (e.g., Fig. 7-38). Nevertheless, the distinctive shape of the polypeptide backbone usually permits it to be traced, which, in turn, allows the positions and orientations of its side chains to be deduced (e.g., Fig. 7-38c). Yet, side chains of comparable size and shape, such as those of Leu, Ile, Thr, and Val, cannot be differentiated with a reasonable degree of confidence (hydrogen atoms, having but one electron, are not visible in protein X-ray structures), so that a

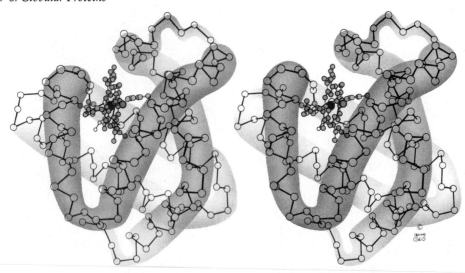

Figure 7-40

A computer-drawn stereo diagram of sperm whale myoglobin in which the C_α atoms are represented by balls and the peptide groups linking them are represented by solid bonds. The 153-residue polypeptide chain is folded into eight α helices (highlighted here by hand-drawn envelopes), connected by short polypeptide links. The protein's bound heme group (*purple*) in complex with an O_2 molecule (*orange sphere*) is shown together with its two closely associated His side chains (*light blue*). Hydrogen atoms have been omitted for the sake of clarity. Instructions for viewing stereo diagrams are given in the appendix to this chapter. [Figure copyrighted © by Irving Geis.]

protein structure cannot be elucidated from its electron density map alone. Rather, the primary structure of the protein must be known, thereby permitting the sequence of amino acid residues to be fitted, by eye, to its electron density map. Mathematical refinement can then reduce the errors in the crystal structure's atomic positions to around 0.1 Å (in contrast, the errors in small molecule X-ray structure determinations may be as little as 0.001 Å).

Most Crystalline Proteins Maintain Their Native Conformations

What is the relationship between the structure of a protein in a crystal and that in solution where most proteins normally function? Several lines of evidence indicate that *crystalline proteins assume very nearly the same structures that they have in solution:*

1. A protein molecule in a crystal is essentially in solution because it is bathed by solvent of crystallization over all of its surface except for the few, generally small patches that contact neighboring protein molecules.

2. A protein may crystallize in one of several forms or "habits," depending on crystallization conditions, that differ in how the protein molecules are arranged in space relative to each other. In the several cases when different crystal forms of the same protein have been independently analyzed, the molecules have virtually identical conformations. Evidently, crystal packing forces do not greatly affect the structures of protein molecules.

3. The most compelling evidence that crystalline proteins have biologically relevant structures, however, is the observation that many enzymes are catalytically active in the crystalline state. The catalytic activity of an enzyme is very sensitive to the relative orientations of the groups involved in binding and catalysis (Chapter 14). Active crystalline enzymes must therefore have conformations that closely resemble their solution conformations.

Protein Molecules Are Most Effectively Illustrated in Simplified Form

The several hundred nonhydrogen atoms of even a small protein makes understanding the detailed structure of a protein a considerable effort. The most instructive method of studying a protein structure is the hands-on examination of its skeletal (ball-and-stick) model. Unfortunately, such models are rarely available and photographs of them are too cluttered to be of much use. A practical alternative is a computer-generated stereo diagram in which the polypeptide backbone is represented only by its C_α atoms and only a few key side chains are included (Fig. 7-40). Another possibility is an artistic rendering of a protein model that has been simplified and slightly distorted to improve its visual clarity (Fig. 7-41). A further level of abstraction may be obtained by representing the protein in a cartoon form that emphasizes its secondary structure (Fig. 7-42; also see Fig. 7-19). Computer-generated drawings of space-filling models, such as Figs. 7-12 and 7-18, may also be employed to illustrate certain features of protein structures.

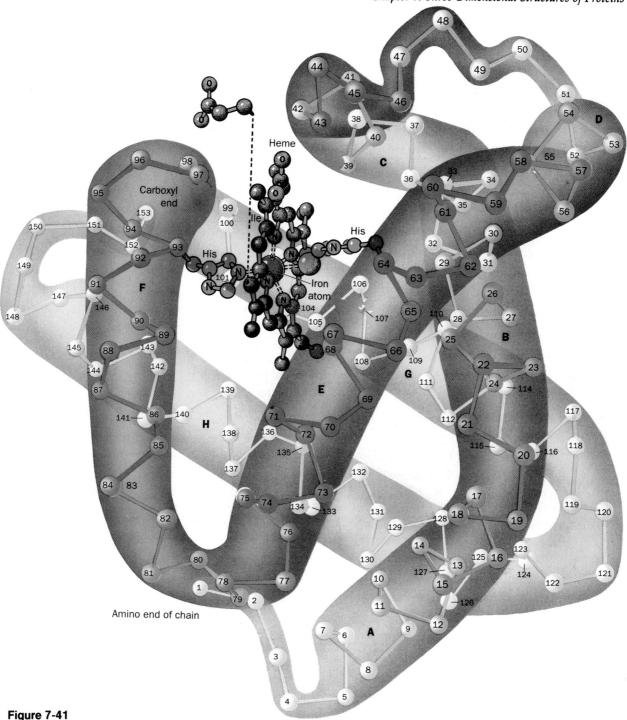

Figure 7-41
An artist's rendering of sperm whale myoglobin analogous to Fig. 7-40. One of the heme group's propionic acid side chains has been displaced for clarity. The amino acid residues are consecutively numbered, starting from the N-terminus, and the eight helices are likewise designated A through H. [Figure copyrighted © by Irving Geis.]

B. Tertiary Structure

The **tertiary structure (3° structure)** of a protein is its three-dimensional arrangement; that is, the folding of its 2° structural elements, together with the spatial dispositions of its side chains. The first protein X-ray structure, that of sperm whale myoglobin, was elucidated in the late 1950s by John Kendrew and coworkers. Its polypeptide chain follows such a tortuous, wormlike path (Figs. 7-40–7-42), that these investigators were moved to indicate their disappointment at its lack of regularity. In the intervening years, well over 200 protein structures have been reported. Each of them is a unique,

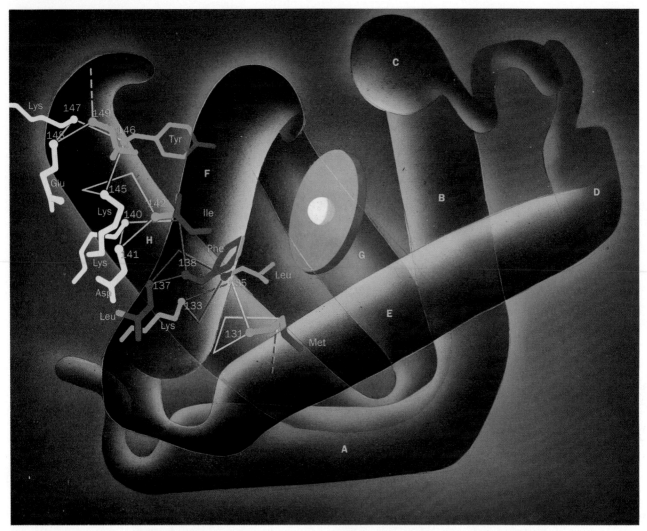

Figure 7-42
A cartoon of sperm whale myoglobin, oriented similarly to Figs. 7-40 and 7-41, which emphasizes the protein's α helical secondary structure (*cylinders*). The pink disk with its central white sphere represents the protein's associated heme group with its bound iron atom. Many of the H helix side chains are shown with polar and nonpolar groups, respectively, colored blue and red. [Figure copyrighted © by Irving Geis.]

highly complicated entity. Nevertheless, their tertiary structures have several outstanding features in common as we shall see below.

Globular Proteins May Contain Both α Helices and β Sheets

The major types of secondary structural elements, α helices and β pleated sheets, commonly occur in proteins but in varying proportions and combinations. Some proteins, such as myoglobin, consist only of α helices spanned by short connecting links that have a coil conformation (Fig. 7-42). Others, such as concanavalin A, have a large proportion of β sheets but are

devoid of α helices (Fig. 7-43). Most proteins, however, have significant amounts of both types of secondary structure (on average, ~27% α helix and 23% β sheet). Human **carbonic anhydrase** (Fig. 7-44) as well as carboxypeptidase and triose phosphate isomerase (Fig. 7-19) are examples of such proteins.

Side Chain Location Varies with Polarity

The primary structures of globular proteins generally lack the repeating or pseudorepeating sequences that are responsible for the regular conformations of fibrous proteins. The amino acid side chains in globular proteins are, nevertheless, spatially distributed according to their polarities:

1. *The nonpolar residues Val, Leu, Ile, Met, and Phe nearly always occur in the interior of a protein, out of contact with the aqueous solvent.* The hydrophobic interactions that promote this distribution, which are largely responsible for the three-dimensional structures of native proteins, are further discussed in Section 7-4C.

2. *The charged polar residues Arg, His, Lys, Asp, and Glu are almost invariably located on the surface of a protein in contact with the aqueous solvent.* This is because the immersion of an ion in the virtually anhydrous interior of a protein results in the uncompensated loss of much of its hydration energy. In the rare instances that these groups occur in the interior of a protein, they usually have a specific chemical function such as promoting catalysis or participating in metal ion binding (e.g., the metal ion-liganding His residues in Figs. 7-41 and 7-44).

3. *The uncharged polar groups Ser, Thr, Asn, Gln, Tyr, and Trp, are usually on the protein surface but frequently occur in the interior of the molecule.* In the latter case, these residues are almost always hydrogen bonded to other groups in the protein. In fact, *virtually all buried hydrogen bond donors form hydrogen bonds with buried acceptor groups;* in a sense, the formation of a hydrogen bond "neutralizes" the polarity of a hydrogen bonding group.

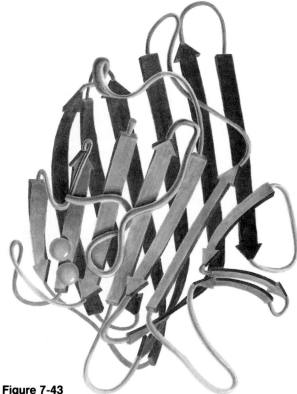

Figure 7-43
The jack bean protein concanavalin A largely consists of extensive regions of antiparallel β pleated sheet, here represented by arrows pointing towards the polypeptide chain's C-terminus. The balls represent protein-bound metal ions. The bottom sheet is shown in a space-filling representation in Fig. 7-18. [After a drawing by Jane Richardson, Duke University.]

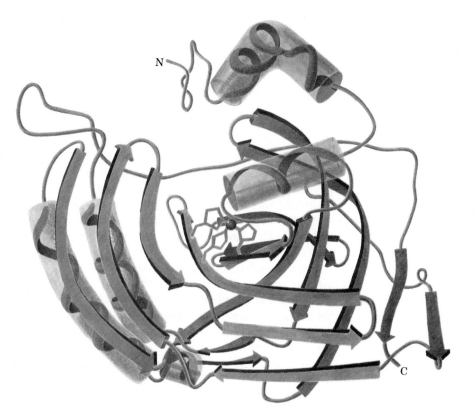

Figure 7-44
Human carbonic anhydrase in which α helices are represented as cylinders and each strand of β sheet is drawn as an arrow pointing towards the polypeptide's C-terminus. The gray ball in the middle represents a Zn^{2+} ion that is coordinated by three His side chains (*blue*). Note that the C-terminus is tucked through the plane of a surrounding loop of polypeptide chain so that carbonic anhydrase is one of the rare native proteins in which a polypeptide chain forms a knot. [After Kannan, K. K., Liljas, A., Waara, I., Bergsten, P.-C., Lovgren, S., Strandberg, B., Bengtsson, J., Carlbom, U., Friedborg, K., Jarup, L., and Petef, M., *Cold Spring Harbor Symp. Quant. Biol.* **36,** 221 (1971).]

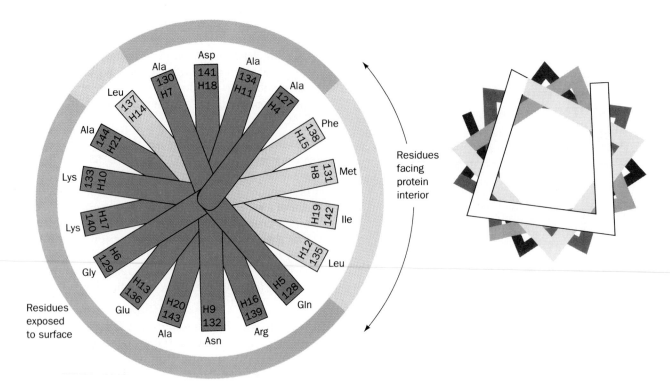

Figure 7-45
A **helical wheel** representation of the sperm whale myoglobin H helix in which side chain positions about the α helix are projected down the helix axis onto a plane. Here each residue is identified both according to its sequence in the polypeptide chain, and according to its position in the H helix. The residues lining the side of the helix facing the protein's interior regions are all nonpolar. The other residues, except Leu 137, which contacts the protein segment linking helices E and F (Figs. 7-40 and 7-41), are exposed to the solvent and are all more or less polar. Compare this diagram with the drawing of the H helix in Fig. 7-42.

This side chain distribution is apparent in Figs. 7-42 and 7-45, which show the surface and interior exposures of the amino acid side chains of myoglobin's H helix.

Globular Protein Cores Are Efficiently Arranged

Globular proteins are quite compact; there is very little space inside them so that water is largely excluded from their interiors. The micellelike arrangement of their side chains (polar groups outside, nonpolar groups inside) has led to their description as "oil drops with polar coats." This generalization, although picturesque, lacks precision. The **packing density** (ratio of the volume enclosed by the van der Waals envelopes of the atoms in a region to the total volume of the region) of the internal regions of globular proteins averages ~0.75, which is in the same range as that of molecular crystals of small organic molecules. In comparison, equal-sized close-packed spheres have a packing density of 0.74, whereas organic liquids (oil drops) have packing densities that are mostly between 0.60 and 0.70. *The interior of a protein is therefore more like a molecular crystal than an oil drop; that is, it is efficiently packed.*

Large Polypeptides Form Domains

Polypeptide chains that consist of more than ~200 residues usually fold into two or more globular clusters known as **domains,** which give these proteins a bi- or multilobal appearance. Most domains consist of 100 to 200 amino acid residues and have an average diameter of ~25 Å. Each subunit of **glyceraldehyde-3-phosphate dehydrogenase,** for example, has two distinct domains (Fig. 7-46). A polypeptide chain wanders back and forth within a domain but neighboring domains are usually connected by one, or less commonly two, polypeptide segments. *Domains are therefore structurally independent units that each have the characteristics of a small globular protein.* Indeed, limited proteolysis of a multidomain protein often liberates its domains without greatly altering their structures. Nevertheless, the domain structure of a protein is not always obvious since its domains may make such extensive contacts with each other that the protein appears to be a single globular entity.

Domains often have a specific function such as the binding of a small molecule. In Fig. 7-46, for example, **nicotinamide adenine dinucleotide (NAD⁺)** binds to the first domain of glyceraldehyde-3-phosphate dehydrogenase. Small molecule-binding sites in multidomain proteins often occur in the clefts between domains; that is, the small molecules are bound by groups from two domains. This arrangement arises, in part, from the need for a flexible interaction between the

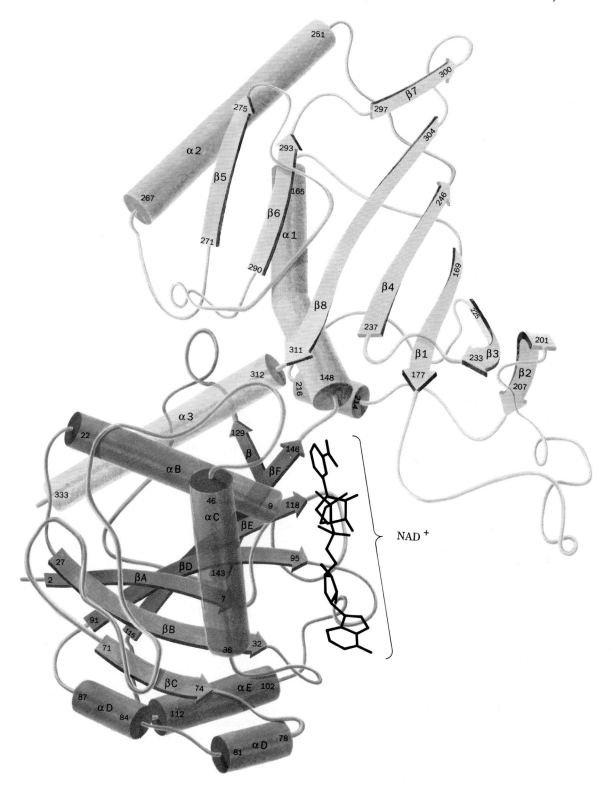

Figure 7-46
One subunit of the enzyme glyceraldehyde-3-phosphate dehydrogenase from *Bacillus stearothermophilus.* The polypeptide folds into two distinct domains. The first domain (*red,* residues 1–146) binds NAD⁺ (*black*) near the C-terminal ends of its parallel β strands, and the second domain (*green*) binds glyceraldehyde-3-phosphate (not shown). [After Biesecker, G., Harris, J. I., Thierry, J. C., Walker, J. E., and Wonacott, A., *Nature* **266,** 331 (1977).]

protein and the small molecule that the relatively pliant covalent connection between the domains can provide.

Supersecondary Structures Have Structural and Functional Roles

Certain groupings of secondary structural elements, named **supersecondary structures,** occur in many unrelated globular proteins:

1. The most common form of supersecondary structure is the **βαβ unit** in which the usually right-handed cross-over connection between two consecutive parallel strands of a β sheet consists of an α helix (Fig. 7-47*a*).

2. Another common supersecondary structure, the **β meander,** consists of an antiparallel β sheet formed by sequential segments of polypeptide chain that are connected by relatively tight reverse turns (Fig. 7-47*b*).

3. In an **αα unit,** two successive antiparallel α helices pack against each other with their axes inclined so as to permit energetically favorable intermeshing of their contacting side chains (Fig. 7-47*c*). Such associations also stabilize the coiled coil conformation of α keratin (Section 7-2A).

4. Extended β sheets are often rolled up to form β barrels (e.g., Fig. 7-19*b*). Three different β barrel topologies (the ways in which the strands and their interconnections are arranged) have been named in analogy with geometric motifs found on Native American and Greek weaving and pottery (Fig. 7-48).

Supersecondary structures may have functional as well as structural significance. A *βαβαβ* unit, for example, in which the β strands form a parallel sheet with right-handed α helical cross-over connections (a double

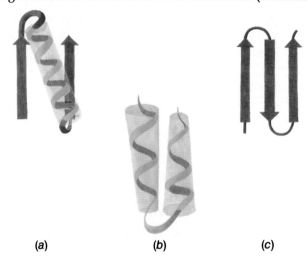

Figure 7-47
Schematic diagrams of (*a*) a βαβ unit, (*b*) a β meander, and (*c*) an αα unit.

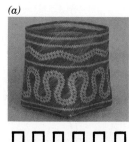

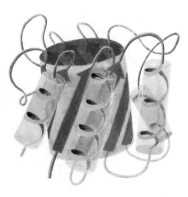

Figure 7-48
Comparisons of the backbone folding patterns of protein β barrels (*right*) with geometric motifs commonly used to decorate Native American and Greek weaving and pottery (*left*). (*a*) Native American polychrome cane basket and the polypeptide backbone of **rubredoxin** from *Clostridium pasteurianum* showing its linked β meanders. [Museum of the American Indian, Heye Foundation.] (*b*) Red figured Greek amphora with its Greek key border area showing Cassandra and Ajax (about 450 B.C.) and the polypeptide backbone of human **prealbumin** with its "Greek key" pattern. [The Metropolitan Museum of Art, Fletcher Fund, 1956.] (*c*) Early Anasazi redware pitcher from New Mexico and the polypeptide backbone of chicken muscle triose phosphate isomerase showing its "lightning" pattern of overlapping βαβ units. This β barrel is also diagrammed in Fig. 7-19*b*. [Museum of the American Indian, Heye Foundation.] [After Richardson, J. S., *Nature* **268,** 498 (1977).]

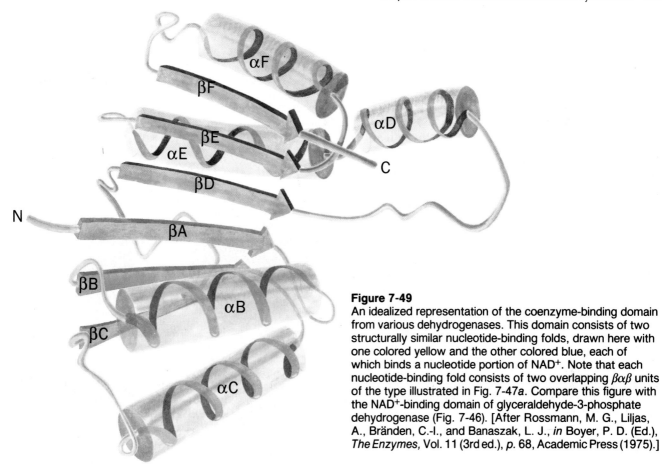

Figure 7-49

An idealized representation of the coenzyme-binding domain from various dehydrogenases. This domain consists of two structurally similar nucleotide-binding folds, drawn here with one colored yellow and the other colored blue, each of which binds a nucleotide portion of NAD⁺. Note that each nucleotide-binding fold consists of two overlapping $\beta\alpha\beta$ units of the type illustrated in Fig. 7-47a. Compare this figure with the NAD⁺-binding domain of glyceraldehyde-3-phosphate dehydrogenase (Fig. 7-46). [After Rossmann, M. G., Liljas, A., Bränden, C.-I., and Banaszak, L. J., *in* Boyer, P. D. (Ed.), *The Enzymes*, Vol. 11 (3rd ed.), *p.* 68, Academic Press (1975).]

$\beta\alpha\beta$ unit), was shown by Michael Rossmann to form the nucleotide-binding site in many enzymes that bind nucleotides (Fig. 7-49). In some enzymes, the second α helix of this **nucleotide-binding fold** or **Rossmann fold** is replaced by a length of nonhelical polypeptide. This occurs, for example, between the βE and βF strands of glyceraldehyde-3-phosphate dehydrogenase (Fig. 7-46).

4. PROTEIN STABILITY

Incredible as it may seem, thermodynamic measurements indicate that *native proteins are only marginally stable entities under physiological conditions.* The free energy required to denature them is ~0.4 kJ·mol⁻¹ of amino acid residues so that 100-residue proteins are typically stable by only around 40 kJ·mol⁻¹. In contrast, the energy required to break a typical hydrogen bond is ~20 kJ·mol⁻¹. The various noncovalent influences to which proteins are subject—electrostatic interactions (both attractive and repulsive), hydrogen bonding (both intramolecular and to water), and hydrophobic forces

—each have energetic magnitudes that may total thousands of kilojoules per mole over an entire protein molecule. Consequently, *a protein structure is the result of a delicate balance among powerful countervailing forces.* In this section we discuss the nature of these forces and end by considering protein denaturation; that is, how these forces can be disrupted.

A. Electrostatic Forces

Molecules are collections of electrically charged particles and hence, to a reasonable degree of approximation, their interactions are determined by the laws of classical electrostatics (more exact calculations require the application of quantum mechanics). The energy of association, U, of two electric charges, q_1 and q_2, that are separated by the distance r, is found by integrating the expression for Coulomb's law, Eq. [2.1], to determine the work necessary to separate these charges by an infinite distance:

$$U = \frac{kq_1q_2}{Dr} \qquad [7.1]$$

Here $k = 9.0 \times 10^9$ J·m·C⁻² and D is the dielectric constant of the medium in which the charges are immersed (recall that $D = 1$ for a vacuum and, for the most part, increases with the polarity of the medium; Table 2-1). The dielectric constant of a molecule-sized region is difficult to estimate. For the interior of a protein, it is usually taken to be in the range 2 to 5 in analogy with the measured dielectric constants of substances that have similar polarities such as benzene and diethyl ether.

Ionic Interactions Are Strong but Do Not Greatly Stabilize Proteins

The association of two ionic protein groups of opposite charge is known as an **ion pair** or **salt bridge**. According to Eq. [7.1], the energy of a typical ion pair, say the carboxyl group of Glu and the ammonium group of Lys, whose charge centers are separated by 4.0 Å in a medium of dielectric constant 4, is -86 kJ·mol^{-1} (one electronic charge $= 1.60 \times 10^{-19}$ C). Free ions in aqueous solution are highly solvated, however, so that the free energy of solvation of two separated ions is about equal to the free energy of formation of their unsolvated ion pair. *Ion pairs therefore contribute little stability towards a protein's native structure.* This accounts for the observations that buried (unsolvated) ion pairs rarely occur in proteins and that ion pairs that are exposed to the aqueous solvent (reasonably common in proteins) are but poorly conserved among homologous proteins.

Dipole–Dipole Interactions Are Weak but Significantly Stabilize Protein Structures

The noncovalent associations between electrically neutral molecules, collectively known as **van der Waals forces,** arise from electrostatic interactions among permanent and/or induced dipoles. These forces are responsible for numerous interactions of varying strengths between nonbonded neighboring atoms. (The hydrogen bond, a special class of dipolar interaction, is considered separately in Section 7-4B.)

The interactions among permanent dipoles are important structural determinants in proteins because many of their groups, such as the carbonyl and amide groups of the peptide backbone, have permanent dipole moments. These interactions are generally much weaker than the charge–charge interactions of ion pairs. Two carbonyl groups, for example, each with dipoles of 4.2×10^{-30} C·m (1.3 debye units) that are oriented in an optimal head-to-tail arrangement (Fig. 7-50a) and separated by 5 Å in a medium of dielectric constant 4, have a calculated attractive energy of only -9.3 kJ·mol^{-1}. Furthermore, these energies vary with r^{-3} so that they rapidly attenuate with distance. In α helices and β sheets, however, the dipolar amide and carbonyl groups of the polypeptide backbone all point in the same direction (Figs. 7-11 and 7-16) so that their interactions are associative and tend to be additive

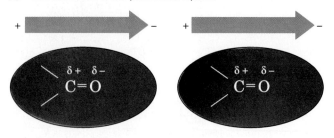

(a) Interactions between permanent dipoles

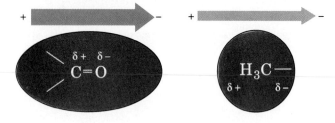

(b) Dipole-induced dipole interactions

(c) London dispersion forces

Figure 7-50
Dipole–dipole interactions. The strength of each dipole is represented by the thickness of the accompanying arrow. (*a*) Interactions between permanent dipoles. These interactions, here represented by carbonyl groups lined up head to tail, may be attractive, as shown here, or repulsive, depending on the relative orientations of the dipoles. (*b*) Dipole-induced dipole interactions. A permanent dipole (here shown as a carbonyl group) induces a dipole in a nearby group (here represented by a methyl group) by electrostatically distorting its electron distribution (*shading*). This always results in an attractive interaction. (*c*) London dispersion forces. The instantaneous charge imbalance (*shading*) resulting from the motions of the electrons in a molecule (*left*) induce a dipole in a nearby group (*right*); that is, the motions of the electrons in neighboring groups are correlated. This always results in an attractive interaction.

(these groups, of course, also form hydrogen bonds but here we are concerned with their residual electric fields). Consequently, *in the low dielectric constant core of a protein, dipole–dipole interactions significantly influence protein folding.*

A permanent dipole also induces a dipole moment on a neighboring group so as to form an attractive interaction (Fig. 7-50b). Such dipole–induced dipole interactions are generally much weaker than are dipole–dipole interactions.

Although nonpolar molecules are nearly electrically

neutral at any instant, they have a small dipole moment resulting from the rapid fluctuating motion of their electrons. This transient dipole moment polarizes the electrons in a neighboring group thereby giving rise to a dipole moment (Fig. 7-50c) such that, near their van der Waals contact distances, the groups are attracted to one another (a quantum mechanical effect that really cannot be explained in terms of only classical physics). These so-called **London dispersion forces** are extremely weak. The 8.2-$kJ \cdot mol^{-1}$ heat of vaporization of CH_4, for example, indicates that the attractive interaction of a nonbonded $H \cdot \cdot \cdot H$ contact between neighboring CH_4 molecules is roughly $-0.3 \ kJ \cdot mol^{-1}$ (in the liquid, a CH_4 molecule touches its 12 nearest neighbors with $\sim 2 \ H \cdot \cdot \cdot H$ contacts apiece).

London forces are only significant for contacting groups because their association energy is proportional to r^{-6}. Nevertheless, *the great numbers of interatomic contacts in protein makes London forces a major influence in determining their conformations.* London forces also provide much of the binding energy in the sterically complementary interactions between proteins and the molecules that they specifically bind.

B. Hydrogen Bonding Forces

Hydrogen bonds $(D—H \cdot \cdot \cdot A)$, as we discussed in Section 2-1A, are predominantly electrostatic interactions between a weakly acidic donor group $(D—H)$ and an acceptor atom (A) that bears a lone pair of electrons. In biological systems, D and A can both be the highly electronegative N and O atoms and occasionally S atoms. Hydrogen bonds, which have association energies in the range -12 to $-30 \ kJ \cdot mol^{-1}$, are much more directional than are van der Waals forces although less so than are covalent bonds. The $D \cdot \cdot \cdot A$ distance is normally in the range 2.7 to 3.1 Å. Hydrogen bonds tend to be linear with the D—H bond pointing along the acceptor's lone pair orbital. Large deviations from this ideal geometry are not unusual, however. For example, in the hydrogen bonds of both α helices (Fig. 7-11) and antiparallel β pleated sheets (Fig. 7-16a), the N—H bonds point approximately along the C=O bonds rather than along an O lone pair orbital, and in parallel β pleated sheets (Fig. 7-16b), the hydrogen bonds depart significantly from linearity.

The internal hydrogen bonding groups of a protein are arranged such that nearly all possible hydrogen bonds are formed (Section 7-3B). Clearly, hydrogen bonding has a major influence on the structures of proteins. An unfolded protein, however, makes all its hydrogen bonds with the water molecules of the aqueous solvent (water, it will be recalled, is a strong hydrogen bonding donor and acceptor). The free energy of stabilization that internal hydrogen bonds confer upon a native protein is therefore equal to the difference in the free energy of hydrogen bonding between the native protein and the unfolded protein. Since the various hydrogen bonds in question, to a first approximation, all have the same free energy, *internal hydrogen bonding cannot significantly stabilize, and, indeed, might even slightly destabilize, the structure of a native protein relative to its unfolded state.* This idea is corroborated by the observation that proteins are more readily denatured in aqueous solutions containing low dielectric constant organic solvents such as ethanol or acetone than they are in the absence of such solvents. Since the electrostatic nature of hydrogen bonding associations makes them stronger as the dielectric constant of the medium is lowered, this is opposite to the result that would be expected if internal hydrogen bonding was the major influence in stabilizing protein structures.

The internal hydrogen bonds of a protein, nevertheless, provide a structural basis for its native folding pattern: If a protein folded in a way that prevented some of its internal hydrogen bonds from forming, their free energy would be lost and such conformations would be less stable than those that are fully hydrogen bonded. This argument also applies to the van der Waals forces discussed in the previous section.

C. Hydrophobic Forces

The **hydrophobic effect** *is the name given to those influences that cause nonpolar substances to minimize their contacts with water, and amphipathic molecules, such as soaps and detergents, to form micelles in aqueous solutions (Section 2-1B).* Since native proteins form a sort of intramolecular micelle in which most of their nonpolar side chains are out of contact with the aqueous solvent, *hydrophobic interactions must be an important determinant of protein structures.*

The hydrophobic effect derives from the special properties of water as a solvent, only one of which is its high dielectric constant. In fact, other polar solvents, such as dimethylsulfoxide (DMSO) and *N,N*-dimethylformamide (DMF), tend to denature proteins. The thermodynamic data of Table 7-4 provide considerable insight as to the origin of the hydrophobic effect because the transfer of a hydrocarbon from water to a nonpolar solvent resembles the transfer of a nonpolar side chain from the exterior of a protein in aqueous solution to its interior. The isothermal Gibbs free energy changes $(\Delta G = \Delta H - T\Delta S)$ for the transfer of a hydrocarbon from an aqueous solution to a nonpolar solvent is negative in all cases, which indicates, as we know to be the case, that such transfers are spontaneous processes (oil and water do not mix). What is perhaps unexpected is that these transfer processes are endothermic (positive ΔH) for aliphatic compounds and athermic $(\Delta H = 0)$ for aromatic compounds; that is, *it is enthalpically more or equally favorable for nonpolar molecules to dissolve in*

Table 7-4

Thermodynamic Changes for Transferring Hydrocarbons from Water to Nonpolar Solvents at 25°C[a]

Process	ΔH (kJ·mol^{-1})	$-T\Delta S_u$ (kJ·mol^{-1})	ΔG_u (kJ·mol^{-1})
CH$_4$ in H$_2$O $\rightleftharpoons$ CH$_4$ in C$_6$H$_6$	11.7	−22.6	−10.9
CH$_4$ in H$_2$O $\rightleftharpoons$ CH$_4$ in CCl$_4$	10.5	−22.6	−12.1
C$_2$H$_6$ in H$_2$O $\rightleftharpoons$ C$_2$H$_6$ in benzene	9.2	−25.1	−15.9
C$_2$H$_4$ in H$_2$O $\rightleftharpoons$ C$_2$H$_4$ in benzene	6.7	−18.8	−12.1
C$_2$H$_2$ in H$_2$O $\rightleftharpoons$ C$_2$H$_2$ in benzene	0.8	−8.8	−8.0
Benzene in H$_2$O $\rightleftharpoons$ liquid benzene[b]	0.0	−17.2	−17.2
Toluene in H$_2$O $\rightleftharpoons$ liquid toluene[b]	0.0	−20.0	−20.0

[a] ΔG_u, the **unitary Gibbs free energy change**, is the Gibbs free energy change, ΔG, corrected for its concentration dependence so that it reflects only the inherent properties of the substance in question and its interaction with solvent. This relationship, according to Eq. [3.13], is

$$\Delta G_u = \Delta G - nRT \ln \frac{[A_f]}{[A_i]}$$

where $[A_i]$ and $[A_f]$ are the initial and final concentrations of the substance under consideration, respectively, and n is the number of moles of that substance. Since the second term in this equation is a purely entropic term (concentrating a substance increases its order), ΔS_u, the **unitary entropy change**, is expressed

$$\Delta S_u = \Delta S + nR \ln \frac{[A_f]}{[A_i]}$$

[b] Data measured at 18°C.

Source: Kauzmann, W., *Adv. Protein Chem.* **14**, 39 (1959).

water than in nonpolar media. In contrast, the entropy component of the unitary free energy change, $-T\Delta S_u$ (see footnote *a* to Table 7-4), is large and negative in all cases. Clearly, *the transfer of a hydrocarbon from an aqueous medium to a nonpolar medium is entropically driven. The same is true of the transfer of a nonpolar protein group from an aqueous environment to the protein's nonpolar interior.*

What is the physical mechanism whereby nonpolar entities are excluded from aqueous solution? Recall that entropy is a measure of the order of a system; it decreases with increasing order (Section 3-2). Thus the decrease in entropy when a nonpolar molecule or side chain is solvated by water (the reverse of the foregoing process) must be due to an ordering process. This is an experimental observation, not a theoretical conclusion. The magnitudes of the entropy changes are too large to be attributed to changes in the conformations of the hydrocarbons; rather, as Henry Frank and Marjorie Evans pointed out in 1945, *these entropy changes can only arise from some sort of ordering of the water structure.*

Liquid water has a highly ordered and extensively hydrogen bonded structure (Section 2-1A). The insinuation of a nonpolar group into this structure disrupts it: A nonpolar group can neither accept nor donate hydrogen bonds so that the water molecules at the surface of the cavity occupied by the nonpolar group cannot hydrogen bond to other molecules in their usual fashion. In order to recover the lost hydrogen bonding energy, these sur-

face waters must orient themselves so as to form a hydrogen bonded network enclosing the cavity (Fig. 7-51). This orientation constitutes an ordering of the water structure since the number of ways that water molecules can form hydrogen bonds about the surface of a nonpolar group is less than the number of ways that they can hydrogen bond in bulk water.

Unfortunately, the complexity of liquid water's basic structure (Section 2-1A) has not yet allowed a detailed structural description of this ordering process. One model that has been proposed is that water forms quasi-crystalline hydrogen bonded cages about the nonpolar groups similar to those of **clathrates** (Fig. 7-52). The magnitudes of the entropy changes that result when nonpolar substances are dissolved in water, however, indicate that the resulting water structures can only be slightly more ordered than bulk water. They also must be quite different from that of ordinary ice, because, for instance, the solvation of nonpolar groups by water causes a large decrease in water volume (e.g., the transfer of CH$_4$ from hexane to water shrinks the water solution by 22.7 mL/mol of CH$_4$), whereas the freezing of water results in a 1.6-mL/mol expansion.

The unfavorable free energy of hydration of a nonpolar substance caused by its ordering of the surrounding water molecules has the net result that *the nonpolar substance is excluded from the aqueous phase.* This is because the surface area of a cavity containing an aggregate of nonpolar molecules is less than the sum of the surface

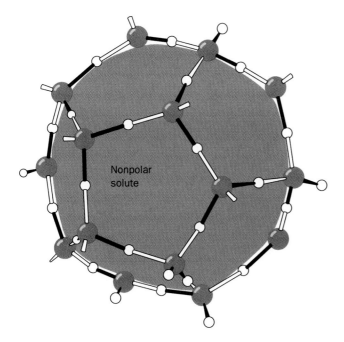

Figure 7-51
The orientational preference of water molecules next to a nonpolar solute. In order to maximize their hydrogen bonding energy, these water molecules tend to straddle the inert solute such that two or three of their tetrahedral directions are tangential to its surface. This permits them to form hydrogen bonds with neighboring water molecules lining the nonpolar surface. This ordering of water molecules extends several layers of water molecules beyond the first hydration shell of the nonpolar solute.

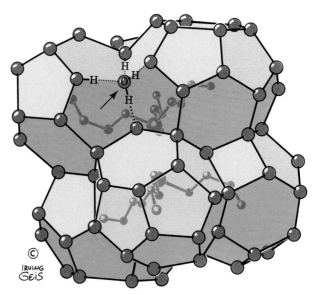

Figure 7-52
The structure of the clathrate (n-C_4H_9)$_3$S$^+$F$^-$·23H$_2$O. Clathrates are crystalline complexes of nonpolar compounds with water (usually formed at low temperatures and high pressures) in which the nonpolar molecules are enclosed, as shown, by a polyhedral cage of tetrahedrally hydrogen bonded water molecules (here represented by only their oxygen atoms). The hydrogen bonding interactions of one such water molecule (*arrow*) are shown in detail. [Figure copyrighted © by Irving Geis.]

areas of the cavities that each of these molecules would individually occupy. The aggregation of the nonpolar groups thereby minimizes the surface area of the cavity and therefore the entropy loss of the entire system. In a sense, the nonpolar groups are squeezed out of the aqueous phase by the hydrophobic interactions. Thermodynamic measurements indicate that the free energy change of removing a —CH_2— group from an aqueous solution is about -3 kJ·mol^{-1}. Although this is a relatively small amount of free energy, *in molecular assemblies involving large numbers of nonpolar contacts, hydrophobic interactions are a potent force.*

Walter Kauzmann pointed out in the 1950s that *hydrophobic forces are a major influence in causing proteins to fold into their native conformations.* Figure 7-53 indicates that the amino acid side chain **hydropathies** (indexes of combined hydrophobic and hydrophilic tendencies; Table 7-5) are, in fact, good predictors of which portions of a polypeptide chain are inside a protein, out of contact with the aqueous solvent, and which portions are outside, in contact with the aqueous solvent. In proteins, the effects of hydrophobic forces are often termed **hydrophobic bonding**, presumably to indicate the specific nature of protein folding under the influence of the hydrophobic effect. You should keep in mind, however,

Table 7-5

Hydropathy Scale for Amino Acid Side Chains

Side Chain	Hydropathy
Ile	4.5
Val	4.2
Leu	3.8
Phe	2.8
Cys	2.5
Met	1.9
Ala	1.8
Gly	-0.4
Thr	-0.7
Ser	-0.8
Trp	-0.9
Tyr	-1.3
Pro	-1.6
His	-3.2
Glu	-3.5
Gln	-3.5
Asp	-3.5
Asn	-3.5
Lys	-3.9
Arg	-4.5

Source: Kyte, J. and Doolittle, R. F., *J. Mol. Biol.* **157,** 110 (1982).

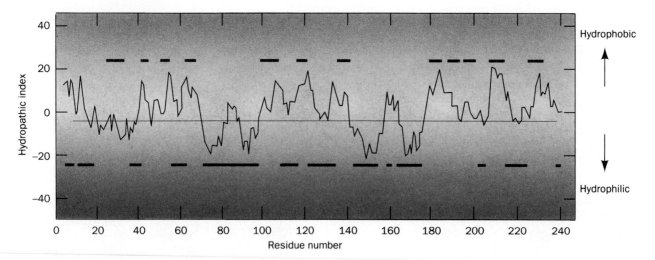

Figure 7-53
The hydropathic index (sum of the hydropathies of nine consecutive residues; see Table 7-5) versus the residue sequence number for bovine **chymotrypsinogen.** A large positive hydropathic index is indicative of a hydrophobic region of the polypeptide chain, whereas a large negative value is indicative of a hydrophilic region. The bars above the midpoint line denote the protein's interior regions, as determined by X-ray crystallography, and the bars below the midpoint line indicate the protein's exterior regions. [After Kyte, J. and Doolittle, R. F., *J. Mol. Biol.* **157**, 111 (1982).]

that hydrophobic bonding does not generate the directionally specific interactions usually associated with the term "bond."

D. Disulfide Bonds

Since disulfide bonds form as a protein folds to its native conformation (Section 8-1B), they function to stabilize its three-dimensional structure. The relatively reducing chemical character of the cytoplasm, however, greatly diminishes the stability of intracellular disulfide bonds. In fact, almost all proteins with disulfide bonds are secreted to more oxidized extracellular destinations where their disulfide bonds are effective in stabilizing protein structures. Apparently, the relative "hostility" of extracellular environments toward proteins (e.g., uncontrolled temperatures and pH's) requires the additional structural stability conferred by disulfide bonds.

E. Protein Denaturation

The low conformational stabilities of native proteins makes them easily susceptible to denaturation by altering the balance of the weak nonbonding forces that maintain the native conformation. When a protein in solution is heated, its conformationally sensitive properties, such as optical rotation (Section 4-2A), viscosity, and UV absorption, change abruptly over a narrow temperature range (Fig. 7-54). Such a nearly discontinuous change indicates that *the native protein structure unfolds in a cooperative manner: Any partial unfolding of the structure destabilizes the remaining structure, which must simultaneously collapse to the random coil.* The temperature at the midpoint of this process is known as the

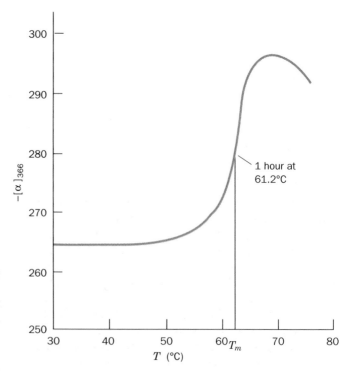

Figure 7-54
The optical rotation, at 366 nm, as a function of temperature, of bovine pancreatic ribonuclease A (RNase A) in 0.15M KCl and 0.013M sodium cacodylate buffer, pH 7. The melting temperature, T_m, is defined as the midpoint of the transition. [After von Hippel, P. H. and Wong, K. Y., *J. Biol. Chem.* **10**, 3911 (1965).]

protein's **melting temperature,** T_m, in analogy with the melting of a solid. Most proteins have T_m values well below 100°C. Among the exceptions to this generalization, however, are the proteins of **thermophilic bacteria,** organisms that inhabit hot springs with temperatures approaching 100°C. Interestingly, the X-ray structures of these heat stable proteins are but subtly different from those of their normally stable homologs.

In addition to high temperatures, proteins are denatured by a variety of other conditions and substances:

1. pH variations alter the ionization states of amino acid side chains (Table 4-2), which changes protein charge distributions and hydrogen bonding requirements.

2. Detergents, some of which significantly perturb protein structures at concentrations as low as $10^{-6}M$, hydrophobically associate with the nonpolar residues of a protein, thereby interfering with the hydrophobic interactions responsible for the protein's native structure.

3. High concentrations of water-soluble organic substances, such as aliphatic alcohols, interfere with the hydrophobic forces stabilizing protein structures through their own hydrophobic interactions with water. Organic substances with several hydroxyl groups, such as ethylene glycol or sucrose,

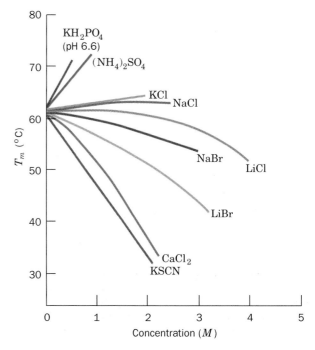

Sucrose

Ethylene glycol

however, are relatively poor denaturants because their hydrogen bonding ability renders them less disruptive of water structure.

The influence of salts is more variable. Figure 7-55 shows the effects of a number of salts on the T_m of bovine pancreatic **ribonuclease A (RNase A).** Some salts, such as $(NH_4)_2SO_4$ and KH_2PO_4, stabilize the native protein structure (raise its T_m); others, such as KCl and NaCl, have little effect, and yet others, such as KSCN and LiBr, destabilize it. The order of effectiveness of the various ions in stabilizing a protein, which is largely independent of the identity of the protein, parallels their capacity to salt out proteins (Section 5-2A). This order is known as the **Hofmeister series:**

Anions $SO_4^{2-} > H_2PO_4^- > CH_3COO^- > Cl^- > Br^- >$
$\qquad I^- > ClO_4^- > SCN^-$

Cations $NH_4^+, Cs^+, K^+, Na^+ > Li^+ > Mg^{2+} > Ca^{2+} >$
$\qquad Ba^{2+}$

Figure 7-55
The melting temperature of RNase A as a function of the concentrations of various salts. All solutions also contain $0.15M$ KCl and $0.013M$ sodium cacodylate buffer, pH 7. [After von Hippel, P. J. and Wong, K. Y., *J. Biol. Chem.* **10,** 3913 (1965).]

The ions in the Hofmeister series that tend to denature proteins, I^-, ClO_4^-, SCN^-, Li^+, Mg^{2+}, Ca^{2+}, and Ba^{2+}, are said to be **chaotropic.** This list should also include the guanidinium ion (Gu^+) and the nonionic urea,

$$H_2N-\overset{\overset{\displaystyle O}{\|}}{C}-NH_2 \qquad\qquad H_2N-\overset{\overset{\displaystyle NH_2^+}{\|}}{C}-NH_2$$

Urea   **Guanidinium ion**

which in concentrations in the range 5 to $10M$, are the most commonly used protein denaturants. The effect of the various ions on proteins is largely cumulative: GuSCN is a much more potent denaturant than the often used GuCl, whereas Gu_2SO_4 stabilizes protein structures.

Chaotropic agents increase the solubility of nonpolar substances in water. Consequently, their effectiveness as denaturing agents stems from their abilities to disrupt hydrophobic interactions although the manner in which they do so is not well understood. Conversely, those substances listed that stabilize proteins strengthen hydrophobic forces thus increasing the tendency of water to expel proteins. This accounts for the correlation between the abilities of an ion to stabilize proteins and to salt them out.

5. QUATERNARY STRUCTURE

Proteins, because of their multiple polar and nonpolar groups, stick to almost anything; anything, that is, but other proteins. This is because the forces of evolution have arranged the surface groups of proteins so as to prevent their association under physiological conditions. If this were not the case, their resulting nonspecific aggregation would render proteins functionally useless (recall, e.g., the consequences of sickle-cell anemia; Section 6-3A). In his pioneering ultracentrifugational studies on proteins, however, The Svedberg discovered that some proteins are composed of more than one polypeptide chain. Subsequent studies established that this is, in fact, true of most proteins, including nearly all those with molecular masses > 100 kD. Furthermore, these polypeptide **subunits** associate in a geometrically specific manner. The spatial arrangement of these subunits is known as a protein's **quaternary structure (4° structure).**

There are several reasons why multisubunit proteins are so common. In large assemblies of proteins, such as collagen fibrils, the advantages of subunit construction over the synthesis of one huge polypeptide chain are analogous to those of using prefabricated components in constructing a building. Defects can be repaired by simply replacing the flawed subunit, the site of subunit manufacture can be different from the site of assembly into the final product, and the only genetic information necessary to specify the entire edifice is that specifying its few different self-assembling subunits. In the case of enzymes, increasing a protein's size tends to better fix the three-dimensional positions of the groups forming the enzyme's active site. Increasing the size of an enzyme through the association of identical subunits is more efficient, in this regard, than increasing the length of its polypeptide chain since each subunit has an active site. More importantly, however, the subunit construction of many enzymes provides the structural basis for the regulation of their activities. Mechanisms for this indispensable function are discussed in Sections 9-4 and 12-4.

In this section we discuss how the subunits of multisubunit proteins associate, what sorts of symmetries they have, how their stoichiometries may be determined, and end by considering the structural characteristics of multienzyme complexes.

A. Subunit Interactions

A multisubunit protein may consist of identical or nonidentical polypeptide chains. Recall that hemoglobin, for example, has the subunit composition $\alpha_2\beta_2$. We shall refer to proteins with identical subunits as **oligomers** and to these identical subunits as **protomers.** A protomer may therefore consist of one polypeptide chain or several unlike polypeptide chains. In this sense, hemoglobin is a **dimer** (oligomer of two protomers) of $\alpha\beta$ protomers (Fig. 7-56).

The contact regions between subunits closely resemble the interior of a single subunit protein. They contain closely packed nonpolar side chains, hydrogen bonds involving the polypeptide backbones and their side chains, and, in some cases, interchain disulfide bonds.

B. Symmetry in Proteins

In the vast majority of oligomeric proteins, the protomers are symmetrically arranged; that is, the protomers occupy geometrically equivalent positions in the oligomer. This implies that each protomer has exhausted its capacity to bind to other protomers; otherwise, higher oligomers would form. As a result of this limited binding capacity, protomers pack about a single point to form a closed shell. Proteins cannot have inversion or mirror symmetry, however, because such symmetry operations convert chiral L-residues to D-residues. Thus, *proteins can only have rotational symmetry.*

Various types of rotational symmetry occur in proteins:

1. Cyclic symmetry

In the simplest type of rotational symmetry, **cyclic symmetry,** subunits are related (brought to coincidence) by a single axis of rotation (Fig. 7-57a). Ob-

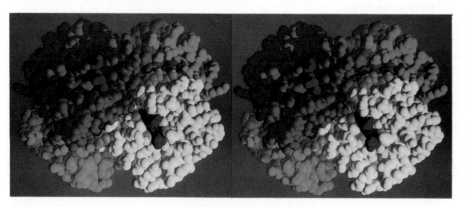

Figure 7-56
A stereo, space-filling drawing showing the quaternary structure of hemoglobin. The α_1, β_1, α_2, and β_2 subunits are colored yellow, light blue, green, and dark blue, respectively. Heme groups are red. The twofold rotation axis relating the $\alpha_1\beta_1$ protomer to the $\alpha_2\beta_2$ protomer is vertical. Instructions for viewing stereo drawings are given in the appendix to this chapter. [Courtesy of Richard J. Feldmann, NIH.]

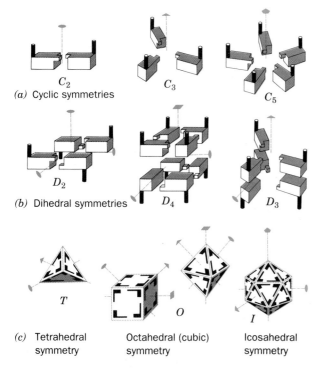

(a) Cyclic symmetries C_2 C_3 C_5

(b) Dihedral symmetries D_2 D_4 D_3

(c) Tetrahedral symmetry T Octahedral (cubic) symmetry O Icosahedral symmetry I

Figure 7-57
Some possible symmetries of proteins with identical protomers. The lenticular shape, the triangle, the square, and the pentagon at the ends of the dashed lines indicate, respectively, the unique twofold, threefold, fourfold, and fivefold rotational axes of the objects shown. (*a*) Assemblies with the cyclic symmetries C_2, C_3, and C_5. (*b*) Assemblies with the dihedral symmetries D_2, D_4, and D_3. In these objects, a twofold axis is perpendicular to the vertical two-, four-, and threefold axes. (*c*) Assemblies with *T*, *O*, and *I* symmetry. Note that the tetrahedron has some but not all of the symmetry elements of the cube, and that the cube and the octahedron have the same symmetry. [Figure copyrighted © by Irving Geis.]

jects with a 2, 3, . . . , or *n*-fold rotational axes are said to have C_2, C_3, . . . , or C_n symmetry, respectively. An oligomer with C_n symmetry consists of *n* protomers that are related by $(360/n)°$ rotations. C_2 symmetry is the most common symmetry in proteins; higher cyclic symmetries are relatively rare.

A common mode of association between protomers related by a twofold rotation axis is the continuation of a β sheet across subunit boundaries. In such cases, the twofold axis is perpendicular to the β sheet so that two symmetry equivalent strands hydrogen bond in an antiparallel fashion. In this manner, the sandwich of two four-stranded β sheets of the **prealbumin** protomer is extended across a twofold axis to form a sandwich of two eight-stranded β sheets (Fig. 7-58).

2. Dihedral symmetry

Dihedral symmetry (D_n), a more complicated type of rotational symmetry, is generated when an *n*-fold

rotation axis and a twofold rotation axis intersect at right angles (Fig. 7-57*b*). An oligomer with D_n symmetry consists of 2*n* protomers. The D_2 symmetry is, by far, the most common type of dihedral symmetry in proteins. Under the proper conditions, many oligomers with D_n symmetry will dissociate into two oligomers, each with C_n symmetry (and which were related by the twofold rotation axis in the D_n oligomer). These, in turn, dissociate to their component protomers under more stringent dissociating conditions.

3. Other rotational symmetries

The only other types of rotationally symmetric objects are those that have the rotational symmetries of a tetrahedron (*T*), a cube or octahedron (*O*) or an icosahedron (*I*), and have 12, 24, and 60 equivalent positions, respectively (Fig. 7-57*c*). The subunit arrangements in the protein coats of the so-called spherical viruses are based on icosahedral symmetry (Section 32-2A).

Under favorable conditions electron microscopy can provide dramatic indications of oligomeric symmetry. Electron microscopy studies suggest, for example, that the 600 kD *E. coli* **glutamine synthetase** has D_6 sym-

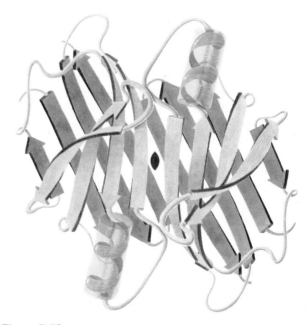

Figure 7-58
A prealbumin dimer viewed down its twofold axis (*red symbol*). Each protomer consists of a sandwich of two four-stranded β sheets. Note how both of these β sheets are continued in an antiparallel fashion in the other protomer to form a sandwich of two eight-stranded β sheets. Two of these dimers associate back to back in the native protein to form a tetramer with D_2 symmetry. [After a drawing by Jane Richardson, Duke University.]

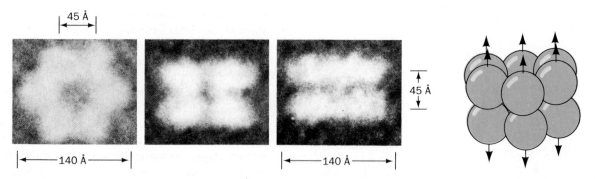

Figure 7-59

Sets of five superimposed electron micrographs (to enhance real detail) of *E. coli* glutamine synthetase molecules in their three characteristic orientations. The mean dimensions are indicated. When the oligomeric molecule rests on its face, it appears to be a hexagonal ring of subunits (*left*). Molecules on edge, however, show two layers of subunits as four spots when viewed exactly between the subunits (*middle*), or as two parallel streaks when viewed in other directions (*right*). This suggests, as the accompanying drawing indicates, that the enzyme molecule has 12 identical subunits organized with D_6 symmetry into two hexagons that are stacked with their subunits in opposition. [Courtesy of Earl Stadtman, NIH.]

metry (Fig. 7-59). Unfortunately, since this technique has insufficient resolution to reveal the relative orientations of the protein subunits (i.e., the directions of the arrows in the interpretive drawing of Fig. 7-59), such symmetry assignments must be taken as tentative; only X-ray crystal structure analysis can unambiguously establish the geometric relationships among protein subunits. In the case of glutamate synthetase, however, X-ray studies have confirmed that it indeed has D_6 symmetry (Section 24-5A).

Helical Symmetry

Some protein oligomers have **helical symmetry** (Fig. 7-60). The chemically identical subunits in a helix are not strictly equivalent because, for instance, those at the end of the helix have a different environment than those in the middle. Nevertheless, the surroundings of all subunits in a long helix, except those near its end, are sufficiently similar that the subunits are said to be **quasi-equivalent.** The subunits of many structural proteins, for example, those of muscle (Section 34-3A), assemble into fibers with helical symmetry.

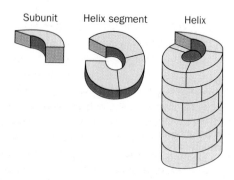

Subunit Helix segment Helix

Figure 7-60

A helical structure composed of a single kind of subunit.

C. Determination of Subunit Composition

The number of different types of subunits in an oligomeric protein may be determined by end group analysis (Section 6-1A). In principle, the subunit composition of a protein may be determined by comparing its molecular mass with those of its component subunits. In practice, however, experimental difficulties, such as the partial dissociation of a supposedly intact protein and uncertainties in molecular mass determinations, often provide erroneous results.

Hybridization Yields Quaternary Structural Information

An alternative procedure may be used if two chemically different and therefore separable species of the protein are available. The species may be proteins with slightly different 1° structures from different organisms or, as is often the case, variants of a protein that occur in the same organism. The two different oligomeric proteins are purified, mixed together, dissociated to their component subunits by exposure to mildly denaturing conditions (e.g., by changing the pH or adding urea), and then allowed to reassemble (e.g., by restoring the pH or dialyzing out the urea). If the native proteins are n-mers, S_n and S'_n, this procedure will yield $(n + 1)$ species of **hybrid molecules** with the mixed subunit compositions $S_n, S_{n-1}S', S_{n-2}S'_2, \ldots, S'_n$, which can be analyzed, for example, by electrophoresis. For instance, vertebrates possess two varieties of the enzyme **lactate dehydrogenase (LDH):** the M type, which predominates in skeletal muscle, and the H type, which predominates in heart tissue. Hybridization of these oligomers, in this case by repeated freezing and thawing, yields five **isozymes** (isoenzymes; catalytically and structurally similar enzymes from the same organism) of LDH that have the subunit compositions M_4, M_3H, M_2H_2, MH_3,

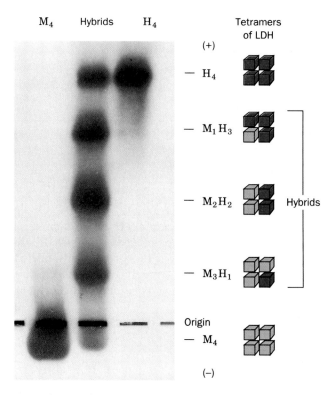

M₄   Hybrids   H₄   Tetramers of LDH

(+)

— H₄

— M₁H₃

— M₂H₂   Hybrids

— M₃H₁

Origin

— M₄

(−)

Figure 7-61
An electrophoretogram of bovine lactate dehydrogenase. The M and H forms of LDH (*outer lanes*) have different electrophoretic mobilities. Upon hybridization of these oligomers, five electrophoretically distinct isozymes are formed (*center lane*), which indicates that LDH is a tetramer. [Courtesy of Clement Markert, North Carolina State University at Raleigh.]

and H₄ (Fig. 7-61). This demonstrates that LDH is a tetramer.

In a related method, a protein subunit may be labeled, for example, by succinylation,

which alters the electrophoretic mobility of a protein by changing its ionic charge. John Gerhart and Howard Schachman used this technique to determine the geometric distribution of subunits in *E. coli* **aspartate trans-**

carbamoylase (ATCase). ATCase has two types of subunits, the catalytic subunit, c, and the regulatory subunit, r (their enzymatic roles are discussed in Section 12-4). Molecular mass measurements (c = 33 kD, r = 17 kD, and ATCase = 300 kD) indicate that ATCase has the subunit composition c_6r_6. This was corroborated by preliminary X-ray studies, which established that ATCase has D_3 symmetry (recall that a protein of D_3 symmetry must have six protomers; Fig. 7-57b). Treatment with organic mercurials such as ***para*-hydroxymercuribenzoate,** which reacts with Cys sulfhydryl groups

causes ATCase to dissociate according to the reaction

$$c_6r_6 \rightarrow 2c_3 + 3r_2$$

The catalytic trimers, c_3, were isolated and succinylated to form c_3^s. When these were mixed with unmodified catalytic trimers and excess regulatory dimers, r_2, under conditions that ATCase reforms, only three products could be electrophoretically distinguished: c_6r_6, $c_3c_3^sr_6$, and $c_6^sr_6$. This indicates that the catalytic subunits were not exchanged between catalytic trimers in ATCase; for example, no $c_4c_2^sr_6$ was formed. The c_3 trimers must therefore be separate entities in the enzyme. Similar studies using succinylated regulatory dimers, r_2^s, established that regulatory dimers likewise maintain their integrity in ATCase. Accordingly, the subunit composition of ATCase is more realistically represented as $(c_3)_2(r_2)_3$. This result was later confirmed by the X-ray crystal structure of ATCase (Fig. 7-62).

Cross-Linking Agents Stabilize Oligomers

Another method for 4° structure analysis, which is especially useful for oligomeric proteins that decompose easily, employs **cross-linking agents,** such as **dimethylsuberimidate** or **glutaraldehyde** (Fig. 7-63). If carried out at sufficiently low protein concentrations to eliminate intermolecular reactions, cross-linking reactions will covalently join only the subunits in a molecule that are no further apart than the length of the cross-link (assuming, of course, that the proper amino acid residues are present). The molecular mass of a cross-linked protein therefore places a lower limit on its number of *(a)*   *(b)*

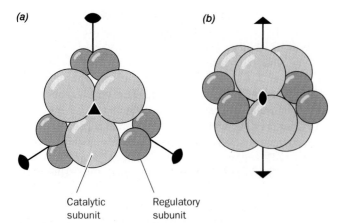

Catalytic   Regulatory
subunit   subunit

Figure 7-62
The quarternary structure of *E. coli* aspartate
transcarbamoylase as established by X-ray structure
analysis. Catalytic subunits and regulatory subunits are
represented, respectively, by large orange spheres and small
purple spheres. The molecule has D_3 symmetry. (*a*) View
along the threefold axis (*triangle*). (*b*) View along a twofold
axis (*lenticular shapes*). [After Kantrowitz, E. R.,
Pastra-Landis, S. C., and Lipscomb, W. N., *Trends Biochem.
Sci.* **5**, 150 (1980).]

subunits. Such studies can also provide some indication
of the distance between subunits, particularly if a series
of cross-linking agents with different lengths are employed.

D. Multienzyme Complexes

   Groups of enzymes that catalyze two or more steps
in a metabolic sequence often associate noncovalently

Table 7-6

**Subunit Composition of the *E. coli* Pyruvate
Dehydrogenase Complex**

Enzyme	Subunit Molecular Mass (kD)	Subunits per Complex
Pyruvate dehydrogenase (E$_1$)	96	24
Dihydrolipoyl transacetylase (E$_2$)	~70	24
Dihydrolipoyl dehydrogenase (E$_3$)	56	12

Source: Reed, L. J., *Acc. Chem. Res.* **7**, 40–46 (1974).

to form **multienzyme complexes.** These symmetric
clusters of enzymes have molecular masses in the
range from several hundred to several thousand kD.
These highly organized assemblies permit efficient
feedthrough of substrates from one enzyme in the metabolic pathway to the next.

   The *E. coli* **pyruvate dehydrogenase complex,** which
has been studied by Lester Reed, is one of the best characterized multienzyme complexes. It consists of three
enzymes, **pyruvate dehydrogenase (E$_1$), dihydrolipoyl transacetylase (E$_2$),** and **dihydrolipoyl dehydrogenase (E$_3$),** that associate in multiple copies to form
an ~4600-kD particle (Table 7-6). The reactions catalyzed by this enzyme system, which mediates an essential step in energy metabolism, are discussed in Section
19-2A.

   Electron micrographs of the entire pyruvate dehydrogenase complex show an ~300 Å in diameter polyhedral particle (Fig. 7-64*a*). Isolated E$_2$ forms a particle

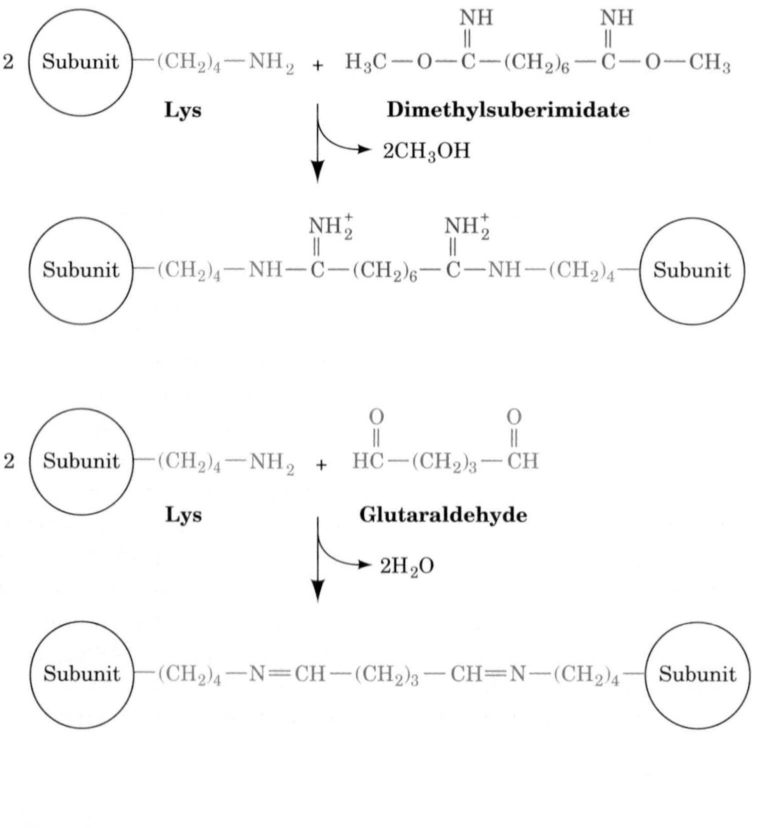

Figure 7-63
Dimethylsuberimidate
and glutaraldehyde are
bifunctional reagents
that react to covalently
cross-link two Lys
residues.

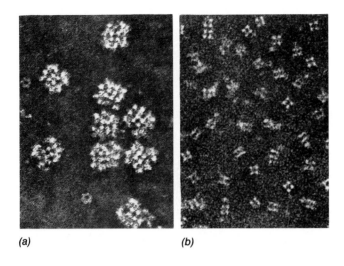

(a) **(b)**

Figure 7-64
Electron micrographs of *E. coli* (*a*) pyruvate dehydrogenase complex, and (*b*) dihydrolipoyl transacetylase (E$_2$) "core" complex. [Courtesy of Lester Reed, Clayton Foundation Biochemical Institute.]

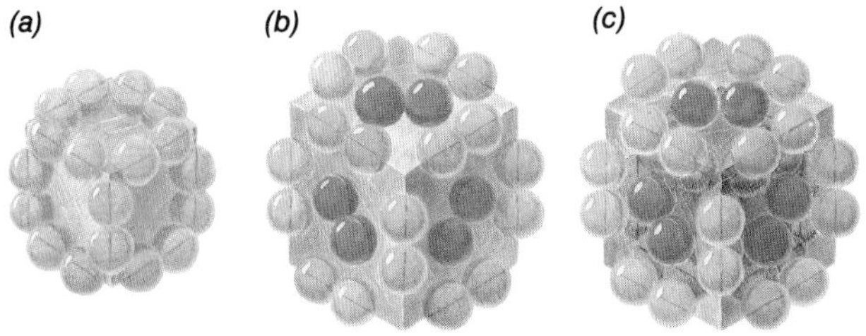

(a) **(b)** **(c)**

Figure 7-65
The *E. coli* pyruvate dehydrogenase multienzyme complex. (*a*) The dihydrolipoyl transacetylase (E$_2$) "core" of the pyruvate dehydrogenase complex. Its 24 subunits (*green spheres*) associate at the corners of a cube to form a particle that has cubic rotational symmetry (*O* symmetry). (*b*) A model of the *E. coli* pyruvate dehydrogenase complex. The 24 pyruvate dehydrogenase (E$_1$) subunits (*orange spheres*) form dimers that associate with the E$_2$ core (*shaded cube*) at the centers of each of its 12 edges. The 12 dihydrolipoyl dehydrogenase (E$_3$) subunits (*purple spheres*) form dimers that attach to the E$_2$ cube at the centers of each of its 6 faces. (*c*) Parts (*a*) and (*b*) combined to form the entire 60-subunit complex.

with 24 identical subunits (polypeptide chains), as it does in the intact complex. Electron micrographs of this E$_2$ particle suggest that it has cubic symmetry (Fig. 7-64*b*). A model of this "core" complex is depicted in Fig. 7-65*a* (compare it with the model for *O* symmetry in Fig. 7-57*c*). The E$_1$ subunits form dimers that associate with the E$_2$ cube at the centers of the cube's 12 edges (Figs. 7-65*b* and *c*). The E$_3$ subunits also form dimers that are located at the centers of the six faces of the E$_2$ cube. Other multienzyme complexes have similarly elaborate sorts of polyhedral symmetries.

Oligomeric proteins and multienzyme complexes are representative of the lowest level of macromolecular structural organization. **Supramolecular structures,** such as ribosomes or the membranous components of the electron-transport chain, are examples of higher levels of macromolecular organization. Indeed, *it is the enormously complex hierarchical organization of individually inanimate molecules that forms the structural basis of life.*

Appendix: Viewing Stereo Pictures

Although we live in a three-dimensional world, the images that we see have been projected onto the two-dimensional plane of our retinas. Depth perception therefore involves binocular vision: The slightly different views perceived by each eye are synthesized by the brain into a single three-dimensional impression.

Two-dimensional pictures of complex three-dimensional objects are difficult to interpret because most of the information concerning the third dimension is suppressed. This information can be recovered by presenting each eye with the image only it would see if the three-dimensional object was actually being viewed. A **stereo pair** therefore consists of two images, one for each eye. Corresponding points of stereo pairs are generally separated by ~6 cm, the average distance between human eyes. Stereo drawings are usually computer generated because of the required precision of the

geometric relationship between the members of a stereo pair.

In viewing a stereo picture, one must overcome the visual habits of a lifetime because each eye must see its corresponding view independently. Viewers are commercially available to aid in this endeavor. However, with some training and practice, equivalent results can be obtained without their use.

In order to train yourself to view stereo pictures, you should become aware that each eye sees a separate image. Hold your finger up about a foot (30 cm) before your eyes while fixing your gaze on some object beyond it. You may realize that you are seeing two images of your finger. If, after some concentration, you are still aware of only one image, try blinking your eyes alternately to ascertain which of your eyes is seeing the image you perceive. Perhaps alternately covering and

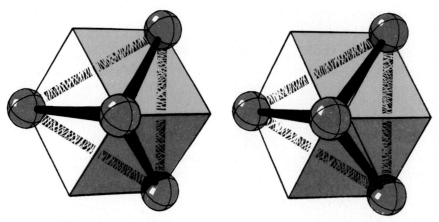

Figure 7-66
A stereo drawing of a tetrahedron inscribed in a cube. When properly viewed, the apex of the tetrahedron should appear to be pointing towards the viewer.

uncovering this dominant eye while staring past your finger will help you become aware of the independent workings of your eyes.

The principle involved in seeing a stereo picture is to visually fuse the left member of the stereo pair seen by the left eye with the right member seen by the right eye. To do this, sit comfortably at a desk, center your eyes about a foot over a stereo drawing such as Fig. 7-66 and stare through it at a point about a foot below the drawing. Try to visually fuse the central members of the four out-of-focus images you see. When you have succeeded, your visual system will "lock onto" it and this fused central image will appear three dimensional. Ignore the outer images. You may have to slightly turn the book, which should be held perfectly flat, or your head in order to bring the two images to the same level. It may help to place the book near the edge of a desk, center your finger about a foot below the drawing and fixate on

your finger while concentrating on the stereo pair. Another trick is to hold your flattened hand or an index card between your eyes so that the left eye sees only the left half of the stereo pair and the right eye sees only the right half and then fuse the two images you see.

The final step in viewing a stereo picture is to focus on the image while maintaining fusion. This may not be easy because our ingrained tendency is to focus on the point at which our gaze converges. It may help to move your head closer to or further from the picture. Most people (including the authors) require a fair amount of practice to become proficient at seeing stereo without a viewer. However, the three-dimensional information provided by stereo pictures, not to mention their esthetic appeal, makes it worth the effort. In any case, the stereo figures used in this text have been selected for their visual clarity without the use of stereo; stereo will simply enhance their impression of depth.

Chapter Summary

The peptide group is constrained by resonance effects to a planar, trans conformation. Steric interactions further limit the conformations of the polypeptide backbone by restricting the torsion angles, ϕ and ψ, of each peptide group to three small regions of the Ramachandran diagram. The α helix, whose conformation angles fall within the allowed regions of the Ramachandran diagram, is held together by hydrogen bonds. The 3_{10} helix, which is more tightly coiled than the α helix, lies in a mildly forbidden region of the Ramachandran diagram. Its infrequent occurrences are most often as single turn terminations of α helices. In the parallel and antiparallel β pleated sheets, two or more almost fully extended polypeptide chains associate such that neighboring chains are hydrogen bonded. These β sheets have a right-handed curl when viewed along

their polypeptide chains. The polypeptide chain often reverses its direction through a β bend. Other arrangements of the polypeptide chain, which are collectively known as coil conformations, are more difficult to describe but are no less ordered than are α or β structures.

The mechanical properties of fibrous proteins can often be correlated with their structures. Keratin, the principal component of hair, horn, and nails, forms protofibrils that consist of two pairs of α helices in which the members of each pair are twisted together into a left-handed coil. The pliability of keratin decreases as the content of disulfide cross-links between the protofibrils increases. Silk fibroin forms flexible but inextensible fibers of great strength. It exists as a semicrystalline array of antiparallel β sheets in which layers of Gly side chains alternate with layers of Ala and Ser side chains. Collagen is the major protein component of connective tissue. Its every third residue is Gly and many of the others are Pro and Hyp. This permits collagen to form a ropelike triple helical structure that has great tensile strength. Collagen molecules aggregate in a staggered array to form fibrils that are covalently cross-linked by groups derived from their His and Lys side chains. Elastin, which has elastic properties, forms a three-dimensional network of fibers that exhibit no regular structure. Its polypeptide strands are cross-linked in a manner similar to that in collagen.

The accuracies of protein X-ray structure determinations are limited by crystalline disorder to resolutions that are mostly in the range 2.0 to 3.5 Å. This requires that a protein's structure be determined by fitting its primary structure to its electron density map. Several lines of evidence indicate that protein crystal structures are nearly identical to their solution structures. A globular protein's 3° structure is the arrangement of its various elements of 2° structure together with the spatial dispositions of its side chains. Its amino acid residues tend to segregate according to residue polarity. Nonpolar residues preferentially occur in the interior of a protein out of contact with the aqueous solvent, whereas charged polar residues are located on its surface. Uncharged polar residues may occur at either location but, if they are internal, they form hydrogen bonds with other protein groups. The interior of a protein molecule resembles a crystal of an organic molecule in its packing efficiency. Larger proteins often fold into two or more domains that may have functionally and structurally independent properties. Certain groupings of secondary structural elements, known as supersecondary structures, repeatedly occur as components of globular proteins. They may have functional as well as structural significance.

Proteins have marginally stable native structures that form as a result of a fine balance among the various noncovalent forces to which they are subject: ionic and dipolar interactions, hydrogen bonding, and hydrophobic forces. Ionic interactions are relatively weak in aqueous solutions due to the solvating effects of water. The various interactions among permanent and induced dipoles, which are collectively referred to as van der Waals forces, are even weaker and are effective only at short range. Nevertheless, because of their large numbers, they cumulatively have an important influence on protein structures. Hydrogen bonding forces are far more directional than are other noncovalent forces. They add little stability to a protein structure, however, because the hydrogen bonds that native proteins form internally are no stronger than those that unfolded proteins form with water. Yet, a protein can only fold stably in ways that almost all of its possible internal hydrogen bonds are formed so that hydrogen bonding is important in specifying the native structure of a protein. Hydrophobic forces arise from the unfavorable ordering of water structure that results from the hydration of nonpolar groups. By folding such that its nonpolar groups are out of contact with the aqueous solvent, a protein minimizes these unfavorable interactions. The fact that most protein denaturants interfere with the hydrophobic effect demonstrates the importance of hydrophobic forces in stabilizing native protein structures. Disulfide bonds often stabilize the native structures of extracellular proteins.

Many proteins consist of noncovalently linked aggregates of subunits in which the subunits may or may not be identical. Most oligomeric proteins are rotationally symmetric. The protomers in many fibrous proteins are related by helical symmetry. The subunit structures of proteins may be elucidated by a variety of techniques, including the hybridization of subunits with their naturally occurring or derivatized variants, cross-linking studies, electron microscopy, and X-ray crystal structure analysis. Many sets of enzymes that catalyze a series of reactions in a metabolic pathway aggregate to form multienzyme complexes. The proximity of different enzymes in these entities, which have polyhedral symmetry and molecular masses ranging up to several thousand kilodaltons, permit the efficient feedthrough of substrates through the metabolic pathway.

References

General

Cantor, C. R. and Schimmel, P. R., *Biophysical Chemistry*, Chapters 2 and 5, Freeman (1980).

Creighton, T. E., *Proteins*, Chapters 4–6, Freeman (1983).

Dickerson, R. E. and Geis, I., *The Structure and Action of Proteins*, Benjamin/Cummings (1969). [A marvelously illustrated exposition of the fundamentals of protein structure.]

Fletterick, R. J., Schroer, T., and Matela, R. J., *Molecular Structure: Macromolecules in Three Dimensions*, Blackwell (1985). [A guide to protein and nucleic acid structures through the construction of molecular models.]

Schultz, G. E. and Schirmer, R. H., *Principles of Protein Structure*, Chapters 2–5 and 7, Springer–Verlag (1979). [An advanced text.]

Wood, W. B., Wilson, J. H., Benbow, R. M., and Hood, L. E., *Biochemistry. A Problems Approach* (2nd ed.), Chapters 4 and 5,

Benjamin/Cummings (1981). [A question-and-answer approach to learning protein structures.]

Secondary Structure

Leszczynski, J. F. and Rose, G. D., Loops in globular proteins: a novel category of secondary structure, *Science* **234,** 849–855 (1986).

Rose, G. D., Gierasch, L. M., and Smith, J. A., Turns in polypeptides and proteins, *Adv. Protein Chem.* **37,** 1–109 (1985).

Salemme, F. R., Structural properties of protein β-sheets, *Prog. Biophys. Mol. Biol.* **42,** 95–133 (1983).

Thornton, J. M., Sibanda, B. L., Edwards, M. S., and Barlow, D. J., Analysis, design and modification of loop regions in proteins, *BioEssays* **8,** 63–69 (1988).

Fibrous Proteins

Bornstein, P. and Traub, W., The chemistry and biology of collagen, *in* Neurath, H. and Hill, R. L. (Eds.), *The Proteins* (3rd ed.), Vol. 4, *pp.* 412–632, Academic Press (1979).

Eyre, D. R., Paz, M. A., and Gallop, P. M., Cross-linking in collagen and elastin, *Annu. Rev. Biochem.* **53**, 717–748 (1984).

Eyre, D. R., Collagen: Molecular diversity in the body's protein scaffold, *Science* **207**, 1315–1322 (1980).

Martin, G. R., Timpl, R., Müller, P. K., and Kühn, K., The genetically distinct collagens, *Trends Biochem. Sci.* **10**, 285–287 (1985).

Nimni, M. E. (Ed.), *Collagen*, Vol. I, CRC Press (1988). [Contains articles on collagen biochemistry.]

Prockop, D. J. and Kivirikko, K. I., Heritable diseases of collagen, *New Engl. J. Med.* **311**, 376–386 (1984).

Steinert, P. M. and Parry, D. A. D., Intermediate filaments, *Annu. Rev. Cell Biol.* **1**, 41–65 (1985). [Discusses the structure of α keratin.]

Globular Proteins

Blundell, T. L. and Johnson, L. N., *Protein Crystallography*, Academic Press (1976). [An advanced treatment of the principles of protein X-ray structure determination.]

Chothia, C., Principles that determine the structures of proteins, *Annu. Rev. Biochem.* **53**, 537–572 (1984).

Cohen, C. and Parry, D. A. D., α-Helical coiled coils — a widespread motif in proteins, *Trends Biochem. Sci.* **11**, 245–248 (1986).

Ptitsyn, O. B. and Finklestein, A. V., Similarities of protein topologies: Evolutionary divergence, functional convergence or principles of folding, *Q. Rev. Biophys.* **3**, 339–386 (1980).

Richards, F. M., Areas, volumes, packing, and protein structure, *Annu. Rev. Biophys. Bioeng.* **6**, 151–176 (1977).

Richardson, J. S., Handedness of crossover connections in β sheets, *Proc. Natl. Acad. Sci.* **73**, 2619–2623 (1976).

Richardson, J. S., The anatomy and taxonomy of protein structures, *Adv. Protein Chem.* **34**, 168–339 (1981). [A detailed discussion of the structural principles governing globular proteins accompanied by an extensive collection of their cartoon representations.]

Rossmann, M. G. and Argos, P., Protein folding, *Annu. Rev. Biochem.* **50**, 497–532 (1981).

Wüthrich, K., Protein structure determination in solution by nuclear magnetic resonance spectroscopy, *Science* **243**, 45–50 (1989).

Wyckoff, H. W., Hirs, C. H. W., and Timasheff, S. N. (Eds.), Diffraction Methods for Biological Macromolecules, Parts A and B, *Methods Enzymol.* **114, 115** (1985). [A series of articles on the theory and practice of X-ray crystallography.]

Protein Stability

Baker, E. N. and Hubbard, R. E., Hydrogen bonding in globular proteins, *Prog. Biophys. Mol. Biol.* **44**, 97–179 (1984).

Burley, S. K. and Petsko, G. A., Weakly polar interactions in proteins, *Adv. Protein Chem.* **39**, 125–189 (1988).

Edsall, J. T. and McKenzie, H. A., Water and proteins, *Adv. Biophys.* **16**, 51–183 (1983).

Franks, F. and Eaglund, D., The role of solvent interactions in protein conformation, *CRC Crit. Rev. Biochem.* **3**, 165–219 (1975).

Hol, W. G. J., The role of the α-helix dipole in protein function and structure, *Prog. Biophys. Mol. Biol.* **45**, 149–195 (1985).

Kauzmann, W., Some factors in the interpretation of protein denaturation, *Adv. Protein Chem.* **14**, 1–63 (1958). [A classic review that first pointed out the importance of hydrophobic bonding in stabilizing proteins.]

Nemethy, G., Pier, W. J., and Scheraga, H. A., Effect of protein-solvent interactions on protein conformation, *Annu. Rev. Biochem.* **10**, 459–497 (1981).

Ramachandran, G. N. and Sasisekharan, V., Conformation of polypeptides and proteins, *Adv. Protein Chem.* **23**, 283–437 (1968).

Saenger, W., Structure and dynamics of water surrounding biomolecules, *Annu. Rev. Biophys. Biophys. Chem.* **16**, 93–114 (1987).

Schellman, J. A., The thermodynamic stability of proteins, *Annu. Rev. Biophys. Biophys. Chem.* **16**, 115–137 (1987).

Tanford, C., The hydrophobic effect and the organization of living matter, *Science* **200**, 1012–1018 (1978).

Tanford, C., *The Hydrophobic Effect: Formation of Micelles and Biological Membranes* (2nd ed.), Chapters 2–4 and 13, Wiley (1980).

Wetlaufer, D. B., Folding of protein fragments, *Adv. Protein Chem.* **34**, 61–92 (1981).

Quaternary Structure

Finch, J. T., Electron microscopy of proteins, *in* Neurath, H. and Hill, R. L. (Eds.), *The Proteins* (3rd ed.), Vol. 1, *pp.* 413–497, Academic Press (1975).

Klotz, I. M., Darnell, D. W., and Langerman, N. R., Quaternary structure of proteins, *in* Neurath, H. and Hill, R. L. (Eds.), *The Proteins* (3rd ed.), Vol. 1, *pp.* 226–411, Academic Press (1975).

Matthews, B. W. and Bernhard, S. A., Structure and symmetry in oligomeric proteins, *Annu. Rev. Biophys. Bioeng.* **6**, 257–317 (1973).

Reed, L. J., Multienzyme complexes, *Acc. Chem. Res.* **7**, 40–46 (1974).

Schachman, H. S., Anatomy and physiology of a regulatory enzyme — aspartate transcarbamylase, *Harvey Lect.* **68**, 67–113 (1974).

Problems

1. What is the length of an α helical section of a polypeptide chain of 20 residues? What is its length when it is fully extended (all trans)?

*2. From an examination of Figs. 7-7 and 7-8, it is apparent that the polypeptide conformation angle ϕ is more constrained than is ψ. By referring to Fig. 7-4, or better yet, by examining a molecular model, indicate the sources of the steric interference that limit the allowed values of ϕ when $\psi = 180°$.

3. For a polypeptide chain made of γ-amino acids, state the nomenclature of the helix analogous to the 3_{10} helix of α-amino acids. Assume the helix has a pitch of 9.9 Å and a rise per residue of 3.2 Å.

*4. Table 7-7 gives the torsion angles, ϕ and ψ, of hen egg white lysozyme for residues 24–73 of this 129-residue protein. (The angle conventions used are those with $\phi = \psi = 180°$ for a fully extended polypeptide chain, as in Fig. 7-7.) (a) What is the secondary structure of residues 26–35? (b) What is the secondary structure of residues

42–53? (c) What is the probable identity of residue 54? (d) What is the secondary structure of residues 56–68? (e) What is the secondary structure of residues 69–71? (f) What additional information besides the torsion angles, ϕ and ψ, of each of its residues are required to define the three-dimensional structure of a protein?

5. Hair splits most easily along its fiber axis whereas fingernails tend to split across the finger rather than along it. What are the directions of the keratin fibrils in hair and in fingernails? Explain your reasoning.

6. What structural features are responsible for the observations that α keratin fibers can stretch to over twice their normal length whereas silk is nearly inextensible?

7. What is the growth rate, in turns per second, of the α helices in a hair that is growing 15 cm·year⁻¹?

8. Can polyproline form a collagenlike triple helix? Explain.

9. As Mother Nature's chief engineer, you have been asked

Table 7-7
Torsion Angles (ϕ, ψ) for Residues 24 to 73 of Hen Egg White Lysozyme

Residue Number	Amino Acid	ϕ (deg)	ψ (deg)	Residue Number	Amino Acid	ϕ (deg)	ψ (deg)
24	Ser	−60	147	49	Gly	95	−75
25	Leu	−49	−32	50	Ser	−18	138
26	Gly	−67	−34	51	Thr	−131	157
27	Asn	−58	−49	52	Asp	−115	130
28	Trp	−66	−32	53	Tyr	−126	146
29	Val	−82	−36	54	xxx	67	−179
30	Cys	−69	−44	55	Ile	−42	−37
31	Ala	−61	−44	56	Leu	−107	14
32	Ala	−72	−29	57	Gln	35	54
33	Lys	−66	−65	58	Ile	−72	133
34	Phe	−67	−23	59	Asn	−76	153
35	Glu	−81	−51	60	Ser	−93	−3
36	Ser	−126	−8	61	Arg	−83	−19
37	Asn	68	27	62	Trp	−133	−37
38	Phe	79	6	63	Trp	−91	−32
39	Asn	−100	109	64	Cys	−151	143
40	Thr	−70	−18	65	Asn	−85	140
41	Glu	−84	−36	66	Asp	133	8
42	Ala	−30	142	67	Gly	73	−8
43	Thr	−142	150	68	Arg	−135	17
44	Asn	−154	121	69	Thr	−122	83
45	Arg	−91	136	70	Pro	−39	−43
46	Asn	−110	174	71	Gly	−61	−11
47	Thr	−66	−20	72	Ser	−45	122
48	Asp	−96	36	73	Arg	−124	146

Source: Imoto, T., Johnson, L. N., North, A. C. T., Phillips, D. C., and Rupley, J. A., *in* Boyer, P. D. (Ed.), *The Enzymes* (3rd ed.), Vol. 7, *pp.* 693–695, Academic Press (1972).

to design a five-turn α helix that is destined to have half its circumference immersed in the interior of a protein. Indicate the helical wheel projection of your prototype α helix and its amino acid sequence (see Fig. 7-45).

10. β-Aminopropionitrile is effective in reducing excessive scar tissue formation after an injury (although its use is contraindicated by side effects). What is the mechanism of action of this lathyrogen?

11. Proteins have been classified as α, β, α/β, or $\alpha + \beta$ proteins depending on whether their tertiary structures, respectively, consist of mostly α helices, mostly β sheets, alternating α helices and β sheets, or some α helices and β sheets that tend to aggregate together rather than alternate along the polypeptide chain. By inspection, classify the following proteins according to this nomenclature and, where possible, identify their supersecondary structures: carboxypeptidase A (Fig. 7-19*a*), triose phosphate isomerase (Fig. 7-19*b*), myoglobin (Fig. 7-42), concanavalin A (Fig. 7-43), carbonic anhydrase (Fig. 7-44), glyceraldehyde-3-phosphate dehydrogenase (Fig. 7-46), and prealbumin (Fig. 7-58).

12. The coat protein of tomato bushy stunt virus consists of 180 chemically identical subunits, each of which is composed of ~ 386 amino acid residues. The probability that a wrong amino acid residue will be biosynthetically incorporated in a polypeptide chain is 1 part in 3000 per residue. Calculate the average number of coat protein subunits that would have to be synthesized in order to produce a perfect viral coat. What would this number be if the viral coat was a single polypeptide chain with the same number of residues that it actually has?

13. State the rotational symmetry of the following objects: (a) a starfish, (b) a square pyramid, (c) a rectangular box, and (d) a trigonal bipyramid.

14. Myoglobin and the subunits of hemoglobin are proteins of similar size and structure. Compare the expected ratio of nonpolar to polar amino acid residues in myoglobin and in hemoglobin.

15. Why are London dispersion forces always attractive?

16. Membrane-bound proteins are generally closely associated with the nonpolar groups of lipid molecules (Section 11-3A). Explain how detergents affect the structural integrity of membrane bound proteins in comparison to their effects on normal globular proteins.

17. Sickle-cell hemoglobin (HbS) differs from normal human adult hemoglobin (HbA) by a single mutational change, Glu $\beta6 \rightarrow$ Val, which causes the HbS molecules to aggregate under proper conditions (Section 6-3A). Under certain conditions, the HbS filaments that form at body temperature disaggregate when the temperature is lowered to 0°C. Explain.

18. Indicate experimental evidence that is inconsistent with the hypothesis that urea and guanidinium ion act to denature proteins by competing for their internal hydrogen bonds.

19. Proteins in solution are often denatured if the solution is shaken violently enough to cause foaming. Indicate the mechanism of this process. (*Hint:* The nonpolar groups of detergents extend into the air at air–water interfaces.)

20. An oligomeric protein in a dilute buffer at pH 7 dissociates to its component subunits when exposed to the following agents. Which of these observations would not support the contention that the quaternary structure of the protein is stabilized exclusively by hydrophobic interactions? Explain. (a) 6*M* guanidinium chloride, (b) 20% ethanol, (c) 2*M* NaCl, (d) temperatures below 0°C, (e) 2-mercaptoethanol, (f) pH 3, and (g) 0.01*M* SDS.

*21. What are the relative amounts of each isozyme formed when equimolar amounts of pure heart and muscle lactate dehydrogenase (H_4 and M_4) are hybridized?

*22. The SDS-polyacrylamide gel electrophoresis of a protein yields two bands corresponding to molecular masses of 10 and 17 kD. After cross-linking this protein with dimethylsuberimidate under sufficient dilution to eliminate intermolecular cross-linking, SDS-polyacrylamide gel electrophoresis of the product yields 12 bands with molecular masses 10, 17, 20, 27, 30, 37, 40, 47, 54, 57, 64, and 74 kD. Assuming that dimethylsuberimidate can only cross-link contacting subunits, diagram the quaternary structure of the protein.

Chapter 8

PROTEIN FOLDING, DYNAMICS, AND STRUCTURAL EVOLUTION

In the preceding chapters, we saw how proteins are constructed from their component parts. This puts us in a similar position to a mechanic who has learned to take apart and put together an automobile engine without any inkling of how the engine works. What we need in order to understand the workings of a protein is knowledge of the types of internal motions it can and must undergo in order to carry out its biological function as well as how it achieves its native structure. Put in terms of our deprived auto mechanic, we need to understand the operations of the "gears" and "levers" with which proteins carry out their function. This is a problem of enormous complexity whose solution we have only glimpsed in outline. We shall see in later chapters, for example, that even though the X-ray structures of many enzymes have been elucidated, few enzyme mechanisms are understood in detail. This is because our comprehension of the ways in which a protein's component groups interact is far from complete. As far as proteins are concerned, we have yet to surpass significantly our hypothetical aborigine's level of understanding.

In this third of four chapters on protein structure, we consider the temporal behavior of proteins. Specifically, we first take up the problem of how random coil polypeptides fold to their native structures. Next we address the dynamic properties of native proteins; that is, the nature and functional significance of their internal mo-

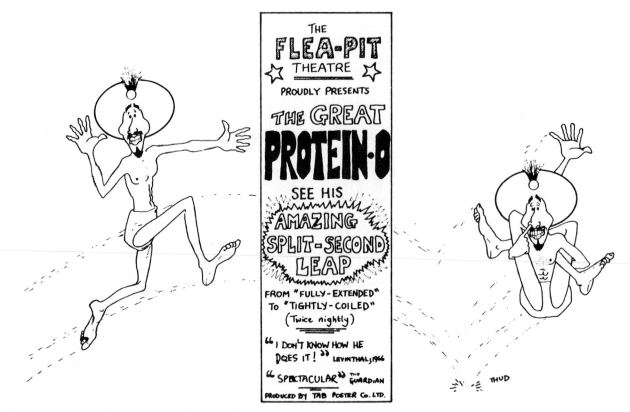

Figure 8-1
[Drawing by T. A. Bramley, *in* Robson, B., *Trends Biochem. Sci.* **1,** 50 (1976). Copyright © Elsevier Biomedical Press, 1976. Used by permission.]

tions. We end by extending the discussions we began in Section 6-3 on protein evolution but do so in terms of the three-dimensional structures of proteins.

1. PROTEIN FOLDING: THEORY AND EXPERIMENT

Early notions of protein folding postulated the existence of "templates" that somehow caused proteins to assume their native conformation. Such an explanation begs the question of how proteins fold because, even if it were true, one would still have to explain how the template achieved its conformation. In fact, *proteins spontaneously fold into their native conformations under physiological conditions.* This implies that *a protein's primary structure dictates its three-dimensional structure.* In general, under the proper conditions, biological structures are **self-assembling** so that they have no need of external templates to guide their formation.

A. Protein Renaturation

Although evidence had been accumulating since the 1930s that proteins could be reversibly denatured, it was not until 1957 that the elegant experiments of Christian Anfinsen on RNase A put **protein renaturation** on a

quantitative basis. RNase A, a 124-residue single-chain protein, is completely unfolded and its four disulfide bonds reductively cleaved in an 8*M* urea solution containing 2-mercaptoethanol (Fig. 8-2). Yet, dialyzing

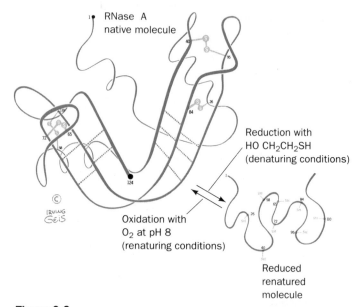

RNase A native molecule

Reduction with
HO CH$_2$CH$_2$SH
(denaturing conditions)

Oxidation with
O$_2$ at pH 8
(renaturing conditions)

Reduced
renatured
molecule

Figure 8-2
The reductive denaturation and oxidative renaturation of RNase A. [Figure copyrighted © by Irving Geis.]

away the urea and exposing the resulting solution to O_2 at pH 8 yields a protein that is virtually 100% enzymatically active and physically indistinguishable from native RNase A. The protein must therefore have spontaneously renatured. Any reservations that this occurs only because RNase A is really not totally denatured by $8M$ urea have been satisfied by the chemical synthesis of enzymatically active RNase A (Section 6-4B).

The renaturation of RNase A demands that its four disulfide bonds reform. The probability of one of the eight Cys residues from RNase A randomly reforming a disulfide bond with its proper (native) mate among the other seven Cys residues is $\frac{1}{7}$; that of one of the remaining six Cys residues then randomly reforming its proper disulfide bond is $\frac{1}{5}$; *etc.* The overall probability of RNase A reforming its four native disulfide links at random is

$$\frac{1}{7} \times \frac{1}{5} \times \frac{1}{3} \times \frac{1}{1} = \frac{1}{105}$$

Clearly, the disulfide bonds from RNase A do not randomly reform under renaturing conditions.

If the RNase A is reoxidized in $8M$ urea so that its disulfide bonds reform while the polypeptide chain is a random coil, then after removal of the urea, the RNase A is, as expected, only ~1% enzymatically active. This "scrambled" RNase A can be made fully active by exposing it to a trace of 2-mercaptoethanol, which, over about a 10-h period, catalyzes disulfide bond interchange reactions until the native structure is achieved (Fig. 8-3). The native state of RNase A under physiological conditions is therefore its thermodynamically most stable "local" conformation (the most stable state without large conformational rearrangements), although it is not certain that this state is its most stable "global"

(overall) conformation. If the protein has a conformation that is more stable than the native state, however, conversion to it must involve such a large activation barrier as to make it kinetically inaccessible (rate processes are discussed in Section 13-1C).

The time for renaturation of "scrambled" RNase A is reduced to ~2 min through the use of an enzyme, **protein-disulfide isomerase,** that catalyzes disulfide interchange reactions. (In fact, the supposition that *in vivo* folding to the native state requires no more than a few minutes prompted the search that led to this enzyme's discovery.) Protein-disulfide isomerase itself contains three Cys residues, one of which must be in the —SH form for enzymatic activity. The enzyme evidently catalyzes the random cleavage and reformation of a protein's disulfide bonds (Fig. 8-3), thereby interchanging them as the protein progressively attains thermodynamically more favorable conformations.

Post-Synthetically Modified Proteins May Not Spontaneously Renature

Many "scrambled" proteins are renatured through the action of protein-disulfide isomerase and are unaffected by it in their native state (their isomerase-cleaved disulfide bonds rapidly reform because these native proteins are in their most stable local conformations). In post-synthetically modified proteins, however, the disulfide bonds may serve to hold the protein in its otherwise unstable native state. For instance, the 51-residue polypeptide hormone **insulin,** which consists of two polypeptide chains joined by two disulfide bonds (Fig. 6-2), is inactivated by the isomerase. This observation led to the discovery that insulin is derived from a single-chain, 84-residue precursor named **proinsulin** (Fig.

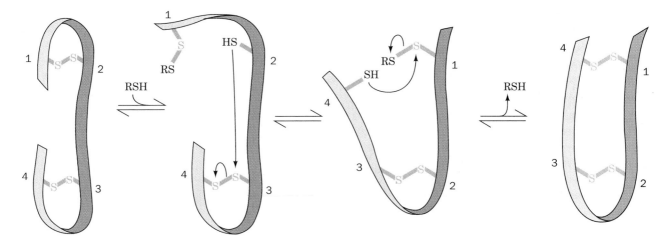

Figure 8-3
A plausible mechanism for the thiol- or enzyme-catalyzed disulfide interchange reaction in a protein. The purple ribbon represents the polypeptide backbone of the protein.

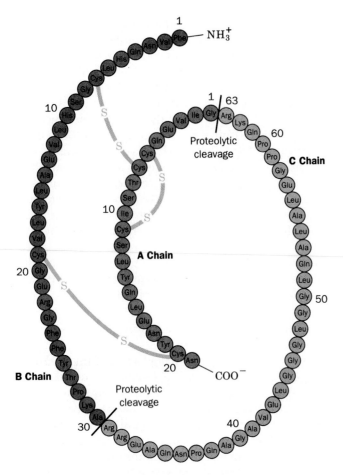

Figure 8-4
The primary structure of porcine proinsulin. Its C chain (*brown*) is proteolytically excised from between its A and B chains to form the mature hormone. [After Chance, R. E., Ellis, R. M., and Brommer, W. W., *Science* **161**, 165 (1968).]

8-4). Only after its disulfide bonds have formed is proinsulin converted to the two-chained active hormone by the specific proteolytic excision of an internal 33-residue segment known as its C chain.

Protein Folding Is Directed Mainly by Internal Residues

Numerous protein modification studies have been aimed at determining the role of various classes of amino acid residues in protein folding. In one particularly revealing study, the free primary amino groups of RNase A (Lys residues and the N-terminus) were derivatized with 8-residue chains of poly-DL-alanine. Surprisingly, these large, water soluble poly-Ala chains could be simultaneously coupled to RNase's 11 free amino groups without significantly altering the protein's native conformation or its ability to refold. Since these free amino groups are all located on the exterior of RNase A, this observation suggests that *it is largely a protein's internal residues that direct its folding to the native conformation*. Similar conclusions have been

reached from studies of protein structure and evolution (Section 8-3): Mutations that change surface residues are accepted more frequently and are less likely to affect protein conformations than are changes of internal residues. It is therefore not surprising that the perturbation of protein folding by limited concentrations of denaturing agents indicates that *protein folding is driven by hydrophobic forces.*

B. Folding Pathways

How does a protein fold to its native conformation? One might guess that this process occurs through the protein's random exploration of all the conformations available to it until it eventually "stumbles" onto the correct one. A "back-of-the-envelope" calculation first made by Cyrus Levinthal, however, convincingly demonstrates that this cannot possibly be the case: Assume that the $2n$ torsional angles, ϕ and ψ, of an n-residue protein each have three stable conformations. This yields $3^{2n} \approx 10^n$ possible conformations for the protein, which is a gross underestimate, if only because the side chains are ignored. If a protein can explore new conformations at the rate that single bonds can reorient, it can find $\sim 10^{13}$ conformations per second which is, no doubt, an overestimate. We can then calculate the time, t, in seconds, required for a protein to explore all the conformations available to it:

$$t = \frac{10^n}{10^{13}} \qquad [8.1]$$

For a rather small protein of $n = 100$ residues, $t = 10^{87}$ s, which is immensely more than the apparent age of the universe (20 billion years $= 6 \times 10^{17}$ s).

It would obviously take even the smallest protein an absurdly long time to explore all its possible conformations. Yet, many proteins fold to their native conformation in less than a second. They therefore must fold by some sort of ordered pathway in which the approach to the native state is accompanied by sharply increasing conformational stability. An analogy to this situation is a ball rolling down a hill to a valley: If the topography of the hill (conformational energy map) was rather featureless, the ball (polypeptide chain) could take many different pathways (conformational sequences) in rolling to the valley (folding to its native state). If, however, the hill (conformational energy map) sloped sharply into a canyon that led to the valley (native state), then the ball (polypeptide chain) would follow a direct pathway in reaching it.

Protein folding is thought to occur via a multistage process (Fig. 8-5):

1. The folding of a random coil polypeptide begins with the random formation of short stretches of 2° structure, such as α helices and β turns, that can act as **nuclei** (scaffolding) for the stabilization of additional

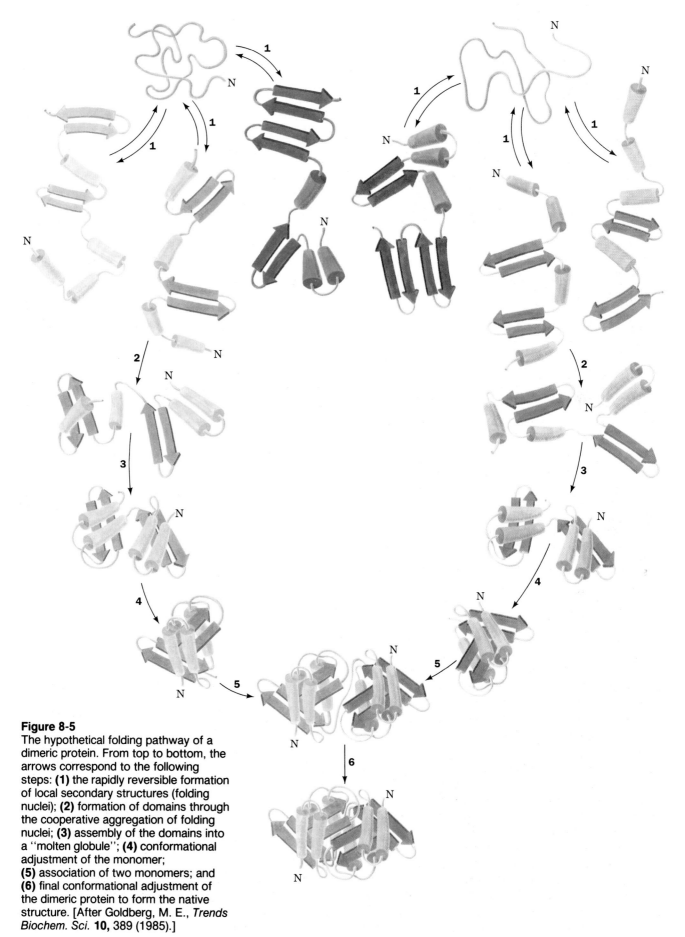

Figure 8-5
The hypothetical folding pathway of a dimeric protein. From top to bottom, the arrows correspond to the following steps: **(1)** the rapidly reversible formation of local secondary structures (folding nuclei); **(2)** formation of domains through the cooperative aggregation of folding nuclei; **(3)** assembly of the domains into a "molten globule"; **(4)** conformational adjustment of the monomer; **(5)** association of two monomers; and **(6)** final conformational adjustment of the dimeric protein to form the native structure. [After Goldberg, M. E., *Trends Biochem. Sci.* **10,** 389 (1985).]

ordered regions of the protein. These nuclei are probably small enough (8–15 residues) to "flicker" in and out of existence in under a millisecond.

2. Nuclei with the proper nativelike structure probably grow by the diffusion, random collision, and adhesion of two or more such nuclei. The stabilities of these ordered regions increase with size so that once they have randomly reached a certain minimum size, they spontaneously grow in a cooperative fashion until they form a nativelike domain. The existence of such a hierarchical folding process in which folding units condense to form larger folding units, *etc.*, is corroborated by the observation that domains are composed of subdomains which, in turn, may consist of sub-subdomains, *etc.* Perhaps the usual upper limit of ~200 residues in a domain reflects the special requirement that it must rapidly fold to its native conformation in an ordered sequence.

3. In multidomain proteins, the domains come together to form a "molten globule" (loosely organized but nativelike subunit) whose hydrophobic side chains remain exposed to solvent.

4. Through a series of relatively small conformational adjustments the polypeptide chain achieves a more compact 3° structure, the native conformation of a single-subunit protein.

5. In a multisubunit protein, the requisite number of subunits assemble in a nativelike 4° structure.

6. Finally, a further series of slight conformational adjustments yields the native protein structure.

BPTI Folds to Its Native Conformation via an Ordered Pathway

The most convincing experimental evidence favoring the above protein folding scheme comes from Thomas Creighton's renaturation studies of **bovine pancreatic trypsin inhibitor (BPTI)**. This 58-residue monomeric protein has three disulfide bonds (Fig. 8-6); it binds to and inactivates trypsin in the pancreas, thereby protecting that secretory organ from self-digestion (Section 14-3). Creighton monitored the folding of BPTI by identifying the order of disulfide bond formation as reduced and unfolded BPTI was oxidized and renatured. *Each distinguishable disulfide-bonded species represents a subset of the conformations that the BPTI polypeptide can assume, so that by following the time course of these various species, the approximate conformational path taken by the renaturing protein was deduced.*

The investigation began with fully reduced BPTI, which is completely unfolded (random coil) under physiological conditions. Disulfide bond formation and subsequent renaturation was induced by reacting the protein with oxidized dithiothreitol (Section 6-1B). As

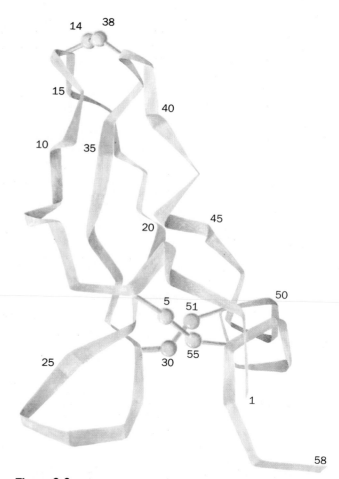

Figure 8-6
The polypeptide backbone and disulfide bonds of native BPTI. [After a drawing by Michael Levitt, *in* Creighton, T. E., *J. Mol. Biol.* **95,** 168 (1975).]

the protein renatured, the intermediates formed were trapped by blocking their remaining free Cys-SH groups with iodoacetate (Section 6-1B). These stabilized intermediates were then chromatographically separated and the positions of their disulfide bonds determined by diagonal electrophoresis (Section 6-1I).

These experiments indicate that *BPTI follows a limited but indirect set of pathways in folding to its native structure* (Fig. 8-7):

1. The six Cys residues of the fully reduced BPTI (R), are equally likely to participate in forming the initial disulfide bond. Yet, after the molecule has equilibrated through a series of rapid internal disulfide interchange reactions, only 2 of the 15 possible one-disulfide intermediates, I_A and I_B, exist in significant quantities. (An intermediate's relative abundance, at equilibrium, is indicative of its thermodynamic stability relative to other intermediates; Section 3-4A.) Of these, only I_B has a disulfide bond that occurs in the native protein (Cys 30-Cys 51) and only this species reacts in significant amounts with sulfhydryl reagents to form a second disulfide bond. NMR studies indicate that I_B is a conformationally fluctuating mol-

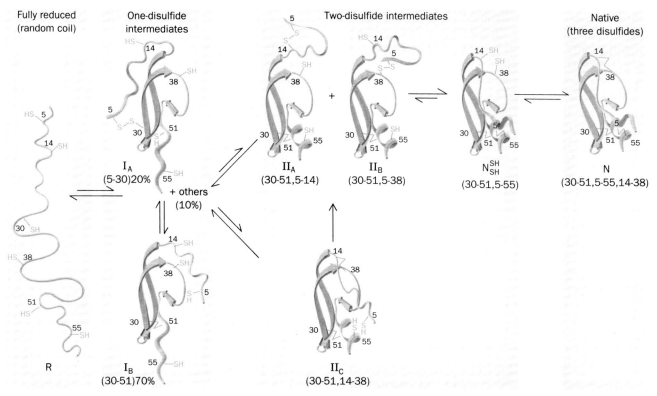

Figure 8-7
The renaturation pathway of BPTI showing the conformations of its polypeptide backbone as deduced from disulfide trapping experiments and NMR measurements (note that these views of the protein differ from that in Fig. 8-6 by a slight rotation about the vertical axis). The fully reduced and native proteins are represented by R and N, respectively. The sequence numbers of the Cys residues involved in each disulfide bond are given in parentheses below the diagram representing each folding intermediate. The two one-disulfide intermediates, I_A and I_B, are in rapid equilibrium. The "+" between intermediates II_A and II_B indicates that both are formed directly from the one-disulfide intermediates, that both convert directly to N_{SH}^{SH}, and that either or both are intermediates in the rearrangement of II_C to N_{SH}^{SH}. [After Creighton, T. E., *J. Mol. Biol.* **113**, 288 (1977) *and* States, D. J., Creighton, T. E., Dobson, C. M., and Karplus, M., *J. Mol. Biol.* **195**, 737 (1987).]

ecule that forms the native protein's β sheet ~ 20% of the time at physiological temperatures but does not detectably contain either of the native protein's two α helices.

2. Of the 45 possible 2-disulfide intermediates, only 3, II_A, II_B, and II_C, all of which contain the original correct disulfide bond, occur in significant quantities even though all four SH groups of I_B are equally reactive. Of these, only II_C contains a second correct disulfide bond (14–38) but, curiously, it is a dead-end species that must first convert to either of the incorrectly disulfide-bridged species, II_A or II_B, in order to fold to the native conformation. The NMR data suggest that this is because II_C, which is certainly more nativelike than I_B, does not have its N-terminal region sufficiently stabilized to permit the formation of the 5–55 disulfide bond.

3. Before any additional disulfide bond formation occurs, II_A and II_B convert to another 2-disulfide intermediate, N_{SH}^{SH}, which also contains a second correct disulfide bond (5–55). This conversion is relatively slow, which is indicative of a large conformational rearrangement. Evidently, the conformation necessary to form the 5–55 disulfide bond is difficult to achieve. NMR studies indicate that N_{SH}^{SH} has a nativelike conformation in that it exhibits all the secondary structural elements of the native protein.

4. This conclusion is corroborated by the very rapid formation of the third disulfide bond of N_{SH}^{SH} (14–38) to yield native BPTI (N).

Protein Primary Structures Determine Their Folding Pathways

Renaturation studies of BPTI and other disulfide-containing proteins are consistent with the hypothesis that *protein folding occurs through a largely ordered, although not necessarily direct, pathway*. Observations on the time course of protein folding made using experimental probes other than disulfide bond formation also support this conclusion. Evidently, *protein primary structures evolved to specify efficient folding pathways as well as stable native conformations.*

Evidence that a protein's folding pathway is, in fact, genetically determined has been obtained by Jonathan King. The tail spikes of **bacteriophage P22** (bacteriophages are viruses that attack bacteria), which are trimers of identical 76-kD polypeptides, denature above 80°C. Several mutant varieties of this protein fail to renature at 39°C. Yet, at 30°C, the mutant proteins fold to structures whose properties, including their T_m values, are indistinguishable from that of the nonmutant tail spike protein. The amino acid changes causing these **temperature-sensitive mutations** apparently act to destabilize intermediate states in the folding process but do not affect the native protein's stability. This observation suggests that *a protein's amino acid sequence dictates its native structure by specifying the series of steps comprising its folding pathway.*

C. Prediction of Protein Structures

Since the primary structure of a protein specifies its three-dimensional structure, it should be possible, at least in principle, to predict the native structure of a protein from a knowledge of only its amino acid sequence. This might be done using theoretical methods based on physicochemical principles, or by empirical methods in which predictive schemes are distilled from the analyses of known protein structures. Theoretical methods, which usually attempt to determine the minimum energy conformation of a protein, are mathematically quite sophisticated and require extensive computations. The enormous difficulty in making such calculations sufficiently accurate and yet computationally tractable has, so far, limited their success. Nevertheless, an understanding of why proteins fold to their native structures must ultimately be based on such theoretical methods.

Empirical methods are usually much easier to apply than are theoretical methods and have had remarkable success. Clearly, certain amino acid sequences limit the conformations available to a polypeptide chain in an easily understood manner; for example, a Pro residue cannot assume an α helical conformation or contribute a backbone hydrogen bond. Likewise, steric interactions between several sequential amino acid residues with side chains branched at C_β (for instance, Ile and Thr) will destabilize an α helix. Furthermore, there are more subtle effects that may not be apparent without a detailed analysis of known protein structures. In what follows, we shall examine two schemes for predicting secondary structures empirically.

The Chou and Fasman Scheme

The first empirical structure prediction scheme we shall consider, which was developed by Peter Chou and

Gerald Fasman, is among the most reliable of those available, and yet, is easy to apply. Its use requires two definitions. The frequency, f_α, with which a given residue occurs in an α helix in a set of protein structures is defined as

$$f_\alpha = \frac{n_\alpha}{n} \qquad [8.2]$$

where n_α is the number of amino acid residues of the given type that occur in α helices and n is the total number of residues of this type in the set. The propensity of a particular amino acid residue to occur in an α helix is defined as

$$P_\alpha = \frac{f_\alpha}{<f_\alpha>} \qquad [8.3]$$

where $<f_\alpha>$ is the average value of f_α for all 20 residues. Accordingly, a value of $P_\alpha > 1$ indicates that a residue occurs with greater than average frequency in an α helix. The propensity, P_β, of a residue to occur in a β sheet is similarly defined.

Table 8-1 contains a list of α- and β-propensities based on the analysis of 29 X-ray structures. In accordance with its value of a given propensity, a residue is classified as a strong former (*H*), former (*h*), weak

Table 8-1

Propensities and Classifications of Amino Acid Residues for α Helical and β Sheet Conformations

Residue	P_α	Helix Classification	P_β	Sheet Classification
Ala	1.42	H_α	0.83	i_β
Arg	0.98	i_α	0.93	i_β
Asn	0.67	b_α	0.89	i_β
Asp	1.01	I_α	0.54	B_β
Cys	0.70	i_α	1.19	h_β
Gln	1.11	h_α	1.10	h_β
Glu	1.51	H_α	0.37	B_β
Gly	0.57	B_α	0.75	b_β
His	1.00	I_α	0.87	h_β
Ile	1.08	h_α	1.60	H_β
Leu	1.21	H_α	1.30	h_β
Lys	1.16	h_α	0.74	b_β
Met	1.45	H_α	1.05	h_β
Phe	1.13	h_α	1.38	h_β
Pro	0.57	B_α	0.55	B_β
Ser	0.77	i_α	0.75	b_β
Thr	0.83	i_α	1.19	h_β
Trp	1.08	h_α	1.37	h_β
Tyr	0.69	b_α	1.47	H_β
Val	1.06	h_α	1.70	H_β

Source: Chou, P. Y. and Fasman, G. D., *Annu. Rev. Biochem.* **47**, 258 (1978).

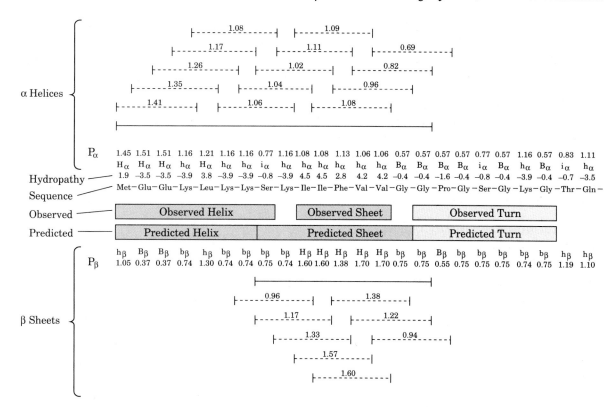

Figure 8-8
The prediction of α helices and β sheets by the method of Chou and Fasman and the prediction of reverse turns by the method of Rose for the N-terminal 24 residues of adenylate kinase. The helix and sheet propensities and classifications are taken from Table 8-1. The solid lines indicate all hexapeptide sequences that can nucleate an α helix (*top*) and all pentapeptide sequences that can nucleate a β sheet (*bottom*) as is explained in the text. The average helix and sheet propensities for each tetrapeptide segment in the helix and sheet regions are given above the corresponding dashed lines. Twelve of 16 residues are observed to have their predicted secondary structures (*middle*) so that the prediction accuracy, in this case, is 75%. Reverse turns are predicted to occur in sequences in which the hydropathy (Table 7-5) is a minimum and which do not occur in helical regions. The region that matches this criterion is observed to have a reverse turn. [After Schultz, G. E. and Schirmer, R. H., *Principles of Protein Structure*, p. 121, Springer–Verlag (1979).]

former (*I*), indifferent former (*i*), breaker (*b*), or strong breaker (*B*) of that secondary structure. Using these data, Chou and Fasman formulated the following empirical rules to predict the secondary structures of proteins:

1. A cluster of four helix-forming residues (H_α or h_α, with I_α counting as one half h_α) out of six contiguous residues will nucleate a helix. The helix segment propagates in both directions until the average value of P_α for a tetrapeptide segment falls below 1.00. A Pro residue, however, can only occur at the N-terminus of an α helix.

2. A cluster of three β sheet formers (H_β or h_β) out of five contiguous residues nucleates a sheet. The sheet is propagated in both directions until the average value of P_β for a tetrapeptide segment falls below 1.00.

3. For regions containing both α and β forming sequences, the overlapping region is predicted to be

helical if its average value of P_α is greater than its average value of P_β; otherwise a sheet conformation is assumed.

These rather simple empirical rules predict the α helix and β sheet strand positions in a protein with an average reliability of ~50% and, in the most favorable cases, ~80% (Fig. 8-8; note that since proteins consist, on average, of ~27% α helix and ~23% β sheet, random predictions of these secondary structures would average 25% correct).

Reverse Turns Are Characterized by a Minimum in Hydrophobicity Along a Polypeptide Chain
The positions of reverse turns can also be predicted by the Chou and Fasman method. However, since a reverse turn usually consists of four consecutive residues, each with a different conformation (Section 7-1D), their prediction algorithm is necessarily more cumbersome than those for sheets and helices.

George Rose has proposed a simpler empirical

method for predicting reverse turns. Reverse turns nearly always occur on the surface of a protein and, in part, define that surface. Since the core of a protein consists of hydrophobic groups and its surface is relatively hydrophilic, reverse turns occur at positions along a polypeptide chain where the hydropathy (Table 7-5) is a minimum. Using this method for partitioning a polypeptide chain, we can deduce the positions of most reverse turns by inspection (Fig. 8-8). Since this method often predicts reverse turns to occur in helical regions (helices are all turns), it should be applied only to regions that are not predicted to be helical.

Secondary Structures Are Partially Dictated by Tertiary Structures

Secondary structures are greatly influenced by tertiary interactions. For instance, in 62 unrelated proteins of known structure comprising some 10,000 amino acid residues, the longest segments with identical amino acid sequences are pentapeptides. In 6 of the 25 such pentapeptide pairs in the sample, the pentapeptide is part of an α helix in one protein and part of a β strand in another. The inability of sophisticated secondary structure prediction schemes to improve significantly on the reliability of simpler schemes is therefore explained by the failure of all of them to take tertiary interactions into account.

Rose has discovered a class of tertiary interactions that appear to be important for helix formation. A backbone C=O group in an α helix accepts a hydrogen bond from the backbone N—H group located four residues further along the polypeptide chain (Section 7-1B). Consequently, an α helix's four N-terminal N—H groups and four C-terminal C=O groups lack intrahelical hydrogen bonding partners. Most α helices in known protein X-ray structures are flanked by residues whose side chains, as conformational analysis indicates, can form hydrogen bonds to these terminal groups. This explains, for example, the observed high preference for an α helix to be preceded by Asn and Pro and to be succeeded by Gly (Asn's side chain C=O and Pro's backbone C=O are, respectively, readily positioned to accept hydrogen bonds from succeeding backbone N—H groups; a Gly at the C-terminus of an α helix, but no other residue, can assume a conformation in which its backbone N—H donates a hydrogen bond to the backbone C=O group three residues down the helix—a 3_{10} helical conformation). The observation that not all of these possible hydrogen bonds occur in native proteins led Rose to postulate that these interactions often function to nucleate helix formation. Such transient interactions may explain the existence of temperature sensitive folding mutants (Section 8-1B), and why identical pentapeptides occur in both α helices and β sheets.

Theoretical calculations of tertiary structures that minimize the conformational energies of polypeptide chains have, so far, yielded results that are not significantly better than random models. The major problem is that polypeptide chains have astronomical numbers of non-nativelike low energy conformations so that it is presently not feasible, even with the fastest available computers, to determine a polypeptide's lowest energy conformation. Conformational energy calculations have been useful, however, in predicting, for example, how a protein of known structure alters its conformation in response to an amino acid residue change or how an enzyme conformationally adjusts to the binding of its substrate.

2. PROTEIN DYNAMICS

The fact that X-ray studies yield time-averaged "snapshots" of proteins may leave the false impression that proteins have fixed and rigid structures. In fact, as is becoming increasing clear, *proteins are flexible and rapidly fluctuating molecules whose structural mobilities have functional significance.* For example, X-ray studies indicate that the heme groups of myoglobin and hemoglobin are so surrounded by protein that there is no clear path for O_2 to approach or escape from its binding pocket. Yet, we know that myoglobin and hemoglobin readily bind and release O_2. These proteins must therefore undergo conformational fluctuations, **breathing motions,** that permit O_2 reasonably free access to their heme groups (Fig. 8-9). The three-dimensional struc-

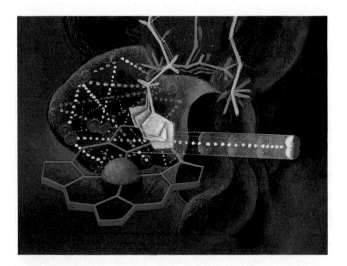

Figure 8-9
An artist's conception of the "breathing" motions in myoglobin that permit the escape of its bound O_2 molecule (*double red spheres*). The dotted lines trace a trajectory an O_2 molecule might take in worming its way through the rapidly fluctuating protein before finally escaping. O_2 binding presumably resembles the reverse of this process. [Figure copyrighted © by Irving Geis.]

tures of myoglobin and hemoglobin undoubtedly evolved the flexibility to facilitate the diffusion of O_2 to its binding pocket.

The intramolecular motions of proteins have been classified into three broad categories according to their coherence:

1. **Atomic fluctuations,** such as the vibrations of individual bonds, which have time periods ranging from 10^{-15} to 10^{-11} s and spatial displacements between 0.01 and 1 Å.

2. **Collective motions,** in which groups of covalently linked atoms, which vary in size from amino acid side chains to entire domains, move as units with time periods ranging from 10^{-12} to 10^{-3} s and spatial displacements between 0.01 and >5 Å. Such motions may occur frequently or infrequently compared with their characteristic time period.

3. **Triggered conformational changes,** in which groups of atoms varying in size from individual side chains to complete subunits move in response to specific stimuli such as the binding of a small molecule, for example, the binding of O_2 to hemoglobin (Sections 9-2A and C). Triggered conformational changes occur over time spans ranging from 10^{-9} to 10^3 s and result in atomic displacements between 0.5 and >10 Å.

In this section, we discuss how these various motions are characterized and their structural and functional significance. We shall mainly be concerned with atomic fluctuations and collective motions; triggered conformational changes are considered in later chapters in connection with specific proteins.

Proteins Have Mobile Structures

X-ray crystallographic analysis is a powerful technique for the analysis of motion in proteins; it reveals not only the average positions of the atoms in a crystal, but also their mean-square displacements from these positions. X-ray analysis indicates, for example, that myoglobin has a rigid core surrounding its heme group and that the regions toward the periphery of the molecule have a more mobile character. A similar analysis of lysozyme has produced the intriguing observation that the enzyme's active site cleft, which undergoes an ~ 1-Å closure upon binding substrate (Section 14-2A), is among the regions of the protein with the greatest mobility.

Theoretical considerations of the internal motions in proteins by Martin Karplus indicate that *a protein's native structure really consists of a large collection of conformational substates that have essentially equal stabilities.* These substates, which each have slightly different atomic arrangements, randomly interconvert at rates that increase with temperature. Computer simulations of protein internal motions (e.g., Fig. 8-10) suggest that the interior of a protein typically has a fluidlike character for structural displacements of up to ~ 2 Å, that is, over excursions that are somewhat larger than a bond distance.

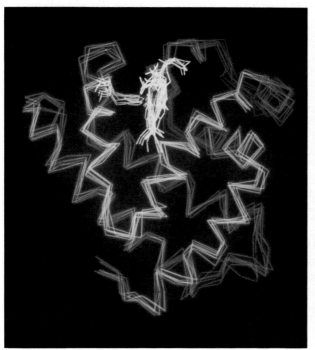

(a)

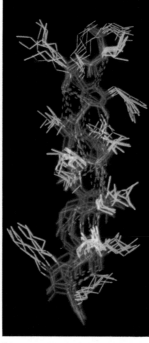

(b)

Figure 8-10
The internal motions of myoglobin as simulated by computerized molecular dynamics calculations. Several "snapshots" of the molecule calculated at intervals of 5×10^{-12} s are superimposed. (a) The C_α backbone and the heme group. The backbone is shown in blue, the heme in yellow, and the proximal His residue in orange. (b) An α helix. The backbone is shown in blue, the side chains in green, and the helix hydrogen bonds as dashed orange lines. Note that the helices tend to move in a coherent fashion so as to retain their shape. [Courtesy of Martin Karplus, Harvard University.]

Protein Core Mobility Is Revealed by Aromatic Ring Flipping

The rate at which an internal Phe or Tyr ring in a protein undergoes 180° "flips" about its C_β—C_γ bonds is indicative of the protein core's rigidity. This is because these bulky asymmetric groups can only move in the close packed interior of a protein when the surrounding groups move aside transiently. The rate that a particular aromatic ring flips is best inferred from an analysis of its NMR spectrum (infrequent motions such as ring flipping are not detected by X-ray crystallography since this technique reveals the average structure of a protein). NMR measurements indicate that the ring flipping rate varies from over 10^6 s^{-1} to one of immobility (<1 s^{-1}) depending on both the protein and the location of the aromatic ring within the protein. Thus, at 4°C, four of BPTI's eight Phe and Tyr rings flip at rates $>5 \times 10^4$ s^{-1}, whereas the remaining four rings flip at rates ranging between 30 and <1 s^{-1}. These ring-flipping rates sharply increase with temperature, as expected.

Infrequent Motions Can Be Detected through Hydrogen Exchange

Conformational changes occurring over time spans of more than several seconds can be chemically characterized through **hydrogen exchange studies.** Weakly acidic protons, such as those of amine and hydroxyl groups (X—H), exchange with those of water as can be demonstrated with the use of tritiated water (HTO).

$$X-H + HTO \rightleftharpoons X-T + H_2O$$

Under physiological conditions, most small organic molecules, such as amino acids and dipeptides, completely exchange such protons in times ranging from milliseconds to seconds.

Proteins bear numerous exchangeable protons such as those of its backbone amide groups. Indeed, tritium exchange studies of native proteins indicate that these protons exchange at rates that vary from under several seconds (the minimum time required for a tritium exchange measurement) to many years (Fig. 8-11). Yet protein interiors, as we have seen, are largely excluded from contact with their surrounding aqueous solvent (Section 7-3B), and, moreover, there is considerable evidence indicating that protons cannot exchange with solvent while they are engaged in hydrogen bonding. The observation that the internal protons of a protein do, in fact, exchange with solvent, must therefore be a consequence of transient local unfolding or "breathing" that physically and chemically exposes these exchangeable protons to the solvent. Hence, *the rate at which a particular proton undergoes hydrogen exchange is a reflection of the conformational mobility of its surroundings.* This hypothesis is corroborated by the observation that the hydrogen exchange rates of proteins decrease as their de-

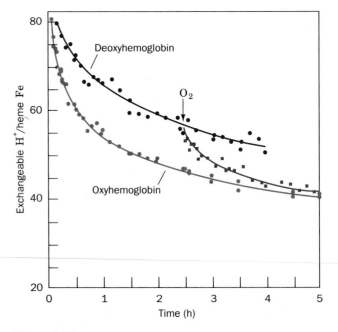

Figure 8-11
The hydrogen–tritium "exchange-out" curve for hemoglobin that has been preequilibrated with tritiated water. The vertical axis expresses the ratio of exchangeable protons to heme Fe atoms. Exchange-out was initiated by replacing the protein's tritiated water solvent with untritiated water through rapid gel filtration (Section 5-3C). As the exchange-out proceeded, additional gel filtration separations were performed and the amount of tritium remaining bound to the protein measured. At the arrow, O_2 was added to exchanging deoxyhemoglobin (hemoglobin lacking bound O_2). The changing slopes of these curves indicate that the hydrogen exchange rates of the ~80 exchangeable protons of each hemoglobin subunit vary by factors of many decades and that O_2 binding increases the exchange rates for ~10 of these protons (the structural changes that O_2 binding induces in hemoglobin are discussed in Section 9-2). [After Englander, S. W. and Mauel, C., *J. Biol. Chem.* **247**, 2389 (1972).]

naturation temperatures increase and that these exchange rates are sensitive to the proteins' conformational states (Fig. 8-11).

3. STRUCTURAL EVOLUTION

Proteins, as we discussed in Section 6-3, evolve through point mutations and gene duplications. Over eons, through processes of natural selection and/or neutral drift, homologous proteins thereby diverge in character and develop new functions. How these primary structure changes affect function, of course, depends on the protein's three-dimensional structure. In this section, we explore the effects of evolutionary change on protein structures.

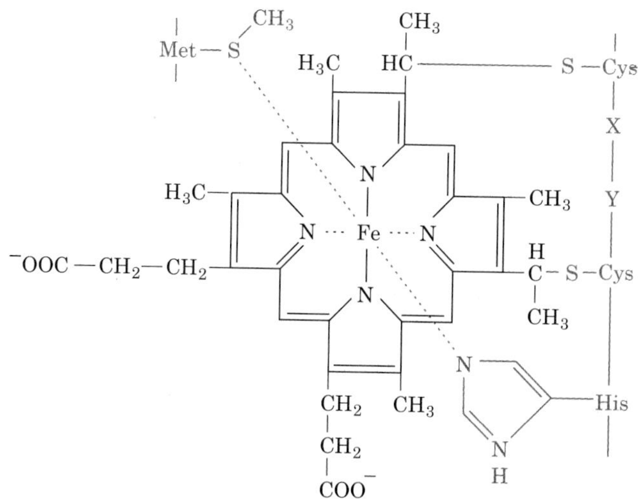

Figure 8-12
The molecular formula of iron–protoporphyrin IX (heme). In c-type cytochromes, the heme is covalently bound to the protein (*red*) by two thioether bonds linking the heme vinyl groups to two Cys residues that occur in the sequence Cys-X-Y-Cys-His. Here X and Y symbolize other amino acids. A fifth and sixth ligand to the Fe atom, both normal to the heme plane, are formed by a nitrogen of the His side chain in this sequence and the sulfur of a Met residue that is further along the polypeptide chain. The iron atom, which is thereby octahedrally liganded, can stably assume either the Fe(II) or the Fe(III) oxidation state. Heme also occurs in myoglobin and hemoglobin but without the thioether bonds or the Met ligand.

A. Structures of Cytochromes *c*

The *c*-type cytochromes are small globular proteins that contain a covalently bound heme (**iron–protoporphyrin IX**) group (Fig. 8-12). The X-ray structures of horse, tuna, and bonito cytochromes *c*, which were elucidated by Richard Dickerson, are quite similar and thus permit the structural significance of cytochrome *c*'s amino acid sequences (Section 6-3B) to be assessed.

The internal residues of cytochrome *c*, particularly those lining its heme pocket, tend to be invariant or conservatively substituted, whereas surface positions have greater variability. This observation is, in part, an indication of the more exacting packing requirements of a protein's internal regions compared to those of its surface (Section 7-3B). Certain invariant or highly conserved residues (Table 6-4) have specific structural and/or functional roles in cytochrome *c*:

1. The invariant Cys 14, Cys 17, His 18, and Met 80 residues form covalent bonds with the heme group (Fig. 8-12).

2. The nine invariant or highly conserved Gly's occupy close-fitting positions in which larger side chains would significantly alter the protein's three-dimensional structure.

3. The highly conserved Lys residues 8, 13, 25, 27, 72, 73, 79, 86, and 87 are distributed in a ring around the exposed edge of the otherwise buried heme group. There is mounting evidence that this unusual constellation of positive charges specifically associates with complementary sets of negative charges on the physiological reaction partners of cytochrome *c*, cytochrome *c* reductase, and cytochrome *c* oxidase (Section 20-2C).

4. The invariant Phe 82, whose aromatic ring lies nearby and parallel to the heme, is thought to form an essential component of the protein's electron conduit in electron-transfer reactions (Section 20-2C).

Prokaryotic *c*-Type Cytochromes Are Structurally Related to Cytochrome *c*

Although cytochrome *c* only occurs in eukaryotes, similar proteins known as *c*-**type cytochromes** are common in prokaryotes, where they function to transfer electrons at analogous positions in a variety of respiratory and photosynthetic electron-transport chains. Unlike the eukaryotic proteins, however, the prokaryotic *c*-type cytochromes exhibit considerable sequence variability among species. For example, the more than 30 bacterial *c*-type cytochromes whose primary structures are known have from 82 to 134 amino acid residues, whereas eukaryotic cytochromes *c* have a narrower range—between 103 and 112 residues. The primary structures of several representative *c*-type cytochromes have few obvious similarities (Fig. 8-13). Yet their X-ray structures closely resemble each other, particularly in their chain folding and side group packing in the regions surrounding the heme group (Fig. 8-14). Furthermore, most of them have aromatic rings in analogous positions and orientations relative to their heme groups as well as similar distributions of positively charged Lys residues about the perimeters of their heme crevices. The major structural differences between these *c*-type cytochromes stem from various loops of polypeptide chain that are located on their surfaces.

The proper alignments of analogous *c*-type cytochrome residues, (thin lines in Fig. 8-13), could not have been made on the basis of only their primary structures: These proteins have diverged so far that their three-dimensional structures were essential guides for this task. Three-dimensional structures are evidently more indicative of the similarities among these distantly related proteins than are primary structures. *It is the essential structural and functional elements of proteins, rather than their amino acid residues, that are conserved during evolutionary change.*

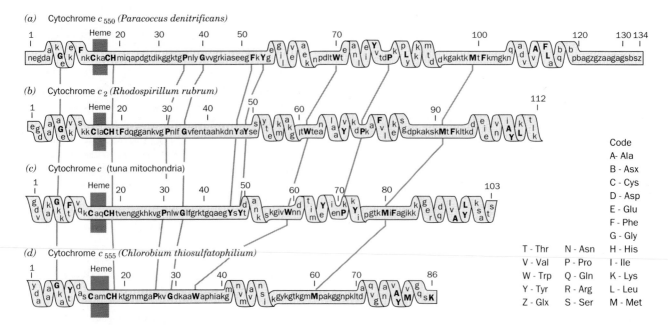

Figure 8-13

The primary structures of some representative *c*-type cytochromes: (*a*) Cytochrome *c₅₅₀* (the subscript indicates the protein's peak absorption wavelength in visible light in nanometers, nm) from *Paracoccus denitrificans,* a respiring bacterium that can use nitrate as an oxidant. (*b*) Cytochrome *c₂* (the subscript only has historical significance) from *Rhodospirillum rubrum,* a purple photosynthetic bacterium. (*c*) Cytochrome *c* from tuna mitochondria. (*d*) Cytochrome *c₅₅₅* from *Chlorobium thiosulfatophilum,* a green photosynthetic bacterium that utilizes H_2S as a hydrogen source. Gray lines connect structurally significant or otherwise invariant residues (capitalized). Helical regions are indicated to facilitate structural comparisons with Fig. 8-14. [After Salemme, F. R., *Annu. Rev. Biochem.* **46,** 307 (1977).]

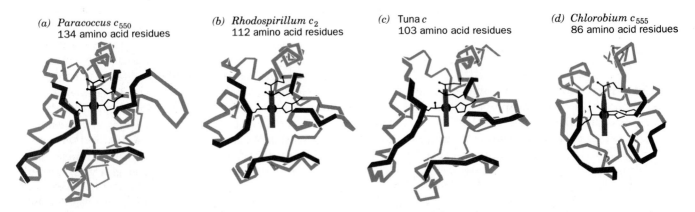

(*a*) *Paracoccus c₅₅₀*
134 amino acid residues

(*b*) *Rhodospirillum c₂*
112 amino acid residues

(*c*) Tuna *c*
103 amino acid residues

(*d*) *Chlorobium c₅₅₅*
86 amino acid residues

Figure 8-14

The three-dimensional structures of the *c*-type cytochromes whose primary structures are displayed in Fig. 8-13. The polypeptide backbones (*blue*) are shown in analogous orientations such that their heme groups (*red*) are viewed edge on. The Cys, Met, and His side chains that covalently link the heme to the protein are also shown. (*a*) Cytochrome *c₅₅₀* from *P. denitrificans.* (*b*) Cytochrome *c₂* from *Rs. rubrum.* (*c*) Tuna cytochrome *c.* (*d*) Cytochrome *c₅₅₅* from *C. thiosulfatophilum.* [Figures copyrighted © by Irving Geis.]

B. Gene Duplication

Gene duplication may promote the evolution of new functions through structural evolution. In over one half of the 50 or so multidomain proteins of known structure, two of the domains are structurally quite similar. Consider, for example, the two domains of the bovine liver enzyme **rhodanese** (Fig. 8-15). It seems highly improbable that its two complex but conformationally similar domains could have independently evolved their present structures (a process known as **convergent evolution**). More likely, they arose through the duplication of the gene specifying an ancestral domain followed by

Figure 8-15   Domain 1   Domain 2
The two structurally similar domains of rhodanese. [After drawings provided by Jane Richardson, Duke University.]

the fusion of the resulting two genes to yield a single gene specifying a polypeptide that folds into two similar domains. The differences between the two domains is therefore due to their **divergent evolution.**

Structurally similar domains occur in proteins whose other domains bear no resemblance to one another. The redox enzymes known as dehydrogenases, for example, each consist of two domains: A coenzyme-binding domain (Fig. 7-49) that is structurally similar in all the dehydrogenases, and a dissimilar substrate-binding domain that determines the specificity and mode of action of each enzyme. Indeed, in some dehydrogenases, such as glyceraldehyde-3-phosphate dehydrogenase (Fig. 7-46), the coenzyme-binding domain occurs at the N-terminal end of the polypeptide chain whereas in others, it occurs at the C-terminal end. Each of these dehydrogenases must have arisen by the fusion of the gene specifying an ancestral coenzyme-binding domain with a gene coding for a proto substrate-binding domain. This must have happened very early in evolutionary history, perhaps in the precellular stage (Section 1-4C), because there are no significant sequence similarities among these coenzyme-binding domains. Evidently, a domain is as much a unit of evolution as it is a unit of structure. *By genetically combining these structural modules in various ways, nature can develop new functions far more rapidly than it can do so by the evolution of completely new structures through point mutations.*

Chapter Summary

Under renaturing conditions, many proteins fold to their native structures in a matter of seconds. Proteins must therefore fold in an ordered manner rather than going through a random search of all their possible conformations. This has been confirmed in the case of BPTI in which the native disulfide bonds reform in a largely specific but indirect sequence. The prediction of protein secondary structures from only amino acid sequences has been reasonably successful using empirical techniques. Techniques for predicting the tertiary structures of proteins, however, are still at a rudimentary stage of development.

Proteins are flexible and fluctuating molecules whose group motions have characteristic periods ranging from 10^{-15} to over 10^3 s. X-ray analysis, which yields the average atomic mobilities in a protein, indicates that proteins tend to be more mobile at their peripheries than in their interiors. Theoretical analysis of protein mobilities suggests that the native protein structures each consist of a large number of closely related and rapidly interconverting conformational substates of nearly equal stabilities. The rates of aromatic ring flipping, as revealed by NMR measurements, indicate that internal group mobilities within proteins vary both with the protein and the position within the protein. Hydrogen exchange studies demonstrate that proteins have a great variety of infrequently occurring internal motions. The exchange of internal protons with solvent probably results from the transient local unfolding of the protein.

The X-ray structures of eukaryotic cytochromes *c* demonstrate that internal residues and those having specific structural and functional roles tend to be conserved during evolution. Prokaryotic *c*-type cytochromes from a variety of organisms structurally resemble each other and those of eukaryotes even though there are few similarities among their amino acid sequences. This indicates that the three-dimensional structures of proteins rather than their amino acid sequences are conserved during evolutionary change. The structural similarities between the domains in many multidomain proteins indicates that these proteins arose through the duplication of a gene specifying an ancestral domain followed by their fusion. Similarly, the structural resemblance between the coenzyme-binding domains of dehydrogenases suggests that these proteins arose by duplication of a primordial coenzyme-binding domain followed by its fusion with a gene specifying a proto-substrate-binding domain. In this manner, proteins with new functions can evolve much faster than by a series of point mutations.

References

Protein Folding

Anfinsen, C. B., Principles that govern the folding of protein chains, *Science* **181**, 223–230 (1973). [A Nobel laureate explains how he got his prize.]

Anfinsen, C. B. and Scheraga, H. A., Experimental and theoretical aspects of protein folding, *Adv. Protein Chem.* **29**, 205–300 (1975).

Cantor, C. R. and Schimmel, P. R., *Biophysical Chemistry,* Chapters 5 and 21, Freeman (1980).

Creighton, T. E., Toward a better understanding of protein folding pathways, *Proc. Natl. Acad. Sci.* **85**, 5082–5086 (1988).

Creighton, T. E., Pathways and mechanisms of protein folding, *Adv. Biophys.* **18**, 1–20 (1984).

Creighton, T. E., *Proteins,* Chapter 7, Freeman (1983).

Creighton, T. E., Experimental studies in protein folding and unfolding, *Prog. Biophys. Mol. Biol.* **33**, 231–297 (1978). [Contains an extensive discussion of the elucidation of the BPTI folding pathway.]

Chou, P. Y. and Fasman, G. D., Empirical predictions of protein structure, *Annu. Rev. Biochem.* **47**, 251–276 (1978) *and* Prediction of the secondary structure of proteins from their amino acid sequence, *Adv. Enzymol.* **47**, 45–148 (1978). [Expositions of one of the most widely used methods for the prediction of protein secondary structures.]

Frauenfelder, H., Parak, F., and Young, R. D., Conformational substates in proteins, *Annu. Rev. Biophys. Biophys. Chem.* **17**, 451–479 (1988).

Gō, N., Theoretical studies of protein folding, *Annu. Rev. Biophys. Bioeng.* **12**, 183–210 (1983).

Goldberg, M. E., The second translation of the genetic message: protein folding and assembly, *Trends Biochem. Sci.* **10**, 388–391 (1985).

Goldenberg, D. P., Genetic studies of protein stability and mechanisms of folding, *Annu. Rev. Biophys. Biophys. Chem.* **17**, 481–507 (1988).

Goldenberg, D. P. and King, J., Trimeric intermediates in the *in vivo* folding and subunit assembly of the tail spike endorhamnosidase of bacteriophage P22, *Proc. Natl. Acad. Sci.* **79**, 3403–3407 (1982). [Evidence that protein folding pathways are genetically controlled.]

Jaenicke, R. (Ed.), *Protein Folding,* Elsevier (1980). [A compendium of papers on various aspects of protein folding.]

Kabsch, W. and Sander, C., On the use of sequence homologies to predict protein structure: identical pentapeptides can have completely different conformations, *Proc. Natl. Acad. Sci.* **81**, 1075–1078 (1984).

Kim, P. S. and Baldwin, R. L., Specific intermediates in the folding reactions of small proteins and the mechanism of protein folding, *Annu. Rev. Biochem.* **51**, 459–489 (1982).

Presta, L. G. and Rose G. D., Helix signals in proteins, *Science* **240**, 1632–1641 (1988).

Richardson, J. S. and Richardson, D. C., Amino acid preferences for specific locations at the ends of α helices, *Science* **240**, 1648–1652 (1988).

Rose, G. D., Prediction of chain turns in globular proteins on a hydrophobic basis, *Nature* **272**, 586–590 (1978).

Schultz, G. E., A critical evaluation of methods for prediction of protein secondary structures, *Annu. Rev. Biophys. Biophys. Chem.* **17**, 1–21 (1988).

Schultz, G. E. and Schirmer, R. H., *Principles of Protein Structure,* Chapters 6 and 8, Springer–Verlag (1979).

States, D. J., Creighton, T. E., Dobson, C. M., and Karplus, M., Conformations of intermediates in the folding of the pancreatic trypsin inhibitor, *J. Mol. Biol.* **195**, 731–739 (1987). [NMR studies of BPTI's folding intermediates.]

Protein Dynamics

Bennett, W. S. and Huber, R., Structural and functional aspects of domain motions in proteins, *CRC Crit. Rev. Biochem.* **15**, 291–384 (1984).

Chothia, C. and Lesk, A. M., Helix movements in proteins, *Trends Biochem. Sci.* **10**, 116–118 (1985).

Huber, R., Flexibility and rigidity of proteins and protein-pigment complexes, *Angew. Chem. Int. Ed. Engl.* **27**, 79–88 (1988).

Karplus, M. and McCammon, J. A., Dynamics of proteins: elements and function, *Annu. Rev. Biochem.* **53**, 263–300 (1983).

Karplus, M. and McCammon, J. A., The dynamics of proteins, *Sci. Am.* **254**(4): 42–51 (1986).

Kossiakoff, A. A., Protein dynamics investigated by neutron diffraction-hydrogen exchange technique, *Nature* **296**, 713–721 (1982).

Levitt, M., Protein conformation, dynamics and folding by computer simulation, *Annu. Rev. Biophys. Bioeng.* **11**, 251–271 (1982).

McCammon, J. A. and Harvey, S. C., *Dynamics of Proteins and Nucleic Acids,* Cambridge University Press (1987).

Ringe, D. and Petsko, G. A., Mapping protein dynamics by X-ray diffraction, *Prog. Biophys. Mol. Biol.* **45**, 197–235 (1985).

Rogero, J. R., Englander, J. J., and Englander, S. W., Measurement and identification of breathing units in hemoglobin by hydrogen exchange, in Sarma R. H. (Ed.), *Biomolecular Stereodynamics,* Vol. 2, pp. 287–298, Adenine Press (1981).

Woodward, C. K. and Hilton, B. D., Hydrogen exchange kinetics and internal motions in proteins and nucleic acids, *Annu. Rev. Biophys. Bioeng.* **8**, 99–127 (1979).

Wüthrich, K. and Wagner, G., Internal motions in globular proteins, *Trends Biochem. Sci.* **3**, 227–230 (1978). [An NMR analysis of motions in BPTI.]

Structural Evolution

Bajaj, M. and Blundell, T., Evolution and the tertiary structure of proteins, *Annu. Rev. Biophys. Bioeng.* **13**, 453–492 (1983).

Dickerson, R. E., The structure and history of an ancient protein, *Sci. Am.* **226**(4): 58–72 (1972) *and* Cytochrome *c* and the evolution of energy metabolism, *Sci. Am.* **242**(3): 137–149 (1980).

Dickerson, R. E., The cytochromes *c*: An exercise in scientific serendipity, *in* Sigman, D. S. and Brazier, M. A. (Eds.), *The*

Evolution of Protein Structure and Function, pp. 172–202, Academic Press (1980).

Dickerson, R. E., Timkovitch, R., and Almassy, R. J., The cytochrome fold and the evolution of bacterial energy metabolism, *J. Mol. Biol.* **100**, 473–491 (1976).

Eventhoff, W. and Rossmann, M., The structures of dehydrogenases, *Trends Biochem. Sci.* **1**, 227–230 (1976).

Mathews, F. S., The structure, function and evolution of proteins, *Prog. Biophys. Mol. Biol.* **45**, 1–56 (1985).

Rossmann, M. G. and Argos, P., The taxonomy of protein structure, *J. Mol. Biol.* **109**, 99–129 (1977).

Salemme, R., Structure and function of cytochromes *c*, *Annu. Rev. Biochem.* **46**, 299–329 (1977).

Schultz, G. E. and Schirmer, R. H., *Principles of Protein Structure*, Chapter 9, Springer–Verlag (1979).

Problems

1. How long should it take the polypeptide backbone of a 6-residue folding nucleus to explore all its possible conformations? Repeat the calculation for 10, 15, and 20-residue folding nuclei. Explain why folding nuclei are thought to be no larger than 15 residues?

*2. Consider a protein with 10 Cys residues. Upon air oxidation, what fraction of the denatured and reduced protein will randomly reform the native set of disulfide bonds if: (a) The native protein has five disulfide bonds? (b) The native protein has three disulfide bonds?

3. Why are β sheets more commonly found in the hydrophobic interiors of proteins rather than on their surfaces so as to be in contact with the aqueous solvent?

4. Under physiological conditions, polylysine assumes a random coil conformation. Under what conditions might it form an α helix?

*5. Predict the secondary structure of the C peptide of proinsulin (Fig. 8-4). Is it likely to have a supersecondary structure?

6. As Mother Nature's chief engineer, now certified as a master helix builder, you are asked to repeat Problem 7-9 with the stipulation that the α helix really be helical. Use Table 8-1.

7. Explain why β sheets are unlikely to form folding nuclei.

8. Folding nuclei are thought to be predominantly stabilized by hydrophobic forces. Why aren't hydrogen bonding forces equally effective?

9. Indicate the probable effects of the following mutational changes on the structure of a protein. Explain your reasoning. (a) Changing a Leu to a Phe, (b) changing a Lys to a Glu, (c) changing a Val to a Thr, (d) changing a Gly to an Ala, and (e) changing a Met to a Pro.

10. Explain why Trp rings are usually completely immobile in proteins that have rapidly flipping Phe and Tyr rings.

*11. Discuss the merits of the hypothesis that the coenzyme-binding domains of the dehydrogenases arose by convergent evolution.

Chapter 9
HEMOGLOBIN: PROTEIN FUNCTION IN MICROCOSM

The existence of hemoglobin, the red blood pigment, is evident to every child who scrapes a knee. Its brilliant red color, widespread occurrence, and ease of isolation have made it an object of inquiry since ancient times. Indeed, the early history of protein chemistry is essentially that of hemoglobin. The observation of crystalline hemoglobin was first reported in 1849, and by 1909 a photographic atlas of hemoglobin crystals from 109 species had been published. In contrast, it was not until 1926 that crystals of an enzyme, those of jack bean **urease,** were first reported. Hemoglobin was one of the first proteins to have its molecular mass accurately determined, the first protein to be characterized by ultracentrifugation and to be associated with a specific physiological function (that of oxygen transport) and, in sickle-cell anemia, the first in which a point mutation was demonstrated to cause a single amino acid change (Section 6-3A). Theories formulated to account for the cooperative binding of oxygen to hemoglobin (Section 9-4) have also been successful in explaining the control of enzyme activity. The first protein X-ray structures to be elucidated were those of hemoglobin and myoglobin. This central role in the development of protein chemistry together with its enzymelike O_2-binding properties have caused hemoglobin to be dubbed an "honorary enzyme."

Hemoglobin is not just a simple oxygen tank. Rather, it is a sophisticated oxygen delivery system that provides the proper amount of oxygen to the tissues under a wide variety of circumstances. In this chapter, we discuss hemoglobin's properties, structure, and mechanism of action, both to understand the workings of this physiologically essential molecule and to illustrate the principles of protein structure that we developed in the preceding chapters. We also consider the properties of abnormal hemoglobins and their relationship to human disease. Finally, we discuss theories of cooperative interactions among proteins, both to better understand the properties of hemoglobin and to set the stage for our later consideration of how enzyme action is regulated.

1. HEMOGLOBIN FUNCTION

Hemoglobin **(Hb)**, as we have seen in Chapters 6 and 7, is a tetrameric protein, $\alpha_2\beta_2$ (alternatively, a dimer of $\alpha\beta$ protomers). The α and β subunits are structurally and evolutionarily related to each other and to myoglobin **(Mb)**, the monomeric oxygen-binding protein of muscle (Section 6-3C).

Hemoglobin transports oxygen from the lungs, gills, or skin of an animal to its capillaries for use in respiration. Very small organisms do not require such a protein because their respiratory needs are satisfied by the simple passive diffusion of O_2 through their bodies. However, since the transport rate of a diffusing substance varies inversely with the square of the distance it must diffuse, the O_2-diffusion rate through tissue thicker than ~ 1 mm is too slow to support life. The evolution of organisms as large and complex as annelids (e.g., earthworms) therefore required the development of circulatory systems that actively transport O_2 and nutrients to the tissues. The blood of such organisms must contain an oxygen transporter such as Hb because the solubility of O_2 in **blood plasma** (the fluid component of blood) is too low ($\sim 10^{-4}M$ under physiological conditions) to carry sufficient O_2 for metabolic needs. In contrast, whole blood, which normally contains ~ 150 g of $Hb \cdot L^{-1}$, can carry O_2 at concentrations as high as $0.01M$, about the same as in air. [Although many invertebrate species have hemoglobin-based oxygen transport systems, others produce one of two alternative types of O_2-binding proteins: (1) **hemocyanin,** a Cu-containing protein that is blue in complex with oxygen and colorless otherwise; or (2) **hemerythrin,** a nonheme Fe-containing protein that is burgundy colored in complex with oxygen and colorless otherwise. Antarctic icefish, the only adult vertebrates that lack hemoglobin — their blood is colorless — are viable because of the relatively high aqueous solubility of O_2 at the $-1.9°C$ temperature of their environment; recall that the solubilities of gases increase with decreasing temperature.]

Although Mb was originally assumed to function only to store oxygen, it is now apparent that *its major physiological role is to facilitate oxygen transport in rapidly respiring muscle.* The rate that O_2 can diffuse from the capillaries to the tissues, and thus the level of respira-

tion, is limited by the oxygen's low solubility in aqueous solution. Mb increases the effective solubility of O_2 in muscle, the most rapidly respiring tissue under conditions of high exertion. Thus, in rapidly respiring muscle, Mb functions as a kind of molecular bucket brigade to facilitate O_2 diffusion. The O_2 storage function of Mb is probably only significant in aquatic mammals such as seals and whales, which have Mb concentrations in their muscles around 10-fold greater than those in terrestrial mammals.

In this section, we begin our discussions of hemoglobin by considering its chemical and physical properties and how they relate to its physiological function.

A. Heme

Mb and each of the four subunits of Hb noncovalently bind a single heme group (Fig. 9-1; spelled "haem" in British English). This is the same group that occurs in the cytochromes (Section 8-3A) and in certain redox enzymes such as **catalase.** Heme is responsible for the characteristic red color of blood and is the site at which each **globin** monomer binds one molecule of O_2 (globins are the heme-free proteins of Hb and Mb). The heterocyclic ring system of heme is a **porphyrin** derivative; it consists of four **pyrrole** rings (numbered in Fig. 9-1) linked by methene bridges. The porphyrin in heme,

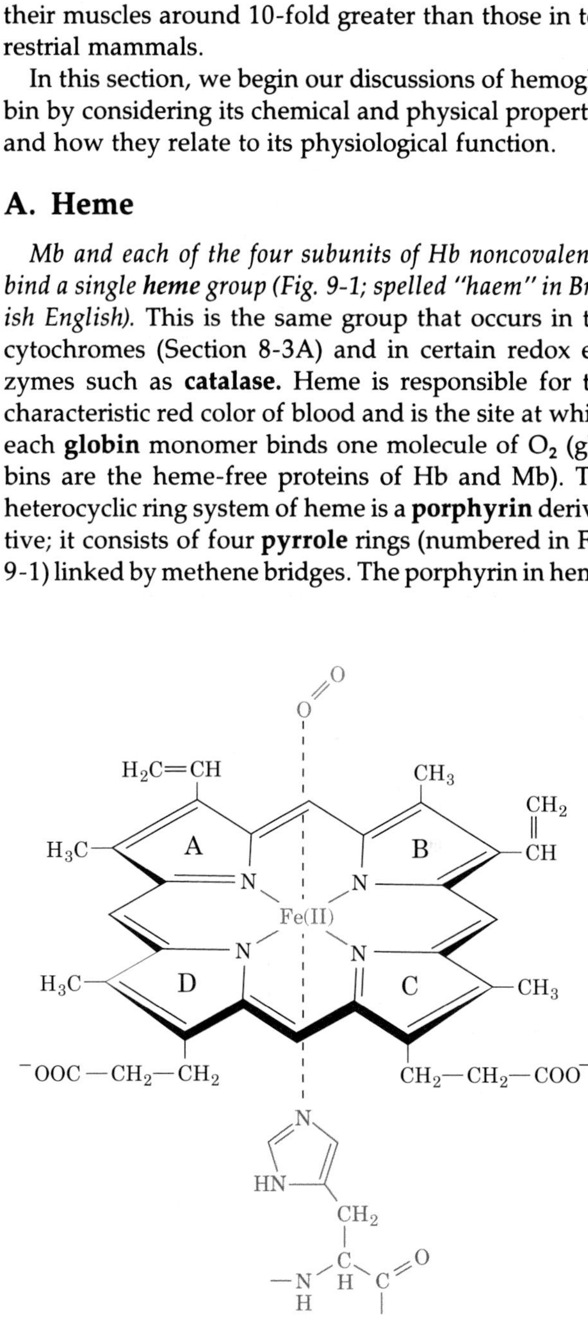

Figure 9-1
Fe(II) heme (ferroprotoporphyrin IX) shown liganded to His and O_2 as it is in oxygenated myoglobin and oxygenated hemoglobin. Note that the heme is a conjugated system so that, although two of its Fe—N bonds are coordinate covalent bonds (bonds in which the bonding electron pair is formally contributed by only one of the atoms forming the bond), all of the Fe—N bonds are equivalent. The pyrrole ring numbering scheme is shown.

with its particular arrangement of four methyl, two propionate, and two vinyl substituents, is known as **protoporphyrin IX**. Heme, then, is protoporphyrin IX with a centrally bound iron atom. In Hb and Mb, the iron atom normally remains in the Fe(II) (ferrous) oxidation state whether or not the heme is oxygenated (binds O_2).

The Fe atom in deoxygenated Hb and Mb is 5-coordinated by a square pyramid of N atoms: four from the porphyrin and one from a His side chain of the protein. Upon oxygenation, the O_2 binds to the Fe(II) on the opposite side of the porphyrin ring from the His ligand so that the Fe(II) is octahedrally coordinated; that is, the ligands occupy the six corners of an octahedron centered on the Fe atom (Fig. 9-1). *Oxygenation changes the electronic state of the Fe(II)-heme as is indicated by the color change of blood from the dark purplish hue characteristic of venous blood to the brilliant scarlet color of arterial blood and blood from a cut finger (Fig. 9-2).*

Certain small molecules, such as CO, NO, and H_2S, coordinate to the sixth liganding position of the Fe(II) in Hb and Mb with much greater affinity than does O_2. This, together with their similar binding to the hemes of cytochromes, accounts for the highly toxic properties of these substances.

The Fe(II) of Hb or Mb can be oxidized to Fe(III) to form **methemoglobin (metHb)** or **metmyoglobin (metMb)**. MetHb does not bind O_2; its Fe(III) is already octahedrally coordinated with an H_2O molecule in the sixth liganding position. The brown color of dried blood and old meat is that of metHb and metMb. Erythrocytes (red blood cells) contain the enzyme **methemoglobin reductase**, which converts the small amount of metHb that spontaneously forms back to the Fe(II) form.

B. Oxygen Binding

The binding of O_2 to myoglobin is described by a simple equilibrium reaction

$$Mb + O_2 \rightleftharpoons MbO_2$$

with dissociation constant

$$K = \frac{[Mb][O_2]}{[MbO_2]} \tag{9.1}$$

(biochemists usually express equilibria in terms of dissociation constants, the reciprocals of the more chemically traditional association constants). The O_2 dissociation of Mb may be characterized by its **fractional saturation, Y_{O_2},** defined as the fraction of O_2-binding sites occupied by O_2.

$$Y_{O_2} = \frac{[MbO_2]}{[Mb] + [MbO_2]} = \frac{[O_2]}{K + [O_2]} \tag{9.2}$$

Since O_2 is a gas, its concentration is conveniently expressed by its partial pressure, pO_2 (also called the **oxygen tension**). Equation [9.2] may therefore be expressed:

$$Y_{O_2} = \frac{pO_2}{K + pO_2} \tag{9.3}$$

Now define p_{50} as the value of pO_2 when $Y_{O_2} = 0.50$, that is, when one half of myoglobin's O_2-binding sites are occupied. Equation [9.3] then indicates that $K = p_{50}$ so that our expression for the fractional saturation of Mb finally becomes:

$$Y_{O_2} = \frac{pO_2}{p_{50} + pO_2} \tag{9.4}$$

Hemoglobin Cooperatively Binds O_2

Myoglobin's O_2-dissociation curve (Fig. 9-3) closely follows the hyperbolic curve described by Eq. [9.4]; its p_{50} is 2.8 torr (1 torr = 1 mmHg at 0°C = 0.133 kPa; 760 torr = 1 atm). Mb therefore gives up little of its bound O_2 over a relatively wide range of pO_2, for example, $Y_{O_2} = 0.97$ at $pO_2 = 100$ torr and 0.88 at 20 torr. In contrast, hemoglobin's O_2-dissociation curve (Fig. 9-3), which has a **sigmoidal** shape (S shape) that Eq. [9.4] cannot describe, indicates the amount of O_2 bound by Hb changes significantly over a relatively small range of pO_2, for example, $Y_{O_2} = 0.95$ at 100 torr and 0.30 at 20 torr in whole blood. Hemoglobin's sigmoidal O_2-dissociation curve is of great physiological importance; it *permits the blood to deliver much more O_2 to the tissues than it could if Hb had a hyperbolic O_2-dissociation curve resembling that of Mb.* A sigmoidal dissociation curve is diagnostic of a **cooperative interaction** between a protein's small molecule binding sites; that is, the binding of one small molecule affects the binding of others. In this case, the binding of O_2 increases the affinity of Hb

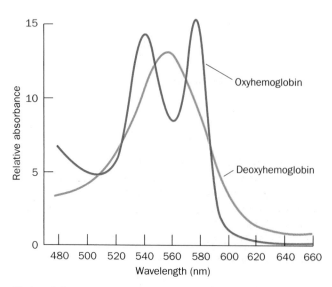

Figure 9-2
The visible absorption spectra of oxygenated and deoxygenated hemoglobins.

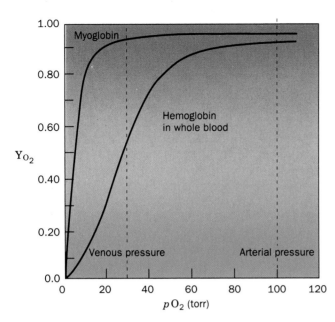

Figure 9-3
The oxygen dissociation curves of Mb and Hb in whole blood. The normal sea level values of human arterial and venous pO_2 values are indicated.

for binding additional O_2. The structural mechanism of hemoglobin cooperativity is described in Section 9-2C.

The Hill Equation Phenomenologically Describes Hemoglobin's O_2-Binding Curve

The earliest attempt to analyze hemoglobin's sigmoidal O_2-dissociation curve was formulated by Archibald Hill in 1910. We shall follow his analysis in general form because it is useful for characterizing the cooperative behavior of oligomeric enzymes as well as that of hemoglobin.

Consider a protein E consisting of n subunits that can each bind a molecule S, which, in analogy with the substituents of metal ion complexes, is known as a **ligand**. Assume that the ligand binds with infinite cooperativity,

$$E + nS \rightleftharpoons ES_n$$

that is, the protein either has all or none of its ligand-binding sites occupied so that there are no observable intermediates ES_1, ES_2, *etc.* The dissociation constant for this reaction is

$$K = \frac{[E][S]^n}{[ES_n]} \qquad [9.5]$$

and, as before, its fractional saturation is expressed:

$$Y_S = \frac{n[ES_n]}{n([E] + [ES_n])} \qquad [9.6]$$

Combining Eqs. [9.5] and [9.6] yields

$$Y_S = \frac{n[E][S]^n/K}{n[E](1 + [S]^n/K)}$$

that upon algebraic rearrangement and cancellation of terms becomes the **Hill equation:**

$$Y_S = \frac{[S]^n}{K + [S]^n} \qquad [9.7]$$

which, in a manner analogous to Eq. [9.4], describes the degree of saturation of a multisubunit protein as a function of ligand concentration.

Infinite ligand-binding cooperativity, as assumed in deriving the Hill equation, is a physical impossibility. Nevertheless, n may be taken to be a nonintegral parameter related to the degree of cooperativity among interacting ligand-binding sites rather than the number of subunits per protein. The Hill equation then becomes a useful empirical curve-fitting relationship rather than an indicator of a particular model of ligand binding. *The quantity n, the **Hill constant**, increases with the degree of cooperativity of a reaction and thereby provides a convenient although simplistic characterization of a ligand-binding reaction.* If $n = 1$, Eq. [9.7] describes a hyperbola as do Eqs. [9.3] and [9.4] for Mb, and the ligand-binding reaction is said to be **noncooperative**. A reaction with $n > 1$ is described as **positively cooperative**: Ligand binding increases the affinity of E for further ligand binding (cooperativity is infinite in the limit that n is equal to the number of ligand-binding sites in E). Conversely, if $n < 1$, the reaction is termed **negatively cooperative**: Ligand binding reduces the affinity of E for subsequent ligand binding.

Hill Equation Parameters May Be Graphically Evaluated

The Hill constant, n, and the dissociation constant, K, that best describe a saturation curve can be graphically determined by rearranging Eq. [9.7] as follows:

$$\frac{Y_S}{1 - Y_S} = \frac{\dfrac{[S]^n}{K + [S]^n}}{1 - \dfrac{[S]^n}{K + [S]^n}} = \frac{[S]^n}{K}$$

and then taking the log of both sides to yield a linear equation:

$$\log\left(\frac{Y_S}{1 - Y_S}\right) = n \log [S] - \log K \qquad [9.8]$$

The linear plot of $\log [Y_S/(1 - Y_S)]$ versus $\log [S]$, the **Hill plot**, has a slope of n and an intercept on the $\log [S]$ axis of $(\log K)/n$ (recall that the linear equation $y = mx + b$ describes a line with a slope of m and an x intercept of $-b/m$).

For Hb, if we substitute pO_2 for $[S]$ as was done for Mb, the Hill equation becomes:

$$Y_{O_2} = \frac{(pO_2)^n}{K + (pO_2)^n} \qquad [9.9]$$

As in Eq. [9.4], let us define p_{50} as the value of pO_2 at

$Y_{O_2} = 0.50$. Then, substituting this value into Eq. [9.9],

$$0.50 = \frac{(p_{50})^n}{K + (p_{50})^n}$$

so that

$$K = (p_{50})^n \qquad [9.10]$$

Substituting this result back into Eq. [9.9] yields

$$Y_{O_2} = \frac{(pO_2)^n}{(p_{50})^n + (pO_2)^n} \qquad [9.11]$$

(*Note:* Eq. [9.4] is a special case of Eq. [9.11] with $n = 1$). Equation [9.8] for the Hill plot of Hb therefore takes the form

$$\log\left(\frac{Y_{O_2}}{1 - Y_{O_2}}\right) = n \log pO_2 - n \log p_{50} \qquad [9.12]$$

so that *this plot has a slope of n and an intercept on the log pO_2 axis of p_{50}.*

Figure 9-4 shows the Hill plots for Mb and Hb. For Mb it is linear with a slope of 1, as expected. Although Hb does not bind O_2 in a single step as is assumed in deriving the Hill equation, its Hill plot is essentially linear for values of Y_{O_2} between 0.1 and 0.9. Its maximum slope, which occurs near $pO_2 = p_{50}$ $[Y_{O_2}/(1 - Y_{O_2}) = 1]$, is normally taken to be the Hill constant. For normal human Hb, the Hill constant is between 2.8 and 3.0, that is, hemoglobin oxygen binding is highly, but not infinitely, cooperative. Many abnormal hemoglobins exhibit smaller Hill constants (Section 9-3A), indicating that they have a less than normal degree of cooperativity. At Y_{O_2} values near 0, when few Hb molecules have bound even one O_2 molecule, the Hill plot of Hb assumes a slope of one (Fig. 9-4, lower asymptote) because the Hb subunits independently compete for O_2 as do

molecules of Mb. At Y_{O_2} values near 1, when at least three of each of hemoglobin's four O_2-binding sites are occupied, the Hill plot also assumes a slope of 1 (Fig. 9-4, upper asymptote) because the few remaining unoccupied sites are on different molecules and therefore bind O_2 independently.

Extrapolating the lower asymptote in Fig. 9-4 to the horizontal axis indicates, according to Eq. [9.11], that $p_{50} = 30$ torr for binding the first O_2 to Hb. Likewise, extrapolating the upper asymptote yields $p_{50} = 0.3$ torr for binding hemoglobin's fourth O_2. Thus *the fourth O_2 to bind to Hb does so with 100-fold greater affinity than the first.* This difference, as we shall see in Section 9-2C, is entirely due to the influence of the globin on the O_2 affinity of heme. It corresponds to a free energy difference of $11.4 \text{ kJ} \cdot \text{mol}^{-1}$ between binding the first and the last O_2 to Hb (Section 3-4A).

More sophisticated mathematical models than the Hill equation have been developed for analyzing the cooperative binding of ligands to proteins. They are examined in Section 9-4.

Globin Prevents Oxyheme from Autooxidizing

Globin not only modulates the O_2-binding affinity of heme, but makes O_2 binding possible. Fe(II) heme by itself is incapable of binding O_2 reversibly. Rather, in the presence of O_2, it autooxidizes irreversibly to the Fe(III) form through the intermediate formation of a complex consisting of an O_2 bridging the Fe atoms of two hemes. This reaction can be inhibited by derivatizing the heme with bulky groups that sterically prevent the close face-to-face approach of two hemes. Such **picket-fence** Fe(II)–porphyrin complexes (Fig. 9-5), which James Collman first synthesized, bind O_2 reversibly. The

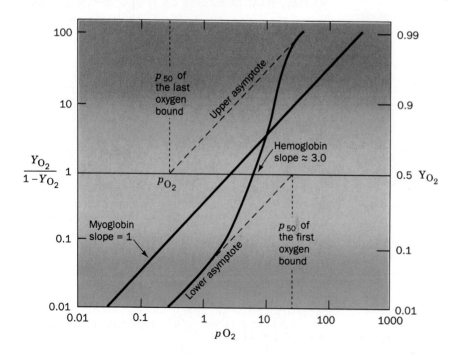

Figure 9-4
Hill plots for Mb and purified Hb. Note that this is a log–log plot. At $pO_2 = p_{50}$, the value of $Y_{O_2}/(1 - Y_{O_2}) = 1$.

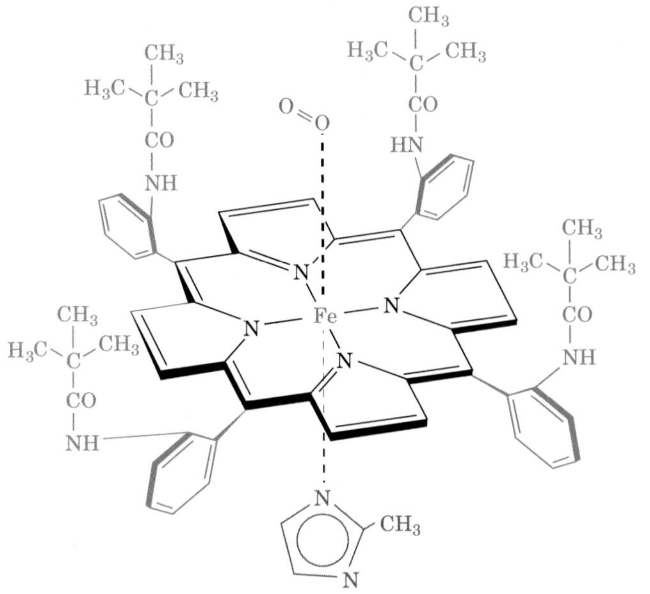

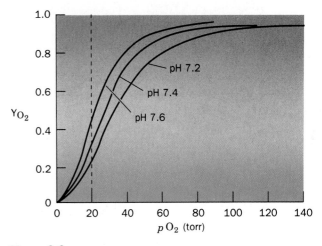

Figure 9-5
A picket-fence Fe(II)–porphyrin complex with bound O_2. [After Collman, J. P., Brauman, J. I., Rose, E., and Suslick, K. S., *Proc. Natl. Acad. Sci.* **75**, 1053 (1978).]

backside of this porphyrin is unhindered and is complexed with a substituted imidazole in a manner similar to that in Mb and Hb. In fact, the O_2 affinity of the picket-fence complex is similar to that of Mb. Thus, the globins of Mb and Hb function to prevent the autooxidation of oxyheme by surrounding it, rather like a hamburger bun surrounds a hamburger, so that only its propionate side chains are exposed to the aqueous solvent (Section 9-2B).

C. Carbon Dioxide Transport and the Bohr Effect

In addition to being an O_2 carrier, *Hb plays an important role in the transport of CO_2 by the blood.* When Hb (but not Mb) binds O_2 at physiological pH's, it undergoes a conformational change (Section 9-2B) that makes it a slightly stronger acid. It therefore releases protons on binding O_2:

$$Hb(O_2)_nH_x + O_2 \rightleftharpoons Hb(O_2)_{n+1} + xH^+$$

where $n = 0, 1, 2,$ or 3 and $x \approx 0.6$ under physiological conditions. Conversely, *increasing the pH, that is, removing protons, stimulates Hb to bind O_2 (Fig. 9-6)*. This phenomenon, whose molecular basis is discussed in Section 9-2E, is known as the **Bohr effect** after Christian Bohr (the father of the pioneering atomic physicist Niels Bohr) who first reported it in 1904.

The Bohr Effect Facilitates O_2 Transport

The ~ 0.8 molecules of CO_2 formed per molecule of O_2 consumed by respiration diffuse from the tissues to the capillaries largely as dissolved CO_2 as a result of the slowness of the reaction forming bicarbonate:

$$CO_2 + H_2O \rightleftharpoons H^+ + HCO_3^-$$

This reaction, however, is catalyzed in the erythrocyte by **carbonic anhydrase** (Fig. 7-44). Accordingly, most of the CO_2 in the blood is carried in the form of bicarbonate (in the absence of carbonic anhydrase, the hydration of CO_2 would equilibrate 100-fold more slowly so that bubbles of the only slightly soluble CO_2 would form in the blood and tissues).

In the capillaries, where pO_2 is low, the H^+ generated by bicarbonate formation is taken up by Hb, which is thereby induced to unload its bound O_2. This H^+ uptake, moreover, facilitates CO_2 transport by stimulating bicarbonate formation. Conversely, in the lungs, where pO_2 is high, O_2 binding by Hb releases the Bohr protons, which drive off the CO_2. These reactions are closely matched so that they cause very little change in blood pH.

The Bohr effect provides a mechanism whereby additional O_2 can be supplied to highly active muscles. Such muscles generate acid (Section 16-3A) so fast that they lower the pH of the blood passing through them from 7.4 to 7.2. At pH 7.2, Hb releases $\sim 10\%$ more O_2 at the <20 torr pO_2 in these muscles than it does at pH 7.4 (Fig. 9-6).

CO_2 and Cl^- Modulate Hemoglobin's O_2 Affinity

CO_2 modulates O_2 binding directly and by combining reversibly with the N-terminal amino groups of blood proteins to form carbamates:

$$R-NH_2 + CO_2 \rightleftharpoons R-NH-COO^- + H^+$$

The conformation of deoxygenated Hb (**deoxyHb**), as we shall see in Section 9-2B, is significantly different from that of oxygenated Hb (**oxyHb**). Consequently,

Figure 9-6
The effect of pH on the O_2-dissociation curve of Hb: The Bohr effect. [After Benesch, R. E. and Benesch, R., *Adv. Protein Chem.* **28**, 212 (1974).]

deoxyHb binds more CO_2 as carbamate than does oxyHb. CO_2, like H^+, is therefore a modulator of hemoglobin's O_2 affinity: A high CO_2 concentration, as occurs in the capillaries, stimulates Hb to release its bound O_2. Note the complexity of this $Hb\text{—}O_2\text{—}CO_2\text{—}H^+$ equilibrium: The protons released by carbamate formation are, in part, taken up through the Bohr effect thereby increasing the amount of O_2 that Hb would otherwise release. Although the difference in CO_2 binding between the oxy and deoxy states of hemoglobin accounts for only ~5% of the total blood CO_2, it is nevertheless responsible for around one half of the CO_2 transported by blood. This is because only ~10% of the total blood CO_2 turns over in each circulatory cycle.

Cl^- is also bound more tightly to deoxyHb than to oxyHb (Section 9-2E). Accordingly, hemoglobin's O_2 affinity also varies with $[Cl^-]$. HCO_3^- freely permeates the erythrocyte membrane (Section 11-3C) so that once formed, it equilibrates with the surrounding plasma. The need for charge neutrality on both sides of the membrane, however, requires that Cl^-, which also freely permeates the membrane, replace the HCO_3^- that leaves the erythrocyte (the erythrocyte membrane is impermeable to cations). Consequently, $[Cl^-]$ in the erythrocyte is greater in the venous blood than it is in the arterial blood. *Cl^- is therefore also a modulator of hemoglobin's O_2 affinity.*

D. Effect of BPG on O_2 Binding

Purified (stripped) hemoglobin has a much greater O_2 affinity than does hemoglobin in whole blood (Fig. 9-7). This observation led Joseph Barcroft, in 1921, to speculate that blood contains some other substance that com-

plexes with Hb so as to reduce its O_2 affinity. In 1967, Reinhold and Ruth Benesch demonstrated that this substance is **D-2,3-bisphosphoglycerate (BPG)**

D-2,3-Bisphosphoglycerate (BPG)

Inositol hexaphosphate (IHP)

Adenosine triphosphate (ATP)

[previously known as **2,3,-diphosphoglycerate (DPG)**]. BPG binds tightly to deoxyHb in a 1:1 mol ratio ($K = 1.5 \times 10^{-5}M$) but only weakly to oxyHb. The presence of BPG therefore decreases hemoglobin's oxygen affinity by keeping it in the deoxy conformation; for example, the p_{50} of stripped hemoglobin is increased from 12 to 22 torr by 4.7 mM BPG, its normal concentration in erythrocytes (similar to that of Hb). Organic polyphosphates, such as **inositol hexaphosphate (IHP)** and **adenosine triphosphate (ATP)**, also have this effect on Hb. In fact, in birds, IHP functionally replaces BPG and ATP does so in fish and most amphibians. The ~2 mM ATP normally present in mammalian erythrocytes is prevented from binding to Hb by its complexation with Mg^{2+}.

BPG has an indispensable physiological function: In arterial blood, where pO_2 is ~100 torr, Hb is ~95% saturated with O_2, but in venous blood where pO_2 is ~30 torr, it is only ~55% saturated (Fig. 9-3). Consequently, in passing through the capillaries, Hb unloads ~40% of its O_2. *In the absence of BPG, little of this O_2 is*

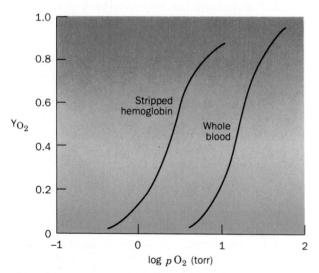

Figure 9-7
Comparison of the O_2-dissociation curves of "stripped" Hb and whole blood in 0.01M NaCl at pH 7.0. [After Benesch, R. E. and Benesch, R., *Adv. Protein Chem.* **28**, 217 (1974).]

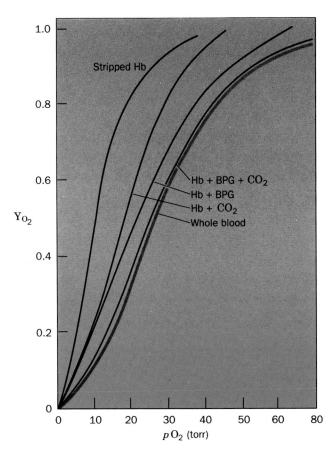

Figure 9-8
The effects of BPG and CO_2, both separately and combined, on hemoglobin's O_2-dissociation curve compared with that of whole blood (*red curve*). In the Hb solutions, which were 0.1*M* KCl and pH 7.22, $pCO_2 = 40$ torr and the BPG concentration was 1.2 times that of Hb. The blood had $pCO_2 = 40$ torr and a plasma pH of 7.40, which corresponds to a pH of 7.22 inside the erythrocyte. [After Kilmartin, J. V. and Rossi-Bernardi, L., *Physiol. Rev.* **53**, 884 (1973).]

released since hemoglobin's O_2-dissociation curve then shifts significantly towards lower pO_2 (Fig. 9-8, left).

CO_2 and BPG independently modulate hemoglobin's O_2 affinity. Figure 9-8 indicates that Hb has the same oxygen-dissociation curve in whole blood and when mixed with CO_2 and BPG in the concentrations found in erythrocytes (the pH and [Cl⁻] are also the same). Hence, *the presence of these four substances in whole blood—BPG, CO_2, H^+, and Cl^-—accounts for the O_2-binding properties of Hb.*

Increased BPG Levels Are Partially Responsible for High Altitude Adaptation

High altitude adaptation is a complex physiological process that involves an increase in the amount of hemoglobin per erythrocyte and in the number of erythrocytes. It normally requires several weeks to complete. Yet, as is clear to anyone who has climbed to high altitude, even a one-day stay there results in a noticeable degree of adaptation. This effect results from a rapid

increase in the erythrocyte BPG concentration (Fig. 9-9; BPG, which cannot pass through the erythrocyte membrane, is synthesized in the erythrocyte; Section 16-2H). The consequent decrease in O_2-binding affinity, as indicated by its elevated p_{50}, increases the amount of O_2 that hemoglobin unloads in the capillaries (Fig. 9-10). Similar increases in BPG concentration occur in individuals suffering from disorders that limit the oxygenation of the blood (**hypoxia**) such as various anemias and cardiopulmonary insufficiency.

Fetal Hemoglobin Has a Low BPG Affinity

The effects of BPG also help supply the fetus with oxygen. A fetus obtains its O_2 from the maternal circulation via the placenta. This process is facilitated if fetal hemoglobin (HbF) has a higher O_2 affinity than does maternal hemoglobin (HbA; recall that HbF has the subunit composition $\alpha_2\gamma_2$ in which the γ subunit is a variant of HbA's β subunit; Section 6-3C). BPG occurs in about the same concentrations in adult and fetal erythrocytes but binds more tightly to deoxyHbA than to deoxyHbF; this accounts for HbF's greater O_2 affinity. In the next section we shall develop the structural rationale for the effect of BPG and for the other aspects of O_2 binding.

2. STRUCTURE AND MECHANISM

The determination of the first protein X-ray structures, those of sperm whale myoglobin by John Kendrew in 1959 and of human deoxyhemoglobin and horse methemoglobin by Max Perutz shortly thereafter, ushered in a revolution in biochemical thinking that has

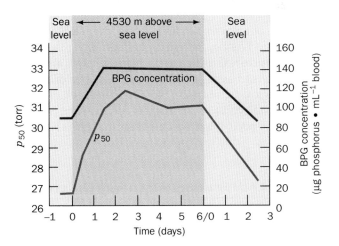

Figure 9-9
The effect of high altitude exposure on the p_{50} and the BPG concentration of blood in sea level-adapted individuals. The region on the right marked "Sea level" indicates the effects of exposure to sea level on high altitude-adapted individuals. [After Lenfant, C., Torrance, J. D., English, E., Finch, C. A., Reynafarje, C., Ramos, J., and Faura, J., *J. Clin. Invest.* **47**, 2653 (1968).]

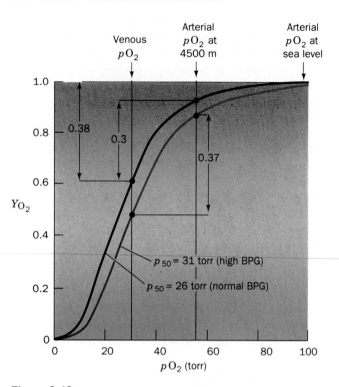

Figure 9-10
The O_2-dissociation curves of blood adapted to sea level (*black*) and to high altitude (*purple*). Between the sea level arterial and venous pO_2 values of 100 and 30 torr, respectively, Hb normally unloads 38% of the O_2 it can maximally carry. However, when the arterial pO_2 drops to 55 torr, as it does at an altitude of 4500 m, this difference is reduced to 30% in nonadapted blood. High altitude adaptation increases the BPG concentration in erythrocytes, which shifts the O_2-dissociation curve of Hb to the right. The amount of O_2 that Hb delivers to the tissues is thereby restored to 37% of its maximum load.

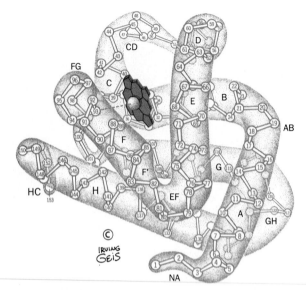

Figure 9-11
The structure of sperm whale myoglobin. Its 153 C_α positions are numbered from the N-terminus and its eight helices are sequentially labeled A through H. The last half of the EF corner is now regarded as a turn of helix and is therefore designated the F' helix. The heme group is shown in red. Also see Figs. 7-40 through 7-42. [Figure copyrighted © by Irving Geis.]

the interactions of complex, structurally well-defined molecules.

The story of hemoglobin's structural determination is a tale of enormous optimism and tenacity. Perutz began this study in 1937 at Cambridge University as a graduate student of John D. Bernal (who, with Dorothy Crowfoot Hodgkin, had taken the first single-crystal X-ray diffraction photographs of a protein in 1934). In 1937, the X-ray crystal structure determination of even the smallest molecule required many months of hand computation and the largest structure yet determined was that of the dye phthalocyanin, which has 40 nonhydrogen atoms. Since hemoglobin has some 4500 nonhydrogen atoms, it must have seemed to Perutz's colleagues that he was pursuing an impossible goal. Nevertheless, the laboratory director, Lawrence Bragg (who in 1912, with his father William Bragg, had solved the first X-ray structure, that of NaCl), realized the tremendous biolog-

reshaped our understanding of the chemistry of life. Before the advent of protein crystallography, macromolecular structures, if they were considered at all, were thought of as having a rather hazy existence of uncertain biological significance. However, as the elucidation of macromolecular structures has continued at an ever quickening pace, it has become clear that *life is based on*

Table 9-1

The Amino Acid Sequences of the α and β Chains of Human Hemoglobin and of Human Myoglobin[a,b]

Helix Boundaries	A1		A16 B1		B16 C1	C7		D1	D7 E1		
Hb α	V−LSPADKTNVKAAWGKVGAHAGEYGAEALERMFLSFPTTKTYFPHF−DLSH−−−−−GSAQVKGHGKKVADALT										
Hb β	VHLTPEEKSAVTALWGKV−−NVDEVGGEALGRLLVVYPWTQRFFESFGDLSTPDAVMGNPKVKAHGKKVLGAFS										
Mb	G−LSDGEWQLVLNVWGKVEADIPGHGQEVLIRLFKGHPETLEKFDKFKHLKSEDEMKASEDLKKHGATVLTALG										

[a] The residues have been aligned in structurally analogous positions. The blue boxes shade the residues that are identical in both Hb chains, the purple boxes shade the residues that are identical in both Hb chains and in Mb, and the dark purple boxes shade

residues that are invariant in all vertebrate Hb and Mb chains (Thr C4, Phe CD1, Leu F4, His F8, and Tyr HC2). The one letter amino acid symbols are defined in Table 4-3.
[b] The first and last residues in helices A−H are indicated, whereas

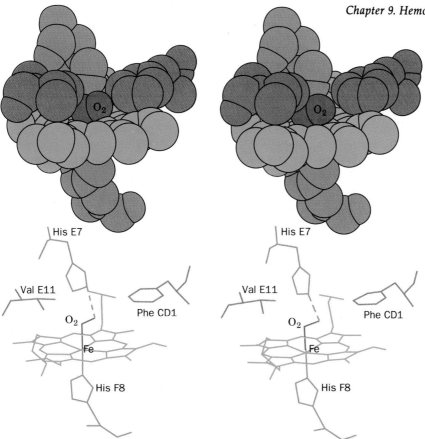

Figure 9-12
Stereo drawings of the heme complex in oxyMb. In the upper drawing, atoms are represented as spheres of van der Waals radii. The lower drawing shows the corresponding skeletal model with a dashed line representing the hydrogen bond between the distal His and the bound O_2. Instructions for viewing stereo drawings are given in the appendix to Chapter 7. [After Phillips, S. E. V., *J. Mol. Biol.* **142**, 544 (1980).]

ical significance of determining a protein structure and supported the project. It was not until 1953 that Perutz finally hit upon the method that would permit him to solve the X-ray structure of hemoglobin, that of isomorphous replacement. Kendrew, a colleague of Perutz, used this technique to solve the X-ray structure of sperm whale myoglobin, first at low resolution in 1957, and then at high resolution in 1959. Hemoglobin's greater complexity delayed its low resolution structural determination until 1959, and it was not until 1968, over 30 years after he had begun the project, that Perutz and his associates obtained the high resolution X-ray structure of horse methemoglobin. Those of human and horse deoxyhemoglobins followed shortly thereafter. Since then, the X-ray structures of hemoglobins from numerous different species, from mutational variants, and with different bound ligands have been elucidated.

This, together with many often ingenious physicochemical investigations have made hemoglobin the most intensively studied, and perhaps the best understood, of proteins.

In this section, we examine the molecular structures of myoglobin and hemoglobin, and consider the structural basis of hemoglobin's oxygen-binding cooperativity, the Bohr effect and BPG binding.

A. Structure of Myoglobin

Myoglobin consists of eight helices (labeled A – H) that are linked by short polypeptide segments to form an ellipsoidal molecule of approximate dimensions 44 × 44 × 25 Å (Fig. 9-11). The helices range in length from 7 to 26 residues and incorporate 121 of myoglobin's 153 residues (Table 9-1). They are largely α helical but with

```
        E19           F1          F9        G1                        G19        H1                         H19      H26
                                                                                                                     H21
        70    75    80    85    90    95    100   105   110   115   120   125   130   135   140
Hb α..N A V A H V D D M P N A L S A L S D L H A H K L R V D P V N F K L L S H C L L V T L A A H L P A E F T P A V H A S L D K F L A S V S T V L T S K Y R
        75    80    85    90    95    100   105   110   115   120   125   130   135   140   145
Hb β...D G L A H L D N L K G T F A T L S E L H C D K L H V D P E N F R L L G N V L V C V L A H H F G K E F T P P V Q A A Y Q K V V A G V A N A L A H K Y H
        75    80    85    90    95    100   105   110   115   120   125   130   135   140   145   150
Mb......G I L K K K G H H E A E I K P L A Q S H A T K H K I P V K Y L E F I S E C I I Q V L Q S K H P G D F G A D A Q G A M N K A L E L F R K D M A S N Y K E L G F Q G
```

the residues between helices constitute the intervening "corners." The refined Hb structure reveals that much of what is designated the EF corner is really helical in both chains. This segment, which encompasses residues EF4-F2, is designated the F' helix.

Source: Dickerson, R. E. and Geis, I., *Hemoglobin*, pp. 68–69, Benjamin/Cummings (1983).

some distortions from this geometry such as a tightening of the final turns of helices A, C, E, and G to form segments of 3_{10} helix. In a helix numbering convention peculiar to Mb and Hb, residues are designated according to their position in a helix or interhelical segment. For example, residue B5 is the 5th residue from the N-terminus of the B helix and residue FG3 is the 3rd residue from the N-terminus in the nonhelical segment (known as a "corner") connecting helices F and G. The nonhelical N- and C-terminal segments are designated NA and HC, respectively. The usual convention of sequentially numbering all amino acid residues from the N-terminal residue of the polypeptide is also used, and often, both conventions are used together. For example, Glu EF6(83) of Mb is the 83rd residue from its N-terminus and the 6th residue in its EF corner.

The heme is tightly wedged in a hydrophobic pocket formed mainly by helices E and F but which includes contacts with helices B, C, G, and H as well as the CD and FG corners. The fifth ligand of the heme Fe(II) is His F8, the **proximal** (near) **histidine.** In oxyMb, the Fe(II) is positioned 0.22 Å out of the heme plane on the side of the proximal His and is coordinated by O_2 with the bent

geometry shown in Fig. 9-12. His E7, the **distal** (distant) **histidine,** hydrogen bonds to the O_2. In deoxyMb, the sixth liganding position of the Fe(II) is unoccupied because the distal His is too far away from the Fe(II) to coordinate with it. Furthermore, the Fe(II) has moved to a point 0.55 Å out of the heme plane. Other structural changes in Mb upon changing oxygenation states consist of small motions of various chain segments and slight readjustments of side chain conformations. *By and large, however, the structures of oxy- and deoxyMb are nearly superimposable.*

B. Structure of Hemoglobin

The hemoglobin tetramer is a spheroidal molecule of dimensions $64 \times 55 \times 50$ Å. Its two $\alpha\beta$ protomers are symmetrically related by a twofold rotation (Figs. 7-56 and 9-13). *The tertiary structures of the α and β subunits are remarkably similar, both to each other and to that of Mb (Figs. 9-11 and 9-13),* even though only 18% of the corresponding residues are identical among these three polypeptides (Table 9-1) and there is no D helix in hemoglobin's α subunit. Indeed, *the α and β subunits in the*

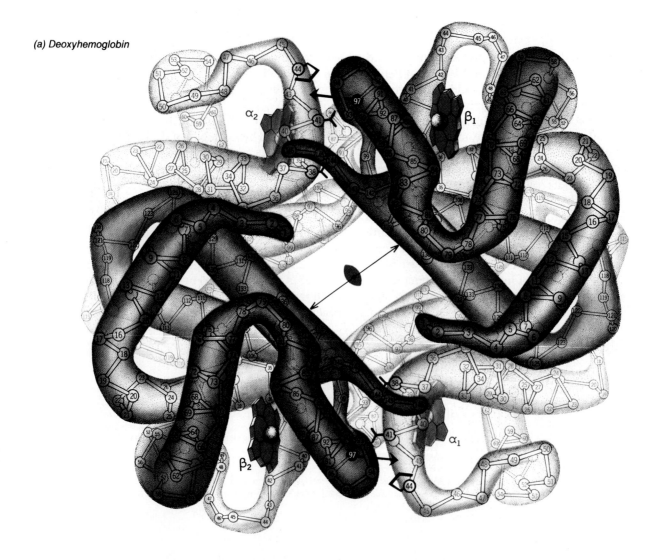

(a) Deoxyhemoglobin

tetramer are related by pseudo- (inexact) twofold rotations so that the subunits occupy the vertices of a tetrahedron (pseudo D_2 symmetry; Section 7-5B).

The polypeptide chains of Hb are arranged such that there are extensive interactions between unlike subunits. The $\alpha_1 - \beta_1$ interface (and its $\alpha_2 - \beta_2$ symmetry equivalent) involves 35 residues, whereas the $\alpha_1 - \beta_2$ (and $\alpha_2 - \beta_1$) interface involves 19 residues. These associations are predominantly hydrophobic in character, although numerous hydrogen bonds and several ion pairs are also involved (see below). In contrast, contacts between like subunits, $\alpha_1 - \alpha_2$ and $\beta_1 - \beta_2$, are few and largely polar in character. This is because like subunits face each other across an ~20 Å in diameter solvent-filled channel that parallels the 50-Å length of the exact twofold axis (Fig. 9-13).

Oxy- and Deoxyhemoglobins Have Different Quaternary Structures

Oxygenation causes such extensive quaternary structural changes to Hb that oxy- and deoxyHb have different crystalline forms; indeed, crystals of deoxyHb shatter upon exposure to O_2. The crystal structures of hemoglobin's oxy and deoxy forms therefore had to be determined independently. *The quaternary structural change preserves hemoglobin's exact twofold symmetry and takes place entirely across its $\alpha_1 - \beta_2$ (and $\alpha_2 - \beta_1$) interface.* The $\alpha_1 - \beta_1$ (and $\alpha_2 - \beta_2$) contact is unchanged, presumably as a result of its more extensive close associations. This contact provides a convenient frame of reference from which the oxy and deoxy conformations may be

Figure 9-13
The structures of (a) deoxyHb and (b) oxyHb, as viewed down their exact twofold axes. The C_α atoms, numbered from each N-terminus, and the heme groups are shown. The Hb tetramer contains a solvent-filled central channel paralleling its twofold axis whose flanking β chains draw closer together upon oxygenation (compare the lengths of double-headed arrows). In the deoxy state, His FG4(97)β (*small single-headed arrow*) fits between Thr C3(38)α and Thr C6(41)α (*lower right* and *upper left*). The relative movements of the two $\alpha\beta$ protomers upon oxygenation (*large gray arrows in Part b*) shift His FG4(97)β to a new position between Thr C6(41)α and Pro CD2(44)α. See Fig. 7-56 for a space-filling model of Hb. [Figure copyrighted © by Irving Geis.]

(b) Oxyhemoglobin

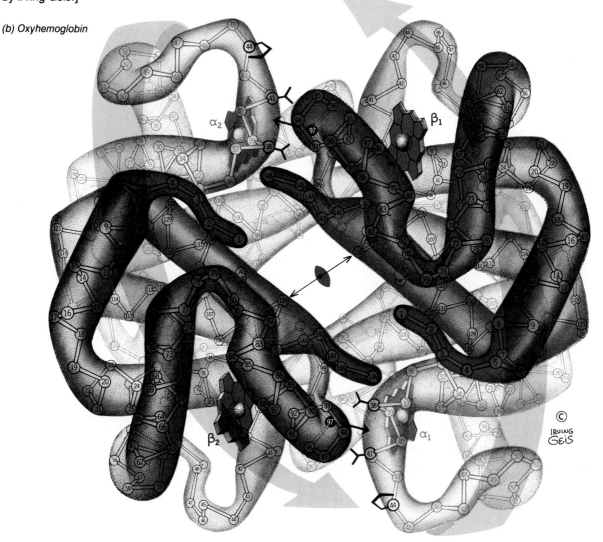

compared. Viewed in this way, oxygenation rotates the $\alpha_1\beta_1$ dimer ~15° with respect to the $\alpha_2\beta_2$ dimer (Fig. 9-14) so that some atoms at the $\alpha_1-\beta_2$ interface shift by as much as 6 Å relative to each other (compare Fig. 9-13a and b).

The quaternary conformation of deoxyHb is named the **T state.** That of oxyHb, which is essentially independent of the ligand used to induce it (e.g., O_2, met, CO, CN^-, and NO hemoglobins all have the same quaternary structure), is called the **R state.** Similarly, the tertiary conformational states for the deoxy and liganded subunits are designated as the **t** and **r states,** respectively. The structural differences between these quaternary and these tertiary conformations are described in the following section in terms of hemoglobin's O_2-binding mechanism.

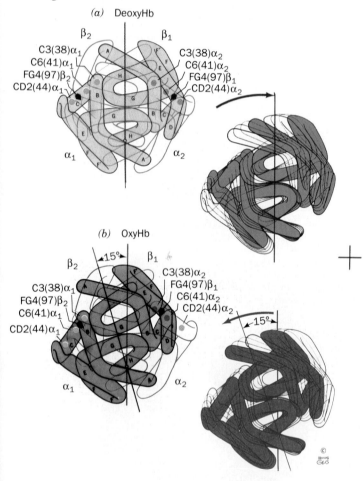

Figure 9-14
The major structural differences between the quaternary conformations of (a) deoxyHb and (b) oxyHb. Upon oxygenation, the $\alpha_1\beta_1$ (*shaded*) and $\alpha_2\beta_2$ (*outline*) protomers move, as indicated on the right, as rigid units such that there is an ~15° off-center rotation of one protomer relative to the other that preserves the molecule's exact twofold symmetry. Note how the position of His FG4β (*pentagons*) changes with respect to Thr C3α, Thr C6α, and Pro CD2α (*yellow dots*) at the $\alpha_1-\beta_2$ and $\alpha_2-\beta_1$ interfaces. The view is from the right side relative to that in Fig. 9-13. [Figure copyrighted © by Irving Geis.]

C. Mechanism of Oxygen-Binding Cooperativity

The positive cooperativity of O_2 binding to Hb arises from the affect of the ligand-binding state of one heme on the ligand-binding affinity of another. Yet, the distances of 25 to 37 Å between the hemes in an Hb molecule are too large for these heme–heme interactions to be electronic in character. Rather, *they are mechanically transmitted by the protein.* The elucidation of how this occurs has motivated much of the structural research on Hb for the past two decades.

X-ray crystal structure analysis has provided "snapshots" of the R and T states of Hb in various states of ligation but does not indicate how the protein changes states. It is difficult to determine the sequence of events that result in such transformations because to do so requires an understanding of the inner workings of proteins that is presently lacking. It is as if you were asked to explain the mechanism of a complicated mechanical clock from its out-of-focus photographs when you had only a hazy notion of how gears, levers, and springs might function. Nevertheless, largely on the basis of the X-ray structures of Hb, Perutz has formulated the following mechanism of Hb oxygenation, the **Perutz mechanism.**

The Movement of Fe(II) into the Heme Plane Triggers the T → R Conformation Shift

In the t state, the Fe(II) is situated ~0.6 Å out of the heme plane on the side of the proximal His because of a pyramidal doming of the porphyrin skeleton and because the Fe—$N_{porphyrin}$ bonds are too long to allow the Fe to lie in the porphyrin plane (Figs. 9-15 and 9-16). The change in the heme's electronic state on binding O_2, however, causes the doming to subside and the Fe—$N_{porphyrin}$ bonds to contract by ~0.1 Å. Consequently, on changing from the t to the r state, the Fe(II) moves to the center of the heme plane (Fig. 9-16) where O_2 can coordinate it without steric interference from the porphyrin. The Fe's movement drags the proximal His along with it, which tilts the attached F helix and translates it ~1 Å across the heme plane (Fig. 9-16). This lateral translation occurs because, in the t state, the imidazole ring of the proximal His is oriented such that its direct movment of 0.6 Å towards the heme plane would cause it to collide with the heme (Figs. 9-15 and 9-16); however, the F helix shift reorients the imidazole ring thereby permitting the Fe(II) to move into the heme plane. Another consideration is that in the β but not the α subunits, Val E11 partially occludes the O_2-binding pocket so that it must be moved aside before O_2 binding can occur.

The $\alpha_1-\beta_2$ and $\alpha_2-\beta_1$ Contacts Have Two Stable Positions

As we saw above, the difference between hemoglo-

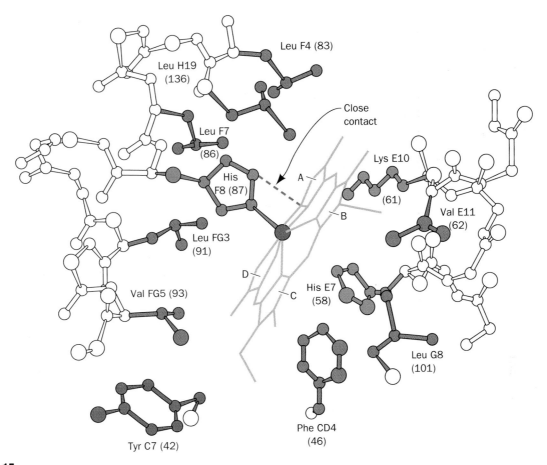

Figure 9-15
The heme group and its environment in the unliganded α chain of Hb. Only selected side chains are shown and the heme 4-propionate group is omitted for clarity. The F helix runs along the left side of the drawing. The close contact between the proximal His and the heme group that inhibits oxygenation of t-state hemes is indicated by a dashed line. [After Gelin, B. R., Lee, A. W.-N., and Karplus, M., *J. Mol. Biol.* **171,** 542 (1983).]

bin's R and T conformations occurs mainly in the $\alpha_1 - \beta_2$ (and the symmetry related $\alpha_2 - \beta_1$) interface, which consists of the C helix and FG corner of α_1, respectively, contacting the FG corner and C helix of β_2. The quaternary change results in a 6-Å relative shift at the α_1C – β_2FG interface (Fig. 9-14). In the T state, His FG4(97)β is in contact with Thr C6(41)α (Fig. 9-13*a* and 9-17, *left*), whereas in the R state it is in contact with Thr C3(38)α, one turn back along the C helix (Figs. 9-13*b* and 9-17, *right*). In both conformations, the "knobs" on one subunit mesh nicely with the "grooves" on the other (Fig. 9-17). An intermediate position, however, would be severely strained because it would bring His FG4(97)β and Thr C6(41)α too close together (knobs on knobs). Hence *these contacts, which are joined by different but equivalent sets of hydrogen bonds in the two states (Figs. 9-18 and 9-19), act as a binary switch that permits only two stable positions of the subunits relative to each other.* In contrast, the quaternary change causes only a 1-Å shift at the α_1FG – β_2C contact so that its side chains maintain the

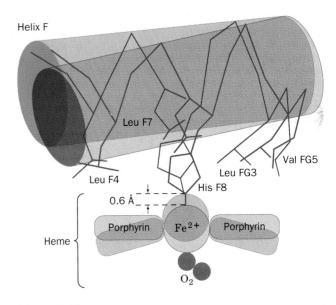

Figure 9-16
The triggering mechanism for the T → R transition in Hb. In the T form (*blue*), the Fe is 0.6 Å above the domed porphyrin ring. Upon assuming the R form (*red*), the Fe moves into the plane of the now undomed porphyrin where it can readily bind O_2 and, in doing so, pulls the proximal His F8 and its attached F helix with it. The Fe—O_2 bond is thereby strengthened because of the relaxation of the steric interference between the O_2 and the heme.

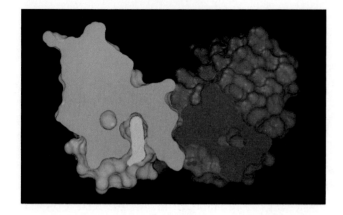
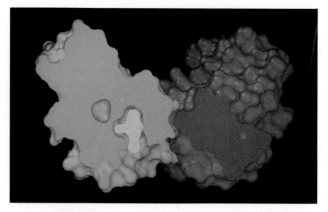

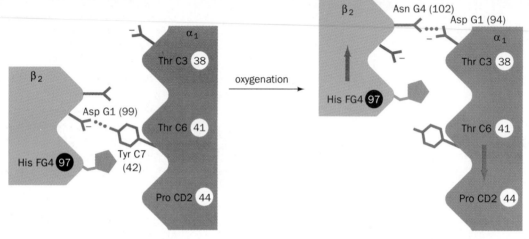

T Form (deoxy)　　　　　　**R Form (oxy)**

Figure 9-17

Surface drawings of the $\alpha_1\beta_2$ dimer of (*a*) T state Hb, and (*b*) R state Hb, that have been sectioned perpendicular to the protein's exact twofold axis so as to show its α_1C–β_2FG interface. Each drawing is accompanied by a corresponding schematic diagram of the α_1–β_2 contact. Upon a T → R transformation, this contact snaps from one position to the other with no stable intermediate. The subunits are joined by different sets of hydrogen bonds in the two quaternary states. Figures 9-13, 9-14 and 9-19 provide additional structural views of these interactions. [Courtesy of Michael Pique, Research Institute of Scripps Clinic.]

same associations throughout the change (Fig. 9-19). *These side chains therefore act as flexible joints or hinges about which the α_1 and β_2 subunits pivot during the quaternary change.*

The T State Is Stabilized by a Network of Salt Bridges That Must Break to Form the R State

The R state is stabilized by ligand binding. But in the absence of ligand, why is the T state more stable than the R state? In the X-ray maps of R state Hb, the C-terminal residues of each subunit (Arg 141α and His 146β) appear as a blur, which suggests that these residues are free to wave about in solution. Maps of the T form, however, show these residues firmly anchored in place via several intersubunit and intrasubunit salt bridges, which evidently help stabilize the T state (Figs. 9-18 and

9-19). *The structural changes accompanying the T → R transition tear away these salt bridges in a process driven by the Fe—O_2 bond's energy of formation.*

Hemoglobin's O_2-Binding Cooperativity Derives from the T → R Conformational Shift

The hemoglobin molecule resembles a finely tooled mechanism that has very little slop. The binding of O_2 requires a series of tightly coordinated movements:

1. The Fe(II) of any subunit cannot move into its heme plane without the reorientation of its proximal His so as to prevent this residue from bumping into the porphyrin ring.

2. The proximal His is so tightly packed by its surrounding groups that it cannot reorient unless this movement is accompanied by the previously described translation of the F helix across the heme plane.

3. The F helix translation is only possible in concert with the quaternary shift that steps the α_1C–β_2 FG contact one turn along the α_1C helix.

4. The inflexibility of the α_1–β_1 and α_2–β_2 interfaces requires that this shift simultaneously occur at both the α_1–β_2 and the α_2–β_1 interfaces.

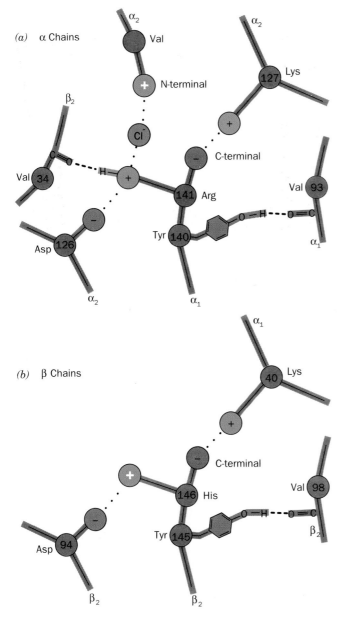

(a) α Chains

(b) β Chains

Figure 9-18
Networks of salt bridges and hydrogen bonds in deoxyHb involving the last two residues of (*a*) the α chains, and (*b*) the β chains of deoxyHb. All of these bonds are ruptured in the T → R transition. The two groups that participate in the Bohr effect by becoming partially deprotonated in the R state are indicated by white plus signs. [Figure copyrighted © by Irving Geis.]

Consequently, *no one subunit or dimer can greatly change its conformation independently of the others. Indeed, the two stable positions of the* $\alpha_1 C - \beta_2 FG$ *contact limits the Hb molecule to only two quaternary forms, R and T.*

We are now in a position to structurally rationalize hemoglobin's O_2-binding cooperativity. Any deoxyHb subunit binding O_2 is constrained to remain in the t state by the T conformation of the tetramer. However, *the t state has reduced O_2 affinity, most probably because its Fe—O_2 bond is stretched beyond its normal length by the steric repulsions between the heme and the O_2, and in the β subunits, by the need to move Val E11 out of the O_2-binding*

site. As more O_2 is bound to the Hb tetramer, this strain, which derives from the Fe—O_2 bond energy, accumulates in the liganded subunits until it is of sufficient strength to snap the molecule into the R conformation. All the subunits are thereby converted to the r state whether or not they are liganded. *Unliganded subunits in the r state have an increased O_2 affinity because they are already in the O_2-binding conformation.* This accounts for the high O_2 affinity of nearly saturated Hb.

Hemoglobin's Sigmoidal O_2-Binding Curve Is a Composite of Its Hyperbolic R- and T-State Curves

The relative stabilities of the T and R states, as indicated by their free energies, vary with fractional saturation (Fig. 9-20*a*). In the absence of ligand, the T state is more stable than the R state, and *vice versa* when all ligand-binding sites are occupied. The formation of Fe—O_2 bonds causes the free energy of both the T and the R states to decrease (become more stable) with oxygenation although the rate of this decrease is smaller for the T state as a result of the strain that liganding imposes on t-state subunits. The R $\rightleftharpoons$ T transformation is, of course, an equilibrium process so that Hb molecules, at intermediate levels of fractional saturation (1, 2, or 3 bound O_2 molecules), continually interconvert between the R and the T states.

The O_2-binding curve of Hb can be understood as a composite of those of its R and T states (Fig. 9-20*b*). For pure states, such as R or T, these curves are hyperbolic because ligand binding at one protomer is unaffected by the state of other protomers in the absence of a quaternary structural change. At low pO_2's, Hb follows the low affinity T-state curve and at high pO_2's, it follows the high affinity R-state curve. At intermediate pO_2's, Hb exhibits an O_2 affinity that changes from T-like to R-like as pO_2 increases. The switchover results in the sigmoidal shape of hemoglobin's O_2-binding curve.

D. Testing the Perutz Mechanism

The Perutz mechanism is a description of the dynamic behavior of Hb that is largely based on the static structures of its R and T end states. Accordingly, without the direct demonstration that Hb actually follows the postulated pathway in changing conformational states, the Perutz mechanism must be taken as being at least partially conjectural. Unfortunately, the physical methods that can follow dynamic changes in proteins are, as yet, incapable of providing detailed descriptions of these changes. Nevertheless, certain aspects of the Perutz mechanism are supported by static measurements as is described below and in Section 9-3.

C-Terminal Salt Bridges Are Required to Maintain the T State

The proposed function of the C-terminal salt bridges in stabilizing the T state has been corroborated by chemically modifying human Hb. Removal of the C-terminal

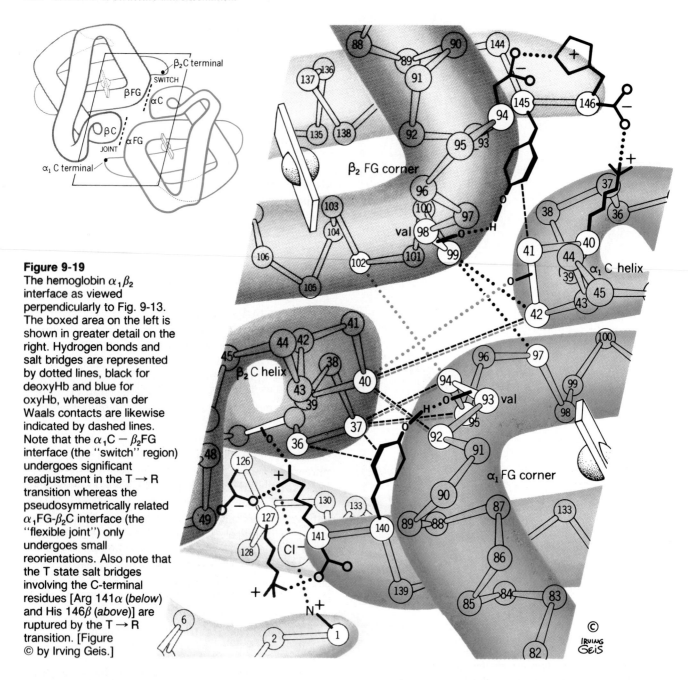

Figure 9-19
The hemoglobin $\alpha_1\beta_2$ interface as viewed perpendicularly to Fig. 9-13. The boxed area on the left is shown in greater detail on the right. Hydrogen bonds and salt bridges are represented by dotted lines, black for deoxyHb and blue for oxyHb, whereas van der Waals contacts are likewise indicated by dashed lines. Note that the α_1C — β_2FG interface (the "switch" region) undergoes significant readjustment in the T → R transition whereas the pseudosymmetrically related α_1FG-β_2C interface (the "flexible joint") only undergoes small reorientations. Also note that the T state salt bridges involving the C-terminal residues [Arg 141α (below) and His 146β (above)] are ruptured by the T → R transition. [Figure © by Irving Geis.]

Arg 141α (by treating isolated α chains with carboxypeptidase followed by reconstitution) drastically reduces the cooperativity of O_2 binding (Hill constant = 1.7; reduced from its normal value of 2.8). It is abolished by the further removal of the other C-terminal residue, His 146β (Hill constant $\approx$ 1.0). Apparently, in the absence of its C-terminal salt bridges, the T form of Hb is unstable. Indeed, human deoxyHb, with its C-terminal residues removed, crystallizes in a form very similar to that of normal human oxyHb.

Fe—O_2 Bond Tension Has Been Spectroscopically Demonstrated

If movement of the Fe into the heme plane upon oxygenation is mechanically coupled via the proximal

His to the T → R transformation, then conversely, forcing Hb into the T form must exert a tension on the Fe, through the proximal His, that tends to pull the Fe out of the heme plane. Perutz demonstrated the existence of this tension as follows. IHP's six phosphate groups cause it to bind to deoxyHb with much greater affinity than does BPG (the structural basis of BPG binding to Hb is discussed in Section 9-2F); the presence of IHP therefore tends to force Hb into the T state. Conversely, nitric oxide (NO) binds to Hb far more strongly than does O_2 and thereby tends to force Hb into the R state. Spectroscopic analysis indicates the consequences of simultaneously binding both NO and IHP to Hb:

1. The NO, as expected, pulls the Fe into the plane of the heme.

(a)

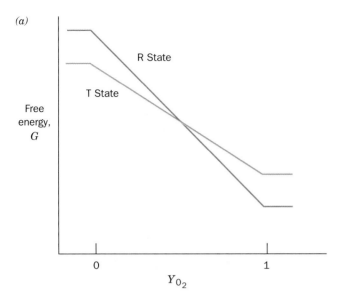

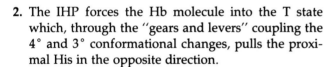

(b)

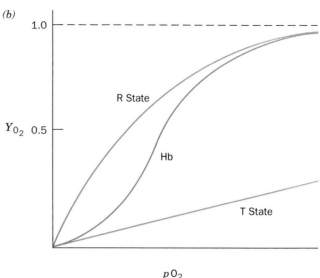

Figure 9-20
(a) The variation of the free energies of hemoglobin's T and R states with their fractional saturation, Y_{O_2}. In the absence of O_2, the T state is more stable and, when saturated with O_2, the R state is more stable. The free energy of both states is reduced with increasing oxygenation as a consequence of O_2 liganding. The Fe(II)—O_2 bonding is more exergonic in the R state than it is in the T state, however, so that the relative stabilities of these two states reverses order at intermediate levels of oxygenation. *(b)* The sigmoid O_2-binding curve of Hb (*purple*) is a composite of its hyperbolic R-state (*red*) and T-state (*blue*) binding curves: It is more T-like at lower pO_2 values and more R-like at higher pO_2 values.

2. The IHP forces the Hb molecule into the T state which, through the "gears and levers" coupling the 4° and 3° conformational changes, pulls the proximal His in the opposite direction.

The bond between the proximal His and the Fe lacks the strength to withstand these two opposing "irresistible" forces; it simply breaks. The spectroscopic observation of this phenomenon therefore confirms the existence of the heme-protein tension predicted by the Perutz mechanism.

E. Origin of the Bohr Effect

The Bohr effect, hemoglobin's release of H⁺ on binding O_2, is also observed when Hb binds other ligands. *It arises from pK changes of several groups caused by changes in their local environments that accompany hemoglobin's T → R transition.* The groups involved have been identified through chemical and structural studies and their quantitative contributions to the Bohr effect have been estimated.

Reaction of the α subunits of Hb with **cyanate** results in the specific **carbamoylation** of its N-terminal amino group (Fig. 9-21). When such carbamoylated α subunits are mixed with normal β subunits, the resulting reconstituted Hb lacks 20 to 30% of the normal Bohr effect. The reason for this is seen on comparing the structure of deoxyHb with that of carbamoylated deoxyHb. In deoxyHb, a Cl⁻ ion binds between the N-terminal amino group of Val $1\alpha_2$ and the guanidino group of Arg

$141\alpha_1$ (the C-terminal residue; Fig. 9-18*a*). This Cl⁻ is absent in carbamoylated deoxyHb. It is also absent in normal R state Hb because its C-terminal residues are not held in place by salt bridges (the origin of the preferential binding of Cl⁻ to deoxyHb; Section 9-1C). N-terminal amino groups of polypeptides normally have pK's near 8.0. On deoxyHb α subunits, however, the N-terminal amino group is electrostatically influenced by its closely associated Cl⁻ to increase its positive charge by binding protons more tightly; that is, to increase its pK. Since at the pH of blood (7.4), N-terminal amino groups are normally only partially charged, this pK shift causes them to bind significantly more protons in the T state than in the R state.

The Hb β chain also contributes to the Bohr effect. Removal of its C-terminal residue, His 146β, reduces the

$$R-NH_2 \; + \; \overset{O}{\underset{\|}{C}}=N^- \; \overset{H^+}{\longrightarrow} \; R-NH-\overset{O}{\underset{\|}{C}}-NH_2$$

Terminal amino group	Cyanate	Carbamoylated terminal amino group

Figure 9-21
The reaction of cyanate with the unprotonated (nucleophilic) forms of primary amino groups. At physiological pH's, N-terminal amino groups, which have pK's near 8.0, readily react with cyanate. Lys ε-amino groups (pK $\approx$ 10.8), however, are fully protonated under these conditions and are therefore unreactive.

Bohr effect by 40%. In normal deoxyHb, the imidazole ring of His 146β associates with the carboxylate of Asp 94β on the same subunit (Figs. 9-18*b* and 9-19) to form a salt bridge that is absent in the R state. Proton NMR measurements indicate that formation of this salt bridge increases the pK of the imidazole group from 7.1 to 8.0. This effect more than accounts for His 146β's share of the Bohr effect.

About 30 to 40% of the Bohr effect remains unaccounted for. It no doubt arises from small contributions of many of the residues whose environments are altered upon hemoglobin's R $\rightarrow$ T transition. A variety of evidence suggests that His 122α, His 143β, and Lys 82β are among these residues.

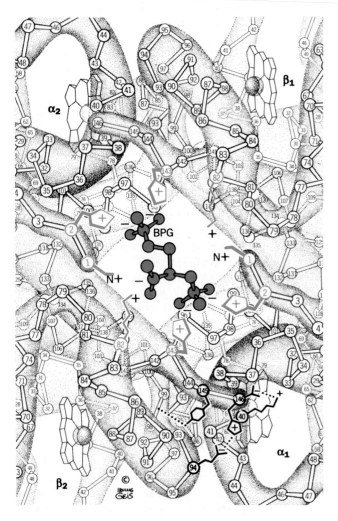

Figure 9-22
The binding of BPG to deoxyHb as viewed down the molecule's exact twofold axis (the same view as in Fig. 9-13*a*). BPG (*red*), with its five anionic groups, binds in the central cavity of deoxyHb where it is surrounded by a ring of eight cationic side chains (*blue*) extending from the two β subunits. In the R state, the central cavity is too narrow to contain BPG (Fig. 9-13*b*). The arrangement of salt bridges and hydrogen bonds between the α_1 and β_2 subunits that partially stabilizes the T state (Figs. 9-18*b* and 9-19) is indicated on the lower right. [Figure copyrighted © by Irving Geis.]

F. Structural Basis of BPG Binding

BPG decreases the oxygen-binding affinity of Hb by preferentially binding to its deoxy state (Section 9-1D). The binding of the physiologically quadruply charged BPG to deoxyHb is weakened by high salt concentrations, which suggests that this association is ionic in character. This explanation is corroborated by the X-ray structure of a BPG–deoxyHb complex, which indicates that BPG binds in the central cavity of deoxyHb on its twofold axis (Fig. 9-22). The anionic groups of BPG are within hydrogen bonding and salt bridging distances of the cationic Lys EF6(82), His H21(143), His NA2(2), and N-terminal amino groups of both β subunits (Fig. 9-22). The T $\rightarrow$ R transformation brings the two β H helices together, which narrows the central cavity (compare Fig. 9-13*a* and *b*) and expels the BPG. It also widens the distance between the β N-terminal amino groups from 16 to 20 Å, which prevents their simultaneous hydrogen bonding with BPG's phosphate groups. BPG therefore stabilizes the T conformation of Hb by cross-linking its β subunits. This shifts the T $\rightleftharpoons$ R equilibrium towards the T state, which lowers hemoglobin's O_2 affinity.

The structure of the BPG–deoxyHb complex also indicates why fetal hemoglobin (HbF) has a reduced affinity for BPG relative to HbA (Section 9-1D). The cationic His H21(143)β of HbA is changed to an uncharged Ser residue in HbF's β-like γ subunit thereby eliminating a pair of ionic interactions stabilizing the BPG–deoxyHb complex (Fig. 9-22).

3. ABNORMAL HEMOGLOBINS

Mutant hemoglobins have provided a unique opportunity to study structure–function relationships in proteins because Hb is the only protein of known structure that has a large number of well-characterized variants. The examination of individuals with physiological disabilities together with the routine electrophoretic screening of human blood samples has led to the discovery of over 400 mutant hemoglobins. Around 95% of these variants result from single amino acid substitutions in a globin polypeptide chain. In this section, we consider the nature of these **hemoglobinopathies.** Hemoglobin diseases characterized by defective globin synthesis, the **thalassemias,** are the subject of Section 33-2G.

A. Molecular Pathology of Hemoglobin

The physiological effect of an amino acid substitution on Hb can, in most cases, be understood in terms of its molecular location:

1. Changes in surface residues
Changes of surface residues are usually innocuous because most of these residues have no specific functional role [although sickle-cell Hb (HbS) is a glaring excep-

tion to this generalization; Section 9-3B]. For example, HbE [Glu B8(26)β → Lys], the most common human Hb mutant after HbS (possessed by up to 10% of the populace in parts of Southeast Asia), has no clinical manifestations in either heterozygotes or homozygotes. About one half of the known Hb mutations are of this type and were only discovered accidentally or through surveys of large populations. It has been estimated that one individual in 800 has a variant hemoglobin.

2. Changes in internally located residues

Changing an internal residue often destabilizes the Hb molecule. The degradation products of these hemoglobins, particularly those of heme, form granular precipitates (known as **Heinz bodies**) that are hydrophobically adsorbed to the erythrocyte cell membrane. The membrane's permeability is thereby increased causing premature cell lysis. Carriers of unstable hemoglobins therefore suffer from **hemolytic anemia** of varying degrees of severity.

The structure of Hb is so delicately balanced that small structural changes may render it nonfunctional. This can occur through the weakening of the heme–globin association or as a consequence of other conformational changes. For instance, the heme group is easily dislodged from its closely fitting hydrophobic binding pocket. This occurs in Hb Hammersmith (Hb variants are often named after the locality of their discovery) in which Phe CD1(42)β, an invariant residue that wedges the heme into its pocket (see Figs. 9-12 and 9-15), is replaced by Ser. The resulting gap permits water to enter the heme pocket, which causes the hydrophobic heme to easily drop out. Similarly, in Hb Bristol, the substitution of Asp for Val E11(67)β, which partially occludes the O_2 pocket, places a polar group in contact with the heme. This weakens the binding of the heme to the protein, probably by facilitating the access of water to the subunit's otherwise hydrophobic interior.

Hb may also be destabilized by the disruption of elements of its 2°, 3°, and/or 4° structures. The instability of Hb Bibba results from the substitution of a helix-breaking Pro for Leu H19(136)α. Likewise, the instability of Hb Savannah is caused by the substitution of Val for the highly conserved Gly B6(24)β, which is located on the B helix where it crosses the E helix with insufficient clearance for side chains larger than an H atom (Fig. 9-13, and Fig. 9-11 where Gly B6 is residue 25). The $\alpha_1-\beta_1$ contact, which does not significantly dissociate under physiological conditions, may do so upon structural alteration. This occurs in Hb Philly in which Tyr C1(35)α, which participates in the hydrogen bonded network that helps knit together the $\alpha_1-\beta_1$ interface, is replaced by Phe.

3. Changes stabilizing methemoglobin

Changes at the O_2-binding site that stabilize the heme in the Fe(III) oxidation state eliminate the binding of O_2 to the defective subunits. Such methemoglobins are designated **HbM** and individuals carrying them are said to have **methemoglobinemia.** These individuals usually have bluish skin, a condition known as **cyanosis,** which results from the presence of deoxyHb in their arterial blood.

All known methemoglobins arise from substitutions that provide the Fe atom with an anionic oxygen atom ligand. In Hb Boston, the substitution of Tyr for His E7(58)α (the distal His) results in the formation of a 5-coordinate Fe(III) complex with the phenolate ion of the mutant Tyr E7 displacing the imidazole ring of His F8(87) as the apical ligand (Fig. 9-23a). In Hb Milwaukee, the γ-carboxyl group of the Glu that replaces Val E11(67)β forms an ion pair with a 5-coordinate Fe(III) complex (Fig. 9-23b). Both the phenolate and glutamate ions in these methemoglobins so stabilize the Fe(III) oxidation state that methemoglobin reductase is ineffective in converting them to the Fe(II) form.

Individuals with HbM are alarmingly cyanotic and have blood that is chocolate brown, even when their normal subunits are oxygenated. In northern Japan, this condition is named "black mouth" and has been known for centuries; it is caused by the presence of HbM Iwate [His F8(87)α → Tyr]. Methemoglobins have Hill constants of ~ 1.2. This indicates a reduced cooperativity in comparison with HbA even though HbM, which can only bind two oxygen molecules, can have a maximum Hill constant of 2 (the unmutated chains remain functional). Surprisingly, heterozygotes with HbM, which have an average of one functional β subunit per Hb molecule, have no apparent physical disabilities. Evidently, the amount of O_2 released in their capillaries is within normal limits. Homozygotes of HbM, however, are unknown; this condition is, no doubt, lethal.

4. Changes at the $\alpha_1-\beta_2$ contact

Changes at the $\alpha_1-\beta_2$ contact often interfere with hemoglobin's quaternary structural changes. Most such hemoglobins have an increased O_2 affinity so that they release less than normal amounts of O_2 in the tissues. Individuals with such defects compensate for it by increasing the concentration of erythrocytes in their blood. This condition, which is named **polycythemia,** often gives them a ruddy complexion. Some amino acid substitutions at the $\alpha_1-\beta_2$ interface instead result in a reduced O_2 affinity. Individuals carrying such hemoglobins are cyanotic.

Amino acid substitutions at the $\alpha_1-\beta_2$ contact may change the relative stabilities of hemoglobin's R and T forms, thereby altering its O_2 affinity. For example, the replacement of Asp G1(99)β by His in Hb Yakima eliminates the hydrogen bond at the $\alpha_1-\beta_2$ contact that stabilizes the T form of Hb (Fig. 9-17a). The interloping imidazole ring also acts as a wedge that

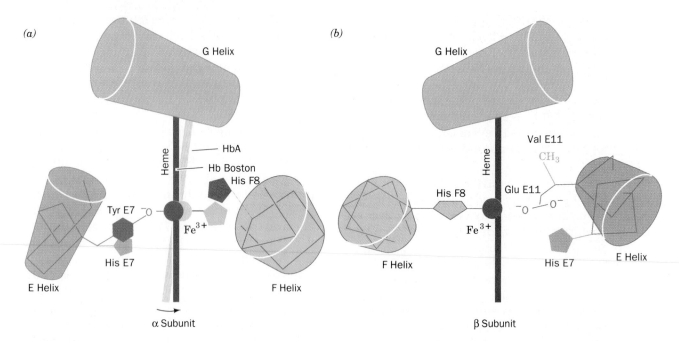

(a) *(b)*

Figure 9-23
Mutations stabilizing the Fe(III) oxidation state of heme: (a) Alterations in the heme pocket of the α subunit on changing from deoxyHbA to Hb Boston [His E7(58)α → Tyr]. The phenolate ion of the mutant Tyr becomes the fifth ligand of the Fe atom thereby displacing the proximal His [F8(87)α]. [After Pulsinelli, P. D., Perutz, M. F., and Nagel, R. L., *Proc. Natl. Acad. Sci.* **70**, 3872 (1973).] (b) The structure of the heme pocket of the β subunit in Hb Milwaukee [Val E11(67)β → Glu]. Here the mutant Glu residue's carboxyl group forms an ion pair with the heme iron atom so as to stabilize its Fe(III) state. [From Perutz, M. F., Pulsinelli, P. D., and Ranney, H. M., *Nature* **237**, 260 (1972).]

pushes the subunits apart and displaces them towards the R state. This change shifts the T $\rightleftharpoons$ R equilibrium almost entirely to the R state, which results in Hb Yakima having an increased O_2 affinity (p_{50} = 12 torr under physiological conditions vs 26 torr for HbA) and a total lack of cooperativity (Hill constant = 1.0). In contrast, the replacement of Asn G4(102)β by Thr in Hb Kansas eliminates the hydrogen bond in the $\alpha_1 - \beta_2$ contact that stabilizes the R state (Fig. 9-17b) so that this Hb variant remains in the T state upon binding O_2. Hb Kansas therefore has a low O_2 affinity (p_{50} = 70 torr) and a low cooperativity (Hill constant = 1.3).

B. Molecular Basis of Sickle-Cell Anemia

Most harmful Hb variants occur in only a few individuals, in many of whom the mutation apparently originated. However, ~10% of American blacks and as many as 25% of African blacks are heterozygotes for **sickle-cell hemoglobin (HbS).** HbS arises, as we have seen (Section 6-3A), from the substitution of a hydrophobic Val residue for the hydrophilic surface residue Glu A3(6)β (Fig. 9-13). The prevalence of HbS results from the protection it affords heterozygotes against ma-

laria. However, homozygotes for HbS, of which there are some 50,000 in the United States, are severely afflicted by hemolytic anemia together with painful, debilitating, and sometimes fatal blood flow blockages caused by the irregularly shaped and inflexible erythrocytes characteristic of the disease.

HbS Fibers Are Stabilized by Intermolecular Contacts Involving Val β6 and Other Residues

The sickling of HbS-containing erythrocytes (Fig. 6-11b) results from the aggregation (polymerization) of deoxy HbS into rigid fibers that extend throughout the length of the cell (Fig. 9-24). Electron microscopy indicates that these

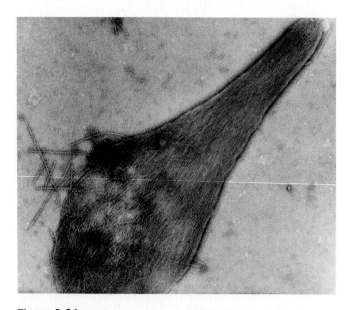

Figure 9-24
An electron micrograph of deoxyHbS fibers spilling out of a ruptured erythrocyte. [Courtesy of Robert Josephs, University of Chicago.]

Figure 9-25
220 Å in diameter fibers of deoxyHbS: (*a*) An electron micrograph of a negatively stained fiber. The accompanying cutaway interpretive drawing indicates the relationship between the inner and outer strands; the spheres represent individual HbS molecules. The fiber has a layer repeat distance of 64 Å and a moderate twist such that it repeats every 350 Å along the fiber axis. [Courtesy of Stuart Edelstein, University of Geneva.] (*b*) A cross-sectional diagram of the fiber indicating a probable pairing and polarity set for its 14 strands; differently shaded strands are antiparallel.

fibers are ∼ 220 Å in diameter elliptical rods consisting of 14 hexagonally packed and helically twisting strands of deoxyHbS molecules that associate in parallel pairs (Figs. 9-25 and 9-26*a*).

The structural relationship among the HbS molecules in the pairs of parallel HbS strands has been established by the X-ray structure analysis of deoxyHbS crystals. When this crystal structure was first determined, it was unclear whether the intermolecular contacts in the crystal resembled those in the fiber. However, the subsequent observation that HbS fibers slowly convert to these crystals with little change in their overall X-ray diffraction pattern indicates that the fibers structurally resemble the crystals. The crystal structure of deoxyHbS consists of double filaments of HbS molecules whose several different intermolecular contacts are diagrammed in Fig. 9-26*b*. Only one of the two Val 6β's per Hb molecule contacts a neighboring molecule. In this contact, the mutant Val side chain occupies a hydrophobic surface pocket on an adjacent molecule's β subunit. This pocket, which is absent in oxyHb, is too small to contain HbA's Glu β6 side chain even if it was not hydrophilic (Fig. 9-26*c*). Other contacts involve residues that also occur in HbA including Asp 73β and Glu 23α (Fig. 9-26*b*). The observation that HbA does not aggregate into fibers, however, even at very high concentrations, indicates that *the contact involving Val 6β is essential for fiber formation.*

The importance of these other intermolecular contacts to the structural integrity of HbS fibers has been demonstrated by studying the effects of other mutant hemoglobins on HbS gelation (polymerization). For example, the doubly mutated Hb Harlem (Glu 6β → Val + Asp 73β → Asn) requires a higher concentration to gel than does HbS (Glu 6β → Val); similarly, mixtures of HbS and Hb Korle-Bu (Asp 73β → Asn) gel less readily than equivalent mixtures of HbS and HbA. This observation suggests that Asp 73β occupies an important intermolecular contact site in HbS fibers (Fig. 9-26*b*). Likewise, the observation that hybrid tetramers consisting of α subunits from Hb Memphis (Glu 23α → Gln) and β subunits from HbS gel less readily than does HbS indicates that Glu 23α also participates in the polymerization of HbS fibers (Fig. 9-26*b*). The other white-lettered residues in Fig. 9-26*b* have been similarly implicated in sickling interactions.

The Initiation of HbS Gelation Is a Complex Process

The gelation of HbS, both in solution and within the red cell, follows an unusual time course. A solution of HbS can be brought to conditions under which it will gel by lowering the pO_2, raising the HbS concentration, and/or raising the temperature. *Upon achieving gelation conditions, there is a reproducible delay that varies according to conditions from milliseconds to days: During this time, no HbS fibers can be detected.* Only after the delay do

(a)

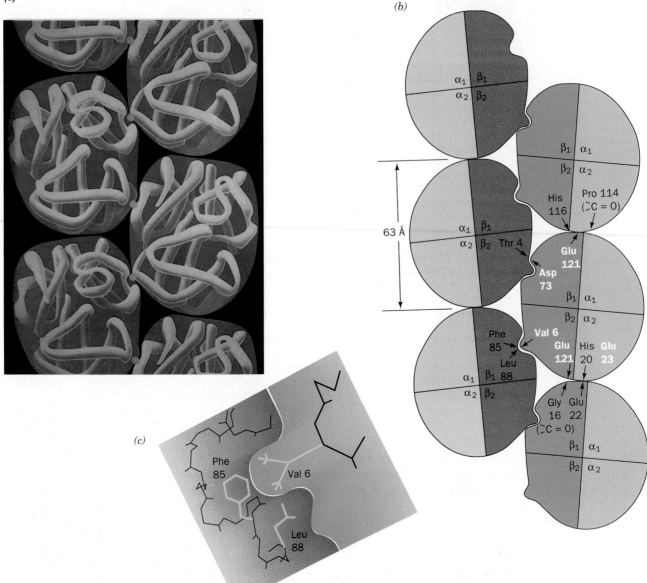

(c)

Figure 9-26
The structure of the deoxyHbS fiber. (*a*) The arrangement of the deoxyHbS molecules in the fiber. [Figure copyrighted © by Irving Geis.] (*b*) A schematic diagram indicating the intermolecular contacts in the crystal structure of deoxyHbS. The white-lettered residues are implicated in forming these contacts. Note that the only intermolecular association in which the mutant residue Val 6β participates involves

subunit β_2; Val 6 of subunit β_1 is free. [After Wishner, B. C., Ward, K. B., Lattman, E. E., and Love, W. E., *J. Mol. Biol.* **98**, 192 (1975).] (*c*) The mutant Val 6β₂ fits neatly into a hydrophobic pocket formed mainly by Phe 85 and Leu 88 of an adjacent β_1 subunit. This pocket, which is located between helices E and F at the periphery of the heme pocket, is absent in oxyHb and is too small to contain the normally occurring Glu 6β side chain. [Figure copyrighted © by Irving Geis.]

fibers first appear and gelation is then completed in about one half the delay time (Fig. 9-27*a*).

William Eaton and James Hofrichter discovered that the delay time, t_d, has a concentration dependence described by

$$\frac{1}{t_d} = k \left(\frac{c_t}{c_s} \right)^n \qquad [9.13]$$

where c_t is the total HbS concentration prior to gelation, c_s is the solubility of HbS measured after gelation is complete, and k and n are constants. Graphical analysis of the data indicates that $k \approx 10^{-7}$ s^{-1} and that n is between 30 and 50 (Fig. 9-27*b*). This is a remarkable result: *No other known solution process even approaches a 30th power concentration dependence.*

A two-stage process accounts for Eq. [9.13]:

1. At first, HbS molecules sequentially aggregate to form a **nucleus** consisting of m HbS molecules (Fig. 9-28*a*):

$$\text{HbS} \rightleftharpoons (\text{HbS})_2 \rightleftharpoons (\text{HbS})_3 \rightleftharpoons \cdots \rightleftharpoons (\text{HbS})_m \longrightarrow \text{Growth}$$

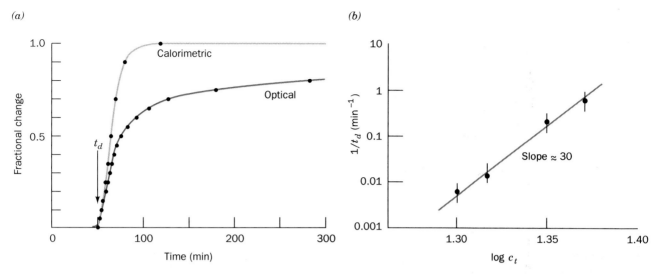

(a)

(b)

Figure 9-27
The time course of deoxyHbS gelation. (a) The extent of gelation as monitored calorimetrically (*yellow*) and optically (*purple*). Gelation of the 0.233 g·mL⁻¹ deoxyHbS solution was initiated by rapidly increasing the temperature from 0°C, where HbS is soluble, to 20°C; t_d is the delay time. (b) A log–log plot showing the concentration dependence of t_d for the gelation of deoxyHbS at 30°C. The slope of this line is ~30. [After Hofrichter, J., Ross, P. D., and Eaton, W. A., *Proc. Natl. Acad. Sci.* **71,** 4865, 4867 (1974).]

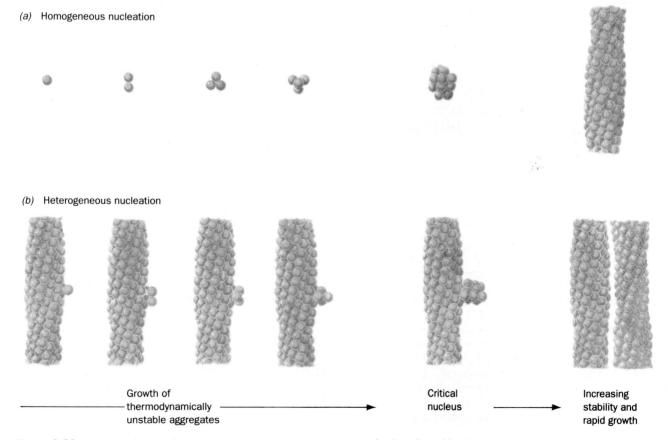

(a) Homogeneous nucleation

(b) Heterogeneous nucleation

Growth of thermodynamically unstable aggregates ⟶ Critical nucleus ⟶ Increasing stability and rapid growth

Figure 9-28
The double nucleation mechanism for deoxyHbS gelation: (a) The initial aggregation of HbS molecules (*circles*) occurs very slowly because this process is thermodynamically unfavorable and hence the intermediates tend to decompose rather than grow. However, once an aggregate reaches a certain size, the **critical nucleus,** its further growth becomes thermodynamically favorable leading to rapid fiber formation. (b) Each fiber, in turn, can nucleate the growth of other fibers leading to the explosive appearance of polymer. [After Ferrone, F. A., Hofrichter, J., and Eaton, W. A., *J. Mol. Biol.* **183,** 614 (1985).]

Prenuclear aggregates are unstable and easily decompose, but once a nucleus has formed, it assumes a stable structure that rapidly elongates to form an HbS fiber.

2. Once a fiber has formed, it can nucleate the growth of other fibers (Fig. 9-28b). These newly formed fibers, in turn, nucleate the growth of yet other fibers, *etc.*, so that this latter process is autocatalytic.

The initial **homogeneous nucleation** process (taking place in solution) accounts for the very high concentration dependence in Eq. [9.13], whereas the secondary **heterogeneous nucleation** process (taking place on a surface — that of a fiber in this case) is responsible for the rapid onset of gelation (Fig. 9-27a).

The foregoing kinetic hypothesis suggests why sickle-cell anemia is characterized by episodic "crises" caused by blood flow blockages. HbS fibers dissolve essentially instantaneously upon oxygenation so that none are present in arterial blood. Erythrocytes take from 0.5 to 2 s to pass through the capillaries where deoxygenation renders HbS insoluble. If the delay time, t_d, for sickling is greater than this transit time, no blood flow blockage occurs (although sickling that occurs in the veins damages the erythrocyte membrane). However, Eq. [9.13] indicates that small increases in HbS concentration, c_t, and/or small decreases in HbS solubility, c_s, caused by conditions known to trigger sickle-cell crises, such as dehydration, O_2 deprivation, and fever, result in significant decreases of t_d. Once a blockage occurs, the resulting lack of O_2 and slow down of blood flow in the area compound the situation.

The kinetic hypothesis of sickling has profound clinical implications for the treatment of sickle-cell anemia. Heterozygotes of HbS, whose blood usually contains ~60% HbA and 40% HbS, rarely show any symptoms of sickling. The t_d for the gelation of their Hb is ~10^6-fold greater than that of homozygotes. Accordingly, a treatment of sickle-cell anemia that increases t_d by this amount, which corresponds to decreasing the ratio c_t/c_s by a factor of ~1.6, would relieve the symptoms of this disease. Three different therapeutic strategies to increase t_d, and thus inhibit HbS gelation, are under investigation:

1. The disruption of intermolecular interactions. Of particular interest are synthetic oligopeptides that have been designed with the aid of the X-ray structure of HbS to bind stereospecifically to its intermolecular contact regions.

2. The use of agents that increase hemoglobin's O_2 affinity. For example, the administration of cyanate carbamoylates the N-terminal amino groups of Hb (Fig. 9-21). This treatment eliminates some of the salt bridges that stabilize the T state (Section 9-2E) and thereby increases the O_2 affinity of Hb. Although cyanate is an effective *in vitro* antisickling agent, its clinical use has been discontinued because of toxic side effects, cataract formation and peripheral nervous system damage, that probably result from the carbamoylation of proteins other than Hb.

3. Lowering the HbS concentration (c_t) in erythrocytes. Agents that alter erythrocyte membrane permeability so as to permit the influx of water have promise in this regard.

Replacing HbS with other Hb molecules is also a promising possibility. Homozygotes for HbS with high levels of HbF in their blood, for example, have a relatively mild form of sickle-cell anemia. This observation has prompted the search for agents that can "switch on" the synthesis of HbF γ subunits in preference to that of mutant HbS β subunits. The use of vasodilators (substances that dilate blood vessels) so as to reduce the entrapment of sickled erythrocytes in the capillaries may also relieve the symptoms of sickle-cell disease.

4. ALLOSTERIC REGULATION

One of the outstanding characteristics of life is the high degree of control exercised in almost all of its processes. Through a great variety of regulatory mechanisms, the exploration of which constitutes a significant portion of this text, an organism is able to respond to changes in its environment, maintain intra- and intercellular communications, and execute an orderly program of growth and development. Regulation is exerted at every organizational level in living systems, from the control of rates of reactions on the molecular level, through the control of expression of genetic information on the cellular level, to the control of behavior on the organismal level. It is therefore not surprising that many, if not most, diseases are caused by aberrations in biological control processes.

Our exploration of the structure and function of hemoglobin continues with a theoretical discussion of the regulation of ligand binding to proteins through **allosteric interactions** (Greek: *allos*, other + *stereos*, solid or space). These cooperative interactions occur when the binding of one ligand at a specific site is influenced by the binding of another ligand, known as an **effector** or **modulator,** at a different (allosteric) site on the protein. If the ligands are identical, this is known as a **homotropic effect,** whereas if they are different, it is described as a **heterotropic effect.** These effects are termed **positive** or **negative** depending on whether the effector increases or decreases the protein's ligand-binding affinity.

Hemoglobin, as we have seen, exhibits both homotropic and heterotropic effects. The binding of O_2 to Hb results in a positive homotropic effect since it increases hemoglobin's O_2 affinity. In contrast, BPG, CO_2, H^+, and Cl^- are negative heterotropic effectors of O_2 binding to Hb because they decrease its affinity for O_2 (negative and are chemically different from O_2 (heterotropic). The O_2 affinity of Hb, as we have seen, depends on its quaternary structure. *In general, allosteric effects result from interactions among subunits of oligomeric proteins.*

Even though hemoglobin catalyzes no chemical reaction, it binds ligands in the same manner as do enzymes. Since an enzyme cannot catalyze a reaction until after it has bound its substrate (the molecule undergoing reaction), the enzyme's catalytic rate varies with its substrate-binding affinity. Consequently, the cooperative binding of O_2 to Hb is taken as a model for the allosteric regulation of enzyme activity. Indeed, in this section, we shall consider several models of allosteric regulation that, for the most part, were formulated to explain the O_2-binding properties of Hb. Following this, we shall compare these models with the realities of Hb behavior.

A. The Adair Equation

The derivation of the Hill equation (Section 9-1B) is predicated on the assumption of all-or-none O_2 binding. The observation of partially oxygenated Hb molecules, however, led Gilbert Adair, in 1924, to propose that the binding of ligands to proteins occurs sequentially with dissociation constants that are not necessarily equal. The expression for the saturation function under this model is straightforwardly derived.

For a protein such as Hb with four ligand-binding sites, the reaction sequence is

$$
\begin{aligned}
E + S &\rightleftharpoons ES & k_1 &= 4K_1 \\
ES + S &\rightleftharpoons ES_2 & k_2 &= \tfrac{3}{2}K_2 \\
ES_2 + S &\rightleftharpoons ES_3 & k_3 &= \tfrac{2}{3}K_3 \\
ES_3 + S &\rightleftharpoons ES_4 & k_4 &= \tfrac{1}{4}K_4
\end{aligned}
$$

where the K_i are the **macroscopic** or **apparent dissociation constants** for binding the ith ligand to the protein,

$$
K_i = \frac{[ES_{i-1}][S]}{[ES_i]} \qquad [9.14]
$$

and the k_i are the **microscopic** or **intrinsic dissociation constants,** that is, the individual dissociation constants for the ligand-binding sites. The intrinsic dissociation constants are equal to the apparent dissociation constants multiplied by **statistical factors,** $4, \tfrac{3}{2}, \tfrac{2}{3},$ and $\tfrac{1}{4},$ that account for the number of ligand-binding sites on the protein molecule. The statistical factor 4 derives from the fact that a tetrameric protein E bears four sites that can bind ligand to form ES (that is, the concentration of ligand-binding sites is 4[E]) but only one site from which ES can dissociate ligand to form E (that is, the concentration of bound ligand is 1[E]); the statistical factor $\tfrac{3}{2}$ is a result of there being three remaining sites on ES that can bind ligand to form ES_2 and two sites from which ES_2 can dissociate ligand to form ES; *etc.* In general, for a protein with n equivalent binding sites:

$$
k_i = \frac{(n-i+1)[ES_{i-1}][S]}{i[ES_i]} = \left(\frac{n-i+1}{i}\right) K_i \quad [9.15]
$$

since $(n - i + 1)[ES_{i-1}]$ is the concentration of free ligand-binding sites in ES_{i-1} and $i[ES_i]$ is the concentration of bound ligand on ES_i. Therefore, solving sequentially for the concentration of each protein–ligand species in a tetrameric protein, we obtain:

$$
\begin{aligned}
[ES] &= [E][S]/K_1 &= 4[E][S]/k_1 \\
[ES_2] &= [ES][S]/K_2 &= \tfrac{3}{2}[ES][S]/k_2 &= 6[E][S]^2/k_1k_2 \\
[ES_3] &= [ES_2][S]/K_3 &= \tfrac{2}{3}[ES_2][S]/k_3 &= 4[E][S]^3/k_1k_2k_3 \\
[ES_4] &= [ES_3][S]/K_4 &= \tfrac{1}{4}[ES_3][S]/k_4 &= [E][S]^4/k_1k_2k_3k_4
\end{aligned}
$$

The fractional saturation of ligand binding, the fraction of occupied ligand-binding sites divided by the total concentration of ligand-binding sites, is expressed

$$
Y_s = \frac{[ES] + 2[ES_2] + 3[ES_3] + 4[ES_4]}{4([E] + [ES] + [ES_2] + [ES_3] + [ES_4])} \quad [9.16]
$$

so that, substituting in the above relationships and cancelling terms, we obtain

$$
Y_s = \frac{\dfrac{[S]}{k_1} + \dfrac{3[S]^2}{k_1k_2} + \dfrac{3[S]^3}{k_1k_2k_3} + \dfrac{[S]^4}{k_1k_2k_3k_4}}{1 + \dfrac{4[S]}{k_1} + \dfrac{6[S]^2}{k_1k_2} + \dfrac{4[S]^3}{k_1k_2k_3} + \dfrac{[S]^4}{k_1k_2k_3k_4}} \quad [9.17]
$$

This is the **Adair equation** for four ligand-binding sites. Equations describing ligand binding to proteins with different numbers of binding sites are similarly derived.

If the microscopic dissociation constants of the Adair equation are not equal, the fractional saturation curve will describe cooperative ligand binding. Decreasing and increasing values of these constants lead to positive and negative cooperativity, respectively. Of course, the values of the microscopic dissociation constants may also alternate so that, for example, $k_1 < k_2 > k_3 < k_4$.

In our discussion of the O_2-dissociation curve of Hb (Section 9-1B), we have seen how its values of k_1 and k_4 may be obtained by extrapolating the lower and upper asymptotes of the Hill plot to the log pO_2 axis. The remaining microscopic dissociation constants can be evaluated by fitting Eq. [9.17] to the Hill plot. The values of these **Adair constants** for Hb are given in Table 9-2. Note that k_4 is relatively insensitive to the presence of

Table 9-2

Adair Constants for Hemoglobin A at pH 7.40

Solution	k_1 (torr)	k_2 (torr)	k_3 (torr)	k_4 (torr)
Stripped	8.8	6.1	0.85	0.25
0.1M NaCl	41.	13.	12.	0.14
2 mM BPG	74.	112.	23.	0.24
0.1M NaCl + 2 mM BPG	97.	43.	119.	0.09

Source: Tyuma, I., Imai, K., and Shimizu, K., *Biochemistry* **12,** 1493, 1495 (1973).

BPG. Hb therefore binds and releases its last O_2 almost independently of the BPG concentration.

Although the Adair equation is the most general relationship describing ligand binding to a protein and is widely used to do so, it provides no physical insight as to why the various microscopic dissociation constants differ from each other. Yet, if the protein consists, as so many do, of identical subunits that are symmetrically related, it is desirable to understand how ligand binding at one site influences the ligand-binding affinity at a seemingly identical site. This need led to the development of models for ligand binding that rationalize how the binding sites of oligomeric proteins can exhibit different affinities. Two of these models are described in the following sections.

B. The Symmetry Model

Perhaps the most elegant model for describing cooperative ligand binding to a protein is the **symmetry model** of allosterism, which was formulated in 1965 by Jacques Monod, Jeffries Wyman, and Jean-Pierre Changeux. This model, alternatively termed the **MWC model**, is defined by the following rules:

1. An allosteric protein is an oligomer of protomers that are symmetrically related (for hemoglobin, we shall assume, for the sake of algebraic simplicity, that all four subunits are functionally identical).

2. Each protomer can exist in (at least) two conformational states, designated T and R; these states are in equilibrium whether or not ligand is bound to the oligomer.

3. The ligand can bind to a protomer in either conformation. *Only the conformational change alters the affinity of the protomer for the ligand.*

4. *The molecular symmetry of the protein is conserved during conformational change.* Protomers must therefore change conformation in a concerted manner, which implies that the conformation of each protomer is constrained by its association with the other protomers; in other words, there are no oligomers that simultaneously contain R- and T-state protomers.

For a ligand S and an allosteric protein consisting of n protomers, these rules imply the following equilibria for conformational conversion and ligand-binding reactions (for the sake of brevity, $T_i \equiv TS_i$ and $R_i \equiv RS_i$).

$$T_0 \rightleftharpoons R_0$$

$$
\begin{array}{ll}
T_0 + S \rightleftharpoons T_1 & R_0 + S \rightleftharpoons R_1 \\
T_1 + S \rightleftharpoons T_2 & R_1 + S \rightleftharpoons R_2 \\
\quad \cdot & \quad \cdot \\
\quad \cdot & \quad \cdot \quad\quad\quad\quad [9.18]\\
\quad \cdot & \quad \cdot \\
T_{n-1} + S \rightleftharpoons T_n & R_{n-1} + S \rightleftharpoons R_n
\end{array}
$$

This is illustrated in Fig. 9-29 for a tetramer.

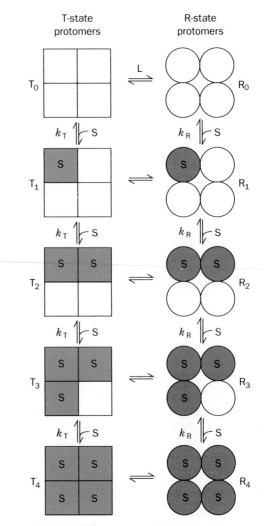

Figure 9-29
The species and reactions permitted under the symmetry model of allosterism. Squares and circles represent T- and R-state protomers, respectively.

The equilibrium constant L for the conformational interconversion of the oligomeric protein in the absence of ligand is expressed

$$L = \frac{[T_0]}{[R_0]} \quad\quad [9.19]$$

The microscopic dissociation constant for the R state, k_R, which according to Rule 3 is independent of the number of ligands bound to R, is expressed according to Eq. [9.15]:

$$k_R = \left(\frac{n-i+1}{i}\right)\frac{[R_{i-1}][S]}{[R_i]} \quad (i = 1, 2, 3, \ldots, n) \quad [9.20]$$

The microscopic dissociation constant for ligand binding to the T state, k_T, is similarly expressed. The fractional saturation, Y_S, for ligand binding is

$$Y_S = \frac{([R_1] + 2[R_2] + \cdots + n[R_n]) + ([T_1] + 2[T_2] + \cdots + n[T_n])}{n\{([R_0] + [R_1] + \cdots + [R_n]) + ([T_0] + [T_1] + \cdots + [T_n])\}} \quad [9.21]$$

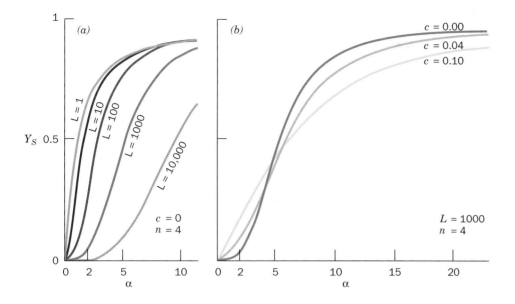

Figure 9-30
Symmetry model saturation function curves for tetramers according to Eq. [9.22]: (a) Their variation with L when $c = 0$. (b) Their variation with c when $L = 1000$. [After Monod, J., Wyman, J., and Changeux, J.-P., *J. Mol. Biol.* **12**, 92 (1965).]

We shall define $\alpha = [S]/k_R$; α may be considered a normalized ligand concentration. We shall also define the ratio of the ligand-binding dissociation constants $c = k_R/k_T$; c increases with the ligand-binding affinity of the T state relative to that of the R state. Then, combining the foregoing relationships as is shown in Section A of the Appendix to this chapter, we obtain the equation describing the symmetry model of allosterism for homotropic interactions:

$$Y_S = \frac{\alpha(1 + \alpha)^{n-1} + Lc\alpha(1 + c\alpha)^{n-1}}{(1 + \alpha)^n + L(1 + c\alpha)^n} \qquad [9.22]$$

Note that this equation depends on three parameters, α, c, and L, which are, respectively, the normalized ligand concentration, the relative affinities of the T and R states for ligand, and the relative stabilities of the T and R states. In contrast, the Hill equation (Section 9-1B) has but two parameters, K and n, whereas the number of parameters in the Adair equation is equal to the number of ligand-binding sites on the protein.

Homotropic Interactions

Let us examine the nature of the symmetry model by plotting Eq. [9.22] for a tetramer ($n = 4$) as a function of α for different values of the parameters L and c (Fig. 9-30). Three major points are evident from an inspection of these plots:

1. The degree of upward curvature exhibited by the initial sections of these sigmoid curves is indicative of their level of cooperativity.

2. When only the R state binds ligand ($c = 0$), the ligand-binding cooperativity increases as the oligomer's conformational preference for the nonligand-bind-

ing T state increases (L increases; Fig. 9-30a). *For high L values, if a single ligand is to bind, it must "force" the protein into its less preferred R state. The requirement that all protomers change their conformational states in a concerted manner causes the remaining three ligand-binding sites to become available.* The binding of the first ligand therefore promotes the binding of subsequent ligands, which is the essence of a positive homotropic effect. Note that cooperativity and ligand-binding affinity are different quantities; in fact, for $c = 0$, curves indicative of high ligand-binding affinity (those with low L) exhibit low cooperativity and *vice versa.*

3. When the T state is highly preferred (L is large), ligand-binding cooperativity increases with the R state's ligand-binding affinity relative to that of the T state (decreasing c; Fig. 9-30b). At low ligand concentrations (low α) the amount of ligand bound (Y_S) increases with the ligand-binding affinity of the T state since the protein is largely in the T state. As α increases, however, the amount of ligand bound to the intrinsically less stable R state eventually surpasses that of the T state thereby resulting in a cooperative effect. This is because *the free energy of ligand-binding stabilizes the R state with respect to the T state.*

Heterotropic Interactions

The symmetry model of allosterism is also capable of accounting for heterotropic effects. This comes about by assuming that each protomer has specific and independent binding sites for the three types of ligands: a substrate, S, that for simplicity let us assume binds only to the R state ($c = 0$); an **activator,** A, that also binds only to the R state; and an **inhibitor,** I, that binds only to the

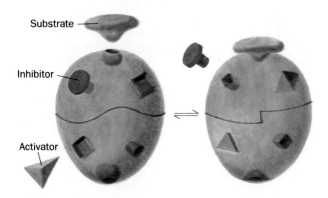

Figure 9-31
In the symmetry model of allosterism, heterotropic effects arise when substrates and activators bind exclusively (or at least preferentially) to the R state (*right*), and inhibitors bind exclusively (or at least preferentially) to the T state (*left*). The binding of substrate and/or activator to the oligomer therefore facilitates the further binding of substrate and activator. Conversely, the binding of inhibitor prevents (or at least inhibits) the oligomer from binding substrate or activator.

T state (Fig. 9-31). Then, through the derivation in Section B of the Appendix to this chapter, we obtain a more general equation for the symmetry model that describes heterotropic interactions as well as homotropic interactions:

$$Y_S = \frac{\alpha(1 + \alpha)^{n-1}}{(1 + \alpha)^n + \dfrac{L(1 + \beta)^n}{(1 + \gamma)^n}} \qquad [9.23]$$

where $\alpha = [S]/k_R$, as before, and analogously, $\beta = [I]/k_I$ and $\gamma = [A]/k_A$. Note that this equation differs from Eq.

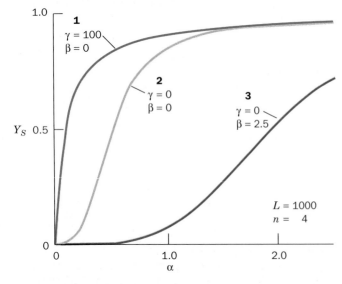

Figure 9-32
The effects of allosteric activator (γ) and inhibitor (β) on the shape of the fractional saturation curve for substrate (α) according to Eq. [9.23] for tetramers. [After Monod, J., Wyman, J., and Changeux, J.-P., *J. Mol. Biol.* **12**, 94 (1965).]

[9.22] for $c = 0$ only in that the second term in the denominator is modulated by terms related to the amounts of activator and inhibitor bound to the oligomer.

Figure 9-32 indicates the consequences of effector binding to a tetramer that follows this model:

1. Activator binding ($\gamma > 0$) increases the concentration of the substrate-binding R state (the second term in the denominator of Eq. [9.23] decreases) because it is the only state capable of binding activator. *The presence of activator therefore increases the protein's substrate-binding affinity* (a positive heterotropic effect) although it decreases the protein's degree of substrate-binding cooperativity (compare Curves 1 and 2 in Fig. 9-32). (*Note:* There is nothing in the derivation of Eq. [9.23] that differentiates the roles of substrate and activator; consequently, the substrate and the activator each bind to the protein with a positive homotropic effect as well as being positive heterotropic effectors of each other.)

2. *The presence of inhibitor ($\beta > 0$), which only binds to the T state, reduces the binding affinity for substrate* (a negative heterotropic effect) by increasing the concentration of the T state (the second term in the denominator of Eq. [9.23] increases). Therefore, since substrate must "work harder" to convert the oligomer to the substrate-binding R state, inhibitor increases the cooperativity of substrate binding (compare Curves 2 and 3 of Fig. 9-32), as well as that for activator binding.

The model derived here is a rather simple one. In a more realistic but algebraically much more complicated symmetry model, all types of ligands would bind to both conformational states of the oligomer. Nevertheless, our model demonstrates that *both homotropic and heterotropic effects can be explained solely by the requirement that the molecular symmetry of the oligomer be conserved rather than by the existence of any direct interactions between ligands.* In Section 9-4D, we compare the theoretical predictions of the symmetry model with our experimentally based model of hemoglobin oxygen binding.

C. The Sequential Model

The symmetry model provides a reasonable rationalization for the ligand-binding properties of many proteins. There are, however, several valid objections to it. Foremost of these is that it is difficult to believe that oligomeric symmetry is invariably preserved in all proteins so that there are never any hybrid conformations such as $R_{n-2}T_2$. Furthermore, there are well-established instances of negative homotropic effects, although the symmetry model, which permits only positive homotropic effects, is unable to account for them.

The symmetry model implicitly assumes Emil Fischer's "lock-and-key" model of ligand binding in

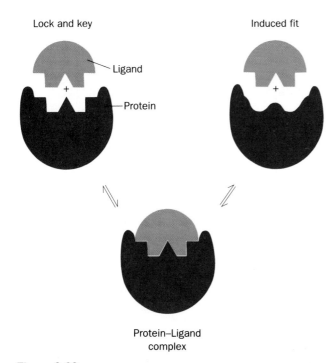

Lock and key   Induced fit

Protein–Ligand
complex

Figure 9-33
In the lock-and-key mechanism of ligand binding (*left*),
proteins are postulated to have preformed ligand-binding
sites that are complementary in shape to their ligand. Under
the induced-fit mechanism, a protein does not have this
complementary-binding site in the absence of ligand (*right*).
Rather, the ligand induces a conformation change at the
binding site that results in the complementary interaction.

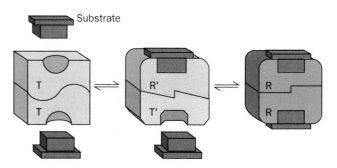

Substrate

Figure 9-34
In the sequential model of allosterism, substrate binding to
the low affinity T state induces conformation changes in
unliganded subunits that gives them ligand-binding affinities
between those of the low affinity T state and the high
affinity R state.

which ligand-binding sites of proteins are rigid and
complementary in shape to their ligand (Fig. 9-33, *left*).
A more sophisticated extension of this "lock-and-key"
model, known as the **induced-fit hypothesis,** postulates that *a flexible interaction between ligand and protein
induces a conformational change in the protein, which results in its increased ligand-binding affinity* (Fig. 9-33,
right). The observation, through X-ray crystal structure
analysis, that such conformational changes occur in several proteins has established the validity of the induced-
fit hypothesis.

Daniel Koshland has adapted the induced-fit hypothesis to explain allosteric effects. *In the resulting **sequential** or **induced-fit model,** ligand binding induces a con-*
*formational change in a subunit; cooperative interactions
arise through the influence that these conformational
changes have on neighboring subunits (Fig. 9-34).* If, for
example, they increase the neighbor's ligand-binding
affinity, then ligand binding is positively cooperative.
*The strengths of these interactions depend on the degree of
mechanical coupling between subunits.* In the limit of very
strong coupling, conformational changes become concerted so that the oligomer maintains its symmetry (the
symmetry model). With looser coupling, however, conformational changes occur sequentially as more and
more ligand is bound (Fig. 9-35). Thus, *the essence of the
sequential model is that a protein's ligand-binding affinity
varies with its number of bound ligands, whereas in the
symmetry model this affinity depends only on the protein's
quaternary state.*

The degree of coupling between oligomer subunits
depends on how these subunits are arranged, that is, on
the protein's symmetry. Consequently, in the sequential
model, the fractional saturation has a different algebraic
form for each oligomeric symmetry. The form of the
Adair equation (Eq. [9.17] for a tetramer) similarly depends on the number of subunits in the protein. In fact,
the sequential model of allosterism may be considered
an extension of the Adair model that provides a physical
rationalization for the values of its microscopic dissociation constants, k_i.

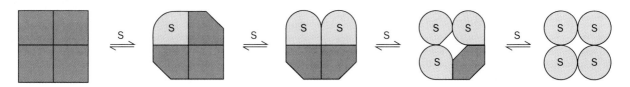

Figure 9-35
The sequential binding of ligand in the sequential model of
allosterism. Ligand-binding progressively induces
conformational changes in the subunits with the greatest
changes occurring in those subunits that have bound ligand.
The coupling between subunits is not necessarily of
sufficient strength to maintain the symmetry of the oligomer
as it is in the symmetry model.

D. Hemoglobin Cooperativity

Hemoglobin's fractional saturation curve is closely approximated by both the symmetry model and the sequential model (Fig. 9-36). Clearly such curves cannot by themselves be used to differentiate between these two models, if, in fact, either is correct. It is of interest, however, to compare these models with the mechanistic model of Hb we developed in Section 9-2C.

Hb, of course, is not composed of identical subunits as the symmetry model demands. At least to a first approximation, however, the functional differences of hemoglobin's closely related α and β subunits may be ignored (although their structural differences are essential to the molecular mechanism of Hb cooperativity). *To this approximation, Hb largely follows the symmetry model although it also exhibits some features of the sequential model* (although to what extent is still a matter of debate). The quaternary $T \rightarrow R$ conformation change is concerted as the symmetry model requires. Yet, ligand binding to the T state does cause small tertiary structural changes as the induced-fit model predicts. This phenomenon is evident in the X-ray structure of human hemoglobin whose α subunits are fully oxygenated and whose β subunits are oxygen free. This partially liganded hemoglobin remains in the T state but its α subunit Fe's are 0.15 Å closer to the still domed porphyrins than they are in deoxyHb (25% of the total distance moved in the $T \rightarrow R$ transition). *Such tertiary structural changes are undoubtedly responsible for the buildup of strain that eventually triggers the $T \rightarrow R$ transition.*

The symmetry and sequential models are often indis-

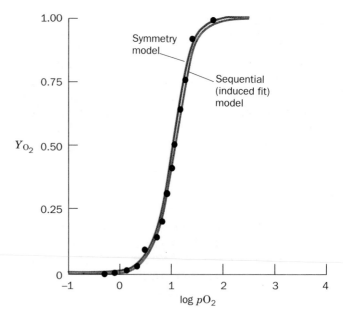

Figure 9-36
The sequential and the symmetry models of allosterism can provide equally good fits to the measured O$_2$-dissociation curve of Hb. [After Koshland, D. E., Jr., Némethy, G., and Filmer, D., *Biochemistry* **5**, 382 (1966).]

pensable guides in rationalizing the allosteric behavior of a protein. If, however, the experience with Hb is any guide, a protein's actual allosteric mechanism is likely to be more complex than either of these idealities. Indeed, recent thermodynamic data suggests that hemoglobin has three rather than two ligand-binding states although the structural basis of this model is obscure.

Appendix: Derivations of Symmetry Model Equations

A. Homotropic Interactions — Equation [9.22]

The fractional saturation Y_S for ligand binding is expressed:

$$Y_S = \frac{([R_1] + 2[R_2] + \cdots + n[R_n]) + ([T_1] + 2[T_2] + \cdots + n[T_n])}{n\{([R_0] + [R_1] + \cdots + [R_n]) + ([T_0] + [T_1] + \cdots + [T_n])\}}$$
[9.21]

Defining $\alpha = [S]/k_R$ and $c = k_R/k_T$, and using Eq. [9.20] to substitute $[R_{n-1}]$ for $[R_n]$, $[R_{n-2}]$ for $[R_{n-1}]$, etc., the terms enclosed by the first parentheses in the numerator of Eq. [9.21] are reduced to

$$[R_0]\left\{n\alpha + \frac{2n(n-1)\alpha^2}{2} + \cdots + \frac{n\,n!\,\alpha^n}{n!}\right\}$$

$$= [R_0]\alpha n\left\{1 + \frac{2(n-1)\alpha}{2} + \cdots + \frac{n(n-1)!\,\alpha^{n-1}}{n(n-1)!}\right\}$$

$$= [R_0]\alpha n(1 + \alpha)^{n-1}$$

and similarly, the terms in the first parentheses of the denominator of Eq. [9.21] become

$$[R_0]\left\{1 + n\alpha + \cdots + \frac{n!\,\alpha^n}{n!}\right\} = [R_0](1 + \alpha)^n$$

Likewise, the terms in the second parentheses of the numerator and the denominator of Eq. [9.21] assume the respective forms

$$[T_0]([S]/k_T)n(1 + [S]/k_T)^{n-1} = L[R_0]c\alpha n(1 + c\alpha)^{n-1}$$

and

$$[T_0](1 + [S]/k_T)^n = L[R](1 + c\alpha)^n$$

Accordingly,

$$Y_S = \frac{[R_0]\alpha n(1 + \alpha)^{n-1} + L[R_0]c\alpha n(1 + c\alpha)^{n-1}}{n\{[R_0](1 + \alpha)^n + L[R_0](1 + c\alpha)^n\}}$$

which, upon cancellation of terms, yields the equation describing the symmetry model for homotropic interactions:

$$Y_S = \frac{\alpha(1 + \alpha)^{n-1} + Lc\alpha(1 + c\alpha)^{n-1}}{(1 + \alpha)^n + L(1 + c\alpha)^n}$$
[9.22]

B. Heterotropic Interactions — Equation [9.23]

For an oligomer that binds activator A and substrate S to only its R state, and inhibitor I to only its T state, the fractional saturation for substrate, Y_S, the fraction of substrate-binding sites occupied by substrate, is expressed:

$$Y_S = \frac{\sum_{i=1}^{n} \sum_{j=0}^{n} i[R_{i,j}]}{n\left(\sum_{i=0}^{n} \sum_{j=0}^{n} [R_{i,j}] + \sum_{k=0}^{n} [T_k]\right)}$$

Here the subscripts i, j, and k indicate the respective numbers of S, A, and I molecules that are bound to one oligomer; that is, $R_{i,j} \equiv RS_iA_j$ and $T_k \equiv TI_k$. Then defining $\alpha = [S]/k_R$ and following the foregoing derivation of Eq. [9.22]:

$$Y_S = \frac{\left(\sum_{j=0}^{n} [R_{0,j}]\right) \alpha n (1+\alpha)^{n-1}}{n\left\{\left(\sum_{j=0}^{n} [R_{0,j}]\right)(1+\alpha)^n + \sum_{k=0}^{n} [T_k]\right\}} = \frac{\alpha(1+\alpha)^{n-1}}{(1+\alpha)^n + L'}$$

where

$$L' = \sum_{k=0}^{n} [T_k] \bigg/ \sum_{j=0}^{n} [R_{0,j}]$$

In analogy with the definition of α, we define $\beta = [I]/k_I$ and $\gamma = [A]/k_A$, and again follow the derivation of Eq. [9.22] to obtain:

$$\sum_{k=0}^{n} [T_k] = [T_0](1+\beta)^n$$

and

$$\sum_{j=0}^{n} [R_{0,j}] = [R_{0,0}](1+\gamma)^n$$

so that

$$L' = \frac{L(1+\beta)^n}{(1+\gamma)^n}$$

The symmetry model equation extended to include heterotropic effects is therefore expressed:

$$Y_S = \frac{\alpha(1+\alpha)^{n-1}}{(1+\alpha)^n + \dfrac{L(1+\beta)^n}{(1+\gamma)^n}} \qquad [9.23]$$

Chapter Summary

The heme group in myoglobin and in each subunit of hemoglobin reversibly binds O_2. In deoxyHb, the Fe(II) is 5 coordinated to the four pyrrole nitrogen atoms of the protoporphyrin IX and to the protein's proximal His. Upon oxygenation, O_2 becomes the sixth ligand of Fe(II). Mb has a hyperbolic fractional saturation curve (Hill constant, $n = 1$). However, that of Hb is sigmoidal ($n \approx 2.8$) as a consequence of its cooperative O_2 binding: Hb binds its fourth O_2 with 100-fold greater affinity than its first O_2. The variation of O_2 affinity with pH, the Bohr effect, causes Hb to release O_2 in the tissues in response to the binding of protons liberated by the hydration of CO_2 to HCO_3^-. Hb facilitates the transport of CO_2, both directly, by binding CO_2 as N-terminal carbamate, and indirectly, by increasing the concentration of HCO_3^- through the Bohr effect. The presence of BPG in erythrocytes, which only binds to deoxyHb, further modulates the O_2 affinity of Hb. Short-term high altitude adaptation results from an increase of BPG concentration in the erythrocytes, which increases the amount of O_2 delivered to the tissues by decreasing hemoglobin's O_2 affinity.

The α and β subunits of Hb consist mostly of seven or eight consecutive helices arranged to form a hydrophobic pocket that almost completely envelops the heme. Oxygen binding moves the Fe(II) from a position 0.6 Å out of the heme plane on the side of the proximal His to the center of the heme thereby relieving the steric interference that would otherwise occur between the bound O_2 and the porphyrin. The Fe(II) pulls the attached proximal His after it in a motion that can only occur if its imidazole ring reorients so as to avoid collision with the heme. In the T → R conformational transition, the symmetry equivalent $\alpha_1 C - \beta_2 FG$ and $\alpha_2 C - \beta_1 FG$ contacts simultaneously shift between two stable positions. Intermediate positions are sterically prevented so that these contacts act as a two-position conformational switch. The Perutz mechanism of O_2 binding proposes that the low O_2 affinity of the T state arises from strain that prevents the Fe(II) from moving into the heme plane to form a strong Fe—O_2 bond. This strain is relieved by the concerted 4° shift of the Hb molecule to the high O_2 affinity R state. The quaternary shift is opposed by a network of salt bridges in the T state that involve the C-terminal carboxyl groups and that are ruptured in the R state. The stability of the R state relative to the T state increases with the degree of oxygenation as a result of the strain of binding O_2 in the T state. The existence of this strain has been demonstrated through the breakage of the Fe(II)—proximal His bond upon hemoglobin's simultaneous binding of IHP, a tight-binding BPG analog that forces Hb into the T state, and NO, a strong ligand that forces Hb into the R state. The Bohr effect results from increases in the pK's of the α N-terminal amino group and His 146β on forming the T-state salt bridges. Lys 82β also participates in the Bohr effect. BPG binding occurs in the central cavity of T-state Hb through several salt bridges.

Around 400 mutant varieties of Hb are known. About one half of them are innocuous because they result in surface residue changes. However, alterations of internal residues often disrupt the structure of Hb, which causes hemolytic anemia. Changes at the O_2-binding site that stabilize the Fe(III) state eliminate O_2 binding to these subunits, which results in cyanosis. Mutations affecting subunit interfaces may stabilize either the R state or the T state, which, respectively, increase and decrease hemoglobin's O_2 affinity. Sickle-cell anemia is caused by the homozygous Hb mutant Glu 6β → Val, which promotes the gelation of the resulting deoxyHbS to form rigid 14-strand fibers that deform erythrocytes. Under gelation conditions, fiber growth occurs via a two-stage nucleation mechanism resulting in a delay time, which varies with the 30 to 50th power of the initial HbS concentration. Agents that increase this delay time to longer than the transit times of erythrocytes through the capillaries should therefore prevent sickling and thus relieve the symptoms of sickle-cell anemia.

The Adair equation rationalizes the O_2-binding cooperativity of Hb by assigning a separate dissociation constant to each O_2 bound. Positive cooperativity results if these constants decrease sequentially. However, the Adair equation offers no physical insight as to why this occurs. The symmetry model proposes that symmetrical oligomers can exist in one of two conformational states, R and T, that differ in ligand-binding affinity. Ligand binding to the high affinity state forces the oligomer to assume this conformation, which facilitates the binding of additional ligand. This homotropic model is extended to heterotropic effects by postulating that activator and substrate can bind only to the R state and inhibitor can bind only to the T state. The binding of activator forces the oligomer into the R state, which facilitates the binding of substrate and additional activator. The binding of inhibitor, however, forces the oligomer into the T state, which prevents substrate and activator binding. The sequential model postulates that an induced fit between ligand and substrate confers conformational strain on the protein that alters its affinity for binding other ligands without requiring the oligomer to maintain its symmetry. The Perutz mechanism for O_2 binding to Hb is largely consistent with the symmetry model but exhibits some elements of the sequential model.

References

General

Antonini, E., Rossi-Bernardi, L., and Chiancone, E. (Eds.), *Methods Enzymol.* **76** (1981). [A collection of articles on techniques of purification and characterization of hemoglobins.]

Bunn, H. F. and Forget, B. G., *Hemoglobin: Molecular, Genetic and Clinical Aspects,* Saunders (1986). [A valuable compendium on normal and abnormal hemoglobins.]

Dickerson, R. E. and Geis, I., *Hemoglobin,* Benjamin/Cummings (1983). [A beautifully written and lavishly illustrated treatise on the structure, function, and evolution of hemoglobin.]

Fermi, G. and Perutz, M. F., *Atlas of Molecular Structures in Biology. 2. Haemoglobin and Myoglobin,* Oxford University Press (1981). [A wide-ranging review of the structural aspects of hemoglobin and myoglobin action that is extensively illustrated with stereo drawings.]

Judson, H. F., *The Eighth Day of Creation,* Chapters 9 and 10, Simon & Schuster (1979). [Includes a fascinating historical account of how our present perception of hemoglobin structure and function came about.]

Perutz, M. F., Hemoglobin structure and respiratory transport, *Sci. Am.* **239**(6): 92–125 (1978). [An instructive discussion of the structure of hemoglobin and its mechanism of oxygen transport.]

Stamatoyannopoulos, G., Nienhuis, A. W., Leder, P., and Majerus, P. W. (Eds.), *The Molecular Basis of Blood Diseases,* Chapters 5 and 6, Saunders (1987). [Authoritative discussions of hemoglobin structure and function and on sickle-cell disease.]

Structures of Myoglobin, Hemoglobin, and Model Compounds

Collman, J. P., Synthetic models for oxygen-binding hemoproteins, *Acc. Chem. Res.* **10**, 265–272 (1977).

Fermi, G., Perutz, M. F., Shaanan, B., and Fourme, R., The crystal structure of human deoxyhaemoglobin at 1.74 Å, *J. Mol. Biol.* **175**, 159–174 (1984).

Jameson, G. B., Molinaro, F. S., Ibers, J. A., Collman, J. P., Brauman, J. I., Rose, E., and Suslick, K. S., Models for the active site of oxygen-binding hemoproteins. Dioxygen binding properties and the structures of (2-methylimidazole)-*meso*-tetra($\alpha,\alpha,\alpha,\alpha$-*o*-pivalamidophenyl)porphinato iron(II)-ethanol and its dioxygen adduct, *J. Am. Chem. Soc.* **102**, 3224–3237 (1980).

Liddington, R., Derewenda, Z., Dodson, G., and Harris, D., Structure of the liganded T state of haemoglobin identifies the origin of cooperative oxygen binding, *Nature* **331**, 725–728 (1988).

Phillips, S. E. V. and Schoenborn, B. P., Neutron diffraction reveals oxygen–histidine hydrogen bond in oxymyoglobin, *Nature* **292**, 81–82 (1982).

Phillips, S. E. V., Structure and refinement of oxymyoglobin at 1.6 Å resolution, *J. Mol. Biol.* **142**, 531–554 (1980).

Shaanan, B., Structure of human oxyhaemoglobin at 2.1 Å resolution, *J. Mol. Biol.* **171**, 31–59 (1983).

Takano, T., Structure of myoglobin refined at 2.0 Å resolution, *J. Mol. Biol.* **110**, 537–568, 569–584 (1977).

Mechanism of Hemoglobin Oxygen Binding

Baldwin, J. and Chothia, C., Haemoglobin: the structural changes related to ligand binding and its allosteric mechanism, *J. Mol. Biol.* **129**, 175–220 (1979). [The exposition of a detailed mechanism of O_2 binding to Hb based on the structures of oxy and deoxyHb.]

Baldwin, J., Structure and cooperativity of haemoglobin, *Trends Biochem. Sci.* **5**, 224–228 (1980). [A review of the previous reference.]

Gelin, B. R., Lee, A. W.-N., and Karplus, M., Haemoglobin tertiary structural change on ligand binding, *J. Mol. Biol.* **171**, 489–559 (1983). [A theoretical study of the dynamics of O_2 binding to Hb.]

Perutz, M. F., Stereochemistry of cooperative effects in haemoglobin, *Nature* **228**, 726–734 (1970). [A landmark paper in which the Perutz mechanism was first proposed. Although many of its details have since been modified, the basic model remains intact.]

Perutz, M. F., Regulation of oxygen affinity of hemoglobin, *Annu. Rev. Biochem.* **48**, 327–386 (1979). [A detailed examination of the Perutz mechanism in light of structural and spectroscopic data.]

Perutz, M. F., Stereochemical mechanism of oxygen transport in haemoglobin, *Proc. R. Soc. London Ser. B* **208**, 135–162 (1980).

Perutz, M. F., Fermi, G., Luisi, B., Shaanan, B., and Liddington, R. C., Stereochemistry of cooperative mechanisms in hemoglobin, *Acc. Chem. Res.* **20**, 309–321 (1987). [The Perutz mechanism in light of refined high resolution X-ray structures.]

Bohr Effect and BPG Binding

Arnone, A., X-ray diffraction study of binding of 2,3-diphosphoglycerate to human deoxyhaemoglobin, *Nature* **237**, 146–149 (1972).

Arnone, A., X-ray studies of the interaction of CO_2 with human deoxyhaemoglobin, *Nature* **247**, 143–145 (1974).

Benesch, R. E. and Benesch, R., The mechanism of interaction of red cell organic phosphates with hemoglobin, *Adv. Protein Chem.* **28**, 211–237 (1974).

Kilmartin, J. V. and Rossi-Bernardi, L., Interactions of hemoglobin with hydrogen ion, carbon dioxide and organic phosphates, *Physiol. Rev.* **53**, 836–890 (1973).

Kilmartin, J. V., Interactions of haemoglobin with protons, CO_2 and 2,3-diphosphoglycerate, *Br. Med. Bull.* **32**, 209–212 (1976).

Kilmartin, J. V., The Bohr effect of human haemoglobin, *Trends Biochem. Sci.* **2**, 247–249 (1977).

Lenfant, C., Torrance, J., English, E., Finch, C. A., Reynafarje, C., Ramos, J., and Faura, J., Effect of altitude on oxygen binding by hemoglobin and on organic phosphate levels, *J. Clin. Invest.* **47**, 2652–2656 (1968).

Perutz, M. F., Kilmartin, J. V., Nishikura, K., Fogg, J. H., and Butler, P. J. G., Identification of residues contributing to the Bohr effect of human haemoglobin, *J. Mol. Biol.* **138**, 649–670 (1980).

Abnormal Hemoglobins

Bellingham, A. J., Haemoglobins with altered oxygen affinity, *Br. Med. Bull.* **32**, 234–238 (1976).

Dykes, G., Crepeau, R. H., and Edelstein, S. J., Three-dimensional reconstruction of the fibers of sickle cell hemoglobin, *Nature* **272**, 506–510 (1978).

Eaton, W. A. and Hofrichter, J., Hemoglobin S gelation and sickle cell disease, *Blood* **70**, 1245–1266 (1987).

Morimoto, H., Lehmann, H., and Perutz, M. F., Molecular pathology of human haemoglobin: stereochemical interpretation of abnormal oxygen affinities, *Nature* **232**, 408–413 (1971).

Noguchi, C. T. and Schechter, A. N., Sickle hemoglobin polymerization in solution and in cells, *Annu. Rev. Biophys. Biophys. Chem.* **14**, 239–263 (1985).

Padlan, E. A. and Love, W. E., Refined crystal structure of deoxyhemoglobin S, *J. Biol. Chem.* **260**, 8280–8291 (1985).

Perutz, M. F. and Lehmann, H., Molecular pathology of human haemoglobin, *Nature* **219**, 902–909 (1968). [A ground-breaking study correlating the clinical symptoms and inferred structural alterations of numerous mutant hemoglobins.]

Winslow, R. M. and Anderson, W. F., The hemoglobinopathies, *in* Stanbury, J. B., Wyngaarden, J. B., Fredrickson, D. S., Goldstein, J. L., and Brown, M. S. (Eds.), *The Metabolic Basis of Inherited Disease* (5th ed.), *pp.* 1666–1710, McGraw–Hill (1983). [A detailed review.]

Allosteric Regulation

Ackers, G. K. and Smith, F. R., The hemoglobin tetramer: a three-state molecular switch for control of ligand affinity, *Annu. Rev. Biophys. Biophys. Chem.* **16**, 583–609 (1987). [Presents thermodynamic arguments that hemoglobin ligand binding is better described by a three-state than a two-state model.]

Baldwin, J. M., A model of co-operative oxygen binding to haemoglobin, *Br. Med. Bull.* **32**, 213–218 (1976).

Cantor, C. R. and Schimmel, P. R., *Biophysical Chemistry*, Chapter 17, Freeman (1980).

Edelstein, S. J., Cooperative interactions of hemoglobin, *Annu. Rev. Biochem.* **44**, 209–232 (1975).

Fersht, A., *Enzyme Structure and Mechanism* (2nd ed.), Chapter 10, Freeman (1985).

Koshland, D. E., Jr., Némethy, G., and Filmer, D., Comparison of experimental binding data and theoretical models in proteins containing subunits, *Biochemistry* **5**, 365–385 (1966). [The formulation of the sequential model of allosteric regulation.]

Monod, J., Wyman, J., and Changeux, J.-P., On the nature of allosteric transitions: a plausible model, *J. Mol. Biol.* **12**, 88–118 (1965). [The exposition of the symmetry model of allosteric regulation.]

Shulman, R. G., Hopfield, J. J., and Ogawa, S., Allosteric interpretation of haemoglobin properties, *Q. Rev. Biophys.* **8**, 325–420 (1975).

Szabo, A. and Karplus, M., A mathematical model for structure-function relations in hemoglobin, *J. Mol. Biol.* **72**, 163–197 (1972). [An analysis of the Perutz mechanism indicating that it is consistent with the symmetry model of allosterism.]

Problems

1. The urge to breathe in humans results from a high blood CO_2 content; there are no direct physiological sensors of blood pO_2. Skindivers often **hyperventilate** (breathe rapidly and deeply for several minutes) just before making a protracted dive in the belief that they will thereby increase the O_2 content of their blood. This belief results from the fact that hyperventilation represses the breathing urge by expelling significant quantities of CO_2 from the blood. In light of what you know about the properties of hemoglobin, is hyperventilation a useful procedure? Is it safe? Explain.

2. Explain why n in the Hill equation can never be larger than the number of ligand-binding sites on the protein.

*3. In the Bohr effect, protonation of the N-terminal amino groups of hemoglobin's α chains is responsible for $\sim 30\%$ of the 0.6 mol of H^+ that combine with Hb upon the release of 1 mol of O_2 at pH 7.4. Assuming that this group has pK 7.0 in oxyHb, what is its pK in deoxyHb?

4. As one of the favorites to win the La Paz, Bolivia marathon, you have trained there for the several weeks it requires to become adapted to its 3700-m altitude. A manufacturer of running equipment who sponsors an opponent has invited you for the weekend to a prerace party at a beach house near Lima, Peru with the assurance that you will be flown back to La Paz at least a day before the race. Is this a token of his respect for you or an underhanded attempt to handicap you in the race? Explain (see Fig. 9-9).

5. In active muscles, the pO_2 may be 10 torr at the cell surface and 1 torr at the mitochondria (the organelles where oxidative metabolism occurs). How does myoglobin ($p_{50} = 2.8$ torr) facilitate the diffusion of O_2 through these cells? Active muscles consume O_2 much faster than do other tissues. Would myoglobin also be an effective O_2-transport protein in other tissues? Explain.

6. Erythrocytes that have been stored for over a week in standard acid–citrate–dextrose medium become depleted in BPG. Discuss the merits of using fresh versus week-old blood in blood transfusions.

7. The following fractional saturation data has been measured for a certain blood sample:

pO_2	Y_{O_2}	pO_2	Y_{O_2}
20	0.14	60	0.59
30	0.26	70	0.66
40	0.39	80	0.72
50	0.50	90	0.76

What are the Hill constant and the p_{50} of this blood sample? Are they normal?

8. An anemic individual, whose blood has only half the normal Hb content, may appear to be in good health. Yet, a normal individual is incapacitated by exposure to sufficient carbon monoxide to occupy half his heme sites (pCO of 1 torr for ~ 1 h; CO binds to Hb with 200 times greater affinity than does O_2). Explain.

*9. The X-ray structure of Hb Rainier (Tyr $145\beta \rightarrow$ Cys) indicates that the mutant Cys residue forms a disulfide bond with Cys 93β of the same subunit. This holds the β subunit's C-terminal residue in a quite different orientation than it assumes in HbA. How would the following quantities for Hb Rainier compare with those of HbA? Explain. (a) The oxygen affinity, (b) the Bohr effect, (c) the Hill constant, and (d) the BPG affinity.

10. The crocodile, which can remain under water without breathing for up to 1 h, drowns its air-breathing prey and then dines at its leisure. An adaptation that aids the crocodile in doing so is that it can utilize virtually 100% of the O_2 in its blood whereas humans, for example, can extract only $\sim 65\%$ of the O_2 in their blood. Crocodile Hb does not bind BPG. However, crocodile deoxyHb preferentially binds HCO_3^-. How does this help the crocodile obtain its dinner?

11. The gelation time of an equimolar mixture of HbA and HbS is less than that of a solution of only HbS in the same concentration that it has in the mixture. What does this observation imply about the participation of HbA in the gelation of HbS?

12. The severely anemic condition of homozygotes for HbS results in an elevated BPG content in their erythrocytes. Discuss whether or not this is a beneficial effect.

13. As organizer of an expedition that plans to climb several very high mountains, it is your responsibility to choose its members. Each of the applicants for one of the positions on the team is a heterozygote for one of the following abnormal hemoglobins: (1) HbS, (2) Hb Hyde Park [His F8(92)$\beta \rightarrow$ Tyr], (3) Hb Riverdale–Bronx [Gly B6(24)$\beta \rightarrow$ Arg], (4) Hb Memphis [Glu B4(23)$\alpha \rightarrow$ Gln], and (5) Hb Cowtown [His HC3(146)$\beta \rightarrow$ Leu]. Assuming that all of these candidates are equal in ability at low altitudes, which one would you choose for the position? Explain your reasoning.

14. Show that the Adair equation for a tetramer reduces to the Hill equation for $k_1 \approx k_2 \approx k_3 \gg k_4$ and to a hyperbolic relationship for $k_1 = k_2 = k_3 = k_4$.

15. Derive the equilibrium constant for the reaction $R_2 \rightleftharpoons T_2$ for a symmetry model n-mer in terms of the parameters L, c, and α.

16. Derive the equation for the fraction of protein molecules in the R state, $\overline{R}$, for the homotropic symmetry model in terms of the parameters n, L, c, and α. Plot this function versus α for $n = 4$, $L = 1000$ and $c = 0$ and discuss its physical significance.

17. In the symmetry model of allosterism, why must an inhibitor (which causes a negative heterotropic effect with the substrate) undergo a positive homotropic effect?

Chapter 10
SUGARS AND POLYSACCHARIDES

Carbohydrates or **saccharides** (Greek: *sakcharon,* sugar) are essential components of all living organisms and are, in fact, the most abundant class of biological molecules. The name carbohydrate, which literally means "carbon hydrate," stems from their chemical composition, which is roughly $(C \cdot H_2O)_n$, where $n \geq 3$. The basic units of carbohydrates are known as **monosaccharides.** Many of these compounds are synthesized from simpler substances in a process named gluconeogenesis (Section 21-1). Others (and ultimately nearly all biological molecules) are the products of photosynthesis (Section 22-3), the light-powered combination of CO_2 and H_2O through which plants and certain bacteria form "carbon hydrates." The metabolic breakdown of monosaccharides (Chapters 16 and 19) provides most of the energy used to power biological processes. Monosaccharides are also principal components of nucleic acids (Section 28-1), as well as important elements of complex lipids (Section 11-1D).

Oligosaccharides consist of a few covalently linked monosaccharide units. They are often associated with proteins (glycoproteins) and lipids (glycolipids) in which they have both structural and regulatory functions. **Polysaccharides** consist of many covalently linked monosaccharide units and have molecular masses ranging well into the millions of daltons. They have indispensable structural functions in all types of organisms but are most conspicuous in plants because **cellulose,** their principal structural material, comprises up to 80% of their dry weight. Polysaccharides such as **starch** in plants and **glycogen** in animals serve as important nutritional reservoirs.

The elucidation of the structures and functions of carbohydrates has lagged well behind those of proteins and nucleic acids. This can be attributed to several factors. Carbohydrate compounds are often heterogeneous, both in size and composition, which greatly complicates their physical and chemical characterization. They are not subject to the types of genetic analysis that have been invaluable in the study of proteins and nucleic acids because saccharide sequences are not genetically specified but rather are built up through the sequential actions of specific enzymes (Section 21-3B). Furthermore, it has been difficult to establish assays for the biological activities of polysaccharides because of their largely passive roles. Nevertheless, recent biochemical progress has made it abundantly clear that carbohydrates are essential elements in many, if not most, biological processes.

In this chapter, we explore the structures, chemistry, and, to a limited extent, the functions of carbohydrates, alone and in association with proteins. Glycolipid structures are considered in Section 11-1D. The biosynthesis of complex carbohydrates is discussed in Section 21-3.

1. MONOSACCHARIDES

Monosaccharides or **simple sugars** are aldehyde or ketone derivatives of straight-chain polyhydroxy alcohols containing at least three carbon atoms. Such substances, for example, D-**glucose** and D-**ribulose,** cannot be hydrolyzed to form simpler saccharides.

Figure 10-1
The stereochemical relationships, shown in Fischer projection, among the D-aldoses with three to six carbon atoms. The configuration about C(2) (*red*) distinguishes the members of each pair.

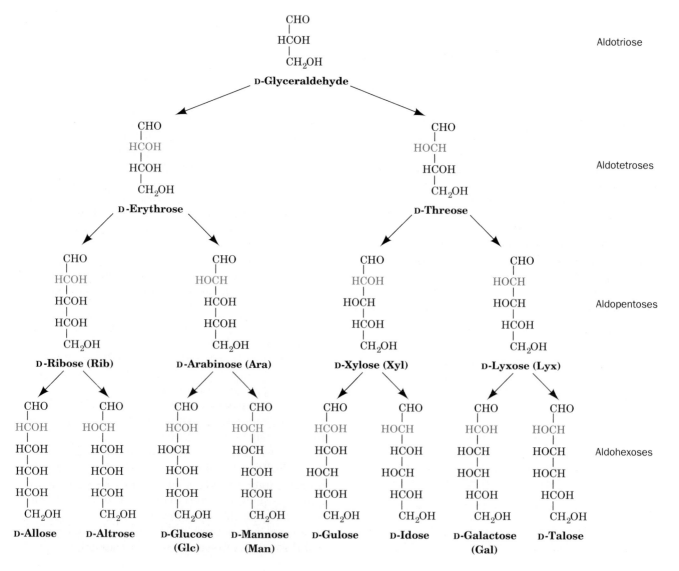

O=C—H
H—²C—OH
HO—³C—H
H—⁴C—OH
H—⁵C—OH
⁶CH₂OH

D-Glucose

¹CH₂OH
²C=O
H—³C—OH
H—⁴C—OH
⁵CH₂OH

D-Ribulose

In this section, the structures of the monosaccharides and some of their biologically important derivatives are discussed.

A. Classification

Monosaccharides are classified according to the chemical nature of their carbonyl group and the number of their C atoms. If the carbonyl group is an aldehyde, as in glucose, the sugar is an **aldose.** If the carbonyl group is a ketone, as in ribulose, the sugar is a **ketose.** The smallest monosaccharides, those with three carbon atoms, are **trioses.** Those with four, five, six, seven, *etc.* C atoms are, respectively, **tetroses, pentoses, hexoses, heptoses,** *etc.* These terms may be combined so that, for example, glucose is an **aldohexose** whereas ribulose is a **ketopentose.**

Examination of D-glucose's molecular formula indicates that all but two of its six C atoms — C(1) and C(6) — are chiral centers so that D-glucose is one of $2^4 = 16$ stereoisomers that comprise all possible aldohexoses. In general, *n*-carbon aldoses have 2^{n-2} stereoisomers. The stereochemistry and names of the D-aldoses are presented in Fig. 10-1. Emil Fischer elucidated these configurations for the aldohexoses in 1896. According to the Fischer convention (Section 4-2B), *D-sugars have the same absolute configuration at the asymmetric center farthest removed from their carbonyl group as does D-glyceraldehyde.* The L-sugars, in accordance with this convention, are mirror images of their D-counterparts as is shown below in Fischer projection for glucose.

O=C—H
H—C—OH
HO—C—H
H—C—OH
H—C—OH
CH₂OH

D-Glucose

O=C—H
HO—C—H
H—C—OH
HO—C—H
HO—C—H
CH₂OH

L-Glucose

Sugars that differ only by the configuration about one C atom are known as **epimers** of one another. Thus D-glu-

cose and **D-mannose** are epimers with respect to C(2), whereas D-glucose and **D-galactose** are epimers with respect to C(4) (Fig. 10-1). However, D-mannose and D-galactose are not epimers of each other because they differ in configuration about two C atoms.

D-Glucose is the only aldose that commonly occurs in nature as a monosaccharide. However, it and several other monosaccharides including D-glyceraldehyde, D-ribose, D-mannose, and D-galactose are important components of larger biological molecules. L-Sugars are biologically much less abundant than D-sugars.

The position of their carbonyl group gives ketoses one less asymmetric center than their isomeric aldoses (e.g., compare D-fructose and D-glucose). *n*-Carbon ketoses therefore have 2^{n-3} stereoisomers. Those with their ketone function at C(2) are the most common form (Fig. 10-2). Note that some of these ketoses are named by the

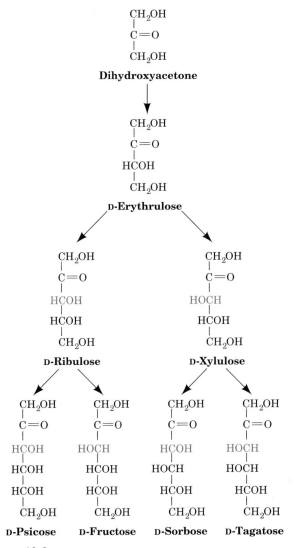

Figure 10-2
The stereochemical relationships among the D-ketoses with three to six carbon atoms. The configuration about C(3) (*red*) distinguishes the members of each pair.

(a)

R—OH + R'—C(H)=O ⇌ [R—O, H / C / R', OH]

Alcohol   Aldehyde   Hemiacetal

(b)

R—OH + R'—C(R'')=O ⇌ [R—O, R'' / C / R', OH]

Alcohol   Ketone   Hemiketal

Figure 10-3
The reactions of alcohols with (*a*) aldehydes to form hemiacetals and (*b*) ketones to form hemiketals.

insertion of *ul* before the suffix *ose* in the name of the corresponding aldose; thus D-xylulose is the ketose corresponding to the aldose D-xylose. Dihydroxyacetone, D-fructose, D-ribulose, and D-xylulose are the biologically most prominent ketoses.

B. Configurations and Conformations

Alcohols react with the carbonyl groups of aldehydes and ketones to form **hemiacetals** and **hemiketals,** respectively (Fig. 10-3). The hydroxyl and either the aldehyde or ketone functions of monosaccharides can likewise react intramolecularly to form cyclic hemiacetals and hemiketals (Fig. 10-4). The configurations of the substituents to each carbon atom of these sugar rings are

conveniently represented by their **Haworth projection formulas.**

A sugar with a 6-membered ring is known as a **pyranose** in analogy with **pyran,** the simplest compound containing such a ring. Similarly, sugars with 5-membered rings are designated **furanoses** in analogy with **furan.** The cyclic forms of glucose and fructose are known therefore as **glucopyranose** and **fructofuranose,** respectively.

Pyran   **Furan**

Cyclic Sugars Have Two Anomeric Forms

The Greek letters preceding the names in Fig. 10-4 still need to be explained. The cyclization of a monosaccharide renders the former carbonyl carbon asymmetric. The resulting pair of diastereomers are known as **anomers** and the hemiacetal or hemiketal carbon is referred to as the **anomeric** carbon. In the α anomer, the

Figure 10-4
The reactions of (*a*) D-glucose in its linear form to yield the cyclic hemiacetal α-D-glucopyranose and (*b*) D-fructose in its linear form to yield the hemiketal α-D-fructofuranose. The cyclic sugars are shown as both Haworth projections and as space-filling models. [Computer graphics models courtesy of Robert Stodola, Fox Chase Cancer Center.]

(a)

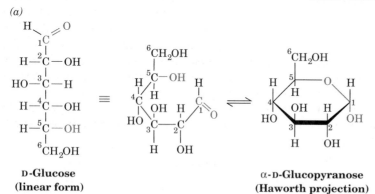

D-Glucose
(linear form)

α-**D-Glucopyranose**
(Haworth projection)

(b)

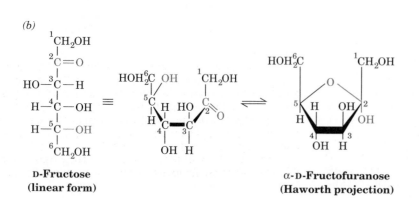

D-Fructose
(linear form)

α-**D-Fructofuranose**
(Haworth projection)

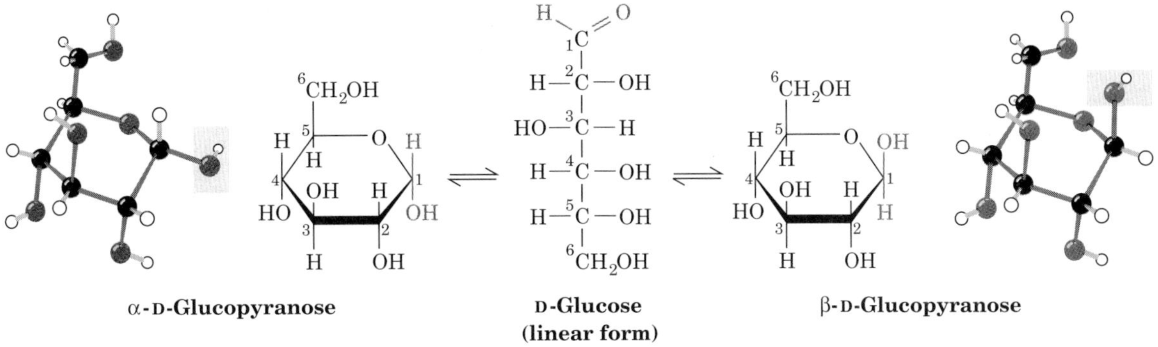

α-D-Glucopyranose   D-Glucose (linear form) β-D-Glucopyranose

Figure 10-5
The anomeric monosaccharides α-D-glucopyranose and β-D-glucopyranose, drawn as both Haworth projections and ball-and-stick models. These pyranose sugars interconvert through the linear form of D-glucose and differ only by the configurations about their anomeric carbon atoms, C(1).

OH substituent to the anomeric carbon is on the opposite side of the sugar ring from the CH_2OH group at the chiral center that designates the D or L configuration [C(5) in hexoses]. The other anomer is known as the β form (Fig. 10-5).

The two anomers of D-glucose, as any pair of diastereomers, have different physical and chemical properties. For example, the values of the specific optical rotation, $[\alpha]_D^{20}$, for α-D-glucose and β-D-glucose are, respectively, $+112.2°$ and $+18.7°$. When either of these pure substances is dissolved in water, however, the specific optical rotation of the solution slowly changes until it reaches an equilibrium value of $[\alpha]_D^{20} = +52.7°$. This phenomenon is known as **mutarotation**; in glucose, it results from the formation of an equilibrium mixture consisting of 63.6% of the β anomer and 36.4% of the α anomer (the optical rotations of separate molecules in solution are independent of each other so that the optical rotation of a solution is the weighted average of the optical rotations of its components). The interconversion between these anomers occurs via the linear form of glucose (Fig. 10-5). Yet, since the linear forms of these monosaccharides are normally present in only minute amounts, these carbohydrates are accurately described as cyclic polyhydroxy hemiacetals or hemiketals.

Sugars Are Conformationally Variable

Hexoses and pentoses may each assume pyranose or furanose forms. The equilibrium composition of a particular monosaccharide depends somewhat on conditions but mostly on the identity of the monosaccharide. For instance, in aqueous solutions, glucose exists almost exclusively in the pyranose form whereas ribose is $\sim 25\%$ furanose and 75% pyranose. Although, in principle, hexoses and larger sugars can form rings of seven or more atoms, such rings are rarely observed because of the greater stabilities of the 5- and 6-membered rings

that these sugars can also form. The internal strain of 3- and 4-membered sugar rings makes them unstable with respect to linear forms.

The use of Haworth formulas may lead to the erroneous impression that furanose and pyranose rings are planar. This cannot be the case, however, because all of the atoms in these rings are tetrahedally (sp^3) hybridized. The pyranose ring, like the cyclohexane ring, may assume a **boat** or a **chair** conformation (Fig. 10-6). The relative stabilities of these various conformations depends on the stereochemical interactions between the substituents on the ring. The boat conformer crowds the substituents on its "bow" and "stern" and eclipses those along its sides so that in cyclohexane it is $\sim 25\ kJ \cdot mol^{-1}$ less stable than the chair conformer. The ring substituents on the chair conformer (Fig. 10-6b) fall into two geometrical classes: the rather close-fitting **axial** groups that extend parallel to the ring's threefold rotational axis, and the staggered, and therefore largely unencumbered, **equatorial** groups. Since the axial and equatorial

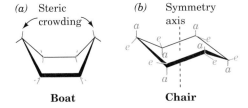

(a) Steric crowding   *(b)* Symmetry axis

Boat   **Chair**

Figure 10-6
The conformations of the cyclohexane ring. (*a*) In the boat conformation, substituents at the "bow" and "stern" (*red*) are sterically crowded while those along its sides (*green*) are eclipsed. (*b*) In the chair conformation, the substituents that extend parallel to the ring's threefold rotation axis are designated axial [*a*] and those that extend roughly outward from this symmetry axis are designated equatorial [*e*]. The equatorial substituents about the ring are staggered so that they alternately extend above and below the mean plane of the ring.

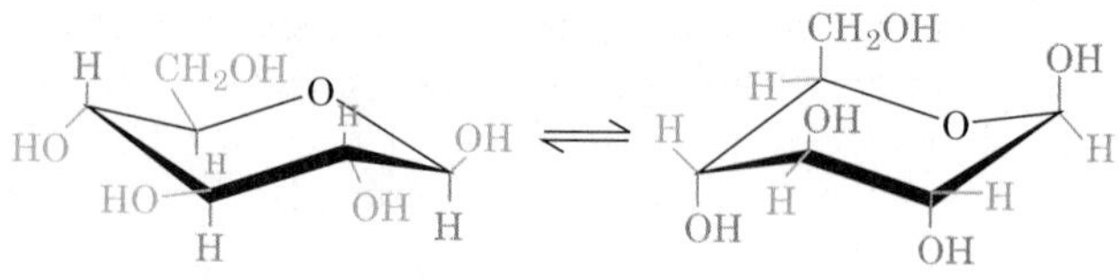

Figure 10-7
The two alternative chair conformations of β-D-glucopyranose. In that on the left, which predominates, the relatively bulky OH and CH_2OH substituents all occupy equatorial positions, whereas on the right (drawn in ball and stick form in Fig. 10-5, *right*) they occupy the more crowded axial positions.

groups on a cyclohexane ring are conformationally interconvertible, a given ring has two alternative chair forms (Fig. 10-7); the one that predominates usually has the lesser crowding among its axial substituents. The conformational situation of a group directly affects its chemical reactivity. For example, equatorial OH groups on pyranoses esterify more readily than do axial OH groups. Note that β-D-glucose is the only D-aldohexose that can simultaneously have all five non-H substituents in the equatorial position (*left side of Fig. 10-7*). Perhaps this is why glucose is the most abundant naturally occurring monosaccharide. The conformational properties of furanose rings are discussed in Section 28-3B in relation to their effects on the conformations of nucleic acids.

C. Sugar Derivatives

Polysaccharides Are Held Together by Glycosidic Bonds

The chemistry of monosaccharides is largely that of their hydroxy and carbonyl groups. For example, in an acid catalyzed reaction, the anomeric hydroxyl group of a sugar reversibly condenses with alcohols to form **α-** and **β-glycosides** [Fig. 10-8 (Greek: *glykys,* sweet)]. The bond connecting the anomeric carbon to the acetal oxygen is termed a **glycosidic bond.** *Polysaccharides are held together by glycosidic bonds between neighboring monosaccharide units.* The glycosidic bond is therefore the carbohydrate analog of the peptide bond in proteins. The hydrolysis of glycosidic bonds is catalyzed by enzymes

known as **glycosidases** that differ in specificity according to the identity and anomeric configuration of the glycoside but are often rather insensitive to the identity of the alcohol residue. Under basic or neutral conditions and in the absence of glycosidases, however, the glycosidic bond is stable so that glycosides do not undergo mutarotation as do monosaccharides. The methylation of the non-anomeric OH groups of monosaccharides requires more drastic conditions than is required for the formation of methyl glycosides, such as treatment with dimethylsulfate.

Oxidation – Reduction Reactions

Because the cyclic and linear forms of aldoses and ketoses interconvert so readily, these sugars undergo reactions typical of aldehydes and ketones. Mild oxidation of an aldose, either chemically or enzymatically, results in the conversion of its aldehyde group to a carboxylic acid function thereby yielding an **aldonic acid** such as **gluconic acid.** Aldonic acids are named by appending the suffix *onic acid* to the root name of the parent aldose.

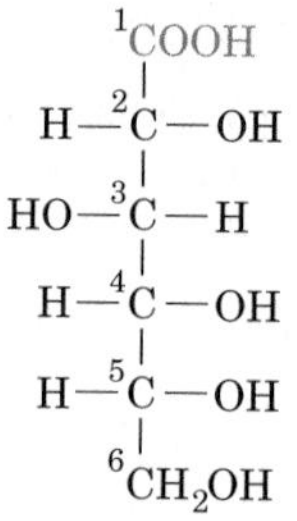

D-**Gluconic acid**

Saccharides bearing anomeric carbon atoms that have not formed glycosides are termed **reducing sugars** because of the facility with which the aldehyde group reduces mild oxidizing agents. A standard test for the presence of a reducing sugar is the reduction of Ag^+ in an ammonia solution **(Tollens' reagent)** to yield a metallic silver mirror lining on the inside of the reaction vessel.

The specific oxidation of the primary alcohol group of aldoses yields **uronic acids,** which are named by ap-

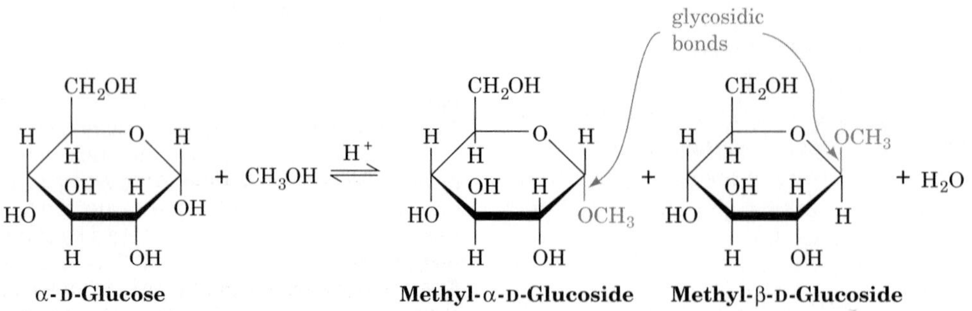

Figure 10-8
The acid catalyzed condensation of α-D-glucose with methanol to form an anomeric pair of methyl-D-glucosides.

pending *uronic acid* to the root name of the parent aldose. **D-Glucuronic acid, D-galacturonic acid,** and **D-mannuronic acid** are important components of many polysaccharides.

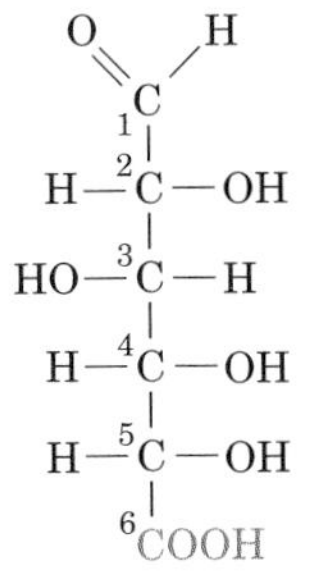

D-Glucuronic acid 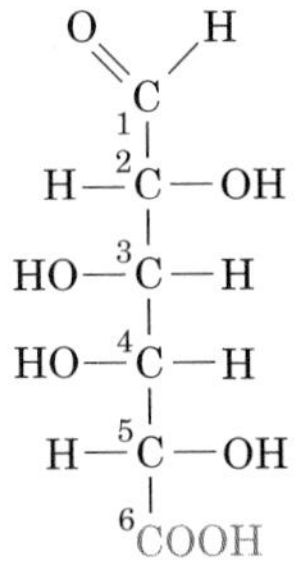   D-Galacturonic acid

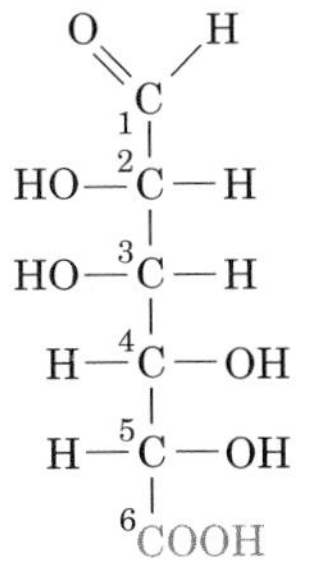

D-Mannuronic acid

Uronic acids can assume the pyranose, furanose, and linear forms.

Both aldonic and uronic acids have a strong tendency to internally esterify so as to form 5- and 6-membered lactones (Fig. 10-9). **Ascorbic acid (vitamin C,** Fig. 10-10) is a γ-lactone that is synthesized by plants and almost all animals except primates and guinea pigs. Its prolonged deficiency in the diet of humans results in the disease known as **scurvy,** which is caused by the impairment of collagen formation (Section 7-2C). Scurvy

generally results from a lack of fresh food. This is because, under physiological conditions, ascorbic acid is reversibly oxidized to **dehydroascorbic acid** which, in turn, is irreversibly hydrolyzed to the vitamin-inactive **diketogulonic acid** (Fig. 10-10).

Aldoses and ketoses may be reduced under mild conditions, for example, by treatment with $NaBH_4$, to yield acyclic polyhydroxy alcohols known as **alditols,** which are named by appending the suffix *itol* to the root name of the parent aldose. **Ribitol** is a component of flavin coenzymes (Section 14-4), and **glycerol** and the cyclic polyhydroxy alcohol *myo***-inositol** are important lipid components (Section 11-1). **Xylitol** is a sweetener that is used in "sugarless" gum and candies.

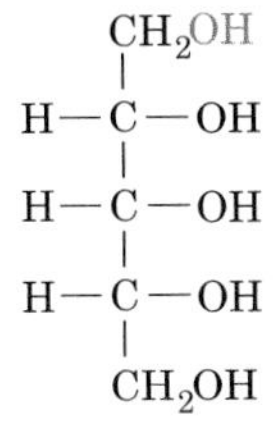

D-Ribitol 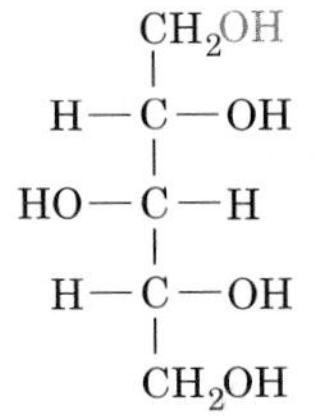   D-Xylitol

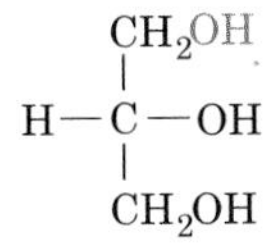

D-Glycerol 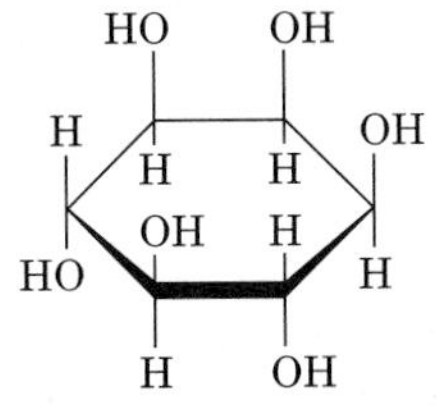   *myo*-Inositol

Other Biologically Important Sugar Derivatives

Monosaccharide units in which an OH group is replaced by H are known as **deoxy sugars.** The biologically most important of these is **D-2-deoxyribose,** the

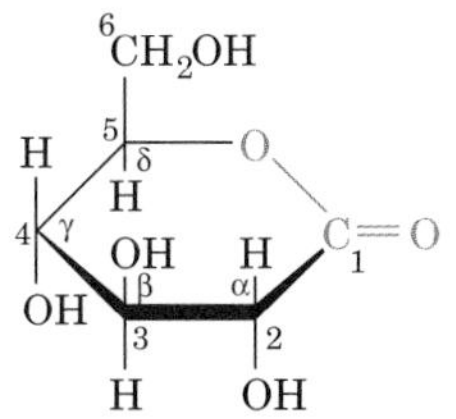

D-Glucono-δ-lactone 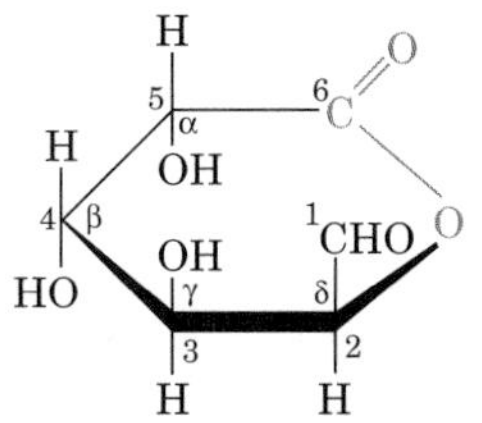   D-Glucurono-δ-lactone

Figure 10-9
D-glucono-δ-lactone and D-glucurono-δ-lactone are, respectively, the lactones of D-gluconic acid and D-glucuronic acid. The "δ" indicates that the O atom closing the lactone ring is also substituent to C_δ.

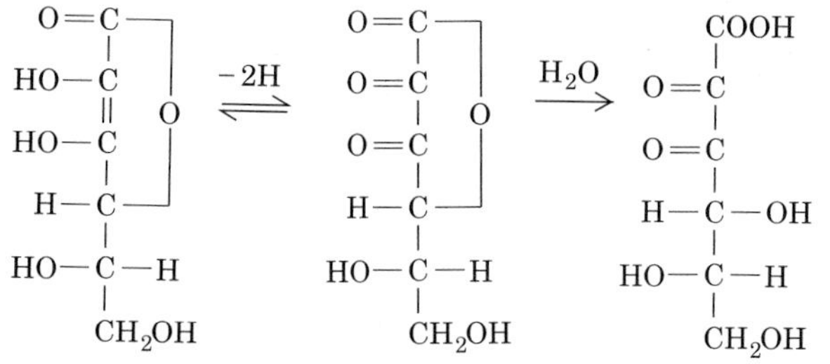

L-Ascorbic acid   L-Dehydroascorbic acid   L-Diketogulonic acid

Figure 10-10
The reversible oxidation of L-ascorbic acid to form L-dehydroascorbic acid followed by the physiologically irreversible hydrolysis of its lactone ring to form L-diketogulonic acid.

sugar component of DNA's sugar–phosphate backbone. **L-Rhamnose** and **L-fucose** are widely occurring polysaccharide components.

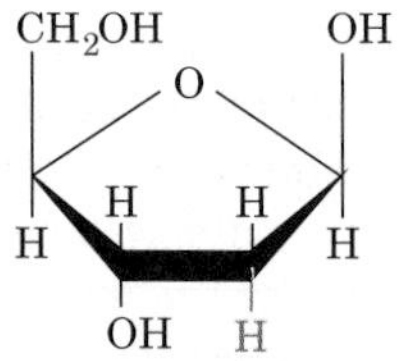

β-D-2-Deoxyribose

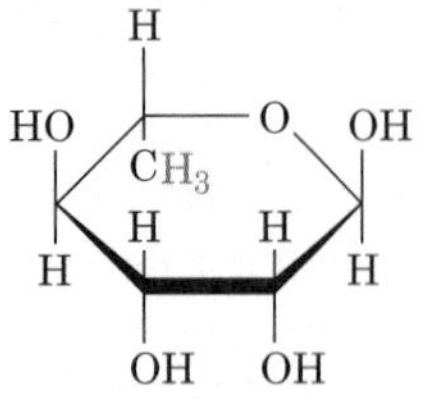

α-L-Rhamnose
(6-deoxy-L-mannose)

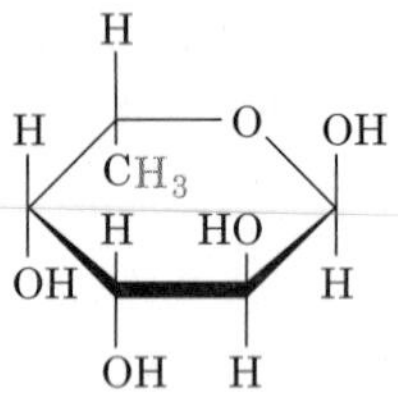

α-L-Fucose
(6-deoxy-L-galactose)

In **amino sugars,** one or more OH groups have been replaced by an often acetylated amino group. **D-Glucosamine** and **D-galactosamine** are components of numerous biologically important polysaccharides.

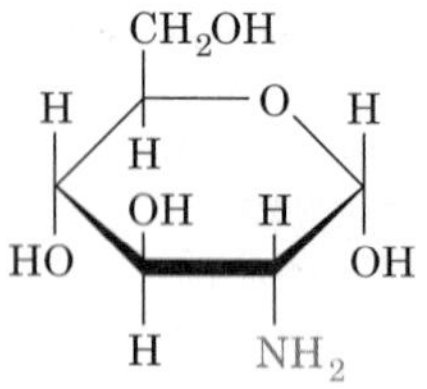

α-D-Glucosamine
(2-amino-2-deoxy-
α-D-glucopyranose)

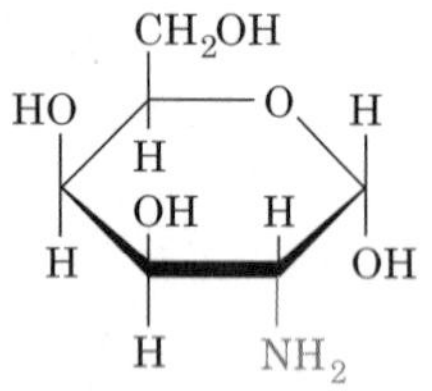

α-D-Galactosamine
(2-amino-2-deoxy-
α-D-galactopyranose)

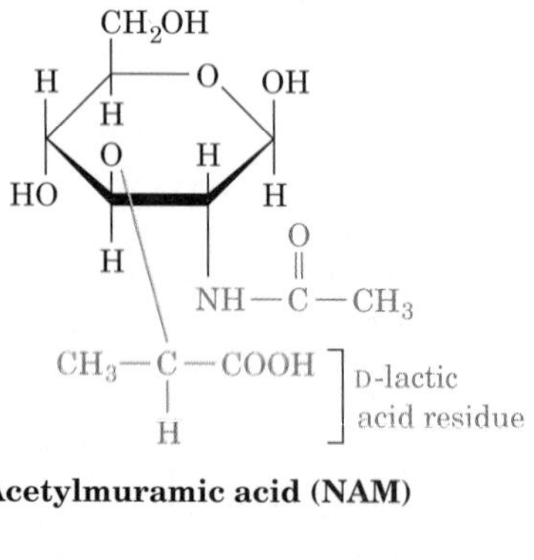

N-Acetylmuramic acid (NAM)

The amino sugar derivative **N-acetylmuramic acid,** which consists of *N*-acetyl-D-glucosamine in an ether linkage with **D-lactic acid,** is a prominent component of bacterial cell walls (Section 10-3B). **N-Acetylneuraminic acid,** which is derived from *N*-acetylmannosamine and pyruvic acid (Fig. 10-11), is an important constituent of glycoproteins (Section 10-3C) and glycolipids (Section 11-1D). *N*-Acetylneuraminic acid and its derivatives are often referred to as **sialic acids.**

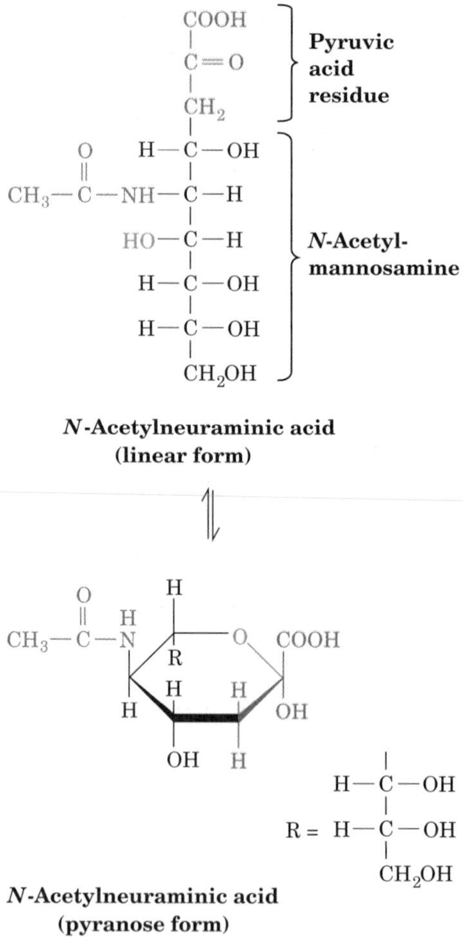

N-Acetylneuraminic acid
(linear form)

N-Acetylneuraminic acid
(pyranose form)

Figure 10-11
N-Acetylneuraminic acid in its linear and pyranose forms. Note that its pyranose ring incorporates the pyruvic acid residue (*blue*) and part of the mannose moiety.

2. POLYSACCHARIDES

Polysaccharides, which are also known as **glycans,** consist of monosaccharides linked together by glycosidic bonds. They are classified as **homopolysaccharides** or **heteropolysaccharides** if they consist of one type or more than one type of monosaccharide, respectively. Homopolysaccharides may be further classified according to the identity of their monomeric unit. For example, **glucans** are polymers of glucose whereas **galactans** are polymers of galactose. Although monosaccharide sequences of heteropolysaccharides can, in principle, be as varied as those of proteins, they are usually composed of only a few types of monosaccharides that alternate in a repetitive sequence.

Polysaccharides, in contrast to proteins and nucleic acids, form branched as well as linear polymers. This is because glycosidic linkages can be made to any of the hydroxyl groups of a monosaccharide. Fortunately for structural biochemists, most polysaccharides are linear and those that branch do so in only a few well-defined ways.

In this section, we discuss the structures of the simplest polysaccharides, disaccharides, and then consider the structures and properties of the most abundant classes of polysaccharides. We begin by outlining how polysaccharide structures are elucidated.

A. Carbohydrate Analysis

The purification of carbohydrates can, by and large, be effected by chromatographic and electrophoretic procedures similar to those used in protein purification (Sections 5-3 and 5-4). Affinity chromatography (Section 5-3D) using immobilized proteins known as **lectins** is a particularly powerful technique in this regard. Lectins are sugar-binding proteins that are usually of plant origin but which also occur in animals and bacteria. Among the best known lectins are jack bean **concanavalin A,** which specifically binds α-D-glucose and α-D-mannose residues, and wheat germ **agglutinin** (so named because it causes cells to agglutinate or clump together), which specifically binds β-N-acetylmuramic acid and α-N-acetylneuraminic acid.

Characterization of an oligosaccharide requires that the identities, anomers, linkages, and order of its component monosaccharides be elucidated. The linkages of the monosaccharides may be determined by **methylation analysis,** a technique pioneered by Norman Ha-

worth in the 1930s: *Methyl ethers not at the anomeric C atom are resistant to acid hydrolysis but glycosidic bonds are not. Consequently, if an oligosaccharide is exhaustively methylated and then hydrolyzed, the free OH groups on the resulting methylated monosaccharides mark the former positions of the glycosidic bonds.* Methylated monosaccharides are usually identified by gas–liquid chromatography (GLC) combined with mass spectrometry (Section 5-3E).

The sequence and anomeric configurations of monosaccharides in an oligosaccharide can be determined through the use of specific **exoglycosidases.** These enzymes, for example, β-galactosidase, specifically remove their corresponding monosaccharides from the nonreducing ends of oligosaccharides (the ends lacking a free anomeric carbon atom) in a manner analogous to the actions of exopeptidases on proteins (Section 6-1A). Comparison of the high resolution proton and/or ^{13}C NMR spectra of oligosaccharides of known and unknown structures can provide similar information and holds much promise for future investigations. Indeed, two-dimensional NMR techniques are providing information on oligosaccharide conformations.

B. Disaccharides

We begin our studies of polysaccharides by considering disaccharides (Fig. 10-12). **Sucrose,** the most abun-

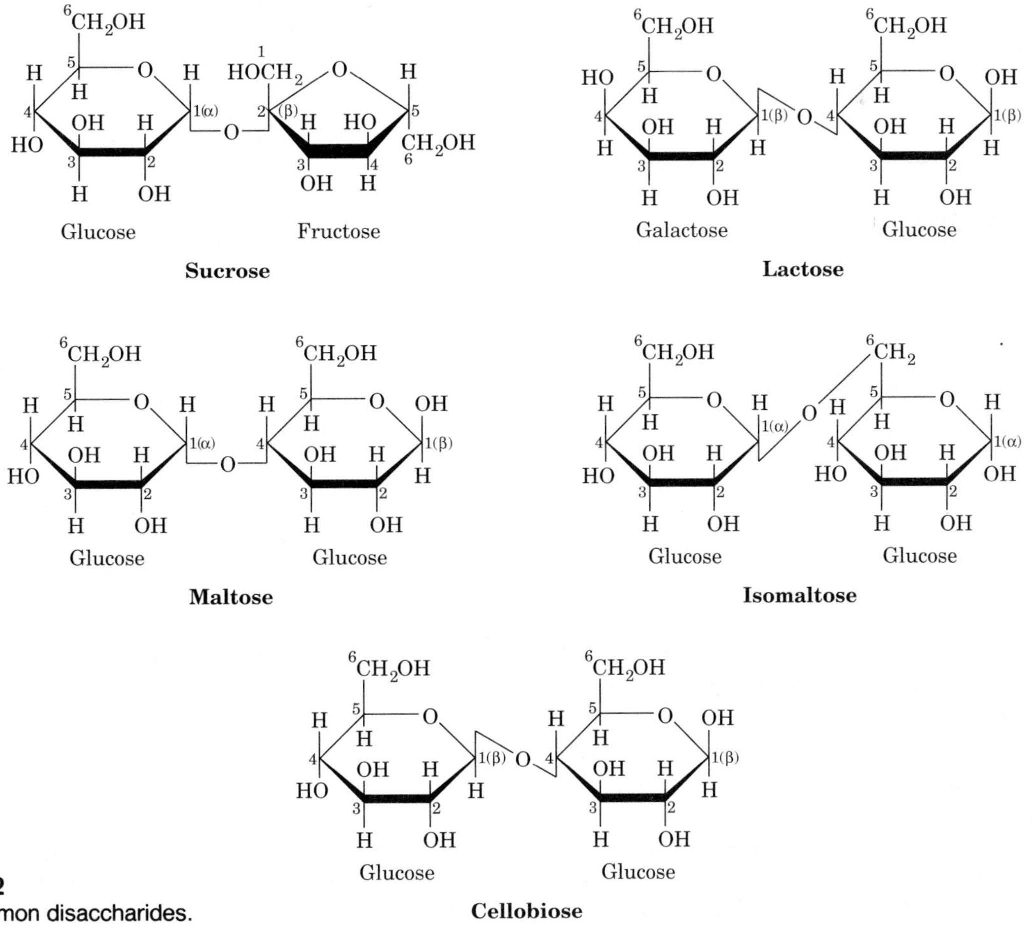

Figure 10-12
Several common disaccharides.

dant disaccharide, occurs throughout the plant kingdom and is familiar to us as common table sugar. Its structure (Fig. 10-12) was established by methylation analysis as described above and was later confirmed by X-ray analysis. To name a polysaccharide systematically, one must specify its component monosaccharides, their ring types, their anomeric forms, and how they are linked together. Sucrose is therefore O-α-D-glucopyranosyl-(1 → 2)-β-D-fructofuranoside, where the symbol (1 → 2) indicates that the glycosidic bond links C(1) of the glucose residue to C(2) of the fructose residue. Note that since these two positions are the anomeric carbon atoms of their respective monosaccharides, sucrose is not a reducing sugar (as the suffix *ide* implies).

The hydrolysis of sucrose to D-glucose and D-fructose is accompanied by a change in optical rotation from *dextro* to *levo*. Consequently, the hydrolyzed sucrose is sometimes called **invert sugar** and the enzyme that catalyzes this process, α-D-**glucosidase**, is archaically named **invertase**.

Lactose [O-β-D-galactopyranosyl-(1 → 4)-D-glucopyranose] or milk sugar occurs naturally only in milk where its concentration ranges from 0 to 7% depending on the species. The free anomeric carbon of its glucose residue makes lactose a reducing sugar.

Infants normally have the intestinal enzyme β-D-galactosidase or **lactase** that hydrolyzes lactose to its component monosaccharides for absorption into the bloodstream. Many adults, however, including most blacks and almost all Orientals, have a low level of this enzyme (as do most adult mammals). Consequently, much of the lactose in any milk they drink moves through their digestive tract to the colon where its bacterial fermentation produces large quantities of CO_2, H_2, and irritating organic acids. This results in an embarrassing and often painful digestive upset termed **milk intolerance.** Perhaps this is why oriental cuisine, which is noted for the wide variety of foodstuffs it employs, is devoid of milk products. Recently, modern food technology has come to the aid of milk lovers who develop milk intolerance: Milk in which the lactose has been hydrolyzed enzymatically is now commercially available. Unfortunately for fastidious bean lovers, there is no equivalent bean product although it has been shown that the well-known gas-producing properties of beans results from their content of indigestible but bacterially fermentable oligosaccharides.

There are several common glucosyl-glucose disaccharides. These include **maltose** [O-α-D-glucopyranosyl-(1 → 4)-D-glucopyranose], an enzymatic hydrolysis product of starch; **isomaltose**, its α(1 → 6) isomer; and **cellobiose**, its β(1 → 4) isomer, the repeating disaccharide of cellulose.

Only a few tri- or higher oligosaccharides occur in significant natural abundance. Not surprisingly, they all occur in plants.

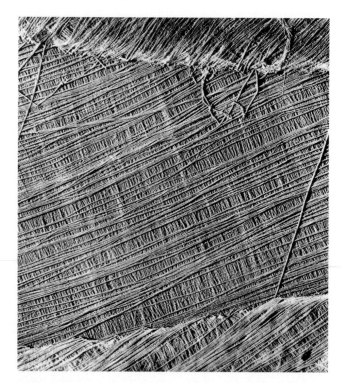

Figure 10-13
Electron micrograph of the cellulose fibers in the cell wall of the alga *Chaetomorpha*. Note that the cell wall consists of layers of parallel fibers. [Biophoto Associates.]

C. Structural Polysaccharides: Cellulose and Chitin

Plants have rigid cell walls (Fig. 1-9) that, in order to maintain their shapes, must be able to withstand osmotic pressure differences between the extracellular and intracellular spaces of up to 20 atm. In large plants, such as trees, the cell walls also have a load-bearing function. Cellulose, the primary structural component of plant cell walls (Fig. 10-13), accounts for over one half of the carbon in the biosphere: $\sim 10^{15}$ kg of cellulose are estimated to be synthesized and degraded annually. Although cellulose is predominantly of vegetable origin, it also occurs in the stiff outer mantles of marine invertebrates known as **tunicates** (urochordates; Fig. 1-11).

The primary structure of cellulose has been determined through methylation analysis. Cellulose is a linear polymer of up to 15,000 D-glucose residues (a glucan) linked by β(1 → 4) glycosidic bonds (Fig. 10-14). As is generally true of large polysaccharides, it has no defined size since, in contrast to proteins and nucleic acids, there is no genetically determined template that directs its synthesis.

X-ray studies of cellulose fibers have led Anatole Sarko to tentatively propose the structure diagrammed in Fig. 10-15. This highly cohesive, hydrogen bonded structure gives cellulose fibers exceptional strength and

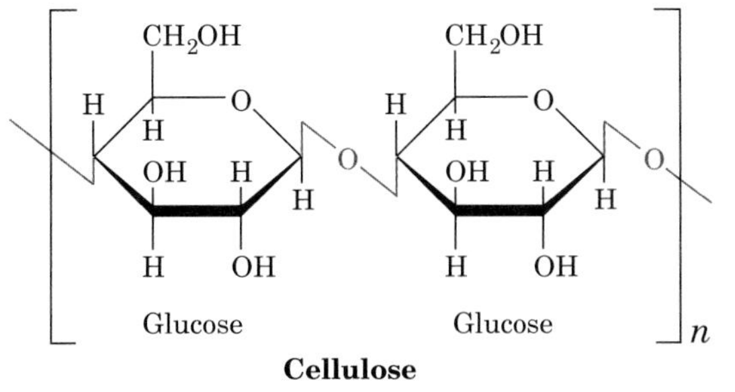

CH₂OH ... (structural formula)

Glucose   Glucose

Cellulose

Figure 10-14
The primary structure of cellulose. Here *n* may be several thousand.

makes them water insoluble despite their hydrophilicity.

In plant cell walls, the cellulose fibers are embedded in and cross-linked by a matrix of several polysaccharides that are composed of glucose as well as other monosaccharides. In wood, this cementing matrix also contains a large proportion of **lignin,** a plastic-like phenolic polymer. One has only to watch a tall tree in a high wind to realize the enormous strength of plant cell walls. In engineering terms, they are "composite materials" as is concrete reinforced by steel rods. Composite materials can withstand large stresses because the matrix evenly distributes the stresses among the reinforcing elements.

Although vertebrates themselves do not possess an enzyme capable of hydrolyzing the $\beta(1 \rightarrow 4)$ linkages of cellulose, the digestive tracts of herbivores contain symbiotic microorganisms that secrete a series of enzymes, collectively known as **cellulase,** that do so. The same is true of termites. Nevertheless, the degradation of cellulose is a slow process because its tightly packed and hydrogen bonded glucan chains are not easily accessible to cellulase and do not separate readily even after many of their glycosidic bonds have been hydrolyzed. The digestion of fibrous plants such as grass by herbivores is therefore a more complex and time-consuming process than is the digestion of meat by carnivores (cows, e.g., have multichambered stomachs and must chew their cud). Similarly, the decay of dead plants by fungi and other organisms, and the consumption of wooden houses by termites, often takes years.

Chitin is the principle structural component of the exoskeletons of invertebrates such as crustaceans, insects, and spiders and is also present in the cell walls of most fungi and many algae. It is therefore almost as abundant as is cellulose. Chitin is a homopolymer of $\beta(1 \rightarrow 4)$-linked *N*-acetyl-D-glucosamine residues (Fig.

Figure 10-15
A proposed structural model of cellulose. Cellulose fibers consist of ~40 parallel glucan chains arranged in an extended fashion. Each of the $\beta(1 \rightarrow 4)$-linked glucose units in a chain is turned over with respect to its preceding residue and is held in this position by intrachain hydrogen bonds (*dashed lines*). The glucan chains line up laterally to form sheets and these sheets stack vertically such that they are staggered by one half the length of a glucose unit. The entire assembly is stabilized by intermolecular hydrogen bonds between glucose units of neighboring chains. Hydrogen atoms not participating in hydrogen bonds have been omitted for clarity.

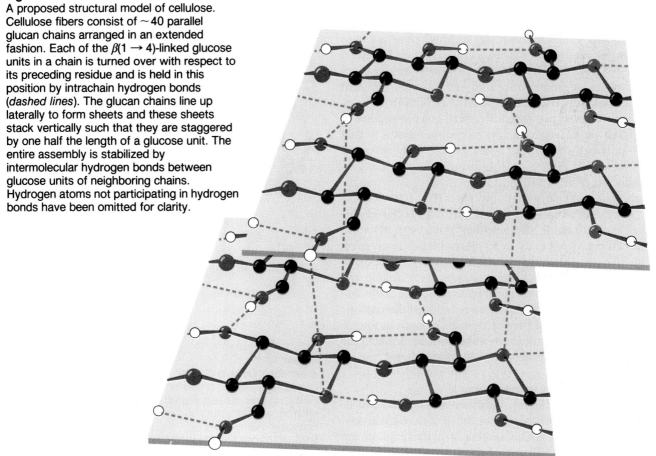

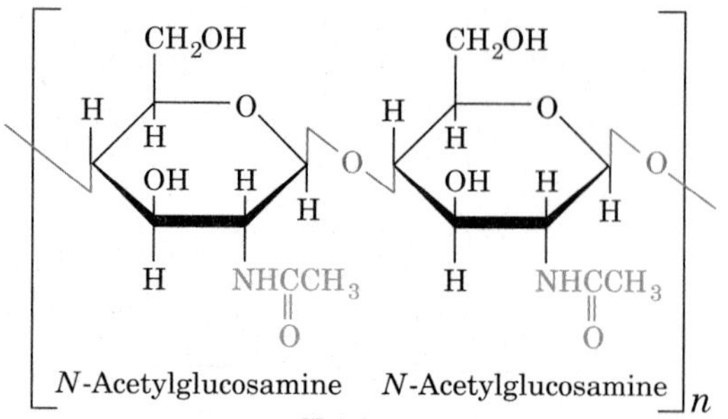

CH₂OH ... CH₂OH

N-Acetylglucosamine ... N-Acetylglucosamine]ₙ

Chitin

Figure 10-16
Chitin is a β(1 → 4)-linked homopolymer of *N*-acetyl-D-glucosamine.

bonds and those next to branches. By the time thoroughly chewed food reaches the stomach, where the acidity inactivates α-amylase, the average chain length of starch has been reduced from several thousand to less than eight glucose units. Starch digestion continues in the small intestine under the influence of pancreatic α-amylase, which is similar to the salivary enzyme. This enzyme degrades starch to a mixture of the disaccharide maltose, the trisaccharide **maltotriose,** which contains three α(1 → 4)-linked glucose residues, and oligosaccharides known as **dextrins** that contain the α(1 → 6) branches. These oligosaccharides are hydrolyzed to their component monosaccharides by specific enzymes contained in the brush border membranes of the intes-

10-16). It differs chemically from cellulose only in that each C(2)-OH group is replaced by an acetamido function. X-ray analysis indicates that chitin and cellulose have similar structures.

D. Storage Polysaccharides: Starch and Glycogen

Starch Is a Food Reserve in Plants and a Major Nutrient for Animals

Starch is a mixture of glucans that plants synthesize as their principal food reserve. It is deposited in the cytoplasm of plant cells as insoluble granules composed of **α-amylose** and **amylopectin.** α-Amylose is a linear polymer of several thousand glucose residues linked by α(1 → 4) bonds (Fig. 10-17*a*). Note that although α-amylose is an isomer of cellulose, it has very different structural properties. This is because cellulose's β-glycosidic linkages cause each successive glucose residue to flip 180° with respect to the preceding residue so that the polymer assumes an easily packed fully extended conformation (Fig. 10-15). In contrast, α-amylose's α-glycosidic bonds cause it to adopt an irregularly aggregating helically coiled conformation (Fig. 10-17*b*).

Amylopectin consists mainly of α(1 → 4)-linked glucose residues but is a branched molecule with α(1 → 6) branch points every 24 to 30 glucose residues on average (Fig. 10-18). Amylopectin molecules contain up to 10⁶ glucose residues, which makes them among the largest molecules occurring in nature. Storage of glucose as starch greatly reduces the large intracellular osmotic pressures that would result from its storage in monomeric form since osmotic pressure is proportional to the number of solute molecules in a given volume.

Starch Digestion Occurs in Stages

The digestion of starch, the main carbohydrate source in the human diet, begins in the mouth. Saliva contains **α-amylase,** which randomly hydrolyzes all the α(1 → 4) glucosidic bonds of starch except its outermost

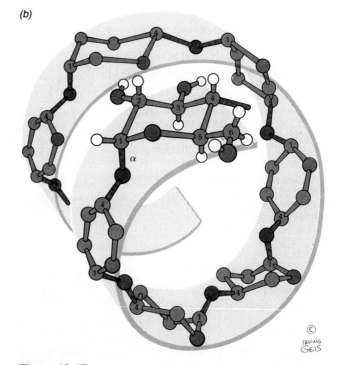

(a)

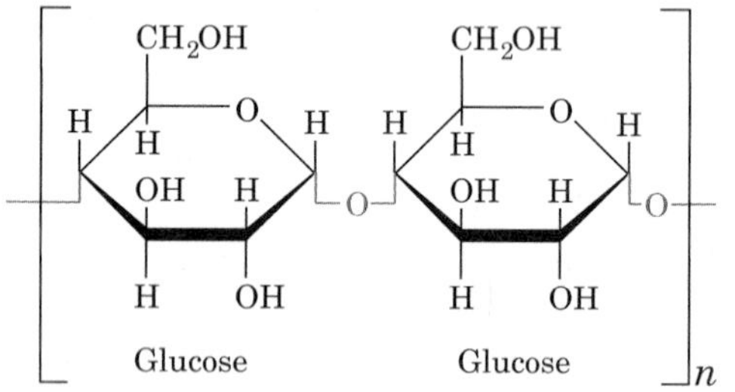

CH₂OH ... CH₂OH

Glucose ... Glucose]ₙ

α-Amylose

(b)

Figure 10-17
α-Amylose. (*a*) Its D-glucose residues are linked by α-(1 → 4) bonds (*red*). Here *n* is several thousand. (*b*) This regularly repeating polymer forms a left-handed helix. Note the great differences in structure and properties that result from changing α-amylose's α(1 → 4) linkages to the β(1 → 4) linkages of cellulose (Fig. 10-15). [Figure copyrighted © by Irving Geis.]

(a)

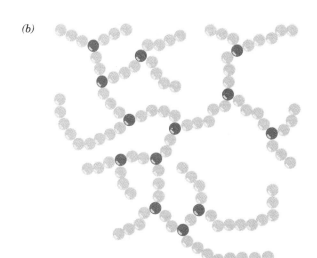

Amylopectin

(b)

Figure 10-18
Amylopectin. (a) its primary structure near one of its
α(1 → 6) branch points (red). (b) Its bushlike structure with
glucose residues at branch points indicated in red. The
actual distance between branch points averages 24 to 30
glucose residues. Glycogen has a similar structure but is
branched every 8 to 12 residues.

tinal mucosa: an **α-glucosidase,** which removes one
glucose residue at a time from oligosaccharides, an **α-
dextrinase** or **debranching enzyme,** which hydrolyzes
α(1 → 6) and α(1 → 4) bonds, a **sucrase** and, at least in
infants, a lactase. The resulting monosaccharides are
absorbed by the intestine and transported to the bloodstream.

Glycogen Is "Animal Starch"

Glycogen, the storage polysaccharide of animals, is
present in all cells but is most prevalent in skeletal muscle and liver where it occurs as cytoplasmic granules
(Fig. 10-19). The primary structure of glycogen resembles that of amylopectin but glycogen is more highly
branched with branch points occurring every 8 to 12
glucose residues. Glycogen's degree of polymerization
is nevertheless similar to that of amylopectin. In the cell,
glycogen is degraded for metabolic use by **glycogen
phosphorylase,** which phosphorolytically cleaves glycogen's α(1 → 4) bonds sequentially inwards from its
nonreducing ends to yield **glucose-1-phosphate.** Gly-

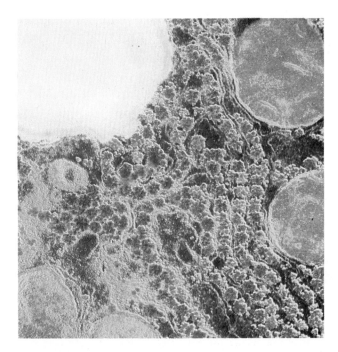

Figure 10-19
A photomicrograph showing the glycogen granules (*pink*)
in the cytoplasm of a liver cell (the greenish objects are
mitochondria and the yellow object is a fat globule). Note
that these granules tend to aggregate. The glycogen
content of liver may reach as high as 10% of its net weight.
[CNRI.]

cogen's highly branched structure, which has many nonreducing ends, permits the rapid mobilization of glucose in times of metabolic need. The $\alpha(1 \rightarrow 6)$ branches of glycogen are cleaved by a debranching enzyme. These enzymes play an important role in glucose metabolism and are discussed further in Section 17-1.

E. Glycosaminoglycans

The extracellular spaces, particularly those of connective tissues such as cartilage, tendon, skin, and blood vessel walls, consist of collagen and elastin fibers (Sections 7-2C and D) embedded in a gel-like matrix known as **ground substance**. Ground substance is composed largely of a **glycosaminoglycans** (alternatively, **mucopolysaccharides**), unbranched polysaccharides of alternating uronic acid and hexosamine residues. Solutions of glycosaminoglycans have a slimy, mucuslike consistency that results from their high viscosity and elasticity. In the following paragraphs, we discuss the structural origin of these important mechanical properties.

Hyaluronic Acid

Hyaluronic acid is an important glycosaminoglycan component of ground substance, synovial fluid (the fluid that lubricates the joints), and the vitreous humor of the eye. It also occurs in the capsules surrounding certain, usually pathogenic, bacteria. Hyaluronic acid molecules are composed of 250 to 25,000 $\beta(1 \rightarrow 4)$-linked disaccharide units that consist of D-glucuronic acid and N-acetyl-D-glucosamine linked by a $\beta(1 \rightarrow 3)$ bond (Fig. 10-20). The anionic character of its glucuronic acid residues causes hyaluronic acid to bind cations such as K^+, Na^+, and Ca^{2+} tightly. X-ray fiber analysis indicates that Ca^{2+} hyaluronate forms an extended left-handed single-stranded helix with three disaccharide units per turn (Fig. 10-21).

Hyaluronate's structural features suit it to its biological function. Its high molecular mass and numerous mutually repelling anionic groups make hyaluronate a rigid and highly hydrated molecule which, in solution, occupies a volume ~ 1000 times that in its dry state. Hyaluronate solutions therefore have a viscosity that is shear

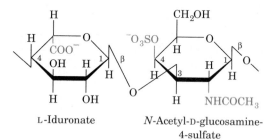

Figure 10-20
The disaccharide repeating units of the common glycosaminoglycans. The anionic groups are shown in red and the *N*-acetylamino groups are shown in blue.

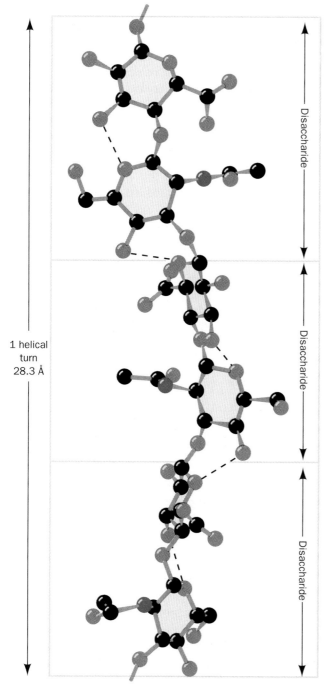

Figure 10-21
The X-ray fiber structure of Ca²⁺ hyaluronate. The hyaluronate polyanion forms an extended left-handed single-stranded helix with three disaccharide units per turn that is stabilized by intramolecular hydrogen bonds (*dashed lines*). H and Ca²⁺ atoms are omitted for clarity. [After Winter, W. T. and Arnott, S., *J. Mol. Biol.* **117,** 777 (1977).]

1 helical turn 28.3 Å

Disaccharide

dependent (an object under shear stress has equal and opposite forces applied across its opposite faces). At low shear rates, hyaluronate molecules form tangled masses that greatly impede flow; that is, the solution is quite viscous. As the shear rate increases, the stiff hyaluronate

molecules tend to line up with the flow and thus offer less resistance to it. This viscoelastic behavior makes hyaluronate solutions excellent biological shock absorbers and lubricants.

Hyaluronic acid and other glycosaminoglycans (see below) are degraded by **hyaluronidase,** which hydrolyzes their β(1 → 4) linkages. Hyaluronidase occurs in a variety of animal tissues, in bacteria (where it presumably expedites their invasion of animal tissue), and in snake and insect toxins.

Other Glycosaminoglycans

Other glycosaminoglycan components of ground substance consist of 50 to 1000 sulfated disaccharide units, which occur in proportions that are both tissue and species dependent. The most prevalent structures of these generally heterogeneous substances are indicated below (Fig. 10-20):

1. **Chondroitin-4-sulfate** (Greek: *chondros,* cartilage), a major component of cartilage and other connective tissue, has *N*-acetyl-D-galactosamine-4-sulfate residues in place of hyaluronate's *N*-acetyl-D-glucosamine residues.

2. **Chondroitin-6-sulfate** is instead sulfated at the C(6) position of its *N*-acetyl-D-galactosamine residues. The two chondroitin sulfates occur separately or in mixtures depending on the tissue.

3. **Dermatan sulfate,** which is so-named because of its prevalence in skin, differs from chondroitin-4-sulfate only by an inversion of configuration about C(5) of the β-D-glucuronate residues to form α-L-iduronate. This results from the enzymatic epimerization of these residues after the formation of chondroitin. The epimerization is usually incomplete so that dermatan sulfate also contains glucuronate residues.

4. **Keratan sulfate** (not to be confused with the protein keratin) contains alternating β(1 → 4)-linked D-galactose and *N*-acetyl-D-glucosamine-6-sulfate residues. It is the most heterogeneous of the major glycosaminoglycans in that its sulfate content is variable and it contains small amounts of fucose, mannose, *N*-acetylglucosamine, and sialic acid.

5. **Heparin** is a variably sulfated glycosaminoglycan that consists predominantly of alternating α(1 → 4)-linked residues of D-glucuronate-2-sulfate and *N*-sulfo-D-glucosamine-6-sulfate. Heparin, in contrast to the above glycosaminoglycans, is not a constituent of connective tissue, but rather, occurs almost exclusively in the intracellular granules of the mast cells that line arterial walls, especially in the liver, lungs, and skin. It inhibits the clotting of blood and its release, through injury, is thought to prevent run-away clot formation (Section 34-1E). Heparin is therefore

in wide clinical use to inhibit blood clotting, for example, in postsurgical patients. **Heparan sulfate,** a ubiquitous cell surface component as well as an extracellular substance in blood vessel walls and brain, resembles heparin but has a far more variable composition with fewer *N*- and *O*-sulfate groups and more *N*-acetyl groups.

3. GLYCOPROTEINS

Until about 1960, carbohydrates were thought to be rather dull compounds that were probably some sort of inert filler. Protein chemists therefore considered them to be a nuisance that complicated protein "purification." Research of the past 20 to 30 years, however, has demonstrated that most proteins are, in fact, **glycoproteins;** that is, they are covalently associated with carbohydrates. Glycoproteins vary in carbohydrate content from <1% to >90% by weight. They occur in all forms of life and have functions that span the entire spectrum of protein activities, including those of enzymes, transport proteins, receptors, hormones, and structural proteins. Their carbohydrate moieties, as we shall see, have several important biological roles, but in many cases their functions remain enigmatic.

The polypeptide chains of glycoproteins, like those of all proteins, are synthesized under genetic control. Their carbohydrate chains, in contrast, are enzymatically generated and covalently linked to the polypeptide without the rigid guidance of nucleic acid templates (Section 21-3B). The processing enzymes are generally not available in sufficient quantities to ensure the synthesis of uniform products. Glycoproteins therefore have variable carbohydrate compositions, a phenomenon known as **microheterogeneity,** that compounds the difficulties in their purification and characterization.

In this section we consider the structures and properties of glycoproteins. In particular, we shall study the glycoproteins of connective tissues, those of bacterial cell walls, and several soluble glycoproteins. We end by discussing the general principles of glycoprotein structure and function.

A. Proteoglycans

Proteins and glycosaminoglycans in ground substance aggregate, covalently and noncovalently, to form a diverse group of macromolecules known as **proteoglycans** (alternatively, **mucoproteins**). Electron micrographs (Fig. 10-22a) together with reconstitution experiments indicate that proteoglycans have a bottle-brushlike molecular architecture (Fig. 10-22b) whose **proteoglycan subunit** "bristles" are noncovalently attached to a filamentous hyaluronic acid "backbone" at intervals of 200 to 300 Å.

*Proteoglycan subunits consist of a **core protein** to which glycosaminoglycans, most often keratan sulfate and chondroitin sulfate, are covalently linked.* There are numerous types of core proteins; those from cartilage have molecular masses of 200 to 300 kD, which makes them among the largest polypeptide chains synthesized by any cell. The N-terminus of the otherwise highly extended core protein forms a globular region of 60 to 70 kD that noncovalently binds to the hyaluronic acid, an attachment that is stabilized by a **link protein** of 40 to 60 kD. The carbohydrates linked to the core protein divide it into three regions (Fig. 10-22b):

1. An N-terminal segment, which includes the globular hyaluronic acid-binding region, binds a relatively few carbohydrate chains. These tend to be oligosaccharides that are covalently linked to the protein via the amide N of specific Asn residues (Section 10-3C).

2. A region rich in oligosaccharides, many of which serve as anchor points for keratan sulfate chains. The oligosaccharides are covalently bonded to side chain O atoms of specific core protein Ser and Thr residues.

3. A C-terminal region rich in chondroitin sulfate, which is *O* linked to specific core protein Ser residues via galactose–galactose–xylose trisaccharides.

Altogether, a central strand of hyaluronic acid, which varies in length from 4000 to 40,000 Å, can have up to 100 associated core proteins, each of which binds ~50 keratan sulfate chains of up to 250 disaccharide units and ~100 chondroitin sulfate chains of up to 1000 disaccharide units. This accounts for the enormous molecular masses of many proteoglycans, which range up to tens of millions of daltons, and for their high degree of polydispersity (range of molecular masses).

Cartilage's Mechanical Properties Are Explained by Its Molecular Structure

Cartilage consists largely of a meshwork of collagen fibrils that is filled in by proteoglycans whose chondroitin sulfate and core protein components specifically interact with the collagen. The tensile strength of cartilage and other connective tissues is, as we have seen (Section 7-2C), a consequence of their collagen content. Cartilage's characteristic resilance, however, results from its high proteoglycan content. Proteoglycan's extended brushlike structure, together with the polyanionic character of its keratan sulfate and chondroitin sulfate components, cause this complex to be highly hydrated. The application of pressure on cartilage squeezes water away from these charged regions until charge–charge repulsions prevent further compression. When the pressure is released, the water returns. Indeed, the cartilage in the joints, which lack blood vessels, is nourished by this flow of liquid brought about by body movements. This explains why long periods of inactivity cause joint cartilage to become thin and fragile.

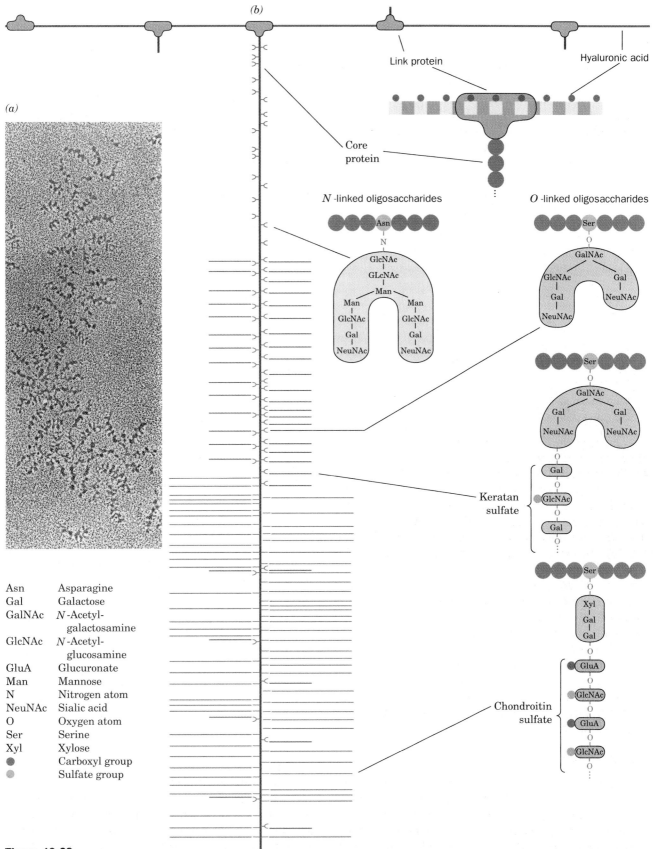

(b)

Link protein

Hyaluronic acid

Core protein

N-linked oligosaccharides

O-linked oligosaccharides

Asn

N

GlcNAc
|
GLcNAc
|
Man

Man   Man
| |
GlcNAc   GlcNAc
| |
Gal   Gal
| |
NeuNAc   NeuNAc

Ser

O

GalNAc

GlcNAc   Gal
| |
Gal   NeuNAc
|
NeuNAc

Ser

O

GalNAc

Gal   Gal
| |
NeuNAc   NeuNAc

Gal

GlcNAc

Gal

Keratan sulfate

Ser

O

Xyl
|
Gal
|
Gal

O

GluA

GlcNAc

GluA

GlcNAc

O

Chondroitin sulfate

Asn	Asparagine
Gal	Galactose
GalNAc	*N*-Acetyl-galactosamine
GlcNAc	*N*-Acetyl-glucosamine
GluA	Glucuronate
Man	Mannose
N	Nitrogen atom
NeuNAc	Sialic acid
O	Oxygen atom
Ser	Serine
Xyl	Xylose
●	Carboxyl group
●	Sulfate group

Figure 10-22

(a) An electron micrograph of a proteoglycan. The central strand of hyaluronic acid, which runs down the field of view, supports numerous projections each consisting of a core protein to which bushy polysaccharide protrusions are linked. [From Caplan, A. I., *Sci. Am.* **251**(4): 87 (1984). Copyright © 1984 Scientific American, Inc. Used by permission.] (b) The bottlebrush model of proteoglycan. The core proteins, one of which is shown extending down through the middle of the diagram, project from the central hyaluronic acid strand. The core is noncovalently anchored to the hyaluronic acid via its globular N-terminal end in an association that is stabilized by link protein. The core has three saccharide-binding regions: the inner region predominantly binds *N*-linked oligosaccharides; the central region binds *O*-linked oligosaccharides, many of which bear keratan sulfate chains; and the outer region binds chondroitin sulfate chains that are *O*-linked to the core protein via a galactose–galactose–xylose linker.

(a) Gram-positive bacteria

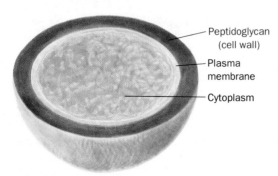

(b) Gram-negative bacteria

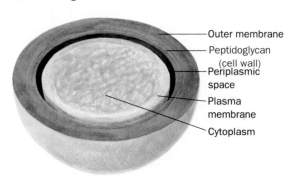

Figure 10-23
A schematic diagram comparing the cell envelopes of (a) gram-positive bacteria and (b) gram-negative bacteria.

B. Bacterial Cell Walls

Bacteria are surrounded by rigid cell walls that give them their characteristic shapes (Fig. 1-1) and permit them to live in hypotonic (less than intracellular salt concentrations) environments that would otherwise cause them to swell osmotically until their plasma (cell) membranes lysed (burst). Bacterial cell walls are of considerable medical significance because they are responsible for bacterial **virulence** (disease-evoking power). In fact, the symptoms of many bacterial diseases can be elicited in animals merely by the injection of bacterial cell walls. Furthermore, the characteristic **antigens** (immunological markers; Section 34-2) of bacteria are components of their cell walls so that injection of bacterial cell wall preparations into an animal often invokes its immunity against these bacteria.

Bacteria are classified as **gram positive** or **gram negative** depending on whether or not they take up gram stain (Section 1-1B). Gram-positive bacteria (Fig. 10-23a) have a thick (~ 250 Å) cell wall surrounding their plasma membrane, whereas gram-negative bacteria (Fig. 10-23b) have a thin (~ 30 Å) cell wall covered by a complex outer membrane.

Bacterial Cell Walls Have a Peptidoglycan Framework

The cell walls of both gram-positive and gram-negative bacteria consist of covalently linked polysaccharide and polypeptide chains, which form a baglike molecule that completely encases the cell. This framework, whose structure was elucidated in large part by Jack Strominger, is known as a **peptidoglycan** or **murein** (Latin: *murus*, wall). Its polysaccharide component consists of linear chains of alternating $\beta(1 \rightarrow 4)$-linked *N*-acetylglucosamine **(NAG)** and *N*-acetylmuramic acid **(NAM)**. The NAM's lactic acid residue forms an amide bond with a D-amino acid-containing tetrapeptide to form the peptidoglycan repeating unit (Fig. 10-24). Neighboring parallel peptidoglycan chains are covalently cross-

linked through their tetrapeptide side chains. In the gram-positive bacterium *Staphylococcus aureus*, whose tetrapeptide has the sequence L-Ala-D-isoglutamyl-L-Lys-D-Ala, this cross-link consists of a pentaglycine chain, that extends from the terminal carboxyl group of one tetrapeptide to the ε-amino group of the Lys in a neighboring tetrapeptide. The bacterial cell wall consists of several concentric layers of peptidoglycan that are probably cross-linked in the third dimension; gram-positive bacteria have up to 20 such layers.

The D-amino acids of peptidoglycans render them resistant to proteases. However, **lysozyme,** which is present in tears, mucus, and other body secretions, as well as in egg whites, catalyzes the hydrolysis of the $\beta(1 \rightarrow 4)$ glycosidic linkage between NAM and NAG. Consequently, treatment of gram-positive bacteria with lysozyme degrades their cell walls, which results in their lysis (gram-negative bacteria are resistant to lysozyme degradation). Lysozyme was discovered in 1922 by the British bacteriologist Alexander Fleming after he noticed that a bacterial culture had dissolved where mucus from a sneeze had landed. It was Fleming's hope that lysozyme would be a universal antibiotic but, unfortunately, it is clinically ineffective against pathogenic bacteria. The structure and mechanism of lysozyme are examined in detail in Section 14-2.

Penicillin Kills Bacteria by Inhibiting Cell Wall Biosynthesis

In 1928, Fleming noticed that the chance contamination of a bacterial culture plate with the mold *Penicillium notatum* lysed nearby bacteria (a clear demonstration of Pasteur's maxim that chance favors a prepared mind). This was caused by the presence of **penicillin** (Fig. 10-25), an antibiotic secreted by the mold. Yet, the difficulties of isolating and characterizing penicillin arising from its instability, led to the passage of over a decade before penicillin was ready for routine clinical use. Penicillin specifically binds to and inactivates enzymes that

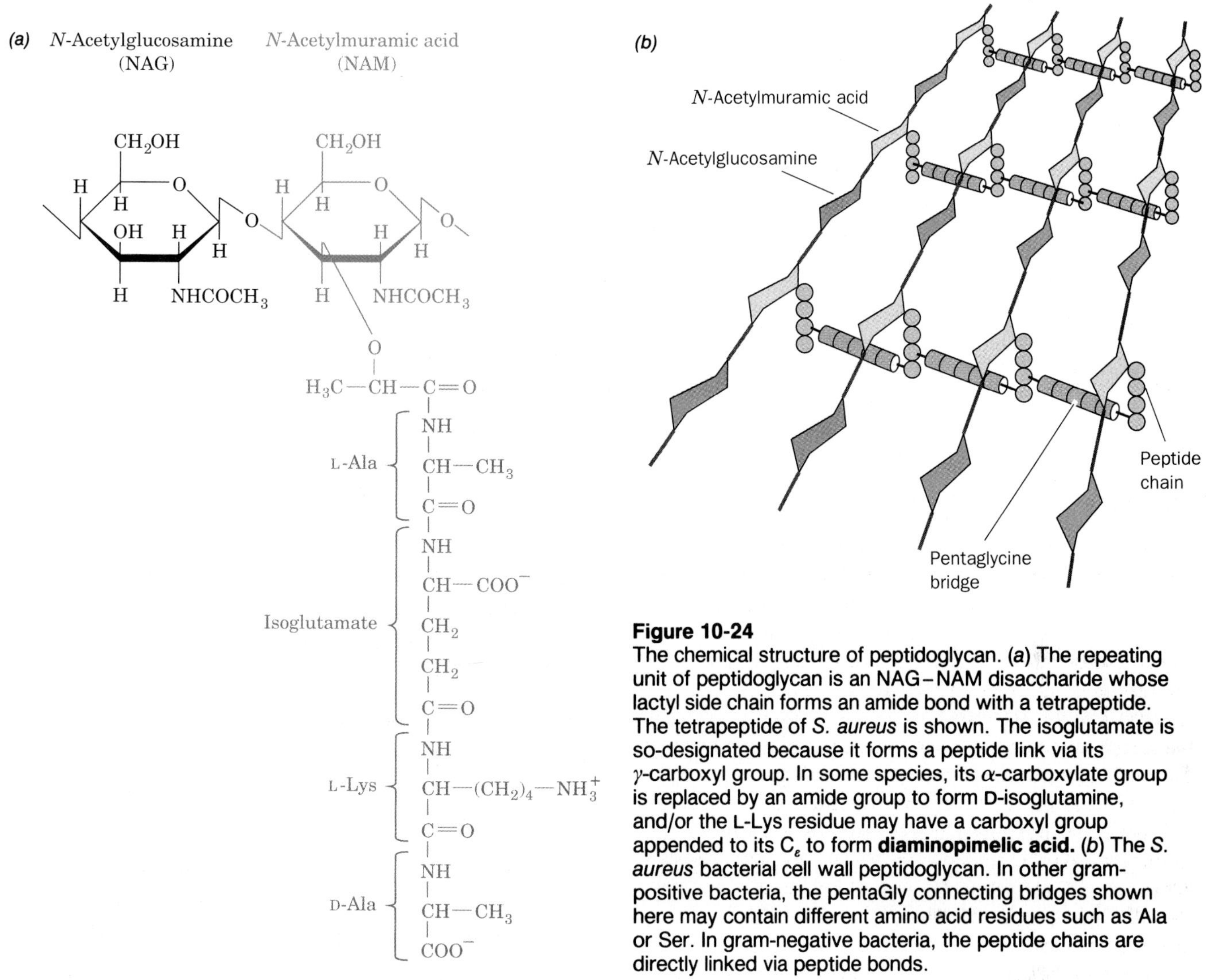

(a) N-Acetylglucosamine   N-Acetylmuramic acid
      (NAG)   (NAM)

(b)

Figure 10-24
The chemical structure of peptidoglycan. (*a*) The repeating
unit of peptidoglycan is an NAG–NAM disaccharide whose
lactyl side chain forms an amide bond with a tetrapeptide.
The tetrapeptide of *S. aureus* is shown. The isoglutamate is
so-designated because it forms a peptide link via its
γ-carboxyl group. In some species, its α-carboxylate group
is replaced by an amide group to form D-isoglutamine,
and/or the L-Lys residue may have a carboxyl group
appended to its C_ε to form **diaminopimelic acid.** (*b*) The *S.
aureus* bacterial cell wall peptidoglycan. In other gram-
positive bacteria, the pentaGly connecting bridges shown
here may contain different amino acid residues such as Ala
or Ser. In gram-negative bacteria, the peptide chains are
directly linked via peptide bonds.

function to cross-link the peptidoglycan strands of bacterial cell walls. Since cell wall expansion also requires
the action of enzymes that degrade cell walls, *exposure of
growing bacteria to penicillin results in their lysis;* that is,
penicillin disrupts the normal balance between cell wall
biosynthesis and degradation. However, since no
human enzyme binds penicillin, it is of low human toxicity, a therapeutic necessity.

> Penicillin treated bacteria remain intact, even though
> they have no cell wall, if they are kept in a hypertonic
> medium. Such bacteria, which are called **protoplasts**
> or **spheroplasts,** are spherical and extremely fragile
> because they are encased by only their plasma mem-
> branes. Protoplasts immediately lyse upon transfer
> to a normal medium.

Most bacteria that are resistant to penicillin secrete
penicillinase, which inactivates penicillin by cleaving

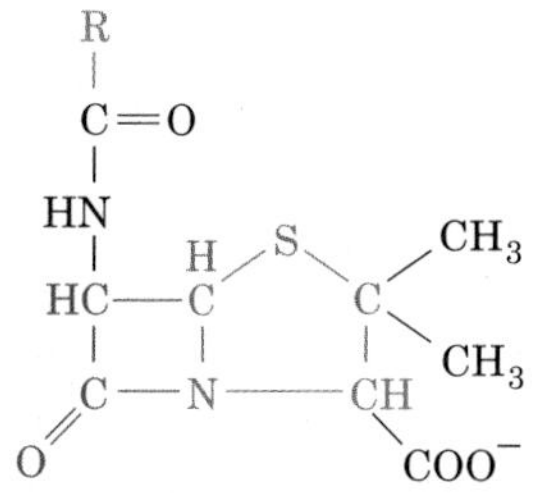

Penicillin

Figure 10-25
Penicillin contains a thiazolidine ring (*red*) fused to a
β-lactam ring (*blue*). A variable R-group is bonded to the
β-lactam ring via a peptide link. In benzyl penicillin (penicillin
G), one of several naturally occurring derivatives that are
clinically effective, R is the benzyl group ($-CH_2\phi$). In
ampicillin, a semisynthetic derivative, R is the aminobenzyl
group [$-CH(NH_2)\phi$].

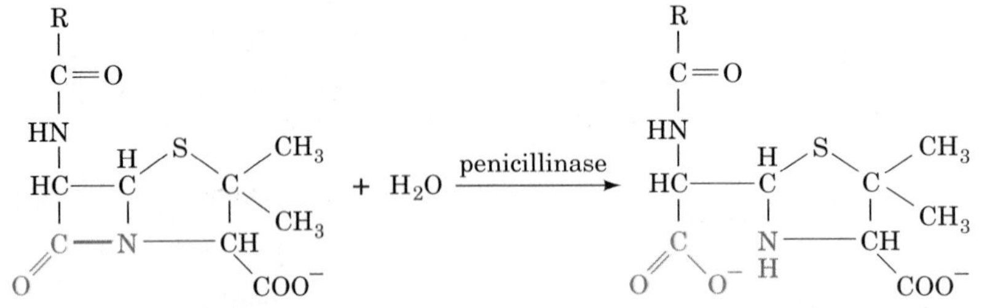

Figure 10-26
Penicillinase inactivates penicillin by catalyzing the hydrolysis of its β-lactam ring to form **penicillinoic acid.**

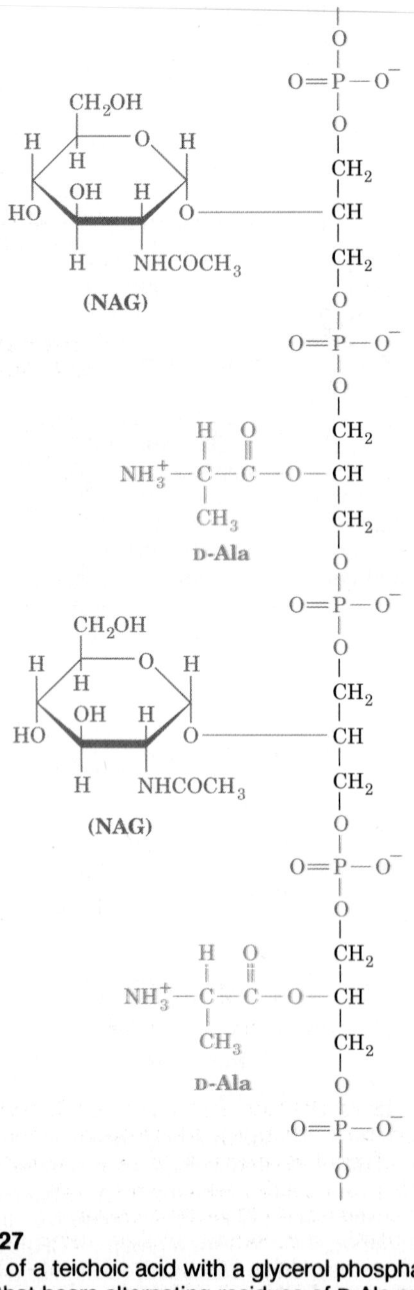

Figure 10-27
A segment of a teichoic acid with a glycerol phosphate backbone that bears alternating residues of D-Ala and NAG.

the amide bond of its β-lactam ring (Fig. 10-26). However, the observation that penicillinase activity varies with the nature of penicillin's R group has prompted the semisynthesis of penicillins, such as **ampicillin** (Fig. 10-25), which are clinically effective against penicillin-resistant strains of bacteria.

Bacterial Cell Walls Are Studded with Antigenic Groups

The surfaces of gram-positive bacteria are covered by **teichoic acids** (Greek: *teichos*, city walls), which account for up to 50% of the dry weight of their cell walls. Teichoic acids are polymers of glycerol or ribitol linked by phosphodiester bridges (Fig. 10-27). The hydroxyl groups of this sugar–phosphate chain are substituted by D-Ala residues and saccharides such as glucose or NAG. Teichoic acids are anchored to the peptidoglycans via phosphodiester bonds to the C(6)-OH groups of their NAG residues. They often terminate in **lipopolysaccharides** (lipids that contain polysaccharides; Section 11-1).

The outer membranes of gram-negative bacteria (Fig. 10-23b) are composed of complex lipopolysaccharides, proteins, and phospholipids that are organized in a complicated manner. The **periplasmic space,** an aqueous compartment that lies between the plasma membrane and the peptidoglycan cell wall, contains proteins that transport sugars and other nutrients. The outer membrane functions as a barrier to exclude harmful substances (such as gram stain). This accounts for the observation that gram-negative bacteria are less affected by lysozyme and penicillin, as well as by other antibiotics, than are gram-positive bacteria.

The outer surfaces of gram-negative bacteria are coated with complex and often unusual polysaccharides known as **O-antigens** that uniquely mark each bacterial strain (Fig. 10-28). The observation that mutant strains of pathogenic bacteria lacking O-antigens are nonpathogenic suggests that O-antigens participate in the recognition of host cells. O-antigens, as their name implies, are also the means by which a host's immunological defense system recognizes invading bacteria as foreign (Section 34-2A). As part of the ongoing biological warfare between pathogen and host, O-antigens are subject

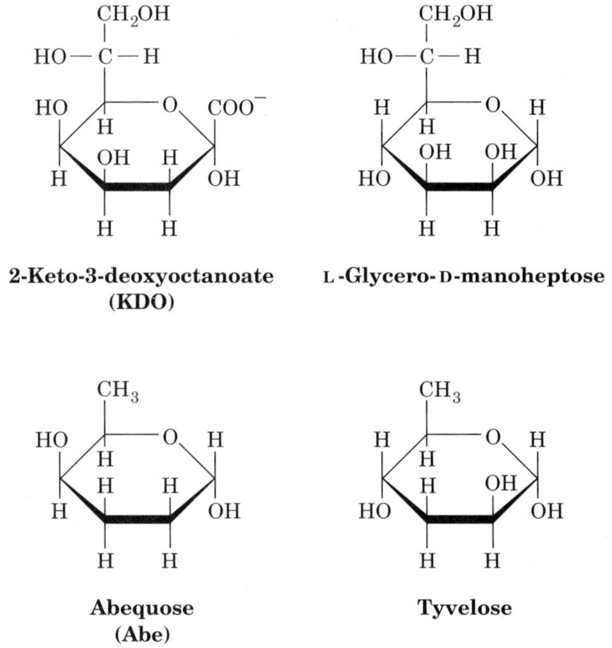

2-Keto-3-deoxyoctanoate
(KDO)

L-Glycero-D-manoheptose

Abequose
(Abe)

Tyvelose

Figure 10-28
Some of the unusual monosaccharides that occur in the O-antigens of gram-negative bacteria. These sugars rarely occur in other organisms.

to rapid mutational alteration so as to generate new bacterial strains that the host does not initially recognize (the mutations are in the genes specifying the enzymes that synthesize the O-antigens).

C. Glycoprotein Structure and Function

Glycoprotein Carbohydrate Chains Are Highly Diverse

Oligosaccharides form two types of attachments to proteins: *N* linked and *O* linked. Sequence analyses of glycoproteins have led to the following generalizations about these attachments.

1. *In N-glycosidic (N-linked) attachments, an NAG is invariably β linked to the amide nitrogen of an Asn in the sequence Asn-X-Ser or Asn-X-Thr, where X is any amino acid residue except possibly Pro or Asp (Fig. 10-29a). The oligosaccharides in these linkages usually have a distinctive **core** (innermost sequence; Fig. 10-29b) whose peripheral mannose residues are linked to either mannose or NAG residues. These latter residues may, in turn, be linked to yet other sugar residues so that an enormous diversity of N-linked oligosaccharides is possible. Some examples of such oligosaccharides are shown in Fig. 10-29c.*

2. *The most common O-glycosidic (O-linked) attachment involves the disaccharide core β-galactosyl-(1 → 3)-α-N-acetylgalactosamine α linked to the OH group of ei-*

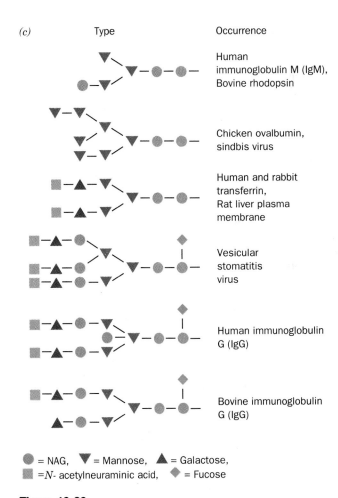

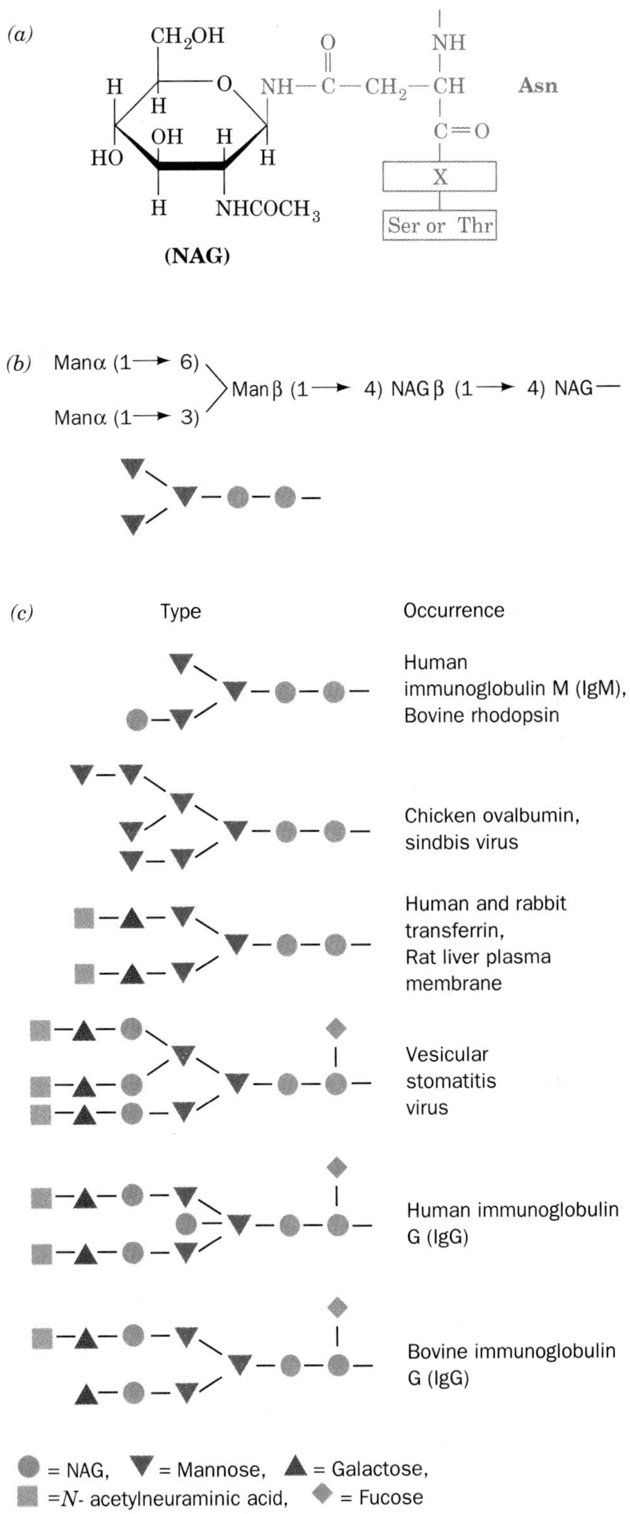

Figure 10-29
N-Linked oligosaccharides. (a) All *N*-glycosidic protein attachments occur through a *β-N*-acetylglucosamino-Asn bond in which the Asn occurs in the sequence Asn-X-Ser/Thr (*red*) where X is any amino acid. (b) *N*-Linked oligosaccharides usually have the branched (mannose)₃(NAG)₂ core shown. (c) Some examples of *N*-linked oligosaccharides. [After Sharon, N. and Lis, H., *Chem. Eng. News* **59**(13): 28 (1981).]

ther Ser or Thr (Fig. 10-30a). Less commonly, galactose, mannose, or xylose form α-O-glycosides with Ser or Thr (Fig. 10-30b). Galactose also forms O-glycosidic bonds to the 5-hydroxylysyl residues of collagen (Section 7-2C). However, there seem to be few, if any, additional generalizations that can be made about O-glycosidically linked oligosaccharides. They vary in size from a single galactose residue in collagen to the chains of up to 1000 disaccharide units in proteoglycans.

3. Oligosaccharides tend to attach to proteins at sequences that form β bends. Taken with their hydrophilic character, this observation suggests that *oligosaccharides extend from the surfaces of proteins rather than participating in their internal structures.* Indeed, the few glycoprotein X-ray structures that have yet been reported, for example, those of **immunoglobulin G** (Section 34-2B) and the influenza virus **hemagglutinin** (Section 32-4B), are consistent with this hypothesis. This accounts for the observation that the protein structures of glycoproteins are unaffected by the removal of their associated oligosaccharides.

Antifreeze Glycoproteins Prevent Antarctic Fish from Freezing

Antarctic fish live in −1.9°C waters, well below the temperature at which their blood is expected to freeze based on its salt and neutral solute content. These fish are prevented from freezing by **antifreeze glycoproteins,** substances that lower water's freezing point about twice as much as an equal weight of NaCl which,

on a molar basis, makes these proteins ~500-fold more effective than NaCl. However, in contrast to NaCl, which is excluded from the ice phase as it forms, antifreeze glycoprotein is included in the ice phase but yet has no effect on this ice's *melting* temperature.

Antifreeze glycoproteins are composed almost entirely of up to 50 repeats of the tripeptide Ala-Ala-Thr in which each Thr residue is glycosylated by the disaccharide β-galactosyl-(1 → 3)-α-N-acetylgalactosamine (Fig. 10-30a for R = CH$_3$). ^{13}C NMR measurements indicate that their large number of closely spaced and rigid disaccharide units cause antifreeze glycoproteins to assume the conformation of a flexible rod. Surprisingly, however, scission of as few as three of these glycoproteins' peptide bonds abolishes their antifreeze properties. Although the antifreeze glycoproteins' mechanism of action is unknown, the observation that ice crystals grown from concentrated solutions of antifreeze glycoproteins have an unusual fibrous form suggests that they may interfere with ice crystal formation and thus stabilize supercooled water.

Some cold water fish similarly inhibit ice formation with **antifreeze polypeptides,** nonglycoproteins that are either Ala-rich, Cys-rich, or neither rich in Ala nor Cys. The X-ray structure of a 37-residue Ala-rich species (22 Ala residues) reveals that it forms a single α helix, which is thought to preferentially bind to ice crystals so as to inhibit their growth. There are also proteins that have the opposite effect: **Ice-nucleation proteins,** which are produced by certain bacteria, presumably to help them invade plants by inducing frost damage. These proteins are used in making artificial snow to cover ski slopes. The repetitive primary structure of one such protein suggests that it forms a template for ice crystal formation in supercooled water.

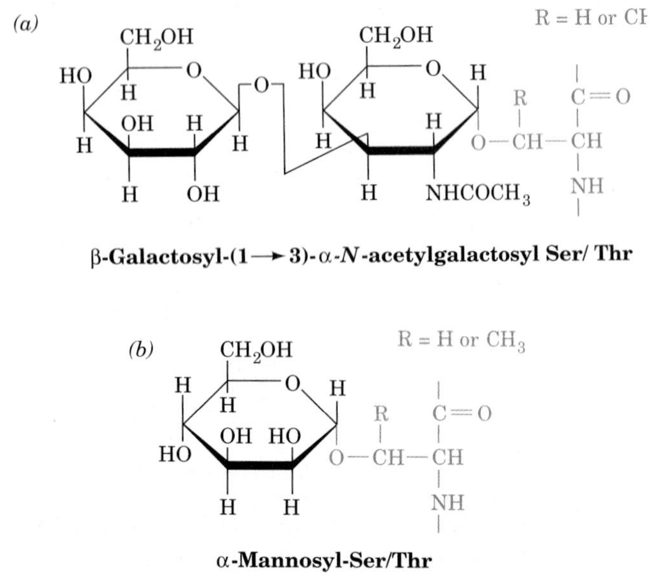

Mucus Contains Mucins

The secretions of mucous membranes have important protective and lubricative functions. Their active principles are glycoproteins, known as **mucins,** that contain numerous negatively charged oligosaccharide chains. For example, each molecule of sheep submaxillary (salivary) gland mucin contains 205 units of α-D-N-acetylneuraminic acid-(2 → 6)-α-N-acetylgalactosamine that are linked to Ser or Thr residues and spaced an average of six peptide units apart. The carbohydrate chains of many mucins are also sulfated. The multiple, mutually repelling and highly hydrated negative charges of mucins give them extended rodlike conformations, which are largely responsible for the characteristic viscoelastic properties of their aqueous solutions.

N-Linked Glycoproteins Exhibit Numerous Glycoforms

There is a growing body of evidence that cells synthesize a large repertoire of a given N-linked glycoprotein in which

β-Galactosyl-(1 → 3)-α-N-acetylgalactosyl Ser/Thr

α-Mannosyl-Ser/Thr

Figure 10-30
Some common O-glycosidic attachments of oligosaccharides to glycoproteins (*red*).

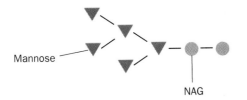

Figure 10-31
The microheterogeneous *N*-linked oligosaccharide of RNase B has the (mannose)₅(NAG)₂ core shown. A sixth mannose residue occurs at various positions on this core.

each variant species **(glycoform)** *has different covalently attached oligosaccharides.* For example, one of the simplest glycoproteins, bovine pancreatic **ribonuclease B (RNase B),** differs from the well-characterized and carbohydrate-free enzyme RNase A (Section 8-1A) only by the attachment of a single *N*-glycosidically linked oligosaccharide chain. The oligosaccharide has the core sequence diagrammed in Fig. 10-31 with considerable microheterogeneity in the position of a sixth mannose residue. The oligosaccharide does not affect the conformation, substrate specificity, or catalytic properties of RNase A. In fact, *carbohydrates generally seem to have little effect on the biochemical properties of proteins such as enzymes, transport proteins, and hormones. Rather, carbohydrates appear to act as recognition markers in various sorts of biological processes as described below. Thus, the species-specific and tissue-specific distribution of glycoforms that each cell synthesizes endows it with a characteristic spectrum of biological properties.*

Oligosaccharide Markers Mediate a Variety of Intercellular Interactions

Glycoproteins are important constituents of plasma membranes (Section 11-3). The location of their carbohydrate moieties can be determined by electron micros-

copy. The glycoproteins are labeled with lectins that have been conjugated (covalently cross-linked) to **ferritin,** an iron-transporting protein that is readily visible in the electron microscope because of its electron-dense iron hydroxide core. Such experiments, with lectins of different specificities and with a variety of cell types, have demonstrated that *the carbohydrate groups of membrane-bound glycoproteins are invariably located on the external surface of the cell membrane.*

A further indication that oligosaccharides function as biological markers is the observation that the carbohydrate content of a glycoprotein may govern its metabolic fate. For example, the excision of sialic acid residues from certain radioactively labeled blood plasma glycoproteins by treatment with **sialidase** greatly increases the rate with which these proteins are removed from the circulation. The proteins are taken up and degraded by the liver in a process that depends on the recognition by liver cell receptors of sugar residues such as galactose and mannose, which are exposed by the sialic acid excision. There is growing evidence that similar *"ticketing" mechanisms govern the compartmentation and degradation of glycoproteins within cells.*

The observation that cancerous cells are more susceptible to agglutination by lectins than are normal cells led to the discovery that *there are significant differences between the cell surface carbohydrate distributions of cancerous and noncancerous cells* (Fig. 10-32). Normal cells stop growing when they touch each other, a phenomenon known as **contact inhibition.** Cancer cells, however, are under no such control and therefore form

Figure 10-32
The surfaces of (a) a normal mouse cell, and (b) a cancerous cell as seen in the electron microscope. Both cells were incubated with the ferritin-labeled lectin concanavalin A. The lectin is evenly dispersed on the normal cell but is aggregated into clusters on the cancerous cell. [Courtesy of Garth Nicholson, University of Texas M. D. Anderson Cancer Center.]

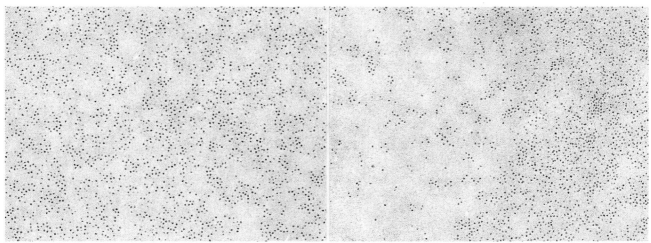

(a)   *(b)*

malignant tumors. Carbohydrates are important mediators of cell–cell recognition and have been implicated in related processes such as fertilization, cellular differentiation, the aggregation of cells to form organs, and the infection of cells by bacteria (Fig. 10-33) and viruses.

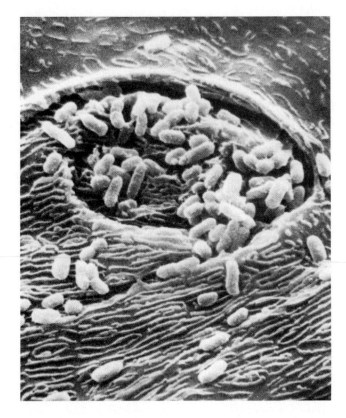

Figure 10-33
A scanning electron micrograph of tissue from the inside of a human cheek. The white cylindrical objects are *E. coli*. The bacteria adhere to mannose residues that are incorporated in the plasma membrane of cheek cells. This is the first step of a bacterial infection. [Courtesy of Fredric Silverblatt and Craig Kuehn, Veterans Administration Hospital, Sepulveda, California.]

Chapter Summary

Carbohydrates are polyhydroxy aldehydes or ketones of approximate composition $(C \cdot H_2O)_n$ that are important components of biological systems. The various monosaccharides, such as ribose, fructose, glucose, and mannose, differ in their number of carbon atoms, the positions of their carbonyl groups, and their diastereomeric configurations. These sugars exist largely as cyclic hemiacetals and hemiketals which, for 5- and 6-membered rings, are, respectively, known as furanoses and pyranoses. The two anomeric forms of these cyclic sugars may interconvert by mutarotation. Pyranose sugars have nonplanar rings with boat and chair conformations similar to those of substituted cyclohexanes. Polysaccharides are held together by glycosidic bonds between neighboring monosaccharide units. Glycosidic bonds do not undergo mutarotation. Monosaccharides can be oxidized to aldonic and glycuronic acids or reduced to alditols. An OH group is replaced by H in deoxy sugars and by an amino group in amino sugars.

Carbohydrates can be purified by electrophoretic and chromatographic procedures. Affinity chromatography using lectins has been particularly useful in this regard. The sequences and linkages of polysaccharides may be determined by methylation analysis and by the use of specific exoglycosidases. Similar information may be obtained through NMR spectroscopy. Cellulose, the structural polysaccharide of plant cell walls, is a linear polymer of $\beta(1 \rightarrow 4)$-linked D-glucose residues. It forms a fibrous hydrogen bonded structure of exceptional strength that in plant cells is embedded in an amorphous matrix. Starch, the food storage polysaccharide of plants, consists of a mixture of the linear $\alpha(1 \rightarrow 4)$-linked glucan α-amylose and the $\alpha(1 \rightarrow 6)$-branched and $\alpha(1 \rightarrow 4)$-linked glucan amylopectin. Glycogen, the animal storage polysaccharide, resembles amylopectin but is more highly branched. Digestion of starch and glycogen is initiated by α-amylase and is completed by specific membrane-bound intestinal enzymes.

Proteoglycans of ground substance are high molecular mass aggregates, many of which structurally resemble a bottlebrush. Their proteoglycan subunits consist of a core protein to which glycosaminoglycans, usually chondroitin sulfate and keratan sulfate, are covalently linked. The rigid framework of a bacterial cell wall consists of chains of alternating $\beta(1 \rightarrow 4)$-linked NAG and NAM that are cross-linked by short polypeptides to form a bag-shaped peptidoglycan molecule that encloses the bacterium. Lysozyme cleaves the glycosidic linkages between NAM and NAG of peptidoglycan. Penicillin specifically inactivates enzymes involved in the cross-linking of peptidoglycans. Both of these substances cause the lysis of susceptible bacteria. Gram-positive bacteria have teichoic acids that are linked covalently to their peptidoglycans. Gram-negative bacteria have outer membranes that bear complex and unusual polysaccharides known as O-antigens. These participate in the recognition of host cells and are important in the immunological recognition of bacteria by the host. Oligosaccharides attach to proteins in only a few ways. In *N*-glycosidic attachments, an NAG is invariably bound to the amide nitrogen of Asn of the sequence Asn-X-Ser(Thr). *O*-Glycosidic attachments are made to Ser or Thr in most proteins and to hydroxylysine in collagen. Oligosaccharides are located on the surfaces of glycoproteins. Glycoproteins have functions that span the entire range of protein activities although the roles of their carbohydrate moieties are not well understood. Antifreeze glycoproteins lower the freezing point of water far more than an equal weight of NaCl, apparently by interfering with ice crystal formation. The viscoelastic properties of mucus largely result from the numerous negatively charged oligosaccharide groups of its component mucins. Ribonuclease B differs from the functionally indistinguishable and carbohydrate-free ribonuclease A only by the attachment of a single oligosaccharide. The carbohydrate moieties of glycoproteins in plasma membranes are invariably located on the external surfaces of the membranes. A glycoprotein's carbohydrate moieties may direct its metabolic fate by governing its uptake by certain cells or cell compartments. Glycoproteins are also important mediators of cell–cell recognition.

References

General

Aspinall, G. O. (Ed.), *The Polysaccharides*, Vols. 1–3, Academic Press (1982, 1983, and 1985).

Cantor, C. R. and Schimmel, P. R., *Biophysical Chemistry*, Chapter 4, Freeman (1980).

El Khadem, H. S., *Carbohydrate Chemistry. Monosaccharides and Their Oligomers*, Academic Press (1988).

Manners, D. J. (Ed.), *Biochemistry of Carbohydrates II, International Review of Biochemistry*, Vol. 16, University Park Press (1978). [Contains several useful reviews.]

Pigman, W. and Horton, D. (Eds.), *The Carbohydrates*, Vol. IA, IB, IIA, and IIB, Academic Press (1970 and 1980). [A series of articles detailing many aspects of carbohydrate chemistry.]

Solomons, T. W. G., *Organic Chemistry* (4th ed.), Chapter 21, Wiley (1988). [A general discussion of carbohydrate nomenclature and chemistry. Other comprehensive organic chemistry texts have similar material.]

Whelan, W. J. (Ed.), *Biochemistry of Carbohydrates, MTP International Review of Science, Biochemistry Series One*, Vol. 5, Butterworths (1975). [Contains a number of useful reviews.]

Oligosaccharide and Polysaccharides

Albersheim, P., The walls of growing plants, *Sci. Am.* **232**(4): 80–95 (1975).

Gray, G. M., Carbohydrate digestion and absorption, *New Engl. J. Med.* **292**, 1225–1230 (1975).

Homans, S. W., Dwek, R. A., and Rademacher, T. W., Solution conformations of *N*-linked oligosaccharides, *Biochemistry*, **26**, 6573–6578 (1987). [Reviews NMR techniques.]

Kretchmer, M., Lactose and lactase, *Sci. Am.* **227**(4): 74–78 (1972).

Lindahl, U. and Höök, M., Glycosaminoglycans and their binding to biological molecules, *Annu. Rev. Biochem.* **47**, 385–417 (1978).

Glycoproteins

Caplan, A. I., Cartilage, *Sci. Am.* **251**(4): 84–94 (1984).

Chain, E., Fleming's contribution to the discovery of penicillin, *Trends Biochem. Sci.* **4**, 143–146 (1979). [An historical account by one of the biochemists who characterized penicillin.]

Comper, W. D. and Laurent, T. C., Physiological function of connective tissue polysaccharides, *Physiol. Rev.* **58**, 255–315 (1978).

DiRenzo, J. M., Nakamura, K., and Inouye, M., The outer membrane proteins of gram-negative bacteria: biosynthesis, assembly and functions, *Annu. Rev. Biochem.* **47**, 481–532 (1978).

Feeney, R. E. and Burcham, T. S., Antifreeze proteins from polar fish blood, *Annu. Rev. Biophys. Biophys. Chem.* **15**, 59–78 (1986).

Fransson, L.-Å., Structure and function of cell-associated proteoglycans, *Trends Biochem. Sci.* **12**, 406–411 (1987).

Gallagher, J. T. and Corfield, A. P., Mucin-type glycoproteins —new perspectives in their structure and synthesis, *Trends Biochem. Sci.* **3**, 38–41 (1978).

Ginsburg, V. and Robbins, P. (Eds.), *Biology of Carbohydrates*, Vols. 1 and 2, Wiley (1981 and 1984). [Contains authoritative articles on glycoproteins.]

Hassell, J. R., Kimura, J. H., and Hascall, V. C., Proteoglycan core protein families, *Annu. Rev. Biochem.* **55**, 539–567 (1986).

Höök, M., Kjellén, L., and Johansson, S., Cell-surface glycosaminoglycans, *Annu. Rev. Biochem.* **53**, 847–869 (1984).

Lennarz, W. P. (Ed.), *The Biochemistry of Glycoproteins and Proteoglycans*, Plenum Press (1980).

Montreuil, J., Primary structure of glycoprotein glycans, *Adv. Carbohydr. Chem. Biochem.* **37**, 157–233 (1980). [A comprehensive review.]

Schauer, R., Sialic acids and their role as biological masks, *Trends Biochem. Sci.* **10**, 357–360 (1985).

Rademacher, T. W., Parekh, R. B., and Dwek, R. A., Glycobiology, *Annu. Rev. Biochem.* **57**, 787–838 (1988). [The role of *N*-linked oligosaccharides in mediating protein-specific biological activity.]

Ruoslahti, E., Structure and biology of proteoglycans, *Annu. Rev. Cell Biol.* **4**, 229–255 (1988).

Sharon, N., *Complex Carbohydrates*, Addison–Wesley (1975).

Sharon, N. and Lis, H., Glycoproteins, *in* Neurath, H. and Hill, R. L. (Eds.), *The Proteins* (3rd ed.), Vol. 5, pp. 125–144, Academic Press (1982).

Sharon, N., Glycoproteins, *Trends Biochem. Sci.* **9**, 198–202 (1984).

Problems

1. The trisaccharide shown is named **raffinose**. What is its systematic name? Is it a reducing sugar?

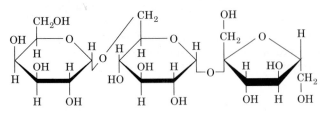

Raffinose

2. The systematic name of **melezitose** is O-α-D-glucopyranosyl-$(1 \rightarrow 3)$-O-β-D-fructofuranosyl-$(2 \rightarrow 1)$-α-D-glucopyranoside. Draw its molecular formula. Is it a reducing sugar?

3. Name the linear form of D-glucose using the (RS) chirality nomenclature system. [See Section 4-2C. *Hint*: The branch towards C(1) has higher priority than the branch towards C(6).]

*4. Draw the α-furanose form of D-talose and the β-pyranose form of L-sorbose.

5. The $NaBH_4$ reduction product of D-glucose may be named L-sorbitol or D-glucitol. Explain.

6. How many different disaccharides of D-glucopyranose are possible? How many trisaccharides?

7. A molecule of amylopectin consists of 1000 glucose residues and is branched every 25 residues. How many reducing ends does it have?

8. Most paper is made by removing the lignin from wood pulp and forming the resulting mass of largely unoriented cellulose fibers into a sheet. Untreated paper loses most of its strength when wet with water but maintains its strength when wet with oil. Explain.

*9. Write a chemical mechanism for the acid-catalyzed mutarotation of glucose.

10. The values of the specific rotation, $[\alpha]_D^{20}$, for the α and β anomers of D-galactose are 150.7° and 52.8°, respectively. A mixture that is 20% α-D-galactose and 80% β-D-galactose is dissolved in water at 20°C. What is its initial specific rotation? After several hours, the specific rotation of this mixture has reached an equilibrium value of 80.2°. What is its anomeric composition?

11. Name the epimers of D-gulose.

12. Exhaustive methylation of a trisaccharide followed by acid hydrolysis yields equimolar quantities of 2,3,4,6-tetra-O-methyl-D-galactose, 2,3,4-tri-O-methyl-D-mannose, and 2,4,6-tri-O-methyl-D-glucose. Treatment of the trisaccharide with β-galactosidase yields D-galactose and a disaccharide. Treatment of this disaccharide with α-mannosidase yields D-mannose and D-glucose. Draw the structure of the trisaccharide and state its systematic name.

13. The enzyme **β-amylase** cleaves successive maltose units from the nonreducing end of $\alpha(1 \rightarrow 4)$ glucans. It will not cleave at glucose residues that have an $\alpha(1 \rightarrow 6)$ bond. The end products of the exhaustive digestion of amylopectin by β-amylase are known as **limit dextrins**. Draw a schematic diagram of an amylopectin molecule and indicate what part(s) of it constitute limit dextrins.

14. One demonstration of P.T. Barnum's maxim that there's a sucker born every minute is that new "reducing aids" regularly appear on the market. A recent eat-all-you-want nostrum, which was touted as a "starch blocker" [and which the Food and Drug Administration (FDA) eventually banned], contains an α-amylase-inhibiting protein extracted from beans. If this substance really worked as advertised, which it does not, what unpleasant side effects would result from its ingestion with a starch-containing meal? Discuss why this substance, which inhibits α-amylase *in vitro*, will not do so in the intestines after oral ingestion?

*15. Treatment of a 6.0-g sample of glycogen with Tollens' reagent followed by exhaustive methylation and then hydrolysis yields 3.1 mmol of 2,3-di-O-methylglucose and 0.0031 mmol of 1,2,3-tri-O-methylgluconic acid as well as other products. (a) What fraction of glucose residues occur at $(1 \rightarrow 6)$ branch points and what is the average number of glucose residues per branch? (b) What are the other products of the methylation–hydrolysis treatment and in what amounts are they formed? (c) What is the average molecular mass of the glycogen?

16. Instilling methyl-α-D-mannoside into the bladder of a mouse prevents the colonization of its urinary tract by *E. coli*. What is the reason for this effect?

Chapter 11
LIPIDS AND MEMBRANES

Membranes function to organize biological processes by compartmentalizing them. Indeed, the cell, the basic unit of life, is essentially defined by its enveloping plasma membrane. Moreover, in eukaryotes, many subcellular organelles, such as nuclei, mitochondria, chloroplasts, the endoplasmic reticulum, and the Golgi apparatus (Fig. 1-5) are likewise membrane bounded.

Biological membranes are organized assemblies of lipids and proteins with small amounts of carbohydrate. Yet they are not impermeable barriers to the passage of materials. Rather, they regulate the composition of the intracellular medium by controlling the flow of nutrients, waste products, ions, etc., into and out of the cell. They do this through membrane-embedded "pumps" and "gates" that transport specific substances against an electrochemical gradient or permit their passage with such a gradient (Chapter 18).

Many fundamental biochemical processes occur on or in a membranous scaffolding. For example, electron transport and oxidative phosphorylation (Chapter 20), processes that oxidize nutrients with the concomitant generation of ATP, are mediated by an organized battery of enzymes that are components of the inner mitochondrial membrane. Likewise, photosynthesis, in which light energy powers the chemical combination of H_2O and CO_2 to form carbohydrates (Chapter 22), occurs in the inner membranes of chloroplasts. The processing of information, such as sensory stimuli or intercellular communications, is generally a membrane-based phenomenon. Thus nerve impulses are mediated by nerve cell membranes (Section 34-4C) and the presence of certain substances such as hormones and nutrients is detected by specific membrane-bound receptors (Section 34-4B).

In this chapter, we examine the compositions and structures of biological membranes and related substances. Specific membrane-based biochemical processes, such as those mentioned above, are dealt with in the later chapters.

1. LIPID CLASSIFICATION

Lipids (Greek: lipos, fat) are substances of biological origin that are soluble in organic solvents such as chloroform but are only sparingly soluble, if at all, in water. Fats, oils, certain vitamins and hormones, and most nonprotein membrane components are lipids. In this section, we discuss the structures and physical properties of the major classes of lipids.

A. Fatty Acids

Fatty acids are carboxylic acids with long-chain hydrocarbon side groups (Fig. 11-1). They are rarely free in nature but, rather, occur in esterified form as the major components of the various lipids described in this chapter. The more common biological fatty acids are listed in Table 11-1. In higher plants and animals, the predominant fatty acid residues are those of the C_{16} and C_{18} species **palmitic, oleic, linoleic,** and **stearic acids.** Fatty acids with <14 or >20 carbon atoms are uncommon. *Most fatty acids have an even number of carbon atoms because they are usually biosynthesized by the concatenation of C_2 units (Section 23-4C).* Over one half of the fatty acid residues of plant and animal lipids are unsaturated (contain double bonds) and are often polyunsaturated (contain two or more double bonds). Bacterial fatty acids are rarely polyunsaturated but are commonly branched, hydroxylated, or contain cyclopropane rings. Unusual fatty acids also occur as components of the oils and **waxes** (esters of fatty acids and long-chain alcohols) produced by certain plants.

Table 11-1

The Common Biological Fatty Acids

Symbol[a]	Common Name	Systematic Name	Structure	mp (°C)
Saturated fatty acids				
12:0	Lauric acid	Dodecanoic acid	$CH_3(CH_2)_{10}COOH$	44.2
14:0	Myristic acid	Tetradecanoic acid	$CH_3(CH_2)_{12}COOH$	52
16:0	Palmitic acid	Hexadecanoic acid	$CH_3(CH_2)_{14}COOH$	63.1
18:0	Stearic acid	Octadecanoic acid	$CH_3(CH_2)_{16}COOH$	69.6
20:0	Arachidic acid	Eicosanoic acid	$CH_3(CH_2)_{18}COOH$	75.4
22:0	Behenic acid	Docosanoic acid	$CH_3(CH_2)_{20}COOH$	81
24:0	Lignoceric acid	Tetracosanoic acid	$CH_3(CH_2)_{22}COOH$	84.2
Unsaturated fatty acids (all double bonds are cis)				
16:1	Palmitoleic acid	9-Hexadecenoic acid	$CH_3(CH_2)_5CH{=}CH(CH_2)_7COOH$	−0.5
18:1	Oleic acid	9-Octadecenoic acid	$CH_3(CH_2)_7CH{=}CH(CH_2)_7COOH$	13.4
18:2	Linoleic acid	9,12-Octadecadienoic acid	$CH_3(CH_2)_4(CH{=}CHCH_2)_2(CH_2)_6COOH$	−9
18:3	α-Linolenic acid	9,12,15-Octadecatrienoic acid	$CH_3CH_2(CH{=}CHCH_2)_3(CH_2)_6COOH$	−17
18:3	γ-Linolenic acid	6,9,12-Octadecatrienoic acid	$CH_3(CH_2)_4(CH{=}CHCH_2)_3(CH_2)_3COOH$	
20:4	Arachidonic acid	5,8,11,14-Eicosatetraenoic acid	$CH_3(CH_2)_4(CH{=}CHCH_2)_4(CH_2)_2COOH$	−49.5
20:5	EPA	5,8,11,14,17-Eicosapentaenoic acid	$CH_3CH_2(CH{=}CHCH_2)_5(CH_2)_2COOH$	−54
24:1	Nervonic acid	15-Tetracosenoic acid	$CH_3(CH_2)_7CH{=}CH(CH_2)_{13}COOH$	39

[a] Number of carbon atoms: Number of double bonds.

Source: Dawson, R. M. C., Elliott, D. C., Elliott, W. H., and Jones, K. M., *Data for Biochemical Research* (2nd ed.), Chapter 11, Clarendon Press (1969).

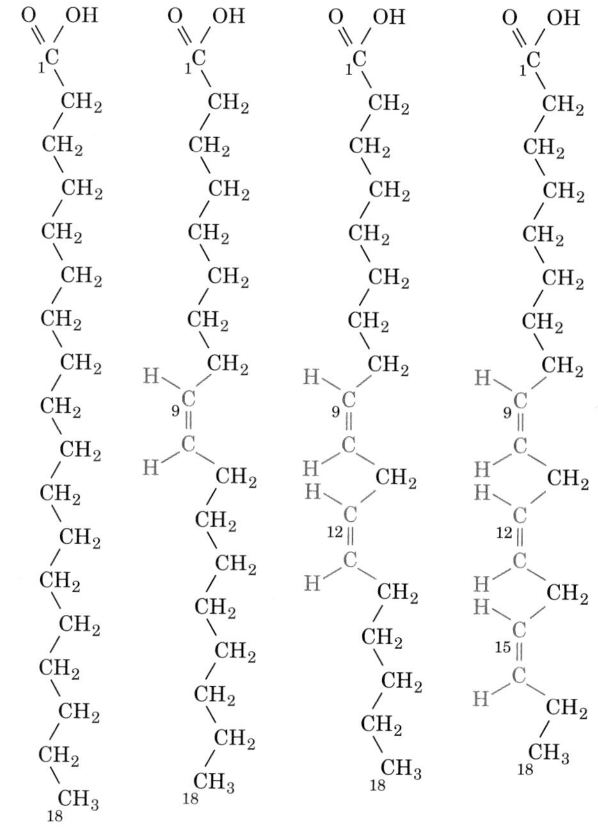

Figure 11-1
The structural formulas of some C_{18} fatty acids. The double bonds all have the cis configuration.

The Physical Properties of Fatty Acids Vary with Their Degree of Unsaturation

Table 11-1 indicates that the first double bond of an unsaturated fatty acid commonly occurs between its C(9) and C(10) atoms counting from the carboxyl C atom (a Δ^9- or 9-double bond). In polyunsaturated fatty acids, the double bonds tend to occur at every third carbon atom towards the methyl terminus of the molecule (such as $-CH=CH-CH_2-CH=CH-$). Double bonds in polyunsaturated fatty acids are almost never conjugated (as in $-CH=CH-CH=CH-$). Triple bonds rarely occur in fatty acids or any other compound of biological origin.

Saturated fatty acids are highly flexible molecules that can assume a wide range of conformations because there is relatively free rotation about each of their $C-C$ bonds. Nevertheless, their fully extended conformation is that of minimum energy because this conformation has the least amount of steric interference between neighboring methylene groups. The melting points of saturated fatty acids, like those of most substances, increase with molecular mass (Table 11-1).

Fatty acid double bonds almost always have the cis configuration (Fig. 11-1). This puts a rigid 30° bend in the hydrocarbon chain of unsaturated fatty acids that interferes with their efficient packing to fill space. The consequent reduced van der Waals interactions cause fatty acid melting points to decrease with their degree of unsaturation (Table 11-1). Lipid fluidity likewise increases with the degree of unsaturation of their component fatty acids residues. This phenomenon, as we shall see in Section 11-3B, has important consequences for membrane properties.

B. Triacylglycerols

The fats and oils that occur in plants and animals consist largely of mixtures of **triacylglycerols** (also referred to as **triglycerides** or **neutral fats**). *These nonpolar, water-insoluble substances are fatty acid triesters of glycerol:*

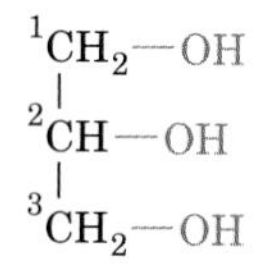

$$\begin{array}{l} ^1CH_2-OH \\ ^2CH-OH \\ ^3CH_2-OH \end{array}$$

Glycerol

$$\begin{array}{l} ^1CH_2-O-\overset{\displaystyle O}{\overset{\|}{C}}-R_1 \\ ^2CH-O-\overset{\displaystyle O}{\overset{\|}{C}}-R_2 \\ ^3CH_2-O-\overset{\displaystyle O}{\overset{\|}{C}}-R_3 \end{array}$$

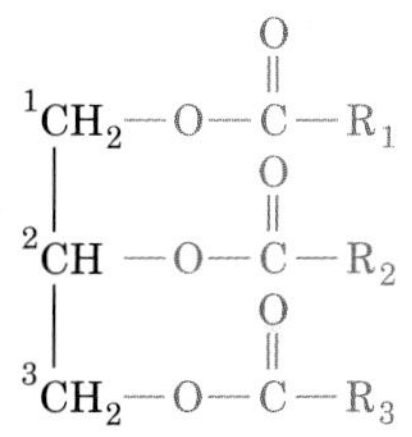

Triacylglycerol

Triacylglycerols function as energy reservoirs in animals and are therefore their most abundant class of lipids even though they are not components of biological membranes.

Triacylglycerols differ according to the identity and placement of their three fatty acid residues. The so-called **simple triacylglycerols** contain one type of fatty acid residue and are named accordingly. For example, **tristearoylglycerol** or **tristearin** contains three stearic acid residues, whereas **trioleoylglycerol** or **triolein** has three oleic acid residues. The more common **mixed triacylglycerols** contain two or three different types of

Stearic acid   Oleic acid   Linoleic acid   α-Linolenic acid fatty acid residues and are named according to their placement on the glycerol moiety.

$$
\begin{array}{ccc}
{}^{1}\text{CH}_2 - {}^{2}\text{CH} - {}^{3}\text{CH}_2 \\
| & | & | \\
\text{O} & \text{O} & \text{O} \\
| & | & | \\
\text{C}_1{=}\text{O} & \text{C}_1{=}\text{O} & \text{C}_1{=}\text{O} \\
| & | & | \\
\text{CH}_2 & \text{CH}_2 & \text{CH}_2 \\
| & | & | \\
\text{CH}_2 & \text{CH}_2 & \text{CH}_2 \\
| & | & | \\
\text{CH}_2 & \text{CH}_2 & \text{CH}_2 \\
| & | & | \\
\text{CH}_2 & \text{CH}_2 & \text{CH}_2 \\
| & | & | \\
\text{CH}_2 & \text{CH}_2 & \text{CH}_2 \\
| & | & | \\
\text{CH}_2 & \text{CH}_2 & \text{CH}_2 \\
| & | & | \\
\text{CH}_2 & \text{CH}_2 & \text{CH}_2 \\
| & | & | \\
\text{CH} & \text{CH} & \text{CH}_2 \\
\|^{9} & \|^{9} & | \\
\text{CH} & \text{CH} & \text{CH}_2 \\
| & | & | \\
\text{CH}_2 & \text{CH}_2 & \text{CH}_2 \\
| & | & | \\
\text{CH}_2 & \text{CH} & \text{CH}_2 \\
| & \|^{12} & | \\
\text{CH}_2 & \text{CH} & \text{CH}_2 \\
| & | & | \\
\text{CH}_2 & \text{CH}_2 & \text{CH}_2 \\
| & | & | \\
\text{CH}_3 & \text{CH}_2 & \text{CH}_2 \\
{}_{16} & | & | \\
& \text{CH}_2 & \text{CH}_2 \\
& | & | \\
& \text{CH}_3 & \text{CH}_3 \\
& {}_{18} & {}_{18}
\end{array}
$$

1-Palmitoleoyl-2-linoleoyl-3-stearoyl-glycerol

Fats and oils (which differ only in that fats are solid and oils are liquid at room temperature) are complex mixtures of simple and mixed triacylglycerols whose fatty acid compositions vary with the organism that produced them. Plant oils are usually richer in unsaturated fatty acid residues than are animal fats as the lower melting points of oils suggest.

Triacylglycerols Function as Energy Reserves

Fats are a highly efficient form in which to store metabolic energy. This is because fats are less oxidized than are carbohydrates or proteins and hence yield significantly more energy on oxidation. Furthermore, fats, being nonpolar substances, are stored in anhydrous form whereas glycogen, for example, binds about twice its weight of water under physiological conditions. Fats therefore provide about six times the metabolic energy of an equal weight of hydrated glycogen.

In animals, **adipocytes** (fat cells; Fig. 11-2) are specialized for the synthesis and storage of triacylglycerols.

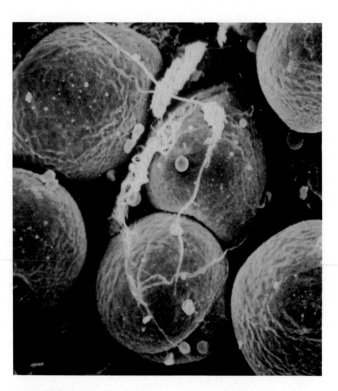

Figure 11-2
A scanning electron micrograph of adipocytes. Each contains a fat globule that occupies nearly the entire cell. [Fred E. Hossler/Visuals Unlimited.]

Whereas other types of cells have only a few small droplets of fat dispersed in their cytosol, adipocytes may be almost entirely filled with fat globules. **Adipose tissue** is most abundant in a subcutaneous layer and in the abdominal cavity. The fat content of normal humans (21% for men, 26% for women) enables them to survive starvation for 2 to 3 months. In contrast, the body's glycogen supply, which functions as a short-term energy store, can provide for the body's metabolic needs for less than a day. The subcutaneous fat layer also provides thermal insulation, which is particularly important for warm-blooded aquatic animals, such as whales, seals, geese, and penguins, which are routinely exposed to low temperatures.

C. Glycerophospholipids

Glycerophospholipids (or phosphoglycerides) are the major lipid components of biological membranes. They consist of *sn*-**glycerol-3-phosphate** (Fig. 11-3*a*) esterified at its C(1) and C(2) positions to fatty acids and at its phosphoryl group to a group, X, to form the class of substances diagrammed in Fig. 11-3*b*. *Glycerophospholipids are therefore amphiphilic molecules with nonpolar aliphatic "tails" and polar phosphoryl-X "heads."* The simplest glycerophospholipids, in which X = H, are **phosphatidic acids;** they are present only in small amounts in biological membranes. *In the glycerophos-*

Figure 11-3
(a) The compound shown in Fischer projection (Section 4-2B) can be equivalently referred to as L-glycerol-3-phosphate or D-glycerol-1-phosphate. However, using **stereospecific numbering** (*sn*), which assigns the 1-position to the group occupying the *pro-S* position of a prochiral center (see Section 4-2C for a discussion of prochirality), the compound is unambiguously named *sn*-glycerol-3-phosphate. (b) The general formula of the glycerophospholipids. R_1 and R_2 are long-chain hydrocarbon tails of fatty acids and X is derived from a polar alcohol (see Table 11-2).

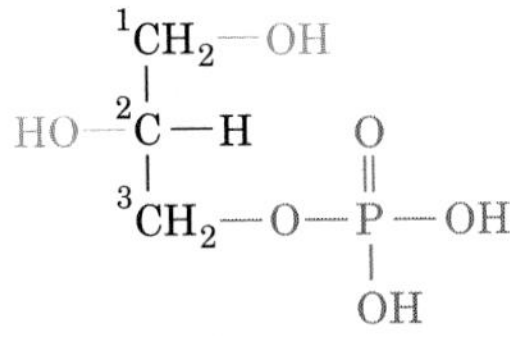

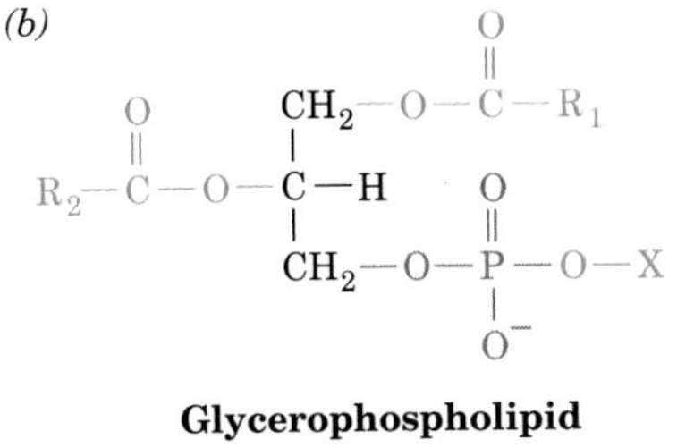

(a)

sn-**Glycerol-3-phosphate**

(b)

Glycerophospholipid

pholipids that commonly occur in biological membranes, the head groups are derived from polar alcohols (Table 11-2). Saturated C_{16} and C_{18} fatty acids usually occur at the C(1) position of glycerophospholipids and the C(2) position is often occupied by an unsaturated C_{16} to C_{20} fatty acid. Glycerophospholipids are, of course, also named according to the identities of these fatty acid

Table 11-2
The Common Classes of Glycerophospholipids

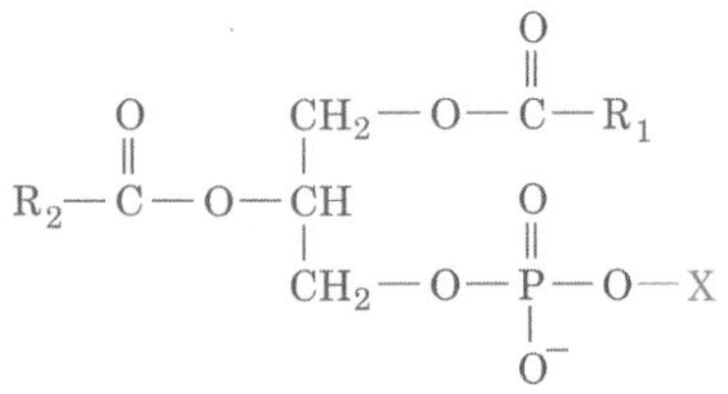

Name of X—OH	Formula of —X	Name of Phospholipid
Water	$-H$	Phosphatidic acid
Ethanolamine	$-CH_2CH_2NH_3^+$	Phosphatidylethanolamine
Choline	$-CH_2CH_2N(CH_3)_3^+$	Phosphatidylcholine (lecithin)
Serine	$-CH_2CH(NH_3^+)COO^-$	Phosphatidylserine
myo-Inositol	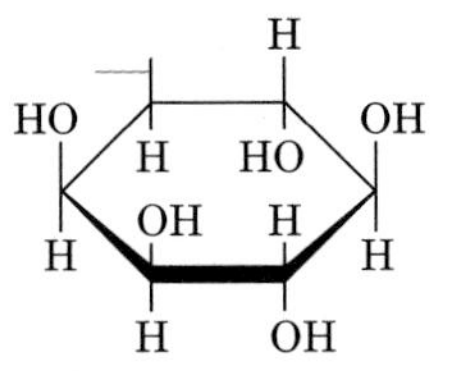	Phosphatidylinositol
Glycerol	$-CH_2CH(OH)CH_2OH$	Phosphatidylglycerol
Phosphatidylglycerol	$-CH_2CH(OH)CH_2-O-\overset{O}{\underset{O^-}{\overset{\|}{P}}}-O-CH_2$ ...	Diphosphatidylglycerol (cardiolipin)

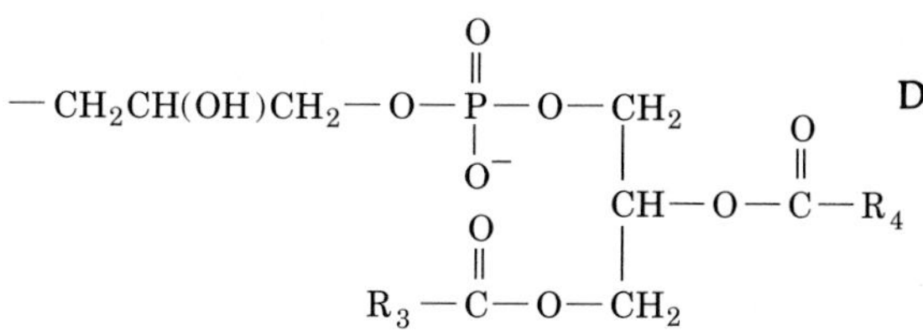

residues (Fig. 11-4). Some glycerophospholipids have common names. For example, phosphatidylcholines are known as **lecithins;** diphosphatidylglycerols, the "double" glycerophospholipids, are known as **cardiolipins** (because they were first isolated from heart muscle).

Plasmalogens are glycerophospholipids in which the C(1) substituent to the glycerol moiety is bonded to it via an α,β-unsaturated ether linkage in the cis configuration rather than through an ester linkage.

A plasmalogen

Ethanolamine, choline, and serine form the most common plasmalogen head groups.

D. Sphingolipids

Sphingolipids, which are also major membrane components, are derivatives of the C_{18} amino alcohols **sphingosine, dihydrosphingosine** (Fig. 11-5), and their C_{16}, C_{17}, C_{19}, and C_{20} homologs. Their *N*-acyl fatty acid derivatives, **ceramides,**

A ceramide

occur only in small amounts in plant and animal tissues but form the parent compounds of more abundant sphingolipids:

1. **Sphingomyelins,** the most common sphingolipids, are ceramides bearing either a phosphocholine (Fig. 11-6) or a phosphoethanolamine moiety so that they can also be classified as **sphingophospholipids.** *Although sphingomyelins differ chemically from phosphatidylcholine and phosphatidylethanolamine, their conformations and charge distributions are quite similar.* The membranous myelin sheath that surrounds and electrically insulates many nerve cell axons (Section 34-4C) is particularly rich in sphingomyelin.

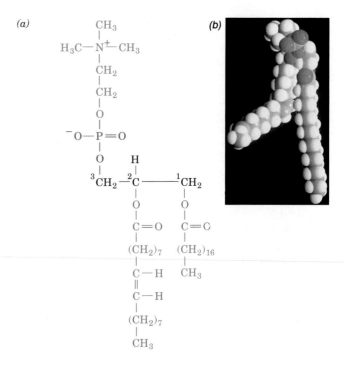

(a) **1-Stearoyl-2-oleoyl-3-phosphatidylcholine**

Figure 11-4
The glycerophospholipid 1-stearoyl-2-oleoyl-3-phosphatidylcholine: (*a*) molecular formula in Fischer projection and (*b*) space-filling model. [Computer graphics courtesy of Richard Pastor, FDA.]

2. **Cerebrosides,** the simplest **sphingoglycolipids,** are ceramides with head groups that consist of a single sugar residue. **Galactocerebrosides,** which are most prevalent in the neuronal cell membranes of the brain, have a β-D-galactose head group.

A galactocerebroside

Glucocerebrosides, which instead have a β-D-glucose residue, occur in the membranes of other tissues. *Cerebrosides, in contrast to phospholipids, lack phos-*

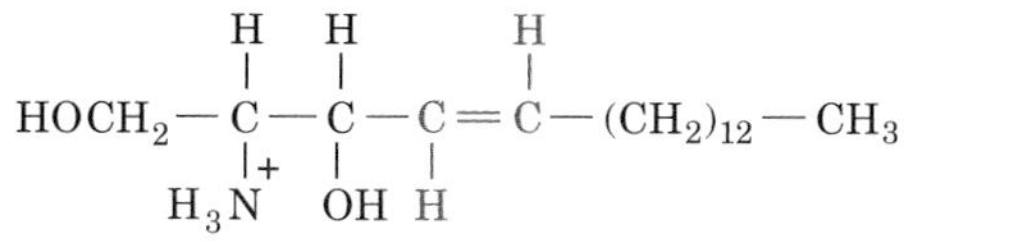

$$HOCH_2 - \underset{\underset{H_3\overset{+}{N}}{|}}{\overset{\overset{H}{|}}{C}} - \underset{\underset{OH}{|}}{\overset{\overset{H}{|}}{C}} - C = \overset{\overset{H}{|}}{C} - (CH_2)_{12} - CH_3$$

Sphingosine

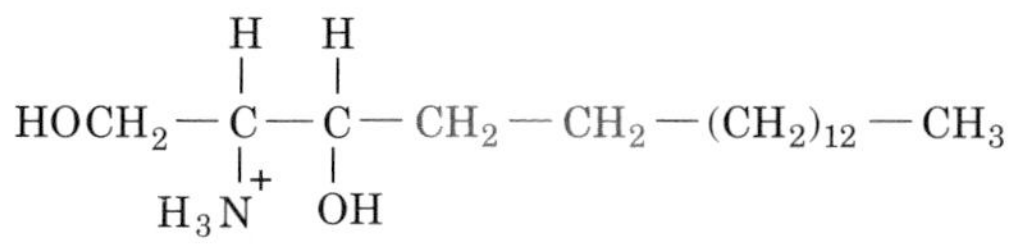

$$HOCH_2 - \underset{\underset{H_3\overset{+}{N}}{|}}{\overset{\overset{H}{|}}{C}} - \underset{\underset{OH}{|}}{\overset{\overset{H}{|}}{C}} - CH_2 - CH_2 - (CH_2)_{12} - CH_3$$

Dihydrosphingosine

Figure 11-5
The chiral centers at C(2) and C(3) of sphingosine and dihydrosphingosine have the configurations shown in Fischer projection. The double bond in sphingosine has the trans configuration.

phate groups and hence are most frequently nonionic compounds. The galactose residues of some galactocerebrosides, however, are sulfated at their C(3) position to form ionic compounds known as **sulfatides.** More complex sphingoglycolipids have unbranched oligosaccharide head groups of up to four sugar residues.

3. **Gangliosides** form the most complex group of sphingoglycolipids. They are ceramide oligosaccharides that include among their sugar groups at least one sialic acid residue (*N*-acetylneuraminic acid and its derivatives; Section 10-1C). The structures of gangliosides G_{M1}, G_{M2}, and G_{M3}, three of the over 60 that are known, are shown in Fig. 11-7. Gangliosides

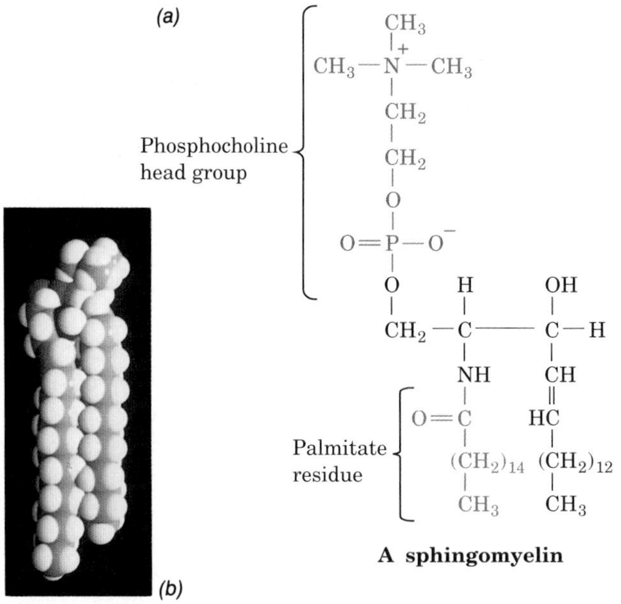

(a)

(b)

A sphingomyelin

Figure 11-6
A sphingomyelin: (*a*) molecular formula in Fischer projection and (*b*) space-filling model. Note its conformational resemblance to glycerophospholipids (Fig. 11-4). [Computer graphics courtesy of Richard Pastor, FDA.]

Figure 11-7
Ganglioside G_{M1}: (*a*) structural formula with its sphingosine residue in Fischer projection and (*b*) space-filling model. Gangliosides G_{M2} and G_{M3} differ from G_{M1} only by the sequential absences of the terminal D-galactose and *N*-acetyl-D-galactosamine residues. Other gangliosides have different oligosaccharide head groups. [Computer graphics courtesy of Richard Venable, FDA.]

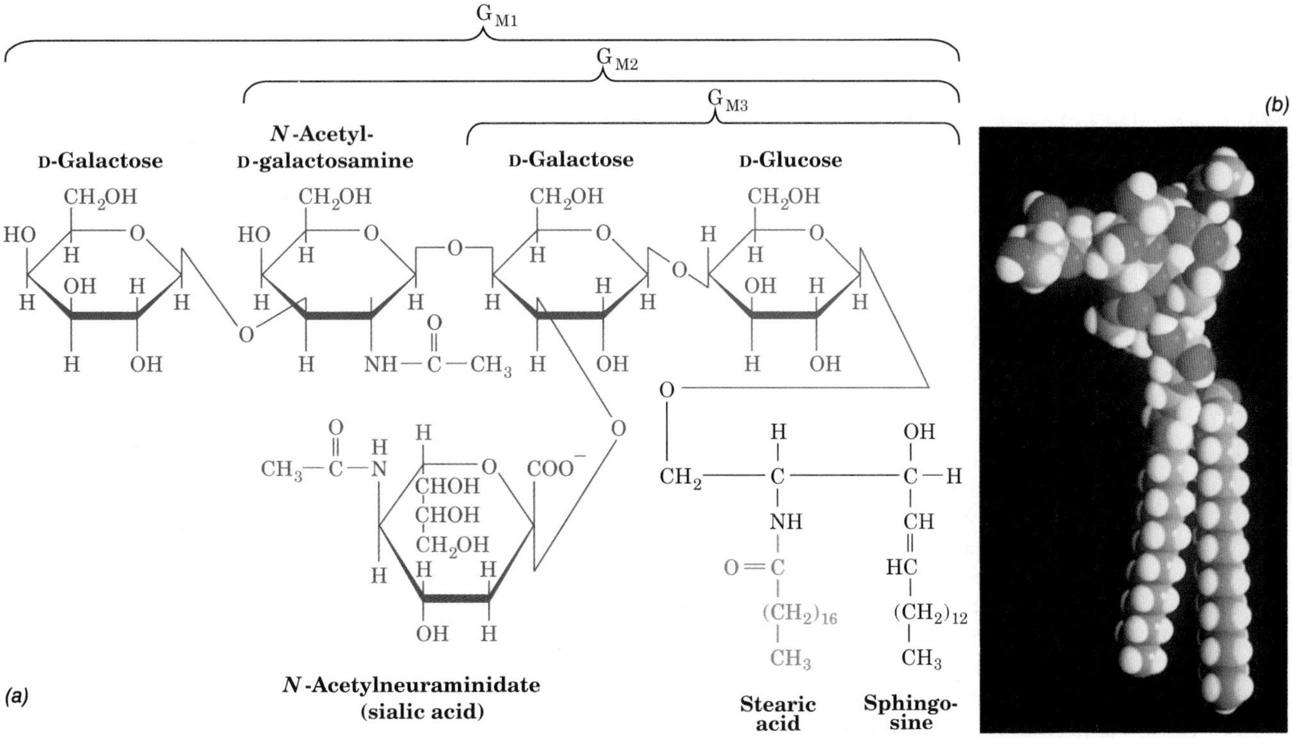

(a)

(b)

N-Acetylneuraminidate
(sialic acid)

Stearic acid

Sphingosine

are primarily components of cell surface membranes and constitute a significant fraction (6%) of brain lipids. Other tissues also contain gangliosides but in lesser amounts.

Gangliosides have considerable physiological and medical significance. Their complex carbohydrate head groups, which extend beyond the surfaces of cell membranes, act as specific receptors for certain pituitary glycoprotein hormones that regulate a number of important physiological functions (Section 34-4A). Gangliosides are also receptors for bacterial protein toxins such as cholera toxin (Section 34-4B). There is considerable evidence that gangliosides are specific determinants of cell–cell recognition so that they probably have an important role in the growth and differentiation of tissues as well as in carcinogenesis. Disorders of ganglioside breakdown are responsible for several hereditary **sphingolipid storage diseases,** such as **Tay-Sachs disease,** which are characterized by an invariably fatal neurological deterioration (Section 23-8C).

E. Cholesterol

Steroids, which are mostly of eukaryotic origin, are derivatives of cyclopentanoperhydrophenanthrene (Fig. 11-8). The much maligned **cholesterol** (Fig. 11-9), the most abundant steroid in animals, is further classified as a **sterol** because of its C(3)-OH group and its branched aliphatic side chain of 8 to 10 carbon atoms at C(17).

Cholesterol is a major component of animal plasma membranes and occurs in lesser amounts in the membranes of their subcellular organelles. Its polar OH group gives it a weak amphiphilic character, whereas its fused ring system provides it with greater rigidity than other membrane lipids. Cholesterol is therefore an important determinant of membrane properties. It is also abundant in blood plasma lipoproteins (Section 11-4) where ~70% of it is esterified to long-chain fatty acids to form **cholesteryl esters.**

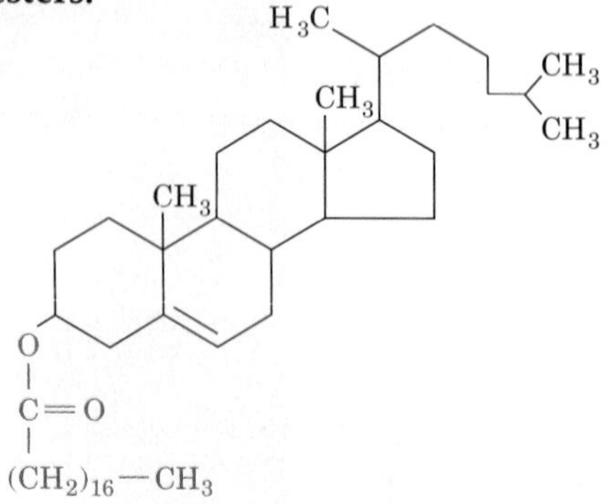

Cholesteryl stearate

Cholesterol is the metabolic precursor of **steroid hormones,** substances that regulate a great variety of physiological functions including sexual development and

Cyclopentanoperhydrophenanthrene

Figure 11-8
Cyclopentanoperhydrophenanthrene, the parent compound of steroids, consists of four fused saturated rings. The standard ring labeling system is indicated.

carbohydrate metabolism (Section 34-4A). The much debated role of cholesterol in heart disease is examined in Section 11-4.

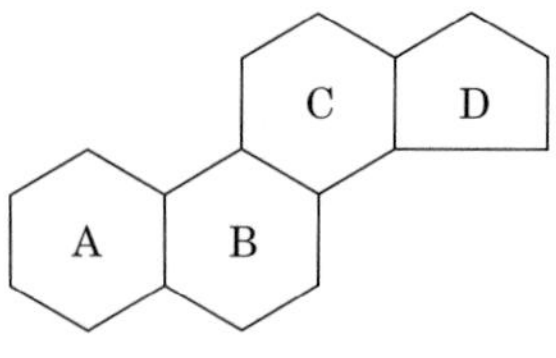

Plants contain little cholesterol. Rather, the most common sterol components of their membranes are **stigmasterol** and **β-sitosterol**

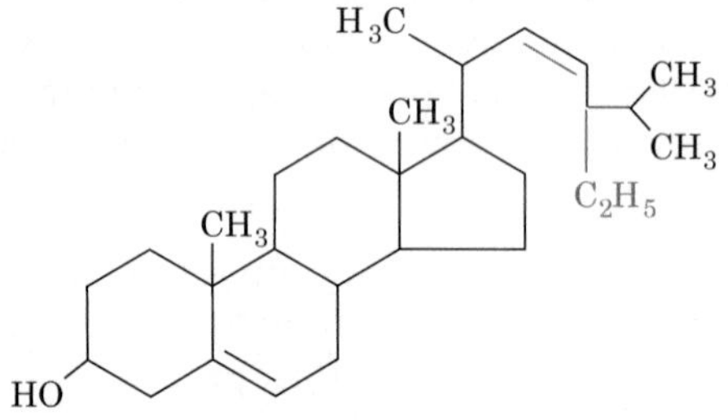

Stigmasterol

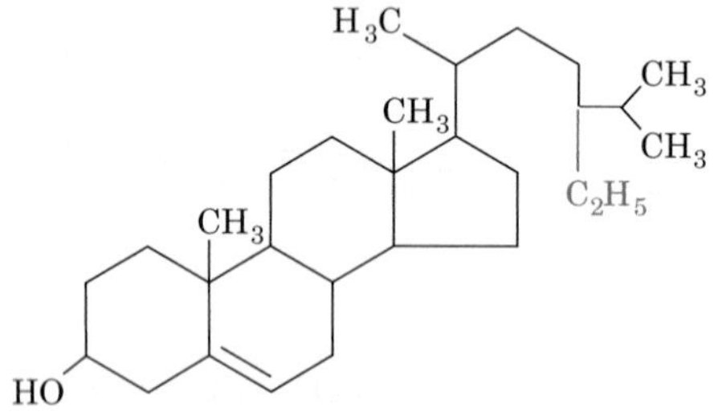

β-Sitosterol

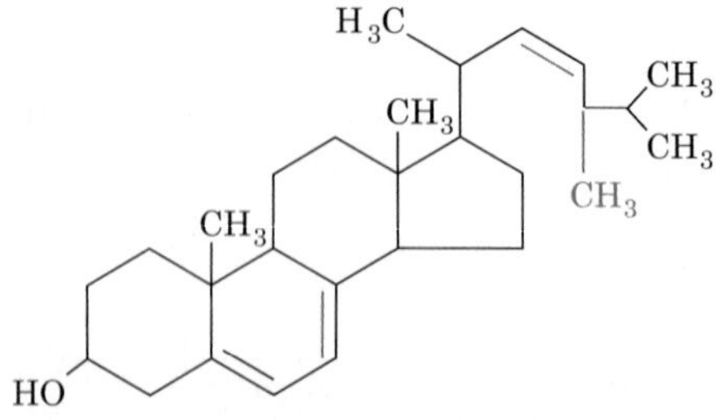

Ergosterol

which differ from cholesterol only in their aliphatic side chains. Yeast and fungi have yet other membrane sterols such as **ergosterol,** which has a C(7) to C(8) double bond. Prokaryotes, with the exception of mycoplasmas (Section 1-1B), contain little, if any, sterol.

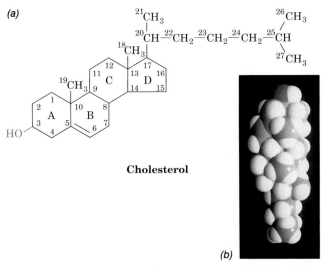

(a)

Cholesterol

(b)

Figure 11-9
Cholesterol: (*a*) structural formula with the standard numbering system and (*b*) space-filling model. Cholesterol's rigid ring system makes it far less conformationally flexible than membrane lipids. [Computer graphics courtesy of Richard Pastor, FDA.]

2. PROPERTIES OF LIPID AGGREGATES

The first recorded experiments on the physical properties of lipids were made in 1774 by the American statesman and scientist Benjamin Franklin. In investigating the well-known (at least among sailors) action of oil in calming waves, Franklin wrote:

> At length being at Clapham [in London] where there is, on the common, a large pond, which I observed to be one day very rough with the wind, I fetched out a cruet of oil [probably olive oil] and dropt a little of it in the water. I saw it spread itself with surprising swiftness upon the surface. . . . I then went to the windward side, where [the waves] began to form; and there the oil, though not more than a teaspoonful, produced an instant calm over a space several yards square, which spread amazingly, and extended itself gradually till it reached the lee side, making all that quarter of the pond, perhaps half an acre, as smooth as a looking glass.

This is sufficient information to permit the calculation of the oil layer's thickness (although there is no indication that Franklin made this calculation, we can; see Problem 4). We now know that oil forms a monomolecular layer on the surface of water in which the polar heads of these amphiphilic molecules are immersed in the water and their hydrophobic tails extend into the air (Fig. 11-10).

The calming effect of oil on rough water is a consequence of a large reduction in the water's surface tension. An oily surface film has the weak intermolecular cohesion characteristic of hydrocarbons rather than the strong intermolecular attractions of water responsible for its normally large surface tension. Oil, nevertheless, calms only smaller waves; it does not, as Franklin later observed, affect the larger swells.

In this section we discuss how lipids aggregate to form micelles and bilayers. We shall also be concerned with the physical properties of lipids in bilayers because these aggregates form the structural basis for biological membranes.

A. Micelles and Bilayers

In aqueous solutions, amphiphilic molecules, such as soaps and detergents, form micelles (globular aggregates whose hydrocarbon groups are out of contact with water; Section 2-1B). This molecular arrangement eliminates unfavorable contacts between water and the hydrophobic tails of the amphiphiles and yet permits the solvation of the polar head groups. Micelle formation is a cooperative process: An assembly of just a few amphiphiles cannot shield its tails from contact with water. Consequently, dilute aqueous solutions of amphiphiles do not form micelles until their concentration surpasses a certain **critical micelle concentration (cmc)**. Above the cmc, almost all the added amphiphile aggregates to form micelles. The value of the cmc depends on the identity of the amphiphile and the solution conditions. For amphiphiles with relatively small single tails, such as dodecyl sulfate, $CH_3(CH_2)_{11}OSO_3^-$, the cmc is ~ 1 mM. Those of biological lipids, most of which have two large hydrophobic tails, are generally $< 10^{-6}M$.

Single-Tailed Lipids Tend to Form Micelles

The approximate size and shape of a micelle can be predicted from geometrical considerations. Single-tailed amphiphiles, such as soap anions, form spheroidal or ellipsoidal micelles because of their tapered shapes (their hydrated head groups are wider than their

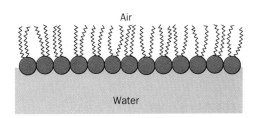

Figure 11-10
At the air–water interface, the hydrophobic tails of a lipid monolayer avoid association with water by extending into the air.

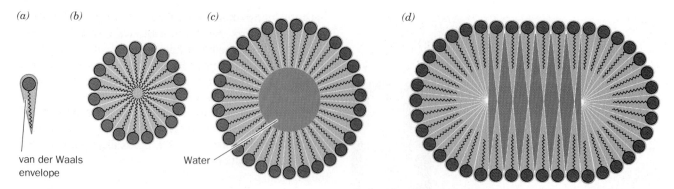

(a)   *(b)*   *(c)*   *(d)*

van der Waals envelope   Water

Figure 11-11
The tapered van der Waals envelope of single-tailed lipids (*a*) permits them to pack efficiently in forming a spheroidal micelle (*b*). The diameter of these micelles and hence their lipid population largely depends on the length of the tails. Spheroidal micelles composed of many more lipid molecules than the optimal number would have an unfavorable water-filled center (*c*). Such micelles could flatten out to collapse the hollow center but as such ellipsoidal micelles become elongated, they also develop water-filled spaces (*d*).

tails; Fig. 11-11*a* and *b*). The number of molecules in such micelles depends on the amphiphile, but for many substances, it is on the order of several hundred. For a given amphiphile, these numbers span a narrow range: Less would expose the hydrophobic core of the micelle to water, whereas more would give the micelle an energetically unfavorable hollow center (Fig. 11-11*c*). Of course, a large micelle could flatten out to eliminate this hollow center but the resulting decrease of curvature at the flattened surfaces would also generate empty spaces (Fig. 11-11*d*).

Glycerophospholipids and Sphingolipids Tend to Form Bilayers

The two hydrocarbon tails of glycerophospholipids and sphingolipids give these amphiphiles more or less rectangular shapes (Fig. 11-12*a*). The steric requirements of packing such molecules together yields large disklike micelles (Fig. 11-12*b*) that are really extended bimolecular leaflets. The existence of such **lipid bilayers** was first proposed in 1925 by E. Gorter and F. Grendel, on the basis of their observation that lipids extracted from erythrocytes cover twice the area when spread as a monolayer at the air–water interface (Fig. 11-10) than in the erythrocyte plasma membrane (the erythrocyte's only membrane). Lipid bilayers have thicknesses of ~ 60 Å, as measured by electron microscopy and X-ray diffraction techniques, the value expected when their hydrocarbon tails are more or less extended. We shall see below that *lipid bilayers form the structural basis of biological membranes.*

B. Liposomes

A suspension of phospholipids in water forms multilamellar vesicles that have an onionlike arrangement of lipid bilayers (Fig. 11-13*a*). Upon **sonication** (agitation by ultrasonic vibrations), these structures rearrange to form **liposomes**—closed, self-sealing, solvent-filled vesicles that are bounded by only a single bilayer (Fig. 11-13*b*). They usually have diameters of several hundred Å and, in a given preparation, are rather uniform in size. Liposomes with diameters ~ 1000 Å are made by injecting an ethanolic solution of phospholipid into water or by dissolving phospholipid in a detergent solution and then dialyzing out the detergent. Once formed, liposomes are quite stable and, in fact, may be separated from the solution in which they reside by dialysis, gel filtration chromatography, or centrifugation. Liposomes with differing internal and external environments can therefore be readily prepared. *Biological membranes consist of lipid bilayers with which proteins are associated (Section 11-3A).* Liposomes composed of synthetic lipids and/or lipids extracted from biological sources (e.g., lecithin from egg yolks) have therefore been extensively studied as models for biological membranes.

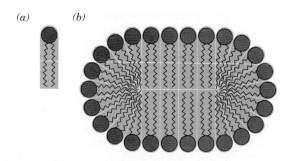

(a)   *(b)*

Figure 11-12
The cylindrical van der Waals envelope of phospholipids (*a*) causes them to form extended disklike micelles (*b*) that are better described as lipid bilayers.

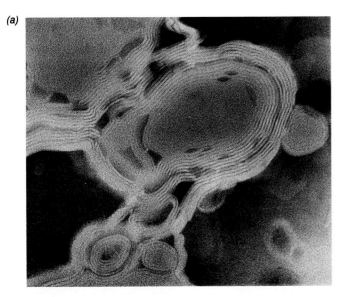

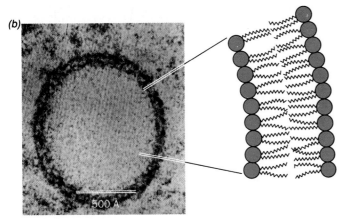

hydration shell and become solvated by the bilayer's hydrocarbon core. Such a process is highly unfavorable for polar molecules so that even the ~30-Å thickness of a lipid bilayer's hydrocarbon core forms an effective barrier for polar substances. However, measurements using tritiated water indicate that lipid bilayers are appreciably permeable to water. Despite the polarity of water, its small molecular size makes it significantly soluble in the hydrocarbon core of lipid bilayers and therefore able to permeate them.

> The stability of liposomes and their impermeability to many substances makes them promising vehicles for the delivery of therapeutic agents, such as drugs and enzymes, to particular tissues. Liposomes are absorbed by many cells through fusion with their plasma membranes. If methods can be developed for targeting liposomes to specific cell populations, then drugs could be directed towards particular tissues through liposome encapsulation. Encouraging progress towards this goal has been made (Fig. 11-14).

Figure 11-13
(a) An electron micrograph of a multilamellar phospholipid vesicle in which each layer is a lipid bilayer. [Courtesy of Alec D. Bangham, Institute of Animal Physiology.] (b) An electron micrograph of a liposome. Its wall, as the accompanying diagram indicates, consists of a bilayer. [Courtesy of Walter Stoekenius, University of California at San Francisco.]

Lipid Bilayers Are Impermeable to Most Polar Substances

Since biological membranes form cell and organelle boundaries, it is important to determine their ability to partition two aqueous compartments. The permeability of a lipid bilayer to a given substance may be determined by forming liposomes in a solution containing the substance, changing the external aqueous solution, and then measuring the rate at which the substance of interest appears in the new external solution. It has been found in this way that *lipid bilayers are extraordinarily impermeable to ionic and polar substances and that the permeabilities of such substances increase with their solubilities in nonpolar solvents.* This suggests that to penetrate a lipid bilayer, a solute molecule must shed its

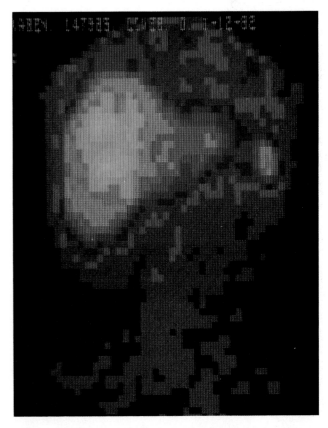

Figure 11-14
γ-Ray scan of a human subject taken 24 h after the intravenous injection of liposomes labeled with a γ-ray emitter. Liposomal uptake (*blue and white squares*) preferentially occurs in the liver (*left*) and the spleen (*right*). [Courtesy of Gabriel Lopez-Berestein, University of Texas M. D. Anderson Cancer Center.]

Lipid Bilayers Are Two-Dimensional Fluids

The transfer of a lipid molecule across a bilayer (Fig. 11-15a), a process termed **transverse diffusion** or a **flip-flop,** is an extremely rare event. This is because a flip-flop requires the polar head group of the lipid to pass through the hydrocarbon core of the bilayer. The flip-flop rates of phospholipids, as measured by several techniques, are characterized by half-times that are minimally several days.

In contrast to their low flip-flop rates, *lipids are highly mobile in the plane of the bilayer* (**lateral diffusion,** Fig.

(a) Transverse diffusion (flip-flop)

very slow

(b) Lateral diffusion

rapid

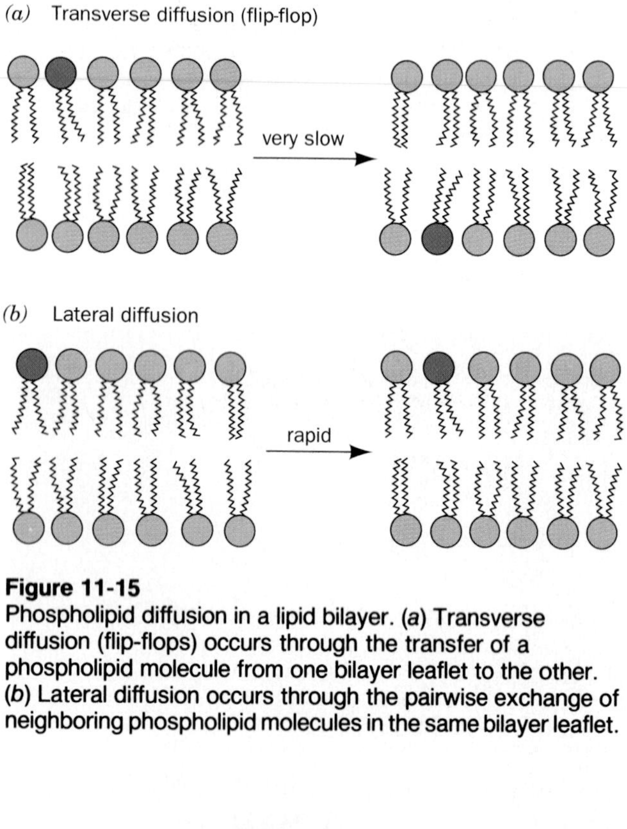

Figure 11-15
Phospholipid diffusion in a lipid bilayer. (a) Transverse diffusion (flip-flops) occurs through the transfer of a phospholipid molecule from one bilayer leaflet to the other. (b) Lateral diffusion occurs through the pairwise exchange of neighboring phospholipid molecules in the same bilayer leaflet.

11-15b). The X-ray diffraction patterns of bilayers at physiological temperatures have a diffuse band centered at a spacing of 4.6 Å whose width is a measure of the distribution of lateral spacings between the hydrocarbon chains in the bilayer plane. This band, which resembles one in the X-ray diffraction patterns of liquid paraffins, is indicative that *the bilayer is a two-dimensional fluid in which the hydrocarbon chains undergo rapid fluxional (continuously changing) motions involving rotations about their C—C bonds.*

The lateral diffusion rate of lipid molecules can be quantitatively determined from the rate of **fluorescence photobleaching recovery** (Fig. 11-16). A fluorescent group (**fluorophore**) is specifically attached to a bilayer component and an intense laser pulse focused on a very small area ($\sim$3 μm^2) is used to destroy (bleach) the fluorophore there. The rate at which the bleached area recovers its fluorescence, as monitored by fluorescence microscopy, indicates the rate at which unbleached and bleached fluorescence-labeled molecules laterally diffuse into and out of the bleached area, respectively. Such observations indicate, as do magnetic resonance measurements, that lipids in bilayers have lateral mobilities similar to those of the molecules in a light machine oil. Lipids in bilayers can therefore diffuse the 1-μm length of a bacterial cell in $\sim$1 s.

Bilayer Fluidity Varies with Temperature

*As a lipid bilayer cools below a characteristic **transition temperature**, it undergoes a sort of phase change, termed an **order–disorder transition**, in which it becomes a gel-like solid* (Fig. 11-17); that is, it loses its fluidity. Below the transition temperature, the 4.6-Å X-ray diffraction band characteristic of the lateral spacing between hydrocarbon chains in a liquid-crystalline bilayer is re-

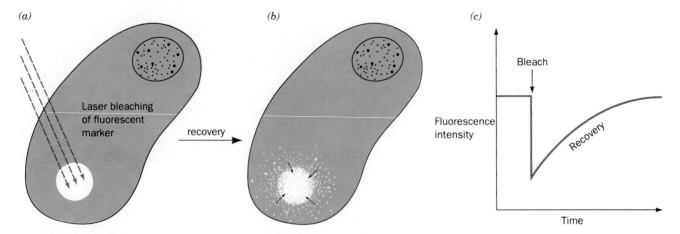

(a)   (b)   (c)

Laser bleaching of fluorescent marker

recovery

Bleach

Fluorescence intensity

Recovery

Time

Figure 11-16
The fluorescence photobleaching recovery technique: (a) An intense laser light pulse bleaches the fluorescent markers (*color*) from a small region of an immobilized cell that has a fluorescence-labeled membrane component. (b) The fluorescence of the bleached area, as monitored by fluorescence microscopy, recovers as the bleached molecules laterally diffuse out of it and intact fluorescence-labeled molecules diffuse into it. (c) The fluorescence recovery rate depends on the diffusion rate of the labeled molecule.

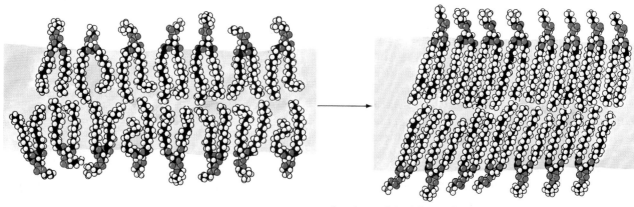

(a) Above transition temperature

(b) Below transition temperature

Figure 11-17
The structure of a lipid bilayer composed of phosphatidyl-choline and phosphatidylethanolamine as the temperature is lowered below the bilayer's transition temperature. *(a)* Above the transition temperature, both the lipid molecules as a whole and their nonpolar tails are highly mobile in the plane of the bilayer. Such a state of matter, which is ordered in some directions but not in others, is called a **liquid crystal.** *(b)* Below the transition temperature, the lipid molecules form a much more orderly array to yield a gel-like solid. [After Robertson, R. N., *The Lively Membranes, pp.* 69–70, Cambridge University Press (1983).]

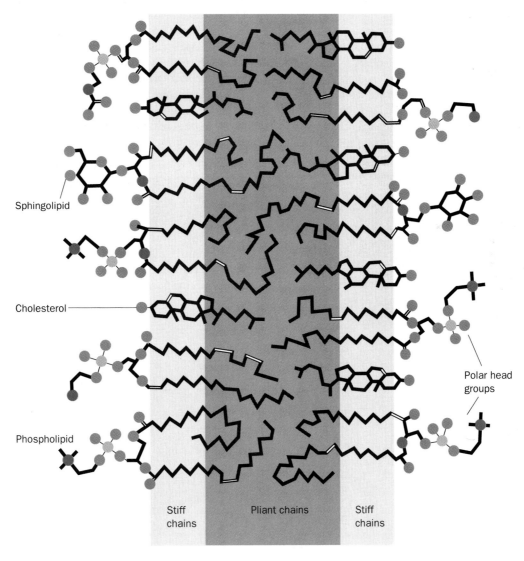

Sphingolipid

Cholesterol

Phospholipid

Polar head groups

Stiff chains | Pliant chains | Stiff chains

Figure 11-18
A lipid bilayer drawn to show the liquidlike packing of its hydrocarbon chains. In the 30 × 30-Å section represented, there are six cholesterol molecules, five glycerophospholipid molecules of three different types, and four sphingolipid molecules of two different types. The cholesterol OH group positions its attached steroid ring system in the exterior of the bilayer's hydrophobic core, thereby stiffening this region. [After Chapman, D., *in* Weissman, G. and Claiborne, R. (Eds.), *Cell Membranes, p.* 22, HP Publishing (1975).]

placed by a sharp 4.2-Å band similar to that exhibited by crystalline paraffins. This indicates that the hydrocarbon chains in a bilayer become fully extended and packed in a hexagonal array as in crystalline paraffins.

The transition temperature of a bilayer increases with the chain length and the degree of saturation of its component fatty acid residues for the same reasons that the melting temperatures of fatty acids increase with these quantities. The transition temperatures of most biological membranes are in the range 10 to 40°C. *Cholesterol, which by itself does not form a bilayer, decreases membrane fluidity because its rigid steroid ring system interferes with the motions of the fatty acid side chains (Fig. 11-18).* It also broadens the temperature range of the order–disorder transition and in high concentrations totally abolishes it. This behavior occurs because cholesterol inhibits the crystallization of fatty acid side chains by fitting in between them. Thus cholesterol functions as a kind of membrane plasticizer.

The fluidity of biological membranes is one of their important physiological attributes since it permits their embedded proteins to interact (Section 11-3B). The transition temperatures of mammalian membranes are well-below body temperatures and hence these membranes all have a fluidlike character. Bacteria and poikilothermic (cold-blooded) animals such as fish modify (through lipid biosynthesis and degradation) the fatty acid compositions of their membrane lipids with the ambient temperature so as to maintain membrane fluidity. For example, the membrane viscosity of *E. coli* at its growth temperature remains constant as the growth temperature is varied from 15 to 43°C.

Gaseous anesthetics, such as diethyl ether, cyclopropane, **halothane** (2-bromo-2-chloro-1,1,1-trifluoroethane), and the inert gas Xe, act by interfering with the transmission of nerve impulses in the central nervous system. Since the body excretes these general anesthetics unchanged, it appears that they do not act by chemical means. Rather, experimental evidence, such as the linear correlation of their anesthetic effectiveness with their lipid solubilities, suggests that these nonpolar substances alter the structures of membranes by dissolving in their hydrocarbon cores. Nerve impulse transmission, which is a membrane based phenomenon (Section 34-4C), is disrupted by these structural changes to which neuronal membranes seem particularly sensitive.

3. BIOLOGICAL MEMBRANES

Biological membranes are composed of proteins associated with a lipid bilayer matrix. Their lipid fractions consist of complex mixtures that vary according to the membrane source (Table 11-3) and, to some extent, with the diet and environment of the organism that produced the membrane. *Membrane proteins carry out the dynamic processes associated with membranes and therefore specific proteins occur only in particular membranes.* Protein to lipid ratios in membranes vary considerably with membrane function as is indicated by Table 11-4, although most membranes are at least one-half protein. The myelin membrane, which functions passively as an insulator around certain nerve fibers (Section 34-4C), is a prominent exception to this generalization in that it contains only 18% protein.

In this section, we discuss the properties of membrane proteins and their behavior in biological membranes. Following this, we examine specific aspects of biological membranes, namely, the erythrocyte membrane skeleton, the nature of blood groups, and gap junctions. Finally, we consider how membranes are assembled and how their component proteins are directed to them.

Table 11-3
Lipid Compositions of Some Biological Membranes[a]

Lipid	Human Erythrocyte	Human Myelin	Beef Heart Mitochondria	*E. coli*
Phosphatidic acid	1.5	0.5	0	0
Phosphatidylcholine	19	10	39	0
Phosphatidylethanolamine	18	20	27	65
Phosphatidylglycerol	0	0	0	18
Phosphatidylinositol	1	1	7	0
Phosphatidylserine	8.5	8.5	0.5	0
Cardiolipin	0	0	22.5	12
Sphingomyelin	17.5	8.5	0	0
Glycolipids	10	26	0	0
Cholesterol	25	26	3	0

[a] The values given are weight percent of total lipid.

Source: Tanford, C., *The Hydrophobic Effect*, p. 109, Wiley (1980).

Table 11-4

Compositions of Some Biological Membranes

Membrane	Protein (%)	Lipid (%)	Carbohydrate (%)	Protein to Lipid Ratio
Plasma membranes:				
Mouse liver cells	46	54	2–4	0.85
Human erythrocyte	49	43	8	1.1
Amoeba	52	42	4	1.3
Rat liver nuclear membrane	59	35	2.0	1.6
Mitochondrial outer membrane	52	48	$(2-4)^a$	1.1
Mitochondrial inner membrane	76	24	$(1-2)^a$	3.2
Myelin	18	79	3	0.23
Gram-positive bacteria	75	25	$(10)^a$	3.0
Halobacterium purple membrane	75	25		3.0

^{a} Deduced from the analyses.

Source: Guidotti, G., *Annu. Rev. Biochem.* **41,** 732 (1972).

A. Membrane Proteins

Membrane proteins are operationally classified according to how tightly they are associated with membranes:

1. *Integral* or *intrinsic proteins* are tightly bound to membranes by hydrophobic forces (Fig. 11-19) and can only be separated from them by treatment with agents that disrupt membranes. These include organic solvents, detergents (e.g., those in Fig. 11-20), and chaotropic agents (ions that disrupt water structure; Section 7-4E). Integral proteins tend to aggregate and precipitate in aqueous solutions in the absence of detergents or water-miscible organic solvents such as butanol or glycerol. Some integral proteins bind lipids so tenaciously that they can only be freed from them under denaturing conditions.

2. *Peripheral* or *extrinsic proteins* are dissociated from membranes by relatively mild procedures that leave the membrane intact, such as exposure to high ionic strength salt solutions (e.g., 1*M* NaCl), metal chelating agents, or pH changes. Peripheral proteins, for example, cytochrome *c*, are stable in aqueous solution and do not bind lipid. They associate with a membrane by binding at its surface, most likely to its integral proteins, through electrostatic and hydrogen bonding interactions.

In the following subsections, we shall concentrate on integral proteins.

Integral Proteins Are Asymmetrically Oriented Amphiphiles

All biological membranes contain integral proteins. Their locations on a membrane may be determined through **surface labeling,** a technique employing

Figure 11-19
Integral membrane proteins in a lipid bilayer are "solvated" by lipids through hydrophobic interactions between the protein and the lipids' nonpolar tails. The polar head groups may also associate with the protein through hydrogen bonding and salt bridges. [After Robertson, R. N., *The Lively Membranes, p.* 56, Cambridge University Press (1983).]

agents that react with proteins but cannot penetrate membranes. For example, an integral protein on the outer surface of an intact cell membrane binds antibodies elicited against it, but a protein on the membrane's inner surface can only do so if the membrane has been ruptured. Membrane-impermeable protein-specific reagents that are fluorescent or radioactively labeled, may

Figure 11-20
A selection of the detergents used in biochemical
manipulations. Note that they may be anionic, cationic,
zwitterionic, or uncharged.

$$CH_3-(CH_2)_n-CH_2-\overset{\overset{\displaystyle CH_3}{|}}{\underset{\underset{\displaystyle CH_3}{|}}{\overset{+}{N}}}-CH_3 \quad Br^-$$

$n = 10$ **Dodecyltriethylammonium bromide (DTAB)**
$n = 15$ **Cetyltrimethylammonium bromide (CTAB)**

$$CH_3-(CH_2)_{11}-OSO_3^- \quad Na^+$$

Sodium dodecyl sulfate (SDS)

$$CH_3-(CH_2)_{11}-(O-CH_2-CH)_n-OH$$

Polyoxyethylenelauryl ether

$n = 4$ **Brij 30**
$n = 25$ **Brij 35**

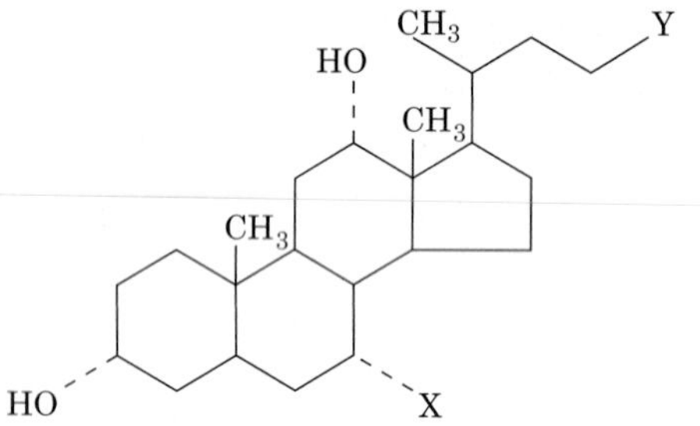

$$CH_3-\overset{\overset{\displaystyle CH_3}{|}}{\underset{\underset{\displaystyle CH_3}{|}}{C}}-CH_2-\overset{\overset{\displaystyle CH_3}{|}}{\underset{\underset{\displaystyle CH_3}{|}}{C}}-\langle\text{phenyl}\rangle-(O-CH_2-CH_2)_n-OH$$

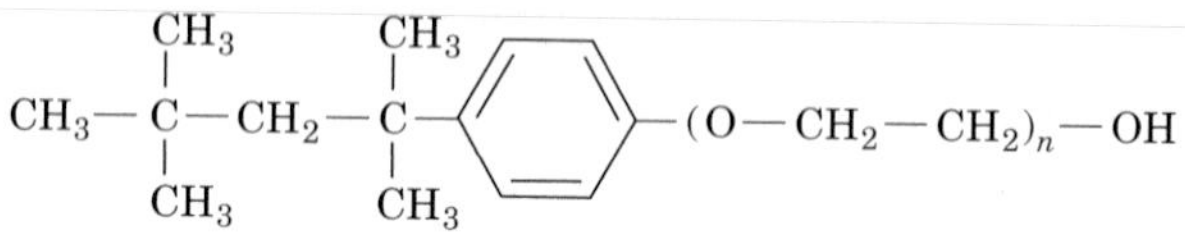

Polyoxyethylene-*p*-isooctylphenyl ether

$n = 5$ **Triton X-20**

$n = 10$ **Triton X-100**

X = H, Y = COO⁻ Na⁺ **Sodium deoxycholate**
X = OH, Y = COO⁻ Na⁺ **Sodium cholate**
X = OH, Y = CO—NH—(CH₂)₃—N⁺(CH₃)₂—SO₃⁻ **CHAPS**

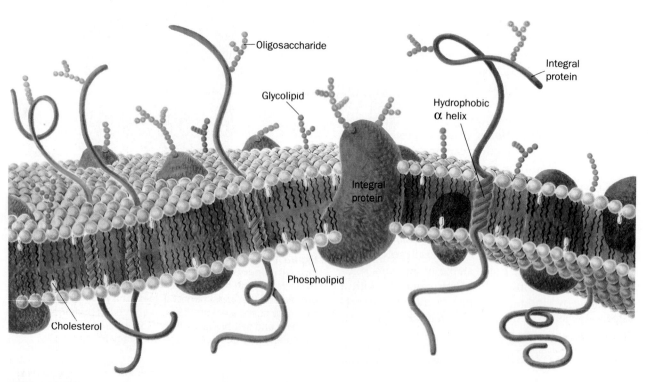

Figure 11-21
A schematic diagram of a plasma membrane. Integral proteins (*orange*) are
embedded in a bilayer composed of phospholipids (*blue spheres with two
wiggly tails;* shown, for clarity, in much greater proportion than they have in
biological membranes) and cholesterol (*yellow*). The carbohydrate
components of glycoproteins (*yellow beaded chains*) and glycolipids (*green
beaded chains*) occur only on the external face of the membrane.

be similarly employed. Using such surface-labeling re-agents, it has been shown that *some integral membrane proteins are exposed to a specific surface of a membrane, whereas others, known as* **trans-membrane proteins,** *span the membrane.* However, no protein is known to be completely buried in a membrane; that is, all have some exposure to the aqueous environment. Such studies have also established that biological membranes are asymmetric in that a particular membrane protein is invariably located on only one particular face of a membrane or, in the case of a trans-membrane protein, oriented in only one direction with respect to the membrane (Fig. 11-21).

Integral proteins are amphiphilic; the protein segments immersed in a membrane's nonpolar interior have predominantly hydrophobic surface residues, whereas those por-

tions that extend into the aqueous environment are by and large sheathed with polar residues. For example, proteolytic digestion and chemical modification studies indicate that the erythrocyte trans-membrane protein **glycophorin A** (Fig. 11-22) has three domains: (1) a 72-residue externally located N-terminal domain that bears 16 carbohydrate chains; (2) a 19-residue sequence, consisting almost entirely of hydrophobic residues, that spans the erythrocyte cell membrane; and (3) a 40-residue cytosplasmic C-terminal domain that has a high proportion of charged and polar residues. The transmembrane domain, as is common in many integral membrane proteins, almost certainly forms an α helix, thereby satisfying the hydrogen bonding requirements of its polypeptide backbone. Indeed, the existence of glycophorin A's single trans-membrane helix is pre-

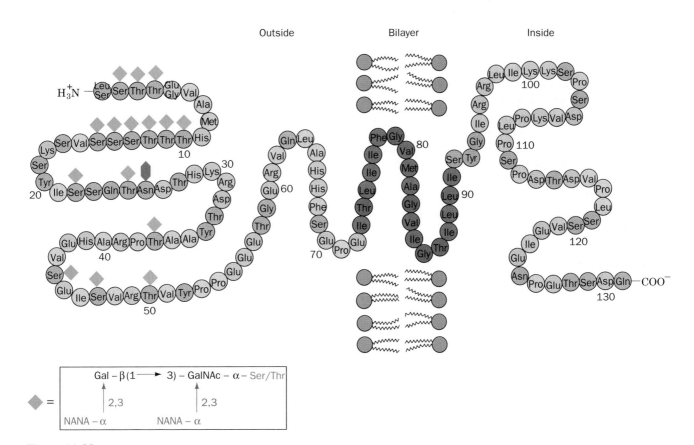

Figure 11-22
The amino acid sequence and membrane location of human erythrocyte glycophorin A. The protein, which is ~60% carbohydrate by weight, bears 15 O-linked oligosaccharides (*diamonds*) and one that is N-linked (*hexagon*). The predominant sequence of the O-linked oligosaccharides is given below. The protein's trans-membrane portion (*bright colors*) consists of 19 sequential predominantly hydrophobic residues. Its C-terminal portion, which is located on the membrane's cytoplasmic face, is rich in anionic (*pink*) and cationic (*blue*) amino acid residues. There are two common genetic variants of glycophorin A: glycophorin A^M has Ser and Gly at positions 1 and 5, respectively, whereas they are Leu and Glu in Glycophorin A^N. (Abbreviations: Gal = galactose, GalNAc = N-acetylgalactosamine, NANA = N-acetylneuraminic acid (sialic acid)). [After Marchesi, V. T., *Semin. Hematol.* **16,** 8 (1979).]

dicted by computing the free energy change in transferring α helically folded polypeptide segments from the nonpolar interior of a membrane to water (Fig. 11-23). Similar computations on other integral membrane proteins have also identified their trans-membrane helices.

In many integral proteins, the hydrophobic segment(s) anchors the active region of the protein to the membrane. For instance, trypsin cleaves the membrane-bound enzyme **cytochrome b₅** into a polar, enzymatically active ∼85-residue N-terminal fragment, and an ∼50-residue C-terminal fragment that remains embedded in the membrane (Fig. 11-24). *The asymmetric orientation of integral membrane proteins in the membrane is maintained by their infinitesimal flip-flop rates, which result from the greater sizes of the membrane protein "head groups" in comparison to those of lipids.* The origin of this asymmetry is discussed in Section 11-3F.

Bacteriorhodopsin Contains a Bundle of Seven Hydrophobic Helical Rods

One of the structurally best characterized integral membrane proteins is **bacteriorhodopsin** from the halophilic (salt loving) bacterium *Halobacter halobium* that inhabits such salty places as the Dead Sea (it grows best in 4.3*M* NaCl and is nonviable below 2.0*M* NaCl; seawater contains 0.6*M* NaCl). Under low O₂ conditions, its cell membrane develops ∼0.5-μm wide patches of **purple membrane** whose only protein component is bacteriorhodopsin. This 247-residue protein is a light-

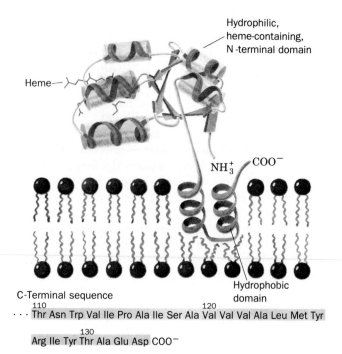

Figure 11-24
Liver cytochrome b₅ in association with a membrane. The protein's enzymatically active N-terminal domain (*purple*), whose X-ray structure has been determined, is anchored in the membrane by a hydrophobic and presumably α helical C-terminal segment (*brown*) that begins and ends with hydrophilic segments (*purple*). The amino acid sequence of the horse enzyme indicates that this hydrophobic anchor consists of a 13-residue segment ending 9 residues from the polypeptide's C-terminus (*below*). [Ribbon diagram of the N-terminal domain after a drawing by Jane Richardson, Duke University. Amino acid sequence from Ozals, J. and Craig, G., *J. Biol. Chem.* **253**, 8549 (1977).]

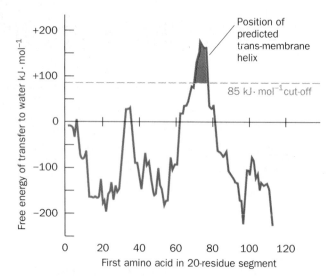

Figure 11-23
A plot, for glycophorin A, of the calculated free energy change in transferring 20-residue long α helical segments from the interior of a membrane to water versus the position of the segment's first residue. Peaks higher than +85 kJ·mol⁻¹ are indicative of a transmembrane helix. [After Engleman, D. M., Steitz, T. A., and Goldman, A., *Annu. Rev. Biophys. Biophys. Chem.* **15**, 343 (1986).]

driven proton pump; it generates a proton concentration gradient across the membrane that powers the synthesis of ATP (by a mechanism discussed in Section 20-3B). Bacteriorhodopsin's light-absorbing element, **retinal,** is covalently bound to its Lys 216 (Fig. 11-25). This **chromophore** (light-absorbing group), which is responsible for the membrane's purple color, is also the light-sensitive element in vision.

The purple membrane, which is 75% protein and 25% lipid, has an unusual structure compared to most other membranes (Section 11-3B): Its bacteriorhodopsin molecules are arranged in a highly ordered two-dimensional array. This crystalline arrangement permitted Richard Henderson and Nigel Unwin to determine the 7-Å resolution structure of bacteriorhodopsin by the combined electron microscopic and electron diffraction analysis of unstained purple membranes.

Bacteriorhodopsin consists largely of a bundle of seven ∼25-residue α helical rods that span the lipid bilayer in directions almost perpendicular to the bilayer plane (Fig. 11-26*a*). The ∼20-Å spaces between the pro-

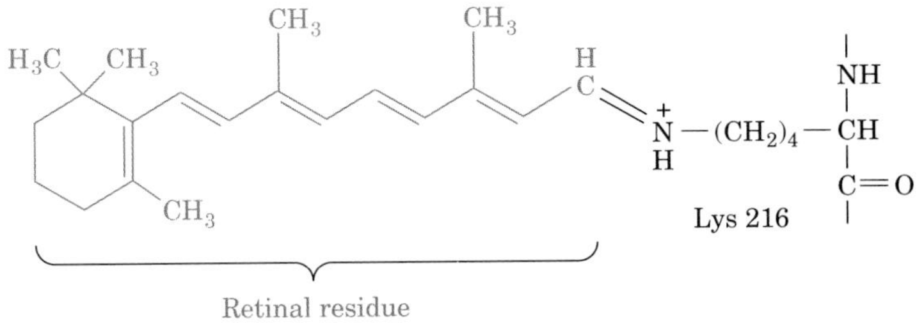

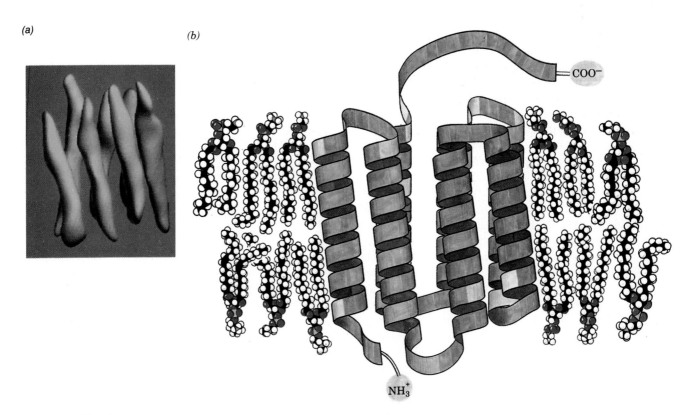

Figure 11-25
Retinal, the prosthetic group of bacteriorhodopsin, forms a Schiff base with Lys 216 of the protein. A similar linkage occurs in **rhodopsin**, the photoreceptor of the eye.

tein molecules in the purple membrane are occupied by this bilayer. Analysis of the proteolytic cleavage pattern of the membrane-embedded protein, together with model-building studies, suggests that adjacent α helices, which are largely hydrophobic in character, are connected in a head-to-tail fashion by short lengths of polypeptide (Fig. 11-26b). This arrangement places the protein's charged residues near the surfaces of the membrane in contact with the aqueous solvent or in a position to form ion pairs with oppositely charged groups. The internal charged residues in this model line the center of the helix bundle so as to form a hydrophilic channel that, it has been suggested, permits the passage of protons. Other membrane pumps and channels (Chapter 18) probably have similar structures.

B. Fluid Mosaic Model of Membrane Structure

The demonstrated fluidity of artificial lipid bilayers suggests that biological membranes have similar properties. This idea was proposed in 1972 by S. Jonathan Singer and Garth Nicholson in their unifying theory of membrane structure known as the **fluid mosaic model.** The theory postulates that integral proteins resemble "icebergs" floating in a two-dimensional lipid "sea" (Fig.

(a)

(b)

Figure 11-26
(*a*) A computer-generated model of bacteriorhodopsin based on its 7-Å resolution electron density map. The seven rodlike segments represent α helices. [Courtesy of Nigel Unwin, Stanford University School of Medicine.] (*b*) A proposed arrangement of bacteriorhodopsin in the lipid bilayer of the purple membrane. The N-terminus is situated on the cell surface side of the membrane and adjacent helices alternately run in and out of the cell so that the C-terminus is inside the cell. The polypeptide chain is colored according to side chain polarity: Cationic residues are blue, anionic residues are red, uncharged polar residues are purple, and nonpolar residues are brown. Note the predominantly hydrophobic character of the protein's periphery as it passes through the membrane's nonpolar core. [After Robertson, R. N., *The Lively Membranes*, p. 34, Cambridge University Press (1983). Based on a model by Engelman, D. M., Henderson, R., McLachlin, A. D., and Wallace, A., *Proc. Natl. Acad. Sci.* **77,** 2024 (1980).]

11-21) and that these proteins freely diffuse laterally in the lipid matrix unless their movements are restricted by associations with other cell components.

The Fluid Mosaic Model Has Been Verified Experimentally

The validity of the fluid mosaic model has been established in several ways. Perhaps the most vivid is an experiment by Michael Edidin (Fig. 11-27). Cultured mouse cells were fused with human cells by treatment with **Sendai virus** to yield a hybrid cell known as a **heterokaryon.** The mouse cells were labeled with mouse protein-specific antibodies to which a green-fluorescing dye had been covalently linked **(immuno-fluorescence).** The proteins on the human cells were similarly labeled with a red-fluorescing marker. Upon cell fusion, the mouse and human proteins, as seen under the fluorescence microscope, were segregated on the two halves of the heterokaryon. After 40 min at 37°C, however, these proteins had thoroughly intermingled. The addition of substances that inhibit metabolism or protein synthesis did not slow this process but lowering the temperature below 15°C did. These observations indicate that the mixing process is independent of both metabolic energy and the insertion into the membrane of newly synthesized proteins. Rather, it is a result of the diffusion of existing proteins throughout the fluid membrane, a process that slows as the temperature is lowered.

Fluorescence photobleaching recovery measurements (Fig. 11-16) indicate that membrane proteins vary in their lateral diffusion rates. Some 30 to 90% of these proteins are freely mobile; they diffuse at rates only an order of magnitude or so slower than those of the much smaller lipids so that they typically take from 10 to 60 min to diffuse the 20-μm length of a eukaryotic cell. Other proteins diffuse more slowly, and some, because of submembrane attachments, are essentially immobile.

The distribution of proteins in membranes may be visualized through electron microscopy using the freeze-fracture and freeze-etch techniques. In the freeze-etch procedure, which was devised by Daniel Branton, a membrane specimen is rapidly frozen to near liquid nitrogen temperatures (-196°C). This immobilizes the sample and thereby minimizes its disturbance by subsequent manipulations. The specimen is then fractured with a cold microtome knife, which often splits the bilayer into monolayers (Fig. 11-28). Since the exposed membrane itself would be destroyed by an electron beam, its metallic replica is made by coating the membrane with a thin layer of carbon, shadowing it (covering it by evaporative deposition under high vacuum) with platinum, and removing the organic matter by treatment with acid. Such a metallic replica can be examined by electron microscopy. In the freeze-etch procedure, the external surface of the membrane adjacent

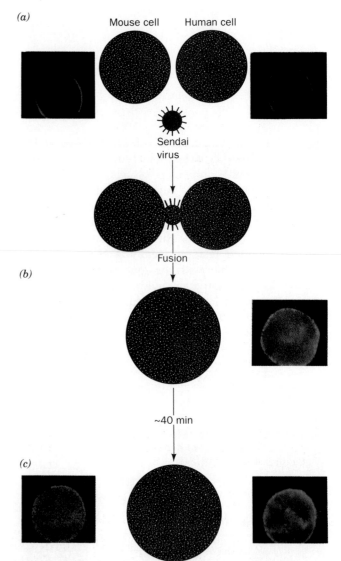

Figure 11-27
Sendai virus-induced fusion of a mouse cell with a human cell and the subsequent intermingling of their cell-surface components as visualized by immunofluorescence. Human and mouse antigens are labeled with red and green fluorescent markers, respectively. (*a*) The membrane-encapsulated Sendai virus specifically binds to cell-surface receptors on both types of cells and subsequently fuses to their cell membranes. (*b*) This results in the formation of a cytoplasmic bridge between the cells that expands so as to form the heterokaryon. (*c*) After 40 min, the red and green markers are fully intermingled. The photomicrographs were taken through filters that allowed only red or green light to reach the camera; that in Part (*b*) is a double exposure. [Immunofluorescence photomicrographs courtesy of Michael Edidin, The Johns Hopkins University.]

to the cleaved area revealed by freeze fracture may also be visualized by first subliming (etching) away, at -100°C, some of the ice in which it is encased (Fig. 11-29).

Freeze-etch electron micrographs of most biological membranes show an inner fracture face that is studded with embedded 50 to 85 Å in diameter globular particles

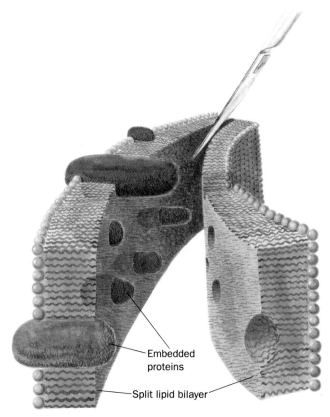

Figure 11-28
A membrane that has been split by freeze fracture, as is schematically diagrammed, exposes the interior of the lipid bilayer and its embedded proteins.

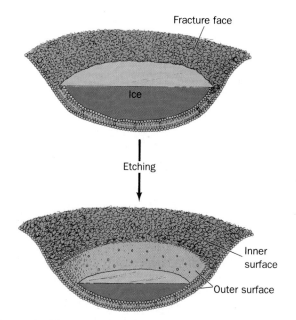

Figure 11-29
In the freeze-etch procedure, the ice that encases a freeze-fractured membrane (*top*) is partially sublimed away so as to expose the outer membrane surface (*bottom*) for electron microscopy.

(Fig. 11-30) that appear to be distributed randomly. These particles correspond to membrane proteins as is demonstrated by their disappearance when the membrane is treated with proteases before its freeze fracture. This is further corroborated by the observation that the myelin membrane, which has a low protein content, as well as liposomes composed of only lipids, have smooth inner fracture faces. Outer membrane surfaces also have a relatively smooth appearance (Fig. 11-30) because integral proteins do not protrude very far beyond them. The distributions of individual external proteins may be visualized by staining procedures, such as the use of ferritin-labeled antibodies, to yield electron micrographs similar in appearance to Fig. 10-32.

Membrane Lipids Are Asymmetrically Distributed

The distribution of lipids in a membrane has been established through the use of phospholipid-hydrolyzing enzymes known as **phospholipases.** Phospholipases cannot pass through membranes so that only phospholipids on the external surfaces of intact cells are susceptible to their action. Such studies indicate that *the lipids in biological membranes, like the proteins, are asymmetrically distributed* (e.g., Fig. 11-31). Carbohydrates,

as we have seen (Section 10-3C), are located exclusively on the external surfaces of plasma membranes.

C. The Erythrocyte Membrane

The erythrocyte membrane's relative simplicity, availability, and ease of isolation have made it the most extensively studied and best understood biological membrane. It is therefore a model for the more complex membranes of other cell types. A mature mammalian

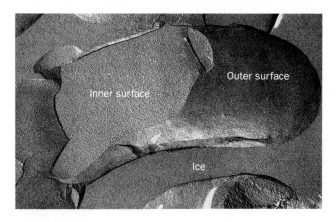

Figure 11-30
A freeze-etch electron micrograph of a human erythrocyte plasma membrane. The exposed interior face of the membrane is studded with numerous globular particles that are integral membrane proteins (see Fig. 11-28). The outer surface of the membrane appears smoother than the inner surface because proteins do not project very far beyond the outer membrane surface. [Courtesy of Vincent Marchesi, Yale University.]

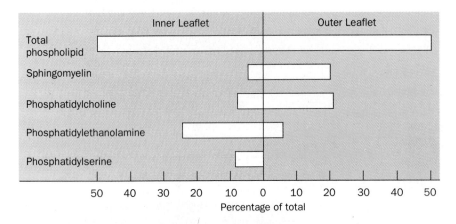

Figure 11-31
The asymmetric distribution of phospholipids in human erythrocyte membranes, expressed in mol %. [After Rothman, J. E. and Lenard, J., *Science* **194**, 1744 (1977).]

erythrocyte is devoid of organelles and carries out few metabolic processes; it is essentially a membranous bag of hemoglobin. Erythrocyte membranes can therefore be obtained by osmotic lysis, which causes the cell contents to leak out. The resultant membranous particles are known as erythrocyte **ghosts** because, upon return to physiological conditions, they reseal to form particles that retain their original shape. Indeed, by transferring sealed ghosts to another medium, their contents can be made to differ from the external solution.

Erythrocyte Membranes Contain a Variety of Proteins

The erythrocyte membrane has a more or less typical plasma membrane composition of about one-half protein, somewhat less lipid, and the remainder carbohydrate (Table 11-4). *Its proteins may be separated by SDS–polyacrylamide gel electrophoresis (Section 5-4C) after first solubilizing the membrane in a 1% SDS solution. The resulting electrophoretogram for a human erythrocyte membrane exhibits seven major and many minor bands when stained with Coomassie brilliant blue (Fig. 11-32).* If the electrophoretogram is instead treated with **periodic acid-Schiff's reagent (PAS)**, which stains carbohydrates, four so-called PAS bands become evident. The polypeptides corresponding to bands 1, 2, 4.1, 4.2, 5, and 6 are readily extracted from the membrane by changes in ionic strength or pH and hence are peripheral proteins. These proteins are located on the inner side of the membrane as is indicated by the observation that they are not altered by the incubation of intact erythrocytes or sealed ghosts with proteolytic enzymes or membrane-impermeable protein labeling reagents. These proteins are altered, however, if "leaky" ghosts are so-treated.

In contrast, bands 3, 7, and all four PAS bands correspond to integral proteins; they can be released from the membrane only by extraction with detergents or organic solvents. Of these, band 3 and PAS bands 1 and 2 correspond to trans-membrane proteins as indicated by their different labeling patterns when intact cells are treated with membrane-impermeable protein-labeling reagents

and when these reagents are introduced inside sealed ghosts.

The transport of CO_2 in blood (Section 9-1C) requires that the erythrocyte membrane be permeable to HCO_3^- and Cl^- (the maintenance of electroneutrality requires that for every HCO_3^- to enter a cell, a Cl^- or some other anion must leave the cell; Section 9-1C). The rapid transport of these and other anions across the erythrocyte membrane is mediated by a specific **anion channel** of which there are ~1 million/cell (comprising >30% of the membrane protein). Band 3 protein (929 residues and 5–8% carbohydrate) specifically reacts with anionic protein-labeling reagents that block the anion channel, thereby indicating that the anion channel is

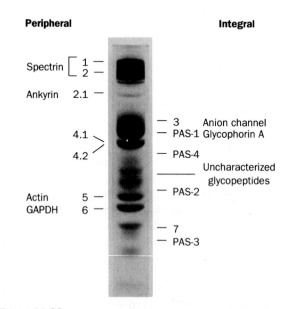

Figure 11-32
An SDS–polyacrylamide gel electrophoretogram of human erythrocyte membrane proteins as stained by Coomassie brilliant blue. The bands designated 4.1 and 4.2 are not separated with the 1% SDS concentration used. The minor bands are not labeled for the sake of simplicity. The positions of the four sialoglycoproteins revealed by PAS staining are indicated. [Courtesy of Vincent Marchesi, Yale University.]

composed of band 3 protein. Furthermore, cross-linking studies with bifunctional reagents (Section 7-5C) demonstrate that the anion channel is at least a dimer. Hemoglobin and the glycolytic (glucose metabolizing) enzymes **aldolase, phosphofructokinase (PFK),** and the band 6 protein **glyceraldehyde-3-phosphate dehydrogenase (GAPDH; Section 16-2)** all specifically and reversibly bind to band 3 protein on the cytoplasmic side of the membrane. The functional significance of this observation is unknown.

The Erythrocyte's Membrane Skeleton Is Responsible for Its Shape

A normal erythrocyte's biconcave disklike shape (Fig. 6-11a) assures the rapid diffusion of O_2 to its hemoglobin molecules by placing them no further than 1 μm from the cell surface. However, the rim and the dimple regions of an erythrocyte do not occupy fixed positions on the cell membrane. This has been demonstrated by anchoring an erythrocyte to a microscope slide by a small portion of its surface and inducing the cell to move laterally with a gentle flow of isotonic buffer. It was observed that a point originally on the rim of the erythrocyte moved across the dimple to the rim on the opposite side of the cell from where it began. Evidently, the membrane rolled across the cell while maintaining its shape, much like the tread of a tractor. This remarkable mechanical property of the erythrocyte membrane results from the presence of a submembranous network of proteins that function as a membrane "skeleton." Indeed, this property is partially duplicated by a mechanical model consisting of a geodesic sphere (a spheroidal cage) that is freely jointed at the intersections of its struts but constrained from collapsing much beyond a flat surface. When placed inside an evacuated plastic bag, this cage also assumes a biconcave disklike shape.

The fluidity and flexibility imparted to an erythrocyte by its membrane skeleton has important physiological consequences. A slurry of solid particles of a size and concentration equal to that of red cells in blood has the flow characteristics approximating that of sand. Consequently, in order for blood to flow at all, much less for its erythrocytes to squeeze through capillary blood vessels smaller in diameter than they are, erythrocyte membranes, with their membrane skeletons, must be fluidlike and easily deformable.

The protein **spectrin,** so-called because it was discovered in erythrocyte ghosts, accounts for ~75% of the erythrocyte membrane skeleton. It is composed of two similar polypeptide chains, band 1 (α subunit; 220 kD) and band 2 (β subunit; 240 kD), which sequence analysis indicates each consist of repeating 106-residue segments that are predicted to fold into triple-stranded α helical coiled coils (Fig. 11-33a). Electron microscopy indicates that these large polypeptides are loosely intertwined to form a flexible wormlike $\alpha\beta$ dimer that is

~1000 Å long (Fig. 11-33b). Two such heterodimers further associate in a head-to-head manner to form an $(\alpha\beta)_2$ tetramer. These tetramers, of which there are ~100,000/cell, are cross-linked at both ends by attachments to bands 4.1 and 5 to form an irregular protein meshwork that underlies the erythrocyte plasma membrane (Fig. 11-33b and c). Band 5, a globular protein that forms filamentous oligomers, has been identified as **actin,** a common cytoskeletal element in other cells (Section 1-2A) and a major component of muscle (Section 34-3A). Spectrin also associates with band 2.1, a 215-kD monomer known as **ankyrin,** which in turn binds to band 3, the anion channel protein. This attachment anchors the membrane skeleton to the membrane. Indeed, upon solubilization of spectrin and actin by low ionic strength solutions, erythrocyte ghosts lose their biconcave shape; their integral proteins that normally occupy fixed positions in the membrane plane, become laterally mobile. Immunochemical studies have recently revealed spectrinlike, ankyrinlike, and band 4.1-like proteins in a variety of tissues.

Hereditary Spherocytosis Arises from Erythrocyte Membrane Skeleton Defects

Individuals with **hereditary spherocytosis** have spheroidal erythrocytes that are relatively fragile and inflexible. These individuals suffer from hemolytic anemia because the spleen, a labyrinthine organ with narrow passages that normally filters out aged erythrocytes (which lose flexibility towards the end of their ~120-day lifetime), prematurely removes spherocytotic erythrocytes. The hemolytic anemia may be alleviated by the spleen's surgical removal. However, the primary defects in spherocytotic cells are reduced synthesis of spectrin, the production of an abnormal spectrin that binds band 4.1 protein with reduced affinity, or the absence of band 4.1 protein.

> The camel, the renowned "ship of the desert," provides a striking example of adaptation involving the erythrocyte membrane. This remarkable animal is still active after loss of 30% of its body weight in water and, upon doing so, can drink sufficient water in a few minutes to become fully rehydrated. Such a rapid rate of water uptake by the blood, which must deliver it to the cells, would lyse the erythrocytes of most animals. Yet, camel erythrocytes, which have the shape of flattened ellipsoids rather than biconcave disks, are resistant to osmotic lysis. Camel spectrin binds to its membrane with particular tenacity, but upon spectrin removal, which requires a strong denaturing agent such as guanidinium chloride, camel erythrocytes assume a spherical shape.

D. Blood Groups

The outer surfaces of erythrocytes and other eukaryotic cells are covered with complex carbohydrates that

(a)

α chain

β chain

N

C

C

N

C

(b)

Figure 11-33

The human erythrocyte membrane skeleton. (*a*) The structure of an $\alpha\beta$ dimer of spectrin. Both of these antiparallel polypeptides contain multiple 106-residue repeats, which are thought to form triple-helical bundles that are flexibly connected by nonhelical segments. Two of these heterodimers join, head to head, to form an $(\alpha\beta)_2$ heterotetramer. [After Speicher, D. W. and Marchesi, V., *Nature* **311,** 177 (1984).] (*b*) An electron micrograph of an erythrocyte membrane skeleton that has been stretched to an area 9 to 10 times greater than that of the native membrane. Stretching makes it possible to obtain clear images of the membrane skeleton which, in its native state, is so densely packed and irregularly flexed that it is difficult to pick out individual molecules and to ascertain how they are interconnected. Note the predominantly hexagonal and pentagonal network composed of spectrin tetramers cross-linked by junctions containing actin and band 4.1. [Courtesy of Daniel Branton, Harvard University.] (*c*) A model of the erythrocyte membrane skeleton. [After Goodman, S. R., Krebs, K. E., Whitfield, C. F., Riederer, B. M., and Zagen, I. S., *CRC Crit. Rev. Biochem.* **23,** 196 (1988).]

(c)

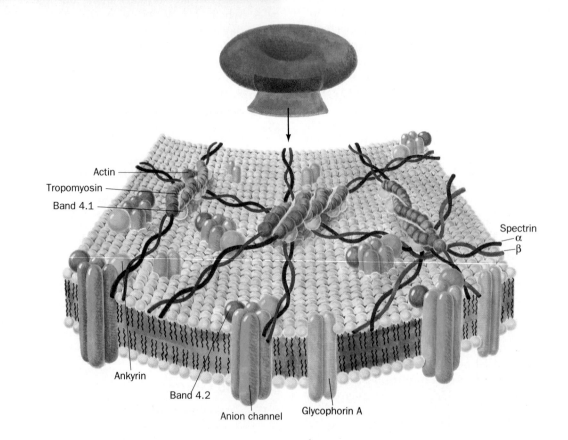

Actin

Tropomyosin

Band 4.1

Spectrin
α
β

Ankyrin

Band 4.2

Anion channel

Glycophorin A

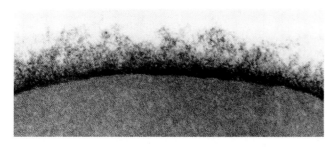

Figure 11-34
The erythrocyte glycocalyx as revealed by electron microscopy using special staining techniques. It is up to 1400 Å thick and composed of closely packed, 12 to 25 Å in diameter oligosaccharide filaments. [Courtesy of Harrison Latta, UCLA.]

are components of plasma membrane glycoproteins and glycolipids. They form a thick, fuzzy cell coating, the **glycocalyx** (Fig. 11-34), which contains numerous identity markers that function in various recognition processes. The human erythrocyte has some 100 known **blood group determinants** that comprise 15 genetically distinct blood group systems. Of these, only two—the **ABO blood group system** (discovered in 1900 by Karl Landsteiner) and the **rhesus (Rh) blood group system**—have major clinical importance. The various blood groups are identified by means of suitable antibodies or by specific plant lectins.

> Knowledge of blood group substances, and of their inheritance according to simple Mendelian laws, has been useful for legal and historical as well as medical purposes. The use of blood types in disproving paternity has even become the stuff of soap operas. Similarly, the analysis of tissue dust from the mummy of Tutankhamen, an Egyptian Pharaoh who reigned from 1334 to 1325 B.C., has indicated his probable relationship to Smenkhkare, another eighteenth dynasty Pharaoh.

ABO Blood Group Substances Are Carbohydrates

The ABO system consists of three blood group substances, the A, B, and H antigens, which are components of erythrocyte surface sphingoglycolipids. [Antigens are characteristic constellations of chemical groups that elicit the production of antibodies when injected into an animal (Section 34-2A). Each antibody molecule can specifically bind to at least two of its corresponding antigen molecules thereby cross-linking them.] Individuals with type A cells carry anti-B antibodies in their serum; those with type B cells carry anti-A antibodies; those with type AB cells, which bear both A and B antigens, carry neither anti-A nor anti-B antibodies; and type O individuals, whose cells bear neither antigen, carry both anti-A and anti-B antibodies. Consequently, the transfusion of type A blood into a type B individual, for example, agglutinates (clumps together) the trans-

fused erythrocytes resulting in an often fatal blockage of blood vessels. The H antigen is discussed below.

The ABO blood group substances are not confined to erythrocytes but also occur in the plasma membranes of many tissues as glycolipids of considerable diversity. In fact, in the ~ 80% of the population known as secretors, these antigens are secreted as *O*-linked components of glycoproteins into various body fluids including saliva, milk, seminal fluid, gastric juice, and urine. These diverse molecules, which are 85% carbohydrate by weight and have molecular masses ranging into thousands of kD, consist of multiple oligosaccharides attached to a polypeptide chain.

The A, B, and H antigens differ only in the sugar residues at their nonreducing ends (Table 11-5). The H antigen occurs in type O individuals; it is also the precursor oligosaccharide of A and B antigens. Type A individuals have an enzyme that specifically adds an *N*-acetylgalactosamine residue to the terminal position of the H antigen; in type B individuals, this enzyme is specific for a galactose residue; and in type O individuals, the enzyme is inactive. The functional differences between these enzyme variants are probably results of single amino acid changes.

MN Blood Groups Arise from Glycophorin A Variants

The **MN blood group system** constitutes another well-characterized set of human blood group determinants. Antigens of this system occur only in the erythrocyte membrane as part of the trans-membrane glycoprotein, glycophorin A (Fig. 11-22). This protein is also

Table 11-5

Structures of the A, B, and H Antigenic Determinants in Erythrocytes

Type	Antigen
H	Galβ(1 → 4)GlcNAc ⋯ ↑1,2 L-Fuc α
A	GalNAc α(1 → 3)Galβ(1 → 4)GlcNAc ⋯ ↑1,2 L-Fuc α
B	Galα(1 → 3)Galβ(1 → 4)GlcNAc ⋯ ↑1,2 L-Fuc α

Abbreviations: Gal = Galactose, GalNAc = *N*-acetylgalactosamine, GlcNAc = *N*-acetylglucosamine, L-Fuc = L-fucose.

the site of the influenza virus receptor (Section 32-4A), as well as a receptor for erythrocyte invasion by the malarial parasite *Plasmodium falciparum* (Section 6-3A). The PAS band 1 protein (Fig. 11-32) is a dimer of glycophorin A, which is formed through an SDS-resistant association between hydrophobic sections of the polypeptide chains; this dimer is presumably the protein's native form. The PAS band 2 protein is the monomeric form of glycophorin A.

Treatment of erythrocytes or glycophorin A with neuraminidase abolishes their reactivity to anti-M or anti-N antibodies as well as destroying their influenza virus receptor activity and reducing the invasion of *P. falciparum*. Thus, sialic acid (*N*-acetylneuraminic acid), which is cleaved by neuraminidase, forms part of the MN antigenic determinants. However, despite initial reports to the contrary, there are no differences between the oligosaccharides of M- and N-specific glycophorins A. Rather, these proteins differ in their amino acid sequence. Glycophorin A^M has a Ser at position 1 (the N-terminal residue) and a Gly at position 5, whereas these residues are Leu and Glu, respectively, in glycophorin A^N (Fig. 11-22). The erythrocytes of heterozygotes with both M and N antigenicity bear both these glycophorin A variants.

It has been suggested that glycophorin A's numerous negatively charged sialic acid residues prevent erythrocytes, which are closely packed in the blood stream, from adhering to one another. Yet, individuals who genetically lack glycophorin A suffer no apparent ill effects. It nevertheless seems unlikely that the million or so glycophorin A molecules per erythrocyte exist solely for the convenience of invading parasites.

E. Gap Junctions

Most eukaryotic cells are in metabolic as well as physical contact with neighboring cells. This contact is brought about by tubular particles, named **gap junctions,** that join discrete regions of neighboring plasma membranes much like hollow rivets (Fig. 11-35). Indeed, these intercellular channels are so widespread that many whole organs are continuous from within. Thus gap junctions are important intercellular communication channels. For example, the synchronized contraction of heart muscle is brought about by flows of ions through gap junctions (heart muscle is not innervated as is skeletal muscle). Likewise, gap junctions serve as conduits for some of the substances that mediate embryonic development; blocking gap junctions with antibodies that bind to them causes developmental abnormalities in species as diverse as hydra, frogs, and mice. Gap junctions also function to nourish cells that are distant from the blood supply, such as bone and lens cells.

Mammalian gap junction channels are 16 to 20 Å in diameter as Werner Loewenstein established by microinjecting single cells with fluorescent molecules of various sizes and observing with a fluorescence microscope whether the fluorescent probe passed into neighboring cells. The molecules and ions that can pass freely between neighboring cells are therefore limited in molecular mass to a maximum of ~1200 D; macromole-

Figure 11-35
Gap junctions between adjacent cells consist of two apposed plasma membrane-embedded hexagonal studs that bridge the gap between the cells. Small molecules and ions, but not macromolecules, can pass between cells via the gap junction's central channel.

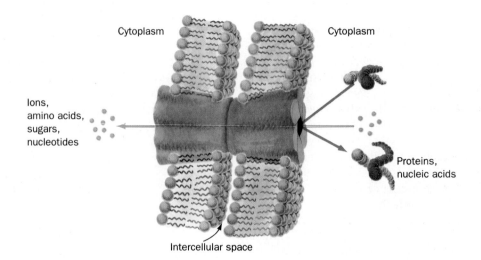

Cytoplasm

Cytoplasm

Ions, amino acids, sugars, nucleotides

Proteins, nucleic acids

Intercellular space

cules such as proteins and nucleic acids cannot leave a cell via this route.

The diameter of a gap junction channel varies with Ca^{2+} concentration: The channels are fully open when the Ca^{2+} level is $< 10^{-7}M$ and narrow as the Ca^{2+} concentration increases until, above $5 \times 10^{-5}M$, they close. This shutter system is thought to protect communities of interconnected cells from the otherwise catastrophic damage that would result from the death of any of their members. Cells generally maintain very low cystosolic Ca^{2+} concentrations ($< 10^{-7}M$) by actively pumping Ca^{2+} out of the cell as well as into their mitochondria and endoplasmic reticulum (Section 18-3B; Ca^{2+} is an important intracellular messenger whose cytosolic concentration is precisely regulated). Ca^{2+} floods back into leaky or metabolically depressed cells, thereby inducing closure of their gap junctions and sealing them off from their neighbors.

Gap Junction Channels Are Formed by Hexagons of Subunits

Purified gap junctions consist of a single type of $\sim$ 32-kD protein subunit. A single gap junction consists of two apposed hexagonal rings of these subunits, one from each of the adjoining plasma membranes (Fig. 11-35). Freeze-etch electron microscopy indicates that membrane-bound gap junctions form rafts of hexagonally packed doughnut-shaped particles of 80 to 90 Å in diameter (Fig. 11-36). This crystal-like packing permitted Unwin to determine their 18-Å resolution structure through electron microscopy in a manner similar to

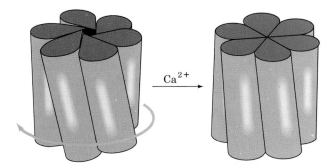

Figure 11-37
A model for the Ca^{2+}-induced closure of the gap junction central channel. [After Unwin, P. T. N. and Zampighi, G., *Nature* **283**, 549 (1980).]

that used with bacteriorhodopsin (Section 11-3A). Gap junction subunits are rods that are 25 Å in diameter and 75 Å long. In the absence of Ca^{2+}, they are inclined with respect to the sixfold axis so as to form a central channel that runs the length of the gap junction (Fig. 11-37, *left*). In 0.05 mM Ca^{2+}, however, the rods are nearly perpendicular to the plane of the junction and their central channel is closed at its cytoplasmic end (Fig. 11-37, *right*). Unwin has proposed that this closure is achieved by relatively slight tilting and twisting motions of the gap junction subunits centered at their bases, which, through the lever arms of their 75 Å lengths, results in large ($\sim$ 9 Å) radial displacements at their cytoplasmic ends. However, X-ray studies by Lee Makowski suggest that the cytoplasmic domains of gap junctions contain gates that close through more localized motions. The resolution of this discrepancy is the object of ongoing research.

F. Membrane Assembly and Protein Targeting

As cells grow and divide, they synthesize new membranes. How are such asymmetric membranes generated? One way in which this might occur is through self-assembly. Indeed, upon removal of the detergent used to disperse a biological membrane, liposomes form in which functional integral proteins are embedded. In most cases, however, these model membranes are symmetrical, both in their lipid distribution between the inner and outer leaflets of the bilayer and in the positions and orientations of their proteins. An alternative hypothesis of membrane assembly is that it occurs on the scaffolding of preexisting membranes; that is, membranes are generated by expansion of old ones rather than by creation of new ones. We shall see below that this is, in fact, the case.

Membrane Lipids Are Synthesized in Membranes

The enzymes involved in the biosynthesis of membrane lipids are mostly integral membrane proteins (Section 23-8).

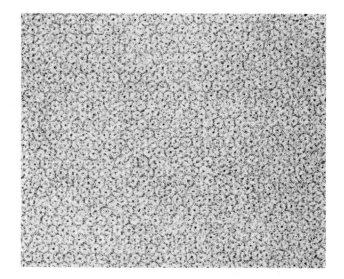

Figure 11-36
An electron micrograph of a gap junction-containing membrane. The gap junctions are arranged in a hexagonal lattice with a repeat distance of 80 to 85 Å. Note the densely stained central hole in each gap junction. [From Unwin, P. T. N. and Zampighi, G., *Nature* **283**, 546 (1980).]

Their substrates and products are themselves membrane components so that membrane lipids are fabricated on site. Eugene Kennedy and James Rothman demonstrated this to be the case in bacteria through the use of selective labeling. They gave growing bacteria a 1-min pulse of $^{32}PO_4^{3-}$ so as to label radioactively the phosphoryl groups of only the newly synthesized phospholipids. **Trinitrobenzenesulfonic acid (TNBS),** a membrane-impermeable reagent that combines with phosphatidyl-ethanolamine (**PE**; Fig. 11-38), was then immediately added to the cell suspension. Analysis of the resulting doubly labeled membrane showed that none of the TNBS-labeled PE was radioactively labeled. This observation indicates that *newly made PE is synthesized on the cytoplasmic face of the membrane (Fig. 11-39, lower left).*

If an interval of only 3 min is allowed to elapse between the $^{32}PO_4^{3-}$ pulse and the TNBS addition, about one half of the ^{32}P-labeled PE is also TNBS labeled (Fig. 11-39, *right*). This observation indicates that the flip-flop rate of PE in the bacterial membrane is $\sim 100,000$ faster than it is in bilayers consisting of only phospholipids (where, it will be recalled, the flip-flop rates have half-times of many days. *Evidently, membranes contain proteins that catalyze flip-flops (flipases). This is how lipids synthesized on one side of a membrane can reach the other side so quickly.* The observed asymmetric distribution of various lipid species between the two faces of a membrane (e.g., Fig. 11-31) is thought to be a consequence of preferential binding of certain lipids by the asymmetrically distributed membrane proteins.

In eukaryotic cells, lipids are synthesized on the cytoplasmic face of the **endoplasmic reticulum (ER),** interconnected membranous vesicles that occupy much of the cytosol (Fig. 1-5); from there, newly synthesized lipids are transported to other membranes. One mechanism of lipid transport is the budding off of membranous vesicles from the ER and their subsequent fusion with other membranes. This mechanism, by itself, does not explain the different lipid compositions of the various membranes in a cell. However, lipids may also be transported between membranes by **phospholipid exchange proteins** that occur in many tissues. These proteins spontaneously transfer specific phospholipids, one molecule at a time, between two membranes separated by an aqueous medium. A membrane's characteristic lipid composition may also be generated by on site remodeling and/or selective degradation of its component lipids (Section 23-8A).

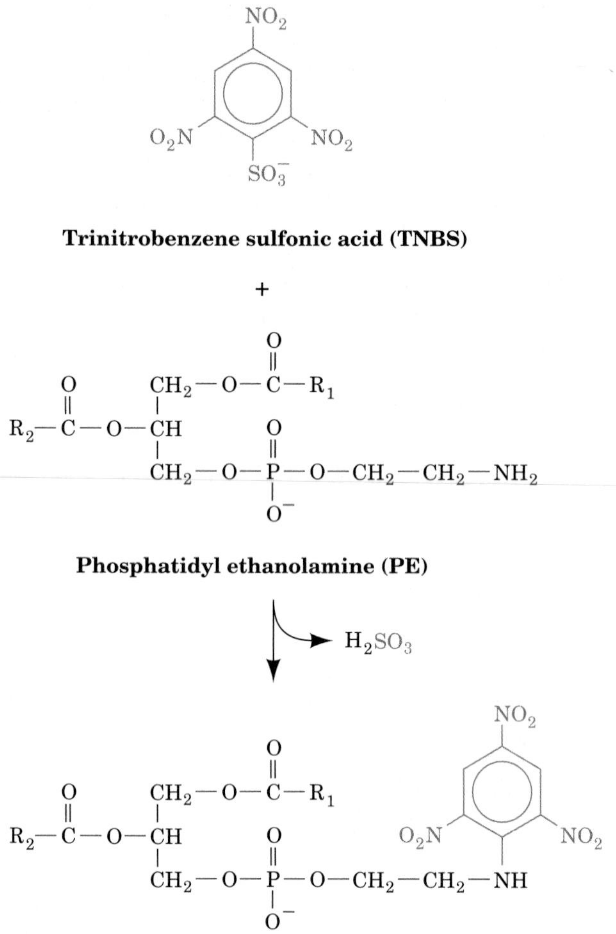

The Signal Hypothesis Accounts for the Targeting of Many Membrane Proteins

Membrane proteins, as are all proteins, are ribosomally synthesized under the direction of messenger RNA templates (a process known as **translation;** Chapter 30). The polypeptide grows from its N-terminus to its C-terminus by the stepwise addition of amino acid resi-

Figure 11-38
The reaction of TNBS with PE.

dues. Cytologists have long noted two classes of eukaryotic ribosomes, those free in the cytosol, and those bound to the ER so as to form the **rough endoplasmic reticulum (RER;** Fig. 1-5; so-called because of the knobby appearance its bound ribosomes give it). Both classes of ribosomes are, nevertheless, structurally identical; they differ only in the nature of the polypeptide they are synthesizing. *Free ribosomes synthesize mostly soluble and mitochondrial proteins, whereas membrane-bound ribosomes manufacture trans-membrane proteins and proteins destined for secretion, operation within the ER, or incorporation into* **lysosomes** (membranous vesicles containing a battery of hydrolytic enzymes that function to degrade and recycle cell components; Section 1-2A). These latter proteins initially appear in the RER.

How are RER-synthesized proteins differentiated from other proteins? And how do these large, relatively polar molecules pass through the RER membrane? The **signal hypothesis,** which was formulated by Günter Blobel, Cesar Milstein, and David Sabatini, partially explains how this occurs (Fig. 11-40):

1. *All secreted ER and lysosomal proteins, as well as many*

trans-membrane proteins, are synthesized with leading (N-terminal) 13 to 36-residue **signal peptides.** These signal peptides consist of a 7 to 13-residue hydrophobic core flanked by several relatively hydrophilic residues that usually include one or more basic residues near the N-terminus (Fig. 11-41). Signal peptides otherwise have little sequence similarity.

2. The signal peptide first protrudes beyond the ribosomal surface after ~80 residues have been linked together. At this point, the **signal recognition particle (SRP),** a 325-kD complex of six different polypeptides and a 300-nucleotide RNA molecule, bind to the ribosome. This arrests further polypeptide growth thereby preventing the RER-destined protein from being released in the cytosol.

3. The SRP–ribosome complex diffuses to the RER surface where it is bound by the **SRP receptor (docking protein),** a trans-membrane heterodimer of 69 and 30-kD subunits. This stimulates the bound ribosome to resume polypeptide synthesis and facilitates the passage of the growing polypeptide's N-terminus through the membrane into the **lumen** (enclosed space) of the RER. How polypeptides penetrate the membrane is discussed below.

4. Shortly after the signal peptide enters the RER lumen, it is specifically cleaved from the growing polypeptide by a membrane-bound **signal peptidase** (polypeptide chains with their signal peptide still attached are known as **preproteins;** signal peptides are alternatively called **presequences).**

5. Other enzymes in the lumen initiate **post-translational modification** of the still growing polypeptide such as by the specific attachments of "core" carbohydrates to form glycoproteins (Section 21-3B).

6. When polypeptide synthesis is completed, the ribosome dissociates from the RER. Secretory, ER, and lysosomal proteins pass completely through the RER membrane into the lumen. Trans-membrane proteins, in contrast, contain a hydrophobic ~20-residue "membrane anchor" or "stop-transfer" sequence that arrests the passage of the growing polypeptide chain through the membrane. *Trans-*

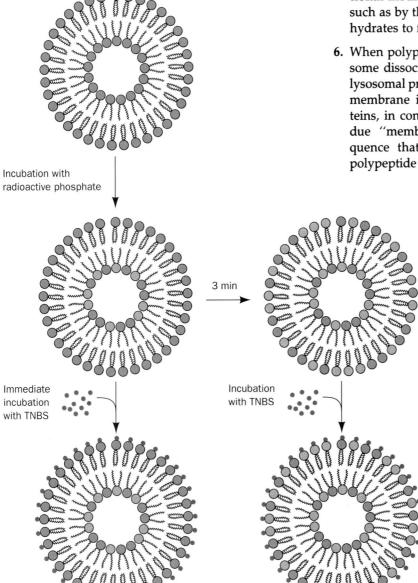

Incubation with
radioactive phosphate

3 min

Immediate
incubation
with TNBS

Incubation
with TNBS

Figure 11-39
The location of lipid synthesis in a bacterial membrane was determined by radioactively labeling newly synthesized PE by a 1-min pulse of $^{32}PO_4^{3-}$ (*orange head groups*) and by independently labeling the PE on the cell surface by treatment with the membrane-impermeable reagent TNBS. If TNBS labeling (*purple circles*) occurred immediately after the ^{32}P pulse, none of the ^{32}P-labeled PE was also TNBS labeled (*lower left*), thereby indicating that the PE is synthesized on the cytoplasmic face of the membrane. If, however, there was even a few minutes delay between the two labeling procedures, much of the TNBS-labeled PE in the external face of the membrane was also ^{32}P labeled (*right*).

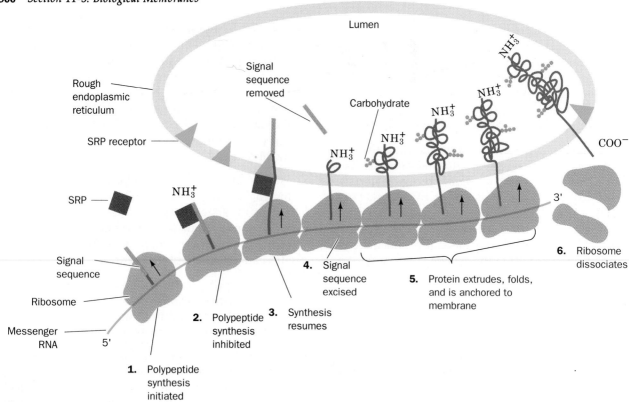

Figure 11-40
The ribosomal synthesis, membrane insertion and initial glycosylation of an integral membrane protein according to the signal hypothesis: **(1)** Protein synthesis is initiated at the N-terminus of the polypeptide, which consists of a 13 to 36 residue signal sequence. **(2)** A signal-recognition particle (SRP) binds to the ribosome as the signal sequence emerges from it thereby arresting polypeptide synthesis. **(3)** The SRP is bound by the trans-membrane SRP receptor, which causes resumption of polypeptide synthesis and facilitates the passage of the growing polypeptide through the membrane. **(4)** Shortly after the entrance of the signal sequence into the lumen of the endoplasmic reticulum, it is proteolytically excised. **(5)** As the growing polypeptide chain is extruded into the lumen, it starts to fold into its native conformation while enzymes initiate its specific glycosylation. Once the protein has folded, it cannot be pulled out of the membrane. At a point determined by its sequence, the protein becomes stuck in the membrane (proteins destined for secretion pass completely into the ER lumen). **(6)** Once polypeptide synthesis is completed, the ribosome dissociates. For the sake of clarity, the ribosome is shown at $\sim\frac{1}{50}$th of its actual size relative to the other cell components.

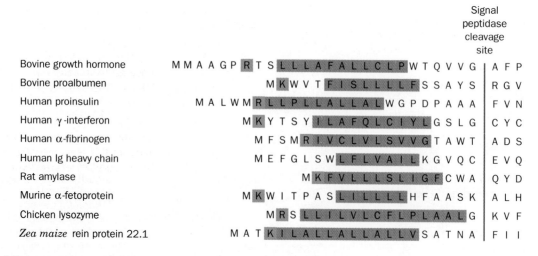

Figure 11-41
The N-terminal sequences of some eukaryotic secretory preproteins. The hydrophobic cores (*brown*) of most signal peptides are preceded by basic residues (*blue*). The one letter code for amino acid residues is given in Table 4-3. [After Watson, M. E. E., *Nucleic Acids Res.* **12,** 5147–5156 (1984).]

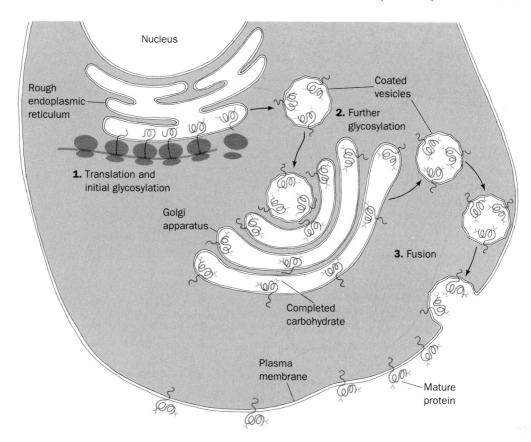

Figure 11-42
The post-translational processing of integral membrane proteins. **(1)** During their ribosomal synthesis, their glycosylation is initiated in the lumen of the endoplasmic reticulum. **(2)** After ribosomal synthesis is completed, coated vesicles containing the protein bud off from the endoplasmic reticulum and move to the Golgi apparatus where protein processing is completed.
(3) Later, coated vesicles containing the mature protein bud off from the Golgi apparatus and fuse to the membrane for which the protein is targeted, here shown as the plasma membrane.

membrane proteins therefore remain embedded in the ER membrane with their C-terminus on its cytoplasmic side.

Sometime after their polypeptide synthesis, the partially processed trans-membrane, secretory, and lysosomal proteins appear in the **Golgi apparatus** (Fig. 1-5), an organelle consisting of a stack of flattened membranous sacs where further post-translational processing occurs (Fig. 11-42; Section 21-3B). There, the proteins are also sorted and packaged into membranous vesicles for transport to their final destinations (see below). The export of soluble proteins that function in the ER is apparently prevented by their common C-terminal sequence, Lys-Asp-Glu-Leu; alteration of this sequence results in their secretion.

Bacterial Membrane Proteins Are Also Preceded by Signal Peptides

The signal hypothesis also applies to bacteria. Proteins that traverse the bacterial plasma membrane have leading signal peptides similar to those of eukaryotes and are synthesized by membrane-bound ribosomes. The importance of signal peptides in directing a protein to its cellular destination was demonstrated through the use of genetically engineered *E. coli.* **Maltose-binding protein,** which is involved in the uptake of the disaccharide maltose, is normally secreted across the plasma membrane into the **periplasmic compartment** (the space between the plasma membrane and the cell wall in gram-negative bacteria; Fig. 10-23*b*). Mutations that change even one hydrophobic residue of maltose-binding protein's signal peptide to a charged residue result in the cytoplasmic accumulation of this protein with the defective signal peptide still attached. Conversely, the use of recombinant DNA techniques (Section 28-8) to add a leading signal peptide to a normally cytoplasmic protein results in transport of the hybrid protein across the cell membrane. Similar manipulations have yielded analogous results in eukaryotes.

The Dispositions of Many Membrane Proteins Require Alternative Explanations

The signal hypothesis accounts for the orientations of numerous integral membrane proteins such as glycophorin *A* (Fig. 11-22) but does not indicate how such relatively polar molecules are able to penetrate biological membranes. Moreover, it does not, for example, explain the following observations:

1. **Influenza virus neuraminidase** (Section 32-4A) is a trans-membrane protein that is oriented with its N-terminus on the cytoplasmic side of the membrane.

2. Bacteriorhodopsin's polypeptide chain traverses the membrane 7 times (Fig. 11-26). Similarly, the erythrocyte anion channel (band 3) protein crosses the membrane at least 12 times and, moreover, is oriented with both its N- and C-termini inside the cell.

3. The integral membrane protein cytochrome b_5 (Fig. 11-24) is synthesized on soluble ribosomes.

4. Most mitochondrial proteins are synthesized in the cytosol and must therefore traverse one or both mitochondrial membranes (Section 1-2A) to reach their final destinations. Similarly, some cytosolically synthesized plant proteins must cross all three chloroplast membranes.

How are proteins inserted into or transported across a membrane? The failure to detect protein "pores" that could mediate the passage of polypeptides through membranes initially led to proposals that the hydrophobic segment of a protein's signal peptide, possibly in association with other nonpolar polypeptide segments, spontaneously inserts into a membrane as an α helix (a helix fully satisfies the main chain hydrogen bonding potential of a polypeptide in the absence of water). Recent observations, however, indicate that *the insertion of*

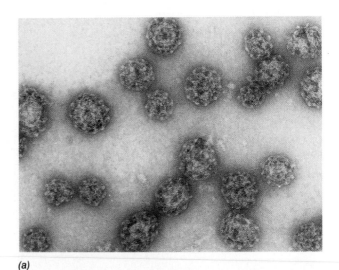

(a)

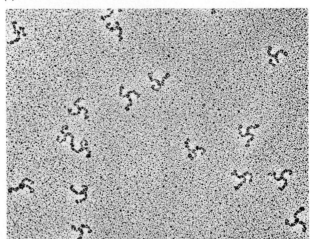

(b)

Figure 11-43
Coated vesicles are membranous sacs encased in polyhedral frameworks of clathrin and its associated proteins. Clathrin cages, which form the structural skeletons of coated vesicles, can be reversibly dissociated to flexible three-legged protein complexes known as **triskelions.** (*a*) An electron micrograph of coated vesicles. [Courtesy of Barbara Pearse, Medical Research Council, U.K.] (*b*) Electron micrograph of triskelions. The variable orientations of their legs is indicative of their flexibility. [Courtesy of Daniel Branton, Harvard University.] (*c*) A three-dimensional map of a clathrin coat generated from electron micrographs. The polyhedral clathrin coat is shown in red, the clathrin terminal domains are green, and an inner shell of accessory proteins is blue. Each vertex of the polyhedron is the center of a triskelion and its edges, which are ~150 Å in length, are formed by the overlapping legs of adjoining triskelions. Such frameworks, which consist of 12 pentagons and a variable number of hexagons (for geometric reasons explained in Section 32-2A), are the most economical way of enclosing spheroidal objects in polyhedral cages. The accessory proteins are thought to bind the membrane-spanning receptors for the specific proteins that the coated vesicle sequesters. [Courtesy of Barbara Pearse, Medical Research Council, U.K.]

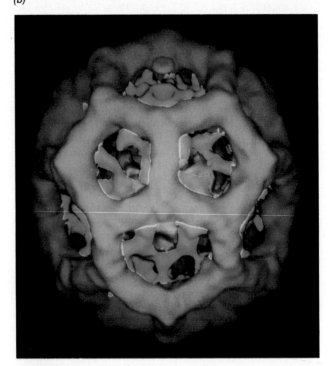

(c)

a protein into a membrane is facilitated by specific proteins in an ATP-driven process. In yeast, some of these proteins belong to a family of 70-kD **heat shock proteins (Hsp;** highly conserved proteins in all organisms whose synthesis, in many cases, is induced by environmental stress such as heat), which are therefore named **Hsp70.** Thus, the genetically engineered shutdown of Hsp70 production in yeast causes these cells to accumulate precursor forms of proteins that are otherwise imported into the ER or the mitochondria.

Evidence is accumulating that *only unfolded proteins can pass through a membrane.* For example, **dihydrofolate reductase (DHFR),** a normally cytosolic enzyme, is imported into yeast mitochondria when it is preceded by the signal peptide of a cytosolically synthesized mitochondrial protein. However, this importation process is arrested by the presence of **methotrexate,** an analog of DHFR's normal substrate, **dihydrofolate** (Section 26-4B). Methotrexate binds to DHFR with such high affinity that it stabilizes the protein's native conformation. Moreover, the rate of Hsp70-facilitated protein translocation across a membrane is further stimulated by the prior denaturation of the protein by urea. *Hsp70 may therefore be an ATP-driven "protein unfoldase."*

Membrane, Secretory, and Lysosomal Proteins Are Transported in Coated Vesicles

The vehicles in which proteins are transported between the RER, the Golgi apparatus, and their final destinations are **coated vesicles** (Fig. 11-43). These membranous sacs are encased on their outer (cytosolic) face by a polyhedral framework of the nonglycosylated protein **clathrin,** which is believed to act as a flexible scaffolding in promoting vesicle formation. A vesicle buds off from its membrane of origin and later fuses to its target membrane. *This process preserves the orientation of the trans-membrane protein (Fig. 11-44), so that the lumens of the ER and Golgi apparatus are topologically equivalent to the outside of the cell. This explains why the carbohydrate moieties of integral membrane glycoproteins only occur on the external surfaces of plasma membranes.*

Proteins Are Directed to the Lysosome by Carbohydrate Recognition Markers

How are proteins in the ER selected for transport to the Golgi apparatus and from there to their respective membranous destinations? A clue as to the nature of this process is provided by the human hereditary defect known as **I-cell disease** which, in homozygotes, is characterized by progressive psychomotor retardation, skeletal deformities, and early death. The lysosomes in the connective tissue of I-cell disease victims contain large inclusions (after which the disease is named) of glycosaminoglycans and glycolipids as a result of the absence of several lysosomal hydrolases. These enzymes are synthesized on the RER with their correct amino acid

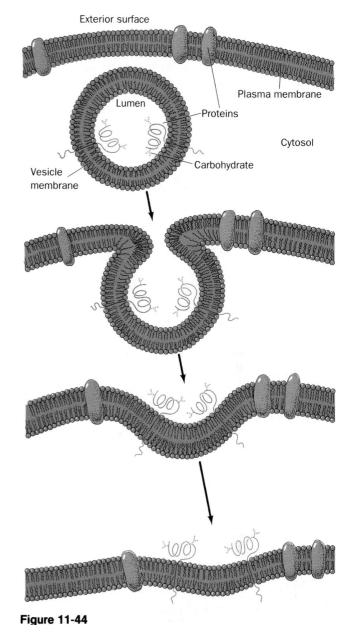

Figure 11-44
The fusion of a vesicle with the plasma membrane preserves the orientation of the integral proteins embedded in the vesicle bilayer. The inside of the vesicle and the exterior of the cell are topologically equivalent because the same side of the protein is always immersed in the cytosol. Note that any soluble proteins contained within the vesicle would be secreted. In fact, proteins destined for secretion are packaged in membranous **secretory vesicles** that eventually fuse with the plasma membrane as shown.

sequences but rather than being dispatched to the lysosomes, are secreted into the extracellular medium. This misdirection results from the absence of a mannose-6-phosphate recognition marker on the carbohydrate moieties of these hydrolases because of a deficiency in an enzyme required for mannose phosphorylation. The mannose-6-phosphate residues are bound by a receptor in the coated vesicles that transport lysosomal hydrolases from the Golgi apparatus to the lysosomes (Section 21-3B). No doubt, other glycoproteins are directed to their intracellular destinations by similar carbohydrate markers. In contrast, nuclear proteins, which have only to pass through the pores in the nuclear membrane (Section 1-2A), have a short, basic, internal sequence that directs their delivery to the nucleus and their retention within it, whereas proteins that normally reside in the lumen of the RER have, as we saw, the C-terminal sequence Lys-Asp-Glu-Leu.

4. LIPOPROTEINS

Proteins that are covalently associated with lipids form a relatively small but rapidly growing list. In most **lipid-linked proteins,** a fatty acid, phospholipid, or glycolipid residue is covalently attached to the protein, often at or near its N- or C-terminus. For example, myristate forms amide linkages with the α-amino groups of N-terminal Gly residues, whereas mainly palmitate forms thioester linkages with Cys residues of the C-terminal sequence CAAX (where C is Cys, A is an aliphatic residue, and X is any C-terminal residue). These lipid residues presumably function to anchor their attached protein to its associated membrane. However, this need for a hydrophobic group does not explain the functional requirements of some of the proteins for specific fatty acid residues.

Lipoproteins, particles that consist of noncovalently

associated lipids and proteins, are presently more familiar than lipid-linked proteins. *Lipoproteins function in the blood plasma as transport vehicles for triacylglycerols and cholesterol.* In this section, we discuss the structure, function, and dysfunction of this interrelated group of complex particles, and how eukaryotic cells take up lipoproteins and other specific proteins from their external medium through **receptor-mediated endocytosis.**

A. Lipoprotein Structure

Plasma lipoproteins form globular micellelike particles that consist of a nonpolar core of triacylglycerols and cholesteryl esters surrounded by an amphiphilic coating of protein, phospholipid, and cholesterol. They have been classified into five broad categories on the basis of their functional and physical properties (Table 11-6):

1. **Chylomicrons,** which transport exogenous (externally supplied; in this case, dietary) triacylglycerols and cholesterol from the intestines to the tissues.

2–4. **Very low density lipoproteins (VLDL), intermediate density lipoproteins (IDL),** and **low density lipoproteins (LDL),** a group of related particles that transport endogenous (internally supplied) triacylglycerols and cholesterol from the liver to the tissues (the liver synthesizes triacylglycerols from excess carbohydrates; Section 23-4).

5. **High density lipoproteins (HDL),** which trans-

Table 11-6

Characteristics of the Major Classes of Lipoproteins in Human Plasma

Lipoprotein Class	Major Lipids[a]	Apoproteins	Density (g·cm^{-3})	Particle Diameter (Å)
Chylomicrons and remnants	Dietary triacylglycerols	A-I, A-II, B-48, C-I, C-II, C-III, E	<0.95	800–5000
VLDL	Endogenous triacylglycerols, cholesteryl esters, cholesterol	B-100, C-I, C-II, C-III, E	0.95–1.006	300–800
IDL	Cholesteryl esters, cholesterol, triacylglycerols	B-100, C-III, E	1.006–1.019	250–350
LDL	Cholesteryl esters, cholesterol, triacylglycerols	B-100	1.019–1.063	180–280
HDL	Cholesteryl esters, cholesterol	A-I, A-II, C-I, C-II, C-III, D, E	1.063–1.210	50–120

[a] Given in order of abundance for substances comprising >5% of the lipid present.

Source: Brown, M. S. and Goldstein, J. L., *in* Braunwald, E., Isselbacher, K. J., Petersdorf, R. G., Wilson, J. D., Martin, J. B., and Fauci, A. S. (Eds.), *Harrison's Principles of Internal Medicine* (11th ed.), *p.* 1651, McGraw–Hill (1987).

port endogenous cholesterol from the tissues to the liver.

Lipoprotein particles undergo continuous metabolic processing so that they have somewhat variable properties and compositions (Table 11-6). Each contains just enough protein, phospholipid, and cholesterol to form an ~20-Å thick monolayer of these substances on the particle surface (Fig. 11-45). Lipoprotein densities increase with decreasing particle diameter because the density of their outer coating is greater than that of their inner core.

The protein components of lipoproteins are known as **apolipoproteins** or just **apoproteins.** They are soluble in water but, like water-soluble membrane proteins, have a tendency to aggregate in aqueous solution. There are at least nine apolipoproteins that are distributed in significant amounts in the different human lipoproteins (Table 11-6). Circular dichroism (CD) measurements indicate that *apolipoproteins have a high helix content, which increases when they are incorporated in lipoproteins.* Apparently the helices are stabilized by a lipid environment, presumably because helices best satisfy the polypeptide backbone's hydrogen bonding potential in a membrane's water-free interior.

Analysis of the 245-residue amino acid sequence of **apoA-I** reveals six similar 22-residue segments of high helix-forming propensity (Section 8-1C). *The sequences of these putative α helices, as well as such helices in the*

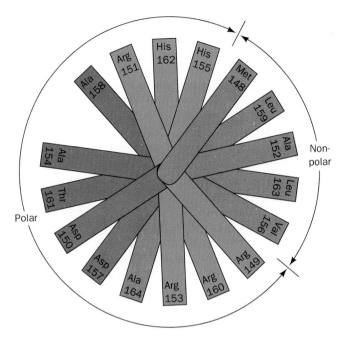

Figure 11-46
A helical wheel projection of the postulated amphipathic α helix constituting residues 148 to 164 of apolipoprotein A-I (in a helical wheel representation, the side chain positions are projected down the helix axis onto a plane). Note the segregation of nonpolar, acidic and basic residues to different sides of the helix. Other apolipoprotein helices have similar polarity distributions. [After Kaiser, E. T., *in* Oxender, D. L. and Fox, C. F. (Eds.), *Protein Engineering*, p. 194, Liss (1987).]

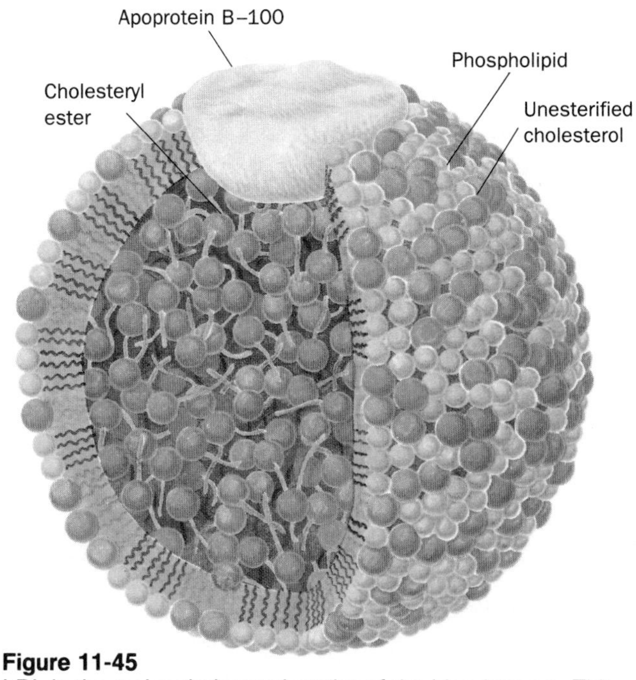

other apolipoproteins, have their hydrophobic and hydrophilic residues on opposite sides of helical cylinders (Fig. 11-46). Furthermore, the polar helix face has a zwitterionic character in that its negatively charged residues project from the center of this face, whereas its positively charged residues are located at its edges. Indeed, a synthetic 22-residue polypeptide of high helix-forming propensity, which was designed by E. Thomas Kaiser to have this polarity distribution but to otherwise have minimal similarity to the repeating apoA-I sequences, behaves much like apoA-I in binding to egg lecithin liposomes. Evidently, the structural role of apoA-I, and probably the other apolipoproteins, is fulfilled by its helical segments rather than by any organized tertiary structure. This suggests that *lipoprotein α helices float on phospholipid surfaces, much like logs on water.* The phospholipids are arrayed with their charged groups bound to oppositely charged residues on the polar face of the helix and with the first few methylene groups of their fatty acid residues in hydrophobic association with the nonpolar face of the helix.

B. Lipoprotein Function

The various lipoproteins have different physiological functions as we discuss below.

Figure 11-45
LDL is the major cholesterol carrier of the bloodstream. This spheroidal particle consists of some 1500 cholesteryl ester molecules surrounded by an amphiphilic coat of 800 phospholipid molecules, 500 cholesterol molecules, and at least one 550-kD molecule of apolipoprotein B-100. [After Brown, M. S. and Goldstein, J. L., *Sci. Am.* **251**(5): 60 (1984). Copyright © 1984 by Scientific American, Inc.]

Chylomicrons Are Delipidated in the Capillaries of Peripheral Tissues

Chylomicrons, which are assembled by the intestinal mucosa, function to keep exogenous triacylglycerols and cholesterol suspended in aqueous solution. These lipoproteins are released into the intestinal lymph (known as **chyle**), which is transported through the lymphatic vessels before draining into the large body veins via the thoracic duct. After a fatty meal, the otherwise clear chyle takes on a milky appearance.

Chylomicrons adhere to binding sites on the inner surface (endothelium) of the capillaries in skeletal muscle and adipose tissue. There, within minutes after entering the bloodstream, the chylomicron's component triacylglycerols are hydrolyzed through the action of **lipoprotein lipase,** an extracellular enzyme that is activated by **apoC-II.** The tissues then take up the liberated

monoacylglycerol and fatty acid hydrolysis products. The chylomicrons shrink as their triacylglycerols are progressively hydrolyzed until they are reduced to cholesterol-enriched **chylomicron remnants.** The chylomicron remnants reenter the circulation by dissociating from the capillary endothelium and are subsequently taken up by the liver as explained below. *Chylomicrons therefore function to deliver dietary triacylglycerols to muscle and adipose tissue and dietary cholesterol to the liver (Fig. 11-47, left).*

VLDL Are Degraded Much Like Chylomicrons

VLDL, which are synthesized in the liver as lipid transport vehicles, are also degraded by lipoprotein lipase (Fig. 11-47, *right*). The VLDL remnants appear in the circulation, first as IDL and then as LDL. In the transformation of VLDL to LDL, almost all their proteins

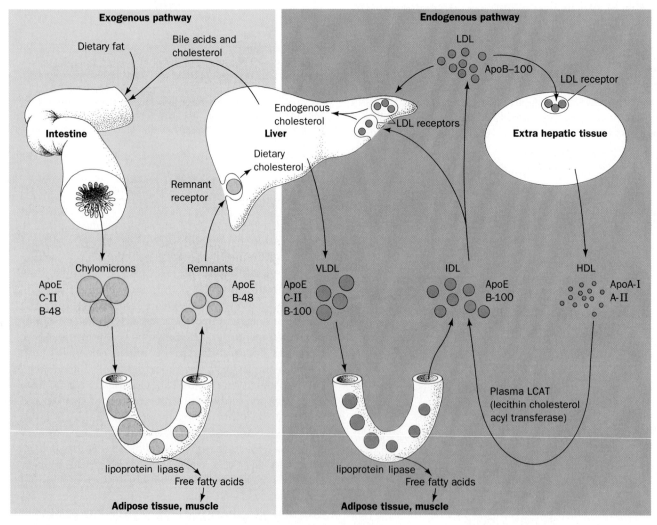

Figure 11-47
A model for plasma triacylglycerol and cholesterol transport in humans.
[After Brown, M. S. and Goldstein, J. L., *in* Brunwald, E., Isselbacher, K. J., Petersdorf, R. G., Wilson, J. D., Martin, J. B., and Fauci, A. S. (Eds.), *Harrison's Principles of Internal Medicine* (11th ed.), *p.* 1652, McGraw–Hill (1987).]

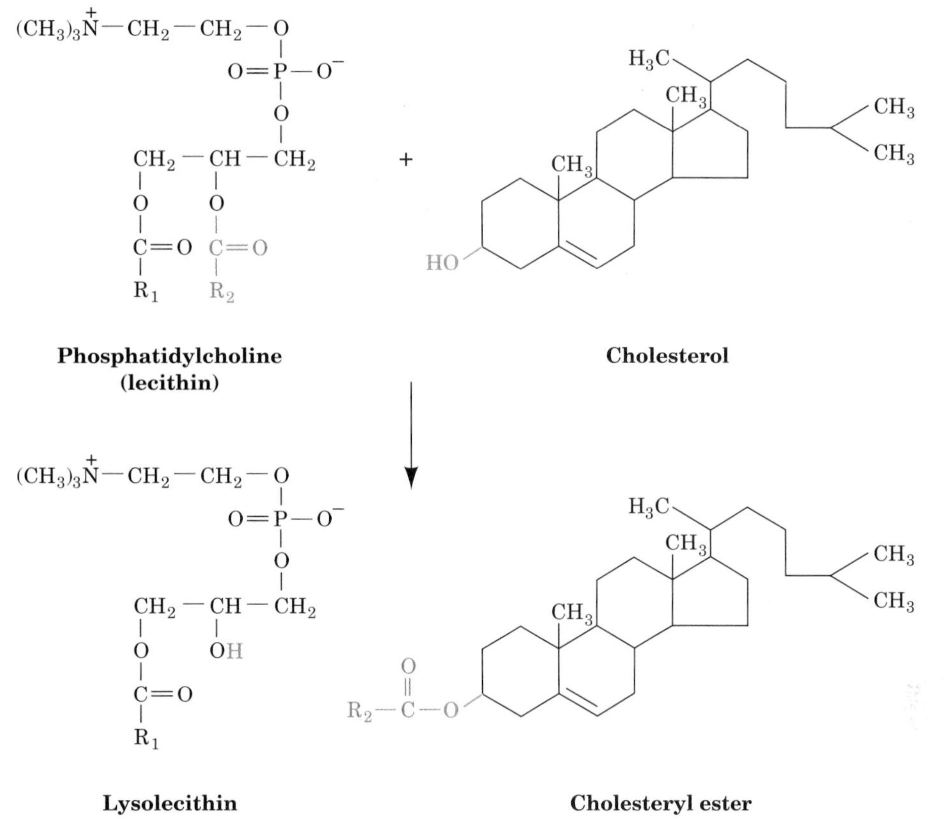

$(CH_3)_3\overset{+}{N}-CH_2-CH_2-O$

$O=P-O^-$

O

$CH_2-CH-CH_2$

$O \quad\quad O$

$C=O \quad C=O$

$R_1 \quad\quad R_2$

**Phosphatidylcholine
(lecithin)**

$+$

Cholesterol

$(CH_3)_3\overset{+}{N}-CH_2-CH_2-O$

$O=P-O^-$

O

$CH_2-CH-CH_2$

$O \quad\quad OH$

$C=O$

R_1

Lysolecithin

$R_2-\overset{O}{\overset{\|}{C}}-O$

Cholesteryl ester

Figure 11-48
The reaction catalyzed by lecithin–cholesterol acyl transferase (LCAT). The transferred acyl group is most often a linoleic acid residue.

but **apoB-100** are removed and much of their cholesterol is esterified by the HDL-associated enzyme **lecithin– cholesterol acyl transferase (LCAT)** as is discussed below. The enzyme transfers a fatty acid residue from the C(2) of lecithin to cholesterol with the concomitant formation of **lysolecithin** (Fig. 11-48).

Cells Take Up Cholesterol through Receptor-Mediated Endocytosis of LDL

Cholesterol, as we have seen, is an essential component of animal cell membranes. The cholesterol may be externally supplied or, if this source is insufficient, internally synthesized (Section 23-6A). Michael Brown and Joseph Goldstein have demonstrated that *cells obtain exogenous cholesterol mainly through the endocytosis (engulfment) of LDL particles in a process that occurs as follows:* The LDL is sequestered by **LDL receptor,** a cell-surface trans-membrane glycoprotein, which specifically binds both apoB-100 and **apoE.** LDL receptors cluster into **coated pits,** which serve to gather the cell-surface receptors that are destined for endocytosis while excluding other cell-surface proteins. The coated pits, which have a clathrin backing (Fig. 11-49), invaginate into the plasma membrane to form coated vesicles that

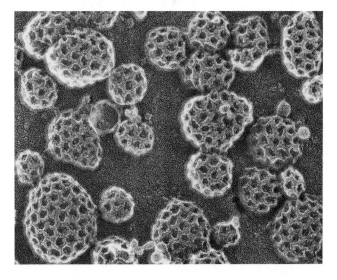

Figure 11-49
A freeze-etch electron micrograph of coated pits on the inner surface of a cultured fibroblast's plasma membrane. Compare this figure with that of coated vesicles (Fig. 11-43a). [Courtesy of John Heuser, Washington University School of Medicine.]

subsequently fuse with lysosomes (Fig. 11-50). *Such receptor-mediated endocytosis (Fig. 11-51) is a general mechanism whereby cells take up large molecules, each through a corresponding specific receptor.* Indeed, the liver takes up chylomicron remnants in this manner through the mediation of a separate **remnant receptor** that specifically binds apoE.

In the lysosome, as demonstrated by radioactive labeling studies, the LDL's apoB-100 is rapidly degraded to its component amino acids (Fig. 11-51). The cholesteryl esters are hydrolyzed by a lysosomal lipase to yield cholesterol, which is subsequently incorporated into the cell membranes. Any excess intracellular cholesterol is reesterified for storage within the cell through the action of **acyl-CoA : cholesterol acyltransferase (ACAT).**

The overaccumulation of cellular cholesteryl esters is prevented by two feedback mechanisms:

1. High intracellular levels of cholesterol suppress the synthesis of LDL receptor thus decreasing the rate of LDL accumulation by endocytosis (although LDL receptor cycles in and out of the cell about every 10 min, this glycoprotein is slowly degraded by the cell such that its half-life is ~ 20 h).

2. Excess intracellular cholesterol inhibits the biosynthesis of cholesterol (Section 23-6B).

HDL Transports Cholesterol from the Tissues to the Liver

HDL has essentially the opposite function of LDL: It removes cholesterol from the tissues. HDL is assembled in the plasma from components largely obtained through the degradation of other lipoproteins. *Circulating HDL probably acquires its cholesterol by extracting it from cell surface membranes and converts it to cholesteryl esters through the action of LCAT, an enzyme that is activated by the HDL component apoA-I. HDL therefore functions as a cholesterol scavenger.* Most evidence indicates that HDL transfers its cholesteryl esters to VLDL in a poorly understood process that is mediated by **apoD,** which is therefore also known as **cholesteryl ester transfer protein.** About one half of the VLDL, after its degradation to IDL and LDL, is taken up by the liver via LDL receptor-mediated endocytosis (Fig. 11-47, *right*). There are also indications that the liver directly takes up HDL through the agency of a specific **HDL receptor.** In any case, the liver is the only organ capable of disposing of significant quantities of cholesterol (as bile acids; Section 23-6C).

C. Lipoprotein Dysfunction in Atherosclerosis

Atherosclerosis, the most common form of **arteriosclerosis** (hardening of the arteries), is characterized by the presence of **atheromas** (Greek: *athere,* mush), arte-

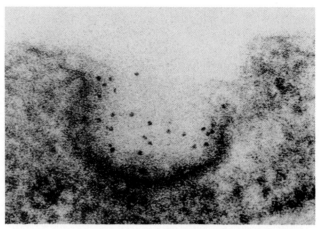

(a)

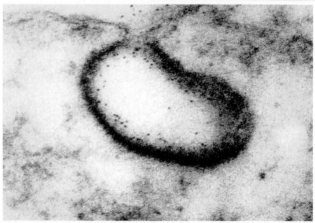

(b)

Figure 11-50
Electron micrographs showing the endocytosis of LDL by cultured human fibroblasts. The LDL was conjugated to ferritin so that it appears as dark dots. (*a*) LDL bound to a coated pit on the cell surface. (*b*) The coated pit invaginates and begins to pinch off from the cell membrane to form a coated vesicle enclosing the bound LDL. [From Anderson, R. G. W., Brown, M. S., and Goldstein, J. L., *Cell* **10,** 356 (1977). Copyright © 1977 by Cell Press.]

rial thickenings that, upon sectioning, exude a pasty yellow deposit of almost pure cholesteryl esters.

Atherosclerosis is a progressive disease that begins as intracellular lipid deposits in the smooth muscle cells of the inner arterial wall. These lesions eventually become fibrous, calcified plaques that narrow and even block the arteries. The resultant roughening of the arterial wall promotes the formation of blood clots, which may also occlude the artery. A blood flow stoppage, known as an **infarction,** causes the death of the deprived tissues. Although atheromas can occur in many different arteries, they are most common in the coronary arteries, the arteries supplying the heart. This results in **myocardial infarctions** or "heart attacks," the most common cause of death in Western man.

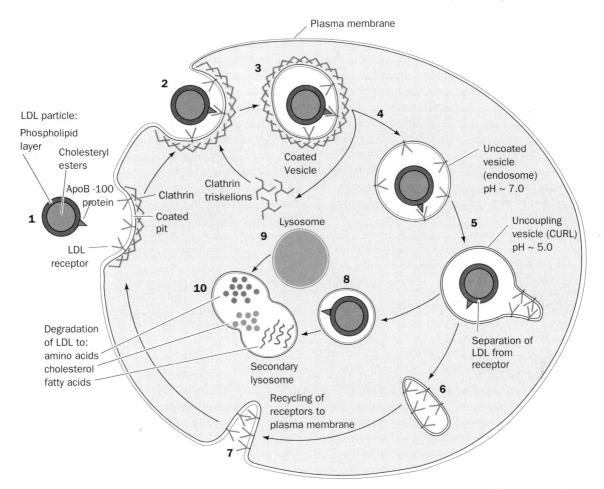

Figure 11-51
The sequence of events in the receptor-mediated endocytosis of LDL. The LDL specifically binds to LDL receptors on coated pits **(1)**. These bud into the cell **(2)** to form coated vesicles **(3)** whose clathrin coats depolymerize as triskelions resulting in the formation of smooth-surfaced vesicles known as **endosomes (4)**. The endosome then fuses with a vesicle called a **CURL (5**; compartment of *un*coupling of *r*eceptor and *l*igand), which has an internal pH of ~5.0. The acidity induces the LDL to dissociate from its receptor. The LDL accumulates in the vesicular portion of the CURL, whereas the LDL receptors concentrate in the membrane of an attached tubular structure, which then separates from the CURL **(6)** and subsequently recycles the LDL receptors to the plasma membrane **(7)**. The vesicular portion of the CURL **(8)** fuses with a lysosome **(9)** yielding a **secondary lysosome (10)** wherein the apoB-100 component of the LDL is degraded to its component amino acids and the cholesteryl esters are hydrolyzed to yield cholesterol and fatty acids.

Atherosclerosis Can Be Caused by Deficient LDL Receptors

The development of atherosclerosis is strongly correlated with the level of plasma cholesterol. This is particularly evident in individuals with **familial hypercholesterolemia.** Homozygotes with this inherited disorder have such high levels of the cholesterol-rich LDL in their plasma that their plasma cholesterol levels are three- to fivefold greater than the average level of ~175 mg/100 mL. This situation results in the deposition of cholesterol in their skin and tendons as yellow nodules known as **xanthomas.** However, far greater damage is caused by the rapid formation of atheromas that in homozygotes cause death from myocardial infarction as early as the age of 5. Heterozygotes, which comprise ~1 person in 500, are less severely afflicted; they develop symptoms of coronary artery disease after the age of 30.

Cells taken from homozygotes of familial hypercholesterolemia completely lack functional LDL receptors, whereas those taken from heterozygotes have about one half of the normal complement. Homozygotes and, to a lesser extent, heterozygotes, are therefore unable to utilize the cholesterol in LDL. Rather, their cells must synthesize most of the cholesterol for their needs. The high level of plasma LDL in these individuals results from two related causes:

1. Its decreased rate of degradation because of the lack of LDL receptors.

2. Its increased rate of synthesis from IDL due to the failure of LDL receptors to take up IDL.

The Incidence of Cardiovascular Disease Is Correlated with High LDL Levels and Low HDL Levels

Atheromas are thought to begin by an injury to the arterial lining through which plasma leaks into the arterial wall at a rate that exceeds the clearance capacity of normal receptor-mediated processes. The smooth muscle cells ingest this foreign material by a nonspecific engulfment process called **phagocytosis**. The lysosomal degradation of the ingested LDL results in an insoluble residue of cholesterol that accumulates in the cells as cholesteryl esters. High plasma levels of LDL, of course, accelerate this process.

If this model of atheroma formation is correct, then *the optimal level of plasma LDL is the lowest concentration that can adequately supply cholesterol to the cells.* Such a level, which is thought to be ~25 mg of cholesterol/100 mL, occurs in various mammalian species that are not naturally susceptible to atherosclerosis as well as in newborn humans. Yet, plasma LDL levels in adult Western man average ~7-fold higher than this supposed optimal level. The reason for these high plasma cholesterol levels is unknown although it is clear that they are affected by diet and by environmental stress. Medical strategies for reducing plasma cholesterol levels are considered in Section 23-6B.

Epidemiological studies indicate that high plasma HDL levels are strongly correlated with a low incidence of cardiovascular disease. Women have HDL levels higher than men and also less heart disease. Many of the factors that decrease the incidence of heart disease also tend to increase HDL levels. These include strenuous exercise, weight loss, certain drugs such as alcohol, and female sex hormones known as **estrogens** (Section 34-4A). Conversely, cigarette smoking is inversely related to HDL concentration. Curiously, in communities that have a very low incidence of coronary artery disease, both the mean HDL and LDL concentrations are low. The reasons for these various effects are unknown.

Chapter Summary

Fatty acids are long-chain carboxylic acids that may have one or more double bonds that are usually cis. Their anions are amphiphilic molecules that form micelles in water. Fatty acids rarely occur free in nature but rather are components of lipids. The most abundant class of lipids, the triacylglycerols or neutral fats, are nonpolar molecules that constitute the major nutritional store of animals. The lipids that occur in membranes are the phospholipids, the sphingolipids, and cholesterol. Sphingolipids such as cerebrosides and gangliosides have complex carbohydrate head groups that act as specific recognition markers in various biological processes.

The molecular shapes of membrane lipids cause them to aggregate in aqueous solution as bilayers. These form closed vesicles known as liposomes that are useful model membranes. Bilayers are essentially impermeable to polar molecules, except for water. Likewise, the flip-flop of a lipid in a bilayer is an extremely rare event. In contrast, bilayers above their transition temperatures behave as two-dimensional fluids in which the individual lipid molecules freely diffuse in the bilayer plane. Cholesterol increases membrane fluidity and broadens the temperature range of its order–disorder transition by interfering with the orderly packing of the lipids' fatty acid side chains.

Biological membranes contain a high proportion of proteins. Integral proteins have nonpolar surface regions that hydrophobically associate with the bilayer core. Peripheral proteins bind to integral proteins in the membrane surface by polar interactions. Specific integral proteins are invariably associated with a particular side of the membrane or, if they are trans-membrane proteins, have only one orientation. According to the fluid mosaic model of membrane structure, integral proteins resemble icebergs floating on a two-dimensional lipid sea. These proteins, as observed by the freeze-fracture and freeze-etch techniques, are randomly distributed in the membrane.

The erythrocyte membrane skeleton is responsible for the shape, flexibility, and fluidity of the red cell. Spectrin, the major constituent of the membrane skeleton, is a wormlike protein dimer that is cross-linked by actin oligomers and band 4.1 protein. The resulting protein meshwork is anchored to the membrane by the association of spectrin with ankyrin which, in turn, binds to band 3 protein, a trans-membrane protein that forms an anion channel.

The erythrocyte surface bears the various blood group antigens. The antigens of the ABO system differ in the sugar at a nonreducing end. The ABO blood group substances occur in the plasma membranes of many cells and in the secretions of many individuals. The M and N blood group antigens differ in the N-terminal sequence of the trans-membrane glycoprotein glycophorin A.

Gap junctions are hexagonal trans-membrane protein tubes that link adjoining cells. The gap junction's central channel, which closes at high intracellular levels of Ca^{2+}, allows small molecules and ions but not macromolecules to pass between cells.

New membranes are generated by the expansion of old ones. Lipids are synthesized by membrane-bound enzymes and are deposited on one side of the membrane. They migrate to the other side by flip-flops that are catalyzed by membrane proteins. In eukaryotes, lipids are transported between different membranes by lipid vesicles or by phospholipid-exchange proteins. Secretory proteins and many trans-membrane proteins are synthesized with a leading signal sequence that causes them to pass into the endoplasmic reticulum as they are extruded from the ribosome. Polypeptide chains are apparently inserted into membranes in unfolded form through an ATP-driven process that is mediated in part by the heat shock protein Hsp70. Once the protein has entered the endoplasmic reticulum, its signal sequence is removed and its glycosylation is initiated. The completed polypeptide is transported via coated vesicles to the Golgi apparatus where its glycosylation is completed. From there, the mature trans-membrane and secretory proteins are transported, again by coated vesicles, to their respective cellular destinations. Proteins may be targeted for their cellular destinations by glycosylation or by specific signal sequences.

Lipids are transported in the blood by plasma lipoproteins. These are essentially droplets of triacylglycerols and cholesteryl esters coated with a monolayer of phospholipids, cholesterol, and apolipoproteins. The amphiphilic apolipoprotein helices float on the lipoprotein surface in hydrophobic contact with its lipid interior. Chylomicrons and VLDL func-

tion to transport triacylglycerols and cholesterol; LDL and HDL carry mostly cholesterol. The triacylglycerols of chylomicrons and VLDL are degraded by lipoprotein lipase that lines the capillaries. LDL, the degradation product of VLDL, binds to cell surface LDL receptors and is taken into the cell by receptor-mediated endocytosis. The presence of excess intracellular cholesterol inhibits the synthesis of both LDL receptor and cholesterol. A major cause of atherosclerosis is an excess of plasma LDL. This disease, however, is also correlated with a low concentration of HDL, which functions as a cholesterol scavenger.

References

General

Bretscher, M., The molecules of the cell membrane, *Sci. Am.* **253**(4): 100–108 (1985).

Cantor, C. R. and Schimmel, P. R., *Biophysical Chemistry*, Chapters 4 and 25, Freeman (1980).

Finean, J. B., Coleman, R., and Michell, R. H., *Membranes and Their Cellular Functions* (3rd ed.), Blackwell (1984).

Harrison, R. and Lunt, G. G., *Biological Membranes, Their Structure and Function* (2nd ed.), Halsted (1980).

Jain, M. K., *Introduction to Biological Membranes* (2nd ed.), Wiley (1988).

Robertson, R. N., *The Lively Membranes*, Cambridge University Press (1983).

Tanford, C., *The Hydrophobic Effect: Formation of Micelles and Biological Membranes* (2nd ed.), Wiley–Interscience (1980). [An exposition of the thermodynamic properties of micelles and membranes.]

Lipids and Bilayers

Fishman, P. H. and Brady, R. O., Biosynthesis and function of gangliosides, *Science* **194**, 906–915 (1976).

Giles, C. H., Franklin's teaspoon of oil, *Chem. Ind.*, 1616–1624 (1969). [An historical account on Benjamin Franklin's investigations of the effect of oil on waves.]

Gurr, A. I. and James, A. T., *Lipid Biochemistry: An Introduction* (3rd ed.), Menthuen (1980).

Hakomori, S., Glycosphingolipids, *Sci. Am.* **254**(5): 44–53 (1986).

Kornberg, R. D. and McConnell, H. M., Inside–outside transitions of phospholipids in vesicle membranes, *Biochemistry* **10**, 1111–1120 (1971). [The first measurements of flip-flop rates.]

Kornberg, R. D. and McConnell, H. M., Lateral diffusion of phospholipids in a vesicle membrane, *Proc. Natl. Acad. Sci.* **68**, 2564–2568 (1971).

Ostro, M. J., Liposomes, *Sci. Am.* **256**(1): 102–111 (1987).

Ostro, M. J. (Ed.), *Liposomes*, Dekker (1983).

Rothman, J. E. and Davidowicz, E. A., Asymmetric exchange of vesicle phospholipids catalyzed by phosphatidylcholine exchange protein. Measurement of inside–outside transitions, *Biochemistry* **14**, 2809–2816 (1975).

Seelig, J. and Seelig, A., Lipid conformation in model membranes and biological membranes, *Q. Rev. Biophys.* **13**, 19–61 (1981).

Storch, J. and Kleinfeld, A. M., The lipid structure of biological membranes, *Trends Biochem. Sci.* **10**, 418–421 (1985).

Membranes and Membrane Proteins

Branton, D., Fracture faces of frozen membranes, *Proc. Natl. Acad. Sci.* **55**, 1048–1056 (1966).

Bretscher, M. S. and Raff, M. C., Mammalian plasma membranes, *Nature* **258**, 43–49 (1975).

Eisenberg, D., Three-dimensional structure of membrane and surface proteins, *Annu. Rev. Biochem.* **53**, 595–623 (1984).

Dawidowicz, E. A., Dynamics of membrane lipid metabolism and turnover, *Annu. Rev. Biochem.* **56**, 43–61 (1987).

Engelman, D. M., Henderson, R., McLachlan, A. D., and Wallace, B. A., Path of the polypeptide in bacteriorhodopsin, *Proc. Natl. Acad. Sci.* **77**, 2023–2027 (1980).

Frye, C. D. and Edidin, M., The rapid intermixing of cell surface antigens after formation of mouse–human heterokaryons, *J. Cell Sci.* **7**, 319–335 (1970).

Henderson, R. and Unwin, P. N. T., Three-dimensional model of purple membrane obtained by electron microscopy, *Nature* **257**, 28–32 (1975). [The low resolution structure of bacteriorhodopsin.]

Oseroff, A. R., Robbins, P. W., and Burger, M. M., The cell surface membrane: biochemical aspects and biophysical probes, *Annu. Rev. Biochem.* **42**, 647–682 (1973).

Singer, S. J. and Nicolson, G. L., The fluid mosaic model of the structure of cell membranes, *Science* **175**, 720–731 (1972). [A landmark paper of membrane structure.]

Unwin, N. and Henderson, R., The structure of proteins in biological membranes, *Sci. Am.* **250**(2): 78–94 (1984).

Webb, W. W., Luminescence measurements of macromolecular mobility, *Annu. NY Acad. Sci.* **366**, 300–314 (1981). [A discussion of the fluorescence photobleaching recovery technique.]

The Red Cell Membrane

Bennett, V., The membrane skeleton of human erythrocytes and its implications for more complex cells, *Annu. Rev. Biochem.* **54**, 273–304 (1985).

Branton, D., Cohen, C. M., and Tyler, J., Interaction of cytoskeletal proteins on the human erythrocyte membrane, *Cell* **24**, 24–32 (1981).

Elgsaeter, A., Stokke, B. T., Mikkelsen, A., and Branton, D., The molecular basis of erythrocyte shape, *Science* **234**, 1217–1223 (1986).

Gratzer, W. B., The red cell membrane and its cytoskeleton, *Biochem. J.* **198**, 1–8 (1981).

Jennings, M. L., Structure and function of the red blood cell anion transport protein, *Annu. Rev. Biophys. Biophys. Chem.* **18**, 397–430 (1989).

Marchesi, V. T., Stabilizing infrastructures of cell membranes, *Annu. Rev. Cell Biol.* **1**, 531–561 (1985).

Marchesi, V. T., Functional proteins of the human red cell membrane, *Semin. Hematol.* **16**, 3–20 (1981).

Rice-Evans, C. A. and Dunn, M. J., Erythrocyte deformability and disease, *Trends Biochem. Sci.* **7**, 282–286 (1982).

Blood Groups

Sharon, N., *Complex Carbohydrates*, Chapters 12 and 13, Addison–Wesley (1975). [A discussion of human blood group antigens.]

Vitala, J. and Järnefelt, J., The red cell surface revisited, *Trends Biochem. Sci.* **10**, 392–395 (1985).

Watkins, H. M., Biochemistry and genetics of the ABO, Lewis and P group systems, *Adv. Human Genet.* **10**, 1–136 (1980).

Gap Junctions

Bennett, M. V. L. and Spray, D. C. (Eds.), *Gap Junctions*, Cold Spring Harbor Laboratory (1985).

Makowsky, L., X-ray diffraction studies of gap junction structure, *Adv. Cell Biol.* **2**, 119–158 (1988).

Unwin, P. T. N. and Zampighi, G., Structure of the junction between communicating cells, *Nature* **283**, 545–549 (1980).

Unwin, P. T. N. and Ennis, P. D., Two configurations of a channel-forming membrane protein, *Nature* **307**, 609–613 (1984).

Membrane Assembly and Protein Targeting

Briggs, M. S. and Gierasch, L. M., Molecular mechanisms of protein secretion: the role of the signal sequence, *Adv. Protein Chem.* **38**, 109–180 (1986).

Brodsky, F. M., Living with clathrin: Its role in intracellular membrane traffic, *Science* **242**, 1396–1402 (1988).

Deshaies, R. J., Koch, B. D., Werner-Washburne, M., Craig, E. A., and Schekman, R., A subfamily of stress proteins facilitates translocation of secretory and mitochondrial precursor polypeptides, *Nature* **332**, 800–805 (1988), *and* Chirico, W. J., Waters, M. G., and Blobel, G., 70K heat shock related proteins stimulate protein translocation into microsomes, *Nature* **332**, 805–810 (1988).

Gierasch, L. M., Signal Sequences, *Biochemistry* **28**, 923–930 (1989).

Hasilik, A. and Neufield, E. F., Biosynthesis of lysosomal enzymes in fibroblasts, Phosphorylation of mannose residues, *J. Biol. Chem.* **255**, 4946–4950 (1980).

Kreil, G., Transfer of proteins across membranes, *Annu. Rev. Biochem.* **50**, 317–348 (1981).

Lodish, H. F. and Rothman, J. E., The assembly of membranes, *Sci. Am.* **240**(1): 48–63 (1979).

Op den Kamp, J. A. F., Lipid asymmetry in membranes, *Annu. Rev. Biochem.* **48**, 47–71 (1979).

Orci, L., Vassalli, J.-D., and Perrelet, A., The insulin factory, *Sci. Am.* **259**(3): 85–94 (1988). [A case study in how a cell produces and exports a secretory protein.]

Pearse, B. M. F. and Crowther, R. A., Structure and assembly of coated vesicles, *Annu. Rev. Biophys. Biophys. Chem.* **16**, 49–68 (1987).

Pearse, B. M. F. and Bretscher, M. S., Membrane recycling by coated vesicles, *Annu. Rev. Biochem.* **50**, 85–101 (1981).

Rose, J. K. and Doms, R. W., Regulation of protein export from the endoplasmic reticulum, *Annu. Rev. Cell Biol.* **4**, 257–288 (1988).

Rothman, J. E. and Fine, R. E., Coated vesicles transport newly synthesized membrane glycoproteins from endoplasmic reticulum to plasma membrane in two successive stages, *Proc. Natl. Acad. Sci.* **77**, 780–784 (1980).

Rothman, J. E. and Leonard, J., Membrane asymmetry, *Science* **195**, 743–753 (1977).

Schekman, R., Protein localization and membrane traffic in yeast, *Annu. Rev. Cell Biol.* **1**, 115–143 (1988).

Verner, K. and Schatz, G., Protein translocation across membranes, *Science* **241**, 1307–1313 (1988).

Walter, P., Gilmore, R., and Blobel, G., Protein translocation across the endoplasmic reticulum, *Cell* **38**, 5–8 (1984).

Walter, P. and Lingappa, V. R., Mechanism of protein translocation across the endoplasmic reticulum membrane, *Annu. Rev. Cell Biol.* **2**, 499–516 (1986).

Wickner, W. T. and Lodish, H. F., Multiple mechanisms of protein insertion into and across membranes, *Science* **230**, 400–407 (1985).

Lipoproteins

Breslow, J. L., Human apolipoprotein molecular biology and genetic variation, *Annu. Rev. Biochem.* **54**, 699–727 (1985).

Brown, M. S. and Goldstein, J. L., A receptor-mediated pathway for cholesterol homeostasis, *Science* **232**, 34–47 (1986). [A Nobel prize address.]

Brown, M. S. and Goldstein, J. L., How LDL receptors influence cholesterol and atherosclerosis, *Sci. Am.* **251**(5): 58–66 (1984).

Fukushima, D., Kupferberg, J. P., Yokoyama, S., Kroon, D. J., Kaiser, E. T., and Kédzy, F. J., A synthetic amphiphilic docosapeptide with the surface properties of plasma apolipoprotein A-I, *J. Am. Chem. Soc.* **101**, 3703–3704 (1979).

Goldstein, J. L., Brown, M. S., Anderson, R. G. W., Russell, D. W., and Schneider, W. J., Receptor-mediated endocytosis: concepts emerging from the LDL system, *Annu. Rev. Cell Biol.* **1**, 1–39 (1985).

Gotto, A. M., Jr. (Ed.), *Plasma Lipoproteins*, Elsevier (1984).

Mahley, R. W., Apolipoprotein E: cholesterol transport protein with expanding role in cell biology, *Science* **240**, 622–630 (1988).

Miller, G. J., High density lipoproteins and atherosclerosis, *Annu. Rev. Med.* **31,** 97–108 (1980).

Nilsson-Ehle, P., Garfinkel, A. S., and Schotz, M. C., Lipolytic enzymes and plasma lipoprotein metabolism, *Annu. Rev. Biochem.* **49,** 667–693 (1980).

Reichl, D. and Miller, N. E., The anatomy and physiology of reverse cholesterol transport, *Clin. Sci.* **70,** 221–231 (1986).

Scanu, A. M., Byrne, R. E., and Mihovilvic, M., Functional roles of plasma high density lipoproteins, *CRC Crit. Rev. Biochem.* **13,** 109–140 (1982).

Schlesinger, M. J., Proteolipids, *Annu. Rev. Biochem.* **50,** 193–206 (1981).

Schultz, A. M., Henderson, L. E., and Oroszlan, S., Fatty acylation of proteins, *Annu. Rev. Cell Biol.* **4,** 611–647 (1988).

Smith, L. C., Parnall, H. J., and Gotts, A. M., Jr., The plasma lipoproteins: structure and metabolism, *Annu. Rev. Biochem.* **47,** 751–777 (1978).

Stanbury, J. B., Wyngaarden, J. B., Fredrickson, D. S., Goldstein, J. L., and Brown, M. S. (Eds.), *The Metabolic Basis of Inherited Disease* (5th ed.), Part 4, McGraw–Hill (1983). [Authoritative discussions of diseases characterized by abnormal lipid metabolism.]

Problems

1. Explain the difference in melting points between *trans*-oleic acid (44.5°C) and *cis*-oleic acid (13.4°C).

2. Why do animals that live in cold climates generally have more polyunsaturated fatty acid residues in their fats than do animals that live in warm climates?

*3. How many different isomers of phosphatidylserine, triacylglycerol, and cardiolipin can be made from four types of fatty acids?

4. Estimate the thickness of the surface layer formed by Benjamin Franklin's teaspoon of oil on Clapham pond (1 teaspoon = 5 mL and 1 acre = 4047 m^2).

5. "Hard water" contains a relatively high concentration of Ca^{2+}. Explain why soap is ineffective for washing in hard water.

6. Explain why pure hydrocarbons do not form monolayers on water.

7. Soap bubbles are inside-out bilayers; that is, the polar head groups of the amphiphiles, together with some water, are in opposition, whereas their hydrophobic tails extend into the air. Explain the physical basis of this phenomenon.

8. Describe the action of detergents in extracting integral proteins from membranes. How do they keep these proteins from precipitating. Why do mild detergents such as Triton X-100 bind only to proteins that form lipid complexes?

*9. Is the trans-membrane portion of glycophorin A α helical (use the Chou and Fasman rules; Section 8-1C)?

10. The symmetries of oligomeric integral membrane proteins are constrained by the requirement that their subunits must all have the same orientation with respect to the plane of the membrane. What symmetries can these proteins have? Explain. (Protein symmetry is discussed in Section 7-5B.)

11. Explain why antibodies against type A blood group antigens are inhibited by *N*-acetylgalactosamine, whereas anti-B antibodies are inhibited by galactose.

12. Individuals with a certain one of the ABO blood types are said to be "universal donors," whereas those with another type are said to be "universal recipients." What are these blood types? Explain.

13. Anti-H antibodies are not normally found in human blood. They may, however, be elicited in animals by the injection of human blood. How would such antibodies be expected to react with tissues from individuals with type A, type B, and type O blood groups?

*14. In a genetically distinct form of familial hypercholesterolemia, LDL binds to the cell surface but fails to be internalized by endocytosis. Electron microscopy reveals that each mutant cell has its normal complement of coated pits but that ferritin-conjugated LDL does not bind to them. Rather, the bound LDL is uniformly distributed about the noncoated regions of the cell surfaces. Apparently the binding properties of the mutant LDL receptors are normal but they are in the wrong place. What does this data suggest about how the LDL receptor is assembled into coated pits?

15. Table 11-6 indicates that the densities of lipoproteins increase as their particle diameters decrease. Explain.

16. Certain types of animal viruses form by budding out from a cell surface much like coated pits bud into the cytoplasm during endocytosis to form coated vesicles. In both cases, the membranous vesicles form on a polyhedral protein scaffolding. Sketch the budding of an animal virus and indicate the location of its membrane relative to its protein shell.

17. **Wolman's disease** is a rapidly fatal homozygous defect characterized by a severe deficiency in **cholesteryl ester hydrolase,** the enzyme that catalyzes the hydrolysis of intracellularly located cholesteryl esters. Describe the microscopic appearance of the cells of victims of Wolman's disease.

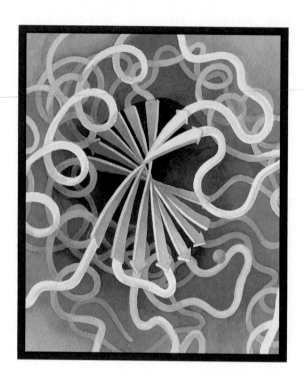

The digestive enzyme carboxypeptidase A showing its central twisted β sheet.

III

Mechanisms of Enzyme Action

Chapter 12
INTRODUCTION TO ENZYMES

The enormous variety of biochemical reactions that comprise life are nearly all mediated by a series of remarkable biological catalysts known as enzymes. Although enzymes are subject to the same laws of nature that govern the behavior of other substances, they differ from ordinary chemical catalysts in several important respects:

1. **Higher reaction rates:** The rates of enzymatically catalyzed reactions are typically factors of 10^6 to 10^{12} greater than those of the corresponding uncatalyzed reactions and are at least several orders of magnitude greater than those of the corresponding chemically catalyzed reactions.

2. **Milder reaction conditions:** Enzymatically catalyzed reactions occur under relatively mild conditions: temperatures below 100°C, atmospheric pressure, and nearly neutral pH's. In contrast, efficient chemical catalysis often requires elevated temperatures and pressures as well as extremes of pH.

3. **Greater reaction specificity:** Enzymes have a vastly greater degree of specificity with respect to both the identities of their **substrates** (reactants) and products than do chemical catalysts; that is, enzymatic reactions rarely have side products. For example, in the enzymatic synthesis of proteins on ribosomes (Section 30-3), polypeptides consisting of well over 1000 amino acid residues are made all but error free. Yet, in the chemical synthesis of polypeptides, side reactions

and incomplete reactions presently limit the lengths of polypeptides that can be accurately produced in reasonable yields to ~50 residues (Section 6-4B).

4. **Capacity for regulation:** The catalytic activities of many enzymes vary in response to the concentrations of substances other than their substrates. The mechanisms of these regulatory processes include allosteric control, covalent modification of enzymes, and variation of the amounts of enzymes synthesized.

Consideration of these remarkable catalytic properties of enzymes leads to one of the central questions of biochemistry: *How do enzymes work?* We address this issue in this part of the text.

In this chapter, following an historical review, we commence our study of enzymes with a discussion of two clear instances of enzyme action: one that illustrates how enzyme specificity is manifested, and a second that exemplifies the regulation of enzyme activity. These are, by no means, exhaustive treatments, but are intended to highlight these all-important aspects of enzyme mechanism. We shall encounter numerous other examples of these phenomena in our study of metabolism (Chapters 15–26). These two expositions are interspersed by a consideration of the role of enzymatic cofactors. The chapter ends with a short synopsis of enzyme nomenclature. In Chapter 13 we take up the formalism of enzyme kinetics because the study of the rates of enzymatically catalyzed reactions provides indispensable mechanistic information. Finally, Chapter 14 is a general discussion of the catalytic mechanisms employed by enzymes followed by an examination of the mechanisms of several specific enzymes.

1. HISTORICAL PERSPECTIVE

The early history of **enzymology,** the study of enzymes, is largely that of biochemistry itself; these disciplines evolved together from nineteenth century investigations of fermentation and digestion. Research on fermentation is widely considered to have begun in 1810 with Joseph Gay-Lussac's determination that ethanol and CO_2 are the principal products of sugar decomposition by yeast. In 1835, Jacob Berzelius, in the first general theory of chemical catalysis, pointed out that an extract of malt known as **diastase** (now known to contain the enzyme α**-amylase;** Section 10-2D) catalyzes the hydrolysis of starch more efficiently than does sulfuric acid. Yet, despite the ability of mineral acids to mimic the effect of diastase, it was the inability to reproduce most other biochemical reactions in the laboratory that led Louis Pasteur, in the mid-nineteenth century, to propose that the processes of fermentation could only occur in living cells. Thus, as was common in his era,

Pasteur assumed that living systems are endowed with a "vital force" that permits them to evade the laws of nature governing inanimate matter. Others, however, notably Justus Liebig, argued that biological processes are caused by the action of chemical substances that were then known as "ferments." Indeed, the name "enzyme" (Greek: *en*, in + *zyme*, yeast) was coined in 1878 by Fredrich Wilhelm Kühne in an effort to emphasize that there is something *in* yeast, as opposed to the yeast itself, that catalyzes the reactions of fermentation. Nevertheless, it was not until 1897 that Eduard Buchner obtained a cell-free yeast extract that could carry out the synthesis of ethanol from glucose (**alcoholic fermentation;** Section 16-3B).

Emil Fischer's discovery, in 1894, that glycolytic enzymes can distinguish between stereoisomeric sugars led to the formulation of his **lock-and-key hypothesis:** *The specificity of an enzyme (the lock) for its substrate (the key) arises from their geometrically complementary shapes.* Yet, the chemical composition of enzymes was not firmly established until well into the twentieth century. In 1926, James Sumner, who crystallized the first enzyme, jack bean **urease,** which catalyzes the hydrolysis of urea to NH_3 and CO_2, demonstrated that these crystals consist of protein. Since Sumner's preparations were somewhat impure, however, the protein nature of enzymes was not generally accepted until the mid-1930s when John Northrop and Moses Kunitz showed that there is a direct correlation between the enzymatic activities of crystalline pepsin, trypsin, and chymotrypsin, and the amounts of protein present. Enzymological experience since then has amply demonstrated that enzymes are proteins (although it has recently been shown that some species of RNA also have catalytic properties; Section 29-4B).

Although the subject of enzymology has a long history, most of our understanding of the nature and functions of enzymes is a product of the latter half of the twentieth century. Only with the advent of modern techniques for separation and analysis (Chapter 5) has the isolation and characterization of an enzyme become less than a monumental task. It was not until 1963 that the first amino acid sequence of an enzyme, that of **bovine pancreatic ribonuclease A** (Section 14-1A), was reported in its entirety, and not until 1965 that the first X-ray structure of an enzyme, that of hen egg white lysozyme (Section 14-2A), was elucidated. In the years since then, nearly 2000 enzymes have been purified and characterized to at least some extent and the pace of this endeavor is rapidly accelerating.

2. SUBSTRATE SPECIFICITY

The noncovalent forces through which substrates and other molecules bind to enzymes are identical in character

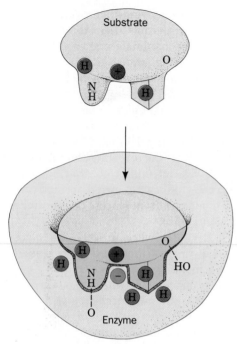

Figure 12-1
An enzyme–substrate complex illustrating both the geometrical and the physical complementarity between enzymes and substrates. Hydrophobic groups are represented by an H in a brown circle and dashed lines represent hydrogen bonds.

to the forces that dictate the conformations of the proteins themselves (Section 7-4): Both involve van der Waals, electrostatic, hydrogen bonding, and hydrophobic interactions. In general, a substrate-binding site consists of an indentation or cleft on the surface of an enzyme molecule that is complementary in shape to the substrate (geometrical complementarity). Moreover, the amino acid residues that form the binding site are arranged to interact specifically with the substrate in an attractive manner (electronic complementarity; Fig. 12-1). Molecules that differ in shape or functional group distribution from the substrate cannot productively bind to the enzyme; that is, they cannot form enzyme–substrate complexes that lead to the formation of products. The substrate-binding site may, in accordance with the lock-and-key hypothesis, exist in the absence of bound substrate or it may, as suggested by the induced fit hypothesis (Section 9-4C), form about the substrate as it binds to the enzyme. *X-ray studies indicate that the substrate-binding sites of most enzymes are largely preformed but that most of them exhibit at least some degree of induced fit upon binding substrate.*

A. Stereospecificity

Enzymes are highly specific both in binding chiral substrates and in catalyzing their reactions. This **stereospecificity** arises because enzymes, by virtue of their inherent chirality (proteins consist of only L-amino acids),

form asymmetric active sites. For example, trypsin readily hydrolyzes polypeptides composed of L-amino acids but not those consisting of D-amino acids. Likewise, the enzymes involved with glucose metabolism (Section 16-2) are specific for D-glucose residues.

Enzymes are absolutely stereospecific in the reactions they catalyze. This was strikingly demonstrated for the case of **yeast alcohol dehydrogenase (YADH)** by Frank Westheimer and Birgit Vennesland. Alcohol dehydrogenase catalyzes the interconversion of ethanol and acetaldehyde according to the reaction:

$$CH_3CH_2OH + NAD^+ \underset{}{\overset{YADH}{\rightleftharpoons}} \underset{}{\overset{O}{\overset{\|}{CH_3CH}}} + NADH + H^+$$

Ethanol   **Acetaldehyde**

The structures of NAD^+ and NADH are presented in Fig. 12-2. Ethanol, it will be recalled, is a prochiral molecule (see Section 4-2C for a discussion of prochirality):

$$H_{pro\text{-}S} - \underset{\underset{CH_3}{|}}{\overset{\overset{OH}{|}}{C}} - H_{pro\text{-}R}$$

Ethanol's two methylene H atoms may be distinguished if *the molecule is held in some sort of asymmetric jig (Fig. 12-3). The substrate-binding sites of enzymes are, of course, just such jigs since they immobilize the reacting groups of the substrate on the enzyme surface.*

Westheimer and Vennesland elucidated the stereospecific nature of the YADH reaction through the following series of experiments:

1. If the YADH reaction is carried out with deuterated ethanol, the product NADH is deuterated:

$$CH_3CD_2OH +$$

NAD⁺

$$CH_3CD \overset{O}{\overset{\|}{}} +$$

NADD $+ H^+$

Note that the nicotinamide ring of NAD^+ is also prochiral.

2. Upon isolating this NADD and using it in the reverse reaction to reduce normal acetaldehyde, the deute-

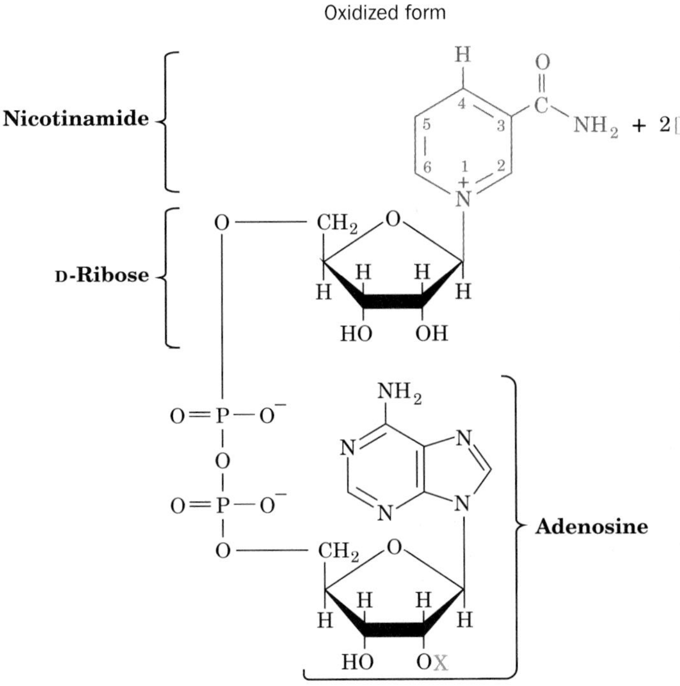

Oxidized form   Reduced form

Nicotinamide

D-Ribose

$$X = H \quad \text{Nicotinamide adenine dinucleotide (NAD}^+\text{)}$$
$$X = PO_3^{2-} \quad \text{Nicotinamide adenine dinucleotide phosphate (NADP}^+\text{)}$$

Figure 12-2
The structures and reactions of **nicotinamide adenine dinucleotide (NAD⁺)** and **nicotinamide adenine dinucleotide phosphate (NADP⁺)**. Their reduced forms are **NADH** and **NADPH**. [In the older literature they are termed **diphosphopyridine nucleotide (DPN⁺)** and **triphosphopyridine nucleotide (TPN⁺)** and their reduced forms are symbolized **DPNH** and **TPNH**.] These substances, which are collectively referred to as the **nicotinamide coenzymes** or **pyridine nucleotides** (nicotinamide is a pyridine derivative) function, as is indicated in later chapters, as intracellular carriers of reducing equivalents (electrons). Note that only the nicotinamide ring is changed in the reaction. Reduction formally involves the transfer of two hydrogen atoms (H·) although the actual reduction may occur via a different mechanism.

rium is quantitatively transferred from the NADD to the product ethanol:

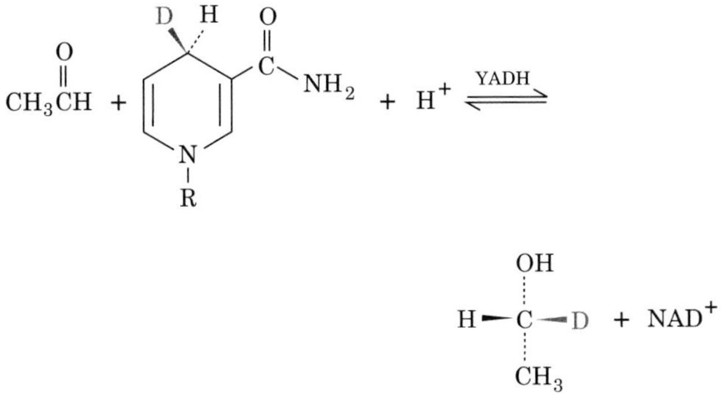

$$CH_3CH + \text{(ring)} + H^+ \xrightleftharpoons{YADH}$$

$$H-\underset{CH_3}{\overset{OH}{C}}-D + NAD^+$$

3. If the enantiomer of the foregoing CH_3CHDOH is made as follows:

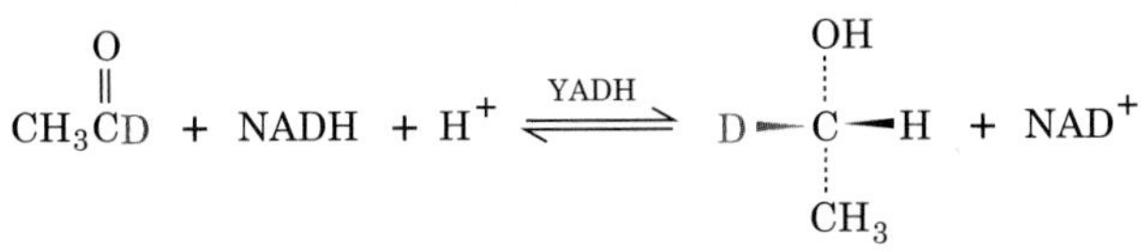

$$CH_3CD + NADH + H^+ \xrightleftharpoons{YADH} D-\underset{CH_3}{\overset{OH}{C}}-H + NAD^+$$

none of the deuterium is transferred from the product ethanol to NAD⁺ in the reverse reaction.

4. If, however, this ethanol is converted to its tosylate and then inverted by S_N2 hydrolysis to yield the

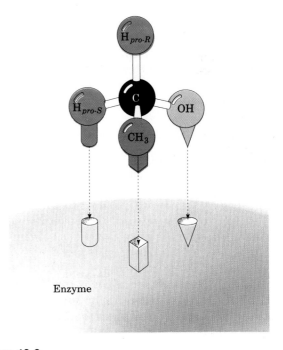

Figure 12-3
The specific attachment of a prochiral center to an enzyme-binding site permits the enzyme to differentiate between prochiral groups.

enantiomorphic ethanol,

**p-Toluenesulfonyl
chloride
(tosyl chloride)**

the deuterium is again quantitatively transferred to NAD^+ in the YADH reaction.

The foregoing data, in addition to showing that there is direct hydrogen atom transfer in the YADH reaction (Experiments 1 and 2), indicates that the enzyme distinguishes between the *pro-S* and *pro-R* hydrogen atoms of ethanol as well as the *si* and *re* faces of the nicotinamide ring of NAD^+ (Experiments 2–4). It was later demonstrated, by stereospecific syntheses, that YADH transfers the *pro-R* hydrogen atom of ethanol to the *re* face of the nicotinamide ring of NAD^+ as is drawn in the preceding diagrams.

The stereospecificity of YADH is by no means unusual. As we consider biochemical reactions we shall find that nearly all enzymes that participate in chiral reactions are absolutely stereospecific.

B. Geometric Specificity

The stereospecificity of enzymes is not particularly surprising in light of the complementarity of an enzymatic binding site for its substrate. A substrate of the wrong chirality will not fit into an enzymatic binding site for much the same reasons that you cannot fit your right hand into your left glove. *In addition to their stereospecificity, however, most enzymes are quite selective about the identities of the chemical groups on their substrates.* Indeed, such **geometric specificity** is a more stringent requirement than is stereospecificity. After all, your left glove will more or less fit left hands that have somewhat different sizes and shapes than your own.

Enzymes vary considerably in their degree of geometric specificity. A few enzymes are absolutely specific for only one compound. Most enzymes, however, catalyze the reactions of a small range of related compounds. For example, YADH catalyzes the oxidation of small primary and secondary alcohols to their corresponding aldehydes or ketones but none so efficiently as that of ethanol. Even methanol or isopropanol, which differ from ethanol only by the deletion or addition of a CH_2 group, are oxidized by YADH at rates that are, respectively, 25-fold and 2.5-fold slower than that for ethanol. Similarly, $NADP^+$, which differs from NAD^+ only by the addition of a phosphoryl group at the 2′ position of its adenosine ribose group (Fig. 12-2), does not bind to YADH. On the other hand, there are many enzymes that bind $NADP^+$ but not NAD^+.

Some enzymes, particularly digestive enzymes, are so permissive in their ranges of acceptable substrates that their geometric specificities are more accurately described as preferences. Carboxypeptidase A, for example, catalyzes the hydrolysis of C-terminal peptide bonds to all residues except Arg, Lys, and Pro if the preceding residue is not Pro (Table 6-1). However, the rate of this enzymatic reaction varies with the identities of the residues in the vicinity of the C-terminus of the polypeptide (see Fig. 6-5). Some enzymes are not even very specific in the type of reaction they catalyze. Thus chymotrypsin, in addition to its ability to mediate peptide bond hydrolysis, also catalyzes ester bond hydrolysis.

Moreover, the acyl group acceptor in the chymotrypsin reaction need not be water; amino acids, alcohols, or ammonia can also act in this capacity. It should be realized, however, that such permissiveness is much more the exception than the rule. Indeed most intracellular enzymes function *in vivo* (in the cell) to catalyze a particular reaction on a specific substrate.

3. COENZYMES

Enzymes catalyze a wide variety of chemical reactions. Their functional groups can facilely participate in acid–base reactions, form certain types of transient covalent bonds, and take part in charge–charge interactions (Section 14-1). They are, however, less suitable for catalyzing oxidation–reduction reactions and many types of group-transfer processes. Although enzymes catalyze such reactions, they can only do so in association with small molecule **cofactors,** which essentially act as the enzymes' "chemical teeth."

Cofactors may be metal ions, such as the Zn^{2+} required for the catalytic activity of carboxypeptidase A, or organic molecules known as **coenzymes** such as the NAD^+ in YADH (Section 12-2A). Some cofactors, for instance NAD^+, are but transiently associated with a given enzyme molecule so that, in effect, they function as cosubstrates. Other cofactors, known as **prosthetic groups,** are essentially permanently associated with their protein, often by covalent bonds. For example, the heme prosthetic group of hemoglobin is tightly bound to its protein through extensive hydrophobic and hydrogen bonding interactions together with a covalent bond between the heme Fe^{2+} ion and His F8 (Sections 9-1A and 2B).

Coenzymes are chemically changed by the enzymatic reactions in which they participate. Thus, in order to complete the catalytic cycle, the coenzyme must be returned to its original state. For prosthetic groups, this can only occur in a separate phase of the enzymatic reaction sequence. For transiently bound coenzymes, such as NAD^+, however, the regeneration reaction may be catalyzed by a different enzyme.

A catalytically active enzyme–cofactor complex is called a **holoenzyme.** The enzymatically inactive protein resulting from the removal of a holoenzyme's cofactor is referred to as an **apoenzyme;** that is,

apoenzyme (*inactive*) + cofactor
$$\rightleftharpoons \text{holoenzyme (active)}$$

Table 12-1 lists the most common coenzymes together with the types of reactions in which they participate. We shall describe the structures of these substances and their reaction mechanisms in the appropriate sections of the text.

Many Vitamins Are Coenzyme Precursors

Many organisms are unable to synthesize certain portions of essential cofactors and therefore these substances must be present in the organism's diet; thus they are **vitamins.** In fact, many coenzymes were discovered as growth factors for microorganisms or substances that cure nutritional deficiency diseases in humans and animals. For example, the NAD^+ component **nicotinamide** (alternatively known as **niacinamide**) or its carboxylic acid analog **nicotinic acid (niacin;** Fig. 12-4), relieve the dietary deficiency disease in humans known as **pellagra.** Pellagra, which is characterized by diarrhea, dermatitis, and dementia, was endemic in the rural Southern United States in the early twentieth century. Most animals, including humans, can synthesize nicotinamide from the amino acid tryptophan (Section 26-6A). The corn-rich diet that was prevalent in the rural South, however, contained little available nicotinamide or tryptophan from which to synthesize it. [Corn actually contains significant quantities of nicotinamide but in a form that requires treatment with base before it can be intestinally absorbed. The Mexican Indians, who are thought to have domesticated the corn plant, customarily soak corn meal in lime water (dilute $Ca(OH)_2$ solution) before using it to bake their staple food, tortillas.]

The vitamins in the human diet that are coenzyme precursors are all **water-soluble vitamins** (Table 12-2). In contrast, the **fat-soluble vitamins,** such as vitamins A and D, are not components of coenzymes, although they are also required in trace amounts in the diets of many higher animals. Man's distant ancestors probably had the ability to synthesize the various vitamins, as do many modern plants and microorganisms. Yet, since vitamins are normally available in the diets of higher animals, which all eat other organisms, or are synthesized by the bacteria that normally inhabit their digestive systems, it is believed that the then superfluous cellular machinery to synthesize them was lost through evolution.

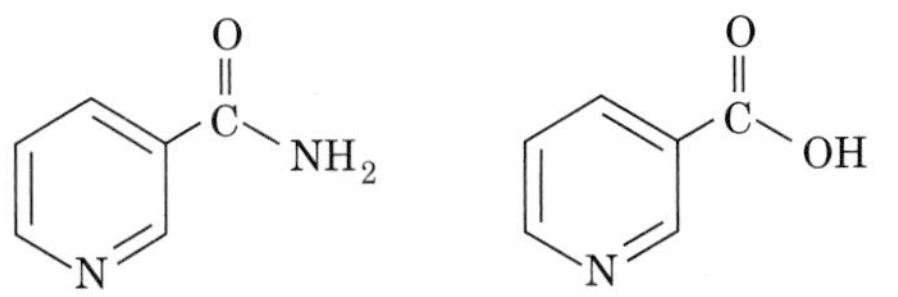

Table 12-1
The Common Coenzymes

Coenzyme	Reaction Mediated	Section Discussed
Biotin	Carboxylation	21-1A
Cobamide (B_{12}) coenzymes	Alkylation	23-2E
Coenzyme A	Acyl transfer	19-2A
Flavin coenzymes	Oxidation–reduction	14-4
Lipoic acid	Acyl transfer	19-2A
Nicotinamide coenzymes	Oxidation–reduction	12-2A
Pyridoxal phosphate	Amino group transfer	24-1A
Tetrahydrofolate	One-carbon group transfer	24-4D
Thiamine pyrophosphate	Aldehyde transfer	16-3B

Figure 12-4
The structures of nicotinamide and nicotinic acid. These vitamins form the redox-active components of the nicotinamide coenzymes NAD^+ and $NADP^+$ (compare with Fig. 12-2).

Table 12-2
Vitamins that Are Coenzyme Precursors

Vitamin	Coenzyme	Human Deficiency Disease
Biotin	Biocytin	a
Cobalamin (B₁₂)	Cobamide (B₁₂) coenzymes	Pernicious anemia
Folic acid	Tetrahydrofolate	Megaloblastic anemia
Nicotinamide	Nicotinamide coenzymes	Pellagra
Pantothenate	Coenzyme A	a
Pyridoxine (B₆)	Pyridoxal phosphate	a
Riboflavin (B₂)	Flavin coenzymes	a
Thiamine (B₁)	Thiamine pyrophosphate	Beriberi

a No specific name; deficiency in humans is rare or unobserved.

4. REGULATION OF ENZYMATIC ACTIVITY

An organism must be able to regulate the catalytic activities of its component enzymes so that it can coordinate its numerous metabolic processes, respond to changes in its environment, and grow and differentiate, all in an orderly manner. There are two ways that this might occur:

1. *Control of enzyme availability: The amount of a given enzyme in a cell depends on both its rate of synthesis and its rate of degradation.* Each of these rates are directly controlled by the cell. For example, *E. coli* grown in the absence of the disaccharide lactose (Fig. 10-12) lack the enzymes to metabolize this sugar. Within minutes of their exposure to lactose, however, these bacteria commence synthesizing the enzymes required to utilize this nutrient (Section 29-1A). Similarly, the various tissues of a higher organism contain different sets of enzymes although most of its cells contain identical genetic information. How cells achieve this control of enzyme synthesis is a major subject of Part IV of this text. The degradation of proteins is discussed in Section 30-6.

2. *Control of enzyme activity: An enzyme's catalytic activity may be directly regulated through conformational or structural alterations.* The rate of an enzymatically catalyzed reaction is directly proportional to the concentration of its enzyme–substrate complex which, in turn, varies with the substrate concentration and the enzyme's substrate-binding affinity (Section 13-2A). The catalytic activity of an enzyme can there-

fore be controlled through the variation of its substrate-binding affinity. Recall that Sections 9-1 and 9-4 detail how hemoglobin's oxygen affinity is allosterically regulated by the binding of ligands such as O_2, CO_2, H^+, and BPG. These homotropic and heterotropic effects (ligand binding that, respectively, alters the binding affinity of the same or different ligands) result in cooperative (sigmoidal) O_2-binding curves such as those of Figs. 9-6 and 9-8. *An enzyme's substrate-binding affinity may likewise vary with the binding of small molecule effectors thereby changing the enzyme's catalytic activity.* In this section we consider the allosteric control of enzymatic activity by examining one particular example—**aspartate transcarbamoylase (ATCase)** from *E. coli.* (The activities of some enzymes are similarly regulated through their reversible covalent modification, usually by the phosphorylation of a Ser residue. We study this form of enzymatic regulation in Section 17-3.)

The Feedback Inhibition of ATCase Regulates Pyrimidine Biosynthesis

Aspartate transcarbamoylase catalyzes the formation of *N*-**carbamoylaspartate** from **carbamoyl phosphate** and aspartate:

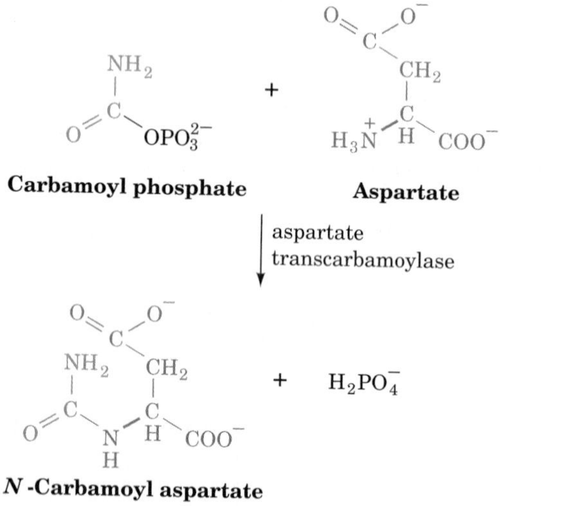

Arthur Pardee demonstrated that this reaction is the first step unique to the biosynthesis of pyrimidines (Section 26-3A), major components of nucleic acids.

The allosteric behavior of *E. coli* ATCase has been investigated by John Gerhart and Howard Schachman who demonstrated that this enzyme exhibits positively homotropic cooperative binding of both its substrates —aspartate and carbamoyl phosphate. Moreover, ATCase is heterotropically inhibited by **cytidine triphosphate (CTP)**, a pyrimidine nucleotide, and is heterotropically activated by **adenosine triphosphate (ATP)**, a purine nucleotide. CTP therefore decreases the enzyme's catalytic rate, whereas ATP increases it (Fig. 12-5).

CTP, a product of the pyrimidine biosynthesis pathway (Fig. 12-6), is a nucleic acid precursor (Section 29-2). Consequently, when rapid nucleic acid biosynthesis has depleted a cell's CTP pool, this effector dissociates from ATCase through mass action thereby deinhibiting the enzyme and increasing the rate of CTP synthesis. Conversely, if the rate of CTP synthesis outstrips its rate of uptake, the resulting excess CTP inhibits ATCase which, in turn, reduces the rate of CTP synthesis. *This is an example of **feedback inhibition,** a common mode of metabolic regulation in which the concentration of a biosynthetic pathway product controls the activity of an enzyme near the beginning of that pathway.*

The metabolic significance of the ATP activation of ATCase is that it tends to coordinate the rates of synthesis of purine and pyrimidine nucleotides, which are required in roughly equal amounts in nucleic acid biosynthesis. For instance, if the ATP concentration is greater than that of CTP, ATCase is activated to synthesize pyrimidines until the concentrations of ATP and CTP become roughly equal. Conversely, if the CTP concentration is greater than that of ATP, CTP inhibition of ATCase permits purine biosynthesis to equalize the ATP and CTP concentrations.

Allosteric Changes Alter ATCase's Substrate-Binding Sites

E. coli ATCase (300 kD) has the subunit composition $c_6 r_6$ where c and r, respectively, represent its catalytic and regulatory subunits (Section 7-5C). The X-ray structure of ATCase (Fig. 12-7), determined by William Lipscomb, reveals that the catalytic subunits are arranged as two sets of trimers (c_3) in complex with three

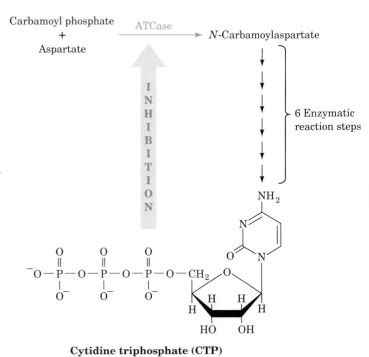

Figure 12-6
A schematic representation of the pyrimidine biosynthesis pathway indicating that CTP, the end product of the pathway, inhibits ATCase, which catalyzes the pathway's first step.

sets of regulatory dimers (r_2) to form a molecule with the rotational symmetry of a trigonal prism (D_3 symmetry; Section 7-5B). Each regulatory dimer joins two catalytic subunits in different c_3 subunits.

Dissociated catalytic trimers retain their catalytic activity, exhibit a noncooperative (hyperbolic) substrate saturation curve, and are unaffected by the presence of either ATP or CTP. The isolated regulatory dimers bind these allosteric effectors but are devoid of enzymatic activity. Evidently, *the regulatory subunits allosterically modulate the activity of the catalytic subunits in the intact enzyme.*

As allosteric theory predicts (Section 9-4), the activator ATP preferentially binds to ATCase's active (R or high substrate affinity) state, whereas the inhibitor CTP preferentially binds to the enzyme's inactive (T or low substrate affinity) state. Similarly, the unreactive bisubstrate analog *N*-**(phosphonacetyl)-L-aspartate (PALA)**

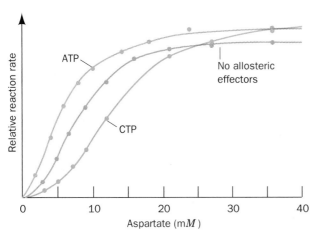

Figure 12-5
The rate of reaction catalyzed by ATCase as a function of aspartate concentration. The rates were measured in the absence of allosteric effectors, in the presence of 0.4 m*M* CTP (inhibition), and in the presence of 2.0 m*M* ATP (activation). [After Kantrowitz, E. R., Pastra-Landis, S. C., and Lipscomb, W. N., *Trends Biochem. Sci.* **5,** 125 (1980).]

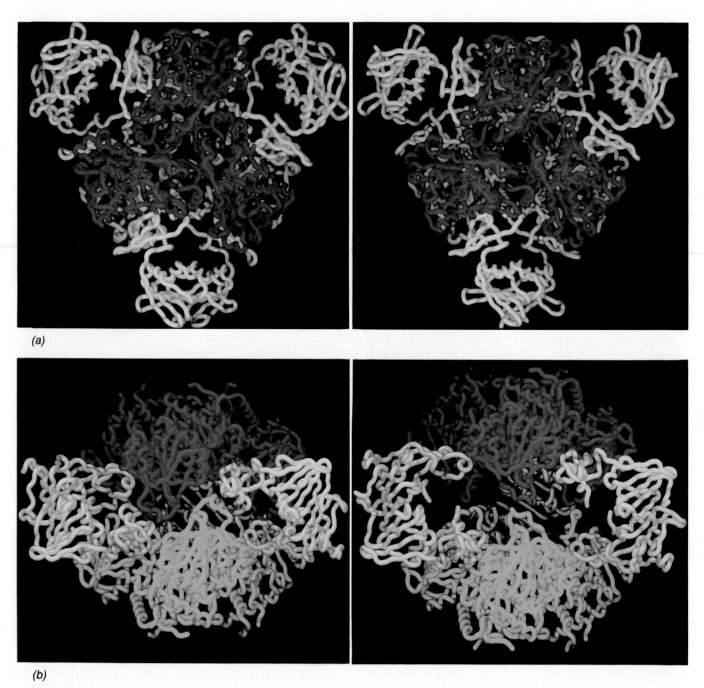

(a)

(b)

Figure 12-7
The X-ray structure of ATCase. The polypeptide backbones of T-state ATCase (*left*) and R-state ATCase (*right*) as viewed (*a*) along the protein's molecular threefold axis of symmetry and (*b*) along a molecular twofold axis of symmetry. The regulatory dimers (*yellow*) join the upper catalytic trimer (*red*) to the lower catalytic trimer (*blue*). [Courtesy of Michael Pique, Research Institute of Scripps Clinic.]

tightly binds to R-state but not to T-state ATCase (the use of unreactive substrate analogs is common in the study of enzyme mechanisms because they form stable complexes that are amenable to structural study rather than rapidly reacting to form products as do true substrates).

The X-ray structures of the T-state ATCase–CTP complex and the R-state ATCase–PALA complex reveal that the T → R transition maintains the protein's D_3 symmetry. The comparison of these two structures (Fig. 12-7) indicates that in the T → R transition, the enzyme's catalytic trimers separate along the molecular threefold axis by 12 Å and reorient about this axis by 10° such that these trimers assume a more nearly eclipsed configuration. In addition, the regulatory dimers rotate by 15° about their twofold axes and separate by ~4 Å along the threefold axis. Such large quaternary shifts are reminiscent of the conformational differences between deoxy- and oxyhemoglobin (Section 9-2B).

ATCase's substrates, carbamoyl phosphate and aspartate, each bind to a separate domain of the catalytic subunit (Fig. 12-8). The binding of PALA to the enzyme, which presumably mimics the binding of both substrates, induces active site closure in a manner that would bring them together so as to promote their reaction. The resulting atomic shifts, up to 8 Å for some residues (Fig. 12-8), trigger ATCase's T → R quaternary shift. Indeed, *ATCases's tertiary and quaternary shifts are so tightly coupled through extensive intersubunit contacts that they cannot occur independently (Fig. 12-9).* The binding of substrate to one catalytic subunit therefore increases the substrate-binding affinity and catalytic activity of the other catalytic subunits and hence accounts for the enzyme's positively cooperative substrate binding—much like occurs in hemoglobin (Section 9-2C). Thus, low levels of PALA actually activate ATCase by promoting its T → R transition: ATCase has such high affinity for this unreactive bisubstrate analog that the binding of one molecule of PALA converts all six of its catalytic subunits to the R state. Evidently, *ATCase closely follows the symmetry model of allosterism (Section 9-4B).*

X-ray studies indicate that CTP and ATP bind to the same region of the regulatory subunit so that these effectors must share at least part of their binding sites even though they have opposite effects on ATCase's catalytic activity. The structural basis of these heterotropic effects has yet to be determined. It is nevertheless expected that the tertiary changes to the regulatory subunits on binding ATP or CTP, respectively, promote or inhibit the T → R shift.

Allosteric enzymes are widely distributed in nature and tend to occupy key regulatory positions in metabolic pathways. The allosteric properties of few of these enzymes have yet been investigated in detail. It seems likely, however, that their regulatory mechanisms will generally resemble the model of ATCase: Inhibitors bind to a less active T-state conformation of the multisubunit enzyme, whereas activators bind to a more active R state. The T → R quaternary transition structurally alters the enzyme's active sites so as to promote substrate binding and facilitate catalysis.

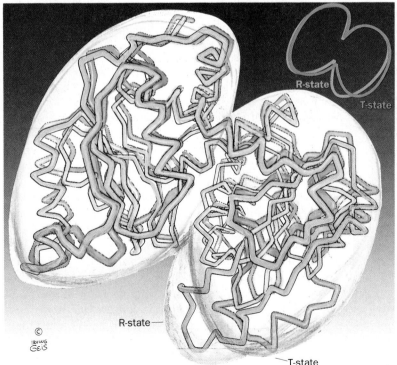

R-state
T-state

Figure 12-8
The comparison of the polypeptide backbones of the ATCase catalytic subunit in the T state (*orange*), and the R state (*blue*). The subunit consists of two domains with the one on the left containing the carbamoyl phosphate binding site and that on the right forming the aspartic acid binding site. The T ⇌ R transition brings the two domains together such that their two bound substrates can react to form product. [Figure copyrighted © by Irving Geis and Eric Gouaux. X-ray structure by William Lipscomb.]

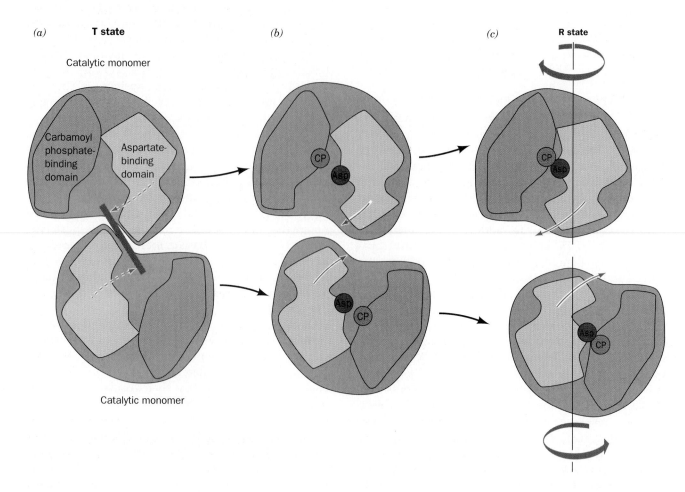

Figure 12-9
A schematic diagram indicating the tertiary and quarternary conformational changes in two vertically interacting catalytic ATCase subunits. (*a*) In the absence of bound substrate the protein is held in the T state because the motions that bring together the two domains of each subunit (*dashed arrows*) are prevented by steric interference (*purple bar*) between the contacting aspartic acid-binding domains. (*b*) The binding of carbamoyl phosphate (CP) followed by aspartic acid (Asp) to their respective binding sites causes the subunits to move apart and rotate with respect to each other so as to permit the T $\rightleftharpoons$ R transition. (*c*) In the R state, the two domains of each subunit come together so as to promote the reaction of their bound substrates to form products. [Figure copyrighted by Irving Geis.]

5. A PRIMER OF ENZYME NOMENCLATURE

Enzymes, as we have seen throughout the text so far, are commonly named by appending the suffix "-ase" to the name of the enzyme's substrate or to a phrase describing the enzyme's catalytic action. Thus "urease" catalyzes the hydrolysis of urea and "alcohol dehydrogenase" catalyzes the oxidation of alcohols to their corresponding aldehydes. Since there were at first no systematic rules for naming enzymes, this practice occasionally resulted in two different names being used for the same enzyme or, conversely, in the same name being used for two different enzymes. Moreover, many enzymes, such as "catalase," which mediates the dis-

mutation of H_2O_2 to H_2O and O_2, were given names that provide no clue as to their function; even such atrocities as "old yellow enzyme" have crept into use. In an effort to eliminate this confusion and to provide rules for rationally naming the rapidly growing number of newly discovered enzymes, a scheme for the systematic functional classification and nomenclature of enzymes was adopted by the International Union of Biochemistry (IUB).

Enzymes are classified and named according to the nature of the chemical reactions they catalyze. There are six major classes of reactions that enzymes catalyze (Table 12-3), as well as subclasses and sub-subclasses within these classes. Each enzyme is assigned two names and a four-number classification. Its **recommended name** is convenient for everyday use and is often an enzyme's pre-

Table 12-3
Enzyme Classification According to Reaction Type

Classification	Type of Reaction Catalyzed
1. Oxidoreductases	Oxidation–reduction reactions
2. Transferases	Transfer of functional groups
3. Hydrolases	Hydrolysis reactions
4. Lyases	Group elimination to form double bonds
5. Isomerases	Isomerization
6. Ligases	Bond formation coupled with ATP hydrolysis

viously used trivial name. Its **systematic name** is used when ambiguity must be minimized; it is the name of its substrate(s) followed by a word ending in "-ase" specifying the type of reaction the enzyme catalyzes according to its major group classification. For example, the

most recent version of *Enzyme Nomenclature* (see references) indicates that the enzyme whose recommended name is carboxypeptidase A (Section 6-1A) has the systematic name **Peptidyl-L-amino acid hydrolase** and the **classification number** EC 3.4.17.1. Here "EC" stands for Enzyme Commission, the first number (3) indicates the enzyme's major class (hydrolases; Table 12-3), the second number (4) denotes its subclass (peptide bonds), the third number (17) designates its sub-subclass (metallocarboxypeptidases; carboxypeptidase A has a bound Zn^{2+} ion that is essential for its catalytic activity), and the fourth number (1) is the enzyme's arbitrarily assigned serial number in its sub-subclass. As another example, the enzyme with the recommended name alcohol dehydrogenase (Section 12-2A) has the systematic name **alcohol:NAD$^+$ oxidoreductase** and the classification number EC 1.1.1.1. In this text, as in general biochemical terminology, we shall most often use the recommended names of enzymes but, when ambiguity must be minimized, we shall refer to an enzyme's systematic name.

Chapter Summary

Enzymes specifically bind their substrates through geometrically and physically complementary interactions. This permits enzymes to be absolutely stereospecific, both in binding substrates and catalyzing reactions. Enzymes vary in the more stringent requirement of geometric specificity. Some are highly specific for the identity of their substrates, whereas others can bind a wide range of substrates and catalyze a variety of related types of reactions.

Enzymatic reactions involving oxidation–reduction reactions and many types of group-transfer processes are mediated by coenzymes. Many vitamins are coenzyme precursors.

Enzymatic activity may be regulated by the allosteric alteration of substrate-binding affinity. For example, the rate of the reaction catalyzed by *E. coli* ATCase is subject to positive ho-

motropic control by substrate, heterotropic inhibition by CTP, and heterotropic activation by ATP. ATCase has the subunit composition c_6r_6. Its isolated catalytic trimers are catalytically active but not subject to allosteric control. The regulatory dimers bind ATP and CTP. Substrate binding induces a tertiary conformational shift in the catalytic subunits, which increases the subunit's substrate-binding affinity and catalytic efficiency. This tertiary shift is strongly coupled to ATCase's large quaternary $T \rightarrow R$ conformational shift thereby accounting for the enzyme's allosteric properties.

Enzymes are systematically classified according to their recommended name, their systematic name, and their classification number, which is indicative of the type of reaction catalyzed by the enzyme.

References

General

Dixon, M. and Webb, E. C., *Enzymes* (3rd ed.), Academic Press (1979). [A treatise on enzymes.]

History

Fruton, J. S., *Molecules and Life, pp. 22–86*, Wiley (1972).

Schlenk, F., Early research on fermentation—a story of missed opportunities, *Trends Biochem. Sci.* **10**, 252–254 (1985).

Substrate Specificity

Fersht, A., *Enzyme Structure and Mechanism* (2nd ed.), Freeman (1985).

Popják, G., Specificity of enzyme reactions, *in* Boyer, P. D. (Ed.), *The Enzymes* (3rd ed.), Vol. 2, pp. 217–279, Academic Press (1970).

Regulation of Enzyme Activity

Allewell, N. M., *Eschericia coli* aspartate transcarbamoylase: structure, energetics, and catalytic and regulatory mechanisms. Annu. Rev. Biophys. Biophys. Chem. **18**, 71–92 (1989).

Kantrowitz, E. R. and Lipscomb, W. N., *Eschericia coli* aspartate transcarbamylase: the relation between structure and function, *Science* **241**, 669–674 (1988).

Kantrowitz, E. R., Pastra-Landis, S. C., and Lipscomb, W. N., *E. coli* aspartate transcarbamylase: Part I: catalytic and regulatory function, *and* Part II: structure and allosteric interactions, *Trends Biochem. Sci.* **5**, 124–128 and 150–153 (1980).

Kim, K. H., Pan, Z., Honzatko, R. B., Ke, H., and Lipscomb, W. N., Structural asymmetry in the CTP-liganded form of aspartate carbamoyltransferase from *Escherichia coli*, *J. Mol. Biol.* **196**, 853–875 (1987).

Krause, K. L., Volz, K. W., and Lipscomb, W. N., 2.5 Å structure of aspartate carbamoyltransferase complexed with the bisubstrate analogue N-(phosphonacetyl)-L-aspartate, *J. Mol. Biol.* **193**, 527–553 (1987).

Schachman, H. K., Can a simple model account for the allosteric transition of aspartate transcarbamylase? *J. Biol. Chem.* **263**, 18583–18586 (1988).

Enzyme Nomenclature

Enzyme Nomenclature, Academic Press (1984). [Recommendations of the Nomenclature Committee of the IUB on the nomenclature and classification of enzymes.]

Dixon, M. and Webb, E. C., *Enzymes* (3rd ed.), Academic Press (1979). [Chapter V contains the rules for enzyme classification, and the Table of Enzymes provides their systematic names and classification numbers.]

Problems

1. Indicate the products of the YADH reaction with normal acetaldehyde and NADH in D_2O solution.

2. Indicate the product(s) of the YADH-catalyzed oxidation of the chiral methanol derivative (R)-TDHCOH.

3. The enzyme **fumarase** catalyzes the hydration of the double bond of **fumarate:**

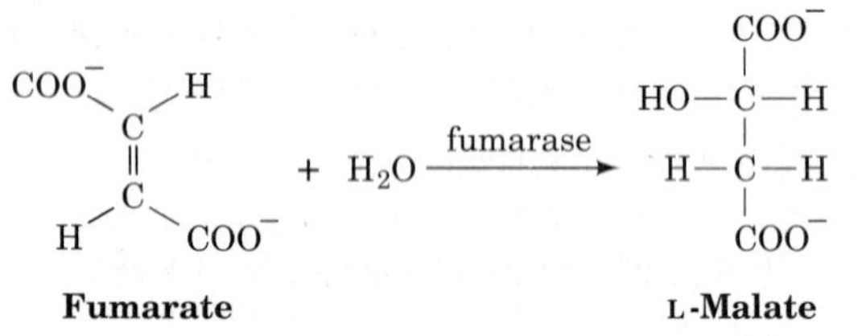

Fumarate   L-**Malate**

Predict the action of fumarase on **maleate,** the cis isomer of fumarate. Explain.

4. Write a balanced equation for the chymotrypsin-catalyzed reaction between an ester and an amino acid.

5. Hominy grits, a regional delicacy of the Southern United States, is made from corn that has been soaked in a weak lye (NaOH) solution. What is the function of this unusual treatment?

6. Which of the curves in Fig. 12-5 exhibits the greatest cooperativity? Explain.

7. What are the advantages of having the final product of a multistep metabolic pathway inhibit the enzyme that catalyzes the first step?

8. Using the references, find the systematic names and classification numbers for the enzymes whose recommended names are catalase, aspartate carbamoyltransferase (aspartate transcarbamoylase), and trypsin.

Chapter 13
RATES OF ENZYMATIC REACTIONS

Kinetics is the study of the rates at which chemical reactions occur. A major purpose of such a study is to gain an understanding of a reaction mechanism; that is, a detailed description of the various steps in a reaction process and the sequence with which they occur. Thermodynamics, as we saw in Chapter 3, tells us whether a given process can occur spontaneously, but provides little indication as to the nature or even the existence of its component steps. In contrast, *the rate of a reaction and how this rate changes in response to different conditions is intimately related to the path followed by the reaction and is therefore indicative of its reaction mechanism.*

In this chapter, we take up the study of **enzyme kinetics**, a subject that is of enormous practical importance in biochemistry because:

1. It is through kinetic studies that the binding affinities of substrates and inhibitors to an enzyme can be determined and that the maximum catalytic rate of an enzyme can be established.

2. By observing how the rate of an enzymatic reaction varies with the reaction conditions and combining this information with that obtained from chemical and structural studies of the enzyme, the enzyme's catalytic mechanism may be elucidated.

3. Most enzymes, as we shall see in later chapters, function as members of metabolic pathways. The study of the kinetics of an enzymatic reaction leads to an understanding of that enzyme's role in an overall metabolic process.

4. Under the proper conditions, the rate of an enzymatically catalyzed reaction is proportional to the amount of the enzyme present and therefore most enzyme assays (measurements of the amount of enzyme present) are based on kinetic studies of the enzyme. Measurements of enzymatically catalyzed reaction rates are therefore among the most commonly employed procedures in biochemical and clinical analyses.

We begin our consideration of enzyme kinetics by reviewing chemical kinetics because enzyme kinetics is based on this formalism. Following that, we derive the basic equations of enzyme kinetics, describe the effects of inhibitors on enzymes, and consider how the rates of enzymatic reactions vary with pH. We end by outlining the kinetics of complex enzymatic reactions.

Kinetics is, by and large, a mathematical subject. Although the derivations of kinetic equations are occasionally rather detailed, the level of mathematical skills it requires should not challenge anyone who has studied elementary calculus. Nevertheless, to prevent mathematical detail from obscuring the underlying enzymological principles, the derivations of all but the most important kinetic equations have been collected in the appendix to this chapter. Those who wish to cultivate a deeper understanding of enzyme kinetics are urged to consult this appendix.

1. CHEMICAL KINETICS

Enzyme kinetics is a branch of chemical kinetics and, as such, shares much of the same formalism. In this section we shall therefore review the principles of chemical kinetics so that, in later sections, we can apply them to enzymatically catalyzed reactions.

A. Elementary Reactions

A reaction of overall stoichiometry

$$A \longrightarrow P$$

may actually occur through a sequence of **elementary reactions** (simple molecular processes) such as

$$A \longrightarrow I_1 \longrightarrow I_2 \longrightarrow P$$

Here A represents reactants, P products, and I_1 and I_2 symbolize **intermediates** in the reaction. *The characterization of the elementary reactions comprising an overall reaction process constitutes its mechanistic description.*

Rate Equations

At constant temperature, elementary reaction rates vary with reactant concentration in a simple manner. Consider the general elementary reaction:

$$aA + bB + \cdots + zZ \longrightarrow P$$

The rate of this process is proportional to the frequency with which the reacting molecules simultaneously come together, that is, to the products of the concentrations of the reactants. This is expressed by the following **rate equation**

$$\text{rate} = k[A]^a[B]^b \cdots [Z]^z \qquad [13.1]$$

where k is a proportionality constant known as a **rate constant**. The **order** of a reaction is defined as $(a + b + \cdots + z)$, the sum of the exponents in the rate equation. For an elementary reaction, the order corresponds to the **molecularity** of the reaction, the number of molecules that must simultaneously collide in the elementary reaction. Thus the elementary reaction $A \rightarrow P$ is an example of a **first-order** or **unimolecular** reaction, whereas the elementary reactions $2A \rightarrow P$ and $A + B \rightarrow P$ are examples of **second-order** or **bimolecular** reactions. Unimolecular and bimolecular reactions are common. **Termolecular** reactions are unusual and fourth- and higher-order reactions are unknown. This is because the simultaneous collision of three molecules is a rare event; that of four or more molecules essentially never occurs.

B. Rates of Reactions

We can experimentally determine the order of a reaction by measuring [A] or [P] as a function of time; that is,

$$v = -\frac{d[A]}{dt} = \frac{d[P]}{dt} \qquad [13.2]$$

where v is the instantaneous rate or **velocity** of the reaction. For the first-order reaction $A \rightarrow P$:

$$v = -\frac{d[A]}{dt} = k[A] \qquad [13.3a]$$

For second-order reactions such as $2A \rightarrow P$:

$$v = -\frac{d[A]}{dt} = k[A]^2 \qquad [13.3b]$$

whereas for $A + B \rightarrow P$, a second-order reaction that is first order in [A] and first order in [B],

$$v = -\frac{d[A]}{dt} = -\frac{d[B]}{dt} = k[A][B] \qquad [13.3c]$$

The rate constants of first- and second-order reactions

must have different units. In terms of units, Eq. [13.3a] is expressed $Ms^{-1} = kM$. Therefore, k must have units of reciprocal seconds (s^{-1}) in order for Eq. [13.3a] to balance. Similarly, for second-order reactions, $Ms^{-1} = kM^2$ so that k has the units $M^{-1}s^{-1}$.

The order of a specific reaction can be determined by measuring the reactant or product concentrations as a function of time and comparing the fit of this data to equations describing this behavior for reactions of various orders. To do this we must first derive these equations.

First-Order Rate Equation

The equation for [A] as a function of time for a first-order reaction, $A \rightarrow P$, is obtained by rearranging Eq. [13.3a]

$$\frac{d[A]}{[A]} \equiv d \ln[A] = -k dt$$

and integrating it from $[A]_o$, the initial concentration of A, to [A], the concentration of A at time t:

$$\int_{[A]_o}^{[A]} d \ln[A] = -k \int_0^t dt$$

This results in

$$\ln[A] = \ln[A]_o - kt \qquad [13.4a]$$

or, by taking the antilogs of both sides,

$$[A] = [A]_o e^{-kt} \qquad [13.4b]$$

Equation [13.4a] is a linear equation in terms of the variables $\ln[A]$ and t as is diagrammed in Fig. 13-1. Therefore, if a reaction is first order, a plot of $\ln[A]$ versus t will yield a straight line whose slope is $-k$, the negative of the first-order rate constant, and whose intercept on the $\ln[A]$ axis is $\ln[A]_o$.

Substances that are inherently unstable, such as radioactive nuclei, decompose through first-order reactions (first-order processes are not just confined to chemical reactions). One of the hallmarks of a first-order reaction is that *the time for one half of the reactant initially present to decompose, its **half-time** or **half-life**, $t_{1/2}$, is a constant and hence independent of the initial concentration of the reactant.* This is easily demonstrated by substituting the relationship $[A] = [A]_o/2$ when $t = t_{1/2}$ into Eq. [13.4a] and rearranging:

$$\ln \left(\frac{[A]_o/2}{[A]_o} \right) = -k t_{1/2}$$

Thus

$$t_{1/2} = \frac{\ln 2}{k} = \frac{0.693}{k} \qquad [13.5]$$

In order to appreciate the course of a first-order reaction, let us consider the decomposition of ^{32}P, a radioac-

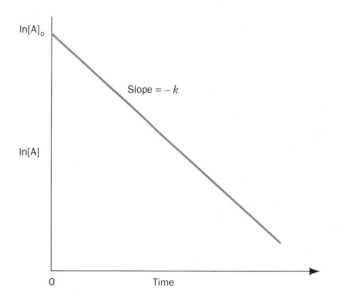

Figure 13-1
A plot of ln[A] versus time for a first-order reaction illustrating the graphical determination of the rate constant k using Eq. [13.4a].

tive isotope that is widely used in biochemical research. It has a half-life of 14 days. Thus, after 2 weeks, one half of the ^{32}P initially present in a given sample will have decomposed; after another 2 weeks, one half of the remainder or three quarters of the original sample will have decomposed; *etc.* The long-term storage of waste ^{32}P therefore presents little problem, since after 1 year (26 half-lives), only 1 part in $2^{26} = 67$ million of the original sample will remain. How much will remain after 2 years? In contrast, ^{14}C, another commonly employed radioactive tracer, has a half-life of 5730 years: Only a small fraction of a given quantity of ^{14}C will decompose over the course of a human lifetime.

Second-Order Rate Equation for One Reactant

In a second-order reaction with one type of reactant, $2A \rightarrow P$, the variation of [A] with time is quite different from that in a first-order reaction. Rearranging Eq. [13.3b] and integrating it over the same limits used for the first-order reaction yields:

$$\int_{[A]_o}^{[A]} -\frac{d[A]}{[A]^2} = k \int_0^t dt$$

so that

$$\frac{1}{[A]} = \frac{1}{[A]_o} + kt \qquad [13.6]$$

Equation [13.6] is a linear equation in terms of the variables $1/[A]$ and t. Consequently, Eqs. [13.4a] and [13.6] may be used to distinguish a first-order from a second-order reaction by plotting $\ln[A]$ versus t and $1/[A]$ versus t and observing which, if any, of these plots is a straight line.

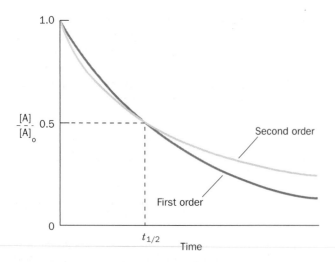

Figure 13-2
A comparison of the progress curves for first- and second-order reactions that have the same value of $t_{1/2}$. [After Tinoco, I., Jr., Sauer, K., and Wang., J. C., *Physical Chemistry. Principles and Applications in Biological Sciences* (2nd ed.), *p*. 291, Prentice–Hall (1985).]

Figure 13-2 compares the different shapes of the progress curves describing the disappearance of A in first- and second-order reactions having the same half-times. Note that before the first half-time, the second-order progress curve descends more steeply than the first-order curve but after this time the first-order progress curve is the more rapidly decreasing of the two. The half-time for a second-order reaction is expressed $t_{1/2} = 1/k[A]_o$ and therefore, in contrast to a first-order reaction, is dependent on the initial reactant concentration.

C. Transition State Theory

The goal of kinetic theory is to describe reaction rates in terms of the physical properties of the reacting molecules. A theoretical framework for doing so, that explicitly considers the structures of the reacting molecules and how they collide, was developed in the 1930s, principally by Henry Eyring. This view of reaction processes, known as **transition state theory** or **absolute rate theory,** is the foundation of much of modern kinetics and has provided an extraordinarily productive framework for understanding how enzymes catalyze reactions.

The Transition State

Consider a bimolecular reaction involving three atoms A, B, and C:

$$A + B—C \longrightarrow A—B + C$$

Clearly atom A must approach the diatomic molecule B—C so that, at some point in the reaction, a high-energy (unstable) complex represented as $A \cdots B \cdots C$ exists in which the A—B covalent bond is in the process of forming while the B—C bond is in the process of breaking.

Let us consider the simplest example of this reaction: that of a hydrogen atom with diatomic hydrogen (H_2) to yield a new H_2 molecule and a different hydrogen atom:

$$H_A + H_B—H_C \longrightarrow H_A—H_B + H_C$$

The potential energy of this triatomic system as a function of the relative positions of its component atoms is plotted in Fig. 13-3a and b. Its shape is of two long and

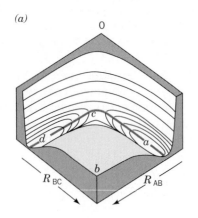

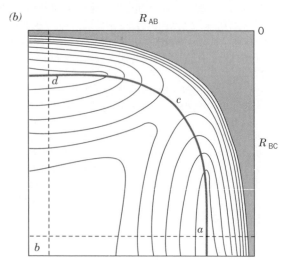

Figure 13-3
The potential energy of the colinear $H + H_2$ system as a function of its internuclear distances, R_{AB} and R_{BC}. The reaction is represented as: (a) A perspective drawing. (b) The corresponding contour diagram. The points *a* and *d* are approaching potential energy minima, *b* is approaching a maximum, and *c* is a saddle point. [After Frost, A. A. and Pearson, R. G., *Kinetics and Mechanism* (2nd ed.), *p*. 80, Wiley (1961).]

deep valleys parallel to the coordinate axes with sheer walls rising towards the axes and less steep ones rising towards a plateau where both coordinates are large (the region of point *b*). The two valleys are joined by a pass or saddle near the origin of the diagram (point *c*). The minimum energy configuration is that of an H_2 molecule and an isolated atom, that is, with one coordinate large and the other at the H_2 covalent bond distance [near points *a* (the reactants) and *d* (the products)]. During a collision, the reactants generally approach one another with little deviation from the minimum energy reaction pathway (line *a—c—d*) because other trajectories would require much greater energy. As the atom and molecule come together, they increasingly repel one another (have increasing potential energy) and therefore usually fly apart. *If, however, the system has sufficient kinetic energy to continue its coalescence, it will cause the covalent bond of the H_2 molecule to weaken until ultimately, if the system reaches the saddle point (point c), there is an equal probability that either the reaction will occur or that the system will decompose back to its reactants.* Therefore, at this saddle point, the system is said to be at its **transition state** and hence to be an **activated complex.** Moreover, since the concentration of the activated complex is small, *the decomposition of the activated complex is postulated to be the rate-determining process of this reaction.*

The minimum energy pathway of a reaction is known as its **reaction coordinate.** Figure 13-4*a*, which is called a **transition state diagram** or a **reaction coordinate diagram,** shows the potential energy of the $H + H_2$ system along the reaction coordinate (line *a—c—d* in Fig. 13-3). It can be seen that the transition state is the point of highest energy on the reaction coordinate. If the atoms in the triatomic system are of different types, as is diagrammed in Fig. 13-4*b*, the transition state diagram is no longer symmetrical because there is an energy difference between reactants and products.

Thermodynamics of the Transition State

The realization that the attainment of the transition state is the central requirement in any reaction process led to a detailed understanding of reaction mechanisms. For example, consider a bimolecular reaction that proceeds along the following pathway:

$$A + B \underset{}{\overset{K^{\ddagger}}{\rightleftharpoons}} X^{\ddagger} \overset{k'}{\longrightarrow} P + Q$$

where $X^{\ddagger}$ represents the activated complex. Therefore, considering the preceding discussion,

$$\frac{d[P]}{dt} = k[A][B] = k'[X^{\ddagger}] \qquad [13.7]$$

where *k* is the ordinary rate constant of the elementary reaction and *k'* is the rate constant for the decomposition of $X^{\ddagger}$ to products.

In contrast to stable molecules, such as A and P, which occur at energy minima, the activated complex occurs at an energy maximum and is therefore only metastable (like a ball balanced on a pin). Transition state theory nevertheless assumes that $X^{\ddagger}$ is in rapid equilibrium with the reactant; that is,

$$K^{\ddagger} = \frac{[X^{\ddagger}]}{[A][B]} \qquad [13.8]$$

where $K^{\ddagger}$ is an equilibrium constant. *This central assumption of transition state theory permits the powerful formalism of thermodynamics to be applied to the theory of reaction rates.*

If $K^{\ddagger}$ is an equilibrium constant it can be expressed as:

$$-RT \ln K^{\ddagger} = \Delta G^{\ddagger} \qquad [13.9]$$

where $\Delta G^{\ddagger}$ is the Gibbs free energy of the activated complex less that of the reactants (Fig. 13-4*b*), *T* is the absolute temperature, and $R \ (= 8.3145 \ J \cdot K^{-1} \ mol^{-1})$ is the gas constant (this relationship between equilibrium

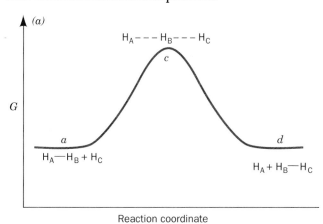

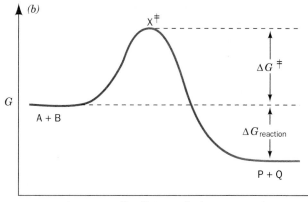

Figure 13-4
Transition state diagrams: (*a*) For the $H + H_2$ reaction. This is a section taken along the *a—c—d* line in Fig. 13-3*b*. (*b*) For a spontaneous reaction, that is, one in which the free energy decreases.

constants and free energy is derived in Section 3-4A). Then combining Eqs. [13.7] through [13.9] yields

$$\frac{d[P]}{dt} = k'e^{-\Delta G^\dagger/RT}[A][B] \qquad [13.10]$$

This equation indicates that the rate of a reaction depends not only on the concentrations of its reactants but also decreases exponentially with $\Delta G^\dagger$. Thus, *the larger the difference between the free energy of the transition state and that of the reactants, that is, the less stable the transition state, the slower the reaction proceeds.*

In order to continue, we must now evaluate k', the rate of passage of the activated complex over the maximum in the transition state diagram (sometimes referred to as the **activation barrier** or the **kinetic barrier** of the reaction). This transition state model permits us to do so (although the following derivation is by no means rigorous). The activated complex is held together by a bond that is associated with the reaction coordinate and which is assumed to be so weak that it flies apart during its first vibrational excursion. Therefore, k' is expressed

$$k' = \kappa v \qquad [13.11]$$

where v is the vibrational frequency of the bond that breaks as the activated complex decomposes to products and κ, the **transmission coefficient,** is the probability that the breakdown of the activated complex, $X^\dagger$, will be in the direction of product formation rather than back to reactants. For most spontaneous reactions in solution, κ is between 0.5 and 1.0; for the colinear $H + H_2$ reaction, we saw that it is 0.5.

We have nearly finished our job of evaluating k'. All that remains is to determine the value of v. Planck's law states that

$$v = \varepsilon/h \qquad [13.12]$$

where, in this case, ε is the average energy of the vibration that leads to the decomposition of $X^\dagger$, and h ($= 6.6261 \times 10^{-34}$ J·s) is Planck's constant. Statistical mechanics tells us that at temperature T, the classical energy of an oscillator is

$$\varepsilon = k_B T \qquad [13.13]$$

where k_B ($= 1.3807 \times 10^{-23}$ J·K^{-1}) is a constant of nature known as the Boltzmann constant and $k_B T$ is essentially the available thermal energy. Combining Eqs. [13.11] through [13.13]

$$k' = \frac{\kappa k_B T}{h} \qquad [13.14]$$

Then assuming, as is done for most reactions, that $\kappa = 1$ (κ can rarely be calculated with any confidence), the combination of Eqs. [13.7] and [13.10] yields the expression for the rate constant k of our elementary reaction:

$$k = \frac{k_B T}{h} e^{-\Delta G^\dagger/RT} \qquad [13.15]$$

This equation indicates that *the rate of reaction decreases as its free energy of activation, $\Delta G^\dagger$, increases.* Conversely, as the temperature rises, so that there is increased thermal energy available to drive the reacting complex over the activation barrier, the reaction speeds up. (Of course, enzymes, being proteins, are subject to thermal denaturation so that the rate of an enzymatically catalyzed reaction falls precipitously with increasing temperature once the enzyme's denaturation temperature has been surpassed.) Keep in mind, however, that transition state theory is an ideal model; real systems behave in a more complicated although qualitatively similar manner.

Multistep Reactions Have Rate-Determining Steps

Since chemical reactions commonly consist of several elementary reaction steps, let us consider how transition state theory treats such reactions. For a multistep reaction such as,

$$A \xrightarrow{k_1} I \xrightarrow{k_2} P$$

where I is an intermediate of the reaction, there is an activated complex for each elementary reaction step; the shape of the transition state diagram for such a reaction reflects the relative rates of the elementary reactions involved. For this reaction, if the first reaction step is slower than the second reaction step ($k_1 < k_2$), then the activation barrier of the first step must be higher than that of the second step, and conversely if the second reaction step is the slower (Fig. 13-5). Since the rate of formation of product P can only be as fast as the slowest elementary reaction, *if one reaction step of an overall reaction is much slower than the other, the slow step acts as a*

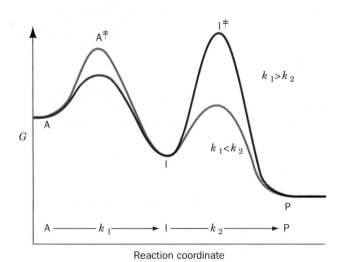

Figure 13-5
The transition state diagram for the two-step overall reaction $A \xrightarrow{k_1} I \xrightarrow{k_2} P$. For $k_1 < k_2$ (*green curve*), the first step is rate determining, whereas if $k_1 > k_2$ (*red curve*), the second step is rate determining.

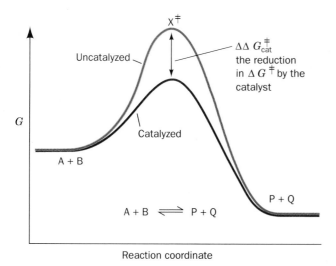

Figure 13-6
A schematic diagram illustrating the effect of a catalyst on the transition state diagram of a reaction. Here $\Delta\Delta G^{\ddagger} = \Delta G^{\ddagger}_{uncat} - \Delta G^{\ddagger}_{cat}$.

"bottleneck" and is therefore said to be the **rate-determining step** *of the reaction.*

Catalysis Reduces $\Delta G^{\ddagger}$

Biochemistry is, of course, mainly concerned with enzyme-catalyzed reactions. Catalysts act by lowering the activation barrier for the reaction being catalyzed (Fig. 13-6). If a catalyst lowers the activation barrier of a reaction by $\Delta\Delta G^{\ddagger}_{cat}$ then, according to Eq. [13.15], the rate of the reaction is enhanced by the factor $e^{\Delta\Delta G^{\ddagger}_{cat}/RT}$. Thus, a 10-fold rate enhancement requires that $\Delta\Delta G^{\ddagger}_{cat} = 5.71$ $kJ \cdot mol^{-1}$, less than one half the energy of a typical hydrogen bond; a millionfold rate acceleration occurs when $\Delta\Delta G^{\ddagger}_{cat} = 34.25$ $kJ \cdot mol^{-1}$, a small fraction of the energy of most covalent bonds. *The rate enhancement is therefore a sensitive function of $\Delta\Delta G^{\ddagger}_{cat}$.*

Note that the kinetic barrier is lowered by the same amount for both the forward and the reverse reactions (Fig. 13-6). Consequently, a catalyst equally accelerates the forward and the reverse reactions so that the equilibrium constant for the reaction remains unchanged. The chemical mechanisms through which enzymes lower the activation barriers of reactions are the subject of Section 14-1. There we shall see that the most potent such mechanism often involves the enzymatic binding of the transition state of the catalyzed reaction in preference to the substrate.

2. ENZYME KINETICS

The chemical reactions of life are mediated by enzymes. These remarkable catalysts, as we saw in Chapter 12, are individually highly specific for particular reactions. Yet, collectively they are extremely versatile in

that the several thousand enzymes now known carry out such diverse reactions as hydrolysis, polymerization, functional group transfer, oxidation–reduction, dehydration, and isomerization to mention only the most common classes of enzymatically mediated reactions. Enzymes are not passive surfaces on which reactions take place but rather, are complex molecular machines that operate through a great diversity of mechanisms. For instance, some enzymes act on only a single substrate molecule; others act on two or more different substrate molecules whose order of binding may or may not be obligatory. Some enzymes form covalently bound intermediate complexes with their substrates; others do not.

Kinetic measurements of enzymatically catalyzed reactions are among the most powerful techniques for elucidating the catalytic mechanisms of enzymes. The remainder of this chapter is therefore largely concerned with the development of the kinetic tools that are most useful in the determination of enzymatic mechanisms. We begin, in this section, with a presentation of the basic theory of enzyme kinetics.

A. The Michaelis–Menten Equation

The study of enzyme kinetics began in 1902 when Adrian Brown reported an investigation of the rate of hydrolysis of sucrose as catalyzed by the yeast enzyme **invertase** (now known as **β-fructofuranosidase**):

$$\text{Sucrose} + H_2O \longrightarrow \text{glucose} + \text{fructose}$$

Brown demonstrated that when the sucrose concentration is much higher than that of the enzyme, the reaction rate becomes independent of the sucrose concentration; that is, the rate is **zeroth-order** with respect to sucrose. He therefore proposed that the overall reaction is composed of two elementary reactions in which the substrate forms a complex with the enzyme that subsequently decomposes to products and enzyme:

$$E + S \underset{k_{-1}}{\overset{k_1}{\rightleftharpoons}} ES \overset{k_2}{\longrightarrow} P + E$$

Here E, S, ES, and P symbolize the enzyme, substrate, **enzyme–substrate complex,** and products, respectively (for enzymes composed of multiple identical subunits, E refers to active sites rather than enzyme molecules). According to this model, *when the substrate concentration becomes high enough to entirely convert the enzyme to the ES form, the second step of the reaction becomes rate limiting and the overall reaction rate becomes insensitive to further increases in substrate concentration.*

The general expression for the **velocity** (rate) of this reaction is

$$v = \frac{d[P]}{dt} = k_2[ES] \qquad [13.16]$$

The overall rate of production of [ES] is the difference between the rates of the elementary reactions leading to its appearance and those resulting in its disappearance:

$$\frac{d[ES]}{dt} = k_1[E][S] - k_{-1}[ES] - k_2[ES] \quad [13.17]$$

This equation cannot be explicitly integrated, however, without simplifying assumptions. Two possibilities are

1. Assumption of equilibrium

In 1913, Leonor Michaelis and Maude Menten, building upon earlier work by Victor Henri, assumed that $k_{-1} \gg k_2$, so that the first step of the reaction achieves equilibrium.

$$K_S = \frac{k_{-1}}{k_1} = \frac{[E][S]}{[ES]} \quad [13.18]$$

Here K_S is the dissociation constant of the first step in the enzymatic reaction. With this assumption, Eq. [13.17] can be integrated. Although this assumption is not often correct, in recognition of the importance of this pioneering work, the noncovalently bound enzyme–substrate complex ES is known as the **Michaelis complex.**

2. Assumption of steady state

Figure 13-7 illustrates the progress curves of the various participants in the preceding reaction model under the physiologically common condition that substrate is in great excess over enzyme. With the exception of the initial stage of the reaction, the so-called **transient phase,** which is usually over within milliseconds of mixing the enzyme and substrate, [ES] remains approximately constant until the substrate is nearly exhausted. Hence, the rate of synthesis of ES must equal its rate of consumption over most of the course of the reaction; that is, [ES] maintains a **steady state.** One can therefore assume with a reasonable degree of accuracy that [ES] is constant; that is,

$$\frac{d[ES]}{dt} = 0 \quad [13.19]$$

This so-called **steady-state assumption** was first proposed in 1925 by G. E. Briggs and James B. S. Haldane.

In order to be of use, kinetic expressions for overall reactions must be formulated in terms of experimentally measurable quantities. The quantities [ES] and [E] are not, in general, directly measurable but the total enzyme concentration

$$[E]_T = [E] + [ES] \quad [13.20]$$

is usually readily determined. The rate equation for our enzymatic reaction is then derived as follows. Combin-

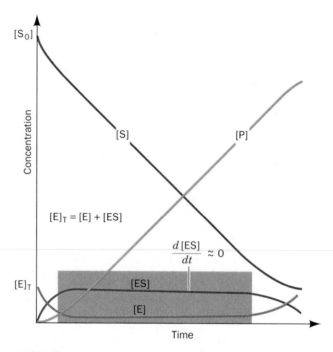

Figure 13-7
The progress curves for the components of a simple Michaelis–Menten reaction. Note that with the exception of the transient phase of the reaction, which occurs before the shaded block, the slopes of the progress curves for [E] and [ES] are essentially zero so long as [S] >> [E]$_T$ (within the shaded block). [After Segel, I. H., *Enzyme Kinetics*, p. 27, Wiley (1975).]

ing Eq. [13.17] with the steady state assumption, Eq. [13.19], and the conservation condition, Eq. [13.20], yields:

$$k_1([E]_T - [ES])[S] = (k_{-1} + k_2)[ES]$$

which upon rearrangement becomes

$$[ES](k_{-1} + k_2 + k_1[S]) = k_1[E]_T[S]$$

Dividing both sides by k_1 and solving for [ES],

$$[ES] = \frac{[E]_T[S]}{K_M + [S]}$$

where K_M, which is known as the **Michaelis constant,** is defined

$$K_M = \frac{k_{-1} + k_2}{k_1} \quad [13.21]$$

The meaning of this important constant is described below.

The **initial velocity** of the reaction can then be expressed in terms of the experimentally measureable quantities [E]$_T$ and [S].

$$v_o = \left(\frac{d[P]}{dt}\right)_{t=0} = k_2[ES] = \frac{k_2[E]_T[S]}{K_M + [S]} \quad [13.22]$$

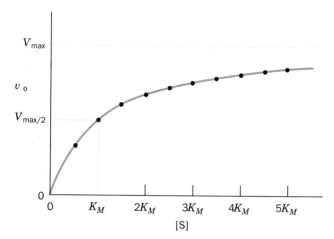

Figure 13-8
A plot of the initial velocity v_o of a simple Michaelis–Menten reaction versus the substrate concentration [S]. Points are plotted in 0.5 K_M intervals of substrate concentration between 0.5 K_M and 5 K_M.

The use of the initial velocity (operationally taken as the velocity measured before more than ~ 10% of the substrate has been converted to product) rather than just the velocity minimizes such complicating factors as the effects of reversible reactions, inhibition of the enzyme by product, and progressive inactivation of the enzyme.

The **maximal velocity** of a reaction, V_{max}, occurs at high substrate concentrations when the enzyme is **saturated,** that is, when it is entirely in the ES form:

$$V_{max} = k_2[E]_T \qquad [13.23]$$

Therefore, combining Eqs. [13.22] and [13.23], we obtain

$$v_o = \frac{V_{max}[S]}{K_M + [S]} \qquad [13.24]$$

This expression, the Michaelis–Menten equation, is the basic equation of enzyme kinetics. It describes a rectangu-

lar hyperbola such as is plotted in Fig. 13-8 (although this curve is rotated by 45° and translated to the origin with respect to the examples of hyperbolas seen in most elementary algebra texts). The saturation function for oxygen binding to myoglobin, Eq. [9.4], has the same functional form.

Significance of the Michaelis Constant

The Michaelis constant, K_M, has a simple operational definition. At the substrate concentration where [S] = K_M, Eq. [13.24] yields $v_o = V_{max}/2$ so that K_M is the *substrate concentration at which the reaction velocity is half-maximal.* Therefore, if an enzyme has a small value of K_M, it achieves maximal catalytic efficiency at low substrate concentrations.

The magnitude of K_M varies widely with the identity of the enzyme and the nature of the substrate (Table 13-1). It is also a function of temperature and pH (see Section 13-4). The Michaelis constant can be expressed as

$$K_M = \frac{k_{-1}}{k_1} + \frac{k_2}{k_1} = K_S + \frac{k_2}{k_1} \qquad [13.25]$$

Since K_S is the dissociation constant of the Michaelis complex, as K_S decreases, the enzyme's affinity for substrate increases. K_M is therefore also a measure of the affinity of the enzyme for its substrate providing k_2/k_1 is small compared with K_S, that is, $k_2 < k_{-1}$.

B. Analysis of Kinetic Data

There are several methods for determining the values of the parameters of the Michaelis–Menten equation. At very high values of [S], the initial velocity v_o asymptotically approaches V_{max}. In practice, however, it is very difficult to accurately assess V_{max} from direct plots of v_o versus [S] such as Fig. 13-8. Even at such high substrate concentrations as [S] = 10 K_M, Eq. [13.24] indicates that v_o is only 91% of V_{max} so that the extrapolated value of

Table 13-1

The Values of K_M, k_{cat}, and k_{cat}/K_M for Some Enzymes and Substrates

Enzyme	Substrate	$K_M(M)$	$k_{cat}(s^{-1})$	$k_{cat}/K_M(M^{-1}\,s^{-1})$
Acetylcholine esterase	Acetylcholine	9.5×10^{-5}	1.4×10^4	1.5×10^8
Carbonic anhydrase	CO_2	1.2×10^{-2}	1.0×10^6	8.3×10^7
	HCO_3^-	2.6×10^{-2}	4.0×10^5	1.5×10^7
Catalase	H_2O_2	2.5×10^{-2}	1.0×10^7	4.0×10^8
Chymotrypsin	N-Acetylglycine ethyl ester	4.4×10^{-1}	5.1×10^{-2}	1.2×10^{-1}
	N-Acetylvaline ethyl ester	8.8×10^{-2}	1.7×10^{-1}	1.9
	N-Acetyltyrosine ethyl ester	6.6×10^{-4}	1.9×10^2	2.9×10^5
Fumarase	Fumarate	5.0×10^{-6}	8.0×10^2	1.6×10^8
	Malate	2.5×10^{-5}	9.0×10^2	3.6×10^7
Urease	Urea	2.5×10^{-2}	1.0×10^4	4.0×10^5

the asymptote will almost certainly be underestimated.

A better method for determining the values of V_{max} and K_M, which was formulated by Hans Lineweaver and Dean Burk, uses the reciprocal of Eq. [13.24]:

$$\frac{1}{v_o} = \left(\frac{K_M}{V_{max}}\right)\frac{1}{[S]} + \frac{1}{V_{max}} \qquad [13.26]$$

This is a linear equation in $1/v_o$ and $1/[S]$. If these quantities are plotted, the so-called **Lineweaver–Burk** or **double reciprocal plot,** the slope of the line is K_M/V_{max}, the $1/v_o$ intercept is $1/V_{max}$, and the extrapolated $1/[S]$ intercept is $-1/K_M$ (Fig. 13-9). A disadvantage of this plot is that most experimental measurements involve relatively high [S] and are therefore crowded onto the left side of the graph. Furthermore, for small values of [S], small errors in v_o lead to large errors in $1/v_o$ and hence to large errors in K_M and V_{max}.

Several other types of plots, each with its advantages and disadvantages, have been formulated for the determination of V_{max} and K_M from kinetic data. With the advent of conveniently available computers, however, kinetic data are commonly analyzed by mathematically sophisticated statistical treatments. Nevertheless, Lineweaver–Burk plots are valuable for the visual presentation of kinetic data as well as being useful in the analysis of kinetic data from enzymes requiring more than one substrate (Section 13-5C).

k_{cat}/K_M Is a Measure of Catalytic Efficiency

An enzyme's kinetic parameters provide a measure of its catalytic efficiency. We may define the **catalytic constant** of an enzyme as

$$k_{cat} = \frac{V_{max}}{[E]_T} \qquad [13.27]$$

This quantity is also known as the **turnover number** of an enzyme because it is the number of reaction processes (turnovers) that each active site catalyzes per unit time. The turnover numbers for a selection of enzymes are given in Table 13-1. Note that these quantities vary by over eight orders of magnitude depending on the identity of the enzyme as well as that of its substrate. Equation [13.23] indicates that for the Michaelis–Menten model, $k_{cat} = k_2$. For enzymes with more complicated mechanisms, k_{cat} may be a function of several rate constants.

When $[S] \ll K_M$, very little ES is formed. Consequently, $[E] \approx [E]_T$, so that Eq. [13.22] reduces to a second-order rate equation:

$$v_o \approx \left(\frac{k_2}{K_M}\right)[E]_T[S] \approx \left(\frac{k_{cat}}{K_M}\right)[E][S] \qquad [13.28]$$

k_{cat}/K_M is the apparent second-order rate constant of the enzymatic reaction; the rate of the reaction varies directly with how often enzyme and substrate encounter one another in solution. *The quantity k_{cat}/K_M is therefore a measure of an enzyme's catalytic efficiency.*

Some Enzymes Have Attained Catalytic Perfection

Is there an upper limit on enzymatic catalytic efficiency? From Eq. [13.21] we find

$$\frac{k_{cat}}{K_M} = \frac{k_2}{K_M} = \frac{k_1 k_2}{k_{-1} + k_2} \qquad [13.29]$$

This ratio is maximal when $k_2 \gg k_{-1}$; that is, when the formation of product from the Michaelis complex, ES, is fast compared to its decomposition back to substrate and enzyme. Then $k_{cat}/K_M = k_1$, the second-order rate constant for the formation of ES. The term k_1, of course, can be no greater than the frequency with which enzyme and substrate molecules collide with each other in solution. This **diffusion-controlled limit** is in the range of 10^8 to $10^9 M^{-1}s^{-1}$. Thus, enzymes with such values of k_{cat}/K_M must catalyze a reaction almost every time they encounter a substrate molecule. Table 13-1 indicates

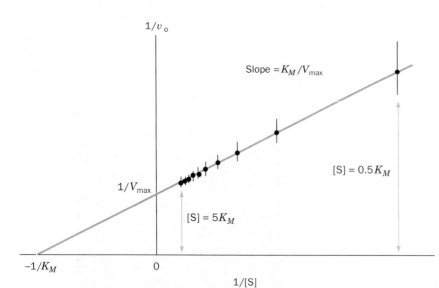

Figure 13-9
A double reciprocal (Lineweaver–Burk) plot with error bars of $\pm 0.05\ V_{max}$. The indicated points are the same as those in Fig. 13-8. Note the large effect of small errors at small [S] (large 1/[S]) and the crowding together of points at large [S].

that several enzymes, namely, catalase, acetylcholine esterase, fumarase, and possibly carbonic anhydrase, have achieved this state of virtual catalytic perfection.

Since the active site of an enzyme generally occupies only a small fraction of its total surface area, how can any enzyme catalyze a reaction every time it encounters a substrate molecule? Although the answer to this question is yet unclear, structural and theoretical evidence is accumulating suggesting that the arrangements of charged groups on the surfaces of enzymes serve to electrostatically guide polar substrates to their enzymes' active sites.

C. Reversible Reactions

The Michaelis–Menten model implicitly assumes that enzymatic reverse reactions may be neglected. Yet, many enzymatic reactions are highly reversible (have a small free energy of reaction) and therefore have products that back react to form substrates at a significant rate. In this section we therefore relax the Michaelis–Menten restriction of no back reaction and, by doing so, discover some interesting and important kinetic principles.

The One-Intermediate Model

Modification of the Michaelis–Menten model to incorporate a back reaction yields the following reaction scheme:

$$E + S \underset{k_{-1}}{\overset{k_1}{\rightleftharpoons}} ES \underset{k_{-2}}{\overset{k_2}{\rightleftharpoons}} P + E$$

(Here ES might just as well be called EP because this model does not specify the nature of the intermediate complex.) The equation describing the kinetic behavior of this model, which is derived in Appendix A of this chapter, is expressed

$$v = \frac{\dfrac{V^f_{max}[S]}{K^S_M} - \dfrac{V^r_{max}[P]}{K^P_M}}{1 + \dfrac{[S]}{K^S_M} + \dfrac{[P]}{K^P_M}} \qquad [13.30]$$

where

$$V^f_{max} = k_2[E]_T \qquad V^r_{max} = k_{-1}[E]_T$$

$$K^S_M = \frac{k_{-1} + k_2}{k_1} \qquad K^P_M = \frac{k_{-1} + k_2}{k_{-2}}$$

and

$$[E]_T = [E] + [ES]$$

This is essentially a Michaelis–Menten equation that works backwards as well as forwards. Indeed, at $[P] = 0$, that is, when $v = v_o$, this equation becomes the Michaelis–Menten equation.

The Haldane Relationship

At equilibrium, $v = 0$ so Eq. [13.30] can be solved to yield

$$K_{eq} = \frac{[P]}{[S]} = \frac{V^f_{max} K^P_M}{V^r_{max} K^S_M} \qquad [13.31]$$

which is known as the **Haldane relationship.** This relationship demonstrates that *the kinetic parameters of a reversible enzymatically catalyzed reaction are not independent of one another. Rather, they are related by the equilibrium constant for the overall reaction which, of course, is independent of the presence of the enzyme.*

Kinetic Data Cannot Unambiguously Establish a Reaction Mechanism

An enzyme that forms a reversible complex with its substrate should likewise form one with its product; that is, have a mechanism such as:

$$E + S \underset{k_{-1}}{\overset{k_1}{\rightleftharpoons}} ES \underset{k_{-2}}{\overset{k_2}{\rightleftharpoons}} EP \underset{k_{-3}}{\overset{k_3}{\rightleftharpoons}} P + E$$

The equation describing the kinetic behavior of this two-intermediate model, whose derivation is analogous to that described in Appendix A for the one-intermediate model, has a form identical to that of Eq. [13.30]. However, its parameters V^f_{max}, V^r_{max}, K^S_M, and K^P_M are defined in terms of the six kinetic constants of the two-intermediate model rather than the four of the one-intermediate model. In fact, the steady state rate equations for reversible reactions with three or more intermediates also have this same form but with yet different definitions of the four parameters.

The values of V^f_{max}, V^r_{max}, K^S_M, and K^P_M in Eq. [13.30] can be determined by suitable manipulations of the initial substrate and product concentrations under steady state conditions. This, however, will not yield the values of the rate constants for our two-intermediate model because there are six such constants and only four equations describing their relationships. Moreover, steady state kinetic measurements are incapable of distinguishing the number of intermediates in a reversible enzymatic reaction because the form of Eq. [13.30] does not change with this number of intermediates.

The functional identities of the equations describing these reaction schemes may be understood in terms of an analogy between our n-intermediate reversible reaction model and a "black box" containing a system of water pipes with one inlet and one drain:

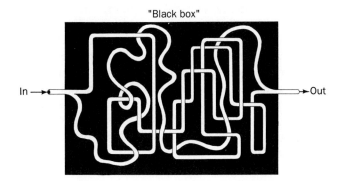

"Black box"

At steady state, that is, after the pipes have filled with water, one can measure the relationship between input pressure and output flow. However, such measurements yield no information concerning the detailed construction of the plumbing connecting the inlet to the drain. This would require additional information such as opening the black box and tracing the pipes. *Likewise, steady state kinetic measurements can provide a phenomenological description of enzymatic behavior, but the nature of the intermediates remains indeterminate. Rather, these intermediates must be detected and characterized by independent means such as by spectroscopic analysis.*

The foregoing discussion brings to light a central principle of kinetic analysis: *The steady state kinetic analysis of a reaction cannot unambiguously establish its mechanism.* This is because no matter how simple, elegant, or rational a mechanism one postulates that fully accounts for kinetic data, there are an infinite number of alternate mechanisms, perhaps complicated, awkward, and seemingly irrational that can account for these kinetic data equally well. Usually it is the simpler and more elegant mechanism that turns out to be correct, but this is not always the case. *If, however, kinetic data are not compatible with a given mechanism, then the mechanism must be rejected.* Therefore, although kinetics cannot be used to ambiguously establish a mechanism without confirming data, such as the physical demonstration of an intermediate's existence, the steady state kinetic analysis of a reaction is of great value because it can be used to eliminate proposed mechanisms.

3. INHIBITION

Many substances alter the activity of an enzyme by combining with it in a way that influences the binding of substrate and/or its turnover number. Substances that reduce an enzyme's activity in this way are known as **inhibitors.**

Many inhibitors are substances that structurally resemble their enzyme's substrate but either do not react or react very slowly compared to substrate. Such inhibitors are commonly used to probe the chemical and conformational nature of a substrate-binding site as part of an effort to elucidate the enzyme's catalytic mechanism. In addition, many enzyme inhibitors are effective chemotherapeutic agents since an "unnatural" substrate analog can block the action of a specific enzyme. For example, **methotrexate** (also called **amethopterin**) chemically resembles **dihydrofolate**. Methotrexate binds tightly to the enzyme **dihydrofolate reductase,** thereby preventing it from carrying out its normal function, the reduction of dihydofolate to **tetrahydrofolate,** an essential cofactor in the biosynthesis of the DNA precursor **thymidylic acid** (Section 26-4B).

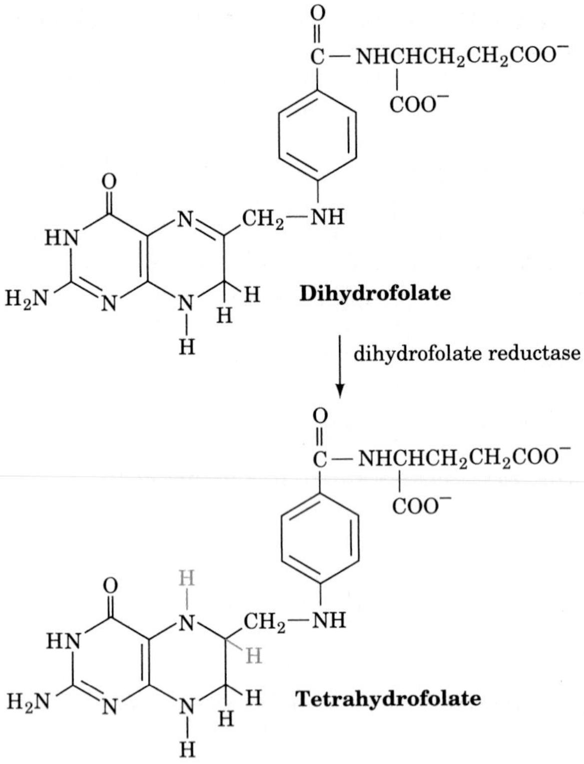

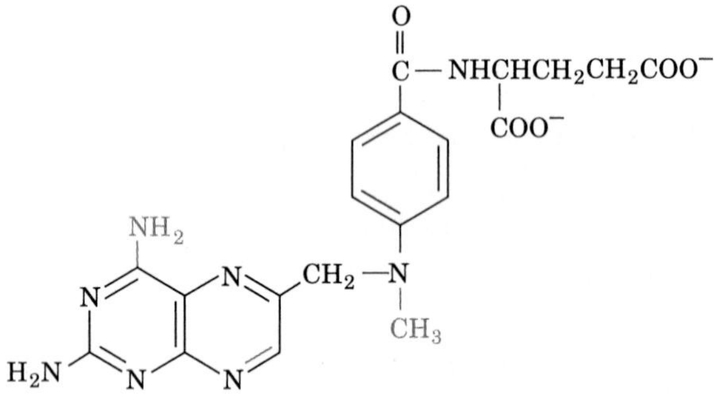

Rapidly dividing cells, such as cancer cells, which are actively engaged in DNA synthesis, are far more susceptible to methotrexate than are slower growing cells such as those of most normal mammalian tissues. Hence, methotrexate, when administered in proper dosage, kills cancer cells without fatally poisoning the host.

There are various mechanisms through which enzyme inhibitors can act. In this section, we discuss several of the simplest such mechanisms and their effects on the kinetic behavior of enzymes that follow the Michaelis–Menten model.

A. Competitive Inhibition

A substance that competes directly with a normal substrate for an enzymatic-binding site is known as a **competitive inhibitor.** Such an inhibitor usually resembles the substrate to the extent that it specifically binds to the active site but differs from it so as to be unreactive. Thus methotrexate is a competitive inhibitor of dihydrofolate reductase. Similarly, **succinate dehydrogenase,** a citric acid cycle enzyme that functions to convert **succinate** to **fumarate** (Section 19-3F), is competitively inhibited by **malonate,** which structurally resembles succinate but cannot be dehydrogenated.

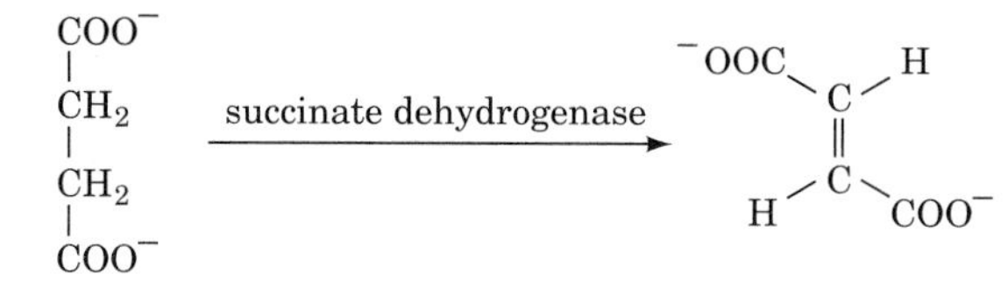

Succinate   **Fumarate**

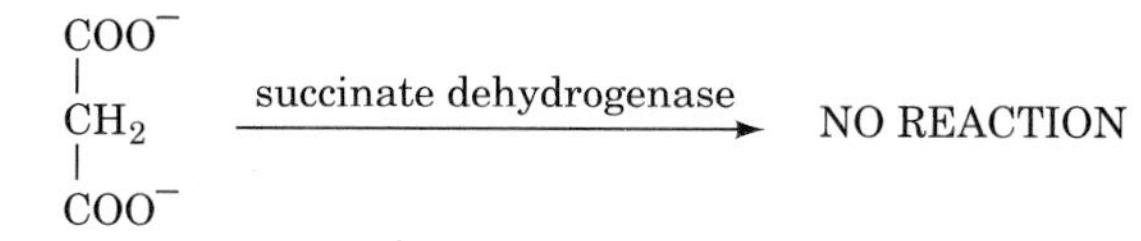

Malonate

The effectiveness of malonate in competitively inhibiting succinate dehydrogenase strongly suggests that the enzyme's substrate-binding site is designed to bind both of the substrate's carboxylate groups, presumably through the influence of two appropriately placed positively charged residues.

The general model for competitive inhibition is given by the following reaction scheme:

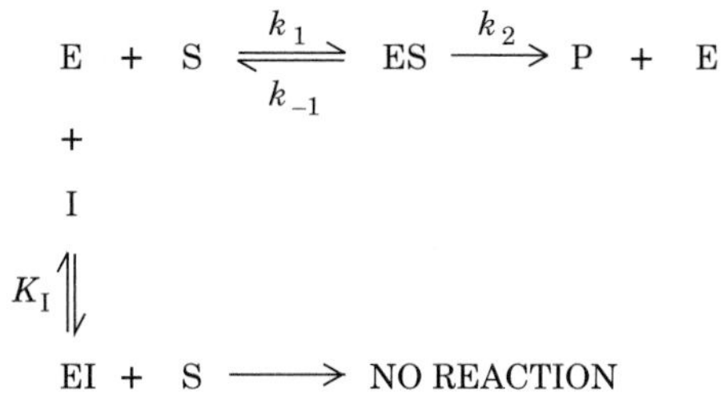

$$E \; + \; S \underset{k_{-1}}{\overset{k_1}{\rightleftharpoons}} ES \xrightarrow{k_2} P \; + \; E$$
$$+$$
$$I$$
$$K_I \Big\Updownarrow$$
$$EI \; + \; S \longrightarrow NO \; REACTION$$

Here it is assumed that I, the inhibitor, binds reversibly to the enzyme and is in rapid equilibrium with it so that

$$K_I = \frac{[E][I]}{[EI]} \qquad [13.32]$$

and EI, the enzyme–inhibitor complex, is catalytically inactive. *A competitive inhibitor therefore acts by reducing the concentration of free enzyme available for substrate binding.*

Our goal, as before, is to express v_o in terms of measurable quantities; in this case $[E]_T$, $[S]$, and $[I]$. We begin, as in the derivation of the Michaelis–Menten equation, with the expression for the conservation condition, which must now take into account the existence of EI.

$$[E]_T = [E] + [EI] + [ES] \qquad [13.33]$$

The enzyme concentration can be calculated by rearranging Eq. [13.17] under the steady state condition:

$$[E] = \frac{K_M[ES]}{[S]} \qquad [13.34]$$

That of the enzyme–inhibitor complex is found by rearranging Eq. [13.32] and substituting Eq. [13.34] into it

$$[EI] = \frac{[E][I]}{K_I} = \frac{K_M[ES][I]}{[S]K_I} \qquad [13.35]$$

Substituting the latter two results into Eq. [13.33] yields

$$[E]_T = [ES]\left\{\frac{K_M}{[S]}\left(1 + \frac{[I]}{K_I}\right) + 1\right\}$$

which can be rearranged to

$$[ES] = \frac{[E]_T[S]}{K_M\left(1 + \dfrac{[I]}{K_I}\right) + [S]}$$

so that, according to Eq. [13.22], the initial velocity is expressed

$$v_o = k_2[ES] = \frac{k_2[E]_T[S]}{K_M\left(1 + \dfrac{[I]}{K_I}\right) + [S]} \qquad [13.36]$$

Then defining

$$\alpha = \left(1 + \frac{[I]}{K_I}\right) \qquad [13.37]$$

and V_{max} as in Eq. [13.23],

$$v_o = \frac{V_{max}[S]}{\alpha K_M + [S]} \qquad [13.38]$$

This is the Michaelis–Menten equation with K_M modulated by α, a function of the inhibitor concentration (which, according to Eq. [13.37], must always be ≥ 1). The value of $[S]$ at $v_o = V_{max}/2$ is therefore αK_M.

Figure 13-10 shows the hyperbolic plot of Eq. [13.38] for various values of α. Note that as $[S] \to \infty$, $v_o \to V_{max}$ for any value of α. The larger the value of α, however, the greater $[S]$ must be to approach V_{max}. Thus, the inhibitor does not affect the turnover number of the enzyme. Rather, the presence of I has the effect of making $[S]$ appear more dilute than it actually is, or alternatively, making K_M appear larger than it really is. Conversely, increasing $[S]$ shifts the substrate-binding equilibrium towards $[ES]$. Hence, there is true competition between I

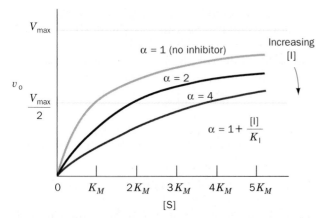

Figure 13-10
A plot of the initial velocity v_o of a simple Michaelis–Menten reaction versus the substrate concentration [S] in the presence of different concentrations of a competitive inhibitor.

and S for the enzyme's substrate-binding site; their binding is mutually exclusive.

Recasting Eq. [13.38] in the double reciprocal form yields

$$\frac{1}{v_o} = \left(\frac{\alpha K_M}{V_{max}}\right)\frac{1}{[S]} + \frac{1}{V_{max}} \qquad [13.39]$$

A plot of this equation is linear and has a slope of $\alpha K_M/V_{max}$, a $1/[S]$ intercept of $-1/\alpha K_M$, and a $1/v_o$ intercept of $1/V_{max}$ (Fig. 13-11). *The double reciprocal plots for a competitive inhibitor at various concentrations of I intersect at $1/V_{max}$; this is diagnostic for competitive inhibition as compared with other types of inhibition (see below).*

By determining the values of α at different inhibitor

concentrations, the value of K_I can be found from Eq. [13.37]. In this way, competitive inhibitors can be used to probe the structural nature of an active site. For example, to ascertain the importance of the various segments of an ATP molecule

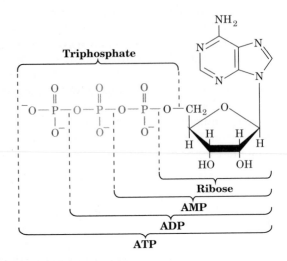

for binding to the active site of an ATP-requiring enzyme, one might determine the K_I, say, for ADP, AMP (adenosine monophosphate), ribose, triphosphate ion, *etc.* Since many of these ATP components are catalytically inactive, inhibition studies are the most convenient means of monitoring their binding to the enzyme.

If the inhibitor binds irreversibly to the enzyme, the inhibitor is classified as an **inactivator** as is any agent that somehow inactivates the enzyme. Inactivators truly reduce the effective level of $[E]_T$ at all values of [S]. Reagents that modify specific amino acid residues, such as those listed in Table 6-3, can act in this manner.

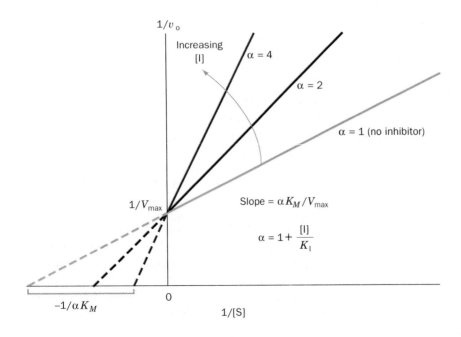

Figure 13-11
A Lineweaver–Burk plot of the competitively inhibited Michaelis–Menten enzyme described by Fig. 13-10. Note that all lines intersect on the $1/v_o$ axis at $1/V_{max}$.

B. Uncompetitive Inhibition

In **uncompetitive inhibition,** the inhibitor binds directly to the enzyme–substrate complex but not to the free enzyme:

$$E \; + \; S \; \underset{k_{-1}}{\overset{k_1}{\rightleftharpoons}} \; ES \; \xrightarrow{k_2} \; P \; + \; E$$

$$+$$

$$I$$

$$K_I' \Big\Vert$$

$$ESI \longrightarrow NO\ REACTION$$

The inhibitor-binding step, which has the dissociation constant

$$K_I' = \frac{[ES][I]}{[ESI]} \qquad [13.40]$$

is assumed to be at equilibrium. The binding of the uncompetitive inhibitor, which need not resemble the substrate, is envisioned to cause structural distortion of the active site thereby rendering the enzyme catalytically inactive. (If the inhibitor binds to enzyme alone, it does so without affecting its affinity for substrate.)

The Michaelis–Menten equation for uncompetitive inhibition, which is derived in Appendix B of this chapter, is

$$v_o = \frac{V_{max}[S]}{K_M + \alpha'[S]} \qquad [13.41]$$

where

$$\alpha' = 1 + \frac{[I]}{K_I'} \qquad [13.42]$$

Inspection of this equation indicates that *at high values of [S], v_o asymptotically approaches V_{max}/α' so that, in contrast to competitive inhibition, the effects of uncompetitive* inhibition *on V_{max} are not reversed by increasing the substrate concentration. However, at low substrate concentrations, that is, when [S] $\ll K_M$, the effect of an uncompetitive inhibitor becomes negligible, again the opposite behavior of a competitive inhibitor.*

When cast in the double reciprocal form, Eq. [13.41] becomes

$$\frac{1}{v_o} = \left(\frac{K_M}{V_{max}}\right)\frac{1}{[S]} + \frac{\alpha'}{V_{max}} \qquad [13.43]$$

The Lineweaver–Burk plot for uncompetitive inhibition is linear with slope K_M/V_{max} as in the uninhibited reaction, and with $1/v_o$ and $1/[S]$ intercepts of α'/V_{max} and $-\alpha'/K_M$, respectively. *A series of Lineweaver–Burk plots at various uncompetitive inhibitor concentrations consists of a family of parallel lines (Fig. 13-12). This is diagnostic for uncompetitive inhibition.*

Uncompetitive inhibition requires that the inhibitor affect the catalytic function of the enzyme but not its substrate binding. For single-substrate enzymes it is difficult to conceive of how this could happen with the exception of small inhibitors such as protons (see Section 13-4) or metal ions. As we discuss in Section 13-5C, however, uncompetitive inhibition is important for multisubstrate enzymes.

C. Mixed Inhibition

If both the enzyme and the enzyme–substrate complex bind inhibitor, the following model results:

$$E \; + \; S \; \underset{k_{-1}}{\overset{k_1}{\rightleftharpoons}} \; ES \; \xrightarrow{k_2} \; P \; + \; E$$

$$+ \qquad\qquad +$$

$$I \qquad\qquad I$$

$$K_I \Big\Vert \qquad\qquad K_I' \Big\Vert$$

$$EI \qquad\qquad ESI \longrightarrow NO\ REACTION$$

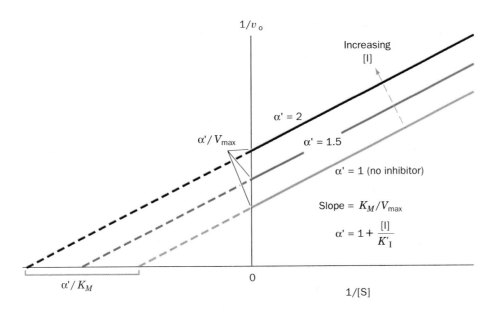

Figure 13-12
A Lineweaver–Burk plot of a simple Michaelis–Menten enzyme in the presence of uncompetitive inhibitor. Note that all lines have identical slopes of K_M/V_{max}.

Both of the inhibitor-binding steps are assumed to be at equilibrium but with different dissociation constants:

$$K_I = \frac{[E][I]}{[EI]} \quad \text{and} \quad K_I' = \frac{[ES][I]}{[ESI]} \quad [13.44]$$

This phenomenon is alternatively known as **mixed inhibition** or **noncompetitive inhibition**. Presumably a mixed inhibitor binds to enzyme sites that participate in both substrate binding and catalysis.

The Michaelis–Menten equation for mixed inhibition, which is derived in Appendix C of this chapter, is

$$v_o = \frac{V_{max}[S]}{\alpha K_M + \alpha'[S]} \quad [13.45]$$

where α and α' are defined in Eqs. [13.37] and [13.42], respectively. It can be seen from Eq. [13.45] that the name mixed inhibition arises from the fact that the denominator has the factor α multiplying K_M as in competitive inhibition, (Eq. [13.38]) and the factor α' multiplying [S] as in uncompetitive inhibition (Eq. [13.41]). Mixed inhibitors are therefore effective at both high and low substrate concentrations.

The Lineweaver–Burk equation for mixed inhibition is

$$\frac{1}{v_o} = \left(\frac{\alpha K_M}{V_{max}}\right)\frac{1}{[S]} + \frac{\alpha'}{V_{max}} \quad [13.46]$$

The plot of this equation consists of lines that have slope $\alpha K_M/V_{max}$ with a $1/v_o$ intercept of α'/V_{max} and a $1/[S]$ intercept of $-\alpha'/\alpha K_M$ (Fig. 13-13). Algebraic manipulation of Eq. [13.46] for different values of [I] reveals that

Table 13-2

The Effects of Inhibitors on the Parameters of the Michaelis–Menten Equation[a]

Type of Inhibition	V_{max}^{app}	K_M^{app}
None	V_{max}	K_M
Competitive	V_{max}	αK_M
Uncompetitive	V_{max}/α'	K_M/α'
Mixed	V_{max}/α'	$\alpha K_M/\alpha'$

[a] $\alpha = 1 + \dfrac{[I]}{K_I}$ and $\alpha' = 1 + \dfrac{[I]}{K_I'}$

this equation describes a family of lines that intersect to the left of the $1/v_o$ axis (Fig. 13-13); for the special case in which $K_I = K_I'(\alpha = \alpha')$, the intersection is, in addition, on the $1/[S]$ axis.

Table 13-2 provides a summary of the preceding results concerning the inhibition of simple Michaelis–Menten enzymes. The quantities K_M^{app} and V_{max}^{app} are the "apparent" values of K_M and V_{max} that would actually be observed in the presence of inhibitor for the Michaelis–Menten equation describing the inhibited enzymes.

4. EFFECTS OF pH

Enzymes, being proteins, have properties that are quite pH sensitive. Most enzymes, in fact, are active only within a narrow pH range, typically 5 to 9. This is a result of the effects of pH on a combination of factors: (1) the binding of substrate to enzyme, (2) the catalytic

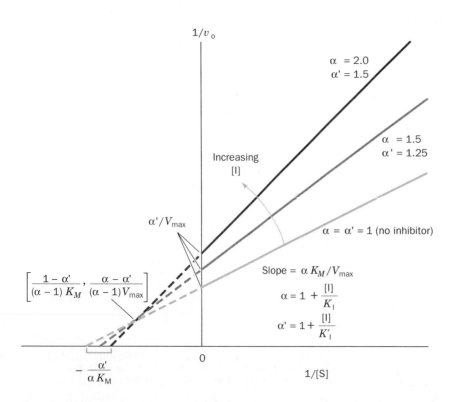

Figure 13-13
A Lineweaver–Burk plot of a simple Michaelis–Menten enzyme in the presence of a mixed inhibitor. Note that the lines all intersect to the left of the $1/v_o$ axis. The coordinates of this intersection point are given in brackets.

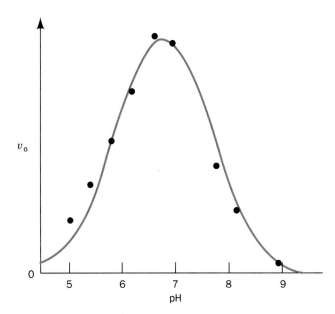

Figure 13-14
The effect of pH on the initial rate of the reaction catalyzed by the enzyme fumarase. [After Tanford, C., *Physical Chemistry of Macromolecules*, p. 647, Wiley (1961).]

activity of the enzyme, (3) the ionization of substrate, and (4) the variation of protein structure (usually significant only at extremes of pH).

pH Dependence of Simple Michaelis–Menten Enzymes

The initial rates for many enzymatic reactions exhibit bell-shaped curves as a function of pH (e.g., Fig. 13-14). These curves reflect the ionizations of certain amino acid residues that must be in a specific ionization state for enzyme activity. The following model can account for such pH effects.

$$
\begin{array}{ccc}
E^- & & E\bar{S} \\
K_{E2} \big\Vert H^+ & & K_{ES2} \big\Vert H^+ \\
EH + S \underset{k_{-1}}{\overset{k_1}{\rightleftharpoons}} ESH & \overset{k_2}{\longrightarrow} & P + EH \\
K_{E1} \big\Vert H^+ & & K_{ES1} \big\Vert H^+ \\
EH_2^+ & & ESH_2^+
\end{array}
$$

In this expansion of the simple one substrate–no back reaction mechanism, it is assumed that only EH and ESH are catalytically active.

The Michaelis–Menten equation for this model, which is derived in Appendix D, is

$$
v_o = \frac{V'_{max}[S]}{K'_M + [S]} \tag{13.47}
$$

Here the apparent Michaelis–Menten parameters are defined

$$
V'_{max} = V_{max}/f_2 \quad \text{and} \quad K'_M = K_M(f_1/f_2)
$$

where

$$
f_1 = \frac{[H^+]}{K_{E1}} + 1 + \frac{K_{E2}}{[H^+]}
$$

$$
f_2 = \frac{[H^+]}{K_{ES1}} + 1 + \frac{K_{ES2}}{[H^+]}
$$

and V_{max} and K_M refer to the active forms of the enzyme, EH and ESH. Note that at any given pH, Eq. [13.47] behaves as a simple Michaelis–Menten equation, but because of the pH dependence of f_1 and f_2, v_o varies with pH in a bell-shaped manner (e.g., Fig. 13-14).

Evaluation of Ionization Constants

The ionization constants of enzymes that obey Eq. [13.47] can be evaluated by the analysis of the curves of $\log V'_{max}$ versus pH, which provides values of K_{ES1} and K_{ES2} (Fig. 13-15a), and of $\log (V'_{max}/K'_M)$ versus pH, which yields K_{E1} and K_{E2} (Fig. 13-15b). This, of course, entails the determination of the enzyme's Michaelis–Menten parameters at each of a series of different pH's.

The measured pK's often provide valuable clues as to the identities of the amino acid residues essential for enzymatic activity. For example, a measured pK of ~ 4 suggests that an Asp or Glu residue is essential to the enzyme. Similarly, pK's of ~ 6 or 10 suggest the participation of a His or a Lys residue, respectively. However, a given acid–base group may vary by as much as several pH units from its expected value as a consequence of the electrostatic influence of nearby charged groups, as well as of the proximity of regions of low polarity. For example, the carboxylate group of a Glu residue forming a salt bridge with a Lys residue is stabilized by the nearby positive charge and therefore has a lower pK than it

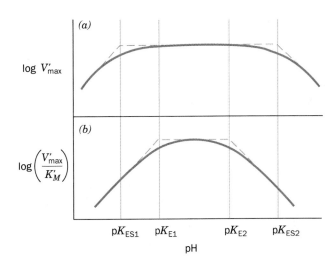

Figure 13-15
The pH dependence of (a) $\log V'_{max}$ and (b) $\log (V'_{max}/K'_M)$ illustrating how the values of the molecular ionization constants can be determined by graphical extrapolation.

would otherwise have; that is, it is more difficult to protonate. Conversely, a carboxylate group immersed in a region of low polarity is less acidic than normal because it attracts protons more strongly than if it were in a region of higher polarity. The identification of a kinetically characterized pK with a particular amino acid residue must therefore be verified by other types of measurements such as the use of group-specific reagents to inactivate a putative essential residue (Section 6-2).

5. BISUBSTRATE REACTIONS

We have heretofore been concerned with reactions involving enzymes that require only a single substrate. Yet, enzymatic reactions involving two substrates and yielding two products

$$A + B \underset{E}{\rightleftharpoons} P + Q$$

account for ~60% of known biochemical reactions. Almost all of these so-called **bisubstrate reactions** are either **transferase** reactions in which the enzyme catalyzes the transfer of a specific functional group, X, from one of the substrates to the other:

$$P-X + B \underset{E}{\rightleftharpoons} P + B-X$$

or oxidation–reduction reactions in which reducing equivalents are transferred between the two substrates. For example, the hydrolysis of a peptide bond by trypsin (Section 6-1E) is the transfer of the peptide carbonyl group from the peptide nitrogen atom to water (Fig. 13-16a). Similarly, in the alcohol dehydrogenase reaction (Section 12-2A), a hydride ion is formally transferred from ethanol to NAD$^+$ (Fig. 13-16b). Although such bisubstrate reactions could, in principle, occur

(a)

$$R_1-\overset{\overset{\displaystyle O}{\|}}{C}-NH-R_2 + H_2O \xrightarrow{\text{trypsin}} R_1-\overset{\overset{\displaystyle O}{\|}}{C}-O^- + H_3\overset{+}{N}-R_2$$

Polypeptide

(b)

$$CH_3-\overset{\overset{\displaystyle H}{|}}{\underset{\underset{\displaystyle H}{|}}{C}}-OH + NAD^+ \xrightarrow[H^+]{\text{alcohol}\atop\text{dehydrogenase}} CH_3-\overset{\overset{\displaystyle O}{\|}}{C}H + NADH$$

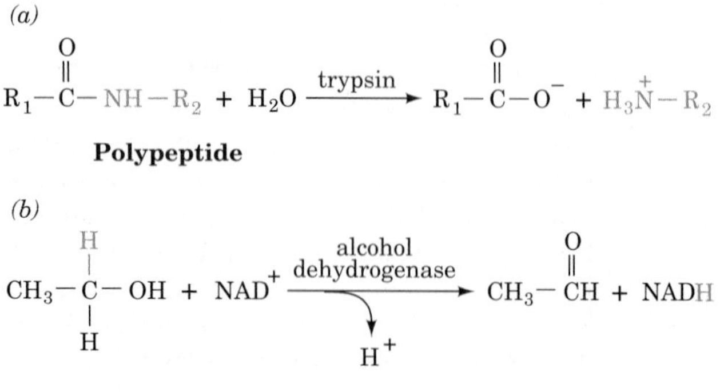

Figure 13-16
Some bisubstrate reactions: (a) In the peptide hydrolysis reaction catalyzed by trypsin, the peptide carbonyl group, with its pendent polypeptide chain, is transferred from the peptide nitrogen atom to a water molecule. (b) In the alcohol dehydrogenase reaction, a hydride ion is formally transferred from ethanol to NAD$^+$.

through a vast variety of mechanisms, only a few types are commonly observed.

A. Terminology

We shall follow the nomenclature system introduced by W. W. Cleland for representing enzymatic reactions:

1. Substrates are designated by the letters A, B, C, and D *in the order that they add to the enzyme.*

2. Products are designated P, Q, R, and S *in the order that they leave the enzyme.*

3. Stable enzyme forms are designated E, F, and G with E being the free enzyme, if such distinctions can be made. A stable enzyme form is defined as one that by itself is incapable of converting to another stable enzyme form (see below).

4. The number of reactants and products in a given reaction are specified, in order, by the terms **Uni** (one), **Bi** (two), **Ter** (three), and **Quad** (four). A reaction requiring one substrate and yielding three products is designated a Uni Ter reaction. In this section, we shall be concerned with reactions that require two substrates and yield two products, that is, Bi Bi reactions. Keep in mind, however, that there are numerous examples of even more complex reactions.

Types of Bi Bi Reactions
Enzyme-catalyzed group-transfer reactions fall under two major mechanistic classifications:

1. Sequential Reactions
Reactions in which all substrates must combine with the enzyme before a reaction can occur and products be released are known as Sequential reactions. In such reactions, the group being transferred, X, is directly passed from A ($=$ P—X) to B yielding P and Q ($=$ B—X). Hence, such reactions are also called **single displacement reactions.**

Sequential reactions can be subclassified into those with a compulsory order of substrate addition to the enzyme, which are said to have an **Ordered mechanism,** and those with no preference for the order of substrate addition, which are described as having a **Random mechanism.** In the Ordered mechanism, the binding of the first substrate is apparently required for the enzyme to form the binding site for the second substrate, whereas for the Random mechanism, both binding sites are present on the free enzyme.

Let us describe enzymatic reactions using Cleland's shorthand notation. The enzyme is represented by a horizontal line and successive additions of substrates and release of products are denoted by vertical arrows. Enzyme forms are placed under the line and rate constants, if given, are to the left of the arrow or on top of the line for forward reactions. An **Ordered Bi Bi** reaction is represented:

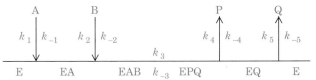

where A and B are said to be the **leading** and **following** substrates, respectively. Here, only minimal details are given concerning the interconversions of intermediate enzyme forms because, as we have seen for reversible single-substrate enzymes, steady state kinetic measurements provide no information concerning the number of intermediates in a given reaction step. Many NAD$^+$ and NADP$^+$ requiring dehydrogenases follow an Ordered Bi Bi mechanism in which the coenzyme is the leading reactant.

A **Random Bi Bi** reaction is diagrammed:

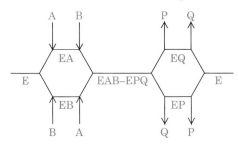

Some dehydrogenases and kinases operate through Random Bi Bi mechanisms.

2. Ping Pong Reactions

*Mechanisms in which one or more products are released before all substrates have been added are known as **Ping Pong reactions**.* The **Ping Pong Bi Bi** reaction is represented by

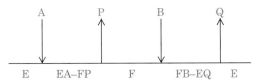

In it, a functional group X of the first substrate A (= P—X) is displaced from the substrate by the enzyme E to yield the first product P and a stable enzyme form F (= E—X) in which X is tightly (often covalently) bound to the enzyme (Ping). In the second stage of the reaction, X is displaced from the enzyme by the second substrate B to yield the second product Q (= B—X) thereby regenerating the original form of the enzyme, E (Pong). Such reactions are therefore also known as **double displacement reactions**. *Note that in Ping Pong Bi Bi reactions, the substrates A and B do not encounter one another on the surface of the enzyme.* Many enzymes, including chymotrypsin (Section 14-3), transaminases and some flavoenzymes, react with Ping Pong mechanisms.

B. Rate Equations

Steady state kinetic measurements can be used to distinguish among the foregoing bisubstrate mechanisms. In order to do so, one must first derive their rate equations. This can be done in much the same manner as for single-substrate enzymes, that is, solving a set of simultaneous linear equations consisting of an equation expressing the steady state condition for each kinetically distinct enzyme complex plus one equation representing the conservation condition for the enzyme. This, of course, is a more complex undertaking for bisubstrate enzymes than it is for single-substrate enzymes.

The rate equations for the above described bisubstrate mechanisms are given below in double reciprocal form.

Ordered Bi Bi

$$\frac{1}{v_o} = \frac{1}{V_{max}} + \frac{K_M^A}{V_{max}[A]} + \frac{K_M^B}{V_{max}[B]} + \frac{K_S^A K_M^B}{V_{max}[A][B]} \quad [13.48]$$

Rapid Equilibrium Random Bi Bi

The rate equation for the general Random Bi Bi reaction is quite complicated. However, in the special case that both substrates are in rapid and independent equilibrium with the enzyme; that is, the EAB–EPQ interconversion is rate determining, the initial rate equation reduces to the following relatively simple form. This mechanism is known as the **Rapid Equilibrium Random Bi Bi** mechanism:

$$\frac{1}{v_o} = \frac{1}{V_{max}} + \frac{K_S^A K_M^B}{V_{max} K_S^B [A]} + \frac{K_M^B}{V_{max}[B]} + \frac{K_S^A K_M^B}{V_{max}[A][B]}$$

$$[13.49]$$

Ping Pong Bi Bi

$$\frac{1}{v_o} = \frac{K_M^A}{V_{max}[A]} + \frac{K_M^B}{V_{max}[B]} + \frac{1}{V_{max}} \quad [13.50]$$

Physical Significance of the Bisubstrate Kinetic Parameters

The kinetic parameters in the equations describing bisubstrate reactions have meanings similar to those for single substrate reactions. The term V_{max} is the maximal velocity of the enzyme obtained when both A and B are present at saturating concentrations, K_M^A and K_M^B are the respective concentrations of A and B necessary to achieve $\frac{1}{2} V_{max}$ in the presence of a saturating concentration of the other, and K_S^A and K_S^B are the respective dissociation constants of A and B from the enzyme, E.

C. Differentiating Bisubstrate Mechanisms

One can discriminate between Ping Pong and Sequential mechanisms from their contrasting properties in linear plots such as those of the Lineweaver–Burk type.

(a)

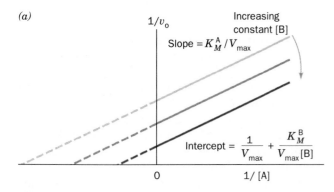

(b)

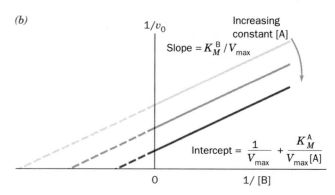

Figure 13-17
Double reciprocal plots for an enzymatic reaction with a Ping Pong Bi Bi mechanism. (*a*) Plots of $1/v_o$ versus $1/[A]$ at various constant concentrations of B. (*b*) Plots of $1/v_o$ versus $1/[B]$ at various constant concentrations of A.

Diagnostic Plot for Ping Pong Bi Bi Reactions

A plot of $1/v_o$ versus $1/[A]$ at constant [B] for Eq. [13.50] yields a straight line of slope K_M^A/V_{max} and an intercept on the $1/v_o$ axis equal to the last two terms in Eq. [13.50]. Since the slope is independent of [B], such plots for different values of [B] yield a family of parallel lines (Fig. 13-17). A plot of $1/v_o$ versus $1/[B]$ for different values of [A] likewise yields a family of parallel lines. *Such parallel lines are diagnostic for a Ping Pong mechanism.*

Diagnostic Plot for Sequential Bi Bi Reactions

The equations representing the Ordered Bi Bi mechanism (Eq. [13.48]) and the Rapid Equilibrium Random Bi Bi mechanism (Eq. [13.49]) have identical functional dependence on [A] and [B].

Equation [13.48] can be rearranged to

$$\frac{1}{v_o} = \frac{K_M^A}{V_{max}}\left(1 + \frac{K_S^A K_M^B}{K_M^A [B]}\right)\frac{1}{[A]} + \frac{1}{V_{max}}\left(1 + \frac{K_M^B}{[B]}\right)$$

$$[13.51]$$

Thus plotting $1/v_o$ versus $1/[A]$ for constant [B] yields a linear plot with an intercept on the $1/v_o$ axis equal to the second term of Eq. [13.51] (Fig. 13-18*a*). Alternatively, Eq. [13.48] can be rearranged to

$$\frac{1}{v_o} = \frac{K_M^B}{V_{max}}\left(1 + \frac{K_S^A}{[A]}\right)\frac{1}{[B]} + \frac{1}{V_{max}}\left(1 + \frac{K_M^A}{[A]}\right) \quad [13.52]$$

which yields a linear plot of $1/v_o$ versus $1/[B]$ for constant [A] with a slope equal to the coefficient of $1/[B]$ in Eq. [13.52] (Fig. 13-18*b*). *The characteristic feature of these plots, which is indicative of a Sequential mechanism, is that the lines intersect to the left of the $1/v_o$ axis.*

Differentiating Random and Ordered Sequential Mechanisms

The Ordered Bi Bi mechanism may be experimentally distinguished from the Random Bi Bi mechanism through **product inhibition studies.** If only one product of the reaction, P or Q, is added to the reaction mixture, the reverse reaction still cannot occur. Never-

(a)

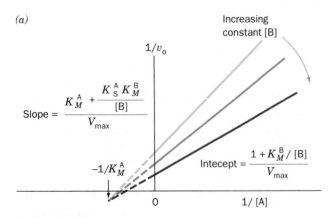

(b)

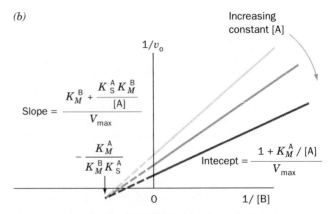

Figure 13-18
Double reciprocal plots of an enzymatic reaction with a Sequential Bi Bi mechanism. (*a*) Plots of $1/v_o$ versus $1/[A]$ at various constant concentrations of B. (*b*) Plots of $1/v_o$ versus $1/[B]$ at various constant concentrations of A. The corresponding plots for Rapid Equilibrium Random Bi Bi reactions have identical appearances; their lines all intersect to the left of the $1/v_o$ axis.

Table 13-3
Patterns of Product Inhibition for Sequential Bisubstrate Mechanisms

Mechanism	Product Inhibitor	A Variable	B Variable
Ordered Bi Bi	P	Mixed	Mixed
	Q	Competitive	Mixed
Rapid Equilibrium Random Bi Bi	P	Competitive	Competitive
	Q	Competitive	Competitive

theless, by binding to the enzyme, this product will inhibit the forward reaction. For an Ordered Bi Bi reaction, Q (= B—X, the second product to be released) directly competes with A (= P—X, the leading substrate) for binding to E and hence is a competitive inhibitor of A when [B] is fixed (the presence of X in Q = B—X interferes with the binding of A = P—X). However, since B combines with EA, not E, Q is a mixed inhibitor of B when [A] is fixed (Q interferes with both the binding of B to enzyme and with the catalysis of the reaction). Similarly, P, which combines only with EQ, is a mixed inhibitor of A when [B] is held constant and of B when [A] is held constant. In contrast, in a Rapid Equilibrium Bi Bi reaction, since both products as well as both substrates can combine directly with E, both P and Q are competitive inhibitors of A when [B] is constant and of B when [A] is constant. These product inhibition patterns are summarized in Table 13-3.

D. Isotope Exchange

Mechanistic conclusions based on kinetic analyses alone are fraught with uncertainties and are easily confounded by inaccurate experimental data. A particular mechanism for an enzyme is therefore greatly corroborated if the mechanism can be shown to conform to experimental criteria other than kinetic analysis.

Sequential (single displacement) and Ping Pong (double displacement) bisubstrate mechanisms may be differentiated through the use of isotope exchange studies. Double displacement reactions are capable of exchanging an isotope from the first product P back to the first substrate A in the absence of the second substrate. Consider an overall Ping Pong reaction catalyzed by the bisubstrate enzyme E

$$P - X + B \underset{E}{\rightleftarrows} P + B - X$$

in which, as usual, A = P—X, Q = B—X, and X is the group that is transferred from one substrate to the other in the course of the reaction. Only the first step of the reaction can take place in the absence of B. If a small amount of isotopically labeled P, denoted P*, is added to

this reaction mixture then, in the reverse reaction, P*—X will form:

Forward reaction	$E + P - X \longrightarrow E - X + P$
Reverse reaction	$E - X + P^* \longrightarrow E + P^* - X$

that is, isotopic exchange will occur.

In contrast, let us consider the first step of a Sequential reaction. Here a noncovalent enzyme–substrate complex forms:

$$E + P - X \rightleftharpoons E \cdot P - X$$

Addition of P* cannot result in an exchange reaction because no covalent bonds are broken in the formation of E·P—X, that is, there is no P released from the enzyme to exchange with P*. The demonstration of isotopic exchange for a bisubstrate enzyme is therefore convincing evidence favoring a Ping Pong mechanism.

Isotope Exchange in Sucrose Phosphorylase and Maltose Phosphorylase

The enzymes **sucrose phosphorylase** and **maltose phosphorylase** provide two clearcut examples of enzymatically catalyzed isotopic exchange reactions. Sucrose phosphorylase catalyzes the overall reaction

Glucose — fructose + phosphate
Sucrose

⇅ E

Glucose-1-phosphate + fructose

If the enzyme is incubated with sucrose and isotopically labeled fructose in the absence of phosphate, it is observed that the label passes into the sucrose:

Glucose — fructose + fructose*
Sucrose

⇅ E

Glucose — fructose* + fructose

For the reverse reaction, if the enzyme is incubated with

glucose-1-phosphate and ^{32}P-labeled phosphate, this label exchanges into the glucose-1-phosphate:

Glucose-1-phosphate + phosphate*

$\Updownarrow$ E

Glucose-1-phosphate* + phosphate

These observations indicate that a tight glucosyl–enzyme complex is formed with the release of fructose, thereby establishing that the sucrose phosphorylase reaction occurs via a Ping Pong mechanism. This finding has been conclusively corroborated by the isolation and characterization of this glucosyl–enzyme complex.

The enzyme **maltose phosphorylase** catalyzes a similar overall reaction:

Glucose—glucose + phosphate

Maltose

$\Updownarrow$ E

Glucose-1-phosphate + glucose

In contrast to sucrose phosphorylase, however, it does not catalyze isotopic exchange between glucose-1-phosphate and [^{32}P]phosphate or between maltose and [^{14}C]glucose. Likewise, a glucosyl–enzyme complex has not been detected. This evidence is consistent with maltose phosphorylase having a sequential mechanism.

Appendix: Derivations of Michaelis–Menten Equation Variants

A. The Michaelis–Menten Equation for Reversible Reactions—Equation [13.30]

The conservation condition for the reversible reaction with one intermediate (Section 13-2C) is

$$[E]_T = [E] + [ES] \qquad [13.A1]$$

The steady state condition is

$$\frac{d[ES]}{dt} = k_1[E][S] + k_{-2}[E][P] - (k_{-1} + k_2)[ES] = 0 \qquad [13.A2]$$

so that

$$[E] = \left(\frac{k_{-1} + k_2}{k_1[S] + k_{-2}[P]}\right)[ES] \qquad [13.A3]$$

Substituting this result into Eq. [13.A1] yields

$$[E]_T = \left(\frac{k_{-1} + k_2}{k_1[S] + k_{-2}[P]} + 1\right)[ES] \qquad [13.A4]$$

The velocity of the reaction is expressed

$$v = -\frac{d[S]}{dt} = k_1[E][S] - k_{-1}[ES] \qquad [13.A5]$$

which can be combined with Eq. [13.A3] to give

$$v = \left(\frac{k_1[S](k_{-1} + k_2)}{k_1[S] + k_{-2}[P]} - k_{-1}\right)[ES] \qquad [13.A6]$$

which, in turn, is combined with Eq. [13.A4] to yield

$$v = \left(\frac{k_1 k_2[S] - k_{-1}k_{-2}[P]}{k_{-1} + k_2 + k_1[S] + k_{-2}[P]}\right)[E]_T \qquad [13.A7]$$

Dividing the numerator and denominator of this equation by $(k_{-1} + k_2)$ results in

$$v = \left(\frac{k_2\left(\dfrac{k_1}{k_{-1} + k_2}\right)[S] - k_{-1}\left(\dfrac{k_{-2}}{k_{-1} + k_2}\right)[P]}{1 + \left(\dfrac{k_1}{k_{-1} + k_2}\right)[S] + \left(\dfrac{k_{-2}}{k_{-1} + k_2}\right)[P]}\right)[E]_T \qquad [13.A8]$$

Then, if we define the following parameters analogously with the constants of the Michaelis–Menten equation (Eqs. [13.23] and [13.21]),

$$V^f_{max} = k_2[E]_T \qquad V^r_{max} = k_{-1}[E]_T$$

$$K^S_M = \frac{k_{-1} + k_2}{k_1} \qquad K^P_M = \frac{k_{-1} + k_2}{k_{-2}}$$

we obtain the Michaelis–Menten equation for a reversible, one intermediate reaction:

$$v = \frac{\dfrac{V^f_{max}[S]}{K^S_M} - \dfrac{V^r_{max}[P]}{K^P_M}}{1 + \dfrac{[S]}{K^S_M} + \dfrac{[P]}{K^P_M}} \qquad [13.30]$$

B. Michaelis–Menten Equation for Uncompetitive Inhibition—Equation [13.41]

For uncompetitive inhibition (Section 13-3B), the inhibitor binds to the Michaelis complex with dissociation constant

$$K'_I = \frac{[ES][I]}{[ESI]} \qquad [13.A9]$$

The conservation condition is

$$[E]_T = [E] + [ES] + [ESI] \qquad [13.A10]$$

Substituting into Eqs. [13.34] and [13.39]

$$[E]_T = [ES]\left(\frac{K_M}{[S]} + 1 + \frac{[I]}{K_I'}\right) \qquad [13.A11]$$

Defining α' similarly to Eq. [13.37] as

$$\alpha' = 1 + \frac{[I]}{K_I'} \qquad [13.A12]$$

and v_o and V_{max} as in Eqs. [13.22] and [13.23], respectively,

$$v_o = k_2[ES] = \frac{V_{max}}{\frac{K_M}{[S]} + \alpha'} \qquad [13.A13]$$

which upon rearrangement yields the Michaelis–Menten equation for uncompetitive inhibition:

$$v_o = \frac{V_{max}[S]}{K_M + \alpha'[S]} \qquad [13.41]$$

C. The Michaelis–Menten Equation for Mixed Inhibition — Equation [13.45]

In mixed inhibition (Section 13-3C), the inhibitor-binding steps have different dissociation constants:

$$K_I = \frac{[E][I]}{[EI]} \quad \text{and} \quad K_I' = \frac{[ES][I]}{[ESI]} \qquad [13.A14]$$

(Here, for the sake of mathematical simplicity, we are making the thermodynamically unsupportable assumption that EI does not react with S to form ESI. Inclusion of this reaction requires a more complex derivation than that given here but leads to results that are substantially the same.) The conservation condition for this reaction scheme is

$$[E]_T = [E] + [EI] + [ES] + [ESI] \qquad [13.A15]$$

so that substituting in Eqs. [13.A14]

$$[E]_T = [E]\left(1 + \frac{[I]}{K_I}\right) + [ES]\left(1 + \frac{[I]}{K_I'}\right) \qquad [13.A16]$$

Defining α and α' as in Eqs. [13.38] and [13.A12], respectively, Eq. [13.A16] becomes

$$[E]_T = [E]\alpha + [ES]\alpha' \qquad [13.A17]$$

Then substituting in Eq. [13.34]

$$[E]_T = [ES]\left(\frac{\alpha K_M}{[S]} + \alpha'\right) \qquad [13.A18]$$

Defining v_o and V_{max} as in Eqs. [13.22] and [13.23] results in the Michaelis–Menten equation for mixed inhibition:

$$v_o = \frac{V_{max}[S]}{\alpha K_M + \alpha'[S]} \qquad [13.45]$$

D. The Michaelis–Menten Equation for Ionizable Enzymes — Equation [13.47]

In the model presented in Section 13-4 to account for the effect of pH on enzymes, the dissociation constants for the ionizations are

$$K_{E2} = \frac{[H^+][E^-]}{[EH]} \qquad K_{ES2} = \frac{[H^+][ES^-]}{[ESH]}$$

$$[13.A19]$$

$$K_{E1} = \frac{[H^+][EH]}{[EH_2^+]} \qquad K_{ES1} = \frac{[H^+][ESH]}{[ESH_2^+]}$$

Protonation and deprotonation are among the fastest known reactions so that, with the exception of the few enzymes with extremely high turnover numbers, it can be reasonably assumed that all acid–base reactions are at equilibrium. The conservation condition is

$$[E]_T = [ESH]_T + [EH]_T \qquad [13.A20]$$

where $[E]_T$ is the total enzyme present in any form,

$$[EH]_T = [EH_2^+] + [EH] + [E^-]$$

$$= [EH]\left(\frac{[H^+]}{K_{E1}} + 1 + \frac{K_{E2}}{[H^+]}\right) = [EH]f_1 \qquad [13.A21]$$

and

$$[ESH]_T = [ESH_2^+] + [ESH] + [ES^-]$$

$$= [ESH]\left(\frac{[H^+]}{K_{ES1}} + 1 + \frac{K_{ES2}}{[H^+]}\right) = [ESH]f_2 \qquad [13.A22]$$

Then making the steady state assumption

$$\frac{d[ESH]}{dt} = k_1[EH][S] - (k_{-1} + k_2)[ESH] = 0 \qquad [13.A23]$$

and solving for [EH]

$$[EH] = \frac{(k_{-1} + k_2)[ESH]}{k_1[S]} = \frac{K_M[ESH]}{[S]} \qquad [13.A24]$$

Therefore, from Eq. [13.A21],

$$[EH]_T = \frac{K_M[ESH]f_1}{[S]} \qquad [13.A25]$$

which, together with Eqs. [13.A20] and [13.A22], yields

$$[E]_T = [ESH]\left(\frac{K_M f_1}{[S]} + f_2\right) \qquad [13.A26]$$

As in the simple Michaelis–Menten derivation, the initial rate is

$$v_o = k_2[ESH] = \frac{k_2[E]_T}{\left(\frac{K_M f_1}{[S]}\right) + f_2} = \frac{(k_2/f_2)[E]_T[S]}{K_M(f_1/f_2) + [S]}$$

$$[13.A27]$$

Then defining the "apparent" values of K_M and $V_{max} = k_2[E]_T$ at a given pH:

$$K'_M = K_M(f_1/f_2) \qquad [13.A28]$$

and

$$V'_{max} = V_{max}/f_2 \qquad [13.A29]$$

the Michaelis–Menten equation modified to account for pH effects is

$$v_o = \frac{V'_{max}[S]}{K'_M + [S]} \qquad [13.47]$$

Chapter Summary

Complicated reaction processes occur through a series of elementary reaction steps defined as having a molecularity equal to the number of molecules that simultaneously collide to form products. The order of a reaction can be determined from the characteristic functional form of its progress curve. Transition state theory postulates that the rate of a reaction depends on the free energy of formation of its activated complex. This complex, which occurs at the free energy maximum of the reaction coordinate, is poised between reactants and products and is therefore also known as the transition state. Transition state theory explains that catalysis results from the reduction of the free energy difference between the reactants and the transition state.

In the simplest enzymatic mechanism, the enzyme and substrate reversibly combine to form an enzyme–substrate complex known as the Michaelis complex, which may irreversibly decompose to form product and the regenerated enzyme. The rate of product formation is expressed by the Michaelis–Menten equation, which is derived under the assumption that the concentration of the Michaelis complex is constant, that is, at a steady state. The Michaelis–Menten equation, which has the functional form of a rectangular hyperbola, has two parameters: V_{max}, the maximal rate of the reaction, which occurs when the substrate concentration is saturating, and K_M, the Michaelis constant, which has the value of the substrate concentration at the half-maximal reaction rate. These parameters may be graphically determined using the Lineweaver–Burk plot. Physically more realistic models of enzyme mechanisms than the Michaelis–Menten model assume the enzymatic reaction to be reversible and to have one or more intermediates. The functional form of the equations describing the reaction rates for these models is independent of their number of intermediates so that the models cannot be differentiated using only steady state kinetic measurements.

Enzymes may be inhibited by competitive inhibitors, which compete with the substrate for the enzymatic-binding site. The effect of a competitive inhibitor may be reversed by increasing the substrate concentration. An uncompetitive inhibitor inactivates a Michaelis complex upon binding to it. The maximal rate of an uncompetitively inhibited enzyme is a function of inhibitor concentration and therefore the effect of an uncompetitive inhibitor cannot be reversed by increasing substrate concentration. In mixed inhibition, the inhibitor binds to both the enzyme and the enzyme–substrate complex to form a complex that is catalytically inactive. The rate equation describing this situation has characteristics of both competitive and uncompetitive reactions.

The rate of an enzymatic reaction is a function of hydrogen ion concentration. At any pH, the rate of a simple enzymatic reaction can be described by the Michaelis–Menten equation. However, its parameters V_{max} and K_M vary with pH. By the evaluation of kinetic rate curves as a function of pH, the pK's of an enzyme's ionizable binding and catalytic groups can be determined, which may help identify these groups.

The majority of enzymatic reactions are bisubstrate reactions in which two substrates react to form two products. Bisubstrate reactions may have Ordered or Random Sequential mechanisms or Ping Pong Bi Bi mechanisms, among others. The initial rate equations for any of these mechanisms involve five parameters, which are analogous to either Michaelis–Menten equation parameters or equilibrium constants. The various bisubstrate mechanisms may be experimentally differentiated according to the forms of their double-reciprocal plots and from the nature of their product inhibition patterns. Isotope exchange reactions provide an additional, nonkinetic method of differentiating bisubstrate mechanisms.

References

Chemical Kinetics

Atkins, P. W., *Physical Chemistry* (3rd. ed.), Chapters 28 and 30, Freeman (1986).

Frost, A. A. and Pearson, R. G., *Kinetics and Mechanism* (2nd ed.), Wiley (1961). [A good introduction to chemical kinetics.]

Hammes, G. G., *Principles of Chemical Kinetics*, Academic Press (1978).

Laidler, K. J., *Theories of Chemical Reaction Rates*, McGraw-Hill (1969).

Enzyme Kinetics

Cleland, W. W., Steady state kinetics, *in* Boyer, P. D. (Ed.), *The Enzymes* (3rd ed.), Vol. 2, pp. 1–65, Academic Press (1970).

Cleland, W. W., Determining the mechanism of enzyme-catalyzed reactions by kinetic studies, *Adv. Enzymol.* **45,** 273 (1977).

Cornish-Bowden, A., *Fundamentals of Enzyme Kinetics*, Butterworths (1979). [A lucid and detailed account of enzyme kinetics.]

Dixon, M. and Webb, E. C., *Enzymes* (3rd ed.). Chapter IV, Academic Press (1979). [An almost exhaustive treatment of enzyme kinetics.]

Fersht, A., *Enzyme Structure and Mechanism*, (2nd ed.), Chapters 2–7, Freeman (1985).

Gutfruend, H., *An Introduction to the Study of Enzymes*, Wiley (1965).

Knowles, J. R., The intrinsic pK_a-values of functional groups in enzymes: Improper deductions from the pH-dependence of steady state parameters, *CRC Crit. Rev. Biochem.* **4,** 165 (1976).

Piszkiewicz, D., *Kinetics of Chemical and Enzyme Catalyzed Reactions*, Oxford University Press (1977). [A highly readable discussion of enzyme kinetics.]

Plowman, K. M., *Enzyme Kinetics*, McGraw-Hill (1972).

Purich, D. L. (Ed.), Enzyme kinetics and mechanisms, *Methods Enzymol.* **63** and **64** (1979). [A collection of articles on advanced topics.]

Segel, I. H., *Enzyme Kinetics*, Wiley (1975). [A detailed and understandable treatise providing full explanations of many aspects of enzyme kinetics.]

Tinoco, I., Jr., Sauer, K., and Wang, J. C., *Physical Chemistry. Principles and Applications for Biological Sciences* (2nd ed.), Chapters 7 and 8, Prentice–Hall (1985).

Westley, J., *Enzymic Catalysis*, Harper & Row (1969). [An informative presentation of enzyme kinetics.]

Wood, W. B., Wilson, J. H., Benbow, R. M., and Hood, L. E., *Biochemistry. A Problems Approach* (2nd ed.), Chapter 8, Benjamin/Cummings (1981). [Contains instructive problems on enzyme kinetics with answers worked out in detail.]

Problems

1. The hydrolysis of sucrose:

$$\text{Sucrose} + H_2O \longrightarrow \text{glucose} + \text{fructose}$$

takes the following time course.

Time (min)	[Sucrose] (M)
0	0.5011
30	0.4511
60	0.4038
90	0.3626
130	0.3148
180	0.2674

Determine the first-order rate constant and the half-life of the reaction. Why does this bimolecular reaction follow a first-order rate law? How long will it take to hydrolyze 99% of the sucrose initially present? How long will it take if the amount of sucrose initially present is twice that given in the table?

2. By what factor will a reaction at 25°C be accelerated if a catalyst reduces the free energy of its activated complex by 1 kJ·mol^{-1}; by 10 kJ·mol^{-1}?

3. For a Michaelis–Menten reaction, $k_1 = 5 \times 10^7 M^{-1}s^{-1}$, $k_{-1} = 2 \times 10^4\ s^{-1}$, and $k_2 = 4 \times 10^2\ s^{-1}$. Calculate K_S and K_M for this reaction. Does substrate binding achieve equilibrium or the steady state?

*4. The following table indicates the rates at which a substrate reacts as catalyzed by an enzyme that follows the Michaelis–Menten mechanism: (1) in the absence of inhibitor; (2) and (3) in the presence of 10 mM concentration, respectively, of each of two inhibitors. Assume $[E]_T$ is the same for all reactions.

[S](mM)	(1) $v_o(\mu M \cdot s^{-1})$	(2) $v_o(\mu M \cdot s^{-1})$	(3) $v_o(\mu M \cdot s^{-1})$
1	2.5	1.17	0.77
2	4.0	2.10	1.25
5	6.3	4.00	2.00
10	7.6	5.7	2.50
20	9.0	7.2	2.86

(a) Determine K_M and V_{max} for the enzyme. For each inhibitor determine the type of inhibition and K_I or K_I'. What additional information would be required to calculate the turnover number of the enzyme? (b) If [S] = 5 mM, what fraction of the enzyme molecules have a bound substrate in the absence of inhibitor, in the presence of 10-mM inhibitor of type (2), and in the presence of 10 mM inhibitor of type (3)?

*5. Ethanol in the body is oxidized to acetaldehyde by liver alcohol dehydrogenase (LADH). Other alcohols are also oxidized by LADH. For example, methanol, which is mildly intoxicating, is oxidized by LADH to the quite toxic product formaldehyde. The toxic effects of ingesting methanol (a component of many commercial solvents) can be reduced by administering ethanol. The ethanol acts as a competitive inhibitor of the methanol by displacing it from LADH. This provides sufficient time for the methanol to be harmlessly excreted by the kidneys. If an individual has ingested 100 mL of methanol (a lethal dose), how much 100 proof whiskey (50% ethanol by volume) must he imbibe to reduce the activity of his LADH towards methanol to 5% of its original value? The adult human body contains ~ 40 L of aqueous fluids throughout which ingested alcohols are rapidly and uniformly mixed. The densities of ethanol and methanol are both 0.79 g$\cdot$cm^{-3}. Assume the K_M values of LADH for ethanol and methanol to be $1.0 \times 10^{-3} M$ and $1.0 \times 10^{-2} M$, respectively, and that $K_I = K_M$ for ethanol.

6. The K_M of a Michaelis–Menten enzyme for a substrate is $1.0 \times 10^{-4} M$. At a substrate concentration of $0.2 M$, $v_o = 43$ $\mu M \cdot$min^{-1} for a certain enzyme concentration. However, with a substrate concentration of $0.02 M$, v_o has the same value. (a) Using numerical calculations, show that this observation is accurate. (b) What is the best range of [S] for measuring K_M?

7. Why are uncompetitive and mixed inhibitors generally considered to be more effective *in vivo* (in a living organism) than competitive inhibitors?

8. Explain why an exact fit to a kinetic model of the experimental parameters describing a reaction does not prove that the reaction follows the model?

9. An enzyme that follows the model for pH effects presented in Section 13-4 has $pK_{ES1} = 4$ and $pK_{ES2} = 8$. What is the pH at which V'_{max} is a maximum for this enzyme? What fraction of V_{max} does V'_{max} achieve at this pH?

10. Derive the initial rate equation for a Rapid Equilibrium Random Bi Bi reaction. Assume the equilibrium constants K_S^A and K_S^B for binding A and B to the enzyme are independent of whether the other substrate is bound (an assumption that constrains $K_M^B = K_S^B$ in Eq. [13.49]).

*11. Consider the following variation of a Ping Pong Bi Bi mechanism.

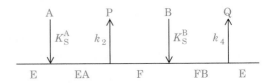

Assume that the substrate-binding reactions are in rapid equilibrium,

$$K_S^A = \frac{[E][A]}{[EA]} \quad \text{and} \quad K_S^B = \frac{[F][B]}{[FB]}$$

that both [A] and [B] $\gg$ [E]$_T$, that neither product release reaction is reversible, and that the steady state approximation is valid. (a) Derive an expression for v_o in terms of K_S^A, K_S^B, k_2, and k_4. (b) Indicate the form of the double reciprocal plots for $1/v_o$ versus $1/[A]$ for various values of [B]. (c) Indicate the form of the double reciprocal plots for $1/v_o$ versus $1/[B]$ for various values of [A].

12. Creatine kinase catalyzes the reaction

$$\text{MgADP}^- + \text{Phosphocreatine} \rightleftharpoons \text{MgATP}^{2-} + \text{creatine}$$

which functions to regenerate ATP in muscle. Rabbit muscle creatine kinase exhibits the following kinetic behavior. In the absence of both products, plots of $1/v_o$ versus $1/[\text{MgADP}^-]$ at different fixed concentrations of phosphocreatine yield lines that intersect to the left of the $1/v_o$ axis. Similarly, plots of $1/v_o$ versus $1/[\text{phosphocreatine}]$ in the absence of product at different fixed concentrations of MgADP$^-$ yield lines that intersect to the left of the $1/v_o$ axis. In the absence of one of the reaction products, MgATP^{2-} or creatine, plots of $1/v_o$ versus $1/[\text{MgADP}^-]$ at different concentrations of the other product intersect on the $1/v_o$ axis. The same is true of the plots of $1/v_o$ versus $1/[\text{phosphocreatine}]$. Indicate a kinetic mechanism that is consistent with this information.

Chapter 14
ENZYMATIC CATALYSIS

Enzymes, as we have seen, cause rate enhancements that are orders of magnitude greater than those of the best chemical catalysts. Yet, they operate under mild conditions and are highly specific as to the identities of both their substrates and their products. These catalytic properties are so remarkable that many nineteenth century scientists concluded that enzymes have characteristics that are not shared by substances of nonliving origin. To this day, there are few enzymes for which we understand in more than cursory detail how they achieve their enormous rate accelerations. Nevertheless, it is now abundantly clear that the catalytic mechanisms employed by enzymes are identical to those used by chemical catalysts. Enzymes are simply better designed.

In this chapter we consider the nature of enzymatic catalysis. We begin by discussing the underlying principles of chemical catalysis as elucidated through the study of organic reaction mechanisms. We then embark on a detailed examination of the catalytic mechanisms of several of the best characterized enzymes: lysozyme, the serine proteases, and glutathione reductase. Their study should lead to an appreciation of the intricacies of these remarkably efficient catalysts as well as of the experimental methods used to elucidate their properties.

1. CATALYTIC MECHANISMS

Catalysis is a process that increases the rate at which a reaction approaches equilibrium. Since, as we discussed in Section 13-1C, the rate of a reaction is a function of its free energy of activation ($\Delta G^{\ddagger}$), a catalyst acts by lowering the height of this kinetic barrier; that is, a catalyst stabilizes the transition state with respect to the uncatalyzed reaction. There is, in most cases, nothing unique about enzymatic mechanisms of catalysis in comparison to nonenzymatic mechanisms. *What apparently make enzymes such powerful catalysts are two related properties: their specificity of substrate binding combined with their optimal arrangement of catalytic groups.* An enzyme's arrangement of binding and catalytic groups is, of course, the product of eons of evolution: nature has had ample opportunity to "fine tune" the performances of most enzymes.

The types of catalytic mechanisms that enzymes employ have been classified as:

1. Acid–base catalysis.

2. Covalent catalysis.

3. Metal ion catalysis.

4. Electrostatic catalysis.

5. Proximity and orientation effects.

6. Preferential binding of the transition state complex.

In this section, we examine these various phenomena. In doing so we shall frequently refer to the organic model compounds that have been used to characterize these catalytic mechanisms.

A. Acid–Base Catalysis

General acid catalysis is a process in which partial proton transfer from a Brønsted acid (a species that can donate protons; Section 2-2A) lowers the free energy of a reaction's transition state. For example, an uncatalyzed keto–enol tautomerization reaction occurs quite slowly as a result of the high energy of its carbanionlike transition state (Fig. 14-1a). Proton donation to the oxygen atom (Fig. 14-1b), however, reduces the carbanion character of the transition state thereby catalyzing the reaction. *A reaction may also be stimulated by general base catalysis if its rate is increased by partial proton abstraction by a Brønsted base (a species that can combine with a proton; Fig. 14-1c). Some reactions may be simultaneously subject to both processes: a concerted general acid–base catalyzed reaction.*

Mutarotation Is Catalyzed by Acids and by Bases

The mutarotation of glucose provides an instructive example of acid–base catalysis. Recall that a glucose molecule can assume either of two anomeric cyclic forms through the intermediacy of its linear form (Sec-

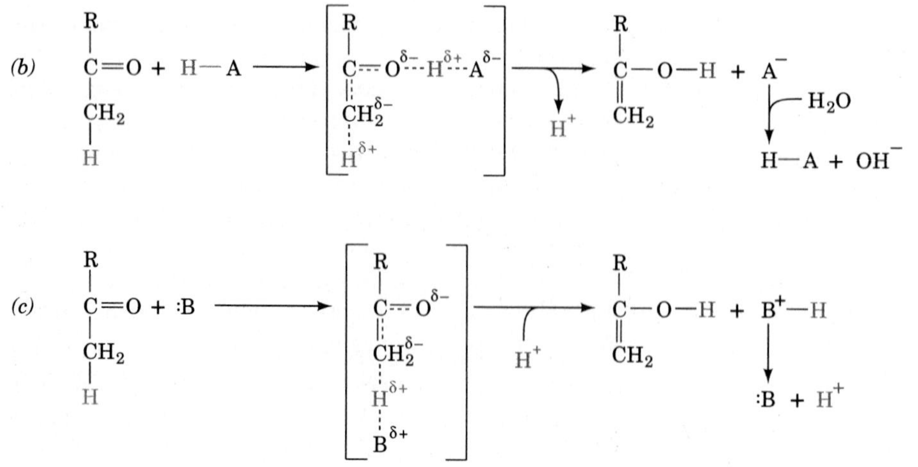

Figure 14-1
Mechanisms of keto–enol tautomerization: (a) uncatalyzed, (b) general acid catalyzed, and (c) general base catalyzed.

tion 10-1B):

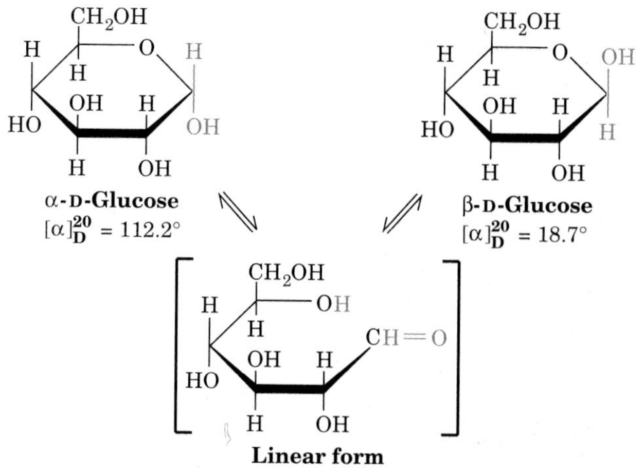

α-D-Glucose
$[\alpha]_D^{20} = 112.2°$

β-D-Glucose
$[\alpha]_D^{20} = 18.7°$

Linear form

In aqueous solvents, the initial rate of mutarotation of α-D-glucose, as monitored by polarimetry (Section 4-2A), is observed to follow the relationship:

$$v = - \frac{d[\alpha\text{-D-glucose}]}{dt} = k_{obs}\,[\alpha\text{-D-glucose}] \quad [14.1]$$

where k_{obs} is the reaction's apparent first-order rate constant. The mutarotation rate increases with the concentrations of general acids and general bases; they are thought to catalyze mutarotation according to the mechanism:

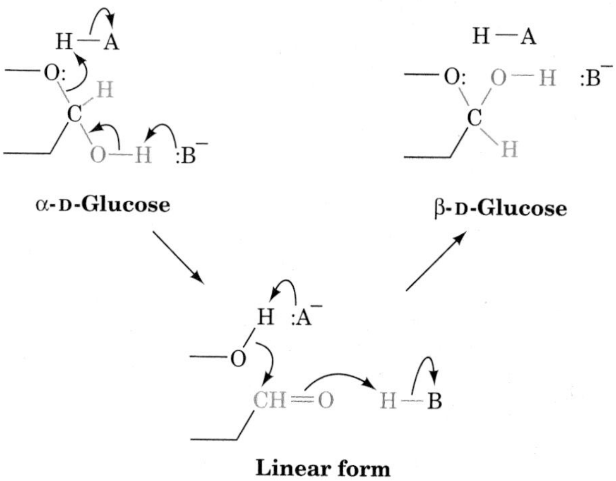

α-D-Glucose

β-D-Glucose

Linear form

This model is consistent with the observation that in aprotic solvents such as benzene, **2,3,4,6-O-tetramethyl-α-D-glucose** (a less polar benzene-soluble analog)

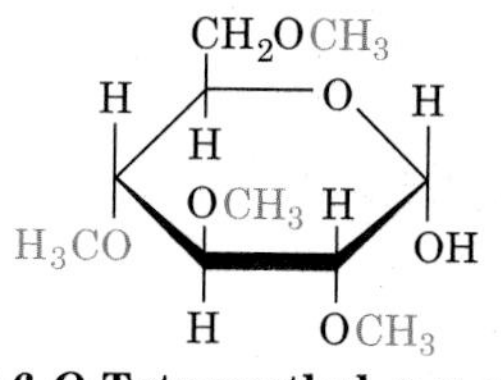

CH_2OCH_3

2,3,4,6-O-Tetramethyl-α-D-glucose

does not undergo mutarotation. Yet, the reaction is catalyzed by the addition of phenol, a weak benzene-soluble acid, together with pyridine, a weak benzene-soluble base, according to the rate equation:

$$v = k[\text{phenol}]\,[\text{pyridine}]\,[\text{tetramethyl-}\alpha\text{-D-glucose}] \quad [14.2]$$

Moreover, in the presence of **α-pyridone,** whose acid and base groups can rapidly interconvert between two tautomeric forms and are situated so that they can simultaneously catalyze mutarotation,

α-Pyridone

Glucose

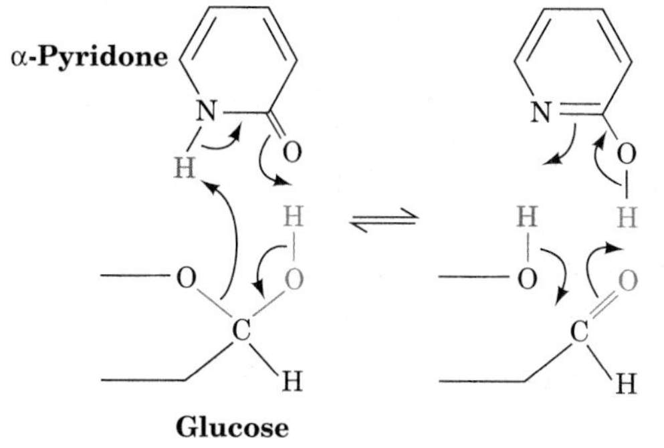

the reaction follows the rate law

$$v = k'[\alpha\text{-pyridone}][\text{tetramethyl-}\alpha\text{-D-glucose}] \quad [14.3]$$

where $k' = 7000M\,k$. This increased rate constant indicates that α-pyridone does, in fact, catalyze mutarotation in a concerted fashion since $1M$ α-pyridone has the same catalytic effect as impossibly high concentrations of phenol and pyridine (e.g., $70M$ phenol and $100M$ pyridine).

Many types of biochemically significant reactions are susceptible to acid and/or base catalysis. These include the hydrolysis of peptides and esters, the reactions of phosphate groups, tautomerizations, and additions to carbonyl groups. The side chains of the amino acid residues Asp, Glu, His, Cys, Tyr, and Lys have pK's in or near the physiological pH range (Table 4-2) which, we shall see, permits them to act in the enzymatic capacity of general acid and/or base catalysts in analogy with known organic mechanisms. Indeed, the ability of enzymes to arrange several catalytic groups about their substrates makes concerted acid–base catalysis a common enzymatic mechanism.

The RNase A Reaction Incorporates General Acid–Base Catalysis

Bovine pancreatic ribonuclease A (RNase A) provides an illuminating example of enzymatically mediated general acid–base catalysis. This digestive enzyme functions to hydrolyze RNA to its component nucleotides. The isolation of **2′,3′-cyclic nucleotides** from RNase A digests of RNA indicates that the enzyme mediates the following reaction sequence:

RNA

2',3'-Cyclic nucleotide +

— H$_2$O

— H$^+$

The RNase A reaction exhibits a pH rate profile that is peaked near pH 6 (Fig. 14-2). Analysis of this curve (Section 13-4), together with chemical derivatization and X-ray studies, indicates that RNase A has two essential His residues, His 12 and His 119, which act in a concerted manner as general acid and base catalysts (the structure of RNase A is sketched in Fig. 8-2). Evidently, the RNase A reaction is a two-step process (Fig. 14-3):

1. His 12, acting as a general base, abstracts a proton from an RNA 2'-OH group thereby promoting its

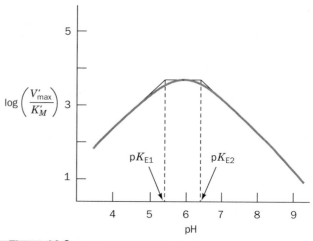

Figure 14-2
The pH dependence of V'_{max}/K'_M in the RNase A-catalyzed hydrolysis of cytidine-2',3'-cyclic phosphate where V'_{max}/K'_M is given in units of $M^{-1}s^{-1}$. Analysis of this curve (Section 13-4) suggests the catalytic participation of groups with pK's of 5.4 and 6.4. [After del Rosario, E. J. and Hammes, G. G., *Biochemistry* **8,** 1887 (1969).]

nucleophilic attack on the adjacent phosphorus atom while His 119, acting as a general acid, promotes bond scission by protonating the leaving group.

2. The 2',3'-cyclic intermediate is hydrolyzed through what is essentially the reverse of the first step in which water replaces the leaving group. Thus His 12 acts as a general acid and His 119 as a general base to yield the hydrolyzed RNA and the enzyme in its original state.

B. Covalent Catalysis

Covalent catalysis involves rate acceleration through the transient formation of a catalyst–substrate covalent bond. The decarboxylation of **acetoacetate,** as chemically catalyzed by primary amines, is an example of such a process (Fig. 14-4). In the first stage of this reaction, the amine nucleophilically attacks the carbonyl group of acetoacetate to form a **Schiff base** (imine bond).

The protonated nitrogen atom of the covalent intermediate then acts as an electron sink (Fig. 14-4, bottom) so as to reduce the otherwise high-energy enolate character of the transition state. The formation and decomposition of the Schiff base occurs quite rapidly so that it is not rate determining in this reaction sequence.

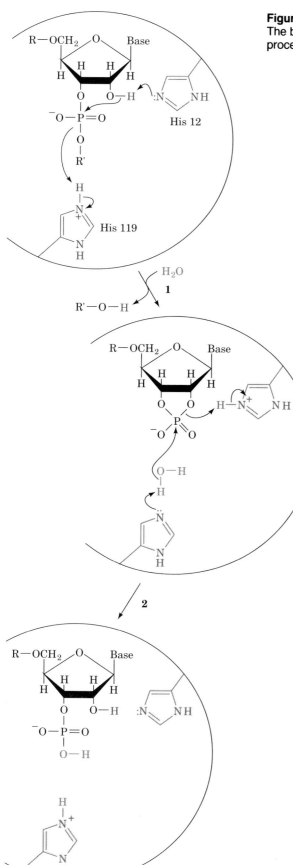

Figure 14-3
The bovine pancreatic RNase A-catalyzed hydrolysis of RNA is a two-step process with the intermediate formation of a 2′,3′-cyclic nucleotide.

Covalent Catalysis Has Both Nucleophilic and Electrophilic Stages

As the preceding example indicates, covalent catalysis may be conceptually decomposed into two stages:

1. The nucleophilic reaction between the catalyst and the substrate to form a covalent bond.

2. The withdrawal of electrons from the reaction center by the now electrophilic catalyst.

Reaction mechanisms are somewhat arbitrarily classified as occurring with either **nucleophilic catalysis** or **electrophilic catalysis** depending on which of these effects provides the greater driving force for the reaction, that is, which catalyzes its rate-determining step. The primary amine-catalyzed decarboxylation of acetoacetate is clearly an electrophilically catalyzed reaction since its nucleophilic phase, Schiff base formation, is not its rate-determining step. In other covalently catalyzed reactions, however, the nucleophilic phase may be rate determining.

The nucleophilicity of a substance is closely related to its basicity. Indeed, the mechanism of nucleophilic catalysis resembles that of general base catalysis except that, rather than abstracting a proton from the substrate, the catalyst nucleophilically attacks it so as to form a covalent bond. Consequently, if covalent bond formation is the rate-determining step of a covalently catalyzed reaction, the reaction rate tends to increase with the covalent catalyst's basicity (pK).

An important aspect of covalent catalysis is that the more stable the covalent bond formed, the less facilely it will decompose in the final steps of a reaction. A good covalent catalyst must therefore combine the seemingly contradictory properties of high nucleophilicity and the ability to form a good leaving group, that is, to easily reverse the bond formation step. Groups with high polarizabilities (highly mobile electrons), such as imidazole and thiol functions, have these properties and hence make good covalent catalysts.

Certain Amino Acid Side Chains and Coenzymes Can Serve as Covalent Catalysts

Enzymes commonly employ covalent catalytic mechanisms as is indicated by the large variety of covalently linked enzyme–substrate reaction intermediates that have been isolated. For example, the enzymatic decarboxylation of acetoacetate proceeds, much as described above, through Schiff base formation with an enzyme Lys residue's ε-amino group. The covalent intermediate,

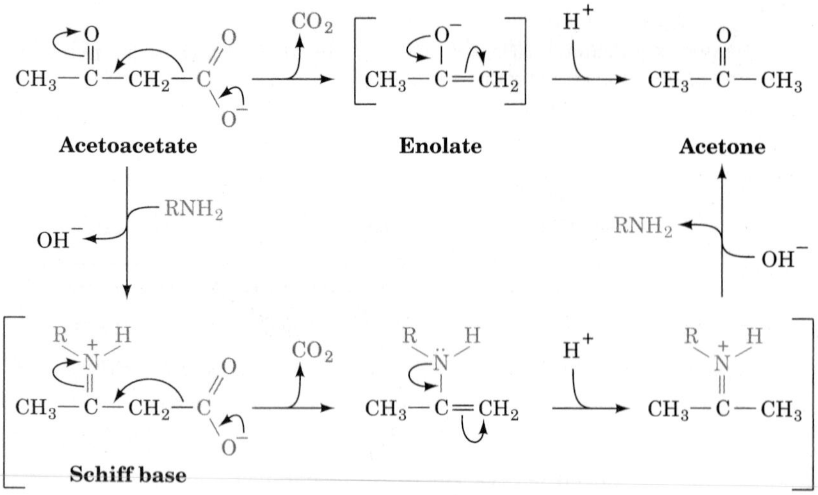

Figure 14-4
The uncatalyzed reaction mechanism for the decarboxylation of acetoacetate (*top*) and the reaction mechanism as catalyzed by primary amines (*bottom*).

in this case, has been isolated through $NaBH_4$ reduction of its imine bond to an amine thereby irreversibly inhibiting the enzyme. Other enzyme functional groups that participate in covalent catalysis include the imidazole moiety of His, the thiol group of Cys, the carboxyl function of Asp, and the hydroxyl group of Ser. In addition, several coenzymes, most notably **thiamine pyrophosphate** (Section 16-3B) and **pyridoxal phosphate** (Section 24-1A), function in association with their holoenzymes mainly as covalent catalysts.

C. Metal Ion Catalysis

Nearly one third of all known enzymes require the presence of metal ions for catalytic activity. There are two classes of metal ion-requiring enzymes that are distinguished by the strengths of their ion–protein interactions:

1. *Metalloenzymes contain tightly bound metal ions,* most commonly transition metal ions such as Fe^{2+}, Fe^{3+}, Cu^{2+}, Zn^{2+}, Mn^{2+}, or Co^{3+}.

2. *Metal-activated enzymes loosely bind metal ions from solution,* usually the alkali and alkaline earth metal ions Na^+, K^+, Mg^{2+}, or Ca^{2+}.

Metal ions participate in the catalytic process in three major ways:

1. By binding to substrates so as to properly orient them for reaction.

2. By mediating oxidation–reduction reactions through reversible changes in the metal ion's oxidation state.

3. By electrostatically stabilizing or shielding negative charges.

In this section we shall be mainly concerned with the third aspect of metal ion catalysis. The other forms of enzyme-mediated metal ion catalysis will be considered

in later chapters in conjunction with discussions of specific enzyme mechanisms.

Metal Ions Promote Catalysis through Charge Stabilization

In many metal ion-catalyzed reactions, the metal ion acts in much the same way as a proton, that is, as a Lewis acid. Yet, *metal ions are often much more effective catalysts than protons because metal ions can be present in high concentrations at neutral pH's and can have charges* $> +1$. Metal ions have therefore been dubbed "superacids."

The decarboxylation of **dimethyloxaloacetate,** as catalyzed by metal ions such as Cu^{2+} and Ni^{2+}, is a nonenzymatic example of catalysis by a metal ion:

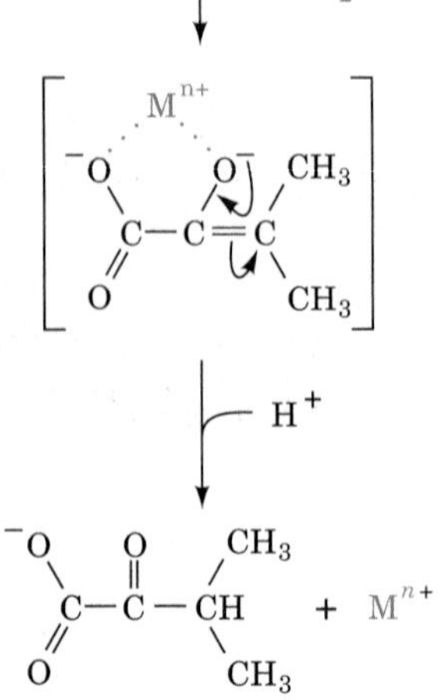

Here the metal ion (M^{n+}), which is chelated by the dimethyloxaloacetate, electrostatically stabilizes the developing enolate ion of the transition state. This mechanism is supported by the observation that acetoacetate, which cannot form such a chelate, is not subject to metal ion-catalyzed decarboxylation. Most enzymes that decarboxylate oxaloacetate require a metal ion for activity.

Metal Ions Promote Nucleophilic Catalysis via Water Ionization

A metal ion's charge makes its bound water molecules more acidic than free H_2O and therefore a source of OH^- ions even below neutral pH's. For example, the water molecule of $(NH_3)_5Co^{3+}(H_2O)$ ionizes according to the reaction:

$$(NH_3)_5Co^{3+}(H_2O) \rightleftharpoons (NH_3)_5Co^{3+}(OH^-) + H^+$$

with a pK of 6.6, which is some 9 pH units below the pK of free H_2O. *The resulting metal ion-bound hydroxyl group is a potent nucleophile.*

An excellent example of this phenomenon occurs in the catalytic mechanism of **carbonic anhydrase** (Section 9-1C), a widely occurring enzyme that catalyzes the reaction:

$$CO_2 + H_2O \rightleftharpoons HCO_3^- + H^+$$

Carbonic anhydrase contains an essential Zn^{2+} ion that is implicated in the enzyme's catalytic mechanism as follows:

1. The crystal structure of human carbonic anhydrase (Fig. 7-44) reveals that its Zn^{2+} is tetrahedrally coordinated by three (evolutionarily invariant) His side chains and a H_2O molecule. This Zn^{2+} polarized H_2O ionizes in a process facilitated through general base catalysis by either Glu 106 or Glu 117 (Fig. 14-5).

2. The resulting Zn^{2+} bound OH^- nucleophilically attacks the nearby enzymatically bound CO_2 thereby converting it to HCO_3^-.

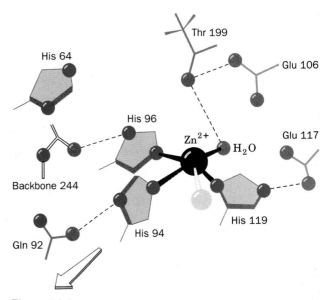

Figure 14-5

The active site of human carbonic anhydrase. The light grey ligand to the Zn^{2+} indicates the probable fifth Zn^{2+} coordination site. The arrow points towards the opening of the active site cavity. [After Sheridan, R. P. and Allen, L. C., *J. Am. Chem. Soc.* **103**, 1545 (1981).]

3. The catalytic site is then regenerated by the binding and ionization of another H_2O to the Zn^{2+}, possibly before the departure of the HCO_3^- ion, so as to transiently form a 5-coordinated Zn^{2+} complex.

Metal Ions Promote Reactions through Charge Shielding

Another important enzymatic function of metal ions is **charge shielding**. For example, the actual substrates of **kinases** (phosphoryl-transfer enzymes utilizing ATP) are Mg^{2+}–ATP complexes such as

$$\text{Adenine—Ribose—O—}\underset{\underset{O}{\|}}{\overset{\overset{O^-}{|}}{P}}\text{—O—}\underset{\underset{O}{\|}}{\overset{\overset{O^-}{|}}{P}}\text{—O—}\underset{\underset{O}{\|}}{\overset{\overset{O^-}{|}}{P}}\text{—O}^-$$

rather than just ATP. Here, the Mg^{2+} ion's role, in addition to its orienting effect, is to electrostatically shield the negative charges of the phosphate groups. Otherwise, these charges would tend to repel the electron pairs of attacking nucleophiles, especially those with anionic character.

D. Electrostatic Catalysis

The binding of substrate generally excludes water from an enzyme's active site. The local dielectric constant of the active site therefore resembles that in an organic solvent where electrostatic interactions are much stronger than they are in aqueous solutions (Section 7-4A). The charge distribution in a medium of low dielectric constant can greatly influence chemical reactivity. Thus, as we have seen, the pK's of amino acid side chains in proteins may vary by several units from their nominal values (Table 4-2) because of the proximity of charged groups.

Although experimental evidence and theoretical analyses on the subject are still sparse, *there are mounting indications that the charge distributions about the active sites of enzymes are arranged so as to stabilize the transition states of the catalyzed reactions.* Such a mode of rate enhancement, which resembles the form of metal ion catalysis discussed above, is termed **electrostatic catalysis.** Moreover, in several enzymes, *these charge distributions apparently serve to guide polar substrates towards their binding sites so that the rates of these enzymatic reactions are greater than their apparent diffusion-controlled limits (Section 13-2B).*

E. Catalysis through Proximity and Orientation Effects

Although enzymes employ catalytic mechanisms that resemble those of organic model reactions, they are far more catalytically efficient than these models. Such efficiency must arise from the specific physical conditions at enzyme catalytic sites that promote the corresponding chemical reactions. The most obvious effects are **proximity** and **orientation:** *Reactants must come together with the proper spatial relationship for a reaction to occur.* For example, in the bimolecular reaction of imidazole with *p*-nitrophenylacetate,

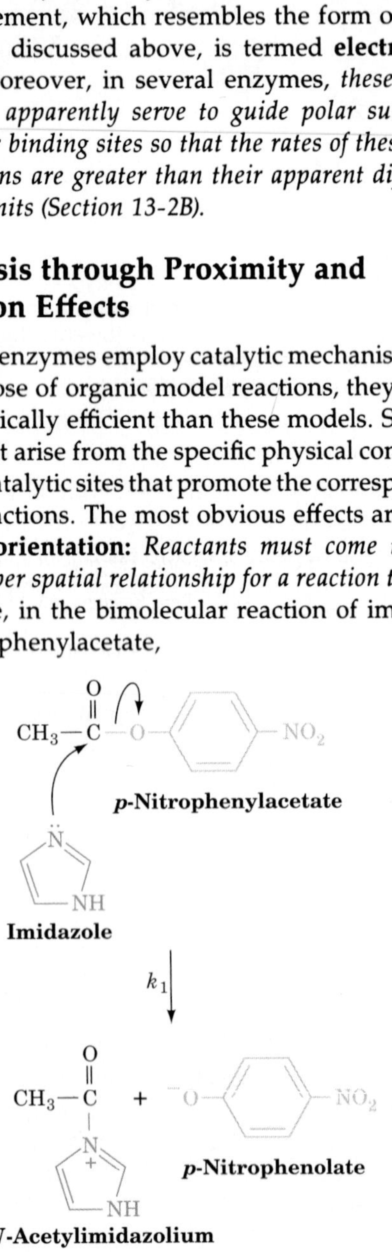

p-Nitrophenylacetate

Imidazole

p-Nitrophenolate

N-Acetylimidazolium

the progress of the reaction is conveniently monitored by the appearance of the intensely yellow *p*-**nitrophenolate** ion:

$$\frac{d\,[p\text{-}NO_2\phi O^-]}{dt} = k_1 [\text{imidazole}][p\text{-}NO_2\phi Ac]$$

$$= k_1' [p\text{-}NO_2\phi Ac] \quad [14.4]$$

Here k_1', the pseudo-first-order rate constant, is 0.0018 s^{-1} when [imidazole] $= 1M (\phi = \text{phenyl})$. However, for the intramolecular reaction

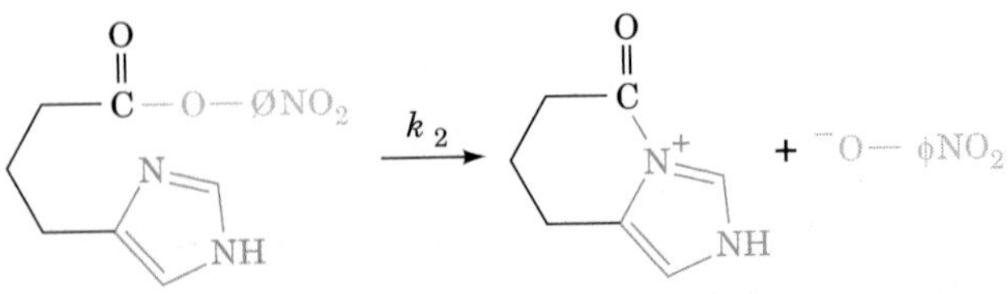

the first-order rate constant $k_2 = 0.043$ s^{-1}; that is, $k_2 = 24\ k_1'$. Thus, when the $1M$ imidazole catalyst is covalently attached to the reactant, it is 24-fold more effective than when it is free in solution, that is, *the imidazole group in the intramolecular reaction behaves as if its concentration is 24M.*

Let us verify this explanation by making a rough calculation as to how the rate of a reaction is affected by the proximity of its reacting groups. Following Daniel Koshland's treatment, we shall make several reasonable assumptions:

1. Reactant species, that is, functional groups, are about the size of water molecules.

2. Each reactant species in solution has 12 nearest-neighbor molecules, as do packed spheres of identical size.

3. Chemical reactions occur only between reactants that are in contact.

4. The reactant concentration in solution is low enough so that the probability of any reactant species being in simultaneous contact with more than one other reactant molecule is negligible.

Then the reaction:

$$A + B \xrightarrow{k_1} A-B$$

obeys the second-order rate equation

$$v = \frac{d\,[A-B]}{dt} = k_1\,[A][B] = k_2\,[A,B]_{pairs} \quad [14.5]$$

where $[A,B]_{pairs}$ is the concentration of contacting molecules of A and B. The value of this quantity is

$$[A,B]_{pairs} = \frac{12\,[A][B]}{55.5M} \quad [14.6]$$

since there are 12 ways that A can be in contact with B, and $[A]/55.5M$ is the probability that a molecule of B will be next to one of A ($[H_2O] = 55.5M$ in dilute aqueous solutions). Combining Eqs. [14.5] and [14.6] yields

$$v = k_1 \left(\frac{55.5}{12}\right) [A,B]_{pairs} = 4.6\ k_1\,[A,B]_{pairs} \quad [14.7]$$

Thus, in the absence of other effects, this model predicts

that for the intramolecular reaction,

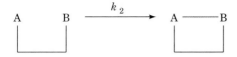

$k_2 = 4.6\,k_1$, which is a rather small rate enhancement. Factors that will increase this value other than proximity alone clearly must be considered.

Arresting Reactants' Relative Motions and Properly Orienting Them Can Result in Large Catalytic Rate Enhancements

The foregoing theory is, of course, quite simple. For example, it does not take into account the motions of the reacting groups with respect to one another. Yet, in the transition state complex, the reacting groups have little relative motion. In fact, as Thomas Bruice demonstrated, the rates of intramolecular reactions are greatly increased by arresting a molecule's internal motions (Table 14-1). Thus, when an enzyme brings two molecules together in a bimolecular reaction, as William Jencks pointed out, not only does it increase their proximity, but it freezes out their relative translational and rotational motions (decreases their entropy), thereby enhancing their reactivity. Table 14-1 indicates that such rate enhancements can be enormous.

Another effect that we have neglected in our treatment of proximity is that of orientation. Molecules do not react from just any direction as Koshland's simple theory assumes. Rather, *they react only if they have the proper relative orientation (Fig. 14-6)*. For example, in an S_N2 (bimolecular nucleophilic substitution) reaction, the incoming nucleophile must attack its target along the direction opposite to that of the bond to the leaving group (backside attack). The approaches of reacting atoms along a trajectory that varies by as little as $10°$ from this optimum direction will result in a significantly reduced reactivity. It has, in fact, been estimated that *properly orientating substrates can increase reaction rates by a factor of up to ~100.*

Enzymes, as we shall see in Sections 14-2 and 14-3, bind substrates in a manner that both immobilizes them and

Table 14-1

Relative Rates of Anhydride Formation for Esters Possessing Different Degrees of Motional Freedom in the Reaction:

Reactants[a]	Relative Rate Constant
$CH_3COO\phi Br$ $+$ CH_3COO^-	1.0
[structure: $COO\phi Br$ / COO^- with curved arrows]	$\sim1 \times 10^3$
[structure: $COO\phi Br$ / COO^-]	$\sim2.2 \times 10^5$
[bicyclic structure: $COO\phi Br$ / COO^-]	$\sim5 \times 10^7$

[a] Curved arrows indicate rotational degrees of freedom.

Source: Bruice, T. C., *Annu. Rev. Biochem.* **45,** 353 (1976).

aligns them so as to optimize their reactivities. The free energy required to do so is derived from the specific binding energy of substrate to enzyme.

F. Catalysis by Transition State Binding

The rate enhancements effected by enzymes are often greater than can be reasonably accounted for by the catalytic mechanisms thus far discussed. However, we have not yet considered one of the most important mechanisms of enzymatic catalysis: *the binding of the transition state to an enzyme with greater affinity than the corresponding substrates or products.* When taken together with the previously described catalytic mechanisms, preferential transition state binding rationalizes the observed rates of enzymatic reactions.

The original concept of transition state binding proposed that enzymes mechanically strain their substrates towards the transition state geometry through binding sites into which undistorted substrates do not properly fit. This so-called **rack mechanism** (in analogy with the medieval torture device) was based on the extensive evidence for the role of strain in promoting organic reac-

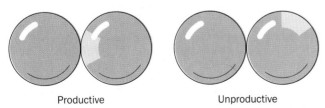

Productive   Unproductive

Figure 14-6
Molecules are only susceptible to chemical attack over limited regions of their surfaces (represented by the colored areas). Without the proper relative orientation (*left*), reactions will not occur (*right*).

tion   example, the rate of the reaction,

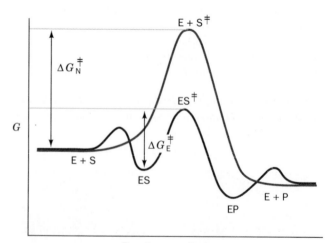

is 315 times faster when R is CH_3 rather than H because of the greater steric repulsions between the CH_3 groups and the reacting groups. Similarly, ring opening reactions are considerably more facile for strained rings such as cyclopropane than for unstrained rings such as cyclohexane. In either process, *the strained reactant more closely resembles the transition state of the reaction than does the corresponding unstrained reactant.* Thus, as was first suggested by Linus Pauling and further amplified by Richard Wolfenden and Gustav Lienhard, *interactions that preferentially bind the transition state increase its concentration and therefore proportionally increase the reaction rate.*

Let us quantitate this statement by considering the kinetic consequences of preferentially binding the transition state of an enzymatically catalyzed reaction involving a single substrate. The substrate S may react to form product P either spontaneously or through enzymatic catalysis:

$$S \xrightarrow{k_N} P$$
$$ES \xrightarrow{k_E} EP$$

Here k_E and k_N are the first-order rate constants for the catalyzed and uncatalyzed reactions, respectively. The relationships between the various states of these two reaction pathways are indicated in the following scheme:

$$E + S \underset{K_N^\ddagger}{\rightleftharpoons} S^\ddagger + E \longrightarrow P + E$$
$$\big\Updownarrow K_R \qquad\quad \big\Updownarrow K_T \qquad\qquad \big\Updownarrow$$
$$ES \underset{K_E^\ddagger}{\rightleftharpoons} ES^\ddagger \longrightarrow EP$$

where

$$K_R = \frac{[ES]}{[E][S]} \qquad\qquad K_T = \frac{[ES^\dagger]}{[E][S^\dagger]}$$

$$K_N^\dagger = \frac{[E][S^\dagger]}{[E][S]} \quad \text{and} \quad K_E^\dagger = \frac{[ES^\dagger]}{[ES]}$$

are all association constants. Consequently,

$$\frac{K_T}{K_R} = \frac{[S][ES^\dagger]}{[S^\dagger][ES]} = \frac{K_E^\dagger}{K_N^\dagger} \qquad [14.8]$$

According to transition state theory, Eqs. [13.7] and [13.14], the rate of the uncatalyzed reaction can be expressed

$$v_N = k_N[S] = \left(\frac{\kappa k_B T}{h}\right)[S^\dagger] = \left(\frac{\kappa k_B T}{h}\right) K_N^\dagger[S] \qquad [14.9]$$

Similarly, the rate of the enzymatically catalyzed reaction is

$$v_E = k_E[ES] = \left(\frac{\kappa k_B T}{h}\right)[ES^\dagger] = \left(\frac{\kappa k_B T}{h}\right) K_E^\dagger[ES] \qquad [14.10]$$

Therefore, combining Eqs. [14.8] to [14.10],

$$\frac{k_E}{k_N} = \frac{K_E^\dagger}{K_N^\dagger} = \frac{K_T}{K_R} \qquad [14.11]$$

This equation indicates that *the more tightly an enzyme binds its reaction's transition state (K_T) relative to the substrate (K_R), the greater the rate of the catalyzed reaction (k_E) relative to that of the uncatalyzed reaction (k_N); that is, catalysis results from the preferential binding and therefore the stabilization of the transition state ($S^\dagger$) relative to that of the substrate (S)* (Fig. 14-7).

According to Eq. [13.15], the ratio of the rates of the catalyzed versus the uncatalyzed reaction is expressed

$$\frac{k_E}{k_N} = \exp\left[(\Delta G_N^\ddagger - \Delta G_E^\ddagger)/RT\right] \qquad [14.12]$$

A rate enhancement factor of 10^6 therefore requires that an enzyme bind its transition state complex with 34.2 kJ·mol^{-1} greater affinity at 25°C than its substrate. This is roughly the free energy of two hydrogen bonds. Consequently, *the enzymatic binding of a transition state ($ES^\dagger$) by two hydrogen bonds that cannot form in the Michaelis complex (ES) should result in a rate enhancement of ~10^6 based on this effect alone.*

Figure 14-7
Reaction coordinate diagrams for a hypothetical enzymatically catalyzed reaction involving a single substrate (*blue*), and the corresponding uncatalyzed reaction (*red*).

It is commonly observed that the specificity of an enzyme is manifested by its turnover number (k_{cat}) rather than by its expressed substrate-binding affinity. In other words, an enzyme binds poor substrates, which have a low reaction rate, as well as or even better than good ones, which have a high reaction rate. Such enzymes apparently use a good substrate's intrinsic binding energy to stabilize the corresponding transition state. The "rack mechanism," which would require converting this binding energy to a mechanical force, seems implausible because theoretical considerations (Sections 8-2 and 14-2C) suggest that proteins lack sufficient structural rigidity to significantly deform their substrates. Rather, it appears that the transition state makes better contacts with the enzyme than S does in the Michaelis complex so that maximum binding energy can only be realized by the transition state; that is, *a good substrate does not necessarily bind to its enzyme with high affinity, but does so upon activation to the transition state.*

Transition State Analogs Are Competitive Inhibitors

If an enzyme preferentially binds its transition state, then it can be expected that **transition state analogs,** *stable molecules that resemble S‡ or one of its components, are potent competitive inhibitors of the enzyme.* For example, the reaction catalyzed by **proline racemase** from *Clostridium sticklandii* is thought to occur via a planar transition state:

L-Proline   **D-Proline**

Planar transition state

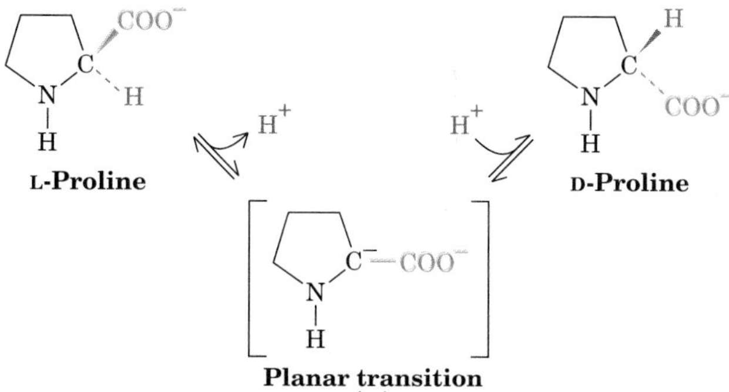

Proline racemase is competitively inhibited by the planar analogs of proline, **pyrrole-2-carboxylate** and **Δ-1-pyrroline-2-carboxylate,**

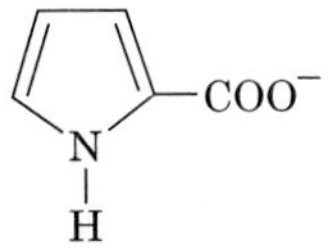

Pyrrole-2-carboxylate 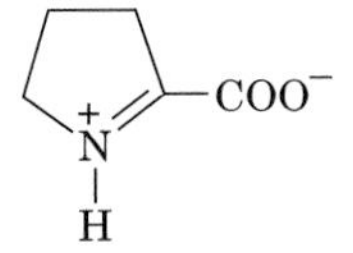   **Δ-1-Pyrroline-2-carboxylate**

both of which bind to the enzyme with 160-fold greater affinity than does proline. These compounds are therefore thought to be analogs of the transition state in the proline racemase reaction. In contrast, **tetrahydro-**

furan-2-carboxylate,

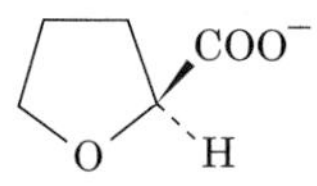

Tetrahydrofuran-2-carboxylate

which more closely resembles the tetrahedral structure of proline, is not nearly as good an inhibitor as these compounds. A 160-fold increase in binding affinity corresponds, according to Eq. [14.9], to a 12.6 kJ·mol^{-1} increase in the free energy of binding. This quantity presumably reflects the additional binding affinity that proline racemase has for proline's planar transition state over that of the undistorted molecule.

Hundreds of transition state analogs for various enzymatic reactions have been reported. Some are naturally occurring antibiotics. Others were designed to investigate the mechanisms of particular enzymes and/or to act as specific enzymatic inhibitors for therapeutic or agricultural use. Indeed, *the theory that enzymes bind transition states with higher affinity than substrates has led to a rational basis of drug design based on the understanding of specific enzyme reaction mechanisms.*

2. LYSOZYME

In the remainder of this chapter we shall investigate the catalytic mechanisms of several well-characterized enzymes. In doing so, we shall see how enzymes apply the catalytic principles described in Section 14-1. You should note that *the great catalytic efficiency of enzymes arises from their simultaneous use of several of these catalytic mechanisms.*

Lysozyme is an enzyme that destroys bacterial cell walls. It does so, as we saw in Section 10-3B, by hydrolyzing the $\beta(1 \to 4)$ glycosidic linkages from **N-acetylmuramic acid (NAM)** to **N-acetylglucosamine (NAG)** in the alternating NAM–NAG polysaccharide component of cell wall peptidoglycans (Fig. 14-8). Lysozyme occurs widely in the cells and secretions of vertebrates. It may function as a bacteriocidal agent although the observation that few pathogenic bacteria are susceptible to lysozyme alone prompted the suggestion that this enzyme helps dispose of bacteria after they have been killed by other means.

Hen egg white (HEW) lysozyme is the most widely studied species of lysozyme and is one of the mechanistically best understood enzymes. It is a rather small protein (14.6 kD) whose single polypeptide chain consists of 129 amino acid residues and is internally cross-linked by four disulfide bonds (Fig. 14-9). HEW lysozyme catalyzes the hydrolysis of its substrate at a rate that is ~10^{10}-fold greater than that of the uncatalyzed reaction.

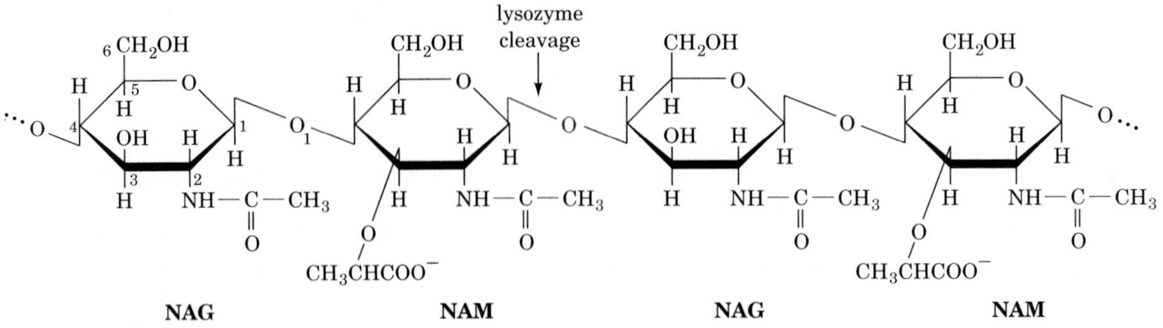

lysozyme
cleavage

NAG   NAM   NAG   NAM

Figure 14-8
The alternating NAG–NAM polysaccharide component of bacterial cell walls showing the position of the lysozyme cleavage site.

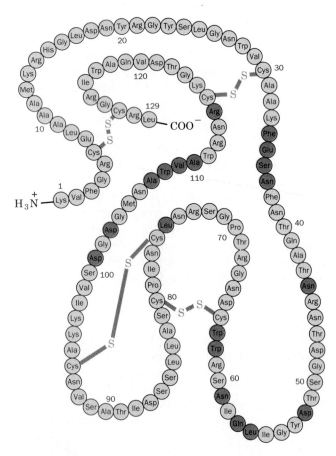

Figure 14-9
The primary structure of HEW lysozyme. The amino acid residues that line the substrate-binding pocket are shown in dark purple.

A. Enzyme Structure

The elucidation of an enzyme's mechanism of action requires a knowledge of the structure of its enzyme–substrate complex. This is because, even if the active site residues have been identified through chemical and physical means, their three-dimensional arrangements relative to the substrate as well as to each other must be known for an understanding of how the enzyme works. However, an enzyme binds its good substrates only transiently before it catalyzes a reaction and releases the products. Consequently, *most of our knowledge of enzyme–substrate complexes derives from X-ray studies of enzymes in complex with inhibitors or poor substrates* that remain stably bound to the enzyme for the day or more required to measure a protein crystal's X-ray diffraction intensities. The large solvent filled channels that occupy much of the volume of most protein crystals (Section 7-3A) often permit the formation of enzyme–inhibitor complexes by the diffusion of inhibitor molecules into crystals of the native protein.

The X-ray structure of HEW lysozyme, which was elucidated by David Phillips in 1965, was the second structure of a protein and the first of an enzyme to be determined at high resolution. The protein molecule is roughly ellipsoidal in shape with dimensions $30 \times 30 \times 45$ Å (Fig. 14-10). *Its most striking feature is a prominent cleft, the substrate-binding site, that traverses one face of the molecule.* The polypeptide chain (Fig. 14-10a) has relatively little regular secondary structure: It contains several more or less helical segments as well as a three-stranded antiparallel β sheet that comprises much of one wall of the binding cleft. As expected, most of the nonpolar side chains are in the interior of the molecule, out of contact with the aqueous solvent.

The Nature of the Binding Site

HEW lysozyme hydrolyzes poly(NAG) (chitin) as well as its natural peptidoglycan substrate. However, NAG oligosaccharides of less than five residues are hydrolyzed very slowly (Table 14-2) although these substrate analogs bind to the enzyme's active site and are thus its competitive inhibitors. The X-ray structure of the (NAG)₃–lysozyme complex reveals that (NAG)₃ is bound on the right side of the enzymatic binding cleft as drawn in Fig. 14-10a for substrate residues A, B, and C.

(a)

Amino end
of chain

Pleated
sheet
region

asn

Substrate
cleavage

Carboxyl
end

Substrate

© IRVING GEIS

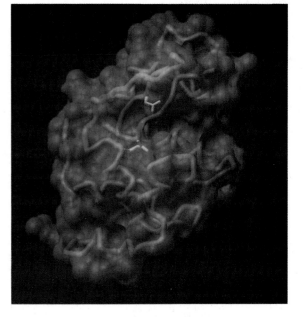

(b)

Figure 14-10

The X-ray structure of HEW lysozyme. (a) The polypeptide chain is shown with a bound (NAG)$_6$ substrate. The positions of the backbone C$_\alpha$ atoms are indicated together with those of the side chains that line the substrate-binding site and form disulfide bonds. The substrate's sugar rings are designated A, at its nonreducing end (*right*), through F, at its reducing end (*left*). Lysozyme catalyzes the hydrolysis of the glycosidic bond between residues D and E. Rings A, B, and C are observed in the X-ray structure of the complex of (NAG)$_3$ with lysozyme; the positions of rings D, E, and F were inferred from model building studies. [Figure copyrighted © by Irving Geis.] (b) A computer-generated model in approximately the same orientation as Part (a) showing the protein's molecular envelope (*purple*) and C$_\alpha$ backbone (*blue*). The side chains of the catalytic residues, Asp 52 (*above*) and Glu 35 (*below*), are colored yellow. Note the enzyme's prominent substrate-binding cleft. [Courtesy of Arthur Olson, Research Institute of Scripps Clinic.]

Table 14-2

Rates of HEW Lysozyme-Catalyzed Hydrolysis of Selected Oligosaccharide Substrate Analogs

Compound	k_{cat} (s^{-1})
(NAG)$_2$	2.5×10^{-8}
(NAG)$_3$	8.3×10^{-6}
(NAG)$_4$	6.6×10^{-5}
(NAG)$_5$	0.033
(NAG)$_6$	0.25
(NAG–NAM)$_3$	0.5

Source: Imoto, T., Johnson, L. N., North, A. C. T., Phillips, D. C., and Rupley, J. A., *in* Boyer, P. D. (Ed.), *The Enzymes* (3rd ed.), Vol. **7**, p. 842, Academic Press (1972).

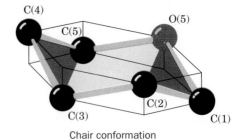

Chair conformation

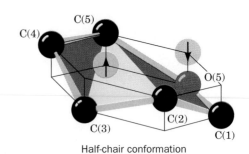

Half-chair conformation

Figure 14-11

Hexose rings normally assume the chair conformation. It is postulated, however, that binding by lysozyme distorts the D ring into the half-chair conformation such that atoms C(1), C(2), C(5), and O(5) are coplanar.

This inhibitor associates with the enzyme through strong hydrogen bonding interactions, some of which involve the acetamido groups of residues A and C, as well as through close fitting hydrophobic contacts. In an example of induced fit ligand binding (Section 9-4C), there is a slight (~1 Å) closing of lysozyme's binding cleft on binding (NAG)$_3$.

Lysozyme's Catalytic Site Was Identified through Model Building

(NAG)$_3$ takes several weeks to hydrolyze under the influence of lysozyme. It is therefore presumed that the complex revealed by X-ray analysis is unproductive; that is, the enzyme's catalytic site occurs at neither the A—B nor the B—C bonds. [Presumably, the rare occasions when (NAG)$_3$ hydrolyzes occur when it binds productively at the catalytic site.]

In order to locate lysozyme's catalytic site, Phillips used model building to investigate how a larger substrate could bind to the enzyme. Lysozyme's active site cleft is long enough to accommodate (NAG)$_6$, which the enzyme rapidly hydrolyzes (Table 14-2). However, the fourth NAG residue (residue D in Fig. 14-10*a*) appears unable to bind to the enzyme because its C(6) and O(6) atoms too closely contact Gln 57, Trp 108, and the acetamido group of NAG residue C. This steric interference can be relieved by distorting the glucose ring from its normal chair conformation to that of a half-chair (Fig. 14-11). *This distortion, which renders atoms C(1), C(2), C(5), and O(5) of residue D coplanar, moves the —C(6)H$_2$OH group from its normal equatorial position to an axial position where it makes no close contacts and can hydrogen bond to the backbone carbonyl group of Gln 57 (Fig. 14-12).* Continuing the model building, Phillips found that residues E and F apparently bind to the enzyme without distortion and with a number of favorable hydrogen bonding and van der Waals contacts.

We are almost in a position to identify lysozyme's catalytic site. In the enzyme's natural substrate, every second residue is an NAM. Model building, however, indicates that a lactyl side chain cannot be accommodated in the binding subsites of either residues C or E. Hence, the NAM residues must bind to the enzyme in subsites B, D, and F.

$$\cdots-\text{NAG}-\text{NAM}-\text{NAG}-\text{NAM}-\text{NAG}-\text{NAM}-\cdots \rightarrow \binom{\text{reducing}}{\text{end}}$$
$$\text{A}\qquad\text{B}\qquad\text{C}\qquad\text{D}\qquad\text{E}\qquad\text{F}$$

The observation that lysozyme hydrolyzes $\beta(1 \rightarrow 4)$ linkages from NAM to NAG implies that bond cleavage occurs either between residues B and C or between residues D and E. Since (NAG)$_3$ is stably bound to but not cleaved by the enzyme while spanning subsites B and C, the probable cleavage site is between residues D and E. This conclusion is supported by John Rupley's observation that lysozyme nearly quantitatively hydrolyzes (NAG)$_6$ between the 2nd and 3rd residues from its reducing terminus [the end with a free C(1)—OH]—just as is expected if the enzyme has six saccharide-binding subsites and cleaves its bound substrate between subsites D and E.

The bond that lysozyme cleaves was identified by carrying out the lysozyme-catalyzed hydrolysis of (NAG)$_3$ in H$_2$^{18}O. The resulting product had ^{18}O bonded to the C(1) atom of its newly liberated reducing terminus

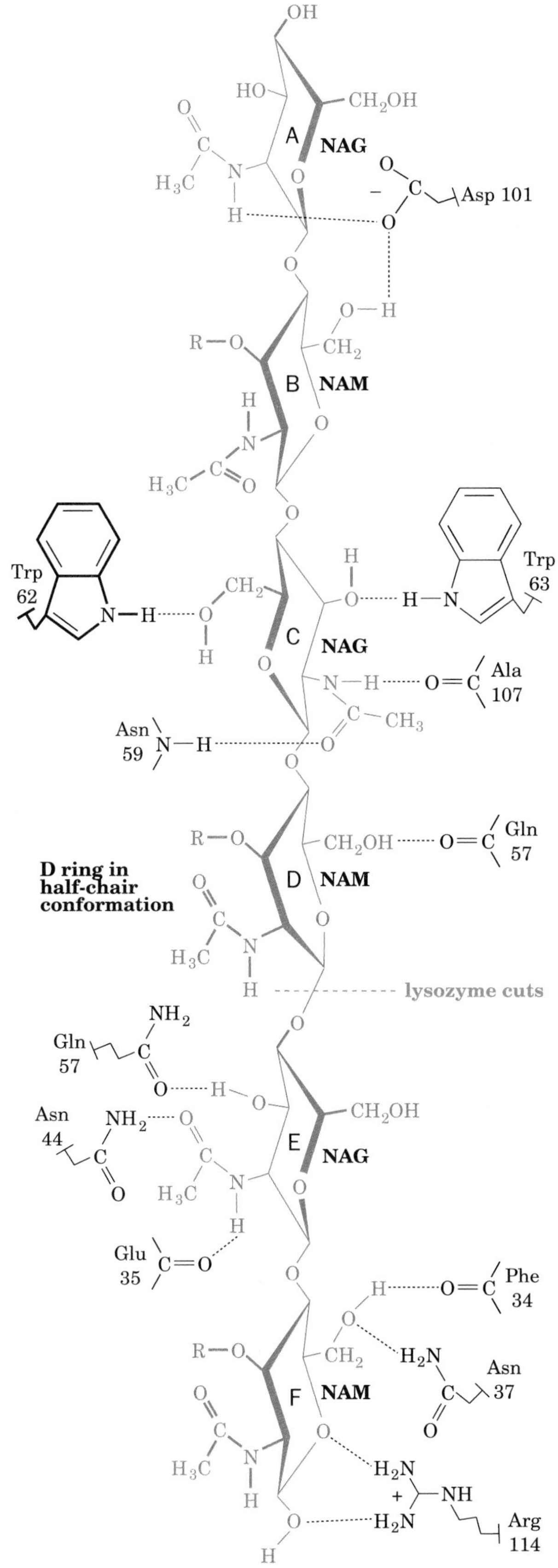

OH
HO
CH₂OH
A NAG

O
C
H₃C N
H
Asp 101
C
O
O
−

O—H
CH₂
R—O
B NAM
H
N
H₃C C=O

Trp 62
N—H
CH₂
O
H
O
C NAG
Trp 63
O···H—N
N—H······O=C Ala 107
C—CH₃

Asn 59
N—H·········O
O
R—O
CH₂OH······O=C Gln 57
D NAM

D ring in half-chair conformation

O
C
H₃C N
H
O ----------- lysozyme cuts
O

NH₂
Gln 57
C
O·····H—O
Asn 44
NH₂···O
C
O
C
CH₃ N
H
E NAG
Glu 35
C=O
O
Phe 34
H······O=C
O
R—O
CH₂
H₂N
Asn 37
F NAM
C
O
O
C
H₃C N
H
H₂N
+ NH
O·········H₂N
Arg 114
H

Figure 14-12
The interactions of lysozyme with its substrate. The view is into the binding cleft with the darker edges of the rings facing the outside of the enzyme and the lighter ones against the bottom of the cleft. [Figure copyrighted © by Irving Geis.]

thereby demonstrating that bond cleavage occurs between C(1) and the bridge oxygen O(1):

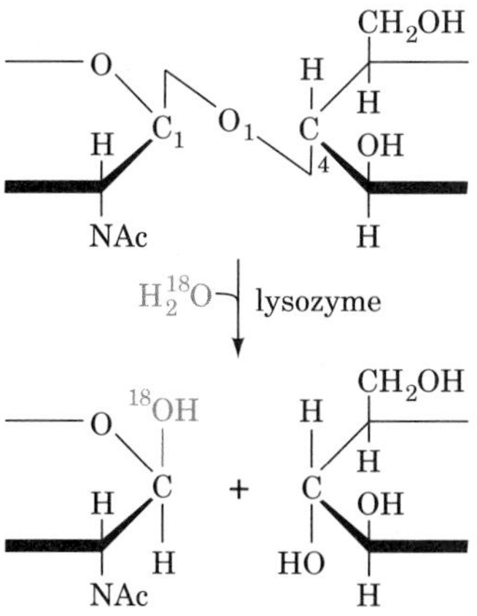

CH₂OH
H
H
O
C₁
O₁
C₄
OH
H
H
NAc
H

$H_2^{18}O$ lysozyme

CH₂OH
H
H
O
¹⁸OH
C
C
OH
H
H
HO
NAc
H

Thus, lysozyme catalyzes the hydrolysis of the C(1)—O(1) bond of a bound substrate's D residue. Moreover, this reaction occurs with retention of configuration so that the D-ring product remains the β anomer.

B. Catalytic Mechanism

It remains to identify lysozyme's catalytic groups. The reaction catalyzed by lysozyme, the hydrolysis of a glycoside, is the conversion of an acetal to a hemiacetal. Nonenzymatic acetal hydrolysis is an acid-catalyzed reaction that involves the protonation of a reactant oxygen atom followed by cleavage of its C—O bond (Fig. 14-13). This results in the formation of a resonance-stabilized carbocation that is called an **oxonium ion.** To attain maximum orbital overlap, and thus resonance stabilization, the oxonium ion's R and R' groups must be coplanar with its C, O, and H atoms. The oxonium ion then adds water to yield the hemiacetal and regenerate the acid catalyst. In searching for catalytic groups on an enzyme that mediates acetal hydrolysis, we should therefore seek a potential acid catalyst and possibly a group that could further stabilize an oxonium ion intermediate.

Glu 35 and Asp 52 Are Lysozyme's Catalytic Residues

The only functional groups in the immediate vicinity of lysozyme's reactive center that have the required catalytic properties are Glu 35 and Asp 52. These residues are disposed to either side of the $\beta(1 \rightarrow 4)$ glycosidic linkage to be cleaved (Fig. 14-10a). The side chains of Asp 52 and Glu 35 have markedly different environments. Asp 52 is surrounded by polar groups and participates in a complex hydrogen bonding network. Asp 52 is therefore predicted to have a normal pK; that is, it should by unprotonated and hence negatively charged throughout the 3 to 8 pH range in which lysozyme is

OR'
|
H—C—O—R" + H⁺ ⇌ H—C—O⁺—R"
| | |
R   R H

Acetal

→ R"OH

$$\left[\begin{array}{ccc} & R' & & R' \\ & O^+ & & O \\ & \| & \longleftrightarrow & | \\ & C & & C^+ \\ H & R & H & R \end{array} \right]$$

**Resonance stabilized
carbocation (oxonium ion)**

H₂O ┐
     └→ H⁺

OR'
|
H—C—OH
|
R

Hemiacetal

Figure 14-13
The mechanism of the nonenzymatic acid-catalyzed hydrolysis of an acetal to a hemiacetal. The reaction involves the protonation of one of the acetal's oxygen atoms followed by cleavage of its C—O bond to form an alcohol (R"OH) and a resonance-stabilized carbocation (oxonium ion). The addition of water to the oxonium ion forms the hemiacetal and regenerates the H⁺ catalyst. Note that the oxonium ion's C, O, H, R, and R' atoms all lie in the same plane.

active. In contrast, *the carboxyl group of Glu 35 is in a predominantly nonpolar region where, as we discussed in Section 14-1D, it is likely to remain protonated at unusually high pH's for carboxyl groups.* The closest approaches between the carboxyl O atoms of both Asp 52 and Glu 35 and the C(1)—O(1) bond of NAG residue D is ~3 Å, which makes them the prime candidates for electrostatic and acid catalysts, respectively.

The Phillips Mechanism

With the foregoing information, Phillips postulated the following enzymatic mechanism for lysozyme (Fig. 14-14):

1. Lysozyme attaches to a bacterial cell wall by binding to a hexasaccharide unit. In the process, *residue D is distorted towards the half-chair conformation* in response to the unfavorable contacts that its —C(6)H₂OH group would otherwise make with the protein.

2. *Glu 35 transfers its proton to the O(1) of the D ring,* the only polar group in its vicinity *(general acid catalysis). The C(1)—O(1) bond is thereby cleaved generating a resonance-stabilized oxonium ion at C(1).*

3. The ionized carboxyl group of Asp 52 acts to *stabilize the developing oxonium ion through charge–charge in-*

teractions (electrostatic catalysis). No covalent bond is formed because of the ~3-Å distance between C(1) and a carboxyl O atom of Asp 52. The bond cleavage reaction is facilitated by the strain in the D ring that distorts it to the planar half-chair conformation. This is a result of the oxonium ion's required planarity; that is, *the initial binding conformation of the D ring resembles that of the reaction's transition state (transition state binding catalysis; Fig. 14-15).*

4. At this point, the enzyme releases the hydrolyzed E ring with its attached polysaccharide yielding a **glycosyl-enzyme intermediate.** This oxonium ion subsequently adds H₂O from solution in a reversal of the

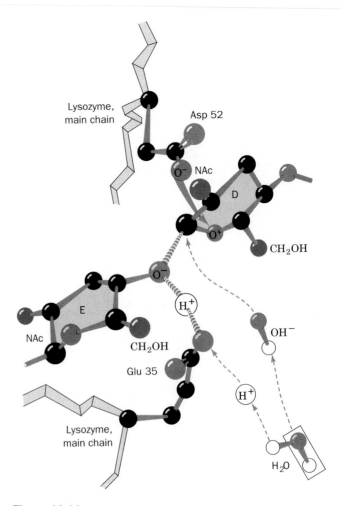

Figure 14-14
The Phillips mechanism for the lysozyme reaction. The cleavage of the glycosidic bond between the substrate D and E rings occurs through protonation of the bridge oxygen atom by Glu 35. The resulting D-ring oxonium ion is stabilized by the proximity of the Asp 52 carboxylate group and the enzyme-induced distortion of the D ring. Once the E ring is released, H₂O from solution provides both an OH⁻ that combines with the oxonium ion and an H⁺ that reprotonates Glu 35. NAc represents the N-acetylamino substituent at C(2) of each glucose ring.

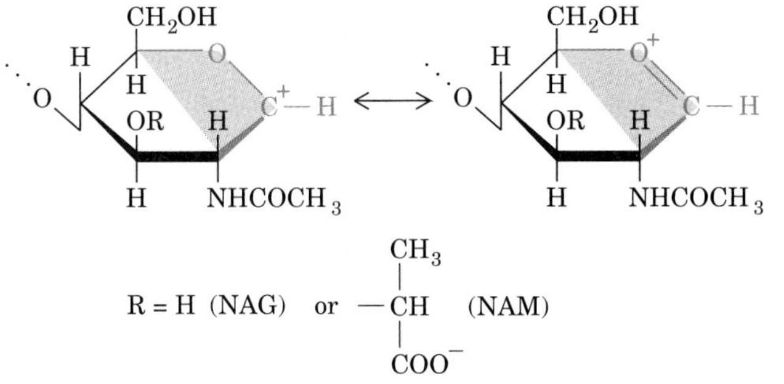

Figure 14-15
The oxonium ion transition state of the D ring in the lysozyme reaction is stabilized by resonance. This requires that atoms C(1), C(2), C(5), and O(5) be coplanar (*shading*); that is, the hexose ring must assume the half-chair conformation.

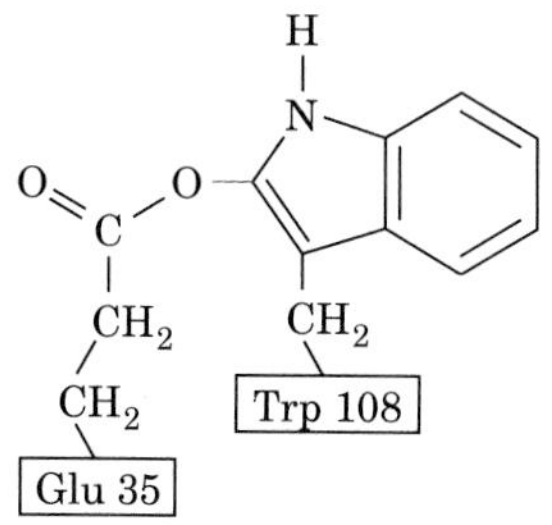

Figure 14-16
The I_2 oxidation of lysozyme results in the formation of a covalent bond between the side chains of Glu 35 and Trp 108.

preceding steps to form product and to reprotonate Glu 35. *The reaction's retention of configuration is dictated by the shielding of one of the oxonium ion's faces by the enzymatic cleft.* The enzyme then releases the D-ring product with its attached saccharide thereby completing the catalytic cycle.

C. Testing the Phillips Mechanism

The Phillips mechanism was formulated largely on the basis of structural investigations of lysozyme and a knowledge of the mechanism of nonenzymatic acetal hydrolysis. A variety of evidence has since been gathered that bears on the validity of this mechanism.

Identification of the Catalytic Residues
Lysozyme's catalytically important groups have been experimentally identified through the use of group specific reagents (Section 6-2):

Asp 52. The ethylating agent **triethoxonium fluoroborate** reacts with carboxylic acids as follows:

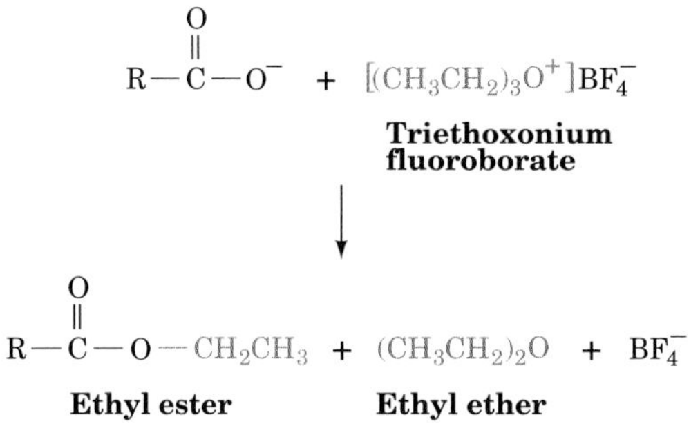

$$R-C(=O)-O^- + [(CH_3CH_2)_3O^+]BF_4^-$$

Triethoxonium fluoroborate

$$\downarrow$$

$$R-C(=O)-O-CH_2CH_3 + (CH_3CH_2)_2O + BF_4^-$$

Ethyl ester   **Ethyl ether**

At pH 4.5, this reagent reacts only with the β carboxyl group of lysozyme's Asp 52. *The resulting monoesterified enzyme binds substrate in a normal manner but is catalytically inactive. Asp 52 must therefore be essential for enzymatic activity.*

Glu 35. In the native enzyme, Glu 35 is in van der Waals contact with Trp 108 (Fig. 14-10*a*). The reaction of lyso-

zyme with I_2 specifically oxidizes Trp 108 (and none of the enzyme's other five Trp residues) to form a modified enzyme that is totally inactive. The X-ray structure of this modified enzyme indicates that its only chemical change is the formation of a covalent bond between Trp 108 and Glu 35 (Fig. 14-16). *Lysozyme's only conformational change upon I_2 oxidation is a reorientation of Glu 35's side chain that does not significantly affect the enzyme's substrate-binding affinity. Glu 35 must therefore also be essential for lysozyme's catalytic activity.*

Verification of Asp 52 and Glu 35 Involvement. The analysis of lysozyme's pH rate profile (Section 13-4) indicates that this enzyme has two catalytically important ionizable groups whose pK's are 3.8 and 6.7. The latter ionization, although abnormally high for a carboxyl group, is attributed to Glu 35 because of this ionization's disappearance upon I_2 oxidation of lysozyme. The only other reasonable possibility for an ionization with this pK is lysozyme's sole His residue but this residue is located far from the active site. The absence of the ionization with pK 3.8 in the triethoxonium fluoroborate-treated enzyme demonstrates that this ionization is due to Asp 52. This observation corroborates the mechanistic postulate that lysozyme is only catalytically active when Asp 52 is ionized and Glu 35 is not.

Noninvolvement of Other Amino Acid Residues. Lysozyme's other carboxyl groups besides Glu 35 and Asp 52 do not participate in the catalytic process as was demonstrated by reacting lysozyme with carboxyl specific reagents in the presence of substrate. This treatment yields an almost fully active enzyme in which all carboxyl groups but Glu 35 and Asp 52 are derivatized. Other group specific reagents that modify, for instance, His, Lys, Met, or Tyr residues, but induce no major protein structure disruptions, cause little change in lysozyme's catalytic efficiency.

Role of Strain

Many of the mechanistic investigations of lysozyme have had the elusive goal of establishing the catalytic role of strain. Not all of these studies, as we shall see, support the Phillips mechanism.

Table 14-3
Binding Free Energies of HEW Lysozyme Subsites

Site	Bound Saccharide	Binding Free Energy (kJ·mol⁻¹)
A	NAG	−7.5
B	NAM	−12.3
C	NAG	−23.8
D	**NAM**	**+12.1**
E	NAG	−7.1
F	NAM	−7.1

Source: Chipman, D. M. and Sharon, N., *Science* **165,** 459 (1969).

Measurements of the binding equilibria of various oligo-saccharides to lysozyme indicate that all saccharide residues except that binding to the D subsite contribute energetically towards the binding of substrate to lysozyme; binding NAM in the D subsite requires a free energy input of 12 kJ·mol⁻¹ (Table 14-3). The Phillips mechanism explains this observation as being indicative of the energy penalty of straining the D ring from its preferred chair conformation towards the half-chair form.

As we discussed in Section 14-1F, an enzyme that catalyzes a reaction by the preferential binding of its transition state has a greater binding affinity for an inhibitor that has the transition state geometry (transition state analog) than it does for its substrate. The δ-lactone analog of (NAG)$_4$ (Fig. 14-17) is a transition state analog of lysozyme since *this compound's lactone ring has the half-chair conformation that geometrically resembles the proposed oxonium ion transition state of the substrate's D ring.* X-ray studies indicate, in accordance with prediction, that this inhibitor binds to lysozyme's A—B—C—D subsites such that the lactone ring occupies the D subsite in a half-chairlike conformation.

Despite the foregoing, *there is considerable evidence that substrate distortion does not play a significant role in lysozyme catalysis.* Theoretical studies by Michael Levitt and Arieh Warshel on substrate binding by lysozyme suggest that *the protein is too flexible to mechanically*

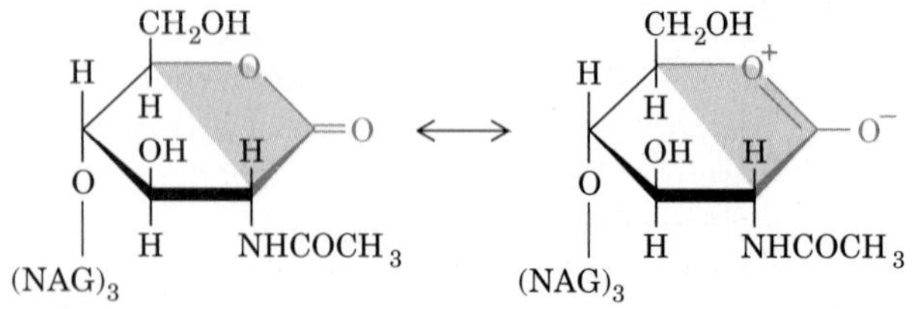

Figure 14-17
The δ-lactone analog of (NAG)$_4$. Its C(1), O(1), C(2), C(5), and O(5) atoms are coplanar (*shading*) because of resonance as is the D ring in the transition state of the lysozyme reaction (compare with Fig. 14-15).

distort the D ring of a bound substrate. Rather, these calculations indicate that transition state stabilization occurs through the displacement by substrate of several tightly bound water molecules from the D subsite. The resulting desolvation of the Asp 52 carboxylate group significantly enhances its capacity to electrostatically stabilize the transition state oxonium ion. This study therefore concludes that "electrostatic strain" rather than steric strain is the more important factor in stabilizing lysozyme's transition state.

Experimental information bearing on the Phillips strain mechanism was gathered by Nathan Sharon and David Chipman who determined the D subsite-binding affinities of several saccharides by comparing the lysozyme-binding affinities of various substrate analogs. The NAG lactone inhibitor binds to the D subsite with 9.2 kJ·mol⁻¹ greater affinity than does NAG. This quantity corresponds, according to Eq. [13.18], to no more than an ∼30-fold rate enhancement of the lysozyme reaction as a result of strain (recall that the difference in binding energy between a transition state analog and a substrate is indicative of the enzyme's rate enhancement arising from the preferential binding of the transition state complex). Such an enhancement is hardly a major portion of lysozyme's ∼10⁸-fold rate enhancement (accounting for only ∼20% of the reaction's $\Delta\Delta G^{\ddagger}_{cat}$; Section 13-1C). Moreover, an **N-acetylxylosamine (XylNAc)** residue,

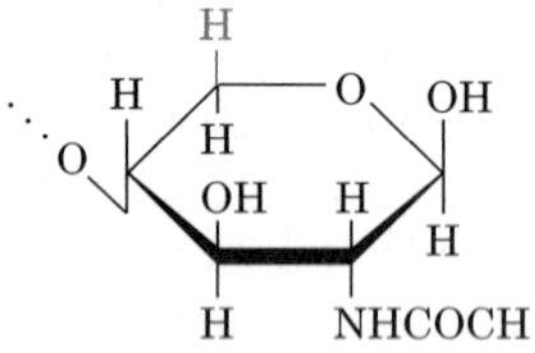

***N*-Acetylxylosamine residue**

which lacks the sterically hindered —C(6)H$_2$OH group of NAM and NAG, has only marginally greater binding affinity for the D subsite (−3.8 kJ·mol⁻¹) than does NAG (−2.5 kJ·mol⁻¹). Yet, recall the Phillips mechanism postulates that it is the unfavorable contacts made by this —C(6)H$_2$OH group that promotes D-ring distortion. The experimental evidence therefore does not support the hypothesis that enzymatically induced distortion of the substrate D ring is an important aspect of lysozyme catalysis. Rather, it seems, *the D ring does not take up the half-chair conformation until it forms the oxonium ion transition state.* Keeping in mind, however, our yet meager understanding of the complex energetics and dynamics of protein molecules, it is possible that lysozyme promotes its reaction by straining the bound D ring towards the transition state without significantly distorting it. At any rate, with the probable exception that the transition state is stabilized by electrostatic strain rather than steric strain, the Phillips mechanism appears to be essentially correct.

3. SERINE PROTEASES

Our next example of enzymatic mechanisms is a diverse group of proteolytic enzymes known as the **serine proteases** (Table 14-4). These enzymes are so-named because they have a common catalytic mechanism characterized by the possession of a peculiarly reactive Ser residue that is essential for their enzymatic activity. The serine proteases are the most thoroughly understood family of enzymes as a result of their extensive examination over a 40-year period by kinetic, chemical, and physical techniques. In this section, we study the best characterized serine proteases, **chymotrypsin, trypsin,** and **elastase.** We also consider how these three enzymes, which are synthesized in inactive forms, are physiologically activated.

A. Kinetics and Catalytic Groups

Chymotrypsin, trypsin, and elastase are digestive enzymes that are synthesized by the pancreatic acinar cells (Fig. 1-10c) and secreted, via the pancreatic duct, into the duodenum (the small intestine's upper loop). All of these enzymes catalyze the hydrolysis of peptide (amide) bonds but with different specificities for the side chains flanking the scissile (to be cleaved) peptide bond (Table 6-2). Together, they form a potent digestive team.

Ester Hydrolysis as a Kinetic Model

That chymotrypsin can act as an esterase as well as a protease is not particularly surprising since the chemical mechanisms of ester and amide hydrolysis are almost identical. The study of chymotrypsin's esterase activity

has led to important insights concerning this enzyme's catalytic mechanism. Kinetic measurements by Brian Hartley of the chymotrypsin-catalyzed hydrolysis of p-nitrophenylacetate

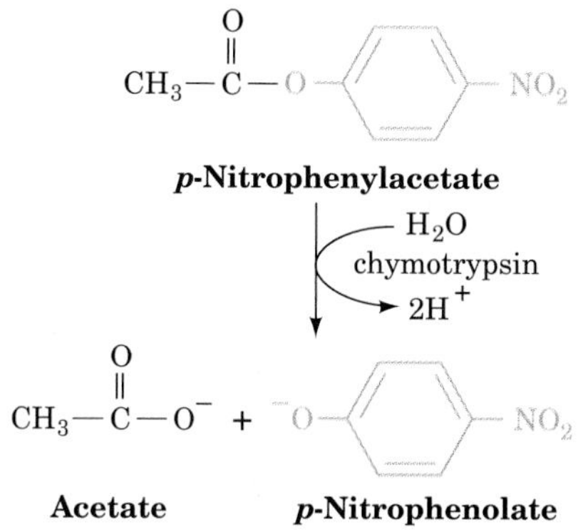

$$CH_3-\overset{\overset{\textstyle O}{\|}}{C}-O-\!\!\!\underset{}{\bigcirc}\!\!\!-NO_2$$

p-Nitrophenylacetate

$$\downarrow \begin{array}{l} -H_2O \\ \text{chymotrypsin} \\ \rightarrow 2H^+ \end{array}$$

$$CH_3-\overset{\overset{\textstyle O}{\|}}{C}-O^- + {}^-O-\!\!\!\underset{}{\bigcirc}\!\!\!-NO_2$$

Acetate   **p-Nitrophenolate**

indicated that the reaction occurs in two phases (Fig. 14-18):

1. The "burst phase," in which the highly colored p-nitrophenolate ion is rapidly formed in amounts stoichiometric with the quantity of active enzyme present.

2. The "steady state phase," in which p-nitrophenolate is generated at a reduced but constant rate that is independent of substrate concentration.

These observations have been interpreted in terms of a two-stage reaction sequence in which the enzyme (1) rapidly reacts with the p-nitrophenylacetate to release p-nitrophenylate ion and form a covalent acyl–enzyme

Table 14-4
A Selection of Serine Proteases

Enzyme	Source	Function
Trypsin	Pancreas	Digestion of proteins
Chymotrypsin	Pancreas	Digestion of proteins
Elastase	Pancreas	Digestion of proteins
Thrombin	Vertebrate serum	Blood clotting
Plasmin	Vertebrate serum	Dissolution of blood clots
Kallikrein	Blood and tissues	Control of blood flow
Complement C1	Serum	Cell lysis in the immune response
Acrosomal protease	Sperm acrosome	Penetration of ovum
Lysosomal protease	Animal cells	Cell protein turnover
Cocoonase	Moth larvae	Dissolution of cocoon after metamorphosis
α-Lytic protease	*Bacillus sorangium*	Possibly digestion
Proteases A and B	*Streptomyces griseus*	Possibly digestion
Subtilisin	*Bacillus subtilus*	Possibly digestion

Source: Stroud, R. M., *Sci. Am.* **231**(1): 86 (1974).

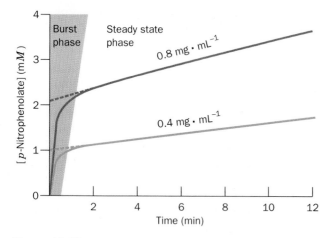

Figure 14-18
The time course of *p*-nitrophenylacetate hydrolysis as
catalyzed by two different concentrations of chymotrypsin.
The enzyme rapidly binds substrate and releases the first
product, *p*-nitrophenolate ion, but the second product,
acetate ion, is released more slowly. Consequently, the rate
of *p*-nitrophenolate generation begins rapidly (burst phase)
but slows as acyl–enzyme complex accumulates until the
rate of *p*-nitrophenolate generation approaches that of
acetate release (steady state). The extrapolation of the
steady state curve to zero time (*dashed lines*) indicates the
initial concentration of active enzyme. [After Hartley, B. S.
and Kilby, B. A., *Biochem. J.* **56**, 294 (1954).]

intermediate that (2) is slowly hydrolyzed to release acetate:

$$CH_3-\overset{\overset{\textstyle O}{\|}}{C}-O-\!\!\!\bigcirc\!\!\!-NO_2 \quad + \quad Enzyme$$

p-Nitrophenylacetate   **Chymotrypsin**

fast

$$^-O-\!\!\!\bigcirc\!\!\!-NO_2$$

p-Nitrophenolate

$$CH_3-\overset{\overset{\textstyle O}{\|}}{C}-Enzyme$$

Acyl–enzyme intermediate

slow H_2O → H^+

$$CH_3-\overset{\overset{\textstyle O}{\|}}{C}-O^- \quad + \quad Enzyme$$

Acetate

Chymotrypsin therefore follows a Ping Pong Bi Bi
mechanism (Section 13-5A). Chymotrypsin-catalyzed
amide hydrolysis has been shown to follow a reaction
pathway similar to that of ester hydrolysis but with the
first step of the reaction, enzyme acylation, being rate
determining rather than the deacylation step.

Identification of the Catalytic Residues

Chymotrypsin's catalytically important groups have
been identified by chemical labeling studies. These are
described below.

Ser 195. A diagnostic test for the presence of the **active
Ser** of serine proteases is its reaction with **diisopropylphosphofluoridate (DIPF):**

$$(Active\ Ser)-CH_2OH \quad + \quad F-\overset{\overset{\textstyle CH(CH_3)_2}{|}\overset{\textstyle O}{|}}{\underset{\overset{\textstyle O}{|}\underset{\textstyle CH(CH_3)_2}{|}}{P}}\!\!=\!\!O$$

**Diisopropylphospho-
fluoridate (DIPF)**

↓

$$(Active\ Ser)-CH_2-O-\overset{\overset{\textstyle CH(CH_3)_2}{|}\overset{\textstyle O}{|}}{\underset{\overset{\textstyle O}{|}\underset{\textstyle CH(CH_3)_2}{|}}{P}}\!\!=\!\!O \quad + \quad HF$$

DIP–Enzyme

which irreversibly inactivates the enzyme. Other Ser
residues, including those on the same protein, do not
react with DIPF. *DIPF reacts only with Ser 195 of chymotrypsin, thereby demonstrating that this residue is the enzyme's active Ser.*

The use of DIPF as an enzyme inactivating agent
came about through the discovery that organophos-
phorus compounds such as DIPF are potent nerve poisons. The neurotoxicity of DIPF arises from its ability to
inactivate **acetylcholinesterase,** a serine esterase that
catalyzes the hydrolysis of **acetylcholine.**

$$(CH_3)_3\overset{+}{N}-CH_2-CH_2-O-\overset{\overset{\textstyle O}{\|}}{C}-CH_3 \quad + \quad H_2O$$

Acetylcholine

acetylcholineesterase

↓

$$(CH_3)_3\overset{+}{N}-CH_2-CH_2-OH \quad + \quad \overset{\overset{\textstyle O}{\|}}{\underset{\overset{\textstyle O}{|}}{C}}-CH_3$$

Choline

Acetylcholine is a **neurotransmitter:** It transmits nerve
impulses across the **synapses** (junctions) between certain types of nerve cells (Section 34-4C). The inactivation of acetylcholinesterase prevents the otherwise
rapid hydrolysis of the acetylcholine released by a nerve
impulse and thereby interferes with the regular sequence of nerve impulses. DIPF is of such great toxicity
to humans that it has been used militarily as a nerve gas.

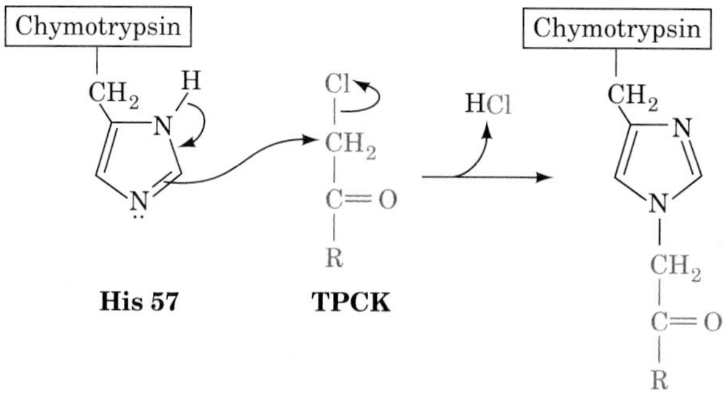

Figure 14-19
The reaction of TPCK with His 57 of chymotrypsin.

Related compounds, such as **parathion** and **malathion,**

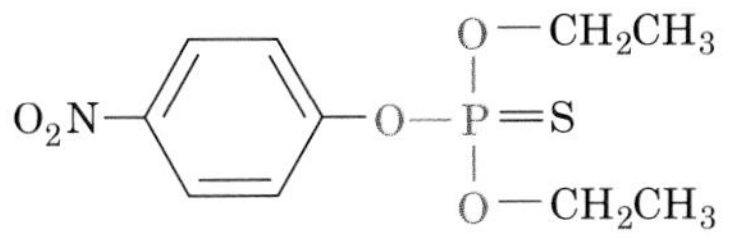

Parathion

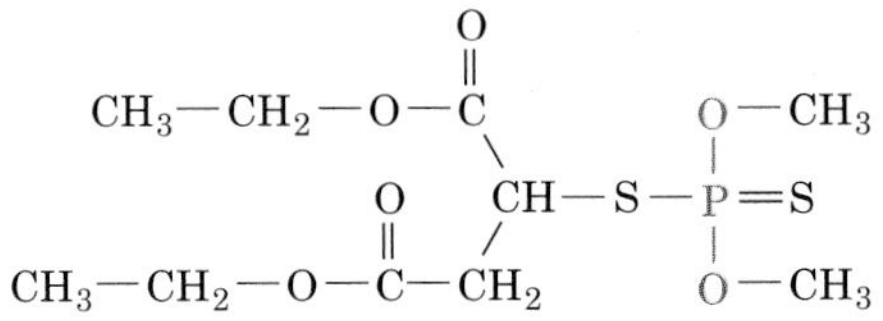

Malathion

are useful insecticides because they are far more toxic to insects than to mammals.

His 57. A second catalytically important residue was discovered through **affinity labeling.** In this technique, a substrate analog bearing a reactive group specifically binds at the enzyme's active site where it reacts to form a stable covalent bond with a nearby susceptible group (these reactive substrate analogs have therefore been described as the "Trojan Horses" of biochemistry). The affinity labeled groups can subsequently be identified by fingerprinting (Section 6-1J). Chymotrypsin specifically binds **tosyl-L-phenylalanine chloromethyl ketone (TPCK),**

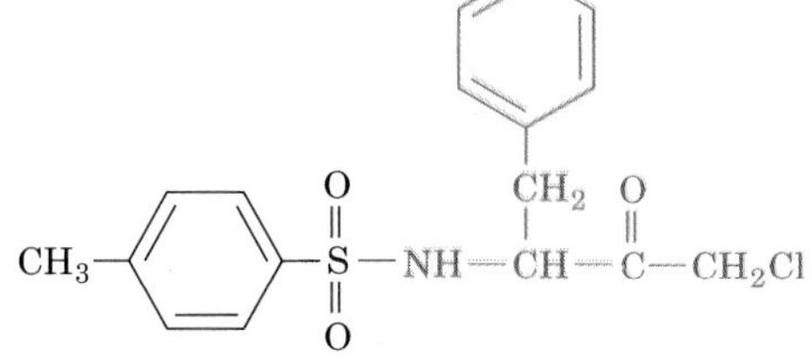

Tosyl-L-phenylalanine chloromethyl ketone (TPCK)

because of its resemblance to a Phe residue (one of chymotrypsin's preferred residues; Table 6-2). Active site-bound TPCK's chloromethyl ketone group is a strong alkylating agent; it reacts only with His 57 (Fig. 14-19) thereby inactivating the enzyme. The TPCK reaction is

inhibited by β-phenylpropionate,

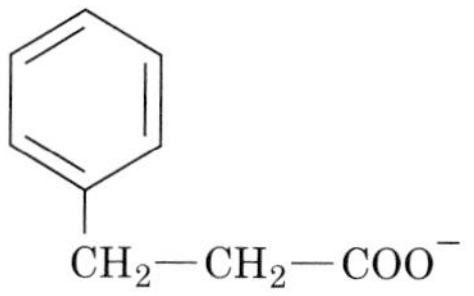

β-Phenylpropionate

a competitive inhibitor of chymotrypsin that presumably competes with TPCK for its enzymatic-binding site. Moreover, the TPCK reaction does not occur in 8*M* urea, a denaturing reagent, or with DIP-chymotrypsin, in which the active site is blocked. These observations establish that *His 57 is an essential active site residue of chymotrypsin.*

B. X-Ray Structures

Bovine chymotrypsin, bovine trypsin, and porcine elastase are strikingly homologous: The primary structures of these ~240-residue monomeric enzymes are ~40% identical and their internal sequences are even more alike (in comparison, the α and β chains of human hemoglobin have a 44% sequence identity). Furthermore, *all of these enzymes have an active Ser and a catalytically essential His as well as similar kinetic mechanisms.* It therefore came as no surprise when their X-ray structures all proved to be closely related.

To most conveniently compare the structures of these three digestive enzymes, they have been assigned the same amino acid residue numbering scheme. Bovine chymotrypsin is synthesized as an inactive 245-residue precursor named **chymotrypsinogen** that is proteolytically converted to chymotrypsin (Section 14-3E). In what follows, the numbering of the amino acid residues in chymotrypsin, trypsin, and elastase will be that of the corresponding residues in bovine chymotrypsinogen.

The X-ray structure of bovine chymotrypsin was elucidated in 1967 by David Blow. This was followed by the determination of the structures of bovine trypsin (Fig. 14-20) by Robert Stroud and Richard Dickerson, and porcine elastase by David Shotton and Herman Watson. Each of these proteins is folded into two domains, both of which have extensive regions of antiparallel β-sheets in a barrel-like arrangement, but contain little helix. *The catalytically essential His 57 and Ser 195 are located at the substrate-binding site together with the invariant (in all serine proteases) Asp 102, which is buried in a solvent inaccessible pocket. These three residues form a hydrogen bonded constellation referred to as the **catalytic triad** (Figs. 14-20 and 14-21).*

Substrate Specificities Can Be Structurally Rationalized

The X-ray structures of the above three enzymes

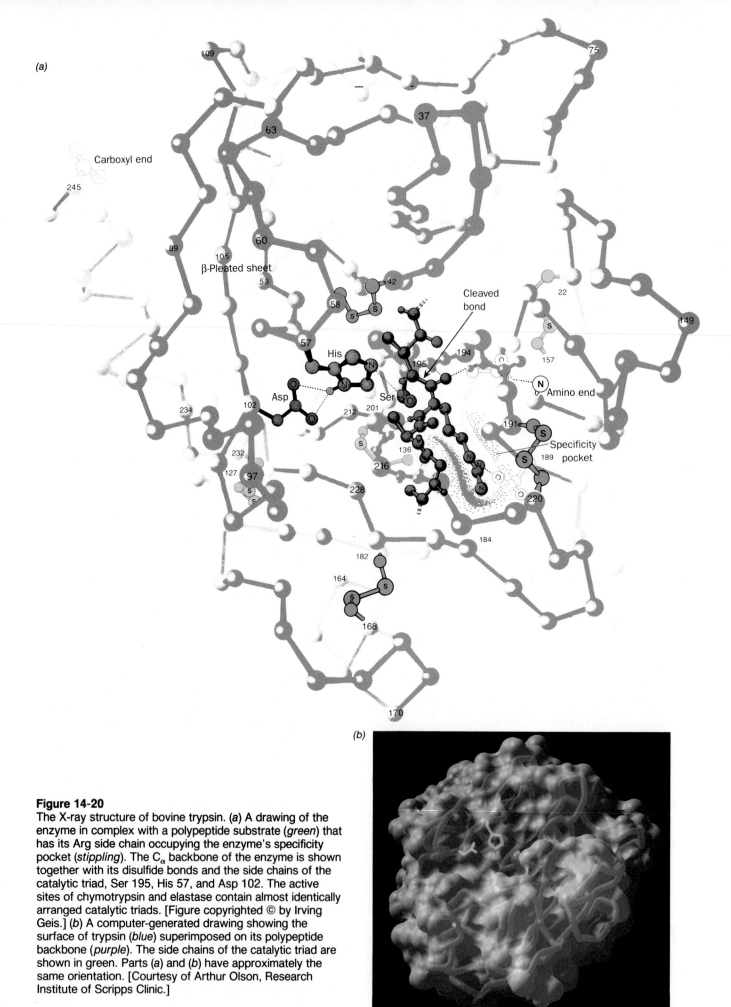

(a)

109

75

Carboxyl end

245

63

37

89

60

105

β-Pleated sheet

53

42

58

S

S

22

S

149

57

His

N

N

194

Cleaved
bond

195

157

N Amino end

Asp

Ser

102

234

212

201

191

S

Specificity
pocket

232

136

S

189

127 97

S
S

216

228

S

220

N

184

182

S

164

S

168

170

Figure 14-20

The X-ray structure of bovine trypsin. (*a*) A drawing of the
enzyme in complex with a polypeptide substrate (*green*) that
has its Arg side chain occupying the enzyme's specificity
pocket (*stippling*). The C_α backbone of the enzyme is shown
together with its disulfide bonds and the side chains of the
catalytic triad, Ser 195, His 57, and Asp 102. The active
sites of chymotrypsin and elastase contain almost identically
arranged catalytic triads. [Figure copyrighted © by Irving
Geis.] (*b*) A computer-generated drawing showing the
surface of trypsin (*blue*) superimposed on its polypeptide
backbone (*purple*). The side chains of the catalytic triad are
shown in green. Parts (*a*) and (*b*) have approximately the
same orientation. [Courtesy of Arthur Olson, Research
Institute of Scripps Clinic.]

(b)

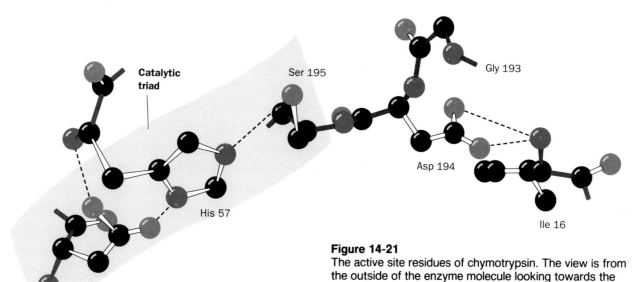

Figure 14-21
The active site residues of chymotrypsin. The view is from the outside of the enzyme molecule looking towards the interior. The catalytic triad consists of Ser 195, His 57, and Asp 102. [After Blow, D. M. and Steitz, T. A., *Annu. Rev. Biochem.* **39**, 86 (1970).]

clearly indicate the basis for their differing substrate specificities (Table 6-2):

1. In chymotrypsin, the bulky aromatic side chain of the preferred Phe, Trp, or Tyr residue that contributes the carbonyl group of the scissile peptide fits snugly into a slitlike hydrophobic pocket located near the catalytic groups.

2. In trypsin, the residue corresponding to chymotrypsin Ser 189, which lies at the back of the binding pocket, is the anionic residue Asp. The cationic side chains of trypsin's preferred residues, Arg or Lys, can therefore form ion pairs with this Asp residue. The rest of chymotrypsin's specificity pocket is preserved in trypsin so that it can accommodate the bulky side chains of Arg and Lys.

3. Elastase is so-named because it rapidly hydrolyzes the otherwise nearly indigestible Ala, Gly, and Val-rich protein elastin (the structure of elastin is discussed in Section 7-2D). Elastase's binding pocket is largely occluded by the side chains of a Val and a Thr residue that replace two Gly's lining the specificity pocket in both chymotrypsin and trypsin. Consequently elastase, whose substrate-binding pocket is better described as a depression, specifically cleaves peptide bonds after small nonpolar residues, particularly Ala. In contrast, chymotrypsin and trypsin hydrolyze such peptide bonds extremely slowly because these small substrates cannot be sufficiently immobilized on the enzyme surface for efficient catalysis to occur (Section 14-1E).

Evolutionary Relationships among Serine Proteases

We have seen that sequence and structural homologies among proteins reveal their evolutionary relationships (Sections 6-3 and 8-3). *The great similarities among chymotrypsin, trypsin, and elastase indicate that these proteins evolved through gene duplications of an ancestral serine protease followed by the divergent evolution of the resulting enzymes (Section 6-3C).*

Two serine proteases of bacterial origin provide further insights into the evolutionary relationships among the serine proteases. ***Streptomyces griseus* protease A (SGPA)** is a bacterial serine protease of chymotryptic specificity whose X-ray structure was determined by Michael James. It exhibits extensive structural similarity, although <20% sequence identity, with the pancreatic serine proteases. The primordial trypsin gene evidently arose before the divergence of prokaryotes and eukaryotes.

Subtilisin (originally isolated from *Bacillus subtilus*) is a serine protease whose primary and tertiary structures bear no discernible relationships to those of chymotrypsin. Yet, as Joseph Kraut demonstrated, the active site groups of these two enzymes are essentially identical and their relative three-dimensional positions are nearly indistinguishable. Since the orders of the corresponding active site residues in the amino acid sequences of the two enzymes are, nevertheless, quite different (Fig. 14-22), *it seems highly improbable that they could have evolved from a common ancestor serine protease. Chymotrypsin and subtilisin apparently constitute a remarkable example of* **convergent evolution**: *Nature seems to have independently discovered the same catalytic mechanism at least twice.*

C. Catalytic Mechanism

The extensive active site homologies among the various serine proteases indicates that they all have the same catalytic mechanism. On the basis of considerable

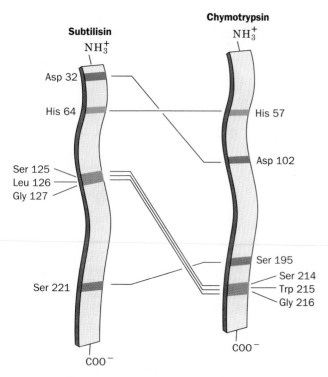

Subtilisin

NH$_3^+$

Asp 32

His 64

Ser 125
Leu 126
Gly 127

Ser 221

COO$^-$

Chymotrypsin

NH$_3^+$

His 57

Asp 102

Ser 195

Ser 214
Trp 215
Gly 216

COO$^-$

Figure 14-22
A diagram indicating the relative positions of the active site residues in the primary structures of subtilisin (*left*) and chymotrypsin (*right*). Ser 221, His 64, and Asp 32 form subtilisin's catalytic triad. The peptide backbones of Ser 214, Trp 215, and Gly 216 in chymotrypsin, and their counterparts in subtilisin, participate in substrate-binding interactions. [After Robertus, J. D., Alden, R. A., Birktoft, J. J., Kraut, J., Powers, J. C., and Wilcox, P. E., *Biochemistry* **11**, 2449 (1972).]

chemical and structural data gathered in many laboratories, the following catalytic mechanism has been formulated for the serine proteases, here given in terms of chymotrypsin:

Rate-Determining Formation of the Tetrahedral Intermediate

After chymotrypsin has bound substrate to form the Michaelis complex,

Michaelis complex

*Ser 195 nucleophilically attacks the scissile peptide's carbonyl group to form the **tetrahedral intermediate** (covalent catalysis).*

Tetrahedral
intermediate

X-ray studies indicate that Ser 195 is ideally positioned to carry out this nucleophilic attack (proximity and orientation effects). The imidazole ring of His 57 takes up the liberated proton thereby forming an imidazolium ion (general base catalysis). This process is aided by the polarizing effect of the unsolvated carboxylate ion of Asp 102, which is hydrogen bonded to His 57 (electrostatic catalysis). Indeed, the replacement of trypsin's Asp 102 with Asn by site-directed mutagenesis (a technique considered in Section 28-8), leaves the enzyme's K_M substantially unchanged at neutral pH but reduces its k_{cat} to $<0.05\%$ of its native value. Neutron diffraction studies, which provide similar information to X-ray diffraction studies but also indicate hydrogen atom positions, have demonstrated that *Asp 102 remains a carboxylate ion rather than abstracting a proton from the imidazolium ion to form an uncharged carboxylic acid group.* The tetrahedral intermediate has a well defined although transient existence. We shall see that *much of chymotrypsin's catalytic power derives from its preferential binding of this transition state (transition state binding catalysis).*

Formation of the Acyl–Enzyme Intermediate

The tetrahedral intermediate decomposes to the **acyl–enzyme intermediate** under the driving force of proton donation from N(3) of His 57 (general acid catalysis):

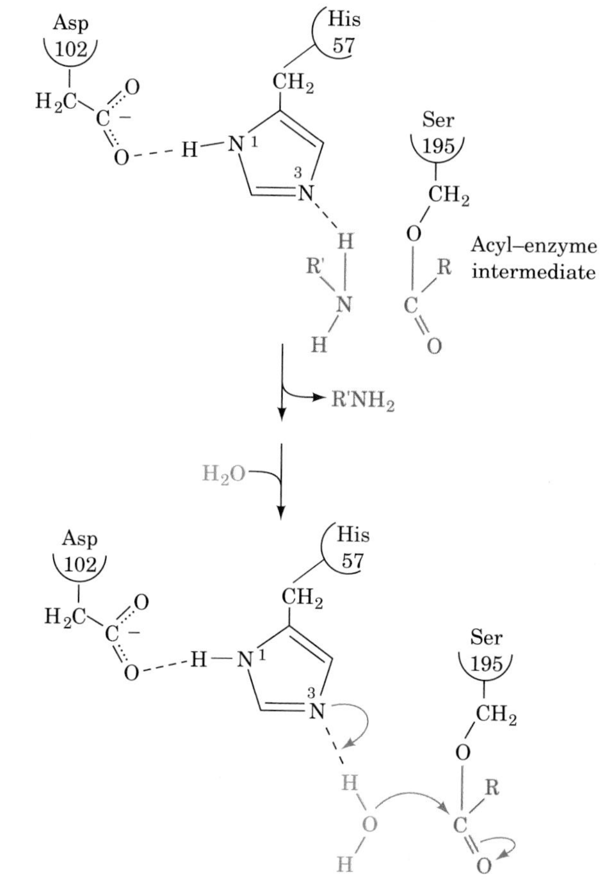

Asp 102 His 57 CH₂ Ser 195 CH₂

H_2C C — O

O - - - H — N¹

Acyl–enzyme intermediate

R'NH₂

H_2O

Asp 102 His 57 CH₂ Ser 195

New C-terminus of cleaved polypeptide chain

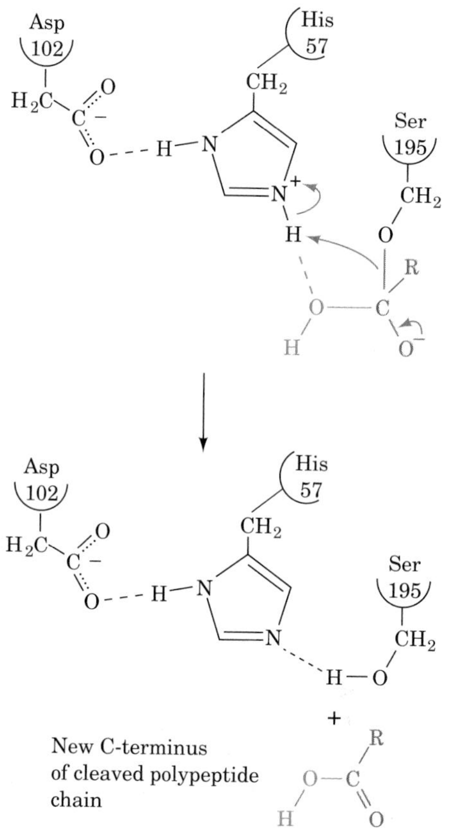

In this process, water is the attacking nucleophile and Ser 195 is the leaving group.

D. Testing the Catalytic Mechanism

The formulation of the foregoing model for catalysis by serine proteases has prompted numerous investigations of its validity. In this section we discuss several of the most revealing of these studies.

The Tetrahedral Intermediate Is Mimicked in a Complex of Trypsin with Trypsin Inhibitor

Perhaps the most convincing structural evidence for the existence of the tetrahedral intermediate was provided by Robert Huber in an X-ray study of the complex between **bovine pancreatic trypsin inhibitor (BPTI)** and trypsin. The 58-residue protein BPTI, whose folding pathway we examined in Section 8-1B, binds to and inactivates trypsin; this interaction prevents any trypsin that is prematurely activated in the pancreas from digesting that organ (see Section 14-3E). BPTI binds to the active site region of trypsin across a tightly packed interface that is cross-linked by a complex network of hydrogen bonds. This complex's $10^{13}M^{-1}$ association constant, the largest of any known protein–protein interaction, emphasizes BPTI's physiological importance.

The portion of BPTI in contact with the trypsin active site resembles bound substrate. The side chain of BPTI Lys 15I (here "I" differentiates BPTI residues from trypsin resi-

The amine leaving group (R'NH₂, the new N-terminal portion of the cleaved polypeptide chain) is released from the enzyme and replaced by water from the solvent. The acyl–enzyme intermediate is extremely unstable to hydrolytic cleavage because of the enzyme's catalytic properties (see below). Despite this instability, the X-ray structure of elastase's acyl–enzyme intermediate has been reported. It was trapped at −55°C at which temperature the rate of the enzymatic reaction is slowed by many orders of magnitude.

The Deacylation Step

The deacylation step proceeds largely through the reversal of the previous steps, with the release of the carboxylate product (the new C-terminal portion of the cleaved polypeptide chain) and the concomitant regeneration of the enzyme:

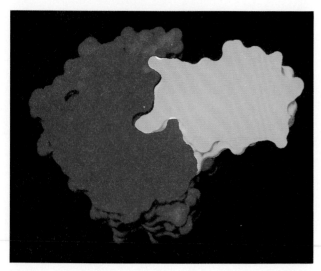

(a)

(b)

Figure 14-23

The trypsin–BPTI complex. (a) The X-ray structure shown as a computer-generated cutaway drawing indicating how trypsin (*red*) binds BPTI (*green*). The green protrusion extending into the red cavity near the center of the figure represents the Lys 15I side chain occupying trypsin's specificity pocket. Note the close complementary fit of these two proteins. [Figure courtesy of Michael Connolly, New York University.] (b) Trypsin Ser 195, the active Ser, is in closer-than-van der Waals contact with the carbonyl carbon of BPTI's scissile peptide, which is pyramidally distorted towards Ser 195. The normal proteolytic reaction is apparently arrested somewhere along the reaction coordinate between the Michaelis complex and the tetrahedral intermediate.

dues) occupies the trypsin specificity pocket (Fig. 14-23*a*) and the peptide bond between Lys 15I and Ala 16I is positioned as if it were the scissile peptide bond (Fig. 14-23*b*). What is most remarkable about this structure is that *its active site complex assumes a conformation well along the reaction coordinate towards the tetrahedral intermediate: The side chain oxygen of trypsin Ser 195, the active Ser, is in closer-than-van der Waals contact (2.6 Å) with the pyramidally distorted carbonyl carbon of BPTI's "scissile" peptide*. Despite this close contact, the proteolytic reaction cannot proceed past this point along the reaction coordinate because of the rigidity of the active site complex and because it is so tightly sealed that the leaving group cannot leave and water cannot enter the reaction site.

Protease inhibitors are common in nature where they have protective and regulatory functions. For exam-

ple, certain plants release protease inhibitors in response to insect bites thereby causing the offending insect to starve by inactivating its digestive enzymes. Protease inhibitors constitute ~10% of the nearly 200 proteins of blood serum. For instance, **α_1-proteinase inhibitor,** which is secreted by the liver, inhibits **leucocyte elastase** (leucocytes are a type of white blood cell; the action of leucocyte elastase is thought to be part of the inflammatory process). Pathological variants of α_1-proteinase inhibitor are associated with **pulmonary emphysema,** a degenerative disease of the lungs resulting from the hydrolysis of its elastic fibers. Smokers also suffer from reduced activity of their α_1-proteinase inhibitor because of the oxidation of its active site Met residue. Full activity of this inhibitor is not regained until several hours after smoking.

Serine Proteases Preferentially Bind the Transition State

Detailed comparisons of the X-ray structures of several serine protease-inhibitor complexes have revealed the structural basis for catalysis in these enzymes (Fig. 14-24):

1. The conformational distortion that occurs with the formation of the tetrahedral intermediate causes the carbonyl oxygen of the scissile peptide to move deeper into the active site so as to occupy a previously unoccupied position—the **oxyanion hole.**

2. *There it forms two hydrogen bonds with the enzyme that cannot form when the carbonyl group is in its normal trigonal conformation.* These two enzymatic hydrogen bond donors were first noted by Kraut to occupy corresponding positions in chymotrypsin and subtilisin. He proposed the existence of the oxyanion hole based on the premise that convergent evolution had made the active sites of these unrelated enzymes functionally identical.

3. The tetrahedral distortion, moreover, permits the formation of an otherwise unsatisfied hydrogen bond between the enzyme and the backbone NH group of the residue preceding the scissile peptide. Consequently, *the enzyme binds the tetrahedral intermediate in preference to either the Michaelis complex or the acyl–enzyme intermediate.*

It is this phenomenon that is largely responsible for the catalytic efficiency of serine proteases. In fact, the reason that DIPF is such an effective inhibitor of serine proteases is because its tetrahedral phosphate group makes this compound a transition state analog of the enzyme.

The Role of the Catalytic Triad

The earlier literature postulated that the Asp 102-polarized His 57 side chain directly abstracts a proton from Ser 195, thereby converting its weakly nucleophilic

(a)

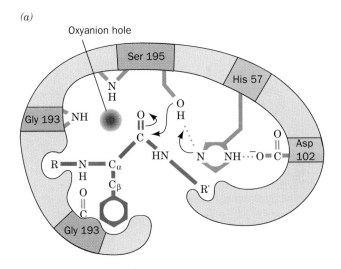

(b)

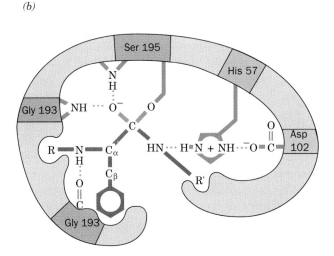

Figure 14-24
Transition state stabilization in the serine proteases: (*a*) In the Michaelis complex, the trigonal carbonyl carbon of the scissile peptide is conformationally constrained from binding in the oxyanion hole (*upper left*). (*b*) In the tetrahedral intermediate, the carbonyl oxygen of the scissile peptide (the oxyanion) has entered the oxyanion hole thereby

hydrogen bonding to the backbone NH groups of Gly 193 and Ser 195. The consequent conformational distortion permits the NH group of the peptide preceding the scissile peptide to form an otherwise unsatisfied hydrogen bond to Gly 193. Serine proteases therefore preferentially bind the tetrahedral intermediate. [After Robertus, J. D., Kraut, J., Alden, R. A., and Birktoft, J. J., *Biochemistry* **11**, 4302 (1972).]

CH$_2$OH group to a highly nucleophilic alkoxide ion, CH$_2$O$^-$.

methylation of His 57 by treating chymotrypsin with **methyl-*p*-nitrobenzene sulfonate,**

"Charge relay system"

In the process, the anionic charge of Asp 102 was thought to be transferred, via a tautomeric shift of His 57, to Ser 195. The catalytic triad was therefore originally named the **charge relay system.** It is now realized, however, that such a mechanism is implausible because an alkoxide ion (p$K \geq 15$) has far greater proton affinity than does His 57 (p$K \approx 7$, as measured by NMR techniques). Ser 195 is therefore unlikely to be significantly activated as a nucleophile by its polarization through the interactions of the catalytic triad. In fact, blocking the action of the catalytic triad through the specific

yields an enzyme that is nevertheless a reasonably good catalyst: It enhances the rate of proteolysis by as much as a factor of 2×10^6 over the uncatalyzed reaction, whereas the native enzyme has a rate enhancement factor of $\sim 10^{10}$. Evidently, the catalytic triad functions largely as an alternate source and sink of protons (general acid–base catalysis). *A major portion of chymotrypsin's rate enhancement must therefore be attributed to its preferential binding of the catalyzed reaction's transition state.*

E. Zymogens

Proteolytic enzymes are usually biosynthesized as somewhat larger inactive precursors known as **zymogens** (enzyme precursors, in general, are known as **proenzymes**). In the case of digestive enzymes, the reason for this is clear: If these enzymes were synthesized in their active forms, they would digest the tissues that synthesized them. Indeed, **acute pancreatitis**, a sometimes fatal condition that can be precipitated by pancreatic trauma, is characterized by the premature activation of the digestive enzymes synthesized by this gland.

Serine Proteases Are Autocatalytically Activated

Trypsin, chymotrypsin, and elastase are activated according to the following pathways:

Trypsin. The activation of **trypsinogen,** the zymogen of trypsin, occurs as a two-stage process when trypsinogen enters the duodenum from the pancreas. **Enteropeptidase** (originally named **enterokinase**), a serine protease that is secreted under hormonal control by the duodenal mucosa, specifically clips off trypsinogen's N-terminal hexapeptide to yield the active enzyme (Fig. 14-25). The proteolytic cleavage that activates trypsinogen occurs between its Lys 15 and Ile 16 residues yielding an active enzyme having Ile 16 at its N-terminus. Since this activating cleavage occurs at a trypsin-sensitive site (recall that trypsin cleaves after Arg and Lys residues), the small amount of trypsin produced by enteropeptidase generates more trypsin, *etc.*; that is, *trypsinogen activation is autocatalytic.*

Chymotrypsin. Chymotrypsinogen is activated by the specific tryptic cleavage of its Arg 15—Ile 16 peptide bond, to form π-chymotrypsin (Fig. 14-26). π-Chymotrypsin subsequently undergoes autolysis (self-digestion) to specifically excise two dipeptides, Ser 14-Arg 15 and Thr 147-Asn 148, thereby yielding the equally active enzyme α-chymotrypsin (heretofore and hereafter referred to as chymotrypsin). The biochemical significance of this latter process, if any, is unknown.

Elastase. Proelastase, the zymogen of elastase, is activated similarly to trypsinogen by a single tryptic cleavage that excises a short N-terminal polypeptide.

Biochemical "Strategies" that Prevent Premature Zymogen Activation

Trypsin activates **procarboxypeptidases A** and **B** and **prophospholipase A_2** (the action of phospholipase A_2 is outlined in Section 23-1) as well as the pancreatic serine proteases. Premature trypsin activation can consequently trigger a series of events that lead to pancreatic self-digestion. Nature has therefore evolved an elaborate defense against such inappropriate trypsin activation. We have already seen (Section 14-3D) that pancreatic trypsin inhibitor binds essentially irreversibly to any trypsin formed in the pancreas so as to inactivate it.

Figure 14-25
The activation of trypsinogen to form trypsin occurs by proteolytic excision of the N-terminal hexapeptide as catalyzed by either enteropeptidase or trypsin. Chymotrypsinogen residue numbering is used here; that is, Val 10 is actually trypsinogen's N-terminus and Ile 16 is trypsin's N-terminus.

Furthermore, the trypsin-catalyzed activation of trypsinogen (Fig. 14-25) occurs quite slowly, presumably because the unusually large negative charge of its highly evolutionarily conserved N-terminal hexapeptide repels the Asp at the back of trypsin's specificity pocket. Finally, pancreatic zymogens are stored in intracellular vesicles called **zymogen granules** whose membranous walls are thought to be resistant to enzymatic degradation.

Zymogens Have Distorted Active Sites

Since the zymogens of trypsin, chymotrypsin, and elastase have all their catalytic residues, why aren't they enzymatically active? The X-ray structures of both trypsinogen and chymotrypsinogen show that upon activation, the newly liberated N-terminal Ile 16 residue moves from the surface of the protein to an internal position where its free cationic amino group forms an ion pair with the invariant anionic Asp 194 (Fig. 14-21). Aside from this change, however, the structures of these zymogens closely resemble those of their corresponding active enzymes. Surprisingly, this resemblance includes their catalytic triads, an observation which led to the discovery that these zymogens are actually enzymatically active, albeit at a very low level. Careful comparisons of the corresponding enzyme and zymogen structures, however, revealed the reason for this low activity: *The zymogens' specificity pockets and oxyanion holes are improperly formed such that, for example, the amide NH of chymotrypsin's Gly 193 points in the wrong direction to form a hydrogen bond with the tetrahedral intermediate (see Fig. 14-24).* Hence, the zymogens' very low enzymatic activity arises from their inability to productively bind substrate and to stabilize the tetrahedral intermediate. These observations provide further structural evidence favoring the role of transition state binding in the catalytic mechanism of serine proteases.

4. GLUTATHIONE REDUCTASE

Lysozyme and the serine proteases all catalyze hy-

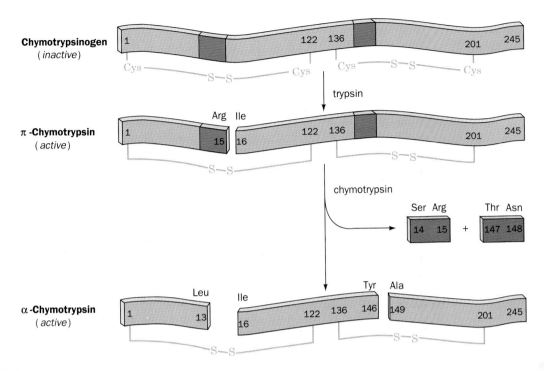

Figure 14-26
The activation of chymotrypsinogen by proteolytic cleavage. Both π- and α-chymotrypsin are enzymatically active.

drolytic reactions. In contrast, the third enzyme that we shall consider in mechanistic detail, **glutathione reductase,** catalyzes an oxidation–reduction reaction. Such reactions are extremely important in metabolic processes. We have chosen to study glutathione reductase, which sequentially catalyzes several electron-transfer processes, because it is one of the few such enzymes in which the pathway of electron flow has been well characterized.

Glutathione reductase is a nearly ubiquitous enzyme that catalyzes the NADPH-dependent reduction of **glutathione disulfide (GSSG)** to **glutathione (GSH):**

$$H_3\overset{+}{N}-CH-CH_2-CH_2-\overset{O}{\overset{\|}{C}}-NH-CH-\overset{O}{\overset{\|}{C}}-NH-CH_2-COO^-$$
$$\underset{COO^-}{|} \qquad\qquad\qquad \underset{CH_2}{|}$$
$$\underset{S}{|}$$
$$\underset{S}{|}$$
$$\underset{COO^-}{|} \qquad\qquad O \qquad \underset{CH_2}{|}\ \ O$$
$$H_3\overset{+}{N}-CH-CH_2-CH_2-\overset{\|}{C}-NH-CH-\overset{\|}{C}-NH-CH_2-COO^-$$

Glutathione disulfide (GSSG)

NADPH + H⁺ ⟍ glutathione
NADP⁺ ⟋ reductase

$$2\ H_3\overset{+}{N}-CH-CH_2-CH_2-\overset{O}{\overset{\|}{C}}-NH-CH-\overset{O}{\overset{\|}{C}}-NH-CH_2-COO^-$$
$$\underset{COO^-}{|} \qquad\qquad\qquad \underset{CH_2}{|}$$
$$\underset{SH}{|}$$

Glutathione (GSH)
(γ-L-Glutamyl-L-cysteinyl glycine)

(the structures of NADP⁺ and NADPH are indicated in Fig. 12-2). This process normally produces a GSH : GSSG ratio of over 100 : 1, which permits GSH to function as an intracellular reducing agent (the thermodynamics of oxidation–reduction reactions is discussed in Section 15-5). For example, the inactivation of proteins (P) that have free SH groups through the spontaneous oxidative formation of mixed disulfides

$$P-SH + P'-SH + \tfrac{1}{2}O_2 \rightleftharpoons P-S-S-P' + H_2O$$

is reversed through disulfide interchange with GSH.

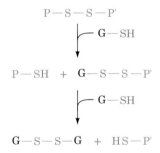

GSH also acts as a coenzyme in several enzymatically catalyzed reductions and plays an important role in the transport of amino acids into certain cells (Section 24-4C).

FAD Is an Essential Redox Coenzyme

Glutathione reductase contains the electron-transfer prosthetic group **flavin adenine dinucleotide (FAD;** Fig. 14-27). **Flavins** (substances that contain the **isoalloxazine** ring) can undergo two sequential one-electron transfers (Fig. 14-27), or a simultaneous two-electron transfer that bypasses the semiquinone state. The gluta-

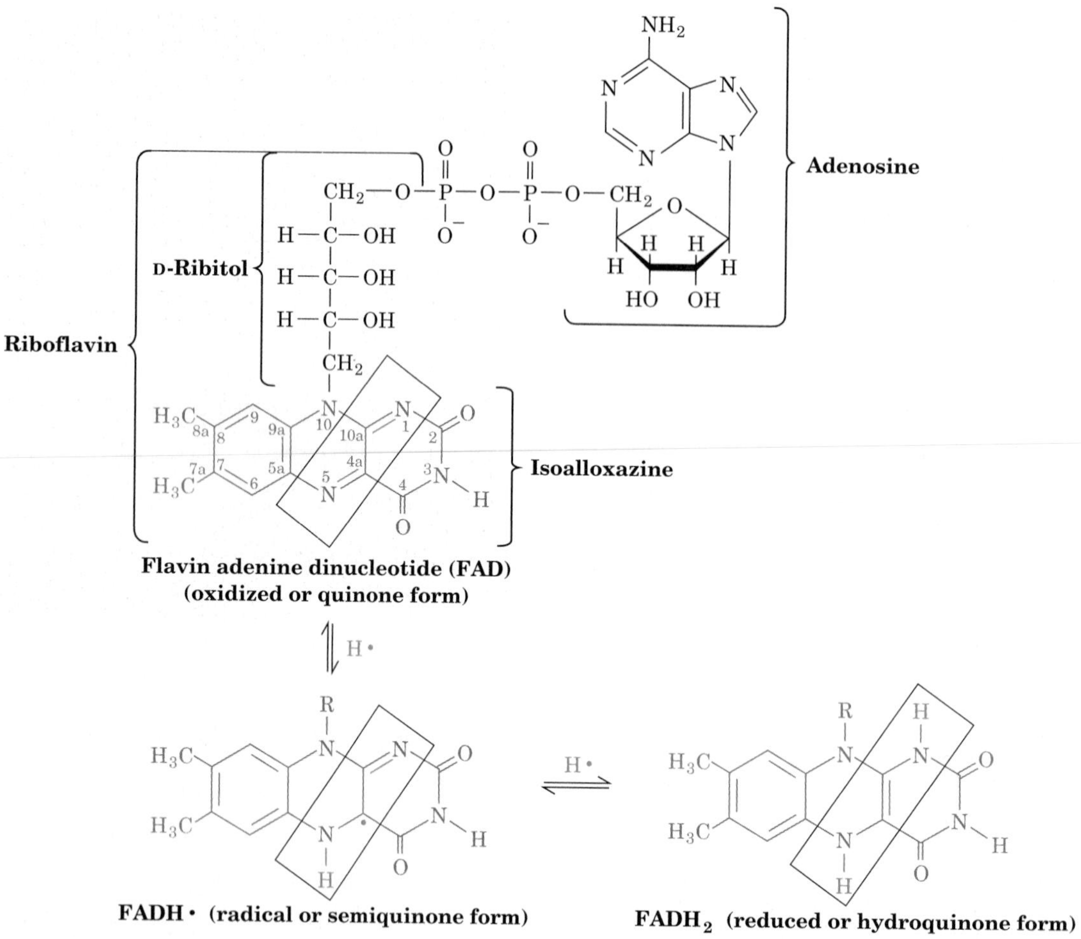

Figure 14-27
The molecular formula and reactions of the coenzyme flavin adenine dinucleotide (FAD). The term "flavin" is synonymous with the isoalloxazine ring system. The D-ribitol residue is derived from the alcohol of the sugar D-ribose. FAD may be half-reduced to the stable radical FADH· or fully reduced to FADH$_2$ (*boxes*). Consequently, different FAD-containing enzymes cycle between different oxidation states of FAD. FAD is usually tightly bound to its enzymes so that this coenzyme normaly is a prosthetic group rather than a cosubstrate as is the case, for example, with NAD$^+$.

thione reductase reaction involves the simultaneous transfer of two electrons so that, in this case, the FAD never assumes its radical form. The oxidation state of the flavin in a **flavoprotein** (flavin-containing protein) is readily established from its characteristic UV–visible spectrum: FAD is an intense yellow whereas FADH$_2$ is pale yellow.

Humans and other higher animals are unable to synthesize the isoalloxazine component of flavins so that they must obtain this substance from their diets, for example, in the form of **riboflavin (vitamin B$_2$)** (Fig. 14-27). Riboflavin deficiency is quite rare in humans, in part because of the tight binding of flavin prosthetic groups to their apoenzymes. The symptoms of riboflavin deficiency, which are associated with general malnutrition or bizarre diets, include an inflamed tongue, lesions in the corners of the mouth, and dermatitis.

Glutathione Reductase Catalyzes a Two-Stage Reaction

Glutathione reductase from human erythrocytes is a dimer of identical 478-residue subunits (52.4 kD per monomer) that are covalently linked by an intersubunit disulfide bond. In the absence of GSSG, the enzyme catalyzes the first stage of a two-stage reaction:

First stage $E + NADPH + H^+ \rightleftharpoons EH_2 + NADP^+$

where E represents fully oxidized glutathione reductase and EH$_2$ is a stable two-electron reduced intermediate whose chemical nature we shall presently discuss. Upon subsequent addition of GSSG, EH$_2$ reacts to form products and complete the catalytic cycle.

Second stage $EH_2 + GSSG \rightleftharpoons E + 2GSH$

The glutathione reductase reaction is more complex than these overall reactions suggest. Vincent Massey and Charles Williams have demonstrated that *oxidized glutathione reductase (E) contains a "redox-active" disulfide bond, which in EH$_2$ has accepted an electron pair through bond cleavage to form a dithiol.*

$$\left[\begin{array}{c} S-S \\ \text{Enzyme} \end{array}\right] \xrightarrow{2e^-} \left[\begin{array}{c} S^- \quad {}^-S \\ \text{Enzyme} \end{array}\right] \xrightarrow{2H^+} \left[\begin{array}{c} SH \quad HS \\ \text{Enzyme} \end{array}\right]$$

Through the use of arsenite, an ion that specifically complexes vicinal (adjacent) dithiols but not disulfide bonds, they obtained the following information:

$$\begin{array}{c} -SH \\ \\ -SH \end{array} + AsO_2^- \longrightarrow \begin{array}{c} -S \\ \quad As-OH \\ -S \end{array} + OH^-$$

Arsenite

1. The spectrum of oxidized glutathione reductase (E) is unaffected by arsenite.

2. When NADPH reacts with the enzyme in the presence of arsenite, the reduced glutathione reductase (EH_2) produced binds arsenite to form an enzymatically inactive species.

3. The spectrum of the inactive EH_2 indicates that its FAD prosthetic group is fully oxidized.

Glutathione reductase must therefore have a second electron acceptor besides FAD; arsenite's known specificity suggests that the acceptor is a disulfide.

Glutathione reductase's amino acid sequence indicates that its redox-active disulfide bond forms between Cys 58 and Cys 63, which occur on a highly conserved segment of the enzyme's polypeptide chain. Thus, in the first stage of the glutathione reductase reaction, NADPH reduces the enzyme's redox-active disulfide and, in the second stage, this reduced disulfide reduces GSSG to 2GSH.

$$\begin{array}{c} S \\ | \ E \\ S \end{array} + NADPH \xrightarrow{H^+} \begin{array}{c} HS \\ \quad E \\ HS \end{array} + NADP^+ \quad \text{(first stage)}$$

$$\begin{array}{c} HS \\ \quad E \\ HS \end{array} + G-S-S-G \longrightarrow \begin{array}{c} S \\ | \ E \\ S \end{array} + 2GSH \quad \text{(second stage)}$$

The X-Ray Structure of Glutathione Reductase

The X-ray structure of human erythrocyte glutathione reductase, which was determined by Georg Schulz and Heiner Schirmer, indicates that the protein's monomer units each contain considerable secondary structure and are folded into five domains (Fig. 14-28). The active site involves at least four of these domains and the GSSG-binding site spans both subunits (Fig. 14-28b). Electrons are passed from NADPH to FAD thereby reducing the flavin which, in turn, reduces the redox-active disulfide and ultimately GSSG. The flavin is almost completely buried in the protein so as to prevent the surrounding solution from interfering with the electron-transfer process ($FADH_2$, but neither NADPH nor disulfide, is rapidly oxidized by O_2).

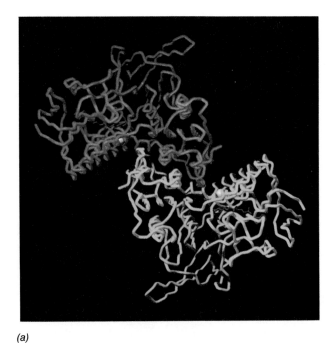

(a)

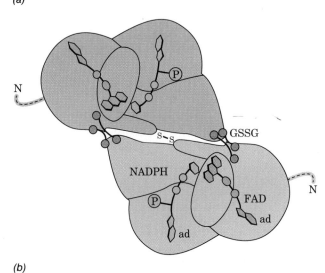

(b)

Figure 14-28
The X-ray structure of the dimeric enzyme glutathione reductase as viewed along the molecule's twofold axis of symmetry. (a) The C_α backbone with the two identical subunits shown in different colors. The S atoms of the redox-active disulfides are represented by yellow spheres and the FAD prosthetic groups are shown in orange (their flavin residues are near the active disulfide groups). [Courtesy of Arthur Olson, Research Institute of Scripps Clinic.] (b) An interpretive diagram of Part (a) showing how each subunit is organized into five domains. The 18-residue N-terminal domain (*dashed lines*) is not visible in the X-ray structure presumably because it is flexibly linked to the rest of the protein. The binding sites of NADPH and GSSG [not shown in Part (a)], as well as those of FAD, are indicated. The two subunits are covalently linked by a disulfide bridge across the molecular twofold axis. [After Pai, E. F. and Schulz, G. E., *J. Biol. Chem.* **258**, 1753 (1983).]

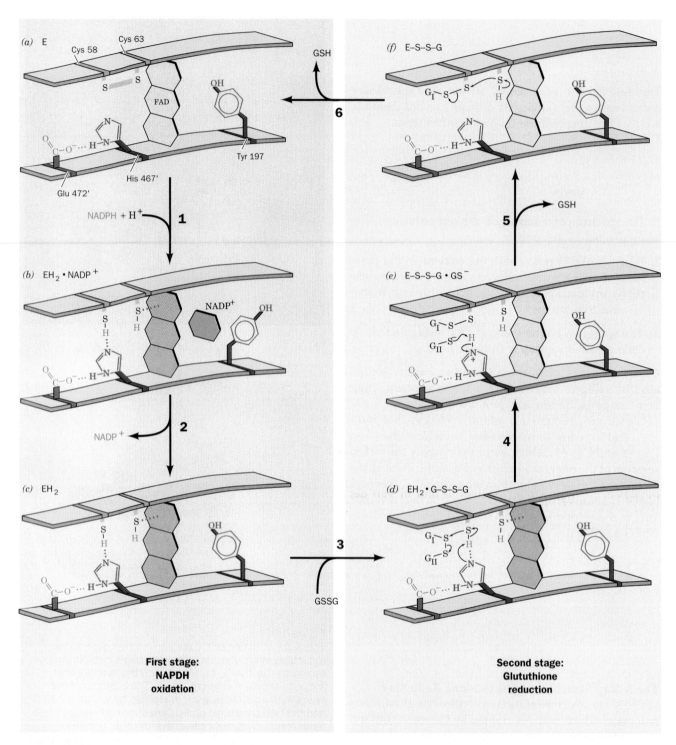

Figure 14-29
The catalytic reaction cycle of glutathione reductase. The catalytic center is surrounded by protein so that the NADPH and the GSSG-binding sites are in deep pockets. (*a*) The oxidized enzyme E. The phenol side chain of Tyr 197 blocks the access of the solvent to the flavin ring. (Primed residues are on a different subunit from unprimed residues.) (*b*) The complex between NADP$^+$ and the reduced enzyme EH$_2$. The phenol side chain of Tyr 197 has been pushed aside by the nicotinamide ring. The redox-active disulfide bond has been reductively cleaved yielding a thiol group on both Cys 58 and Cys 63. That on Cys 63 forms a charge-transfer complex with the flavin ring (*dashed line*). (*c*) The stable reduced enzyme, EH$_2$. Tyr 197 again blocks access to the FAD thereby preventing the oxidation of the enzyme by O$_2$. (*d*) The EH$_2$·GSSG complex. (*e*) The mixed disulfide between GSH I and Cys 58 has formed through the attack of the Cys 58 thiol on GSSG yielding GSH II, which is subsequently released. (*f*) The Cys 63 thiol group attacks the mixed disulfide so as to reform the redox-active disulfide and release GSH I. [After Douglas, K. T., *Adv. Enzymol.* **59**, 150 (1987).]

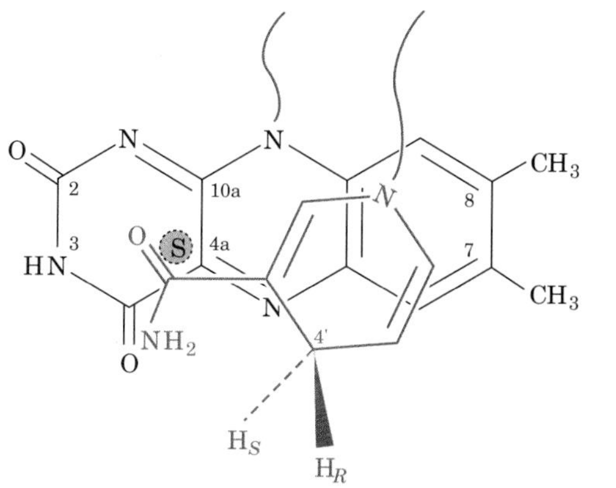

scopic studies as well as the chemistry of model compounds suggest that rapid electron transfer between the flavin and the redox-active disulfide bond occurs through the transient formation of a covalent bond between the sulfur of Cys 63 and flavin atom C(4a). This releases Cys 58, which acquires a solvent proton yielding a thiol group (Fig. 14-31; Step 1). The cova-

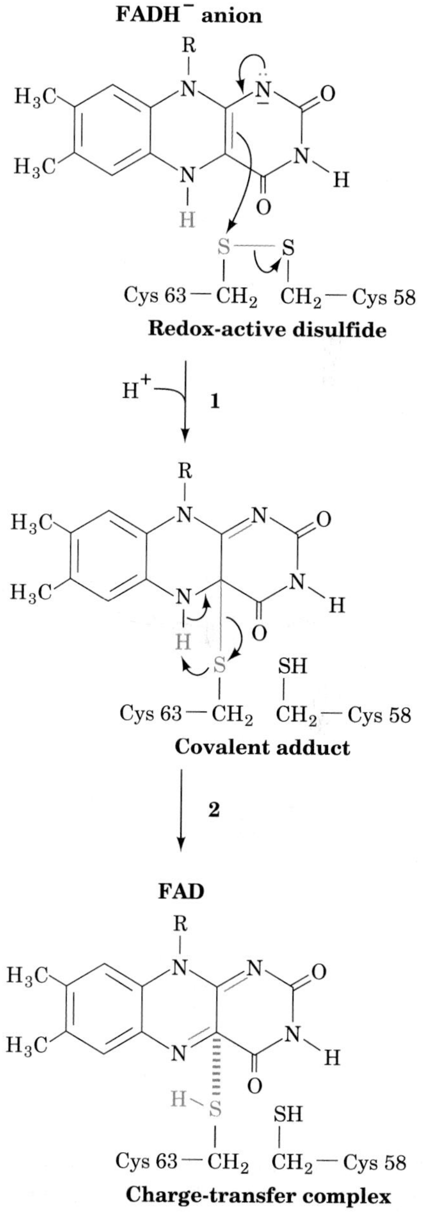

Figure 14-30
The complex between the flavin (*black*) and nicotinamide rings (*red*) observed in the X-ray structure of glutathione reductase. The two planar heterocycles are parallel and in van der Waals contact. The position of the Cys 63 S atom below the flavin ring is indicated by the dashed circle. This atom, which is a member of the redox-active disulfide, is in van der Waals contact with the flavin ring. [After Pai, E. F. and Schulz, G. E., *J. Biol. Chem.* **253**, 1755 (1983).]

The arrangement of the groups in glutathione reductase's catalytic center and its reaction sequence is diagrammed in Fig. 14-29. In the absence of NADPH, the phenol side chain of Tyr 197 covers the nicotinamide-binding pocket so as to shield the flavin from contact with the solution (Fig. 14-29*a*). This side chain moves aside in the presence of NADPH thereby permitting the nicotinamide ring to bind parallel to and in van der Waals contact with the flavin ring of the fully oxidized enzyme E (Fig. 14-30). The H_S substituent to reduced nicotinamide's prochiral C(4) atom, the H atom that is known to be lost in the glutathione reductase reaction, lies near flavin atom N(5), the position through which reducing equivalents often enter a flavin ring upon its reduction. This positioning is particularly significant in view of the catalytic mechanism for glutathione reductase described below.

Catalytic Mechanism

Protein X-ray structures neither reveal H atom positions nor indicate pathways of electron transfer. The electron-transfer pathway in the glutathione reductase reaction has nevertheless been inferred from the X-ray structures of a series of stable enzymatic reaction intermediates as augmented with a variety of enzymological data (Fig. 14-29):

First Stage of the Catalytic Cycle: NADPH Reduction

1. NADPH binds to the oxidized enzyme, E, and immediately reduces the flavin yielding NADP+. The resulting reduced flavin anion (FADH−) has but transient existence. The redox-active disulfide bond linking Cys 58 and Cys 63 lies between the GSSG-binding site and the flavin ring on the opposite side of the flavin from the nicotinamide pocket. Spectro-

Figure 14-31
The reaction transferring electrons from the reduced FADH− anion in glutathione reductase to the enzyme's redox-active disulfide. **(1)** The formation of a covalent bond from flavin atom C(4a) to the sulfur atom of Cys 63 thereby releasing the thiol group of Cys 58. **(2)** The collapse of the C(4a)—S bond to form a charge-transfer complex (*dashed line*) between the resulting Cys 63 thiol group and the FAD. Cys 63's S atom is located out of the plane of the flavin as Fig. 14-30 indicates.

lent Cys 63–flavin adduct rapidly collapses to a **charge-transfer complex** (a noncovalent interaction in which an electron pair is partially transferred from a donor, in this case the Cys 63 thiol, to an acceptor, in this case the oxidized flavin ring; Fig. 14-31, Step 2).

2. The release of NADP$^+$ yields reduced glutathione reductase, EH$_2$. This model is corroborated by the X-ray structure of EH$_2$, which indicates that the redox-active disulfide bridge has opened such that the resulting Cys 63 thiol group is in contact with the flavin ring near its C(4a) position (Fig. 14-30). The red color of EH$_2$ crystals is indicative of this charge-transfer complex (the yellow color of crystals of oxidized glutathione reductase, E, is indicative of oxidized FAD).

Second Stage of the Catalytic Cycle: Glutathione Reduction

3. The second stage of the glutathione reductase reaction begins with the binding of GSSG to EH$_2$.

4. S$_{58}$ (the Cys 58 thiol) nucleophilically attacks S$_I$ (the S on the GS unit nearest Cys 58) yielding the mixed disulfide and GS$^-$ II. His 467′ promotes this reaction by abstracting the proton on S$_{58}$ (general base catalysis; primed residues are members of the other subunit). Indeed, Glu 472′, His 467′, and Cys 58 are arranged much like the catalytic triad of serine proteases (Section 14-3B) with the Cys SH replacing Ser OH.

5. GS$^-$ II is protonated by His 467′ (general acid catalysis) and the product GSH is released by the enzyme.

6. In the final step of the glutathione reductase reaction, the Cys 63 thiol group nucleophilically attacks S$_{58}$ to reform the redox-active disulfide and split the mixed disulfide. His 467′ may also function as a proton sink and source in this process. The second GSH is then released from the enzyme, completing the catalytic cycle.

In this chapter we have discussed the various mechanisms that enzymes utilize in catalyzing reactions and have shown in detail how these mechanisms are applied in three of the best understood systems. Throughout the text we shall continue our discussion of catalytic mechanisms, using the principles we have set out here as they are understood to apply to the various enzymes we encounter along the way.

Chapter Summary

Most enzymatic mechanisms of catalysis have ample precedent in organic catalytic reactions. Acid–base catalyzed reactions occur, respectively, through the donation or abstraction of a proton to or from a reactant so as to stabilize the reaction's transition state complex. Enzymes often employ ionizable amino acid side chains as general acid–base catalysts. Covalent catalysis involves nucleophilic attack of the catalyst on the substrate to transiently form a covalent bond followed by the electrophilic stabilization of a developing negative charge in the reaction's transition state. Various protein side chains as well as certain coenzymes can act as covalent catalysts. Metal ions, which are common enzymatic components, catalyze reactions by stabilizing developing negative charges in a manner resembling general acid catalysis. Metal ion-bound water molecules are potent sources of OH$^-$ ions at neutral pH's. Metal ions also facilitate enzymatic reactions through the charge shielding of bound substrates. The arrangement of charged groups about an enzymatic active site of low dielectric constant in a manner that stabilizes the transition state complex results in the electrostatic catalysis of the enzymatic reaction. Enzymes catalyze reactions by bringing their substrates into close proximity in reactive orientations. The enzymatic binding of the substrates in a bimolecular reaction arrests their relative motions resulting in a rate enhancement. The preferential enzymatic binding of the transition state of a catalyzed reaction over the substrate is an important rate enhancement mechanism. Transition state analogs are potent competitive inhibitors because they bind to the enzyme more tightly than does the corresponding substrate.

Lysozyme catalyzes the hydrolysis of β (1 → 4)-linked poly(NAG–NAM), the bacterial cell wall polysaccharide, as well as that of poly(NAG). According to the Phillips mechanism, lysozyme binds a hexasaccharide so as to distort its D ring towards the half-chair conformation of the planar oxonium ion transition state. This is followed by cleavage of the C(1)—O(1) bond between the D and E rings as promoted by proton donation from Glu 35. Finally, the resulting oxonium ion transition state is electrostatically stabilized by the nearby carboxyl group of Asp 52 so that the E ring can be replaced by OH$^-$ to form the hydrolyzed product. The roles of Glu 35 and Asp 52 in lysozyme catalysis have been verified through chemical modification studies. However, theoretical and inhibitor binding studies suggest that strain is not of major catalytic importance in lysozyme. Rather, it seems that electrostatic stabilization of the transition state is lysozyme's most important rate enhancement factor.

Serine proteases constitute a widespread class of proteolytic enzymes that are characterized by the possession of a reactive Ser residue. The pancreatically synthesized digestive enzymes trypsin, chymotrypsin, and elastase are sequentially and structurally related but have different side chain specificities for their substrates. All have the same catalytic triad, Asp 102, His 57, and Ser 195, at their active sites. Subtilisin is an unrelated serine protease that has essentially the same active site geometry as do the pancreatic enzymes. The catalytic process in serine proteases is initiated by the nucleophilic attack of the active Ser on the carbonyl carbon atom of the scissile peptide to form the tetrahedral intermediate transition state. The tetrahedral intermediate, which is stabilized by its preferential binding to the enzyme's active site, then decomposes to the acyl–enzyme intermediate under the impetus of proton donation from the Asp 102-polarized His 57. After the replacement of the leaving group by solvent H_2O, the catalytic process is reversed to yield the second product and the regenerated enzyme. The Asp 102–His 57 couple therefore functions in the reaction as a proton shuttle. The active Ser is not unusually reactive but is ideally situated to nucleophilically attack the activated scissile peptide. The X-ray structure of the trypsin–BPTI complex indicates the existence of the tetrahedral intermediate. The pancreatic serine proteases are synthesized as zymogens to prevent pancreatic self-digestion. Trypsinogen is activated by a single proteolytic cleavage by enteropeptidase. The resulting trypsin similarly activates trypsinogen as well as chymotrypsinogen, proelastase, and other pancreatic digestive enzymes. Trypsinogen's catalytic triad is structurally intact. The zymogen's low catalytic activity arises from a distortion of its specificity pocket and oxyanion hole so that it is unable to productively bind substrate or preferentially bind the tetrahedral intermediate.

Glutathione reductase is a nearly ubiquitous FAD-containing enzyme that catalyzes the NADPH reduction of GSSG to GSH. The first stage of the reaction is the formation of a stable two-electron reduced intermediate, EH_2. The electron pair is passed from the reduced nicotinamide ring, through the parallel flavin ring, to the enzyme's redox-active disulfide. The latter stage occurs via the transient formation of a covalent bond between the S atom of Cys 63 and flavin C(4a). In EH_2, the Cys 63 thiol group forms a charge-transfer complex with the oxidized flavin ring. In the second step of the glutathione reductase reaction, EH_2 reacts with GSSG, through disulfide interchange reactions, to regenerate the oxidized enzyme and 2GSH.

References

General

Bender, M. L., Bergeron, R. J., and Komiyama, M., *The Bioorganic Chemistry of Enzymatic Catalysis,* Wiley (1984).

Fersht, A., *Enzyme Structure and Mechanism* (2nd ed.), Freeman (1985).

Hammes, G. G., *Enzyme Catalysis and Regulation,* Academic Press (1982).

Jencks, W. P., *Catalysis in Chemistry and Enzymology,* McGraw–Hill (1969).

Walsh, C., *Enzymatic Reaction Mechanisms,* Freeman (1979).

Enzymatic Catalysis

Bruice, T. C., Proximity effects and enzyme catalysis, *in* Boyer, P. D., (Ed.), *The Enzymes* (3rd ed.), Vol. 2, *pp.* 217–279, Academic Press (1970).

Bruice, T. C., Some pertinent aspects of mechanism as determined with small molecules, *Annu. Rev. Biochem.* **45,** 331–373 (1976).

Jencks, W. P., Binding energy, specificity, and enzymatic catalysis: the Circe effect, *Adv. Enzymol.* **43,** 219–410 (1975).

Kirsch, J. F., Mechanism of enzyme action, *Annu. Rev. Biochem.* **42,** 205–234 (1973).

Kraut, J., How do enzymes work? *Science* **242,** 533–540 (1988).

Lienhard, G. E., Enzymatic catalysis and transition state theory, *Science* **180,** 149–154 (1973).

Lipscomb, W. N., Acceleration of reactions by enzymes, *Acc. Chem. Res.* **15,** 232–238 (1982).

Lipscomb, W. N., Structure and catalysis of enzymes, *Annu. Rev. Biochem.* **52,** 17–34 (1983).

Mildvan, A. S., Metals in enzyme catalysis, *in* Boyer, P. D. (Ed.), *The Enzymes* (3rd ed.), Vol. 2, *pp.* 446–536, Academic Press (1970).

Page, M. I., Entropy, binding energy, and enzyme catalysis, *Angew. Chem. Int. Ed. Engl.* **16,** 449–459 (1977).

Wolfenden, R., Analogue approaches to the structure of the transition state in enzyme reactions, *Acc. Chem. Res.* **5,** 10–18 (1972).

Lysozyme

Beddell, C. R., Blake, C. C. F., and Oatley, S. J., An X-ray study of the structure and binding properties of iodine-inactivated lysozyme, *J. Mol. Biol.* **97,** 643–654 (1975).

Blake, C. C. F., Johnson, L. N., Mair, G. A., North, A. C. T., Phillips, D. C., and Sarma, V. R., Crystallographic studies of the activity of hen egg-white lysozyme, *Proc. R. Soc. London Ser. B* **167,** 378–388 (1967).

Chipman, D. M. and Sharon, N., Mechanism of lysozyme action, *Science* **165,** 454–465 (1969).

Ford, L. O., Johnson, L. N., Machin, P. A., Phillips, D. C., and Tijan, R., Crystal structure of a lysozyme–tetrasaccharide lactone complex, *J. Mol. Biol.* **88,** 349–371 (1974).

Imoto, T., Johnson, L. N., North, A. C. T., Phillips, D. C., and Rupley, J. A., Vertebrate lysozymes, *in* Boyer, P. D. (Ed.), *The Enzymes* (3rd ed.), Vol. 7, *pp.* 665–868, Academic Press (1972). [An exhaustive review.]

Kelly, J. A., Sielecki, A. R., Sykes, B. D., James, M. N. G., and Phillips, D. C., X-ray crystallography of the binding of the bacterial cell wall trisaccharide NAM–NAG–NAM to lysozyme, *Nature* **282,** 875–878 (1979).

Kirby, A. J., Mechanism and stereoelectronic effects in the lysozyme reaction, *CRC Crit. Rev. Biochem.* **22,** 283–315 (1987).

Parsons, S. M. and Raftery, M. A., The identification of aspartic acid 52 as being critical to lysozyme activity, *Biochemistry* **8**, 4199–4205 (1969).

Phillips, D. C., The three-dimensional structure of an enzyme molecule, *Sci. Am.* **215**(5): 75–80 (1966). [A marvelously illustrated article on the structure and mechanism of lysozyme.]

Secemski, I. I., Lehrer, S. S., and Lienhard, G. E., A transition state analogue for lysozyme, *J. Biol. Chem.* **247**, 4740–4748 (1972). [Binding studies on the lactone derivative of (NAG)$_4$.]

Schindler, M., Assaf, Y., Sharon, N., and Chipman, D. M., Mechanism of lysozyme catalysis: role of ground-state strain in subsite D in hen egg-white and human lysozymes, *Biochemistry* **16**, 423–431 (1977). [Experimental indications that lysozyme catalysis does not occur through D ring strain.]

Warshel, A. and Levitt, M., Theoretical studies of enzymatic reactions; dielectric, electrostatic and steric stabilization of the carbonium ion in the reaction of lysozyme, *J. Mol. Biol.* **103**, 227–249 (1976). [Theoretical indications that lysozyme catalysis occurs through electrostatic rather than steric strain.]

Neurath, H., Evolution of proteolytic enzymes, *Science* **224**, 350–357 (1984).

Sprang, S., Standing, T., Fletterick, R. J., Stroud, R. M., Finer-Moore, J., Xuong, N.-H., Hamlin, R., Rutter, W. J., and Craik, C. S., The three-dimensional structure of Asn^{102} mutant of trypsin: role of Asp^{102} in serine protease catalysis, *Science* **237**, 905–909 (1987) *and* Craik, C. S., Roczniak, S., Largman, C., and Rutter, W. J., The catalytic role of the active site aspartic acid in serine proteases, *Science* **237**, 909–913 (1987).

Steitz, T. A. and Shulman, R. G., Crystallographic and NMR studies of the serine proteases, *Annu. Rev. Biophys. Bioeng.* **11**, 419–444 (1982).

Stroud, R. M., A family of protein-cutting proteins, *Sci. Am.* **231**(1): 74–88 (1974).

Stroud, R. M., Kossiakoff, A. A., and Chambers, J. L., Mechanism of zymogen activation, *Annu. Rev. Biophys. Bioeng.* **6**, 177–193 (1977).

Warshel, A., Naray-Szabo, G., Sussman, F., and Hwang, J.-K., How do serine proteases really work? *Biochemistry* **28**, 3629–3637 (1989). [Theoretical arguments that catalysis in serine proteases arises from the enzyme's electronic complementarity to the changes in charge distribution accompanying the catalyzed reaction.]

Serine Proteases

Blow, D. M., Structure and mechanism of chymotrypsin, *Acc. Chem. Res.* **9**, 145–152 (1976).

Boyer, P. D. (Ed.), *The Enzymes*, Vol. 3 (3rd ed.), Academic Press (1971). [Contains reviews on many serine proteases.]

Huber, R. and Bode, W., Structural basis of the activation and action of trypsin, *Acc. Chem. Res.* **11**, 114–122 (1978).

James, M. N. G., Sielecki, A. R., Brayer, G. D., Delbaere, L. T. J., and Bauer, C. A., Structure of product and inhibitor complexes of *Streptomyces griseus* protease A at 1.8 Å resolution, *J. Mol. Biol.* **144**, 45–88 (1980).

Kossiakoff, A. A., Catalytic properties of trypsin, *in* Jurnak, F. A. and McPherson, A., *Biological Macromolecules and Assemblies*, Vol. 3, pp. 369–412, Wiley (1987).

Kraut, J., Serine proteases: structure and mechanism of catalysis, *Annu. Rev. Biochem.* **46**, 331–358 (1977).

Laskowski, M., Jr., and Kato, I., Protein inhibitors of proteinases, *Annu. Rev. Biochem.* **49**, 593–626 (1980).

Glutathione Reductase

Douglas, K. T., Mechanism of action of glutathione-dependent enzymes, *Adv. Enzymol.* **59**, 103–167 (1986). [Section VIII discusses glutathione reductase.]

Karplus, P. A. and Schultz, G. E., Refined structure of glutathione reductase at 1.54 Å resolution, *J. Mol. Biol.* **195**, 701–729 (1987).

Schulz, G. E. and Pai, E. F., The catalytic mechanism of glutathione reductase as derived from X-ray diffraction analyses of reaction intermediates, *J. Biol. Chem.* **258**, 1752–1757 (1983).

Thieme, R., Pai, E. F., Schirmer, R. H., and Schulz, G. E., Three-dimensional structure of glutathione reductase at 2 Å resolution, *J. Mol. Biol.* **152**, 763–782 (1981).

Williams, C. H., Jr., Flavin-containing dehydrogenases, *in* Boyer, P. D. (Ed.), *The Enzymes* (3rd ed.), Vol. 13, pp. 89–173, Academic Press (1976).

Problems

1. Explain why γ-pyridone is not nearly as effective a catalyst for glucose mutarotation as is α-pyridone. What about β-pyridone?

2. RNA is rapidly hydrolyzed in alkaline solution to yield a mixture of nucleotides whose phosphate groups are bonded to either the 2′ or the 3′ positions of the ribose residues. DNA, which lacks RNA's 2′-OH groups, is resistant to alkaline degradation. Explain.

3. Carboxypeptidase A, a Zn^{2+}-containing enzyme, hydrolyzes the C-terminal peptide bonds of polypeptides (Section 6-1A). In the enzyme–substrate complex, the Zn^{2+} ion is coordinated to three enzyme side chains, the carbonyl oxygen of the scissile peptide bond, and a water molecule. A plausible model for the enzyme's reaction mechanism that is consistent with X-ray and enzymological data is diagrammed in Fig. 14-32. What are the roles of the Zn^{2+} ion and Glu 270 in this mechanism?

4. In the following lactonization reaction,

the relative reaction rate when $R = CH_3$ is 3.4×10^{11} times that when $R = H$. Explain.

*5. Derive the analog of Eq. [14.11] for an enzyme that catalyzes the reaction:

$$A + B \rightarrow P$$

O
||
Glu 270 — C
O⁻
... H
O
... H
Zn²⁺

CO₂⁻
|
CHR
|
NH   R'
     \ /
      C=O

Michaelis complex

attack of water

O
||
Glu 270 — C
O
... H
O
... H
Zn²⁺

CO₂⁻
|
CHR
|
NH   R'
     \ /
      C
     / \
    O⁻ O⁻

Tetrahedral intermediate

scissile bond scission

CO₂⁻
|
CHR
|
N⁺
/|\
H H H

O
||
Glu 270 — C
O⁻

+

R'
\
C=O
O
... H
Zn²⁺

Enzyme–product complex

Figure 14-32
The mechanism of carboxypeptidase A.

Assume the enzyme must bind A before it can bind B.

$$E + A + B \rightleftharpoons EA + B \rightleftharpoons EAB \rightarrow EP$$

6. Suggest a transition state analog for proline racemase that differs from those discussed in the text. Justify your suggestion.

7. Wolfenden has stated that it is meaningless to distinguish between the "binding sites" and the "catalytic sites" of enzymes. Explain.

8. Explain why oxalate ($^-$OOCCOO$^-$) is an inhibitor of oxaloacetate decarboxylase.

9. In light of the information given in this chapter, why are enzymes such large molecules? Why are active sites almost always located in clefts or depressions in enzymes rather than on protrusions?

10. Predict the effects on lysozyme catalysis of changing Phe 34, Ser 36, and Trp 108 to Arg, assuming that this change does not significantly alter the structure of the protein.

*11. The incubation of (NAG)$_4$ with lysozyme results in the slow formation of (NAG)$_6$ and (NAG)$_2$. Propose a mechanism for this reaction. What aspect of the Phillips mechanism is established by this reaction?

12. Why does the following $\beta\,(1 \rightarrow 4)$-linked tetrasaccharide

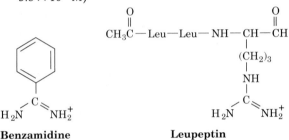

bind to lysozyme with 100 times greater affinity than does NAG–NAM–NAG–NAM?

13. A major difficulty in investigating the properties of the pancreatic serine proteases is that these enzymes, being proteins themselves, are self-digesting. This problem is less severe, however, for solutions of chymotrypsin than it is for solutions of trypsin or elastase. Explain.

14. The comparison of the active site geometries of chymotrypsin and subtilisin under the assumption that their similarities have catalytic significance has led to greater mechanistic understanding of both these enzymes. Discuss the validity of this strategy.

15. **Benzamidine** ($K_I = 1.8 \times 10^{-5} M$) and **leupeptin** ($K_I = 3.8 \times 10^{-7} M$)

O
||
CH₃C — Leu — Leu — NH — CH — CH
                         |
                        (CH₂)₃
                         |
                        NH
                         |
                         C
                        / \\
                     H₂N   NH₂⁺

C
/ \\
H₂N   NH₂⁺

Benzamidine   **Leupeptin**

are both specific competitive inhibitors of trypsin. Explain their mechanisms of inhibition. Design leupeptin analogs that inhibit chymotrypsin and elastase.

16. Trigonal boronic acid derivatives have a high tendency to form tetrahedral adducts. **2-Phenylethyl boronic acid**

CH₂ — CH₂ — B
                OH
                OH

2–Phenylethyl boronic acid

is an inhibitor of subtilisin and chymotrypsin. Indicate the structure of these enzyme–inhibitor complexes.

17. Tofu (bean curd), a high protein soybean product that is widely consumed in the Far East, is prepared in such a way as to remove the trypsin inhibitor present in soybeans. Explain the reason(s) for this treatment?

18. Explain why chymotrypsin is not self-activating as is trypsin.

19. The reaction of glutathione reductase with an excess of NADPH in the presence of arsenite yields a nonphysiological four-electron reduced form of the enzyme. What is the chemical nature of this catalytically inactive species?

20. Two-electron reduced glutathione reductase (EH₂), but not the oxidized enzyme (E), reacts with iodoacetate (ICH₂COO$^-$) to yield an inactive enzyme. Explain.

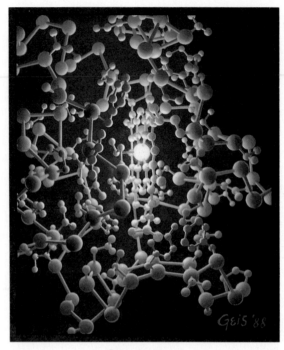

Cytochrome *c*

Cytochrome *c* illuminated by its single iron atom.
The interior amino acid side-chains are shown in red.

IV
Metabolism

Chapter 15

INTRODUCTION TO METABOLISM

Living organisms are not at equilibrium. Rather, they require a continuous influx of free energy to maintain order in a universe bent on maximizing disorder. **Metabolism** is the overall process through which living systems acquire and utilize the free energy they need to carry out their various functions. *They do so by coupling the exergonic reactions of nutrient oxidation to the endergonic processes required to maintain the living state* such as the performance of mechanical work, the active transport of molecules against concentration gradients, and the biosynthesis of complex molecules. How do living things acquire this necessary free energy? And what is the nature of the energy-coupling process? **Phototrophs** (plants and certain bacteria) acquire free energy from the sun through **photosynthesis,** a process in which light energy powers the endergonic reaction of CO_2 and H_2O to form carbohydrates and O_2 (Chapter 22). **Chemotrophs** obtain their free energy by oxidizing organic compounds (carbohydrates, lipids, proteins) obtained from other organisms, ultimately phototrophs. This free energy is most often coupled to endergonic reactions through the intermediate synthesis of "high-energy" phosphate compounds such as **adenosine triphosphate (ATP;** Section 15-4). In addition to being completely oxidized, nutrients are broken down in a series of metabolic reactions to common intermediates that are used as precursors in the synthesis of other biological molecules.

A remarkable property of living systems is that, despite the complexity of their internal processes, they maintain a steady state. This is strikingly demonstrated by the observation that, over a 40-year time span, a

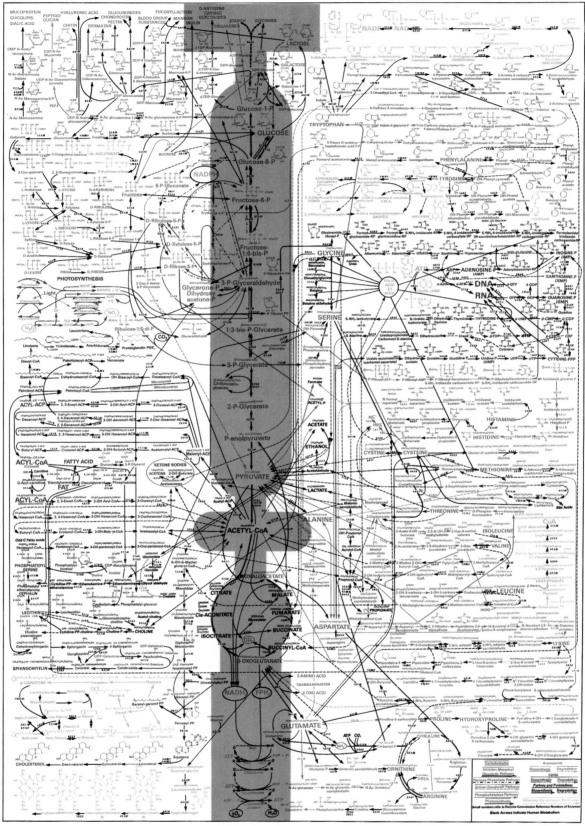

Figure 15-1
A map of the major metabolic pathways in a typical cell. The main pathways
of glucose metabolism are shaded. [Designed by D. E. Nicholson. Published
by BDH Ltd., Pool 2, Dorset, England.]

normal human adult consumes literally tons of nutrients and imbibes over 20,000 L of water, but does so without significant weight change. This steady state is maintained by a sophisticated set of metabolic controls. In this introductory chapter to metabolism, we outline the general characteristics of metabolic pathways and their regulation, study the main types of chemical reactions that comprise these pathways, and consider the experimental techniques that have been most useful in their elucidation. We then discuss the free energy changes associated with reactions of phosphate compounds and oxidation–reduction reactions. Finally, we consider the thermodynamic nature of biological processes; that is, what are the properties of life that are responsible for its self-sustaining character.

1. METABOLIC PATHWAYS

Metabolic pathways are series of consecutive enzymatic reactions that produce specific products. Their reactants, intermediates, and products are referred to as **metabolites.** Since an organism utilizes many metabolites, it has many metabolic pathways. Figure 15-1 shows a metabolic map for a typical cell with many of its interconnected pathways. Each reaction on the map is catalyzed by a distinct enzyme of which there are more than 2000 known. At first glance, this network seems incomprehensibly complex. Yet, by focusing on its major areas in the following chapters, for example, the major pathways of glucose oxidation (the shaded areas of Fig. 15-1), we shall become familiar with its most important avenues and their interrelationships.

The reaction pathways that comprise metabolism are often divided into two categories: those involved in degradation **(catabolism)** and those involved in biosynthesis **(anabolism).** In catabolic pathways, complex metabolites are exergonically broken down to simpler products. The free energy released during these processes is conserved by the synthesis of ATP from ADP and phosphate or by the reduction of the coenzyme NADP$^+$ to NADPH (Fig. 12-2). ATP and NADPH are the major free energy sources for anabolic pathways (Fig. 15-2).

A striking characteristic of degradative metabolism is that *it converts a large number of diverse substances (carbohydrates, lipids, and proteins) to common intermediates.* These intermediates are then further metabolized in a central oxidative pathway that terminates in a few end products. Figure 15-3 outlines the breakdown of various foodstuffs, first to their monomeric units, and then to the common intermediate, **acetyl-coenzyme A (acetyl-CoA)** (Section 19-2; structural formula in Fig. 19-2). This is followed by the oxidation of the acetyl group to CO_2 and H_2O by the sequential actions of the **citric acid cycle** (Chapter 19), the **electron-transport chain,** and **oxidative phosphorylation** (Chapter 20).

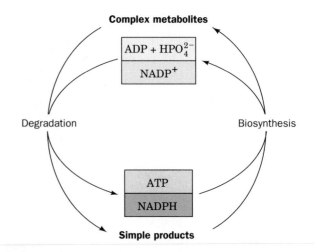

Figure 15-2
ATP and NADPH are the sources of free energy for biosynthetic reactions. They are generated through the degradation of complex metabolites.

Biosynthesis carries out the opposite process. *Relatively few metabolites, mainly pyruvate, acetyl-CoA, and the citric acid cycle intermediates, serve as starting materials for a host of varied biosynthetic products.* In the next several chapters we discuss many degradative and biosynthetic pathways in detail. For now, let us consider some general characteristics of these processes.

Four principal characteristics of metabolic pathways stem from their function of producing products for use by the cell:

1. Metabolic pathways are irreversible

They are highly exergonic (have large negative free energy changes) so that their reactions go to completion. This characteristic provides the pathway with direction. Consequently, *if two metabolites are metabolically interconvertible, the pathway from the first to the second must differ from the pathway from the second back to the first:*

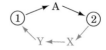

The reason for this difference is that if the route from the first metabolite to the second is exergonic, free energy must be supplied in order to bring it "back up the hill." This requires a different pathway for at least some of the reaction steps. *The existence of independent interconversion routes, as we shall see, is an important property of metabolic pathways because it allows independent control of the rates of the two processes.* If metabolite 2 is required by the cell, it is necessary to "turn off" the pathway from 2 to 1 while "turning on" the pathway from 1 to 2. Such independent control would be impossible without different pathways.

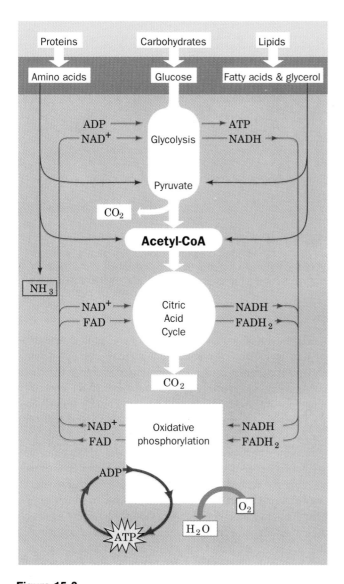

Figure 15-3
Complex metabolites such as carbohydrates, lipids, and proteins are degraded first to their monomeric units, chiefly glucose, fatty acids, glycerol, and amino acids, and then to the common intermediate, acetyl coenzyme A (acetyl-CoA). The acetyl group is then oxidized to CO_2 via the citric acid cycle with concomitant reduction of NAD^+ and FAD. Reoxidation of these latter coenzymes by O_2 via the electron-transport chain and oxidative phosphorylation yields H_2O and ATP.

2. Every metabolic pathway has a first committed step

Although metabolic pathways are irreversible, most of their component reactions function close to equilibrium. Early in each pathway, however, there is generally an irreversible (exergonic) reaction that "commits" the intermediate it produces to continue down the pathway.

3. All metabolic pathways are regulated

In order to exert control on the flux of metabolites through a metabolic pathway, it is necessary to regulate its rate-limiting step. The first committed step, being irreversible, functions too slowly to permit its substrates and products to equilibrate. Since most of the other reactions in a pathway function close to equilibrium, the first committed step is often its rate-limiting step. Most metabolic pathways are therefore controlled by regulating the enzymes that catalyze their first committed steps. This is the most efficient way to exert control because it prevents the unnecessary synthesis of metabolites further along the pathway when they are not required. Specific aspects of such flux control are discussed in Section 16-4A.

4. Metabolic pathways in eukaryotic cells occur in specific cellular locations

The synthesis of metabolites in specific membrane-bounded subcellular compartments makes their transport between these compartments a vital component of eukaryotic metabolism. Biological membranes are selectively permeable to metabolites because of the presence in membranes of specific transport proteins. For example, ATP is generated in the mitochondria but utilized in the cytosol. The transport protein that facilitates the passage of ATP through the mitochondrial membrane is discussed in Section 18-4C, along with the characteristics of membrane transport processes in general. The synthesis and utilization of acetyl-CoA is also compartmentalized. This metabolic intermediate is utilized in the cytosolic synthesis of fatty acids, but is synthesized in mitochondria. Yet there is no transport protein for acetyl-CoA in the mitochondrial membrane. How cells solve this fundamental problem is discussed in Section 23-4D.

2. ORGANIC REACTION MECHANISMS

Almost all of the reactions that occur in metabolic pathways are enzymatically catalyzed organic reactions. Section 14-1 details the various mechanisms enzymes have at their disposal for catalyzing reactions: acid–base catalysis, covalent catalysis, metal ion catalysis, electrostatic catalysis, proximity and orientation effects, and transition state binding. Yet, few enzymes alter the chemical mechanisms of these reactions so that *much can be learned about enzymatic mechanisms from the study of nonenzymatic model reactions*. We therefore begin our study of metabolic reactions by outlining the types of reactions we shall encounter and the mechanisms by which they have been observed to proceed in nonenzymatic systems.

Christopher Walsh has classified biochemical reactions into four categories: (1) **group-transfer reactions;**

(2) **oxidations and reductions;** (3) **eliminations, iso-merizations, and rearrangements;** and (4) **reactions that make or break carbon–carbon bonds.** Much is known about the mechanisms of these reactions and about the enzymes that catalyze them. The discussions in the next several chapters focus on these mechanisms as they apply to specific metabolic interconversions. In this section we outline the four reaction categories and discuss how our knowledge of their reaction mechanisms derives from the study of model organic reactions. We begin by briefly reviewing the chemical logic used in analyzing these reactions.

A. Chemical Logic

A covalent bond consists of an electron pair shared between two atoms. In breaking such a bond, the electron pair can either remain with one of the atoms **(heterolytic bond cleavage)** or separate such that one electron accompanies each of the atoms **(homolytic bond cleavage)** (Fig. 15-4). Homolytic bond cleavage, which usually produces unstable radicals, occurs mostly in oxidation–reduction reactions. Heterolytic C—H bond cleavage involves either carbanion and proton (H$^+$) formation or carbocation (carbonium ion) and hydride ion (H$^-$) formation. Since hydride ions are highly reactive species and carbon atoms are slightly more electronegative than hydrogen atoms, bond cleavage in which the electron pair remains with the carbon atom is the predominant mode of C—H bond breaking in biochemical systems. Hydride ion abstraction occurs only if the hydride is transferred directly to an acceptor such as NAD$^+$ or NADP$^+$.

Compounds participating in reactions involving heterolytic bond cleavage and bond formation are categorized into two broad classes: electron rich and electron deficient. Electron-rich compounds, which are called **nucleophiles** (nucleus lovers), are negatively charged or contain unshared electron pairs that easily form covalent bonds with electron-deficient centers. Biologically important nucleophilic groups include amino, hydroxyl, imidazole, and sulfhydryl functions (Fig. 15-5). The nucleophilic forms of these groups are also their basic forms. Indeed, nucleophilicity and basicity are closely related properties (Section 14-1B): A compound acts as a base when it forms a covalent bond with H$^+$, whereas it acts as a nucleophile when it forms a covalent bond with an electron-deficient center other than H$^+$, usually an electron-deficient carbon atom.

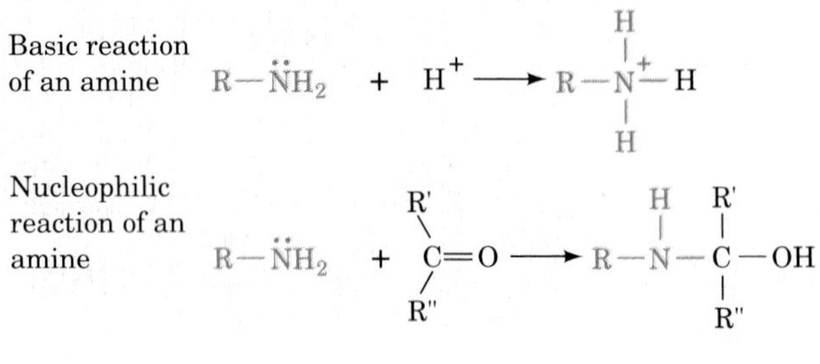

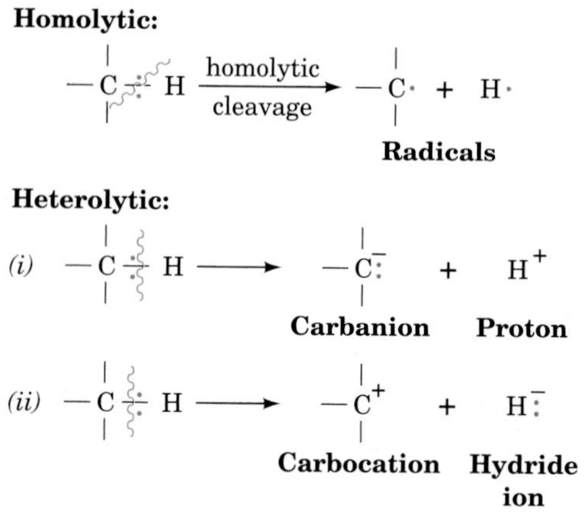

Homolytic:

$$-\overset{|}{\underset{|}{C}} \overset{\curvearrowright}{:} H \xrightarrow[\text{cleavage}]{\text{homolytic}} -\overset{|}{\underset{|}{C}} \cdot \ + \ H\cdot$$

Radicals

Heterolytic:

(i) $\quad -\overset{|}{\underset{|}{C}} \overset{\curvearrowright}{\underset{}{:}} H \longrightarrow -\overset{|}{\underset{|}{C}}\overset{..}{:}^- \ + \ H^+$

Carbanion   Proton

(ii) $\quad -\overset{|}{\underset{|}{C}} \overset{\curvearrowright}{\underset{}{:}} H \longrightarrow -\overset{|}{\underset{|}{C}}^+ \ + \ H\overset{..}{:}^-$

Carbocation   Hydride ion

Figure 15-4
Modes of C—H bond breaking. Homolytic cleavage yields radicals, whereas heterolytic cleavage yields either (*i*) a carbanion and a proton or (*ii*) a carbocation and a hydride ion.

Electron-deficient compounds are called **electrophiles** (electron lovers). They may be positively charged, contain an unfilled valence electron shell, or contain an electronegative atom. The most common electrophiles in biochemical systems are H$^+$, metal ions, the carbon atoms of carbonyl groups, and cationic imines (Fig. 15-6).

Reactions are best understood if the electron pair rearrangements involved in going from reactants to products can be traced. In illustrating these rearrangements we shall use the **curved arrow convention** in which the movement of an electron pair is symbolized by a curved arrow emanating from the electron pair and pointing to the electron-deficient center being "attacked" by the electron pair. When you use this convention, think of the nucleophile as the attacking reagent. For example, imine formation, a biochemically important reaction between an amine and an aldehyde or ketone, is repre-

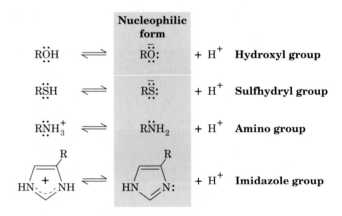

Figure 15-5
Biologically important nucleophilic groups. Nucleophiles are the conjugate bases of weak acids.

Basic reaction of an amine $\quad R-\overset{..}{N}H_2 \ + \ H^+ \longrightarrow R-\overset{\overset{\textstyle H}{|}}{\underset{\underset{\textstyle H}{|}}{N}}{}^+\!-H$

Nucleophilic reaction of an amine $\quad R-\overset{..}{N}H_2 \ + \ \overset{R'}{\underset{R''}{\diagdown}}C{=}O \longrightarrow R-\overset{\overset{\textstyle H}{|}}{N}-\overset{\overset{\textstyle R'}{|}}{\underset{\underset{\textstyle R''}{|}}{C}}-OH$

sented:

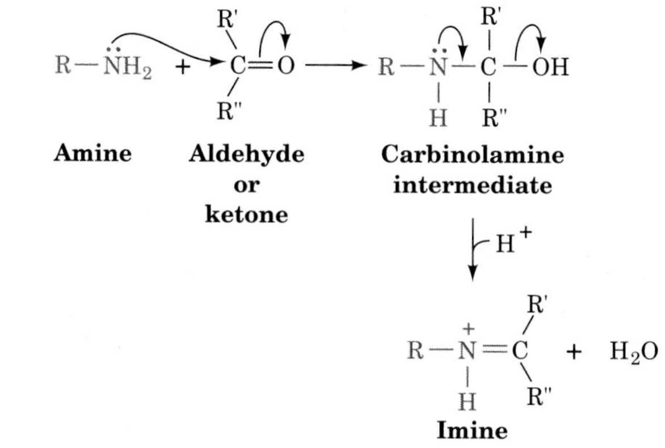

In the first reaction step, the amine's unshared electron pair attacks the electron-deficient carbonyl carbon atom thereby driving an electron pair from its C=O double bond onto the oxygen atom. In the second step, the unshared electron pair on the nitrogen atom attacks the electron-deficient carbon atom so as to eliminate water. *At all times, the rules of chemical reason apply to the system:* For example, there are never five bonds to a carbon atom or two bonds to a hydrogen atom.

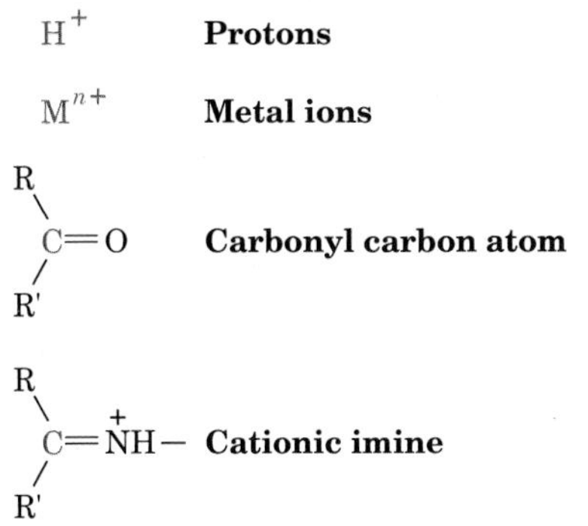

Figure 15-6
Biologically important electrophiles. Electrophiles contain an electron-deficient atom *(red)*.

B. Group-Transfer Reactions

The group transfers that occur in biochemical systems involve the transfer of an electrophilic group from one nucleophile to another.

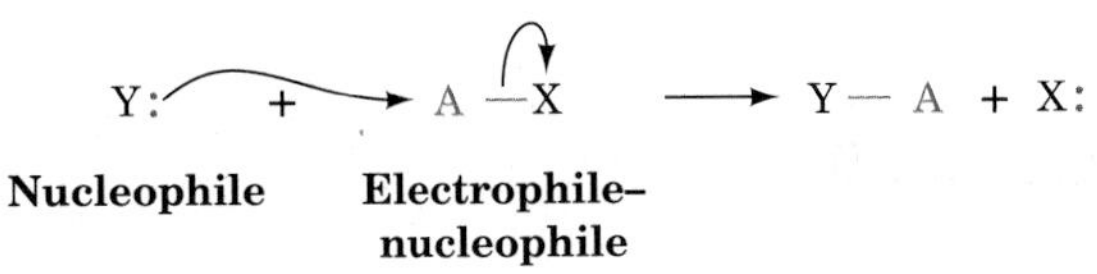

They could equally well be called nucleophilic substitution reactions. The most commonly transferred groups in biochemical reactions are acyl groups, phosphoryl groups, and glycosyl groups (Fig. 15-7):

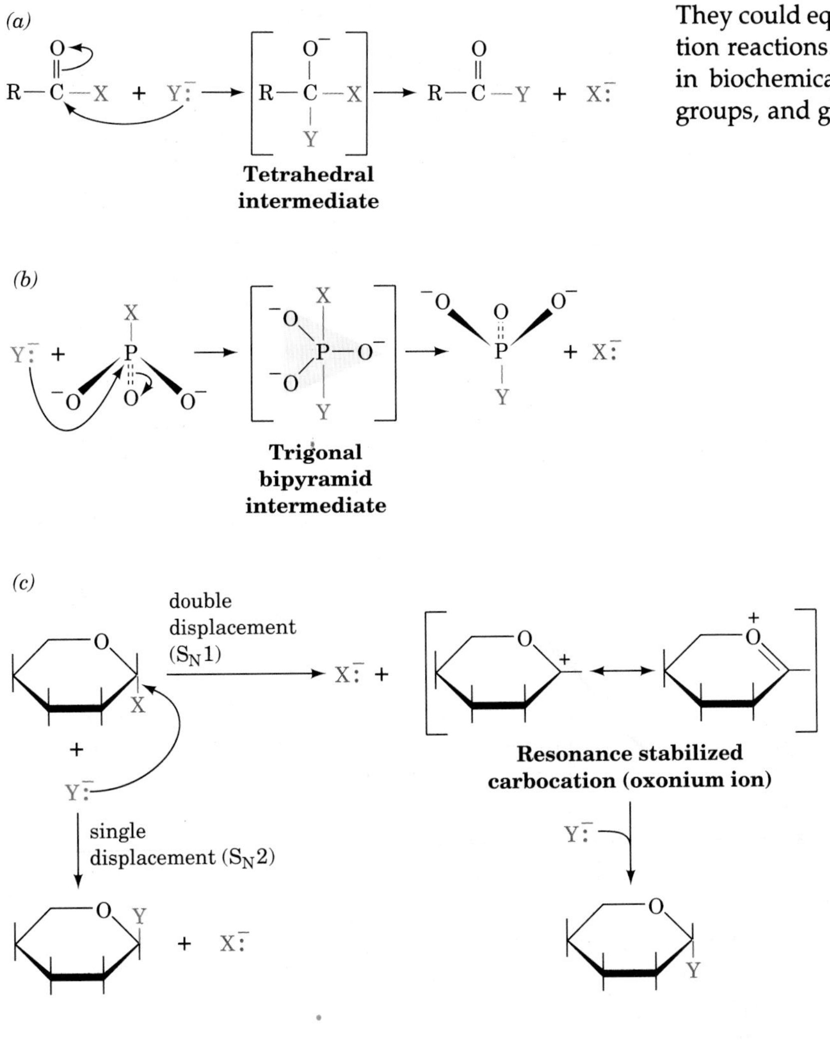

Figure 15-7
Types of metabolic group-transfer reactions: (*a*) Acyl group transfer involves attack of a nucleophile (Y) on the electrophilic carbon atom of an acyl compound to form a tetrahedral intermediate. The original acyl carrier (X) is then expelled to form a new acyl compound. (*b*) Phosphoryl group transfer involves attack of a nucleophile (Y) on the electrophilic phosphorus atom of a tetrahedral phosphoryl group. This yields a trigonal bipyramidal intermediate whose apical positions are occupied by the leaving group (X) and the attacking group (Y). Elimination of the leaving group (X) to complete the transfer reaction results in the phosphoryl group's inversion of configuration. (*c*) Glycosyl group transfer involves the substitution of one nucleophilic group for another at C(1) of a sugar ring. This reaction usually occurs via a double displacement mechanism in which the elimination of the original glycosyl carrier (X) is accompanied by the intermediate formation of a resonance-stabilized carbocation (oxonuim ion) followed by the addition of the attacking nucleophile (Y). The reaction also may occur via a single displacement mechanism in which Y directly displaces X with inversion of configuration.

1. **Acyl group transfer** from one nucleophile to another almost invariably involves the attack of a nucleophile on the acyl carbonyl carbon atom so as to form a tetrahedral intermediate (Fig. 15-7a). Peptide bond hydrolysis, as catalyzed, for example, by chymotrypsin (Section 14-3C), is a familiar example of such a reaction.

2. **Phosphoryl group transfer** proceeds via the attack of a nucleophile on a phosphoryl phosphorus atom to yield a trigonal bipyramidal intermediate whose apexes are occupied by the attacking and leaving groups (Fig. 15-7b). The overall reaction results in the tetrahedral phosphoryl group's inversion of configuration. Indeed, chiral phosphoryl compounds have been shown to undergo just such an inversion. For example, Jeremy Knowles has synthesized ATP made chiral at its γ-phosphoryl group by isotopic substitution and demonstrated that this group is inverted upon its transfer to glucose in the reaction catalyzed by **hexokinase** (Fig. 15-8).

3. **Glycosyl group transfer** involves the substitution of one nucleophilic group for another at C(1) of a sugar ring (Fig. 15-7c). This is the central carbon atom of an acetal. Chemical models of acetal reactions generally proceed via acid catalyzed cleavage of the first bond to form a resonance stabilized carbocation at C(1) (an oxonium ion). The lysozyme-catalyzed hydrolysis of bacterial cell wall polysaccharides (Section 14-2B) is such a reaction.

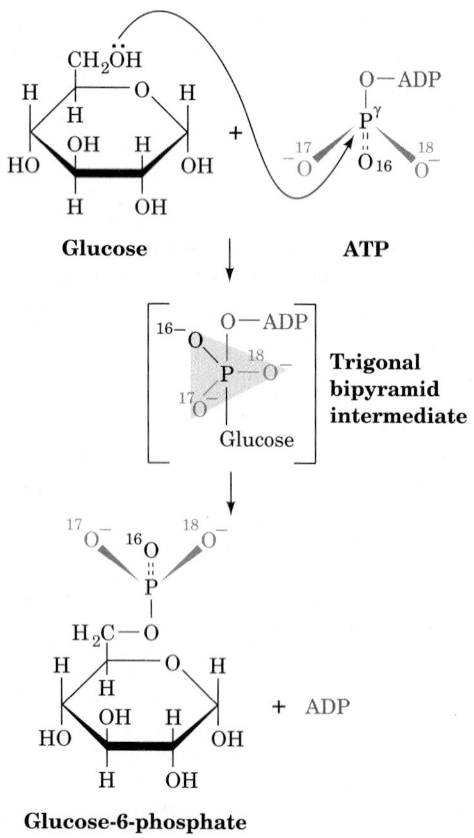

C. Oxidations and Reductions

Oxidation–reduction (redox) reactions involve the loss or gain of electrons. The thermodynamics of these reactions is discussed in Section 15-5. Many of the redox reactions that occur in metabolic pathways involve C—H bond cleavage with the ultimate loss of two bonding electrons by the carbon atom. These electrons are transferred to an electron acceptor such as NAD^+ (Fig. 12-2). Whether these reactions involve homolytic or heterolytic bond cleavage has not always been rigorously established. In most instances heterolytic cleavage is assumed when radical species are not observed. It is useful, however, to visualize redox C—H bond cleavage reactions as hydride transfers as diagrammed below for the oxidation of an alcohol by NAD^+:

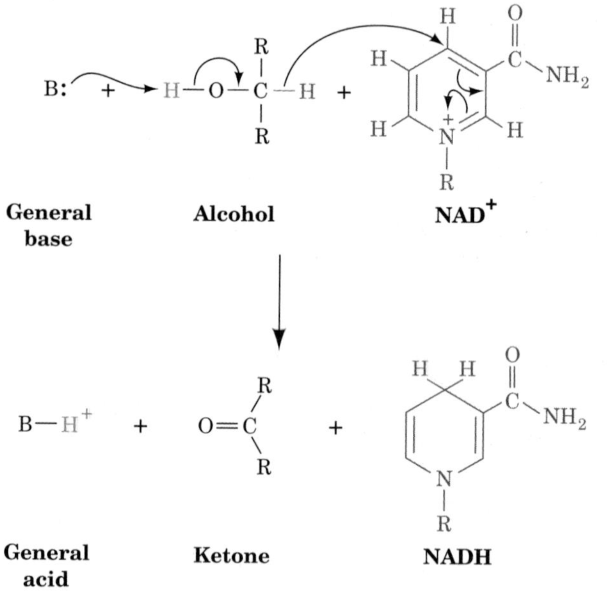

Figure 15-8
In the phosphoryl-transfer reaction catalyzed by hexokinase, the γ-phosphoryl group of ATP made chiral by isotopic substitution undergoes inversion of configuration.

The terminal acceptor for the electron pairs removed from metabolites by their oxidation is, for aerobic organisms, molecular oxygen (O_2). Recall that this molecule is a ground state diradical species whose unpaired electrons have parallel spins. The rules of electron pairing (the Pauli exclusion principle) therefore dictate that O_2 can only accept unpaired electrons; that is, electrons must be transferred to O_2 one at a time (in contrast to redox processes in which electrons are transferred in pairs). Electrons that are removed from metabolites as pairs must therefore be passed to O_2 via the electron-transport chain one at a time. This is accomplished through the use of conjugated coenzymes that have

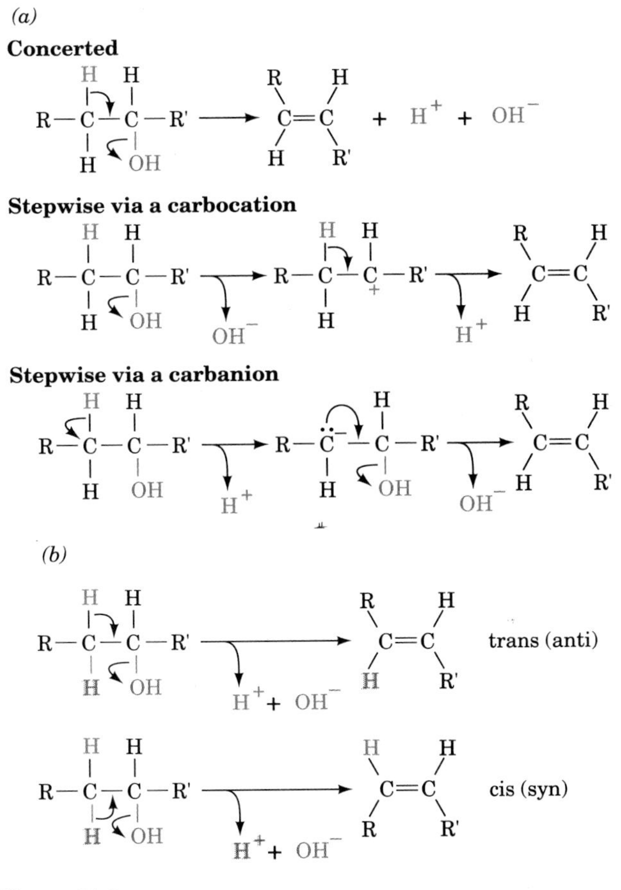

(a)
Concerted

Stepwise via a carbocation

Stepwise via a carbanion

(b)

Figure 15-9
Possible elimination reaction mechanisms using dehydration as an example. Reactions may be *(a)* either concerted, stepwise via a carbocation intermediate, or stepwise via a carbanion intermediate; and *(b)* occur with either trans (anti) or cis (syn) stereochemistry.

stable radical oxidation states and can therefore undergo both $1e^-$ and $2e^-$ redox reactions. One such coenzyme is FAD (whose structure and oxidation states are indicated in Fig. 14-27).

D. Eliminations, Isomerizations, and Rearrangements

Elimination Reactions Form Carbon–Carbon Double Bonds

Elimination reactions result in the formation of a double bond between two previously single-bonded saturated centers. The substances eliminated may be H_2O, NH_3, an alcohol (ROH), or a primary amine (RNH_2). The dehydration of an alcohol, for example, is an elimination reaction:

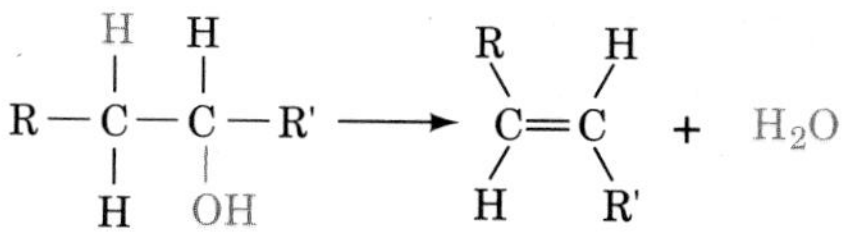

Bond breaking and bond making in this reaction may proceed via either of three mechanisms (Fig. 15-9a): (1) concerted; (2) stepwise with the C—O bond breaking first to form a carbocation; or (3) stepwise with the C—H bond breaking first to form a carbanion.

Enzymes catalyze dehydration reactions by either of two simple mechanisms: (1) protonation of the OH group by an acidic group (acid catalysis), or (2) abstraction of the proton by a basic group (base catalysis). Moreover, in a stepwise reaction, the charged intermediate may be stabilized by an oppositely charged active site group. The glycolytic enzyme **enolase** (Section 16-2I) and the citric acid cycle enzyme **fumarase** (Section 19-3G) catalyze such dehydration reactions.

Elimination reactions may take one of two possible stereochemical courses (Fig. 15-9b): (1) trans (anti) eliminations, the most prevalent biochemical mechanism, and (2) cis (syn) eliminations, which are biochemically less common.

Biochemical Isomerizations Involve Intramolecular Hydrogen Atom Shifts

Biochemical **isomerization reactions** involve the intramolecular shift of a hydrogen atom so as to change the location of a double bond. In such a process, a proton is removed from one carbon atom and added to another. The metabolically most prevalent isomerization reaction is the **aldose–ketose interconversion,** a base-catalyzed reaction that occurs via **enediolate anion** intermediates (Fig. 15-10). The glycolytic enzyme **phosphoglucose isomerase** catalyzes such a reaction (Section 16-2B).

Racemization is an isomerization reaction in which a hydrogen atom shifts its stereochemical position at a molecule's only chiral center so as to invert that chiral center. Such an isomerization is called an **epimerization** in a molecule with more than one chiral center.

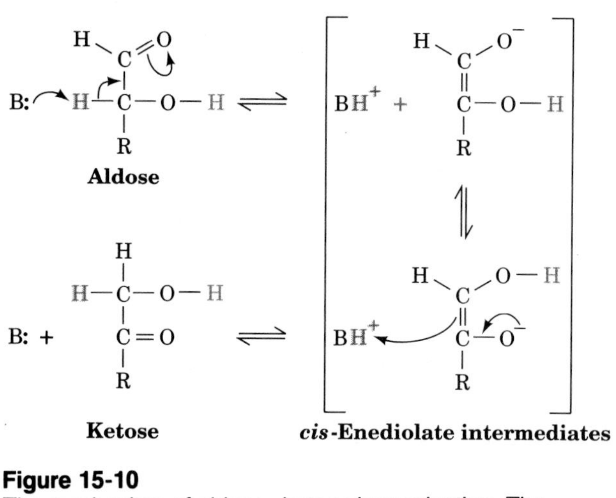

Figure 15-10
The mechanism of aldose–ketose isomerization. The reaction occurs with acid–base catalysis and proceeds via *cis*-enediolate intermediates.

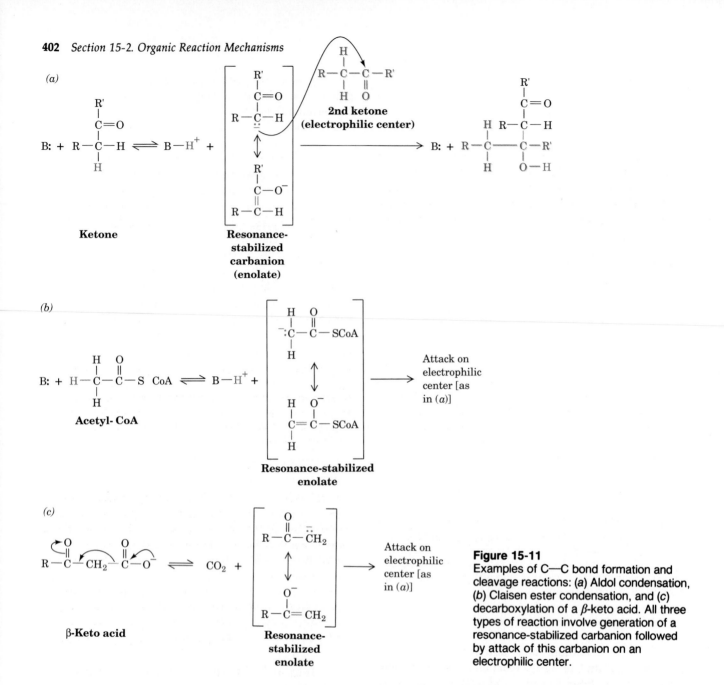

Figure 15-11
Examples of C—C bond formation and cleavage reactions: (a) Aldol condensation, (b) Claisen ester condensation, and (c) decarboxylation of a β-keto acid. All three types of reaction involve generation of a resonance-stabilized carbanion followed by attack of this carbanion on an electrophilic center.

Rearrangements Produce Altered Carbon Skeletons

Rearrangement reactions break and reform C—C bonds so as to rearrange a molecule's carbon skeleton. There are few such metabolic reactions. One is the conversion of L-methylmalonyl-CoA to succinyl-CoA by methylmalonyl-CoA mutase, an enzyme whose prosthetic group is a vitamin B$_{12}$ derivative:

This reaction is involved in the oxidation of fatty acids with an odd number of carbon atoms (Section 23-2E) and several amino acids (Section 24-3E).

E. Reactions that Make and Break Carbon – Carbon Bonds

Reactions that make and break carbon–carbon bonds form the basis of both degradative and biosynthetic metabolism. The breakdown of glucose to CO$_2$ involves five such cleavages while its synthesis involves the reverse process. Such reactions, considered from the synthetic direction, involve attack of a nucleophilic carbanion on an electrophilic carbon atom. The most common electrophilic carbon atoms in such reactions are the sp^2-hybridized carbonyl carbon atoms of aldehydes, ketones, esters, and CO_2.

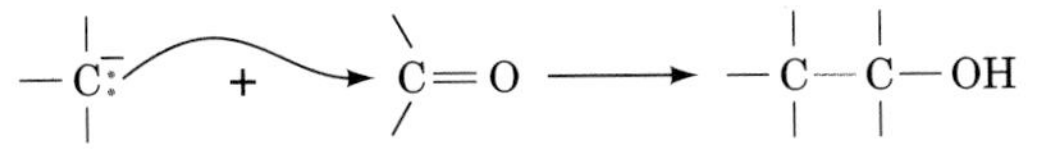

$$-\overset{|}{\underset{|}{C}}{:}^{-} + \quad \overset{\diagdown}{\diagup}C{=}O \quad \longrightarrow \quad -\overset{|}{\underset{|}{C}}{-}\overset{|}{\underset{|}{C}}{-}OH$$

Stabilized carbanions must be generated to attack these electrophilic centers. Three examples are the **aldol condensation** (catalyzed, e.g., by **aldolase;** Section 16-2D), **Claisen ester condensation (citrate synthase;** Section 19-3A), and the decarboxylation of a β-keto acid (**isocitrate dehydrogenase;** Section 19-3C and **fatty acid synthase;** Section 23-4C). In nonenzymatic systems, both the aldol condensation and Claisen ester condensation involve the base-catalyzed generation of a carbanion α to a carbonyl group (Fig. 15-11*a* and *b*). The carbonyl group is electron withdrawing and thereby provides resonance stabilization (Fig. 15-12*a*). In many enzymatic systems, resonance stabilization is enhanced through conversion of the carbonyl group to a protonated Schiff base (covalent catalysis; Fig. 15-12*b*) or by its coordination to a metal ion (metal ion catalysis; Fig. 15-12*c*). Decarboxylation of a β-keto acid does not require base catalysis for the generation of the resonance-stabilized carbanion; the highly exergonic formation of CO_2 provides its driving force (Fig. 15-11*c*).

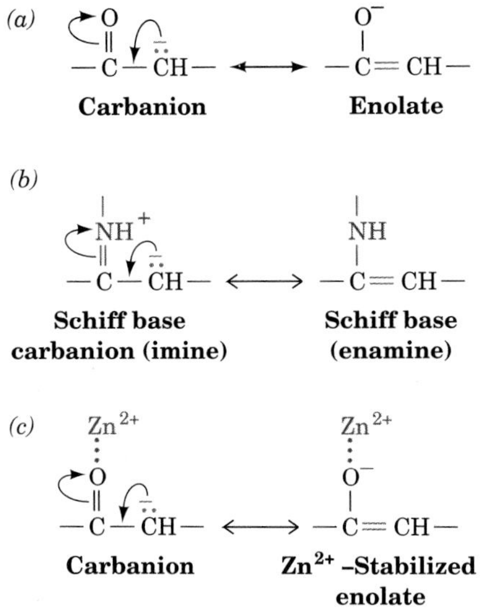

Figure 15-12
The stabilization of carbanions: (*a*) Carbanions adjacent to carbonyl groups are stabilized by the formation of enolates. (*b*) Carbanions adjacent to protonated imines (Schiff bases) are stabilized by the formation of enamines. (*c*) Metal ions stabilize carbanions adjacent to carbonyl groups by the electrostatic stabilization of the enolate.

3. EXPERIMENTAL APPROACHES TO THE STUDY OF METABOLISM

A metabolic pathway can be understood at several levels:

1. In terms of the sequence of reactions by which a specific nutrient is converted to end products, and the energetics of these conversions.

2. In terms of the mechanisms by which each intermediate is converted to its successor. Such an analysis requires the isolation and characterization of the specific enzymes that catalyze each reaction.

3. In terms of the control mechanisms that regulate the flow of metabolites through the pathway. An exquisitely complex network of regulatory processes renders metabolic pathways remarkably sensitive to the needs of the organism; the output of a pathway is generally only as great as required.

As you might well imagine, the elucidation of a metabolic pathway on all of these levels is a complex process involving contributions from a variety of disciplines. Most of the techniques used to do so involve somehow perturbing the system and observing the perturbation's effect on growth or on the production of metabolic intermediates. One such technique is the use of metabolic inhibitors that block metabolic pathways at specific enzymatic steps. Another is the study of genetic abnormalities that interrupt specific metabolic pathways. Techniques have also been developed for the dissection of organisms into their component organs, tissues, cells, and subcellular organelles, and for the purification and identification of metabolites as well as the enzymes that catalyze their interconversions. The use of isotopic tracers to follow the paths of specific atoms and molecules through the metabolic maze has become routine. This section outlines the use of these various techniques.

A. Metabolic Inhibitors, Growth Studies, and Biochemical Genetics

Pathway Intermediates Accumulate in the Presence of Metabolic Inhibitors

The first metabolic pathway to be completely traced was the conversion of glucose to ethanol in yeast by a process known as **glycolysis** (Section 16-1A). In the course of these studies, certain substances, called **metabolic inhibitors,** were found to block the pathway at specific points thereby causing preceding intermediates to buildup. For instance, iodoacetate causes yeast extracts to accumulate fructose-1,6-bisphosphate, whereas fluoride causes the buildup of two phosphate esters, 3-phosphoglycerate and 2-phosphoglycerate. The isolation and characterization of these intermediates was vital to the elucidation of the glycolytic pathway: Chemical intuition combined with this information led to the prediction of the pathway's intervening steps. Each of the proposed reactions was eventually shown to occur *in vitro* (in the "test tube") as catalyzed by a purified enzyme.

Genetic Defects Also Cause Metabolic Intermediates to Accumulate

Archibald Garrod's realization, in the early 1900s, that human genetic diseases are the consequence of deficiencies in specific enzymes also contributed to the elucidation of metabolic pathways. For example, upon the ingestion of either phenylalanine or tyrosine, individuals with the largely harmless inherited condition known as **alcaptonuria,** but not normal subjects, excrete **homogentisic acid** in their urine (Sections 24-3H and 27-1C). This is because the liver of alcaptonurics lacks an enzyme that catalyzes the breakdown of homogentisic acid. Another genetic disease, **phenylketonuria** (Section 24-3H), results in the accumulation of **phenylpyruvate** in the urine (which, if untreated, causes severe mental retardation in infants). Ingested phenylalanine and phenylpyruvate appear as phenylpyruvate in the

urine of affected subjects while tyrosine is metabolized normally. The effects of these two abnormalities suggested the pathway for phenylalanine metabolism diagrammed in Fig. 15-13. However, the supposition that phenylpyruvate but not tyrosine occurs on the normal pathway of phenylalanine metabolism because phenylpyruvate accumulates in the urine of phenylketonurics has proven to be incorrect. This indicates the pitfalls of relying solely on metabolic blocks and the consequent buildup of intermediates as indicators of a metabolic pathway. In this case, phenylpyruvate formation was later shown to arise from a normally minor pathway that becomes significant only when the phenylalanine concentration is abnormally high as in phenylketonurics.

Metabolic Blocks Can Be Generated by the Genetic Manipulation of Microorganisms

Early metabolic studies led to the astounding discovery that *the basic metabolic pathways in most organisms are essentially identical.* This metabolic uniformity has greatly facilitated the study of metabolic reactions. Thus, although a mutation that inactivates or deletes an enzyme in a pathway of interest may be unknown in higher organisms, it can be readily generated in rapidly reproducing microorganisms through the use of **mutagens** (chemical agents that induce genetic changes; Section 30-1A), X-rays, or, more recently, through genetic engineering techniques (Section 28-8). Desired mutants are identified by their requirement of the pathway's end product for growth. For example, George Beadle and Edward Tatum proposed a pathway of arginine biosynthesis in the mold *Neurospora crassa* based on their analysis of three arginine-requiring **auxotrophic mutants** (mutants requiring a specific nutrient for growth),

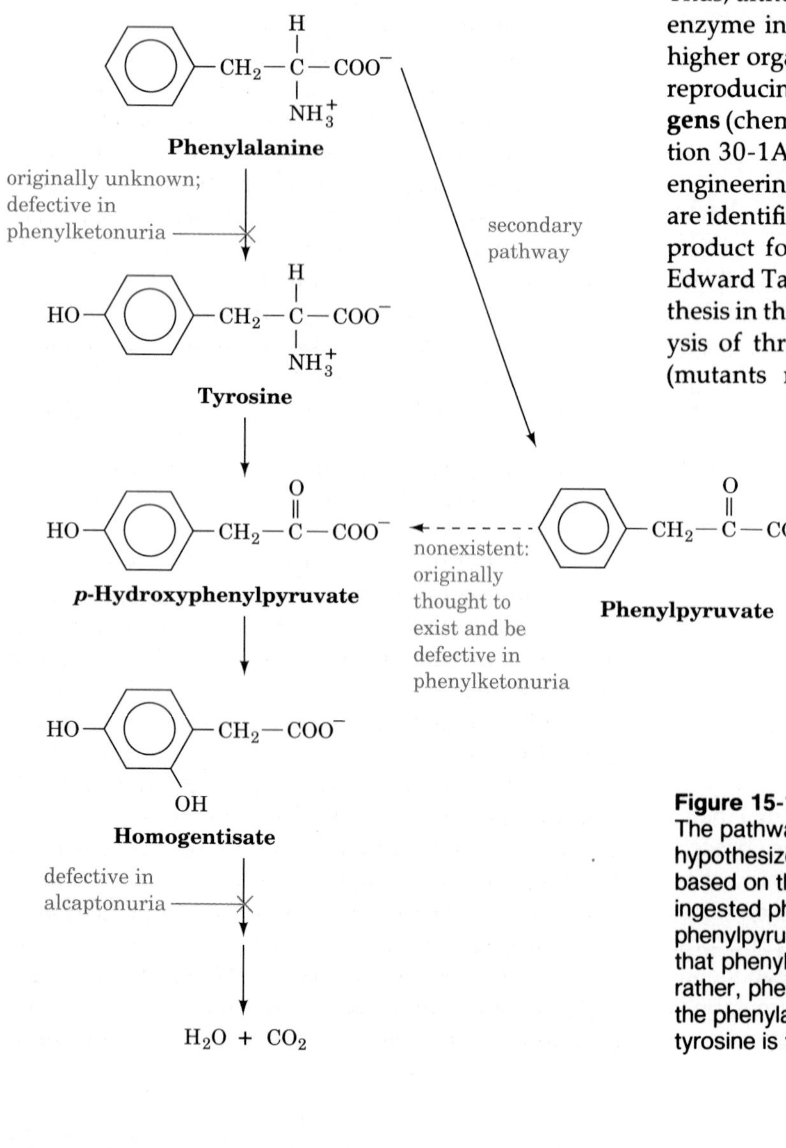

Figure 15-13
The pathway for phenylalanine degradation. It was originally hypothesized that phenylpyruvate is a pathway intermediate based on the observation that phenylketonurics excrete ingested phenylalanine and phenylpyruvate as phenylpyruvate. Further studies, however, demonstrated that phenylpyruvate is not a homogentisate precursor but, rather, phenylpyruvate production is only significant when the phenylalanine concentration is abnormally high. Instead, tyrosine is the normal product of phenylalanine degradation.

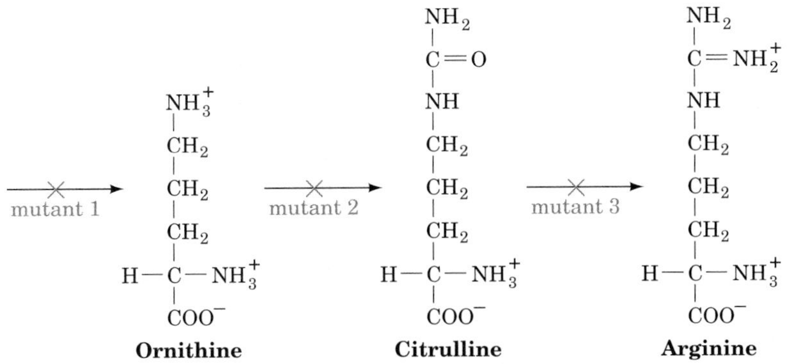

Figure 15-14
The pathway of arginine biosynthesis
indicating the positions of genetic blocks.

which were isolated after X-irradiation (Fig. 15-14). All of these mutants grow in the presence of arginine, but mutant 1 also grows in the presence of the (nonstandard) α-amino acids **citrulline** or **ornithine** and mutant 2 grows in the presence of citrulline. This is because in mutant 1, an enzyme leading to the production of ornithine is absent while enzymes further along the pathway are normal. In mutant 2, the enzyme catalyzing citrulline production is defective while in mutant 3 an enzyme involved in the conversion of citrulline to arginine is lacking. This landmark study also conclusively demonstrated that enzymes are specified by genes (Section 27-1C).

B. Isotopes in Biochemistry

The specific labeling of metabolites such that their interconversions can be traced is an indispensable technique for elucidating metabolic pathways. Franz Knoop formulated this technique in 1904 in order to study fatty acid oxidation. He fed dogs fatty acids chemically labeled with phenyl groups and isolated the phenyl-substituted end products from their urine. From the differences in these products when the phenyl-substituted starting material contained odd and even numbers of carbon atoms he deduced that fatty acids are degraded in C_2 units (Section 23-2).

Isotopes Specifically Label Molecules Without Altering Their Chemical Properties

Chemical labeling has the disadvantage that the chemical properties of labeled metabolites differ from those of normal metabolites. This problem is eliminated by labeling molecules of interest with **isotopes** (atoms with the same number of protons but a different number of neutrons in their nuclei). Recall that the chemical properties of an element are a consequence of its electron configuration which, in turn, is determined by its atomic number, not its atomic mass. The metabolic fate of a specific atom in a metabolite can therefore be elucidated by isotopically labeling that position and following its progress through the metabolic pathway of interest. The advent of isotopic labeling and tracing techniques in the 1940s therefore revolutionized the study of metabolism. (**Isotope effects,** which are changes in reaction rates arising from the mass differences between isotopes, are in most instances, negligible. Where they are significant, most noticably between hydrogen and its isotopes deuterium and tritium, they have been used to gain insight into enzymatic reaction mechanisms.)

The Detection of Isotopes

All elements have isotopes. For example, the atomic mass of naturally occurring Cl is 35.45 D because, at least on earth, it is a mixture of 55% ^{35}Cl and 45% ^{36}Cl (other isotopes of Cl are only present in trace amounts). Stable isotopes are generally identified and quantitated by mass spectrometry or NMR techniques. Many isotopes, however, are unstable; they undergo **radioactive decay,** a process that involves the emission from the radioactive nuclei of subatomic particles such as helium nuclei (**α particles**), electrons (**β particles**), and/or photons (**γ radiation**). Radioactive nuclei emit radiation with characteristic energies. For example, ^{3}H, ^{14}C, and ^{32}P all emit β particles but with respective energies of 0.018, 0.155, and 1.71 MeV. The radiation from ^{32}P is therefore highly penetrating, whereas that from ^{3}H and ^{14}C is not. (^{3}H and ^{14}C, as all radioactive isotopes, must, nevertheless, be handled with great caution because they can cause genetic damage upon ingestion.)

Radiation can be detected by a variety of techniques. Those most commonly used in biochemical investigations are **proportional counting** (known in its simplest form as **Geiger counting**), **liquid scintillation counting,** and **autoradiography.** Proportional counters electronically detect the ionizations in a gas caused by the passage of radiation. Moreover, they can also discriminate between particles of different energies and thus simultaneously determine the amounts of two or more different isotopes present.

Although proportional counters are quite simple to use, the radiation from two of the most widely used isotopes in biochemical analysis, ^{3}H and ^{14}C, have insufficient penetrating power to enter a proportional counter's detection chamber with reasonable efficiency.

This limitation is circumvented through liquid scintillation counting. In this technique, a radioactive sample is dissolved or suspended in a solution containing fluorescent substances that emit a pulse of light when struck by radiation. The light is detected electronically so that the number of light pulses can be counted. The emitting nucleus can also be identified because the intensity of a light pulse is proportional to the radiation energy (the number of fluorescent molecules excited by a radioactive particle is proportional to the particle's energy).

In autoradiography, radiation is detected by its blackening of photographic film. The radioactive sample is laid on, or in some cases mixed with, the photographic emulsion and after sufficient exposure time (from minutes to months) the film is developed. Autoradiography is widely used to locate radioactive substances in polyacrylamide gels (e.g., Fig. 5-29).

Isotopes Have Characteristic Half-Lives

Radioactive decay is a random process whose rate for a given isotope depends only on the number of radioactive atoms present. It is therefore a simple first-order process whose half-life, $t_{1/2}$, is a function only of the

Table 15-1
Some Isotopes of Biochemical Importance

Stable Isotopes	
Nucleus	Natural Abundance (%)
^2H	0.015
^{13}C	1.1
^{15}N	0.37
^{18}O	0.20

Radioactive Isotopes		
Nucleus	Radiation Type	Half-Life
^3H	β	12.26 years
^{14}C	β	5730 years
^{22}Na	γ	2.60 years
^{32}P	β	14.3 days
^{35}S	β	88 days
^{45}Ca	β	165 days
^{60}Co	γ	5.26 years
^{125}I	γ	60 days
^{131}I	β,γ	8.07 days

Source: Heath, R. L., *in* Weast, R. C. (Ed.), *Handbook of Chemistry and Physics* (69th ed.), *pp.* B232–B448, CRC Press (1988).

Figure 15-15
The metabolic origin of the nitrogen atoms in heme. Only [^{15}N]glycine, of many ^{15}N-labeled metabolites, is an ^{15}N-labeled heme precursor.

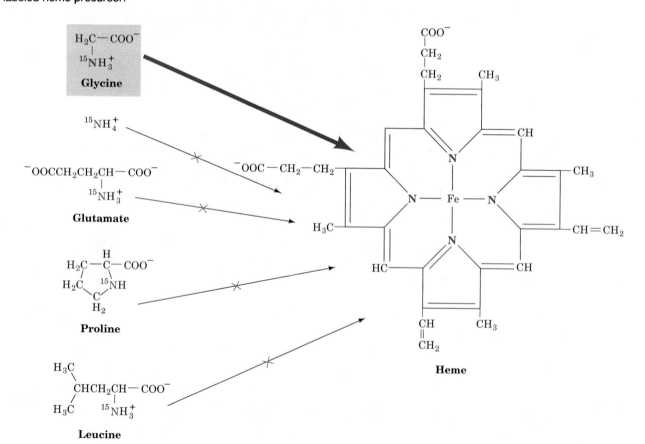

rate constant, k, for the decay process (Section 13-1B):

$$t_{1/2} = \frac{\ln 2}{k} = \frac{0.693}{k} \quad [13.5]$$

Because k is different for each radioactive isotope, each has a characteristic half-life. The properties of some isotopes in common biochemical use are listed in Table 15-1.

Isotopes Are Indispensable for Establishing the Metabolic Origins of Complex Metabolites and Precursor–Product Relationships

The metabolic origins of complex molecules such as heme, cholesterol, and phospholipids may be determined by administering isotopically labeled starting materials to animals and isolating the resulting products. One of the early advances in metabolic understanding resulting from the use of isotopic tracers was the demonstration, by David Shemin and David Rittenberg in 1945, that the nitrogen atoms of heme are derived from glycine rather than from ammonia, glutamic acid, proline, or leucine (Section 24-4A). They did so by feeding rats these ^{15}N-labeled nutrients, isolating the heme in their blood, and analyzing it for ^{15}N content. Only when the rats were fed [^{15}N]glycine did the heme contain ^{15}N (Fig. 15-15). This technique was also used to demonstrate that all of cholesterol's carbon atoms are derived from acetyl-CoA (Section 23-6A).

Isotopic tracers are also useful in establishing the order of appearance of metabolic intermediates, their so-called **precursor–product relationships.** An example of such an analysis concerns the biosynthesis of the complex phospholipids called **plasmalogens** and **alkylacylglycerophospholipids** (Section 23-8A). Alkylacylglycerophospholipids are ethers while the closely related plasmalogens are vinyl ethers. Their similar structures brings up the interesting question of their biosynthetic relationship: Which is the precursor and which is the product? Two possible modes of synthesis can be envisioned (Fig. 15-16):

I. The starting material is converted to the vinyl ether (plasmalogen), which is then reduced to yield the ether (alkylacylglycerophospholipid). Accordingly, the vinyl ether would be the precursor and the ether the product.

II. The ether is formed first and then oxidized to yield the vinyl ether. The ether would then be the precursor and the vinyl ether the product.

Precursor–product relationships can be most easily sorted out through the use of radioactive tracers. A pulse of the labeled starting material is administered to an organism and the specific radioactivities of the resulting metabolic products are followed with time (Fig. 15-17):

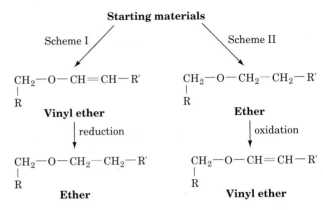

Figure 15-16
Two possible pathways for the biosynthesis of ether- and vinyl ether-containing phospholipids. **(I)** The vinyl ether is the precursor and the ether is the product. **(II)** The ether is the precursor and the vinyl ether is the product.

$$\text{Starting material*} \longrightarrow \text{A*} \longrightarrow \text{B*} \longrightarrow \text{later products*}$$

(here the * represents the radioactive label). Metabolic pathways, as we shall see in Section 15-6B, normally operate in a steady state; that is, the throughput of metabolites in each of its reaction steps is equal. Moreover, the rates of most metabolic reactions are first order for a given substrate. Making these assumptions, we note that the rate of change of B's radioactivity, [B*], is equal to the rate of passage of label from A* to B* less the rate of passage of label from B* to the pathway's next product:

$$\frac{d[\text{B*}]}{dt} = k[\text{A*}] - k[\text{B*}] = k([\text{A*}] - [\text{B*}]) \quad [15.1]$$

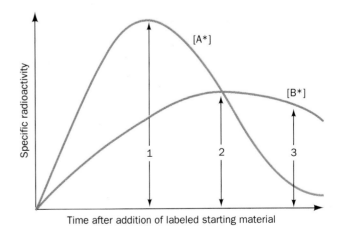

Figure 15-17
The flow of a pulse of radioactivity from precursor (A*, *orange*) to product (B*, *purple*). At point 1, product radioactivity is increasing and is less than that of its precursor; at point 2, product radioactivity is maximal and is equal to that of its precursor; and at point 3, product radioactivity is decreasing and is greater than that of its precursor.

where k is the pseudo-first-order rate constant for both the conversion of A to B and the conversion of B to its product, and t is time. Inspection of this equation indicates the criteria that must be met to establish that A is the precursor of B (Fig. 15-17):

1. Before the radioactivity of the product [B*] is maximal, $d[B^*]/dt > 0$ so that [A*] > [B*]; that is, *while the radioactivity of a product is rising, it should be less than that of its precursor.*

2. When [B*] is maximal, $d[B^*]/dt = 0$ so that [A*] = [B*]; that is, *when the radioactivity of a product is at its peak, it should be equal to that of its precursor.* This result also implies that *the radioactivity of a product peaks after that of its precursor.*

3. After [B*] begins to decrease, $d[B^*]/dt < 0$ so that [A*] < [B*]; that is, *after the radioactivity of a product has peaked, it should remain greater than that of its precursor.*

Such a determination of the precursor–product relationship between alkylacylglycerophospholipid and plasmalogen, using ^{14}C-starting materials, indicated that the ether is the precursor and the vinyl ether is the product (Fig. 15-16, Scheme II).

C. Isolated Organs, Cells, and Subcellular Organelles

In addition to understanding the chemistry and catalytic events that occur at each step of a metabolic pathway, it is important to learn where a given pathway occurs within an organism. Early workers studied metabolism in whole animals. For example, the role of the pancreas in diabetes was established by Frederick Banting and George Best in 1921 by surgically removing that organ from dogs and observing that these animals then developed the disease.

The metabolic products produced by a particular organ can be studied by **organ perfusion** or in **tissue slices.** In organ perfusion, a specific organ is surgically removed from an animal and the organ's arteries and veins connected to an artificial circulatory system. The composition of the material entering the organ can thereby be controlled and its metabolic products monitored. Metabolic processes can be similarly studied in slices of tissue thin enough to be nourished by free diffusion in an appropriate nutrient solution. Otto Warburg pioneered the tissue slice technique in the early twentieth century, through his studies of respiration in which he used a manometer to measure the changes in gas volume above tissue slices as a consequence of their O_2 consumption.

A given organ or tissue generally contains several cell types. **Cell sorters** are devices that can separate cells

Table 15-2
Metabolic Functions of Eukaryotic Organelles

Organelle	Function
Mitochondrion	Citric acid cycle, oxidative phosphorylation, fatty acid oxidation, amino acid breakdown
Cytosol	Glycolysis, pentose phosphate pathway, fatty acid biosynthesis, many reactions of gluconeogenesis
Lysosomes	Enzymatic digestion of cell components and ingested matter
Nucleus	DNA replication and transcription, RNA processing
Golgi apparatus	Post-translational processing of membrane and secretory proteins; formation of plasma membrane and secretory vesicles
Rough endoplasmic reticulum	Synthesis of membrane-bound and secretory proteins
Smooth endoplasmic reticulum	Lipid and steroid biosynthesis
Peroxisomes (glyoxisomes in plants)	Oxidative reactions catalyzed by amino acid oxidases and catalase; glyoxylate cycle reactions in plants

according to type once they have been treated with the enzymes trypsin and collagenase to destroy the intercellular matrix that binds them into a tissue. This technique allows further localization of metabolic function. A single cell type may also be grown in **tissue culture** for study. Although culturing cells often results in their loss of differentiated function, techniques have been developed for maintaining several cell types that still express their original characteristics.

Metabolic pathways in eukaryotes are compartmentalized in various subcellular organelles as Table 15-2 indicates (these organelles are described in Section 1-2A). For example, oxidative phosphorylation occurs in the mitochondrion whereas glycolysis and fatty acid biosynthesis occur in the cytosol. Such observations are made by breaking cells open and fractionating their components by differential centrifugation (Section 5-1B), possibly followed by zonal ultracentrifugation through a sucrose density gradient or by equilibrium density gradient ultracentrifugation in a CsCl density gradient, which, respectively, separate particles according to their size and density (Section 5-5B). The cell fractions are then analyzed for biochemical function.

4. THERMODYNAMICS OF PHOSPHATE COMPOUNDS

The endergonic processes that maintain the living state are driven by the exergonic reactions of nutrient oxidation. This coupling is most often mediated through the synthesis of a few types of "high-energy" intermediates whose exergonic consumption drives the endergonic processes. These intermediates therefore form a sort of universal free energy "currency" through which free energy producing reactions "pay for" the free energy consuming processes in biological systems.

Adenosine triphosphate (ATP; Fig. 15-18), which occurs in all known life forms, is the "high-energy" intermediate that constitutes the most common cellular energy currency. Its central role in energy metabolism was first recognized in 1941 by Fritz Lipmann and Herman Kalckar. ATP consists of an **adenosine** moiety to which three **phosphoryl groups** ($-PO_3^{2-}$) are sequentially linked via a **phosphoester bond** followed by two **phosphoanhydride bonds. Adenosine diphosphate (ADP)** and **5'-adenosine monophosphate (AMP)** are similarly constituted but with only two and one phosphoryl units, respectively.

In this section we consider the nature of phosphoryl-transfer reactions, discuss why some of them are so exergonic, and outline how the cell consumes and regenerates ATP.

A. Phosphoryl-Transfer Reactions

Phosphoryl-transfer reactions:

$$R_1-O-PO_3^{2-} + R_2-OH \rightleftharpoons R_1-OH + R_2-O-PO_3^{2-}$$

are of enormous metabolic significance. Some of the most important reactions of this type involve the synthesis and hydrolysis of ATP:

$$ATP + H_2O \rightleftharpoons ADP + P_i$$
$$ATP + H_2O \rightleftharpoons AMP + PP_i$$

where P_i and PP_i, respectively, represent **orthophosphate** (PO_4^{3-}) and **pyrophosphate** ($P_2O_7^{4-}$) in any of their ionization states. *These highly exergonic reactions are coupled to numerous endergonic biochemical processes so as to drive them to completion. Conversely, ATP is regenerated by coupling its formation to a more highly exergonic metabolic process* (the thermodynamics of coupled reactions is discussed in Section 3-4C).

To illustrate these concepts, let us consider two examples of phosphoryl-transfer reactions. The initial step in the metabolism of glucose is its conversion to glucose-6-phosphate (Section 16-2A). Yet, the direct reaction of

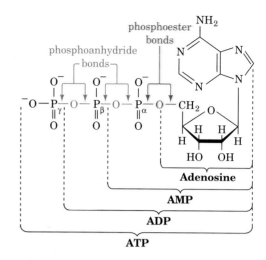

Figure 15-18
The structure of ATP indicating its relationship to ADP, AMP, and adenosine. The phosphoryl groups, starting with that on AMP, are referred to as the α, β, and γ phosphates. Note the differences between phosphoester and phosphoanhydride bonds.

glucose and P_i is thermodynamically unfavorable (Fig. 15-19a). In biological systems, however, this reaction is coupled to the exergonic hydrolysis of ATP so that the overall reaction is thermodynamically favorable. ATP can be similarly regenerated by coupling its synthesis from ADP and P_i to the even more exergonic hydrolysis of **phosphoenolpyruvate** (Fig. 15-19b; Section 16-2J).

The bioenergetic utility of phosphoryl-transfer reactions stems from their kinetic stability to hydrolysis combined with their capacity to transmit relatively large amounts of free energy. The $\Delta G°'$ values of hydrolysis of several phosphorylated compounds of biochemical importance are tabulated in Table 15-3. The negatives of these values are often referred to as **phosphate group-transfer potentials;** they are a measure of the tendency of phosphorylated compounds to transfer their phosphoryl groups to water. Note that ATP has an intermediate phosphate group-transfer potential. Under standard conditions, the compounds above ATP in Table 15-3 can spontaneously transfer a phosphoryl group to ADP to form ATP, which can, in turn, spontaneously transfer a phosphoryl group to the hydrolysis products (ROH form) of the compounds below it.

ΔG of ATP Hydrolysis Varies with pH, Divalent Metal Ion Concentration, and Ionic Strength

The ΔG of a reaction varies with the total concentrations of its reactants and products and thus with their ionic states (Eq. [3.15]). The ΔG's of hydrolysis of phos-

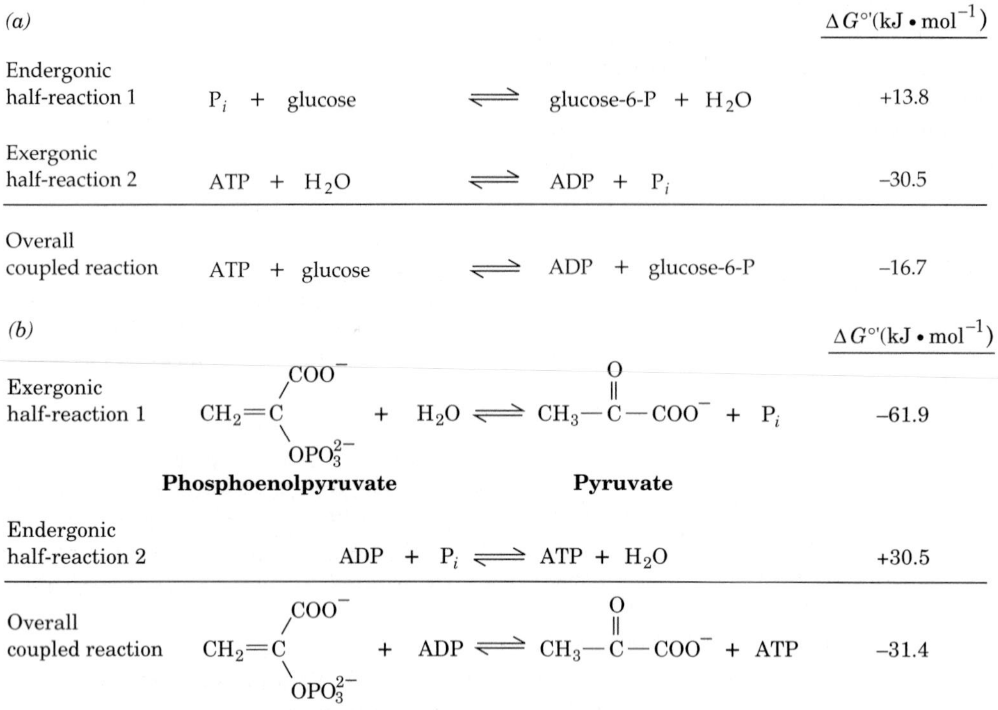

(a) $\Delta G^{\circ\prime}(\text{kJ} \cdot \text{mol}^{-1})$

Endergonic half-reaction 1	P_i + glucose	$\rightleftharpoons$ glucose-6-P + H_2O	+13.8
Exergonic half-reaction 2	ATP + H_2O	$\rightleftharpoons$ ADP + P_i	−30.5
Overall coupled reaction	ATP + glucose	$\rightleftharpoons$ ADP + glucose-6-P	−16.7

(b) $\Delta G^{\circ\prime}(\text{kJ} \cdot \text{mol}^{-1})$

Exergonic half-reaction 1

$$\underset{\textbf{Phosphoenolpyruvate}}{CH_2\!\!=\!\!\overset{\displaystyle COO^-}{\underset{\displaystyle OPO_3^{2-}}{C}}} + H_2O \rightleftharpoons \underset{\textbf{Pyruvate}}{CH_3\!-\!\overset{\displaystyle O}{\overset{\|}{C}}\!-\!COO^-} + P_i \qquad -61.9$$

Endergonic half-reaction 2   ADP + P_i $\rightleftharpoons$ ATP + H_2O +30.5

Overall coupled reaction

$$CH_2\!\!=\!\!\overset{\displaystyle COO^-}{\underset{\displaystyle OPO_3^{2-}}{C}} + ADP \rightleftharpoons CH_3\!-\!\overset{\displaystyle O}{\overset{\|}{C}}\!-\!COO^- + ATP \qquad -31.4$$

Figure 15-19
Some overall coupled reactions involving ATP: *(a)* The phosphorylation of glucose to form glucose-6-phosphate and ADP. *(b)* The phosphorylation of ADP by phosphoenolpyruvate to form ATP and pyruvate. Each reaction has been conceptually decomposed into a direct phosphorylation step (half-reaction **1**) and a step in which ATP is hydrolyzed (half-reaction **2**). Both half-reactions proceed in the direction in which the overall reaction is exergonic ($\Delta G < 0$).

Table 15-3

Standard Free Energies of Phosphate Hydrolysis of Some Compounds of Biological Interest

Compound	$\Delta G^{\circ\prime}$ (kJ·mol^{-1})
Phosphoenol pyruvate	−61.9
1,3-Bisphosphoglycerate	−49.4
Acetyl phosphate	−43.1
Phosphocreatine	−43.1
PP$_i$	−33.5
ATP ($\longrightarrow$ AMP + PP$_i$)	−32.2
ATP ($\longrightarrow$ ADP + P$_i$)	−30.5
Glucose-1-phosphate	−20.9
Fructose-6-phosphate	−13.8
Glucose-6-phosphate	−13.8
Glycerol-3-phosphate	−9.2

Source: Jencks, W. P., *in* Fasman, G. D. (Ed.), *Handbook of Biochemistry and Molecular Biology* (3rd ed.), Physical and Chemical Data, Vol. I, *pp.* 296–304, CRC Press (1976).

phorylated compounds are therefore highly dependent on pH, divalent metal ion concentration (divalent metal ions such as Mg^{2+} have high phosphate-binding affinities), and ionic strength. Reasonable estimates of the intracellular values of these quantities as well as of [ATP], [ADP], and [P$_i$] (which are generally on the order of millimolar) indicate that ATP hydrolysis under physiological conditions has $\Delta G \approx -50 \text{ kJ} \cdot \text{mol}^{-1}$ rather than the −30.5-kJ·mol^{-1} value of its $\Delta G^{\circ\prime}$. Nevertheless, for the sake of consistency in comparing reactions, we shall usually refer to the latter value.

The above situation for ATP is not unique. It is important to keep in mind that *within a given cell, the concentrations of most substances vary both with location and time. Indeed, the concentrations of many ions, coenzymes, and metabolites commonly vary by several orders of magnitude across membranous organelle boundaries.* Unfortunately, it is usually quite difficult to obtain an accurate measurement of the concentration of any particular chemical species in a specific cellular compartment. The ΔG's for most *in vivo* (in the cell) reactions are therefore little more than estimates.

B. Rationalizing the "Energy" in "High-Energy" Compounds

Bonds whose hydrolysis proceeds with large negative values of $\Delta G°'$ (customarily more negative than -25 kJ $\cdot$ mol^{-1}) are often referred to as **"high-energy"** **bonds** or **"energy-rich"** **bonds** and are frequently symbolized by the squiggle ($\sim$). Thus ATP may be represented as AR—P $\sim$ P $\sim$ P where A, R, and P symbolize adenyl, ribosyl, and phosphoryl groups, respectively. Yet, the phosphoester bond joining the adenosyl group of ATP to its α-phosphoryl group appears to be not greatly different in electronic character from the so-called "high-energy" bonds bridging its β and γ phosphoryl groups. In fact, none of these bonds have any unusual properties so that the term "high-energy" bond is somewhat of a misnomer. (In any case, it should not be confused with the term "bond energy," which is defined as the energy required to break, not hydrolyze, a covalent bond.) Why then, should the phosphoryl-transfer reactions of ATP be so exergonic? The answer comes from the comparison of the stabilities of the reactants and products of these reactions.

Several different factors appear to be responsible for the "high-energy" character of phosphoanhydride bonds such as those in ATP (Fig. 15-20):

1. The resonance stabilization of a phosphoanhydride bond is less than that of its hydrolysis products. This is because a phosphoanhydride's two strongly electron-withdrawing phosphoryl groups must compete for the π electrons of its bridging oxygen atom, whereas this competition is absent in the hydrolysis products. In other words, the electronic requirements of the phosphoryl groups are less satisfied in a phosphoanhydride than in its hydrolysis products.

2. Of perhaps greater importance is the destabilizing effect of the electrostatic repulsions between the charged groups of a phosphoanhydride in comparison to that of its hydrolysis products. In the physiological pH range, ATP has three to four negative charges whose mutual electrostatic repulsions are partially relieved by ATP hydrolysis.

3. Another destabilizing influence, which is difficult to assess, is the smaller solvation energy of a phosphoanhydride in comparison to that of its hydrolysis products. Some estimates suggest that this factor provides the dominant thermodynamic driving force for the hydrolysis of phosphoanhydrides.

A further property of ATP that suits it to its role as an energy intermediate stems from the relative kinetic stability of phosphoanhydride bonds to hydrolysis. Most types of anhydrides are rapidly hydrolyzed in aqueous

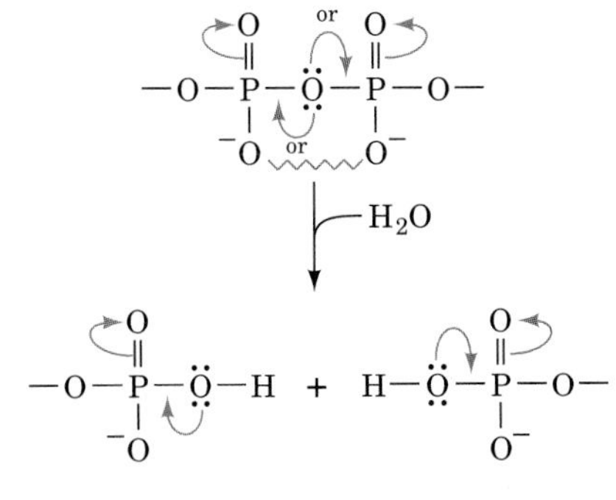

Figure 15-20
The competing resonances (*curved arrows* from central O) and charge–charge repulsions (*zigzag line*) between the phosphoryl groups of a phosphoanhydride decrease its stability relative to its hydrolysis products.

solution. Phosphoanhydride bonds, however, have unusually large free energies of activation. Consequently, ATP is reasonably stable under physiological conditions but is readily hydrolyzed in enzymatically mediated reactions.

Other "High-Energy" Compounds

The compounds in Table 15-3 with phosphate group-transfer protentials significantly greater than that of ATP have additional destabilizing influences:

1. **Acyl phosphates**

   The hydrolysis of **acyl phosphates** (mixed phosphoric-carboxylic anhydrides), such as **acetyl phosphate** and **1,3-bisphosphoglycerate**,

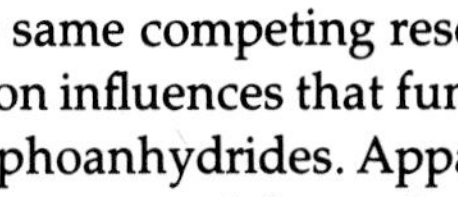

Acetyl phosphate

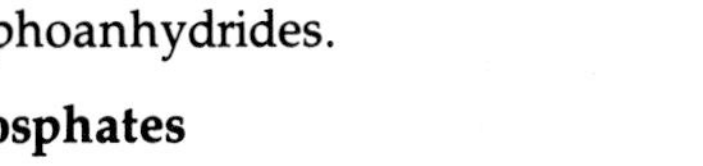

1,3-Bisphosphoglycerate

   is driven by the same competing resonance and differential solvation influences that function in the hydrolysis of phosphoanhydrides. Apparently these effects are more pronounced for acyl phosphates than for phosphoanhydrides.

2. **Enol phosphates**

   The high phosphate group-transfer potential of **enol phosphates** such as phosphoenolpyruvate (Fig. 15-19*b*), derives from their **enol** hydrolysis product being less stable than its **keto** tautomer. Consider the hydrolysis reaction of an enol phosphate as occurring

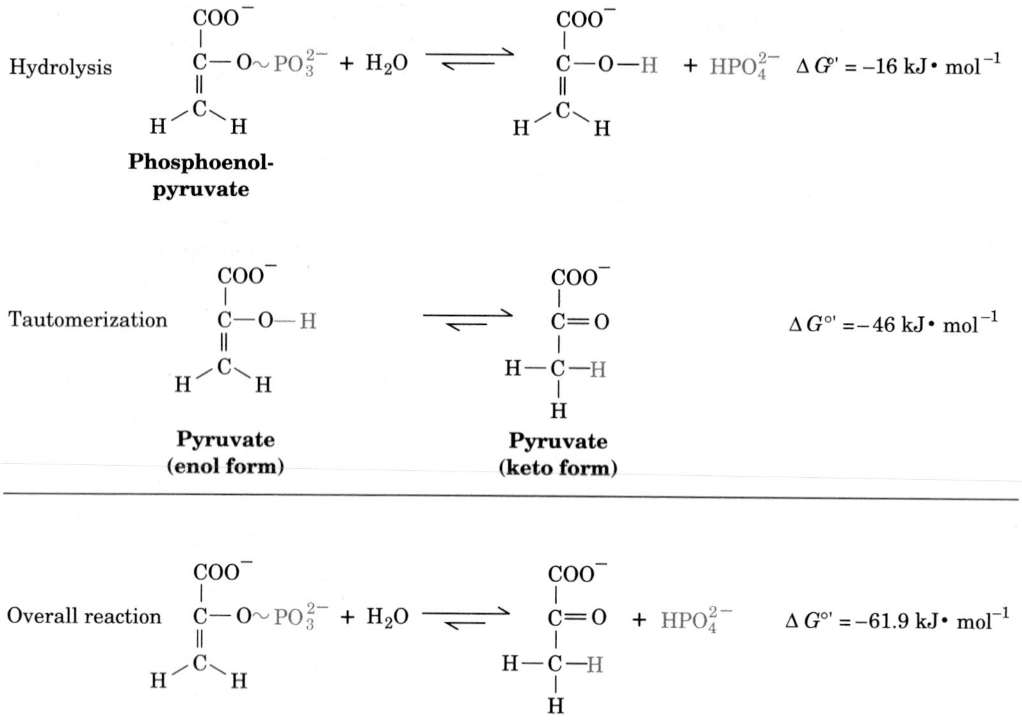

Hydrolysis

Phosphoenol-
pyruvate

$$\text{Phosphoenolpyruvate} + H_2O \rightleftharpoons \text{(pyruvate enol)} + HPO_4^{2-} \qquad \Delta G^{\circ\prime} = -16 \text{ kJ} \cdot \text{mol}^{-1}$$

Tautomerization

Pyruvate
(enol form)

Pyruvate
(keto form)

$$\Delta G^{\circ\prime} = -46 \text{ kJ} \cdot \text{mol}^{-1}$$

Overall reaction

$$\Delta G^{\circ\prime} = -61.9 \text{ kJ} \cdot \text{mol}^{-1}$$

Figure 15-21
The hydrolysis of phosphoenolpyruvate. The reaction is broken down into two steps, hydrolysis and tautomerization.

in two steps (Fig. 15-21). The hydrolysis step is subject to the driving forces discussed above. *It is therefore the highly exergonic enol–keto conversion that provides phosphoenolpyruvate with the added thermodynamic impetus to phosphorylate ADP to form ATP.*

3. Phosphoguanidines

The high phosphate group-transfer potentials of **phosphoguanidines**, such as **phosphocreatine** and **phosphoaginine**, largely result from the competing resonances in their **guanidino** group, which are even more pronounced than they are in the phosphate group of phosphoanhydrides (Fig. 15-22). Consequently, phosphocreatine can phosphorylate ADP (see Section 15-4C).

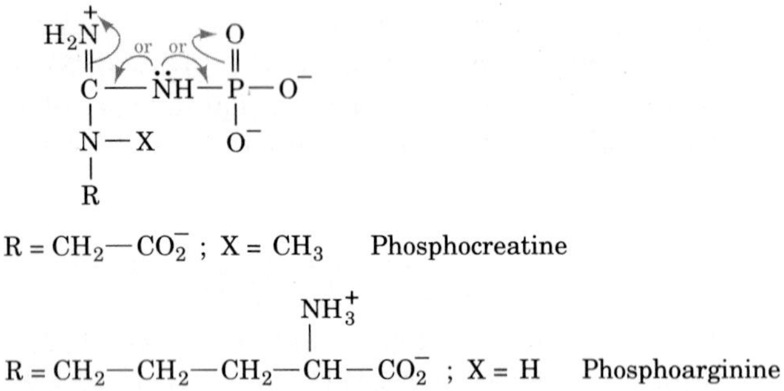

$R = CH_2 - CO_2^-$; $X = CH_3$ Phosphocreatine

$R = CH_2 - CH_2 - CH_2 - CH - CO_2^-$; $X = H$ Phosphoarginine

with NH_3^+ on the CH.

Figure 15-22
The competing resonances in phosphoguanidines.

Compounds such as **glucose-6-phosphate** or **glycerol-3-phosphate**,

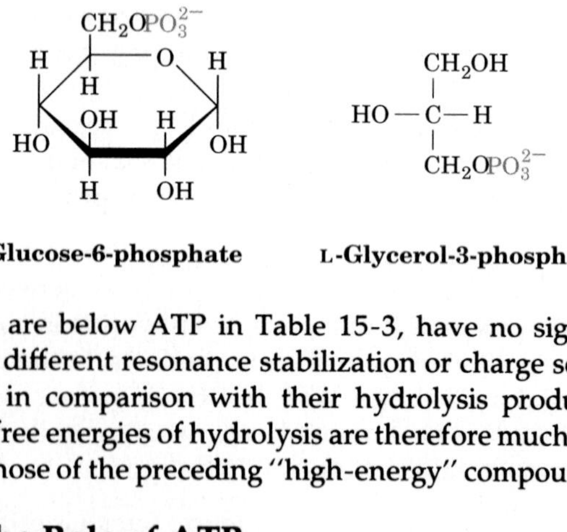

α- D-**Glucose-6-phosphate**   L-**Glycerol-3-phosphate**

which are below ATP in Table 15-3, have no significantly different resonance stabilization or charge separation in comparison with their hydrolysis products. Their free energies of hydrolysis are therefore much less than those of the preceding "high-energy" compounds.

C. The Role of ATP

As Table 15-3 indicates, in the thermodynamic hierarchy of phosphoryl-transfer agents, ATP occupies the middle rank. This enables ATP to serve as an energy conduit between "high-energy" phosphate donors and "low-energy" phosphate acceptors (Fig. 15-23). Let us examine the general biochemical scheme of how this occurs.

In general, the highly exergonic phosphoryl-transfer reactions of nutrient degradation are coupled to the formation of ATP from ADP and P_i through the auspices of

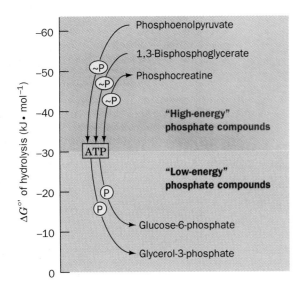

Figure 15-23
The flow of phosphoryl groups from high-energy phosphate donors, via the ATP–ADP system, to low-energy phosphate acceptors.

various enzymes known as **kinases;** these enzymes catalyze the transfer of phosphoryl groups between ATP and other molecules. Consider the two reactions in Fig. 15-19*b*. If carried out independently, these reactions would not influence each other. In the cell, however, the enzyme **pyruvate kinase** couples the two reactions by catalyzing the transfer of the phosphoryl group of phosphoenolpyruvate directly to ADP to result in an overall exergonic reaction.

Consumption of ATP

In its role as the universal energy currency of living systems, ATP is consumed in a variety of ways:

1. Early stages of nutrient breakdown

The exergonic hydrolysis of ATP to ADP may be enzymatically coupled to an endergonic phosphorylation reaction to form "low-energy" phosphate compounds. We have seen one example of this in the hexokinase-catalyzed formation of glucose-6-phosphate (Fig. 15-19*a*). Another example is the **phosphofructokinase**-catalyzed phosphorylation of

fructose-6-phosphate to form **fructose-1,6-bisphosphate** (Fig. 15-24). Both of these reactions occur in the first stage of glycolysis (Section 16-2).

2. Interconversion of nucleoside triphosphates

Many biosynthetic processes, such as the synthesis of proteins and nucleic acids, require nucleoside triphosphates other than ATP. These include the ribonucleoside triphosphates CTP, GTP, and UTP (Section 1-3C) which, together with ATP, are utilized, for example, in the biosynthesis of RNA (Section 29-2) and the deoxyribonucleoside triphosphate DNA precursors dATP, dCTP, dGTP, and dTTP (Section 31-1A). All these **nucleoside triphosphates (NTPs)** are synthesized from ATP and the corresponding **nucleoside diphosphate (NDP)** in reactions catalyzed by the nonspecific enzyme **nucleoside diphosphate kinase:**

$$ATP + NDP \rightleftharpoons ADP + NTP$$

The $\Delta G°'$ values for these reactions are nearly zero as might be expected from the structural similarities among the NTPs. These reactions are driven by the depletion of the NTPs through their exergonic hydrolysis in the biosynthetic reactions in which they participate (Section 3-4C).

3. Physiological processes

The hydrolysis of ATP to ADP and P_i energizes many essential endergonic physiological processes such as muscle contraction and the transport of molecules and ions against concentration gradients. In general, these processes result from conformational changes in proteins (enzymes) that occur in response to their binding ATP. This is followed by the exergonic hydrolysis of ATP and release of ADP and P_i, thereby causing these processes to be unidirectional (irreversible).

4. Additional phosphoanhydride cleavage in highly endergonic reactions

Although many reactions involving ATP yield ADP and P_i **(orthophosphate cleavage),** others yield AMP and PP_i **(pyrophosphate cleavage).** In these latter cases, the PP_i is rapidly hydrolyzed to $2P_i$ by **inorganic pyrophosphatase** ($\Delta G°' = -33.5$ kJ·mol^{-1}) so that *the pyrophosphate cleavage of ATP ultimately results in the hydrolysis of two "high-energy" phosphoanhydride bonds.* The first step in the oxidation of

Figure 15-24
The phosphorylation of fructose-6-phosphate by ATP to form fructose-1,6-bisphosphate and ADP.

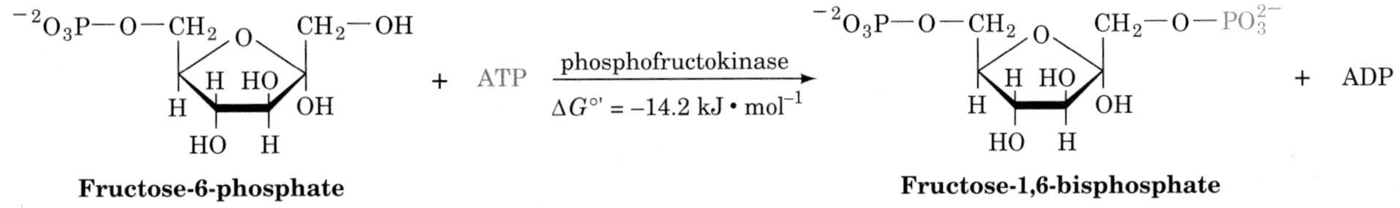

$$^{-2}O_3P-O-CH_2 \quad CH_2-OH \quad + \quad ATP \quad \xrightarrow[\Delta G°' = -14.2 \text{ kJ · mol}^{-1}]{\text{phosphofructokinase}} \quad ^{-2}O_3P-O-CH_2 \quad CH_2-O-PO_3^{2-} \quad + \quad ADP$$

Fructose-6-phosphate   **Fructose-1,6-bisphosphate**

fatty acids (Fig. 15-25 and Section 23-2A) provides an example of this process. Pyrophosphate cleavage alone is insufficiently exergonic to drive the fatty acid-activation reaction to completion. This reaction is made irreversible, however, by the additional thermodynamic impetus of PP$_i$ hydrolysis. Nucleic acid biosynthesis from the appropriate NTPs also releases PP$_i$ (Sections 29-2 and 31-1A). The free energy changes of these vital reactions are around zero so that the subsequent hydrolysis of PP$_i$ is essential for the synthesis of nucleic acids.

Formation of ATP

To complete its intermediary metabolic function, ATP must be replenished. This is accomplished through three types of processes:

1. **Substrate-level phosphorylation**

   ATP may be formed, as is indicated in Fig. 15-19*b* from phosphoenolpyruvate, by direct transfer of a phosphoryl group from a "high-energy" compound to ADP. Such reactions, which are referred to as **substrate-level phosphorylations,** most commonly occur in the early stages of carbohydrate metabolism (Section 16-2).

2. **Oxidative phosphorylation and photophosphorylation**

   Both oxidative metabolism and photosynthesis act to generate a proton (H$^+$) concentration gradient across a membrane (Sections 20-3 and 22-2D). Discharge of this gradient is enzymatically coupled to formation of ATP from ADP and P$_i$ (the reverse of ATP hydrolysis). In oxidative metabolism, this process is called **oxidative phosphorylation,** whereas in photosynthesis it is termed **photophosphorylation.** Most of the ATP produced by respiring and photosynthesizing organisms is generated in this manner.

3. **Adenylate kinase reaction**

   The AMP resulting from pyrophosphate cleavage reactions of ATP is converted to ADP in a reaction catalyzed by the enzyme **adenylate kinase:**

$$\text{AMP} + \text{ATP} \rightleftharpoons 2\,\text{ADP}$$

   The ADP is subsequently converted to ATP through substrate-level phosphorylation, oxidative phosphorylation, or photophosphorylation.

Rate of ATP Turnover

The cellular role of ATP is that of a free energy transmitter rather than a free energy reservoir. The amount of ATP in a cell is typically only enough to supply its free energy needs for a minute or two. Hence, ATP is continually being hydrolyzed and regenerated. Indeed, ^{32}P-labeling experiments indicate that the metabolic half-life of an ATP molecule varies from seconds to minutes depending on the cell type and its metabolic activity. For in-

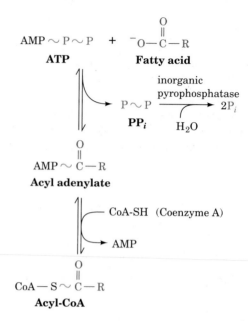

Figure 15-25
The activation of a fatty acid involves its conversion to a thioester. In the first reaction step, the fatty acid is **adenylylated** by reaction with ATP. In the second step, coenzyme A (CoA-SH), a complicated organic molecule bearing a sulfhydryl group (Fig. 19-2), displaces the AMP moiety to form an acyl-CoA adduct. For these two steps, $\Delta G^{\circ\prime} = +4.6$ kJ·mol^{-1} so that, even under physiological conditions, the reaction is unfavorable. However, the hydrolysis of the product PP$_i$, which has $\Delta G^{\circ\prime} = -33.5$ kJ·mol^{-1}, drives the activation reaction to completion ($\Delta G^{\circ\prime} = -33.5 + 4.6 = -28.9$ kJ·mol^{-1}).

stance, brain cells have only a few seconds supply of ATP (which, in part, accounts for the rapid deterioration of brain tissue by oxygen deprivation). *An average person at rest consumes and regenerates ATP at a rate of ~ 3 mol (1.5 kg)·h^{-1} and as much as an order of magnitude faster during strenuous activity.*

Phosphocreatine Provides a "High-Energy" Reservoir for ATP Formation

Muscle and nerve cells, which have a high ATP turnover, have a free energy reservoir that functions to regenerate ATP rapidly. In vertebrates, phosphocreatine (Fig. 15-22) functions in this capacity. It is synthesized by the reversible phosphorylation of creatine by ATP as catalyzed by **creatine kinase:**

$$\text{ATP} + \text{Creatine} \rightleftharpoons \text{phosphocreatine} + \text{ADP}$$
$$\Delta G^{\circ\prime} = +12.6 \text{ kJ·mol}^{-1}$$

Note that under standard conditions this reaction is endergonic; however, the intracellular concentrations of its reactants and products are such that it operates close to equilibrium ($\Delta G \approx 0$). Accordingly, when the cell is in a resting state, so that [ATP] is relatively high, the reaction proceeds with net synthesis of phosphocreatine, whereas at times of high metabolic activity, when [ATP] is low, the equilibrium shifts so as to yield net synthesis of ATP. *Phosphocreatine thereby acts as an ATP "buffer" in cells that contain creatine kinase.* A resting vertebrate skeletal muscle normally has sufficient phosphocreatine to supply its free energy needs for several minutes (but for only a few seconds at maximum exertion). In the muscles of some invertebrates, such as lobsters, phosphoarginine performs the same function. These phosphoguanidines are collectively named **phosphagens.**

5. OXIDATION – REDUCTION REACTIONS

Oxidation – reduction reactions, processes involving the transfer of electrons, are of immense biochemical significance; living things derive most of their free energy from them. In photosynthesis (Chapter 22), CO_2 is **reduced** (gains electrons) and H_2O is **oxidized** (loses electrons) to yield carbohydrates and O_2 in an otherwise endergonic process that is powered by light energy. In aerobic metabolism, which is carried out by all eukaryotes and many prokaryotes, the overall photosynthetic reaction is essentially reversed so as to harvest the free energy of oxidation of carbohydrates and other organic compounds in the form of ATP (Chapter 20). Anaerobic metabolism generates ATP, although in lower yields, through intramolecular oxidation–reductions of various organic molecules; for example, glycolysis (Chapter 16), or in certain anaerobic bacteria, through the use of non-O_2 oxidizing agents such as sulfate or nitrate. In this section we outline the thermodynamics of oxidation–reduction reactions in order to understand the quantitative aspects of these crucial biological processes.

A. The Nernst Equation

Oxidation – reduction reactions (also known as **redox** or **oxido – reduction reactions**) resemble other types of chemical reactions in that they involve group transfer. For instance, hydrolysis transfers a functional group to water. In oxidation–reduction reactions, the "groups" transferred are electrons, which are passed from an **electron donor (reductant** or **reducing agent)** to an **electron acceptor (oxidant** or **oxidizing agent).** For example, in the reaction

$$Fe^{3+} + Cu^+ \rightleftharpoons Fe^{2+} + Cu^{2+}$$

Cu^+, the reductant, is oxidized to Cu^{2+} while Fe^{3+}, the oxidant, is reduced to Fe^{2+}.

Redox reactions may be divided into two **half-reactions** or **redox couples,** such as

$$Fe^{3+} + e^- \rightleftharpoons Fe^{2+} \quad \text{(reduction)}$$
$$Cu^+ \rightleftharpoons Cu^{2+} + e^- \quad \text{(oxidation)}$$

whose sum is the above whole reaction. These half-reactions occur during oxidative metabolism in the vital mitchondrial electron transfer mediated by **cytochrome *c* oxidase** (Section 20-2C). Note that for electrons to be transferred, both half-reactions must simultaneously occur. In fact, the electrons are the two half-reactions' common intermediate.

Electrochemical Cells

A half-reaction consists of an electron donor and its conjugate electron acceptor; in the oxidation half-reaction shown above, Cu^+ is the electron donor and Cu^{2+} is its conjugate electron acceptor. Together these constitute a **conjugate redox pair** analogous to the conjugate acid – base pair (HA and A^-) of a Brønsted acid (Section 2-2A). An important difference between redox pairs and acid – base pairs, however, is that *the two half-reactions of a redox reaction, each consisting of a conjugate redox pair, may be physically separated so as to form an electrochemical cell* (Fig. 15-26). In such a device, each half-reaction takes place in its separate **half-cell,** and electrons are passed between half-cells as an electric current in the wire connecting their two electrodes. A salt bridge is necessary to complete the electrical circuit by providing a conduit for ions to migrate in the maintenance of electrical neutrality.

The free energy of an oxidation – reduction reaction is particularly easy to determine through a simple measurement of the voltage difference between its two half-cells. Consider the general redox reaction:

$$A_{ox}^{n+} + B_{red} \rightleftharpoons A_{red} + B_{ox}^{n+}$$

in which n electrons per mole of reactants are transferred from reductant (B_{red}) to oxidant (A_{ox}^{n+}). The free energy of this reaction is expressed, according to Eq. [3.15], as

$$\Delta G = \Delta G° + RT \ln \left(\frac{[A_{red}][B_{ox}^{n+}]}{[A_{ox}^{n+}][B_{red}]} \right) \qquad [15.2]$$

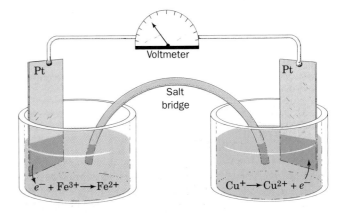

Figure 15-26
An example of an electrochemical cell. The half-cell undergoing oxidation (here $Cu^+ \rightarrow Cu^{2+} + e^-$) passes the liberated electrons through the wire to the half-cell undergoing reduction (here $e^- + Fe^{3+} \rightarrow Fe^{2+}$). Electroneutrality in the two half-cells is maintained by the transfer of ions through the electrolyte-containing salt bridge.

Equation [3.12] indicates that, under reversible conditions,

$$\Delta G = -w' = -w_{el} \qquad [15.3]$$

where w', the nonpressure–volume work, is, in this case, w_{el}, the electrical work required to transfer the n moles of electrons through the electric potential difference $\Delta\mathscr{E}$. This, according to the laws of electrostatics, is

$$w_{el} = n\mathscr{F}\Delta\mathscr{E} \qquad [15.4]$$

where $\mathscr{F}$, the **faraday,** is the electrical charge of one mole of electrons ($1\ \mathscr{F} = 96{,}494\ \mathrm{C\cdot mol^{-1}} = 96{,}494\ \mathrm{J\cdot V^{-1}\,mol^{-1}}$). Thus, substituting Eq. [15.4] into Eq. [15.3],

$$\Delta G = -n\mathscr{F}\Delta\mathscr{E} \qquad [15.5]$$

Combining Eqs. [15.2] and [15.5], and making the analogous substitution for $\Delta G°$, yields the **Nernst equation:**

$$\Delta\mathscr{E} = \Delta\mathscr{E}° - \frac{RT}{n\mathscr{F}} \ln\left(\frac{[A_{red}][B_{ox}^{n+}]}{[A_{ox}^{n+}][B_{red}]}\right) \qquad [15.6]$$

which was originally formulated in 1881 by Walther Nernst. Here $\Delta\mathscr{E}$, the **electromotive force (emf)** or **redox potential,** may be described as the "electron pressure" that the electrochemical cell exerts. The quantity $\Delta\mathscr{E}°$, the redox potential when all components are in their standard states, is called the **standard redox potential.** If these standard states refer to biochemical standard states (Section 3-4B), then $\Delta\mathscr{E}°$ is replaced by $\Delta\mathscr{E}°'$. Note that a postive $\Delta\mathscr{E}$ in Eq. [15.5] results in a negative ΔG; in other words, *a positive $\Delta\mathscr{E}$ is indicative of a spontaneous reaction, one that can do work.*

B. Measurements of Redox Potentials

The free energy change of a redox reaction may be determined, as Eq. [15.5] indicates, by simply measuring its redox potential with a volt meter (Fig. 15-26). Consequently, voltage measurements are commonly employed to characterize the sequence of reactions comprising a metabolic electron-transport pathway (such as mediates, e.g., oxidative metabolism; Chapter 20).

Any redox reaction can be divided into its component half-reactions:

$$A_{ox}^{n+} + ne^- \rightleftharpoons A_{red}$$
$$B_{ox}^{n+} + ne^- \rightleftharpoons B_{red}$$

where, by convention, both half-reactions are written as reductions. These half-reactions can be assigned **reduction potentials,** $\mathscr{E}_A$ and $\mathscr{E}_B$, in accordance with the Nernst equation:

$$\mathscr{E}_A = \mathscr{E}_A° - \frac{RT}{n\mathscr{F}} \ln\left(\frac{[A_{red}]}{[A_{ox}^{n+}]}\right) \qquad [15.7a]$$

$$\mathscr{E}_B = \mathscr{E}_B° - \frac{RT}{n\mathscr{F}} \ln\left(\frac{[B_{red}]}{[B_{ox}^{n+}]}\right) \qquad [15.7b]$$

For the redox reaction of any two half-reactions:

$$\Delta\mathscr{E}° = \mathscr{E}°_{(e^-\,acceptor)} - \mathscr{E}°_{(e^-\,donor)} \qquad [15.8]$$

Thus, when the reaction proceeds with A as the electron acceptor and B as the electron donor, $\Delta\mathscr{E}° = \mathscr{E}_A° - \mathscr{E}_B°$ and similarly for $\Delta\mathscr{E}$.

Reduction potentials, like free energies, must be defined with respect to some arbitrary standard. By convention, **standard reduction potentials** are defined with respect to the **standard hydrogen half-reaction**

$$2H^+ + 2e^- \rightleftharpoons H_2(g)$$

in which H^+ at pH 0, 25°C, and 1 atm is in equilibrium with $H_2(g)$ that is in contact with a Pt electrode. This half-cell is arbitrarily assigned a standard reduction potential $\mathscr{E}°$ of 0 V (1 V = 1 $\mathrm{J\cdot C^{-1}}$). Similarly, $\mathscr{E}°'$ for the hydrogen half-reaction at pH 7 is defined as 0 V [the standard hydrogen half-cell (pH 0) has $\mathscr{E}' = -0.421$ V]. When $\Delta\mathscr{E}$ is positive, ΔG is negative (Eq. [15.5]), indicating a spontaneous process. In combining two half-reactions under standard conditions, the direction of spontaneity therefore involves the reduction of the redox couple with the more positive standard reduction potential. In other words, *the more positive the standard reduction potential, the higher the electron affinity of the redox couple's oxidized form.*

Biochemical Half-Reactions Are Physiologically Significant

The biochemical standard reduction potentials ($\mathscr{E}°'$) of some biochemically important half-reactions are listed in Table 15-4. The oxidized form of a redox couple with a large positive standard reduction potential has a high electron affinity and is a strong electron acceptor (oxidizing agent), whereas its conjugate reductant is a weak electron donor (reducing agent). For example, O_2 is the strongest oxidizing agent in Table 15-4 whereas H_2O, which tightly holds its electrons, is the table's weakest reducing agent. The converse is true of half-reactions with large negative standard reduction potentials. Since electrons spontaneously flow from low to high reduction potentials, they are transferred, under standard conditions, from the reduced products in any half-reaction in Table 15-4 to the oxidized reactants of any half-reaction above it (although this may not occur at a measurable rate in the absence of a suitable enzyme). Note that Fe^{3+} ions of the various cytochromes tabulated in Table 15-4 have significantly different redox potentials. This indicates that *the protein components of redox enzymes play active roles in electron-transfer reactions by modulating the redox potentials of their bound redox-active centers.*

Electron-transfer reactions are of great biological importance. For example, in the mitochondrial electron-transport chain (Section 20-2), the primary source of ATP in eukaryotes, electrons are passed from NADH

Table 15-4
Standard Reduction Potentials of Some Biochemically Important Half-Reactions

Half-Reaction	$\mathscr{E}°'$(V)
$\frac{1}{2}O_2 + 2H^+ + 2e^- \rightleftharpoons H_2O$	0.815
$SO_4^{2-} + 2H^+ + 2e^- \rightleftharpoons SO_3^{2-} + H_2O$	0.48
$NO_3^- + 2H^+ + 2e^- \rightleftharpoons NO_2^- + H_2O$	0.42
Cytochrome a_3 (Fe^{3+}) + $e^- \rightleftharpoons$ cytochrome a_3 (Fe^{2+})	0.385
$O_2(g) + 2H^+ + 2e^- \rightleftharpoons H_2O_2$	0.295
Cytochrome a (Fe^{3+}) + $e^- \rightleftharpoons$ cytochrome a (Fe^{2+})	0.29
Cytochrome c (Fe^{3+}) + $e^- \rightleftharpoons$ cytochrome c (Fe^{2+})	0.254
Cytochrome c_1 (Fe^{3+}) + $e^- \rightleftharpoons$ cytochrome c_1 (Fe^{2+})	0.22
Cytochrome b (Fe^{3+}) + $e^- \rightleftharpoons$ cytochrome b (Fe^{2+}) *(mitochondrial)*	0.077
Ubiquinone + $2H^+ + 2e^- \rightleftharpoons$ ubiquinol	0.045
Fumarate$^-$ + $2H^+ + 2e^- \rightleftharpoons$ succinate$^-$	0.031
FAD + $2H^+ + 2e^- \rightleftharpoons FADH_2$ *(in flavoproteins)*	~0.
Oxaloacetate$^-$ + $2H^+ + 2e^- \rightleftharpoons$ malate$^-$	−0.166
Pyruvate$^-$ + $2H^+ + 2e^- \rightleftharpoons$ lactate$^-$	−0.185
Acetaldehyde + $2H^+ + 2e^- \rightleftharpoons$ ethanol	−0.197
FAD + $2H^+ + 2e^- \rightleftharpoons FADH_2$ *(free coenzyme)*	−0.219
S + $2H^+ + 2e^- \rightleftharpoons H_2S$	−0.23
Lipoic acid + $2H^+ + 2e^- \rightleftharpoons$ dihydrolipoic acid	−0.29
$NAD^+ + H^+ + 2e^- \rightleftharpoons$ NADH	−0.315
$NADP^+ + H^+ + 2e^- \rightleftharpoons$ NADPH	−0.320
Cystine + $2H^+ + 2e^- \rightleftharpoons$ 2 cysteine	−0.340
Acetoacetate$^-$ + $2H^+ + 2e^- \rightleftharpoons \beta$-hydroxybutyrate$^-$	−0.346
$H^+ + e^- \rightleftharpoons \frac{1}{2}H_2$	−0.421
Acetate$^-$ + $3H^+ + 2e^- \rightleftharpoons$ acetaldehyde + H_2O	−0.581

Source: Mostly from Loach, P. A., *in* Fasman, G. D. (Ed.), *Handbook of Biochemistry and Molecular Biology* (3rd ed.), Physical and Chemical Data, Vol. I, pp. 123–130, CRC Press (1976).

(Fig. 12-2) along a series of electron acceptors of increasing reduction potential (many of which are listed in Table 15-4), to O_2. ATP is generated from ADP and P_i by coupling its synthesis to this free energy cascade. *NADH thereby functions as an energy-rich electron-transfer coenzyme.* In fact, the oxidation of one NADH to NAD^+ supplies sufficient free energy to generate three ATPs. The NAD^+/NADH redox couple functions as the electron acceptor in many exergonic metabolite oxidations. In serving as the electron donor in ATP synthesis, it fulfills its cyclic role as a free energy conduit in a manner analogous to ATP. The metabolic roles of redox coenzymes are further discussed in succeeding chapters.

C. Concentration Cells

A concentration gradient has a lower entropy (greater order) than the corresponding uniformly mixed solution and therefore requires the input of free energy for its formation. Consequently, discharge of a concentration gradient is an exergonic process that may be harnessed to drive an ender- gonic reaction. For example, discharge of a proton concentration gradient (generated by the reactions of the electron-transport chain) across the inner mitochondrial membrane, drives the enzymatic synthesis of ATP from ADP and P_i (Section 20-3). Likewise, nerve impulses, which require electrical energy, are transmitted through the discharge of $[Na^+]$ and $[K^+]$ gradients that nerve cells generate across their cell membranes (Section 34-4C). Quantitation of the free energy contained in a concentration gradient is accomplished by use of the concepts of electrochemical cells.

The reduction potential and free energy of a half-cell vary with the concentrations of its reactants. An electrochemical cell may therefore be constructed from two half-cells that contain the same chemical species but at different concentrations. The overall reaction for such an electrochemical cell may be represented

$$A_{ox}^{n+}(\text{half-cell 1}) + A_{red}(\text{half-cell 2}) \rightleftharpoons$$
$$A_{ox}^{n+}(\text{half-cell 2}) + A_{red}(\text{half-cell 1}) \quad [15.9]$$

and, according to the Nernst equation, since $\Delta\mathscr{E}°$ vanishes when the same reaction occurs in both cells

$$\Delta\mathscr{E} = -\frac{RT}{n\mathscr{F}} \ln\left(\frac{[A_{ox}^{n+}(\text{half-cell 2})][A_{red}(\text{half-cell 1})]}{[A_{ox}^{n+}(\text{half-cell 1})][A_{red}(\text{half-cell 2})]}\right)$$

Such **concentration cells** are capable of generating electrical work until they reach equilibrium. This occurs when the concentrations in the half-cells become equal ($K_{eq} = 1$). The reaction, in effect, constitutes a mixing of the two half-cells; the free energy generated is a reflection of the entropy of this mixing.

6. THERMODYNAMICS OF LIFE

One of the last refuges of **vitalism,** the doctrine that biological processes are not bound by the physical laws that govern inanimate objects, was the belief that living things can somehow evade the laws of thermodynamics. This view was partially refuted by elaborate calorimetric measurements on living animals that are entirely consistent with the energy conservation predictions of the first law of thermodynamics. However, the experimental verification of the second law of thermodynamics in living systems is more difficult. It has not been possible to measure the entropy of living matter since the heat, q_P, of a reaction at constant T and P is only equal to $T\Delta S$ if the reaction is carried out reversibly (Eq. [3.8]). Obviously, the dismantling of a living organism to its component molecules for such a measurement would invariably result in its irreversible death. Consequently, the present experimentally verified state of knowledge is that the entropy of living matter is less than that of the products to which it decays.

In this section we consider the special aspects of the thermodynamics of living systems. Knowledge of these matters, which is by no means complete, has enhanced our understanding of how metabolic pathways are regulated, how cells respond to stimuli, and how organisms grow and change with time.

A. Living Systems Cannot Be at Equilibrium

Classical or **equilibrium thermodynamics** (Chapter 3) applies largely to reversible processes in closed systems. The fate of any closed system, as we discussed in Section 3-4A, is that it must inevitably reach equilibrium. For example, if its reactants are in excess, the forward reaction will proceed faster than the reverse reaction until equilibrium is attained ($\Delta G = 0$). In contrast, open systems may remain in a nonequilibrium state as long as they are able to acquire free energy from their surroundings in the form of reactants, heat or work. While classical thermodynamics provides invaluable information concerning open systems by indicating if a given process can occur spontaneously, further thermo-

dynamic analysis of open systems requires the application of the relatively recently elucidated principles of **nonequilibrium** or **irreversible thermodynamics.** In contrast to classical thermodynamics, this theory explicitly takes time into account.

Living organisms are open systems and therefore can never be at equilibrium. As indicated above, they continuously ingest high-enthalpy, low-entropy nutrients, which they convert to low-enthalpy, high-entropy waste products. The free energy resulting from this process is used to do work and to produce the high degree of organization characteristic of life. If this process is interrupted, the organism ultimately reaches equilibrium, which for living things is synonymous with death. For example, one theory of aging holds that senescence results from the random but inevitable accumulation in cells of genetic defects that interfere with and ultimately disrupt the proper functioning of living processes. [The theory does not, however, explain how single-celled organisms or the germ cells of multicellular organisms (sperm and ova), which are in effect immortal, are able to escape this so-called **error catastrophe.**]

Living systems must maintain a nonequilibrium state for several reasons:

1. Only a nonequilibrium process can perform useful work.

2. The intricate regulatory functions characteristic of life require a nonequilibrium state because a process at equilibrium cannot be directed (similarly, a ship that is dead in the water will not respond to its rudder).

3. The complex cellular and molecular systems that conduct biological processes can only be maintained in the nonequilibrium state. Living systems are inherently unstable because they are degraded by the very biochemical reactions to which they give rise. Their regeneration, which must occur almost simultaneously with their degradation, requires the continuous influx of free energy. For example, the ATP-generating consumption of glucose (Section 16-2), as has been previously mentioned, occurs with the initial consumption of ATP through its reactions with glucose to form glucose-6-phosphate and with fructose-6-phosphate to form fructose-1,6-bisphosphate. Consequently, if metabolism is suspended long enough to exhaust the available ATP supply, glucose metabolism cannot be resumed. Life therefore differs in a fundamental way from a complex machine such as a computer. Both require a throughput of free energy to be active. However, the function of the machine is based on a static structure so that the machine can be repeatedly switched on and off. Life, in contrast, is based on a self-destructing but self-renewing process which, once interrupted, cannot be reinitiated.

B. Nonequilibrium Thermodynamics and the Steady State

In a nonequilibrium process, something (such as matter, electrical charge, or heat) must flow, that is, change its spatial distribution. In classical mechanics, the acceleration of mass occurs in response to force. *Similarly, flow in a thermodynamic system occurs in response to a thermodynamic force (driving force), which results from the system's nonequilibrium state.* For example, the flow of matter in diffusion is motivated by the thermodynamic force of a concentration gradient; the migration of electrical charge (electric current) occurs in response to a gradient in an electric field (a voltage difference); the transport of heat results from a temperature gradient; and a chemical reaction results from a difference in chemical potential. Such flows are said to be **conjugate** to their thermodynamic force.

A thermodynamic force may also promote a **nonconjugate flow** under the proper conditions. For example, a gradient in the concentration of matter can give rise to an electric current (a concentration cell), heat (such as occurs upon mixing H_2O and HCl), or a chemical reaction (the mitochondrial production of ATP through the dissipation of a proton gradient). Similarly, a gradient in electrical potential can motivate a flow of matter (electrophoresis), heat (resistive heating), or a chemical reaction (the charging of a battery). When a thermodynamic force stimulates a nonconjugate flow, the process is called an **energy transduction.**

Living Things Maintain the Steady State

*Living systems are, for the most part, characterized by being in a **steady state**.* By this it is meant that all flows in the system are constant so that the system does not change with time. Some environmental steady state processes are schematically illustrated in Fig. 15-27. Ilya Prigogine, a pioneer in the development of irreversible thermodynamics, has shown that a steady state system produces the maximum amount of useful work for a given energy expenditure under the prevailing conditions. *The steady state of an open system is therefore its state of maximum thermodynamic efficiency.* Furthermore, in analogy with Le Chatelier's principle, slight perturbations from the steady state give rise to changes in flows that counteract these perturbations so as to return the system to the steady state. *The steady state of an open system is therefore analogous to the equilibrium state of a closed system; both are stable states.*

In the following chapters it will be seen that many biological regulatory mechanisms function to maintain a steady state. For example, the flow of reaction intermediates through a metabolic pathway is often inhibited by an excess of final product and stimulated by an excess of starting material through the allosteric regulation of its key enzymes (Section 12-4). Living things

have apparently evolved so as to take maximum thermodynamic advantage of their environment.

C. Thermodynamics of Metabolic Control

Enzymes Selectively Catalyze Required Reactions

Biological reactions are highly specific; only reactions that lie on metabolic pathways take place at significant rates despite the many other thermodynamically favorable reactions that are also possible. As an example, let us consider the reactions of ATP, glucose, and water. Two thermodynamically favorable reactions that ATP can undergo are phosphoryl transfer to form ADP and glucose-6-phosphate, and hydrolysis to form ADP and P_i (Fig. 15-19a). The free energy profiles of these reactions are diagrammed in Fig. 15-28. ATP hydrolysis is thermodynamically favored over the phosphoryl transfer to glucose. However, their relative rates are determined by their free energies of activation to their transition states ($\Delta G^{\ddagger}$ values; Section 13-1C) and the relative concentrations of glucose and water. The larger $\Delta G^{\ddagger}$, the slower the reaction. In the absence of enzymes, $\Delta G^{\ddagger}$ for the phosphoryl-transfer reaction is greater than that for hydrolysis so that the hydrolysis reaction predominates (although neither reaction occurs at a biologically significant rate).

The free energy barriers of both of the nonenzymatic reactions are far higher than that of the enzyme-catalyzed phosphoryl transfer to glucose. Hence enzymatic formation of glucose-6-phosphate is kinetically favored over the nonenzymatic hydrolysis of ATP. *It is the role of an enzyme, in this case hexokinase, to selectively reduce the free energy of activation of a chemically coupled reaction so that it approaches equilibrium faster than the more thermodynamically favored uncoupled reaction.*

Many Enzymatic Reactions Are Near Equilibrium

Although metabolism as a whole is a nonequilibrium process, many of its component reactions function close to equilibrium. The reaction of ATP and creatine to form phosphocreatine (Section 15-4C) is an example of such a reaction. The ratio [creatine]/[phosphocreatine] depends on [ATP] because creatine kinase, the enzyme catalyzing this reaction, has sufficient activity to equilibrate the reaction rapidly. The net rate of such an equilibrium reaction is effectively regulated by varying the concentrations of its reactants and/or products.

Pathway Throughput Is Controlled by Enzymes Operating Far from Equilibrium

Other biological reactions function far from equilibrium. For example, the phosphofructokinase reaction (Fig. 15-24) has an equilibrium constant of $K'_{eq} = 300$ but under physiological conditions in rat heart muscle has the mass action ratio [fructose-1,6-bisphosphate][ADP]/[fructose-6-phosphate][ATP] = 0.03, which cor-

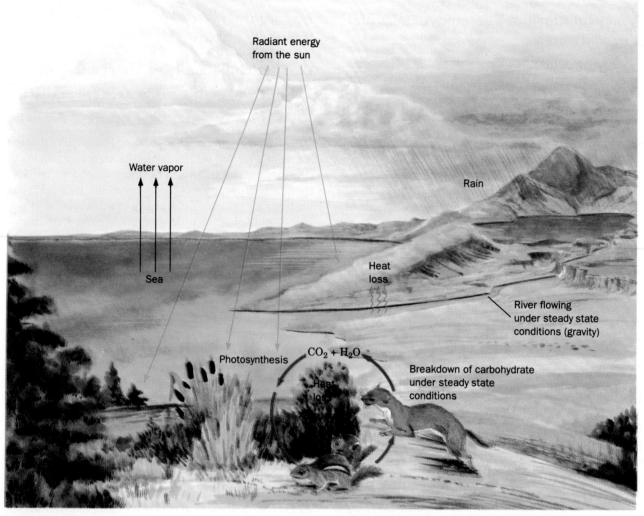

Figure 15-27

Two examples of open systems in a steady state. A constant flow of water in the river occurs under the influence of the force of gravity (*background*). The water level in the reservoir is maintained by rain, the major source of which is the evaporation of seawater. Hence the entire cycle is ultimately powered by the sun. The steady state of the biosphere (*foreground*) is similarly maintained by the sun. Plants harness the sun's radiant energy to synthesize carbohydrates from CO_2 and H_2O. The eventual metabolism of the carbohydrates by the plants or the animals that eat them results in the release of their stored free energy and the return of the CO_2 and H_2O to the environment to complete the cycle.

responds to $\Delta G = -25.7 \text{ kJ} \cdot \text{mol}^{-1}$ (Eq. [3.15]). This situation arises from a buildup of reactants resulting from insufficient phosphofructokinase activity to equilibrate the reaction. Changes in substrate concentrations therefore have relatively little effect on the rate of the phosphofructokinase reaction; the enzyme is essentially saturated. Only changes in the activity of the enzyme, through allosteric interactions, for example, can significantly alter this rate. An enzyme, such as phosphofructokinase, is therefore analogous to a dam on a river. It controls substrate **flux** (rate of flow) by varying its activity (allosterically or by other means), much as a dam controls the flow of a river by varying the opening of its flood gates.

Understanding of how reactant flux in a metabolic

pathway is controlled requires knowledge of which reactions are functioning near equilibrium and which are far from it. Most enzymes in a metabolic pathway operate near equilibrium and therefore have net rates that vary with their substrate concentrations. However, as we shall see in the following chapters (particularly Section 16-4), *certain allosteric enzymes, which are strategically located in a metabolic pathway, operate far from equilibrium. The relative insensitivity of the rates of the reactions catalyzed by such "flux-generating" enzymes to variations in the concentrations of their substrates permits the establishment of a steady state flux of metabolites through the pathway.* This situation, as we have seen, maximizes the pathway's thermodynamic efficiency and allows the flux to be allosterically controlled.

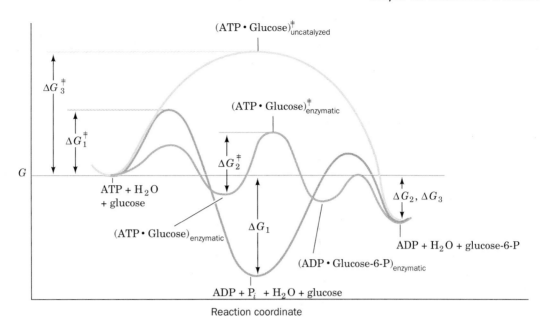

Figure 15-28
Reaction coordinate diagrams for (1) the reaction of ATP and water (*purple curve*), and the reaction of ATP and glucose (2) in the presence (*orange curve*), and (3) in the absence (*yellow curve*) of an appropriate enzyme. Although the hydrolysis of ATP is a more exergonic reaction than the phosphorylation of glucose (ΔG_1 more negative than ΔG_2), the latter reaction is predominant in the presence of a suitable enzyme because it is kinetically favored ($\Delta G_2^\ddagger < \Delta G_1^\ddagger$).

Chapter Summary

Metabolic pathways are series of consecutive enzymatic reactions that produce specific products for use by an organism. The free energy released by degradation (catabolism) is, through the intermediacy of ATP and NADPH, used to drive the endergonic processes of biosynthesis (anabolism). Carbohydrates, lipids, and proteins are all converted to the common intermediate acetyl-CoA whose acetyl group is then converted to CO_2 and H_2O through the action of the citric acid cycle and oxidative phosphorylation. A relatively few metabolites serve as starting materials for a host of biosynthetic products. Metabolic pathways have four principal characteristics: (1) metabolic pathways are irreversible so that if two metabolites are interconvertible, the synthetic route from the first to the second must differ from the route from the second to the first; (2) every metabolic pathway has an exergonic first committed step; (3) all metabolic pathways are regulated, usually at the first committed step; and (4) metabolic pathways in eukaryotes occur in specific subcellular compartments.

Almost all metabolic reactions fall into four categories: (1) group-transfer reactions; (2) oxidation–reduction reactions; (3) eliminations, isomerizations, and rearrangements; and (4) reactions that make or break carbon–carbon bonds. Most of these reactions involve heterolytic bond cleavage or formation occurring through attack of nucleophiles on electrophilic carbon atoms. Group-transfer reactions therefore involve transfer of an electrophilic group from one nucleophile to another. The main electrophilic groups transferred are acyl groups, phosphoryl groups, and glycosyl groups. The most common nucleophiles are amino, hydroxyl, imidazole, and sulfhydryl groups. Oxidation–reduction reactions involve loss or gain of electrons. Oxidation at carbon usually involves C—H bond cleavage with the ultimate loss by C of the two bonding electrons through their transfer to an electron acceptor such as NAD^+. The terminal electron acceptor in aerobes is O_2. Elimination reactions are those in which a C=C double bond is created from two saturated carbon centers with the loss of H_2O, NH_3, ROH, or RNH_2. Dehydration reactions are the most common eliminations. Isomerizations involve shifts of double bonds within molecules. Rearrangements are biochemically uncommon reactions in which intramolecular C—C bonds are broken and reformed to produce new carbon skeletons. Reactions that make and break C—C bonds form the basis of both degradative and biosynthetic metabolism. In the synthetic direction, these reactions involve attack of a nucleophilic carbanion on an electrophilic carbon atom. The most common electrophilic carbon atom is the carbonyl carbon while carbanions are usually generated by removal of a proton from a carbon atom adjacent to a carbonyl group or by decarboxylation of a β-keto acid.

Experimental approaches employed in elucidation of metabolic pathways include the use of metabolic inhibitors, growth studies, and biochemical genetics. Metabolic inhibitors block pathways at specific enzymatic steps. Identification of the resulting intermediates indicates the course of the pathway. Mutations, which occur naturally in genetic diseases or can be induced in microorganisms by mutagens, X-rays, or genetic engineering, may also result in the absence or inactivity of an enzyme. When isotopic labels are incorporated into metabolites and allowed to enter a metabolic system, their paths may be traced from the distribution of label in the intermediates.

Studies on isolated organs, tissue slices, cells, and subcellular organelles have contributed enormously to our knowledge of the localization of metabolic pathways.

Free energy is supplied to endergonic metabolic processes by the ATP produced via exergonic metabolic processes. ATP's -30.5 kJ $\cdot$ mol^{-1} $\Delta G^{\circ\prime}$ of hydrolysis, is intermediate between those of "high-energy" metabolites such as phosphoenol pyruvate and "low-energy" metabolites such as glucose-6-phosphate. The "high-energy" phosphoryl groups are enzymatically transferred to ADP and the resulting ATP, in a separate reaction, phosphorylates "low-energy" compounds. ATP may also undergo pyrophosphate cleavage to yield PP$_i$ whose subsequent hydrolysis adds further thermodynamic impetus to the reaction. ATP is present in too short a supply to act as an energy reservoir. This function, in vertebrate nerve and muscle cells, is carried out by phosphocreatine, which under low ATP conditions, readily transfers its phosphoryl group to ADP to form ATP.

The half-reactions of redox reactions may be physically separated to form an electrochemical cell. The reduction potential for the reduction of A by B,

$$A_{ox}^{n+} + B_{red} \rightleftharpoons A_{red} + B_{ox}^{n+}$$

in which n electrons are transferred, is given by the Nernst equation

$$\Delta \mathscr{E} = \Delta \mathscr{E}^{\circ} - \frac{RT}{n\mathscr{F}} \ln \left(\frac{[A_{red}][B_{ox}^{n+}]}{[A_{ox}^{n+}][B_{red}]} \right)$$

The change in reduction potential of such a reaction is related to the reduction potentials of its component half-reactions, $\mathscr{E}_A$ and $\mathscr{E}_B$, by

$$\Delta \mathscr{E} = \mathscr{E}_A - \mathscr{E}_B$$

If $\mathscr{E}_A > \mathscr{E}_B$, then A_{ox}^{n+} has a greater electron affinity than does B_{ox}^{n+}. The reduction potential scale is defined by arbitrarily setting the reduction potential of the standard hydrogen half-cell to zero. Redox reactions are of great metabolic importance. For example, the oxidation of NADH yields three ATPs through the mediation of the electron-transport chain.

Living organisms are open systems and therefore cannot be at equilibrium. They must continuously dissipate free energy in order to carry out their various functions and to preserve their highly ordered structures. The study of nonequilibrium thermodynamics has indicated that the steady state, which living processes maintain, is the state of maximum efficiency under the constraints governing open systems. Control mechanisms that regulate biological processes preserve the steady state by regulating the activities of enzymes that are strategically located in metabolic pathways.

References

Metabolic Studies

Beadle, G. W., Biochemical genetics, *Chem. Rev.* **37,** 15–96 (1945). [A classical review summarizing the "one gene–one enzyme" hypothesis.]

Cooper, T. G., *The Tools of Biochemistry*, Chapter 3, Wiley–Interscience (1977). [A presentation of radiochemical techniques.]

Freifelder, D., *Biophysical Chemistry* (2nd ed.), Chapters 5 and 6, Freeman (1982). [A discussion of the principles of radioactive counting and autoradiography.]

Fruton, J. S. and Simmons, S., *General Biochemistry*, Chapter 16, Wiley (1958). [Outlines the classical methods for the study of intermediate metabolism.]

Hevesy, G., Historical sketch of the biological application of tracer elements, *Cold Spring Harbor Symp. Quant. Biol.* **13,** 129–150 (1948).

Shemin, D. and Rittenberg, D., The biological utilization of glycine for the synthesis of the protoporphyrin of hemoglobin, *J. Biol. Chem.* **166,** 621–625 (1946).

Suckling, K. E. and Suckling, C. J., *Biological Chemistry*, Cambridge University Press (1980). [Presents the organic chemistry of biochemical reactions.]

Walsh, C., *Enzymatic Reaction Mechanisms*, Chapter 1, Freeman (1979). [A discussion of the types of biochemical reactions.]

Westheimer, F. H., Why nature chose phosphates, *Science* **235,** 1173–1178 (1987).

Bioenergetics

Alberty, R. A., Standard Gibbs free energy, enthalpy and entropy changes as a function of pH and pMg for reactions involving adenosine phosphates, *J. Biol. Chem.* **244,** 3290–3302 (1969).

Caplan, S. R., Nonequilibrium thermodynamics and its application to bioenergetics, *Curr. Top. Bioenerg.* **4,** 1–79 (1971).

Crabtree, B. and Taylor, D. J., Thermodynamics and metabolism, *in* Jones, M. N. (Ed.), *Biochemical Thermodynamics*, pp. 333–378, Elsevier (1979).

Dickerson, R. E., *Molecular Thermodynamics*, Chapter 7, Benjamin (1969). [An interesting chapter on the thermodynamics of life.]

Henley, H. J. M., An introduction to nonequilibrium thermodynamics, *J. Chem. Educ.* **41,** 647–655 (1964).

Katchelsky, A. and Curran, P. F., *Nonequilibrium Thermodynamics in Biophysics*, Harvard University Press (1965).

Lehninger, A. L., *Bioenergetics* (2nd ed.), Benjamin (1972). [An introductory work by one of the field's originators.]

Morowitz, H. J., *Foundations of Bioenergetics*, Academic Press (1978).

Problems

1. Glycolysis (glucose breakdown) has the overall stoichiometry:

$$\text{Glucose} + 2\text{ADP} + 2\text{P}_i + 2\text{NAD}^+ \longrightarrow$$
$$2\text{pyruvate} + 2\text{ATP} + 2\text{NADH} + 2\text{H}^+ + 2\text{H}_2\text{O}$$

whereas that of gluconeogenesis (glucose synthesis) is

$$2\text{Pyruvate} + 4\text{ATP} + 2\text{NADH} + 2\text{H}^+ + 4\text{H}_2\text{O} \longrightarrow$$
$$\text{glucose} + 4\text{ADP} + 4\text{P}_i + 2\text{NAD}^+$$

What is the overall stoichiometry of the glycolytic breakdown of 1 mol of glucose followed by its gluconeogenic synthesis? Explain why it is necessary that the pathways of these two processes be independently controlled and why they must differ by at least one reaction.

2. It has been postulated that a trigonal bipyrimidal pentacovalent phosphorus intermediate can undergo a vibrational deformation process known as **pseudorotation** in which its apical ligands exchange with two of its equatorial ligands via a tetragonal pyrimidal transition state:

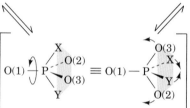

Trigonal bipyramid
[X and Y apical]

Trigonal bipyramid
[O(2) and O(3) apical]

Tetragonal pyramidal
transition state

In a nucleophilic substitution reaction, would two cycles of pseudorotation, so as to place the leaving group (X) in an apical position and the attacking group (Y) in an equatorial position, lead to retention or inversion of configuration on the departure of the leaving group?

3. One **Curie (Ci)** of radioactivity is defined as 3.70×10^{10} disintegrations per second, the number that occurs in 1 g of pure ^{226}Ra. A sample of $^{14}\text{CO}_2$ has a specific radioactivity of 5 $\mu\text{Ci} \cdot \mu\text{mol}^{-1}$. What percentage of its C atoms are ^{14}C?

4. In the hydrolysis of ATP to ADP and P_i, the equilibrium concentration of ATP is too small to enable it to be measured accurately. A better way of determining K'_{eq}, and hence $\Delta G^{\circ\prime}$ of this reaction, is to break it up into two steps whose values of $\Delta G^{\circ\prime}$ can be accurately determined. This has been done using the following pair of reactions (the first being catalyzed by **glutamine synthetase**):

(1) $\text{ATP} + \text{Glutamate} + \text{NH}_3^+ \rightleftharpoons \text{ADP} + \text{P}_i +$
  $\text{glutamine} + \text{H}^+ \quad \Delta G_1^{\circ\prime} = -16.3 \text{ kJ} \cdot \text{mol}^{-1}$

(2) $\text{Glutamate} + \text{NH}_3^+ \rightleftharpoons \text{glutamine} + \text{H}_2\text{O} + \text{H}^+$
  $\Delta G_2^{\circ\prime} = 14.2 \text{ kJ} \cdot \text{mol}^{-1}$

What is the $\Delta G^{\circ\prime}$ of ATP hydrolysis according to this data?

*5. Consider the reaction catalyzed by hexokinase:

$$\text{ATP} + \text{Glucose} \rightleftharpoons \text{ADP} + \text{glucose-6-phosphate}$$

A mixture containing 40 mM ATP and 20 mM glucose was incubated with hexokinase at pH 7 and 25°C. Calculate the equilibrium concentrations of the reactants and products (see Table 15-3).

6. In aerobic metabolism, glucose is completely oxidized in the reaction

$$\text{Glucose} + 6\,\text{O}_2 \rightleftharpoons 6\text{CO}_2 + 6\text{H}_2\text{O}$$

with the coupled generation of 38ATP molecules from $38\text{ADP} + 38\text{P}_i$. Assuming the ΔG for the hydrolysis of ATP to ADP and P_i under intracellular conditions is $-50 \text{ kJ} \cdot \text{mol}^{-1}$ and that for the combustion of glucose is $-2823.2 \text{ kJ} \cdot \text{mol}^{-1}$, what is the efficiency of the glucose oxidation reaction in terms of the free energy sequestered in the form of ATP?

7. Typical intracellular concentrations of ATP, ADP, and P_i in muscles are 5.0, 0.5, and 1.0 mM, respectively. At 25°C and pH 7: (a) What is the free energy of hydrolysis of ATP at these concentrations? (b) Calculate the equilibrium concentration ratio of phosphocreatine to creatine in the creatine kinase reaction:

$$\text{Creatine} + \text{ATP} \rightleftharpoons \text{phosphocreatine} + \text{ADP}$$

if ATP and ADP have the above concentrations. (c) What concentration ratio of ATP to ADP would be required under the foregoing conditions to yield an equilibrium concentration ratio of phosphocreatine to creatine of 1? Assuming the concentration of P_i remained 1.0 mM, what would the free energy of hydrolysis of ATP be under these latter conditions?

*8. Assuming the intracellular concentrations of ATP, ADP, and P_i, are those given in Problem 7: (a) Calculate the concentration of AMP at pH 7 and 25°C under the condition that the adenylate kinase reaction:

$$2\text{ADP} \rightleftharpoons \text{ATP} + \text{AMP}$$

is at equilibrium. (b) Calculate the equilibrium concentration of AMP when the free energy of hydrolysis of ATP to ADP and P_i is $-55 \text{ kJ} \cdot \text{mol}^{-1}$. Assume $[\text{P}_i]$ and $([\text{ATP}] + [\text{ADP}])$ remain constant.

9. Using the data in Table 15-4, list the following substances in order of their increasing reducing power: (a) fumarate$^-$, (b) cystine, (c) O_2, (d) NADP$^+$, (e) cytochrome c (Fe^{3+}), and (f) lipoic acid.

10. Calculate the equilibrium concentrations of reactants and products for the reactions:

Acetoacetate$^-$ + NADH + H$^+$ $\rightleftharpoons$
$$\beta\text{-hydroxybutyrate}^- + \text{NAD}^+$$

when the initial concentrations of acetoacetate$^-$ and NADH are 0.01 and 0.005M, respectively, and β-hydroxybutyrate and NAD$^+$ are initially absent. Assume the reaction takes place at 25°C and pH 7.

11. In anaerobic bacteria, the final metabolic electron acceptor is some molecule other than O$_2$. A major requirement for any redox pair utilized as a metabolic free energy source is that it provides sufficient free energy to generate ATP from ADP and P$_i$. Indicate which of the following redox pairs are sufficiently exergonic to enable a properly equipped bacterium to utilize them as a major energy source. Assume that redox reactions forming ATP require two electrons and that $\Delta\mathscr{E} = \Delta\mathscr{E}°'$.

(a) Ethanol + NO$_3^-$ (c) H$_2$ + S

(b) Fumarate$^-$ + SO$_3^{2-}$ (d) Acetaldehyde + acetaldehyde

12. Calculate $\Delta G°'$ for the following pairs of half-reactions at pH 7 and 25°C. Write a balanced equation for the overall reaction and indicate the direction in which it occurs spontaneously under standard conditions.

(a) (H$^+$/$\frac{1}{2}$H$_2$)   and   ($\frac{1}{2}$O$_2$ + H$_2$/H$_2$O)

(b) (pyruvate$^-$ + 2H$^+$/lactate$^-$)   and (NAD$^+$ + H$^+$/NADH)

*13. The chemiosmotic hypothesis (Section 20-3) postulates that ATP is generated in the two-electron reaction:

ADP + P$_i$ + 2H$^+$(low pH) $\rightleftharpoons$
$$\text{ATP} + \text{H}_2\text{O} + 2\text{H}^+(\text{high pH})$$

which is driven by a metabolically generated pH gradient in the mitochondria. What is the magnitude of the pH gradient required for net synthesis of ATP at 25°C and pH 7, if the steady state concentrations of ATP, ADP, and AMP are 0.01, 10, and 10 mM, respectively?

14. Gastric juice is 0.15M HCl. The blood plasma, which is the source of this H$^+$ and Cl$^-$, is 0.10M in Cl$^-$ and has a pH of 7.4. Calculate the free energy necessary to produce the HCl in 0.1 L of gastric juice at 37°C.

At this point we commence our discussions of specific metabolic pathways by considering **glycolysis** (Greek: *glykos*, sweet; *lysis*, loosening), the pathway by which **glucose** is converted via **fructose-1,6-bisphosphate** to **pyruvate** with the generation of 2 mol of ATP/mol of glucose. This sequence of 10 enzymatic reactions, which is probably the most completely understood biochemical pathway, plays a key role in energy metabolism by providing a significant portion of the energy utilized by most organisms and by preparing glucose, as well as other carbohydrates, for oxidative degradation.

In our study of glycolysis, and indeed of all of metabolism, we shall attempt to understand the pathway on four levels:

1. The chemical interconversion steps, that is, the sequence of reactions by which glucose is converted to the pathway's end products.

2. The mechanism of the enzymatic conversion of each pathway intermediate to its successor.

3. The energetics of the conversions.

4. The mechanisms controlling the flux (rate of flow) of metabolites through the pathway.

The flux of metabolites through a pathway is remarkably sensitive to the needs of the organism for the products of the pathway. Through an exquisitely complex network of control mechanisms, flux through a pathway is only as great as required.

1. THE GLYCOLYTIC PATHWAY

An overview of glucose metabolism is diagrammed in Fig. 16-1. *Under aerobic conditions, the pyruvate formed by glycolysis is further oxidized by the citric acid cycle (Chapter 19) and oxidative phosphorylation (Chapter 20) to CO_2 and water. Under anaerobic conditions, however, the pyruvate is instead converted to a reduced end product, which is **lactate** in muscle (**homolactic fermentation;** a fermentation is an anaerobic biological reaction process) and ethanol + CO_2 in yeast (**alcoholic fermentation**).*

A. Historical Perspective

The fermentation of glucose to ethanol and CO_2 by yeast (Fig. 16-2) has been a useful process since before the dawn of recorded history. Winemaking and baking both exploit this process. Yet, the scientific investigation of the mechanism of glycolysis began only in the latter half of the nineteenth century.

In the years 1854 to 1864, Louis Pasteur established

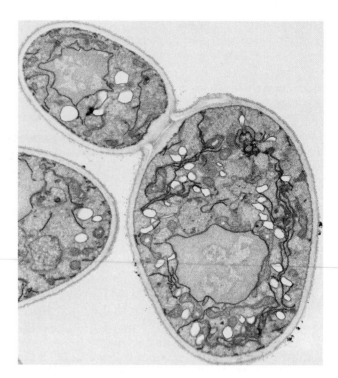

Figure 16-2
An electron micrograph of yeast cells. [Biophoto Associates.]

that fermentation is caused by microorganisms. It was not until 1897, however, that Eduard Buchner demonstrated that cell-free yeast extracts can also carry out this process. This discovery refuted the then widely held belief that fermentation, and every other biological process, was mediated by some "vital force" inherent in living matter, and thereby brought glycolysis within the province of chemistry. This was a major step in the development of biochemistry as a science. Although, in principle, the use of cell-free extracts enabled a systematic "dissection" of the reactions involved in the pathway, the complete elucidation of the glycolytic pathway was still a long-range project because analytical techniques for the isolation and identification of intermediates and enzymes had to be developed concurrently.

In the years 1905 to 1910, Arthur Harden and William Young made two important discoveries:

1. Inorganic phosphate is required for fermentation and is incorporated into fructose-1,6-bisphosphate, an intermediate in the process.

2. A cell-free yeast extract can be separated, by dialysis, into two fractions that are both required for fermentation: a nondialyzable heat-labile fraction they named **zymase;** and a dialyzable, heat-stable fraction they called **cozymase.** It was later shown by others that zymase is a mixture of enzymes and that cozymase is a mixture of cofactors: coenzymes such as NAD^+, ATP, and ADP, as well as metal ions.

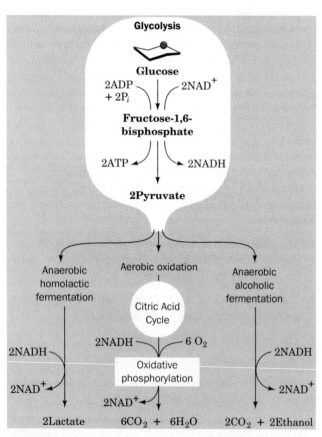

Figure 16-1
Glycolysis converts glucose to pyruvate while generating two ATPs. Under anaerobic conditions, alcoholic fermentation of pyruvate occurs in yeast while homolactic fermentation occurs in muscle. Under aerobic conditions, pyruvate is oxidized to H_2O and CO_2 via the citric acid cycle (Chapter 19) and oxidative phosphorylation (Chapter 20).

In their efforts to identify pathway intermediates, the early investigators of glycolysis developed a general technique of metabolic investigation that is still widely used today: *Reagents are found that inhibit the production of pathway products thereby causing the buildup of metabolites that can then be identified as pathway intermediates.* Over years of investigation that attempted to identify glycolytic intermediates, various reagents were found that inhibit the production of ethanol from glucose in yeast extracts. The use of different inhibitors results in the accumulation of different intermediates. For example, the addition of iodoacetate to fermenting yeast extracts causes the buildup of fructose-1,6-bisphosphate while addition of fluoride ion induces the accumulation of **2-phosphoglycerate** and **3-phosphoglycerate**.

3-Phosphoglycerate   **2-Phosphoglycerate**

The mechanisms by which these inhibitors act are discussed in Sections 16-2F and 16-2I, respectively.

One remarkable finding of these studies was that the same intermediates and enzyme activities could be isolated not only from yeast, but from a great variety of other organisms. With few exceptions (see Problem 10), *living things all metabolize glucose by identical pathways. In spite of their enormous diversity, they share a common biochemistry.*

By 1940, the efforts of many investigators had come to fruition with the elucidation of the complete pathway of glycolysis. Three of these individuals, Gustav Embden, Otto Meyerhof, and Jacob Parnas, have been honored in that glycolysis is alternatively known as the **Embden–Meyerhof–Parnas pathway.** Other major contributors to the elucidation of this pathway were Carl and Gerti Cori, Carl Neuberg, Robert Robison, and Otto Warburg.

B. Pathway Overview

Before beginning our detailed discussion of the enzymes of glycolysis let us first take a moment to survey the overall pathway as it fits in with animal metabolism as a whole. Glucose usually arises in the blood as a result of the breakdown of higher polysaccharides (Sections 10-2B, 10-2C, and 17-1), or from its synthesis from noncarbohydrate sources (gluconeogenesis; Section 21-1). The fate of nonglucose hexoses is discussed in Section 16-5. Glucose enters most cells by a specific carrier that transports it from the exterior of the cell into the cytosol (Section 18-2). *The enzymes of glycolysis are located in the cytosol, where they are only loosely associated, if at all, with cell structures such as membranes, and apparently form no organized complexes with each other.*

Glycolysis converts glucose to two C_3 units (pyruvate) of lower free energy in a process that harnesses the released free energy to synthesize ATP from ADP and P_i. This process requires a pathway of chemically coupled phosphoryl-transfer reactions (Sections 15-4 and 15-6). Thus the chemical strategy of glycolysis is

1. Add phosphoryl groups to the glucose.

2. Chemically convert phosphorylated intermediates into compounds with high phosphate group-transfer potentials.

3. Chemically couple the subsequent hydrolysis of reactive substances to ATP synthesis.

The 10 enzyme-catalyzed reactions of glycolysis are diagrammed in Fig. 16-3. Note that ATP is used early in the pathway to synthesize phosphoryl compounds (Reactions 1 and 3) but is later resynthesized (Reactions 7 and 10). Glycolysis may therefore be considered to occur in two stages:

Stage I (Reactions 1–5): A preparatory stage in which the hexose glucose is phosphorylated and cleaved to yield two molecules of the triose **glyceraldehyde-3-phosphate.** This process utilizes two ATPs in a kind of energy investment.

Stage II (Reactions 6–10): The two molecules of glyceraldehyde-3-phosphate are converted to pyruvate with concomitant generation of four ATPs. Glycolysis therefore has a net "profit" of two ATPs per glucose: Stage I consumes two ATPs; Stage II produces four ATPs.

The overall reaction is

Glucose + 2NAD$^+$ + 2ADP + 2P$_i$ $\longrightarrow$
2NADH + 2pyruvate + 2ATP + 2H$_2$O + 4H$^+$

The enzymes of glycolysis, their subunit compositions, molecular masses, and cofactors, are listed in Table 16-1.

The Oxidizing Power of NAD$^+$ Must Be Recycled

NAD$^+$ is the primary oxidizing agent of glycolysis. The NADH produced by this process (Fig. 16-3, Reaction 6) must be continually reoxidized to keep the pathway supplied with NAD$^+$. There are three common ways that this occurs (Fig. 16-1, *bottom*):

1. Under anaerobic conditions in muscle, NAD$^+$ is regenerated when NADH reduces pyruvate to lactate (homolactic fermentation; Section 16-3A).

2. Under anaerobic conditions in yeast, pyruvate is decarboxylated to acetaldehyde, which is then reduced

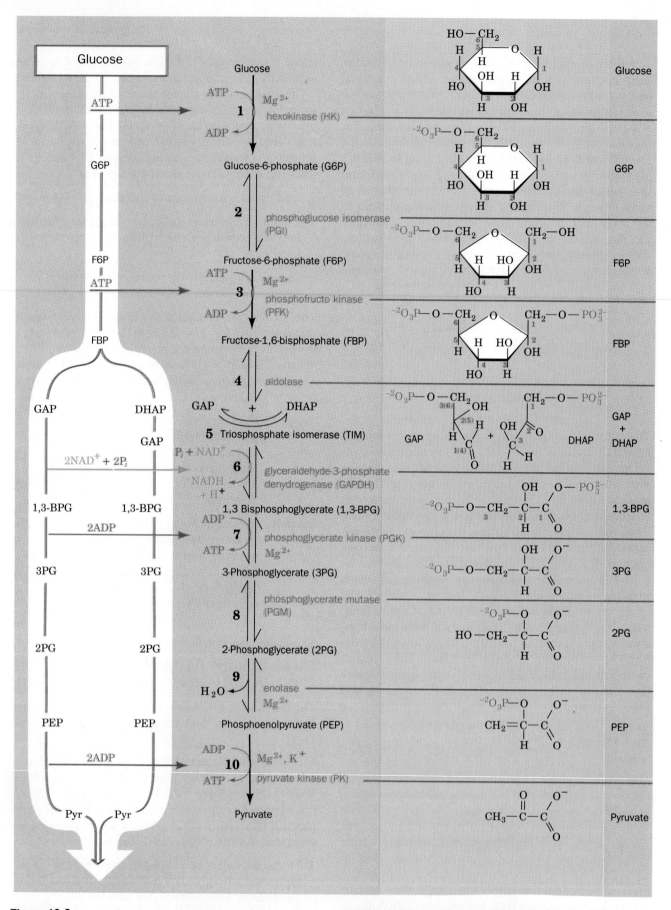

Figure 16-3

The degradation of glucose via the glycolytic pathway. Glycolysis may be considered to occur in two stages. Stage I (Reactions **1–5**): Glucose is phosphorylated and cleaved to form two molecules of the triose glyceraldehyde-3-phosphate. This requires the expenditure of two ATPs in an "energy investment" (Reactions **1** and **3**). Stage II (Reactions **6–10**): The two molecules of glyceraldehyde-3-phosphate are converted to pyruvate with the concomitant generation of four ATPs (Reactions **7** and **10**).

Table 16-1
Enzymes of Glycolysis and Fermentation

Enzyme	Source	Oligomeric Composition	Subunit Molecular Mass (kD)	Cofactors
Hexokinase	Mammalian tissue	Monomer	100	
Glucokinase	Mammalian liver	Monomer	50	
Phosphoglucose isomerase (PGI)	Rabbit muscle	Dimer	61	
Phosphofructokinase (PFK)	Rabbit muscle	Tetramer	78	
Aldolase	Rabbit muscle	Tetramer	40	
Triose phosphate isomerase (TIM)	Chicken muscle	Dimer	27	
Glyceraldehyde-3-phosphate dehydrogenase (GAPDH)	Rabbit muscle	Tetramer	37	NAD^+
Phosphoglycerate kinase (PGK)	Rabbit muscle	Monomer	64	Mg^{2+}
Phosphoglycerate mutase (PGM)	Rabbit muscle	Dimer	27	2,3-BPG
Enolase	Rabbit muscle	Dimer	41	Mg^{2+}
Pyruvate kinase (PK)	Rabbit muscle	Tetramer	57	Mg^{2+}, K^+
Lactate dehydrogenase (LDH)	Crab muscle	Tetramer	35	NAD^+
Pyruvate decarboxylase	Yeast	Tetramer	44	TPP
Yeast alcohol dehydrogenase (YADH)	Yeast	Dimer	37	NAD^+
Liver alcohol dehydrogenase (LADH)	Mammalian liver	Dimer	41	NAD^+, Zn^{2+}

by NADH to ethanol (alcoholic fermentation; Section 16-3B).

3. Under aerobic conditions, the mitochondrial oxidation of each NADH yields three ATPs (Section 20-2A).

Thus in aerobic glycolysis, NADH may be thought of as a ''high-energy'' compound, whereas in anaerobic glycolysis its free energy of oxidation is dissipated as heat.

2. THE REACTIONS OF GLYCOLYSIS

In this section we examine the reactions of glycolysis more closely, describing the properties of the individual enzymes and their mechanisms. In Section 16-3 we consider the anaerobic fate of pyruvate. Finally, in Section 16-4 we consider the thermodynamics of the entire process and address the problem of how the flux (rate of flow) of metabolites through the pathway is controlled. As we study the individual glycolytic enzymes we shall encounter many organic reaction mechanisms (Section 15-2). Indeed, the study of organic reaction mechanisms has been invaluable in understanding the mechanisms by which enzymes catalyze reactions.

A. Hexokinase: First ATP Utilization

Reaction 1 of glycolysis is the transfer of a phosphoryl

group from ATP to glucose to form **glucose-6-phosphate (G6P)** in a reaction catalyzed by **hexokinase.**

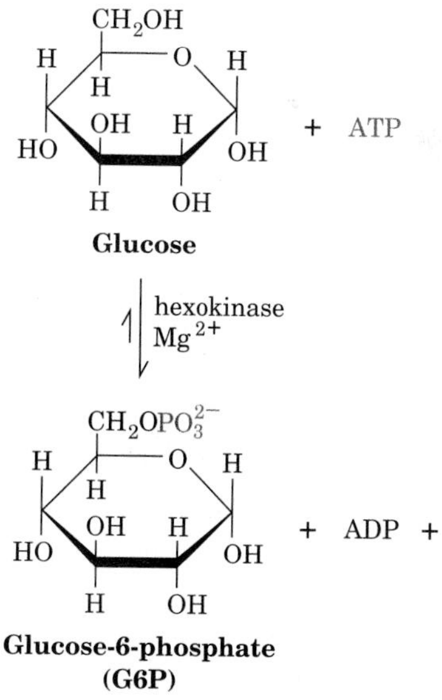

Glucose

hexokinase
Mg^{2+}

Glucose-6-phosphate (G6P)

A **kinase** is an enzyme that transfers phosphoryl groups between ATP and a metabolite (Section 15-4C). The metabolite that serves as the phosphoryl group acceptor for a specific kinase is identified in the prefix of the kinase name. Hexokinase is a relatively nonspecific enzyme contained in all cells that catalyzes the phosphorylation of hexoses such as D-glucose, D-mannose, and D-fructose. Liver cells also contain **glucokinase,** which

Figure 16-4
The nucleophilic attack of the C(6)-OH group of glucose on the γ-phosphate of an Mg^{2+}–ATP complex. The position of the Mg^{2+} ion is shown as an example; its actual position(s) has not been conclusively established. In any case, the Mg^{2+} functions to shield the negatively charged groups of ATP and thereby facilitates the nucleophilic attack.

catalyzes the same reaction but which is specific for glucose and is involved in the maintenance of blood glucose levels (Section 17-3F). The second substrate for hexokinase, as with other kinases, is an Mg^{2+}–ATP complex. In fact, uncomplexed ATP is a potent competitive inhibitor of hexokinase. In what follows, we shall rarely refer to this Mg^{2+} requirement but keep in mind that it is essential for enzymatic activity (other divalent metal ions such as Mn^{2+} often satisfy the metal ion requirements of kinases but Mg^{2+} is the normal physiological species).

Kinetics and Mechanism of the Hexokinase Reaction

Hexokinase has a Random Bi Bi mechanism in which the enzyme forms a ternary complex with glucose and Mg^{2+}–ATP before the reaction occurs. The Mg^{2+}, by complexing with the phosphate oxygen atoms, is thought to shield their negative charges, making the

phosphorus atom more accessible for the nucleophilic attack of the C(6)-OH group of glucose (Fig. 16-4).

An important mechanistic question is why does the kinase catalyze the transfer of a phosphoryl group from ATP to glucose to yield G6P, but not to water to yield $ADP + P_i$ (ATP hydrolysis)? Water is certainly small enough to fit into the phosphoryl acceptor group's enzymatic-binding site. Furthermore, phosphoryl transfer from ATP to 55.5M water, is more exergonic than that to a hexose (Table 15-3). Yet, hexokinase catalyzes phosphoryl transfer to glucose 40,000 times faster than it does to water.

The answer was provided by Thomas Steitz' X-ray structural studies of yeast hexokinase. Comparison of the X-ray structures of hexokinase and the glucose–hexokinase complex indicates that *glucose induces a large conformational change in hexokinase (Fig. 16-5). The two lobes that form its active site cleft swing together by up to 8 Å so as to engulf the glucose in a manner that suggests the closing of jaws. This movement places the ATP in close proximity to the —C(6)H₂OH group of glucose and excludes water from the active site (catalysis by proximity effects; Section 14-1E).* If the catalytic and reacting groups were in the proper position for reaction while the enzyme was in the open position (Fig. 16-5a), ATP hydrolysis would almost certainly be the dominant reaction. This conclusion is confined by the observation that **xylose,** which differs from glucose only by the lack of the —C(6)H₂OH group,

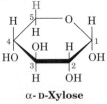

α- D-**Xylose**

(a)   *(b)*

Figure 16-5
A space-filling model of a subunit of (*a*) yeast hexokinase and (*b*) its complex with glucose (*purple*). Note the prominent bilobal appearance of the free enzyme (the C atoms in the small lobe are shaded green, whereas those in the large lobe are light gray). In the enzyme–substrate complex these lobes have swung together so as to engulf the substrate. [Courtesy of Robert Stodola, Fox Chase Cancer Center.]

greatly enhances the rate of ATP hydrolysis by hexokinase (presumably xylose induces the activating conformational change, while water occupies the binding site of the missing hydroxymethyl group). Clearly, *this substrate-induced conformational change in hexokinase is responsible for the enzyme's specificity.* In addition, the active site polarity is reduced by exclusion of water, thereby expediting the nucleophilic reaction process. Other kinases have the same deeply clefted structure as hexokinase (Section 16-2G), and undergo conformational changes upon binding their substrates. This suggests that all kinases have similar mechanisms for maintaining specificity.

B. Phosphoglucose Isomerase

Reaction 2 of glycolysis is the conversion of G6P to **fructose-6-phosphate (F6P)** by **phosphoglucose isomerase (PGI).** This is the isomerization of an aldose to a ketose:

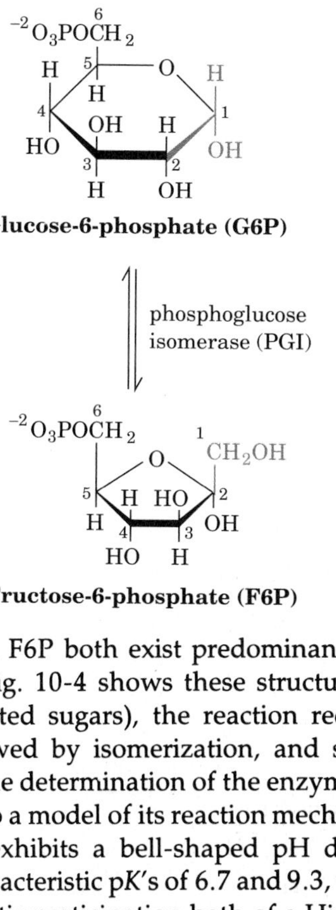

Glucose-6-phosphate (G6P)

phosphoglucose isomerase (PGI)

Fructose-6-phosphate (F6P)

Since G6P and F6P both exist predominantly in their cyclic forms (Fig. 10-4 shows these structures for the unphosphorylated sugars), the reaction requires ring opening, followed by isomerization, and subsequent ring closure. The determination of the enzyme's pH dependence led to a model of its reaction mechanism. The catalytic rate exhibits a bell-shaped pH dependence curve with characteristic pK's of 6.7 and 9.3, which suggests the catalytic participation both of a His and a Lys (Section 13-4).

A proposed reaction mechanism for the phosphoglucose isomerase reaction involves general acid–base catalysis by the enzyme (Fig. 16-6):

Step 1   An acid, presumably the Lys ε-amino group, catalyzes ring opening.

Step 2   A base, presumably the His imidazole ring, abstracts the acidic proton from C(2) to form a

cis-enediolate intermediate (this proton is acidic because it is α to a carbonyl group).

Step 3   The proton is replaced on C(1) in an overall proton transfer. Protons abstracted by bases are labile and exchange rapidly with solvent protons. Nevertheless, Irwin Rose confirmed this step by demonstrating that 2-[³H]G6P is occasionally converted to 1-[³H]F6P by intramolecular proton transfer before the ³H has a chance to exchange with the medium.

Step 4   Ring closure to form the product.

Phosphoglucose isomerase, like most enzymes, catalyzes reactions with absolute stereospecificity. To appreciate this, let us compare the proposed enzymatic reaction mechanism with that in the nonenzymatic base-catalyzed isomerization of glucose, fructose, and mannose (Fig. 16-7). Glucose and mannose differ with respect to their configuration at C(2). In the enediolate intermediate, as well as in fructose, C(2) has no chirality. Therefore, in nonenzymatic systems, base catalyzed isomerization of glucose also results in racemization of C(2) with the production of mannose. In the phosphoglucose isomerase reaction, however, no mannose-6-phosphate is ever formed because the face of the enediolate to which H⁺ must be added to form mannose-6-phosphate is shielded by the enzyme.

C. Phosphofructokinase: Second ATP Utilization

In Reaction 3 of glycolysis, **phosphofructokinase (PFK)** phosphorylates F6P to yield fructose-1,6-bisphosphate [**FBP** or **F1,6P**; previously known as fructose-1,6-diphosphate (FDP)].

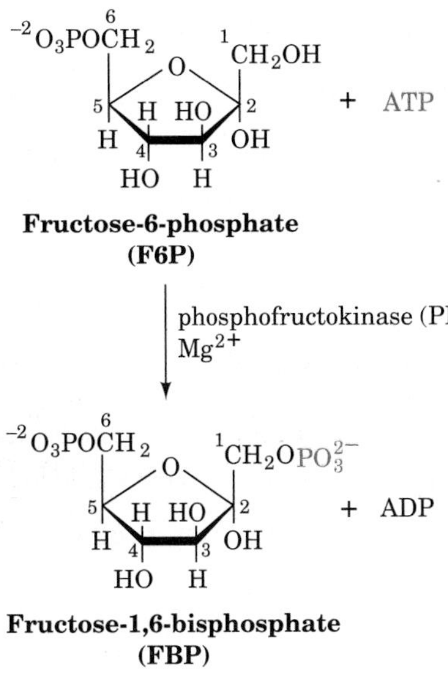

Fructose-6-phosphate (F6P)

phosphofructokinase (PFK)
Mg^{2+}

Fructose-1,6-bisphosphate (FBP)

This reaction is similar to the hexokinase reaction (Reac-

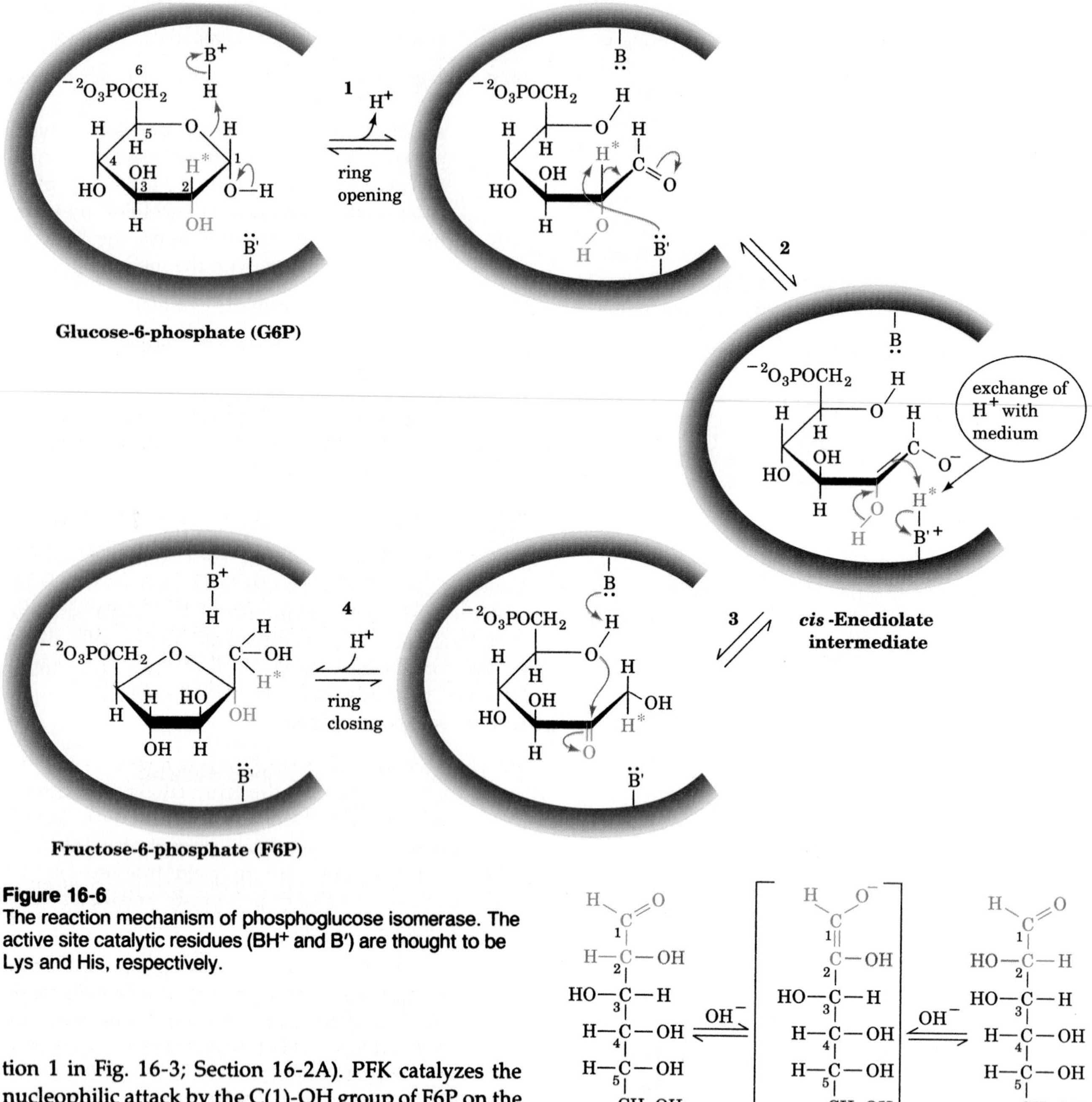

Glucose-6-phosphate (G6P)

Fructose-6-phosphate (F6P)

cis-**Enediolate intermediate**

Figure 16-6
The reaction mechanism of phosphoglucose isomerase. The active site catalytic residues (BH+ and B') are thought to be Lys and His, respectively.

tion 1 in Fig. 16-3; Section 16-2A). PFK catalyzes the nucleophilic attack by the C(1)-OH group of F6P on the electrophilic γ-phosphorus atom of the Mg^{2+}–ATP complex.

PFK plays a central role in control of glycolysis because it catalyzes one of the pathway's rate-determining reactions. In many organisms the activity of PFK is enhanced allosterically by several substances, including AMP, and inhibited allosterically by several other substances, including ATP and citrate. The regulatory properties of

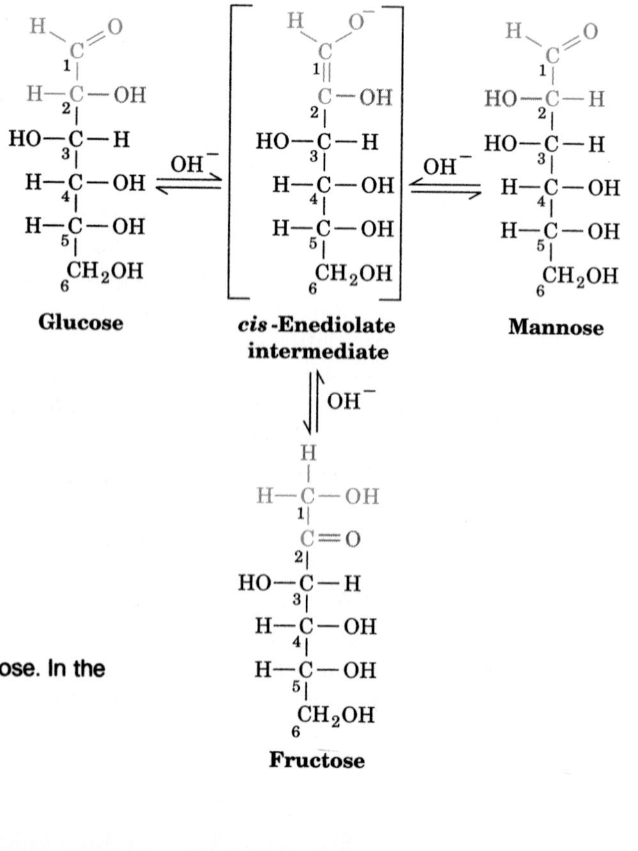

Glucose

cis-**Enediolate intermediate**

Mannose

Fructose

Figure 16-7
The base-catalyzed isomerization of glucose, mannose, and fructose. In the absence of enzyme, this reaction is nonstereospecific.

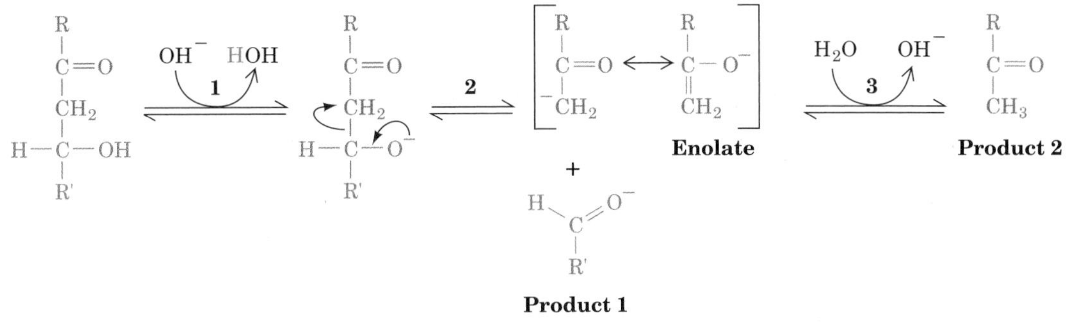

R | C=O | CH$_2$ | H—C—OH | R' → (OH⁻, HOH, 1) R | C=O | CH$_2$ | H—C—O⁻ | R' → (2) [R | C=O | CH$_2$ ↔ R | C—O⁻ || CH$_2$] **Enolate** → (H$_2$O, OH⁻, 3) R | C=O | CH$_3$ **Product 2**

+

H—C(=O⁻) | R' **Product 1**

Figure 16-8
The mechanism for base-catalyzed aldol cleavage. Aldol condensation occurs by the reverse mechanism.

PFK are exquisitely complex; the mechanism by which it exerts control over the glycolytic pathway is examined in Section 16-4B.

D. Aldolase

Aldolase catalyzes Reaction 4 of glycolysis, the cleavage of FBP to form the two trioses **glyceraldehyde-3-phosphate (GAP)** and **dihydroxyacetone phosphate (DHAP).**

Dihydroxyacetone phosphate (DHAP)

CH$_2$OPO$_3^{2-}$ (1) | C=O (2) | HO—CH$_2$ (3)

Fructose 1,6-bisphosphate (FBP)

CH$_2$OPO$_3^{2-}$ (1) | C=O (2) | HO—C—H (3) | H—C—OH (4) | H—C—OH (5) | CH$_2$OPO$_3^{2-}$ (6)

→ aldolase ⇌

+

H—C(=O) (4) | H—C—OH (5) | CH$_2$OPO$_3^{2-}$ (6)

Glyceraldehyde-3-phosphate (GAP)

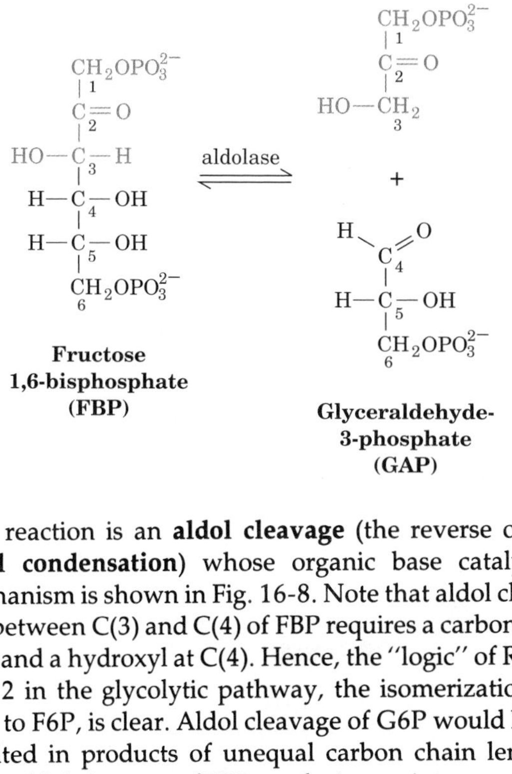

This reaction is an **aldol cleavage** (the reverse of an **aldol condensation**) whose organic base catalyzed mechanism is shown in Fig. 16-8. Note that aldol cleavage between C(3) and C(4) of FBP requires a carbonyl at C(2) and a hydroxyl at C(4). Hence, the "logic" of Reaction 2 in the glycolytic pathway, the isomerization of G6P to F6P, is clear. Aldol cleavage of G6P would have resulted in products of unequal carbon chain length, while aldol cleavage of FBP results in two interconvertible C(3) compounds that can therefore enter a common degradative pathway. The enolate intermediate in the aldol cleavage reaction is stabilized by resonance, as shown, as a result of the electron-withdrawing character of the carbonyl oxygen atom.

There Are Two Mechanistic Classes of Aldolases

Aldol cleavage is catalyzed by stabilizing its enolate intermediate through increased electron delocalization. There are two types of aldolases that are classified according to the chemistry they employ to stabilize the enolate. In Class I aldolases, which occur in animals and plants, the reaction occurs as follows (Fig. 16-9):

Step 1 Substrate binding.

Step 2 Reactions of the FBP carbonyl group with the ε-amino group of the active site Lys to form an iminium cation, that is, a protonated **Schiff base.**

Step 3 C(3)—C(4) bond cleavage resulting in enamine formation and the release of GAP. The iminium ion, as we saw in Section 15-2E, is a better electron-withdrawing group than is the oxygen atom of the precursor carbonyl group. Thus, *catalysis occurs, because the enamine intermediate (Fig. 16-9, Step 3) is more stable than the corresponding enolate intermediate of the base-catalyzed aldol cleavage reaction (Fig. 16-8, Step 2).*

Step 4 Protonation of the enamine to an iminium cation.

Step 5 Hydrolysis of this imine to release DHAP, with regeneration of the free enzyme.

The catalytic participation of the Cys and His residues as acids and bases that facilitate proton transfers (Fig. 16-9), was inferred through the use of appropriate group specific reagents (Section 6-2) that inactivate the enzyme by reacting with these residues. Proof for the formation of the Schiff base was provided by "trapping" ¹⁴C-labeled DHAP on the enzyme by reacting it

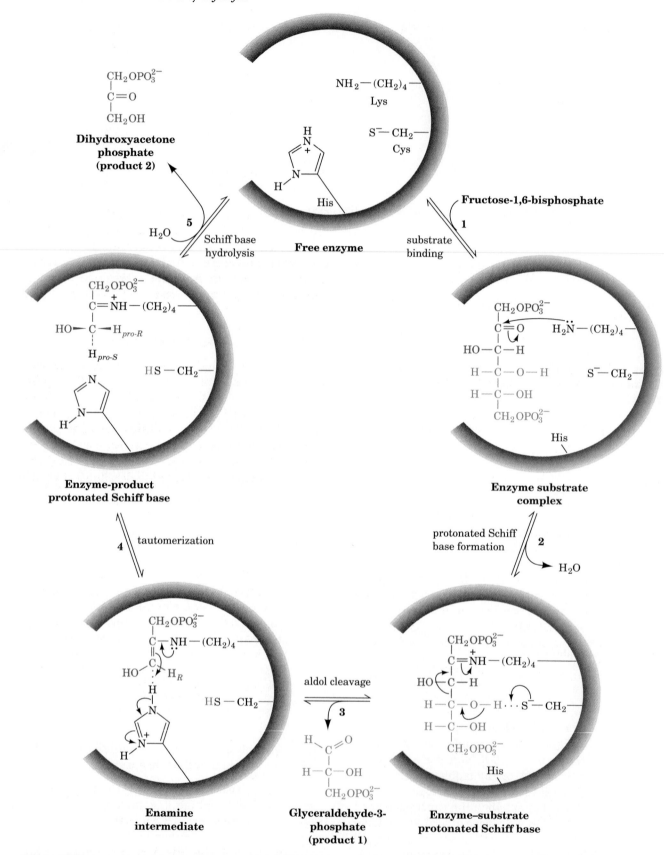

Figure 16-9
The enzymatic mechanism of Class I aldolase. The reaction involves **(1)** substrate binding; **(2)** Schiff base formation between the enzyme's active site Lys residue and FBP; **(3)** aldol cleavage to form an enamine intermediate of the enzyme and DHAP with release of GAP (shown with its *re* face up); **(4)** tautomerization to the imine form of the Schiff base; and **(5)** hydrolysis of the Schiff base with release of DHAP.

with NaBH$_4$, which reduces imines to amines:

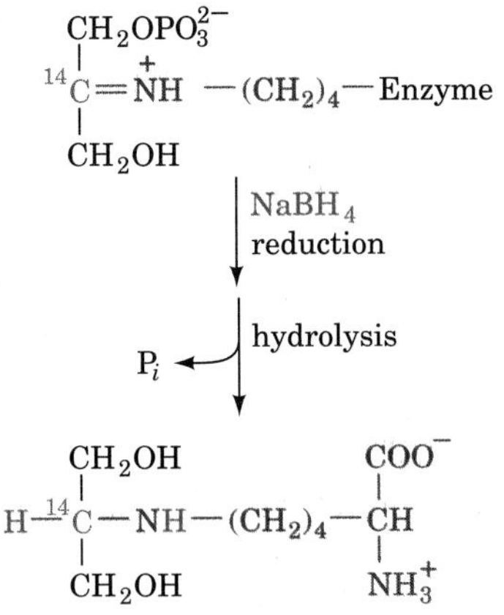

$$^{14}C = \overset{+}{NH} - (CH_2)_4 - Enzyme$$

with side groups $CH_2OPO_3^{2-}$ and CH_2OH

$$\downarrow \text{NaBH}_4 \text{ reduction}$$

$$P_i \downarrow \text{hydrolysis}$$

$$H - ^{14}C - NH - (CH_2)_4 - CH$$

with groups CH_2OH, CH_2OH (left) and COO^-, NH_3^+ (right)

N^6-β-glyceryl lysine

The radioactive product was hydrolyzed and identified as N^6-β-glyceryl lysine.

Class II aldolases, which occur in fungi, algae, and some bacteria, do not form a Schiff base with the substrate. Rather, a divalent cation, usually Zn^{2+}, polarizes the carbonyl oxygen of the substrate to stabilize the enolate intermediate of the reaction:

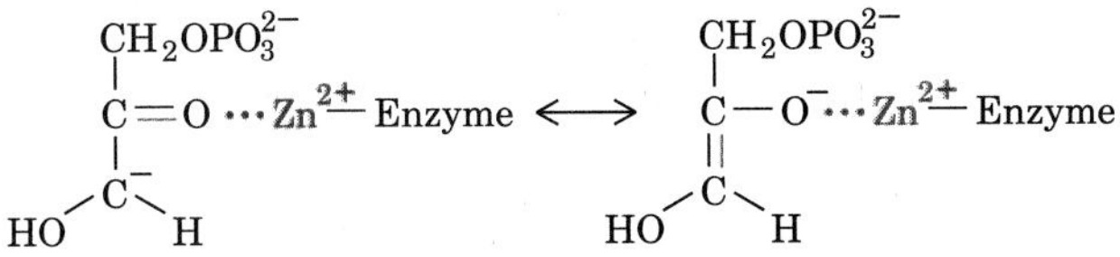

$$\begin{array}{ccc} CH_2OPO_3^{2-} & & CH_2OPO_3^{2-} \\ C=O \cdots Zn^{2+} \text{ Enzyme} & \longleftrightarrow & C-O^- \cdots Zn^{2+} \text{ Enzyme} \\ HO \overset{C}{} H & & HO \overset{C}{} H \end{array}$$

Both classes of aldolases exhibit the Uni Bi kinetics implicit in these mechanisms.

Aldolase Is Stereospecific

The aldolase reaction provides another example of the extraordinary stereospecificity of enzymes. In the nonenzymatic aldol condensation to form hexose-1,6-bisphosphate from DHAP and GAP, there are four possible products depending on whether the *pro-R* or *pro-S* hydrogen at C(3) of DHAP is removed and whether the resulting carbanion attacks GAP on its *re* or *si* face:

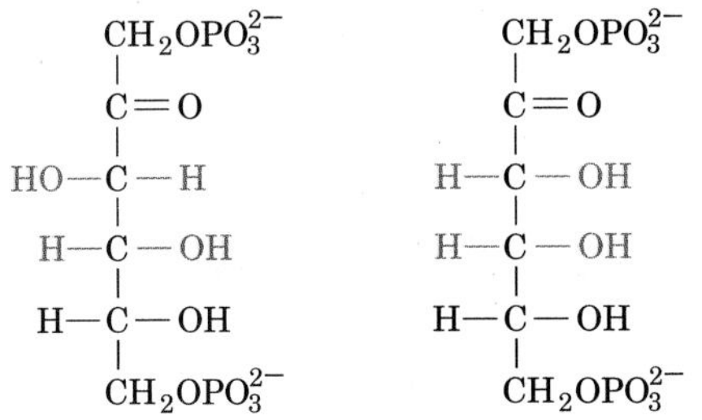

$$\begin{array}{cc} CH_2OPO_3^{2-} & CH_2OPO_3^{2-} \\ C=O & C=O \\ HO-C-H & H-C-OH \\ H-C-OH & H-C-OH \\ H-C-OH & H-C-OH \\ CH_2OPO_3^{2-} & CH_2OPO_3^{2-} \end{array}$$

D-Fructose-1,6-bisphosphate   **D-Psicose-1,6-bisphosphate**

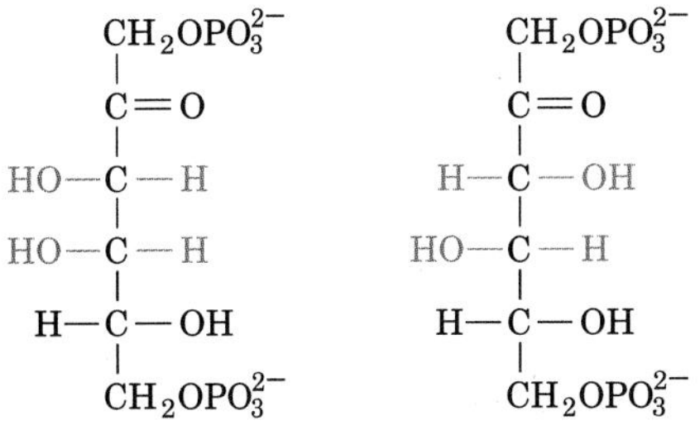

$$\begin{array}{cc} CH_2OPO_3^{2-} & CH_2OPO_3^{2-} \\ C=O & C=O \\ HO-C-H & H-C-OH \\ HO-C-H & HO-C-H \\ H-C-OH & H-C-OH \\ CH_2OPO_3^{2-} & CH_2OPO_3^{2-} \end{array}$$

D-Tagatose-1,6-bisphosphate   **D-Sorbose-1,6-bisphosphate**

In the enzymatic aldol condensation (Fig. 16-9 in reverse), carbanion formation from the enzyme-DHAP iminium ion (Fig. 16-9, Step 4 in reverse) occurs with removal of only the *pro-S* hydrogen. Attack of this carbanion occurs only on the *si* face of the enzyme-bound GAP carbonyl group so that only FBP is formed (Fig. 16-9, Step 3 in reverse).

E. Triose Phosphate Isomerase

Only one of the products of the aldol cleavage reaction, GAP, continues along the glycolytic pathway (Fig. 16-3). However, DHAP and GAP are ketose–aldose isomers just as are F6P and G6P. Interconversion of GAP and DHAP is therefore possible via an enediolate intermediate in analogy with the phosphoglucomutase reaction (Fig. 16-6). **Triose phosphate isomerase (TIM; Fig. 7-19b)** catalyzes this process in Reaction 5 of glycolysis, the final reaction of Stage I.

$$\begin{array}{cc} \overset{1}{C}\!\!\nearrow^{O}_{\searrow H} & \overset{OH}{H-\overset{1}{C}-H} \\ H-\overset{2}{C}-OH & \overset{2}{C}=O \\ \overset{3}{CH_2OPO_3^{2-}} & \overset{3}{CH_2OPO_3^{2-}} \end{array}$$

Glyceraldehyde-3-phosphate (an aldose)   **Dihydroxyacetone phosphate (a ketose)**

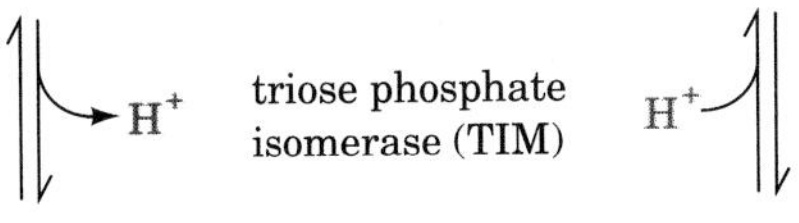

$\rightarrow H^+$ triose phosphate isomerase (TIM) H^+

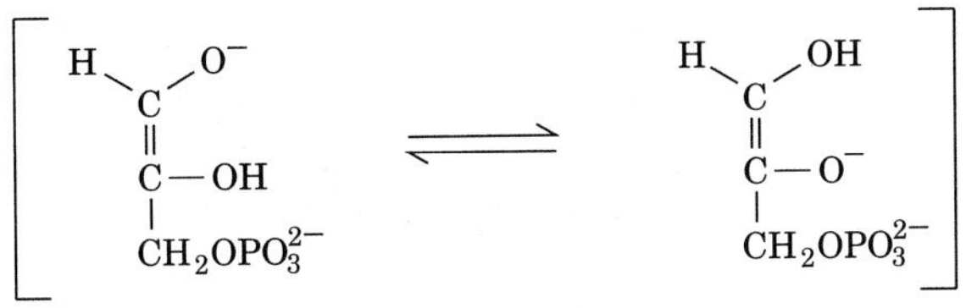

$$\left[\begin{array}{ccc} H\!\!\diagdown_{C}\!\!\diagup^{O^-} & & H\!\!\diagdown_{C}\!\!\diagup^{OH} \\ \| & \rightleftharpoons & \| \\ C-OH & & C-O^- \\ CH_2OPO_3^{2-} & & CH_2OPO_3^{2-} \end{array} \right]$$

Enediolate intermediate

Identification of Active Site Amino Acids

The pH dependence of the TIM reaction is a bell-shaped curve with pK's of 6.5 and 9.5. The similarity of these pK's to the corresponding quantities of the phosphoglucose isomerase reaction suggests the participation of both an acid and a base in the TIM reaction. pH studies alone are difficult to interpret in terms of specific amino acid residues, however, since the environment of the active site may alter the pK of an acidic or basic group (Section 13-4).

Affinity labeling reagents have been employed in an effort to identify the base at the active site of TIM. Both **bromohydroxyacetone phosphate** and **glycidol phosphate**

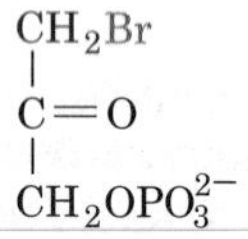

Bromohydroxyacetone phosphate 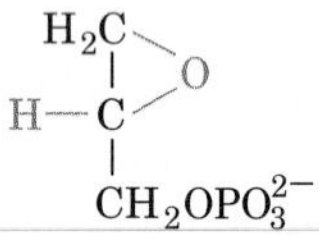   **Glycidol phosphate**

inactivate TIM by forming esters of a specific Glu that X-ray structural studies indicate is in the proper region of the active site to interact with the substrate. Assuming that the carboxyl group of this Glu is, in fact, the general base responsible for the observed pH dependence of TIM activity, its pK is drastically altered from the 4.1 value of the free amino acid to the observed 6.5 value. This provides a striking example of the effect of the environment on the properties of amino acid side chains.

The similarity of the TIM and phosphoglucose isomerase reactions suggests that both enzymes operate by a reaction mechanism that involves the participation of an enediolate intermediate. Support for this idea comes from the use of the transition state analog **phosphoglycohydroxamate,** a stable compound whose geometric structure resembles that of the proposed intermediate:

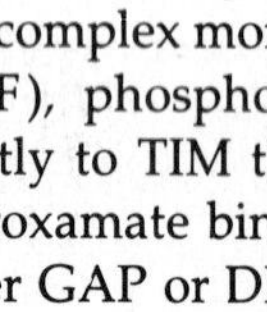

Phospho-glycohydroxamate   **Proposed enediolate intermediate**

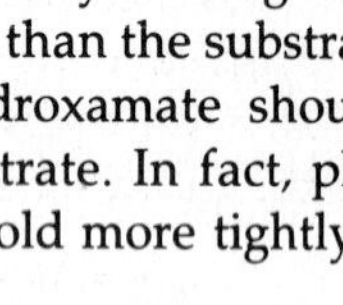

Since enzymes catalyze reactions by binding the transition state complex more tightly than the substrate (Section 14-1F), phosphoglycohydroxamate should bind more tightly to TIM than substrate. In fact, phosphoglycohydroxamate binds 155-fold more tightly to TIM than either GAP or DHAP.

TIM Is a Perfect Enzyme

TIM, as Jeremy Knowles demonstrated, has achieved catalytic perfection in that the rate of bimolecular reaction between enzyme and substrate is diffusion controlled; that is, product formation occurs as rapidly as enzyme and substrate can collide in solution so that any increase in TIM's catalytic efficiency would not increase

the reaction rate (Section 13-2B). Because of the high interconversion efficiency of GAP and DHAP, these two metabolites are maintained in equilibrium: $K = [GAP]/[DHAP] = 4.73 \times 10^{-2}$; that is, [DHAP] is $\gg$[GAP] at equilibrium. However, *as GAP is utilized in the succeeding reaction of the glycolytic pathway, more DHAP is converted to GAP so that these compounds maintain their equilibrium ratio.* One common pathway therefore accounts for the metabolism of both products of the aldolase reaction.

Let us now take stock of where we are in our travels down the glycolytic pathway. At this point, the glucose, which has been transformed into two GAPs, has completed the preparatory stage of glycolysis. This process has required the expenditure of two ATPs. However, this investment has resulted in the conversion of one glucose to two C_3 units, each of which has a phosphoryl group that, with a little chemical artistry, can be converted to a "high-energy" compound (Section 15-14B) whose free energy of hydrolysis can be coupled to ATP synthesis. *This energy investment will be doubly repaid in the final stage of glycolysis in which the two phosphorylated C_3 units are transformed to two pyruvates with the coupled synthesis of four ATPs per glucose.*

F. Glyceraldehyde-3-Phosphate Dehydrogenase: First "High-Energy" Intermediate Formation

Reaction 6 of glycolysis, involves the oxidation and phosphorylation of GAP by NAD^+ and P_i as catalyzed by **glyceraldehyde-3-phosphate dehydrogenase (GAPDH;** Fig. 7-46). This is the first instance of chemical artistry alluded to above. *In this reaction, aldehyde oxidation, an exergonic reaction, drives the synthesis of the acyl phosphate 1,3-bisphosphoglycerate (1,3-BPG; also called 1,3-diphosphoglycerate).* Recall that acyl phosphates are compounds with high phosphate group-transfer potential (Section 15-4B).

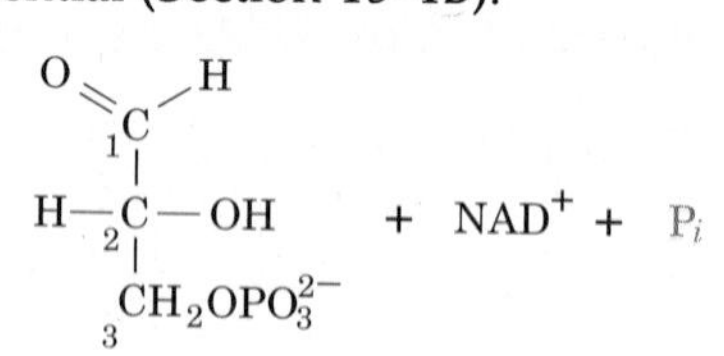

Glyceraldehyde-3-phosphate (GAP)

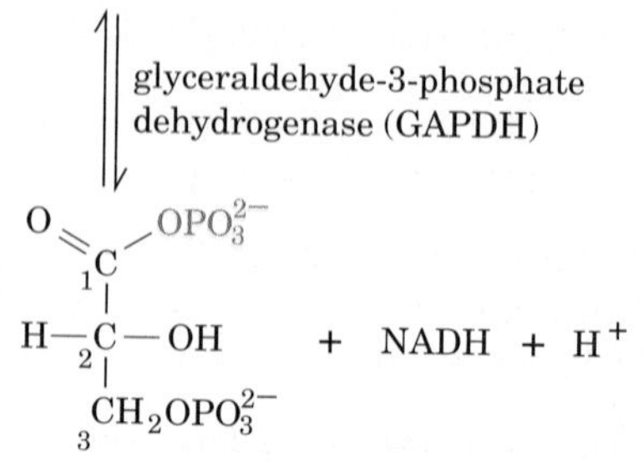

glyceraldehyde-3-phosphate dehydrogenase (GAPDH)

1,3-Bisphoglycerate (1,3-BPG)

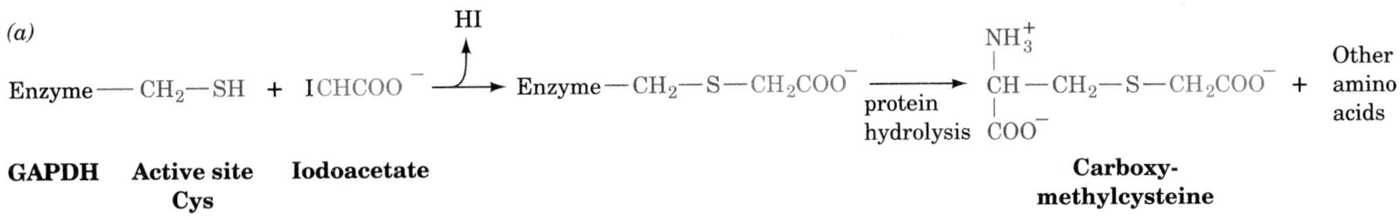

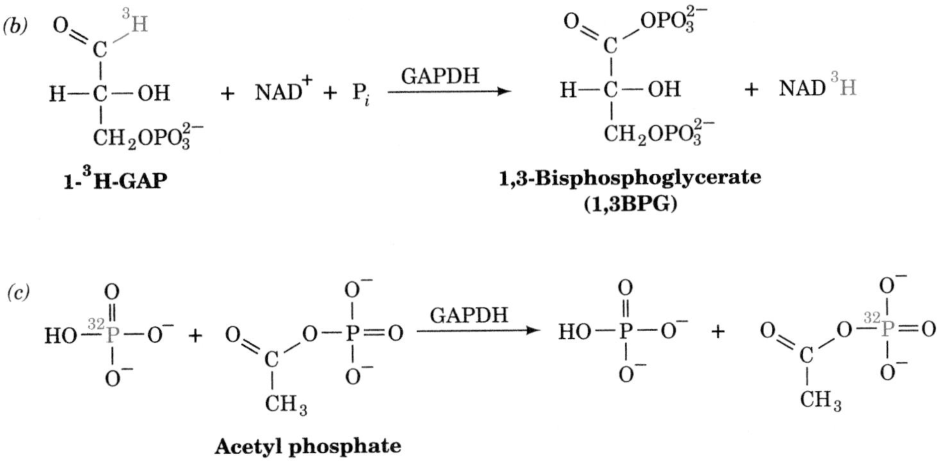

(a)

$$\text{Enzyme}-\text{CH}_2-\text{SH} + \text{ICHCOO}^- \xrightarrow{\text{HI}} \text{Enzyme}-\text{CH}_2-\text{S}-\text{CH}_2\text{COO}^- \xrightarrow[\text{hydrolysis}]{\text{protein}} \underset{\underset{\text{COO}^-}{|}}{\overset{\overset{\text{NH}_3^+}{|}}{\text{CH}}}-\text{CH}_2-\text{S}-\text{CH}_2\text{COO}^- + \begin{array}{l}\text{Other}\\\text{amino}\\\text{acids}\end{array}$$

GAPDH   **Active site**   **Iodoacetate**
         Cys

                                                 Carboxy-
                                             methylcysteine

(b)

$$\underset{\underset{\text{CH}_2\text{OPO}_3^{2-}}{|}}{\underset{\underset{\text{OH}}{|}}{\overset{\overset{\text{O}\diagdown~~\diagup^3\text{H}}{\text{C}}}{\text{H}-\text{C}}}} + \text{NAD}^+ + \text{P}_i \xrightarrow{\text{GAPDH}} \underset{\underset{\text{CH}_2\text{OPO}_3^{2-}}{|}}{\underset{\underset{\text{OH}}{|}}{\overset{\overset{\text{O}\diagdown~~\diagup\text{OPO}_3^{2-}}{\text{C}}}{\text{H}-\text{C}}}} + \text{NAD}{}^3\text{H}$$

       1-³H-GAP   **1,3-Bisphosphoglycerate**
                                             (1,3BPG)

(c)

$$\underset{\underset{\text{O}^-}{|}}{\overset{\overset{\text{O}}{||}}{\text{HO}-\overset{32}{\text{P}}-\text{O}^-}} + \underset{\underset{\text{CH}_3}{|}}{\overset{}{\text{O}\diagdown~\underset{\text{C}}{}~\diagup\overset{\text{O}^-}{}}}\text{O}-\underset{\underset{\text{O}^-}{|}}{\text{P}}{=}\text{O} \xrightarrow{\text{GAPDH}} \underset{\underset{\text{O}^-}{|}}{\overset{\overset{\text{O}}{||}}{\text{HO}-\text{P}-\text{O}^-}} + \underset{\underset{\text{CH}_3}{|}}{\overset{}{\text{O}\diagdown~\underset{\text{C}}{}~\diagup\overset{\text{O}^-}{}}}\text{O}-\underset{\underset{\text{O}^-}{|}}{\overset{32}{\text{P}}}{=}\text{O}$$

         Acetyl phosphate

Figure 16-10
Some reactions employed in elucidating the enzymatic mechanism of GAPDH. (*a*) The reaction of iodoacetate with an active site Cys residue. (*b*) Quantitative tritium transfer from substrate to NAD⁺. (*c*) The enzyme-catalyzed exchange of ³²P from phosphate to acetyl phosphate.

Mechanistic Studies

Several key enzymological experiments have contributed to the elucidation of the GAPDH reaction mechanism:

1. GAPDH is inactivated by alkylation with stoichiometric amounts of iodoacetate. The presence of carboxymethylcysteine in the hydrolysate of the resulting alkylated enzyme (Fig. 16-10*a*) suggests that GAPDH has an active site Cys sulfhydryl group. [GAPDH inactivation results in FBP accumulation (Section 16-1A) as a consequence of its equilibration with GAP by the TIM and aldolase reactions.]

2. GAPDH quantitatively transfers ³H from C(1) of GAP to NAD⁺ (Fig. 16-10*b*) thereby establishing that this reaction occurs via direct hydride transfer.

3. GAPDH catalyzes exchange of ³²P between [³²P]–P$_i$ and the product analog acetyl phosphate (Fig. 16-10*c*). Such isotope exchange reactions are indicative of an acyl–enzyme intermediate (Section 13-5D).

David Trentham has proposed a mechanism for GAPDH based on this information and the results of kinetic studies (Fig. 16-11):

Step 1   GAP binds to the enzyme.

Step 2   The essential sulfhydryl group, acting as a nucleophile, attacks the aldehyde to form a **thiohemiacetal.**

Step 3   The thiohemiacetal undergoes oxidation to an **acyl thioester** by direct transfer of a hydride to NAD⁺. This intermediate, which has been iso-

lated, has a high group-transfer potential. *The energy of aldehyde oxidation has not been dissipated but has been conserved through the synthesis of the thioester, and the reduction of NAD⁺ to NADH.*

Step 4   Another molecule of NAD⁺ replaces NADH.

Step 5   The thioester intermediate undergoes nucleophilic attack by P$_i$ to regenerate free enzyme and form 1,3-BPG. This "high-energy" mixed anhydride generates ATP from ADP in the next reaction of glycolysis.

Hydride Transfers Involving NAD⁺ or NADP⁺ Are Stereospecific

Dehydrogenases, like most enzymes, bind substrates and catalyze reactions with absolute stereospecificity. The hydride transfers to and from NAD⁺ or NADP⁺ catalyzed by these enzymes are therefore to a specific face of the coenzyme's nicotinamide ring although the identity of that face varies with the enzyme. For example, deuterium-labeling studies have demonstrated that GAPDH transfers hydride ion from C(1) of GAP to the *si* face of the nicotinamide ring of NAD⁺:

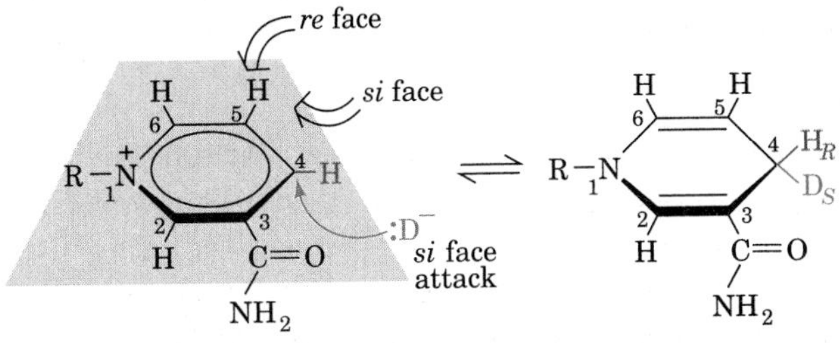

Hence for GAPDH, the transferred H atom is the *pro-S*

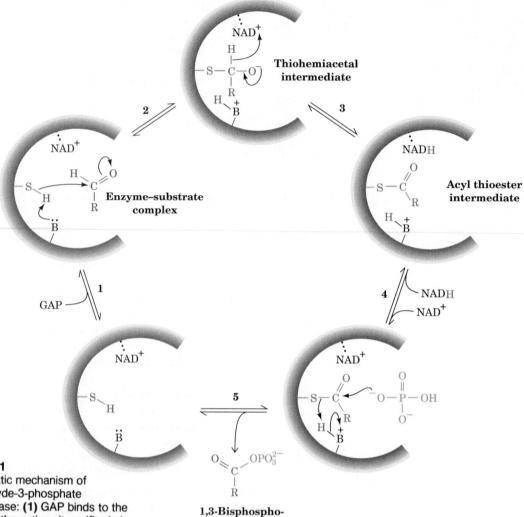

Figure 16-11
The enzymatic mechanism of glyceraldehyde-3-phosphate dehydrogenase: **(1)** GAP binds to the enzyme; **(2)** the active site sulfhydryl group forms a thiohemiacetal with the substrate; **(3)** NAD⁺ oxidizes the thiohemiacetal to a thioester; **(4)** the newly formed NADH is replaced on the enzyme by NAD⁺; and **(5)** P_i attacks the thioester forming the acyl phosphate product, 1,3-BPG, and regenerating the active enzyme.

substituent to the C(4) atom of NADH. In contrast, both lactate dehydrogenase (Section 16-3A) and alcohol dehydrogenase (Section 16-3B) catalyze transfer of the *pro-R* hydrogen of NADH.

Before the absolute stereochemistry of hydride ion addition to NAD⁺ was known, the "relative" faces of the molecule were labeled A and B. It has since been found that A-side addition is to the *re* face *(pro-R)* of NAD⁺ and B-side addition is to the *si* face.

G. Phosphoglycerate Kinase: First ATP Generation

Reaction 7 of the glycolytic pathway results in the first formation of ATP together with **3-phosphoglycerate**

(3PG) in a reaction catalyzed by **phosphoglycerate kinase (PGK):**

$$
\begin{array}{c}
\underset{1}{\text{C}} \overset{\text{O}}{\underset{}{\diagup}} \text{OPO}_3^{2-} \\
\text{H}-\underset{2}{\text{C}}-\text{OH} \qquad + \text{ ADP}\\
\underset{3}{\text{CH}_2}\text{OPO}_3^{2-}
\end{array}
$$

**1,3-Bisphosphoglycerate
(1,3-BPG)**

$$\text{Mg}^{2+} \Bigg\downarrow \begin{array}{l}\text{phosphoglycerate}\\\text{kinase (PGK)}\end{array}$$

$$
\begin{array}{c}
{}^{-}\text{O} \diagdown \underset{1}{\text{C}} \diagup \text{O} \\
\text{H}-\underset{2}{\text{C}}-\text{OH} \qquad + \text{ ATP}\\
\underset{3}{\text{CH}_2}\text{OPO}_3^{2-}
\end{array}
$$

**3-Phosphoglycerate
(3PG)**

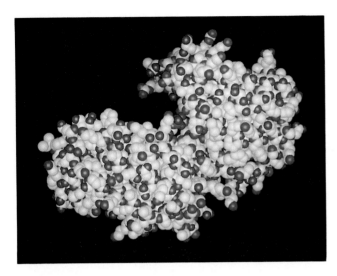

Figure 16-12
A space-filling model of yeast phosphoglycerate kinase showing its deeply clefted bilobal structure. The substrate-binding site is at the bottom of the cleft as marked by the P atom (*purple*) of 3PG. Compare this structure with that of hexokinase (Fig. 16-5a). [Courtesy of Robert Stodola, Fox Chase Cancer Center. X-ray structure determined by Herman Watson.]

(*Note:* The name "kinase" is given to any enzyme that transfers a phosphoryl group between ATP and a metabolite. Nothing is implied as to the exergonic direction of transfer.)

PGK (Fig. 16-12) is conspicuously bilobal in appearance. The Mg^{2+}–ADP binding site is located on one domain, ~ 10 Å from the 1,3-BPG binding site, which is on the other domain. Physical measurements suggest that, *upon substrate binding, the two domains of PGK swing together so as to permit the substrates to react in a water-free environment as occurs with hexokinase (Section 16-2A).* Indeed, the appearance of PGK is remarkably similar to that of hexokinase (Fig. 16-5a), even though the structures of these proteins are otherwise unrelated.

Figure 16-13 indicates a reaction mechanism for PGK

that is consistent with its observed sequential kinetics. The terminal phosphoryl oxygen of ADP nucleophilically attacks the C(1) phosphorus atom of 1,3-BPG to form the reaction product.

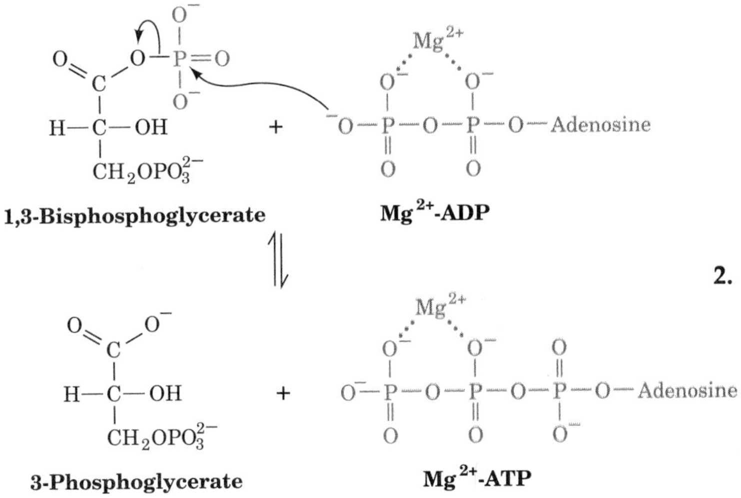

H. Phosphoglycerate Mutase

In Reaction 8 of glycolysis, 3PG is converted to **2-phosphoglycerate (2PG)** by **phosphoglycerate mutase (PGM)**:

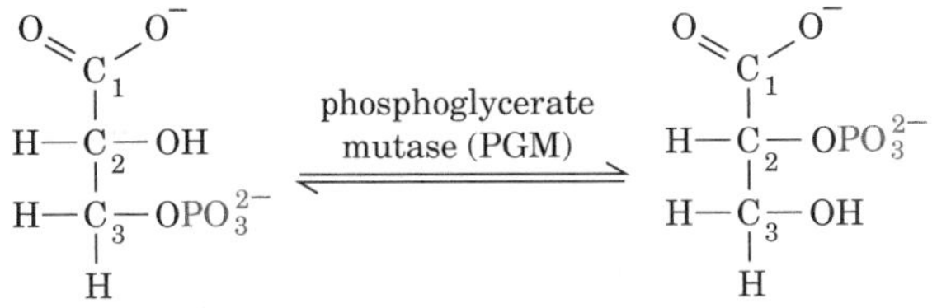

$$\underset{\substack{\textbf{3-Phosphoglycerate}\\ \textbf{(3PG)}}}{\ce{O=C(O^-)-CH(OH)-CH(H)(OPO_3^{2-})}} \xrightleftharpoons[\text{}]{\text{phosphoglycerate}\atop \text{mutase (PGM)}} \underset{\substack{\textbf{2-Phosphoglycerate}\\ \textbf{(2PG)}}}{\ce{O=C(O^-)-CH(OPO_3^{2-})-CH(H)(OH)}}$$

A **mutase** catalyzes the transfer of a functional group from one position to another on a molecule. This reaction is necessary preparation for the next reaction in glycolysis, which generates a "high-energy" phosphoryl compound for use in ATP synthesis.

Reaction Mechanism of PGM

At first sight, the reaction catalyzed by PGM appears to be a simple intramolecular phosphoryl transfer. This is not the case, however. *The active enzyme has a phosphoryl group at its active site, which it transfers to the substrate to form a bisphospho intermediate. This intermediate then rephosphorylates the enzyme to form the product and regenerate the active phosphoenzyme.* The following experimental data permitted the elucidation of PGM's enzymatic mechanism:

1. Catalytic amounts of **2,3-bisphosphoglycerate (2,3-BPG, also known as 2,3-diphosphoglycerate)**

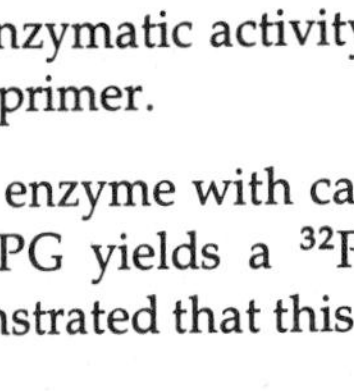

$$\underset{\substack{\textbf{2,3-Bisphosphoglycerate}\\ \textbf{(2,3-BPG)}}}{\ce{O=C(O^-)-CH(OPO_3^{2-})-CH(H)(OPO_3^{2-})}}$$

are required for enzymatic activity; that is, 2,3-BPG acts as a reaction primer.

2. Incubation of the enzyme with catalytic amounts of ^{32}P-labeled 2,3-BPG yields a ^{32}P-labeled enzyme. Zelda Rose demonstrated that this was a result of the

Figure 16-13
The mechanism of the PGK reaction. The Mg^{2+} positions are shown as examples; their actual binding sites are unknown.

1,3-Bisphosphoglycerate Mg^{2+}-ADP

3-Phosphoglycerate Mg^{2+}-ATP

phosphorylation of a His residue:

Enzyme—CH₂

Phospho His residue

3. The enzyme's X-ray structure shows His at the active site (Fig. 16-14). In the active enzyme, His 8 is phosphorylated.

These data are consistent with a mechanism in which the active enzyme contains a phospho-His residue at the active site (Fig. 16-15):

Step 1   3PG binds to the phosphoenzyme in which His 8 is phosphorylated.

Step 2   This phosphoryl group is transferred to the substrate, resulting in an intermediate 2,3-BPG · enzyme complex.

Steps 3 and 4   The complex decomposes to form the product 2PG with regeneration of the phosphoenzyme.

The phosphoryl group on 3PG therefore ends up on the C(2) of the next 3PG to undergo reaction.

Occasionally, 2,3-BPG dissociates from the enzyme (Fig. 16-15; Step 5) leaving it in an inactive form. Trace amounts of 2,3-BPG must therefore always be available to regenerate the active phosphoenzyme by the reverse reaction.

Glycolysis Influences Oxygen Transport

2,3-BPG specifically binds to deoxyhemoglobin and thereby alters the oxygen affinity of hemoglobin (Section 9-1D). The concentration of 2,3-BPG in erythrocytes is much higher (~ 5 mM) than the trace amounts required for its use as a primer of PGM. Erythrocytes synthesize and degrade 2,3-BPG by a detour from the glycolytic pathway diagrammed in Fig. 16-16. **Bisphosphoglycerate mutase** catalyzes the transfer of a phosphoryl group from C(1) to C(2) of 1,3-BPG. The resulting 2,3-BPG is hydrolyzed to 3PG by **2,3-bisphosphoglycerate phosphatase**.

The rate of glycolysis affects the oxygen affinity of hemoglobin through the mediation of 2,3-BPG. Consequently, inherited defects of glycolysis in erythrocytes alter the capacity of the blood to transport oxygen (Fig. 16-17). For example, the concentration of glycolytic intermediates in hexokinase-deficient erythrocytes is less than normal because hexokinase catalyzes the first reaction of glycolysis. This results in a diminished 2,3-BPG concentration and therefore in increased hemoglobin oxygen affinity. Conversely, pyruvate kinase deficiency decreases hemoglobin oxygen affinity through the in-

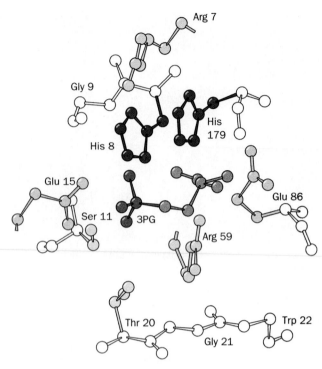

Figure 16-14
The active site region of yeast phosphoglycerate mutase (dephospho form) showing the substrate, 3-phosphoglycerate, and some of the side chains that approach it. His 8 is phosphorylated in the active enzyme. [After Winn, S. I., Watson, H. I., Harkins, R. N., and Fothergill, L. A., *Philos. Trans. R. Soc. London Ser. B* **293**, 126 (1981).]

crease of 2,3-BPG resulting from the blockade of the last reaction in glycolysis.

I. Enolase: Second "High-Energy" Intermediate Formation

In Reaction 9 of glycolysis, 2PG is dehydrated to **phosphoenolpyruvate (PEP)** in a reaction catalyzed by **enolase**:

2-Phosphoglycerate   **Phosphoenolpyruvate**
(2PG)   **(PEP)**

The enzyme forms a complex with a divalent cation such as Mg^{2+} before the substrate is bound. As is mentioned in Section 16-1A, fluoride ion inhibits glycolysis with the accumulation of 2PG and 3PG. It does so by strongly inhibiting enolase in the presence of P_i. The inhibitory species is **fluorophosphate ion (FPO_3^{3-})**, which probably complexes the enzyme-bound Mg^{2+}

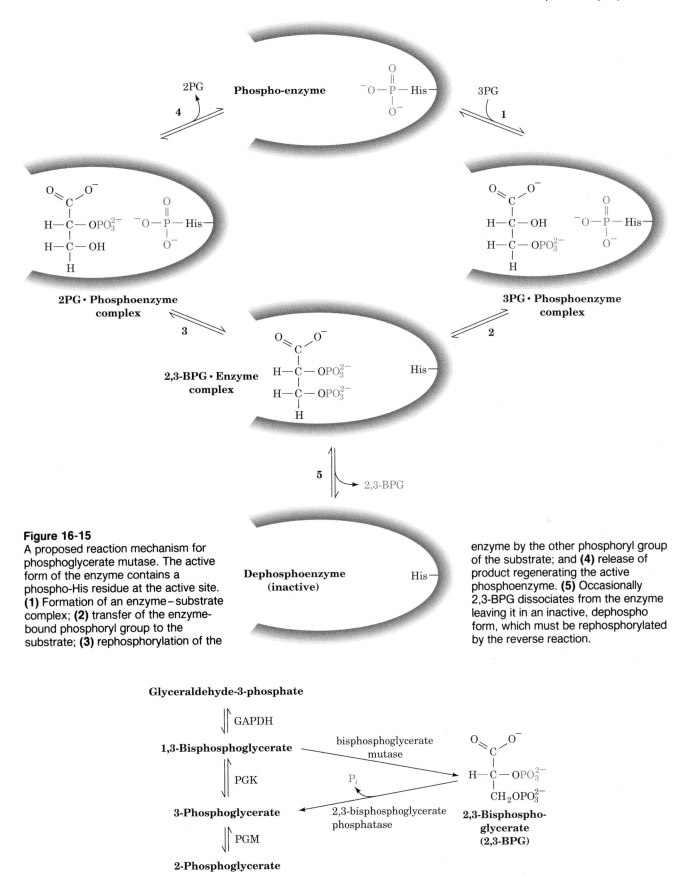

Figure 16-15
A proposed reaction mechanism for phosphoglycerate mutase. The active form of the enzyme contains a phospho-His residue at the active site. **(1)** Formation of an enzyme–substrate complex; **(2)** transfer of the enzyme-bound phosphoryl group to the substrate; **(3)** rephosphorylation of the enzyme by the other phosphoryl group of the substrate; and **(4)** release of product regenerating the active phosphoenzyme. **(5)** Occasionally 2,3-BPG dissociates from the enzyme leaving it in an inactive, dephospho form, which must be rephosphorylated by the reverse reaction.

Figure 16-16
The pathway for the synthesis and degradation of 2,3-BPG in erythrocytes is a detour from the glycolytic pathway.

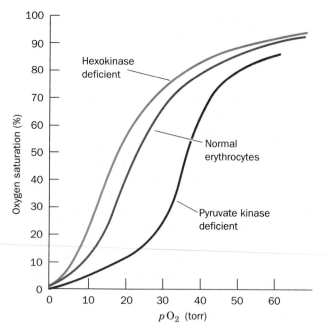

Figure 16-17
The oxygen-saturation curves of hemoglobin in normal
erythrocytes (*red curve*) and those from patients with
hexokinase deficiency (*green*) and with pyruvate kinase
deficiency (*purple*). [After Delivoria-Papadopoulos, M., Oski,
F. A., and Gottlieb, A. J., *Science* **165**, 601 (1969).]

thereby inactivating the enzyme. Enolase's substrate
2PG, therefore builds up and as it does so, is equilibrated
with 3PG by PGM.

Catalytic Mechanism

The dehydration (elimination of H_2O) catalyzed by
enolase might occur in one of three ways (Fig. 15-9*a*): (1)
the C(3)-OH group can leave first generating a carbocation at C(3); (2) the C(2) proton can leave first, generating
a carbanion at C(2); or (3) the reaction can be concerted.
Isotope exchange studies by Paul Boyer demonstrated
that the C(2) proton of 2PG exchanges with solvent 12
times faster than the rate of PEP formation. However,
the C(3) oxygen exchanges with solvent at a rate roughly
equivalent with the overall reaction rate. This suggests
the following mechanism (Fig. 16-18):

Step 1 Rapid carbanion formation at C(2) faciliated by
a general base on the enzyme. The abstracted
proton can readily exchange with the solvent
accounting for its observed rapid exchange
rate.

Step 2 Rate-limiting elimination of the C(3)-OH
group. This is consistent with the slow rate of
exchange of this hydroxyl group with solvent.

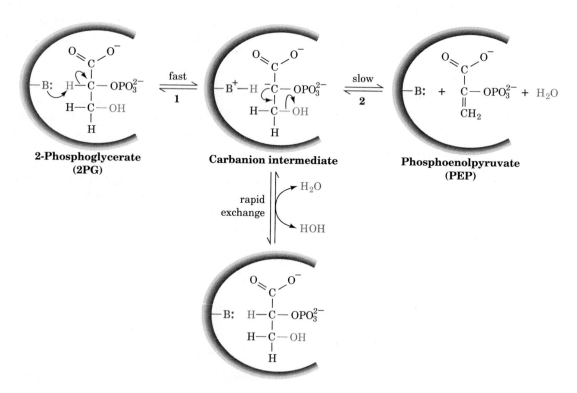

Figure 16-18
The proposed reaction mechanism of enolase: **(1)** Rapid formation of a
carbanion by removal of a proton at C(2); this proton can rapidly exchange
with the solvent. **(2)** Slow elimination of an OH group to form
phosphoenolpyruvate; ^{18}O can only exchange with solvent as rapidly as this
step occurs.

J. Pyruvate Kinase: Second ATP Generation

In Reaction 10 of glycolysis, the final reaction, **pyruvate kinase (PK)** couples the free energy of PEP hydrolysis to the synthesis of ATP to form **pyruvate**:

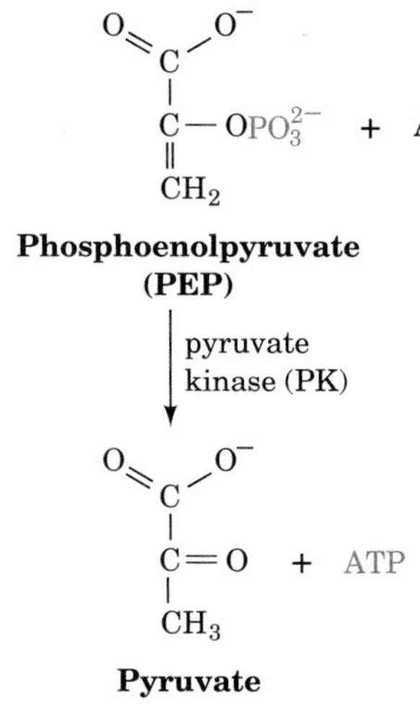

Phosphoenolpyruvate
(PEP)

pyruvate
kinase (PK)

Pyruvate

Catalytic Mechanism of PK

The PK reaction occurs as follows (Fig. 16-19):

Step 1 A β-phosphoryl oxygen of ADP nucleophilically attacks the PEP phosphorus atom thereby displacing **enol pyruvate** and forming ATP. This reaction conserves the free energy of PEP hydrolysis.

Step 2 Enol pyruvate converts to pyruvate. This enol–keto tautomerization is sufficiently exergonic to drive the coupled endergonic synthesis of ATP (Section 15-4C).

We can now see the "logic" of the enolase reaction. The standard free energy of hydrolysis of 2PG is only $\Delta G°' = -17.6 \text{ kJ} \cdot \text{mol}^{-1}$, which is insufficient to drive ATP synthesis ($\Delta G°' = 30.5 \text{ kJ} \cdot \text{mol}^{-1}$ for ATP synthesis from ADP and P_i). The dehydration of 2PG results in the formation of a "high-energy" compound capable of such synthesis [the standard free energy of hydrolysis of

PEP is $-61.9 \text{ kJ} \cdot \text{mol}^{-1}$ (Table 3-4 and Section 3-4B)]. In other words, PEP is a "high-energy" compound, 2PG is not.

Metabolic Pathways Probably Evolved Backwards through Gene Duplication

We have seen the ingenious manner in which the free energy of degradation of glucose to pyruvate is utilized to synthesize ATP. This process involves an investment of ATP to form a phosphoryl compound (FBP), which is then cleaved to two C_3 units. The free energy of oxidation of GAP is then utilized to synthesize an acyl phosphate, a "high-energy" intermediate (1,3-BPG) that is used to phosphorylate ADP to ATP. The second "high-energy" compound of the pathway, PEP, which is produced from 2PG, also phosphorylates ADP to ATP. The overall reaction of glycolysis is therefore

$$\text{Glucose} + 2\text{ADP} + 2\text{P}_i + 2\text{NAD}^+ \longrightarrow$$
$$2\text{pyruvate} + 2\text{ATP} + 2\text{NADH} + 4\text{H}^+ + 2\text{H}_2\text{O}$$

How might such a complex pathway have evolved? It seems highly unlikely that it simply arose through the random appearance of the required enzymatic activities followed by evolutionary fine tuning. Rather, as the existence of the nucleotide-binding fold (Section 7-3B) in a number of nucleotide-binding enzymes suggests, pathways probably arose through gene duplication followed by divergent evolution. Indeed, since the substrate of a given enzyme in a pathway is the product of the preceding enzyme, it has been postulated that metabolic pathways evolved backwards with each enzyme giving rise to the preceding enzyme through gene duplication and divergent evolution. This notion is consistent with the observation that the successive glycolytic enzymes enolase and PK both contain C-terminal PEP-binding β barrels (Section 7-3B) whose backbone atoms are nearly superimposable but which bear lesser structural resemblance with other β barrels of known structure.

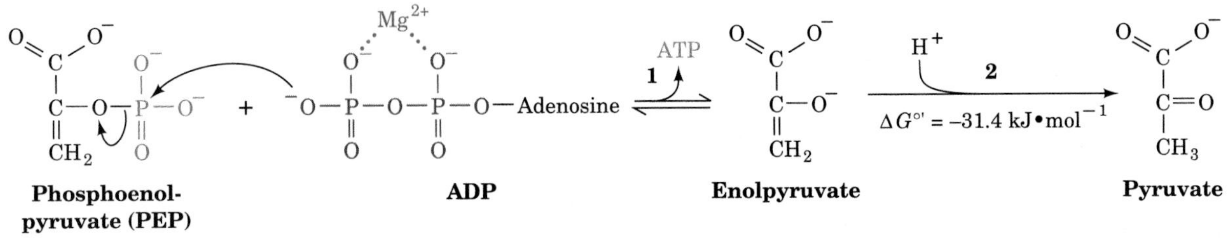

Phosphoenol-
pyruvate (PEP)

ADP

Enolpyruvate

Pyruvate

$\Delta G°' = -31.4 \text{ kJ} \cdot \text{mol}^{-1}$

Figure 16-19
The mechanism of the reaction catalyzed by pyruvate kinase:
(1) Nucleophilic attack of an ADP β-phosphoryl oxygen atom on the phosphorous atom of PEP to form ATP and phosphoenolpyruvate; and
(2) tautomerization of enolpyruvate to pyruvate.

3. FERMENTATION: THE ANAEROBIC FATE OF PYRUVATE

For glycolysis to continue, NAD$^+$, which cells have in limited quantities, must be recycled after its reduction to NADH by GAPDH (Fig. 16-3; Reaction 6). In the presence of oxygen, the reducing equivalents of NADH are passed into the mitochondria for reoxidation (Chapter 20). Under anaerobic conditions, on the other hand, the NAD$^+$ is replenished by the reduction of pyruvate in an extension of the glycolytic pathway. Two processes for the anaerobic replenishment of NAD$^+$ are homolactic and alcoholic fermentation, which occur in muscle and yeast, respectively.

A. Homolactic Fermentation

In muscle, particularly during vigorous activity when the demand for ATP is high and oxygen has been depleted, **lactate dehydrogenase (LDH)** catalyzes the reduction of NADH by pyruvate to yield NAD$^+$ and **lactate.** This reaction is often classified as Reaction 11 of glycolysis:

Pyruvate   **NADH**

lactate dehydrogenase (LDH)

L-Lactate   **NAD$^+$**

LDH, as do other NAD$^+$-requiring enzymes, catalyzes its reaction with absolute stereospecificity: The *pro-R* (A-side) hydrogen at C(4) of NADH is stereospecifically transferred to the *re* face of pyruvate at C(2) to form L-(or S-) lactate. This regenerates NAD$^+$ for participation in the GAPDH reaction. Note that the hydride transfer to pyruvate is from the opposite face of the nicotinamide ring as that to GAP in the GAPDH reaction (Section 16-2F). Figure 16-20 is a superposition of the NAD$^+$ cofactors bound to GAPDH and LDH in their respective crystal structures. The coenzyme conformations in the two enzymes are similar except that the orientations of the nicotinamide rings differ by ~180°.

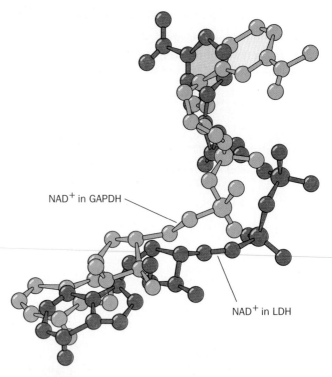

NAD$^+$ in GAPDH

NAD$^+$ in LDH

Figure 16-20
The superposition of the NAD$^+$ molecules in the crystal structures of LDH (*red*) and GAPDH (*green*). The nicotinamide rings (*shaded*) of the two coenzymes face in opposite directions. [After Rossmann, M. G., Liljas, A., Bränden, C.-I., and Banaszak, L. J., *in* Boyer, P. D. (Ed.), *The Enzymes* (3rd ed.), Vol. 11, *p*. 85, Academic Press (1975).]

Mammals have two different types of LDH subunits, the M type and the H type, which together form five tetrameric isozymes: M$_4$, M$_3$H, M$_2$H$_2$, MH$_3$, and H$_4$ (Section 7-5C). Although these hybrid forms occur in most tissues, the H-type subunit predominates in aerobic tissues such as heart muscle, while the M-type subunit predominates in tissues that are subject to anaerobic conditions such as skeletal muscle and liver. H$_4$ LDH has a low K_M for pyruvate and is allosterically inhibited by high levels of this metabolite, whereas the M$_4$ isozyme has a higher K_M for pyruvate and is not inhibited by it. The other isozymes have intermediate properties that vary with the ratio of their two types of subunits. It has therefore been proposed, although not without disagreement, that H-type LDH is better adapted to function in the oxidation of lactate to pyruvate while M-type LDH is more suited to catalyze the reverse reaction.

The X-ray structure of dogfish M$_4$ LDH was elucidated by Michael Rossmann. The complex of LDH with a synthetic adduct of NAD$^+$ and pyruvate (Fig. 16-21) suggests a mechanism for pyruvate reduction (Fig. 16-22): *The hydride is transferred from C(4) of NADH to C(2) of pyruvate with concomitant transfer of a proton from the imidazolium moiety of His 195. The latter interaction also serves to orient the substrate as does the salt bridge

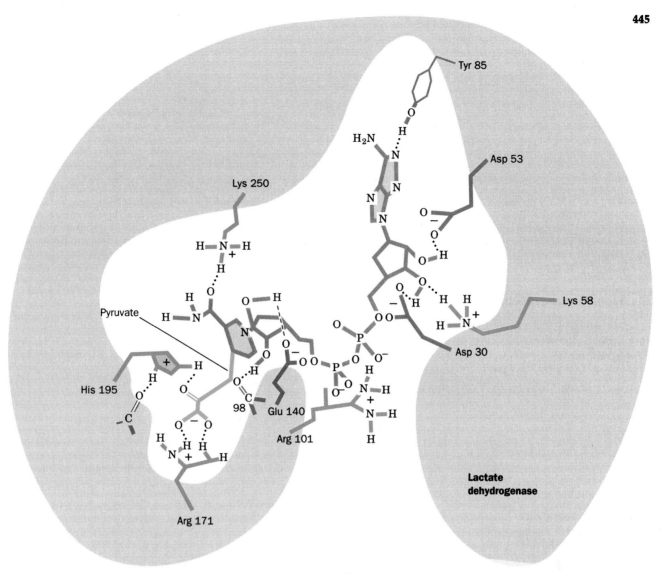

Figure 16-21
The binding of the synthetic covalent NAD–pyruvate adduct in the active site of lactate dehydrogenase. The pyruvate residue is drawn in green and the NAD⁺ is brown. [After Holbrook, J. J., Liljas, A., Steindel, S. J., and Rossmann, M. G., *in* Boyer, P. D. (Ed.), *The Enzymes* (3rd ed.), Vol. 11, p. 240, Academic Press (1975).]

that the substrate carboxyl group forms with Arg 171.

The overall process of anaerobic glycolysis in muscle can be represented:

$$\text{Glucose} + 2\text{ADP} + 2\text{P}_i \longrightarrow 2\text{lactate} + 2\text{ATP} + 2\text{H}^+$$

Much of the lactate, the end product of anaerobic glycolysis, is exported from the muscle cell and carried by the blood to the liver, where it is reconverted to glucose (Section 21-1C). Contrary to the widely held belief, it is not lactate buildup in the muscle *per se* that causes muscle fatigue and soreness but rather the accumulation of glycolytic-

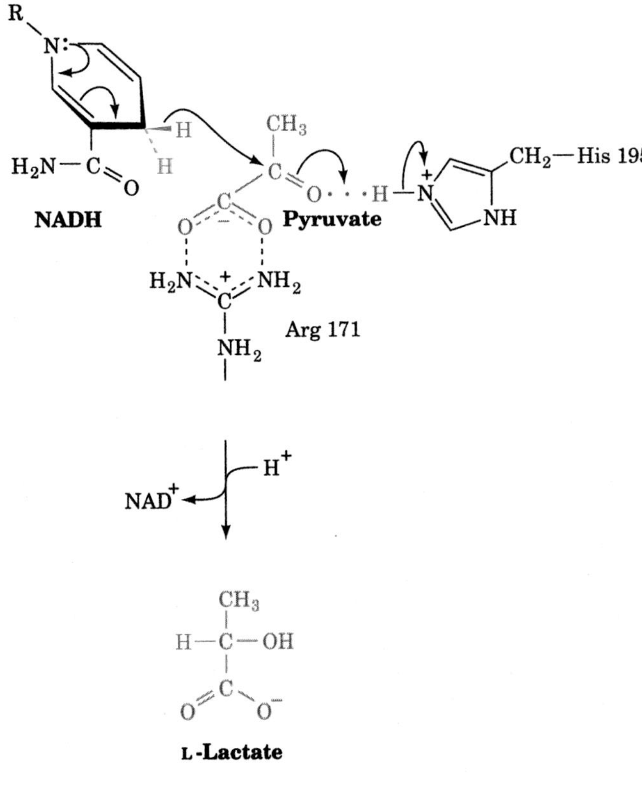

Figure 16-22
The reaction mechanism of lactate dehydrogenase involves direct hydride transfer from NADH to pyruvate's carbonyl carbon atom.

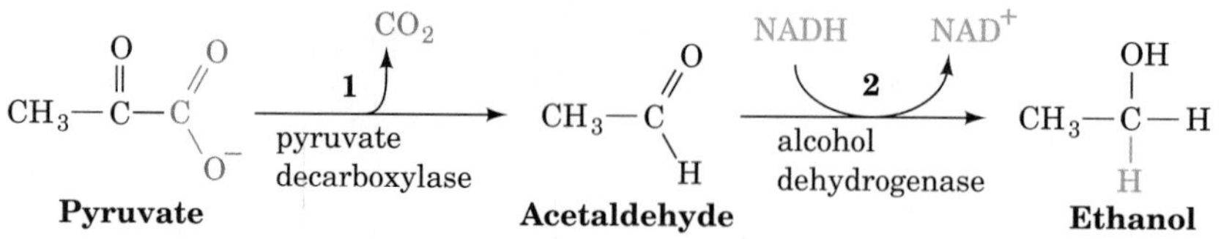

Figure 16-23
The two reactions of alcoholic fermentation are (1) decarboxylation of pyruvate to form acetaldehyde, followed by (2) reduction to ethanol by NADH.

ally generated acid (muscles can maintain their work load in the presence of high lactate concentrations if the pH is kept constant). Indeed, it is well known among hunters that the meat of an animal that has run to exhaustion before being killed has a sour taste. This is a result of lactic acid buildup in the muscles.

B. Alcoholic Fermentation

Under anaerobic conditions in yeast, NAD^+ is regenerated in a manner that has been of importance to mankind for thousands of years: the conversion of pyruvate to ethanol and CO_2. Ethanol is, of course, the active ingredient of wine and spirits; CO_2 so-produced leavens bread.

TPP Is an Essential Cofactor of Pyruvate Decarboxylase

Yeast produces ethanol and CO_2 via two reactions (Fig. 16-23). The first reaction is the decarboxylation of pyruvate to form acetaldehyde and CO_2 as catalyzed by **pyruvate decarboxylase** (an enzyme not present in animals). This enzyme contains the coenzyme **thiamine pyrophosphate** (**TPP;** Fig. 16-24) which it binds tightly but noncovalently. The coenzyme is employed because decarboxylation of an α-keto acid such as pyruvate requires the buildup of negative charge on the carbonyl carbon atom in the transition state, an unstable situation:

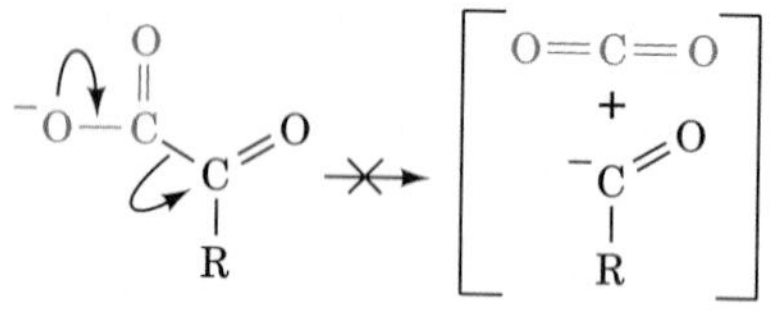

This transition state may be stabilized by delocalization of the developing negative charge into a suitable "electron sink." The amino acid residues of proteins function poorly in this capacity but TPP does so easily.

*The "business" end of TPP is the **thiazolium ring*** (Fig. 16-24). Its C(2)–H group is relatively acidic because of the adjacent positively charged quaternary nitrogen atom which electrostatically stabilizes the carbanion formed on dissociation of the proton. This dipolar carbanion (or **ylid**) is the active form of the coenzyme. The mechanism of pyruvate decarboxylase catalysis is as follows (Fig. 16-25):

Step 1 Nucleophilic attack by the ylid form of TPP on the carbonyl carbon of pyruvate.

Step 2 Departure of CO_2 to generate a resonance-stabilized carbanion adduct in which the thiazolium ring of the coenzyme acts as an electron sink.

Step 3 Protonation of the carbanion.

Step 4 Elimination of the TPP ylid to form acetaldehyde and regenerate the active enzyme.

This mechanism has been corroborated by the isolation of the **hydroxyethylthiamine pyrophosphate** intermediate (Fig. 16-25).

Beriberi Is a Thiamine Deficiency Disease

The ability of TPP's thiazolium ring to add to carbonyl groups and act as an "electron sink" makes it the coenzyme most utilized in α-keto acid decarboxylations. TPP is also involved in decarboxylation reactions that we shall encounter in other metabolic pathways. Consequently, thiamine (**vitamin B_1**), which is neither synthesized nor stored in significant amounts by the tissues of most vertebrates, is required in their diets. Its deficiency in man results in an ultimately fatal condition known as **beriberi** that is characterized by neurological disturbances causing pain, paralysis and atrophy (wasting) of the limbs, and/or cardiac failure resulting in edema (the accumulation of fluid in tissues and body cavities). Beriberi was particulary prevalent in the rice-consuming areas of the Orient because of the custom of polishing this staple grain to remove its coarse but thiamine-containing outer layers. Beriberi frequently develops in chronic alcoholics as a consequence of their penchant for drinking but not eating.

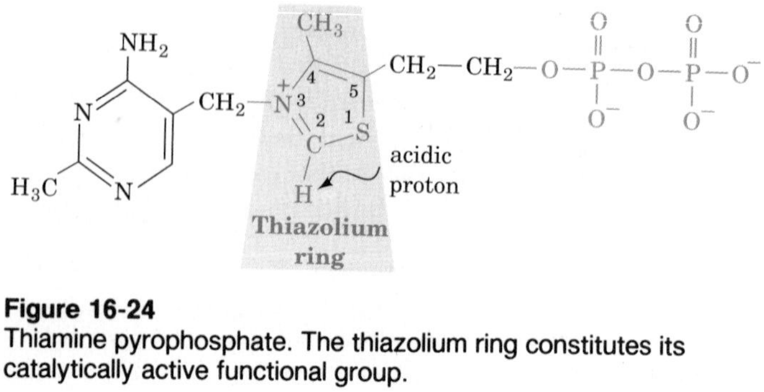

Figure 16-24
Thiamine pyrophosphate. The thiazolium ring constitutes its catalytically active functional group.

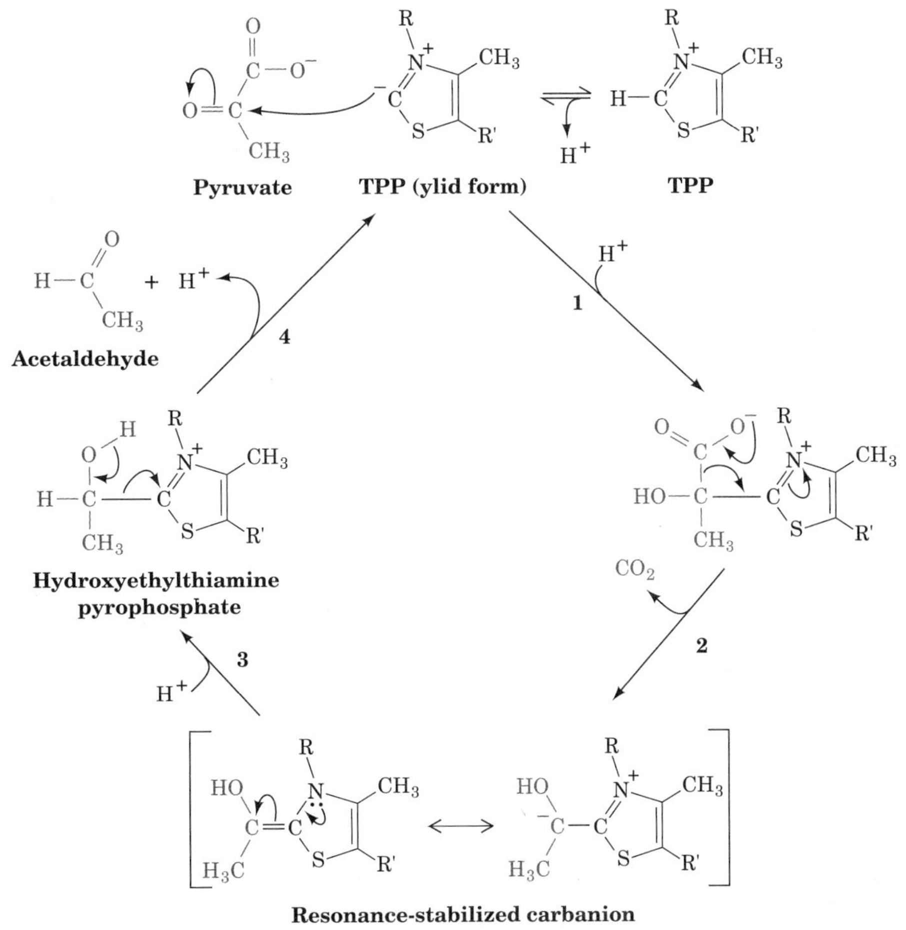

Figure 16-25
The reaction mechanism of pyruvate decarboxylase:
(1) nucleophilic attack by the ylid form of TPP on the
carbonyl carbon of pyruvate; **(2)** departure of CO_2 to
generate a resonance stabilized carbanion; **(3)** protonation
of the carbanion; and **(4)** elimination of the TPP ylid and
release of product.

Reduction of Acetaldehyde and Regeneration of NAD$^+$

The acetaldehyde formed by the decarboxylation of
pyruvate is reduced to ethanol by NADH in a reaction
catalyzed by **alcohol dehydrogenase (ADH).** Each subunit of the tetrameric yeast ADH (YADH) binds one
NADH and one Zn^{2+} ion. The Zn^{2+} ion functions to
polarize the carbonyl group of acetaldehyde (Fig.
16-26), so as to stabilize the developing negative charge
in the transition state of the reaction (the role of metal
ions in enzymes is discussed in Section 14-1C). This
facilitates the transfer of NADH's *pro-R* hydrogen (the
same atom that LDH transfers) to acetaldehyde's *re* face
forming ethanol with the transferred hydrogen in the
pro-R position (Section 12-1A).

Both homolactic and alcoholic fermentation have the
same function: the anaerobic regeneration of NAD$^+$ for
continued glycolysis. Their main difference is in their
metabolic products.

Mammalian liver ADH **(LADH)** functions to metabolize the alcohols anaerobically produced by intestinal

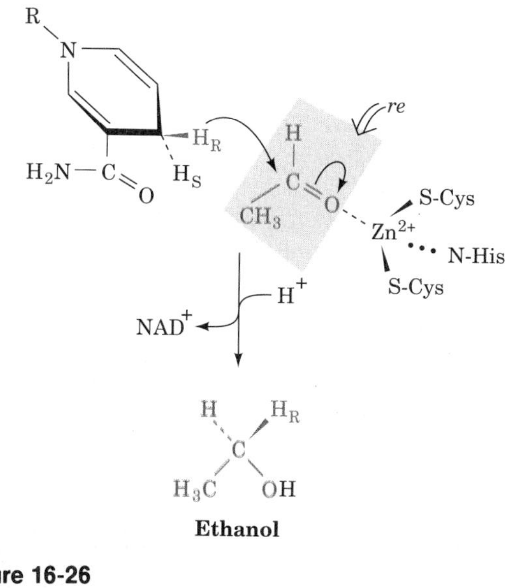

Figure 16-26
The reaction mechanism of alcohol dehydrogenase involves
direct hydride transfer of the *pro-R* hydrogen of NADH to
the *re* face of acetaldehyde.

flora as well as those from external sources (the direction of the ADH reaction varies with the relative concentrations of ethanol and acetaldehyde). Each subunit of this dimeric enzyme binds one NAD^+ and two Zn^{2+} ions although only one of these ions participates directly in catalysis. There is significant amino acid sequence similarity between YADH and LADH so that it is commonly assumed that both enzymes have the same general mechanism.

C. Energetics of Fermentation

Thermodynamics permits us to dissect the process of fermentation into its component parts and to account for the free energy changes that occur. This enables us to calculate the efficiency with which the free energy of degradation of glucose is utilized in the synthesis of ATP. The overall reaction of homolactic fermentation is

$$Glucose \longrightarrow 2lactate + 2H^+$$
$$\Delta G°'(pH\ 7) = -196\ kJ \cdot mol^{-1}\ of\ glucose$$

($\Delta G°'$ is calculated from the data in Table 3-4 using Eqs. [3.19] and [3.21] adapted for $2H^+$ ions.) For alcoholic fermentation, the overall reaction is

$$Glucose \longrightarrow 2CO_2 + 2ethanol$$
$$\Delta G°' = -235\ kJ \cdot mol^{-1}\ of\ glucose$$

Each of these reactions is coupled to the net formation of two ATPs, which requires $\Delta G°' = +61\ kJ \cdot mol^{-1}$ of glucose consumed (Table 15-3). Dividing the $\Delta G°'$ of ATP formation by that of lactate formation indicates that homolactic fermentation is 31% "efficient"; that is, 31% of the free energy released by this process under standard biochemical conditions is sequestered in the form of ATP. The rest is dissipated as heat, thereby making the process irreversible. Likewise, alcoholic fermentation is 26% efficient under biochemical standard state conditions. Actually, *under physiological conditions, where the concentrations of reactants and products differ from those of the standard state, these reactions have a free energy efficiency of > 50%.*

Glycolysis Is Used for Rapid ATP Production

Anaerobic fermentation utilizes glucose in a profligate manner compared to oxidative phosphorylation: Fermentation results in the production of 2ATPs per glucose, whereas oxidative phosphorylation yields 38ATPs per glucose (Chapter 20). This accounts for Pasteur's observation that yeast consumes far more sugar when growing anaerobically than when growing aerobically (the **Pasteur effect;** Section 20-4C). However, *the rate of ATP production by anaerobic glycolysis can be up to 100 times faster than that of oxidative phosphorylation. Consequently, when tissues such as muscle are rapidly consuming ATP, they regenerate it almost entirely by anaerobic glycolysis.* (Homolactic fermentation does not really "waste" glucose since the lactate so-produced is reconverted to glucose by the liver; Section 21-1C.)

Skeletal muscles consist of both **slow-twitch** (Type I) and **fast-twitch** (Type II) **fibers.** Fast-twitch fibers, so-called because they predominate in muscles capable of short bursts of rapid activity, are nearly devoid of mitochondria so that they must obtain nearly all of their ATP through glycolysis. Muscles designed to contract slowly and steadily, in contrast, are enriched in slow-twitch fibers that are rich in mitochondria and obtain most of their ATP through oxidative phosphorylation. (Fast- and slow-twitch fibers were originally known as white and red fibers, respectively, because otherwise pale colored muscle tissue, when enriched with mitochondria, takes on the red color characteristic of their heme-containing cytochromes. However, fiber color has been shown to be an imperfect indicator of muscle physiology.) In a familiar example, the flight muscles of migratory birds such as ducks and geese, which need a continuous energy supply, are rich in slow-twitch fibers and therefore such birds have dark breast meat. In contrast, the flight muscles of less ambitious fliers such as chickens and turkeys, which are used only for short bursts (often to escape danger), consist mainly of fast-twitch fibers that form white meat. In humans, the muscles of sprinters are relatively rich in fast-twitch fibers, whereas distance runners have a greater proportion of slow-twitch fibers (although their muscles have the same color). World class distance runners have a remarkably high capacity to aerobically generate ATP. This was demonstrated by the noninvasive ^{31}P NMR monitoring of the ATP, P_i, phosphocreatine, and pH levels in their exercising but untrained forearm muscles. These observations suggest that the muscles of these athletes are better endowed genetically for endurance exercise than those of "normal" individuals.

4. CONTROL OF METABOLIC FLUX

Living organisms, as we saw in Section 15-6, are thermodynamically open systems that tend to maintain a steady state rather than reaching equilibrium (death for living things). Thus the *flux (rate of flow) of intermediates through a metabolic pathway is constant; that is, the rates of synthesis and breakdown of each pathway intermediate maintain it at a constant concentration.* Such a state, it will be recalled, is one of maximum thermodynamic efficiency (Section 15-6B).

The concentrations of intermediates and the level of metabolic flux at which a pathway is maintained varies with the needs of the organism through a highly responsive system of precise controls. Such pathways are analogous to rivers that have been dammed to provide a means of generating electricity. Although water is continually flowing in and out of the lake formed by the dam, a relatively constant water level is maintained. The rate of water outflow from the lake is precisely controlled at the dam, and is varied in response to the need for electrical power. In this section, we examine the mechanisms by which metabolic pathways in general, and the glycolytic pathway in particular, are controlled in response to biological energy needs.

A. Flux Generation

Since a metabolic pathway is a series of enzyme-catalyzed reactions, it is easiest to describe the flux of metabolites through the pathway by considering its reaction steps individually. The flux of metabolites, J, through each reaction step is the rate of the forward reaction, v_f, less that of the reverse reaction, v_r:

$$J = v_f - v_r \qquad [16.1]$$

At equilibrium, by definition, there is no flux ($J = 0$), although v_f and v_r may be quite large. At the other extreme, in reactions that are far from equilibrium, $v_f \gg v_r$, so that the flux is essentially equal to the rate of the forward reaction, $J \approx v_f$. *The flux throughout a steady state pathway is constant and is set (generated) by the pathway's rate-determining step (or steps). Consequently, control of flux through a metabolic pathway requires: (1) that the flux through this **flux-generating step** vary in response to the organism's metabolic requirements, and (2) that this change in flux be communicated throughout the pathway.* We begin our discussion with the second process, the communication of flux changes in the rate-determining step of a pathway to the other enzymes of the pathway.

The Rates of Enzymatic Reactions Vary with Flux

Let us consider how a constant flux is maintained throughout a metabolic pathway by analyzing the response of an enzyme-catalyzed reaction to a change in the flux of the reaction preceding it. In the following steady state pathway:

$$S \xrightarrow[\substack{\text{rate-determining} \\ \text{step}}]{J} A \underset{v_r}{\overset{v_f}{\rightleftarrows}} B \xrightarrow{J} P$$

the flux, J, through the reaction $A \rightleftharpoons B$, which must be identical to the flux through the rate-determining step, is expressed by Eq. [16.1]. If the flux of the rate-determining step increases by the amount ΔJ, the increase must be communicated to the next reaction step in the pathway by an increase in v_f (Δv_f) in order to reestablish the steady state. Thus

$$\Delta J = \Delta v_f \qquad [16.2]$$

Dividing Eq. [16.2] by J, multiplying the right side by v_f/v_f, and substituting in Eq. [16.1] yields

$$\frac{\Delta J}{J} = \frac{\Delta v_f}{v_f} \frac{v_f}{J} = \frac{\Delta v_f}{v_f} \frac{v_f}{(v_f - v_r)} \qquad [16.3]$$

which relates $\Delta J/J$, the fractional change in flux through the rate-determining step, and $\Delta v_f/v_f$, the fractional

change in v_f, the forward rate of the next reaction in the pathway.

In Section 13-2A, we discussed the relationship between substrate concentration and the rates of an enzymatic reaction as expressed by the Michaelis–Menten equation:

$$v_f = \frac{V_{max}^f[A]}{K_M + [A]} \qquad [13.24a]$$

In the simplest and physiologically most common situation, $[A] \ll K_M$ so that

$$v_f = \frac{V_{max}^f[A]}{K_M} \qquad [16.4]$$

and

$$\Delta v_f = \frac{V_{max}^f \Delta[A]}{K_M} \qquad [16.5]$$

Hence,

$$\frac{\Delta v_f}{v_f} = \frac{\Delta[A]}{[A]} \qquad [16.6]$$

that is, the fractional change in reaction rate is equal to the fractional change in substrate concentration. Then, by substituting Eq. [16.6] into Eq. [16.3], we find that

$$\frac{\Delta J}{J} = \frac{\Delta[A]}{[A]} \frac{v_f}{(v_f - v_r)} \qquad [16.7]$$

This equation relates the fractional change in flux through a metabolic pathway's rate-determining step to the fractional change in substrate concentration necessary to communicate that change to the following reaction steps. *The quantity $v_f/(v_f - v_r)$ is a measure of the sensitivity of a reaction's fractional change in flux to its fractional change in substrate concentration.* This quantity is also a measure of the reversibility of the reaction; that is, how close it is to equilibrium:

1. In an irreversible reaction, v_r approaches 0 and $v_f/(v_f - v_r)$ approaches 1. The reaction therefore responds to a fractional increase in flux by a nearly equal fractional increase in its substrate concentration.

2. As a reaction approaches equilibrium, v_r approaches v_f and $v_f/(v_f - v_r)$ approaches infinity. The reaction therefore responds to a fractional increase in flux by a much smaller fractional increase in its substrate concentration.

Consequently, *the ability of a reaction to communicate a change in flux increases as the reaction approaches equilibrium.* A series of sequential reactions that are all near equilibrium therefore have the same flux.

Rate-Determining Step in the Control of Pathway Flux

The metabolic flux through an entire pathway is determined by its rate-determining step (or steps) which, by definition, is much slower than the following reaction step(s). The product(s) of the rate-determining step is therefore removed before it can equilibrate with reactant so that the rate-determining step functions far from equilibrium and has a large negative free energy change. In an analogous manner, the flow of a river can only be controlled by a dam, which creates a difference in water levels between its upstream and downstream sides, a situation that also has a large negative free energy change, in this case resulting from the hydrostatic pressure head. Yet, as we have just seen, the fractional change in flux, $\Delta J/J$, of a nonequilibrium reaction ($v_f \gg v_r$), is not very sensitive to the fractional change in its substrate concentration, $\Delta[A]/[A]$; for example, its substrate concentration must double (in the absence of other controlling effects) in order to double the reaction flux (Eq. [16.7]). Yet, some pathway fluxes vary by factors that are much greater than can be explained by changes in substrate concentrations. For example, glycolytic fluxes are known to vary by factors of 100 or more while variations of substrate concentrations over such a large range are unknown. Consequently, while changes in substrate concentration can communicate a change in flux at the rate-determining step to the other (near equilibrium; $v_f \approx v_r$) reaction steps of the pathway, there must be other mechanisms that control the flux of the rate-determining step.

The flux through the rate-determining step of a pathway may be altered by several mechanisms:

1. **Allosteric control:** Many enzymes are allosterically regulated (Section 12-4) by effectors that are often substrates, products, or coenzymes in the pathway but not necessarily of the enzyme in question (feedback regulation). One such enzyme is PFK, an important glycolytic control enzyme (Section 16-4B).

2. **Covalent modification (enzymatic interconversion):** Many enzymes that control pathway fluxes have specific sites that may be enzymatically phosphorylated and dephosphorylated or covalently modified in some other way. Such enzymatic modification processes, which are themselves subject to control, greatly alter the activities of the modified enzymes. This flux control mechanism is discussed in Section 17-3.

3. **Substrate cycles:** If v_f and v_r in Eq. [16.7] represent the rates of two opposing nonequilibrium reactions that are catalyzed by different enzymes, v_f and v_r may be independently varied. The flux through such a substrate cycle, as we shall see in the next section, is more sensitive to the concentrations of allosteric effectors than is the flux through a single unopposed nonequilibrium reaction.

4. **Genetic control:** Enzyme concentrations, and hence enzyme activities, may be altered by protein synthesis in response to metabolic needs. Genetic control of enzyme concentrations is a major concern of Part V of this text.

Mechanisms 1 to 3 can respond rapidly (within seconds or minutes) to external stimuli and are therefore classified as "short-term" control mechanisms. Mechanism 4 responds more slowly to changing conditions (within hours or days in higher organisms) and is therefore referred to as a "long-term" control mechanism.

B. Control of Glycolysis in Muscle

Elucidation of the flux control mechanisms of a given pathway involves the determination of the pathway's regulatory enzymes controlling the rate-determining steps together with the identification of the modulators of these enzymes and their mechanism(s) of modulation. A hypothesis may then be formulated that can be tested *in vivo*. A common procedure for establishing control mechanisms involves three steps.

1. Identification of the rate-determining step(s) of the pathway. One way to do so is to measure the *in vivo* ΔG's of all the reactions in the pathway to determine how close to equilibrium they function. Those that operate far from equilibrium are potential control points; the enzymes catalyzing them may be regulated by one or more of the mechanisms listed above. Another way of establishing the rate-determining step(s) of a pathway is to measure the effect of a known inhibitor on a specific reaction step and on the flux through the pathway as a whole. The ratio of the fractional change in the activity of the inhibited enzyme to the fractional change in the total flux will vary between 0 and 1. The closer the ratio is to 1, the more control is exerted by that particular enzyme on the total flux through the pathway.

2. *In vitro* identification of allosteric modifiers of the enzymes catalyzing the rate-determining reactions. The mechanisms by which these compounds act are determined from their effects on the enzyme's kinetics. From this information, a model of the allosteric control mechanisms for the pathway may be formulated.

3. Measurement of the *in vivo* levels of the proposed regulators under various conditions to establish whether these concentration changes are consistent with the proposed control mechanism.

Free Energy Changes in the Reactions of Glycolysis

Let us examine the thermodynamics of glycolysis with an eye towards understanding its control mechanisms. This must be done separately for each type of

Table 16-2

Standard Free Energy Changes ($\Delta G^{\circ\prime}$), and Physiological Free Energy Changes (ΔG) in Heart Muscle, of the Reactions of Glycolysis[a]

Reaction	Enzyme	$\Delta G^{\circ\prime}$ (kJ·mol^{-1})	ΔG (kJ·mol^{-1})
1	Hexokinase	-20.9	-27.2
2	PGI	$+2.2$	-1.4
3	PFK	-17.2	-25.9
4	Aldolase	$+22.8$	-5.9
5	TIM	$+7.9$	$+4.4$
6+7	GAPDH + PGK	-16.7	-1.1
8	PGM	$+4.7$	-0.6
9	Enolase	-3.2	-2.4
10	PK	-23.0	-13.9

[a] Calculated from data in Newsholme, E. A. and Start, C., *Regulation in Metabolism*, p. 97, Wiley (1973).

tissue in question because different tissues control glycolysis in different ways. We shall confine ourselves to muscle tissue. First we establish the pathway's possible control points through the identification of its nonequilibrium reactions. Table 16-2 lists the standard free energy change ($\Delta G^{\circ\prime}$) and the actual physiological free energy change (ΔG) associated with each reaction in the pathway. It is important to realize that the free energy changes associated with the reactions under standard conditions may differ dramatically from those in effect under physiological conditions. For example, the $\Delta G^{\circ\prime}$ for aldolase is $+22.8$ kJ·mol^{-1} while under physiological conditions in heart muscle it is close to zero, indicating that the *in vivo* activity of aldolase is sufficient to equilibrate its substrates and products. The same is true of the GAPDH + PGK reaction series.

Only three reactions, those catalyzed by hexokinase, phosphofructokinase, and pyruvate kinase, function with large negative free energy changes in heart muscle under physiological conditions. These nonequilibrium reactions of glycolysis are the candidates for the flux-control points. The other glycolytic reactions function near equilibrium: Their forward and reverse rates are much faster than the actual flux through the pathway. Consequently, these equilibrium reactions are very sensitive to changes in the concentration of pathway intermediates and rapidly communicate any changes in flux generated at the rate-determining step(s) throughout the rest of the pathway.

Phosphofructokinase Is the Major Flux-Controlling Enzyme of Glycolysis in Muscle

In vitro kinetic studies of hexokinase, phosphofructokinase, and pyruvate kinase indicate that each is controlled by a variety of compounds, some of which are listed in Table 16-3. Yet, when the G6P source for gly-

Table 16-3

Some Effectors of the Nonequilibrium Enzymes of Glycolysis

Enzyme	Inhibitors	Activators[a]
Hexokinase	G6P	
PFK	ATP, citrate	ADP, AMP, cAMP, FBP, F2,6P, F6P, NH$_4^+$, P$_i$
PK (muscle)	ATP	

[a] The activators for PFK are better described as deinhibitors of ATP because they reverse the effect of inhibitory concentrations of ATP.

colysis is glycogen, rather than glucose, as is often the case in skeletal muscle (Section 17-1), the hexokinase reaction is not required. *PFK, an elaborately regulated enzyme functioning far from equilibrium, evidently is the major control point for glycolysis in muscle under most conditions.*

PFK (Fig. 16-27) is a tetrameric enzyme with two conformational states, R and T, that are in equilibrium. ATP is both a substrate and an allosteric inhibitor of PFK. Each subunit has two binding sites for ATP: a substrate site and an inhibitor site. The substrate site binds ATP

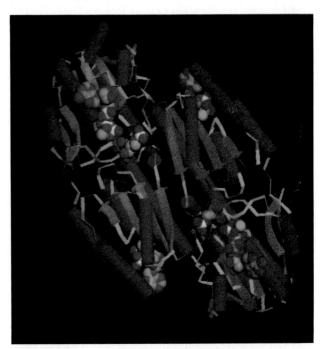

Figure 16-27

A computer-drawn ribbon diagram of PFK from *Bacillus stearothermophilus*. Two subunits of the tetrameric molecule are shown (related by a twofold axis perpendicular to the page through the center of the figure). Each of the two subunits in the protein is in association with its substrates F6P (*near the center of each subunit*) and ATP·Mg^{2+} (*lower right and upper left;* the green balls represent Mg^{2+}), together with the activator ADP·Mg^{2+} (*top right and lower left*). [Courtesy of Arthur Lesk, Cambridge University and EMBL. X-ray structure determined by Phillip Evans.]

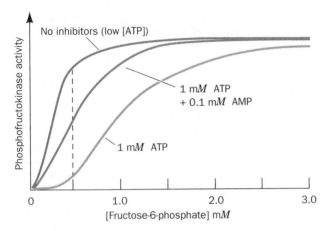

Figure 16-28
PFK activity versus F6P concentration under various conditions: blue, no inhibitors (low, noninhibitory [ATP]); yellow-green, 1 mM ATP (inhibitory); and red, 1 mM ATP + 0.1 mM AMP. [After data from Mansour, T. E. and Ahlfors, C. E., *J. Biol. Chem.* **243**, 2523–2533 (1968).]

equally well in either conformation but the inhibitor site binds ATP almost exclusively in the T state. The other substrate of PFK, F6P, preferentially binds to the R state. Consequently, at high concentrations, ATP acts as a heterotropic allosteric inhibitor of PFK by binding to the T state, thereby shifting the T ⇌ R equilibrium in favor of the T state and thus decreasing PFK's affinity for F6P (this is similar to the action of 2,3-BPG in decreasing the affinity of hemoglobin for O_2; Section 9-2F). In graphical terms, at high concentrations of ATP, the hyperbolic (noncooperative) curve of PFK activity versus [F6P] is converted to the sigmoidal (cooperative) curve characteristic of allosteric enzymes (Fig. 16-28; cooperative and noncooperative processes are discussed in Section 9-1B). For example, when [F6P] = 0.5 mM (the dashed line in Fig. 16-28), the enzyme is maximally active, but in the presence of 1 mM ATP, the activity drops to 15% of its original level (a nearly sevenfold decrease). [Actually, the most potent allosteric effector of PFK is **fructose-2,6-bisphosphate (F2,6P)**. We discusss the role of F2,6P in controlling PFK activity when we study the mechanism by which the liver maintains blood glucose concentrations (Section 17-3F).]

Direct allosteric regulation of PFK by ATP may superficially appear to be the means by which glycolytic flux is controlled. After all, when [ATP] is high as a result of low metabolic demand, PFK is inhibited and flux through the pathway is low; conversely, when [ATP] is low, flux through the pathway is high and ATP is synthesized to replenish the pool. Consideration of the physiological variation in ATP concentration, however, indicates that the situation must be more complex. The metabolic flux through glycolysis may vary by 100-fold or more, depending on the metabolic demand for ATP. However, measurements of [ATP] *in vivo* at various levels of metabolic activity indicate that [ATP] varies

<10% between rest and vigorous exertion. Yet, *there is no known allosteric mechanism that can account for a 100-fold change in flux of a nonequilibrium reaction with only 10% change in effector concentration.* Thus some other mechanism, or mechanisms, must be responsible for controlling glycolytic flux.

AMP Overcomes the ATP Inhibition of PFK

The inhibition of PFK by ATP is relieved by AMP. This results from AMP's preferential binding to the R state of PFK. If a PFK solution containing 1 mM ATP and 0.5 mM F6P is brought to 0.1 mM in AMP, the activity of PFK rises from 15 to 50% of its maximal activity, a threefold increase (Fig. 16-28).

[ATP] decreases by only 10% in going from a resting state to one of vigorous activity because it is buffered by the action of two enzymes: creatine kinase (Section 15-4C) and, of particular importance to this discussion, **adenylate kinase** (also known as **myokinase**). Adenylate kinase catalyses the reaction

$$2ADP \rightleftharpoons ATP + AMP \qquad K = \frac{[ATP][AMP]}{[ADP]^2} = 0.44$$

which rapidly equilibrates the ADP resulting from ATP hydrolysis in muscle contraction with ATP and AMP.

In muscle, [ATP] is ~50 times [AMP] and ~10 times [ADP] so that, *as a result of the adenylate kinase reaction, a 10% decrease in [ATP] will cause over a fourfold increase in [AMP] (see Problem 11).* Consequently, a metabolic signal consisting of a decrease in [ATP] too small to relieve PFK inhibition is amplified significantly by the adenylate kinase reaction, which increases [AMP] by an amount sufficient to produce a much larger increase in PFK activity.

Substrate Cycling Increases Flux Sensitivity

Even though a mechanism exists for amplifying the effect of a small change in [ATP] by producing a larger change in [AMP], a fourfold increase in [AMP] would only allosterically increase the activity of PFK by ~10-fold, an amount insufficient to account for the observed 100-fold increase in glycolysis flux. Small changes in effector concentration (and therefore v_f) can only cause relatively large changes in the flux through a reaction $(v_f - v_r)$ if the reaction is functioning close to equilibrium. The reason for this high sensitivity is that for such reactions, the term $v_f/(v_f - v_r)$ in Eq. [16.7] is large, that is, the reverse reaction contributes significantly to the value of the net flux. This is not the case for the PFK reaction.

Such equilibriumlike conditions may be imposed on a nonequilibrium reaction if a second enzyme catalyzes the regeneration of substrate from product in a thermodynamically favorable manner. Then v_r is no longer negligible compared to v_f (although increasing v_r forces the forward reaction even further out of equilibrium than it would otherwise be). This situation requires that the forward process (formation of FBP from F6P) and reverse process (breakdown of FBP to F6P) be accomplished by different reactions since the laws of thermodynamics would otherwise be violated. In the following paragraphs, we discuss the nature of such **substrate cycles.**

Under physiological conditions, the reaction catalyzed by PFK:

Fructose-6-phosphate + ATP ⟶
$\qquad$ fructose-1,6-bisphosphate + ADP

is highly exergonic ($\Delta G = -25.9$ kJ·mol^{-1}, Table 16-2). Consequently, the back reaction has a negligible rate compared to the forward reaction. **Fructose-1,6-bis-phosphatase (FBPase),** however, which is present in many mammalian tissues, catalyzes the exergonic hydrolysis of FBP ($\Delta G = -8.6$ kJ·mol^{-1}):

Fructose-1,6-bisphosphate + H$_2$O ⟶
$\qquad$ fructose-6-phosphate + P$_i$

Note that the combined reactions catalyzed by PFK and FBPase result in net ATP hydrolysis:

$$ATP + H_2O \rightleftharpoons ADP + P_i$$

Such a set of opposing reactions is known as a substrate cycle because it cycles a substrate to an intermediate and back again. When this set of reactions was discovered, it was referred to as a **futile cycle** since its net result seemed to be the useless consumption of ATP. In fact, when it was found that activators of PFK, such as AMP, inhibit FBPase and *vice versa*, it was suggested that only one of these enzymes was functional in a cell under any given set of conditions. It was subsequently demonstrated, however, that both enzymes often function simultaneously at significant rates.

Substrate Cycling Can Account for Glycolytic Flux Variation

Eric Newsholme has proposed that substrate cycles are not at all "futile," but rather, have a regulatory function. The *in vivo* activities of enzymes and concentrations of metabolites are extremely difficult to measure so that their values are rarely known accurately. However, let us make the physiologically reasonable assumption that a fourfold increase in [AMP], resulting from the adenylate kinase reaction, causes PFK activity (v_f) to increase from 10 to 90% of its maximum and FBPase activity (v_r) to decrease from 90 to 10% of its maximum. The maximum activity of muscle PFK is known from *in vitro* studies to be ~ 10-fold greater than that of muscle FBPase. Hence, if we assign full activity of PFK to be 100 arbitrary units, then full activity of FBPase is 10 such units. The flux through the PFK reaction in glycolysis under conditions of low [AMP] is

$$J_{low} = v_f(low) - v_r(low) = 10 - 9 = 1$$

where v_f is catalyzed by PFK and v_r by FBPase. The flux under conditions of high [AMP] is

$$J_{high} = v_f(high) - v_r(high) = 90 - 1 = 89$$

Substrate cycling could therefore amplify the effect of changes in [AMP] on the net rate of phosphorylation of F6P. Without the substrate cycle, a fourfold increase in [AMP] increases the net flux by about ninefold whereas, with the cycle, the same increase in [AMP] causes a $J_{high}/J_{low} = 89/1 \approx 90$-fold increase in net flux. Consequently, under the above assumptions, *a 10% change in [ATP] could stimulate a 90-fold change in flux through the glycolytic pathway by a combination of the adenylate kinase reaction and substrate cycles.*

Physiological Impact of Substrate Cycling

Substrate cycling, if it has a regulatory function, does not increase the maximum flux through a pathway. On the contrary, it functions to decrease its minimum flux. In a sense, the substrate is put into a "holding pattern." In the case described above, *the cycling of substrate is the energetic "price" that a muscle must pay to be able to change rapidly from a resting state, in which substrate cycling is maximal, to one of sustained high activity.* However, the rate of substrate cycling may itself be under hormonal or nervous control so as to increase the sensitivity of the metabolic system under conditions when high activity (fight or flight) is anticipated (we address the involvement of hormones in metabolic regulation in Sections 17-3E and F).

In some tissues, substrate cycles function to produce heat. For example, many insects require a thoracic temperature of 30°C to be able to fly. Yet bumblebees are capable of flight at ambient temperatures as low as 10°C. Bumblebee flight muscle FBPase has a maximal activity similar to that of its PFK (10-fold greater than our example for mammalian muscle); furthermore, unlike all other known muscle FBPases, it is not inhibited by AMP. This permits the FBPase and PFK of bumblebee flight muscle to be highly active simultaneously so as to generate heat. Since the maximal rate of FBP cycling possible in bumblebee flight muscle generates only 10 to 15% of the required heat, however, other mechanisms of thermogenesis must also be operative. Nevertheless, FBP cycling must be signficant because, unlike bumblebees, honeybees, which have no FBPase activity in their flight muscles, cannot fly when the temperature is low.

For those individuals with a rare hereditary condition known as **malignant hyperthermia,** the administration of halogenated anesthetic gases causes an extremely rapid rise in body temperature that is fatal in the absence of aggressive treatment. This condition, in a strain of "stress-prone" pigs, is caused by uncontrolled substrate cycling of FBP in the muscles. The mechanism by which the anesthetic causes this loss of control is unknown although it has been postulated that it somehow deranges the normal hormonal regulatory systems.

5. METABOLISM OF HEXOSES OTHER THAN GLUCOSE

While glucose is the primary end product of the digestion of starch and glycogen (Sections 10-2B and C), three other hexoses are also prominent digestion products: **fructose**, obtained from fruits and from the hydrolysis of sucrose (table sugar); **galactose**, obtained from the hydrolysis of lactose (milk sugar); and **mannose**, obtained from the digestion of polysaccharides and glycoproteins. After digestion, these monosaccharides enter the bloodstream, which carries them to various tissues. *The metabolism of fructose, galactose, and mannose proceeds by their conversion to glycolytic intermediates, from which point they are broken down identically to glucose.*

A. Fructose

Fructose is a major fuel source in diets that contain large amounts of sucrose (a disaccharide of fructose and glucose). There are two pathways for the metabolism of fructose; one occurs in muscle and the other occurs in liver. This dichotomy results from the different enzymes present in these various tissues.

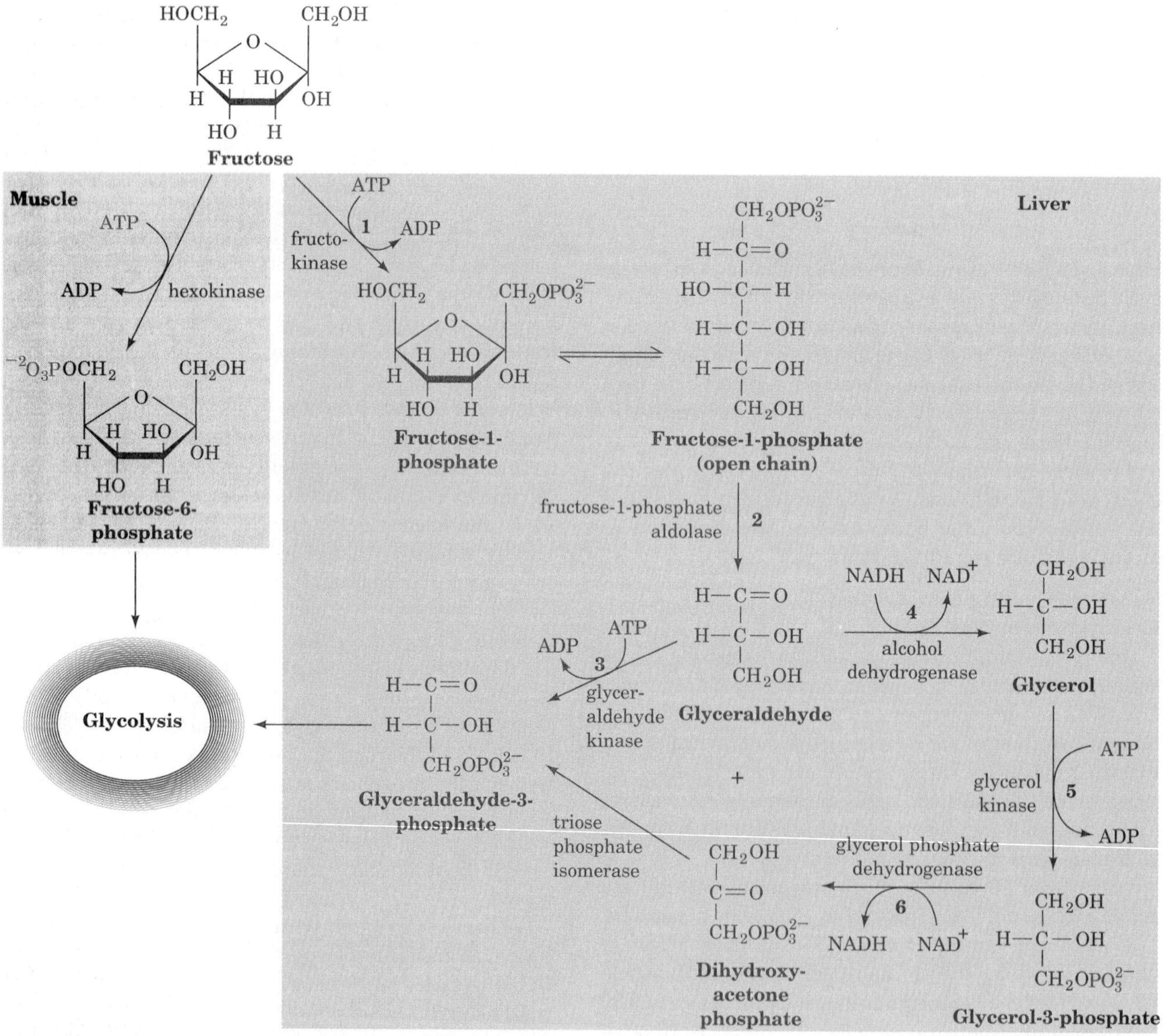

Figure 16-29
The metabolism of fructose. In muscle (*left*), the conversion of fructose to the glycolytic intermediate F6P involves only one enzyme, hexokinase. In liver (*right*), six enzymes participate in the conversion of fructose to glycolytic intermediates: **(1)** fructokinase, **(2)** fructose-1-phosphate aldolase, **(3)** glyceraldehyde kinase, **(4)** alcohol dehydrogenase, **(5)** glycerol kinase, and **(6)** glycerol phosphate dehydrogenase.

Fructose metabolism in muscle differs little from that of glucose. Hexokinase (Section 16-2A), which converts glucose to G6P on entry into muscle cells, also phosphorylates fructose yielding F6P (Fig. 16-29, *left*). The entry of fructose into glycolysis therefore involves only one reaction step:

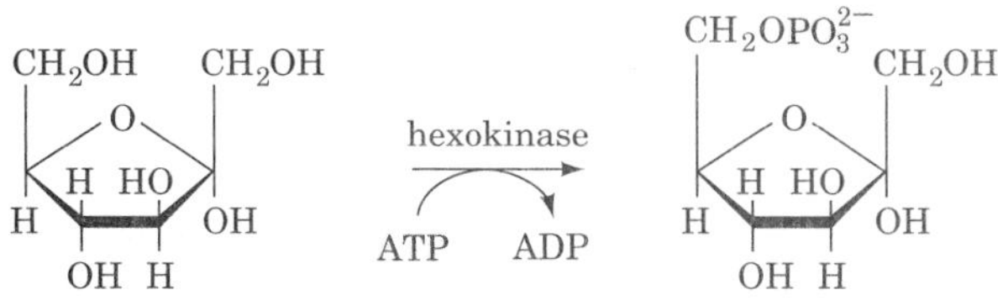

Liver contains little hexokinase; rather it contains glucokinase, which phosphorylates only glucose (Section 16-2A). Fructose metabolism in liver must therefore differ from that in muscle. In fact, liver converts fructose to glycolytic intermediates through a pathway that involves six enzymes (Fig. 16-29, *right*):

1. **Fructokinase** catalyzes the phosphorylation of fructose by ATP at C(1) to form **fructose-1-phosphate.** *Neither hexokinase nor phosphofructokinase can phosphorylate fructose-1-phosphate at C(6) to form the glycolytic intermediate fructose-1,6-bisphosphate.*

2. Class I aldolase (Section 16-2D) has several isoenzymic forms. Muscle contains Type A aldolase, which is specific for fructose-1,6-bisphosphate. Liver, however, contains Type B aldolase, which also utilizes fructose-1-phosphate as a substrate (Type B aldolase is sometimes called **fructose-1-phosphate aldolase**). In liver, fructose-1-phosphate therefore undergoes an aldol cleavage (Section 16-2D):

   Fructose-1-phosphate $\rightleftharpoons$
   dihydroxyacetone phosphate + glyceraldehyde

   The glyceraldehyde thus formed is converted to glyceraldehyde-3-phosphate by Reaction 3, or to dihydroxyacetone phosphate by a combination of Reactions 4 to 6 (Fig. 16-29).

3. Direct phosphorylation of glyceraldehyde by ATP through the action of **glyceraldehyde kinase** forms the glycolytic intermediate glyceraldehyde-3-phosphate.

4-6. Alternatively, glyceraldehyde is converted to the glycolytic intermediate dihydroxyacetone phosphate by reduction to glycerol by NAD$^+$ as catalyzed by alcohol dehydrogenase (Reaction 4), phosphorylation to glycerol-3-phosphate by ATP through the action of **glycerol kinase** (Reaction 5), and reoxidation by NADH to dihydroxyacetone phosphate as mediated by glycerol phosphate dehydrogenase (Reaction 6).

As this complex series of reactions suggests, the liver has an enormous repertory of enzymes. This is because the liver is involved in the breakdown of a great variety of metabolites. Efficiency in metabolic processing dictates that many of these substances be converted to glycolytic intermediates. The liver, in fact, contains many of the enzymes necessary to do so.

Excessive Fructose Depletes Liver P$_i$

Until recently, fructose was thought to have advantages over glucose for intravenous feeding. The liver, however, encounters metabolic problems when the blood concentration of this sugar is too high (higher than can be attained by simply eating fructose-containing foods). When the fructose concentration is high, fructose-1-phosphate may be produced faster than Type B aldolase can cleave it. Intravenous feeding of large amounts of fructose may therefore result in high enough fructose-1-phosphate accumulation to severely deplete the liver's store of P$_i$. Under these conditions [ATP] drops, thereby activating glycolysis and lactate production. The lactate concentration in the blood under such conditions can reach life-threatening levels.

Fructose intolerance, a genetic disease in which ingestion of fructose causes the same fructose-1-phosphate accumulation as with its intravenous feeding, results from a deficiency of Type B aldolase. This condition appears to be self-limiting: Individuals with fructose intolerance rapidly develop a strong distaste for anything sweet.

B. Galactose

Galactose comprises half of the milk sugar lactose, and is thus a major fuel constituent of dairy products. Galactose and glucose are epimers that differ only in their configuration about C(4).

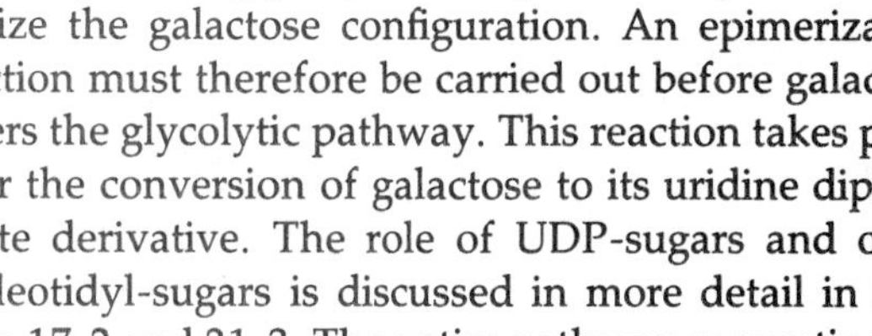

The enzymes of glycolysis are specific; they do not recognize the galactose configuration. An epimerization reaction must therefore be carried out before galactose enters the glycolytic pathway. This reaction takes place after the conversion of galactose to its uridine diphosphate derivative. The role of UDP-sugars and other nucleotidyl-sugars is discussed in more detail in Sections 17-2 and 21-3. The entire pathway converting galactose to a glycolytic intermediate involves five reactions (Fig. 16-30):

1. Galactose is phosphorylated at C(1) by **galactokinase.**

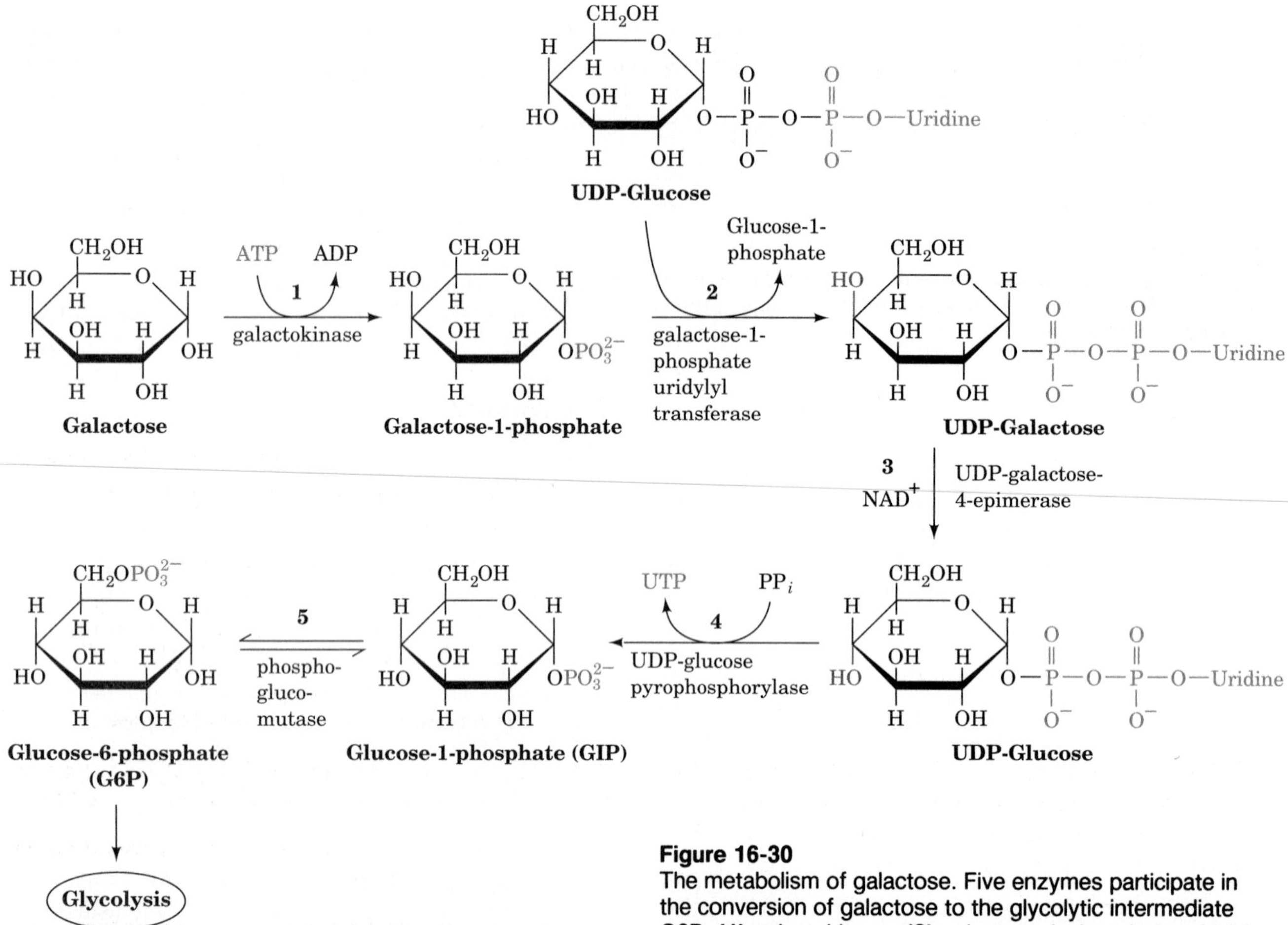

UDP-Glucose

Galactose → (ATP, ADP) **1** galactokinase → **Galactose-1-phosphate** → (Glucose-1-phosphate) **2** galactose-1-phosphate uridylyl transferase → **UDP-Galactose**

3 NAD$^+$ UDP-galactose-4-epimerase

Glucose-6-phosphate (G6P) ⇌ **5** phosphoglucomutase **Glucose-1-phosphate (GIP)** ← (UTP, PP$_i$) **4** UDP-glucose pyrophosphorylase ← **UDP-Glucose**

→ Glycolysis

Figure 16-30
The metabolism of galactose. Five enzymes participate in the conversion of galactose to the glycolytic intermediate G6P: **(1)** galactokinase, **(2)** galactose-1-phosphate uridylyl transferase, **(3)** UDP-galactose-4-epimerase, **(4)** UDP-glucose pyrophosphorylase, and **(5)** phosphoglucomutase.

2. **Galactose-1-phosphate uridylyl transferase** transfers the uridylyl group of **UDP-glucose** to galactose-1-phosphate to form **UDP-galactose** by the reversible cleavage of UDP-glucose's pyrophosphoryl bond.

3. **UDP-galactose-4-epimerase** interconverts UDP-galactose and UDP-glucose. This enzyme has an associated NAD$^+$, which suggests that the reaction involves the sequential oxidation and reduction of the hexose C(4) atom:

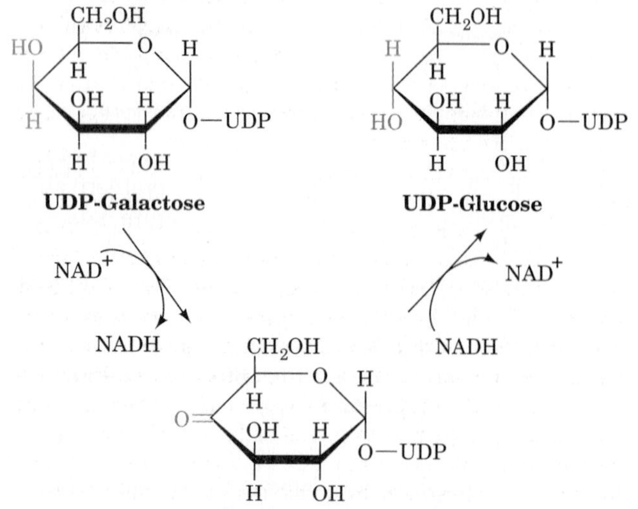

UDP-Galactose (NAD$^+$ → NADH) ⇌ ⇌ (NADH → NAD$^+$) **UDP-Glucose**

4. UDP-glucose is converted to G1P by **UDP-glucose pyrophosphorylase** (Section 17-2A).

5. G1P is converted to the glycolytic intermediate G6P by the action of **phosphoglucomutase** (Section 17-1B).

Galactosemia

Galactosemia is a genetic disease characterized by an inability to convert galactose to glucose. Its symptoms include failure to thrive, mental retardation and, in some instances, death from liver damage. Most cases of galactosemia involve a deficiency in the enzyme catalyzing Reaction 2 of the interconversion, galactose-1-phosphate uridylyl transferase. Formation of UDP-galactose from galactose-1-phosphate is thus prevented, leading to a buildup of toxic metabolic byproducts. For example, the increased galactose concentration in the blood results in a higher galactose concentration in the

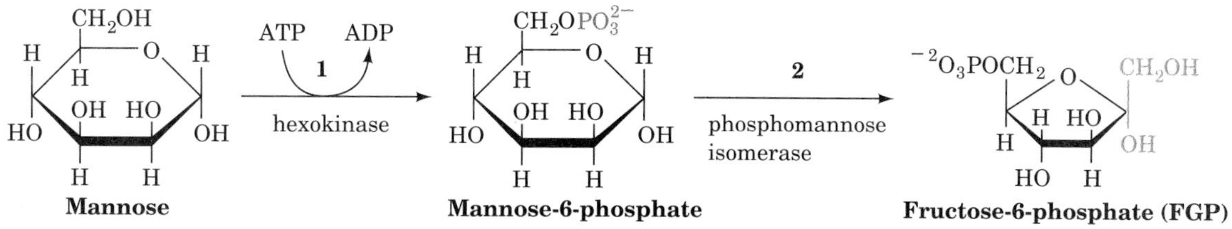

Figure 16-31
The metabolism of mannose. Two enzymes are required to convert mannose to the glycolytic intermediate F6P: **(1)** hexokinase and **(2)** phosphomannose isomerase.

lens of the eye where this sugar is reduced to **galactitol.**

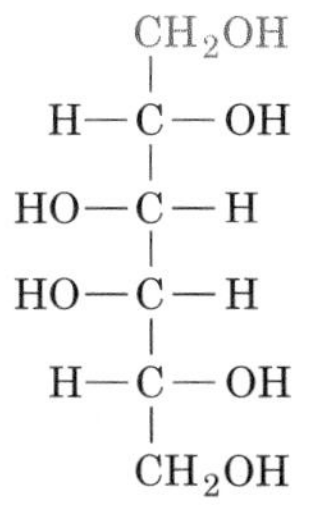

D-Galactitol

The presence of this sugar alcohol in the lens eventually causes cataract formation (clouding of the lens).

Galactosemia is treated by a galactose-free diet. Except for the mental retardation, this reverses all symptoms of the disease. The galactosyl units that are essential for the synthesis of glycoproteins (Section 10-3C) and glycolipids (Section 11-1D) may be synthesized from glucose by a reversal of the epimerase reaction. These syntheses therefore do not require dietary galactose.

C. Mannose

Mannose, a common component of glycoproteins (Section 10-3C), and glucose are C(2) epimers:

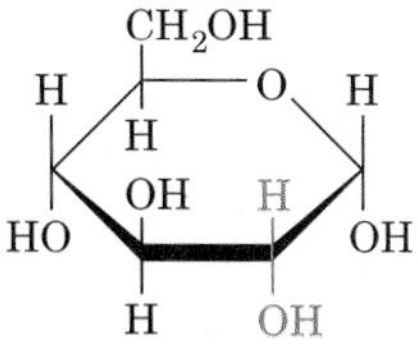

α-D-Glucose 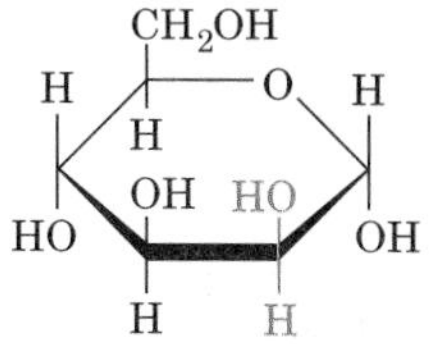   α-D-Mannose

Mannose enters the glycolytic pathway after its conversion to F6P via a two-reaction pathway (Fig. 16-31):

1. **Hexokinase** (Section 16-2A) converts mannose to mannose-6-phosphate.

2. **Phosphomannose isomerase** then converts this aldose to the ketose F6P. The mechanism of the phosphomannose isomerase reaction resembles that catalyzed by phosphoglucose isomerase (Section 16-2B); it involves an enediolate intermediate.

Chapter Summary

Glycolysis is the metabolic pathway by which most life forms degrade glucose to two molecules of pyruvate with the concomitant net generation of two ATPs. The overall reaction:

Glucose + 2NAD$^+$ + 2ADP + 2P$_i$ $\longrightarrow$
$\qquad$ 2NADH + 2pyruvate + 2ATP + 2H$_2$O + 4H$^+$

occurs in 10 enzymatically catalyzed reactions.

In the preparatory stage of glycolysis, which encompasses its first five reactions, glucose reacts with 2ATPs, in an "energy investment," to form fructose-1,6-bisphosphate, which is subsequently converted to two molecules of glyceraldehyde-3-phosphate. In the second stage of glycolysis, the "payoff" stage, which comprises its last five reactions, glyceraldehyde-3-phosphate reacts with NAD$^+$ and P$_i$ to form the "high-energy" compound 1,3-bisphosphoglycerate. This compound reacts in the last four reactions of the pathway with 2ADPs to form pyruvate and 2ATPs/molecule. The mechanisms of most of the 10 glycolytic enzymes have been eluci-

dated through chemical and kinetic measurements combined with X-ray structural studies. The glycolytic enzymes exhibit stereospecificity in the reactions that they catalyze. In at least two kinases, phosphoryl transfer from substrate to water is prevented by substrate-induced conformational changes that form the active site and exclude water from it.

The NAD$^+$ consumed in the formation of 1,3-BPG must be regenerated if glycolysis is to continue. In the presence of O$_2$, NAD$^+$ is regenerated by oxidative phosphorylation in the mitochondria. Under anaerobic conditions in muscle, pyruvate is reduced by NADH yielding lactate and NAD$^+$ in a reaction catalyzed by lactate dehydrogenase. In many muscles, particularly during strenuous activity, the process of homolactic fermentation is a major free energy source. In anaerobic yeast, NAD$^+$ is regenerated by alcoholic fermentation in two reactions. First pyruvate is decarboxylated to acetaldehyde by pyruvate decarboxylase, an enzyme that requires thiamine pyrophosphate as a cofactor. The acetaldehyde is then reduced by NADH to form ethanol and NAD$^+$ in a reaction catalyzed by alcohol dehydrogenase.

The flux through a reaction that is close to equilibrium is very sensitive to changes in substrate concentration. Hence, the steady state flux through a metabolic pathway can only be controlled by a nonequilibrium reaction. Nonequilibrium reactions are regulated by allosteric interactions, substrate cycles, covalent modification, and genetic (long-term) control mechanisms. In muscle glycolysis, phosphofructokinase (PFK) catalyzes the flux-generating step. Although PFK is inhibited by high concentrations of one of its substrates, ATP, the 10% variation of [ATP] over the range of metabolic activity has insufficient influence on PFK activity to account for the observed 100-fold range in glycolytic flux. [AMP] has a four-fold variation in response to the 10% variation of [ATP] through the action of adenylate kinase. Although AMP relieves the ATP inhibition of PFK, its concentration variation is also insufficient to account for the observed glycolytic flux range. However, the product of the PFK reaction, fructose-1,6-bisphosphate, is hydrolyzed to F6P by FBPase, which is inhibited by AMP. The substrate cycle catalyzed by these two enzymes confers, at least in principle, the necessary sensitivity of the glycolytic flux to variations in [AMP].

Digestion of carbohydrates yields glucose as the primary product. Other prominent products are fructose, galactose, and mannose. These monosaccharides are metabolized through their conversion to glycolytic intermediates.

References

General

Cunningham, E. B., *Biochemistry. Mechanisms of Metabolism*, Chapter 9, McGraw–Hill (1978).

Fersht, A., *Enzyme Structure and Mechanism* (2nd ed.), Freeman (1985).

Fruton, J. S., *Molecules and Life: Historical Essays on the Interplay of Chemistry and Biology*, Wiley–Interscience (1974). [Includes a detailed historical account of the elucidation of fermentation.]

Saier, M. H., Jr., *Enzymes in Metabolic Pathways*, Chapter 5, Harper & Row (1987).

Walsh, C., *Enzymatic Reaction Mechanisms*, Freeman (1979).

Enzymes of Glycolysis

Structure and function of consecutive glycolytic enzymes, *Biochem. Soc. Trans.* **5**, 642–659 (1977). [A series of papers on the structures of TIM, GAPDH, phosphoglycerate kinase, phosphoglycerate mutase, and PK.]

The Enzymes of Glycolysis: Structure, Activity and Evolution, *Philos. Trans. R. Soc. London Ser. B* **293**, 1–214 (1981). [A collection of authoritative discussions on the enzymes of glycolysis. Also available from the Royal Society as a bound volume.]

Anderson, C. M., Zucker, F. H., and Steitz, T. A., Space-filling models of kinase clefts and conformation changes, *Science* **204**, 375–380 (1979). [The examination of computer-generated space-filling models demonstrates that kinases have deep active site clefts that close upon binding substrate.]

Banks, R. D., Blake, C. C. F., Evans, P. R., Rice, D. W., Hardy, G. W., Merritt, M., and Phillips, A. W., Sequence, structure and activity of phosphoglycerate kinase: A possible hinge bending enzyme, *Nature* **279**, 773–777 (1979).

Bennett, W. S. Jr., and Steitz, T. A., Glucose-induced conformational change in yeast hexokinase, *Proc. Natl. Acad. Sci.* **75**, 4848–4852 (1978).

Biesecker, G., Harris, J. I., Thierry, J. C., Walker, J. E., and Wonacott, A. J., Sequence and structure of D-glyceraldehyde-3-phosphate dehydrogenase from *Bacillus stearothermophilus*, *Nature* **266**, 328–333 (1977).

Boyer, P. D. (Ed.) *The Enzymes* (3rd ed.,), Vols. 5–9 and 13, Academic Press (1972–1976). [Contains detailed reviews of the various glycolytic enzymes.]

Buehner, M., Ford, G. C., Moras, D., Olsen, K. W., and Rossmann, M. G., The three-dimensional structure of D-glyceraldehyde-3-phosphate dehydrogenase, *J. Mol. Biol.* **90**, 25–49 (1974). [The structure of the lobster enzyme.]

Delivoria-Papadopoulos, M., Oska, F. A., and Gottlieb, A. J., Oxygen–hemoglobin dissociation curves: effect of inherited enzyme defects of the red cell, *Science* **165**, 601–602 (1969).

Evans, P. R. and Hudson, P. J., Structure and control of phosphofructokinase from *Bacillus stearothermophilus*, *Nature* **279**, 500–504 (1979).

Knowles, J. R. and Albery, W. J., Perfection in enzyme catalysis: The energetics of triosephosphate isomerase, *Acc. Chem. Res.* **10**, 105–111 (1977).

Lebioda, L. and Stec, B., Crystal structure of enolase indicates that enolase and pyruvate kinase evolved from a common ancestor, *Nature* **333**, 683–686 (1988).

Muirhead, H., Pyruvate kinase, *in* Jurnak, F. A. and McPherson, A. (Eds.), *Biological Macromolecules and Assemblies*, Vol. 3., *pp.* 141–186, Wiley (1987).

Muirhead, H., Triose phosphate isomerase, pyruvate kinase and other α/β-barrel enzymes, *Trends Biochem. Sci.* **8**, 326–330 (1983).

Rose, Z. B., The enzymology of 2,3-bisphosphoglycerate, *Adv. Enzymol.* **51**, 211–253 (1980).

Sygusch, J., Beaudry, D., and Allaire, M., Molecular architecture of rabbit skeletal muscle aldolase at 2.7-Å resolution, *Proc. Natl. Acad. Sci.* **84**, 7846–7850 (1987).

Enzymes of Anaerobic Fermentation

Adams, M. J., Buehner, M., Chandrasekhar, K., Ford, G. C., Hackert, M. L., Liljas, A., Rossmann, M. G., Smiley, I. E., Allison, W. S., Everse, J., Kaplan, N. O., and Taylor, S., Structure-function relationships in lactate dehydrogenase, *Proc. Natl. Acad. Sci.* **70**, 1968–1972 (1973).

Boyer, P. D. (Ed.), *The Enzymes* (3rd ed.), Vol. 11, Academic Press (1975). [Contains authoritative reviews on alcohol dehydrogenase, lactate dehydrogenase, and the evolutionary and structural relationships among the dehydrogenases.]

Park, J. H., Brown, R. L., Park, C. R., Cohn, M., and Chance, B., Energy metabolism in the untrained muscle of elite runners as observed by ^{31}P magnetic resonance spectroscopy: evidence suggesting a genetic endowment for endurance exercise. *Proc. Natl. Acad. Sci.* **85**, 8780–8785 (1988).

Control of Flux

Boscá, L. and Corredor, C., Is phosphofructokinase the rate-limiting step of glycolysis? *Trends Biochem. Sci.* **9**, 372–373 (1984).

Hoffman, E., Phosphofructokinase — a favorite of enzymologists and students of metabolic regulation, *Trends Biochem. Sci.* **3**, 145–147 (1978).

Katz, J. and Rognstad, R., Futile cycling in glucose metabolism, *Trends Biochem. Sci.* **3**, 171–174 (1978).

Newsholme, E. A., Challiss, R. A. J., and Crabtree, B., Substrate cycles: their role in improving sensitivity in metabolic control, *Trends Biochem. Sci.* **9**, 277–280 (1984).

Newsholme, E. A., Substrate cycles: their metabolic, energetic and thermic consequences in man, *Biochem. Soc. Symp.* **43**, 183–205 (1978).

Newsholme, E. A. and Crabtree, B., Substrate cycles in metabolic regulation and heat generation, *Biochem. Soc. Symp.* **41**, 61–110 (1976).

Newsholme, E. A. and Start, C., *Regulation in Metabolism*, Chapters 1 and 3, Wiley–Interscience (1973). [Contains a detailed discussion of the control of glycolysis in muscle.]

Pettigrew, D. W. and Frieden, C., Rabbit muscle phosphofructokinase. A model for regulatory kinetic behavior, *J. Biol. Chem.* **254**, 1896–1901 (1979).

Problems

1. Write out the reactions of the glycolytic pathway from glucose to lactate using structural formulas for all intermediates. Learn the names of these intermediates and the enzymes that catalyze the reactions.

2. $\Delta G^{\circ\prime}$ for the aldolase reaction is $+22.8$ kJ·mol^{-1}. In the cell, at 37°C, the mass action ratio [DHAP]/[GAP] = 5.5. Calculate the equilibrium ratio of [FBP]/[GAP] when [GAP] is (a) $2 \times 10^{-5}M$ and (b) $10^{-3}M$.

3. Iodoacetic acid inhibits glycolysis by inactivating glyceraldehyde-3-phosphate dehydrogenase (GAPDH). Fructose-1,6-bisphosphate accumulates under these conditions. Why does this metabolite accumulate rather than glyceraldehyde-3-phosphate and dihydroxyacetone phosphate, the products of the reaction immediately preceding GAPDH?

4. Arsenate, a structural analog of phosphate, can act as a substrate for any reaction in which phosphate is a substrate. Arsenate esters, unlike phosphate esters, are kinetically as well as thermodynamically unstable and hydrolyze almost instantaneously. Write a balanced overall equation for conversion of glucose to pyruvate in the presence of ATP, ADP, NAD$^+$, and either (a) phosphate or (b) arsenate. (c) Why is arsenate a poison?

5. When glucose is degraded anaerobically via glycolysis there is no overall oxidation or reduction of the substrate. The fermentation reaction is therefore said to be "balanced." The free energy required for ATP formation is nevertheless obtained from favorable electron-transfer reactions. Which metabolic intermediate is the electron donor and which is the electron acceptor when glucose is degraded by a balanced glycolytic fermentation: (a) in muscle and (b) in yeast?

6. In which carbon atoms of pyruvate would radioactivity be found if glucose metabolized by the glycolytic pathway were labeled with ^{14}C at: (a) C(1) and (b) C(4)?

(*Note:* Assume that triose phosphate isomerase is able to equilibrate dihydroxyacetone phosphate and glyceraldehyde-3-phosphate.)

*7. The following reaction is catalyzed by an enzyme very similar to Class I aldolases:

Fructose-6-phosphate + **Erythrose-4-phosphate**

⇅ transaldolase

Glyceraldehyde-3-phosphate + **Sedoheptulose-7-phosphate**

Write a plausible mechanism for this reaction using curved arrows to indicate the electron flow.

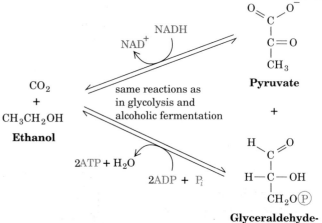

$$\textcircled{P} \equiv -PO_3^{2-}$$

Figure 16-32
The Entner–Doudoroff pathway for glucose breakdown.

8. The half-reactions involved in the LDH reaction and their standard reduction potentials are

$$\text{Pyruvate} + 2H^+ + 2e \longrightarrow \text{lactate} \qquad \mathscr{E}°' = -0.185 \text{ V}$$
$$NAD^+ + 2H^+ + 2e^- \longrightarrow NADH + H^+ \qquad \mathscr{E}°' = -0.315 \text{ V}$$

Calculate ΔG for the reaction under the following conditions:

(a) [lactate]/[pyruvate] = 1
   [NAD$^+$]/[NADH] = 1

(b) [lactate]/[pyruvate] = 160
   [NAD$^+$]/[NADH] = 160

(c) [lactate]/[pyruvate] = 1000
   [NAD$^+$]/[NADH] = 1000

(d) Under what conditions will the reaction spontaneously favor NADH oxidation?

(e) In order for the free energy change of the glyceraldehyde-3-phosphate dehydrogenase reaction to favor glycolysis, the [NAD$^+$]/[NADH] ratio must be maintained close to 10^3. Under anaerobic conditions in mammalian muscle, lactate dehydrogenase performs this function. How high can the [lactate]/[pyruvate] ratio become in muscle cells before the LDH-catalyzed reaction ceases to be favorable in the direction of NAD$^+$ production while maintaining the above [NAD$^+$]/[NADH] level?

*9. Based on the involvement of thiamine pyrophosphate (TPP) in the pyruvate decarboxylase reaction, which of the following reactions, if any, might be expected to utilize TPP as a cofactor?

(a)

(b)

Write hypothetical mechanisms for each reaction showing where TPP is involved or why it is unnecessary.

10. The glycolytic pathway for glucose breakdown is almost universal. Some bacteria, however, utilize an alternate route called the **Entner–Doudoroff pathway** (Fig. 16-32). Like the glycolytic pathway in yeast, the final product is ethanol. (a) Write balanced equations for the conversion of glucose to ethanol and CO_2 via the Entner–Doudoroff pathway and the yeast alcoholic fermentation. (b) Infer from your stoichiometries why the glycolytic pathway rather than the Entner–Doudoroff pathway is almost universal.

*11. The hydrolysis of ATP to ADP in the cell results in a concomitant change in [AMP] as mediated by adenylate kinase. (a) Assuming that [ATP] $\gg$ [AMP] and that the total adenine nucleotide concentration in the cell, $A_T = [AMP] + [ADP] + [ATP]$ is constant, derive an expression for [AMP] in terms of [ATP] and A_T. (b) Assuming an initial [ATP]/[ADP] of 10 and $A_T = 5$ mM, calculate the ratio of the final to initial values of [AMP] upon a 10% decrease of [ATP].

12. The symptoms of malignant hyperthermia include muscle rigidity as well as a dramatic rise in body temperature. The meat of stress-prone pigs that have died of this syndrome is highly acidic. Explain.

Chapter 17
GLYCOGEN METABOLISM

> Everything should be made as simple as possible but not simpler.
> *Albert Einstein*

Glucose, a major metabolic fuel source, is degraded via glycolysis to produce ATP (Chapter 16). Higher organisms protect themselves from potential fuel shortage by polymerizing excess glucose for storage as high molecular mass glucans (glucose polysaccharides) that may be readily mobilized in times of metabolic need. In plants, this glucose-storage substance is starch, a mixture of the $\alpha(1 \rightarrow 4)$-linked glucan α-amylose (Fig. 10-17), and amylopectin, which differs from α-amylose by the presence of $\alpha(1 \rightarrow 6)$ branches every 24 to 30 residues (Fig. 10-18). In animals, the storage glucan is **glycogen** (Fig. 17-1), which differs from amylopectin only in that its branches occur every 8 to 12 residues. Glycogen occurs in 100 to 400-Å diameter cytoplasmic granules that are especially prominent in the cells that make the greatest use of it, muscle (maximally 1–2% by weight) and liver cells (maximally 10% by weight, an ~12-h energy supply for the body; Fig. 10-19). Glycogen granules also contain the enzymes that catalyze glycogen synthesis and degradation as well as some of the enzymes that regulate these processes.

As we shall see in this chapter, glycogen's glucose units are mobilized by their sequential removal from the glucan chains' nonreducing ends [ends lacking a C(1)-OH group]. *Glycogen's highly branched structure is*

(a)

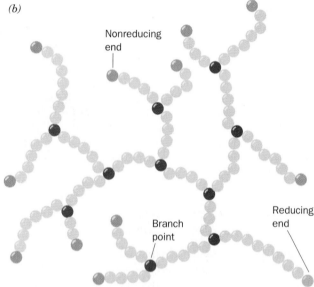

$\alpha(1 \longrightarrow 6)$ linkage

Reducing end

Nonreducing ends

Branch point

$\alpha(1 \longrightarrow 4)$ linkage

(b)

Nonreducing end

Branch point

Reducing end

Figure 17-1
The structure of glycogen. (a) Molecular formula. In the actual molecule the chains are much longer than shown. (b) Schematic diagram illustrating its branched structure. Branch points in the actual molecule are separated by 8 to 12 glucosyl units. Note that the molecule, no matter how big, has but one reducing end.

not adequately maintain essential blood glucose levels (Section 17-3F).

As with all metabolic processes, there are several levels on which glycogen metabolism may be understood. We shall examine this process in order to understand the pathway's thermodynamics and the reaction mechanisms of its individual steps but will emphasize the mechanisms by which glycogen synthesis and breakdown rates are controlled. We began our consideration of metabolic control mechanisms in Section 16-4 with a discussion of the role of allosteric interactions and substrate cycles in the regulation of glycolysis. Glycogen metabolism's more complex control systems provide us with examples of several additional regulatory processes: covalent modification of enzymes and enzyme cascades. In addition, we shall consider glycogen metabolism as a model for the role of hormones in the overall regulatory process. We end the chapter by discussing the consequences of genetic defects in various enzymes of glycogen metabolism.

1. GLYCOGEN BREAKDOWN

Liver and muscle are the two major storage tissues for glycogen. In muscle, the need for ATP results in the conversion of glycogen to glucose-6-phosphate (G6P) for entry into glycolysis. In liver, low blood glucose concentration triggers glycogen breakdown to G6P which, in this case, is hydrolyzed to glucose and released into the bloodstream to reverse this situation.

Glycogen breakdown requires the actions of three enzymes:

therefore physiologically significant: It permits glycogen's rapid degradation through the simultaneous release of the glucose units at the end of every branch.

Why does the body go to such metabolic effort to use glycogen for energy storage when fat, which is far more abundant in the body, seemingly serves the same purpose. The answer is threefold:

1. Muscles cannot mobilize fat as rapidly as glycogen.

2. The fatty acid residues of fat cannot be metabolized anaerobically (Section 23-2).

3. Animals cannot convert fatty acids to glucose precursors (Section 21-1) so that fat metabolism alone can-

1. **Glycogen phosphorylase** (or simply **phosphorylase**) catalyzes glycogen **phosphorolysis** (bond cleavage by the substitution of a phosphate group) to yield **glucose-1-phosphate (G1P).**

$$\text{Glycogen} + P_i \rightleftharpoons \text{glycogen} + \text{G1P}$$
$$(n \text{ residues}) \qquad\qquad (n-1 \text{ residues})$$

This enzyme will only release a glucose unit that is at least five units from a branch point.

2. **Glycogen debranching enzyme** removes glycogen's branches thereby permitting the glycogen phosphorylase reaction to go to completion. Consequently, 90% of glycogen's glucose residues are converted to G1P. The remaining 10%, those at the branch points, are converted to glucose.

3. **Phosphoglucomutase** converts G1P to G6P which, as we have seen (Section 16-2A), is also formed in the first step of glycolysis through the action of either hexokinase or glucokinase. G6P can either continue along the glycolytic pathway (in muscle) or be hydrolyzed to glucose (in liver).

In this section, we discuss the structures and mechanisms of action of these three enzymes.

A. Glycogen Phosphorylase

Glycogen phosphorylase is a dimer of identical 842-residue (97 kD) subunits that catalyzes the controlling step in glycogen breakdown. It is regulated by both allosteric interactions and by covalent modification. *The enzyme-catalyzed modification/demodification process yields two forms of phosphorylase: **phosphorylase a**, which has a phosphoryl group esterified to Ser 14 in each of its subunits, and **phosphorylase b**, which lacks these phosphoryl groups. Phosphorylase's allosteric inhibitors, ATP, G6P and glucose, and its allosteric activator, AMP (to name only the enzyme's most prominent effectors), interact differently with the phospho- and dephosphoenzymes resulting in an extremely sensitive regulation process. We study this process in Section 17-3C.*

Kinetics and Reaction Mechanism

The phosphorylase reaction results in the cleavage of the C(1)—O(1) bond from a nonreducing terminal glucosyl unit of glycogen yielding G1P. This reaction proceeds with retention of configuration, which suggests that the phosphorolysis occurs via a double displacement mechanism (two sequential nucleophilic substitutions each occurring with inversion of configuration; Section 13-5D) involving a covalent glucosyl-enzyme intermediate. Yet, phosphorylase exhibits Rapid Equilibrium Random Bi Bi kinetics (Section 13-5), not Ping Pong kinetics as would be expected for a double displacement mechanism. Furthermore, all attempts to establish the existence of the putative covalent intermediate have been unsuccessful. An alternative mechanism (Fig. 17-2) involves the formation of a ternary $P_i \cdot$ enzyme $\cdot$ glycogen complex followed by the generation of an intermediate shielded oxonium ion similar to that in the lysozyme reaction (which also in-

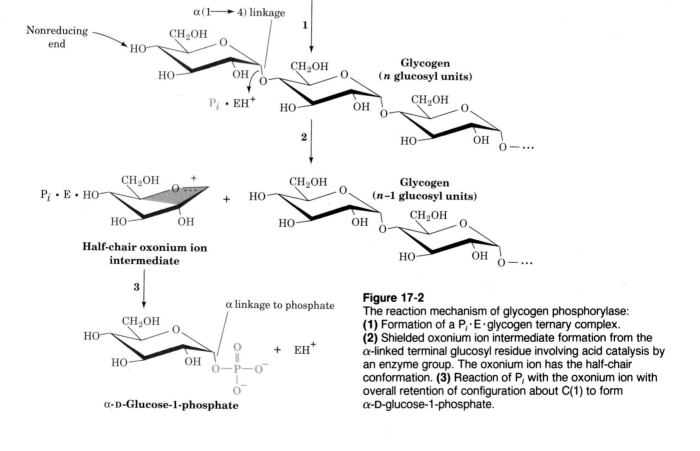

Figure 17-2
The reaction mechanism of glycogen phosphorylase:
(1) Formation of a $P_i \cdot$ E $\cdot$ glycogen ternary complex.
(2) Shielded oxonium ion intermediate formation from the α-linked terminal glucosyl residue involving acid catalysis by an enzyme group. The oxonium ion has the half-chair conformation. **(3)** Reaction of P_i with the oxonium ion with overall retention of configuration about C(1) to form α-D-glucose-1-phosphate.

volves polysaccharide bond cleavage; Section 14-2B). *Bond cleavage, with its consequent oxonium ion formation, results from acid catalysis by an enzymatic group. The oxonium ion then reacts with P_i to form the product G1P with retention of configuration.*

Support for the oxonium ion mechanism comes from the observation that **1,5-gluconolactone**

1,5-Gluconolactone

is a potent inhibitor of phosphorylase. 1,5-Gluconolactone has the same half-chair conformation as the proposed oxonium ion, suggesting that it is a transition state analog that mimics the oxonium ion at the active site of phosphorylase (Section 14-1F).

Structural Domains and Binding Sites

The high resolution X-ray structures of phosphorylase *a* and phosphorylase *b* were determined by Robert Fletterick and Louise Johnson, respectively. The structure of phosphorylase *b*, despite its lack of a Ser-linked phosphate, is very similar to that of phosphorylase *a* (Fig. 17-3). Both structures have two domains, an N-terminal domain (residues 1–484; the largest known domain), and a C-terminal domain (residues 484–842). The N-terminal domain is further divided into an interface subdomain (residues 1–315), which includes the covalent modification site (Ser 14), the allosteric effector site, and all the intersubunit contacts in the dimer; and a glycogen-binding subdomain (residues 316–484), which contains the "glycogen storage site" (see below). The catalytic site is located at the center of the subunit where these two subdomains come together with the C-terminal domain.

Glycogen forms a left-handed helix with 6.5 glucose residues per turn (Fig. 10-17*b*). An ~30-Å long crevice on the surface of the phosphorylase monomer that has the same radius of curvature as glycogen, connects the glycogen storage site, which binds glycogen, to the active site, which phosphorylizes it. *Since this crevice can accommodate four or five sugar residues in a chain but is too narrow to admit branched oligosaccharides, it provides a clear physical rationalization for the inability of phosphorylase to cleave glycosyl residues closer than five units from a branch point.* Presumably the glycogen storage site increases the catalytic efficiency of phosphorylase by permitting it to phosphorylize many glucose residues on the same glycogen particle without having to dissociate and reassociate completely between catalytic cycles.

Pyridoxal Phosphate Is an Essential Cofactor for Phosphorylase

Phosphorylase contains **pyridoxal phosphate (PLP)**

Pyridoxal phosphate (PLP)

PLP covalently bound to phosphorylase via a Schiff base to Lys 679

and requires it for activity. This vitamin B_6 derivative is covalently linked to phosphorylase via a Schiff base to Lys 679. PLP is similarly linked to a variety of enzymes involved in amino acid metabolism where it is an essential cofactor in transamination reactions (Section 24-1A). The mechanism of PLP participation in the phosphorylase reaction must differ from that in these other enzymes because, for example, reduction of the Schiff base with $NaBH_4$ ($-HC=N- \rightarrow -H_2C-NH-$) has no effect on the activity of phosphorylase while it inactivates the PLP-requiring enzymes of amino acid metabolism. This is an intriguing example of nature's opportunism in using the same cofactor to perform different chemistries.

Extensive studies on phosphorylase using PLP analogs in which various parts of this molecule are missing or modified indicate that only its phosphate group participates in the catalytic process. Indeed, the above X-ray structures reveal that only PLP's phosphate group is near phosphorylase's active site. The role of this phosphoryl group in phosphorylase's incompletely characterized catalytic mechanism has not been convincingly established but it most probably functions as a proton donor (acid catalyst).

B. Phosphoglucomutase

Phosphorylase converts the glucosyl units of glycogen to G1P, which, in turn, is converted by phosphoglucomutase to G6P either for entry into glycolysis in muscle or hydrolysis to glucose in liver. This phospho-

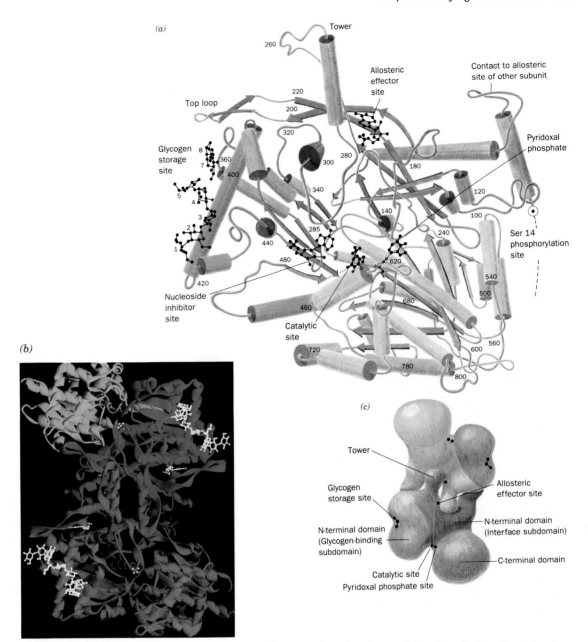

Figure 17-3

The X-ray structure of rabbit muscle glycogen phosphorylase. (*a*) Ribbon diagram of a phosphorylase *b* subunit. It consists of an N-terminal domain, which is subdivided into an interface subdomain (residues 1–315) and a glycogen-binding subdomain (residues 316–484), and a C-terminal domain (residues 485–841). AMP is shown bound at both the allosteric effector site and the nucleoside inhibitor site. G1P is shown bound at the catalytic site. The pyridoxal phosphate, which is partially hidden from view, is bound at Lys 679 in the C-terminal domain. Maltoheptose is bound to the glycogen storage site. Residues 1–18, which do not appear in the electron density map, are drawn as a dashed line. Ser 14 is the site of enzymatic phosphorylation. [After McLaughlin, P. J., Stuart, D. I., Klein, H. W., Oikonomakos, N. G., and Johnson, L. N., *Biochemistry* **23**, 5865 (1984).] (*b*) A computer graphics drawing of the glycogen phosphorylase *a* dimer viewed along its molecular twofold axis of symmetry [this view is related to that in Part (*a*) by an ~ 45° rotation about the vertical axis; the structural differences between the enzyme's two forms are relatively small]. The bottom subunit is colored orange whereas the top subunit's N-terminal and C-terminal domains are colored blue and green, respectively. The various bound ligands are white: The phosphate group at the center of each subunit marks the enzyme's catalytic sites (the Ser 14 phosphate groups in both subunits are hidden in this drawing), two chains of maltoheptose are bound at each glycogen storage site, and the AMPs at the "back" of the protein identify the allosteric effector sites. [Courtesy of Stephen Sprang, University of Texas, Southwestern Medical Center.] (*c*) An interpretive drawing of Part (*b*) showing the enzyme's various ligand binding sites.

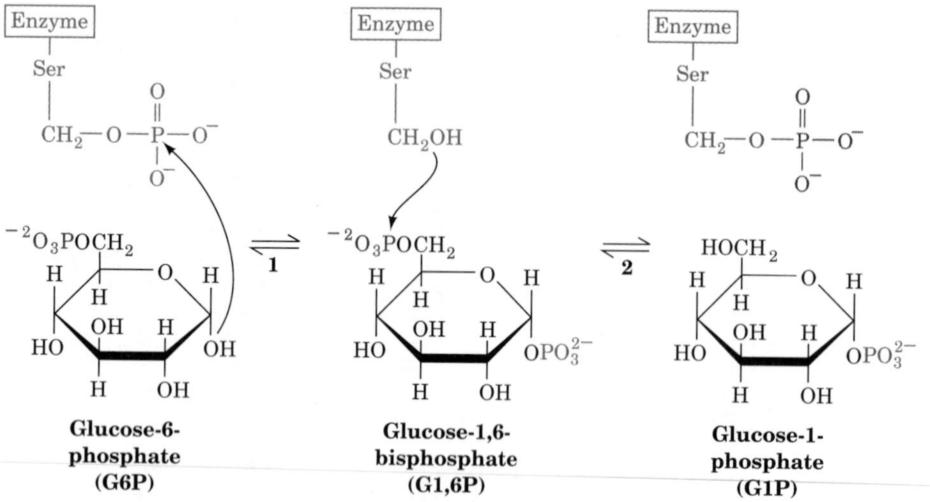

Figure 17-4
The mechanism of action of phosphoglucomutase showing the involvement of both G1,6P and a phosphoenzyme intermediate: **(1)** The OH group at C(1) of G6P attacks the phosphoenzyme to form a dephosphoenzyme–G1,6P intermediate. **(2)** The Ser-OH group on the dephosphoenzyme attacks the phosphoryl group at C(6) to regenerate the phosphoenzyme with the formation of G1P.

glucomutase reaction is similar to that catalyzed by phosphoglycerate mutase (Section 16-2H). A phosphoryl group is transferred from the active phosphoenzyme to G1P forming **glucose-1,6-bisphosphate (G1,6P)** which then rephosphorylates the enzyme to yield G6P (Fig. 17-4). An important difference between this enzyme and phosphoglycerate mutase is that the phosphoryl group in phosphoglucomutase is covalently bound to a Ser hydroxyl group rather than to a His residue.

G1,6P occasionally dissociates from phosphoglucomutase resulting in the inactivation of this enzyme. The presence of small amounts of G1,6P is therefore necessary to keep phosphoglucomutase fully active. This product is provided by **phosphoglucokinase,** which catalyzes the phosphorylation of the C(6)-OH group of G1P by ATP.

C. Glycogen Debranching Enzyme

Glycogen debranching enzyme acts as an $\alpha(1 \rightarrow 4)$ transglycosylase (glycosyl transferase) by transferring an $\alpha(1 \rightarrow 4)$-linked trisaccharide unit from a "limit branch" of glycogen to the nonreducing end of another branch (Fig. 17-5). This reaction forms a new $\alpha(1 \rightarrow 4)$ linkage with three more units available for phosphorylase-catalyzed phosphorolysis. The $\alpha(1 \rightarrow 6)$ bond linking the remaining glycosyl residue in the branch to the main chain is hydrolyzed (not phosphorylized) by the same debranching enzyme to yield glucose and debranched glycogen. Thus *debranching enzyme has active sites for both the transferase reaction and the $\alpha(1 \rightarrow 6)$-glucosidase reaction.* The presence of two independent catalytic activities on the same enzyme no doubt improves the efficiency of the debranching process.

The maximal rate of the glycogen phosphorylase reaction is much greater than that of the glycogen debranching reaction. Consequently, the outermost branches of glycogen, which comprise nearly one half

of its residues, are degraded in muscle in a few seconds under conditions of high metabolic demand. Glycogen degradation beyond this point requires debranching and hence occurs more slowly. This, in part, accounts for the fact that a muscle can sustain its maximum exertion for only a few seconds.

D. Thermodynamics of Glycogen Metabolism: The Need for Separate Pathways of Synthesis and Breakdown

The $\Delta G°'$ (ΔG under standard biochemical conditions) for the phosphorylase reaction is $+3.1 \text{ kJ} \cdot \text{mol}^{-1}$, so that, as Eq. [3.15] indicates, this reaction is at equilibrium ($\Delta G = 0$) when $[P_i]/[G1P] = 3.5$. In the cell, however, this concentration ratio varies between 30 and 100, which places ΔG in the range -5 to $-8 \text{ kJ} \cdot \text{mol}^{-1}$; that is, *under physiological conditions, glycogen breakdown is exergonic.* The synthesis of glycogen from G1P under physiological conditions is therefore thermodynamically unfavorable without free energy input. Consequently, *glycogen biosynthesis and breakdown must occur by separate pathways. Thus we encounter a recurrent metabolic strategy: Biosynthetic and degradative pathways of metabolism are almost always different* (Section 15-1). There are two important reasons for this. The first, as we have seen, is that both pathways may be required under similar *in vivo* metabolite concentrations. This situation is thermodynamically impossible if one pathway is just the reverse of the other. The second reason is equally important: Reactions catalyzed by different enzymes can be independently regulated, which permits very fine flux control. We have seen this principle in operation in the glycolytic conversion of fructose-6-phosphate (F6P) to fructose-1,6-bisphosphate (F1,6P) by phosphofructokinase (PFK; Section 16-4B). The reverse process in that case (hydrolysis of F1,6P) is catalyzed by fructose bisphosphatase (FBPase). Independent regula-

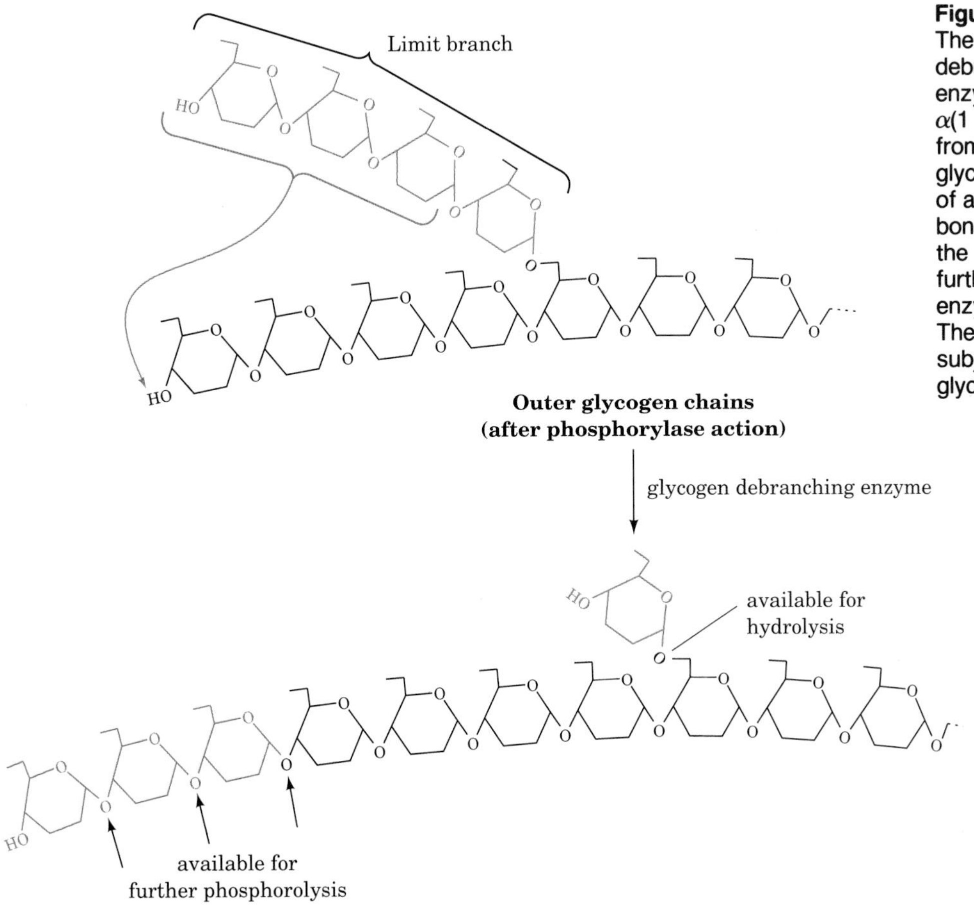

Figure 17-5
The reaction catalyzed by debranching enzyme. The enzyme transfers three $\alpha(1 \rightarrow 4)$-linked glucose residues from a "limit branch" of glycogen to the nonreducing end of another branch. The $\alpha(1 \rightarrow 6)$ bond of the residue remaining at the branch point is hydrolyzed by further action of debranching enzyme to yield free glucose. The newly elongated branch is subject to degradation by glycogen phosphorylase.

tion of those two enzymes provides precise control of glycolytic flux.

Glycogen metabolism, like glycolysis, is exquisitely regulated by the independent control of its synthetic and degradative pathways. In the next section we examine the pathway of glycogen synthesis and, in Section 17-3, we explore the regulatory process.

2. GLYCOGEN SYNTHESIS

Although the thermodynamic arguments presented in Section 17-1D demonstrate that glycogen synthesis and breakdown must occur by separate pathways, it was not thermodynamic arguments that led to the general acceptance of this idea. Rather, it was the elucidation of the cause of **McArdle's disease,** a rare inherited glycogen storage disease that results in painful muscle cramps upon strenuous exertion (Section 17-4). The muscle tissue from individuals with McArdle's disease exhibits no glycogen phosphorylase activity and is therefore incapable of glycogen breakdown. Their muscles, nevertheless, contain moderately high quantities of normal glycogen. Clearly, there must be separate pathways for glycogen synthesis and breakdown.

Since the direct conversion of G1P to glycogen and P_i

is thermodynamically unfavorable (positive ΔG) under all physiological P_i concentrations, glycogen biosynthesis requires an additional exergonic step. This is accomplished, as Luis Leloir discovered in 1957, by combining G1P with uridine triphosphate (UTP) to form **uridine diphosphate glucose (UDP-glucose** or **UDPG):**

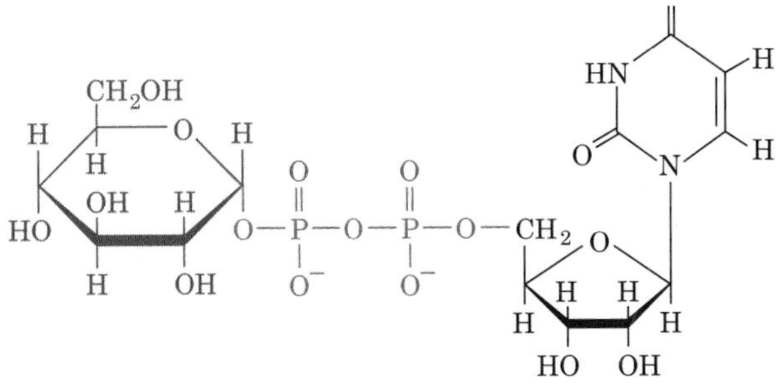

Uridine diphosphate glucose
(UDPG)

UDPG's "high-energy" status permits it to spontaneously donate glucosyl units to the growing glycogen chain.

The enzymes catalyzing the three steps involved in the glycogen synthesis pathway are **UDP-glucose pyrophosphorylase, glycogen synthase,** and **glycogen branching enzyme.** In this section, we examine the re-

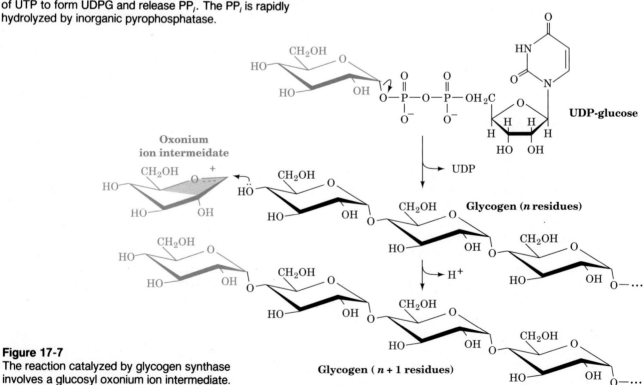

actions catalyzed by these enzymes. Discussion of how these enzymes are regulated is reserved for Section 17-3.

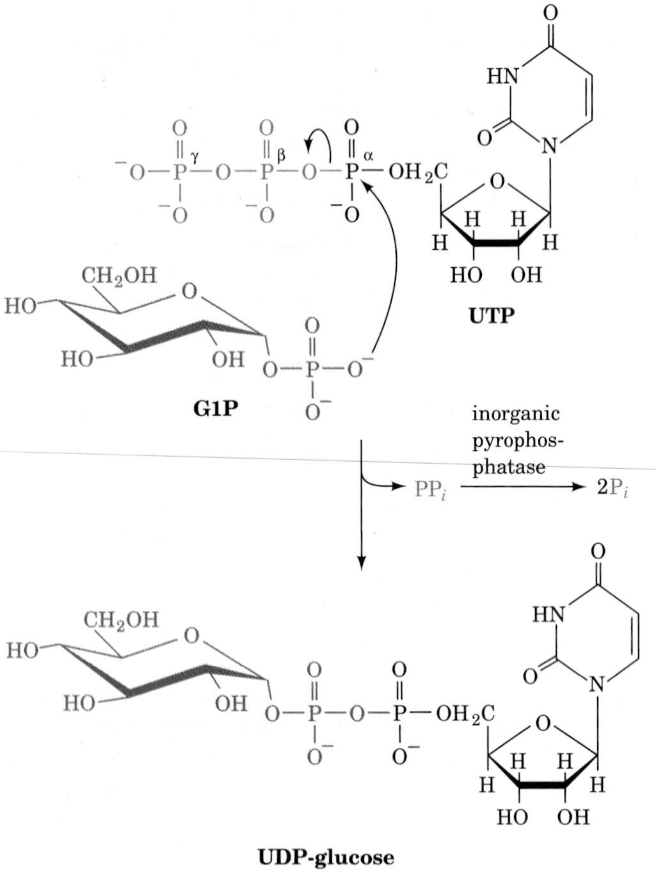

A. UDP-Glucose Pyrophosphorylase

UDP-glucose pyrophosphorylase catalyzes the reaction of UTP and G1P (Fig. 17-6). In this reaction, the phosphoryl oxygen of G1P attacks the α phosphorus atom of UTP to form UDPG and release PP_i. The $\Delta G°'$ of this phosphoanhydride exchange is, as expected, nearly zero. However, the PP_i formed is hydrolyzed in a highly exergonic reaction by the omnipresent enzyme **inorganic pyrophosphatase**. The overall reaction for the formation of UDPG is therefore also highly exergonic:

	$\Delta G°'(kJ\ mol^{-1})$
$G1P\ +\ UTP \rightleftharpoons UDPG\ +\ PP_i$	~ 0
$H_2O\ +\ PP_i \longrightarrow 2P_i$	−31
Overall $G1P\ +\ UTP \longrightarrow UDPG\ +\ 2P_i$	−31

The cleavage of a nucleoside triphosphate to form PP_i is a common biosynthetic strategy. The free energy of PP_i hydrolysis can then be utilized together with the free energy of nucleoside triphosphate hydrolysis to drive an otherwise endergonic reaction to completion (Section 15-4C).

Figure 17-6
The reaction catalyzed by UDP-glucose pyrophosphorylase. The reaction is a phosphoanhydride exchange in which the phosphoryl oxygen of G1P attacks the α phosphorus atom of UTP to form UDPG and release PP_i. The PP_i is rapidly hydrolyzed by inorganic pyrophosphatase.

Figure 17-7
The reaction catalyzed by glycogen synthase involves a glucosyl oxonium ion intermediate.

B. Glycogen Synthase

In the next step of glycogen synthesis, the glycogen synthase reaction, the glucosyl unit of UDPG is transferred to the C(4)-OH group on one of glycogen's nonreducing ends to form an $\alpha(1 \rightarrow 4)$-glycosidic bond (Fig. 17-7). The glycogen synthase reaction, like those of glycogen phosphorylase and lysozyme, is thought to involve a glucosyl oxonium ion transition state since it is also inhibited by 1,5-gluconolactone, an analog that mimics the oxonium ion's half-chair geometry.

The $\Delta G°'$ for the glycogen synthase reaction is -13.4 $kJ \cdot mol^{-1}$ making the overall reaction spontaneous under the same conditions that glycogen breakdown by glycogen phosphorylase is also spontaneous. The rates of both reactions may then be independently controlled. There is, however, an energetic price for doing so. In this case, *for each molecule of G1P that is converted to glycogen and then regenerated, one molecule of UTP is hydrolyzed to UDP and P_i. The cyclic synthesis and breakdown of glycogen is therefore not a perpetual motion "machine" but, rather, is an "engine" that is powered by UTP hydrolysis.* The UTP is replenished through a phosphate-transfer reaction mediated by **nucleoside diphosphate kinase** (Section 26-3B):

$$UDP + ATP \rightleftharpoons UTP + ADP$$

so that UTP hydrolysis is energetically equivalent to ATP hydrolysis.

Glycogen synthase cannot simply link together two glucose residues; it can only extend an already existing $\alpha(1 \rightarrow 4)$-linked glucan chain. How, then, is glycogen synthesis initiated? The answer is that a glycogen synthase can append a glucose residue onto a specific Tyr-OH group of a protein named **glycogenin** which thereby acts as a "primer" for the initiation of glycogen synthesis.

C. Glycogen Branching

Glycogen synthase catalyzes only $\alpha(1 \rightarrow 4)$-linkage formation to yield α-amylose. Branching to form glycogen is accomplished by a separate enzyme, **amylo-(1,4 $\rightarrow$ 1,6)-transglycosylase (branching enzyme)**, which is distinct from glycogen debranching enzyme. Debranching (Section 17-1C) involves breaking and reforming $\alpha(1 \rightarrow 4)$-glycosidic bonds and only the hydrolysis of $\alpha(1 \rightarrow 6)$-glycosidic bonds; branching, on the other hand, involves breaking $\alpha(1 \rightarrow 4)$-glycosidic bonds and reforming $\alpha(1 \rightarrow 6)$ linkages. Branches are created by the transfer of terminal chain segments consisting of ~ 7 glucosyl residues to the C(6)-OH groups of glucose residues on the same or another glycogen chain (Fig. 17-8). Each transferred segment must come from a chain of at least 11 residues and the new branch point must be at least 4 residues away from other branch points.

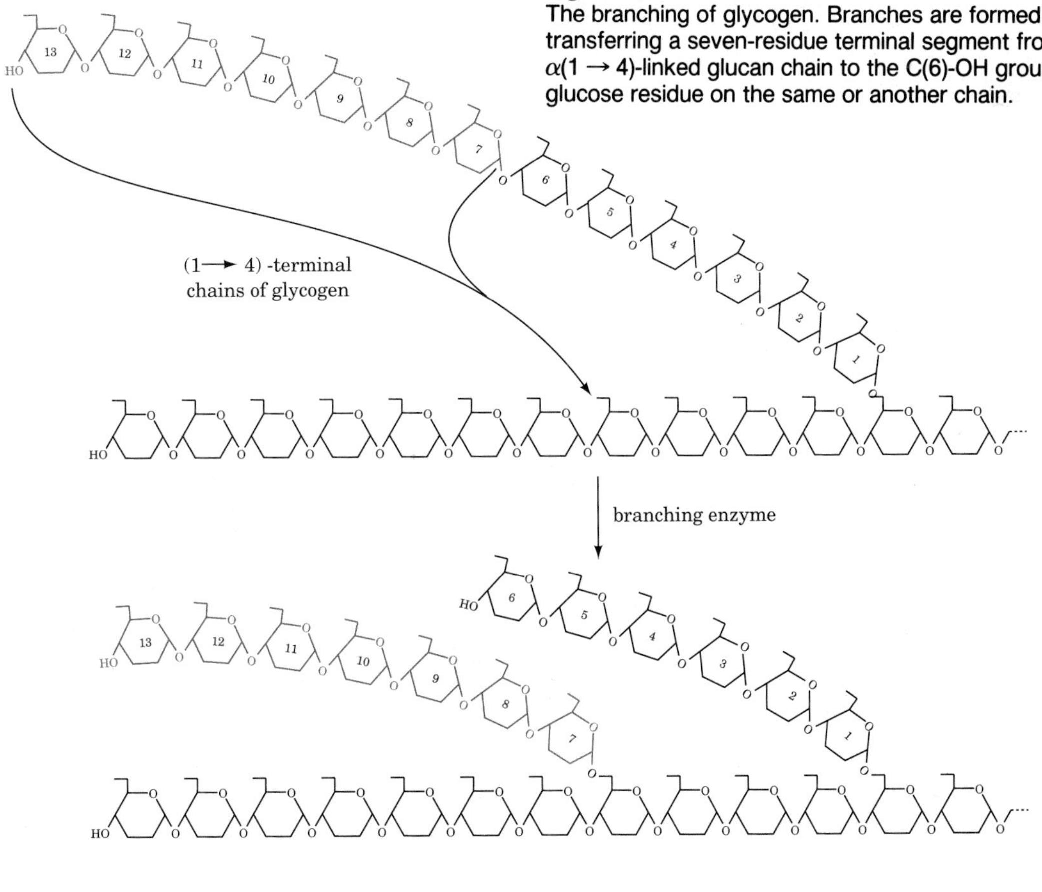

Figure 17-8
The branching of glycogen. Branches are formed by transferring a seven-residue terminal segment from an $\alpha(1 \rightarrow 4)$-linked glucan chain to the C(6)-OH group of a glucose residue on the same or another chain.

(1 $\rightarrow$ 4)-terminal chains of glycogen

branching enzyme

3. CONTROL OF GLYCOGEN METABOLISM

We have just seen that both glycogen synthesis and breakdown are exergonic under the same physiological conditions. If both pathways operate simultaneously, however, all that is achieved is wasteful hydrolysis of UTP. This situation is similar to that of the phosphofructokinase–fructose bisphosphatase substrate cycle (Section 16-4B). Glycogen phosphorylase and glycogen synthase therefore must be under stringent control such that glycogen is either synthesized or utilized according to cellular needs. The astonishing mechanism of this control is the next topic of our discussion. It involves not only allosteric regulation and substrate cycles but enzyme-catalyzed covalent modification of both glycogen synthase and glycogen phosphorylase. The covalent modification reactions are themselves ultimately under hormonal control through an enzymatic cascade.

A. Direct Allosteric Control of Glycogen Phosphorylase and Glycogen Synthase

As we saw in Section 16-4A, the net flux of reactants, J, through a step in a metabolic pathway is the difference between the forward and reverse reaction velocities, v_f and v_r. The variation in the flux through any step in a pathway with a change in substrate concentration maximizes as that reaction step approaches equilibrium

($v_f \approx v_r$; Eq. [16.7]). The flux through a near-equilibrium reaction is therefore all but uncontrollable. As we have seen for the case of PFK and FBPase, however, *precise flux control of a pathway is possible when an enzyme functioning far from equilibrium is opposed by a separately controlled enzyme. Then, v_f and v_r vary independently. In fact, under these circumstances, even the flux direction is controlled if v_r can be made larger than v_f.* Exactly this situation occurs in glycogen metabolism through the opposition of the glycogen phosphorylase and glycogen synthase reactions. The rates of both of these reactions are under allosteric control by effectors that include ATP, G6P, and AMP. In muscle, glycogen phosphorylase is activated by AMP and inhibited by ATP and G6P. Glycogen synthase, on the other hand, is activated by G6P. When there is high demand for ATP (low [ATP], low [G6P], and high [AMP]), glycogen phosphorylase is stimulated while glycogen synthase is inhibited so that flux through this pathway favors glycogen breakdown. When [ATP] and [G6P] are high, the reverse is true and glycogen synthesis is favored.

For instance, in terms of the symmetry model of allosterism (Section 9-4B), AMP promotes phosphorylase's T (*inactive*) → R (*active*) conformational shift by binding to the R state of the enzyme at its allosteric effector site (Fig. 17-9). In doing so, AMP's adenine, ribose, and phosphate groups bind to separate segments of the polypeptide chain so as to link the active site, the subunit interface, and the N-terminal region (Fig. 17-10). Curiously, ATP also binds to the allosteric effector site

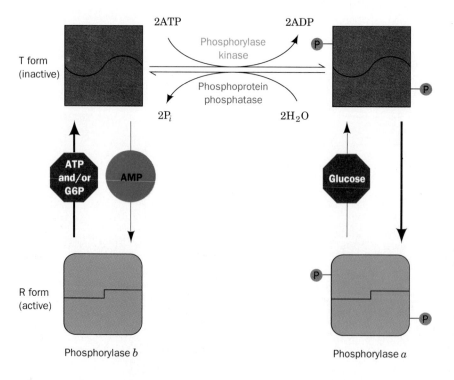

Figure 17-9
The control of glycogen phosphorylase activity. The enzyme may assume the enzymatically inactive T conformation (*above*) or the catalytically active R form (*below*). The conformation of phosphorylase *b* is allosterically controlled by effectors such as AMP, ATP, and G6P and is mostly in the T state under physiological conditions. In contrast, the modified form of the enzyme, phosphorylase *a*, is unresponsive to these effectors and is largely in the R state unless there is a high level of glucose. Under usual physiological conditions, the enzymatic activity of glycogen phosphorylase is essentially determined by its rates of modification and demodification. Note that only the T form enzyme is subject to phosphorylation and dephosphorylation so that effector binding influences the rates of these modification/demodification events.

but in the T state so that it inhibits rather than promotes the T → R conformational shift. This is because, as structural analysis indicates, the β and γ phosphate groups of ATP bind to the enzyme so as to displace its ribose and α phosphate groups relative to those of AMP and thus destabilize the R state. The inhibitory action of ATP on phosphorylase is therefore simply understood: It competes with AMP for binding to phosphorylase and, in doing so, prevents the relative motions of the three polypeptide segments required for phosphorylase activation.

The above allosteric interactions are superimposed on an even more sophisticated control system involving covalent modifications of glycogen phosphorylase and glycogen synthase. These modifications alter the structures of the enzymes so as to change their responses to allosteric regulators. We shall therefore discuss the general concept of covalent modification and how it increases the sensitivity of a metabolic system to effector concentration changes. We subsequently consider the functions of such modifications in glycogen metabolism. Only then will we be in a position to take up the detailed consideration of allosteric regulation in glycogen metabolism.

B. Covalent Modification of Enzymes by Cyclic Cascades: Effector "Signal" Amplification

*Glycogen synthase and glycogen phosphorylase can each be enzymatically interconverted between two forms with different kinetic and allosteric properties through a complex series of reactions known as a **cyclic cascade**. The interconversion of these different enzyme forms involves distinct, enzyme-catalyzed **covalent modification and demodification reactions**.*

Compared with other regulatory enzymes, enzymatically interconvertible enzyme systems:

1. Can respond to a greater number of allosteric stimuli.

2. Exhibit greater flexibility in their control patterns.

3. Possess enormous amplification potential in their responses to variations in effector concentrations.

This is because *the enzymes that modify and demodify a target enzyme are themselves under allosteric control. It is therefore possible for a small change in concentration of an allosteric effector of a modifying enzyme to cause a large change in the concentration of an active, modified target*

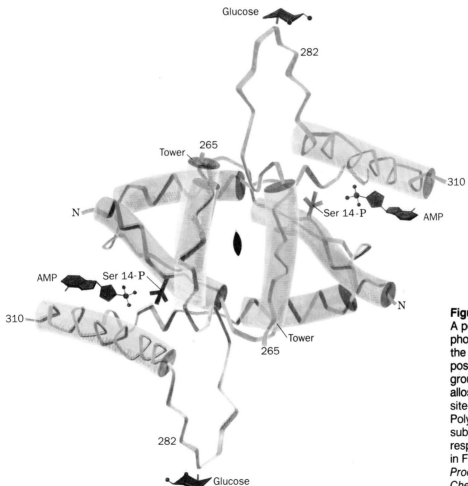

Figure 17-10
A portion of the glycogen phosphorylase *a* dimer in the vicinity of the dimer interface indicating the positions of the Ser 14 phosphoryl groups, the AMPs bound in the allosteric effector sites, and the active site-bound glucose molecules. Polypeptide segments from the two subunits are colored blue and green, respectively. The view is similar to that in Fig. 17-3*b*. [After Fletterick, R. J., *Proc. Robert A. Welch Found. Conf. Chem. Res.* **27,** 197 (1984).]

(a)

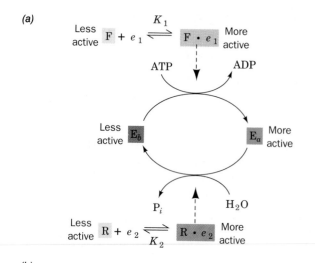

(b)

$$ATP + F \cdot e_1 + E_b \underset{}{\overset{K_f}{\rightleftharpoons}} E_b \cdot ATP \cdot F \cdot e_1 \overset{k_f}{\longrightarrow} F \cdot e_1 + E_a + ADP$$

$$R \cdot e_2 + E_a \underset{}{\overset{K_r}{\rightleftharpoons}} E_a \cdot R \cdot e_2 \overset{k_r}{\longrightarrow} R \cdot e_2 + E_b + P_i$$

Figure 17-11
(a) A monocyclic enzyme cascade. F and R are, respectively, the modifying and demodifying enzymes. These are allosterically converted from their inactive to their active conformations upon binding their respective effectors, e_1 and e_2. The target enzyme, E, is more active in the modified form (E_a) and less active in the unmodified form (E_b). Dashed arrows symbolize catalysis of the indicated reactions. (b) Chemical equations for the interconversion of the target enzyme's unmodified and modified forms E_b and E_a.

enzyme. Such a cyclic cascade is diagrammed in Fig. 17-11.

Description of a General Cyclic Cascade

Figure 17-11*a* shows a general scheme for a cyclic cascade where, by convention, the more active target enzyme form has the subscript "*a*" while the less active form has the subscript "*b*." Here, modification, in this case, phosphorylation, activates the enzyme. Note that the modifying enzymes, F and R, are only active when they are binding their respective allosteric effectors e_1 and e_2. The kinetic mechanisms for the interconversion of the unmodified and modified forms of the target enzyme, E_b and E_a, are indicated in Fig. 17-11*b*.

In the steady state, the fraction of E in the active form, $[E_a]/[E]_T$ (where $[E]_T = [E_a] + [E_b]$ is the total enzyme concentration) determines the rate of the reaction catalyzed by E. This fraction is a function of the total concentrations of the modifying enzymes, $[F]_T$ and $[R]_T$, the concentrations of their allosteric effectors, e_1 and e_2, the dissociation constants of these effectors, K_1 and K_2, and the rate constants, k_f and k_r, as well as the dissociation constants, K_f and K_r, of the interconversions themselves (Fig. 17-11). This relationship is obviously quite com-

plex. Nevertheless, as is shown in the appendix to this chapter, it can be reduced to simple terms. The appendix demonstrates that in a cyclic cascade, a relatively small change in the concentration of e_1, the allosteric effector of the modifying enzyme F, can result in a much larger change in $[E_a]/[E]_T$, the fraction of E in the active form. In other words, *the cascade functions to amplify the sensitivity of the system to an allosteric effector.*

We have so far considered the covalent modification of only one enzyme, a **monocyclic cascade.** Imagine a **bicyclic cascade** involving the covalent modification of one of the modifying enzymes (F), as well as the metabolic target enzyme (E) (Fig. 17-12). As you might expect, the amplification potential of a "signal", e_1, as well as the control flexibility of such a system, is enormous.

The activities of both glycogen phosphorylase and glycogen synthase are controlled by bicyclic cascades. Let us now examine the enzymatic interconversions involved in these bicyclic cascades. We shall specifically focus on the covalent modifications of glycogen phosphorylase and glycogen synthase, the structural effects of these covalent modifications, and how these structural changes affect the interactions of their allosteric effectors. We shall then consider the cyclic cascades as a whole, studying the various modification enzymes involved and their "ultimate" allosteric effectors. Finally, we shall see how the various cyclic cascades of glycogen metabolism function in different physiological situations.

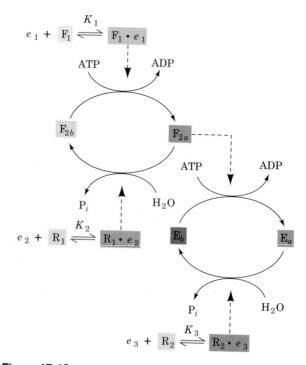

Figure 17-12
A bicyclic enzyme cascade. See the legend of Fig. 17-11 for symbol definitions. In a bicyclic cascade, one of the modifying enzymes (F_2) is also subject to chemical modification. It is active in the modified state (F_{2a}) and inactive in the unmodified state (F_{2b}).

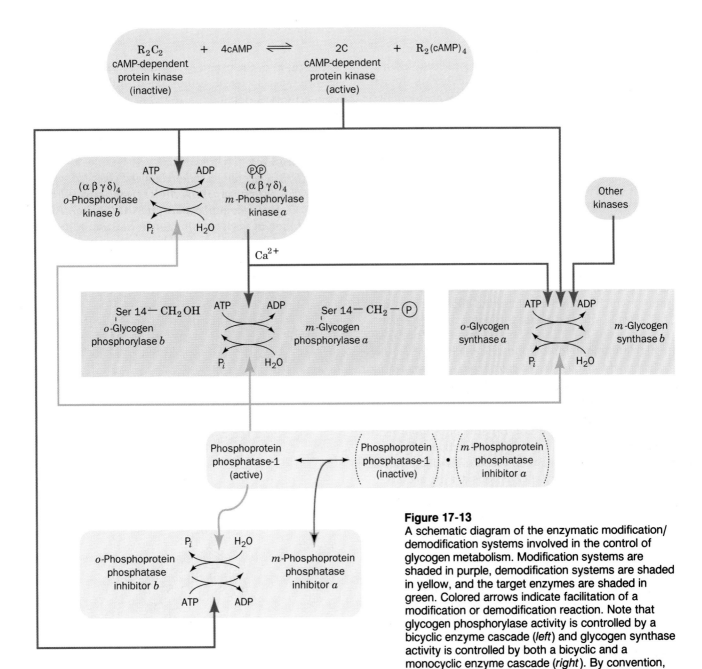

$$R_2C_2 + 4cAMP \rightleftharpoons 2C + R_2(cAMP)_4$$

cAMP-dependent protein kinase (inactive) ⟶ cAMP-dependent protein kinase (active)

$(\alpha\beta\gamma\delta)_4$ *o*-Phosphorylase kinase *b* → ATP, ADP, P_i, H_2O → $(\alpha\beta\gamma\delta)_4$ *m*-Phosphorylase kinase *a*

Other kinases

Ca^{2+}

Ser 14—CH_2OH *o*-Glycogen phosphorylase *b* → ATP, ADP, P_i, H_2O → Ser 14—CH_2—Ⓟ *m*-Glycogen phosphorylase *a*

o-Glycogen synthase *a* → ATP, ADP, P_i, H_2O → *m*-Glycogen synthase *b*

Phosphoprotein phosphatase-1 (active) ⟷ Phosphoprotein phosphatase-1 (inactive) · *m*-Phosphoprotein phosphatase inhibitor *a*

o-Phosphoprotein phosphatase inhibitor *b* → P_i, H_2O, ATP, ADP → *m*-Phosphoprotein phosphatase inhibitor *a*

Figure 17-13
A schematic diagram of the enzymatic modification/demodification systems involved in the control of glycogen metabolism. Modification systems are shaded in purple, demodification systems are shaded in yellow, and the target enzymes are shaded in green. Colored arrows indicate facilitation of a modification or demodification reaction. Note that glycogen phosphorylase activity is controlled by a bicyclic enzyme cascade (*left*) and glycogen synthase activity is controlled by both a bicyclic and a monocyclic enzyme cascade (*right*). By convention, the modified form of the enzyme bears the prefix *m* and the "original" (unmodified) form bears the prefix *o*. The most active and least active forms of the enzymes are identified by the suffixes *a* and *b*, respectively.

C. Glycogen Phosphorylase Bicyclic Cascade

In 1938, Carl and Gerti Cori found that glycogen phosphorylase exists in two forms, the *b* form that requires AMP for activity, and the *a* form that is active without AMP. It nevertheless took 20 years for the development of the protein chemistry techniques through which Edwin Krebs and Edmund Fischer demonstrated, in 1959, that phosphorylases *a* and *b* correspond to forms of the protein in which a specific residue, Ser 14, is enzymatically phosphorylated or dephosphorylated, respectively.

Glycogen Phosphorylase: The Cascade's Target Enzyme

The activity of glycogen phosphorylase is allosterically controlled, as we saw, through AMP activation and ATP, G6P, and glucose inhibition (Section 17-3A). Superimposed upon this allosteric control is control by enzymatic interconversion through the actions of three enzymes (Fig. 17-13, *left*):

1. **Phosphorylase kinase,** which specifically phosphorylates Ser 14 of glycogen phosphorylase *b*.

2. **cAMP-dependent protein kinase,** which phosphorylates and thereby activates phosphorylase kinase.

3. **Phosphoprotein phosphatase-1,** which dephosphorylates and thereby deactivates both glycogen phosphorylase *a* and phosphorylase kinase.

In an interconvertible enzyme system, the "modified"

form of the enzyme bears the prefix *m* and the "original" (unmodified) form bears the prefix *o*, whereas the enzyme's most active and least active forms are identified by the suffixes *a* and *b*, respectively. In this case, *o*-phosphorylase *b* (unmodified, least active) is the form under allosteric control by AMP, ATP, and G6P (Fig. 17-9). Phosphorylation to yield *m*-phosphorylase *a* (modified, most active) all but removes the effects of these allosteric modulators. In terms of the symmetry model of allosterism (Section 9-4B), *the phosphorylation of Ser 14 shifts the enzyme's T (inactive) $\rightleftharpoons$ R (active) equilibrium in favor of the R state (Fig. 17-9)*. Indeed, *phosphorylase a's Ser 14-phosphoryl group is analogous to an allosteric activator: It promotes the binding of the protein's otherwise disordered N-terminal segment to the subunit interface in a manner which resembles that of the allosteric activator AMP (Fig. 17-10) and hence facilitates the T $\rightarrow$ R shift*.

In the resting cell, the concentrations of ATP and G6P are high enough to inhibit phosphorylase *b*. *The level of phosphorylase activity is therefore largely determined by the fraction of the enzyme present as phosphorylase a*. As Eq. [17.A1] indicates, the steady state fraction of phosphorylated enzyme (E_a) depends on the relative activities of phosphorylase kinase (F) and phosphoprotein phosphatase-1 (R; not to be confused with an allosteric state). This interrelationship is remarkably elaborate for glycogen phosphorylase. Let us consider the actions of these enzymes.

Phosphorylase Kinase: Coordination of Enzyme Activation with [Ca²⁺]

Phosphorylase kinase, which converts phosphorylase b to phosphorylase a, is activated by Ca²⁺ concentrations as low as 10⁻⁷M, and is itself subject to covalent modification (Fig. 17-13). For phosphorylase kinase to be fully active, Ca²⁺ must be present and the protein must be phosphorylated. This 1200-kD enzyme contains four nonidentical subunits that form the active oligomer $(\alpha\beta\gamma\delta)_4$. The isolated γ subunit has full catalytic activity (ability to convert phosphorylase *b* to phosphorylase *a*), whereas the α, β, and δ subunits are inhibitors of the catalytic reaction. The δ subunit, which is known as **calmodulin,** confers Ca²⁺ sensitivity on the complex. When Ca²⁺ binds to any of calmodulin's four Ca²⁺-binding sites, this ubiquitous eukaryotic regulatory protein undergoes an extensive conformational change (Section 18-3B) that activates phosphorylase kinase. Glycogen phosphorylase therefore becomes phosphorylated and the rate of glycogen breakdown increases. *The physiological significance of this Ca²⁺ activation process is that muscle contraction is triggered by a transient increase in the cytosolic Ca²⁺ level (Section 34-3C). The rate of glycogen breakdown is therefore linked to the rate of muscle contraction. This is an important link because glycogen breakdown in muscle provides fuel for glycolysis which, in turn, generates the ATP required for muscle contraction.*

Sites on both the α and β subunits of phosphorylase kinase may also be phosphorylated. Indeed, full enzyme activity is obtained in the presence of Ca²⁺ only when both these subunits are phosphorylated.

cAMP-Dependent Protein Kinase: A Crucial Regulatory Link

In both the glycogen phosphorylase and glycogen synthase cascades, the primary intracellular signal, e_1, is adenosine-3',5'-cyclic monophosphate (cAMP). The cAMP concentration in a cell is a function of the ratio of its rate of synthesis from ATP by **adenylate cyclase,** and its rate of breakdown to AMP by a specific **phosphodiesterase.**

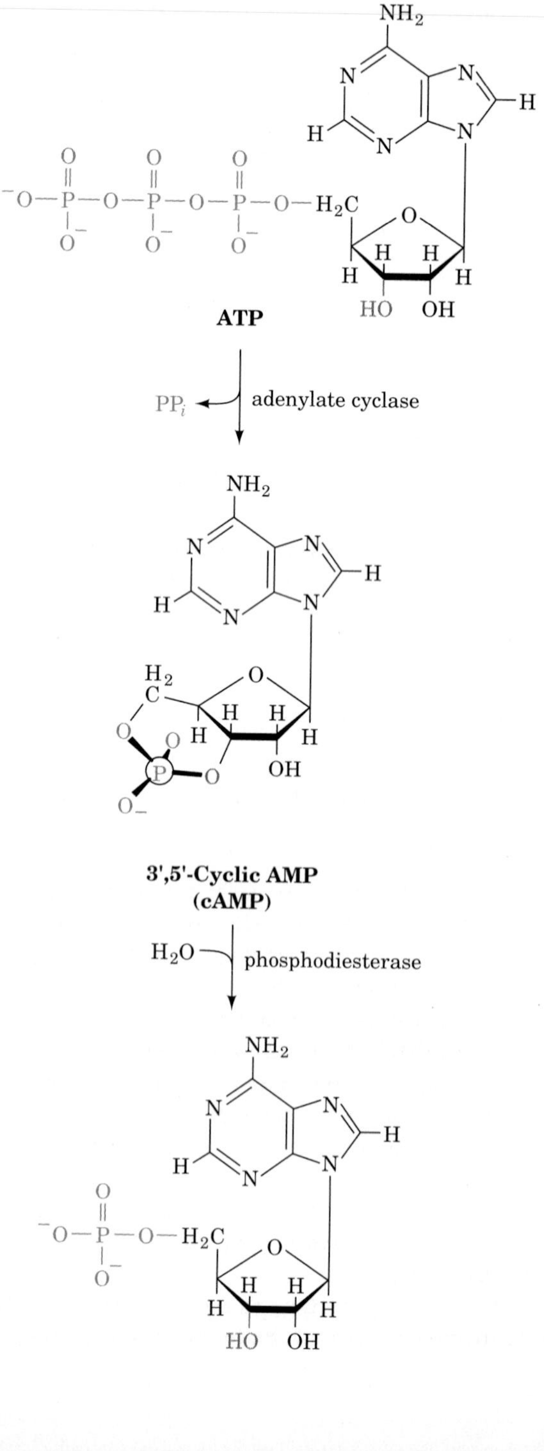

ATP

PP$_i$ ←| adenylate cyclase

3',5'-Cyclic AMP (cAMP)

H$_2$O →| phosphodiesterase

*cAMP is absolutely required for the activity of **cAMP-dependent protein kinase**, an enzyme that phosphorylates specific Ser and/or Thr residues of numerous cellular proteins including phosphorylase kinase (both its α and β subunits) and glycogen synthase.* In the absence of cAMP, the enzyme is an inactive tetramer consisting of two regulatory and two catalytic subunits, R_2C_2. The cAMP binds to the regulatory subunits so as to cause the dissociation of active catalytic monomers (Fig. 17-13; top). *The intracellular concentration of cAMP therefore determines the fraction of the protein kinase in its active form and thus the rate at which it phosphorylates its substrates.* In fact, in all known eukaryotic cases, the physiological effects of cAMP are exerted through the activation of specific protein kinases.

Phosphoprotein Phosphatase-1

The steady state phosphorylation levels of most enzymes involved in cyclic cascades are maintained by the opposition of kinase-catalyzed phosphorylations and the hydrolytic dephosphorylations catalyzed by phosphoprotein phosphatase-1. This enzyme hydrolyses the phosphoryl groups from *m*-glycogen phosphorylase *a*, both α and β subunits of phosphorylase kinase and, as is discussed below, two other proteins involved in glycogen metabolism.

Phosphoprotein phosphatase-1 is inhibited by its binding to the protein **phosphoprotein phosphatase inhibitor**. This protein provides yet another example of control by enzymatic interconversion: It too is modified by cAMP-dependent protein kinase and demodified by phosphoprotein phosphatase-1 (Fig. 17-13, bottom) although, in this case, a Thr, not a Ser, is phosphorylated/dephosphorylated. The protein is a functional inhibitor only when it is phosphorylated. *The concentration of cAMP therefore controls the fraction of an enzyme in its phosphorylated form, not only by increasing the rate at which it is phosphorylated, but by decreasing the rate at which it is dephosphorylated. In the case of glycogen phosphorylase, an increase in [cAMP] results not only in an increase in the rate of activation of the enzyme, but also in a decrease in its rate of deactivation.*

The activity of phosphoprotein phosphatase-1 is also controlled by its binding to *m*-phosphorylase *a*. X-ray studies indicate that phosphorylase undergoes extensive conformational changes in converting from the T to the R states. Among the major structural changes in this conformational shift is the movement of the Ser 14-phosphoryl group from the surface of the T state (inactive) enzyme to a position buried a few angstroms beneath the protein's surface in the R state (active) enzyme (Fig. 17-10). *Both the R and T forms of phosphorylase a strongly bind phosphoprotein phosphatase-1, but only in the T state enzyme is the Ser 14-phosphoryl group accessible for hydrolysis, which converts phosphorylase a to phosphorylase b. Consequently, when phosphorylase a is in its active R form, it effectively removes phosphoprotein phos-*

phatase-1 *from circulation.* However, under the conditions that phosphorylase *a* converts to the T state (Section 17-3G), phosphoprotein phosphatase-1 hydrolyzes the now exposed Ser 14-phosphoryl group thereby converting *m*-phosphorylase *a* to *o*-phosphorylase *b*, which has only a low affinity for binding phosphoprotein phosphatase-1. One effect of phosphorylase *a* demodification, therefore, is to relieve the inhibition of phosphoprotein phosphatase-1 by releasing it and thus allowing it to excise the phosphoryl groups of other susceptible phosphoproteins. Since glycogen phosphorylase is in ~10-fold greater concentration than is phosphoprotein phosphatase-1, this release only occurs when more than ~90% of the glycogen phosphorylase is in the *o*-phosphorylase *b* form. Glycogen synthase is among the proteins that are dephosphorylated by phosphoprotein phosphatase-1 when it is released from phosphorylase. In contrast to phosphorylase, dephosphorylation activates glycogen synthase. This enzyme is involved in its own bicyclic cascade whose properties we shall now examine.

D. Glycogen Synthase Bicyclic Cascade

Like glycogen phosphorylase, glycogen synthase exists in two enzymatically interconvertible forms:

1. The modified (*m*; phosphorylated) form that is inactive under physiological conditions (the *b* form).

2. The original (*o*; dephosphorylated) form that is active (the *a* form).

m-Glycogen synthase *b* is under allosteric control; it is strongly inhibited by physiological concentrations of ATP, ADP, and P_i. This inhibition may be overcome by G6P but only at nonphysiological concentrations above 10 mM. Since the physiological concentration of G6P in muscle tissue is only 0.2 to 0.4 mM, the modified enzyme is almost totally inactive *in vivo*. The activity of the unmodified enzyme is essentially independent of these effectors so that the cell's glycogen synthase activity varies with the fraction of the enzyme in its unmodified form.

The mechanistic details of the interconversion of modified and unmodified forms of glycogen synthase are particularly complex and are therefore not as well understood as those of glycogen phosphorylase. It has been clearly established that the fraction of unmodified glycogen synthase is, in part, controlled by a bicyclic cascade involving phosphorylase kinase and phosphoprotein phosphatase-1, enzymes that are also involved in the glycogen phosphorylase bicyclic cascade (Fig. 17-13, right). However, 6 other protein kinases are known to at least partially deactivate human muscle glycogen synthase by phosphorylating this homodimer at 1 or more of 9 Ser residues on its 737-residue subunits. These enzymes include cAMP-dependent protein kinase so that glycogen synthase deactivation may also be considered to occur via a monocyclic cascade; **cal-modulin-dependent protein kinase,** which is activated by the presence of Ca^{2+}; **protein kinase C,** which responds to the extracellular presence of certain hormones via a mechanism described in Section 34-4B; and **glycogen synthase kinase-3** whose mode of activation is unknown. Why glycogen synthase deactivation is so elaborately regulated compared to its activation or the activation/deactivation of glycogen phosphorylase is unclear although, whatever the reasons, it closely monitors the organism's metabolic state.

E. Integration of Glycogen Metabolism Control Mechanisms

Whether there is net synthesis or degradation of glycogen and at what rate depends on the relative balance of the active forms of glycogen synthase and glycogen phosphorylase. This, in turn, largely depends on the rates of the phosphorylation and dephosphorylation reactions of the two bicyclic cascades. These cascades, one controlling the rate of glycogen breakdown and the other controlling the rate of glycogen synthesis, are intimately related. They are linked by cAMP-dependent protein kinase and phosphorylase kinase which, through phosphorylation, activate phosphorylase as they inactivate glycogen synthase. The cascades are also linked by phosphoprotein phosphatase-1, which is inhibited by phosphorylase *a* and therefore unable to activate (dephosphorylate) glycogen synthase unless it first inactivates (also by dephosphorylation) phosphorylase *a*.

Hormones Trigger Glycogen Metabolism

Glycogen metabolism in the liver is ultimately controlled by the polypeptide hormone **glucagon.**

$$\overset{+}{H_3N}-His-Ser-Glu-Gly-Thr-Phe-Thr-Ser-Asp-Tyr-\quad 10$$

$$Ser-Lys-Tyr-Leu-Asp-Ser-Arg-Arg-Ala-Gln-\quad 20$$

$$Asp-Phe-Val-Gln-Trp-Leu-Met-Asn-Thr-COO^-\quad 29$$

Glucagon

In muscles and various tissues, control is exerted by the adrenal hormones **epinephrine (adrenalin)** and **norepinephrine (noradrenalin).**

X = CH_3 Epinephrine
X = H   Norepinephrine

Earl Sutherland found that these hormones act at cell surfaces to stimulate adenylate cyclase, thus increasing [cAMP], which then acts inside the cell as a **second messenger,** that is, as the intracellular mediator of the hormonal message (the mechanism of adenylate cyclase activation is elaborated in Section 34-4B). Following this discovery, it was realized that cAMP, which is present in all forms of life, is an essential control element in many biological processes.

When hormonal stimulation by glucagon or epinephrine increases the cAMP concentration, the cAMP-dependent protein kinase activity increases, increasing the rates of phosphorylation of many proteins and decreasing their dephosphorylation rates as well. A decrease in dephosphorylation rates, as previously noted, increases the phosphorylation level of phosphoprotein phosphatase inhibitor which, in turn, inhibits phosphoprotein phosphatase-1. An increase in the concentration of phosphorylase *a* also contributes to the inhibition of phosphoprotein phosphatase-1.

Because of the amplifying properties of the cyclic cascades, a small change in [cAMP] results in a large change in the fraction of enzymes in their phosphorylated forms. When a large fraction of the glycogen metabolism enzymes are present in their phosphorylated forms, the metabolic flux is in the direction of glycogen breakdown since glycogen phosphorylase is active and glycogen synthase is inactive. When [cAMP] decreases, phosphorylation rates decrease, dephosphorylation rates increase, and the fraction of enzymes in their dephospho forms increases. The resultant activation of glycogen synthase and the inhibition of glycogen phosphorylase causes a change in the flux direction towards net glycogen synthesis.

F. Maintenance of Blood Glucose Levels

An important function of the liver is to maintain the blood concentration of glucose, the brain's primary fuel source, at ~5 m*M*. When blood [glucose] decreases beneath this level, usually during exercise or well after meals have been digested, the liver releases glucose into the blood stream. The process is mediated by the hormone glucagon as follows:

1. Low concentrations of blood glucose cause the pancreatic α cells to secrete glucagon into the blood stream.

2. Glucagon receptors on liver cell surfaces respond to the presence of glucagon by activating adenylate cyclase, thereby increasing the [cAMP] inside these cells.

3. The [cAMP] increase, as described above, triggers an increase in the rate of glycogen breakdown leading to increased intracellular [G6P].

4. G6P, in contrast to glucose, cannot pass through the

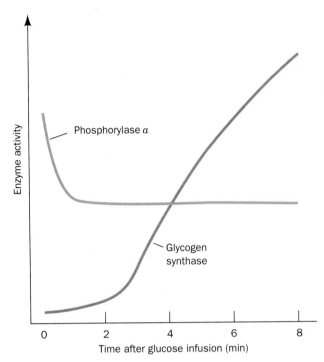

Figure 17-14
The enzymatic activities of phosphorylase *a* and glycogen synthase in mouse liver in response to an infusion of glucose. Phosphorylase *a* is rapidly inactivated and, somewhat later, glycogen synthase is activated. [After Stalmans, W., De Wulf, H., Hue, L., and Hers, H.-G., *Eur. J. Biochem.* **41**, 129 (1974).]

cell membrane. However, in liver, which does not employ glucose as a major energy source, the enzyme **glucose-6-phosphatase** hydrolyzes G6P:

$$G6P + H_2O \longrightarrow Glucose + P_i$$

The resulting glucose enters the bloodstream thereby increasing the blood glucose concentration. Muscle and brain cells, however, lack glucose-6-phosphatase so that they retain their G6P.

How does this delicately balanced system respond to an increase in blood [glucose]? When blood sugar is high, normally immediately after meals have been digested, glucagon levels decrease and the polypeptide hormone **insulin** (Fig. 6-2) is released from the pancreatic β cells. *The rate of glucose transport across many cell membranes increases in response to insulin (Section 18-2D).* Although the mechanism of insulin action is quite complex and poorly understood (Section 34-4B), it is thought that glucose itself may be the messenger to which the glycogen metabolism system responds. *Glucose inhibits phosphorylase by binding only to the active site of the enzyme's inactive T state, but in a manner different from that of substrate.* The presence of glucose therefore shifts phosphorylase's T ⇌ R equilibrium towards the T state (Fig. 17-9). This conformational shift, as we

saw in Section 17-3C, exposes the Ser 14 phosphoryl group to phosphoprotein phosphatase-1, resulting in the demodification of phosphorylase *a*. An increase in glucose concentration therefore promotes inactivation of glycogen phosphorylase *a* through the enzyme's conversion to phosphorylase *b* (Fig. 17-14). The concomitant release of phosphoprotein phosphatase-1 (recall that it specifically binds to phosphorylase *a*), moreover, results in the activation (dephosphorylation) of *m*-glycogen synthase *b*. Above a glucose concentration of 7 m*M*, this process reverses the flux of glycogen metabolism. The liver can thereby store the excess glucose as glycogen.

Glucokinase Forms G6P at a Rate Proportional to the Glucose Concentration

The liver's function in "buffering" the blood [glucose] is made possible because this organ contains a variant of hexokinase (the first glycolytic enzyme) known as **glucokinase**. The hexokinase in most cells has a high glucose affinity ($K_M < 0.1$ m*M*) and is inhibited by its reaction product, G6P. Glucokinase, in contrast, has a much lower glucose affinity ($K_M \approx 10$ m*M*) so that *its activity is proportional to the blood [glucose] over the normal physiological range (Fig. 17-15). Glucokinase, moreover, is not inhibited by G6P.* Consequently, the higher the blood [glucose], the faster the liver converts glucose to G6P (liver cells, unlike most cells, are freely permeable to glucose; their glucose transport rate is unresponsive to insulin). Thus at low blood [glucose], the liver does not compete with other tissues for the available glucose supply whereas at high blood [glucose], when the glucose needs of these tissues are met, the liver converts the excess glucose to glycogen.

Phosphoglucomutase, which has a high enough ac-

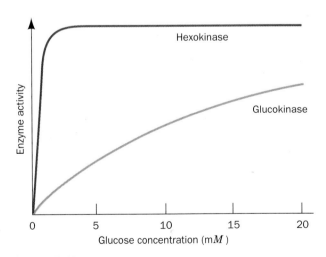

Figure 17-15
A comparison of the relative enzymatic activities of hexokinase and glucokinase over the physiological blood glucose range. The K_M for the glucokinase·glucose interaction is much higher than that for the hexokinase·glucose interaction.

tivity to equilibrate its substrate and product and therefore functions in either direction, transforms G6P to G1P, which is then converted to glycogen. Some of the G6P is also reconverted to glucose by the action of glucose-6-phosphatase in what amounts to a "futile" cycle. This is apparently the energetic price of effective glucose "buffering" of the blood.

Fructose-2,6-biphosphate Activates Glycolysis

β-D-Fructose-2,6-bisphosphate (F2,6P)

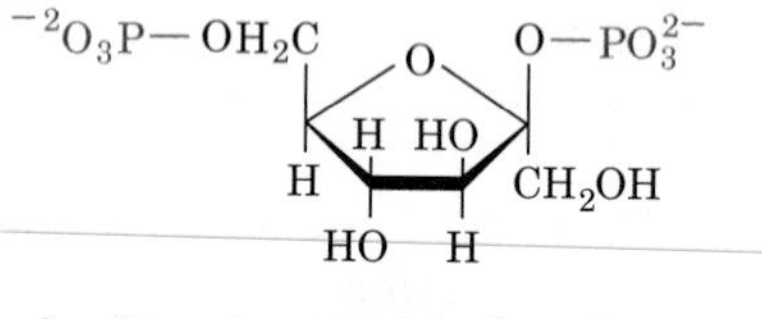

$$^{-2}O_3P-OH_2C \quad O \quad O-PO_3^{2-}$$

**β- D-Fructose-2,6-bisphosphate
(F2,6P)**

is also an important factor in the liver's maintenance of blood [glucose]. *F2,6P, which is not a glycolytic metabolite, is an extremely potent allosteric activator of animal phosphofructokinase (PFK) and an inhibitor of fructose bisphosphatase (FBPase).* F2,6P, which was discovered in 1980 by Emile Van Schaftingen and Henri-Géry Hers and independently by Kosaku Uyeda, therefore stimulates glycolytic flux (the F6P–FBP substrate cycle is discussed in Section 16-4B).

The concentration of F2,6P in the cell depends on the balance between its rates of synthesis and degradation by phosphofructokinase-2 (PFK-2) and fructose bisphosphatase-2 (FBPase-2), respectively (Fig. 17-16). These enzyme activities, which are both located on the same ~100-kD homodimeric protein, are subject to phosphorylation/dephosphorylation by cAMP-dependent protein kinase and phosphoprotein phosphatase-1. Phosphorylation of the liver enzyme at a specific Ser residue inhibits its PFK-2 activity and activates its FBPase-2 activity. Thus, the pancreatic α cell's release of glucagon in response to low blood [glucose] results,

through an increase in liver [cAMP], in a decreased liver [F2,6P]. This situation, in turn, decreases the PFK activity thereby inhibiting glycolysis. Hence, the G6P resulting from the concurrent stimulation of glycogen degradation is converted to glucose and secreted as described above rather than being metabolized. Simultaneously, the deinhibition of FBPase by the decrease of [F2,6P] stimulates **gluconeogenesis,** the formation of glucose from nonglucose precursors such as amino acids by a pathway that effectively reverses glycolytic flux (Section 21-1). This process provides a second means of glucose production. Conversely, when the blood [glucose] is high, cAMP levels decrease, the liver PFK-2/FPase-2 is dephosphorylated by phosphoprotein phosphatase-1 activating PFK-2 which, in turn, causes a rise in [F2,6P]. PFK is therefore activated, FBPase is inhibited, and the net glycolytic flux changes from gluconeogensis to glycolysis.

The F2,6P control system in heart muscle functions quite differently from that in liver. In heart muscle, increased glycogen breakdown is coordinated with increased glycolysis rather than increased glucose secretion. This is because phosphorylation of the heart muscle PFK-2/FBPase-2 isozyme, which differs from that in liver, activates rather than inhibits PFK-2. Consequently, hormones that stimulate glycogen breakdown also increase heart muscle [F2,6BP] thereby stimulating glycolysis as well.

G. Response to Stress

Epinephrine and norepinephrine, which are often called the "fight or flight" hormones, are released into the bloodstream by the adrenal glands in response to stress. Epinephrine receptors (known as **β-adrenergic receptors;** Section 34-4A) present on the surfaces of liver and muscle cells respond to these hormones just as glucagon receptors respond to the presence of glucagon; they activate adenylate cyclase thereby increasing intracellular [cAMP]. Indeed, epinephrine also stimulates the

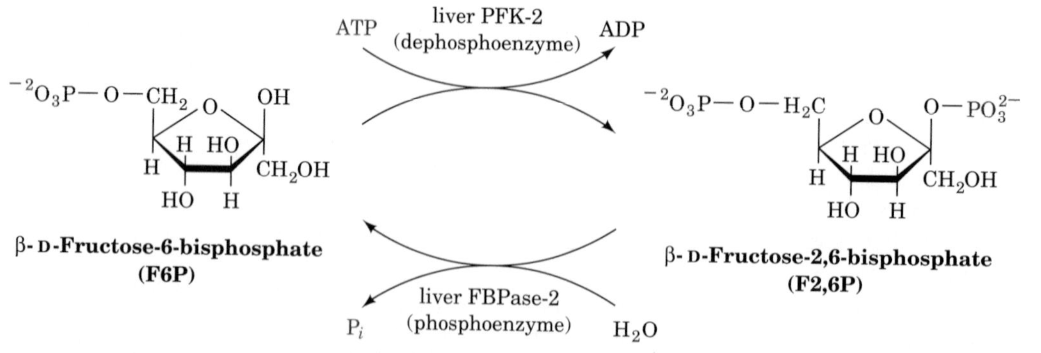

Figure 17-16
The formation and degradation of β-D-fructose-2,6-bisphosphate as catalyzed by PFK-2 and FBPase-2, two enzyme activities that occur on the same protein. The dephosphorylation of the liver enzyme activates PFK-2 but deactivates FBPase-2.

pancreatic α cells to release glucagon, which further increases liver [cAMP]. The G6P produced by the consequent glycogen breakdown in muscle enters the glycolytic pathway thereby generating ATP and helping the muscles cope with the stress that triggered the epinephrine release. The liver simultaneously releases glucose into the bloodstream to further fuel these muscles.

4. GLYCOGEN STORAGE DISEASES

With glycogen metabolism being such a finely controlled system, it is not surprising that genetically determined enzyme deficiencies result in disease states. The study of these disease states and the enzyme deficiencies that cause them has provided insights into the system's balance. In this sense, genetic diseases are valuable research tools. Conversely, the biochemical characterization of the pathways affected by a genetic disease often leads, as we shall see, to useful strategies for its treatment. Many diseases have been characterized that result from inherited deficiencies of one or another of the enzymes of glycogen metabolism. These defects are listed in Table 17-1 and discussed below.

Type I: Glucose-6-Phosphatase Deficiency (von Gierke's Disease)

Glucose-6-phosphatase catalyzes the final step leading to the release of glucose into the bloodstream by the liver. Deficiency of this enzyme results in an increase of intracellular [G6P], which leads to a large accumulation of glycogen of normal structure in the liver and kidney (recall that G6P inhibits glycogen phosphorylase and activates glycogen synthase) and an inability to increase blood glucose concentration in response to glucagon or epinephrine. Its symptoms include massive liver enlargement, severe **hypoglycemia** (low blood sugar) after a few hour fast, and a general failure to thrive. Treatment of the disease has included drug-induced inhibition of glucose uptake by the liver to increase blood [glucose], continuous intragastric feeding overnight, again to increase blood [glucose], and surgical transposition of the portal vein, which ordinarily feeds the liver directly from the intestines, so as to allow this glucose-rich blood to reach peripheral tissues before it reaches the liver. This latter treatment has the double benefit of allowing the tissues to receive more glucose while decreasing the storage of this glucose as liver glycogen. Liver transplantation has also been successful in the few patients in which this treatment has been tried.

Type II: α-1,4-Glucosidase Deficiency (Pompe's Disease)

This is the most devastating glycogen storage disease. It results in a large accumulation of glycogen of normal structure in the lysosomes of all cells and causes death by cardiorespiratory failure, usually before the age of 1 year. We have not discussed **α-1,4-glucosidase** in the sections on the pathways of glycogen synthesis and breakdown since it is not among those enzymes. It occurs in lysosomes where it functions to hydrolyze maltose (Section 10-2B) and other linear oligosaccharides, as well as the outer branches of glycogen, thereby yielding free glucose. The reason why lysosomes normally take up and degrade glycogen granules is unknown.

Type III: Amylo-1,6-Glucosidase (Debranching Enzyme) Deficiency (Cori's Disease)

In this disease, glycogen of abnormal structure containing very short outer chains accumulates in both liver

Table 17-1
Hereditary Glycogen Storage Diseases[a]

Type	Enzyme Deficiency	Tissue	Common Name	Glycogen Structure
I	Glucose-6-phosphatase	Liver	von Gierke's disease	Normal
II	α-1,4-Glucosidase	All lysosomes	Pompe's disease	Normal
III	Amylo-1,6-glucosidase (debranching enzyme)	All organs	Cori's disease	Outer chains missing or very short
IV	Amylo-$(1,4 \rightarrow 1,6)$-transglycosylase (branching enzyme)	Liver, probably all organs	Andersen's disease	Very long unbranched chains
V	Glycogen phosphorylase	Muscle	McArdle's disease	Normal
VI	Glycogen phosphorylase	Liver	Hers' disease	Normal
VII	Phosphofructokinase	Muscle		Normal
VIII	Phosphorylase kinase	Liver	Tarui's disease	Normal
IX	Glycogen synthase	Liver		Normal, deficient in quantity

[a] All types but Type VIII are autosomal recessive; Type VIII is sex linked.

and muscle since, in the absence of debranching enzyme, the glycogen cannot be further degraded. Its hypoglycemic symptoms are similar to, but not as severe as, those of von Gierke's Disease (Type I). The low blood sugar, which in this case is a result of the decreased efficiency of glycogen breakdown, is treated with frequent feeding and a high protein diet [in response to low blood sugar, the liver, through gluconeogenesis (Section 21-1), synthesizes glucose from protein]. For unknown reasons, the symptoms of Cori's disease often disappear at puberty.

Type IV: Amylo-(1,4 → 1,6)-transglycosylase (Branching Enzyme) Deficiency (Andersen's Disease)

This is one of the most severe glycogen storage diseases; victims rarely survive past the age of 4 years because of liver dysfunction. Glycogen concentration in liver is not increased but its structure is abnormal, with very long unbranched chains resulting from the lack of branching enzyme. This decreased branching greatly reduces the solubility of glycogen. It has been suggested that the liver dysfunction may be due to a "foreign body" immune reaction to the abnormal glycogen.

Type V: Muscle Phosphorylase Deficiency (McArdle's Disease)

We have mentioned this condition in connection with the realization that glycogen synthesis and breakdown must occur by different pathways (Section 17-2). Its major symptom, which is most severely manifested in early adulthood, is painful muscle cramps upon the onset of exertion. This situation is a result of the inability of the glycogen breakdown system to provide sufficient fuel for glycolysis to keep up with the metabolic demand for ATP. Studies, by ^{31}P NMR, on human forearm muscle have noninvasively corroborated this conclusion by demonstrating that exercise in individuals with McArdle's disease leads to elevated muscle ADP levels compared to those of normal individuals (Fig. 17-17). Curiously, if McArdle's victims continue their exertions, their cramps subside. This "second wind" effect has been attributed to vasodilation, which gives the muscles increased access to the glucose and fatty acids in the blood for use as alternative fuels to glycogen. Liver glycogen phosphorylase is normal in these individuals implying the presence of different glycogen phosphorylase **isozymes** in muscle and liver (isozymes are structurally similar enzymes that catalyze the same reaction but which are specified by different genes).

Type VI: Liver Phosphorylase Deficiency (Hers' Disease)

Patients with a deficiency of liver phosphorylase have symptoms similar to those with mild forms of Type I glycogen storage disease. The hypoglycemia in this case is due to the inability of glycogen phosphorylase to respond to the need for glucose production by the liver.

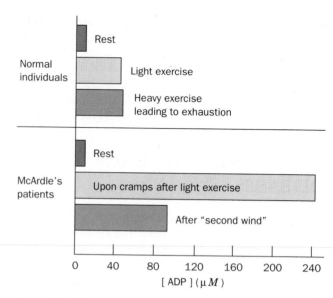

Figure 17-17
The ADP concentration in human forearm muscles during rest and following exertion in normal individuals and those with McArdle's disease. The ADP concentration was determined from ^{31}P NMR measurements on intact forearms. [After Radda, G. K., *Biochem. Soc. Trans.* **14**, 522 (1986).]

Type VII: Muscle Phosphofructokinase Deficiency

The result of a deficiency of the glycolytic enzyme PFK in muscle is an abnormal buildup of the glycolytic metabolites G6P and F6P. High concentrations of G6P increase the activities of glycogen synthase (G6P activates glycogen synthase and inactivates glycogen phosphorylase) and UDP-glucose pyrophosphorylase (G6P is in equilibrium with G1P, a substrate for the enzyme) so that glycogen accumulates in muscle. Other symptoms are similar to those of Type V glycogen storage disease, muscle phosphorylase deficiency, since PFK deficiency prevents glycolysis from keeping up with the ATP demand of muscle contraction.

Type VIII: Liver Phosphorylase Kinase Deficiency (Tarui's Disease)

Some individuals with symptoms of Type VI glycogen storage disease have liver phosphorylase of normal structure. Rather, they have a defective phosphorylase kinase, which results in their inability to convert phosphorylase *b* to phosphorylase *a*.

Type IX: Liver Glycogen Synthase Deficiency

This is the only disease of glycogen metabolism in which there is a deficiency rather than an overabundance of glycogen. Although the activity of liver glycogen synthase is extremely low in individuals with this disease, the primary lesion may not be in the synthase itself. Other metabolic defects may lead to an imbalance of the glycogen synthase cyclic cascade. The root cause of Type IX glycogen storage disease is still under investigation.

Appendix: Kinetics of a Cyclic Cascade

In the general monocyclic cascade of Fig. 17-11, the fraction of the target enzyme E in the more active state E_a may be expressed:

$$\frac{[E_a]}{[E]_T} = \frac{\left[\dfrac{\dfrac{k_f[F]_T}{K_f}}{\dfrac{k_r[R]_T e_2}{K_r(K_2 + e_2)} + \dfrac{k_f[F]_T}{K_f}}\right] e_1}{\left[\dfrac{\dfrac{k_r[R]_T e_2}{K_r(K_2 + e_2)}}{\dfrac{k_r[R]_T e_2}{K_r(K_2 + e_2)} + \dfrac{k_f[F]_T}{K_f}} K_1\right] + e_1} \qquad [17.A1]$$

where $[E]_T = [E_a] + [E_b]$ is the total concentration of E; $[F]_T$ and $[R]_T$ are the total concentrations of the modifying enzymes, F and R; e_1 and e_2 are the concentrations of their respective allosteric effectors that they bind with dissociation constants K_1 and K_2; k_f and k_r are the rate constants of the enzymatic modifications; and K_f and K_r are the corresponding dissociation constants. Here, for simplicity, we have ignored the participation of ATP in the modification.

This all but incomprehensible relationship is reproduced to demonstrate that even very complex behavior can be reduced to simple terms so as to permit us to make predictions. The mathematical form of Eq. [17.A1] is that of the Michaelis–Menten equation (Eq. [13.24]):

$$\frac{v_0}{[E]_T} = \frac{k_2[S]}{K_M + [S]}$$

which, it will be recalled, describes the effect of substrate concentration on the rate of an enzymatic reaction. Therefore, in analogy with the Michaelis–Menten equation, where K_M is the concentration of substrate required for half-maximal velocity, the apparent K_M for Eq. [17.A1] (the first term in the denominator) represents the concentration of e_1 required for one half of E to be in its active form; that is

$$K_M^{app} = \frac{\dfrac{k_r[R]_T e_2}{K_r(K_2 + e_2)}}{\dfrac{k_r[R]_T e_2}{K_r(K_2 + e_2)} + \dfrac{k_f[F]_T}{K_f}} K_1 \qquad [17.A2]$$

Equation [17.A2] indicates that the concentration of e_1 necessary for one half of E to be in its active form is the product of a complex of constants times K_1, the dissociation constant for e_1 from F (the e_1 concentration required to half convert F to its active form; $e_1 = K_1$ when $[F \cdot e_1] = [F]$). We wish to understand from this analysis whether cyclic cascades are able to amplify the sensitivity of a system, in this case the reaction catalyzed by E, to the change in concentration of an allosteric effector, in this case e_1.

How can we define amplification? *The effect of an allosteric effector (e_1) has been amplified if a small effect on one enzyme (F) causes a larger effect on another enzyme (E).* That is, if it takes less e_1 to convert one half of E to its active form than it takes to convert one half of F to its active form, we can say that the effect of e_1 has been amplified. Let us define an amplification factor, A, as the ratio of K_1 (the e_1 concentration required to half convert F to its active form) to K_M^{app} (the concentration of e_1 required for one half of E to be in its active form). Then, using the relationship in Eq. [17.A2] and rearranging:

$$A = \frac{K_1}{K_M^{app}} = 1 + \frac{k_f[F]_T K_r(K_2 + e_2)}{K_f k_r[R]_T e_2} \qquad [17.A3]$$

We now have a relationship that enables us to evaluate the effectiveness of a cyclic cascade in amplifying the effect of an allosteric modifier. Let us see what the amplification factor would be if, for the sake of argument, we assign the following arbitrary, but not unrealistic values to the rate constants, equilibrium constants, and enzyme concentrations: $k_f = 2k_r$, $K_r = 2K_f$, $[F]_T = 2[R]_T$, and $K_2 = 2e_2$. Upon substitution of these values into Eq. [17.A3] we find that A = 25. Thus, with our choice of parameters, there is a 25-fold amplification of the effect of e_1 on F by the time it reaches E because of this cyclic cascade. A larger difference between the forward and reverse parameters would provide an even larger amplification.

Chapter Summary

In animals, when glucose is not needed as a source of metabolic energy, it is stored, predominantly in liver and muscle cells, as glycogen, an $\alpha(1 \rightarrow 4)$-linked glucan with $\alpha(1 \rightarrow 6)$ branches every 8 to 12 residues. Glycogen breakdown to glucose-6-phosphate (G6P) is a two-step process. Glycogen phosphorylase catalyzes the phosphorolysis of the glycosidic linkage of a terminal glucosyl residue to form glucose-1-phosphate (G1P). Phosphoglucomutase interconverts G1P and G6P. A debranching enzyme allows complete degradation of glycogen by catalyzing the transfer of three-residue chains onto the nonreducing ends of other chains and catalyzing the hydrolysis of the remaining $\alpha(1 \rightarrow 6)$-linked glucosyl unit to glucose.

Glycogen is synthesized from G6P by a pathway different from that of glycogen breakdown. G6P is converted to G1P under the influence of phosphoglucomutase. UDP-glucose pyrophosphorylase utilizes UTP to convert G1P to UDP-glucose, the activated intermediate in glycogen synthesis. The hydrolysis of the PP_i product by inorganic pyrophosphatase drives the reaction to completion. Glucosyl units are transferred from UDP-glucose to the C(4)-OH group of a terminal residue on a growing glycogen chain by glycogen synthase. Branching occurs through the action of a branching enzyme, which transfers ~7-residue sections of $\alpha(1 \rightarrow 4)$-linked chains to the C(6)-OH group of a glucosyl residue on the same or another glycogen chain.

The rates at which glycogen is synthesized by glycogen phosphorylase and degraded by glycogen synthase are controlled by the levels of their allosteric effectors such as ATP, AMP, G6P, and glucose. Superimposed on this allosteric control is control by the phosphorylation/dephosphorylation of these enzymes. The kinases that catalyze these modifications are part of amplifying cascades that are ultimately controlled by the hormones glucagon and epinephrine. These hormones stimulate glycogen breakdown by causing an increase in the intracellular [cAMP]. cAMP activates cAMP-dependent protein kinase which, through its activation of phosphorylase kinase, results in the phosphorylation of both glycogen phosphorylase and glycogen synthase. Phosphorylation activates glycogen phosphorylase but inactivates glycogen synthase. A decrease in [cAMP] results in dephosphorylation of these enzymes by phosphoprotein phosphatase-1. The concentration in liver of F2,6P, an activator of PFK and an inhibitor of FBPase, is also dependent on the rates of cAMP-dependent phosphorylation and dephosphorylation.

Glycogen storage diseases are caused by a genetic deficiency of one or another of the enzymes of glycogen metabolism. Nine different deficiencies of varying severity have been reported in humans.

References

General

Boyer, P. D. and Krebs, E. G. (Eds.), *The Enzymes* (3rd ed.), Vol. 17, Academic Press (1986). [Contains detailed articles on the enzymes of glycogen metabolism and their control.]

Cunningham, E. B., *Biochemistry: Mechanisms of Metabolism*, Chapter 9, McGraw–Hill (1978).

Newsholme, E. A. and Start, C., *Regulation in Metabolism*, Chapter 4, Wiley (1973).

Walsh, C., *Enzymatic Reaction Mechanisms*, Freeman (1979).

Glycogen Metabolism

Chock, P. B., Rhee, S. G., and Stadtman, E. R., Interconvertible enzyme cascades in cellular regulation, *Annu. Rev. Biochem.* **49**, 813–843 (1980).

Cohen, P., The role of cAMP-dependent protein kinase in the regulation of glycogen metabolism in mammalian skeletal muscle, *Curr. Top. Cell. Regul.* **14**, 117–196 (1978).

Cohen, P., The role of protein phosphorylation in neural and hormonal control of cellular activity, *Nature* **296**, 613–620 (1982).

Cohen, P., Burchell, A., Foulkes, J. G., Cohen, P. T. W., Vanaman, T. C., and Nairn, A. C., Identification of the Ca^{2+}-dependent modulator protein as the fourth subunit of rabbit skeletal muscle phosphorylase kinase, *FEBS Lett.* **92**, 287–293 (1978). [The discovery of calmodulin as a subunit of phosphorylase kinase.]

Edelman, A. M., Blumenthal, D. K., and Krebs, E. G., Protein serine/threonine kinases, *Annu. Rev. Biochem.* **56**, 567–613 (1987).

Fletterick, R. J., Glycogen phosphorylase: plasticity and specificity in ligand binding, *Proc. Robert A. Welch Found. Conf. Chem. Res.* **27**, 172–220 (1984).

Fletterick, R. J. and Madsen, N. B., Structures and related functions of phosphorylase *a*, *Annu. Rev. Biochem.* **49**, 31–61 (1980).

Fletterick, R. J., Sprang, S., and Madsen, N. B., Analysis of the surface topography of glycogen phosphorylase *a*: implications for the metabolic interconversion and regulatory mechanisms, *Can. J. Biochem.* **57**, 789–797 (1979).

Hers, H.-G., Hue, L., and van Schaftingen, E., Fructose 2,6-bisphosphate, *Trends Biochem. Sci.* **7**, 329–331 (1982).

Jenkins, J. A., Johnson, L. N., Stuart, D. I., Stura, E. A., Wilson, K. S., and Zanotti, G., Phosphorylase: control and activity, *Philos. Trans. R. Soc. London Ser. B* **293**, 23–41 (1981).

Krebs, E. G., Phosphorylation and dephosphorylation of glycogen phosphorylase: a prototype for reversible covalent enzyme modification, *Curr. Top. Cell. Regul.* **18**, 401–419 (1981).

Madsen, N. B., Kravinsky, P. J., and Fletterick, R. J., Allosteric transitions of phosphorylase *a* and regulation of glycogen metabolism, *J. Biol. Chem.* **253**, 9097–9101 (1978).

Pilkus, S. J., Claus, T. H., Kountz, P. D., and El-Maghrabi, M. R., Enzymes of the fructose 6-phosphate–fructose 1,6-bisphosphatase substrate cycle, *in* Boyer, P. D. and Krebs, E. G.

(Eds.), *The Enzymes* (3rd ed.), Vol. 19, *pp.* 3–46, Academic Press (1987).

Roach, P. J., Glycogen synthase and glycogen synthase kinases, *Curr. Top. Cell. Regul.* **20**, 45–105 (1981).

Sprang, S. R., Acharya, K. R., Goldsmith, E. J., Stuart, D. I., Varvill, K., Fletterick, R. J., Madsen, N. B., and Johnson, L. N., Structural changes in glycogen phosphorylase induced by phosphorylation, *Nature* **336**, 215–221 (1988).

Stadtman, E. R. and Chock, P. B., Superiority of interconvertible enzyme cascades in metabolic regulation: analysis of monocyclic systems, *Proc. Natl. Acad. Sci.* **74**, 2761–2765 (1977).

Stalmans, W., The role of liver in the homeostasis of blood glucose, *Curr. Top. Cell. Regul.* **17**, 51–97 (1976).

Taylor, S. S., cAMP-dependent protein kinase, *J. Biol. Chem.* **264**, 8443–8446 (1989).

Van Schaftingen, E., Fructose 2,6-bisphosphate, *Adv. Enzymol.* **59**, 315–395 (1987).

Glycogen Storage Diseases

Radda, G. K., Control of bioenergetics: from cells to man by phosphorus nuclear-magnetic-resonance spectroscopy, *Biochem. Soc. Trans.* **14**, 517–525 (1986). [Discusses the noninvasive diagnosis of McArdle's disease by ^{31}P NMR.]

Stanbury, J. B., Wyngaarden, J. B., Frederickson, D. S., Goldstein, J. L., and Brown, M. S. (Eds.), *The Metabolic Basis of Inherited Disease* (5th ed.), Chapters 5–7, McGraw–Hill (1983).

Problems

1. A glycogen molecule consisting of 10,000 glucose residues is branched, on average, every 10 residues. How many reducing ends does it have?

2. The complete metabolic oxidation of glucose to CO_2 and O_2 yields 38ATPs (Chapter 20). What is the fractional energetic cost of storing glucose as glycogen and later metabolizing the glycogen rather than directly metabolizing the glucose? (Recall that glycogen's branched structure results in its degradation to 90% G1P and 10% glucose.)

3. What are the effects of the following on the rates of glycogen synthesis and glycogen degradation: (a) increasing the Ca^{2+} concentration, (b) increasing the ATP concentration, (c) inhibiting adenylate cyclase, (d) increasing the epinephrine concentration, and (e) increasing the AMP concentration?

4. Demonstrate that hexokinase activity but not glucokinase activity is insensitive to blood [glucose] over the physiological range. Calculate the ratio of glucokinase to hexokinase activities when [glucose] is 2 mM (hypoglycemic), 5 mM (normal), and 25 mM (diabetic). Assume that both enzymes have the same V_{max}.

5. What is the amplification factor in a monocyclic cascade if $k_f = 3k_r$, $K_r = 3K_f$, $[F]_T = 3[R]_T$, and $K_2 = 3e_2$?

6. Compare the properties of a bicyclic cascade with those of a monocyclic cascade.

7. The V_{max} of muscle glycogen phosphorylase is much larger than that of liver. Discuss the functional significance of this phenomenon.

*8. A complication of glycogen metabolism that we have not discussed is that many protein kinases, including phosphorylase kinase, are autophosphorylating; that is, they can specifically phosphorylate and thereby activate themselves. Discuss how this phenomenon affects glycogen metabolism taking into consideration the possibilities that phosphorylase kinase autophosphorylation may be an intramolecular or an intermolecular process.

9. Explain the symptoms of von Gierke's disease.

10. A sample of glycogen from a patient with liver disease is incubated with P_i, normal glycogen phosphorylase, and normal debranching enzyme. The ratio of glucose-1-phosphate to glucose formed in this reaction mixture is 100. What is the patient's most likely enzymatic deficiency? What is the probable structure of the patient's glycogen?

Chapter 18
TRANSPORT THROUGH MEMBRANES

Metabolism occurs within cells that are separated from their environments by plasma membranes. Eukaryotic cells, in addition, are compartmentalized by intracellular membranes that form the boundaries and internal structures of their various organelles. The nonpolar cores of biological membranes make them highly impermeable to most ionic and polar substances so that *these substances can traverse membranes only through the action of specific* **transport proteins.** Such proteins are therefore required to mediate all trans-membrane movements of ions, such as Na^+, K^+, Ca^{2+}, and Cl^-, as well as metabolites such as pyruvate, amino acids, sugars, and nucleotides. Transport proteins are also responsible for all biological electrochemical phenomena. In this chapter, we discuss the thermodynamics, kinetics, and chemical mechanisms of these membrane transport systems.

1. THERMODYNAMICS OF TRANSPORT

As we saw in Section 3-4A, the free energy of a solute, A, varies with its concentration:

$$\overline{G}_A - \overline{G}_A^{\circ\prime} = RT \ln [A] \qquad [18.1]$$

where $\overline{G}_A$ is the **chemical potential** (partial molar free energy) of A (the bar indicates quantity per mole) and $\overline{G}_A^{\circ\prime}$ is the chemical potential of its standard state. Strictly speaking, this equation applies only to ideal solutions; for nonideal (real) solutions, molar concentrations must be replaced by activities (Appendix to Chapter 3). This added complication is unnecessary for our purposes, however, because at the low physiological concentrations of biological substances, activities closely approach their corresponding molar concentrations in value.

The diffusion of a substance between two sides of a membrane

$$A \ (out) \rightleftharpoons A \ (in)$$

thermodynamically resembles a chemical equilibration. A difference in the concentrations of the substance on two sides of a membrane generates a chemical potential difference:

$$\Delta \overline{G}_A = \overline{G}_A(in) - \overline{G}_A(out) = RT \ln \left(\frac{[A]_{in}}{[A]_{out}} \right) \quad [18.2]$$

Consequently, if the concentration of A outside the membrane is greater than that inside, $\Delta \overline{G}_A$ for the transfer of A from outside to inside will be negative and the spontaneous net flow of A will be inward. If, however, [A] is greater inside than outside, $\Delta \overline{G}_A$ is positive and an inward net flow of A can only occur if an exergonic process, such as ATP hydrolysis, is coupled to it to make the overall free energy change negative.

Membrane Potentials Arise from Trans-Membrane Concentration Differences of Ionic Substances

The permeabilities of biological membranes to ions such as H^+, Na^+, K^+ Cl^-, Ca^{2+}, *etc.*, are controlled by specific membrane-embedded transport systems that we shall discuss in later sections. *The resulting charge differences across a biological membrane generate an electric potential difference,* $\Delta \Psi = \Psi(in) - \Psi(out)$, *where* $\Delta \Psi$ *is termed the* **membrane potential.** Consequently, if A is ionic, Eq. [18.2] must be amended to include the electrical work required to transfer a mole of A across the membrane from outside to inside:

$$\Delta \overline{G}_A = RT \ln \left(\frac{[A]_{in}}{[A]_{out}} \right) + Z_A \mathscr{F} \, \Delta \Psi \quad [18.3]$$

where Z_A is the ionic charge of A; $\mathscr{F}$, the Faraday constant, is the charge on a mole of electrons (96,494 $C \cdot mol^{-1}$); and, $\overline{G}_A$ is now termed the **electrochemical potential** of A.

Membrane potentials in living cells can be directly measured through the use of microelectrodes. $\Delta \Psi$ values of -100 mV (inside negative) are not uncommon (note that $1 \text{ V} = 1 \text{ J} \cdot C^{-1}$). Thus the last term of Eq. [18.3] is often significant for ionic substances.

2. KINETICS AND MECHANISMS OF TRANSPORT

Thermodynamics indicates whether a given transport process will be spontaneous but, as we saw for chemical and enzymatic reactions, provides no indication of the rates of these processes. Kinetic analyses of transport processes together with mechanistic studies have nevertheless permitted these processes to be characterized. There are two types of transport processes: **nonmediated transport** and **mediated transport.** Nonmediated transport occurs through simple diffusion. In contrast, *mediated transport occurs through the action of specific carrier proteins* that are variously called **permeases, porters, translocases, translocators,** and **transporters.** Mediated transport is further classified into two categories depending on the thermodynamics of the system:

1. **Passive-mediated transport** or **facilitated diffusion** in which specific molecules flow from high concentration to low concentration so as to equilibrate their concentration gradients.

2. **Active transport** in which specific molecules are transported from low concentration to high concentration, that is, against their concentration gradients. Such an endergonic process must be coupled to a sufficiently exergonic process to make it favorable.

In this section, we consider the nature of nonmediated transport and then compare it to passive-mediated transport as exemplified by the erythrocyte glucose transporter and ionophores. Active transport is examined in succeeding sections.

A. Nonmediated Transport

The driving force for the nonmediated flow of a substance A through a medium is A's electrochemical potential gradient. This relationship is expressed by the **Nernst–Planck equation:**

$$J_A = -[A] \, U_A \, (d\overline{G}_A/dx) \quad [18.4]$$

where J_A is the flux (rate of passage per unit area) of A, x is distance, $d\overline{G}_A/dx$ is the electrochemical potential gradient of A, and U_A is its **mobility** (velocity per unit force) in the medium. If we assume, for simplicity, that A is an uncharged molecule so that $\overline{G}_A$ is given by Eq. [18.1], the Nernst–Planck equation reduces to

$$J_A = -D_A \, (d\,[A]/dx) \quad [18.5]$$

where $D_A \equiv RTU_A$ is the **diffusion coefficient** of A in the medium of interest. This is **Fick's first law of diffusion,** which states that *a substance diffuses in the direction that eliminates its concentration gradient,* $d\,[A]/dx$, *at a rate proportional to the magnitude of this gradient.*

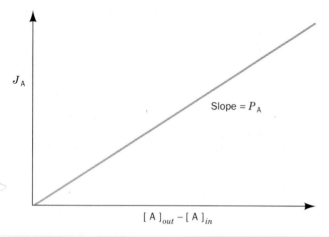

Figure 18-1
The linear relationship between diffusional flux (J_A) and ($[A]_{out} - [A]_{in}$ across a semipermeable membrane; see Eq. [18.6].

For a membrane of thickness x, Eq. [18.5] is approximated by

$$J_A = \frac{D_A}{x}([A]_{out} - [A]_{in}) = P_A([A]_{out} - [A]_{in})\ [18.6]$$

where D_A is the diffusion coefficient of A inside the membrane and $P_A = D_A/x$ is termed the membrane's **permeability coefficient** for A. The permeability coefficient is indicative of the solute's tendency to transfer from the aqueous solvent to the membrane's nonpolar core. It should therefore vary with the ratio of the solute's solubility in a nonpolar solvent resembling the membrane's core (e.g., olive oil) to that in water, a quantity known as the solute's **partition coefficient** between the two solvents. Indeed, the fluxes of many nonelectrolytes across erythrocyte membranes vary linearly with their concentration differences across the membrane as predicted by Eq. [18.6] (Fig. 18-1). Moreover, their permeability coefficients, as obtained from the

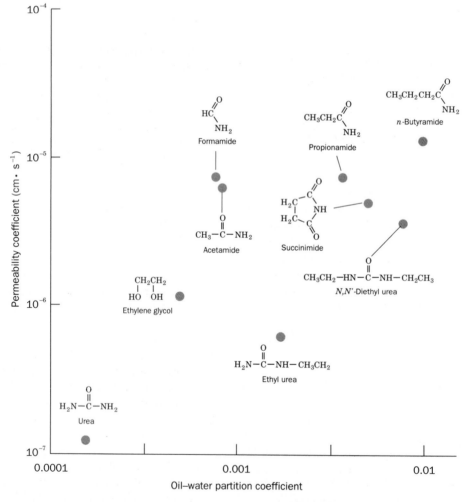

Figure 18-2
The permeability coefficients of various organic molecules in plasma membranes from the alga *Nitella mucronata* versus their partition coefficients between olive oil and water (a measure of a molecule's polarity). This more or less linear log–log plot indicates that the rate-limiting step for the nonmediated entry of a molecule into a cell is its passage through the membrane's hydrophobic core. [Based on data from Collander, R., *Physiol. Plant.* **7**, 433–434 (1954).]

slopes of plots such as Fig. 18-1, correlate rather well with their measured partition coefficients between non-polar solvents and water (Fig. 18-2).

B. Kinetics of Mediated Transport: Glucose Transport into Erythrocytes

Despite the success of the foregoing model in predicting the rates at which many molecules pass through membranes, there are numerous combinations of solutes and membranes that do not obey Eq. [18.6]. The flux in such a system is not linear with the solute concentration difference across the corresponding membrane (Fig. 18-3) and, furthermore, the solute's permeability coefficient is much larger than is expected on the basis of its partition coefficient. Such behavior indicates that *these solutes are conveyed across membranes in complex with carrier molecules; that is, they undergo mediated transport.*

The system that transports glucose across the erythrocyte membrane provides a well-characterized example of passive-mediated transport: It invariably transports glucose down its concentration gradient but not at the rate predicted by Eq. [18.6]. Indeed, the **erythrocyte glucose transporter** exhibits four characteristics that differentiate mediated from nonmediated transport: (1) *speed and specificity,* (2) *saturation kinetics,* (3) *susceptibility to competitive inhibition,* and (4) *susceptibility to chemical inactivation.* In the following paragraphs we

Table 18-1

Permeability Coefficients of Natural and Synthetic Membranes to D-Glucose and D-Mannitol at 25° C

Membrane Preparation	Permeability Coefficients (cm · s^{-1})	
	D-Glucose	D-Mannitol
Synthetic lipid bilayer	2.4×10^{-10}	4.4×10^{-11}
Calculated nonmediated diffusion	4×10^{-9}	3×10^{-9}
Intact human erythrocyte	2.0×10^{-4}	5×10^{-9}

Source: Jung, C. Y., *in* Surgenor, D. (Ed.), *The Red Blood Cell,* Vol. 2, *p.* 709, Academic Press (1975).

shall see how the erythrocyte glucose transporter exhibits these qualities.

Speed and Specificity

Table 18-1 indicates that the permeability coefficients of D-glucose and D-mannitol in synthetic bilayers, and that of D-mannitol in the erythrocyte membrane, are in reasonable agreement with the values calculated from the diffusion and partition coefficients of these sugars in olive oil. However, the experimentally determined permeability coefficient for D-glucose in the erythrocyte membrane is four orders of magnitude greater than its predicted value. *The erythrocyte membrane must therefore contain a system that rapidly transports glucose and that can distinguish D-glucose from D-mannitol.*

Saturation Kinetics

The concentration dependence of glucose transport indicates that its flux obeys the relationship:

$$J_A = \frac{J_{max}[A]}{K_M + [A]} \qquad [18.7]$$

This **saturation function** has a familiar hyperbolic form (Fig. 18-3). We have seen it in the equation describing the binding of O_2 to myoglobin (Eq. [9.2]) and in the Michaelis–Menten equation describing the rates of enzymatic reactions (Eq. [13.24]). Here, as before, K_M may be defined operationally as the concentration of glucose when the transport flux is one half of its maximal rate, $J_{max}/2$. *This observation of* **saturation kinetics** *for glucose transport was the first evidence that a specific saturable number of sites on the membrane were involved in the transport of any substance.*

The transport process can be described by a simple four-step kinetic scheme involving binding, transport, dissociation, and recovery (Fig. 18-4). Its binding and dissociation steps are analogous to the recognition of a substrate and the release of product by an enzyme. The mechanisms of transport and recovery are under active investigation and are discussed in Section 18-2D.

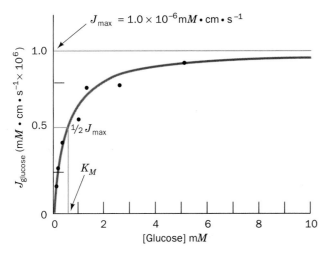

Figure 18-3
The variation of glucose flux into human erythrocytes with the external glucose concentration at 5°C. The black dots are experimentally determined data points while the solid line is computed from Eq. [18.7] with $J_{max} = 1.0 \times 10^{-6}$ $mM \cdot cm \cdot s^{-1}$ and $K_M = 0.5$ mM. The nonmediated glucose flux increases linearly with [glucose] (Fig. 18-1) but would not visibly depart from the baseline on the scale of this drawing. [Based on data from Stein, W. D., *Movement of Molecules Across Membranes,* p. 134, Academic Press (1967).]

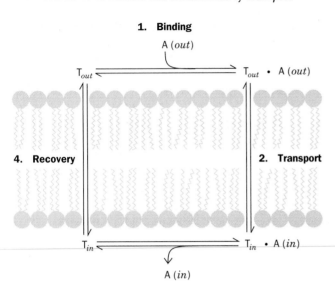

1. **Binding**

4. **Recovery**

2. **Transport**

3. **Dissociation**

Figure 18-4
A general kinetic scheme for membrane transport involving four steps: binding, transport, dissociation, and recovery. T is the transport protein whose binding site for solute A is located on either the inner or the outer side of the membrane at any one time.

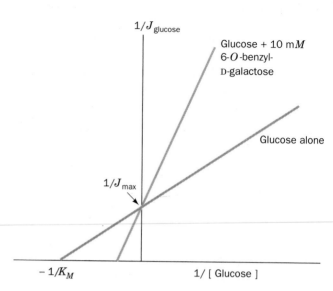

$1/J_{glucose}$

Glucose + 10 mM 6-O-benzyl-D-galactose

Glucose alone

$1/J_{max}$

$-1/K_M$

$1/$[Glucose]

Figure 18-5
Double-reciprocal plots for the net flux of glucose into erythrocytes in the presence and absence of 6-O-benzyl-D-galactose. The pattern is that of competitive inhibition. [After Barnett, J. E. G., Holman, G. D., Chalkley, R. A., and Munday, K. A., *Biochem. J.* **145**, 422 (1975).]

Susceptibility to Competitive Inhibition

Many compounds structurally similar to D-glucose inhibit glucose transport. A double-reciprocal plot (Section 13-2B) for the flux of glucose into erythrocytes in the presence or absence of 6-O-benzyl-D-galactose (Fig. 18-5) shows behavior typical of competitive inhibition of glucose transport (competitive inhibition of enzymes is discussed in Section 13-3A). *Susceptibility to competitive inhibition indicates that there is a limited number of sites available for mediated transport.*

Susceptibility to Chemical Inactivation

Treatment of erythrocytes with $HgCl_2$, which reacts with protein sulfhydryl groups (Section 6-2) and thus inactivates many enzymes, causes the rapid, saturable flux of glucose to disappear so that its permeability constant approaches that of mannitol. *The erythrocyte glucose transport system's susceptibility to such protein modifying agents indicates that it, in fact, is a protein.*

All of the above observations indicate that *glucose transport across the erythrocyte membrane is mediated by a limited number of protein carriers.* Before we discuss the mechanism of this transport system, however, we shall examine some simpler models of facilitated diffusion.

C. Ionophores

Our understanding of mediated transport has been enhanced by the study of **ionophores,** substances that vastly increase the permeability of membranes to particular ions.

Ionophores May Be Carriers or Channel Formers

Ionophores are organic molecules of diverse types, many of which are antibiotics of bacterial origin. Cells and organelles actively maintain concentration gradients of various ions across their membranes (Section 18-3A). The antibiotic properties of ionophores arise from their tendency to discharge these vital concentration gradients.

There are two types of ionophores:

1. *Carriers, which increase the permeabilities of membranes to their selected ion by binding it, diffusing through the membrane, and releasing the ion on the other side (Fig. 18-6a). For net transport to occur, the uncomplexed ionophore must then return to the original side of the membrane ready to repeat the process. Carriers therefore share the common property that their ionic complexes are soluble in nonpolar solvents.*

2. *Channel formers, which form trans-membrane channels or pores through which their selected ions can diffuse (Fig. 18-6b).*

Both types of ionophores transport ions at a remarkable rate. For example, a single molecule of the carrier antibiotic **valinomycin** transports up to 10^4 K^+ ions/s across a membrane. Channel formers have an even greater ion throughput; for example, each membrane channel composed of the antibiotic **gramicidin A** permits the passage of over 10^7 K^+ ions/s. Clearly, the presence of either type of ionophore, even in small amounts, greatly increases the permeability of a membrane towards the specific ions transported. However,

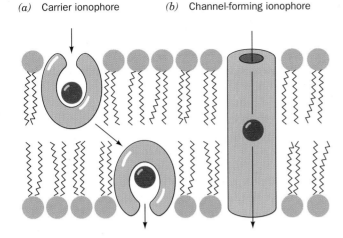

(a) Carrier ionophore   *(b)* Channel-forming ionophore

Figure 18-6
The ion transport modes of ionophores: (*a*) Carrier ionophores transport ions by diffusing through the lipid bilayer. (*b*) Channel-forming ionophores span the membrane with a channel through which ions can diffuse.

since ionophores passively permit ions to diffuse across a membrane in either direction, their effect can only be to equilibrate the concentrations of their selected ions across the membrane.

Carriers and channel formers are easily distinguished experimentally through differences in the temperature dependence of their action. Carriers depend on their ability to freely diffuse across the membrane. Consequently, cooling a membrane below its transition temperature (the temperature below which it becomes a gel-like solid; Section 11-2B) essentially eliminates its ionic permeability in the presence of carriers. In contrast, membrane permeability in the presence of channel formers is rather insensitive to temperature because,

once in place, channel formers need not move to mediate ion transport.

The K⁺–Valinomycin Complex Has a Polar Interior and Hydrophobic Exterior

Valinomycin, which is perhaps the best characterized carrier ionophore, specifically binds K^+ and the biologically unimportant Rb^+. It is a **cyclic depsipeptide** that has both D- and L-amino acid residues (Fig. 18-7; a depsipeptide contains ester linkages as well as peptide bonds). The X-ray structure of valinomycin's K^+ complex (Fig. 18-8a) indicates that the K^+ is octahedrally

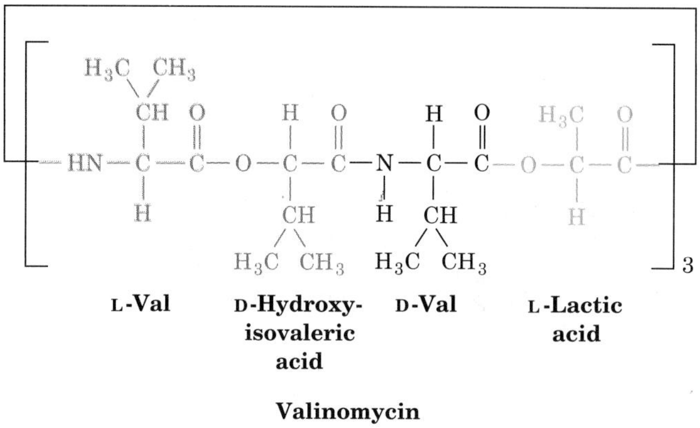

Figure 18-7
Valinomycin is a cyclic depsipeptide (has both ester and amide bonds) that contains both D- and L-amino acids.

Figure 18-8
Valinomycin X-ray structures. (*a*) The K⁺ complex. The six oxygen atoms that octahedrally complex the K⁺ ion are darker red than the other atoms. [After Neupert-Laves, K. and Dobler, M., *Helv. Chim. Acta* **58,** 439 (1975).] (*b*) Uncomplexed valinomycin. [After Smith, G. D., Duax, W. L., Langs, D. A., DeTitta, G. T., Edmonds, R. C., Rohrer, D. C., and Weeks, C. M., *J. Am. Chem. Soc.* **97,** 7242 (1975).] Hydrogen atoms are not shown.

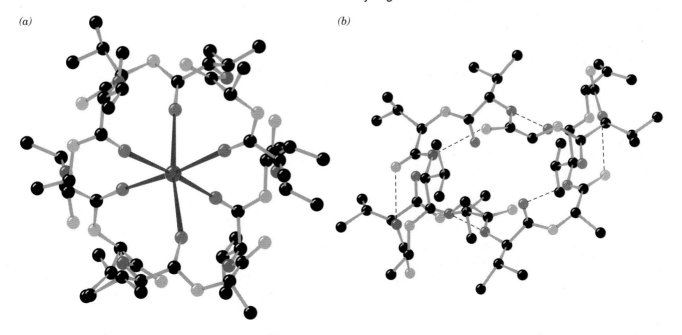

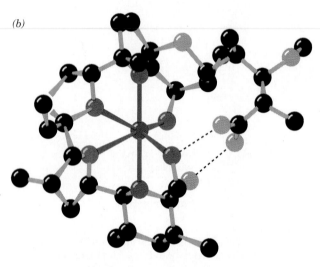

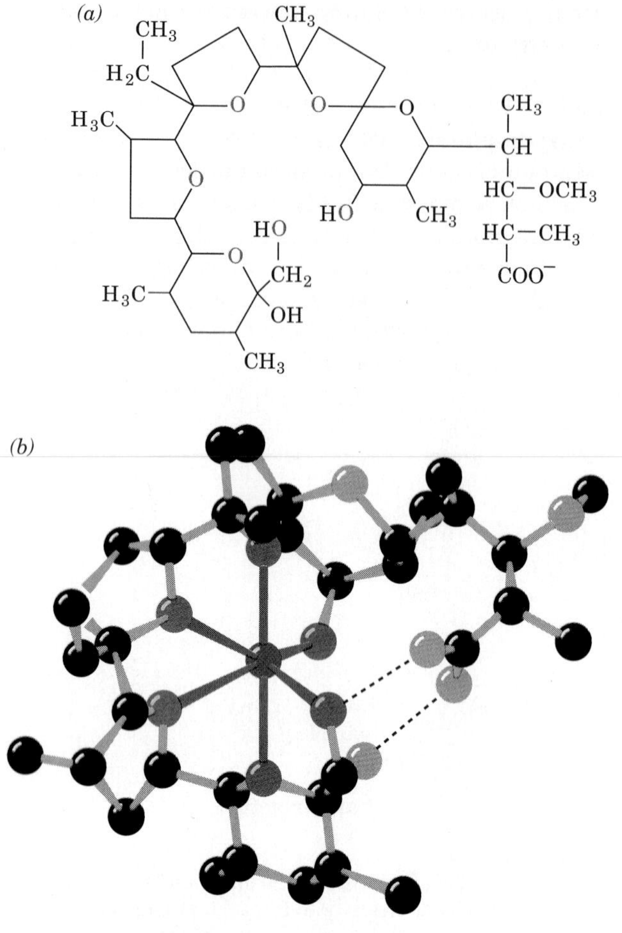

Figure 18-9
Monensin. (a) The structural formula with the six oxygen atoms that octahedrally complex Na^+ indicated in red. (b) The X-ray structure of the Na^+ complex (hydrogen atoms not shown). [After Duax, W. L., Smith, G. D., and Strong, P. D., *J. Am. Chem. Soc.* **102**, 6728 (1980).]

ions. Complexes of these ions with water are therefore energetically more favorable than their complexes with valinomycin. This accounts for valinomycin's 10,000-fold greater binding affinity for K^+ over Na^+. Indeed, no other known substance discriminates so acutely between Na^+ and K^+.

The Na^+-binding ionophore **monensin** (Fig. 18-9a), a linear polyether carboxylic acid, is chemically different from valinomycin. Nevertheless, X-ray analysis reveals that monensin's Na^+ complex has the same general features as valinomycin's K^+ complex in that monensin octahedrally coordinates Na^+ so as to wrap it in a nonpolar jacket (Fig. 18-9b). Other carrier ionophores have similar characteristics.

Gramicidin A Forms Helical Trans-Membrane Channels

Gramicidin A is a channel-forming ionophore from *Bacillus brevis* that permits the passage of protons and alkali cations but is blocked by Ca^{2+}. It is a 15-residue linear polypeptide of alternating L and D residues that is chemically blocked at its amino terminus by formylation and at its carboxyl terminus by an amide bond with ethanolamine (Fig. 18-10). Note that all of its residues are hydrophobic as is expected for a small trans-membrane polypeptide. NMR and X-ray crystallographic evidence indicate that *gramicidin A dimerizes in a head-to-head fashion to form a trans-membrane channel* (Fig. 18-11). This is corroborated by the observation that two gramicidin A molecules with their N-terminal amino groups covalently cross-linked form a functional ion channel.

The gramicidin A channel cannot be α helical because α helices lack a central channel and cannot consist of alternating L and D residues. Dan Urry has proposed that gramicidin A forms a novel helix that he named the β **helix** because it resembles a rolled up parallel β sheet.

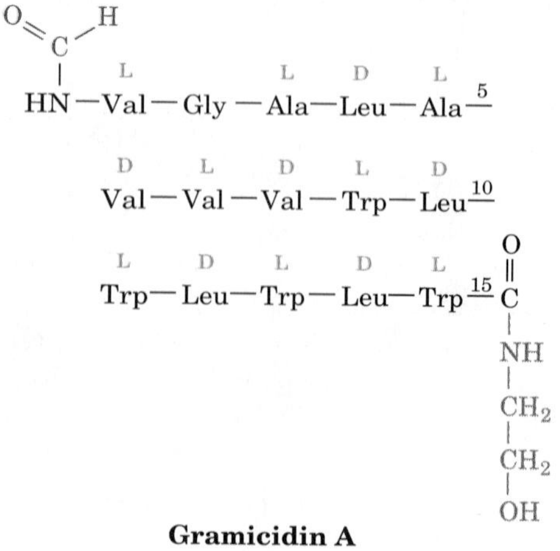

coordinated by the carbonyl groups of its 6 Val residues, which also form its ester linkages. The cyclic, intramolecularly hydrogen bonded valinomycin backbone follows a zigzag path that surrounds the K^+ coordination shell with a sinuous molecular bracelet. *Its methyl and isopropyl side chains project outward from the bracelet to provide the spheroidal complex with a hydrophobic exterior that makes it soluble in nonpolar solvents and in the hydrophobic cores of lipid bilayers.* Uncomplexed valinomycin (Fig. 18-8b) has a more open conformation than its K^+ complex, which presumably facilitates the rapid binding of K^+.

K^+ (ionic radius, $r = 1.33$ Å) and Rb^+ ($r = 1.49$ Å) fit snugly into valinomycin's coordination site. However, the rigidity of the valinomycin complex makes this site too large to accommodate Na^+ ($r = 0.95$ Å) or Li^+ ($r = 0.60$ Å) properly; that is, valinomycin's six carbonyl oxygen atoms cannot simultaneously coordinate these

Gramicidin A

Figure 18-10
Gramicidin A consists of 15 alternating D- and L-amino acid residues and is blocked at both its N- and C-termini.

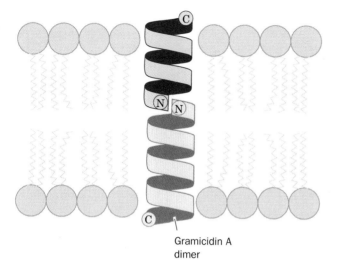

Figure 18-11
A schematic diagram of the trans-membrane channel formed by two molecules of gramicidin A. The molecules presumably dimerize by a hydrogen bonding association between their *N*-formyl ends (N).

Successive backbone N—H groups in this model alternately point up and down the helix to hydrogen bond with backbone carbonyl groups (Fig. 18-12). As a consequence of its alternating D and L residues, the side chains of the β helix festoon its periphery to form the channel's required hydrophobic exterior (recall that in a β sheet of all L-amino acid residues, the side chains alternately extend to opposite sides of the sheet; Section 7-1C). The polar backbone groups thus line the central channel and thereby facilitate the passage of ions. The model is consistent with the results of spectroscopic and NMR measurements on gramicidin A in lipid bilayers.

D. Mechanism of Passive-Mediated Glucose Transport

Integral membrane proteins are either exposed only at one surface of the membrane or, in the case of transmembrane proteins, oriented in only one direction with respect to the membrane (Section 11-3A). Since protein flip-flop rates are negligible, *the mobile carrier model that describes the mechanism of ionophores such as valinomycin (Fig. 18-6a) is not applicable to protein-mediated transport; rather, some sort of channel or pore mechanism appears likely.*

Glucose Transport Occurs via a Gated Pore Mechanism

The erythrocyte glucose transporter is a 55-kD glycoprotein which, according to sequence analysis, has three major domains: (1) a bundle of 12 membrane-spanning

α helices that are thought to form a hydrophobic cylinder surrounding a hydrophilic channel through which the glucose is transported; (2) a large, highly charged, cytoplasmic domain; and (3) a smaller, carbohydrate-bearing, external domain. The glucose transporter accounts for 2% of erythrocyte membrane proteins and runs as band 4.5 in SDS–PAGE gels of erythrocyte

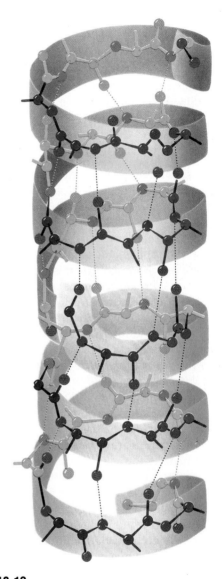

Figure 18-12
A proposed model for the gramicidin A trans-membrane channel. Each gramicidin A molecule of the head-to-head dimer forms a left-handed helix in which the hydrogen bonding pattern resembles that in a parallel β pleated sheet. The novel hydrogen bonding arrangement of this 26-Å long "β helix" is made possible because the alternating D and L configurations of the amino acid residues permit both successive NH and successive CO groups to point in opposite directions along the helix. The helix's bore diameter of 4 Å is sufficient to permit passage of alkali metal cations. [After Dobler, M., *Ionophores and Their Structures*, p. 215, Wiley–Interscience (1981). Based on a model proposed by D. W. Urry.]

membranes (Section 11-3C; it is not visible on the gel depicted in Fig. 11-32 because the heterogeneity of its oligosaccharide components makes the protein band diffuse).

Two observations support the hypothesis that the erythrocyte glucose transporter is asymmetrically disposed in the membrane:

1. **Galactose oxidase** oxidizes the galactose units of the glucose transporter's carbohydrate moieties only when the oxidase is outside the erythrocyte. The transporter's galactose units must therefore be located on the erythrocyte cell surface.

2. Trypsin only disrupts glucose transport when it acts from within an erythrocyte ghost. The transporter's trypsin-sensitive amino acid residues must therefore be located only on the cytoplasmic surface of the erythrocyte plasma membrane.

The glucose-binding sites on the two sides of the erythrocyte membrane have different steric requirements as well. John Barnett showed that addition of a propyl group to glucose C(1) prevents glucose binding to the outer surface of the membrane while addition of a propyl group to C(6) prevents binding to the inner surface. He therefore proposed that this trans-membrane protein has two alternate conformations: one with the glucose site facing the external cell surface, requiring O(1) contact, and leaving O(6) free; the other with the glucose site facing the internal cell surface, requiring O(6) contact, and leaving O(1) free (Fig. 18-13). *Transport apparently takes place by binding glucose to the protein on one face of the membrane, followed by a conformational change that closes the first site while exposing the other.* Glucose can then dissociate from the protein having been translocated across the membrane. The transport cycle is completed by the reversion of the glucose transporter to its initial conformation in the absence of bound glucose. Since this cycle can occur in either direction, the direction of net glucose transport is from high to low glucose concentrations. The glucose transporter thereby provides a means of equilibrating the glucose concentration across the erythrocyte membrane without any accompanying leakage of small molecules or ions.

The mechanism of glucose transport across erythrocyte membranes is a general one, often referred to as a **gated pore.** Indeed, *all known transport proteins appear to be asymmetrically situated trans-membrane proteins that alternate between two conformational states in which the ligand-binding sites are exposed, in turn, to alternate sides of the membrane.*

Cellular Glucose Uptake Is Regulated through the Insulin-Sensitive Exocytosis/Endocytosis of Glucose Transporters

Insulin stimulates fat and muscle cells to take up glucose. Within ~ 15 min after the administration of insulin, the

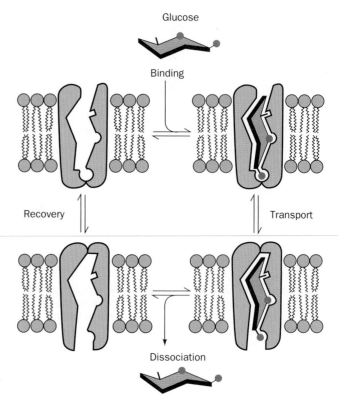

Figure 18-13
The alternating conformation model for glucose transport. Such a system is also known as a "gated pore." [After Baldwin, S. A., and Lienhard, G. E., *Trends Biochem. Sci.* **6,** 210 (1981).]

J_{max} for passive-mediated glucose transport into these cells increases by 6- to 12-fold, whereas the K_M remains constant. Upon withdrawal of the insulin, the rate of glucose uptake returns to its basal level within 20 min to 2 h depending on conditions. Neither the increase nor the decrease in the rate of glucose transport is affected by the presence of protein synthesis inhibitors so that these observations cannot be a consequence of the synthesis of new glucose transporter nor of a protein that inhibits it. How, then, does insulin regulate glucose transport?

Basal state cells are thought to store most of their glucose transporters in internal membranous vesicles. Upon insulin stimulation, these vesicles fuse with the plasma membrane in a process known as **exocytosis** *(Fig. 18-14).* The consequent increased number of cell-surface glucose transporters results in a proportional increase in the cell's glucose uptake rate. Upon insulin withdrawal, the process is reversed through the endocytosis of plasma membrane-embedded glucose transporters. Apparently insulin speeds up the basal rate of glucose transporter vesicle exocytosis and/or slows the basal rate of endocytosis, although how insulin does so is unknown.

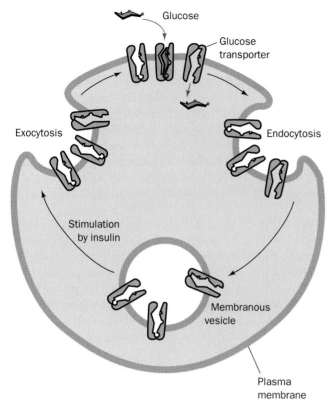

Figure 18-14
The regulation of glucose uptake in muscle and fat cells by the insulin-stimulated exocytosis (the opposite of endocytosis; Section 11-4C) of membranous vesicles containing glucose transporters (*left*). Upon insulin withdrawal, the process reverses itself through endocytosis (*right*).

2. **Electrogenic** if the transport process results in a charge separation across the membrane.

Since the glucose concentration in blood plasma is generally higher than that in cells, the erythrocyte glucose transporter normally transports glucose into the erythrocyte where it is metabolized via glycolysis. Many substances, however, are available on one side of a membrane in lower concentrations than are required on the other side of the membrane. Such substances must be actively and selectively transported across the membrane.

Active transport is an endergonic process that is often coupled to the hydrolysis of ATP. How is this coupling accomplished? In biosynthetic reactions, it often occurs through the direct phosphorylation of a substrate by ATP; for example, the formation of UTP in the synthesis of glycogen (Section 17-2). Membrane transport, however, is usually a physical rather than a chemical process; the transported molecule is not chemically altered. Determining the mechanism by which the free energy of ATP hydrolysis is coupled to endergonic physical processes has therefore been a challenging problem. In this section, we discuss animal cell transport proteins that are, in fact, directly phosphorylated by ATP during the transport process. We also examine a bacterial active transport process in which the molecules transported are concomitantly phosphorylated. In the next section, we study transport systems that couple the exergonic discharge of electrochemical potential gradients to the transport of ions and neutral molecules against their concentration gradients.

3. ATP-DRIVEN ACTIVE TRANSPORT

Mediated transport is categorized according to the stoichiometry of the transport process (Fig. 18-15):

1. A **uniport** involves the movement of a single molecule at a time. The erythrocyte glucose transporter is a uniport system.

2. A **symport** simultaneously transports two different molecules in the same direction.

3. An **antiport** simultaneously transports two different molecules in opposite directions.

The electrical character of ion transport is further specified as:

1. **Electroneutral** (electrically silent) if there is simultaneous charge neutralization, either by symport of oppositely charged ions or antiport of similarly charged ions.

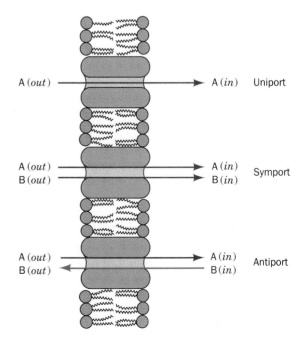

Figure 18-15
Uniport, symport, and antiport translocation systems.

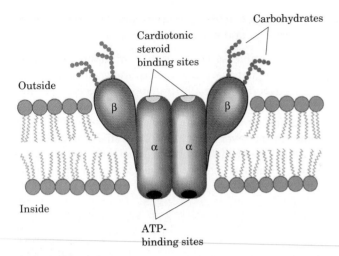

Figure 18-16
The putative dimeric structure of the (Na$^+$-K$^+$)-ATPase indicating its orientation in the plasma membrane.

A. (Na$^+$-K$^+$)-ATPase of Plasma Membranes

One of the most thoroughly studied active transport systems is the **(Na$^+$-K$^+$)-ATPase** of plasma membranes. This trans-membrane protein was first isolated in 1957 by Jens Skou. It consists of two types of subunits: A 110-kD nonglycosylated α subunit that contains the enzyme's catalytic activity and ion-binding sites, and an ~55-kD glycoprotein β subunit of unknown function. The protein is thought to have the subunit composition $(\alpha\beta)_2$ (Fig. 18-16), but it is unclear whether this dimeric structure is a functional necessity. The enzyme is often called the **(Na$^+$-K$^+$) pump** because *it pumps Na$^+$ out of and K$^+$ into the cell with the concomitant hydrolysis of intracellular ATP.* The overall stoichiometry of the (Na$^+$-K$^+$)-ATPase reaction is

$$3\text{Na}^+(in) + 2\text{K}^+(out) + \text{ATP} \rightleftharpoons$$
$$3\text{Na}^+(out) + 2\text{K}^+(in) + \text{ADP} + \text{P}_i$$

The (Na$^+$-K$^+$)-ATPase is therefore an electrogenic antiport: Three positive charges exit the cell for every two that enter. This extrusion of Na$^+$ enables animal cells to control their water content osmotically; without functioning (Na$^+$-K$^+$) pumps, animal cells, which lack cell walls, would swell and burst (recall that lipid bilayers are permeable to H$_2$O; Section 11-2B). Moreover, the electrochemical potential gradient generated by the (Na$^+$-K$^+$) pump is responsible for the electrical excitability of nerve cells (Section 34-4C), and provides the free energy for the active transport of glucose and amino acids into some cells (Section 18-4A). In fact, *all cells expend a large fraction of the ATP they produce (up to 70% in nerve cells) to maintain their required cytosolic Na$^+$ and K$^+$ concentrations.*

ATP Phosphorylates an Essential Asp During the Transport Process

The free energy of ATP hydrolysis powers the endergonic transport of Na$^+$ and K$^+$ against an electrochemical gradient. In coupling these two processes, a kinetic barrier must somehow be erected against the "downhill" transport of Na$^+$ and K$^+$ along their ion concentration gradients, while simultaneously facilitating their "uphill" transport. In addition, futile ATP hydrolysis must be prevented in the absence of uphill transport. How the enzyme does so is by no means well understood, although many of its mechanistic aspects have been elucidated.

A key discovery was that the protein is phosphorylated by ATP in the presence of Na$^+$ during the transport process. The use of chemical trapping techniques demonstrated that this phosphorylation occurs on an Asp residue to form a highly reactive **aspartyl phosphate** intermediate. For instance, sodium borohydride reduces acyl phosphates to their corresponding alcohols. In the case of an aspartyl phosphate residue, the alcohol is homoserine. By use of [^{3}H]NaBH$_4$ to reduce the phosphorylated enzyme, radioactive homoserine was, in fact, isolated from the acid hydrolysate (Fig. 18-17).

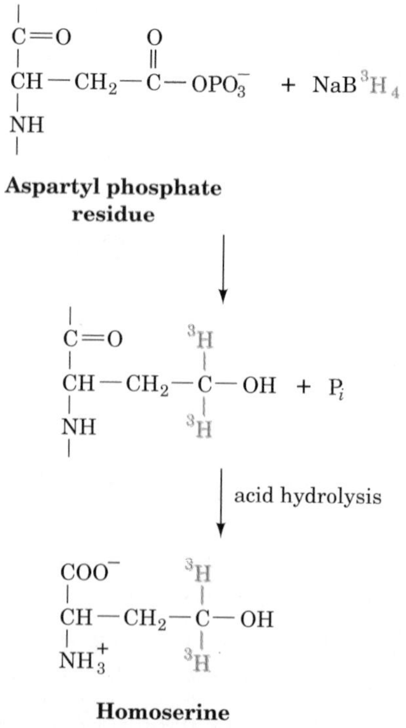

Figure 18-17
Reaction of [^{3}H]NaBH$_4$ with phosphorylated (Na$^+$-K$^+$)-ATPase. The isolation of [^{3}H]homoserine following acid hydrolysis of the protein indicates that the original phosphorylated amino acid residue is Asp.

The (Na⁺-K⁺)-ATPase Has Two Conformational States

The observations that ATP only phosphorylates the (Na⁺-K⁺)-ATPase in the presence of Na⁺, while the aspartyl phosphate residue is only subject to hydrolysis in the presence of K⁺, led to the realization that *the enzyme has two conformational states, E_1 and E_2.* These states have different tertiary structures, different catalytic activities, and different ligand specificities:

1. E_1 has an inward-facing high affinity Na⁺-binding site ($K_M = 0.2$ mM, well below the intracellular [Na⁺]) and reacts with ATP to form the activated product $E_1 \sim P$ only when Na⁺ is bound.

2. E_2—P has an outward-facing high affinity K⁺-binding site ($K_M = 0.05M$, well below the extracellular [K⁺]), and hydrolyzes to form $P_i + E_2$ only when K⁺ is bound.

An Ordered Sequential Kinetic Reaction Mechanism Accounts for the Coupling of Active Transport with ATP Hydrolysis

The (Na⁺-K⁺)-ATPase is thought to operate in accordance with the following ordered sequential reaction scheme (Fig. 18-18):

1. $E_1 \cdot 3Na^+$, which acquired its Na⁺ inside the cell, binds ATP to yield the ternary complex $E_1 \cdot ATP \cdot 3Na^+$.

2. The ternary complex reacts to form the "high-energy" aspartyl phosphate intermediate $E_1 \sim P \cdot 3Na^+$.

3. This "high-energy" intermediate relaxes to its "low-energy" conformation, E_2—$P \cdot 3Na^+$, and releases its

bound Na⁺ outside the cell, that is, Na⁺ is transported through the membrane.

4. E_2—P binds 2K⁺ from outside the cell to form E_2—$P \cdot 2K^+$.

5. The phosphate group is hydrolyzed yielding $E_2 \cdot 2K^+$.

6. $E_2 \cdot 2K^+$ changes conformation, releases its 2K⁺ inside the cell, and replaces it with 3Na⁺, thereby completing the transport cycle.

The obligatory order of the reaction requires that ATP can only be hydrolyzed as Na⁺ is transported "uphill." Conversely, Na⁺ can only be transported "downhill" if ATP is concomitantly synthesized. Consequently, although each of the above reaction steps is, in fact, individually reversible, the cycle, as is diagrammed in Fig. 18-18, circulates only in the clockwise direction under normal physiological conditions. The enzyme is thought to have only one set of cation-binding sites, which apparently changes both its orientation and its affinity during the course of the transport cycle.

Mutual Destabilization Accounts for the Rate of Na⁺ and K⁺ Transport

The above ordered kinetic mechanism accounts only for the coupling of active transport with ATP hydrolysis. *In order to maintain a reasonable rate of transport, the free energies of all its intermediates must be roughly equal. If one intermediate were more stable than another, the most stable would accumulate with a severe reduction in the overall transport rate.* For example, in order for Na⁺ to be transported out of the cell, uphill, its binding to E_1 must

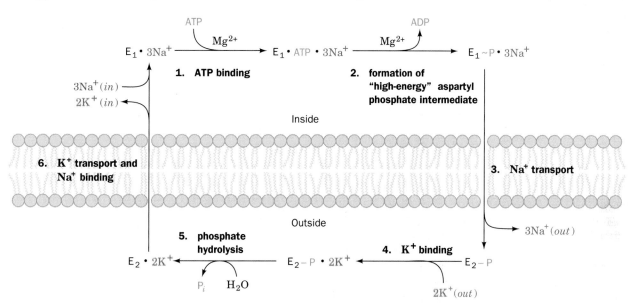

Figure 18-18
A kinetic scheme for the active transport of Na⁺ and K⁺ by (Na⁺-K⁺)-ATPase.

be strong on the inside and weak to E_2 on the outside. Strong binding means greater stability and a potential bottleneck. This difficulty is counteracted by the phosphorylation of $E_1 \cdot 3Na^+$ and its subsequent conformational change to yield the low Na^+ affinity E_2—P (Steps 2 and 3, Fig. 18-18). Likewise, the strong binding of K^+ to E_2—P on the outside is attenuated by its dephosphorylation and conformational change to yield the low K^+ affinity E_1 (Steps 5 and 6, Fig. 18-18). It is these mutual destabilizations that permit Na^+ and K^+ to be transported at a rapid rate.

Cardiac Glycosides Specifically Inhibit the (Na^+-K^+)-ATPase

Study of the (Na^+-K^+)-ATPase has been greatly facilitated by the use of **cardiac glycosides** (also called **cardiotonic steroids**), natural products that increase the intensity of heart muscle contraction. Indeed, **digitalis**, an extract of purple foxglove leaves (Fig. 18-19a), which contains a mixture of cardiac glycosides, has been used to treat congestive heart failure for centuries. Two of these steroids, **digitoxigenin** and **ouabain** (pronounced wa-bane; Fig. 18-19b), are still among the most commonly prescribed cardiac drugs. These substances inhibit the (Na^+-K^+)-ATPase by binding strongly to an externally exposed portion of the enzyme (the drugs are ineffective when injected inside cells) so as to block Step 3 in Fig. 18-18. The resultant increase in intracellular $[Na^+]$ causes an increased activity of the cardiac (Na^+-Ca^{2+}) antiport system, which pumps Na^+ out of and Ca^{2+} into the cell (Section 20-1B). Since Ca^{2+} triggers muscle contraction (Section 34-3C), the resulting increase in intracellular $[Ca^{2+}]$ increases the force of cardiac muscle contraction.

B. Ca^{2+}-ATPase

Ca^{2+} often acts as a second messenger in a manner similar to cAMP. Transient increases in cytosolic $[Ca^{2+}]$ trigger numerous cellular responses including muscle contraction (Section 34-3C), release of neurotransmitters (Section 34-4C) and, as we have seen, glycogen breakdown (Section 17-3C). Moreover, Ca^{2+} is an important activator of oxidative metabolism (Section 19-4).

The $[Ca^{2+}]$ in extracellular spaces ($\sim$1500 μM) is four orders of magnitude higher than in the cytosol ($\sim$0.1 μM). This large concentration gradient is maintained by the active transport of Ca^{2+} across the plasma membrane, the endoplasmic reticulum (the sarcoplasmic reticulum in muscle), and the mitochondrial inner membrane. We discuss the mitochondrial system in Section 20-1B. Plasma membrane and endoplasmic reticulum each contain a **Ca^{2+}-ATPase** that actively pumps Ca^{2+} out of the cytosol at the expense of ATP hydrolysis. Their kinetic mechanisms (Fig. 18-20) are very similar to that of the (Na^+-K^+)-ATPase (Fig. 18-18).

Figure 18-19
(a) The leaves of the purple foxglove plant are the source of the heart muscle stimulant digitalis. [Derek Fell.] (b) Two of the major components of digitalis are the cardiac glycosides ouabain and digitoxigenin.

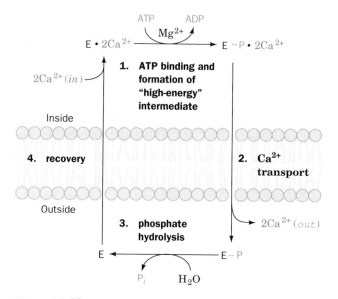

$$E \cdot 2Ca^{2+} \xrightarrow[\text{Mg}^{2+}]{\text{ATP} \quad \text{ADP}} E \sim P \cdot 2Ca^{2+}$$

1. ATP binding and formation of "high-energy" intermediate

$2Ca^{2+}$ (in)

Inside

4. recovery

Outside

2. Ca^{2+} transport

3. phosphate hydrolysis

$2Ca^{2+}$ (out)

$$E \leftarrow E-P$$

$P_i \qquad H_2O$

Figure 18-20
The kinetic mechanism of Ca²⁺-ATPase. Here (*in*) refers to the cytosol and (*out*) refers to the outside of the cell for plasma membrane Ca²⁺-ATPase or the lumen of the endoplasmic reticulum (sarcoplasmic reticulum) for the Ca²⁺-ATPase of that membrane.

In fact, there is significant sequence similarity among these various proteins suggesting that they arose from a common ancestor.

Calmodulin Regulates the Ca²⁺ Pump

For a cell to maintain its proper physiological state, it must regulate the activities of its ion pumps precisely. *The regulation of the Ca²⁺ pump in the plasma membrane is controlled by the level of Ca²⁺ through the mediation of calmodulin (CaM).* This ubiquitous eukaryotic Ca²⁺-binding protein participates in numerous cellular regulatory processes including, as we have seen, the control of glycogen metabolism (Section 17-3C). CaM's highly conserved 148-residue sequence has the distinctive feature that Lys 115 is trimethylated so that this quaternary amine is invariably positively charged. The protein, whose X-ray structure was determined by Charles Bugg, has a curious dumbbell-shaped structure consisting of two globular domains connected by a seven-turn α helix (Fig. 18-21).

Calmodulin has four high affinity Ca²⁺-binding sites, two on each of its globular domains. All of these sites are formed by nearly superimposable helix–loop–helix motifs known as **EF hands** (Fig. 18-22), which also occur in several other Ca²⁺-sensing proteins of known structure. The Ca²⁺ ion in each of these sites is octahedrally coordinated by oxygen atoms from the backbone and side chains of the loop as well as from a protein-associated water molecule. The binding of Ca²⁺ to any of these four sites in CaM triggers a large conformational change in that domain, which is thought to expose an

otherwise inaccessible nonpolar surface. The hydrophobic interactions of calmodulin's target proteins with one or both of these nonpolar surfaces probably mediate calmodulin's Ca²⁺-dependent regulatory properties.

Ca²⁺·calmodulin activates the Ca²⁺-ATPase of plasma membranes. The activation, as deduced from the study of the isolated ATPase, results in a decrease in its K_M for Ca²⁺ from 20 to 0.5 μM. Now we can see how Ca²⁺ regulates its own cytoplasmic concentration: At Ca²⁺ levels below calmodulin's ~1-μM dissociation constant for Ca²⁺, the Ca²⁺-ATPase is relatively inactive. If, however, the [Ca²⁺] rises to this level, Ca²⁺ binds to calmodulin which, in turn, binds to and activates the Ca²⁺ pump:

$$Ca^{2+} + CaM \rightleftharpoons$$
$$Ca^{2+} \cdot CaM^* + \text{pump (inactive)} \rightleftharpoons$$
$$Ca^{2+} \cdot CaM^* \cdot \text{pump (active)}$$

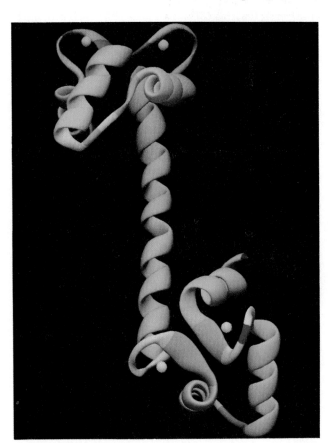

Figure 18-21
The X-ray structure of rat testes calmodulin. This monomeric 148-residue protein contains two remarkably similar globular domains separated by a seven-turn α helix. The residues are color coded according to their backbone conformation angles (φ and ψ; Fig. 7-7): light blue, α helical angles; green, β sheet angles; yellow, between helix and sheet; and purple, left-handed helix. The Gly residues are white and the N-terminus is dark blue. The two Ca²⁺-binding sites in each domain are represented by white spheres. [Courtesy of Michael Carson, University of Alabama at Birmingham. X-ray structure determined by Charles Bugg.]

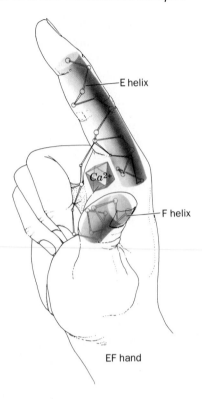

Figure 18-22
The Ca^{2+}-binding sites in many proteins that function to sense the level of Ca^{2+} are formed by helix–loop–helix motifs named EF hands. [After Kretsinger, R. H. *Annu. Rev. Biochem.* **45**, 241 (1976).]

(CaM* indicates activated calmodulin). This interaction decreases the pump's K_M for Ca^{2+} to below the ambient $[Ca^{2+}]$ thereby causing Ca^{2+} to be pumped out of the cytosol. When the $[Ca^{2+}]$ decreases sufficiently, Ca^{2+} dissociates from calmodulin and this series of events reverses itself, thereby inactivating the pump. The entire system is therefore analogous to a basement sump pump, that is automatically activated by a float when the water reaches a preset level.

C. (H⁺-K⁺)-ATPase of Gastric Mucosa

Parietal cells of the mammalian gastric mucosa secrete HCl at a concentration of $0.15M$ (pH 0.8). Since the cytosolic pH of these cells is 7.4, this represents a pH difference of 6.6 units, the largest known in eukaryotic cells. The secreted protons are derived from the intracellular hydration of CO_2 by carbonic anhydrase:

$$CO_2 + H_2O \rightleftharpoons HCO_3^- + H^+$$

The secretion of H^+ involves the participation of an **(H⁺-K⁺)-ATPase,** an electroneutral antiport with properties similar to that of (Na⁺-K⁺)-ATPase. Like the related (Na⁺-K⁺)- and Ca^{2+}-ATPases, it is phosphorylated during the transport process. In this case, however, the K^+, which enters the cell as H^+ is pumped out, is subse-

quently externalized by its electroneutral cotransport with Cl^-. HCl is therefore the overall transported product.

For many years, effective treatment of peptic ulcers, a frequently fatal condition caused by the attack of stomach acid on the gastric mucosa, often required the surgical removal of the affected portions of the stomach. The discovery, by James Black, of **cimetidine,**

$$H_3C \quad CH_2-S-CH_2-CH_2-NH-\overset{\displaystyle N-C\equiv N}{\underset{\displaystyle \|}{C}}-NH-CH_3$$

Cimetidine

$$CH_2-CH_2-NH_3^+$$

Histamine

which inhibits stomach acid secretion, has almost entirely eliminated the need for this dangerous and debilitating surgery. The (H⁺-K⁺)-ATPase of the gastric mucosa is activated by histamine stimulation of a cell surface receptor in a process mediated by cAMP. Cimetidine (trade name: Tagamet) and its analogs, which competitively inhibit the binding of histamine to this receptor, are presently the most commonly prescribed drugs in the United States.

D. Group Translocation

Group translocation is a variation of ATP-driven active transport that gram-negative bacteria use to import certain sugars; *it differs from active transport in that the molecules transported are simultaneously modified chemically.* The most extensively studied example of group translocation is the **phosphoenolpyruvate-dependent phosphotransferase system (PTS)** of *E. coli* discovered by Saul Roseman in 1964. Phosphoenolpyruvate (PEP) is the phosphoryl donor for this system (recall that PEP is the "high-energy" phosphoryl donor for ATP synthesis in the pyruvate kinase reaction of glycolysis; Section 16-2J). *The PTS simultaneously transports and phosphorylates sugars. Since the cell membrane is impermeable to sugar phosphates, once they enter the cell, they remain there.* The sugars transported are listed in Table 18-2.

The PTS system involves two soluble cytoplasmic proteins, **Enzyme I (E I)** and **HPr,** which participate in the transport of all sugars (Fig. 18-23). In addition, for each sugar the system transports, there is a specific trans-membrane transport protein **E II,** and in some cases, a sugar-specific protein **E III** that is membrane bound for some sugars and cytoplasmic for others. Glucose transport, for example, requires the participation of **E II_g** and **E III_g.**

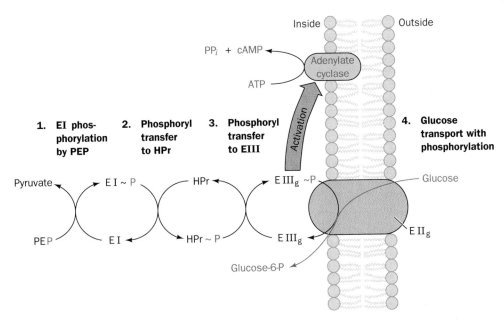

Figure 18-23
Transport of glucose by the PEP-dependent phosphotransferase system (PTS). HPr and E I are cytoplasmic proteins common to all sugars transported. E II$_g$ and E III$_g$ are proteins specific for glucose. Adenylate cyclase is activated by the presence of E III$_g$ ~ P.

Glucose transport, which resembles that of other sugars, occurs as follows (Fig. 18-23):

1. PEP phosphorylates E I to form a reactive phosphohistidine adduct.

$$\underset{\text{Phosphohistidine}}{H_3\overset{+}{N}-\underset{\underset{COO^-}{|}}{\overset{\overset{H}{|}}{C}}-CH_2-\underset{4}{\underset{5}{\overset{3}{\overset{N}{\diagup}}}}\overset{2}{\underset{1}{\underset{N}{\diagdown}}}C}$$

Phosphohistidine

2. The phosphoryl group is transferred to N(1) of a His residue on HPr. His is apparently a favored phosphoryl group acceptor in phosphoryl transfer reactions. It also participates in the phosphoglycerate mutase reaction of glycolysis (Section 16-2H).

3. HPr ~ P continues the phosphoryl-transfer chain by phosphorylating E III$_g$.

4. The phosphoryl group is finally transferred from E III$_g$ to glucose, which, in the process, is transported across the membrane by E II$_g$. Glucose is released into the cytoplasm only after it has been phosphorylated to glucose-6-phosphate (G6P).

Thus the transport of glucose is driven by its indirect, exergonic phosphorylation by PEP. The PTS is an energy-efficient system since only one ATP-equivalent is required to both transport and phosphorylate glucose. When the active transport and phosphorylation steps occur separately, as they do in many cells, two ATPs are hydrolyzed per glucose processed.

Bacterial Sugar Transport Is Genetically Regulated

The PTS is more complex than the other transport systems we have encountered, probably because it is part of a complicated regulatory system governing sugar transport. When any of the sugars transported by the PTS is abundant, the active transport of sugars which enter the cell via other transport systems is inhibited. This inhibition, called **catabolite repression,** is mediated through the cAMP concentration (Section 29-3C). cAMP activates the transcription of genes that encode various sugar transport proteins, including **lactose permease** (Section 18-4B). The presence of glucose results in a decrease in [cAMP] which, in turn, represses the synthesis of these other sugar transport proteins.

Table 18-2

Sugars Transported by the PEP-Dependent Phosphotransferase System (PTS) of *E. coli*

Glucose	Mannitol
Fructose	Sorbitol
Mannose	Galactitol
N-Acetylglucosamine	Lactose

The mechanism for control of [cAMP] is thought to reside in E III$_g$, which is transiently phosphorylated in Step 3 of the PTS transport process (Fig. 18-23). In the absence of glucose, this enzyme accumulates in its phosphorylated form. E III$_g$ ~ P is thought to activate adenylate cyclase, which results in elevated cAMP concentrations. *When glucose is plentiful, the steady state concentration of E III$_g$ ~ P decreases as it forms G6P during the PTS transport process. Adenylate cyclase is therefore deactivated, the cAMP concentration decreases, and the synthesis of lactose permease, as well as other proteins involved in lactose metabolism, slows down.* This is a form of energy conservation for the cell. Why synthesize the proteins required for the transport and metabolism of all sugars when the metabolism of only one sugar at a time will do?

4. ION GRADIENT-DRIVEN ACTIVE TRANSPORT

Systems such as the (Na$^+$-K$^+$)-ATPase discussed above utilize the free energy of ATP hydrolysis to generate electrochemical potential gradients across membranes. Conversely, *the free energy stored in an electrochemical potential gradient may be harnessed to power various endergonic physiological processes.* Indeed, ATP synthesis by mitochondria and chloroplasts is powered by the dissipation of proton gradients generated through electron transport and photosynthesis (Sections 20-3C and 22-2D). In this section we discuss active transport processes that are driven by the dissipation of ion gradients. We consider three examples: intestinal uptake of glucose by the **Na$^+$-glucose symport,** uptake of lactose by *E. coli* **lactose permease,** and the mitochondrial **ADP–ATP transporter.**

A. Na$^+$-Glucose Symport

Nutritionally derived glucose is actively concentrated in **brush border cells** of the intestinal epithelium by a Na$^+$-dependent symport (Fig. 18-24). It is transported from these cells to the circulatory system via a passive-mediated glucose uniport located on the capillary side of the cell and which is similar to that of the erythrocyte membrane (Section 18-2B). Note that *although the immediate energy source for glucose transport from the intestine is the Na$^+$ gradient, it is really the free energy of ATP hydrolysis that powers this process through the maintenance of the Na$^+$ gradient by the (Na$^+$-K$^+$)-ATPase.*

Active and Passive Glucose Transporters Exhibit Differential Drug Susceptibilities

The two glucose transport systems are inhibited by different drugs:

1. **Phlorizin** inhibits Na$^+$-dependent glucose transport.

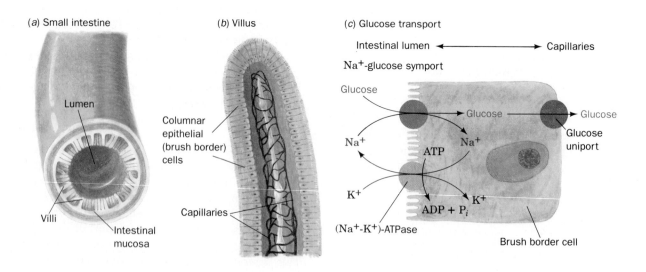

(a) Small intestine

Lumen

Villi

Intestinal mucosa

(b) Villus

Columnar epithelial (brush border) cells

Capillaries

(c) Glucose transport

Intestinal lumen ⟷ Capillaries

Na$^+$-glucose symport

Glucose

Na$^+$

ATP

Glucose

Na$^+$

Glucose

Glucose uniport

K$^+$

ADP + P$_i$

K$^+$

(Na$^+$-K$^+$)-ATPase

Brush border cell

Figure 18-24

Glucose transport in the intestinal epithelium. The brushlike villi lining the small intestine greatly increase its surface area thereby facilitating the absorption of nutrients. The brush border cells from which the villi are formed actively concentrate glucose from the intestinal lumen in symport with Na$^+$, a process that is driven by the (Na$^+$-K$^+$)-ATPase. The glucose is exported to the bloodstream via a separate passive-mediated uniport system like that in the erythrocyte.

2. Cytochalasin B inhibits Na⁺-independent glucose transport.

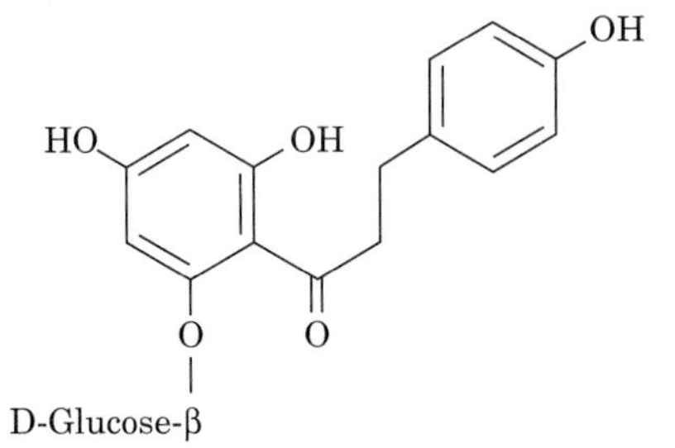

Phlorizin

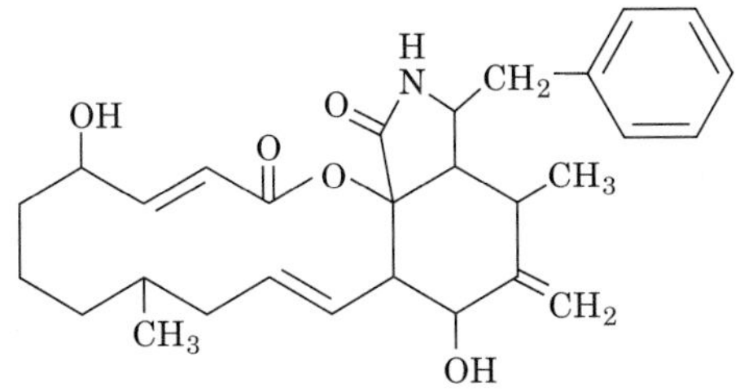

Cytochalasin B

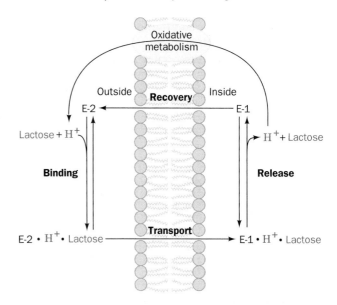

Figure 18-26
Kinetic mechanism of lactose permease in *E. coli*. Lactose and H⁺ are bound and released in random order from E-2 outside the cell and from E-1 inside the cell. E-2 must bind both lactose and H⁺ in order to change conformation to E-1, thereby cotransporting these substances into the cell. E-1 changes conformation to E-2 when neither lactose nor H⁺ are bound, thus completing the transport cycle.

Both inhibitors bind only to the external surfaces of their respective target proteins, thereby further indicating that these proteins are asymmetrically inserted into membranes. The use of these inhibitors permits the actions of the two glucose transporters to be studied separately in intact cells.

Kinetic studies indicate that the Na⁺-glucose symport binds its substrates, Na⁺ and glucose, in random order (Fig. 18-25). Only when both substrates are bound, however, does the protein change its conformation to expose the binding sites to the inside of the cell. This requirement for concomitant Na⁺ and glucose transport prevents the wasteful dissipation of the Na⁺ gradient.

B. Lactose Permease

Gram-negative bacteria such as *E. coli* contain several active transport systems for concentrating sugars. We have already discussed the PTS system. Another extensively studied system, **lactose permease** (also known as **galactoside permease**), *utilizes the proton gradient across the bacterial cell membrane to cotransport H⁺ and lactose (Fig. 18-26)*. The proton gradient is metabolically generated through oxidative metabolism in a manner similar to that in mitochondria (Section 20-3B). The electrochemical potential gradient created by both these systems is used mainly to drive the synthesis of ATP.

How do we know that lactose transport requires the presence of a proton gradient? Ronald Kaback has established the requirement for this gradient through the following observations:

1. The rate of lactose transport into bacteria is increased

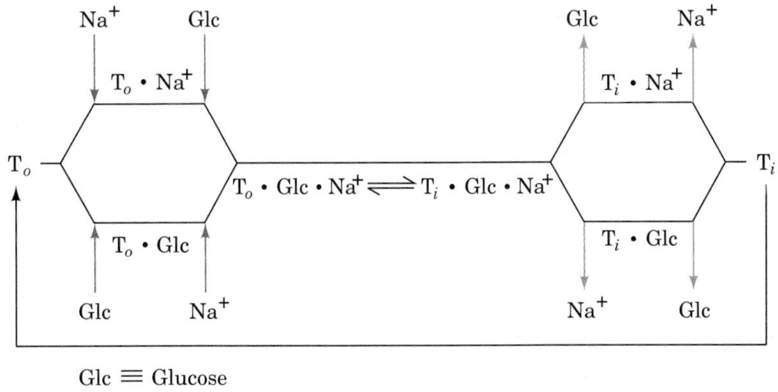

Figure 18-25
The Na⁺-glucose symport system represented as a Random Bi Bi kinetic mechanism. T_o and T_I, respectively, represent the transport protein with its binding sites exposed to the outer and inner surfaces of the membrane. [After Crane, R. K., and Dorando, F. C., *in* Martonosi, A. N. (Ed.), *Membranes and Transport*, Vol. 2, p. 154, Plenum Press (1982).]

$$T_o \cdot Glc \cdot Na^+ \rightleftharpoons T_i \cdot Glc \cdot Na^+$$

Glc ≡ Glucose

enormously by the addition of D-lactate, an energy source for trans-membrane proton gradient generation. Conversely, inhibitors of oxidative metabolism, such as cyanide, block both the formation of the proton gradient and lactose transport.

2. 2,4-Dinitrophenol, a proton ionophore that dissipates trans-membrane proton gradients (Section 20-3D), inhibits lactose transport into both intact bacteria and membrane vesicles.

3. The fluorescence of **dansylaminoethylthiogalactoside,**

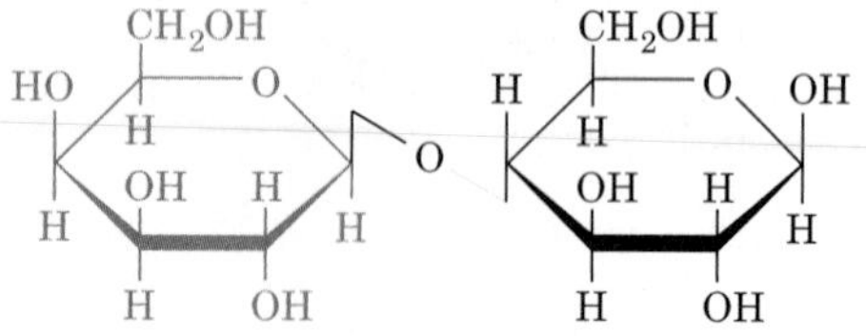

Lactose

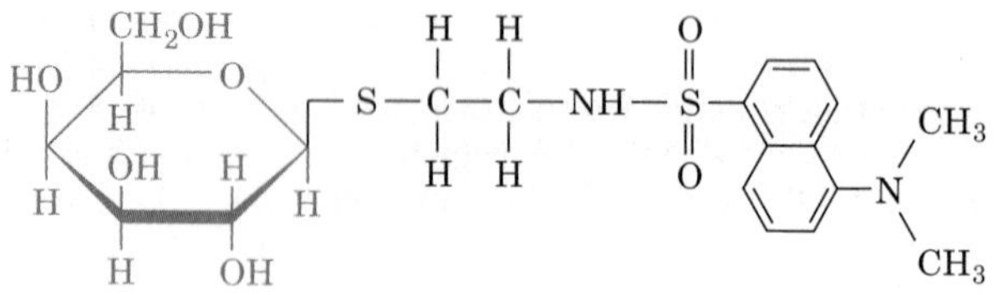

Dansylaminoethylthiogalactoside

a competitive inhibitor of lactose transport, is sensitive to the polarity of its environment and thus changes when it binds to lactose permease. Fluorescence measurements indicate that it does not bind to membrane vesicles that contain lactose permease in the absence of a trans-membrane proton gradient.

Lactose Permease Has Two Conformational States

Lactose permease, like the (Na^+-K^+)-ATPase, is a trans-membrane protein with two conformational states (Fig. 18-26):

1. E-1, which has a low affinity lactose-binding site facing the interior of the cell.

2. E-2, which has a high affinity lactose-binding site facing the exterior of the cell.

E-1 and E-2 can only interconvert when their H^+ and lactose-binding sites are either both filled or both empty. This prevents both dissipation of the H^+ gradient without cotransport of lactose into the cell, and transport of lactose out of the cell without cotransport of H^+ against its concentration gradient.

C. ADP–ATP Translocator

ATP generated in the mitochondrial matrix (its inner compartment; Section 1-2A) through oxidative phos-

phorylation (Section 20-3C) is largely utilized in the cytosol to drive such endergonic processes as biosynthesis, active transport, and muscle contraction. The inner mitochondrial membrane contains a system that transports ATP out of the matrix in exchange for ADP produced in the cytosol by ATP hydrolysis. This antiport is electrogenic since it exchanges ADP^{3-} for ATP^{4-}.

Several natural products inhibit ADP–ATP translocation. **Atractyloside** and its derivative **carboxyatractyloside** inhibit the process only from the external surface of the inner mitochondrial membrane; **bongkrekic acid** only exerts its effects on the internal surface.

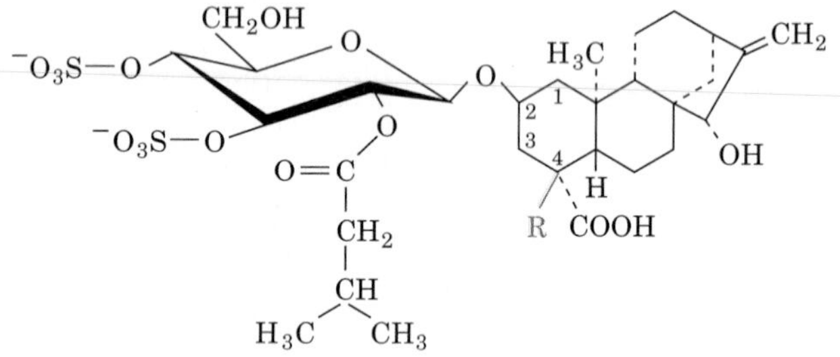

R = H **Atractyloside**
R = COOH **Carboxyatractyloside**

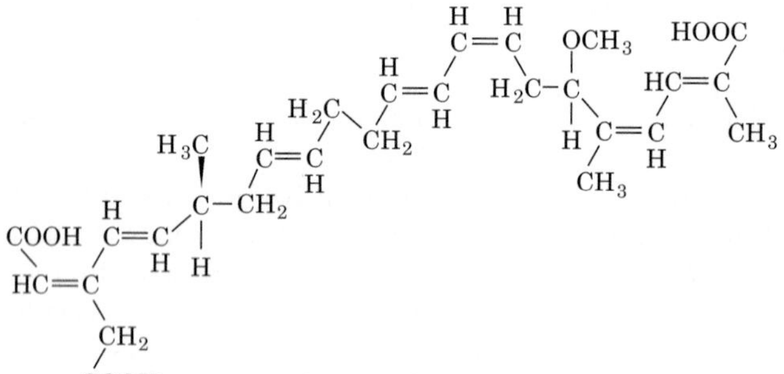

Bongkrekic acid

These differentially acting inhibitors have been valuable tools in the isolation of the transport protein and in the elucidation of its mechanism of action. For example, the translocator has been purified by affinity chromatography (Section 5-3D) using atractyloside derivatives as affinity ligands. Atractyloside binding is also a convenient means of identifying the translocator.

The **ADP–ATP translocator,** a dimer of identical 30-kD subunits, has characteristics similar to those of other transport proteins. It has one binding site for which ADP and ATP compete. It has two conformations, one with its ATP–ADP binding site facing the inside of the mitochondrion, and the other with the site facing outward (Fig. 18-27). The translocator is an antiport because it must bind ligand to change from one conformational state to the other at a physiologically reasonable rate.

The ADP–ATP translocator is not itself an active trans-

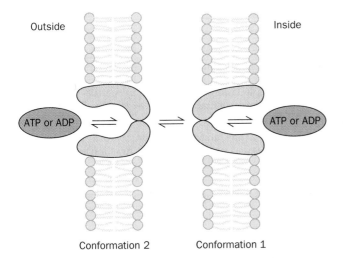

Outside

Inside

ATP or ADP

ATP or ADP

Conformation 2   Conformation 1

Figure 18-27
Conformational mechanism of the ADP–ATP translocator.
An adenine nucleotide-binding site located in the intersubunit
contact area of the translocator dimer is alternately exposed
to the two sides of the membrane. In contrast to the
glucose transporter (Fig. 18-13), the ADP–ATP translocator
can only change its conformation when bound to ADP or ATP.

port system. However, its electrogenic export of one negative charge per transport cycle in the direction of ATP export–ADP import is driven by the membrane potential difference, $\Delta\Psi$, across the inner mitochondrial membrane (positive outside). This results in the formation of gradients in ATP and ADP across the membrane.

Chapter Summary

Polar molecules and ions are transported across biological membranes by specific trans-membrane transport proteins. The free energy change of the species transported depends on the ratio of its concentrations on the two sides of the membrane and, if the species is charged, on the membrane potential $\Delta\Psi$. The rate of nonmediated diffusion across a membrane is a linear function of the difference in concentration of the species on the two sides of the membrane as governed by Fick's first law of diffusion. The rate of mediated transport is characterized by rapid saturation kinetics and specificity for the substance transported. It is also subject to competitive inhibition and chemical inactivation. Ionophores transport ions through membranes. Carrier ionophores, such as valinomycin, do so by wrapping a specific ion in a hydrophobic, membrane-soluble coat that can freely diffuse through the membrane. Channel-forming ionophores, such as gramicidin A, form a trans-membrane pore through which selected ions can diffuse. Glucose transport across erythrocyte membranes is mediated by a dimeric trans-membrane glycoprotein that can assume two conformations: One with a glucose-binding site facing the external cell surface and the other with the glucose site facing the cytosol. Transport occurs by the binding of glucose to the protein on one face of the membrane, followed by a conformational change that closes this site and exposes the other (a gated pore).

Active transport of molecules or ions against a concentration gradient requires an input of free energy. The free energy of ATP hydrolysis is coupled to the transport of three Na^+ ions

out of and two K^+ ions into the cell by the $(Na^+\text{-}K^+)$-ATPase. This electrogenic process involves phosphorylation of an Asp residue in the presence of Na^+ and its dephosphorylation in the presence of K^+. Phosphorylation and dephosphorylation are accompanied by conformational changes that ensure rapid interconversion of all intermediates along the transport pathway. ATP-driven transport of Ca^{2+} by the $Ca^{2+}\cdot$calmodulin-activated Ca^{2+}-ATPase and of H^+ by $(H^+\text{-}K^+)$-ATPase occur by similar phosphorylation–dephosphorylation mechanisms. Bacteria transport sugars by group translocation, a process in which the transported substance is chemically modified. The PTS system phosphorylates sugars as they are transported by utilizing phosphoenolpyruvate as a phosphoryl donor.

Active transport may be driven by the free energy stored in ion gradients. Glucose is transported into intestinal epithelial cells against its concentration gradient by a Na^+-glucose symport. This process is ultimately powered by the free energy of ATP hydrolysis since the Na^+ gradient is constantly being replenished via the $(Na^+\text{-}K^+)$-ATPase. The system conforms to a Random Bi Bi kinetic mechanism implying that both Na^+ and glucose must be bound for the transport-producing conformational change to occur. Lactose is transported into *E. coli* by lactose permease, an H^+-lactose symport. This process is driven by the cell's electrochemical H^+ gradient which is, in turn, maintained by a proton pump coupled with oxidative metabolism. The mitochondrial ADP-ATP antiport system also interacts with the membrane potential in the asymmetric transport of ATP out of and ADP into the mitochondrion.

References

General

Finean, J. B., Coleman, R., and Michell, R. H., *Membranes and their Cellular Functions* (3rd ed.), Chapters 3 and 4, Blackwell Scientific Publications (1984).

Franklin, H. M., *The Vital Force: A Study of Bioenergetics,* Chapters 9 and 10, Freeman (1986).

Martonosi, A. N. (Ed.), *The Enzymes of Biological Membranes* (2nd ed.), Vols. 3 and 4, Plenum Press (1985). [Contains authoritative reviews on various transport proteins.]

Martonosi, A. N. (Ed.), *Membranes and Transport,* Vols. 1 and 2, Plenum Press (1982). [A collection of short reviews on most of the topics discussed in this chapter.]

Stein, W. D., *Transport and Diffusion across Cell Membranes,* Academic Press (1986).

Tosteson, D. C., (Ed.), *Membrane Transport in Biology,* Vol. 2, *Transport Across Single Biological Membranes,* Springer–Verlag (1979).

Kinetics and Mechanism of Transport

Dobler, M., *Ionophores and Their Structures,* Wiley–Interscience (1981).

Hall, J. L., and Baker, D. A., *Cell Membranes and Ion Transport,* Longmans, Green, (1977).

Jencks, W. P., The utilization of binding energy in coupled vectorial processes, *Adv. Enzymol.* **51,** 75–106 (1980).

Schultz, S. G., *Basic Principles of Membrane Transport,* Cambridge University Press (1980).

Tanford, C., Mechanism of free energy coupling in active transport, *Annu. Rev. Biochem.* **52,** 379–409 (1983).

Glucose Transport

Baldwin, S. A. and Lienhard, G. E., Glucose transport across plasma membranes: facilitated diffusion systems, *Trends Biochem. Sci.* **6,** 208–211 (1981).

Barnett, J. E. G., Holman, G. D., Chalkley, R. A., and Munday, K. A., Evidence for two asymmetric conformational states in the human erythrocyte sugar transport system, *Biochem. J.* **145,** 417–429 (1975).

Jung, C. Y., Carrier mediated glucose transport across human red cell membranes, *in* Surgenor, D. (Ed.), *The Red Blood Cell,* Vol. 2, *pp.* 705–751, Academic Press (1975).

Lienhard, G., Regulation of cellular membrane transport by the exocytotic insertion and endocytotic retrieval of transporters, *Trends Biochem. Sci.* **8,** 125–127 (1983).

Walmsley, A. R., The dynamics of the glucose transporter, *Trends Biochem. Sci.* **13,** 226–231 (1988).

Widdas, W. F., Old and new concepts of the membrane transport for glucose in cells, *Biochim. Biophys. Acta* **947,** 385–404 (1988).

(Na⁺-K⁺)-ATPase

Cantley, L., Ion transport systems sequenced, *Trends Neurosci.* **9,** 1–3 (1986).

Cantley, L. C., Carilli, C. T., Smith, R. L., and Perlman, D., Conformational changes of Na,K-ATPase necessary for transport, *Curr. Top. Membr. Transp.* **19,** 315–322 (1983).

Gadsby, D. C., The Na/K pump of cardiac cells, *Annu. Rev. Biophys. Bioeng.* **13,** 373–398 (1984).

Langer, G. A., Relationship between myocardial contractility and the effects of digitalis on ion exchange, *Fed. Proc.* **36,** 2231–2234 (1977).

Pedersen, P. L. and Carafoli, E., Ion motive ATPases. I. Ubiquity, properties and significance to cell function *and* II. Energy coupling and work output, *Trends Biochem. Sci.* **12,** 146–150, 186–189 (1987).

Sweadner, K. J. and Goldin, S. M., Active transport of sodium and potassium ions: mechanism, function and regulation, *New Engl. J. Med.* **302,** 777–783 (1980).

Ca²⁺-ATPase

Babu, Y. S., Sack, J. S., Greenough, T. J., Bugg, C. E., Means, A. R., and Cook, W. J., Three-dimensional structure of calmodulin, *Nature* **315,** 37–40 (1985).

Carafoli, E. and Penniston, J. T., The calcium signal, *Sci. Am.* **253**(5): 70–78 (1985).

Carafoli, E. and Zurini, M., The Ca²⁺-pumping ATPase of plasma membranes: purification, reconstitution and properties, *Biochim. Biophys. Acta* **683,** 279–301 (1982). [Contains a discussion of calmodulin regulation of Ca²⁺-ATPase.]

Klee, C. B. and Vaenaman, T. C., Calmodulin, *Adv. Protein Chem.* **35,** 213–321 (1982).

Pickhart, C. M. and Jencks, W. P., Energetics of the calcium-transporting ATPase, *J. Biol. Chem.* **259,** 1629–1643 (1984).

(H⁺-K⁺)-ATPase

Sachs, G., Wallmark, B., Saccomani, G., Rabon, E., Stewart, H. B., DiBona, D. R., and Berglindh, T., The ATP-dependent component of gastric acid secretion, *Curr. Top. Membr. Transp.* **16,** 135–159 (1982).

Lactose Permease

Kaback, H. R., Active transport in *Escherichia coli*: passage to permease, *Annu. Rev. Biophys. Biophys. Chem.* **15,** 279–319 (1986).

Kaback, H. R., From membrane to molecule with the *lac* permease of *Escherichia coli, Curr. Top. Cell. Regul.* **26,** 137–148 (1985).

Overath, P. and Wright, K. J., Lactose permease: a carrier on the move, *Trends Biochem. Sci.* **8,** 404–408 (1983).

ADP–ATP Translocator

LaNoue, K., Mizani, S. M., and Kilingenberg, M., Electrical imbalance of adenine nucleotide transport across the mitochondrial membrane, *J. Biol. Chem.* **253,** 191–198 (1978).

Vignais, P. V., Molecular and physiological aspects of adenine nucleotide transport in mitochondria, *Biochim. Biophys. Acta* **456,** 1–38, (1976).

PEP-Dependent Phosphotransferase System

Saier, M. H., Jr., Bacterial phosphoenolpyruvate: sugar phosphotransferase systems: structural, functional and evolutionary interrelationships, *Bacteriol. Rev.* **41,** 856–871 (1977).

Problems

1. If the glucose concentration outside a cell is 10 mM but that inside a cell is 0.1 mM, what is glucose's chemical potential difference across the membrane at 37°C?

*2. If a solution of an ionic macromolecule is equilibrated with a salt solution from which it is separated by a membrane through which the salt ions but not the macromolecule can pass, a membrane potential is generated across the membrane. This so-called **Donnan equilibrium** arises because the impermeability of the membrane to some ions but not others prevents the equalization of the ionic concentrations on the two sides of the membrane. To demonstrate this effect, assume that the Cl^- salt of a monocationic protein, P^+, is dissolved in water to the extent that $[Cl^-] = 0.1M$ and is separated by a membrane impermeable to the protein but not NaCl from an equal volume of $0.1M$ NaCl solution. Assuming no volume change in either compartment, what are the concentrations of the various ionic species on either side of the membrane after the system has equilibrated? What is the membrane potential across the membrane? (*Hint:* Mass is conserved and the solution on each side of the membrane must be electrically neutral. At equilibrium, $\Delta G_{Na^+} + \Delta G_{Cl^-} = 0$.)

3. How long would it take one molecule of gramicidin A to transport enough Na$^+$ to change the concentration inside an erythrocyte of volume 80 μm^3 by 10 mM. Assume the erythrocyte's Na$^+$ pumps are inoperative.

4. The (Na$^+$-K$^+$)-ATPase is inhibited by nanomolar concentrations of vanadate, which forms a pentavalent ion, VO_5^{5-}, with trigonal bipyrimidal symmetry. Explain the mechanism of this inhibition. (*Hint:* See Section 15-2B.)

5. The (H$^+$-K$^+$)-ATPase secretes H$^+$ at a concentration of $0.18M$ from cells that have an internal pH of 7. What is the ΔG required for the transport of 1 mol of H$^+$ under these conditions? Assuming that the ΔG for ATP hydrolysis is -31.5 kJ · mol^{-1} under these conditions, and that the membrane potential is 0.06 V, inside negative, how much ATP must be hydrolyzed per mole of H$^+$ transported in order to make this transport exergonic?

6. A 100-Å wide membrane has a membrane potential of 100 mV. What is the magnitude of this potential difference in V · cm^{-1}? Comment on the magnitude of this potential field in macroscopic terms.

7. The resting membrane potential ($\Delta\Psi$) of a neuron is -60 mV (inside negative). If the free energy change associated with the transport of one Na$^+$ ion from outside to inside is -1.9 kJ · mol^{-1}, and [Na$^+$] outside the cell is 260 mM, what is [Na$^+$] inside the cell?

8. You have isolated a new strain of bacteria and would like to know whether leucine and ethylene glycol enter the cells by mediated diffusion or only by a nonmediated route. To do this you measure the initial rates of uptake of these molecules as a function of external concentration and obtain the data below. Which compound(s) enters by a mediated route? What criteria did you use for this decision?

Compound	Concentration (M)	Initial Uptake Rate (arbitrary units)
Leucine	1×10^{-6}	110
	2×10^{-6}	220
	5×10^{-6}	480
	1×10^{-5}	830
	3×10^{-5}	1700
	1×10^{-4}	2600
	5×10^{-4}	3100
	1×10^{-3}	3200
Ethylene glycol	1×10^{-3}	1
	5×10^{-3}	5
	0.01	10
	0.05	50
	0.1	100
	0.5	500
	1.0	1000

9. Draw the structures of the following compounds and predict whether they can cross a membrane without mediation or will require facilitation. Indicate the criteria you used to make these predictions? (a) Ethanol, (b) glycine, (c) cholesterol, and (d) ATP.

10. Write a kinetic scheme for the (H$^+$-K$^+$)-ATPase that provides for coupled ATP hydrolysis with H$^+$ transport. Discuss the order of substrate addition required for coupling. Identify the steps in which mutual destabilization results in reasonable rates of transport.

Chapter 19

CITRIC ACID CYCLE

In this chapter we continue our metabolic explorations by examining the **citric acid cycle,** the common mode of oxidative degradation in eukaryotes and prokaryotes. This cycle, which is alternatively known as the **tricarboxylic acid (TCA) cycle** or the **Krebs cycle,** marks the "hub" of the metabolic system: *It accounts for the major portion of carbohydrate, fatty acid, and amino acid oxidation and generates numerous biosynthetic precursors.* The citric acid cycle is therefore **amphibolic,** that is, it operates both catabolically and anabolically.

We begin our study of the citric acid cycle with an overview of its component reactions and a historical synopsis of its elucidation. Next, we explore the origin of the cycle's starting compound, **acetyl-coenzyme A (acetyl-CoA),** the common intermediate formed by the breakdown of most metabolic fuels. Then, after discussing the reaction mechanisms of the enzymes that catalyze the cycle, we consider the various means by which it is regulated. Finally, we deal with the citric acid cycle's amphibolic nature by examining its interrelationships with other metabolic pathways.

1. CYCLE OVERVIEW

The citric acid cycle (Fig. 19-1) is an ingenious series of reactions that oxidizes the acetyl group of acetyl-CoA to two molecules of CO_2 in a manner that conserves the liberated

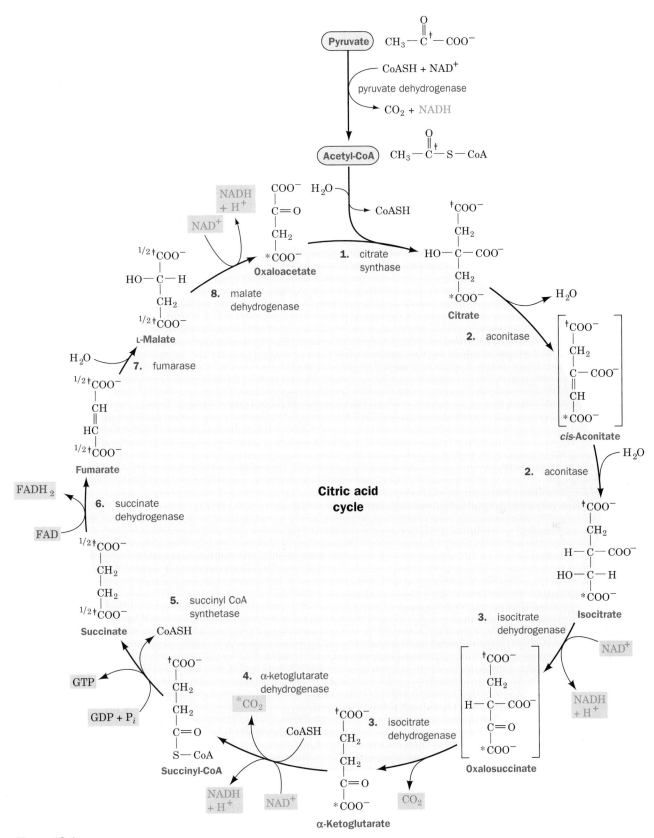

Figure 19-1

The reactions of the citric acid cycle. The reactants and products of this catalytic cycle are boxed. The pyruvate → acetyl-CoA reaction (*top*) supplies the cycle's substrate via carbohydrate metabolism but is not considered to be part of the cycle. The bracketed compounds are enzyme-bound intermediates. An isotopic label at C(4) of oxaloacetate (*) becomes C(1) of α-ketoglutarate and is released as CO_2 in Reaction 4. An isotopic label at C (1) of acetyl-CoA (‡) becomes C(5) of α-ketoglutarate and is scrambled between C(1) and C(4) of succinate (1/2‡).

free energy for utilization in ATP generation. Before we study these reactions in detail, let us consider the cycle's chemical strategy by "walking" through the cycle and noting the fate of the acetyl group at each step. Following this preview, we shall consider some of the major discoveries that led to our present understanding of the citric acid cycle.

A. Reactions of the Cycle

The eight enzymes of the citric acid cycle (Fig. 19-1) catalyze a series of well-known organic reactions that cumulatively oxidize an acetyl group to two CO_2 molecules with the concomitant generation of three NADHs, one $FADH_2$, and one GTP:

1. **Citrate synthase** catalyzes the condensation of acetyl-CoA and **oxaloacetate** to yield **citrate,** giving the cycle its name.

2. The strategy of the cycle's next two steps is to rearrange citrate to a more easily oxidized isomer and then oxidize it. **Aconitase** converts citrate, a not readily oxidized tertiary alcohol, to the easily oxidized secondary alcohol **isocitrate.** The reaction sequence involves a dehydration, producing enzyme-bound *cis*-**aconitate,** followed by a hydration so that citrate's hydroxyl group is, in effect, transferred to an adjacent carbon atom.

3. **Isocitrate dehydrogenase** oxidizes isocitrate to the β-keto acid intermediate **oxalosuccinate** with the coupled reduction of NAD^+ to NADH; oxalosuccinate is then decarboxylated yielding α-**ketoglutarate.** This is the first step in which oxidation is coupled to NADH production and also the first CO_2-generating step.

4. The multienzyme complex α-**ketoglutarate dehydrogenase** oxidatively decarboxylates α-ketoglutarate to **succinyl-coenzyme A.** The reaction involves the reduction of a second NAD^+ to NADH and the generation of a second molecule of CO_2. At this point in the cycle, two molecules of CO_2 have been produced so that the net oxidation of the acetyl group is complete. Note, however, that it is not the carbon atoms of the entering acetyl-CoA that have been oxidized.

5. **Succinyl-CoA synthetase** converts succinyl-coenzyme A to **succinate.** The free energy of the thioester bond is conserved in this reaction by the formation of "high-energy" GTP from GDP + P_i.

6. The remaining reactions of the cycle serve to oxidize succinate back to oxaloacetate in preparation for another round of the cycle. **Succinate dehydrogenase** catalyzes the oxidation of succinate's central single bond to a trans double bond yielding **fumarate** with the concomitant reduction of the redox coenzyme FAD to $FADH_2$.

7. **Fumarase** then catalyzes the hydration of fumarate's double bond to yield **malate.**

8. Finally, **malate dehydrogenase** reforms oxaloacetate by oxidizing malate's secondary alcohol group to the corresponding ketone with concomitant reduction of a third NAD^+ to NADH.

Acetyl groups are thereby completely oxidized to CO_2 with the following stoichiometry:

$$3NAD^+ + FAD + GDP + \text{acetyl-CoA} + P_i \longrightarrow$$
$$3NADH + FADH_2 + GTP + CoA + 2CO_2$$

The citric acid cycle functions catalytically as a consequence of its regeneration of oxaloacetate: An endless number of acetyl groups can be oxidized through the agency of a single oxaloacetate molecule.

NADH and $FADH_2$ are vital products of the citric acid cycle. Their reoxidation by O_2 through the mediation of the electron-transport chain and oxidative phosphorylation (Chapter 20) completes the breakdown of metabolic fuel in a manner that drives the synthesis of ATP. Other functions of the cycle are discussed in Section 19-5.

B. Historical Perspective

The citric acid cycle was proposed in 1937 by Hans Krebs, a contribution that ranks as one of the most important achievements of metabolic chemistry. We therefore outline the intellectual history of this cycle's discovery.

By the early 1930s, significant progress had been made in elucidating the glycolytic pathway (Section 16-1A). Yet, the mechanism of glucose oxidation and its relationship to cellular respiration (oxygen uptake) was still a mystery. Nevertheless, the involvement of several metabolites in cellular oxidative processes was recognized. It was well known, for example, that in addition to lactate and acetate, the dicarboxylates succinate, malate, and α-ketoglutarate, as well as the tricarboxylate citrate, are rapidly oxidized by muscle tissue during respiration. It had also been shown that **malonate** (Section 19-3F), a potent inhibitor of succinate oxidation to fumarate, also inhibits cellular respiration thereby suggesting that succinate plays a central role in oxidative metabolism rather than being just another metabolic fuel.

In 1935, Albert Szent-Györgyi demonstrated that cellular respiration is dramatically accelerated by catalytic amounts of succinate, fumarate, malate, or oxaloacetate; that is, *the addition of any of these substances to minced muscle tissue stimulates O_2 uptake and CO_2 production far in excess of that required to oxidize the added dicarboxylic acid.* Szent-Györgyi further showed that these compounds were interconverted according to the reaction sequence:

$$\text{Succinate} \rightarrow \text{fumarate} \rightarrow \text{malate} \rightarrow \text{oxaloacetate}$$

Shortly afterward, Carl Martins and Franz Knoop demonstrated that citrate is rearranged, via *cis*-aconitate, to isocitrate and then dehydrogenated to α-ketoglutarate. α-Ketoglutarate was already known to undergo oxidative decarboxylation to succinate and CO_2. This extended the proposed reaction sequence to

$$\text{Citrate} \longrightarrow \textit{cis-}\text{aconitate} \longrightarrow \text{isocitrate} \longrightarrow$$
$$\alpha\text{-ketoglutarate} \longrightarrow \text{succinate} \longrightarrow \text{fumarate}$$
$$\longrightarrow \text{malate} \longrightarrow \text{oxaloacetate}$$

What was necessary to close the circle so as to make the system catalytic was to establish that oxaloacetate is converted to citrate. In 1936, Martins and Knoop demonstrated that citrate could be formed nonenzymatically from oxaloacetate and pyruvate by treatment with hydrogen peroxide under basic conditions. Krebs used this chemical model as the point of departure for the biochemical experiments that led to his proposal of the citric acid cycle.

Krebs' hypothesis was based on his investigations, starting in 1936, on respiration in minced pigeon breast muscle (which has a particularly high rate of respiration). The idea of a catalytic cycle was not new to him: In 1932, he and Kurt Henseleit had elucidated the outlines of the **urea cycle,** a process in which ammonia and CO_2 are converted to urea (Section 24-2). The most important observations Krebs made in support of the existence of the citric acid cycle were

1. Succinate is formed from fumarate, malate, or oxaloacetate in the presence of the metabolic inhibitor malonate. Since malonate inhibits the direct reduction of fumarate to succinate, the succinate must be formed by an oxidative cycle.

2. Pyruvate and oxaloacetate can form citrate enzymatically. Krebs therefore suggested that the metabolic cycle is closed with the reaction:

$$\text{Pyruvate} + \text{oxaloacetate} \longrightarrow \text{citrate} + CO_2$$

3. The interconversion rates of the cycle's individual steps are sufficiently rapid to account for observed respiration rates so that it must be (at least) the major pathway for pyruvate oxidation in muscle.

Although Krebs had established the existence of the citric acid cycle, some major gaps still remained in its complete elucidation. The mechanism of citrate formation did not become clear until Nathan Kaplan and Fritz Lipmann discovered **coenzyme A** in 1945 (Section 19-2), and Severo Ochoa and Feodor Lynen established, in 1951, that acetyl-CoA is the intermediate that condenses with oxaloacetate to form citrate. Oxidative decarboxylation of α-ketoglutarate to succinate was also shown to involve coenzyme A with succinyl-CoA as an intermediate.

The elucidation of the citric acid cycle was a major

achievement and, like all achievements of this magnitude, required the efforts of numerous investigators. Indeed, many biochemists are still working to understand the cycle on a molecular and enzymatic level. We shall study the eight enzymes that catalyze the cycle after first discussing the cycle's major fuel, acetyl-CoA, and its formation from pyruvate.

2. METABOLIC SOURCES OF ACETYL-COENZYME A

*Acetyl groups enter the citric acid cycle as **acetyl-coenzyme A (acetyl-SCoA or acetyl-CoA; Fig. 19-2), the** common product of carbohydrate, fatty acid, and amino acid breakdown.* **Coenzyme A (CoASH or CoA)** consists of a β-mercaptoethylamine group bonded through an amide linkage to the vitamin **pantothenic acid,** which, in turn, is attached to a 3-phosphoadenosine moiety via a pyrophosphate bridge. The acetyl group of acetyl-CoA is bonded as a thioester to the sulfhydryl portion of the β-mercaptoethylamine group. *CoA thereby functions as a carrier of acetyl and other acyl groups (the A of CoA stands for "Acetylation").*

Acetyl-CoA is a "high-energy" compound: The $\Delta G°'$ for

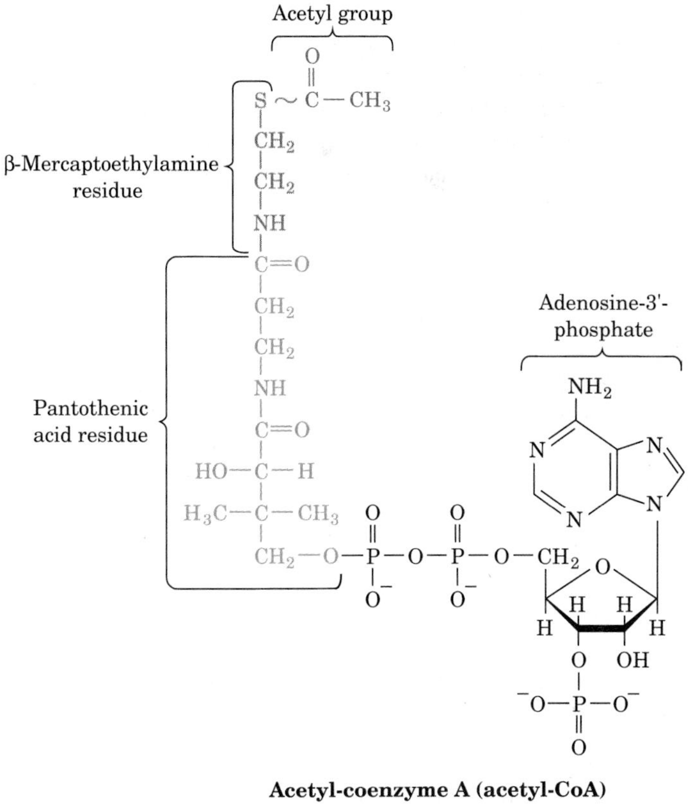

Figure 19-2
The chemical structure of acetyl-CoA. The thioester bond is drawn with a $\sim$ to indicate that it is a "high-energy" bond. In CoA, the acetyl group is replaced by hydrogen.

the hydrolysis of its thioester bond is -31.5 kJ·mol^{-1}, which makes this reaction slightly (1 kJ·mol^{-1}) more exergonic than that of ATP hydrolysis (Section 15-4B). The formation of this thioester bond in a metabolic intermediate therefore conserves a portion of the free energy of oxidation of a metabolic fuel.

A. Pyruvate Dehydrogenase Multienzyme Complex

The immediate precursor to acetyl-CoA from carbohydrate sources is the glycolytic product pyruvate. As we saw in Section 16-3, under anaerobic conditions the NADH produced by glycolysis is reoxidized with concomitant reduction of pyruvate to lactate (in muscle) or ethanol (in yeast). Under aerobic conditions, however, NADH is reoxidized by the mitochondrial electron-transport chain (Section 20-2) so that pyruvate, which enters the mitochondrion via a specific pyruvate-H$^+$ symport, can undergo further oxidation. (The formation of acetyl-CoA from fatty acids and amino acids is discussed in Sections 23-2 and 24-3.)

Acetyl-CoA is formed from pyruvate through oxidative decarboxylation by the multienzyme complex **pyruvate dehydrogenase**. The pyruvate dehydrogenase multienzyme complex contains three catalytic activities: **pyruvate dehydrogenase (E$_1$), dihydrolipoyl transacetylase (E$_2$),** and **dihydrolipoyl dehydrogenase (E$_3$).** The *E. coli* complex has been studied extensively; its elaborate polyhedral structure is described in Section 7-5D. The mammalian complex is located in the mitochondrion and differs in structure from that of *E. coli* although both catalyze the same reactions by similar mechanisms. The 8400-kD bovine heart complex con-

sists of 30 E$_1$ $\alpha_2\beta_2$ tetramers and 6 E$_3$ dimers surrounding a core of 60 E$_2$ monomers.

Multienzyme complexes are a step forward in the evolution of catalytic efficiency; they offer the following mechanistic advantages:

1. Enzymatic reaction rates are limited by the frequency that enzymes collide with their substrates (Section 13-2B). If a series of reactions occurs within a multienzyme complex, the distance that substrates must diffuse between active sites is minimized thereby achieving a rate enhancement.

2. Complex formation provides the means for channeling metabolic intermediates between successive enzymes in a metabolic pathway thereby minimizing side reactions.

3. The reactions catalyzed by a multienzyme complex may be coordinately controlled.

Acetyl-CoA Formation Occurs in Five Reactions

The pyruvate dehydrogenase multienzyme complex catalyzes five sequential reactions (Fig. 19-3) with the overall stoichiometry:

$$\text{Pyruvate} + \text{CoA} + \text{NAD}^+ \rightarrow$$
$$\text{acetyl-CoA} + \text{CO}_2 + \text{NADH}$$

The coenzymes and prosthetic groups required in this reaction sequence are thiamine pyrophosphate (TPP; Fig. 16-24), flavin adenine dinucleotide (FAD; Fig. 14-27), nicotinamide adenine dinucleotide (NAD$^+$); (Fig. 12-2), and **lipoamide** (Fig. 19-4); their functions are listed in Table 19-1. Lipoamide consists of **lipoic acid** joined in amide linkage to the ε-amino group of a

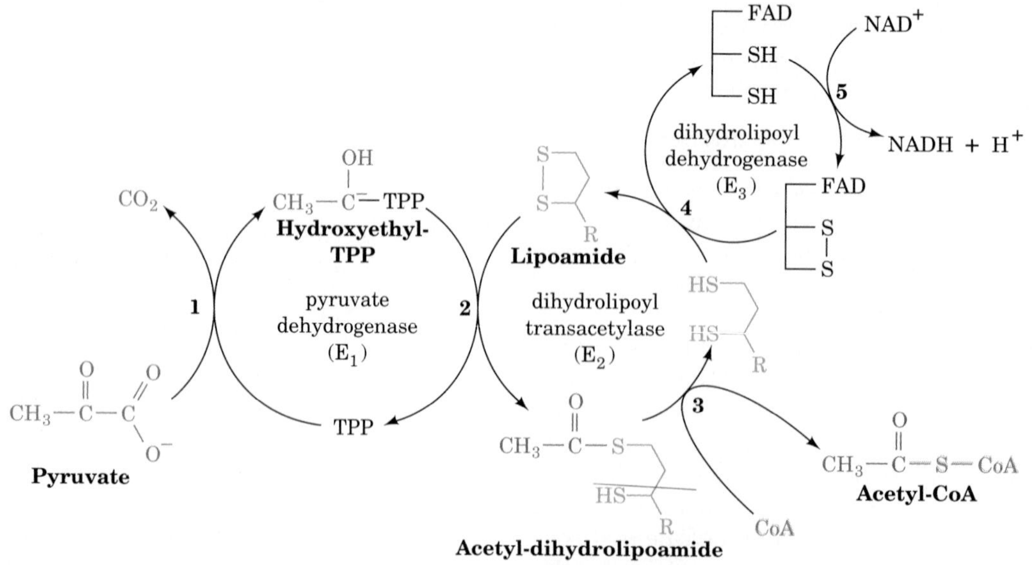

Figure 19-3
The five reactions of the pyruvate dehydrogenase multienzyme complex. E$_1$ (pyruvate dehydrogenase) contains TPP and catalyzes Reactions 1 and 2. E$_2$ (dihydrolipoyl transacetylase) contains lipoamide and catalyzes Reaction 3. E$_3$ (dihydrolipoyl dehydrogenase) contains FAD and a redox active disulfide and catalyzes Reactions 4 and 5.

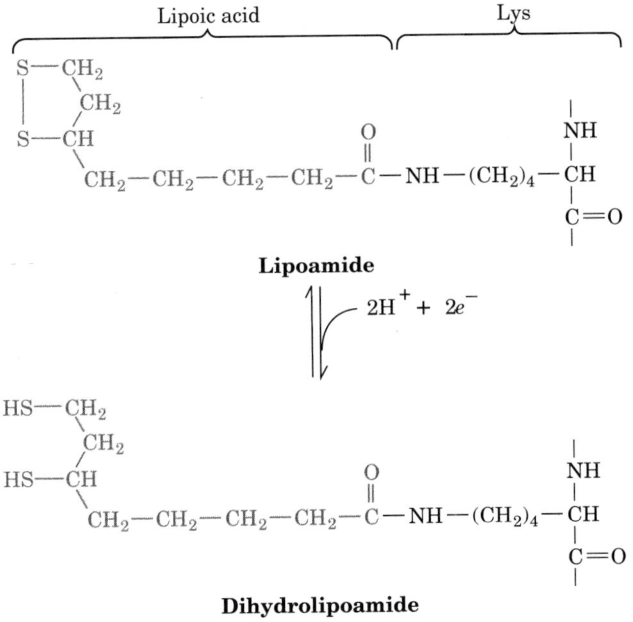

Figure 19-4
The chemical structures and interconversion of lipoamide and dihydrolipoamide (lipoic acid covalently joined to the ε-amino group of a Lys residue via an amide linkage).

Lys residue. Reduction of its cyclic disulfide to a dithiol, **dihydrolipoamide,** and its reoxidation (Fig. 19-4) are the "business" of this prosthetic group.

The five reactions catalyzed by the pyruvate dehydrogenase multienzyme complex are (Fig. 19-3):

1. Pyruvate dehydrogenase (E_1), a TPP-requiring enzyme, decarboxylates pyruvate with the intermediate formation of hydroxyethyl-TPP. This reaction is identical with that catalyzed by yeast pyruvate decarboxylase (Section 16-3B).

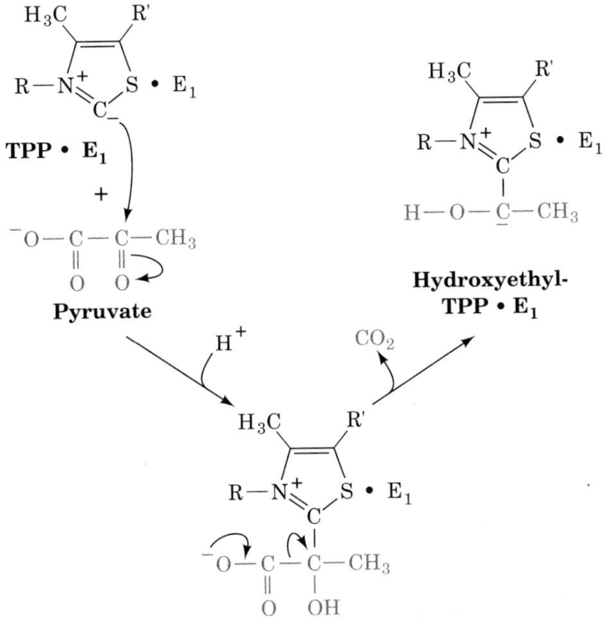

Unlike the yeast enzyme, however, pyruvate dehydrogenase does not convert the hydroxyethyl-TPP intermediate into acetaldehyde and TPP, but channels it to the next enzyme in the multienzyme sequence, dihydrolipoyl transacetylase (E_2).

2. E_1 also catalyzes the oxidation of hydroxyethyl-TPP by lipoamide-E_2 to form **acetyl-dihydrolipoamide-E_2** and eliminate TPP. The reaction occurs by attack of the hydroxyethyl group carbanion on the lipoamide disulfide followed by the elimination of TPP from the intermediate adduct.

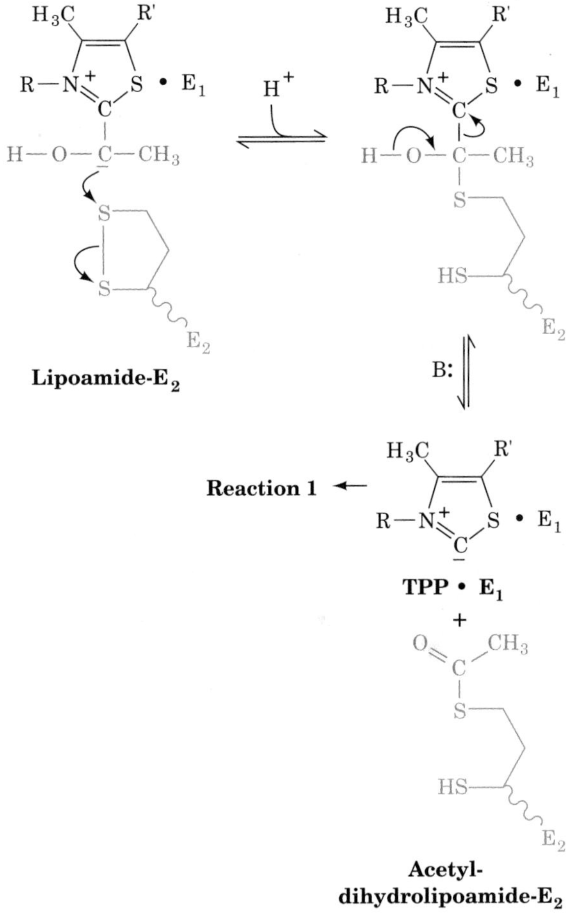

3. E_2 catalyzes the transfer of the acetyl group to CoA yielding acetyl-CoA and dihydrolipoamide-E_2.

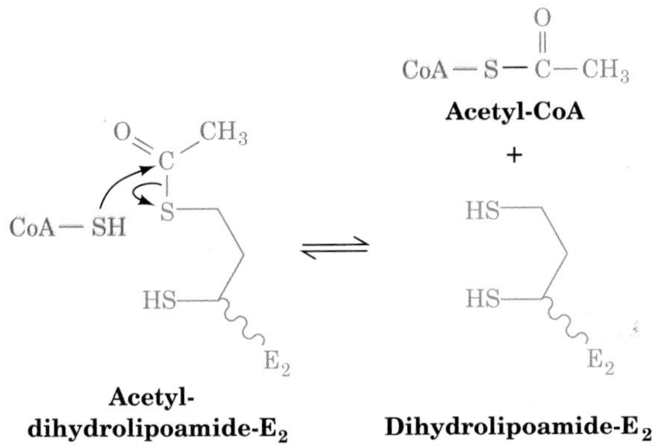

Table 19-1

The Coenzymes and Prosthetic Groups of Pyruvate Dehydrogenase

Cofactor	Location	Function
Thiamine pyrophosphate (TPP)	Bound to E_1	Decarboxylates pyruvate yielding a hydroxyethyl-TPP carbanion
Lipoic acid	Covalently linked to a Lys on E_2 (lipoamide)	Accepts the hydroxyethyl carbanion from TPP as an acetyl group
Coenzyme A (CoA)	Substrate for E_2	Accepts the acetyl group from lipoamide
Flavin adenine dinucleotide (FAD)	Bound to E_3	Reduced by lipoamide
Nicotinamide adenine dinucleotide (NAD⁺)	Substrate for E_3	Reduced by $FADH_2$

This is a transesterification in which the sulfhydryl group of CoA attacks the acetyl group of acetyl-dihydrolipoamide-E_2 to form a tetrahedral intermediate (not shown), which decomposes to acetyl-CoA and dihydrolipoamide-E_2.

4. Dihydrolipoyl dehydrogenase (E_3; also named **lipoamide dehydrogenase**) reoxidizes dihydrolipoamide thereby completing the catalytic cycle of E_2.

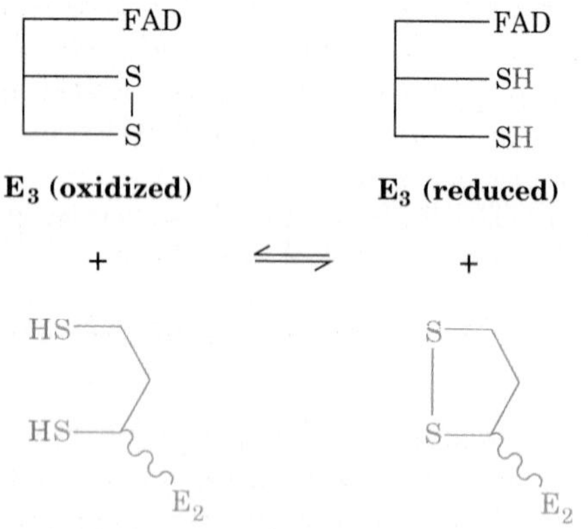

E_3 (oxidized)   **E_3 (reduced)**

Oxidized E_3 contains a reactive disulfide group and a tightly bound FAD. The oxidation of dihydrolipoamide is a disulfide interchange reaction (Section 8-2A): The lipoamide disulfide bond forms with concomitant reduction of E_3's reactive disulfide to two sulfhydryl groups. The catalytic mechanism of E_3 is

essentially identical to that of glutathione reductase acting in reverse (Section 14-4). Indeed, the amino acid sequences of pig heart dihydrolipoyl dehydrogenase and human erythrocyte glutathione reductase are ~40% identical (the presence of both enzymes in *E. coli* indicates that their divergence preceded the appearance of eukaryotes).

5. Reduced E_3 is reoxidized by NAD⁺.

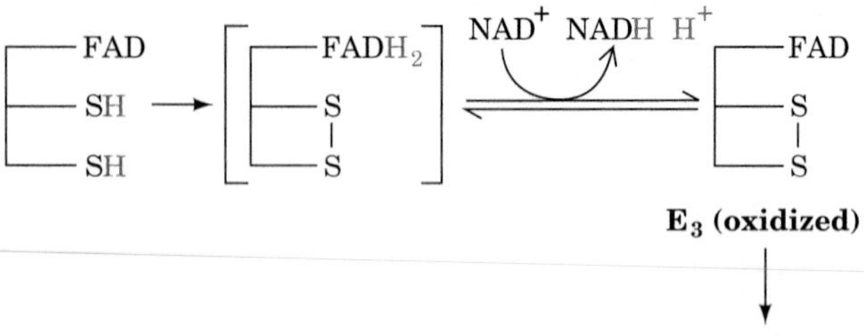

E_3 (oxidized)

Reaction 4

The enzyme's active sulfhydryl groups are reoxidized by the enzyme-bound FAD, which is thereby reduced to $FADH_2$. The $FADH_2$ is then reoxidized to FAD by NAD⁺, producing NADH.

The Lipoyllysyl Arm Transfers Intermediates between Enzyme Subunits

How are reaction intermediates channeled between the enzymes of the pyruvate dehydrogenase multienzyme complex? The grouping between the lipoamide disulfide bond and the E_2 polypeptide backbone, the so-called **lipoyllysyl arm,** has a fully extended length of 14 Å.

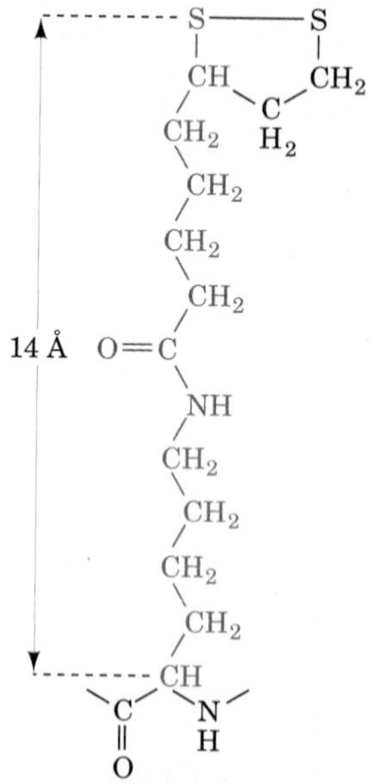

Lipollysyl arm (fully extended)

This suggests that *the lipoyllysyl arm acts as a long tether that swings the disulfide together with its reduced acetylated reaction product (Reaction 2) between E_1 and E_3.* Indeed, spectroscopic evidence indicates that the section of E_2 that bears the lipoyllysyl arm is flexibly linked to the rest of the subunit. Moreover, there is rapid interchange of acetyl groups among the lipoyl groups of the E_2 core (24 lipoyl groups in *E. coli*, 60 in mammals); the tethered arms evidently also swing among themselves exchanging both acetyl groups and disulfides:

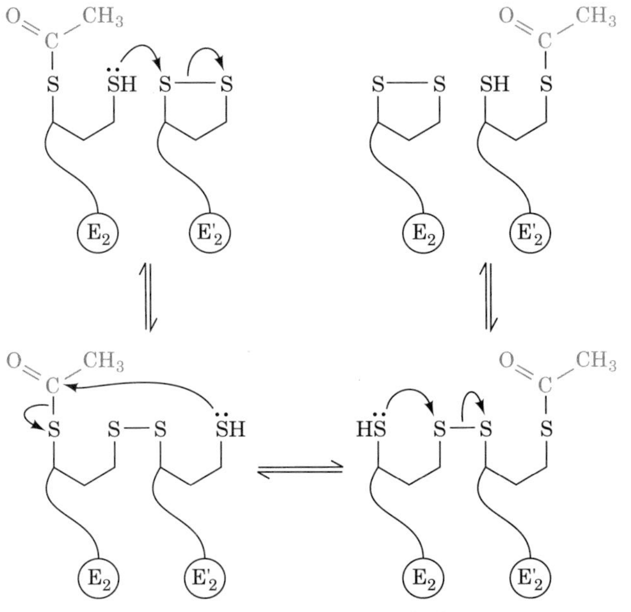

Intersubunit transacetylation

One E_1 subunit can therefore acetylate numerous E_2 subunits while one E_3 subunit can reoxidize several dihydrolipoamide groups.

Arsenic Compounds Are Poisonous because They Sequester Lipoamide

Arsenic has been known to be a poison since ancient times. As(III) compounds, such as **arsenite** (AsO_3^{3-}) and organic arsenicals, are toxic because of their ability to covalently bind sulfhydryl compounds. This is particularly true of sulfhydryls such as lipoamide that can form bidentate adducts.

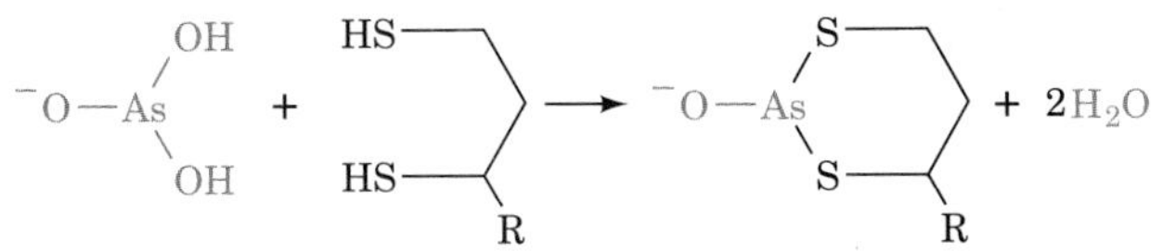

Arsenite   Dihydro-lipoamide

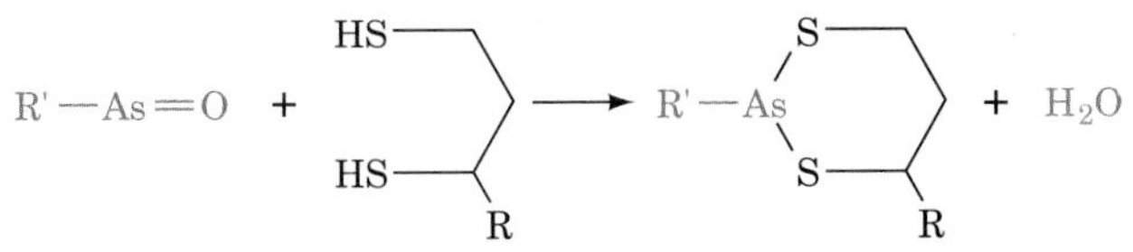

Organic arsenical

The resultant inactivation of lipoamide-containing enzymes, especially pyruvate dehydrogenase and α-

ketoglutarate dehydrogenase (Section 19-3D), brings respiration to a halt.

Organic arsenicals are more toxic to microorganisms than to humans, apparently because of differences in the sensitivities of their various enzymes to these compounds. This differential toxicity is the basis for the early twentieth century use of these organic arsenicals in the treatment of syphilis (now superseded by penicillin) and trypanosomiasis (typanosomes are parasitic protozoa). These compounds were really the first antibiotics although, not surprisingly, they had severe side effects.

Arsenic is often suspected as a poison in untimely deaths. In fact, it is thought that Napoleon Bonaparte was murdered by arsenic poisoning while in exile on the island of St. Helena. This suspicion, and the chemical analyses it sparked, makes a fascinating chemical anecdote. The finding that a lock of Napoleon's hair indeed contains high levels of arsenic strongly suggests that he died of arsenic poisoning. However, arsenic-containing dyes were used in wallpaper at the time and in damp weather, it was eventually determined, fungi convert the arsenic to a volatile compound. Samples of the wallpaper in Napoleon's room have survived and their analysis indicates that they contain arsenic. Napoleon's arsenic poisoning may therefore have been unintentional.

Retrospective detective work also suggests that Charles Darwin was a victim of chronic arsenic poisoning. For most of his life after he returned from his epic voyage, Darwin complained of numerous ailments including eczema, vertigo, headaches, arthritis, gout, palpitations, and nausea, all symptoms of arsenic poisoning. Fowler's solution, a common nineteenth century tonic, contained 10 mg of arsenite · mL^{-1}. Many patients, quite possibly Darwin himself, took this "medication" for years.

B. Control of Pyruvate Dehydrogenase

The pyruvate dehydrogenase multienzyme complex regulates the entrance of acetyl units derived from carbohydrate sources into the citric acid cycle. The decarboxylation of pyruvate by E_1 is irreversible and, since there are no other pathways in mammals for the synthesis of acetyl-CoA from pyruvate, it is crucial that the reaction be carefully controlled. Two regulatory systems are employed:

1. Product inhibition by NADH and acetyl-CoA (Fig. 19-5*a*)

2. Covalent modification by phosphorylation/dephosphorylation of the pyruvate dehydrogenase (E_1) subunit (Fig. 19-5*b*; enzymatic regulation by covalent modification is discussed in Section 17-3B).

Control by Product Inhibition

NADH and acetyl-CoA compete with NAD⁺ and CoA for binding sites on their respective enzymes. They also drive the reversible transacetylase (E_2) and dihydrolipoyl

(a)

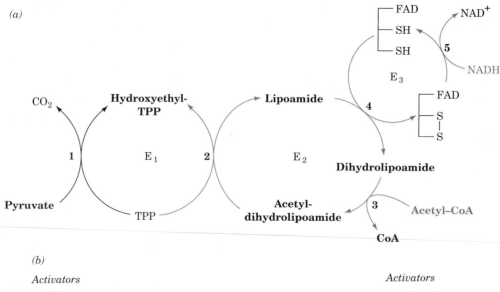

(b)

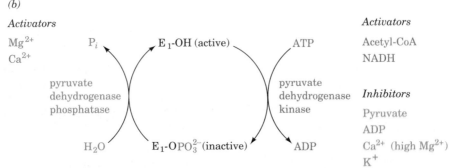

Figure 19-5
The factors controlling the activity of the pyruvate dehydrogenase multienzyme complex. (*a*) Product inhibition. NADH and acetyl-CoA, respectively, compete with NAD⁺ and CoA in Reactions **3** and **5** of the pyruvate dehydrogenase reaction sequence. When the relative concentrations of NADH and acetyl-CoA are high, the reversible reactions catalyzed by E_2 and E_3 are driven backwards (*red arrows*) thereby inhibiting further formation

of acetyl-CoA. (*b*) Covalent modification. Pyruvate dehydrogenase (E_1) is inactivated by the specific phosphorylation of one of its Ser residues in a reaction catalyzed by pyruvate dehydrogenase kinase (*right*). This phosphoryl group is hydrolyzed through the action of pyruvate dehydrogenase phosphatase (*left*), thereby reactivating E_1. The activators and inhibitors of the kinase are listed on the right and the activators of the phosphatase are listed on the left.

dehydrogenase (E₃) reactions backwards (Fig. 19-5a). High ratios of [NADH]/[NAD⁺] and [acetyl-CoA]/[CoA] therefore maintain E_2 in the acetylated form, incapable of accepting the hydroxyethyl group from the TPP on E_1. This, in turn, ties up the TPP on the E_1 subunit in its hydroxyethyl form, decreasing the rate of pyruvate decarboxylation.

Control by Phosphorylation/Dephosphorylation

Control by phosphorylation/dephosphorylation occurs only in eukaryotic enzyme complexes. These complexes contain **pyruvate dehydrogenase kinase** and **pyruvate dehydrogenase phosphatase** bound to the dihydrolipoyl transacetylase core. *The kinase inactivates the pyruvate dehydrogenase (E_1) subunit by phosphorylation of a specific dehydrogenase Ser residue by ATP (Fig. 19-5b). Hydrolysis of this phospho-Ser residue by the phosphatase reactivates the complex.*

The products of the reaction, NADH and acetyl-CoA, in addition to their direct effects on the pyruvate dehy-

drogenase multienzyme complex, activate pyruvate dehydrogenase kinase. The resultant phosphorylation inactivates the complex just as the products themselves inhibit it. Various other activators and inhibitors regulate the pyruvate dehydrogenase system (Fig. 19-5b), but in contrast to the glycogen metabolism control system (Section 17-3), it is unaffected by cAMP.

3. ENZYMES OF THE CITRIC ACID CYCLE

In this section we discuss the reaction mechanisms of the eight citric acid cycle enzymes. Our knowledge of these mechanisms rests on an enormous amount of experimental work; as we progress, we shall pause to examine some of these experimental details. Table 19-2 lists the subunit compositions, molecular masses, and cofactors of the citric acid cycle enzymes. Consideration

Table 19-2
Enzymes of the Citric Acid Cycle

Enzyme	Subunit Composition	Subunit Mass (kD)	Cofactors
Citrate synthase (pig heart)	α_2	49	
Aconitase (pig heart)	α_2	44.5	[4Fe–4S] cluster
Isocitrate dehydrogenase (mammalian mitochondria)	$\alpha_2\beta\gamma$		Mg^{2+} or Mn^{2+}, NAD^+
α-Ketoglutarate dehydrogenase multienzyme complex (*E. coli*):	$(E_1)_6(E_2)_{24}(E_3)_6$		
$\quad$ α-Ketoglutarate dehydrogenase (E_1)	α_2	95	TPP
$\quad$ Dihydrolipoyl transsuccinylase (E_2)	Monomer	42	Lipoamide, CoA
$\quad$ Dihydrolipoyl dehydrogenase (E_3)	α_2	56	FAD, NAD^+
Succinyl-CoA synthetase (pig heart)	$\alpha\beta$	34.5, 42.5	GDP
Succinate dehydrogenase (ox heart)	$\alpha\beta$	70, 27	FAD (covalent), Fe–S clusters
Fumarase (pig heart)	α_4	48.5	
Malate dehydrogenase (pig heart)	α_2	35	NAD^+

of how this cycle is regulated and its relationship to cellular metabolism are the subjects of the following sections.

A. Citrate Synthase

*Citrate synthase (originally named **citrate condensing enzyme**) catalyzes the condensation of acetyl-CoA and oxaloacetate (Reaction 1 of Fig. 19-1).* This initial reaction of the citric acid cycle is the point at which carbon atoms are "fed into the furnace" as acetyl-CoA. The citrate synthase reaction proceeds with an ordered sequential kinetic mechanism (Section 13-5B), oxaloacetate adding to the enzyme before acetyl-CoA.

The X-ray structure of the free dimeric enzyme, one of the few enzymes of the cycle for which an X-ray crystal structure is known, shows it to adopt an "open" form in which the two domains of each subunit form a deep cleft that contains the oxaloacetate-binding site (Fig. 19-6a). Upon binding oxaloacetate, however, the smaller do-

Figure 19-6
A space-filling drawing of citrate synthase in (*a*) the open conformation, and (*b*) the closed, substrate-binding conformation. The C atoms of the small domain in each subunit of the enzyme are colored green and those of the large domain are colored purple. The view is along the dimeric protein's twofold rotation axis. The large conformational shift between the open and closed forms entails relative interatomic movements of up to 15 Å. [Courtesy of Anne Dallas, University of Pennsylvania; and Helen Berman, Fox Chase Cancer Center. X-Ray structures determined by Stephen Remington and Robert Huber.]

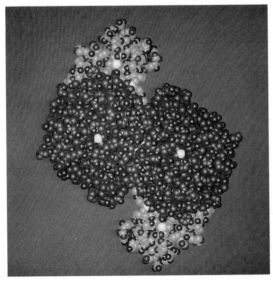

(*a*)

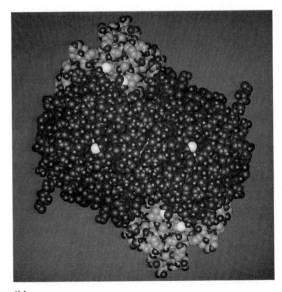

(*b*)

main undergoes a remarkable 18° rotation relative to the larger domain, which closes the cleft (Fig. 19-6*b*). **Acetonyl-CoA** (an inhibitory analog of acetyl-CoA),

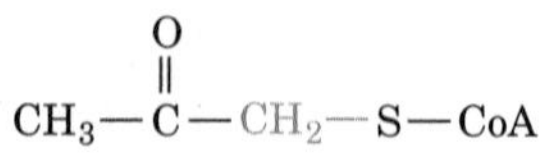

$$CH_3 - \overset{\displaystyle O}{\overset{\displaystyle \|}{C}} - CH_2 - S - CoA$$

Acetonyl–CoA

binds to the enzyme in this "closed" form, thereby identifying the acetyl-CoA binding site. The existence of these "open" and "closed" forms explains the enzyme's ordered sequential kinetic behavior: *The conformational change induced by oxaloacetate binding generates the acetyl-CoA binding site while sealing off the solvent's access to the bound oxaloacetate.* This is a classic example of the induced-fit model of substrate binding (Section 9-4C). Hexokinase exhibits similar behavior (Section 16-2A).

The citrate synthase reaction is a mixed aldol–Claisen ester condensation, which occurs in three steps (Fig. 19-7):

1. The carbanion of acetyl-CoA is generated by an enzyme His residue acting as a base. The presence of the thioester bond to CoA permits the enolization stabilizing the carbanion intermediate; enolization of acetate alone would require the highly unfavorable generation of a double negative charge on the carboxyl group.

2. The acetyl-CoA carbanion nucleophilically attacks the carbonyl group of oxaloacetate. The reaction product, **citryl-CoA,** remains bound to the enzyme.

3. Citryl-CoA is hydrolyzed to citrate and CoA. This hydrolysis provides the reaction's thermodynamic driving force ($\Delta G^{\circ \prime} = -31.5 \text{ kJ} \cdot \text{mol}^{-1}$). We shall see presently why the reaction requires such a large, seemingly wasteful, driving force.

Enzyme-catalyzed reactions, as we have previously noted, are stereospecific. The aldol–Claisen condensation that occurs here involves attack of the acetyl-CoA carbanion exclusively at the *si* face of oxaloacetate's carbonyl carbon atom thereby yielding *S*-citryl-CoA (chirality nomenclature is presented in Section 4-2C). The acetyl group of acetyl-CoA thereby only forms citrate's *pro-S* carboxymethyl group.

B. Aconitase

Aconitase catalyzes the reversible isomerization of citrate and isocitrate with cis-aconitate as an intermediate (Reaction 2 of Fig. 19-1). Although citrate has a plane of symmetry and is therefore not optically active, it is, nevertheless, prochiral; aconitase can distinguish between citrate's *pro-R* and *pro-S* carboxymethyl groups. Formation of the *cis*-aconitate intermediate involves a dehydration in which an enzyme base abstracts the *pro-R* proton at C(2) of citrate's *pro-R* carboxymethyl group.

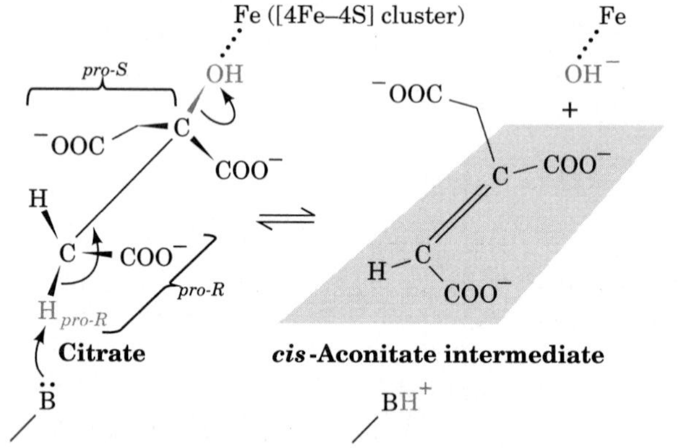

This is followed by loss of the OH group at C(3) in a trans elimination to form the *cis*-aconitate intermediate. Aconitase contains a covalently bound **[4Fe–4S] iron–sulfur cluster,** which is required for catalytic activity (the properties of iron–sulfur clusters are discussed in Section 20-2C). One of the Fe(II) atoms in this cluster is

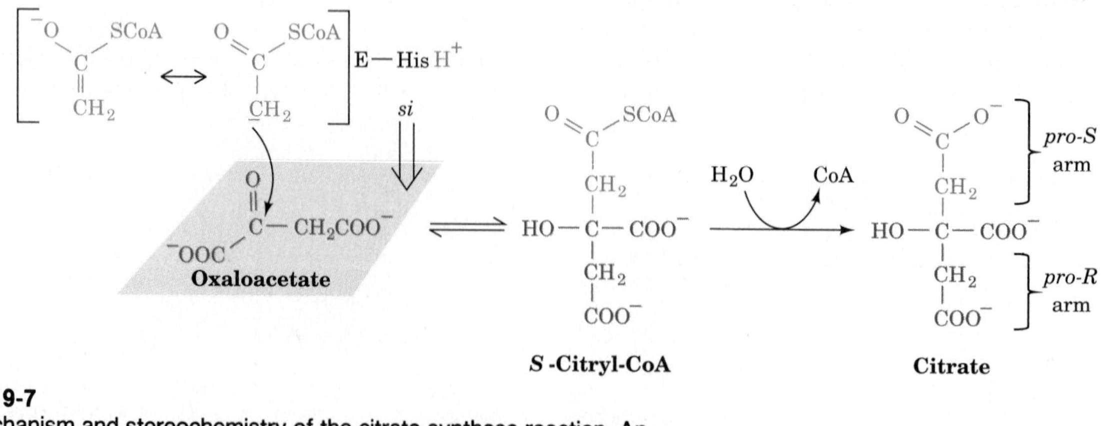

Figure 19-7
The mechanism and stereochemistry of the citrate synthase reaction. An acetyl-CoA carbanion, formed through proton abstraction by an enzyme His residue, nucleophilically attacks the *si* face of oxaloacetate's carbonyl carbon (*left*). The resulting intermediate, *S*-citryl-CoA, is hydrolyzed to yield citrate (*right*).

thought to coordinate the OH group of the substrate so as to facilitate its elimination. Iron–sulfur clusters are almost always associated with redox processes although, curiously, not in the case of aconitase.

The second stage of the aconitase reaction is rehydration of *cis*-aconitate's double bond to form isocitrate (Fig. 19-1). The nonenzymatic addition of H_2O across the double bond of *cis*-aconitate yields four stereoisomers. Aconitase, however, catalyzes the stereospecific trans addition of OH^- and H^+ across the double bond to form only (2R,3S)-isocitrate in the forward reaction and citrate in the reverse reaction.

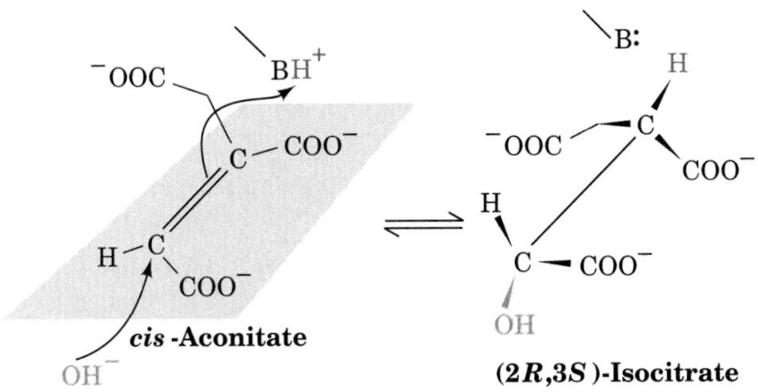

Although citrate's OH group is lost to the solvent in the aconitase reaction, the abstracted H^+ is retained by the enzyme's basic group. Remarkably, it adds to the opposite faces of *cis*-aconitate's double bond in forming citrate and isocitrate. Since the *cis*-aconitate does not dissociate from the enzyme, it must expose a different face to the sequestered H^+ for the formation of isocitrate.

Fluorocitrate Inhibits Aconitase

Fluoroacetate, one of the most toxic small molecules known ($LD_{50} = 0.2$ mg·kg^{-1} of body weight in rats), occurs in the leaves of certain African, Australian, and South American poisonous plants. Interestingly, fluoroacetate itself has little toxic effect on cells; rather, cells enzymatically convert it first to fluoroacetyl-CoA and then to **(2R,3R)-fluorocitrate,** which specifically inhibits aconitase.

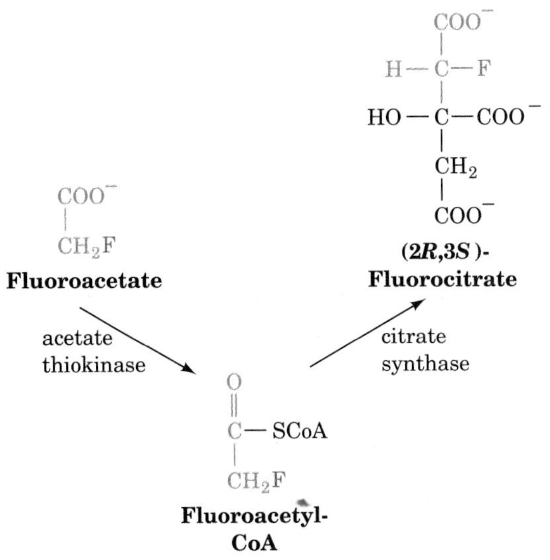

It is not clear, however, that the inhibition of aconitase fully accounts for fluorocitrate's high toxicity. Indeed, fluorocitrate also inhibits the transport of citrate across the mitochondrial membrane.

C. NAD$^+$-Dependent Isocitrate Dehydrogenase

Isocitrate dehydrogenase catalyzes the oxidative decarboxylation of isocitrate to α-ketoglutarate to produce the citric acid cycle's first CO_2 and NADH (Reaction 3 of Fig. 19-1). Mammalian tissues contain two different forms of this enzyme. One form participates in the citric acid cycle, is located entirely in the mitochondrion, and utilizes NAD$^+$ as a cofactor. The other form occurs in both the mitochondrion and the cytosol and utilizes NADP$^+$ as a cofactor.

NAD$^+$-dependent isocitrate dehydrogenase, which requires an Mn^{2+} or Mg^{2+} cofactor, is thought to catalyze the oxidation of a secondary alcohol (isocitrate) to a ketone (oxalosuccinate) followed by the decarboxylation of the carboxyl group β to the ketone (Fig. 19-8). In

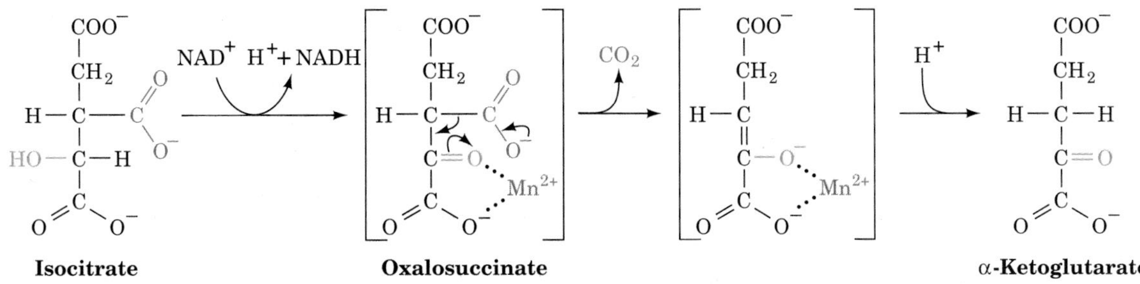

Figure 19-8
The probable reaction mechanism of isocitrate dehydrogenase. Oxalosuccinate is shown in brackets because it does not dissociate from the enzyme.

this sequence, the keto group β to the carboxyl group facilitates the decarboxylation by acting as an electron sink. The oxidation occurs with the stereospecific reduction of NAD^+ at its *re* face (A-side addition; Section 16-2F). Mn^{2+} coordinates the newly formed carbonyl group so as to polarize its electronic charge.

Although the intermediate formation of oxalosuccinate is a logical chemical prediction, evidence for it is sparse. Oxalosuccinate is decarboxylated by the $NADP^+$-dependent enzyme in the absence of $NADP^+$ but at a much lower rate than the overall reaction. Yet, intermediates should be converted to products at least as fast as the overall reaction. The specific radioactivity $(cpm \cdot mmol^{-1})$ of the α-ketoglutarate formed from $[^{14}C]$isocitrate is not diluted by the addition of unlabeled oxalosuccinate. This indicates that if $[^{14}C]$oxalosuccinate is, in fact, a reaction intermediate, it does not dissociate from the enzyme to exchange with unlabeled material before being converted to product.

D. α-Ketoglutarate Dehydrogenase

α-Ketoglutarate dehydrogenase catalyzes the oxidative decarboxylation of an α-keto acid (α-ketoglutarate), releasing the citric acid cycle's second CO_2 and NADH (Reaction 4 of Fig. 19-1). The overall reaction closely resembles that catalyzed by the pyruvate dehydrogenase multienzyme complex (Section 19-2A). It involves an analogous multienzyme complex consisting of **α-ketoglutarate dehydrogenase (E_1), dihydrolipoyl transsuccinylase (E_2),** and **dihydrolipoyl dehydrogenase (E_3).** The dihydroli-

poyl dehydrogenase subunit is, in fact, identical to that of the pyruvate dehydrogenase complex.

Individual reactions catalyzed by the complex occur by mechanisms identical to those of the pyruvate dehydrogenase reaction (Section 19-2A), the product likewise being a "high-energy" thioester, in this case succinyl-CoA. There are no covalent modification enzymes in the α-ketoglutarate dehydrogenase complex, however.

E. Succinyl-CoA Synthetase

*Succinyl-CoA synthetase (also called **succinate thiokinase**) hydrolyzes the "high-energy" compound succinyl-CoA with the coupled synthesis of a "high-energy" nucleoside triphosphate (Reaction 5 of Fig. 19-1).* (*Note:* Enzyme names can refer to either the forward or the reverse reaction; in this case succinyl-CoA synthetase and succinate thiokinase refer to the reverse reaction.) GTP is synthesized from $GDP + P_i$ by the mammalian enzyme; plant and bacterial enzymes utilize $ADP + P_i$ to form ATP. These reactions are nevertheless equivalent since ATP and GTP are rapidly interconverted through the action of nucleoside diphosphate kinase (Section 15-4C):

$$GTP + ADP \rightleftharpoons GDP + ATP \qquad \Delta G°' = 0$$

The Succinyl-CoA Thioester Bond Energy Is Preserved through the Formation of a Series of "High-Energy" Phosphates

How does succinyl-CoA synthetase couple the exer-

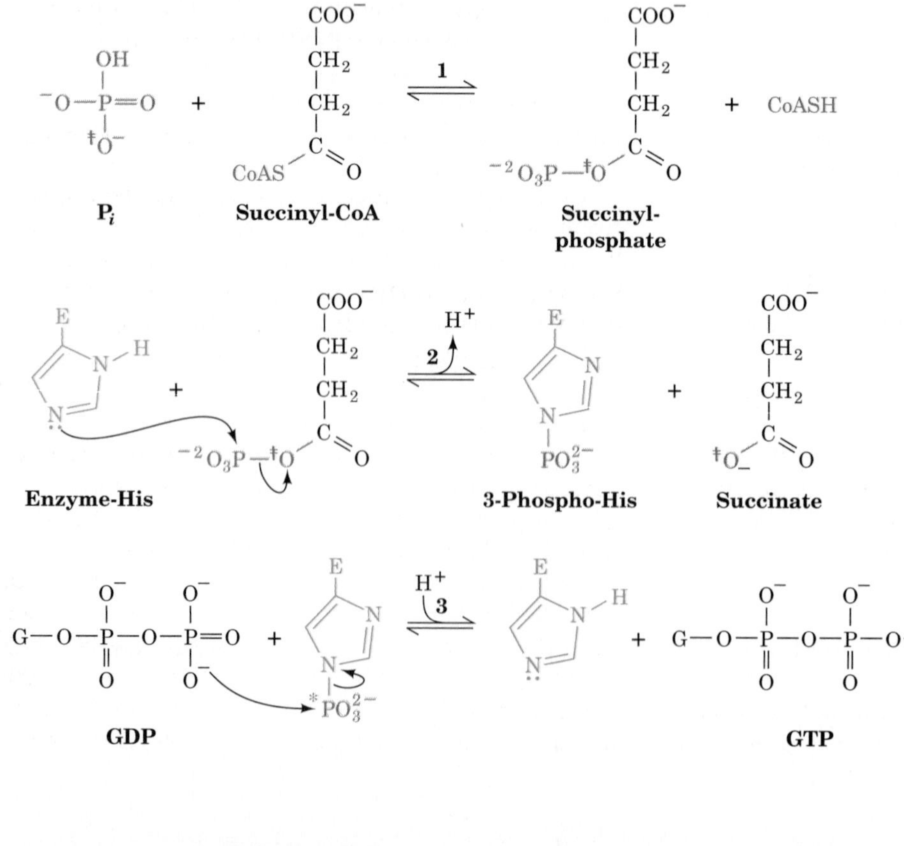

Figure 19-9
The reactions catalyzed by succinyl-CoA synthetase. **(1)** Formation of succinyl phosphate, a "high-energy" mixed anhydride. **(2)** Formation of phosphoryl-His, a "high-energy" intermediate. **(3)** Transfer of the phosphoryl group to GDP, forming GTP. The symbols † and *, respectively, represent ^{18}O and ^{32}P in isotopic labeling reactions.

gonic hydrolysis of succinyl-CoA ($\Delta G°' = -32.6$ kJ·mol^{-1}) to the endergonic formation of a nucleoside triphosphate ($\Delta G°' = 30.5$ kJ·mol^{-1})? This question was answered through the creative use of isotope tracers. In the absence of succinyl-CoA, the spinach enzyme (which utilizes adenine nucleotides), catalyzes the transfer of ATP's γ-phosphoryl group to ADP. Such an exchange reaction suggests the participation of a phosphoryl-enzyme intermediate that mediates the reaction sequence:

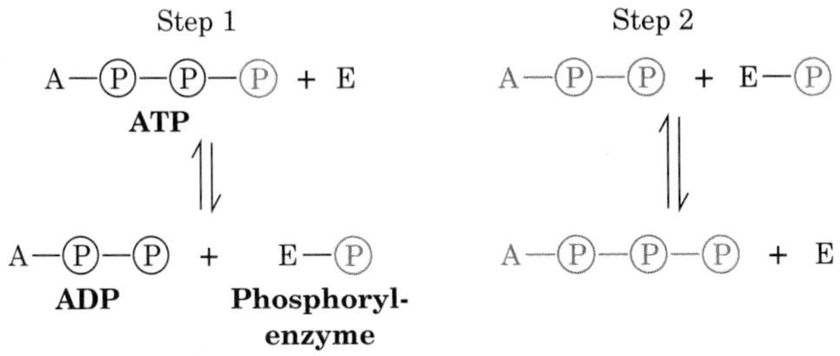

Step 1

$$A\!-\!\circledP\!-\!\circledP\!-\!\circledP + E$$
ATP

$$\Updownarrow$$

$$A\!-\!\circledP\!-\!\circledP + E\!-\!\circledP$$
ADP   **Phosphoryl-
enzyme**

Step 2

$$A\!-\!\circledP\!-\!\circledP + E\!-\!\circledP$$

$$\Updownarrow$$

$$A\!-\!\circledP\!-\!\circledP\!-\!\circledP + E$$

Indeed, this information led to the isolation of a kinetically active phosphoryl-enzyme in which the phosphoryl group is covalently linked to the N(3) position of a His residue.

When the succinyl-CoA synthetase reaction, which is freely reversible, is run in the direction of succinyl-CoA synthesis (opposite to its direction in the citric acid cycle) using [^{18}O]succinate as a substrate, ^{18}O is transferred from succinate to phosphate. Evidently, succinyl phosphate, a "high-energy" mixed anhydride, is transiently formed during the reaction.

These observations suggest the following three-step sequence for the mammalian succinyl-CoA synthetase reaction (Fig. 19-9):

1. Succinyl-CoA reacts with P$_i$ to form succinyl phosphate and CoA (accounting for the ^{18}O-exchange reaction).

2. Succinyl phosphate's phosphoryl group is transferred to an enzyme His residue, releasing succinate (accounting for the 3-phospho-His residue).

3. The phosphoryl group on the enzyme is transferred to GDP, forming GTP (accounting for the ^{32}P-exchange reaction).

Note how in each of these steps *the "high-energy" succinyl-CoA's free energy of hydrolysis is conserved through the successive formation of "high-energy" compounds: first succinyl phosphate, then a 3-phospho-His residue, and finally GTP.* The process is reminiscent of passing a hot potato.

A Pause for Perspective

Up to this point in the cycle, one acetyl equivalent has been completely oxidized to CO_2. Two NADHs and one GTP (in equilibrium with ATP) have also been generated. In order to complete the cycle, succinate must be

converted back to oxaloacetate. This is accomplished by the cycle's remaining three reactions.

F. Succinate Dehydrogenase

Succinate dehydrogenase catalyzes stereospecific dehydrogenation of succinate to fumarate (Reaction 6 of Fig. 19-1).

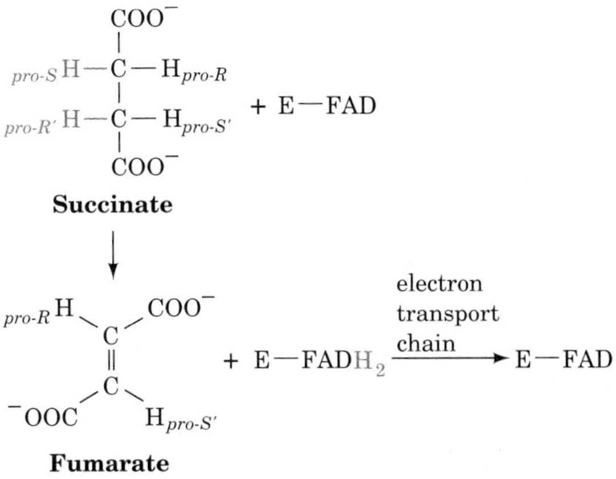

Succinate

Fumarate

The enzyme is strongly inhibited by **malonate,** a structural analog of succinate and a classic example of a competitive inhibitor:

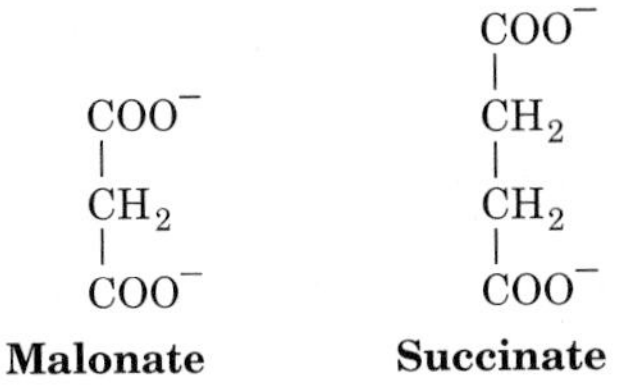

Malonate   **Succinate**

Recall that malonate inhibition of cellular respiration was one of the observations that led Krebs to hypothesize the citric acid cycle (Section 19-1B).

Succinate dehydrogenase contains an FAD, the reaction's electron acceptor. In general, FAD functions biochemically to oxidize alkanes to alkenes while NAD$^+$ oxidizes alcohols to aldehydes or ketones. This is because the oxidation of an alkane (such as succinate) to an alkene (such as fumarate) is sufficiently exergonic to reduce FAD to FADH$_2$ but not to reduce NAD$^+$ to NADH. Alcohol oxidation, in contrast, can reduce NAD$^+$ (Table 15-4). Succinate dehydrogenase's FAD is covalently bound via its C(8a) atom to an enzyme His residue (Fig. 19-10). A covalent link between FAD and a protein is unusual; in most cases FAD is noncovalently although tightly bound to its associated enzyme.

How does succinate dehydrogenase's FADH$_2$ become reoxidized? Being permanently linked to the enzyme, this prosthetic group cannot function as a metabolite as does NADH. Rather, *succinate dehydrogenase is reoxidized by the electron-transport chain,* an aspect of its function that we discuss in Section 20-2C. This rationalizes why succinate dehydrogenase, which is embedded in the inner mitochondrial membrane, is the only mem-

Figure 19-10
The covalent attachment of FAD to a His residue of succinate dehydrogenase.

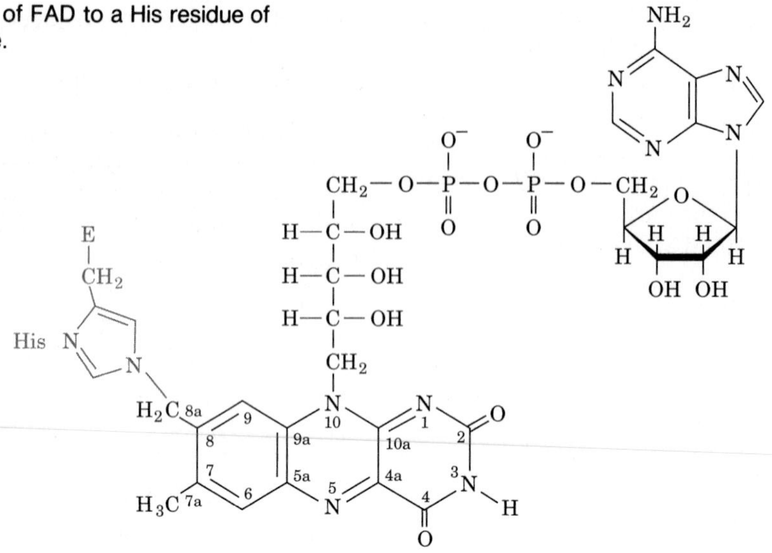

FAD

brane-bound citric acid cycle enzyme. The others are all dissolved in the mitochondrial matrix (mitochondrial anatomy is described in Section 20-1A).

G. Fumarase

Fumarase (fumarate hydratase) catalyzes the hydration of fumarate's double bond to form S-malate (L-malate) (Reaction 7 of Fig. 19-1). Consideration of experiments that have contributed to our understanding of the fumarase mechanism illustrates the role played by independent investigations.

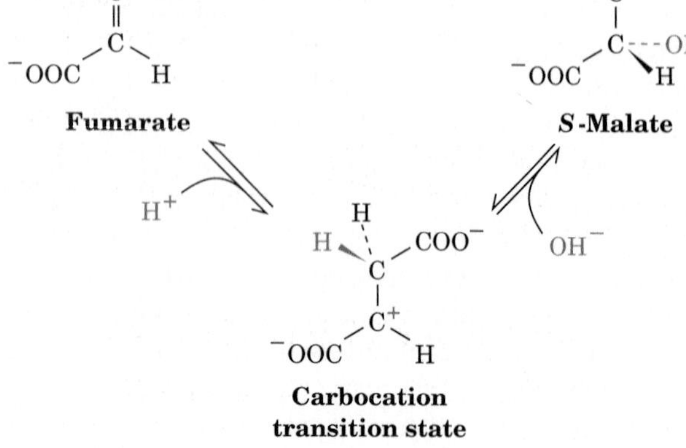

Figure 19-11
Possible mechanisms for the hydration of fumarate as catalyzed by fumarase.

Conflicting Mechanistic Evidence: What is the Sequence of H⁺ and OH⁻ Addition?

Experiments designed to establish whether the fumarase reaction occurs by a carbanion (OH⁻ addition first) or carbocation (H⁺ addition first) mechanism have provided contradictory information (Fig. 19-11). Evidence favoring the carbocation mechanism was obtained by studying the dehydration of *S*-malate (the fumarase reaction run in reverse) in $H_2^{18}O$. [^{18}O]Malate appears in the reaction mixture more rapidly than it would if the ^{18}O were incorporated via a back reaction of the newly formed fumarate. This suggests the rapid formation of a carbocation intermediate at C(2), from which OH⁻ could exchange with $^{18}OH^-$, followed by slow hydrogen removal at C(3).

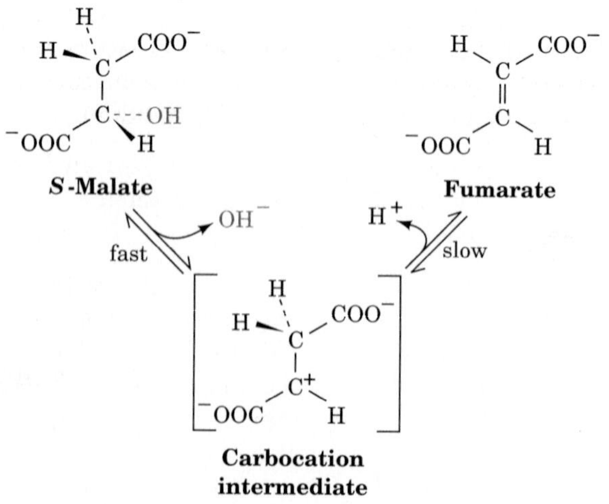

Other observations, however, indicate that the reaction occurs via the formation of a carbanion intermediate at C(3). David Porter synthesized **3-nitro-2-S-hydroxypropionate,** which sterically resembles *S*-malate.

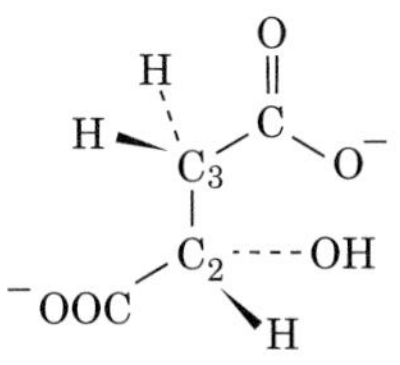

S-Malate

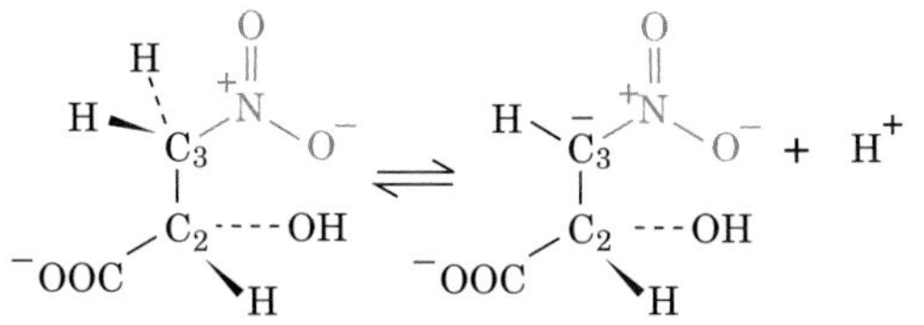

**3-Nitro-2-
S-hydroxypropionate**

The nitro group's electron-withdrawing character renders the C(3) protons relatively acidic (pK ≈ 10). The resulting anion is an analog of the postulated C(3) carbanion transition state of the fumarase reaction but not of the C(2) carbocation transition state (Fig. 19-11; transition state analogs are discussed in Section 14-1F). This anion is, in fact, an excellent inhibitor of fumarase: It has an 11,000-fold higher binding affinity for the enzyme than does S-malate.

If the fumarase reaction proceeds by a carbanion mechanism, how can the rapid OH^- exchange be explained? Conversely, if it has a carbocation mechanism, why is the nitro anion such an effective inhibitor? This contradictory set of observations makes only one thing clear: *When studying enzyme reaction mechanisms, it is always necessary to approach the problem from several different directions. One set of experiments should never be taken as proof, and interpretation should never be taken as fact.* Indeed, reinterpretation of the ^{18}O exchange experiment makes it consistent with a carbanion mechanism. It turns out that product release is the enzyme's rate-determining step:

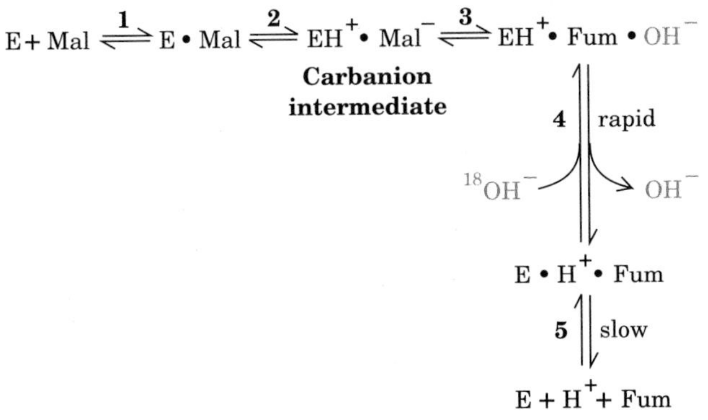

Malate binds to fumarase (1), forms a carbanion (2), eliminates OH^- to form fumarate (3), and rapidly releases OH^- from the enzyme surface (4). Release of the other products (5) is slow. $^{18}OH^-$ can therefore exchange

with OH^- to produce [^{18}O]malate more rapidly than the overall rate of the fumarase reaction.

H. Malate Dehydrogenase

Malate dehydrogenase catalyzes the final reaction of the citric acid cycle, the regeneration of oxaloacetate (Reaction 8 of Fig. 19-1). This occurs through the oxidation of S-malate's hydroxyl group to a ketone in an NAD$^+$-dependent reaction.

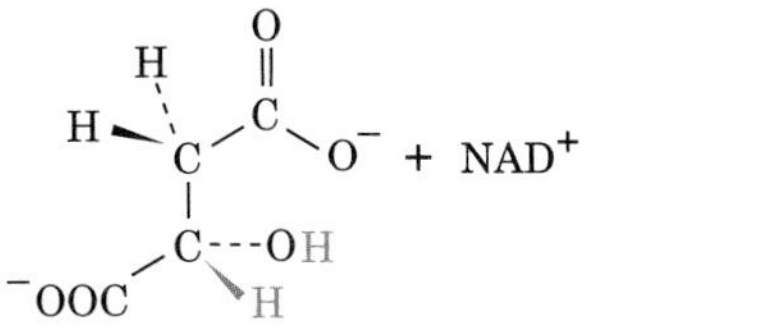

S-Malate

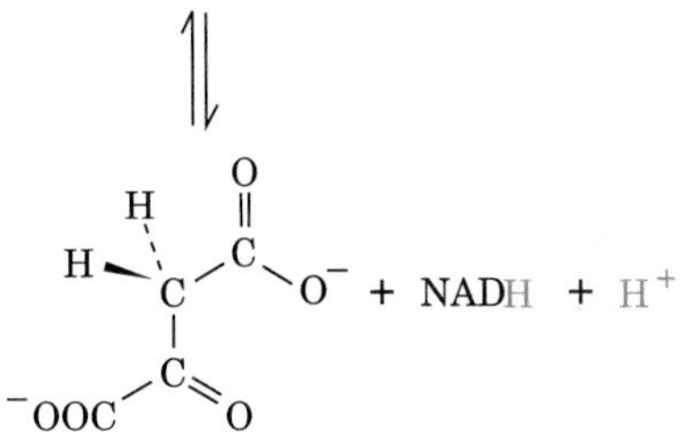

Oxaloacetate

The hydride ion released by the alcohol in this reaction is transferred to the *re* face of NAD$^+$, the same face acted upon by lactate dehydrogenase (Section 16-3A) and alcohol dehydrogenase (Section 16-3B). In fact, *X-ray crystallographic comparisons of the NAD$^+$-binding domains of these three dehydrogenases indicate that they are remarkably similar and have led to the proposal that all NAD$^+$-binding domains have evolved from a common ancestor (Section 8-3B).*

The $\Delta G^{\circ\prime}$ for the malate dehydrogenase reaction is $+29.7$ kJ·mol^{-1}; the concentration of oxaloacetate formed at equilibrium is consequently very low. Recall, however, that the reaction catalyzed by citrate synthase, the first enzyme in the cycle, is highly exergonic ($\Delta G^{\circ\prime} = -31.5$ kJ·mol^{-1}; Section 19-3A) because of the hydrolysis of citryl-CoA's thioester bond. We can now understand the necessity for such a seemingly wasteful process. It allows citrate formation to be exergonic at even low physiological concentrations of oxaloacetate with the resultant initiation of another turn of the cycle.

I. Integration of the Citric Acid Cycle

The Impact of the Citric Acid Cycle on ATP Production

The foregoing discussion indicates that one turn of the citric acid cycle results in the following chemical transformations (Fig. 19-1):

1. One acetyl group is oxidized to two molecules of CO_2, a four-electron pair process (although, as we discuss below, it is not the carbon atoms of the entering acetyl group that are oxidized).

2. Three molecules of NAD^+ are reduced to NADH, which accounts for three of the electron pairs.

3. One molecule of FAD is reduced to $FADH_2$, which accounts for the fourth electron pair.

4. One "high-energy" phosphate group is produced as GTP (or ATP).

The eight electrons abstracted from the acetyl group in the citric acid cycle subsequently pass into the electron-transport chain where they ultimately reduce two molecules of O_2 to H_2O. The three electron pairs from NADH each produce 3ATPs via oxidative phosphorylation while the one electron pair of $FADH_2$ produces 2ATPs. One turn of the citric acid cycle therefore results in the ultimate generation of 12ATPs. Electron transport and oxidative phosphorylation are the subject of Chapter 20.

Isotopic Tests of Citric Acid Cycle Stereochemistry

The reactions of the citric acid cycle have been confirmed through the use of radioactive tracer experiments that became possible in the late 1930s and early 1940s. At that time, compounds could be synthesized enriched with the stable isotope ^{13}C (detectable at the time by mass spectrometry and now by NMR) or with the radioactive isotope ^{11}C, which has a half-life of only 20 min; ^{14}C, whose use was pioneered in the late 1940s by Samuel Ruben and Martin Kamen, has the advantage over ^{11}C that its half-life is 5570 years.

In one landmark experiment, 4-[^{11}C]oxaloacetate was generated from $^{11}CO_2$ and pyruvate,

$$^*CO_2 + CH_3 - \overset{\overset{\displaystyle O}{\|}}{C} - CO_2^-$$

Pyruvate

$$\overset{\text{ATP}}{\underset{\text{ADP} + P_i}{\searrow}} \Bigg| \begin{array}{l} \text{pyruvate} \\ \text{carboxylase} \end{array}$$

$$^-OOC \overset{*}{-} CH_2 - \overset{\overset{\displaystyle O}{\|}}{C} - CO_2^-$$

Oxaloacetate

utilized in the citric acid cycle of metabolizing muscle cells, and the resulting cycle intermediates isolated. Identification of the labeled position in the isolated α-ketoglutarate caused a furor. As we have seen in Sections 19-3A and B, citrate synthase and aconitase catalyze stereospecific reactions in which citrate synthase can distinguish between the two faces of oxaloacetate's carbonyl group and aconitase can distinguish between

citrate's *pro-R* and *pro-S* carboxymethyl groups. In the early 1940s, however, the concept of prochirality had not been established; it was assumed that the two halves of citrate were indistinguishable (in nonenzymatic systems, this is, in effect, the case). It was therefore assumed that the radioactivity originally located at C(4) in oxaloacetate (* in Fig. 19-1) would be scrambled in citrate so that its C(1) and C(6) atoms would be equally labeled, resulting in α-ketoglutarate labeled at both C(1) and C(5). In fact, only C(1), the carboxyl group α to the keto group of α-ketoglutarate, was found to be radioactive (Fig. 19-1). This result threw the identity of the condensation product of oxaloacetate and acetyl-CoA into doubt. How could it be the symmetrical citrate molecule in light of such "conclusive" labeling experiments? This problem of which tricarboxylic acid was the cycle's original condensation product resulted in a name change from the citric acid cycle (proposed by Krebs) to the tricarboxylic acid (TCA) cycle. In 1948, Alexander Ogston pointed out that citric acid, while symmetrical, is prochiral and thus can interact asymmetrically with the surface of aconitase (Section 19-3B). Even though citrate is now accepted as a cycle intermediate, the duality of the cycle's name remains.

In following the fate of isotopically labeled carbon (*) through the citric acid cycle, we can see (Fig. 19-1) that it is lost as CO_2 at the α-ketoglutarate dehydrogenase reaction. While the net reaction of the cycle is oxidation of the carbon atoms of acetyl units to CO_2, the CO_2 lost in a given turn of the cycle is derived from the carbon skeleton of oxaloacetate.

Experiments have been performed using 1-[^{14}C]acetate which cells convert to 1-[^{14}C]acetyl-CoA:

$$CH_3 - {}^*COO^- + CoASH \xrightarrow[\substack{\text{acetate} \\ \text{thiokinase}}]{\overset{\text{ATP} \quad \text{AMP} + PP_i}{\overset{\curvearrowright}{}}} CH_3 - \overset{\overset{\displaystyle O}{\|}}{\underset{*}{C}} - SCoA$$

Acetate   **Acetyl–CoA**

Tracing the path of this label (‡ in Fig. 19-1) allows the reader to ascertain that the label becomes scrambled (1/2‡ in Fig. 19-1) during the cycle, but not until the formation of succinate, the cycle's first rotationally symmetric (nonprochiral) intermediate.

4. REGULATION OF THE CITRIC ACID CYCLE

In this section we consider how metabolic flux through the citric acid cycle is regulated. In our discussions of metabolic flux control (Section 16-4), we established that to understand how a metabolic pathway is controlled, we must identify the enzyme(s) that catalyze its rate-determining step(s), the *in vitro* effectors of these enzymes, and the *in vivo* concentrations of these substances. *A proposed mechanism of flux control must dem-*

onstrate that an increase or decrease in flux is correlated with an increase or decrease in the concentration of the proposed effector.

Citrate Synthase, Isocitrate Dehydrogenase, and α-Ketoglutarate Dehydrogenase Are the Citric Acid Cycle's Rate-Controlling Enzymes

Establishing the rate-determining steps of the citric acid cycle is more difficult than it is for glycolysis because most of the cycle's metabolites are present in both mitochondria and cytosol and we do not know their distribution between these two compartments [recall that identifying a pathway's rate-determining step(s) requires determining the ΔG of each of its reactions from the concentrations of its substrates and products]. If, however, we assume equilibrium between the two compartments, we can use the total cell contents of these substances to estimate their mitochondrial concentrations. Table 19-3 gives the standard free energy changes for the eight citric acid cycle enzymes and estimates of the physiological free energy changes for the reactions in heart muscle or liver tissue. We can see that three of the enzymes are likely to function far from equilibrium under physiological conditions (negative ΔG): citrate synthase, NAD^+-dependent isocitrate dehydrogenase, and α-ketoglutarate dehydrogenase. We shall therefore focus our discussion on how these enzymes are regulated (Fig. 19-12).

The Citric Acid Cycle Is Largely Regulated by Substrate Availability, Product Inhibition, and Inhibition by Other Cycle Intermediates

In heart muscle, where the citric acid cycle functions mainly to generate ATP for use in muscle contraction, the enzymes of the cycle almost always act as a func-

Table 19-3

Standard Free Energy Changes ($\Delta G°'$) and Physiological Free Energy Changes (ΔG) of Citric Acid Cycle Reactions

Reaction	Enzyme	$\Delta G°'$ $(kJ \cdot mol^{-1})$	ΔG $(kJ \cdot mol^{-1})$
1	Citrate synthase	−31.5	Negative
2	Aconitase	~5	~0
3	Isocitrate dehydrogenase	−21	Negative
4	α-Ketoglutarate dehydrogenase multienzyme complex	−33	Negative
5	Succinyl-CoA synthetase	−2.1	~0
6	Succinate dehydrogenase	+6	~0
7	Fumarase	−3.4	~0
8	Malate dehydrogenase	+29.7	~0

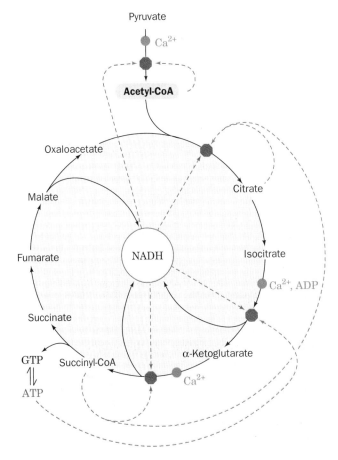

Figure 19-12
A diagram of the citric acid cycle and the pyruvate dehydrogenase reaction indicating their points of inhibition (*red octagons*) and the pathway intermediates that function as inhibitors (*dashed red arrows*). ADP and Ca²⁺ (*green dots*) are activators.

tional unit with their metabolic flux proportional to the rate of cellular oxygen consumption. *Since oxygen consumption, NADH reoxidation, and ATP production are tightly coupled (Section 20-5), the citric acid cycle must be regulated by feedback mechanisms that coordinate its NADH production with energy expenditure.* Unlike the rate-limiting enzymes of glycolysis and glycogen metabolism, which utilize elaborate systems of allosteric control, substrate cycles and covalent modification as flux control mechanisms, the regulatory enzymes of the citric acid cycle seem to be controlled almost entirely in three simple ways: (1) substrate availability, (2) product inhibition, and (3) competitive feedback inhibition by intermediates further along the cycle. We shall encounter several examples of these straightforward mechanisms in the following discussion.

Perhaps the most crucial regulators of the citric acid cycle are its substrates, acetyl-CoA and oxaloacetate, and its product NADH. Both acetyl-CoA and oxaloacetate are present in mitochondria at concentrations that do not saturate citrate synthase. The metabolic flux through the enzyme therefore varies with substrate concentration and is subject to control by substrate availability. The production of acetyl-CoA from pyruvate is regulated by the activity of pyruvate dehydrogenase (Section 19-2B). Oxaloacetate is in equilibrium with malate, its concentration fluctuating with the [NADH]/[NAD$^+$] ratio according to the equilibrium expression

$$K = \frac{[\text{oxaloacetate}][\text{NADH}]}{[\text{malate}][\text{NAD}^+]}$$

In the transition from low to high work and respiration rates, mitochondrial [NADH] decreases. The consequent increase in [oxaloacetate] stimulates the citrate synthase reaction, which controls the rate of citrate formation.

The observation that [citrate] invariably falls as the work load increases indicates that the rate of citrate removal increases more than its rate of formation. The rate of citrate removal is governed by NAD$^+$-dependent isocitrate dehydrogenase (aconitase functions close to equilibrium), which is strongly inhibited *in vitro* by NADH (product inhibition). Citrate synthase is also inhibited by NADH. Evidently, NAD$^+$-dependent isocitrate dehydrogenase is more sensitive to [NADH] changes than citrate synthase.

The decrease in [citrate] that occurs upon transition from low to high work and respiration rates results in a domino effect:

1. Citrate is a competitive inhibitor of oxaloacetate for citrate synthase (product inhibition); the fall in [citrate] caused by increased isocitrate dehydrogenase activity increases the rate of citrate formation.

2. α-Ketoglutarate dehydrogenase is also strongly inhibited by its products, NADH and succinyl-CoA. Its activity therefore increases when [NADH] decreases.

3. Succinyl-CoA also competes with acetyl-CoA in the citrate synthase reaction (competitive feedback inhibition).

This interlocking system serves to keep the citric acid cycle coordinately regulated.

ADP, ATP, and Ca^{2+} Are Allosteric Regulators of Citric Acid Cycle Enzymes

In vitro studies on the enzymes of the citric acid cycle have identified a few allosteric activators and inhibitors. Increased workload is accompanied by increased [ADP] resulting from the consequent increased rate of ATP hydrolysis. ADP acts as an allosteric activator of isocitrate dehydrogenase by decreasing its apparent K_M for isocitrate. ATP, which builds up when muscle is at rest, inhibits this enzyme.

Ca^{2+}, among its many biological functions, is an essential metabolic regulator. It stimulates glycogen breakdown (Section 17-3C), triggers muscle contraction (Section 34-3C), and mediates many hormonal signals

as a second messenger (Section 34-4B). Ca^{2+} also plays an important role in the regulation of the citric acid cycle (Fig. 19-12). It activates pyruvate dehydrogenase phosphatase which, in turn, activates the pyruvate dehydrogenase complex to produce acetyl-CoA. In addition, Ca^{2+} activates both isocitrate dehydrogenase and α-ketoglutarate dehydrogenase. Thus, the same signal stimulates muscle contraction and the production of the ATP to fuel it.

In the liver, the role of the citric acid cycle is more complex than in heart muscle. The liver synthesizes many substances required by the body including glucose, fatty acids, cholesterol, amino acids, and porphyrins. Reactions of the citric acid cycle play a part in many of these biosynthetic pathways in addition to their role in energy metabolism. In the next section, we discuss the contribution of the citric acid cycle to these processes.

5. THE AMPHIBOLIC NATURE OF THE CITRIC ACID CYCLE

Ordinarily one thinks of a metabolic pathway as being either catabolic with the release (and conservation) of free energy, or anabolic with a requirement for free energy. The citric acid cycle is, of course, catabolic because it involves degradation and is a major free energy conservation system in most organisms. Cycle intermediates are only required in catalytic amounts to maintain the degradative function of the cycle. However, several biosynthetic pathways utilize citric acid cycle intermediates as starting materials (anabolism). The citric acid cycle is therefore amphibolic (both anabolic and catabolic).

All of the biosynthetic pathways that utilize citric acid cycle intermediates also require free energy. Consequently, the catabolic function of the cycle cannot be interrupted; *cycle intermediates that have been siphoned off must be replaced.* Although the mechanistic aspects of the enzymes involved in the pathways that utilize and replenish citric acid cycle intermediates are discussed in subsequent chapters, it is useful to mention these metabolic interconnections briefly (Fig. 19-13).

Pathways that Utilize Citric Acid Cycle Intermediates

1. **Glucose biosynthesis (gluconeogenesis;** Section 21-1), which occurs in the cytosol, utilizes malate that has been transported across the mitochondrial membrane.

2. **Lipid biosynthesis,** which includes **fatty acid biosynthesis** (Section 23-4) and **cholesterol biosynthesis** (Section 23-6A), are cytosolic processes that re-

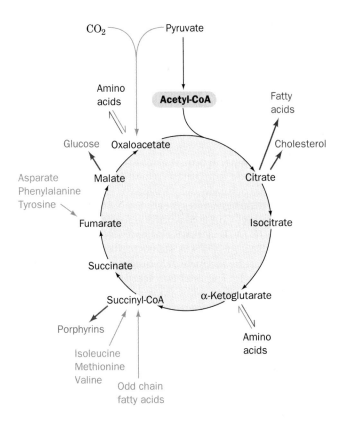

Figure 19-13
A diagram of the citric acid cycle indicating the positions at which intermediates are drawn off for use in anabolic pathways (*red arrows*) and the points where anaplerotic reactions replenish depleted cycle intermediates (*green arrows*). Reactions involving amino acid transamination and deamination are reversible so that their direction varies with metabolic demand.

quire acetyl-CoA. Acetyl-CoA is generated in the mitochondrion and is not transported across the inner mitochondrial membrane. Cytosolic acetyl-CoA is therefore generated by the breakdown of citrate, which can cross the inner mitochondrial membrane, in a reaction catalyzed by **ATP-citrate lyase** (Section 23-4D):

$$\text{Citrate} + \text{ATP} + \text{CoA} \rightleftharpoons$$
$$\text{ADP} + P_i + \text{oxaloacetate} + \text{acetyl-CoA}$$

3. **Amino acid biosynthesis** utilizes citric acid cycle intermediates in two ways. α-Ketoglutarate is used to synthesize glutamate in a reductive amination reaction involving either NAD^+ or $NADP^+$ catalyzed by **glutamate dehydrogenase** (Section 24-1):

$$\alpha\text{-Ketoglutarate} + \text{NAD(P)H} + NH_4^+ \rightleftharpoons$$
$$\text{glutamate} + \text{NAD(P)}^+ + H_2O$$

α-Ketoglutarate and oxaloacetate are also used to synthesize glutamate and aspartate in transamination reactions (Section 24-1):

$$\alpha\text{-Ketoglutarate} + \text{alanine} \rightleftharpoons$$
$$\text{glutamate} + \text{pyruvate}$$

and

$$\text{Oxaloacetate} + \text{alanine} \rightleftharpoons \text{aspartate} + \text{pyruvate}$$

4. **Porphyrin biosynthesis** (Section 24-4A) utilizes succinyl-CoA as a starting material.

Reactions that Replenish Citric Acid Cycle Intermediates

Reactions that replenish citric acid cycle intermediates are called **anaplerotic reactions** (filling up, Greek: *ana*, up + *plerotikos*, to fill). The main reaction of this type is catalyzed by **pyruvate carboxylase**, which produces oxaloacetate (Section 21-1A):

$$\text{Pyruvate} + CO_2 + \text{ATP} + H_2O \rightleftharpoons$$
$$\text{oxaloacetate} + \text{ADP} + P_i$$

This enzyme "senses" the need for more citric acid cycle intermediates through its activator, acetyl-CoA. Any decrease in the rate of the cycle caused by insufficient oxaloacetate or other cycle intermediates results in an increased level of acetyl-CoA because of its underutilization. This activates pyruvate carboxylase, which replenishes oxaloacetate, increasing the rate of the cycle. Of course, if the citric acid cycle is inhibited at some other step, by high NADH concentration, for example, increased oxaloacetate concentration will not activate the cycle. The excess oxaloacetate instead equilibrates with malate, which is transported out of the mitochondria for use in gluconeogenesis.

Degradative pathways generate citric acid cycle intermediates:

1. Oxidation of odd-chain fatty acids (Section 23-2E) leads to the production of succinyl-CoA.

2. Breakdown of the amino acids isoleucine, methionine, and valine (Section 24-3E) also leads to the production of succinyl-CoA.

3. Transamination and deamination of amino acids leads to the production of α-ketoglutarate and oxaloacetate. These reactions are reversible and, depending on metabolic demand, serve to remove or replenish these citric acid cycle intermediates.

The citric acid cycle is truly at the center of metabolism. Its reduced products, NADH and $FADH_2$, are reoxidized by the electron-transport chain during oxidative phosphorylation and the free energy released is coupled to the biosynthesis of ATP. Citric acid cycle intermediates are utilized in the biosynthesis of many vital cellular constituents. In the next few chapters we shall explore the interrelationships of these pathways in more detail.

Chapter Summary

The citric acid cycle, the common mode of oxidative metabolism in most organisms, is mediated by eight enzymes that collectively convert one acetyl-CoA to two CO_2 molecules so as to yield three NADHs, one $FADH_2$, and one GTP (or ATP). The NADH and $FADH_2$ are oxidized by O_2 in the electron-transport chain with the concomitant synthesis of eleven ATPs.

Pyruvate, the end product of glycolysis under aerobic conditions, is converted to acetyl-CoA by the pyruvate dehydrogenase multienzyme complex, a symmetrical cluster of three enzymes: pyruvate dehydrogenase, dihydrolipoyl transacetylase, and dihydrolipoyl dehydrogenase. The pyruvate dehydrogenase subunit catalyzes the conversion of pyruvate to CO_2 and a hydroxyethyl-TPP intermediate. The latter is channeled to dihydrolipoyl transacetylase, which oxidizes the hydroxyethyl group to acetate and transfers it to CoA to form acetyl-CoA. The lipoamide prosthetic group, which is reduced to the dihydro form in the process, is reoxidized by dihydrolipoamide dehydrogenase in a reaction involving bound FAD that reduces NAD^+ to NADH. Dihydrolipoamide transacetylase is inactivated by the formation of a covalent adduct between lipoate and As(III) compounds. The activity of the pyruvate dehydrogenase complex varies with the [NADH]/[NAD^+] and [acetyl-CoA]/[CoA] ratios. In eukaryotes, the pyruvate dehydrogenase subunit is also inactivated by phosphorylation of a specific Ser residue and is reactivated by its removal. These modifications are mediated, respectively, by pyruvate dehydrogenase kinase and pyruvate dehydrogenase phosphatase, which are components of the multienzyme complex and respond to the levels of metabolic intermediates such as NADH and acetyl-CoA.

Citrate is formed by the condensation of acetyl-CoA and oxaloacetate by citrate synthase. The citrate is dehydrated to *cis*-aconitate and then rehydrated to isocitrate in a stereospecific reaction catalyzed by aconitase. This enzyme is specifically inhibited by (2R,3R)-fluorocitrate, which is enzymatically synthesized from fluoroacetate and oxaloacetate. Isocitrate is oxidatively decarboxylated to α-ketoglutarate by isocitrate dehydrogenase, which produces NADH and CO_2. The α-ketoglutarate, in turn, is oxidatively decarboxylated by α-ketoglutarate dehydrogenase, a multienzyme complex that resembles the pyruvate dehydrogenase multienzyme complex. This reaction generates the second NADH and CO_2. The resulting succinyl-CoA is converted to succinate with the generation of GTP (ATP in plants and bacteria) by succinyl-CoA synthetase. The succinate is stereospecifically dehydrogenated to fumarate by succinate dehydrogenase in a reaction that generates $FADH_2$. The final two reactions of the citric acid cycle, which are catalyzed by fumarase and malate dehydrogenase, in turn hydrate fumarate to *S*-malate and oxidize this alcohol to its corresponding ketone, oxaloacetate, with concomitant production of the pathway's third and final NADH.

The enzymes of the citric acid cycle act as a functional unit that keeps pace with the metabolic demands of the cell. The flux-controlling enzymes appear to be citrate synthase, isocitrate dehydrogenase, and α-ketoglutarate dehydrogenase. Their activities are controlled by substrate availability, product inhibition, inhibition by cycle intermediates, and activation by Ca^{2+}.

Several anabolic pathways utilize citric acid cycle intermediates as starting materials. These essential substances are replaced by anaplerotic reactions of which the major one is synthesis of oxaloacetate from pyruvate and CO_2 by pyruvate carboxylase.

References

General

Cunningham, E. B., *Biochemistry: Mechanisms of Metabolism*, Chapter 10, McGraw–Hill (1978).

Kornberg, H. L., Tricarboxylic acid cycles, *BioEssays* **7**, 236–238 (1987). [An historical synopsis of the intellectual background leading to the discovery of the citric acid cycle.]

Krebs, H. A., The history of the tricarboxylic acid cycle, *Perspect. Biol. Med.* **14**, 154–170 (1970).

Lowenstein, J. M. (Ed.), *The Citric Acid Cycle: Control and Compartmentation*, Dekker (1969).

Lowenstein, J. M., The tricarboxylic cycle, *in* Greenberg, D. M. (Ed.), *Metabolic Pathways* (3rd ed.), Vol. 1, *pp.* 146–270, Academic Press (1967).

Enzyme Mechanisms

Angelides, J. K. and Hammes, G. G., Mechanism of action of the pyruvate dehydrogenase multienzyme complex from *E. coli*, *Proc. Natl. Acad. Sci.* **75**, 4877–4880 (1978).

Emptage, M. H., Kent, T. A., Kennedy, M. C., Beinert, H., and Münck, E., Mössbauer and EPR studies of activated aconitase: development of a localized valence state at a subsite of the [4Fe–4S] cluster on binding citrate, *Proc. Natl. Acad. Sci.* **80**, 4674–4678 (1983).

Kent, T. A., Emptage, M. H., Merkle, H., Kennedy, M. C., Beinert, H., and Münck, E., Mössbauer studies of aconitase: substrate and inhibitor binding, reaction intermediates and hyperfine interactions of reduced 3Fe and 4Fe clusters, *J. Biol. Chem.* **260**, 6871–6881 (1985).

Banaszak, L. J. and Bradshaw, R. A., Malate dehydrogenases, in Boyer, P. D. (Ed.), *The Enzymes* (3rd ed.), Vol. 11, *pp.* 369–396, Academic Press (1975).

Porter, D. J. T. and Bright, H. J., 3-Carbanionic substrate analogues bind very tightly to fumarase and aspartase, *J. Biol. Chem.* **255**, 4772–4780 (1980).

Reed, L. J., Pettit, F., and Yeaman, S., Pyruvate dehydrogenase complex: structure, function and regulation, *in* Srere, P. A. and Estabrook, R. W. (Eds.), *Microenvironments and Metabolic Compartmentation*, *pp.* 305–321, Academic Press (1978).

Srere, P. A., The enzymology of the formation and breakdown of citrate, *Adv. Enzymol.* **43**, 57–101 (1975).

Walsh, C., *Enzymatic Reaction Mechanisms*, Freeman (1979).

[Contains discussions of the mechanisms of various citric acid cycle enzymes.]

Weitzman, P. D. J. and Danson, M. J., Citrate synthase, *Curr. Top. Cell. Regul.* **10**, 161–204 (1976).

Wiegand, G. and Remington, S. J., Citrate synthase: structure, control and mechanism, *Annu. Rev. Biophys. Biophys. Chem.* **15**, 97–117 (1986).

Weiland, O. H., The mammalian pyruvate dehydrogenase complex: structure and regulation, *Rev. Physiol. Biochem. Pharmacol.* **96**, 123–170 (1983).

Wilkinson, K. D. and Williams, C. H., Jr., Evidence for multiple electronic forms of two-electron-reduced lipoamide dehydrogenase from *E. coli, J. Biol. Chem.* **254**, 852–862 (1979).

Metabolic Poisons

Arsenic, Committee on the Medical and Biological Effects of Environmental Pollutants, Subcommittee on Arsenic, National Research Council, National Academy of Sciences (1977).

Gibble, G. W., Fluoroacetate toxicity, *J. Chem. Educ.* **50**, 460–462 (1973).

Jones, D. E. H. and Ledingham, K. W. D., Arsenic in Napoleons's wallpaper, *Nature* **299**, 626–627 (1982).

Winslow, J. H., *Darwin's Victorian Malady,* American Philosophical Society (1971).

Control Mechanisms

Denton, R. M. and Halestrap, A. P., Regulation of pyruvate metabolism in mammalian tissues, *Essays Biochem.* **15**, 37–77 (1979).

Hansford, R. G., Control of mitochondrial substrate oxidation, *Curr. Top. Bioenerget.* **10**, 217–278 (1980).

Reed, L. J., Damuni, Z., and Merryfield, M. L., Regulation of mammalian pyruvate and branched-chain α-keto-acid dehydrogenase complexes by phosphorylation and dephosphorylation, *Curr. Top. Cell. Regul.* **27**, 41–49 (1985).

Williamson, J. R., Mitochondrial metabolism and cell regulation, *in* Packer, L. and Gómez-Puyon, A. (Eds.), *Mitochondria,* pp. 79–107, Academic Press (1976).

Williamson, J. R., Mitochondrial function in the heart, *Annu. Rev. Physiol.* **41**, 485–506 (1979).

Problems

1. Trace the course of the radioactive label in 2-[^{14}C]glucose through glycolysis and the citric acid cycle. At what point(s) in the cycle will the radioactivity be released as $^{14}CO_2$? How many turns of the cycle will be required for complete conversion of the radioactivity to CO_2? Repeat this problem for pyruvate that is ^{14}C labeled at its methyl group.

2. Given the following information, calculate the physiological ΔG of the isocitrate dehydrogenase reaction at 25°C and pH 7.0: [NAD$^+$]/[NADH] = 8; [α-ketoglutarate] = 0.1 mM; [isocitrate] = 0.02 mM; assume standard conditions for CO_2 ($\Delta G°'$ is given in Table 19-3). Is this reaction a likely site for metabolic control? Explain.

3. The oxidation of acetyl-CoA to two molecules of CO_2 involves the transfer of four electron pairs to redox coenzymes. In which of the cycle's reactions do these electron transfers occur? Identify the redox coenzyme in each case. For each reaction, draw the structural formulas of the reactants, intermediates, and products and show, using curved arrows, how the electrons are transferred.

4. The citrate synthase reaction has been proposed to proceed via the formation of an acetyl-CoA carbanion. How, then, would you account for the observation that ^{3}H is not incorporated into acetyl-CoA when acetyl-CoA is incubated with citrate synthase in ^3H_2O?

5. Malonate is a competitive inhibitor of succinate in the succinate dehydrogenase reaction. Sketch the graphs that would be obtained upon plotting $1/v$ versus $1/$[succinate] at three different malonate concentrations. Label the lines for low, medium, and high [malonate].

6. Krebs found that malonate inhibition of the citric acid cycle could be overcome by raising the oxaloacetate concentration. Explain the mechanism of this process in light of your findings in Problem 5.

7. Which of the following metabolites undergo net oxidation by the citric acid cycle: (a) α-ketoglutarate, (b) succinate, (c) citrate, and (d) acetyl-CoA?

8. Although there is no net synthesis of intermediates by the citric acid cycle, citric acid cycle intermediates are used in biosynthetic reactions such as the synthesis of porphyrins from succinyl-CoA. Give a reaction for the net synthesis of succinyl-CoA from pyruvate.

9. Oxaloacetate and α-ketoglutarate are precursors of the amino acids aspartate and glutamate as well as being catalytic intermediates in the citric acid cycle. Describe the net synthesis of α-ketoglutarate from pyruvate in which no citric acid cycle intermediates are depleted.

10. Lipoic acid is bound to enzymes that catalyze oxidative decarboxylation of α-keto acids. (a) What is the chemical mode of attachment of lipoic acid to enzymes? (b) Using chemical structures, show how lipoic acid participates in the oxidative decarboxylation of α-keto acids.

11. **British anti-lewisite (BAL),** which was designed to counter the effects of the arsenical war gas **lewisite,** is useful in treating arsenic poisoning. Explain.

$$CH_2—SH$$
$$|$$
$$CH—SH \qquad\qquad Cl—CH=CH—AsCl_2$$
$$|$$
$$CH_2—OH$$

British anti-lewisite   **Lewisite**
(BAL)

Chapter 20
ELECTRON TRANSPORT AND OXIDATIVE PHOSPHORYLATION

In 1789, Armand Séguin and Antoine Lavoisier (the father of modern chemistry) wrote:

> . . . in general, respiration is nothing but a slow combustion of carbon and hydrogen, which is entirely similar to that which occurs in a lamp or lighted candle, and that, from this point of view, animals that respire are true combustible bodies that burn and consume themselves.

Lavoisier had by this time demonstrated that living animals consume oxygen and generate carbon dioxide. It was not until the early twentieth century, however, after the rise of enzymology, that it was established, largely through the work of Otto Warburg, that biological oxidations are catalyzed by intracellular enzymes. As we have seen, glucose is completely oxidized to CO_2 through the enzymatic reactions of glycolysis and the citric acid cycle. In this chapter we shall examine the fate of the electrons that are removed from glucose by this oxidation process.

The complete oxidation of glucose by molecular oxygen is described by the following redox equation:

$$C_6H_{12}O_6 + 6\,O_2 \longrightarrow 6CO_2 + 6H_2O$$
$$\Delta G^{\circ\prime} = -2823 \text{ kJ} \cdot \text{mol}^{-1}$$

To see more clearly the transfer of electrons, this equation may be broken down into two half-reactions. In the first reaction the glucose carbon atoms are oxidized:

$$C_6H_{12}O_6 + 6H_2O \longrightarrow 6CO_2 + 24H^+ + 24e^-$$

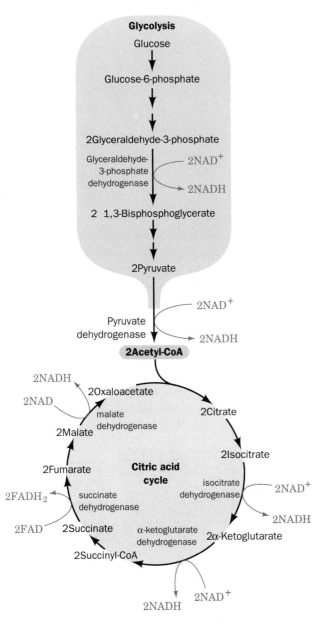

Figure 20-1
The sites of electron transfer that form NADH and FADH$_2$ in glycolysis and the citric acid cycle.

and in the second, molecular oxygen is reduced:

$$6\,O_2 + 24H^+ + 24e^- \longrightarrow 12H_2O$$

In living systems, the electron-transfer process connecting these half-reactions occurs through a multistep pathway that harnesses the liberated free energy to form ATP.

The 12 electron pairs involved in glucose oxidation are not transferred directly to O_2. Rather, as we have seen, *they are transferred to the coenzymes NAD$^+$ and FAD to form 10NADH + 2FADH$_2$ (Fig. 20-1)* in the reactions catalyzed by the glycolytic enzyme glyceraldehyde-3-phosphate dehydrogenase (Section 16-2F), pyruvate dehydrogenase (Section 19-2A), and the citric acid cycle

enzymes isocitrate dehydrogenase, α-ketoglutarate dehydrogenase, succinate dehydrogenase (the only FAD reduction), and malate dehydrogenase (Section 19-3). *The electrons then pass into the electron-transport chain where, through reoxidation of NADH and FADH$_2$, they participate in the sequential oxidation–reduction of over 10 redox centers before reducing O_2 to H_2O. In this process, protons are expelled from the mitochondrion. The free energy stored in the resulting pH gradient drives the synthesis of ATP from ADP and P$_i$ through* **oxidative phosphorylation.** Reoxidation of each NADH results in the synthesis of 3ATPs, and reoxidation of FADH$_2$ yields 2ATPs for a total of 38ATPs/glucose completely oxidized to CO_2 and H_2O.

In this chapter we explore the mechanisms of electron transport and oxidative phosphorylation and their regulation. We begin with a discussion of mitochondrial structure and transport systems.

1. THE MITOCHONDRION

The mitochondrion (Section 1-2A) is the site of eukaryotic oxidative metabolism. It contains, as Albert Lehninger and Eugene Kennedy demonstrated in 1948, the enzymes that mediate this process, including pyruvate dehydrogenase, the citric acid cycle enzymes, the enzymes catalyzing fatty acid oxidation (Section 23-2C), and the enzymes and redox proteins involved in electron transport and oxidative phosphorylation. It is therefore with good reason that the mitochondrion is described as the cell's "power plant."

A. Mitochondrial Anatomy

Mitochondria vary considerably in size and shape depending on their source and metabolic state. They are typically ellipsoids of ~ 0.5 μm in diameter and 1 μm in length (about the size of a bacterium; Fig. 20-2). The mitochondrion is bounded by a smooth outer membrane and contains an extensively invaginated inner membrane. The number of invaginations, called **cristae,** varies with the respiratory activity of the particular type of cell. This is because the proteins mediating electron transport and oxidative phosphorylation are bound to the inner mitochondrial membrane so that the respiration rate varies with membrane surface area. Liver, for example, which has a relatively low respiration rate, contains mitochondria with relatively few cristae whereas those of heart muscle contain many. Nevertheless, the aggregate area of the inner mitochondrial membranes in a liver cell is ~ 15-fold greater than that of its plasma membrane. The inner mitochondrial compartment consists of a gel-like substance of $<50\%$ water, named the **matrix,** which contains remarkably high concentrations of the soluble enzymes of oxidative metabolism (e.g., citric acid cycle enzymes), as well as

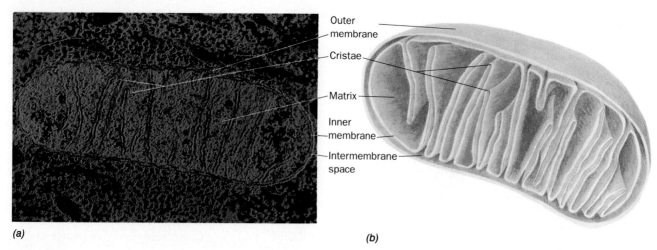

(a)

(b)

Figure 20-2
Mitochondria. (a) An electron micrograph of an animal mitochondrion. [Secchi-Lacaque/Roussel-UCLAF/CNRI.] (b) Cut-away diagram of a mitochondrion.

substrates, nucleotide cofactors, and inorganic ions. The matrix also contains the mitochondrial genetic machinery—DNA, RNA, and ribosomes—that generate several (but by no means all) mitochondrial proteins (Section 1-2A).

The Inner Mitochondrial Membrane Compartmentalizes Metabolic Functions

The outer mitochondrial membrane contains **porin,** a protein that forms nonspecific pores that permit free diffusion of up to 10-kD molecules. The inner membrane, which contains ~75% protein by weight, is considerably richer in proteins than the outer membrane (Fig. 20-3). It is freely permeable only to O_2, CO_2, and H_2O and contains, in addition to respiratory chain proteins, numerous transport proteins that control the passage of metabolites such as ATP, ADP, pyruvate, Ca^{2+}, and phosphate (see below). *This controlled impermeability of the inner mitochondrial membrane to most ions, metabolites, and low molecular mass compounds permits the generation of ionic gradients across this barrier and results in the compartmentalization of metabolic functions between cytosol and mitochondria.*

B. Mitochondrial Transport Systems

The inner mitochondrial membrane is impermeable to most hydrophilic substances. It must therefore contain specific transport systems to permit the following processes:

1. Glycolytically produced cytosolic NADH must gain access to the electron-transport chain for aerobic oxidation.

2. Mitochondrially produced metabolites such as oxaloacetate and acetyl-CoA, the respective precursors for cytosolic glucose and fatty acid biosynthesis, must reach their metabolic destinations.

3. Mitochondrially produced ATP must reach the cytosol where most ATP-utilizing reactions take place,

while ADP and P_i, the substrates for oxidative phosphorylation, must enter the mitochondrion.

We have already studied the ADP–ATP translocator (Section 18-4C). The export of oxaloacetate and acetyl-CoA from the mitochondrion are, respectively, discussed in Sections 21-1A and 23-4D. In the remainder of this section we examine the mitochondrial transport systems for P_i and Ca^{2+} and shuttle systems for NADH.

P_i Transport

ATP is generated from ADP + P_i in the mitochondrion but is utilized in the cytosol. The P_i produced is returned to the mitochondrion via an electroneutral

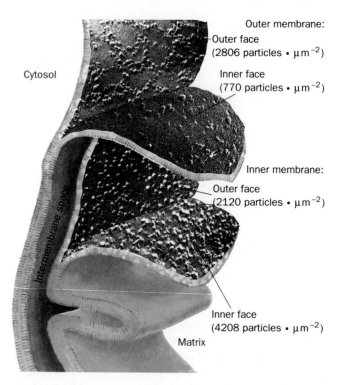

Figure 20-3
Freeze-fracture and freeze-etch electron micrographs of the inner and outer mitochondrial membranes. The inner membrane contains about twice the density of embedded particles as does the outer membrane. [Courtesy of Lester Packer, University of California at Berkeley.]

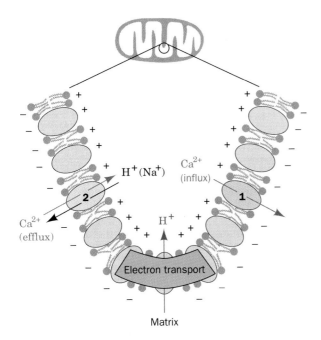

Figure 20-4
The two mitochondrial Ca^{2+}-transport systems. System 1 mediates Ca^{2+} influx to the matrix in response to the membrane potential (negative inside). System 2 mediates Ca^{2+} efflux in exchange for H^+ (Na^+ in heart muscle).

P_i—H^+ symport that is driven by ΔpH. The electrochemical potential gradient generated by the redox-driven proton pumps of electron transport (Section 20-3B) is therefore responsible for maintaining high mitochondrial ADP and P_i concentrations in addition to providing the free energy for ATP synthesis.

Ca^{2+} Transport

Since Ca^{2+}, like cAMP, functions as a second messenger (Section 18-3B), its cytosolic concentration must be precisely controlled. The mitochondrion, endoplasmic reticulum, and the extracellular spaces act as Ca^{2+} storage tanks. We studied the Ca^{2+}-ATPases of the plasma membrane and endoplasmic reticulum in Section 18-3B. Here we consider the mitochondrial Ca^{2+}-transport systems.

Mitochondrial inner membrane systems separately mediate the influx and the efflux of Ca^{2+} (Fig. 20-4). The Ca^{2+} influx is driven by the inner mitochondrial membrane's membrane potential (negative inside), which attracts positively charged ions. The rate of influx varies with the external $[Ca^{2+}]$ because the K_M for Ca^{2+} transport by this system is greater than the cytosolic Ca^{2+} concentration. Ca^{2+} efflux is independently driven by the electron-transport-generated H^+ gradient across the inner mitochondrial membrane or, in heart mitochondria, by the Na^+ gradient. Ca^{2+} exits the matrix only in exchange for H^+ (or Na^+) so that this system is an antiport. This exchange process normally operates at its maximal velocity. *Mitochondria therefore act as a "buffer" for cytosolic Ca^{2+} (Fig. 20-5):* If cytosolic $[Ca^{2+}]$ rises, the rate of mitochondrial Ca^{2+} influx increases while that of Ca^{2+} efflux remains constant causing the mitochondrial $[Ca^{2+}]$ to increase while the cytosolic $[Ca^{2+}]$ decreases to its original level (its set-point). Conversely, a decrease in cytosolic $[Ca^{2+}]$ reduces the influx rate causing net efflux of $[Ca^{2+}]$ and an increase of cytosolic $[Ca^{2+}]$ back to the set-point.

Cytoplasmic Shuttle Systems "Transport" NADH Across the Inner Mitochondrial Membrane

Although most of the NADH generated by glucose oxidation is formed in the mitochondrial matrix via the citric acid cycle, that generated by glycolysis occurs in the cytosol. Yet, the inner mitochondrial membrane lacks an NADH transport protein. *Only the electrons from cytosolic NADH are transported into the mitochondrion by one of several ingenious "shuttle" systems.* In the **glycerophosphate shuttle** (Fig. 20-6) of insect flight muscle (the tissue with the largest known sustained power output—about the same power-to-weight ratio as a small automobile engine), 3-phosphoglycerol dehydrogenase catalyzes the oxidation of cytosolic NADH by dihydroxyacetone phosphate to yield NAD^+, which reenters glycolyis. The electrons of the resulting 3-phosphoglycerol are transferred to **flavoprotein dehydrogenase** to form $FADH_2$. This enzyme, which is situated on the inner mitochondrial membrane's outer surface,

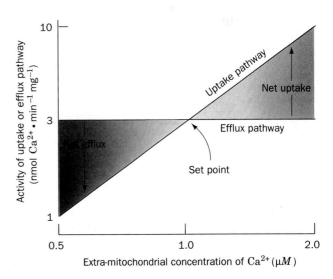

Figure 20-5
The regulation of cytosolic $[Ca^{2+}]$. The efflux pathway operates at a constant rate independent of $[Ca^{2+}]$, whereas the activity of the influx pathway varies with $[Ca^{2+}]$. At the set-point, the activities of the two pathways are equal and there is no net Ca^{2+} flux. An increase in cytosolic $[Ca^{2+}]$ results in net influx while a decrease in cytosolic $[Ca^{2+}]$ results in net efflux. Both effects lead to the restoration of the cytosolic $[Ca^{2+}]$. [After Nicholls, D., *Trends Biochem. Sci.* **6**, 37 (1981).]

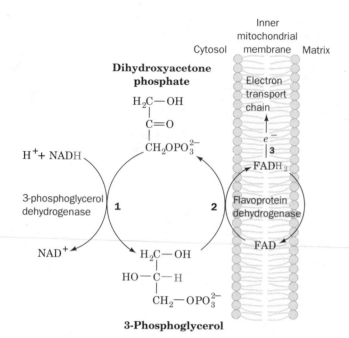

Figure 20-6
The glycerophosphate shuttle. The electrons of cytosolic NADH are transported to the mitochondrial electron-transport chain in three steps (shown in red as hydride transfers): **(1)** Cytosolic oxidation of NADH by dihydroxy-acetone phosphate catalyzed by 3-phosphoglycerol dehydrogenase. **(2)** Oxidation of 3-phosphoglycerol by flavoprotein dehydrogenase with reduction of FAD to $FADH_2$. **(3)** Reoxidation of $FADH_2$ with passage of electrons into the electron-transport chain.

supplies electrons to the electron-transport chain in a manner similar to that of succinate dehydrogenase (Section 20-2C). *The glycerophosphate shuttle therefore results in the synthesis of 2ATPs for every cytoplasmic NADH reoxidized.*

The **malate–aspartate shuttle** (Fig. 20-7) of mammalian systems is more complex but more energy-efficient than the glycerophosphate shuttle. Mitochondrial NAD^+ is reduced by cytosolic NADH through the intermediate reduction and subsequent regeneration of oxaloacetate. This process occurs in two phases of three reactions each:

Phase A (transport of electrons into the matrix):

1. NADH is reoxidized by cytosolic oxaloacetate through the action of cytosolic malate dehydrogenase.

2. The malate–α-ketoglutarate carrier transports the malate formed in Reaction 1 into the mitochondrial matrix in exchange for α-ketoglutarate.

3. In the mitochondrial matrix, NADH is regenerated from NAD^+ through oxidation of malate to oxaloacetate by mitochondrial malate dehydrogenase (Section 19-3H).

Phase B (regeneration of cytosolic oxaloacetate):

4. A transaminase (Section 24-1A) converts mitochondrial oxaloacetate to aspartate with concomitant conversion of glutamate to α-ketoglutarate.

5. Aspartate is transported from the matrix to the cytosol by the **glutamate–aspartate carrier** in exchange for cytosolic glutamate.

6. Cytosolic aspartate is converted to oxaloacetate by a transaminase in conjunction with α-ketoglutarate conversion to glutamate.

The electrons of cytosolic NADH are thereby transferred to mitochondrial NADH, which is subject to reoxidation via the electron-transport chain. *The malate–aspartate shuttle yields three ATPs for every cytosolic NADH, one more than the glycerophosphate shuttle.*

2. ELECTRON TRANSPORT

In the electron-transport process, the free energy of electron transfer from NADH and $FADH_2$ to O_2 via protein-bound redox centers is coupled to ATP synthesis. We begin our study of this process by considering its thermodynamics. We then examine the path of electrons through the redox centers of the system and discuss experiments used to unravel this path. Finally, we study the four complexes that comprise the electron-transport chain. In the next section we discuss how the free energy released by the electron-transport process is coupled to ATP synthesis.

A. Thermodynamics of Electron Transport

We can estimate the thermodynamic efficiency of electron transport through knowledge of standard reduction potentials. As we have seen in our thermodynamic considerations of oxidation–reduction reactions (Section 15-5), an oxidized substrate's electron affinity increases with its standard reduction potential, $\mathscr{E}°'$ [the voltage generated by the reaction of the half-cell under standard biochemical conditions ($1M$ reactants and products with $[H^+]$ defined as 1 at pH 7) relative to the standard hydrogen electrode; Table 15-4 lists the standard reduction potentials of several half-reactions of biochemical interest]. The standard reduction potential difference, $\Delta\mathscr{E}°'$, for a redox reaction involving any two half-reactions is therefore expressed:

$$\Delta\mathscr{E}°' = \mathscr{E}°'_{(e^-\text{-acceptor})} - \mathscr{E}°'_{(e^-\text{-donor})}$$

NADH Oxidation Is a Highly Exergonic Reaction

The half-reactions for O_2 oxidation of NADH are (Table 15-4):

$$NAD^+ + H^+ + 2e^- \rightleftharpoons NADH \qquad \mathscr{E}°' = -0.315 \text{ V}$$

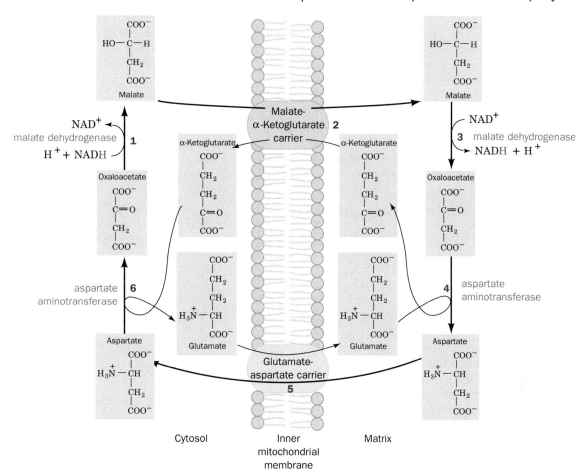

Figure 20-7
The malate–aspartate shuttle. The electrons of cytosolic NADH are transported to mitochondrial NADH (shown in red as hydride transfers) in Steps **1** to **3**. Steps **4** to **6** then serve to regenerate cytosolic oxaloacetate.

and

$$\tfrac{1}{2}O_2 + 2H^+ + 2e^- \rightleftharpoons H_2O \qquad \mathscr{E}°' = 0.815 \text{ V}$$

Since the O_2/H_2O half-reaction has the greater standard reduction potential and therefore the higher electron affinity, the NADH half-reaction is reversed so that NADH is the electron donor in this couple and O_2 the electron acceptor. The overall reaction is

$$\tfrac{1}{2} O_2 + NADH + H^+ \rightleftharpoons H_2O + NAD^+$$

so that

$$\Delta\mathscr{E}°' = 0.815 - (-0.315) = 1.130 \text{ V}$$

The standard free energy change for the reaction can then be calculated from Eq. [15.7]:

$$\Delta G°' = -n\mathscr{F}\Delta\mathscr{E}°'$$

where $\mathscr{F}$, the Faraday constant, is 96,494 $C \cdot mol^{-1}$ of electrons and n is the number of electrons transferred per mole of reactants. Thus, for NADH oxidation:

$$\Delta G°' = -2 \frac{mol \; e^-}{mol \; reactant} \times 96,494 \frac{C}{mol \; e^-} \times 1.13 \text{ J} \cdot C^{-1}$$

$$= -218 \text{ kJ} \cdot mol^{-1}$$

since $1 \text{ V} = 1 \text{ J} \cdot C^{-1}$. In other words, the oxidation of 1 mol of NADH by O_2 (the transfer of $2e^-$) under standard biochemical conditions is associated with the release of 218 kJ of free energy.

Electron Transport Is Thermodynamically Efficient

The standard free energy required to synthesize 1 mol of ATP from ADP + P_i is 30.5 $kJ \cdot mol^{-1}$. The standard free energy of oxidation of NADH by O_2, if coupled to ATP synthesis, is therefore sufficient to drive the formation of several moles of ATP. This coupling, as we shall see, is achieved by an electron-transport chain in which electrons are passed through three protein complexes containing redox centers with progressively greater electron affinity (increasing standard reduction potentials) instead of directly to O_2. *This allows the large overall free energy change to be broken up into three smaller packets, each of which is coupled with ATP synthesis in a process called* **oxidative phosphorylation.** *Oxidation of one NADH therefore results in the synthesis of three ATPs.* (Oxidation of $FADH_2$, whose entrance into the electron-transport chain is regulated by a fourth protein complex, is similarly coupled to the synthesis of two ATPs.) The thermodynamic efficiency of oxidative phosphorylation is therefore 3×30.5 kJ $\cdot$ mol$^{-1} \times 100/218$ kJ $\cdot$ mol$^{-1} = 42\%$ under standard biochemical conditions. However, under physiological conditions in active mitochondria (where the reactant and product concentrations as well as the pH deviate from standard conditions), this thermodynamic efficiency is thought to be $\sim 70\%$. In comparison, the energy efficiency of a typical automobile engine is $<30\%$.

B. The Sequence of Electron Transport

The free energy necessary to generate ATP is extracted from the oxidation of NADH and FADH$_2$ by the electron-transport chain, a series of four protein complexes through which electrons pass from lower to higher standard reduction potentials (Fig. 20-8). Electrons are carried from **Complexes I** and **II** to **Complex III** by **coenzyme Q** (**CoQ** or **ubiquinone;** so named because of its ubiquity in respiring organisms), and from Complex III to **Complex IV** by the peripheral membrane protein **cytochrome c** (Sections 6-3B and 8-3A).

Complex I catalyzes oxidation of NADH by CoQ:

$$\text{NADH} + \text{CoQ (oxidized)} \longrightarrow \text{NAD}^+ + \text{CoQ (reduced)}$$
$$\Delta\mathscr{E}^{\circ\prime} = 0.360 \text{ V} \qquad \Delta G^{\circ\prime} = -69.5 \text{ kJ} \cdot \text{mol}^{-1}$$

Complex III catalyzes oxidation of CoQ (reduced) by cytochrome c.

$$\text{CoQ (reduced)} + \text{cytochrome } c \text{ (oxidized)} \longrightarrow$$
$$\text{CoQ (oxidized)} + \text{cytochrome } c \text{ (reduced)}$$
$$\Delta\mathscr{E}^{\circ\prime} = 0.190 \text{ V} \qquad \Delta G^{\circ\prime} = -36.7 \text{ kJ} \cdot \text{mol}^{-1}$$

Complex IV catalyzes oxidation of cytochrome c (reduced) by O$_2$, the terminal electron acceptor of the electron-transport process.

$$\text{Cytochrome } c \text{ (reduced)} + \tfrac{1}{2} \text{ O}_2 \longrightarrow$$
$$\text{cytochrome } c \text{ (oxidized)} + \text{H}_2\text{O}$$
$$\Delta\mathscr{E}^{\circ\prime} = 0.580 \text{ V} \qquad \Delta G^{\circ\prime} = -112 \text{ kJ} \cdot \text{mol}^{-1}$$

The changes in standard reduction potential of an electron pair as it successively traverses Complexes I, III, and IV corresponds, at each stage, to sufficient free energy to power the synthesis of an ATP molecule.

Complex II catalyzes the oxidation of FADH$_2$ by CoQ.

$$\text{FADH}_2 + \text{CoQ (oxidized)} \longrightarrow \text{FAD} + \text{CoQ (reduced)}$$
$$\Delta\mathscr{E}^{\circ\prime} = 0.015 \text{ V} \qquad \Delta G^{\circ\prime} = -2.9 \text{ kJ} \cdot \text{mol}^{-1}$$

This redox reaction does not release sufficient free energy to synthesize ATP; it functions only to inject the electrons from FADH$_2$ into the electron-transport chain.

The Workings of the Electron-Transport Chain Have Been Elucidated through the Use of Inhibitors

Our understanding of the sequence of events in electron transport is largely based on the use of specific inhibitors. This sequence has been corroborated by measurements of the standard reduction potentials of

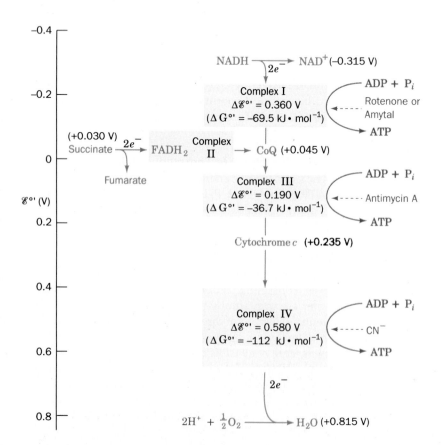

Figure 20-8
The mitochondrial electron-transport chain. The standard reduction potentials of its most mobile components (*green*) are indicated as are the points where sufficient free energy is harvested to synthesize ATP (*blue*) and the sites of action of several respiratory inhibitors (*red*).

Figure 20-9
The oxygen electrode consists of an Ag/AgCl reference electrode and a Pt electrode, both immersed in a KCl solution and in contact with the sample chamber through an O_2-permeable Teflon membrane. O_2 is reduced to H_2O at the Pt electrode, thereby generating a voltage with respect to the Ag/AgCl electrode that is proportional to the O_2 concentration in the sealed sample chamber. [After Cooper, T. G., *The Tools of Biochemistry*, p. 69, Wiley (1977).]

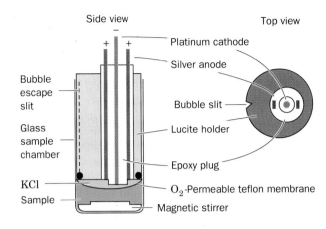

the redox components of each of the complexes as well as by determining the stoichiometry of the electron transport and coupled ATP synthesis.

The rate at which O_2 is consumed by a suspension of mitochondria is a sensitive measure of the functioning of the electron-transport chain. It is conveniently measured with an **oxygen electrode** (Fig. 20-9). Compounds that inhibit electron transport, as judged by their effect on O_2 disappearance in such an experimental system, have been invaluable experimental probes in tracing the path of electrons through the electron-transport chain and in determining the points of entry of electrons from various substrates. Among the most useful such substances are **rotenone** (a plant toxin used by Amazonian Indians to poison fish and which is also used as an insecticide), **amytal** (a barbiturate), **antimycin A** (an antibiotic), and **cyanide**.

Rotenone

Amytal

Cyanide

Antimycin A

The following experiment illustrates the use of these inhibitors:

A buffered solution containing excess ADP and P_i is equilibrated in the reaction vessel of an oxygen electrode. Reagents are then injected into the chamber and the O_2 consumption recorded (Fig. 20-10):

1. Mitochondria and **β-hydroxybutyrate** are injected into the chamber. Mitochondria mediate the NAD^+-linked oxidation of β-hydroxybutyrate (Section 23-3).

β-Hydroxybutyrate

Acetoacetate

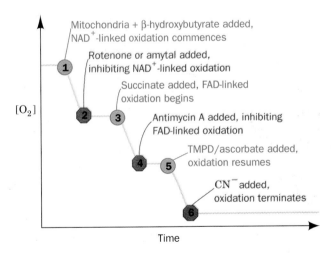

Figure 20-10
Oxygen electrode trace of a mitochondrial suspension containing excess ADP and P_i. At the numbered points, the indicated reagents are injected into the sample chamber and the resulting changes in $[O_2]$ are recorded. The numbers refer to the discussion in the text. [After Nicholls, D. G., *Bioenergetics*, p. 110, Academic Press (1982).]

As the resulting NADH is oxidized by the electron-transport chain with O_2 as the terminal electron acceptor, the O_2 concentration in the reaction mixture decreases.

2. Addition of rotenone or amytal completely stops the β-hydroxybutyrate oxidation.

3. Addition of succinate, which undergoes FAD-linked oxidation, causes the $[O_2]$ to resume its decrease. Electrons from $FADH_2$ are therefore still able to reduce O_2 in the presence of rotenone; that is, *electrons from $FADH_2$ enter the electron-transport chain after the rotenone-blocked step.*

4. Addition of antimycin A inhibits electron transport from $FADH_2$.

5. Although NADH and $FADH_2$ are the electron-transport chain's two physiological electron donors, nonphysiological reducing agents can also be used to probe the flow of electrons. **Tetramethyl-*p*-phenylenediamine (TMPD)** is an ascorbate-reducible redox carrier that transfers electrons directly to cytochrome *c*.

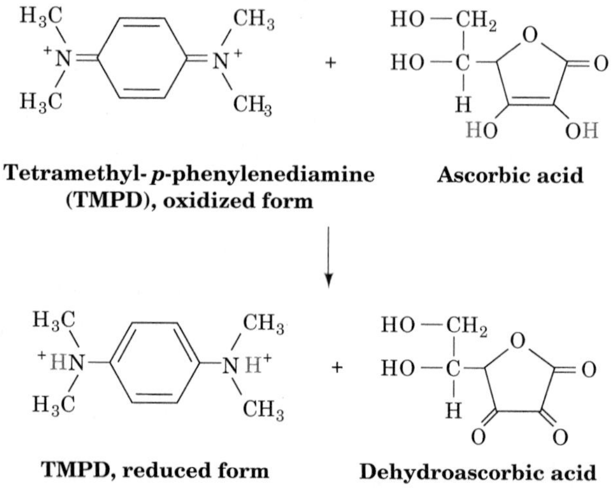

Tetramethyl-*p*-phenylenediamine (TMPD), oxidized form   **Ascorbic acid**

TMPD, reduced form   **Dehydroascorbic acid**

Addition of TMPD and ascorbate to the antimycin A-inhibited reaction mixture results in resumption of oxygen consumption; *evidently there is a third point for electrons to enter the electron-transport chain.*

6. The addition of CN^- completely inhibits oxidation of all three electron donors, indicating that it blocks the electron-transport chain after the third point of entry of electrons.

Experiments such as these established the order of electron flow through the electron-transport chain complexes and the positions blocked by various electron-transport inhibitors (Fig. 20-8). This order was confirmed and extended by observations that the standard reduction potentials of the redox carriers forming the electron-transport chain complexes are very close to the standard reduction potentials of their electron donor

Table 20-1

Standard Reduction Potentials of Electron-Transport Chain Components in Resting Mitochondria

Component	$\mathscr{E}^{\circ\prime}$ (V)
NADH	−0.315
Complex I (NADH-CoQ reductase; 850 kD, 26 subunits):	
FMN	?
(Fe–S)N-1a	−0.380
(Fe–S)N-1b	−0.250
(Fe–S)N-2	−0.030
(Fe–S)N-3,4	−0.245
(Fe–S)N-5,6	−0.270
Succinate	0.030
Complex II (succinate-CoQ reductase; 127 kD, 5 subunits):	
FAD	−0.040
(Fe–S)S-1	−0.030
(Fe–S)S-2	−0.245
(Fe–S)S-3	0.060
Cytochrome b_{560}	−0.080
Coenzyme Q	0.045
Complex III (CoQ-cytochrome *c* reductase; 280 kD, 10 subunits):	
Cytochrome b_K	0.030
Cytochrome b_T	−0.030
(Fe–S)	0.280
Cytochrome c_1	0.215
Cytochrome *c*	0.235
Complex IV (cytochrome *c* oxidase; 160–170 kD, 7–8 subunits):	
Cytochrome *a*	0.210
Cu_A	0.245
Cu_B	0.340
Cytochrome a_3	0.385
O_2	0.815

Source: Wilson, D. F., Erecińska, M., and Dutton, P. L., *Annu. Rev. Biophys. Bioeng.* **3**, 205 and 208 (1974) *and* Wilson, D. F., in Bittar, E. E. (Ed.), *Membrane Structure and Function*, Vol. 1, p. 160, Wiley (1980).

substrates (Table 20-1). *The three jumps in reduction potential between NADH, CoQ, cytochrome c, and O_2 are each of sufficient magnitude to drive ATP synthesis.* Indeed, these redox potential jumps correspond to the points of inhibition of rotenone (or amytal), antimycin A, and CN^-.

Phosphorylation and Oxidation Are Rigidly Coupled

The foregoing thermodynamic studies suggest that oxidation of NADH, $FADH_2$, and ascorbate by O_2 are

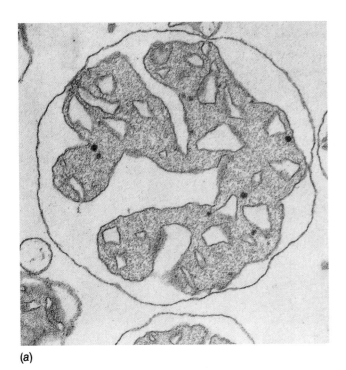

(a)

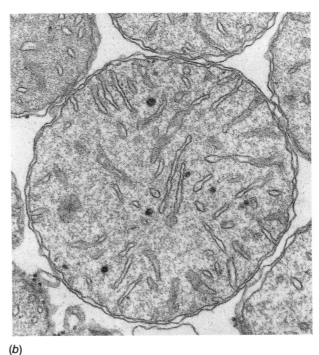

(b)

Figure 20-11
Electron micrographs of mouse liver mitochondria in (a) the actively respiring state and (b) the resting state. The cristae in actively respiring mitochondria are far more condensed than they are in resting mitochondria. [Courtesy of Charles Hackenbrock, University of North Carolina Medical School.]

associated with the synthesis of three, two, and one ATPs, respectively. This stoichiometry, called the **P/O ratio,** has been confirmed experimentally through measurements of O_2 uptake by resting and active mitochondria. In a typical experiment, a suspension of mitochondria (isolated by differential centrifugation after cell disruption; Section 5-1B) containing an excess of P_i but no ADP was incubated in an oxygen electrode reaction chamber. *Oxidation and phosphorylation are closely coupled in well-functioning mitochondria so that electron transport can only occur if ADP is being phosphorylated (Section 20-3).* Indeed, mitochondrial metabolism is so tightly regulated that even the appearances of actively respiring and resting mitochondria are greatly different (Fig. 20-11). Since no ADP was present in the reaction mixture, the mitochondria were resting and the O_2 consumption rate was minimal (Fig. 20-12, Region 1). The system was then manipulated as follows:

(a) 90 μmol of ADP and an excess of β-hydroxybutyrate (an NAD^+-linked substrate) were added. The mitochondria immediately entered the active state and the rate of oxygen consumption increased (Fig. 20-12, Region 2) and maintained this elevated level until all the ADP was phosphorylated. The mitochondria then returned to the resting state (Fig. 20-12, Region 3). Phosphorylation of 90 μmol of ADP under these conditions consumed 15 μmol of O_2. Since the oxidation of NADH by O_2 consumes

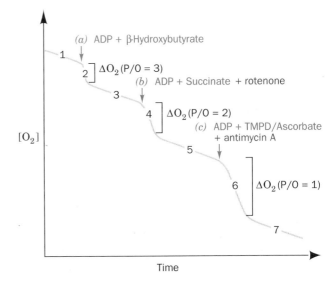

Figure 20-12
Stoichiometry of coupled oxidation and phosphorylation (the P/O ratio) with different electron donors. Mitochondria are incubated in excess phosphate buffer in the sample chamber of an oxygen electrode. (a) 90 μmol of ADP and excess β-hydroxybutyrate are added. Respiration continues until all the ADP is phosphorylated. ΔO_2 in Region 2 is 15 μmol, corresponding to 30 μmol of NADH oxidized; P/O = 90/30 = 3. (b) 90 μmol of ADP and excess succinate are added together with rotenone to inhibit electron transfer from NADH. ΔO_2 in Region 4 is 22.5 μmol, corresponding to 45 μmol of $FADH_2$ oxidized; P/O = 90/45 = 2. (c) 90 μmol of ADP and excess TMPD/ascorbate are added with antimycin A to inhibit electron transfer from $FADH_2$. ΔO_2 in Region 6 is 45-μmol, corresponding to 90 μmol of ascorbate oxidized; P/O = 90/90 = 1.

twice as many moles of NADH as O_2, the P/O ratio for NADH reoxidation at Region 2 is 90 μmol of ADP/(2 × 15 μmol of O_2) = 3; that is, *3 mol of ADP are phosphorylated per mole of NADH oxidized.*

(b) The experiment was continued by inhibiting electron transfer from NADH by rotenone and adding an additional 90 μmol of ADP (Fig. 20-12, Region 4), this time together with an excess of the FAD-linked substrate succinate. Oxygen consumption again continued until all the ADP was phosphorylated, and the system again returned to the resting state (Fig. 20-12, Region 5). Calculation of the P/O ratio for $FADH_2$ oxidation yielded the value 2; that is, *2 mol of ADP are phosphorylated per mole of $FADH_2$ oxidized.*

(c) In the same manner, *the oxidation of ascorbate/TMPD yielded a P/O ratio of 1 (Fig. 20-12; Regions 6 and 7).*

These conclusions agree with the inhibitor studies indicating that there are three entry points for electrons into the electron-transport chain and with the standard reduction potential measurements exhibiting three potential jumps, each sufficient to provide the free energy for ATP synthesis (Fig. 20-8). How the free energy of electron transport is actually coupled to ATP synthesis, a subject of active research, is discussed in Section 20-3. We first examine the structures of the four respiratory complexes in order to understand how they are related to the function of the electron-transport chain. Keep in mind, however, that as in most areas of biochemistry, this field is under intense investigation and much of the information we need for a complete understanding of these relationships has yet to be uncovered.

C. Components of the Electron-Transport Chain

Many of the proteins embedded in the inner mitochondrial membrane are organized into the four respiratory complexes of the electron-transport chain. Each complex consists of several protein components that are associated with a variety of redox-active prosthetic groups with successively increasing reduction potentials (Table 20-1). The complexes are all laterally mobile within the inner mitochondrial membrane; they do not appear to form any stable higher structures. Indeed, they are not present in equimolar ratios. In the following paragraphs, we examine their structures and the agents that transfer electrons between them. Their relationships are summarized in Fig. 20-13.

1. Complex I (NADH–Coenzyme Q Reductase)

Complex I passes electrons from NADH to CoQ. This probably largest protein component of the inner mitochondrial membrane (850 kD) contains one molecule of

flavin mononucleotide (**FMN;** a redox-active prosthetic group that differs from FAD only by the absence of the AMP group) and six to seven **iron–sulfur clusters** that participate in the electron-transport process (Table 20-1).

Iron–Sulfur Clusters Are Redox Active

Three types of iron–sulfur clusters are known to occur as prosthetic groups of **iron–sulfur proteins (nonheme iron proteins):**

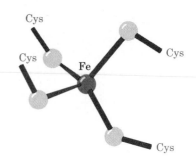

[Fe–S]

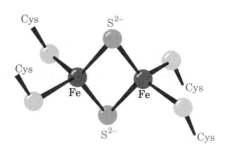

[2Fe–2S]

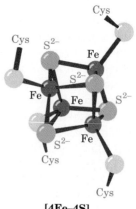

[4Fe–4S]

The two most common types, designated **[2Fe–2S]** and **[4Fe–4S] clusters,** consist of equal numbers of iron and sulfide ions and are both coordinated to four protein Cys sulfhydryl groups. One means of identifying these clus-

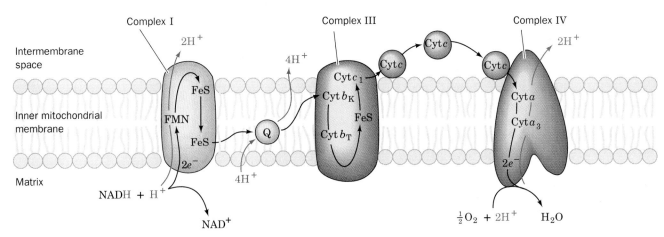

Figure 20-13
A diagram of the mitochondrial electron-transport chain indicating the pathway of electron transfer (*black*) and proton pumping (*red*). Electrons are transferred between Complexes I and III by the membrane-soluble CoQ and between Complexes III and IV by the peripheral membrane protein cytochrome *c*. Complex II (not shown) transfers electrons from succinate to CoQ.

ters utilizes the fact that their sulfide ions are acid labile: They are released as H_2S near pH 1. The **[Fe–S] cluster,** which has only been found in bacteria, consists of a single Fe atom liganded to four Cys residues. Note that the Fe atoms in all three types of clusters are each coordinated by four S atoms, which are more or less tetrahedrally disposed around the Fe. (A **[3Fe–4S] cluster,** which has been found in only one bacterial species, resembles a [4Fe–4S] cluster that lacks an Fe atom.) *The oxidized and reduced states of all iron–sulfur clusters differ by one formal charge regardless of their number of Fe atoms.* This is because the Fe atoms in each cluster form a conjugated system and thus can have oxidation states between the +2 and +3 values possible for individual Fe atoms.

Iron–sulfur proteins also occur in the photosynthetic electron-transport chains of plants and bacteria (Section 22-2); indeed, photosynthetic electron-transport chains are thought to be the evolutionary precursors of oxidative electron-transport chains (Section 1-4C). The X-ray structure of a bacterial iron–sulfur protein, **ferredoxin,** which contains two [4Fe–4S] clusters, reveals how the clusters are attached to the polypeptide chain (Fig. 20-14).

The Coenzymes of Complex I

FMN and CoQ, the coenzymes of Complex I, can each adopt three oxidation states (Fig. 20-15). Although NADH can only participate in a two-electron transfer, both FMN and CoQ are capable of accepting and donating either one or two electrons because their semiquinone forms are stable. In contrast, the cytochromes of Complex III (see below), to which reduced CoQ passes its electrons, are only capable of one-electron reductions. *FMN and CoQ therefore provide an electron conduit between the two-electron donor NADH and the one-electron acceptors, the cytochromes.*

CoQ's hydrophobic tail makes it soluble in the inner mitochondrial membrane's lipid bilayer. In mammals, this tail consists of 10 C_5 isoprenoid units and hence the coenzyme is designated Q_{10}. In other organisms, CoQ may have only 6 (Q_6) or 8 (Q_8) isoprenoid units.

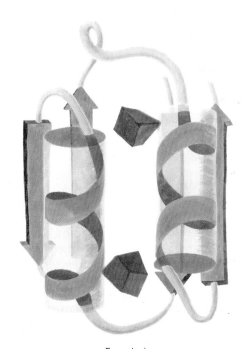

Ferredoxin

Figure 20-14
Structure of ferredoxin from *Peptococcus aerogenes*, a 54-residue protein containing two [4Fe–4S] clusters (*red*). [After a drawing provided by Jane Richardson, Duke University. X-ray structure determined by Elinor Adman, Larry Sieker, and Lyle Jensen.]

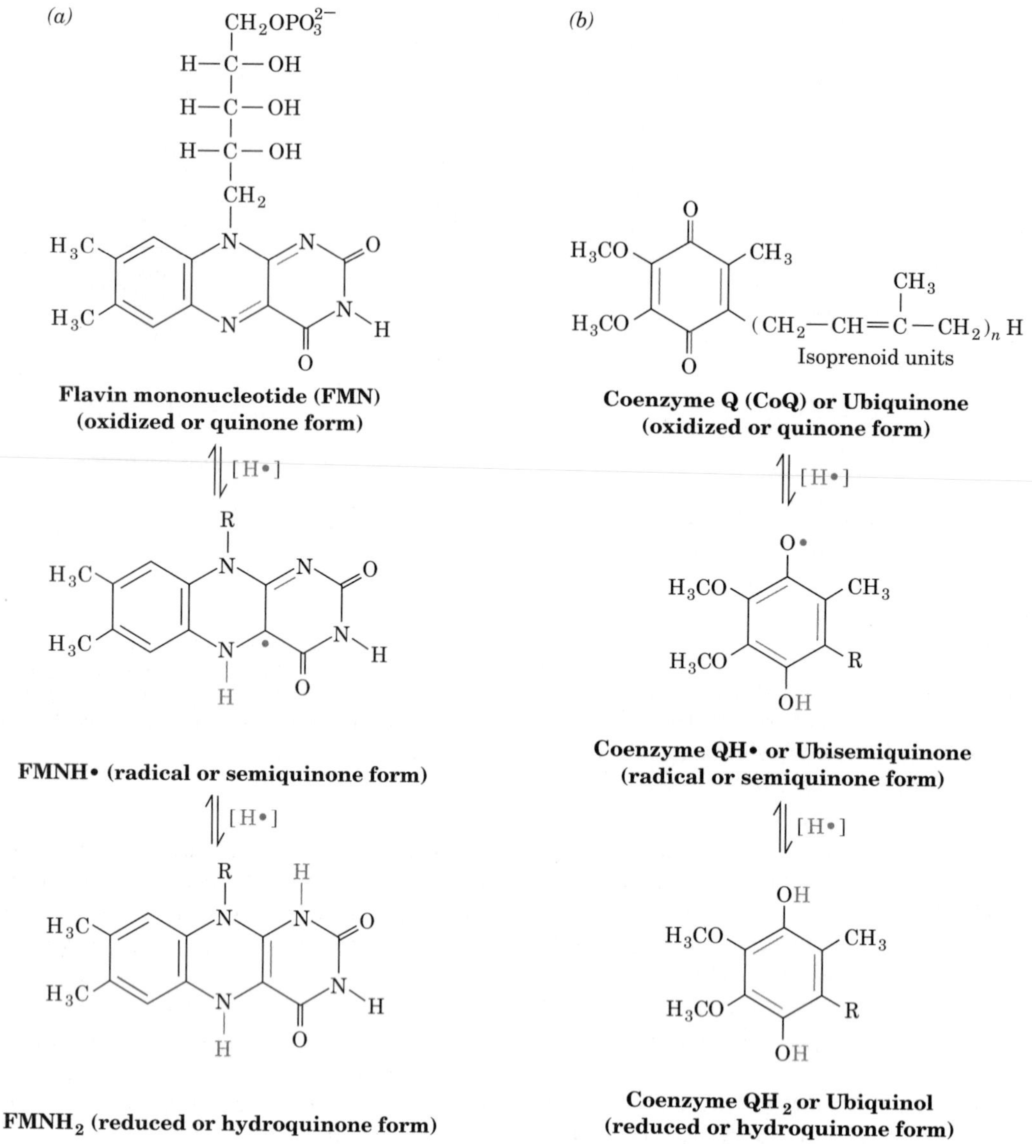

(a)

$$CH_2OPO_3^{2-}$$

Flavin mononucleotide (FMN)
(oxidized or quinone form)

⇅ [H•]

FMNH• (radical or semiquinone form)

⇅ [H•]

FMNH$_2$ (reduced or hydroquinone form)

(b)

Coenzyme Q (CoQ) or Ubiquinone
(oxidized or quinone form)

Isoprenoid units

⇅ [H•]

Coenzyme QH• or Ubisemiquinone
(radical or semiquinone form)

⇅ [H•]

Coenzyme QH$_2$ or Ubiquinol
(reduced or hydroquinone form)

Figure 20-15
The oxidation states of *(a)* FMN and *(b)* CoQ. Both coenzymes form stable semiquinone free radical states.

2. Complex II (Succinate–Coenzyme Q Reductase)

Complex II, which contains the dimeric citric acid cycle enzyme succinate dehydrogenase (Section 19-3F) and three other small hydrophobic subunits, passes electrons from succinate to CoQ. It does so with the participation of a covalently bound FAD, one [4Fe–4S] cluster, two [2Fe–2S] clusters, and one cytochrome b_{560} (Table 20-1). We discuss the structures of the cytochromes in connection with that of Complex III below. [One of Complex II's iron–sulfur clusters has a standard reduction potential that is too negative (-0.245 V) to accept electrons from succinate; its function is unknown.]

The standard redox potential for electron transfer from succinate to CoQ (Fig. 20-8) is insufficient to pro-

vide the free energy necessary to drive ATP synthesis. The complex is, nevertheless, important because it allows these relatively high-potential electrons to enter the electron-transport chain.

3. Complex III (Coenzyme Q–Cytochrome *c* Reductase)

*Complex III passes electrons from reduced CoQ to cytochrome c. It contains two **b-cytochromes**, one **cytochrome c_1**, and one [2Fe–2S] cluster (Table 20-1).*

Cytochromes: Electron-Transport Heme Proteins

Cytochromes, whose function was elucidated in 1925 by David Keilin, are redox-active proteins that occur in all organisms except a few types of obligate anaerobes.

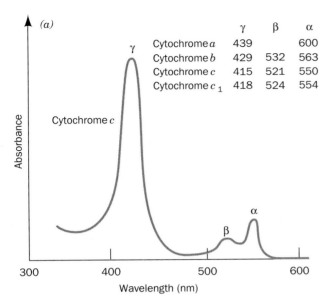

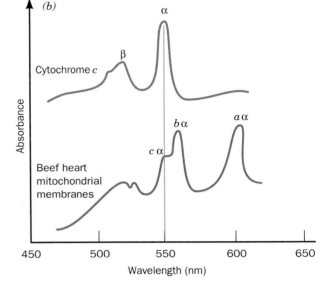

Figure 20-16
Visible absorption spectra of cytochromes. (a) Absorption
spectrum of reduced cytochrome c showing its characteristic
α, β, and γ (Soret) absorption bands. The absorption
maxima for cytochromes a, b, c, and c_1 are listed. (b) The
three separate α bands in the visible absorption spectra of

beef heart mitochondrial membranes (*below*) indicate the
presence of cytochromes a, b, and c. The spectrum of
purified cytochrome c (*above*) is provided for reference.
[After Nicholls, D. G., *Bioenergetics*, p. 104, Academic Press
(1982).]

These proteins contain heme groups, which reversibly
alternate between their Fe(II) and Fe(III) oxidation states
during electron transport.

The heme groups of the reduced [Fe(II)] cytochromes
have prominent visible absorption spectra consisting of
three peaks: the α, β, and γ (**Soret**) bands (Fig. 20-16a).
The wavelength of the α peak, which varies characteristically with the particular reduced cytochrome species (it
is absent in oxidized cytochromes), is useful in differentiating the various cytochromes. Accordingly, the spectra of mitochondrial membranes (Fig. 20-16b) indicate
that they contain three cytochrome species, cytochromes a, b, and c.

Within each group of cytochromes, different heme
group environments may be characterized by slightly
different α peak wavelengths. For example, Complex III
has two b-type cytochrome hemes: That absorbing
maximally at 562 nm is referred to as b_K or b_{562}, whereas
that absorbing maximally at 566 nm is referred to as b_T
or b_{566}. (The second type of nomenclature is a recent
adoption that identifies a cytochrome with the wavelength at which its α band absorbance is maximal. Previously, cytochromes were identified nondescriptively
with either numbers or letters.) Complex II has previously been noted to contain cytochrome b_{560}.

Each group of cytochromes contains a differently
substituted porphyrin ring (Fig. 20-17a) coordinated
with the redox-active iron atom. The b-type cytochromes contain **protoporphyrin IX,** which also
occurs in hemoglobin (Section 9-1A). The heme group

of c-type cytochromes differs from protoporphyrin IX in
that its vinyl groups have added Cys sulfhydryls across
their double bonds to form thioether linkages to the
protein. Heme a contains a long hydrophobic tail of
isoprene units attached to the porphyrin, as well as a
formyl group in place of a methyl substituent. The axial
ligands of the heme iron also vary with the cytochrome
type. In cytochromes a and b, both ligands are His residues, whereas in cytochrome c, one is His and the other
is Met (Fig. 20-17b).

Complex III is arranged asymmetrically in the inner
mitochondrial membrane as judged from various chemical labeling studies. Both cytochrome c_1 and the nonheme iron protein (often called the **Rieske iron–sulfur
protein** after its discoverer, John Rieske) are located on
the membrane's outer surface while cytochrome b is a
trans-membrane protein. Cytochrome b is a particularly
interesting protein because it contains both b-type cytochrome hemes, b_{562} and b_{566}, associated with a single
polypeptide chain.

Since few X-ray structures of integral membrane proteins have yet been determined, investigators have attempted to predict the gross structures of such proteins
from their amino acid sequences. The amino acid sequence of cytochrome b, which is encoded by mitochondrial DNA, has been deduced from the base sequence of this DNA (Section 30-1E). The polypeptide
chains from various species are 380 to 385 amino acids
long and exhibit considerable sequence homology. They
have nine > 20 residue stretches of predominantly hy-

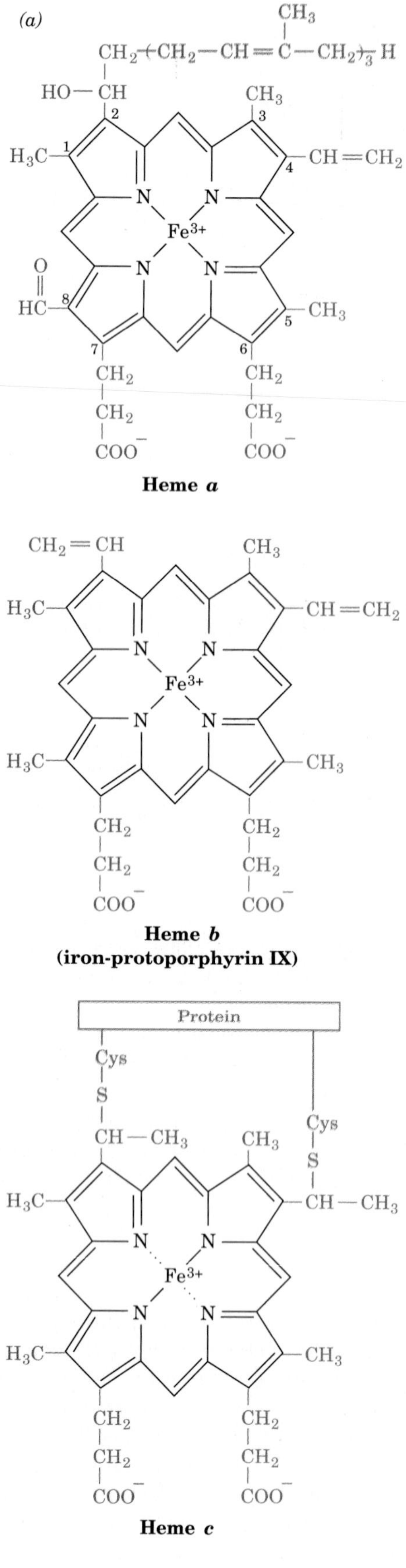

(a)

Heme a

Heme b
(iron-protoporphyrin IX)

Protein

Heme c

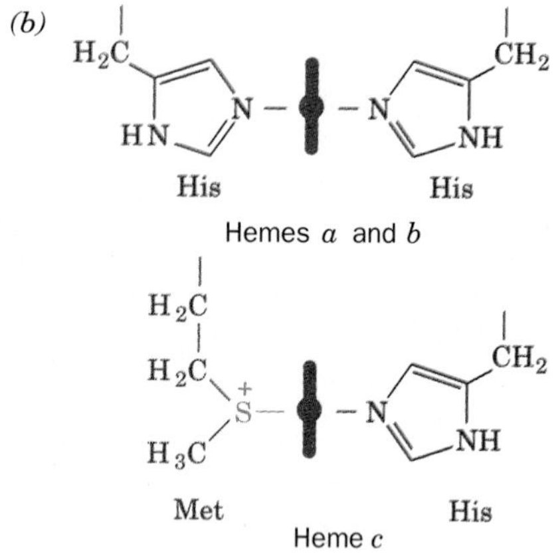

(b)

Hemes *a* and *b*

Heme *c*

Figure 20-17
The (a) chemical structures and (b) axial liganding of the heme groups contained in cytochromes *a*, *b*, and *c*.

drophobic residues that are predicted to form stable helices that can span the membrane bilayer. Therefore, it is likely that the polypeptide chain of cytochrome *b* spans the membrane nine times (Fig. 20-18). The two heme groups are postulated to be coordinated to four invariant His residues, located on Helices II and V at positions 82 and 198, and 96 and 183. As more aspects of this structure are revealed, the mechanism by which Complex III preserves the free energy of electron transfer from $CoQH_2$ to cytochrome *c* for ATP synthesis will presumably be revealed.

4. Cytochrome *c*

Cytochrome *c* is a peripheral membrane protein of known crystal structure (Fig. 8-14c) that is loosely bound to the outer surface of the inner mitochondrial membrane. *It alternately binds to cytochrome c_1 of Complex III and to cytochrome c oxidase (Complex IV) and thereby functions to shuttle electrons between them.* Reduced hemes are highly reactive entities; they can transfer electrons over distances of 10 to 20 Å at physiologically significant rates. Hence cytochromes, in a sense, have the opposite function of enzymes: Rather than persuading unreactive substrates to react, they must prevent their hemes from transferring electrons nonspecifically to other cellular components. This, no doubt, is why these hemes are almost entirely enveloped by protein. However, cytochromes must also provide a path for electron transfer to an appropriate partner.

Binding Site Structure and Electron-Transfer Function

Cytochrome *c*'s binding site involves several invariant Lys residues that lie in a ring around the exposed

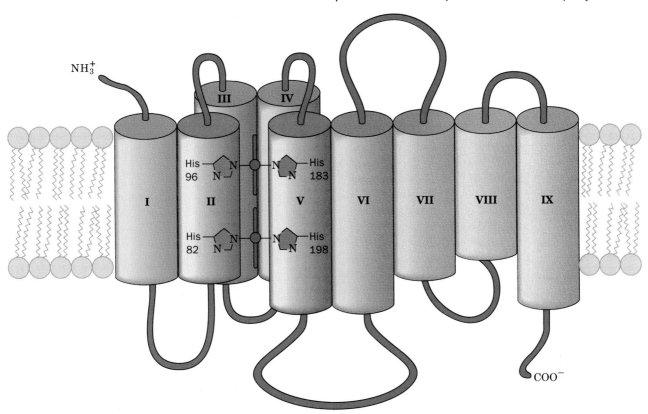

Figure 20-18
Predicted secondary structure of mitochondrial cytochrome *b* based on its amino acid sequences in several species. The polypeptide has nine stretches of nonpolar amino acids that can form helices long enough to traverse the membrane (represented as cylinders). The invariant His residues 96 and 183, as well as 82 and 198, form the axial ligands to the protein's two heme groups, thereby bridging Helices II and V. [After Barber, J., *Trends Biochem. Sci.* **9,** 210 (1984).]

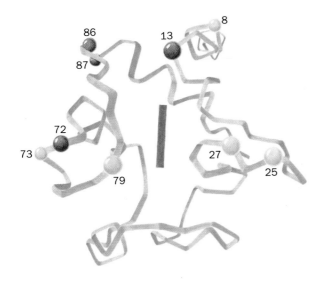

Figure 20-19
Ribbon diagram of cytochrome *c* showing the Lys residues involved in intermolecular complex formation with cytochrome *c* oxidase or reductase as inferred from chemical modification studies. Dark and light blue balls, respectively, mark the positions of Lys residues whose ε-amino groups are strongly and less strongly protected by cytochrome *c* oxidase and reductase against acetylation by acetic anhydride. Note that these Lys residues form a ring around the heme (*solid bar*) on one face of the protein [After Mathews, F. S., *Prog. Biophys. Mol. Biol.* **45,** 45 (1986).]

edge of its otherwise buried heme group (Fig. 20-19). This binding site has been identified by **differential labeling:** Treatment of cytochrome *c* with acetic anhydride (which acetylates Lys residues) in the presence and absence of cytochrome c_1 demonstrated that cytochrome c_1 completely shields these cytochrome *c* Lys residues. The reactivities of other cytochrome *c* Lys residues that are distant from the exposed heme edge are unaffected by complex formation. Nearly identical results were obtained when cytochrome c_1 was replaced by cytochrome *c* oxidase. Evidently, both these proteins have negatively charged sites that are complementary to the ring of positively charged Lys residues on cytochrome *c* (see below).

The mechanism of electron transfer between proteins and the role of the protein in this process are areas of active research. Thomas Poulos and Joseph Kraut have proposed an intriguing structural model for this electron-transfer process based on the known crystal structures of cytochrome *c* and yeast **cytochrome *c* peroxidase (CCP).** CCP, a heme-containing protein, catalyzes the reduction of organic hydroperoxides (ROOH) in a three-step reaction cycle that results in the oxidation of two molecules of cytochrome *c*:

$$CCP + ROOH + 2H^+ \longrightarrow CCP(I) + ROH + H_2O$$

$$CCP(I) + cyt\ c\ (Fe^{2+}) \longrightarrow CCP(II) + cyt\ c\ (Fe^{3+})$$

$$CCP(II) + cyt\ c\ (Fe^{2+}) \longrightarrow CCP + cyt\ c\ (Fe^{3+})$$

Here CCP(I) represents a $2e^-$ oxidized state and CCP(II) represents a $1e^-$ oxidized state of CCP.

The atomic models of CCP and yeast cytochrome *c* (which closely resembles that of the highly homologous tuna cytochrome *c*; Section 8-3A) fit together with remarkable precision. A convincing feature of this hypothetical complex is that CCP bears a ring of negatively charged Asp residues that, as we hypothesized above, is in complementary association with the ring of positively charged Lys residues that surrounds cytochrome *c*'s exposed heme edge. Of particular mechanistic interest is that the hemes, whose closest mutual approach is ~18 Å (as close as possible without severe conformational changes in the proteins), are parallel to each other as well as to the intervening rings of CCP His 181 and the invariant Phe 82 of cytochrome *c* (Fig. 20-20). His 181 is also involved in a hydrogen bonding network with the CCP heme propionate, Arg 48 and Asp 37, one of the residues proposed to form the CCP–cytochrome *c* interface. Poulos and Kraut propose that this assembly of four parallel conjugated rings and hydrogen bonds provides a conduit through which electrons are efficiently transferred between the two proteins. Indeed, Brian Hoffman has shown that changing cytochrome *c* Phe 82 to an aliphatic residue such as Gly, Ser, or Leu by genetic engineering techniques reduces the rate that cytochrome *c* transfers an electron to CCP by a factor of 10^4. Presumably similar interactions guide the flow of electrons between cytochrome *c* and Complexes III and IV.

5. Complex IV (Cytochrome *c* Oxidase)

Cytochrome c oxidase catalyzes the one-electron oxidations of four consecutive reduced cytochrome c molecules and the concomitant four-electron reduction of one O_2 molecule:

$$4\text{Cytochrome } c^{2+} + 4H^+ + O_2 \longrightarrow$$
$$4\text{cytochrome } c^{3+} + 2H_2O$$

Mammalian Complex IV is an ~200-kD trans-membrane protein composed of 6 to 13 subunits (Fig. 20-21*a*) whose largest and most hydrophobic subunits, I, II, and III, are encoded by mitochondrial DNA. These subunits collectively have 18 hydrophobic segments that probably form membrane-spanning helices similar to those of cytochrome *b* (Fig. 20-18). Subunits I and II of this complex contain all four of its redox-active centers: two *a*-type hemes (*a* and a_3; Fig. 20-17) and two Cu atoms that alternate between their +1 and +2 oxidation states

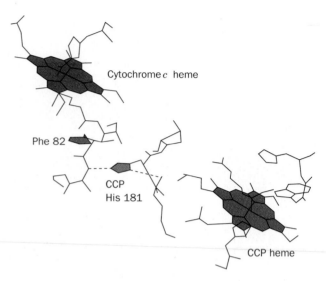

Figure 20-20
The proposed electron-transfer conduit in the hypothetical cytochrome *c* · cytochrome *c* peroxidase complex. The two hemes, together with the intervening cytochrome *c* Phe 82 and CCP His 181, form a system of parallel conjugated rings (*shaded*) and hydrogen bonds (*dashed lines*) that are postulated to guide the transfer of an electron between the two heme irons. [After Poulos, T. L. and Kraut, J., *J. Biol. Chem.* **255**, 10327 (1980).]

(Table 20-1). Heme *a* and the Cu atom designated Cu_A (alternatively, Cu_a) are of low potential (~0.24 V), whereas heme a_3 and Cu_B (alternatively, Cu_{a3}) are of higher potential (~0.34 V; Table 20-1). A variety of spectroscopic evidence indicates that Cu_A is liganded to Subunit II through two Cys and one His residue and a fourth ligand (also a nitrogen), and is part of the cytochrome *c*-binding site. Heme a_3 and Cu_B are bridged, most probably by an S atom, to form a binuclear complex that comprises the O_2-binding site (Fig. 20-21*b*).

Cytochrome *c* oxidase's interaction site with cytochrome *c* presumably has several Asp and/or Glu residues that interact with the above-discussed ring of Lys residues surrounding cytochrome *c*'s heme crevice. Indeed, differential labeling of cytochrome *c* oxidase's carboxyl groups in the presence and absence of cytochrome *c*, demonstrated that cytochrome *c* shields the invariant residues Asp 112, Glu 114, and Glu 198 of cytochrome oxidase Subunit II. Glu 198 is located between the two Cys residues of Subunit II that are thought to ligand Cu_A. This observation supports the spectroscopic evidence that places the cytochrome *c*-binding site on Subunit II in close proximity to Cu_A.

3. OXIDATIVE PHOSPHORYLATION

The endergonic synthesis of ATP from ADP and P_i in mitochondria, which, as we shall see, is catalyzed by

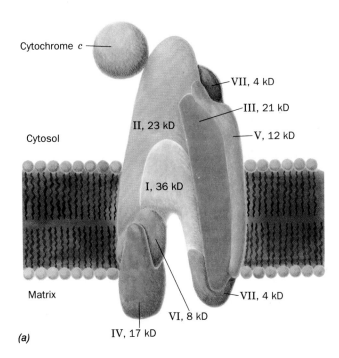

(a)

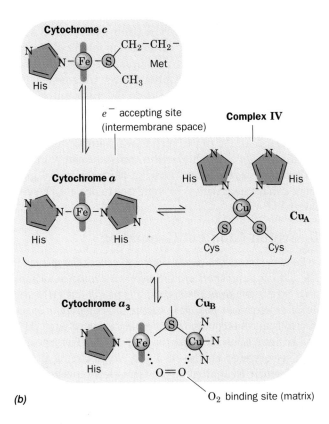

(b)

Figure 20-21
Cytochrome *c* oxidase. (*a*) The structure and orientation of cytochrome *c* oxidase in relation to the inner mitochondrial membrane. Only one monomer unit of the dimeric protein complex is shown. The overall Y shape of the monomer was elucidated by electron microscopy combined with image reconstruction techniques. Spatial relationships among the subunits were deduced from the binding of specific antibodies together with the results of cross-linking studies. The complex, evidently, contains two differently located copies of subunit VII. The subunit molecular masses are indicated. (*b*) The flow of electrons from cytochrome *c* through the four redox-active centers of cytochrome *c* oxidase to O_2. The most probable ligands of the heme Fe and Cu ions are indicated although they have not been determined unambiguously in every case. O_2 binds to the cytochrome a_3–Cu_B binuclear complex where, in a poorly understood process, it is reduced to $2H_2O$ by the stepwise transfer of four electrons from cytochrome *a* and Cu_A together with the acquisition of four protons. The entire reaction is extremely fast; it goes to completion in ∼1 ms at room temperature.

ATP synthase (Complex V), is driven by the electron-transport process. Yet, since Complex V is physically distinct from the proteins mediating electron transport (Complexes I–IV), *the free energy released by electron transport must be conserved in a form that ATP synthase can utilize.* Such energy conservation is referred to as **energy coupling** or **energy transduction.**

The physical characterization of energy coupling has proven to be surprisingly elusive; many sensible and often ingenious ideas have failed to withstand the test of experimental scrutiny. In this section we first examine some of the hypotheses that have been formulated to explain the coupling of electron transport and ATP synthesis. We shall then explore the coupling mechanism that has garnered the most experimental support, analyze the mechanism by which ATP is synthesized by ATP synthase, and finally, discuss how electron transport and ATP synthesis can be uncoupled.

A. Energy Coupling Hypotheses

In the more than 50 years that electron transport and oxidative phosphorylation have been studied, numerous mechanisms have been proposed to explain how these processes are coupled. In the following paragraphs, we examine the mechanisms that have received the greatest experimental attention:

1. The chemical coupling hypothesis

In 1953, Edward Slater formulated the **chemical-coupling hypothesis,** in which he proposed that electron transport yields reactive intermediates whose subsequent breakdown drives oxidative phosphorylation. We have seen, for example, that such a mechanism is responsible for ATP synthesis in glycolysis (Sections 16-2F and G). Thus, the exergonic oxidation of glyceraldehyde-3-phosphate by NAD^+ yields 1,3-bisphosphoglycerate, a reactive ("high-energy") acyl phosphate whose phosphoryl group is then transferred to ADP to form ATP in the phosphoglycerate kinase reaction. The difficulty with such a mechanism for oxidative phosphorylation,

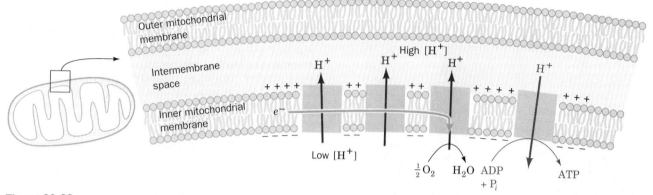

Figure 20-22

The coupling of electron transport (*green arrow*) and ATP synthesis by the generation of a proton electrochemical gradient across the inner mitochondrial membrane. H$^+$ is pumped out of the mitochondrion during electron transport (*blue arrows*) and its exergonic return powers the synthesis of ATP (*red arrows*).

which has largely caused it to be abandoned, is that despite intensive efforts in numerous laboratories over many years, no appropriate reactive intermediates have been identified.

2. The conformational coupling hypothesis

The **conformational coupling hypothesis,** which Paul Boyer formulated in 1964, proposes that electron transport causes proteins of the inner mitochondrial membrane to assume "activated" or "energized" conformational states. These proteins are somehow associated with ATP synthase such that their relaxation back to the deactivated conformation drives ATP synthesis. As with the chemical coupling hypothesis, the conformational coupling hypothesis has found little experimental support. However, conformational coupling of a different sort is probably involved in ATP synthesis (Section 20-3C).

3. The chemiosmotic hypothesis

The **chemiosmotic hypothesis,** proposed in 1961 by Peter Mitchell, has spurred considerable controversy as well as research, and now appears to be the model most consistent with the experimental evidence. It postulates that *the free energy of electron transport is conserved by pumping H$^+$ from the mitochondrial matrix to the intermembrane space so as to create an electrochemical H$^+$ gradient across the inner mitochondrial membrane. The electrochemical potential of this gradient is harnessed to synthesize ATP (Fig. 20-22).*

Several key observations are explained by the chemiosmotic hypothesis:

(a) Oxidative phosphorylation requires an intact inner mitochondrial membrane.

(b) The inner mitochondrial membrane is impermeable to ions such as H$^+$, OH$^-$, K$^+$, and Cl$^-$, whose free diffusion would discharge an electrochemical gradient.

(c) Electron transport results in the transport of H$^+$

out of intact mitochondria thereby creating a measureable electrochemical gradient across the inner mitochondrial membrane.

(d) Compounds that increase the permeability of the inner mitochondrial membrane to protons, and thereby dissipate the electrochemical gradient, allow electron transport (from NADH and succinate oxidation) to continue but inhibit ATP synthesis; that is, they "uncouple" electron transport from oxidative phosphorylation. Conversely, increasing the acidity outside the inner mitochondrial membrane stimulates ATP synthesis.

In the remainder of this section we examine the mechanisms through which electron transport can result in proton translocation and how an electrochemical gradient can interact with ATP synthase to drive ATP synthesis.

B. Proton Gradient Generation

Electron transport, as we shall see, causes Complexes I, III, and IV to transport protons across the inner mitochondrial membrane from the matrix, a region of low [H$^+$] and negative electrical potential, to the intermembrane space (which is in contact with the cytosol), a region of high [H$^+$] and positive electrical potential (Fig. 20-13). The free energy sequestered by the resulting electrochemical gradient [which, in analogy to the term electromotive force (emf), is called **proton-motive force (pmf)**] powers ATP synthesis.

Proton Pumping Is an Endergonic Process

The free energy change of transporting a proton out of the mitochondrion against an electrochemical gradient is expressed by Eq. [18.3] which, in terms of pH, is

$$\Delta G = 2.3RT\,[\text{pH }(in) - \text{pH }(out)] + Z\mathscr{F}\Delta\Psi \quad [20.1]$$

where Z is the charge on the proton (including sign), $\mathscr{F}$ is the faraday constant, and $\Delta\Psi$ is the membrane potential. The sign convention for $\Delta\Psi$ is that when an ion is transported from negative to positive, $\Delta\Psi$ is positive. Since pH (*out*) is less than pH (*in*), the export of protons from the mitochondrial matrix (against the proton gradient) is an endergonic process. In addition, *proton transport out of the matrix makes the inner membrane's internal surface more negative than its external surface.* Outward transport of a positive ion is consequently associated with a positive $\Delta\Psi$ and an increase in free energy (endergonic process), whereas the outward transport of a negative ion yields the opposite result. Clearly, it is always necessary to describe membrane polarity when specifying a membrane potential.

The measured membrane potential across the inner membrane of a liver mitochondrion, for example, is 0.168 V (which corresponds to an $\sim 210,000\text{-V} \cdot \text{cm}^{-1}$ electric field across its ~ 80-Å thickness). The pH of its matrix is 0.75 units higher than that of its intermembrane space. ΔG for proton transport out of this mitochondrial matrix is therefore 21.5 kJ $\cdot$ mol^{-1}.

The Passage of Two to Three Protons Is Required to Synthesize One ATP

An ATP molecule's estimated physiological free energy of synthesis, around $+40$ to $+50$ kJ $\cdot$ mol^{-1}, is too large to be driven by the passage of a single proton back into the mitochondrial matrix; at least two protons are required. This number is difficult to measure precisely, in part because transported protons tend to leak back across the mitochondrial membrane. Estimates range from about two to three protons passed per ATP synthesized.

Two Mechanisms of Proton Transport Have Been Proposed

Three of the four electron-transport complexes, Complexes I, III, and IV, are involved in proton translocation. Two mechanisms have been entertained that would couple the free energy of electron transport with the active transport of protons — the **redox loop mechanism** and the **proton pump mechanism.**

1. The Redox Loop Mechanism

This mechanism, proposed by Mitchell, requires that the redox centers of the respiratory chain (FMN, CoQ, cytochromes, and iron–sulfur clusters) be so arranged in the membrane that reduction would involve a redox center simultaneously accepting e^- and H$^+$ from the matrix side of the membrane. Reoxidation of this redox center by the next center in the chain would involve release of H$^+$ on the cytosolic side of the membrane together with the transfer of electrons back to the matrix side (Fig. 20-23*a*). Electron flow from one center to the next would therefore yield net translocation of H$^+$ and the creation of an electrochemical gradient ($\Delta\Psi$ and ΔpH).

The redox loop mechanism requires that the first redox carrier contain more hydrogen atoms in its reduced state than in its oxidized state while the second redox carrier have no difference in its hydrogen atom content between its reduced and oxidized states. Is this requirement met in the electron-transport chain? Some of the redox carriers, FMN and CoQ, in fact, contain more hydrogen atoms in their reduced state than in their oxidized state and thus can qualify as proton carriers as well as electron carriers.

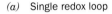

(a) Single redox loop

(b) Redox loop mechanism

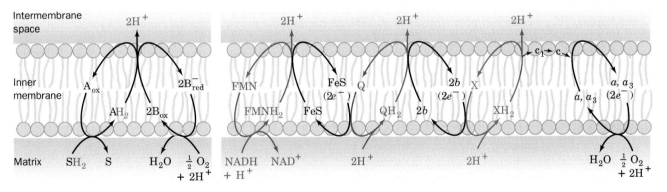

Figure 20-23

The Redox Loop mechanism. (*a*) A single redox loop for electron-transport-linked H$^+$ translocation. (*b*) The proposed Redox Loop mechanism of electron-transport-linked proton translocation in mitochondria. FMN and CoQ function as (H$^+$ + e^-) carriers while the iron–sulfur clusters and the cytochromes function as pure e^- carriers. These components are so arranged as to require that electron transport be accompanied by H$^+$ translocation. Note that a mystery (H$^+$ + e^-) carrier, X, is included to account for a third H$^+$ translocation site.

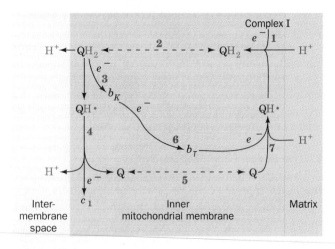

Figure 20-24
The Q cycle, an electron-transport cycle in Complex III proposed to account for H^+ translocation during the transport of electrons from cytochrome b to cytochrome c: **(1)** Coenzyme $QH \cdot$ semiquinone is reduced to QH_2 by Complex I on the matrix side of the membrane. **(2)** QH_2 diffuses to the cytosolic side of the membrane. **(3)** QH_2 reduces cytochrome b_K with the release of H^+. **(4)** A second H^+ is translocated when the semiquinone reduces cytochrome c_1 on the cytosolic side releasing a second H^+. **(5)** Q diffuses to the matrix side. **(6)** Cytochrome b_K reduces cytochrome b_T. **(7)** Q is reduced to $QH \cdot$ semiquinone by cytochrome b_T with the concomitant absorption of matrix H^+. The overall cycle involves translocation of H^+ both as coenzyme Q is reduced by Complex I and as Complex III reduces cytochrome c_1. [After Nicholls, D. G., *Bioenergetics*, p. 122, Academic Press (1982).]

If these centers were spatially alternated with pure electron carriers (cytochromes and iron–sulfur clusters), such a mechanism could well be accommodated (Fig. 20–23b).

The main difficulty with the redox loop mechanism involves the deficiency of $(H^+ + e^-)$ carriers that can alternate with pure e^- carriers. While there are as many as 15 pure e^- carriers (up to 8 iron–sulfur proteins, 5 cytochromes, and 2 Cu atoms), only two $(H^+ + e^-)$ carriers are known. The fact that there are three sites for ATP synthesis suggests the need for at least three proton-transport sites. This problem is emphasized in Fig. 20-23b by showing X as an unknown $(H^+ + e^-)$ carrier.

In an attempt to solve this problem, Mitchell has described a way coenzyme Q might be involved twice in proton translocation in Complex III (CoQ-cytochrome c reductase). In the so-called **Q cycle**, CoQ is proposed to undergo a two-stage reduction involving the semiquinone as a stable intermediate (Fig. 20-24). *The result of this sequence in Complex III would be the transport of two protons for each electron transferred from Complex I to cytochrome c_1.* The Q cycle cannot operate in Complex IV (cytochrome oxidase), however, because Complex IV contains no $(H^+ + e^-)$ carriers even though it pumps protons from the matrix to the cytosol during electron transport.

2. The Proton Pump Mechanism

The proton pump mechanism does not require that the redox centers themselves be hydrogen carriers. In this model, *the transfer of electrons results in conformational changes to the complex. The translocation of protons occurs as a result of the influence of these conformational changes on the pK's of amino acid side chains and their alternate exposure to the internal and external side of the membrane (Fig. 20-25).* We have already seen that conformation can influence pK. The Bohr effect in hemoglobin, for example, results from conformational changes induced by O_2 binding which induces pK changes in protein acid–base groups (Section 9-2E). If such a protein were located in a membrane and if, in addition to pK changes, the conformational changes altered the side of the membrane to which the affected amino acid side chains were exposed, the result would be H^+ transport and the system would be a proton pump.

One documented proton pump is the intrinsic membrane protein **bacteriorhodopsin** of *Halobacter halobium.* This protein, which has seven membrane-spanning helical segments forming a polar channel (Section 11-3A), obtains the free energy required for pumping protons through the absorption of light by its retinaldehyde–Schiff base prosthetic group. The proposed mechanism for light-induced H^+ transport in bacteriorhodopsin is diagrammed in Fig. 20-26. A similar mechanism is thought to operate in cytochrome oxidase.

Proton-Transport Mechanisms May Be Distinguished Experimentally

If the electrochemical H^+ gradient across the mitochondrial membrane was generated by a proton pump such as bacteriorhodopsin, the stoichiometry of proton transport would depend on the number of groups whose pK's changed in going from the oxidized to the reduced conformation and the magnitudes of these pK

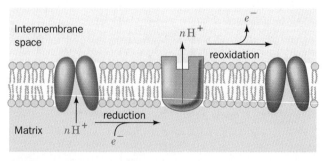

Figure 20-25
The proton pump mechanism of electron-transport-linked proton translocation. At each H^+ translocation site, *n* protons bind to amino acid side chains on the matrix side of the membrane. Reduction causes a conformational change that decreases the pK's of these side chains and exposes them to the cytosolic side of the membrane where the protons dissociate. Reoxidation results in a conformational change that restores the pump to its original conformation.

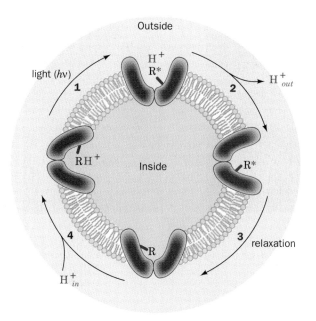

Figure 20-26
The proton pump of bacteriorhodopsin: **(1)** In its resting state, the retinaldehyde–Schiff base prosthetic group (R) is protonated and faces the inside of the membrane. Upon the absorption of light (*hv*), the pump undergoes a conformational change that causes R*H$^+$ to face the outside of the membrane (R* is the light-excited form of R). **(2)** The conformational change lowers the pK of R* compared to R, which causes H$^+$ to dissociate from R* on the outside of the membrane. **(3)** The pump relaxes back to its resting state so that R again faces the inside of the membrane. **(4)** The pK increase accompanying relaxation causes R to be protonated by an H$^+$ from the inside of the membrane. The pump has thereby recovered its original state with the only change to the system being the translocation of a proton from the inside to the outside of the membrane.

changes. The redox loop mechanism, in contrast, requires that one proton be translocated for each electron transported in Complex I and two protons transported for each electron in Complex III if the Q cycle is employed. [There are no (H$^+$ + e^-) carriers in Complex IV so the redox loop mechanism cannot operate there.] It might therefore be possible to distinguish between the redox loop and proton pump mechanisms if the actual ratio of protons translocated to electrons transported were known. Of course, since three complexes translocate protons, there is also the possibility that different mechanisms may operate at different points in the electron-transport chain. This complicated situation is currently receiving much research attention.

C. Mechanism of ATP Synthesis

The free energy of the electrochemical proton gradient across the mitochondrial membrane is harnessed in the synthesis of ATP by **proton-translocating ATP synthase** *(also known as* **proton pumping ATPase** *and* **F$_0$F$_1$-ATPase).** In the following subsections we discuss the location and structure of this ATP synthase and the mechanism by which it harnesses proton flux to drive ATP synthesis. As with many other aspects of oxidative phosphorylation, this mechanism is by no means well established and is an area of active investigation.

Proton-Translocating ATP Synthase Is a Multisubunit Trans-Membrane Protein

Proton-translocating ATP synthase is the most complex structure in the inner mitochondrial membrane. It contains two major substructures and several different subunits (Table 20-2). Electron micrographs of mitochondria show lollipop-shaped structures studding the matrix surface of the inner mitochondrial membrane (Fig. 20-27*a*). Similar entities have been observed to line the inner surface of the bacterial plasma membrane. Sonication of the inner mitochondrial membrane yields sealed vesicles, **submitochondrial particles,** from which the "lollipops" project (Fig. 20-27*b*) and which can carry out ATP synthesis.

Efraim Racker discovered that the proton-translocating ATP synthase from submitochondrial particles is comprised of two functional units, **F$_0$** and **F$_1$**. F$_0$ is a water-insoluble trans-membrane protein composed of four or five types of subunits that contains a channel for proton translocation. F$_1$ is a water-soluble peripheral membrane protein composed of five types of subunits that is easily dissociated from F$_0$ by treatment with urea. Submitochondrial particles from which F$_1$ has been re-

Table 20-2

Components of Mitochondrial Proton-Translocating Synthase (F$_0$ F$_1$-ATPase)^a

Component	Subunit Composition	Functions
F$_1$	$\alpha_3\beta_3\gamma\delta\varepsilon$	β Contains the ATP synthase site; δ forms the gate coupling the F$_0$ proton channel with F$_1$
F$_0$	4–5 Types of subunit including 6–10 copies of DCCD-binding proteolipid	DCCD-binding proteolipid oligomer forms the proton channel
Stalk	One copy each of OSCP and F$_6$	Required to bind F$_0$ to F$_1$
Associated polypeptides	IF$_1$	Inhibits ATP hydrolysis; binds to the F$_1$ β subunit
	F$_B$	

^{a} Total molecular mass = 450 kD.

(a)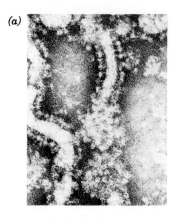

(b)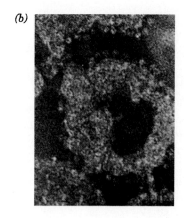

(c)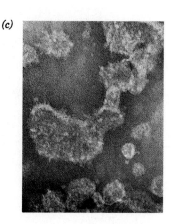

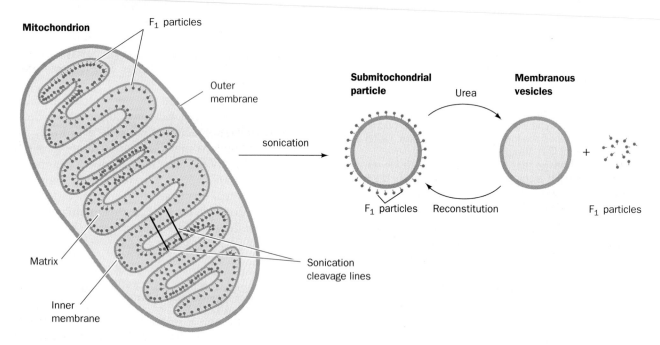

Figure 20-27

Electron micrographs and their interpretive drawings (*below*) of the mitochondrial membrane at various stages of dissection: (*a*) Cristae from intact mitochondria showing their F_1 "lollipops" projecting into the matrix. [From Parsons, D. F., *Science* **140,** 985 (1963). Copyright © 1963 American Association for the Advancement of Science. Used by permission.] (*b*) Submitochondrial particles showing their outwardly projecting F_1 lollipops. Submitochondrial particles are prepared by the sonication (ultrasonic disruption) of inner mitochondrial membranes. [Courtesy of Efraim Racker, Cornell University.] (*c*) Submitochondrial particles after treatment with urea. [Courtesy of Efraim Racker, Cornell University.]

(a)   (b)

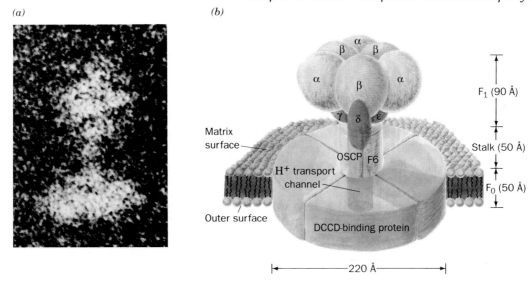

Figure 20-28
(a) An electron micrograph of reconstituted mitochondrial proton-translocating ATP synthase (F$_o$F$_1$-ATPase) with (b) an accompanying interpretive drawing indicating the postulated positions of its component subunits. [Electron micrograph courtesy of Peter Pedersen, The Johns Hopkins University School of Medicine.]

moved by urea treatment no longer exhibit the lollipops in their electron micrographs (Fig. 20-27c) or the ability to synthesize ATP. If, however, F$_1$ is added back to these F$_o$-containing submitochondrial particles, their ability to synthesize ATP is restored and their electron micrographs again exhibit the lollipops. Thus *the lollipops are the F$_1$ particles.* Electron micrographs of reconstituted F$_o$F$_1$ particles clearly show their dumbbell-shaped structure in which F$_o$ and F$_1$ are connected by a 50-Å stalk (Fig. 20-28). This stalk is composed of two proteins, **oligomycin-sensitivity-conferring protein (OSCP)** and **coupling factor 6 (F$_6$)**.

F$_1$'s catalytic site for ATP synthesis is contained on the β subunit of this $\alpha_3\beta_3\gamma\delta\varepsilon$ multimer. The δ subunit is required for binding of F$_1$ to F$_o$. **Oligomycin,** a *Streptomyces*-produced antibiotic,

inhibits ATP synthase by binding to a subunit of F$_o$ (not OSCP) so as to interfere with H$^+$ transport through F$_o$.
Dicyclohexylcarbodiimide (DCCD),

**Dicyclohexylcarbodiimide
(DCCD)**

a lipid-soluble carboxyl reagent, also inhibits proton transport through F$_o$ by reacting (analogously with EDAC; Table 6-3) with a single Glu residue on one of the subunits of mammalian F$_o$ (Asp in *E. coli*). Reaction with DCCD usually implies that a carboxylic acid group is located in a lipid environment; that is, "buried" in the membrane. Mammalian F$_o$ contains six copies of this DCCD-binding protein (also known as **DCCD-binding proteolipid**), which are thought to associate like staves of a barrel so as to form the polar H$^+$-transport channel that contains the buried Glu residues (Fig. 20-28).

Proton-Translocating ATP Synthase Is Driven by Conformational Changes

The mechanism of ATP synthesis by proton-translocating ATP synthase can be conceptually broken down into three phases:

1. Translocation of protons carried out by F$_o$.

2. Catalysis of formation of the phosphoanhydride bond of ATP carried out by F$_1$.

3. Coupling of the dissipation of the proton gradient with ATP synthesis, which requires interaction of F$_1$ and F$_o$.

Oligomycin B

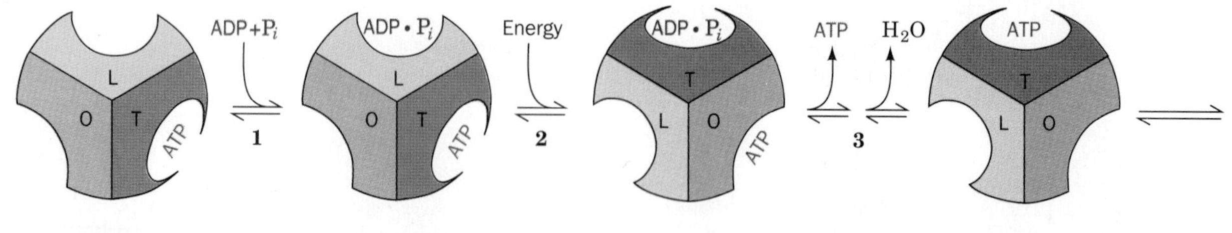

Figure 20-29

The energy-dependent binding change mechanism for ATP synthesis by proton-translocating ATP synthase. F_1 has three chemically identical but conformationally distinct interacting subunits: O, the open conformation, has very low affinity for ligands and is catalytically inactive; L has loose binding for ligands and is catalytically inactive; T has tight binding for ligands and is catalytically active. ATP synthesis occurs in three steps: **(1)** Binding of ADP and P_i to site L. **(2)** Energy-dependent conformational change converting binding site L to T, T to O, and O to L. **(3)** Synthesis of ATP at site T and release of ATP from site O. The enzyme returns to its initial state after two more passes of this reaction sequence. [After Cross, R. L., *Annu. Rev. Biochem.* **50**, 687 (1980).]

The available evidence supports a mechanism for ATP formation proposed by Boyer, which resembles the conformational coupling hypothesis of oxidative phosphorylation (Section 20-3A). However, the conformational changes in the ATP synthase that power ATP formation are generated by proton translocation rather than by direct electron transfer as proposed in the original formulation of the conformational coupling hypothesis. F_1 is proposed to have three interacting catalytic subunits, each in a different conformational state: one that binds substrates and products loosely (L state), one tightly (T state), and one not at all (open or O state). The reaction involves three steps (Fig. 20-29):

1. Binding of ADP and P_i to the "loose" (L) binding-site.

2. An energy-dependent conformational change that converts the L site to a "tight" (T)-binding site that catalyzes the formation of ATP. Exergonic proton translocation presumably occurs here to provide the necessary free energy, thus dissipating the electrochemical gradient generated by electron transport. This step also involves conformational changes of the other two subunits that convert the ATP-containing T site to an "open" (O) site and convert the O site to an L site.

3. ATP is synthesized at the T site on one subunit while ATP dissociates from the O site on another subunit.

The "tight coupling" between electron transport and ATP synthesis in the mitochondrion depends on the impermeability of the inner mitochondrial membrane. This impermeability allows an electrochemical gradient to be established across this membrane during the H^+ translocation associated with electron transport. The only way for H^+ to reenter the matrix is through the F_0 portion of the proton-translocating ATP synthase. The electrochemical gradient therefore builds until the free energy required to transport H^+ balances the free energy of electron transport. Electron transport must then cease.

ATP synthesis, by dissipating the electrochemical gradient, allows electron transport to continue.

D. Uncoupling of Oxidative Phosphorylation

Electron transport (the oxidation of NADH and $FADH_2$ by O_2) and oxidative phosphorylation (the synthesis of ATP) are normally tightly coupled. In the resting state, when oxidative phosphorylation is minimal, the electrochemical gradient across the inner mitochondrial membrane builds up to the extent that it prevents further proton pumping and therefore inhibits electron transport. Over the years, however, many compounds, including **2,4-dinitrophenol (DNP)** and **carbonylcyanide-*p*-trifluoromethoxyphenylhydrazone (FCCP)**, have been found to "uncouple" these processes. The chemiosmotic hypothesis has provided a rationale for understanding the mechanism by which these uncouplers act.

The presence in the inner mitochondrial membrane of an agent that increases its permeability to H^+ uncouples oxidative phosphorylation from electron transport by providing a route for the dissipation of the proton electrochemical gradient that does not require ATP synthesis. Uncoupling therefore allows electron transport to proceed unchecked even when ATP synthesis is inhibited. DNP and FCCP are lipophilic weak acids that therefore readily pass through membranes. In a pH gradient, they bind protons on the acidic side of the membrane, diffuse through, and release them on the alkaline side, thereby dissipating the gradient (Fig. 20-30). Thus, *such uncouplers are proton-transporting ionophores.*

Even before the mechanism of uncoupling was known, it was recognized that metabolic rates were increased by such compounds. Studies at Stanford University in the early part of the twentieth century documented an increase in respiration and weight loss

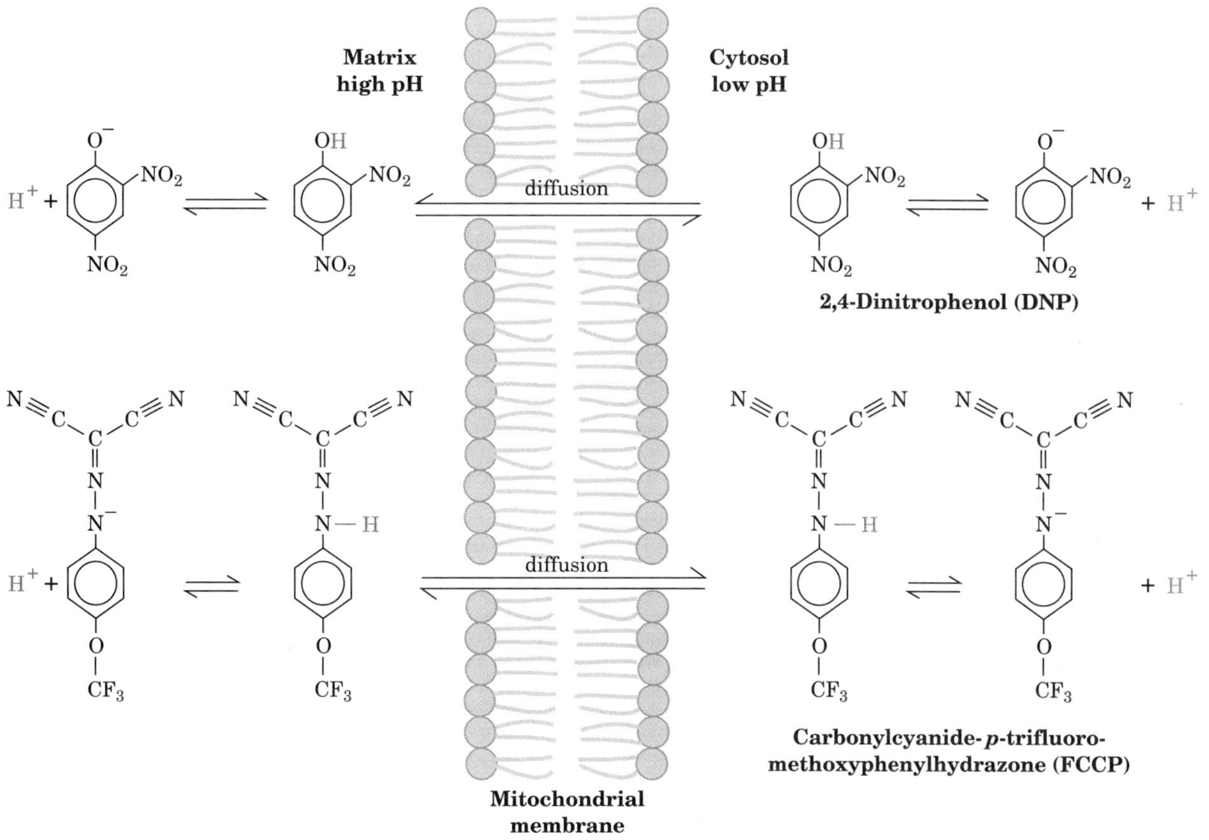

Figure 20-30
The proton-transporting ionophores DNP and FCCP uncouple oxidative phosphorylation from electron transport

by discharging the electrochemical proton gradient generated by electron transport.

caused by DNP. The compound was even used as a "diet pill" for several years. In the words of Efraim Racker (*A New Look at Mechanisms in Bioenergetics, p. 155*):

> *In spite of warnings from the Stanford scientists, some enterprising physicians started to administer dinitrophenol to obese patients without proper precautions. The results were striking. Unfortunately in some cases the treatment eliminated not only the fat but also the patients, and several fatalities were reported in the Journal of the American Medical Association in 1929. This discouraged physicians for a while . . .*

Hormonally Controlled Uncoupling in Brown Adipose Tissue Functions to Generate Heat

The dissipation of an electrochemical H$^+$ gradient, which is generated by electron transport and uncoupled from ATP synthesis, produces heat. Heat generation is the physiological function of **brown adipose tissue (brown fat).** This tissue is unlike typical (white) adipose tissue in that, besides containing large amounts of triacylglycerols, it contains numerous mitochondria whose cytochromes cause its brown color. Newborn mammals that lack fur,

such as humans, as well as hibernating mammals, contain brown fat in their neck and upper back that functions in **nonshivering thermogenesis,** that is, as a "biological heating pad." (The ATP hydrolysis that occurs during the muscle contractions of shivering—or any other movement—also produces heat.)

The mechanism of heat generation in brown fat involves the regulated uncoupling of oxidative phosphorylation in their mitochondria. These mitochondria contain **thermogenin,** a protein dimer of 32-kD subunits that is absent in the mitochondria of other tissues, which acts as a channel to control the permeability of the inner mitochondrial membrane to protons. In cold-adapted animals, thermogenin constitutes up to 15% of brown fat inner mitochondrial membrane proteins. The flow of protons through this channel protein is inhibited by physiological concentrations of purine nucleotides (ADP, ATP, GDP, GTP) but this inhibition can be overcome by free fatty acids. The components of this system interact under hormonal control.

Thermogenesis in brown fat mitochondria is activated by free fatty acids. These counteract the inhibitory effects of purine nucleotides thereby stimulating the flux through the proton channel and uncoupling electron transport from oxidative phosphorylation. *The concentration of fatty acids in brown adipose tissue is controlled by the hormone* **norepinephrine (noradrenaline)**

HO

OH

HO—C—CH₂—NH₃⁺

H

Norepinephrine

with cAMP acting as a second messenger (Section 17-3). Under norepinephrine stimulation (Fig. 20-31), the adenylate cyclase component of the norepinephrine receptor system synthesizes cAMP as described in Section 34-4B. The cAMP, in turn, allosterically activates cAMP-dependent protein kinase, which activates **hor-**

mone-sensitive triacylglycerol lipase by phosphorylating it (Section 23-5). Finally, the activated lipase hydrolyzes triacylglycerols to yield the free fatty acids that open the proton channel.

4. CONTROL OF ATP PRODUCTION

An adult woman requires some 1500 to 1800 kcal (6300–7500 kJ) of metabolic energy per day. This corresponds to the free energy of hydrolysis of over 200 mol of ATP to ADP and P_i. Yet the total amount of ATP present in the body at any one time is <0.1 mol; obviously, this sparse supply of ATP must be contin-

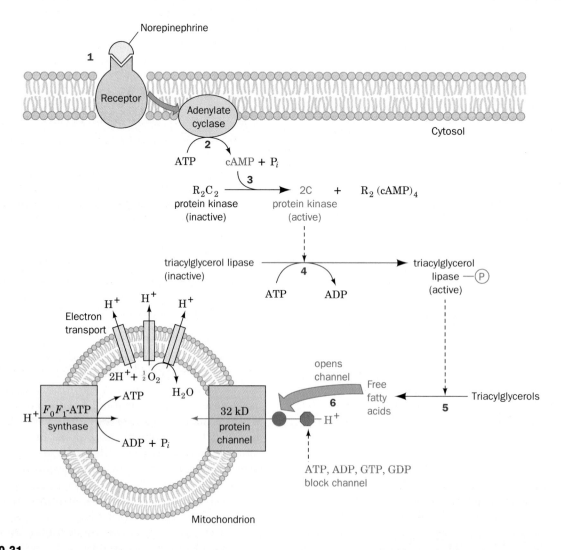

Figure 20-31
The mechanism of hormonally induced uncoupling of oxidative phosphorylation in brown fat mitochondria:
(1) Norepinephrine binds to its cell-surface receptor. **(2)** The norepinephrine-receptor complex stimulates adenylate cyclase thereby causing cAMP levels to rise. **(3)** cAMP binding activates cAMP-dependent protein kinase.

(4) cAMP-dependent protein kinase phosphorylates hormone-sensitive triacylglycerol lipase thereby activating it.
(5) Triacylglycerols are hydrolyzed yielding free fatty acids.
(6) Free fatty acids overcome the purine nucleotide block of the 32-kD proton channel, allowing H⁺ to enter the mitochondrion uncoupled from ATP synthesis.

ually recycled. As we have seen, when carbohydrates serve as the energy supply and aerobic conditions prevail, this recycling involves glycogenolysis, glycolysis, the citric acid cycle, and oxidative phosphorylation.

Of course the need for ATP is not constant. There is a 100-fold change in ATP utilization between sleep and vigorous activity. *The activities of the pathways that produce ATP are under strict coordinated control so that ATP is never produced more rapidly than necessary.* We have already discussed the control mechanisms of glycolysis, glycogenolysis, and the citric acid cycle (Sections 16-4, 17-3, and 19-4). In this section we discuss the mechanisms through which oxidative phosphorylation is controlled and observe how all four systems are synchronized to produce ATP at precisely the rate required at any particular moment.

A. Control of Oxidative Phosphorylation

In our discussion of the control of glycolysis, we saw that most of the reactions in a metabolic pathway function close to equilibrium. *The few irreversible reactions constitute the potential control points of the pathway and usually are catalyzed by regulatory enzymes that are under allosteric control.* In the case of oxidative phosphorylation, the pathway from NADH to cytochrome c functions near equilibrium ($\Delta G' \approx 0$):

$$\tfrac{1}{2}\text{NADH} + \text{cytochrome } c^{3+} + \text{ADP} + P_i \rightleftharpoons$$
$$\tfrac{1}{2}\text{NAD}^+ + \text{cytochrome } c^{2+} + \text{ATP}$$

for which

$$K_{eq} = \left(\frac{[\text{NAD}^+]}{[\text{NADH}]}\right)^{\frac{1}{2}} \frac{[c^{2+}]}{[c^{3+}]} \frac{[\text{ATP}]}{[\text{ADP}][P_i]} \quad [20.2]$$

This pathway is therefore readily reversed by the addition of ATP. *In the cytochrome oxidase reaction, however, the terminal step of the electron-transport chain is irreversible and is therefore a prime candidate as the control site of the pathway.* Cytochrome oxidase, in contrast to most regulatory enzyme systems, appears to be controlled exclusively by the availability of one of its substrates, reduced cytochrome c (c^{2+}). Since this substrate is in equilibrium with the rest of the coupled oxidative phosphorylation system, (Eq. [20.2]), its concentration ultimately depends on the intramitochondrial [NADH]/[NAD$^+$] ratio and the **ATP mass action ratio** ([ATP]/[ADP] [P$_i$]). By rearranging Eq. [20.2], the ratio of reduced to oxidized cytochrome c is expressed

$$\frac{[c^{2+}]}{[c^{3+}]} = \left(\frac{[\text{NADH}]}{[\text{NAD}^+]}\right)^{\frac{1}{2}} \left(\frac{[\text{ADP}][P_i]}{[\text{ATP}]}\right) \frac{1}{K_{eq}} \quad [20.3]$$

Consequently, the higher the [NADH]/[NAD$^+$] ratio and the lower the ATP mass action ratio, the higher [c^{2+}] (reduced cytochrome c) and thus the higher the cytochrome oxidase activity.

How is this system affected by changes in physical activity? In an individual at rest, ATP hydrolysis to ADP and P$_i$ is minimal and the ATP mass action ratio is high; the concentration of reduced cytochrome c is therefore low and oxidative phosphorylation is minimal. Increased activity results in hydrolysis of ATP to ADP and P$_i$, thereby decreasing the ATP mass action ratio and increasing the concentration of reduced cytochrome c. This results in an increase in the electron-transport rate and its coupled phosphorylation. Such control of oxidative phosphorylation by the ATP mass action ratio is called **acceptor control** because the rate of oxidative phosphorylation increases with the concentration of ADP, the phosphoryl group acceptor.

The compartmentalization of the cell into mitochondria, where ATP is synthesized, and cytosol, where ATP is utilized, presents an interesting control problem: Is it the ATP mass action ratio in the cytosol or in the mitochondrial matrix that ultimately controls oxidative phosphorylation? Clearly the ATP mass action ratio that exerts direct control must be that of the mitochondrial matrix where ATP is synthesized. However, the inner mitochondrial membrane, which is impermeable to adenine nucleotides and P$_i$, depends on specific transport systems to maintain communication between the two compartments (Section 18-4C). This organization makes it possible for the transport of adenine nucleotides or P$_i$ to be the rate-limiting step in oxidative phosphorylation. Martin Klingenberg has proposed just such a control function for the ADP–ATP translocator. David Wilson and Maria Erecińska assert that there is equilibration of ATP, ADP, and P$_i$ between cytosol and mitochondria and that the cytosolic ATP mass action ratio ultimately controls mitochondrial oxidative phosphorylation. As in most areas of investigation, this sort of controversy is what spurs research.

B. Coordinated Control of ATP Production

Glycolysis, the citric acid cycle, and oxidative phosphorylation constitute the major pathways for cellular ATP production. Control of oxidative phosphorylation by the ATP mass action ratio depends, of course, on an adequate supply of electrons to fuel the electron-transport chain. This aspect of the system's control is, in turn, dependent on the [NADH]/[NAD$^+$] ratio (Eq. [20.3]), which is maintained high by the combined action of glycolysis and the citric acid cycle in converting 10 molecules of NAD$^+$ to NADH per molecule of glucose oxidized (Fig. 20-1). It is clear, therefore, that coordinated control is necessary for the three processes. This is provided by the regulation of each of the control points of glycolysis (phosphofructokinase; PFK) and the citric acid cycle (pyruvate dehydrogenase, citrate synthase, isocitrate dehydrogenase, and α-ketoglutarate dehydrogenase) by adenine nucleotides or NADH or both as well as by certain metabolites (Fig. 20-32).

Citrate Inhibits Glycolysis

The main control points of glycolysis and the citric acid cycle are regulated by several effectors besides adenine nucleotides or NADH (Fig. 20-32). This is an extremely complex system with complex demands. Its many effectors, which are involved in various aspects of metabolism, increase its regulatory sensitivity. One particularly interesting regulatory effect is the inhibition of PFK by citrate. When demands for ATP decrease, [ATP] increases and [ADP] decreases. The citric acid cycle slows down at its isocitrate dehydrogenase (activated by ADP) and α-ketoglutarate dehydrogenase (inhibited by ATP) steps, thereby causing the citrate concentration to build up. Citrate can leave the mitochondrion via a specific transport system and, *once in the cytosol, acts to restrain further carbohydrate breakdown by inhibiting PFK.*

C. Physiological Implications of Aerobic versus Anaerobic Metabolism

In 1861, Louis Pasteur observed that *when yeast are exposed to aerobic conditions, their glucose consumption and ethanol production drops precipitously* (the **Pasteur effect**; alcoholic fermentation in yeast to produce ATP, CO_2, and ethanol are discussed in Section 16-3B). An analogous effect is observed in mammalian muscle; the concentration of lactic acid, the anaerobic product of muscle glycolysis, drops dramatically when cells switch to aerobic metabolism. (This situation is of opposite concern in yeast and muscle since alcohol production is the desired pathway in yeast, at least for winemakers, while lactic acid accumulation in muscle — or more precisely, the associated decrease in pH — leads to soreness and fatigue.)

Aerobic ATP Production Is Far More Efficient than Anaerobic ATP Production

One reason for the decrease in glucose consumption on switching from anaerobic to aerobic metabolism is clear from an examination of the stoichiometries of anaerobic and aerobic breakdown of glucose ($C_6H_{12}O_6$).

Anaerobic glycolysis:

$$C_6H_{12}O_6 + 2ADP + 2P_i \longrightarrow$$
$$2 lactate + 2H^+ \ 2H_2O + 2ATP$$

Figure 20-32
A schematic diagram representing the coordinated control of glycolysis and the citric acid cycle by ATP, ADP, AMP, P_i, Ca^{2+}, and the [NADH]/[NAD$^+$] ratio (the vertical arrows indicate increases in this ratio). Here a green dot signifies activation and a red octagon represents inhibition. [After Newsholme, E. A. and Leech, A. R., *Biochemistry for the Medical Sciences*, pp. 316, 320, Wiley (1983).]

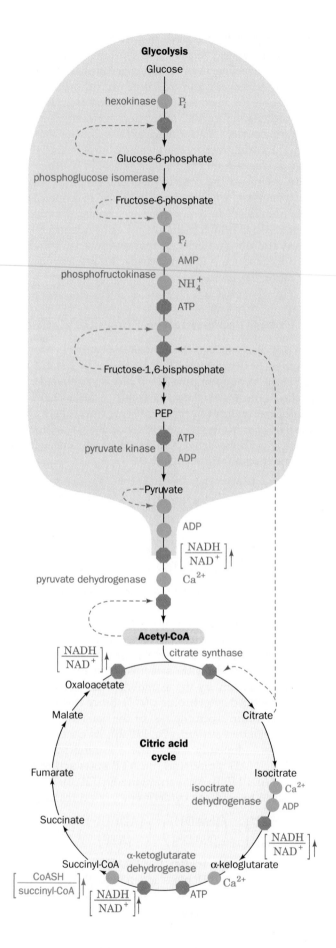

Aerobic metabolism of glucose:

$$C_6H_{12}O_6 + 38ADP + 38P_i + 6\ O_2 \longrightarrow$$
$$6CO_2 + 44H_2O + 38ATP$$

(3ATPs for each of the 10NADHs generated per glucose oxidized, 2ATPs for each of the 2FADH$_2$s generated, 2ATPs produced in glycolysis, and 2GTP $\rightleftharpoons$ 2ATP produced in the citric acid cycle.) Thus *aerobic metabolism is 19 times more efficient than anaerobic glycolysis in producing ATP.* The switch to aerobic metabolism therefore rapidly increases the ATP mass action ratio. As the ATP mass action ratio increases, the rate of electron transport decreases, which has the effect of increasing the [NADH]/[NAD$^+$] ratio. The increases in [ATP] and [NADH] inhibit their target enzymes in the citric acid cycle and in the glycolytic pathway. *The activity of PFK, the citrate- and adenine nucleotide-regulated rate-controlling enzyme of glycolysis, decreases many-fold on switching from anaerobic to aerobic metabolism. This accounts for the dramatic decrease in glycolysis.*

Anaerobic Glycolysis Has Advantages As Well As Limitations

Animals can only sustain anaerobic glycolysis for short periods of time. This is because PFK, which cannot function effectively much below pH 7, is inhibited by the acidification arising from lactic acid production. Despite this limitation and the low efficiency of glycolytic ATP production, *the enzymes of glycolysis are present in such great concentrations that when they are not inhibited, ATP can be produced much more rapidly than through oxidative phosphorylation.*

The different characteristics of aerobic and anaerobic metabolism permit us to understand certain aspects of cancer cell metabolism and cardiovascular disease.

Cancer Cell Metabolism

As Warburg first noted in 1926, certain cancer cells produce more lactic acid under aerobic conditions than do normal cells. This is because the glycolytic pathway in these cells produces pyruvate more rapidly than the citric acid cycle can accommodate. How can this happen given the interlocking controls on the system? One explanation is that these controls have broken down in cancer cells. Another is that their ATP utilization occurs at rates too rapid to be replenished by oxidative phosphorylation. This would alter the ratios of adenine nucleotides so as to relieve the inhibition of PFK. Attempts to understand the metabolic differences between cancer cells and normal cells may one day provide a clue to the treatment of certain forms of this devastating disease.

Cardiovascular Disease

Oxygen deprivation of certain tissues resulting in cardiovascular disease is of major medical concern. For example, two of the most common causes of human death, **myocardial infarction** (heart attack) and **stroke,** are caused by interruption of the blood (O$_2$) supply to a portion of the heart or the brain, respectively. It seems obvious why this should result in a cessation of cellular activity but why does it cause cell death?

In the absence of O$_2$, a cell, which must then rely only on glycolysis for ATP production, rapidly depletes its stores of phosphocreatine (a source of rapid ATP production; Section 15-4C) and glycogen. As the rate of ATP production falls below the level required by membrane ion pumps for the maintenance of proper intracellular ionic concentrations, the osmotic balance of the system is disrupted so that the cell and its membrane-enveloped organelles begin to swell. The resulting over-stretched membranes become permeable thereby leaking their enclosed contents. [In fact, a useful diagnostic criterion for myocardial infarction is the presence in the blood of heart-specific enzymes, such as the H-type isozyme of lactate dehydrogenase (Section 7-5C), which leak out of necrotic (dead) heart tissue.] Moreover, the decreased intracellular pH that accompanies anaerobic glycolysis (because of lactatic acid production) permits the released lysosomal enzymes (which are only active at acidic pH's) to degrade the cell contents. Thus, the cessation of metabolic activity results in irreversible cell damage. Rapidly respiring tissues, such as those of heart and brain, are particularly susceptible to such damage.

Chapter Summary

Oxidative phosphorylation is the process through which the NADH and FADH$_2$ produced by nutrient oxidation are oxidized with the concomitant formation of ATP. The process takes place in the mitochondrion, an ellipsoidal organelle that is bounded by a permeable outer membrane and contains an impermeable and highly invaginated inner membrane that encloses the matrix. Enzymes of oxidative phosphorylation are embedded in the inner mitochondrial membrane. P$_i$ is imported into the mitochondrion by a specific transport protein, whereas Ca^{2+} import and Ca^{2+} export proteins operate to maintain a constant cytosolic [Ca^{2+}]. NADH's electrons are imported into the mitochondrion by one of several shuttle systems such as the glycerophosphate shuttle or the malate–aspartate shuttle.

The standard free energy change for the oxidation of NAD$^+$ by O$_2$ is $\Delta G°' = -218$ kJ $\cdot$ mol^{-1}, whereas that for the synthesis of ATP from ADP and P$_i$ is $\Delta G°' = 30.5$ kJ $\cdot$ mol^{-1}. Consequently, the molar free energy of oxidation of NADH by O$_2$ is sufficient to power the synthesis of several moles of ATP under standard conditions. The electrons generated by oxidation of NADH and FADH$_2$ pass through four protein complexes, the electron-transport chain, with the coupled synthesis of ATP. Complexes I, III, and IV participate in the oxidation of NADH producing three ATPs per NADH whereas FADH$_2$ oxidation, which involves Complexes II, III, and IV, produces only two ATPs per FADH$_2$. Thus, the ratio of moles of ATP produced per mole of coenzyme oxidized by O$_2$, the P/O ratio, is three for NADH oxidation and two for FADH$_2$ oxidation. The route taken by electrons through the electron-transport chain was elucidated, in part, through the use of electron-transport inhibitors. Rotenone and amytal inhibit Complex I, antimycin A inhibits Complex III, and CN$^-$ inhibits Complex IV. Also involved were measurements of the reduction potentials of the electron-carrying prosthetic groups contained in the electron-transport complexes. Complex I contains FMN and six to seven iron–sulfur clusters in a 26-subunit membrane protein complex. This complex passes electrons from NADH to CoQ, a nonpolar small molecule that diffuses freely within the membrane. Complex II contains the citric acid cycle enzyme succinate dehydrogenase and also passes electrons to CoQ, in this case from succinate through FAD and an iron–sulfur cluster. CoQ passes electrons to Complex III, which contains two b-type cytochromes, one iron–sulfur cluster, and cytochrome c_1. Electrons from cytochrome c_1 of Complex III are passed to cytochrome a of Complex IV (cytochrome c oxidase) via the peripheral membrane protein cytochrome c. This protein alternately associates with Complexes III and IV such that electrons are transferred between the two hemes of the resulting specific complex through a postulated conduit of parallel conjugated rings and hydrogen bonds. Complex IV passes electrons through cytochrome a, two Cu atoms, and cytochrome a_3 to O$_2$ which, in a four-electron process, is reduced to H$_2$O.

The mechanism by which the free energy released by the electron-transport chain is stored and utilized in ATP synthesis is best described by the chemiosmotic hypothesis. This hypothesis states that the free energy released by electron transport is conserved by generation of an electrochemical proton gradient across the inner mitochondrial membrane (outside acidic), which is harnessed to synthesize ATP. The proton gradient is created and maintained by the obligatory outward translocation of H$^+$ across the inner mitochondrial membrane as electrons travel through Complexes I, III, and IV. Transport of 2e^- through one of these complexes creates a proton gradient sufficient for synthesis of one ATP. The mechanism of this H$^+$ translocation is a subject of active research. The energy stored in the electrochemical proton gradient is utilized by proton-translocating ATP synthase (proton-pumping ATPase, F$_0$F$_1$-ATPase) in the synthesis of ATP by coupling this process to the exergonic transport of H$^+$ back into the mitochondrial matrix. Proton-translocating ATPase contains two oligomeric components: F$_1$, a peripheral membrane protein that appears as "lollipops" in electron micrographs of the inner mitochondrial membrane, and F$_0$, an integral membrane protein that contains the proton channel. Compounds such as 2,4-dinitrophenol are uncouplers of oxidative phosphorylation because they carry H$^+$ across the mitochondrial membrane thereby dissipating the proton gradient and allowing electron transport to continue without concomitant ATP synthesis. Brown fat mitochondria contain a regulated uncoupling system that, under hormonal control, generates heat instead of ATP.

Under aerobic conditions, the rate of ATP synthesis by oxidative phosphorylation is regulated, in a phenomenon known as acceptor control, by the ATP mass action ratio. ATP synthesis is tightly coupled to the oxidation of NADH and FADH$_2$ by the electron-transport chain. Glycolysis and the citric acid cycle are coordinately controlled so as to produce NADH and FADH$_2$ only at a rate required to meet the system's demand for ATP.

References

Historical Overview

Ernster, L. and Schatz, G., Mitochondria: a historical review, *J. Cell Biol.* **91**, 227s–255s (1981).

Fruton, J. S., *Molecules and Life*, pp. 262–396, Wiley–Interscience (1972).

Racker, E., *A New Look At Mechanisms in Bioenergetics*, Academic Press (1976). [A fascinating personal account by one of the outstanding contributors to the field.]

General

Boyer, P. D., Chance, B., Ernster, L., Mitchell, P., Racker, E., and Slater, E. C., Oxidative phosphorylation and photophosphorylation, *Annu. Rev. Biochem.* **46**, 955–1026 (1977). [A collection of articles by many of the leaders in the field.]

Ernster, L. (Ed.), *Bioenergetics*, Elsevier (1984).

Harold, F. M., *The Vital Force: A Study of Bioenergetics*, Chapter 7, Freeman (1986).

Hatefi, Y., The mitochondrial electron transport chain and oxidative phosphorylation system, *Annu. Rev. Biochem.* **54**, 1015–1069 (1985).

Hinkle, P. C. and McCarty, R. E., How cells make ATP, *Sci. Am.* **238**(3): 104–123 (1978).

Martonosi, A. N. (Ed.), *The Enzymes of Biological Membranes* (2nd ed.), Vol. 4, *Bioenergetics of Electron and Proton Transport*, Plenum Press (1985).

Newsholme, E. and Leech, T., *The Runner*, Fitness Books (1983). [A delightful book on the physiology and biochemistry of running.]

Nicholls, D. G., *Bioenergetics*, Academic Press (1982). [An authoritative monograph devoted entirely to the mechanism of oxidative phosphorylation and the techniques used to elucidate it.]

Mitochondrial Structure

Lindén M., Gellerfors, P., and Nelson, B. D., Purification of a protein having pore forming activity from the rat liver mitochondrial outer membrane, *Biochem. J.* **208**, 77–82 (1982).

Mannella, C. A., Structure of the outer mitochondrial membrane: ordered arrays of porelike subunits in outer-membrane fractions from *Neurospora crassa* mitochondria, *J. Cell Biol.* **94**, 680–687 (1982).

Electron Transport

Barber, J., Further evidence for the common ancestry of cytochrome *b*–*c* complexes, *Trends Biochem. Sci.* **9**, 209–211 (1984).

Beinert, H. and Albracht, S. P. J., New insights and unanswered questions concerning iron–sulfur clusters in mitochondria, *Biochim. Biophys. Acta* **683**, 245–277 (1982).

Brunori, M. and Wilson, M. T., Cytochrome oxidase, *Trends Biochem. Sci.* **7**, 295–299 (1982).

Capaldi, R. A., Arrangement of proteins in the mitochondrial inner membrane, *Biochim. Biophys. Acta* **695**, 291–306 (1982).

Capaldi, R. A., Malatesta, F., and Darley-Usmar, V. M., Struc-ture of cytochrome *c* oxidase, *Biochim. Biophys. Acta* **726**, 135–148 (1983).

Lee, C. P. (Ed.), *Curr. Top. Bioenerg.* **15** (1987). [Contains articles on the structures of the components of the electron-transport chain.]

Mathews, F. S., The structure, function and evolution of cytochromes, *Prog. Biophys. Mol. Biol.* **45**, 1–56 (1985).

Palmer, G., Cytochrome oxidase: a perspective, *Pure Appl. Chem.* **59**, 749–758 (1987).

Poulos, T. L. and Kraut, J., A hypothetical model of the cytochrome *c* peroxidase · cytochrome *c* electron transfer complex, *J. Biol. Chem.* **255**, 10322–10330 (1980).

Rieder, R. and Bosshard, H. R., Comparison of the binding sites on cytochrome *c* for cytochrome *c* oxidase, cytochrome bc_1 and cytochrome c_1, *J. Biol. Chem.* **255**, 4732–4739 (1980).

Scott, R. A., X-ray absorption spectroscopic investigations of cytochrome *c* oxidase structure and function, *Annu. Rev. Biophys. Biophys. Chem.* **18**, 137–158 (1989).

Smith, H. T., Ahmed, A. J., and Millett, F., Electrostatic interaction of cytochrome *c* with cytochrome c_1 and cytochrome oxidase, *J. Biol. Chem.* **256**, 4984–4990 (1981).

Sweeney, W. V. and Rabinowitz, J. C., Proteins containing 4Fe–4S clusters: an overview, *Annu. Rev. Biochem.* **49**, 139–161 (1980).

von Jagow, G., *b*-Type cytochromes, *Annu. Rev. Biochem.* **49**, 281–314 (1980).

Oxidative Phosphorylation

Alexandre, A., Reynafarje, B., and Lehninger, A. L., Stoichiometry of vectorial H^+ movements coupled to electron transport and to ATP synthesis in mitochondria, *Proc. Natl. Acad. Sci.* **75**, 5296–5300 (1978).

Amzel, L. M. and Pedersen, P. L., Proton ATPase: structure and function, *Annu. Rev. Biochem.* **52**, 801–824 (1983).

Cross, R. L., The mechanism and regulation of ATP synthesis by F_1 ATPases, *Annu. Rev. Biochem.* **50**, 681–714 (1981).

Ferguson, S. J. and Sorgato, M. C., Proton electrochemical gradients and energy transduction processes, *Annu. Rev. Biochem.* **51**, 185–217 (1982).

Fillingame, R. H., The proton-translocating pump of oxidative phosphorylation, *Annu. Rev. Biochem.* **49**, 1079–1113 (1980).

Futai, M. and Kanazawa, H., Role of subunits in proton-translocating ATPase (F_0-F_1), *Curr. Top. Bioenerg.* **10**, 181–215 (1980).

Lehninger, A. L., Reynafarje, B., Alexandre, A., and Villalobo, A., Respiration-coupled H^+ ejection by mitochondria, *Ann. N.Y. Acad. Sci.* **341**, 585–592 (1980).

Mitchell, P., Vectorial chemistry and the molecular mechanics of chemiosmotic coupling: power transmission by proticity, *Biochem. Soc. Trans.* **4**, 398–430 (1976).

Nicholls, D. G. and Rial, E., Brown fat mitochondria, *Trends Biochem. Sci.* **9**, 489–491 (1984).

Ovchinnikov, Yu. A., Abdulaev, N. G., and Modyanov, N. N., Structural basis of proton-translocating function, *Annu. Rev. Biophys. Bioeng.* **11**, 445–463 (1982).

Senior, A. E., Secondary and tertiary structure of membrane proteins involved in proton translocation, *Biochim. Biophys. Acta* **726**, 81–95 (1983).

Slater, E. C., The Q cycle, an ubiquitous mechanism of electron transfer, *Trends Biochem. Sci.* **8**, 239–242 (1983).

Stoeckenius, W. and Bogomolni, R. A., Bacteriorhodopsin and related pigments of halobacteria, *Annu. Rev. Biochem.* **51**, 587–616 (1982).

Tanford, C., Mechanism of free energy coupling in active transport, *Annu. Rev. Biochem.* **52**, 379–409 (1983).

Wang, J. H., Coupling of proton flux to the hydrolysis and synthesis of ATP, *Annu. Rev. Biophys. Bioeng.* **12**, 21–34 (1983).

Wikström, M., Identification of the electron transfers in cytochrome oxidase that are coupled to proton pumping, *Nature* **338**, 776–778 (1989).

Wikström, M., Krab, K., and Saraste, M., Proton-translocating cytochrome complexes, *Annu. Rev. Biochem.* **50**, 623–655 (1981).

Wilson, D. F., Energy transduction in biological membranes, *in* Bittar, E. E. (Ed.), *Membrane Structure and Function*, Vol. 1, *pp.* 153–195, Wiley-Interscience (1980).

Control of ATP Production

Erecińska, M. and Wilson, D. F., Regulation of cellular energy metabolism, *J. Membr. Biol.* **70**, 1–14 (1982).

Klingenberg, M., The ADP,ATP shuttle of the mitochondrion, *Trends Biochem. Sci.* **4**, 249–252 (1979).

Williamson, J. R., Mitochondrial metabolism and cell regulation, *in* Packer, L. and Gómez-Puyou, A. (Eds.). *Mitochondria*, *pp.* 79–107, Academic Press (1976).

Williamson, J. R., Mitochondrial function in the heart, *Annu. Rev. Physiol.* **41**, 485–506 (1979).

Problems

1. Rank the following redox-active coenzymes and prosthetic groups of the electron-transport chain in order of increasing electron affinity: cytochrome a, CoQ, FAD, cytochrome c, NAD^+.

2. Why is the oxidation of succinate to fumarate only associated with the production of two ATPs during oxidative phosphorylation while the oxidation of malate to oxaloacetate is associated with the production of three ATPs?

3. What is the thermodynamic efficiency of oxidizing $FADH_2$ so as to synthesize two ATPs under standard biochemical conditions?

4. Sublethal cyanide poisoning may be reversed by the administration of nitrites. These substances oxidize hemoglobin, which has a relatively low affinity for CN^-, to methemoglobin, which has a relatively high affinity for CN^-. Why is this treatment effective?

5. Match the compound with it's behavior: (1) rotenone, (2) dinitrophenol, and (3) antimycin A. (a) Inhibits oxidative phosphorylation when the substrate is pyruvate but not when the substrate is succinate. (b) Inhibits oxidative phosphorylation when the substrate is either pyruvate or succinate. (c) Allows pyruvate to be oxidized by mitochondria even in the absence of ADP.

*6. **Nigericin** is an ionophore (Section 18-2C) that exchanges K^+ for H^+ across membranes. Explain how the treatment of functioning mitochondria with nigericin uncouples electron transport from oxidative phosphorylation. Does valinomycin, an ionophore that transports K^+ but not H^+, do the same? Explain.

7. The difference in pH between the internal and external surfaces of the inner mitochondrial membrane is 1.4 pH units (external side acidic). If the membrane potential is assumed to be 0.06 V (inside negative) what is the free energy released upon transporting 1 mol of protons back across the membrane? How many protons must be transported to provide enough free energy for the synthesis of 1 mol of ATP (assume standard biochemical conditions)?

8. Explain why: (a) Submitochondrial particles from which F_1 has been removed are permeable to protons. (b) Addition of oligomycin to F_1-depleted submitochondrial particles decreases this permeability severalfold.

9. Oligomycin and cyanide both inhibit oxidative phosphorylation when the substrate is either pyruvate or succinate. Dinitrophenol can be used to distinguish between these inhibitors. Explain.

10. For the oxidation of a given amount of glucose, does nonshivering thermogenesis by brown fat or shivering thermogenesis by muscle produce more heat?

11. How does atractyloside affect mitochondrial respiration? (*Hint:* see Section 18-4C.)

12. Certain unscrupulous operators offer, for a fee, to freeze recently deceased individuals in liquid nitrogen until medical science can cure the disease from which they died. What is the biochemical fallacy of this procedure?

Chapter 21

OTHER PATHWAYS OF CARBOHYDRATE METABOLISM

Heretofore, we have dealt with many aspects of carbohydrate metabolism. We have seen how the free energy of glucose oxidation is sequestered in ATP through glycolysis, the citric acid cycle, and oxidative phosphorylation. We have also studied the mechanism by which glucose is stored as glycogen for future use and how glycogen metabolism is controlled in response to the needs of the organism. In this chapter, we examine several other carbohydrate metabolism pathways of importance:

1. **Gluconeogenesis,** through which noncarbohydrate precursors such as lactate, pyruvate, glycerol, and amino acids are converted to glucose.

2. The **glyoxylate pathway,** through which plants convert acetyl-CoA to glucose.

3. Oligosaccharide and glycoprotein biosynthesis, through which oligosaccharides are synthesized and added to specific amino acid residues of proteins.

4. The **pentose phosphate pathway,** an alternate pathway of glucose degradation, which generates **NADPH,** the source of reducing equivalents in reductive biosynthesis, and **ribose-5-phosphate,** the sugar precursor of the nucleic acids.

This chapter completes our study of carbohydrate metabolism in animals; photosynthesis, which occurs only in plants and certain bacteria, is the subject of Chapter 22.

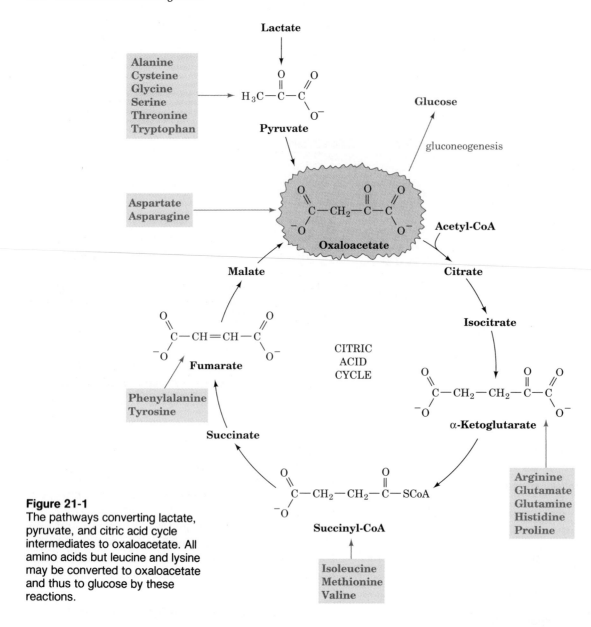

Figure 21-1
The pathways converting lactate, pyruvate, and citric acid cycle intermediates to oxaloacetate. All amino acids but leucine and lysine may be converted to oxaloacetate and thus to glucose by these reactions.

1. GLUCONEOGENESIS

Glucose occupies a central role in metabolism, both as a fuel and as a precursor of essential structural carbohydrates and other biomolecules. The brain and red blood cells are almost completely dependent on glucose as an energy source. Yet the liver's capacity to store glycogen is only sufficient to supply the brain with glucose for about half a day under fasting or starvation conditions. Thus, *when fasting, most of the body's glucose needs must be met by gluconeogenesis (literally, new glucose synthesis), the biosynthesis of glucose from noncarbohydrate precursors.* Gluconeogenesis occurs in liver and, to a smaller extent, in kidney.

The noncarbohydrate precursors that can be converted to glucose include the glycolysis products lactate and pyruvate, citric acid cycle intermediates, and most amino acids. First, however, all these substances must be converted to oxaloacetate, the starting material for gluconeogenesis (Fig. 21-1). The only amino acids that cannot be converted to oxaloacetate in animals are leucine and lysine because their breakdown yields only acetyl-CoA (Section 24-3F) *which animals are unable to convert to oxaloacetate.* Likewise, fatty acids cannot serve as glucose precursors in animals because most fatty acids are degraded completely to acetyl-CoA (Section 23-2C). In plants, however, the glyoxylate cycle (Section 21-2) generates oxaloacetate from acetyl-CoA so that lipids can serve as a plant cell's only carbon source. Glycerol, a triacylglycerol breakdown product, is converted to glucose via synthesis of the glycolytic intermediate dihydroxyacetone phosphate as described in Section 23-1.

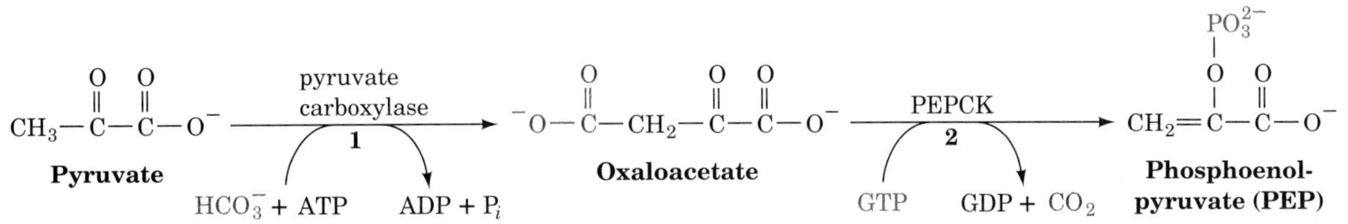

Figure 21-2
The conversion of pyruvate to oxaloacetate and then to phosphoenolpyruvate involves **(1)** pyruvate carboxylase and **(2)** PEP carboxykinase (PEPCK).

A. The Gluconeogenesis Pathway

Gluconeogenesis utilizes glycolytic enzymes. Yet, three of these enzymes, hexokinase, phosphofructokinase (PFK), and pyruvate kinase, catalyze reactions with large negative free energy changes in the direction of glycolysis. These reactions must therefore be replaced in gluconeogenesis by reactions that make glucose synthesis thermodynamically favorable. Here, as in glycogen metabolism (Section 17-1D), we see the recurrent theme that *biosynthetic and degradative pathways differ in at least one reaction. This not only permits both directions to be thermodynamically favorable under the same physiological conditions but allows the pathways to be independently controlled so that one direction can be activated while the other is inhibited.*

Pyruvate Is Converted to Oxaloacetate before Conversion to Phosphoenolpyruvate

The formation of phosphoenolpyruvate (PEP) from pyruvate, the reverse of the pyruvate kinase reaction, is endergonic and therefore requires free energy input. This is accomplished by first converting the pyruvate to oxaloacetate. Oxaloacetate is a "high-energy" intermediate whose exergonic decarboxylation provides the free energy necessary for PEP synthesis. The process requires the participation of two enzymes (Fig. 21-2):

1. **Pyruvate carboxylase** catalyzes the ATP-driven formation of oxaloacetate from pyruvate and HCO_3^-.

2. **PEP carboxykinase (PEPCK)** converts oxaloacetate to PEP in a reaction that uses GTP as a phosphorylating agent.

Pyruvate Carboxylase Has a Biotin Prosthetic Group

Pyruvate carboxylase, discovered in 1959 by Merton Utter, is a tetrameric protein of identical ~120-kD subunits, each of which has a **biotin** prosthetic group. *Biotin (Fig. 21-3a) functions as a CO_2 carrier by forming a carboxyl substituent at its ureido group (Fig. 21-3b).* Biotin is covalently bound to the enzyme by an amide linkage between the carboxyl group of its valerate side chain and the ε-amino group of an enzyme Lys residue to form a **biocytin** (alternatively, **biotinyllysine**) residue (Fig.

21-3*b*). The biotin ring system is therefore at the end of a 14-Å long flexible arm, much like that of the lipoic acid prosthetic group in the pyruvate dehydrogenase complex (Section 19-2A).

Biotin, which was first identified in 1935 as a growth factor in yeast, is an essential human nutrient. Its nutritional deficiency is rare, however, because it occurs in many foods and is synthesized by intestinal bacteria. Human biotin deficiency almost always results from the consumption of large amounts of raw eggs. This is because egg whites contain a protein, **avidin**, that binds

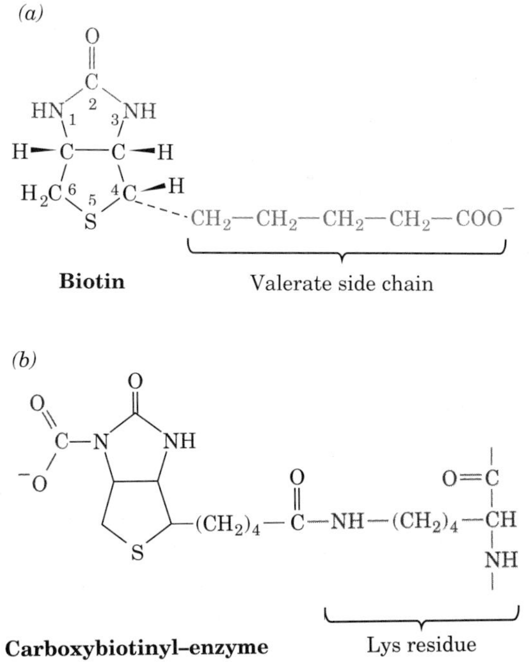

Figure 21-3
(a) Biotin consists of an imidazoline ring that is cis fused to a tetrahydrothiophene ring bearing a valerate side chain. The chirality at each of its three asymmetric centers is indicated. Positions 1, 2, and 3 constitute a ureido group. (b) Carboxybiotinyl–enzyme: N(1) of the biotin ureido group is the carboxylation site. Biotin is covalently attached to carboxylases by an amide linkage between its valeryl carboxyl group and an ε-amino group of an enzyme Lys side chain.

biotin so tightly (dissociation constant, $K = 10^{-15}M$) as to prevent its intestinal absorption (cooked eggs do not cause this problem because cooking denatures avidin). The presence of avidin in eggs is thought to inhibit the growth of microorganisms in this highly nutritious environment.

The Pyruvate Carboxylase Reaction

The pyruvate carboxylase reaction occurs in two steps (Fig. 21-4):

1. Biotin is carboxylated at its N(1′) atom by bicarbonate ion in a reaction driven by the concomitant hydrolysis of ATP to ADP and P_i. The resulting carboxyl group is "activated," since $\Delta G°'$ for its cleavage is $-19.7 \text{ kJ} \cdot \text{mol}^{-1}$. This carboxyl group can therefore be transferred without further free energy input.

2. The activated carboxyl group is transferred from carboxybiotin to pyruvate to form oxaloacetate.

These two reaction steps occur on different subsites of the same enzyme; the 14-Å arm of biocytin serves to transfer the biotin ring between the two sites.

Acetyl-CoA Regulates Pyruvate Carboxylase

Oxaloacetate synthesis is an anaplerotic (filling up) reaction that increases citric acid cycle activity (Section 19-4). Accumulation of the citric acid cycle substrate acetyl-CoA therefore signals the need for more oxaloacetate. Indeed, acetyl-CoA is a powerful allosteric activator of pyruvate carboxylase; the enzyme is all but inactive without bound acetyl-CoA. *If, however, the citric acid cycle is inhibited (by ATP and NADH whose presence in high concentrations indicates a satisfied demand for oxidative phosphorylation; Section 19-4), oxaloacetate instead undergoes gluconeogenesis.*

PEP Carboxykinase

PEPCK, a monomeric 74-kD enzyme, catalyzes the GTP-driven decarboxylation of oxaloacetate to form PEP and GDP (Fig. 21-5). Note that the CO_2 that carboxylates pyruvate to yield oxaloacetate is eliminated in the formation of PEP. Oxaloacetate may therefore be considered as "activated" pyruvate with CO_2 and biotin facilitating the activation at the expense of ATP hydrolysis. Acetyl-CoA is similarly activated for fatty acid biosynthesis through such a carboxylation–decarboxylation process (Section 23-4B).

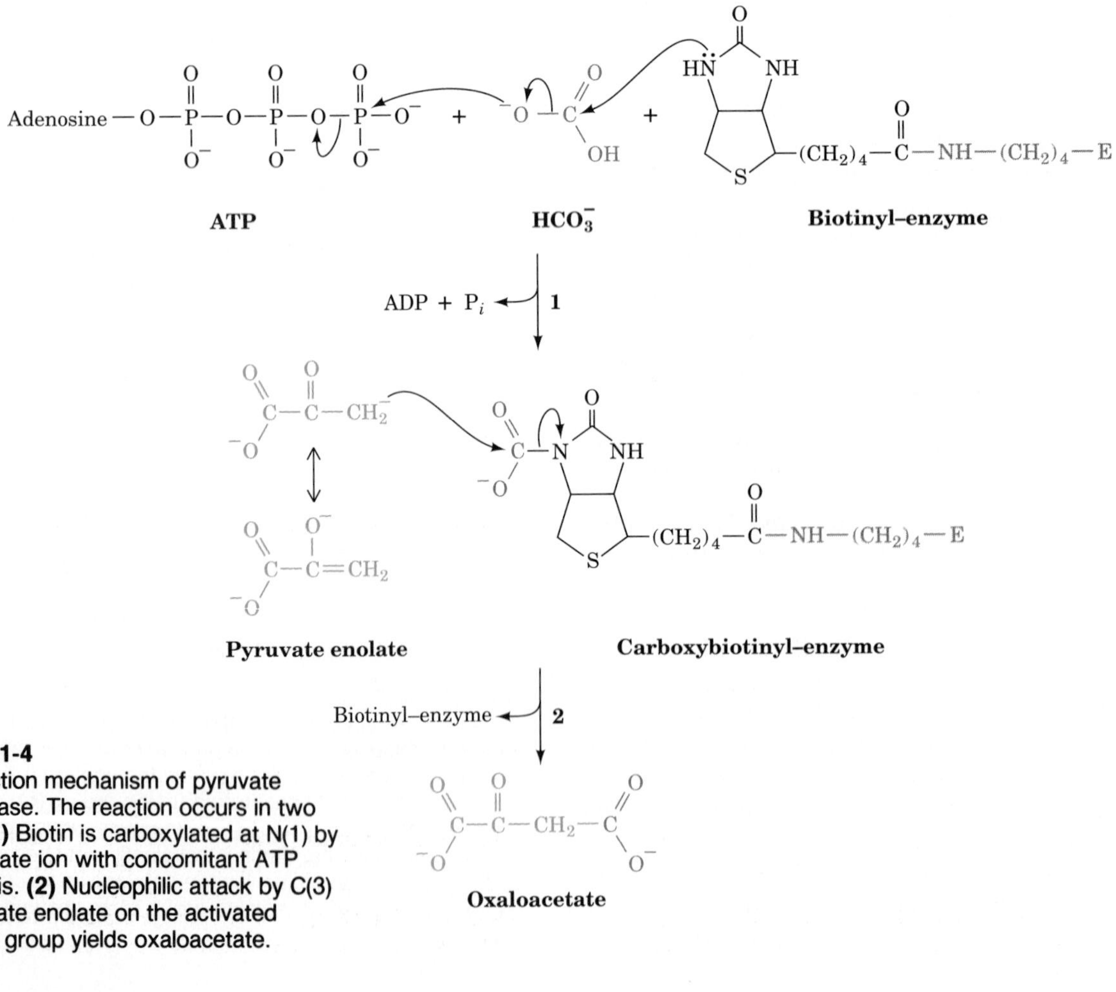

Figure 21-4
The reaction mechanism of pyruvate carboxylase. The reaction occurs in two steps: **(1)** Biotin is carboxylated at N(1) by bicarbonate ion with concomitant ATP hydrolysis. **(2)** Nucleophilic attack by C(3) of pyruvate enolate on the activated carboxyl group yields oxaloacetate.

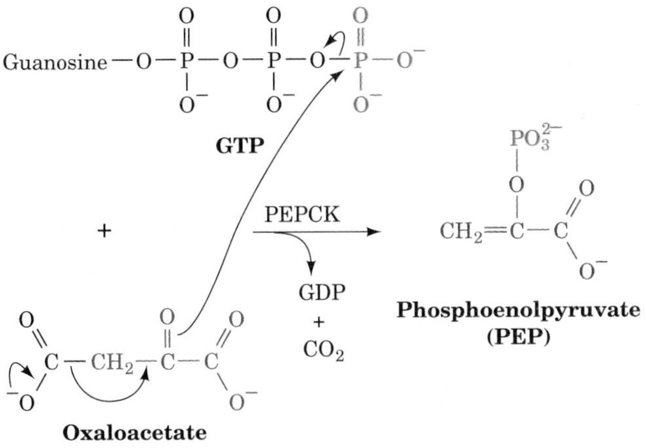

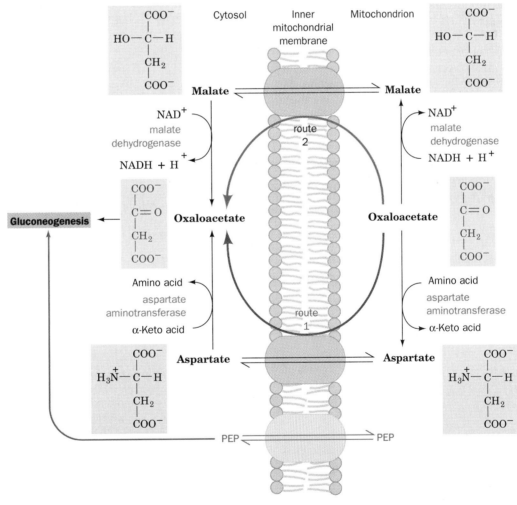

Figure 21-5
The PEPCK mechanism. Decarboxylation of oxaloacetate (a β-keto acid) forms a resonance-stabilized enolate anion whose oxygen atom attacks the γ-phosphoryl group of GTP forming PEP and GDP.

Gluconeogenesis Requires Metabolite Transport between Mitochondria and Cytosol

The generation of oxaloacetate from pyruvate or citric acid cycle intermediates occurs only in the mitochondrion, whereas the enzymes that convert PEP to glucose are cytosolic. The cellular location of PEPCK varies with the species. In mouse and rat liver it is located almost exclusively in the cytosol, in pigeon and rabbit liver it is mitochondrial, and in guinea pig and humans it is more or less equally distributed between both compartments. In order for gluconeogenesis to occur, either oxaloacetate must leave the mitochondrion for conversion to PEP, or the PEP formed there must enter the cytosol.

PEP is transported across the mitochondrial membrane by specific membrane transport proteins. There is, however, no such transport system for oxaloacetate. It must first be converted either to aspartate (Fig. 21-6, Route 1) or to malate (Fig. 21-6, Route 2) for which

Figure 21-6
The transport of PEP and oxaloacetate from the mitochondrion to the cytosol. PEP is directly transported between these compartments. Oxaloacetate, however, must first be converted to either aspartate through the action of

aspartate aminotransferase (Route 1) or to malate by malate dehydrogenase (Route 2). Route 2 involves the mitochondrial oxidation of NADH followed by the cytosolic reduction of NAD⁺ and therefore also transfers NADH reducing equivalents from the mitochondrion to the cytosol.

mitochondrial transport systems exist (Section 20-1B). The difference between these two routes involves the transport of NADH reducing equivalents. The **malate dehydrogenase** route (Route 2) results in the transport of reducing equivalents from the mitochondrion to the cytosol, since it utilizes mitochondrial NADH and produces cytosolic NADH. The **aspartate aminotransferase** route (Route 1) does not involve NADH. Cytosolic NADH is required for gluconeogenesis so, under most conditions, the route through malate is a necessity. If, however, lactate is the gluconeogenic precursor (Section 21-1C), its oxidation to pyruvate generates cytosolic NADH so that either transport route may then be used. Of course, as we have seen, the two routes may alternate (with Route 2 reversed) to form the malate–aspartate shuttle, which transports NADH reducing equivalents into the mitochondrion during oxidative metabolism (Section 20-1B).

Hydrolytic Reactions Bypass PFK and Hexokinase

The opposing pathways of gluconeogenesis and glycolysis utilize many of the same enzymes (Fig. 21-7). However, the free energy change is highly unfavorable in the gluconeogenic direction at two other points in the pathway in addition to the pyruvate kinase reaction: the PFK reaction and the hexokinase reaction. At these points, instead of generating ATP by reversing the glycolytic reactions, FBP and G6P are hydrolyzed, releasing P_i in exergonic processes catalyzed by **fructose-1,6-bisphosphatase (FBPase)** and **glucose-6-phosphatase**, respectively. *Glucose-6-phosphatase is unique to liver and kidney, permitting them to supply glucose to other tissues.*

Because of the presence of separate gluconeogenic enzymes at the three irreversible steps in the glycolytic conversion of glucose to pyruvate, both glycolysis and gluconeogenesis are rendered thermodynamically favorable. This is accomplished at the expense of the free energy of hydrolysis of two molecules each of ATP and GTP per molecule of glucose synthesized by gluconeogenesis in addition to that which would be consumed by the direct reversal of glycolysis.

Glycolysis:

$$\text{Glucose} + 2\text{NAD}^+ + 2\text{ADP} + 2P_i \longrightarrow$$
$$2\text{pyruvate} + 2\text{NADH} + 4\text{H}^+ + 2\text{ATP} + 2\text{H}_2\text{O}$$

Gluconeogenesis:

$$2\text{Pyruvate} + 2\text{NADH} + 4\text{H}^+ + \mathbf{4\text{ATP}} + \mathbf{2\text{GTP}} + 6\text{H}_2\text{O}$$
$$\longrightarrow \text{glucose} + 2\text{NAD}^+ + 4\text{ADP} + 2\text{GDP} + 6P_i$$

Overall:

$$2\text{ATP} + 2\text{GTP} + 4\text{H}_2\text{O} \longrightarrow 2\text{ADP} + 2\text{GDP} + 4P_i$$

Such free energy losses in a cyclic process are thermodynamically inescapable. They are the energetic price that

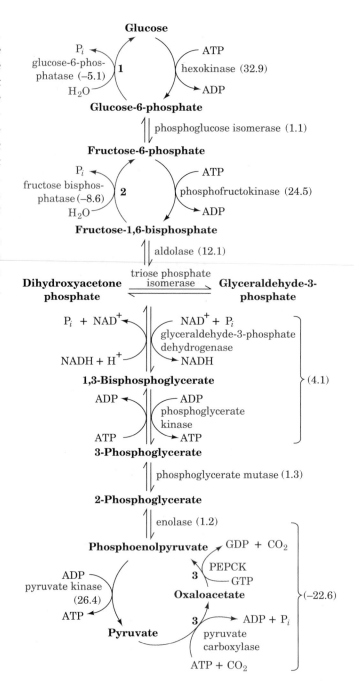

Figure 21-7
The pathways of gluconeogenesis and glycolysis. The three numbered steps, which are catalyzed by different enzymes in gluconeogenesis, have red arrows. The ΔG's for the reactions in the direction of gluconeogenesis under physiological conditions in liver are given in parentheses in kJ·mol^{-1}. [ΔG's obtained from Newsholme, E. A. and Leech, A. R., *Biochemistry for the Medical Sciences*, p. 448, Wiley (1983).]

Table 21-1
Regulators of Gluconeogenic Enzyme Activity

Enzyme	Allosteric Inhibitors	Allosteric Activators	Enzyme Phosphorylation	Protein Synthesis
PFK	ATP, citrate	AMP, F2,6P		
FBPase	AMP, F2,6P			
PK	Alanine	F1,6P	Inactivates	
Pyruvate carboxylase		Acetyl-CoA		
PEPCK				Stimulated by glucagon
PFK-2	Citrate	AMP, F6P, P_i	Inactivates	
FBPase-2	F6P	Glycerol-3-P	Activates	

must be paid to maintain independent regulation of the two pathways.

B. Regulation of Gluconeogenesis

If both glycolysis and gluconeogenesis were to proceed in an uncontrolled manner, the net effect would be a futile cycle wastefully hydrolyzing ATP and GTP. This does not occur. Rather, *these pathways are reciprocally regulated so as to meet the needs of the organism.* In the fed state, when the blood glucose level is high, the liver is geared toward fuel conservation: Glycogen is synthesized and the glycolytic pathway and pyruvate dehydrogenase are activated, breaking down glucose to acetyl-CoA for fatty acid biosynthesis and fat storage. In the fasted state, however, the liver maintains the blood glucose level both by stimulating glycogen breakdown and by reversing the flux through glycolysis toward gluconeogenesis (using mainly protein degradation products).

Glycolysis and Gluconeogenesis Are Controlled by Allosteric Interactions and Covalent Modifications

The rate and direction of glycolysis and gluconeogenesis are controlled at the points in these pathways where the forward and reverse directions can be independently regulated: the reactions catalyzed by (1) hexokinase/glucose-6-phosphatase, (2) PFK/FBPase, and (3) pyruvate kinase/pyruvate carboxylase–PEPCK (Fig. 21-7). Table 21-1 lists these regulatory enzymes and their regulators. The dominant mechanisms are allosteric interactions and cAMP-dependent covalent modifications (phosphorylation/dephosphorylation; Section 17-3). cAMP-dependent covalent modification renders this system sensitive to control by glucagon and other hormones that alter cAMP levels.

One of the most important allosteric effectors involved in the regulation of glycolysis and gluconeogenesis is fructose-2,6-bisphosphate (F2,6P), which activates PFK and inhibits FBPase (Section 17-3F). The concentration of F2,6P is controlled by its rates of synthesis and breakdown by phosphofructokinase-2 (PFK-2) and fructose bisphos-

phatase-2 (FBPase-2), respectively. Control of the activities of PFK-2 and FBPase-2 is therefore an important aspect of gluconeogenic regulation even though these enzymes do not catalyze reactions of the pathway. PFK-2 and FBPase-2 activities, which both occur on the same bifunctional enzyme, are subject to allosteric regulation as well as control by covalent modifications (Table 21-1). Low levels of blood glucose result in hormonal activation of gluconeogenesis through regulation of [F2,6P] (Fig. 21-8).

Activation of gluconeogenesis in liver also involves inhibition of glycolysis at the level of pyruvate kinase. *Liver pyruvate kinase is inhibited, both allosterically by alanine (a pyruvate precursor; Section 24-1A) and by phosphorylation.* Glycogen breakdown, in contrast, is stimulated by phosphorylation (Section 17-3C). Both pathways then flow towards G6P, which is converted to

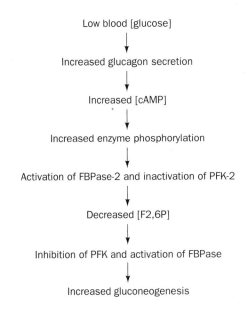

Figure 21-8
Hormonal regulation of [F2,6P] activates gluconeogenesis in liver in response to low blood [glucose].

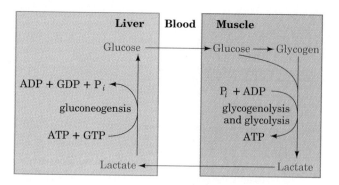

Figure 21-9
The Cori cycle. Lactate produced by muscle glycolysis is transported by the bloodstream to the liver where it is converted to glucose by gluconeogenesis. The bloodstream carries the glucose back to the muscles where it may be stored as glycogen.

glucose for export to muscle and brain. Muscle pyruvate kinase, an isozyme of the liver enzyme, is not subject to these controls. Indeed, such controls would be counterproductive in muscle since this tissue lacks the ability to synthesize glucose via gluconeogenesis.

C. The Cori Cycle

Muscle contraction is powered by hydrolysis of ATP, which is then regenerated through oxidative phosphorylation in the mitochondria of slow-twitch (red) muscle fibers and by glycolysis yielding lactate in fast-twitch (white) muscle fibers. Slow-twitch fibers also produce lactate when ATP demand exceeds oxidative flux. The lactate is transferred, via the bloodstream, to the liver where it is reconverted to pyruvate by lactate dehydrogenase and then to glucose by gluconeogenesis. Thus, through the intermediacy of the bloodstream, liver and muscle participate in a metabolic cycle known as the **Cori Cycle** (Fig. 21-9) in honor of Carl and Gerti Cori who first described it. This is the same ATP-consuming glycolysis/gluconeogenesis "futile cycle" we discussed above. Here, however, instead of occurring in the same cell, the two pathways occur in different organs. Liver ATP is used to resynthesize glucose from lactate produced in muscle. The resynthesized glucose is returned to the muscle, where it is stored as glycogen and used, on demand, to generate ATP for muscle contraction. The ATP utilized by the liver for this process is regenerated by oxidative phosphorylation. After vigorous exertion, it often takes at least 30 min for the oxygen consumption rate to return to its resting level, a phenomenon known as **oxygen debt**.

2. THE GLYOXYLATE PATHWAY

Plants, but not animals, possess enzymes that mediate the net conversion of acetyl-CoA to oxaloacetate. This is accomplished via the **glyoxylate pathway** (Fig.

21-10), a route involving enzymes of both the mitochondrion and the **glyoxysome** (a membranous plant organelle; Section 1-2A). Mitochondrial oxaloacetate is converted to aspartate by aspartate aminotransferase and transported to the glyoxysome where it is reconverted to oxaloacetate (Fig. 21-10, Reactions 1). The oxaloacetate is then condensed with acetyl-CoA to form citrate, which is isomerized to isocitrate as in the citric acid cycle (Fig. 21-10, Reactions 2 and 3). Glyoxysomal **isocitrate lyase** then cleaves isocitrate to succinate and glyoxylate (hence, the pathway's name; Fig. 21-10, Reaction 4). Succinate is transported to the mitochondrion where it enters the citric acid cycle for conversion to oxaloacetate. *The glyoxylate pathway therefore results in the net conversion of acetyl-CoA to glyoxylate instead of to two molecules of CO_2 as occurs in the citric acid cycle.*

Glyoxylate is converted to oxaloacetate in two reactions (Fig. 21-10):

Reaction 5. Malate synthase, a glyoxysomal enzyme, condenses glyoxylate with a second molecule of acetyl-CoA to form malate.

Reaction 6. Cytosolic malate dehydrogenase catalyzes the oxidation of malate to oxaloacetate by NAD^+.

The overall reaction of the glyoxylate cycle is therefore the formation of oxaloacetate from two molecules of acetyl-CoA.

$$2\text{Acetyl-CoA} + 2\text{NAD}^+ + \text{FAD} \longrightarrow$$
$$\text{oxaloacetate} + 2\text{CoASH} + 2\text{NADH} + \text{FADH}_2 + 2\text{H}^+$$

Isocitrate lyase and malate synthase, the only enzymes of the glyoxylate pathway unique to plants, enable germinating seeds to convert their stored triacylglycerols, through acetyl-CoA, to glucose.

3. BIOSYNTHESIS OF OLIGOSACCHARIDES AND GLYCOPROTEINS

Oligosaccharides consist of monosaccharide units joined together by glycosidic bonds [linkages between C(1), the anomeric carbon, of one unit and an OH group of a second unit; Section 10-1C]. About 80 different kinds of naturally occurring glycosidic linkages are known, most of which involve mannose, *N*-acetylglucosamine, *N*-acetylmuramic acid, glucose, fucose (6-deoxygalactose), galactose, *N*-acetylneuraminic acid (sialic acid), and *N*-acetylgalactosamine (Section 10-1). Glycosidic linkages also occur to lipids (glycosphingolipids; Section 11-1D), and proteins (glycoproteins; Section 10-3C).

Glycosidic bond formation requires free energy input under physiological conditions ($\Delta G°' = 16$ kJ·mol^{-1}). This free energy, as we have seen in the case of glycogen synthesis (Section 17-2B), is acquired through the con-

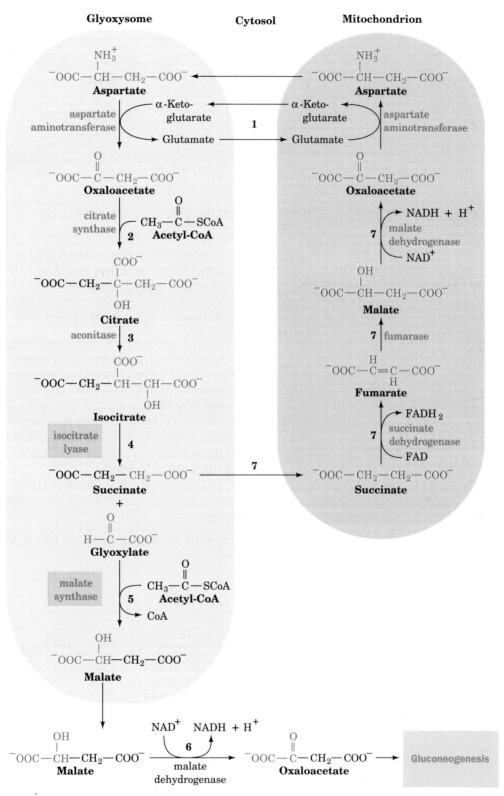

Figure 21-10
The glyoxylate pathway involves the participation of both mitochondrial and glyoxysomal enzymes. Isocitrate lyase and malate synthase, enzymes unique to plant glyoxysomes, are boxed. The pathway results in the net conversion of two acetyl-CoA to oxaloacetate. **(1)** Mitochondrial oxaloacetate is converted to aspartate, transported to the glyoxysome, and reconverted to aspartate. **(2)** Oxaloacetate is condensed with acetyl-CoA to form citrate. **(3)** Aconitase catalyzes the conversion of citrate to isocitrate. **(4)** Isocitrate lyase catalyzes the cleavage of isocitrate to succinate and glyoxylate. **(5)** Malate synthase catalyzes the condensation of glyoxylate with acetyl-CoA to form malate. **(6)** After transport to the cytosol, malate dehydrogenase catalyzes the oxidation of malate to oxaloacetate, which can then be used in gluconeogenesis. **(7)** Succinate is transported to the mitochondrion where it is reconverted to oxaloacetate via the citric acid cycle.

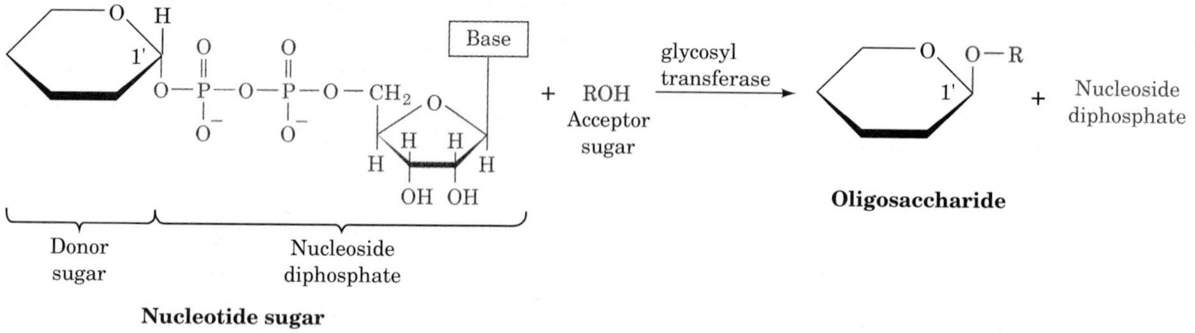

Figure 21-11
Nucleotide sugars are glycosyl donors in oligosaccharide biosynthesis catalyzed by glycosyl transferases.

Table 21-2

Sugar Nucleotides and Their Corresponding Monosaccharides in Glycosyl Transferase Reactions

UDP	GDP	CMP
N-Acetylgalactosamine	Fucose	Sialic acid
N-Acetylglucosamine	Mannose	
N-Acetylmuramic acid		
Galactose		
Glucose		
Glucuronic acid		
Xylose		

version of monosaccharide units to nucleotide sugars. A nucleotide at a sugar's anomeric carbon atom is a good leaving group and thereby facilitates formation of a glycosidic bond to a second sugar unit via reactions catalyzed by **glycosyl transferases** (Fig. 21-11). The nucleotides that participate in monosaccharide transfers are UDP, GDP, and CMP; a given sugar is associated with only one of these nucleotides (Table 21-2).

A. Lactose Synthesis

Several disaccharides are synthesized for future use as metabolic fuels. Typical of these is lactose [β-galactosyl-(1 → 4)-glucose; milk sugar], which is synthesized in the mammary gland by **lactose synthase** (Fig. 21-12). The donor sugar is UDP–galactose, which is formed by epimerization of UDP–glucose (Section 16-5B). The acceptor sugar is glucose.

Lactose synthase consists of two subunits:

1. **Galactosyl transferase,** the catalytic subunit, occurs in many tissues where it catalyzes the reaction of UDP–galactose and N-acetylglucosamine to yield N-acetyllactosamine, a constituent of many complex oligosaccharides (see, e.g., Fig. 21-16, Reaction 10).

2. **α-Lactalbumin,** a mammary gland protein with no catalytic activity, alters the specificity of galactosyl

transferase such that it utilizes glucose as an acceptor, rather than N-acetylglucosamine, to form lactose instead of N-acetyllactosamine.

B. Glycoprotein Synthesis

Proteins destined for secretion, incorporation into membranes, or localization inside membranous organelles, contain carbohydrates and are therefore classified as glycoproteins. *Glycosylation and oligosaccharide processing play an indispensable role in the sorting and the distribution of these proteins to their proper cellular destinations.* Their polypeptide components are ribosomally synthesized and processed by addition and modification of oligosaccharides.

The oligosaccharide portions of glycoproteins, as we have seen in Section 10-3C, are classified into two groups:

1. **N-Linked oligosaccharides,** which are attached to

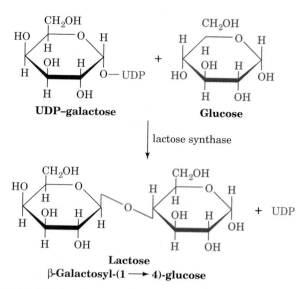

Figure 21-12
Lactose synthase catalyzes the formation of lactose from UDP–galactose and glucose.

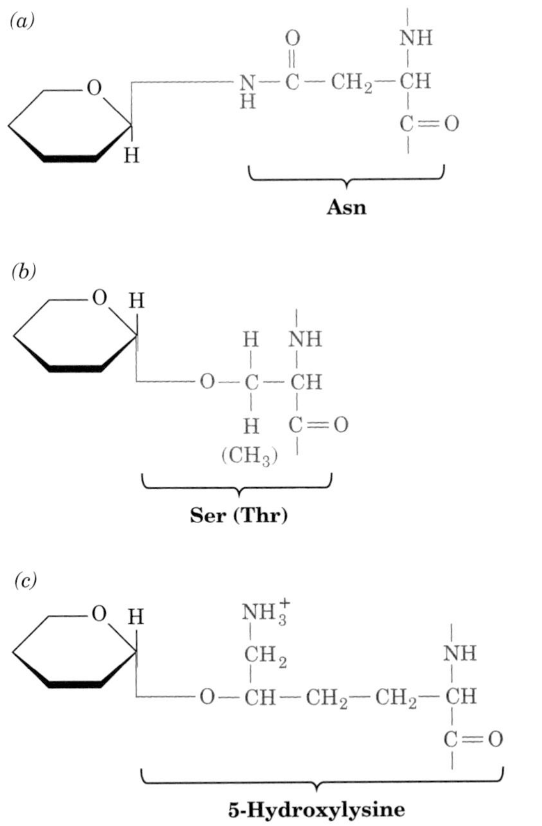

Figure 21-13
Types of glycosidic bonds in glycoproteins. (*a*) *N*-Linked glycosidic bond to an Asn residue in the sequence Asn-X-Ser/Thr. (*b*) *O*-Linked glycosidic bond to a Ser (or Thr) residue. (*c*) *O*-Linked glycosidic bond to a 5-hydroxylysine residue in collagen.

their polypeptide chain by a β-*N*-glycosidic bond to an Asn residue in the sequence Asn-X-Ser or Asn-X-Thr, where X is any amino acid residue except Pro or perhaps Asp (Fig. 21-13*a*).

2. ***O*-Linked oligosaccharides,** which are attached to their polypeptide chain through an α-*O*-glycosidic bond to Ser or Thr (Fig. 21-13*b*) or, only in collagens, to 5-hydroxylysine residues (Fig. 21-13*c*).

We shall consider the synthesis of these two types of oligosaccharides separately.

N-Linked Glycoproteins Are Synthesized in Four Stages

N-Linked glycoproteins are formed in the endoplasmic reticulum and further processed in the Golgi apparatus. Synthesis of their carbohydrate moieties occurs in four stages:

1. Synthesis of a lipid-linked oligosaccharide precursor.

2. Transfer of this precursor to the NH_2 group of an Asn residue on a growing polypeptide.

3. Removal of some of the precursor's sugar units.

4. Addition of sugar residues to the remaining core oligosaccharide.

We shall discuss these stages in order.

N-Linked Oligosaccharides Are Constructed on Dolichol Carriers

N-Linked oligosaccharides are initially synthesized as lipid-linked precursors. The lipid component in this process is **dolichol,** a long-chain polyisoprenol of 14 to 24 isoprene units (17–21 units in animals and 14–24 units in fungi and plants; isoprene units are C_5 units with the carbon skeleton of isoprene; Section 23-6A), which is linked to the oligosaccharide precursor via a pyrophosphate bridge (Fig. 21-14). Dolichol apparently anchors the growing oligosaccharide to the endoplasmic reticulum membrane. Involvement of lipid-linked oligosaccharides in *N*-linked glycoprotein synthesis was first demonstrated in 1972 by Armando Parodi and Luis Leloir who showed that, when a lipid-linked oligosaccharide containing [^{14}C]glucose is incubated with rat liver microsomes (vesicular fragments of isolated endoplasmic reticulum), the radioactivity becomes associated with protein.

N-Linked Glycoproteins Have a Common Oligosaccharide Core

The pathway of dolichol-PP-oligosaccharide synthesis involves stepwise addition of monosaccharide units to the growing glycolipid by specific glycosyl transferases to form a common "core" structure. Each monosaccharide unit is

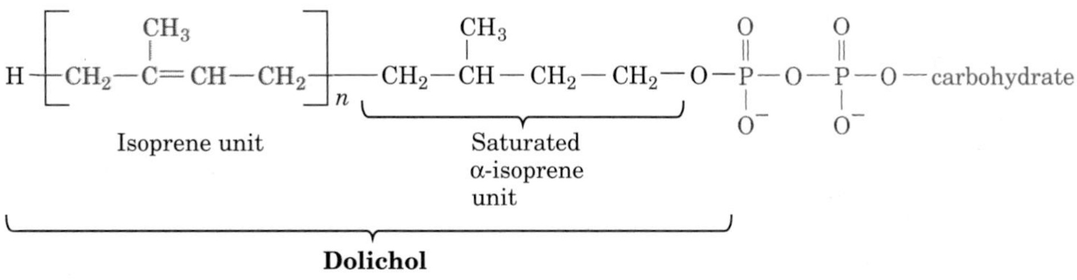

Figure 21-14
The carbohydrate precursors of *N*-linked glycosides are synthesized as dolichol pyrophosphate glycosides. Dolichols are long-chain polyisoprenols ($n = 14$–24) in which the α-isoprene unit is saturated.

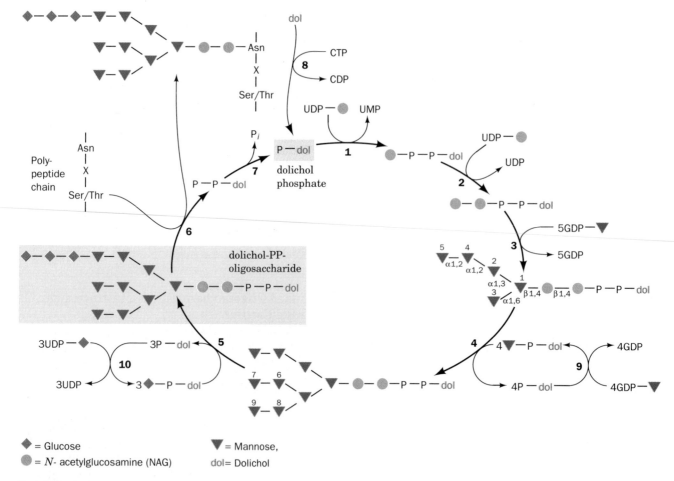

= Glucose

= N- acetylglucosamine (NAG)

= Mannose,

dol= Dolichol

Figure 21-15
The pathway of dolichol-PP-oligosaccharide synthesis:
(1) Addition of *N*-acetylglucosamine-1-P to dolichol-P.
(2) Addition of a second *N*-acetylglucosamine. **(3)** Addition of five mannosyl residues from GDP-mannose in reactions catalyzed by five different mannosyl transferases. **(4)** Addition of four mannosyl residues from dolichol-P-mannose in reactions catalyzed by four different mannosyl transferases. **(5)** Addition of three glucosyl residues from

dolichol-P-glucose. **(6)** Transfer of the oligosaccharide from dolichol-PP to the polypeptide chain at an Asn residue in the sequence Asn-X-Ser/Thr, releasing dolichol-PP.
(7) Hydrolysis of dolichol-PP to dolichol-P. **(8)** Dolichol-P can also be formed by phosphorylation of dolichol by CTP.
(9) Synthesis of dolichol-P-mannose from GDP-mannose and dolichol-P. **(10)** Synthesis of dolichol-P-glucose from UDPG and dolichol-P.

added by a unique glycosyl transferase (Fig. 21-15). For example, in Reaction 3 of Fig. 21-15, five mannosyl units are added through the action of five different mannosyl transferases, each with a different oligosaccharide-acceptor specificity. The oligosaccharide core, the product of Reaction 5 in Fig. 21-15, has the composition $(N\text{-acetylglucosamine})_2$ $(\text{mannose})_9$ $(\text{glucose})_3$.

Although nucleotide sugars are the most common monosaccharide donors in glycosyl transferase reactions, *several mannosyl and glucosyl residues are transferred to the growing dolichol-PP-oligosaccharide from their corresponding dolichol-P derivatives*. The requirement for **dolichol-P-mannose** was discovered by Stuart Kornfeld who found that mutant mouse lymphoma cells (lymphoma is a type of cancer) unable to synthesize the normal lipid-linked oligosaccharides formed a de-

fective, smaller glycolipid. These cells contain all the requisite glycosyl transferases but are unable to synthesize dolichol-P-mannose (Reaction 9 in Fig. 21-15 is blocked). When this substance is supplied to the mutant cells, mannosyl units are added to the defective dolichol-PP-oligosaccharide.

Asparagine-Linked Oligosaccharides Are Cotranslationally Added to Proteins

Vesicular-stomatitis virus (VSV), which infects cattle producing influenzalike symptoms, provides an excellent model system for studying *N*-linked glycoprotein processing. The VSV coat consists of host–cell membrane in which a single viral glycoprotein, the **VSV G protein,** is embedded. Since a viral infection almost totally usurps an infected cell's protein synthesizing ma-

Figure 21-16
Schematic pathway of oligosaccharide processing on newly synthesized vesicular-stomatitis virus glycoprotein. The reactions are catalyzed by: **(1)** membrane-bound oligosaccharide-transferring enzyme **(2)** α-glucosidase I, **(3)** α-glucosidase II, **(4)** ER α-1,2-mannosidase, **(5)** Golgi α-mannosidase I, **(6)** *N*-acetylglucosaminyltransferase I, **(7)** Golgi α-mannosidase II, **(8)** *N*-acetylglucosaminyltransferase II, **(9)** fucosyltransferase, **(10)** galactosyltransferase, and **(11)** sialyltransferase. Lysosomal proteins are modified by: **(I)** *N*-acetylglucosaminyl phosphotransferase and **(II)** *N*-acetylglucosamine-1-phosphodiester α-*N*-acetylglucosaminidase. The transfer of intermediates from RER to cis to medial to trans Golgi cisternae occurs via membranous vesicles. [Modified from Kornfeld, R. and Kornfeld, S., *Annu. Rev. Biochem.* **54**, 640 (1985).]

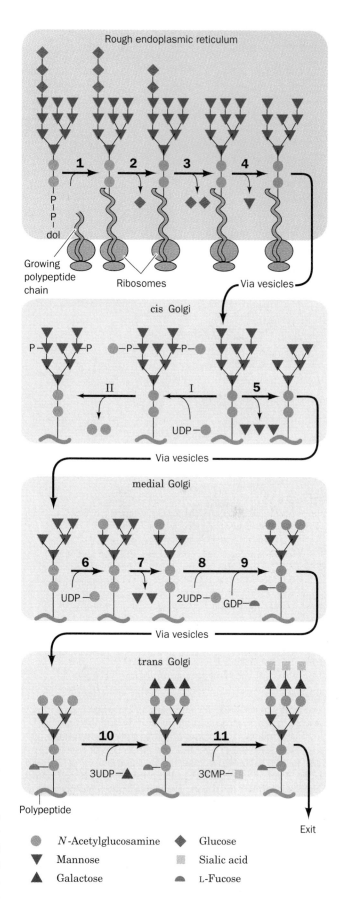

chinery, a VSV-infected cell's Golgi apparatus, which normally contains hundreds of different types of glycoproteins, contains virtually no other glycoprotein but G protein. Consequently, the maturation of the G protein is relatively easy to follow.

Study of VSV-infected cells indicated that the *transfer of the lipid-linked oligosaccharide to a polypeptide chain occurs while the polypeptide chain is still being synthesized.* The G protein is *N*-glycosylated by **membrane-bound oligosaccharide-transferring enzyme,** which recognizes the amino acid sequence Asn-X-Ser/Thr (Fig. 21-15, Reaction 6; Fig. 21-16, Reaction 1). Yet, only about one third of the Asn-X-Ser/Thr sites of eukaryotic proteins are actually *N*-glycosylated. The application of structure prediction algorithms (Section 8-1C) to the amino acid sequences flanking known *N*-glycosylation sites, together with glycosylation studies of model polypeptides, suggests that these sites occur at β turns or loops in which the Asn peptide N—H group is hydrogen bonded to the Ser/Thr hydroxyl O atom. This explains why Pro cannot occupy the X position; it would prevent Asn-X-Ser/Thr from assuming the putative required hydrogen bonded conformation.

Glycoprotein Processing Begins in the Endoplasmic Reticulum and Is Completed in the Golgi Apparatus

Processing of primary glycoproteins begins in the endoplasmic reticulum by the enzymatic trimming (removal) of their three glucose residues (Fig. 21-16, Reactions 2 and 3) and one of their mannose residues (Fig. 21-16, Reaction 4). The glycoproteins are then transported, in membranous vesicles, to the Golgi apparatus for further processing.

The Golgi apparatus consists of a stack of 5 to 20 (depending on the species) flattened membranous sacs (Fig. 21-17; an electron micrograph of the Golgi apparatus is presented in Fig. 1-5). The Golgi contains at least three different types of sacs, the **cis, medial,** and **trans cisternae,** each of which, as shown by James Rothman and Kornfeld, contains different sets of glycoprotein

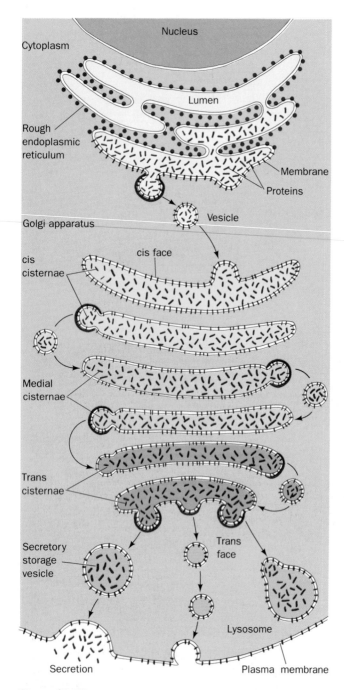

Figure 21-17
Most proteins destined for secretion or insertion into a membrane are synthesized by ribosomes attached to the rough endoplasmic reticulum (*top*). As they are synthesized, the proteins are either injected into the lumen of the endoplasmic reticulum or inserted into its membrane (Section 11-3F). After initial processing, the proteins are encapsulated in vesicles formed from endoplasmic reticulum membrane, which subsequently fuse with the cis cisternae of the Golgi apparatus. The proteins are progressively processed (Fig. 21-16), according to their cellular destinations, in the cis, medial, and trans cisternae of the Golgi between which they are transported by other membranous vesicles. In the trans cisternae (*bottom*), the completed glycoproteins are sorted for delivery to their final destinations, for example, lysosomes, the plasma membrane, or secretory granules, to which they are transported by yet other vesicles.

processing enzymes. As a glycoprotein traverses the Golgi stack, from the cis to the medial to the trans cisternae, mannose residues are trimmed and *N*-acetylglucosamine, galactose, fucose, and sialic acid residues are added to complete its processing (Fig. 21-16; Reactions 5–11). The glycoproteins are then sorted in the trans cisternae for transfer to their respective cellular destinations. The glycoproteins are transported between these various locations in membranous vesicles (Section 11-3F).

There is enormous diversity among the different oligosaccharides of *N*-linked glycoproteins as is indicated, for example, in Fig. 10-29c. Indeed, *even glycoproteins with a given polypeptide chain exhibit considerable microheterogeneity* (Section 10-3C), presumably as a consequence of incomplete glycosylation and lack of absolute specificity on the part of glycosyl transferases and glycosylases.

The processing of all *N*-linked oligosaccharides appears to be identical through Reaction 4 of Fig. 21-16 so that all of them have a common $(N\text{-acetylglucosamine})_2$ $(\text{mannose})_3$ core (five "noncore" mannose residues are subsequently trimmed from VSV glycoprotein in Reactions 5 and 7). The diversity of the *N*-linked oligosaccharides therefore arises through divergence from this sequence after Reaction 7. The resulting oligosaccharides are classified into three groups:

1. **High mannose oligosaccharides** (Fig. 21-18a), which contain 2 to 9 mannose residues appended to the common pentasaccharide core (red residues in Fig. 21-18).

2. **Complex oligosaccharides** (Fig. 21-18b), which contain variable numbers of *N*-acetylgalactosamine units as well as sialic acid and/or fucose residues linked to the core.

3. **Hybrid oligosaccharides** (Fig. 21-18c), which contain elements of both high mannose and complex chains.

It is unclear how different types of oligosaccharides are related to the functions and/or final cellular locations of their glycoproteins. Lysosomal glycoproteins, however, appear to be of the high mannose variety.

Inhibitors Have Aided the Study of *N*-Linked Glycosylation

Elucidation of the events in the glycosylation process has been greatly facilitated through the use of inhibitors that block specific glycosylation enzymes. Two of the most useful are the antibiotics **tunicamycin** (Fig. 21-19a), a hydrophobic analog of UDP-*N*-acetylglucosamine, and **bacitracin** (Fig. 21-20), a cyclic polypeptide. Both were discovered because of their ability to inhibit bacterial cell wall biosynthesis, a process that also involves the participation of lipid-linked oligosac-

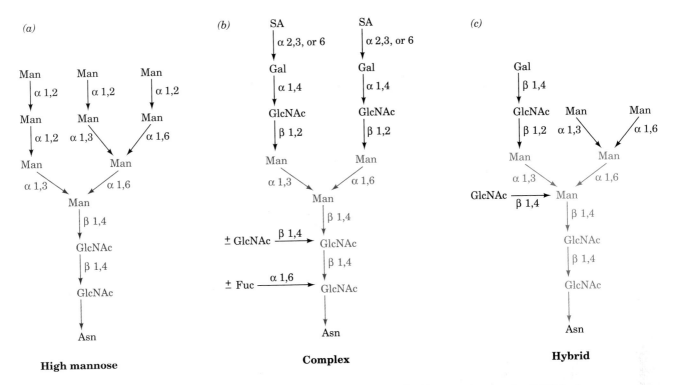

Figure 21-18
Typical primary structures of (a) high mannose, (b) complex, and (c) hybrid N-linked oligosaccharides. The pentasaccharide core common to all N-linked oligosaccharides is indicated in red. [After Kornfield, R. and Kornfield, S., *Annu. Rev. Biochem.* **54,** 633 (1985).]

charides. Tunicamycin blocks the formation of dolichol-PP-oligosaccharides by inhibiting the synthesis of dolichol-PP-N-acetylglucosamine from dolichol-P and UDP-N-acetylglucosamine (Fig. 21-15, Reaction 1). Tunicamycin resembles an adduct of these reactants (Fig. 21-19b) and, in fact, binds to the enzyme with a dissociation constant of $7 \times 10^{-9}M$.

Bacitracin forms a complex with dolichol-PP that inhibits its dephosphorylation (Fig. 21-15, Reaction 7), thereby preventing glycoprotein synthesis from lipid-linked oligosaccharide precursors. Bacitracin is clinically useful because it destroys bacterial cell walls but does not affect animal cells because it cannot cross cell membranes (bacterial cell wall biosynthesis is an extracellular process).

O-Linked Oligosaccharides Are Post-Translationally Formed

The study of the biosynthesis of **mucin,** an O-linked glycoprotein secreted by the submaxillary salivary gland indicates that *O-linked oligosaccharides are synthesized in the Golgi apparatus by serial addition of monosaccharide units to a completed polypeptide chain (Fig. 21-21).* Synthesis starts with the transfer of N-acetylgalactosamine (GalNAc) from UDP-GalNAc to a Ser or Thr residue on the polypeptide by **GalNAc transferase.** In contrast to N-linked oligosaccharides, which are transferred to an Asn in a specific amino acid sequence, the O-glycosylated Ser and Thr residues are not members of

any common sequence. Rather, it is thought that the location of glycosylation sites is specified only by the secondary or tertiary structure of the polypeptide. Glycosylation continues with stepwise addition of galactose, sialic acid, N-acetylglucosamine, and fucose by the corresponding glycosyl transferases.

Oligosaccharides on Glycoproteins Act as Recognition Sites

Glycoproteins synthesized in the endoplasmic reticulum and processed in the Golgi apparatus are targeted for secretion, insertion into cell membranes, or incorporation into cellular organelles such as lysosomes (Fig. 21-17). This suggests that *oligosaccharides serve as recognition markers for this sorting process.* For example, the study of I-cell disease (Section 11-3F) demonstrated that in glycoprotein enzymes destined for the lysosome, a mannose residue is converted to mannose-6-phosphate (M6P) in the cis cisternae of the Golgi. The process involves two enzymes (Fig. 21-16, Reactions I and II), although how they recognize lysosomal protein precursors is unknown. In the trans cisternae, M6P-bearing glycoproteins are sorted into lysosome-bound coated vesicles through their specific binding to a 215-kD membrane glycoprotein called the **M6P receptor.** Individuals with I-cell disease lack the enzyme catalyzing mannose phosphorylation (Fig. 21-16, Reaction I), resulting in the secretion of the normally lysosome-resident enzymes.

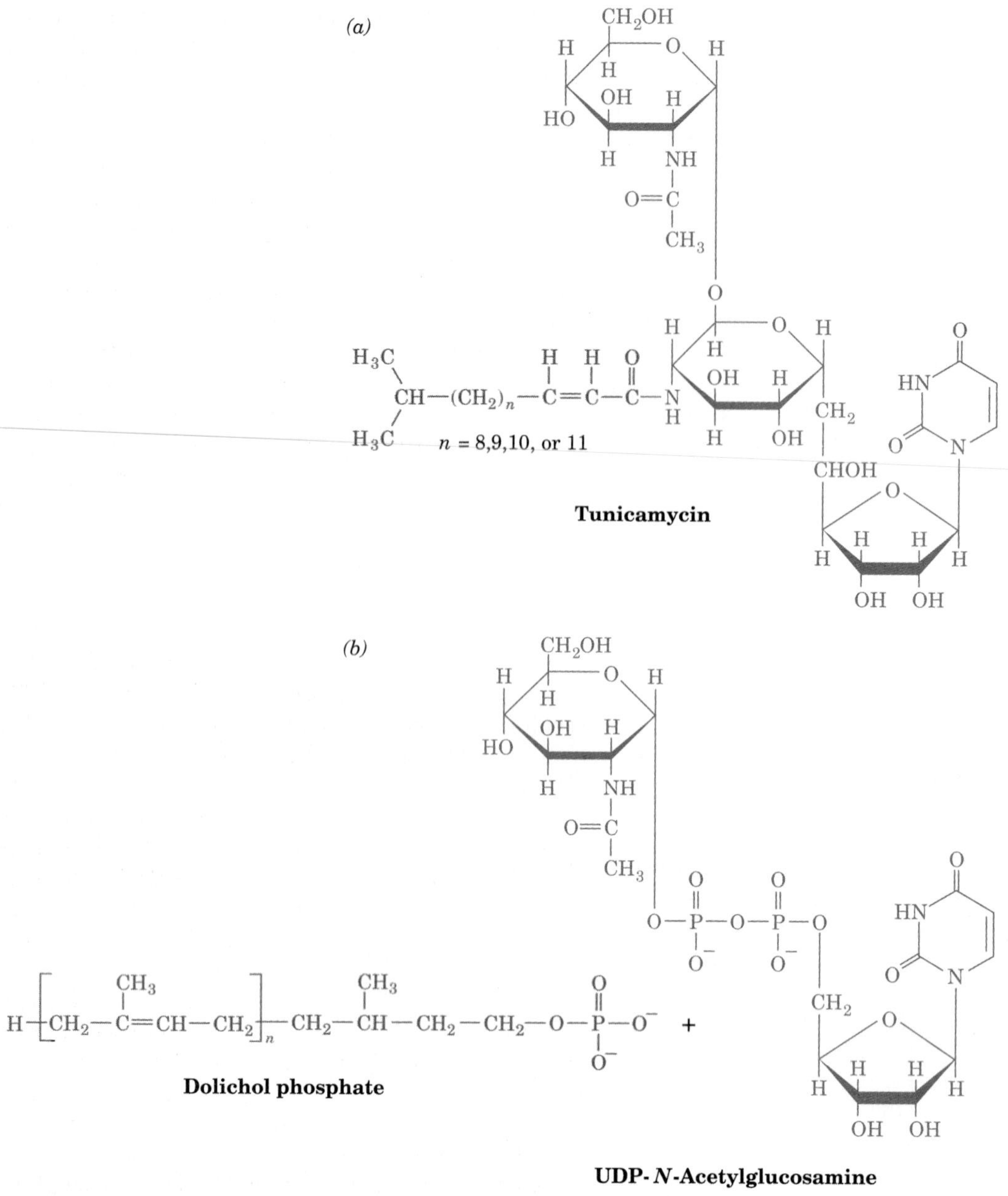

(a)

Tunicamycin

n = 8,9,10, or 11

(b)

Dolichol phosphate

UDP-*N*-Acetylglucosamine

Figure 21-19
Comparison of the chemical structures of (a) tunicamycin and (b) dolichol-P + UDP-*N*-acetylglucosamine.

H$_3$N—CH—C≡N—CH—C—Leu—D-Glu—Ile—Lys—D-Orn—Ile—D-Phe—His—Asp—D-Asn—COO$^-$

Ile   Cys   (CH$_2$)$_4$ CH$_2$

NH ——————————————— C=O

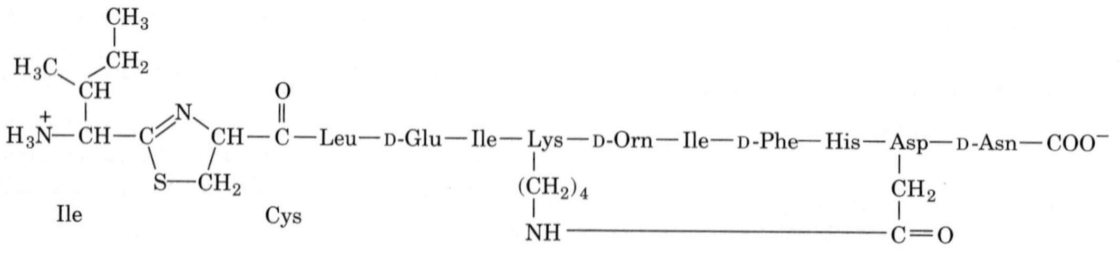

Bacitracin

Figure 21-20
The chemical structure of bacitracin. Note that this dodecapeptide has four D-amino acid residues and two unusual intrachain linkages. "Orn" represents the nonstandard amino acid residue ornithine.

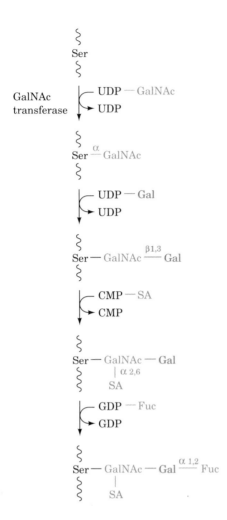

Figure 21-21
The proposed synthesis pathway for the carbohydrate moiety of an *O*-linked oligosaccharide chain of canine submaxillary mucin.

ABO blood group antigens (Section 11-3D) are *O*-linked glycoproteins. Their characteristic oligosaccharides are components of both cell-surface lipids and of proteins that occur in various secretions such as saliva. These oligosaccharides form antibody recognition sites.

Glycoproteins are believed to mediate cell–cell recognition. For example, an *O*-linked oligosaccharide on a glycoprotein that coats the mouse ovum surface (zona pellucida) acts as the sperm receptor. Even when this oligosaccharide is separated from its protein, it retains the ability to bind mouse sperm.

4. THE PENTOSE PHOSPHATE PATHWAY

ATP is the cell's "energy currency"; its exergonic hydrolysis is coupled to many otherwise endergonic cell functions. *Cells have a second currency, reducing power.*

Many endergonic reactions, notably the reductive biosynthesis of fatty acids (Section 23-4) and cholesterol (Section 23-6A), as well as photosynthesis (Section 22-3A), require NADPH in addition to ATP. Despite their close chemical resemblance, *NADPH and NADH are not metabolically interchangeable* (recall that these coenzymes differ only by a phosphate group at the 2'-OH group of NADPH's adenosine moiety; Fig. 12-2). While NADH participates in utilizing the free energy of metabolite oxidation to synthesize ATP (oxidative phosphorylation), *NADPH is involved in utilizing the free energy of metabolite oxidation for otherwise endergonic reductive biosynthesis.* This differentiation is possible because the dehydrogenase enzymes involved in oxidative and reductive metabolism exhibit a high degree of specificity towards their respective coenzymes. Indeed, cells normally maintain their $[NAD^+]/[NADH]$ ratio near 1000, which favors metabolite oxidation, while keeping their $[NADP^+]/[NADPH]$ ratio near 0.01, which favors metabolite reduction.

NADPH is generated by the oxidation of G6P via an alternative pathway to glycolysis, the **pentose phosphate pathway** *[also called the* **hexose monophosphate (HMP) shunt** *or the* **phosphogluconate pathway**; *Fig. 21-22].* The first evidence of this pathway's existence was obtained in the 1930s by Otto Warburg who discovered $NADP^+$ through his studies on the oxidation of G6P to 6-phosphogluconate. Further indications came from the observation that tissues continue to respire in the presence of high concentrations of fluoride ion which, it will be recalled, blocks glycolysis by inhibiting enolase (Section 16-2I). It was not until the 1950s, however, that the pentose phosphate pathway was elucidated by Frank Dickens, Bernard Horecker, Fritz Lipmann, and Efraim Racker. Tissues most heavily involved in fatty acid and cholesterol biosynthesis (liver, mammary gland, adipose tissue, and adrenal cortex), are rich in pentose phosphate pathway enzymes. Indeed, some 30% of the glucose oxidation in liver occurs via the pentose phosphate pathway.

The overall reaction of the pentose phosphate pathway is

$$3G6P + 6NADP^+ + 3H_2O \rightleftharpoons$$
$$6NADPH + 6H^+ + 3CO_2 + 2F6P + GAP$$

However, the pathway may be considered to have three stages:

1. Oxidative reactions (Fig. 21-22, Reactions 1–3), which yield NADPH and **ribulose-5-phosphate (Ru5P)**.

$$3G6P + 6NADP^+ + 3H_2O \longrightarrow$$
$$6NADPH + 6H^+ + 3CO_2 + 3Ru5P$$

2. Isomerization and epimerization reactions (Fig. 21-22, Reactions 4 and 5), which transform Ru5P

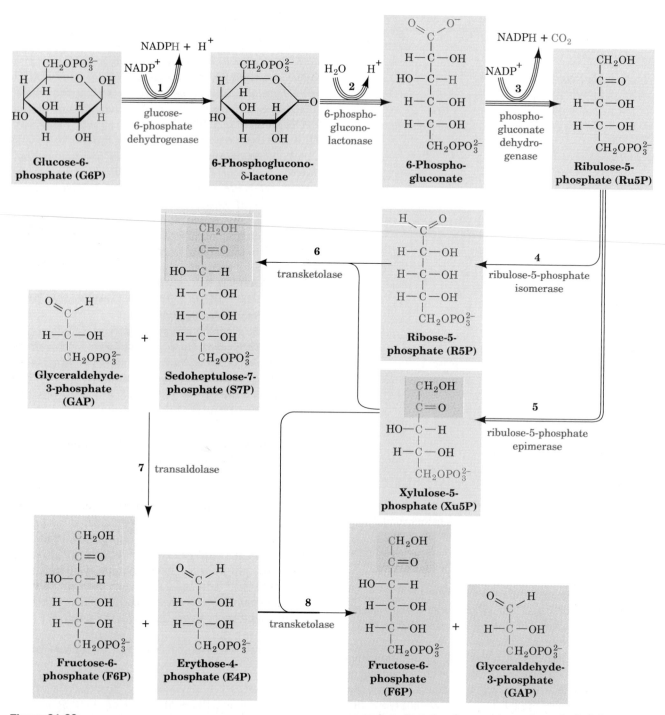

Figure 21-22
The pentose phosphate pathway. The number of lines in an arrow represents the number of molecules reacting in one turn of the pathway so as to convert three G6Ps to three CO_2s, two F6Ps and one GAP. For the sake of clarity, sugars from Reaction 3 onward are shown in their linear forms. The carbon skeleton of R5P and the atoms derived from it are drawn in red. The C_2 units transferred by transketolase are shaded in green and the C_3 units transferred by transaldolase are shaded in blue.

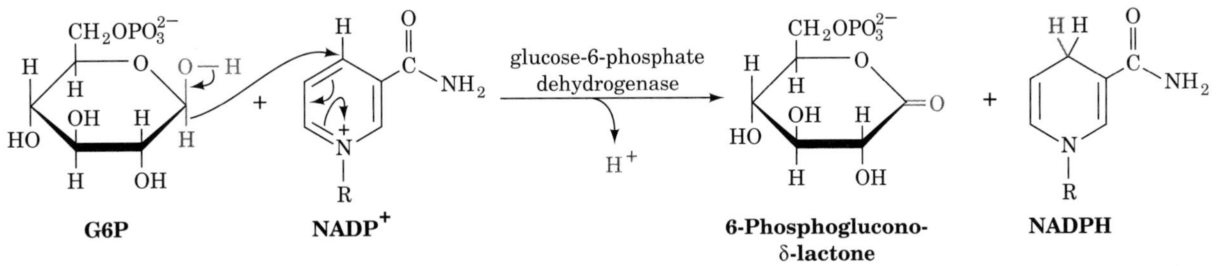

Figure 21-23
The glucose-6-phosphate dehydrogenase reaction.

either to **ribose-5-phosphate (R5P)** or to **xylulose-5-phosphate (Xu5P).**

$$3\text{Ru5P} \rightleftharpoons \text{R5P} + 2\text{Xu5P}$$

3. A series of C—C bond cleavage and formation reactions (Fig. 21-22, Reactions 6–8) that convert two molecules of Xu5P and one molecule of R5P to two molecules of fructose-6-phosphate (F6P) and one of glyceraldehyde-3-phosphate (GAP).

$$\text{R5P} + 2\text{Xu5P} \rightleftharpoons 2\text{F6P} + \text{GAP}$$

The reactions of Stages 2 and 3 are freely reversible so that the products of the pathway vary with the needs of the cell. For example, when R5P is required for nucleotide biosynthesis, Stage 3 produces less F6P and GAP. In this section, we discuss the three stages of the pentose phosphate pathway and how this pathway is controlled. We close by considering the consequences of one of its abnormalities.

A. Oxidative Reactions of NADPH Production

Only the first three reactions of the pentose phosphate pathway are involved in NADPH production.

1. **Glucose-6-phosphate dehydrogenase** catalyzes net transfer of a hydride ion to NADP$^+$ from C(1) of G6P to form **6-phosphoglucono-δ-lactone** (Fig. 21-23). G6P, a cyclic hemiacetal with C(1) in the aldehyde oxidation state, is thereby oxidized to a cyclic ester (lactone). The enzyme is specific for NADP$^+$ and is strongly inhibited by NADPH.

2. **6-Phosphogluconolactonase** increases the rate of hydrolysis of 6-phosphoglucono-δ-lactone to **6-phosphogluconate** (the nonenzymatic reaction occurs at a significant rate), the substrate of the next oxidative enzyme in the pathway.

3. **Phosphogluconate dehydrogenase** catalyzes the oxidative decarboxylation of 6-phosphogluconate, a β-hydroxy acid, to Ru5P and CO$_2$ (Fig. 21-24). The reaction is similar to that catalyzed by the citric acid cycle enzyme isocitrate dehydrogenase (Section 19-3C).

Formation of Ru5P completes the oxidative portion of the pentose phosphate pathway. *It generates two molecules of NADPH for each molecule of G6P that enters the pathway.* The product Ru5P must subsequently be converted to R5P or Xu5P for further use.

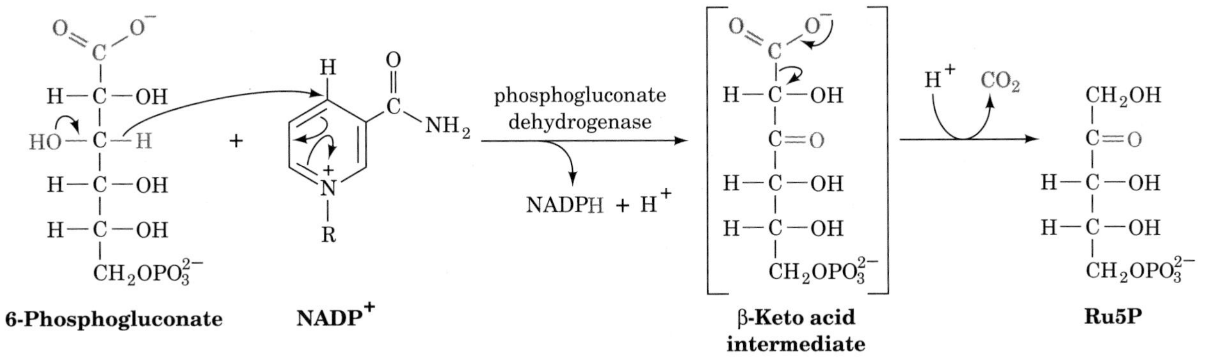

Figure 21-24
The phosphogluconate dehydrogenase reaction. Oxidation of the OH group forms an easily decarboxylated β-keto acid (although the proposed intermediate has never been isolated).

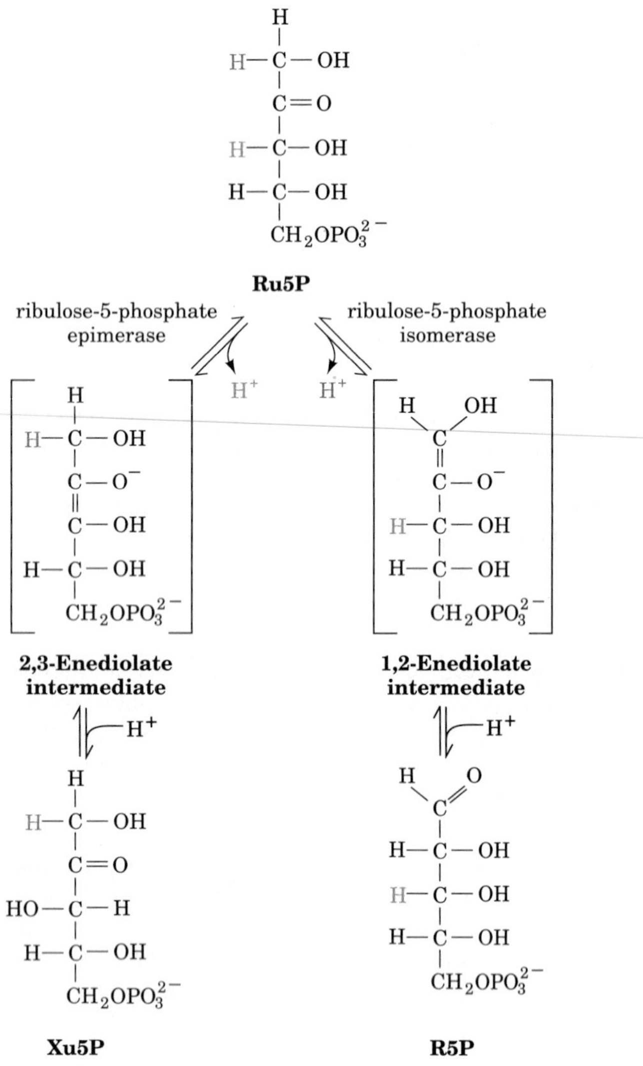

Figure 21-25
The ribulose-5-phosphate isomerase and ribulose-5-phosphate epimerase reactions both involve enediolate intermediates. In the isomerase reaction (*right*), a base on the enzyme removes a proton from C(1) of Ru5P to form a 1,2-enediolate, and then adds a proton at C(2) to form R5P. In the epimerase reaction (*left*), a base on the enzyme removes a C(3) proton to form a 2,3-enediolate. A proton is then added to the same carbon atom but with inversion of configuration to yield Xu5P.

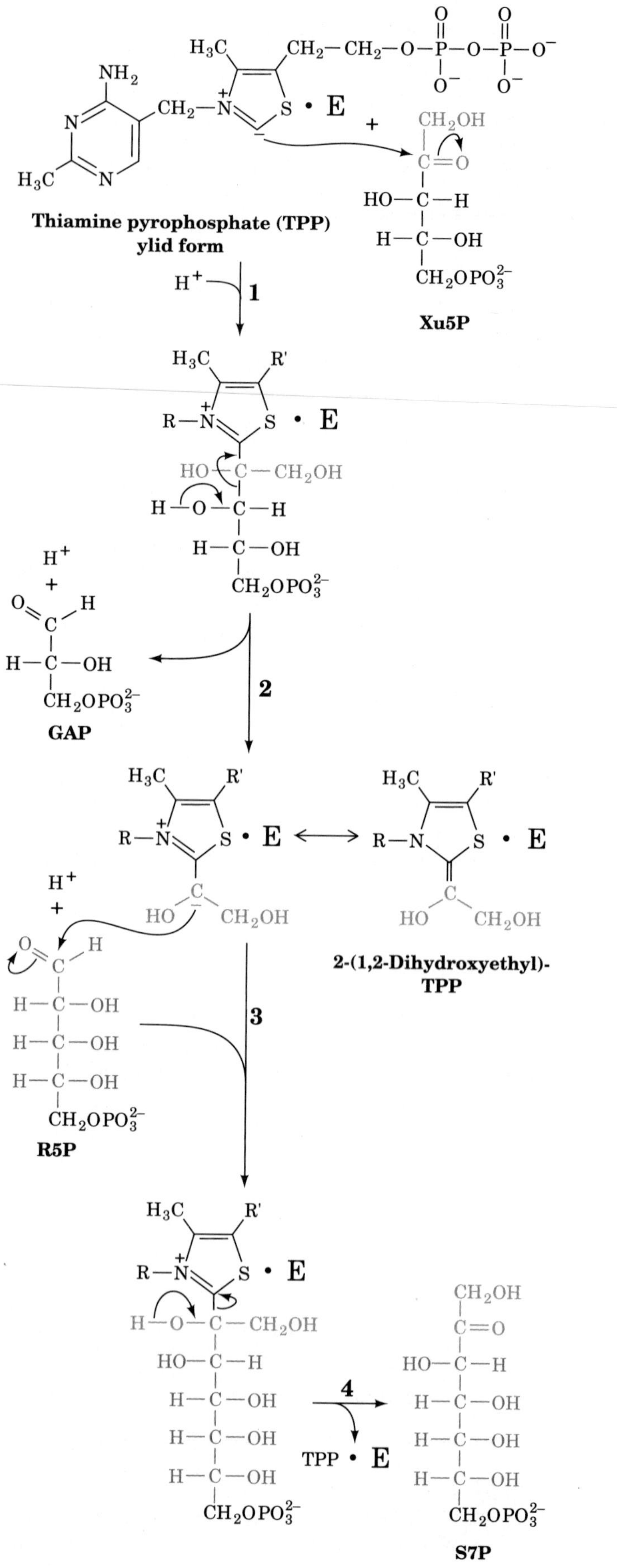

B. Isomerization and Epimerization of Ribulose-5-phosphate

*Ru5P is converted to R5P by **ribulose-5-phosphate isomerase** (Fig. 21-22, Reaction 4) and to Xu5P by **ribulose-5-phosphate epimerase** (Fig. 21-22, Reaction 5).* These isomerization and epimerization reactions, as discussed in Section 15-2B, are both thought to occur via enediolate intermediates (Fig. 21-25).

R5P is an essential precursor in the biosynthesis of nucleotides (Sections 26-2, 3, and 6). If, however, more R5P is

Figure 21-26 (*opposite*)
Transketolase utilizes the coenzyme thiamine pyrophosphate to stabilize the carbanion formed on cleavage of the C(2)—C(3) bond of Xu5P. The reaction occurs as follows: **(1)** The TPP ylid attacks the carbonyl group of Xu5P. **(2)** C(2)—C(3) bond cleavage yields GAP and enzyme-bound 2-(1,2-dihydroxyethyl)-TPP, a resonance stabilized carbanion. **(3)** The C(2) carbanion attacks the aldehyde carbon of R5P forming an S7P–TPP adduct. **(4)** TPP is eliminated yielding S7P and the regenerated TPP–enzyme.

formed than the cell needs, the excess, along with Xu5P, is converted to the glycolytic intermediates F6P and GAP as is described below.

C. Carbon–Carbon Bond Cleavage and Formation Reactions

The conversion of three C_5 sugars to two C_6 sugars and one C_3 sugar involves a remarkable "juggling act" catalyzed by two enzymes, **transaldolase** and **transketolase**. As we discussed in Section 15-2E, enzymatic reactions that make or break carbon–carbon bonds usually have mechanisms that involve generation of a stabilized carbanion and its addition to an electrophilic center such as an aldehyde. This is the dominant theme of both the transaldolase and the transketolase reactions.

Transketolase Catalyzes the Transfer of C_2 Units

Transketolase, which has a thiamine pyrophosphate cofactor (TPP; Section 16-3B), catalyzes the transfer of a C_2 unit from Xu5P to R5P yielding GAP and **sedoheptulose-7-phosphate (S7P)** (Fig. 21-22, Reaction 6). The reaction involves the intermediate formation of a covalent adduct between Xu5P and TPP (Fig. 21-26).

Transaldolase Catalyzes the Transfer of C_3 Units

Transaldolase catalyzes the transfer of a C_3 unit from S7P to GAP yielding **erythrose-4-phosphate (E4P)** and F6P (Fig. 21-22, Reaction 7). The reaction occurs by aldol cleavage (Section 16-2D), which begins with the formation of a Schiff base between an ε-amino group of an essential enzyme Lys residue and the carbonyl group of S7P (Fig. 21-27).

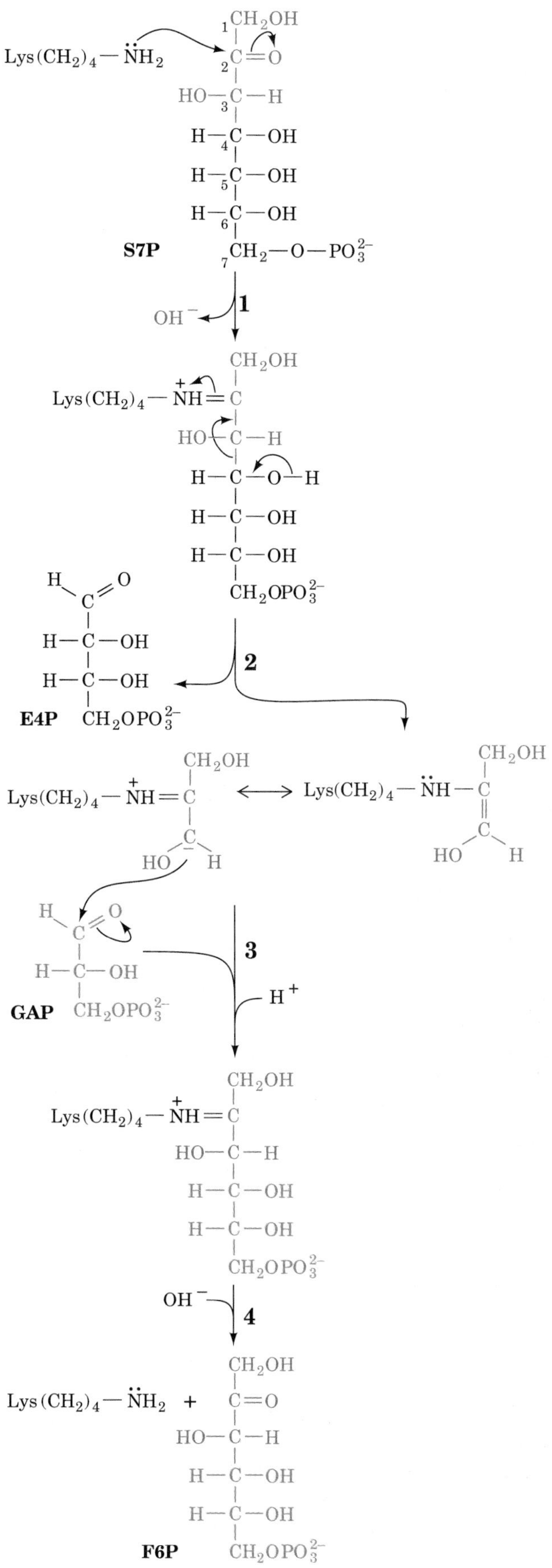

Figure 21-27
Transaldolase contains an essential Lys residue that forms a Schiff base with S7P to facilitate an aldol cleavage reaction. The reaction occurs as follows: **(1)** The ε-amino group of an essential Lys residue forms a Schiff base with the carbonyl group of S7P. **(2)** A Schiff base-stabilized C(3) carbanion is formed in an aldol cleavage reaction between C(3) and C(4) that eliminates E4P. **(3)** The enzyme-bound resonance-stabilized carbanion adds to the carbonyl C atom of GAP forming F6P linked to the enzyme via a Schiff base. **(4)** The Schiff base hydrolyzes, regenerating active enzyme and releasing F6P.

(6) $C_5 + C_5 \rightleftharpoons C_7 + C_3$

(7) $C_7 + C_3 \rightleftharpoons C_6 + C_4$

(8) $\underline{C_5 + C_4 \rightleftharpoons C_6 + C_3}$

(Sum) $3C_5 \rightleftharpoons 2C_6 + C_3$

Figure 21-28
The carbon–carbon bond formations and cleavages that convert three C_5 sugars to two C_6 and one C_3 sugar in the pentose phosphate pathway. The number to the left of each reaction is keyed to the corresponding reaction in Fig. 21-22.

A Second Transketolase Reaction Yields GAP and a Second F6P Molecule

In a second transketolase reaction, a C_2 unit is transferred from a second molecule of Xu5P to E4P to form GAP and another molecule of F6P (Fig. 21-22, Reaction 8). The third phase of the pentose phosphate pathway thus transforms two molecules of Xu5P and one of R5P to two molecules of F6P and one molecule of GAP. These carbon skeleton transformations (Fig. 21-22, Reactions 6–8) are summarized in Fig. 21-28.

D. Control of the Pentose Phosphate Pathway

The principal products of the pentose phosphate pathway are R5P and NADPH. The transaldolase and transketolase reactions serve to convert excess R5P to glycolytic intermediates when the metabolic need for NADPH exceeds that of R5P in nucleotide biosynthesis. The resulting GAP and F6P can be consumed through glycolysis and oxidative phosphorylation or recycled by gluconeogenesis to form G6P. *In the latter case, 1 molecule of G6P can be converted, via 6 cycles of the pentose phosphate pathway and gluconeogenesis, to 6CO$_2$ molecules with the concomitant generation of 12NADPH molecules.* When the need for R5P outstrips that for NADPH, F6P and GAP can be diverted from the glycolytic pathway for use in the synthesis of R5P by reversal of the transaldolase and transketolase reactions.

Flux through the pentose phosphate pathway and thus the rate of NADPH production is controlled by the rate of the glucose-6-phosphate dehydrogenase reaction (Fig. 21-22, Reaction 1). The activity of this enzyme, which catalyzes the pentose phosphate pathway's first committed step ($\Delta G = -17.6\ \text{kJ}\cdot\text{mol}^{-1}$ in liver), is regulated by the NADP$^+$ concentration (substrate availability). When the cell consumes NADPH, the NADP$^+$ concentration rises, increasing the rate of the glucose-6-phosphate dehydrogenase reaction, thereby stimulating NADPH regeneration.

E. Glucose-6-phosphate Dehydrogenase Deficiency

NADPH is required for several reductive processes in addition to biosynthesis. For example, erythrocyte membrane integrity requires a plentiful supply of reduced glutathione (GSH), a Cys-containing tripeptide (Section 14-4). A major function of GSH in the erythrocyte is to eliminate H_2O_2 and organic hydroperoxides. H_2O_2, a toxic product of various oxidative processes, reacts with double bonds in the fatty acid residues of the erythrocyte cell membrane to form organic hydroperoxides. These, in turn, react to cleave fatty acid C—C bonds thereby damaging the membrane. In erythrocytes, the unchecked buildup of peroxides results in premature cell lysis. Peroxides are eliminated through the action of **glutathione peroxidase,** one of the handful of enzymes with a selenium cofactor, yielding glutathione disulfide (GSSG).

$$2\text{GSH} + \text{R}-\text{O}-\text{O}-\text{H} \xrightarrow{\text{glutathione peroxidase}} \text{GSSG} + \text{ROH} + \text{H}_2\text{O}$$

Organic hydroperoxide

GSH is subsequently regenerated by the NADPH reduction of GSSG by glutathione reductase (Section 14-4).

$$\text{GSSG} + \text{NADPH} + \text{H}^+ \xrightarrow{\text{glutathione reductase}} 2\text{GSH} + \text{NADP}^+$$

A steady supply of NADPH is therefore vital for erythrocyte integrity.

Primaquine Causes Hemolytic Anemia in Glucose-6-phosphate Dehydrogenase Mutants

A genetic defect, common in African, Asian, and Mediterranean populations, results in severe hemolytic anemia on administration of certain drugs including the antimalarial agent **primaquine**

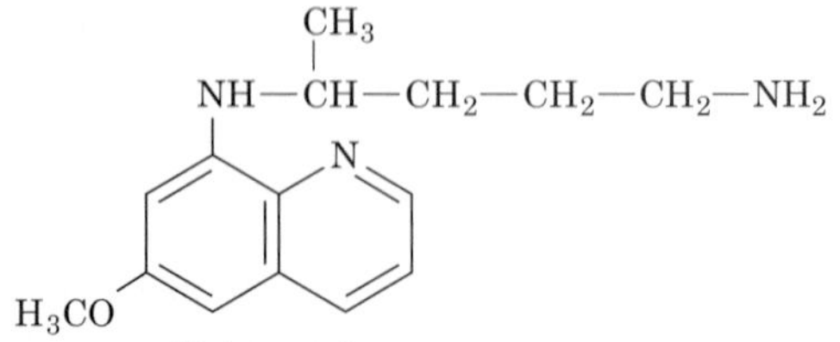

Primaquine

This sex-linked trait has been traced to an altered gene for glucose-6-phosphate dehydrogenase. Under most conditions, mutant erythrocytes have sufficient enzyme activity for normal function. Primaquine, however, stimulates peroxide formation thereby increasing the demand for NADPH to a level that mutant cells cannot meet.

The major reason for low enzymatic activity in affected cells appears to be an accelerated rate of breakdown of the mutant enzyme (protein degradation is discussed in Section 30-6). This explains why patients with glucose-6-phosphate dehydrogenase deficiency react to primaquine with hemolytic anemia but recover within a week despite continued primaquine treatment. Mature erythrocytes lack a nucleus and protein synthesizing machinery and therefore cannot synthesize new enzyme molecules to replace degraded ones (they likewise cannot synthesize new membrane components which is why they are so sensitive to membrane damage in the first place). The initial primaquine treatments result in the lysis of old red blood cells whose defective glucose-6-phosphate dehydrogenase has been largely degraded. Lysis products stimulate the release of young cells that contain more enzyme and are therefore better able to cope with primaquine stress.

Over 100 glucose-6-dehydrogenase mutations are known, several of which occur with high incidence. For example, the so-called type A⁻ deficiency, which exhibits ∼ 10% of the normal glucose-6-phosphate dehydrogenase activity, has an incidence of 11% among black Americans. This, together with the high prevalence of defective glucose-6-phosphate dehydrogenase in malarial areas of the world, suggests that such mutations confer resistance to the malarial parasite, *Plasmodium falciparum*. Indeed, erythrocytes with glucose-6-phosphate dehydrogenase deficiency appear to be less suitable hosts for *Plasmodia* than normal cells because the parasite requires the products of the pentose phosphate pathway and/or because the erythrocyte is lysed before the parasite has had a chance to mature. Thus, like the sickle-cell trait (Section 6-3A), *a defective glucose-6-phosphate dehydrogenase confers a selective advantage on individuals living where malaria is endemic.*

Chapter Summary

Lactate, pyruvate, citric acid cycle intermediates, and many amino acids may be converted, by gluconeogenesis, to glucose via the formation of oxaloacetate. For this to occur, the three irreversible steps of glycolysis must be bypassed. The pyruvate kinase reaction is bypassed by converting pyruvate to oxaloacetate in an ATP-driven reaction catalyzed by the biotinyl–enzyme pyruvate carboxylase. The oxaloacetate is subsequently phosphorylated by GTP and decarboxylated to PEP in a reaction catalyzed by PEPCK. For this to happen in species in which PEPCK is a cytosolic enzyme, the oxaloacetate must be transported from the mitochondrion to the cytosol via its interim conversion to either malate or aspartate. Conversion to malate concomitantly transports reducing equivalents to the cytosol in the form of NADH. The two other irreversible steps of glycolysis, the PFK reaction and the hexokinase reaction, are bypassed by simply hydrolyzing their products, FBP and G6P, by FBPase and glucose-6-phosphatase, respectively. A glucose molecule may therefore be synthesized from pyruvate at the expense of four ATPs more than is generated by the reverse process. Glycolysis and gluconeogenesis are reciprocally regulated so as to consume glucose when the demand for ATP is high and synthesize it when the demand is low. The control points in these processes are at pyruvate kinase/pyruvate carboxylase–PEPCK, PFK/FBPase, and hexokinase/glucose-6-phosphatase. Regulation of these enzymes is exerted largely through allosteric interactions and cAMP-dependent enzyme modifications. Muscle, which is incapable of gluconeogenesis, transfers much of the lactate it produces to the liver via the blood for conversion to glucose and return to the muscle. This Cori cycle shifts the metabolic burden of oxidative ATP generation for gluconeogenesis from muscle to liver.

Animals cannot convert fatty acids to glucose because they lack the enzymes necessary to synthesize oxaloacetate from acetyl-CoA. Plants, however, can do so via the glyoxylate cycle, a glyoxysomal process that converts two molecules of acetyl-CoA to one molecule of oxaloacetate via the intermediate formation of glyoxylate.

Glycosidic bonds are formed by transfer of the monosaccharide unit of a sugar nucleotide to a second sugar unit. Such reactions occur in the synthesis of disaccharides such as lactose and of the carbohydrate components of glycoproteins. In *N*-linked glycoproteins, the carbohydrate component is attached to the protein via an *N*-glycosidic bond to an Asn residue in the sequence Asn-X-Ser/Thr. In *O*-linked glycoproteins, the carbohydrate attachment is an *O*-glycosidic bond to Ser or Thr, or in collagens, to 5-hydroxylysine. Synthesis of *N*-linked oligosaccharides begins in the endoplasmic reticulum with the multistep formation of a lipid-linked precursor consisting of dolichol pyrophosphate bonded to a common 14-residue core oligosaccharide. The carbohydrate is then transferred to an Asn residue of a growing polypeptide chain. The immature *N*-linked glycoprotein is subsequently transferred, via a membranous vesicle, to the cis cisternae of the Golgi apparatus. Processing is completed by trimming of mannose residues followed by attachment of a variety of other monosaccharides as catalyzed by specific enzymes in the cis, medial, and trans Golgi cisternae. Completed *N*-linked glycoproteins are sorted in the trans Golgi cisternae according to the identities of their carbohydrate components for transport, via membranous vesicles, to their final cellular destinations. Three major types of *N*-linked oligosaccharides have been identified, high mannose, complex, and hybrid oligosaccharides, all of which contain a common pentasaccharide core. Studies of glycoprotein formation have been facilitated by the use of antibiotics, such as tunicamycin and bacitracin, which inhibit specific enzymes involved in the synthesis of these oligosaccharides. *O*-Linked oligosaccharides are synthesized in the Golgi apparatus by sequential attachments of specific monosaccharide units to certain Ser or Thr residues. Carbohydrate components of glycoproteins are thought to act as recognition markers for the transport of glycoproteins to their proper cellular destinations and for cell–cell and antibody recognition.

The cell uses NAD$^+$ in oxidative reactions, and employs NADPH in reductive biosynthesis. NADPH is synthesized by the pentose phosphate pathway, an alternate mode of glucose oxidation. This pathway also synthesizes R5P for use in nucleotide biosynthesis. The first three reactions of the pentose phosphate pathway involve oxidation of G6P to Ru5P with release of CO_2 and formation of two NADPH molecules. This is followed by reactions that either isomerize Ru5P to R5P or epimerize it to Xu5P. Each molecule of R5P not required for nucleotide biosynthesis, together with two Xu5P, is converted to two molecules of F6P and one molecule of GAP via the sequential actions of transketolase, transaldolase and, again, transketolase. The products of the pentose phosphate pathway depend on the needs of the cell. The F6P and GAP may be metabolized through glycolysis and the citric acid cycle or recycled via gluconeogenesis. If NADPH is in excess, the latter portion of the pentose phosphate pathway may be reversed to synthesize R5P from glycolytic intermediates. The pentose phosphate pathway is controlled at its first committed step, the glucose-6-phosphate dehydrogenase reaction, by the NADP$^+$ concentration. A genetic deficiency in glucose-6-phosphate dehydrogenase leads to hemolytic anemia on administration of the antimalarial drug primaquine. This deficiency, which results from the accelerated degradation of the mutant enzyme, provides resistance against malaria.

References

Gluconeogenesis

Hers, H. G. and Hue, L., Gluconeogenesis and related aspects of glycolysis, *Annu. Rev. Biochem.* **52**, 617–653 (1983).

Hue, L., The role of futile cycles in the regulation of carbohydrate metabolism in the liver, *Adv. Enzymol.* **52**, 247–330 (1981).

Krauss-Friedman, N., *Hormonal Control of Gluconeogenesis*, Vols. I and II, CRC Press (1986).

Pilkus, S. J., Mahgrabi, M. R., and Claus, T. H., Hormonal regulation of hepatic gluconeogenesis and glycolysis, *Annu. Rev. Biochem.* **57**, 755–783 (1988).

Wood, H. G. and Barden, R. E., Biotin enzymes, *Annu. Rev. Biochem.* **46**, 385–413 (1977).

The Glyoxylate Pathway

Tolbert, N. E., Metabolic pathways in peroxisomes and glyoxysomes, *Annu. Rev. Biochem.* **50**, 133–157 (1981).

Oligosaccharide Biosynthesis

Elbein, A. D., Inhibitors of the biosynthesis and processing of N-linked oligosaccharide chains, *Annu. Rev. Biochem.* **56**, 497–534 (1987).

Elbein, A. D., The role of lipid-linked saccharides in the biosynthesis of complex carbohydrates, *Annu. Rev. Plant Physiol.* **30**, 239–272 (1979).

Florman, H. M. and Wasserman, P. M., *O*-Linked oligosaccharides of mouse egg ZP3 account for its sperm receptor activity, *Cell* **41**, 313–324 (1985).

Hanover, J. A. and Lennarz, W. J., Transmembrane assembly of membrane and secretory glycoproteins, *Arch. Biochem. Biophys.* **211**, 1–19 (1981).

Hirschberg, C. B. and Snider, M. D., Topography of glycosylation in the rough endoplasmic reticulum and the Golgi apparatus, *Annu. Rev. Biochem.* **56**, 63–87 (1987).

Hubbard, S. C. and Ivatt, R. J., Synthesis and processing of asparagine-linked oligosaccharides, *Annu. Rev. Biochem.* **50**, 555–583 (1981).

Li, Y.-T. and Li, S.-C., Biosynthesis and catabolism of glyco-sphingolipids, *Adv. Carbohydr. Chem. Biochem.* **40**, 235–286 (1982).

Keller, R. K., Boon, D. Y., and Crum, F. C., *N*-Acetylglucosamine-1-phosphate transferase from hen oviduct: solubilization, characterization and inhibition by tunicamycin, *Biochemistry* **18**, 3946–3952 (1979).

Kornfeld, R. and Kornfeld, S., Assembly of asparagine-linked oligosaccharides, *Annu. Rev. Biochem.* **54**, 631–664 (1985).

Lodish, H. F., Transport of secretory and membrane glycoproteins from the rough endoplasmic reticulum to the Golgi, *J. Biol. Chem.* **263**, 2107–2110 (1988).

Parodi, A. J. and Leloir, L. F., The role of lipid intermediates in the glycosylation of proteins in the eucaryotic cell, *Biochim. Biophys. Acta* **559**, 1–37 (1979).

Pfeffer, S. R. and Rothman, J. E., Biosynthetic protein transport and sorting by the endoplasmic reticulum and Golgi, *Annu. Rev. Biochem.* **56**, 829–852 (1987).

Presper, K. A. and Heath, E. C., Glycosylated lipid intermediates involved in glycoprotein biosynthesis, *in* Boyer, P. D. (Ed.), *The Enzymes* (3rd ed.), Vol. 16, pp. 449–488, Academic Press (1983).

Rothman, J. E., The compartmental organization of the Golgi apparatus, *Sci. Am.* **253**(3): 74–89 (1985).

Schwartz, R. T. and Datema, R., The lipid pathway of protein glycosylation and its inhibitors: the biological significance of protein-bound carbohydrates, *Adv. Carbohydr. Chem. Biochem.* **40**, 287–379 (1982).

Schwartz, R. T. and Datema, R., Inhibitors of trimming: new tools in glycoprotein research, *Trends Biochem. Sci.* **9**, 32–34 (1984).

von Figura, K. and Hasilik, A., Lysosomal enzymes and their receptors, *Annu. Rev. Biochem.* **55**, 167–193 (1986).

The Pentose Phosphate Pathway

Beutler, E., Glucose-6-phosphate dehydrogenase deficiency, *in* Stanbury, J. B., Wyngaarden, J. B., Fredrickson, D. S., Goldstein, J. L., and Brown, M. S. (Eds.), *The Metabolic Basis of Inherited Disease* (5th ed.), *pp.* 1629–1653, McGraw-Hill (1983).

Wood, T., *The Pentose Phosphate Pathway*, Academic Press (1985).

Problems

1. Compare the relative energetic efficiencies, in ATPs per mole of glucose oxidized, of glucose oxidation via glycolysis + the citric acid cycle versus glucose oxidation via the pentose phosphate pathway + gluconeogenesis. Assume that NADH and NADPH are each energetically equivalent to three ATPs.

2. Although animals cannot synthesize glucose from acetyl-CoA, if a rat is fed ^{14}C-labeled acetate, some of the label will appear in the glycogen extracted from its muscles. Explain.

3. Substances that inhibit specific trimming steps in the processing of *N*-linked glycoproteins have been useful tools in elucidating the pathway of this process. Explain.

4. Through clever genetic engineering, you have developed an unregulatable enzyme that can interchangeably use NAD$^+$ or NADP$^+$ in a redox reaction. What would be the physiological consequence(s) on an organism of having such an enzyme?

5. What is the free energy change of the reaction

$$NADH + NADP^+ \rightleftharpoons NAD^+ + NADPH$$

under physiological conditions? Assume that $\Delta G^{\circ\prime} = 0$ for this reaction and that $T = 37°C$.

6. If G6P is ^{14}C labeled at its C(2) position, what is the distribution of the radioactive label in the products of the pentose phosphate pathway after one turnover of the pathway? What is the distribution of the label after passage of these products through gluconeogenesis followed by a second round of the pentose phosphate pathway?

*7. The relative metabolic activities in an organism of glycolysis + the citric acid cycle versus pentose phosphate pathway + gluconeogenesis can be measured by comparing the rates of $^{14}CO_2$ generation upon administration of glucose labeled with ^{14}C at C(1) with that of glucose labeled at C(6). Explain.

8. In light of the finding that an otherwise benign or even advantageous mutation leads to abnormal primaquine sensitivity combined with the fact that human beings have enormous genetic complexity, comment on the possibility of developing drugs that exhibit no atypical side effects in any individual.

Chapter 22

PHOTOSYNTHESIS

Life on earth depends on the sun. *Plants and cyanobacteria chemically sequester light energy through* **photosynthesis**, *a light-driven process in which CO_2 is "fixed" to yield carbohydrates (CH_2O).*

$$CO_2 + H_2O \xrightarrow{\text{light}} (CH_2O) + O_2$$

This process, in which both CO_2 and H_2O are reduced to yield carbohydrate and O_2, is essentially the reverse of oxidative carbohydrate metabolism. Photosynthetically produced carbohydrates therefore serve as an energy source for the organism that produced them as well as for nonphotosynthetic organisms that directly or indirectly consume photosynthetic organisms. In fact, even modern industry is highly dependent on the products of photosynthesis because coal, oil, and gas (the so-called fossil fuels) are the remains of ancient organisms. It is estimated that photosynthesis annually fixes $\sim 10^{11}$ tons of carbon, which represents the storage of over 10^{18} kJ of energy. Moreover, photosynthesis, over the eons, has produced the O_2 in the earth's atmosphere.

The notion that plants obtain nourishment from such insubstantial things as light and air took nearly two centuries to evolve. In 1648, the Flemish physician Jean-Baptiste von Helmont reported that growing a potted willow tree from a shoot caused an insignificant change in the weight of the soil in which the tree had been rooted. Although another century was to pass before the law of conservation of matter was formulated, van Helmont attributed the tree's weight gain to the water it had taken up. This idea was extended in 1727 by Stephen

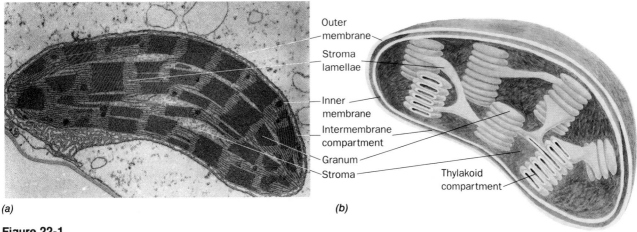

Figure 22-1
An electron micrograph (*a*) and a schematic diagram (*b*) of a chloroplast from corn. [Electron micrograph courtesy of Lester Shumway, College of Eastern Utah.]

Hales who proposed that plants extract some of their matter from the air.

The first indication that plants produce oxygen was found by the English clergyman and pioneering chemist Joseph Priestley who reported:

> Finding that candles burn very well in air in which plants had grown a long time, and having some reason to think, that there was something attending vegetation, which restored air that had been injured by respiration, I thought it was possible that the same process might also restore the air that had been injured by the burning of candles. Accordingly, on the 17th of August, 1771, I put a sprig of mint into a quantity of air, in which a wax candle had burned out, and found that, on the 27th of the same month, another candle burned perfectly well in it.

Although Priestley later discovered oxygen, which he named "dephlogisticated air," it was Antoine Lavosier who elucidated its role in combustion and respiration. Nevertheless, Priestley's work inspired the Dutch physician Jan Ingen-Housz who in 1779 demonstrated that the "purifying" power of plants resides in the influence of sunlight upon their green parts. In 1782, the Swiss pastor Jean Senebier showed that CO_2, which he called "fixed air," is taken up during photosynthesis. His compatriot, Théodore de Saussure found, in 1804, that the combined weights of the organic matter produced by plants and the oxygen they evolve is greater than the weight of the CO_2 they consume. He therefore concluded that water, the only other substance he added to his system, was also necessary for photosynthesis. The final ingredient in the overall photosynthetic recipe was established in 1842 by the German physiologist Robert Mayer, one of the formulators of the first law of thermodynamics, who concluded that plants convert light energy to chemical energy.

1. CHLOROPLASTS

*The site of photosynthesis in eukaryotes (algae and higher plants) is the **chloroplast**, a membranous subcellular organelle (Section 1-2A).* The first indication that chloroplasts function in this manner was Theodor Englemann's observation, in 1882, that small, motile, O_2-seeking bacteria congregate at the surface of the alga *Spirogyra* overlying its single chloroplast, but only while the chloroplast is illuminated. Chloroplasts must therefore be the site of light-induced O_2 evolution, that is, photosynthesis. Chloroplasts, of which there are 1 to 1000 per cell, vary considerably in size and shape but are typically ~5-μm long ellipsoids. Like mitochondria, which they resemble in many ways, chloroplasts have a highly permeable outer membrane and a nearly impermeable inner membrane separated by a narrow intermembrane space (Fig. 22-1). The inner membrane encloses the **stroma**, a concentrated solution of enzymes that also contains the DNA, RNA, and ribosomes involved in the synthesis of several chloroplast proteins —much like the mitochondrial matrix. The stroma, in turn, surrounds a third membranous compartment, the **thylakoid** (Greek: *thylakos*, a sac or pouch). The thylakoid is probably a single highly folded vesicle although in most organisms it appears to consist of stacks of disklike sacs named **grana**, which are interconnected by unstacked **stroma lamellae**. A chloroplast usually contains 10 to 100 grana. Thylakoid membranes arise from invaginations in the inner membrane of developing chloroplasts and therefore resemble mitochondrial cristae.

The lipids of the thylakoid membrane have a distinctive composition. They consist of only ~10% phospholipids; the majority, ~80%, are uncharged **mono-** and **digalactosyl diacylglycerols,** and the remaining ~10%

In the figure labels: Outer membrane, Stroma lamellae, Inner membrane, Intermembrane compartment, Granum, Stroma, Thylakoid compartment.

are the sulfolipids **sulfoquinovosyl diacylglycerols.**

X = OH   Galactosyl diacylglycerol

$X =$ Digalactosyl diacylglycerol

$X = SO_3^-$ Sulfoquinovosyl diacylglycerol

The acyl chains of these lipids have a high degree of unsaturation, which gives the thylakoid membrane a highly fluid character.

Photosynthesis occurs in two distinct phases:

1. The **light reactions,** which use light energy to generate NADPH and ATP.

2. The **dark reactions,** actually light-independent reactions, which use NADPH and ATP to drive the synthesis of carbohydrate from CO_2 and H_2O.

The light reactions occur in the thylakoid membrane and involve processes that resemble mitochondrial electron transport and oxidative phosphorylation (Sections 20-2 and 3). In photosynthetic prokaryotes, which lack chloroplasts, the light reactions take place in the

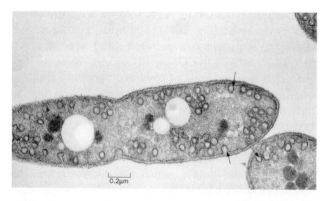

Figure 22-2
An electron micrograph of a section through the purple photosynthetic bacterium *Rhodobacter sphaeroides.* Its plasma membrane invaginates to form externally connected tubules known as chromatophores (*arrows*; seen here in circular cross-section) that are the sites of photosynthesis. [Courtesy of Gerald A. Peters, Virginia Commonwealth University.]

cell's plasma (inner) membrane or in highly invaginated structures derived from it called **chromatophores** (e.g., Fig. 22-2; recall that chloroplasts evolved from cyanobacteria that assumed a symbiotic relationship with a nonphotosynthetic eukaryote; Section 1-2A). In eukaryotes, the dark reactions occur in the stroma through a cyclic series of enzyme-catalyzed reactions. In the following sections, we consider the light and dark reactions in detail.

2. LIGHT REACTIONS

In the first decades of this century, it was generally assumed that light, as absorbed by photosynthetic pigments, directly reduces CO_2 which, in turn, combines with water to form carbohydrate. In this view, CO_2 is the source of the O_2 generated by photosynthesis. In 1931, however, Cornelis van Niel showed that green photosynthetic bacteria, anaerobes that use H_2S in photosynthesis, generate sulfur:

$$CO_2 + 2H_2S \xrightarrow{\text{light}} (CH_2O) + 2S + H_2O$$

The chemical similarity between H_2S and H_2O led van Niel to propose that the general photosynthetic reaction is

$$CO_2 + 2H_2A \xrightarrow{\text{light}} (CH_2O) + 2A + H_2O$$

where H_2A is H_2O in green plants and cyanobacteria and H_2S in photosynthetic sulfur bacteria. This suggests that photosynthesis is a two-stage process in which light energy is harnessed to oxidize H_2A (the light reactions):

$$2H_2A \xrightarrow{\text{light}} 2A + 4[H]$$

and the resulting reducing agent [H] subsequently reduces CO_2 (the dark reactions):

$$4[H] + CO_2 \longrightarrow (CH_2O) + H_2O$$

Thus, in aerobic photosynthesis, H_2O, not CO_2, is photolyzed (split by light).

The validity of van Niel's hypothesis was established unequivocally by two experiments. In 1937, Robert Hill discovered that when isolated chloroplasts that lack CO_2 are illuminated in the presence of an artificial electron acceptor such as ferricyanide [$Fe(CN)_6^{3-}$], O_2 is evolved with concomitant reduction of the acceptor [to ferrocyanide, $Fe(CN)_6^{4-}$, in our example]. This so-called **Hill reaction** demonstrates that CO_2 does not participate directly in the O_2-producing reaction. It was discovered eventually that the natural photosynthetic electron acceptor is $NADP^+$ (Section 22-2C) whose reduction product, NADPH, is utilized in the dark reactions to reduce CO_2 to carbohydrate (Section 22-3A). In 1941, when the oxygen isotope ^{18}O became available, Samuel Ruben and Martin Kamen directly demonstrated that the source of the O_2 formed in photosynthesis is H_2O:

$$H_2{}^{18}O + CO_2 \xrightarrow{\text{light}} (CH_2O) + {}^{18}O_2$$

This section is a discussion of the major aspects of the light reactions.

A. Absorption of Light

The principal photoreceptor in photosynthesis is **chlorophyll.** This cyclic tetrapyrrole, like the heme group of globins and cytochromes (Sections 9-1A and 20-2E), is derived biosynthetically from protoporphyrin IX. Chlorophyll, however, differs from heme in four major respects (Fig. 22-3):

1. Its central metal ion is Mg^{2+} rather than Fe(II) or Fe(III).

2. It has a cyclopentanone ring, Ring V, fused to pyrrole Ring III.

3. Pyrrole Ring IV is partially reduced in **chlorophyll a (Chl a)** and **chlorophyll b (Chl b),** the two major chlorophyll varieties in eukaryotes and cyanobacteria, whereas in **bacteriochlorophyll a (BChl a)** and **bacteriochlorophyll b (BChl b),** the principal chlorophylls of photosynthetic bacteria, Rings II and IV, are partially reduced.

4. The propionyl side chain of Ring IV is esterified to a tetraisoprenoid alcohol. In Chls a and b as well as in BChl b it is **phytol** but in BChl a it is either phytol or **geranylgeraniol** depending on the bacterial species.

Chl b has a formyl group in place of the methyl substituent to Ring II of Chl a. Similarly, BChl a and BChl b have different substituents to atom C(4).

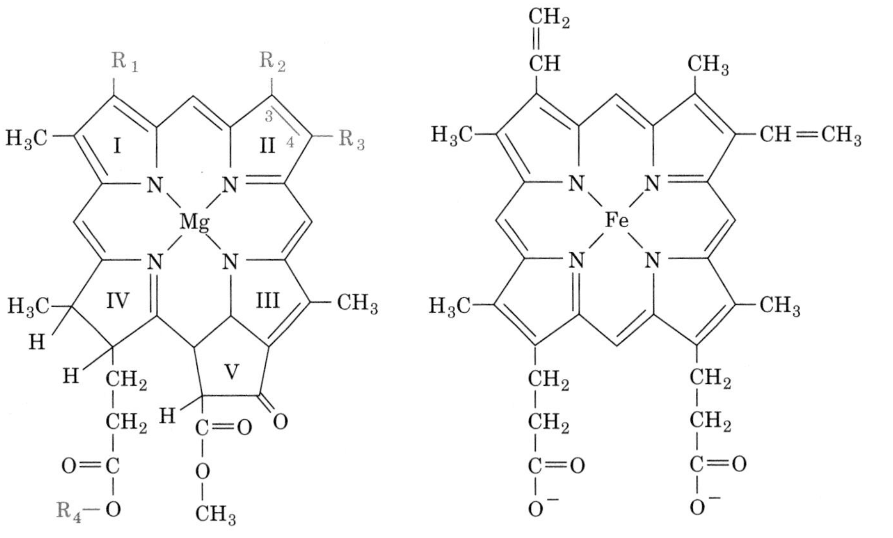

Chlorophyll   **Iron Protoporphyrin IX**

	R_1	R_2	R_3	R_4
Chlorophyll a	$-CH=CH_2$	$-CH_3$	$-CH_2-CH_3$	P
Chlorophyll b	$-CH=CH_2$	$-\overset{\displaystyle O}{\overset{\|}{C}}-H$	$-CH_2-CH_3$	P
Bacteriochlorophyll a	$-\overset{\displaystyle O}{\overset{\|}{C}}-CH_3$	$-CH_3{}^a$	$-CH_2-CH_3{}^a$	P or G
Bacteriochlorophyll b	$-\overset{\displaystyle O}{\overset{\|}{C}}-CH_3$	$-CH_3{}^a$	$=CH-CH_3{}^a$	P

^{a} No double bond between positions C(3) and C(4).

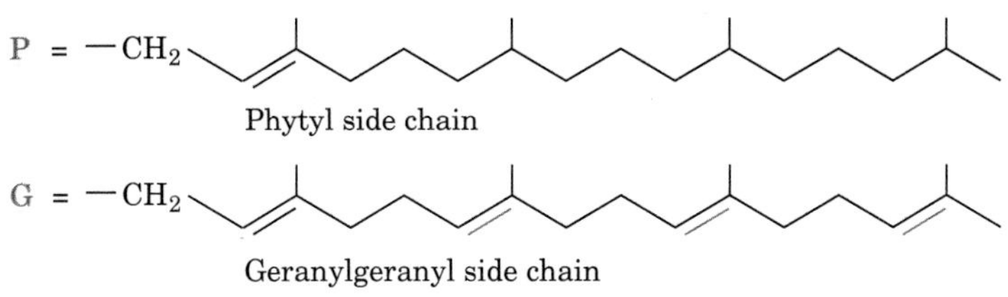

P = $-CH_2$
Phytyl side chain

G = $-CH_2$
Geranylgeranyl side chain

Figure 22-3
The molecular formulas of chlorophylls *a* and *b* and bacteriochlorophylls *a* and *b* compared to that of iron protoporphyrin IX (heme). The isoprenoid phytyl and geranylgeranyl tails presumably increase the chlorophylls' solublity in nonpolar media.

Light and Matter Interact in Complex Ways

As photosynthesis is a light-driven process, it is worthwhile reviewing how light and matter interact. Electromagnetic radiation is propagated as discrete **quanta (photons)** whose energy E is given by **Planck's law:**

$$E = h\nu = \frac{hc}{\lambda} \qquad [22.1]$$

where h is **Planck's constant** $(6.626 \times 10^{-34}$ J·s$)$, c is the speed of light $(2.998 \times 10^{8}$ m·s^{-1} in a vacuum$)$, ν is the frequency of the radiation, and λ is its wavelength (visible light ranges in wavelength from 400–700 nm). Thus red light with $\lambda = 700$ nm has an energy of 171 kJ·einstein^{-1} (an **einstein** is a mole of photons).

Molecules, like atoms, have numerous electronic quantum states of differing energies. Moreover, because molecules contain more than one nucleus, each of their electronic states has an associated series of vibrational and rotational substates that are closely spaced in en-

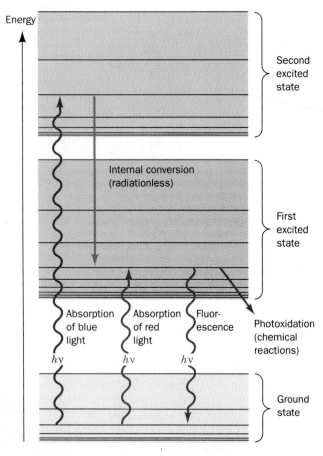

Figure 22-4
An energy diagram schematically indicating the electronic states of chlorophyll and their most important modes of interconversion. The wiggly arrows represent the absorption of photons or their fluorescent emission. Excitation energy may also be dissipated in radiationless processes such as internal conversion (heat production) or chemical reactions.

ergy (Fig. 22-4). Absorption of light by a molecule usually occurs through the promotion of an electron from its ground (lowest energy) state molecular orbital to one of higher energy. However, *a given molecule can only absorb photons of certain wavelengths because, as is required by the law of conservation of energy, the energy difference between the two states must exactly match the energy of the absorbed photon.*

The amount of light absorbed by a substance at a given wavelength is described by the **Beer–Lambert law:**

$$I = I_0 \, 10^{-\varepsilon c\ell} \qquad [22.2]$$

where I_0 and I are, respectively, the intensities of incident and transmitted light, c is the molar concentration of the sample, ℓ is the length of the light path through the sample in cm, and ε is the molecule's **molar extinction coefficient.** Consequently, a plot of ε versus λ for a given molecule, its **absorption spectrum** (Fig. 22-5), is indicative of its electronic structure.

The various chlorophylls are highly conjugated molecules (Fig. 22-3). It is just such molecules that strongly absorb visible light (the spectral band in which the solar radiation reaching the earth's surface is of peak intensity; Fig. 22-5). In fact, the peak molar extinction coefficients of the various chlorophylls, over 10^5 M^{-1} cm^{-1}, are among the highest known for organic molecules. Yet, the relatively small chemical differences among the various chlorophylls greatly affect their absorption spectra. These spectral differences, as we shall see, are functionally significant.

An electronically excited molecule can dissipate its excitation energy in many ways. Those modes with the greatest photosynthetic significance are (Fig. 22-4):

1. **Internal conversion,** a common mode of decay in which electronic energy is converted to the kinetic energy of molecular motion, that is, to heat. This process occurs very rapidly, being complete in $<10^{-11}$ s. Many molecules relax in this manner to their ground states. Chlorophyll molecules, however, usually relax only to their lowest excited states. Therefore, *the photosynthetically applicable excitation energy of a chlorophyll molecule that has absorbed a photon in its short wavelength band, which corresponds to its second excited state, is no different than if it had absorbed a photon in its less energetic long wavelength band.*

2. **Fluorescence,** in which an electronically excited molecule decays to its ground state by emitting a photon. Such a process requires $\sim 10^{-8}$ s so that it occurs much more slowly than internal conversion. Consequently, a fluorescently emitted photon generally has a longer wavelength (lower energy) than that initially absorbed. Fluorescence accounts for the dissipation of only 3 to 6% of the light energy absorbed by living plants. However, chlorophyll in solution,

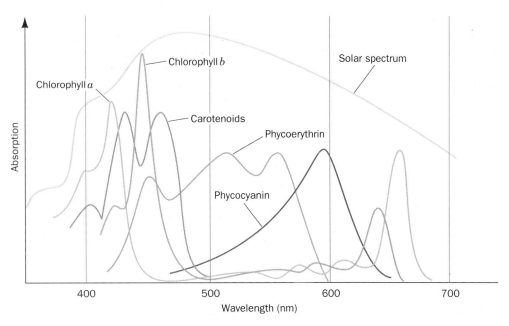

Figure 22-5
The absorption spectra of various photosynthetic pigments. The chlorophylls have two absorption bands, one in the red and one in the blue. Phycoerythrin absorbs blue and green light, whereas phycocyanin absorbs yellow light. Together, these pigments absorb most of the visible light in the solar spectrum.

where of course the photosynthetic uptake of this energy cannot occur, has an intense red fluorescence.

3. **Exciton transfer** (also known as **resonance energy transfer**), in which an excited molecule directly transfers its excitation energy to nearby unexcited molecules with similar electronic properties. This process occurs through interactions between the molecular orbitals of the participating molecules in a manner analogous to the interactions between mechanically coupled pendulums of similar frequencies. An exciton (excitation) may be serially transferred between members of a group of molecules or, if their electronic coupling is strong enough, the entire group may act as a single excited "supermolecule." We shall see that *exciton transfer is of particular importance in funneling light energy to photosynthetic reaction centers.*

4. **Photooxidation,** in which a light-excited donor molecule is oxidized by transferring an electron to an acceptor molecule, which is thereby reduced. This process occurs because the transferred electron is less tightly bound to the donor in its excited state than it is to the ground state. In photosynthesis, excited chlorophyll (Chl*) is such a donor. *The energy of the absorbed photon is thereby chemically transferred to the photosynthetic reaction system.* Photooxidized chlorophyll, Chl⁺, a cationic free radical, eventually returns to its ground state by oxidizing some other molecule.

Light Absorbed by Antenna Chlorophylls and Accesory Pigments Is Transferred to Photosynthetic Reaction Centers

The primary reactions of photosynthesis, as is explained in Sections 22-2B and C, take place at **photosynthetic reaction centers.** Yet, *photosynthetic organelles contain far more chlorophyll molecules than reaction centers.* This was demonstrated in 1932 by Robert Emerson and William Arnold in their studies of O_2 production by the green alga *Chlorella* (a favorite experimental subject), which had been exposed to repeated brief (10 μs) flashes of light. The amount of O_2 generated per flash was maximal when the interval between flashes was at least 20 ms. Evidently, this is the time required for a single turnover of the photosynthetic reaction cycle. Emerson and Arnold then measured the variation of O_2 yield with flash intensity when the flash interval was the optimal 20 ms. With weak flashes, the O_2 increased linearly with flash intensity such that about one molecule of O_2 was generated per eight photons absorbed (Fig. 22-6). With increasing flash intensity the efficiency of this process fell off, no doubt because the number of photons began to approach the number of photochemical units. What was unanticipated, however, was that each flash of saturating intensity produced only 1 molecule of O_2 per ~2400 molecules of chlorophyll present. Since at least eight photons must be sequentially absorbed to liberate one O_2 molecule (Section 22-2C), these results suggest that the photosyn-

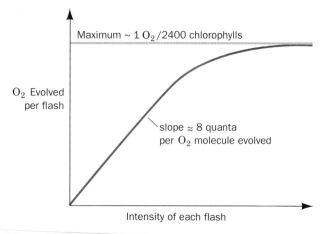

Figure 22-6
The amount of O_2 evolved by *Chlorella* algae versus the intensity of light flashes that are separated by dark intervals > 20 ms.

timizes the efficiency of exciton transfer throughout the LHC.

Most LHCs contain organized arrays of other light-absorbing substances besides chlorophyll. These **accessory pigments** function to fill in the absorption spectra of the antenna complexes in spectral regions where chlorophylls do not absorb strongly (Fig. 22-5). **Carotenoids,**

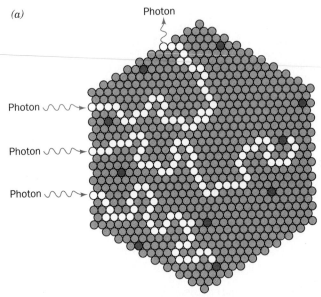

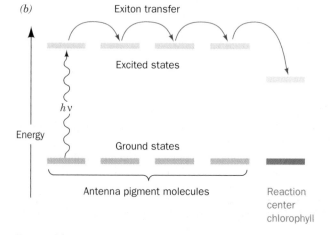

Figure 22-7
The flow of energy through a photosynthetic antenna complex. (*a*) The excitation resulting from photon absorption randomly migrates by exciton transfer among the molecules of the antenna complex (*light green circles*) until it is either trapped by a reaction center chlorophyll (*dark green circles*) or, less frequently, fluorescently reemitted. (*b*) The excitation is trapped by the reaction center chlorophyll because its lowest excited state has a lower energy than those of the antenna pigment molecules.

thetic apparatus contains ~ 2400/8 = 300 chlorophyll molecules per photosynthetic reaction center.

With such a great excess of chlorophyll molecules per reaction center, it seems unlikely that all participate directly in photochemical reactions. Rather, as subsequent experiments have shown, *most chlorophylls function to gather light; that is, they act as light-harvesting antennas.* These **antenna chlorophylls** pass the energy of an absorbed photon, by exciton transfer, from molecule to molecule until the excitation reaches a photosynthetic reaction center (Fig. 22-7*a*). There, the excitation is trapped because reaction center chlorophylls, although chemically identical to antenna chlorophylls, have slightly lower excited state energies because of their different environments (Fig. 22-7*b*).

Transfer of energy from the antenna system to a reaction center occurs in < 10^{-10} s with an efficiency of > 90%. This high efficiency depends on the chlorophyll molecules having appropriate spacings and relative orientations. Although the structures of the **light-harvesting complexes (LHCs)** are not well understood, it appears that they consist of arrays of membrane-bound hydrophobic proteins that each contain several pigment molecules. For example, an antenna protein from the green photosynthetic bacterium *Prosthecochloris aestuarii*, whose X-ray structure was determined by Brian Matthews (Fig. 22-8), is a trimer of identical 47-kD subunits, each of which contains 7 BChl *a* molecules. Each subunit largely consists of 15 strands of β sheet wrapped around a chlorophyll-containing core. This protein therefore has been described as a "string bag" for holding pigment molecules. The porphyrin rings are not in van der Waals contact and exhibit a complex pattern of relative orientations. This arrangement presumably op-

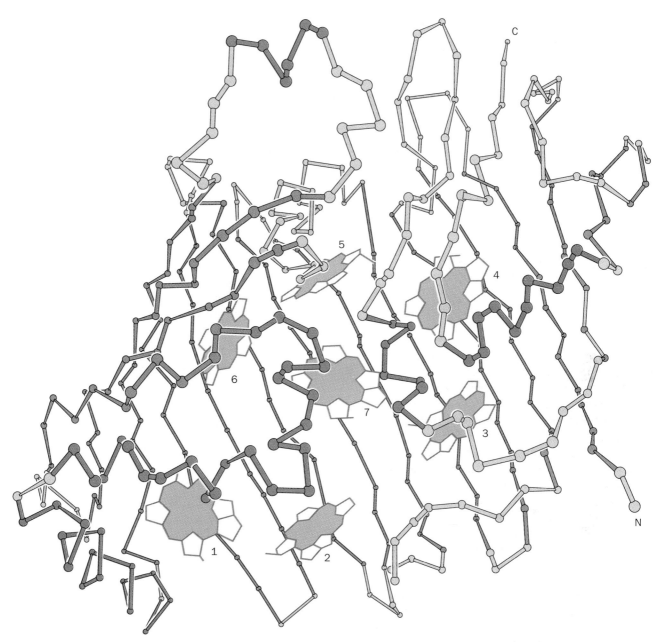

Figure 22-8
The structure of a subunit of bacteriochlorophyll *a* protein from *P. aestuarii*. The polypeptide backbone, only the C$_\alpha$ positions of which are shown, is folded into a structure that consists almost entirely of 15 strands of β sheet (*orange*).

The phytyl side chains and the other substituents to the subunit's 7 BChl *a* molecules (*green*) have been omitted for clarity. [After Matthews, B. W., Fenna, R. E., Bolognesi, M. C., Schmid, M. F., and Olson, J. M., *J. Mol. Biol.* **131,** 272 (1979).]

which are linear polyenes such as **β-carotene,**

β-Carotene

are components of all green plants and many photosyn-

thetic bacteria and are therefore the most common accessory pigments (they are largely responsible for the brilliant fall colors of deciduous trees as well as for the orange color of carrots, after which they are named). Water-dwelling photosynthetic organisms, which are responsible for nearly one half of the photosynthesis on earth, additionally contain other types of accessory pigments. This is because light outside the wavelengths 450 to 550 nm (blue and green light) is absorbed almost completely by passage through more than 10 m of water. In red algae and cyanobacteria, Chl *a* therefore is replaced as an antenna pigment by a series of linear tetrapyrroles, notably the red **phycoerythrobilin** and the blue **phycocyanobilin** (spectra in Fig. 22-5).

Phycoerythrobilin

The lowest excited states of these various **bilins** have higher energies than those of the chlorophylls, thereby facilitating energy transfer to the photosynthetic reaction centers. The bilins are covalently linked to **phycobiliproteins** which are, in turn, arranged in organized high molecular mass particles called **phycobilisomes**. The phycobilisomes are bound to the outer faces of photosynthetic membranes so as to funnel excitation energy to reaction centers over long distances with >90% efficiency.

B. Electron Transport in Photosynthetic Bacteria

Photosynthesis is a process in which electrons from excited chlorophyll molecules are passed through a series of acceptors that convert electronic energy to chemical energy. Thus two questions arise: (1) What is the mechanism of energy transduction; and (2) How do photooxidized chlorophyll molecules regain their lost electrons? We shall see that photosynthetic bacteria solve these problems somewhat differently from cyanobacteria and plants. We first discuss these mechanisms in photosynthetic bacteria where they are simpler and better understood. Electron transport in cyanobacteria and plants is the subject of Section 22-2C.

The Photosynthetic Reaction Center Is a Trans-Membrane Protein Containing a Variety of Chromophores

The first indication that chlorophyll undergoes direct photooxidation during photosynthesis was obtained by Louis Duysens in 1952. He observed that illumination of membrane preparations from the purple photosynthetic bacterium *Rhodospirillum* (*Rs.*) *rubrum* caused a slight (~2%) bleaching of their absorbance at 870 nm, which returned to their original levels in the dark. Duysens suggested that this bleaching is caused by photooxidation of a bacteriochlorophyll complex that he named

P870 (P for pigment and 870 nm is the position of the major long wave absorption band of BChl *a*; photosynthetic bacteria tend to inhabit murky stagnant ponds so that they require a near infrared-absorbing species of chlorophyll). The ability to detect the presence of P870 eventually led to the purification and characterization of the photosynthetic reaction centers to which it is bound.

Reaction center particles from several species of purple photosynthetic bacteria have similar compositions. That from *Rhodopseudomonas* (*Rps.*) *viridis* consists of three hydrophobic subunits: H (258 residues), L (273 residues), and M (323 residues). The L and M subunits of this membrane-spanning particle collectively bind four molecules of BChl *b* (which maximally absorbs light at 960 nm), two molecules of **bacteriopheophytin** *b* (**BPheo** *b*; BChl *b* in which the Mg^{2+} is replaced by two protons), one nonheme/non-Fe–S Fe(II) ion, one molecule of the redox coenzyme ubiquinone (Section 20-2B), and one molecule of the related **menaquinone**

Menaquinone

(**vitamin** K_2, a substance required for proper blood clotting; Section 34-1B). In many purple photosynthetic bacteria, however, the BChl *b*, BPheo *b*, and menaquinone are replaced by BChl *a*, BPheo *a*, and a second ubiquinone, respectively.

The photosynthetic reaction center of *Rps. viridis*, whose X-ray structure was determined by Johann Deisenhofer, Robert Huber, and Harmut Michel in 1984, was the first trans-membrane protein to be described in atomic detail (Fig. 22-9). *The protein's membrane-spanning portion consists of 11 α helices that form a 45-Å long cylinder with the expected hydrophobic surface.* A *c*-type cytochrome containing four hemes, which is an integral constituent of the reaction center complex in only some photosynthetic bacteria, binds to the reaction center on the external side of the plasma membrane (the photosynthetic reaction center from another species, *Rhodobacter* (*Rb.*) *sphaeroides*, whose X-ray structure is nearly identical to that of *Rps. viridis*, lacks such a bound cytochrome).

The most striking aspect of the reaction center is that its chromophoric prosthetic groups are arranged with nearly perfect twofold symmetry (Fig. 22-10a). This symmetry arises because the L and M subunits, with which these prosthetic groups are exclusively associated, have homologous sequences and similar folds. Two of the BChl *b* molecules, the so-called **"special pair,"** are closely associated; they are nearly parallel and have an Mg—Mg distance of ~7 Å. The "special pair" occupies

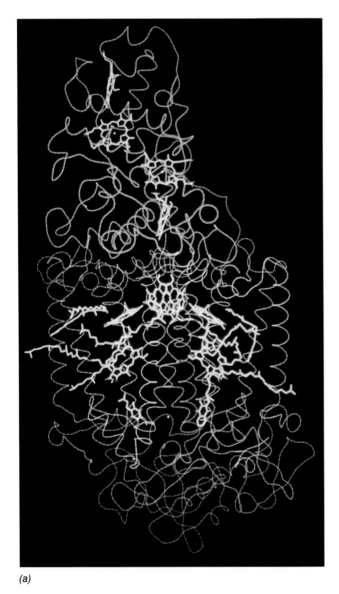

(a)

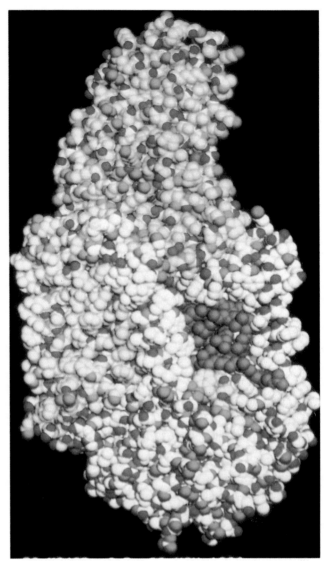

(b)

Figure 22-9
The X-ray structure of the photosynthetic reaction center of *Rps. viridis*. (a) A diagram in which only the C_α backbone and the prosthetic groups (*yellow*) are shown. The H, M, and L subunits (*pink, blue,* and *orange,* respectively) collectively have 11 trans-membrane helices. The 4-heme *c*-type cytochrome (*green*), which does not occur in all species of photosynthetic bacteria, is bound to the external face of the complex. (b) A space-filling model in which nitrogen atoms are blue, oxygen atoms are red, sulfur atoms are yellow, and the carbon atoms of the H, M, L, and cytochrome subunits are tinted pink, blue, orange, and green, respectively. Exposed portions of prosthetic groups are brown. Note how few polar groups (nitrogen and oxygen atoms) are externally exposed in the region surrounding the protein's midsection. This hydrophobic belt forms the reaction center's trans-membrane segment. [From Deisenhofer, J. and Michel, H., *Les Prix Nobel* (1989).]

a predominantly hydrophobic region of the protein and each of its Mg^{2+} ions has a His side chain as a fifth ligand [much like the Fe(II) in deoxyhemoglobin]. Each member of the "special pair" is in contact with another His-liganded BChl *b* molecule which, in turn, is associated with a BPheo *b* molecule. The menaquinone is in close association with the L subunit BPheo *b* (Fig. 22-10*a*, *right*), whereas the ubiquinone, which is but loosely bound to the protein, associates with the M subunit BPheo *b* (Fig. 22-10*a*, *left*). These various chromophores are closely associated with a number of protein aromatic rings, which are therefore also thought to participate in the electron-transfer process described below. The Fe(II) is positioned between the menaquinone and ubiquinone rings and is octahedrally liganded by four His side chains and two carboxyl oxygen atoms of a Glu side chain. Curiously, the two symmetry related groups of chromophores are not functionally equivalent; electrons, as we shall see, are only transferred through the L subunit along the arm that ends in menaquinone.

Figure 22-10
The sequence of excitations in the bacterial photosynthetic reaction center. The reaction center chromophores are shown in the same view as in Fig. 22-9. Note that their rings, but not their aliphatic side chains, are arranged with close to twofold symmetry. (*a*) At zero time, a photon is absorbed by the "special pair" of BChl *a* molecules thereby collectively raising them to an excited state [in each step, the excited molecule(s) is shown in red]. (*b*) Within 4 ps, an excited electron has been passed to the BPheo *a* of the L subunit (right arm of the system) without becoming closely associated with the accessory BChl *a*. The special pair is thereby left with a positive charge. (*c*) Some 200 ps later, the excited electron has transferred to the menaquinone (Q_A). (*d*) Within the next 100 μs, the *c*-type cytochrome reduces the "special pair" thereby eliminating its positive charge while the excited electron migrates to the ubiquinone (Q_B). After a second such electron has been transferred to Q_B, it picks up two protons from solution and exchanges with the membrane-bound ubiquinone pool.

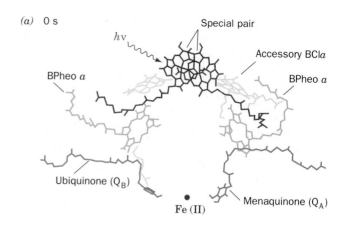

(*a*) 0 s

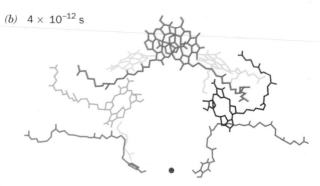

(*b*) 4×10^{-12} s

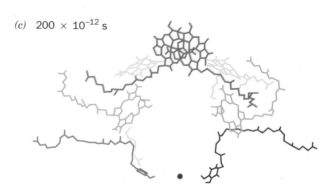

(*c*) 200×10^{-12} s

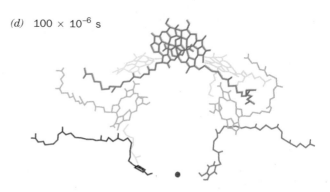

(*d*) 100×10^{-6} s

The Electronic States of Molecules Undergoing Fast Reactions Can Be Monitored by EPR and Laser Spectroscopy Techniques

The turnover time of a photosynthetic reaction cycle, as we have seen, is only a few milliseconds. Its sequence of reactions can therefore only be traced by measurements that can follow extremely rapid electronic changes in molecules. Two techniques are well suited to this task:

1. **Electron paramagnetic resonance (EPR) spectroscopy** [also called **electron spin resonance (ESR) spectroscopy**], which detects the spins of unpaired electrons in a manner analogous to the detection of nuclear spins in NMR spectroscopy. A molecular species with unpaired electrons, such as an organic radical or a transition metal ion, has a characteristic EPR spectrum because its unpaired electrons interact with the magnetic fields generated by the nuclei and the other electrons of the molecule. Paramagnetic species as short lived as 10^{-11} s can exhibit definitive EPR spectra.

2. Optical spectroscopy using pulsed lasers. Laser flashes shorter than 1 femtosecond (fs, 10^{-15} s) have been generated. By monitoring the bleaching (disappearance) of certain absorption bands and the emergence of others, laser spectroscopy can track the time course of a fast reaction process.

Photon Absorption Rapidly Photooxidizes the "Special Pair"

The sequence of photochemical events mediated by the photosynthetic reaction center is diagrammed in Fig. 22-10:

(**a**) *The primary photochemical event of bacterial photosynthesis is absorption of a photon by the special pair* (P870 or **P960** depending on whether it consists of

BChl *a* or *b*; here, for argument's sake, we assume it to be P870). This event is nearly instantaneous; it occupies the < 1-fs oscillation time of a light wave. EPR measurements established that P870 is, in fact, a pair of BChl *a* molecules, and indicated that the excited electron is delocalized over both of them.

(b) P870*, the excited state of P870, has but a fleeting existence. Laser spectroscopy has demonstrated that within 4 picosecond (ps; 10^{-12} s) after its formation, P870* has transferred an electron to the BPheo a on the right in Fig. 22-10b to yield P870$^+$ BPheo a^-. In forming this radical pair, the transferred electron must pass near but seems not to reduce the intervening BChl a (which is therefore termed an accessory chlorophyll), although its position strongly suggests that it has an important role in conveying electrons.

(c) By some 200 ps later, the electron has further migrated to the menaquinone (or, in many species, the second ubiquinone), designated Q_A, to form the anionic semiquinone radical Q_A^-. All these electron transfers, as diagrammed in Fig. 22-11, are to progressively lower energy states, which makes this process all but irreversible.

Rapid removal of the excited electron from the vicinity of P870$^+$ is an essential feature of the photosynthetic reaction center; this prevents back reactions that would return the electron to P870$^+$ so as to provide the time required for the wasteful internal conversion of its excitation energy to heat. In fact, *this sequence of electron transfers is so efficient that its overall **quantum yield** (ratio of molecules reacted to photons absorbed) is virtually 100%.* No man-made device has yet to approach this level of efficiency.

Electrons Are Returned to the Photooxidized Special Pair via an Electron-Transport Chain

The remainder of the photosynthetic electron-transport process occurs on a much slower time scale. Within ~ 100 μs after its formation, Q_A^-, which occupies a hydrophobic pocket in the protein, transfers its excited electron to the more solvent-exposed ubiquinone, Q_B, to form Q_B^- (Fig. 22-10d). The nonheme Fe(II), which is not reduced by the passing electron, is implicated in this process because its removal prevents the formation of Q_B^-. Q_A never becomes fully reduced; it shuttles between its oxidized and semiquinone forms. Moreover, the lifetime of Q_A^- is so short that it never becomes protonated. In contrast, once the reaction center again becomes excited, it transfers a second electron to Q_B to form the fully reduced Q_B^{2-}. This anionic quinol takes up two protons from the solution on the cytoplasmic side of the plasma membrane to form Q_BH_2. Thus Q_B is a *molecular transducer that converts two light-driven one-electron excitations to a two-electron chemical reduction.*

The electrons taken up by Q_BH_2 are eventually returned to P870$^+$ via a complex electron-transport chain. The details of this process, which are much more species dependent than the preceding, are not well understood. In *Rb. sphaeroides*, the available redox carriers include a membrane-bound pool of ubiquinone molecules, two or three b-type cytochromes, and three iron–sulfur proteins. This collection of redox carriers is strikingly similar to those present in the proton-translocating Complex III of mitochondria (Section 20-2C). It is thought that the pathway (or possibly parallel pathways) of electron transport lead from Q_BH_2 through the ubiquinone pool (with which Q_BH_2 exchanges), the b-type cytochromes and then the iron–sulfur proteins. These carriers ultimately transfer their electrons to **cytochrome c_2** on the cytoplasmic side of the plasma membrane. Cytochrome c_2, which, as its name implies, closely resembles mitochondrial cytochrome c, reacts with the membrane-spanning reaction center to transfer an electron to P870$^+$ (the structures of several c-type cytochromes, including that of cytochrome c_2 from *Rs. rubrum*, are diagrammed in Fig. 8-13). In *Rps. viridis*, the c-type cytochrome

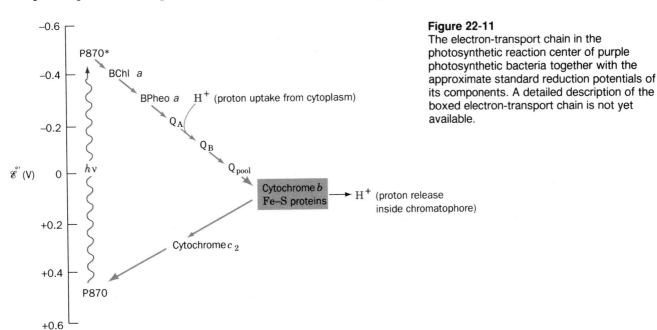

Figure 22-11
The electron-transport chain in the photosynthetic reaction center of purple photosynthetic bacteria together with the approximate standard reduction potentials of its components. A detailed description of the boxed electron-transport chain is not yet available.

bound to the reaction center complex on the external side of the plasma membrane (Fig. 22-9) also participates in this process. Note that one of its hemes is positioned to reduce the photooxidized special pair. The reaction center is thereby prepared to absorb another photon.

Photosynthetic Electron Transport Drives the Formation of a Proton Gradient

Since electron transport in purple photosynthetic bacteria is a cyclic process (Fig. 22-11), *it results in no net oxidation–reduction. Rather, it functions to translocate the cytoplasmic protons acquired by Q_BH_2 across the plasma membrane, thereby making the cell alkaline relative to its environment.* The mechanism of this process is unknown although it is thought to resemble proton transport in mitochondria (Section 20-3A); that is, protons may be translocated through a redox loop mechanism such as a Q cycle or by a proton pump. *Synthesis of ATP, a process known as* **photophosphorylation,** *is driven by the dissipation of the resulting pH gradient in a manner that closely resembles ATP synthesis in oxidative phosphorylation (Section 20-3C).* The mechanism of photophosphorylation is further discussed in Section 22-2D.

Photosynthetic bacteria use photophosphorylation-generated ATP to drive their various endergonic processes. However, unlike cyanobacteria and plants, which generate their required reducing equivalents by light-driven oxidation of H_2O (see below), photosynthetic bacteria must obtain their reducing equivalents from the environment. Various substances, such as H_2S, S, $S_2O_3^{2-}$, H_2 and many organic compounds, function in this capacity, depending on the bacterial species.

Modern photosynthetic bacteria are thought to resemble the original photosynthetic organisms. These presumably arose very early in the history of cellular life when environmentally supplied sources of high-energy compounds were dwindling but reducing agents were still plentiful (Section 1-4C). During this era, photosynthetic bacteria were no doubt the dominant form of life. However, their very success eventually caused them to exhaust the available reductive resources. The ancestors of modern cyanobacteria adapted to this situation by evolving a photosynthetic system with sufficient electromotive force to abstract electrons from H_2O. The gradual accumulation of the resulting toxic waste product, O_2, forced photosynthetic bacteria, which cannot photosynthesize in the presence of O_2 (although some species have evolved the ability to respire), into the narrow ecological niches to which they are presently confined.

C. Two-Center Electron Transport

Plants and cyanobacteria use the reducing power generated by light-driven oxidation of H_2O to produce NADPH.

The component half-reactions of this process, together with their standard reduction potentials, are

$$O_2 + 4e^- + 4H^+ \rightleftharpoons 2H_2O \qquad \mathscr{E}°' = +0.815 \text{ V}$$

and

$$NADP^+ + H^+ + 2e^- \rightleftharpoons NADPH \qquad \mathscr{E}°' = -0.320 \text{ V}$$

Hence, the overall four-electron reaction and its standard redox potential is

$$2NADP^+ + 2H_2O \rightleftharpoons 2NADPH + O_2 + 2H^+$$
$$\Delta\mathscr{E}°' = -1.135 \text{ V}$$

This latter quantity corresponds (Eq. [15.5]), to a standard free energy change of $\Delta G°' = 438 \text{ kJ} \cdot \text{mol}^{-1}$, which Eq. [22.1] indicates is the energy of an einstein of 223-nm photons (UV light). Clearly, *even if photosynthesis were 100% efficient, which it is not, it would require more than one photon of visible light to generate a molecule of O_2. In fact, experimental measurements indicate that algae minimally require 8 to 10 photons of visible light to produce one molecule of O_2.* In the following subsections, we discuss how plants and cyanobacteria manage this multiphoton process.

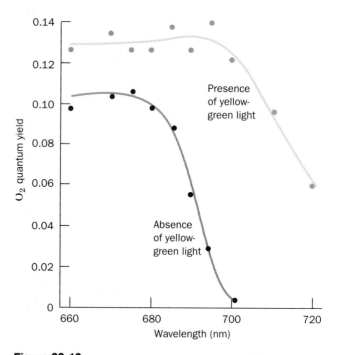

Figure 22-12

The quantum yield for O_2 production by *Chlorella* algae as a function of the wavelength of the incident light in the absence (*lower curve*) and the presence (*upper curve*) of supplementary yellow-green light. The upper curve has been corrected for the amount of O_2 production stimulated by the supplementary light alone. Note that the lower curve falls off precipitously above 680 nm (the red drop). However, the supplementary light greatly increases the quantum yield in the wavelength range above 680 nm (far-red) in which the algae absorb light. [After Emerson, R., Chalmers, R., and Cederstrand, C., *Proc. Natl. Acad. Sci.* **49**, 137 (1957).]

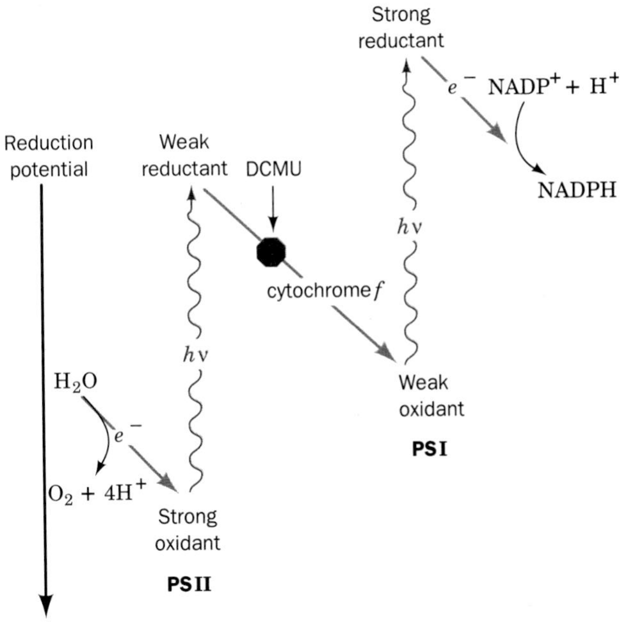

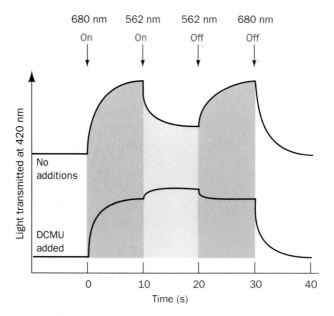

Figure 22-13
The Z-scheme for photosynthesis in plants and cyanobacteria. Two photosystems, PSI and PSII, function to drive electrons from H_2O to NADPH. The reduction potential increases downwards so that electron flow occurs spontaneously in this direction. The herbicide DCMU (see below) blocks photosynthetic electron transport from PSII to cytochrome *f*.

Photosynthetic O₂ Production Requires Two Sequential Photosystems

Two seminal observations led to the elucidation of the basic mechanism of photosynthesis in plants:

1. The quantum yield for O_2 evolution by *Chlorella pyrenoidosa* varies little with the wavelength of the illuminating light between 400 and 675 nm but decreases precipitously above 680 nm (Fig. 22-12, lower curve). This phenomenon, the "red drop," was unexpected because Chl *a* absorbs such far-red light (Fig. 22-5).

2. Shorter wavelength light, such as yellow-green light, enhances the photosynthetic efficiency of 700-nm light well in excess of the energy content of the short wavelength light; that is, *the rate of O₂ evolution by both lights is greater than the sum of the rates for each light acting alone (Fig. 22-12, upper curve).* Moreover, this enhancement still occurs if the yellow-green light is switched off several seconds before the red light is turned on and vice versa.

These observations clearly indicate that two processes are involved. They are explained by a mechanistic model, the **Z-scheme,** which postulates that *O₂-producing photosynthesis occurs through the actions of two photosynthetic reaction centers that are connected essentially in series (Fig. 22-13):*

1. **Photosystem I (PSI),** generates a strong reductant capable of reducing NADP$^+$, and concomitantly, a weak oxidant.

2. **Photosystem II (PSII),** generates a strong oxidant capable of oxidizing H_2O, and concomitantly, a weak reductant.

The weak reductant reduces the weak oxidant so that *PSI and PSII form a two-stage electron "energizer." Both photosystems must therefore function for photosynthesis (electron transfer from H₂O to NADPH) to occur.*

The red drop is explained in terms of the Z-scheme by the observation that PSII is only poorly activated by 680-nm light. In the presence of only this far-red light, PSI is activated but is unable to obtain more than a few of the electrons it is capable of energizing. Yellow-green light, however, efficiently stimulates PSII to supply these electrons. The observation that the far-red and yellow-green lights can be alternated indicates that both photosystems remain activated for a time after the light is switched off.

The validity of the Z-scheme was established as follows. The oxidation state of **cytochrome *f*,** a *c*-type cytochrome of the electron-transport chain connecting PSI and PSII (see below), can be spectroscopically monitored. Illumination of algae with 680 nm (far-red) light results in the oxidation of cytochrome *f* (Fig. 22-14). However, the additional imposition of a 562-nm (yellow-green) light results in this protein's partial rereduc-

Figure 22-14
The oxidation state of cytochrome *f* in *Porphyridium cruentum* algae as monitored by a weak beam of 420 nm (blue-violet) light. An increase in the transmitted light signals the oxidation of cytochrome *f*. In the upper curve, strong light at 680 nm (far-red) causes the oxidation of the cytochrome *f* but the superposition of 562-nm (yellow-green) light causes its partial rereduction. In the lower curve, the presence of the herbicide DCMU, which inhibits photosynthetic electron transport, causes 562-nm light to further oxidize rather than reduce the cytochrome *f*.

tion. In the presence of the herbicide **3-(3,4-dichloro-phenyl)-1,1-dimethylurea (DCMU),**

3-(3,4-Dichlorophenyl)-1,1-dimethylurea (DCMU)

which abolishes photosynthetic oxygen production, 680-nm light still oxidizes cytochrome *f* but simultaneous 562-nm light only oxidizes it further. The explanation for these effects is that 680-nm light, which efficiently activates only PSI, causes it to withdraw electrons from (oxidize) cytochrome *f*. The 562-nm light also activates PSII, which thereby transfers electrons to (reduces) cytochrome *f*. DCMU blocks electron flow

from PSII to cytochrome *f* (Fig. 22-13), so that an increased intensity of light, whatever its wavelength, only serves to activate PSI further.

O₂-Producing Photosynthesis Is Mediated by Three Trans-Membrane Protein Complexes Linked by Mobile Electron Carriers

The pathway of electron transport in the chloroplast has been traced in broad outline but not in the same detail as in the simpler photosynthetic bacterial systems. *The components involved in the electron transport from H_2O to NADPH are largely organized into three thylakoid membrane-bound particles (Fig. 22-15): (1) PSII, (2) cytochrome b_6–f complex, and (3) PSI.* As in oxidative phosphorylation, electrons are transferred between these complexes via mobile electron carriers. The ubiquinone analog **plastoquinone (Q),** via its reduction to

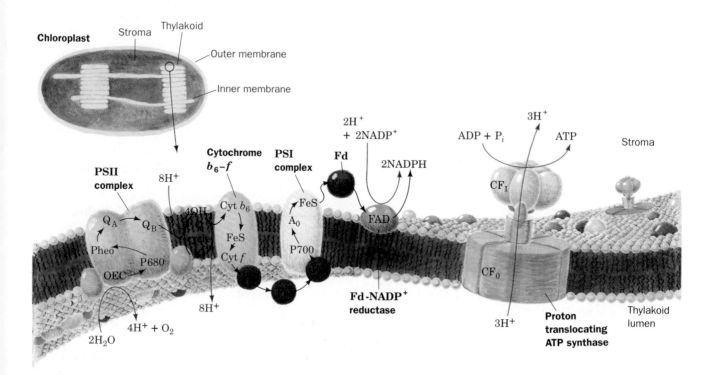

Figure 22-15
A schematic representation of the thylakoid membrane showing the components of its electron-transport chain. The system consists of three protein complexes: PSII, the cytochrome b_6–f complex, and PSI, which are electrically "connected" by the diffusion of the electron carriers plastoquinol (Q) and plastocyanin (PC). Light-driven transport of electrons (*black arrows*) from H_2O to NADPH motivates the transport of protons (*red arrows*) into the thylakoid

space. Additional protons are split off from water by the oxygen-evolving complex (OEC) yielding O_2. The resulting proton gradient powers the synthesis of ATP by the CF_0–CF_1 ATP synthase. The membrane also contains light-harvesting complexes whose component chlorophylls and other chromophores transfer their excitations to PSI and PSII. [After Ort. D. R. and Good, N. E., *Trends Biochem. Sci.* **13**, 469 (1988).]

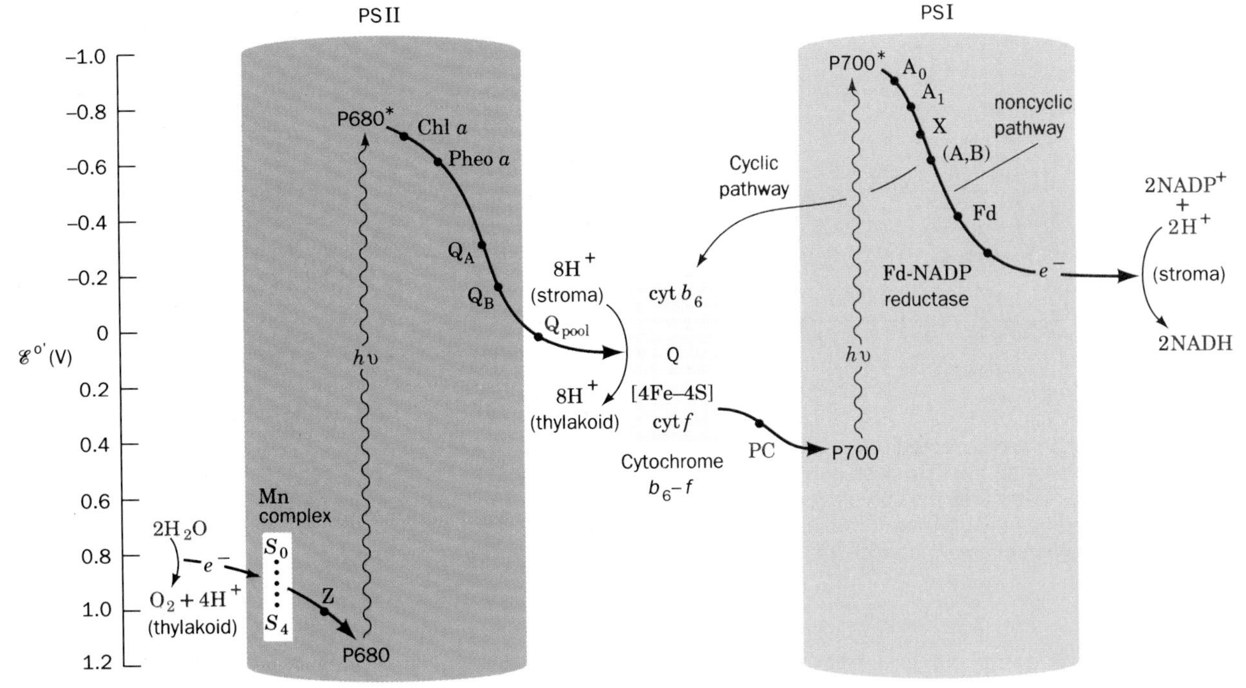

A detailed diagram of the Z-scheme of photosynthesis. Electrons ejected from P680 by the absorption of photons are replaced with electrons abstracted from H_2O by an Mn complex (OEC), thereby forming O_2 and $4H^+$. Each ejected electron is passed through a chain of electron carriers to a pool of plastoquinone molecules (Q). The resulting plastoquinol, in turn, reduces the cytochrome $b_6 - f$ particle (*yellow*) that transfers electrons, via a poorly characterized pathway, to plastocyanin (PC) with the concomitant translocation of protons into the thylakoid space. The plastocyanin regenerates photooxidized P700. The electron ejected from P700, through the intermediacy of a chain of electron carriers, reduces $NADP^+$ to NADPH in noncyclic electron transport. Alternatively, the electron may be returned to the cytochrome $b_6 - f$ complex in a cyclic process that translocates protons into the thylakoid space.

plastoquinol (QH_2),

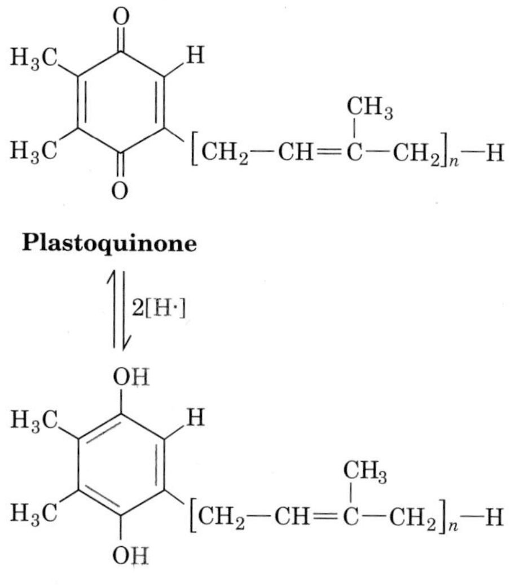

Plastoquinone

Plastoquinol

links PSII to the cytochrome b_6–f complex which, in turn, interacts with PSI through the mobile protein **plastocyanin (PC).** In what follows, we trace the electron pathway through this chloroplast system, insofar as it is known, from H_2O to NADPH (Fig. 22-16).

O_2 Is Generated in a Five-Stage Water-Splitting Reaction Mediated by an Mn-Containing Protein Complex

The oxidation of two molecules of H_2O to form one molecule of O_2 requires four electrons. Since transfer of a single electron from H_2O to $NADP^+$ requires two photochemical events, this accounts for the observed minimum of 8 to 10 photons absorbed per molecule of O_2 produced.

Must the four electrons necessary to produce a given O_2 molecule be removed by a single photosystem or can they be extracted by several different photosystems? This question was answered by monitoring the rate at which dark-adapted chloroplasts produce O_2 when exposed to a series of short flashes. O_2 was evolved with a peculiar oscillatory pattern (Fig. 22-17). There is virtually no O_2 evolved by the first two flashes. The third flash results in the maximum O_2 yield. Thereafter, the amount of O_2 produced peaks with every fourth flash

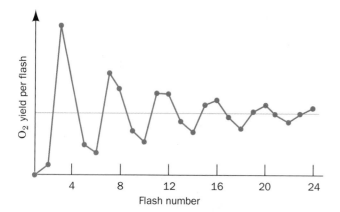

Figure 22-17
The O_2 yield per flash in dark-adapted spinach chloroplasts. Note that the yield peaks on the third flash and then on every fourth flash thereafter until the curve eventually damps out to its average value. [After Forbush, B., Kok, B., and McGloin, M. P., *Photochem. Photobiol.* **14**, 309 (1971).]

until the oscillations damp out to a steady state. This periodicity indicates that each O_2-evolving center cycles through five different states, S_0 through S_4 (Fig. 22-18). Each of the transitions between S_0 and S_4 is a photon-driven redox reaction; that from S_4 to S_0 results in the release of O_2. Thus, *each O_2 molecule must be produced by a single photosystem.* The observation that O_2 evolution peaks at the third rather than the fourth flash indicates that the oxygen-evolving center's resting state is predominantly S_1 rather than S_0. The oscillations gradually damp out because a small fraction of the reaction centers fail to be excited or become doubly excited by a given flash of light so that the reaction centers eventually lose synchrony. The five reaction steps release a total of four water-derived protons into the inner thylakoid space in a step-wise manner (Fig. 22-18).

Since the S states function to abstract electrons from H_2O, their standard reduction potentials must average more than the 0.815-V value of the O_2/H_2O half-reaction. PSII has the remarkable capacity of stabilizing these highly reactive intermediates for extended periods (typically minutes) in close proximity to water. We are just beginning to understand how this occurs. PSII contains four protein-bound Mn ions which, upon excitation of chloroplasts with short flashes of light, exhibit EPR signals that have a four-flash periodicity similar to that of O_2 production (Fig. 22-17). These Mn ions evidently form a catalytically active complex, the **oxygen-evolving complex (OEC)**, which binds two H_2O molecules so as to facilitate O_2 formation. The OEC cycles through a series of oxidation states [the S states, which probably involve various combinations of Mn(III), Mn(IV), and Mn(V)] while abstracting protons and electrons from the H_2O molecules, and finally releases O_2 into the inner thylakoid space. In a model consistant with the EPR data, the S_0, S_1, and S_2 states are adamantane-like Mn_4O_6 complexes and the S_3 and S_4 states are cubane-like Mn_4O_4 complexes (Fig. 22-19). In the $S_4 \rightarrow S_0$ transition, the adamantane-like structure changes to the cubane-like structure with the release of O_2.

The next link in the PSII electron transport chain is a substance known as Z (Fig. 22-16), which relays electrons from the Mn-protein water-splitting complex to the reaction center of PSII. The existence of Z is signaled by a transient EPR spectrum of illuminated chloroplasts, that parallels the S-state transitions. The change in this spectrum upon feeding deuterated tyrosine to cyanobacteria under conditions that they incorporate this amino acid in their proteins indicates that Z^+ is a tyrosine radical (EPR spectra reflect the nuclear spins of the atoms with which the unpaired electrons interact).

The PSII Reaction Center Resembles That of Photosynthetic Bacteria

PSII reaction center's photon-absorbing species is named **P680,** after the wavelength of its absorption maximum. Spectroscopic analysis of P680 indicates that it consists of Chl *a* but it has not been definitively established whether it is a "special pair" of Chl *a* molecules, similar to P870 of purple photosynthetic bacteria (Section 22-2B), or a monomer. The P680$^+$ formed by light excitation, which is among the most powerful biological oxidants known, abstracts electrons from H_2O via the intermediacy of Z and the S states.

The chain of electron carriers on the reducing side of P680 bears a remarkable resemblance to the bacterial photosynthetic reaction center (Section 22-2B) even though the two systems operate over different ranges of reduction potentials (compare Figs. 22-11 and 22-16). Indeed, *the two sets of proteins have similar amino acid sequences indicating that they arose from a common ancestor.* A single electron is transferred, as diagrammed in the central portion of Fig. 22-16, from P680* to a mole-

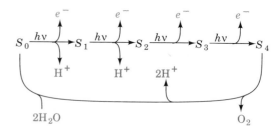

Figure 22-18
The schematic mechanism of O_2 generation in chloroplasts. Four electrons are stripped, one at a time in light-driven reactions, from two bound H_2O molecules (*top*). In the recovery step (*bottom*), which is light independent, the resulting O_2 is released and two more H_2O molecules are bound. Three of these five steps release protons into the thylakoid space.

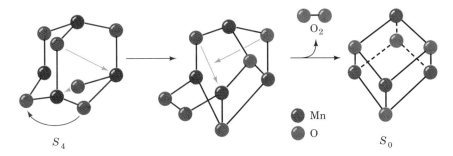

S_4 Mn   O S_0

Figure 22-19

A proposed structure of PSII's water splitting Mn complex (OEC). In the $S_4 \rightarrow S_0$ transition, the complex changes from the adamantane-like Mn_4O_6 complex to the cubane-like Mn_4O_4 complex with the release of O_2 (adamantane and cubane are saturated hydrocarbons with C atoms in the positions occupied by the Mn and O atoms in these complexes). The Mn complex retains the cubane-like structure through its S_1 and S_2 states, switching back to the adamantane-like structure in its S_3 state. [After Brudvig, G. W. and Crabtree, R. H., *Proc. Natl. Acad. Sci.* **83**, 4586 (1986).]

cule of **pheophytin** *a* (**Pheo** *a*; Chl *a* with its Mg^{2+} replaced by two protons), probably via a Chl *a* molecule, and then to a plastoquinone–Fe(II) complex designated Q_A. Subsequently, two electrons are transferred, one at a time, to a second plastoquinone molecule, Q_B, which takes up two protons at the stromal surface of the thylakoid membrane. The resulting plastoquinol, Q_BH_2, then exchanges with a membrane-bound pool of plastoquinone molecules. DCMU, as well as many other commonly used herbicides, compete with plastoquinone for the Q_B-binding site on PSII, which explains how they inhibit photosynthesis.

Electron Transport through the Cytochrome b_6–f Complex Generates a Proton Gradient

From the plastoquinone pool, electrons pass through the cytochrome b_6–f complex. This integral membrane assembly, which closely resembles both its bacterial counterpart (Section 22-2B) and Complex III of the mitochondrial electron-transport chain (Section 20-2C), contains one molecule of cytochrome *f*, a two heme-containing **cytochrome b_6,** one [2Fe–2S] iron–sulfur protein, and one bound plastoquinol. *The cytochrome b_6–f complex transports protons as well as electrons from the outside to the inside of the thylakoid membrane.* This proton translocation, it has been proposed, occurs through a Q cycle (Section 20-3B) in which plastoquinone is the $(H^+ + e^-)$ carrier. However, the roles of the various available e^- carriers in this process have not been sorted out. The Q cycle mechanism predicts that two protons are translocated across the thylakoid membrane for every electron transported but the experimental difficulties of measuring this ratio have precluded its unambiguous determination. It is, nevertheless, clear that *electron transport, via the cytochrome b_6–f complex, generates much of the electrochemical proton gradient that drives the synthesis of ATP (see below).*

Electron transfer between cytochrome b_6–f complex and PSI is mediated by plastocyanin, a peripheral membrane protein located on the thylakoid luminal surface (Fig. 22-15). The Cu-containing redox center of this mobile 10.5-kD monomer cycles between its Cu(I) and Cu(II) oxidation states. The X-ray structure of plastocyanin from poplar leaves, determined by Hans Freeman, shows the Cu atom coordinated with distorted tetrahedral geometry by a Cys, a Met, and two His residues (Fig. 22-20). Cu(II) complexes with four ligands normally adopt a square planar coordination geometry, whereas those of Cu(I) are generally tetrahedral. Evidently, the strain of Cu(II)'s protein-imposed tetrahedral coordination in plastocyanin promotes its reduction to Cu(I). This hypothesis accounts for plastocyanin's high standard reduction potential (0.370 V) compared to that of the normal Cu(II)/Cu(I) half-reaction (0.158 V). This is an example of how proteins modulate the reduction potentials of their redox centers so as to

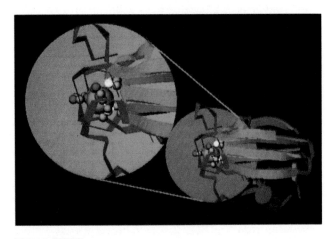

Figure 22-20

A computer-generated ribbon diagram of plastocyanin. The expanded view shows its Cu atom (*large orange sphere*), which alternates between its Cu(I) and Cu(II) oxidation states, in complex with its four ligands, His 37, Cys 84, His 87, and Met 92. [Courtesy of Arthur Lesk, Cambridge University and EMBL.]

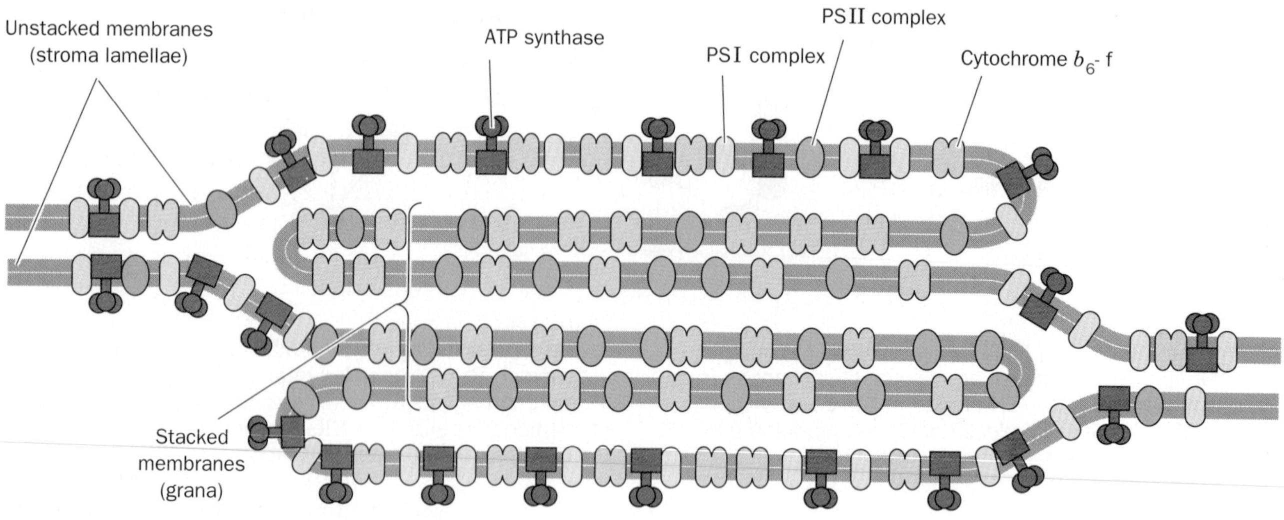

Figure 22-21
The distribution of photosynthetic protein complexes between the stacked (grana) and the unstacked (stroma exposed) regions of the thylakoid membrane. [After Anderson, J. M. and Anderson, B., *Trends Biochem. Sci.* **7**, 291 (1982).]

match them to their function—in the case of plastocyanin, the efficient transfer of electrons from the cytochrome b_6–f complex to PSI.

PSI-Activated Electrons May Reduce NADP$^+$ or Motivate Proton Gradient Formation

PSI, the chloroplast's lower reduction potential photosystem, differs greatly from both PSII and the reaction centers of photosynthetic bacteria. The photon-absorbing center of PSI, **P700,** probably consists of a dimer of chlorophyll *a* molecules. Photooxidation of P700 yields P700$^+$, a weak oxidant that subsequently accepts an electron directly from plastocyanin. On the reducing side of P700, the analysis of light-induced EPR changes indicates that the electron passes through a chain of electron carriers of the increasing reduction potential (right side of Fig. 22-16). The first of these carriers, designated A$_0$, appears to be a Chl *a* monomer, whereas the second carrier, A$_1$, is probably **phylloquinone (vitamin K$_1$;** note that its phytyl side chain is the same as that of the chlorophylls; Fig. 22-3).

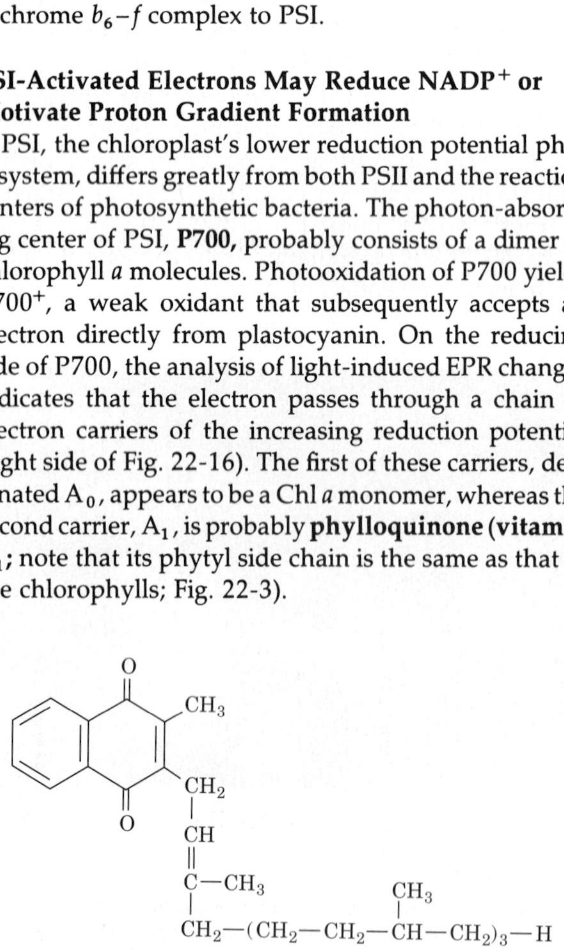

Phylloquinone

The electron finally proceeds through three membrane-bound **ferredoxins,** X, A, and B, which probably all contain [4Fe–4S] clusters (Section 20-2C; note that the terminal electron carriers of the other photosystems are all quinones).

Electrons ejected from PSI may follow either of two alternative pathways:

1. Most electrons follow a noncyclic pathway by passing to an 11-kD, [2Fe–2S]-containing soluble ferredoxin, Fd, that is located in the stroma. Reduced Fd, in turn, reduces NADP$^+$ in a reaction mediated by the 37 kD, FAD-containing **ferredoxin-NADP$^+$ reductase,** to yield the final product of the chloroplast light reaction, NADPH.

2. Some electrons are returned from PSI, via cytochrome b_6, to the plastoquinone pool, thereby traversing a cyclic pathway that translocates protons across the thylakoid membrane (Fig. 22-16). This accounts for the observation that chloroplasts absorb more than eight photons per O$_2$ molecule evolved. Note that the cyclic pathway is independent of the action of PSII and hence does not result in the evolution of O$_2$. PSI, in this way, functionally resembles the photosynthetic bacterial system. It therefore came as a surprise when it was discovered that PSII, but not PSI, is related genetically to bacterial photosystems.

The cyclic electron flow presumably functions to increase the amount of ATP produced relative to that of NADPH and thus permits the cell to adjust the relative amounts of these two substances produced according to its needs. However, the mechanism that apportions electrons between the cyclic and noncyclic pathways is unknown.

PSI and PSII Occupy Different Parts of the Thylakoid Membrane

Freeze-fracture electron microscopy (Section 11-3B) has revealed that the protein complexes of the thylakoid membrane have characteristic distributions (Fig. 22-21):

1. PSI occurs mainly in the unstacked stroma lamellae, in contact with the stroma where it has access to $NADP^+$.

2. PSII is located almost exclusively between the closely stacked grana, out of direct contact with the stroma.

3. Cytochrome b_6–f is uniformly distributed throughout the membrane.

The high mobilities of plastoquinone and plastocyanin, the electron carriers that shuttle electrons between these particles, permits photosynthesis to proceed at a reasonable rate.

What function is served by the segregation of PSI and PSII? If these two photosystems were in close proximity, the higher excitation energy of PSII (P680 vs P700) would cause it to pass a large fraction of its absorbed photons to PSI via exciton transfer; that is, PSII would act as a light-harvesting antenna for PSI (Fig. 22-7b). The separation of these particles by around 100 Å eliminates this difficulty.

The physical separation of PSI and PSII also permits the chloroplast to respond to changes in illumination. The relative amounts of light absorbed by the two photosystems vary with how the light-harvesting complexes (LHCs) are distributed between the stacked and unstacked portions of the thylakoid membrane. Under high illumination (normally direct sunlight, which contains a high proportion of short wavelength blue light), all else being equal, PSII absorbs more light than PSI. PSI is then unable to take up electrons as fast as PSII can supply them so the plastoquinone is predominantly in its reduced state. The reduced plastoquinone activates a protein kinase to phosphorylate specific Thr residues of the LHCs which, in response, migrate to the unstacked regions of the thylakoid membrane where they bind to PSI. A greater fraction of the incident light is thereby funneled to PSI. Under low illumination (normally shady light, which contains a high proportion of long wavelength red light), PSI takes up electrons faster than PSII can provide them so that plastoquinone predominantly assumes its oxidized form. The LHCs are consequently dephosphorylated and migrate to the stacked portions of the thylakoid membrane where they drive PSII. The chloroplast therefore maintains the balance between its two photosystems by a light-activated feedback mechanism.

D. Photophosphorylation

Chloroplasts generate ATP in much the same way as mitochondria, that is, by coupling the dissipation of a proton gradient to the enzymatic synthesis of ATP (Section 20-3C). This was clearly demonstrated by the imposition of an artificially produced pH gradient across the thylakoid membrane. Chloroplasts were soaked, in the dark, for several hours in a succinic acid solution of pH 4 so as to bring the thylakoid space to this pH (the thylakoid membrane is permeable to un-ionized succinic acid). The abrupt transfer of these chloroplasts to an ADP + P_i-containing buffer at pH 8 resulted in an impressive burst of ATP synthesis: About 100 ATPs were synthesized per molecule of cytochrome f present. Moreover, the amount of ATP synthesized was unaffected by the presence of electron-transport inhibitors such as DCMU. This, together with the observations that photophosphorylation requires an intact thylakoid membrane and that proton translocators such as 2,4-dinitrophenol (Section 20-3D) uncouple photophosphorylation from light-driven electron transport, provide convincing evidence favoring Peter Mitchell's chemiosmotic hypothesis (Section 20-3A).

Chloroplast Proton-Translocating ATP Synthase Resembles That of Mitochondria

Electron micrographs of thylakoid membrane stromal surfaces and bacterial plasma membrane inner surfaces reveal lollipop-shaped structures (Fig. 22-22). These closely resemble the F_1 units of the proton-translocating ATP synthase studding the matrix surfaces of inner mitochondrial membranes (Fig. 20-27a). In fact, the chlor-

Figure 22-22
Electron micrographs of thylakoids showing the CF_1 "lollipops" of their ATP synthases projecting from their stromal surfaces. Compare this with Fig. 20-27a and b. [Courtesy of Efraim Racker, Cornell University.]

oplast ATP synthase, which is termed **CF$_0$CF$_1$ complex** (C for chloroplast), has remarkably similar properties to the mitochondrial F$_0$F$_1$ complex (Section 20-3C). For example,

1. Both F$_0$ and CF$_0$ units are hydrophobic trans-membrane proteins that contain a proton translocating channel.

2. Both F$_1$ and CF$_1$ are hydrophilic peripheral membrane proteins of subunit composition $\alpha_3\beta_3\gamma\delta\varepsilon$, of which β is a reversible ATPase, and γ forms the gate controlling proton flow from (C)F$_1$ to (C)F$_0$.

3. Both ATP synthases are inhibited by oligomycin and by dicyclohexylcarbodiimide.

Clearly, proton-translocating ATP synthases must have evolved very early in the history of cellular life. Note, however, that whereas chloroplast ATP synthase translocates protons out of the thylakoid space (Fig. 22-15), mitochondrial ATP synthase conducts them into the matrix space (Section 20-3A). Chloroplast ATP synthase is located in the unstacked portions of the thylakoid membrane, in contact with the stroma, where there is room for the bulky CF$_1$ globule and access to ADP (Fig. 22-21).

Photosynthesis with Noncyclic Electron Transport Produces around 1.25 ATPs per Absorbed Photon

At saturating light intensities, chloroplasts generate proton gradients of ~3.5 pH units across their thylakoid membranes. This, as we have seen, arises from two sources:

1. The evolution of a molecule of O$_2$ from 2H$_2$O molecules releases 4 protons into the thylakoid space.

2. The transport of the liberated 4 electrons through the cytochrome b_6–f complex occurs with the translocation of what is estimated to be 8 protons from the stroma to the thylakoid space.

Altogether ~12 protons are translocated per molecule of O$_2$ produced by noncyclic electron transport.

The thylakoid membrane, in contrast to the inner mitochondrial membrane, is permeable to ions such as Mg^{2+} and Cl$^-$. Translocation of protons and electrons across the thylakoid membrane is consequently accompanied by the passage of these ions so as to maintain electrical neutrality (Mg^{2+} out and Cl$^-$ in). This all but eliminates the membrane potential, $\Delta\Psi$ (Eq. [20.1]). *The electrochemical gradient in chloroplasts is therefore almost entirely a result of the pH gradient.*

Chloroplast ATP synthase, according to most estimates, produces one ATP for every three protons it transports out of the thylakoid space. Noncyclic electron transport in chloroplasts therefore results in the production of ~$\frac{12}{3}$ = 4 molecules of ATP per molecule of O$_2$

evolved (although this quantity is subject to revision) or around one half of an ATP per photon absorbed. Cyclic electron transport is a more productive ATP generator since it yields two thirds of an ATP (2 protons) per absorbed photon. The noncyclic process, of course, also yields NADPH, each molecule of which has the free energy to produce three ATPs (Section 20-2A), for a total of six more ATP equivalents per O$_2$ produced. Consequently, the energetic efficiency of the noncyclic process is $\frac{4}{8} + \frac{6}{8} = 1.25$ ATP equivalents per absorbed photon.

3. DARK REACTIONS

In the previous section we saw how light energy is harnessed to generate ATP and NADPH. In this section we discuss how these products are used to synthesize carbohydrates and other substances from CO$_2$.

A. The Calvin Cycle

The metabolic pathway by which plants incorporate CO$_2$ into carbohydrates was elucidated between 1946 and 1953 by Melvin Calvin, James Bassham, and Andrew Benson. They did so by tracing the metabolic fate of the radioactive label from ^{14}CO$_2$ as it passed through a series of photosynthetic intermediates. The basic experimental strategy they used was to expose growing cultures of algae, such as *Chlorella,* to ^{14}CO$_2$ for varying times and under differing illumination conditions and then to drop the cells into boiling alcohol so as to disrupt them while preserving their labeling pattern. The radioactive products were subsequently separated and identified (an often difficult task) through the use of the then recently developed technique of two-dimensional paper chromatography (Section 5-3B) coupled with autoradiography. The overall pathway, diagrammed in Fig. 22-23, is known as the **Calvin cycle** or the **reductive pentose phosphate cycle.**

Some of Calvin's earliest experiments indicated that algae exposed to ^{14}CO$_2$ for a minute or more had synthesized a complex mixture of labeled metabolic products including sugars and amino aicds. By inactivating the algae within 5 s of their exposure to ^{14}CO$_2$, however, it was shown that *the first stable radioactively labeled compound formed is 3-phosphoglycerate (3PG), which is ini-*

Figure 22-23 *(opposite)*
The Calvin cycle. The number of lines in an arrow indicates the number of molecules reacting in that step for a single turn of the cycle that converts three CO$_2$ molecules to one GAP molecule. For the sake of clarity, the sugars are all shown in their linear forms although the hexoses and heptoses predominantly exist in their cyclic forms (Section 10-1B). The ^{14}C-labeling patterns generated in one turn of the cycle through the use of ^{14}CO$_2$ are indicated in red. Note that two of the Ru5Ps are labeled only at C(3), whereas the third Ru5P is equally labeled at C(1), C(2), and C(3).

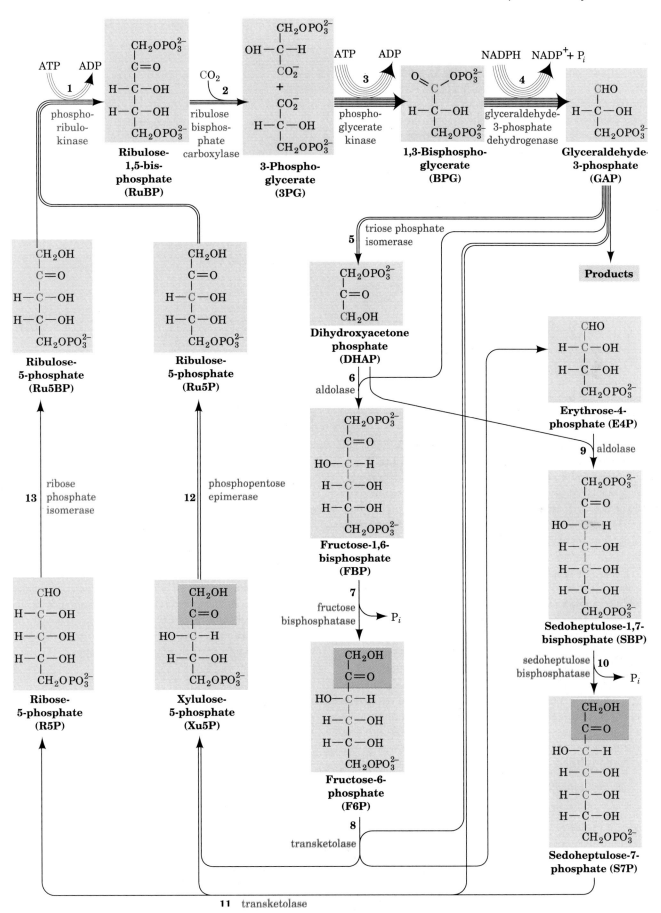

tially labeled only in its carboxyl group. This result immediately suggested, in analogy with most biochemical experience, that the 3PG was formed by the carboxylation of a C_2 compound. Yet, the failure to find any such precursor eventually forced this hypothesis to be abandoned. The actual carboxylation reaction was discovered through an experiment in which illuminated algae had been exposed to $^{14}CO_2$ for ~10 min so that the levels of their labeled photosynthetic intermediates had reached a steady state. The CO_2 was then withdrawn. As expected, the carboxylation product, 3PG, decreased in concentration (Fig. 22-24) because it was depleted by reactions further along the pathway. The concentration of **ribulose-5-phosphate (Ru5P),**

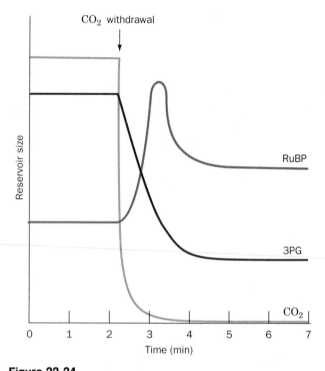

$$
\begin{array}{c}
CH_2OH \\
| \\
C=O \\
| \\
H-C-OH \\
| \\
H-C-OH \\
| \\
CH_2OPO_3^{2-}
\end{array}
$$

Ribulose-5-phosphate (Ru5P)

however, simultaneously increased. Evidently, Ru5P is the Calvin cycle's carboxylation substrate. If so, the resulting C_6 carboxylation product must split into two C_3 compounds, one of which is 3PG (Fig. 22-23, Reaction 2). A consideration of the oxidation states of Ru5P and CO_2 indicates that, in fact, both C_3 compounds must be 3PG and that the carboxylation reaction requires no external redox source.

While the search for the carboxylation substrate was going on, several other photosynthetic intermediates had been identified and, through chemical degradation studies, their labeling patterns had been elucidated. For example, the hexose fructose-1,6-bisphosphate (FBP) is initially labeled only at its C(3) and C(4) positions (Fig. 22-23) but later becomes labeled to a lesser degree at its other atoms. Similarly, a series of tetrose, pentose, hexose, and heptose phosphates were isolated that had the identities and initial labeling patterns indicated in Fig. 22-23. A consideration of the flow of the labeled atoms through these various intermediates led, in what is a milestone of metabolic biochemistry, to the deduction of the Calvin cycle as is diagrammed in Fig. 22-23. The existence of many of its postulated reactions was eventually confirmed by *in vitro* studies using purified enzymes.

The Calvin Cycle Generates GAP from CO_2 via a Two-Stage Process

The Calvin cycle may be considered to have two stages:

Stage 1 The production phase (top line of Fig. 22-23), in which three molecules of Ru5P react with three molecules of CO_2 to yield six molecules

Figure 22-24
The time course of the levels of 3PG (*purple curve*) and RuBP (*green curve*) in steady state $^{14}CO_2$-labeled, illuminated algae during a period in which the CO_2 (*orange curve*) is abruptly withdrawn. In the absence of CO_2, the 3PG concentration rapidly decreases because it is taken up by the reactions of the Calvin cycle but cannot be replenished by them. Conversely, the RuBP concentration transiently increases as it is synthesized from the residual pool of Calvin cycle intermediates but, in the absence of CO_2, cannot be used for their regeneration.

of glyceraldehyde-3-phosphate (GAP) at the expense of nine ATP and six NADPH molecules. *The cyclic nature of the pathway makes this process equivalent to the synthesis of one GAP from three CO_2 molecules.* Indeed, at this point, one GAP can be bled off from the cycle for use in biosynthesis (see Stage 2).

Stage 2 The recovery phase (bottom lines of Fig. 22-23), in which the carbon atoms of the remaining five GAPs are shuffled in a remarkable series of reactions, similar to those of the pentose phosphate pathway (Section 21-4), to reform the three Ru5Ps with which the cycle began. Indeed, the elucidation of the pentose phosphate pathway at about the same time that the Calvin cycle was being worked out provided much of the biochemical evidence in support of the Calvin cycle. This stage can be conceptually decomposed into four sets of reactions (with the numbers keyed to the corresponding reactions in Fig. 22-23):

6. $C_3 + C_3 \longrightarrow C_6$

8. $C_3 + C_6 \longrightarrow C_4 + C_5$

9. $C_3 + C_4 \longrightarrow C_7$

11. $C_3 + C_7 \longrightarrow C_5 + C_5$

The overall stoichiometry for this process is therefore

$$5C_3 \longrightarrow 3C_5$$

Note that this stage of the Calvin cycle occurs without further input of free energy (ATP) or reducing power (NADPH).

Most Calvin Cycle Reactions Also Occur in Other Metabolic Pathways

The types of reactions that comprise the Calvin cycle are all familiar, with the exception of the carboxylation reaction. Thus the first stage of the Calvin cycle begins with the phosphorylation of Ru5P by **phosphoribulokinase** to form **ribulose-1,5-bisphosphate (RuBP).** Following the carboxylation step, which is discussed below, the resulting 3PG is converted first to 1,3-bisphosphoglycerate (BPG) and then to GAP. This latter sequence is the reverse of two consecutive glycolytic reactions (Sections 16-2G and F) except that the Calvin cycle reaction involves NADPH rather than NADH.

The second stage of the Calvin cycle begins with the reverse of a familiar glycolytic reaction, the isomerization of GAP to dihydroxyacetone phosphate (DHAP) by triose phosphate isomerase (Section 16-2E). Following this, DHAP is directed along two analogous paths (Fig. 22-23): Reactions 6–8 or Reactions 9–11. Reactions 6 and 9 are aldolase-catalyzed aldol condensations in which DHAP is linked to an aldehyde (aldolase is specific for DHAP but accepts a variety of aldehydes). Reaction 6 is also the reverse of a glycolytic reaction (Section 16-2D). Reactions 7 and 10 are phosphate hydrolysis reactions that are catalyzed, respectively, by fructose bisphosphatase (FBPase, which we previously encountered in our discussion of glycolytic futile cycles and gluconeogenesis; Sections 16-4B and 21–1A), and **sedoheptulose bisphosphatase.** The remaining Calvin cycle reactions are catalyzed by enzymes that also participate in the pentose phosphate pathway. In Reactions 8 and 11, both catalyzed by **transketolase,** a C_2 keto unit (shaded in green in Fig. 22-23) is transferred from a ketose to GAP to form **xylulose-5-phosphate (Xu5P)** and leave the aldoses **erythrose-4-phosphate (E4P)** in Reaction 8 and **ribose-5-phosphate (R5P)** in Reaction 11. The E4P produced by Reaction 8 feeds into Reaction 9. The Xu5Ps produced by Reactions 8 and 11 are converted to Ru5P by **phosphopentose epimerase** in Reaction 12. The R5P from Reaction 11 is also converted to Ru5P by **ribose phosphate isomerase** in Reaction 13, thereby completing a turn of the Calvin cycle. Thus only three of the 11 Calvin cycle enzymes, phosphoribulo-

kinase, the carboxylation enzyme **ribulose bisphosphate carboxylase,** and SBPase, have no equivalents in animal tissues.

RuBP Carboxylase Catalyzes CO₂ Fixation in an Exergonic Process

The enzyme that catalyzes CO_2 fixation, ribulose bisphosphate carboxylase **(RuBP carboxylase),** is arguably the world's most important enzyme since nearly all life on earth ultimately depends on its action. This protein comprises some 15% of chloroplast protein and is therefore the most abundant protein in the biosphere (it is estimated to be synthesized at the rate of 4×10^{13} g·yr^{-1}). RuBP carboxylase from higher plants and most photosynthetic microorganisms consists of eight large (L) subunits (56 kD) encoded by chloroplast DNA, and eight small (S) subunits (14 kD) specified by a nuclear gene (the RuBP carboxylase from certain photosynthetic bacteria is an L_2 dimer whose L subunit has 30% sequence identity and is structurally similar to that of the L_8S_8 enzyme). X-ray studies by David Eisenberg demonstrated that this protein has the symmetry of a square prism (Fig. 22-25). The large subunit contains the

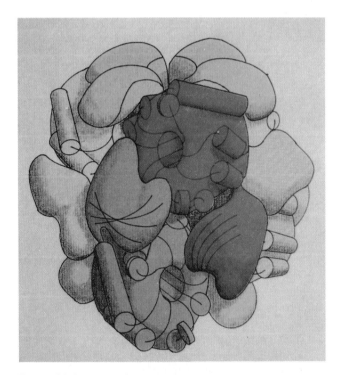

Figure 22-25
The X-ray structure of tobacco RuBP carboxylase drawn to show the quaternary structure of this L_8S_8 protein. The S subunits (six visible) are blue and each of the bilobal L subunits (four visible) is a different color. The protein, which has D_4 symmetry (the symmetry of a square prism; Section 7-5B), is viewed here along one of its twofold axes (which relates the red and green L subunits). The fourfold rotation axis, which is nearly vertical in this view, relates the members of the S_4 tetramer visible on the top of the complex. [Courtesy of David Eisenberg, UCLA.]

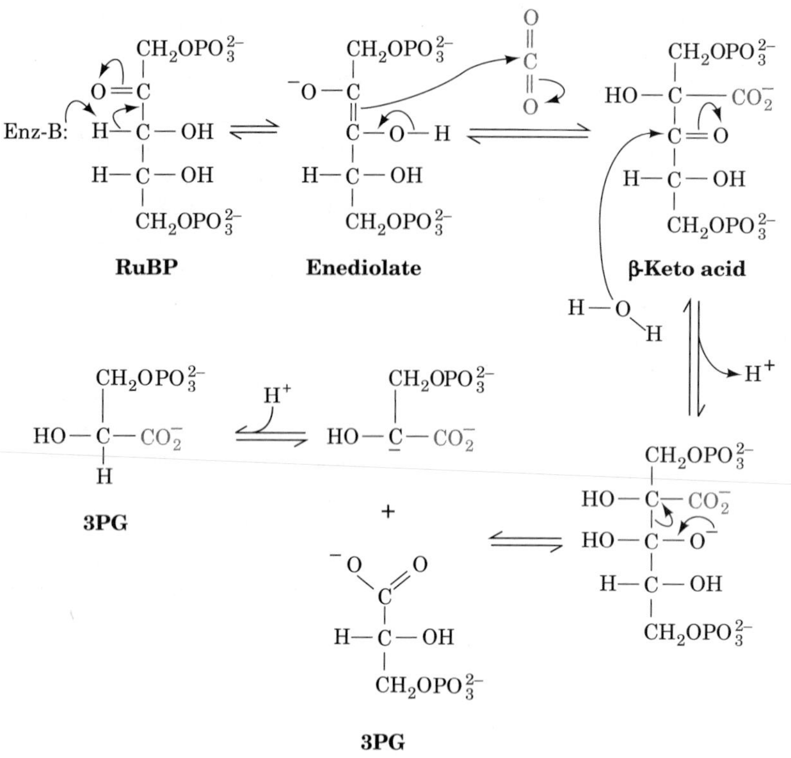

Figure 22-26
The probable reaction mechanism of the carboxylation reaction catalyzed by RuBP carboxylase. The reaction proceeds via an enediolate intermediate that nucleophilically attacks CO_2 to form a β-keto acid. This intermediate reacts with water to yield two molecules of 3PG.

enzyme's catalytic site as is demonstrated by its enzymatic activity in the absence of the small subunit. The function of the small unit is unknown; attempts to show that it has a regulatory role, in analogy with other enzymes, have been unsuccessful.

The accepted mechanism of RuBP carboxylase, which was largely formulated by Calvin, is indicated in Fig. 22-26. Abstraction of the C(3) proton of RuBP, the reaction's rate-determining step, generates an enediolate that nucleophilically attacks CO_2 (not HCO_3^-). The resulting β-keto acid is rapidly attacked at its C(3) position by H_2O to yield an adduct that splits, by a reaction similar to aldol cleavage, to yield the two product 3PG molecules. Evidence favoring this mechanism is

1. The C(3) proton of enzyme-bound RuBP exchanges with solvent, an observation compatible with the existence of the enediolate intermediate.

2. The C(2) and C(3) oxygen atoms remain attached to their respective C atoms, which eliminates mechanisms involving a covalent adduct such as a Schiff base between RuBP and the enzyme.

3. The trapping of the proposed β-keto acid intermediate by borohydride reduction, and the tight enzymatic binding of its analog such as **2-carboxyarabin-**

itol-1-phosphate,

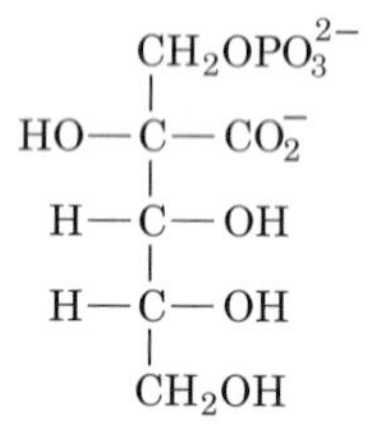

2-Carboxyarabinitol-1-phosphate

provide strong evidence for the existence of this intermediate.

Enzyme activity requires a bound divalent metal ion such as Mg^{2+}, which probably acts to stabilize developing negative charges during catalysis. *The driving force for the overall reaction, which is highly exergonic ($\Delta G°' = -35.1$ kJ · mol^{-1}), is provided by the cleavage of the β-keto acid intermediate to yield an additional resonance-stabilized carboxylate group.*

GAP Is the Precursor of Glucose-1-phosphate and Other Biosynthetic Products

The overall stoichiometry of the Calvin cycle is

$$3CO_2 + 9ATP + 6NADPH \longrightarrow$$
$$GAP + 9ADP + 8P_i + 6NADP^+$$

GAP, the primary product of photosynthesis, is used in a variety of biosynthetic pathways, both inside and outside the chloroplast. For example, it can be converted to fructose-6-phosphate by the further action of Calvin cycle enzymes and then to glucose-1-phosphate (G1P) by phosphoglucose isomerase and phosphoglucomutase (Section 17-1B). *G1P is the precursor of the higher carbohydrates characteristic of plants.* These most notably include sucrose (Section 10-2B), their major transport sugar for delivering carbohydrates to nonphotosynthesizing cells; starch (Section 10-2D), their chief storage polysaccharide; and cellulose (Section 10-2C), the primary structural component of their cell walls. In the synthesis of all these substances, G1P is activated by the formation of either ADP–, CDP–, GDP–, or UDP–glucose (Section 17-2), depending on the species and the pathway. Its glucose unit is then transferred to the nonreducing end of a growing polysaccharide chain much as occurs in the synthesis of glycogen (Section 17-2B). In the case of sucrose synthesis, the acceptor is the reducing end of F6P with the resulting **sucrose-6-phosphate** being hydrolyzed to sucrose by a phosphatase. Fatty acids and amino acids are synthesized from GAP as is described, respectively, in Sections 23-4 and 24-5.

B. Control of the Calvin Cycle

During the day, plants satisfy their energy needs via the light and dark reactions of photosynthesis. At night, however, like other organisms, they must use their nutritional reserves to generate their required ATP and NADPH through glycolysis, oxidative phosphorylation, and the pentose phosphate pathway. Since the stroma contains the enzymes of glycolysis and the pentose phosphate pathway as well as those of the Calvin cycle, *plants must have a light-sensitive control mechanism to prevent the Calvin cycle from consuming this catabolically produced ATP and NADPH in a wasteful futile cycle.*

As we saw in Section 16-4A, the control of flux in a metabolic pathway occurs at enzymatic steps that are far from equilibrium; that is, they have a large negative value of ΔG. Inspection of Table 22-1 indicates that the three best candidates for flux control in the Calvin cycle are the reactions catalyzed by RuBP carboxylase, FBPase, and SBPase (Reactions 2, 7, and 10, Fig. 22-23). In fact, the catalytic efficiencies of these three enzymes all vary, *in vivo*, with the level of illumination.

The activity of RuBP carboxylase responds to four light-dependent factors:

1. It varies with pH. Upon illumination, the pH of the stroma increases from around 7.0 to about 8.0 as protons are pumped from the stroma into the thylakoid space. RuBP carboxylase has a sharp pH optimum near pH 8.0.

2. It is stimulated by Mg^{2+}. Recall that the light-induced influx of protons to the thylakoid space is accompanied by the efflux of Mg^{2+} to the stroma (Section 22-2D).

Table 22-1

Standard and Physiological Free Energy Changes for the Reactions of the Calvin Cycle

Step[a]	Enzyme	$\Delta G°'$ (kJ·mol⁻¹)	ΔG (kJ·mol⁻¹)
1	Phosphoribulokinase	−21.8	−15.9
2	Ribulose bisphosphate carboxylase	−35.1	−41.0
3 + 4	Phosphoglycerate kinase + glyceraldehyde-3-phosphate dehydrogenase	+18.0	−6.7
5	Triose phosphate isomerase	−7.5	−0.8
6	Aldolase	−21.8	−1.7
7	Fructose bisphosphatase	−14.2	−27.2
8	Transketolase	+6.3	−3.8
9	Aldolase	−23.4	−0.8
10	Sedoheptulose bisphosphatase	−14.2	−29.7
11	Transketolase	+0.4	−5.9
12	Phosphopentose isomerase	+0.8	−0.4
13	Ribose phosphate isomerase	+2.1	−0.4

[a] Refer to Fig. 22-23.

Source: Bassham, J. A. and Buchanan, B. B., *in* Govindjee (Ed.), *Photosynthesis,* Vol. II, *p.* 155, Academic Press (1982).

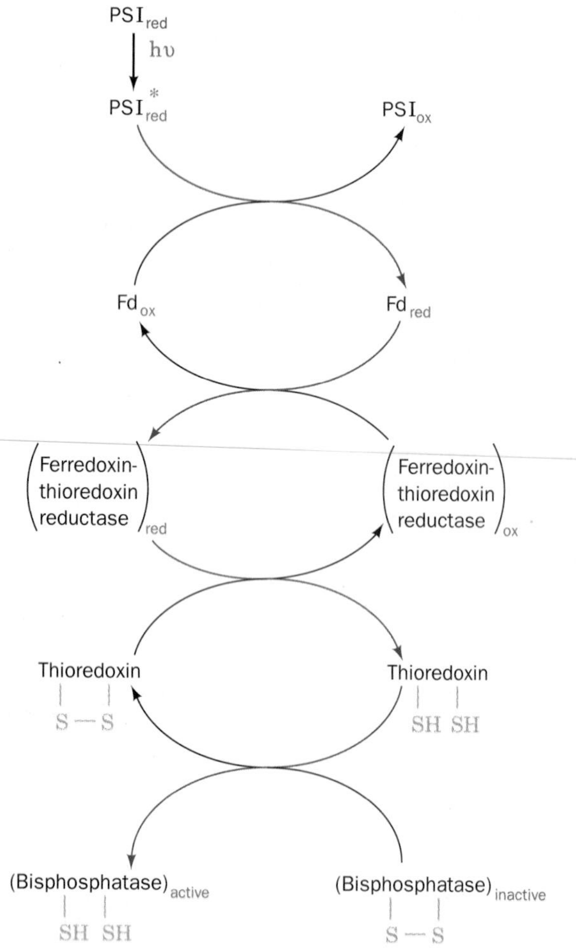

Figure 22-27
The light-activation mechanism of FBPase and SBPase. Photoactivated PSI reduces soluble ferredoxin (Fd), which reduces ferredoxin–thioredoxin reductase which, in turn, reduces the disulfide linkage of thioredoxin. Reduced thioredoxin reacts with the inactive bisphosphatases by disulfide exchange, thereby activating these flux-generating Calvin cycle enzymes.

3. It is allosterically activated by NADPH, which is produced by illuminated PSI (Section 22-2C).

4. It is strongly inhibited by 2-carboxyarabinitol-1-phosphate (Section 22-3A), which many plants synthesize only in the dark.

FBPase and SBPase are also activated by increased pH, Mg^{2+}, and NADPH. The action of these factors is complemented by a second regulatory system that responds to the redox potential of the stroma. **Thioredoxin,** a 12-kD protein that occurs in many types of cells, contains a reversibly reducable cystine disulfide group. Reduced thioredoxin activates both FBPase and SBPase by a disulfide exchange reaction (Fig. 22-27). This explains why these Calvin cycle enzymes are activated by reduced disulfide reagents such as dithiothreitol. The redox level of thioredoxin is maintained by a second enzyme, **ferredoxin-thioredoxin reductase,** which directly responds to the redox state of the soluble ferredoxin in the stroma. This in turn varies with the illumination level. The thioredoxin system also deactivates phosphofructokinase (PFK), the main flux-generating enzyme of glycolysis (Section 16-4B). Thus in plants, *light stimulates the Calvin cycle while deactivating glycolysis, whereas darkness has the opposite effect* (that is, the so-called dark reactions do not occur in the dark).

C. Photorespiration and the C_4 Cycle

It has been known since the 1960s that *illuminated plants consume O_2 and evolve CO_2 in a pathway distinct*

Figure 22-28
The probable mechanism of the oxygenase reaction catalyzed by RuBP carboxylase–oxygenase. Note the similarity of this mechanism to that of the carboxylase reaction catalyzed by the same enzyme (Fig. 22-26).

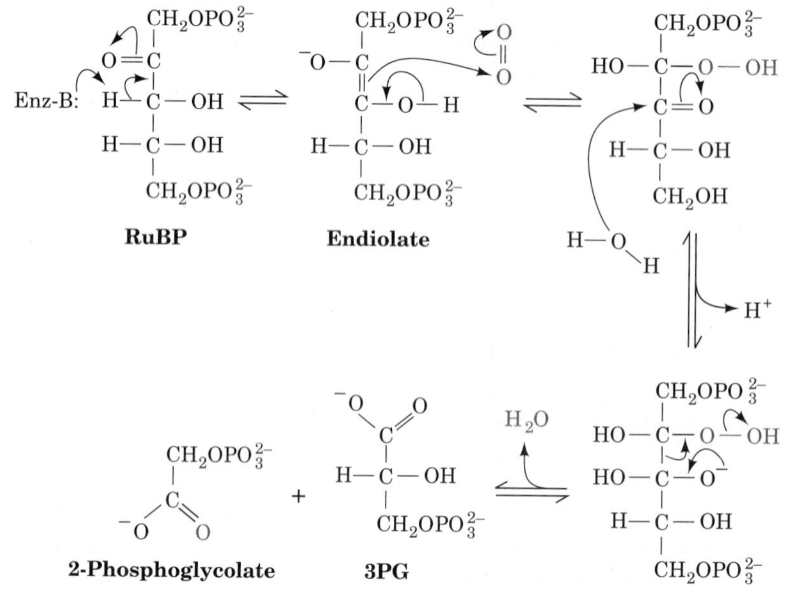

from oxidative phosphorylation. In fact, at low CO₂ and high O₂ levels, this **photorespiration** *process can outstrip photosynthetic CO₂ fixation.* The basis of photorespiration was unexpected: *O₂ competes with CO₂ as a substrate for RuBP carboxylase* (RuBP carboxylase is therefore also called **RuBP carboxylase–oxygenase** or **Rubisco**). In the oxygenase reaction, O₂ reacts with Rubisco's second substrate, RuBP, to form 3PG and **2-phosphoglycolate** (Fig. 22-28). The 2-phosphoglycolate is hydrolyzed to **glycolate** by **glycolate phosphatase** and, as described below, is partially oxidized to yield CO₂ by a series of enzymatic reactions that occur in the peroxisome and the mitochondrion. Thus photorespiration is a seemingly wasteful process that undoes some of the work of photosynthesis. In the following subsections we discuss the biochemical basis of photorespiration, its significance, and how certain plants manage to evade its deleterious effects.

Photorespiration Dissipates ATP and NADPH

The photorespiration pathway is outlined in Fig. 22-29. Glycolate is exported from the chloroplast to the peroxisome (also called the glyoxisome, Section 1-2A), where it is oxidized by **glycolate oxidase** to **glyoxylate** and H₂O₂. The H₂O₂, a powerful and potentially harmful oxidizing agent, is disproportionated to H₂O and O₂ in the peroxisome by the heme-containing enzyme **catalase.** Some of the glyoxylate is further oxidized by glycolate oxidase to oxalate. The remainder is converted to glycine in a **transamination reaction,** as discussed in Section 24-1A, and exported to the mitochondrion. There, two molecules of glycine are converted to one molecule of serine and one of CO₂ by a reaction described in Section 24-1B. *This is the origin of the CO₂ generated by photorespiration.* The serine is transported back to the peroxisome where a transamination reaction converts it to **hydroxypyruvate.** This substance is reduced to **glycerate** and phosphorylated in the cytosol to 3PG, which reenters the chloroplast where it is reconverted to RuBP in the Calvin cycle. *The net result of this complex photorespiration cycle is that some of the ATP and NADPH generated by the light reactions is uselessly dissipated.*

Although photorespiration has no known metabolic function, the RuBP carboxylases from the great variety of photosynthetic organisms so far tested all exhibit oxygenase activity. Yet, over the eons, the forces of evolution must have optimized the function of this important enzyme. It is thought that photosynthesis evolved at a time when the earth's atmosphere contained large quantities of CO₂ and very little O₂ so that photorespiration was of no consequence. It has therefore been suggested that the RuBP carboxylation reaction has an obligate intermediate that is inherently autooxidizable. Another possibility is that photorespiration protects the photosynthetic apparatus from photooxidative damage

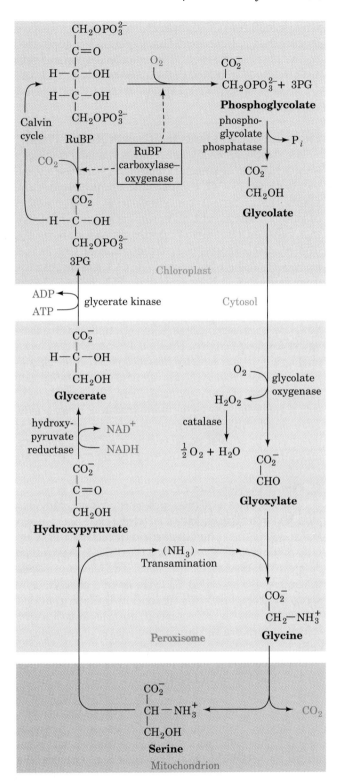

Figure 22-29

The photorespiration pathway for the metabolism of the phosphoglycolate produced by the RuBP carboxylase-catalyzed oxidation of RuBP. The reactions occur, as indicated, in the chloroplast, the peroxisome, the mitochondrion, and the cytosol. Note that two glycines are required to form serine + CO₂ (Section 24-3B).

when insufficient CO_2 is available to otherwise dissipate its absorbed light energy. This hypothesis is supported by the observation that when chloroplasts or leaf cells are brightly illuminated in the absence of both CO_2 and O_2, their photosynthetic capacity is rapidly and irreversibly lost.

Photorespiration Limits the Growth Rate of Plants

The steady state CO_2 concentration attained when a photosynthetic organism is illuminated in a sealed system is named its **CO_2 compensation point.** For healthy plants, this is the CO_2 concentration at which the rates of photosynthesis and photorespiration are equal. For many species it is ~ 40 to 70 ppm (parts per million) CO_2 (the normal atmospheric concentration of CO_2 is 330 ppm) so that their photosynthetic CO_2 fixation usually dominates their photorespiratory CO_2 release. However, the CO_2 compensation point increases with temperature because the oxygenase activity of RuBP carboxylase increases more rapidly with temperature than its carboxylase activity. Thus, *on a hot bright day, when photosynthesis has depleted the level of CO_2 at the chloroplast and raised that of O_2, the rate of photorespiration may approach that of photosynthesis. This phenomenon is, in fact, a major limiting factor in the growth of many plants. The control of photorespiration is therefore an important unsolved agricultural problem that is presently being attacked through genetic engineering studies (Section 28-8).*

C_4 Plants Concentrate CO_2

Certain species of plants, such as sugar cane, corn, and most important weeds, have a metabolic cycle that concentrates CO_2 in their photosynthetic cells thereby almost totally preventing photorespiration (their CO_2 compensation points are in the range 2 to 5 ppm). The leaves of plants that have this so-called **C_4 cycle** have a characteristic anatomy. Their fine veins are concentrically surrounded by a single layer of so-called **bundle-sheath cells,** which in turn are surrounded by a layer of **mesophyll cells.**

The C_4 cycle (Fig. 22-30) was elucidated in the 1960s by Marshall Hatch and Rodger Slack. It begins with the uptake of atmospheric CO_2 by the mesophyll cells which, lacking RuBP carboxylase in their chloroplasts, do so by condensing it as HCO_3^- with phosphoenolpyruvate (PEP) to yield oxaloacetate. The oxaloacetate is reduced by NADPH to **malate,** which is exported to the bundle-sheath cells (the name C_4 refers to these four-carbon acids). There the malate is oxidatively decarboxylated by $NADP^+$ to form CO_2, pyruvate, and NADPH. The CO_2, which has been concentrated by this process, enters the Calvin cycle. The pyruvate is returned to the mesophyll cells where it is phosphorylated to again form PEP. The enzyme that mediates this reaction, **pyruvate-phosphate dikinase,** has the unusual action of activating a phosphate group through the hydrolysis of ATP to AMP + PP_i. This PP_i is further hydrolyzed to two P_i, which is tantamount to the consumption of a second ATP. *CO_2 is thereby concentrated in the bundle-sheath cells at the expense of two ATPs per CO_2. Photosynthesis in C_4 plants therefore consumes a total of five ATPs per CO_2 fixed versus the three ATPs required by the Calvin cycle alone.*

C_4 plants occur largely in tropical regions because they grow faster under hot and sunny conditions than other, so called **C_3 plants** (so-named because they initially fix CO_2 in the form of three-carbon acids). In cooler climates, where photorespiration is less of a burden, C_3 plants have the advantage because they require less energy to fix CO_2.

CAM Plants Store CO_2 through a Variant of the C_4 Cycle

A variant of the C_4 cycle that separates CO_2 acquisition and the Calvin cycle in time rather than in space occurs in many desert-dwelling succulent plants. If, as most plants, they opened their **stomata** (the pores leading to their internal leaf spaces) by day to acquire CO_2, they would simultaneously transpire (lose by evaporation) what for them would be unacceptable amounts of water. To minimize this loss, these succulents only absorb CO_2 at night when the temperature is relatively cool. They store this CO_2, in a process known as **Crassulacean acid metabolism (CAM;** so-named because it was first discovered in plants of the family *Crassulaceae*), by the synthesis of malate through the reactions of the C_4 pathway (Fig. 22-30). The large amount of PEP necessary to store a day's supply of CO_2 is obtained by the breakdown of starch via glycolysis. During the course of the day, this malate is broken down to CO_2, which enters the Calvin cycle, and pyruvate, which is used to resynthesize starch. CAM plants are able, in this way, to carry out photosynthesis with minimal water loss.

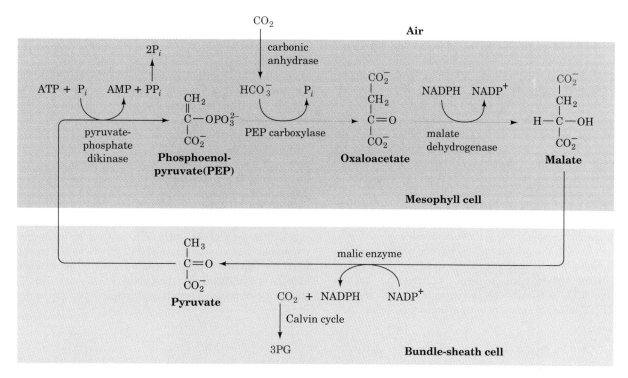

Figure 22-30
The C_4 pathway for concentrating CO_2 in the mesophyll cells and transporting it to the bundle-sheath cells for entry into the Calvin cycle.

Chapter Summary

Photosynthesis is the light-driven fixation of CO_2 to form carbohydrates and other biological molecules. In plants, photosynthesis takes place in the chloroplast, which consists of an inner and outer membrane surrounding the stroma, a concentrated enzyme solution, in which the thylakoid membrane system is immersed. Photosynthesis occurs in two stages, the so-called light reactions in which light energy is harnessed to synthesize ATP and NADPH, and the dark reactions in which these products are used to drive the synthesis of carbohydrates from CO_2 and H_2O. The thylakoid membrane is the site of the photosynthetic light reactions, whereas the dark reactions take place in the stroma. The counterpart of the thylakoid in photosynthetic bacteria is a portion of the plasma membrane termed the chromatophore.

Chlorophyll is the principal photoreceptor of photosynthesis. Light is absorbed initially by a light-harvesting antenna system consisting of chlorophyll and accessory pigments. The resulting excitation then migrates via exciton transfer until it reaches the reaction center chlorophyll where it is trapped.

In purple photosynthetic bacteria, the reaction center is a particle that consists of three subunits and several redox-active small molecules. The primary photon absorbing species of the bacterial reaction center is a "special pair" of BChl a molecules known as P870. By rapid measurement techniques it has been determined that the electron ejected by P870* passes by a third BChl a to a BPheo a molecule and then sequentially to a

menaquinone (Q_A) and a ubiquinone (Q_B). The resulting Q_B^- is subsequently further reduced in a second one-electron transfer process and then takes up two protons from the cytosol to form Q_BH_2. The electrons taken up by this species are returned to P870 via a series of b-type cytochromes, iron–sulfur proteins, and finally cytochrome c_2. This cyclic electron-transport process functions to translocate protons, probably through a Q cycle, from the cytoplasm to the outside of the cell. The resulting proton gradient, in a process known as photophosphorylation, drives the synthesis of ATP. Since bacterial photosynthesis does not generate the reducing equivalents needed in many biosynthetic processes, photosynthetic bacteria require an outside source of reducing agents such as H_2S.

In plants and cyanobacteria, the light reactions occur in two reaction centers, PSI and PSII, that are electrically "connected" in series. This enables the system to generate sufficient electromotive force to form NADPH by oxidizing H_2O in a noncyclic pathway known as the Z-scheme. PSII contains an Mn complex that oxidizes $2H_2O$ to $4H^+$ and O_2 in 4 one-electron steps. The electrons are passed singly, through a poorly characterized carrier named Z, to photooxidized P680, the reaction center's photon-absorbing species, which consists of one or two Chl a molecules. The electron previously ejected from P680* passes through a series of carriers similar in character to those of the bacterial reaction center to a pool of plastoquinone molecules. The electrons then enter the cytochrome b_6-f complex, which transports protons, probably via a Q cycle, from the stroma to the thylakoid space. These electrons are transferred individually, by a plastocyanin carrier, directly to PSI's photooxidized photon-absorbing pigment, P700, which is a single molecule of Chl *a*. The electron that had been previously released by P700* migrates through a chain of Chl *a* molecules and then through a chain of ferredoxin molecules. The electron may be returned cyclically, via cytochrome b_6, to the plastoquinone pool so as to translocate protons across the thylakoid membrane. Alternatively, it may act to reduce $NADP^+$ in a noncyclic process mediated by ferredoxin-$NADP^+$ reductase. ATP is synthesized by the CF_0CF_1–ATP synthase, which closely resembles the analogous mitochondrial complex, in a reaction driven by the dissipation of the proton gradient across the thylakoid membrane.

CO_2 is fixed in the photosynthetic dark reactions of plants and cyanobacteria by reactions of the Calvin cycle. The first stage of the Calvin cycle, in sum, mediates the reaction $3RuBP + 3CO_2 \rightarrow 6GAP$ with the consumption of $9ATP + 6NADPH$. The second stage reshuffles the atoms of five GAPs to reform the three RuBPs with which the cycle began, a process that requires no further input of free energy or reduction equivalents. The sixth GAP, the product of the Calvin cycle, is used to synthesize carbohydrates, amino acids, and fatty acids. The flux-controlling enzymes of the Calvin cycle are activated in the light through variations in the pH, the Mg^{2+} and NADPH concentrations, and by the redox level of thioredoxin. The central enzyme of the Calvin cycle, RuBP carboxylase, catalyzes both a carboxylase and an oxygenase reaction with RuBP. The latter reaction is the first step in the photorespiration cycle that liberates CO_2. The rate of photorespiration increases with temperature and decreases with CO_2 concentration so that photorespiration constitutes a significant energetic drain on most plants on hot bright days. C_4 plants, which are most common in the tropics, have a system for concentrating CO_2 in their photosynthetic cells so as to minimize the effects of photorespiration but at the cost of two ATPs per CO_2 fixed. Certain desert plants conserve water by absorbing CO_2 at night and releasing it to the Calvin cycle by day. This Crassulacean acid metabolism occurs through a process similar to the C_4 cycle.

References

General

Danks, S. M., Evans, E. H., and Whittaker, P. A., *Photosynthetic Systems*, Wiley (1983).

Foyer, C. H., *Photosynthesis*, Wiley (1984).

Govindjee (Ed.), *Photosynthesis*, Vols. I and II, Academic Press (1982). [A comprehensive series of authoritative articles. Volume I deals with light reactions and Volume II reviews dark reactions.]

Chloroplasts

Bogorad, L., Chloroplasts, *J. Cell Biol.* **91**, 256s–270s (1981).

Hoober, J. K., *Chloroplasts*, Plenum Press (1984).

Light Reactions

Amesz, J., The role of manganese in photosynthetic oxygen evolution, *Biochim. Biophys. Acta* **726**, 1–12 (1983).

Anderson, J. M., Photoregulation of the composition, function and structure of thylakoid membranes, *Annu. Rev. Plant Physiol.* **37**, 93–136 (1986).

Andréasson, L.-E. and Vänngard, T., Electron transport in photosystems I and II, *Annu. Rev. Plant Physiol. Plant Mol. Biol.* **39**, 379–411 (1988).

Barber, J., Rethinking the structure of the photosystem two reaction centre, *Trends Biochem. Sci.* **12**, 123–124 (1987).

Beck, W. F. and dePaula, J. C., Mechanism of photosynthetic water oxidation, *Annu. Rev. Biophys. Biophys. Chem.* **18** 25–46 (1989).

Brudvig, G. W., Beck, W. F., and de Paula, J. C., Mechanism of photosynthetic water oxidation, *Annu. Rev. Biophys. Biophys. Chem.* **18**, 25–46 (1989).

Clayton, R. K., *Photosynthesis*, Cambridge University Press (1980). [A concise and informative monograph on light reactions.]

Cramer, W. A., Widger, W. R., Herrmann, R. G., and Trebst, A., Topography and function of thylakoid membrane proteins, *Trends Biochem. Sci.* **10**, 125–129 (1985).

Deisenhofer, J., Epp, O., Miki, K., Huber, R., and Michel, H., Structure of the protein subunits in the photosynthetic reaction centre of *Rhodopseudomonas viridis* at 3 Å resolution, *Nature* **318**, 618–624 (1985).

Deisenhofer, J., Michel, H., and Huber, R., The structural basis of photosynthetic light reactions in bacteria, *Trends Biochem. Sci.* **10**, 243–248 (1985).

Feher, G., Allen, J. P., Okamura, M. Y., and Rees, D. C., Structure and function of bacterial photosynthetic reaction centres, *Nature* **339**, 111–116 (1989).

Glazer, A. N. and Melis, H., Photochemical reaction center: structure, organization and function, *Annu. Rev. Plant Physiol.* **38**, 11–45 (1987).

Golbeck, J.H., Structure, function and organization of the Photosystem I reaction center complex, *Biochim. Biophys. Acta* **895**, 167–204

Govindjee and Govindjee, R., The primary events of photosynthesis, *Sci. Am.* **231**(6): 68–82 (1974).

Haehnel, W., Photosynthetic electron transport in higher plants, *Annu. Rev. Plant Physiol.* **35**, 659–693 (1984).

Hunter, C. N., van Grondelle, R., and Olsen, J. D., Photosynthetic antenna proteins: 100 ps before photochemistry starts, *Trends Biochem. Sci.* **14**, 72–76 (1989).

Knaff, D. B., The photosystem I reaction centre, *Trends Biochem. Sci.* **13**, 460–461 (1988).

Michel, H. and Deisenhofer, J., Relevance of the photosynthetic reaction center from purple bacteria to the structure of photosystem II, *Biochemistry* **27**, 1–7 (1988).

Parsons, W. W., Photosynthetic reaction centers, *Annu. Rev. Biophys. Bioeng.* **11**, 57–80 (1982).

Stanier, R. Y., Ingraham, J., Wheelis, M. L., and Painter, P. R., *The Microbiol World* (5th ed.), Chapter 15, Prentice–Hall (1986). [The biology of photosynthetic eubacteria.]

Staehelin, J. K. and Arntzen, C. J. (Eds.), *Encyclopedia of Plant Physiology*, Vol. 19, Photosynthesis III, Springer–Verlag (1986). [Authoritative articles on the major aspects of light reactions.]

Strotmann, H. and Bickel-Sandkötter, S., Structure, function, and regulation of chloroplast ATPase, *Annu. Rev. Plant Physiol.* **35**, 97–120 (1984).

Thorber, P. J. and Markwell, J. P., Photosynthetic pigment-protein complexes in plant and bacterial membranes, *Trends Biochem. Sci.* **6**, 122–125 (1981).

Youvain, D. C. and Marrs, B. L., Molecular mechanisms of photosynthesis, *Sci. Am.* **256**(6): 42–48 (1987).

Dark Reactions

Buchanan, B. B., Role of light in the regulation of chloroplast enzymes, *Annu. Rev. Plant Physiol.* **31**, 341–374 (1980).

Edwards, G. and Walker, D., C_3, C_4: *mechanisms, and cellular and environmental regulation, of photosynthesis*, University of California Press (1983).

Hatch, M. D., C_4 photosynthesis: a unique blend of modified biochemistry, anatomy and ultrastructure, *Biochim. Biophys. Acta* **895**, 81–106 (1987).

Heber, U. and Krause, G. H., What is the physiological role of photorespiration? *Trends Biochem. Sci.* **5**, 32–34 (1980).

Miziorko, H. M. and Lorimer, G. H., Ribulose-1,5-bisphosphate carboxylase–oxygenase, *Annu. Rev. Biochem.* **52**, 507–535 (1983).

Ogren, W. L., Photorespiration: pathways, regulation, and modification, *Annu. Rev. Plant Physiol.* **35**, 415–442 (1984).

Ting, I. P., Crassulacean acid metabolism, *Annu. Rev. Plant Physiol.* **36**, 595–622 (1985).

Walker, D. A., Leegood, R. C., and Sivak, M. N., Ribulose bisphosphate carboxylase–oxygenase: its role in photosynthesis, *Phil. Trans. R. Soc. London Ser. B* **313**, 305–324 (1986).

Problems

1. Why is chlorophyll green in color when it absorbs in the red and the blue regions of the spectrum (Fig. 22-5)?

2. Indicate, where appropriate, the analogous components in the photosynthetic electron-transport chains of purple photosynthetic bacteria and chloroplasts.

3. Antimycin A inhibits photosynthesis in chloroplasts. Indicate its most likely site of action and explain your reasoning.

4. Calculate the energy efficiency of cyclic and noncyclic photosynthesis in chloroplasts using 680-nm light. What would this efficiency be with 500-nm light? Assume that ATP formation requires $59 \text{ kJ} \cdot \text{mol}^{-1}$ under physiological conditions.

*5. What is the minimum pH gradient required to synthesize ATP from ADP + P_i? Assume $[ATP]/([ADP][P_i]) = 10^3$, $T = 25°C$, and that three protons must be translocated per ATP generated. (See Table 15-3 for useful thermodynamic information.)

6. Indicate the average Calvin cycle labeling pattern in ribulose-5-phosphate after two rounds of exposure to $^{14}CO_2$.

7. Chloroplasts are illuminated until the levels of their Calvin cycle intermediates reach a steady state. The light is then turned off. How do the levels of RuBP and 3PG vary after this time?

8. What is the energy efficiency of the Calvin cycle combined with glycolysis and oxidative phosphorylation; that is, what percentage of the input energy can be metabolically recovered in synthesizing starch from CO_2 using photosynthetically produced NADPH and ATP rather than somehow directly storing these "high-energy" intermediates? Assume that each NADPH is energetically equivalent to three ATPs and that starch synthesis and breakdown are energetically equivalent to glycogen synthesis and breakdown.

9. If a C_3 plant and a C_4 plant are placed together in a sealed illuminated box with sufficient moisture, the C_4 plant thrives while the C_3 plant sickens and eventually dies. Explain.

10. The leaves of some species of desert plants taste sour in the early morning but, as the day wears on, they become tasteless and then bitter. Explain.

Chapter 23
LIPID METABOLISM

Lipids play indispensable roles in cell structure and metabolism. For example, triacylglycerols are the major storage form of metabolic energy in animals; cholesterol is a vital component of cell membranes and a precursor of the steroid hormones and bile acids; arachidonate is an unsaturated fatty acid that serves as the precursor of the prostaglandins, prostacyclins, thromboxanes, and leukotrienes, potent intercellular mediators that control a variety of complex processes; and complex glycolipids and phospholipids are major components of biological membranes. We discussed the structures of simple and complex lipids in Section 11-1. In the first half of this chapter, we consider the metabolism of fatty acids and triacylglycerols including their digestion, oxidation, and biosynthesis. We then consider how cholesterol is synthesized and utilized, and how arachidonate is converted to prostaglandins, prostacyclins, thromboxanes, and leukotrienes. We end by studying how complex glycolipids and phospholipids are synthesized from their simpler lipid and carbohydrate components.

1. LIPID DIGESTION, ABSORPTION, AND TRANSPORT

*Triacylglycerols (also called **fats, triglycerides,** and **depot lipids**), constitute both ~90% of the dietary lipid and the major form of metabolic energy storage in humans.* Triacylglycerols consist of glycerol triesters of fatty acids such as palmitic and oleic acids

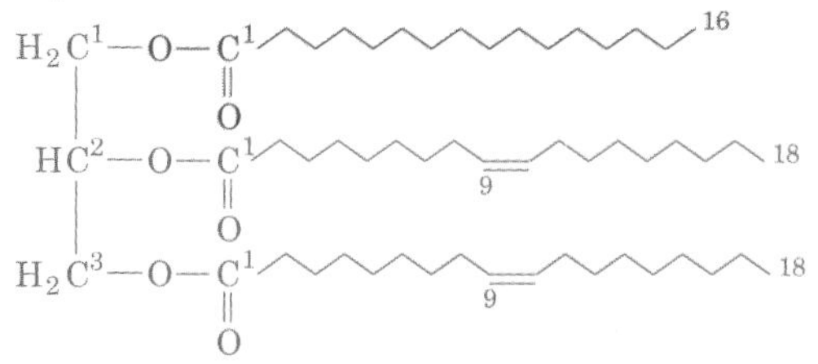

1-Palmitoyl-2,3-dioleoyl-glycerol

(the names and structural formulas of some biologically common fatty acids are listed in Table 11-1). Like glucose, they are metabolically oxidized to CO_2 and H_2O. Yet, since most carbon atoms of triacylglycerols have lower oxidation states than those of glucose, *the oxidative metabolism of fats yields over twice the energy of an equal weight of dry carbohydrate or protein (Table 23-1).* Moreover, fats, being nonpolar, are stored in an anhydrous state whereas glycogen, the storage form of glucose, is polar, and is consequently stored in a hydrated form that contains about twice its dry weight of water. Fats therefore provide up to six times the metabolic energy of an equal weight of hydrated glycogen.

Table 23-1

Energy Content of Food Constituents

Constituent	ΔH (kJ $\cdot$ g^{-1} dry weight)
Carbohydrate	16
Fat	37
Protein	17

Source: Newsholme, E. A. and Leech, A. R., *Biochemistry for the Medical Sciences, p.* 16, Wiley (1983).

Lipid Digestion Occurs at Lipid–Water Interfaces

Since triacylglycerols are water-insoluble whereas digestive enzymes are water-soluble, *triacylglycerol digestion takes place at lipid–water interfaces.* The rate of triacylglycerol digestion therefore depends on the surface area of the interface, a quantity that is greatly increased by the churning peristaltic movements of the intestine combined with the emulsifying action of **bile acids.** The bile acids are powerful digestive detergents which, as we shall see in Section 23-6C, are synthesized by the liver and secreted via the gallbladder into the small intestine where lipid digestion and absorption mainly take place.

Pancreatic **lipase** catalyzes the hydrolysis of triacylglycerols at their 1 and 3 positions to sequentially form **1,2-diacylglycerols** and **2-acylglycerols** together with the Na$^+$ and K$^+$ salts of fatty acids (soaps). These soaps, being amphipathic, aid in the lipid emulsification process. Lipase, as are many proteins, is rapidly denatured at interfaces including lipid–water interfaces. **Colipase,** a pancreatic protein that forms a 1:1 complex with lipase, inhibits the surface denaturation of lipase and anchors it to the lipid–water interface.

Phospholipids are degraded by pancreatic **phospholipase A$_2$,** which hydrolytically excises the fatty acid residue at C(2) to yield the corresponding **lysophospholipids** (Fig. 23-1), which are also powerful detergents. Indeed, the phospholipid lecithin (phosphatidyl-

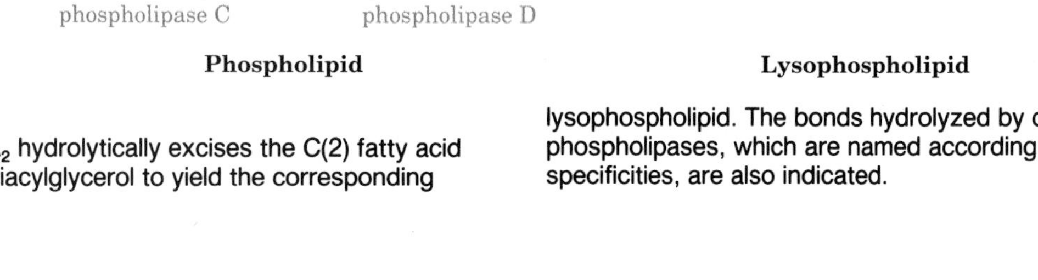

Figure 23-1
Phospholipase A$_2$ hydrolytically excises the C(2) fatty acid residue from a triacylglycerol to yield the corresponding lysophospholipid. The bonds hydrolyzed by other types of phospholipases, which are named according to their specificities, are also indicated.

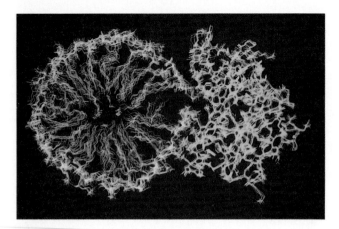

Figure 23-2
A hypothetical model of phospholipase A_2 in complex with a micelle of lysophosphatidyl ethanolamine as shown in cross-section. The protein is drawn in blue-green, the phospholipid head groups are yellow, and their hydrocarbon tails are blue. The calculated atomic motions of the assembly are indicated through a series of superimposed images taken at 5-picosecond intervals. [Courtesy of Raymond Salemme, E. I. du Pont de Nemours & Company.]

choline) is secreted in the bile, presumably to aid in lipid digestion.

Pancreatic phospholipase A_2, as does pancreatic lipase, preferentially catalyzes reactions at interfaces. The X-ray structure of the bovine enzyme, determined by Jan Drenth, suggests the structural basis for this preference. The enzyme's active site, which resembles that of chymotrypsin (Section 14-3B) occupies a depression surrounded by a ring and two flanking patches of 21 surface residues whose side chains extend toward and presumably bind phospholipid aggregates. This grouping is thought to position the active site over a phospholipid molecule (Fig. 23-2), which NMR measurements indicate assumes a conformation in the micelle that is complementary to the active site.

Bile Acids Facilitate Intestinal Absorption of Lipids

The mixture of fatty acids, mono-, and diacylglycerols produced by lipid digestion is absorbed by the cells lining the small intestine (the intestinal mucosa) in a pro-

cess facilitated by bile acids. The micelles formed by the bile acids take up the nonpolar lipid degradation products so as to permit their transport across the unstirred aqueous boundary layer at the intestinal wall. The importance of this process is demonstrated in individuals with obstructed bile ducts: They absorb little of their dietary lipids, but rather, eliminate them in hydrolyzed form in the feces (**steatorrhea**). Evidently, *bile acids are not only an aid to lipid digestion but are essential for the absorption of lipid digestion products.* Bile acids are likewise required for the efficient intestinal absorption of the lipid-soluble vitamins A, D, E, and K.

Lipids Are Transported in Lipoprotein Complexes

The lipid digestion products absorbed by the intestinal mucosa are converted by these tissues to triacylglycerols (Section 23-4F) and then packaged into lipoprotein particles called **chylomicrons.** These, in turn, are released into the bloodstream via the lymph system for delivery to the tissues. Similarly, triacylglycerols synthesized by the liver are packaged into **very low density lipoproteins (VLDL)** and released directly into the blood. These lipoproteins, whose origins, structures, and functions we discussed in Section 11-4, maintain their otherwise insoluble lipid components in aqueous solution.

The triacylglycerol components of chylomicrons and VLDL are hydrolyzed to free fatty acids and glycerol in the capillaries of adipose tissue and skeletal muscle by **lipoprotein lipase** (Section 11-4B). The resulting free fatty acids are taken up by these tissues while the glycerol is transported to the liver or kidneys. There it is converted to the glycolytic intermediate dihydroxyacetone phosphate by the sequential actions of **glycerol kinase** and **glycerol-3-phosphate dehydrogenase** (Fig. 23-3).

Mobilization of triacylglycerols stored in adipose tissue involves their hydrolysis to glycerol and free fatty acids by **hormone-sensitive triacylglycerol lipase** (Section 23-5). The free fatty acids are released into the bloodstream where they bind to **albumin,** a soluble 66.5-kD monomeric protein, which comprises about one half of the blood serum protein. In the absence of

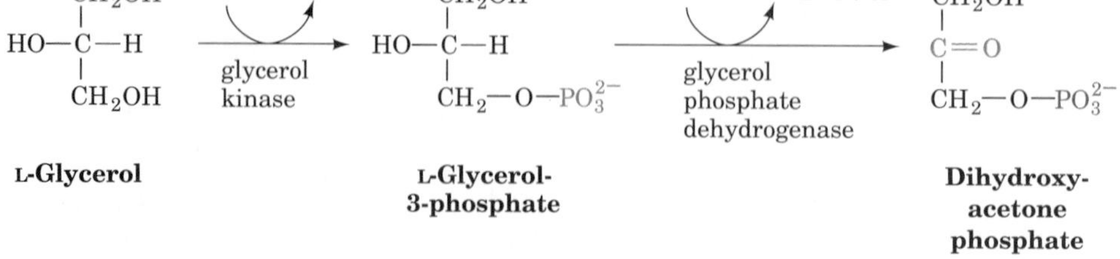

Figure 23-3
The conversion of glycerol to the glycolytic intermediate dihydroxyacetone phosphate.

albumin, the maximum solubility of free fatty acids is $\sim 10^{-6}M$. Above this concentration, free fatty acids form micelles that act as detergents to disrupt protein and membrane structure and would therefore be toxic. However, the effective solubility of fatty acids in fatty acid–albumin complexes is as much as 2 mM. Nevertheless, those rare individuals with **analbuminemia** (severely depressed levels of albumin) suffer no apparent adverse symptoms; evidently, their fatty acids are transported in complex with other serum proteins.

2. FATTY ACID OXIDATION

The biochemical strategy of fatty acid oxidation was understood long before the advent of modern biochemical techniques involving enzyme purification or the use of radioactive tracers. In 1904, Franz Knoop, in the first use of chemical labels to trace metabolic pathways, fed dogs fatty acids labeled at their ω (last)-carbon atom by a benzene ring and isolated the phenyl-containing metabolic products from their urine. Dogs fed labeled odd-chain fatty acids excreted **hippuric acid,** the glycine amide of **benzoic acid,** whereas those fed labeled even-chain fatty acids excreted **phenylaceturic acid,** the glycine amide of **phenylacetic acid** (Fig. 23-4). Knoop therefore deduced that the oxidation of the carbon atom β to the carboxyl group is involved in fatty acid breakdown. Otherwise, the phenylacetic acid would be further oxidized to benzoic acid. Knoop proposed that this breakdown occurs by a mechanism known as β **oxidation** in which the fatty acid's C_β atom is oxidized. It was not until after 1950, following the discovery of

coenzyme A, that the enzymes of fatty acid oxidation were isolated and their reaction mechanisms elucidated. This work confirmed Knoop's hypothesis.

A. Fatty Acid Activation

Before fatty acids can be oxidized, they must be "primed" for reaction in an ATP-dependent acylation reaction to form fatty acyl-CoA. This "activation" process is catalyzed by a family of at least three **acyl-CoA synthetases** (also called **thiokinases**) that differ according to their chain-length specificities. These enzymes, which are associated with either the endoplasmic reticulum or the outer mitochondrial membrane, all catalyze the reaction

$$\text{Fatty acid} + \text{CoA} + \text{ATP} \rightleftharpoons \text{acyl-CoA} + \text{AMP} + \text{PP}_i$$

In the activation of ^{18}O-labeled palmitate by a long-chain acyl-CoA synthetase, both the AMP and the acyl-CoA products become ^{18}O labeled. This observation indicates that the reaction has an acyladenylate mixed anhydride intermediate that is attacked by the sulfhydryl group of CoA to form the thioester product (Fig. 23-5). The reaction involves both the cleavage and the synthesis of bonds with large negative free energies of hydrolysis so that the free energy change associated with the overall reaction is close to zero. The reaction is driven to completion in the cell by the highly exergonic hydrolysis of the product pyrophosphate (PP$_i$) catalyzed by the ubiquitous **inorganic pyrophosphatase.** Thus, as commonly occurs in metabolic pathways, *a reaction forming a "high-energy" bond through the hydrolysis of one of ATP's phosphoanhydride bonds is driven to completion by the hydrolysis of its second such bond.*

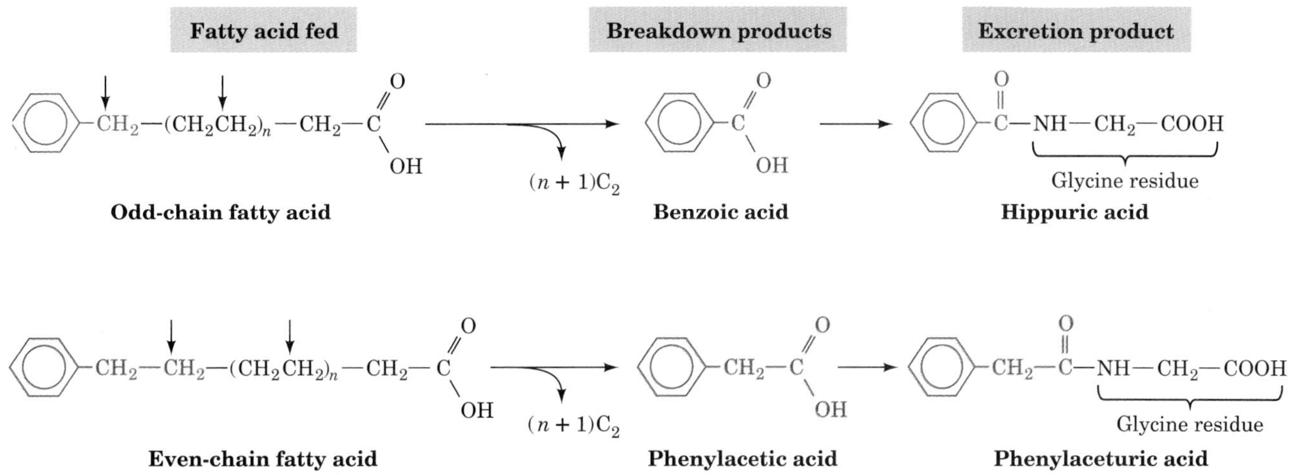

| Fatty acid fed | Breakdown products | Excretion product |

Figure 23-4
Franz Knoop's classic experiment indicating that fatty acids are metabolically oxidized at their β-carbon atom. ω-Phenyl-labeled fatty acids containing an odd number of carbon atoms are oxidized to the phenyl-labeled C_1 product benzoic acid, whereas those with an even number of carbon atoms are oxidized to the phenyl-labeled C_2 product phenylacetic acid. These products are excreted as their respective glycine amides hippuric, and phenylaceturic acids. The vertical arrows indicate the deduced sites of carbon oxidation. The intermediate C_2 products are oxidized to CO_2 and H_2O and were therefore not isolated.

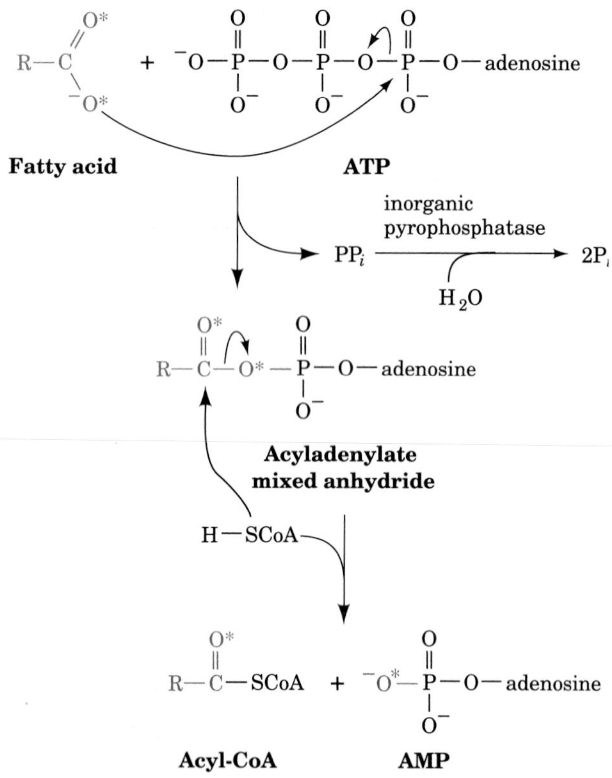

Figure 23-5
The mechanism of fatty acid "activation" catalyzed by acyl-CoA synthase. Experiments utilizing ^{18}O-labeled fatty acids (*) demonstrate that the formation of acyl-CoA involves an intermediate acyladenylate mixed anhydride.

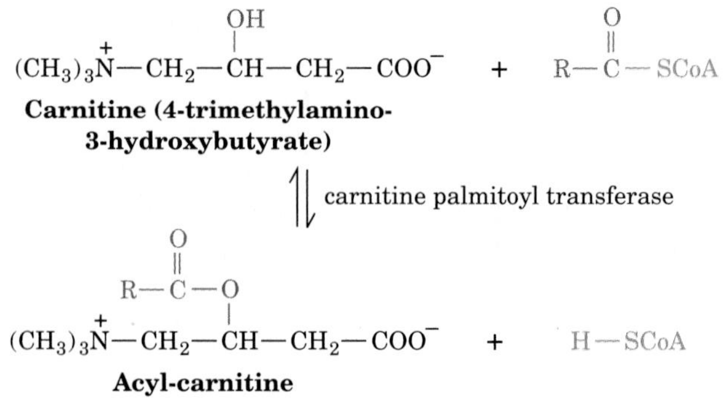

Figure 23-6
The acylation of carnitine is catalyzed by carnitine palmitoyl transferase.

B. Transport Across the Mitochondrial Membrane

Although fatty acids are activated for oxidation in the cytosol, they are, as Eugene Kennedy and Albert Lehninger established in 1950, oxidized in the mitochondrion. We must therefore consider how fatty acyl-CoA is transported across the inner mitochondrial membrane. A long-chain fatty acyl-CoA cannot directly cross the inner mitochondrial membrane. Rather, its acyl por-

tion is first transferred to **carnitine** (Fig. 23-6), a compound that occurs in both plant and animal tissues. This transesterification reaction has an equilibrium constant close to 1, which indicates that the O-acyl bond of acyl-carnitine has a free energy of hydrolysis similar to that of the thioester. **Carnitine palmitoyl transferases I and II**, which can transfer a variety of acyl groups, are located, respectively, on the external and internal surfaces of the inner mitochondrial membrane. The translocation process itself is mediated by a specific carrier protein that transports acyl-carnitine into the mitochondrion while transporting free carnitine in the opposite direction. Acyl-CoA transport therefore occurs via four reactions (Fig. 23-7):

1. The acyl group of a cytosolic acyl-CoA is transferred to carnitine thereby releasing the CoA to its cytosolic pool.

2. The resulting acyl-carnitine is transported into the mitochondrial matrix by the transport system.

3. The acyl group is transferred to a CoA molecule from the mitochondrial pool.

4. The product carnitine is returned to the cytosol.

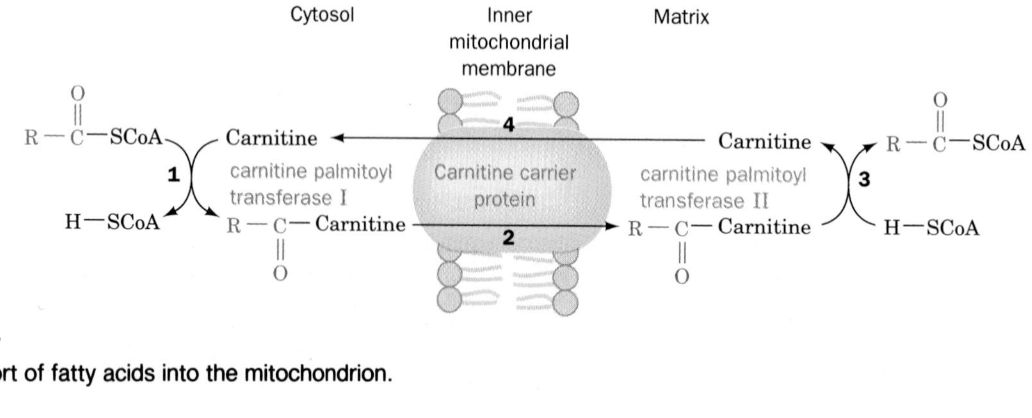

Figure 23-7
The transport of fatty acids into the mitochondrion.

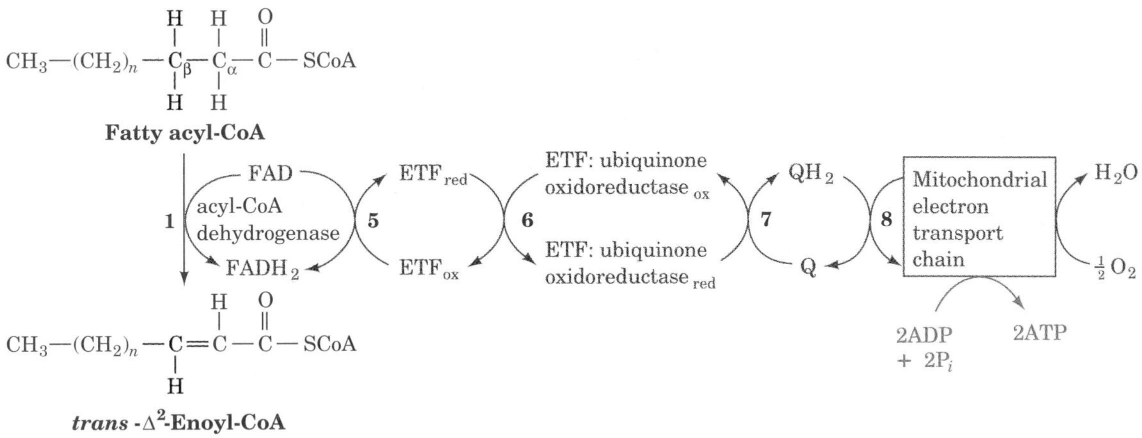

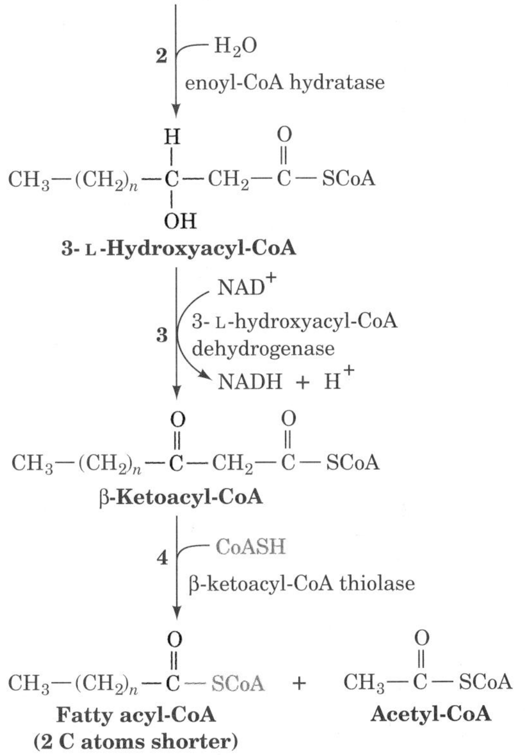

$$CH_3-(CH_2)_n-\underset{\underset{H}{|}}{\overset{\overset{H}{|}}{C_\beta}}-\underset{\underset{H}{|}}{\overset{\overset{H}{|}}{C_\alpha}}-\overset{\overset{O}{\|}}{C}-SCoA$$

Fatty acyl-CoA

1 acyl-CoA dehydrogenase: FAD → FADH$_2$

5 ETF$_{red}$ / ETF$_{ox}$

6 ETF: ubiquinone oxidoreductase$_{ox}$ / ETF: ubiquinone oxidoreductase$_{red}$

7 QH$_2$ / Q

8 Mitochondrial electron transport chain → H$_2$O, ½O$_2$; 2ADP + 2P$_i$ → 2ATP

$$CH_3-(CH_2)_n-\underset{\underset{H}{|}}{C}=C-\overset{\overset{O}{\|}}{C}-SCoA$$

***trans*-Δ^2-Enoyl-CoA**

2 enoyl-CoA hydratase $-$H$_2$O

$$CH_3-(CH_2)_n-\underset{\underset{OH}{|}}{\overset{\overset{H}{|}}{C}}-CH_2-\overset{\overset{O}{\|}}{C}-SCoA$$

3-L-Hydroxyacyl-CoA

3 3-L-hydroxyacyl-CoA dehydrogenase: NAD$^+$ → NADH + H$^+$

$$CH_3-(CH_2)_n-\overset{\overset{O}{\|}}{C}-CH_2-\overset{\overset{O}{\|}}{C}-SCoA$$

β-Ketoacyl-CoA

4 β-ketoacyl-CoA thiolase $-$CoASH

$$CH_3-(CH_2)_n-\overset{\overset{O}{\|}}{C}-SCoA \;+\; CH_3-\overset{\overset{O}{\|}}{C}-SCoA$$

Fatty acyl-CoA **Acetyl-CoA**
(2 C atoms shorter)

Figure 23-8
The β-oxidation pathway of fatty acyl-CoA.

1. Formation of a *trans*-α, β double bond through dehydrogenation by the flavoenzyme **acyl-CoA dehydrogenase** (Fig. 23-9).

2. Hydration of the double bond by **enoyl-CoA hydratase** to form a **3-L-hydroxyacyl-CoA**.

3. Dehydrogenation of this β-hydroxyacyl-CoA by **3-L-hydroxyacyl-CoA dehydrogenase** to form the corresponding β-ketoacyl-CoA.

4. C$_\alpha$ — C$_\beta$ cleavage in a thiolysis reaction with CoA as catalyzed by **β-ketoacyl-CoA thiolase** (or just **thiolase**) to form acetyl-CoA and a new acyl-CoA containing two less C atoms than the original one.

The first three steps of this process chemically resemble the citric acid cycle reactions that convert succinate to

The cell thereby maintains separate cytosolic and mitochondrial pools of CoA. The mitochondrial pool functions in the oxidative degradation of pyruvate (Section 19-2A) and certain amino acids (Sections 24-3E – G) as well as fatty acids, whereas the cytosolic pool supplies fatty acid biosynthesis (Section 23-4). The cell similarly maintains separate cytosolic and mitochondrial pools of ATP and NAD$^+$.

C. β Oxidation

Fatty acids are dismembered through the β oxidation of fatty acyl-CoA, a process that occurs in four reactions (Fig. 23-8):

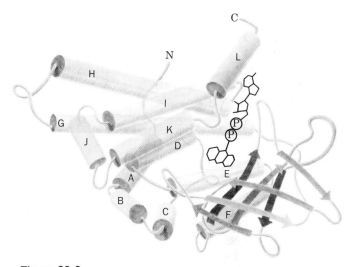

Figure 23-9
The X-ray structure of a subunit of medium-chain acyl-CoA dehydrogenase from pig liver mitochondria. The enzyme is a tetramer of four identical 43-kD subunits, each of which binds an FAD in an extended conformation. [After Kim. J.-J. P. and Wu, J., *Proc. Natl. Acad. Sci.* **85**, 6679 (1988).]

oxaloacetate (Sections 19-3F–H).

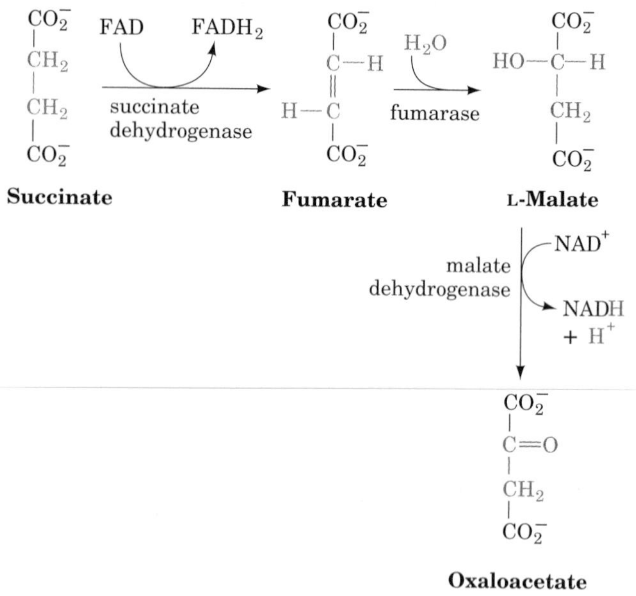

Succinate   **Fumarate**   L-**Malate**

Oxaloacetate

Acyl-CoA Dehydrogenase Is Reoxidized via the Electron-Transport Chain

Mitochondria contain three acyl-CoA dehydrogenases, with specificities for short, medium, and long chain fatty acyl-CoAs. The reaction catalyzed by these enzymes is thought to involve removal of a proton at C_α and transfer of a hydride ion equivalent from C_β to FAD (Fig. 23-8, Reaction 1). The resulting $FADH_2$ is reoxidized by the mitochondrial electron-transport chain through the intermediacy of a series of electron-transfer reactions. **Electron-transfer flavoprotein (ETF)** transfers an electron pair from $FADH_2$ to the flavo-iron–sulfur protein **ETF:ubiquinone oxidoreductase** which, in turn, transfers an electron pair to the mitochondrial electron-transport chain by reducing coenzyme Q (CoQ; Fig. 23-8, Reactions 5–8). Reduction of O_2 to H_2O by the electron-transport chain beginning at the CoQ stage results in the synthesis of two ATPs per electron pair transferred (Section 20-2B).

Acyl-CoA Dehydrogenase Deficiency Has Fatal Consequences

The unexpected death of an apparently healthy infant, often overnight, has been, for lack of any real explanation, termed **sudden infant death syndrome (SIDS)**. **Medium-chain acyl-CoA dehydrogenase (MCAD)** has been shown to be deficient in up to 10% of these infants, making this genetic disease more prevalent than **phenylketonuria (PKU)** (Section 24-3H), a genetic defect in phenylalanine degradation for which babies born in the United States are routinely tested. Glucose is the principal energy metabolism substrate just after eating, but when the glucose level later decreases, the rate of fatty acid oxidation must correspondingly increase. The sudden death in infants lacking MCAD may be caused by the imbalance between glucose and fatty acid oxidation.

Deficiency of acyl-CoA dehydrogenase has also been implicated in **Jamaican vomiting sickness,** whose victims suffer violent vomiting followed by convulsions, coma, and death. Severe hypoglycemia is observed in most cases. This condition results from eating unripe **ackee fruit,** which contains **hypoglycin A,** an unusual amino acid, which is metabolized to **methylenecyclopropylacetyl-CoA (MCPA-CoA)**. MCPA-CoA, a substrate for acyl-CoA dehydrogenase, is thought to undergo the first step of the reaction that this enzyme catalyzes, removal of a proton from C_α, to form a reactive intermediate that covalently modifies the enzyme's FAD prosthetic group (Fig. 23-10). Since a normal step in the enzyme's reaction mechanism generates the reactive intermediate, MCPA-CoA is said to be a **mechanism-based inhibitor.**

The Thiolase Reaction Occurs via Claisen Ester Cleavage

Enoyl-CoA hydratase catalyzes stereospecific addition of H_2O to its substrate's trans-α, β double bond to form 3-L-(S)-hydroxyacyl-CoA. 3-L-Hydroxyacyl-CoA dehydrogenase oxidizes this secondary alcohol to a ketone utilizing NAD^+ as its oxidizing agent.

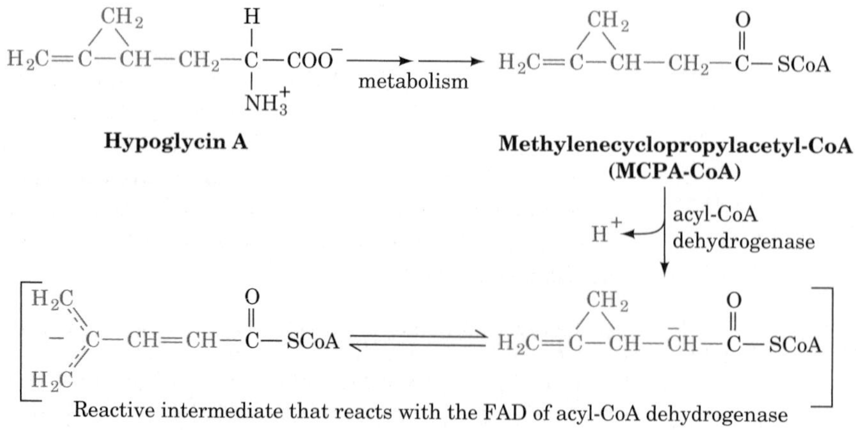

Hypoglycin A

Methylenecyclopropylacetyl-CoA (MCPA-CoA)

Reactive intermediate that reacts with the FAD of acyl-CoA dehydrogenase

Figure 23-10
Metabolic conversions of hypoglycin A to yield a product that inactivates acyl-CoA dehydrogenase. Spectral changes suggest that the enzyme's FAD prosthetic group has been modified although the purported adduct has not yet been characterized.

The final stage of the fatty acid β-oxidation process, the thiolase reaction, forms acetyl-CoA and a new acyl-CoA, which is two carbon atoms shorter than the one that began the cycle (Fig. 23-11):

1. The first step of the thiolase reaction involves formation of a thioester bond to the substrate by an active site thiol group, a deduction based on the observation that [^{14}C]acetyl-CoA labels a specific enzyme Cys residue.

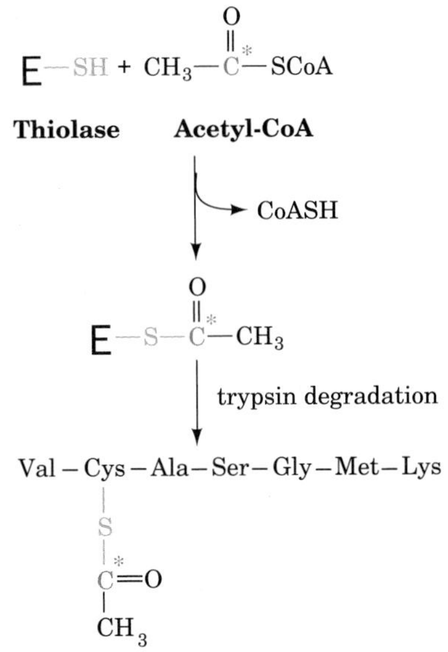

E—SH + CH$_3$—C(=O)—SCoA

Thiolase   Acetyl-CoA

→ CoASH

E—S—C(=O)—CH$_3$

trypsin degradation

Val—Cys—Ala—Ser—Gly—Met—Lys
         |
         S
         |
         C=O
         |
         CH$_3$

2. The second step involves carbon–carbon bond cleavage to form an acetyl-CoA carbanion intermediate that is stabilized by electron withdrawal into this thioester's carbonyl group. This type of reaction is known as a Claisen ester cleavage (the reverse of a Claisen condensation). The citric acid cycle enzyme citrate synthase also catalyzes a reaction that involves an acetyl-CoA carbanion intermediate (Section 19-3A).

3. The acetyl-CoA carbanion intermediate is protonated by an enzyme acid group yielding acetyl-CoA.

4. Finally, CoA displaces the enzyme thiol group from the enzyme–thioester intermediate yielding acyl-CoA.

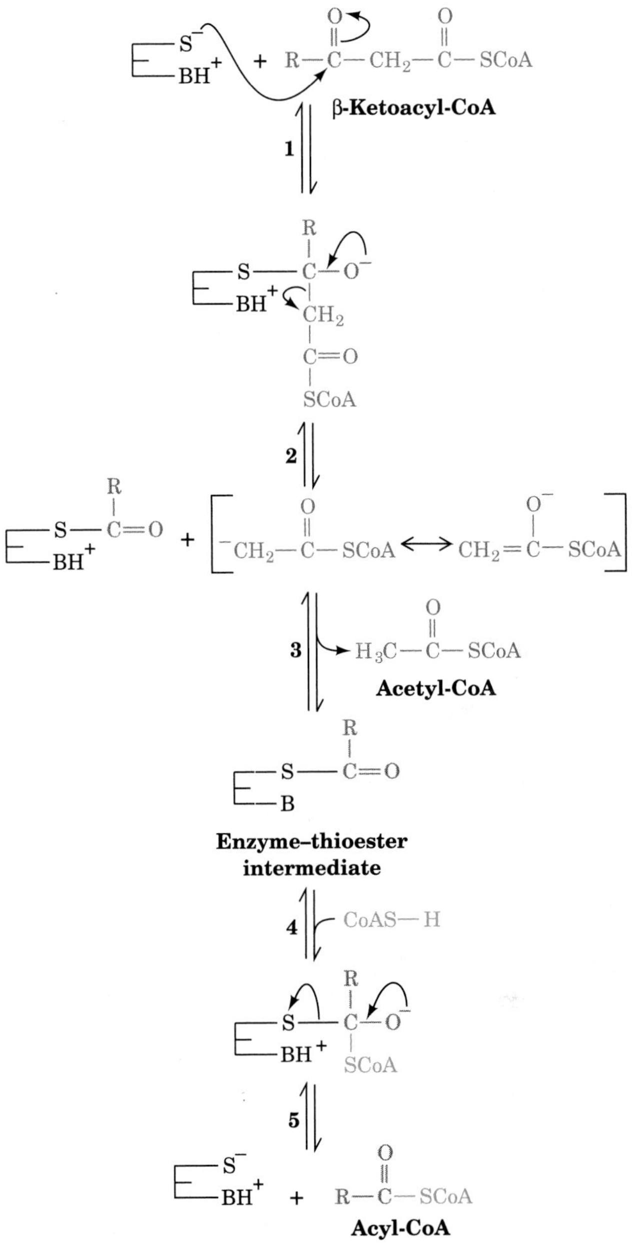

Fatty Acid Oxidation Is Highly Exergonic

The function of fatty acid oxidation is, of course, to generate metabolic energy. Each round of β oxidation produces one NADH, one FADH$_2$, and one acetyl-CoA. Oxidation of acetyl-CoA via the citric acid cycle generates additional FADH$_2$ and NADH which are reoxidized through oxidative phosphorylation to form ATP. Complete oxidation of a fatty acid molecule is therefore a highly exergonic process, which yields numerous ATPs. For example, oxidation of palmitoyl-CoA (which has a C$_{16}$ fatty acyl group) involves seven rounds of β oxidation yielding 7FADH$_2$, 7NADH, and 8acetyl-CoA.

Figure 23-11
The mechanism of action of β-ketoacyl-CoA thiolase. An active site Cys residue participates in the formation of an enzyme thioester intermediate.

Oxidation of the 8acetyl-CoA, in turn, yields 8GTP, 24NADH, and 8FADH$_2$. Since oxidative phosphorylation of the 31NADH molecules yields 93ATP and that of the 15FADH$_2$ yields 30ATPs, subtracting the 2ATP equivalents required for fatty acyl-CoA formation (Section 23-2A), *the oxidation of one palmitate molecule has a net yield of 129ATP.*

D. Oxidation of Unsaturated Fatty Acids

Almost all unsaturated fatty acids of biological origin (Section 11-1A) contain only cis double bonds, which most

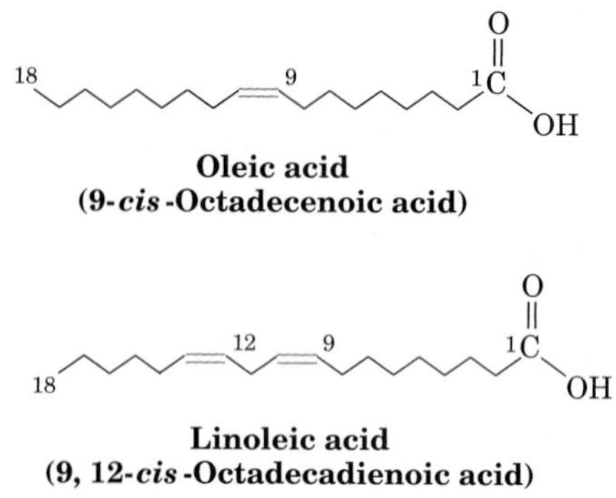

Oleic acid
(9-*cis*-Octadecenoic acid)

Linoleic acid
(9, 12-*cis*-Octadecadienoic acid)

Figure 23-12
The structures of two common unsaturated fatty acids. Most unsaturated fatty acids contain unconjugated cis-double bonds.

often begin between C(9) and C(10) (referred to as a Δ^9 or 9-double bond; Table 11-1). Additional double bonds, if any, occur at three-carbon intervals and are therefore never conjugated. Two examples of unsaturated fatty acids are oleic acid and linoleic acid (Fig. 23-12). Note that one of the double bonds in linoleic acid is at an odd numbered carbon atom and the other is at an even numbered carbon atom. Double bonds at these positions in fatty acids pose two problems for the β-oxidation pathway that are solved through the actions of three additional enzymes (Fig. 23-13):

Problem 1: A β, γ Double Bond
The first enzymatic difficulty occurs after the third round of β oxidation: The resulting cis-β, γ double bond-containing enoyl-CoA is not a substrate for enoyl-CoA hydratase. **Enoyl-CoA isomerase,** however, mediates conversion of the cis Δ^3-double bond to the more stable, ester-conjugated trans-Δ^2 form:

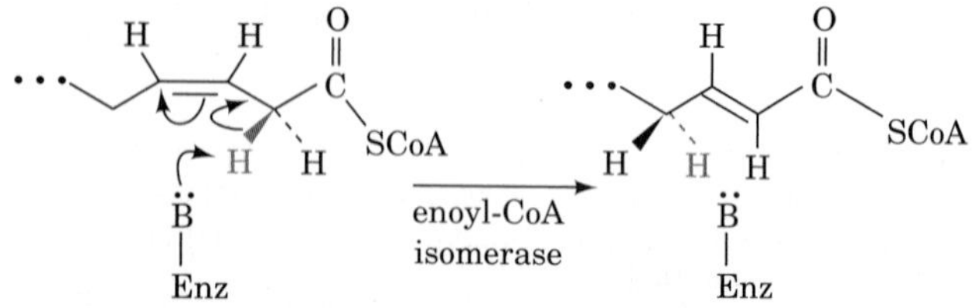

Such compounds are normal substrates of enoyl-CoA hydratase so that β oxidation can then continue.

Problem 2: A Δ^4 Double Bond Inhibits Hydratase Action
The next difficulty arises in the fifth round of β oxidation. Presence of a double bond at an even numbered carbon atom results in the formation of 2,4-dienoyl-CoA, which is a poor substrate for enoyl-CoA hydratase. However, NADPH-dependent **2,4-dienoyl-CoA reductase** reduces the Δ^4 double bond. The *E. coli* reductase produces *trans*-2-enoyl-CoA, a normal substrate of β oxidation. The mammalian reductase, how-

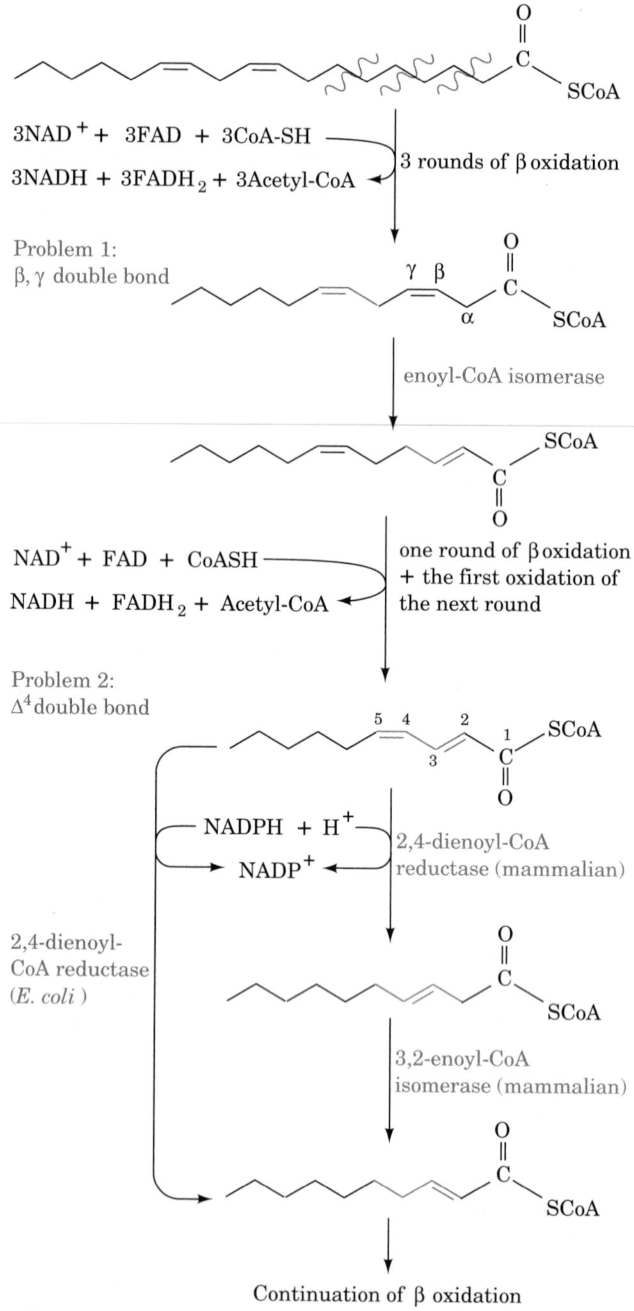

Figure 23-13
Problems in the oxidation of unsaturated fatty acids and their solutions. Linoleic acid is used as an example. The first problem, the presence of a β,γ double bond, is solved by the bond's enoyl-CoA isomerase-catalyzed conversion to a trans-α,β double bond. The second problem, that a 2,4-dienoyl-CoA is not a substrate for enoyl-CoA hydratase, is eliminated by the NADPH-dependent reduction of the 4-double bond by 2,4-dienoyl-CoA-4-reductase to yield the β-oxidation substrate *trans*-2-enoyl-CoA in *E. coli* but *trans*-3-enoyl-CoA in mammals. Mammals therefore also have 3,2-enoyl-CoA isomerase, which converts the *trans*-3-enoyl-CoA to *trans*-2-enoyl-CoA.

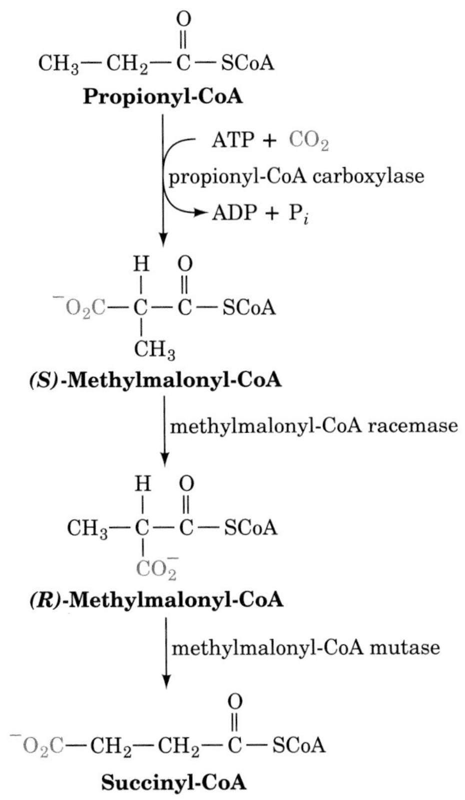

$$CH_3-CH_2-\overset{\overset{\displaystyle O}{\|}}{C}-SCoA$$

Propionyl-CoA

ATP + CO_2

propionyl-CoA carboxylase

ADP + P_i

$$^-O_2C-\overset{\overset{\displaystyle H}{|}}{\underset{\underset{\displaystyle CH_3}{|}}{C}}-\overset{\overset{\displaystyle O}{\|}}{C}-SCoA$$

(S)-Methylmalonyl-CoA

methylmalonyl-CoA racemase

$$CH_3-\overset{\overset{\displaystyle H}{|}}{\underset{\underset{\displaystyle CO_2^-}{|}}{C}}-\overset{\overset{\displaystyle O}{\|}}{C}-SCoA$$

(R)-Methylmalonyl-CoA

methylmalonyl-CoA mutase

$$^-O_2C-CH_2-CH_2-\overset{\overset{\displaystyle O}{\|}}{C}-SCoA$$

Succinyl-CoA

Figure 23-14
The conversion of propionyl-CoA to succinyl-CoA.

ever, yields *trans*-3-enoyl-CoA, which, to proceed along the β-oxidation pathway, must first be isomerized to *trans*-2-enoyl-CoA by **3,2-enoyl-CoA isomerase.**

> Until recently, it was generally accepted that 2,4-dienoyl-CoA is a substrate for enoyl-CoA hydratase yielding, after a further cycle of β oxidation, *cis*-Δ²-enoyl CoA which, in turn, is converted by enoyl-CoA hydratase to 3-D-hydroxyacyl-CoA. This D-isomer is not a β-oxidation substrate but was thought to be converted to its normally metabolizable L-stereoisomer by **3-hydroxyacyl-CoA epimerase.** The finding that this epimerase is, in fact, a peroxisomal rather than a mitochondrial enzyme, combined with the discovery of 2,4-dienoyl-CoA reductase led to the formulation of the revised pathway diagrammed at the bottom of Fig. 23-13.

E. Oxidation of Odd-Chain Fatty Acids

Most fatty acids have even numbers of carbon atoms and are therefore completely converted to acetyl-CoA. Some plants and marine organisms, however, synthesize fatty acids with an odd number of carbon atoms. *The final round of β oxidation of these fatty acids forms propionyl-CoA, which, as we shall see, is converted to succinyl-CoA for entry into the citric acid cycle.* Propionate or propionyl-CoA is also produced by oxidation of the amino acids isoleucine, valine, and methionine (Section 24-3E). Furthermore, ruminant animals such as cattle

derive most of their caloric intake from the acetate and propionate produced in their rumen (stomach) by bacterial fermentation of carbohydrates. These products are absorbed by the animal and metabolized after conversion to the corresponding acyl-CoA.

Propionyl-CoA Carboxylase Has a Biotin Prosthetic Group

The conversion of propionyl-CoA to succinyl-CoA involves three enzymes (Fig. 23-14). The first reaction is that of **propionyl-CoA carboxylase,** a tetrameric enzyme that contains a biotin prosthetic group (Section 21-1A). The reaction occurs in two steps (Fig. 23-15):

1. Carboxylation of biotin at N(1') by bicarbonate ion as in the reaction catalyzed by pyruvate carboxylase (Fig. 21-4). This step, which is driven by the concomitant hydrolysis of ATP to ADP and P_i, activates the resulting carboxyl group for transfer without further free energy input.

2. Stereospecific transfer of the activated carboxyl group from carboxybiotin to propionyl-CoA to form *(S)*-methylmalonyl-CoA. This step occurs via nucleophilic attack on carboxybiotin by a carbanion at C(2) of propionyl-CoA (see below).

These two reaction steps occur at different catalytic sites on propionyl-CoA carboxylase. It has therefore been proposed that the biotinyllysine linkage attaching the biotin ring to the enzyme forms a flexible tether that permits the efficient transfer of the biotin ring between these two active sites as postulated for the biotin enzyme pyruvate carboxylase (Section 21-1A).

Formation of the C(2) carbanion in the second stage of the propionyl-CoA carboxylase reaction involves removal of a proton α to a thioester. This proton is relatively acidic since, as we have seen in Section 23-2C, the negative charge on a carbanion α to a thioester can be delocalized into the thioester's carbonyl group. This may explain the relatively convoluted path taken in the conversion of propionyl-CoA to succinyl-CoA (Fig. 23-14). It would seem simpler, at least on paper, for this process to occur in one step, with carboxylation occurring on C(3) of propionyl-CoA so as to form succinyl-CoA directly. Yet, the C(3) carbanion required for such a carboxylation would be extremely unstable. Nature has instead chosen a more facile, albeit less direct route, which carboxylates propionyl-CoA at a more reactive position and then rearranges the C_4 skeleton to form the desired product.

Methylmalonyl-CoA Mutase Contains a Coenzyme B_{12} Prosthetic Group

Methylmalonyl-CoA mutase, which catalyzes the third reaction of the propionyl-CoA to succinyl-CoA conversion (Fig. 23-14), is specific for *(R)*-methylmalonyl-CoA even though propionyl-CoA carboxylase stereospecifically synthesizes *(S)*-methylmalonyl-

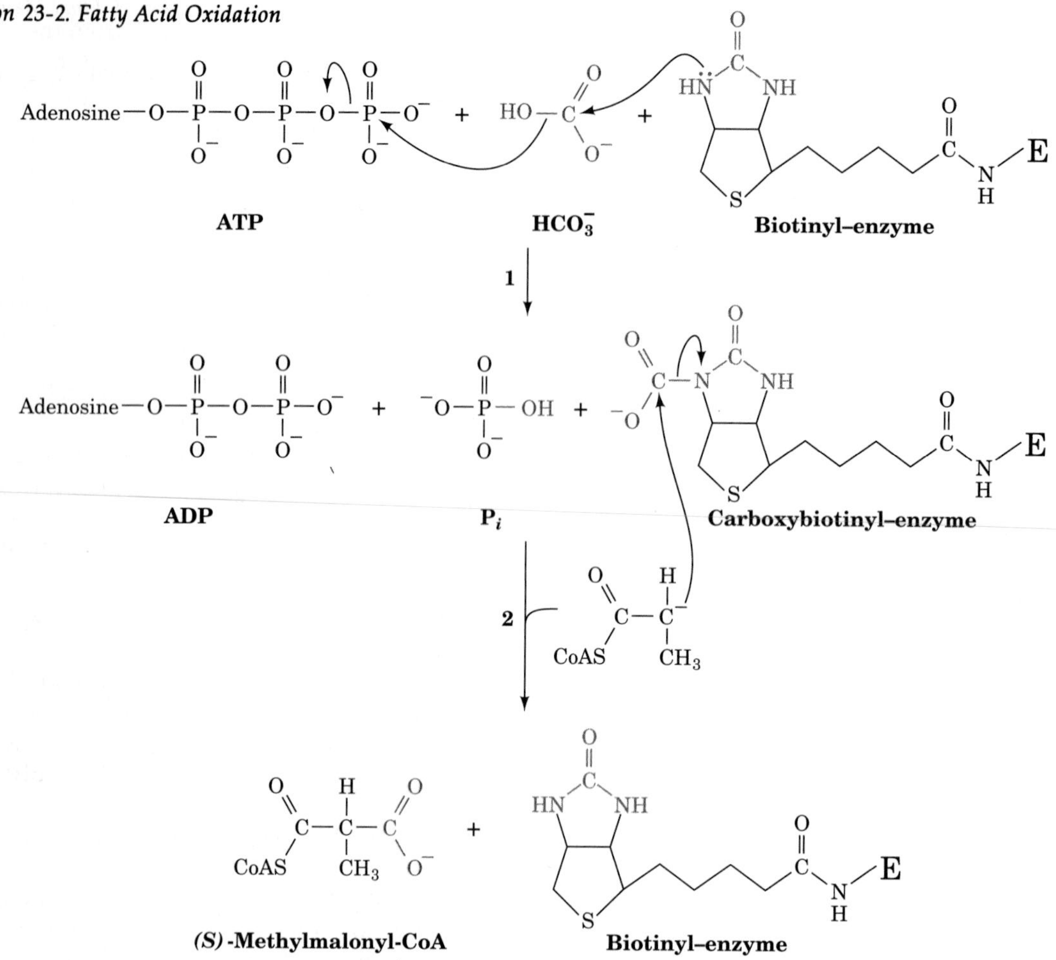

Figure 23-15
The propionyl-CoA carboxylase reaction involves: **(1)** the carboxylation of biotin with the concomitant hydrolysis of ATP; followed by **(2)** the carboxylation of a propionyl-CoA carbanion by its attack on carboxybiotin.

CoA. This diversion is rectified by **methylmalonyl-CoA racemase,** which interconverts the *(R)* and *(S)* configurations of methylmalonyl-CoA, presumably by promoting the reversible dissociation of its acidic α-H via formation of a resonance-stabilized carbanion intermediate.

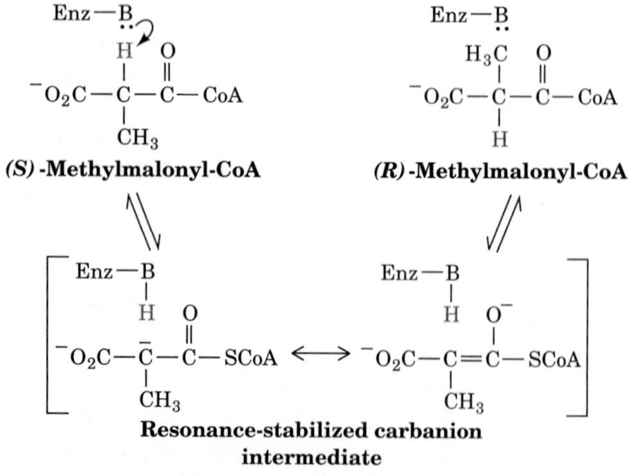

Methylmalonyl-CoA mutase, which catalyzes an unusual carbon skeleton rearrangement (Fig. 23-16), utilizes a **5'-deoxyadenosylcobalamin** prosthetic group (also called **coenzyme B$_{12}$**). Dorothy Hodgkin determined the structure of this complex molecule (Fig. 23-17) in 1956, a landmark achievement, through X-ray crystallographic analysis combined with chemical degradation studies. 5'-Deoxyadenosylcobalamin contains a hemelike **corrin** ring whose four pyrrole N atoms each ligand a 6-coordinate Co ion. The fifth Co ligand is an N atom of a **5,6-dimethylbenzimidazole (DMB)** nucleotide that is covalently linked to the corrin D ring. The sixth ligand is a 5'-deoxyadenosyl group in which the

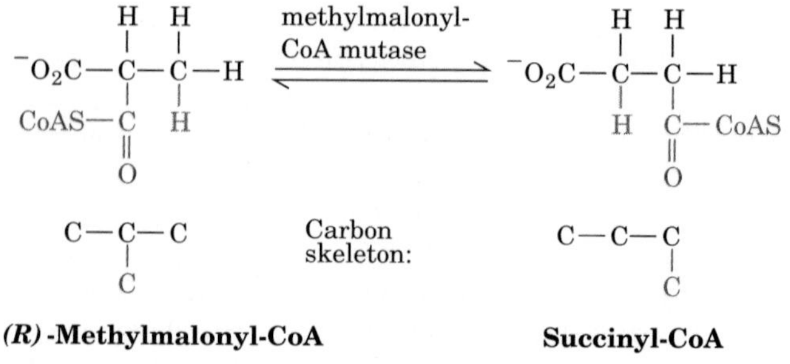

Figure 23-16
The rearrangement catalyzed by methylmalonyl-CoA mutase.

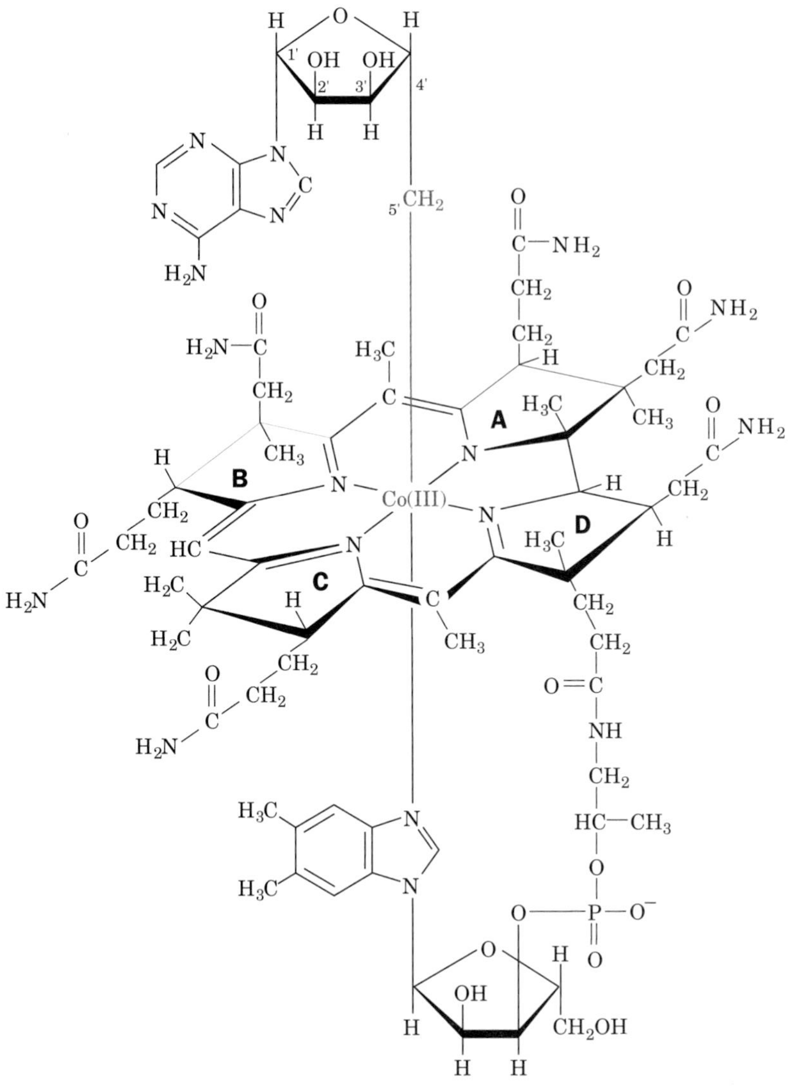

5'-**Deoxyadenosylcobalamin (coenzyme B₁₂)**

Figure 23-17
The structure of 5'-deoxyadenosylcobalamin (coenzyme B_{12}).

deoxyribose C(5') atom forms a covalent C—Co bond, *the only carbon–metal bond known in biology.* In some enzymes, the sixth ligand instead is a CH_3 group that likewise forms a C—Co bond.

Coenzyme B_{12}'s reactive C—Co bond participates in two types of enzyme-catalyzed reactions:

1. Rearrangements in which a hydrogen atom is directly transferred between two adjacent carbon atoms with concomitant exchange of the second substituent, X:

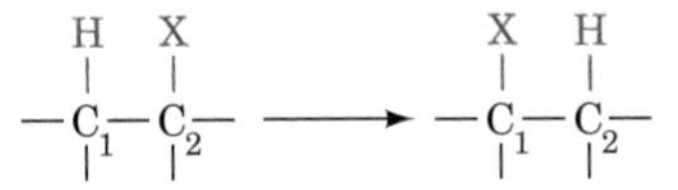

where X may be a carbon atom with substituents, an oxygen atom of an alcohol, or an amine.

2. Methyl group transfers between two molecules.

There are about a dozen known cobalamin-dependent enzymes. Only two occur in mammalian systems:

methylmalonyl-CoA mutase, which catalyzes a carbon skeleton rearrangement (the X group in the rearrangement is —COSCoA; Fig. 23-16) and **homocysteine methyltransferase,** a methyl transfer enzyme that participates in methionine biosynthesis (Sections 24-3E and 5B).

The proposed methylmalonyl-CoA mutase reaction mechanism (Fig. 23-18) begins with **homolytic cleavage** of the cobalamin C—Co(III) bond (the C and Co atoms each acquire one of the electrons that formed the cleaved electron pair bond). The Co ion therefore fluctuates between its Co(III) and Co(II) oxidation states [the two states are spectroscopically distingishable: Co(III) is red and diamagnetic (no unpaired electrons), whereas Co(II) is yellow and paramagnetic (unpaired electrons)]. Hence, *the role of coenzyme B_{12} in the catalytic process is that of a reversible free radical generator.* The C—Co(III) bond is well suited to this function because it is inherently weak (dissociated energy = 109 kJ · mol⁻¹) and appears to be further weakened through steric interactions with the enzyme. Note that a homolytic cleavage reaction is unusual in biology; most other biological bond cleavage reactions occur via **heterolytic cleavage** (in which the electron pair forming the cleaved bond is fully acquired by one of the separating atoms).

Succinyl-CoA Cannot Be Directly Consumed by the Citric Acid Cycle

Methylmalonyl-CoA mutase catalyzes the conversion of a metabolite to a citric acid cycle intermediate other than acetyl-CoA. You should note, however, that the route of succinyl-CoA oxidation is not as simple as it may first appear. The citric acid cycle regenerates all of its C_4 intermediates so that these compounds are really catalysts, not substrates. Consequently, succinyl-CoA cannot undergo net degradation by citric acid cycle enzymes alone. Rather, *in order for a metabolite to undergo net oxidation by the citric acid cycle, it must first be converted either to pyruvate or directly to acetyl-CoA.* Net degradation of succinyl-CoA begins with its conversion, via the citric acid cycle, to malate. At high concentrations, malate is transported, by a specific transport protein, to the cytosol where it may be oxidatively decarboxylated to pyruvate and CO_2, by **malic enzyme (malate dehydrogenase, decarboxylating);**

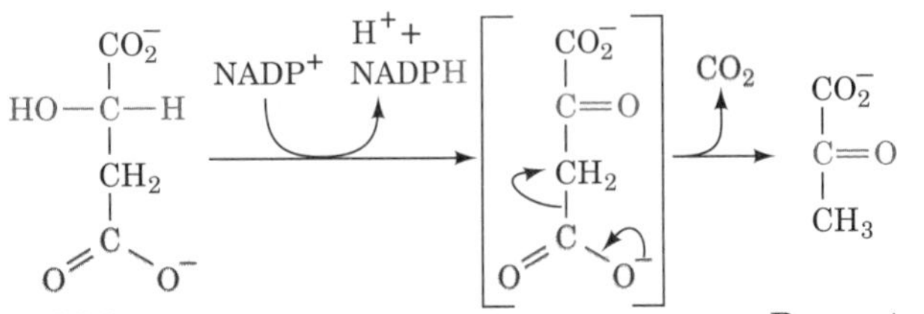

(an enzyme we previously encountered in the C_4 cycle of photosynthesis; Fig. 22-30). Pyruvate is then completely oxidized via pyruvate dehydrogenase and the citric acid cycle.

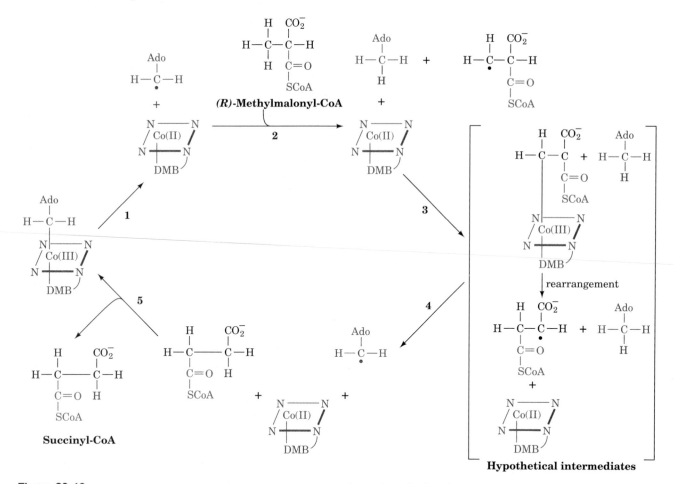

Figure 23-18
The proposed mechanism of methylmalonyl-CoA mutase:
(1) The homolytic cleavage of the C—Co(III) bond yielding a 5'-deoxyadenosyl radical and cobalamin in its Co(II) oxidation state. **(2)** Abstraction of a hydrogen atom from the methylmalonyl-CoA by the 5'-deoxyadenosyl radical thereby generating a methylmalonyl-CoA radical. **(3)** Hypothetical formation of a C—Co bond between the methylmalonyl-CoA radical and the coenzyme followed by carbon skeleton rearrangement to form a succinyl-CoA radical. **(4)** Abstraction of a hydrogen atom from 5'-deoxyadenosine by the succinyl-CoA radical to regenerate the 5'-deoxyadenosyl radical. **(5)** Release of succinyl-CoA and reformation of the coenzyme.

Pernicious Anemia Results from Vitamin B$_{12}$ Deficiency

The existence of **vitamin B$_{12}$** came to light in 1926 when George Minot and William Murphy discovered that **pernicious anemia,** an often fatal disease of the elderly characterized by decreased numbers of red blood cells, low hemoglobin levels, and progressive neurological deterioration, can be treated by the daily consumption of large amounts of raw liver (a treatment that some patients considered worse than the disease). It was not until 1948, however, after a bacterial assay for antipernicious anemia factor had been developed, that vitamin B$_{12}$ was isolated.

Vitamin B$_{12}$ is synthesized neither by plants nor animals but only by a few species of bacteria. Herbivores obtain their vitamin B$_{12}$ from the bacteria that inhabit their gut (in fact, some animals, such as rabbits, must periodically eat some of their feces to obtain sufficient amounts of this essential substance). Humans, however, obtain almost all their vitamin B$_{12}$ directly from their diet, particularly from meat. The vitamin is specifically bound in the intestine by the glycoprotein **intrinsic factor** that is secreted by the stomach. This complex is absorbed by a specific receptor in the intestinal mucosa where the complex is dissociated and the liberated vitamin B$_{12}$ transported to the bloodstream. There it is bound by at least three different plasma globulins called **transcobalamins,** which facilitate its uptake by the tissues.

Pernicious anemia is not usually a dietary deficiency disease but, rather, results from insufficient secretion of intrinsic factor. The normal human requirement for cobalamin is very small, ~3 μg·day^{-1}, and the liver stores a 3 to 5-year supply of this vitamin. This accounts for the insidious onset of pernicious anemia and the fact that true dietary deficiency of vitamin B$_{12}$, even among strict vegetarians, is extremely rare.

F. Peroxisomal β Oxidation

The β oxidation of fatty acids occurs in the peroxisome as well as in the mitochondrion. Peroxisomal β oxidation functions to shorten long-chain fatty acids so as to facilitate their degradation by the mitochondrial β-oxidation system. The peroxisomal pathway results in the same chemical changes to fatty acids as does the mitochondrial pathway, although the enzymes in these two organelles are different. There is no carnitine requirement for transport of fatty acyl-CoA into the peroxisome. Rather, long-chain fatty acids diffuse into this compartment, are activated by a peroxisomal acyl-CoA synthetase, and are oxidized directly. The shorter chain acyl products of this β-oxidation process are then linked to carnitine for transport to mitochondria for further oxidation.

The β-oxidation process in peroxisomes involves three enzymatic reactions:

1. The **acyl-CoA oxidase** reaction:

Fatty acyl-CoA $+ O_2 \longrightarrow$

$$trans\text{-}\Delta^2\text{-enoyl-CoA} + H_2O_2$$

This reaction involves participation of an FAD cofactor but differs from its mitochondrial counterpart in that the abstracted electrons are transferred directly to O_2 rather than passing through the electron-transport chain with its concomitant oxidative phosphorylation (Fig. 23-8). Peroxisomal fatty acid oxidation is therefore less efficient than the mitochondrial process by two ATPs for each C_2 cycle. The H_2O_2 produced disproportionates to H_2O and O_2 through the action of peroxisomal catalase (Section 1-2A).

2. Peroxisomal enoyl-CoA hydratase and 3-L-hydroxy-acyl-CoA dehydrogenase, activities that occur on a single polypeptide and therefore join the growing list of multifunctional enzymes. The reactions catalyzed are identical to those of the mitochondrial system (Fig. 23-8).

3. Peroxisomal thiolase, which has a different chain-length specificity than its mitochondrial counterpart. It is almost inactive with acyl-CoAs of length C_8 or less so that fatty acids are incompletely oxidized by peroxisomes.

Although peroxisomal β oxidation is not dependent on the transport of acyl groups into the peroxisome as their carnitine esters, the peroxisome contains both a carnitine–acetyl transferase and a transferase specific for longer-chain acyl groups. Acyl-CoAs that have been chain-shortened by peroxisomal β oxidation are thereby converted to their carnitine esters. These substances, for the most part, passively diffuse out of the peroxisome to the mitochondrion where they are oxidized further.

G. Minor Pathways of Fatty Acid Oxidation

β Oxidation is blocked by an alkyl group at the C_β of a fatty acid, and thus at any odd-numbered carbon atom. One such branched-chain fatty acid, a common dietary component, is **phytanic acid**. This metabolic breakdown product of chlorophyll's phytyl side chain (Section 22-2A) is present in dairy products and ruminant fats although, surprisingly, chlorophyll itself is but a poor dietary source of phytanic acid for humans. The oxidation of branched-chain fatty acids such as phytanic acid is facilitated by **α oxidation** (Fig. 23-19). In this process, the fatty acid C_α is hydroxylated and the

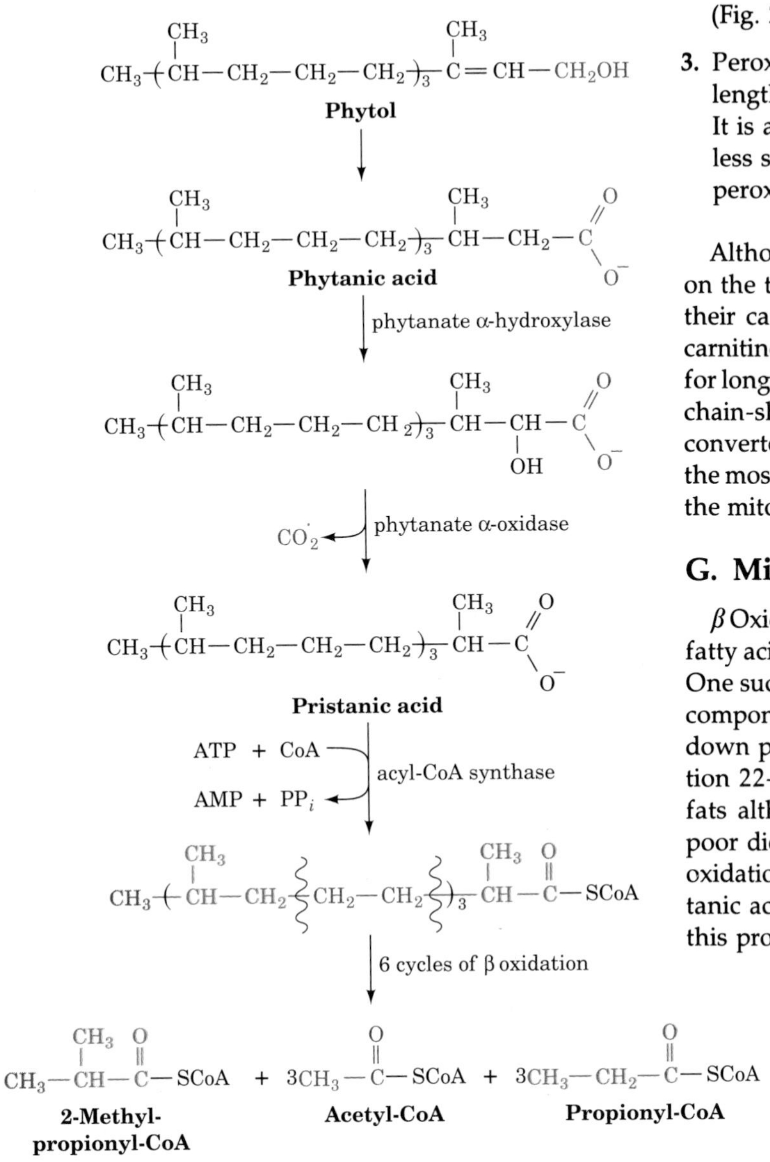

Figure 23-19
Phytanic acid, a degradation product of the phytol side chain of chlorophyll, is metabolized through α oxidation to **pristanic acid** followed by β oxidation.

resulting product is oxidatively decarboxylated to yield a new fatty acid with an unsubstituted C_β. Further degradation of the molecule can then continue via six cycles of normal β oxidation to yield three propionyl-CoAs, three acetyl-CoAs, and one 2-methylpropionyl-CoA, (which is converted to succinyl-CoA).

A rare genetic defect, **Refsum's disease** or **phytanic acid storage syndrome,** results from the accumulation of this metabolite throughout the body. The disease, which is characterized by progressive neurological difficulties such as tremors, unsteady gait, and poor night vision, results from a greatly reduced α-hydroxylation activity. Its symptoms can therefore be attenuated by a diet that restricts the intake of phytanic acid-containing foods.

Medium- and long-chain fatty acids are converted to dicarboxylic acids through ω **oxidation** (oxidation of the last carbon atom). This process, which is catalyzed by enzymes of the endoplasmic reticulum (microsomes), involves hydroxylation of a fatty acid's C_ω atom by **cytochrome P_{450},** a monooxygenase that utilizes NADPH and O_2. The OH group is then oxidized to a carboxyl group, converted to a CoA derivative at either end, and oxidized via the β-oxidation pathway. ω Oxidation is probably of only minor significance in fatty acid oxidation.

3. KETONE BODIES

Acetyl-CoA produced by oxidation of fatty acids in liver mitochondria can be further oxidized via the citric acid cycle as is discussed in Chapter 19. A significant fraction of this acetyl-CoA has another fate, however. *By a process known as* **ketogenesis,** *which occurs primarily in liver mitochondria, acetyl-CoA is converted to* **acetoacetate** *or* **D-β-hydroxybutyrate.** *These compounds, which together with* **acetone** *are somewhat inaccurately referred to as* **ketone bodies,**

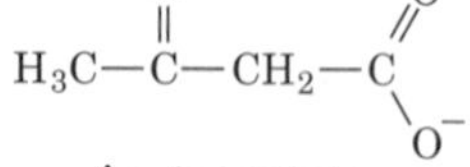

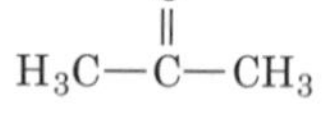

serve as important metabolic fuels for many peripheral tissues, particularly heart and skeletal muscle. The brain, under normal circumstances, uses only glucose as its energy source (fatty acids are unable to pass the blood–brain barrier) but during starvation, ketone bodies become the brain's major fuel source (Section 25-3A). *Ketone bodies are water soluble equivalents of fatty acids*

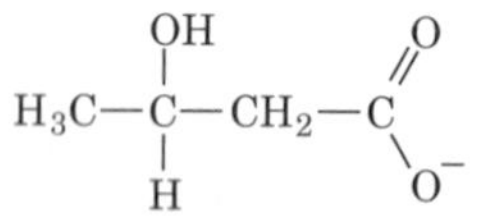

Acetoacetate formation occurs in three reactions (Fig. 23-20):

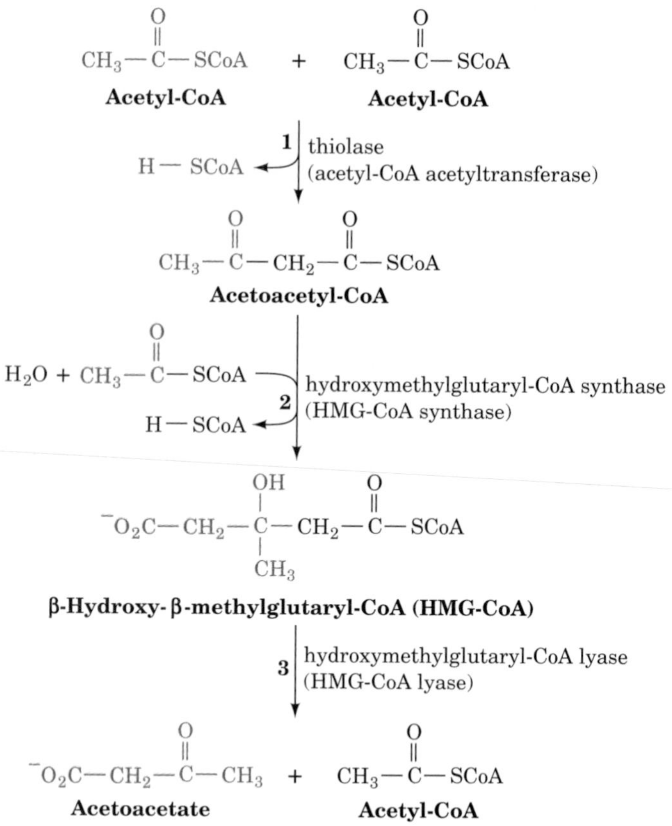

Figure 23-20
The enzymatic reactions forming acetoacetate from acetyl-CoA. **(1)** Two molecules of acetyl-CoA condense to form acetoacetyl-CoA in a thiolase-catalyzed reaction. **(2)** A Claisen ester condensation of the acetoacetyl-CoA with a third acetyl-CoA to form β-hydroxy-β-methylglutaryl-CoA (HMG-CoA) as catalyzed by HMG-CoA synthase. **(3)** The degradation of HMG-CoA to acetoacetate and acetyl-CoA in a mixed aldol–Claisen ester cleavage catalyzed by HMG-CoA lyase.

1. Two molecules of acetyl-CoA are condensed to **acetoacetyl-CoA** by thiolase (also called **acetyl-CoA acetyltransferase**) working in the reverse direction from the way it does in the final step of β oxidation (Section 23-2C).

2. Condensation of the acetoacetyl-CoA with a third acetyl-CoA by **HMG-CoA synthase** forms β-**hydroxy-β-methylglutaryl-CoA (HMG-CoA).** The mechanism of this reaction resembles the reverse of the thiolase reaction (Fig. 23-11) in that an active site thiol group forms an acyl-thioester intermediate.

3. Degradation of HMG-CoA to acetoacetate and acetyl-CoA in a mixed aldol–Claisen ester cleavage by **HMG-CoA lyase.** The mechanism of this reaction is analogous to the reverse of the citrate synthase reaction (Section 19-3A). (HMG-CoA is also a precursor in cholesterol biosynthesis and hence may be diverted to this purpose as is discussed in Section 23-6A.)

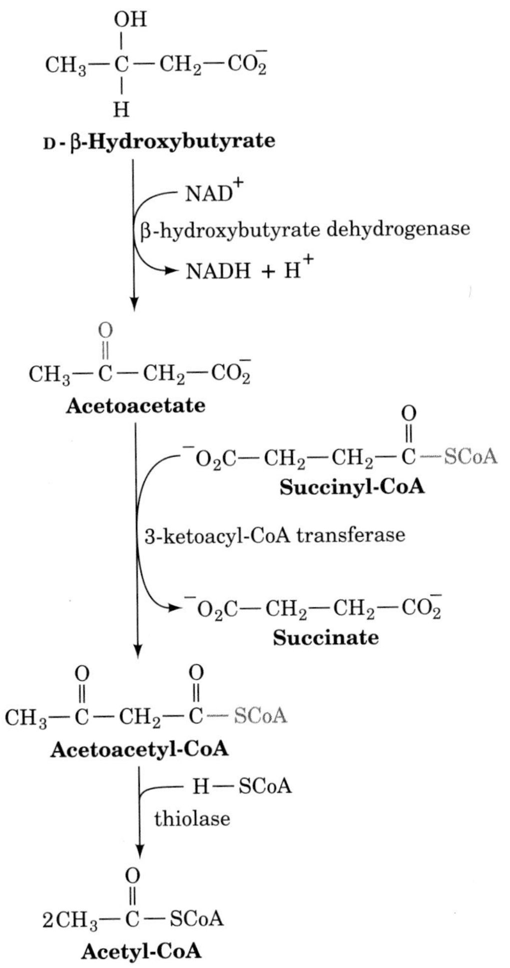

Figure 23-21
The metabolic conversion of ketone bodies to acetyl-CoA.

The overall reaction catalyzed by HMG-CoA synthase and HMG-CoA lyase is

Acetoacetyl-CoA + $H_2O \longrightarrow$ acetoacetate + CoA

One may well ask why this apparently simple hydrolysis reaction occurs in such an indirect manner. The answer is unclear but may lie in the regulation of the process.

Acetoacetate may be reduced to D-β-hydroxybutyrate by **β-hydroxybutyrate dehydrogenase:**

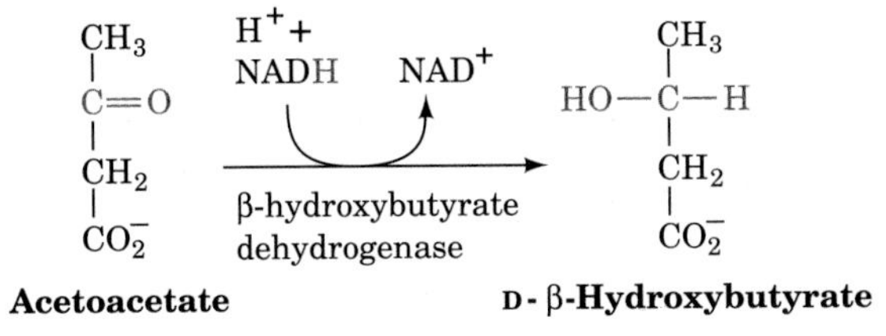

Note that this product is the stereoisomer of the L-β-hydroxyacyl-CoA that occurs in the β-oxidation pathway. Acetoacetate, being a β-keto acid, also undergoes relatively facile nonenzymatic decarboxylation to ace-

tone and CO_2. Indeed, the breath of individuals with **ketosis,** a pathological condition in which acetoacetate is produced faster than it can be metabolized (a symptom of diabetes; Section 25-3B), has the characteristic sweet smell of acetone.

The liver releases acetoacetate and β-hydroxybutyrate, which are carried by the bloodstream to the peripheral tissues for use as alternative fuels. There, these products are converted to acetyl-CoA as is diagrammed in Fig. 23-21. The proposed reaction mechanism of **3-ketoacyl-CoA-transferase** (Fig. 23-22), which catalyzes this pathway's second step, involves the participation of an active site carboxyl group both in an enzyme–CoA thioester intermediate and in an unstable anhydride. Succinyl-CoA, which acts as the CoA donor in this reaction, can also be converted to succinate with the coupled synthesis of GTP in the succinyl-CoA synthase reaction of the citric acid cycle (Section 19-3E). The "activation" of acetoacetate bypasses this step and

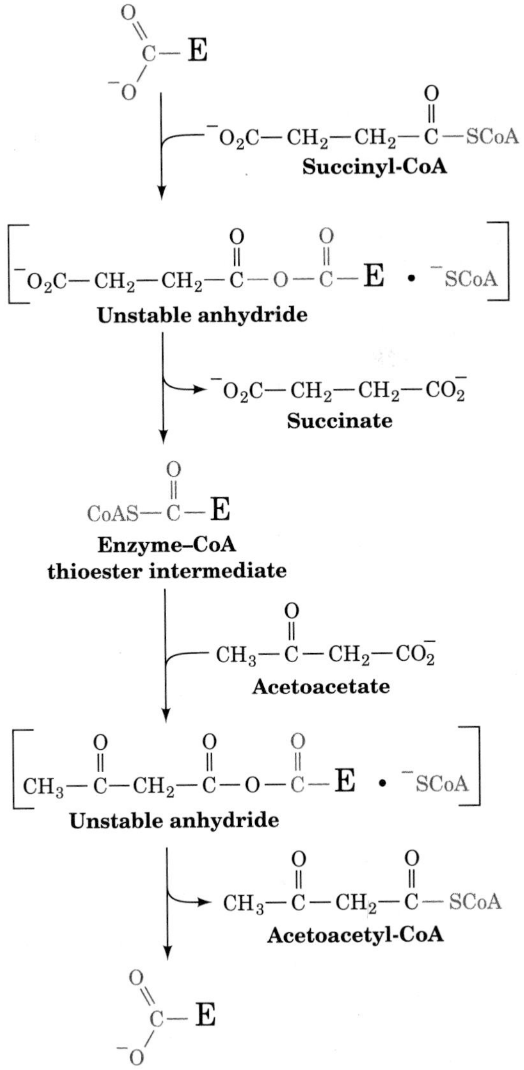

Figure 23-22
The proposed mechanism of 3-ketoacyl-CoA transferase involves an enzyme-CoA thioester intermediate.

therefore "costs" the free energy of GTP hydrolysis. The liver lacks 3-ketoacyl-CoA transferase, which permits it to supply ketone bodies to other tissues.

4. FATTY ACID BIOSYNTHESIS

Fatty acid biosynthesis occurs through condensation of C_2 units, the reverse of the β-oxidation process. Through isotopic labeling techniques, David Rittenberg and Konrad Bloch demonstrated, in 1945, that these condensation units are derived from acetic acid. Acetyl-CoA was soon proved to be a precursor of the condensation reaction but its mechanism remained obscure until the late 1950s when Salih Wakil discovered a requirement for bicarbonate in fatty acid biosynthesis and malonyl-CoA was shown to be an intermediate. In this section, we discuss the reactions of fatty acid biosynthesis.

A. Pathway Overview

The pathway of fatty acid synthesis differs from that of fatty acid oxidation. This situation, as we saw in Section 17-1D, is typically the case of opposing biosynthetic and degradative pathways because it permits them both to be thermodynamically favorable and independently regulated under similar physiological conditions. Figure 23-23 outlines fatty acid oxidation and synthesis with

emphasis on the differences between these pathways. While fatty acid oxidation occurs in the mitochondrion and utilizes fatty acyl-CoA esters, fatty acid biosynthesis occurs in the cytosol, as Roy Vagelos discovered, with the growing fatty acids esterified to **acyl-carrier protein** (**ACP**; Fig. 23-24). ACP, like CoA, contains a phosphopantetheine group that forms thioesters with acyl groups. The phosphopantetheine phosphoryl group is esterified to the OH group of an ACP Ser residue, whereas in CoA it is esterified to AMP. In *E. coli*, ACP is a 10-kD polypeptide while in animals it is part of a large multifunctional protein.

The redox coenzymes of the fatty acid oxidative and biosynthetic pathways differ (NAD^+ and FAD for oxidation; NADPH for biosynthesis) as does the stereochemistry of their intermediate steps, but their main difference is the manner in which C_2 units are removed from or added to the fatty acyl thioester chain. In the oxidative pathway, β-ketothiolase catalyzes the cleavage of the C_α—C_β bond of β-ketoacyl-CoA so as to produce acetyl-CoA and a new fatty acyl-CoA, which is shorter by a C_2 unit. The $\Delta G°'$ of this reaction is very close to zero so that it can also function in the reverse direction (ketone body formation). In the biosynthetic pathway, the condensation reaction is coupled to the hydrolysis of ATP thereby driving the reaction to completion. This process involves two steps: (1) the ATP-dependent carboxylation of acetyl-CoA by **acetyl-CoA carboxylase** to form **malonyl-CoA,** and (2) the exer-

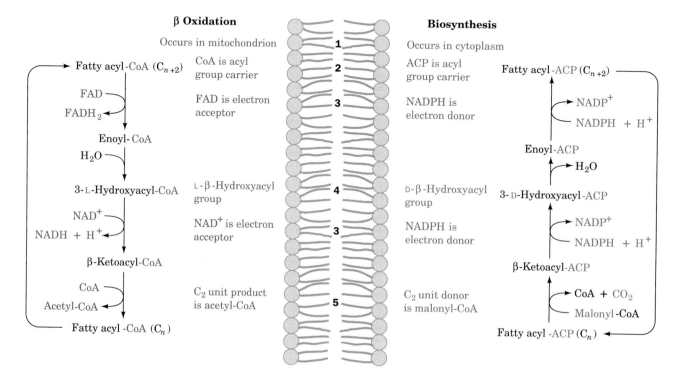

Figure 23-23
Differences between the pathways of fatty acid β oxidation and fatty acid biosynthesis with respect to: **(1)** cellular location; **(2)** acyl group carrier; **(3)** electron acceptor/donor; **(4)** stereochemistry of the hydration/dehydration reaction; and **(5)** the form in which C_2 units are produced/donated.

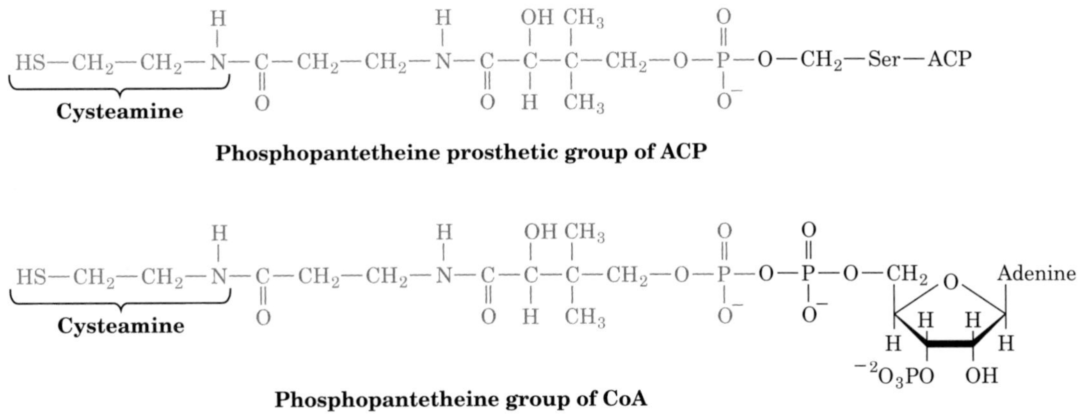

$$HS-CH_2-CH_2-\overset{H}{\underset{}{N}}-\overset{}{\underset{O}{C}}-CH_2-CH_2-\overset{H}{\underset{}{N}}-\overset{}{\underset{O}{C}}-\overset{OH}{\underset{H}{C}}-\overset{CH_3}{\underset{CH_3}{C}}-CH_2-O-\overset{O}{\underset{O^-}{P}}-O-CH_2-Ser-ACP$$

$$\underbrace{\qquad\qquad}_{\textbf{Cysteamine}}$$

Phosphopantetheine prosthetic group of ACP

$$HS-CH_2-CH_2-\overset{H}{\underset{}{N}}-\overset{}{\underset{O}{C}}-CH_2-CH_2-\overset{H}{\underset{}{N}}-\overset{}{\underset{O}{C}}-\overset{OH}{\underset{H}{C}}-\overset{CH_3}{\underset{CH_3}{C}}-CH_2-O-\overset{O}{\underset{O^-}{P}}-O-\overset{O}{\underset{O^-}{P}}-O-CH_2$$

$$\underbrace{\qquad\qquad}_{\textbf{Cysteamine}}$$

Adenine

$^{-2}O_3PO$ OH

Phosphopantetheine group of CoA

Figure 23-24
The phosphopantetheine group in acyl carrier protein (ACP) and in CoA.

gonic decarboxylation of the malonyl group in the condensation reaction catalyzed by **fatty acid synthase.** The mechanisms of these enzymes are described in the next section.

B. Acetyl-CoA Carboxylase

Acetyl-CoA carboxylase catalyzes the first committed step of fatty acid biosynthesis and one of its rate-controlling steps. The mechanism of this biotin-dependent enzyme is very similar to that of propionyl-CoA carboxylase (Section 23-2E) in that it occurs in two steps, a CO_2-activation and a carboxylation:

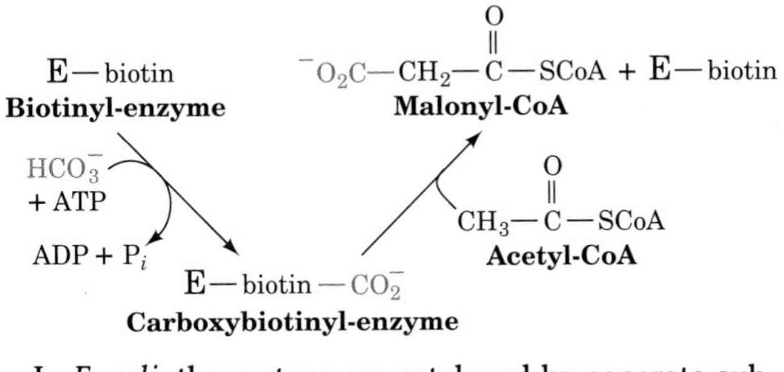

$$E-biotin \qquad\qquad \overset{O}{\underset{\|}{^-O_2C-CH_2-C-SCoA}} + E-biotin$$
Biotinyl-enzyme   **Malonyl-CoA**

$$HCO_3^- + ATP$$

$$ADP + P_i$$

$$\overset{O}{\underset{\|}{CH_3-C-SCoA}}$$
Acetyl-CoA

$$E-biotin-CO_2^-$$
Carboxybiotinyl-enzyme

In *E. coli*, these steps are catalyzed by separate subunits, known as **biotin carboxylase** and **transcarboxylase,** respectively. In addition, the biotin is bound as a biocytin residue to a third subunit termed **biotin carboxyl carrier protein.** The mammalian and avian enzymes contain both enzymatic activities as well as the biotin carboxyl carrier on a single 230-kD polypeptide chain.

Avian and Mammalian Acetyl-CoA Carboxylase Are Regulated through Enzyme Polymerization

Electron microscopy reveals that the flat rectangular protomers of both avian and mammalian acetyl-CoA carboxylases associate to form long filaments with molecular masses in the range 4000 to 8000 kD (Fig. 23-25). *This polymeric form of the enzyme is catalytically active but*

the protomer is not. The rate of fatty acid biosynthesis is therefore controlled by the position of the equilibrium between these forms:

$$\text{Protomer } (inactive) \rightleftharpoons \text{polymer } (active)$$

Metabolites that most affect the position of this equilib-

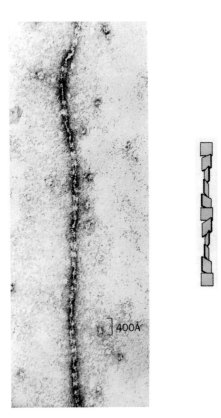

Figure 23-25
An electron micrograph with an accompanying interpretive drawing indicating that filaments of avian liver acetyl-CoA carboxylase consist of linear chains of flat rectangular protomers. [Courtesy of Malcolm Lane, The Johns Hopkins University School of Medicine.]

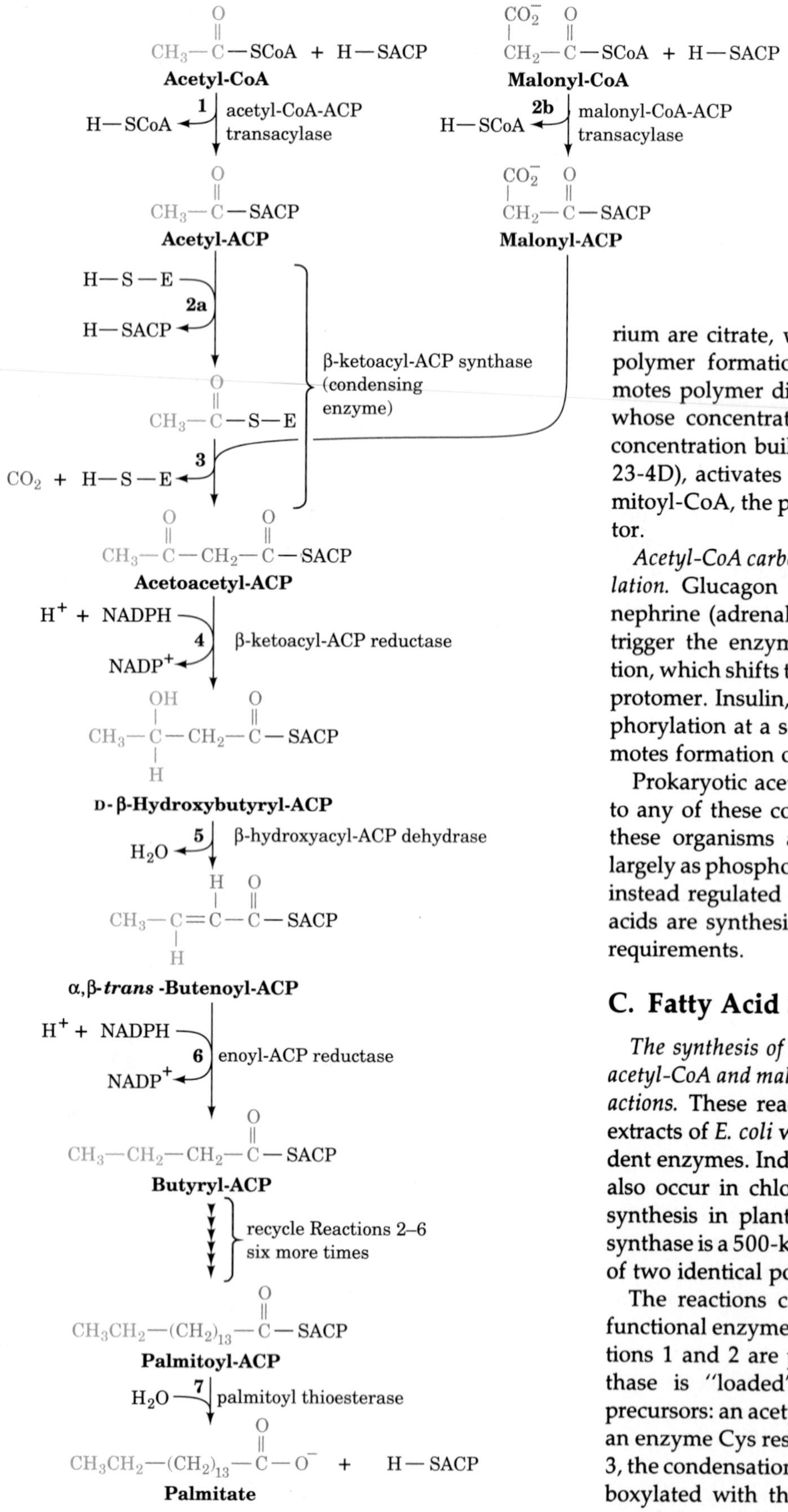

Figure 23-26
The reaction sequence for the biosynthesis of fatty acids. In forming palmitate, the pathway is repeated for seven cycles of C_2 elongation followed by a final hydrolysis step.

rium are citrate, which shifts the equilibrium towards polymer formation, and palmitoyl-CoA, which promotes polymer disaggregation. Thus, cytosolic citrate, whose concentration increases when the acetyl-CoA concentration builds up in the mitochondrion (Section 23-4D), activates fatty acid biosynthesis, whereas palmitoyl-CoA, the pathway product, is a feedback inhibitor.

Acetyl-CoA carboxylase is also subject to hormonal regulation. Glucagon as well as epinephrine and norepinephrine (adrenalin and noradrenalin; Section 17-3E) trigger the enzyme's cAMP-dependent phosphorylation, which shifts the equilibrium in favor of the inactive protomer. Insulin, on the other hand, stimulates phosphorylation at a separate site on the enzyme that promotes formation of the active polymer.

Prokaryotic acetyl-CoA carboxylases are not subject to any of these controls. This is because fatty acids in these organisms are not stored as fats but function largely as phospholipid precursors. The *E. coli* enzyme is instead regulated by guanine nucleotides so that fatty acids are synthesized in response to the cell's growth requirements.

C. Fatty Acid Synthase

The synthesis of fatty acids, mainly palmitic acid, from acetyl-CoA and malonyl-CoA involves seven enzymatic reactions. These reactions were first studied in cell-free extracts of *E. coli* where they are catalyzed by independent enzymes. Individual enzymes with these activities also occur in chloroplasts (the only site of fatty acid synthesis in plants). In animals, however, fatty acid synthase is a 500-kD multifunctional enzyme consisting of two identical polypeptide chains.

The reactions catalyzed by the mammalian multifunctional enzyme are diagrammed in Fig. 23-26. Reactions 1 and 2 are priming reactions in which the synthase is "loaded" with the condensation reaction precursors: an acetyl group linked by a thioester bond to an enzyme Cys residue, and malonyl-ACP. In Reaction 3, the condensation reaction, the malonyl-ACP is decarboxylated with the resulting carbanion attacking the acetyl-thioester to form a β-ketoacyl-ACP. Reactions 4–6 are the reductions and dehydration that convert this ketone to an alkyl group. The coenzyme in both reductive steps is NADPH, whereas in β oxidation, the analogs of Reactions 4 and 6, respectively, use NAD^+ and FAD (Fig. 23-23). Moreover, Reaction 5 requires a D-β-hydroxyacyl substrate whereas the analogous reaction in β oxidation forms the corresponding L-isomer.

All of the enzyme activities but those catalyzing Reactions 2a and 3, remain functional when the native dimeric enzyme is dissociated into monomers. Electron microscopy of these monomers indicates that they consist of a linear chain of at least four 50-Å in diameter lobes. Moreover, fragments resulting from the limited proteolysis of fatty acid synthase exhibit many of the enzymatic activities of the intact protein. Thus, *contiguous stretches of its polypeptide chain fold to form a series of autonomous domains, each with a specific but different catalytic activity.* Several other enzymes, such as mammalian acetyl-CoA carboxylase (Section 23-4B), exhibit similar multifunctionality but none have as many separate catalytic activities as does animal fatty acid synthase.

Since the condensation reaction requires the juxtaposition of the sulfhydryl groups of the ACP phosphopantetheine and an enzyme Cys residue, it has been proposed that these groups are on separate subunits that interact in a head-to-tail manner (Fig. 23-27). The mechanism of palmitate synthesis by this dimer is therefore thought to occur as is diagrammed in Fig. 23-28 with the long flexible phosphopantetheine chain of ACP (Fig. 23-24) functioning to transport the substrate between the enzyme's various catalytic sites:

1. The "priming" of the condensing enzyme (Fig. 23-26, Reaction 1) followed by formation of an acetyl-Cys residue (Fig. 23-26, Reaction 2a).

2. The "loading" of the condensing enzyme by the formation of malonyl-ACP (Fig. 23-26, Reaction 2b).

3. The coupling of the acetyl group to the C_β of the malonyl group on the other subunit with the latter's accompanying decarboxylation so as to form aceto-

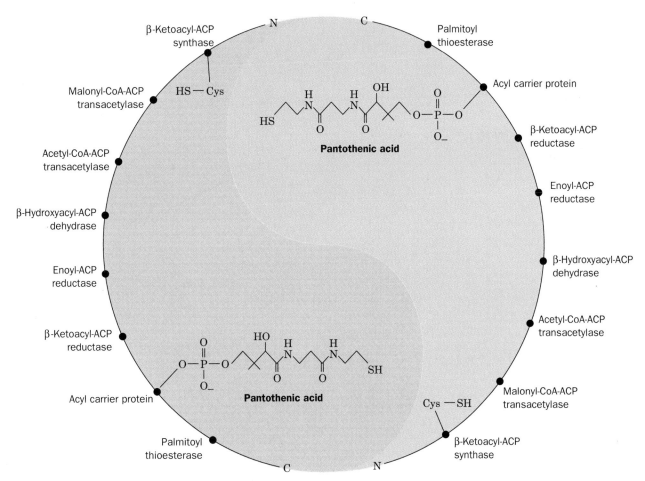

Figure 23-27
Schematic representation of two multifunctional subunits of animal fatty acid synthase in head-to-tail association to form the active dimer. [After Wakil, S. J., Stoops, J. K., and Joshi, V. C., *Annu. Rev. Biochem.* **52,** 556 (1983).]

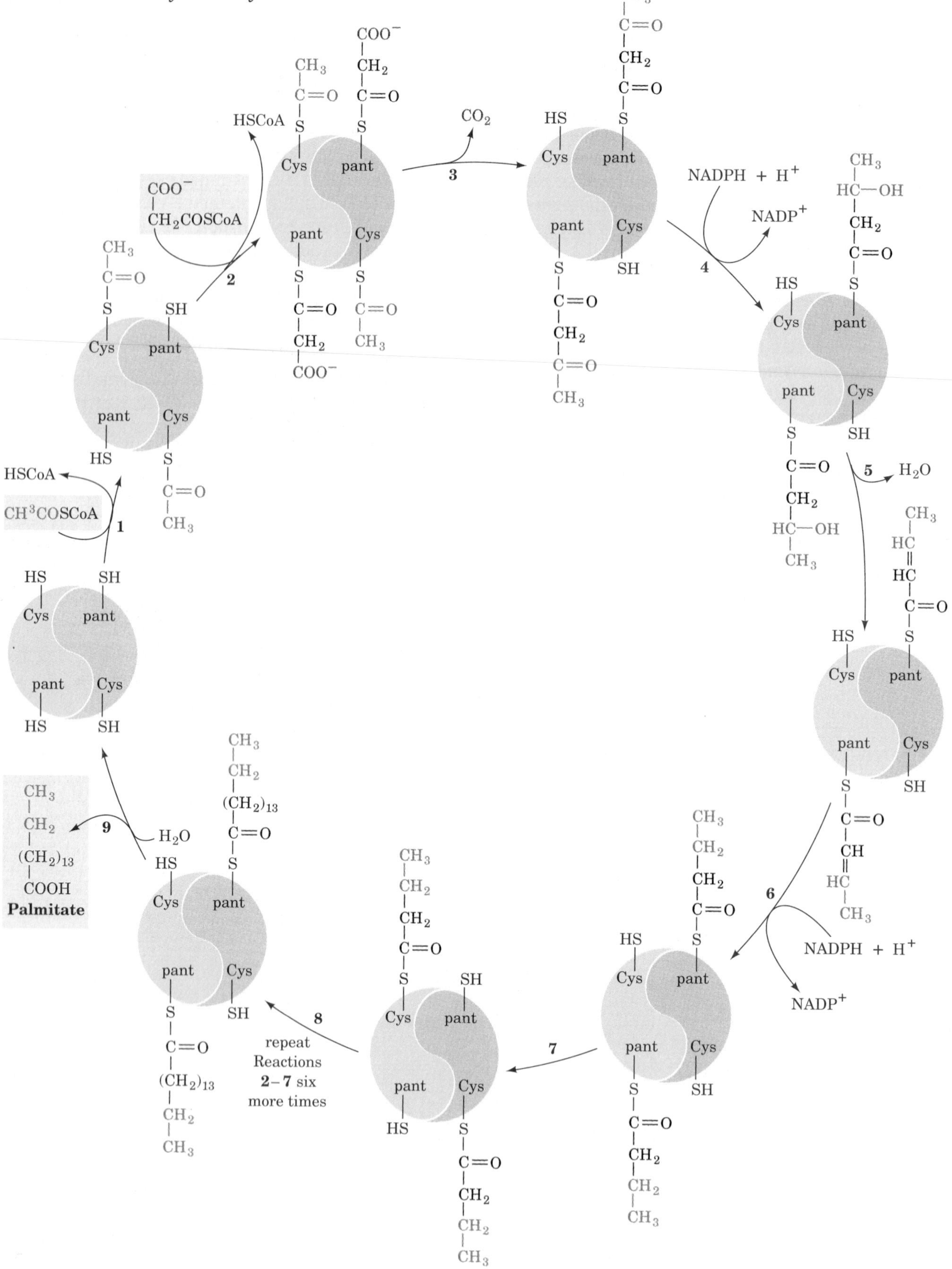

Figure 23-28
Proposed mechanism of palmitate synthesis as mediated by animal fatty acid synthase. The circles represent the multifunctional dimer of fatty acid synthase. The Cys-SH represents the active cysteine sulfhydryl of the β-ketoacyl-ACP synthase and pant-SH represents the pantetheine sulfhydryl of the acyl carrier protein. The other catalytic domains indicated in Fig. 23-27 are not shown but are present in both subunits of the dimer. [After Wakil, S. J., Stoops, J. K., and Joshi, V. C., *Annu. Rev. Biochem.* **52**, 568 (1983).]

acetyl-ACP and free the active site Cys-SH group (Fig. 23-26, Reaction 3). Consequently, the CO_2 taken up in the acetyl-CoA carboxylase reaction (Section 23-4B) does not appear in the product fatty acid. Rather, the decarboxylation functions to drive the condensation reaction which, through the acetyl-CoA carboxylase reaction, is coupled to ATP hydrolysis.

4–6. The reduction, dehydration, and further reduction (Fig. 23-26, Reactions 4–6) of the acetoacetyl-ACP to form butyryl-ACP followed by the transfer of this group to the Cys-SH of the first subunit. Thus the acetyl group with which the system was initially primed has been elongated by a C_2 unit.

7. The transfer of the butyryl group to the Cys-SH of the first subunit (Fig. 23-26, repeat of Reaction 2a).

The ACP group is "reloaded" with a malonyl group followed by another cycle of C_2 elongation. This process occurs altogether seven times to form palmitoyl-ACP (Fig. 23-28, Reaction 8). Its thioester bond is then hydrolyzed by **palmitoyl thioesterase** (Fig. 23-26, Reaction 7) yielding palmitate, the normal product of the fatty acid synthase pathway, and regenerating the enzyme for a new round of synthesis. The stoichiometry of palmitate synthesis therefore is

$$\text{Acetyl-CoA} + 7\text{malonyl-CoA}$$
$$+ 14\text{NADPH} + 7\text{H}^+ \longrightarrow \text{palmitate} + 7\text{CO}_2$$
$$+ 14\text{NADP}^+ + 8\text{CoA} + 6\text{H}_2\text{O}$$

Since the 7malonyl-CoA are derived from acetyl-CoA as follows:

$$7\text{Acetyl-CoA} + 7\text{CO}_2 + 7\text{ATP} \longrightarrow$$
$$7\text{malonyl-CoA} + 7\text{ADP} + 7\text{P}_i + 7\text{H}^+$$

the overall stoichiometry for palmitate biosynthesis is

$$8\text{Acetyl-CoA} + 7\text{ATP} + 14\text{NADPH} \longrightarrow$$
$$\text{palmitate} + 14\text{NADP}^+ + 8\text{CoA} + 6\text{H}_2\text{O} + 7\text{ADP} + 7\text{P}_i$$

We next consider the means of transport of mitochondrial acetyl-CoA to the cytosol, the site of fatty acid synthesis. Following that, we examine the reactions by which fatty acids are elongated and desaturated.

D. Transport of Mitochondrial Acetyl-CoA into the Cytosol

Acetyl-CoA is generated in the mitochondrion by the oxidative decarboxylation of pyruvate as catalyzed by pyruvate dehydrogenase (Section 19-2A) as well as by the oxidation of fatty acids. When the need for ATP synthesis is low, so that the oxidation of acetyl-CoA via the citric acid cycle and oxidative phosphorylation is minimal, this mitochondrial acetyl-CoA may be stored for future use as fat. Fatty acid biosynthesis occurs in the cytosol but the mitochondrial membrane is essentially impermeable to acetyl-CoA. *Acetyl-CoA enters the cyto-*

sol in the form of citrate via the **tricarboxylate transport system** *(Fig. 23-29).* Cytosolic **ATP-citrate lyase** then catalyzes the reaction.

$$\text{ATP} + \text{citrate} + \text{CoA} \Longrightarrow$$
$$\text{acetyl-CoA} + \text{oxaloacetate} + \text{ADP} + \text{P}_i$$

which resembles the reverse of the citrate synthase reaction (Section 19-3A) except that ATP hydrolysis is required to drive the intermediate synthesis of the "high-energy" citryl-CoA (whose hydrolysis drives the citrate synthase reaction to completion). ATP hydrolysis is therefore required in the ATP–citrate lyase reaction to power the resynthesis of this thioester bond. Oxaloacetate is reduced to malate by **malate dehydrogenase.** Malate may be oxidatively decarboxylated to pyruvate by **malic enzyme** and, in this form, returned to the mitochondrion. The malic enzyme reaction resembles that of isocitrate dehydrogenase in which a β-hydroxy acid is oxidized to a β-keto acid, whose decarboxylation is strongly favored (Section 19-3C). Malic enzyme's coenzyme is $NADP^+$ so that, when this route is used, NADPH is produced for use in the reductive reactions of fatty acid biosynthesis.

Citrate transport out of the mitochondrion must be balanced by anion transport into the mitochondrion. Malate, pyruvate, and P_i can act in this capacity. Malate may therefore also be transported directly back to the mitochondrion without generating NADPH. As we have seen in Section 23-4C, synthesis of each palmitate ion requires 8 molecules of acetyl-CoA and 14 molecules of NADPH. As many as 8 of these NADPH molecules may be supplied with the 8 molecules of acetyl-CoA if all the malate produced in the cytosol is oxidatively decarboxylated. The remaining NADPH is provided through the pentose phosphate pathway (Section 21-4).

E. Elongases and Desaturases

Palmitate (16:0; recall that the symbol n:m indicates a C_n *fatty acid with m double bonds), the normal product of the fatty acid synthase pathway, is the precursor of longer-chain saturated and unsaturated fatty acids through the actions of* **elongases** *and* **desaturases.** Elongases are present in both the mitochondrion and the endoplasmic reticulum but the mechanisms of elongation at the two sites differ. Mitochondrial elongation (a process independent of the fatty acid synthase pathway) occurs by successive addition and reduction of acetyl units in a reversal of fatty acid oxidation; the only difference between these two pathways occurs in the final reduction step in which NADPH takes the place of $FADH_2$ as the terminal redox coenzyme (Fig. 23-30). Elongation in the endoplasmic reticulum involves the successive condensations of malonyl-CoA with acyl-CoA. These reactions are each followed by NADPH-associated reductions similar to those catalyzed by fatty acid synthase, the

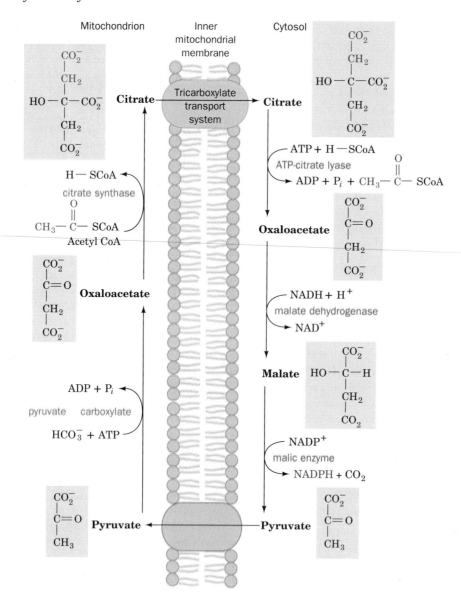

Figure 23-29
The transfer of acetyl-CoA from mitochondrion to cytosol via the tricarboxylate transport system.

only difference being that the fatty acid is elongated as its CoA derivative rather than as its ACP derivative.

Unsaturated fatty acids are produced by **terminal desaturases.** Mammalian systems contain four terminal desaturases of broad chain-length specificities designated Δ^9-, Δ^6-, Δ^5-, and Δ^4-**fatty acyl-CoA desaturases.** These nonheme iron-containing enzymes catalyze the general reaction:

$$CH_3-(CH_2)_x-\overset{\overset{H}{|}}{\underset{\underset{H}{|}}{C}}-\overset{\overset{H}{|}}{\underset{\underset{H}{|}}{C}}-(CH_2)_y-\overset{O}{\overset{\|}{C}}-SCoA + NADH + H^+ + O_2$$

$$\downarrow$$

$$CH_3-(CH_2)_x-\underset{\underset{H}{|}}{C}=\underset{\underset{H}{|}}{C}-(CH_2)_y-\overset{O}{\overset{\|}{C}}-SCoA + 2H_2O + NAD^+$$

where x is at least five and where $(CH_2)_x$ can contain one or more double bonds. The $(CH_2)_y$ portion of the substrate is always saturated. Double bonds are inserted between existing double bonds in the $(CH_2)_x$ portion of the substrate and the CoA group such that the new double bond is three carbon atoms closer to the CoA group than the next double bond (not conjugated to an existing double bond) and, in animals, never at positions beyond C(9).

A variety of unsaturated fatty acids may be synthesized by combinations of elongation and desaturation reactions. However, since palmitic acid is the shortest available fatty acid in animals, the above rules preclude the formation of the Δ^{12} double bond of linoleic acid ($\Delta^{9,12}$-octadecadienoic acid), a required precursor of **prostaglandins** (Section 23-7). *Linoleic acid must conse-*

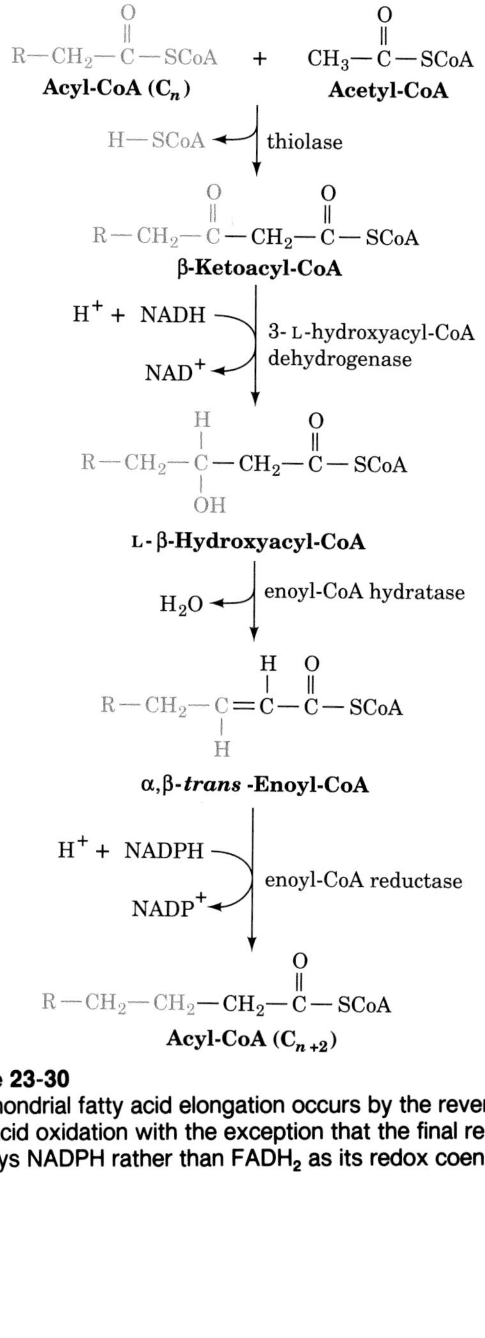

Figure 23-30
Mitochondrial fatty acid elongation occurs by the reversal of fatty acid oxidation with the exception that the final reaction employs NADPH rather than $FADH_2$ as its redox coenzyme.

Figure 23-31
The electron-transfer reactions mediated by the Δ^9-fatty acyl-CoA desaturase complex. Its three proteins, desaturase, cytochrome b_5, and NADH–cytochrome b_5 reductase, are situated in the endoplasmic reticulum membrane. [After Jeffcoat, R., *Essays Biochem.* **15**, 19 (1979).]

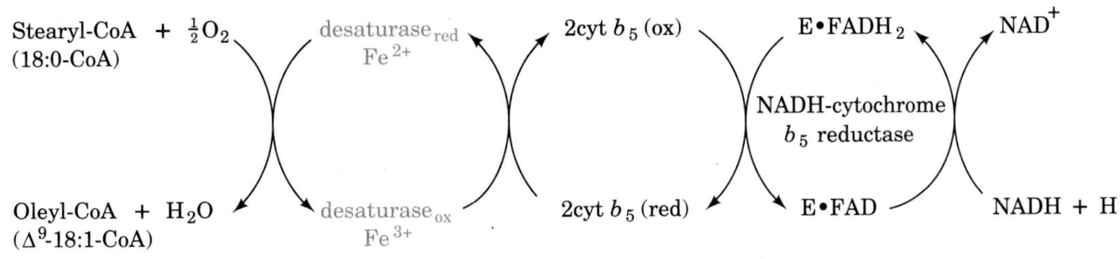

quently be obtained in the diet (*ultimately from plants which have Δ^{12}- and Δ^{15}-desaturases*) and is therefore an *essential fatty acid*. Indeed, animals maintained on a fat-free diet develop an ultimately fatal condition that is initially characterized by poor growth, poor wound healing, and dermatitis. Linoleic acid is also an important constituent of epidermal sphingolipids that function as the skin's water-permeability barrier.

Mammalian terminal desaturases are components of mini-electron-transport systems that contain two other proteins; **cytochrome b_5** and **NADH–cytochrome b_5 reductase**. The electron-transfer reactions mediated by these complexes occur at the inner surface of the endoplasmic reticulum membrane (Fig. 23-31) and are therefore not associated with oxidative phosphorylation.

F. Synthesis of Triacylglycerols

Triacylglycerols are synthesized from fatty acyl-CoA esters and glycerol-3-phosphate or dihydroxyacetone phosphate (Fig. 23-32). The initial step in this process is catalyzed either by **glycerol-3-phosphate acyltransferase** in mitochondria and endoplasmic reticulum, or by **dihydroxyacetone phosphate acyltransferase** in endoplasmic reticulum or peroxisomes. In the latter case, the product acyl-dihydroxyacetone phosphate is reduced to the corresponding **lysophosphatidic acid** by an NADPH-dependent reductase. Lysophosphatidic acid is converted to triacylglycerol by the successive actions of **1-acylglycerol-3-phosphate acyltransferase, phosphatidic acid phosphatase**, and **diacylglycerol acyltransferase**. The intermediate phosphatidic acid and diacylglycerol can also be converted to phospholipids by the pathways described in Section 23-8. The acyltransferases are not completely specific for particular fatty acyl-CoAs, either in chain length or degree of unsaturation, but in human adipose tissue triacylglycerols, palmitate tends to be concentrated at position 1 and oleate at position 2.

5. REGULATION OF FATTY ACID METABOLISM

Discussions of metabolic control are usually concerned with the regulation of metabolite flow through a pathway in response to the differing energy needs and dietary states of an organism. For example, the differ-

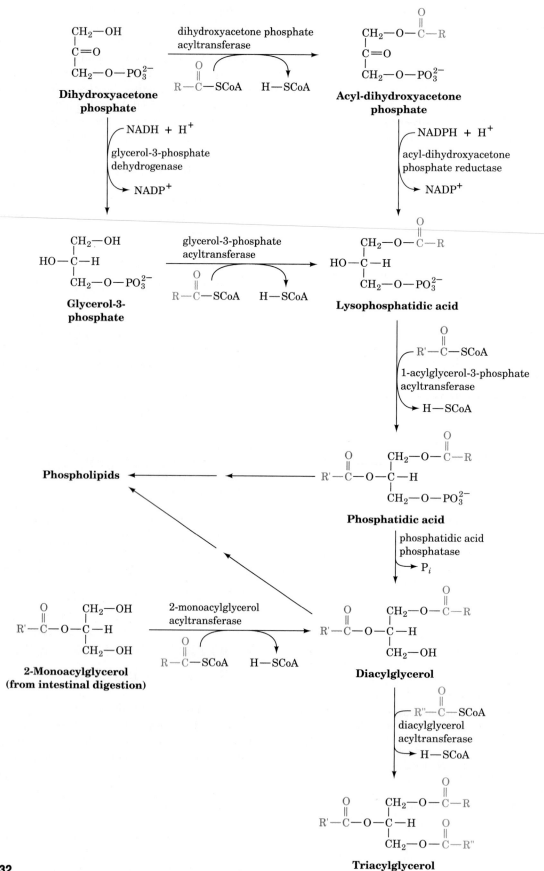

Figure 23-32
The reactions of triacylglycerol biosynthesis.

ence in the energy requirement of muscle between rest and vigorous exertion may be as much as 100-fold. Such varying demands may be placed on the body when it is in either a fed or a fasted state. For instance, Eric Newsholme, an authority on the biochemistry of exercise, enjoys a two-hour run before breakfast. Others might wish for no greater exertion than the motion of hand to mouth. In both individuals, glycogen and triacylglycerols serve as primary fuels for energy requiring-processes, and are synthesized in times of quiet plenty for future use.

Synthesis and breakdown of glycogen and triacylglycerols, as detailed in Chapter 17 and above, are processes that concern the whole organism, with its organs and tissues forming an interdependent network connected by the bloodstream. The blood carries the metabolites responsible for energy production: triacylglycerols in the form of chylomicrons and VLDL (Section 11-4A), fatty acids as their albumin complexes (Section 23-1), ketone bodies, amino acids, lactate, and glucose. The pancreatic α and β cells sense the organism's dietary and energetic state mainly through the glucose concentration in the blood. The α cells respond to the low blood glucose concentration of the fasting and energy-demanding states by secreting glucagon. The β cells respond to the high blood glucose concentration of the fed and resting states by secreting insulin. We have previously discussed (Sections 17-3E and F) how these hormones are involved in glycogen metabolism. *They also regulate the rates of the opposing pathways of lipid metabolism and therefore control whether fatty acids will be oxidized or synthesized.* Their targets are the regulatory (flux-generating) enzymes of fatty acid synthesis and breakdown in specific tissues (Fig. 23-33).

We are already familiar with most of the mechanisms by which the catalytic activities of regulatory enzymes may be controlled: substrate availability, allosteric interactions, and covalent modification (phosphorylation). These are examples of **short-term regulation**, regulation that occurs with a response time of minutes or less. *Fatty acid synthesis is controlled, in part, by short-term regulation.* Acetyl-CoA carboxylase, which catalyzes the first committed step of this pathway, is inhibited by palmitoyl-CoA and by glucagon-stimulated cAMP-dependent phosphorylation, and is activated by citrate and by insulin-stimulated phosphorylation (Section 23-4B).

Another mechanism exists for controlling a pathway's regulatory enzymes: alteration of the amount of enzyme present by changes in the rates of protein synthesis and/or breakdown. This process requires hours or days and is therefore called **long-term regulation** (the control of protein synthesis and breakdown is discussed in Chapters 29 and 30). *Lipid biosynthesis is also controlled by long-term regulation,* with insulin stimulating and starvation inhibiting the synthesis of acetyl-

CoA carboxylase and fatty acid synthase. The presence in the diet of polyunsaturated fatty acids also decreases the concentrations of these enzymes. The amount of adipose tissue lipoprotein lipase, the enzyme that initiates the entry of lipoprotein-packaged fatty acids into adipose tissue for storage (Section 11-4B), is also increased by insulin and decreased by starvation. In contrast, the concentration of heart lipoprotein lipase, which controls the entry of fatty acids from lipoproteins into heart tissue for oxidation rather than storage, is decreased by insulin and increased by starvation. *Starvation and/or regular exercise, by decreasing the glucose concentration in the blood, change the body's hormone balance. This situation results in long-term increases in the levels of fatty acid oxidation enzymes accompanied by long-term decreases in those of lipid biosynthesis.*

Hormones Regulate Fatty Acid Metabolism

Fatty acid oxidation is regulated largely by the concentration of fatty acids in the blood, which is, in turn, controlled by the hydrolysis rate of triacylglycerols in adipose tissue by **hormone-sensitive triacylglycerol lipase.** This enzyme is so-named because it is susceptible to regulation by phosphorylation and dephosphorylation in response to hormonally controlled cAMP levels. Epinephrine and norepinephrine, as does glucagon, act to increase adipose tissue cAMP concentrations. cAMP allosterically activates cAMP-dependent protein kinase which, in turn, increases the phosphorylation levels of susceptible enzymes. Phosphorylation activates hormone-sensitive lipase, thereby stimulating lipolysis in adipose tissue, raising blood fatty acid levels, and ultimately activating the β-oxidation pathway in other tissues such as liver and muscle. In liver, this process leads to the production of ketone bodies that are secreted into the bloodstream for use as an alternative fuel to glucose by peripheral tissues. cAMP-dependent protein kinase also inactivates acetyl-CoA carboxylase (Section 23-4B), one of the rate-determining enzymes of fatty acid synthesis, so that *cAMP-dependent phosphorylation simultaneously stimulates fatty acid oxidation and inhibits fatty acid synthesis.*

Insulin has the opposite effect of glucagon and epinephrine: It stimulates the formation of glycogen and triacylglycerols. This protein hormone, which is secreted in response to high blood glucose concentrations, decreases cAMP levels. This situation leads to the dephosphorylation and thus the inactivation of hormone-sensitive lipase thereby reducing the amount of fatty acid available for oxidation. Insulin also stimulates the activity of certain **cAMP-independent protein kinases,** which phosphorylate proteins at different sites than cAMP-dependent kinases. Acetyl-CoA carboxylase, for instance, is activated by cAMP-independent (insulin-dependent) phosphorylation (Section 23-4B), the opposite effect of cAMP-dependent phosphorylation. *The*

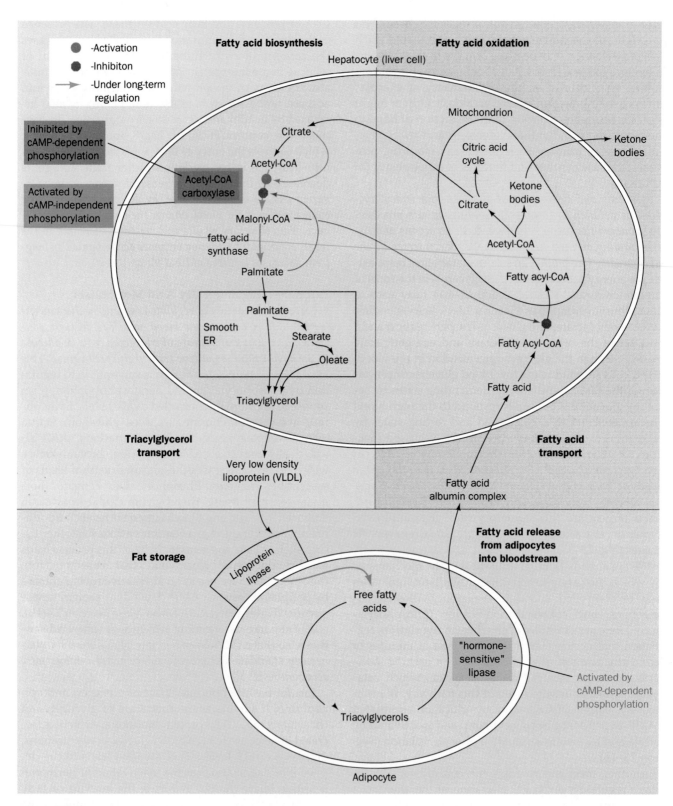

Figure 23-33
Sites of regulation of fatty acid metabolism.

glucagon – insulin ratio is therefore of prime importance in determining the rate and direction of fatty acid metabolism.

Another control point that inhibits fatty acid oxidation when fatty acid synthesis is stimulated is the inhibition of carnitine palmitoyl transferase I by malonyl-CoA. This inhibition keeps the newly synthesized fatty acids out of the mitochondrion (Section 23-1) and thus away from the β-oxidation system.

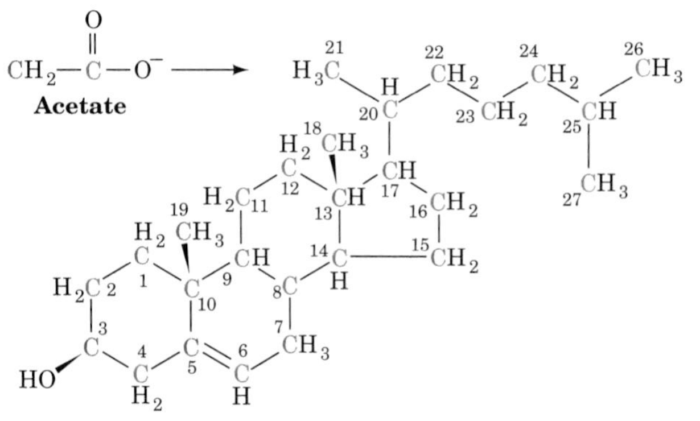

6. CHOLESTEROL METABOLISM

Cholesterol is a vital constituent of cell membranes and the precursor of steroid hormones and bile acids. It is clearly essential to life, yet its deposition in arteries has been associated with heart disease and atherosclerosis, two leading causes of death in humans. In a healthy organism, an intricate balance is maintained between the biosynthesis, utilization, and transport of cholesterol, keeping its harmful deposition to a minimum. In this section, we study the pathways of cholesterol biosynthesis and transport and how they are controlled. We also examine how cholesterol is utilized in the biosynthesis of steroid hormones and bile acids.

A. Cholesterol Biosynthesis

All of the carbon atoms of cholesterol are derived from acetate (Fig. 23-34). Observation of their pattern of incorporation into cholesterol led Konrad Bloch to propose that acetate was first converted to **isoprene units,** C_5 units that have the carbon skeleton of **isoprene.**

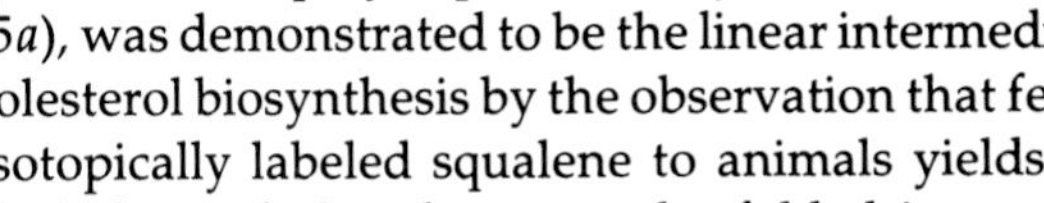

$$CH_2\!=\!\underset{\underset{CH_3}{|}}{C}\!-\!CH\!=\!CH_2 \qquad\qquad C\!-\!\underset{\underset{C}{|}}{C}\!-\!C\!-\!C$$

Isoprene · **An isoprene unit**
(2-methyl-1,3-butadiene)

Isoprene units are condensed to form a linear precursor to cholesterol, and then cyclized.

Squalene, a polyisoprenoid hydrocarbon (Fig. 23-35*a*), was demonstrated to be the linear intermediate in cholesterol biosynthesis by the observation that feeding isotopically labeled squalene to animals yields labeled cholesterol. Squalene may be folded in several ways that would enable it to cyclize to the four-ring

Figure 23-34
All of cholesterol's carbon atoms are derived from acetate.

sterol nucleus (Section 11-1E). The folding pattern proposed by Bloch and Robert B. Woodward (Fig. 23-35*b*) proved to be correct.

Bloch's outline for the major stages of cholesterol biosynthesis was

Acetate $\longrightarrow$ isoprenoid intermediate $\longrightarrow$
  squalene $\longrightarrow$ cyclization product $\longrightarrow$ cholesterol

This pathway has been experimentally verified and its details elaborated. It is now known to be part of a branched pathway (Fig. 23-36) that produces several other essential isoprenoids in addition to cholesterol, namely, ubiquinone (CoQ; Fig. 20-15*b*), dolichol (Fig. 21-14), and isopentenyl adenosine (a modified base of tRNA; Fig. 30–13). We shall examine in detail the portion of this pathway that synthesizes cholesterol.

HMG-CoA Is a Key Cholesterol Precursor

Acetyl-CoA is converted to isoprene units by a series of reactions that begins with formation of hydroxymethylglutaryl-CoA (HMG-CoA; Fig. 23-20), a compound we previously encountered as an intermediate in ketone body biosynthesis (Section 23-3). HMG-CoA synthesis requires the participation of two enzymes: thiolase and HMG-CoA synthase. The enzymes forming the HMG-CoA leading to ketone bodies occur in the mitochondria while those responsible for the synthesis of the HMG-CoA that is destined for cholesterol biosyn-

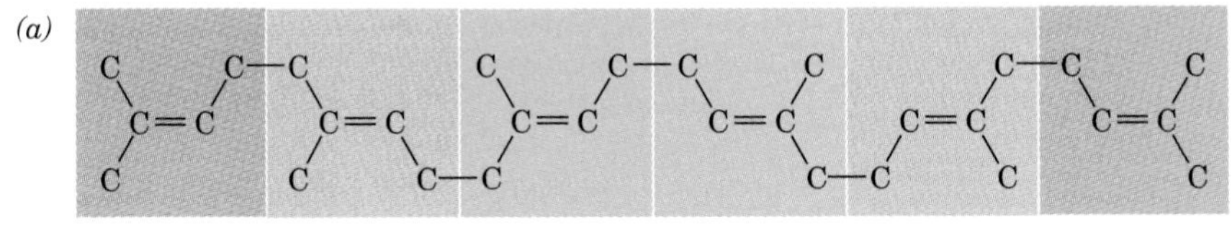

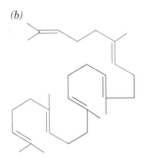

Figure 23-35
Squalene. (*a*) Extended conformation. Each box contains one isoprene unit.
(*b*) Folded in preparation for cyclization as predicted by Bloch and Woodward.

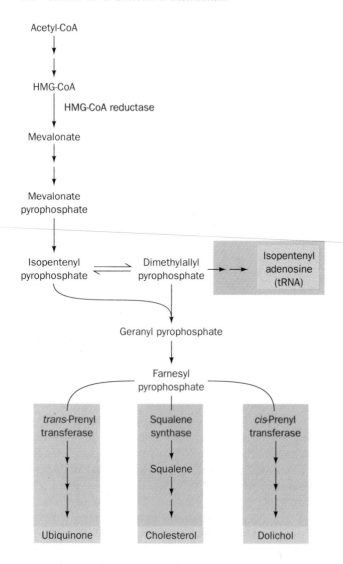

Figure 23-36
The branched pathway of isoprenoid metabolism in mammalian cells. The pathway produces ubiquinone, dolichol, and isopentenyl adenosine, a modified tRNA base, in addition to cholesterol.

Figure 23-37
The formation of isopentenyl pyrophosphate from HMG-CoA.

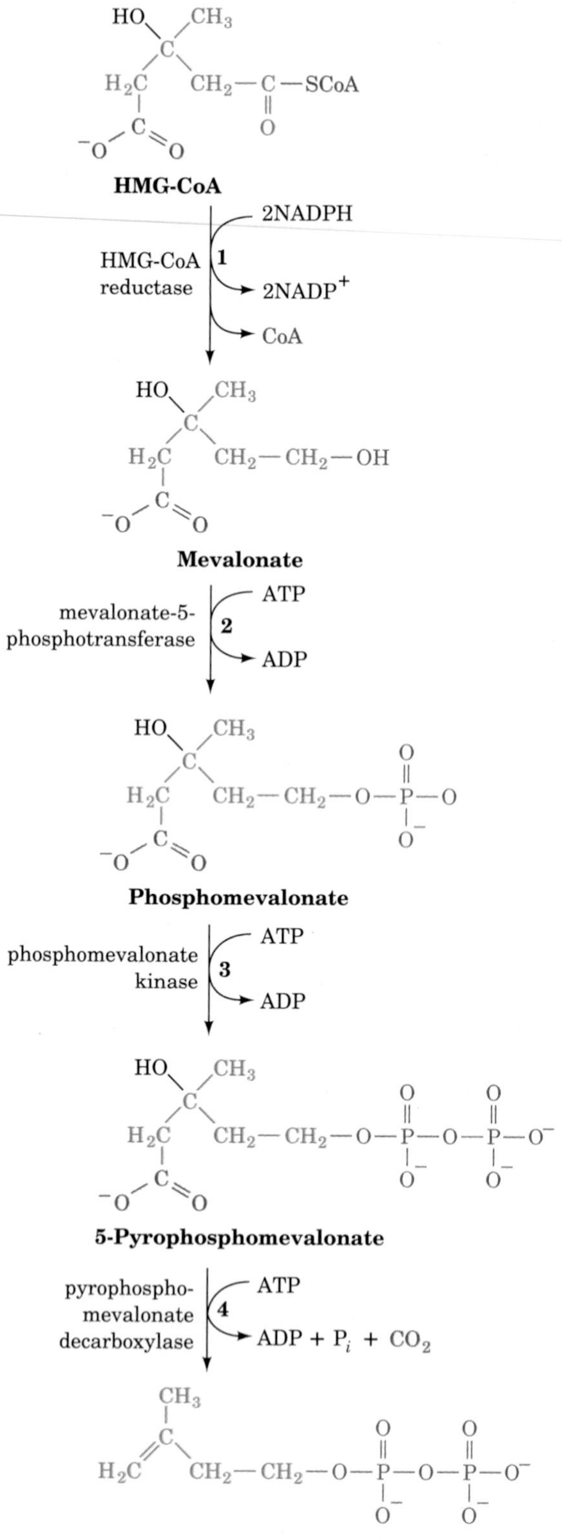

thesis are located in the cytosol. Their catalytic mechanisms, however, are identical.

HMG-CoA is the precursor of two isoprenoid intermediates, **isopentenyl pyrophosphate** and **dimethylallyl pyrophosphate.**

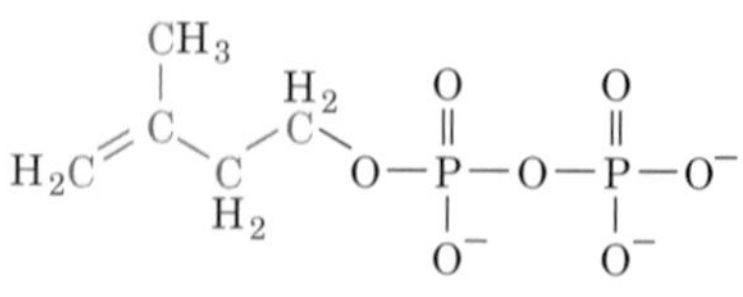

Isopentenyl pyrophosphate

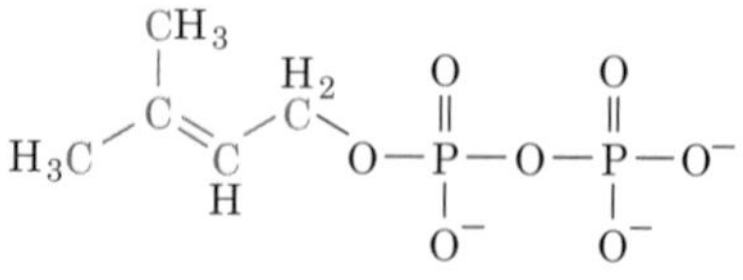

Dimethylallyl pyrophosphate

HMG-CoA

Mevalonate

Phosphomevalonate

5-Pyrophosphomevalonate

Isopentenyl pyrophosphate

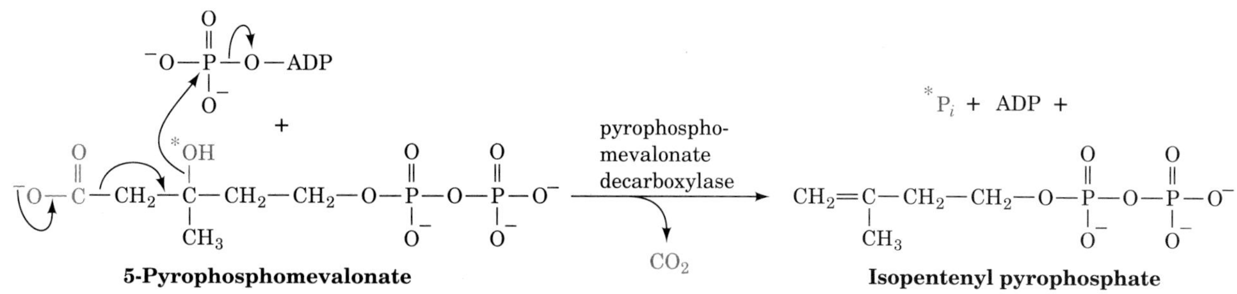

Figure 23-38
Pyrophosphomevalonate decarboxylase catalyzes an ATP-dependent concerted dehydration–decarboxylation of pyrophosphomevalonate yielding isopentenyl pyrophosphate.

The formation of isopentenyl pyrophosphate involves four reactions (Fig. 23-37):

1. The CoA thioester group of HMG-CoA is reduced to an alcohol in an NADPH-dependent four-electron reduction catalyzed by **HMG-CoA reductase,** yielding **mevalonate.**

2. The new OH group is phosphorylated by **mevalonate-5-phosphotransferase.**

3. The phosphate group is converted to a pyrophosphate by **phosphomevalonate kinase.**

4. The molecule is decarboxylated and the resulting alcohol dehydrated by **pyrophosphomevalonate decarboxylase.**

HMG-CoA reductase mediates the rate-determining step of cholesterol biosynthesis and is the most elaborately regulated enzyme of this pathway. It is regulated, as we shall see in Section 23-6B, by competitive and allosteric mechanisms, phosphorylation/dephosphorylation, and long-term regulation. Cholesterol itself is an important feedback regulator of the enzyme.

Pyrophosphomevalonate Decarboxylase and Isopentenyl Pyrophosphate Isomerase Both Catalyze Apparently Concerted Reactions

5-Pyrophosphomevalonate is converted to isopentenyl pyrophosphate by an ATP-dependent dehydration/decarboxylation reaction catalyzed by **pyrophosphomevalonate decarboxylase** (Fig. 23-38). When

[3-^{18}O]-5-pyrophosphomevalonate (*O in Fig. 23-38) is used as a substrate, the labeled oxygen appears in P_i. This observation suggests that 3-phospho-5-pyrophosphomevalonate is a reaction intermediate. Since all attempts to isolate this intermediate have failed, however, it has been proposed that phosphorylation, the α, β elimination of CO_2, and the elimination of P_i occur in a concerted reaction.

The equilibration between isopentenyl pyrophosphate and dimethylallyl pyrophosphate is catalyzed by **isopentenyl pyrophosphate isomerase.** The reaction is thought to occur via a concerted protonation/deprotonation reaction (Fig. 23-39).

Squalene Is Formed by the Condensation of Six Isoprene Units

Four isopentenyl pyrophosphates and two dimethylallyl pyrophosphates condense to form the C_{30} cholesterol precursor squalene in three reactions catalyzed by two enzymes (Fig. 23-40):

1. **Prenyl transferase (farnesyl pyrophosphate synthase)** catalyzes the head-to-tail (1′–4) condensation of dimethylallyl pyrophosphate and isopentenyl pyrophosphate to yield **geranyl pyrophosphate.**

2. **Prenyl transferase** catalyzes a second head-to-tail condensation of geranyl pyrophosphate and isopentenyl pyrophosphate to yield **farnesyl pyrophosphate.**

Figure 23-39
Isopentenyl pyrophosphate isomerase interconverts isopentenyl pyrophosphate and dimethylallyl pyrophosphate by a concerted protonation/deprotonation reaction.

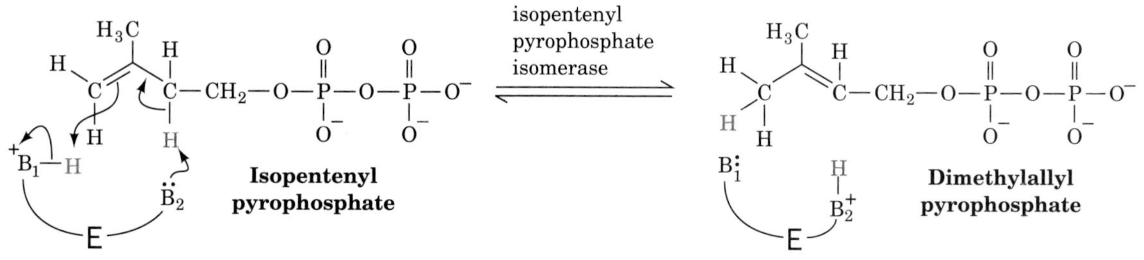

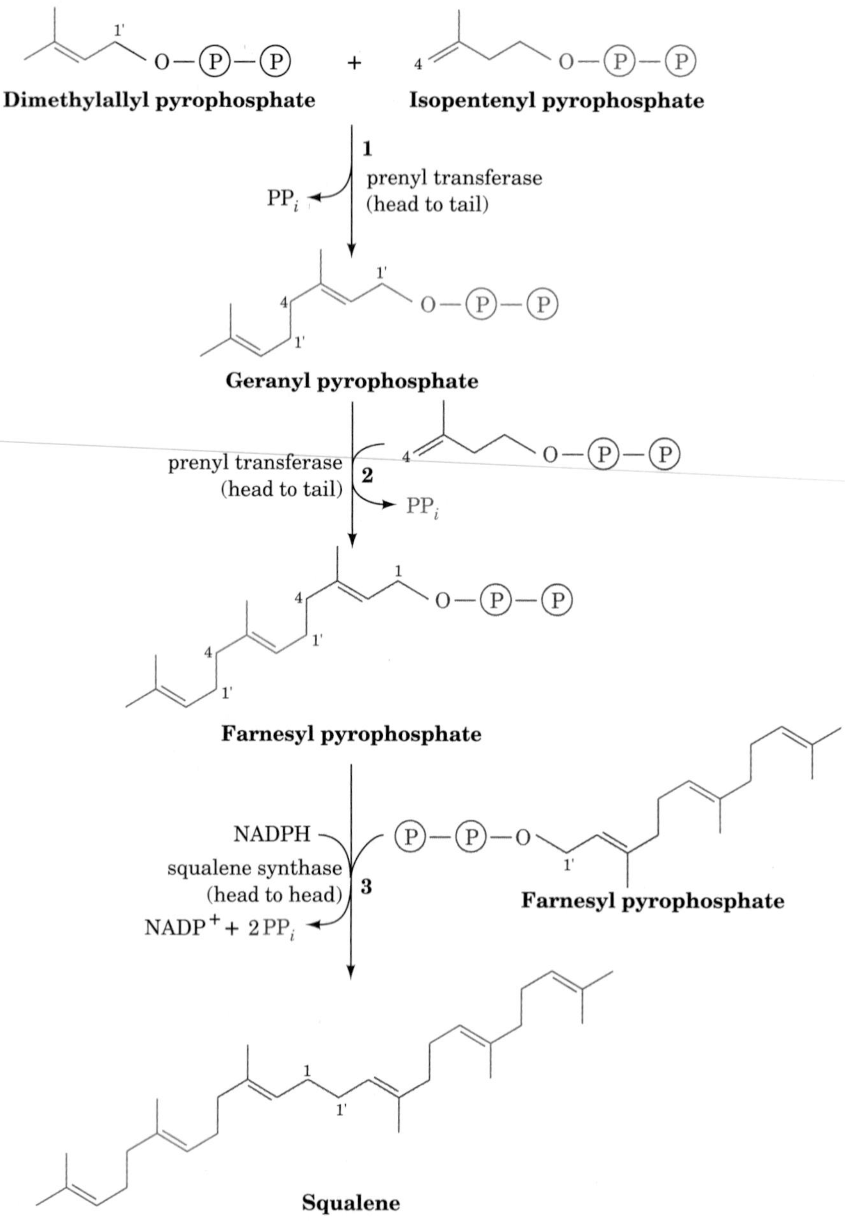

Figure 23-40
The formation of squalene from isopentenyl pyrophosphate and dimethylally pyrophosphate. The pathway involves two head-to-tail condensations catalyzed by prenyl transferase and a head-to-head condensation catalyzed by squalene synthase.

3. **Squalene synthase** then catalyzes the head-to-head (1–1′) condensation of two farnesyl pyrophosphate molecules to form squalene. Farnesyl pyrophosphate is also a precursor to diolichol and ubiquinone (Fig. 23-36).

Prenyl transferase catalyzes the condensation of isopentenyl pyrophosphate with an allylic pyrophosphate. It is specific for isopentenyl pyrophosphate, but can use either the 5-carbon dimethylallyl pyrophosphate or the 10-carbon **geranyl pyrophosphate** as its allylic substrate. The prenyl transferase-catalyzed condensation mechanism is particularly interesting since it is one of the very few known enzyme-catalyzed reactions that

proceed via a carbocation intermediate. Two possible condensation mechanisms can be envisioned (Fig. 23-41):

Scheme I   An S_N1 mechanism in which an allylic carbocation forms by the elimination of PP_i. Isopentenyl pyrophosphate then condenses with this carbocation forming a new carbocation that eliminates a proton to form product.

Scheme II   An S_N2 reaction in which the allylic PP_i is displaced in a concerted manner. In this case, an enzyme nucleophile, X, assists in the reaction. This group is eliminated in the

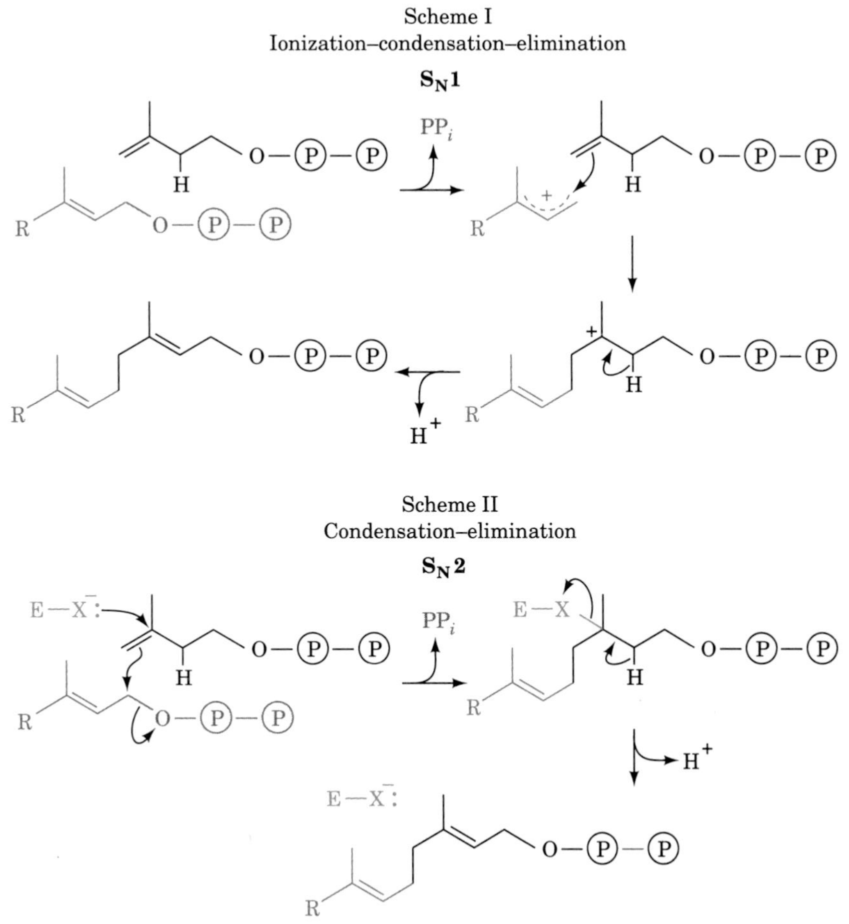

Scheme I
Ionization–condensation–elimination

S_N1

Scheme II
Condensation–elimination

S_N2

Figure 23-41
Two possible mechanisms for the prenyl transferase reaction. Scheme I involves the formation of a carbocation intermediate, whereas Scheme II involves the participation of an enzyme nucleophile, X.

second step with the loss of a proton to form product.

C. Dale Poulter and Hans Rilling used chemical logic to differentiate between these two mechanisms. Capitalizing on the observation that S_N1 reactions are much more sensitive to electron-withdrawing groups than S_N2 reactions, they synthesized a geranyl pyrophosphate derivative in which the H at C(3) is replaced by the electron-withdrawing group F. This allylic substrate for the second (1'–4) condensation catalyzed by prenyl transferase, not surprisingly, has the same K_M as the natural substrate (F and H have similar atomic radii).

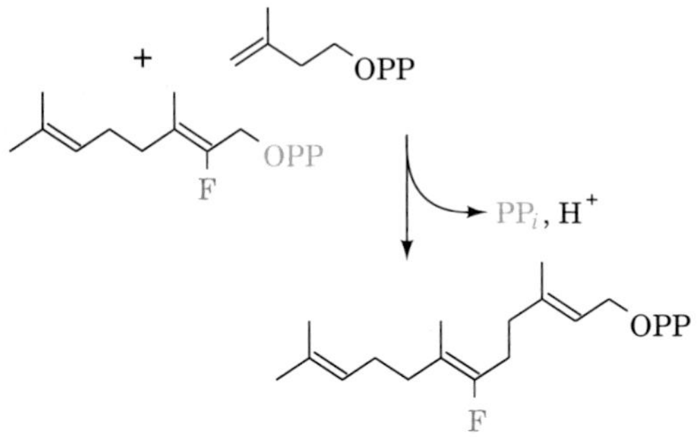

It is, however, the V_{max} of this reaction that tells the story. If the reaction is an S_N2 displacement, the fluoro derivative should react at a rate similar to that of the natural substrate. If, instead, the reaction has an S_N1 mechanism, the fluoro derivative should react orders of magnitude more slowly than the natural substrate. In fact, 3-fluorogeranyl pyrophosphate forms product at <1% of the rate of the natural substrate, strongly supporting an S_N1 mechanism with a carbocation intermediate.

Squalene, the immediate sterol precursor, is formed by the head-to-head condensation of two farnesyl pyrophosphate molecules by **squalene synthase.** The reaction is not a simple head-to-tail condensation, as might be expected, but rather, proceeds via a complex two-step reaction (Fig. 23-42):

Step I The insertion of C(1) of one farnesyl pyrophosphate molecule into the C(2)—C(3) double bond of the second molecule, eliminating PP_i and forming **presqualene pyrophosphate,** a cyclopropylcarbinyl pyrophosphate.

Step II Rearrangement and reduction of presqualene

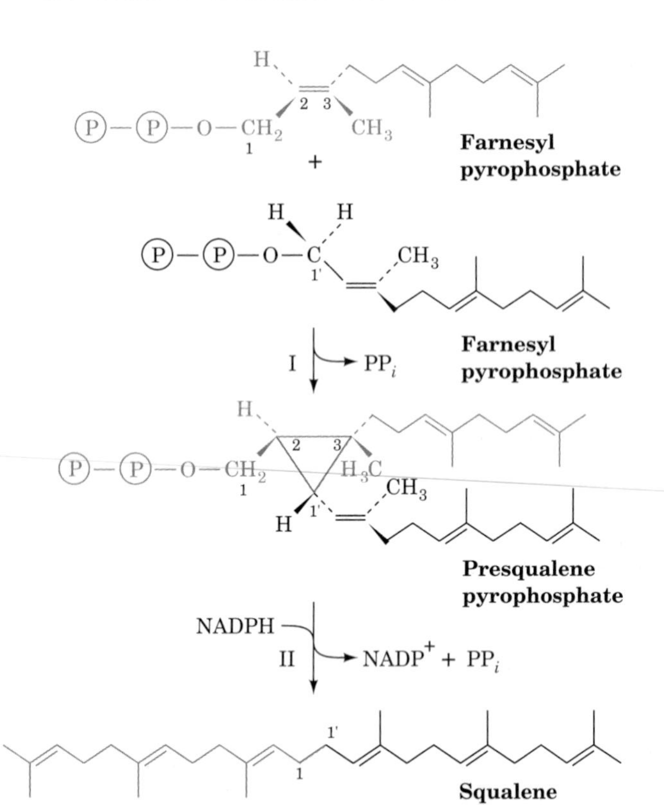

Figure 23-42
Squalene synthase catalyzes the head-to-head condensation of two farnesyl pyrophosphate molecules to form squalene.

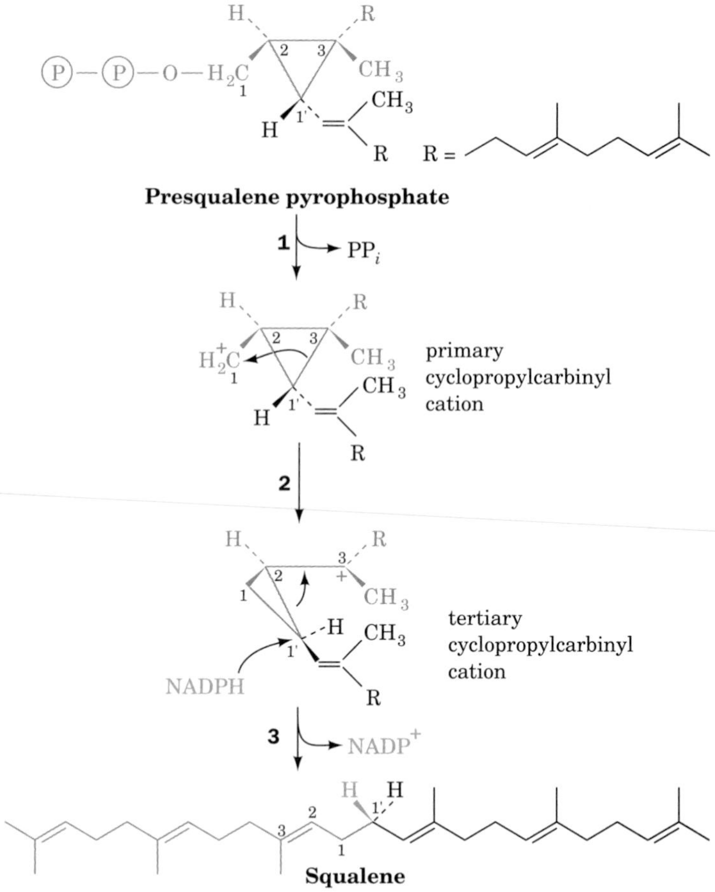

Figure 23-43
The mechanism of rearrangement and reduction of presqualene pyrophosphate to squalene as catalyzed by squalene synthase: **(1)** Presqualene's pyrophosphate group leaves, yielding a primary carbocation at C(1). **(2)** The electrons forming the C(1')—C(3) bond migrate to C(1) forming squalene's C(1)—C(1') bond and a tertiary carbocation at C(3). **(3)** The process is completed by the addition of an NADPH-supplied hydride ion to C(1') and the formation of the C(2)—C(3) double bond.

pyrophosphate by NADPH to form squalene. This reaction involves the formation and rearrangement of a cyclopropylcarbinyl cation in a complex reaction sequence called a **1'–2–3 process** (Fig. 23-43).

Lanosterol Is Produced by Squalene Cyclization

Squalene, an open-chain C_{30} hydrocarbon, is cyclized to form the tetracyclic steroid skeleton in two steps. **Squalene epoxidase** catalyzes oxidation of squalene to form **2,3-oxidosqualene** (Fig. 23-44). **Squalene oxidocyclase** converts this epoxide to **lanosterol,** the sterol precursor of cholesterol. The reaction is a complex process involving cyclization of 2,3-oxidosqualene to a **protosterol** cation and rearrangement of this cation to lanosterol by a series of 1,2 hydride and methyl shifts (Fig. 23-45).

Cholesterol Is Synthesized from Lanosterol

Conversion of lanosterol to cholesterol (Fig. 23-46) is a 19-step process that we shall not explore in detail. It involves an oxidation and loss of three methyl groups. The first methyl group is removed as formate while the other two are eliminated as CO_2 in reactions that all require NADPH and O_2. The enzymes involved in this process are embedded in the endoplasmic reticulum membrane.

Cholesterol Is Transported in the Blood and Taken Up by Cells in Lipoprotein Complexes

Transport and cellular uptake of cholesterol is described in Section 11-4. To recapitulate, cholesterol synthesized by the liver is either converted to bile acids for use in the digestive process (Section 23-1) or esterified by **acyl-CoA:cholesterol acyl transferase (ACAT)** to form **cholesteryl esters**

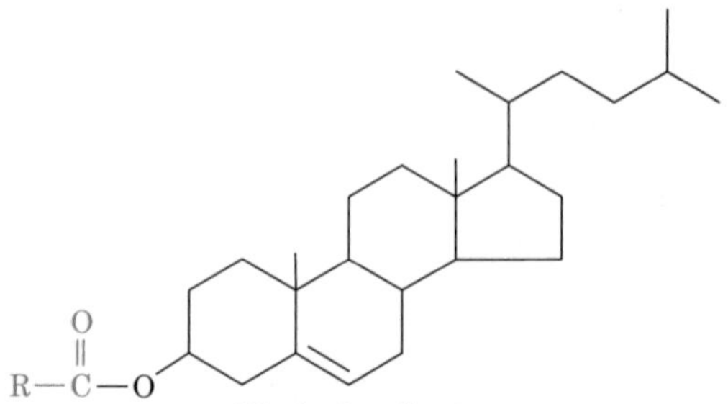

Cholesteryl ester

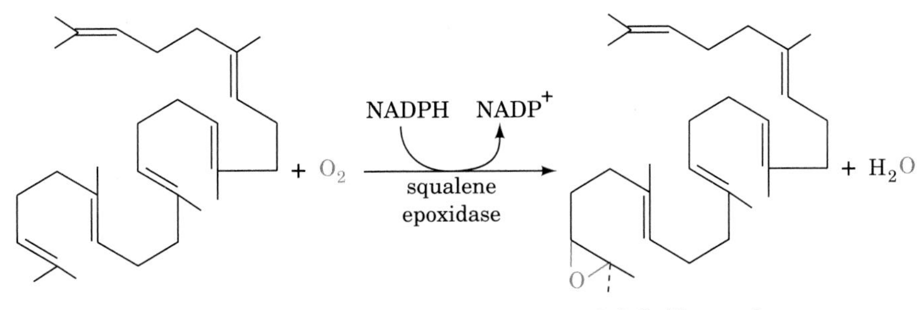

Figure 23-44
The squalene epoxidase reaction.

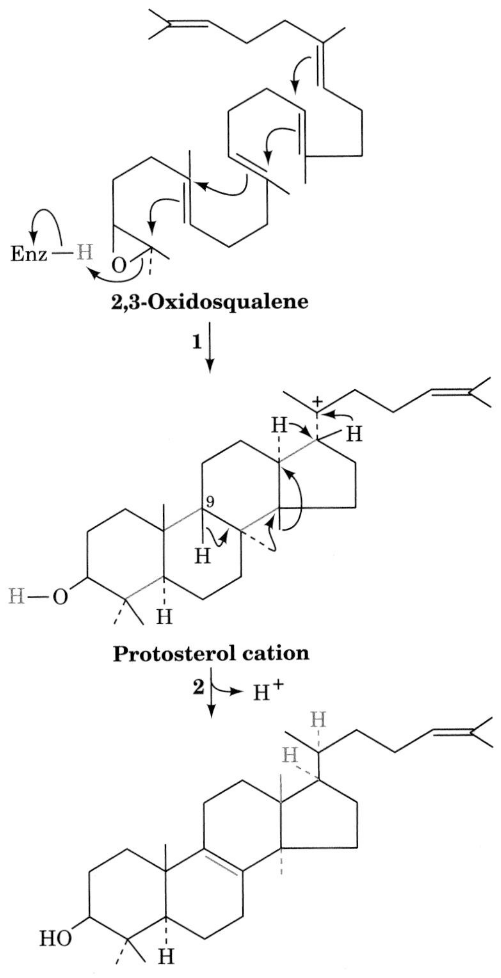

Figure 23-45
The squalene oxidocyclase reaction: **(1)** 2,3-Oxidosqualene is cyclized to the protosterol cation in a process that is initiated by the enzyme-mediated protonation of the squalene epoxide oxygen while this extended molecule is folded in the manner predicted by Bloch and Woodward. The opening of the epoxide leaves an electron-deficient center whose migration drives the series of cyclizations that form the protosterol cation. **(2)** The elimination of a proton from C(9) of the sterol to form a double bond accompanied by a series of methyl and hydride migrations yields neutral lanosterol.

which are secreted into the bloodstream as part of the lipoprotein complexes called **very low density lipoproteins (VLDL)**. As the VLDL circulate, their component triacylglycerols and most types of their **apolipoproteins** (Table 11-6) are removed in the capillaries of muscle and adipose tissues, sequentially converting the VLDL to **intermediate density lipoproteins (IDL)** and then to **low density lipoproteins (LDL)**. Peripheral tissues normally obtain most of their exogenous cholesterol from LDL by receptor-mediated endocytosis (Fig. 23-47; Section 11-4B). Inside the cell, cholesteryl esters are hydrolyzed by a lysosomal lipase to free cholesterol, which is either incorporated into cell membranes or reesterified by ACAT for storage as cholesteryl ester droplets.

Dietary cholesterol, cholesteryl esters, and triacylglycerols are transported in the blood by intestinally synthesized lipoprotein complexes called **chylomicrons**. After removal of their triacylglycerols at the peripheral tissues, the resulting **chylomicron remnants** bind to specific liver cell remnant receptors and are taken up by receptor-mediated endocytosis in a manner similar to that of LDL. In the liver, dietary cholesterol is either used in bile acid biosynthesis (Section 23-6C) or packaged into VLDL for export. *Liver and peripheral tissues therefore have two ways of obtaining cholesterol: They may either synthesize it from acetyl-CoA by the de novo pathway we have just discussed, or they may obtain it from the bloodstream by receptor-mediated endocytosis.* A small amount of cholesterol also enters cells by a nonreceptor mediated pathway.

Cholesterol actually circulates back and forth between the liver and peripheral tissues. While LDL transports cholesterol from the liver, cholesterol is transported back to the liver by **high density lipoproteins (HDL)**. Surplus cholesterol is disposed of by the liver as bile acids, thereby protecting the body from an overaccumulation of this water-insoluble substance.

B. Control of Cholesterol Biosynthesis and Transport

Cholesterol biosynthesis and transport must be tightly regulated. There are three ways in which the cellular cholesterol supply is maintained:

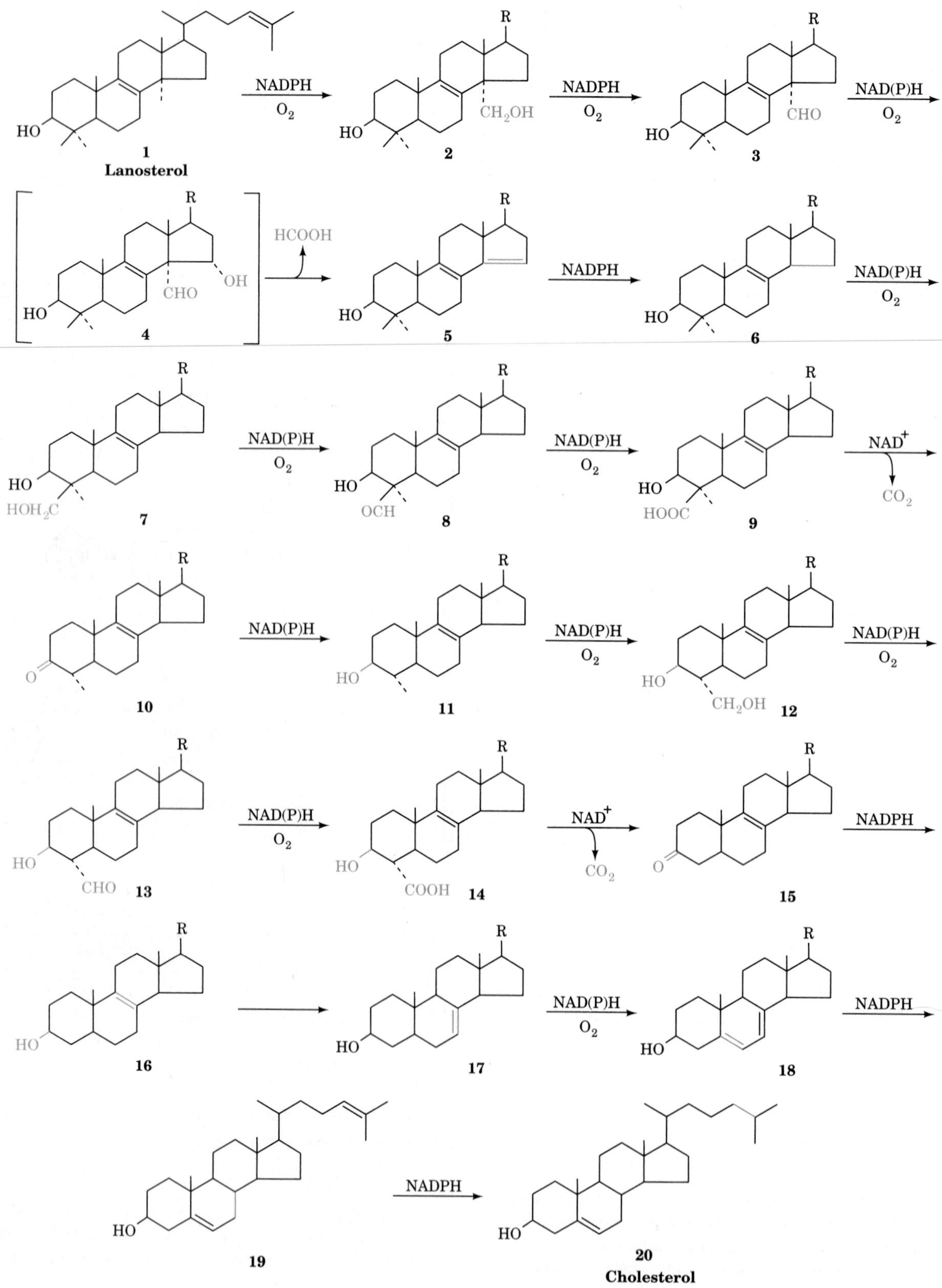

Figure 23-46
The 19-reaction conversion of lanosterol to cholesterol. [After Rilling, H. C. and Chayet, L. T., *in* Danielsson, H. and Sjövall, J. (Eds.), *Sterols and Bile Acids, p.* 33, Elsevier (1985).]

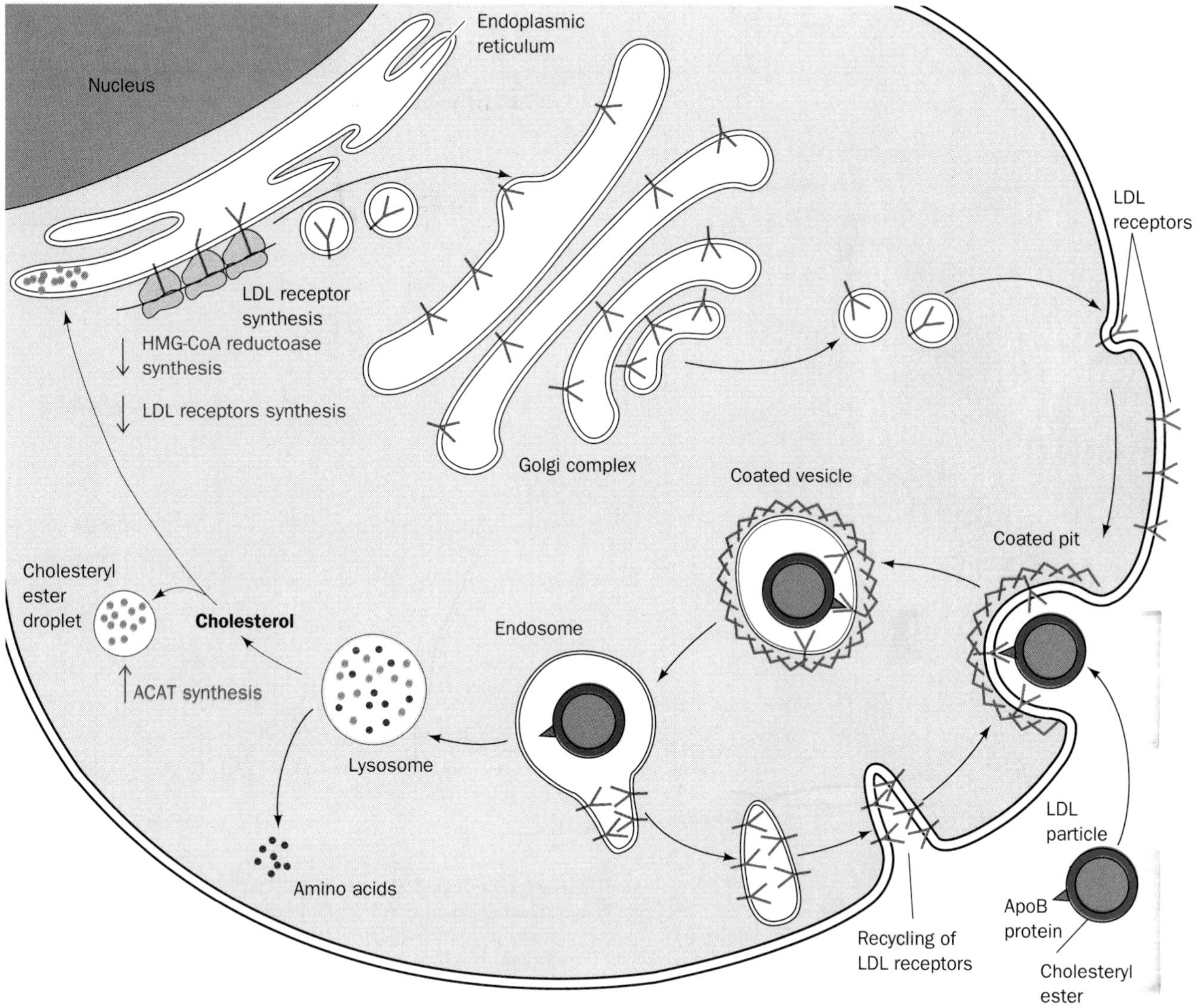

Figure 23-47
LDL receptor-mediated endocytosis in mammalian cells. LDL receptor is synthesized on the endoplasmic reticulum, processed in the Golgi complex, and inserted into the plasma membrane as a component of coated pits. LDL is specifically bound by the receptor on the coated pit and brought into the cell in endosomes that deliver LDL to lysosomes while recycling LDL receptor to the plasma membrane (Section 11-4B). Lysosomal degradation of LDL releases cholesterol, whose presence decreases the rate of synthesis of HMG-CoA reductase and LDL receptors (*down arrows*) while increasing that of acyl-CoA:cholesterol acyl transferase (ACAT; *up arrow*). [After Brown, M. S. and Goldstein, J. L., *Curr. Top. Cell. Regul.* **26,** 7 (1985).]

1. By regulating the activity of HMG-CoA reductase, the enzyme catalyzing the rate-limiting step in the *de novo* pathway. This is accomplished in two ways:
   (i) Short-term regulation of the enzyme's catalytic activity by (a) competitive inhibition (b) allosteric effects, and (c) covalent modification involving reversible phosphorylation.
   (ii) Long-term regulation of the enzyme's concentration by modulating its rates of synthesis and degradation.

2. By regulating the rate of LDL receptor synthesis. High intracellular concentrations of cholesterol sup-press LDL receptor synthesis, whereas low cholesterol concentrations stimulate it.

3. By regulating the rate of esterification and hence the removal of free cholesterol. ACAT, the enzyme that catalyzes intracellular cholesterol esterification, is regulated by reversible phosphorylation and by long-term control.

HMG-CoA Reductase Is the Primary Control Site for Cholesterol Biosynthesis

HMG-CoA reductase is the rate-limiting enzyme in cholesterol biosynthesis and, as therefore might be ex-

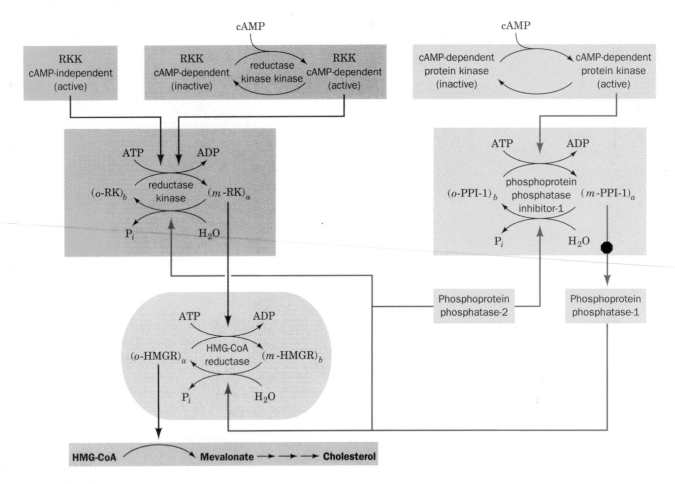

Figure 23-48
The modulation of HMG-CoA reductase activity by a bicyclic phosphorylation/dephosphorylation cascade. Colored arrows indicate catalysis of a reaction. Note that there are two types of phosphoprotein phosphatases: phosphoprotein phosphatase-1, which is inhibited by modified phosphoprotein phosphatase inhibitor-1; and phosphoprotein phosphatase-2, which is not so-inhibited.

pected, constitutes the pathway's main regulatory site. The pathway branches after this reaction, however (Fig. 23-36); ubiquinone, dolichol, and isopentenyl adenosine are also essential, albeit minor products. HMG-CoA is therefore subject to "multivalent" control in order to coordinate the synthesis of all of these products. Full suppression of the enzyme requires at least two regulators: cholesterol, normally derived from LDL, and a nonsterol product normally synthesized endogenously from mevalonate (next subsection). These regulators exert both short-term and long-term control over HMG-CoA reductase.

HMG-CoA Reductase Is Regulated by Reversible Phosphorylation

HMG-CoA reductase exists in interconvertible more active and less active forms as do glycogen phosphorylase (Section 17-3C), glycogen synthase (Section 17-3D), and pyruvate dehydrogenase (Section 19-2B), among others. The unmodified form of HMG-CoA reductase is more active; the phosphorylated form is less active. HMG-CoA reductase is phosphorylated (inactivated) in a bi-

cyclic cascade system by the covalently modifiable enzyme **HMG-CoA reductase kinase (RK)** (Fig. 23-48). The cascade is initiated by phosphorylation of unmodified (*o*riginal) and less active reductase kinase (*o*-RK)$_b$ to its *m*odified and more active form (*m*-RK)$_a$ by a cAMP-independent **reductase kinase kinase (RKK)** (this nomenclature is explained in Section 17-3C). In turn, (*m*-RK)$_a$ phosphorylates HMG-CoA reductase (HMGR), thereby converting the more active (*o*-HMGR)$_a$ to the less active (*m*-HMGR)$_b$. The reactivation of (*m*-HMGR)$_b$ and the inactivation of (*m*-RK)$_a$ are both catalyzed by **phosphoprotein phosphatase-1** (Fig. 23-48). The cAMP concentration also plays a role in this system due to the presence of a cAMP-dependent RKK in addition to the cAMP-independent RKK. Moreover, cAMP-dependent protein kinase phosphorylates, and thereby activates, **phosphoprotein phosphatase inhibitor-1** resulting in an increased phosphorylation level (decreased activity level) of HMG-CoA reductase (a process similar to the activation of glycogen phosphorylase; Section 17-3C).

The relationship between the phosphorylation state of

*HMG-CoA reductase and [cAMP] places HMG-CoA re-
ductase activity and thus cholesterol biosynthesis under
hormonal control.* Insulin, which decreases [cAMP],
stimulates cholesterol biosynthesis. Glucagon, which
increases [cAMP], inhibits it. These effects are consistent
with the actions of insulin and glucagon on other metabolic pathways in liver, such as glycolysis, glycogen synthesis and breakdown, gluconeogenesis, and fatty acid
biosynthesis and breakdown (Sections 17-3F and 23-5).
The modification/demodification system is also influenced by the presence of LDL–cholesterol and **mevalonolactone** (an internal ester of mevalonate that is
hydrolyzed to mevalonate and metabolized in the cell),

Mevalonolactone

which inhibit HMG-CoA reductase activity by stimulating its phosphorylation. Research is actively underway
to determine how this process occurs.

LDL Receptor Activity Controls Cholesterol Homeostasis

LDL receptors clearly play an important role in the
maintenance of plasma LDL–cholesterol levels. In normal individuals, about one half of the IDL formed from
the VLDL reenters the liver through LDL receptor–
mediated endocytosis (IDL and LDL both contain apolipoproteins that specifically bind to the LDL receptor;
Section 11-4B). The remaining IDL are converted to LDL
(Fig. 23-49a). *The serum concentration of LDL therefore
depends on the rate that liver removes IDL from the circulation which, in turn, depends on the number of functioning
LDL receptors on the liver cell surface.*

High blood cholesterol **(hypercholesterolemia),**
which results from the overproduction and/or underutilization of LDL, is known to be caused by two metabolic irregularities: (1) the genetic disease **familial hypercholesterolemia (FH);** or (2) the consumption of a

Figure 23-49
Liver LDL receptors control plasma LDL production and
uptake. (a) In normal human subjects, VLDL is secreted by
the liver and converted to IDL in the capillaries of the
peripheral tissues. About one half of the plasma IDL particles
bind to the LDL receptor and are taken up by the liver. The
remainder are converted to LDL at the peripheral tissues.
(b) In individuals with familial hypercholesterolemia (FH), liver
LDL receptors are diminished or eliminated because of a
genetic defect. (c) In normal individuals who ingest a high
cholesterol diet, the liver is filled with cholesterol, which
represses the rate of LDL receptor production. Receptor
deficiency, whether of genetic or dietary cause, raises the
plasma LDL level by increasing the rate of LDL production
and decreasing the rate of LDL uptake. [After Goldstein J. L.
and Brown, M. S., *J. Lipid Res.* **25,** 1457 (1984).]

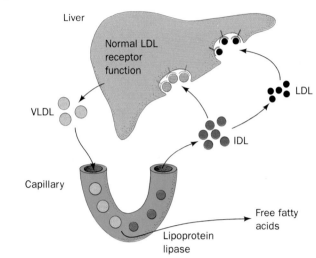

(a) Normal

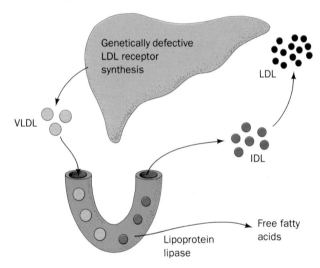

(b) Familial hypercholesterolemia

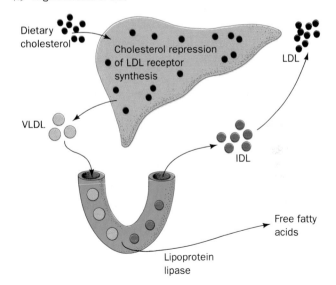

(c) High cholesterol diet

high cholesterol diet. FH is a dominant genetic defect that results in a deficiency of functional LDL receptors (Section 11-4C). Homozygotes for this disorder lack functional LDL receptors so that their cells can absorb neither IDL nor LDL by receptor–mediated endocytosis. The increased concentration of IDL in the bloodstream leads to a corresponding increase in LDL which is, of course, underutilized since it cannot be taken up by the cells (Fig. 23-49b). FH homozygotes therefore have plasma LDL–cholesterol levels three to five times higher than average. FH heterozygotes, which are far more common, have about one half of the normal number of functional LDL receptors and plasma LDL–cholesterol levels of about twice the average.

The ingestion of a high cholesterol diet has an effect similar, although not as extreme, as FH (Fig. 23-49c). Excessive dietary cholesterol enters the liver cells in chylomicron remnants and represses the synthesis of LDL–receptor protein. The resulting insufficiency of LDL receptors on the liver cell surface has consequences similar to those of FH.

LDL receptor deficiency, whether of genetic or dietary origin, raises the LDL level by two mechanisms: (1) increased LDL production resulting from decreased IDL uptake; and (2) decreased LDL uptake. Two strategies for reversing these conditions (besides maintaining a low cholesterol diet) have been tested:

1. *Ingestion of resins that bind bile acids thereby preventing their intestinal absorption.* Bile acids are normally efficiently recycled by the liver (Section 23-6C). Elimination of resin-bound cholesterol in the feces forces the liver to convert more cholesterol to bile acids than normal. The consequent decrease in the serum cholesterol concentration induces synthesis of LDL receptors (of course, not in FH homozygotes). Unfortunately, the decreased serum cholesterol level also induces the synthesis of HMG-CoA reductase, which increases the rate of cholesterol biosynthesis. Ingestion of bile acid-binding resins therefore provides only a 15 to 20% drop in serum cholesterol levels.

2. *Treatment with competitive inhibitors of HMG-CoA reductase, notably the fungal products* **compactin** *and* **lovastatin** *(also called* **mevinolin**; *Fig. 23-50), so as to decrease the rate of cholesterol biosynthesis.* Indeed, lovastatin has recently received clinical approval for the treatment of hypercholesterolemia. The resulting decreased cholesterol supply is again met by induction of LDL receptors and HMG-CoA reductase. Lovastatin-treated FH heterozygotes nevertheless routinely show a serum cholesterol decrease of 30%.

The combined use of these agents, moreover, results in a clinically dramatic 50 to 60% decrease in serum cholesterol levels.

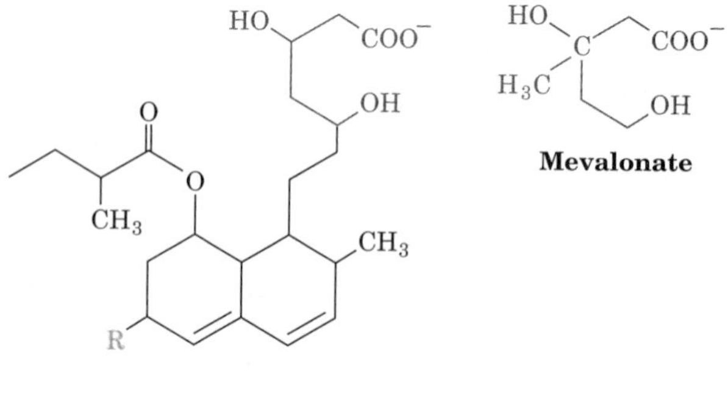

Figure 23-50
Compactin and lovastatin, two potent inhibitors of HMG-CoA reductase. The structure of mevalonate is shown for comparison.

C. Cholesterol Utilization

Cholesterol is the precursor of **steroid hormones** *and bile acids.* Steroid hormones are grouped into five categories: **progestins, glucocorticoids, mineralocorticoids, androgens,** *and* **estrogens.** These hormones, as described in Section 34-4A, mediate a wide variety of vital physiological functions. All contain the four-ring structure of the sterol nucleus and are remarkably similar in structure, considering the enormous differences in their physiological effects. A simplified biosynthetic scheme (Fig. 23-51) indicates their structural similarities and differences. We shall not discuss the details of these pathways.

The quantitatively most important pathway for the excretion of cholesterol in mammals is the formation of bile acids (also called **bile salts**). The major bile acids, **cholic acid** and **chenodeoxycholic acid,** are synthesized in the liver and secreted as glycine or **taurine** conjugates (Fig. 23-52) into the gallbladder. From there, they are secreted into the small intestine where they act as emulsifying agents in the digestion and absorption of fats and fat-soluble vitamins (Section 23-1). An efficient recycling system allows the bile acids to reenter the bloodstream and return to the liver for reuse several times each day. The <1 g $\cdot$ day^{-1} of bile acids that normally escape this recycling system are further metabolized by microorganisms in the large intestine and excreted. *This is the body's only route for cholesterol excretion.*

Comparison of the structures of cholesterol and the bile acids (Figs. 23-34 and 23-52) indicates that biosynthesis of bile acids from cholesterol involves (1) saturation of the 5,6-double bond, (2) epimerization of the 3β-OH group, (3) introduction of OH groups into the 7α and 12α positions, (4) oxidation of C(24) to a carboxylate, and (5) conjugation of this side chain carboxylate with glycine or taurine.

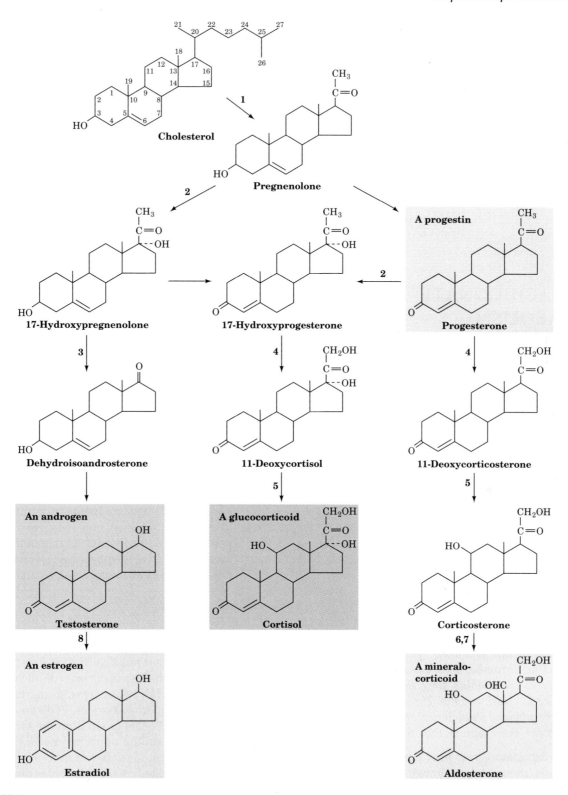

Figure 23-51
A simplified scheme of steroid biosynthesis. The enzymes involved are **(1)** the cholesterol side chain cleavage enzyme; **(2)** steriod C(17) hydroxylase; **(3)** steroid C(17),C(20) lyase; **(4)** steroid C(21) hydroxylase; **(5)** steroid 11β-hydroxylase; **(6)** steroid C(18) hydroxylase; **(7)** 18-hydroxysteroid oxidase; and **(8)** aromatase.

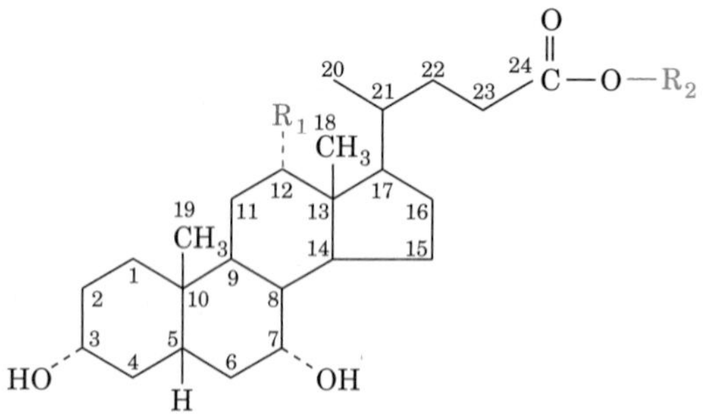

Figure 23-52
Structures of the major bile acids and their glycine and taurine conjugates.

	$R_1 = OH$	$R_1 = H$
$R_2 = H$	**Cholic acid**	**Chenodeoxycholic acid**
$R_2 = NH—CH_2—COOH$	**Glycocholic acid**	**Glycochenodeoxycholic acid**
$R_2 = NH—CH_2—CH_2—SO_3H$	**Taurocholic acid**	**Taurochenodeoxycholic acid**

7. ARACHIDONATE METABOLISM: PROSTAGLANDINS, PROSTACYCLINS, THROMBOXANES, AND LEUKOTRIENES

Prostaglandins (PG's) were first identified in human semen by Ulf von Euler in the early 1930s through their ability to stimulate uterine contractions and lower blood pressure. von Euler thought that these compounds originated in the prostate gland (hence their name) but they were later shown to be synthesized in the seminal vesicles. By the time the mistake was realized, the name was firmly entrenched. In the mid-1950s, crystalline materials were isolated from biological fluids and called PGE (ether-soluble) and PGF (phosphate buffer-soluble; *fosfat* in Swedish). This began an explosion of work on these potent substances.

Almost all mammalian cells except red blood cells produce prostaglandins and their related compounds, the prostacyclins, thromboxanes, and leukotrienes (known collectively as eicosanoids since they are all C$_{20}$ compounds; Greek: eikosi, twenty). The eicosanoids, like hormones, have profound physiological effects at extremely low concentrations. For example, they mediate:

1. The inflammatory response, notably as it involves the joints (rheumatoid arthritis), skin (psoriasis), and eyes.

2. The production of pain and fever.

3. The regulation of blood pressure.

4. The induction of blood clotting.

5. The control of several reproductive functions such as the induction of labor.

6. The regulation of the sleep/wake cycle.

The enzymes that synthesize these compounds and the receptors to which they bind are therefore the targets of intensive pharmacological research.

The eicosanoids are also hormonelike in that many of their effects are intracellularly mediated by cAMP. Unlike hormones, however, they are not transported in the bloodstream to their sites of action. Rather, these chemically and biologically unstable substances (some decompose within minutes or less *in vitro*) are local mediators; that is, *they act in the same environment in which they are synthesized.*

In this section, we discuss the structures of the eicosanoids and outline their biosynthetic pathways and modes of action. As we do so, note the great diversity of their structures and functions, a phenomenon that makes the elucidation of the physiological roles of these potent substances a challenging research area.

A. Background

*Prostaglandins are all derivatives of the hypothetical C$_{20}$ fatty acid **prostanoic acid** in which carbon atoms 8 to 12 comprise a cyclopentane ring* (Fig. 23-53a). Prostaglandins A through I differ in the substituents on the cyclopentane ring (Fig. 23-53b): PGAs are α, β-unsaturated ketones, PGEs are β-hydroxy ketones, PGFs are 1,3-diols, *etc.* In PGF$_\alpha$, the C(9)-OH group is on the same side of the ring as R$_1$ and on the opposite side in PGF$_\beta$. The numerical subscript in the name refers to the number of double bonds contained on the side chains of the cyclopentane ring (Fig. 23-53c).

*In humans, the most important prostaglandin precursor is **arachidonic acid (5,8,11,14-eicosatetraenoic acid)**, a C$_{20}$ polyunsaturated fatty acid that has four nonconjugated double bonds.* The double bond at C(14) is six carbon atoms from the terminal carbon atom (the ω-carbon atom), making arachidonic acid an ω − 6 fatty acid. Arachidonic acid is synthesized from linoleic acid (also

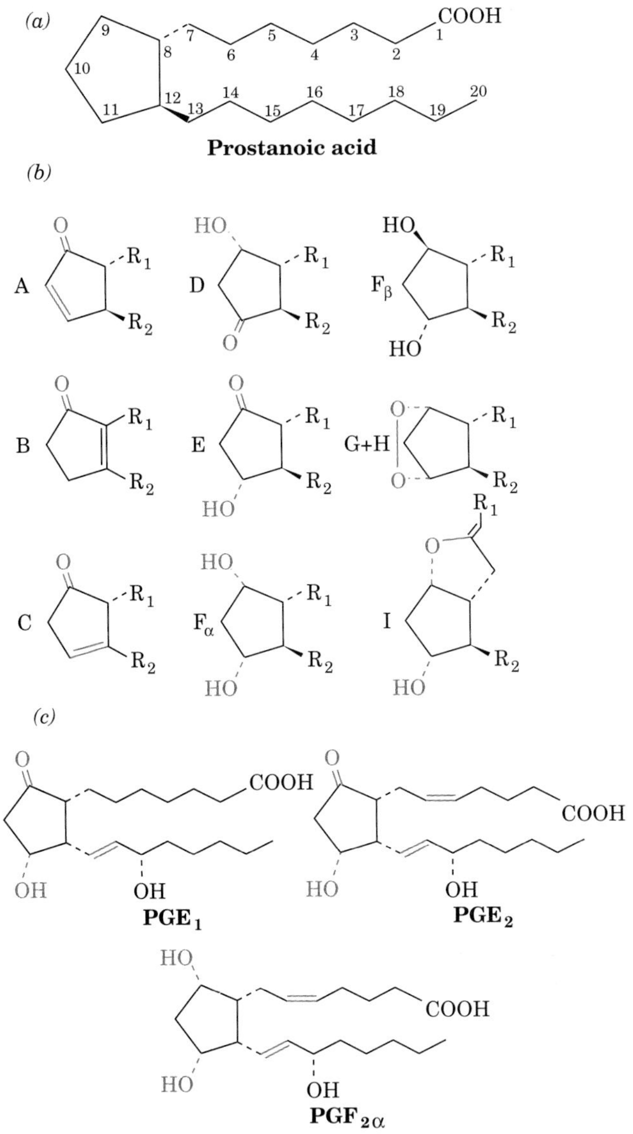

Figure 23-53
Prostaglandin structures: (*a*) the carbon skeleton of prostanoic acid, the prostaglandin parent compound. (*b*) Structures of prostaglandins A through I. (*c*) Structures of prostaglandins E₁, E₂, and F₂ₐ (the first prostaglandins to be identified).

an ω-6 fatty acid) by elongation and desaturation (Fig. 23-54; Section 23-4E). Prostaglandins with the subscript 1 (the "series-1" prostaglandins) are synthesized from **8,11,14-eicosatrienoic acid,** whereas "series-2" prostaglandins are synthesized from arachidonic acid. **α-Linolenic acid** is a precursor of **5,8,11,14,17-eicosapentaenoic acid (EPA)** and the "series-3" prostaglandins. Since arachidonate is the primary prostaglandin precursor in humans, we shall mostly refer to the series-2 prostaglandins in our examples.

Arachidonate Is Generated by Phospholipid Hydrolysis

Arachidonate is stored in cell membranes esterified at glycerol C(2) of phosphatidylinositol and other phospholipids. The production of arachidonate metabolites is controlled by the rate of arachidonate release from these phospholipids through three alternative pathways (Fig. 23-55):

1. **Phospholipase A₂** hydrolyzes acyl groups at C(2) of phospholipids (Fig. 23-55*b, left*).

Figure 23-54
The synthesis of prostaglandin precursors. The linoleic acid derivatives 8,11,14-eicosatrienoic acid and arachidonic acid are the respective precursors of the series-1 and series-2 prostaglandins. The *γ*-linolenic acid derivative 5,8,11,14,17-eicosapentaenoic acid is the series-3 prostaglandin precursor.

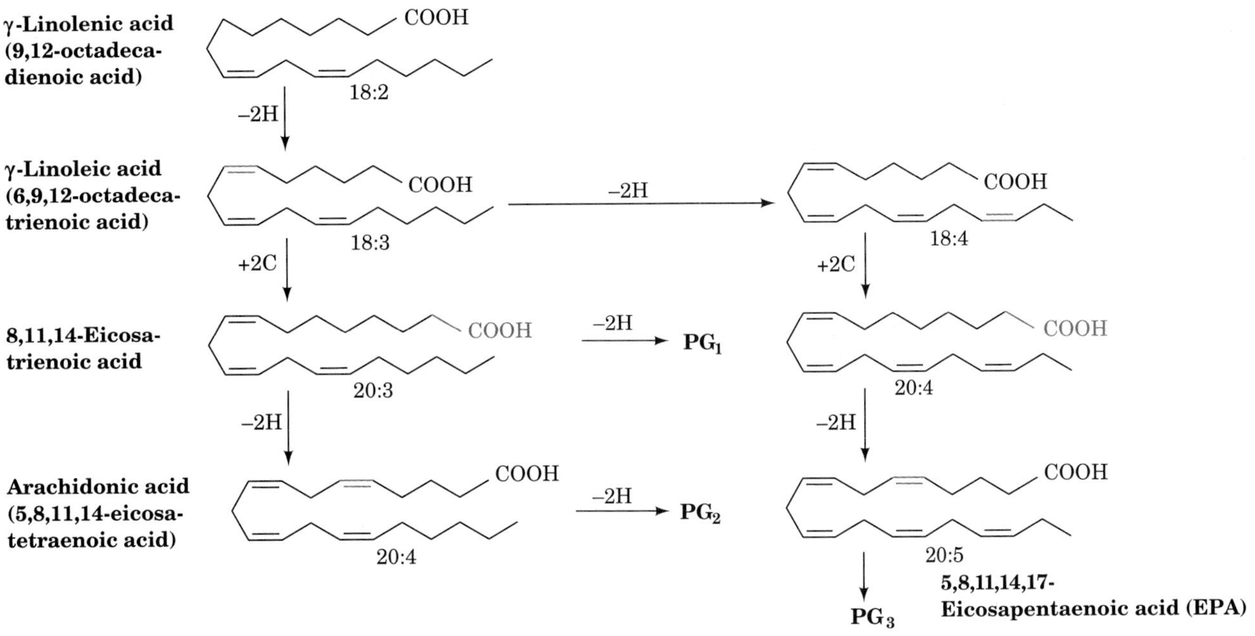

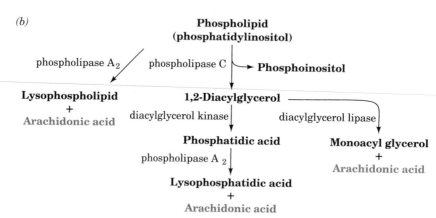

Figure 23-55
(*a*) The sites of hydrolytic cleavage mediated by phospholipases A$_2$ and C. The polar head group, X, is often inositol. (*b*) Pathways of arachidonic acid liberation from phospholipids.

2. **Phospholipase C** specifically hydrolyzes the phosphatidylinositol head group to yield a 1,2-diacylglycerol, which is phosphorylated by **diglycerol kinase** to phosphatidic acid, a phospholipase A$_2$ substrate (Fig. 23-55*b*, *center*).

3. The 1,2-diacylglycerol also may be hydrolyzed directly by **diacylglycerol lipase** (Fig. 23-55*b*, *right*).

Corticosteroids are used as anti-inflammatory agents because they inhibit phospholipase A$_2$, reducing the rate of arachidonate production.

Aspirin Inhibits Prostaglandin Synthesis

The use of **aspirin** as an analgesic (pain reliever), antipyretic (fever-reducing), and anti-inflammatory agent has been widespread since the nineteenth century. Yet, it was not until 1971 that John Vane discovered its mechanism of action. *Aspirin, as well as other "nonsteroidal anti-inflammatory drugs," inhibit the synthesis of prostaglandins from arachidonic acid (Section 23-7B).* These inhibitors have therefore proven to be valuable tools in the elucidation of prostaglandin biosynthesis pathways as well as providing a starting point for the rational synthesis of new anti-inflammatory drugs.

Arachidonic Acid Is a Precursor of Leukotrienes, Thromboxanes, and Prostacyclins

Arachidonic acid also serves as a precursor to compounds whose synthesis is not inhibited by aspirin. In

fact, there are two main pathways of arachidonate metabolism. The so-called "cyclic pathway," which is inhibited by "nonsteroidal anti-inflammatory drugs," forms prostaglandin's characteristic cyclopentane ring, whereas the so-called "linear pathway," which is not inhibited by these agents, leads to the formation of the **leukotrienes** and **HPETEs** (Fig. 23-56; Section 23-7C).

Studies using nonsteroidal anti-inflammatory drugs helped demonstrate that two structurally related and highly short-lived classes of compounds, the prostacyclins and the thromboxanes (Fig. 23-57), are also products of the cyclic pathway of arachidonic acid metabolism. The specific products produced by this branched pathway depend on the tissue involved. For example, blood platelets (thrombocytes) produce thromboxanes almost exclusively; vascular endothelial cells, which make up the walls of veins and arteries, predominantly synthesize the prostacyclins; and heart muscle makes PGI$_2$, PGE$_2$, and PGF$_{2\alpha}$ in more or less equal quantities. In the remainder of this section, we study the cyclic and the linear pathways of arachidonate metabolism.

B. The Cyclic Pathway of Arachidonate Metabolism: Prostaglandins, Prostacyclins, and Thromboxanes

The first step in the cyclic pathway of arachidonic acid metabolism is catalyzed by **prostaglandin endoperox-**

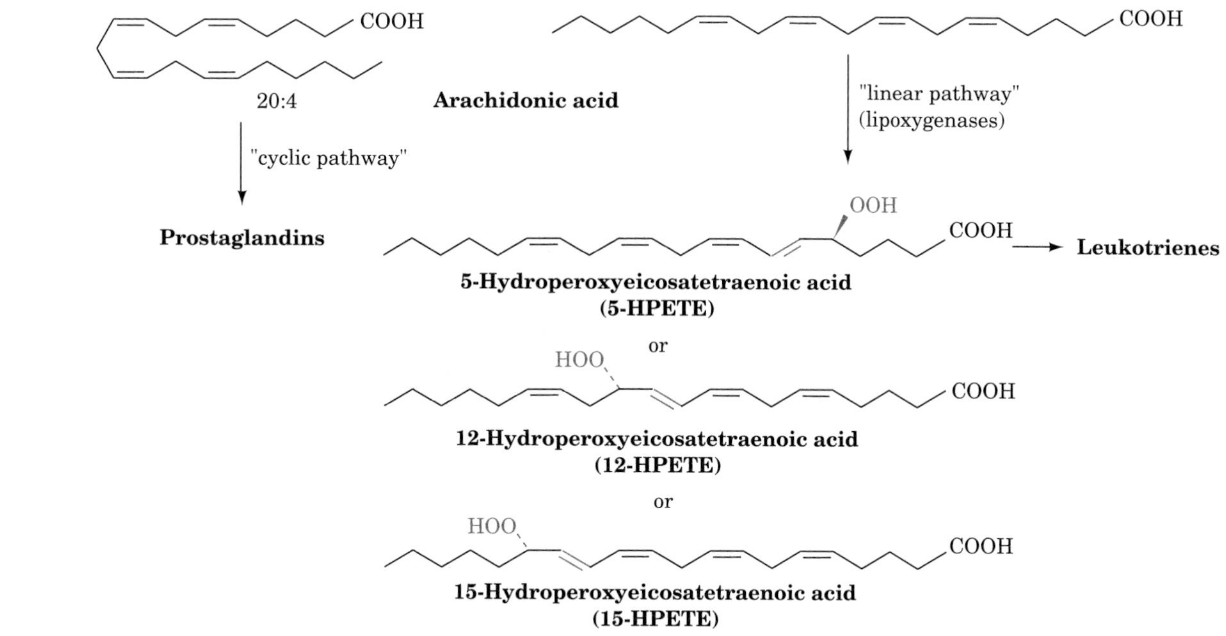

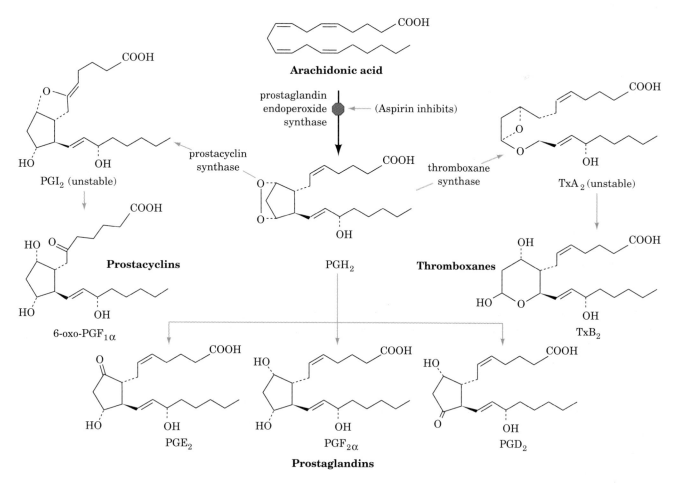

Figure 23-56
The cyclic and linear pathways of arachidonic acid metabolism.

Figure 23-57
The cyclic pathway of arachidonic acid metabolism is branched,
leading to prostaglandins, prostacyclins, and thromboxanes.

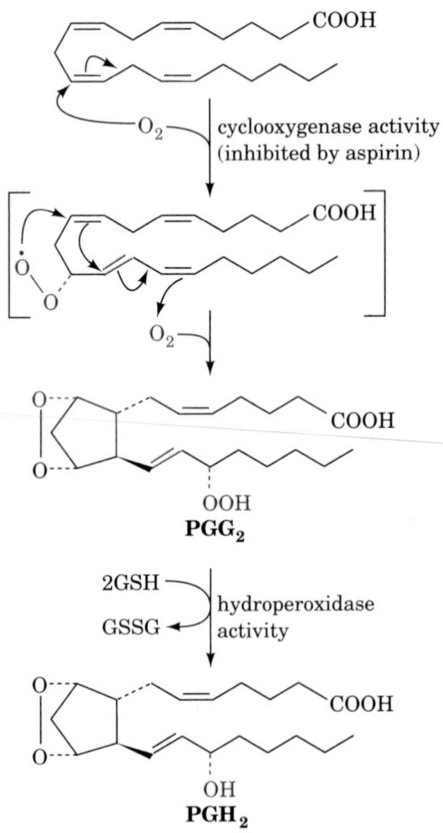

Figure 23-58
The reactions catalyzed by prostaglandin endoperoxide synthase. The enzyme, which is inhibited by aspirin, contains two activities: a cyclooxygenase and a glutathione-dependent hydroperoxidase.

ide synthase (Fig. 23-58). This enzyme contains two catalytic activities: a cyclooxygenase activity and a hydroperoxidase activity. The former catalyzes the addition of two molecules of O_2 to arachidonic acid, forming PGG_2. The latter mediates a glutathione-dependent reaction that converts the hydroperoxy function of PGG_2 to an OH group (PGH_2). *PGH_2 is the immediate precursor of all series-2 prostaglandins, prostacyclins, and thromboxanes (Fig. 23-57).*

The fate of PGH_2 depends on the relative activities of the enzymes catalyzing the specific interconversions (Fig. 23-57). Platelets contain **thromboxane synthase,** which mediates formation of **thromboxane A_2 (TxA_2),** a vasoconstrictor and stimulator of platelet aggregation (an initial step in blood clotting; Section 34-1). Vascular endothelial cells contain **prostacyclin synthase,** which catalyzes the synthesis of **prostacyclin I_2 (PGI_2),** a vasodilator and inhibitor of platelet aggregation. These two substances act in opposition, maintaining a balance in the cardiovascular system.

Nonsteroidal anti-inflammatory drugs (Figs. 23-59 and 23-60) inhibit the synthesis of the prostaglandins, prostacyclins, and thromboxanes by inhibiting or inactivating the cyclooxygenase activity of prostaglandin

endoperoxide synthase. Aspirin **(acetylsalicylic acid),** for example, acetylates this enzyme: If [^{14}C-acetyl]-acetylsalicylic acid is incubated with the enzyme, radioactivity becomes irreversibly associated with the inactive enzyme (Fig. 23-59). Other nonsteroidal anti-inflammatory drugs are thought to act either by competing with arachidonic acid at the enzyme's active site, or by assuming a conformation resembling the proposed peroxy radical intermediate in the reaction (Fig. 23-58).

Low doses of aspirin, say one tablet every 2 days, have been reported to significantly reduce the long-term incidence of heart attacks and strokes. Such low doses selectively inhibit platelet aggregation and thus blood clot formation because these enucleated cells cannot resynthesize their inactivated enzymes as can most other tissues.

C. The Linear Pathway of Arachidonate Metabolism: Leukotrienes

Leukotrienes are synthesized by a variety of white blood cells, mast cells (connective tissue cells derived from the blood-forming tissues that secrete substances that mediate inflammatory and allergic reactions) as well as lung, spleen, brain, and heart. **Peptidoleukotrienes** are now recognized to be the components of the **slow reacting substances of anaphylaxis (SRS-A;** anaphalaxis is a violent and potentially fatal allergic reaction) released from sensitized lung after immunological challenge. These substances act at very low concentrations (as little as $10^{-10}M$) to contract vascular, respiratory, and intestinal smooth muscle. Peptidoleukotrienes, for example, are ~10,000-fold more potent than histamine, a noted stimulant of allergic reactions. In the

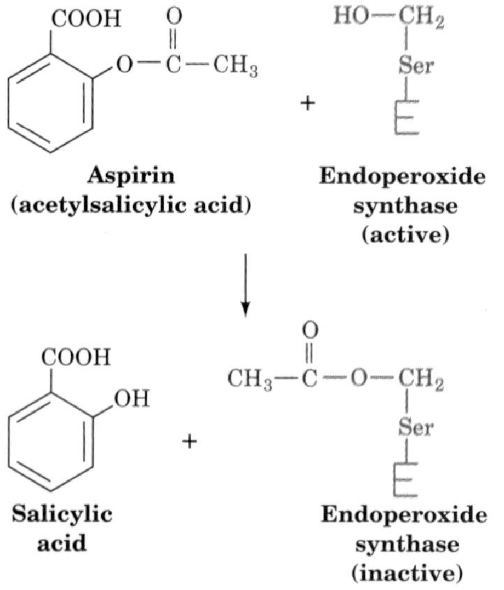

Figure 23-59
Aspirin acetylates a Ser residue of prostaglandin endoperoxide synthase thereby blocking the enzyme's cyclooxygenase activity.

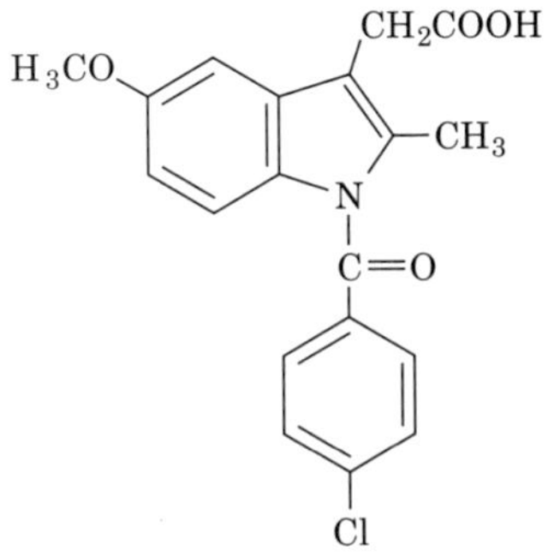

CH₂COOH
CH_2COOH

Indomethacin

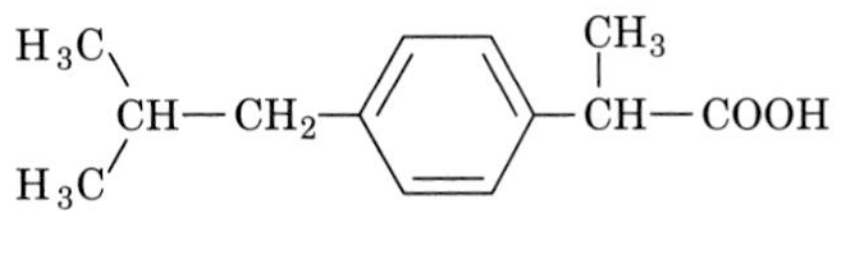

Ibuprofen

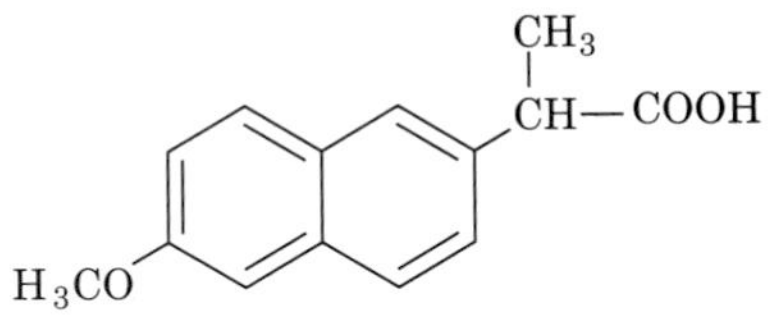

Naproxen

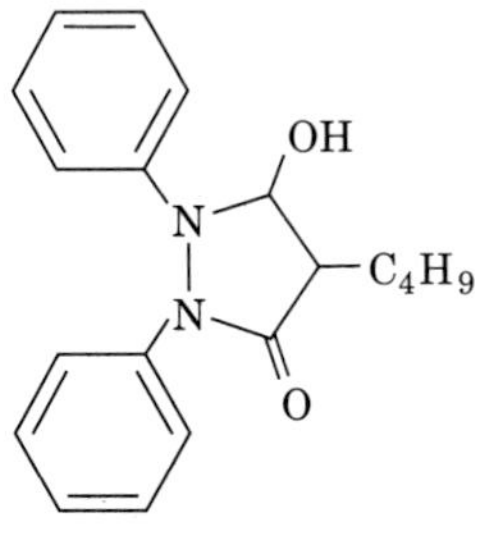

Phenylbutazone

Figure 23-60
Some nonsteroidal anti-inflammatory drugs.

respiratory system, they constrict bronchi, especially the smaller airways, increase mucus secretion, and are thought to be the mediators in asthma. They are also implicated in immediate hypersensitivity (allergic) reactions, inflammatory reactions, and heart attacks.

The first reaction in conversion of arachidonate to leukotrienes is the **lipoxygenase**-catalyzed oxidation at its 5, 12, or 15 position to form **hydroperoxyeicosatetraenoic acids (HPETEs;** Fig. 23-56), substances that, in themselves, are not physiological mediators. The reaction is thought to be a hydrogen atom abstraction from a 1,4-pentadiene unit followed by addition of O_2 and then the readdition of the hydrogen atom (Fig. 23-61). Different types of cells contain lipoxygenases with characteristic specificities.

5-HPETE, the product of the **5-lipoxygenase**-catalyzed oxidation of arachidonic acid (Fig. 23-56), is converted to peptidoleukotrienes by first forming an unstable epoxide, **leukotriene A₄ (LTA₄** Fig. 23-62; the subscript indicates the number of carbon–carbon double bonds in the molecule as well as the series to which the leukotriene belongs). **Glutathione-S-transferase** then catalyzes the addition of the glutathione sulfhydryl group to the epoxide, forming the first of the peptidoleukotrienes, **leukotriene C₄(LTC₄).** *γ***-Glutamyltransferase** removes glutamic acid, converting LTC₄ to **leukotriene D₄(LTD₄).** LTD₄ is converted to **leukotriene E₄ (LTE₄)** by a dipeptidase that removes glycine. LTA₄ can also be converted to **leukotriene B₄ (LTB₄),** a potent chemotactic agent (a substance that attracts motile cells) involved in attracting certain types of white blood cells to fight infection.

Diets Rich in Marine Lipids May Decrease Cholesterol, Prostaglandin, and Leukotriene Levels

Greenland Eskimos have a very low incidence of coronary heart disease and thrombosis despite their high dietary intake of cholesterol and fat. Their consumption of marine animals provides them with a higher proportion of unsaturated fats than the typical American diet.

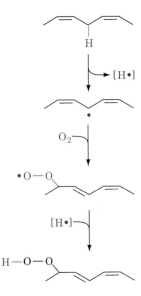

Figure 23-61
The lipoxygenase-catalyzed oxidation of arachidonate (Fig. 23-56) involves hydrogen abstraction from a 1,4-pentadiene unit, addition of O_2, and readdition of the hydrogen atom.

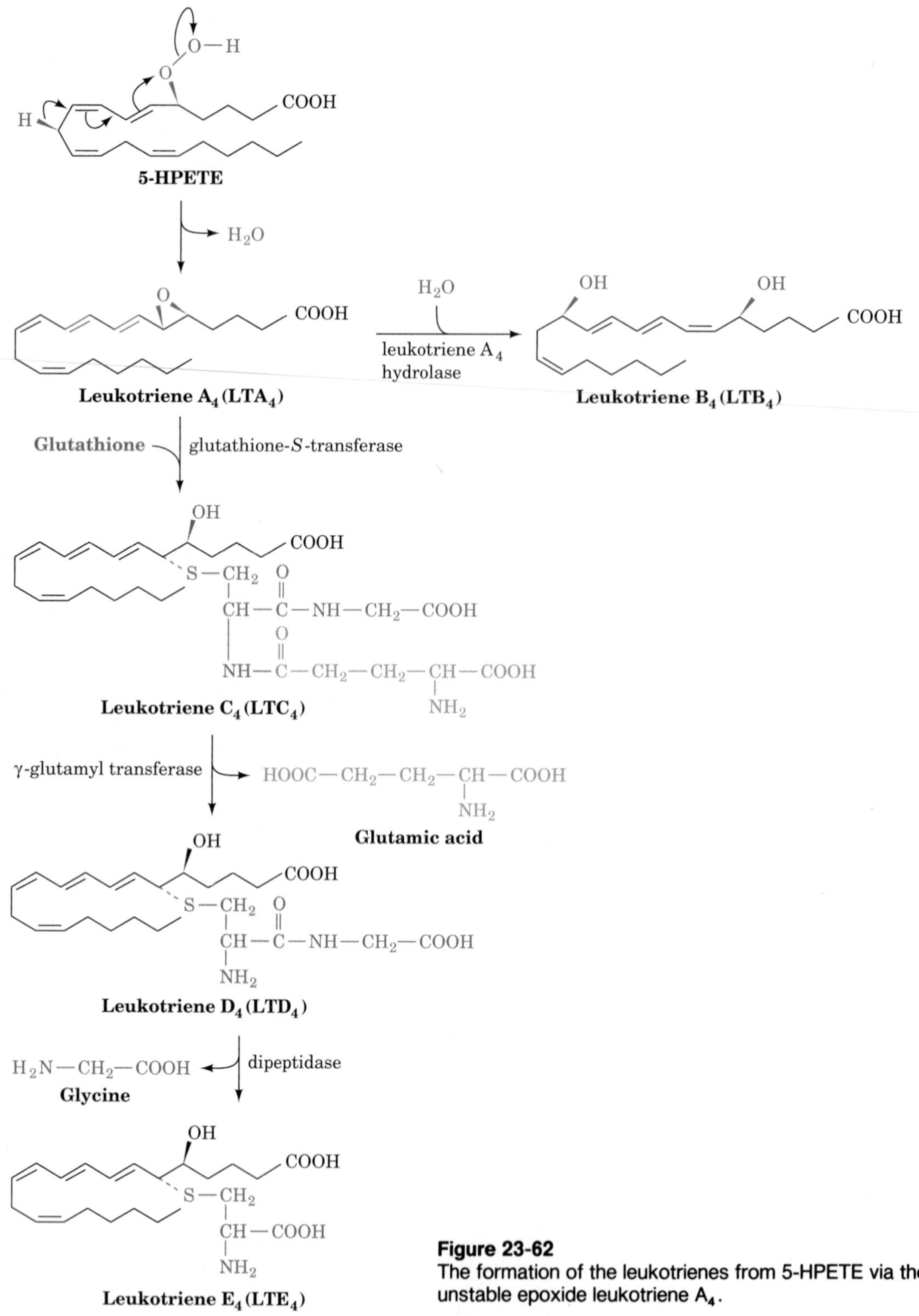

Figure 23-62
The formation of the leukotrienes from 5-HPETE via the unstable epoxide leukotriene A₄.

The major unsaturated component of marine lipids is 5,8,11,14,17-eicosapentaenoic acid (EPA; Fig. 23-54), an $\omega - 3$ fatty acid, rather than the arachidonic acid precursor linoleic acid, an $\omega - 6$ fatty acid. EPA inhibits formation of TxA_2 (Fig. 23-57) and is a precursor of the **series-5 leukotrienes,** compounds with substantially lower physiological activities than their arachidonate-derived (series-4) counterparts. This suggests that a diet containing marine lipids should decrease the extent of prostaglandin- and leukotriene-mediated inflammatory responses. Indeed, dietary enrichment with EPA inhibits the *in vitro* chemotactic and aggregating activities of neutrophils (a type of white blood cell). Moreover, an EPA-rich diet decreases the cholesterol and triacylglyc-

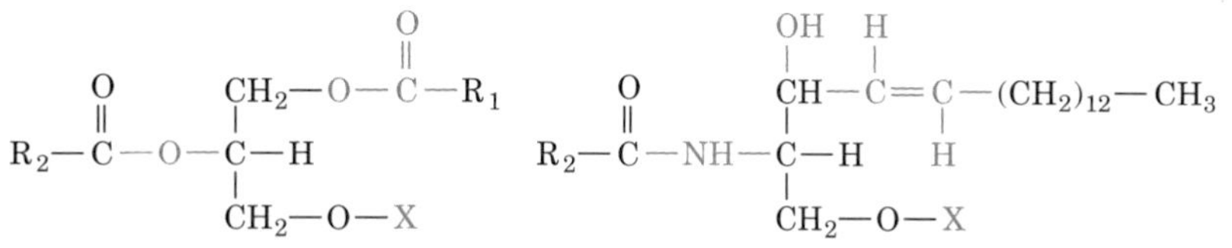

Glycerolipid **Sphingolipid**

X = H	**1,2-Diacyl-*sn*-glycerol**	***N*-Acylsphingosine (ceramide)**
X = Carbohydrate	**Glyceroglycolipid**	**Sphingoglycolipid**
X = Phosphate ester	**Glycerophospholipid**	**Sphingophospholipid**

Figure 23-63
The glycerolipids and sphingolipids. The structures of the common head groups, X, are presented in Table 11-2.

erol levels in the plasma of hypertriacylglycerolemic patients.

These are indeed exciting times in the study of arachidonate metabolism and its physiological manifestations. As the mechanisms of action of the prostaglandins, prostacyclins, thromboxanes, and leukotrienes are becoming better understood, they are providing the insights required for the development of new and improved therapeutic agents.

8. PHOSPHOLIPID AND GLYCOLIPID METABOLISM

The "complex lipids" are dual-tailed amphipathic molecules composed of either 1,2-diacyl-sn-glycerol, or N-acylsphingosine (ceramide) linked to a polar head group that is either a carbohydrate or a phosphate ester (Fig. 23-63; Sections 11-C and D; *sn* stands for stereospecific numbering, which assigns the 1-position to the group occupying the *pro-S* position of a prochiral center). Hence, there are two categories of phospholipids, **glycerophospholipids** and **sphingophospholipids,** and two categories of glycolipids, **glyceroglycolipids** and **sphingoglycolipids.** In this section we describe the biosynthesis of the complex lipids from their simplier components. We shall see that the great variety of these substances is matched by the numerous enzymes required for their specific syntheses. Note also that these substances are synthesized in membranes, mostly on the cytosolic face of the endoplasmic reticulum, and from there are transported to their final cellular destinations as indicated in Section 11-3F.

A. Glycerophospholipids

Glycerophospholipids have significant asymmetry in their C(1)- and C(2)-linked fatty acyl groups: C(1) substituents are mostly saturated fatty acids, whereas those at C(2) are by and large unsaturated fatty acids. We shall examine the major pathways of biosynthesis and metabolism of the glycerophospholipids with an eye toward understanding the origin of this asymmetry.

Biosynthesis of Diacylglycerophospholipids

The triacyglycerol precursor 1,2-diacyl-sn-glycerol is also the precursor of certain glycerophospholipids (Fig. 23-63). Activated phosphate esters of the polar head groups (Table 11-2) react with the C(3)-OH group of 1,2-diacyl-*sn*-glycerol to form the phospholipid's phosphodiester bond. The mechanism of activated phosphate ester formation is the same for both the polar head groups **ethanolamine** and **choline** (Fig. 23-64):

1. ATP first phosphorylates the OH group of choline or ethanolamine.
2. The phosphoryl group of the resulting **phosphoethanolamine** or **phosphocholine** then attacks CTP, displacing PP_i, to form the corresponding CDP derivatives, which are activated phosphate esters of the polar head group.
3. Finally, the C(3)-OH group of 1,2-diacyl-*sn*-glycerol attacks the phosphoryl group of the activated CDP–ethanolamine or CDP–choline, displacing CMP to yield the corresponding glycerophospholipid.

The liver also converts phosphatidylethanolamine to phosphatidylcholine by trimethylating its amino group, using ***S*-adenosylmethionine** (Section 24-3E) as the methyl donor.

Phosphatidylserine is synthesized from phosphatidylethanolamine by a head group exchange reaction catalyzed by **phosphatidylethanolamine:serine transferase** in which serine's OH group attacks the donor's phosphoryl group (Fig. 23-65). The original head group is then eliminated forming phosphatidylserine.

In the synthesis of **phosphatidylinositol** and **phosphatidylglycerol,** the hydrophobic tail is activated

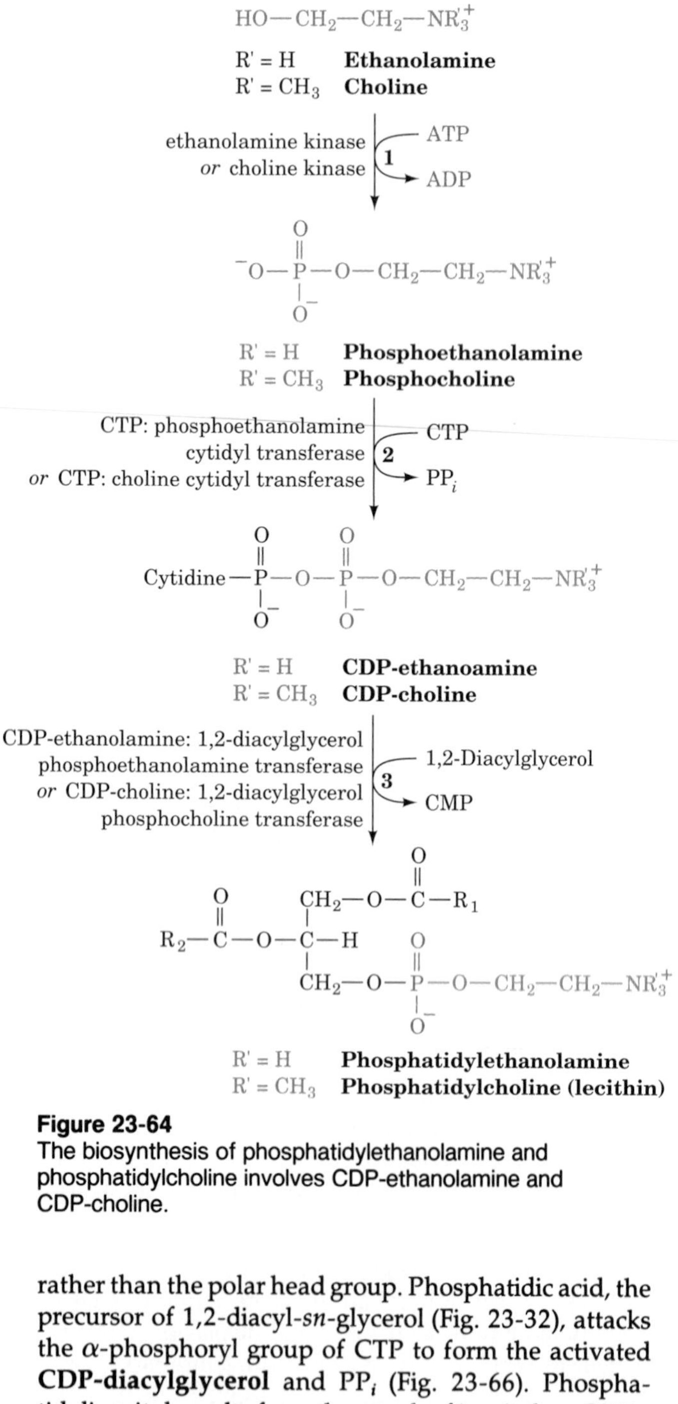

$$HO-CH_2-CH_2-NR_3'^+$$

R' = H   **Ethanolamine**
R' = CH$_3$ **Choline**

ethanolamine kinase
or choline kinase   **1**   ATP → ADP

$$^-O-\overset{\overset{\displaystyle O}{\|}}{\underset{\underset{\displaystyle O^-}{|}}{P}}-O-CH_2-CH_2-NR_3'^+$$

R' = H   **Phosphoethanolamine**
R' = CH$_3$ **Phosphocholine**

CTP: phosphoethanolamine
cytidyl transferase   **2**   CTP → PP$_i$
or CTP: choline cytidyl transferase

$$Cytidine-\overset{\overset{\displaystyle O}{\|}}{\underset{\underset{\displaystyle O^-}{|}}{P}}-O-\overset{\overset{\displaystyle O}{\|}}{\underset{\underset{\displaystyle O^-}{|}}{P}}-O-CH_2-CH_2-NR_3'^+$$

R' = H   **CDP-ethanoamine**
R' = CH$_3$ **CDP-choline**

CDP-ethanolamine: 1,2-diacylglycerol
phosphoethanolamine transferase   **3**   1,2-Diacylglycerol → CMP
or CDP-choline: 1,2-diacylglycerol
phosphocholine transferase

$$R_2-\overset{\overset{\displaystyle O}{\|}}{C}-O-\underset{\underset{\displaystyle CH_2-O-\overset{\overset{\displaystyle O}{\|}}{\underset{\underset{\displaystyle O^-}{|}}{P}}-O-CH_2-CH_2-NR_3'^+}{\overset{\overset{\displaystyle CH_2-O-\overset{\overset{\displaystyle O}{\|}}{C}-R_1}{|}}{C}}-H$$

R' = H   **Phosphatidylethanolamine**
R' = CH$_3$ **Phosphatidylcholine (lecithin)**

Figure 23-64
The biosynthesis of phosphatidylethanolamine and phosphatidylcholine involves CDP-ethanolamine and CDP-choline.

Enzymes that synthesize phosphatidic acid have a general preference for saturated fatty acids at C(1) and for unsaturated fatty acids at C(2). Yet, this general preference cannot account, for example, for the observations that ~80% of brain phosphatidylinositol has a stearoyl group (18 : 0) at C(1) and an arachidonoyl group (20 : 4) at C(2), and that ~40% of lung phosphatidylcholine has palmitoyl groups (16 : 0) at both positions (this latter substance is the major component of the surfactant that prevents the lung from collapsing when air is expelled; its deficiency is responsible for **respiratory distress syndrome** in premature infants). William Lands showed that *such side chain specificity results from "remodeling" reactions in which specific acyl groups of individual glycerophospholipids are exchanged by specific phospholipases and acyl transferases.*

Biosynthesis of Plasmalogens and Alkylacylglycerophospholipids

Eukaryotic membranes contain significant amounts of two other types of glycerophospholipids:

1. **Plasmalogens,** which contain a hydrocarbon chain linked to glycerol C(1) via a vinyl ether linkage.

2. **Alkylacylglycerophospholipids,** in which the alkyl substituent at glycerol C(1) is attached via an ether

rather than the polar head group. Phosphatidic acid, the precursor of 1,2-diacyl-*sn*-glycerol (Fig. 23-32), attacks the α-phosphoryl group of CTP to form the activated **CDP-diacylglycerol** and PP$_i$ (Fig. 23-66). Phosphatidylinositol results from the attack of inositol on CDP–diacylglycerol. Phosphatidylglycerol is formed in two reactions: (1) attack of the C(1)-OH group of *sn*-glycerol-3-phosphate on CDP-diacylglycerol yielding **phosphatidylglycerol phosphate;** and (2) hydrolysis of the phosphoryl group to form phosphatidylglycerol.

Cardiolipin, an important phospholipid first isolated from heart tissue, is synthesized from two molecules of phosphatidylglycerol (Fig. 23-67). The reaction occurs by the attack of the C(1)-OH group of one of the phosphatidylglycerol molecules on the phosphoryl group of the other, displacing a molecule of glycerol.

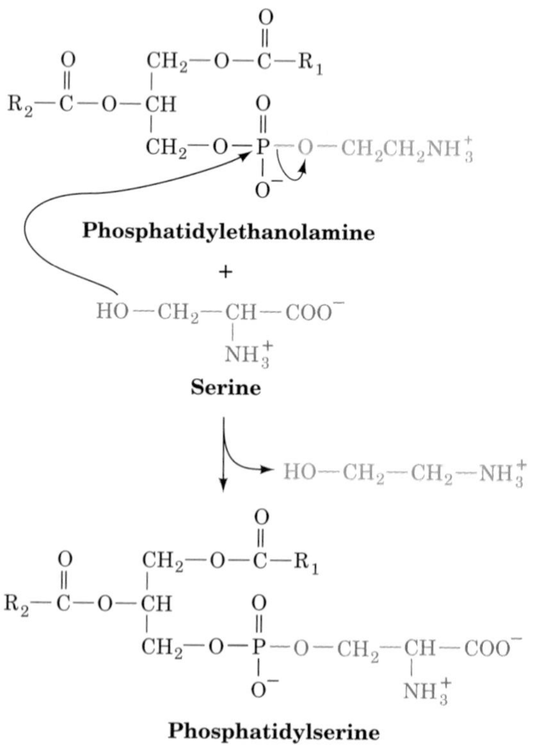

$$R_2-\overset{\overset{\displaystyle O}{\|}}{C}-O-CH$$
$$CH_2-O-\overset{\overset{\displaystyle O}{\|}}{C}-R_1$$
$$CH_2-O-\overset{\overset{\displaystyle O}{\|}}{\underset{\underset{\displaystyle O^-}{|}}{P}}-O-CH_2CH_2NH_3^+$$

Phosphatidylethanolamine

+

$$HO-CH_2-\underset{\underset{\displaystyle NH_3^+}{|}}{CH}-COO^-$$

Serine

→ $HO-CH_2-CH_2-NH_3^+$

$$R_2-\overset{\overset{\displaystyle O}{\|}}{C}-O-CH$$
$$CH_2-O-\overset{\overset{\displaystyle O}{\|}}{C}-R_1$$
$$CH_2-O-\overset{\overset{\displaystyle O}{\|}}{\underset{\underset{\displaystyle O^-}{|}}{P}}-O-CH_2-\underset{\underset{\displaystyle NH_3^+}{|}}{CH}-COO^-$$

Phosphatidylserine

Figure 23-65
Phosphatidylserine synthesis from phosphatidylethanolamine occurs by a head group exchange reaction.

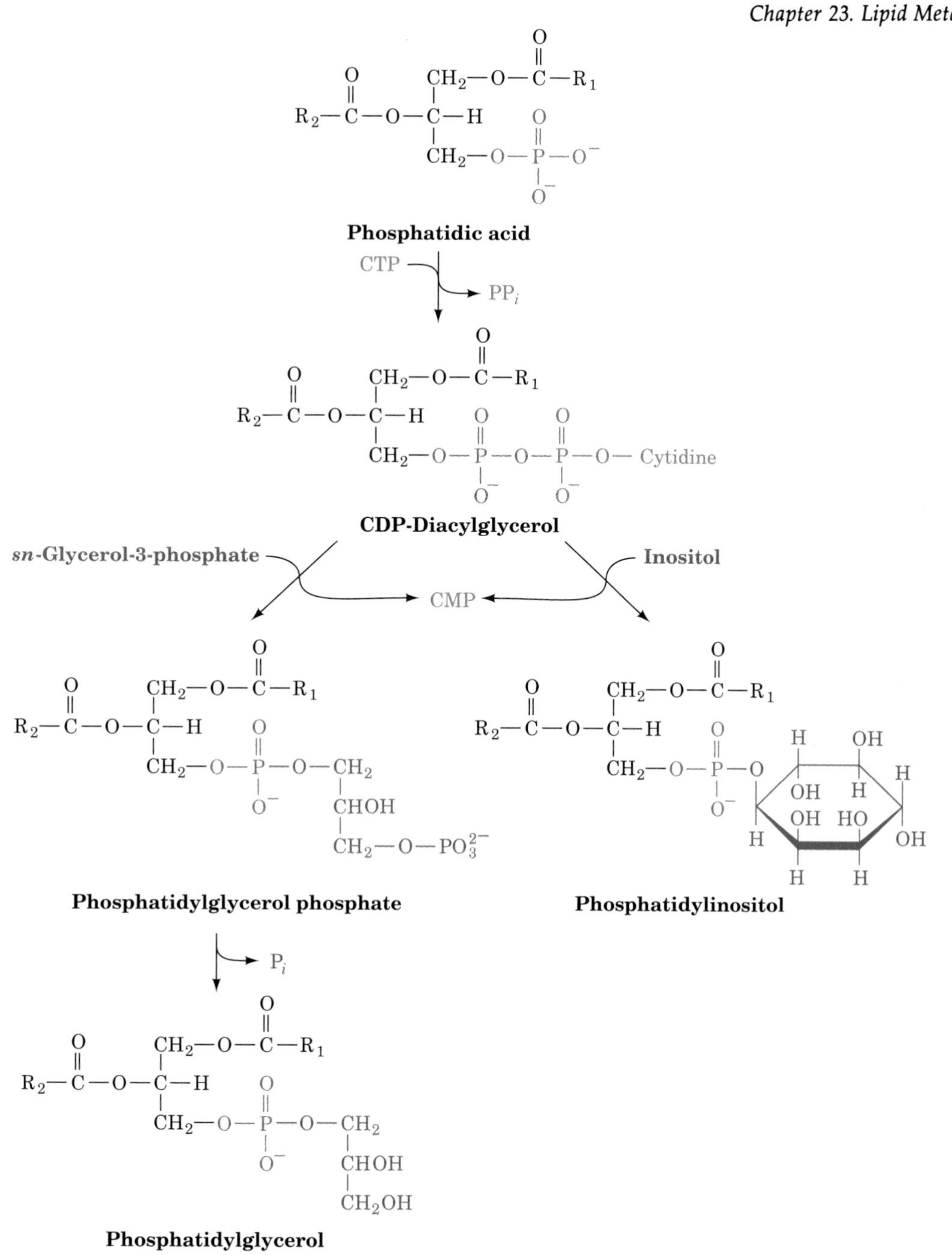

Phosphatidic acid

CDP-Diacylglycerol

sn-Glycerol-3-phosphate

Inositol

CMP

Phosphatidylglycerol phosphate

Phosphatidylinositol

P$_i$

Phosphatidylglycerol

Figure 23-66
The biosynthesis of phosphatidylinositol and phosphatidylglycerol involves a CDP–diacylglycerol intermediate.

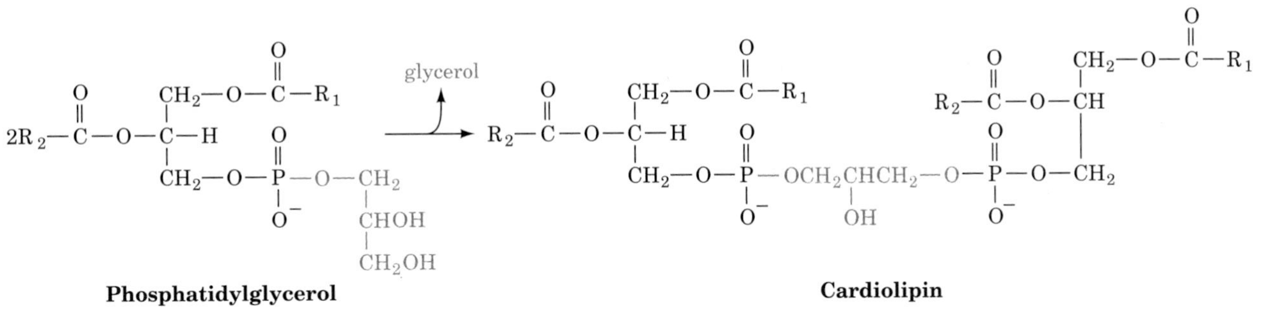

glycerol

Phosphatidylglycerol

Cardiolipin

Figure 23-67
The formation of cardiolipin.

linkage.

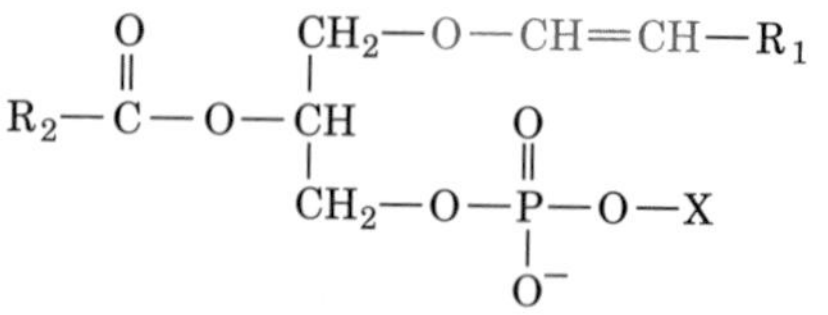

A plasmalogen

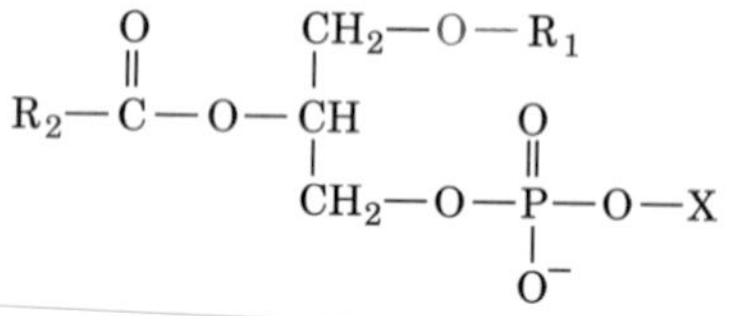

**An alkylacyl
glycerophospholipid**

About 20% of mammalian glycerophospholipids are plasmalogens. The exact percentage varies both from species to species and from tissue to tissue within a given organism. While plasmalogens comprise only 0.8% of the phospholipids in human liver, they account for 23% of those in human nervous tissue. The alkylacylglycer-

ophospholipids are less abundant than the plasmalogens; for instance, 59% of the ethanolamine glycerophospholipids of human heart are plasmalogens, whereas only 3.6% are alkylacylglycerophospholipids. However, in bovine erythrocytes, 75% of the ethanolamine glycerophospholipids are of the alkylacyl type.

The pathway forming ethanolamine plasmalogens and alkylacylglycerophospholipids involves several reactions (Fig. 23-68):

1. Exchange of the acyl group of **1-acyl-dihydroxyacetone phosphate** for an alcohol.

2. Reduction of the ketone to **1-aklyl-*sn*-glycerol-3-phosphate**.

3. Acylation of the resulting C(2)-OH group by acyl-CoA.

4. Hydrolysis of the phosphoryl group to yield an alkylacylglycerol.

5. Attack by the new OH group of alkylacylglycerol on CDP–ethanolamine to yield **1-alkyl-2-acyl-*sn*-gly-**

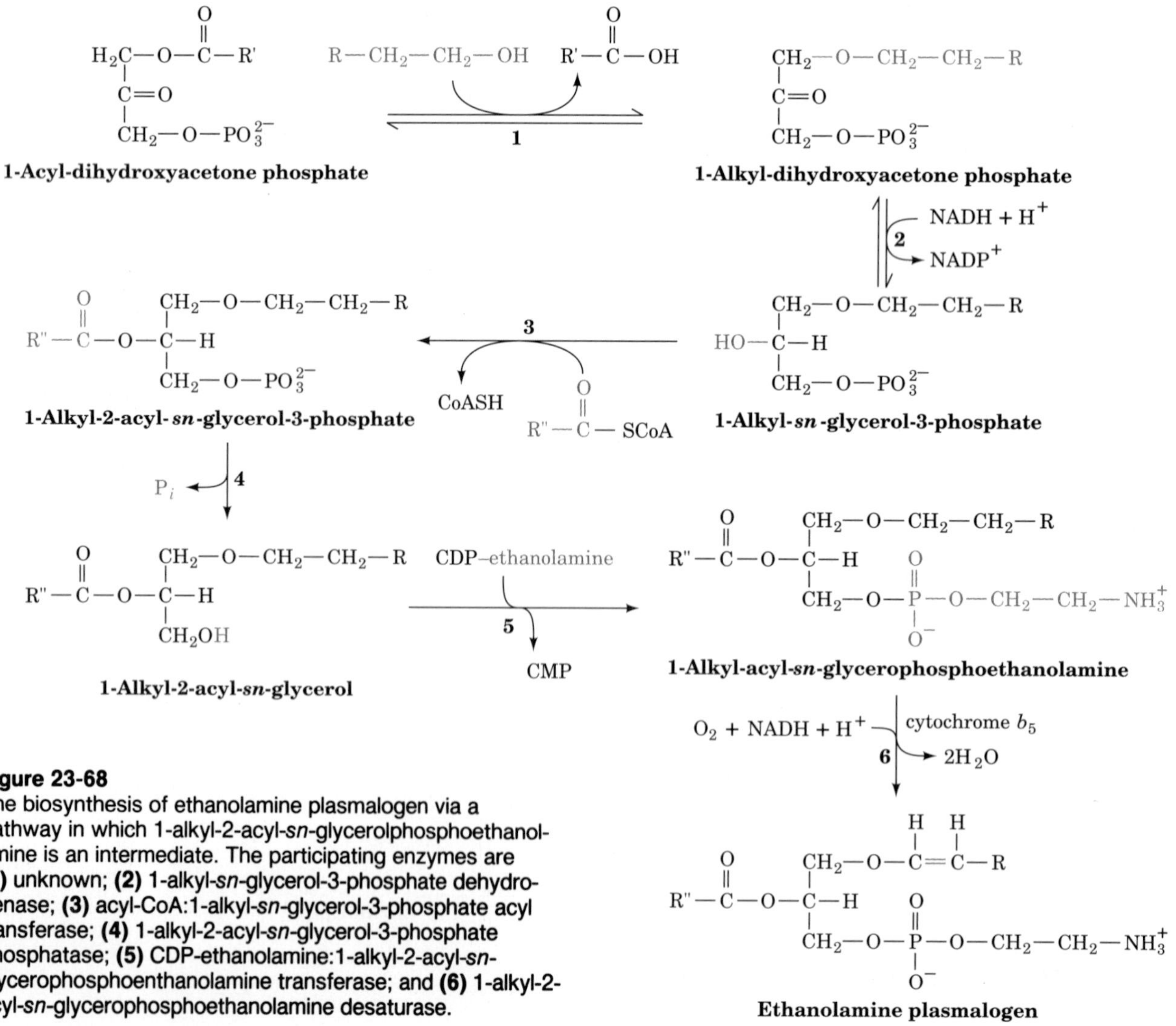

Figure 23-68
The biosynthesis of ethanolamine plasmalogen via a pathway in which 1-alkyl-2-acyl-*sn*-glycerolphosphoethanolamine is an intermediate. The participating enzymes are **(1)** unknown; **(2)** 1-alkyl-*sn*-glycerol-3-phosphate dehydrogenase; **(3)** acyl-CoA:1-alkyl-*sn*-glycerol-3-phosphate acyl transferase; **(4)** 1-alkyl-2-acyl-*sn*-glycerol-3-phosphate phosphatase; **(5)** CDP-ethanolamine:1-alkyl-2-acyl-*sn*-glycerophosphoenthanolamine transferase; and **(6)** 1-alkyl-2-acyl-*sn*-glycerophosphoethanolamine desaturase.

$$\underset{\text{Ceramide (}N\text{-Acylsphingosine)}}{R-\overset{\displaystyle O}{\overset{\|}{C}}-NH-\underset{\underset{\displaystyle CH_2OH}{|}}{\overset{\overset{\displaystyle OH \quad H}{|\qquad|}}{\underset{|}{\overset{\displaystyle CH-C=C-(CH_2)_{12}-CH_3}{\underset{\displaystyle H}{}}}}}-H} \qquad + \qquad \underset{\text{CDP-Choline}}{\text{Cytidine}-O-\overset{\displaystyle O}{\overset{\|}{\underset{\underset{\displaystyle O^-}{|}}{P}}}-O-\overset{\displaystyle O}{\overset{\|}{\underset{\underset{\displaystyle O^-}{|}}{P}}}-O-CH_2-CH_2-\overset{+}{N}(CH_3)_3}$$

CDP-choline: ceramide
choline phosphotransferase $\longrightarrow$ CMP

$$\underset{\text{Sphingomyelin}}{R-\overset{\displaystyle O}{\overset{\|}{C}}-NH-\overset{\overset{\displaystyle OH \quad H \quad H}{|\qquad|\quad|}}{\underset{\underset{\displaystyle CH_2-O-\overset{\overset{\displaystyle O}{\|}}{\underset{\underset{\displaystyle O^-}{|}}{P}}-O-CH_2-CH_2-\overset{+}{N}(CH_3)_3}{|}}{CH-C=C-(CH_2)_{12}-CH_3}}}-H}$$

Figure 23-69
The synthesis of sphingomyelin from
N-acylsphingosine and CDP-choline.

cerophosphoethanolamine.

6. Introduction of a double bond into the alkyl group to form the plasmalogen by a desaturase having the same cofactor requirements as the fatty acid desaturases (Section 23-4E).

Recall that the precursor–product relationship between the alkylacylglycerophospholipid and the plasmalogen was established through studies using [¹⁴C]ethanolamine (Section 15-3B).

B. Sphingophospholipids

Only one major phospholipid contains ceramide (*N*-acylsphingosine) as its hydrophobic tail: **sphingomyelin** (*N*-acylsphingosine phosphocholine; Section 11-D), an important structural lipid of nerve cell membranes. The molecule is synthesized from *N*-acylsphingosine and CDP-choline (Fig. 23-69). An alternate route of sphingomyelin synthesis occurs through donation of the phosphocholine group of phosphatidylcholine to *N*-acylsphingosine.

The most prevalent acyl groups of sphingomyelin are palmitoyl (16:0) and stearoyl (18:0) groups. Longer-chain fatty acids such as nervonic acid (24:1), and behenic acid (22:0) occur with lesser frequency in sphingomyelins.

C. Sphingoglycolipids

Most sphingolipids are sphingoglycolipids; that is, their polar head groups consist of carbohydrate units (Section 11-1D). The principal classes of sphingoglycolipids, as indicated in Fig. 23-70, are **cerebrosides** (cer-

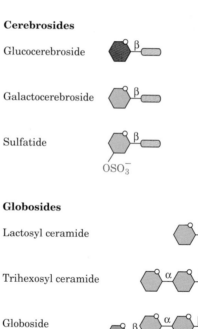

Cerebrosides

Glucocerebroside

Galactocerebroside

Sulfatide

Globosides

Lactosyl ceramide

Trihexosyl ceramide

Globoside

Gangliosides

G_{M3}

G_{M2}

G_{M1}

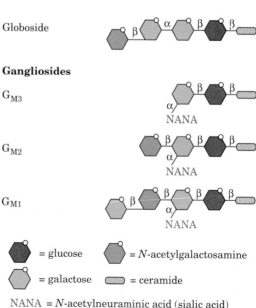

= glucose   = *N*-acetylgalactosamine

= galactose   = ceramide

NANA = *N*-acetylneuraminic acid (sialic acid)

Figure 23-70
Diagrammatic representation of the principal classes of sphingoglycolipids. The ganglioside structures are presented in greater detail in Fig. 11-7.

amide monosaccharides), **sulfatides** (ceramide monosaccharide sulfates), **globosides** (neutral ceramide oligosaccharides), and **gangliosides** (acidic, sialic acid-containing ceramide oligosaccharides). The carbohydrate unit is glycosidically attached to the *N*-acylsphingosine at its C(1)-OH group (Fig. 23-63). In the following subsections, we discuss the biosynthesis and breakdown of *N*-acylsphingosine and sphingoglycolipids and consider the diseases caused by deficiencies in their degradative enzymes.

Biosynthesis of Ceramide (*N*-Acylsphingosine)

Biosynthesis of *N*-acylsphingosine occurs in four reactions from the precursors palmitoyl-CoA and serine (Fig. 23-71):

1. **3-Ketosphinganine synthase,** a pyridoxal phosphate-dependent enzyme, catalyzes condensation of palmitoyl-CoA with serine yielding **3-ketosphinganine** (pyridoxal phosphate-dependent reactions are discussed in Section 24-1A).

2. **3-Ketosphinganine reductase** catalyzes the NADPH-dependent reduction of 3-ketosphinganine's keto group to form **sphinganine (dihydrosphingosine).**

3. **Dihydroceramide** is formed by transfer of an acyl group from an acyl-CoA to the sphinganine's 2-amino group forming an amide bond.

4. **Dihydroceramide reductase** converts dihydroceramide to ceramide by an FAD-dependent oxidation reaction.

Biosynthesis of Cerebrosides

Galactocerebroside (1-*β*-galactoceramide) and glucocerebroside (1-*β*-glucoceramide) are the two most common cerebrosides. In fact, the term cerebroside is often used synonymously with galactocerebroside. Both are synthesized from ceramide by addition of a glycosyl unit from the corresponding UDP-hexose (Fig. 23-72). Galactocerebroside is a common component of brain lipids. Glucocerebroside, although relatively uncommon, is the precursor of globosides and gangliosides.

Biosynthesis of Sulfatides

Sulfatides (galactocerebroside-3-sulfate) account for 15% of the lipids of white matter in the brain. They are formed by transfer of an "activated" sulfate group from 3′-phosphoadenosine-5′-phosphosulfate (PAPS) to the C(3)-OH group of galactose in galactocerebroside (Fig. 23-73).

Biosynthesis of Globosides and Gangliosides

Biosynthesis of both globosides (neutral ceramide oligosaccharides) and gangliosides (acidic, sialic acid-containing ceramide oligosaccharides) is catalyzed by a

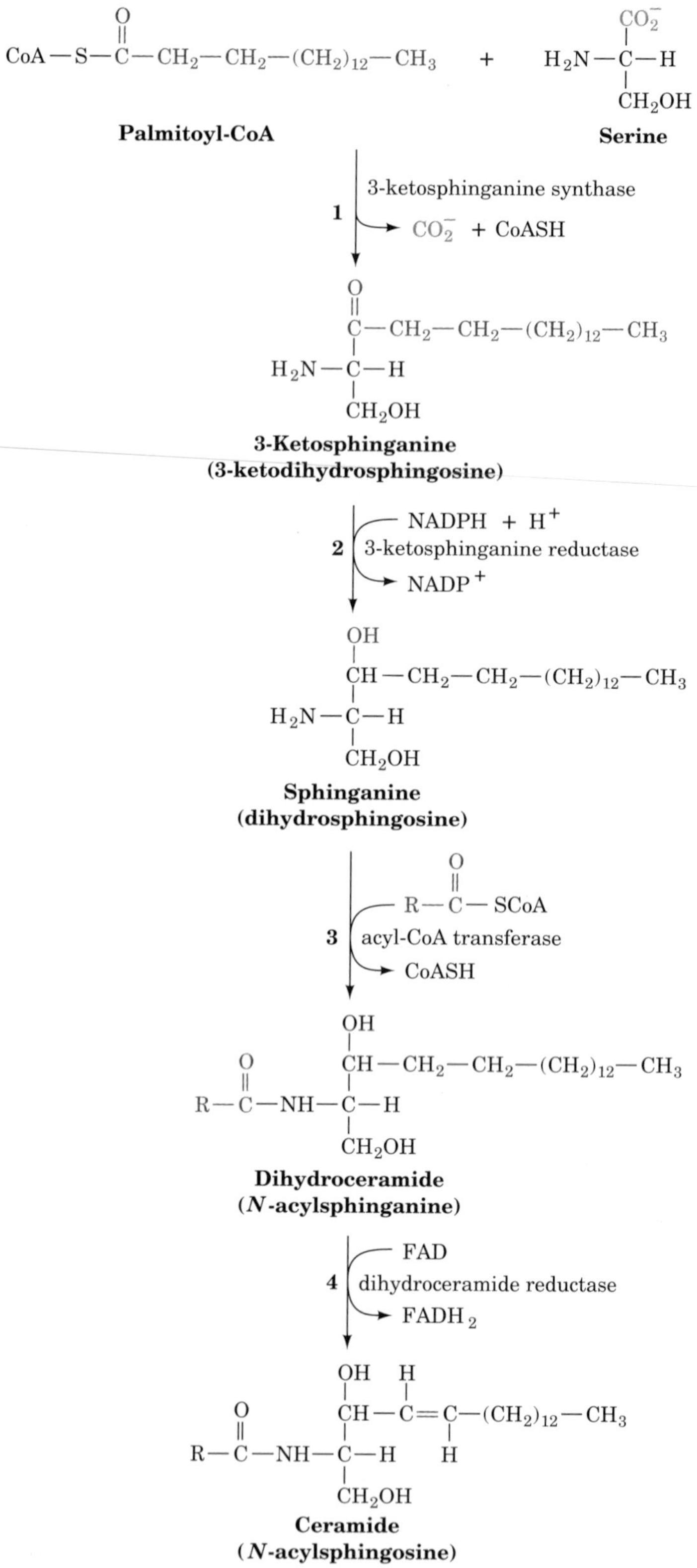

Figure 23-71
The biosynthesis of ceramide (*N*-acylsphingosine).

Glucocerebroside (1-β-D-glucoceramide)

Ceramide

Galactocerebroside (1-β-D-galactoceramide)

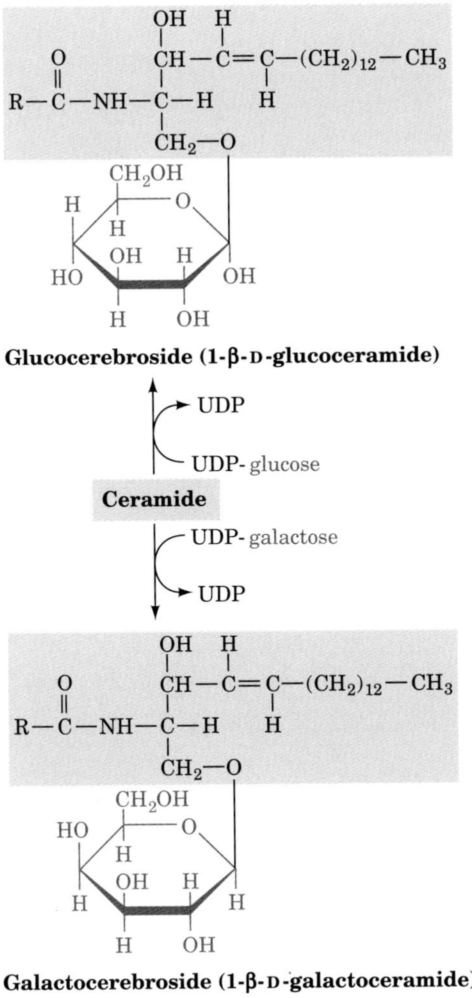

Figure 23-72
The biosynthesis of cerebrosides.

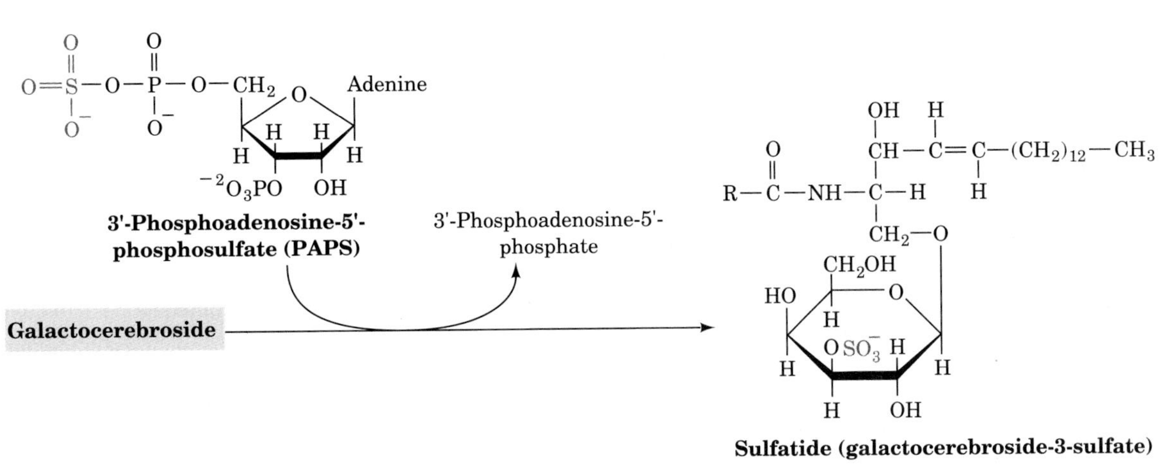

series of **glycosyl transferases.** While the reactions are chemically similar, each is catalyzed by a specific enzyme. The pathways begin with transfer of a galactosyl unit from UDP-Gal to glucocerebroside to form a $\beta(1 \rightarrow 4)$ linkage (Fig. 23-74). Since this bond is the same as that linking glucose and galactose in lactose, this glycolipid is often referred to as **lactosyl ceramide.** Lactosyl ceramide is the precursor of both globosides and gangliosides. To form a globoside, one galactosyl and one N-acetylgalactosyl unit are sequentially added to lactosyl ceramide from UDP-Gal and UDP-GalNAc, respectively. Gangliosides are formed by addition of **N-acetylneuraminic acid (NANA, sialic acid)**

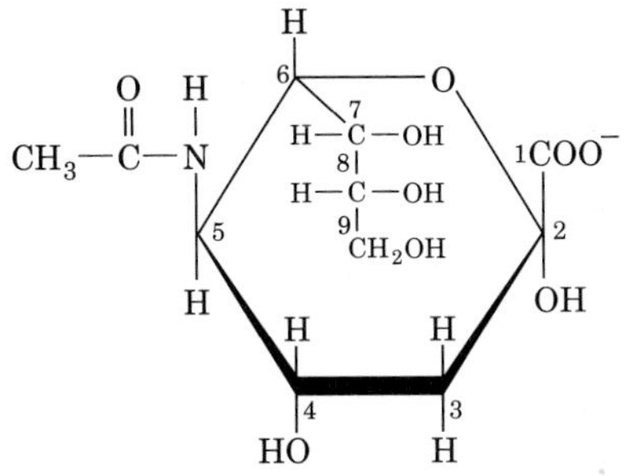

N-Acetylneuraminic acid
(NANA, sialic acid)

from CMP-NANA to lactosyl ceramide in $\alpha(2 \rightarrow 3)$ linkage yielding G_{M3}. The sequential addition to G_{M3} of the N-acetylgalactosamine and galactose units from UDP-GalNAc and UDP-Gal yield gangliosides G_{M2} and G_{M1}.

3'-Phosphoadenosine-5'-phosphosulfate (PAPS)

Galactocerebroside

3'-Phosphoadenosine-5'-phosphate

Sulfatide (galactocerebroside-3-sulfate)

Figure 23-73
The biosynthesis of sulfatides.

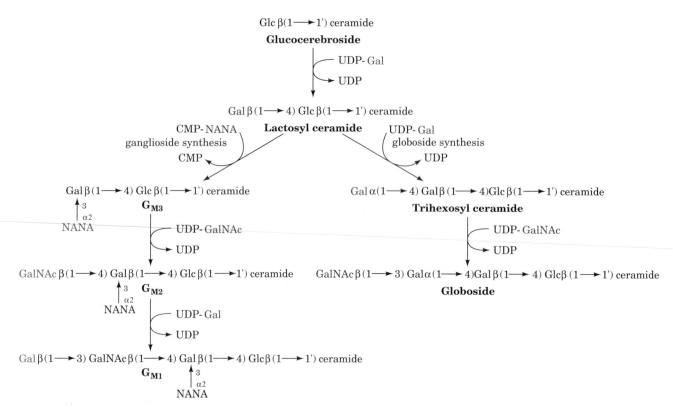

Figure 23-74
The biosynthesis of globosides and gangliosides.

Sphingoglycolipid Degradation and Lipid Storage Diseases

Sphingoglycolipids are lysosomally degraded by a series of enzymatically mediated hydrolytic reactions (Fig. 23-75). The hereditary absence of one of these enzymes results in a **sphingolipid storage disease** (Table 23-2). The most common such condition is **Tay-Sachs disease,** an autosomal recessive deficiency in **hexosaminidase A,** which hydrolyzes *N*-acetylgalactosamine from ganglioside G$_{M2}$. The absence of hexosaminidase A activity results in the neuronal accumulation of G$_{M2}$ as shell-like inclusions (Fig. 23-76).

Although infants born with Tay-Sachs disease at first appear normal, by ~1 year of age, when sufficient G$_{M2}$ has accumulated to interfere with neuronal function, they become progressively weaker, retarded, and blinded until they die, usually by the age of 3 years. It is possible, however, to screen potential carriers of this disease by a simple serum assay. It is also possible to detect the disease *in utero* by assay of amniotic fluid or amniotic cells obtained by amniocentesis. The assay involves use of an artificial hexosaminidase substrate, **4 - methylumbelliferyl - β - D - N - acetylglucosamine,** which yields a fluorescent product on hydrolysis.

4-Methylumbelliferyl- β-D-*N*-acetylglucosamine

hexosaminidase

4-Methylumbelliferone (fluorescent in alkaline medium)

Since this substrate is also recognized by **hexosaminidase B,** which is unaffected in Tay-Sachs disease, the hexosaminidase B is first heat-inactivated since it is more heat-labile than hexosaminidase A. As a result of mass screening efforts, the tragic consequences of this genetic enzyme deficiency are being averted. The other sphingolipid storage diseases, while less common, have similar consequences (Table 23-2).

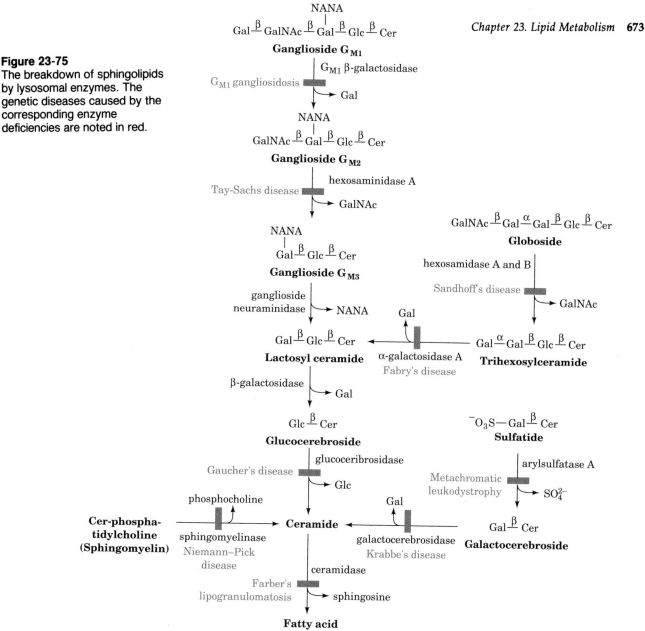

Figure 23-75
The breakdown of sphingolipids by lysosomal enzymes. The genetic diseases caused by the corresponding enzyme deficiencies are noted in red.

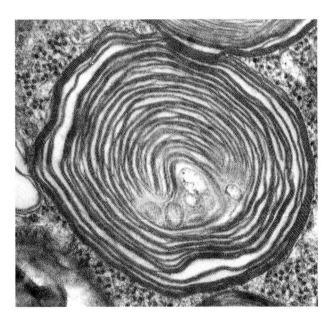

Figure 23-76
Cytoplasmic membranous body in a neuron affected by Tay-Sachs disease. [Courtesy of John S. O'Brien, University of California at San Diego Medical School.]

Table 23-2
Sphingolipid Storage Diseases

Disease	Enzyme Deficiency	Principal Storage Substance	Major Symptoms
G_{M1} Gangliosidosis	G_{M1} β-Galactosidase	Ganglioside G_{M1}	Mental retardation, liver enlargement, skeletal involvement, death by age 2
Tay-Sachs disease	Hexosaminidase A	Ganglioside G_{M2}	Mental retardation, blindness, death by age 3
Fabry's disease	α-Galactosidase A	Trihexosylceramide	Skin rash, kidney failure, pain in lower extemities
Sandhoff's disease	Hexosamindases A and B	Ganglioside G_{M2} and globoside	Similar to Tay-Sachs disease but more rapidly progressing
Gaucher's disease	Glucocerebrosidase	Glucocerebroside	Liver and spleen enlargement, erosion of long bones, mental retardation in infantile form only
Niemann–Pick disease	Sphingomyelinase	Sphingomyelin	Liver and spleen enlargement, mental retardation
Farber's lipogranulomatosis	Ceramidase	Ceramide	Painful and progressively deformed joints, skin nodules, death within a few years
Krabbe's disease	Galactocerebrosidase	Galactocerebroside	Loss of myelin, mental retardation, death by age 2
Sulfatide lipidosis	Arylsulfatase A	Sulfatide	Mental retardation, death in first decade

Chapter Summary

Triacylglycerols, the storage form of metabolic energy in animals, provide up to six times the metabolic energy of an equal weight of hydrated glycogen. Dietary lipids are digested at the lipid–water interface by pancreatic digestive enzymes such as lipase and phospholipase A_2 that are active at the lipid–water interface of bile acid-stabilized emulsions. Bile acids are also essential for the intestinal absorption of dietary lipids. Dietary triacylglycerols and those synthesized by the liver are transported in the blood as chylomicrons and VLDL, respectively. Triacylglycerols present in these lipoproteins are hydrolyzed by lipoprotein lipase outside the cells and enter them as free fatty acids. Fatty acids resulting from hydrolysis of adipose tissue triacylglycerols by hormone-sensitive lipase are transported in the bloodstream as fatty acid–albumin complexes.

Before fatty acids are oxidized, they are converted to their acyl-CoA derivatives by acyl-CoA synthase in an ATP-requiring process, transported into mitochondria as carnitine esters, and reconverted inside the mitochondrial matrix to acyl-CoA. β Oxidation of fatty acyl-CoA occurs in 2-carbon increments so as to convert even-chain fatty acyl-CoAs completely to acetyl-CoA. The pathway involves FAD-dependent dehydrogenation of an alkyl group, hydration of the resulting double bond, NAD$^+$-dependent oxidation of this alcohol to a ketone, and C—C bond cleavage to form acetyl-CoA and a new fatty acyl-CoA with two fewer carbon atoms. The process then repeats itself. Complete oxidation of the acetyl-CoA, NADH, and FADH$_2$ is achieved by the citric acid cycle and oxidative phosphorylation. Oxidation of unsaturated fatty acids and odd-chain fatty acids also occur by β oxidation but require the participation of additional enzymes. Odd-chain fatty acid oxidation generates propionyl-CoA, whose further metabolism requires the participation of (1) propionyl-CoA carboxylase,

which has a biotin prosthetic group, (2) methylmalonyl-CoA racemase, and (3) methylmalonyl-CoA mutase, which contains coenzyme B_{12}. β Oxidation of fatty acids takes place in the peroxisomes in addition to the mitochondrion. The peroxisomal pathway differs from the mitochondrial pathway in that the FADH$_2$ produced in the first step is directly oxidized by O_2 to produce H_2O_2 rather than generating ATP by oxidative phosphorylation. Peroxisomal enzymes are specific for long chain fatty acids and are thought to function in a chain-shortening process. The resultant intermediate chain-length products are transferred to the mitochondrion for complete oxidation.

A significant fraction of the acetyl-CoA produced by fatty acid oxidation in the liver is converted to acetoacetate and D-β-hydroxybutyrate, which, together with acetone, are referred to as ketone bodies. The first two compounds serve as important fuels for the peripheral tissues.

Fatty acid biosynthesis differs from fatty acid oxidation in several respects. While fatty acid oxidation occurs in the mitochondrion utilizing fatty acyl-CoA esters, fatty acid biosynthesis occurs in the cytosol with the growing fatty acids esterified to acyl carrier protein (ACP). The redox coenzymes differ (FAD and NAD$^+$ for oxidation; NADPH for biosynthesis) as does the stereochemistry of the pathway's intermediate steps. Oxidation produces acetyl-CoA while malonyl-CoA is the immediate precursor in biosynthesis. HCO$_3^-$ is required for biosynthesis but not for oxidation. Acetyl-CoA is transferred from the mitochondrion to the cytosol as citrate via the tricarboxylate transport system and citrate cleavage, a process that also generates some of the NADPH required for biosynthesis. Palmitate is the primary product of fatty acid biosynthesis in animals. Longer-chain fatty acids and unsaturated fatty acids are synthesized from palmitate by elongation and desaturation reactions. Triacylglycerols are synthesized from fatty acyl-CoA esters and glycerol-3-phosphate.

Fatty acid metabolism is regulated through the allosteric control of hormone-sensitive triacylglycerol lipase and acetyl-CoA carboxylase, phosphorylation/dephosphorylation and/or changes in the rates of protein synthesis and breakdown. This regulation is mediated by the hormones glucagon, epinephrine, and norepinephrine which activate degradation, and by insulin which activates biosynthesis. These hormones interact to control the cAMP concentration which, in turn, controls phosphorylation/dephosphorylation ratios.

Cholesterol is a vital constituent of cell membranes and is the precursor of the steroid hormones and bile acids. Its biosynthesis, transport, and utilization are rigidly controlled. Cholesterol is synthesized in the liver from acetate in a pathway that involves formation of HMG-CoA from three molecules of acetate followed by reduction, phosphorylation, decarboxylation, and dehydration to the isoprene units isopentenyl pyrophosphate and dimethylallyl pyrophosphate. These isoprene units are then condensed to form squalene, which, in turn, undergoes a cyclization reaction to form lanosterol, the sterol precursor to cholesterol. The pathway's major control point is at HMG-CoA reductase. This enzyme is regulated by competitive and allosteric mechanisms, phosphorylation/dephosphorylation, and long-term control of the rates of enzyme synthesis and degradation. The liver secretes cholesterol into the bloodstream in esterified form as part of the VLDL. This complex is sequentially converted to IDL and then to LDL. LDL, which is brought into the cells by receptor-mediated endocytosis, carries the major portion of cholesterol to peripheral tissues for utilization. Excess cholesterol is returned to the liver from peripheral tissues by HDL. The cellular supply of cholesterol is controlled by three mechanisms: (1) long-and short-term regulation of HMG-CoA reductase; (2) control of LDL receptor synthesis by cholesterol concentration; and (3) long- and short-term regulation of acyl-CoA:cholesterol acyl transferase (ACAT), which mediates cholesterol esterification. Cholesterol is the precursor to the steroid hormones which are classified as progestins, glucocorticoids, mineralocorticoids, androgens, and estrogens. The quantitatively most important pathway for the excretion of cholesterol in mammals is the formation of bile acids.

Prostaglandins, prostacyclins, thromboxanes, and leuko-trienes are products of arachidonate metabolism. These highly unstable compounds have profound physiological effects at extremely low concentrations. They are involved in the inflammatory response, the production of pain and fever, the regulation of blood pressure, and many other important physiological processes. Arachidonate is synthesized from linoleic acid, an essential fatty acid, and stored as phosphatidylinositol and other phospholipids. Prostaglandins, prostacyclins, and thromboxanes are synthesized via the "cyclic pathway" while leukotrienes are synthesized via the "linear pathway." Aspirin and other nonsteroidal anti-inflammatory drugs inhibit the cyclic pathway but not the linear pathway. Peptidoleukotrienes have been identified as the Slow Reacting Substances of Anaphylaxis (SRS-A) released from sensitized lung after immunological challenge.

Complex lipids have either a phosphate ester or a carbohydrate as their polar head group and either 1,2-diacyl-*sn*-glycerol or ceramide (*N*-acylsphingosine) as their hydrophobic tail. Phospholipids are either glycerophospholipids or sphingophospholipids, whereas glycolipids are either glyceroglycolipids or sphingoglycolipids. The polar head groups of glycerophospholipids, which are phosphate esters of either ethanolamine, serine, choline, inositol, or glycerol, are attached to 1,2-diacyl-*sn*-glycerol's C(3)-OH group by means of CTP-linked transferase reactions. The specific long-chain fatty acids found at the C(1) and C(2) positions are incorporated by "remodeling reactions" after the addition of the polar head group. Plasmalogens and alkylacylglycerophospholipids, respectively, contain a long chain alkyl group in a vinyl-ether linkage or an ether linkage to glycerol's C(1)-OH group. The only major sphingophospholipid is sphingomyelin (*N*-acylsphingosine phosphocholine), an important structural lipid of nerve cell membranes. Most sphingolipids contain polar head groups composed of carbohydrate units and are therefore referred to as sphingoglycolipids. The principal classes of sphingoglycolipids are cerebrosides, sulfatides, globosides, and gangliosides. Their carbohydrate units, which are attached to *N*-acylsphingosine's C(1)-OH group by glycosidic linkages, are formed by stepwise addition of activated monosaccharide units. Several lysosomal sphingolipid storage diseases, including Tay-Sachs disease, result from deficiencies in the enzymes that degrade sphingoglycolipids.

References

General

Boyer, P. D. (Ed.), *The Enzymes* (3rd ed.), Vol. 16, Academic Press (1983). [An excellent collection of reviews on lipid enzymology. Section 1 deals with fatty acid biosynthesis; Section 2 covers glyceride synthesis and degradation; Sections 3–5 review phospholipid, sphingolipid, and glycolipid metabolism; and Section 6 deals with aspects of cholesterol metabolism.]

Newsholme, E. A. and Leech, A. R., *Biochemistry for the Medical Sciences*, Wiley (1983). [Chapters 6–8 contain a wealth of information on the control of fatty acid metabolism and its integration into the overall scheme of metabolism.]

Numa, S. (Ed.), *Fatty Acid Metabolism and its Regulation*, Elsevier (1984).

Stanbury, J. B., Wyngaarden, J. B., Fredrickson, D. S., Goldstein, J. L., and Brown, M. S. (Eds.), *The Metabolic Basis of Inherited Disease* (5th ed.), McGraw–Hill (1983). [Chapter 23 deals with disorders of propionate and methylmalonate metabolism, Chapter 33 discusses hypercholesterolemia, Chapter 35 describes Refsum's disease, and Chapters 40–46 are concerned with diseases associated with sphingolipid degradation.]

Vance, D. E. and Vance, J. E. (Eds.), *Biochemistry of Lipids and Membranes*, Benjamin/Cummings (1985). [An advanced textbook.]

Lipid Digestion

Borgström, B., Barrowman, J. A., and Lindström, M., Roles of bile acids in intestinal lipid digestion and absorption, *in* Danielsson, H. and Sjövall, J. (Eds.), *Sterols and Bile Acids, pp.* 405–425, Elsevier (1985).

Drenth, J., Dijkstra, B. W., and Renetseder, R., Catalysis by phospholipases A2, *in* Jurnak, F. A. and McPherson, A., *Biological Macromolecules and Assemblies*, Vol. 3., *pp.* 287–312, Wiley (1987).

Fatty Acid Oxidation

Bieber, L. L., Carnitine, *Annu. Rev. Biochem.* **88**, 261–283 (1988).

Halpern, J., Mechanisms of coenzyme B$_{12}$ rearrangements, *Science* **277**, 869–875 (1985).

Masters, C. and Crane, D., The role of peroxisomes in lipid metabolism, *Trends Biochem. Sci.* **9**, 314–319 (1984).

Osumi, T. and Hashimoto, T., The inducible fatty acid oxidation system in mammalian peroxisomes, *Trends Biochem. Sci.* **9**, 317–319 (1984).

Schulz, H. and Kunau, W.-H., Beta-oxidation of unsaturated fatty acids: a revised pathway, *Trends Biochem. Sci.* **12**, 403–406 (1987).

Sudden infant death and inherited disorders of fat oxidation, *Lancet*, 1073–1075, Nov. 8, 1986.

Tanaka, K., Kean, E. A., and Johnson, B., Jamaican vomiting sickness, biochemical investigation of two cases, *New Eng. J. Med.* **295**, 461–467 (1976).

Walsh, C., *Enzymatic Reaction Mechanisms, pp.* 644–655, Freeman (1979). [An informative description of coenzyme B$_{12}$-dependent rearrangements.]

Wenz, A., Thorpe, C., and Ghisla, S., Inactivation of general acyl-CoA dehydrogenase from pig kidney by a metabolite of hypoglycin A, *J. Biol. Chem.* **256**, 9809–9812 (1981).

Fatty Acid Biosynthesis

Brownsey, R. W. and Denton, R. M., Acetyl-coenzyme A carboxylase, *in* Boyer, P. D. and Krebs, E. G. (Eds.), *The Enzymes* (3rd ed.), Vol. 18, *pp.* 123–146, Academic Press (1987).

Jeffcoat, R., The biosynthesis of unsaturated fatty acids and its control in mammalian liver, *Essays Biochem.* **15**, 1–36 (1979).

Lane, D. M., Moss, J., and Polakis, S. E., Acetyl coenzyme A carboxylase, *Curr. Top. Cell. Regul.* **8**, 139–195 (1974).

McCarthy, A. D. and Hardie, D. G., Fatty acid synthase—an example of protein evolution by gene fusion, *Trends Biochem. Sci.* **9**, 60–63 (1984).

Thorpe, C., Green enzymes and suicide substrates: a look at acyl-CoA dehydrogenases in fatty acid oxidation, *Trends Biochem. Sci.* **14**, 148–151 (1989).

Wakil, S. J., Fatty acid synthase, a proficient multifunctional enzyme, *Biochemistry* **28**, 4523–4530 (1989).

Wakil, S. J., Stoops, J. K., and Joshi, V. C., Fatty acid synthesis and its regulation, *Annu. Rev. Biochem.* **52**, 537–579 (1983).

Regulation of Fatty Acid Metabolism

McGarry, J. D. and Foster, D. W., Regulation of hepatic fatty acid oxidation and ketone body production, *Annu. Rev. Biochem.* **49**, 395–420 (1980).

Kim, K.-H., Regulation of acetyl-CoA carboxylase, *Curr. Top. Cell. Regul.* **22**, 143–176 (1983).

Saggerson, E. D., Regulation of lipid metabolism in adipose tissue and liver cells, *in* Bittar, E. E. (Ed.), *Biochemistry of Cellular Regulation*, Vol. 2, *pp.* 207–256, CRC Press (1980).

Stralfors, P., Olsson, H., and Belfrage, P., Hormone-sensitive lipase, *in* Boyer, P. D. and Krebs, E. G. (Eds.), *The Enzymes* (3rd ed.), Vol. 18, *pp.* 147–177, Academic Press (1987).

Cholesterol Metabolism

Beg, Z. H. and Brewer, H. B., Jr., Regulation of liver 3-hydroxy-3-methylglutaryl-coenzyme A reductase, *Curr. Top. Cell. Regul.* **20**, 139–184 (1981).

Billheimer, D. W., Grundy, S. M., Brown, M. S., and Goldstein, J. L., Mevinolin and colestipol stimulate receptor-mediated clearance of low density lipoprotein from plasma in familial hypercholesterolemia heterozygotes, *Proc. Natl. Acad. Sci.* **80**, 4124–4128 (1983).

Bloch, K., The biological synthesis of cholesterol, *Science* **150**, 19–28 (1965).

Brown, M. S. and Goldstein, J. L., A receptor-mediated pathway for cholesterol homeostasis, *Science* **232**, 34–47 (1986).

Brown, M. S. and Goldstein, J. L., The LDL receptor and HMG-CoA reductase—two membrane molecules that regulate cholesterol homeostasis, *Curr. Top. Cell. Regul.* **26**, 3–15 (1985).

Brown, M. S. and Goldstein, J. L., How LDL receptors influence cholesterol and atherosclerosis, *Sci. Am.* **251**(5): 58–66 (1984).

Cohen, D. C., Massoglia, S. L., and Gospodarowicz, D., Feedback regulation of 3-hydroxy-3-methylglutaryl-coenzyme A reductase in vascular endothelial cells: separate sterol and non-sterol components, *J. Biol. Chem.* **257**, 11106–11112 (1982).

Gibson, D. M. and Parker, R. A., Hydroxymethylglutaryl-coenzyme A reductase, *in* Boyer, P. D. and Krebs, E. G. (Eds.), *The Enzymes*, (3rd ed.), Vol. 18, *pp.* 179–215, Academic Press (1987).

Poulter, C. D. and Rilling, H. C., The prenyl transfer reaction, Enzymatic and mechanistic studies of the 1′-4 coupling reaction in the terpene biosynthetic pathway, *Acc. Chem. Res.* **11**, 307–313 (1978).

Rilling, H. C. and Chayet, L. T., Biosynthesis of cholesterol, *in* Danielsson, H. and Sjövall, J. (Eds.), *Sterols and Bile Acids, pp.* 1–40, Elsevier (1985).

Arachidonate Metabolism

Hayaishi, O., Sleep–wake regulation by prostaglandins D$_2$ and E$_2$, *J. Biol. Chem.* **263**, 14593–14596 (1988).

Johnson, M., Carey, F., and McMillan, R. M., Alternative pathways of arachidonate metabolism: prostaglandins, thromboxane and leukotrienes, *Essays Biochem.* **19**, 40–141 (1983).

Lee, T. H., Hoover, R. L., Williams, J. D. Sperling, R. I., Ravelese, J., III, Spur, B. W., Robinson, D. R., Corey, E. J., Lewis, R. A., and Austen, K. F., Effect of dietary enrichment with eicosapentaenoic and docosahexaenoic acids on in vitro neutrophil and monocyte leukotriene generation and neutrophil function, *New Engl. J. Med.* **19**, 1217–1224 (1985).

Needleman, P., Turk, J., Jakschik, B. A., Morrison, A. R., and

Lefkowith, J. B., Arachidonic acid metabolism, *Annu. Rev. Biochem.* **55**, 69–102 (1986).

Pace-Asciak, C. and Grandström, E. (Eds.), Prostaglandins and related subjects, *New Comprehensive Biochemistry*, Vol. 5, Elsevier (1983).

Padley, F. B. and Podmore, J. (Eds.), The role of fats in human nutrition, Ellis Horwood, Chichester (1985).

Phillipson, B. E., Rothrock, D. W., Conner, W. E., Harris, W. S., and Illingworth, D. R., Reduction of plasma lipids, lipoproteins and apoproteins by dietary fish oils in patients with hypertriglyceridemia, *New Engl. J. Med.* **19**, 1210–1216 (1985).

Samuelsson, B., Dahlén, S.-E., Lindgren, J. Å., Rouzier, C. A., and Serhan, C. N., Leukotrienes and lipoxins: structures, biosynthesis, and biological effects, *Science* **237**, 1171–1176 (1987).

Taylor, G. W. and Clarke, S. R., The leukotriene biosynthetic pathway: a target for pharmacological attack, *Trends Pharm. Sci.* **7**, 100–103 (1986).

Phospholipid and Glycolipid Metabolism

Conzelmann, E. and Sandhoff, K., Glycolipid and glycoprotein degradation, *Adv. Enzymol.* **60**, 89–216 (1987).

Hawthorne, J. N. and Ansell, G. B., (Eds.), Phospholipids, *New Comprehensive Biochemistry*, Vol. 4, Elsevier (1982). [A summary of the properties, biosynthesis, and effects of phospholipids.]

Neufield, E. F., Natural history and inherited disorders of a lysosomal enzyme, β-hexosaminidase, *J. Biol. Chem.* **264**, 10927–10930 (1989).

Robinson, M., Blank, M. L., and Snyder, F., Acylation of lysophospholipids by rabbit alveolar macrophages: specificities of CoA-dependent and CoA-independent reactions, *J. Biol. Chem.* **260**, 7889–7895 (1985).

Wiegandt, H. (Ed.), Glycolipids, *New Comprehensive Biochemistry*, Vol. 10, Elsevier (1982).

Problems

1. The venoms of many poisonous snakes, including rattlesnakes, contain a phospholipase A_2 that causes tissue damage that is seemingly far out of proportion to the small amount of enzyme injected. Explain.

2. Why are the livers of Jamaican vomiting sickness victims usually depleted of glycogen?

3. Compare the metabolic efficiencies, in moles of ATP produced per gram, of completely oxidized fat (tripalmitoyl glycerol) versus glucose derived from glycogen. Assume that the fat is anhydrous and the glycogen is stored with twice its weight in water.

4. A fasting animal is fed palmitic acid that has a ^{14}C-labeled carboxyl group. (a) After allowing sufficient time for fatty acid breakdown and resynthesis, what would be the ^{14}C-labeling pattern in the animal's palmitic acid residues? (b) The animal's liver glycogen becomes ^{14}C labeled although there is no net increase in the amount of this substance present. Indicate the sequence of reactions whereby the glycogen becomes labeled. Why is there no net glycogen synthesis?

5. What is the ATP yield from the complete oxidation of a molecule of (a) α-linolenic acid (9,12,15-octadecatrienoic acid, 18:3), and (b) **margaric acid** (heptadecanoic acid, 17:0)? Which has the greater amount of available biological energy on a per carbon basis?

*6. The role of coenzyme B_{12} in mediating hydrogen transfer was established using the coenzyme B_{12}-dependent bacterial enzyme **dioldehydrase,** which catalyzes the reaction:

$$CH_3-\underset{\underset{OH}{|}}{CH}-\underset{\underset{H}{|}}{CH}-OH \longrightarrow CH_3-\underset{\underset{H}{|}}{CH}-\underset{\underset{OH}{|}}{CH}-OH$$

1,2-Propanediol

$$\downarrow \rightarrow H_2O$$

$$CH_3-CH_2-\overset{\overset{O}{\|}}{CH}$$

Propionaldehyde

The enzyme converts $[1\text{-}^3H_2]1,2$-propanediol to $[1,2\text{-}^3H]$propionaldehyde with the incorporation of tritium into both C(5') positions of 5'-deoxyadenosylcobalamin's 5'-deoxyadenosyl residue. Suggest the mechanism of this reaction. What would be the products of the dioldehydrase reaction if the enzyme was supplied with $[5'\text{-}^3H]$deoxyadenosylcobalamin and unlabeled 1,2-propanediol?

7. What is the energetic price, in ATP equivalents, of breaking down palmitic acid to acetyl-CoA and then resynthesizing it?

8. What is the energetic price, in ATP equivalents, of synthesizing cholesterol from acetyl-CoA?

9. What would be the ^{14}C-labeling pattern in cholesterol if it was synthesized from HMG-CoA that was ^{14}C labeled (a) at C(5), its carboxyl carbon atom or (b) C(1), its thioester carbon atom?

*10. A child suffering from severe abdominal pain is admitted to the hospital several hours after eating a meal consisting of hamburgers, fried potatoes, and ice cream. Her blood has the appearance of "creamed tomato soup" and upon analysis is found to contain massive quantities of chylomicrons. As attending physician, what is your diagnosis of the patient's difficulty (the cause of the abdominal pain is unclear)? What treatment would you prescribe to alleviate the symptoms of this inherited disease?

11. Although linoleic acid is an essential fatty acid in animals, it is not required by animal cells in tissue culture. Explain.

Chapter 24
AMINO ACID METABOLISM

α-Amino acids, in addition to their role as protein monomeric units, are energy metabolites and precursors of many biologically important nitrogen-containing compounds, notably heme, physiologically active amines, glutathione, nucleotides, and nucleotide coenzymes. Amino acids are classified into two groups: **essential** and **nonessential**. Mammals synthesize the nonessential amino acids from metabolic precursors but must obtain the essential amino acids from their diet. Excess dietary amino acids are neither stored for future use nor excreted. Rather, they are converted to common metabolic intermediates such as pyruvate, oxaloacetate, and α-ketoglutarate. Consequently, amino acids are also precursors of glucose, fatty acids, and ketone bodies and are therefore metabolic fuels.

In this chapter, we consider the pathways of amino acid breakdown, synthesis, and utilization. We begin by examining the three common stages of amino acid breakdown:

1. **Deamination** (amino group removal), whereby amino groups are converted to either ammonia or to the amino group of aspartate.
2. Incorporation of ammonia and aspartate nitrogen atoms into urea for excretion.
3. Conversion of amino acid carbon skeletons (the α-keto acids produced by deamination) to common metabolic intermediates.

Many of these reactions are similar to those we have considered in other pathways. Others employ enzyme cofactors we have not previously encountered. One of our goals in studying amino acid metabolism is to understand the mechanisms of action of these cofactors.

After our discussion of amino acid breakdown, we examine the pathways by which amino acids are utilized in the biosynthesis of heme, physiologically active amines, and glutathione (the synthesis of nucleotides and nucleotide coenzymes is the subject of Chapter 26). Next, we study amino acid biosynthesis pathways. The chapter ends with a discussion of nitrogen fixation, a process that converts atmospheric N_2 to ammonia and is therefore the ultimate biological source of metabolically useful nitrogen.

1. AMINO ACID DEAMINATION

The first reaction in the breakdown of an amino acid is almost always removal of its α-amino group with the object of excreting excess nitrogen and degrading the remaining carbon skeleton. Urea, the predominant nitrogen excretion product in terrestrial mammals, is synthesized from ammonia and aspartate. Both of these latter substances are derived mainly from glutamate, a product of most deamination reactions. In this section we examine the routes by which α-amino groups are incorporated into glutamate and then into aspartate and ammonia. In Section 24-2, we discuss urea biosynthesis from these precursors.

Most amino acids are deaminated by **transamination,** the transfer of their amino group to an α-keto acid to yield the α-keto acid of the original amino acid and a new amino acid, in reactions catalyzed by **aminotransferases** (alternatively, **transaminases**). The predominant amino group acceptor is α-ketoglutarate, producing glutamate as the new amino acid:

Amino acid + α-ketoglutarate $\rightleftharpoons$
$$\alpha\text{-keto acid} + \text{glutamate}$$

Glutamate's amino group, in turn, is transferred to oxaloacetate in a second transamination reaction yielding aspartate:

Glutamate + oxaloacetate $\rightleftharpoons$
$$\alpha\text{-ketoglutarate} + \text{aspartate}$$

Transamination, of course, does not result in any net deamination. Deamination occurs largely through the oxidative deamination of glutamate by **glutamate dehydrogenase** yielding ammonia. The reaction requires NAD^+ or $NADP^+$ as an oxidizing agent and regenerates α-ketoglutarate for use in additional transamination reactions:

Glutamate + $NAD(P)^+$ + H_2O $\rightleftharpoons$
$$\alpha\text{-ketoglutarate} + NH_3 + NAD(P)H$$

The mechanisms of transamination and oxidative deamination are the subjects of this section. We also consider other means of amino group removal from specific amino acids.

A. Transamination

Aminotransferase reactions occur in two stages:

1. The amino group of an amino acid is transferred to the enzyme, producing the corresponding keto acid and the aminated enzyme.

Amino acid + enzyme $\rightleftharpoons$
$$\alpha\text{-keto acid} + \text{enzyme}-NH_2$$

2. The amino group is transferred to the keto acid acceptor (e.g., α-ketoglutarate) forming the amino acid product (e.g., glutamate) and regenerating the enzyme.

α-Ketoglutarate + enzyme$-NH_2$ $\rightleftharpoons$
$$\text{enzyme} + \text{glutamate}$$

In order to carry the amino group, aminotransferases require participation of pyridoxal-5'-phosphate (PLP), a derivative of pyridoxine (vitamin B_6; Fig. 24-1). The amino group is accommodated by conversion of this coenzyme to **pyridoxamine-5'-phosphate (PMP;** Fig. 24-1). PLP is covalently attached to the enzyme via a Schiff base (imine) linkage formed by the condensation of its aldehyde group with the ε-amino group of an enzymatic Lys residue.

This Schiff base, which is conjugated to the coenzyme's pyridinium ring, is the focus of the coenzyme's activity.

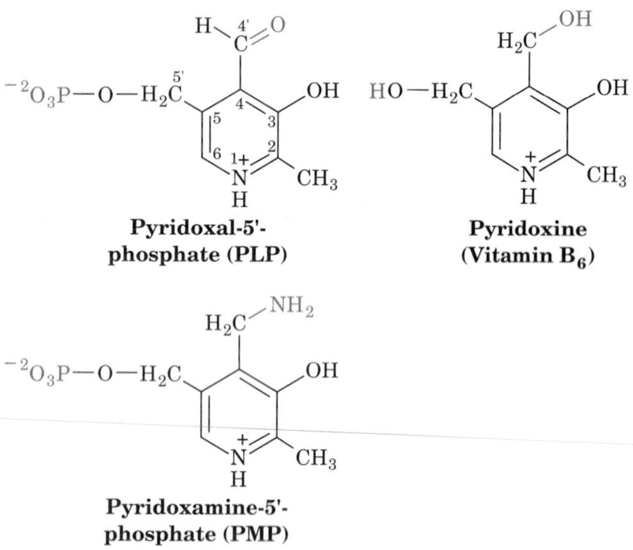

Figure 24-1
The coenzymes pyridoxal-5'-phosphate (PLP) and pyridoxamine-5'-phosphate (PMP) are derived from pyridoxine (vitamin B_6).

Esmond Snell, Alexander Braunstein, and David Metzler demonstrated that the aminotransferase reaction occurs via a Ping Pong Bi Bi mechanism whose two stages consist of three steps each (Fig. 24-2):

Stage One: Conversion of an Amino Acid to a Keto Acid

1. The amino acid's nucleophilic amino group attacks the enzyme–PLP Schiff base carbon atom in a **transimination (trans-Schiffization)** reaction to form an amino acid–PLP Schiff base (aldimine) with concomitant release of the enzyme's Lys amino group.

2. The amino acid–PLP Schiff base tautomerizes to an α-keto acid–PMP Schiff base by removal of the amino acid α hydrogen and protonation of PLP atom C(4') via a resonance-stabilized carbanion intermediate. This resonance stabilization facilitates the cleavage of the C_α—H bond.

3. The α-keto acid–PMP Schiff base is hydrolyzed to PMP and an α-keto acid.

Stage Two: Conversion of an α-Keto Acid to an Amino Acid

To complete the aminotransferase's catalyic cycle, the coenzyme must be converted from PMP back to the enzyme–PLP Schiff base. This involves the same three steps as above, but in reverse order:

3'. PMP reacts with an α-keto acid to form a Schiff base.

2'. The α-keto acid–PMP Schiff base tautomerizes to form an amino acid–PLP Schiff base.

1'. The ε-amino group of a Lys residue attacks the amino acid–PLP Schiff base in a transimination reaction to regenerate the active enzyme–PLP Schiff base with release of the newly formed amino acid.

The reaction's overall stoichiometry therefore is

Amino acid$_1$ + α-keto acid$_2$ $\rightleftharpoons$
α-keto acid$_1$ + amino acid$_2$

Examination of the amino acid–PLP Schiff base's structure (Fig. 24-2, Step 1) reveals why this system is called "an electron-pusher's delight." *Cleavage of any of the amino acid C_α atom's three bonds (labeled a, b, and c) produces a resonance-stabilized C_α carbanion whose electrons are delocalized all the way to the coenzyme's protonated pyridinium nitrogen atom; that is, PLP functions as an electron sink.* For transamination reactions, this electron-withdrawing capacity facilitates removal of the α proton (a bond cleavage) in the tautomerization of the Schiff base. PLP-dependent reactions involving *b* bond cleavage (amino acid decarboxylation) and *c* bond labilization are discussed in Sections 24-4B and 24-3B and G, respectively.

Aminotransferases differ in their specificity for amino acid substrates in the first stage of the transamination reaction thereby producing the correspondingly different α-keto acid products. Most aminotransferases, however, accept only α-ketoglutarate or (to a lesser extent) oxaloacetate as the α-keto acid substrate in the second stage of the reaction thereby yielding glutamate or aspartate as their only amino acid products. *The amino groups of most amino acids are consequently funneled into the formation of glutamate or aspartate, which are themselves interconverted by* **glutamate–aspartate aminotransferase:**

Glutamate + oxaloacetate $\rightleftharpoons$
α-ketoglutarate + aspartate

Oxidative deamination of glutamate (Section 24-1B) yields ammonia and regenerates α-ketoglutarate for another round of transamination reactions. Ammonia and aspartate are the two amino group donors in the synthesis of urea.

The Glucose–Alanine Cycle Transports Nitrogen to the Liver

An important exception to the foregoing is a group of muscle aminotransferases that accept pyruvate as their α-keto acid substrate. The product amino acid, alanine, is released into the bloodstream and transported to the liver where it undergoes transamination to yield pyruvate for use in gluconeogenesis (Section 21-1A). The resulting glucose is returned to the muscles where it is glycolytically degraded to pyruvate. This is the **glucose–alanine cycle.** The amino group ends up in either ammonia or aspartate for urea biosynthesis. Evi-

Steps 1 & 1': Transimination:

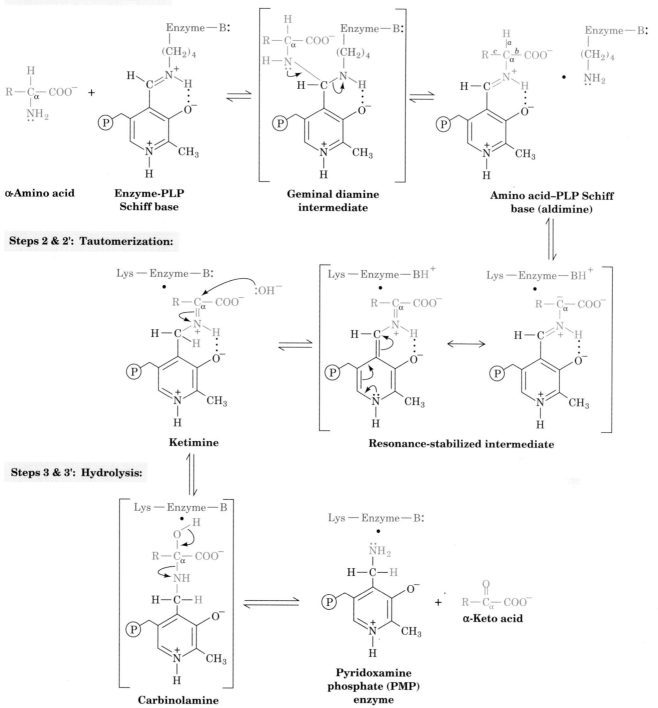

α-Amino acid   Enzyme-PLP Schiff base   Geminal diamine intermediate   Amino acid–PLP Schiff base (aldimine)

Steps 2 & 2': Tautomerization:

Ketimine   Resonance-stabilized intermediate

Steps 3 & 3': Hydrolysis:

Carbinolamine   Pyridoxamine phosphate (PMP) enzyme   α-Keto acid

Figure 24-2

The mechanism of PLP dependent enzyme-catalyzed transamination. The first stage of the reaction, in which the α-amino group of an amino acid is transferred to PLP yielding an α-keto acid and PMP, consists of three steps: **(1)** transimination, **(2)** tautomerization, and **(3)** hydrolysis.

The second stage of the reaction, in which the amino group of PMP is transferred to a different α-keto acid to yield a new α-amino acid and PLP, is essentially the reverse of the first stage: Steps 3', 2', and 1' are, respectively, the reverse of Steps 3, 2, and 1.

dently, the glucose–alanine cycle functions to transport nitrogen from muscle to liver.

B. Oxidative Deamination: Glutamate Dehydrogenase

Glutamate is oxidatively deaminated in the mitochondrion by glutamate dehydrogenase, the only known enzyme which, in at least some organisms, can accept either NAD^+ or $NADP^+$ as its redox coenzyme. Oxidation is thought to occur with transfer of a hydride ion from glutamate's C_α to $NAD(P)^+$, thereby forming α-iminoglutarate, which is hydrolyzed to α-ketoglutarate and ammonia (Fig. 24-3).

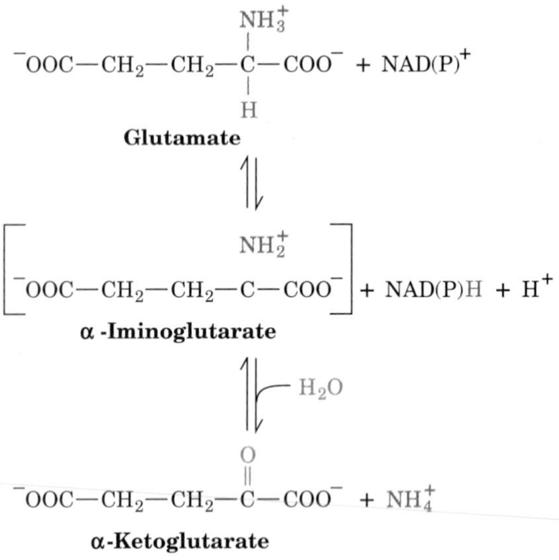

Glutamate dehydrogenase is inhibited by GTP and activated by ADP *in vitro* suggesting that these nucleotides regulate the enzyme *in vivo*. Studies of the cellular substrate and product concentrations nevertheless indicate that the enzyme functions close to equilibrium ($\Delta G \approx 0$) *in vivo*. Changes in glutamate dehydrogenase activity resulting from allosteric interactions are therefore unlikely to result in flux changes. Most probably the flux is controlled by the concentrations of substrates and products (Section 16-4A). The equilibrium position greatly favors glutamate formation over ammonia formation ($\Delta G°' \approx 30$ kJ·mol^{-1} for the reaction as written in Fig. 24-3). As high concentrations of ammonia are toxic, this equilibrium position is physiologically important; it helps maintain low ammonia concentrations. The ammonia produced is converted to urea (Section 24-2).

C. Other Deamination Mechanisms

Two nonspecific amino acid oxidases, **L-amino acid oxidase** and **D-amino acid oxidase,** catalyze the oxidation of L- and D-amino acids, utilizing FAD as their redox coenzyme [rather than $NAD(P)^+$]. The resulting $FADH_2$ is reoxidized by O_2.

$$Amino\ acid + FAD + H_2O \longrightarrow$$
$$\alpha\text{-keto acid} + NH_3 + FADH_2$$

$$FADH_2 + O_2 \longrightarrow FAD + H_2O_2$$

D-Amino acid oxidase occurs mainly in kidney. Its function is enigmatic since D-amino acids are associated mostly with bacterial cell walls (Section 10-3B). A few amino acids, such as serine and histidine, are deaminated nonoxidatively (Sections 24-3B and D).

2. THE UREA CYCLE

Living organisms excrete the excess nitrogen resulting from the metabolic breakdown of amino acids in one of three ways. Many aquatic animals simply excrete ammonia. Where water is less plentiful, however, processes have evolved that convert ammonia to less toxic waste products that therefore require less water for ex-

Figure 24-3
The oxidation of glutamate by glutamate dehydrogenase involves the intermediate formation of α-iminoglutarate.

cretion. One such product is urea, which is excreted by most terrestrial vertebrates; another is **uric acid,** which is excreted by birds and terrestrial reptiles.

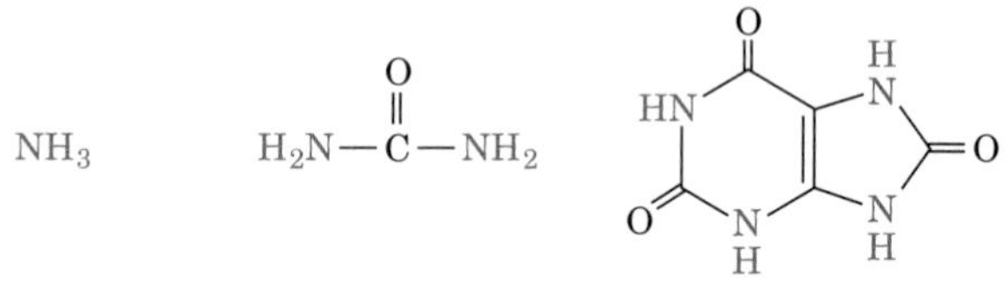

Accordingly, living organisms are classified as being either **ammonotelic** (ammonia excreting), **ureotelic** (urea excreting), or **uricotelic** (uric acid excreting). Some animals can shift from ammonotelism to ureotelism or uricotelism if their water supply becomes restricted. Here we focus our attention on urea formation. Uric acid biosynthesis is discussed in Section 26-5A.

Urea is synthesized in the liver by the enzymes of the urea cycle. It is then secreted into the bloodstream and sequestered by the kidneys for excretion in the urine. The urea cycle was elucidated in outline in 1932 by Hans Krebs and Kurt Henseleit (the first known metabolic cycle; Krebs did not elucidate the citric acid cycle until 1937). Its individual reactions were later described in detail by Sarah Ratner and Philip Cohen. The overall urea cycle reaction is

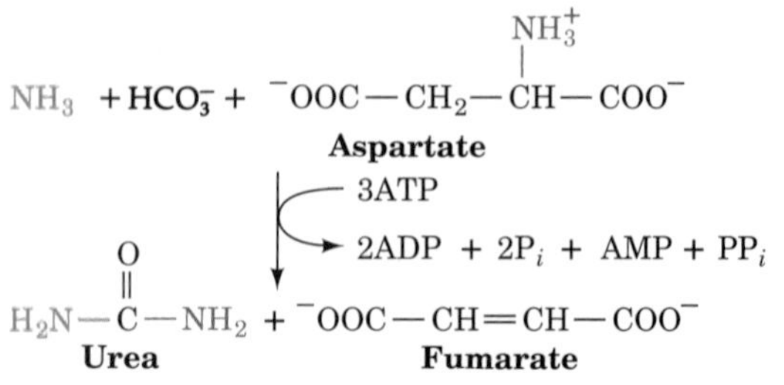

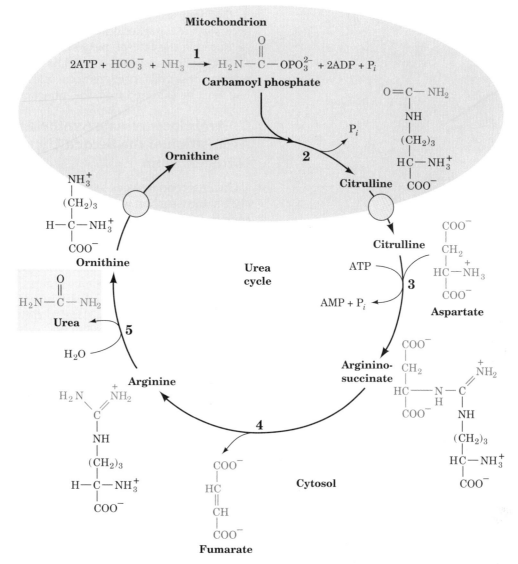

Figure 24-4
The urea cycle occurs partially in the mitochondrion and partially in the cytosol with ornithine and citrulline being transported across the mitochondrial membrane by specific transport systems. Five enzymes participate in the urea cycle: **(1)** carbamoyl phosphate synthetase, **(2)** ornithine transcarbamoylase, **(3)** argininosuccinate synthetase, **(4)** argininosuccinase, and **(5)** arginase.

Thus, the two urea nitrogen atoms are contributed by ammonia and aspartate while the carbon atom comes from HCO_3^-. Five enzymatic reactions are involved in the urea cycle, two of which are mitochondrial and three cytosolic (Fig. 24-4). In this section, we examine the mechanisms of these reactions and their regulation.

A. Carbamoyl Phosphate Synthetase: Acquisition of the First Urea Nitrogen Atom

Carbamoyl phosphate synthetase (CPS) is technically not a member of the urea cycle. It catalyzes the condensation and activation of NH_4^+ and HCO_3^- to form **carbamoyl phosphate,** the first of the cycle's two nitrogen-containing substrates, with the concomitant hydrolysis of two ATPs. Eukaryotes have two forms of CPS:

1. Mitochondrial **CPS I** uses ammonia as its nitrogen donor and participates in urea biosynthesis.

2. Cytosolic **CPS II** uses glutamine as its nitrogen donor and is involved in pyrimidine biosynthesis (Section 26-3A).

The reaction catalyzed by CPS I is thought to involve three steps (Fig. 24-5):

1. Activation of HCO_3^- by ATP to form **carbonyl phosphate** and ADP.

2. Attack of ammonia on carbonyl phosphate, displacing the phosphate to form **carbamate** and P_i.

3. Phosphorylation of carbamate by the second ATP to form carbamoyl phosphate and ADP.

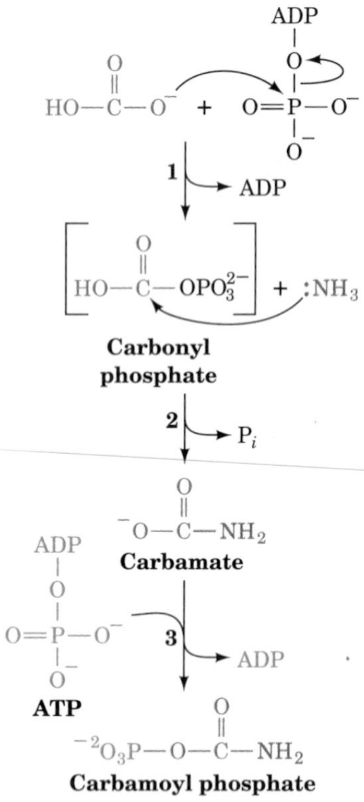

Figure 24-5
The mechanism of action of CPS I: **(1)** activation of HCO_3^- by phosphorylation to form the postulated intermediate, carbonyl phosphate, **(2)** attack on carbonyl phosphate by NH_3 to form carbamate, and **(3)** phosphorylation of carbamate by ATP yielding carbamoyl phosphate.

The reaction is essentially irreversible and is the rate-limiting step of the urea cycle. CPS I is subject to allosteric activation by **N-acetylglutamate** as is discussed in Section 24-2F.

B. Ornithine Transcarbamoylase

Ornithine transcarbamoylase transfers the carbamoyl group of carbamoyl phosphate to **ornithine**, yielding **citrulline** (Fig. 24-4, Reaction 2; note that both of these compounds are "nonstandard" α-amino acids

in that they do not occur in proteins). The reaction occurs in the mitochondrion so that ornithine, which is produced in the cytosol, must enter the mitochondrion via a specific transport system. Likewise, since the remaining urea cycle reactions occur in the cytosol, citrulline must be exported from the mitochondrion.

C. Argininosuccinate Synthetase: Acquisition of the Second Urea Nitrogen Atom

Urea's second nitrogen atom is introduced in the urea cycle's third reaction when the citrulline ureido group is condensed with an aspartate amino group by **argininosuccinate synthetase** (Fig. 24-6). The ureido oxygen atom is activated as a leaving group through formation of a citrullyl–AMP intermediate, which is subsequently displaced by the aspartate amino group. Support for the existence of the citrullyl–AMP intermediate comes from experiments using ^{18}O-labeled citrulline (* in Fig. 24-6). The label was isolated in the AMP produced by the reaction, demonstrating that at some stage of the reaction, AMP and citrulline are linked covalently through the ureido oxygen atom.

D. Argininosuccinase

With formation of argininosuccinate, all of the urea molecule components have been assembled. However, the amino group donated by aspartate is still attached to the aspartate carbon skeleton. This situation is remedied by the **argininosuccinase**-catalyzed elimination of arginine from the aspartate carbon skeleton forming fumarate (Fig. 24-4, Reaction 4). Arginine is urea's immediate precursor. Note that the urea cycle and the citric acid cycle are linked via production of fumarate in the argininosuccinase reaction and the transamination of oxaloacetate to aspartate, which is used in the argininosuccinate synthetase reaction (Fig. 24-7).

E. Arginase

The urea cycle's fifth and final reaction is the **arginase**-catalyzed hydrolysis of arginine to yield urea and

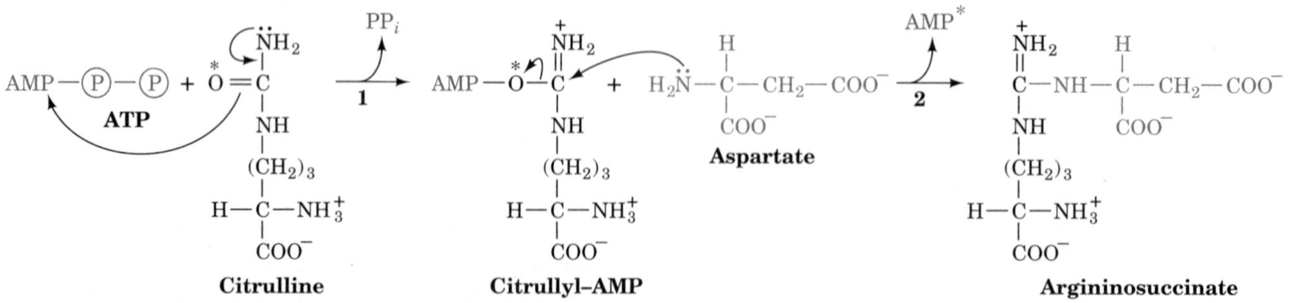

Figure 24-6
The mechanism of action of argininosuccinate synthetase: **(1)** activation of the ureido oxygen of citrulline through the formation of citrullyl–AMP, and **(2)** displacement of AMP by the α-amino group of aspartate. The asterisk (*) traces the fate of ^{18}O originating in citrulline's ureido group.

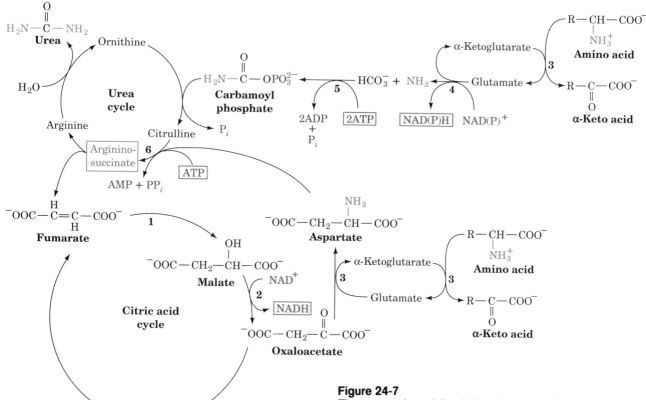

regenerate ornithine (Fig. 24-4). Ornithine is then returned to the mitochondrion for another round of the cycle. The urea cycle thereby converts two amino groups, one from ammonia and one from aspartate, and a carbon atom from HCO_3^- to the relatively nontoxic excretion product, urea, at the cost of four "high-energy" phosphate bonds (three ATP hydrolyzed to two ADP, two P_i, AMP, and PP_i followed by rapid PP_i hydrolysis). This energetic cost is more than recovered, however, by the energy released upon the formation of urea cycle substrates. The ammonia released by the glutamate dehydrogenase reaction is accompanied by NADH formation as is the reconversion of fumarate through oxaloacetate to aspartate (Fig. 24-7). Mitochondrial reoxidation of this NADH yields six ATPs.

F. Regulation of the Urea Cycle

Carbamoyl phosphate synthetase I, the mitochondrial enzyme that catalyzes the first committed step of the urea cycle, is allosterically activated by **N-acetylglutamate.**

$$
\begin{array}{c}
COO^- \\
| \\
(CH_2)_2 \quad O \\
| \qquad\quad \| \\
H-C-N-C-CH_3 \\
| \quad\; | \\
{}^-OOC \quad H
\end{array}
$$

N-Acetylglutamate

Figure 24-7
The urea cycle and the citric acid cycle are linked via the formation and breakdown of argininosuccinate. The enzymes (1) fumarase and (2) malate dehydrogenase are citric acid cycle enzymes (Sections 19-2 G and H). Oxaloacetate is diverted from the citric acid cycle to form aspartate through the action of (3) an aminotransferase. ATP is hydrolyzed at reactions (5) carbamoyl phosphate synthetase I, and (6) argininosuccinate synthetase. This ATP is regenerated by oxidative phosphorylation from the NAD(P)H produced in (4) the glutamate dehydrogenase reaction, and (2) the malate dehydrogenase reaction.

This metabolite is synthesized from glutamate and acetyl-CoA by **N-acetylglutamate synthase,** and hydrolyzed by a specific hydrolase. The rate of urea production by the liver is, in fact, correlated with the N-acetylglutamate concentration. Increased urea synthesis is required when amino acid breakdown rates increase, generating excess nitrogen that must be excreted. Increases in these breakdown rates are signaled by an increase in glutamate concentration through transamination reactions (Section 24-1). This situation, in turn, causes an increase in N-acetylglutamate synthesis, stimulating carbamoyl phosphate synthetase and thus the entire urea cycle.

The remaining enzymes of the urea cycle are controlled by the concentrations of their substrates. Thus, inherited deficiencies in urea cycle enzymes other than arginase do not result in significant decreases in urea production (the total lack of any urea cycle enzyme results in death shortly after birth). Rather, the deficient enzyme's substrate builds up, increasing the rate of the deficient reaction to normal. The anomalous substrate buildup is not without cost, however. The substrate concentrations become elevated all the way back up the cycle to ammonia, resulting in **hyperammonemia** (elevated levels of ammonia in the blood). Although the root cause of ammonia toxicity is not completely understood, a high ammonia concentration puts an enormous strain on the ammonia-clearing system, especially in the brain (symptoms of urea cycle enzyme deficiencies include mental retardation and lethargy). This clearing system involves glutamate dehydrogenase (working in reverse) and **glutamine synthetase,** which decrease the α-ketoglutarate and glutamate pools (Sections 24-1 and 24-5A). The brain is most sensitive to the depletion of these pools. Depletion of α-ketoglutarate decreases the rate of the energy-generating citric acid cycle, while glutamate is both a neurotransmitter and a precursor to γ-aminobutyrate (GABA), another neurotransmitter (Section 34-4C).

3. METABOLIC BREAKDOWN OF INDIVIDUAL AMINO ACIDS

The degradation of amino acids converts them to citric acid cycle intermediates or their precursors so that they can be metabolized to CO_2 and H_2O or used in gluconeogenesis. Indeed, oxidative breakdown of amino acids typically accounts for 10 to 15% of the metabolic energy generated by animals. In this section we consider how amino acid carbon skeletons are catabolized. The 20 "standard" amino acids (the amino acids of proteins) have widely differing carbon skeletons, so their conversions to citric acid cycle intermediates follow correspondingly diverse pathways. We shall not describe all of the many reactions involved in detail. Rather, we shall consider how these pathways are organized and focus on a few reactions of chemical and/or medical interest.

A. Amino Acids Can Be Glucogenic, Ketogenic, or Both

"Standard" amino acids are degraded to one of seven metabolic intermediates: pyruvate, α-ketoglutarate, succinyl-CoA, fumarate, oxaloacetate, acetyl-CoA, or acetoacetate (Fig. 24-8). The amino acids may therefore be divided into two groups based on their catabolic pathways (Fig. 24-8):

1. **Glucogenic amino acids,** whose carbon skeletons are degraded to pyruvate, α-ketoglutarate, succinyl-CoA, fumarate, or oxaloacetate and are therefore glucose precursors (Section 21-1A).

2. **Ketogenic amino acids,** whose carbon skeletons are broken down to acetyl-CoA or acetoacetate and can thus be converted to fatty acids or ketone bodies (Section 23-3).

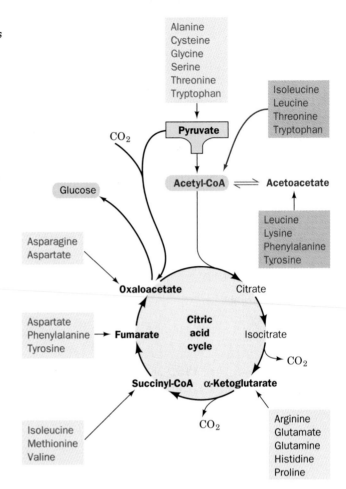

Figure 24-8
Amino acids are degraded to one of seven common metabolic intermediates. Glucogenic and ketogenic degradations are indicated in green and red, respectively.

For example, alanine is glucogenic because its transamination product, pyruvate (Section 24-1A), can be converted to glucose via gluconeogenesis (Section 21-1A). Leucine, on the other hand, is ketogenic; its carbon skeleton is converted to acetyl-CoA and acetoacetate (Section 24-3F). Since animals lack any metabolic pathway for the net conversion of acetyl-CoA or acetoacetate to gluconeogenic precursors, no net synthesis of carbohydrates is possible from leucine, or from lysine, the only other purely ketogenic amino acid. Isoleucine, phenylalanine, threonine, tryptophan, and tyrosine, however, are both glucogenic and ketogenic; isoleucine, for example, is broken down to succinyl-CoA and acetyl-CoA and hence is a precursor of both carbohydrates and ketone bodies (Section 24-3E). The remaining 13 amino acids are purely glucogenic.

In studying the specific pathways of amino acid breakdown, we shall organize the amino acids into groups that are degraded into each of the seven metabolic intermediates mentioned above: pyruvate, oxalo-

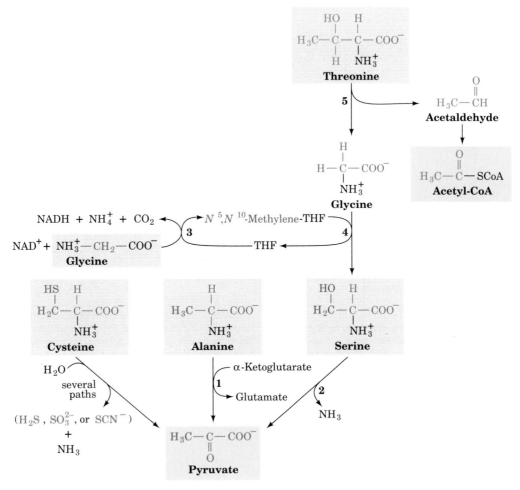

Figure 24-9
The pathways converting alanine, cysteine, glycine, serine, and threonine to pyruvate. The enzymes involved are **(1)** alanine aminotransferase, **(2)** serine dehydratase, **(3)** glycine cleavage system, **(4)** and **(5)** serine hydroxymethyl transferase.

acetate, α-ketoglutarate, succinyl-CoA, fumarate, acetyl-CoA, and acetoacetate.

B. Alanine, Cysteine, Glycine, Serine, and Threonine Are Degraded to Pyruvate

Five amino acids, alanine, cysteine, glycine, serine, and threonine, are broken down to yield pyruvate (Fig. 24-9). Tryptophan should also be included in this group since one of its breakdown products is alanine (Section 24-3G), which, as we have seen, is transaminated to pyruvate.

Serine is converted to pyruvate through dehydration by **serine dehydratase.** This PLP–enzyme, like the aminotransferases (Section 24-1), functions by forming a PLP–amino acid Schiff base so as to facilitate the removal of the amino acid's α-hydrogen atom. In the serine dehydratase reaction, however, the C_α carbanion breaks down with the elimination of the amino acid's C_β—OH, rather than with tautomerization (Fig. 24-2, Step 2), so that the substrate undergoes α, β elimination

of H_2O rather than deamination (Fig. 24-10). The product of the dehydration, the enamine **aminoacrylate,** tautomerizes nonenzymatically to the corresponding imine, which spontaneously hydrolyzes to pyruvate and ammonia.

Cysteine may be converted to pyruvate via several routes in which the sulfhydryl group is released as H_2S, SO_3^{2-}, or SCN^-.

Glycine is converted to serine by the enzyme **serine hydroxymethyl transferase,** another PLP-containing enzyme (Fig. 24-9; Reaction 4). This enzyme utilizes N^5,N^{10}-methylene-tetrahydrofolate (N^5,N^{10}-methylene-THF) as a cofactor to provide the C_1 unit necessary for this conversion. We shall defer discussion of reactions catalyzed by THF cofactors until Section 24-4D.

The methylene group of the N^5,N^{10}-methylene-THF utilized in conversion of glycine to serine is obtained through a second glycine degradation (Fig. 24-9, Reaction 3) catalyzed by the **glycine cleavage system** (also called **glycine synthase** when acting in the reverse direction; Section 24-5A). The glycine cleavage system, a

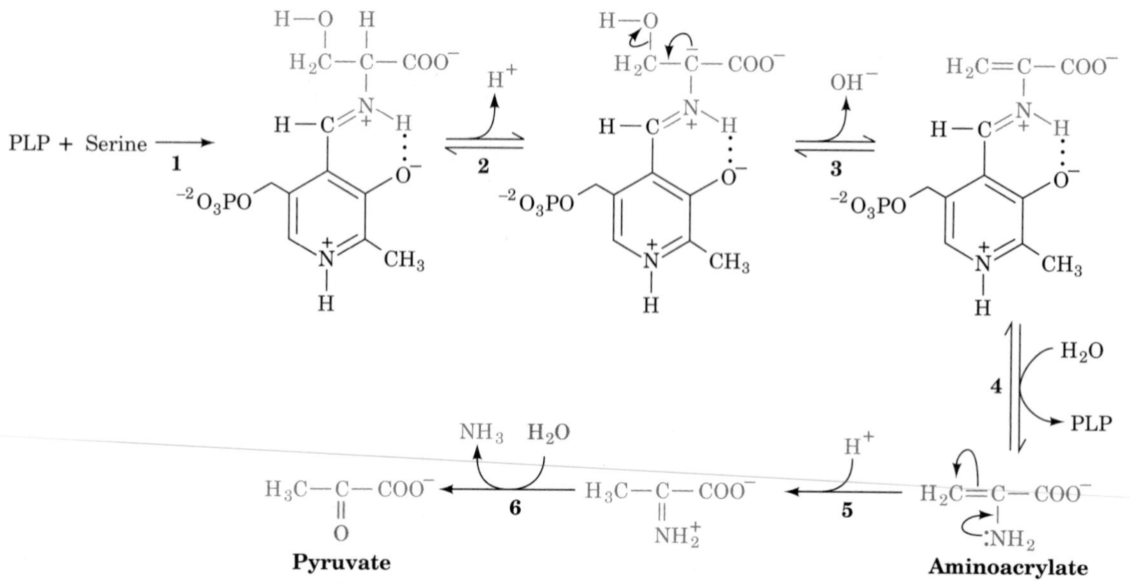

Figure 24-10
Serine dehydratase, a PLP-dependent enzyme, catalyzes the elimination of water from serine. The steps in the reaction are (1) formation of a serine–PLP Schiff base, (2) removal of the α-H atom of serine to form a resonance-stabilized carbanion, (3) β elimination of OH^-, (4) hydrolysis of the Schiff base to yield the PLP–enzyme and aminoacrylate, (5) nonenzymatic tautomerization to the imine, and (6) nonenzymatic hydrolysis to form pyruvate and ammonia.

multienzyme complex that resembles pyruvate dehydrogenase (Section 19-2A), contains four protein components (Fig. 24-11):

1. A PLP-dependent glycine decarboxylase (P protein).

2. A lipoamide-containing aminomethyltransferase, which carries the aminomethyl group remaining after glycine decarboxylation (H protein).

3. An N^5,N^{10}-methylene-THF synthesizing enzyme (T protein), which accepts a methylene group from the aminomethyltransferase (the amino group is released as ammonia).

4. An NAD^+-dependent FAD-requiring lipoamide dehydrogenase (L protein).

Two observations indicate that this pathway is the major route of glycine degradation in mammalian tissues:

1. The serine isolated from an animal that has been fed [2-^{14}C]glycine is ^{14}C labeled at both C(2) and C(3). This observation indicates that the methylene group of the N^5,N^{10}-methylene-THF utilized by serine hydroxymethyl transferase is derived from glycine C(2).

2. The inherited human disease **nonketotic hyperglycinemia**, which is characterized by mental retardation and accumulation of large amounts of glycine in body fluids, results from the absence of the glycine cleavage system.

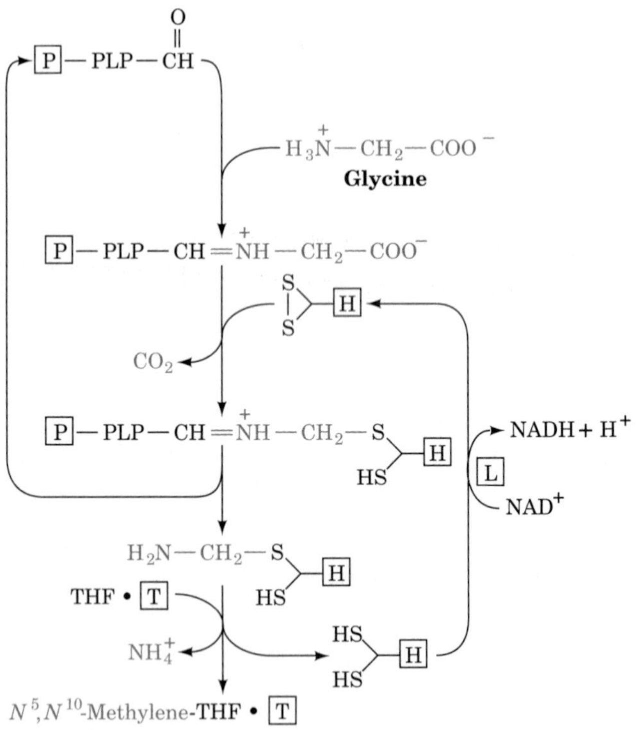

Figure 24-11
The reactions catalyzed by the glycine cleavage system, a multienzyme complex (also called glycine synthase). The enzymes involved are (1) a PLP-dependent glycine decarboxylase (P protein), (2) a lipoamide-containing protein (H protein), (3) a THF-requiring enzyme (T protein), and (4) an NAD^+-dependent, FAD-requiring lipoamide dehydrogenase (L protein).

Threonine is both glucogenic and ketogenic, since one of its degradation routes produces both pyruvate and acetyl-CoA (Fig. 24-9, Reaction 5). Serine hydroxymethyl transferase (acting in the reverse direction from that in Reaction 4) cleaves threonine to acetaldehyde and glycine. This reaction does not require THF as an aldehyde acceptor; the acetaldehyde is directly released. The glycine may be converted, through serine, to pyruvate, whereas the acetaldehyde is oxidized to acetyl-CoA.

Serine Hydroxymethyl Transferase Catalyzes PLP-Dependent C_α—C_β Bond Cleavage

We have heretofore considered PLP-catalyzed reactions that begin with the cleavage of an amino acid's C_α—H bond (Fig. 24-2). Degradation of threonine to glycine and acetaldehyde by serine hydroxymethyl transferase demonstrates that PLP also facilitates cleavage of an amino acid's C_α—C_β bond by delocalizing the electrons of the resulting carbanion into the conjugated PLP ring.

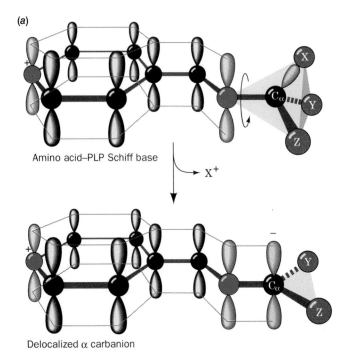

(a)

Amino acid–PLP Schiff base

$\longrightarrow X^+$

Delocalized α carbanion

How can the same amino acid–PLP Schiff base be involved in the cleavage of the different bonds to an amino acid C_α in different enzymes? The answer to this puzzle was suggested by Harmon Dunathan. For electrons to be withdrawn into the conjugated ring system of PLP, the π-orbital system of PLP must overlap with the bonding orbital containing the electron pair being delocalized. This is only possible if the bond being broken lies in the plane perpendicular to the plane of the PLP π-orbital system (Fig. 24-12a). Different bonds to C_α can be placed in this plane by rotation about the C_α—N bond. Indeed, the X-ray structure of aspartate aminotransferase reveals that the C_α—H of its aspartate substrate assumes just this conformation (Fig. 24-12b). Evidently, *each enzyme specifically cleaves its corresponding bond because the enzyme binds the amino acid–PLP Schiff base adduct with this bond in the plane perpendicular to that of the PLP ring.*

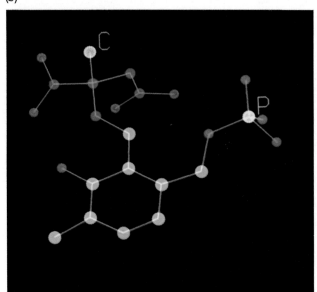

(b)

Figure 24-12
(a) The π-orbital framework of a PLP-amino acid Schiff base. The bond to C_α in the plane perpendicular to the PLP π-orbital system (from X in the illustration) is labile as a consequence of its overlap with the π system, which permits the broken bond's electron pair to be delocalized over the conugated molecule. (b) The Schiff base complex of the inhibitor α-methylaspartate with pyridoxal phosphate in the X-ray structure of porcine aspartate aminotransferase. Here, the methyl C atom (*yellow ball marked* C) occupies the position of the H atom that the enzyme normally excises from aspartate. Note that the bond linking the methyl C atom to aspartate is in the plane perpendicular to the pyridoxal ring and is thus ideally oriented for bond cleavage. [Courtesy of Craig Hyde, NIH.]

C. Asparagine and Aspartate Are Degraded to Oxaloacetate

Transamination of aspartate leads directly to oxaloacetate:

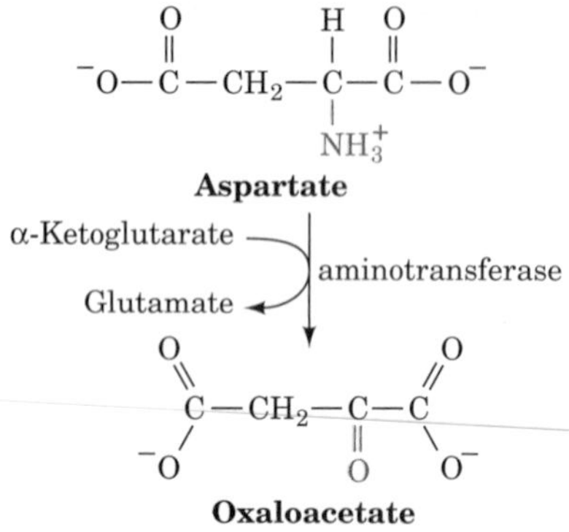

Aspartate

α-Ketoglutarate ⟶
⟵ Glutamate ⟍ aminotransferase

Oxaloacetate

Asparagine is also converted to oxaloacetate in this manner after its hydrolysis to aspartate by **asparaginase:**

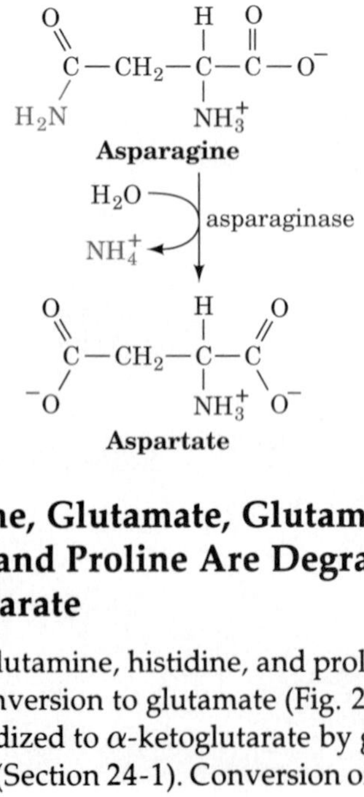

Asparagine

H_2O ⟶
NH_4^+ ⟵ ⟍ asparaginase

Aspartate

D. Arginine, Glutamate, Glutamine, Histidine, and Proline Are Degraded to α-Ketoglutarate

Arginine, glutamine, histidine, and proline are all degraded by conversion to glutamate (Fig. 24-13), which, in turn, is oxidized to α-ketoglutarate by glutamate dehydrogenase (Section 24-1). Conversion of glutamine to glutamate involves only one reaction: hydrolysis by **glutaminase.** Histidine's conversion to glutamate is more complicated: It is nonoxidatively deaminated, hydrated, and its imidazole ring cleaved to form **N-formiminoglutamate.** The formimino group is then transferred to tetrahydrofolate forming glutamate and N^5-**formimino-tetrahydrofolate** (Section 24-4D). Both arginine and proline are converted to glutamate through the intermediate formation of **glutamate-5-semialdehyde.**

E. Isoleucine, Methionine, and Valine Are Degraded to Succinyl-CoA

Isoleucine, methionine, and valine have complex degradative pathways that all yield propionyl-CoA. Pro-

pionyl-CoA, which is also a product of odd-chain fatty acid degradation, is converted, as we have seen, to succinyl-CoA by a series of reactions involving the participation of biotin and coenzyme B_{12} (Section 23-2E).

Methionine Breakdown Involves Synthesis of S-Adenosylmethionine and Cysteine

Methionine degradation (Fig. 24-14) begins with its reaction with ATP to form **S-adenosylmethionine (SAM).** *This sulfonium ion's highly reactive methyl group makes it an important biological methylating agent.* For instance, we have already seen that SAM is the methyl donor in the synthesis of phosphatidylcholine from phosphatidylethanolamine (Section 23-8A). It is also the methyl donor in the conversion of norepinephrine to epinephrine (Section 24-4B).

Methylation reactions involving SAM yield **S-adenosylhomocysteine** in addition to the methylated acceptor. The former product is hydrolyzed to adenosine and **homocysteine** in the next reaction of the methionine degradation pathway. The homocysteine may be methylated to form methionine via a reaction in which N^5-**methyl-THF** is the methyl donor. Alternatively, the homocysteine may combine with serine to yield **cystathionine,** which subsequently forms cysteine (cysteine biosynthesis) and **α-ketobutyrate.** The α-ketobutyrate continues along the degradative pathway to propionyl-CoA and then succinyl-CoA.

Branched-Chain Amino Acid Degradation Pathways Contain Themes Common to All Acyl-CoA Oxidations

Degradation of the branched-chain amino acids isoleucine, leucine, and valine, begins with three reactions that employ common enzymes (Fig. 24-15, *top*): (1) transamination to the corresponding α-keto acid, (2) oxidative decarboxylation to the corresponding acyl-CoA, and (3) dehydrogenation by FAD to form a double bond.

The remainder of the isoleucine degradation pathway (Fig. 24-15, *left*) is identical to that of fatty acid oxidation (Section 23-2C): (4) double-bond hydration, (5) dehydrogenation by NAD^+, and (6) thiolytic cleavage yielding acetyl-CoA and propionyl-CoA, which is subsequently converted to succinyl-CoA. Valine degradation is a variation on this theme (Fig. 24-15, *center*): following (7) double-bond hydration, (8) the CoA thioester bond is hydrolyzed before (9) the second dehydrogenation reaction. The thioester bond is then regenerated as propionyl-CoA in the sequence's last reaction (10), an oxidative decarboxylation rather than a thiolytic cleavage.

Maple Syrup Urine Disease Results from a Defect in Branched-Chain Amino Acid Degradation

α-Ketoisovalerate dehydrogenase, which catalyzes Reaction 2 of branched-chain amino acid degradation

(Fig. 24-15), is a multienzyme complex that employs the coenzymes TPP, lipoamide, and FAD in addition to its terminal oxidizing agent, NAD⁺. A genetic deficiency in this enzyme causes **maple syrup urine disease,** so-named because the consequent buildup of branched-chain α-keto acids imparts the urine with the characteristic odor of maple syrup. Unless promptly treated by a diet low in branched-chain amino acids, maple syrup urine disease is rapidly fatal.

F. Leucine and Lysine Are Degraded to Acetoacetate and/or Acetyl-CoA

Leucine is oxidized by a combination of reactions used in β oxidation and ketone body synthesis (Fig. 24-15, *right*). The first dehydrogenation and the hydra-

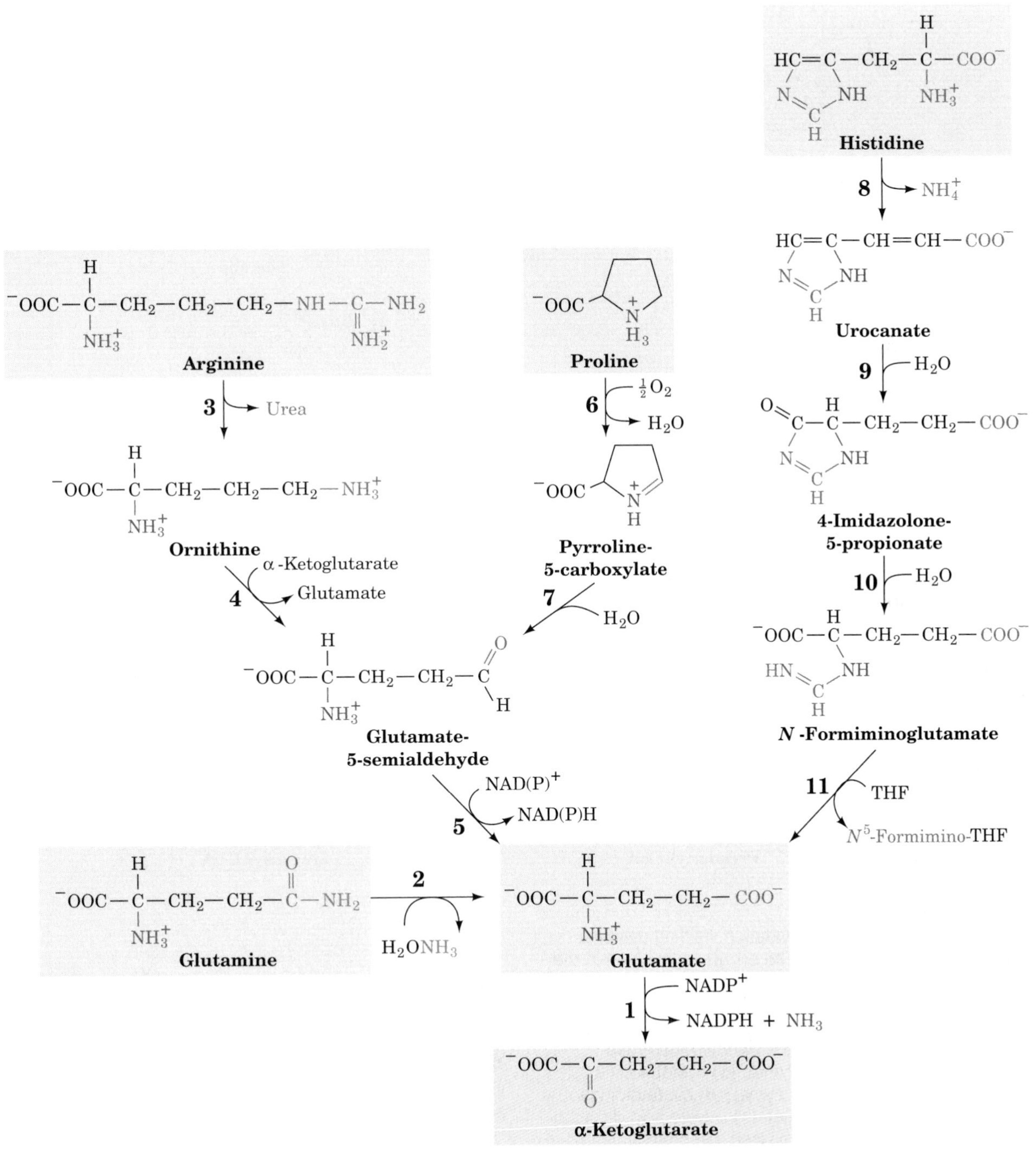

Figure 24-13

Degradation pathways of arginine, glutamate, glutamine, histidine, and proline to α-ketoglutarate. The enzymes catalyzing the reactions are **(1)** glutamate dehydrogenase, **(2)** glutaminase, **(3)** arginase, **(4)** ornithine-δ-amino- transferase, **(5)** glutamate semialdehyde dehydrogenase, **(6)** proline oxidase, **(7)** spontaneous, **(8)** histidine ammonia lyase, **(9)** urocanate hydratase, **(10)** imidazolone propionase, and **(11)** glutamate formimino transferase.

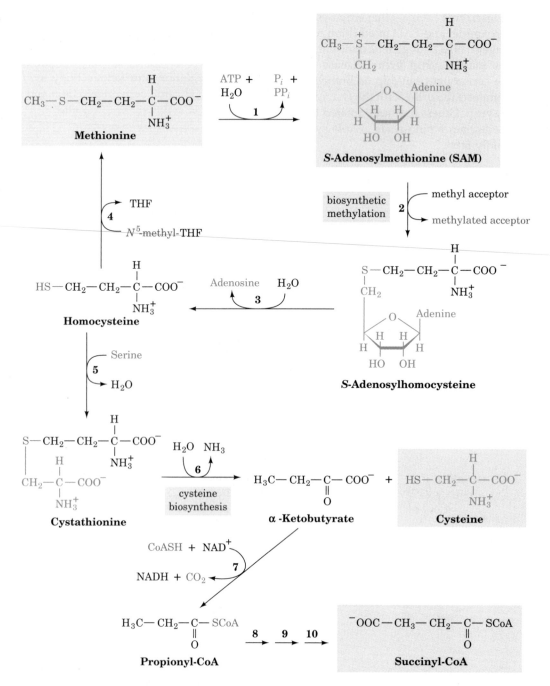

Figure 24-14
The pathway of methionine degradation yielding cysteine and succinyl-CoA as products. The enzymes involved in the pathway are **(1)** methionine adenosyl transferase in a reaction that yields the biological methylating agent *S*-adenosylmethionine (SAM), **(2)** methylase, **(3)** adenosyl-homocysteinase, **(4)** homocysteine methyltransferase (a coenzyme B_{12}-dependent enzyme), **(5)** cystathionine β-synthase (a PLP-dependent enzyme), **(6)** cystathionine γ-lyase (a PLP-dependent enzyme), **(7)** α-keto acid dehydrogenase, **(8)** propionyl-CoA carboxylase, **(9)** methylmalonyl-CoA racemase, and **(10)** methylmalonyl-CoA mutase (a coenzyme B_{12}-dependent enzyme; Reactions 8–10 are discussed in Section 23-2E).

tion reactions are interspersed by (11) a carboxylation reaction catalyzed by a biotin-containing enzyme. The hydration reaction (12) then produces **β-hydroxy-β-methylglutaryl-CoA (HMG-CoA),** which is cleaved by HMG-CoA lyase to form acetyl-CoA and the ketone body acetoacetate (13) (which, in turn, may be converted to 2 acetyl-CoA; Section 23-3).

Although there are several pathways for lysine degradation, the one that proceeds via formation of the α-ketoglutarate–lysine adduct **saccharopine** predominates in mammalian liver (Fig. 24-16). This pathway is

Figure 24-15

The degradation of the branched-chain amino acids (A) isoleucine, (B) valine, and (C) leucine. The first three reactions of each pathway utilize the common enzymes **(1)** branched-chain amino acid aminotransferase, **(2)** α-ketoisovalerate dehydrogenase, and **(3)** acyl-CoA dehydrogenase. Isoleucine degradation then continues (*left*) with **(4)** enoyl-CoA hydratase, **(5)** β-hydroxyacyl-CoA dehydrogenase, and **(6)** acetyl-CoA acetyl transferase to yield acetyl-CoA and the succinyl-CoA precursor propionyl-CoA. Valine degradation (*center*) continues with **(7)** enoyl-CoA hydratase, **(8)** β-hydroxyisobutyryl-CoA hydrolase, **(9)** β-hydroxyisobutyrate dehydrogenase, and **(10)** methylmalonate semialdehyde dehydrogenase to also yield propionyl-CoA. Leucine degradation (*right*) continues with **(11)** β-methylcrotonyl-CoA carboxylase (a biotin-dependent enzyme), **(12)** β-methylglutaconyl-CoA hydratase, and **(13)** HMG-CoA lyase to yield acetyl-CoA and acetoacetate.

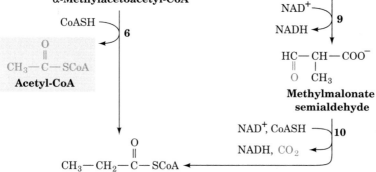

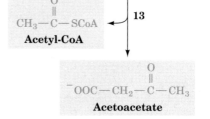

(A) Isoleucine : $R_1 = CH_3—$, $R_2 = CH_3—CH_2—$
(B) Valine : $R_1 = CH_3—$, $R_2 = CH_3—$
(C) Leucine : $R_1 = H—$, $R_2 = (CH_3)_2CH—$

(A) α-Keto-β-methylvalerate
(B) α-Ketoisovalerate
(C) α-Ketoisocaproic acid

(A) α-Methylbutyryl-CoA
(B) Isobutyryl-CoA
(C) Isovalyryl-CoA

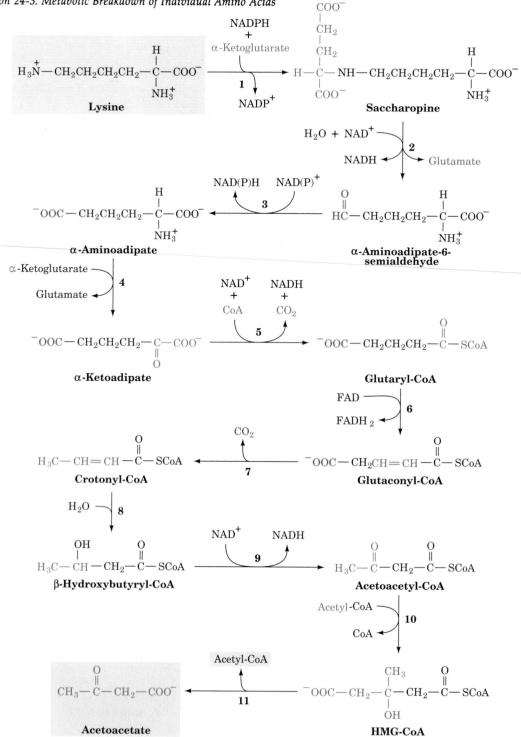

Figure 24-16

The pathway of lysine degradation in mammalian liver. The enzymes involved are **(1)** saccharopine dehydrogenase (NADP⁺, lysine forming), **(2)** saccharopine dehydrogenase (NAD⁺, glutamate forming), **(3)** aminoadipate semialdehyde dehydrogenase, **(4)** aminoadipate aminotransferase (a PLP enzyme), **(5)** α-keto acid dehydrogenase, **(6)** glutaryl-CoA dehydrogenase, **(7)** decarboxylase, **(8)** enoyl-CoA hydratase, **(9)** β-hydroxyacyl-CoA dehydrogenase, **(10)** HMG-CoA synthase, and **(11)** HMG-CoA lyase (Reactions 10 and 11 are discussed in Section 23-3).

of interest because we have encountered 7 of its 11 reactions in other pathways. Reaction 4 is a PLP-dependent transamination. Reaction 5 is the oxidative decarboxylation of an α-keto acid by a multienzyme complex similar to pyruvate dehydrogenase and α-ketoglutarate dehydrogenase (Sections 19-2A and 19-3D). Reactions 6, 8, and 9 are standard reactions of fatty acyl-CoA oxidation: dehydrogenation by FAD, hydration, and dehydrogenation by NAD⁺. Reactions 10 and 11 are

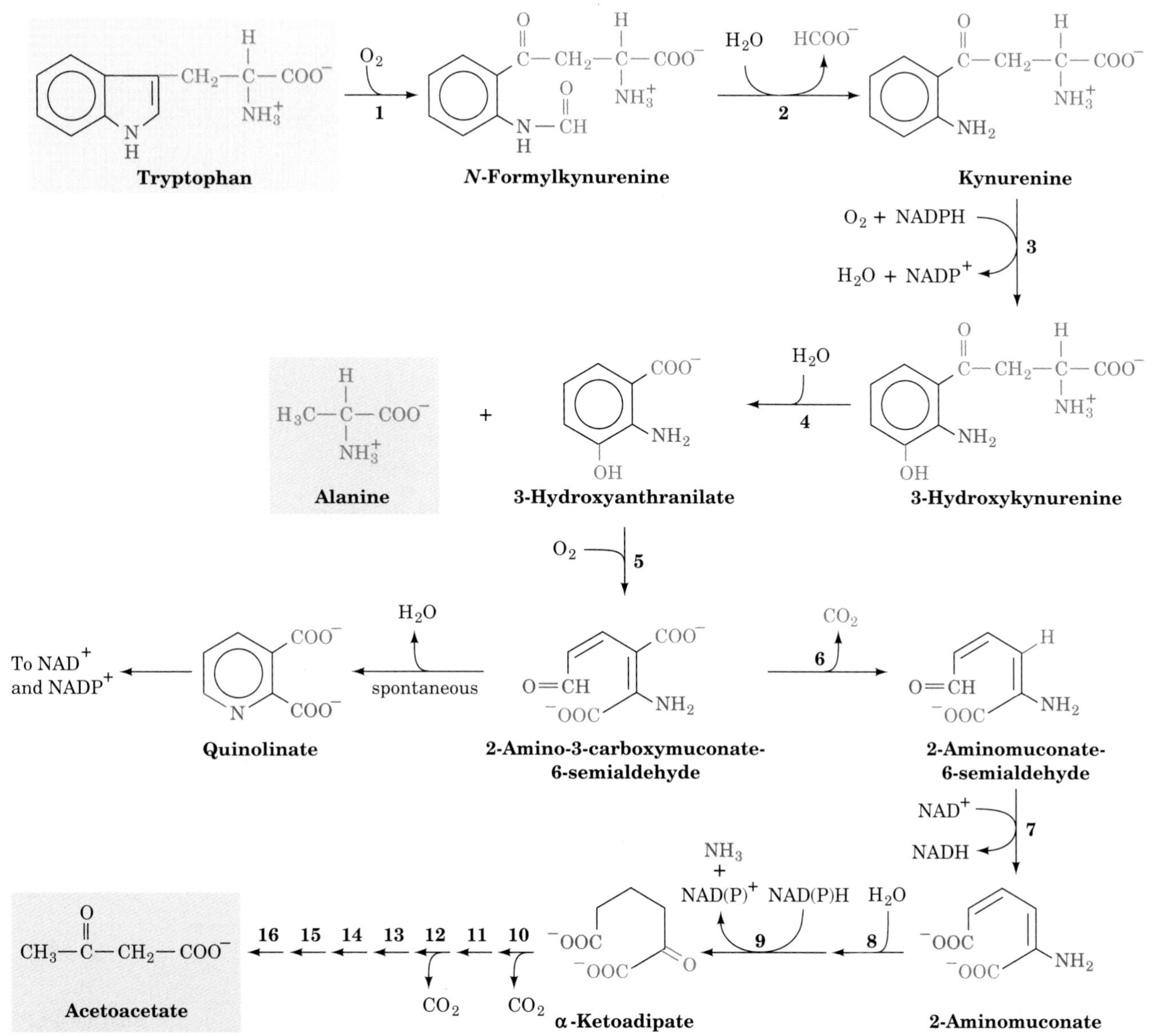

Figure 24-17
The pathway of tryptophan degradation. The enzymes involved are **(1)** tryptophan-2,3-dioxygenase, **(2)** formamidase, **(3)** kynurenine-3-monooxygenase, **(4)** kynureninase (PLP dependent), **(5)** 3-hydroxyanthranilate-3,4-dioxygenase, **(6)** amino carboxymuconate semialdehyde decarboxylase, **(7)** aminomuconate semialdehyde dehydrogenase, **(8)** hydratase, **(9)** dehydrogenase, **(10–16)** Reactions 5 to 11 in lysine degradation (Fig. 24-16). 2-Amino-3-carboxymuconate-6-semialdehyde, in addition to undergoing Reaction 6, spontaneously forms **quinolinate,** an NAD$^+$ and NADP$^+$ precursor (Section 26-6A).

standard reactions in ketone body formation. Two moles of CO_2 are produced at Reactions 5 and 7 of the pathway.

The saccharopine pathway is thought to predominate in mammals because a genetic defect in the enzyme that catalyzes Reaction 1 in the sequence results in **hyperly-**

sinemia and **hyperlysinuria** (elevated levels of lysine in the blood and urine, respectively) along with mental and physical retardation. This is yet another example of how the study of rare inherited disorders has helped to trace metabolic pathways.

Leucine's carbon skeleton, as we have seen, is converted to one molecule each of acetoacetate and acetyl-CoA, whereas that of lysine is converted to one molecule of acetoacetate and two of CO_2. Since neither acetoacetate nor acetyl-CoA can be converted to glucose in animals, leucine and lysine are purely ketogenic amino acids.

G. Tryptophan Is Degraded to Alanine and Acetyl-CoA

The complexity of the major tryptophan degradation pathway (Fig. 24-17) precludes the detailed discussion of all of its reactions. However, one reaction in the path-

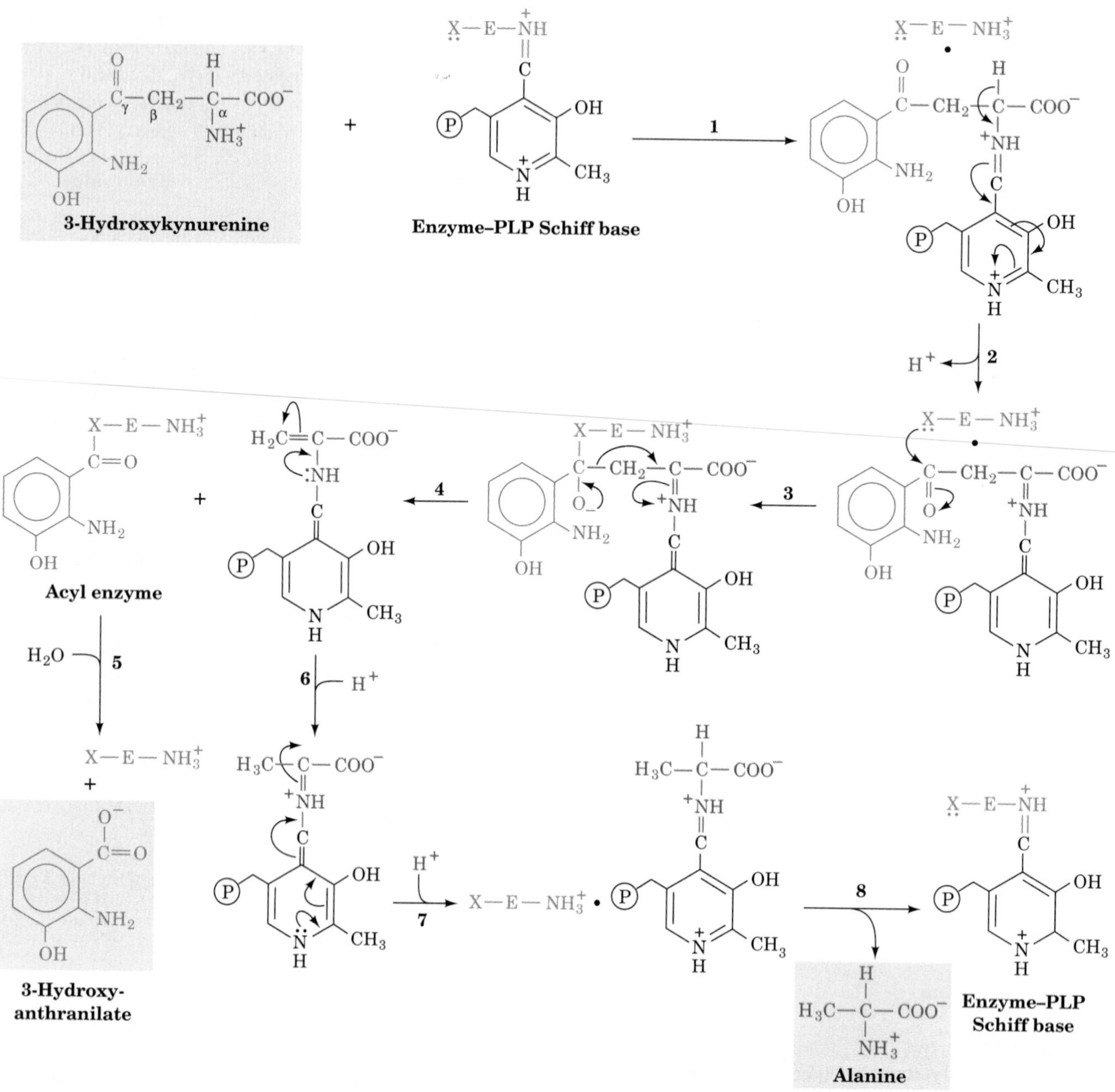

Figure 24-18
The proposed mechanism for the PLP-dependent kynureninase-catalyzed
C_β—C_γ bond cleavage of 3-hydroxykynurenine occurs in eight steps:
(1) transimination, **(2)** tautomerization, **(3)** attack of an enzyme nucleophile,
(4) C_β—C_γ bond cleavage with formation of an acyl–enzyme intermediate,
(5) acyl–enzyme hydrolysis, **(6)** and **(7)** tautomerization, and
(8) transimination.

way is of particular interest. Reaction 4, cleavage of
3-hydroxykynurenine to alanine and **3-hydroxy-
anthranilate,** is catalyzed by **kynureninase,** a PLP-dependent enzyme. The reaction further demonstrates the
enormous versatility of PLP. We have seen how PLP can
labilize an α-amino acid's C_α—H bond. Here we see the
facilitation of C_β—C_γ bond cleavage. The reaction follows the same steps as transamination reactions but

does not hydrolyze the tautomerized Schiff base (Fig.
24-18). The proposed reaction mechanism involves an
attack of an enzyme nucleophile on the carbonyl carbon
(C_γ) of the tautomerized 3-hydroxykynurenine–PLP
Schiff base (Fig. 24-18, Step 3). This is followed by
C_β—C_γ bond cleavage to generate an acyl–enzyme intermediate together with a tautomerized alanine–PLP
adduct (Fig. 24-18, Step 4). Hydrolysis of the acyl–

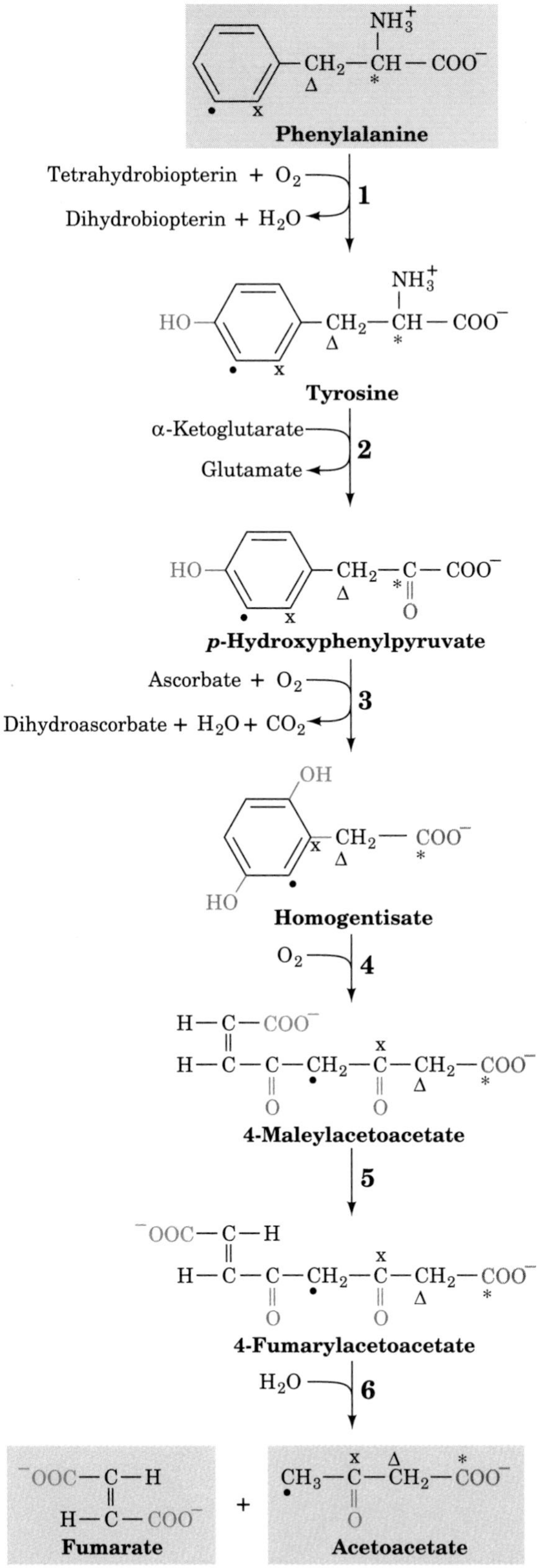

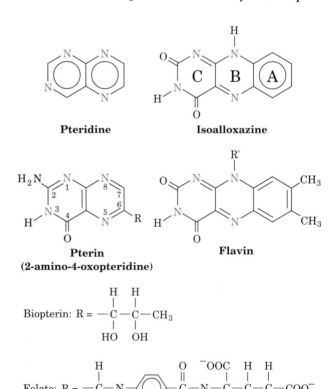

Phenylalanine

Tetrahydrobiopterin + O_2

Dihydrobiopterin + H_2O

1

Tyrosine

α-Ketoglutarate

Glutamate

2

***p*-Hydroxyphenylpyruvate**

Ascorbate + O_2

Dihydroascorbate + H_2O + CO_2

3

Homogentisate

O_2

4

4-Maleylacetoacetate

5

4-Fumarylacetoacetate

H_2O

6

Fumarate + **Acetoacetate**

Figure 24-19
The pathway of phenylalanine degradation. The enzymes involved are **(1)** phenylalanine hydroxylase, **(2)** aminotransferase, **(3)** *p*-hydroxyphenylpyruvate dioxygenase, **(4)** homogentisate dioxygenase, **(5)** maleylacetoacetate isomerase, and **(6)** fumarylacetoacetase. The symbols labeling the various carbon atoms serve to indicate the group migration that occurs in Reaction 3 of the pathway (see Fig. 24-23).

enzyme then yields 3-hydroxyanthranilate whose further degradation yields **α-ketoadipate** (Fig. 24-17, Reactions 4–9). α-Ketoadipate is also an intermediate in lysine breakdown (Fig. 24-16, Reaction 4) so that the last seven reactions in the degradation of both these amino acids are identical, forming acetoacetate and two molecules of CO_2.

H. Phenylalanine and Tyrosine Are Degraded to Fumarate and Acetoacetate

Since the first reaction in phenylalanine degradation is its hydroxylation to tyrosine, a single pathway (Fig. 24-19) is responsible for the breakdown of both of these amino acids. The final products of the six-reaction degradation are fumarate, a citric acid cycle intermediate, and acetoacetate, a ketone body.

Pterins Are Redox Cofactors

The hydroxylation of phenylalanine by the Fe(III)-containing enzyme **phenylalanine hydroxylase** requires the participation of a cofactor we have not yet encountered: **biopterin**, a **pterin** derivative. Pterins are compounds that contain the **pteridine** ring (Fig. 24-20). Note the resemblance between the pteridine ring and the isoalloxazine ring of the flavin coenzymes; the posi-

Pteridine   **Isoalloxazine**

Pterin
(2-amino-4-oxopteridine)   **Flavin**

Biopterin: R =

Folate: R =

Figure 24-20
The pteridine ring is the nucleus of biopterin and folate. Note the similar structures of pteridine and the isoalloxazine ring of flavin coenzymes.

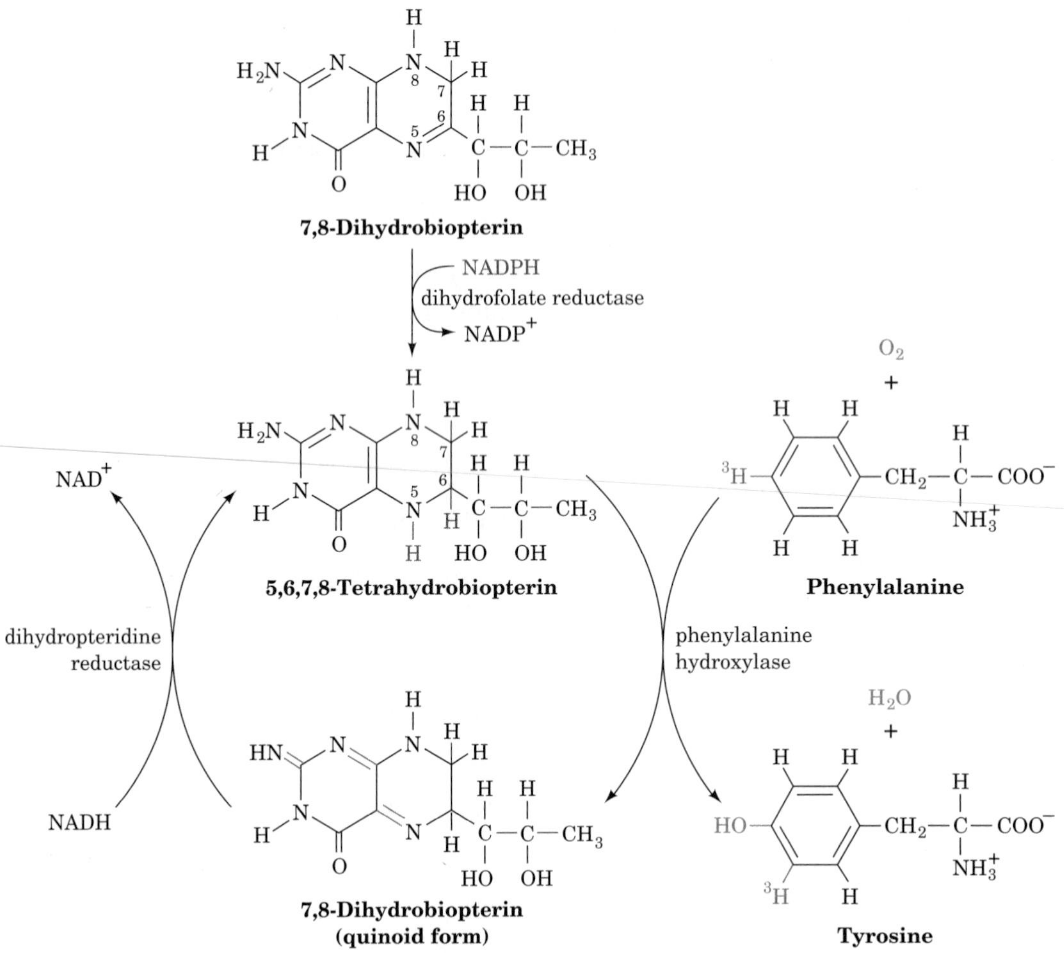

Figure 24-21
The formation, utilization, and regeneration of 5,6,7,8-tetrahydrobiopterin in the phenylalanine hydroxylase reaction.

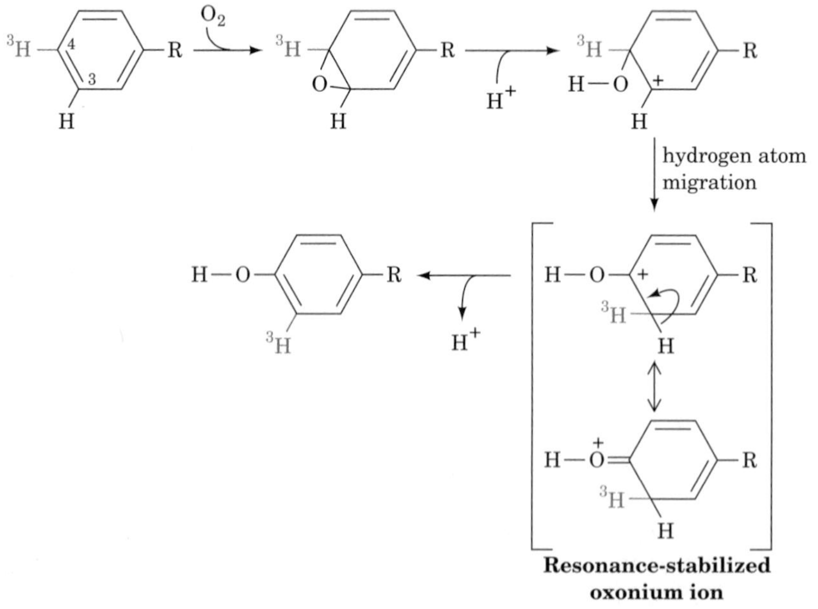

Figure 24-22
The proposed mechanism of the NIH shift. The rearrangement is driven by the formation of a resonance-stabilized oxonium ion.

tions of the nitrogen atoms in pteridine are identical with those of the B and C rings of isoalloxazine. Folate derivatives also contain the pterin ring (Section 24-4D).

Pterins, like flavins, participate in biological oxidations. The active form of biopterin is the fully reduced form, **5,6,7,8-tetrahydrobiopterin.** It is produced from **7,8-dihydrobiopterin** and NADPH, in what may be considered a priming reaction, by **dihydrofolate reductase** (Fig. 24-21). In the phenylalanine hydroxylase reaction, 5,6,7,8-tetrahydrobiopterin is oxidized to **7,8-dihydrobiopterin (quinoid form).** This quinoid is subsequently reduced by the NADH-requiring enzyme **dihydropteridine reductase** to regenerate the active cofactor.

The NIH Shift

An unexpected aspect of the foregoing reaction is that a ^3H atom, which begins on C(4) of phenylalanine's phenyl ring, ends up on C(3) of this ring in tyrosine (Fig. 24-21, *right*) rather than being lost to the solvent by replacement with the OH group. The mechanism postulated to account for this **NIH shift** (so-called because it was first characterized by chemists at the National Institutes of Health) involves the activation of oxygen to form an epoxide across the phenyl ring's 3,4 bond, followed by epoxide opening to form a carbocation at C(3) (Fig. 24-22). Migration of a hydride from C(4) to C(3) forms a more stable carbocation (an oxonium ion). This migration is followed by ring aromatization to form tyrosine.

Reaction 3 (Fig. 24-19) in the phenylalanine degradation pathway provides a second example of an NIH shift. This reaction, which is catalyzed by the Cu-containing *p*-**hydroxyphenylpyruvate dioxygenase,** involves the oxidative decarboxylation of an α-keto acid as well as ring hydroxylation. In this case, the NIH shift involves migration of an alkyl group rather than of a hydride ion to form a more stable carbocation (Fig. 24-23). This shift, which has been demonstrated through isotope-labeling studies (represented by the different symbols in Figs. 24-19 and 24-23), accounts for the observation that C(3) is bonded to C(4) in *p*-**hydroxyphenylpyruvate** but to C(5) in **homogentisate.**

Alcaptonuria and Phenylketonuria Result from Defects in Phenylalanine Degradation

Archibald Garrod realized in the early 1900s that human genetic diseases result from specific enzyme deficiencies. We have repeatedly seen how this realization has contributed to the elucidation of metabolic pathways. The first such disease to be recognized was **alkaptonuria,** which, Garrod observed, resulted in the excretion of large quantities of homogentisic acid. This condition results from deficiency of **homogentisate dioxygenase** (Fig. 24-19, Reaction 4). Alcaptonurics suffer

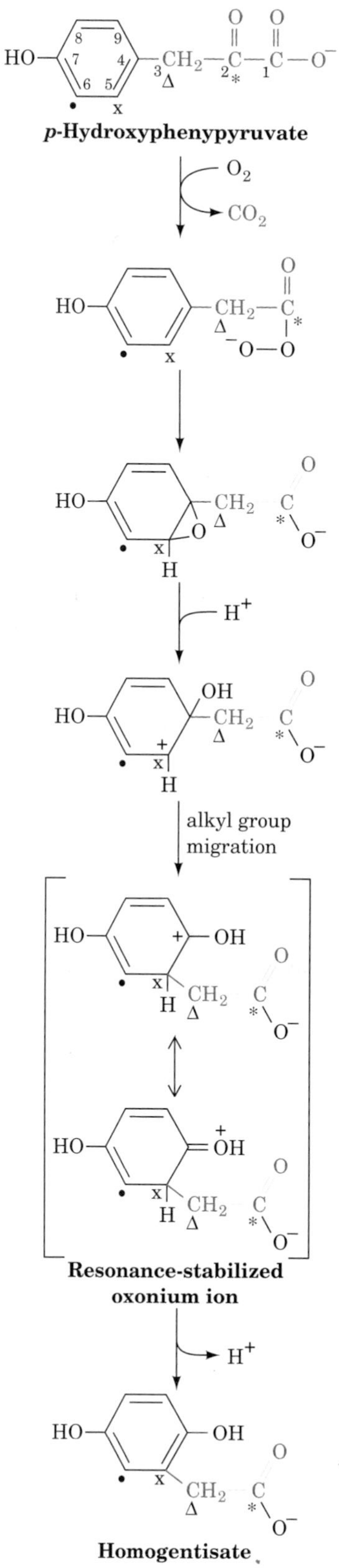

Figure 24-23
The NIH shift in the *p*-hydroxyphenylpyruvate dioxygenase reaction. Carbon atoms are labeled as an aid to following the group migration constituting the shift.

no ill effects other than arthritis later in life (although their urine darkens alarmingly resulting from the rapid air oxidation of the homogentisate they excrete).

Individuals suffering from **phenylketonuria (PKU)** are not so fortunate. Severe mental retardation occurs within a few months of birth if the disease is not detected and treated immediately (see below). Indeed, ~1% of the patients in mental institutions are phenylketonurics. PKU is caused by the inability to hydroxylate phenylalanine (Fig. 24-19, Reaction 1) and therefore results in increased blood levels of phenylalanine **(hyperphenylalaninemia).** The excess phenylalanine is transaminated to **phenylpyruvate**

$$\text{\Large$\bigcirc$}-CH_2-\overset{\overset{\displaystyle O}{\|}}{C}-COO^-$$

Phenylpyruvate

by an otherwise minor pathway. The "spillover" of phenylpyruvate (a phenylketone) into the urine was the first observation connected with the disease and gave the disease its name. All babies born in the United States are now screened for PKU immediately after birth by testing for elevated levels of phenylalanine in the blood.

Classical PKU results from a deficiency in phenylalanine hydroxylase. When this was established in 1947, it was the first human inborn error of metabolism whose basic biochemical defect had been identified. Since all of the tyrosine breakdown enzymes are normal, treatment consists in providing the patient with a low phenylalanine diet and monitoring the blood level of phenylalanine to ensure that it remains within normal limits for the first 5 to 10 years of life (the adverse effects of hyperphenylalaninemia seem to disappear after that age). Phenylalanine hydroxylase deficiency also accounts for another common symptom of PKU: Its victims have lighter hair and skin color than their siblings. This is because tyrosine hydroxylation, the first reaction in the formation of the black skin pigment **melanin** (Section 24-4B), is inhibited by elevated phenylalanine levels.

Other causes of hyperphenylalaninemia have been discovered since the introduction of infant screening techniques. These are caused by deficiencies in the enzymes catalyzing the formation or regeneration of 5,6,7,8-tetrahydrobiopterin, the phenylalanine hydroxylase cofactor (Fig. 24-21). In such cases, patients must also be supplied with **L-3,4-dihydroxyphenylalanine (L-DOPA)** and **5-hydroxytryptophan,** metabolic precursors of the neurotransmitters **norepinephrine** and **serotonin,** respectively, since the enzymes that produce these physiologically active amines also require 5,6,7,8-tetrahydrobiopterin (Section 24-4B). Simply adding 5,6,7,8-tetrahydrobiopterin to the diet of affected individuals is not sufficient because this compound is unstable and cannot cross the blood–brain barrier.

4. AMINO ACIDS AS BIOSYNTHETIC PRECURSORS

Certain amino acids, in addition to their major function as protein building blocks, are essential precursors of a variety of important biomolecules including nucleotides and nucleotide coenzymes, heme, various hormones and neurotransmitters, and glutathione. In this section, we therefore consider the pathways producing some of these substances. We begin by discussing the biosynthesis of heme from glycine and succinyl-CoA. We then examine the pathways by which tyrosine, tryptophan, glutamate, and histidine are converted to various neurotransmitters and study certain aspects of glutathione biosynthesis and the involvement of this tripeptide in amino acid transport and other processes. Finally, we consider the role of folate derivatives in the biosynthetic transfer of C_1 units. The biosynthesis of nucleotides and nucleotide coenzymes is the subject of Chapter 26.

A. Heme Biosynthesis and Degradation

Heme (Fig. 24-24), as we have seen, is an Fe-containing prosthetic group that is an essential component of many proteins, notably hemoglobin, myoglobin, and the cytochromes. The initial reactions of heme biosynthesis are common to the formation of other tetrapyr-

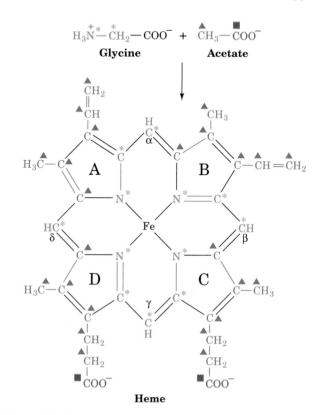

Figure 24-24
Heme's C and N atoms are derived from those of glycine and acetate.

roles including chlorophyll in plants and bacteria (Section 22-1A) and coenzyme B_{12} in bacteria (Section 23-2E).

Porphyrins Are Derived from Succinyl-CoA and Glycine

Elucidation of the heme biosynthesis pathway involved some interesting detective work. David Shemin and David Rittenberg, who were among the first to use isotopic tracers in the elucidation of metabolic pathways, demonstrated, in 1945, that *all of heme's C and N atoms can be derived from acetate and glycine.* Only glycine, out of a variety of ^{15}N-labeled metabolites they tested, including ammonia, glutamate, leucine, and proline, yielded ^{15}N-labeled heme in the hemoglobin of experimental subjects to whom these metabolites were administered. Similar experiments, using acetate labeled with ^{14}C in its methyl or carboxyl groups, or $[^{14}C_\alpha]$glycine, demonstrated that 24 of heme's 34 carbon atoms are derived from acetate's methyl carbon, 2 from acetate's carboxyl carbon, and 8 from glycine's C_α atom (Fig. 24-24). None of the heme atoms are derived from glycine's carboxyl carbon atom.

Figure 24-24 indicates that heme C atoms derived from acetate methyl groups occur in groups of three linked atoms. Evidently, acetate is first converted to some other metabolite that has this labeling pattern. Shemin and Rittenbenberg postulated that this metabolite is succinyl-CoA based on the following reasoning (Fig. 24-25):

1. Acetate is metabolized via the citric acid cycle (Section 19-3 I).

2. Labeling studies indicate that atom C(3) of the citric acid cycle intermediate succinyl-CoA is derived from acetate's methyl C atom while atom C(4) comes from acetate's carboxyl C atom.

3. After many turns of the citric acid cycle, C(1) and C(2) of succinyl-CoA likewise become fully derived from acetate's methyl C atom.

We shall see that this labeling pattern indeed leads to that of heme.

The first phase of heme biosynthesis is a condensation of succinyl-CoA with glycine followed by decarboxylation to form **δ-aminolevulinic acid (ALA)** as catalyzed by the PLP-dependent enzyme **δ-aminolevulinate synthase** (Fig. 24-26). The carboxyl group lost in the decarboxylation (Fig. 24-26, Reaction 5) originates in glycine, which is why heme contains no label from this group.

The Pyrrole Ring Is the Product of Two ALA Molecules

The pyrrole ring is formed in the next phase of the pathway through linkage of two molecules of ALA to

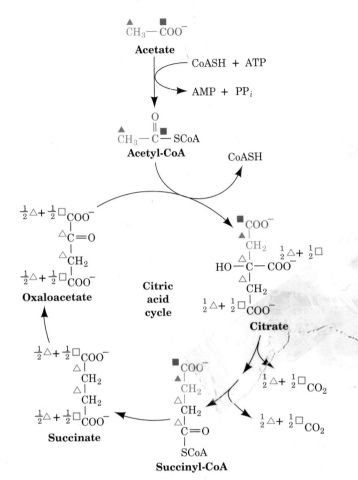

Figure 24-25
The origin of the C atoms of succinyl-CoA as derived from acetate via the citric acid cycle. C atoms labeled with triangles and squares are derived, respectively, from acetate's methyl and carboxyl C atoms. Filled symbols label atoms derived from acetate in the present round of the citric acid cycle, whereas open symbols label atoms derived from acetate in previous rounds of the citric acid cycle. Note that the C(1) and C(4) atoms of succinyl-CoA are scrambled on forming the twofold symmetric succinate.

yield **porphobilinogen (PBG).** The reaction is catalyzed by the Zn-requiring enzyme **porphobilinogen synthase** (alternatively, **δ-aminolevulinic acid dehydratase),** and involves Schiff base formation of one of the substrate molecules with an enzyme amine group. One possible mechanism of this condensation–elimination reaction involves formation of a second Schiff base between the ALA–enzyme Schiff base and the second ALA molecule (Fig. 24-27). At this point, if we continue tracing the acetate and glycine labels through the PBG synthase reaction (Fig. 24-27), we can begin to see how heme's labeling pattern arises.

Inhibition of PBG synthase by lead is one of the major manifestations of acute lead poisoning. Indeed, it has been suggested that the accumulation, in the blood, of ALA, which resembles the neurotransmitter **γ-amino-**

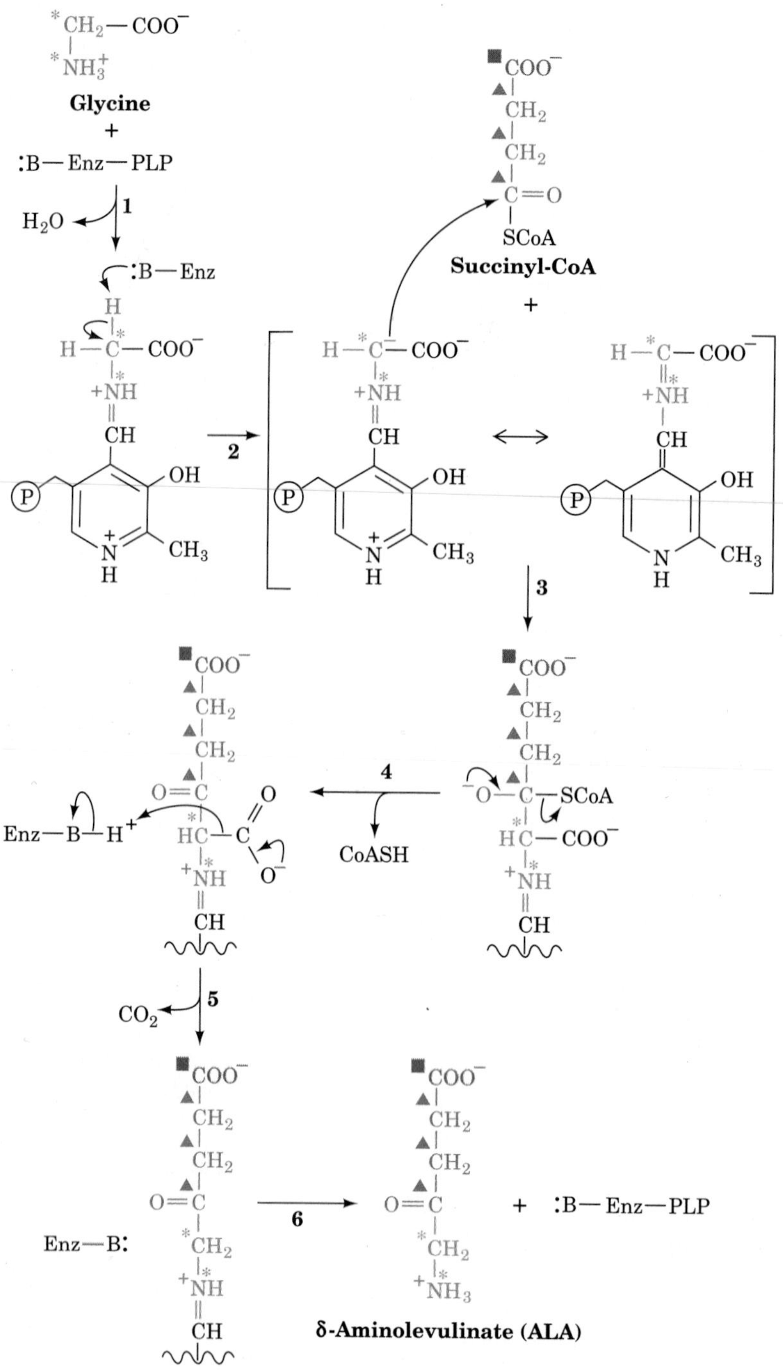

Figure 24-26
The mechanism of action of the PLP-dependent enzyme,
δ-aminolevulinate synthase. The reaction steps are
(1) transimination, (2) PLP-stabilized carbanion formation,
(3) C—C bond formation, (4) CoA elimination,
(5) decarboxylation facilitated by the PLP–Schiff base, and
(6) transimination yielding ALA and regenerating the
PLP–enzyme.

butyric acid (Section 24-4B), is responsible for the psychosis that often accompanies lead poisoning.

The Porphyrin Ring Is Formed from Four PBG Molecules

The next phase of heme biosynthesis is the condensation of four PBG molecules to form **uroporphyrinogen** **III**, the porphyrin nucleus, in a series of reactions catalyzed by **uroporphyrinogen synthase** (alternatively, **porphobilinogen deaminase**) and **uroporphyrinogen III cosynthase**. The reaction (Fig. 24-28), whose mechanism was elucidated by Alan Battersby, begins with the enzyme's displacement of the amino group in PBG to form a covalent adduct. A second, third, and fourth PBG

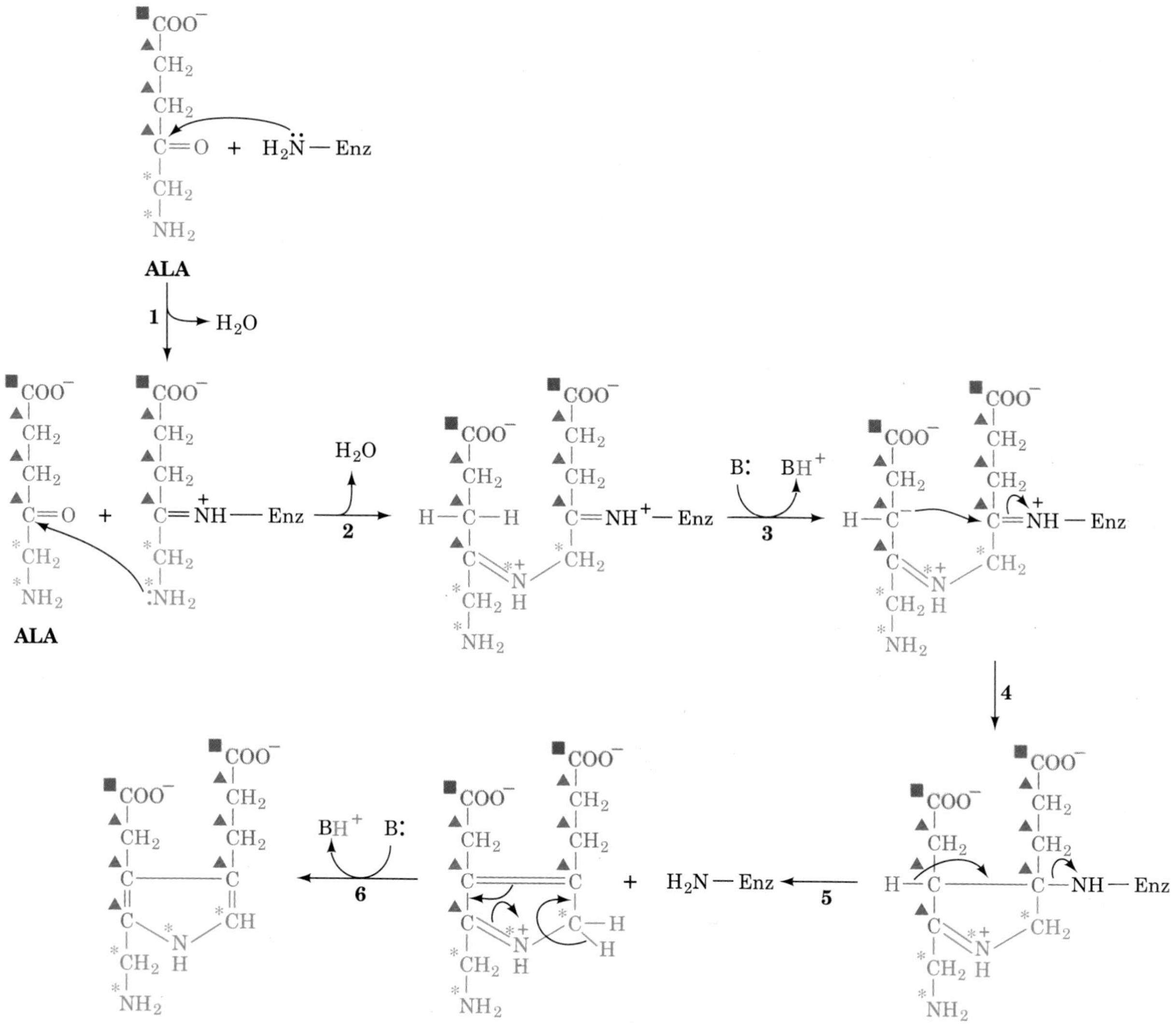

Figure 24-27

A possible mechanism for porphobilinogen synthase. The reaction involves: **(1)** Schiff base formation, **(2)** second Schiff base formation, **(3)** formation of a carbanion α to a Schiff base, **(4)** cyclization by an aldol-type condensation, **(5)** elimination of the enzyme NH_2 group, and **(6)** tautomerization.

then sequentially add through the displacement of the primary amino group on one PBG by a carbon atom on the pyrrole ring of the succeeding PBG to yield the linear tetrapyrrole **hydroxymethylbilane.**

Cyclization requires the participation of uroporphyrinogen III cosynthase (Fig. 24-28). In the absence of the cosynthase, hydroxymethylbilane is released from the synthase and rapidly cyclizes nonenzymatically to the symmetric **uroporphyrinogen I.** Heme, however, is an asymmetric molecule; the methyl substituent of pyrrole ring D has an inverted placement compared to those of rings A, B, and C (Fig. 24-24). This ring reversal to yield uroporphyrinogen III is thought to proceed through attachment of the methylenes from rings A and C to the same carbon of ring D so as to form a spiro compound (a bicyclic compound with a carbon atom common to both rings; Fig. 24-28).

Heme biosynthesis takes place partly in the mitochondrion and partly in the cytosol (Fig. 24-29). ALA is mitochondrially synthesized and is transported to the cytosol for conversion to PBG and then to uroporphyrinogen III. **Protoporphyrin IX,** to which Fe is added to form heme, is produced from uroporphyrinogen III in a series of reactions catalyzed by (1) **uroporphyrinogen decarboxylase,** which decarboxylates all four acetate side chains (A) to form methyl groups (M); (2) **coproporphyrinogen oxidase,** which oxidatively decarboxylates two of the propionate side chains (P) to vinyl groups (V); and (3) **protoporphyrinogen oxidase,** which oxidizes the methylene groups linking the pyrrole rings to

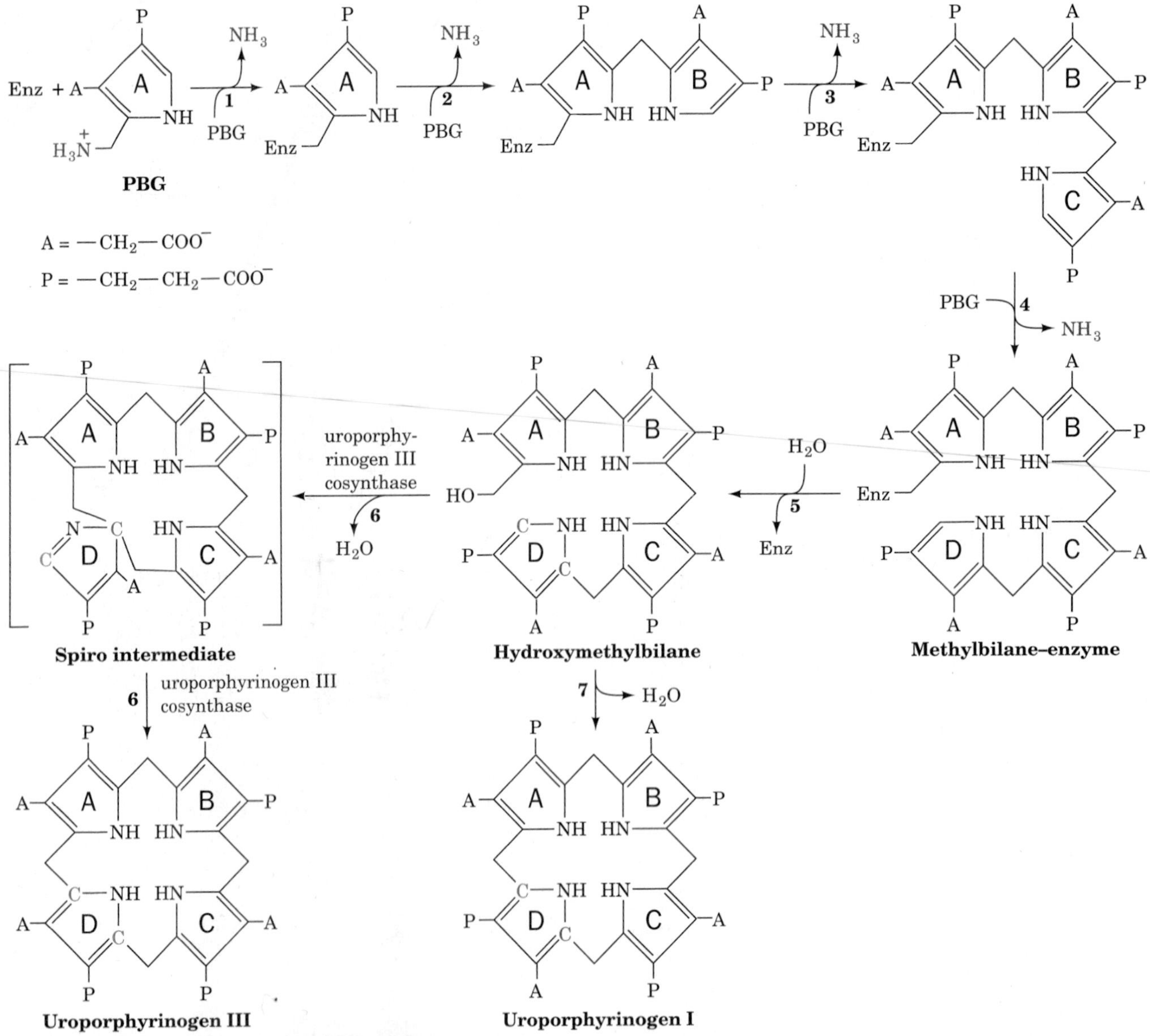

Figure 24-28

The synthesis of uroporphyrinogen III from PBG as catalyzed by uroporphyrinogen synthase and uroporphyrinogen III cosynthase: **(1)** Enzyme displacement of a PBG's amino group to form a covalent adduct. **(2–4)** Sequential addition of a second, third, and fourth PBG through the displacement of the primary amino group on one PBG by a pyrrole ring carbon atom on the succeeding PBG. **(5)** Hydrolysis of the methylbilane–enzyme to form hydroxymethylbilane. **(6)** Synthesis of uroporphyrinogen III via a spiro intermediate by uroporphyrinogen synthase and uroporphyrinogen III cosynthase. **(7)** Spontaneous cyclization of hydroxymethyl-bilane in the absence of uroporphyrinogen III cosynthase. A and P, respectively, represent acetyl and propionyl groups.

methenyl groups. Altogether, six carboxyl groups originally from carboxyl-labeled acetate are lost as CO_2. The only remaining C atoms from carboxyl-labeled acetate are the carboxyl groups of heme's two propionate side chains (P). During the coproporphyrinogen oxidase reaction, the macrocycle is transported back into the mitochondrion for the pathway's final reactions. Protoporphyrin IX is converted to heme by the insertion of Fe(II) into the tetrapyrrole nucleus by **ferrochelatase.**

Heme Biosynthesis Is Regulated Differently in Erythroid and Liver Cells

The two major sites of heme biosynthesis are erythroid cells, which synthesize ~85% of the body's heme groups, and the liver, which synthesizes most of the remainder. An important function of heme in liver is as the prosthetic group of **cytochrome P$_{450}$,** an oxidative enzyme involved in detoxification, which is required throughout the liver cell's lifetime in amounts that vary

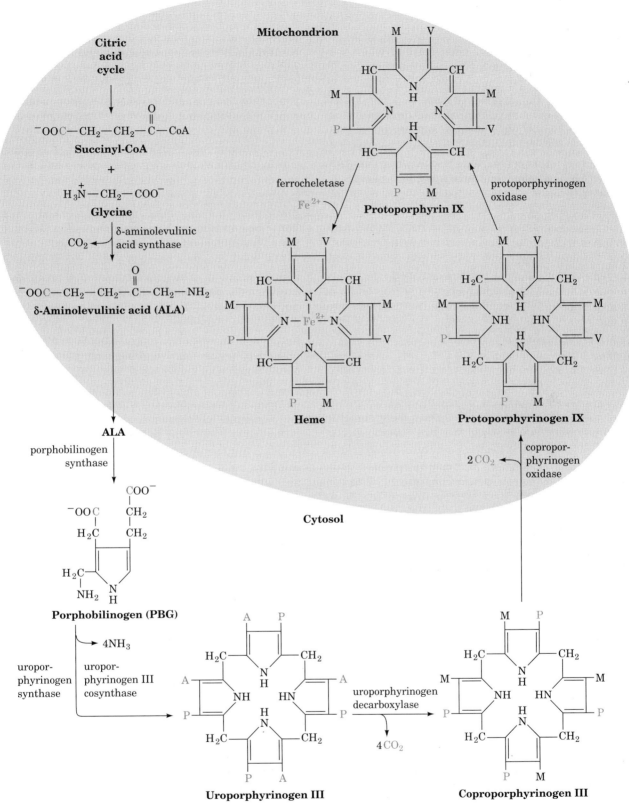

Figure 24-29

The overall pathway of heme biosynthesis. δ-Aminolevulinic acid (ALA) is synthesized in the mitochondrion by ALA-synthase. ALA (*left*) leaves the mitochondrion and is converted to PBG, four molecules of which condense to form a porphyrin ring. The next three reactions involve oxidation of the pyrrole ring substituents yielding protoporphyrinogen IX whose formation is accompanied by its transport back into the mitochondrion. After oxidation of the methylene groups linking the pyrroles to yield protoporphyrin IX, ferrochelatase catalyzes the insertion of Fe^{2+} to yield heme. A, P, M, and V, respectively, represent acetyl, propionyl, methyl, and vinyl ($-CH_2=CH_2$) groups. C atoms originating in the carboxyl group of acetate are red.

with conditions. In contrast, erythroid cells, where heme is, of course, a hemoglobin component, engage in heme synthesis only upon differentiation when they synthesize hemoglobin in vast quantities. This is a one-time synthesis; the heme must last the erythrocyte's lifetime (normally 120 days) since heme and hemoglobin synthesis stop upon red cell maturation (protein synthesis stops because of a loss of nuclei and ribosomes). The different ways that heme biosynthesis is regulated in liver and erythroid cells reflect these different demands: in liver, heme biosynthesis must really be "controlled," whereas in erythroid cells, the process is more like breaking a dam.

In liver, the main control target in heme biosynthesis is ALA-synthase, the enzyme catalyzing the pathway's first committed step. Heme, or its Fe(III) oxidation product **hemin,** controls this enzyme's activity through three mechanisms: (1) feedback inhibition, (2) inhibition of the transport of ALA-synthase from its site of synthesis in the cytosol to its site of action in the mitochondrion (Fig. 24-29), and (3) repression of ALA-synthase synthesis.

In erythroid cells, heme exerts quite a different effect on its biosynthesis. Heme stimulates, rather than represses, protein synthesis in reticulocytes (immature erythrocytes; Section 30-4A). Although the vast majority of the protein synthesized by reticulocytes is globin, there is evidence that heme also induces these cells to synthesize the enzymes of the heme biosynthesis pathway. Moreover, the rate-determining step of heme biosynthesis in erythroid cells may not be the ALA-synthase reaction. Experiments on various systems of differentiating erythroid cells implicate ferrochelatase, the enzyme catalyzing iron insertion, and uroporphyrinogen synthase in the control of heme biosynthesis in these cells. There are also indications that cellular uptake of iron may be rate limiting. Iron is transported in the plasma complexed with the iron transport protein **transferin.** The rate at which the iron–transferrin complex enters most cells, including those of liver, is controlled by receptor-mediated endocytosis (Section 11-4B). Nevertheless, lipid–soluble iron complexes that diffuse directly into reticulocytes stimulate *in vitro* heme biosynthesis. The existence of several control points supports the supposition that when erythroid heme biosynthesis is "switched on," all of its steps function at their maximal rates rather than any one step limiting the flow through the pathway. Heme-stimulated synthesis of globin also ensures that heme and globin are synthesized in the correct ratio for assembly into hemoglobin (Section 30-4A).

Porphyrias Have Bizarre Symptoms

Several genetic defects in heme biosynthesis, in liver or erythroid cells, are recognized. All involve the accumulation of prophyrin and/or its precursors and are therefore known as **porphyrias.** Two such defects are known to affect erythroid cells: uroporphyrinogen III cosynthase deficiency **(congenital erythropoietic porphyria)** and ferrochelatase deficiency **(erythropoietic protoporphyria).** The former results in accumulation of uroporphyrinogen I and its decarboxylation product **coproporphyrinogen I.** Excretion of these compounds colors the urine red, their deposition in the teeth turns them a fluorescent reddish-brown, and their accumulation in the skin renders it extremely photosensitive such that it ulcerates and forms disfiguring scars. Increased hair growth is also observed in afflicted individuals such that fine hair may cover much of their faces and extremities. These symptoms have prompted speculation that the werewolf legend has a biochemical basis.

The most common porphyria that primarily affects liver is uroporphyrinogen synthase deficiency **(acute intermittent porphyria).** This disease is marked by intermittent attacks of abdominal pain and neurological dysfunction. Excessive amounts of ALA and PBG are excreted in the urine during and after such attacks. The urine may become red resulting from the excretion of excess porphyrins synthesized from PBG in nonhepatic cells although the skin does not become unusually photosentitive. King George III, who ruled England during the American Revolution, and who has been widely portrayed as being mad, in fact had attacks characteristic of acute intermittent porphyria, was reported to have urine the color of port wine, and had several descendants who were diagnosed as having this disease. American history might have been quite different had George III not inherited this metabolic defect.

Heme Is Degraded to Bile Pigments

At the end of their lifetime, red cells are removed from the circulation and their components degraded. Heme catabolism (Fig. 24-30) begins with oxidative cleavage of the porphyrin between rings A and B to form **biliverdin,** a green linear tetrapyrrole. Biliverdin's central methenyl bridge (between rings C and D) is then reduced to form the red-orange **bilirubin.** The changing colors of a healing bruise are a visible manifestation of heme degradation.

In the reaction forming biliverdin, the methenyl bridge carbon between porphyrin rings A and B is released as CO which, we have seen, is a tenacious heme ligand (with 200-fold greater affinity for hemoglobin than O_2; Section 9-1A). Consequently, ~1% of hemoglobin's O_2-binding sites are blocked by CO, even in the absence of air pollution. This amount would be much greater were it not for the presence of hemoglobin's distal His residue (E7, the His residue that hydrogen bonds to bound O_2; Section 9-2). The distal His sterically strains CO from its preferred linear geometry in liganding heme Fe(II) towards the bent geometry favored by O_2. This reduces heme's affinity for CO by

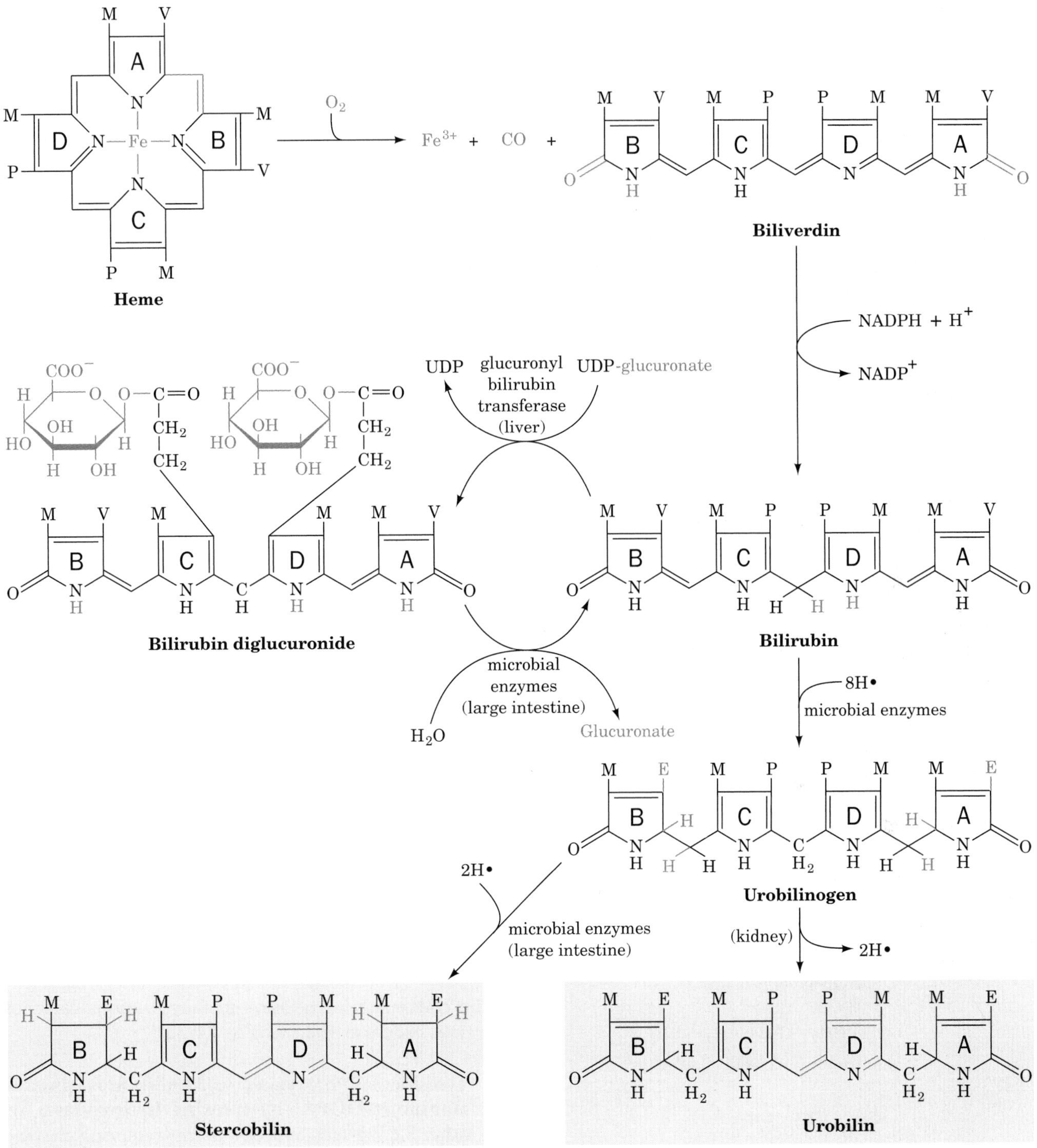

Figure 24-30
The heme degradation pathway. M, V, P, and E, respectively, represent methyl, vinyl, propionyl, and ethyl groups.

over 100-fold thus permitting the CO to be exhaled slowly. Indeed, in the mutant **Hb Zurich** [His E7(63)$\beta \rightarrow$ Arg], around 10% of the hemes carry CO.

The highly lipophilic bilirubin is insoluble in aqueous solutions. Like other lipophilic metabolites, such as free fatty acids, it is transported in the blood in complex with serum albumin. In the liver, its aqueous solubility is

increased by esterification of its two propionate side groups with glucuronic acid yielding **bilirubin diglucuronide,** which is secreted into the bile. Bacterial enzymes in the large intestine hydrolyze the glucuronic acid groups and, in a multistep process, convert bilirubin to several products, most notably **urobilinogen.** Some urobilinogen is reabsorbed and transported via the bloodstream to the kidney where it is converted to the yellow **urobilin** and excreted, thus giving urine its characteristic color. Most of the urobilinogen, however, is microbially converted to the deeply red-brown **stercobilin,** the major pigment of feces.

When the blood contains excessive amounts of bilirubin, the deposition of this highly insoluble substance colors the skin and the whites of the eyes yellow. This condition, called **jaundice** (French: *jaune,* yellow), signals either an abnormally high rate of red cell destruction, liver dysfunction, or bile duct obstruction. Newborn infants, particularly when premature, often become jaundiced because their livers do not yet make sufficient **bilirubin glucuronyl transferase** to glucuronidate the incoming bilirubin. Jaundiced infants are treated by bathing them with light from a fluorescent lamp; this photochemically converts bilirubin to more soluble isomers that the infant can degrade and excrete.

B. Biosynthesis of Physiologically Active Amines

Epinephrine, norepinephrine, dopamine, serotonin (5-hydroxytryptamine), γ-aminobutyric acid (GABA), and histamine

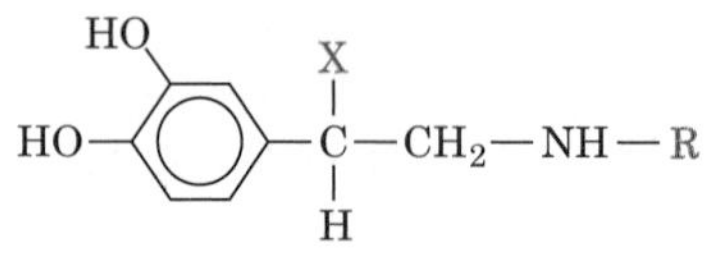

X = OH, R = CH$_3$ **Epinephrine**
X = OH, R = H   **Norepinephrine**
X = H,   R = H   **Dopamine**

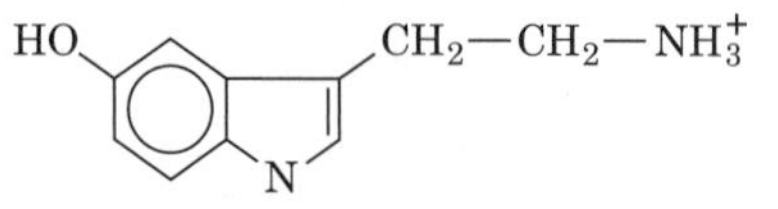

Serotonin
(5-hydroxytryptamine)

$$^-OOC-CH_2-CH_2-CH_2-NH_3^+$$

γ-Aminobutyric acid (GABA)

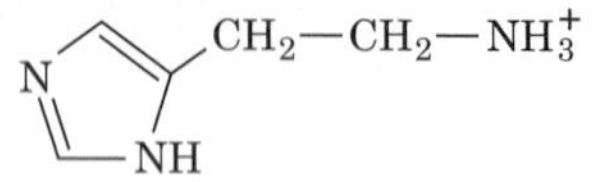

Histamine

are hormones and/or neurotransmitters derived from amino acids. For instance, epinephrine, as we have seen, activates muscle adenylate cyclase thereby stimulating glycogen breakdown (Section 17-3E); deficiency in dopamine production is associated with **Parkinson's disease,** a degenerative condition causing "shaking palsy"; serotonin causes smooth muscle contraction; GABA is one of the brain's major inhibitory neurotrans-

mitters, being released at 30% of its synapses; and histamine is involved in allergic responses (as allergy sufferers who take antihistamines will realize), as well as in the control of acid secretion by the stomach (Section 18-3C).

The biosynthesis of each of these physiologically active amines involves decarboxylation of the corresponding precursor amino acid. Amino acid decarboxylases are PLP-dependent enzymes that form a PLP–Schiff base with the substrate so as to stabilize the C$_\alpha$ carbanion formed upon C$_\alpha$—COO$^-$ bond cleavage (Section 24-1A).

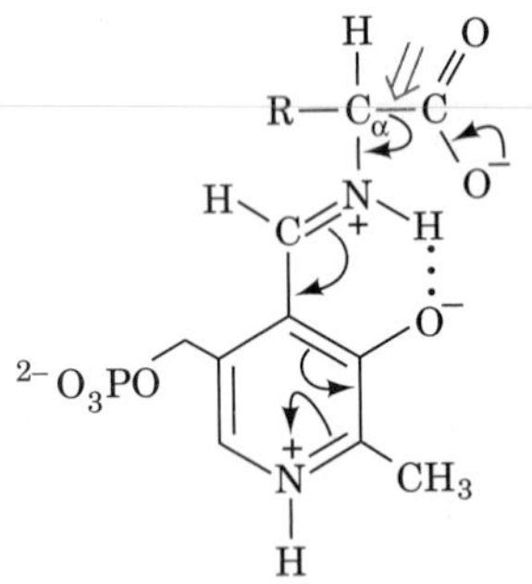

Formation of histamine and GABA are one-step processes (Fig. 24-31). In the synthesis of serotonin from tryptophan, the decarboxylation is preceded by a hydroxylation (Fig. 24-32) by **tryptophan hydroxylase,** one of three mammalian enzymes that has a 5,6,7,8-tetrahydrobiopterin cofactor (Section 24-3H). Dopamine, norepinephrine, and epinephrine are all termed **catecholamines** because they are amine derivatives of **catechol.**

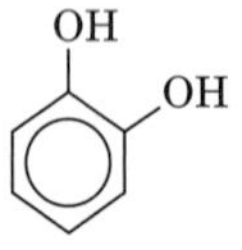

Catechol

The conversion of tyrosine to these various catecholamines occurs as follows (Fig. 24-33):

1. Tyrosine is hydroxylated to **3,4-dihydroxyphenylalanine (L-DOPA)** by **tyrosine hydroxylase,** another 5,6,7,8-tetrahydrobiopterin-requiring enzyme.

2. L-DOPA is decarboxylated to dopamine.

3. A second hydroxylation yields norepinephrine.

4. Methylation of norepinephrine's amino group by S-adenosylmethionine (Section 24-3E) produces epinephrine.

The specific catecholamine that a cell produces depends on which enzymes of the pathway are present. In adrenal medulla, which functions to produce hormones (Section 34-4A), epinephrine is the predominant prod-

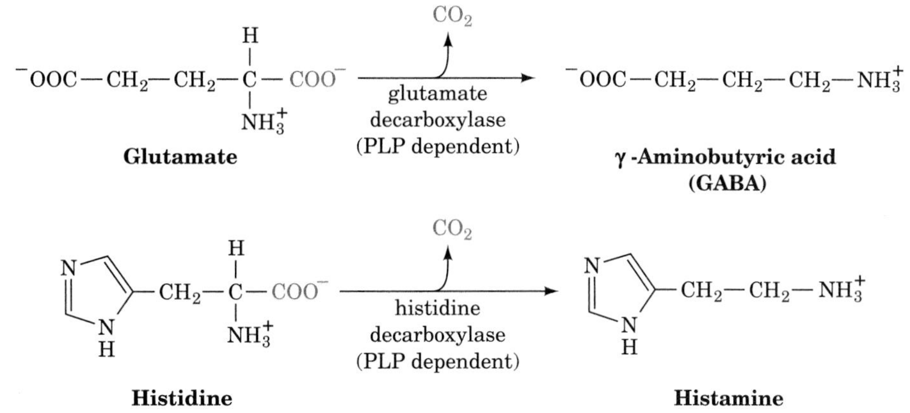

Figure 24-31
The decarboxylation reactions forming γ-aminobutyric acid (GABA) from glutamate and histamine from histidine.

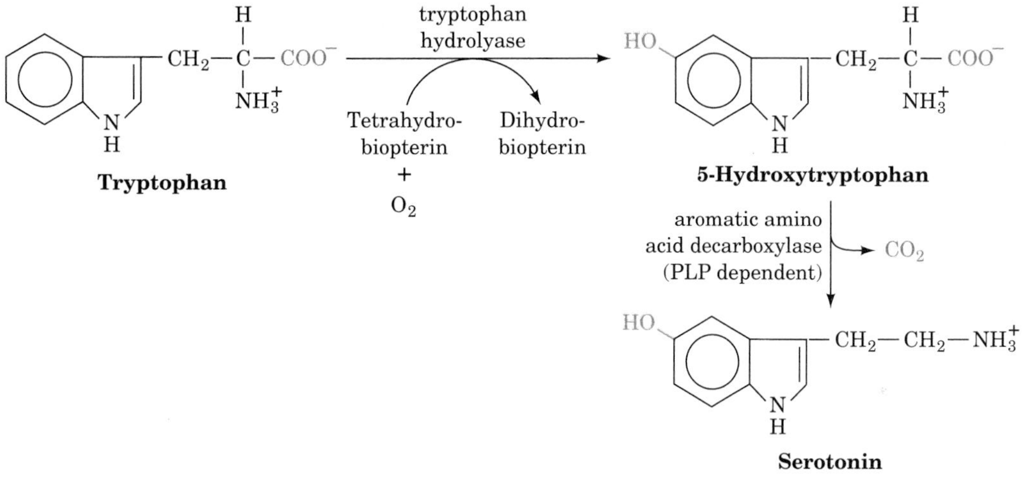

Figure 24-32
Serotonin is formed by the hydroxylation and subsequent decarboxylation of tryptophan.

uct. In some areas of the brain, norepinephrine is more common. In other areas, most prominently the *substantia nigra*, the pathway stops at dopamine synthesis. Indeed, Parkinson's disease, which is caused by degeneration of the *substantia nigra*, has been treated with some success by the administration of L-DOPA, dopamine's immediate precursor. Dopamine itself is ineffective because it cannot cross the blood–brain barrier. L-DOPA, however, is able to get to its sites of action where it is decarboxylated to dopamine. The enzyme catalyzing this reaction, **aromatic amino acid decarboxylase,** decarboxylates all aromatic amino acids and is therefore also responsible for serotonin formation. In a recently developed approach to the treatment of Parkinsonism, a portion of the patient's adrenal medulla is surgically transplanted to his/her brain. Presumably, the dopamine and L-DOPA released by this tissue serve to replace that lost via degeneration of the *substantia*

nigra. L-DOPA is also a precursor of the black skin pigment melanin.

C. Glutathione

Glutathione (GSH; γ-glutamylcysteinylglycine),

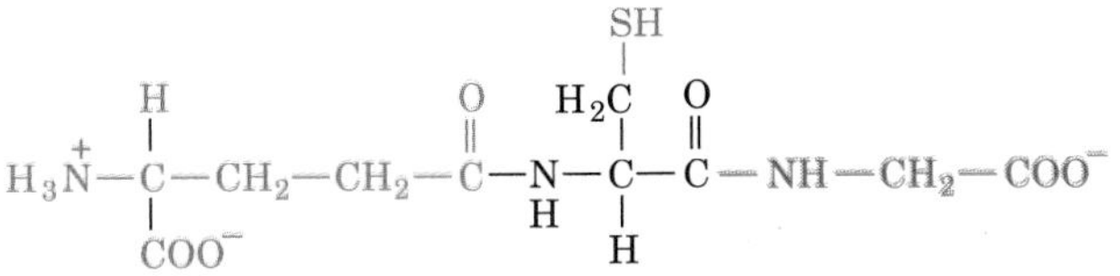

γ-**Glutamylcysteinylglycine**
glutathione (GSH)

a tripeptide that contains an unusual γ-amide bond, participates in a variety of detoxification, transport, and metabolic processes (Fig. 24-34). For instance, it is a substrate for peroxidase reactions, helping to destroy peroxides generated by oxidases; it is involved in leuko-

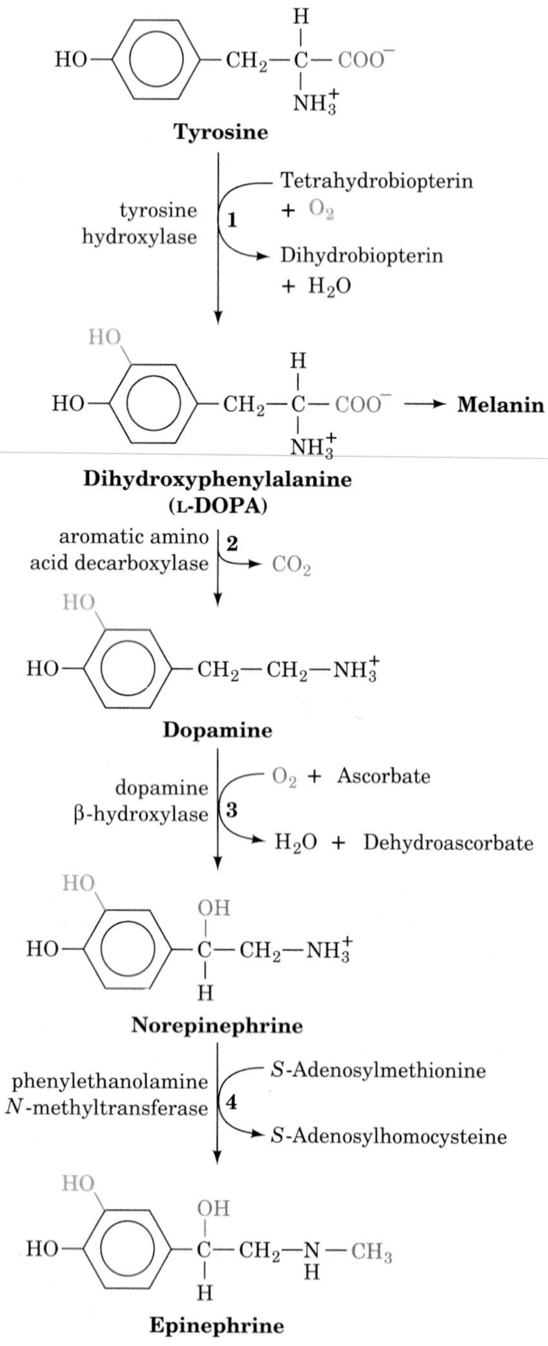

Tyrosine

tyrosine hydroxylase | **1**
— Tetrahydrobiopterin + O_2
→ Dihydrobiopterin + H_2O

Dihydroxyphenylalanine (L-DOPA) → **Melanin**

aromatic amino acid decarboxylase | **2** → CO_2

Dopamine

dopamine β-hydroxylase | **3**
— O_2 + Ascorbate
→ H_2O + Dehydroascorbate

Norepinephrine

phenylethanolamine *N*-methyltransferase | **4**
— *S*-Adenosylmethionine
→ *S*-Adenosylhomocysteine

Epinephrine

Figure 24-33
The sequential synthesis of L-DOPA, dopamine, norepinephrine, and epinephrine from tyrosine. L-DOPA is also the precursor of the black skin pigment melanin, an oxidized polymeric material.

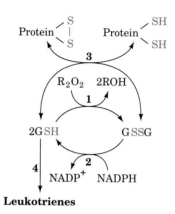

Leukotrienes

Figure 24-34
Some reactions involving glutathione: **(1)** peroxide detoxification by **glutathione peroxidase, (2)** regeneration of GSH from GSSG by glutathione reductase (Section 14-4), **(3)** thiol transferase modulation of protein thiol–disulfide balance, and **(4)** leukotriene biosynthesis by glutathione-*S*-transferase.

triene biosynthesis (Section 23-7C); and the balance between its reduced (GSH) and oxidized (GSSG) forms maintains the sulfhydryl groups of intracellular proteins in their correct oxidation states.

The **γ-glutamyl cycle,** which was elucidated by Alton Meister, *provides a vehicle for the energy-driven transport of amino acids into cells through the synthesis and break-* down of GSH (Fig. 24-35). GSH is synthesized from glutamate, cysteine, and glycine by the consecutive action of **γ-glutamylcysteine synthetase** and **GSH synthetase** (Fig. 24-35, Reactions 1 and 2). ATP hydrolysis provides the free energy for each reaction. The carboxyl group is activated for peptide bond synthesis by formation of an acyl phosphate intermediate.

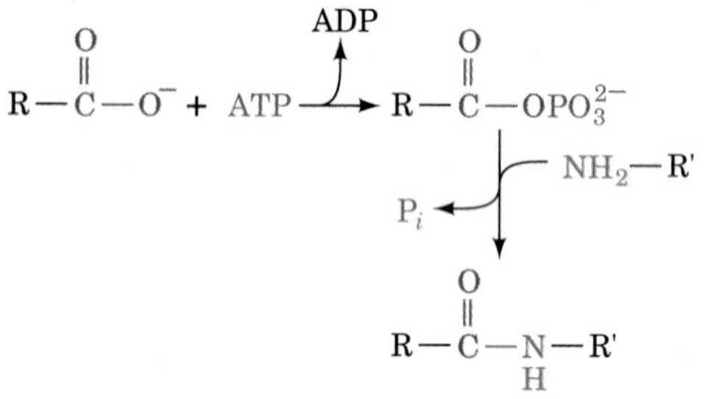

$$R-\overset{\overset{\displaystyle O}{\|}}{C}-O^- + ATP \xrightarrow{\text{ADP}} R-\overset{\overset{\displaystyle O}{\|}}{C}-OPO_3^{2-}$$

$$\xrightarrow[P_i]{NH_2-R'} R-\overset{\overset{\displaystyle O}{\|}}{C}-\underset{\underset{\displaystyle H}{|}}{N}-R'$$

The breakdown of GSH is catalyzed by **γ-glutamyl transpeptidase, γ-glutamyl cyclotransferase, 5-oxoprolinase,** and an intracellular protease (Fig. 24-35, Reactions 3–6).

Amino acid transport occurs because whereas GSH is synthesized intracellularly and is located largely within the cell, γ-glutamyl transpeptidase, which catalyzes GSH breakdown (Fig. 24-35, Reaction 3), is situated on the cell membrane's external surface and accepts amino acids, notably cysteine and methionine. GSH is first transported to the external surface of the cell membrane where the transfer of the γ-glutamyl group from GSH to an external amino acid occurs. The γ-glutamyl amino acid is then transported back into the cell and converted to glutamate by a two-step process in which the transported amino acid is released and **5-oxoproline** is formed as an intermediate. The last step in the cycle, the hydrolysis of 5-oxoproline, requires ATP hydrolysis.

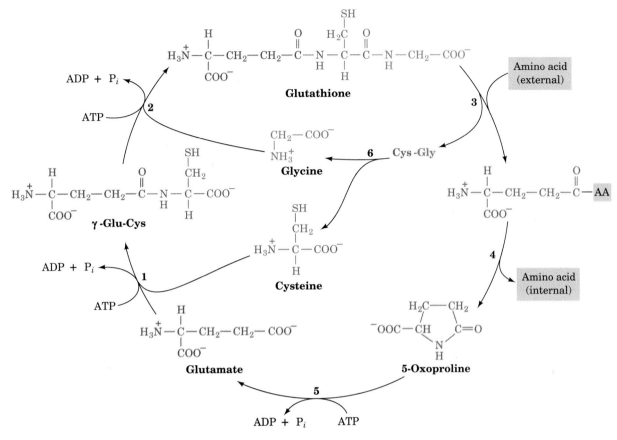

Figure 24-35
Glutathione synthesis as part of the γ-glutamyl cycle of glutathione metabolism. The cycle's reactions are catalyzed by: **(1)** γ-glutamylcysteine synthetase, **(2)** glutathione synthetase, **(3)** γ-glutamyl transpeptidase, **(4)** γ-glutamyl cyclotransferase, **(5)** 5-oxoprolinase, and **(6)** an intracellular protease.

This surprising observation (amide bond hydrolysis is almost always an exergonic process) is a consequence of 5-oxoproline's unusually stable internal amide bond.

D. Tetrahydrofolate Cofactors: The Metabolism of C_1 Units

Many biosynthetic processes involve the addition of a C_1 unit to a metabolic precursor. A familiar example is carboxylation. For instance, gluconeogenesis from pyr-

uvate begins with the addition of a carboxyl group to form oxaloacetate (Section 21-1A). The coenzyme involved in this and most other carboxylation reactions is biotin (Section 21-1A). In contrast, *S*-adenosylmethionine functions as a methylating agent (Section 24-2E).

Tetrahydrofolate (THF) is more versatile than the above cofactors in that it functions to transfer C_1 units in several oxidation states. THF is a 6-methylpterin derivative linked in sequence to *p*-aminobenzoic acid and Glu residues (Fig. 24-36). Up to five additional Glu residues

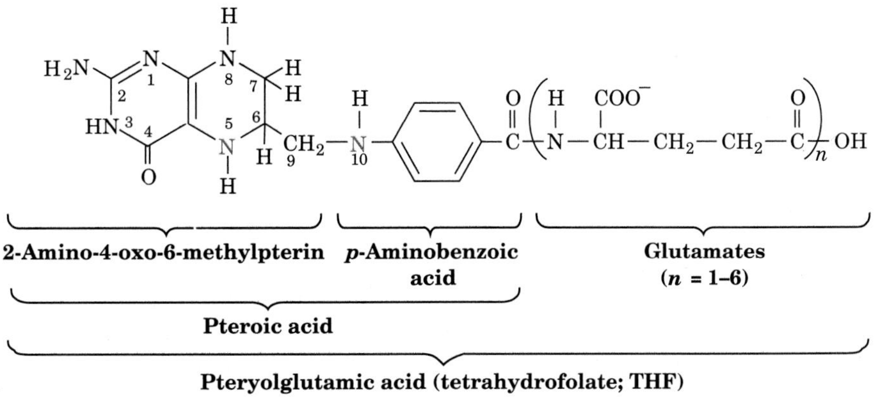

Figure 24-36
Tetrahydrofolate (THF).

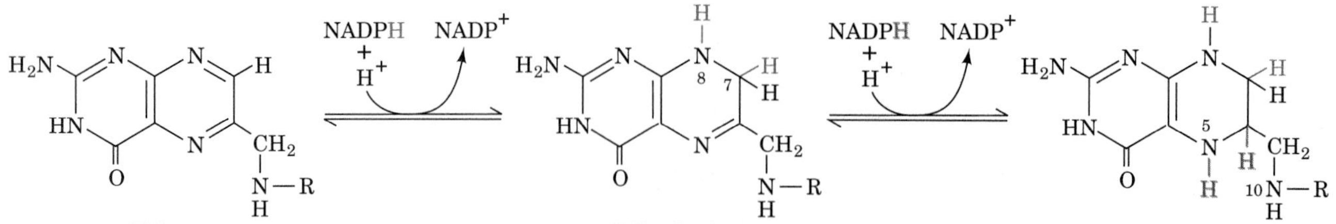

Figure 24-37
The two-stage reduction of folate to THF. Both reactions are catalyzed by dihydrofolate reductase (DHFR).

may be linked to the first glutamate via isopeptide bonds to form a polyglutamyl tail.

THF is derived from **folic acid** (Latin: *folium,* leaf), a doubly oxidized form of THF which must be enzymatically reduced before it becomes an active coenzyme (Fig. 24-37). Both reductions are catalyzed by **dihydrofolate reductase (DHFR)**. Mammals cannot synthesize folic acid so it must be provided in the diet or by intestinal microorganisms.

C_1 units are covalently attached to THF at its positions N(5), N(10), or both N(5) and N(10). These C_1 units, which may be at the oxidation levels of formate, formaldehyde, or methanol (Table 24-1), are all interconvertible by enzymatic redox reactions (Fig. 24-38).

The main entry of C_1 units into the THF pool is as N^5,N^{10}-**methylene-THF** through the conversion of serine to glycine by serine hydroxymethyl transferase (Sections 24-3B and 24-5A) and the cleavage of glycine by glycine synthase (the glycine cleavage system; Section 24-3B, Fig. 24-11). Histidine also contributes C_1 units through its degradation with formation of N^5-**form-imino-THF** (Fig. 24-13, Reaction 11).

A C_1 unit in the THF pool can have several fates (Fig. 24-39):

1. It may be used directly as N^5,N^{10}-methylene-THF in the conversion of the deoxynucleotide dUMP to dTMP by **thymidylate synthase** (Section 26-4B).

2. It may be reduced to N^5-**methyl-THF** for the synthesis of methionine from homocysteine (Section 24-3E).

3. It may be oxidized through N^5,N^{10}-methenyl-THF to

N^{10}-**formyl-THF** for use in the synthesis of purines (Section 26-2A). Since the purine ring of ATP is involved in histidine biosynthesis in microorganisms and plants (Section 24-5B), N^{10}-formyl-THF is indirectly involved in this pathway as well. Prokaryotes use N^{10}-formyl-THF in a formylation reaction yielding formylmethionyl-tRNA, which they require for the initiation of protein synthesis (Section 30-3C).

Sulfonamides (sulfa drugs) such as **sulfanilamide** are antibiotics that are structural analogs of the *p*-aminobenzoic acid constituent of THF.

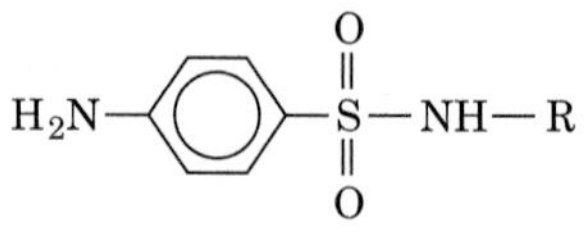

Sulfonamides   *p*-Aminobenzoic acid
(R = H sulfanilamide)

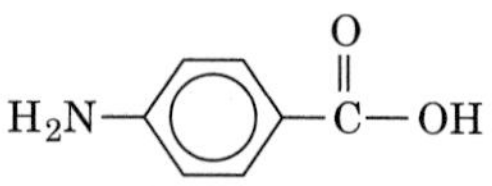

They competitively inhibit bacterial synthesis of THF at the *p*-aminobenzoic acid incorporation step, thereby blocking the above THF-requiring reactions. The inability of mammals to synthesize folic acid leaves them unaffected by sulfonamides, which accounts for the medical utility of these antibacterial agents.

5. AMINO ACID BIOSYNTHESIS

Many amino acids are synthesized by pathways that are only present in plants and microorganisms. Since mammals must obtain these amino acids in their diets, these substances are known as **essential amino acids**. The

Table 24-1

Oxidation Levels of C_1 Groups Carried by THF

Oxidation Level	Group Carried	THF Derivative(s)
Methanol	Methyl (—CH_3)	N^5-Methyl-THF
Formaldehyde	Methylene (—CH_2—)	N^5,N^{10}-Methylene-THF
Formate	Formyl (—CH=O)	N^5-Formyl-THF, N^{10}-formyl-THF
	Formimino (—CH=NH)	N^5-Formimino-THF
	Methenyl (—CH=)	N^5,N^{10}-Methenyl-THF

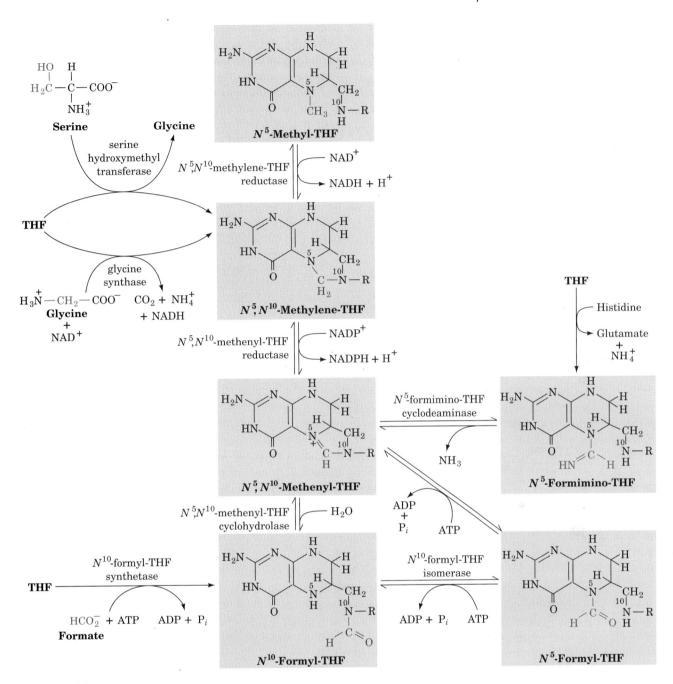

Figure 24-38
Interconversion of the C_1 units carried by THF.

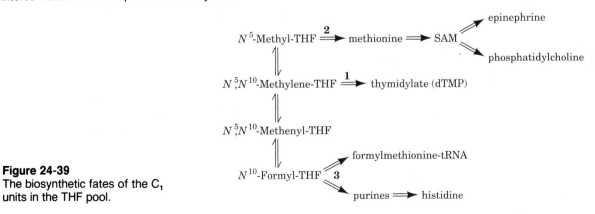

Figure 24-39
The biosynthetic fates of the C_1
units in the THF pool.

Table 24-2

Essential and Nonessential Amino Acids in Humans

Essential	Nonessential
Arginine[a]	Alanine
Histidine	Asparagine
Isoleucine	Aspartate
Leucine	Cysteine
Lysine	Glutamate
Methionine	Glutamine
Phenylalanine	Glycine
Threonine	Proline
Tryptophan	Serine
Valine	Tyrosine

[a] Although mammals synthesize arginine, they cleave most of it to form urea (Sections 24-2D and E).

other amino acids, which can be synthesized by mammals from common intermediates, are termed **nonessential amino acids**. The essential and nonessential amino acids for humans are listed in Table 24-2. Argi-

nine is classified as essential, even though it is synthesized by the urea cycle (Section 24-2D), because it is required in greater amounts than can be produced by this route during the normal growth and development of children (but not adults).

The essential amino acids occur in animal and vegetable proteins. Different proteins, however, contain different proportions of the essential amino acids. Milk proteins, for example, contain them all in the proportions required for proper human nutrition. Bean protein, on the other hand, contains an abundance of lysine but is deficient in methionine, whereas wheat is deficient in lysine but contains ample methionine. A balanced protein diet therefore must contain a variety of different protein sources that complement each other to supply the proper proportions of all the essential amino acids.

In this section we study the pathways involved in the formation of the nonessential amino acids. We also briefly consider such pathways for the essential amino acids as they occur in plants and microorganisms. You should note, however, that although we discuss some of the most common pathways for amino acid biosynthe-

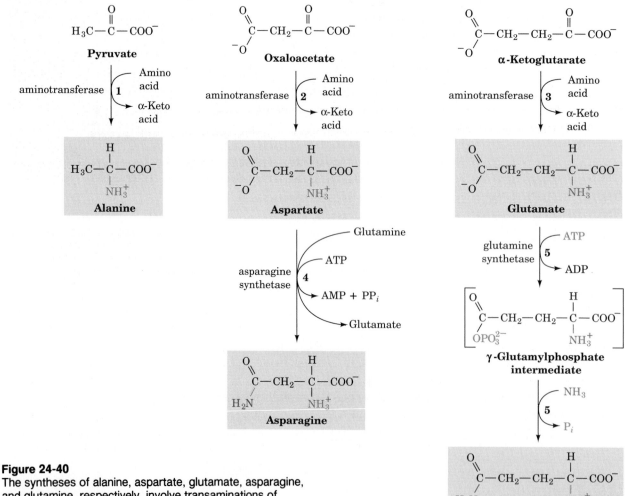

Figure 24-40
The syntheses of alanine, aspartate, glutamate, asparagine, and glutamine, respectively, involve transaminations of (1) pyruvate, (2) oxaloacetate, and (3) α-ketoglutarate, and amidation of (4) aspartate and (5) glutamate.

sis, there is considerable variation in these pathways among different species. In contrast, as we have seen, the basic pathways of carbohydrate and lipid metabolism are all but universal.

A. Biosynthesis of the Nonessential Amino Acids

All the nonessential amino acids except tyrosine are synthesized by simple pathways leading from one of four common metabolic intermediates: pyruvate, oxaloacetate, α-ketoglutarate, and 3-phosphoglycerate. Tyrosine, which is really misclassified as nonessential, is synthesized by the one-step hydroxylation of the essential amino acid phenylalanine (Section 24-3H). Indeed, the dietary requirement for phenylalanine reflects the need for tyrosine as well. The presence of dietary tyrosine therefore decreases the need for phenylalanine.

Alanine, Asparagine, Aspartate, Glutamate, and Glutamine Are Synthesized from Pyruvate, Oxaloacetate, and α-Ketoglutarate

Pyruvate, oxaloacetate, and α-ketoglutarate are the keto acids that correspond to alanine, aspartate, and glutamate, respectively. Indeed, as we have seen (Section 24-1), the synthesis of each of these amino acids is a one-step transamination reaction (Fig. 24-40, Reactions 1–3). Asparagine and glutamine are, respectively, synthesized from aspartate and glutamate by amidation (Fig. 24-40, Reactions 4 and 5). **Glutamine synthetase** catalyzes the formation of glutamine in a reaction in which NH_3 is the amino group donor and ATP is hydrolyzed to ADP and P_i via the intermediacy of **γ-glutamylphosphate** (Fig. 24-40, Reaction 5). Curiously, aspartate amidation by **asparagine synthetase** to form asparagine follows a different route; it utilizes glutamine as its amino group donor and hydrolyzes ATP to AMP + PP_i (Fig. 24-40, Reaction 4).

Glutamine Synthetase Is a Central Control Point in Nitrogen Metabolism

Glutamine is the amino group donor in the formation of many biosynthetic products (see below), as well as being a storage form of ammonia. Glutamine synthetase's consequent pivotal position in nitrogen metabolism makes this enzyme an excellent candidate for controlling the metabolic flux of nitrogenous compounds. Indeed, mammalian glutamine synthetases are activated by α-ketoglutarate, the product of glutamate's oxidative deamination. This control presumably prevents the accumulation of the ammonia produced by that reaction.

Bacterial glutamine synthetase, as Earl Stadtman showed, has a much more elaborate control system. This enzyme, which consists of 12 identical 51.6-kD subunits arranged at the corners of a hexagonal prism (Fig. 24-41), is regulated by several allosteric effectors as well as by covalent modification. Although a complete description of this complex enzyme is not given here, sev-

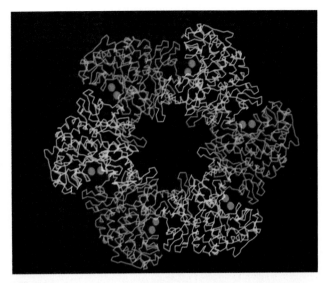

(a)

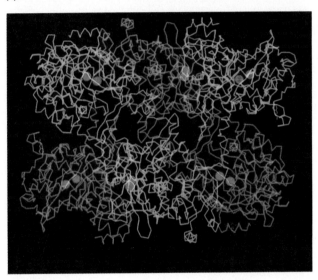

(b)

Figure 24-41
The X-ray structure of glutamine synthetase from *Salmonella typhimurium*. The enzyme consists of 12 identical subunits arranged with D_6 symmetry (the symmetry of a hexagonal prism). (a) View down the sixfold axis of symmetry showing only the six subunits of the upper ring in alternating blue and green. The subunits of the lower ring are roughly directly below those of the upper ring. The protein, including its side chains (not shown), has a radius of 143 Å. The six active sites shown are marked by the pairs of Mn^{2+} ions (*red spheres;* a divalent metal ion, physiologically Mg^{2+}, is required for enzymatic activity). Each adenylylation site, Tyr 397 (*red*), lies between two subunits at a higher radius than the corresponding active site. (b) Side view along one of the twofold axes showing only the six nearest subunits. The molecule extends 103 Å along the sixfold axis, which is vertical in this view. Compare these diagrams with electron micrographs of glutamine synthetase (Fig. 7-59). [Courtesy of David Eisenberg, UCLA.]

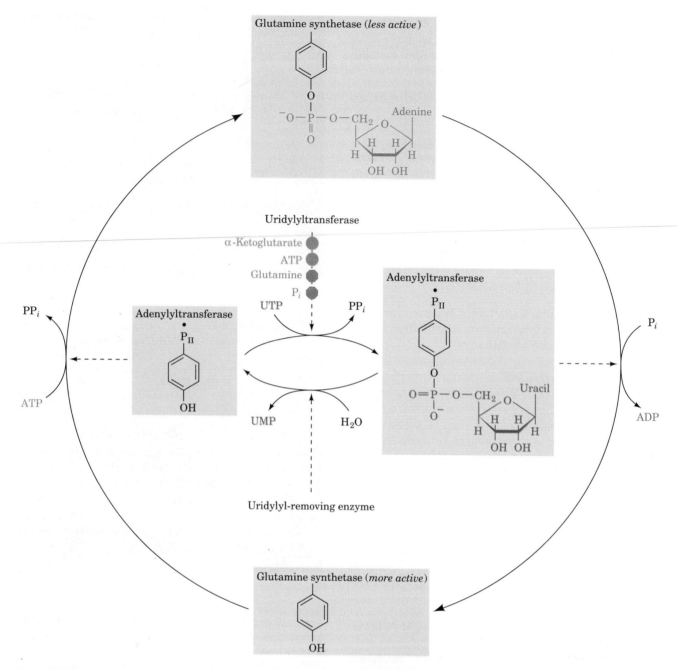

Figure 24-42
The regulation of bacterial glutamine synthetase by the adenylylation/deadenylylation of a specific Tyr residue. The adenylylation level of glutamine synthetase is controlled by the level of uridylylation of a specific adenylyltransferase·P_{II} Tyr residue. This uridylylation level, in turn, is controlled by the relative activities of uridylyltransferase, which is sensitive to the levels of a variety of nitrogen metabolites, and uridylyl-removing enzyme, whose activity is independent of these metabolite levels.

eral aspects of its control system bear note. *Nine allosteric feedback inhibitors, each with its own binding site, control the activity of bacterial glutamine synthetase in a cumulative manner:* histidine, tryptophan, carbamoyl phosphate (as synthesized by carbamoyl phosphate synthetase II), glucosamine-6-phosphate, AMP, and CTP are all end products of pathways leading from glutamine whereas alanine, serine, and glycine reflect the cell's nitrogen level.

E. coli glutamine synthetase is covalently modified by adenylylation of a specific Tyr residue (Fig. 24-42). The enzyme's susceptibility to cumulative feedback inhibition increases and thus its activity decreases with its degree of adenylylation. The level of adenylylation is controlled by a complex metabolic cascade which is conceptually similar to that controlling glycogen phosphorylase (although the type of covalent modification differs in that glycogen phosphorylase is phosphorylated at a specific Ser residue; Sections 17-3B and C). Both adenylylation and deadenylylation of glutamine synthetase are catalyzed by **adenylyl transferase** in complex with a tetrameric regulatory protein, P_{II}. This complex deadenylylates glutamine synthetase when P_{II} is uridylylated (also at a Tyr residue) and adenylylates glutamate synthetase when P_{II} lacks UMP residues. The level of P_{II} uridylylation, in turn, depends on the relative activities of two enzymatic activities located on the same protein: a **uridylyl transferase** that uridylylates P_{II} and a **uridylyl-removing enzyme** that hydrolytically excises the attached UMP groups of P_{II} (Fig. 24-42). The uridylyl transferase is activated by α-ketoglutarate and ATP and inhibited by glutamine and P_i, whereas uridylyl-removing enzyme is insensitive to these metabolites. This complex metabolic cascade therefore renders the activity of *E. coli* glutamine synthetase extremely responsive to the cell's nitrogen requirements.

Glutamate Is the Precursor of Proline, Ornithine, and Arginine

Conversion of glutamate to proline (Fig. 24-43, Reactions 1–4) involves the reduction of the γ-carboxyl group to an aldehyde followed by the formation of an internal Schiff base whose further reduction yields proline. Reduction of the glutamate γ-carboxyl group to an aldehyde is an endergonic process that is facilitated by the carboxyl group's prior phosphorylation by **γ-glutamyl kinase**. The unstable product, **glutamate-5-phosphate,** has not been isolated from reaction mixtures but is presumed to be the substrate for the reduction that follows. The resulting **glutamate-5-semialdehyde** cyclizes spontaneously to form the internal Schiff base **Δ^1-pyrroline-5-carboxylate.** The final reduction to proline is catalyzed by **pyrroline-5-carboxylate reductase.** Whether the enzyme requires NADH or NADPH is unclear.

The *E. coli* pathway from glutamate to ornithine and hence to arginine likewise involves the ATP-driven reduction of the glutamate γ-carboxyl group to an aldehyde (Fig. 24-43, Reactions 6 and 7). Spontaneous cyclization of this intermediate, **N-acetylglutamate-5-semialdehyde,** is prevented by prior acetylation of its amino group by *N*-acetylglutamate synthase to form *N*-acetylglutamate (Fig. 24-43, Reaction 5). *N*-acetylglutamate-5-semialdehyde, in turn, is converted to the corresponding amine by transamination (Fig. 24-43, Reaction 8). Hydrolysis of the acetyl protecting group finally yields ornithine which, as we have seen (Section 24-2), is converted to arginine via the urea cycle. In humans, however, the pathway to ornithine is more direct. The *N*-acetylation of glutamate that protects it from cyclization does not occur. Rather, glutamate-5-semialdehyde, which is in equilibrium with Δ^1-pyrroline-5-carboxylate, is directly transaminated to yield or-

nithine in a reaction catalyzed by **ornithine-δ-aminotransferase** (Fig. 24-43, Reaction 10).

Serine, Cysteine, and Glycine Are Derived from 3-Phosphoglycerate

Serine is formed from the glycolytic intermediate 3-phosphoglycerate in a three-reaction pathway (Fig. 24-44):

1. Conversion of 3-phosphoglycerate's 2-OH group to a ketone yielding **3-phosphohydroxypyruvate,** serine's phosphorylated keto acid analog.

2. Transamination of 3-phosphohydroxypyruvate to phosphoserine.

3. Hydrolysis of phosphoserine to yield serine.

Serine participates in glycine synthesis in two ways (Section 24-3B):

1. Direct conversion of serine to glycine by serine hydroxymethyl transferase in a reaction that also yields N^5,N^{10}-methylene-THF (Fig. 24-9; Reaction 4 in reverse).

2. Condensation of the N^5,N^{10}-methylene-THF with CO_2 and NH_4^+ by glycine synthase (Fig. 24-9; Reaction 3 in reverse).

We have already discussed the synthesis of cysteine from serine and homocysteine, a breakdown product of methionine (Section 24-3E). Homocysteine combines with serine to yield cystathionine, which subsequently forms cysteine and α-ketobutyrate (Fig. 24-14, Reactions 5 and 6). Since cysteine's sulfhydryl group is derived from the essential amino acid methionine, cysteine is really an essential amino acid.

B. Biosynthesis of the Essential Amino Acids

Essential amino acids, like nonessential amino acids, are synthesized from familiar metabolic precursors. Their synthetic pathways are only present in microorganisms and plants, however, and usually involve more steps than those of the nonessential amino acids. For example, lysine, methionine, and threonine are all synthesized from aspartate in pathways whose common first reaction is catalyzed by **aspartokinase,** an enzyme that is only present in plants and microorganisms. Similarly, valine and leucine are formed from pyruvate, isoleucine is formed from pyruvate and α-ketobutyrate, and tryptophan, phenylalanine, and tyrosine are formed from phosphoenolpyruvate and erythrose-4-phosphate. The enzymes that synthesize essential amino acids have apparently been lost early in animal evolution, possibly because of the ready availability of these amino acids in the diet.

Time and space prevent a detailed discussion of the

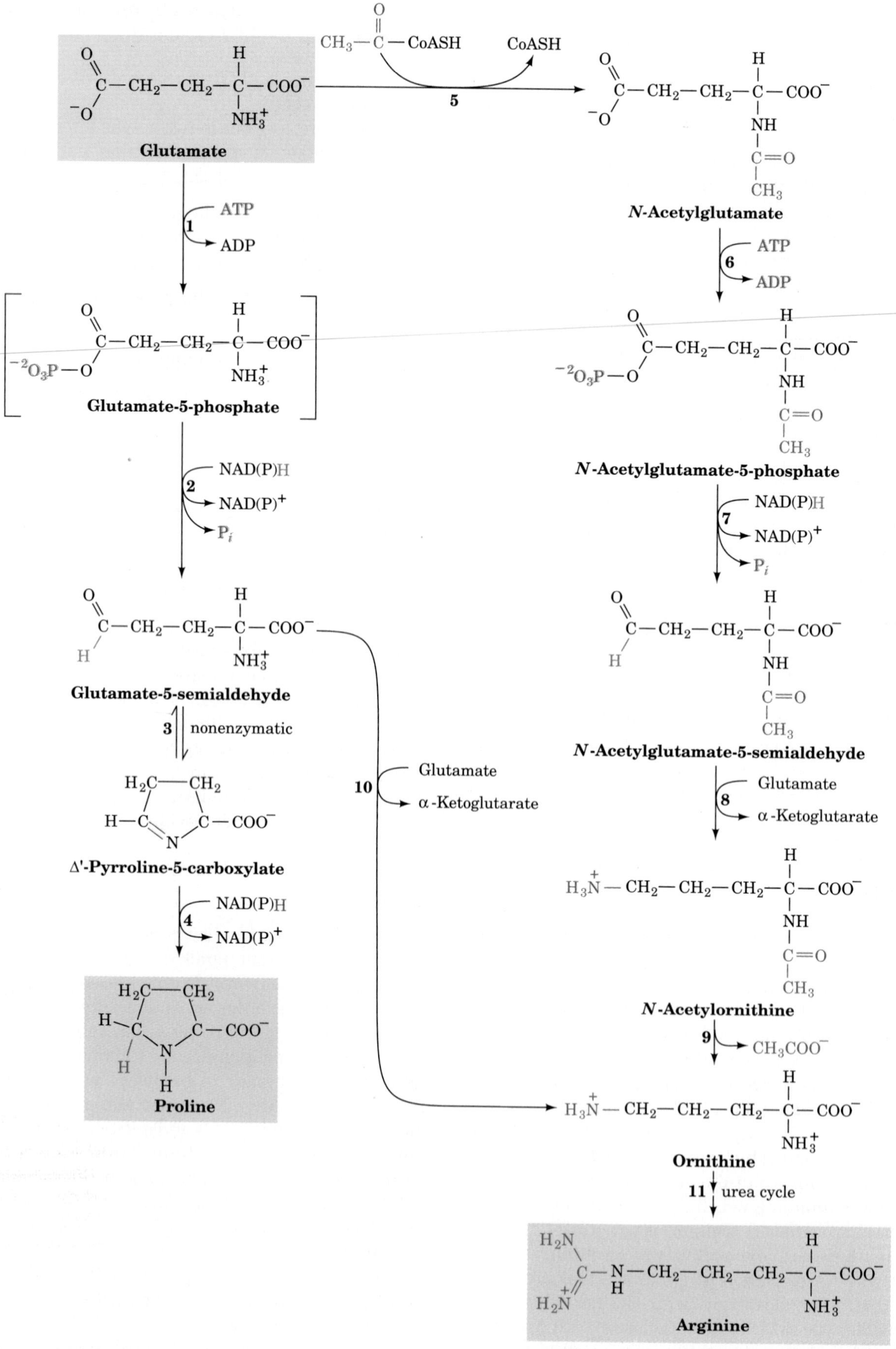

Glutamate
CoASH
5
N-Acetylglutamate
ATP
1
ADP
ATP
6
ADP
Glutamate-5-phosphate
N-Acetylglutamate-5-phosphate
NAD(P)H
2
NAD(P)+
Pi
NAD(P)H
7
NAD(P)+
Pi
Glutamate-5-semialdehyde
N-Acetylglutamate-5-semialdehyde
3
nonenzymatic
Glutamate
10
α-Ketoglutarate
Glutamate
8
α-Ketoglutarate
Δ'-Pyrroline-5-carboxylate
NAD(P)H
4
NAD(P)+
Proline
N-Acetylornithine
9
Ornithine
11
urea cycle
Arginine

Figure 24-43 (*opposite*)
The biosynthesis of the "glutamate family" of amino acids: arginine, ornithine, and proline. The enzymes catalyzing proline biosynthesis are (**1**) γ-glutamyl kinase, (**2**) dehydrogenase, (**3**) nonenzymatic, and (**4**) pyrroline-5-carboxylate reductase. The enzymes catalyzing ornithine biosynthesis are (**5**) *N*-acetylglutamate synthase, (**6**) acetylglutamate kinase, (**7**) *N*-acetyl-γ-glutamyl phosphate dehydrogenase, (**8**) *N*-acetylornithine-δ-aminotransferase, and (**9**) acetylornithine deacetylase. An alternate pathway to ornithine is though reaction (**10**), ornithine-δ-aminotransferase. Ornithine is converted to arginine (**11**) via the urea cycle (Fig. 24-4; Reactions 2–4).

many interesting reactions that occur in these pathways. The biosynthetic pathways of the aspartate family of amino acids, the pyruvate family, the aromatic family, and histidine are presented in Figs. 24-45 to 24-48 and 24-50 together with lists of the enzymes involved. Several recently developed agriculturally useful herbicides are specific inhibitors of some of these enzymes. Such herbicides have little toxicity towards animals and hence pose minimal risk to human health and the environment.

The Aspartate Family: Lysine, Methionine, and Threonine

In bacteria, aspartate is the common precursor of lysine, methionine, and threonine (Fig. 24-45). The biosynthesis of these essential amino acids all begin with the aspartokinase-catalyzed phosphorylation of aspartate to yield **aspartyl-β-phosphate.** We have seen that the control of metabolic pathways commonly occurs at the first committed step of the pathway. One might therefore expect that lysine, methionine, and threonine biosynthesis would be controlled as a group. Each of these pathways is, in fact, independently controlled. *E. coli* does so via three isozymes of aspartokinase that respond differently to the three amino acids in terms both of feedback inhibition of enzyme activity and repression of enzyme synthesis. Table 24-3 summarizes this differential control. In addition, the pathway direction is controlled by feedback inhibition at the branch

Table 24-3
Differential Control of Aspartokinase Isoenzymes in *E. Coli*

Enzyme	Feedback Inhibitor	Corepressor(s)[a]
Aspartokinase I	Threonine	Threonine and isoleucine
Aspartokinase II	None	Methionine
Aspartokinase III	Lysine	Lysine

[a] Compounds whose presence results in the repression of enzyme synthesis (Section 29-3E).

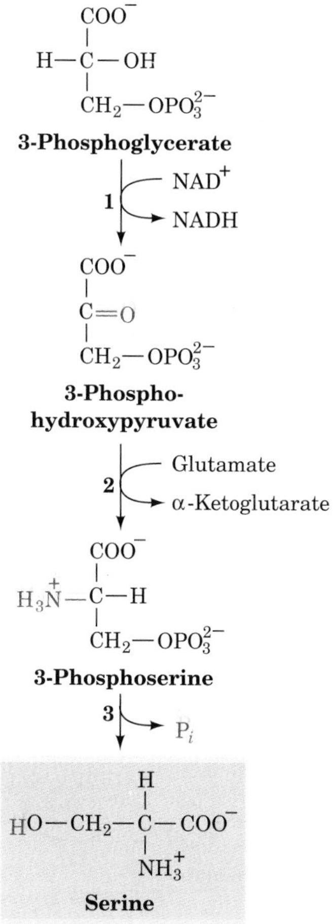

Figure 24-44
The conversion of 3-phosphoglycerate to serine. The pathway enzymes are (**1**) 3-phosphoglycerate dehydrogenase, (**2**) a PLP-dependent aminotransferase, and (**3**) phosphoserine phosphatase.

points by the individual amino acids. Thus methionine inhibits the *O*-acylation of homoserine (Fig. 24-45; Reaction 6), and lysine inhibits dihydrodipicolinate formation (Fig. 24-45; Reaction 10).

The Pyruvate Family: Leucine, Isoleucine, and Valine

Valine and isoleucine are both synthesized via the same five-step pathway (Fig. 24-46), the only difference being in the first step of the series. In this thiamine pyrophosphate-dependent reaction, which resembles those catalyzed by pyruvate decarboxylase (Section 16-3B) and transketolase (Section 21-4C), pyruvate forms an adduct with TPP, which is decarboxylated to hydroxyethyl-TPP. This resonance-stabilized carbanion adds to either the keto group of a second pyruvate to form **acetolactate** on the way to valine, or to the keto group of threonine-derived **α-ketobutyrate** to form **α-aceto-α-hydroxybutyrate** on the way to isoleucine. The leucine biosynthetic pathway branches off from the valine pathway at α-ketoisovalerate (Fig. 24-46, Reac-

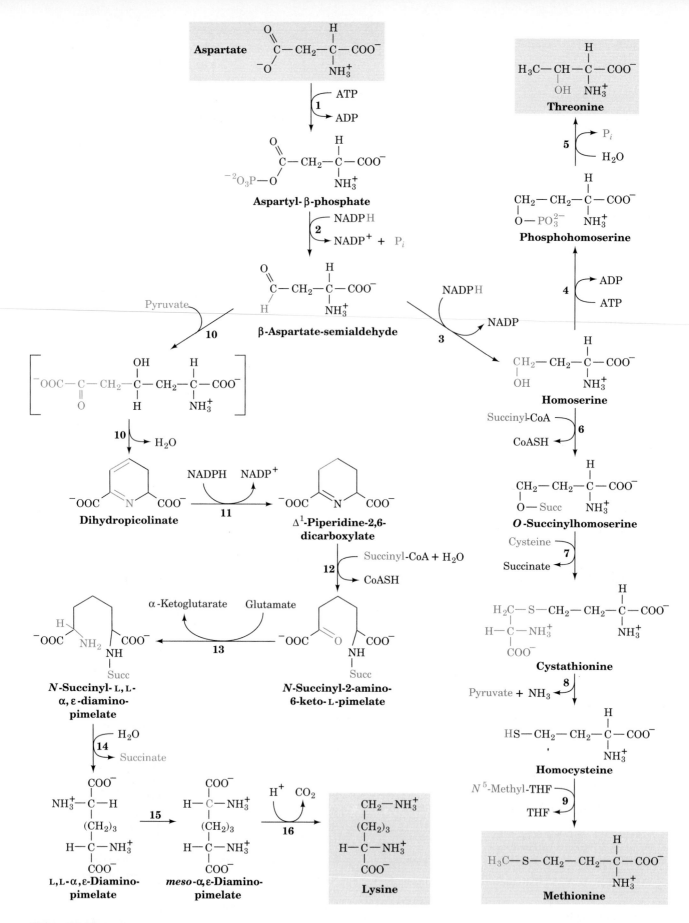

Figure 24-45

The biosynthesis of the "aspartate family" of amino acids: lysine, methionine, and threonine. The pathway enzymes are **(1)** aspartokinase, **(2)** β-aspartate semialdehyde dehydrogenase, **(3)** homoserine dehydrogenase, **(4)** homoserine kinase, **(5)** threonine synthase (a PLP enzyme), **(6)** homoserine acyltransferase, **(7)** cystathionine-γ-synthase, **(8)** cystathionine-β-lyase, **(9)** homocysteine methyl transferase (which also occurs in mammals; Section 24-3E), **(10)** dihydropicolinate synthase, **(11)** Δ¹-piperidine-2,6-dicarboxylate dehydrogenase, **(12)** N-succinyl-2-amino-6-ketopimelate synthase, **(13)** succinyl–diaminopimelate aminotransferase (a PLP enzyme), **(14)** succinyl–diaminopimelate desuccinylase, **(15)** diaminopimelate epimerase, and **(16)** diaminopimelate decarboxylase.

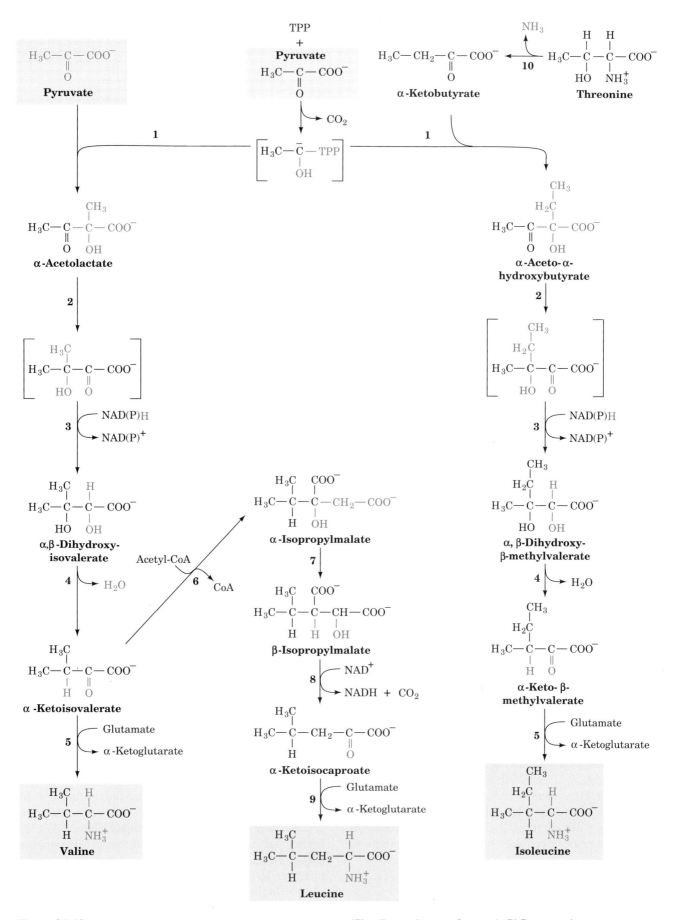

Figure 24-46
The biosynthesis of the "pyruvate family" of amino acids: isoleucine, leucine, and valine. The pathway enzymes are **(1)** acetolactate synthase (a TPP enzyme), **(2)** acetolactate mutase, **(3)** reductase, **(4)** dihydroxy acid dehydratase, **(5)** valine aminotransferase, (a PLP enzyme), **(6)** α-isopropylmalate synthase, **(7)** α-isopropylmalate dehydratase, **(8)** isopropylmalate dehydrogenase, **(9)** leucine aminotransferase, (a PLP enzyme), and **(10)** threonine deaminase (serine dehydratase, a PLP enzyme).

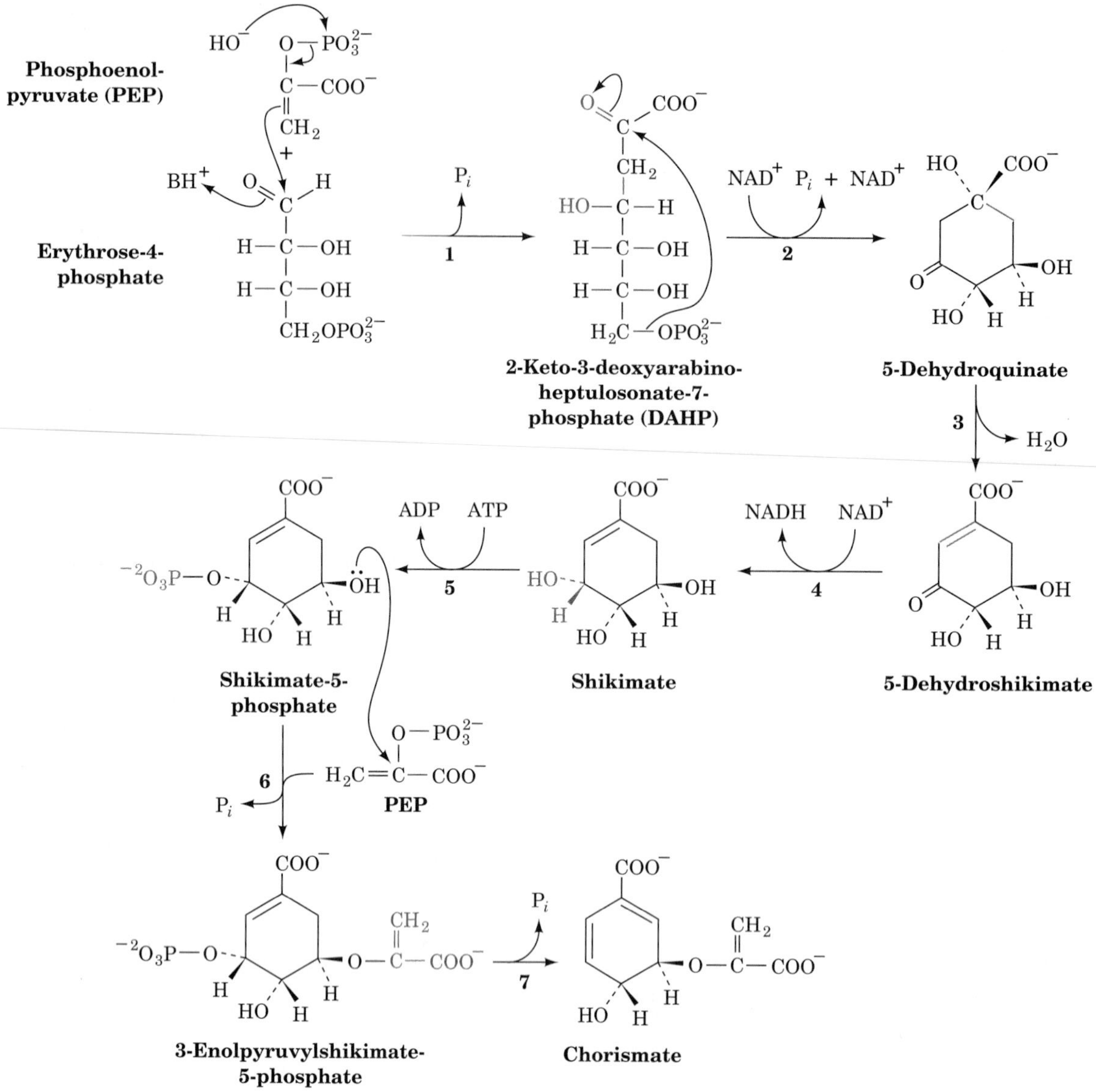

Figure 24-47
The biosynthesis of chorismate, the aromatic amino acid precursor. The pathway enzymes are **(1)** 2-keto-3-deoxy-D-arabinoheptulosonate-7-phosphate synthase, **(2)** dehydroquinate synthase (an NAD$^+$-requiring reaction that yields an unchanged NAD$^+$ product and is thereby indicative of an oxidized intermediate as similarly occurs in the UDP–galactose-4-epimerase reaction: Section 16-5B), **(3)** 5-dehydroquinate dehydratase, **(4)** shikimate dehydrogenase, **(5)** shikimate kinase, **(6)** 3-enolpyruvylshikimate-5-phosphate synthase, and **(7)** chorismate synthase.

tion 6). Reactions 6 to 8 in Fig. 24-46 are reminiscent of the first three reactions of the citric acid cycle (Sections 19-3A–C). Here, acetyl-CoA condenses with **α-keto-isovalerate** to form **α-isopropylmalate,** which then undergoes a dehydration/hydration reaction followed by oxidative decarboxylation and transamination to yield leucine.

The Aromatic Amino Acids: Phenylalanine, Tyrosine, and Tryptophan

The precursors to the aromatic amino acids are the glycolytic intermediate phosphoenolpyruvate (PEP) and erythrose-4-phosphate (an intermediate in the pentose phosphate pathway; Section 21-4C). Their condensation forms **2-keto-3-deoxy-D-arabinoheptulosonate-7-phosphate,** a C$_7$ compound that cyclizes and is ultimately converted to **chorismate** (Fig. 24-47), the branch point for tryptophan biosynthesis. Chorismate is either converted to anthranilate and then on to tryptophan, or to **prephenate** and on to either tyrosine or phenylalanine (Fig. 24-48). Although mammals synthesize tyrosine by the hydroxylation of phenylalanine (Section 24-3H), many microorganisms synthesize it directly from prephenate.

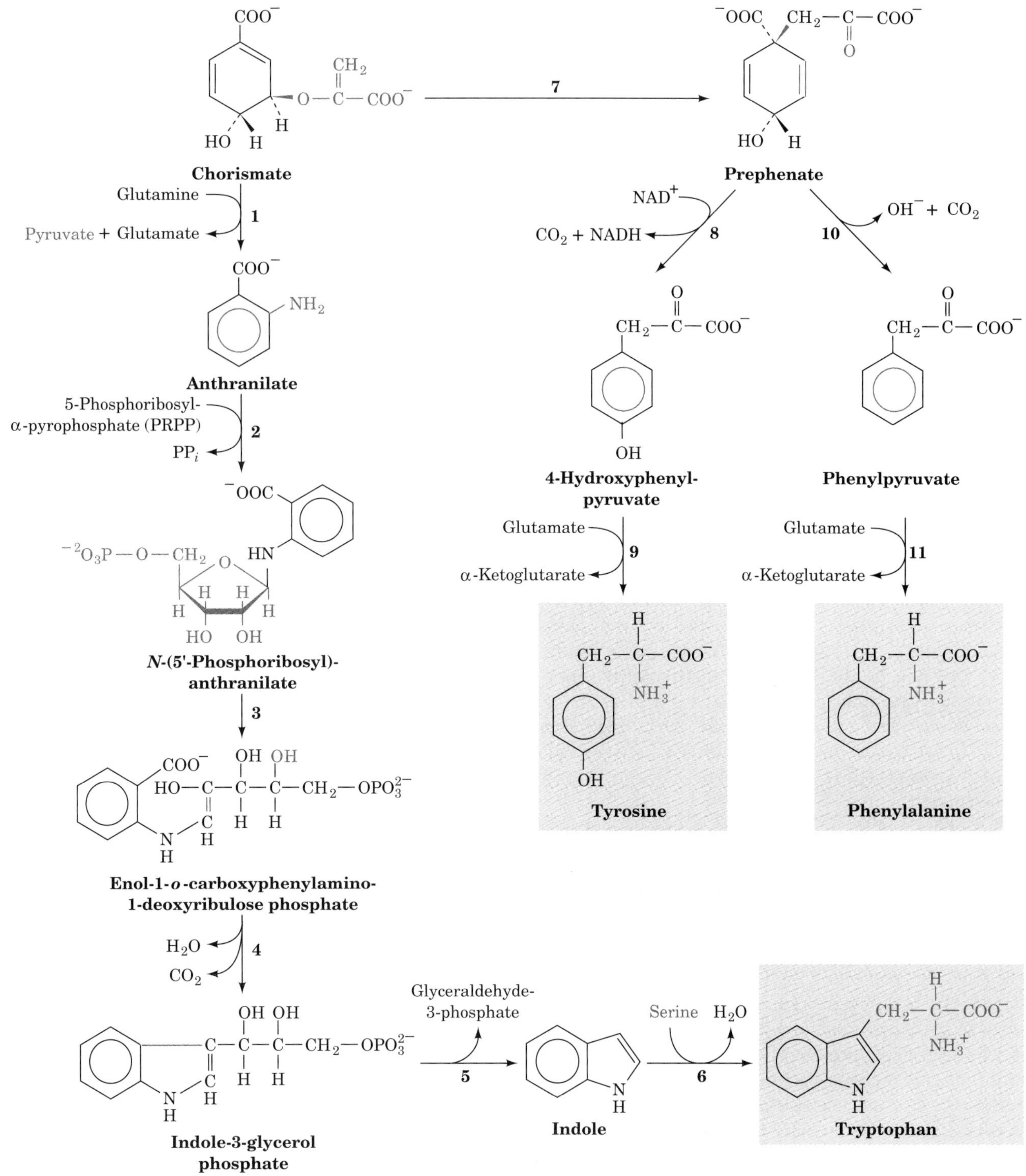

Figure 24-48
The biosynthesis of phenylalanine, tryptophan, and tyrosine from chorismate. The pathway enzymes are **(1)** anthranilate synthase, **(2)** anthranilate–phosphoribosyl transferase, **(3)** *N*-(5′-phosphoribosyl)-anthranilate isomerase, **(4)** indole-3-glycerol phosphate synthase **(5)** tryptophan synthase, α subunit **(6)** tryptophan synthase, β subunit, **(7)** chorismate mutase, **(8)** prephenate dehydrogenase, **(9)** aminotransferase, **(10)** prephenate dehydratase, and **(11)** aminotransferase.

A Protein Tunnel Channels the Intermediate Product of Tryptophan Synthase between Two Active Sites

The final two reactions in tryptophan biosynthesis, Reactions 5 and 6 in Fig. 24-48, are both catalyzed by **tryptophan synthase:**

1. The α subunit (29 kD) of this $\alpha_2\beta_2$ bifunctional enzyme cleaves **indole-3-glycerol phosphate** yielding

indole and glyceraldehyde-3-phosphate (Reaction 5).

2. The β subunit (43 kD) joins indole with L-serine in a PLP-dependent reaction to form L-tryptophan (Reaction 6).

Either subunit alone is enzymatically active but, when joined in the $\alpha_2\beta_2$ tetramer, the rates of both reactions and their substrate affinities are increased by 1 to 2 orders of magnitude. Indole, the intermediate product, does not appear free in solution; the enzyme apparently sequesters it.

The X-ray structure of tryptophan synthase from *Salmonella typhimurium*, determined by Craig Hyde, Edith Miles, and David Davies, explains the latter observation. The protein forms a 150-Å long, twofold symmetric α-β-β-α complex (Fig. 24-49) in which the active sites of neighboring α and β subunits are separated by ~25 Å. *These active sites are joined by a solvent-filled tunnel that is wide enough to contain the intermediate substrate, indole.* This unprecedented structure suggests the following series of events. The indole-3-glycerol phosphate substrate binds to the α subunit through an opening into its active site, its "front door," and the glyceraldehyde-3-phosphate product leaves via the same route. Similarly, the β subunit active site has a "front door" opening to the solvent through which serine enters and tryptophan leaves. Both active sites also have "back doors" that are connected by the tunnel. The indole intermediate presumably diffuses between the two active sites via the tunnel and hence does not escape to the solvent.

This phenomenon, in which the intermediate of two reactions is directly transferred from one enzyme active site to another, is called **channeling** (the term "tunneling" is reserved for certain quantum mechanical phenomena). Channeling increases the rate of a metabolic pathway by preventing the loss of its intermediate products as well as protecting this intermediate from degradation. We have seen a similar phenomenon in the series of reactions catalyzed by fatty acid synthase in which the growing product is kept in the vicinity of the multifuntional enzyme's active site by covalent attachment to the enzyme's flexible phosphopantetheine arm (Section 23-4C). Channeling is also implicated in the multistep biosyntheses of purines and pyrimidines (Sections 26-3A).

Histidine Biosynthesis

Five of histidine's six C atoms are derived from **5-phosphoribosyl-α-pyrophosphate (PRPP;** Fig. 24-50), an intermediate also involved in the biosynthesis of tryptophan, (Fig. 24-48, Reaction 2), purine nucleotides (Section 26-2A), and pyrimidine nucleotides (Section 26-3A). The histidine's sixth carbon originates from ATP. The ATP atoms that are not incorporated into histidine are eliminated as **5-aminoimidazole-4-car-**

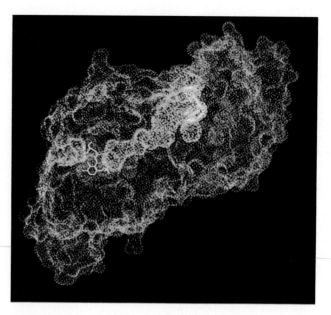

Figure 24-49
The X-ray structure of the bifunctional enzyme tryptophan synthase from *S. typhimurium* as represented by its molecular surface. Only one $\alpha\beta$ unit of this twofold symmetric $\alpha_2\beta_2$ heterotetramer is shown: blue dots outline the surface of the α subunit (268 residues), turquoise dots outline the surface of the β subunits (397 residues), and yellow dots outline the interface between adjacent β subunits. The active site of the α subunit is located by its bound competitive inhibitor **indolepropanol phosphate** (*red dot surface on right*), whereas that of the β subunit is marked by its PLP coenzyme (*yellow with red dot surface on left*). The "tunnel" connecting the active sites of the α and β subunits, through which the indole product of the α subunit reaction is believed to pass to the β subunit active site, is highlighted in green. [Courtesy of Craig Hyde, Edith Miles, and David Davies, NIH.]

boxamide ribonucleotide (Fig. 24-50, Reaction 5), which is also an intermediate in purine biosynthesis (Section 26-2A).

The unusual biosynthesis of histidine from a purine has been cited as evidence supporting the hypothesis that life was originally RNA-based (Section 1-4C). His residues, as we have seen, are often components of enzyme active sites where they act as nucleophiles and/or general acid–base catalysts. The recent discovery that RNA can have catalytic properties (Section 29-4C) therefore suggests that the imidazole moiety of purines plays a similar role in these RNA enzymes **(ribozymes).** This further suggests that the histidine biosynthesis pathway is a "fossil" of the transition to more efficient protein-based life forms.

6. NITROGEN FIXATION

The most prominent chemical elements in living systems are O, H, C, N, and P. The elements O, H, and P, occur widely in metabolically available forms (H_2O, O_2,

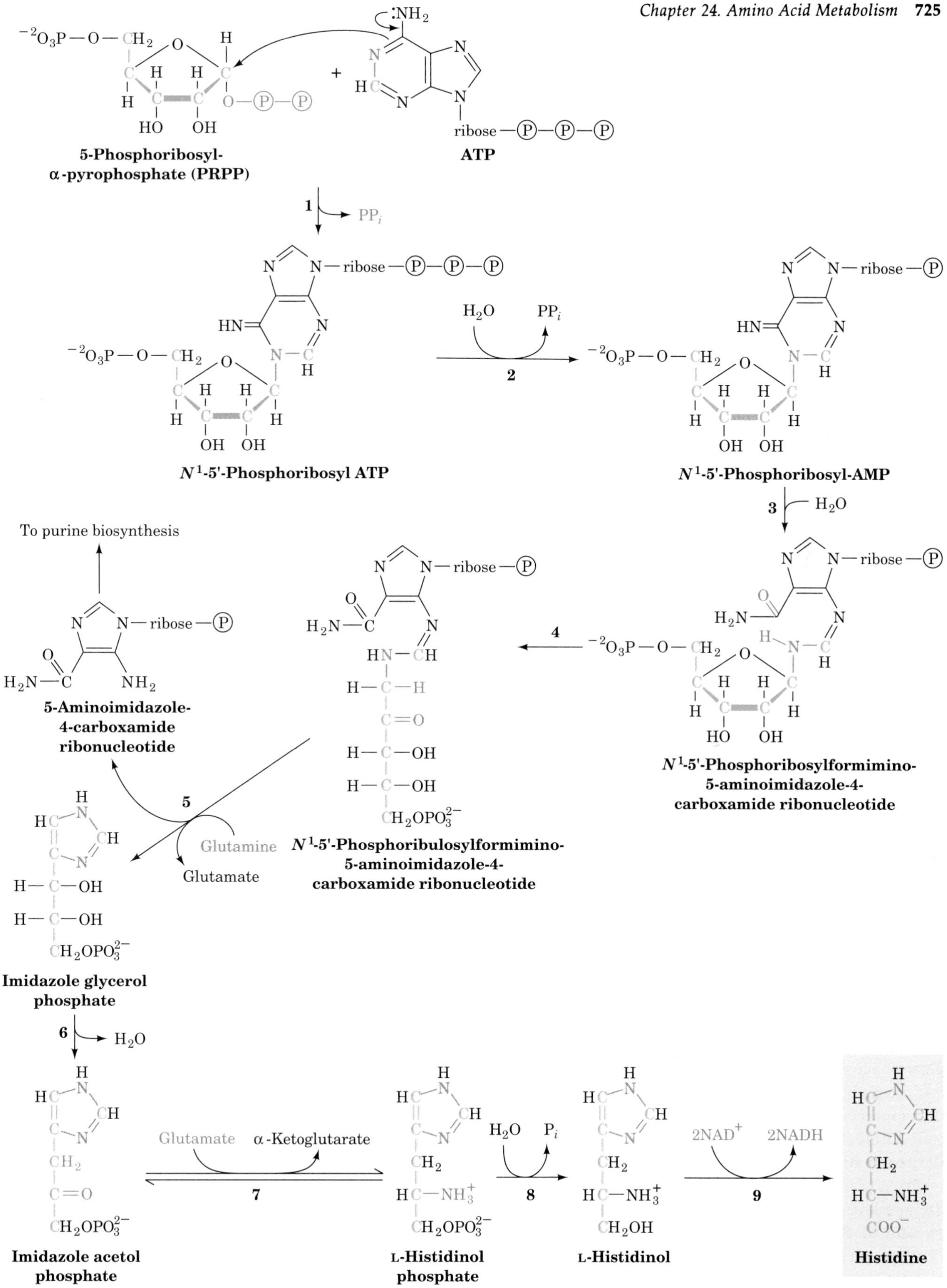

Figure 24-50
The biosynthesis of histidine. The pathway enzymes are
(1) ATP phosphoribosyl transferase,
(2) pyrophosphohydrolase, **(3)** phosphoribosyl-AMP
cyclohydrolase, **(4)** phosphoribosylformimino-5-
aminoimidazole carboxamide ribonucleotide isomerase,
(5) glutamine amidotransferase, **(6)** imidazole glycerol
phosphate dehydratase, **(7)** L-histidinol phosphate
aminotransferase, **(8)** histidinol phosphate phosphatase,
(9) histidinol dehydrogenase.

and P_i). However, the major available forms of C and N, CO_2 and N_2, are extremely stable (unreactive); for example, the $N \equiv N$ triple bond has a bond energy of $945 \text{ kJ} \cdot \text{mol}^{-1}$ (vs $351 \text{ kJ} \cdot \text{mol}^{-1}$ for a $C—O$ single bond). CO_2, with only minor exceptions, is metabolized (fixed) only by photosynthetic organisms (Chapter 22). *N_2 fixation is even less common; this element is only converted to metabolically useful forms by a few strains of bacteria.*

Nitrogen fixing bacteria of the genus *Rhizobium* live in symbiotic relationship with root nodule cells of legumes (plants belonging to the pea family including beans, clover, and alfalfa) where they convert N_2 to NH_3.

$$N_2 + 8H^+ + 8e^- + 16ATP \longrightarrow$$
$$2NH_3 + H_2 + 16ADP + 16P_i$$

The NH_3 thus formed can be either incorporated into glutamate by glutamate dehydrogenase (Section 24-1) or into glutamine by glutamine synthetase (Section 24-5A). This nitrogen fixing system produces more metabolically useful nitrogen than the legume needs; the excess is excreted into the soil, enriching it. It is therefore common agricultural practice to plant a field with alfalfa every few years to build up the supply of usable nitrogen in the soil for later use in growing other crops.

Nitrogenase Contains Several Redox Centers

Nitrogenase, which catalyzes the reduction of N_2 to NH_3, consists of two proteins:

1. The **MoFe protein,** an ~ 220-kD protein of subunit structure $\alpha_2 \beta_2$ that contains Fe and Mo.

2. The **Fe protein,** an ~ 64-kD dimer of identical subunits that contains Fe.

Much of the Fe in these proteins is contained in [4Fe–4S] clusters (Section 20-2C). The Fe protein contains one [4Fe–4S] cluster in each dimer and two ATP-binding sites, whereas the MoFe protein contains four [4Fe–4S] clusters together with two Mo-containing clusters of unknown structure but thought to have the composition [6Fe–Mo–6S].

Nitrogenase is rapidly inactivated by O_2 so that the enzyme must be protected from this reactive substance. Cyanobacteria (photosynthetic oxygen-evolving bacteria; Section 1-1A) provide protection by carrying out nitrogen fixation in specialized nonphotosynthetic cells called **heterocysts,** which have Photosystem I but lack Photosytem II (Section 22-2C). In the root nodules of legumes (Fig. 24-51), however, protection is afforded by the symbiotic synthesis of **leghemoglobin.** The globin portion of this monomeric oxygen-binding protein is synthesized by the plant (an evolutionary curiosity since globins are otherwise known to occur only in animals) while the heme is synthesized by the *Rhizobium*. Leghemoglobin has a very high O_2 affinity, keeping the pO_2 low enough to protect the nitrogenase while providing passive O_2 transport for the aerobic bacterium.

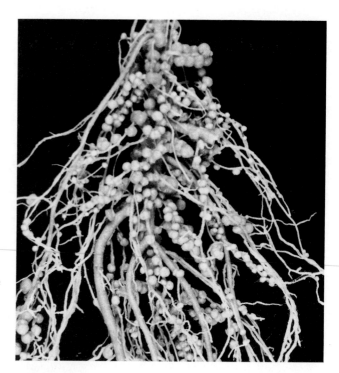

Figure 24-51
A photograph showing the root nodules of the legume Bird's Foot Trefoil. [The Nitragin Co., Inc.]

N_2 Reduction Is Energetically Costly

Nitrogen fixation requires two participants in addition to N_2 and nitrogenase: (1) a source of electrons and (2) ATP. Electrons are generated either oxidatively or photosynthetically, depending on the organism. These electrons are transferred to ferredoxin, a [4Fe–4S]-containing electron carrier that transfers an electron to the Fe protein of nitrogenase, beginning the nitrogen fixation process (Fig. 24-52). Two molecules of ATP bind to the reduced Fe protein and are hydrolyzed as the electron is passed from the Fe protein to the MoFe protein. ATP hydrolysis is thought to cause a conformational change in the Fe protein that alters its redox potential from -0.29 to -0.40 V, making the electron capable of N_2 reduction ($\mathscr{E}^{\circ\prime}$ for the half-cell $N_2 + 6H^+ \rightleftharpoons 2NH_3$ is -0.34 V).

The actual reduction of N_2 occurs on the MoFe protein in three discrete steps, each involving an electron pair:

$$N \equiv N \xrightarrow{2H^+ + 2e^-} H—N=N—H \xrightarrow{2H^+ + 2e^-} \underset{\text{Hydrazine}}{\overset{H}{\underset{H}{N}}—\overset{H}{\underset{H}{N}}} \xrightarrow{2H^+ + 2e^-} 2NH_3$$

$$\underset{\text{Diimine}}{}$$

An electron transfer must occur six times per N_2 molecule fixed so that a total of 12ATPs are required to fix one N_2 molecule. However, nitrogenase also reduces H_2O to H_2 which, in turn, reacts with **diimine** to reform N_2.

$$HN=NH + H_2 \longrightarrow N_2 + 2H_2$$

This futile cycle is favored when the ATP level is low

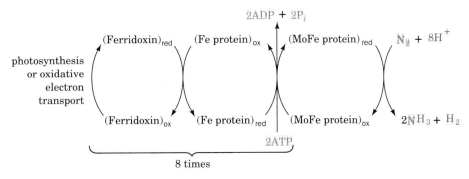

Figure 24-52
The flow of electrons in the nitrogenase-catalyzed reduction of N_2.

and/or the reduction of the Fe protein is sluggish. Even when ATP is plentiful, however, the cycle cannot be entirely suppressed beyond about one H_2 molecule produced per N_2 reduced. The total cost of N_2 reduction is therefore 8 electrons transferred and 16ATPs hydrolyzed. Hence nitrogen fixation is an energetically expensive process; indeed, the nitrogen fixing bacteria in the root nodules of pea plants consume nearly 20% of the ATP the plant produces.

While atmospheric N_2 is the ultimate nitrogen source for all living things, most plants do not support the symbiotic growth of nitrogen fixing bacteria. They must therefore depend on a source of "prefixed" nitrogen such as nitrate or ammonia. These nutrients come from decaying organic matter in the soil or from fertilizer applied to it. The Haber process, which was invented by Fritz Haber in 1910, is a chemical process for N_2 fixation

that is still widely used in fertilizer manufacture. This direct reduction of N_2 by H_2 to form NH_3 requires a temperature of 500°C, a pressure of 300 atm, and an iron catalyst. One of the major long-term goals of genetic engineering (Section 28-8) is to induce agriculturally useful nonleguminous plants to fix their own nitrogen, a complex undertaking in which the plant must be made to provide a hospitable environment for nitrogen fixation as well as acquire the enzymatic machinery to do so. This would free farmers, particularly those in developing countries, from the need to either purchase fertilizers, periodically let their fields lie fallow (giving legumes the opportunity to grow), or follow the slash-and-burn techniques that are rapidly destroying the world's tropical forests and contributing significantly to the greenhouse effect (atmospheric CO_2 pollution causing long-term global warming).

Chapter Summary

Amino acids are the precursors for numerous nitrogen-containing compounds such as heme, physiologically active amines, and glutathione. Excess amino acids are converted to common metabolic intermediates for use as fuels. The first step in amino acid breakdown is removal of the α-amino group by transamination. Transaminases require pyridoxal phosphate (PLP) and convert amino acids to their corresponding α-keto acids. The amino group is transferred to α-ketoglutarate to form glutamate, oxaloacetate to form aspartate, or pyruvate to form alanine. Glutamate is subsequently oxidatively deaminated to form ammonia and regenerate α-ketoglutarate.

In the urea cycle, amino groups from ammonia and aspartate combine with HCO_3^- to form urea. This pathway takes place in the liver, partially in the mitochondrion, and partially in the cytosol. It begins with the ATP-dependent condensation of NH_4^+ and HCO_3^- by carbamoyl phosphate synthetase. The resulting carbamoyl phosphate then combines with ornithine to yield citrulline, which combines with aspartate to form argininosuccinate which, in turn, is cleaved to fumarate and arginine. The arginine is then hydrolyzed to urea, which is

excreted, and ornithine, which reenters the urea cycle. N-Acetylglutamate regulates the urea cycle by activating carbamoyl phosphate synthetase allosterically.

The α-keto acid products of transamination reactions are degraded to citric acid cycle intermediates or their precursors. The amino acids leucine and lysine are ketogenic in that they are converted only to the ketone body precursors acetyl-CoA and acetoacetate. The remaining amino acids are, at least in part, glucogenic in that they are converted to the glucose precursors pyruvate, oxaloacetate, α-ketoglutarate, succinyl-CoA, and fumarate. Alanine, cysteine, glycine, serine, and threonine are converted to pyruvate. Serine hydroxymethyl transferase catalyzes the PLP-dependent C_α—C_β bond cleavage of serine to form glycine. The reaction requires tetrahydrofolate (THF) to act as a C_1 unit acceptor. Asparagine and aspartate are converted to oxaloacetate. α-Ketoglutarate is a product of arginine, glutamate, glutamine, histidine, and proline degradation. Methionine, isoleucine, and valine are degraded to succinyl-CoA. Methionine breakdown involves the synthesis of S-adenosylmethionine (SAM), a sulfonium ion that acts as a methyl donor in many biosynthetic reactions.

Maple syrup urine disease is caused by an inherited defect in branched-chain amino acid degradation. Branched-chain amino acid degradation pathways contain reactions common to all acyl-CoA oxidations. Tryptophan is degraded to alanine and acetyl-CoA. Phenylalanine and tyrosine are degraded to fumarate and acetoacetate. Most individuals with the hereditary disease phenylketonuria lack phenylalanine hydroxylase, which converts phenylalanine to tyrosine.

Heme is synthesized from glycine and succinyl-CoA. These precursors condense to from δ-aminolevulinic acid (ALA), which cyclizes to form the pyrrole, porphobilinogen (PBG). Four molecules of PBG condense to form uroporphyrinogen III, which then goes on to form heme. Heme is degraded to form linear tetrapyrroles, which are subsequently excreted as bile pigments. The hormones and neurotransmitters L-DOPA, epinephrine, norepinephrine, serotonin, γ-aminobutyric acid (GABA), and histamine are all synthesized from amino acid precursors. Glutathione, a tripeptide that is synthesized from glutamate, cysteine, and glycine, is involved in a variety of

protective, transport, and metabolic processes. Tetrahydrofolate is a coenzyme that participates in the transfer of C_1 units.

Amino acids are required for many vital functions of an organism. Those that mammals can synthesize are known as nonessential amino acids; those that mammals must obtain from their diet are called essential amino acids. The biosynthesis of nonessential amino acids involves relatively simple pathways, whereas those forming the essential amino acids are generally more complex.

Although the ultimate source of nitrogen for amino acid biosynthesis is atmospheric N_2, this nearly inert gas must first be reduced to a metabolically useful form, NH_3, by nitrogen fixation. This process only occurs in certain types of bacteria, one species of which occurs in symbiotic relationship with legumes. N_2 is fixed in these organisms by an oxygen-sensitive enzyme, nitrogenase, that consists of two proteins in which [4Fe–4S] clusters as well as [6Fe–Mo–6S] clusters function as electron carriers.

References

General

Bender, David A., *Amino Acid Metabolism,* Wiley (1985).

Stanbury, J. B., Wyngaarden, J. B., Fredrickson, D. S., Goldstein, J. L., and Brown, M. S. (Eds.), *The Metabolic Basis of Inherited Disease* (5th ed.), Chapters 11–28, 60, and 61, McGraw–Hill (1983).

Walsh, C., *Enzymatic Reaction Mechanisms,* Chapters 24 and 25, Freeman (1979). [Discusses reactions involving PLP, THF, and SAM cofactors.]

Amino Acid Deamination and the Urea Cycle

Braunstein, A. E., Amino group transfer, *in* Boyer, P. D. (Ed.), *The Enzymes* (3rd ed.), Vol. 9, pp. 379–482, Academic Press (1973).

Cohen, P. P., The ornithine–urea cycle: biosynthesis and regulation of carbamyl phosphate synthetase I and ornithine transcarbamylase, *Curr. Top. Cell. Regul.* **18**, 1–19 (1981). [An interesting historical review of the discovery of urea and the urea cycle, as well as a discussion of the cycle's regulation.]

Martell, A. E., Vitamin B_6 catalyzed reactions of α-amino and α-keto acids: model systems, *Acc. Chem. Res.* **22**, 115–124 (1989).

Smith, E. L., Austen, B. M., Blumenthal, K. M., and Nyc, J. F., Glutamate dehydrogenases, *in* Boyer, P. D. (Ed.), *The Enzymes* (3rd ed.), Vol. 11, pp. 293–367, Academic Press (1975).

Torchinsky, Yu. M., Transamination: its discovery, biological and chemical aspects (1937–1987), *Trends Biochem. Sci.* **12**, 115–117 (1987).

Degradation and Biosynthesis of Amino Acids

Adams, E. and Frand, L., Metabolism of proline and the hydroxyprolines, *Annu. Rev. Biochem.* **49**, 1005–1061 (1980).

Almassey, R. J., Janson, C. A., Hamlin, R., Xuong, N.-H., and Eisenberg, D., Novel subunit–subunit interactions in the structure of glutamine synthetase, *Nature* **323**, 304–309 (1986).

Cooper, A. J. L., Biochemistry of sulfur-containing amino acids, *Annu. Rev. Biochem.* **52**, 187–222 (1983).

Hyde, C. C., Ahmed, S. A., Padlan, E. A., Miles, E. W., and Davies, D. R., Three-dimensional structure of the tryptophan synthase $\alpha_2\beta_2$ multienzyme complex from *Salmonella typhimurium, J. Biol. Chem.* **263**, 17857–17871 (1988).

Herrmann, K. M. and Somerville, R. L. (Eds.), *Amino Acids: Biosynthesis and Genetic Regulation,* Addison–Wesley (1983).

Jansonius, J. N. and Vincent, M. G., Structural basis for catalysis by aspartate aminotransferase, *in* Jurnak, F. A. and McPherson, A. (Eds.), *Biological Macromolecules and Assemblies,* Vol. 3, pp. 187–285, Wiley (1987).

Kishore, G. M. and Shah, D. M., Amino acid biosynthesis inhibitors as herbicides, *Annu. Rev. Biochem.* **57**, 627–663 (1988). [Discusses the biosynthesis of the essential amino acids.]

Meister, A., Asparagine synthesis *and* Glutamine synthetase of mammals, *in* Boyer, P. D. (Ed.), *The Enzymes* (3rd ed.), Vol. 10, pp. 561–580 *and* 699–754, Academic Press (1974).

Nichol, C. A., Smith, G. K., and Duch, D. S., Biosynthesis and metabolism of tetrahydrobiopterin and molybdopterin, *Annu. Rev. Biochem.* **54**, 729–764 (1985).

Stadtman, E. R. and Ginsburg, A., The glutamine synthetase of *E. coli:* structure and control, *in* Boyer, P. D. (Ed.), *The Enzymes* (3rd ed.), Vol. 10, pp. 755–807, Academic Press (1974).

Tyler, B., Regulation of the assimilation of nitrogen compounds, *Annu. Rev. Biochem.* **47**, 1127–1162 (1978).

Umbarger, H. E., Amino acid biosynthesis and its regulation, *Annu. Rev. Biochem.* **47**, 533–606 (1978).

Wellner, D. and Meister, A., A survey of inborn errors of amino acid metabolism and transport in man, *Annu. Rev. Biochem.* **50**, 911–968 (1981).

Amino Acids as Biosynthetic Precursors

Battersby, A. R., Fookes, C. J. R., Matcham, G. W. J., and McDonald, E., Biosynthesis of the pigments of life: formation of the macrocycle, *Nature* **285**, 17–21 (1980).

Beru, N. and Goldwasser, E., The regulation of heme biosynthesis during erythropoietin-induced erythroid differentiation, *J. Biol. Chem.* **260**, 9251–9257 (1985).

Cooper, J. R., Bloom, F. E., and Roth, R. H., *The Biochemical Basis of Neuropharmacology* (4th ed.), Oxford University Press, Chapters 6–9 and 12 (1982).

Grandchamp, B., Beaumont, C., de Verneuil, H., and Nordmann, Y., Accumulation of porphobilinogen deaminase, uroporphyrinogen decarboxylase, and α- and β-globin mRNAs during differentiation of mouse erythroleukemic cells: effects of succinylacetone, *J. Biol. Chem.* **260**, 9630–9635 (1985).

Jordan, P. M. and Gibbs, P. N. B., Mechanism of action of 5-aminolevulinate dehydratase from human erythrocytes, *Biochem. J.* **227**, 1015–1021 (1985).

Macalpine, I. and Hunter, R., Porphyria and King George III, *Sci. Am.* **221**(1): 38–46 (1969).

Meister, A., Glutathione metabolism and its selective modification, *J. Biol. Chem.* **263**, 17205–17208 (1988).

Meister, A., Glutamine synthetase of mammals, *in* Boyer, P. D. (Ed.), *The Enzymes* (3rd ed.), Vol. 10, *pp.* 699–754, Academic Press (1974).

Meister, A. and Anderson, M. E., Glutathione, *Annu. Rev. Biochem.* **52**, 711–760 (1983).

Ponka, P. and Schulman, H. M., Acquisition of iron from transferrin regulated reticulocyte heme synthesis, *J. Biol. Chem.* **260**, 14717–14721 (1985).

Rutherford, T., Thompson, G. G., and Moore, M. R., Heme biosynthesis in Friend erythroleukemia cells: control by ferrochelatase, *Proc. Natl. Acad. Sci.* **76**, 833–836 (1979).

Perutz, M. F., Myoglobin and hemoglobin: role of distal residues in reactions with haem ligands, *Trends Biochem. Sci.* **14**, 42–44 (1989).

Nitrogen Fixation

Dilworth, M. and Glenn, A., How does a legume nodule work? *Trends Biochem. Sci.* **9**, 519–523, (1984).

Haaker, H. and Veeger, C., Enzymology of nitrogen fixation, *Trends Biochem. Sci.* **9**, 188–192 (1984).

Mortenson, L. E. and Thorneley, R. N. F., Structure and function of nitrogenase, *Annu. Rev. Biochem.* **48**, 387–418 (1979).

Orme-Johnson, W. H., Molecular basis of nitrogen fixation, *Annu. Rev. Biophys. Biophys. Chem.* **14**, 419–459 (1985).

Shah, V. K., Ugalde, R. A., Imperial, J., and Brill, W. J., Molybdenum in nitrogenase, *Annu. Rev. Biochem.* **53**, 231–257 (1984).

Problems

1. The symptoms of the partial deficiency of a urea cycle enzyme may be attenuated by a low protein diet. Explain.

2. Why are people on a high protein diet instructed to drink lots of water?

3. A student on a particular diet expends 10,000 kJ·day^{-1} while excreting 40 g of urea. Assuming that protein is 16% N by weight and that its metabolism yields 18 kJ·g^{-1}, what percentage of the student's energy requirement is met by protein.

*4. Among the many eat-all-you-want-and-lose-weight diets that have been popular for a time is one that eliminates all carbohydrates but permits the consumption of all the protein and fat desired. Would such a diet be effective? (*Hint:* Individuals on such a diet often complain that they have bad breath.)

5. Why are phenylketonurics warned against eating products containing the artificial sweetener **aspartame (NutraSweet®**; chemical name L-aspartyl-L-phenylalanine methyl ester)?

6. Demonstrate that the synthesis of heme from PBG as labeled in Fig. 24-27 results in the heme-labeling pattern given in Fig. 24-24.

7. Explain why certain drugs and other chemicals can precipitate an attack of acute intermittent porphyria.

8. One of the symptoms of **kwashiorkor,** the dietary protein deficiency disease in children, is the depigmentation of the skin and hair. Explain the biochemical basis of this symptom.

9. What are the metabolic consequences of a defective uridylyl-removing enzyme in *E. coli*?

10. Figure 24-45, Reaction 9 indicates that methionine is synthesized in microorganisms by the methylation of homocysteine in a reaction in which N^5-methyl-THF is the methyl donor. Yet, in the breakdown of methionine (Fig. 24-14), its demethylation occurs in three steps in which SAM is an intermediate. Discuss why this reaction does not occur via the simpler one-step reverse of the methylation reaction.

*11. In the glucose–alanine cycle, glycolytically derived pyruvate is transaminated to alanine and exported to the liver for conversion to glucose and return to the cell. Explain how a muscle cell is able to participate in this cycle under anaerobic (vigorously contracting) conditions. (*Hint:* The breakdown of many amino acids yields NH_3.)

12. Suggest a reason why the nitrogen fixing heterocysts of cyanobacteria have lost Photosystem II but retain Photosystem I.

Chapter 25

ENERGY METABOLISM: INTEGRATION AND ORGAN SPECIALIZATION

At this point in our narrative we have studied all of the major pathways of energy metabolism. Consequently, we are now in a position to consider how organisms, mammals in particular, orchestrate the metabolic symphony to meet their energy needs. This chapter therefore begins with a recapitulation of the major metabolic pathways and their control systems, then considers how these processes are apportioned among the various organs of the body, and ends with a discussion of how the body deals with the metabolic challenge of starvation and how it responds to the loss of control resulting from diabetes mellitus.

1. MAJOR PATHWAYS AND STRATEGIES OF ENERGY METABOLISM: A SUMMARY

Figure 25-1 indicates the interrelationships among the major pathways involved in energy metabolism. Let us review these pathways and their control mechanisms.

1. Glycolysis (Chapter 16)

The metabolic degradation of glucose begins with its conversion to two molecules of pyruvate with the net generation of two molecules each of ATP and

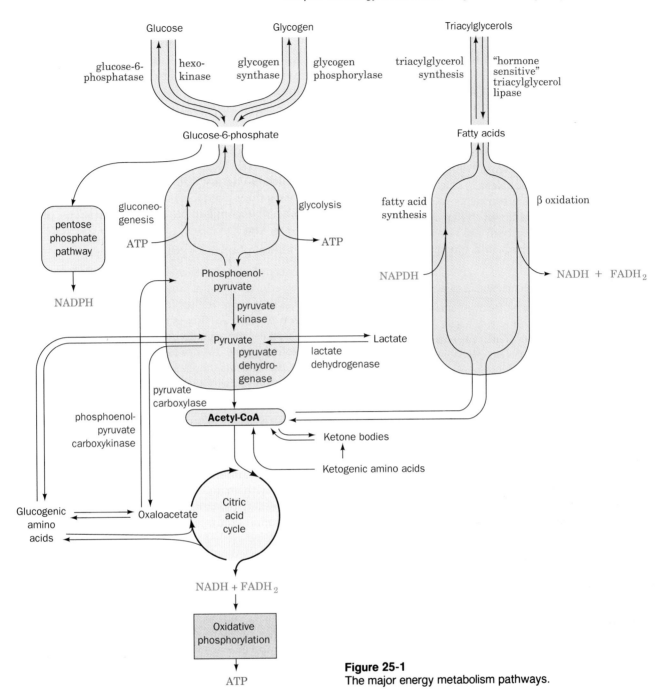

Figure 25-1
The major energy metabolism pathways.

NADH. Under anaerobic conditions, pyruvate is converted to lactate (or, in yeast, to ethanol) so as to recycle the NADH. Under aerobic conditions, however, when glycolysis serves to prepare glucose for further oxidation, the NAD$^+$ is regenerated through oxidative phosphorylation (see below). The flow of metabolites through the glycolytic pathway is largely controlled by the activity of phosphofructokinase (PFK). This enzyme is activated by AMP and ADP, whose concentrations rise as the need for energy metabolism increases, and is inhibited by ATP and citrate, whose concentrations increase when the de-

mand for energy metabolism has slackened. PFK is also activated by fructose-2,6-bisphosphate whose concentration is regulated by the levels of glucagon, epinephrine, and norepinephrine through the intermediacy of cAMP (Section 17-3F). Liver and muscle F2,6P levels are regulated oppositely: a [cAMP] increase causes an [F2,6P] decrease in liver and an [F2,6P] increase in muscle.

2. Gluconeogenesis (Section 21-1)

Mammals can synthesize glucose from a variety of precursors, including pyruvate, lactate, glycerol, and glucogenic amino acids, through pathways that occur mainly in liver and kidney. Many of these precursors are converted to oxaloacetate which, in turn, is converted to phosphoenolpyruvate and then, through a series of reactions that largely reverse the path of glycolysis, to glucose. The irreversible steps of glycolysis, those catalyzed by PFK and hexokinase, are bypassed in gluconeogenesis by hydrolytic reactions catalyzed, respectively, by fructose-1,6-bisphosphatase (FBPase) and glucose-6-phosphatase. FBPase and PFK may both be at least partially active simultaneously, creating a substrate cycle. This cycle, and the reciprocal regulation of PFK and FBPase, are important in regulating both the rate and direction of flux through glycolysis and gluconeogenesis (Sections 16-4 and 21-1B).

3. **Glycogen degradation and synthesis** (Chapter 17)

Glycogen, the storage form of glucose in animals, occurs mostly in liver and muscle. Its conversion to glucose-6-phosphate (G6P) for entry into glycolysis is catalyzed, in part, by glycogen phosphorylase whereas the opposing synthetic pathway is mediated by glycogen synthase. These enzymes are reciprocally regulated through phosphorylation/dephosphorylation reactions as catalyzed by amplifying cascades that respond to the levels of the hormones glucagon and epinephrine through the intermediacy of cAMP.

4. **Fatty acid degradation and synthesis** (Sections 23-1–23-5)

Fatty acids are broken down in increments of C_2 units through β oxidation to form acetyl-CoA. They are synthesized from this compound via a separate pathway. The activity of the β-oxidation pathway varies with the fatty acid concentration. This, in turn, depends on the activity of "hormone-sensitive" triacylglycerol lipase in adipose tissue that is stimulated, through cAMP-regulated phosphorylation/dephosphorylation reactions, by glucagon and epinephrine but inhibited by insulin. The fatty acid synthesis rate varies with the activity of acetyl-CoA carboxylase, which is activated by citrate and inhibited by the pathway product palmitoyl-CoA. Fatty acid synthesis is also subject to long-term regulation through alterations in the rates of synthesis of the enzymes mediating this process as stimulated by insulin and inhibited by fasting.

5. **Citric acid cycle** (Chapter 19)

The citric acid cycle oxidizes acetyl-CoA, the common degradation product of glucose, fatty acids and ketogenic amino acids, to CO_2 and H_2O with the concomitant production of NADH and $FADH_2$. Many glucogenic amino acids can also be oxidized via the citric acid cycle through their breakdown to one of its intermediates (Section 24-3). The activities of the citric acid cycle regulatory enzymes, citrate synthase, isocitrate dehydrogenase, and α-ketoglutarate dehydrogenase, are controlled by substrate availability and feedback inhibition by cycle intermediates.

6. **Oxidative phosphorylation** (Chapter 20)

This mitochondrial pathway oxidizes NADH and $FADH_2$ to NAD^+ and FAD with the coupled synthesis of ATP. The rate of oxidative phosphorylation, which is tightly coordinated with the metabolic fluxes through glycolysis and the citric acid cycle, is largely dependent on the concentrations of ATP, ADP, and P_i.

7. **Pentose phosphate pathway** (Section 21-4)

This pathway functions to generate NADPH for use in reductive biosynthesis, as well as the nucleotide precursor ribose-5-phosphate, through the oxidation of G6P. Its flux-generating step is catalyzed by glucose-6-phosphate dehydrogenase, which is controlled by the level of $NADP^+$. *The ability of enzymes to distinguish between NADH, which is mainly utilized in energy metabolism, and NADPH, permits energy metabolism and biosynthesis to be regulated independently.*

8. **Amino acid degradation and synthesis** (Sections 24-1–24-5)

Excess amino acids may be degraded to common metabolic intermediates. Most of these pathways begin with an amino acid's transamination to its corresponding α-keto acid with the eventual transfer of the amino group to urea via the urea cycle. Leucine and lysine are ketogenic amino acids in that they can be converted only to acetyl-CoA or acetoacetate and hence cannot be glucose precursors. The other amino acids are glucogenic in that they may be, at least in part, converted to one of the glucose precursors— pyruvate, oxaloacetate, α-ketoglutarate, succinyl-CoA, or fumarate. Five amino acids are both ketogenic and glucogenic. Essential amino acids are those that an animal cannot synthesize itself; they must be obtained from plant and microbial sources. Nonessential amino acids can be synthesized by animals via pathways that are generally simpler than those synthesizing essential amino acids.

Two compounds lie at the crossroads of the foregoing metabolic pathways: acetyl-CoA and pyruvate (Fig. 25-1). Acetyl-CoA is the common degradation product of most metabolic fuels including polysaccharides, lipids, and proteins. Its acetyl group may be oxidized to CO_2 and H_2O via the citric acid cycle and oxidative phosphorylation or used to synthesize fatty acids. Pyruvate is the product of glycolysis, the dehydrogenation of lactate, and the breakdown of certain glucogenic amino acids. It may be oxidatively decarboxylated to yield acetyl-CoA, thereby committing its atoms to either oxidation or to the biosynthesis of fatty acids. Alternatively, it may be carboxylated via the pyruvate carboxylase reaction to form oxaloacetate which, in turn, either replenishes citric acid cycle intermediates or enters gluconeogenesis via phosphoneolpyruvate, thereby bypassing an irreversible step in glycolysis. Pyruvate is therefore a precursor of several amino acids as well as of glucose.

The foregoing pathways occur in specific cellular compartments. Glycolysis, glycogen synthesis and degradation, fatty acid synthesis, and the pentose phosphate pathway are largely or entirely cytosolically based whereas fatty acid degradation, the citric acid cycle, and oxidative phosphorylation occur largely in the mitochondrion. Different phases of gluconeogenesis and amino acid degradation occur in each of these compartments. *The flow of metabolites across compartment membranes is mediated, in most cases, by specific carriers that are also subject to regulation.*

The enormous number of enzymatic reactions that simultaneously occur in every cell must be coordinated and strictly controlled to meet the cell's needs. Such regulation occurs on many levels. Intercellular communications regulating metabolism occur via certain hormones including epinephrine, norepinephrine, gluca-gon, and insulin, as well as through a series of steroid hormones known as **glucocorticoids,** (whose actions are discussed in Section 34-4A). These hormonal signals trigger a variety of cellular responses including the synthesis of second messengers such as cAMP in the short term and the modulation of protein synthesis rates in the long term. On the molecular level, the enzymatic reaction rates are controlled by phosphorylation/dephosphorylation via amplifying reaction cascades, by allosteric responses to the presence of effectors, which are usually precursors or products of the reaction pathway being controlled, and by substrate availability. The regulatory machinery of opposing catabolic and anabolic pathways is generally arranged such that these pathways are reciprocally regulated.

2. ORGAN SPECIALIZATION

In this section we consider how the special needs of the mammalian body organs are met and how their metabolic capabilities are coordinated to meet these needs. In particular, we discuss brain, muscle, adipose tissue, and liver (Fig. 25-2).

Figure 25-2
The metabolic interrelationships among brain, adipose tissue, muscle, and liver. The red arrows indicate pathways that predominate in the well-fed state when glucose, amino acids, and fatty acids are directly available from the intestines.

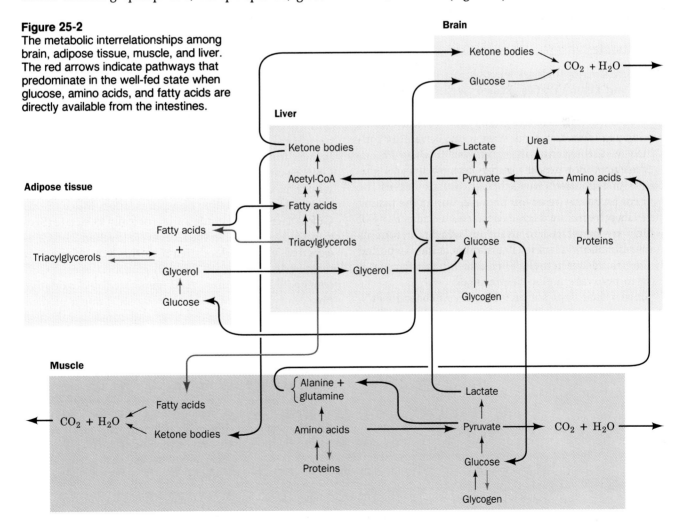

A. Brain

Brain tissue has a remarkably high respiration rate. For instance, the human brain only constitutes some 2% of the adult body mass but is responsible for ~20% of its resting O_2 consumption. This consumption, moreover, is independent of the state of mental activity; it varies little between sleep and the intense concentration required of, say, the study of biochemistry. Most of the brain's energy production serves to power the plasma membrane (Na^+-K^+)-ATPase (Section 18-3A), which maintains the membrane potential required for nerve impulse transmission (Section 34-4C). In fact, the respiration of brain slices is over 50% reduced by the (Na^+-K^+)-ATPase inhibitor ouabain (Section 18-3A).

Under usual conditions, glucose serves as the brain's only fuel (although, with extended fasting, it gradually switches to ketone bodies; Section 25-3A). Indeed, since brain cells store very little glycogen, they require a steady supply of glucose from the blood. A blood glucose concentration of less than one half of the normal value of ~5 mM results in brain dysfunction. Levels much below this, for example, caused by severe insulin overdose, result in coma, irreversible damage, and ultimately death. One of liver's major functions, therefore, is to maintain the blood glucose level (Section 25-2D).

B. Muscle

Muscle's major fuels are glucose from glycogen, fatty acids, and ketone bodies. Rested, well-fed muscle, in contrast to brain, synthesizes a glycogen store comprising 1 to 2% of its mass. The glycogen serves muscle as a readily available fuel depot since it can be rapidly converted to G6P for entry into glycolysis (Section 17-1).

Muscle cannot export glucose because it lacks glucose-6-phosphatase. Nevertheless, muscle serves the body as an energy reservoir because, during the fasting state, its proteins are degraded to amino acids, many of which are converted to pyruvate which, in turn, is transaminated to alanine. The alanine is then exported via the bloodstream to the liver, which transaminates it back to pyruvate, a glucose precursor.

Since muscle does not participate in gluconeogenesis, it lacks the machinery that regulates this process in such gluconeogenic organs as liver and kidney. Muscle does not have receptors for glucagon, which it will be recalled, stimulates an increase in blood glucose levels (Section 17-3F). However, muscle possesses epinephrine receptors, which through the intermediacy of cAMP, control the phosphorylation/dephosphorylation cascade system that regulates glycogen breakdown and synthesis (Section 17-3). This is the same cascade system that controls the competition between glycolysis and gluconeogenesis in liver in response to glucagon. The cAMP-controlled enzyme system that synthesizes and degrades F2,6P in muscle acts oppositely to that in liver. The concentration of F2,6P, the principle PFK activator and FBPase inhibitor, rises in heart muscle but falls in liver in response to an increase in [cAMP]. Moreover the muscle isozyme of pyruvate kinase, which it will be recalled, catalyzes the final step of glycolysis, is not subject to phosphorylation/dephosphorylation as is the liver isozyme (Section 21-1B). Thus, *whereas an increase in liver cAMP stimulates glycogen breakdown and gluconeogenesis resulting in glucose export, an increase in muscle cAMP activates glycogen breakdown and glycolysis resulting in glucose consumption. Consequently, epinephrine, which prepares the organism for action (fight or flight), acts independently of glucagon which, in concert with insulin, regulates the general level of blood glucose.*

Muscle Contraction Is Anaerobic Under Conditions of High Exertion

Muscle contraction is driven by ATP hydrolysis (Section 34-3B) and is therefore ultimately dependent on respiration. Skeletal muscle at rest utilizes ~30% of the O_2 consumed by the human body. A muscle's respiration rate may increase in response to a heavy work load by as much as 25-fold. Yet, its rate of ATP hydrolysis can increase by a much greater amount. The ATP is initially regenerated by the reaction of phosphocreatine (Section 15-4C):

$$\text{Phosphocreatine} + \text{ADP} \rightleftharpoons \text{creatine} + \text{ATP}$$

(phosphocreatine is resynthesized in resting muscle by the reversal of this reaction). Under conditions of maximum exertion, however, such as occurs in a sprint, a muscle has only about a 4-s supply of phosphocreatine. It must then shift to ATP production via glycolysis of G6P resulting from glycogen breakdown, a process whose maximum flux greatly exceeds those of the citric acid cycle and oxidative phosphorylation. Much of this G6P is therefore degraded anaerobically to lactate (Section 16-3A) which, in the Cori cycle (Section 21-1C), is exported via the bloodstream to the liver where it is reconverted to glucose through gluconeogenesis. Gluconeogenesis requires ATP generated by oxidative phosphorylation. Muscles thereby shift much of their respiratory burden to the liver and consequently also delay the O_2-consumption process, a phenomenon known as oxygen debt.

Muscle Fatigue Has a Protective Function

Muscle fatigue, defined as the inability of a muscle to maintain a given power output, occurs in ~20 s under conditions of maximum exertion. Such fatigue is not caused by the exhaustion of the muscle's glycogen supply. Rather, it results from glycolytic proton generation that can drop the intramuscular pH from its resting value of 7.0 to as low as 6.4 (fatigue does not, as is widely believed, result from the buildup of lactate itself as is demonstrated by the observation that muscles can sustain a large power output under high lactate concentrations if the pH is maintained near 7.0). The mechanism whereby increased acidity causes muscle fatigue is obscure. A reasonable possibility is that PFK, the enzyme controlling glycolytic flux, has reduced activity at low pH's. Whatever the case, it seems likely that this phenomenon is an adaptation that prevents muscle cells from commiting suicide by exhausting their ATP supply (recall that glycolysis and other ATP-generating pathways must be primed by ATP).

The Heart Is a Largely Aerobic Organ

The heart is a muscular organ but one that must maintain continuous rather than intermittent activity. Thus heart muscle, except for short periods of extreme exertion, relies entirely on aerobic metabolism. It is therefore richly endowed with mitochondria; they comprise up to 40% of its cytoplasmic space, whereas some types of skeletal muscle are nearly devoid of mitochondria. The heart can metabolize fatty acids, ketone bodies, glucose, pyruvate, and lactate. Fatty acids are the resting heart's fuel of choice but, upon the imposition of a heavy work load, the heart greatly increases its rate of consumption of glucose, which is derived mostly from its relatively limited glycogen store.

C. Adipose Tissue

Adipose tissue, which consists of cells known as adipocytes (Fig. 11-2), is widely distributed about the body but occurs most prominently under the skin, in the abdominal cavity, in skeletal muscle, around blood vessels, and in mammary gland. The adipose tissue of a normal 70-kg man contains ~15 kg of fat. This amount represents some 590,000 kJ of energy (141,000 dieter's Calories), which is sufficient to maintain life for ~3 months. Yet, adipose tissue is by no means just a passive storage depot. In fact, it is second in importance only to liver in the maintenance of metabolic homeostasis.

Adipose tissue obtains most of its fatty acids from the liver or from the diet as described in Section 23-1. Fatty acids are activated by the formation of the corresponding fatty acyl-CoA and then esterified with glycerol-3-phosphate to form the stored triacylglycerols (Section 23-4F). The glycerol-3-phosphate arises from the reduction of dihydroxyacetone phosphate, which must be glycolytically generated from glucose because adipocytes lack a kinase that phosphorylates endogenous glycerol.

Adipocytes hydrolyze triacylglycerols to fatty acids and glycerol in response to the levels of glucagon, epinephrine, and insulin through a reaction catalyzed by "hormone-sensitive" lipase (Section 23-5). If glycerol-3-phosphate is abundant, many of the fatty acids so-formed are reesterified to triacylglycerols. Indeed, the average turnover time for triacylglycerols in adipocytes is only a few days. If, however, glycerol-3-phosphate is in short supply, the fatty acids are released into the bloodstream. *The rate of glucose uptake by adipocytes, which is regulated by insulin as well as by glucose availability, is therefore also a controlling factor in triacylglycerol formation and mobilization.*

Obesity Results from Aberrant Metabolic Control

Obesity is one of the major health-related problems in the developed world. Most obese people (those who are at least 20% above their desirable weights) find it inordinately difficult to lose weight, or having done so, to keep it off. Yet most animals, including humans, tend to have stable weights; that is, if they are given free access to food, they eat just enough to maintain this so-called "set point" weight. The nature of the regulatory machinery that controls the set point, which in obese individuals seems to be aberrantly high, is just beginning to come to light.

Formerly grossly obese individuals who have lost at least 100 kg to reach their normal weights exhibit some of the metabolic symptoms of starvation: they are obsessed with food, have low heart rates, are cold intolerant, and require 25% less caloric intake than normal individuals of similar heights and weights. In both normal and obese individuals, some 50% of the fatty acids liberated by the hydrolysis of triacylglycerols are reesterified before they can leave the adipocytes. In formerly obese subjects, this reesterification rate is only 35 to 40%, a level similar to that observed in normal individuals after a several day fast. The fat cells in normal and obese individuals, moreover, are of roughly the same size; obese people just have more of them. In fact, adipocyte precursor cells from massively obese individuals proliferate excessively in tissue culture compared to those from normal or even moderately obese subjects (adipocytes themselves do not replicate). Since fat cells, once gained, are never lost, this suggests that adipocytes, although highly elastic in size, tend to maintain a certain fixed volume and in doing so influence the metabolism and thus the appetite. This insight, unfortunately, has not yet led to a method for lowering the set points of individuals with a tendency towards obesity.

D. Liver

The liver is the body's central metabolic clearing house. It functions to maintain the proper levels of nutrients in the blood for use by the brain, muscles, and other tissues. The liver is uniquely situated to carry out this task because all the nutrients absorbed by the intestines except fatty acids are released into the portal vein, which drains directly into the liver.

One of the liver's major functions is to act as a blood glucose "buffer." It does so by taking up or releasing glucose in response to the levels of glucagon, epinephrine, and insulin as well as to the concentration of glucose itself. After a carbohydrate-containing meal, when the blood glucose concentration reaches ~6 mM, the liver takes up glucose by converting it to G6P. The process is catalyzed by glucokinase (Section 17-3F), which differs from hexokinase, the analogous glycolytic enzyme in other cells, in that glucokinase has a much lower affinity for glucose ($K_M \approx 10$ mM for glucokinase vs <0.1 mM for hexokinase) and is not inhibited by G6P. Liver cells, in contrast to muscle and adipose cells, are permeable to glucose and thus insulin has no direct effect on their glucose uptake. Since the blood glucose concentration is normally less than glucokinase's K_M, the rate of glucose phosphorylation in the liver is more or less proportional to the blood glucose concentration. The other intestinally absorbed sugars, mostly fructose, galactose, and mannose, are also converted to G6P in the liver (Section 16-5). After an overnight fast, the blood glucose level drops to ~4 mM. The liver keeps it from dropping below this level by releasing glucose into the blood as is described below. In addition, lactate, the product of anaerobic glucose metabolism in the muscle, is taken up by the liver for use in gluconeogenesis and lipogenesis as well as in oxidative phosphorylation.

The Fate of Glucose Varies with Metabolic Requirements

G6P is at the crossroads of carbohydrate metabolism; it can have several alternative fates depending on the glucose demand (Fig. 25-1):

1. G6P can be converted to glucose by the action of glucose-6-phosphatase for transport via the bloodstream to the peripheral organs.

2. G6P can be converted to glycogen (Section 17-1) when the body's demand for glucose is low. Yet, increased glucose demand, as signaled by higher levels of glucagon and/or epinephrine, reverses this process (Section 17-2).

3. G6P can be converted to acetyl-CoA via glycolysis and the action of pyruvate dehydrogenase (Chapter 16). Most of this glucose-derived acetyl-CoA is used in the synthesis of fatty acids (Section 23-4), whose fate is described below, and in the synthesis of phospholipids (Section 23-8) and cholesterol (Section 23-6A). Cholesterol, in turn, is a precursor of bile acids, which are produced by the liver (Section 23-6C) for use as emulsifying agents in the intestinal digestion and absorption of fats (Section 23-1).

4. G6P may be degraded via the pentose phosphate pathway (Section 21-4) to generate the NADPH required for fatty acid biosynthesis and the liver's many other biosynthetic functions.

The Liver Can Synthesize or Degrade Triacylglycerols

Fatty acids are also subject to alternative metabolic fates in the liver (Fig. 25-1):

1. When the demand for metabolic fuels is high, fatty acids are degraded to acetyl-CoA and then to ketone bodies (Section 23-3) for export via the bloodstream to the peripheral tissues.

2. When the demand for metabolic fuels is low, fatty acids are used to synthesize triacylglycerols that are secreted into the bloodstream as VLDL for uptake by adipose tissue. Fatty acids may also be incorporated into phospholipids (Section 23-8).

Since the rate of fatty acid oxidation varies only with fatty acid concentration (Section 23-5), it might be expected that fatty acids produced by the liver are subject to reoxidation before they can be exported. Such a futile cycle is prevented by the compartmentation of fatty acid oxidation in the mitochondrion and fatty acid synthesis in the cytosol. Carnitine palmitoyl transferase I, a component of the system that transports fatty acids into the mitochondrion (Section 23-2B), is inhibited by malonyl-CoA, the key intermediate in fatty acid biosynthesis (Section 23-4A). Hence, when the demand for metabolic fuels is low so that fatty acids are being synthesized, they cannot enter the mitochondrion for conversion to acetyl-CoA. Rather, the liver's biosynthetic demand for acetyl-CoA is met through the degradation of glucose.

When the demand for metabolic fuel rises so as to inhibit fatty acid biosynthesis, however, fatty acids are transported into the liver mitochondria for conversion to ketone bodies. Under such conditions of low blood glucose concentrations, glucokinase has reduced activity so that there is net glucose export (there is, however, always a futile cycle between the reactions catalyzed by glucokinase and glucose-6-phosphatase; Section 17-3F). The liver cannot use ketone bodies for its own metabolic purposes because liver cells lack 3-ketoacyl-CoA transferase (Section 23-3). Fatty acids rather than glucose or ketone bodies are therefore the liver's major acetyl-CoA source under conditions of high metabolic demand. The liver generates its ATP from this acetyl-CoA through the citric acid cycle and oxidative phosphorylation.

Amino Acids Are Important Metabolic Fuels

The liver degrades amino acids to a variety of metabolic intermediates (Section 24-3). These pathways mostly begin with amino acid transamination to yield the corresponding α-keto acid (Section 24-1A) with the amino group being ultimately converted, via the urea cycle (Section 24-2), to the subsequently excreted urea. Glucogenic amino acids can be converted in this manner to pyruvate or citric acid cycle intermediates such as oxaloacetate and are thereby gluconeogenic precursors (Section 21-1). Ketogenic amino acids, many of which are also glucogenic, may be converted to ketone bodies.

The liver's glycogen store is insufficient to supply the body's glucose needs for more than ~6 h after a meal. After that, glucose is supplied through gluconeogenesis from amino acids arising mostly from muscle protein degradation to alanine and glutamine (the transport form of ammonia; Section 24-1B). (Animals cannot convert fat to glucose because they lack a pathway for the net conversion of acetyl-CoA to oxaloacetate; Section 21-2). Thus proteins, in addition to their structural and functional roles, are important fuel resources.

The Liver Is the Body's Major Metabolic Processing Unit

The liver has numerous specialized biochemical functions in addition to those already mentioned. Prominent among them are the synthesis of blood plasma proteins, the degradation of porphyrins (Section 24-4A) and nucleic acid bases (Section 26-5), the storage of iron, and the detoxification of biologically active substances such as drugs, poisons, and hormones by a variety of oxidation, reduction, hydrolysis, conjugation, and methylation reactions.

3. METABOLIC ADAPTATION

In this section we consider the body's responses to two metabolically abnormal situations: (1) starvation and (2) the disease diabetes mellitus.

A. Starvation

Glucose is the metabolite of choice of both brain and working muscle. Yet, the body stores less than a day's supply of carbohydrate (Table 25-1). Thus, the low blood sugar resulting from even an overnight fast results, through an increase in glucagon secretion and a decrease in insulin secretion, in the mobilization of fatty acids from adipose tissue (Section 23-5). The diminished insulin level also inhibits glucose uptake by muscle tissue. Muscles therefore switch from glucose to fatty acid metabolism for energy production. The brain, however, still remains heavily dependent on glucose.

In animals, glucose cannot be synthesized from fatty acids. This is because neither pyruvate nor oxaloacetate, the precursors of glucose in gluconeogenesis (Section 21-1), can be synthesized from acetyl-CoA (the oxaloacetate in the citric acid cycle is derived from acetyl-CoA

Table 25-1
Fuel Reserves of a Normal 70-kg Man

Fuel	Mass (kg)	Calories[a]
Tissues		
Fat (adipose triacylglycerols)	15	141,000
Protein (mainly muscle)	6	24,000
Glycogen (muscle)	0.150	600
Glycogen (liver)	0.075	300
Circulating fuels		
Glucose (extracellular fluid)	0.020	80
Free fatty acids (plasma)	0.0003	3
Triacylglycerols (plasma)	0.003	30
Total		166,000

[a]1 (dieters) Calorie = 1 kcal = 4.184 kJ.

Source: Cahill, G. F., *New Engl. J. Med.* **282,** 669 (1970).

but the cyclic nature of this process requires that the oxaloacetate be consumed as fast as it is synthesized; Section 19-1A). During starvation, glucose must therefore be synthesized from the glycerol product of triacylglycerol breakdown and, more importantly, from the amino acids derived from the proteolytic degradation of proteins, the major source of which is muscle. Yet, the continued breakdown of muscle during prolonged starvation would ensure that this process became irreversible since a large muscle mass is essential for an animal to move about in search of food. The organism must therefore make alternative metabolic arrangements.

After several days of starvation, gluconeogenesis has so depleted the liver's oxaloacetate supply that this organ's ability to metabolize acetyl-CoA via the citric acid cycle is greatly diminished. Rather, the liver converts the acetyl-CoA to ketone bodies (Section 23-3), which it releases into the blood. The brain gradually adapts to using ketone bodies as fuel through the synthesis of the appropriate enzymes: After a 3-day fast, only about one third of the brain's energy requirements are satisfied by ketone bodies but after 40 days of starvation, ~70% of its energy needs are so-met. The rate of muscle breakdown during prolonged starvation consequently decreases to ~25% of its rate after a several-day fast. The survival time of a starving individual is therefore much more dependent on the size of his fat reserves than it is on his muscle mass. Indeed, highly obese individuals can survive for over a year without eating (and have occasionally done so in clinically supervised weight reduction programs).

B. Diabetes Mellitus

The polypeptide hormone insulin acts mainly on muscle, liver, and adipose tissue cells to stimulate the synthesis of glycogen, fats, and proteins while inhibiting the breakdown of these metabolic fuels. In addition, insulin stimulates the uptake of glucose by most cells with the notable exception of brain and liver cells. Together with glucagon, which has largely opposite effects, insulin acts to maintain the proper level of blood glucose.

In the disease **diabetes mellitus,** which is the third leading cause of death in the United States after heart disease and cancer, insulin is either not secreted in sufficient amounts or does not efficiently stimulate its target cells. As a consequence, blood glucose levels become so elevated that the glucose "spills over" into the urine providing a convenient diagnostic test for the disease. Yet, despite these high blood glucose levels, cells "starve" since insulin-stimulated glucose entry into cells is impaired. Triacylglycerol hydrolysis, fatty acid oxidation, gluconeogenesis, and ketone body formation are accelerated and, in a condition termed **ketosis,** ketone body levels in the blood become abnormally high. Since ketone bodies are acids, their high concentration puts a strain on the buffering capacity of the blood and on the kidney, which controls blood pH by excreting the excess H^+ into the urine. This H^+ excretion is accompanied by Na^+, K^+, P_i, and H_2O excretion, causing severe dehydration (which compounds the dehydration resulting from the osmotic effect of the high glucose concentration in the blood; excessive thirst is a classical symptom of diabetes) and a decrease in blood volume; ultimately life-threatening situations.

Their are two major forms of diabetes mellitus:

1. **Insulin-dependent** or **juvenile-onset diabetes,** which most often strikes suddenly in childhood.

2. **Noninsulin-dependent** or **maturity-onset diabetes,** which usually develops rather gradually after the age of 40.

In insulin-dependent diabetes, insulin is absent or nearly so because the pancreas lacks or has defective β cells. This condition results, in genetically susceptible individuals, from a drug or virus-triggered autoimmune response that selectively destroys their β cells. Individuals with insulin-dependent diabetes, as Frederick Banting and George Best first demonstrated in 1921, require daily insulin injections to survive and must follow carefully balanced diet and exercise regimens. Their lifespans are, nevertheless, reduced by up to one third as a result of degenerative complications such as kidney malfunction, nerve impairment, and cardiovascular disease, as well as blindness, that apparently arise from the imprecise metabolic control provided by periodic insulin injections. Perhaps newly developed systems that monitor blood glucose levels and continuously deliver insulin in the required amounts will rectify this situation.

Noninsulin-dependent diabetes, which accounts for over 90% of the diagnosed cases of diabetes, usually occurs in obese individuals with a genetic predisposition for this condition (although one that differs from that associated with insulin-dependent diabetes). These individuals have normal or even greatly increased insulin levels. Their symptoms arise from a paucity of **insulin receptors** on normally insulin-responsive cells. Perhaps the increased insulin production resulting from overeating (obesity is almost always the consequence of overeating) eventually suppresses the synthesis of insulin receptor (a plasma membrane-bound glycoprotein; Section 34-4B). This hypothesis accounts for the observation that diet alone is often sufficient to control this type of diabetes.

Chapter Summary

The complex network of processes involved in energy metabolism are distributed among different compartments within cells and in different organs of the body. These processes function to generate ATP "on demand," to generate and store glucose, triacylglycerols, and proteins in times of plenty for use when needed, and to keep the concentration of glucose in the blood at the proper level for use by organs such as the brain whose sole fuel source, under normal conditions, is glucose. The major energy metabolism pathways include glycolysis, glycogen degradation and synthesis, gluconeogenesis, the pentose phosphate pathway, and triacylglycerol and fatty acid synthesis, which are cytosolically based, in addition to fatty acid oxidation, the citric acid cycle, and oxidative phosphorylation, which are confined to the mitochondrion. Amino acid degradation occurs, in part, in both compartments. The mediated membrane transport of metabolites therefore also plays an essential metabolic role.

The main organs involved in energy metabolism are brain, muscle, adipose tissue, and liver. The brain normally consumes large amounts of glucose. Muscle, under intense ATP demand such as in sprinting, degrades glucose and glycogen anaerobically, thereby producing lactate, which is exported via the blood to the liver for reconversion to glucose through gluconeogenesis. During moderate activity, muscle generates ATP by oxidizing glucose from glycogen, fatty acids, and ketone bodies completely to CO_2 and H_2O via the citric acid cycle and oxidative phosphorylation. Adipose tissue stores triacylglycerols and releases fatty acids into the bloodstream in response to the organism's metabolic needs. These metabolic needs are communicated to adipose tissue by means of the hormones insulin, which indicates a fed state in which storage is appropriate, and glucagon, epinephrine, and norepinephrine, which signal a need for fatty acid release to provide fuel for other tissues. The liver, the body's central metabolic clearing house, maintains blood glucose concentrations by storing glucose as glycogen in times of plenty and releasing glucose in times of need both by glycogen breakdown and gluconeogenesis. It also converts fatty acids to ketone bodies for use by peripheral tissues. During a fast, it breaks down amino acids resulting from protein degradation, to metabolic intermediates that can be used to generate glucose.

During prolonged starvation, the brain slowly adapts from the use of glucose as its sole fuel source to the use of ketone bodies, thereby shifting the metabolic burden from protein breakdown to fat breakdown. Diabetes mellitus is a disease in which insulin is either not secreted or does not efficiently stimulate its target tissues. Cells "starve" in the midst of plenty since they cannot absorb blood glucose and their hormonal signals remain those of starvation. Abnormally high production of ketone bodies is one of the most dangerous effects of uncontrolled diabetes.

References

Part IV of this text.

Cahill, G. F., Jr., Starvation in man, *New Engl. J. Med.* **282,** 668–675 (1970).

Niven, N. J. and Hitman, G. A., The molecular genetics of diabetes mellitus, *Biosci. Rep.* **6,** 501–512 (1986).

Notkins, A. L., The causes of diabetes, *Sci. Am.* **241**(5): 62–73 (1979).

Problems

1. Describe the metabolic effects of liver failure.

2. What is the basis of the hypothesis that athletes' muscles are more heavily buffered than those of normal individuals?

3. Explain why urea output is vastly decreased during starvation.

4. Explain why the breath of an untreated diabetic smells of acetone.

Chapter 26
NUCLEOTIDE METABOLISM

Nucleotides are biologically ubiquitous substances that participate in nearly all biochemical processes:

1. They form the monomeric units of nucleic acids. Indeed, nucleic acids are synthesized directly from nucleoside triphosphates, the activated form of nucleotides (Section 31-1A).

2. Nucleoside triphosphates, most conspicuously ATP, are the "energy-rich" end products of most energy-releasing pathways whose utilization drives most energy-requiring processes. Several activated intermediates, such as UDP-glucose in glycogen synthesis (Section 17-2A), contain nucleotide components.

3. Most metabolic pathways are regulated, at least in part, by the levels of nucleotides such as ATP, ADP, and AMP. Similarly, many hormonal signals, such as those controlling glycogen metabolism (Section 17-3), are mediated intracellularly by cAMP or its guanine analog **cGMP**.

4. Adenine nucleotides are components of the coenzymes NAD⁺, NADP⁺, FMN, FAD, and coenzyme A.

The importance of nucleotides in cellular metabolism is indicated by the observation that nearly all cells can synthesize them both *de novo* (anew) and from the degradation products of nucleic acids. In this chapter, we consider the nature of these biosynthetic pathways. In doing so, we shall examine how they are regulated and the consequences of their blockade, both by genetic defects and through the administration of chemotherapeutic agents. We then discuss how nucleotides are degraded. Finally, we outline the biosynthesis of the nucleotide coenzymes. We begin with a presentation of the chemical nomenclature of nucleotides and their components.

1. CHEMICAL STRUCTURES OF NUCLEOTIDES, NUCLEOSIDES, AND BASES

Nucleotides are phosphate esters of pentoses (C_5 sugars) in which a nitrogenous base is linked to C(1') of the sugar residue. In **ribonucleotides** (Fig. 26-1a), the pentose is the D-ribose residue while in **deoxyribonucleotides** (or just **deoxynucleotides;** Fig. 26-1b), which occur in DNA, the sugar is 2'-deoxy-D-ribose (note that the "primed" numbers refer to the atoms of the ribose residue; "unprimed" numbers refer to the nitrogenous base). The phosphate group may be bonded to the C(3') or C(5') of a pentose to form its 3'-nucleotide or its 5'-nucleotide, respectively. If the phosphate group is absent, the compound is known as a **nucleoside**. A 5'-nucleotide, for example, may therefore be referred to as a nucleoside-5'-phosphate. In all naturally occurring nucleotides and nucleosides, the *N*-glycosidic bond linking the nitrogenous base to the pentose C(1') atom has the β configuration [extending from the same side of the furanose ring as the C(4')—C(5') bond; Section 10-1B]. Note that nucleotide phosphate groups are doubly ionized at physiological pH's; that is, *nucleotides are moderately strong acids.*

The nitrogenous bases are planar, aromatic, heterocyclic molecules which, for the most part, are derivatives of either **purine** *or* **pyrimidine**.

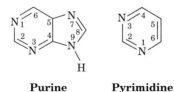

Purine   **Pyrimidine**

The structures, names, and abbreviations of the common bases, nucleosides, and nucleotides are given in Table 26-1. The major purine components of nucleic acids are **adenine** and **guanine** residues; the major pyrimidine residues are those of **cytosine, uracil** (which occurs in RNA), and **thymine** (which occurs mainly in DNA). The purines form glycosidic bonds to ribose via their N(9) atoms, whereas pyrimidines do so through their N(1) atoms (note that purines and pyrimidines have dissimilar atom numbering schemes).

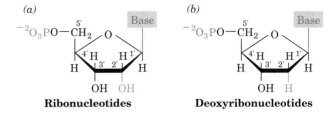

Ribonucleotides   **Deoxyribonucleotides**

Figure 26-1
The chemical structures of (a) ribonucleotides and (b) deoxyribonucleotides.

2. SYNTHESIS OF PURINE RIBONUCLEOTIDES

In this section we commence our considerations of how nucleic acids and their components are synthesized by describing the synthesis of purine ribonucleotides. In 1948, John Buchanan obtained the first clues as to how this process occurs *de novo* by feeding a variety of isotopically labeled compounds to pigeons and chemically determining the positions of the labeled atoms in their excreted **uric acid** (a purine).

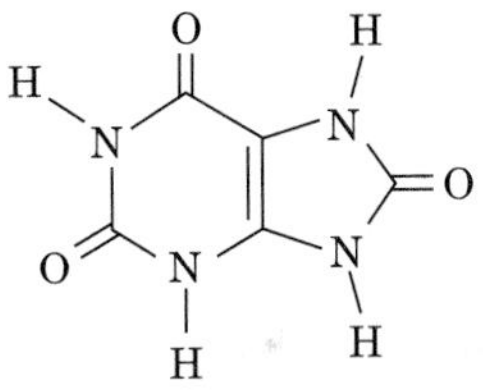

Uric acid

He used birds in these experiments because they excrete waste nitrogen almost entirely as uric acid, a water insoluble and therefore easily isolated substance. The results of his studies, which are summarized in Fig. 26-2, demonstrated that N(1) of purines arises from the amine group of aspartate, C(2) and C(8) originate from formate, N(3) and N(9) are contributed by the amide group of glutamine, C(4), C(5), and N(7) are derived from gly-

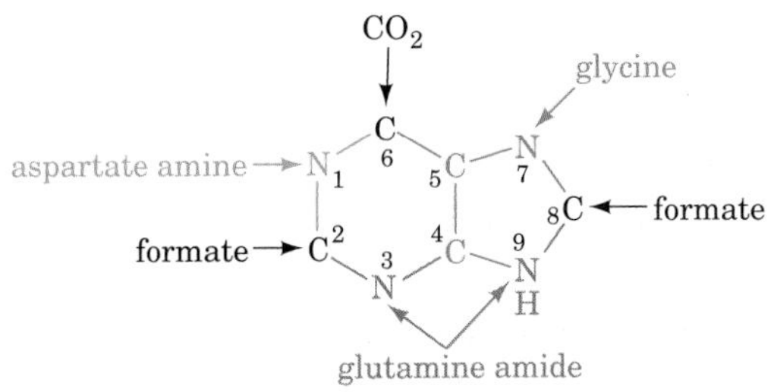

Figure 26-2
The biosynthetic origins of purine ring atoms. Note that C(4), C(5), and N(7) come from a single glycine molecule but each of the other atoms are derived from independent precursors.

Table 26-1

Names and Abbreviations of Nucleic Acid Bases, Nucleosides, and Nucleotides

Base Formula	Base X = H	Nucleoside X = ribose[a]	Nucleotide[b] X = ribose phosphate[a]
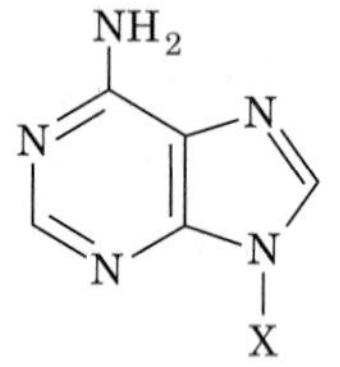 *(adenine structure)*	Adenine Ade A	Adenosine Ado A	Adenylic acid Adenosine monophosphate AMP
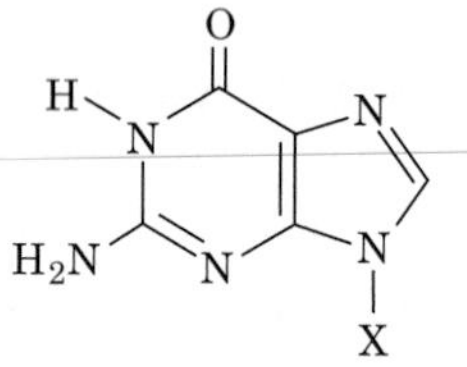 *(guanine structure)*	Guanine Gua G	Guanosine Guo G	Guanylic acid Guanosine monophosphate GMP
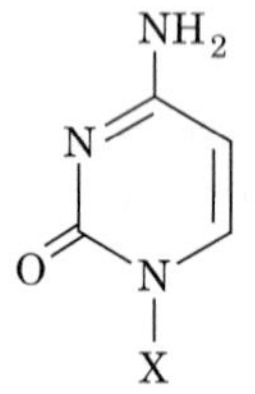 *(cytosine structure)*	Cytosine Cyt C	Cytidine Cyd C	Cytidylic acid Cytosine monophosphate CMP
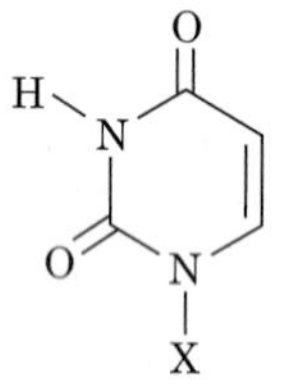 *(uracil structure)*	Uracil Ura U	Uridine Urd U	Uridylic acid Urdine monophosphate UMP
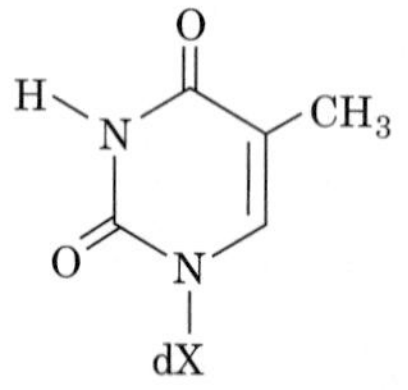 *(thymine structure)*	Thymine Thy T	Deoxythymidine dThd dT	Deoxythymidylic acid Deoxythymidine monophosphate dTMP

[a] The presence of a 2′-deoxyribose unit in place of ribose, as occurs in DNA, is implied by the prefixes "deoxy" or "d." For example, the deoxynucleoside of adenine is deoxyadenosine or dA. However, for thymine-containing residues, which rarely occur in RNA, the prefix is redundant and may be dropped. The presence of a ribose unit may be explicitly implied by the prefixes "ribo" or "r." Thus the ribonucleotide of thymine is ribothymidine or rT.

[b] The position of the phosphate group in a nucleotide may be explicitly specified as in, for example, 3′-AMP and 5′-GMP.

cine (strongly suggesting that this molecule is wholly incorporated into the purine ring), and C(6) comes from CO_2.

The actual pathway by which these precursors are incorporated into the purine ring, the subject of Section 26-2A, was elucidated in subsequent investigations performed largely by Buchanan and by G. Robert Greenberg. These investigations showed that the initially synthesized purine derivative is **inosine monophosphate (IMP),**

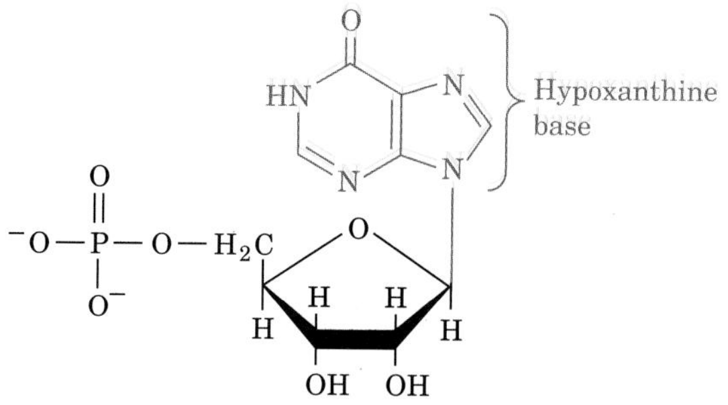

Inosine monophosphate (IMP)

the nucleotide of the base **hypoxanthine**. AMP and GMP are subsequently synthesized from this intermediate via separate pathways (Section 26-2B). Thus, contrary to naive expectation, purines are initially formed as ribonucleotides rather than as free bases. Additional studies have demonstrated that such widely divergent organisms as *E. coli*, yeast, pigeons, and humans have virtually identical pathways for the biosynthesis of purine nucleotides thereby further demonstrating the biochemical unity of life.

A. Synthesis of Inosine Monophosphate

IMP is synthesized in a pathway comprised of 11 reactions (Fig. 26-3):

1. Activation of ribose-5-phosphate

The starting material for purine biosynthesis is α-D-ribose-5-phosphate, a product of the pentose phosphate pathway (Section 21-4). In the first step of purine biosynthesis, **ribose phosphate pyrophosphokinase** activates this compound by reacting it with ATP to form **5-phosphoribosyl-α-pyrophosphate (PRPP)**. This reaction, which occurs via the nucleophilic attack of the ribose C(1)-OH group on the P_β of ATP, is unusual in that a pyrophosphoryl group is directly transferred from ATP to C(1) of ribose-5-phosphate and that the product has the α configuration about C(1). PRPP is also a precursor in the biosynthesis of pyrimidines (Section 26-3A) and the amino acids histidine and tryptophan (Section 24-5B). Thus, as is expected for an enzyme at such an important biosynthetic crossroad, the activity of ribose phosphate pyrophosphokinase varies with the concentrations of numerous metabolites including PP_i and 2,3-bisphosphoglycerate, which are activators, and ADP and GDP, which are mixed inhibitors (Section 13-3C). The regulation of purine nucleotide biosynthesis is further discussed in Section 26-2C.

2. Acquisition of purine atom N(9)

In the first reaction unique to purine biosynthesis,

amidophosphoribosyl transferase catalyzes the displacement of PRPP's pyrophosphate group by glutamine's amide nitrogen. The reaction occurs with inversion of configuration about ribose C(1) thereby forming β-5-phosphoribosylamine and establishing the anomeric form of the future nucleotide. This reaction, which is driven to completion by the subsequent hydrolysis of the released PP_i, is the pathway's flux-generating step. Not surprisingly, therefore, amidophosphoribosyl transferase, is subject to feedback inhibition by purine nucleotides (Section 26-2C).

3. Acquisition of purine atoms C(4), C(5), and N(7)

Glycine's carboxyl group forms an amide with the amino group of phosphoribosylamine yielding **glycinamide ribotide (GAR)** in a reaction facilitated by the intermediate phosphorylation of glycine's carboxyl group. The reaction, which is reversible despite its concomitant hydrolysis of ATP to ADP + P_i, is the only step of the purine biosynthesis pathway in which more than one purine ring atom is acquired.

4. Acquisition of purine atom C(8)

GAR's free α-amino group is formylated to yield **formylglycinamide ribotide (FGAR)**. The formyl donor in this reaction is N^{10}-formyltetrahydrofolate. This cofactor transfers C_1 units from such donors as serine, glycine, and formate to various acceptors in biosynthetic reactions (Section 24-4D).

5. Acquisition of purine atom N(3)

The amide amino group of a second glutamine is transferred to the growing purine ring to form **formylglycinamidine ribotide (FGAM)**. This reaction, which is driven by the coupled hydrolysis of ATP to ADP + P_i, is thought to proceed by the mechanism diagrammed in Fig. 26-4. Here the oxygen of the FGAR isoamide form reacts with ATP to yield a phosphoryl ester intermediate. This intermediate then reacts with "NH_3," the glutamine amide nitrogen as labelized through the transient formation of an enzyme thiol ester, to form a tetrahedral adduct. The adduct then eliminates P_i to yield the imine product, FGAM. Such reactions, in which a carboxamide oxygen is replaced by an imino group, are common in the biosynthesis of nucleotides. For example, Reaction 6 of this pathway, and the reaction converting IMP to AMP (Section 26-2B), and UTP to CTP (Section 26-3B), follow similar mechanisms; that is, conversion of a carboxamide oxygen to a phosphoryl ester that is nucleophilically attacked by an amine nitrogen atom to yield a tetrahedral adduct that, in turn, expels P_i to form product.

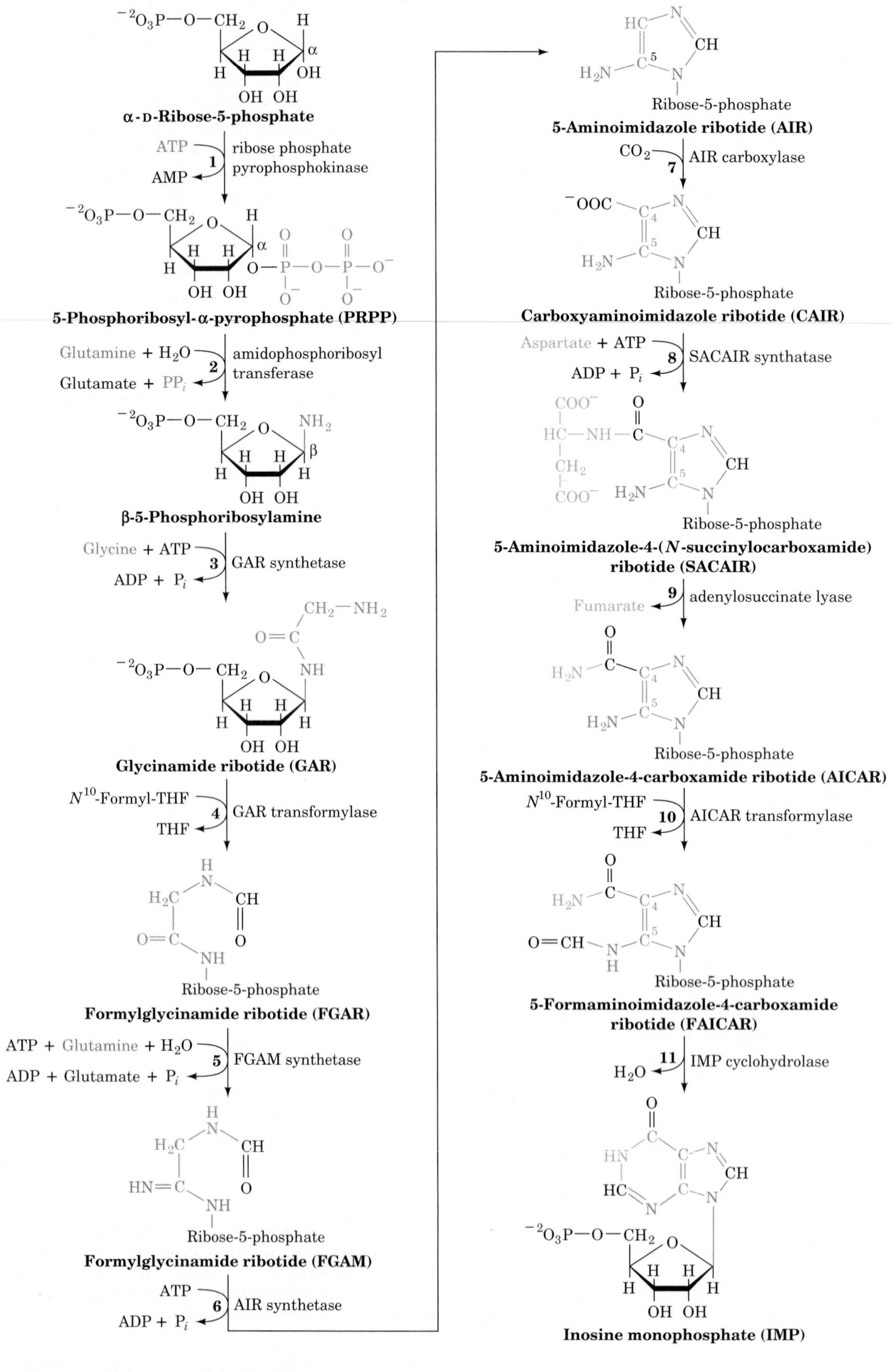

α-D-Ribose-5-phosphate

ATP ⟶ | **1** ribose phosphate pyrophosphokinase
AMP ⟵

5-Phosphoribosyl-α-pyrophosphate (PRPP)

Glutamine + H₂O ⟶ | **2** amidophosphoribosyl transferase
Glutamate + PP$_i$ ⟵

β-5-Phosphoribosylamine

Glycine + ATP ⟶ | **3** GAR synthetase
ADP + P$_i$ ⟵

Glycinamide ribotide (GAR)

N^{10}-Formyl-THF ⟶ | **4** GAR transformylase
THF ⟵

Formylglycinamide ribotide (FGAR)

ATP + Glutamine + H₂O ⟶ | **5** FGAM synthetase
ADP + Glutamate + P$_i$ ⟵

Formylglycinamide ribotide (FGAM)

ATP ⟶ | **6** AIR synthetase
ADP + P$_i$ ⟵

5-Aminoimidazole ribotide (AIR)

CO₂ ⟶ | **7** AIR carboxylase

Carboxyaminoimidazole ribotide (CAIR)

Aspartate + ATP ⟶ | **8** SACAIR synthatase
ADP + P$_i$ ⟵

5-Aminoimidazole-4-(N-succinylocarboxamide) ribotide (SACAIR)

Fumarate ⟵ | **9** adenylosuccinate lyase

5-Aminoimidazole-4-carboxamide ribotide (AICAR)

N^{10}-Formyl-THF ⟶ | **10** AICAR transformylase
THF ⟵

5-Formaminoimidazole-4-carboxamide ribotide (FAICAR)

H₂O ⟶ | **11** IMP cyclohydrolase

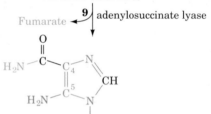

Inosine monophosphate (IMP)

Figure 26-3 *(opposite)*
The metabolic pathway for the *de novo* biosynthesis of IMP. Here the purine residue is built up on a ribose ring in 11 enzymatically catalyzed reactions.

6. Formation of the purine imidazole ring

The purine imidazole ring is closed in an ATP-requiring intramolecular condensation that yields **5-aminoimidazole ribotide (AIR)**. The aromatization of the imidazole ring is facilitated by the tautomeric shift of the reactant from its imine to its enamine form.

7. Acquisition of C(6)

Purine C(6) is introduced as CO_2 in a carboxylation reaction that yields **carboxyaminoimidazole ribotide (CAIR)**. The reaction is unusual in that it neither utilizes biotin (a carboxylation cofactor; Section 21-1A) nor requires an energy source such as ATP. The reaction, which, as might be expected, has an unfavorable equilibrium, is driven by its coupling to the exergonic reactions further along the pathway.

8. Acquisition of N(1)

Purine atom N(1) is contributed by aspartate in an amide-forming condensation reaction yielding **5-aminoimidazole-4-(N-succinylocarboxamide) ribotide (SACAIR)**. The reaction, which is driven by the hydrolysis of ATP to ADP + P_i, chemically resembles Reaction 3.

9. Elimination of fumarate

SACAIR is cleaved with the release of fumarate yielding **5-aminoimidazole-4-carboxamide ribotide (AICAR)**. Reactions 8 and 9 chemically resemble the reactions in the urea cycle in which citrulline is aminated to form arginine (Sections 24-2C and D). In both pathways, aspartate's amino group is transferred to an acceptor through an ATP-driven coupling reaction followed by the elimination of the aspartate carbon skeleton as fumarate. In plants and microorganisms, AICAR is also formed in the biosynthesis of histidine (Section 24-5B) but since, in that process, the AICAR is derived from ATP, it provides for no net purine biosynthesis.

10. Acquisition of C(2)

The final purine ring atom is acquired through formylation by N^{10}-formyltetrahydrofolate yielding **5-formaminoimidazole-4-carboxamide ribotide (FAICAR)**. In bacteria, this reaction and that of Reaction 4 are indirectly inhibited by sulfonamides which, it will be recalled, prevent the synthesis of folate by competing with its *p*-aminobenzoate component (Section 24-4D). Animals, including humans, must acquire folate through the diet, since they are incapable of synthesizing it. They are there-fore unaffected by sulfonamides. The antibiotic properties of sulfonamides are therefore largely a result of their inhibition of nucleic acid biosynthesis in susceptible bacteria.

11. Cyclization to form IMP

The final reaction in the pathway, ring closure to form IMP, occurs through the elimination of water. In contrast to Reaction 6, the cyclization that forms the imidazole ring, this reaction does not entail ATP hydrolysis.

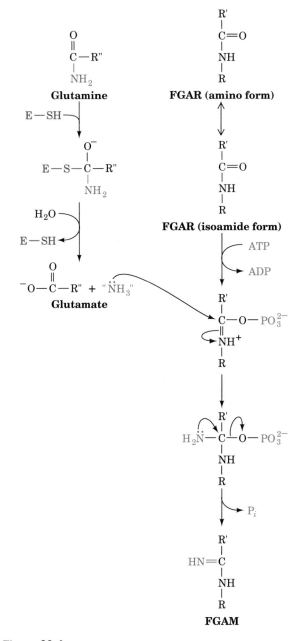

Figure 26-4
The mechanism for the ATP-driven replacement of a carboxamide oxygen with an imino group in nucleotide biosynthesis. Here, "NH₃" indicates that the covalent state of the glutamine amide-derived nitrogen, which does not exchange with solvent, is uncertain.

B. Synthesis of Adenine and Guanine Ribonucleotides from Inosine Monophosphate

IMP does not accumulate in the cell but, rather, is rapidly converted to AMP and GMP. AMP, which differs from IMP only in the replacement of its 6-keto group by an amino group, is synthesized in a two-reaction pathway (Fig. 26-5, *left*). In the first reaction, aspartate's amino group is linked to IMP in a reaction powered by the hydrolysis of GTP to GDP + P_i, to yield **adenylosuccinate**. In the second reaction, **adenylosuccinate lyase** elminates fumarate from adenylosuccinate to form AMP. This enzyme also catalyzes Reaction 9 of the IMP pathway (Fig. 26-3).

GMP is also synthesized from IMP in a two-reaction pathway (Fig. 26-5, *right*). In the first reaction, IMP is dehydrogenated via the reduction of NAD$^+$ to form **xanthosine monophosphate (XMP;** the ribonucleotide of the base **xanthine**). XMP is then converted to GMP by the transfer of the glutamine amide nitrogen in a reaction driven by the hydrolysis of ATP to AMP + PP$_i$ (and subsequently to 2P$_i$).

Nucleoside Diphosphates and Triphosphates Are Synthesized by the Phosphorylation of Nucleoside Monophosphates

In order to participate in nucleic acid synthesis, nucleoside monophosphates must first be converted to the corresponding nucleoside triphosphates. In the first of the two sequential phosphorylation reactions that do so, nucleoside diphosphates are synthesized from the corresponding nucleoside monophosphates by base-specific **nucleoside monophosphate kinases.** For example, adenylate kinase (Section 15-4C) catalyzes the phosphorylation of AMP to ADP:

$$AMP + ATP \rightleftharpoons 2ADP$$

Similarly, GDP is produced by a guanine-specific enzyme:

$$GMP + ATP \rightleftharpoons GDP + ADP$$

These nucleoside monophosphate kinases do not discriminate between ribose and deoxyribose in the substrate.

Nucleoside diphosphates are converted to the corresponding triphosphates by **nucleoside diphosphate ki-**

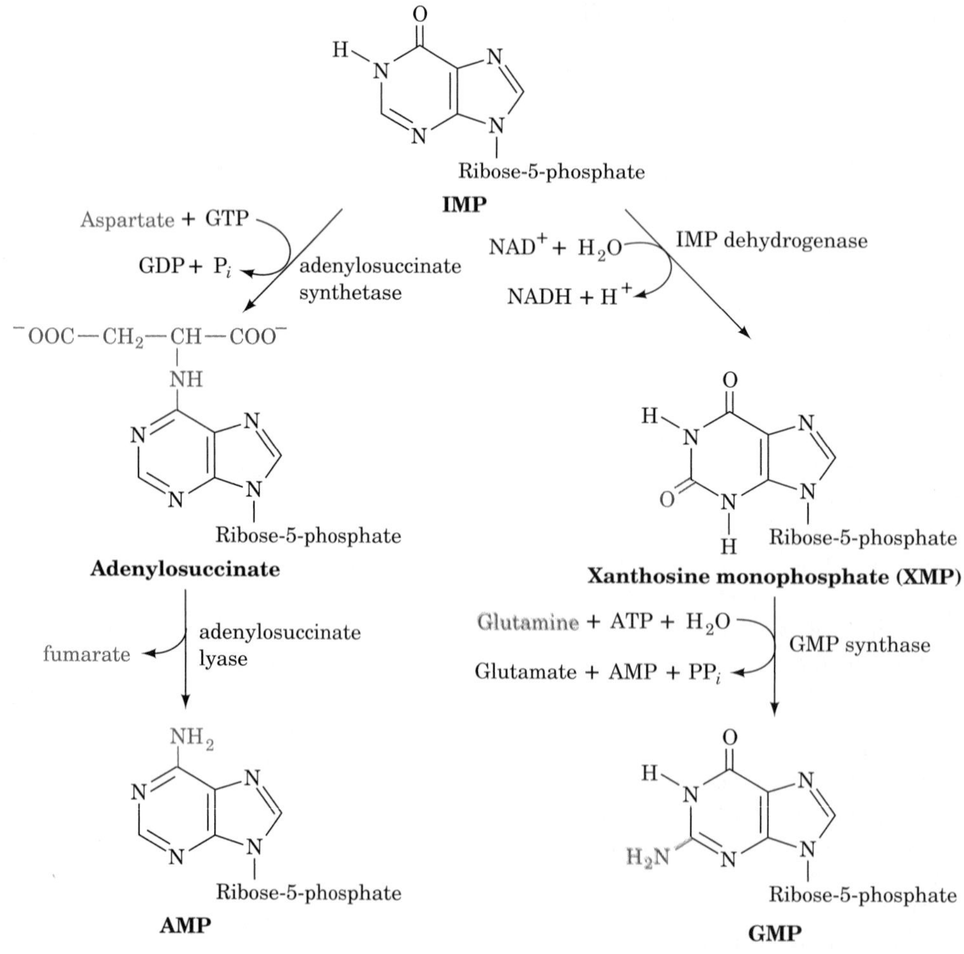

Figure 26-5
IMP is converted to AMP or GMP in separate two-reaction pathways.

nase; for instance:

$$GDP + ATP \rightleftharpoons GTP + ADP$$

Although this reaction is written with ATP as the phosphoryl donor and GDP as the acceptor, nucleoside diphosphate kinase is nonspecific as to the bases on either of its substrates and as to whether their sugar residues are ribose or deoxyribose. These reactions, as might be expected from the nearly identical structures of their substrates and products, normally operate close to equilibrium ($\Delta G \approx 0$). ADP is, of course, also converted to ATP by a variety of energy releasing reactions such as those of glycolysis and oxidative phosphorylation. Indeed, it is these reactions that ultimately drive the foregoing kinase reactions.

C. Regulation of Purine Nucleotide Biosynthesis

The pathways involved in nucleic acid metabolism are tightly regulated as is evidenced, for example, by the increased rates of nucleotide synthesis during cell proliferation. In fact, the pathways synthesizing IMP, ATP, and GTP are individually regulated in most cells so as not only to control the total amounts of purine nucleotides produced but to also coordinate the relative amounts of ATP and GTP. This control network is diagrammed in Fig. 26-6.

The IMP pathway is regulated at its first two reactions: those catalyzing the synthesis of PRPP and 5-phosphoribosylamine. We have already seen that ribose phosphate pyrophosphokinase, the enzyme catalyzing Reaction 1 of the IMP pathway, is inhibited by both ADP and GDP (Section 26-2A). Amidophosphoribosyl transferase, the enzyme catalyzing the first committed step of the IMP pathway (Reaction 2), is likewise subject to feedback inhibition. In this case, however, the enzyme binds ATP, ADP, and AMP at one inhibitory site and GTP, GDP, and GMP at another. *The rate of IMP production is consequently independently but synergistically controlled by the levels of adenine nucleotides and guanine nucleotides.* Moreover, amidophosphoribosyl transferase is allosterically stimulated by PRPP (feedforward activation).

A second level of regulation occurs immediately below the branch point leading from IMP to AMP and GMP. AMP and GMP are each competitive inhibitors of IMP in their own synthesis so that excessive buildup of these products is impeded. In addition, the synthesis rates of adenine and guanine nucleotides are coordinated. Recall that GTP powers the synthesis of AMP from IMP, whereas ATP powers the synthesis of GMP from IMP (Section 26-2B). This reciprocity serves to balance the production of AMP and GMP (which are required in roughly equal amounts in nucleic acid biosyn-

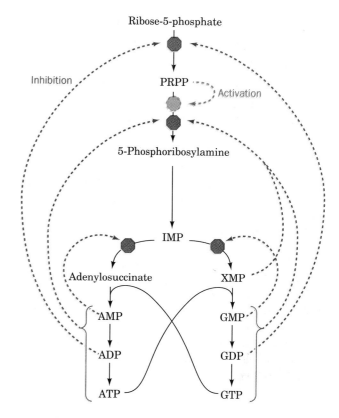

Figure 26-6
The control network for the purine biosynthesis pathway. Red octagons and green dots indicate control points. Feedback inhibition is indicated by dashed red arrows and feedforward activation is represented by dashed green arrows.

thesis): *The rate of synthesis of GMP increases with [ATP], whereas that of AMP increases with [GTP].*

D. Salvage of Purines

Most cells have an active turnover of many of their nucleic acids (particularly some types of RNA) which, through degradative processes described in Section 26-5A, result in the release of adenine, guanine, and hypoxanthine. These free purines are reconverted to their corresponding nucleotides through **salvage pathways.** In contrast to the *de novo* purine nucleotide synthesis pathway, which is virtually identical in all cells, salvage pathways are diverse in character and distribution. In mammals, purines are, for the most part, salvaged by two different enzymes. **Adenine phosphoribosyltransferase (APRT)** mediates AMP formation through the transfer of adenine to PRPP with release of PP_i:

$$Adenine + PRPP \rightleftharpoons AMP + PP_i$$

Hypoxanthine-guanine phosphoribosyltransferase (HGPRT) catalyzes the analogous reaction for both hypoxanthine and guanine:

$$Hypoxanthine + PRPP \rightleftharpoons IMP + PP_i$$
$$Guanine + PRPP \rightleftharpoons GMP + PP_i$$

Lesch–Nyhan Syndrome Results from HGPRT Deficiency

The symptoms of **Lesch–Nyhan syndrome,** which is caused by a severe HGPRT deficiency, indicate that purine salvage reactions have functions other than conservation of the energy required for *de novo* purine biosynthesis. This sex-linked congenital defect (affects almost only males) results in excessive uric acid production (uric acid is a purine degradation product; Section 26-5A) and neurological abnormalities such as spasticity, mental retardation, and highly aggressive and destructive behavior including a bizarre compulsion towards self-mutilation. For example, many children with Lesch–Nyhan syndrome have such an irresistible urge to bite their lips and fingers that they must be restrained. If the restraints are removed, communicative patients will plead that the restraints be replaced even as they attempt to injure themselves.

The excessive uric acid production in patients with Lesch–Nyhan syndrome is readily explained. The lack of HGPRT activity leads to an accumulation of the PRPP that would normally be used in the salvage of hypoxanthine and guanine. The excess PRPP activates amidophosphoribosyl transferase (which catalyzes Reaction 2 of the IMP biosynthesis pathway) thereby greatly increasing the rate of synthesis of purine nucleotides and consequently that of their degradation product, uric acid. Yet, the physiological basis of the associated neurological abnormalities remains obscure. That a defect in a single enzyme can cause such profound but well-defined behavioral changes nevertheless has important psychiatric implications.

3. SYNTHESIS OF PYRIMIDINE RIBONUCLEOTIDES

The biosynthesis of pyrimidines is a simpler process than that of purines. Isotopic labeling experiments have shown that atoms N(1), C(4), C(5), and C(6) of the pyrimidine ring are all derived from aspartic acid, C(2) arises from HCO_3^-, and N(3) is contributed by glutamine (Fig. 26-7). In this section we discuss the pathways for pyrimidine ribonucleotide biosynthesis and how these processes are regulated.

A. Synthesis of UMP

The major breakthrough in the determination of the

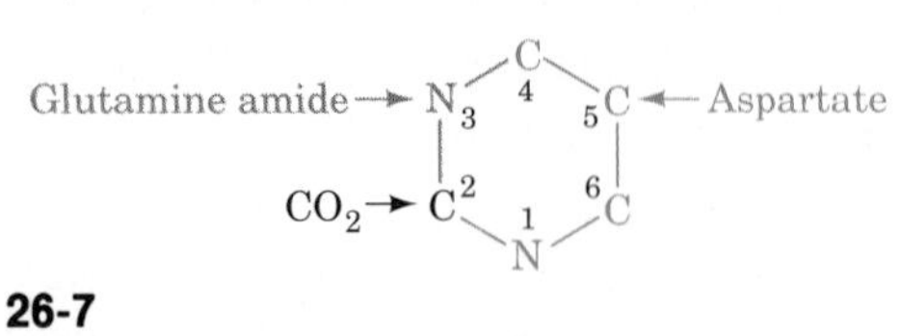

Figure 26-7
The biosynthetic origins of pyrimidine ring atoms.

pathway for the *de novo* biosynthesis of pyrimidine ribonucleotides was the observation that mutants of the bread mold *Neurospora crassa,* which are unable to synthesize pyrimidines and therefore require both cytosine and uracil in their growth medium, grow normally when supplied instead with the pyrimidine **orotic acid** (uracil-6-carboxylic acid).

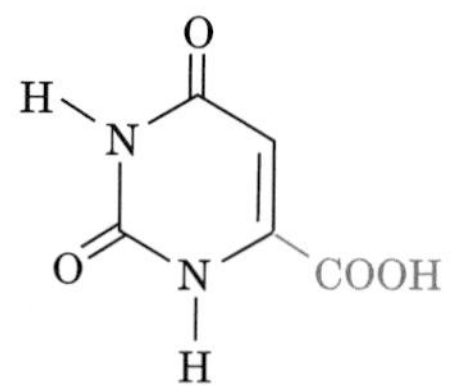

Orotic acid (Uracil-6-carboxylic acid)

This observation led to the elucidation of the following six-reaction pathway for the biosynthesis of UMP (Fig. 26-8). Note that, in contrast to the case for purine nucleotides, the pyrimidine ring is coupled to the ribose-5-phosphate moiety *after* it has been synthesized.

1. **Synthesis of carbamoyl phosphate**

   The first reaction of pyrimidine biosynthesis is synthesis of **carbamoyl phosphate** from HCO_3^- and the amide nitrogen of glutamine by the cytosolic enzyme **carbamoyl phosphate synthetase II.** This reaction is unusual in that it does not use biotin and consumes two molecules of ATP: one provides a phosphate group and the other energizes the reaction. We have previously discussed the synthesis of carbamoyl phosphate in connection with the formation of arginine (Section 24-2A). When arginine is synthesized via the urea cycle, the carbamoyl phosphate that is used in this process is synthesized by a separate mitochondrial enzyme, **carbamoyl phosphate synthetase I,** which uses ammonia as its nitrogen source. Prokaryotes only have one carbamoyl phosphate synthetase, which supplies both pyrimidine and arginine biosynthesis and utilizes glutamine.

2. **Synthesis of carbamoyl aspartate**

   Condensation of carbamoyl phosphate with aspartate to form **carbamoyl aspartate,** is catalyzed by **aspartate transcarbamoylase (ATCase).** This reaction, the pathway's flux-generating step, occurs without need of ATP because carbamoyl phosphate is intrinsically activated. The structure and regulation of *E. coli* ATCase is discussed in Section 12-4.

3. **Ring closure to form dihydroorotate**

   The third reaction of the pathway was elucidated by Arthur Kornberg following his observation that microorganisms forced to exist on orotic acid as a carbon source first reduce it to **dihydroorotate.** The reaction forming the pyrimidine ring yields dihydroorotate in

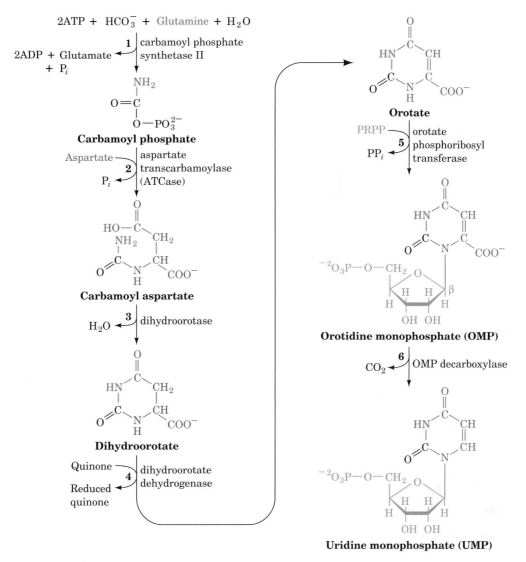

Figure 26-8
The metabolic pathway for the *de novo* synthesis of UMP consists of six enzymatically catalyzed reactions. Note that, in contrast to the case for purine biosynthesis (Fig. 26-3), the pyrimidine ring is formed before its attachment to a ribose ring.

an intramolecular condensation catalyzed by **dihydroorotase.**

4. Oxidation of dihydroorotate

Dihydroorotate is irreversibly oxidized to orotate by **dihydroorotate dehydrogenase.** The eukaryotic enzyme, which contains FMN and nonheme Fe, is associated with the inner mitochondrial membrane where quinones supply its oxidizing power. (Many bacterial dihydroorotate dehydrogenases are NAD^+-linked flavoproteins that contain FMN, FAD, and nonheme Fe. These enzymes normally function degradatively, that is, in the direction orotate $\rightarrow$ dihydroorotate, thereby permitting these bacteria to metabolize orotate and accounting for Kornberg's observation.)

5. Acquisition of the ribose phosphate moiety

Orotate reacts with PRPP to yield **orotidine-5'-monophosphate (OMP)** in a reaction catalyzed by **orotate phosphoribosyl transferase** and which is driven by hydrolysis of the eliminated PP_i. This reaction fixes the anomeric form of pyrimidine nucleotides in the β configuration. Orotate phosphoribosyl transferase also acts to salvage other pyrimidine bases, such as uracil and cytosine, by converting them to their corresponding nucleotides.

6. Decarboxylation to form UMP

The final reaction of the pathway is the decarboxylation of OMP by **OMP decarboxylase** to form UMP. This is an unusual reaction in that it requires no cofactors.

In bacteria, the six enzymes of UMP biosynthesis occur as independent proteins. In animals, however, as Mary Ellen Jones demonstrated, the first three enzymatic activities of the pathway, carbamoyl phosphate synthetase II, ATCase, and dihydroorotase, occur on a single 210-kD polypeptide chain. Similarly, Reactions 5 and 6 of the animal pyrimidine pathway are catalyzed by a single polypeptide (although this observation is disputed) as are Reactions 3, 4, and 6, Reactions 7 and 8, and Reactions 10 and 11 in animal purine biosynthesis (Section 26-2A). *The intermediate products of these multifunctional enzymes are not readily released to the medium but rather, are* **channeled** *to the succeeding enzymatic activities of the pathway. Channeling increases the overall rate of multistep process and protects the intermediates from degradation by other cellular enzymes.* We have previously seen that all seven enzymatic activities catalyzing fatty acid synthesis in animals also occur on a single protein molecule (Section 23-4C), and that the indole intermediate in the tryptophan synthase reaction passes between the bifunctional enzyme's two active sites via a protein tunnel (Section 24-5B). It is becoming increasingly apparent that the association of functionally related enzymes is a widespread phenomenon.

B. Synthesis of UTP and CTP

The synthesis of UTP from UMP is analogous to the synthesis of purine nucleotide triphosphates (Section 26-2B). The process occurs by the sequential actions of a nucleoside monophosphate kinase and nucleoside diphosphate kinase:

$$UMP + ATP \rightleftharpoons UDP + ADP$$
$$UDP + ATP \rightleftharpoons UTP + ADP$$

CTP is formed by amination of UTP by **CTP synthase** (Fig. 26-9). In animals, the amino group is donated by glutamine, whereas in bacteria it is supplied directly by ammonia.

C. Regulation of Pyrimidine Nucleotide Biosynthesis

In bacteria, the pyrimidine biosynthesis pathway is primarily regulated at Reaction 2, the ATCase reaction (Fig. 26-10a). In *E. coli*, control is exerted there through the allosteric stimulation of ATCase by ATP and its inhibition by CTP (Section 12-4). In many bacteria, however, UTP is the major ATCase inhibitor.

In animals, ATCase is not a regulatory enzyme. Rather, pyrimidine biosynthesis is controlled by the activity of carbamoyl phosphate synthetase II, which is inhibited by UDP and UTP and activated by ATP and PRPP (Fig. 26-10b). A second level of control in the mammalian pathway occurs at OMP decarboxylase for which UMP and to a lesser extent CMP are competitive inhibitors.

In all organisms, the rate of OMP production varies

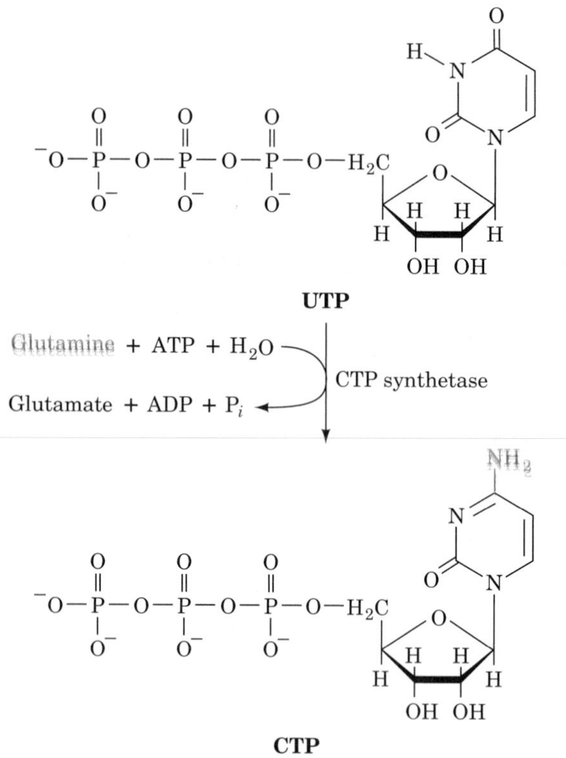

Figure 26-9
The synthesis of CTP from UTP.

with the availability of its precursor, PRPP. The PRPP level, it will be recalled, depends on the activity of ribose phosphate pyrophosphokinase, which is inhibited by ADP and GDP (Section 26-2A).

Orotic Aciduria Results from an Inherited Enzyme Deficiency

Orotic aciduria, an inherited human disease, is characterized by the excretion of large amounts of orotic acid in the urine, retarded growth, and severe anemia. It results from a deficiency in the bifunctional enzyme catalyzing Reactions 5 and 6 of pyrimidine nucleotide biosynthesis. Consideration of the biochemistry of this situation led to its effective treatment: the administration of uridine and/or cytidine. The UMP formed through the phosphorylation of these nucleosides, besides replacing that normally synthesized, inhibits carbamoyl phosphate synthetase II so as to attenuate the rate of orotic acid synthesis.

4. FORMATION OF DEOXYRIBONUCLEOTIDES

DNA differs chemically from RNA in two major respects: (1) its nucleotides contain 2'-deoxyribose residues rather than ribose residues, and (2) it contains the base thymine (5-methyluracil) rather than uracil. In this section we consider the biosynthesis of these DNA components.

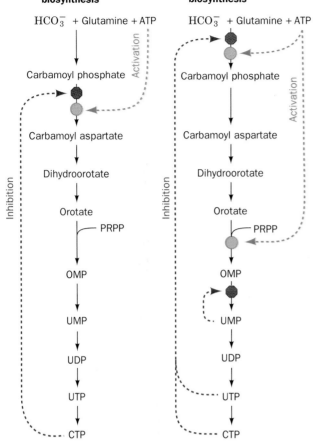

(a) **E. coli pyrimidine biosynthesis**

(b) **Animal pyrimidine biosynthesis**

Figure 26-10
The control networks for pyrimidine biosynthesis in
(a) E. coli, and *(b)* animals. Red octagons and green dots
indicate control points. Feedback inhibition is represented by
dashed red arrows and activation is indicated by dashed
green arrows.

A. Production of Deoxyribose Residues

Deoxyribonucleotides are synthesized from their corresponding ribonucleotides by the reduction of their C(2') position rather than by their de novo synthesis from deoxyribose-containing precursors.

NDP

dNDP

This pathway was established through Irwin Rose's
study of how rats metabolize cytidine that is ^{14}C labeled

in both its base and ribose components. The dCMP recovered from the rats' DNA had the same labeling ratio in its cytosine and deoxyribose residues as had the original cytidine indicating that the DNA's components remained linked during DNA synthesis. If the cytosine and the ribose residues had become separated, dilution of the labeled cytosine and ribose residues with unlabeled residues, which are present in the rat's tissues in different amounts, would have altered this ratio.

Enzymes that catalyze the formation of deoxyribonucleotides by the reduction of the corresponding ribonucleotides are named **ribonucleotide reductases.** Three types of ribonucleotide reductases are known: (1) enzymes that contain nonheme Fe(III) at their active sites; (2) enzymes that utilize 5'-deoxyadenosylcobalamin cofactors (coenzyme B_{12}; Section 23-2E); and (3) Mn-dependent enzymes. The chemistries of the reactions catalyzed by at least the first two types of these enzymes are, nevertheless, surprisingly similar; both replace the 2'-OH group of ribose with H via a free radical mechanism in which the immediate source of reducing equivalents is a pair of enzyme sulfhydryl groups. The first two types of enzyme are widely distributed among prokaryotes; some species have the Fe(III) enzyme while other, sometimes related species, have the coenzyme B_{12} enzyme (the Mn-dependent enzyme has only recently been discovered in a few prokaryotes and is largely uncharacterized). All eukaryotes except a few unicellular species, however, have the Fe-containing variety. We shall discuss only the Fe-containing enzyme.

E. Coli Ribonucleotide Reductase

Fe-containing ribonucleotide reductases reduce ribonucleoside diphosphates to the corresponding deoxyribonucleoside diphosphates:

$$NDP \longrightarrow dNDP$$

E. coli ribonucleotide reductase, as Peter Reichard demonstrated, consists of two copies each of two different subunits, B1 (160 kD) and B2 (78 kD), which together form the enzyme's active site (Fig. 26-11). Each B1 subunit contains a redox-active sulfhydryl pair as well as two independent effector-binding sites that control both the enzyme's catalytic activity and its substrate specificity (see below). The B2 subunits' nonheme Fe(III) ions form an Fe^{3+}—O—Fe^{3+} group that interacts with Tyr 122 on one of the B2 subunits to form, as EPR measurements indicate, an unusual tyrosyl free radical. The dimeric enzyme therefore has only one catalytically active site.

E. coli ribonucleotide reductase is inhibited by **hydroxyurea,**

Hydroxyurea

8-Hydroxyquinoline

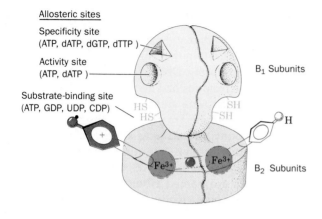

Figure 26-11

E. coli ribonucleotide reductase. Note that although the enzyme consists of two identical pairs of subunits, it contains only one tyrosyl radical (*red*) and one Fe^{3+}—O—Fe^{3+} group and therefore has only one active site. [After Thelander, L. and Reichard, P., *Annu. Rev. Biochem.* **48**, 126 (1979).]

which specifically quenches (destroys) the tyrosyl radical, and by **8-hydroxyquinoline,** which chelates Fe^{3+} ions. Mammalian ribonucleotide reductases have similar characteristics to the *E. coli* enzyme. Indeed, hydroxyurea is in clinical use as an antitumor agent.

If *E. coli* ribonucleotide reductase is incubated with [3'-³H]UDP, a small but reproducible fraction of the ³H is released as ³H₂O. This observation, together with kinetic measurements, led JoAnne Stubbe to propose the following catalytic mechanism for *E. coli* ribonucleotide reductase (Fig. 26-12):

1. Ribonucleotide reductase's free radical abstracts an H atom from C(3') of the substrate in the reaction's rate-determining step.

2 and 3. Acid-catalyzed cleavage of the C(2')—OH bond releases H_2O to yield a radical–cation intermediate. The radical mediates the stabilization of the C(2') cation by the 3'-OH group's unshared electron pair, thereby accounting for the radical's catalytic role.

4. The radical–cation intermediate is reduced by the enzyme's redox-active sulfhydryl pair to yield a 3'-deoxynucleotide radical and a protein disulfide group.

5. The 3'-radical reabstracts an H atom from the protein to yield the product deoxynucleoside diphosphate and restore the enzyme to its radical state. A small fraction of the originally abstracted H atom exchanges with solvent before it can be replaced, thus accounting for the release of ³H upon reduction of [3'-³H]UDP.

Thioredoxin and Glutaredoxin Are Ribonucleotide Reductase's Physiological Reducing Agents

The final step in the ribonucleotide reductase catalytic cycle is reduction of the enzyme's newly formed disulfide bond to reform its redox-active sulfhydryl pair. Dithiols such as dithiothreitol (Section 6-1B) can serve as the reducing agent for this process *in vitro* through a disulfide exchange reaction. One of the enzyme's physiological reducing agents, however, is **thioredoxin,** a ubiquitous 12-kD monomeric protein that has a pair of closely proximal Cys residues (we have previously encountered thioredoxin in our study of the light-induced activation of the Calvin cycle; Section 22-3B). Thioredoxin reduces oxidized ribonucleotide reductase via disulfide exchange.

The X-ray structure of thioredoxin (Fig. 26-13) reveals that its redox-active disulfide group is located on a molecular protrusion making this protein the only known example of a "male" enzyme. Oxidized thioredoxin is, in turn, reduced by NADPH in a reaction mediated by the flavoprotein **thioredoxin reductase.** NADPH therefore serves as the terminal reducing agent in the ribonucleotide reductase-mediated reduction of NDPs to dNDPs (Fig. 26-14).

The existence of a viable *E. coli* mutant devoid of thioredoxin indicates that this protein is not the only substance capable of reducing oxidized ribonucleotide reductase *in vivo.* This observation led to the discovery of **glutaredoxin,** a monomeric disulfide-containing 11-kD protein that can also reduce ribonucleotide reductase (mutants devoid of both thioredoxin and glutaredoxin are nonviable). Oxidized glutaredoxin is reduced, via disulfide exchange, by the Cys-containing tripeptide glutathione which, in turn, is reduced by NADPH as catalyzed by the flavoprotein glutathione reductase (glutathione and glutathione reductase are described in Section 14-4). The relative importance of thioredoxin and glutaredoxin in the reduction of ribonucleoside diphosphates remains to be established.

Ribonucleotide Reductase Radical Generation Involves Superoxide Dismutase and Catalase

One of the most remarkable aspects of *E. coli* ribonucleotide reductase is its ability to maintain its normally highly reactive radical state indefinitely. Yet, quenching the radical, say by hydroxyurea treatment, inactivates

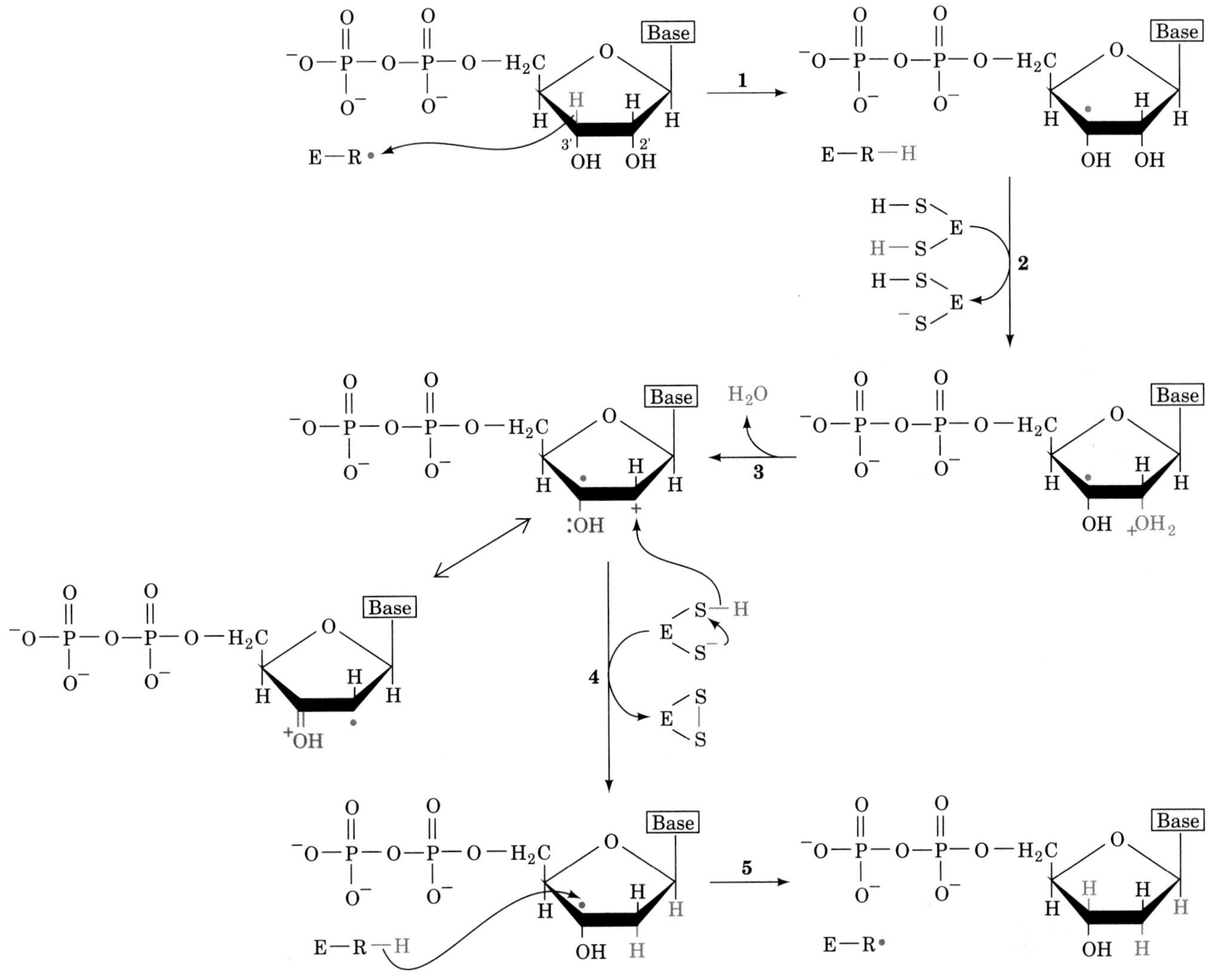

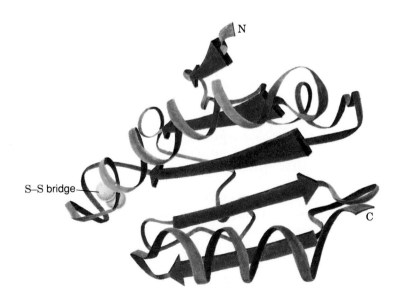

Figure 26-12
The enzymatic mechanism of ribonucleotide reductase. The reaction occurs via a free radical-mediated process in which reducing equivalents are supplied by the formation of an enzyme disulfide bond. [After Stubbe, J. A., Ator, M., and Krenitsky, T., *J. Biol. Chem.* **258**, 1630 (1983).]

Figure 26-13
The X-ray structure of *E. coli* thioredoxin in its oxidized (disulfide) state. The two yellow balls in the protrusion on the left represent the protein's redox-active disulfide group. [After a drawing by B. Furugren *in* Holmgren, A., Söderberg, B.-O., Eklund, H., and Brändén, C.-I., *Proc. Natl. Acad. Sci.* **72**, 2307 (1975).]

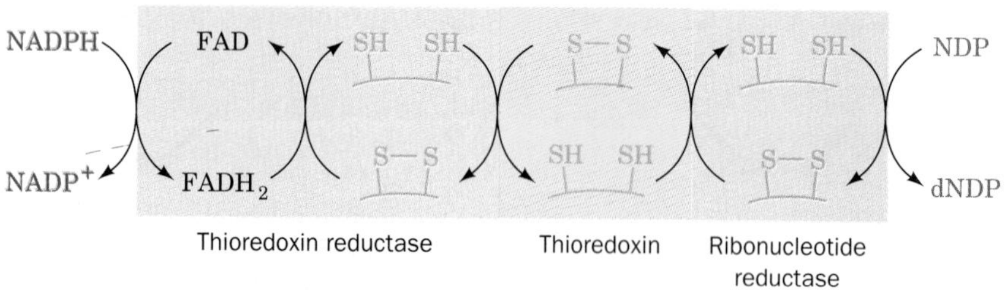

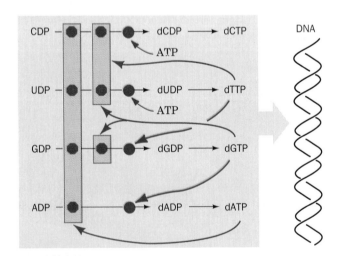

Figure 26-14
An electron-transfer pathway for nucleoside diphosphate (NDP) reduction. NADPH provides the reducing equivalents for this process through the intermediacy of thioredoxin reductase, thioredoxin, and ribonucleotide reductase.

the enzyme. How, then, is the radical generated in the first place? The radical may be restored *in vitro* by simply treating the inactive enzyme with Fe(II) and ascorbic acid in the presence of O_2. *In vivo*, however, radical generation requires four proteins and O_2:

1. An NAD(P)H:flavin oxidoreductase, which participates in tyrosine radical generation and releases **superoxide ion, $O_2^- \cdot$** (which is also a minor product of oxidative metabolism).

2. **Superoxide dismutase (SOD),** which eliminates the highly reactive and hence destructive superoxide radical:

$$2 \ O_2^- \cdot + 2H^+ \longrightarrow H_2O_2 + O_2$$

3. Catalase, which completes the oxygen radical detoxification process by disproportionating the H_2O_2 to H_2O and O_2 (Section 1-2A).

4. A protein whose function is unknown but that can be replaced, *in vitro*, by Fe(II).

Ribonucleotide Reductase Is Regulated by a Complex Feedback Network

The synthesis of the four dNTPs in the amounts required for DNA synthesis is accomplished through feedback control. The maintenance of the proper intracellular ratios of dNTPs is essential for normal growth. Indeed, *a disturbance in this balance is mutagenic because the probability that a given dNTP will be erroneously incorporated into a growing DNA strand increases with its concentration relative to those of the other dNTPs.*

The activities of both *E. coli* and mammalian ribonucleotide reductases are remarkably responsive to the levels of the various nucleotides present. *E. coli* ribonucleotide reductase's B1 subunit has two independent allosteric sites that control both the enzyme's catalytic activity and its substrate specificity (Fig. 26-11):

1. The binding of ATP to the allosteric site controlling the enzyme's overall activity (the activity site) activates the enzyme towards substrates determined by the effector bound at the allosteric site controlling the enzyme's specificity (the specificity site). The binding of dATP to the activity site inhibits the enzyme towards all substrates.

2. At the substrate specificity site, ATP or dATP binding

stimulates CDP and UDP reduction, dTTP binding stimulates GDP reduction but inhibits CDP and UDP reduction, and dGTP binding stimulates ADP reduction but inhibits CDP, UDP, and GDP reduction. In the absence of any of these effectors, ribonucleotide reductase is inactive.

These allosteric effects suggest the following sequence of events for the reduction of ribonucleotides (Fig. 26-15). In the presence of a mixture of NDPs, ribonucleotide reductase commences dNDP production by the ATP-stimulated reduction of CDP and UDP. The resulting dUDP is converted to dTTP, as described in Section 26-4B, which inhibits further CDP and UDP reduction but stimulates dGDP production. This product, after its phosphorylation to dGTP, inhibits the reduction of CDP, UDP, and GDP but stimulates production of dADP and thus dATP. As the dATP accumulates, it binds to the activity site thereby inhibiting all NDP reduction unless the ATP level is sufficiently high to displace the dATP. The proper intracellular balance between dCTP and dTTP is not controlled by ribonucleotide reductase but is maintained by **deoxycytidine de-**

Figure 26-15
The control network for the regulation of deoxyribonucleotide biosynthesis of both *E. coli* and mammalian ribonucleotide reductases. Green dots and arrows represent activation; red octagons and arrows indicate inhibition. [After Thelender, L. and Reichard, P., *Annu. Rev. Biochem.* **48,** 153 (1978).]

aminase which yields dUMP, the precursor of dTTP. This enzyme is activated by dCTP and inhibited by dTTP. Although this scheme is no doubt an oversimplified description of a dynamic process, it accounts for the observed ability of cells to synthesize deoxynucleotides in the amounts of each required for DNA synthesis.

dNTPs Are Produced by Phosphorylation of dNDPs

The final step in the production of all dNTPs is the phosphorylation of the corresponding dNDPs:

$$\text{dNDP} + \text{ATP} \rightleftharpoons \text{dNTP} + \text{ADP}$$

This reaction is catalyzed by nucleoside diphosphate kinase, the same enzyme that phosphorylates NDPs (Section 26-2B). As before, the reaction is written with ATP as the phosphoryl donor although any NTP or dNTP can function in this capacity.

B. Origin of Thymine

The dTMP component of DNA is synthesized, as we discuss below, by methylation of dUMP. The dUMP is generated through the hydrolysis of dUTP by **dUTP diphosphohydrolase:**

$$\text{dUTP} + \text{H}_2\text{O} \rightleftharpoons \text{dUMP} + \text{PP}_i$$

The apparent reason for this energetically wasteful process (dTMP, once it is formed, is rephosphorylated to dTTP) is that cells must minimize their concentration of dUTP in order to prevent incorporation of uracil into their DNA. This is because, as we discuss in Section 31-5B, the enzyme system that synthesizes DNA from dNTPs does not discriminate between dUTP and dTTP efficiently.

Thymidylate Synthase

dTMP is synthesized from dUMP by **thymidylate synthase** *with* N^5,N^{10}*-methylenetetrahydrofolate (*N^5,N^{10}*-methylene-THF) as the methyl donor:*

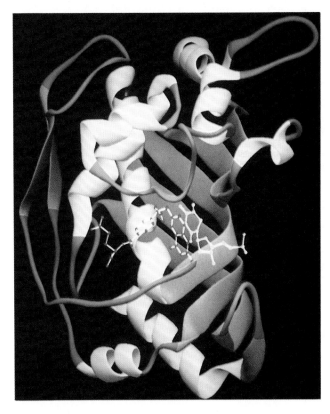

(THF cofactors are discussed in Section 24-4D). Note that the transferred methylene group (in which the carbon has the oxidation state of formaldehyde) is reduced to a methyl group (which has the oxidation state of methanol) at the expense of the oxidation of the THF cofactor to dihydrofolate (DHF).

The catalytic mechanism of thymidylate synthase, a highly conserved 70-kD dimeric protein (Fig. 26-16), has been extensively investigated. Upon incubation of the enzyme with N^5,N^{10}-methylene-6-[³H]THF and dUMP, the ³H is quantitatively transferred to the methyl group of the product dTMP. When 5-[³H]dUMP is the substrate, however, the ³H is released into the aqueous solvent. Such information, together with the knowledge

Figure 26-16
The X-ray structure of a subunit of the dimeric *E. coli* enzyme thymidylate synthase in complex with 5-fluorodeoxyuridylate (FdUMP; *yellow balls*) and the nonreactive folic acid analog **10-propargyl-5,8-dideazafolic acid** (*red balls*). The polypeptide backbone is colored such that the helices are green, the β sheets are brown, and the remaining segments are blue. [Courtesy of David Matthews, Agouron Pharmaceuticals.]

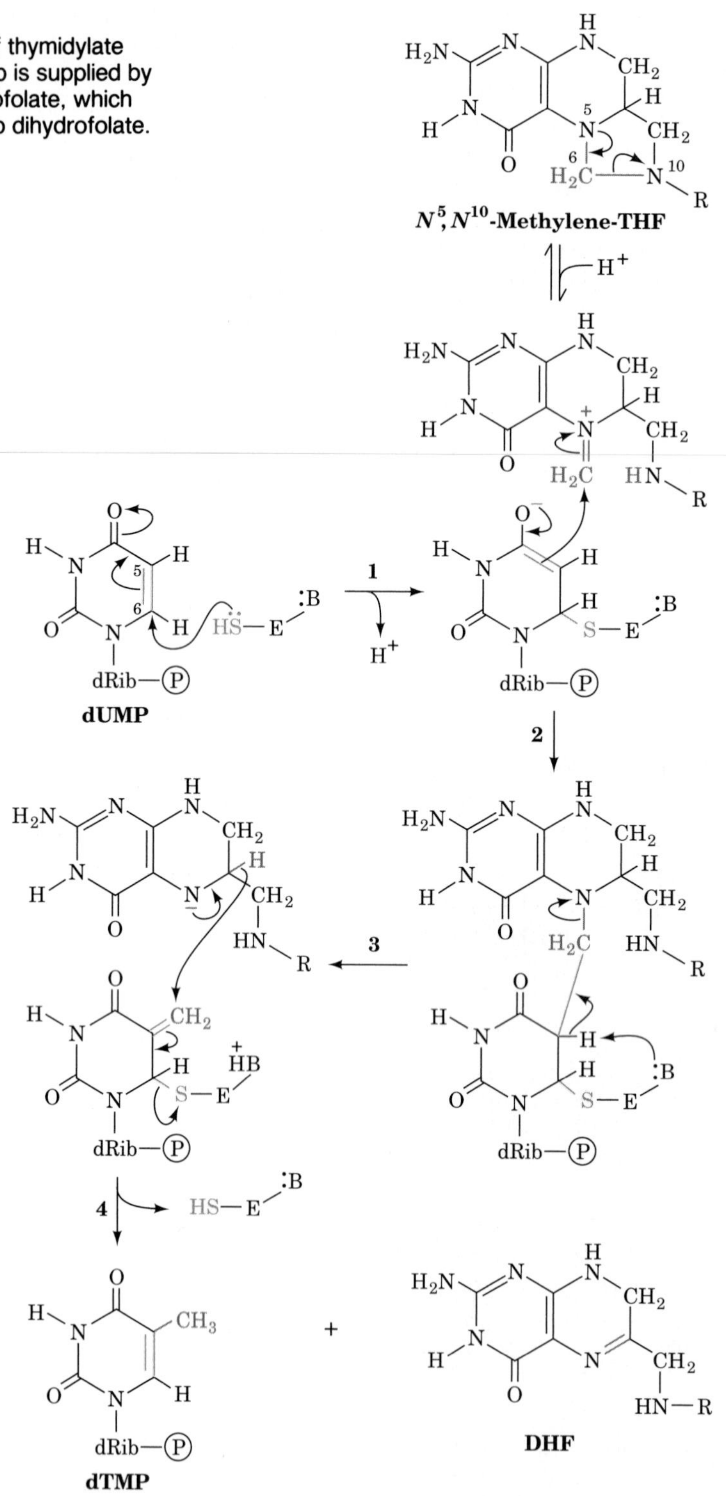

Figure 26-17
The catalytic mechanism of thymidylate synthase. The methyl group is supplied by N^5,N^{10}-methylenetetrahydrofolate, which is concomitantly oxidized to dihydrofolate.

that uracil C(6), being the β position of an α,β-unsaturated ketone, is susceptible to nucleophilic attack, led Daniel Santi to propose the following mechanistic scheme (Fig. 26-17):

1. An enzyme nucleophile, identified as the sulfhydryl group of Cys 198, attacks the C(6) position of dUMP to form a covalent adduct.

2. C(5) of the resulting enolate ion attacks the CH$_2$ group of the iminium cation in equilibrium with N^5,N^{10}-methylene-THF to form a ternary enzyme–dUMP–THF covalent complex.

3. An enzyme base abstracts the acidic proton at the C(5) position of the enzyme-bound dUMP forming an exocyclic methylene group and eliminating the

THF cofactor. The abstracted proton subsequently exchanges with solvent.

4. The redox change occurs via the migration of the N(6)-H atom of THF as a hydride ion to the exocyclic methylene group, converting it to a methyl group (thus accounting for the above described transfer of ^{3}H) and yielding DHF. This reduction promotes displacement of the Cys sulfhydryl group from the intermediate so as to release product, dTMP, and reform active enzyme.

5-Fluorodeoxyuridylate Is a Potent Antitumor Agent

The above mechanism is supported by the observation that **5-fluorodeoxyuridylate (FdUMP)**

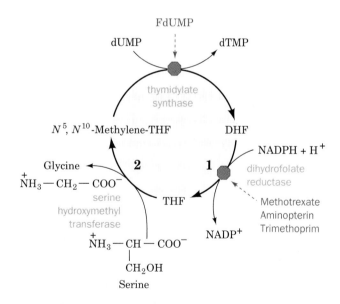

5-Fluorodeoxyuridylate (FdUMP)

is an irreversible inhibitor of thymidylate synthase. This substance, like dUMP, binds to the enzyme (an F atom is about the same size as an H atom) and undergoes the first two steps of the normal enzymatic reaction. In Step 3, however, the enzyme cannot abstract the F atom as F^+ (recall that F is the most electronegative element) so that the enzyme is all but permanently immobilized as an enzyme–FdUMP–THF ternary complex. Enzymatic inhibitors such as FdUMP, which inactivate an enzyme only after undergoing part or all of its normal catalytic reaction, are called **mechanism-based inhibitors** (alternatively, **suicide substrates** because they cause the enzyme to "commit suicide"). *Mechanism-based inhibitors, being targeted for particular enzymes, are among the most powerful, specific, and therefore useful enzyme inactivators.*

The strategic position of thymidylate synthase in DNA biosynthesis has led to the clinical use of FdUMP as an antitumor agent. Rapidly proliferating cells, such as cancer cells, require a steady supply of dTMP in order to survive and are therefore killed by treatment with FdUMP. In contrast, most normal mammalian cells, which grow slowly if at all, have a lesser requirement for dTMP so that they are relatively insensitive to FdUMP (some exceptions are the bone marrow cells that comprise the blood-forming tissues and much of the immune system, the intestinal mucosa, and hair follicles). **5-Fluorouracil** and **5-fluorodeoxyuridine** are also ef-

fective antitumor agents since they are converted to FdUMP through salvage reactions.

N^5,N^{10}-Methylene-THF Is Regenerated in Two Reactions

The thymidylate synthase reaction is biochemically unique in that it oxidizes THF to DHF; no other enzymatic reaction employing a THF cofactor alters this coenzyme's net oxidation state. The DHF product of the thymidylate synthase reaction is recycled to the enzyme's N^5,N^{10}-methylene-THF cofactor through two sequential reactions (Fig. 26-18):

1. DHF is reduced to THF by NADPH as catalyzed by **dihydrofolate reductase (DHFR;** Section 24-40).

2. Serine hydroxymethyl transferase (Section 24-3B) transfers the hydroxymethyl group of serine to THF yielding N^5,N^{10}-methylene-THF and glycine.

Antifolates Are Anticancer Agents

Inhibition of DHFR quickly results in all of a cell's limited supply of THF being converted to DHF by the thymidylate synthase reaction. Inhibition of DHFR therefore not only prevents dTMP synthesis (Fig. 26-18), but also blocks all other THF-dependent biological reactions such as the synthesis of purines (Section 26-2A), histidine, and methionine (Section 24-5B). DHFR (Fig. 26-19) therfore offers an attractive target for chemotherapy.

Figure 26-18
The N^5,N^{10}-methylenetetrahydrofolate that is converted to DHF in the thymidylate synthase reaction is regenerated by the sequential actions of (1) dihydrofolate reductase and (2) serine hydroxymethyl transferase. Thymidylate synthase is inhibited by FdUMP, whereas dihydrofolate reductase is inhibited by the antifolates methotrexate, aminopterin, and trimethoprim.

Methotrexate (amethopterin), aminopterin, and tri-methoprim

R = H   Aminopterin
R = CH$_3$ Methotrexate (amethopterin)

Trimethoprim

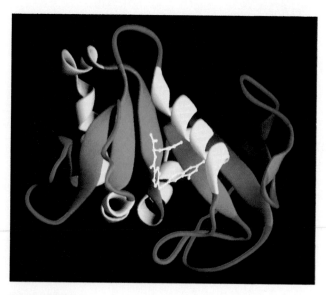

Figure 26-19
The X-ray structure of human dihydrofolate reductase (*ribbon*) in complex with folate. The helices of this monomeric enzyme are drawn in yellow, the β sheets in brown, and the other polypeptide segments in blue. [Courtesy of Jay F. Davies, II and Joseph Kraut, University of California at San Diego.]

are DHF analogs that competitively although nearly irreversibly bind to DHFR with an ~1000-fold greater affinity than does DHF. These **antifolates** (substances that interfere with the action of folate cofactors) are effective anticancer agents, particularly against childhood leukemias. In fact, a successful chemotherapeutic strategy is to treat a cancer victim with a lethal dose of methotrexate and some hours later "rescue" the patient (but hopefully not the cancer) by administering massive doses of 5-formyl-THF and/or thymidine. Trimethoprim, which was discovered by George Hitchings and Gertrude Elion, binds much more tightly to bacterial DHFRs than to those of mammals and is therefore a clinically useful antibacterial agent.

5. NUCLEOTIDE DEGRADATION

Most foodstuffs, being of cellular origin, contain nucleic acids. Dietary nucleic acids survive the acid medium of the stomach; they are degraded to their component nucleotides, mainly in the duodenum, by pancreatic nucleases and intestinal phosphodiesterases. These ionic compounds, which cannot pass through cell membranes, are then hydrolyzed to nucleosides by a variety of group-specific nucleotidases and nonspecific phosphatases. Nucleosides may be directly absorbed by the intestinal mucosa or first undergo further degradation to free bases and ribose or ribose-1-phosphate through the action of **nucleosidases** and **nucleoside phosphorylases:**

$$\text{Nucleoside} + \text{H}_2\text{O} \xrightarrow{\text{nucleosidase}} \text{base} + \text{ribose}$$

$$\text{Nucleoside} + \text{P}_i \xrightarrow[\text{phosphorylase}]{\text{nucleoside}} \text{base} + \text{ribose-1-P}$$

Radioactive labeling experiments have demonstrated that only a small fraction of the bases of ingested nucleic acids are incorporated into tissue nucleic acids. Evidently, the *de novo* pathways of nucleotide biosynthesis largely satisfy an organism's need for nucleotides. Consequently, ingested bases, for the most part, are degraded and excreted. Cellular nucleic acids are also subject to degradation as part of the continual turnover of nearly all cellular components. In this section we outline these catabolic pathways and discuss the consequences of several of their inherited defects.

A. Catabolism of Purines

The major pathways of purine nucleotide and deoxynucleotide catabolism in animals are diagrammed in Fig. 26-20. Other organisms may have somewhat different pathways among these various intermediates (including adenine) but all of these pathways lead to uric acid. Of course, the intermediates in these processes may instead be reused to form nucleotides via salvage reactions. In addition, **ribose-1-phosphate**, a product of the reaction catalyzed by **purine nucleotide phosphorylase (PNP;** Fig. 26-21), is isomerized by **phosphoribomutase** to the PRPP precursor ribose-5-phosphate.

Adenosine and deoxyadenosine are not degraded by mammalian PNP. Rather, adenine nucleosides and nucleotides are deaminated by **adenosine deaminase** and **AMP deaminase,** to their corresponding inosine derivatives which, in turn, may be further degraded.

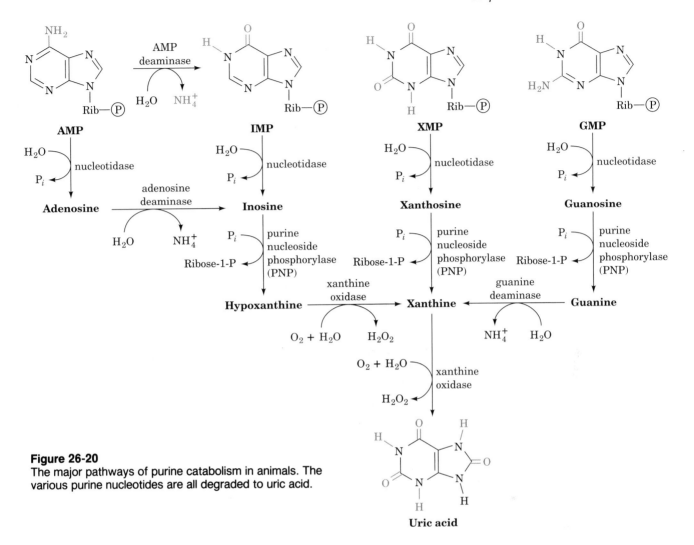

Figure 26-20
The major pathways of purine catabolism in animals. The various purine nucleotides are all degraded to uric acid.

Uric acid

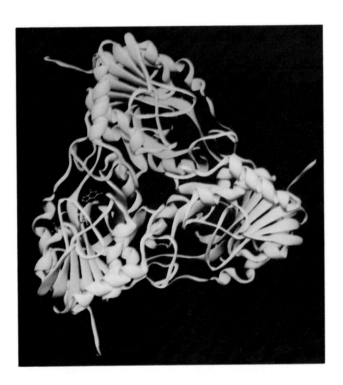

Defective Adenosine Deaminase Results in Severe Combined Immunodeficiency Disease

A genetic lack of adenosine deaminase activity causes an abnormality in purine nucleoside metabolism that selectively kills lymphocytes (a type of white blood cell). The consequent lack of lymphocytes, which mediate the immune response (Section 34-2A), results in **severe combined immunodeficiency disease (SCID)** that, without special protective measures, is invariably fatal in infancy because of overwhelming infection. Biochemical considerations provide a plausible explanation of SCID's etiology (causes). In the absence of active adenosine deaminase, deoxyadenosine is phosphorylated to yield levels of dATP that are 50-fold greater than

Figure 26-21
The X-ray structure of human erythrocyte purine nucleoside phosphorylase as viewed along this trimeric enzyme's threefold axis. Each identical subunit is differently colored with that in yellow shown in complex with a guanine molecule and two phosphate ions. [Courtesy of Michael Carson, University of Alabama at Birmingham; X-ray structure determined by Stephen Ealick and Charles Bugg.]

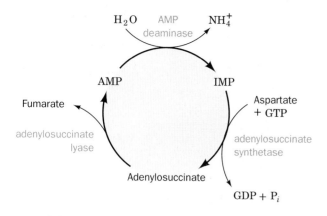

Figure 26-22
The purine nucleotide cycle. This pathway functions, in muscle, to prime the citric acid cycle by generating fumarate.

normal. This high concentration of dATP inhibits ribonucleotide reductase (Section 26-4A), thereby preventing the synthesis of the other dNTPs, choking off DNA synthesis, and thus cell proliferation. The tissue-specific effect of this enzyme defect on the immune system may be explained by the observation that lymphoid tissue is particularly active in deoxyadenosine phosphorylation.

The Purine Nucleotide Cycle

The deamination of AMP to IMP, when combined with the synthesis of AMP from IMP (Fig. 26-5, *left*), has the effect of deaminating aspartate to yield fumarate

(Fig. 26-22). John Lowenstein demonstrated that this **purine nucleotide cycle** has an important metabolic role in skeletal muscle. An increase in muscle activity requires an increase in the activity of the citric acid cycle. This process usually occurs through the generation of additional citric acid cycle intermediates (Section 19-4). Muscles, however, lack most of the enzymes that catalyze these anaplerotic (filling up) reactions in other tissues. Rather, muscle replenishes its citric acid cycle intermediates as fumarate generated in the purine nucleotide cycle. The importance of the purine nucleotide cycle in muscle metabolism is indicated by the observation that the activities of the three enzymes involved are all severalfold higher in muscle than in other tissues. In fact, individuals with an inherited deficiency in muscle AMP deaminase (**myoadenylate deaminase deficiency**) are easily fatigued and usually suffer from cramps after exercise.

Xanthine Oxidase Is a Mini-Electron-Transport Protein

Xanthine oxidase converts hypoxanthine to xanthine, and xanthine to uric acid (Fig. 26-20, *bottom*). In mammals, this enzyme occurs almost exclusively in the liver and the small intestinal mucosa. It is a dimeric protein of identical 130-kD subunits, each of which contain an entire "zoo" of electron-transfer agents: an FAD, a Mo complex that cycles between its Mo(VI) and Mo(IV) oxidation states, and two different Fe–S clusters. The final electron acceptor is O_2, which is converted to H_2O_2, a potentially harmful oxidizing agent that is subsequently disproportionated to H_2O and O_2 by catalase.

Xanthine oxidase hydroxylates xanthine at its C(8)

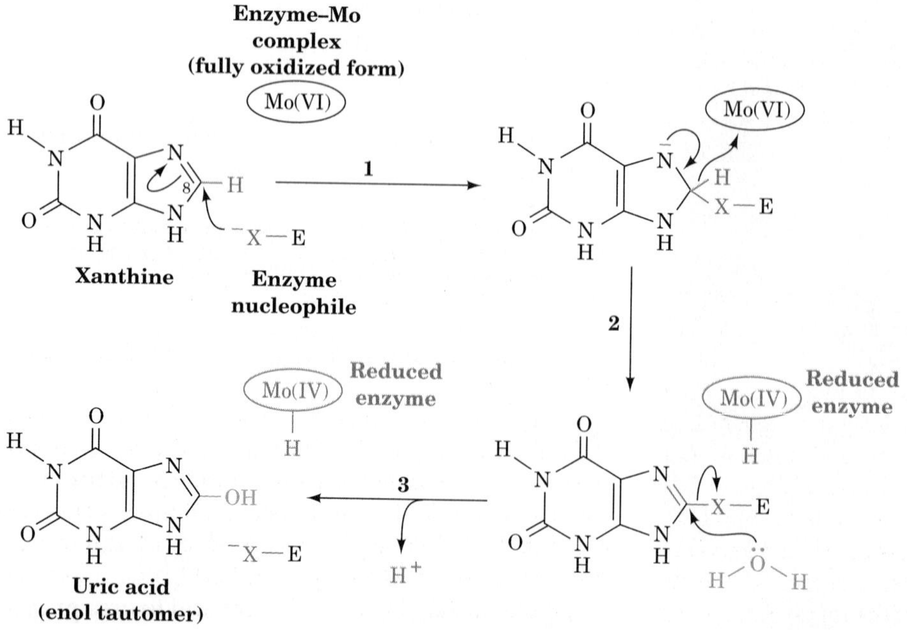

Figure 26-23
The mechanism of xanthine oxidase. The reduced enzyme is subsequently reoxidized by O_2 yielding H_2O_2.

position [and hypoxanthine at its C(2) position] yielding a product that tautomerizes to the more stable keto form:

Uric acid (enol tautomer)   **Uric acid (keto tautomer)**

$pK = 5.4$

Urate

(its enol form ionizes with a pK of 5.4; hence, the name uric *acid*). ^{18}O-Labeling experiments have demonstrated that the C(8)-keto oxygen of uric acid is derived from H_2O, whereas the oxygen atoms of H_2O_2 come from O_2. Chemical and spectroscopic studies suggest that the enzyme has the following mechanism (Fig. 26-23):

1. The reaction is initiated by the attack of an enzyme nucleophile, X, on the C(8) position of xanthine oxidase.

2. The C(8)-H atom is eliminated as a hydride ion that combines with the Mo(VI) complex thereby reducing it to the Mo(IV) state.

3. Water displaces the enzyme nucleophile producing uric acid.

In the second stage of the reaction, the now reduced enzyme is reoxidized to its original Mo(VI) state by reaction with O_2. This complex process, not surprisingly, is but poorly understood. EPR measurements indicate that electrons are funneled from the Mo(IV) through the enzyme's other redox centers and are ultimately passed from the flavin to O_2 yielding H_2O_2 and regenerated enzyme.

B. Fate of Uric Acid

In humans and other primates, the final product of purine degradation is uric acid, which is excreted in the urine. The same is true of birds, terrestrial reptiles, and many insects, but these organisms, which do not excrete urea, also catabolize their excess amino acid nitrogen to uric acid via purine biosynthesis. This complicated system of nitrogen excretion has a straightforward function: *It conserves water.* Uric acid is only sparingly soluble in water so that its excretion as a paste of uric acid crystals is accompanied by very little water. In contrast, the excretion of an equivalent amount of the much more water

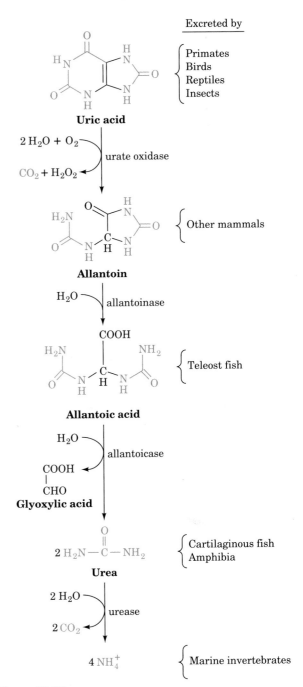

Figure 26-24
The degradation of uric acid to ammonia. The process is arrested at different stages in the indicated species and the resulting nitrogen-containing product is excreted.

soluble urea osmotically sequesters a significant amount of water.

In all other organisms, uric acid is further processed before excretion (Fig. 26-24). Mammals other than primates oxidize it to their excretory product, **allantoin,** in a reaction catalyzed by the Cu-containing enzyme **urate oxidase.** A further degradation product, **allantoic acid,** is excreted by teleost (bony) fish. Cartilaginous fish and amphibia further degrade allantoic acid to urea prior to

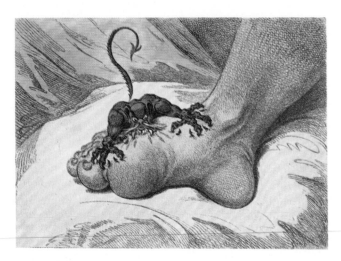

Figure 26-25
Gout, a cartoon by James Gilroy. [Yale University Medical
Historical Library.]

excretion. Finally, marine invertebrates decompose urea
to their nitrogen excretory product, NH_4^+.

Gout Is Caused by an Excess of Uric Acid

Gout is a disease characterized by elevated levels of uric acid in body fluids. Its most common manifestation is excruciatingly painful arthritic joint inflammation of sudden onset, most often in the big toe (Fig. 26-25), caused by deposition of nearly insoluble crystals of sodium urate. Sodium urate and/or uric acid may also precipitate in the kidneys and ureters as stones, resulting in renal damage and urinary tract obstruction. Gout, which affects ~3 per 1000 persons, predominantly males, has been traditionally, although inaccurately, associated with overindulgent eating and drinking. The probable origin of this association is that in previous centuries, when wine was often contaminated with lead during its manufacture and storage, heavy drinking resulted in chronic lead poisoning which, among other things, decreases the kidney's ability to excrete uric acid.

The most prevalent cause of gout is impaired uric acid excretion (although usually for other reasons than lead poisoning). Gout may also result from a number of metabolic insufficiencies, most of which are not well characterized. One well understood cause is HGPRT deficiency (Lesch–Nyhan syndrome in severe cases), which leads to excessive uric acid production through PRPP accumulation (Section 26-2D). Uric acid overproduction is also caused by glucose-6-phosphatase deficiency (von Gierke's glycogen storage disease; Section 17-4): The increased availability of glucose-6-phosphate stimulates the pentose phosphate pathway (Section 21-4), increasing the rate of ribose-5-phosphate production and consequently that of PRPP, which in turn stimulates purine biosynthesis.

Gout may be treated by administration of the xan-

thine oxidase inhibitor **allopurinol,** a hypoxanthine analog with interchanged N(7) and C(8) positions.

Allopurinol   **Hypoxanthine**

Xanthine oxidase hydroxylates allopurinol, as it does hypoxanthine, yielding **alloxanthine,**

Alloxanthine

which remains tightly bound to the reduced form of the enzyme, thereby inactivating it. Allopurinol consequently alleviates the symptoms of gout by decreasing the rate of uric acid production while increasing the levels of the more soluble hypoxanthine and xanthine. While allopurinol controls the gouty symptoms of Lesch–Nyhan syndrome, it has no effect on its neurological symptoms.

C. Catabolism of Pyrimidines

Animal cells degrade pyrimidine nucleotides to their component bases (Fig. 26-26, *top*). These reactions, like those of purine nucleotides, occur through dephosphorrylation, deamination, and glycosidic bond cleavages. The resulting uracil and thymine are then broken down in the liver through reduction (Fig. 26-26, *middle*) rather than by oxidation as occurs in purine catabolism. The end products of pyrimidine catabolism, **β-alanine** and **β-aminoisobutyrate,** are amino acids and are metabolized as such. They are converted, through transamination and activation reactions to malonyl-CoA and methylmalonyl-CoA (Fig. 26-26, *bottom left*) for further utilization (Sections 23-4A and 2E).

6. BIOSYNTHESIS OF NUCLEOTIDE COENZYMES

In this section we outline the assembly, in animals, of the nucleotide coenzymes NAD^+ and $NADP^+$, FMN and FAD, and coenzyme A, from their vitamin precursors. These vitamins are synthesized *de novo* only by plants and microorganisms.

A. Nicotinamide Coenzymes

The nicotinamide moiety of the nicotinamide coenzymes (NAD^+ and $NADP^+$) is derived, in humans, from

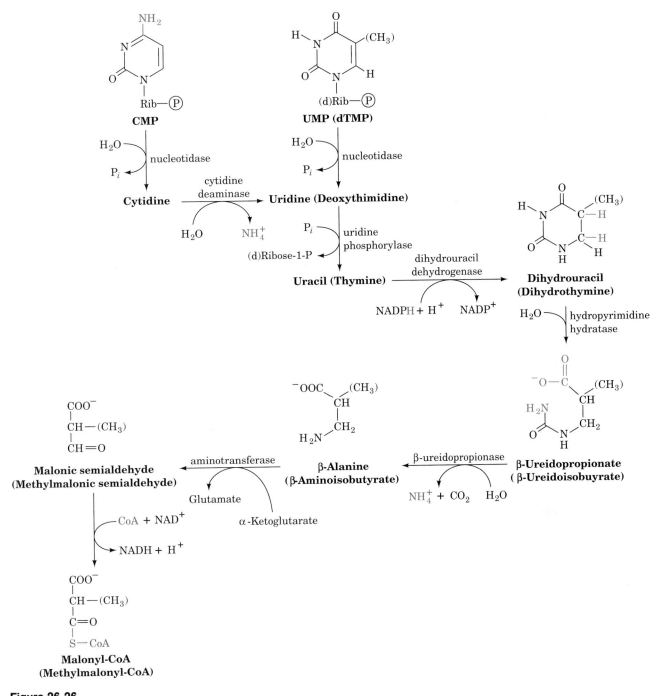

Figure 26-26
The major pathways of pyrimidine catabolism in animals. The amino acid
products of these reactions are taken up in other metabolic processes.
UMP and dTMP are degraded by the same enzymes; the pathway for
dTMP degradation is given in parentheses.

Figure 26-27
The pathways for the biosynthesis of NAD⁺ and
NADP⁺ from their vitamin precursors, nicotinate
and nicotinamide, and from the tryptophan
degradation product, quinolinate.

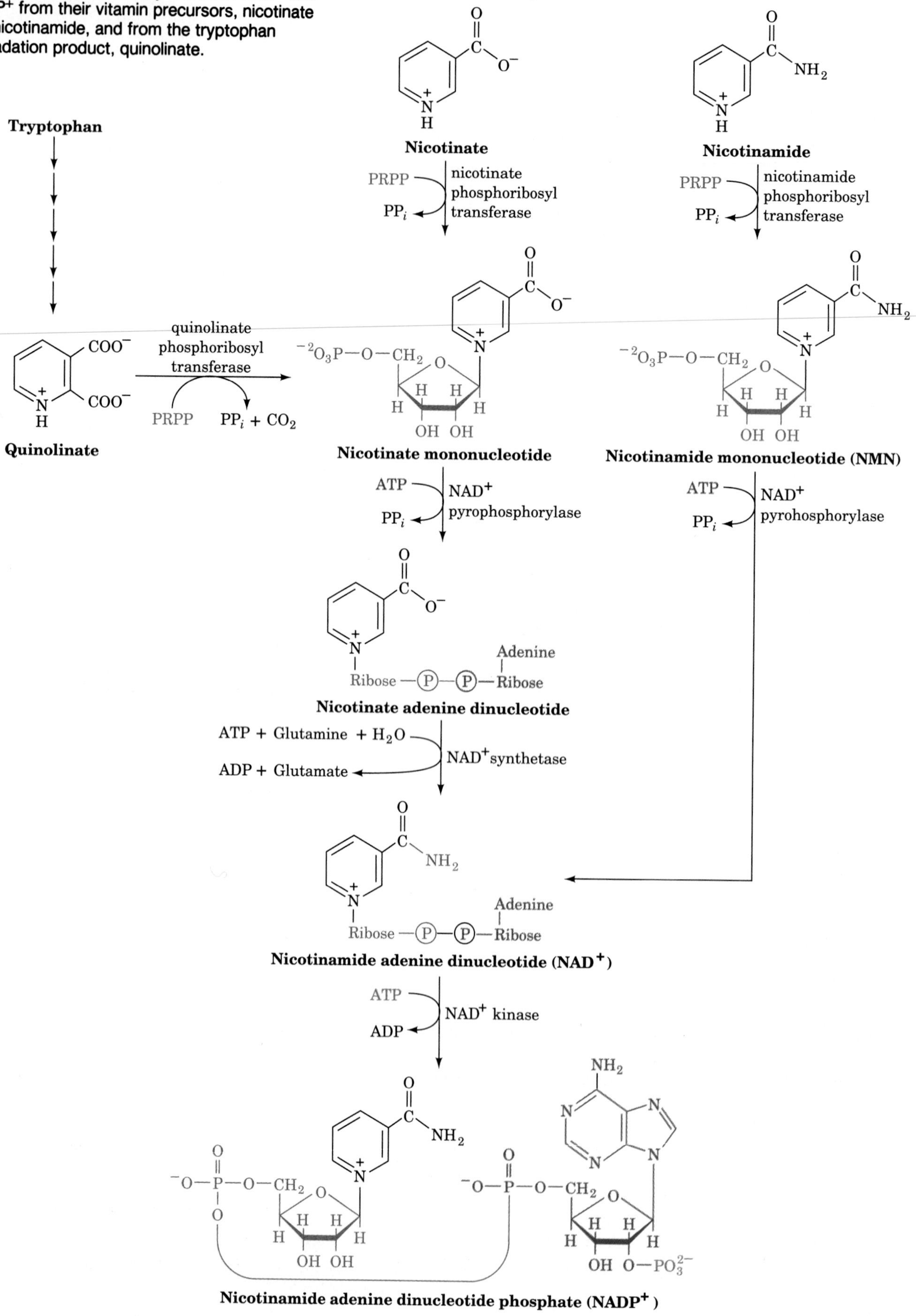

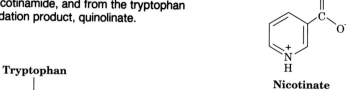

Tryptophan

Nicotinate

Nicotinamide

nicotinate
phosphoribosyl
transferase

PRPP
PP$_i$

nicotinamide
phosphoribosyl
transferase

PRPP
PP$_i$

quinolinate
phosphoribosyl
transferase

PRPP   PP$_i$ + CO$_2$

Quinolinate

Nicotinate mononucleotide

Nicotinamide mononucleotide (NMN)

ATP   NAD⁺
PP$_i$ pyrophosphorylase

ATP   NAD⁺
PP$_i$ pyrohosphorylase

Nicotinate adenine dinucleotide

Ribose—(P)—(P)—Ribose
Adenine

ATP + Glutamine + H$_2$O

ADP + Glutamate

NAD⁺synthetase

Nicotinamide adenine dinucleotide (NAD⁺)

Ribose—(P)—(P)—Ribose
Adenine

ATP   NAD⁺ kinase
ADP

Nicotinamide adenine dinucleotide phosphate (NADP⁺)

dietary nicotinamide, nicotinic acid, or the essential amino acid tryptophan (Fig. 26-27). **Nicotinate phosphoribosyl transferase,** which occurs in most mammalian tissues, catalyzes the formation of **nicotinate mononucleotide** from nicotinate and PRPP. This intermediate may also be synthesized from **quinolinate,** a degradation product of tryptophan (Section 24-3G), in a reaction mediated by **quinolinate phosphoribosyl transferase,** which occurs mainly in liver and kidney. A poor diet, nevertheless, may result in pellagra (nicotinic acid deficiency; Section 12-3), since, under such conditions, tryptophan will be almost entirely utilized in protein biosynthesis. Nicotinate mononucleotide is linked via a pyrophosphate linkage to an AMP residue by **NAD⁺ pyrophosphorylase** to yield **nicotinate adenine dinucleotide (desamido NAD⁺).** Finally, **NAD⁺ synthetase** converts this intermediate to NAD⁺ by a transamidation reaction in which glutamine is the NH₂ donor.

NAD⁺ may also be synthesized from nicotinamide. This vitamin is converted to NMN by **nicotinamide phosphoribosyl transferase,** a widely occurring enzyme distinct from nicotinate phosphoribosyl transferase. However, NAD⁺ is synthesized from NMN and PRPP by NAD pyrophosphorylase, the same enzyme that synthesizes nicotinate adenine dinucleotide.

NADP⁺ is formed via the phosphorylation of the NAD⁺ adenosine residue's C(2′)-OH group by **NAD⁺ kinase.**

B. Flavin Coenzymes

FAD is synthesized from riboflavin in a two-way reaction pathway (Fig. 26-28). First, the 5′-OH group of riboflavin's ribityl side chain is phosphorylated by **flavokinase** yielding flavin mononucleotide (FMN; not a true nucleotide since the ribityl residue is not a true sugar). FAD may then be formed by the coupling of FMN and AMP in a pyrophosphate linkage in a reaction catalyzed by **FAD pyrophosphorylase.** Both of these enzymes are widely distributed in nature.

C. Coenzyme A

Coenzyme A is synthesized in mammalian cells according to the pathway diagrammed in Fig. 26-29. Pantothenate, an essential vitamin, is phosphorylated by **pantothenate kinase** and then coupled to cysteine, the future business end of CoA, by **phosphopantothenoylcysteine synthetase.** After decarboxylation by **phosphopantothenoylcysteine decarboxylase,** the resulting **4′-phosphopantethiene** is coupled to AMP in a pyrophosphate linkage by **dephospho-CoA pyrophosphorylase** and then phosphorylated at its adenosine 3′-OH group by **dephospho-CoA kinase** to form CoA. These latter two enzymatic activities occur on a single protein.

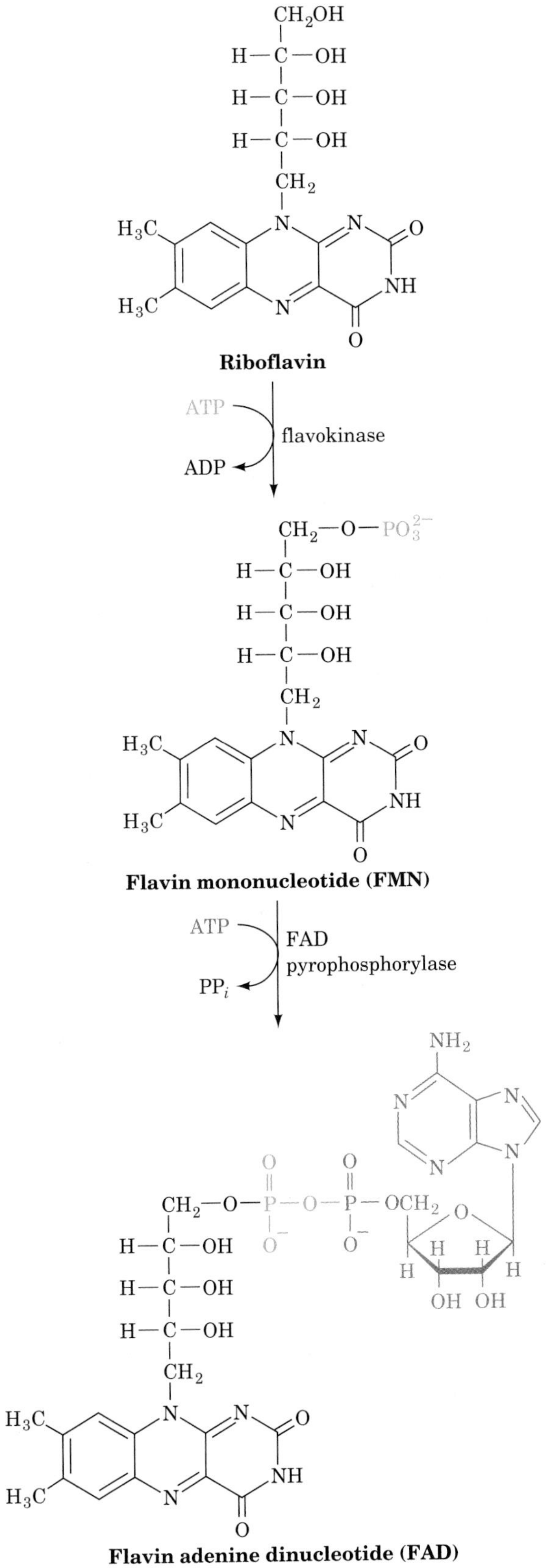

Figure 26-28
The biosynthesis of FMN and FAD from the vitamin precursor riboflavin.

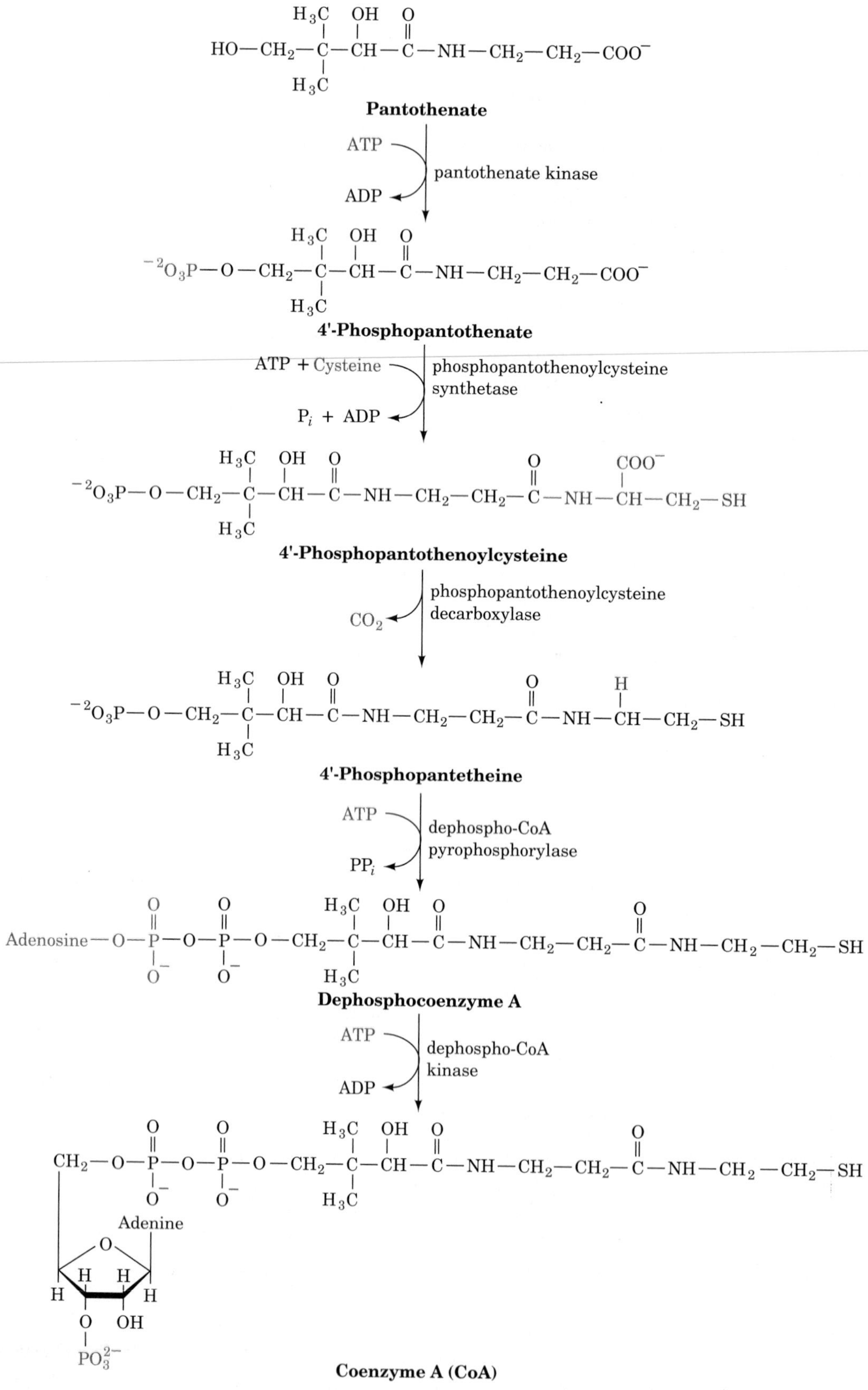

Figure 26-29
The biosynthesis of coenzyme A from pantothenate, its vitamin precursor.

Chapter Summary

Almost all cells synthesize purine nucleotides *de novo* via similar metabolic pathways. The purine ring is constructed in an 11-step reaction sequence that yields IMP. AMP and GMP are then synthesized from IMP in separate pathways. Nucleoside diphosphates and triphosphates are sequentially formed from these products via phosphorylation reactions. The rates of synthesis of these various nucleotides are interrelated through feedback inhibition mechanisms that monitor their concentrations. Purine nucleotides may also be synthesized from free purines salvaged from nucleic acid degradation processes. The importance of these salvage reactions is demonstrated, for example, by the devastating and bizarre consequences of Lesch–Nyhan syndrome.

Cells also synthesize pyrimidines *de novo* but, in this six-step process, a free base is formed before it is converted to a nucleotide, UMP. UTP is then formed by phosphorylation of UMP, and CTP is synthesized by the amination of UTP. Pyrimidine biosynthesis is regulated by feedback inhibition as well as by the concentrations of purine nucleotides.

Deoxynucleotides are formed by reduction of the corresponding ribonucleotides. There are three known types of ribonucleotide reductase: one, which occurs in nearly all eukaryotes and many prokaryotes, contains an Fe^{3+}—O—Fe^{3+} group and a tyrosyl free radical at its active site; the others, which occur only in prokaryotes, contain either a coenzyme B_{12} cofactor or Mn. The oxidized Fe(III) enzyme from *E. coli* is reduced to its original state by electron-transport chains involving either thioredoxin, thioredoxin reductase and NADPH, or glutaredoxin, glutathione, glutathione reductase, and NADPH. Ribonucleotide reductase has two independent regulatory sites that control its substrate specificity as well as its catalytic activity, thereby generating deoxynucleotides in the amounts required for DNA synthesis. Thymine is synthesized by methylation of dUMP by thymidylate synthase to form dTMP. The reaction's methyl source, N^5,N^{10}-methylene-THF is oxidized in the reaction to yield dihydrofolate. N^5,N^{10}-Methylene-THF is subsequently regenerated through the sequential actions of dihydrofolate reductase and serine hydroxymethyl transferase. Since this sequence of reactions is required for DNA biosynthesis, it presents an excellent target for chemotherapy. FdUMP, a mechanism-based inhibitor of thymidylate synthase, and methotrexate, an antifolate that essentially irreversibly inhibits dihydrofolate reductase, are both highly effective anticancer agents.

Purine nucleotides are catabolized to yield uric acid. Depending on the species, the uric acid is either directly excreted or first degraded to simpler nitrogen-containing substances. Overproduction or underexcretion of uric acid in humans causes gout. Pyrimidines are catabolized in animal cells to amino acids.

The nucleotide coenzymes, NAD^+ and $NADP^+$, FMN and FAD, and coenzyme A are synthesized in animals from vitamin precursors.

References

Ashley, G. W. and Stubbe, J. A., Current ideas on the chemical mechanism of ribonucleotide reductases, *Pharm. Ther.* **30**, 301–329 (1986).

Benkovic, S. J., On the mechanism of action of folate- and biopterin-requiring enzymes, *Annu. Rev. Biochem.* **49**, 227–251 (1980).

Boss, G. R. and Seegmiller, J. E., Genetic defects in human purine and pyrimidine metabolism, *Annu. Rev. Genet.* **16**, 297–328 (1982).

Eliasson R., Jörnvall, H., and Reichard, P., Superoxide dismutase participates in the enzymatic formation of the tyrosine radical of ribonucleotide reductase from *Escherichia coli*, *Proc. Natl. Acad. Sci.* **83**, 2373–2377 (1986).

Hardy, L. W., Finer-Moore, J. S., Montfort, W. R., Jones, M. O., Santi, D. V., and Stroud, R. M., Atomic structure of thymidylate synthase: target for rational drug design, *Science* **235**, 448–455 (1987). [The X-ray structure of the *Lactobacillus casei* enzyme.]

Henderson, J. F. and Paterson, A. R. P., *Nucleotide Metabolism,* Academic Press (1973).

Henikoff, S., Multifunctional polypeptides for purine *de novo* synthesis, *BioEssays* **6**, 8–13 (1987).

Holmgren, A., Thioredoxin, *Annu. Rev. Biochem.* **54**, 237–271 (1985).

Holmgren, A., Regulation of ribonucleotide reductase, *Curr. Top. Cell. Regul.* **19**, 47–76 (1981).

Jones, M. E., Pyrimidine nucleotide biosynthesis in animals: genes, enzymes, and regulation of UMP biosynthesis, *Annu. Rev. Biochem.* **49**, 253–279 (1980).

Kellems, R. E., Yeung, C.-Y., and Ingolia, D. E., Adenosine deaminase deficiency and severe combined immunodeficiencies, *Trends Genet.* **1**, 278–283 (1985).

Kornberg, A., *DNA Replication,* Chapter 2, and *1982 Supplement to DNA Replication,* Chapter S2, Freeman (1980 and 1982).

Kraut, J. and Matthews, D. A., Dihydrofolate reductase, *in* Jurnak, F. A. and McPherson, A. (Eds.), *Biological Macromolecules and Assemblies,* Vol. 3, pp. 1–71, Wiley (1987).

Stanbury, J. B., Wyngaarden, J. B., Fredrickson, D. S., Goldstein, J. L., and Brown, M. S. (Eds.), *The Metabolic Basis of Inherited Disease* (5th ed.), Chapters 50–56, McGraw–Hill (1983).

Reichard, P., Interactions between deoxyribonucleotide and DNA synthesis, *Annu. Rev. Biochem.* **57**, 349–374 (1988).

Reichard, P., Regulation of deoxyribonucleotide synthesis, *Biochemistry* **26**, 3245–3248 (1987).

Reichard, P. and Ehrenberg, A., Ribonucleotide reductase—a radical enzyme, *Science* **221**, 514–519 (1983).

Thelander, L. and Reichard, P., Reduction of nucleotides, *Annu. Rev. Biochem.* **48**, 133–158 (1979).

Watts, R. W. E., Some regulatory and integrative aspects of purine nucleotide biosynthesis and its control: an overview, *Adv. Enzyme Regul.* **21**, 33–51 (1983).

Wyngaarden, J. B., Regulation of purine biosynthesis and turnover, *Adv. Enzyme Regul.* **14**, 25–42 (1977).

Problems

1. **Azaserine (*O*-diazoacetyl-L-serine) and 6-diazo-5-oxo-L-norleucine (DON)**

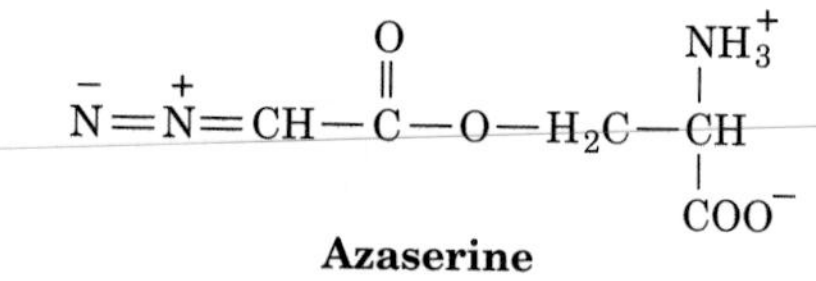

$$\overset{-}{N}=\overset{+}{N}=CH-\overset{\overset{O}{\|}}{C}-O-H_2C-\overset{\overset{+}{NH_3}}{\underset{\underset{COO^-}{|}}{CH}}$$

Azaserine

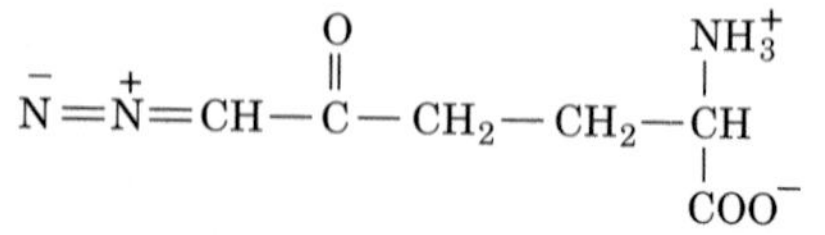

$$\overset{-}{N}=\overset{+}{N}=CH-\overset{\overset{O}{\|}}{C}-CH_2-CH_2-\overset{\overset{+}{NH_3}}{\underset{\underset{COO^-}{|}}{CH}}$$

6-Diazo-5-oxo-L-norleucine (DON)

are glutamine analogs. They form covalent bonds to nucleophiles at the active sites of enzymes that bind glutamine, thereby irreversibly inactivating these enzymes. Identify the nucleotide biosynthesis intermediates that accumulate in the presence of either of these glutamine antagonists.

2. Suggest a mechanism for the AIR synthetase reaction (Fig. 26-3, Reaction 6).

*3. What is the energetic price, in ATPs, of synthesizing the hypoxanthine residue of IMP from CO_2 and NH_4^+?

4. Why is deoxyadenosine toxic to mammalian cells?

5. Indicate which of the following substances are mechanism-based inhibitors and explain your reasoning. (a) Tosyl-L-phenylanine chloromethylketone with chymotrypsin (Section 14-3A). (b) Trimethoprim with bacterial dihydrofolate reductase. (c) The δ-lactone analog of $(NAG)_4$ with lysozyme (Section 14-2C). (d) Allopurinol with xanthine oxidase.

6. Why do individuals who are undergoing chemotherapy with cytotoxic (cell killing) agents such as FdUMP or methotrexate temporarily go bald?

7. Normal cells die in a nutrient medium containing thymidine and methotrexate that supports the growth of mutant cells defective in thymidylate synthase. Explain.

8. FdUMP and methotrexate, when taken together, are less effective chemotherapeutic agents than when either drug is taken alone. Explain.

9. Why is gout more prevalent in populations that eat relatively large amounts of meat?

10. Gout resulting from the *de novo* overproduction of purines can be distinguished from gout caused by impaired excretion of uric acid by feeding a patient ^{15}N-labeled glycine and determining the distribution of ^{15}N in his/her excreted uric acid. What isotopic distributions are expected for each type of defect?

11. **6-Mercaptopurine,**

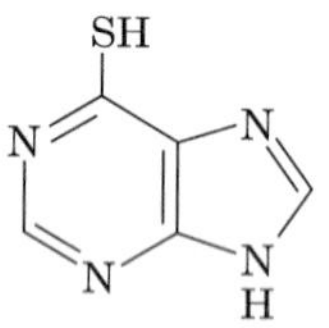

6-Mercaptopurine

after conversion to the corresponding nucleotide through salvage reactions, is a potent competitive inhibitor of IMP in the pathways for AMP and GMP biosynthesis. It is therefore a clinically useful anticancer agent. The chemotherapeutic effectiveness of 6-mercaptopurine is enhanced when it is administered with allopurinol. Explain the mechanism of this enhancement.

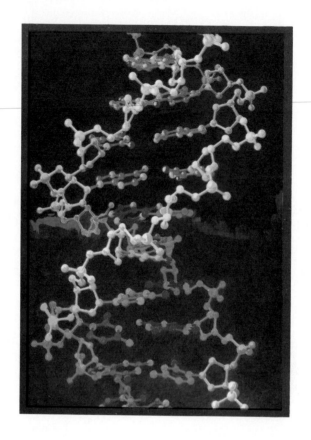

The structure of a specific sequence of B-DNA,
derived from x-ray crystal structure analysis.

The Expression and Transmission of Genetic Information

Chapter 27
DNA: THE VEHICLE OF INHERITANCE

DNA, as is now common knowledge, is the carrier of genetic information in all cellular life as well as in many viruses. Yet, a period of over 75 years passed from the time the laws of inheritance were discovered until the biological role of DNA was discovered. Even now, many details of how genetic information is expressed and transmitted to future generations are still unclear, particularly in eukaryotes, and are subjects of intense research. Thus, our ideas of how genetic information is expressed and transmitted have been attained through a slow evolutionary process that was only occasionally punctuated by incisive experiments or brilliant insights.

In this chapter, we commence our study of **molecular genetics;** that is, how genetic information is transmitted and expressed on the molecular level. We begin by reviewing "classical" genetics, whose understanding is prerequisite for assimilating molecular genetics. We then consider how we have come to know that DNA is the carrier of genetic information. This chapter has been written with an historical perspective in order to illustrate how these ideas developed, in particular, and how scientific concepts evolve, in general. The major aspects of molecular genetics are considered in detail in subsequent chapters.

1. GENETICS: A REVIEW

One has only to note the resemblance between parent and child to realize that physical traits are inherited. Yet, the mechanism of inheritance has, until recent decades, been unknown. The theory of **pangenesis,** which originated with the ancient Greeks, held that semen, which clearly has something to do with procreation, consists of representative particles from all over the body **(pangenes).** This idea was extended in the late eighteenth century by Jean Baptiste de Lamarck who, in a theory known as **Lamarckism,** hypothesized that an individual's acquired characteristics, such as large muscles resulting from exercise, would be transmitted to his/her offspring. Pangenesis and at least some aspects of Lamarckism were accepted by most nineteenth century biologists, including Charles Darwin.

The realization, in the mid-nineteenth century, that all organisms are derived from single cells set the stage for the development of modern biology. In his **germ plasm theory,** August Weismann pointed out that sperm and ova, the **germ cells** (whose primordia are set aside early in embryonic development), are directly descended from the germ cells of the previous generation and that other cells of the body, the **somatic cells,** although derived from germ cells, do not give rise to them. He refuted the ideas of pangenesis and Lamarckism by demonstrating that the progeny of many successive generations of mice whose tails had been cut off had tails of normal length.

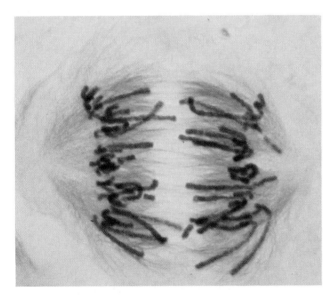

Figure 27-1
A photomicrograph of a plant cell (*Scadoxus katherinae* Bak.) during anaphase of mitosis showing its chromosomes being pulled to opposite poles of the cell by the mitotic spindle. The microtubules forming the mitotic spindle are stained red and the chromosomes are blue. [Courtesy of Andrew S. Bajer, University of Oregon.]

A. Chromosomes

In the 1860s, eukaryotic cell nuclei were observed to contain linear bodies that were named **chromosomes** (Greek: *chromos,* color; *soma,* body) because they are strongly stained by certain basic dyes (Fig. 27-1). There are normally two copies of each chromosome **(homologous pairs)** present in every somatic cell. The number of unique chromosomes (*N*) in such a cell is known as its **haploid number** and the total number of chromosomes (*2N*) is its **diploid number.** Different species differ in their haploid number of chromosomes (Table 27-1).

Somatic Cells Divide by Mitosis

The division of somatic cells, a process known as **mitosis** (Fig. 27-2), is preceded by the duplication of each chromosome to form a cell with *4N* chromosomes. During cell division, each chromosome attaches by its **centromere** to the **mitotic spindle** such that the members of each duplicate pair line up across the equatorial plane of the cell. The members of each duplicate pair are then pulled to opposite poles of the dividing cell by the action of the spindle to yield diploid daughter cells that each have the same *2N* chromosomes as the parent cell.

Germ Cells Are Formed by Meiosis

The formation of germ cells, a process known as **meiosis** (Fig. 27-3), requires two consecutive cell divisions. Prior to the first meiotic division, each chromosome replicates but the resulting sister **chromatids** remain attached at their centromere. The homologous pairs of the doubled chromosomes then line up across the equatorial plane of the cell in zipperlike fashion, which permits an exchange of the corresponding sections of homologous chromosomes in a process known as **crossing-over** (Section 27-1B). The spindle then

Table 27-1
Number of Chromosomes (*2N*) in Some Eukaryotes

Organism	Chromosomes
Humans	46
Dog	78
Rat	42
Turkey	82
Frog	26
Fruit fly	8
Hermit crab	~254
Garden pea	14
Potato	48
Yeast	34
Green alga	~20

Source: Ayala, F. J. and Kiger, J. A., Jr., *Modern Genetics* (2nd ed.), *p.* 9, Benjamin/Cummings (1984).

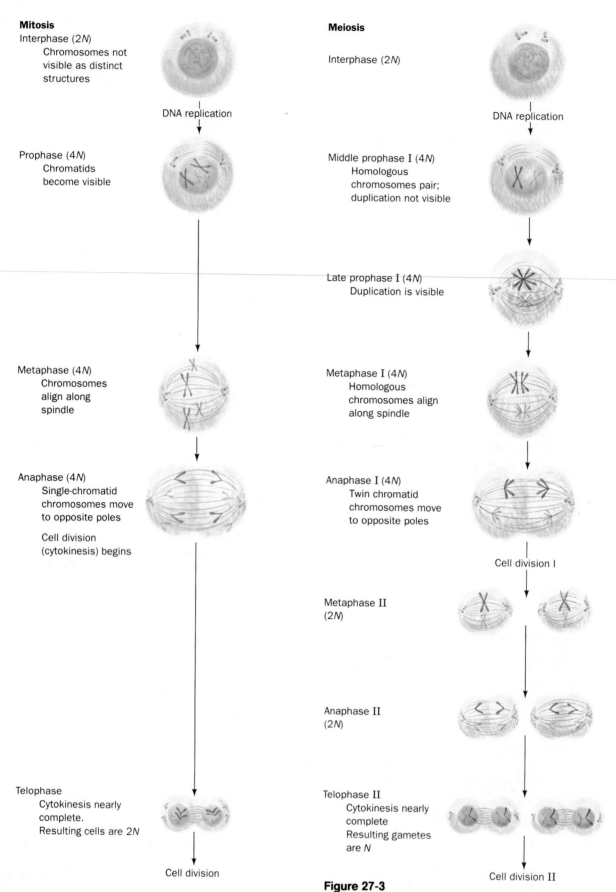

Mitosis

Interphase (2N)
  Chromosomes not
  visible as distinct
  structures

DNA replication

Prophase (4N)
  Chromatids
  become visible

Metaphase (4N)
  Chromosomes
  align along
  spindle

Anaphase (4N)
  Single-chromatid
  chromosomes move
  to opposite poles

  Cell division
  (cytokinesis) begins

Telophase
  Cytokinesis nearly
  complete.
  Resulting cells are 2N

Cell division

Meiosis

Interphase (2N)

DNA replication

Middle prophase I (4N)
  Homologous
  chromosomes pair;
  duplication not visible

Late prophase I (4N)
  Duplication is visible

Metaphase I (4N)
  Homologous
  chromosomes align
  along spindle

Anaphase I (4N)
  Twin chromatid
  chromosomes move
  to opposite poles

Cell division I

Metaphase II
(2N)

Anaphase II
(2N)

Telophase II
  Cytokinesis nearly
  complete
  Resulting gametes
  are N

Cell division II

Figure 27-2
Mitosis, the usual form of cell division in eukaryotes, yields
two daughter cells, each with the same chromosomal
complement as the parental cell.

Figure 27-3
Meiosis, which leads to the formation of gametes (sex cells),
comprises two consecutive cell divisions to yield four
daughter cells, each with one half of the chromosomal
complement of the parental cell.

moves the members of each homologous pair to opposite poles of the cell so that, after the first meiotic division, each daughter cell contains *N* doubled chromosomes. In the second meiotic division, the sister chromatids separate to form chromosomes and move to opposite poles of the dividing cell to yield a total of four haploid cells that are known as **gametes.** Fertilization consists of the fusion of a male gamete (sperm) with a female gamete (ovum) to yield a diploid cell known as a **zygote** that has received *N* chromosomes from each of its parents.

B. Mendelian Inheritance

The basic laws of inheritance were reported in 1866 by Gregor Mendel. They were elucidated by the analysis of a series of **genetic crosses** between true-breeding strains (producing progeny that have the same characteristics as the parents) of garden peas, *Pisum sativum,* that differ in certain well-defined traits such as seed shape (round vs wrinkled), seed color (yellow vs green), or flower color (purple vs white). Mendel found that in crossing parents (*P*) that differ in a single trait, say seed shape, the progeny (*F₁*; first filial generation) all have the trait of only one of the parents, in this case round seeds (Fig. 27-4). The trait appearing in *F₁* is said to be **dominant** whereas the alternative trait is called **recessive.** In *F₂*, the progeny of *F₁*, three quarters have the dominant trait and one quarter have the recessive trait. Those peas with the recessive trait breed true; that is, self-crossing recessive *F₂*'s results in progeny (*F₃*) that also have the recessive trait. The *F₂*'s exhibiting the dominant trait, however, fall into two categories: one third of them breed true, whereas the remainder have progeny with the same 3 : 1 ratio of dominant to recessive traits as do the members of *F₂*.

Mendel accounted for his observations by hypothesizing that *the various pairs of contrasting traits each result from a factor (now called a **gene**) that has alternative forms (**alleles**). Every plant contains a pair of genes governing a particular trait, one inherited from each of its parents.* The alleles for seed shape are symbolized *R* for round seeds and *r* for wrinkled seeds (gene symbols are generally given in italics). The pure breeding plants with round and wrinkled seeds, respectively, have *RR* and *rr* **genotypes** (genetic composition) and are both said to be **homozygous** in seed shape. Plants with the *Rr* genotype are **heterozygous** in seed shape and have the round seed **phenotype** (appearance or character) because *R* is dominant over *r*. *The two alleles do not blend or mix in any way in the plant and are independently transmitted through gametes to progeny* (Fig. 27-5).

Mendel also found that *different traits are independently inherited.* For example, crossing peas that have round yellow seeds (*RRYY*) with peas that have wrinkled green seeds (*rryy*) results in *F₁* progeny (*RrYy*) that have round yellow seeds (yellow seeds are dominant

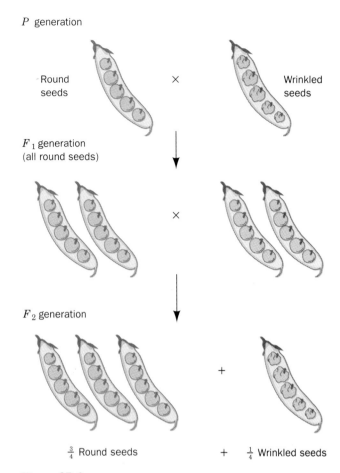

P generation

Round seeds × Wrinkled seeds

F₁ generation (all round seeds)

×

F₂ generation

+

$\frac{3}{4}$ Round seeds + $\frac{1}{4}$ Wrinkled seeds

Figure 27-4
Crossing a pea plant that has round seeds with one that has wrinkled seeds yields *F₁* progeny that all have round seeds. Crossing these *F₁* peas yields an *F₂* generation, of which three quarters have round seeds and one quarter have wrinkled seeds.

over green seeds). The *F₂* phenotypes appear in the ratio 9 round yellow: 3 round green: 3 wrinkled yellow: 1 wrinkled green. This result indicates that there is no tendency for the genes from any parent to assort together (Fig. 27-6). It was later shown, however, that *only genes that occur on different chromosomes exhibit such independence.*

The dominance of one trait over another is a common but not universal phenomenon. For example, crossing a pure-breeding red variety of the snapdragon *Antirrhinum* with a pure-breeding white variety results in pink colored *F₁* progeny. The *F₂* progeny have red, pink, and white flowers in 1 : 2 : 1 ratio because the flowers of homozygotes for the red color (*AA*) contain more red pigment than do the heterozygotes (*Aa*; Fig. 27-7). The red and white traits are therefore said to be **codominant.** In the case of codominance, the phenotype reveals the genotype.

A given gene may have multiple alleles. A familiar example is the human ABO blood group system (Section

P generation

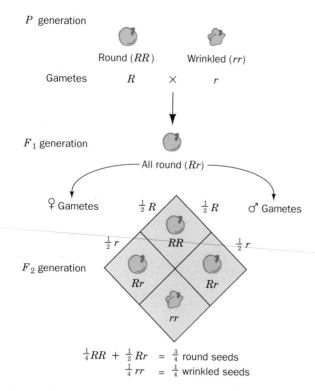

Round (*RR*)   Wrinkled (*rr*)

Gametes R × r

F_1 generation

All round (*Rr*)

♀ Gametes $\frac{1}{2}R$ $\frac{1}{2}R$ ♂ Gametes

F_2 generation

$$\frac{1}{4}RR + \frac{1}{2}Rr = \frac{3}{4} \text{ round seeds}$$
$$\frac{1}{4}rr = \frac{1}{4} \text{ wrinkled seeds}$$

Figure 27-5
In a genetic cross between peas with round seeds and peas with wrinkled seeds the F_1 generation has the round seed phenotype because round seeds are dominant over wrinkled seeds. The F_2 generation's seeds are $\frac{3}{4}$ round and $\frac{1}{4}$ wrinkled because the genes for these alleles are independently transmitted by haploid gametes.

11-3D). The A and B antigens are specified by the co-dominant I^A and I^B alleles, respectively, and the O phenotype is homozygous for the recessive i allele. The different blood types, it will be recalled, arise from the action of a glycosyltransferase that, if specified by an I^A, I^B, or i allele, respectively, transfers an N-acetylgalactos-amine residue, a galactose residue, or is inactive.

C. Chromosomal Theory of Inheritance

Mendel's theory of inheritance was almost universally ignored by his contemporaries. This was partially because in analyzing his data he used probability theory, an alien subject to most biologists of the time. The major reason his theory was ignored, however, is that it was ahead of its time: Contemporary knowledge of anatomy and physiology provided no basis for its understanding. For instance, mitosis and meiosis had yet to be discovered. Yet, after Mendel's work was rediscovered in 1900, it was shown that his principles explained inheritance in animals as well as in plants. In 1903, as a result of the realization that chromosomes and genes behaved in a parallel fashion, Walter Sutton formulated the **chromosomal theory of inheritance** in which he hypothesized that genes are parts of chromosomes.

The first trait to be assigned a chromosomal location was that of sex. *In most eukaryotes, the cells of females each contain two copies of the* **X** *chromosome* (XX), *whereas male cells contain one copy of X and a morphologically distinct* **Y** *chromosome* (XY; Fig. 27-8). Ova must therefore contain a single X chromosome while sperm contain either an X or a Y chromosome (Fig. 27-8). Fertilization by an X-bearing sperm therefore results in a female zygote and by a Y-bearing sperm yields a male zygote. This explains the observed 1:1 ratio of males to females in most species. The X and Y chromosomes are referred to as **sex chromosomes;** the others are known as **autosomes.**

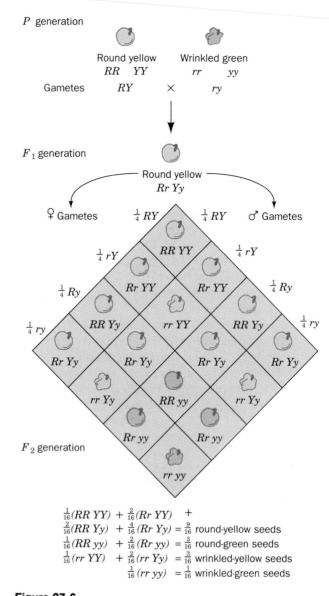

$$\frac{1}{16}(RR\ YY) + \frac{2}{16}(Rr\ YY)\ +$$
$$\frac{2}{16}(RR\ Yy) + \frac{4}{16}(Rr\ Yy) = \frac{9}{16} \text{ round-yellow seeds}$$
$$\frac{1}{16}(RR\ yy) + \frac{2}{16}(Rr\ yy) = \frac{3}{16} \text{ round-green seeds}$$
$$\frac{1}{16}(rr\ YY) + \frac{2}{16}(rr\ Yy) = \frac{3}{16} \text{ wrinkled-yellow seeds}$$
$$\frac{1}{16}(rr\ yy) = \frac{1}{16} \text{ wrinkled-green seeds}$$

Figure 27-6
The genes for round (*R*) versus wrinkled (*r*) and yellow (*Y*) versus green (*y*) pea seeds assort independently. The F_2 progeny consist of nine genotypes comprising the four possible phenotypes.

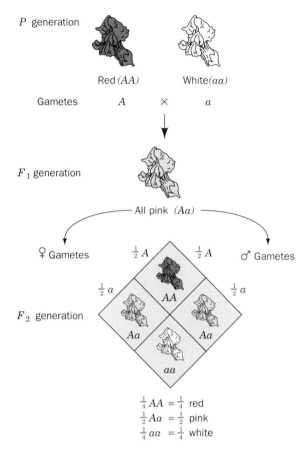

Figure 27-7
In a cross between snapdragons with red (*AA*) and white (*aa*) flowers, the F_1 generation is pink (*Aa*), which demonstrates that the two alleles, *A* and *a*, are codominant. The F_2 flowers are red, pink, and white in 1 : 2 : 1 ratio.

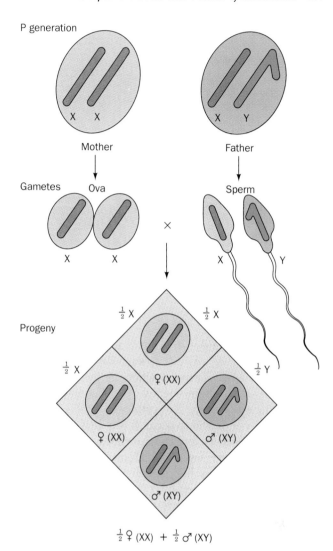

Figure 27-8
The independent segregation of the sex chromosomes, X and Y, results in a 1 : 1 ratio of males to females.

Fruit Flys Are Favorite Genetic Subjects

The pace of genetic research greatly accelerated after Thomas Hunt Morgan began using the fruit fly *Drosophila melanogaster* as an experimental subject. This small prolific insect (Fig. 27-9), which is often seen hovering around ripe fruit in summer and fall, is easily maintained in the laboratory where it produces a new generation every 14 days. With *Drosophila*, the results of genetic crosses can be determined some 25 times faster than they can with peas. *Drosophila* is presently the genetically best characterized higher organism.

The first known mutant strain of *Drosophila* had white eyes rather than the red eyes of the **wild-type** (occurring in nature). Through genetic crosses of the white eye strain with the wild-type, Morgan showed that the distribution of the white eye gene (*wh*) parallels that of the X chromosome. This indicates that the *wh* gene is located on the X chromosome and that the Y chromosome does not contain it. The *wh* gene is therefore said to be **sex linked**.

Genetic Maps Can Be Constructed from an Analysis of Cross-Over Rates

In succeeding years, the chromosomal locations of many *Drosophila* genes were determined. Those genes

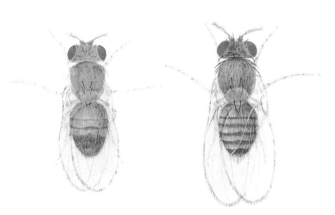

Figure 27-9
The fruit fly *Drosophila melangoster.* The male (*left*) and the female (*right*) are shown in their relative sizes; they are actually ~2 mm long and weigh ~1 mg.

(a)

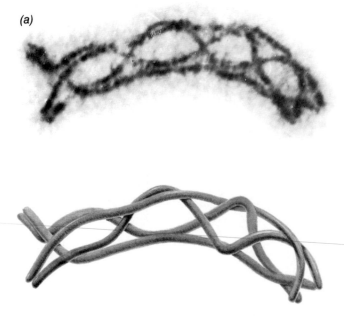

(b)

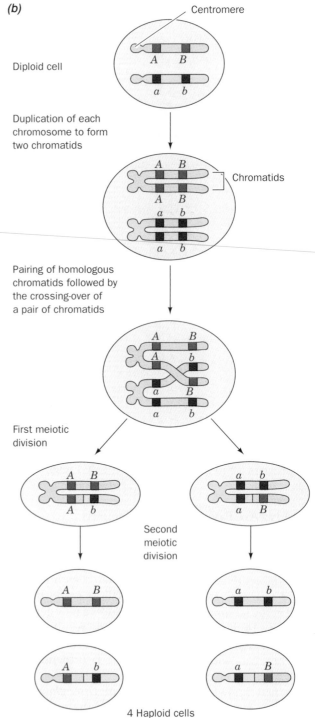

that reside on the same chromosome do not assort independently. However, any pair of such **linked** genes **recombine** (exchange relative positions with their allelic counterparts on the homologous chromosome) with a characteristic frequency. The cytological basis of this phenomenon was found to occur at the start of meiosis when the homologous doubled chromosomes line up in parallel (Metaphase I; Fig. 27-3). Homologous chromatids are observed to exchange equivalent sections in a process known as **crossing-over** (Fig. 27-10). The chromosomal location of the cross-over point varies nearly randomly from event to event. Consequently, *the cross-over frequency of a pair of linked genes varies directly with their physical separation along the chromosome.* Morgan and Alfred Sturtevant made use of this phenomenon to **map** (locate) the relative positions of genes on *Drosophila*'s four unique chromosomes. Such studies have demonstrated that *chromosomes are linear unbranched structures.* We now know that such **genetic maps** (Fig. 27-11) parallel the corresponding base sequences of the DNA within the chromosomes.

Nonallelic Genes Complement One Another

Whether or not two recessive traits that affect similar functions are allelic (different forms of the same gene) can be determined by a **complementation test.** In this test, a homozygote for one of the traits is crossed with a homozygote for the other. If the two traits are nonallelic, the progeny will have the wild-type phenotype because each of the homologous chromosomes supplies the wild-type function that the other lacks; that is, they complement each other. For example, crossing a *Drosophila* that is homozygous for an eye color mutation known as purple (*pr*) with a homozygote for another eye color mutation known as brown (*bw*) yields progeny with wild-type eye color, thereby demonstrating that

Figure 27-10

Crossing-over. (*a*) An electron micrograph, together with an interpretive drawing, of two homologous pairs of chromatids during meiosis in the grasshopper *Chorthippus parallelus*. Nonsister chromatids (*different colors*) may recombine at any of the points where they cross-over. [Courtesy of Bernard John, The Australian National University.] (*b*) A diagram showing the recombination of pairs of allelic genes (*A, B*) and (*a, b*) during crossover.

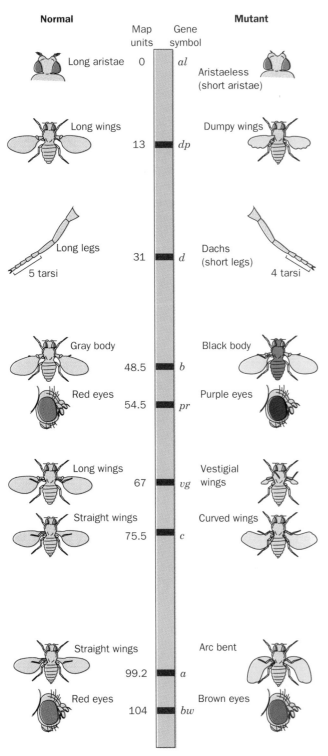

Figure 27-11
A portion of the genetic map of chromosome 2 of *Drosophila*. The positions of the genes are given in map units. Two genes separated by *m* map units recombine with a frequency of *m*%.

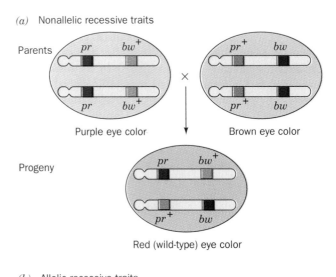

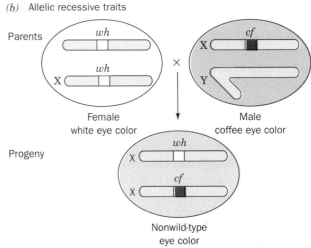

Figure 27-12
The complementation test indicates whether two recessive traits are allelic. Two examples in *Drosophila* are (*a*) Crossing a homozygote for purple eye color (*pr*) with a homozygote for brown eye color (*bw*) yields progeny with wild-type eye color. This indicates that *pr* and *bw* are nonallelic. Here the superscript "+" indicates the wild-type allele. (*b*) In crossing a female that is homozygous for the sex-linked white eye color gene *wh* with a male bearing the sex-linked coffee eye color gene *cf*, the female progeny do not have the wild-type eye color. The *wh* and *cf* genes must therefore be allelic.

these two genes are not allelic (Fig. 27-12*a*). In contrast, in crossing a female *Drosophila* that is homozygous for the sex-linked white eye color allele (*wh*) with a male carrying the sex-linked coffee eye color allele (*cf*), the female progeny do not have wild-type eye color (Fig. 27-12*b*). The *wh* and *cf* genes must therefore be allelic.

Genes Direct Protein Expression

The question of how genes control the characteristics of organisms took some time to be answered. Archibald Garrod was the first to suggest a specific connection between genes and enzymes. Individuals with **alkaptonuria** produce urine that darkens alarmingly on exposure to air, a consequence of the oxidation of the **homogentisate** they excrete (Section 15-3A). In 1902, Garrod showed that this rather benign metabolic disorder (its only adverse effect is arthritis in later life) results from a recessive trait that is inherited in a Mendelian fashion. He further demonstrated that alcaptonurics are unable to metabolize the homogentisate fed to them and therefore concluded that *they lack an enzyme that metabolizes this substance.* This enzyme is now known to be homogentisate oxidase, which is involved in the degradation of phenylalanine and tyrosine (Section 24-3H). Garrod described alcaptonuria and several other inherited human diseases he had studied as **inborn errors of metabolism.**

Beginning in 1940, George Beadle and Edward Tatum, in a series of investigations that mark the beginning of biochemical genetics, showed that *there is a one-to-one correspondence between a mutation and the lack of a specific enzyme.* The wild-type mold *Neurospora* grows on a "minimal medium" in which the only sources of carbon and nitrogen are glucose and NH_3. Certain mutant varieties of *Neurospora* that were generated by means of irradiation with X-rays, however, require an additional substance, such as arginine or thiamine, in order to grow. Beadle and Tatum demonstrated, in several cases, that the mutants lack a normally present enzyme that participates in the biosynthesis of the required substance (Section 15-3A). This resulted in their famous maxim **one gene–one enzyme.** Today we know this principle to be only partially true since many genes specify proteins that are not enzymes and many proteins consist of several independently specified subunits (in humans, e.g., the α and β subunits of hemoglobin are specified by genes that reside on different chromosomes). A more accurate dictum might be **one gene–one polypeptide.** Yet, even this is not completely accurate because RNAs with structural and functional roles, such as transfer RNA and ribosomal RNA, are also genetically specified.

D. Bacterial Genetics

Bacteria offer several advantages for genetic study. Foremost of these is that *under favorable conditions, many have generation times of under 20 min. Consequently, the results of a genetic experiment with bacteria can be ascertained in a matter of hours rather than the weeks or years required for an analogous study with higher organisms. The tremendous number of bacteria that can be quickly grown* ($\sim 10^{10} \ mL^{-1}$) *permits the observation of extremely rare biological events.* For example, an event that occurs with a frequency of 1 per million can be readily detected in bacteria, with only a few minutes work. To do so in *Drosophila* would be an enormous and probably futile

effort. Moreover, bacteria are usually haploid so that their phenotype indicates their genotype. Nevertheless, the basic principles of genetics were elucidated from the study of higher plants and animals. This is because bacteria do not reproduce sexually in the manner of higher organisms so that the basic technique of classical genetics, the genetic cross, is not normally applicable to bacteria (but see below). In fact, before it was shown that DNA is the carrier of hereditary information, it was not altogether clear that bacteria had chromosomes.

The study of bacterial genetics effectively began in the 1940s when procedures were developed for isolating bacterial mutants. Since bacteria have few easily recognized morphological features, *their mutants are usually detected (selected for) by their ability or inability to grow under certain conditions.* For example, wild-type *E. coli* can grow on a medium in which glucose is the only carbon source. Mutants that are unable to synthesize methionine, for example, require the presence of methionine in their growth media. Mutants that are resistant

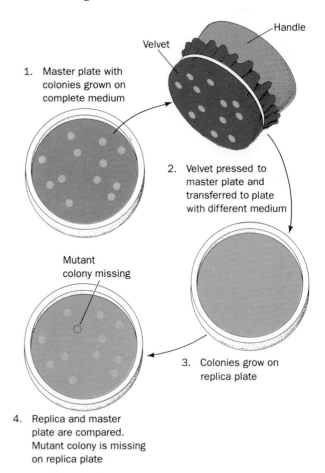

1. Master plate with colonies grown on complete medium

2. Velvet pressed to master plate and transferred to plate with different medium

3. Colonies grow on replica plate

Mutant colony missing

4. Replica and master plate are compared. Mutant colony is missing on replica plate

Figure 27-13
Replica plating is a technique for rapidly and conveniently transferring colonies from a "master" culture plate (Petri dish) to a different medium on another culture plate. Since the colonies on the master plate and on the replicas should have the same spatial distribution, it is easy to identify the desired mutants.

to an antibiotic, say ampicillin, can grow in the presence of this antibiotic whereas the wild-type cannot. Mutants in which an essential protein has become temperature sensitive grow at 30 but not 42°C, whereas the wild-type grows at either temperature. By using a suitable screening protocol, a bacterial colony containing a particular mutation or combination of mutations can be selected. This is conveniently done by the method of **replica plating** (Fig. 27-13).

Bacterial Chromosomes Have Been Mapped through Interrupted Mating

In 1946, Joshua Lederberg and Tatum discovered *that some bacteria can transfer genetic information to others through a process known as* **conjugation.** The ability to conjugate ("mate") is conferred on otherwise indifferent bacteria by a **plasmid** (a DNA molecule distinct from the bacterial chromosome that is replicated by the cell; Section 28-8A) called an **F factor** (for fertility). Bacteria that possess an F factor (designated F$^+$ or male) are covered by hairlike projections known as **F pili.** These bind to cell-surface receptors on bacteria that lack the F factor (F$^-$ or female), which leads to the formation of a cytoplasmic bridge between the cells (Fig. 27-14). The F factor then replicates and, as the newly replicated single strand is formed, it passes through the cytoplasmic bridge to the F$^-$ cell where the complementary strand is synthesized (Fig. 27-15). This converts the F$^-$ cell to F$^+$ so that the F factor is an infectious agent (a bacterial venereal disease?).

On very rare occasions, the F factor spontaneously integrates into the chromosome of the F$^+$ cell (plasmids with this capability are termed **episomes**). In the resulting **Hfr** (for *high frequency of recombination*) cells, the F factor behaves much as it does in the autonomous state. Its replication commences at a specific internal point in the F factor, and the replicated section passes through a cytoplasmic bridge to the F$^-$ cell where its complemen-

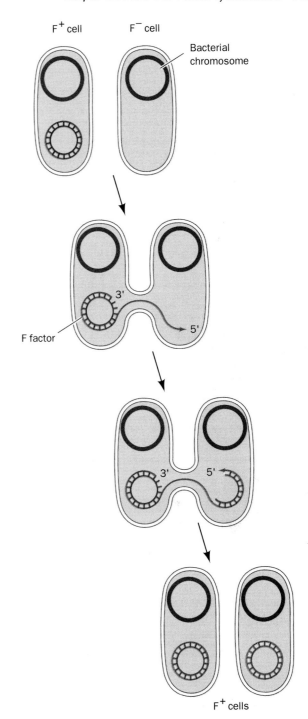

Figure 27-15
A diagram showing how an F$^-$ cell acquires an F factor from an F$^+$ cell. A single strand of the F factor is replicated, via the rolling circle mode (Section 31-3B), and is transferred to the F$^-$ cell where its complementary strand is synthesized to form a new F factor.

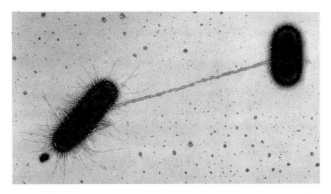

Figure 27-14
An electron micrograph of an F$^+$ (*left*) and an F$^-$ (*right*) *E. coli* engaged in sexual conjugation. The F pili have been made more visible by the addition of male-specific bacteriophages, which infect F$^+$ *E. coli* by adsorbing to their F pili. [Courtesy of Charles Brinton, University of Pittsburgh.]

tary strand is synthesized. In this case, however, *the replicated chromosome of the Hfr cell is also transmitted to the F⁻ cell (Fig. 27-16)*. Usually, only part of the Hfr bacterial chromosome is transferred during sexual conjugation because the cytoplasmic bridge almost always breaks off sometime during the ~90 min required to complete the transfer process. In the resulting **merozygote** (a partially diploid bacterium), the chromosomal fragment, which lacks a complete F factor, neither transforms the F⁻ cell to Hfr nor is subsequently replicated. However, *the transferred chromosomal fragment*

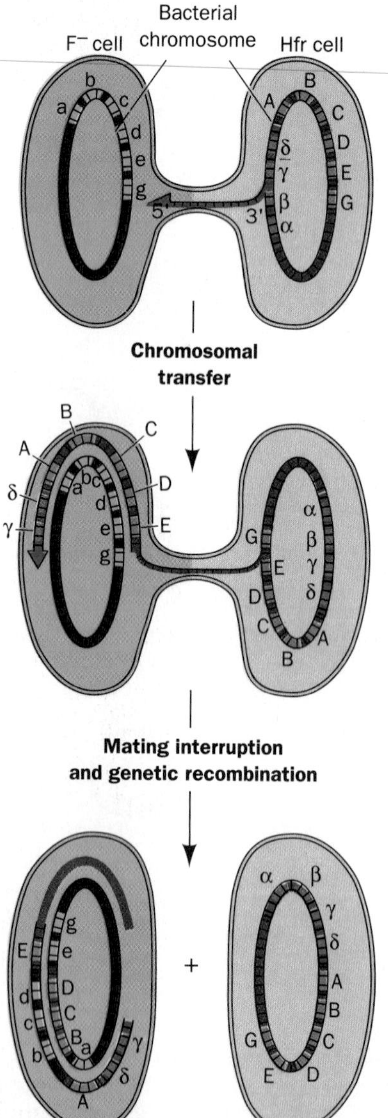

Figure 27-16
The transfer of the bacterial chromosome from an Hfr cell to an F⁻ cell and its subsequent recombination with the F⁻ chromosome. Here, Greek letters represent F factor genes, upper case Roman letters represent bacterial genes from the Hfr cell, and lower case Roman letters represent the corresponding alleles in the F⁻ cell. Since chromosomal transfer, which begins within the F factor, is rarely complete, the entire F factor is seldom transferred. Hence the recipient cell usually remains F⁻.

recombines with the chromosome of the F⁻ cell in a manner similar to chromosomal crossing-over in eukaryotes, thereby permanently endowing the F⁻ cell with some of the traits of the Hfr strain.

Bacterial genes are transferred from the Hfr cell to the F⁻ cell in fixed order. This is because the F factor in a given Hfr strain is integrated into the bacterial chromosome at a specific site and because only a particular strand of the Hfr chromosomal DNA is replicated and transferred to the F⁻ cell. The bacterial chromosome can therefore be mapped by the following procedure. Hfr and F⁻ strains containing different alleles of the genes to be mapped are mixed, permitted to conjugate, and, after a certain time, the conjugation is interrupted by violent agitation in a kitchen blender so as to break the cytoplasmic bridges between conjugating cells (**interrupted mating;** Fig. 27-17). Subsequent screening procedures reveal which allelic genes have entered the F⁻ cell and recombined with its chromosome. *By interrupting the conjugation after various times, the order of the allelic genes can be determined* (Fig. 27-18). The difficulty in determining the order of genes located towards the end of the rarely completely transferred chromosome is circumvented by using several Hfr strains that differ according to the point at which the F factor is integrated into the chromosome. This mapping procedure demonstrated that the *E. coli* chromosome is circular as is its DNA (Fig. 27-18). Other bacterial chromosomes have been similarly mapped.

The integrated F factor in an Hfr cell occasionally undergoes spontaneous excision to yield an F⁺ cell. In rare instances, the F factor is aberrantly excised such that a portion of the adjacent bacterial chromosome is incorporated in the subsequently autonomously replicating F factor. Bacteria carrying such a so-called **F′ factor** are permanently diploid for its bacterial genes.

E. Viral Genetics

*Viruses are infectious particles consisting of a nucleic acid molecule enclosed by a protective **capsid** (coat) that consists largely or entirely of protein.* A virus specifically adsorbs to a susceptible cell into which it insinuates its nucleic acid. Over the course of the infection (Fig. 27-19), the viral chromosome redirects the cell's metabolism so as to produce new viruses. A viral infection usually culminates in the lysis of the host cell thereby releasing large numbers (tens to thousands) of mature virus particles that can each initiate a new round of infection. Viruses, having no metabolism of their own, are the ultimate parasites. They are not living organisms since, in the absence of their host, they are as biologically inert as any other large molecule.

The Fine Details of a Bacterial Genetic Map Can Be Elucidated through Transductional Mapping

Certain species of **bacteriophages** (viruses infecting

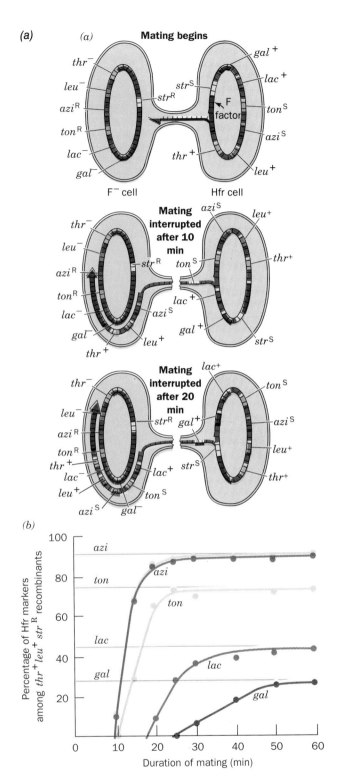

Figure 27-17
An example of the mapping of a bacterial chromosome by interrupted mating. (a) The ordered transfer of an Hfr chromosome to an F⁻ cell. Bacterial mating is interrupted at various times by agitation in a blender. Bacteria carrying the mutant alleles *thr⁻* and *leu⁻*, respectively, require the amino acids threonine and leucine in their growth media; those with *gal⁻* and *lac⁻* are, respectively, unable to grow on media containing the sugars galactose or lactose as their only carbon source; *azi*ᴿ confers resistance to azide; *ton*ᴿ confers resistance to **bacteriophage T1;** and *str*ᴿ confers resistance to the antibiotic streptomycin. The superscripts +, R and S indicate wild-type, "resistance" and "sensitivity," respectively. (b) The frequencies of occurence of the genetic markers *azi, ton, lac,* and *gal* in recombinants as a function of mating time. After their mating was interrupted, the bacteria were grown on a medium containing streptomycin with glucose as the only carbon source to select for recombinants containing the *thr⁺, leu⁺,* and *str*ᴿ alleles. These recombinants were scored for their sensitivity to azide, bacteriophage T1, or their ability to grow on lactose or galactose as their only carbon source. Extrapolation to zero of the frequency of bacterial colonies on the various restrictive media indicates the earliest times that the corresponding alleles became available for recombination with the F⁻ chromosome. [After Jacob, F. and Wollman, E. L., *Sexuality and the Genetics of Bacteria,* p. 135, Academic Press (1961).]

bacteria, **phages** for short; Greek: *phagein,* to eat) have been useful in elucidating bacterial genetics. In an infection by such bacteriophages, about one in every thousand progeny particles contains a segment of the bacterial chromosome in place of the viral chromosome. These defective phage particles can inject their DNA into another bacterial cell but this does not kill the bacterium since the viral genome is absent. The transferred chromosomal segment can, however, recombine with homologous portions of the bacterium's chromosome. This phage-mediated recombinational process is known as **transduction.**

A transducing phage can contain no more than a capsid full of DNA [typically 50,000 base pairs **(bp)**] so that it can only transduce genes that are no further than this distance apart (maximally 2 min) on a bacterial chromosome. Consequently, *the relative frequency with which closely linked bacterial genes are cotransduced accurately reflects their separation on the bacterial chromosome.* The finer details of the *E. coli* genetic map shown in Fig. 27-18 were elucidated by such **transductional mapping** using **bacteriophage P1.**

Viruses Are Subject to Complementation and Recombination

The genetics of viruses can be studied in much the same way as that of cellular organisms. Since viruses have no metabolism, however, their presence is usually detected by their ability to kill their host. The presence of

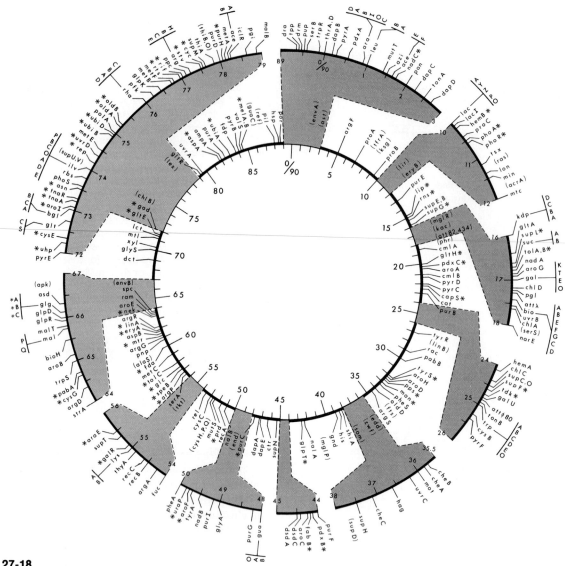

Figure 27-18
The genetic map of *E. coli* based on the results of interrupted mating experiments with the finer details determined by transductional mapping procedures. The inner circle indicates gene transfer time, in minutes, with the *thrA* locus arbitrarily placed at 0. The outer circle displays crowded sections of the map in expanded form. The 310 genes mapped account for ~10% of the potential information content of the *E. coli* genome. Since the time this map was made (1970), many hundreds more *E. coli* genes have been discovered and mapped. [Courtesy of Austin L. Taylor, University of Colorado.]

viable bacteriophages is conveniently indicated by **plaques** (clear spots) on a "lawn" of bacteria on a culture plate (Fig. 27-20). Plaques mark the spots where single phage particles had multiplied with the resulting lysis of the bacteria in the area. A mutant phage, which can produce progeny under certain **permissive conditions,** is detected by its inability to do so under other **restrictive conditions** in which the wild-type phage is viable. These conditions usually involve differences in the strain of the bacterial host employed or in the temperature.

Viruses are subject to complementation. Simultaneous infection of a bacterium by two different mutant varieties of a phage may yield progeny under conditions in which neither variety by itself can reproduce. If this occurs, then each mutant phage must have supplied a function that could not be supplied by the other. Each such mutation is said to belong to a different **complementation group,** a term synonomous for gene.

Viral chromosomes are also subject to recombination. This occurs when a single cell is simultaneously infected by two mutant strains of a virus (Fig. 27-21). The dynamics of viral recombination differ from those in eukaryotes or bacteria because the viral chromosome undergoes recombination throughout the several rounds of DNA replication that occur during the viral life cycle.

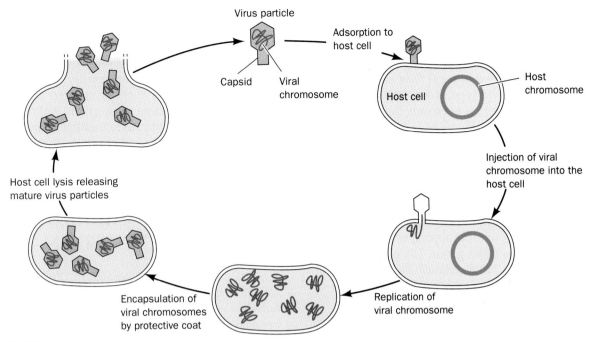

Figure 27-19
The life cycle of a virus.

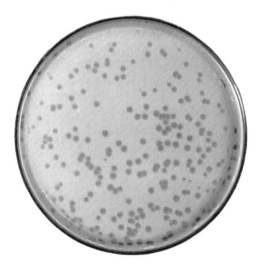

Figure 27-20
A culture plate covered with a lawn of *E. coli* on which bacteriophage T4 has formed plaques. [Bruce Iverson.]

Recombinant viral progeny therefore consist of many if not all of the possible recombinant types.

The Recombinational Unit Is a Base Pair

The enormous rate at which bacteriophages reproduce permits the detection of recombinational events that occur with a frequency of as little as one in 10^8. In the 1950s, Seymour Benzer carried out high resolution genetic studies of the *rII* region of the **bacteriophage T4** chromosome. This ~ 4000 bp region, which represents ~ 2%

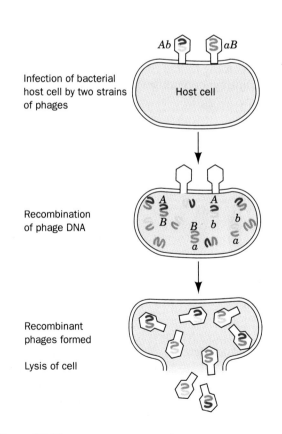

Figure 27-21
Recombination of bacteriophage chromosomes occurs on simultaneous infection of a bacterial host by two phage strains carrying the genes *Ab* and *aB*.

of the T4 chromosome, consists of two adjacent complementation groups designated *rIIA* and *rIIB*. In a permissive host, *E. coli* B, a mutation that inactivates the product of either gene causes the formation of plaques that are easily identified because they are much larger than those of the wild-type phage (the designation *r* stands for rapid lysis). However, only the wild-type will lyse the restrictive host, *E. coli* K12(λ). The presence of plaques in an *E. coli* K12(λ) culture plate that had been simultaneously infected with two different *rII* mutants in the same complementation group demonstrated that *recombination can take place within a gene.* This refuted a then widely held model of the chromosome in which genes were thought to be discrete entities, rather like beads on a string, such that recombination could take place only between intact genes. The genetic mapping of mutations at over 300 distinguishable sites in the *rIIA* and *rIIB* regions indicated that *genes, as are chromosomes, are linear unbranched structures.*

Benzer also demonstrated that a complementation test between two mutations on the same complementation group yields progeny in the restrictive host when the two mutations are in the **cis** configuration (on the same chromosome; Fig. 27-22*a*), but fails to do so when they are in the **trans** configuration (on physically different chromosomes; Fig. 27-22*b*). This is because only when both mutations physically occur in the same gene will the other gene be functionally intact. The term **cistron** was coined to mean a functional genetic unit defined according to this **cis–trans test.** This word has since become synonomous with gene or complementation group.

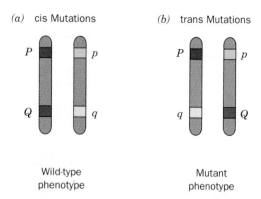

(a)   cis Mutations   (b)   trans Mutations

Wild-type
phenotype   Mutant
                             phenotype

Figure 27-22
The cis–trans test. Consider a chromosome that is present in two copies in which two positions on the same gene, *P* and *Q*, have defective (recessive) mutants, *p* and *q*, respectively. (a) If the two mutations are cis (physically on the same chromosome), one gene will be wild-type so that the organism will have a wild-type phenotype. (b) If the mutations are trans (on physically different chromosomes), both genes will be defective and the organism will have a mutant phenotype.

The recombination of pairs of *rII* mutants was observed to occur at frequencies as low as 0.01% (although frequencies as low as 0.0001% could, in principle, be detected). Since a recombination frequency in T4 of 1% corresponds to a 240 bp separation of mutation sites, the unit of recombination can be no larger than $0.01 \times 240 = 2.4$ bp. For reasons having to do with the mechanism of recombination, this is an upper limit estimate. On the basis of high resolution genetic mapping, it was therefore concluded that *the unit of recombination is about the size of a single base pair.*

2. DNA IS THE CARRIER OF GENETIC INFORMATION

Nucleic acids were first isolated in 1869 by Friedrich Miescher and so-named because he found them in the nuclei of **leukocytes** (pus cells) from discarded surgical bandages. The presence of nucleic acids in other cells was demonstrated within a few years but it was not until some 75 years after their discovery that their biological function was elucidated. Indeed, in the 1930s and 1940s, it was widely held, in what was termed the **tetranucleotide hypothesis,** that nucleic acids have a monotonously repeating sequence of all four bases so that they were not suspected of having a genetic function. Rather, it was generally assumed that genes were proteins since proteins were the only biochemical entities that, at that time, seemed capable of the required specificity. In this section, we outline the experiments that established DNA's genetic role.

A. Transforming Principle Is DNA

The virulent form of pneumococcus (*Diplococcus pneumoniae*), a bacterium that causes pneumonia, is encapsulated by a gelatinous polysaccharide coating that contains the O antigens through which it recognizes the cells it infects (Section 10-3B). Mutant pneumococci that lack this coating, because of a defect in an enzyme involved in its formation, are not pathogenic. The virulent and nonpathogenic pneumococci are known as the S and R forms, respectively, because of the smooth and rough appearances of their colonies in culture (Fig. 27-23).

In 1928, Frederick Griffith made a startling discovery. He injected mice with a mixture of live R and heat-killed S pneumococci. This experiment resulted in the death of most of the mice. More surprising yet was that the blood of the dead mice contained live S pneumococci. The dead S pneumococci initially injected into the mice had somehow **transformed** the otherwise innocuous R pneumococci to the virulent S form. Furthermore, the progeny of the transformed pneumococci were also S; the transformation was permanent. Eventually, it was shown that the transformation could also be made *in*

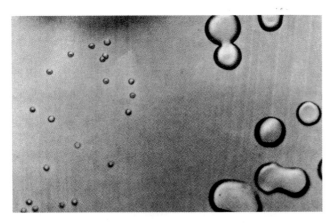

Figure 27-23
The large glistening colonies are virulent S-type pneumococci that resulted from the transformation of nonpathogenic R-type pneumococci (smaller colonies) by DNA from heat-killed S pneumococci. [From Avery, O. T., MacLeod, C. M., and McCarty, M., *J. Exp. Med.* **79,** 153 (1944). Copyright © 1944 by Rockefeller University Press.]

vitro by mixing R cells with a cell-free extract of S cells. The question remained: What is the nature of the **transforming principle?**

In 1944, Oswald Avery, Colin MacLeod, and Maclyn McCarty, after a 10-year investigation, reported that *transforming principle is DNA.* The conclusion was based on the observations that the laboriously purified (few modern fractionation techniques were then available) transforming principle had all the physical and chemical properties of DNA, contained no detectable protein, was unaffected by trypsin, chymotrypsin or ribonuclease, and was totally inactivated by treatment with DNase. *DNA must therefore be the carrier of genetic information.*

Avery's discovery was another idea whose time had not yet come. This seminal advance was initially greeted with skepticism and then largely ignored. Indeed, even Avery did not directly state that DNA is the hereditary material but merely that it has "biological specificity." His work, however, influenced several biochemists, including Erwin Chargaff whose subsequent accurate determination of DNA base ratios (Section 28-1) refuted the tetranucleotide hypothesis and thereby indicated that DNA could be a complex molecule.

It was eventually demonstrated that eukaryotes are also subject to transformation by DNA. Thus DNA, which cytological studies had shown resides in the chromosomes, must also be the hereditary material of eukaryotes. In a spectacular demonstration of eukaryotic transformation, Ralph Brinster, in 1982, microinjected DNA bearing the gene for rat **growth hormone** (a polypeptide) into the nuclei of fertilized mouse eggs and implanted these eggs into the uteri of foster mothers. The resulting "supermice" (Fig. 27-24), which had high levels of rat growth hormone in their serum, grew to nearly twice the weight of their normal litter mates. More recently, a surprisingly simple method for transforming eukaryotes has been reported. The live sperm of mice take up foreign DNA. Up to 30% of the mouse eggs that are fertilized with these "transformed" sperm develop into adult organisms into whose chromosomes the DNA is permanently (heritably) incorporated. Such genetically altered organisms are said to be **transgenic.**

B. The Hereditary Molecule of Bacteriophage Is DNA

Electron micrographs of phage-infected bacteria show empty-headed phage "ghosts" attached to the bacterial surface (Fig. 27-25). This observation led Roger

Figure 27-24
The gigantic mouse (*left*) grew from a fertilized ovum that had been microinjected with DNA bearing the rat growth hormone gene. His normal littermate (*right*) is shown for comparison. [Courtesy of Ralph Brinster, University of Pennsylvania.]

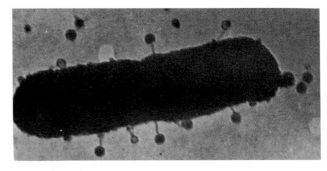

Figure 27-25
Electron micrograph of an *E. coli* cell to which **bacteriophage T5** are adsorbed by their tails. (Courtesy of Thomas F. Anderson, Fox Chase Cancer Center.)

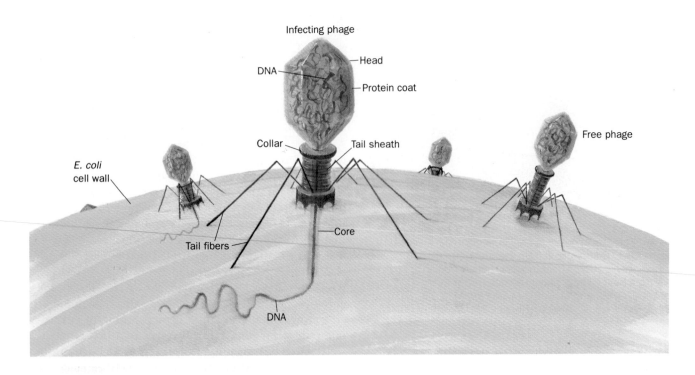

Figure 27-26
A diagram of T2 bacteriophage injecting its DNA into an *E. coli* cell.

Herriott to suggest "that the virus may act like a little hypodermic needle full of transforming principle," which it injects into the bacterial host (Fig. 27-26). This proposal was tested in 1952 by Alfred Hershey and Martha Chase as is diagrammed in Fig. 27-27. **Bacteriophage T2** was grown on *E. coli* in a medium containing the radioactive isotopes ^{32}P and ^{35}S. This labeled the phage capsid, which contains no P, with ^{35}S, and its DNA, which contains no S, with ^{32}P. These phages were added to an unlabeled culture of *E. coli* and, after allowing sufficient time for the phages to infect the bacterial cells, the culture was agitated in a kitchen blender so as to shear the phage ghosts from the bacterial cells. This rough treatment neither injured the bacteria nor altered the course of the phage infection. When the phage ghosts were separated from the bacteria by centrifugation, the ghosts were found to contain most of the ^{35}S whereas the bacteria contained most of the ^{32}P. Furthermore, 30% of the ^{32}P appeared in the progeny phages but only 1% of the ^{35}S did so. Hershey and Chase therefore concluded that only the phage DNA was essential

for the production of progeny. *DNA therefore must be the hereditary material.* In later years it was shown that, in a process known as **transfection**, purified phage DNA can, by itself, induce a normal phage infection in a properly treated bacterial host (transfection differs from transformation in that the latter results from the recombination of the bacterial chromosome with a fragment of homologous DNA).

In 1952, the state of knowledge of biochemistry was such that Hershey's discovery was much more readily accepted than Avery's identification of the transforming principle had been some 8 years earlier. Within a few months, the first speculations arose as to the nature of the genetic code, and James Watson and Frances Crick were inspired to investigate the structure of DNA. In 1955, it was shown that the somatic cells of eukaryotes have twice the DNA of the corresponding germ cells. When this observation was proposed to be a further indicator of DNA's genetic role, there was little comment even though the same could be said of any other chromosomal component.

Chapter Summary

Eukaryotic cells contain a characteristic number of homologous pairs of chromosomes. In mitosis, each daughter cell receives a copy of each of these chromosomes but in meiosis each resulting gamete receives only one member of each homologous pair. Fertilization is the fusion of two haploid ga-

metes to form a diploid zygote.

The Mendelian laws of inheritance state that alternative forms of true-breeding traits are specified by different alleles of the same gene. Alleles may be dominant, codominant, or recessive depending on the phenotype of the heterozygote.

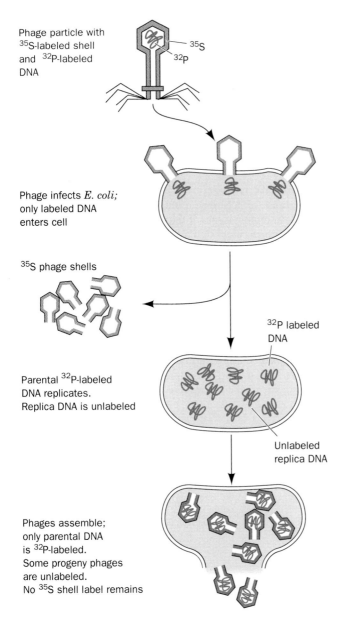

Phage particle with
^{35}S-labeled shell
and ^{32}P-labeled
DNA

^{35}S

^{32}P

Phage infects *E. coli;*
only labeled DNA
enters cell

^{35}S phage shells

^{32}P labeled
DNA

Parental ^{32}P-labeled
DNA replicates.
Replica DNA is unlabeled

Unlabeled
replica DNA

Phages assemble;
only parental DNA
is ^{32}P-labeled.
Some progeny phages
are unlabeled.
No ^{35}S shell label remains

Figure 27-27
A diagram of the Hershey–Chase experiment demonstrating that only the nucleic acid component of bacteriophages enters the bacterial host during phage infection.

Different genes assort independently unless they are on the same chromosome. The linkage between genes on the same chromosome, however, is never complete because of crossing-over among homologous chromosomes during meiosis. The rate that genes recombine varies with their physical separation because crossing-over occurs essentially at random. This permits the construction of genetic maps. Whether two recessive traits are allelic may be determined by the complementation test. The nature of genes is largely defined by the dictum "one gene–one polypeptide."

The rapid reproduction rate of bacteria and bacteriophages permits the detection of extremely rare genetic events. F^+ strains of bacteria transfer a copy of their F factor to F^- cells through conjugation. In Hfr cells, the F factor is integrated into the bacterial chromosome. Hfr cells transfer the leading section of their F factor together with a portion of the attached bacterial chromosome, in fixed order, to the F^- cell where the chromosomal fragment recombines with the F^- chromosome. This permits the genetic mapping of bacterial chromosomes by interrupting the mating process. The fine structures of these maps may be deduced through transductional mapping.

Mutant varieties of bacteriophages are detected by their ability to kill their host under various restrictive conditions. The fine structure analysis of the *rII* region of the bacteriophage T4 chromosome has revealed that recombination may take place within a gene, that genes are linear, unbranched structures, and that the unit of mutation is ~1 bp.

Extracts of virulent S-type pneumococci transform nonpathogenic R-type pneumococci to the S form. The transforming principle is DNA. Similarly, radioactive labeling demonstrated that the genetically active substance of bacteriophage T2 is its DNA. The viral capsid serves only to protect its enclosed DNA and to inject it into the bacterial host. This demonstrates that DNA is the hereditary molecule.

References

Genetics

Ayala, F. J. and Kiger, J. A. Jr., *Modern Genetics* (2nd ed.), Benjamin/Cummings (1984).

Benzer, S., The fine structure of the gene, *Sci. Am.* **206**(1): 70–84 (1962).

Cairns, J., Stent, G. S., and Watson, J. (Eds.), *Phage and the Origins of Molecular Biology,* Cold Spring Harbor Laboratory of Quantitive Biology (1966). [A series of scientific memoirs by many of the pioneers of molecular biology.]

Stent, G. S. and Calender, R., *Molecular Genetics* (2nd. ed.), Freeman (1978).

Suzuki, D. T., Griffiths, A. J. F., Miller, J. H., and Lewontin, R. C., *An Introduction To Genetic Analysis* (4th ed.), Freeman (1989).

The Role of DNA

Avery, O. T., MacLeod, C. M., and McCarty, M., Studies on the chemical nature of the substance inducing transformation of pneumococcal types, *J. Exp. Med.* **79**, 137–158 (1944). [The milestone report identifying transforming principle as DNA.]

Hershey, A. D. and Chase, M., Independent functions of viral proteins and nucleic acid in growth of bacteriophage, *J. Gen. Physiol.* **36**, 39–56 (1952).

Lavitrano, M., Camaioni, A., Fazio, V. M., Dolci, S., Farace, M. G., and Spadafora, C., Sperm cells as vectors for introducing foreign DNA into eggs: genetic transformation of mice, *Cell* **57**, 717–723 (1989).

McCarty, M., *The Transforming Principle,* Norton (1985). [A chronicle of the discovery that genes are DNA.]

Palmiter, R. D., Brinster, R. L., Hammer, R. E., Trumbauer, M. E., Rosenfeld, M. G., Birmberg, N. C., and Evans, R. M., Dramatic growth of mice that develop from eggs microinjected with metallothionein-growth hormone fusion genes, *Nature* **300**, 611–615 (1982).

Stent, G. S., Prematurity and uniqueness in scientific discovery, *Sci. Am.* **227**(6): 84–93 (1972). [A fascinating philosophical discourse on what it means for discoveries such as Avery's to be "ahead of their time" and on the nature of creativity in science.]

Problems

1. One method that Mendel used to test his laws is known as a **testcross**. In it, F_1 hybrids are crossed with their recessive parent. What is the expected distribution of progeny and what are their phenotypes in a testcross involving peas with different color seeds? What is it for snapdragons with different flower colors (use the white parent in this testcross)?

2. The disputed paternity of a child can often be decided on the basis of blood tests. The M, N, and MN blood groups (Section 11-3D) result from two alleles, L^M and L^N; the Rh$^+$ blood group arises from a dominant allele, R. Both sets of alleles occur on a different chromosome from each other and from the alleles responsible for the ABO blood groups. The following table gives the blood types of three children, their mother, and their two possible fathers. Indicate, where possible, each child's paternity and justify your answer.

Child 1	B	M	Rh$^-$
Child 2	B	MN	Rh$^+$
Child 3	AB	MN	Rh$^+$
Mother	B	M	Rh$^+$
Male 1	B	MN	Rh$^+$
Male 2	AB	N	Rh$^+$

3. The most common form of color blindness, red–green color blindness, afflicts almost only males. What are the genotypes and phenotypes of the children and grandchildren of a red–green color blind man and a woman with no genetic history of color blindness? Assume the children mate with individuals who also have no history of color blindness.

4. How might F′ cells be useful in the genetic analysis of bacteria?

5. Hfr strains of *E. coli* differ both in their origins of replication and in their directions of chromosomal transfer to F$^-$ recipients. The following table presents the transfer order of genes near the replication origin, reading left to right, in several Hfr strains. Use this data to construct a genetic map of the bacterial chromosome.

Hfr Strain	Order of Gene Transfer
1	*met-thi-thr-leu-azi-ton-pro*
2	*thi-met-ile-mtl-xyl-mal-str-his*
3	*ton-pro-lac-ade-gal-trp*
4	*xyl-mal-str-his-trp-gal-ade-lac*
5	*thi-thr-leu-azi-ton*

Chapter 28
NUCLEIC ACID STRUCTURES AND MANIPULATION

*There are two classes of nucleic acids, **deoxyribonucleic acid (DNA)** and **ribonucleic acid (RNA)**. DNA is the hereditary molecule in all cellular life forms, as well as in many viruses. It has but two functions:*

1. To direct its own replication during cell division.

2. To direct the **transcription** of complementary molecules of RNA.

RNA, in contrast, has more varied biological functions:

1. The RNA transcripts of DNA sequences that specify polypeptides, **messenger RNA (mRNA),** direct the ribosomal synthesis of these polypeptides in a process known as **translation.**

2. The RNAs of ribosomes, which are about two-thirds RNA and one-third protein, probably have functional as well as structural roles.

3. During protein synthesis, amino acids are delivered to the ribosome by molecules of **transfer RNA (tRNA).**

4. Certain RNAs are associated with specific proteins to form **ribonucleoproteins** that participate in the post-transcriptional processing of other RNAs.

5. In many viruses, RNA, not DNA, is the carrier of hereditary information.

In this chapter we examine the structures of nucleic acids with emphasis on DNA (the structure of RNA is detailed in Section 30-2A), and discuss methods of purifying, sequencing, and chemically synthesizing nucleic acids. We end by outlining how recombinant DNA technology, which has revolutionized the study of biochemistry, is used to manipulate, synthesize, and express DNA.

1. CHEMICAL STRUCTURE AND BASE COMPOSITION

The chemical structures of the nucleic acids were elucidated by the early 1950s largely through the efforts of Phoebus Levine followed by those of Alexander Todd. *Nucleic acids are, with few exceptions, linear polymers of nucleotides whose phosphates bridge the 3' and 5' positions of successive sugar residues (e.g., Fig. 28-1). The phosphates of these* **polynucleotides,** *the* **phosphodiester** *groups, are acidic so that, at physiological pH's, nucleic acids are polyanions.*

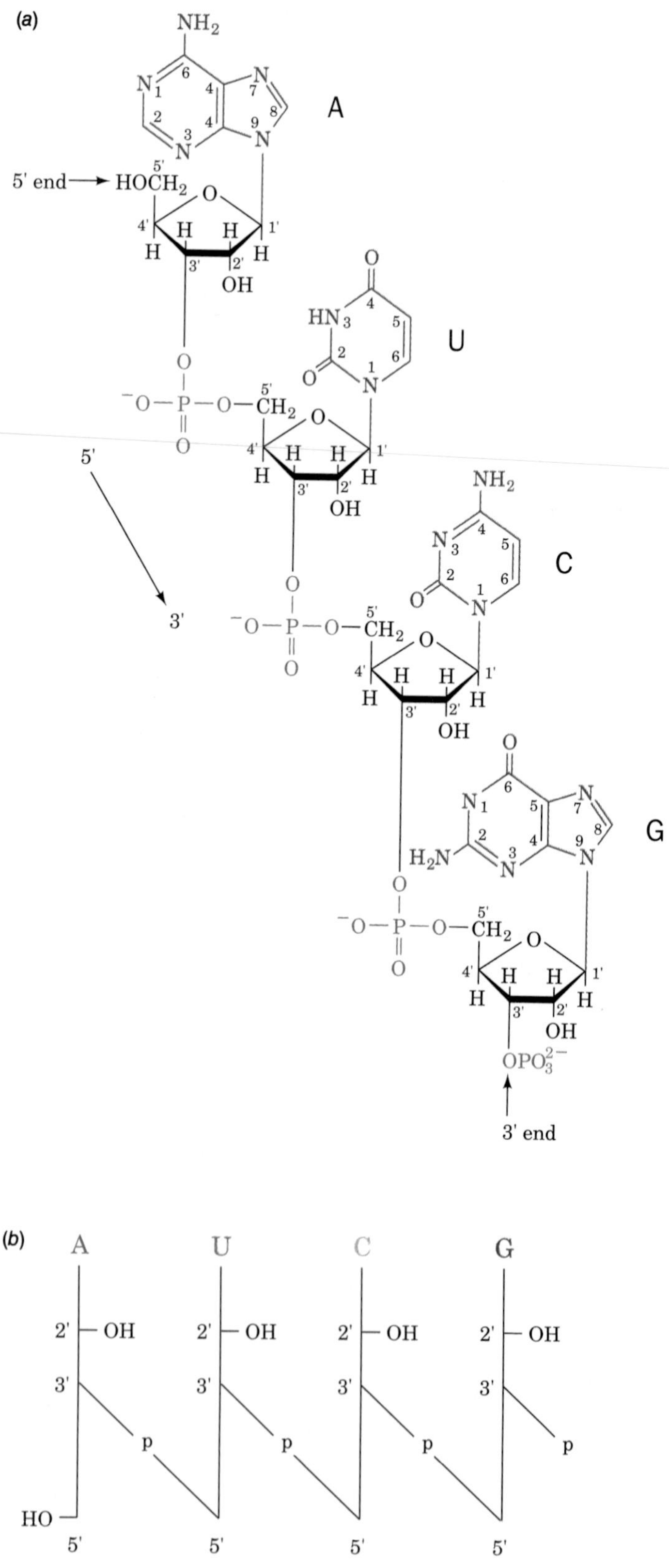

Figure 28-1

(a) The tetranucleotide adenyl-3',5'-uridyl-3',5'-cytidyl-3',5'-guanylyl-3'-phosphate. The sugar atom numbers are primed to distinguish them from the atomic positions of the bases. By convention, polynucleotide sequences are written with their 5' end at the left and their 3' end to the right. Thus, reading left to right, the phosphodiester bond links neighboring ribose residues in the 5' → 3' direction. The above sequence may be abbreviated ApUpCpGp or just AUCGp (where a "p" to the left and/or right of a nucleoside symbol indicates a 5' and/or a 3' phosphoryl bond, respectively; see Table 26-1 for other symbol definitions). The corresponding deoxytetranucleotide is abbreviated d(ApUpCpGp) or d(AUCGp). (b) A schematic representation of AUCGp. Here a vertical line denotes a ribose residue, its attached base is indicated by the corresponding one letter abbreviation and a diagonal line flanking an optional "p" represents a phosphodiester bond. The atomic numbering of the ribose residues, which is indicated here, is usually omitted. The equivalent representation of deoxypolynucleotides differ only by the absence of the 2'-OH groups.

DNA's Base Composition Is Governed by Chargaff's Rules

DNA has equal numbers of adenine and thymine residues (A = T) and equal numbers of guanine and cytosine residues (G = C). These relationships, known as **Chargaff's rules,** were discovered in the late 1940s by Erwin Chargaff who first devised reliable quantitative methods for the separation (by paper chromatography) and analysis of DNA hydrolysates. Chargaff also found that the base composition of DNA from a given organism is characteristic of that organism; that is, it is independent of the tissue from which the DNA is taken as well as the age of the organism, its nutritional state or any other environmental factor. The structural basis of Chargaff's rules derives from DNA's double-stranded character (Section 28-2A).

DNA's base composition varies widely among different organisms. It ranges from ~25 to 75% G + C in different species of bacteria. It is, however, more or less constant among related species; for example, in mammals G + C ranges from 39 to 46%.

RNA, which usually occurs as a single-stranded molecule, has no apparent constraints on its base composition. However, double-stranded RNA, which comprises the genetic material of several viruses, obeys Chargaff's rules. Conversely, single-stranded DNA, which occurs in certain viruses, does not obey Chargaff's rules. Upon entering its host organism, however, such DNA is replicated to form a double-stranded molecule, which then obeys Chargaff's rules.

Nucleic Acid Bases May Be Modified

Some DNAs contain bases that are chemical derivatives of the standard set. For example, dA and dC in the DNAs of many organisms are partially replaced by N^6-**methyl-dA** and **5-methyl-dC**, respectively.

N^6-Methyl-dA   **5-Methyl-dC**

The altered bases are generated by the sequence specific enzymatic modification of normal DNA (Sections 28-6A and 31-7). The modified DNAs obey Chargaff's rules if the derivatized bases are taken as equivalent to their parent bases. Likewise, many bases in RNA and, in particular, in tRNA (Section 30-2), are derivatized.

RNA but Not DNA Is Susceptible to Base-Catalyzed Hydrolysis

RNA is highly susceptible to base-catalyzed hydrolysis by the reaction mechanism diagrammed in Fig. 28-2 so as to yield a mixture of 2' and 3' nucleotides. In

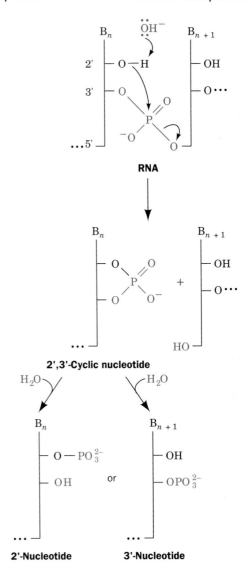

Figure 28-2
The mechanism of base-catalyzed RNA hydrolysis. The base-induced deprotonation of the 2'-OH group facilitates its nucleophilic attack on the adjacent phosphorus atom thereby cleaving the RNA backbone. The resultant 2',3'-cyclic phosphate group subsequently hydrolyzes to either the 2' or 3' phosphate. Note that the RNase-catalyzed hydrolysis of RNA follows a nearly identical reaction sequence (Section 14-1A).

contrast, DNA, which lacks 2'-OH groups, is resistant to base catalyzed hydrolysis and is therefore much more chemically stable than RNA. This is probably why DNA rather than RNA evolved to be the cellular genetic archive.

2. DOUBLE HELICAL STRUCTURES

The determination of the structure of DNA by James Watson and Francis Crick in 1953 is often said to mark

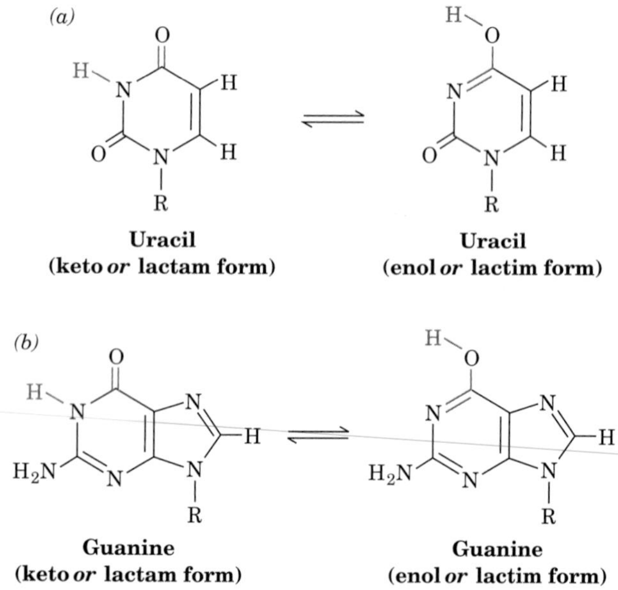

(a)

Uracil
(keto *or* lactam form)   ⇌   Uracil
(enol *or* lactim form)

(b)

Guanine
(keto *or* lactam form)   ⇌   Guanine
(enol *or* lactim form)

Figure 28-3
Some possible tautomeric conversions for (*a*) uracil and (*b*)
guanine residues. Cytosine and adenine residues can
undergo similar proton shifts.

the birth of modern molecular biology. The **Watson –
Crick structure** of DNA is of such importance because,
in addition to providing the structure of what is arguably the central molecule of life, it suggested the molecular mechanism of heredity. Watson and Crick's accom-
plishment, which is ranked as one of science's major
intellectual achievements, tied together the less than
universally accepted results of several diverse studies:

1. **Chargaff's rules.** At the time, these relationships
   were quite obscure because their significance was not
   apparent. In fact, even Chargaff did not emphasize
   them.

2. **The correct tautomeric forms of the bases.** X-ray,
   NMR, and spectroscopic investigations have firmly
   established that the nucleic acid bases are overwhelmingly in the keto tautomeric forms shown in
   Fig. 28-1. In 1953, however, this was not generally
   appreciated. Indeed, guanine and uracil were widely
   believed to be in their enol forms (Fig. 28-3) because
   it was thought that the resonance stability of these
   aromatic molecules would thereby be maximized.
   Knowledge of the dominant tautomeric forms, which
   was prerequisite for the prediction of the correct hydrogen bonding associations of the bases, was provided by Jerry Donohue, an office mate of Watson
   and Crick and an expert on the X-ray structures of
   small organic molecules.

3. **Information that DNA is a helical molecule.** This was
   provided by an X-ray diffraction photograph of a

DNA fiber taken by Rosalind Franklin (Fig. 28-4;
DNA, being a threadlike molecule, does not crystallize but, rather, can be drawn out in fibers consisting
of parallel bundles of molecules; Section 7-2). A description of the photograph enabled Crick, an X-ray
crystallographer by training who had earlier derived
the equations describing diffraction by helical molecules, to deduce that DNA is (*a*) a helical molecule,
and (*b*) that its planar aromatic bases form a stack of
parallel rings that is parallel to the fiber axis.

This information only provided a few crude landmarks
that guided the elucidation of the DNA structure; it
mostly sprang from Watson and Crick's imaginations
through model building studies. Once the Watson –
Crick model had been published, however, its basic simplicity combined with its obvious biological relevance
led to its rapid acceptance. Later investigations have
confirmed the essential correctness of the Watson –
Crick model although its details have been modified.

It is now realized that double helical DNA and RNA
can assume several distinct structures that vary with
such factors as the humidity and the identities of the
cations present, as well as with base sequence. In this
section, we describe these various structures.

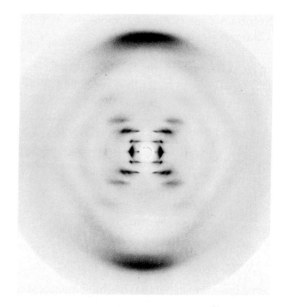

Figure 28-4
An X-ray diffraction photograph of a vertically oriented Na+
DNA fiber in the B conformation. This is the photograph that
provided key information for the elucidation of the Watson –
Crick structure. The central X-shaped pattern of spots is
indicative of a helix, whereas the heavy black arcs on the top
and bottom of the diffraction pattern correspond to a
distance of 3.4 Å and indicate that the DNA structure largely
repeats every 3.4 Å along the fiber axis. [Courtesy of
Maurice Wilkins, King's College, London.]

Table 28-1
Structural Features of Ideal A, B, and Z-DNA

	A	B	Z
Helical sense	Right handed	Right handed	Left handed
Diameter	~26 Å	~20 Å	~18 Å
Base pairs per helical turn	11	10	12 (6 dimers)
Helical twist per base pair	33°	36°	60° (per dimer)
Helix pitch (rise per turn)	28 Å	34 Å	45 Å
Helix rise per base pair	2.6 Å	3.4 Å	3.7 Å
Base tilt normal to the helix axis	20°	6°	7°
Major groove	Narrow and deep	Wide and Deep	Flat
Minor groove	Wide and shallow	Narrow and deep	Narrow and deep
Sugar pucker	C(3')-*endo*	C(2')-*endo*	C(2')-*endo* for pyrimidines; C(3')-*endo* for purines
Glycosidic bond	Anti	Anti	Anti for pyrimidines; syn for purines

A. The Watson–Crick Structure: B-DNA

Fibers of DNA assume the so-called B conformation, as indicated by their X-ray diffraction patterns, when the counterion is an alkali metal such as Na^+ and the relative humidity is 92%. *B-DNA is regarded as the native form because its X-ray pattern resembles that of the DNA in intact sperm heads.*

The Watson–Crick structure of B-DNA has the following major features (Table 28-1):

1. *It consists of two polynucleotide strands that wind about a common axis with a right-handed twist to form an ~20 Å in diameter double helix (Fig. 28-5). The two strands are antiparallel (run in opposite directions) and* wrap around each other such that they cannot be separated without unwinding the helix (a phenomenon known as **plectonemic coiling**). The bases occupy the core of the helix while its sugar–phosphate chains are coiled about its periphery thereby minimizing the repulsions between charged phosphate groups.

2. The planes of the bases are nearly perpendicular to the helix axis. *Each base is hydrogen bonded to a base on the opposite strand to form a planar **base pair** (Fig. 28-5).* It is these hydrogen bonding interactions, a phenomenon known as **complementary base pairing,** that result in the specific association of the two chains of the double helix.

3. The "ideal" B-DNA helix has 10 base pairs (bp) per turn (a helical twist of 36° per bp) and, since the aromatic bases have van der Waals thicknesses of 3.4 Å and are partially stacked on each other (**base stacking;** Fig. 28-5b), the helix has a pitch (rise per turn) of 34 Å.

The most remarkable feature of the Watson–Crick structure is that *it can accommodate only two types of base pairs: Each adenine residue must pair with a thymine residue and vice versa, and each guanine residue must pair with a cytosine residue and vice versa. The geometries of these A·T and G·C base pairs, the so-called* **Watson–Crick base pairs,** are shown in Fig. 28-6. It can be seen that *both of these base pairs are interchangeable in that they can replace each other in the double helix without altering the positions of the sugar–phosphate backbone's C(1') atoms. Likewise, the double helix is undisturbed by exchanging the partners of a Watson–Crick base pair, that is, by changing a G·C to a C·G or an A·T to a T·A.* In contrast, any other combination of bases would significantly distort the double helix since the formation of a non-Watson–Crick base pair would require considerable reorientation of the sugar–phosphate chain.

The two deep grooves that wind about the outside of B-DNA between the sugar–phosphate chains are of unequal size (Fig. 28-5a) because: (1) the top edge of each base pair, as drawn in Fig. 28-6, is structurally distinct from the bottom edge; and (2) the deoxyribose residues are asymmetric. The **minor groove** is that in which the

(a)

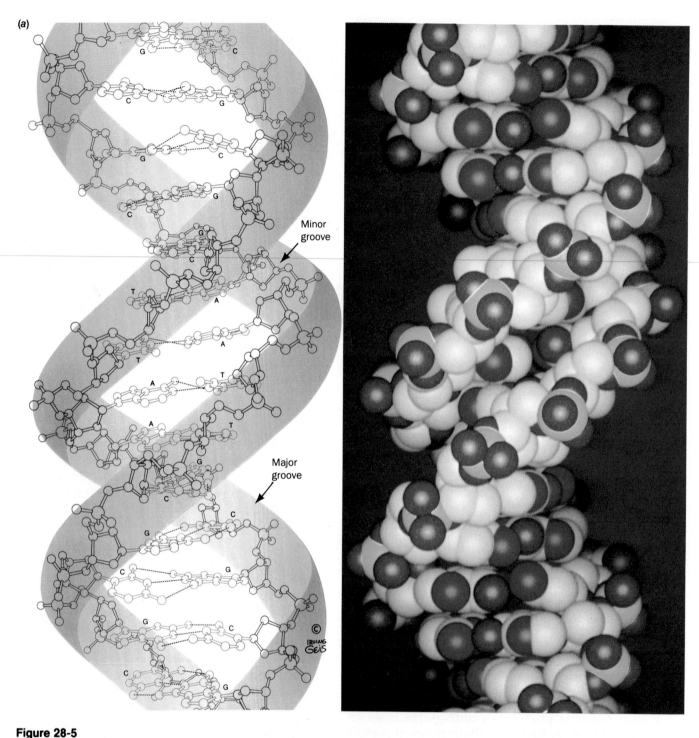

Minor groove

Major groove

Figure 28-5

The structure of B-DNA as represented by ball-and-stick drawings and the corresponding computer-generated space-filling models. The repeating helix is based on the X-ray structure of the self-complementary dodecamer d(CGCGAATTCGCG) determined by Richard Dickerson and Horace Drew. (a) View perpendicular to the helix axis. In the drawing, the sugar–phosphate backbones, which wind about the periphery of the molecule, are blue, and the bases, which occupy its core, are red. In the space-filling model, C, N, O, and P atoms are white, blue, red, and green, respectively. H atoms have been omitted for clarity in both drawings. Note that the two sugar–phosphate chains run in opposite directions. (b) (opposite) View along the helix axis. In the drawing, the ribose ring O atoms are red and the nearest base pair is white. Note that the helix axis passes through the base pairs so that the helix has a solid core. [Drawings copyrighted © by Irving Geis. Computer graphics courtesy of Robert Stodola, Fox Chase Cancer Center.]

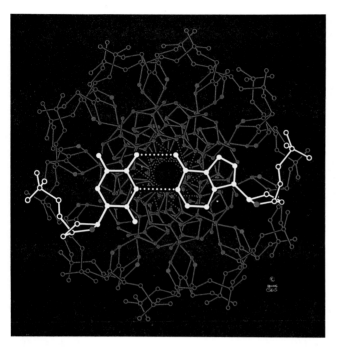 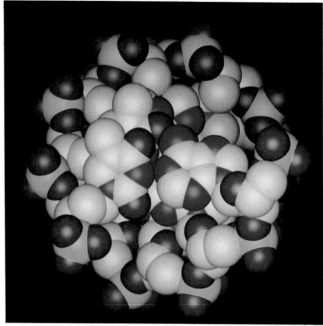

Figure 28-5 (*b*)

C(1′)-helix axis-C(1′) angle is < 180° (opening towards the bottom in Fig. 28-6; the helix axis passes through the middle of each base pair in B-DNA), whereas the **major groove** opens towards the opposite edge of each base pair (Fig. 28-6).

The Watson–Crick structure can accommodate any sequence of bases on one polynucleotide strand if the opposite strand has the complementary base sequence. This immediately accounts for Chargaff's rules. More importantly, *it suggests that hereditary information is encoded in the sequence of bases on either strand.*

Real DNA Deviates from the Ideal
Watson–Crick Structure

By the late 1970s, advances in nucleic acid chemistry permitted the synthesis and crystallization of ever longer oligonucleotides of defined sequences (Section 28-7). Consequently, some 25 years after the Watson–Crick structure had been formulated. the X-ray crystal structures of DNA fragments were clearly visualized for the first time (fiber diffraction studies provide only crude low resolution images in which the base pair elec-

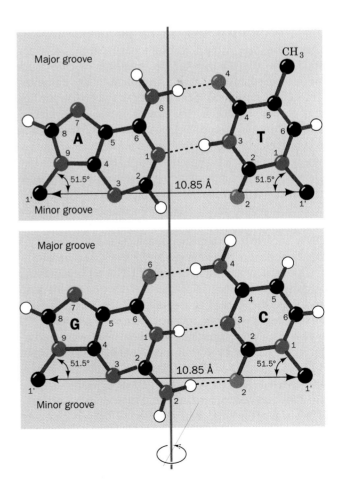

Figure 28-6
The Watson–Crick base pairs. The line joining the C(1′) atoms is the same length in both base pairs and makes equal angles with the glycosidic bonds to the bases. This gives DNA a series of pseudo-twofold symmetry axes (often referred to as **dyad axes**) that pass through the center of each base pair (*red line*) and are perpendicular to the helix axis. Note that A·T base pairs associate via two hydrogen bonds, whereas C·G base pairs are joined by three hydrogen bonds. [After Arnott, S., Dover, S. D., and Wonacott, A. J., *Acta Cryst.* **B25**, 2196 (1969).]

tron density is the average electron density of all the base pairs in the fiber). Richard Dickerson and Horace Drew have shown that the self-complementary dodecamer d(CGCGAATTCGCG) crystallizes in the B-conformation. The molecule has an average rise per residue of 3.4 Å and has 10.1 bp per turn (a helical twist of 35.6° per bp), which is nearly equal to that of ideal B-DNA. Nevertheless, *individual residues significantly depart from this average conformation in a manner that appears to be sequence dependent (Fig. 28-5).* For example, the helical twist per base pair in this dodecamer ranges from 28 to 42°. Each base pair further deviates from its ideal conformation by such distortions as propeller twisting (the opposite rotation of paired bases about the base pair's long axis; in the above dodecamer these values range from 10 to 20°) and base pair roll (the tilting of a base pair as a whole about its long axis). Indeed, rapidly accumulating X-ray and NMR studies of other double helical DNA oligomers have amply demonstrated that *the structure of DNA is surprisingly irregular in a sequence-specific manner. This phenomenon, as we shall see (Sections 29-3C and E) is important for the sequence-specific binding to DNA of proteins that process genetic information.*

DNA Is Semiconservatively Replicated

The Watson–Crick structure also suggests how DNA can direct its own replication. Each polynucleotide strand can act as a template for the formation of its complementary strand through base pairing interactions. The two strands of the parent molecule must therefore separate so that a complementary daughter strand may be enzymatically synthesized on the surface of each parent strand. This results in two molecules of **duplex** (double stranded) DNA, each consisting of one polynucleotide strand from the parent molecule and a newly synthesized complementary strand (Fig. 1-16). Such a mode of replication is termed **semiconservative** in contrast with **conservative** replication which, if it occurred, would result in a newly synthesized duplex copy of the original DNA molecule with the parent DNA molecule remaining intact. The mechanism of DNA replication is the main subject of Chapter 31.

The semiconservative nature of DNA replication was elegantly demonstrated in 1958 by Matthew Meselson and Franklin Stahl. The density of DNA was increased by labeling it with ^{15}N, a heavy isotope of nitrogen (^{14}N is the naturally abundant isotope). This was accomplished by growing *E. coli* for 14 generations in a medium that contained ^{15}NH$_4$Cl as its only nitrogen source. The labeled bacteria were then abruptly transferred to an ^{14}N-containing medium and the density of their DNA was monitored as a function of bacterial growth by equilibrium density gradient ultracentrifugation (Section 5-5B; a technique Meselson, Stahl, and Jerome Vinograd had developed for the purpose of distinguishing ^{15}N-labeled DNA from unlabeled DNA).

The results of the Meselson–Stahl experiment are displayed in Fig. 28-7. After one generation (doubling of the cell population), all of the DNA had a density exactly halfway between the densities of fully ^{15}N-labeled DNA and unlabeled DNA. This DNA must therefore contain equal amounts of ^{14}N and ^{15}N as is expected after one generation of semiconservative replication. Conservative DNA replication, in contrast, would result in the preservation of the parental DNA, so that it maintained its original density, and the generation of an equal amount of unlabeled DNA. After two generations, one half of the DNA molecules were unlabeled and the remainder were ^{14}N—^{15}N hybrids. This is also in accord with the predictions of the semiconservative replication model and in disagreement with the conservative replication model. In succeeding generations, the amount of unlabeled DNA increased relative to the amount of hybrid DNA although the hybrid never totally disappeared. This is again in harmony with semiconservative replication but at odds with conservative replication, which predicts that the fully labeled parental DNA will always be present and that hybrid DNA never forms.

Meselson and Stahl also demonstrated that DNA is double stranded. DNA from ^{15}N-labeled *E. coli* that were grown for one generation in an ^{14}N medium was heat denatured at 100°C (which causes strand separation; Section 28-3A) and then subjected to density gradient ultracentrifugation. Two bands were observed; one at the density of fully ^{15}N-labeled DNA and the other at the density of unlabeled DNA. Moreover the molecular masses of the DNA in these bands, as estimated from their peak shapes, was one half that of undenatured DNA (the peak width varies with molecular mass). Native DNA must therefore be composed of two equal-sized strands that separate upon heat denaturation.

B. Other Nucleic Acid Helices

Double-stranded DNA is a conformationally variable molecule. In the following subsections we discuss its major conformational states besides B-DNA and also those of double-stranded RNA.

A-DNA's Base Pairs Are Inclined to the Helix Axis

When the relative humidity is reduced to 75%, B-DNA undergoes a reversible conformational change to the so-called A form. Fiber X-ray studies indicate that *A-DNA forms a wider and flatter right-handed helix than does B-DNA* (Fig. 28-8; Table 28-1). A-DNA has 11 bp per turn and a pitch of 28 Å which gives A-DNA an axial hole (Fig. 28-8b). The most striking feature of A-DNA, however, is that the planes of its base pairs are tilted 20° with respect to the helix axis. A-DNA therefore has a deep major groove and a very shallow minor groove; it can be described as a flat ribbon wound around a 6 Å in diameter cylindrical hole. Most self-

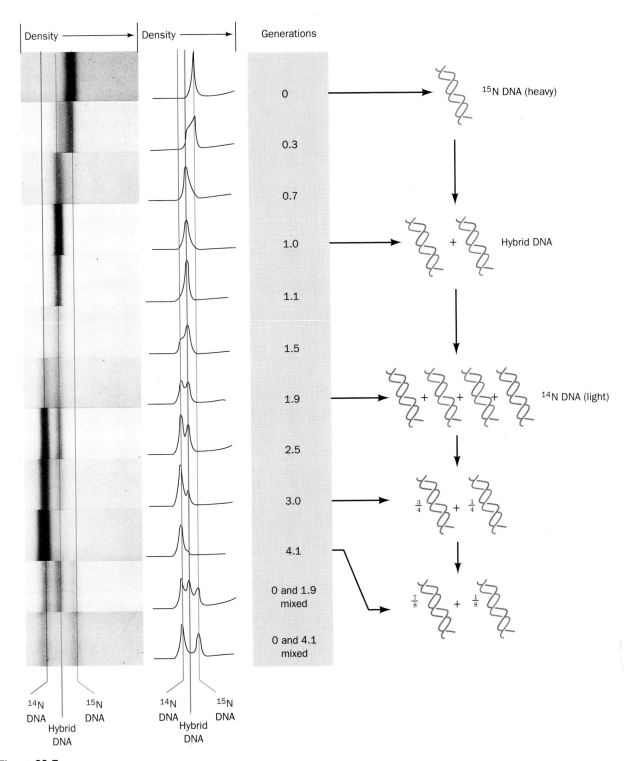

Figure 28-7

The demonstration of the semiconservative nature of DNA replication in *E. coli.* DNA in a CsCl solution of density 1.71 g·cm⁻³ was subjected to equilibrium density gradient ultracentrifugation at 140,000 *g* in an analytical ultracentrifuge (a device in which the spinning sample can be optically observed). The enormous centrifugal acceleration caused the CsCl to form a density gradient in which DNA migrated to its position of buoyant density. The left panels are UV absorption photographs of ultracentrifuge cells (DNA strongly absorbs UV light) and are arranged such that regions of equal density have the same horizontal positions. The middle panels are microdensitometer traces of the corresponding photographs in which the vertical displacement is proportional to the DNA concentration. The buoyant density of DNA increases with its ^{15}N content. The bands furthest to the right (greatest radius and density) arise from DNA that is fully ^{15}N labeled, whereas unlabeled DNA, which is 0.014 g·cm⁻³ less dense, forms the leftmost bands. The bands in the intermediate position result from duplex DNA in which one strand is ^{15}N labeled and the other strand is unlabeled. The accompanying interpretive drawings (*right*) indicate the relative numbers of DNA strands at each generation donated by the original parents (*blue,* ^{15}N labeled) and synthesized by succeeding generations (*red,* unlabeled). [From Meselson, M. and Stahl, F. W., *Proc. Natl. Acad. Sci.* **44,** 674 (1958).]

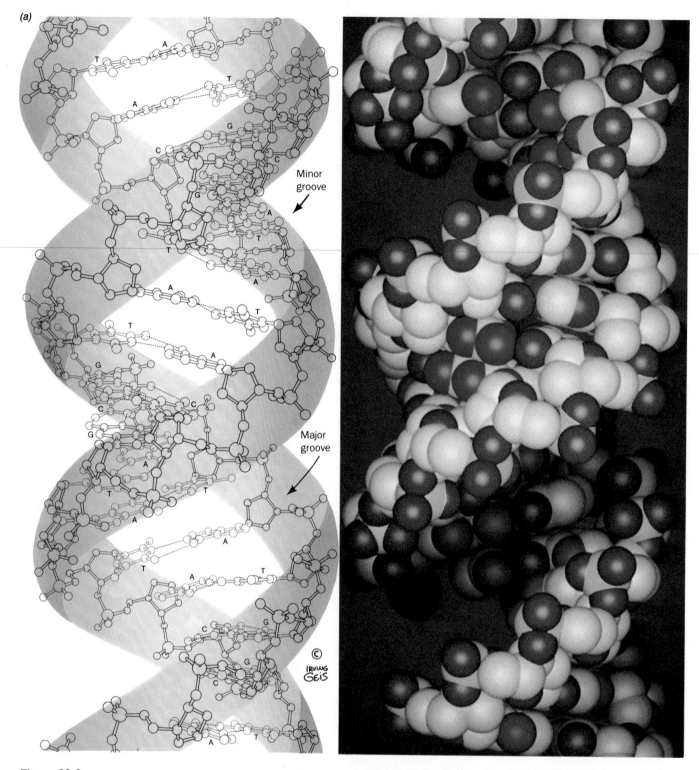

(a)

Minor groove

Major groove

IRUIAG GEIS ©

Figure 28-8

Ball-and-stick drawings and the corresponding space-filling models of A-DNA as viewed (a) perpendicular to the helix axis, and (b) (opposite) along the helix axis. The color codes are given in Fig. 28-5. The repeating helix was generated by Richard Dickerson based on the X-ray structure of the self-complementary octamer d(GGTATACC) determined by Olga Kennard, Dov Rabinovitch, Zippora Shakked, and Mysore Viswamitra. Note that the base pairs are inclined to the helix axis and that the helix has a hollow core. Compare this figure with Fig. 28-5. [Drawings copyrighted © by Irving Geis. Computer graphics courtesy of Robert Stodola, Fox Chase Cancer Center.]

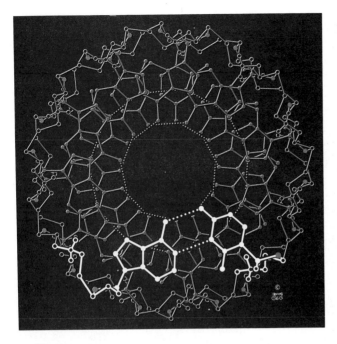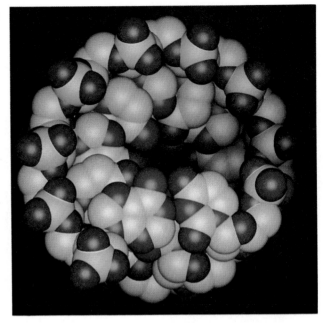

Figure 28.8 (*b*)

complementary oligonucleotides of < 10 base pairs; for example, d(GGCCGGCC) and d(GGTATACC), crystallize in the A-DNA conformation. Like B-DNA, these molecules exhibit considerable sequence-specific conformational variation. It has not been established that A-DNA exists *in vivo* although a few experimental observations suggest that certain DNA segments normally assume the A conformation.

Z-DNA Forms a Left-Handed Helix

Occasionally, a seemingly well understood or at least familiar system exhibits quite unexpected properties. Over 25 years after the discovery of the Watson–Crick structure, the crystal structure determination of d(CGCGCG) by Andrew Wang and Alexander Rich revealed, quite surprisingly, *a left-handed double helix (Fig. 28-9; Table 28-1).* A similar helix is formed by d(CGCATGCG). *This helix, which has been dubbed Z-DNA, has 12 Watson–Crick base pairs per turn, a pitch of 45 Å and, in contrast to A-DNA, a deep minor groove and no discernable major groove.* Z-DNA therefore resembles a left-handed drill bit in appearance. The base pairs in Z-DNA are flipped 180° relative to those in B-DNA (Fig. 28-10) through conformational changes discussed in Section 28-3B. As a consequence, the repeating unit of Z-DNA is a dinucleotide, d(XpYp), rather than a single nucleotide as it is in the other DNA helices. Here, X is usually a pyrimidine residue and Y is usually a purine residue because the purine nucleotide assumes a conformation that would be sterically unfavorable in the pyrimidine nucleotide. The line joining successive phosphate groups on a polynucleotide strand of Z-DNA therefore follows a zigzag path around the helix (Fig.

28-9*a*; hence the name Z-DNA) rather than a smooth curve as it does in A- and B-DNAs (Figs. 28-5*a* and 28-8*a*).

Fiber diffraction and NMR studies have shown that complementary polynucleotides with alternating purines and pyrimidines, such as poly d(GC)·poly d(GC) or poly d(AC)·poly d(GT), take up the Z-DNA conformation at high salt concentrations. Evidently, *the Z-DNA conformation is most readily assumed by DNA segments with alternating purine–pyrimidine base sequences (for structural reasons explained in Section 28-3B).* A high salt concentration stabilizes Z-DNA relative to B-DNA by reducing the otherwise increased electrostatic repulsions between closest approaching phosphate groups on opposite strands (8 Å in Z-DNA vs 12 Å in B-DNA). The methylation of cytosine residues at C(5), a common biological modification (Section 31-7), also promotes Z-DNA formation since a hydrophobic methyl group in this position is less exposed to solvent in Z-DNA than it is in B-DNA.

Does Z-DNA have any biological significance? Rich has proposed that the reversible conversion of specific segments of B-DNA to Z-DNA under appropriate circumstances acts as a kind of switch in regulating genetic expression. Yet, the *in vivo* existence of Z-DNA has been difficult to prove. A major problem is demonstrating that a particular probe for detecting Z-DNA, a Z-DNA-specific antibody, for example, does not in itself cause what would otherwise be B-DNA to assume the Z conformation — a kind of biological uncertainty principle (the act of measurement inevitably disturbs the system being measured). Recently, however, Z-DNA has been shown to be present in *E. coli* by employing an *E. coli* enzyme that methylates a specific base sequence

(a)

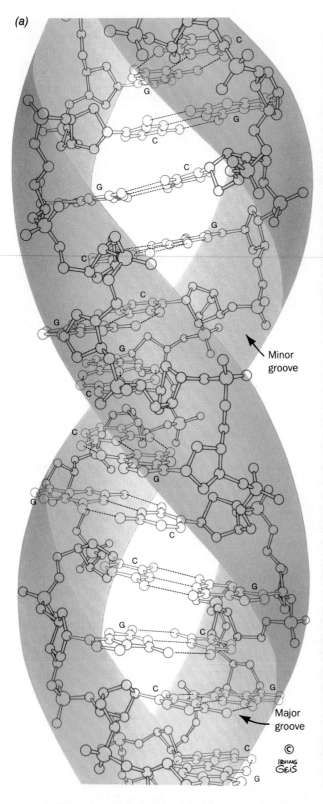

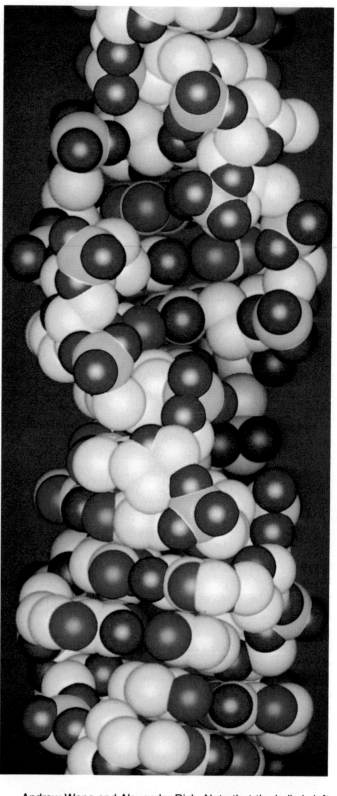

Figure 28-9

Ball-and-stick drawings and the corresponding space-filling models of Z-DNA as viewed (a) perpendicular to the helix axis and (b) (*opposite*) along the helix axis. The color codes are given in Fig. 28-5. The repeating helix was generated by Richard Dickerson based on the X-ray structure of the self-complementary hexamer d(CGCGCG) determined by Andrew Wang and Alexander Rich. Note that the helix is left handed and that the sugar–phosphate chains follow a zigzag course (alternate ribose residues lie at different radii in Part *b*) indicating that the Z-DNA's repeating motif is a dinucleotide. Compare this figure with Figs. 28-5 and 28-8. [Drawings copyrighted © by Irving Geis. Computer graphics courtesy of Robert Stodola, Fox Chase Cancer Center.]

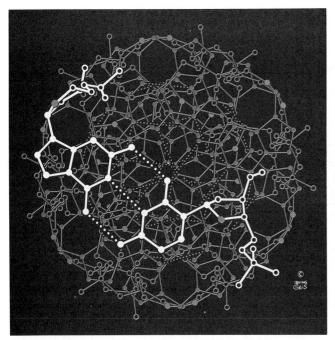
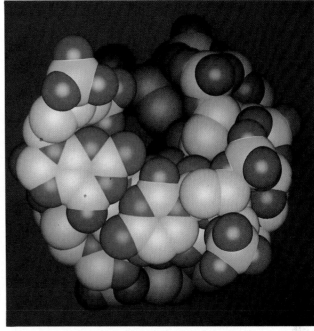

Figure 28-9 (*b*)

in vitro when the DNA is in the B form but not when it is in the Z form. The *in vivo* methylation of this base sequence is inhibited when it is cloned in *E. coli* (by techniques discussed in Section 28-8) within or adjacent to a DNA segment that can form Z-DNA. Moreover, there is a balance between the *in vivo* B and Z forms of these DNAs that is thought to be influenced by environmental factors such as salt concentration and protein binding. Nevertheless, the biological function of Z-DNA, if any, remains unknown.

RNA-11 and RNA–DNA Hybrids Have an A-DNA-Like Conformation

Double helical RNA is unable to assume a B-DNA-like conformation because of steric clashes involving its 2′-OH groups. Rather, it usually assumes a conformation resembling A-DNA (Fig. 28-8), known as **A-RNA** or **RNA-11,** which has 11 bp per helical turn, a pitch of 30 Å, and its base pairs inclined to the helix axis by ~14°. Many RNAs, for example, transfer and ribosomal RNAs (whose structures are detailed in Sections 30-2A

Figure 28-10
The conversion of B-DNA to Z-DNA, here represented by a 4 bp segment, involves a 180° flip of each base pair (*curved arrows*) relative to the sugar–phosphate chains. Here, the different faces of the base pairs are colored red and green. [After Rich, A., Nordheim, A., and Wang, A. H.-J., *Annu. Rev. Biochem.* **53,** 799 (1984).]

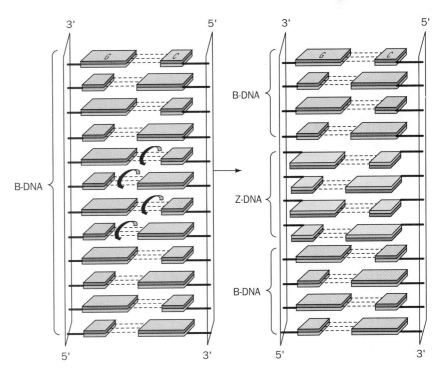

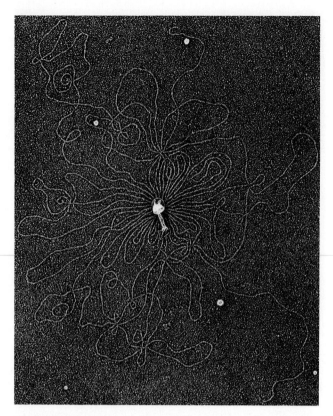

Figure 28-11
An electron micrograph of a T2 bacteriophage that had been osmotically lysed in distilled water so that its DNA spilled out. Without special treatment, duplex DNA, which is only 20 Å in diameter, is difficult to visualize in the electron microscope. In the **Kleinschmidt procedure,** DNA is fattened to ~200 Å in diameter by coating it with denatured cytochrome *c* or some other basic protein. The preparation is rendered visible in the electron microscope by shadowing it with platinum. [From Kleinschmidt, A. K., Lang, D., Jacherts, D., and Zahn, R. K., *Biochim. Biophys. Acta* **61,** 861 (1962).]

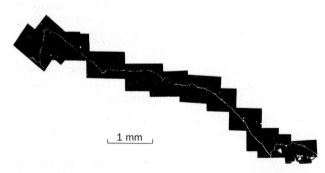

⊢ 1 mm ⊣

Figure 28-12
An autoradiograph of *Drosophila melanogaster* DNA. Lysates of *D. melanogaster* cells that had been cultured with [³H]thymidine were spread on a glass slide and covered with a photographic emulsion that was developed after a 5-month exposure. The measured contour length of the DNA is 1.2 cm. [From Kavenoff, R., Klotz, L. C., and Zimm, B. H., *Cold Spring Harbor Symp. Quant. Biol.* **38,** 4 (1973). Copyright © 1973 by Cold Spring Harbor Laboratory.]

and 30-3A), contain complementary sequences that form double helical stems. Hybrid double helices, which consist of one strand each of RNA and DNA, also have an A-DNA-like conformation. Small segments of RNA · DNA hybrid helices must occur in both the transcription of RNA on DNA templates (Section 29-2D) and in the initiation of DNA replication by short lengths of RNA (Section 31-1D).

C. The Size of DNA

DNA molecules are generally enormous (Fig. 28-11). The molecular mass of DNA has been determined by a variety of techniques including hydrodynamic methods (Section 5-5), length measurements by electron microscopy, and autoradiography [Fig. 28-12; a base pair of Na⁺ B-DNA has an average molecular mass of 660 D and a length (thickness) of 3.4 Å]. The number of base pairs and the **contour lengths** (the end-to-end lengths of the stretched out native molecules) of the DNAs from a selection of organisms of increasing complexity are presented in Table 28-2. Not surprisingly, an organism's haploid quantity (unique amount) of DNA varies more or less with its complexity (although there are notable exceptions to this generalization such as the last entry in Table 28-2).

The visualization of DNAs from prokaryotes has demonstrated that their entire **genome** (complement of genetic information) is contained on a single, usually circular, length of DNA. Similarly, Bruno Zimm demonstrated that the *largest chromosome of the fruit fly Drosophila melanogaster contains a single molecule of DNA* by comparing the molecular mass of this DNA with the cytologically measured amount of DNA contained in the chromosome. Presumably other eukaryotic chromosomes also contain only single molecules of DNA.

The highly elongated shape of duplex DNA (recall B-DNA is only 20 Å in diameter), together with its stiffness, make it extremely susceptible to mechanical damage outside the cell's protective environment (for instance, if the *Drosophila* DNA of Fig. 28-12 were expanded by a factor of 500,000, it would have the shape and some of the mechanical properties of a 6-km long strand of uncooked spaghetti). The hydrodynamic shearing forces generated by such ordinary laboratory manipulations as stirring, shaking, and pipetting, break DNA into relatively small pieces so that the isolation of an intact molecule of DNA requires extremely gentle handling. Before 1960, when this was first realized, the measured molecular masses of DNA were no higher than 10 million D. DNA fragments of uniform molecular mass and as small as a few hundred base pairs may be generated by **shear degrading** DNA in a controlled manner; for instance, by pipetting, through the use of a high speed blender, or by **sonication** (exposure to high frequency sound waves).

Table 28-2
Sizes of Some DNA Molecules

Organism	Number of Base Pairs (kb)[a]	Contour Length (μm)
Viruses		
Polyoma, SV40	5.1	1.7
Bacteriophage λ	48.6	17
T2,T4,T6 bacteriophage	166	55
Fowlpox	280	193
Bacteria		
Mycoplasma hominis	760	260
Escherichia coli	4,000	1,360
Eukaryotes		
Yeast (*in 17 haploid chromosomes*)	13,500	4,600
Drosophila (*in 4 haploid chromosomes*)	165,000	56,000
Human (*in 23 haploid chromosomes*)	2,900,000	990,000
Lungfish (*in 19 haploid chromosomes*)	102,000,000	34,700,000

[a] kb = kilo base pair = 1000 base pairs (bp).

Source: Kornberg, A., *DNA Replication, p.* 20, Freeman (1980).

3. FORCES STABILIZING NUCLEIC ACID STRUCTURES

DNA does not exhibit the structural complexity of proteins because it has only a limited repertoire of secondary structures and no comparable tertiary or quaternary structures. This is perhaps to be expected since there is a far greater range of chemical and physical properties among the 20 amino acid residues of proteins than there is among the four DNA bases. As we discuss in Sections 30-2B and 3A, however, many RNAs have well-defined tertiary structures.

In this section we examine the forces that give rise to the structures of nucleic acids. These forces are, of course, much the same as those that are responsible for the structures of proteins (Section 7-4) but, as we shall see, the way they combine gives nucleic acids properties that are quite different from those of proteins.

A. Denaturation and Renaturation

When a solution of duplex DNA is heated above a characteristic temperature, its native structure collapses and its two complementary strands separate and assume the random coil conformation (Fig. 28-13). This denaturation process is accompanied by a qualitative change in the DNA's physical properties. For instance, the characteristic high viscosity of native DNA solutions, which

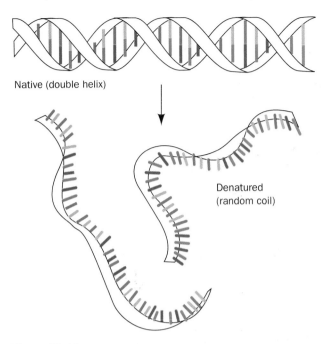

Native (double helix)

Denatured (random coil)

Figure 28-13
A schematic representation of DNA denaturation.

arises from the resistance to deformation of its rigid and rodlike duplex molecules, drastically decreases when the DNA decomposes to relatively freely jointed single strands.

DNA Denaturation Is a Cooperative Process

The most convenient way of monitoring the native state of DNA is by its ultraviolet (UV) absorbance spectrum. When DNA denatures, its UV absorbance, which is almost entirely due to its aromatic bases, increases by $\sim 40\%$ at all wavelengths (Fig. 28-14). This phenomenon, which is known as the **hyperchromic effect** (Greek; *hyper,* above; *chroma,* color), results from the disruption of the electronic interactions among nearby bases. DNA's hyperchromic shift, as monitored at a particular wavelength (usually 260 nm), occurs over a narrow temperature range (Fig. 28-15). This indicates that the denaturation of DNA is a cooperative phenomenon in which the collapse of one part of the structure destabilizes the remainder. The denaturation of DNA may be described as the melting of a one-dimensional solid so that Fig. 28-15 is referred to as a **melting curve** and the temperature at its midpoint is known as its **melting temperature,** T_m.

The stability of the DNA double helix, and hence its T_m, depends on several factors including the nature of the solvent, the identities and concentrations of the ions in solution, and the pH. T_m also increases linearly with the mole fraction of $G \cdot C$ base pairs (Fig. 28-16), which indicates that triply hydrogen bonded $G \cdot C$ base pairs

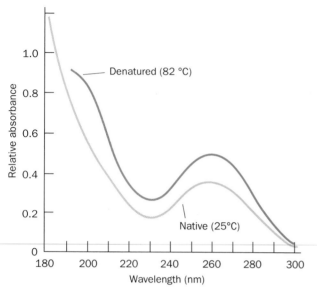

Figure 28-14
The UV absorbance spectra of native and heat denatured *E. coli* DNA. Note that denaturation does not change the general shape of the absorbance curve but only increases its intensity. [After Voet, D., Gratzer, W. B., Cox, R. A., and Doty, P., *Biopolymers* **1**, 205 (1963).]

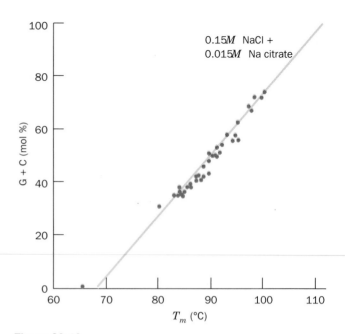

Figure 28-16
The variation of the melting temperatures, T_m, of various DNAs with their G + C content. The DNAs were dissolved in a solution containing 0.15M NaCl and 0.015M Na citrate. [After Marmur, J. and Doty, P., *J. Mol. Biol.* **5**, 113 (1962).]

are more stable than doubly hydrogen bonded A · T base pairs.

Denatured DNA Can Be Renatured

If a solution of denatured DNA is rapidly cooled below its T_m, the resulting DNA will be only partially base paired (Fig. 28-17) because the complementary strands will not have had sufficient time to find each other before the partially base paired structures become effectively "frozen in." If, however, the temperature is maintained ~25°C below the T_m, enough thermal energy is available for short base paired regions to rearrange by melting and reforming but not so much as to melt out long complementary stretches. Under such **annealing conditions,** as Julius Marmur discovered in 1960, denatured DNA eventually completely renatures. Likewise, complementary strands of RNA and DNA, in a process known as **hybridization,** form RNA–DNA hybrid double helices that are only slightly less stable than the corresponding DNA double helices.

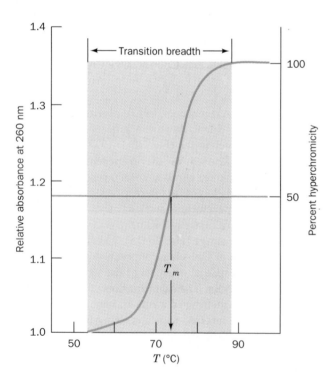

Figure 28-15
An example of a DNA melting curve. The relative absorbance is the ratio of the absorbance (customarily measured at 260 nm) at the indicated temperature to that at 25°C. The melting temperature, T_m, is the temperature at which one half of the maximum absorbance increase is attained.

B. Sugar–Phosphate Chain Conformations

The conformation of a nucleotide unit, as Fig. 28-18 indicates, is specified by the six torsion angles of the sugar–phosphate backbone and the torsion angle describing the orientation of the base about the glycosidic bond [the bond joining C(1′) to the base]. It would seem that these seven degrees of freedom per nucleotide would render polynucleotides highly flexible. Yet, as we shall see, these torsion angles are subject to a variety of

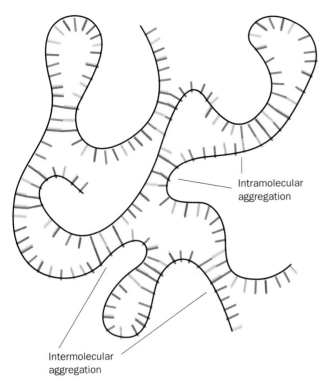

Figure 28-17
A schematic representation of the imperfectly base paired structures assumed by DNA that has been heat denatured and then rapidly cooled. Note that both intramolecular and intermolecular aggregation may occur.

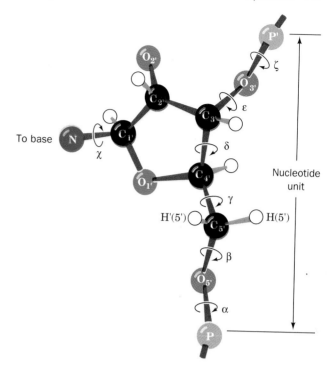

Figure 28-18
The conformation of a nucleotide unit is determined by the seven indicated torsional angles.

internal constraints that greatly restrict their conformational freedom.

Torsion Angles about Glycosidic Bonds Have One or Two Stable Positions

The rotation of a base about its glycosidic bond is greatly hindered, as is best seen by the manipulation of a space-filling molecular model. Purine residues have two sterically permissible orientations relative to the sugar known as the **syn** (Greek: with) and **anti** (Greek: against) conformations (Fig. 28-19). For pyrimidines, only the anti conformation is easily formed because, in the syn conformation, the sugar residue sterically interferes with the pyrimidine's C(2) substituent. In most

double helical nucleic acids, all bases are in the anti conformation. The exception is Z-DNA (Section 28-2B), in which the alternating pyrimidine and purine residues are anti and syn, respectively. *This explains Z-DNA's pyrimidine – purine alternation.* Indeed, the base pair flips that convert B-DNA to Z-DNA (Fig. 28-10) are brought about by rotating each purine base about its glycosidic bond from the anti to syn conformations, whereas the sugars rotate in the pyrimidine nucleotides thereby maintaining their anti conformations.

Sugar Ring Pucker Is Limited to Only a Few of Its Possible Arrangements

The ribose ring has a certain amount of flexibility that significantly affects the conformation of the sugar – phosphate backbone. The vertex angles of a regular pentagon are 108°, a value quite close to the tetrahedral

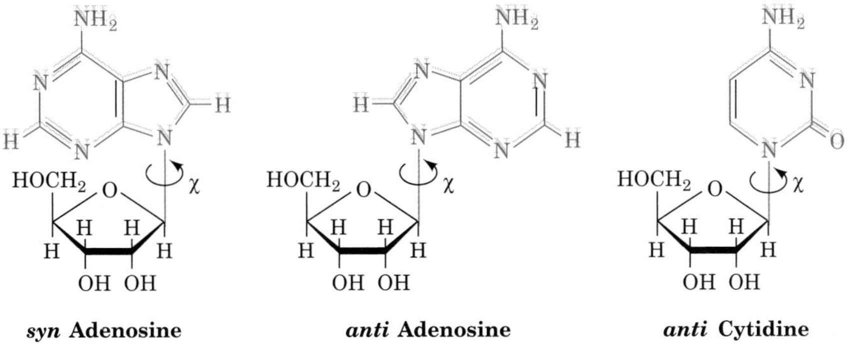

Figure 28-19
The sterically allowed orientations of purine and pyrimidine bases with respect to their attached ribose units.

syn Adenosine   *anti* Adenosine   *anti* Cytidine

(a)

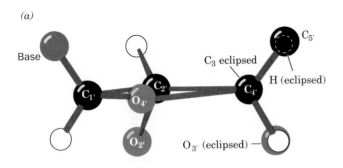

(b)

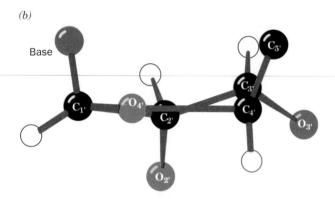

Figure 28-20
The substituents to (a) a planar ribose ring [here viewed down the C(3′)—C(4′) bond] are all eclipsed. The resulting steric strain is partially relieved by ring puckering such as in (b), a half-chair conformation in which C(3′) is the out-of-plane atom.

angle (109.5°), so that one might expect the ribofuranose ring to be nearly flat. However, the ring substituents are eclipsed when the ring is planar. To relieve the resultant crowding, which even occurs between hydrogen atoms, the ring **puckers;** that is, it becomes slightly nonplanar, so as to reorient the ring substituents (Fig. 28-20; this is readily observed by the manipulation of a skeletal molecular model).

One would, in general, expect only three of a ribose ring's five atoms to be coplanar since three points define a plane. Nevertheless, in the great majority of the > 50 nucleoside and nucleotide crystal structures that have been reported, four of the ring atoms are coplanar to within a few hundreths of an Å and the remaining atom is out of this plane by several tenths of an Å (the **half-chair** conformation). If the out-of-plane atom is displaced to the same side of the ring as atom C(5′), it is said to have the **endo** conformation (Greek: *endon,* within), whereas displacement to the opposite side of the ring from C(5′) is known as the **exo** conformation (Greek: *exo,* out of). In the great majority of known nucleoside and nucleotide structures, the out-of-plane atom is either C(2′) or C(3′) (Fig. 28-21). C(2′)-*endo* is the most frequently occurring ribose pucker with C(3′)-*endo* and -*exo* also being common. Other ribose conformations are rare.

The ribose pucker is conformationally important in nu-

cleic acids because it governs the relative orientations of the phosphate substituents to each ribose residue. For instance, it is difficult to build a model of a double helical nucleic acid unless the sugars are either C(2′)-*endo* or C(3′)-*endo*. In fact, B-DNA has the C(2′)-*endo* conformation whereas A-DNA and RNA-11 are C(3′)-*endo*. In Z-DNA, the purine nucleotides are all C(3′)-*endo* and the pyrimidine nucleotides are C(2′)-*endo*, which is another reason why the repeating unit of Z-DNA is a dinucleotide. Note that the most common sugar puckers of independent nucleosides and nucleotides, molecules that are subject to few of the conformational constraints of double helices, are the same as those of double helices.

The Sugar–Phosphate Backbone Is Conformationally Constrained

If the torsion angles of the sugar–phosphate chain (Fig. 28-18) were completely free to rotate, there could probably be no stable nucleic acid structure. However, the comparison, by Muttaiya Sundaralingam, of some

(a)

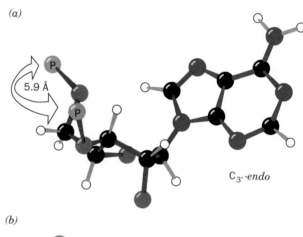

$C_{3'}$-*endo*

(b)

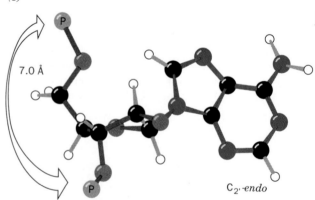

$C_{2'}$-*endo*

Figure 28-21
Nucleotides in (a) the C(3′)-*endo* conformation [on the same side of the sugar ring as C(5′)], and (b) the C(2′)-*endo* conformation which occur, respectively, in A-DNA and B-DNA. The distances between adjacent P atoms in the sugar–phosphate backbone are indicated. [After Saenger, W., *Principles of Nucleic Acid Structure,* p. 237, Springer–Verlag (1983).]

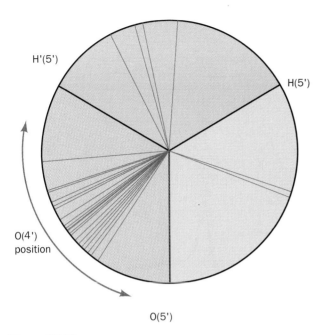

Figure 28-22
A conformational wheel showing the distribution of the torsion angle about the C(4')—C(5') bond (γ in Fig. 28-18) in 33 X-ray structures of nucleosides, nucleotides and polynucleotides. Each radial line represents the position of the C(4')—O(4') bond in a single structure relative to the substituents of C(5') as viewed from C(5') to C(4'). Note that most of the observed torsion angles fall within a relatively narrow range. [After Sundaralingam, M., *Biopolymers* **7**, 838 (1969).]

40 nucleoside and nucleotide crystal structures revealed that these angles are really quite restricted. For example, the torsion angle about the C(4')—C(5') bond (γ in Fig. 28-18) is rather narrowly distributed such that O(4') usually has a gauche conformation with respect to O(5') (Fig. 28-22). This is because the presence of the ribose ring together with certain noncovalent interactions of the phosphate group stiffens the sugar–phosphate chain by restricting its range of torsion angles. These restrictions are even greater in polynucleotides because of steric interference between residues.

The sugar–phosphate conformational angles of the various double helices are all reasonably strain free. *Double helices are therefore conformationally relaxed arrangements of the sugar–phosphate backbone.* Nevertheless, the sugar–phosphate backbone is by no means a rigid structure so that, upon strand separation, it assumes a random coil conformation.

C. Base Pairing

Base pairing is apparently a "glue" that holds together double-stranded nucleic acids. Only Watson–Crick pairs occur in the crystal structures of self-complementary oligonucleotides. It is therefore important to understand how Watson–Crick base pairs differ from other doubly hydrogen bonded arrangements of the bases that have reasonable geometries (e.g., Fig. 28-23).

Unconstrained A·T Base Pairs Assume Hoogsteen Geometry

When monomeric adenine and thymine derivatives are cocrystallized, the A·T base pairs that form invariably have adenine N(7) as the hydrogen bonding acceptor (**Hoogsteen geometry;** Fig. 28-23b) rather than N(1) (Watson–Crick geometry; Fig. 28-6). This suggests that Hoogsteen geometry is inherently more stable for A·T pairs than is Watson–Crick geometry. Apparently steric and other environmental influences make Watson–Crick geometry the preferred mode of base pairing in double helices. A·T pairs with Hoogsteen geometry are nevertheless of biological importance; for example, they help stabilize the tertiary structures of tRNAs (Section

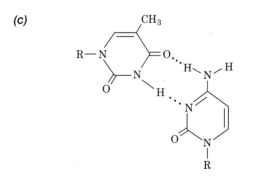

Figure 28-23
Some non-Watson–Crick base pairs. (a) the pairing of adenine residues in the crystal structure of 9-methyladenine. (b) The Hoogsteen pairing between adenine and thymine residues in the crystal structure of 9-methyladenine·1-methylthymine. (c) A hypothetical pairing between cytosine and thymine residues.

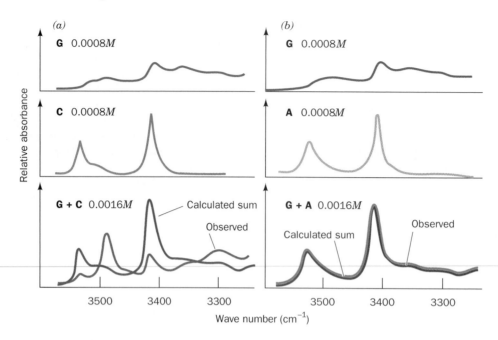

Figure 28-24

The IR spectra, in the N—H stretch region, of guanine, cytosine, and adenine derivatives, both separately and in the indicated mixtures. The solvent, CDCl₃, does not hydrogen bond with the bases and is relatively transparent in the frequency range of interest. (*a*) G + C. The line in the lower spectrum, which is the sum of the two upper spectra, is the calculated spectrum of G + C for noninteracting molecules.

The band near 3500 cm⁻¹ in the observed G + C spectrum is indicative of a specific hydrogen bonding association between G and C. (*b*) G + A. The close match between the calculated and observed spectra of the G + A mixture indicates that G and A do not significantly interact. [After Kyogoku, Y., Lord, R. C., and Rich, A., *Science* **154**, 5109 (1966).]

30-2B). In contrast, monomeric G·C pairs always co-crystallize with Watson–Crick geometry as a consequence of their triply hydrogen bonded structures.

Non-Watson–Crick Base Pairs Are of Low Stability

The bases of a double helix, as we have seen (Section 28-2A), associate such that any base pair position may interchangeably be A·T, T·A, G·C, or C·G without affecting the conformations of the sugar–phosphate chains. One might reasonably suppose that this requirement of **geometric complementarity** of the Watson–Crick base pairs, A with T and G with C, is the only reason that other base pairs do not occur in a double helical environment. In fact, this was precisely what was believed for many years after the DNA double helix was discovered.

Eventually, the failure to detect pairs of different bases in nonhelical environments other than A with T (or U) and G with C led Richard Lord and Rich to demonstrate, through spectroscopic studies, that *only the bases of Watson–Crick pairs have a high mutual affinity*. Figure 28-24*a* shows the infrared (IR) spectrum in the N—H stretch region of guanine and cytosine derivatives, both separately and in a mixture. The band in the spectrum of the G + C mixture that is not present in the spectra of either of its components is indicative of a specific hydrogen bonding interaction between G and C. Such an association, which can occur between like as

well as unlike molecules, may be described by ordinary mass action equations.

$$B_1 + B_2 \rightleftharpoons B_1 \cdot B_2 \qquad K = \frac{[B_1 \cdot B_2]}{[B_1][B_2]} \qquad [28.1]$$

From the analyses of IR spectra such as Fig. 28-24, the values of K for the various base pairs have been determined. The self-association constants of the Watson–Crick bases are given in the top of Table 28-3 (the hydrogen bonded association of like molecules is indicated

Table 28-3

Association Constants for Base Pair Formation

Base Pair	$K(M^{-1})^a$
Self-Association	
A·A	3.1
U·U	6.1
C·C	28
G·G	$10^3 - 10^4$
Watson–Crick Base Pairs	
A·U	100
G·C	$10^4 - 10^5$

[a] Data measured in deuterochloroform at 25°C.

Source: Kyogoku, Y., Lord, R. C., and Rich, A., *Biochim. Biophys. Acta* **179**, 10 (1969).

by the appearance of new IR bands as the concentration of the molecule is increased). The bottom of Table 28-3 lists the association constants of the Watson–Crick pairs. Note that each of these latter quantities is larger than the self-association constants of both their component bases so that Watson–Crick base pairs preferentially form from their constituents. In contrast, the non-Watson–Crick base pairs, A·C, A·G, C·U, and G·U, whatever their geometries, have association constants that are negligible compared with the self-pairing association constants of their constituents (e.g., Fig. 28-24*b*). *Evidently, a second reason that non-Watson–Crick base pairs do not occur in DNA double helices is that they have relatively little stability.* Conversely, the exclusive presence of Watson–Crick base pairs in DNA results, in part, from an **electronic complementarity** matching A to T and G to C. The theoretical basis of this electronic complementarity, which is an experimental observation, is obscure. This is because the approximations inherent in present day theoretical treatments make them unable to accurately account for the few $kJ \cdot mol^{-1}$ energy differences between specific and nonspecific hydrogen bonding associations. The double helical segments of many RNAs, however, contain occasional non-Watson–Crick base pairs, most often G·U, which have functional as well as structural significance (e.g., Sections 30-2B and D).

Hydrogen Bonds Do Not Stabilize DNA

It is clear that hydrogen bonding is required for the specificity of base pairing in DNA that is ultimately responsible for the enormous fidelity required to replicate DNA with almost no error (Section 31-3D). Yet, as is also true for proteins (Section 7-4B), *hydrogen bonding contributes little to the stability of the double helix.* For instance, adding the relatively nonpolar ethanol to an aqueous DNA solution, which strengthens hydrogen bonds, destabilizes the double helix as is indicated by its decreased T_m. This is because hydrophobic forces, which are largely responsible for DNA's stability (see Section 28-3D), are disrupted by nonpolar solvents. In contrast, *the hydrogen bonds between the base pairs of native DNA are replaced in denatured DNA by energetically more or less equivalent hydrogen bonds between the bases and water.*

D. Base Stacking and Hydrophobic Interactions

Purines and pyrimidines tend to form extended stacks of planar parallel molecules. This has been observed in the structures of nucleic acids (Figs. 28-5, 8, and 9) and in the several hundred reported X-ray crystal structures that contain nucleic acid bases. The bases in these structures are usually partially overlapped (e.g., Fig. 28-25). In fact, crystal structures of chemically related bases often exhibit similar stacking patterns. Apparently

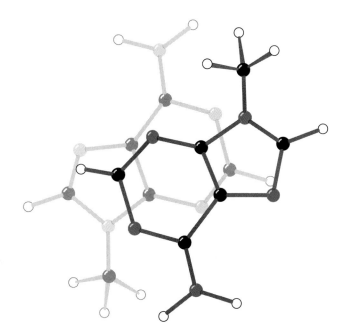

Figure 28-25
The stacking of adenine rings in the crystal structure of 9-methyladenine. The partial overlap of the rings is typical of the association between bases in crystal structures and in double helical nucleic acids. [After Stewart, R. F. and Jensen, L. H., *J. Chem. Phys.* **40**, 2071 (1964).]

stacking interactions, which in the solid state are a form of van der Waals interaction (Section 7-4A), have some specificity although certainly not as much as base pairing.

Nucleic Acid Bases Stack in Aqueous Solution

Bases aggregate in aqueous solution as has been demonstrated by the variation of osmotic pressure with concentration. The van't Hoff law of osmotic pressure is

$$\pi = RTm \qquad [28.2]$$

where π is the osmotic pressure, m is the molality of the solute (mol solute/kg solvent), R is the gas constant, and T is the temperature. The molecular mass, M, of an ideal solute can be determined from its osmotic pressure since $M = c/m$, where $c =$ g solute/kg solvent.

If the species under investigation is of known molecular mass but aggregates in solution, Eq. [28.2] must be rewritten:

$$\pi = \phi RTm \qquad [28.3]$$

where ϕ, the **osmotic coefficient,** indicates the solute's degree of association. ϕ varies from 1 (no association) to 0 (infinite association). The variation of ϕ with m for nucleic acid bases in aqueous solution (e.g., Fig. 28-26) is consistent with a model in which the bases aggregate in successive steps:

$$A + A \rightleftharpoons A_2 + A \rightleftharpoons A_3 + A \rightleftharpoons \cdots \rightleftharpoons A_n$$

where n is at least 5 (if the reaction goes to completion,

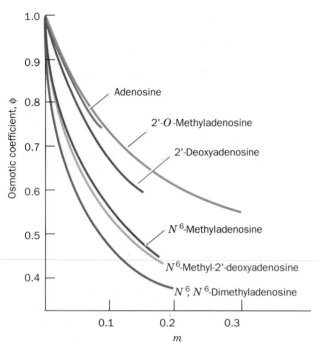

Figure 28-26
The variation of the osmotic coefficient ϕ with the molal concentrations m of adenosine derivatives in H_2O. The decrease of ϕ with increasing m indicates that these derivatives aggregate in solution. [After Broom, A. D., Schweizer, M. P., and Ts'o, P. O. P., *J. Am. Chem. Soc.* **89**, 3613 (1967).]

$\phi = 1/n$). This association cannot be a result of hydrogen bonding since N^6,N^6-**dimethyladenosine,**

N^6,N^6-**Dimethyladenosine**

which cannot form interbase hydrogen bonds, has a greater degree of association than does adenosine (Fig. 28-26). Apparently *the aggregation arises from the formation of stacks of planar molecules.* This model is corroborated by proton NMR studies: The directions of the aggregates' chemical shifts are compatible with a stacked but not a hydrogen bonded model. The stacking associations of monomeric bases are not observed in nonaqueous solutions.

Single-stranded polynucleotides also exhibit stacking interactions. For example, poly(A) shows a broad increase of UV absorbance with temperature (Fig. 28-27a). This hyperchromism is independent of poly(A) concentration so that it cannot be a consequence of intermolecular aggregation. Likewise, it is not due to intramolecular hydrogen bonding because poly(N^6,N^6-dimethyl A)

has a greater degree of hyperchromism than does poly(A). The hyperchromism must therefore arise from some sort of stacking associations within a single strand that melt out with increasing temperature. This is not a very cooperative process as is indicated by the broadness of the melting curve and the observation that short polynucleotides, including dinucleoside phosphates such as ApA, exhibit similar melting curves (Fig. 28-27b).

Nucleic Acid Structures Are Stabilized by Hydrophobic Forces

Stacking associations in aqueous solutions are largely stabilized by hydrophobic forces. One might reasonably suppose that hydrophobic interactions in nucleic acids are similar in character to those that stabilize protein structures. However, closer examination reveals that these two types of interactions are qualitatively different in character. Thermodynamic analysis of dinucleoside phosphate melting curves in terms of the reaction

Dinucleoside phosphate *(unstacked)* $\rightleftharpoons$
dinucleoside phosphate *(stacked)*

(Table 28-4) indicates that *base stacking is enthalpically driven and entropically opposed.* Thus the hydrophobic interactions responsible for the stability of base stacking associations in nucleic acids are diametrically opposite in character to those that stabilize protein structures (which are enthalpically opposed and entropically driven; Section 7-4C). This is reflected in the differing structural properties of these interactions. For example, the aromatic side chains of proteins are almost never stacked and the crystal structures of aromatic hydrocarbons such as benzene, which resemble these side chains, are characteristically devoid of stacking interactions.

Hydrophobic forces in nucleic acids are but poorly understood. The observation that they are different in character from the hydrophobic forces that stabilize proteins

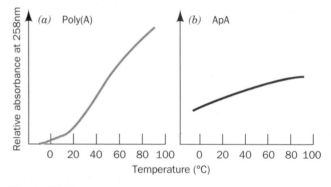

Figure 28-27
The broad temperature range of hyperchromic shifts at 258 nm of (a) poly(A) and (b) ApA is indicative of noncooperative conformational changes in these substances. Compare this figure with Fig. 28-15. [After Leng, M. and Felsenfeld, G., *J. Mol. Biol.* **15,** 457 (1966).]

Table 28-4
Thermodynamic Parameters for the Reaction

Dinucleoside phosphate $\rightleftharpoons$ *dinucleoside phosphate*
(unstacked)   (stacked)

Dinucleoside Phosphate	$\Delta H_{stacking}$ (kJ·mol^{-1})	$-T\Delta S_{stacking}$ (kJ·mol^{-1} at 25°C)
ApA	-22.2	24.9
ApU	-35.1	39.9
GpC	-32.6	34.9
CpG	-20.1	21.2
UpU	-32.6	36.2

Source: Davis, R. C. and Tinoco, I., Jr., *Biopolymers* **6**, 230 (1968).

is nevertheless not surprising because the nitrogenous bases are considerably more polar than the hydrocarbon residues of proteins that participate in hydrophobic bonding. There is, however, no theory available that adequately explains the nature of hydrophobic forces in nucleic acids (our understanding of hydrophobic forces in proteins, it will be recalled, is similarly incomplete). They are complex interactions of which base stacking is probably a significant component. Whatever their origins, hydrophobic forces are of central importance in determining nucleic acid structures.

E. Ionic Interactions

Any theory of the stability of nucleic acid structures must take into account the electrostatic interactions of their charged phosphate groups. Unfortunately, the theory of polyelectrolytes is, as yet, incapable of making reliable predictions of molecular conformations. We can, however, make experimental observations.

The melting temperature of duplex DNA increases with the cation concentration because these ions electrostatically shield the anionic phosphate groups from each other. The observed relationship for Na$^+$ is

$$T_m = 41.1\, X_{G+C} + 16.6 \log[\text{Na}^+] + 81.5 \quad [28.4]$$

where X_{G+C} is the mole fraction of G·C base pairs (recall that T_m increases with the G + C content); the equation is valid in the ranges $0.3 < X_{G+C} < 0.7$ and $10^{-3}M < [\text{Na}^+] < 1.0M$. Other monovalent cations such as Li$^+$ and K$^+$ have similar nonspecific interactions with phosphate groups. Divalent cations, such as Mg^{2+}, Mn^{2+}, and Co^{2+}, in contrast, specifically bind to phosphate groups so that *divalent cations are far more effective shielding agents for nucleic acids than are monovalent cations*. For example, an Mg^{2+} ion has an influence on the DNA double helix comparable to that of 100 to 1000 Na$^+$ ions. Indeed, enzymes that mediate reactions with nucleic acids or just nucleotides (e.g., ATP) usually require Mg^{2+} for activity.

4. NUCLEIC ACID FRACTIONATION

In Chapter 5 we considered the most commonly used procedures for isolating and, to some extent, characterizing proteins. Most of these methods, often with some modification, are also regularly used to fractionate nucleic acids according to size, composition, and sequence. There are also many techniques that are applicable only to nucleic acids. In this section we shall outline some of the most useful of the separation procedures that are specific for nucleic acids.

A. Solution Methods

Nucleic acids are invariably associated with proteins. Once cells have been broken open (Section 5-1B), the nucleic acids must be deproteinized. This may be accomplished by shaking (very gently if high molecular mass DNA is being isolated) the protein–nucleic acid mixture with a phenol solution and/or a CHCl$_3$–isoamyl alcohol mixture so that the protein precipitates and can be removed by centrifugation. Alternatively, the protein can be dissociated from the nucleic acids by detergents, guanidinium chloride, or high salt concentrations, or it can be enzymatically degraded by proteases such as pronase. In all cases, the nucleic acids, a mixture of RNA and DNA, can then be isolated by precipitation with ethanol. The RNA can be recovered from such precipitates by treating them with pancreatic DNase to eliminate the DNA. Conversely, the DNA can be freed of RNA by treatment with RNase. Alternatively, RNA and DNA may be separated by ultracentrifugation (Section 28-4D).

In all these and subsequent manipulations, the nucleic acids must be protected from degradation by nucleases that occur both in the experimental materials and on human hands. Nucleases may be inhibited by the presence of chelating agents such as EDTA, which sequester the divalent metal ions that nucleases require for activity. In cases where no nuclease activity can be tolerated, all glassware must be autoclaved to heat denature the nucleases and the experimenter should wear plastic gloves. Nevertheless, nucleic acids are generally easier to handle than proteins because their lack of a tertiary structure, in most cases, makes them relatively tolerant of extreme conditions.

B. Chromatography

Many of the chromatographic techniques that are used to separate proteins (Section 5-3) are also applicable to nucleic acids. Paper chromatography and thin layer chromatography are useful in fractionating oligonucleotides. They have been largely replaced, however, by the more powerful techniques of HPLC, particularly those using reverse-phase chromatography. Larger nucleic acids are often separated by procedures that include ion exchange chromatography and gel filtration chromatography.

Hydroxyapatite Binds Double-Stranded DNA More Tightly Than Single-Stranded DNA

Hydroxyapatite (a form of calcium phosphate; Section 5-3E) is particularly useful in the chromatographic purification and fractionation of DNA. Double-stranded DNA binds to hydroxyapatite more tightly than do most other molecules. Consequently DNA can be rapidly isolated by passing a cell lysate through a hydroxyapatite column, washing the column with a phosphate buffer of concentration low enough to release only the RNA and proteins, and then eluting the DNA with a concentrated phosphate solution.

Single-stranded DNA elutes from hydroxyapatite at a lower phosphate concentration than does double-stranded DNA (Fig. 28-28). This phenomenon forms the basis of a technique, known as **thermal chromatography,** for separating DNA according to its base composition. A hydroxyapatite column to which double-stranded DNA is bound is eluted with a phosphate buffer that releases only single-stranded DNA while the temperature of the column is gradually increased. As the DNA melts and is converted to the single-stranded form it is eluted from the column. Since the T_m of a duplex DNA varies with its G + C content (Eq. [28.4]), thermal chromatography permits the fractionation of double-stranded DNA according to its base composition.

Messenger RNAs Can Be Isolated by Affinity Chromatography

Affinity chromatography is useful in isolating specific nucleic acids. For example, most eukaryotic messenger RNAs (mRNAs) have a poly(A) sequence at their 3' ends (Section 28-4A). They can be isolated on agarose or cellulose to which poly(U) is covalently attached. The poly(A) sequences specifically bind to the complementary poly(U) in high salt and at low temperatures and can later be released by altering these conditions. Moreover, if the (partial) sequence of an mRNA is known (e.g., as deduced from the corresponding protein's amino acid sequence), the complementary DNA strand may be synthesized (via methods discussed in Section 28-7) and used to isolate that particular mRNA.

C. Electrophoresis

Nucleic acids of a given type may be separated by polyacrylamide gel electrophoresis (Sections 5-4B and C) because their electrophoretic mobilities in such gels vary inversely with their molecular masses. However, DNAs of more than a few thousand base pairs cannot penetrate even a weakly cross-linked polyacrylamide gel. This difficulty is partially overcome through the use

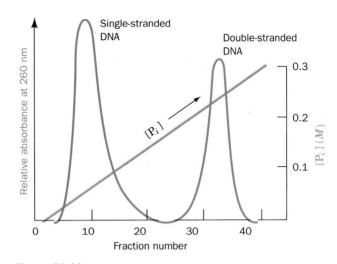

Figure 28-28
The chromatographic separation of single-stranded and duplex DNAs on hydroxyapatite by elution with a solution of increasing phosphate concentration.

of agarose gels. By using gels with an appropriately low agarose content, relatively large DNAs in various size ranges may be fractionated. In this manner, **plasmids** (small, autonomously replicating DNA molecules that occur in bacteria and yeast), for example, may be separated from the larger chromosomal DNA of bacteria.

Very Large DNAs Are Separated by Pulsed-Field Gel Electrophoresis

The sizes of the DNAs that can be separated by conventional gel electrophoresis are limited to ~100,000 bp, even when gels containing as little as 0.1% agarose (which makes an extremely fragile gel) are used. However, the recent development of **pulsed-field gel electrophoresis (PFG)** by Charles Cantor and Cassandra Smith has extended this limit to DNAs with up to 10 million bp (6.6 million kD). The electrophoresis apparatus used in PFG has two or more pairs of electrodes arrayed around the periphery of an agarose slab gel. The different electrode pairs are sequentially pulsed for times varying from 0.1 to 1000 s depending on the sizes of the DNAs being separated. Gel electrophoresis of DNA requires that these elongated molecules worm their way through the gel's labyrinthine channels more or less in the direction from the cathode to the anode. If the direction of the electric field abruptly changes, these DNAs must reorient their long axes along the new direction of the field before they can continue their passage through the gel. The time required to reorient very long gel-embedded DNA molecules evidently increases with their size. Consequently, a judicious choice of electrode distribution and pulse lengths causes shorter DNAs to migrate through the gel faster than longer DNAs, thereby effecting their separation.

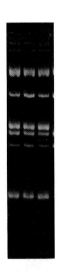

Figure 28-29
An agarose gel electrophoretogram of double helical DNA. After electrophoresis, the gel was soaked in a solution of ethidium bromide, washed, and photographed under UV light. The fluorescence of the ethidium cation is strongly enhanced by binding to DNA so that each fluorescent band marks a different sized DNA fragment. The three parallel lanes contain identical DNA samples so as to demonstrate the technique's reproducibility. [Photo by Elizabeth Levine. From Freifelder, D., *Biophysical Chemistry. Applications to Biochemistry and Molecular Biology* (2nd ed.), p. 294, W. H. Freeman (1982). Used by permission.]

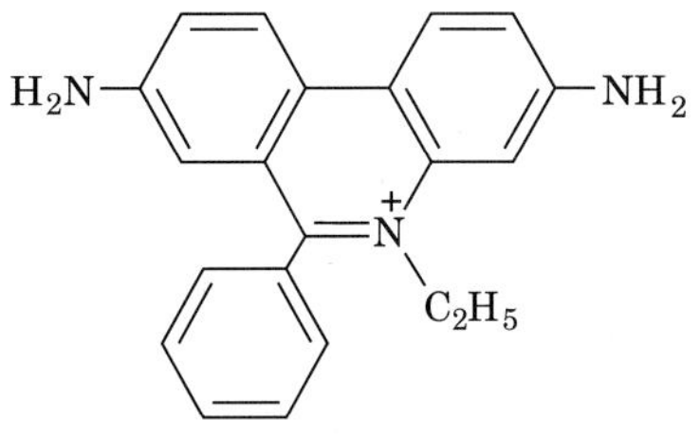

Ethidium

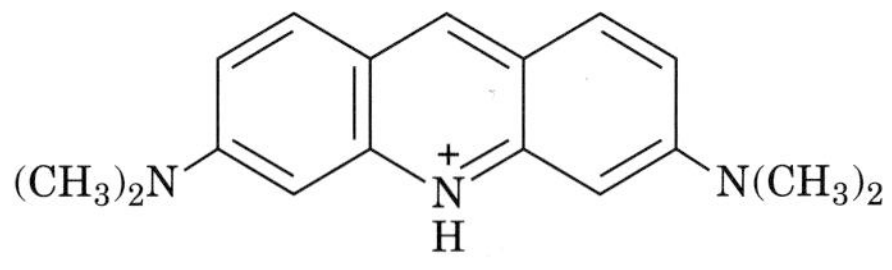

Acridine orange

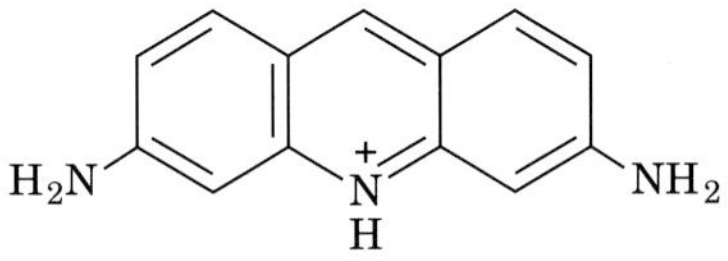

Proflavin

These dyes bind to duplex DNA by **intercalation** (slipping in between the stacked base pairs) where they exhibit a fluorescence under UV light that is far more intense than that of the free dye. As little as 50 ng of DNA may be detected in a gel by staining it with ethidium bromide (Fig. 28-29). Single-stranded DNA and RNA also stimulate the fluorescence of ethidium but to a lesser extent than does duplex DNA.

Duplex DNA Is Detected by Selectively Staining It with Intercalation Agents

The various DNA bands in a gel must be detected if they are to be isolated. Double-stranded DNA is readily stained by planar aromatic cations such as **ethidium ion, acridine orange,** or **proflavin.**

Southern Blotting Identifies DNAs with Specific Sequences

DNA with a specific base sequence may be identified through a procedure developed by Edwin Southern known as the **Southern transfer technique** or more colloquially as **Southern blotting** (Fig. 28-30). This pro-

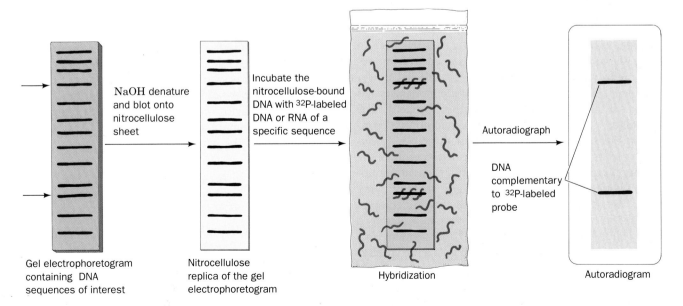

Gel electrophoretogram containing DNA sequences of interest

NaOH denature and blot onto nitrocellulose sheet

Nitrocellulose replica of the gel electrophoretogram

Incubate the nitrocellulose-bound DNA with 32P-labeled DNA or RNA of a specific sequence

Hybridization

Autoradiograph

DNA complementary to 32P-labeled probe

Autoradiogram

Figure 28-30
The detection of DNAs containing specific base sequences by the Southern transfer technique.

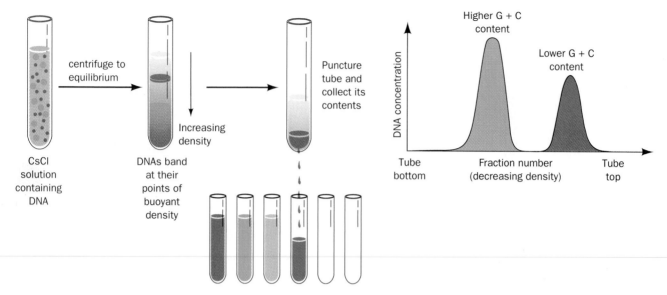

Figure 28-31
The separation of DNAs according to base composition by equilibrium density gradient ultracentrifugation in CsCl solution. An initially 8M CsCl solution forms a density gradient that varies linearly from ~1.80 g·cm^{-3} at the bottom of the centrifuge tube to ~1.55 g·cm^{-3} at the top. The amount of DNA in each fraction is estimated from its UV absorbance, usually at 260 nm.

cedure takes advantage of the valuable property of nitrocellulose that it tenaciously binds single-stranded but not duplex DNA. Following the gel electrophoresis of double-stranded DNA, the gel is soaked in 0.5M NaOH solution, which converts the DNA to the single-stranded form. The gel is then overlaid by a sheet of nitrocellulose paper which, in turn, is covered by a thick layer of paper towels and the entire assembly is compressed by a heavy plate. The liquid in the gel is thereby forced (blotted) through the nitrocellulose so that the single-stranded DNA binds to it at the same position it had in the gel (the transfer to nitrocellulose can alternatively be accomplished by an electrophoretic process named **electroblotting**). After vacuum drying the nitrocellulose at 80°C, which permanently fixes the DNA in place, the nitrocellulose sheet is moistened with a minimal quantity of solution containing ^{32}P-labeled single-stranded DNA or RNA that is complementary in sequence to the DNA of interest (the "probe"). The moistened filter is held at a suitable renaturation temperature for several hours to permit the probe to hybridize to its target sequence(s), washed to remove the unbound radioactive probe, dried, and then autoradiographed by placing it for a time over a sheet of X-ray film. The positions of the molecules that are complementary to the radioactive sequences are indicated by a blackening of the developed film. A DNA segment containing a particular base sequence (e.g., a gene specifying a certain protein) may, in this manner, be detected and isolated. Specific DNAs may likewise be detected by linking the probe to an enzyme that generates a colored or fluorescent deposit on the blot. Such nonradioactive detection techniques are desirable in a clinical setting because of the health hazards, disposal

problems, and the more cumbersome nature of autoradiographic methods.

Northern and Western Blotting, Respectively, Detect RNAs and Proteins

Variations of Southern transfer, which are punningly called **northern transfer (northern blotting)** and **western transfer (western blotting),** respectively detect specific RNAs and proteins. In northern blotting, RNA is immobilized on nitrocellulose paper and detected through the use of complementary radiolabeled RNA or DNA probes. In western blotting, a protein mixture is bound to nitrocellulose paper and specific proteins identified by their binding of antibodies raised against them. Nitrocellulose, however, binds proteins so tenaciously that, in some cases, it may interfere with their immunochemical identification. In such cases, the nitrocellulose paper can be replaced by paper derivatized with diazobenzyloxymethyl groups, which react with the proteins' primary amino groups so as to covalently couple them to the paper.

Paper—O—CH$_2$—O—CH$_2$—

Diazobenzyloxymethyl group
Protein

N$_2$

Paper—O—CH$_2$—O—CH$_2$—

Protein

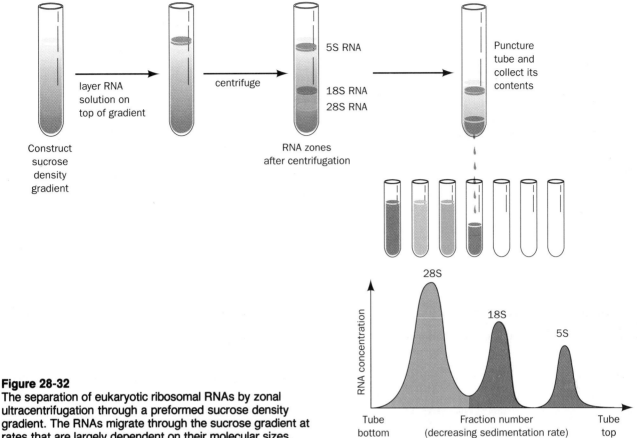

Figure 28-32
The separation of eukaryotic ribosomal RNAs by zonal ultracentrifugation through a preformed sucrose density gradient. The RNAs migrate through the sucrose gradient at rates that are largely dependent on their molecular sizes.

D. Ultracentrifugation

Equilibrium density gradient ultracentrifugation (Fig. 28-31; Section 5-5B) in CsCl constitutes one of the most commonly used DNA separation procedures. The bouyant density, ρ, of double-stranded Cs^+ DNA depends on its base composition:

$$\rho = 1.660 + 0.098\, X_{G+C} \qquad [28.5]$$

so that a CsCl density gradient fractionates DNA according to its base composition. For example, eukaryotic DNAs often contain minor fractions that band separately from the major species. Some of these **satellite bands** consist of mitochondrial and chloroplast DNAs. Another important class of satellite DNA is composed of **repetitive sequences** that are short segments of DNA tandemly (one behind the other) repeated hundreds, thousands, and in some cases, millions of times in a chromosome (Section 33-2B). Likewise, plasmids may be separated from bacterial chromosomal DNA by equilibrium density gradient ultracentrifugation.

Single-stranded DNA is $\sim 0.015\ \text{g} \cdot \text{cm}^{-3}$ denser than the corresponding double-stranded DNA so that the two may be separated by equilibrium density gradient ultracentrifugation. RNA is too dense to band in CsCl but does so in Cs_2SO_4 solutions. RNA–DNA hybrids will band in CsCl but at a higher density than the corresponding duplex DNA.

RNA may be fractionated by zonal ultracentrifugation through a sucrose gradient (Fig. 28-32; Section 5-5B). RNAs are separated by this technique largely on the basis of their size. In fact, ribosomal RNA, which constitutes the major portion of cellular RNA, is classified according to its sedimentation rate; for example, the RNA of the *E. coli* small ribosomal subunit is known as 16S RNA (Section 30-3A).

5. SUPERCOILED DNA

The circular genetic maps of viruses and bacteria implies that their chromosomes are likewise circular. This conclusion has been confirmed by electron micrographs in which circular DNAs are seen (Fig. 28-33). Some of these circular DNAs have a peculiar twisted appearance, a phenomenon that is known equivalently as **supercoiling, supertwisting,** or **superhelicity.** Supercoiling arises from a biologically important topological property of covalently closed circular duplex DNA that is the subject of this section. It is occasionally referred to as DNA's tertiary structure.

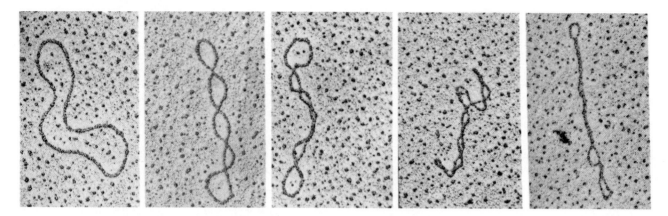

Figure 28-33
Electron micrographs of circular duplex DNAs that vary in their conformations from no supercoiling (*left*) to tightly supercoiled (*right*). [Electron micrographs by Laurien Polder. From Kornberg, A., *DNA Replication*, p. 29, W. H. Freeman (1980). Used by permission.]

A. Superhelix Topology

Consider a double helical DNA molecule in which both strands are covalently joined to form a circular duplex molecule as is diagrammed in Fig. 28-34 (each strand can only be joined to itself because the strands are antiparallel). *A geometric property of such an assembly is that its number of coils cannot be altered without first cleaving at least one of its polynucleotide strands.* You can easily demonstrate this to yourself with a buckled belt in which each edge of the belt represents a strand of DNA. The number of times the belt is twisted before it is buckled cannot be changed without unbuckling or cutting the belt (cutting a polynucleotide strand).

This phenomenon is mathematically expressed

$$L = T + W \qquad [28.6]$$

in which:

1. L, the **linking number,** is the number of times that one DNA strand winds about the other. This integer quantity is most easily counted when the molecule's duplex axis is constrained to lie in a plane (see below). However, *the linking number is invariant no matter how the circular molecule is twisted or distorted so long as both its polynucleotide strands remain covalently intact; the linking number is therefore a topological property of the molecule.*

2. T, the **twist,** is the number of complete revolutions that one polynucleotide strand makes about the duplex axis in the particular conformation under consideration. By convention, T is positive for right-handed duplex turns so that, for B-DNA in solution, the twist is normally the number of base pairs divided by 10.5 (the number of base pairs per turn of the B-DNA double helix under physiological conditions; see Section 28-5B).

3. W, the **writhing number,** is the number of turns that the duplex axis makes about the superhelix axis in the conformation of interest. *It is a measure of the DNA's superhelicity.* The difference between writhing and twisting is illustrated by the familiar example in Fig. 28-35. $W = 0$ when DNA's duplex axis is constrained to lie in a plane (e.g., Fig. 28-34); then $L = T$ so that L may be evaluated by counting the DNA's duplex turns.

The two DNA conformations diagrammed on the right of Fig. 28-36 are topologically equivalent; that is, they have the same linking number, L, but differ in their twists and writhing numbers. Note that T and W need not be integers, only L.

Since L is constant in an intact duplex DNA circle, for

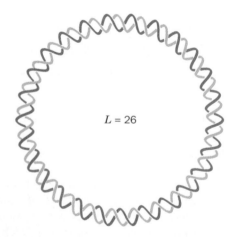

$$L = 26$$

Figure 28-34
A schematic diagram of covalently closed circular duplex DNA that has 26 double helical turns. Its two polynucleotide strands are said to be **topologically bonded** to each other because, although they are not covalently linked, they cannot be separated without breaking covalent bonds.

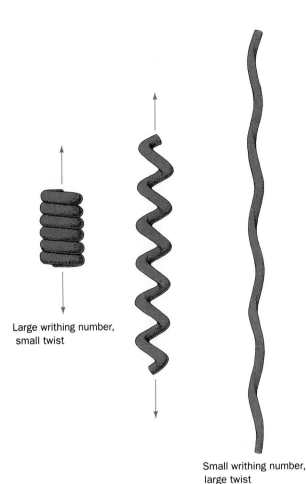

Large writing number, small twist

Small writhing number, large twist

Figure 28-35

The difference between writhing and twist as demonstrated by a coiled telephone cord. In its relaxed state (*left*), the cord is in a helical form that has a large writhing number and a small twist. As the coil is pulled out (*middle*) until it is nearly straight (*right*), its writhing number becomes small as its twist becomes large.

every new double helical twist, ΔT, there must be an equal and opposite superhelical twist; that is, $\Delta W = -\Delta T$. For example, a closed circular DNA without supercoils (Fig. 28-36, *upper right*) can be converted to a negatively supercoiled conformation (Fig. 28-36, *lower right*) by winding the duplex helix the same number of positive (right handed) turns.

Supercoils May Be Toroidal or Interwound

A supercoiled duplex may assume two topologically equivalent forms:

1. **A toroidal helix** in which the duplex axis is wound as if about a cylinder (Fig. 28-37a).

2. An **interwound helix** in which the duplex axis is twisted around itself (Fig. 28-37b).

Note that these two interconvertible superhelical forms have opposite handedness. Since left-handed toroidal turns may be converted to left-handed duplex turns (see Fig. 28-35), left-handed toroidal turns and right-handed

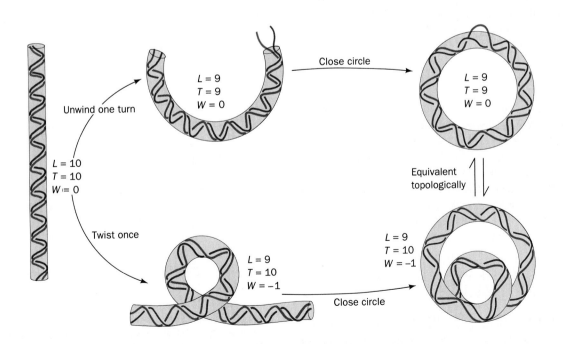

Figure 28-36

Two ways of introducing one supercoil into a DNA with 10 duplex turns. The two closed circular forms shown (*right*) are topologically equivalent; that is, they are interconvertible without breaking any covalent bonds. The linking number L, twist T, and writhing number W are indicated for each form. Strictly speaking, the linking number is only defined for a covalently closed circle.

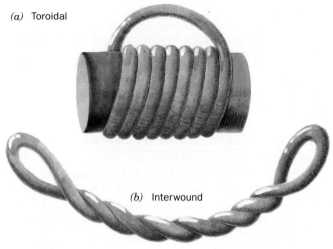

(a) Toroidal

(b) Interwound

Figure 28-37
A rubber tube that has been (a) toroidally coiled around a cylinder with its ends joined such that it has no twist, jumps to (b) an interwound helix with the opposite handedness when the cylinder is removed. Neither the linking number, twist, nor writing number are changed in this transformation.

interwound turns both have negative writing numbers. Thus an underwound duplex ($T <$ number of bp/10.5), for example, will tend to develop right-handed interwound or left-handed toroidal superhelical turns when the constraints causing it to be underwound are released (the molecular forces in a DNA double helix promote its winding to its normal number of helical turns).

Supercoiled DNA Is Relaxed by Nicking One Strand

Supercoiled DNA may be converted to **relaxed circles** (as appears in the left-most panel of Fig. 28-33) by treatment with **pancreatic DNase I,** an **endonuclease** (an enzyme that cleaves phosphodiester bonds within a polynucleotide strand), which cleaves only one strand of a duplex DNA. *One single-strand nick is sufficient to relax a supercoiled DNA.* This is because the sugar–phosphate chain opposite the nick is free to swivel about its backbone bonds (Fig. 28-18) so as to change the molecule's linking number and thereby alter its superhelicity. Supercoiling builds up elastic strain in a DNA circle, much as it does in a rubber band. This is why the relaxed state of a DNA circle is not supercoiled.

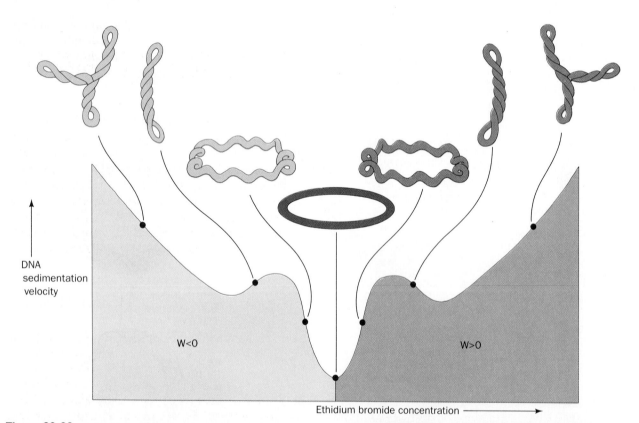

DNA sedimentation velocity

W<0

W>0

Ethidium bromide concentration ⟶

Figure 28-38
The sedimentation rate of closed circular duplex DNA as a function of ethidium bromide concentration. The intercalation of ethidium between the base pairs locally unwinds the double helix which, since the linking number of the circle is constant, is accompanied by an equivalent increase in the writing number. As the superhelix unwinds, it becomes less compact and sediments more slowly. At the low point on the curve, the DNA circles have bound sufficient ethidium to become fully relaxed. As the ethidium concentration is further increased, the DNA supercoils in the opposite direction. The supertwisted appearances of the depicted DNAs have been verified by electron microscopy. [After Bauer, W. R., Crick, F. H. C., and White, J. H., *Sci. Am.* **243**(1): 129 (1980). Copyright © 1980 by Scientific American, Inc.]

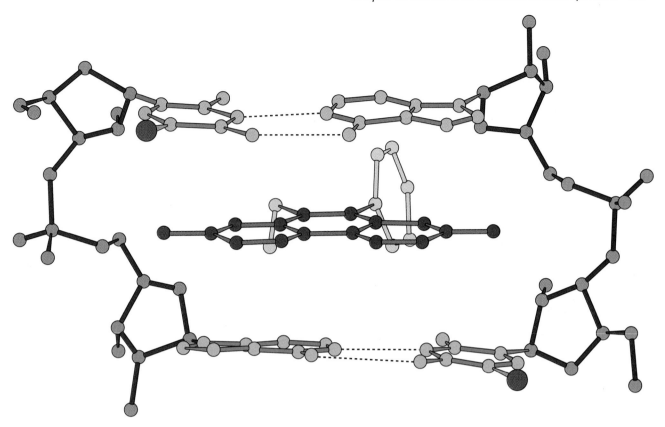

Figure 28-39
The X-ray structure of a complex of ethidium with 5-iodo UpA. Ethidium (*red*) intercalates between the base pairs of the double helically paired dinucleoside phosphate and thereby provides a model for the binding of ethidium to duplex DNA. [After Tsai, C.-C., Jain, S. C., and Sobell, H. M., *Proc. Natl. Acad. Sci.* **72,** 629 (1975).]

B. Measurements of Supercoiling

Supercoiled DNA, far from being just a mathematical curiosity, has been widely observed in nature. In fact, its discovery in polyoma virus DNA by Jerome Vinograd stimulated the elucidation of the topological properties of superhelices rather than *vice versa*.

Intercalating Agents Control Supercoiling by Unwinding DNA

All naturally occurring DNA circles are underwound; that is, their linking numbers are less than those of their corresponding relaxed circles. This phenomenon has been established by observing the effect of ethidium binding on the sedimentation rate of circular DNA (Fig. 28-38). Intercalating agents such as ethidium alter a circular DNA's degree of superhelicity because they cause the DNA double helix to unwind by $\sim 26°$ at the site of the intercalated molecule (Fig. 28-39). $W < 0$ in an unconstrained underwound circle because of the tendency of a duplex DNA to maintain its normal twist of 1 turn/10.5 bp. The titration of a DNA circle by ethidium unwinds the duplex (decreases T), which must be accompanied by a compensating increase in W. This, at first, lessens the superhelicity of an underwound circle. However, as the circle binds more and more ethidium, its value of W passes through zero (relaxed circles) and then becomes positive so that the circle again becomes superhelical. Thus the sedimentation rate of underwound DNAs, which is a measure of their compactness and therefore their superhelicity, passes through a minimum as the ethidium concentration increases. This is what is observed with native DNAs (Fig. 28-38). In contrast, the sedimentation rate of an overwound circle would only increase with increasing ethidium concentration.

DNAs Are Separated According to Their Linking Number by Gel Electrophoresis

Gel electrophoresis also separates similar molecules on the basis of their compactness so that the rate of migration of a circular duplex DNA increases with its degree of superhelicity. The agarose gel electrophoresis pattern of a population of chemically identical DNA molecules with different linking numbers therefore consists of a series of discrete bands (Fig. 28-40). The molecules in a given band all have the same linking number and differ from those in adjacent bands by $\Delta L = \pm 1$.

Comparison of the electrophoretic band patterns of **simian virus 40 (SV40)** DNA that had been enzymatic-

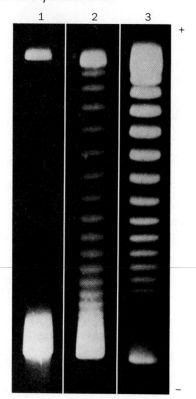

Figure 28-40
The agarose gel electrophoresis pattern of SV40 DNA. Lane 1 contains the negatively supercoiled native DNA (*lower band*). In lanes 2 and 3, the DNA has been exposed for 5 and 30 min, respectively, to an enzyme, known as a Type I topoisomerase (Section 28-5C), that relaxes the supercoils one at a time. Neighboring bands contain DNAs that differ by $\Delta L = \pm 1$. [From Keller, W., *Proc. Natl. Acad. Sci.* **72**, 2553 (1975).]

C. Topoisomerases

The normal biological functioning of DNA occurs only if it is in the proper topological state. In such basic biological processes as RNA transcription, DNA replication and genetic recombination, the recognition of a base sequence requires the local separation of complementary polynucleotide strands. The negative supercoiling of naturally occurring DNAs results in a torsional strain that promotes such separations since it tends to unwind the duplex helix (an increase in W must be accompanied by a decrease in T). *If DNA lacks the proper superhelical tension, the above vital processes occur quite slowly, if at all.*

The supercoiling of DNA is controlled by a remarkable group of enzymes known as **topoisomerases.** They are so named because they alter the topological state (linking

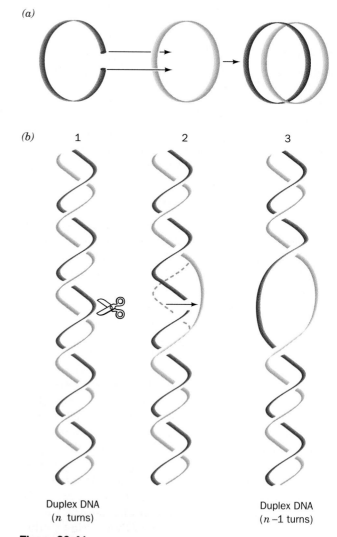

Figure 28-41
By cutting a single-stranded DNA, passing a loop of it through the break and then resealing the break, Type I topoisomerase can (a) catenate two single-stranded circles or (b) unwind duplex DNA by one turn.

ally relaxed to varying degrees and then resealed (Fig. 28-40) reveals that 26 bands separate native from fully relaxed SV40 DNAs. Native SV40 DNA therefore has $W = -26$ (although it is somewhat heterogeneous in this quantity). Since SV40 DNA consists of 5243 bp, it has 1 superhelical turn per ~ 19 duplex turns. Such a **superhelix density** is typical of circular DNAs from various biological sources.

DNA in Physiological Solution Has 10.5 Base Pairs Per Turn

The insertion, using genetic engineering techniques (Section 28-8B), of an additional x base pairs into a superhelical DNA will increase its linking number by $x/h°$, where $h°$ is the number of base pairs per duplex turn. Such an insertion will shift the position of a band in the DNA's gel electrophoretic pattern by $x/h°$ of the distance to the preceding band. By measuring the effects of several such insertions James Wang established that $h° = 10.5 \pm 0.1$ bp for B-DNA in solution under physiological conditions.

number) of circular DNA but not its covalent structure. There are two classes of topoisomerases:

1. **Type I topoisomerases** act by creating transient single-strand breaks in DNA.

2. **Type II topoisomerases** act by making transient double-strand breaks in DNA.

Type I Topoisomerases Incrementally Relax Supercoiled DNA

Type I topoisomerases, which are also known as **nicking–closing enzymes,** are monomeric proteins of 100 to 120 kD that are widespread in both prokaryotes and eukaryotes. *They catalyze the relaxation of negative supercoils in DNA by increasing its linking number in increments of one turn.* The exposure of a negatively supercoiled DNA to nicking–closing enzyme sequentially increases its linking number until the supercoil is entirely relaxed. A clue to the mechanism of action of this enzyme was provided by the observation that it reversibly **catenates** (interlinks) single-stranded circles (Fig. 28-41*a*). Apparently the enzyme operates by cutting a single strand, passing a single-strand loop through the resulting gap, and then resealing the break (Fig. 28-41*b*) thereby twisting double helical DNA by one turn. In support of this hypothesis, the denaturation of prokaryotic nicking–closing enzyme that has been incubated with single-stranded circular DNA yields a linear DNA that has its 5′-terminal phosphoryl group linked to the enzyme via a phosphotyrosine diester linkage.

Denatured eukaryotic nicking–closing enzymes are instead linked to the 3′ end of DNA in a like manner. *By forming such covalent enzyme-DNA intermediates, the free energy of the cleaved phosphodiester bond is preserved so that no energy input is required to reseal the nick.*

Type II Topoisomerases Supercoil DNA at the Expense of ATP Hydrolysis

Prokaryotic Type II topoisomerases, which are also known as **DNA gyrases,** are ~400-kD proteins that consist of two pairs of subunits designated *A* and *B.* *These enzymes catalyze the stepwise negative supercoiling of DNA with the concomitant hydrolysis of an ATP to ADP + P_i.* In the absence of ATP, DNA gyrase relaxes negatively supercoiled DNA but at a relatively slow rate. It can also tie knots in double-stranded circles as well as catenate them. Eukaryotic Type II topoisomerases only relax supercoils; they neither generate them nor hydrolyze ATP. DNA supercoiling in eukaryotes is generated somewhat differently (Section 33-1B).

Prokaryotic DNA gyrases are specifically inhibited by two classes of antibiotic. One of these classes includes the *Streptomyces* derived **novobiocin** and the other contains the clinically useful antibacterial agent **oxolinic acid.**

Novobiocin

Oxolinic acid

Both classes of antibiotic profoundly inhibit bacterial DNA replication and RNA transcription thereby demonstrating the importance of supercoiled DNA in these processes. Studies using antibiotic resistant *E. coli* mutants demonstrated that oxolinic acid associates with DNA gyrase's *A* subunit and novobiocin binds to its *B* subunit.

The gel electrophoretic pattern of duplex circles that have been exposed to DNA gyrase, with or without ATP, show a band pattern in which the linking numbers differ by increments of two rather than one as occurs with nicking–closing enzymes. This observation is strong evidence that *DNA gyrase acts by cutting both strands of a duplex, passing the duplex through the break and resealing it* (Fig. 28-42). This hypothesis is corroborated by the observation that when DNA gyrase is incubated with DNA and oxolinic acid, and subsequently denatured with guanidinium chloride, its *A* subunits remain covalently linked to the 5′ ends of both cut strands through phosphotyrosine linkages. Apparently oxolinic acid interferes with gyrase action by blocking the strand breaking–rejoining process. Novobiocin, on the other hand, prevents ATP from binding to the enzyme.

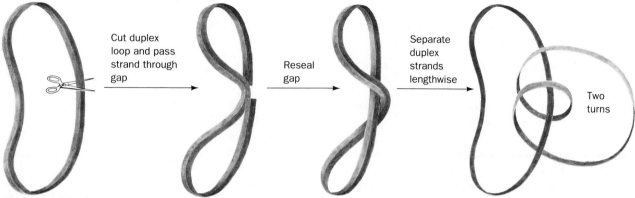

Figure 28-42

A demonstration, in which DNA is represented by a ribbon, that cutting a duplex circle, passing the strand through the resulting gap, and then resealing the break changes the linking number by two. Separating the resulting strands (slitting the ribbon along its length; *right*), indicates that one strand makes two complete revolutions about the other.

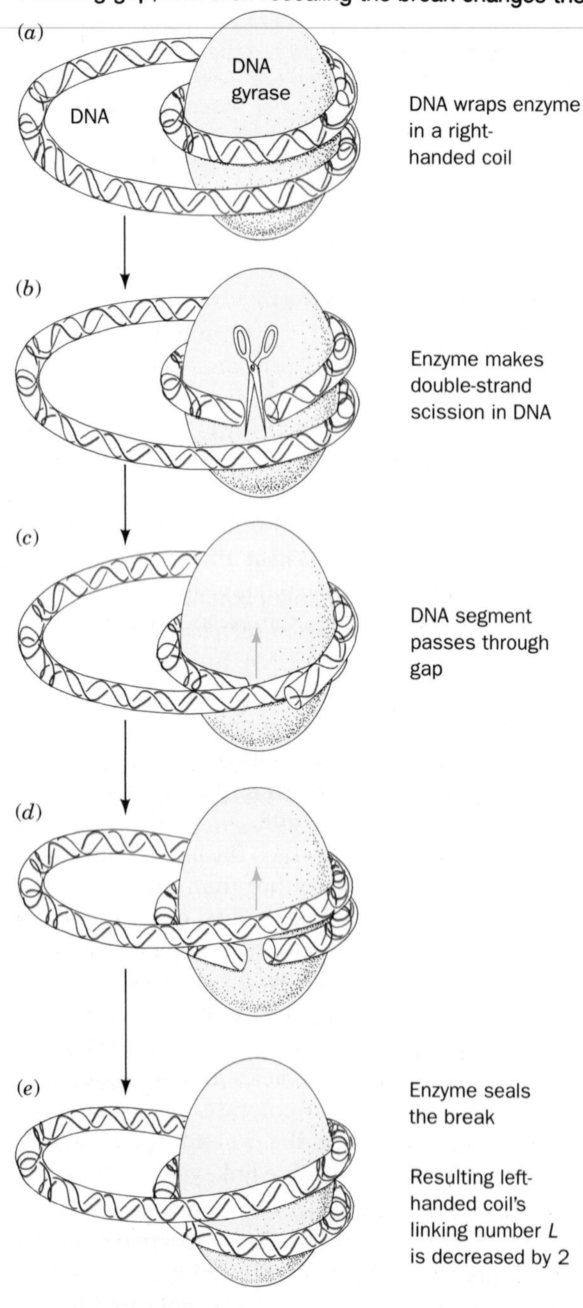

(a)

DNA gyrase

DNA

DNA wraps enzyme in a right-handed coil

(b)

Enzyme makes double-strand scission in DNA

(c)

DNA segment passes through gap

(d)

(e)

Enzyme seals the break

Resulting left-handed coil's linking number *L* is decreased by 2

The exposure of a gyrase–DNA complex to *Staphylococcal* nuclease protects the DNA from nuclease degradation in a 140 bp fragment that is roughly centered on the gyrase cleavage site. The length of this protected fragment suggests that the DNA is wrapped around the enzyme. This observation led Nicholas Cozzarelli to propose the mechanism of gyrase–DNA action diagrammed in Fig. 28-43. It is named the **sign inversion** mechanism because it converts a right-handed toroidal supercoil to a left-handed toroidal supercoil.

6. NUCLEIC ACID SEQUENCING

The basic strategy of nucleic acid sequencing is identical to that of protein sequencing (Section 6-1). It involves:

1. The specific degradation and fractionation of the polynucleotide of interest to fragments small enough to be fully sequenced.

2. The sequencing of the individual fragments.

3. The ordering of the fragments by repeating the preceding steps using a degradation procedure that yields a set of polynucleotide fragments that overlap the cleavage points in the first such set.

Before about 1975, however, nucleic acid sequencing techniques lagged far behind those of protein sequencing largely because there were no available endonucle-

Figure 28-43

The sign inversion mechanism of DNA gyrase action. The duplex DNA is initially wrapped about the enzyme in a right-handed toroidal coil **(1)**. The enzyme then makes a double-strand scission in the DNA **(2)**, passes a DNA segment through the gap **(3, 4)**, and reseals the break **(5)**. This changes the handedness of the coil to the left-handed form so that the DNA's linking number *L* is decreased by 2.

ases that were specific for sequences greater than a nucleotide. Rather, nucleic acids were cleaved into relatively short fragments by partial digestion with enzymes such as **ribonuclease T1** (from *Aspergillus oryzae),* which cleaves RNA after guanine residues, or **pancreatic ribonuclease A,** which does so after pyrimidine residues. Moreover, there is no reliable polynucleotide reaction analogous to the Edman degradation for proteins (Section 6-1A). Consequently, the polynucleotide fragments were sequenced by their partial digestion with either of two **exonucleases** (enzymes that sequentially cleave nucleotides from the end of a polynucleotide strand): **snake venom phosphodiesterase,** which removes residues from the 3′ end of polynucleotides (Fig. 28-44), or **spleen phosphodiesterase,** which does so from the 5′ end. The resulting oligonucleotide fragments were identified from their chromatographic and electrophoretic mobilities. Sequencing RNA in this manner is a lengthy and painstaking procedure.

The first biologically significant nucleic acid to be sequenced was that of yeast alanine tRNA (Section 30-2A). The sequencing of this 76-nucleotide molecule by Robert Holley, a labor of 7 years, was completed in 1965, some 12 years after Frederick Sanger had determined the amino acid sequence of insulin. This was followed, at an accelerating pace, by the sequencing of numerous species of tRNAs and the 5S ribosomal RNAs (Section 30-3A) from several organisms. The art of RNA sequencing by these techniques reached its zenith in 1976 with the sequencing, by Walter Fiers, of the entire 3569 nucleotide genome of the **bacteriophage MS2.** In comparison, DNA sequencing was in a far more primitive state because of the lack of available DNA endonucleases with any sequence specificity.

Since 1975 there has been dramatic progress in nucleic acid sequencing technology. This has been made possible by three advances:

1. The discovery of **restriction endonucleases,** enzymes that cleave duplex DNA at specific sequences.

2. The development of DNA sequencing techniques.

3. The development of **molecular cloning** techniques (Section 28-8), which permit the acquisition of any identifiable DNA segment in the amounts required for sequencing. Their use is necessary because most specific DNA sequences are normally present in a genome in only a single copy.

These procedures are largely responsible for the present "revolution" in molecular biology that is discussed in succeeding chapters. The use of restriction endonucleases and DNA sequencing techniques are the subject of this section.

The pace of nucleic acid sequencing has become so rapid that directly determining a protein's amino acid sequence is far more difficult than determining the base

```
G C A C U U G A
         |  snake venom
         |  phosphodiesterase
         ↓
G C A C U U G A
G C A C U U G
G C A C U U
G C A C U
G C A C
G C A
G C   + Mononucleotides
```

Figure 28-44
The sequence determination of an oligonucleotide by partial digestion with snake venom phosphodiesterase. This enzyme sequentially cleaves the nucleotides from the 3′ end of a polynucleotide that has a free 3′-OH group. Partial digestion of an oligonucleotide with snake venom phosphodiesterase yields a mixture of fragments of all lengths, as indicated, that may be chromatographically separated. Comparison of the base compositions of pairs of fragments that differ in length by one nucleotide establishes the identity of the 3′-terminal nucleotide of the larger fragment. In this way the base sequence of the oligonucleotide may be elucidated.

sequence of its corresponding gene (although amino acid and base sequences provide complementary information; Section 6-1K). There has been such a flood of new DNA sequences—so far ~40 million bases and increasing at the rate of 10 million bases per year—that only computers can keep track of them. A recent high point in the sequencer's art was the determination of the entire 172,282 bp sequence of **Epstein–Barr virus** (human **herpesvirus**) DNA. Indeed, preparations are under way to sequence the 2.9 billion bp human genome (although the magnitude of this project is such that if the DNA sequencing rate can be increased, as it is hoped, to 1 million bp/day, the project will still take nearly 10 years to complete).

A. Restriction Endonucleases

Bacteriophages that propagate efficiently on one bacterial strain, such as *E. coli* K12, have a very low rate of infection (~0.001%) in a related bacterial strain such as *E. coli* B. However, the few viral progeny of this latter infection propagate efficiently in the new host but only poorly in the original host. What is the molecular basis of this **host-specific modification** system? Werner Arber showed that it results from a **restriction-modification system** in the bacterial host that consists of a **restriction endonuclease** and a matched **modification methylase.** *The restriction endonuclease recognizes a specific base sequence of four to eight bases in double-stranded DNA and cleaves both strands of the duplex.* The modification methylase methylates a specific base (usually at the amino group of an adenine residue or the 5-position of a cytosine) in the same base sequence recognized by the restriction enzyme. The restriction enzyme does not cleave such a modified DNA. A newly replicated strand of bacterial DNA, which is protected from degradation by the methylated parent strand with which it forms a duplex, is modified before the next cycle of replication. *The restriction-modification system is therefore thought to protect the bacterium against invasion by foreign (usually viral) DNAs* which, once they have been cleaved by a restriction endonuclease, are further degraded by bacterial exonucleases. Invading DNAs are only rarely modified before being attacked by restriction enzymes. Once a viral genome becomes modified, however, it is able to reproduce in its new host. Its progeny, however, are no longer modified in the way that permits them to propagate in the original host.

There are three known types of restriction endonucleases. **Type I** and **Type III** restriction enzymes each carry both the endonuclease and the methylase activity on a single protein molecule. Type I restriction enzymes cleave the DNA at a possibly random site located at least 1000 bp from the recognition sequence, whereas Type III enzymes do so 24 to 26 bp distant from the recognition sequence. However, **Type II** restriction enzymes, which were discovered and characterized by Hamilton

Smith and Daniel Nathans in the late 1960s, are separate entities from their corresponding modification methylases. *They cleave DNAs at specific sites within the recognition sequence, a property that makes Type II restriction enzymes indispensible biochemical tools for DNA manipulation.* In the remainder of this section we discuss only Type II restriction enzymes.

Over 500 species of Type II restriction enzymes with > 100 differing specificities and from a variety of bacteria have been characterized. Several of the more widely used species are listed in Table 28-5. A restriction endonuclease is named by the first letter of the genus of the bacterium that produced it and the first two letters of its species, followed by its serotype or strain designation, if any, and a roman numeral if the bacterium contains more than one type of restriction enzyme. For example, *Eco*RI is produced by *E. coli* strain RY13.

Most Restriction Endonucleases Recognize Palindromic DNA Sequences

Most restriction enzyme recognition sites possess exact twofold rotational symmetry as is diagrammed in Fig. 28-45. Such sequences are known as **palindromes.**

> A palindrome is a word, verse, or sentence that reads the same backwards or forwards. Two examples are ''Madam, I'm Adam'' and ''Sex at noon taxes.''

Many restriction enzymes, such as *Eco*RI (Fig. 28-45a), catalyze the cleavage of the two DNA strands at positions that are symmetrically staggered about the center of the palindromic recognition sequence. This yields restriction fragments with complementary single-stranded ends that are from one to four nucleotides in length. Restriction fragments with such **cohesive** or **sticky ends** can associate by complementary base pairing with other restriction fragments generated by the

Table 28-5
Recognition and Cleavage Sites of Some Type II Restriction Enzymes

Enzyme	Recognition Sequence[a]	Microorganism
*Alu*I	AG↓CT	*Arthrobacter luteus*
*Bam*HI	G↓GATC*C	*Bacillus amyloliquefaciens* H
*Bgl*I	GCCNNNN↓ NGCC	*Bacillus globigii*
*Bgl*II	A↓GATCT	*Bacillus globigii*
*Eco*RI	G↓AA*TTC	*Escherichia coli* RY13
*Eco*RII	↓CC*(^A_T)GG	*Escherichia coli* R245
*Fnu*DI	GG↓CC	*Fusobacterium nucleatum* D
*Hae*II	PuGCGC↓Py	*Haemophilus aegyptius*
*Hae*III	GG↓C*C	*Haemophilus aegyptius*
*Hind*III	A*↓AGCTT	*Haemophilus influenzae* R_d
*Hpa*II	C↓C*GG	*Haemophilus parainfluenzae*
*Pst*I	CTGCA↓G	*Providencia stuartii* 164
*Sal*I	G↓TCGAC	*Streptomyces albus* G
*Taq*I	T↓CGA*	*Thermus aquaticus*
*Xho*I	C↓TCGAG	*Xanthomonas holcicola*

[a] The recognition sequence is abbreviated so that only one strand, reading 5' to 3', is given. The cleavage site is represented by an arrow (↓) and the modified base, where it is known, is indicated by an asterisk (A* is N^6-methyladenine and C* is 5-methylcytosine). Pu, Py, and N represent purine nucleotide, pyrimidine nucleotide, and any nucleotide, respectively.

Source: Roberts, R. J., *Methods Enzymol.* **68,** 27–41 (1979).

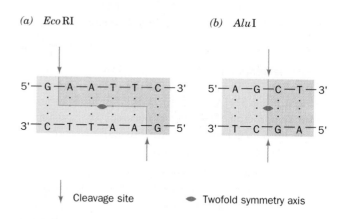

↓ Cleavage site   ● Twofold symmetry axis

Figure 28-45
The recognition sequences of the restriction endonucleases (a) *Eco*RI and (b) *Alu*I showing their twofold (palindromic) symmetry and indicating their cleavage sites.

same restriction enzyme. Some restriction cuts, such as that of *Alu*I (Fig. 28-45*b*), pass through the twofold axis of the palindrome to yield restriction fragments with fully base paired **blunt ends.** Since a given base has a one fourth probability of occurring at any nucleotide position (assuming the DNA has equal proportions of all bases), a restriction enzyme with an *n*-base pair recognition site produces restriction fragments that are, on average, 4^n base pairs long. Thus *Alu*I (4 bp recognition sequence) and *Eco*RI (6 bp recognition sequence) restriction fragments should average $4^4 = 256$ and $4^6 = 4096$ bp in length, respectively.

The X-Ray Structure of the *Eco*RI · DNA Complex Reveals the Molecular Basis of Its Recognition Specificity

The X-ray structure of *Eco*RI endonuclease in complex with a segment of B-DNA containing the enzyme's recognition site was determined by John Rosenberg. The DNA binds in the twofold symmetric cleft between the two identical 276-residue subunits of the dimeric enzyme (Fig. 28-46) thereby accounting for the DNA's palindromic recognition sequence. The protein induces the DNA to kink in three places in a manner that partially unwinds the DNA so as to widen the major groove at the recognition site. Recognition specificity is provided by a tight complementary association of the protein with the major groove of the DNA involving 12 hydrogen bonds between the side chains of Glu 144, Arg 145, and Arg 200 on both protein subunits and the purine bases of the palindromic recognition site.

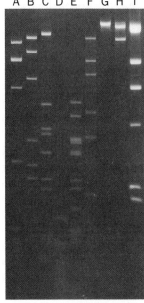

Figure 28-47
Agarose gel electrophoretograms of restriction digests of *Agrobacterium radiobacter* plasmid pAgK84 with (A) *Bam*HI, (B) *Pst*I, (C) *Bgl*II, (D) *Hae*III, (E) *Hinc*II, (F) *Sac*I, (G) *Xba*I, and (H) *Hpa*I. Lane (I) contains λ phage DNA digested with *Hind*III as a standard since these fragments have known sizes. [From Slota, J. E. and Farrand, S. F., *Plasmid* **8,** 180 (1982). Copyright © 1982 by Academic Press.]

Restriction Maps Provide a Means of Characterizing a DNA Molecule

The treatment of DNA with a restriction endonuclease produces a series of precisely defined fragments that can be separated according to size by gel electrophoresis (Fig. 28-47). Complementary single strands can be sepa-

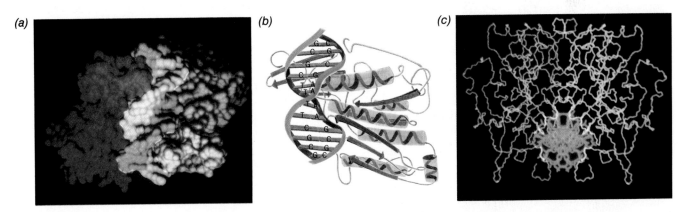

Figure 28-46
The X-ray structure of the *Eco*RI endonuclease · DNA complex. (*a*) Space-filling model showing the duplex DNA bound to the dimeric protein as viewed along the complex's twofold axis of symmetry. The two subunits of the dimeric protein are shown in yellow and orange while the DNA's complementary strands are shown in green and blue. (*b*) Ribbon drawing of one *Eco*RI endonuclease subunit interacting with the DNA's major groove. The view is ∼90° away from that in Part (*a*). (*c*) A skeletal model of the complex as viewed down the DNA's helical axis showing the protein's polypeptide backbone and the DNA's nonhydrogen atoms. The two protein subunits are drawn in yellow and pink while the DNA is drawn in blue. [Parts (*a*) and (*c*) Courtesy of John M. Rosenberg, University of Pittsburgh. Part (*b*) after Rosenberg, J. M., McClarin, J. A., Frederick, C. A., Wang, B.-C., Grable, J., Boyer, H. W., and Greene, P., *Trends Biochem. Sci.* **12,** 396 (1987).]

rated either by melting the DNA and subjecting it to gel electrophoresis, or by density gradient ultracentrifugation in alkaline CsCl. The single strands can be sequenced by one of the methods described below. If a DNA segment is too long to sequence, it may be further fragmented with a second, *etc.*, restriction enzyme before its strands are separated.

A diagram of a DNA molecule showing the relative positions of the cleavage sites of various restriction enzymes is known as its **restriction map.** Such a map is generated by subjecting the DNA to digestion with two or more restriction enzymes, both individually and in mixtures. By comparing the lengths of the fragments in the various digests, as determined, for instance, by their electrophoretic mobilities, a restriction map can be constructed. For example, consider the 4-kilobase pair **(kb)** linear DNA molecule that *Bam*HI, *Hind*III, and their mixture cut to fragments of the lengths indicated in Fig. 28-48*a*. This information is sufficient to deduce the positions of the restriction sites in the intact DNA and hence to construct the restriction map diagrammed in Fig. 28-48*b*. The restriction map of the SV40 chromosome is shown in Fig. 28-49. The restriction sites are physical reference points on a DNA molecule that are easily located. *Restriction maps therefore constitute a convenient framework for locating particular base sequences on*

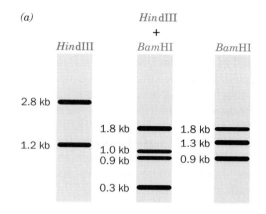

(a)

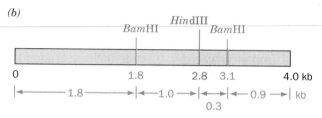

(b)

Figure 28-48
(*a*) The gel electrophoretic patterns of digests of a hypothetical DNA molecule with *Hind*III, *Bam*HI, and their mixture. The lengths of the various fragments are indicated. (*b*) The restriction map of the DNA resulting from the information in Part (*a*). This map is equivalent to one that has been reversed, right to left.

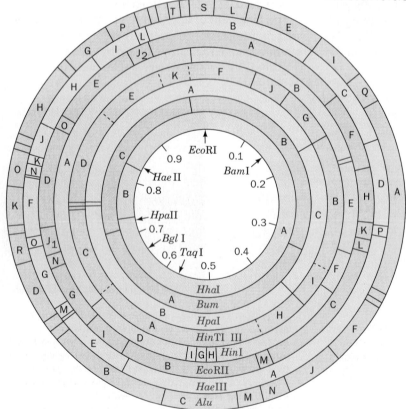

Figure 28-49
A restriction map for the 5243 bp circular DNA of SV40. The central circle indicates the fractional map coordinates with respect to the single *Eco*RI restriction site. The letters A,B,C, . . . in each ring represent the various restriction fragments of the corresponding restriction enzyme in order of decreasing length. [After Nathans, D., *Science* **206**, 905 (1979).]

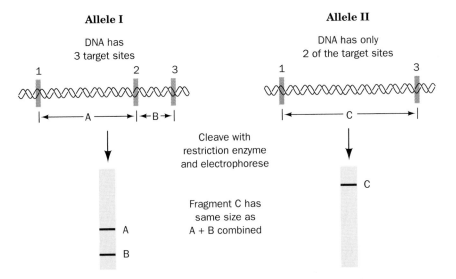

Figure 28-50
A mutational change that affects a restriction site in a DNA segment alters the number and sizes of its restriction fragments.

a chromosome and for estimating the degree of difference between related chromosomes.

Restriction-Fragment Length Polymorphisms Provide Markers for Characterizing Genes

Individuality in humans and other species derives from their high degree of genetic polymorphism; homologous human chromosomes differ in sequence, on average, every 200 to 500 bp. These genetic differences create or eliminate restriction sites. Restriction enzyme digests of the corresponding segments from homologous chromosomes therefore contain fragments with different lengths; that is, these DNAs exhibit **restriction-fragment length polymorphisms (RFLPs; Fig. 28-50).**

RFLPs are useful markers for identifying chromosomal differences (Fig. 28-51). They are particularly valuable for diagnosing inherited diseases for which the molecular defect is unknown. If a particular RFLP is so closely linked to a defective gene that there is little chance the two will recombine from generation to generation (recall that the probability of recombination between two genes increases with their physical separation on a chromosome; Section 27-1C), then the detection of that RFLP in an individual is indicative that the individual has also inherited the defective gene. For example, **Huntington's chorea,** a progressive and invariably fatal neurological deterioration, whose symptoms first appear around age 40, is caused by a dominant but unknown genetic defect. The identification of an RFLP that is closely linked to the defective Huntington's gene has permitted the children of Huntington's chorea victims (50% of whom inherit this devastating condition) to make informed decisions in ordering their lives.

By the same token, the identification of RFLPs associated with the genetic defects causing **cystic fibrosis** (a debilitating and often fatal autosomal recessive disease; heterozygotes, who comprise 5% of the Caucasian population, are asymptomatic), and **Duchenne muscular dystrophy** (an X-linked degenerative disease of muscle that is invariably fatal by around age 25) have permitted

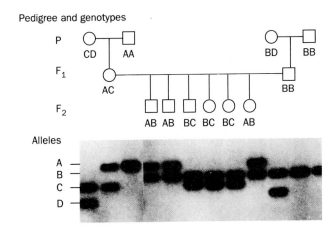

Figure 28-51
RFLPs are inherited according to the rules of Mendelian genetics. Four alleles of a particular gene, each characterized by different restriction markers, can occur in all possible pairwise combinations and segregate independently in each generation (circles represent females and squares represent males). In the P (parental) generation, two individuals are heterozygous (CD and BD) and the other two are homozygous (AA and BB) for the gene in question. Their children, the F_1 generation, are AC and BB. Consequently, every individual in the F_2 generation (grandchildren) inherited either an A or a C from their mother and a B from their father. [Courtesy of Ray White, University of Utah Medical School.]

the *in utero* diagnoses of these diseases. (Note that the availability of fetal testing has actually increased the number of births because many couples who knew they had a high risk of conceiving a genetically defective child previously chose not to have children.)

RFLPs are also valuable markers for isolating and thus sequencing their closely linked but unknown genes. Indeed, the first phase in sequencing the human genome, which is already well underway, is to identify a series of ~ 100 equally spaced markers on each of the 23 human chromosomes.

B. Chemical Cleavage Method

After 1975, several methods were developed for the rapid sequencing of long stretches of DNA. Two of them, the **chemical cleavage** method of Allan Maxam and Walter Gilbert (Fig. 28-52), and the **chain-terminator** procedure of Frederick Sanger (the same individual who pioneered the amino acid sequencing of proteins), are widely used and are largely responsible for the vast number of DNA sequences that have been elucidated. In the remainder of this section, we discuss the chemical

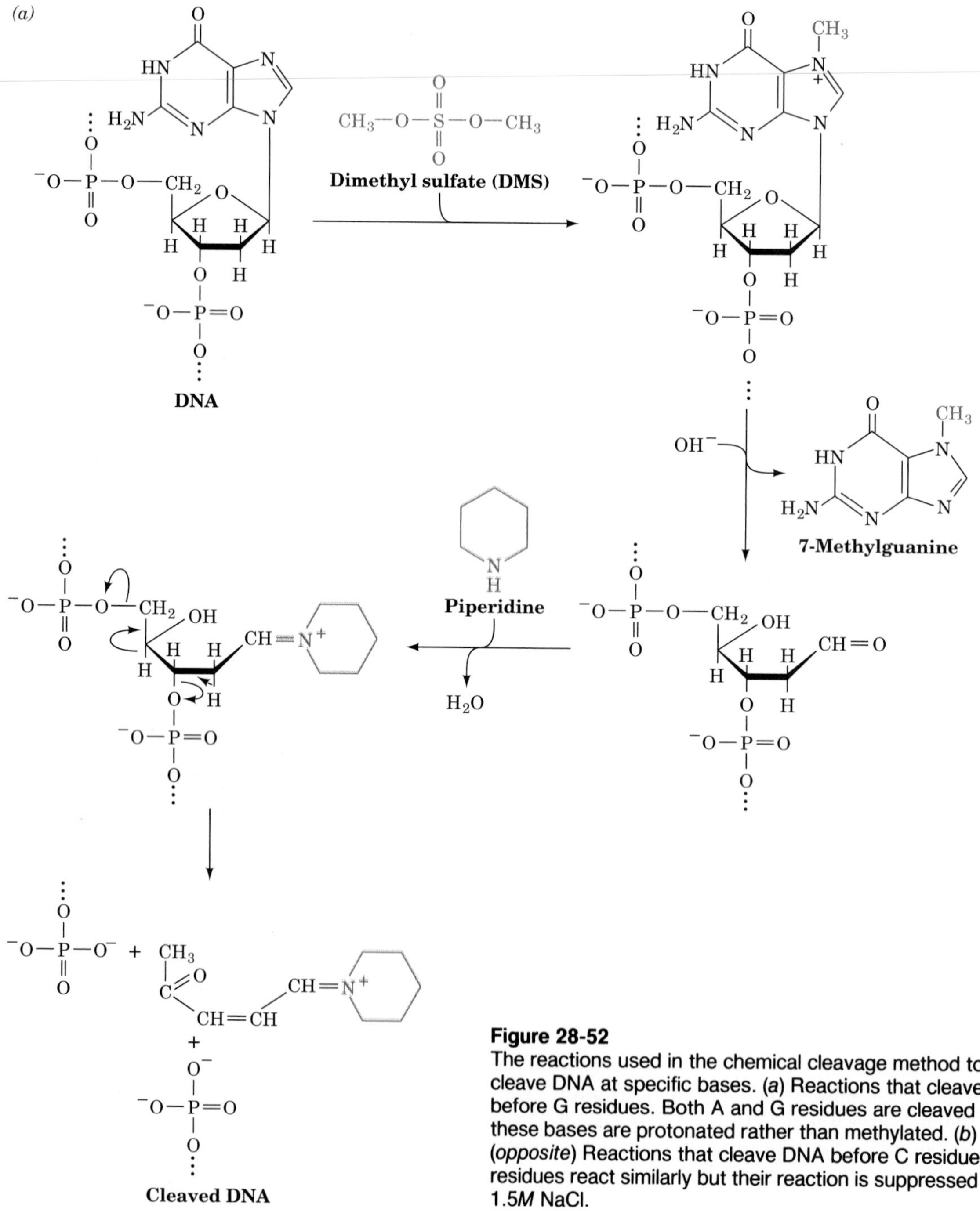

Figure 28-52
The reactions used in the chemical cleavage method to cleave DNA at specific bases. (a) Reactions that cleave DNA before G residues. Both A and G residues are cleaved if these bases are protonated rather than methylated. (b) (*opposite*) Reactions that cleave DNA before C residues. T residues react similarly but their reaction is suppressed in 1.5*M* NaCl.

cleavage and chain-terminator methods as well as methods for sequencing RNA.

One End of the DNA Must Be Radioactively Labeled

The first step in the chemical cleavage method is to radioactively label one end of the DNA, usually the 5′ end, with ^{32}P. If the DNA already has a 5′ phosphate group, this first must be removed by treatment with **alkaline phosphatase** from *E. coli*.

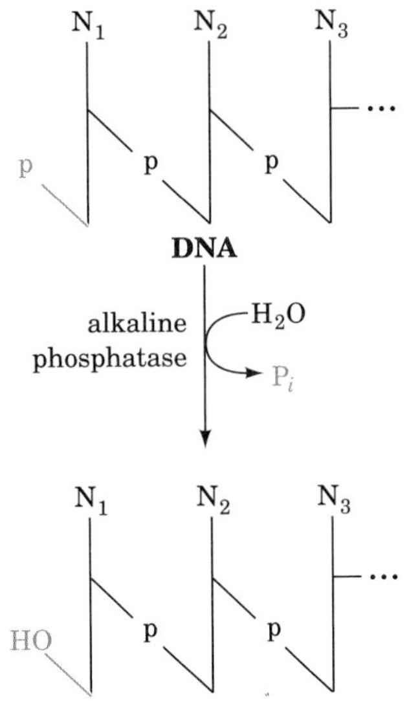

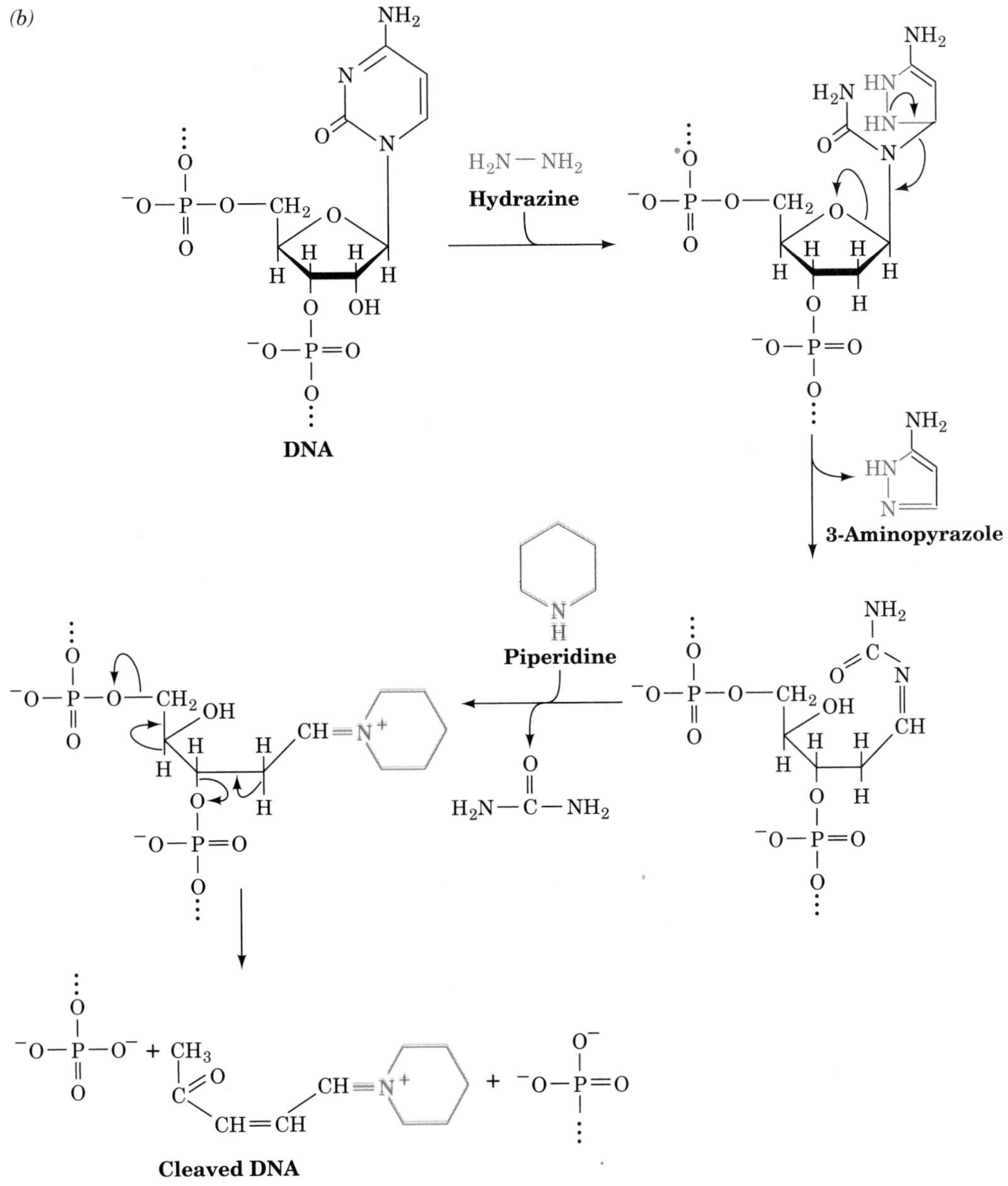

Then the 5′ terminus is labeled in a reaction with [γ-^{32}P]ATP as catalyzed by **polynucleotide kinase** from *E. coli* infected with bacteriophage T4.

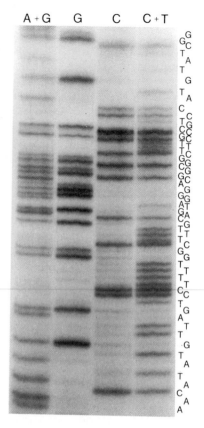

Figure 28-53

An autoradiograph of a sequencing gel containing fragments of a DNA segment that was treated according to the chemical cleavage method of sequence analysis. The DNA was ^{32}P labeled at its 5′ end. The DNA's deduced sequence is written beside the gel. Since the shorter fragments, which have the larger spacing, are at the bottom of the gel, the 5′ → 3′ direction in the sequence corresponds to the upward direction in the gel. [Courtesy of David Dressler, Harvard University Medical School.]

The DNA Is Cleaved in a Base-Specific Manner

The basic strategy of the chemical cleavage method is to specifically cleave the end-labeled DNA at only one type of nucleotide under conditions such that each molecule is broken at an average of one randomly located susceptible bond. This produces a set of radioactive fragments whose members extend from the ^{32}P-labeled end to one of the positions occupied by the chosen base. For example, if the DNA to be sequenced is

^{32}P-TGTAGGAGCT

cleavage on the 5′ side of the G residues, for instance, would produce the following set of 5′-labeled fragments:

^{32}P-TGTAGGA
^{32}P-TGTAG
^{32}P-TGTA
^{32}P-T

Polyacrylamide gel electrophoresis separates these fragments according to size. Hence *the positions of the G residues in the DNA may be identified from the relative positions on the gel of their corresponding ^{32}P-labeled fragments as revealed by autoradiography.* (The unlabeled cleavage fragments are, of course, not observed in this procedure.) In order for this method to work, the gel must be of sufficient resolving power to unambiguously separate fragments that differ in length by only one nucleotide.

The DNA to be sequenced may be cleaved at specific bases by subjecting it, in separate aliquots, to four different treatments:

1. G only

The DNA is reacted with **dimethyl sulfate (DMS),** which methylates G residues at N(7), thereby ren-

dering the glycosidic bond of the methylated residue susceptible to hydrolysis (Fig. 28-52*a*). Subsequent treatment by **piperidine** cleaves the polynucleotide chain before the depurinated residue.

2. A + G

DMS preferentially methylates A residues at N(3) rather than N(7) and hence the above treatment cleaves DNA at A residues at only about one fifth the rate it does at G residues. If, instead, the DNA is treated with acid, both A and G are released at comparible rates to yield the same depurinated product indicated in Fig. 28-52*a*. Piperidine treatment then causes strand cleavage before both A and G residues. The A residues are identified by comparing the positions of the G and the A + G cleavages.

3. C + T

The reaction of DNA with **hydrazine** (NH_2—NH_2) followed by piperidine treatment cleaves DNA before both its C and T residues (Fig. 28-52*b*).

4. C only

If DNA is treated with the hydrazine in 1.5*M* NaCl, only its C residues react appreciably. Then, as with the purines, the comparison of the C and the C + T cleavage positions identifies the T residues.

In all four reactions, the conditions are adjusted so that the strands are cleaved at an average of one randomly located position each.

Cleavage Fragments Are Separated According to Size

The four differently fragmented samples of the DNA, A + G, G, C, and C + T, are simultaneously electrophoresed in parallel lanes on a **sequencing gel.** This is a long, thin (as little as 0.1 mm × up to 200 cm) polyacrylamide slab. It contains ~8M urea and is run at ~70°C so as to eliminate all hydrogen bonding associations. *These conditions ensure that the DNA fragments separate only according to their size. The sequence of the DNA can then be directly read off an autoradiogram of the sequencing gel* as is indicated in Fig. 28-53. Indeed, computerized devices are available to aid in doing so. However, a single gel is incapable of resolving much more than 100 consecutive fragments. This limitation is circumvented by electrophoresing three sets of the four differently cleaved samples for successively longer

times so as to best resolve the shortest, intermediate length and longest fragments, respectively. In this manner, the base sequence of a 200 to 300 nucleotide DNA fragment can normally be determined from one set of sequencing reactions (although technical advances are steadily increasing this number).

Since the base-specific cleavages destroy the corresponding nucleotide, there is no fragment corresponding to the 5'-terminal nucleotide. Furthermore, the mononucleotide identifying the second base is usually not detected on a gel. The identities of these two nucleotides may be determined by sequencing the complementary strand which, just as importantly, verifies the sequence of the first strand.

C. Chain-Terminator Method

The chain-terminator method utilizes the E. coli enzyme DNA polymerase I to make complementary copies of the single-stranded DNA being sequenced. Under the direction of the strand being replicated (the **template strand**), DNA polymerase I assembles the four deoxy-

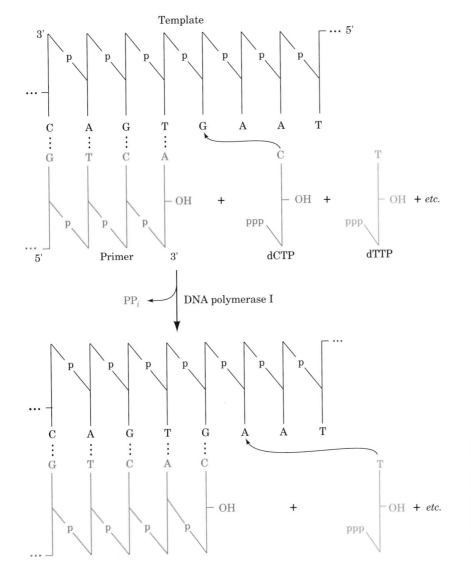

Figure 28-54
The replication of DNA as catalyzed by *E. coli* DNA polymerase I. Under the direction of the template strand, the primer is elongated by the stepwise addition of complementary nucleotides in the 5' → 3' direction on the growing polynucleotide.

nucleoside triphosphates, dATP, dCTP, dGTP, and dTTP, into a complementary polynucleotide chain that it elongates in the 5′ → 3′ direction (Fig. 28-54). DNA polymerase I can only sequentially add deoxyribonucleotides to the 3′ end of a polynucleotide. Hence, to initiate replication, it requires the presence of the 5′ end of the chain being synthesized (a **primer**) in a stable base paired complex with the template. If the DNA being sequenced is a restriction fragment, as it usually is, it begins and ends with a restriction site. The primer can therefore be a short DNA segment containing this restriction fragment annealed to the strand being replicated. The role of DNA polymerase I in DNA replication is examined in Section 31-2A.

DNA polymerase I has a 5′ → 3′ exonuclease activity (degrades DNA one nucleotide at a time from its 5′ end), which is catalyzed by a separate active site from that which mediates the polymerization reaction. This is demonstrated by the observation that upon proteolytic cleavage of the enzyme into two fragments, the larger fragment, which is known as the **Klenow fragment,** possesses the full polymerase activity of the enzyme whereas the smaller fragment has the 5′ → 3′ exonuclease activity. Only the Klenow fragment is used in DNA sequencing to ensure that all replicated chains have the same 5′ terminus.

The Synthesis of Labeled DNA by DNA Polymerase Is Terminated After Specific Bases

In the chain-terminator technique (Fig. 28-55), *the DNA to be sequenced is incubated with the Klenow fragment of DNA polymerase I, a suitable primer and the four deoxynucleoside triphosphates, of which at least one (usually dATP) is [α-^{32}P]-labeled. In addition, a small amount of the 2′,3′-dideoxynucleoside triphosphate*

$$(P)-(P)-(P)-OCH_2 \quad O \quad \boxed{Base}$$

2′,3′-Dideoxynucleoside triphosphate

of one of the bases is added to the reaction mixture. When the dideoxy analog is incorporated in the growing polynucleotide in place of the corresponding normal nucleotide, chain growth is terminated because of the absence of a 3′-OH group. By using only a small amount of the dideoxy analog, a series of truncated chains are generated that are each terminated by the dideoxy analog at one of the positions occupied by the corresponding base. Sequence gel electrophoresis separates these chains according to their lengths and therefore indicates the positions at which that base occurs.

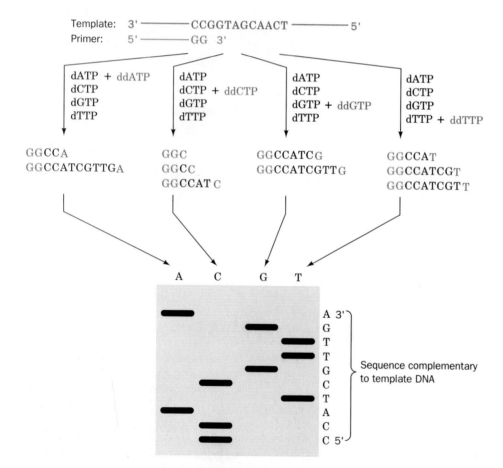

Figure 28-55
A flow diagram of the chain-terminator method of DNA sequencing. The symbol ddATP represents dideoxyadenosine triphosphate, *etc.*

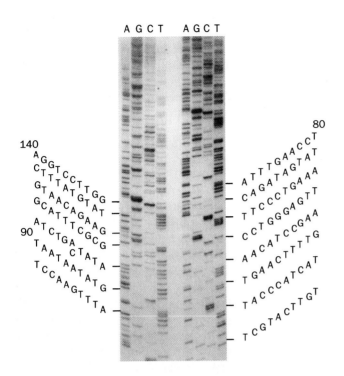

A G C T   A G C T

140

80

Figure 28-56
An autoradiograph of a sequencing gel
containing DNA fragments produced by the
chain-terminator method of DNA
sequencing. A second loading of the gel
(*right*) was made 90 min after the initial
loading. The deduced sequence of 140
nucleotides is written along side. [From
Hindley, J., DNA Sequencing, *In* Work,
T. S. and Burdon, R. H. (Eds.), *Laboratory
Techniques in Biochemistry and Molecular
Biology*, Vol. 10, *p.* 82, Elsevier (1983).
Used by permission.]

Each of the dideoxy analogs of the four bases are
reacted in separate vessels and the resulting ^{32}P-labeled
product mixtures are subjected to sequence gel electrophoresis in parallel lanes. The sequence of the replicated
strand can then be directly read from an autoradiogram
of the gel (Fig. 28-56), much like that in the chemical
cleavage method.

Both the chain-terminator and the chemical cleavage
procedures are widely used for DNA sequencing. With a
few hours effort by a skilled operator, either method can
sequence a DNA of several hundred nucleotides. Indeed, the major obstacle to sequencing a very long DNA
molecule is ensuring that all of its fragments are cloned
(by methods discussed in Section 28-8C) rather than
sequencing them once they have been obtained. The
chemical cleavage method is somewhat easier to set up
for occasional use while the chain-terminator method is
generally chosen for routine use. Note that the sequence
obtained by the chain-terminator method is complementary to the DNA strand being sequenced, whereas
the sequence obtained by the chemical cleavage method
is that of the original DNA strand.

The Chain-Terminator Method Is Readily Automated

If large DNA segments such as entire chromosomes
are to be sequenced, then existing sequencing methods
must be greatly accelerated, that is, automated. The
chain-terminator method has been adapted to computerized procedures. Rather than use radiolabeled nucleotides (with their inherent health hazards and storage
problems), each dideoxynucleoside triphosphate is covalently linked to a differently fluorescing dye. The
chain-extension reaction is carried out in a single vessel
containing all four fluorescent dideoxy analogs and thus
yielding a series of increasingly longer polynucleotides,
each with a fluorescence spectrum characteristic of its
3'-terminal nucleotide. The reaction mixture is then
subject to sequence gel electrophoresis in a single lane
yielding a series of bands, each with the fluorescence
spectrum indicative of a successive base in the DNA
being sequenced (Fig. 28-57). The gel fluorescence detection system is computer-controlled and hence data
acquisition is automated. This device can identify
~ 10,000 bases per day in contrast to the ~ 50,000 bases
per year that a skilled operator can identify using the
above-described manual methods (note that with the
use of only one such device, it would still take nearly
1000 years to sequence the human genome).

D. RNA Sequencing

RNA may be rapidly sequenced by only a slight modification of DNA sequencing procedures. The RNA to be
sequenced is transcribed into a complementary strand of
DNA (**cDNA**) through the action of **RNA-directed
DNA polymerase** (also known as **reverse transcriptase**). This enzyme, which is produced by certain RNA-
containing viruses (Section 31-4C), uses an RNA tem-

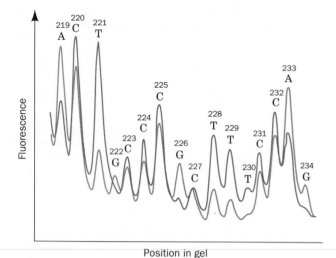

Figure 28-57
The detection, in a sequencing gel, of fluorescent
terminator-labeled DNA fragments generated by the
automated chain-terminator technique. The ratio of the
intensities of the laser-excited dye fluorescence as
separately measured in two wavelength bands (*blue*, short
wavelength; *red*, long wavelength) unambiguously identifies
the fluorescent 3′-terminal base in each gel band as A, T, G,
or C. The number above each band indicates its position in
the DNA segment being sequenced. [After Prober, J. M.,
Trainor, G. L., Dam, R. J., Hobbs, F. W., Robertson, C. W.,
Zagursky, R. J., Cocuzza, A. J., Jensen, M. A., and
Baumeister, K., *Science* **238**, 340 (1987).]

plate but is otherwise similar in its action to DNA
polymerase I. The resulting cDNA may then be sequenced by either the chemical cleavage or the chain-
terminator methods. Alternatively, RNA may be directly sequenced by a chemical cleavage method similar
to that of DNA sequencing, which employs reactions
that cleave RNA after specific bases.

7. CHEMICAL SYNTHESIS OF OLIGONUCLEOTIDES

Molecular cloning techniques (Section 28-8) have
permitted the genetic manipulation of organisms in
order to investigate their cellular machinery, change
their characteristics, and produce scarce or specifically
altered proteins in large quantities. *The ability to chemically synthesize DNA oligonucleotides of specified base sequences is an indispensable part of this powerful technology.* For example, suppose we wished to obtain the gene
specifying a protein whose amino acid sequence is at
least partially known. Reference to the **genetic code** (the
correspondence between an amino acid sequence and
the base sequence of the gene specifying it; Section
30-1) permits the synthesis of a short (~ 15 nucleotide)
[32]P-labeled oligonucleotide that is complementary to a
segment of the gene of interest. The oligonucleotide is

used as a probe in the Southern transfer procedure (Section 28-4C) on restriction enzyme-digested DNA from
the organism that produced the protein. The probe specifically labels the required gene and thereby permits its
isolation.

Synthetic oligonucleotides are also required to specifically alter genes through **site-directed mutagenesis.**
An oligonucleotide containing a short gene segment
with the desired altered base sequence is used as a
primer in the DNA polymerase I replication of the gene
of interest. Such a primer will hybridize to the corresponding wild-type sequence if there are only a few
mismatched base pairs, and its extension, by DNA
polymerase I (Section 28-6C), yields the desired altered
gene (Fig. 28-58). The altered gene can then be inserted
in a suitable organism via techniques discussed in Section 28-8 and grown (cloned) in quantity.

Oligonucleotides Are Valuable Diagnostic Tools

*The use of synthetic oligonucleotides as probes in South-
ern transfer analysis has great promise for the diagnosis and
prenatal detection of genetic diseases. These diseases often
result from a specific change in a single gene such as a base
substitution, deletion, or insertion.* The temperature at

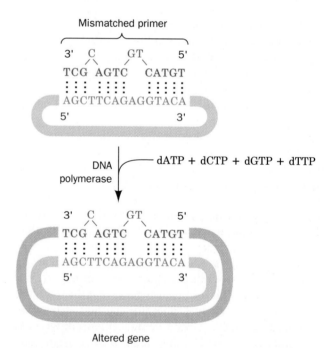

Figure 28-58
Site-directed mutagenesis. A chemically synthesized
oligonucleotide incorporating the desired base changes is
hybridized to the DNA encoding the gene to be altered. The
mismatched primer is then extended by DNA polymerase I
thereby generating the mutated gene. The mutated gene
can subsequently be inserted into a suitable host organism
so as to yield the mutant DNA, or the corresponding RNA, in
quantity, produce a specifically altered protein, and/or
generate a mutant organism.

which probe hybridization is carried out may be adjusted so that only an oligonucleotide that is perfectly complementary to a length of DNA will hybridize to it. Even a single base mismatch, under appropriate conditions, will result in a failure to hybridize. For example, sickle-cell anemia (Section 9-3B) arises from a single base change that causes the amino acid substitution Glu $\beta6 \rightarrow$ Val in hemoglobin. A 19-residue oligonucleotide that is complementary to the sickle-cell gene's mutated segment hybridizes, at the proper temperature, to DNA from homozygotes for the sickle-cell gene but not to DNA from normal individuals. An oligonucleotide that is complementary to the normal Hbβ gene gives opposite results. DNA from sickle-cell heterozygotes hybridizes to both probes but in reduced amounts relative to the DNAs from homozygotes. The oligonucleotides may consequently be used in the prenatal diagnosis of sickle-cell disease. DNA probes are also rapidly replacing the much slower and less accurate culturing techniques for the identification of pathogenic bacteria.

Oligonucleotides Are Synthesized in a Stepwise Manner

The basic strategy of oligonucleotide synthesis is analogous to that of polypeptide synthesis (Section 6-4): *A suitably protected nucleotide is coupled to the growing end of the oligonucleotide chain, the protecting group is removed, and the process is repeated until the desired oligonucleotide has been synthesized.* The first practical technique for DNA synthesis, the **phosphodiester method,** which was developed by H. Gobind Khorana in the 1960s, is a laborious process in which all reactions are carried out in solution and the products must be isolated at each stage of the multistep synthesis. Khorana, nevertheless, used this method, in combination with enzymatic techniques, to synthesize a 126-nucleotide tRNA gene, a project that required several years of intense effort by numerous skilled chemists.

The Phosphoramidite Method

By the early 1980s, these difficult and time consuming processes had been replaced by much faster solid phase methodologies that permitted oligonucleotide synthesis to be automated. The presently most widely used chemistry, which was formulated by Robert Letsinger and further developed by Marvin Caruthers, is known as the **phosphoramidite method.** This nonaqueous reaction sequence adds a single nucleotide to a growing oligonucleotide chain as follows (Fig. 28-59):

1. The **dimethoxytrityl (DMTr)** protecting group at the 5' end of the growing oligonucleotide chain (which is anchored via a linking group at its 3' end to a solid support, S) is removed by treatment with acid.

2. The newly liberated 5' end of the oligonucleotide is coupled to the 3'-phosphoramidite derivative of the

next deoxynucleoside to be added to the chain. The coupling agent in this reaction is **tetrazole.**

3. Any unreacted 5' end (the coupling reaction has a yield of over 99%) is capped by acetylation so as to block its extension in subsequent coupling reactions. This prevents the extension of erroneous oligonucleotides.

4. The phosphite triester group resulting from the coupling step is oxidized to the phosphotriester thereby yielding a chain that has been lengthened by one nucleotide.

The above reaction sequence, in commercially available automated synthesizers, can be routinely repeated at least 50 times with a cycle time of 40 min or less. Once an oligonucleotide of desired sequence has been synthesized, it is released from its support and its various blocking groups, including those on the bases, are removed. The product can then be purified by HPLC and/or gel electrophoresis.

8. MOLECULAR CLONING

A major problem in almost every area of biochemical research is obtaining sufficient quantities of the substance of interest. For example, a 10-L culture of *E. coli* grown to its maximum titer of $\sim 10^{10}$ cells $\cdot$ mL^{-1} contains, at most, 7 mg of DNA polymerase I, and many of its proteins are present in far lesser amounts. Yet, it is rare that as much as one half of any protein originally present in an organism can be recovered in pure form. Eukaryotic proteins may be even more difficult to obtain because many eukaryotic tissues, whether acquired from an intact organism or grown in tissue culture, are available in only small quantities. As far as the amount of DNA is concerned, our 10-L *E. coli* culture would contain $\sim$ 0.1 mg of any 1000 bp length of chromosomal DNA (a length sufficient to contain most prokaryotic genes) but its purification in the presence of the rest of the chromosomal DNA would be an all but impossible task. These difficulties have been largely eliminated in recent years through the development of **molecular cloning** techniques (a **clone** is a collection of identical organisms that are derived from a single ancestor). These methods, which are also referred to as **genetic engineering** and **recombinant DNA** technology, deserve much of the credit for the enormous progress in biochemistry since the mid-1970s.

*The main idea of molecular cloning is to insert a DNA segment of interest into an autonomously replicating DNA molecule, a so-called **cloning vector** or *vehicle*, so that the DNA segment is replicated with the vector.* Cloning such a **chimeric vector** (*chimera:* A monster in Greek mythology that has a lion's head, a goat's body, and a serpent's tail) in a suitable **host organism** such as *E. coli* or yeast

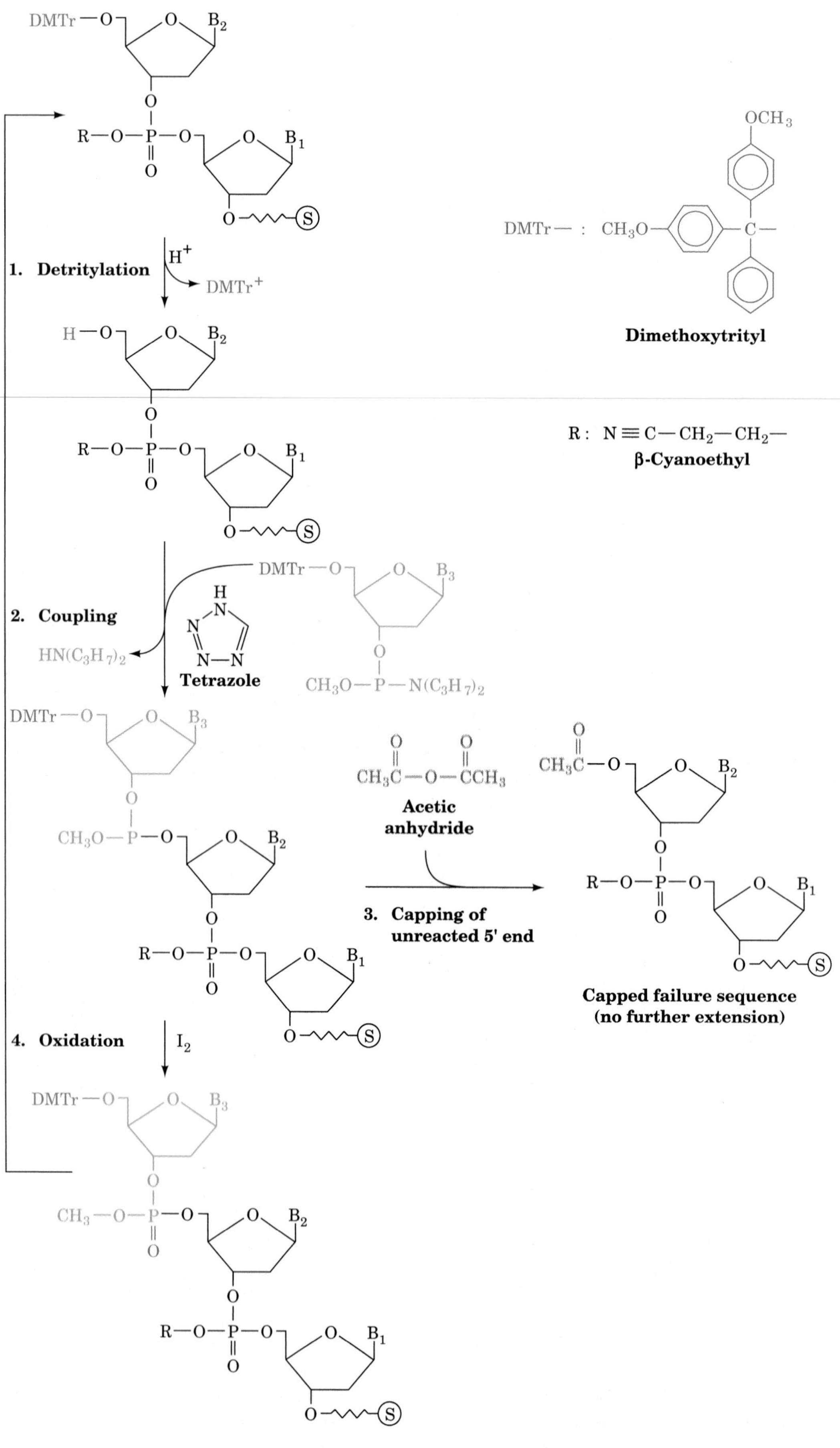
DMTr—O
B2
R—O—P—O
B1
1. Detritylation
H+
DMTr+
H—O
Dimethoxytrityl
DMTr— : CH3O—
OCH3
R: N≡C—CH2—CH2—
β-Cyanoethyl
2. Coupling
HN(C3H7)2
Tetrazole
CH3O—P—N(C3H7)2
B3
CH3O—P—O
Acetic anhydride
3. Capping of unreacted 5' end
CH3C—O
Capped failure sequence (no further extension)
4. Oxidation
I2
CH3—O—P—O

Figure 28-59 (*opposite*)
The reaction cycle in the phosphite-triester method of oligonucleotide synthesis. Here B₁, B₂, and B₃ represent protected bases, and S represents an inert solid phase support such as controlled-pore glass.

results in the production of large amounts of the inserted DNA segment. If a cloned gene is flanked by the properly positioned control sequences for RNA and protein synthesis (Chapters 29 and 30), the host may also produce large quantities of the RNA and protein specified by that gene. The techniques of genetic engineering are outlined in this section.

A. Cloning Vectors

Both plasmids, bacteriophages, and yeast artificial chromosomes are used as cloning vectors in genetic engineering.

Plasmid-Based Cloning Vectors

Plasmids are circular DNA duplexes of 1 to 200 kb that contain the requisite genetic machinery, such as a **replication origin** (a site at which DNA replication is initiated; Section 31-2), to permit their autonomous propagation in a bacterial host or in yeast. Plasmids may be considered molecular parasites but in many instances they benefit their host by providing functions, such as resistance to an antibiotic, that the host lacks. Indeed, the widespread appearance, since antibiotics came into use, of antibiotic-resistant pathogens is a result of the rapid proliferation among these organisms of plasmids containing genes that confer resistance to antibiotics.

Some types of plasmids, which are present in one or a few copies per cell, replicate once per cell division as does the bacterial chromosome; their replication is said to be under **stringent control**. The plasmids used in molecular cloning, however, are under **relaxed control**; they are normally present in 10 to 200 copies per cell. Moreover, if protein synthesis in the bacterial host is inhibited, for example, by the antibiotic **chloramphenicol** (Section 30-3G), the copy number of these plasmids may increase to several thousand per cell (about one half of the cell's total DNA). The plasmids that have been constructed (by genetic engineering techniques) for use in molecular cloning are relatively small, carry genes specifying resistance to several antibiotics, and contain a number of conveniently located restriction endonuclease sites into which the DNA to be cloned may be inserted (via techniques described in Section 28-8B). The *E. coli* plasmid designated **pBR322** (Fig. 28-60) is among the most widely used cloning vectors.

The expression of a chimeric plasmid in a bacterial host was first demonstrated in 1973 by Herbert Boyer and Stanley Cohen. The host bacterium takes up a plasmid when the two are mixed together in a process that is greatly enhanced by the presence of Ca²⁺, (which is

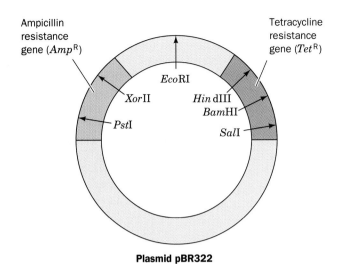

Plasmid pBR322

Figure 28-60
A restriction map of plasmid pBR322 indicating the positions of its antibiotic resistance genes.

thought to increase membrane permeability). An absorbed plasmid vector becomes permanently established in its bacterial host (transformation) with an efficiency of ~0.1%.

Plasmid vectors cannot be used to clone DNAs of more than ~ 10 kb. This is because the time required for plasmid replication increases with plasmid size. Hence intact plasmids with large unessential (to them) inserts are lost through the faster proliferation of plasmids that have eliminated these inserts by random deletions.

Bacteriophage-Based Cloning Vectors

Bacteriophage λ (Fig. 28-61) is an alternative cloning

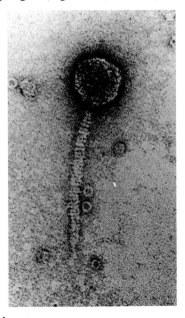

Figure 28-61
An electron micrograph of bacteriophage λ. [Courtesy of A. F. Howatson. From Lewin, B., *Gene Expression*, Vol. 3, Fig. 5.23, John Wiley & Sons Inc. (1977).]

vehicle that can be used to clone DNAs of up to 16 kb. The central third of this virus' 48.5-kb genome is not required for phage infection (Section 32-3A) and can therefore be replaced by foreign DNAs of up to slightly greater size using techniques discussed in Section 28-8B. The chimeric phage DNA can then be introduced into the host cells by infecting them with phages formed from the DNA by an *in vitro* packaging system (Section 32-3B). The use of phages as cloning vectors has the additional advantage that the chimeric DNA is produced in large amounts and in easily purified form.

λ Phages can be used to clone even longer DNA inserts. The viral apparatus that packages DNA into phage heads requires only that the DNA have a specific 14 bp sequence known as a *cos* **site** located at both ends and that these ends be 36 to 51 kb apart (Section 32-3B). Placing two *cos* sites the proper distance apart on a plasmid vector yields, via an *in vitro* packaging system, a so-called **cosmid** vector, which can contain foreign DNA of up to ~49 kb. Cosmids have no phage genes and hence, upon introduction into a host cell via phage infection, reproduce as plasmids.

The **filamentous bacteriophage M13** (Fig. 28-62) is also a useful cloning vector. It has a single-stranded circular DNA that is contained in a protein tube composed of ~2700 helically arranged identical protein subunits. This number is controlled, however, by the length of the phage DNA being coated; insertion of foreign DNA in a nonessential region of the M13 chromosome results in the production of longer phage particles. Although M13 cloning vectors cannot stably maintain DNA inserts of >1 kb, they are widely used in the production of DNA for sequence analysis by the chain-terminator method (Section 28-6C) because these phages directly produce the ~300-nucleotide single-stranded DNA that this technique requires. Furthermore, since the DNA to be sequenced is always inserted at the same point in the viral chromosome (a restriction site; Section 28-8B), an ~15 base synthetic oligonucleotide (the so-called "universal primer") that is complementary to the viral DNA on the 3' side of the cloning site may be used as the primer for any DNA segment sequenced by this method.

YAC Vectors

DNA segments larger than those that can be carried by cosmids may be cloned in **yeast artificial chromosomes (YACs).** YACs are linear DNA segments that contain all the molecular paraphernalia required for replication in yeast: a replication origin [known as an **autonomously replicating sequence (ARS)**], a centromere (the chromosomal segment attached to the spindle during mitosis and meiosis), and telomeres (the ends of linear chromosomes that permit their replication). DNAs of several hundred kb have been spliced into YACs and successfully cloned.

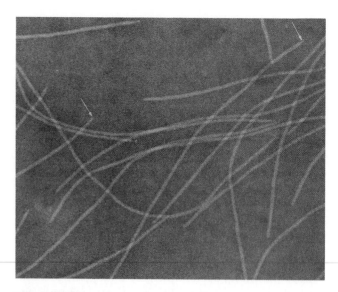

Figure 28-62
An electron micrograph of the filamentous bacteriophage M13. Note that some filaments appear to be pointed at one end (*arrows*). [Courtesy of Robley Williams, Stanford University, Emeritus and Harold Fisher, University of Rhode Island.]

B. Gene Splicing

A DNA to be cloned is, in many cases, obtained as a defined fragment through the application of restriction endonucleases (for M13 vectors, the restriction enzymes' requirement of duplex DNA necessitates the use of this phage DNA in double-stranded form). Recall that most restriction endonucleases cleave duplex DNA at specific palindromic sites so as to yield single-stranded ends that are complimentary to each other (Section 28-6A). Therefore, as Janet Mertz and Ron Davis first demonstrated in 1972, *a restriction fragment may be inserted into a cut made in a cloning vector by the same restriction enzyme* (Fig. 28-63). *The complimentary (cohesive) ends of the two DNAs specifically associate under annealing conditions and are covalently joined (spliced) through the action of an enzyme named DNA ligase* (Section 31-2C; DNA ligase produced by **bacteriophage T4** must be used for blunt-ended restriction cuts such as those generated by *Alu*I or *Hae*III; Table 28-5). *A great advantage of using a restriction enzyme to construct a chimeric vector is that the DNA insert can be precisely excised from the cloned vector by cleaving it with the same restriction enzyme.*

If the foreign DNA and cloning vector have no common restriction sites at innocuous positions, they may still be spliced, using a procedure pioneered by Dale Kaiser and Paul Berg, through the use of **terminal deoxynucleotidyl transferase (terminal transferase).** This mammalian enzyme, which has been implicated in the generation of antibody diversity (Section 34-2C), adds nucleotides to the 3'-terminal OH group of a DNA chain; it is the only known DNA polymerase that does not require a template. Terminal transferase and dTTP,

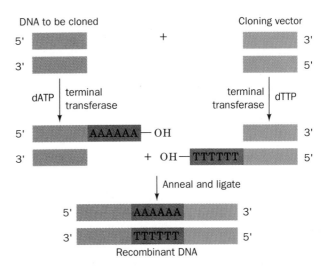

Figure 28-64
Two DNA fragments may be joined through the generation of complementary homopolymer tails. The poly(dA) and poly(dT) tails shown in this example may be replaced by poly(dC) and poly(dG) tails.

nates the restriction sites that were used to generate the foreign DNA insert and to cleave the vector. It may therefore be difficult to recover the insert from the cloned vector. This difficulty can be circumvented by appending to both ends of the foreign DNA a chemically synthesized palindromic "linker" which has a restriction site matching that of the cloning vector. The linker is attached to the foreign DNA by blunt end ligation with T4 ligase and then cleaved with the appropriate restriction enzyme to yield the correct cohesive ends for ligation to the vector (Fig. 28-65).

Properly Transformed Cells Must Be Selected

How can one select only those host organisms that contain a properly constructed vector? In the case of plasmid transformation, this is usually done through the use of antibiotics. For example, an *E. coli* transformed by a pBR322 plasmid (Fig. 28-60) containing a foreign DNA insert in its *Bam*HI site is **tetracycline**-sensitive (*tet⁻*; tetracycline is an antibiotic that inhibits bacterial protein synthesis; Section 30-3G) because of the interruption of its *tet* gene by the insert, but is **ampicillin**-resistant (*amp⁺*; ampicillin is a penicillin derivative) as conferred by the plasmid's intact *amp* gene. Bacterial colonies (clones) can therefore be grown on culture plates containing ampicillin to select for bacteria that have been transformed by this plasmid. Of these colonies, the ones with plasmids containing the foreign DNA can be detected, through replica plating (Section 27-1D), by their failure to grow on a tetracycline-containing medium.

Genetically engineered λ phage variants contain restriction sites that flank the dispensable central third of the phage genome (Section 28-8A). This segment may

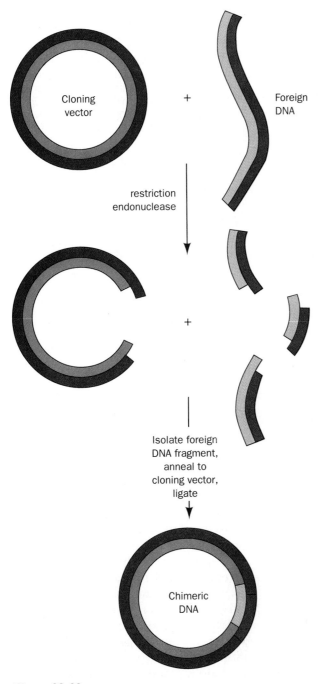

Figure 28-63
The construction of a recombinant DNA molecule by the insertion of a restriction fragment in a cloning vector's corresponding restriction cut.

for example, can build up poly(dT) tails of ∼100 residues on the 3′ ends of the DNA segment to be cloned (Fig. 28-64). The cloning vector is enzymatically cleaved at a specific site and the 3′ ends of the cleavage site are similarly extended with poly(dA) tails. The complimentary homopolymer tails are annealed, any gaps resulting from differences in their lengths filled in by DNA polymerase I, and the strands joined by DNA ligase.

A disadvantage of the above technique is that it elimi-

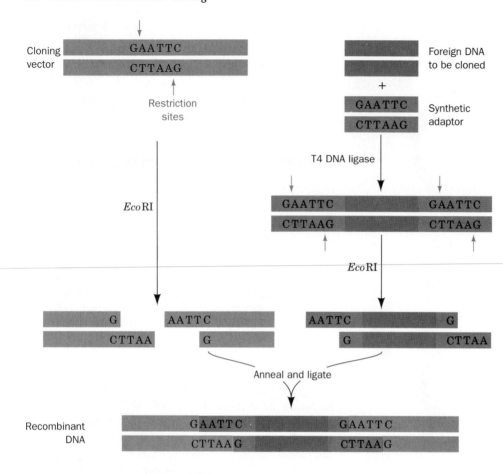

Figure 28-65
The construction of a recombinant DNA molecule through the use of synthetic oligonucleotide adaptors. In this example, the adaptor and the cloning vector have *Eco*RI restriction sties (*red arrows*). *arrows*).

therefore be replaced, as is described above, by a foreign DNA insert (Fig. 28-66). DNA is only packaged in λ phage heads if its length is from 75 to 105% of the 48.5-kb wild-type λ genome. Consequently, λ phage vectors that have failed to acquire a foreign DNA insert are unable to propagate because they are too short to form infectious phage particles. Cosmid vectors are subject to the same limitation. Moreover, cloned cosmids are harvested by repackaging them into phage particles. Hence, any cosmids that have lost sufficient DNA through random deletion to make them shorter than the above limit are not recovered. This is why cosmids can support the proliferation of large DNA inserts, whereas other types of plasmids cannot.

C. Genomic Libraries

In order to clone a particular DNA fragment, it must first be obtained in relatively pure form. The magnitude of this task may be appreciated when it is realized that, for example, a 1-kb fragment of human DNA represents only 0.000035% of the 2.9 billion bp human genome. A DNA fragment might be identified by Southern blotting of a restriction digest of the genomic DNA under investigation (Section 28-4C). The radioactive probe used in this procedure could be the corresponding mRNA if it is produced in sufficient quantity to be isolated (e.g., reticulocytes, which produce little protein besides hemoglobin, are rich in globin mRNAs). Alternatively, in cases

where the amino acid sequence of the protein encoded by the gene is known, the probe could be a mixture of the various synthetic oligonucleotides that might be complimentary to a segment of the gene's inferred base sequence (Section 30-1E).

In practice, it is usually more difficult to identify a particular gene from an organism and then clone it than it is to clone the organism's entire genome as DNA fragments and then identify the clone(s) containing the sequences(s) of interest. Such a set of cloned fragments is known as a **genomic library.** A genomic library of a particular organism need only be made once since it can be perpetuated for use whenever a new probe becomes available.

Genomic libraries are generated according to a procedure known as **shotgun cloning.** The chromosomal DNA of the organism of interest is isolated, cleaved to fragments of clonable size, and inserted in a cloning vector by the methods described in Section 28-8B. The DNA is fragmented by partial rather than exhaustive restriction digestion so that the genomic library contains intact representatives of all the organism's genes, including those whose sequences contain restriction sites. Shear fragmentation by rapid stirring of a DNA solution can also be used but requires further treatment of the fragments to insert them into cloning vectors. Genomic libraries have been established for a number of organisms including yeast, *Drosophila,* and humans.

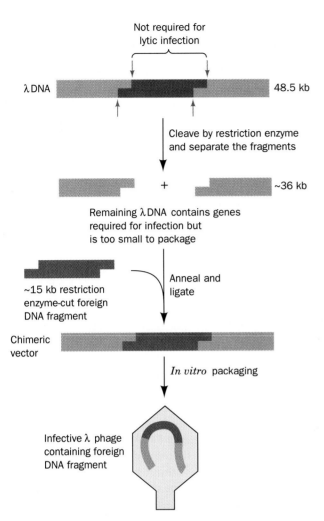

Figure 28-66
The cloning of foreign DNA in λ phages.

Many Clones Must Be Screened to Obtain a Gene of Interest

The number of random cleavage fragments that must be cloned to ensure a high probability that a given sequence is represented at least once in the genomic library is calculated as follows: The probability P that a set of N clones contains a fragment that constitutes a fraction f, in bp, of the organism's genome is

$$P = 1 - (1 - f)^N \qquad [28.7]$$

Consequently,

$$N = \ln(1 - P)/\ln(1 - f) \qquad [28.8]$$

Thus, in order for $P = 0.99$ for fragments averaging 10 kb in length, $N = 1840$ for the 4000-kb *E. coli* chromosome and 76,000 for the 165,000-kb *Drosophila* genome. The recent development of YAC-based genomic libraries therefore promises to greatly reduce the effort needed to obtain a given gene segment from a large genome.

Since a genomic library lacks an index, it must be screened for the presence of a particular gene. This is done

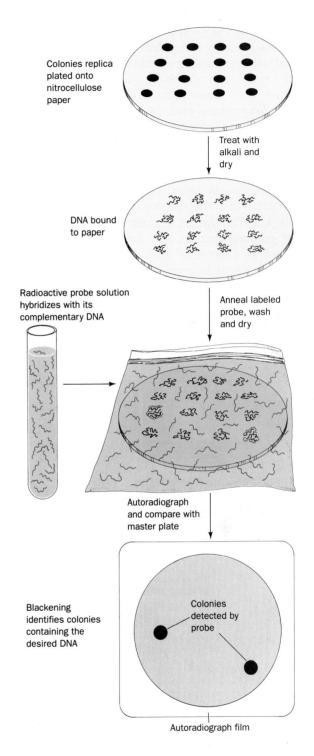

Figure 28-67
Colony (*in situ*) hybridization identifies the clones containing a DNA of interest.

by a process known as **colony** or *in situ* **hybridization** (Fig. 28-67; Latin: *in situ*, in position). The cloned yeast colonies, bacterial colonies, or phage plaques to be tested are transferred, by replica plating, from a master plate, to a nitrocellulose filter. The filter is treated with

NaOH, which lyses the cells/phages and denatures the DNA so that it binds to the nitrocellulose (recall that single-stranded DNA is preferentially bound to nitrocellulose). The filter is then dried to fix the DNA in place, treated under annealing conditions with a radioactive probe for the gene of interest, washed, and autoradiographed. *Only those colonies/plaques containing the sought-after gene will bind the probe and thereby blacken the film.* The corresponding clones can then be retrieved from the master plate. Using this technique, even an ~ 1 million clone human genomic library can be readily screened for the presence of a particular DNA segment.

Many eukaryotic genes and gene clusters span enormous tracts of DNA (Section 33-2); some consist of > 1000 kb. With the use of plasmid, phage, or cosmid-based genomic libraries, such long DNAs can only be obtained as a series of overlapping fragments (Fig. 28-68): Each gene fragment that has been isolated is, in turn, used as a probe to identify a successive but partially overlapping fragment of that gene, a process called **chromosome walking.** The use of YACs, however, greatly reduces the need for this laborious and error-prone process.

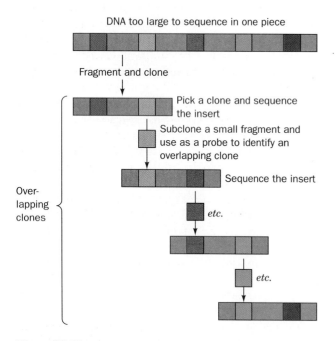

Figure 28-68
Chromosome walking. A DNA segment too large to sequence in one piece is fragmented and cloned. A clone is picked and the DNA insert it contains is sequenced. A small fragment of the insert near one end is subcloned (cloned from a clone) and used as a probe to select a clone containing an overlapping insert which, in turn, is sequenced. The process is repeated so as to "walk" down the chromosome. Chromosome walking can, of course, extend in both directions.

D. DNA Amplification by the Polymerase Chain Reaction

Although molecular cloning techniques are indispensible to modern biochemical research, the use of the **polymerase chain reaction (PCR)** offers a more convenient method of amplifying a specific DNA segment of up to 6 kb. In this technique, a denatured DNA sample is incubated with DNA polymerase and two oligonucleotide primers that direct the DNA polymerase to synthesize new complimentary strands. Multiple cycles of this process, each approximately doubling the amount of DNA present, exponentially amplify the DNA starting from as little as a single gene copy. In each cycle, the two strands of the duplex DNA are separated by heat denaturation, the primers are annealed to their complimentary segments on the DNA, and the DNA polymerase directs the synthesis of the complimentary strands (Section 28-6C). The use of a heat-stable DNA polymerase from the thermophilic bacterium *Thermus aquaticus* eliminates the need to add fresh enzyme after each heat denaturation step. Hence, each amplification cycle is controlled by simply varying the temperature.

Twenty-five cycles of PCR amplification increase the amount of the target sequence by around a millionfold with high specificity. Indeed, the method has been shown to amplify a target DNA present only once in a sample of 10^5 cells thereby demonstrating that the method can be used without prior DNA purification. The amplified DNA can be characterized by the various techniques we have discussed: Southern blotting, RFLP analysis, and direct sequencing. PCR amplification is therefore a form of "cell-free molecular cloning" that can accomplish in an automated 3 to 4 h *in vitro* reaction what would otherwise take days or weeks via the cloning techniques discussed above.

The use of PCR amplification holds great promise for a variety of applications. Clinically, it can be used for the rapid diagnosis of infectious diseases and the detection of rare pathological events such as chromosomal translocations. Forensically, the DNA from a single hair or sperm can be used to unambiguously identify the donor. RNA may also be amplified by the PCR method by first converting it to cDNA through the use of reverse transcriptase (Section 28-6D).

E. Production of Proteins

One of the greatest potential uses of recombinant DNA technology is in the production of large quantities of scarce and/or novel proteins. This is a relatively straightforward procedure for bacterial proteins: A cloned structural gene must be accompanied by the properly positioned transcriptional and translational control sequences for its expression. With the use of a relaxed control plasmid and an efficient **promoter** (a type of transcriptional control element; Section 29-3A), the production of a protein of interest may reach 30% of the host's total cellular protein. Such genetically engineered organisms are called **overproducers.** Bacterial cells often sequester such large amounts of useless (to the bacterium) protein as insoluble and denatured inclusions. Protein extracted from these inclusions must therefore be renatured, usually by dissolving it in a guanidinium chloride or urea solution (Section 8-1A) and then dialyzing away the denaturant.

The synthesis of a eukaryotic protein in a prokaryotic host presents several problems not encountered with prokaryotic proteins:

1. The eukaryotic control elements for RNA and protein synthesis are not recognized by bacterial hosts.

2. Most eukaryotic genes contain one or more internal unexpressed sequences called **introns,** which are specifically excised from the gene's RNA transcript to form the mature mRNA (Section 29-4A). Bacterial genes lack introns and hence, bacteria are unable to excise them.

3. Bacteria lack the enzyme systems to carry out the specific post-translational processing that many eukaryotic proteins require for biological activity (Section 30-5). Most conspicuously, bacteria do not glycosylate proteins (although, in many cases, glycosylation does not seem to affect protein function).

4. Eukaryotic proteins may be preferentially degraded by bacterial proteases (Section 30-6A).

The problem of nonrecognition of eukaryotic control elements can be eliminated by inserting the protein-encoding portion of a eukaryotic gene into a vector containing correctly placed bacterial control elements. The need to excise introns can be circumvented by cloning the cDNA of the protein's mRNA. Alternatively, genes encoding small proteins of known sequence can be chemically synthesized (Section 28-7). Neither of these strategies is universally applicable, however, because few mRNAs are sufficiently abundant to be isolated and many eukaryotic proteins are large (although the maximum available size of synthetic polynucleotides is increasing rapidly). Likewise, no general approach has been developed for the post-translational modification of eukaryotic proteins although polypeptide cleavage by treatment with trypsin or cyanogen bromide (Section 6-1E) has been successfully employed in the *in vitro* activation of some eukaryotic proenzymes. Lastly, the preferential bacterial proteolysis of certain eukaryotic proteins has been prevented by inserting the eukaryotic gene within a bacterial gene. The resulting hybrid protein has an N-terminal polypeptide of bacterial origin that, in some cases, prevents bacterial proteases from recognizing the eukaryotic segment as being foreign. However, the development of cloning vectors that prop-

agate in eukaryotic hosts, such as yeast or cultured animal cells, has led to the elimination of many of these problems (although post-translational processing may vary among different eukaryotes). Indeed, **shuttle vectors** are available that can propagate in both yeast and *E. coli* and thus transfer (shuttle) genes between these two types of cells.

The ability to synthesize a given protein in large quantities has enormous medical, agricultural, and industrial potential. Human insulin and human growth hormone, to mention but two, are already in widespread clinical use, and many others are under development. *Of equal importance is the ability to tailor proteins to specific applications through site-directed mutagenesis* (Section 28-7). For many purposes, however, it will be preferable to tailor an intact organism rather than just its proteins—true genetic engineering. For example, if nitrogen fixing bacteria can be persuaded to associate with agriculturally important plants besides legumes (a complicated process whose requirements are by no means understood), the need for nitrogenous fertilizers to grow these plants in high yield will perhaps be entirely eliminated.

F. Social Considerations

In the early 1970s, when strategies for genetic engineering were first being discussed, it was realized that little was known about the safety of the proposed experiments. Certainly it would be foolhardy to attempt experiments such as introducing the gene for **diphtheria toxin** (Section 30-3G) into *E. coli* so as to convert this human symbiont into a deadly pathogen. But what biological hazards would result, for example, from cloning tumor virus genes in *E. coli* (a useful technique for analyzing these viruses)? Consequently, in 1975, molecular biologists declared a voluntary moratorium on molecular cloning experiments until these risks could be assessed. There ensued a spirited debate, at first among molecular biologists and later in the public arena, between two camps: those who thought that the enormous potential benefits of recombinant DNA research warranted its continuation once adequate safety precautions had been instituted, and those who felt that its potential dangers were so great that it should not be pursued under any circumstances.

The former viewpoint eventually prevailed with the promulgation, in 1976, of a set of U.S. government regulations for recombinant DNA research. Experiments that are obviously dangerous were forbidden. In other experiments, the escape of laboratory organisms was to be prevented by both physical and biological containment. By biological containment it is meant that vectors will only be cloned in host organisms with biological defects that prevent their survival outside the laboratory. For example, $\chi 1776$, the first approved "safe"

Figure 28-69

[Drawing by T. A. Bramley, *in* Andersen, K., Shanmugam, K. T., Lim, S. T., Csonka, L.N., Tait, R., Hennecke, H., Scott, D. B., Hom, S. S. M., Haury, J. F., Valentine, A., and Valentine, R. C., *Trends Biochem. Sci.* **5**, 35 (1980). Copyright © Elsevier Biomedical Press, 1980. Used by permission.]

strain of *E. coli,* has among its several defects the requirement for diaminopimelic acid, an intermediate in lysine biosynthesis (Section 24-5B), which is neither present in human intestines nor commonly available in the environment.

As experience with recombinant DNA research accumulated, it became evident that the foregoing reservations were largely groundless. No genetically altered organism yet reported has caused an unexpected health hazard. Indeed, recombinant DNA techniques have, in many cases, eliminated the health hazards of studying dangerous pathogens such as the virus causing AIDS. Consequently, since 1979, the regulations governing recombinant DNA research have been gradually relaxed.

There are other social, ethical, and legal considerations that will have to be faced as new genetic engineering techniques become available (Fig. 28-69). Bacterially produced human insulin is now routinely prescribed to treat diabetes and few would dispute the use of "gene therapy," if it can be developed, to cure such genetic defects as sickle-cell anemia (Section 9-3B) and Lesch-Nyhan syndrome (Section 26-2D). If, however, it becomes possible to alter complex traits such as athletic ability or intelligence, which changes would be considered desirable, under what circumstances would they be made, and who would decide whether to make them? If it becomes easy to determine an individual's genetic makeup, should this information be used, for example, in evaluating applications for educational and employment opportunities, or in assessing a person's eligibility for health insurance? The U.S. Supreme Court has affirmed that novel life forms developed in the laboratory may be patented. But to what extent will such proprietory rights impede the free exchange of ideas and information that has heretofore permitted the rapid development of recombinant DNA technology?

Chapter Summary

Nucleic acids are linear polymers of nucleotides containing either ribose residues in RNA or deoxyribose residues in DNA that are linked by $3' \rightarrow 5'$ phosphodiester bonds. In double helical DNAs and RNAs, the base compositions obey Chargaff's rules: A = T and G = C. RNA, but not DNA, is susceptible to base-catalyzed hydrolysis.

B-DNA consists of a right-handed double helix of antiparallel sugar–phosphate chains with ∼ 10 bp per turn of 34 Å and with the bases all perpendicular to the helix axis. Bases on opposite strands hydrogen bond in a geometrically complementary manner to form A·T and G·C Watson–Crick base pairs. DNA replicates in a semiconservative manner as has been demonstrated by the Meselson–Stahl experiment. At low humidity, B-DNA undergoes a reversible transformation to a wider, flatter right-handed double helix known as A-DNA. Z-DNA, which is formed at high salt concentrations by polynucleotides of alternating purine and pyrimidine base sequences, is a left-handed double helix. Double-helical RNA and RNA·DNA hybrids have A-DNA-like structures. DNA occurs in nature as molecules of enormous lengths which, because they are also quite stiff, are easily mechanically cleaved by laboratory manipulations.

When heated past its melting temperature, T_m, DNA denatures and undergoes strand separation. This process may be monitored by the hyperchromism of the DNA's UV spectrum. The orientations about the glycosidic bond and the various torsion angles in the sugar–phosphate chain are sterically constrained in nucleic acids. Likewise, only a few of the possible sugar pucker conformations are commonly observed. Watson–Crick base pairing is both geometrically and electronically complementary. Yet, hydrogen bonding interactions do not significantly stabilize nucleic acid structures. Rather, they are largely stabilized by hydrophobic interactions. Nevertheless, the hydrophobic forces in nucleic acids are qualitatively different in character from those that stabilize proteins. Electrostatic interactions between charged phosphate groups are also important structural determinants of nucleic acids.

Nucleic acids are fractionated by many of the techniques that are used to separate proteins. Hydroxyapatite chromatography separates single-stranded from double-stranded DNA. Polyacrylamide or agarose gel electrophoresis separates DNA largely on the basis of size. Very large DNAs can be separated by pulsed-field gel electrophoresis on agarose gels. Specific base sequences may be detected in DNA with the Southern transfer technique and in RNA by the similar northern transfer technique. DNA may be fractionated according to base composition by CsCl density gradient ultracentrifugation. Different species of RNA are separated by rate-zonal ultracentrifugation through a sucrose gradient.

The linking number of a covalently closed circular DNA is topologically invariant. Consequently, any change in the twist of a circular duplex must be balanced by an equal and opposite change in its writhing number, which indicates its degree of supercoiling. Supercoiling can be induced by intercalation agents. The gel electrophoretic mobility of DNA increases with its degree of superhelicity. Naturally occurring DNAs are all negatively supercoiled and must be so in order to partici-

pate in DNA replication, RNA transcription, and genetic recombination. Type I topoisomerases (nicking–closing enzymes) relax negatively supercoiled DNAs, one supertwist at a time, by creating a single-strand break, passing a single-strand loop through the gap, and resealing it. Type II topoisomerases (gyrases) generate negative supertwists at the expense of ATP hydrolysis. They do so, two supertwists at a time, by making a double-strand scisson in the DNA, passing the duplex through the break, and resealing it.

Nucleic acids may be sequenced by the same basic strategy used to sequence proteins. Defined DNA fragments are generated by Type II restriction endonucleases, which cleave DNA at specific and usually palindromic sequences of four to six bases. Restriction maps provide easily located physical reference points on a DNA molecule. In the chemical cleavage method of DNA sequencing, a defined fragment of DNA is ^{32}P-labeled at one end and subjected to a chemical cleavage process that randomly cleaves it after a particular type of base. The electrophoresis of the four differently cleaved DNA samples in parallel lanes of a sequencing gel resolves fragments that differ in size by one nucleotide. The base sequence of the DNA can be directly read from an autoradiogram of the gel. In the chain-terminator method, the DNA to be sequenced is replicated by DNA polymerase I in the presence of a $[\alpha\text{-}^{32}\text{P}]$-labeled deoxynucleoside triphosphate and a small amount of the dideoxy analog of one of the nucleoside triphosphates. This results in a series of ^{32}P-labeled chains that are terminated after the various positions occupied by the corresponding base. An autoradiograph of the sequencing gel containing the four sets of fragments, each terminated after a different type of base, indicates the DNA's base sequence. RNA may be sequenced by determining the sequence of its corresponding cDNA or by directly sequencing it by a variation of the chemical cleavage method.

Oligonucleotides are indispensible to recombinant DNA technology; they are used to identify normal and mutated genes and to specifically alter genes through site-directed mutagenesis. Oligonucleotides of defined sequence are efficiently synthesized by the phosphite-triester method, a cyclic, nonaqueous, solid phase process that has been automated.

A DNA fragment may be produced in large quantities by inserting it, using recombinant DNA techniques, into a suitable cloning vector. These may be genetically engineered plasmids, bacteriophages, cosmids, or yeast artificial chromosomes (YACs). The DNA to be cloned is usually obtained as a restriction fragment so that it can be specifically ligated into a corresponding restriction cut in the cloning vector. Gene splicing may also occur through the generation of complementary homopolymer tails on the DNA fragment and the cloning vector or through the use of synthetic palindromic linkers containing restriction sequences. Introduction of a recombinant cloning vector into a suitable host organism permits the foreign DNA segment to be produced in nearly unlimited quantities. A particular gene may be isolated through the screening of a genomic library of the organism producing the gene. Genetic engineering techniques may also be used to produce otherwise scarce or specifically altered proteins in large quantities.

References

General

Bloomfield, V. A., Crothers, D. M., and Tinoco, I. Jr., *Physical Chemistry of Nucleic Acids,* Harper & Row (1974).

Cantor, C. R. and Schimmel, P. R., *Biophysical Chemistry,* Chapters 3, 5, 22–24, Freeman (1980).

Landegren, R., Kaiser, R., Caskey, C. T., and Hood, L., DNA diagnostics—molecular techniques and automation, *Science* **242,** 230–237 (1988).

Saenger, W., *Principles of Nucleic Acid Structure,* Springer–Verlag (1984). [A detailed and authoritative exposition.]

Structures of DNA, Cold Spring Harbor Symp. Quant. Biol. **47** (1983). [A cornucopia of structural information on DNA.]

Wood, W. B., Wilson, J. H., Benbow, R. M., and Hood, L. E., *Biochemistry, A Problems Approach* (2nd ed.), Chapters 17 and 19, Benjamin/Cummings (1980).

Structures and Stabilities of Nucleic Acids

Dickerson, R. E., Drew, H. R., Conner, B. N., Wing, R. M., Fratini, A. V., and Kopka, M. L., The anatomy of A-, B- and Z-DNA, *Science* **216,** 475–485 (1982).

Dickerson, R. E., The DNA helix and how it is read, *Sci. Am.* **249**(6): 94–111 (1983).

Drew, H. R., McCall, M. J., and Calladine, C. R., Recent studies of DNA in the crystal, *Annu. Rev. Cell Biol.* **4,** 1–20 (1988).

Jaworski, A., Hsieh, W.-T., Blaho, J. A., Larson, J. E., and Wells, R. D., Left-handed DNA in vivo, *Science* **238,** 773–777 (1988).

Jurnak, F. A. and McPherson, A. (Eds.), *Biological Macromolecules and Assemblies,* Vol. 2: *Nucleic Acids and Interactive Proteins,* Wiley (1985). [Chapters 1–3 are authoritative reviews on the structures of A-, B-, and Z-DNAs, respectively.]

Meselson, M. and Stahl, F. W., The replication of DNA in Escherichia coli, *Proc. Natl. Acad. Sci.* **44,** 671–682 (1958). [The classic paper establishing the semiconservative nature of DNA replication.]

Rich, A., Nordheim, A., and Wang, A. H.-J., The chemistry and biology of left-handed Z-DNA, *Annu. Rev. Biochem.* **53,** 791–846 (1984).

Shakked, Z. and Rabinovich, D., The effect of the base sequence on the fine structure of the DNA double helix, *Prog. Biophy. Mol. Biol.* **47,** 159–195 (1986).

Sundaralingam, M., Stereochemistry of nucleic acids and their constituents. IV. Allowed and preferred conformations of nucleosides, nucleoside mono-, di-, tri-, and tetraphosphates, nucleic acids and polynucleotides, *Biopolymers* **7,** 821–860 (1969).

Voet, D. and Rich, A., The crystal structures of purines, pyrimidines and their intermolecular structures, *Prog. Nucleic Acid Res. Mol. Biol.* **10,** 183–265 (1970).

Watson, J. D. and Crick, F. H. C., Molecular structure of nucleic acids, *Nature* **171,** 737–738 (1953) *and* Genetical implications of the structure of deoxyribonucleic acid, *Nature* **171,** 964–967 (1953). [The seminal papers that are widely held to mark the origin of modern molecular biology.]

Wells, R. D., Unusual DNA structures, *J. Biol. Chem.* **263,** 1095–1098 (1988).

Wing, R., Drew, H., Takano, T., Broka, C., Tanaka, S., Itakura, K., and Dickerson, R. E., Crystal structure analysis of a complete turn of B-DNA, *Nature* **287,** 755–758 (1980).

Zimmerman, S. B., The three-dimensional structures of DNA, *Annu. Rev. Biochem.* **51,** 395–427 (1982).

Fractionation of Nucleic Acids

Anand, R., Pulsed field gel electrophoresis: a technique for fractionating large DNA molecules, *Trends Genet.* **2,** 278–283 (1986).

Cantor, C. R., Smith, C. L., and Mathew, M. K., Pulsed-field gel electrophoresis of very large molecules, *Annu. Rev. Biophys. Biophys. Chem.* **17,** 287–304 (1988).

Freifelder, D., *Physical Biochemistry. Applications to Biochemistry and Molecular Biology* (2nd ed.), Freeman (1982).

Gould, H. and Matthews, H. R., Separation methods for nucleic acids and oligonucleotides, *in* Work, T. S. and Work E. (Eds.), *Laboratory Techniques in Biochemistry and Molecular Biology,* Vol. 4, Part III, North–Holland (1976).

Rickwood, D. and Hames, B. D. (Eds.), *Gel Electrophoresis of Nucleic Acids. A Practical Approach,* IRL Press (1982).

Schleif, R. F. and Wensink, P. C., *Practical Methods in Molecular Biology,* Chapter 5, Springer–Verlag (1981).

Walker, J. M. (Ed.), *Methods in Molecular Biology,* Vol. 2, *Nucleic Acids,* Humana Press (1984).

Supercoiled DNA

Bauer, W. R., Crick, F. H. C., and White, J. H., Supercoiled DNA, *Sci. Am.* **243**(1): 118–133 (1980). [The topology of supercoiling.]

Cozzarelli, N. R., DNA gyrase and the supercoiling of DNA, *Science* **207,** 953–960 (1980).

Gellert, M., DNA topoisomerases, *Annu. Rev. Biochem.* **50,** 879–910 (1981).

Maxwell, A. and Gellert, M., Mechanistic aspects of DNA topoisomerases, *Adv. Protein Chem.* **38,** 69–107 (1986).

Wang, J. C., DNA topoisomerases, *Sci. Am.* **247**(1): 94–109 (1982).

Wang, J. C., DNA topoisomerases, *Annu. Rev. Biochem.* **54,** 665–697 (1985).

Nucleic Acid Sequencing

Brownlee, G. G., Determination of sequences in RNA, *in* Work, T. S. and Work, E. (Eds.), *Laboratory Techniques in Biochemistry and Molecular Biology,* Vol. 3, Part I, North–Holland (1972). [A review of the sequencing procedures used before the advent of restriction enzymes and fast DNA sequencing techniques.]

Fiers, W., Contreras, R., Duerinck, F., Haegeman, G., Iserantant, D., Merregaert, J., Min Jou, W., Molemans, F., Raeymaekers, A., Van den Berghe, A., Volckaert, G., and Ysebaert, M., Complete nucleotide sequence of bacteriophage MS2 RNA: primary and secondary structure of the replicase gene, *Nature* **260,** 500–507 (1976).

Frederick, C. A., Wang, B.-C., Greene, P., Boyer, H. W., Grable, J., and Rosenberg, J. M., Kinked DNA in crystalline complex with *EcoRI* endonuclease, *Science* **234,** 1526–1551 (1987).

Ghosh, P. K., Reddy, V. B., Piatak, M., Lebowitz, P., and Weissman, S. M., Determination of RNA sequences by primer directed synthesis and sequencing of their cDNA transcripts, *Methods Enzymol.* **65,** 580–595 (1980).

Gusella, J. F., DNA polymorphism and human disease, *Annu. Rev. Biochem.* **55,** 831–854 (1986).

Hindley, J., DNA sequencing, *in* Work, T. S. and Burdon, R. S. (Eds.), *Laboratory Techniques in Molecular Biology,* Vol. 10, North–Holland (1983).

Maxam, A. M. and Gilbert, W., Sequencing end-labeled DNA with base-specific chemical cleavages, *Methods Enzymol.* **65,** 499–560 (1980).

Nathans, D. and Smith, M. O., Restriction endonucleases in the analysis and restructuring of DNA, *Annu. Rev. Biochem.* **44,** 273–293 (1975).

Peattie, D. M., Direct chemical method for sequencing RNA, *Proc. Natl. Acad. Sci.* **76,** 1760–1764 (1979).

Prober, J. M., Trainor, G. L., Dam, R. J., Hobbs, F. W., Robertson, C. W., Zagursky, R. J., Cocuzza, A. J., Jensen, M. A., and Baumeister, K., A system for rapid DNA sequencing with fluorescent chain-terminating dideoxynucleotides, *Science* **238,** 336–341 (1987).

Rosenberg, J. M., McClarin, J. A., Frederick, C. A., Wang, B.-C., Grable, J., Boyer, H. W., and Greene, P., Structure and recognition mechanism of *EcoRI* endonuclease, *Trends Biochem. Sci.* **12,** 395–398 (1987).

Sanger, F., Sequences, sequences, and sequences, *Annu. Rev. Biochem.* **57,** 1–28 (1988). [A scientific memoir.]

Smith, A. J. H., DNA sequence analysis by primed synthesis, *Methods Enzymol.* **65,** 560–580 (1980). [The chain-terminator method.]

Weissman, S. M., *Methods of DNA and RNA Sequencing,* Preager (1983).

Wells, R. D., Klein, R. D., and Singleton, C. K., Type II restriction enzymes, *in* Boyer, P. D. (Ed.), *The Enzymes* (3rd ed.), Vol. 14, *pp.* 137–156, Academic Press (1981).

White, R. and Lalouel, J.-M., Chromosome mapping with DNA markers, *Sci. Am.* **258**(2): 40–48 (1988). [Describes the use of RFLPs.]

Yuan, R., Structure and mechanism of multifunctional restriction endonucleases, *Annu. Rev. Biochem.* **50,** 285–315 (1981). [A discussion of Types I and III restriction endonucleases.]

Chemical Synthesis of Oligonucleotides

Caruthers, M. H., Barone, A. D., Beaucage, S. L., Dodds, D. R., Fisher, E. F., McBride, L. J., Matteucci, M., Stabinsky, Z., and Tang, J. Y., Chemical synthesis of deoxyoligonucleotides, *Methods Enzymol.* **154,** 287–313 (1987).

Conner, B. J., Reyes, A. A., Morin, C., Itakura, K., Teplitz, R. L., and Wallace, R. B., Detection of sickle cell β^S-globin allele by hybridization with synthetic oligonucleotides, *Proc. Natl. Acad. Sci.* **80,** 278–282 (1983).

Gait, M. J. (Ed.), *Oligonucleotide Synthesis. A Practical Approach,* IRL Press (1984).

Itakura, K., Rossi, J. J., and Wallace, B. R., Synthesis and use of synthetic oligonucleotides, *Annu. Rev. Biochem.* **53,** 323–356 (1984).

Khorana, H. G., Total synthesis of a gene, *Science* **203,** 614–625 (1979). [The use of the classical phosphodiester method of oligonucleotide synthesis.]

Molecular Cloning

Berger, S. L. and Kimmel, A. R. (Eds.), Guide to Molecular Cloning Techniques, *Methods Enzymol.* **152** (1987). [A "cookbook" describing the basic techniques of molecular biology.]

Burke, D. T., Carle, G. F., and Olso, M. V., Cloning of large segments of exogenous DNA into yeast by means of artificial chromosome vectors, *Science* **236,** 806–812 (1987).

Glover, D. M. (Eds.), *DNA Cloning. A Practical Approach,* Vols. 1 and 2, IRL Press (1985).

Johnson, I. S., Human insulin from recombinant DNA technology, *Science* **219,** 632–637 (1983).

Leatherbarrow, R. J. and Fersht, A. R., Protein engineering, *Protein Eng.* **1,** 7–16 (1986).

Maniatis, T., Fritsch, E. F., and Sambrook, J., *Molecular Cloning. A Laboratory Manual,* Cold Spring Harbor Laboratory (1982). [Contains extensive data and descriptions of techniques for preparing recombinant DNA.]

Saiki, R. K., Gelfand, D. H., Stoffel, S., Scharf, S. J., Higuchi, R., Horn, G. T., Mullis, K. B., and Erlich, H. A., Primer-directed enzymatic amplification of DNA with a thermostable DNA polymerase, *Science* **239,** 487–494 (1988).

Watson, J. D. and Tooze, J., *The DNA Story,* Freeman (1981). [A scrapbooklike account of the recombinant DNA debate of the 1970s.]

Watson, J. D., Tooze, J., and Kurtz, D. T., *Recombinant DNA. A Short Course,* Scientific American Books (1983).

Wu, R., Grossman, L., and Moldave, K., (Eds.), Recombinant DNA, Parts A–F, *Methods Enzymol.* **68, 100, 101,** and **153–155** (1979, 1983, and 1987).

Historical Aspects

Crick, F., *What Mad Pursuit,* Basic Books (1988). [A scientific autobiography.]

Judson, H. F., *The Eighth Day of Creation,* Part I, Simon & Schuster (1979). [A fascinating narration of the discovery of the DNA double helix.]

Olby, R., *The Path to the Double Helix,* Macmillan (1974).

Portugal, F. H. and Cohen, J. S., *A Century of DNA,* MIT Press (1977).

Sayre, A., *Rosalind Franklin and DNA,* Norton (1975). [A biographical work which argues that Rosalind Franklin, who died in 1958, deserves far more credit than is usually accorded her for the discovery of the structure of DNA.]

Schlenk, F., Early nucleic acid chemistry, *Trends Biochem. Sci.* **13,** 67–68 (1988).

Watson, J. D., *The Double Helix,* Atheneum (1968). [A lively but controversial autobiographical account of the discovery of the DNA structure.]

Problems

1. Non-Watson–Crick base pairs are of biological importance. For example: (a) **Hypoxanthine** (6-oxopurine) is often one of the bases of the anticodon of tRNA (the three consecutive nucleotides that base pair with mRNA). With what base on mRNA is hypoxanthine likely to pair? Draw the structure of this base pair. (b) tRNA often makes a G·U base pair with mRNA. Draw a plausible structure for such a base pair. (c) Many species of tRNA contain a hydrogen bonded U·A·U assembly. Draw two plausible structures for this assembly in which each U forms at least two hydrogen bonds with the A. (d) Mutations may arise during DNA replication when mispairing occurs as a result of the transient formation of a rare tautomeric form of a base. Draw the structure of a base pair with proper Watson–Crick geometry that contains a rare tautomeric form of adenine. What base sequence change would be caused by such mispairing?

2. What is the molecular mass and contour length of a section of B-DNA that specifies a 40-kD protein? Each amino acid is specified by three contiguous bases on a single strand of DNA (Section 30-1).

*3. The antiparallel orientation of complementary strands in duplex DNA was elegantly demonstrated in 1960 by Arthur Kornberg by **nearest-neighbor analysis.** In this technique, DNA is synthesized by DNA polymerase I from one [α-^{32}P]-labeled and three unlabeled deoxynucleoside triphosphates. The resulting product is hydrolyzed by a DNase that cleaves phosphodiester bonds on the 3' sides of all deoxynucleotides.

$$ppp^*A + pppC + pppG + pppT$$

$$PP_i \swarrow | \text{ DNA polymerase}$$

$$\cdots pCpTp^*ApCpCp^*ApGp^*Ap^*ApTp\cdots$$

$$H_2O \searrow | \text{ DNase I}$$

$$\cdots + Cp + Tp^* + Ap + Cp + Cp^* + Ap + Gp^* + Ap^* + Ap + Tp + \cdots$$

In this example, the relative frequencies of occurrence of ApA, CpA, GpA, and TpA in the DNA can be determined by measuring the relative amounts of Ap*, Cp*, Gp*, and Tp*, respectively, in the product. The relative frequencies with which the other 12 dinucleotides occur may likewise be determined by labeling, in turn, the other 3 nucleoside triphosphates in the above reactions. There are equivalencies between the amounts of certain pairs of dinucleotides. However, the identities of these equivalencies depend on whether the DNA consists of parallel or antiparallel strands. What are these equivalences in both cases?

4. What would be the effect of the following agents on the melting curve of an aqueous solution of duplex DNA? Explain. (a) Decreasing the ionic strength of the solution. (b) Squirting the DNA solution, at high pressure, through a very narrow orifice. (c) Bringing the solution to 0.1M adenine. (d) Heating the solution to 25°C above the DNA's melting point and then rapidly cooling it to 25°C below the DNA's melting point. (e) Adding a small amount of ethanol to the DNA solution.

5. What is the mechanism of alkaline denaturation of DNA?

*6. At Na$^+$ concentrations $> 10M$, the T_m of DNA decreases with increasing [Na$^+$]. Explain this behavior. (*Hint:* Consider the solvation requirements of Na$^+$.)

*7. Why are the most commonly observed conformations of the ribose ring those in which either atom C(2') or C(3') is out of the plane of the other four ring atoms? (*Hint:* In puckering a planar ring such that one atom is out of the plane of the other four, the substituents about the bond opposite the out-of-plane atom remain eclipsed. This is best observed with a ball-and-stick model.)

8. Polyoma virus DNA can be separated by sedimentation at neutral pH into three components that have sedimentation coefficients of 20, 16, and 14.5S and which are known as Types I, II, and III DNAs, respectively. These DNAs all have identical base compositions and molecular masses. In 0.15M NaCl, both Types II and III DNAs have melting curves of normal cooperativity and a T_m of 88°C. Type I DNA, however, exhibits a very broad melting curve and a T_m of 107°C. At pH 13, Types I and III DNAs have sedimentation coefficients of 53 and 16S, respectively, and Type II separates into two components with sedimentation coefficients of 16S and 18S. How do Types I, II, and III DNAs differ from one another? Explain their different physical properties.

9. A closed circular duplex DNA has a 100 bp segment of alternating C and G residues. Upon transfer to a solution containing a high salt concentration, this segment undergoes a transition from the B conformation to the Z conformation. What is the accompanying change in its linking number, writhing number, and twist?

10. You have discovered an enzyme secreted by a particularly virulent bacterium that cleaves the C(2')—C(3') bond in the deoxyribose residues of duplex DNA. What is the effect of this enzyme on supercoiled DNA?

11. SV40 DNA is a circular molecule of 5243 bp that is 40% G + C. In the absence of sequence information, how many restriction cuts would *Taq*I, *Eco*RII, *Pst*I, and *Hae*II be expected to make in SV40 DNA? (Figure 28-49 indicates the number of restriction cuts that these enzymes actually make.)

12. A bacterial chromosome consists of a protein–DNA complex in which its single DNA molecule appears to be supercoiled as demonstrated by ethidium bromide titration. However, in contrast to the case with naked circular duplex DNA, the single-strand nicking of chromosomal DNA does not abolish this supercoiling. What does this indicate about the structure of the bacterial chromosome, that is, how do its proteins constrain its DNA?

13. Which of the restriction endonucleases listed in Table 28-5 produce blunt ends? Which sets of them are **isoschizomers** (enzymes that have the same recognition se-

quence but do not necessarily cleave at the same sites; Greek: *isos,* equal; *schizein,* to cut); which of them are **isocaudamers** (enzymes that produce identical sticky ends: Latin: *cauda,* tail)?

14. In investigating a newly discovered bacterial species that inhabits the sewers of Berkeley, you isolate a plasmid which you suspect carries genes that confer resistance to several antibiotics. To characterize this plasmid you decide to make its restriction map. The sizes of the plasmid's restriction fragments, as determined from their electrophoretic mobilities on agarose gels, are given in the following table. From the data, construct the restriction map of the plasmid.

Sizes of Restriction Fragments from a Plasmid DNA

Restriction Enzymes	Fragment Sizes (kb)
*Eco*RI	5.4
*Hind*III	2.1, 1.9, 1.4
*Sal*I	5.4
*Eco*RI + *Hind*III	2.1, 1.4, 1.3, 0.6
*Eco*RI + *Sal*I	3.2, 2.2
*Hind*III + *Sal*I	1.9, 1.4, 1.2, 0.9

15. Figure 28-70 pictures an autoradiograph of the sequencing gel of a *Hae*III restriction fragment from the *E. coli* K12 gene that codes for dihydrofolate reductase. The DNA was treated according to the chemical cleavage method of DNA sequencing after being ^{32}P labeled at its 3' end. Read the sequence of the first 50 bases from the bottom of the gel.

16. How many yeast DNA fragments of average length 5 kb must be cloned in order to be 90, 99, and 99.9% certain that a genomic library contains a particular segment? The yeast chromosome consists of 13,500 kb.

17. Many of the routine operations in genetic engineering are carried out using commercially available "kits." Genbux Inc., a prospective manufacturer of such kits, has asked your advice on the feasibility of supplying a kit of intact λ phage cloning vectors with the nonessential central section of their DNA already removed. Presumably a "gene jockey" could then grow the required amount of phage, isolate its DNA, and restriction cleave it without having to go to the effort of separating out the central section. What advice would you give the company?

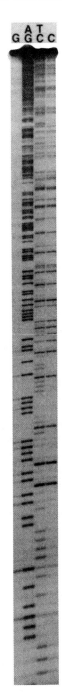

Figure 28-70
[From Smith, D. R. and Calvo, J. M., *Nuclei Acids Res.* **8,** 2268 (1980).]

Chapter 29
TRANSCRIPTION

There are three major classes of RNA, all of which participate in protein synthesis: **ribosomal RNA (rRNA), transfer RNA (tRNA),** and **messenger RNA (mRNA).** All of these RNAs are synthesized under the direction of DNA templates, a process known as **transcription.**

RNA's involvement in protein synthesis became evident in the late 1930s through investigations by Torbjörn Caspersson and Jean Brachet. Caspersson, using microscopic techniques, found that DNA is confined almost exclusively to the eukaryotic cell nucleus, whereas RNA occurs largely in the cytosol. Brachet, who had devised methods for fractionating cellular organelles, came to similar conclusions based on direct chemical analyses. He found, in addition, that the cytosolic RNA-containing particles are also protein rich. Both investigators noted that the concentration of these RNA-protein particles (which were later named ribosomes) is correlated with the rate that a cell synthesizes protein, inferring a relationship between RNA and protein synthesis. Indeed, Brachet even suggested that *the RNA-protein particles are the site of protein synthesis.*

Brachet's suggestion was shown to be valid when radioactively labeled amino acids became available in the 1950s. A short time after injecting a rat with a labeled amino acid, most of the label that had been incorporated in proteins was associated with ribosomes. This experiment also established that *protein synthesis is not immediately directed by DNA because, at least in eukaryotes, DNA and ribosomes are never in contact.*

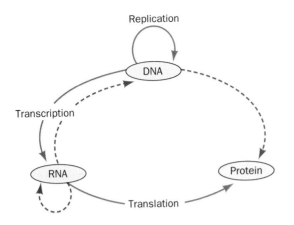

Figure 29-1
The central dogma of molecular biology. Solid arrows indicate the types of genetic information transfers that occur in all cells. Special transfers are indicated by the dashed arrows: RNA-directed RNA polymerase occurs both in certain RNA viruses and in some plants (where it is of unknown function); RNA-directed DNA polymerase (reverse transcriptase) occurs in other RNA viruses; and DNA directly specifying a protein is unknown but does not seem beyond the realm of possibility. However, the missing arrows are information transfers the central dogma postulates never occur: protein specifying either DNA, RNA, or protein. In other words, *proteins can only be recipients of genetic information.* [After Crick, F., *Nature* **227**, 562 (1970).]

In 1958, Francis Crick summarized the then dimly perceived relationships among DNA, RNA, and protein in a flow scheme he described as the **central dogma** of molecular biology: *DNA directs its own replication and its transcription to RNA which, in turn, directs its translation to proteins (Fig. 29-1).*

> The peculiar use of the word "dogma," one definition of which is a religious doctrine that the true believer cannot doubt, stemmed from a misunderstanding. When Crick formulated the central dogma, he was under the impression that dogma meant "an idea for which there was no reasonable evidence."

We begin this chapter by discussing experiments that led to the elucidation of mRNA's central role in protein synthesis. We then study the mechanism of transcription and its control in prokaryotes. Finally, in the last section, we consider post-transcriptional processing of RNA in both prokaryotes and eukaryotes. Translation is the subject of Chapter 30.

1. THE ROLE OF RNA IN PROTEIN SYNTHESIS

Proteins are specified by mRNA and synthesized on ribosomes. This idea arose from the study of **enzyme induction,** a phenomenon in which bacteria vary the synthesis rates of specific enzymes in response to environmental changes. We shall see below that *enzyme induction occurs as a consequence of the regulation of mRNA synthesis by proteins that specifically bind to the mRNA's DNA templates.*

A. Enzyme Induction

E. coli can synthesize an estimated 3000 different polypeptides (Section 27-1D). There is, however, enormous variation in the amounts of these different polypeptides that are produced. For instance, the various ribosomal proteins may each be present in over 10,000 copies per cell, whereas certain regulatory proteins (see below) normally occur in <10 copies per cell. Many enzymes, particularly those involved in basic cellular "housekeeping" functions, are synthesized at a more or less constant rate; they are called **constitutive enzymes.** Other enzymes, termed **adaptive** or **inducible enzymes,** are synthesized at rates that vary with the cell's circumstances.

Lactose-Metabolizing Enzymes Are Inducible

Bacteria, as has been recognized since 1900, adapt to their environments by producing enzymes that metabolize certain nutrients, for example, lactose, only when those substances are available. *E. coli* grown in the absence of lactose are initially unable to metabolize this disaccharide. To do so they require the presence of two proteins: **β-galactosidase,** which catalyzes the hydrolysis of lactose to its component monosaccharides;

$$\text{Lactose} \xrightarrow{\beta\text{-galactosidase}} \text{Galactose} + \text{Glucose}$$

and **galactoside permease** (also known as **lactose permease;** Section 18-4B), which transports lactose into the cell. *E. coli* grown in the absence of lactose contain only a few molecules of these proteins. Yet, a few minutes after lactose is introduced into their medium, *E. coli* increase the rate at which they synthesize these proteins by ~1000-fold and maintain this pace until lactose is no longer available. The synthesis rate then returns to its original miniscule level (Fig. 29-2). *This ability to produce a series of proteins only when the substances they metabo-*

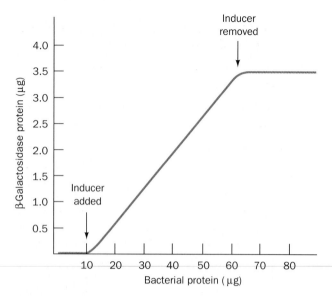

Figure 29-2
The induction kinetics of β-galactosidase in *E. coli*. [After Cohn, M., *Bacteriol. Rev.* **21**, 156 (1957).]

lize are present permits bacteria to adapt to their environment without the debilitating need to continuously synthesize large quantities of otherwise unnecessary substances.

Lactose or one of its metabolic products must somehow trigger the synthesis of the above proteins. Such a substance is known as an **inducer**. The physiological inducer of the lactose system, the lactose isomer **1,6-allolactose,**

1,6-Allolactose

arise's from lactose's occasional transglycosylation by β-galactosidase. Most studies of the lactose system use **isopropylthiogalactoside (IPTG),**

Isopropylthiogalactoside (IPTG)

a potent inducer that structurally resembles allolactose but which is not degraded by β-galactosidase.

Lactose system inducers also stimulate the synthesis of **thiogalactoside transacetylase,** an enzyme that, *in vitro,* transfers an acetyl group from acetyl-CoA to the C(6)-OH group of a β-thiogalactoside such as IPTG. Since lactose fermentation procedes normally in the absence of thiogalactoside transacetylase, however, this enzyme's physiological role is unknown.

lac System Genes Form an Operon

The genes specifying wild-type β-galactosidase, galactoside permease, and thiogalactoside transacetylase are designated Z^+, Y^+, and A^+, respectively. Genetic mapping of the defective mutants Z^-, Y^-, and A^- indicated that these *lac* **structural genes** (genes that specify polypeptides) are contiguously arranged on the *E. coli* chromosome (Fig. 29-3; genetic mapping is reviewed in Section 27-1). *These genes, together with the control elements P and O, form a genetic unit called an* **operon,** *specifically the* **lac operon.** The nature of the control elements is discussed below. The role of operons in prokaryotic gene expression is examined in Section 29-3.

lac Repressor Inhibits the Synthesis of *lac* Operon Proteins

An important clue as to how *E. coli* synthesizes protein was provided by a mutation that causes the proteins of the *lac* operon to be synthesized in large amounts in the absence of inducer. This so-called **constitutive mutation** occurs in a gene, designated *I*, that is distinct from although closely linked to the genes specifying the *lac* enzymes (Fig. 29-3). What is the nature of the *I* gene product? This riddle was solved through an ingeneous experiment performed by Arthur Pardee, Francois Jacob, and Jacques Monod. Hfr bacteria of genotype I^+Z^+ were mated to an F^- strain of genotype I^-Z^- in the absence of inducer while the β-galactosidase activity of the culture was monitored (Fig. 29-4; bacterial mating is described in Section 27-1D). At first, as expected, there was no β-galactosidase activity because the Hfr donors lacked inducer and the F^- recipients were unable to produce active enzyme (only DNA passes through the cytoplasmic bridge connecting mating bacteria). About 1 h after conjugation began, however, when the I^+Z^+ genes had just entered the F^- cells, β-galactosidase synthesis began and only ceased after about another hour. The explanation for these observations is that the donated Z^+ gene, upon entering the cytoplasm of the I^- cell, directs the synthesis of β-galactosidase in a constitutive manner. Only after the donated I^+ gene has had sufficient time to be expressed is it able to repress β-galactosidase synthesis. *The I^+ gene must therefore give rise*

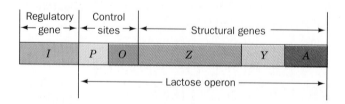

Figure 29-3
A genetic map of the *E. coli lac* operon, that is, genes encoding the proteins mediating lactose metabolism and the genetic sites that control their expression. The Z, Y, and A genes, respectively, specify β-galactosidase, galactoside permease, and thiogalactoside transacetylase.

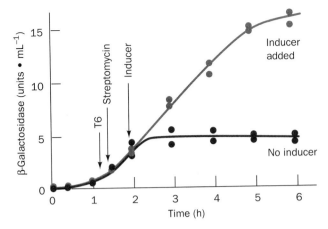

Figure 29-4
The demonstration of the existence of *lac* repressor through the appearance of β-galactosidase in the transient merozygotes (partial diploids) formed by mating *I⁺Z⁺* Hfr donors with a *I⁻Z⁻ F⁻* recipients. The F⁻ strain was also resistant to both bacteriophage T6 and streptomycin, whereas the Hfr strain was sensitive to these agents. Both types of cells were grown and mated in the absence of inducer. After sufficient time had passed for the transfer of the *lac* genes, the Hfr cells were selectively killed by the addition of T6 phage and streptomycin. In the absence of inducer (*lower curve*), β-galactosidase synthesis commenced at around the time that the *lac* genes had entered the F⁻ cells but stopped ~1 h later. If inducer was added shortly after the Hfr donors had been killed (*upper curve*), enzyme synthesis continued unabated. This demonstrates that the cessation of β-galactosidase synthesis in uninduced cells is not due to the intrinsic loss of the ability to synthesize this enzyme but to the production of a repressor specified by the *I⁺* gene. [After Pardee, A. B., Jacob, F., and Monod, J., *J. Mol. Biol.* **1**, 173 (1959).]

*to a diffusible product, the **lac repressor**, which inhibits the synthesis of β-galactosidase (and the other lac proteins).* Inducers such as IPTG temporarily inactivate *lac* repressor, whereas *I⁻* cells constitutively synthesize *lac* enzymes because they lack a functional repressor. *Lac* repressor, as we shall see in Section 29-3B, is a protein.

B. Messenger RNA

The nature of the *lac* repressor's target molecule was deduced in 1961 through a penetrating genetic analysis by Jacob and Monod. A second type of constitutive mutation in the lactose system, designated *Oᶜ* (for **operator constitutive**), which complementation analysis indicated to be independent of the *I* gene, maps between the *I* and *Z* genes (Fig. 29-3). In the partially diploid F′ strain *Oᶜ Z⁻/F O⁺Z⁺*, β-galactosidase activity is inducible by IPTG whereas the strain *Oᶜ Z⁺/F O⁺Z⁻* constitutively synthesizes this enzyme (in F′ bacteria, the F factor plasmid contains a segment of the bacterial chromosome, in this case a portion of the *lac* operon; Section 27-1D). *An O⁺ gene can therefore only control the expression of a Z gene on the same chromosome.* The same is true with the *Y⁺* and *A⁺* genes.

Jacob and Monod's observations led them to conclude the proteins are synthesized in two-stage process:

1. The structural genes on DNA are transcribed onto complementary strands of **messenger RNA (mRNA)**.

2. The mRNAs transiently associate with ribosomes, which they direct in polypeptide synthesis.

This hypothesis explains the behavior of the *lac* system (Fig. 29-5). *In the absence of inducer, the lac repressor specifically binds to the O gene (the **operator**) so as to physically block the enzymatic transcription of mRNA. Upon binding inducer, the repressor dissociates from the operator thereby permitting the transcription and subsequent translation of the lac enzymes.* The operator–repressor–inducer system thereby acts as a molecular switch so that the *lac* operator can only control the expression of *lac* enzymes on the same chromosome. The *Oᶜ* mutants constitutively synthesize *lac* enzymes because they are unable to bind repressor. The **coordinate** (simultaneous) expression of all three *lac* enzymes under the control of a single operator site arises, as Jacob and Monod theorized, from the transcription of the *lac*

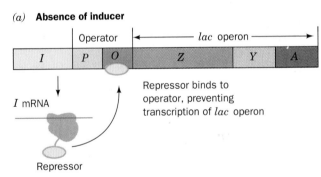

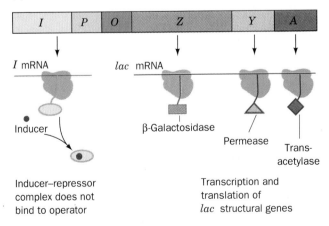

Figure 29-5
The expression of the *lac* operon. (*a*) In the absence of inducer, the repressor, the product of the *I* gene, binds to the operator thereby preventing transcription of the *lac* operon. (*b*) Upon binding inducer, the repressor dissociates from the operator, which permits the transcription and subsequent translation of the *lac* structural genes to proceed.

operon as a single **polycistronic mRNA** which directs the ribosomal synthesis of each of these proteins. This transcriptional control mechanism is further discussed in Section 29-3. [Pairs of DNA sequences, which are on the same DNA molecule, are said to be in cis (Latin: on this side) while those on different DNA molecules are said to be in trans (Latin: across). Control sequences such as the *O* gene, which are only active on the same DNA molecule as the genes they control, are called **cis-acting elements**. Those such as *lacI*, which specify the synthesis of diffusible products and can therefore be located on a different DNA molecule from the genes they control, are said to direct the synthesis of **trans-acting factors**.]

mRNAs Have Their Predicted Properties

The kinetics of enzyme induction, as indicated, for example, in Figs. 29-2 and 29-4, requires that the postulated mRNA be both rapidly synthesized and rapidly degraded. An RNA with such quick turnover had, in fact, been observed in T2-infected *E. coli*. Moreover, the base composition of this RNA fraction resembles that of the viral DNA rather than that of the bacterial RNA. Ribosomal RNA, which comprises up to 90% of a cell's RNA, turns over much more slowly than mRNA. Ribosomes are therefore not permanently committed to the synthesis of a particular protein (a once popular hypothesis). Rather, *ribosomes are nonspecific protein synthesizers that produce the polypeptide specified by the mRNA with which they are transiently associated.* A bacterium can therefore respond within a few minutes to changes in its environment.

Evidence favoring the Jacob and Monod model rapidly accumulated. Sydney Brenner, Jacob, and Matthew Meselson carried out experiments designed to characterize the RNA that *E. coli* synthesized after T4 phage infection. *E. coli* were grown in a medium containing ^{15}N and ^{13}C so as to label all cell constituents with these heavy isotopes. The cells were then infected with T4 phages and immediately transferred to an unlabeled medium (which contained only the light isotopes ^{14}N and ^{12}C) so that cell components synthesized before and after phage infection could be separated by equilibrium density gradient ultracentrifugation in CsCl solution. No "light" ribosomes were observed, which indicates, in agreement with the above mentioned T2 phage results, that no new ribosomes are synthesized after phage infection.

The growth medium also contained either ^{32}P or ^{35}S so as to radioactively label the newly synthesized and presumably phage-specific RNA and protein, respectively. Much of the ^{32}P-labeled RNA was associated, as was postulated for mRNA, with the preexisting "heavy" ribosomes (Fig. 29-6). Likewise, the ^{35}S-labeled proteins were transiently associated with, and therefore synthesized by, these ribosomes.

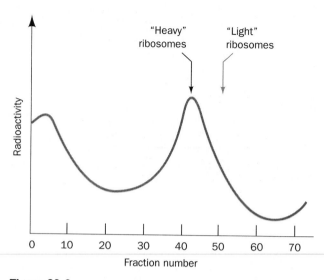

Figure 29-6
The distribution, in a CsCl density gradient, of ^{32}P-labeled RNA that had been synthesized by *E. coli* after T4 phage infection. Free RNA, being relatively dense, bands at the bottom of the centrifugation cell (*left*). Much of the RNA, however, is associated with the [15]N- and ^{13}C-labeled "heavy" ribosomes that had been synthesized before the phage infection. The predicted position of unlabeled "light" ribosomes, which are not synthesized by phage-infected cells, is also indicated. [After Brenner, S., Jacob, F., and Meselson, M., *Nature* **190**, 579 (1961).]

Sol Spiegelman developed the RNA–DNA hybridization technique (Section 28-3A) in 1961 to characterize the RNA synthesized by T2-infected *E. coli*. He found that this phage-derived RNA hybridizes with T2 DNA (Fig. 29-7) but neither does so with DNAs from unrelated phage nor with the DNA from uninfected *E. coli*. This RNA must therefore be complementary to T2 DNA in agreement with Jacob and Monod's prediction; that is, the phage-specific RNA is a messenger RNA. Hybridization studies have likewise shown that mRNAs from uninfected *E. coli* are complementary to portions of *E. coli* DNA. In fact, other RNAs, such as transfer RNA and ribosomal RNA, have corresponding complementary sequences on DNA from the same organism. Thus, *all cellular RNAs are transcribed from DNA templates.*

2. RNA POLYMERASE

RNA polymerase, the enzyme responsible for the DNA-directed synthesis of RNA, was discovered independently in 1960 by Samuel Weiss and Jerard Hurwitz. *The enzyme couples together the ribonucleoside triphosphates ATP, CTP, GTP, and UTP, on DNA templates in a reaction that is driven by the release and subsequent hydrolysis of PP$_i$:*

$$(RNA)_{n \text{ residues}} + \underset{\substack{\text{Nucleoside} \\ \text{triphosphate}}}{NTP} \rightleftharpoons (RNA)_{n+1 \text{ residues}} + PP_i$$

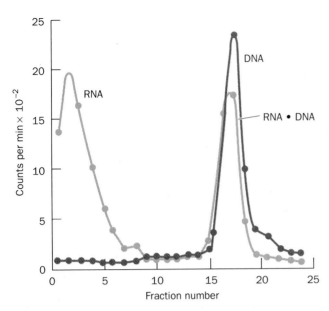

Figure 29-7
The hybridization of ^{32}P-labeled RNA produced by T2-infected *E. coli* with ^{3}H-labeled T2 DNA. Upon radioactive decay, ^{32}P and ^{3}H emit β particles with characteristically different energies so that these isotopes can be independently detected. Although free RNA (*left*) in a CsCl density gradient is denser than DNA, much of the RNA bands with the DNA (*right*). This indicates that the two polynucleotides have hybridized and are therefore complementary in sequence. [After Hall, B. D. and Spiegelman, S., *Proc. Natl. Acad. Sci.* **47**, 141 (1961).]

All cells contain RNA polymerase. In bacteria, one species of this enzyme synthesizes all of the cell's RNA except the short RNA primers employed in DNA replication (Section 31-1D). Various bacteriophages generate RNA polymerases that synthesize only phage-specific RNAs. Eukaryotic cells contain four or five RNA polymerases, that each synthesize a different class of RNA. In this section we first concentrate on the properties of the *E. coli* enzyme because it is the best characterized RNA polymerase; other bacterial RNA polymerases have similar properties. We then consider the eukaryotic enzymes.

A. Enzyme Structure

E. coli RNA polymerase's so-called **holoenzyme** is an ~480-kD protein with subunit composition $\alpha_2\beta\beta'\sigma$. Once RNA synthesis has been initiated, however, the σ subunit (also called **σ factor**) dissociates from the **core enzyme**, $\alpha_2\beta\beta'$, which carries out the actual polymerization process (see below). The β' subunit contains two atoms of Zn^{2+} which are thought to participate in the enzyme's catalytic function. The active enzyme also requires the presence of Mg^{2+}.

The holoenzyme, which is among the largest known soluble enzymes, is ~100 Å in diameter, which renders it visible in electron micrographs (Fig. 29-8); these clearly indicate that RNA polymerase binds to DNA as a protomer. The large size of the holoenzyme is presumably required by its several complex functions that include (1) template binding, (2) RNA chain initiation, (3) chain elongation, and (4) chain termination. We discuss these various functions below.

B. Template Binding

RNA synthesis is normally initiated only at specific sites on the DNA template. This was first demonstrated through hybridization studies of **bacteriophage φX174** DNA with the RNA produced by φX174-infected *E. coli*. Bacteriophage φX174 carries a single strand of DNA known as the "plus" strand. Upon its injection into *E. coli*, the plus strand directs the synthesis of the complementary "minus" strand with which it combines to form a circular duplex DNA known as the **replicative form** (Section 31-3B). The RNA produced by φX174-infected *E. coli* does not hybridize with DNA from intact phage but does so with the replicative form. Thus only the minus strand of φX174 DNA, the so-called **sense strand,** is transcribed, that is, acts as a template; the plus strand, the **antisense strand,** does not do so. Similar studies indicate that in larger phages, such as T4 and λ, the two viral DNA strands are the sense strands for different sets of genes. The same appears to be true of cellular organisms.

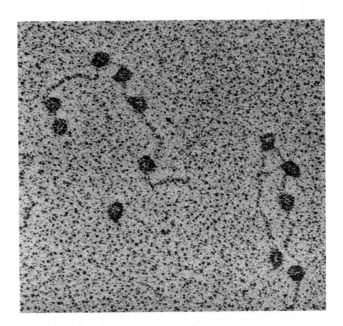

Figure 29-8
An electron micrograph of *E. coli* RNA polymerase holoenzyme attached to various promoter sites on bacteriophage T7 DNA. [From Williams, R. C., *Proc. Natl. Acad. Sci.* **74**, 2313 (1977).]

Holoenzyme Specifically Binds to Promoters

RNA polymerase binds to its initiation sites through base sequences known as **promoters** *that are recognized by the corresponding σ factor.* The existence of promoters was first recognized through mutations that enhance or diminish the transcription rates of certain genes including those of the *lac* operon. *Genetic mapping of such mutations indicated that the promoter consists of an ~40 bp sequence that is located on the 5′ side of the transcription start site.* [By convention, the sequence of template DNA is represented by its antisense (nontemplate) strand so that it will have the same directionality as the transcribed RNA. A base pair in a promoter region is assigned a positive or negative number that indicates its position, upstream or downstream in the direction of RNA polymerase travel, from the first nucleotide that is transcribed to RNA; this start site is +1 and there is no 0.] RNA, as we shall see, is synthesized in the 5′→ 3′ direction (Section 29-2D). Consequently, the promoter lies on the "upstream" side of the RNA's starting nucleotide. Sequencing studies indicate that the *lac* promoter (*lacP*) overlaps the *lac* operator (Fig. 29-3).

The holoenzyme forms tight complexes with promoters (dissociation constant $K \approx 10^{-14} M$) and thereby protects the bound DNA segments from digestion by DNase I. The region from about −20 to +20 is protected against exhaustive DNase I degradation. The region extending upstream to about −60 is also protected but to a lesser extent, presumably because it binds holoenzyme less tightly.

Sequence determinations of the protected regions from numerous *E. coli* and phage genes have revealed the "consensus" sequence of *E. coli* promoters (Fig. 29-9). *Their most conserved sequence is a hexamer centered at about the −10 position known as the* **Pribnow box** *(after David Pribnow who pointed out its existence in 1975).* It has a consensus sequence of TATAAT in which the leading TA and final T are highly conserved. *Upstream sequences around position −35 also have a region of sequence similarity,* TCTTGACAT, which is most evident in efficient promoters. The initiating (+1) nucleotide, which is nearly always A or G, is centered in a poorly conserved CAT or CGT sequence located 5 to 8 bp downstream from the Pribnow box. Most promoter sequences vary considerably from the consensus sequence (Fig. 29-9). Nevertheless, a mutation in one of the partially conserved regions can greatly increase or decrease a promoter's initiation efficiency. *The rates at which genes are transcribed, which span a range of at least 1000, varies directly with the rate that their promoters form stable initiation complexes with the holoenzyme.*

Initiation Requires the Formation of an Open Complex

The promoter regions in contact with the holoenzyme have been identified by determining where the enzyme

Operon	−35 region · Pribnow box (−10 region) · Initiation site (+1)
lac	ACCCCAGGCTTTACACTTTATGCTTCCGGCTCGTATGTTGTGTGGAATTGTGAGCGG
lacI	CCATCGAATGGCGCAAAACCTTTCGCGGTATGGCATGATAGCGCCCGGAAGAGAGTC
galP2	ATTTATTCCATGTCACACTTTTCGCATCTTTGTTATGCTATGGTTATTTCATACCAT
araBAD	GGATCCTACCTGACGCTTTTTATCGCAACTCTCTACTGTTTCTCCATACCCGTTTTT
araC	GCCGTGATTATAGACACTTTTGTTACGCGTTTTTGTCATGGCTTTGGTCCCGCTTTG
trp	AAATGAGCTGTTGACAATTAATCATCGAACTAGTTAACTAGTACGCAAGTTCACGTA
bioA	TTCCAAAACGTGTTTTTTGTTGTTAATTCGGTGTAGACTTGTAAACCTAAATCTTTT
bioB	CATAATCGACTTGTAAACCAAATTGAAAAGATTTAGGTTTACAAGTCTACACCGAAT
*t*RNA[Tyr]	CAACGTAACACTTTACAGCGGCGCGTCATTTGATATGATGCGCCCCGCTTCCCGATA
rrnD1	CAAAAAAATACTTGTGCAAAAAATTGGGATCCCTATAATGCGCCTCCGTTGAGACGA
rrnE1	CAATTTTTCTATTGCGGCCTGCGGAGAACTCCCTATAATGCGCCTCCATCGACACGG
rrnA1	AAAATAAATGCTTGACTCTGTAGCGGGAAGGCGTATTATGCACACCCCGCGCCGCTG

Concensus sequence:	−35 region		Pribnow box		Initiation site
	T C T T G A C A T	... 11–15 bp ...	T A T A A T	... 5–8 bp ...	A (51), C (55), T (48), G (42)
	42 38 82 84 79 64 53 45 41		79 95 44 59 51 96		

Figure 29-9
The noncoding strand nucleotide sequences of selected *E. coli* promoters. The Pribnow box (*red shading*), a 6 bp region centered around the −10 position, and an 8 to 12 bp sequence around the −35 region (*blue shading*) are both conserved. The transcription initiation sites (+1), which in most promoters occurs at a single purine nucleotide, are shaded in green. The bottom row shows the consensus sequence of 112 promoters with the number below each base indicating its percent occurrence. [After Rosenberg, M. and Court, D., *Annu. Rev. Genet.* **13,** 321−323 (1979). Consensus sequence from Hawley, D. K. and McClure, W. R., *Nucleic Acids Res.* **11,** 2244 (1983).]

alters the susceptibility of the DNA to alkylation by agents such as dimethyl sulfate (DMS), a procedure named **footprinting** (Section 33-3B). These experiments demonstrated that the holoenzyme contacts the promoter only around the Pribnow box and the -35 region. Model building indicates that these protected sites are both on the same side of the double helix which suggests that RNA polymerase binds to only one face of the double helical promoter.

DMS, in addition to methylating G residues at N(7) and A residues at N(3) (Section 28-6B), methylates N(1) of A and N(3) of C. Since these latter positions participate in base pairing interactions, however, they can only react with DMS in single-stranded DNA. This differential methylation of single- and double-standed DNAs provides a sensitive test for DNA strand separation or "melting." Footprinting studies indicate that the binding of holoenzyme "melts out" the promoter in an 11 bp region extending from the middle of the Pribnow box to just past the initiation site (-9 to $+2$). The need to form this "open complex" explains why promoter efficiency tends to decrease with the number of G·C base pairs in the Pribnow box; this presumably increases the difficulty in opening the double helix as is required for chain initiation (G·C pairs, it will be recalled, are stronger than A·T pairs).

Core enzyme, which does not specifically bind promoter, tightly binds duplex DNA (the complex's dissociation constant is $K \approx 5 \times 10^{-12}M$ and its half-life is ~60 min). Holoenzyme, in contrast, binds to nonpromoter DNA comparatively loosely ($K \approx 10^{-7}M$ and half-life >1 s). Apparently, the σ subunit allows holoenzyme to move rapidly along a DNA strand in search of the σ subunit's corresponding promoter. Once transcription has been initiated and the σ subunit jettisoned, the tight binding of core enzyme to DNA apparently stabilizes the ternary enzyme–DNA–RNA complex.

C. Chain Initiation

The 5′-terminal base of prokaryotic RNAs is almost always a purine with A occurring more often than G. The initiating reaction of transcription is the coupling of two nucleoside triphosphates in the reaction

$$\text{pppA} + \text{pppN} \rightleftharpoons \text{pppApN} + \text{PP}_i$$

Bacterial RNAs therefore have 5′-triphosphate groups as was demonstrated by the incorporation of radioactive label into RNA when it was synthesized with $[\gamma\text{-}^{32}\text{P}]$ATP. Only the 5′ terminus of the RNA can retain the label because the internal phosphodiester groups of RNA are derived from the α-phosphate groups of nucleoside triphosphates.

Once holoenzyme has initiated RNA transcription, the σ factor dissociates from the core–DNA–RNA complex and can join with another core to form a new

initiation complex. This is demonstrated by a burst of RNA synthesis upon the addition of core enzyme to a transcribing reaction mixture that initially contained holoenzyme.

Rifamycins Inhibit Prokaryotic Transcription Initiation

Two related antibiotics, **rifamycin B,** which is produced by *Streptomyces mediterranei,* and its semisynthetic derivative **rifampicin,**

Rifamycin B $R_1 = CH_2COO^-$; $R_2 = H$

Rifampicin $R_1 = H$; $R_2 = CH = N$ ⬡ $N - CH_3$

specifically inhibit transcription by prokaryotic, but not eukaryotic, RNA polymerases. This selectivity and their high potency (bacterial RNA polymerase is 50% inhibited by $2 \times 10^{-8}M$ rifampicin) has made them medically useful bacteriocidal agents against gram-positive bacteria and tuberculosis. The isolation of rifamycin resistant mutants whose β subunits have altered electrophoretic mobilities indicates that this subunit contains the rifamycin-binding site. Rifamycins neither inhibit the binding of RNA polymerase to the promoter nor the formation of the first phosphodiester bond, but they prevent further chain elongation. The inactivated RNA polymerase remains bound to the promoter thereby blocking its initiation by uninhibited enzyme. Once RNA chain initiation has occurred, however, rifamycins have no effect on the subsequent elongation process. The rifamycins are useful research tools because they permit the transcription process to be dissected into its initiation and its elongation phases.

D. Chain Elongation

What is the direction of RNA chain elongation; that is, does it occur by the addition of incoming nucleotides to the 3′ end of the nascent (growing) RNA chain (5′ → 3′ growth; Fig. 29-10a), or by their addition to its 5′ terminus (3′ → 5′ growth; Fig. 29-10b)? This question was answered by determining the rate that the radioactive label from $[\gamma\text{-}^{32}\text{P}]$GTP is incorporated into RNA. The

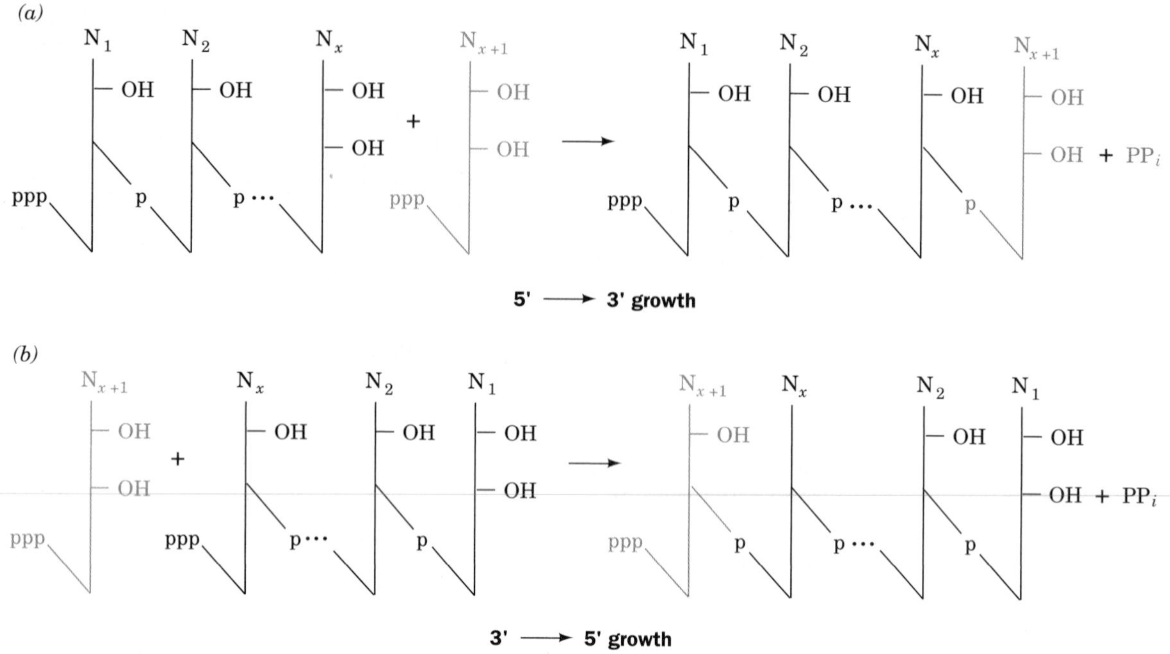

(a)

5' ⟶ 3' growth

(b)

3' ⟶ 5' growth

Figure 29-10
The two possible modes of RNA chain growth: (a) by the addition of nucleotides to the 3' end, and (b) by the addition of nucleotides to the 5' end. RNA polymerase catalyzes the former reaction.

ratio of ^{32}P to total nucleotide was highest just after chain initiation and decreased with time thereby indicating that the 5'-terminal pppG was incorporated into the RNA chain first rather than last. *Chain growth must therefore occur in the 5' → 3' direction (Fig. 29-10a).* This conclusion is corroborated by the observation that the antibiotic **cordycepin,**

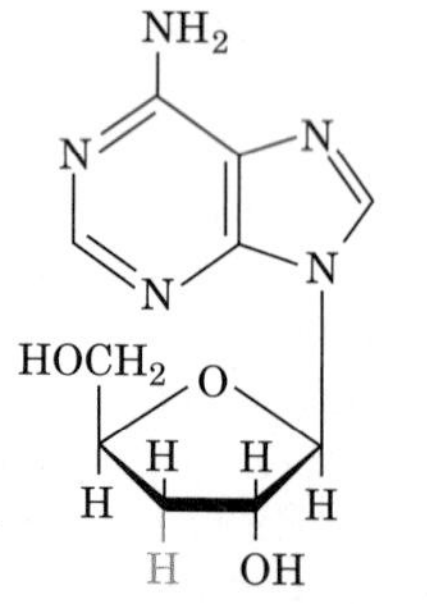

Cordycepin
(3'-deoxyadenosine)

an adenosine analog that lacks a 3'-OH group, inhibits bacterial RNA synthesis. Its addition to the 3' end of RNA, as is expected for 5' → 3' growth, prevents the RNA chain's further elongation. Cordycepin would not have this effect if chain growth occurred in the opposite direction because it cannot be appended to an RNA's 5' end.

Transcription Probably Supercoils DNA

RNA chain elongation requires that the double-stranded DNA template be opened up at the point of RNA synthesis so that the sense strand can be tran-

scribed onto its complementary RNA strand. In doing so, the RNA chain only transiently forms a short length of RNA–DNA hybrid duplex as is indicated by the observation that transcription leaves the template duplex intact and yields single-stranded RNA. The unpaired "bubble" of DNA in the open initiation complex apparently travels along the DNA with the RNA polymerase. There are two ways this might occur (Fig. 29-11):

1. If the RNA polymerase follows the template strand in its helical path around the DNA, the DNA would build up little supercoiling because the DNA duplex would never be unwound by more than about a turn. However, the RNA transcript would wrap around the DNA, once per duplex turn. This model is implausible since it is unlikely that its DNA and RNA could be readily untangled: The RNA would not spontaneously unwind from the long and often circular DNA in any reasonable time, and no topoisomerase is known to accelerate this process.

2. If the RNA polymerase moves in a straight line while the DNA rotates, the RNA and DNA would not become entangled. Rather, the DNA's helical turns would be pushed ahead of the advancing transcription bubble so as to more tightly wind the DNA ahead of the bubble (which promotes positive supercoiling) while the DNA behind the bubble would be equivalently unwound (which promotes negative supercoiling, although note that the linking number of the entire DNA remains unchanged). This model is supported by the observations that the transcription of plasmids in *E. coli* causes their positive supercoil-

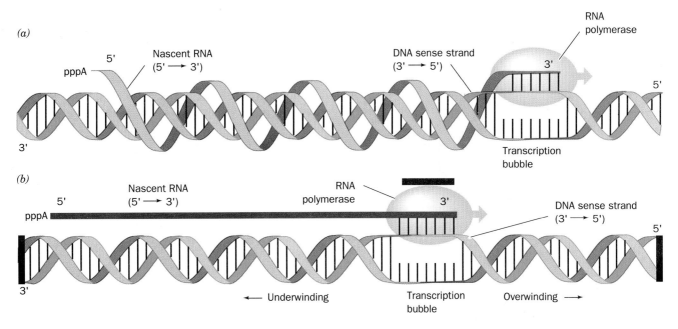

(a)

5' Nascent RNA (5' ⟶ 3')
pppA
3'

DNA sense strand (3' ⟶ 5')
RNA polymerase
3'
5'
Transcription bubble

(b)

5' Nascent RNA (5' ⟶ 3')
pppA
3'

RNA polymerase
3'
DNA sense strand (3' ⟶ 5')
5'

⟵ Underwinding   Transcription bubble   Overwinding ⟶

Figure 29-11

RNA chain elongation by RNA polymerase. In the region being transcribed, the DNA double helix is unwound by about a turn to permit the DNA's sense strand to form a short segment of DNA–RNA hybrid double helix with the RNA's 3' end. As the RNA polymerase advances along the DNA template (here to the right), the DNA unwinds ahead of the RNA's growing 3' end and rewinds behind it thereby stripping the newly synthesized RNA from the sense strand. (a) One way this might occur is by the RNA polymerase following the path of the sense strand about the DNA double helix in which case the transcript becomes wrapped about the DNA once per duplex turn. (b) A second, and more plausible possibility, is that the RNA moves in a straight line while the DNA rotates beneath it. In this case the RNA would not wrap around the DNA but the DNA would become overwound ahead of the advancing transcription bubble and unwound behind it (consider the consequences of placing your finger between the twisted DNA strands in this model and pushing towards the right). The model presumes that the ends of the DNA as well as the RNA polymerase, are prevented from rotating by attachments within the cell (*black bars*). [After Futcher, B., *Trends Genet.* **4,** 271, 272 (1988).]

ing in gyrase mutants (which cannot relax positive supercoils; Section 28-5C) and their negative supercoiling in topoisomerase I mutants (which cannot relax negative supercoils).

Whatever the case, recall that inappropriate superhelicity halts transcription (Section 28-5C). Perhaps the torsional tension in the DNA generated by negative superhelicity behind the transcription bubble is required to help drive the transcriptional process, whereas too much such tension prevents the opening and maintenance of the transcription bubble.

Transcription Occurs Rapidly and Accurately

The *in vivo* rate of transcription is 20 to 50 nucleotides/s at 37°C as indicated by the rate that *E. coli* incorporate ³H-labeled nucleosides into RNA (cells cannot take up nucleoside triphosphates from the medium). Once an RNA polymerase molecule has initiated transcription and moved away from the promoter, another RNA polymerase can follow suit. The synthesis of RNAs that are needed in large quantities, ribosomal RNAs, for example, are initiated as often as is sterically possible, about once per second (Fig. 29-12).

The error frequency in RNA synthesis, as estimated

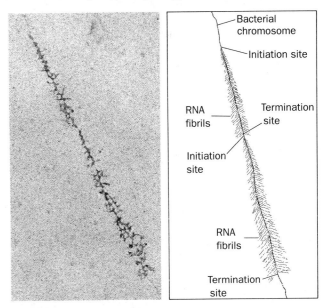

Bacterial chromosome
Initiation site
RNA fibrils
Termination site
Initiation site
RNA fibrils
Termination site

Figure 29-12

An electron micrograph and its interpretive drawing of two contiguous *E. coli* ribosomal genes undergoing transcription. The "arrowhead" structures result from the increasing lengths of the nascent RNA chains as the RNA polymerase molecules synthesizing them move from the initiation site on the DNA to the termination site. [Courtesy of Oscar L. Miller, Jr., University of Virginia.]

(a)

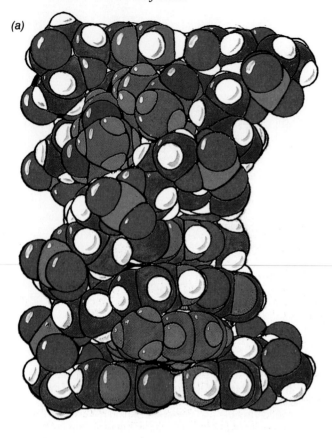

from the analysis of transcripts of simple templates such as poly[d(AT)]·poly[d(AT)], is one wrong base incorporated for every ~10^4 transcribed. This rate is tolerable because of the repeated transcription of most genes, because the genetic code contains numerous synonyms (Section 30-1E), and because amino acid substitutions in proteins are often functionally innocuous.

Intercalating Agents Inhibit Both RNA and DNA Polymerases

Daunomycin and the closely related **adriamycin**,

Daunomycin: R = H
Adriamycin: R = OH

(b)

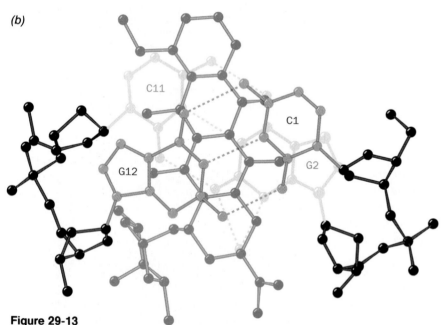

Figure 29-13
The structure of a complex of daunomycin with the self-complementary hexanucleotide d(CGTACG). Each double helical fragment binds two daunomycin molecules by intercalation between its G·C pairs to form a complex with twofold rotational symmetry. (*a*) A space-filling representation of the complex showing the upper daunomycin's amino sugar extending into the minor groove of the double helix and the edge of the lower daunomyucin's intercalated ring. The daunomycin molecules are colored green with purple oxygen atoms. [After a drawing provided by Andrew Wang, University of Illinois.] (*b*) A view perpendicular to the bases indicating the stacking of the intercalated daunomycin ring system (*green*) with its surrounding G·C pairs. The C1·G12 base pair is closer to the viewer than the daunomycin ring system while the G2·C11 base pair is farther away. [After Wang, A. H.-J., Ughetto, G., Quigley, G. J., and Rich, A., *Biochemistry* **26**, 1155, 1157 (1987).]

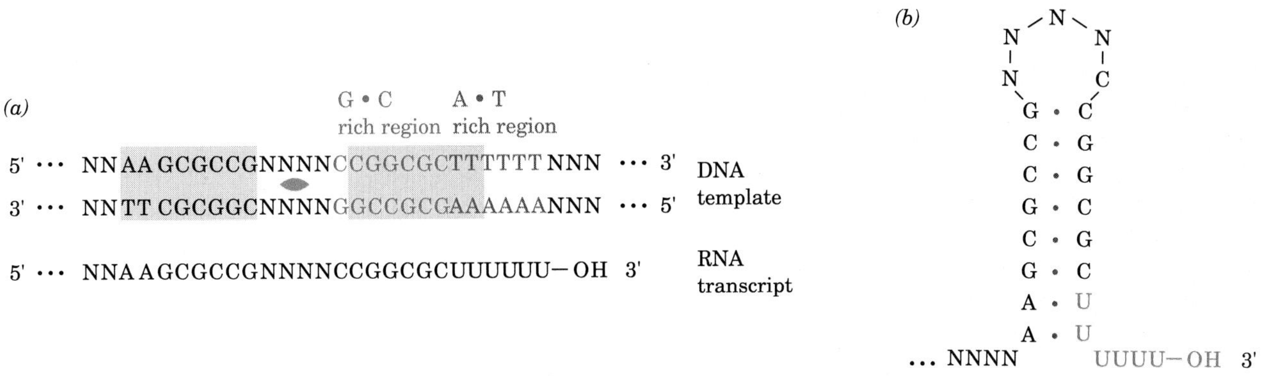

(a)

G • C   A • T
rich region   rich region

5' ··· NN AA GCGCCG NNNNC CGGCGC TTTTTT NNN ··· 3'   DNA
3' ··· NN TT CGCGGC NNNNG GCCGCG AAAAAA NNN ··· 5'   template

5' ··· NNA AGCGCCG NNNNCCGGCGC UUUUUU—OH 3'   RNA transcript

(b)

$$
\begin{array}{ccc}
 & N \diagdown N & \\
N & & N \\
| & & | \\
N & & C \\
G & \cdot & C \\
C & \cdot & G \\
C & \cdot & G \\
G & \cdot & C \\
C & \cdot & G \\
G & \cdot & C \\
A & \cdot & U \\
A & \cdot & U \\
\end{array}
$$

··· NNNN   UUUU—OH 3'

Figure 29-14
The base sequence of a hypothetical strong (efficient) terminator as deduced from the sequences of several transcripts. (a) The DNA sequence together with its corresponding RNA. The A · T-rich and G · C-rich sequences are shown in blue and red, respectively. The twofold symmetry axis (*lenticular symbol*) relates the flanking shaded segments that form an inverted repeat. (b) The RNA hairpin structure and poly(U) tail that triggers transcription termination. [After Pribnow, D., *in* Goldberger, R. F. (Ed.), *Biological Regulation and Development*, Vol. 1, *p.* 253, Plenum Press (1979).]

which are valuable chemotherapeutic agents in the treatment of certain human cancers, specifically bind to duplex DNA so as to inhibit both its transcription and its replication. These antibiotics presumably act by interfering with the passage of both RNA polymerase and DNA polymerase. The X-ray crystal structure of a complex of daunomycin with the self-complementary hexanucleotide d(CGTACG) reveals that daunomycin's planar aromatic ring system (rings B–D) is intercalated between the G · C pairs at both ends of the double helical fragment (Fig. 29-13). The nonplanar A ring extends into the minor groove where its side groups stabilize the complex through hydrogen bonding interactions with the DNA.

Actinomycin D,

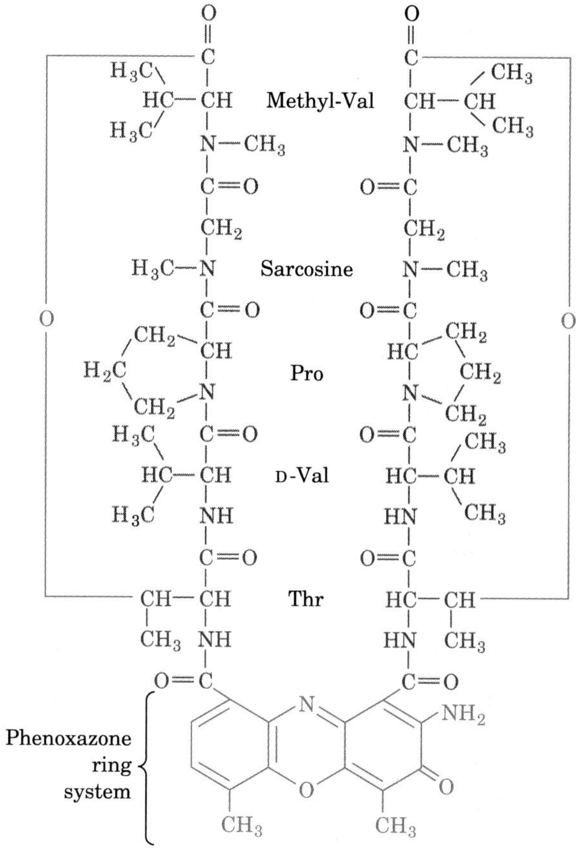

H₃C—N ... Sarcosine ... Phenoxazone ring system

Actinomycin D

an antibiotic produced by *Streptomyces antibioticus*, is also a potent inhibitor of nucleic acid synthesis. It acts by intercalating its **phenoxazone ring** between two successive G · C pairs of duplex DNA in a manner similar to daunomycin. Actinomycin's two identical cyclic pentapeptide groups, which have an unusual composition, stabilize this interaction through specific contacts with the double helix. Other intercalating agents, ethidium and proflavin (Section 28-4C), for example, also inhibit nucleic acid synthesis, presumably by similar mechanisms.

E. Chain Termination

Electron micrographs such as Fig. 29-12 suggest that DNA contains specific sites at which transcription is terminated. The transcriptional termination sequences of several *E. coli* genes share two common features (Fig. 29-14a):

1. A series of 4 to 10 consecutive A · T's with the A's on the template strand. The transcribed RNA is terminated in or just past this sequence.

2. A G + C-rich region with a palindromic (twofold symmetric) sequence that immediately precedes the series of A · T's.

The RNA transcript of this region can therefore form a self-complementary "hairpin" structure that is terminated by several U residues (Fig. 29-14b).
The stability of a terminator's G + C-rich hairpin and the weak base pairing of its oligo(U) tail to template DNA appear to be important factors in ensuring proper chain

termination. Indeed, model studies have shown that oligo(dA·rU) forms a particularly unstable hybrid helix although oligo(dA·dT) forms a helix of normal stability. The formation of the G + C-rich hairpin causes RNA polymerase to pause for several seconds at the termination site. This, it has been proposed, induces a conformational change in the RNA polymerase, which permits the noncoding DNA strand to displace the weakly bound oligo(U) tail from the template strand thereby terminating transcription. Consistent with this notion is the observation that mutations that alter the strengths of these associations reduce the efficiency of chain termination and often eliminate it. Termination is similarly diminished when *in vitro* transcription is carried out with GTP replaced by **inosine triphosphate (ITP).**

Inosine triphosphate (ITP)

I·C pairs are weaker than those of G·C because the hypoxanthine base of I, which lacks the 2-amino group of G, can only make two hydrogen bonds to C thereby decreasing the hairpin's stability. UTP replacement by 5-bromo-UTP also diminishes chain termination because 5Br-U forms stronger base pairs with A than does U itself thus inhibiting the nascent RNA's displacement from the template DNA strand.

Termination Often Requires the Assistance of Rho Factor

The foregoing termination sequences induce the spontaneous termination of transcription. Other termination sites, however, lack any obvious similarities and are unable to form strong hairpins; *they require the participation of a protein known as* **rho factor** *to terminate transcription.* The existence of rho factor was suggested by the observation that *in vivo* transcripts are often shorter than the corresponding *in vitro* transcripts. Rho factor, a hexamer of identical 419-residue subunits, enhances the termination efficiency of spontaneously terminating transcripts as well as inducing the termination of nonspontaneously terminating transcripts.

Several key observations have led to a model of rho-dependent termination:

1. Rho factor is an enzyme that catalyzes the unwinding of RNA–DNA and RNA–RNA double helices. This process is powered by the hydrolysis of nucleoside triphosphates (NTPs) to nucleoside diphosphates + P_i with little preference for the identity of the base.

NTPase activity is required for rho-dependent termination as is demonstrated by its *in vitro* inhibition when the NTPs are replaced by their β,γ-imido analogs,

β,γ-**Imido nucleoside triphosphate**

substances that are RNA polymerase substrates but cannot be hydrolyzed by rho factor.

2. Genetic manipulations indicate that rho-dependent termination requires the presence of a specific recognition sequence upstream of the termination site. The recognition sequence must be on the nascent RNA rather than the DNA as is demonstrated by rho's inability to terminate transcription in the presence of pancreatic RNase A. The essential features of this termination site have not been fully elucidated; the construction of synthetic termination sites indicate that it consists of 80 to 100 nucleotides which lack a stable secondary structure and probably contain multiple C-rich regions.

These observations suggest that rho factor attaches to nascent RNA at its recognition sequence and then migrates along the RNA in the 5' → 3' direction until it encounters an RNA polymerase paused at the termination site (without the pause, rho might not be able to overtake the RNA polymerase). There, rho unwinds the RNA–DNA duplex forming the transcription bubble thereby releasing the RNA transcript.

F. Eukaryotic RNA Polymerases

Eukaryotic nuclei contain three distinct types of RNA polymerases that differ in the RNAs they synthesize:

1. **RNA polymerase I,** which is located in the nucleoli (dense granular bodies in the nuclei that contain the ribosomal genes; Section 29-4B), synthesizes precursors of most ribosomal RNAs.

2. **RNA polymerase II,** which occurs in the nucleoplasm, synthesizes mRNA precursors.

3. **RNA polymerase III,** which also occurs in the nucleoplasm, synthesizes the precursors of 5S ribosomal RNA, the tRNAs, and a variety of other small nuclear and cytosolic RNAs.

In addition to these nuclear enzymes (which are also known as **RNA polymerases A, B,** and **C),** eukaryotic cells contain separate mitochondrial and chloroplast RNA polymerases.

Eukaryotic RNA polymerases, whose molecular masses vary between 500 and 700 kD, are characterized by subunit compositions of Byzantine complexity. Each type of enzyme contains two nonidentical "large" (>100 kD) subunits and an array of up to 12 different "small" (<50 kD) subunits. Some of the small subunits occur in 2 or all 3 of the nuclear RNA polymerases. As yet, little is known about their functions or interactions although, intriguingly, the largest subunits of yeast RNA polymerases II and III exhibit extensive homology to each other and to the largest (β') subunit of *E. coli* RNA polymerase.

The RNA Polymerase I Promoter Consists of Nested Control Regions

The RNA polymerase I promoter has been identified by determining the transcription rates of a series of mutant rRNA genes from *Xenopus laevis* (an African clawed frog) with increasingly longer deletions from either their 5' or their 3' ends. (It is not possible to deduce the RNA polymerase I promoter from the sequence homologies common to the genes it transcribes because, as we shall see in Section 29-4B, there is only one type of rRNA gene.) Optimal rRNA expression requires the presence of the rRNA gene segment extending from -142 to $+6$. The minimal base sequence required for accurate initiation, however, extends between nucleotides -7 and $+6$. It therefore appears that this latter promoter element acts to guide RNA polymerase I to its proper initiation site, whereas the rest of the promoter functions to bind proteins known as **transcription factors** (see below).

RNA Polymerase II Promoters Are Complex and Diverse

The promoters recognized by RNA polymerase II, which are considerably longer and more diverse than those of prokaryotic genes, have, as yet, been only su-

perficially described. The structural genes expressed in all tissues, the so-called "housekeeping" genes, which are thought to be constituitively transcribed, have one or more copies of the sequence GGGCGG or its complement (the **GC box**) located upstream from their transcription start sites. The analysis of deletion and point mutations in eukaryotic viruses such as SV40 indicates that GC boxes function analogously to prokaryotic promoters. On the other hand, structural genes that are selectively expressed in one or a few types of cells often lack these GC-rich sequences. Rather, *they contain a conserved AT-rich sequence located 25 to 30 bp upstream from their transcription start sites (Fig. 29-15).* Note that this so-called **TATA** or **Goldberg–Hogness box** (after Michael Goldberg and David Hogness who deduced its existence in 1978) resembles the prokaryotic Pribnow box (TATAAT) although they differ in their locations relative to the transcription start site (-27 vs -10). The functions of these two promoter elements are not strictly analogous, however, since the deletion of the TATA box does not necessarily eliminate transcription. Rather, TATA box deletion or mutation generates heterogeneities in the transcriptional start site thereby indicating that the TATA box participates in selecting this site.

The gene region extending between about -50 and -110 also contains promoter elements. For instance, many eukaryotic structural genes, including those encoding the various globins, have a conserved sequence of consensus CCAAT (the **CCAAT box**) located between about -70 and -80 whose alteration greatly reduces the gene's transcription rate. Globin genes have, in addition, a conserved **CACCC box** upstream from the CCAAT box that has also been implicated in transcriptional initiation. Evidently, the promoter sequences upstream of the TATA box form the initial DNA-binding sites for RNA polymerase II and the other proteins involved in transcriptional initiation (see below).

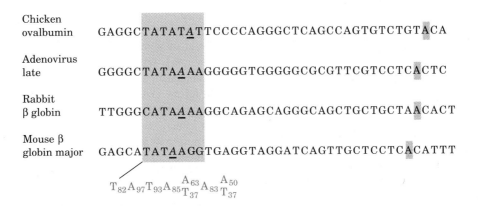

Figure 29-15
The promoter sequences of selected eukaryotic structural genes. The homologous segment, the TATA box, is shaded in red with the base at position -27 underlined and the initial nucleotide to be transcribed ($+1$) shaded in green. The bottom row indicates the consensus sequence of several such promoters with the subscripts indicating the percent occurrence of the corresponding base. [After Gannon, F., O'Hare, K., Perrin, F., Le Pennec, J. P., Benoist, C., Cochet, M., Breathnach, R., Royal, A., Garapin, A., Cami, B., and Chambon, P., *Nature* **278**, 433 (1978).)

Enhancers Are Transcriptional Activators That Can Have Variable Positions and Orientations

Perhaps the most suprising aspect of eukaryotic transcriptional control elements is that some of them need not have fixed positions and orientations relative to their corresponding transcribed sequences. For example, the SV40 genome, in which such elements were first discovered, contains two repeated sequences of 72 bp each that are located upstream from the promoter for early gene expression. Transcription is unaffected if one of these repeats is deleted but is nearly eliminated when both are absent. The analysis of a series of SV40 mutants containing only one of these repeats demonstrated that its ability to stimulate transcription from its corresponding promoter is all but independent of its position and orientation. Indeed, transcription is unimpaired when this segment is several thousand base pairs upstream or downstream from the transcription start site. Gene segments with such properties are named **enhancers** to indicate that they differ from promoters, with which they must be associated in order to trigger site-specific and strand-specific transcription initiation (although the characterization of numerous promoters and enhancers indicates that their functional properties are similar). Enhancers occur both in eukaryotic viruses and cellular genes.

Enhancers are required for the full activities of their cognate promoters. But how do they act? Two not mutually exclusive possibilities are given the most credence:

1. Enhancers are "entry points" on DNA for RNA polymerase II, perhaps through a lack of binding affinity for the histones that normally coat eukaryotic DNA, so as to (as seems likely) block RNA polymerase II binding (Section 33-1A). Alternatively, enhancers may alter DNA's local conformation in a way that favors RNA polymerase II binding. In fact, some enhancers contain a segment of alternating purines and pyrimidines which, we have seen, is just the type of sequence most likely to form Z-DNA (Section 28-2B).

2. Enhancers are recognized by specific proteins called **transcription factors** that stimulate RNA polymerase II to bind to a nearby promoter.

All cellular enhancers that have yet been identified are associated with genes that are selectively expressed in specific tissues. It therefore seems, as we discuss in Section 33-3B, that *enhancers mediate much of the selective gene expression in eukaryotes.*

RNA Polymerase III Promoters Can Be Located Downstream from Their Transcription Start Sites

The promoters of genes transcribed by RNA polymerase III can be located entirely within the genes' transcribed regions. Donald Brown established this through the construction of a series of deletion mutants of a *Xenopus borealis* 5S RNA gene. Deletions of base sequences that start from outside one or the other end of the transcribed portion of the 5S gene only prevent transcription if they extend into the segment between nucleotides +40 and +80. Indeed, a fragment of the 5S gene consisting of only nucleotides 41 to 87, when cloned in a bacterial plasmid, is sufficient to direct specific initiation by RNA polymerase III at an upstream site. This is because, as was subsequently demonstrated, the sequence contains the binding site for a transcription factor that stimulates the upstream binding of RNA polymerase III. Further studies have shown, however, that the promoters of other RNA polymerase III-transcribed genes may lie partially or even entirely upstream of their start sites.

Amatoxins Specifically Inhibit RNA Polymerases II and III

The poisonous mushroom *Amanita phalloides* (death cap), which is responsible for the majority of fatal mushroom poisonings, contains several types of toxic substances including a series of unusual bicyclic octapeptides known as **amatoxins**. α-Amanitin,

α-Amanitin

which is representative of the amatoxins, forms a tight 1:1 complex with RNA polymerase II ($K = 10^{-8}M$) and a looser one with RNA polymerase III ($K = 10^{-6}M$), so as to specifically block their elongation steps. α-Amanitin is therefore a useful tool for mechanistic studies of these enzymes. RNA polymerase I as well as mitochondrial, chloroplast, and prokaryotic RNA polymerases are insensitive to α-amanitin.

Despite the amatoxins' high toxicity (5–6 mg, which occur in ~40 g of fresh mushrooms, are sufficient to kill a human adult), they act slowly. Death, usually from liver dysfunction, occurs no earlier than several days after mushroom ingestion (and after recovery from the effects of other mushroom toxins). This, in part, reflects the slow turnover rate of eukaryotic mRNAs and proteins.

3. CONTROL OF TRANSCRIPTION IN PROKARYOTES

Prokaryotes respond to sudden environmental changes, such as the influx of nutrients, by inducing the synthesis of the appropriate proteins. This process takes only minutes because transcription and translation in prokaryotes are closely coupled: *Ribosomes commence translation near the 5' end of a nascent mRNA soon after it is extruded from RNA polymerase (Fig. 29-16). Moreover, most prokaryotic mRNAs are enzymatically degraded within 1 to 3 min of their synthesis,* thereby eliminating the wasteful synthesis of unneeded proteins after a change in conditions (protein degradation is discussed in Section 30-6). In fact, the 5' ends of some mRNAs are degraded before their 3' ends have been synthesized.

In contrast, the induction of new proteins in eukaryotic cells frequently takes hours or days because transcription takes place in the nucleus and the resulting mRNAs must be transported to the cytoplasm where translation occurs. However, eukaryotic cells, particularly those of multicellular organisms, have relatively stable environments; changes in their transcriptional patterns usually occur only during cell differentiation.

In this section we examine some of the ways in which prokaryotic gene expression is regulated through transcriptional control. Eukaryotes, being vastly more complex creatures than are prokaryotes, have a correspondingly more complicated transcriptional control system whose general outlines are just coming into focus. We therefore defer discussion of eukaryotic transcriptional control until Section 33-3 where it can be considered in light of what we know about the structure and organization of the eukaryotic chromosome.

A. Promoters

In the presence of high concentrations of inducer, the *lac* operon is rapidly transcribed. In contrast, the *lacI* gene is transcribed at such a low rate that a typical *E. coli* cell contains < 10 molecules of the *lac* repressor. Yet, the *I* gene has no repressor. Rather, it has such an inefficient promoter that it is transcribed an average of about once per bacterial generation. *Genes that are transcribed at high rates have efficient promoters.* In general, the more efficient a promoter, the more closely its sequence resembles that of the corresponding consensus sequence.

Gene Expression in Certain Phages Is Controlled by a Succession of σ Factors

The processes of development and differentiation involve the temporally ordered expression of sets of genes according to genetically specified programs. Phage infections are among the simplest examples of developmental processes. Typically, only a subset of the phage genome, often referred to as *early* genes, are expressed in the host immediately after phage infection. As time passes, *middle* genes start to be expressed and the *early* genes as well as the bacterial genes are turned off. In the final stages of phage infection, the *middle* genes give way to the *late* genes. Of course some phage types express more than three sets of genes and some genes may be expressed in more than one stage of an infection.

One way in which families of genes are sequentially expressed is through "cascades" of σ factors. In the infection of *Bacillus subtilus* by **bacteriophage SP01,** for example, the *early* gene promoters are recognized by the bacterial RNA polymerase holoenzyme. Among these *early* genes is gene 28 whose gene product is a new σ subunit, designated σ^{sp28}, that displaces the bacterial σ subunit from the core enzyme. This reconstituted holoenzyme recognizes only the phage *middle* gene promoters, which all have similar −35 and −10 (Pribnow box) regions, but bear little resemblance to the corresponding regions of bacterial and phage *early* genes.

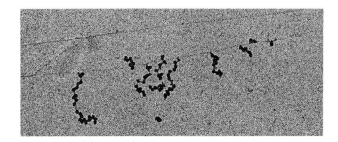

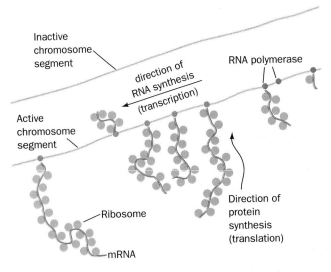

Figure 29-16
An electron micrograph and its interpretive drawing showing the simultaneous transcription and translation of an *E. coli* gene. RNA polymerase molecules are transcribing the DNA from right to left while ribosomes are translating the nascent RNAs (mostly from bottom to top). [Courtesy of Oscar L. Miller, Jr., University of Virginia.]

Inactive chromosome segment

direction of RNA synthesis (transcription)

RNA polymerase

Active chromosome segment

Ribosome

mRNA

Direction of protein synthesis (translation)

The *early* genes therefore become inactive once their corresponding mRNAs have been degraded. The phage *middle* genes include genes 33 and 34, which together specify yet another σ factor, $\sigma^{gp33/34}$ which, in turn, permits the transcription of only *late* phage genes.

Several bacteria, including *E. coli* and *B. subtilus*, likewise have several different σ factors. These are not utilized in a sequential manner. Rather, those that differ from the predominant or primary σ factor control the transcription of coordinately expressed groups of special purpose genes whose promoters are quite different from those recognized by the primary σ factor.

B. *lac* Repressor

In 1966, Beno Müller-Hill and Walter Gilbert isolated *lac* repressor on the basis of its ability to bind ^{14}C-labeled IPTG and demonstrated that it is a protein. This was an exceedingly difficult task because *lac* repressor comprises only ~0.002% of the protein in wild-type *E. coli*. Now, however, *lac* repressor is available in quantity through the application of molecular cloning techniques (Section 28-8D).

lac Repressor Finds Its Operator by Sliding Along DNA

The *lac* repressor is a tetramer of identical 360-residue subunits arranged with three mutually perpendicular twofold axes (D_2 symmetry; Section 7-5B). Each subunit is capable of binding one IPTG molecule with a dissociation constant of $K = 10^{-6}M$. In the absence of inducer, the repressor tetramer nonspecifically binds duplex DNA with a dissociation constant of $K \approx 10^{-4}M$. However, it specifically binds to the *lac* operator with far greater affinity: $K \approx 10^{-13}M$. Limited proteolysis of *lac* repressor with trypsin splits a 58-residue N-terminal peptide from each subunit. The remaining "core" tetramer binds IPTG but is unable to bind DNA. Apparently the DNA and inducer binding regions of each subunit occupy separate domains.

The observed rate constant for the binding of *lac* repressor to *lac* operator is $k_f \approx 10^{10}M^{-1}s^{-1}$. This "on" rate is much greater than that calculated for the diffusion-controlled process in solution: $k_f = 10^7M^{-1}s^{-1}$ for molecules the size of *lac* repressor. Since it is impossible for a reaction to proceed faster than its diffusion-controlled rate, the *lac* repressor must not encounter operator from solution in a random three-dimensional search. Rather, *it appears that lac repressor finds operator by nonspecifically binding to DNA and diffusing along it in a far more efficient one-dimensional search.*

lac Operator Has a Nearly Palindromic Sequence

The availability of large quantities of *lac* repressor made it possible to characterize the *lac* operator. *E. coli* DNA that had been sonicated to small fragments was mixed with *lac* repressor and passed through a nitro-

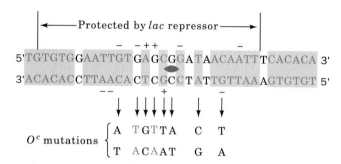

Figure 29-17
The base sequence of the *lac* operator. The symmetry related regions (*red*), comprise 28 of its 35 bp. A "+" denotes positions at which repressor binding enhances methylation by dimethyl sulfate [which methylates G at N(7) and A at N(3)] and a "−" indicates where this footprinting reaction is inhibited. The bottom row indicates the positions and identities of different point mutations that prevent *lac* repressor binding (O^c mutants). Those in color increase the operator's symmetry. [After Sobell, H.M., *in* Goldberger, R. F. (Ed.), *Biological Regulation and Development*, Vol. 1, p. 193, Plenum Press (1979).]

cellulose filter. Protein, with or without bound DNA, sticks to nitrocellulose whereas duplex DNA, by itself, does not. The DNA was released from the filter-bound protein by washing it with IPTG solution, recombined with *lac* repressor, and the resulting complex treated with DNase I. The DNA fragment that *lac* repressor protects from nuclease degradation consists of a run of 26 bp that is embedded in a nearly twofold symmetric sequence of 35 bp (Fig. 29-17; *top*). *Such palindromic symmetry is a common feature of DNAs that are specifically bound by proteins;* recall that restriction endonuclease recognition sites are also palindromic (Section 28-6A).

It has been suggested that the *lac* operator's symmetry matches that of its repressor; that is, operator binds to repressor in a twofold symmetric cleft between two subunits much like *Eco*RI restriction endonuclease binds to its recognition site (Section 28-6A). Methylation protection experiments, however, do not support this contention. There is an asymmetric pattern of differences between free and repressor-bound operator in the susceptibility of its bases to reaction with DMS (Fig. 29-17). Furthermore, point mutations in the operator that render it operator-constitutive (O^c), and which invariably weaken the binding of repressor to operator, may increase as well as decrease the operator's twofold symmetry (Fig. 29-17).

lac Repressor Prevents RNA Polymerase from Forming a Productive Initiation Complex

Operator occupies positions −7 through +28 of the *lac* operon relative to the transcription start site (Fig. 29-18). Nuclease protection studies, it will be recalled, indicate that, in the initiation complex, RNA polymerase tightly binds to the DNA between positions −20 and

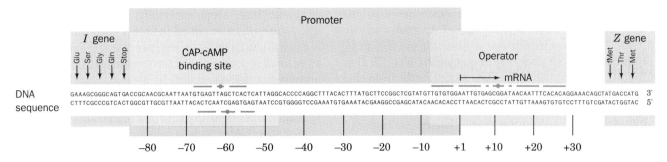

Figure 29-18

The nucleotide sequence of the *E. coli lac* promoter–operator region extending from the C-terminal region of *lacI* (*left*) to the N-terminal region of *lacZ* (*right*). The palindromic sequences of the operator and the CAP-binding site (Section 29-3C) are overscored or underscored. [After Dickson, R. C., Abelson, J., Barnes, W. M., and Reznikoff, W. A., *Science* **187,** 32 (1975).]

+20 (Section 29-2B). Thus, *the lac operator and promoter sites overlap.* This suggests that repressor binding and RNA polymerase binding are mutually exclusive. However, both proteins simultaneously bind to the *lac* operon, at least *in vitro*, to form a transcriptionally inactive complex. Evidently, operator-bound *lac* repressor prevents RNA polymerase from forming a productive initiation complex although how it does so is unknown.

C. Catabolite Repression: An Example of Gene Activation

Glucose is E. coli's metabolite of choice; the availability of adequate amounts of glucose prevents the full expression of genes specifying proteins involved in the fermentation of numerous other catabolites, including lactose (Fig. 29-19), arabinose and galactose, even when they are present in high concentrations. This phenomenon, which is known as **catabolite repression,** prevents the wasteful duplication of energy-producing enzyme systems.

cAMP Signals the Lack of Glucose

The first indication of the mechanism of catabolite repression was the observation that, in *E. coli,* the level of cAMP, which was known to be a second messenger in animal cells (Section 17-3E), is greatly diminished in the presence of glucose. This observation led to the finding that the addition of cAMP to *E. coli* cultures overcame catabolite repression by glucose. Recall that, in *E. coli,* adenylate cyclase is activated by a phosphorylated enzyme (E III$_g$), which is dephosphorylated upon the transport of glucose across the cell membrane (Section 18-3D). *The presence of glucose, therefore, normally lowers the cAMP level in E. coli.*

CAP–cAMP Complex Stimulates the Transcription of Catabolite Repressed Operons

Certain *E. coli* mutants, in which the absence of glu-

cose does not relieve catabolite repression, are missing a cAMP-binding protein that is synonymously named **catabolite gene activator protein (CAP)** or **cAMP receptor protein (CRP).** CAP is a dimeric protein of identical 210-residue subunits that undergoes a large conformational change upon binding cAMP. Its function was elucidated by Ira Pastan who showed that *CAP–cAMP complex, but not CAP itself, binds to the lac operon (among others) and stimulates transcription from its otherwise low efficiency promoter in the absence of repressor.* CAP is therefore a **positive regulator** (turns on transcription), in contrast to *lac* repressor, which is a **negative regulator** (turns off transcription).

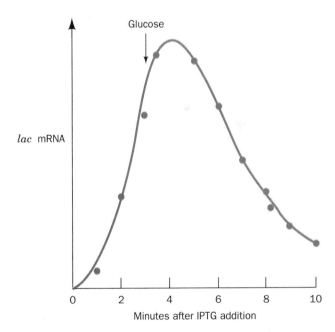

Figure 29-19

The kinetics of *lac* operon mRNA synthesis following its induction with IPTG, and of its degradation after glucose addition. *E. coli* were grown on a medium containing glycerol as their only carbon-energy source and ^{3}H-labeled uridine. IPTG was added to the medium at the beginning of the experiment to induce the synthesis of the *lac* enzymes. After 3 min, glucose was added to stop the synthesis. The amount of ^{3}H-labeled *lac* RNA was determined by hybridization with DNA containing the *lacZ* and *lacY* genes. [After Adesnik, M. and Levinthal, C., *Cold Spring Harbor Symp. Quant. Biol.* **35,** 457 (1970).]

Why is CAP–cAMP complex necessary to stimulate the transcription of its target operons? And how does it do so? The *lac* repressor has a weak (low efficiency) promoter; its −10 and −35 sequences (TATGTT and CTTTACACT; Fig. 29-18) differ significantly from the corresponding consensus sequences of strong (high efficiency) promoters (TATAAT and TCTTGACAT; Fig. 29-9). Such weak promoters evidently require some sort of help for efficient transcriptional initiation. There are two plausible (and not mutually exclusive) ways that CAP–cAMP could provide such help:

1. CAP–cAMP may stimulate transcriptional initiation through direct interaction with RNA polymerase. This hypothesis is supported by the observation that the *lac* operon fragment that CAP–cAMP complex protects from DNase I digestion contains two overlapping pseudopalindromic sequences that are located in the *lac* promoter's upstream segment (Fig. 29-18).

2. The binding of CAP–cAMP complex to promoter may conformationally alter this DNA. For example, it may induce the formation of the open RNA polymerase initiation complex (Section 29-2B). This idea is corroborated by the observation that negative supercoiling, which tends to unwind B-DNA, promotes the *in vitro* CAP stimulation of *lac* operon transcription. The binding of CAP to promoter, however, does not alter DNA's superhelicity so that CAP does not, by itself, unwind promoter. Another possibility is that CAP binding bends the DNA so as to store elastic energy for subsequent use in transcription. Indeed, the anomalously low polyacrylamide gel electrophoretic mobility of CAP in complex with its ∼30 bp target sequence indicates that CAP binding induces at least a 90° bend in this DNA segment.

Many Prokaryotic Repressors and Activators Bind Their Operators in a Similar Fashion

Since genetic expression is controlled by proteins such as CAP and *lac* repressor, an important issue in the study of gene regulation is how do these proteins interact with DNA. The X-ray crystal structure of CAP, determined by Thomas Steitz, reveals that each monomer of this dimeric protein consists of two flexibly linked domains (Fig. 29-20a). The N-terminal domains bind cAMP and form the intersubunit contacts. The C-terminal domains form the DNA-binding site as is demonstrated by the observation that their excision by limited proteolysis results in a dimeric cAMP-binding protein that does not bind DNA.

The CAP dimer's two symmetrically disposed F helices protrude from the protein surface in such a way that, according to model building studies, they fit into successive major grooves of B-DNA (Fig. 29-20b). *CAP's E and F helices form a **helix–turn–helix** supersecondary*

structure that conformationally resembles analogous *helix–turn–helix motifs in the other repressors of known X-ray structures: the E. coli* **trp repressor** *(Section 29-3E) and the* **cI repressors** *and* **Cro proteins** *from* **bacteriophages** *λ and* **434** *(Section 32-3D).*

Specific Protein–DNA Interactions Arise from Mutual Conformational Accommodations

Model building, such as that indicated in Fig. 29-20b, and, more importantly, the direct visualization of protein–DNA complexes (see below), indicates that *these DNA-binding proteins associate with their target base pairs mainly via the side chains extending from the second helix of the helix–turn–helix motif, the so-called "recognition" helix* (helix F in CAP, E in *trp* repressor, and α3 in the phage proteins). Indeed, replacing the outward-facing residues of the 434 repressor's "recognition" helix with the corresponding residues of the related **bacteriophage P22** (using the gentic engineering techniques described in Section 28-8) yields a hybrid repressor that binds to P22 operators but not to those of 434. Moreover, the ∼20-residue helix–turn–helix motifs in all these proteins have amino acid sequences that are similar to each other and to polypeptide segments in numerous other prokaryotic DNA-binding proteins, including *lac* repressor. Evidently, *these proteins are evolutionarily related and bind their target DNAs in a similar manner* (but in a way that differs from that of *Eco*RI restriction endonuclease; Section 28-6A).

How does the "recognition" helix recognize its target sequence? Each base pair presents a different and presumably readily differentiated constellation of hydrogen bonding groups in DNA's major groove (see Fig. 28-6). It has therefore been proposed that there is a simple correspondence, analogous to Watson–Crick base pairing, between the amino acid residues of the "recognition" helix and the bases they contact in forming sequence-specific associations. The above X-ray structures, however, indicate this proposal to be incorrect. Rather, base sequence recognition arises from complex structural interactions. For instance:

1. The X-ray structures of 434 repressor and 434 Cro protein in complex with the identical 20 bp target DNA (434 phage expression is regulated through the differential binding of these proteins to the same DNA segments; Section 32-3D) were both determined by Stephen Harrison. Both dimeric proteins, as predicted for CAP (Fig. 29-20b), associate with the DNA in a twofold symmetric manner with their "recognition" helices bound in successive turns of the DNA's major groove (Figs. 29-21 and 29-22). In both complexes, the protein closely conforms to the DNA surface and interacts with its paired bases and sugar–phosphate chains through elaborate systems of hydrogen bonds, salt bridges, and van der Waals contacts. Nevertheless, the detailed geometries of

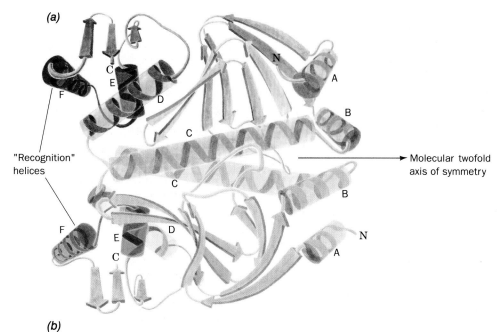

(a)

"Recognition" helices

Molecular twofold axis of symmetry

(b)

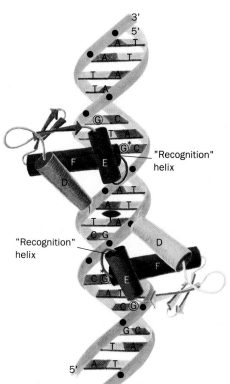

"Recognition" helix

"Recognition" helix

Figure 29-20

The structure and interactions of CAP. (*a*) A ribbon diagram of the CAP dimer. The cAMP-binding N-terminal domains, which contain the dimer contacts, are colored green and yellow whereas, the C-terminal domains are colored blue and purple with their DNA-binding helix–turn–helix domains colored in darker shades. The helices are labeled alphabetically starting from the N-terminus. [Based on a drawing by Jane Richardson, Duke University.] (*b*) The proposed association, based on model building, between CAP's DNA-binding domains and their binding site on the *lac* operon as viewed down the protein's twofold axis of symmetry. Note how the dimeric protein's two symmetry related "recognition" helices are spaced to fit into successive turns of the DNA's major groove. The DNA-binding site was identified through chemical, enzymatic, and mutagenic modification studies. Dots mark the phosphates whose ethylation prevents CAP binding, circled G's are protected from methylation when CAP binds, and * indicates the *lac* mutation sites that decrease CAP affinity. [After Weber, I. T. and Steitz, T. A., *Proc. Natl. Acad. Sci.* **81,** 3975 (1984).]

these associations are significantly different. In the repressor–DNA complex (Fig. 29-21), the DNA bends around the protein in an arc of radius ~65 Å so as to compress the minor groove by ~2.5 Å near its center (between the two protein monomers) and widen it by ~2.5 Å towards its ends [the phosphate–phosphate distance across the minor groove in canonical (ideal) B-DNA is 11.5 Å]. In contrast, the DNA in complex with Cro (Fig. 29-22), although also bent, is nearly straight at its center and has a less compressed minor groove (compare Figs. 29-21a and 29-22a). This explains why the simultaneous replacement of three residues in the repressor "recognition" helix with those occurring in Cro does not cause the resulting hybrid protein to bind DNA with Cro-like affinity: *The different conformations of the DNA in the*

(a) (b) (c)

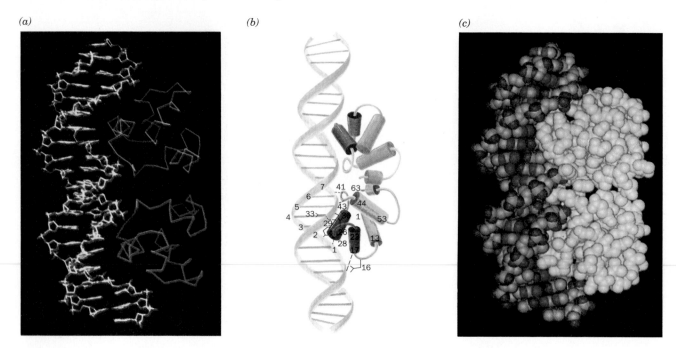

Figure 29-21

The X-ray structure of 434 phage repressor (actually only the repressor's 69-residue N-terminal domain) in complex with a 20 bp fragment of its target sequence [one strand of which has the sequence d(TATACAAGAAAGTTTGTACT)] as viewed perpendicularly to the complex's twofold axis of symmetry. (a) A skeletal model with the DNA on the left and with the protein's two identical subunits (C_α backbone only) shown in red and blue. Only the first 63 residues of the protein are visible. [Courtesy of Aneel Aggarwal, John Anderson, and Stephen Harrison, Harvard University.] (b) An interpretive drawing showing how the helix–turn– helix motif (*darker shading*) interacts with the DNA. In the lower protein monomer, the side chains important for interaction with the DNA are indicated and the numbers of the first and last residues of the helix are given. Note how the dimer's two "recognition" helices bind in successive major grooves of the DNA. [After Anderson, J. E., Ptashne, M., and Harrison, S. C., *Nature* **326,** 847 (1987).] (c) A space-filling model corresponding to Part (a). All of the protein's non-H atoms are drawn in yellow. [Courtesy of Aneel Aggarwal, John Anderson, and Stephen Harrison, Harvard University.]

repressor and Cro complexes prevents any particular side chain from interacting identically with the DNA in the two complexes.

2. Paul Sigler determined the X-ray structure of *E. coli trp* repressor in complex with an 18 bp palindromic DNA that closely resembles *trp* operator (Section 29-3E). The dimeric protein contacts the relatively straight DNA via 24 direct and 6 solvent-mediated (water bridged) hydrogen bonds to the DNA's phosphate groups (Fig. 29-23). Astoundingly, however, *there are no direct hydrogen bonds or nonpolar contacts that can explain the repressor's specificity for its operator* (the few such contacts in the structure are with bases that are tolerant to mutation). Evidently, *trp* repressor recognizes its operator via "indirect readout": The operator's sequence permits the DNA to assume a conformation that makes favorable contacts with the repressor. Model building indicates that canonical B-DNA can only make a small fraction of the contacts that operator makes to repressor. Other DNA sequences could conceivably assume repressor-bound operator's conformation but at too

high an energy cost to form a stable complex with repressor (*trp* repressor's measured 10^4-fold preference for its operator over other DNAs implies an ∼23 kJ · mol^{-1} difference in their binding free energies). Thus, *specificity arises here from sequence-specific conformational variations in DNA rather than from sequence-specific hydrogen bonding interactions between DNA and protein.*

It therefore appears that *there are no simple rules governing how particular amino acid residues interact with bases. Rather, sequence specificity results from an ensemble of mutually favorable interactions between a protein and its target DNA.*

D. *araBAD* Operon: Positive and Negative Control by the Same Protein

Humans neither metabolize nor intestinally absorb the plant sugar L-arabinose. Hence, the *E. coli* that normally inhabit the human gut are periodically presented with a banquet of this pentose. Three of the five *E. coli*

(a)

(b)

(c)

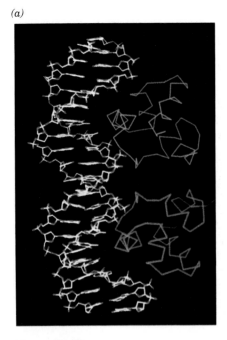

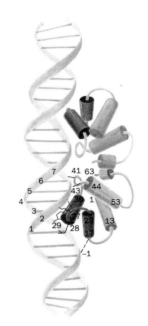

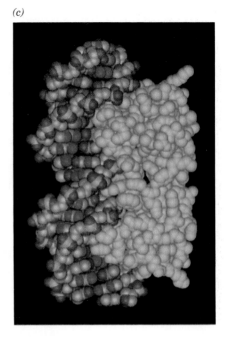

Figure 29-22
The X-ray structure of the 72-residue 434 Cro protein in complex with the same 20 bp DNA shown in Fig. 29-21 as viewed perpendicularly to the complex's twofold axis of symmetry. Only the first 64 residues of the protein are visible. Parts (a), (b), and (c) correspond to those in Fig. 29-21 with the protein in Part (c) shown in light blue. Note the close but not identical correspondence between the two structures. [Parts (a) and (c) courtesy of Alfonso Mondragon, Cynthia Wolberger, and Stephen Harrison, Harvard University. Part (b) after Wolberger, C., Dong, Y., Ptashne, M., and Harrison, S. C., *Nature* **335,** 791 (1988).]

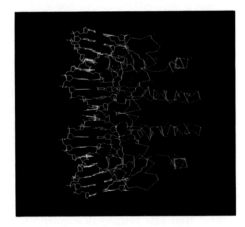

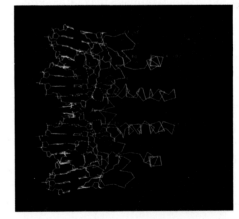

Figure 29-23
The X-ray structure of a *E. coli trp* repressor–operator complex as viewed, in stereo, perpendicular to the molecular twofold axis of symmetry. The protein's C$_\alpha$ backbone is shown (*blue*) together with the side chains (*green*) that make hydrogen bonds (*dashed lines*) to the 18 bp palindromic operator (*yellow*). The protein only binds its operator if L-tryptophan (*red*) is simultaneously bound to the protein. Note that the protein's "recognition" helices bind, as expected, in successive major grooves of the DNA but extend perpendicularly to the DNA duplex axis. In contrast, the "recognition" helices of 434 repressor and Cro proteins bind parallel to the major groove of their DNA (Figs. 29-21 and 29-22). Instructions for viewing stereo diagrams are given in the appendix to Chapter 7. [Courtesy of Paul Sigler, Yale University.]

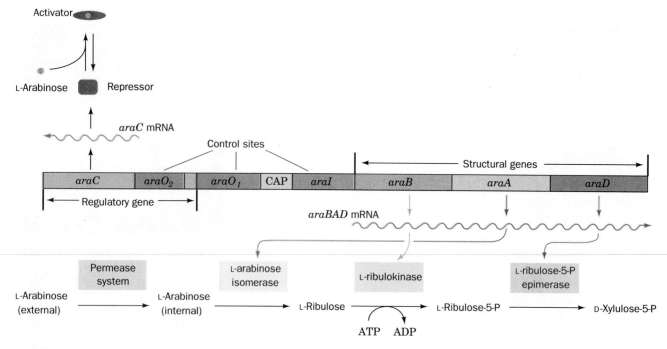

Figure 29-24

A genetic map of the *E. coli araC* and *araBAD* operons indicating the proteins they encode and the reactions in which these proteins participate. The permease system, which transports arabinose into the cell, is the product of the *araE* and *araF* genes, which occur in two independent operons. The pathway product, xylulose-5-phosphate, is converted, via the transketolase reaction, to the glycolytic intermediate fructose-6-phosphate (Section 21-4C). [After Lee, N., *in* Miller. J.H. and Rezinkoff, W. S. (Eds.), *The Operon, pp.* 390, Cold Spring Harbor Laboratory (1979).]

enzymes that metabolize L-arabinose are products of the catabolite repressible ***araBAD* operon** (Fig. 29-24).

The transcription of the araBAD operon is regulated by both CAP–cAMP and the L-arabinose-binding protein, **AraC** (the *araC* gene product; proteins may be assigned the name of the gene specifying them but in roman letters with the first letter capitalized; Fig. 29-25):

1. In the absence of AraC, RNA polymerase initiates transcription of the *araC* gene in the direction away from its upstream neighbor, *araBAD*. The *araBAD* operon remains repressed.

2. When AraC is present, with or without L-arabinose, but not CAP–cAMP (high glucose), AraC binds to three different gene sites: *araI*, which just precedes the *araBAD* promoter; *araO_1*, which overlaps the *araC* promoter; and *araO_2*, which, surprisingly, is located in a noncoding upstream region of the *araC* gene, around position -280 relative to the *araBAD* start site. *araO_1* is the operator for the *araC* gene; its association with AraC blocks *araC* transcription so that this process is autoregulatory. A series of deletion mutations indicate that both *araO_2* and *araI* must be present for *araBAD* to be repressed in the presence of AraC. The remarkably large separation between *araO_2* and the *araBAD* promoter therefore suggests

that the DNA is looped such that a single molecule or molecular complex of AraC protein simultaneously binds to both *araO_2* and *araI*. This cooperative arrangement is required for the AraC-mediated repression of *araBAD* (negative control).

3. When the cAMP level is high (low glucose), CAP–cAMP binds to a site between *araO_1* and *araI*. When L-arabinose is also present, it binds to AraC causing it to release *araO_2* so as to open the DNA loop. This combined influence of CAP–cAMP and AraC–arabinose, which is probably mediated through a direct interaction between these two complexes, activates RNA polymerase to transcribe the *araBAD* operon (positive control). The observation that *araO_2* deletion permits AraC–arabinose to activate *araBAD* in the absence of CAP–cAMP indicates that CAP–cAMP stimulates AraC–arabinose to release *araO_2* and that this release is required to convert AraC–arabinose to an activator. *araC* remains repressed by AraC.

The function of DNA loop formation is obscure although it has been demonstrated to occur in numerous bacterial and eukaryotic systems. Perhaps it permits several regulatory proteins and/or regulatory sites on one protein to simultaneously influence transcription

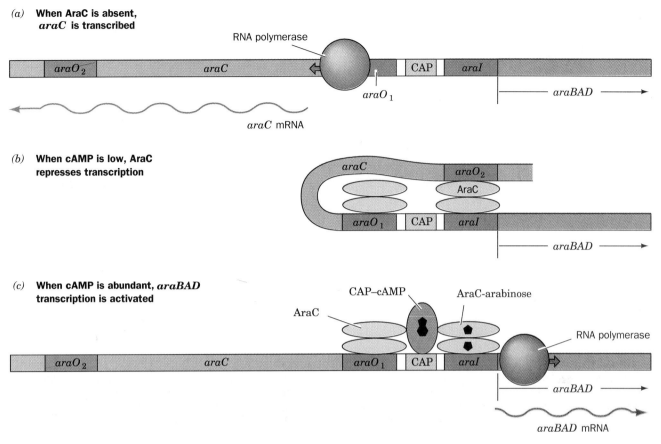

Figure 29-25
The proposed mechanism for *araBAD* regulaion: (*a*) In the absence of AraC, RNA polymerase initiates the transcription of *araC* but not *araBAD*. (*b*) When AraC is present, with or without L-arabinose, and the cAMP level is low, AraC binds to *araO₁* and links together *araO₂* and *araI* to form a DNA loop, thereby repressing both *araC* and *araBAD*. (*c*) When AraC and L-arabinose are both present and cAMP is abundant, CAP-cAMP stimulates the AraC–arabinose complex to release *araO₂* but to remain bound to *araI* where it activates *araBAD* transcription. *araC* remains repressed.

initiation by RNA polymerase. In fact, as recent studies have shown, the *lac* operon contains a second, relatively weak operator located 400 bp downstream from the transcription start site (within the *lacZ* gene). This secondary operator (O_2) cooperates with the primary operator (now called O_1) to form a repression complex that is stronger than with either operator alone. It is thought that during severe repression, both operators bind to a single *lac* repressor tetramer to form a DNA loop-containing complex.

E. *trp* Operon: Attenuation

In the following paragraphs we discuss a sophisticated transcriptional control mechanism named **attenuation** through which bacteria regulate the expression of certain operons involved in amino acid biosynthesis. This mechanism was discovered through the study of the *E. coli trp operon* (Fig. 29-26) which encodes five polypeptides comprising three enzymes that mediate the synthesis of tryptophan from chorismate (Section 24-5B). Charles Yanofsky established that the *trp* operon genes are coordinately expressed under the control of *trp* **repressor,** a dimeric protein of identical 107-residue subunits that is the product of the *trpR* gene (which forms an independent operon). *The trp repressor binds L-tryptophan, the pathway's end product, to form a complex that specifically binds to trp operator (trpO; Fig. 29-27), so as to reduce the rate of trp operon transcription by 70-fold.* The X-ray structure of the *trp* repressor–operator complex (Section 29-3C) indicates that tryptophan binding allosterically orients *trp* repressor's two symmetry related helix–turn–helix "DNA reading heads" so that they can simultaneously bind to *trpO* (Fig. 29-28; also see Fig. 29-23). Moreover, the bound tryptophan forms a hydrogen bond to a DNA phosphate group, thereby strengthening the repressor–operator association. Tryptophan therefore acts as a **corepressor;** its presence prevents what is then super-

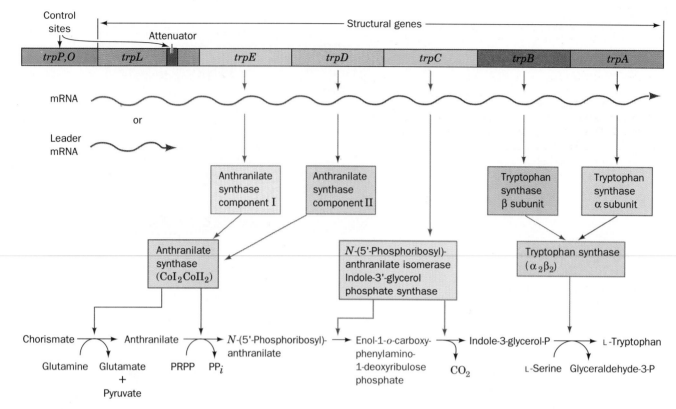

Figure 29-26
A genetic map of the *E. coli trp* operon indicating the enzymes it specifies and the reactions they catalyze. The gene product of *trpC* catalyzes two sequential reactions in the synthesis of tryptophan. [After Yanofsky, C., *J. Am. Med. Assoc.* **218,** 1027 (1971).]

fluous tryptophan biosynthesis. The *trp* repressor also controls the synthesis of at least two other operons: the **trpR operon** and the **aroH operon** (which encodes one of three isozymes that catalyze the initial reaction of aromatic amino acid biosynthesis: Section 24-5B).

Tryptophan Biosynthesis Is Also Regulated by Attenuation

The *trp* repressor–operator system was at first thought to fully account for the regulation of tryptophan biosynthesis in *E. coli.* However, the discovery of *trp* deletion mutants located downstream from *trpO* that increase *trp* operon expression sixfold indicated the existence of an additional transcriptional control element.

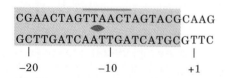

Figure 29-27
The base sequence of the *trp* operator. The nearly palindromic sequence is boxed and the Pribnow box is overscored.

Sequence analysis established that *trpE,* the *trp* operon's leading structural gene, is preceded by a 162-nucleotide **leader sequence** *(trpL).* Genetic analysis indicated that the new control element is located in *trpL,* ~30 to 60 nucleotides upstream of *trpE* (Fig. 29-26).

When tryptophan is scarce, the entire 6720-nucleotide polycistronic *trp* mRNA, including the *trpL* sequence, is synthesized. As the tryptophan concentration increases, the rate of *trp* transcription decreases as a result of the *trp* repressor–corepressor complex's consequent greater abundance. Of the *trp* mRNA that is transcribed, however, an increasing proportion consists of only a 140-nucleotide segment corresponding to the 5' end of *trpL. The availability of tryptophan therefore results in the premature termination of trp operon transcription.* The control element responsible for this effect is consequently termed an **attenuator.**

The *trp* Attenuator's Transcription Terminator Is Masked When Tryptophan Is Scarce

What is the mechanism of attenuation? The attenuator transcript contains four complementary segments that can form one of two sets of mutually exclusive based paired hairpins (Fig. 29-29). *Segments 3 and 4 together with the succeeding residues comprise a normal transcription terminator (Section 29-2E):* a G + C-rich se-

Figure 29-28
The structure of the *trp* repressor–tryptophan complex in association with its operator. The "recognition" helix (*blue*) of the dimeric protein's helix–turn–helix motif binds in the major groove of its operator DNA (see Fig. 29-23). Comparison of the X-ray structures of the *trp* repressor with and without bound tryptophan (*red*) indicates that, upon tryptophan dissociation, the "recognition" helices swing inwards (*arrows*) so that they can no longer simultaneously engage the DNA's major groove. [After Robertson, M., *Nature* **327**, 465 (1987).]

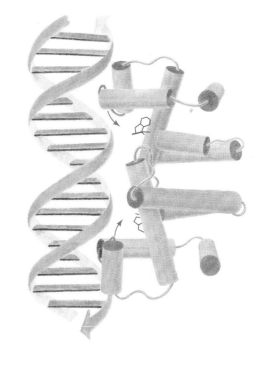

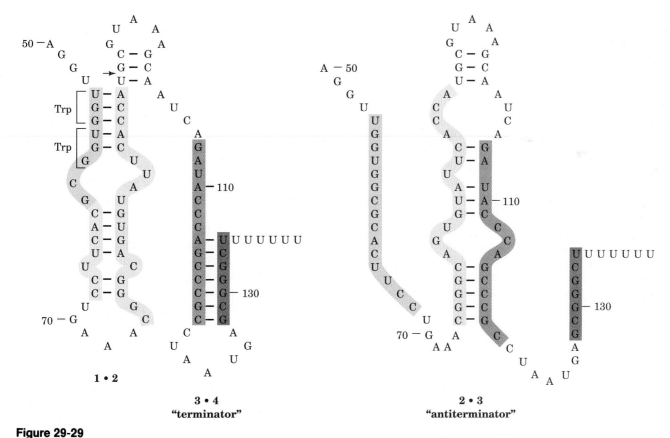

Figure 29-29
The alternative secondary structures of *trpL* mRNA. The formation of the base paired 2·3 (antiterminator) hairpin (*right*) precludes the formation of the 1·2 and 3·4 (terminator) hairpins (*left*) and *vice versa*. Attenuation results in the premature termination of transcription immediately after nucleotide 140 when the 3·4 hairpin is present. The arrow indicates the mRNA site past which RNA polymerase pauses until approached by an active ribosome. [After Fisher, R. F. and Yanofsky, C., *Proc. Natl. Acad. Sci.* **258**, 8147 (1983).]

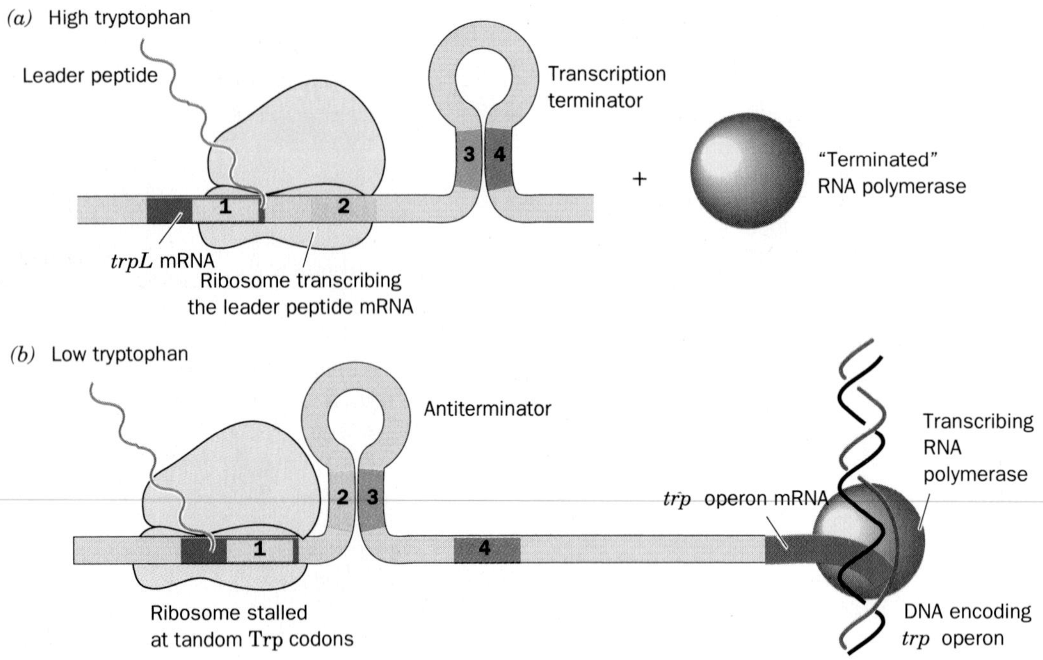

Figure 29-30
Attenuation in the *trp* operon. (*a*) When tryptophanyl–tRNA^{Trp} is abundant, the ribosome translates *trpL* mRNA. The presence of the ribosome on segment 2 prevents the formation of the base paired 2·3 hairpin. The 3·4 hairpin, an essential component of the transcriptional terminator, can thereby form thus aborting transcription. (*b*) When tryptophanyl–tRNA^{Trp} is scarce, the ribosome stalls on the tandem Trp codons of segment 1. This situation permits the formation of the 2·3 hairpin which, in turn, precludes the formation of the 3·4 hairpin. RNA polymerase therefore transcribes through this unformed terminator and continues *trp* operon transcription.

quence that can form a self-complementary hairpin structure followed by several sequential U's (compare with Fig. 29-14). *Transcription rarely proceeds beyond this termination site unless tryptophan is in short supply.*

A section of the leader sequence, which includes segment 1 of the attenuator, is translated to form a 14-residue polypeptide that contains two consecutive Trp residues (Fig. 29-29, *left*). The position of this particularly rare dipeptide segment (only ~1% of the residues in *E. coli* proteins are Trp) provided an important clue to the mechanism of attenuation. An additional essential aspect of this mechanism is that ribosomes commence the translation of a prokaryotic mRNA shortly after its 5′ end has been synthesized.

The above considerations led Yanofsky to propose the following model of attenuation (Fig. 29-30). An RNA polymerase that has escaped repression initiates *trp* operon transcription. Soon after the ribosomal initiation site of the *trpL* gene has been transcribed, a ribosome attaches to it and begins translation of the leader peptide. When tryptophan is abundant, so that there is a plentiful supply of tryptophanyl–tRNA^{Trp} (the transfer RNA specific for Trp with an attached Trp residue; Section 30-2C), the ribosome follows closely behind the transcribing RNA polymerase so as to sterically block the formation of the 2·3 hairpin. Indeed, RNA polymer-

ase pauses past position 92 of the transcript and only continues transcription upon the approach of a ribosome, thereby ensuring the proximity of these two entities at this critical position. The prevention of 2·3 hairpin formation permits the formation of the 3·4 hairpin, the transcription terminator pause site, which results in the termination of transcription (Fig. 29-30*a*). When tryptophan is scarce, however, the ribosome stalls at the tandem UGG codons (the three sequential nucleotides specifying Trp; Section 30-1E) because of the lack of tryptophanyl–tRNA^{Trp}. As transcription continues, the newly synthesized segments 2 and 3 form a hairpin because the stalled ribosome prevents the otherwise competitive formation of the 1·2 hairpin (Fig. 29-30*b*). The formation of the transcriptional terminator's 3·4 hairpin is thereby preempted for sufficient time for RNA polymerase to transcribe through it and consequently through the remainder of the *trp* operon. The cell is thus provided with a regulatory mechanism that is responsive to tryptophanyl–tRNA^{Trp} level, which, in turn, depends on the protein synthesis rate as well as the tryptophan supply.

There is considerable evidence supporting the preceding model of attentuation. The *trpL* transcript is resistant to limited RNase T1 digestion indicating that it has extensive secondary structure. The significance of the tandem Trp codons in the *trpL* transcript is corroborated by their presence in *trp* leader regions of several other bacterial species. Moreover, the leader peptides of the five other amino acid-biosynthesizing operons known to be regulated by attenuation (most exclusively so) are all rich in their corresponding amino acid residues (Table 29-1). For example, the *E. coli his* **operon,** which specifies enzymes synthesizing histidine, has seven tandem His residues in its leader peptide; similarly, the *ilv* **operon,** which specifies enzymes participating in isoleucine, leucine, and valine biosynthesis, has five Ile's, three Leu's, and six Val's in its leader peptide. Finally, the leader transcripts of these operons resemble that of the *trp* operon in their capacity to form two alternative secondary structures, one of which contains a trailing termination structure.

F. Regulation of Ribosomal RNA Synthesis: The Stringent Response

E. coli cells growing under optimal conditions divide every 20 min. Such cells contain nearly 10,000 ribosomes. Yet, RNA polymerase can initiate the transcription of an rRNA gene no faster than about once every second. If the *E. coli* genome contained only one copy of each of the three types of rRNA genes (those specifying the so-called 23S, 16S, and 5S rRNAs; Section 30-3A), there could be no more than ~1200 ribosomes/cell. However, *the E. coli chromosome contains seven separately located rRNA operons, all of which contain one nearly identical copy of each type of rRNA gene,* thereby accounting for the observed rRNA synthesis rate.

Cells have the remarkable ability to coordinate the rates at which their thousands of components are synthesized. For example, *E. coli* adjust their ribosome content to match the rate that they can synthesize proteins under the prevailing growth conditions. The rate of rRNA synthesis is therefore proportional to the rate of protein synthesis. One mechanism by which this occurs is known as the **stringent response:** *A shortage of any species of amino acid-charged tRNA (usually a result of "stringent" or poor growth conditions) that limits the rate of protein synthesis triggers a sweeping metabolic readjustment.* A major facet of this change is an abrupt 10-to 20-fold reduction in the rate of rRNA and tRNA synthesis. This **stringent control,** moreover, depresses numerous metabolic processes (including DNA replication and the biosynthesis of carbohydrates, lipids, nucleotides, proteoglycans, and glycolytic intermediates) while stimulating others (such as amino acid biosynthesis). The cell is thereby prepared to withstand nutritional deprivation.

ppGpp Mediates the Stringent Response

*The stringent response is correlated with a rapid intracellular accumulation of the unusual nucleotide **ppGpp** and its prompt decay when amino acids become available.* The observation that mutants, designated *relA⁻*, which do not exhibit the stringent response (they are said to have **relaxed control**), lack ppGpp suggests that this substance mediates the stringent response. This idea was corroborated by *in vitro* studies demonstrating, for example, that ppGpp inhibits the transcription of rRNA genes but stimulates the transcription of the *trp* and *lac* operons as does the stringent response *in vivo*. It therefore seems that ppGpp acts by somehow altering RNA polymerase's promoter specificity at stringently controlled operons, an hypothesis that is supported by the isolation of RNA polymerase mutants that exhibit reduced responses to ppGpp.

Experiments with cell-free *E. coli* extracts have established that the protein encoded by wild-type *relA* gene, named **stringent factor,** catalyzes the reaction

$$\text{ATP} + \text{GDP} \rightleftharpoons \text{AMP} + \text{ppGpp}$$

Stringent factor is only active in association with a ribo-

Table 29-1

Amino Acid Sequences of Some Leader Peptides in Operons Subject to Attenuation

Operon	Amino Acid Sequence[a]
trp	Met-Lys-Ala-Ile-Phe-Val-Leu-Lys-Gly-TRP-TRP-Arg-Thr-Ser
pheA	Met-Lys-His-Ile-Pro-PHE-PHE-PHE-Ala-PHE-PHE-PHE-Thr-PHE-Pro
his	Met-Thr-Arg-Val-Gln-Phe-Lys-HIS-HIS-HIS-HIS-HIS-HIS-HIS-Pro-Asp
leu	Met-Ser-His-Ile-Val-Arg-Phe-Thr-Gly-LEU-LEU-LEU-LEU-Asn-Ala-Phe-Ile-Val-Arg-Gly-Arg-Pro-Val-Gly-Gly-Ile-Gln-His
thr	Met-Lys-Arg-ILE-Ser-THR-THE-ILE-THR-THR-THR-ILE-THR-ILE-THR-THR-Gln-Asn-Gly-Ala-Gly
ilv	Met-Thr-Ala-LEU-LEU-Arg-VAL-ILE-Ser-LEU-VAL-VAL-ILE-Ser-VAL-VAL-VAL-ILE-ILE-ILE-Pro-Pro-Cys-Gly-Ala-Ala-Leu-Gly-Arg-Gly-Lys-Ala

[a] Upper case residues are synthesized in the pathway catalyzed by the operon's gene products.

Source: Yanofsky, C., *Nature* **289,** 753 (1981).

some that is actively engaged in translation. ppGpp synthesis occurs at a maximal rate when a ribosome binds its mRNA-specified but uncharged (lacking an amino acid residue) tRNA. The binding of a specified and charged tRNA greatly reduces the rate of ppGpp synthesis. *The ribosome apparently signals the shortage of an amino acid by stimulating the synthesis of ppGpp which, acting as an intracellular messenger, influences the rates at which a great variety of operons are transcribed.*

ppGpp degradation is catalyzed by the *spoT* gene product. The *spoT⁻* mutants show a normal increase in ppGpp level upon amino acid starvation but an abnormally slow decay of ppGpp to basal levels when amino acids again become available. The *spoT⁻* mutants therefore exhibit a sluggish recovery from the stringent response. *The ppGpp level is apparently regulated by the countervailing activities of stringent factor and the spoT gene product.*

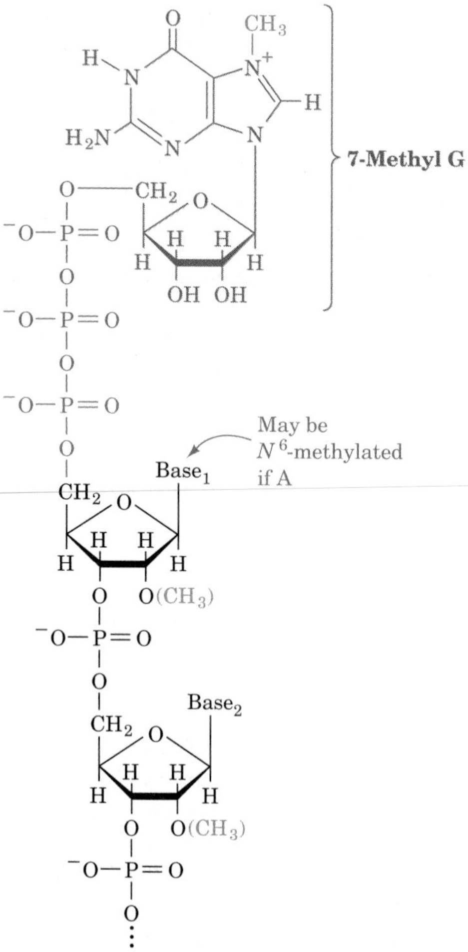

4. POST-TRANSCRIPTIONAL PROCESSING

The immediate products of transcription, the **primary transcripts,** are not necessarily functional entities. In order to acquire biological activity, many of them must be specifically altered in several ways: (1) by the exo and endonucleolytic removal of polynucleotide segments; (2) by appending nucleotide sequences to their 3′ and 5′ ends; and (3) by the modification of specific nucleosides. The three major classes of RNAs, mRNA, rRNA, and tRNA, are altered in different ways in prokaryotes and in eukaryotes. In this section we shall outline these **post-transcriptional modification** processes.

A. Messenger RNA Processing

In prokaryotes, most primary mRNA transcripts function in translation without further modification. Indeed, as we have seen, ribosomes in prokaryotes usually commence translation on nascent mRNAs. In eukaryotes, however, mRNAs are synthesized in the cell nucleus while translation occurs in the cytosol. Eukaryotic mRNA transcripts can therefore undergo extensive post-transcriptional processing while still in the nucleus.

Eukaryotic mRNAs Are Capped

Eukaryotic mRNAs have a peculiar enzymatically appended cap structure consisting of a 7-methylguanosine residue joined to the transcript's initial (5′) nucleoside via a 5′–5′ triphosphate bridge (Fig. 29-31). The cap, which a specific guanylyltransferase adds to the growing transcript before it is > 20-nucleotides long, defines the eu-

Figure 29-31
The structure of the 5′ cap of eukaryotic mRNAs. It is known as cap-0, cap-1, or cap-2, respectively, if it has no further modifications, if the leading nucleoside of the transcript is $O(2')$-methylated, or if its first two nucleosides are $O(2')$-methylated.

karyotic translational start site (Section 30-3C). A cap may be $O(2')$-methylated at the transcript's leading nucleoside (**cap-1,** the predominant cap in multicellular organisms), at its first two nucleosides (**cap-2),** or at neither of these positions (**cap-0,** the predominant cap in unicellular eukaryotes). If the leading nucleoside is adenosine (it is usually a purine), it may also be N^6-methylated.

Eukaryotic mRNAs Have Poly(A) Tails

Eukaryotic mRNAs, in contrast to those of prokaryotes, are invariably monocistronic. Yet, the sequences signaling transcriptional termination in eukaryotes have not been identified. This is largely because the termination process is imprecise; that is, the primary transcripts of a given structural gene have heterogeneous 3′ sequences. Nevertheless, mature eukaryotic mRNAs have well-defined 3′ ends; *almost all of them have 3′-poly(A) tails of 100 to 200 nucleotides.* The poly(A) tails are enzymatically appended to the primary transcripts in two reactions:

1. A transcript is cleaved 10 to 30 nucleotides past a highly conserved AAUAAA sequence, whose mutation abolishes cleavage and polyadenylation, and within 50 nucleotides before a less conserved U-rich or GU-rich sequence.

2. The poly(A) tail is subsequently generated from ATP through the stepwise action of **poly(A) polymerase.**

The precision of the cleavage reaction has apparently eliminated the need for accurate transcriptional termination; to put things another way, all's well that ends well.

In vitro studies indicate that a poly(A) tail is not required for mRNA translation. Rather, the observations that an mRNA's poly(A) tail shortens as it ages in the cytosol and that unadenylated mRNAs have abbreviated cytosolic lifetimes suggest that poly(A) tails have a protective role. In fact, the only mature mRNAs that generally lack poly(A) tails, those of histones (which, with few exceptions, lack the AAUAAA cleavage–polyadenylation signal), have lifetimes of <30 min in the cytosol, whereas most other mRNAs last hours or days.

Eukaryotic Genes Consist of Alternating Expressed and Unexpressed Sequences

The most striking difference between eukaryotic and prokaryotic structural genes is that the coding sequences of most eukaryotic genes are interspersed with unexpressed regions. Early investigations of eukaryotic structural gene transcription found, quite surprisingly, that primary transcripts are quite heterogeneous in length (from ~2000 to well over 20,000 nucleotides) and are much larger than is expected from the known sizes of eukaryotic proteins. Rapid labeling experiments demonstrated that little of this so-called **heterogeneous nuclear RNA (hnRNA)** is ever transported to the cytosol; most of it is quickly turned over in the nucleus. Yet, the hnRNA's 5′ caps and 3′ tails eventually appear in cytosolic mRNAs. *The straightforward explanation of these observations, that pre-mRNAs are processed by the excision of internal sequences, seemed so bizarre that it came as a great suprise in 1977 when it was independently demonstrated in several laboratories that this is actually the case.* In fact, the pre-mRNA's noncoding **intervening sequences (IVSs or introns)** are usually of greater length than their flanking **expressed sequences (exons).** This situation is graphically illustrated in Fig. 29-32, which is an electron micrograph of chicken **ovalbumin** mRNA hybridized to the sense strand of the ovalbumin gene (ovalbumin is the major protein component of egg white). The lengths of introns in vertebrate genes ranges from ~65 to over 100,000 nucleotides with no obvious periodicity. In-

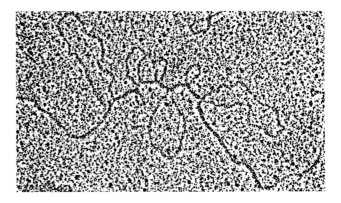

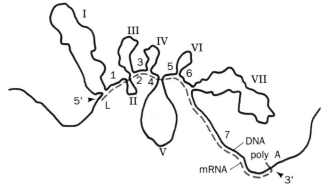

Figure 29-32
An electron micrograph and its interpretive drawing of a hybrid between the sense strand of the chicken ovalbumin gene (as obtained by molecular cloning methods; Section 31-5) and its corresponding mRNA. The complementary segments of the DNA (*purple line in drawing*) and mRNA (*red line*) have annealed to reveal the exon positions (L, 1–7). The looped-out segments (I–VII), which have no complementary sequences in the mRNA, are the introns. [From Chambon, P., *Sci. Am.* **244**(5): 61 (1981)].

deed, the corresponding introns from genes in two vertebrate species can vary extensively in both length and sequence so as to bear little resemblance to one another.

Further investigations established that the formation of eukaryotic mRNA begins with the transcription of an entire structural gene, including its introns, to form pre-mRNA (Fig. 29-33). Then, following capping and perhaps polyadenylation, the introns are excised and their flanking exons are connected, a process called **gene splicing,** to yield the mature mRNA. *The most striking aspect of gene splicing is its precision; if one base too few or too many were excised, the resulting mRNA could not be translated properly (Section 30-1B). Moreover, exons are never shuffled; their order in the mature mRNA is exactly the same as that in the gene from which it is derived.* In the following subsections we discuss the mechanism of this remarkable splicing process.

Exons Are Spliced in a Two-Stage Reaction

Sequence comparisons of exon–intron junctions from a diverse group of eukaryotes indicate that they

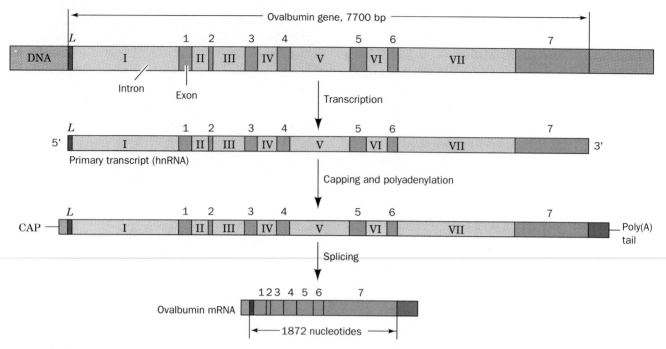

Figure 29-33

The sequence of steps in the production of mature eukaryotic mRNA as shown for the chicken ovalbumin gene. Following transcription, the primary transcript is capped and polyadenylated. The introns are then excised and the exons spliced together to form the mature mRNA.

have a high degree of homology (Fig. 29-34), including, as Richard Breathnach and Pierre Chambon first pointed out, *an invariant GU at the intron's 5′ boundary and an invariant AG at its 3′ boundary. These sequences are necessary and sufficient to define a splice junction:* Mutations that alter the sequences interfere with splicing, whereas mutations that change a nonjunction to a consensuslike sequence can generate a new splice junction.

Investigations of both cell free and *in vivo* splicing systems by Argiris Efstradiadis, Michael Rosbash, Phillip Sharp, and Tom Maniatis have established that intron excision occurs in two reactions (Fig. 29-35):

1. The formation of a 2′,5′ phosphodiester bond between an intron adenosine residue and its 5′-terminal phosphate group with the concommitant release of

the 5′ exon. *The intron thereby assumes a novel lariat structure.* The adenosine residue at the lariat branch has been identified as the *A* in the sequence CURAY [where R represents purines (A or G) and Y represents pyrimidines (C or U)], which is highly conserved in vertebrate mRNAs and is typically located 20 to 50 residues upstream of the 3′ splice site (yeast have a similar UACUAAC sequence that occurs ~50-residues upstream from all its 3′ splice sites). Mutations that change this branch point A residue abolish splicing at that site.

2. The now free 3′-OH group of the 5′ exon forms a phosphodiester bond with the 5′-terminal phosphate of the 3′ exon yielding the spliced product. The intron is thereby eliminated in its lariat form and, *in vivo*, is rapidly degraded. Mutations that alter the conserved

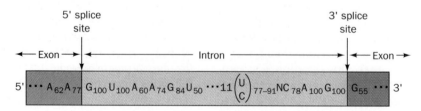

Figure 29-34

The consensus sequence at the exon–intron junctions of eukaryotic pre-mRNAs. The subscripts indicate the percent of pre-mRNAs in which specified base(s) occurs. Note that the 3′ splice site is preceded by a tract of 11 predominantly pyrimidine nucleotides. [Based on data from Padgett, R. A., Grabowski, P. J., Konarska, M. M., Seiler, S. S., and Sharp, P. A., *Annu. Rev. Biochem.* **55**, 1123 (1986).]

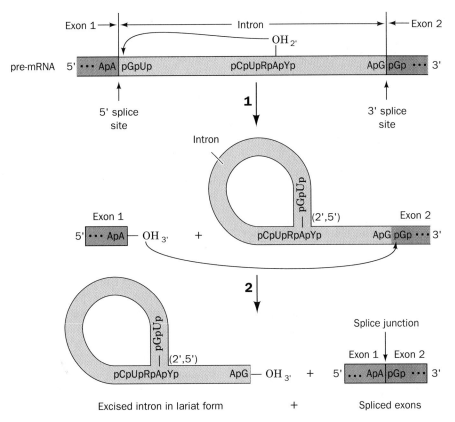

Figure 29-35
The sequence of transesterification reactions that splice together the exons of eukaryotic pre-mRNAs (the exons and introns are drawn in blue and orange; R and Y represent purine and pyrimidine residues): **(1)** The 2'-OH group of a specific intron A residue nucleophilically attacks the 5' phosphate at the 5' intron boundary to yield an unusual 2',5'-phosphodiester bond and thus form a lariat structure. **(2)** The liberated 3'-OH group forms a 3',5'-phosphodiester bond with the 5' terminal residue of the 3' exon, thereby splicing the two exons together and releasing the intron in lariat form.

AG at the 3' splice junction block this second step although they do not interfere with lariat formation.

Note that the splicing process proceeds without free energy input; its transphosphorylation reactions preserve the free energy of each cleaved phosphodiester bond through the concomitant formation of a new one.

Splicing Is Mediated by snRNPs

How are splice junctions recognized and how are the two exons to be joined brought together in the splicing process? Part of the answer to this question was established by Joan Steitz going on the assumption that one nucleic acid is best recognized by another. The eukaryotic nucleus, as has been known since the 1960s, contains numerous copies of several highly conserved 60 to 300 nucleotide RNAs called **small nuclear RNAs (snRNAs),** which form protein complexes termed **small nuclear ribonucleoproteins (snRNPs;** pronounced "snurps"). Steitz recognized that the 5' end of one of these snRNAs, **U1-snRNA** (so-called because it is a member of a U-rich subfamily of snRNAs), is partially complementary to the 5' consensus sequence of mRNA splice junctions. The consequent hypothesis, that *U1-snRNP recognizes the 5' splice junction,* was corroborated by the observations that splicing is inhibited by the selective destruction of the U1-snRNP sequences that are complementary to the 5' splice junction and by the presence of anti-U1-snRNP antibodies (produced by patients suffering from **systemic lupus erythematosus,** an often fatal autoimmune disease). Similar studies have implicated **U2-snRNP** in recognizing the intron region that forms the lariat branch point and **U5-snRNP** in recognizing the 3' splice junction. Altogether, ~65 pre-mRNA nucleotides participate in this recognition process, which rationalizes why introns are minimally 65 nucleotides in length.

Splicing takes place in an as yet poorly characterized 50S to 60S particle dubbed the splicosome (Fig. 29-36). The splicosome brings together a pre-mRNA, the foregoing snRNPs, **U4–U6-snRNP** (in which U4 and U6 snRNAs associate by base pairing and which binds to the other snRNPs rather than directly to pre-mRNA), and a vari-

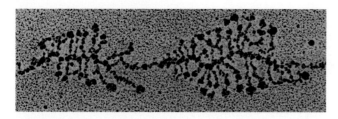

Figure 29-36
An electron micrograph of splicosomes in action. The splicosomes are the large beads on the pre-mRNAs extending above and below the horizontal DNA. [From Steitz, J. A., *Sci. Am.* **258**(6): 59 (1988). Electron micrograph by Yvonne N. Osheim.]

ety of pre-mRNA binding proteins (U4 – U6-snRNP has also been implicated in the previously described polyadenylation reaction). Note that the splicosome is a large particle; the similarly sized large ribosomal subunit of E. coli consists of 3004 nucleotides and 31 polypeptides and has a particle mass of 1.6 million daltons (Section 30-3A). The biochemical significance of splicing is discussed in Section 33-2F.

mRNA Is Methylated at Certain Adenylate Residues

During or shortly after the synthesis of vertebrate pre-mRNAs, ~0.1% of their A residues are methylated at N(6). These m^6A's tend to occur in the sequence RRm^6ACX, where X is rarely G. Although the functional significance of these methylated A's is unknown, it should be noted that a large fraction of them are components of the corresponding mature mRNAs.

B. Ribosomal RNA Processing

The seven *E. coli* rRNA operons all contain one (nearly identical) copy of each of the three types of rRNA genes (Section 29-3F). Their polycistronic primary transcripts, which are >5500 nucleotides in length, contain 16S rRNA at their 5' ends followed by the transcripts for 1 or 2 tRNAs, 23S rRNA, 5S rRNA and, in some rRNA operons, 1 or 2 more tRNAs at the 3' end (Fig. 29-37). The steps in processing these primary transcripts to mature rRNAs (Fig. 29-37) were elucidated with the aid of mutants defective in one or more of the processing enzymes.

The initial processing, which yields products known as **pre-rRNAs**, commences while the primary transcript is still being synthesized. It consists of specific endonucleolytic cleavages by **RNase III, RNase P, RNase E,** and **RNase F** at the sites indicated in Fig. 29-37. The base sequence of the primary transcript suggests the existence of several base paired stems. The RNase III cleavages occur in a stem consisting of complementary sequences flanking the 5' and 3' ends of the 23S segment (Fig. 29-38) as well as that of the 16S segment. Presumably certain features of these stems constitute the RNase III recognition site.

The 5' and 3' ends of the pre-rRNA's are trimmed away in secondary processing steps (Fig. 29-37) through the action of **RNAses M16, M23,** and **M5** to produce the mature rRNAs. These final cleavages only occur after the pre-rRNAs become associated with ribosomal proteins.

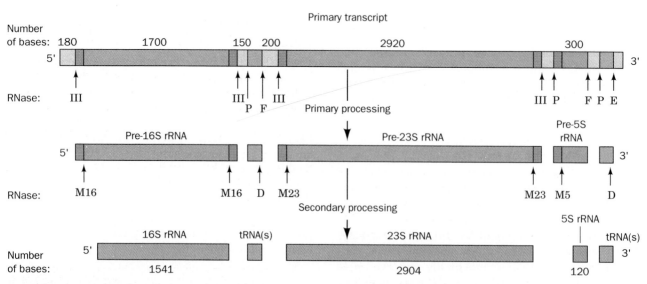

Figure 29-37
The post-transcriptional processing of *E. coli* rRNA. The transcriptional map is shown approximately to scale. The labeled arrows indicate the positions of the various nucleolytic cuts and the nucleases that generate them.

[After Apiron, D., Ghora, B. K., Plantz, G., Misra, T. K., and Gegenheimer, P., *in* Söll, D., Abelson, J. N., and Schimmel P. R. (Eds.), *Transfer RNA: Biological Aspects, p.* 148, Cold Spring Harbor Laboratory (1980).]

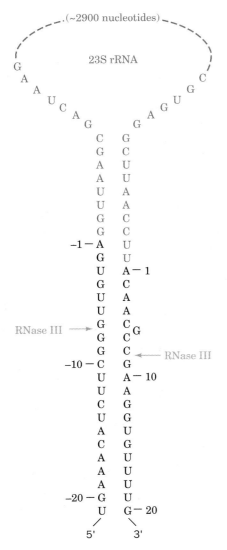

Figure 29-38
The proposed stem-and-giant-loop secondary structure in the 23S region of the *E. coli* primary rRNA transcript. The RNase III cleavage sites are indicated. [After Young. R. R., Bram, R. J., and Steitz, J. A., *in* Söll, D., Abelson, J. N., and Schimmel, P. R. (Eds.), *Transfer RNA: Biological Aspects*, p. 102, Cold Spring Harbor Laboratory (1980).]

Ribosomal RNAs Are Methylated

During ribosomal assembly, the 16S and 23S rRNAs are methylated at a total of 24 specific nucleosides. The methylation reactions, which employ S-adenosylmethionine (Section 24-3E) as a methyl donor, yield N^6,N^6-dimethyladenine and $O^{2'}$-methylribose residues. $O^{2'}$-methyl groups are thought to protect adjacent phosphodiester bonds from degradation by intracellular RNases (the mechanism of RNase hydrolysis involves utilization of the free 2'-OH group of ribose to eliminate the substituent on the 3'-phosphoryl group via the formation of a 2',3'-cyclic phosphate intermediate; Section 28-1). However, the function of base methylation is unknown.

Eukaryotic rRNA Processing Resembles That of Prokaryotes

The eukaryotic genome typically has several hundred tandemly repeated copies of rRNA genes that are contained in small dark-staining nuclear bodies known as nucleoli (the site of rRNA transcription, processing, and ribosomal subunit assembly; Fig. 1-5). The primary rRNA transcript is an ~7500-nucleotide 45S RNA that contains, starting from the 5' end, the 18S, 5.8S, and 28S rRNAs separated by spacer sequences (Fig. 29-39). In the first stage of its processing, 45S RNA is specifically methylated at ~110 sites that occur mostly in its rRNA sequences. About 80% of these modifications yield $O^{2'}$-methylribose residues and the remainder form methylated bases such as N^6,N^6-dimethyladenine and 2-methylguanine. The subsequent cleavage and trimming of the 45S RNA superfically resembles that of prokaryotic rRNAs. In fact, enzymes exhibiting RNAse III and RNase P-like activities occur in eukaryotes. The 5S eukaryotic rRNA is separately processed in a manner resembling that of tRNA (Section 29-4C).

Some Eukaryotic rRNA Genes Are Self-Splicing

Only a few eukaryotic rRNA genes contain introns. Nevertheless, Thomas Cech's study of how such genes are spliced in the ciliated protozoan *Tetrahymena thermophila* led to an astonishing discovery: *RNA can act as an enzyme. When the isolated pre-rRNA of this organism is incubated with guanosine or a free guanine nucleotide (GMP, GDP, or GTP), but in the absence of protein, its single 413-nucleotide intron excises itself and splices together its flanking exons; that is, this pre-rRNA is self-splicing.* The three-step reaction sequence of this process (Fig. 29-40) resembles that of mRNA splicing:

1. The 3'-OH group of the guanosine forms a phosphodiester bond with the intron's 5' end.

2. The 3'-terminal OH group of the newly liberated 5' exon forms a phosphodiester bond with the 5'-terminal phosphate of the 3' exon thereby splicing together the two exons and releasing the intron.

3. The 3'-terminal OH group of the intron forms a phosphodiester bond with the phosphate of the nucleotide 15 residues from the intron's 5' end, yielding the 5'-terminal fragment with the remainder of the intron in cyclic form.

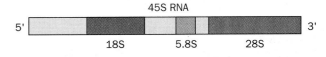

Figure 29-39
The organization of the 45S primary transcript of eukaryotic rRNA.

Figure 29-40
The sequence of reactions in the self-splicing of *Tetrahymena* pre-rRNA: **(1)** The 3'-OH group of a guanine nucleotide attacks the intron's 5'-terminal phosphate so as to form a phosphodiester bond and release the 5' exon. **(2)** The newly generated 3'-OH group of the 5' exon attacks the 5'-terminal phosphate of the 3' exon thereby splicing the two exons and releasing the intron. **(3)** The 3'-OH group of the intron attacks the phosphate of the nucleotide that is 15 residues from the 5' end so as to cyclize the intron and release its 5'-terminal fragment. Throughout this process, the RNA maintains a folded, internally hydrogen bonded conformation that permits the precise excision of the intron.

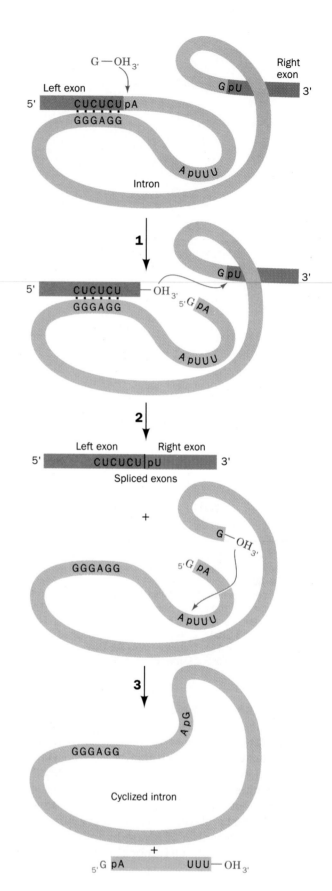

This self-splicing process, which similarly occurs in fungal mitochondrial rRNA, consists of a series of transesterifications and therefore does not require free energy input. Cech further established the enzymatic properties of the *Tetrahymena* intron, which presumably stem from its three-dimensional structure, by demonstrating that it catalyzes the *in vitro* cleavage of poly(C) with an enhancement factor of 10^{10} over the rate of spontaneous hydrolysis. Indeed, this RNA catalyst even exhibits Michaelis–Menton kinetics ($K_M = 42 \ \mu M$ and $k_{cat} = 0.033 \ \text{s}^{-1}$ for C_5). Such RNA enzymes have been named **ribozymes**.

Although the idea that an RNA can have enzymatic properties may be unorthodox, *there is no fundamental reason why an RNA, or any other macromolecule, cannot have catalytic activity.* Of course, in order to be an efficient catalyst, a macromolecule must be able to assume a stable structure but, as we shall see in Sections 30-2B and 3A, RNAs in the form of tRNA and most probably rRNA do just that. The chemical similarities of the mRNA and rRNA splicing reactions therefore suggest that splicosomes are ribozymal systems that evolved from primordial self-splicing RNAs and that their protein components merely serve to fine tune the ribozymes' structure and function. Similarly, the RNA components of ribosomes, which are more than one-half RNA and the rest protein, probably have catalytic functions in addition to the structural and recognition roles usually attributed to them (Section 30-3). Thus, the observations that cells contain batteries of enzymes for manipulating DNA but few for processing RNA, and that many coenzymes are ribonucleotides (e.g., ATP, NAD+, and CoA), led to the hypothesis that *RNAs were the original biological catalysts in precellular times and that the chemically more versatile proteins were relative latecomers in macromolecular evolution* (Section 1-4C).

C. Transfer RNA Processing

tRNAs, as we discuss in Section 30-2A, consist of ~80 nucleotides that assume a secondary structure with four base paired stems known as the **cloverleaf structure**

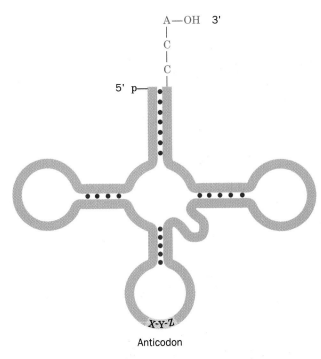

Figure 29-41
A schematic diagram of the tRNA cloverleaf secondary structure. Each dot indicates a base pair in the hydrogen bonded stems. The position of the anticodon triplet and the 3'-terminal —CCA are indicated.

(Fig. 29-41). All tRNAs have a large fraction of modified bases (whose structure, function, and synthesis is also considered in Section 30-2A) and each has the 3'-terminal sequence —CCA to which the corresponding amino acid is appended in the amino acid-charged tRNA. The **anticodon** (which is complementary to the codon specifying the tRNA's corresponding amino acid) occurs in the loop of the cloverleaf structure opposite the stem containing the terminal nucleotides.

The *E. coli* chromosome contains ~60 tRNA genes. Some of them are components of rRNA operons (Section 29-4A); the others are distributed, often in clusters, throughout the chromosome. The primary tRNA transcripts, which contain from one to as many as four or five identical tRNA species, have extra nucleotides at the 3' and 5' ends of each tRNA sequence. The excision and trimming of these tRNA sequences resembles that for *E. coli* rRNAs (Section 29-4B) in that both processes employ some of the same nucleases.

RNase P Is a Ribozyme

RNase P, which generates the 5' ends of all *E. coli* tRNAs (Fig. 29-37), is a particularly interesting enzyme because it has a 377-nucleotide RNA component (~125 kD vs 14 kD for its protein subunit) that is essential for enzymatic activity. The enzyme's RNA was, quite understandably, first proposed to function in recognizing the substrate RNA through base pairing and to thereby guide the protein subunit, which was presumed to be the actual nuclease, to the cleavage site. However, Sidney Altman has shown that *the RNA component of RNase P is, in fact, the enzyme's catalytic subunit* by demonstrating that protein-free RNase P RNA catalyzes the cleavage of substrate RNA at high salt concentrations. RNase P protein, which is basic, evidently functions at physiological salt concentrations to electrostatically reduce the repulsions between the polyanionic ribozyme and substrate RNAs. The argument that trace quantities of RNase P protein are really responsible for the RNase P reaction was disposed of by showing that catalytic activity is exhibited by RNase P RNA that has been transcribed in a cell-free system. Thus we now have two independent examples of ribozymes.

Many Eukaryotic Pre-tRNAs Have Introns

Eukaryotic genomes contain from several hundred to several thousand tRNA genes. Many eukaryotic primary tRNA transcripts, for example, yeast tRNA^Tyr (Fig. 29-42), contain a small intron adjacent to their anticodons as well as extra nucleotides at their 5' and 3' ends. Note that this intron is unlikely to disrupt the tRNA's cloverleaf structure. Eukaryotic tRNA transcripts lack the obligatory —CCA sequence at their 3' end. This is appended to the immature tRNAs by the enzyme **tRNA nucleotidyltransferase,** which sequentially adds two C's and an A to tRNA using CTP and ATP as substrates. This enzyme also occurs in prokaryotes although, at least in *E. coli,* the tRNA genes all encode a —CCA terminus. The *E. coli* tRNA nucleotidyltransferase is therefore thought to function in the repair of degraded tRNAs.

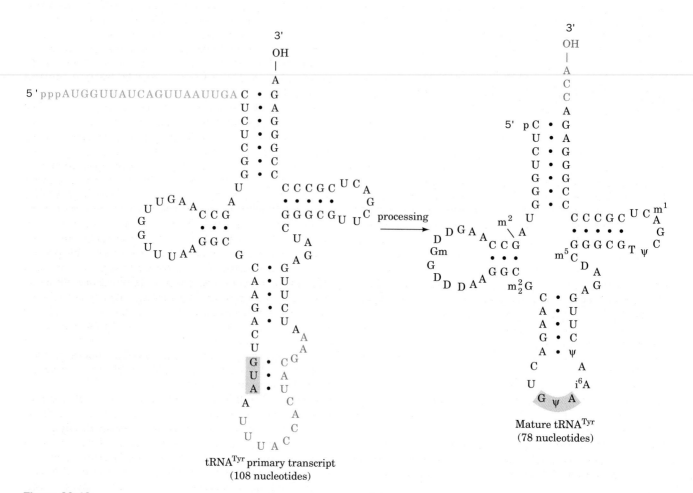

tRNA^Tyr primary transcript
(108 nucleotides)

Mature tRNA^Tyr
(78 nucleotides)

Figure 29-42
The post-transcriptional processing of yeast tRNA^Tyr. A 14-nucleotide intervening sequence and a 19-nucleotide 5'-terminal sequence are excised from the primary transcript, a —CCA is appended to the 3' end and several of the bases are modified (their symbols are defined in Fig. 30-13) to form the mature tRNA. The anticodon is shaded. [After DeRobertis, E. M. and Olsen, M. V., *Nature* **278,** 142 (1989).]

Chapter Summary

The central dogma of molecular biology states that "DNA makes RNA makes protein" (although RNA can also "make" DNA). There is, however, enormous variation among the rates that the various proteins are made. Certain enzymes, such as those of the *lac* operon, are synthesized only when the substances they metabolize are present. The *lac* operon consists of the control sequences *lacP* and *lacO* followed by the tandemly arranged genes for β-galactosidase *(lacZ)*, galactoside permease *(lacY)*, and thiogalactoside transacetylase *(lacA)*. In the absence of inducer, physiologically allolactose, the *lac* repressor, the product of the *lacI* gene, binds to operator *(lacO)* so as to prevent the transcription of the *lac* operon by RNA polymerase. The binding of inducer causes the repressor to release the operator that allows the *lac* structural genes to be transcribed onto a single polycistronic mRNA. The mRNAs transiently associate with ribosomes so as to direct them to synthesize the encoded polypeptides.

The holoenzyme of *E. coli* RNA polymerase has the subunit structure $\alpha_2 \beta \beta' \sigma$. It initiates transcription on the sense strand of a gene at a position designated by its promoter. The most conserved region of the promoter is the Pribnow box, which is centered at about the -10 position and has the consensus sequence TATAAT. The -35 region is also conserved in efficient promoters. Methylation protection studies indicate that holoenzyme forms an "open" initiation complex with the promoter. After the initiation of RNA synthesis, the σ subunit dissociates from the core enzyme, which then autonomously catalyzes chain elongation in the $5' \rightarrow 3'$ direction. RNA synthesis is terminated by a segment of the transcript that forms a G + C-rich hairpin with an oligo(U) tail that spontaneously dissociates from the DNA. Termination sequences that lack these sequences require the assistance of rho factor for proper chain termination. In the nuclei of eukaryotic cells, RNA polymerases I, II, and III, respectively, synthesize rRNA precursors, hnRNA, and tRNAs + 5S RNA. The minimal RNA polymerase I promoter extends between nucleotides -7 and $+6$. Many RNA polymerase II promoters contain a conserved TATAAAA sequence, the TATA box, located around position -27. Enhancers are transcriptional activators that can have variable positions and orientations relative to the transcription start site. RNA polymerase III promoters are located within the transcribed regions of their gene between positions $+40$ and $+80$.

Prokaryotes can respond rapidly to environmental changes, in part, because the translation of mRNAs commences during their transcription and because most mRNAs are degraded within 1 to 3 min of their synthesis. The temporally ordered expression of sets of genes in some bacteriophages is controlled by cascades of σ factors. The *lac* repressor is a tetrameric protein of identical subunits that, in the absence of inducer, nonspecifically binds to duplex DNA but binds much more tightly to *lac* promoter. The promoter sequence that *lac* repressor protects from nuclease digestion has nearly palindromic symmetry. Yet, methylation protection and mutational studies indicate that repressor is not symmetrically bound to promoter. Repressor and RNA polymerase compete for the same promoter-binding sites.

The presence of glucose represses the transcription of operons specifying certain catabolic enzymes through the mediation of cAMP. Upon binding cAMP, which is only formed in the absence of glucose, catabolite gene activator protein (CAP) binds to the promoters of certain operons, such as the *lac* operon, thereby activating their transcription. CAP's two symmetry equivalent DNA-binding domains each bind in the major groove of their target DNA via a helix–turn–helix motif that occurs in numerous prokaryotic repressors. The binding between these repressors and their target DNAs is mediated by mutually favorable associations between these macromolecules rather than any specific interactions between base pairs and amino acid side chains analogous to Watson–Crick base pairing. *araBAD* transcription is controlled by CAP–cAMP and AraC through a remarkable complex of AraC to two-binding sites, *araO_2* and *araI*, that forms a DNA loop. In this system, AraC also regulates its own synthesis by binding to the *araO_1* site so as to repress the transcription of the *araC* gene. The expression of the *E. coli trp* operon is regulated both by attenuation as well as repression. Upon binding tryptophan, its corepressor, *trp* repressor binds to the *trp* operator thereby blocking *trp* operon transcription. When tryptophan is available, much of the *trp* transcript that has escaped repression is prematurely terminated in the *trpL* sequence because its transcript contains a segment that forms a normal terminator structure. When tryptophanyl–tRNA^{Trp} is scarce, ribosomes stall at the transcript's two tandem Trp codons. This permits the newly synthesized RNA to form a base paired stem and loop that prevents the formation of the terminator structure. Several other operons are similarly regulated by attenuation. The stringent response is another mechanism by which *E. coli* match the rate of transcription to charged tRNA availability. When a specified charged tRNA is scarce, stringent factor on active ribosomes synthesizes ppGpp, which inhibits the transcription of rRNA and some mRNAs while stimulating the transcription of other mRNAs.

Prokaryotic mRNA transcripts require no additional processing. However, eukaryotic mRNAs have an enzymatically appended 5' cap and, in most cases, an enzymatically generated poly(A) tail. Moreover, the introns of eukaryotic mRNA primary transcripts (hnRNAs) are precisely excised and their flanking exons are spliced together to form mature mRNAs in a snRNP-mediated process that takes place in spliceosomes. The primary transcript of *E. coli* rRNAs contains all three rRNAs together with some tRNAs. These are excised and trimmed by specific endonucleases and exonucleases. The rRNAs are also modified by the methylation of specific nucleosides. The eukaryotic 18S, 5.8S, and 28S rRNAs are similarly transcribed as a 45S precursor which is processed in a manner resembling that of *E. coli* rRNAs. The intron of *Tetrahymena* pre-rRNA is removed in an RNA-catalyzed self-splicing reaction. Prokaryotic tRNAs are excised from their primary transcripts and trimmed in much the same manner as rRNAs. In RNase P, one of the enzymes mediating this process, the catalytic subunit is an RNA. Eukaryotic tRNA transcripts also require the excision of a short intron and the enzymatic addition of a 3'-terminal —CCA to form the mature tRNA.

References

General

Lewin, B., *Genes* (3rd ed.), Chapters 8–11 and 22–24, Wiley (1987).

Watson, J. D., Hopkins, N. H., Roberts, J. W., Steitz, J. A., and Weiner, A. M., *Molecular Biology of the Gene* (4th ed.), Chapters 16 and 20, Benjamin/Cummings (1987).

The Genetic Role of RNA

Brenner, S., Jacob, F., and Meselson, M., An unstable intermediate carrying information from genes to ribosomes for protein synthesis, *Nature* **190**, 576–581 (1960). [The experimental verification of mRNA's existence.]

Brachet, J., Reminiscences about nucleic acid cytochemistry and biochemistry, *Trends Biochem. Sci.* **12**, 244–246 (1987).

Crick, F., Central dogma of molecular biology, *Nature* **227**, 561–563 (1970).

Hall, B. D. and Spiegelman, S., Sequence complementarity of T2-DNA and T2-specific RNA, *Proc. Natl. Acad. Sci.* **47**, 137–146 (1964). [The first use of RNA–DNA hybridization.]

Jacob, F. and Monod, J., Genetic regulatory mechanisms in the synthesis of proteins, *J. Mol. Biol.* **3**, 318–356 (1961). [The classic paper postulating the existence of mRNA and operons and explaining how the transcription of operons is regulated.]

Spiegelman, S., Hybrid nucleic acids, *Sci. Am.* **210**(5): 48–56 (1964).

RNA Polymerase and mRNA

Adhya, S. and Gottesman, M., Control of transcription termination, *Annu. Rev. Biochem.* **47**, 967–996 (1978).

Bear, D. G. and Peabody, D. S., The *E. coli* rho protein: an ATPase that terminates transcription, *Trends Biochem. Sci.* **13**, 343–347 (1988).

Berman, H. M. and Young, P. R., The interaction of intercalulating drugs with nucleic adids, *Annu. Rev. Biophys. Bioeng.* **10**, 87–114 (1981).

Chamberlain, M. J., Bacterial DNA-dependent RNA polymerases, *in* Boyer, P. D. (Ed.), *The Enzymes* (3rd ed.), Vol. 15, *pp.* 61–86, Academic Press (1982).

Futcher, B., Supercoiling and transcription, or vice versa? *Trends Genet.* **4**, 271–272 (1988).

Gannan, F., O'Hare, K., Perrin, F., LePennec, J. P., Benoist, C., Cochet, M., Breathnach, R., Royal, A., Garapin, A., Cami, B., and Chambon, P., Organization and sequences of the 5' end of a cloned complete ovalbumin gene, *Nature* **278**, 428–434 (1979).

Gale, E. F., Cundliffe, E., Reynolds, P. E., Richmond, M. H., and Waring, M. J., *The Molecular Basis of Antibiotic Action* (2nd ed.), Chapter 5, Wiley (1981).

Geiduschek, E. P. and Tocchini-Valentini, G. P., Transcription by RNA polymerase III, *Annu. Rev. Biochem.* **57**, 873–914 (1988).

Hansen, U. and Sharp, P. A., Transcription by RNA polymerase II, *in* Fraenkel-Conrat, H. and Wagner, R. R. (Eds.), *Comprehensive Virology*, Vol. 19, *pp.* 65–97, Plenum Press (1984).

Khoury, G. and Gruss, P., Enhancer elements, *Cell* **33**, 313–314 (1983).

Lewis, M. K. and Burgess, R. R., Eukaryotic RNA polymerases, *in* Boyer, P. D. (Ed.), *The Enzymes* (3rd ed.), Vol. 15, *pp.* 110–153, Academic Press (1982).

Losick, R. and Chamberlain, M. (Eds.), *RNA Polymerase*, Cold Spring Harbor Laboratory (1976). [A series of authorative articles on various aspects of RNA polymerase.]

McKnight, S. L. and Kingsbury, R., Transcriptional control signals of a eukaryotic protein-coding gene, *Science* **217**, 316–324 (1982).

Paule, M. R. Comparitive subunit composition of eukaryotic nuclear RNA polymerases, *Trends Biochem. Sci.* **6**, 128–131 (1981).

Platt, T., Transcription termination and the regulation of gene expression, *Annu. Rev. Biochem.* **55**, 339–372 (1986).

Pribnow, D., Genetic control signals in DNA, *in* Goldberger, R. F. (Ed.), *Biological Regulation and Development*, Vol. 1, *pp.* 217–277, Plenum Press (1979).

Rosenberg, M. and Court, D., Regulatory sequences involved in the promotion and termination of RNA transcription, *Annu. Rev. Genet.* **13**, 319–353 (1979).

Sakonju, S., Bogenhagen, D. F., and Brown, D. D., A control region in the center of the 5S RNA gene directs specific initiation of transcription: I. The 5' border of the region; *and* II. The 3' border of the region, *Cell* **19**, 13–25, 27–35 (1980).

Sentenac, A., Eukaryotic RNA polymerases, *CRC Crit. Rev. Biochem.* **18**, 31–90 (1985).

Siebenlist, U., RNA polymerase unwinds an 11-base pair segment of phage T7 promoter, *Nature* **279**, 651–652 (1979).

Sollner-Webb, B., Wilkinson, J. A. K., Roan, J., and Reeder, R. H., Nested control regions promote Xenopus ribosomal RNA synthesis by RNA polymerase I, *Cell* **35**, 199–206 (1983).

Wang, A. H.-J., Ughetto, G., Quigley, G. J., and Rich, A., Interactions between an anthracycline antibiotic and DNA: molecular structure of daunomycin complexed to d(CpGpTpApCpG) at 1.2 Å resolution, *Biochemistry* **26**, 1152–1163 (1987).

Control of Transcription

Aggarwal, A. K., Rogers, D. W., Drottar, M., Ptashne, M., and Harrison, S. C., Recognition of the DNA operator by the repressor of phage 434: a view at high resolution, *Science* **242**, 899–907 (1988); *and* Anderson, J. E., Ptashne, M., and Harrison, S. C., The structure of the repressor–operator complex of bacteriophage 434, *Nature* **326**, 846–852 (1987).

Brennan, R. G. and Matthews, B. W., The helix–turn–helix DNA binding motif, *J. Biol. Chem.* **264**, 1903–1906 (1989).

de Crombrugghe, B., Busby, S., and Buc, H., Cyclic AMP receptor protein: role in transcription activation, *Science* **244**, 831–838 (1984).

Dickson, R. C., Abelson, J., Barnes, W. M., and Reznikoff, W. S., Genetic regulation: The lac control region, *Science* **187**, 27–35 (1975).

Dunn, T. M., Hahn, S., Ogden, S., and Schleif, R. F., An operator at −280 base pairs that is required for repression of *araBAD* operon promoter: addition of DNA helical turns between the operator and promoter cyclically hinders repression, *Proc. Natl. Acad. Sci.* **81**, 5017–5020 (1984).

Friedman, D. I., Imperiale, M. J., and Adhya, S. L., RNA 3′ end formation in the control of gene expression, *Annu. Rev. Genet.* **21**, 453–488 (1987).

Gallant, J. A., Stringent control in *E. coli*, *Annu. Rev. Genet.* **13**, 393–415 (1979).

Gilbert, W. and Müller-Hill, B., Isolation of the lac repressor, *Proc. Natl. Acad. Sci.* **56**, 1891–1898 (1966).

Gralla, J.D., Specific repression in the *lac* repressor — the 1988 version, *in* Gralla, J.D. (Ed.), *DNA – Protein Interactions in Transcription*, pp. 3–10, Liss (1989).

Helmann, J. D. and Chamberlin, M. J., Structure and function of bacterial sigma factors, *Annu. Rev. Biochem.* **57**, 839–872 (1988).

Kolter, R. and Yanofsky, C., Attenuation in amino acid biosynthetic operons, *Annu. Rev. Genet.* **16**, 113–134 (1982).

Lamond, A. I. and Travers, A. A., Stringent control of bacterial transcription, *Cell* **41**, 6–8 (1985).

Losick, R. and Pero. J., Cascades of sigma factors, *Cell* **25**, 582–584 (1981).

McClure, W. R., Mechanism and control of transcription initiation in prokaryotes, *Annu. Rev. Biochem.* **54**, 171–204 (1985).

Miller, J. H. and Reznikoff, W. S. (Eds.), *The Operon*, Cold Spring Harbor Laboratory (1978). [An informative collection of reviews on the *lac* operon as well as other operons.]

Otwinowski, Z., Schevitz, R. W., Zhang, R. -G., Lawson, C. L., Joachimiak, A., Marmorstein, R. Q., Luisi, B. F., and Sigler, P. B., Crystal structure of *trp* repressor/operator complex at atomic resolution, *Nature* **335**, 321–329 (1988).

Pabo C. O. and Sauer, R. T., Protein – DNA recognition, *Annu. Rev. Biochem.* **53**, 293–321 (1984).

Pastan, I. and Adhya, S., Cyclic adenosine 5′-monophosphate in *Escherichia coli*, *Bacteriol. Rev.* **40**, 527–551 (1976).

Reznikoff, W. S., Siegele, D. A., Cowing, D. W., and Gross, C. A., The regulation of transcription initiation in bacteria, *Annu. Rev. Genet.* **19**, 355–387 (1985).

Raibaud, O. and Schwartz, O., Positive control of transcription in bacteria, *Annu. Rev. Genet.* **18**, 173–206 (1984).

Steitz, T. A., Ohlendorf, D. H., McKay, D. B., Anderson, W. F., and Matthews, B. W., Structural similarity in the DNA-binding domains of catabolite gene activator and *cro* repressor proteins, *Proc. Natl. Acad. Sci.* **79**, 3097–3100 (1982).

von Hippel, P. H., Bear, D. G., Morgan, W. D., and McSwiggen, J. A., Protein-nucleic acid interactions in transcription: a molecular analysis, *Annu. Rev. Biochem.* **53**, 389–346 (1984).

Weber, I. T. and Steitz, T. A., Model of specific complex between catabolite gene activator protein and B-DNA suggested by electrostatic complementarity, *Proc. Natl. Acad. Sci.* **81**, 3973–3977 (1984).

Wolberger, C., Dong, Y., Ptashne, M., and Harrison, S. C., Structure of phage 434 Cro/DNA complex, *Nature* **335**, 789–795 (1988).

Yanofsky, C., Transcription attenutation, *J. Biol. Chem.* **263**, 609–612 (1988).

Yanofsky, C., Attenuation in the control of expression of bacterial operons, *Nature* **289**, 751–758 (1981).

Post-Transcriptional Processing

Altman, S., Baer, M., Guerrier-Takada, C., and Vioque, A., Enzymatic cleavage of RNA by RNA, *Trends Biochem. Sci.* **11**, 515–518 (1986). [Discusses RNase P.]

Banerjee, A. K., 5′-Terminal cap structure in eukaryotic messenger ribonucleic acids, *Microbiol. Rev.* **44**, 175–205 (1980).

Birnstiel, M. L., Busslinger, M., and Strub, K., Transcription termination and 3′ processing: the end is in sight! *Cell* **41**, 349–359 (1985).

Boyer, P. D. (Ed.), *The Enzymes* (3rd ed.), Vol. 15, Academic Press (1982). [Contains articles on RNA processing enzymes.]

Breathnach, R. and Chambon, P., Organization and expression of eucaryotic split genes coding for proteins, *Annu. Rev. Biochem.* **50**, 349–383 (1981).

Cech, T. R., The chemistry of self-splicing RNA and RNA enzymes, *Science* **236**, 1532–1539 (1987).

Cech, T. R. and Bass, B. L., Biological catalysis by RNA, *Annu. Rev. Biochem.* **55**, 599–629 (1986).

Cech, T. R., RNA as an enzyme, *Sci. Am.* **255**(5): 64–75 (1986).

Chambon, P., Split genes, *Sci. Am.* **244**(5): 60–71 (1981).

Darnell, J. E., Jr., RNA, *Sci. Am.* **253**(4): 68–78 (1985).

Darnell, J. E., Jr., The processing of RNA, *Sci. Am.* **249**(4): 90–100 (1983).

Deutscher, M. P., The metabolic role of RNases, *Trends Biochem. Sci.* **13**, 137–138 (1988).

Gegenheimer, P. and Apiron, D., Processing of prokaryotic ribonucleic acid, *Microbiol. Rev.* **45**, 502–541 (1981).

Guthrie, C. and Patterson, B., Spliceosomal snRNAs, *Annu. Rev. Genet.* **22**, 387–419 (1988).

Humphrey, T. and Proudfoot, N. J., A beginning to the biochemistry of polyadenylation, *Trends Genet.* **4**, 243–245 (1988).

Littauer, U. Z. and Soreq, H., The regulatory function of poly(A) and adjacent 3′ sequences in translated RNA, *Prog. Nucleic Acid Res. Biol.* **27**, 53–83 (1982).

Maniatis, T. and Reed, R., The role of small ribonucleoprotein particles in pre-mRNA splicing, *Nature* **325**, 673–678 (1987).

Mattaj, I. W., snRNAs: from gene architecture to RNA processing, *Trends Biochem. Sci.* **9**, 435–437 (1984).

Mowry, K. L. and Steitz, J. A., snRNP mediators of 3′ end processing: functional fossils?, *Trends Biochem. Sci.* **13**, 447–451 (1988).

Ogden, R. C., Knapp, G., Peebles, C. L., Johnson, J., and Abelson, J., The mechanism of tRNA splicing, *Trends Biochem. Sci.* **6**, 154–158 (1981).

Padgett, R. A., Grabowski, P. J., Konarska, M. M., Seiler, S., and Sharp, P. A., Splicing of messenger RNA precursors, *Annu. Rev. Biochem.* **55**, 1119–1150 (1986).

Proudfoot, N. J., How RNA polymerase II terminates transcription in higher eukaryotes, *Trends Biochem. Sci.* **14**, 105–110 (1989).

Sharp, P. A., Splicing of messenger RNA precursors, *Science* **235**, 766–771 (1987).

Söll, D., Abelson, J. N., and Schimmel, P. R. (Eds.), *Transfer RNA: Biological Aspects*, Cold Spring Harbor Laboratory (1980). [Contains several articles on the processing of tRNA and rRNA.]

Steitz, J. A., "Snurps", *Sci. Am.* **258**(6): 56–63 (1988).

Problems

1. Indicate the phenotypes of the following *E. coli lac* partial diploids in terms of inducibility and active enzymes synthesized.
   (a) $I^-P^+O^+Z^+Y^-/I^+P^-O^+Z^+Y^+$
   (b) $I^-P^+O^c Z^+Y^-/I^+P^+O^+Z^-Y^+$
   (c) $I^-P^+O^c Z^+Y^+/I^-P^+O^+Z^+Y^+$
   (d) $I^+P^-O^c Z^+Y^+/I^-P^+O^c Z^-Y^-$

2. **Superrepressed** mutants, I^s, encode *lac* repressors that bind operator but do not respond to the presence of inducer. Indicate the phenotypes of the following genotypes in terms of inducibility and enzyme production.
   (a) $I^sO^+Z^+$ (b) $I^sO^cZ^+$ (c) $I^+O^+Z^+/I^sO^+Z^+$

*3. Why do *lacZ⁻ E. coli* fail to show galactoside permease activity after the addition of lactose in the absence of glucose? Why do *lac Y⁻* mutants lack β-galactosidase activity under the same conditions?

4. What is the experimental advantage of using IPTG instead of 1,6-allolactose as an inducer of the *lac* operon?

5. Indicate the Pribnow box, -35 region and initiating nucleotide on the antisense strand of the *E. coli* tRNA^Tyr promoter shown below.

   5′ CAACGTAACACTTTACAGCGGCGCGTCATTTGATATGATGCGCCCCGCTTCCCGATA 3′

   3′ GTTGCATTGTGAAATGTCGCCGCGCAGTAAACTATACTACGCGGGGCGAAGGGCTAT 5′

*6. Why are *E. coli* that are diploid for rifamycin resistance and rifamycin sensitivity *(rif^R/rif^S)* sensitive to rifamycin?

7. What is the probability that the 4026-nucleotide DNA sequence coding for the β subunit of *E. coli* RNA polymerase will be transcribed with the correct base sequence. Perform the calculations for the probabilities of 0.0001, 0.001, and 0.01 that each base is incorrectly transcribed.

8. What is the probability that the symmetry of the *lac* operator is merely accidental?

9. Why does the inhibition of DNA gyrase in *E. coli* inhibit the expression of catabolite sensitive operons?

10. Describe the transcription of the *trp* operon in the absence of active ribosomes and tryptophan.

11. Why can't eukaryotic transcription be regulated by attenuation?

12. Charles Yanofsky and his associates have synthesized a 15-nucleotide RNA that is complementary to segment 1 of *trpL* mRNA (but only partially complementary to segment 3). What is its effect on the *in vitro* transcription of *trp* operon? What is its effect if the *trpL* gene contains a mutation in segment 2 that destablizes the $2 \cdot 3$ stem and loop?

13. Why are *relA⁻* mutants defective in the *in vivo* transcription of the *his* and *trp* operons?

14. Why aren't primary rRNA transcripts observed in wild-type *E. coli?*

TRANSLATION

In this chapter we consider **translation,** the mRNA-directed biosynthesis of polypeptides. Although peptide bond formation is a relatively simple reaction, the complexity of the translational process, which involves the coordinated participation of over 100 macromolecules, is mandated by the need to link 20 different amino acid residues accurately in the order specified by a particular mRNA.

We begin by considering the **genetic code,** the correspondence between nucleic acid sequences and polypeptide sequences. Next, we examine the structures and properties of **tRNAs,** the amino acid-bearing entities that mediate the translation process. Following this, we take up what is known about **ribosomes,** the complex molecular machines that catalyze peptide bond formation between the mRNA-specified amino acids. Peptide bond formation, however, does not necessarily yield a functional protein; many polypeptides must first be post-translationally modified as we discuss in the subsequent section. We then study how cells degrade proteins, a process that must balance protein synthesis, and finally, consider the nonribosomal synthesis of certain small and unusual polypeptides.

1. THE GENETIC CODE

How does DNA encode genetic information? According to the one gene–one polypeptide hypothesis, the genetic message dictates the amino acid sequences of proteins. Since the base sequence of DNA is the only variable element in this otherwise monotonously repeating polymer, the amino acid sequence of a protein must somehow be specified by the base sequence of the corresponding segment of DNA.

A DNA base sequence might specify an amino acid sequence in many conceivable ways. With only 4 bases to code for 20 amino acids, a group of several bases, termed a **codon,** is necessary to specify a single amino acid. A triplet code, that is, one with 3 bases per codon, is minimally required since there are $4^3 = 64$ different triplets of bases whereas there can be only $4^2 = 16$ different doublets, which is insufficient to specify all the amino acids. In a triplet code, as many as 44 codons might not code for amino acids. On the other hand, many amino acids could be specified by more than one codon. Such a code, in a term borrowed from mathematics, is said to be **degenerate.**

Another mystery was, how does the polypeptide synthesizing apparatus group DNA's continuous sequence of bases into codons? For example, the code might be overlapping; that is, in the sequence

ABCDEFGHIJ $\cdot\cdot\cdot$

ABC might code for one amino acid, BCD for a second, CDE for a third, *etc.* Alternatively, the code might be nonoverlapping so that ABC specifies one amino acid, DEF a second, HIJ a third, *etc.* The code might also contain internal "punctuation" such as in the nonoverlapping triplet code

ABC,DEF,GHI, $\cdot\cdot\cdot$

in which the commas represent particular bases or base sequences. A related question is how does the genetic code specify the beginning and the end of a polypeptide chain.

The genetic code is, in fact, a nonoverlapping, comma-free, degenerate, triplet code. How this was determined and how the genetic code dictionary was elucidated is the subject of this section.

A. Chemical Mutagenesis

The triplet character of the genetic code, as we shall see below, was established through the use of **chemical mutagens,** substances that induce mutations. We therefore precede our study of the genetic code with a discussion of these substances. There are two major classes of mutations:

1. **Point mutations,** in which one base pair replaces another. These are subclassified as:

    (a) **Transitions,** in which one purine (or pyrimidine) is replaced by another.
    (b) **Transversions,** in which a purine is replaced by a pyrimidine or *vice versa.*

2. **Insertion/deletion mutations,** in which one or more nucleotide pairs are inserted in or deleted from DNA.

A mutation in any of these three categories may be reversed by a subsequent mutation of the same but not another category.

Point Mutations Are Generated by Altered Bases

Point mutations can result from the treatment of an organism with base analogs or substances that chemically alter bases. For example, the base analog **5-bromouracil (5BU)** sterically resembles thymine (5-methyluracil) but, through the influence of its electronegative Br atom, frequently assumes a tautomeric form that base pairs with guanine instead of adenine (Fig. 30-1). Consequently, when 5BU is incorporated into DNA in place of thymine, as it usually is, it occasionally induces an $A \cdot T \rightarrow G \cdot C$ transition in subsequent rounds of DNA replication. Occasionally, 5BU is also incorporated into DNA in place of cytosine, which instead generates a $G \cdot C \rightarrow A \cdot T$ transition.

The adenine analog **2-aminopurine (2AP),** normally base pairs with thymine (Fig. 30-2a) but occasionally forms an undistorted but singly hydrogen bonded base pair with cytosine (Fig. 30-2b). Thus 2AP also generates $A \cdot T \rightarrow G \cdot C$ and $G \cdot C \rightarrow A \cdot T$ transitions.

In aqueous solutions, **nitrous acid** (HNO_2) oxidatively deaminates aromatic primary amines so that it converts cytosine to uracil (Fig. 30-3a) and adenine to the guanine-like **hypoxanthine** (which forms two of guanine's three hydrogen bonds with cytosine; Fig.

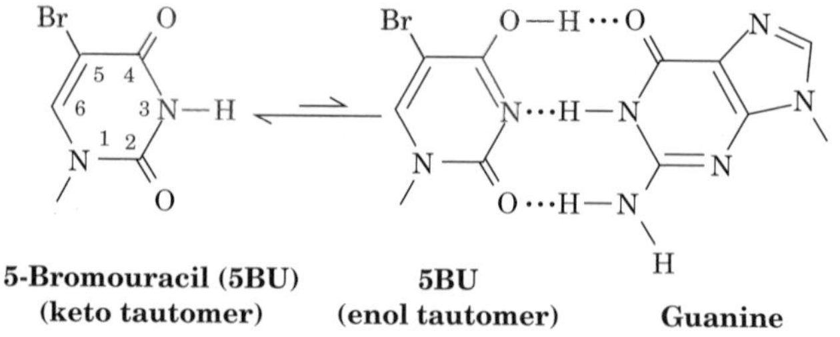

5-Bromouracil (5BU)   **5BU**
(keto tautomer)   **(enol tautomer)**   **Guanine**

Figure 30-1
The keto form of 5-bromouracil (*left*) is its most common tautomer. However, it frequently assumes the enol form (*right*), which base pairs with guanine.

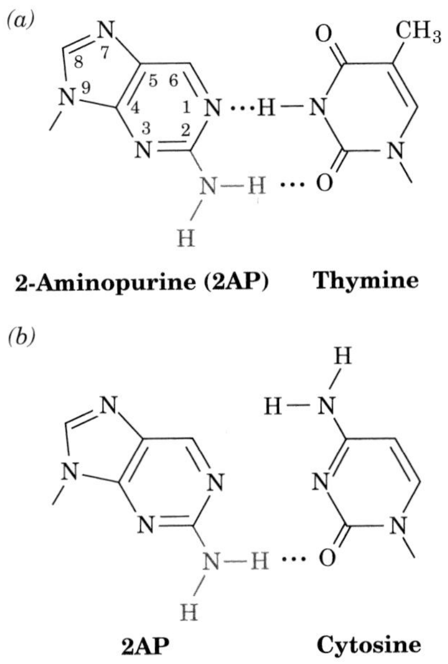

(a)

2-Aminopurine (2AP)   Thymine

(b)

2AP   Cytosine

Figure 30-2
The adenine analog 2-aminopurine normally base pairs with (a) thymine but occasionally also does so with (b) cytosine.

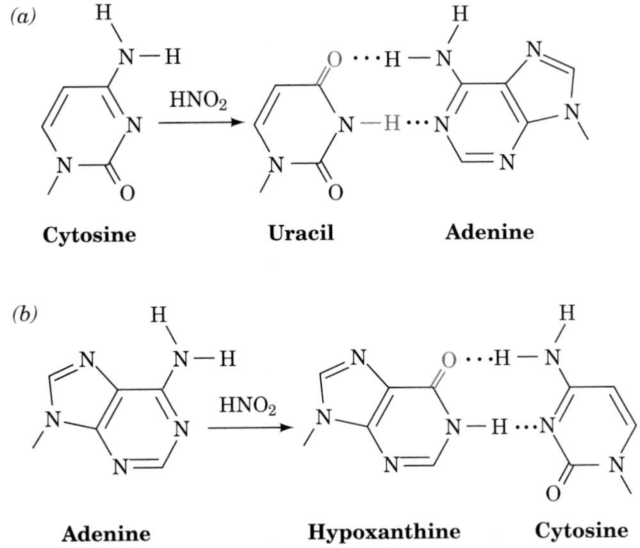

(a)

Cytosine   Uracil   Adenine

(b)

Adenine   Hypoxanthine   Cytosine

Figure 30-3
Reaction with nitrous acid converts (a) cytosine to uracil which base pairs with adenine; and (b) adenine to hypoxanthine, a guanine derivative (it lacks guanine's 2-amino group) which base pairs with cytosine.

30-3b). Hence, treatment of DNA with nitrous acid, or compounds such as **nitrosamines**

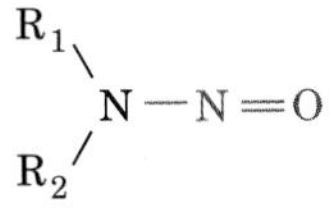

Nitrosoamines

that react to form nitrous acid, results in both $A \cdot T \rightarrow G \cdot C$ and $G \cdot C \rightarrow A \cdot T$ transitions.

> **Nitrite,** the conjugate base of nitrous acid, has long been used as a preservative of prepared meats such as frankfurters. However, the observation that many mutagens are also carcinogens (Section 31-5E) suggests that the consumption of nitrite-containing meat is harmful to humans. Proponents of nitrite preservation nevertheless argue that to stop it would result in far more fatalities. This is because lack of such treatment would greatly increase the incidence of **botulism,** an often fatal form of food poisoning caused by the ingestion of protein neurotoxins secreted by the anaerobic bacterium *Clostridium botulinum* (Section 34-4C).

Hydroxylamine (NH_2OH) also induces $G \cdot C \rightarrow A \cdot T$ transitions by specifically reacting with cytosine to convert it to a compound that base pairs with adenine (Fig. 30-4). The use of alkylating agents such as dimethyl sulfate, **nitrogen mustard,** and **ethylnitrosourea**

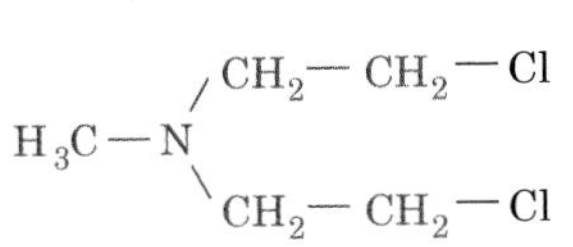

Nitrogen mustard 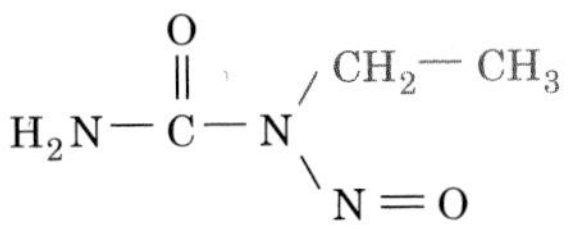   Ethylnitrosourea

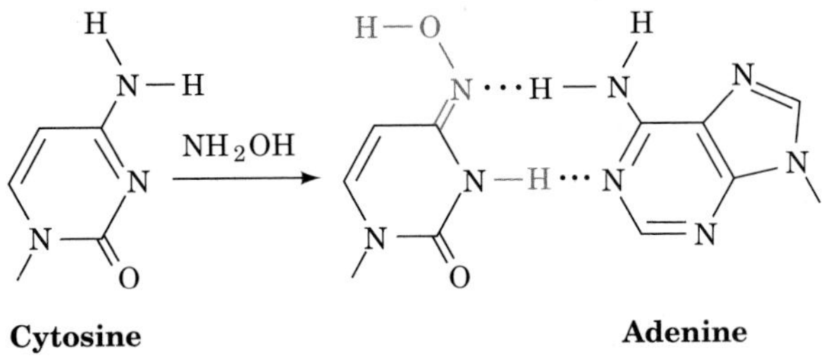

Cytosine   Adenine

Figure 30-4
Reaction with hydroxylamine converts cytosine to a derivative which base pairs with adenine.

often generates transversions. The alkylation of the N(7) position of a purine nucleotide causes its subsequent depurination in a reaction similar to that diagrammed in Fig. 28-52a. The resulting gap in the sequence is filled in by an error-prone enzymatic repair system (Section 31-5B). Transversions arise when the missing purine is replaced by a pyrimidine. The enzymatic repair of DNA that has been damaged by UV radiation may also generate transversions.

Insertion/Deletion Mutations Are Generated by Intercalating Agents

Insertion/deletion mutations may arise from the treatment of DNA with intercalating agents such as acridine orange or proflavin (Section 28-4C). The distance between two consecutive base pairs is doubled by the intercalation of such a molecule between them. The replication of such a distorted DNA occasionally results in the insertion or deletion of one or more nucleotides in the newly synthesized polynucleotide. (Insertions and deletions of large DNA segments generally arise from aberrant cross-over events; Section 33-2C.)

B. Codons Are Triplets

In 1961, Francis Crick and Sydney Brenner, through genetic invesitgations into the previously unknown character of proflavin-induced mutations, determined the triplet character of the genetic code. In bacteriophage T4, a particular proflavin-induced mutation, designated *FC0*, maps in the *rIIB* cistron (Section 27-1E). The growth of this mutant phage on a permissive host (*E. coli* B) resulted in the occasional spontaneous appearance of phenotypically wild-type phages as was demonstrated by their ability to grow on a restrictive host [*E. coli* K12(λ); recall that *rIIB* mutants form characteristically large plaques on *E. coli* B but cannot lyse *E. coli* K12(λ)]. Yet, these doubly mutated phages are not genotypically wild-type; the simultaneous infection of a permissive host by one of them and true wild-type phage yielded recombinant progeny that have either the *FC0* mutation or a new mutation designated *FC1*. Thus the phenotypically wild-type phage is a double mutant that actually contains both *FC0* and *FC1*. *These two genes are therefore* **suppressors** *of one another; that is, they cancel each other's mutant properties.* Furthermore, since they map together in the *rIIB* cistron, they are mutual **intragenic suppressors** (suppressors in the same gene).

The treatment of *FC1* in a manner identical to that described for *FC0* provided similar results: the appearance of a new mutant, *FC2*, that is an intragenic suppressor of *FC1*. By proceeding in this iterative manner, Crick and Brenner collected a series of different *rIIB* mutants, *FC3, FC4, FC5, etc.*, in which each mutant *FC(n)* is an intragenic suppressor of its predecessor, *FC(n − 1)*. Recombination studies showed, moreover, that odd numbered mutations are intragenic suppressors of even numbered mutations but neither pairs of different odd numbered mutations nor pairs of different even numbered mutations suppress each other. However, recombinants containing three odd-numbered mutations or three even-numbered mutations all are phenotypically wild-type.

Crick and Brenner accounted for these observations by the following set of assumptions:

1. The proflavin-induced mutation *FC0* is either an insertion or a deletion of one nucleotide pair from the *rIIB* cistron. If it is a deletion then *FC1* is an insertion, *FC2* is a deletion, *etc.*, and vice versa.

2. *The code is read in a sequential manner starting from a fixed point in the gene.* The insertion or deletion of a nucleotide shifts the **frame** (grouping) in which succeeding nucleotides are read as codons (insertions or deletions of nucleotides are therefore also known as

frameshift mutations). Thus the code has no internal punctuation that indicates the reading frame; that is, *the code is comma-free.*

3. *The code is a triplet code.*

4. All or nearly all of the 64 triplet codons code for an amino acid; that is, *the code is degenerate.*

These principles are illustrated by the following analogy. Consider a sentence (gene) in which the words (codons) each consist of 3 letters (bases).

THE   BIG   RED   FOX   ATE   THE   EGG

(Here the spaces separating the words have no physical significance; they are only present to indicate the reading frame.) The deletion of the 4th letter, which shifts the reading frame, changes the sentence to

THE   IGR   EDF   OXA   TET   HEE   GG

so that all words past the point of deletion are unintelligible (specify the wrong amino acids). An insertion of any letter, however, say an X in the 9th position,

THE   IGR   EDX   FOX   ATE   THE   EGG

restores the original reading frame. Consequently, only the words between the two changes (mutations) are altered. As in this example, such a sentence might still be intelligible (the gene could still specify a functional protein), particularly if the changes are close together. Two deletions or two insertions, no matter how close together, would not suppress each other but just shift the reading frame. However, three insertions, say X, Y, and Z in the 5th, 8th, and 12th positions, respectively, would change the sentence to

THE   BXI   GYR   EDZ   FOX   ATE   THE   EGG

which, after the third insertion, restores the original reading frame. The same would be true of three deletions. As before, if all three changes were close together, the sentence might still retain its meaning.

Crick and Brenner did not unambiguously demonstrate that the genetic code is a triplet code because they had no proof that their insertions and deletions involved only single nucleotides. Strictly speaking, they showed that a codon consists of *3r* nucleotides where *r* is the number of nucleotides in an insertion or deletion. Although it was generally assumed at the time that $r = 1$, proof of this assertion had to await the elucidation of the genetic code (Section 30-1D).

C. Genes Are Colinear with Their Specified Polypeptides

In the early 1960s, Charles Yanofsky demonstrated the colinearity of genes and polypeptides. He did so by isolating a number of mutants of the 268-residue α

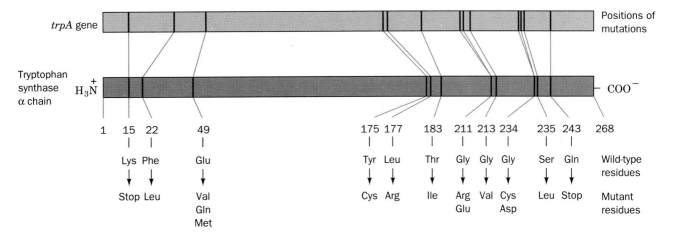

Figure 30-5
The colinearity of the *E. coli trpA* gene with the polypeptide it specifies, the tryptophan synthase α chain. The gene mutation positions, as determined by transductional mapping, have the same order as the corresponding amino acid changes in the polypeptide as determined by fingerprinting.

chain of *E. coli* **tryptophan synthase** (specified by the *trpA* gene; Section 29-3E). The genetic map of these mutants was elucidated by transductional mapping (Section 27-1E) and the amino acid changes to which they give rise were established by fingerprinting (Section 6-1J). The order of the mutants in the gene is the same as the order of the corresponding amino acid changes in the protein (Fig. 30-5). *The E. coli trpA gene is therefore colinear with the polypeptide it specifies.*

D. Deciphering the Genetic Code

In order to understand how the genetic code dictionary was elucidated we must first preview how proteins are synthesized. The mRNAs cannot directly recognize amino acids. Rather, *they specifically bind molecules of tRNA that each carry a corresponding amino acid* (Fig. 30-6). Each tRNA contains a trinucleotide sequence, its **anticodon,** which is complementary to an mRNA codon specifying the tRNA's amino acid. An amino acid is covalently linked to its corresponding tRNA through the action of a specific enzyme that recognizes both of these molecules (a process called "charging" the tRNA). During translation, the mRNA passes through the ribosome such that each codon, in turn, binds its corresponding charged tRNA (Fig. 30-7). As this occurs, the ribosome transfers the tRNA's appended amino acid to the end of the growing polypeptide chain.

UUU Specifies Phe

The genetic code could, in principle, be determined by simply comparing the base sequence of an mRNA with the amino acid sequence of the polypeptide it specifies. In the 1960s, however, techniques for isolating and sequencing mRNAs had not yet been developed. The elucidation of the genetic code dictionary therefore proved to be a difficult task.

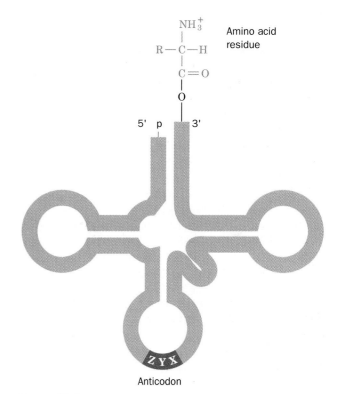

Figure 30-6
Transfer RNA in its "cloverleaf" form showing its covalently linked amino acid residue (*top*) and its anticodon (*bottom*; a trinucleotide segment that base pairs with the complementary mRNA codon during translation).

The major breakthrough in deciphering the genetic code came in 1961 when Marshall Nirenberg and Heinrich Matthaei established that UUU is the codon specifying Phe. They did so by demonstrating that the addition of poly(U) to a cell-free protein synthesizing system stimulates only the synthesis of poly(Phe). The

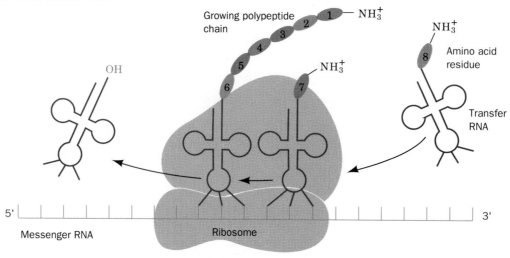

Figure 30-7

A schematic diagram of the processes of translation (ribosomal synthesis of a polypeptide from an mRNA template).

direction of ribosome movement on mRNA

cell-free protein synthesizing system was prepared by gently breaking open *E. coli* cells by grinding them with powdered alumina and centrifuging the resulting cell sap to remove the cell walls and membranes. This extract contained DNA, mRNA, ribosomes, enzymes, and other cell constituents necessary for protein synthesis. When fortified with ATP, GTP, and amino acids, the system synthesized small amounts of proteins. This was demonstrated by the incubation of the system with ^{14}C-labeled amino acids followed by the precipitation of its proteins by the addition of trichloroacetic acid. The precipitate proved to be radioactive.

A cell-free protein synthesizing system, of course, produces proteins specified by the cell's DNA. Upon addition of DNase, however, protein synthesis stops within a few minutes because the system can no longer synthesize mRNA while that originally present is rapidly degraded. Nirenberg found that crude mRNA-containing fractions from other organisms were highly active in stimulating protein synthesis in a DNase-treated protein synthesizing system. This system, is likewise responsive to synthetic mRNAs.

The synthetic mRNAs that Nirenberg used in subsequent experiments were synthesized by the *Azotobacter vinelandii* enzyme **polynucleotide phosphorylase**. This enzyme, which was discovered by Severo Ochoa and Marianne Grunberg-Manago, links together nucleotides in the reaction

$$(RNA)_n + NDP \rightleftharpoons (RNA)_{n+1} + P_i$$

where NDP represents a ribonucleoside diphosphate. In contrast to RNA polymerase, however, polynucleotide phosphorylase does not utilize a template. Rather, it randomly links together the available NDPs so that the base composition of the product RNA reflects that of the reactant NDP mixture.

Nirenberg and Matthaei demonstrated that poly(U) stimulates the synthesis of poly(Phe) by incubating poly(U) and a mixture of 1 radioactive and 19 unlabeled amino acids in a DNase treated protein synthesizing system. Significant radioactivity appeared in the protein precipitate only when phenylalanine was labeled. *UUU must therefore be the codon specifying Phe.* In similar experiments using poly(A) and poly(C), it was found that poly(Lys) and poly(Pro), respectively, were synthesized. Thus AAA *specifies* Lys *and* CCC *specifies* Pro. [Poly(G) cannot function as a synthetic mRNA because, even under denaturing conditions, it aggregates to form what is thought to be a four-stranded helix. A mRNA must be single stranded to direct its transcription; Section 30-2D.]

Nirenberg and Ochoa independently employed ribonucleotide copolymers to further elucidate the genetic code. For example, in a poly(UG) composed of 76% U and 24% G, the probability of a given triplet being UUU is $0.76 \times 0.76 \times 0.76 = 0.44$. Likewise, the probability of a triplet consisting of 2U's and 1G; that is, UUG, UGU, or GUU is $0.76 \times 0.76 \times 0.24 = 0.14$. The use of the poly(UG) as a mRNA therefore indicated the base compositions, but not the sequences of the codons specifying several amino acids (Table 30-1). Through the use of copolymers containing 2, 3, and 4 bases, the base compositions of codons specifying each of the 20 amino acids were inferred. Moreover, *these experiments demonstrated that the genetic code is degenerate since, for example, poly(UA), poly(UC), and poly(UG) all direct the incorporation of Leu into a polypeptide.*

The Genetic Code Was Elucidated through Triplet Binding Assays and the Use of Polyribonucleotides with Known Sequences

In the absence of GTP, which is necessary for protein synthesis, trinucleotides but not dinucleotides are almost as effective as mRNAs in promoting the ribosomal

Table 30-1

Amino Acid Incorporation Stimulated by a Random Copolymer of U and G in Mole Ratio 0.76:0.24

Codon	Probability of Occurrence	Relative Incidence[a]	Amino Acid	Relative Amount of Amino Acid Incorporated
UUU	0.44	100	Phe	100
UUG	0.14	32	Leu	36
UGU	0.14	32	Cys	35
GUU	0.14	32	Val	37
UGG	0.04	9	Trp	14
GUG	0.04	9		
GGU	0.04	9	Gly	12
GGG	0.01	2		

[a] Relative incidence is defined here as 100 × probability of occurrence/0.44.

Source: Matthaei, J. H., Jones, O. W., Martin, R. G., and Nirenberg, M., *Proc. Natl. Acad. Sci.* **48**, 666 (1962).

binding of specific tRNAs. This phenomenon, which Nirenberg and Philip Leder discovered in 1964, permitted the various codons to be identified by a simple binding assay. Ribosomes, together with their bound tRNAs, are retained by a nitrocellulose filter but free tRNA is not. The bound tRNA was identified by using charged tRNA mixtures in which only one of the pendent amino acid residues was radioactively labeled. For instance, it was found, as expected, that UUU stimulates the ribosomal binding of only Phe tRNA. Likewise, UUG, UGU, and GUU stimulate the binding of Leu, Cys, and Val tRNAs, respectively. Hence UUG, UGU, and GUU must be codons that specify Leu, Cys, and Val, respectively. In this way, the amino acids specified by some 50 codons were identified. For the remaining codons, the binding assay was either negative (no tRNA bound) or ambiguous.

The genetic code dictionary was completed and previous results confirmed through H. Gobind Khorana's synthesis of polyribonucleotides with specified repeating sequences. In a cell-free protein synthesizing system, UCUCUCUC · · ·, for example, is read

UCU CUC UCU CUC UCU C···

so that it specifies a polypeptide chain of two alternating amino acid residues. In fact, it was observed that this mRNA stimulated the production of

Ser — Leu — Ser — Leu — Ser — Leu — ···

which indicates that either UCU or CUC specifies Ser and the other specifies Leu. This information, together with the tRNA-binding data, permitted the conclusion that UCU codes for Ser and CUC codes for Leu. These data also proved that codons consist of an odd number

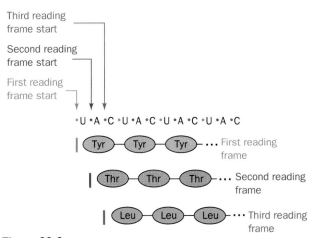

Figure 30-8
An mRNA might be read in any of three reading frames, each of which yields a different polypeptide.

of nucleotides thereby relieving any residual suspicions that codons consist of six rather than three nucleotides.

Alternating sequences of three nucleotides, such as poly(UAC), specify three different homopolypeptides because ribosomes may initiate polypeptide synthesis on these synthetic mRNAs in any of the three possible reading frames (Fig. 30-8). Analyses of the polypeptides specified by various alternating sequences of two and three nucleotides confirmed the identity of many codons and filled out missing portions of the genetic code.

mRNAs Are Read in the 5' → 3' Direction

The use of repeating tetranucleotides indicated the reading direction of the code and identified the chain termination codons. Poly(UAUC) specifies, as expected, a polypeptide with a tetrapeptide repeat:

$$5' \text{ UAU CUA UCU AUC UAU CUA } \cdots \, 3'$$

Tyr — Leu — Ser — Ile — Tyr — Leu — ···

The amino acid sequence of this polypeptide indicates that the mRNA's 5' end corresponds to the polypeptide's N-terminus; that is, *the mRNA is read in the 5' → 3' direction.*

UAG, UAA, and UGA Are Stop Codons

In contrast to the above results, poly(AUAG) yields only dipeptides and tripeptides. This is because *UAG is a signal to the ribosome to terminate protein synthesis:*

AUA GAU AGA UAG AUA GAU ···

Ile — Asp — Arg Stop Ile — Asp — ···

Likewise, poly(GUAA) yields dipeptides and tripeptides because UAA is also a chain termination signal:

GUA AGU AAG UAA GUA AGU ···

Val — Ser — Lys Stop Val — Ser — ···

UGA is a third stop signal. These stop codons, whose existence was first inferred from genetic experiments, are known, somewhat inappropriately, as **nonsense codons** because they are the only codons that do not specify amino acids. UAG, UAA, and UGA are often referred to as *amber, ochre,* and *opal* codons. [They were so named as the result of a laboratory joke: The German word for amber is Bernstein, the name of an individual who helped discover *amber* mutations (mutations that change some other codon to UAG); *ochre* and *opal* are puns on *amber.*]

AUG and GUG Are Chain Initiation Codons

The codons AUG, and less frequently GUG, form part of the chain initiation sequence (Section 30-3C). However, they also specify the amino acid residues Met and Val, respectively, at internal positions of polypeptide chains. (Nirenberg and Matthaei's discovery that UUU specifies Phe was only possible because ribosomes indiscriminately initiate polypeptide synthesis on a mRNA when the Mg^{2+} concentration is unphysiologically high as it was, serendipitously, in their experiments.)

E. The Nature of the Code

The genetic code dictionary, as elucidated by the above methods, is presented in Table 30-2. Examination of this table indicates that the genetic code has several remarkable features:

1. *The code is highly degenerate.* Three amino acids, Arg, Leu, and Ser are each specified by six codons, and most of the rest are specified by either four, three, or two codons. Only Met and Trp are represented by a single codon. Codons that specify the same amino acid are termed **synonyms.**

2. *The arrangement of the code table is nonrandom.* Most synonyms occupy the same box in Table 30-2; that is, they differ only in their third nucleotide. The only exceptions are Arg, Leu, and Ser which, having six codons each, must occupy more than one box. XYU and XYC always specify the same amino acid; XYA and XYG do so in all but two cases. Moreover, changes in the first codon position tend to specify similar (if not the same) amino acids, whereas codons with second position pyrimidines encode mostly hydrophobic amino acids, and those with second position purines encode mostly polar amino acids. Apparently *the code evolved so as to minimize the deleterious effects of mutations.*

Many of the mutations causing amino acid substitutions in a protein can be rationalized, according to the genetic code, as a single point mutation. For instance, all but one of the amino acid substitutions indicated in Fig. 30-5 for the α chain of tryptophan synthase result from

Table 30-2
The "Standard" Genetic Code

First position (5' end)	Second position				Third position (3' end)
	U	**C**	**A**	**G**	
U	UUU ⎱ Phe ⎰ UUC UUA ⎱ Leu ⎰ UUG	UCU UCC ⎱ Ser UCA ⎰ UCG	UAU ⎱ Tyr ⎰ UAC UAA Stop UAG Stop	UGU ⎱ Cys ⎰ UGC UGA Stop UGG Trp	U C A G
C	CUU CUC ⎱ Leu CUA ⎰ CUG	CCU CCC ⎱ Pro CCA ⎰ CCG	CAU ⎱ His ⎰ CAC CAA ⎱ Gln ⎰ CAG	CGU CGC ⎱ Arg CGA ⎰ CGG	U C A G
A	AUU AUC ⎱ Ile AUA ⎰ AUG Met[a]	ACU ACC ⎱ Thr ACA ⎰ ACG	AAU ⎱ Asn ⎰ AAC AAA ⎱ Lys ⎰ AAG	AGU ⎱ Ser ⎰ AGC AGA ⎱ Arg ⎰ AGG	U C A G
G	GUU GUC ⎱ Val GUA ⎰ GUG	GCU GCC ⎱ Ala GCA ⎰ GCG	GAU ⎱ Asp ⎰ GAC GAA ⎱ Glu ⎰ GAG	GGU GGC ⎱ Gly GGA ⎰ GGG	U C A G

[a] AUG forms part of the initiation signal as well as coding for internal Met residues.

single base changes. *As a consequence of the genetic code's degeneracy, however, many point mutations at a third codon position are phenotypically silent; that is, the mutated codon specifies the same amino acid as the wild-type.* Degeneracy may account for as much as 33% of the 25 to 75% range in the G + C content among the DNAs of different organisms (Section 28-1). The frequent occurrence of Arg, Ala, Gly, and Pro also tends to give a high G + C content, whereas Asn, Ile, Lys, Met, Phe, and Tyr contribute to a low G + C content.

Some Phage DNA Segments Contain Overlapping Genes in Different Reading Frames

Since any nucleotide sequence may have three reading frames, it is possible, at least in principle, for a polynucleotide to encode two or even three different polypeptides. This idea was never seriously entertained, however, because it seemed that the constraints on even two overlapping genes in different reading frames would be too great for them to evolve so that both could specify sensible proteins. It therefore came as a great surprise, in 1976, when Frederick Sanger reported that the DNA of bacteriophage φX174 contains two genes that are completely contained within larger genes of different reading frames (Fig. 30-9). Moreover, the end of the overlapping D and E genes contains the control

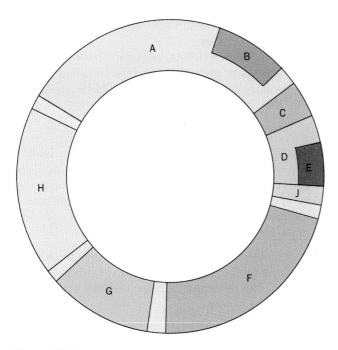

Figure 30-9
The genetic map of bacteriophage φX174 as determined by DNA sequence analysis. Genes are labeled A, B, C, *etc.* Note that gene B is wholly contained within gene A and gene E is wholly contained within gene D. These pairs of genes are read in different reading frames and therefore specify unrelated proteins. The unlabeled regions correspond to untranslated control sequences.

Table 30-3

Mitochondrial Deviations from the "Standard" Genetic Code

Mitochondrion	UGA	AUA	CUN[a]	AG^A_C	CGG
Mammalian	Trp	Met[b]		Stop	
Baker's yeast	Trp	Met[b]	Thr		?
Neurospora crassa	Trp				?
Drosophila	Trp	Met[b]		Ser[c]	
Protozoan	Trp				
Plant					Trp
"Standard" code	Stop	Ile	Leu	Arg	Arg

[a] N represents any of the four nucleotides.
[b] Also acts as part of an initiation signal.
[c] AGA only; no AGG codons occur in *Drosophila* mitochondrial DNA.

Source: Breitenberger, C. A. and RajBhandary, U. L., *Trends Biochem. Sci.* **10**, 481 (1985).

sequence for the ribosomal initiation of the J gene so that this short DNA segment performs triple duty. Bacteria also exhibit such coding economy; the ribosomal initiation sequence of one gene in a polycistronic mRNA often overlaps the end of the preceding gene. Nevertheless, completely overlapping genes have only been found in small single-stranded DNA phages, which presumably must make maximal use of the little DNA that they can pack inside their capsids.

The "Standard" Genetic Code Is Widespread but Not Universal

For many years it was thought that the "standard" genetic code (that given in Table 30-2) is universal. This assumption was, in part, based on the observations that one kind of organism (e.g., *E. coli*), can accurately translate the genes from quite different organisms, (e.g., humans). This phenomenon is, in fact, the basis of genetic engineering. Once the "standard" genetic code had been established, presumably during the time of prebiotic evolution (Section 1-4B), any mutation that would alter the way the code is translated would result in numerous, mostly deleterious, protein sequence changes. Undoubtedly there is strong selection against such mutations. DNA sequencing studies in 1981 nevertheless revealed that *the genetic codes of certain mitochondria (mitochondria contain their own genes and protein* synthesizing systems that produce 10 to 20 mitochondrial proteins) are variants of the "standard" genetic code (Table 30-3). For example, in mammalian mitochondria, AUA, as well as the standard AUG, is a Met/initiation codon, UGA specifies Trp rather than "Stop," and AGA and AGG are "Stop" rather than Arg. Note that all mitochondrial genetic codes except those of plants simplify the "standard" code by increasing its degeneracy. For example, in the mammalian mitochondrial code, each amino acid is specified by at least two codons that differ only in their third nucleotide. Apparently the constraints preventing alterations of the genetic code are eased by the small sizes of mitochondrial genomes. More recent studies, however, have revealed that in ciliated protozoa, the codons UAA and UAG specify Gln rather than "Stop." Perhaps UAA and UAG were sufficiently rare codons in a primordial ciliate (which molecular phylogenetic studies indicate branched off very early in eukaryotic evolution) to permit the code change without unacceptable deleterious effects. At any rate, *the "standard" genetic code, although very widely utilized, is not universal.*

2. TRANSFER RNA

The establishment of the genetic function of DNA led to the realization that cells somehow "translate" the language of base sequences into the language of polypeptides. Yet, nucleic acids do not specifically bind amino acids. In 1955, Francis Crick, in what became known as the **adaptor hypothesis**, hypothesized that translation occurs through the mediation of "adaptor" molecules. Each adaptor was postulated to carry a specific enzymatically appended amino acid and to recog-

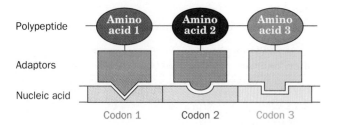

Figure 30-10
The adaptor hypothesis postulates that the genetic code is read by molecules that recognize a particular codon and carry the corresponding amino acid.

nize the corresponding codon (Fig. 30-10). Crick suggested that these adaptors contain RNA because codon recognition could then occur by complementary base pairing. At about this time, Paul Zamecnik and Mahlon Hoagland discovered that in the course of protein synthesis, ^{14}C-labeled amino acids became transiently bound to a low molecular mass fraction of RNA. Further investigations indicated that these RNAs, which at first were called "soluble RNA" or "sRNA" but are now known as transfer RNA (tRNA), are, in fact, Crick's putative adaptor molecules.

A. PRIMARY AND SECONDARY STRUCTURES

In 1965, after a seven year effort, Robert Holley reported the first known base sequence of a biologically significant nucleic acid, that of yeast alanine tRNA (tRNA^{Ala}; Fig. 30-11). To do so Holley had to overcome several major obstacles:

1. All organisms contain many species of tRNAs (at least one for each of the 20 amino acids) which, because of their nearly identical properties (see below), are not easily separated. Preparative techniques had to be developed to provide the gram or so of pure yeast tRNA^{Ala} Holley required for its sequence determination.

2. Holley had to invent the methods that were initially used to sequence RNA (Section 28-6).

3. Ten of the 76 bases of yeast tRNA^{Ala} are modified (see below). Their structural formulas had to be elucidated although they were never available in more than milligram quantities.

Since 1965, the techniques for tRNA purification and sequencing have vastly improved. A tRNA may now be sequenced in a few days time with only ~1 μg of material. Presently, the base sequences of ~300 tRNAs from a great variety of organisms are known (many from their corresponding DNA sequences). They vary in length

from 60 to 95 nucleotides (18–28 kD) although most have ~76 nucleotides.

Almost all known tRNAs, as Holley first recognized, may be schematically arranged in the so-called cloverleaf secondary structure (Fig. 30-12). Starting from the 5' end, they have the following common features:

1. A 5' terminal phosphate group.

2. A 7 bp stem that includes the 5'-terminal nucleotide and which may contain non-Watson–Crick base pairs such as G·U. This assembly is known as the **acceptor** or **amino acid stem** because the amino acid residue carried by the tRNA is appended to its 3'-terminal OH group (Section 30-2C).

3. A 3 or 4 bp stem ending in a loop that frequently contains the modified base **dihydrouridine** (D; see below). This stem and loop are therefore collectively termed the **D arm**.

4. A 5 bp stem ending in a loop that contains the **anticodon**, the triplet of bases that is complementary to the codon specifying the tRNA. These features are known as the **anticodon arm**.

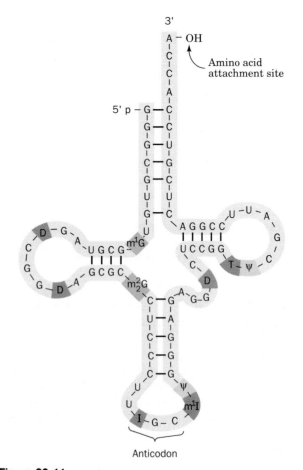

Figure 30-11
The base sequence of yeast tRNA^{Ala} drawn in the cloverleaf form. The symbols for the modified nucleosides (*color*), are explained in Fig. 30-13.

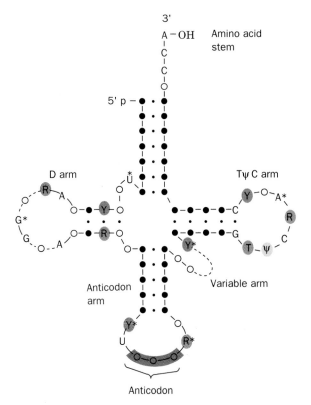

Figure 30-12

The cloverleaf secondary structure of tRNA. Filled circles connected by dots represent Watson–Crick base pairs and open circles in the double helical regions indicate bases involved in non-Watson–Crick base pairing. Invariant positions are indicated: R and Y represent invariant purines and pyrimidines, respectively, ψ signifies pseudouracil. The starred nucleosides are often modified. The dashed regions in the D and variable arms contain different numbers of nucleotides in the various tRNAs.

5. A 5 bp stem ending in a loop that usually contains the sequence TψC (where ψ is the symbol for **pseudouridine;** see below). This assembly is called the **TψC or T arm.**

6. All tRNAs terminate in the sequence CCA with a free 3'-OH group. The —CCA may be genetically specified or enzymatically appended to immature tRNA (Section 29-4C).

7. There are 13 invariant positions (always have the same base) and 8 **semiinvariant** positions (only a purine or only a pyrimidine) that occur mostly in the loop regions. These regions also contain **correlated invariants;** that is, pairs of nonstem nucleotides that are base paired in all tRNAs. The purine on the 3' side of the anticodon is invariably modified. The structural significance of these features is examined in Section 30-2B.

The site of greatest variability among the known tRNAs occurs in the so-called **variable arm.** It has from 3 to 21 nucleotides and may have a stem consisting of up to 7 bp. The D loop also varies in length from 5 to 7 nucleotides.

tRNAs Have Numerous Modified Bases

One of the most striking characteristics of tRNAs is their large proportion, up to 20%, of post-translationally modified or hypermodified bases. A few of the >50 such bases, together with their standard abbreviations, are indicated in Fig. 30-13. Hypermodified nucleosides, such as i^6A, are usually adjacent to the anticodon's 3' nucleotide when it is A or U. Their low polarities probably strengthen the otherwise relatively weak pairing associations of these bases with the codon thereby increasing translational fidelity. Conversely, certain methylations block base pairing and hence prevent inappropriate structures from forming. Yet, neither of these modifications are essential for maintaining a tRNA's structural integrity (see below), for its proper binding to the ribosome, nor, with one known exception (Section 30-2C), for binding the enzyme that attaches the correct amino acid. The functions of most modified bases therefore remain unknown although mutant bacteria unable to form certain modified bases compete poorly against the corresponding normal bacteria.

B. Tertiary Structure

The earliest physicochemical investigations of tRNA indicated that it has a well-defined conformation. Yet, despite numerous hydrodynamic, spectroscopic, and chemical cross-linking studies, its three-dimensional structure remained an enigma until 1974. In that year, the 2.5-Å resolution X-ray crystal structure of yeast tRNA[Phe] was separately elucidated by Alexander Rich in collaboration with Sung Hou Kim and, in a different crystal form, by Aaron Klug. *The molecule assumes an L-shaped conformation in which one leg of the L is formed by the acceptor and T stems folded into a continuous A-RNA-like double helix (Section 28-2B) and the other leg is similarly composed of the D and anticodon stems* (Fig. 30-14). Each leg of the L is ~ 60 Å long and the anticodon and amino acid acceptor sites are at opposite ends of the molecule, some 76-Å apart. The narrow 20 to 25-Å width of native tRNA is essential to its biological function: During protein synthesis, two RNA molecules must simultaneously bind in close proximity at adjacent codons on mRNA (Section 30-3D).

tRNA's Complex Tertiary Structure Is Maintained by Hydrogen Bonding and Stacking Interactions

The structural complexity of yeast tRNA[Phe] is reminiscent of that of a protein. Although only 42 of its 76 bases occur in double helical stems, *71 of them participate in stacking associations* (Fig. 30-15). The structure also contains 9 base pairing interactions that cross-link its tertiary structure (Figs. 30-14a and 30-15). Remarkably, all but one of these tertiary interactions, which appear to be

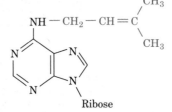

Uracil derivatives

Pseudouridine (ψ)   Dihydrouridine (D)   Ribothymidine (T)   4-Thiouridine (s⁴U)

Cytosine derivatives

3-Methylcytidine (m³C) N^4-Acetylcytidine (ac⁴C)   Lysidine (L)

Adenine derivatives

1-Methyladenosine (m¹A) N^6-Isopentenyladenosine (i⁶A)   Inosine (I)

Guanine derivatives

N^7-Methylguanosine (m⁷G) N^2,N^2-Dimethylguanosine (m₂²G)   R = H   Wyosine (Wyo)

R = CH₂CH₂CH(COCH₃)₂   **Y**

Figure 30-13

A selection of the modified nucleosides that occur in tRNAs together with their standard abbreviations. Note that although inosine chemically resembles guanosine, it is biochemically derived from adenosine. Nucleosides may also be methylated at their ribose 2′ positions to form residues symbolized, for instance, by Cm, Gm, and Um.

(a)

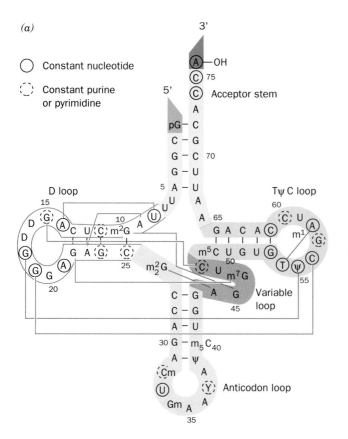

○ Constant nucleotide

◌ Constant purine or pyrimidine

Figure 30-14

The structure of yeast tRNA^Phe. (*a*) The base sequence drawn in cloverleaf form. Base pairing interactions are represented by thin red lines connecting the participating bases. Bases that are conserved or semiconserved in all tRNAs are circled by solid and dashed lines, respectively. The 5′ terminus is colored bright green, the acceptor stem is yellow, the D arm is white, the anticodon arm is light green,

(b)

the variable arm is orange, the TψC arm is light blue, and the 3′ terminus is red. (*b*) The X-ray structure drawn to show how its base paired stems are arranged form the L-shaped molecule. The sugar–phosphate backbone is represented by a ribbon with the same color scheme as that in Part a. [Courtesy of Michael Carson, University of Alabama at Birmingham.]

the mainstays of the molecular structure, are non-Watson–Crick associations. Moreover, most of the bases involved in these interactions are either invariant or semiinvariant, which strongly suggests that all tRNAs have similar conformations (see below). The structure is also stabilized by several unusual hydrogen bonds between bases and either phosphate groups or the 2′-OH groups of ribose residues.

The compact structure of yeast tRNA^Phe results from its large number of intramolecular associations, which renders most of its bases inaccessible to solvent. The most notable exceptions to this are the anticodon bases and those of the amino acid-bearing —CCA terminus. No doubt both of these groupings must be accessible in order to carry out their biological functions.

The observation that the molecular structures of yeast tRNA^Phe in two different crystal forms are essentially identical lends much credence to the supposition that its crystal structure closely resembles its solution structure. Transfer RNAs other than yeast tRNA^Phe have, unfortunately, been notoriously difficult to crystallize. The

crystal structures of only three other native species of tRNA, all at resolutions of 3.0 Å or greater, have thus far been reported. The molecular structures of these tRNAs closely resemble that of yeast tRNA^Phe. The major structural differences among them result from an apparent flexibility in the anticodon loop and the —CCA terminus as well as from a hingelike mobility between the two legs of the L that gives, for instance, yeast tRNA^Asp a boomeranglike shape. Such observations are in accord with the expectation that all tRNAs fit into the same ribosomal cavities.

C. Aminoacyl-tRNA Synthetases

Accurate translation requires two equally important recognition steps: (1) the choice of the correct amino acid for covalent attachment to a tRNA; and (2) the selection of the amino acid-charged tRNA specified by mRNA. The first of these steps, which is catalyzed by amino acid-specific enzymes known as **aminoacyl-tRNA synthetases,** appends an amino acid to the 3′-terminal ribose residue of its cognate tRNA to form an **aminoacyl-tRNA** (Fig.

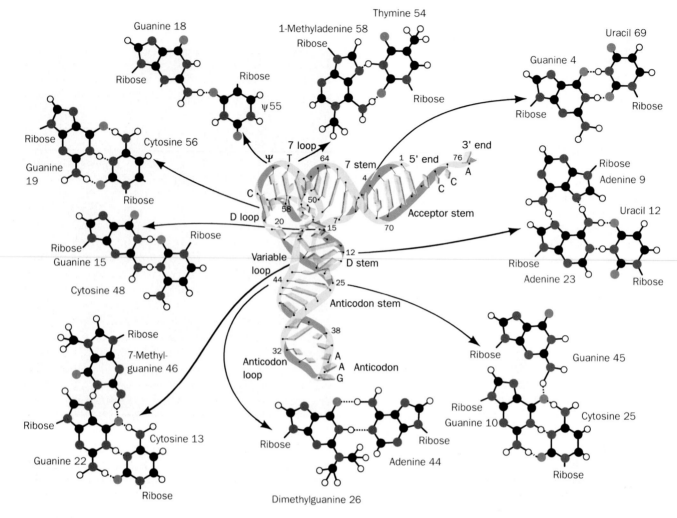

Figure 30-15
The nine tertiary base pairing interactions in yeast tRNA[Phe]. Note that all but one involve non-Watson–Crick pairs and that they are all located near the corner of the L. [After Kim,

S. H., *in* Schimmel, P. R., Söll, D., and Abelson, J. N. (Eds.), *Transfer RNA: Structure, Properties and Recognition, p. 87,* Cold Spring Harbor Laboratory (1979). Drawing of tRNA copyrighted © by Irving Geis.]

30-16). This otherwise unfavorable process is driven by the hydrolysis of ATP in two sequential reactions that are catalyzed by a single enzyme.

1. The amino acid is first "activated" by reaction with ATP to form an **aminoacyl-adenylate**:

$$
\begin{array}{c}
\text{R} - \underset{\underset{\text{NH}_3^+}{|}}{\overset{\overset{\text{H}}{|}}{\text{C}}} - \overset{\overset{\text{O}}{\parallel}}{\text{C}} \overset{\diagup}{\underset{\diagdown}{\,}} {}^{\text{O}}_{\text{O}^-} \quad + \quad \text{ATP}
\end{array}
$$

Amino acid

$$
\begin{array}{c}
\text{R} - \underset{\underset{\text{NH}_3^+}{|}}{\overset{\overset{\text{H}}{|}}{\text{C}}} - \overset{\overset{\text{O}}{\parallel}}{\text{C}} - \text{O} - \underset{\underset{\text{O}^-}{|}}{\overset{\overset{\text{O}}{\parallel}}{\text{P}}} - \text{O} - \text{Ribose} - \text{Adenine} + \text{PP}_i
\end{array}
$$

**Aminoacyl-adenylate
(Aminoacyl-AMP)**

Aminoacyl-tRNA

Figure 30-16
In aminoacyl-tRNAs, the amino acid residue is esterified to the tRNA's 3′-terminal nucleoside at either its 3′-OH group, as shown here, or its 2′-OH group.

which, with most aminoacyl-tRNA synthetases in the absence of tRNA, may be isolated although it normally remains tightly bound to the enzyme.

2. This mixed anhydride then reacts with tRNA to form the aminoacyl-tRNA:

Aminoacyl-AMP + tRNA $\rightleftharpoons$
$$\text{aminoacyl-tRNA} + \text{AMP}$$

Some aminoacyl-tRNA synthetases exclusively append an amino acid to the terminal 2′-OH group of their cognate tRNAs, others do so at the 3′-OH group, and yet others do so at either such position. This selectivity or its absence was established with the use of chemically modified tRNAs that lack either the 2′- or 3′-OH group of their 3′-terminal ribose residue. The use of these derivatives was necessary because, in solution, the aminoacyl group rapidly equilibrates between the 2′ and 3′ positions.

The overall aminoacylation reaction is

Amino acid + tRNA + ATP $\rightleftharpoons$
$$\text{aminoacyl-tRNA} + \text{AMP} + \text{PP}_i$$

These reaction steps are readily reversible because the free energies of hydrolysis of the bonds formed in both the aminoacyl-adenylate and the aminoacyl-tRNA are comparable to that of ATP hydrolysis. The overall reaction is driven to completion by the inorganic pyrophosphatase-catalyzed hydrolysis of the PP$_i$ generated in the first reaction step. Amino acid activation therefore chemically resembles fatty acid activation (Section 23-2A); the major difference between these two processes, which were both elucidated by Paul Berg, is that tRNA is the acyl acceptor in amino acid activation, whereas CoA performs this function in fatty acid activation.

Different Aminoacyl-tRNA Synthetases Are No More Than Distantly Related

Cells must have at least one aminoacyl-tRNA synthetase for each of the 20 amino acids. The similarity of the reactions catalyzed by these enzymes and the structural resemblance of all tRNAs suggests that all aminoacyl-tRNA synthetases evolved from a common ancestor and should therefore be structurally related. This is not the case, however. In fact, *the aminoacyl-tRNA synthetases form a diverse group of enzymes.* The over 100 such enzymes that have been characterized each have one of four different types of subunit structures, α, α_2, α_4, and $\alpha_2\beta_2$, with subunit sizes ranging from 334 to >1000 residues. Moreover, although synthetases specific for a given amino acid exhibit considerable sequence homology from organism to organism, there is little sequence similarity among synthetases specific for different amino acids. Quite possibly, aminoacyl-tRNA synthetases arose very early in evolution, before the development of the modern protein synthesis apparatus other than tRNAs.

Tyrosyl-tRNA Synthetase Operates via Transition State Binding

The X-ray structure of tyrosyl-tRNA synthetase from *Bacillus stearothermophilus,* determined by David Blow, is illustrated in Fig. 30-17. The 419-residue subunit of this α_2 dimer contains a region of β sheet reminiscent of the dinucleotide-binding fold (Section 7-3B), which forms the tyrosyl adenylate-binding site. This region is remarkably similar in structure to the ATP-binding region of *E. coli* methionyl-tRNA synthetase, the only other aminoacyl-tRNA synthetase of known structure. The C-terminal 99 residues of tyrosyl-tRNA synthetase, as well as three other short segments of its polypeptide chain, are not visible in the crystal structure and are therefore presumed to be conformationally disordered. Each of these segments has several Lys and Arg residues that are implicated in the binding of the polyanionic tRNA molecule. Indeed, the N-terminal 320 residues alone, as generated via protein engineering, catalyzes tyrosine adenylate formation with unchanged k_{cat} and K_M, but neither aminoacylates nor binds tRNA^{Tyr}. Most nucleic acid-binding proteins of known structure, as we shall see, have conformationally mobile regions that interact with their corresponding nucleic acid. It has therefore been suggested that these disordered regions function to bind their nuclei acid through flexible interactions.

Although the way in which tyrosyl-tRNA synthetase interacts with tRNA^{Tyr} remains obscure, model building coupled with protein engineering studies have revealed how this enzyme adenylylates tyrosine. Chemical stud-

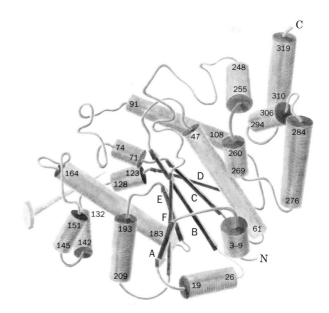

Figure 30-17
The X-ray structure of residues 1 to 320 of tyrosyl-tRNA synthetase. The position of the molecular twofold axis of this dimeric protein is indicated on the lower left. [After Blow, D. M. and Brick, P., *in* Jurnak, F. A. and McPherson, A., *Biological Macromolecules and Assembly*, Vol 2: *Nucleic Acids and Interactive Proteins*, p. 448, Wiley (1985).]

ies have demonstrated that this reaction proceeds via inversion of configuration at ATP's α phosphorus. This observation implies that the reaction involves a single displacement in which the tyrosyl carboxylate group is the nucleophile and PP_i is the leaving group (Fig. 30-18*a*). Model building studies by Alan Fersht and Greg Winter based on this premise, together with the X-ray structure of tyrosyl-tRNA synthetase's tyrosyl adenylate complex, indicate that the enzyme operates by preferentially binding the transition state (Section 14-1F): The γ phosphate in the reaction's pentacoordinate transition state, but not its reactants or products, hydrogen bonds to the enzyme's Thr 40 and His 45 side chains (Fig. 30-18*b*). Fersht and Winter confirmed this conclusion through protein engineering studies in which they replaced Thr 40 with Ala and/or His 45 (which is evolutionarily conserved in aminoacyl-tRNA synthetases) with Gly. All of these mutant enzymes have greatly reduced catalytic activities (a 3×10^5-fold reduction in k_{cat} in the double mutant) even though they all bind both tyrosine and ATP with nearly undiminished affinities. Note that the interactions stabilizing the transition state occur at some distance from the α phosphorous reaction site. Moreover, the enzyme has no catalytically active functional groups, such as general acids or bases, in the vicinity of the reaction site. Evidently, the inherent reactivities of the nucleophilic tyrosyl carboxyl group and ATP's activated PP_i leaving group are sufficient to drive the reaction at a satisfactorily high rate (an $\sim 10^9$-fold increase over the uncatalyzed reaction) with only transition state binding combined with reactant proximity and orientation effects (Section 14-1E).

The Structural Features Recognized by Aminoacyl-tRNA Synthetases May Be Quite Simple

Considerable effort has been expended in elucidating the manner in which aminoacyl-tRNA synthetases recognize their corresponding tRNAs. The methods used to do so include the use of specific tRNA fragments, mutationally altered tRNAs, and chemical cross-linking agents. The most common synthetase contact sites on tRNA occur on the inner (concave) face of the L. Other than that, there appears to be little regularity in how the various tRNAs recognize their cognate synthetases. Indeed, the smaller synthetases appear to recognize only the acceptor region of their tRNAs, whereas the larger enzymes contact much of their tRNA's inner surface. Thus, *the anticodon does not necessarily participate in this recognition process.*

The foregoing suggests that the features of a tRNA recognized by its cognate aminoacyl-tRNA synthetase are idiosyncratic. Genetic manipulations by Paul Schimmel revealed that these features, for at least one type of tRNA, are surprisingly simple. Numerous sequence alterations of *E. coli* tRNA^{Ala} do not appreciably affect its capacity to be aminoacylated with alanine. Yet,

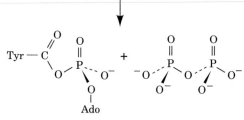

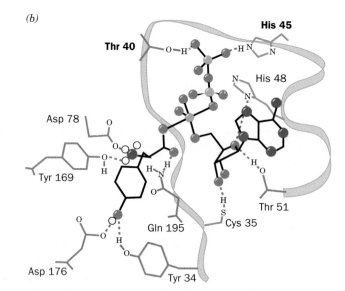

Figure 30-18

The mechanism of tyrosyl adenylate formation as catalyzed by tyrosyl-tRNA synthetase. (*a*) In the reaction's chemical mechanism, the tyrosyl carboxylate group nucleophilically attacks ATP's α phosphorus (*top*) in a single displacement reaction that proceeds via a trigonal bipyramidal transition state (*middle*) to yield tyrosyl adenylate and PP_i (*bottom*). (*b*) Model building studies based on the X-ray structure of tyrosyl-tRNA synthetase in complex with tyrosyl adenylate (a stable complex in the absence of tRNA^{Tyr}) indicate that the γ phosphate of the reacting ATP hydrogen bonds to Thr 40 and His 45 (*top*) only in the reaction's transition state. Tyrosyl adenylate also makes 12 hydrogen bonds with the enzyme (several of which are indicated here by dashed lines) that apparently are not significantly disturbed in the transition state. [After Leatherbarrow, R. J., Fersht, A. R., and Winter, G., *Proc. Natl. Acad. Sci.* **82**, 7841 (1985).]

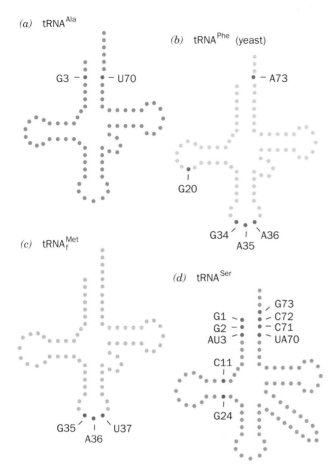

(a) tRNA^{Ala}

(b) tRNA^{Phe} (yeast)

G3 — U70

— A73

G20

G34 — A36
A35

(c) tRNA_f^{Met}

(d) tRNA^{Ser}

G73
G1 — C72
G2 — C71
AU3 — UA70

C11

G24

G35 — U37
A36

Figure 30-19
Major identity elements in four tRNAs. Each base in the tRNA is represented by a circle. Red circles indicate positions that have been shown to be identity elements for the recognition of the tRNA by its cognate aminoacyl-tRNA synthetase. In each case, other identity elements may yet be discovered. [After Schulman, L. H. and Abelson, J., *Science* **240**, 1592 (1988).]

most base substitutions in the $G3 \cdot U70$ base pair located in the tRNA's acceptor stem (Fig. 30-19a) greatly diminish this reaction. Moreover, the introduction of a $G \cdot U$ base pair into the analogous position of tRNA^{Cys} and tRNA^{Phe} causes them to be aminoacylated with alanine even though there are few other sequence similarities between these mutant tRNAs and tRNA^{Ala} (e.g., Fig. 30-20). In fact, *E. coli* alanyl-tRNA synthetase even efficiently aminoacylates a 24-nucleotide "microhelix" derived from only the $G3 \cdot U70$-containing acceptor stem of *E. coli* tRNA^{Ala}. Since the only known *E. coli* tRNAs that normally have a $G3 \cdot U70$ base pair are the tRNA^{Ala}, and this base pair is also present in the tRNA^{Ala} from many organisms including yeast (Fig. 30-11), the foregoing observations strongly suggest that *the $G3 \cdot U70$ base pair is a major feature recognized by alanyl-tRNA synthetases.* These enzymes presumably recognize the distorted shape of the $G \cdot U$ base pair (Fig. 30-15), an idea corroborated by the observation that base changes

at $G3 \cdot U70$, which least affect the acceptor identity of tRNA^{Ala} yield base pairs that structurally resemble $G \cdot U$.

The elements of three other tRNAs, which are recognized by their cognate tRNA synthetases, are indicated in Fig. 30-19. As with tRNA^{Ala}, these identifiers appear to comprise only a few bases. Note that the anticodon is an identifier in two of these tRNAs. In another example of an anticodon identifier, the *E. coli* tRNA^{Ile} specific for the codon AUA has the anticodon LAU where L is **lysidine,** a modified cytosine whose 2-keto group is replaced by the amino acid lysine (Fig. 30-13). The L in this context pairs with A rather than G, a unique case of base modification altering base pairing specificity. The replacement of this L with unmodified C, as expected, yields a tRNA which recognizes the Met codon AUG (codons bind anticodons in an antiparallel fashion). Surprisingly, however, this altered tRNA^{Ile} is also a much better substrate for methionyl-tRNA synthetase than it is for isoleucyl-tRNA synthetase. Thus, both the codon and the amino acid specificity of this tRNA are changed by a single post-transcriptional modification.

Proofreading Enhances the Fidelity of Amino Acid Attachment to tRNA

The charging of a tRNA with its cognate amino acid is a remarkably accurate process. Experimental measure-

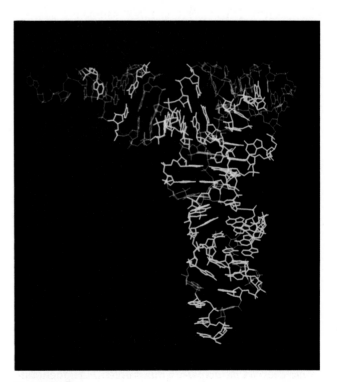

Figure 30-20
A three-dimensional model of *E. coli* tRNA^{Ala} based on the X-ray structure of yeast tRNA^{Phe} (Fig. 30-14) in which the nucleotides that are different in *E. coli* tRNA^{Cys} are highlighted in blue-white and the $G3 \cdot U70$ base pair is highlighted in ivory. [Courtesy of Ya-Ming Hou, MIT.]

ments indicate, for example, that, at equal concentrations of isoleucine and valine, isoleucyl-tRNA synthetase transfers ~ 50,000 isoleucines to tRNA^{Ile} for every valine it so-transfers. Yet, *there are insufficient structural differences between Val and Ile to permit such a high degree of accuracy in the direct generation of aminoacyl-tRNAs.* It seems likely that isoleucyl-tRNA synthetase has a binding site of sufficient size to admit isoleucine but which would exclude larger amino acids. On the other hand, valine, which differs from isoleucine by only the lack of a single methylene group, fits into the isoleucine-binding site. The binding free energy of a methylene group is estimated to be ~ 12 kJ · mol^{-1}. Equation [3.16] indicates that the ratio f of the equilibrium constants, K_1 and K_2, with which two substances bind to a given binding site is given by

$$f = \frac{K_1}{K_2} = \frac{e^{-\Delta G_1^{o'}/RT}}{e^{-\Delta G_2^{o'}/RT}} = e^{-\Delta\Delta G^{o'}/RT} \quad [30.1]$$

where $\Delta\Delta G^{o'} = \Delta G_1^{o'} - \Delta G_2^{o'}$ is the difference between the free energies of binding of the two substances. It is therefore estimated that isoleucyl-tRNA synthetase could discriminate between isoleucine and valine by no more than a factor of ~ 100.

Berg resolved this apparent paradox by demonstrating that, in the presence of tRNA^{Ile}, isoleucyl-tRNA synthetase catalyzes the quantitative hydrolysis of valine-adenylate to valine + AMP rather than forming Val-tRNA^{Ile}. Thus, *isoleucyl-tRNA synthetase subjects aminoacyl-adenylates to a* **proofreading** *or editing step that has been shown to occur at a separate catalytic site.* This site presumably binds Val residues but excludes the larger Ile residues. *The enzymes's overall selectivity is therefore the product of the selectivities of its adenylation and proofreading steps, thereby accounting for the high fidelity of translation.* Many other synthetases discriminate against noncognate amino acids in a similar fashion. However, synthetases that have adequate selectivity for their corresponding amino acid (e.g., tyrosyl-tRNA synthetase discriminates between tyrosine and phenylalanine through hydrogen bonding with the tyrosine-OH group), lack editing functions. Note that *editing occurs at the expense of ATP hydrolysis, the thermodynamic price of high fidelity (increased order).*

D. Codon – Anticodon Interactions

In protein synthesis, the proper tRNA is selected only through codon – anticodon interactions; the aminoacyl group does not participate in this process. This phenomenon was demonstrated as follows. Cys-tRNA^{Cys}, in which the Cys residue was ^{14}C labeled, was reductively desulfurized with Raney nickel so as to convert the Cys residue to Ala:

$$\text{HS}-\text{CH}_2-\underset{\underset{\text{NH}_3^+}{|}}{\overset{\overset{\text{H}}{|}}{\text{C}}}-\overset{\overset{\text{O}}{\|}}{\text{C}}-\text{O}-\text{tRNA}^{Cys} \;+\; \text{Ni}(\text{H})_x$$

Cys-tRNA^{Cys}   **Raney nickel**

$\big\Updownarrow$

$$\text{H}-\text{CH}_2-\underset{\underset{\text{NH}_3^+}{|}}{\overset{\overset{\text{H}}{|}}{\text{C}}}-\overset{\overset{\text{O}}{\|}}{\text{C}}-\text{O}-\text{tRNA}^{Cys} \;+\; \text{H}_2\text{S} \;+\; \text{Ni}$$

Ala-tRNA^{Cys}

The resulting ^{14}C-labeled hybrid, Ala-tRNA^{Cys}, was added to a cell-free protein synthesizing system extracted from rabbit reticulocytes. The product hemoglobin α chain's only radioactive tryptic peptide was the one that normally contains the subunit's only Cys. No radioactivity was found in the peptides that normally contain Ala but no Cys. Evidently, *only the anticodons of aminoacyl-tRNAs are involved in codon recognition.*

Genetic Code Degeneracy Is Largely Mediated by Variable Third Position Codon – Anticodon Interactions

One might naively guess that each of the 61 codons specifying an amino acid would be read by a different tRNA. Yet, even though most cells contain several groups of **isoaccepting tRNAs** (different tRNAs that are specific for the same amino acid), *many tRNAs bind to two or three of the codons specifying their cognate amino acids.* For example, yeast tRNA^{Phe}, which has the anticodon GmAA, recognizes the codons UUC and UUU (remember that the anticodon pairs with the codon in an antiparallel fashion),

```
                3'              5'   3'              5'
Anticodon:   — A — A — Gm —       — A — A — Gm —
                :   :   :            :   :   :
             5' :   :   : 3'      5' :   :   : 3'
Codon:       — U — U — C —        — U — U — U —
```

and yeast tRNA^{Ala}, which has the anticodon IGC, recognizes the codons GCU, GCC, and GCA.

```
                3'              5'   3'              5'
Anticodon:   — C — G — I —        — C — G — I —
                :   :   :            :   :   :
             5' :   :   : 3'      5' :   :   : 3'
Codon:       — G — C — U —        — G — C — C —


                   3'              5'
Anticodon:      — C — G — I —
                   :   :   :
                5' :   :   : 3'
Codon:          — G — C — A —
```

It therefore seems that non-Watson–Crick base pairing can occur at the third codon–anticodon position (the anticodon's first position is defined as its 3' nucleotide), the site of most codon degeneracy (Table 30-2). Note also that the third (5') anticodon position commonly contains a modified base such as Gm or I.

The Wobble Hypothesis Structurally Accounts for Codon Degeneracy

By combining structural insight with logical deduction, Crick proposed, in what he named the **wobble hypothesis,** how a tRNA can recognize several degenerate codons. He assumed that the first two codon–anticodon pairings have normal Watson–Crick geometry. The structural constraints that this places on the third codon–anticodon pairing ensure that its conformation does not drastically differ from that of a Watson–Crick pair. Crick then proposed that there could be a small amount of play or "wobble" in the third codon position which allows limited conformational adjustments in its pairing geometry. This permits the formation of several non-Watson–Crick pairs such as $U \cdot G$ and $I \cdot A$ (Fig. 30-21a). The allowed "wobble" pairings are indicated in Fig. 30-21b. Then, by analyzing the known pattern of codon–anticodon pairing, Crick deduced the most plausible sets of pairing combinations in the third codon–anticodon position (Table 30-4). Thus, an anticodon with C or A in its third position can only pair with its Watson–Crick complementary codon. If U, G, or I occupies the third anticodon position, two, two, or three codons are recognized, respectively.

No prokaryotic or eukaryotic cytoplasmic tRNA is known to participate in a nonwobble pairing combination. There is, however, no known instance of such a tRNA with an A in its third anticodon position which suggests that the consequent $U \cdot A$ pair is not permitted. The structural basis of wobble pairing is poorly understood although it is clear that it is influenced by base modifications.

A consideration of the various wobble pairings indicates that at least 31 tRNAs are required to translate all 61 coding triplets of the genetic code (there are 32 tRNAs in the minimal set because translational initiation requires a separate tRNA; Section 30-3C). Most

Table 30-4

Allowed Wobble Pairing Combinations in the Third Codon–Anticodon Position

5'-Anticodon Base	3'-Codon Base
C	G
A	U
U	A or G
G	U or C
I	U, C, or A

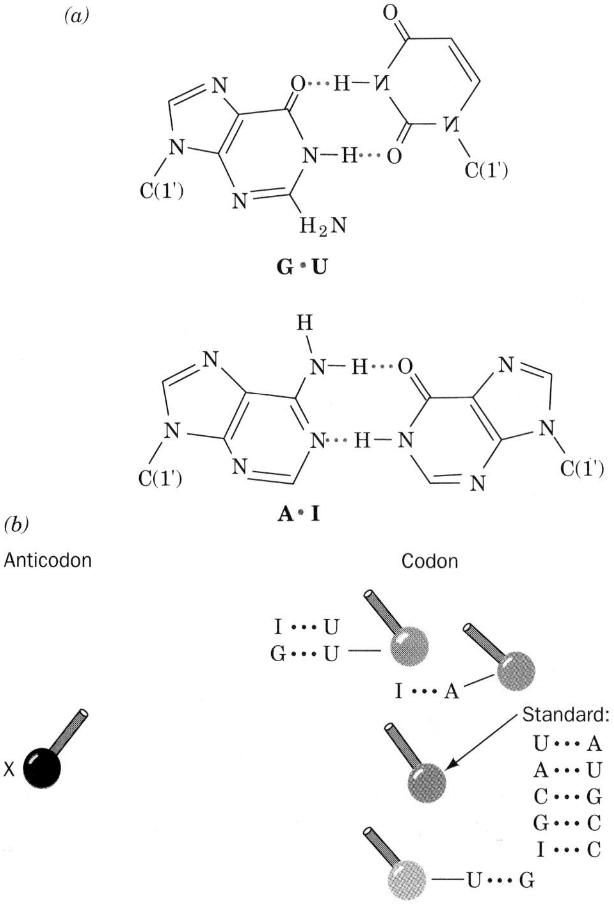

Figure 30-21

Wobble pairing. (*a*) $U \cdot G$ and $I \cdot A$ wobble pairs. (*b*) The geometry of wobble pairing. The spheres and their attached bonds represent the positions of ribose C(1') atoms with their accompanying glycosidic bonds. X (*left*) designates the nucleoside at the 5' end of the anticodon (tRNA). The positions on the right are those of the 3' nucleoside of the codon (mRNA) in the indicated wobble pairings. [After Crick, F. H. C., *J Mol. Biol.* **19,** 55 (1966).]

cells have >32 tRNAs, some of which have identical anticodons. Nevertheless, *all isoaccepting tRNAs in a cell are recognized by a single aminoacyl-tRNA synthetase.*

Some Mitochondrial tRNAs Have More Permissive Wobble Pairings Than Other tRNAs

The codon recognition properties of mitochondrial tRNAs must reflect the fact that mitochondrial genetic codes are variants of the "standard" genetic code (Table 30-3). For instance, the human mitochondrial genome, which consists of only 16,569 bp, encodes 22 tRNAs (together with 2 ribosomal RNAs and 13 proteins). Fourteen of these tRNAs each read one of the synonymous pairs of codons indicated in Tables 30-2 and 30-3 (MNX, where X is either C or U or else A or G) according to normal $G \cdot U$ wobble rules: The tRNAs have either a G or a modified U in their third anticodon position that, respectively, permits them to pair with codons having $X = C$ or U or else $X = A$ or G. The remaining 8 tRNAs, which, contrary to wobble rules, each recognize 1 of the groups of 4 synonymous codons (MNY, where $Y = A$, C, G, or U), all have anticodons with a U in their third position. Either this U can somehow pair with any of the 4 bases or these tRNAs read only the first two codon positions and ignore the third. Thus, not surprisingly, many mitochondrial tRNAs have unusual structures in which, for example, the GTψCRA sequence (Fig. 30-12) is missing, or, in the most bizarre case, a tRNA^{Ser} lacks the entire D arm.

Frequently Used Codons Are Complementary to the Most Abundant tRNA Species

The analysis of the base sequences of several highly expressed structural genes of baker's yeast, *Saccharomyces cerevisiae*, has revealed a remarkable bias in their codon usage. Only 25 of the 61 coding triplets are commonly used. *The preferred codons are those that are most nearly complementary, in the Watson–Crick sense, to the anticodons in the most abundant species in each set of isoaccepting tRNAs.* Furthermore, codons that bind anticodons with two consecutive $G \cdot C$ pairs or three $A \cdot U$ pairs are avoided so that the preferred codon–anticodon complexes all have approximately the same binding free energies. A similar phenomenon occurs in *E. coli* although several of its 22 preferred codons differ from those in yeast. The degree with which the preferred codons occur in a given gene is strongly correlated, in both organisms, with the gene's level of expression. This, it has been proposed, permits the mRNAs of proteins that are required in high abundance to be rapidly and smoothly translated.

Selenocysteine Is Specified by a tRNA

Although it is widely stated, even in this text, that proteins are synthesized from the 20 "standard" amino acids, that is, those specified by the "standard" genetic code, some organisms, in fact, use a 21st amino acid, **selenocysteine,** in synthesizing a few of their proteins.

$$\begin{array}{c} | \\ NH \\ | \\ CH - CH_2 - Se - H \\ | \\ C = O \\ | \end{array}$$

The selenocysteine
residue

Selenium, a biologically essential trace element, is a component of several enzymes in both prokaryotes and eukaryotes. *E. coli* contains two selenoproteins, both **formate dehydrogenases,** which each contain a selenocysteine residue. The selenocysteine residues are ribosomally incorporated into these proteins by a unique tRNA bearing a UCA anticodon that is specified by a particular (in the mRNA) UGA codon (normally the *opal* stop codon). How the ribosomal system differentiates this UGA from normal *opal* stop codons is unknown although mRNA context effects and the physiological state of the cell are probably involved. The selenocysteinyl-tRNA is synthesized by the aminoacylation of its tRNA with L-serine by the same aminoacyl-tRNA synthetase that charges tRNA^{Ser}, followed by the enzymatic selenylation of the resulting Ser residue.

E. Nonsense Suppression

Nonsense mutations are usually lethal when they prematurely terminate the synthesis of an essential protein. An organism with such a mutation may nevertheless be "rescued" by a second mutation on another part of the genome. For many years after their discovery, the existence of such **intergenic suppressors** was quite puzzling. It is now known, however, that they usually arise from mutations in a tRNA gene that cause the tRNA to recognize a nonsense codon. Such a **nonsense suppressor** tRNA appends its amino acid (which is the same as that carried by the corresponding wild-type tRNA) to a growing polypeptide in response to the recognized stop codon thereby preventing chain termination. For example, the *E. coli amber* suppressor known as *su3* is a tRNA^{Tyr} whose anticodon has mutated from the wild-type GUA (which reads the Tyr codons UAU and UAC) to CUA (which recognizes the *amber* stop codon UAG). An *su3*$^+$ *E. coli* with an otherwise lethal *amber* mutation in a gene coding for an essential protein would be viable if the replacement of the wild-type amino acid residue by Tyr does not inactivate the protein.

There are several well-characterized examples of *amber* (UAG), *ochre* (UAA), and *opal* (UGA) suppressors in *E. coli* (Table 30-5). Most of them, as expected, have mutated anticodons. UGA-1 tRNA, however, differs from the wild-type only by a G $\rightarrow$ A mutation in its D stem, which changes a $G \cdot U$ pair to a stronger $A \cdot U$ pair.

Table 30-5

Some *E. coli* Nonsense Suppressors

Name	Codon Suppressed	Amino Acid Inserted
*su*1	UAG	Ser
*su*2	UAG	Gln
*su*3	UAG	Tyr
*su*4	UAA, UAG	Tyr
*su*5	UAA, UAG	Lys
*su*6	UAA	Leu
*su*7	UAA	Gln
UGA-1	UGA	Trp
UGA-2	UGA	Trp

Source: Körner, A. M., Feinstein, S. I., and Altman, S., *in* Altman, S. (Ed.), *Transfer RNA, p.* 109, MIT Press (1978).

This mutation apparently alters the conformation of the tRNA's CCA anticodon so that it can form an unusual wobble pairing with UGA as well as with its normal codon, UGG. Nonsense suppressors also occur in yeast.

Suppressor tRNAs Are Mutants of Minor tRNAs

How do cells tolerate a mutation that both eliminates a normal tRNA and prevents the termination of polypeptide synthesis? They survive because the mutated tRNA is usually a minor member of a set of isoaccepting tRNAs and because nonsense suppressor tRNAs must compete for stop codons with the protein factors that mediate the termination of polypeptide synthesis (Section 30-E3). Consequently, the rate of suppressor-mediated synthesis of active proteins with either UAG or UGA nonsense mutations rarely exceeds 50% of the wild-type rate whereas mutants with UAA, the most common termination codon, have suppression efficiencies of <5%. Many mRNAs, moreover, have two tandem stop codons so that even if their first stop codon was suppressed, termination could occur at the second. Nevertheless, many suppressor-rescued mutants grow relatively slowly because they cannot make an otherwise prematurely terminated protein as efficiently as do wild-type cells.

Other types of suppressor tRNAs are also known. **Missense suppressors** act similarly to nonsense suppressors but substitute one amino acid in place of another. **Frameshift suppressors** have eight nucleotides in their anticodon loops rather than the normal seven. They read a four base codon beyond a base insertion thereby restoring the wild-type reading frame.

3. RIBOSOMES

Ribosomes were first seen in cellular homogenates by dark field microscopy in the late 1930s by Albert Claude who referred to them as "microsomes". It was not until the mid-1950s, however, that George Palade observed them in cells by electron microscopy thereby disposing of the contention that they are merely artifacts of cell disruption. The name ribosome derives from the fact that these particles in *E. coli* consist of ~⅔ RNA and ⅓ protein. (**Microsomes** are now defined as the artifactual vesicles formed by the endoplasmic reticulum upon cell disruption. They are easily isolated by differential centrifugation and are rich in ribosomes.) The correlation between the amount of RNA in a cell and the rate at which it synthesizes protein led to the suspicion that ribosomes are the site of protein synthesis. This hypothesis was confirmed in 1955 by Paul Zamecnik who demonstrated that ^{14}C-labeled amino acids are transiently associated with ribosomes before they appear in free proteins. Further research showed that ribosomal polypeptide synthesis has three distinct phases: (1) chain initiation, (2) chain elongation, and (3) chain termination.

In this section we examine the structure of the ribosome, insofar as it is known, and then outline the ribosomal mechanism of polypeptide synthesis. In doing so we shall compare the properties of ribosomes from prokaryotes (mostly *E. coli*) with those of eukaryotes (mostly rat liver cytoplasm).

A. Ribosome Structure

The *E. coli* ribosome, which has a particle mass of ~2.5×10^6 D and a sedimentation coefficient of 70S, is a spheroidal particle that is ~250 Å across in its largest dimension. It may be dissociated, as James Watson discovered, into two unequal subunits (Table 30-6). The small (30S) subunit consists of a 16S rRNA molecule and 21 different polypeptides, whereas the large (50S) subunit contains a 5S and a 23S rRNA together with 32

Table 30-6
Components of *E. coli* Ribosomes

	Ribosome	Small Subunit	Large Subunit
Sedimentation coefficient	70S	30S	50S
Mass (kD)	2520	930	1590
RNA			
Major		16S, 1542 nucleotides	23S, 2904 nucleotides
Minor			5S, 120 nucleotides
RNA mass (kD)	1664	560	1104
Proportion of mass	66%	60%	70%
Proteins		21 polypeptides	32 polypeptides
Protein mass (kD)	857	370	487
Proportion of mass	34%	40%	30%

Source: Lewin, B., *Genes* (3rd ed.), p. 145, Wiley (1987).

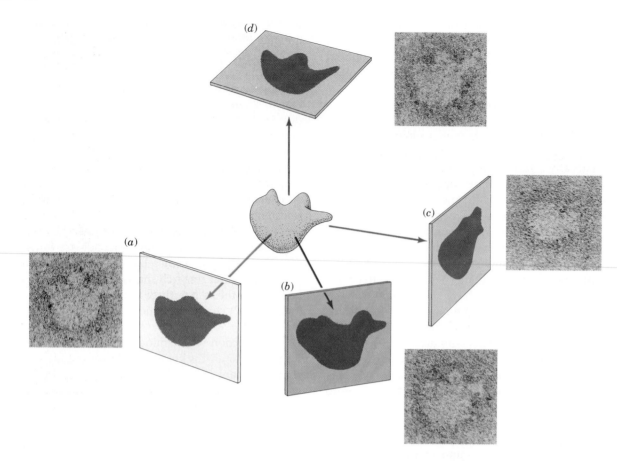

Figure 30-22

The three-dimensional model of the large ribosomal subunit was deduced by mathematically combining its two-dimensional electron microscope images as viewed from different directions. The model of the small subunit was similarly determined. [Courtesy of James Lake, UCLA.]

different polypeptides. The up to 20,000 ribosomes in an *E. coli* cell account for ~80% of its RNA content and 10% of its protein.

Although the ribosome has recently been crystallized by Ada Yonath, it is such a complex entity that it will be many years before its structure is known in molecular detail. However, the low resolution structures of the ribosome and its subunits have been determined through image reconstruction techniques pioneered by Aaron Klug in which electron micrographs of a single particle or ordered sheets of particles taken from several directions are combined to yield its three-dimensional image (Fig. 30-22). The small subunit is a roughly mitten-shaped particle, whereas the large subunit is spheroidal with three protuberances on one side (Fig. 30-23). The large subunit also contains a tunnel, up to 25 Å in diameter and 100 to 120 Å long, that extends from a cleft between the subunit's three protuberances and is postulated to provide the nascent polypeptide's exit path (Fig. 30-24).

Ribosomal RNAs Have Evolutionarily Conserved Secondary Structures

The *E. coli* 16S rRNA, which was sequenced by Harry Noller, consists of 1542 nucleotides. A computerized search of this sequence for stable double helical segments yielded many plausible but often mutually exclusive secondary structures. However, the comparison of the sequences of 16S rRNAs from several prokaryotes, under the assumption that their structures have been evolutionarily conserved, led to the flowerlike secondary structure for 16S rRNA proposed in Fig. 30-25. This four-domain structure, which is 46% base paired, is reasonably consistent with the results of nuclease digestion and chemical modification studies. Its double helical stems tend to be short (<8 bp) and many of them are imperfect. Intriguingly, electron micrographs of the 16S rRNA resemble those of the complete 30S subunit, thereby suggesting that the 30S subunit's overall shape is largely determined by the 16S rRNA.

The large ribosomal subunit's 5S and 23S rRNAs,

(a)

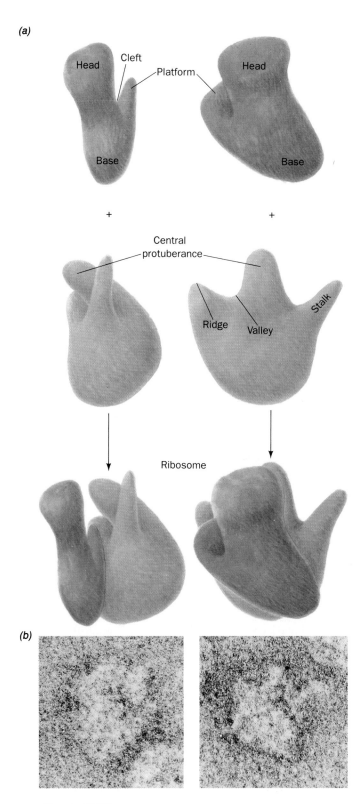

(b)

Figure 30-23
(*a*) A three-dimensional model of the *E. coli* ribosome deduced as indicated in Fig. 30-22. The small subunit (*top*) combines with the large subunit (*middle*) to form the complete ribosome (*bottom*). The two views of the ribosome match those seen in (*b*) the electron micrographs. [Courtesy of James Lake, UCLA.]

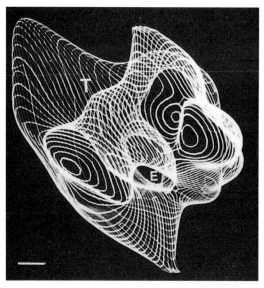

Figure 30-24
A computer-generated image of the large ribosomal subunit from *Bacillus stearothermophilus* as determined by electron micrographic image reconstruction of oriented two-dimensional arrays of particles. An ~25 Å in diameter tunnel extends ~100 Å from the cleft between the subunit's three protrusions (T) to the nascent polypeptide's probable exit site (E). The bar is 20 Å long. [Courtesy of Ada Yonath, Weizmann Institute of Science.]

which consist of 120 and 2904 nucleotides, respectively, have also been sequenced. As with the 16S rRNA, they appear to have extensive secondary structures. That proposed for 5S rRNA is shown in Fig. 30-26. Of course, as we have seen for tRNA, the secondary structure of an RNA provides little indication of its three-dimensional structure (but see below).

Ribosomal Proteins Have Been Partially Characterized

Ribosomal proteins are difficult to separate because most of them are insoluble in ordinary buffers. By convention, ribosomal proteins from the small and large subunits are designated with the prefixes S and L, respectively, followed by a number indicating their position, from upper left to lower right, on a two-dimensional gel electrophoretogram (roughly in order of decreasing molecular mass; Fig. 30-27). Only protein S20/L26 is common to both subunits. It is apparently located at the interface between the two subunits. One of the large subunit proteins is partially acetylated at its N-terminus so that it gives rise to two electrophoretic spots (L7/L12). Four copies of this protein are present in the large subunit. Moreover, these four copies of L7/L12 aggregate with L10 to form a stable complex that was initially thought to be a unique protein, "L8."

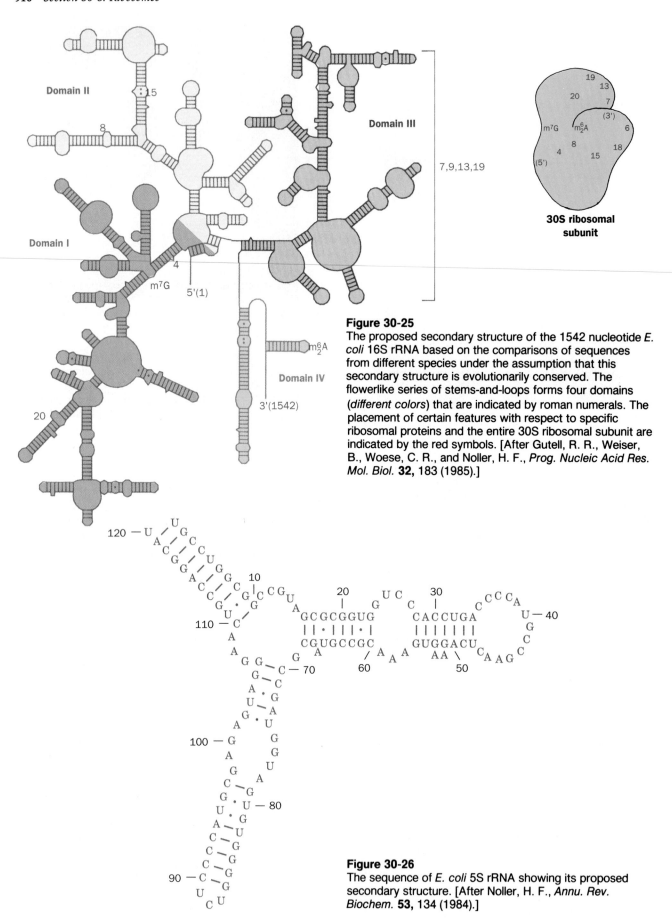

Figure 30-25
The proposed secondary structure of the 1542 nucleotide *E. coli* 16S rRNA based on the comparisons of sequences from different species under the assumption that this secondary structure is evolutionarily conserved. The flowerlike series of stems-and-loops forms four domains (*different colors*) that are indicated by roman numerals. The placement of certain features with respect to specific ribosomal proteins and the entire 30S ribosomal subunit are indicated by the red symbols. [After Gutell, R. R., Weiser, B., Woese, C. R., and Noller, H. F., *Prog. Nucleic Acid Res. Mol. Biol.* **32,** 183 (1985).]

Figure 30-26
The sequence of *E. coli* 5S rRNA showing its proposed secondary structure. [After Noller, H. F., *Annu. Rev. Biochem.* **53,** 134 (1984).]

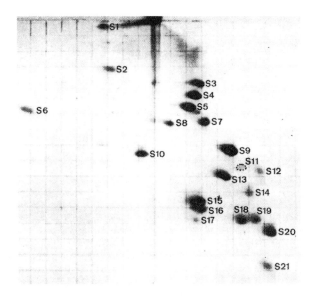

Figure 30-27
A two-dimensional gel electrophoretogram of *E. coli* small subunit proteins. First dimension (*vertical*): 8% acrylamide, pH 8.6; second dimension (*horizontal*): 18% acrylamide; pH 4.6. [From Kaltschmidt, E. and Wittmann, H. G., *Proc. Natl. Acad. Sci.* **67**, 1277 (1970).]

All the other ribosomal proteins occur in only one copy per subunit.

The amino acid sequences of all 52 *E. coli* ribosomal proteins have been elucidated, mainly by Heinz-Günter Wittmann and Brigitte Wittmann-Liebold. They range in size from 46 residues for L34 to 557 residues for S1. Most of these proteins, which exhibit little sequence similarity with one another, are rich in the basic amino acids Lys and Arg and contain few aromatic residues as expected for proteins that are closely associated with polyanionic RNA molecules. The X-ray structure of only two ribosomal proteins have so far been reported: those of L30 and a C-terminal segment of L7/L12 (Fig. 30-28). These proteins have remarkably similar structures despite their only 14% amino acid sequence identity.

Ribosomal Subunits Are Self-Assembling

Ribosomal subunits form, under proper conditions, from mixtures of their numerous macromolecular components. *Ribosomal subunits are therefore self-assembling entities.* Masayasu Nomura determined how this occurs through partial reconstitution experiments. If one macromolecular component is left out of an otherwise self-assembling mixture of proteins and RNA, the other components that fail to bind to the resulting partially assembled subunit must somehow interact with the omitted component. Through the analysis of a series of such partial reconstitution experiments, Nomura constructed an assembly map of the small subunit (Fig. 30-29). This map indicates that the first steps in small subunit assembly are the independent binding of cer-

(a)   (b)

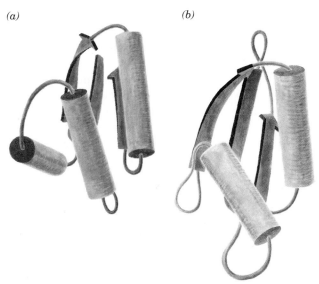

Figure 30-28
The X-ray structures of two ribosomal proteins: (*a*) The 74-residue C-terminal fragment of *E. coli* L7/L12. (*b*) *Bacillus stearothermophilus* L30 (61 residues). The two protein molecules are oriented so as to show their closely similar backbone comformations. [After Leijonmarck, M., Appelt, K., Badger, J., Liljas, A., Wilson, K. S., and White, S. W., *Proteins* **3**, 244 (1988).]

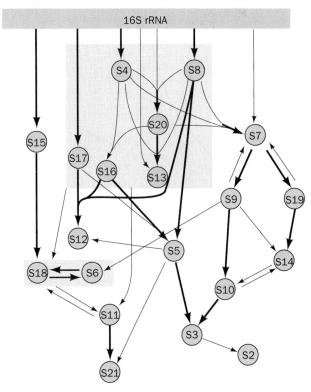

Figure 30-29
The assembly map of the *E. coli* small subunit. Thick and thin arrows between components indicate strong and weak facilitation of binding, respectively. For example, the thick arrow from 16S rRNA to S4 indicates that S4 binds directly to 16S rRNA in the absence of other proteins, whereas the thin arrows from 16S rRNA, S4, S8, S9, S19, and S20 to S7 indicate that the former components all participate in binding S7. [After Held, W. A., Ballou, B., Mizushima, S., and Nomura, M., *J. Biol. Chem.* **249**, 3109 (1974).]

tain proteins to 16S rRNA. The resulting assembly intermediates provide the molecular scaffolding for binding other proteins. At one stage of the assembly process, an intermediate particle must undergo a marked conformational change before assembly can continue. The large subunit self-assembles in a similar manner. The observation that similar assembly intermediates occur *in vivo* and *in vitro* suggests that *in vivo* and *in vitro* assembly processes are much alike.

Ribosomal Architecture Has Been Deduced through Immune Electron Microscopy and Neutron Diffraction Studies

The positions of most ribosomal components have been determined through a variety of physical and chemical techniques. Many proteins have been located by James Lake and by Georg Stöffler through **immune electron microscopy.** Rabbit antibodies [**immunoglobulin G (IgG);** Section 34-2A] raised against a specific ribosomal protein bind to this protein where it is exposed on the surface of its subunit. Electron microscopy of the ribosomal subunit · IgG complex indicates the point of attachment of the IgG and hence the site of the ribosomal protein to which it binds (Figs. 30-30 and 30-31). These results have been confirmed and extended through neutron diffraction measurements of 30S subunits conducted by Donald Engleman and Peter Moore (Fig. 30-32*a*) and by similar studies on 50S subunits by Knud Nierhaus. The protein positions indicated in Figs. 30-31 and 30-32*a* are consistent with the subunit assembly map shown in Fig. 30-29 in that pairs of proteins that must interact for proper subunit assembly (al-

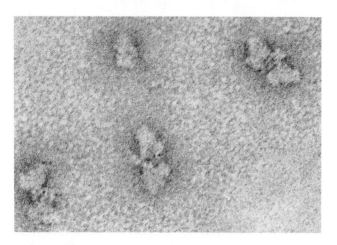

Figure 30-30
Immune electron microscopy reveals the positions of ribosomal proteins. Immunoglobin G (IgG) raised against a particular ribosomal protein, here S6, is mixed with ribosomes. The IgG, which is a Y-shaped protein (Section 34-2B), binds to its corresponding antigen at the ends of the two short prongs of the Y, thereby binding together two ribosomes. The position of the protein on the surface of the ribosome is indicated by the point of attachment of the IgG. [Courtesy of James Lake, UCLA.]

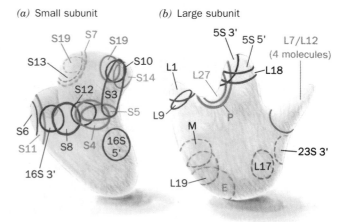

Figure 30-31
Maps of the *E. coli* ribosomal (*a*) small and (*b*) large subunits indicating the locations of some of their component proteins as determined by immune electron microscopy. Sites that are dashed are located on the back side of the subunit. On the small subunit, the symbols 16S 3' and 16S 5' mark the ends of the 16S RNA. On the large subunit, P indicates the peptidyl transferase site, E marks the site where the nascent polypeptide emerges from the ribosome (the end of the tunnel in Fig. 30-24), M specifies the ribosome's membrane anchor site, and 5S 3' and 5S 5' mark the ends of the 5S rRNA. [After Lake, J. A., *Annu. Rev. Biochem.* **54**, 512 (1985).]

though not necessarily by direct contact) are in close proximity.

Many of the secondary structural elements of the 16S rRNA have been located on the small subunit (Figs. 30-25 and 30-32*b*). Their positions were indirectly established from the known positions of proteins that nuclease protection and RNA–protein cross-linking experiments indicate bind to these elements. Thus, we now have a complete, albeit crude, model of the *E. coli* 30S ribosomal subunit.

Affinity Labeling Has Helped Identify the Ribosome's Functional Components

Considerable effort has gone into identifying the ribosome's functional components such as the peptidyl transferase center that catalyzes peptide bond formation (Section 30-3D). Many of these investigations have involved **affinity labeling,** a technique in which a reactive group is attached to a natural ligand of the system of interest such as an antibiotic that binds to the ribosome (Section 30-3G). The reactive group, which may be spontaneously reactive or photolabile so that it only reacts upon UV illumination (**photoaffinity labeling),** is carried to the ligand-binding site where it reacts to cross-link the ligand to the surrounding groups. Dissociation of the resulting particle permits the identification of the components with which the usually radioactive affinity label has reacted.

The results of affinity labeling the ribosome have often been difficult to interpret because its various functions each appear to involve several ribosomal components. For example, mRNA binding apparently involves

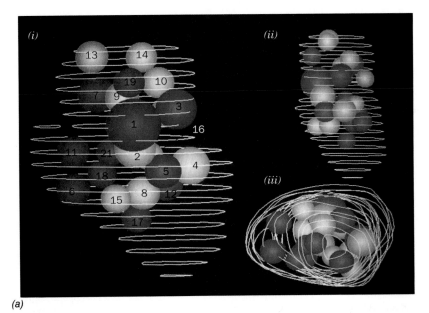

(a)

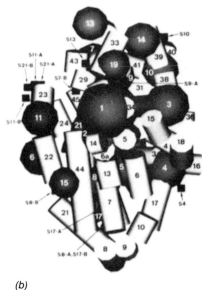

(b)

Figure 30-32

The structure of the 30S ribosomal subunit. (a) The relative positions of all 21 proteins of the 30S ribosomal subunit superimposed on its surface outline. Calling Part (*i*) the front view, then Parts (*ii*) and (*iii*) are the left side and bottom views, respectively. The proteins are assigned their standard numbers in Part (*i*) in which S20 is directly behind S3 (the different colors of spheres are only a viewing aid). The distances between pairs of these proteins were determined from the neutron scattering of concentrated solutions of reconstituted 30S subunits in which the two proteins of interest were heavily deuterated while all other subunit components were normally protonated (deuterons scatter neutrons quite differently from protons). Such measurements on many different pairs of proteins permitted the construction of this map in which the volume of each sphere is proportional to the corresponding protein's mass and its position marks the protein's center of mass. Compare this map with Fig. 30-31a. [Courtesy of Peter Moore, Yale University and Malcolm Capel, Brookhaven National Laboratory.] (b) The locations of the double helical elements of the 16S RNA (*cylinders*) relative to the 30S subunit proteins (*spheres*) as deduced from protein-RNA cross-linking studies. The view is the same as in Figs. 30-31a and 30-32(a,i). [From Schüler, D. and Brinacombe, R., *EMBO J.* **7,** 1512 (1988).]

proteins S1, S3, S4, S5, S9, S12, and S18 as well as the 16S rRNA, whereas proteins L2, L11, L15, L16, L18, L23, and L27, and the 23S RNA are implicated in the peptidyl transferase function. To further confuse matters, studies with mutants deficient in various ribosomal proteins have revealed that the absence of any of at least 15 of the 52 *E. coli* ribosomal proteins does not greatly affect the ribosome's translational ability. Nevertheless, the following functionalities have been located (Figs. 30-23 and 30-31):

1. The 3' end of the 16S rRNA, which is known to participate in mRNA binding (Section 30-3C), is located on the small subunit's "platform." The locations of the proteins implicated in ribosomal mRNA binding, together with the observation that the ribosome protects an ~40 nucleotide mRNA segment from RNase digestion, indicates that mRNA binds to the small subunit across the region connecting its "head" to its "base."

2. The anticodon-binding sites occur in the small subunit's "cleft" region.

3. The four L7/L12 subunits forming the large subunit's "stalk" participate in the ribosome's various GTPase reactions.

4. The peptidyl transferase function (P) occupies the "valley" between the large subunit's other two protuberances.

5. The site that binds ribosomes to membranes (M; Section 11-3F), occurs on the large subunit adjacent to the polypeptide exit tunnel, E.

Thus, *the large subunit appears to be mainly involved in mediating biochemical tasks such as catalyzing the reactions of polypeptide elongation, whereas the small subunit is the major actor in ribosomal recognition processes such as mRNA and tRNA binding (although the large subunit is also implicated in tRNA binding).* Note that rRNA probably has a major functional role in many, if not all, ribosomal processes (recall that RNA has demonstrated catalytic properties; Sections 29-4A and C).

Eukaryotic Ribosomes Are Larger and More Complex Than Prokaryotic Ribosomes

Although eukaryotic and prokaryotic ribosomes resemble each other in both structure and function, they

Table 30-7
Components of Rat Liver Cytoplasmic Ribosomes

	Ribosome	Small Subunit	Large Subunit
Sedimentation coefficient	80S	40S	60S
Mass (kD)	4220	1400	2820
RNA			
Major		18S, 1874 nucleotides	28S, 4718 nucleotides
Minor			5.8S, 160 nucleotides
			5S, 120 nucleotides
RNA mass (kD)	2520	700	1820
Proportion of mass	60%	50%	65%
Proteins		33 polypeptides	49 polypeptides
Protein mass (kD)	1700	700	1000
Proportion of mass	40%	50%	35%

Source: Lewin, B., *Genes* (3rd ed.), *p.* 146, Wiley (1987).

differ in nearly all details. Eukaryotic ribosomes have particle masses in the range 3.9 to 4.5×10^6 D and have a nominal sedimentation coefficient of 80S. They dissociate into two unequal subunits that have compositions that are distinctly different from those of prokaryotes (Table 30-7; compare with Table 30-6). The small (40S) subunit of the rat liver cytoplasmic ribosome, the most well-characterized eukaryotic ribosome, consists of 33 unique polypeptides and an 18S rRNA. Its large (60S) subunit contains 49 different polypeptides and three

rRNAs of 28S, 5.8S, and 5S. Electron microscopy indicates that these subunits, as well as the intact ribosome, have shapes that are similar to those of their prokaryotic counterparts.

Sequence comparisons of the corresponding rRNAs from various species indicates that evolution has conserved their secondary structures rather than their base sequences (Figs. 30-25 and 30-33). For example, a $G \cdot C$ in a base paired stem of *E. coli* 16S rRNA has been replaced by an $A \cdot U$ in the analogous stem of yeast 18S

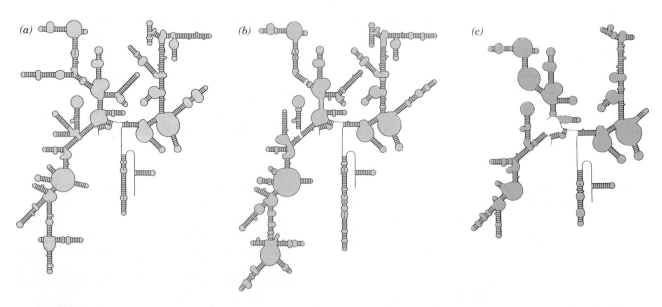

Figure 30-33
The predicted secondary structures of evolutionarily distant 16S-like rRNAs from (*a*) archaebacteria (*Halobacterium volcanii*), (*b*) eukaryotes (baker's yeast), and (*c*) mammalian mitochondria (bovine). Compare them with Fig. 30-25, the predicted secondary structure of 16S RNA from eubacteria

(*E. coli*). Note the close similarities of these assemblies; they differ mostly by insertions and deletions of stem-and-loop structures. The 23S-like rRNAs from a variety of species likewise have similar secondary structures. [After Gutell, R. R., Weiser, B., Woese, C. R., and Noller, H. F., *Prog. Nucleic Acid Res. Mol. Biol.* **32**, 183 (1985).]

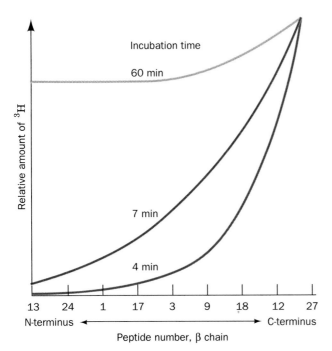

Figure 30-34
Distribution of [^{3}H]Leu among the tryptic peptides from the β subunit of soluble rabbit hemoglobin after the incubation of rabbit reticulocytes with [^{3}H]leucine for the indicated times. [After Dintzis, H. M., *Proc. Natl. Acad. Sci.* **47,** 255 (1961).]

rRNA. The 5.8S rRNA, which occurs in the large eukaryotic subunit in base paired complex with 28S rRNA, is homologous in sequence to the 5′ end of prokaryotic 23S rRNA. Apparently 5.8S RNA arose through mutations that altered rRNA's post-transcriptional processing producing a fourth rRNA.

B. Polypeptide Synthesis: An Overview

Before we commence our detailed discussion of polypeptide synthesis, it will be helpful to outline some of its major features.

Polypeptide Synthesis Proceeds from N-Terminus to C-Terminus

The direction of ribosomal polypeptide synthesis was established, in 1961, by Howard Dintzis through radioactive labeling experiments. He exposed reticulocytes that were actively synthesizing hemoglobin to ^{3}H-labeled leucine for times less than that required to make an entire polypeptide. The extent that the tryptic peptides from the soluble (completed) hemoglobin molecules were labeled increased with their proximity to the C-terminus (Fig. 30-34). Incoming amino acids must therefore be appended to a growing polypeptide's C-terminus; that is, *polypeptide synthesis proceeds from N-terminus to C-terminus.*

Ribosomes Read mRNA in the 5′ → 3′ Direction

The direction that the ribosome reads mRNAs was determined through the use of a cell-free protein synthesizing system in which the mRNA was poly(A) with a 3′-terminal C.

$$5'\quad A-A-A-\cdots-A-A-A-C\quad 3'$$

Such a system synthesizes a poly(Lys) that has a C-terminal Asn.

$$\overset{+}{H_3N}-Lys-Lys-Lys-\cdots-Lys-Lys-Asn-COO^-$$

This, together with the knowledge that AAA and AAC code for Lys and Asn and the polarity of polypeptide synthesis, indicates that *the ribosome reads mRNA in the 5′ → 3′ direction.* Since mRNA is synthesized in the 5′ → 3′ direction, this accounts for the observation that, in prokaryotes, ribosomes initiate translation on nascent mRNAs (Section 29-3).

Active Translation Occurs on Polyribosomes

Electron micrographs reveal that ribosomes engaged in protein synthesis are tandemly arranged on mRNAs like beads on a string (Figs. 30-35 and 29-16). The indi-

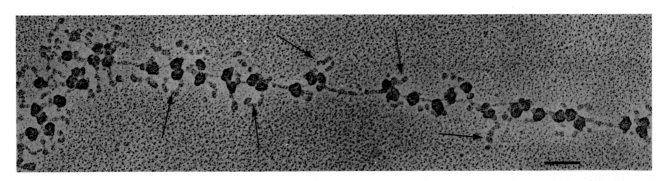

Figure 30-35
Electron micrographs of polysomes from silk gland cells of the silkworm *Bombyx mori.* The 3′ end of the mRNA is on the right. Arrows point to the silk fibroin polypeptides. The bar represents 0.1 μm. [Courtesy of Oscar L. Miller, Jr., University of Virginia.]

vidual ribosomes in these **polyribosomes (polysomes)** are separated by gaps of 50 to 150 Å so that they have a maximum density on mRNA of ~1 ribosome per 80 nucleotides. Polysomes arise because once an active ribosome has cleared its initiation site, a second ribosome can initiate at that site.

Chain Elongation Occurs by the Linkage of the Growing Polypeptide to the Incoming tRNA's Amino Acid Residue

During polypeptide synthesis, amino acid residues are sequentially added to the C-terminus of the nascent, ribosomally bound polypeptide chain. If the growing polypeptide is released from the ribosome by treatment with high salt concentrations, its C-terminal residue is invariably esterified to a tRNA molecule as a **peptidyl-tRNA.**

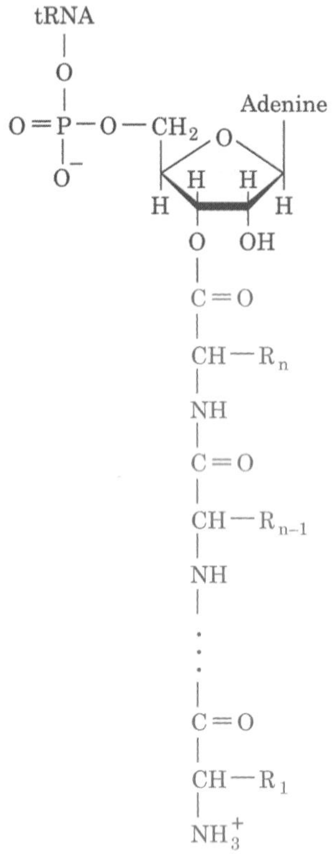

Peptidyl-tRNA

The nascent polypeptide must therefore grow by being transferred from the peptidyl-tRNA to the incoming amino-acyl-tRNA to form a peptidyl-tRNA with one more residue (Fig. 30-36). Apparently, the ribosome has at least two tRNA-binding sites: the so-called **P site,** which binds the peptidyl-tRNA, and the **A site,** which binds the incoming aminoacyl-tRNA (Fig. 30-36). Consequently, after the formation of a peptide bond, the newly deacylated P-site tRNA must be released and replaced by the newly formed peptidyl-tRNA from the A site thereby

permitting a new round of peptide bond formation. The recent finding that each ribosome can bind up to three deacylated tRNAs but only two aminoacyl-tRNAs indicates, however, that the ribosome has a third tRNA-binding site: the **exit or E site,** which transiently binds the outgoing tRNA.

The details of the chain elongation process are discussed in Section 30-3D. Chain initiation and chain termination, which are special processes, are examined in Sections 30-3C and 30-3E, respectively. In all of these sections we shall first consider the process of interest in *E. coli* and then compare it with the analogous eukaryotic activity.

C. Chain Initiation

fMet Is the N-Terminal Residue of Prokaryotic Polypeptides

The first indication that the initiation of translation requires a special codon, since identified as AUG (and, in prokaryotes, occasionally GUG), was the observation that almost one half of the *E. coli* proteins begin with the otherwise uncommon amino acid Met. This was followed by the discovery of a peculiar form of Met-tRNA^{Met} in which the Met residue is *N*-formylated.

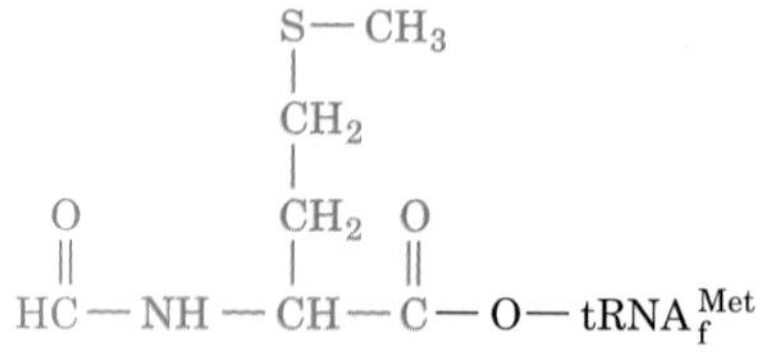

Formylmethionine-tRNA$_f^{Met}$
(fMet-tRNA$_f^{Met}$)

The *N*-formylmethionine residue **(fMet),** which already has an amide bond, can therefore only be the N-terminal residue of a polypeptide. In fact, polypeptides synthesized in an *E. coli* derived cell-free protein synthesizing system always have a leading fMet residue. *fMet must therefore be E. coli's initiating residue.*

The tRNA that recognizes the initiation codon, **tRNA$_f^{Met}$** (Fig. 30-37), differs from the tRNA that carries internal Met residues, **tRNA$_m^{Met}$,** although they both recognize the same codon. In *E. coli,* uncharged (deacylated) tRNA$_f^{Met}$ is first aminoacylated with Met by the same aminoacyl-tRNA synthetase that charges tRNA$_m^{Met}$. The resulting Met-tRNA$_f^{Met}$ is specifically *N*-formylated to yield fMet-tRNA$_f^{Met}$ in an enzymatic reaction which employs N^{10}-formyltetrahydrofolate (Section 24-4D) as its formyl donor. The formylation enzyme does not recognize Met-tRNA$_m^{Met}$. The X-ray structures of *E. coli* tRNA$_f^{Met}$ and yeast tRNA^{Phe} (Fig. 30-14) are largely similar but differ conformationally in

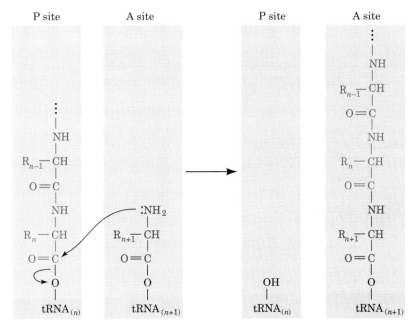

Figure 30-36
The ribosomal peptidyl transferase reaction forming a peptide bond. The amino group of the aminoacyl-tRNA in the A site nucleophilically displaces the tRNA of the peptidyl-tRNA in the P site thereby transferring the nascent polypeptide to the A site tRNA.

their acceptor stems and anticodon loops. Perhaps these structural differences permit tRNA$_f^{Met}$ to be distinguished from tRNA$_m^{Met}$ in the reactions of chain initiation and elongation (see Section 30-3D).

E. coli proteins are post-translationally modified by deformylation of their fMet residue and, in many proteins, by the subsequent removal of the resulting N-terminal Met. This processing usually occurs on the nascent polypeptide, which accounts for the observation that *E. coli* proteins all lack fMet.

Base Pairing between mRNA and the 16S rRNA Helps Select the Translational Initiation Site

AUG codes for internal Met residues as well as the initiating Met residue of a polypeptide. Moreover, mRNAs usually contain many AUGs (and GUGs) in different reading frames. Clearly, *a translational initiation site must be specified by more than just an initiation codon.*

In *E. coli,* the 16S rRNA contains a pyrimidine-rich sequence at its 3′ end. This sequence, as John Shine and Lynn Dalgarno pointed out in 1974, is partially complementary to a purine-rich tract of 3 to 10 nucleotides, the **Shine–Dalgarno sequence,** that is centered ~10 nucleotides upstream from the start codon of nearly all

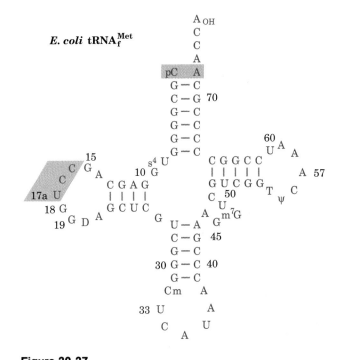

Figure 30-37
The nucleotide sequence of *E. coli* tRNA$_f^{Met}$ shown in cloverleaf form. The shaded boxes indicate the significant differences between this initiator tRNA and noninitiator tRNAs such as yeast tRNA^{Ala} (Fig. 30-11). [After Woo, N. M., Roe, B. A., and Rich, A., *Nature* **286,** 346 (1980).]

Initiation
codon

araB	– U U U G G A U G G A G U G A A A C G A U G G C G A U U–
galE	– A G C C U A A U G G A G C G A A U U A U G A G A G U U–
LacI	– C A A U U C A G G G U G G U G A U U G U G A A A C C A–
lacZ	– U U C A C A C A G G A A A C A G C U A U G A C C A U G–
Q β phage replicase	– U A A C U A A G G A U G A A A U G C A U G U C U A A G–
φX174 phage A protein	– A A U C U U G G A G G C U U U U U U A U G G U U C G U–
R17 phage coat protein	– U C A A C C G G G G U U U G A A G C A U G G C U U C U–
Ribosomal S12	– A A A A C C A G G A G C U A U U U A A U G G C A A C A–
Ribosomal L10	– C U A C C A G G A G C A A A G C U A A U G G C U U U A–
trpE	– C A A A A U U A G A G A A U A A C A A U G C A A A C A–
trp leader	– G U A A A A A G G G U A U C G A C A A U G A A A G C A–

3' end of 16S rRNA   3' ₕₒA U U C C U C C A C U A G– 5'

Figure 30-38
Some translational initiation sequences recognized by *E. coli* ribosomes. The mRNAs are aligned according to their initiation codons (*blue shading*). Their Shine–Dalgarno sequences (*red shading*) are complementary, counting G·U pairs, to a portion of the 16S rRNA's 3' end (*below*). [After Steitz, J. A., *in* Chambliss, G., Craven, G. R., Davies, J., Davis, K., Kahan, L., and Nomura, M. (Eds.), *Ribosomes. Structure, Function and Genetics, pp.* 481–482, University Park Press (1979).]

known prokaryotic mRNAs (Fig. 30-38). *Base pairing interactions between a mRNA's Shine–Dalgarno sequence and the 16S rRNA apparently permit the ribosome to select the proper initiation codon.* Thus ribosomes with mutationally altered anti-Shine–Dalgarno sequences often have greatly reduced ability to recognize natural mRNAs, although they efficiently translate mRNAs whose Shine–Dalgarno sequences have been made complementary to the altered anti-Shine–Dalgarno sequences. Moreover, treatment of ribosomes with the bacteriocidal protein **colicin E3** (produced by *E. coli* strains carrying the E3 plasmid), which specifically cleaves a 49-nucleotide fragment from the 3' terminus of 16S rRNA, yields ribosomes that cannot initiate new polypeptide synthesis but can complete the synthesis of a previously initiated chain. In fact, when ribosomes that have bound a fragment of **R17 phage** mRNA containing the initiation sequence for its so-called A protein are treated with colicin E3 and then dissociated in 1% SDS, the mRNA fragment is released in complex with the 49-nucleotide rRNA fragment (Fig. 30-39).

Initiation Is a Three-Stage Process that Requires the Participation of Soluble Protein Initiation Factors

Intact ribosomes do not directly bind mRNA so as to initiate polypeptide synthesis. Rather, *initiation is a complex process in which the two ribosomal subunits and fMet–tRNA$_f^{Met}$ assemble on a properly aligned mRNA to form a complex that is competent to commence chain elongation. This assembly process also requires the participation of protein initiation factors that are not permanently associated with the ribosome.* Initiation in *E. coli* involves three initiation factors designated **IF-1**, **IF-2**, and **IF-3** (Table 30-8). Their existence was discovered when it was found that washing small ribosomal subunits with 1*M* ammonium chloride solution, which removes the initiation factors but not the "permanent" ribosomal proteins, prevents initiation.

The initiation sequence in *E. coli* ribosomes has three stages (Fig. 30-40):

1. Upon completing a cycle of polypeptide synthesis, the 30S and 50S subunits remain associated as inactive 70S ribosomes. IF-3 binds to the 30S subunit so as to promote the dissociation of this complex. IF-1 increases this dissociation rate, perhaps by assisting the binding of IF-3.

2. GTP, mRNA, and a complex of IF-2 with fMet–tRNA$_f^{Met}$ subsequently bind to the 30S subunit in unknown order. Hence, fMet–tRNA$_f^{Met}$ recognition must not be mediated by a codon–anticodon interaction; it is the only tRNA–ribosome association not to require one. This interaction, nevertheless, helps

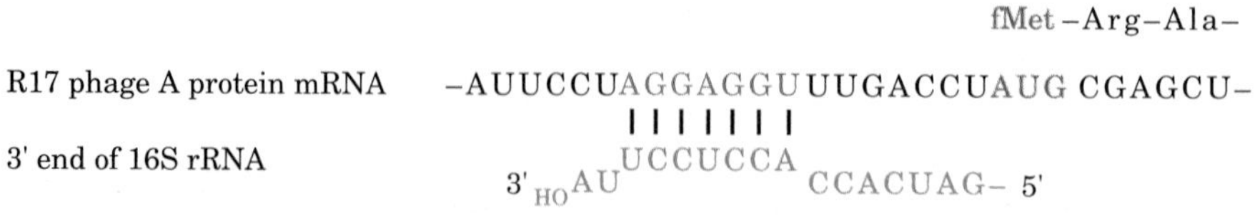

fMet–Arg–Ala–

R17 phage A protein mRNA   –A U U C C U A G G A G G U U U G A C C U A U G C G A G C U–

‖ ‖ ‖ ‖ ‖ ‖

3' end of 16S rRNA   3' ₕₒA U ᵁᶜᶜᵁᶜᶜᴬ C C A C U A G– 5'

Figure 30-39
Base pairing interactions between the colicin E3 fragment of *E. coli* 16S rRNA and the R17 phage A protein initiator region. [After Steitz, J. A. and Jakes, K., *Proc. Natl. Acad. Sci.* **72**, 4735 (1975).]

Table 30-8
The Soluble Protein Factors of *E. coli* Protein Synthesis

Factor	Mass (kD)	Function
Initiation factors		
IF-1	9	Assists IF-3 binding
IF-2	97	Binds initiator tRNA and GTP
IF-3	22	Releases 30S subunit from inactive ribosome and aids mRNA binding
Elongation factors		
EF-Tu	43	Binds aminoacyl-tRNA and GTP
EF-Ts	74	Displaces GDP from EF-Tu
EF-G	77	Promotes translocation by binding GTP to the ribosome
Release factors		
RF-1	36	Recognizes UAA and UAG Stop codons
RF-2	38	Recognizes UAA and UGA Stop codons
RF-3	46	Binds GTP and stmulates RF-1 and RF-2 binding

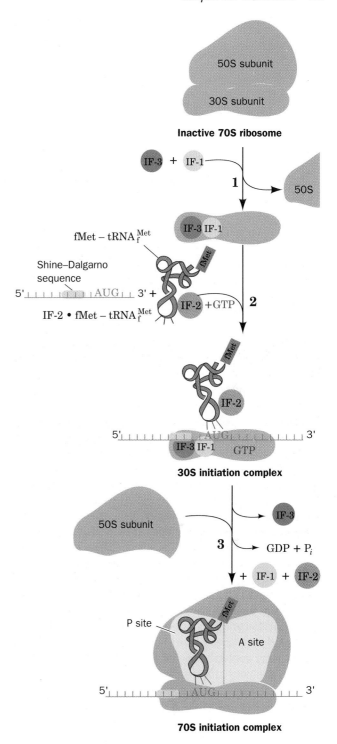

Figure 30-40
The initiation pathway in *E. coli* ribosomes.

bind fMet-tRNA$_f^{Met}$ to the ribosome. IF-3 also functions in this stage of the initiation process: It assists the 30S subunit in binding the mRNA's Shine–Dalgarno sequence.

3. Lastly, in a process that is preceded by IF-3 release, the 50S subunit joins the 30S initiation complex in a manner that hydrolyzes its bound GTP to GDP + P$_i$. This irreversible reaction conformationally rearranges the 30S subunit and releases IF-1 and IF-2 for participation in further initiation reactions.

Initiation results in the formation of an fMet–tRNA$_f^{Met}$·mRNA·ribosome complex in which the fMet–tRNA$_f^{Met}$ occupies the ribosome's P site while its A site is poised to accept an incoming aminoacyl-tRNA (an arrangement analogous to that at the conclusion of a round of elongation: Section 30-3D). This arrangement was established through the use of the antibiotic **puromycin** as is discussed in Section 30-3D. Note that tRNA$_f^{Met}$ is the only tRNA that directly enters the P site. All other tRNAs must do so via the A site during chain elongation (Section 30-3D).

Eukaryotic Initiation Resembles that of Prokaryotes

Eukaryotic initiation resembles the overall prokaryotic process but differs from it in detail. Ribosomes have a far more extensive "zoo" of initiation factors (designated eIF-*n*; "e" for eukaryotic) than do prokaryotes. Over 10, many with multiple subunits, occur in some eukaryotic systems although they are more difficult to distinguish from ribosomal proteins than are prokaryotic initiation factors.

The most striking difference between eukaryotic and prokaryotic ribosomal initiation occurs in the second

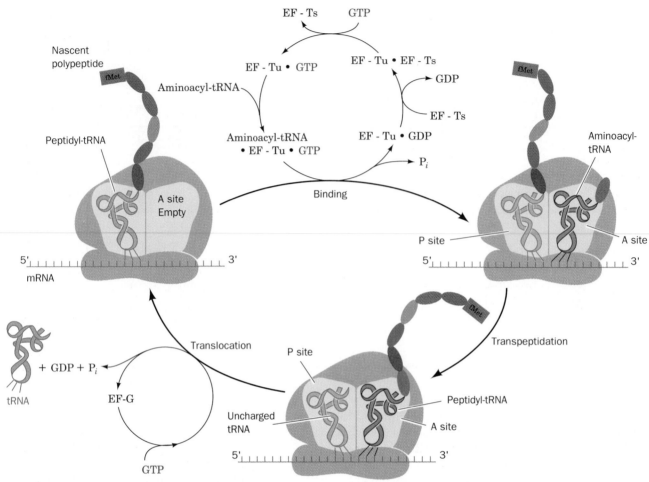

Figure 30-41
The elongation cycle in *E. coli* ribosomes. Eukaryotic elongation follows a similar cycle but EF-Tu and EF-Ts are replaced by a single multisubunit protein, eEF-1, and EF-G is replaced by eEF-2.

stage of the process, the binding of mRNA and a complex of **eIF-2**, GTP, and **Met-tRNA$_i^{Met}$** to the 40S ribosomal subunit (here the subscript "i" distinguishes eukaryotic initiator tRNA, whose appended Met residue is never *N*-formylated, from that of prokaryotes; both species are, nevertheless, readily interchangeable *in vitro*). Eukaryotic mRNAs lack the complementary sequences to bind 18S rRNA in the Shine–Dalgarno manner. Rather, *translation of eukaryotic mRNAs, which are invariably monocistronic, almost always starts at their first AUG*. This, together with the observations that (1) prokaryotic but not eukaryotic ribosomes can inititiate on circular RNAs, and (2) a subunit of **eIF-4F** is a **cap binding protein,** suggests that the 40S subunit binds at or near eukaryotic mRNA's 5' cap (Section 29-4A) and migrates downstream until it encounters the first AUG. This hypothesis explains the greatly reduced initiation rates of improperly capped mRNAs.

D. Chain Elongation

Ribosomes elongate polypeptide chains in a three-stage reaction cycle that adds amino acid residues to a growing *polypeptide's C-terminus (Fig. 30-41)*. This process, which occurs at a rate of up to 40 residues/s, involves the participation of several nonribosomal proteins known as **elongation factors** (Table 30-8).

Aminoacyl-tRNA Binding

In the "binding" stage of the *E. coli* elongation cycle, a binary complex of GTP with the elongation factor **EF-Tu** combines with an aminoacyl-tRNA. The resulting ternary complex binds to the ribosome and, in a reaction that hydrolyzes the GTP to GDP + P$_i$, the aminoacyl-tRNA is bound in a codon–anticodon complex to the ribosomal A site and EF-Tu·GDP + P$_i$ is released. In the remainder of this stage, which serves to regenerate the EF-Tu·GTP complex, GDP is displaced from EF-Tu·GDP by the elongation factor **EF-Ts** which, in turn, is displaced by GTP.

Aminoacyl-tRNAs can bind to the ribosomal A site without the mediation of EF-Tu but at a rate too slow to support cell growth. The importance of EF-Tu is indicated by the fact that it is the most abundant *E. coli* protein; it is present in ~100,000 copies per cell (>5%

of the cell's protein), which is approximately the number of tRNA molecules in the cell. Consequently, *the cell's entire complement of aminoacyl-tRNAs is essentially sequestered by EF-Tu.*

EF-Tu binds neither formylated nor unformylated Met-tRNA$_f^{Met}$, which is why the initiator tRNA never reads internal AUG or GUG codons. What is the structural basis of this discrimination? *E. coli* tRNA$_f^{Met}$ differs from other *E. coli* tRNAs by the absence of a base pair at the end of its amino acid stem (Fig. 30-37). The conversion of its 5'-terminal C residue to U by bisulfite treatment, which reestablishes the missing base pair as U · A, allows EF-Tu binding. Evidently, EF-Tu recognizes the amino acid stem of noninitiator tRNAs. However, the initiator tRNAs from several other sources do have fully base paired amino acid stems.

Transpeptidation

The peptide bond is formed in the second stage of the elongation cycle through the nucleophilic displacement of the P-site tRNA by the amino group of a 3'-linked aminoacyl-tRNA in the A site (Fig. 30-36). The nascent polypeptide chain is thereby lengthened at its C-terminus by one residue and transferred to the A-site tRNA, a process called **transpeptidation.** The reaction occurs without the need of activating cofactors such as ATP because the ester linkage between the nascent polypeptide and the P-site tRNA is a "high-energy" bond. The peptidyl transferase center that catalyzes peptide bond formation is located entirely on the large subunit as is demonstrated by the observation that in high concentrations of organic solvents such as ethanol, the large subunit alone catalyzes peptide bond formation. The organic solvent apparently distorts the large subunit in a way that mimics the effect of small subunit binding. Peptidyl transferase activity seems to arise from the juxtaposition of several polypeptide chains in the large subunit together with the 23S RNA (Section 30-3A).

Translocation

In the final stage of the elongation cycle, the now uncharged P-site tRNA (at first tRNA$_f^{Met}$ but subsequently a noninitiator tRNA) is expelled (or, perhaps, transferred to the E site and expelled in the next binding reaction) and, in a process known as **translocation,** *the peptidyl-tRNA in the A site, together with its bound mRNA, is moved to the P site.* This prepares the ribosome for the next elongation cycle. The maintenance of the peptidyl-tRNA's codon–anticodon association is no longer necessary for amino acid specification. Rather, it probably acts as a placekeeper that permits the ribosome to precisely step off the three nucleotides along the mRNA required to preserve the reading frame. Indeed, the observation that frameshift suppressor tRNAs induce a four nucleotide translocation (Section 30-2E) indicates that mRNA movement is directly coupled to tRNA movement.

The translocation process requires the participation of an elongation factor, **EF-G,** that binds to the ribosome together with GTP and is only released upon hydrolysis of the GTP to GDP + P$_i$. EF-G release is prerequiste for beginning the next elongation cycle because the ribosomal binding of EF-G and EF-Tu are mutually exclusive. Translocation is clearly a highly complex mechanical process and, unfortunately, one that is but poorly understood.

Puromycin Is an Aminoacyl-tRNA Analog

The ribosomal elongation cycle was originally characterized through the use of the antibiotic **puromycin** (Fig. 30-42). *This substance, which resembles the 3' end of Tyr-tRNA, causes the premature termination of polypeptide chain synthesis.* Puromycin, in competition with the specified aminoacyl-tRNA but without the need of elongation factors, binds to the ribosomal A site which, in turn, catalyzes a normal transpeptidation reaction to form peptidyl-puromycin. Yet, the ribosome cannot catalyze the transpeptidation reaction in the next elonga-

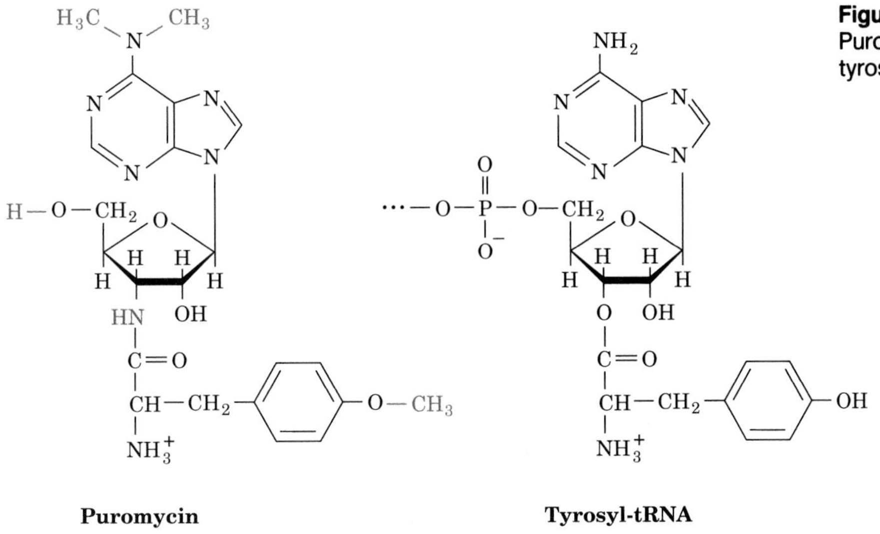

Figure 30-42
Puromycin (*left*) resembles the 3' terminus of tyrosyl-tRNA (*right*).

Puromycin

Tyrosyl-tRNA

tion cycle because puromycin's "amino acid residue" is linked to its "tRNA" via an amide rather than an ester bond. Polypeptide synthesis is therefore aborted and the peptidyl-puromycin is released.

In the absence of EF-G and GTP, an active ribosome cannot bind puromycin because its A site is already occupied by a peptidyl-tRNA. A newly initiated ribosome, however, violates this rule; it catalyzes fMet-puromycin formation. *These observations demonstrated the functional existence of the ribosomal P and A sites and established that fMet-tRNA$_f^{Met}$ binds directly to the P site, whereas other aminoacyl-tRNAs must first enter the A site.*

The Eukaryotic Elongation Cycle Resembles that of Prokaryotes

The eukaryotic elongation cycle closely resembles that of prokaryotes. In eukaryotes, the functions of EF-Tu and EF-Ts are assumed by two different subunits of the eukaryotic elongation factor **eEF-1.** Likewise **eEF-2** functions in a manner analogous to EF-G. However, the corresponding eukaryotic and prokaryotic elongation factors are not interchangable.

E. Chain Termination

Polypeptide synthesis under the direction of synthetic mRNAs such as poly(U) terminates with a peptidyl-tRNA in association with the ribosome. However, *the translation of natural mRNAs, which contain the termination codons UAA, UGA, or UAG, results in the production of free polypeptides (Fig. 30-43).* In *E. coli,* the termination codons, the only codons that normally have no corresponding tRNAs, are recognized by protein **release factors** (Table 30-8): **RF-1** recognizes UAA and UAG, whereas **RF-2** recognizes UAA and UGA. Neither of these release factors can bind to the ribosome simultaneously with EF-G. A third release factor, **RF-3,** which binds GTP, stimulates the ribosomal binding of RF-1 and RF-2. The release factors act at the ribosomal A site as is indicated by the observation that they compete with suppressor tRNAs for termination codons.

The binding of a release factor to the appropriate termination codon induces the ribosomal peptidyl transferase to transfer the peptidyl group to water rather than to an aminoacyl-tRNA (Fig. 30-44). The consequent uncharged tRNA subsequently dissociates from the ribosome and the release factors are expelled with the concomitant hydrolysis of GTP to GDP + P_i. The resulting inactive ribosome then releases its bound mRNA preparatory for a new round of polypeptide synthesis.

Termination in eukaryotes resembles that in prokaryotes but requires only a single release factor, **eRF,** that binds to the ribosome together with GTP. This GTP is hydrolyzed to GDP + P_i in a reaction that is thought to trigger eRF's dissociation from the ribosome.

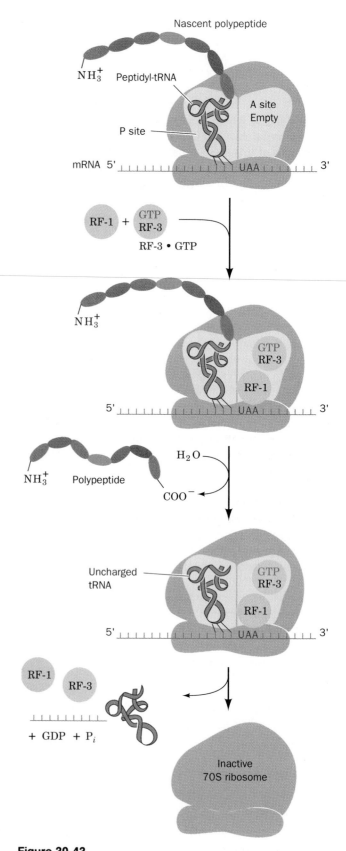

Figure 30-43

The termination sequence in *E. coli* ribosomes. RF-1 recognizes the termination codons UAA and UAG, whereas RF-2 recognizes UAA and UGA. Eukaryotic termination follows an analogous pathway but requires only a single release factor, eRF, that recognizes all three termination codons.

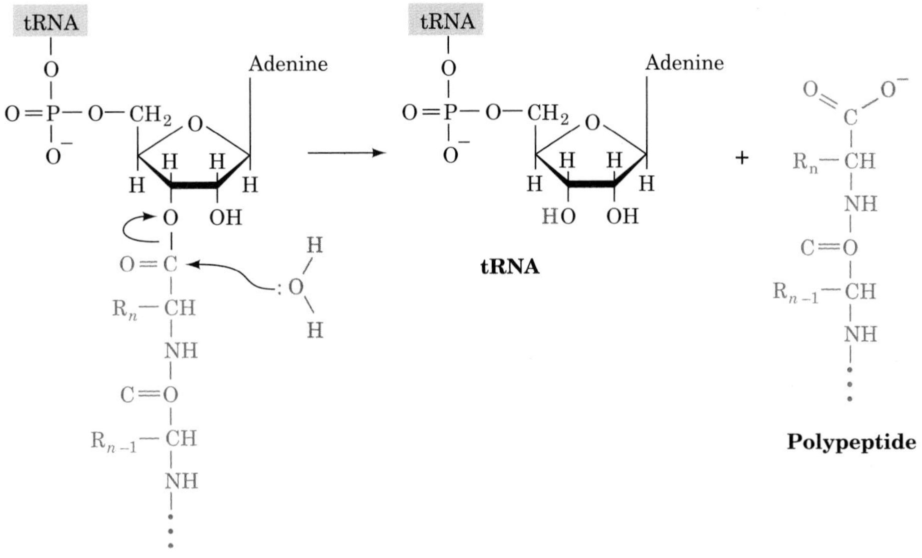

tRNA

$O=P-O-CH_2$ Adenine

Peptidyl-tRNA

tRNA

$O=P-O-CH_2$ Adenine

tRNA

Polypeptide

Figure 30-44
The ribosome catalyzed hydrolysis of peptidyl-tRNA to form a polypeptide and free tRNA.

GTP Hydrolysis Speeds Up Ribosomal Processes

What is the role of the various GTP hydrolysis reactions that are essential for normal ribosomal function? Translation occurs in the absence of GTP, albeit extremely slowly, so that the free energy of the transpeptidation reaction is sufficient to drive the entire translational process. Moreover, none of the GTP hydrolysis reactions yield a "high-energy" covalent intermediate as does, say ATP hydrolysis in numerous biosynthetic reactions. It is therefore thought that GTP binding allosterically causes ribosomal components to change their conformations in a way that facilitates a particular process such as translocation. This conformational change also catalyzes GTP hydrolysis which, in turn, permits the ribosome to relax to its initial conformation with the concomitant release of products including GDP + P_i. *The high rate and irreversibility of the GTP hydrolysis reaction therefore ensures that the various complex ribosomal processes to which it is coupled, initiation, elongation, and termination, will themselves be fast and irreversible.* GTP hydrolysis also facilitates translational accuracy (see below).

F. Translational Accuracy

The genetic code is normally translated with remarkable fidelity. We have already seen that transcription and tRNA aminoacylation both proceed with high accuracy (Sections 29-2D and 30-2C). The accuracy of ribosomal mRNA decoding was estimated from the rate of misincorporation of ^{35}S-Cys into highly purified **flagel-lin,** an *E. coli* protein (Section 34-3G) that normally lacks Cys. These measurements indicated that the mistranslation rate is $\sim 10^{-4}$ errors/per codon. This rate is greatly increased in the presence of **streptomycin,** an antibiotic that increases the rate of ribosomal misreading (Section 30-3G). From the types of reading errors that streptomycin is known to induce, it was concluded that the mistranslation arose almost entirely from the confusion of the Arg codons CGU and CGC for the Cys codons UGU and UGC. The above error rate is therefore largely caused by mistakes in ribosomal decoding.

Aminoacyl-tRNAs are selected by the ribosome only according to their anticodon. Yet, the binding energy loss arising from a single base mismatch in a codon–anticodon interaction is estimated to be ~ 12 kJ·mol^{-1} which, according to Eq. [30.1], cannot account for a ribosomal decoding accuracy of less than $\sim 10^{-2}$ errors per codon. Evidently, the ribosome has some sort of proofreading mechanism that increases its overall decoding accuracy.

How might a ribosome proofread a codon–anticodon interaction? Two types of mechanisms can be envisaged: (1) a selective binding mechanism, such as those of aminoacyl-tRNA synthetases (Section 30-2C); and (2) a kinetic mechanism. The problem with a ribosomal selective binding mechanism is that there is little evidence indicating the existence of a second aminoacyl-tRNA binding site that functions to exclude improper codon–anticodon interactions. Evidence is accumulating, however, that is consistent with a **kinetic proofreading** mechanism.

Kinetic Proofreading Requires a Branched Reaction Path

Kinetic proofreading models of tRNA selection require only one-binding site. John Hopfield theorized that such a process can occur via a branched reaction mechanism of polypeptide chain elongation such as is diagrammed in Fig. 30-45:

1. The initial-binding reaction discriminates, as discussed, between cognate (specified) and noncognate (unspecified) tRNAs according to their codon–anticodon binding energies.

2. Following this, the bound GTP irreversibly hydrolyzes yielding an activated intermediate as is indicated by the star. This complex can then react in one of two ways:

   (a) EF-Tu·GDP can dissociate from the ribosome with rate constant k_3 thereby committing the ribosome to form a peptide bond.

   (b) Alternatively, the aminoacyl-tRNA can dissociate from the ribosome with rate constant k_4 thereby aborting the elongation step. EF-Tu·GDP subsequently dissociates from the

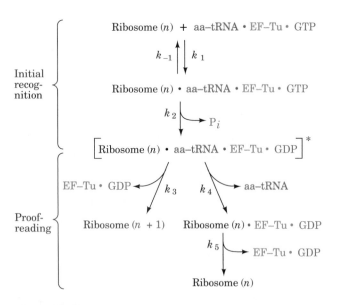

Figure 30-45
A kinetic proofreading mechanism for selecting a correct codon–anticodon interaction. The initial recognition reaction screens the aminoacyl-tRNA (aa-tRNA) for the correct codon–anticodon interaction. The resulting complex converts, in a GTP-driven process, to a "high-energy" intermediate (*) which, in turn, either releases EF-Tu·GDP preparatory to forming a peptide bond or releases aminoacyl-tRNA before EF-Tu·GDP is released. If k_4/k_3 is greater for a codon–anticodon mismatch than it is for a match, then these latter steps constitute a proofreading mechanism for proper tRNA binding.

Table 30-9
Some Ribosomal Inhibitors

Inhibitor	Action
Chloramphenicol	Inhibits peptidyl transferase on the prokaryotic large subunit
Cycloheximide	Inhibits peptidyl transferase on the eukaryotic large subunit
Erythromycin	Inhibits translocation by the prokaryotic large subunit
Fusidic acid	Inhibits elongation in prokaryotes by preventing EF-G·GDP dissociation from the large subunit
Puromycin	An aminoacyl-tRNA analog that causes premature chain termination in prokaryotes and eukaryotes
Streptomycin	Causes mRNA misreading and inhibits chain initiation in prokaryotes
Tetracycline	Inhibits the binding of aminoacyl-tRNAs to the prokaryotic small subunit
Diphtheria toxin	Catalytically inactivates eEF-2 by ADP-ribosylation
Ricin/Abrin	Poisonous plant proteins that catalytically inactivate the eukaryotic large subunit

ribosome permitting it to reinitiate the elongation step.

If the ratio k_4/k_3 is greater for noncognate than for cognate tRNAs, then a second screening will have occurred. It is thought that the physical basis of this second screening is that k_3 is independent of the tRNA's identity, whereas k_4 is larger for a relatively weakly bound noncognate tRNA than it is for a cognate tRNA. The rate of EF-Tu·GDP dissociation therefore provides a countdown clock against which the ribosome measures the rate of tRNA dissociation: A noncognate tRNA usually dissociates from the ribosome before EF-Tu·GDP does (an average period of several milliseconds), whereas a cognate tRNA usually remains bound. The activated intermediate is essential for this process because otherwise its tRNA dissociation step (that characterized by k_4) would be identical to that of the initial recognition step (that characterized by k_{-1}). GTP hydrolysis therefore provides the second context necessary for proofreading.

The kinetic proofreading model is supported by the following observations:

1. More GTP is hydrolyzed per peptide bond formed with noncognate than with cognate tRNAs (although this observation is also consistent with selective binding models).

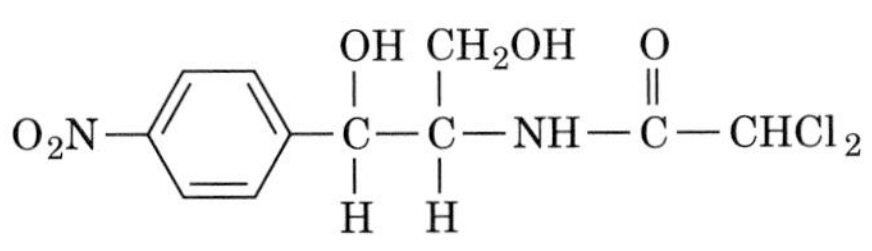

Chloramphenicol

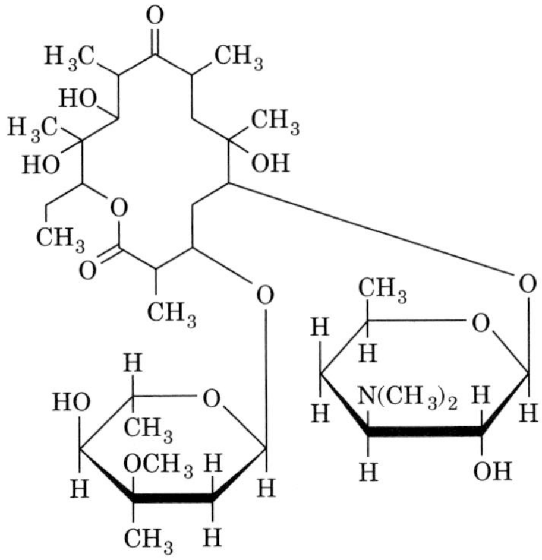

Erythromycin

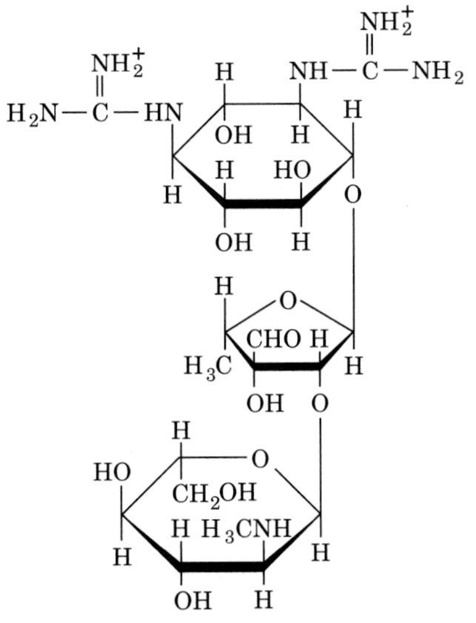

Streptomycin

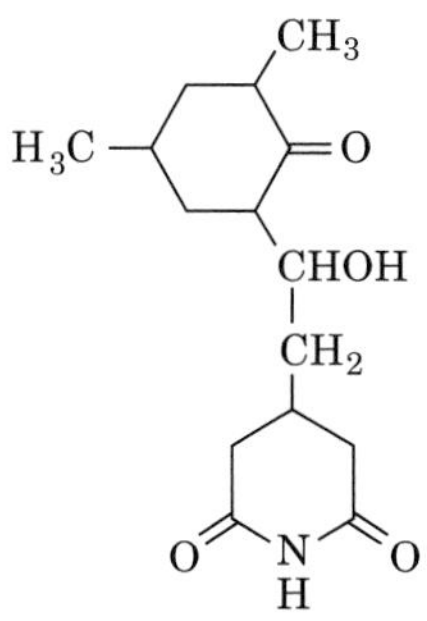

Cycloheximide

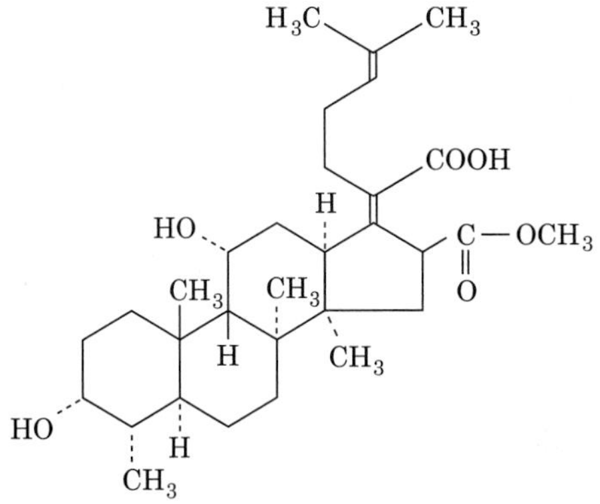

Fusidic acid

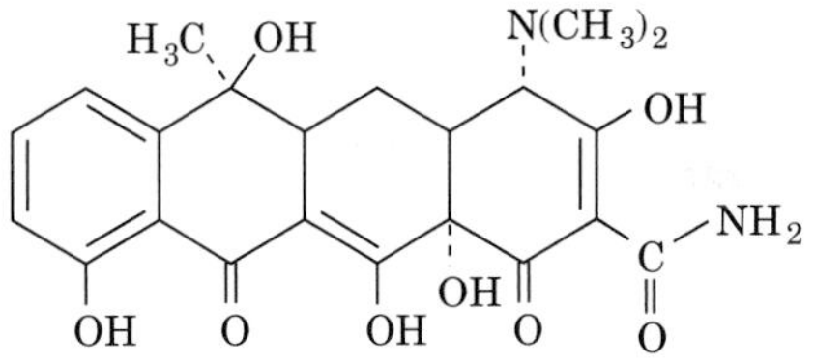

Tetracycline

Figure 30-46
A selection of antibiotics that act as translational inhibitors.

2. The rate of EF-Tu · GDP dissociation from a ribosome is, in fact, independent of its bound aminoacyl-tRNA's identity.

G. Protein Synthesis Inhibitors: Antibiotics

Antibiotics are bacterially or fungally produced substances that inhibit the growth of other organisms. Antibiotics are known to inhibit a variety of essential biological processes including DNA replication (e.g., novobiocin, Section 28-5C), transcription (e.g., rifamycin B; Section 29-2C), and bacterial cell wall synthesis (e.g., penicillin; Section 10-3B). However, *the majority of known antibiotics, including a great variety of medically useful substances, block translation.* This situation is presumably a consequence of the translational machinery's enormous complexity, which makes it vulnerable to disruption in many ways. Antibiotics have also been useful in analyzing ribosomal mechanisms because, as we have seen for puromycin (Section 30-3D), the blockade of a specific function often permits its biochemical dissection into its component steps. Table 30-9 and Fig. 30-46 present several medically significant and/or biochemically useful translational inhibitors. We study the mechanisms of a few of the best characterized of them below.

Streptomycin

Streptomycin, which was discovered in 1944 by Selman Waksman, is a medically important member of a family of antibiotics known as **aminoglycosides** that inhibit prokaryotic ribosomes in a variety of ways. At low concentrations, streptomycin induces the ribosome to characteristically misread mRNA: One pyrimidine may be mistaken for the other in the first and second codon positions and either pyrimidine may be mistaken for adenine in the first position. This inhibits the growth of susceptible cells but does not kill them. At higher concentrations, however, streptomycin prevents proper chain initiation and thereby causes cell death.

Certain streptomycin resistant mutants (str^R) have ribosomes with an altered protein S12 compared with streptomycin sensitive bacteria (str^S). Intriguingly, a change in base C912 of 16S rRNA (which lies in the central loop in Fig. 30-25) also confers streptomycin resistance. (Some mutant bacteria are not only resistant to streptomycin but dependent on it; they require it for growth.) In partial diploid bacteria that are heterozygous for streptomycin resistance (str^R/str^S), streptomycin sensitivity is dominant. This puzzling observation is explained by the finding that, in the presence of streptomycin, str^S ribosomes remain bound to initiation sites thereby excluding str^R ribosomes from these sites. Moreover, the mRNAs in these blocked complexes are degraded after a few minutes, which alows the str^S ribosomes to bind to newly synthesized mRNAs as well.

Chloramphenicol

Chloramphenicol, the first of the "broad-spectrum" antibiotics, inhibits the peptidyl transferase activity on the large subunit of prokaryotic ribosomes. However, its clinical uses are limited to only severe infections because of its toxic side effects which are caused, at least in part, by the chloramphenicol sensitivity of mitochondrial ribosomes. Binding experiments with reconstituted 50S subunits that are missing one or another component suggest that protein L16 is necessary for chloramphenicol binding. This is corroborated by affinity labeling experiments indicating that L16, as well as several other large subunit proteins and the 23S RNA, are in proximity to bound chloramphenicol. The 23S RNA is also implicated in chloramphenicol resistance by the observation that some of its mutants are chloramphenicol resistant. Chloramphenicol's binding-site must lie near the large subunit's A site since chloramphenicol competes for binding with puromycin and the 3' end of aminoacyl-tRNAs but not with peptidyl-tRNAs. This observation suggests that chloramphenicol inhibits peptidyl transfer by interfering with the interactions of ribosomes with A site-bound aminoacyl-tRNAs.

Tetracycline

Tetracycline and its derivatives are broad-spectrum antibiotics that bind to the small subunit of prokaryotic ribosomes where they inhibit aminoacyl-tRNA binding. Tetracycline also blocks the stringent response (Section 29-3F) by inhibiting ppGpp synthesis. This indicates that deacylated tRNA must bind to the A site in order to activate stringent factor.

Tetracycline-resistant bacterial strains have become quite common thereby precipitating a serious clinical problem. Most often, however, resistance is conferred by a decrease in bacterial cell membrane permeability to tetracycline rather than any alteration of ribosomal components.

Diphtheria Toxin

Diphtheria is a disease that results from bacterial infection by *Corynebacterium diphtheriae* that harbor the bacteriophage **corynephage β.** Diphtheria was a leading cause of childhood death until early in this century when immunization became prevalent. Although the bacterial infection is usually confined to the upper respiratory tract, the bacteria secrete a phage-encoded protein, known as **diphtheria toxin,** that is responsible for the disease's lethal effects. *Diphtheria toxin specifically inactivates the eukaryotic elongation factor eEF-2 thereby inhibiting eukaryotic protein synthesis.*

The pathogenic effects of diphtheria are prevented, as was discovered in the 1880s, by immunization with **toxoid,** formaldehyde inactivated toxin. Individuals who have contracted diphtheria are treated with antitoxin from horse serum, which binds to and thereby inactivates diphtheria toxin, as well as with antibiotics to combat the bacterial infection.

Diphtheria toxin acts in a particularly interesting way. It is a monomeric 58-kD protein that is readily cleaved by trypsin and trypsin-like enzymes into two fragments, A and B. The B domain of intact toxin binds to an unknown receptor on the plasma membrane of susceptible cells. The toxin is then proteolytically cleaved whereupon the B fragment facilitates the A fragment's cytosolic uptake via receptor-mediated endocytosis (free fragment A is devoid of toxic activity).

Within the cytosol, the A fragment catalyzes the **ADP-ribosylation** of eEF-2 by NAD^+,

$$\begin{array}{c} \text{eEF-2} + \text{NAD}^+ \\ \text{(active)} \\ \downarrow \text{diphtheria toxin} \\ \text{ADP-ribosyl-eEF-2} + \text{Nicotinamide} + \text{H}^+ \\ \text{(inactive)} \end{array}$$

thereby inactivating this elongation factor. Since the A fragment acts catalytically, *one molecule is sufficient to ADP-ribosylate all of a cell's eEF-2s, which halts protein synthesis and kills the cell.* Only a few micrograms of diphtheria toxin are therefore sufficient to kill an unimmunized individual.

Diphtheria toxin specifically ADP-ribosylates a modified His residue on eEF-2 known as **diphthamide:**

ADP-Ribosylated diphthamide

Diphthamide occurs only in eEF-2 (not even in its bacterial counterpart, EF-G), which accounts for the specificity of diphtheria toxin in exclusively modifying eEF-2. The observation that diphthamide occurs in all eukaryotic eEF-2's suggests that this residue is essential to eEF-2 activity. Yet, certain mutant cultured animal cells, which have unimpaired capacity to synthesize proteins, lack the enzymes that post-translationally modify His to diphthamide. Diphthamide's normal biological role is therefore a mystery.

4. CONTROL OF EUKARYOTIC TRANSLATION

The rates of ribosomal initiation on prokaryotic mRNAs vary by factors of up to 100. For example, the proteins specified by the *E. coli lac* operon, β-galactosidase, galactose permease, and thiogalactoside transacetylase, are synthesized in molar ratios of 10:5:2. This variation is probably a consequence of their different Shine–Dalgarno sequences. Alternatively, ribosomes may attach to *lac* mRNA only at its β-galactosidase gene and occasionally detach in response to a chain termination signal (thereby accounting for the decreasing translational rates along the operon). At any rate, there is no evidence that prokaryotic translation rates are responsive to environmental changes. *Genetic expression in prokaryotes is therefore almost entirely transcriptionally controlled (Section 29-3).* Of course, since their mRNAs have lifetimes of only a few minutes, it would seem that prokaryotes have little need of translational controls.

Eukaryotic transcriptional control, although far more complex than that in prokaryotes, is largely reserved for regulating cell differentiation (Section 33-3). There are,

however, increasing indications that eukaryotic cells respond to their needs, at least in part, through translational control. This is feasible because the lifetimes of eukaryotic mRNAs are generally hours or days. In this section, we examine the two best-characterized eukaryotic translational control mechanisms: (1) the regulation of hemoglobin synthesis by heme, and (2) the effects of virus-induced proteins known as **interferons.** We also consider the phenomenon of mRNA masking.

A. Translational Control by Heme

Reticulocytes synthesize protein, almost exclusively hemoglobin, at an exceedingly high rate and are therefore a favorite subject for the study of eukaryotic translation. Hemoglobin synthesis in fresh reticulocyte lysates proceeds normally for several minutes but then abruptly stops because of the inhibition of translational initiation and the consequent polysome disaggregation. This effect is prevented by the addition of heme [a mitochondrial product (Section 24-4A) that this *in vitro* system cannot synthesize] thereby indicating that *globin synthesis is regulated by heme availability.* The inhibition of globin translational initiation is also reversed by the addition of the eukaryotic initiation factor eIF-2 and by high levels of GTP.

In the absence of heme, reticulocyte lysates accumulate a protein, **heme-controlled inhibitor (HCI),** that phosphorylates a specific Ser residue on the α subunit of eIF-2 (eIF-2 is an αβγ trimer that carries GTP and Met-tRNA$_i^{Met}$ to the 40S ribosomal subunit; Section 30-3C). HCI is generated, in the absence of heme, from a pre-existing proinhibitor by a poorly characterized process that probably involves at least one other protein.

Phosphorylated eIF-2 can participate in the ribosomal initiation process in much the same way as unphosphorylated eIF-2. This puzzling situation was clarified by the discovery that GDP does not spontaneously dissociate from eIF-2 at the completion of initiation as it does from IF-2 in the corresponding prokaryotic process (Fig. 30-40). Rather, eIF-2 exchanges its GDP for GTP in a reaction mediated by another initiation factor, **eIF-2B** (Fig. 30-47). It turns out that phosphorylated eIF-2 forms a much tighter complex with eIF-2B than does unphosphorylated eIF-2. This sequesters eIF-2B (Fig.

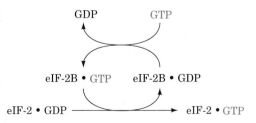

Figure 30-47
The eIF-2·GDP product of eukaryotic ribosomal initiation is regenerated by GDP–GTP interchange with eIF-2B·GTP.

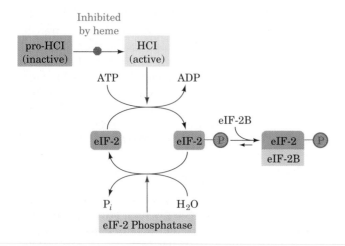

Figure 30-48
A model for heme-controlled protein synthesis in reticulocytes.

30-48), which is present in lesser amounts than is eIF-2, thereby preventing regeneration of the eIF-2·GTP required for translational initiation. The presence of heme reverses translational inhibition by inhibiting HCI. The phosphorylated eIF-2 molecules are reactivated through the action of **eIF-2 phosphatase,** which is unaffected by heme. Reticulocytes, in addition, contain a recently discovered 67-kD protein that protects eIF-2 from HCI-catalyzed phosphorylation.

B. Interferon

Interferons are glycoproteins that are secreted by virus infected vertebrate cells. Upon binding to surface receptors of other cells, interferons convert them to an antiviral state, which impairs the replication of a wide variety of RNA and DNA viruses. Indeed, the discovery of interferon in the 1950s arose from the observation that virus-infected individuals are resistant to infection by a second type of virus.

There are three families of interferons: **type α or leucocyte interferon** (leucocytes are white blood cells), the related **type β or fibroblast interferon** (fibroblasts are connective tissue cells), and **type γ or lymphocyte interferon** (lymphocytes are immune system cells). *Interferon synthesis is induced by double-stranded RNA (dsRNA),* which is probably generated during infection by both DNA and RNA viruses, as well as by the synthetic dsRNA poly(I)·poly(C). Interferons are effective antiviral agents in concentrations as low as $3 \times 10^{-14}M$, which makes them among the most potent biological substances. Moreover, they have far wider specificities than antibodies raised against a particular virus. They have therefore elicited great medical interest, particularly since some cancers are virally induced (Section 33-4C). Indeed, they are in clinical use against certain tumors and viral infections. These treatments are made possible by the production of large quantities of these otherwise

quite scarce proteins through molecular cloning techniques (Section 28-8).

Interferons prevent viral proliferation largely by inhibiting protein synthesis in infected cells (lymphocyte interferon also modulates the immune response). They do so in two independent ways (Fig. 30-49):

1. Interferons induce the production of a protein kinase that, in the presence of dsRNA, phosphorylates the eIF-2 α subunit identically to the action of HCI in reticulocytes, thereby inhibiting ribosomal initiation. This observation suggests that eIF-2 phosphorylation may be a general mechanism of eukaryotic translational control.

2. Interferons also induce the synthesis of (2',5')-

(a)

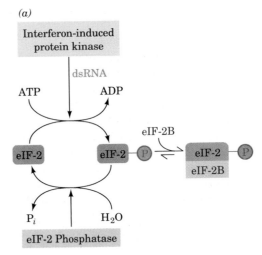

Inhibition of Translation

(b)

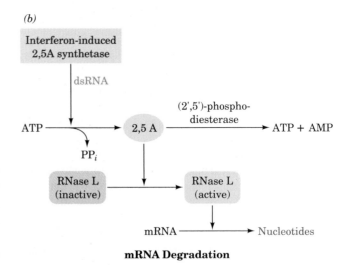

mRNA Degradation

Figure 30-49
In interferon treated cells, the presence of dsRNA, which normally results from a viral infection, causes (*a*) the inhibition of translational initiation, and (*b*) the degradation of mRNA, thereby blocking translation and preventing virus replication.

oligoadenylate synthetase. In the presence of dsRNA, this enzyme catalyzes the synthesis from ATP of the unusual oligonucleotide **pppA(2'p5'A)$_n$** where $n = 1$ to 10. *This compound, 2,5-A, activates a preexisting endonuclease, RNase L, to degrade mRNA thereby inhibiting protein synthesis.* 2,5-A is itself rapidly degraded by an enzyme named **(2',5')-phosphodiesterase** so that it must be continually synthesized to maintain its effect.

The independence of the 2,5-A and the interferon-induced protein kinase systems is demonstrated by the observation that the effect of 2,5-A on protein synthesis is reversed by added mRNA but not by eIF-2.

C. mRNA Masking

It has been known since the previous century that early embryonic development in organisms such as sea urchins is governed almost entirely by information present in the egg before fertilization. Indeed, sea urchin embryos exposed to sufficient actinomycin D (Section 29-2D) to inhibit RNA synthesis without blocking DNA synthesis develop normally through their early stages without a change in their protein synthesis program. This is because *an unfertilized egg contains large quantities of mRNA that is "masked" by associated proteins so as to prevent its association with the ribosomes that are also present. Upon fertilization, this mRNA is somehow "unmasked" in a controlled fashion and commences directing protein synthesis.* Development of the embryo can therefore start immediately upon fertilization rather than waiting for the generation of paternally specified mRNAs.

The cytoplasms of many eukaryotic cells contain large amounts of protein-complexed mRNAs that are not associated with ribosomes. It remains to be seen, however, whether mRNA masking is used for translational control in nonembryonic tissues.

5. POST-TRANSLATIONAL MODIFICIATION

To become mature proteins, polypeptides must fold to their native conformations, their disulfide bonds, if any, must form, and, in the case of multisubunit proteins, the subunits must properly combine. Moreover, as we have seen throughout this text, many proteins are modified in enzymatic reactions that proteolytically cleave certain peptide bonds and/or derivatize specific residues. In this section we shall review some of these **post-translational modifications.**

A. Proteolytic Cleavage

Proteolytic cleavage is the most common type of post-translational modification. Probably all mature proteins have been so-modified, if by nothing else than the endoproteolytic removal of their leading Met (or fMet) residue shortly after it emerges from the ribosome. Many proteins, which are involved in a wide variety of biological processes, are synthesized as inactive precursors that are activated under proper conditions by limited proteolysis. Some examples of this phenomenon that we have encountered are the conversion of trypsinogen and chymotrypsinogen to their active forms by tryptic cleavages of specific peptide bonds (Section 14-3E), and the formation of active insulin from the 84-residue proinsulin by excision of an internal 33-residue polypeptide (Section 8-1A). Inactive proteins that are activated by removal of polypeptides are called **proproteins,** whereas the excised polypeptides are termed **propeptides.**

Propeptides Direct Collagen Assembly

Collagen biosynthesis is illustrative of many facets of post-translational modification. Recall that collagen, a major extracellular component of connective tissue, is a fibrous triple helical protein whose polypeptides largely consist of the repeating amino acid sequence (Gly-X-Y)$_n$ where X is often Pro, Y is often 4-hydroxyproline (Hyp), and $n \approx 340$ (Section 7-2C). The polypeptides of **procollagen** (Fig. 30-50) differ from those of the mature protein by the presence of both N- and C-terminal propeptides of ~ 100 residues whose sequences, for the

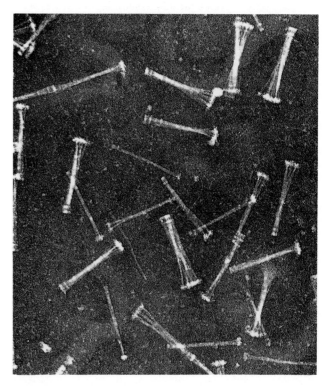

Figure 30-50
An electron micrograph of procollagen aggregates that have been secreted into the extracellular medium. [Courtesy of Jerome Gross, Harvard Medical School.]

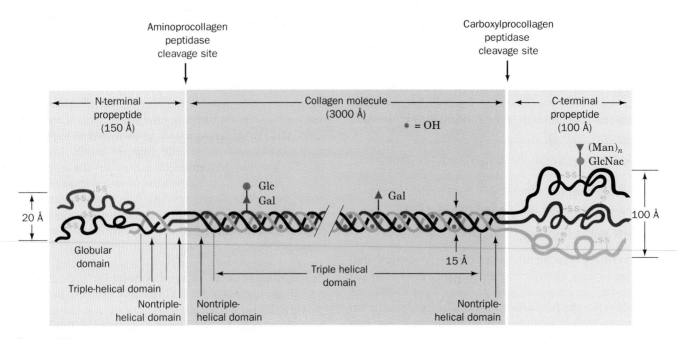

Figure 30-51
A schematic representation of the procollagen molecule. Gal, Glc, GlcNac, and Man, respectively, denote galactose, glucose, *N*-acetylglucosamine, and mannose residues. Note that the N-terminal propeptide has intrachain disulfide bonds while the C-terminal propeptide has both intrachain and interchain disulfide bonds. [After Prockop, D. J., Kivirikko, K. I., Tuderman, L., and Guzman, N. A., *New Engl. J. Med.* **301**, 16 (1979).]

most part, are unlike those of mature collagen. The procollagen polypeptides rapidly assemble, *in vitro* as well as *in vivo*, to form a collagen triple helix. In contrast, polypeptides extracted from mature collagen will only reassemble over a period of days, if at all. *The collagen propeptides are apparently necessary for proper procollagen folding.*

The N- and C-terminal propeptides of procollagen are removed by **amino-** and **carboxylprocollagen peptidases** (Fig. 30-51), which may also be specific for the different collagen types. An inherited defect of aminoprocollagen peptidase in cattle and sheep results in a bizarre condition, **dermatosparaxis,** that is characterized by extremely fragile skin. An analogous disease in man, **Ehler–Danlos syndrome VII,** is caused by a mutation in one of the procollagen polypeptides that inhibits the enzymatic removal of its aminopropeptide. Collagen molecules normally spontaneously aggregate to form collagen fibrils (Figs. 7-33 and 7-34). However, electron micrographs of dermatosparaxic skin show sparse and disorganized collagen fibrils. *The retention of collagen's aminopropeptides apparently interferes with proper fibril formation.* (The dermatosparaxis gene was bred into some cattle herds because heterozygotes produce tender meat.)

Signal Peptides Are Removed from Nascent Proteins by a Signal Peptidase

Many trans-membrane proteins or proteins that are destined to be secreted are synthesized with an N-terminal **signal peptide** of 13 to 36 predominantly hydrophobic residues. According to the **signal hypothesis** (Section 11-3F), a signal peptide is recognized by a **signal recognition particle (SRP).** The SRP binds a ribosome synthesizing a signal peptide to a receptor on the membrane [the rough endoplasmic reticulum (RER) in eukaryotes and the plasma membrane in bacteria] and conducts the signal peptide and the following nascent polypeptide through it.

Proteins bearing a signal peptide are known as **preproteins** or, if they also contain propeptides, as **preproproteins.** Once the signal peptide has passed through the membrane, it is specifically cleaved from the nascent polypeptide by a **signal peptidase.** Both insulin and collagen are secreted proteins and are therefore synthesized with leading signal peptides in the form of **preproinsulin** and **preprocollagen.** These and many other proteins are therefore subject to three sets of sequential proteolytic cleavages: (1) the deletion of their initiating Met residue, (2) the removal of their signal peptides, and (3) the excision of their propeptides.

Polyproteins

Some proteins are synthesized as segments of **polyproteins,** polypeptides that contain the sequences of two or more proteins. Examples include most polypeptide hormones (Section 33-3C), the proteins synthesized by many viruses including those causing polio (Section 32-2C) and AIDS, and **ubiquitin,** a highly conserved eukaryotic protein involved in protein degradation

(Section 30-6B). Specific proteases post-translationally cleave polyproteins to their component proteins, presumably through the recognition of the cleavage site sequences. Some of these proteases are conserved over remarkable evolutionary distances. For instance, ubiquitin is synthesized as several tandem repeats (polyubiquitin) that *E. coli* properly cleave even though prokaryotes lack ubiquitin.

B. Covalent Modification

Proteins are subject to specific chemical derivatizations, both at the functional groups of their side chains and at their terminal amino and carboxyl groups. Over 150 different types of side chain modifications, involving all side chains but those of Ala, Gly, Ile, Leu, Met, and Val, are known (Section 4-3A). These include acetylations, glycosylations, hydroxylations, methylations, nucleotidylylations, phosphorylations, and ADP-ribosylations as well as numerous "miscellaneous" modifications.

Some protein modifications, such as the phosphorylation of glycogen phosphorylase (Section 17-1A) and the ADP-ribosylation of eEF-2 (Section 30-3G), modulate protein activity. Several side chain modifications convalently bond cofactors to enzymes, presumably to increase their catalytic efficiency. Examples of linked cofactors that we have encountered are N^ε-lipoyllysine in dihydrolipoyl transacetylase (Section 19-2A) and 8α-histidylflavin in succinate dehydrogenase (Section 19-3F). The attachment of complex carbohydrates, which occur in almost infinite variety, alter the structural properties of proteins and form recognition markers in various types of targeting and cell–cell interactions (Sections 10-3C, 11-3D, and 21-3B). Modifica-

tions that cross-link proteins, such as occur in collagen and elastin (Sections 7-2C and D), stabilize supramolecular aggregates. The functions of most side chain modifications, however, remain enigmatic.

Collagen Assembly Requires Chemical Modification

Collagen biosynthesis (Fig. 30-52) is illustrative of protein maturation through chemical modification. As the nascent procollagen polypeptides pass into the RER of the fibroblasts that synthesized them, the Pro and Lys residues are hydroxylated to Hyp, 3-hydroxy-Pro, and 5-hydroxy-Lys. The enzymes that do so are sequence specific: **prolyl 4-hydroxylase** and **lysyl hydroxylase** act only on the Y residues of the Gly-X-Y sequences, whereas **prolyl 3-hydroxylase** acts on the X residues but only if Y is Hyp. Glycosylation, which also occurs in the RER, subsequently attaches sugar residues to 5-hydroxy-Lys residues (Section 7-2C). The folding of three polypeptides into the collagen triple helix must follow hydroxylation and glycosylation because the hydroxylases and glycosylases do not act on helical substrates. Moreover, the collagen triple helix denatures below physiological temperatures unless stabilized by hydrogen bonding interactions involving Hyp residues (Section 7-2C). Folding is also preceded by the formation of specific interchain disulfide bonds between the carboxylpropeptides. This observation bolsters the previously discussed conclusion that collagen propeptides help select and align the three collagen polypeptides for proper folding.

The procollagen molecules pass into the Golgi apparatus where they are packaged into secretory granules (Sections 11-3F and 21-3B) and secreted into the extracellular spaces of connective tissue. The aminopropeptides are excised just after procollagen leaves the cell and

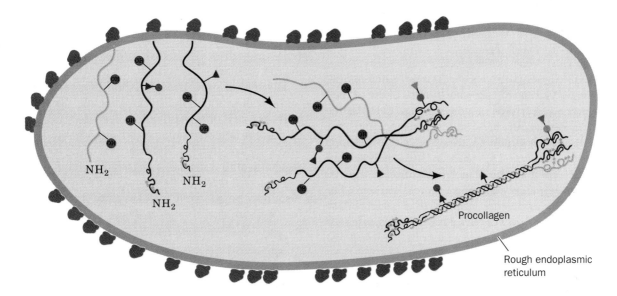

Figure 30-52
A schematic representation of procollagen biosynthesis. The diagram does not indicate the removal of signal peptides.

[After Prockop, D. J., Kivirikko, K. I., Tuderman, L., and Guzman, N. A., *New Engl. J. Med.* **301,** 18 (1979).]

the carboxylpropeptides are removed sometime later. The collagen molecules then spontaneously assemble into fibrils, which suggests that an important propeptide function is to prevent intracellular fibril formation. Finally, after the action of the extracellular enzyme lysyl oxidase, the collagen molecules in the fibrils spontaneously cross-link (Fig. 7-35).

6. PROTEIN DEGRADATION

The pioneering work of Henry Borsook and Rudolf Schoenheimer around 1940 demonstrated that the components of living cells are constantly turning over. Proteins have lifetimes that range from as short as a few minutes to weeks or more. In any case, *cells continuously synthesize proteins from and degrade them to their component amino acids.* The function of this seemingly wasteful process is twofold: (1) to eliminate abnormal proteins whose accumulation would be harmful to the cell, and (2) to permit the regulation of cellular metabolism by eliminating superfluous enzymes and regulatory proteins. Indeed, since the level of an enzyme depends on its rate of degradation as well as its rate of synthesis, *controlling a protein's rate of degradation is as important to the cellular economy as is controlling its rate of synthesis.* In this section we consider the processes of intracellular protein degradation and their consequences.

A. Degradation Specificity

Cells selectively degrade abnormal proteins. For example, hemoglobin that has been synthesized with the valine analog α-amino-β-chlorobutyrate

α-Amino-β-Chorobutyrate   Valine

has a half-life in reticulocytes of ~ 10 min, whereas normal hemoglobin lasts the 120-day lifetime of the red cell (which makes it perhaps the longest lived cytoplasmic protein). Likewise, unstable mutant hemoglobins are degraded soon after their synthesis which, for reasons explained in Section 9-3A, results in the hemolytic anemia characteristic of these molecular disease agents. Bacteria also selectively degrade abnormal proteins. For instance, *amber* and *ochre* mutants of β-galactosidase have half-lives in *E. coli* of only a few minutes, whereas the wild-type enzyme is almost indefinitely stable. Most abnormal proteins, however, probably arise from the chemical modification and/or spontaneous denaturation of these fragile molecules in the cell's reactive environment rather than by mutations or the rare errors in transcription or translation. *The ability to eliminate damaged proteins selectively is therefore an essential recycling*

Table 30-10
Half-Lives of Some Rat Liver Enzymes

Enzyme	Half-Life (h)
Short-lived enzymes	
Ornithine decarboxylase	0.2
RNA polymerase I	1.3
Tyrosine aminotransferase	2.0
Serine hydratase	4.0
PEP carboxylase	5.0
Long-lived enzymes	
Aldolase	118
GAPDH	130
Cytochrome *b*	130
LDH	130
Cytochrome *c*	150

Source: Dice, J. F. and Goldberg, A. L., *Arch. Biochem. Biophys.* **170**, 214 (1975).

mechanism that prevents the buildup of substances that would otherwise interfere with cellular processes.

Normal intracellular proteins are eliminated at rates that depend on their identities. A given protein is eliminated with first-order kinetics indicating that the molecules being degraded are chosen at random rather than according to their age. The half-lives of different enzymes in a given tissue vary substantially as is indicated for rat liver in Table 30-10. Remarkably, *the most rapidly degraded enzymes all occupy important metabolic control points, whereas the relatively stable enzymes have nearly constant catalytic activities under all physiological conditions. The susceptibilities of enzymes to degradation have evidently evolved together with their catalytic and allosteric properties so that cells can efficiently respond to environmental changes and metabolic requirements.* The criteria through which native proteins are selected for degradation are considered in Section 30-6B.

The rate of protein degradation in a cell also varies with its nutritional and hormonal state. Under conditions of nutritional deprivation, cells increase their rate of protein degradation so as to provide the necessary nutrients for indispensible metabolic processes. The mechanism that increases degradative rates in *E. coli* is the stringent response (Section 29-3F). A similar mechanism may be operative in eukaryotes since, as happens in *E. coli*, increased rates of degradation are prevented by antibiotics that block protein synthesis.

B. Degradation Mechanisms

Eukaryotic cells have dual systems for protein degradation, a lysosomal mechanism and an ATP-dependent cytosolically based mechanism. We consider both mechanisms below.

Lysosomes Degrade Proteins Nonselectively

Lysosomes are membrane-encapsulated organelles (Section 1-2A) that contain ~50 hydrolytic enzymes, including a variety of proteases known as **cathepsins.** The lysosome maintains an internal pH of ~5 and its enzymes have acidic pH optima. This situation presumably protects the cell against accidental lysosomal leakage since lysosomal enzymes are largely inactive at cytosolic pH's.

Lysosomes recycle intracellular constituents by fusing with membrane-enclosed bits of cytoplasm known as **autophagic vacuoles** and subsequently breaking down their contents. They similarly degrade substances that the cell takes up via endocytosis (Section 11-4B). The existence of these processes has been demonstrated through the use of lysosomal inhibitors. For example, the antimalarial drug **chloroquine**

$$Cl \underset{}{\overset{}{\bigotimes}} N$$

$$NH-CH-CH_2-CH_2-CH_2-N(C_2H_5)_2$$
$$\underset{CH_3}{|}$$

Chloroquine

is a weak base that freely penetrates the lysosome in uncharged form where it accumulates in charged form thereby increasing the intralysosomal pH and inhibiting lysosomal function. The treatment of cells with chloroquine reduces their rate of protein degradation. Similar effects arise from treatment of cells with cathepsin inhibitors such as the polypeptide antibiotic **antipain.**

$$^-OOCCHNH-\overset{O}{\overset{||}{C}}-NHCHC-\overset{O}{\overset{||}{}}NHCHC-\overset{O}{\overset{||}{}}NHCHC-\overset{O}{\overset{||}{}}H$$
$$\quad\underset{Phe}{|}\qquad\qquad\underset{Arg}{|}\quad\underset{Val}{|}\quad\underset{Arg}{|}$$

Lysosomal protein degradation appears to be nonselective. Lysosomal inhibitors do not affect the rapid degradation of abnormal proteins or short-lived enzymes. However, they prevent the acceleration of protein breakdown upon starvation.

Many normal and pathological processes are associated with increased lysosomal activity. **Diabetes mellitus** (Section 25-3B) stimulates the lysosomal breakdown of proteins. Similarly, muscle wastage caused by disuse, denervation, or traumatic injury arises from increased lysosomal activity. The regression of the uterus after childbirth, in which this muscular organ reduces its mass from 2 kg to 50 g in 9 days, is a striking example of this process. Many chronic inflammatory diseases, such as **rheumatoid arthritis,** involve the extracellular release of lysosomal enzymes which break down the surrounding tissues.

Ubiquitin Marks Proteins Selected for Degradation

It was initially assumed that protein degradation in eukaryotic cells is primarily a lysosomal process, Yet, reticulocytes, which lack lysosomes, selectively degrade abnormal proteins. The observation that protein breakdown is inhibited under anaerobic conditions led to the discovery of a cytosolically based ATP-dependent proteolytic system that is independent of the lysosomal system. This phenomenon was thermodynamically unexpected since peptide hydrolysis is an exergonic process.

Analysis of a cell-free rabbit reticulocyte system has demonstrated that **ubiquitin,** a protein of previously unknown function (Fig. 30-53), is required for ATP-dependent protein degradation. *This 76-residue monomeric protein, so-named because it is ubiquitous as well as abundant in eukaryotes, is the most highly conserved protein known:* It is identical in such diverse organisms as humans, toad, trout, and *Drosophila,* and differs in only three residues between humans and yeast. Evidently, ubiquitin is all but uniquely suited for some essential cellular process.

Proteins that are selected for degradation are so-marked by covalently linking them to ubiquitin. This process, which is reminiscent of amino acid activation (Section 30-2C), occurs in three steps (Fig. 30-54):

1. In an ATP-requiring reaction, ubiquitin's terminal carboxyl group is conjugated, via a thioester bond, to **ubiquitin-activating enzyme (E1),** a 105-kD dimer of identical subunits.

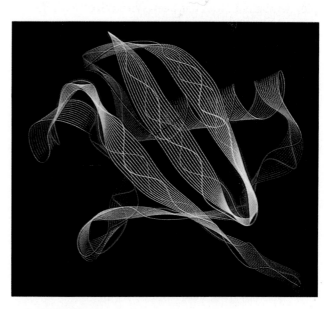

Figure 30-53
The X-ray structure of ubiquitin. The white ribbon represents the polypeptide backbone and the red and blue curves, respectively, indicate the directions of the carbonyl and amide groups. [Courtesy of Michael Carson, University of Alabama at Birmingham. X-ray structure determined by Charles Bugg.]

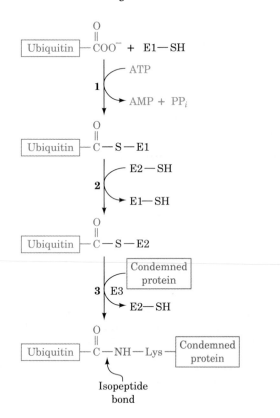

Figure 30-54
The reactions involved in the attachment of ubiquitin to a protein. In the first part of the process, ubiquitin's terminal carboxyl group is joined, via a thioester linkage, to E1 in a reaction driven by ATP hydrolysis. The activated ubiquitin is subsequently transferred to a sulfhydryl group of E2 and then, in a reaction catalyzed by E3, to a Lys ε-amino group on a condemned protein thereby flagging the protein for proteolytic degradation by UCDEN.

Table 30-11

The Half-Lives of Cytoplasmic Enzymes as a Function of Their N-Terminal Residues

N-Terminal Residue	Half-Life
Stabilizing	
Met	>20 h
Ser	
Ala	
Thr	
Val	
Gly	
Destabilizing	
Ile	~30 min
Glu	
Tyr	~10 min
Gln	
Highly destabilizing	
Phe	~3 min
Leu	
Asp	
Lys	
Arg	~2 min

Source: Bachmair, A., Finley, D., and Varshavsky, A., *Science* **234**, 180 (1986).

2. The ubiquitin is then transferred to a sulfhydryl group of one of several small proteins (25–70 kD) named **ubiquitin-carrier proteins (E2's)**.

3. Finally, **ubiquitin-protein ligase (E3;** ~180 kD**)** transfers the activated ubiquitin from E2 to a Lys ε-amino group of a previously bound protein thereby forming an **isopeptide bond**. E3 therefore appears to have a key role in selecting the protein to be degraded. Usually, several ubiquitin molecules are so-linked to this condemned protein. In addition, as many as 20 ubiquitin molecules may be tandemly linked to a target protein to form a multiubiquitin chain in which Lys 48 of each ubiquitin forms an isopeptide bond with the C-terminal carboxyl group of the following ubiquitin.

The ubiquitinated protein is proteolytically degraded in an ATP-dependent process mediated by a large (≥1000 kD) but otherwise poorly characterized multiprotein complex named **ubiquitin-conjugate degrading enzyme** *(***UCDEN***). This protease only degrades ubiquitin-linked proteins.*

A Protein's Half-Life Is Partially Determined by Its N-Terminal Residue

The structural features that E3 uses to select at least native proteins for destruction may be remarkably simple. *The half-life of a cytoplasmic protein varies with the identity of its N-terminal residue (Table 30-11).* Indeed, in a selection of 208 cytoplasmic proteins known to be long lived, all have a "stabilizing" residue, Met, Ser, Ala, Thr, Val, Gly, or Cys, at their N-termini. This is true for both eukaryotes and prokaryotes, which suggests the system that selects proteins for degradation is conserved in eukaryotes and prokaryotes, even though prokaryotes lack ubiquitin. Nevertheless, there are clear indications that other, more complex signals are also important in the selection of proteins for degradation. For instance, proteins with segments rich in Pro (P), Glu (E), Ser (S), and Thr (T) residues are rapidly degraded although how these so-called **PEST proteins** are recognized is unknown. Likewise, the criteria by which cells select defective proteins for degradation are unknown.

7. NONRIBOSOMAL POLYPEPTIDE SYNTHESIS

Several hundred polypeptide antibiotics, such as actinomycin D (Section 29-2D) and gramicidin A (Section 18-2C), have been characterized. These often cyclic molecules consist of rarely more than 20, often unusual, amino acid residues. *Many polypeptide antibiotics are synthesized by soluble enzymes rather than ribosomally from mRNA templates.* The synthesis of these substances is, consequently, unaffected by ribosomal inhibitors such as chloramphenicol (Section 30-3G) that arrest protein synthesis. In this section, we consider the mechanism of biosynthesis of the channel-forming ionophore **gramicidin S,** which is representative of the synthesis of many other polypeptide antibiotics.

Gramicidin S, a product of *Bacillus brevis,* is a cyclic decapeptide that consists of two identical pentapeptides joined head to tail (Fig. 30-55). Fritz Lipmann demonstrated that this antibiotic is synthesized by two enzymes, E_I (280 kD) and E_{II} (100 kD) that activate the amino acids indicated in Fig. 30-56. Each of the amino acids of gramicidin S is activated by the ATP-driven linkage of the amino acid via a thioester bond to its corresponding enzyme. E_{II} binds only a D-Phe residue, whereas E_I simultaneously binds the other four gramicidin S residues.

The polymerization process begins when E_{II} transfers its D-Phe residue to the E_I-bound Pro residue to form a dipeptide (Fig. 30-57a). The growing oligopeptide is then sequentially transferred to the remaining amino acid residues of the pentapeptide (Fig. 30-57b). The absence of any amino acid from the *in vitro* reaction mixture results in premature termination of the reaction at

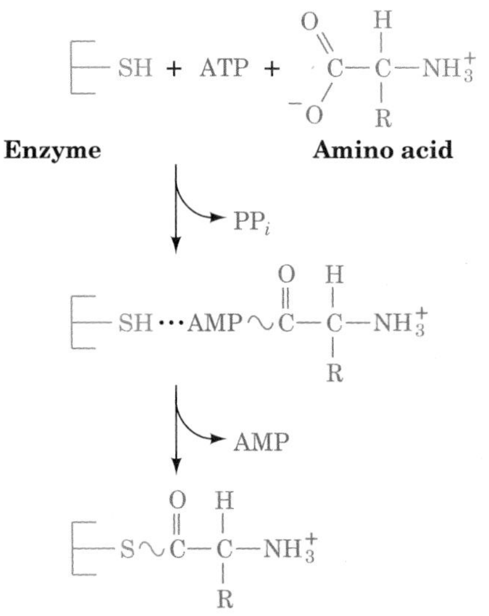

Figure 30-56
The activation of amino acids by the enzymes that synthesize gramicidin S. In the first step of the reaction an enzyme bound aminoacyl-adenylate is formed as it is in the aminoacyl-tRNA synthetase reaction (Section 30-2C). In the second reaction step, however, the amino acid residue is linked to the enzyme via a thioester bond rather than to a tRNA.

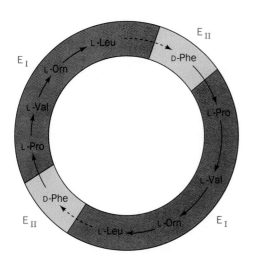

Figure 30-55
The amino acid sequence of gramicidin S. The amino acids activated by E_I and E_{II} are shaded red and green. Dashed arrows indicate the points of cyclization. [After Lipmann, F., *Acc. Chem. Res.* **11,** 363 (1971).]

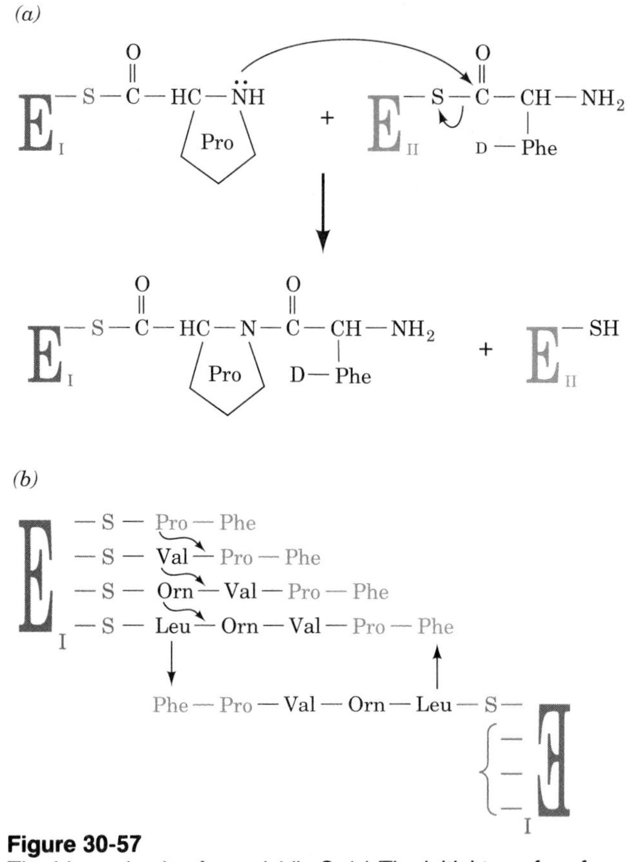

Figure 30-57
The biosynthesis of gramicidin S. (*a*) The initial transfer of Phe to an E_I-linked Pro residue to form a peptide bond. (*b*) The elongation (*top*) and cyclization (*bottom*) reactions on E_I (here the arrows indicate group transfer, not electron transfer). [After Lipmann, F., *Science* **173,** 878 (1971).]

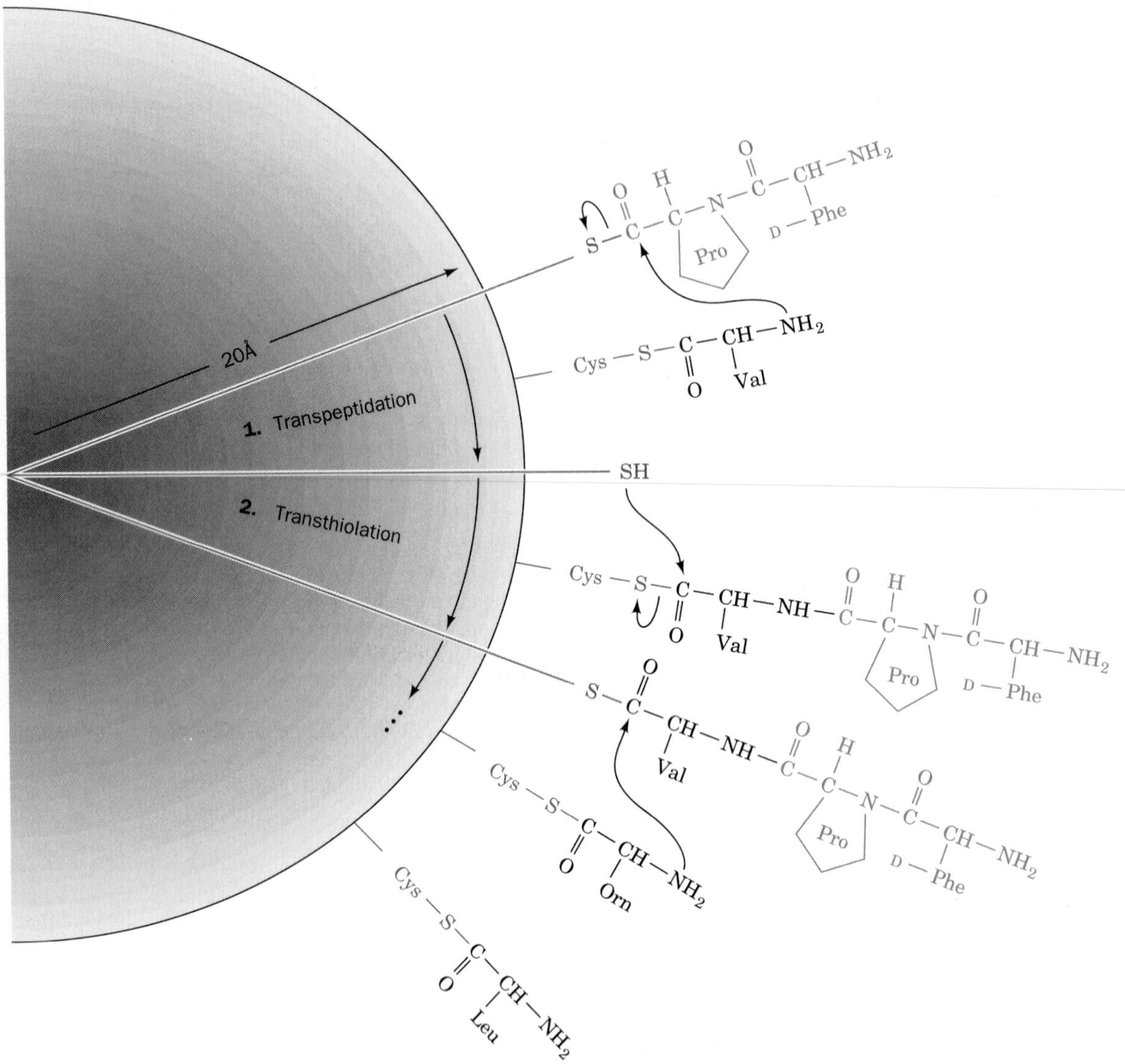

Figure 30-58
The proposed scheme for the participation of the pantetheine residue (*blue*) in the biosynthesis of gramicidin S. The circular arrows indicate the movement of the pantetheine in collecting the amino acid residues that are linked to the enzyme via Cys thioester bonds. The transpeptidation and transthiolation reactions alternate to synthesize the pentapeptide. [After Lipmann, F., *Acc. Chem. Res.* **6**, 366 (1973).]

that point. Note that chain elongation proceeds towards the C-terminus as it does in ribosomal polypeptide synthesis. The final enzyme-linked pentapeptide reacts in a head-to-tail fashion with a second such assembly to form the decapeptide product (Fig. 30-55).

The resemblance of the above reaction sequence to that of fatty acid synthesis led Lipmann to propose that phosphopantetheine is a cofactor in polypeptide synthesis as it is in acyl carrier protein (Section 23-4A). In fact, E_I contains a single Ser-linked phosphopantetheine.

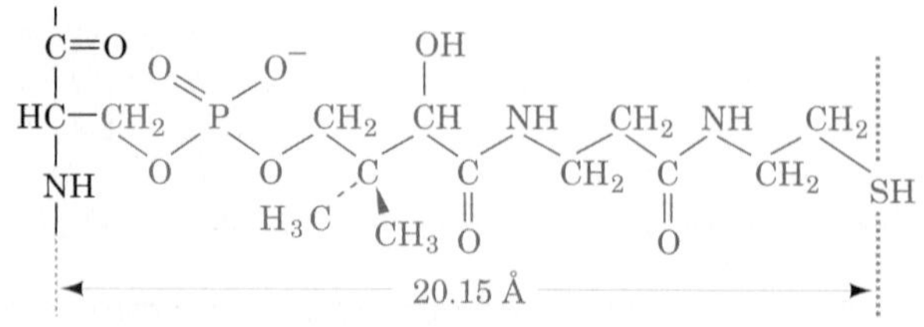

This 20-Å long residue is thought to sequentially collect the enzyme-linked amino acids on the growing oligopeptide through alternating transpeptidation and transthiolation reactions (Fig. 30-58). Transthiolation in polypeptide synthesis is therefore analogous to translocation in ribosomal chain elongation (Section 30-3D).

Chapter Summary

Point mutations are caused by either base analogs that mispair during DNA replication or by substances that react with bases to form products that mispair. Insertion/deletion (frameshift) mutations arise from the association of DNA with intercalating agents that distort the DNA structure. The analysis of a series of frameshift mutations that supressed one another established that the genetic code is an unpunctuated triplet code. Fine structure genetic mapping combined with amino acid sequence analyses demonstrated that genes are colinear with the polypeptides they specify. In a cell-free protein synthesizing system, poly(U) directs the synthesis of poly(Phe) thereby demonstrating that UUU is the codon specifying Phe. The genetic code was established through the use of polynucleotides of known composition but random sequence, the ability of defined triplets to promote the ribosomal binding of tRNAs bearing specific amino acids, and through the use of synthetic mRNAs of known alternating sequences. The latter investigations also demonstrated that the 5' end of mRNA corresponds to the N-terminus of the polypeptide it specifies and established the sequences of the stop codons. Degenerate codons differ mostly in the identities of their third base. Small single-stranded DNA phages such as ϕX174 contain overlapping genes in different reading frames. The genetic code used by mitochondria differs in several respects from the "standard" genetic code.

Transfer RNAs consist of 60 to 95 nucleotides that can be arranged in the cloverleaf secondary structure. As many as 10% of a tRNA's bases may be modified. Yeast tRNA^{Phe} forms a narrow L-shaped three-dimensional structure that resembles that of other tRNAs. Most of the bases are involved in stacking and base pairing associations including nine tertiary interactions that appear to be essential for maintaining the molecule's native conformation. Amino acids are appended to their cognate tRNAs in a two-stage reaction catalyzed by the corresponding aminoacyl-tRNA synthetase. The great accuracy of tRNA charging arises from the proofreading of the bound amino acid by aminoacyl-tRNA synthetase at the expense of ATP hydrolysis. Ribosomes select tRNAs solely on the basis of their anticodons. Sets of degenerate codons are read by a single tRNA through wobble pairing. Nonsense mutations may be suppressed by tRNAs whose anticodons have mutated to recognize a Stop codon.

The ribosome consists of a small and a large subunit whose complex shapes have been revealed by electron microscopy. The three RNAs and 52 proteins comprising the *E. coli* ribosome self-assemble under proper conditions. The positions of many ribosomal components relative to the subunit surfaces have been largely determined by immune electron microscopy and neutron diffraction measurements. Affinity labeling experiments have identified the ribosomal components in the vicinity of various ribosomal binding sites and catalytic centers. Ribosomal polypeptide synthesis proceeds by the addition of amino acid residues to the C-terminal end of the nascent polypeptide. The mRNAs are read in the 5' → 3' direction. mRNAs are usually simultaneously translated by several ribosomes in the form of polysomes. The ribosome has at least three tRNA-binding sites: the P site, which binds the peptidyl-tRNA, the A site, which binds the incoming amino-

acyl-tRNA, and the E site, which transiently binds the outgoing tRNA. During polypeptide synthesis, the nascent polypeptide is transferred to the aminoacyl-tRNA thereby lengthening the nascent polypeptide by one residue. The discharged tRNA is then released and the new peptidyl-tRNA, with its associated codon, is translocated to the P site. In prokaryotes, the initiation sites on mRNA are recognized through their Shine–Dalgarno sequences and by their initiating codon. Prokaryotic initiating codons specify fMet–tRNA$_f^{Met}$. Initiation is a complex process involving the participation of three initiation factors that induce the assembly of the ribosomal subunits with mRNA and fMet–tRNA$_f^{Met}$. The eukaryotic initiation site is usually the first AUG downstream from the 5'-terminal cap and this AUG specifies unformylated Met-tRNA$_i^{Met}$. Polypeptides are elongated in a three-part cycle, aminoacyl-tRNA binding, transpeptidation, and translocation, that requires the participation of elongation factors and is vectorially driven by GTP hydrolysis. Termination codons bind release factors that induce the peptidyl transferase to hydrolyze the peptidyl-tRNA bond. The high accuracy of translation indicates that the ribosome proofreads the codon–anticodon interaction, most probably via a kinetic mechanism. Ribosomal inhibitors, many of which are antibiotics, are medically important and biochemically useful in elucidating ribosomal function. Streptomycin causes mRNA misreading and inhibits prokaryotic chain initiation, chloramphenicol inhibits prokaryotic peptidyl transferase, tetracycline inhibits aminoacyl-tRNA binding to the prokaryotic small subunit, and diphtheria toxin ADP-ribosylates eEF-2.

Several mechanisms of translational control have been elucidated in eukaryotes. Hemoglobin synthesis in reticulocytes is inhibited in the absence of heme by heme-controlled inhibitor. This enzyme catalyzes the phosphorylation of eIF-2, which then tightly binds eIF-2B, thereby blocking translational initiation. In the presence of dsRNA, cells treated with interferon are translationally inhibited. This happens through two independent mechanisms: the induction of a protein kinase that phosphorylates eIF-2, and the induction of 2,5-A synthetase whose product 2,5-A, activates an endonuclease that degrades mRNA. Translation is also inhibited by mRNA masking, at least in certain embryos.

Proteins may be post-translationally modified in a variety of ways. Protelytic cleavages, usually by specific peptidases, activate proproteins. The signal peptides of preproteins are removed by signal peptidases. Covalent modifications alter many types of side chains in a variety of ways that modulate the catalytic activities of enzymes, provide recognition markers, and stabilize protein structures.

Proteins in living cells are continually turning over. This controls the level of regulatory enzymes and disposes of abnormal proteins that would otherwise interfere with cellular processes. Proteins are degraded by lysosomes in a nonspecific process that is stimulated during starvation, as well as by various pathological and normal states. A cytosolically based ATP-dependent system degrades normal as well as abnormal proteins in a process that flags these proteins by the covalent attachment of ubiquitin.

The biosynthesis of gramicidin S, which is representative of the synthesis of many other polypeptide antibiotics, is mediated by soluble enzymes rather than by ribosomes. The proper amino acids are linked as thioesters to E_I and E_{II} in reactions driven by ATP hydrolysis. Chain elongation occurs by the pantetheine-mediated transfer of the growing oligopeptides to the activated amino acid residues. The reaction of two enzyme-linked pentapeptides yields the product cyclic decapeptide.

References

General

Lewin, B., *Genes* (3rd ed.), Chapters 5–7, Wiley (1987).

Watson, J. D., Hopkins, N. H., Roberts, J. W., Steitz, J. A., and Weiner, A. M., *Molecular Biology of the Gene* (4th ed.), Chapters 14 and 15, Benjamin/Cummings (1987).

The Genetic Code

Attardi, G., Animal mitochondrial DNA: an extreme example of genetic economy, *Int. Rev. Cytol.* **93**, 93–145 (1985).

Benzer, S., The fine structure of the gene, *Sci. Am.* **206**(1): 70–84 (1962).

The Genetic Code, Cold Spring Harbor Symp. Quant. Biol. **31** (1966). [A collection of papers describing the establishment of the genetic code. See especially the articles by Crick, Nirenberg, and Khorana.]

Crick, F. H. C., Burnett, L., Brenner, S., and Watts-Tobin, R. J., General nature of the genetic code for proteins, *Nature* **192**, 1227–1232 (1961).

Crick, F. H. C., The genetic code, *Sci. Am.* **207**(4): 66–74 (1962). [The structure of the code as determined by phage genetics.]

Crick, F. H. C., The genetic code: III, *Sci. Am.* **215**(4): 55–62 (1966). [A description of the nature of the code after its elucidation was almost complete.]

Fox, T. D., Natural variation in the genetic code, *Annu. Rev. Genet.* **21**, 67–91 (1987).

Garen, A., Sense and nonsense in the genetic code, *Science* **160**, 149–159 (1968). [A discussion of the genetic aspects of nonsense mutations.]

Judson, J. F., *The Eighth Day of Creation*, Part II, Simon & Schuster (1979). [A fascinating historical narrative on the elucidation of the genetic code.]

Khorana, H. G., Nucleic acid synthesis in the study of the genetic code, *Nobel Lectures in Molecular Biology, 1933–1975, pp.* 303–331, Elsevier (1977).

Nirenberg, M. W. and Matthaei, J. H., The dependence of cell-free protein synthesis in *E. Coli* upon naturally occurring or synthetic polyribonucleotides, *Proc. Natl. Acad. Sci.* **47**, 1588–1602 (1961). [The landmark paper reporting the finding that poly(U) stimulates the synthesis of poly(Phe).]

Nirenberg, M. W., The genetic code: II, *Sci. Am.* **208**: 80–94 (1963). [Discusses the use of synthetic mRNAs to analyze the genetic code.]

Nirenberg, M. and Leder, P., RNA code words and protein synthesis, *Science* **145**, 1399–1407 (1964). [The determination of the genetic code by the ribosomal binding of tRNAs using specific trinucleotides.]

Nirenberg, M., The genetic code, *Nobel Lectures in Molecular Biology, 1933–1975, pp.* 335–360, Elsevier (1977).

Singer, B. and Kuśmierek, J. T., Chemical mutagenesis, *Annu. Rev. Biochem.* **51**, 655–693 (1982).

Yanofsky, C., Carlton, B. C., Guest, J. R., Helinski, D. R., and Henning, U., On the colinearity of gene structure and protein structure, *Proc. Natl. Acad. Sci.* **51**, 266–272 (1964).

Yanofsky, C., Gene structure and protein structure, *Sci. Am.* **216**(5): 80–94 (1967).

Transfer RNA

Altman, S. (Ed.), *Transfer RNA,* MIT Press (1978). [A series of general reviews on tRNAs.]

Björk, G. R., Ericson, J. U., Gustafsson, C. E. D., Hagervall, T. G., Jösson, Y. H., and Wikström, P. M., Transfer RNA modification, *Annu. Rev. Biochem.* **56**, 263–287 (1987).

Blow, D. M. and Brick, P., Aminoacyl-tRNA synthetases, *in* Jurnak, F. A. and McPherson, A. (Eds.), *Biological Macromolecules and Assembly*, Vol. 2: *Nucleic Acids and Interactive Proteins, pp.* 441–469, Wiley (1985).

Crick, F. H. C., Codon–anticodon pairing: the wobble hypothesis, *J. Mol. Biol.* **19**, 548–555 (1966).

Bennetzen, J. L. and Hall, B. D., Codon selection in yeast, *J. Biol. Chem.* **257**, 3026–3031 (1982).

Freist, W., Isoleucyl-tRNA synthetase: an enzyme with several catalytic cycles displaying variation in specificity and energy consumption, *Angew. Chem. Int. Ed. Engl.* **27**, 773–788 (1988).

Kersten, H., On the biological significance of modified nucleosides in tRNA, *Prog. Nucleic Acid Res. Mol. Biol.* **31**, 59–114 (1984).

Kim, S. H., Suddath, F. L., Quigley, G. J., McPherson, A., Sussman, J. L., Wang, A. M. J., Seeman, N. C., and Rich, A., Three-dimensional tertiary structure of yeast phenylalanine transfer RNA, *Science* **185**, 435–440 (1974); *and* Robertus, J. D., Ladner, J. E., Finch, J. T., Rhodes, D., Brown, R. S., Clark, B. F. C., and Klug, A., Structure of yeast phenylalanine tRNA at 3 Å resolution, *Nature* **250**, 546–551 (1974). [The landmark papers describing the high resolution structure of a tRNA.]

Kline, L. K. and Söll, D., Nucleotide modifications in RNA, *in* Boyer, P. D. (Ed.), *The Enzymes* (3rd ed.), Vol. 15, *pp.* 567–582, Academic Press (1982).

Leatherbarrow, R. J., Fersht, A. R., and Winter, G., Transition-state stabilization in the mechanisim of tyrosyl-tRNA synthetase revealed by protein engineering, *Proc. Natl. Acad. Sci.* **82**, 7840–7844 (1985).

Leinfelder, W., Zehelein, E., Mandrand-Berthelot, M.-A., and Böck, A., Gene for a novel tRNA species that accepts L-serine and cotranslationally inserts selenocysteine, *Nature* **331**, 723–725 (1988).

Muramatsu, T., Nishakawa, K., Nemoto, F., Kuchino, Y., Nishamura, S., Miyazawa, T., and Yokoyama, S., Codon and amino-acid specificities of a transfer RNA are both converted by a single post-transcriptional modification, *Nature* **336,** 179–181 (1988).

Rich, A. and Kim, S. H., The three-dimensional structure of transfer RNA, *Sci. Am.* **238**(1): 52–62 (1978).

Schimmel, P., Parameters for the molecular recognition of tRNAs, *Biochemistry* **28,** 2747–2759 (1989).

Schimmel, P. R., Aminoacyl tRNA synthetases: general scheme of structure-function relationships in the polypeptides and recognition of transfer RNAs, *Annu. Rev. Biochem.* **56,** 125–158 (1987).

Schimmel, P. R., Söll, D., and Abelson, J. N. (Eds.), *Transfer RNA,* Cold Spring Harbor Laboratory (1979). [A two volume treatise on tRNA. Part 1 deals with structure, properties, and recognition; Part 2 treats biological aspects.]

Schulman, L. H. and Abelson, J., Recent excitement in understanding transfer RNA identity, *Science* **240,** 1591–1592 (1988).

Sprinzl, M., Moll, J., Meissner, F., and Hartmann, T., Compilation of tRNA sequences, *Nucleic Acids Res.* **13,** r1–r49 (1985).

Steege, D. A. and Söll, D. G., Suppression, *in* Goldberger, R. F. (Ed.), *Biological Regulation and Development,* Vol. 1., pp. 433–485, Plenum Press (1979).

von Ehrenstein, G., Weisblum, B., and Benzer, S., The function of sRNA as amino acid adapter in the synthesis of hemoglobin, *Proc. Natl. Acad. Sci.* **49,** 669–675 (1963).

Ribosomes

Bielka, H. (Ed.), *The Eukaryotic Ribosome,* Springer–Verlag (1982). [A detailed treatise.]

Brinacombe, R., The emerging three-dimensional structure and function of 16S ribosomal RNA, *Biochemistry* **27,** 4207–4214 (1988).

Capel, M. S., Engelman, D. M., Freeborn, B. R., Kjeldgaard, M., Langer, J. A., Ramakrishnan, V., Schindler, D. G., Schneider, D. K., Schoenborn, B. P., Siller, I.-Y., Yabuki, S., and Moore, P. B., A complete mapping of the proteins in the small ribosomal subunit of *Escherichia coli, Science* **238,** 1403–1406 (1987).

Caskey, C. Th., Peptide chain termination, *Trends Biochem. Sci.* **5,** 234–237 (1980).

Chambliss, G., Craven, G. R., Davies, J., Davis, K., Kahan, L., and Nomura, M. (Eds.), *Ribosomes. Structure, Function and Genetics,* University Park Press (1979). [An informative series of reviews.]

Clark, B., The elongation step of protein biosynthesis, *Trends Biochem. Sci.* **5,** 207–210 (1980).

Dahlberg, A. E., The functional role of ribosomal RNA in protein synthesis, *Cell* **57,** 525–529 (1989).

Dintzis, H. M., Assembly of the peptide chains of hemoglobin, *Proc. Natl. Acad. Sci.* **47,** 247–261 (1961). [The determination of the direction of polypeptide biosynthesis.]

Edelmann, P. and Gallant, J., Mistranslation in E. coli, *Cell* **10,** 131–137 (1977).

Fersht. A., *Enzyme Structure and Mechanism* (2nd ed.), Chapter 13, Freeman (1985). [A discussion of enzymatic specificity and editing mechanisms.]

Gale, E. F., Cundliffe, E., Reynolds, P. E., Richmond, M. H., and Waring, M. J., *The Molecular Basis of Antibiotic Action,* Chapter 6, Wiley (1981).

Gutell, R. R., Weiser, B., Woese, C. R., and Noller, H. F., Comparitive anatomy of 16-S-like ribosomal RNA, *Prog. Nucleic Acid Res. Mol. Biol.* **32,** 155–216 (1985).

Hardesty, B. and Kramer, G. (Eds.), *Structure, Function, and Genetics of Ribosomes,* Springer–Verlag (1985). [A series of useful reviews on current aspects of ribosomology.]

Held, W. A., Ballou, B., Mizushima, S., and Nomura, M., Assembly mapping of 30S ribosomal proteins from *Escherichia coli, J. Biol. Chem.* **249,** 3103–3111 (1974).

Hunt, T., The initiation of protein synthesis, *Trends Biochem. Sci.* **5,** 178–181 (1980).

Kurland, C. G. and Ehrenberg, M., Optimization of translation, *Prog. Nucleic Acid Res. Mol. Biol.* **31,** 191–219 (1984).

Lake, J. A., Evolving ribosome structure: domains in archaebacteria, eubacteria, eocytes and eukaryotes, *Annu. Rev. Biochem.* **54,** 507–530 (1985).

Lake, J. A., The ribosome, *Sci. Am.* **245**(2): 84–97 (1981).

Liljas, A., Structural studies of ribosomes, *Prog. Biophys. Mol. Biol.* **40,** 161–228 (1982).

Maitra, U., Stringer, E. A., and Chaudhuri, A., Initiation factors in protein biosynthesis, *Annu. Rev. Biochem.* **51,** 869–900 (1982).

Moldave, K., Eukaryotic protein synthesis, *Annu. Rev. Biochem.* **54,** 109–149 (1983).

Moore, P. B. and Capel, M. S., Structure-function correlations in the small ribosomal subunit from *Escherichia coli, Annu. Rev. Biophys. Biophys. Chem.* **17,** 349–367 (1988).

Moore, P. B., The ribosome returns, *Nature* **331,** 223–227 (1988).

Nagano, K. and Harel, M., Approaches to a three-dimensional model of the *E. coli* ribosome, *Prog. Biophys. Mol. Biol.* **48,** 67–101 (1986).

Nierhaus, K. H., Structure, function and assembly of ribosomes, *Curr. Top. Microbiol. Immunol.* **97,** 81–155 (1982). [A thorough and enlightening review.]

Nierhaus, K. H. and Wittman, H. G., Ribosomal function and its inhibition by antibiotics in prokaryotes, *Naturwissenschaften* **67,** 234–250 (1980).

Noller, H. F., Jr., and Moldave, K., (Eds.), Ribosomes, *Methods Enzymol.* **164** (1988). [Contains numerous articles on ribosome methodology.]

Noller, H. F., Structure of ribosomal RNA, *Annu. Rev. Biochem.* **53,** 119–162 (1984).

Pappenheimer, A. M., Jr., Diphtheria toxin, *Annu. Rev. Biochem.* **46,** 69–94 (1977).

Prince, J. B., Gutell, R. R., and Garrett, R. A., A consensus model of the *E. coli* ribosome, *Trends Biochem. Sci.* **8,** 359–363 (1983).

Rané, H. A., Klootwijk, J., and Musters, W., Evolutionary conservation of structure and function of high molecular weight ribosomal RNA, *Prog. Biophys. Mol. Biol.* **51,** 77–129 (1988).

Rhoads, R. E., Cap recognition and the entry of mRNA into the protein initiation cycle, *Trends Biochem. Sci.* **13,** 52–56 (1988).

Schüler, D. and Brinacombe, R., The *Escherichia coli* 30S ribosomal subunit; an optimized three-dimensional fit between the ribosomal proteins and the 16S RNA, *EMBO J.* **7**, 1509–1513 (1988).

Shatkin, A. J., mRNA cap binding proteins: essential factors for initiating translation, *Cell* **40**, 223–224 (1985).

Shine, J. and Dalgarno, L., The 3'-terminal sequence of *Escherichia coli* 16S ribosomal RNA: complementarity to nonsense triplets and ribosome binding sites, *Prog. Natl. Acad. Sci.* **71**, 1342–1346 (1974).

Spirin, A. S., Ribosomal translocation: facts and models, *Prog. Nucleic. Res. Mol. Biol.* **32**, 75–114 (1985).

Steitz, J. A. and Jakes, K., How ribosomes select initiator regions in mRNA: base pair formation between the 3' terminus of 16S RNA and the mRNA during initiation of protein synthesis in *Escherichia coli*, *Proc. Natl. Acad. Sci.* **72**, 4734–4738 (1975).

Stern, S., Powers, T., Changchien, L.-I., and Noller, H. F., RNA–protein interactions in 30S ribosomal subunits: folding and function of 16S rRNA, *Science* **244**, 783–790 (1989).

Stöffler, G. and Stöffler-Meilicke, M., Immunoelectron microscopy of ribosomes, *Annu. Rev. Biophys. Bioeng.* **13**, 303–330 (1984).

Thompson, R. C., EFTu provides an internal kinetic standard for translational accuracy, *Trends Biochem. Sci.* **13**, 91–93 (1988).

Control of Translation

Jagus, R., Anderson, W. F., and Safer, B., The regulation of initiation of mammalian protein synthesis, *Prog. Nucleic Acid Res. Mol. Biol.* **25**, 127–183 (1981).

Kaempfer, R., Regulation of eukaryotic translation, *in* Fraenkel-Conrat, H. and Wagner, R. R., (Eds.), *Comprehensive Virology*, Vol. 19, *pp.* 99–175, Plenum Press (1984).

Lengyel, P., Biochemistry of interferons and their actions, *Annu. Rev. Biochem.* **51**, 251–282 (1982).

Pestka, S., Langer, J. A., Zoon, K. C., and Samuel, C. E., Interferons and their actions, *Annu. Rev. Biochem.* **56**, 757–777 (1987).

Proud, C. G., Guanine nucleotides, protein phosphorylation and control of translation, *Trends Biochem. Sci.* **11**, 73–77 (1986).

Raff, R. A., Masked messenger RNA and the regulation of protein synthesis in eggs and embryos, *in* Prescott, D. M., and Goldstein, L. (Eds.), *Cell Biol.* Vol. 4, *pp.* 107–136, Academic Press (1980).

Williams, B. R. G. and Kerr, I. M., The 2–5A(pppA^{2}p^{5}A^{2}p^{5}A) system in interferon-treated and control cells, *Trends Biochem. Sci.* **5**, 138–140 (1980).

Post-Translational Modification

Fessler, J. H. and Fessler, L. I., Biosynthesis of procollagen, *Annu. Rev. Biochem.* **47**, 129–162 (1978).

Prockop, D. J., Kivirikko, K. I., Tuderma, L., and Guzman, N. A., The biosynthesis of collagen and its disorders, *New Engl. J. Med.* **301**, 13–23 (1979).

Wold, F., In vivo chemical modification of proteins, *Annu. Rev. Biochem.* **50**, 783–814 (1981).

Wold, F. and Moldave, K. (Eds.), Posttranslational Modifications, Parts A and B, *Methods Enzymol.* **106** and **107** (1984). [Contains extensive descriptions of the amino acid "zoo".]

Protein Degradation

Bachmair, A., Finley, D., and Varshavsky, A., In vivo half-life of a protein is a function of its amino terminal residue, *Science* **234**, 179–186 (1986).

Chau, V., Tobias, J. W., Bachmair, A., Marriott, D., Ecker, D. J., Gonda, D. J., and Varshavsky, A., A multiubiquitin chain is confined to a specific lysine in a targeted short-lived protein. *Science* **243**, 1576–1583 (1989).

Finley, D. and Varshavsky, A., The ubiquitin system: functions and mechanisms, *Trends Biochem. Sci.* **10**, 343–347 (1985).

Goldberg, A. L. and Dice, J. F., *and* Goldberg, A. L. and St. John, A. C., Intracellular protein degradation in mammalian and bacterial cells: parts I *and* II, *Annu. Rev. Biochem.* **42**, 835–869 (1974) *and* **45**, 747–803 (1976).

Hershko, A., Ubiquitin-mediated protein degradation, *J. Biol. Chem.* **262**, 15237–15240 (1988).

Hershko, A. and Ciechanover, A., The ubiquitin pathway for the degradation of intracellular proteins, *Prog. Nucleic Acid Res. Mol. Biol.* **33**, 19–56 (1986).

Rechsteiner, M., *Ubiquitin,* Plenum Press (1988). [A series of authoritative articles on ubiquitin structure and function.]

Rechsteiner, M., Rogers, S., and Rote, K., Protein structure and intracellular stability, *Trends Biochem. Sci.* **12**, 390–394 (1987).

Senahdi, V.-J., Bugg, C. E., Wilkinson, K. D., and Cools, W. J., Three-dimensional structure of ubiquitin at 2.8 Å resolution, *Prog. Natl. Acad. Sci.* **82**, 3582–3585 (1985).

Waxman, L., Fagan, J. M., and Goldberg, A. L., Demonstration of two distinct high molecular weight proteases in rabbit reticulocytes, one of which degrades ubiquitin conjugates, *J. Biol. Chem.* **262**, 2451–2457 (1987).

Nonribosomal Polypeptide Synthesis

Kleinkauf, H. and Döhren, H., Nonribosomal polypeptide formation on multifunctional proteins, *Trends Biochem. Sci.* **8**, 281–283 (1983).

Lipmann, F., Attempts to map a process evolution of peptide biosynthesis, *Science* **173**, 875–884 (1971).

Lipmann, F., Nonribosomal peptide synthesis on polyenzyme templates, *Acc. Chem. Res.* **11**, 361–367 (1973).

Problems

1. What is the product of reacting guanine with nitrous acid? Is the reaction mutagenic? Explain.

2. What is the polypeptide specified by the following DNA antisense strand? Assume translation starts after the first initiation codon.

    5'-TCTGACTATTGAGCTCTCTGGCACATAGCA-3'

*3. The fingerprint of a protein from a phenotypically revertant mutant of bacteriophage T4 indicates the presence of an altered tryptic peptide with respect to the wild-type. The wild-type and mutant peptides have the following sequences:

    Wild-type   Cys-Glu-Asp-His-Val-Pro-Gln-Tyr-Arg
    Mutant   Cys-Glu-Thr-Met-Ser-His-Ser-Tyr-Arg

    Indicate how the mutant could have arisen and give the base sequences, as far as possible, of the mRNAs specifying the two peptides. Comment on the function of the peptide in the protein.

4. Explain why the various classes of mutations can reverse a mutation of the same class but not a different class.

5. Which amino acids are specified by codons that can be changed to an *amber* codon by a single point mutation?

6. The mRNA specifying the α chain of human hemoglobin contains the base sequence

    · · · UCCAAAUACCGUUAAGCUGGA · · ·

    The C-terminal tetrapeptide of the normal α chain, which is specified by part of this sequence, is

    -Ser-Lys-Tyr-Arg

    In hemoglobin Constant Spring, the corresponding region of the α chain has the sequence

    -Ser-Lys-Tyr-Arg-Gln-Ala-Gly- · · ·

    Specify the mutation that causes hemoglobin Constant Spring.

7. Explain why a minimum of 32 tRNAs are required to translate the "standard" genetic code.

8. Draw the wobble pairings not in Fig. 30-21*a*.

9. A colleague of yours claims that by exposing *E. coli* to HNO_2 she has mutated a tRNA^{Gly} to an *amber* suppressor. Do you believe this claim? Explain.

*10. Deduce the anticodon sequences of all suppressors listed in Table 30-5 except UGA-1 and indicate the mutations that caused them.

11. How many different types of macromolecules must be minimally contained in a cell-free protein synthesizing system from *E. coli* ? Count each type of ribosomal component as a different macromolecule.

12. Why do oligonucleotides containing Shine–Dalgarno sequences inhibit translation in prokaryotes? Why don't they do so in eukaryotes?

13. Why does m^{7}GTP inhibit translation in eukaryotes? Why doesn't it do so in prokaryotes?

14. What would be the distribution of radioactivity in the completed hemoglobin chains upon exposing reticulocytes to ^{3}H-labeled leucine for a short time followed by a chase with unlabeled leucine?

15. Design an mRNA with the necessary prokaryotic control sites that codes for the octapeptide Lys-Pro-Ala-Gly-Thr-Glu-Asn-Ser.

16. Indicate the translational control sites in and the amino acid sequence specified by the following prokaryotic mRNA.

    5'-CUGAUAAGGAUUUAAAUUAUGUGUCAAUCA-
    CGAAUGCUAAUCGAGGCUCCAUAAUAACACUU-
    CGAC-3'

17. What is the energetic cost, in ATPs, for the *E. coli* synthesis of a polypeptide chain of 100 residues starting from amino acids and mRNA? Assume that no losses are incurred as a result of proofreading.

*18. It has been suggested that Gly-tRNA synthetase does not require an editing mechanism. Why?

19. An antibiotic named fixmycin, which was isolated from a fungus growing on ripe passion fruit, is effective in curing many types of veneral disease. In characterizing fixmycin's mode of action, you have found that it is a bacterial translational inhibitor that binds exclusively to the large subunit of *E. coli* ribosomes. The initiation of protein synthesis in the presence of fixmycin results in the generation of dipeptides that remain associated with the ribosome. Suggest a mechanism of fixmycin action.

20. Heme inhibits protein degradation in reticulocytes by allosterically regulating ubiquitin-activating enzyme. What physiological function might this serve?

21. Genbux Inc., a genetic engineering firm, has cloned the gene encoding an industrially valuable enzyme into *E. coli* such that the enzyme is produced in large quantities. However, since the firm wishes to produce the enzyme in ton quantities, the expense of isolating it would be greatly reduced if the bacterium could be made to secrete it. As a high priced consultant, what general advice would you offer to solve this problem?

Chapter 31
DNA REPLICATION, REPAIR, AND RECOMBINATION

People are DNA's way of making more DNA.

Anon

In this chapter we consider how DNA is synthesized and maintained. The need for nearly error-free transmission of genetic information to succeeding generations, and yet to proliferate advantageous mutations, has led to the evolution of elaborate mechanisms for the replication, repair, and recombination of DNA in even the simplest organisms. The general outlines of most of these processes have been elucidated although, as will become apparent, the details are largely obscure.

1. DNA REPLICATION: AN OVERVIEW

Watson and Crick's seminal paper describing the DNA double helix ended with the statement: "It has not escaped our notice that the specific pairing we have postulated immediately suggests a possible copying mechanism for the genetic material." In a succeeding

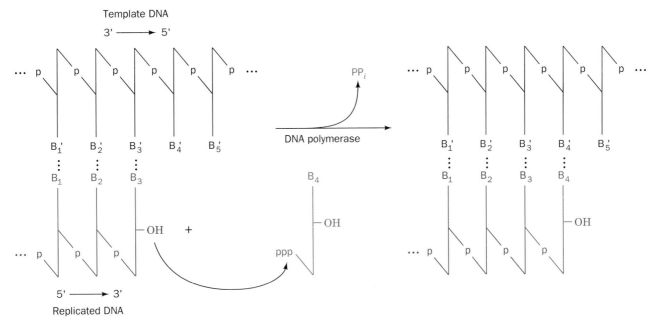

Figure 31-1
DNA polymerases assemble incoming deoxynucleoside triphosphates on single-stranded
DNA templates such that the growing strand is elongated in its 5′ → 3′ direction.

paper they expanded on this rather cryptic remark by
pointing out that a DNA strand could act as a template to
direct the synthesis of its complementary strand. Although Meselson and Stahl demonstrated, in 1958, that
DNA is, in fact, semiconservatively replicated (Section
28-2A), it was not until some 20 years later that the
mechanism of DNA replication in prokaryotes was understood in reasonable detail. This is because, as we
shall see in this chapter, the DNA replication process
rivals translation in complexity but is mediated by often
loosely associated protein assemblies that are present in
only a few copies per cell. *The surprising intricacy of DNA
replication compared to the chemically similar transcription process (Section 29-3) arises from the need for extreme
accuracy in DNA replication so as to preserve the integrity
of the genome from generation to generation.*

A. Replication Forks

*DNA is replicated by enzymes known as DNA-directed
DNA polymerases or simply DNA polymerases.* These
enzymes utilize single-stranded DNA as templates on
which to catalyze the synthesis of the complementary
strand from the appropriate deoxynucleoside triphosphates (Fig. 31-1). The incoming nucleotides are selected by their ability to form Watson–Crick base pairs
with the template DNA so that the newly synthesized
DNA strand forms a double helix with the template
strand. *Nearly all known DNA polymerases can only add a
nucleotide donated by a nucleoside triphosphate to the free
3′-OH group of a base paired polynucleotide so that DNA
chains are extended only in the 5′ → 3′ direction.* DNA

polymerases are discussed further in Sections 31-2A,
2B, and 4B.

Duplex DNA Replicates Semiconservatively at Replication Forks

John Cairns obtained the earliest indications of how
chromosomes replicate through the autoradiography of
replicating DNA. Autoradiograms of circular chromosomes grown in a medium containing [³H]thymine
show the presence of replication "eyes" or "bubbles"
(Fig. 31-2). These so-called θ **structures** (after their resemblance to the Greek letter theta) indicate that *duplex
DNA replicates by the progressive separation of its two*

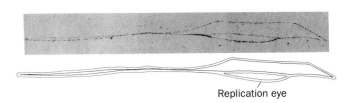

Replication eye

Figure 31-2
An autoradiogram and its interpretive drawing of a replicating
E. coli chromosome. The bacterium had been grown for
somewhat more than one generation in a medium containing
[³H]thymine, thereby labeling the subsequently synthesized
DNA so that it appears as a line of dark grains in the
photographic emulsion (*red lines in the interpretive drawing*).
The size of the replication eye indicates that the circular
chromosome is about one sixth duplicated in the present
round of replication. [Courtesy of John Cairns, Cold Spring
Harbor Laboratory.]

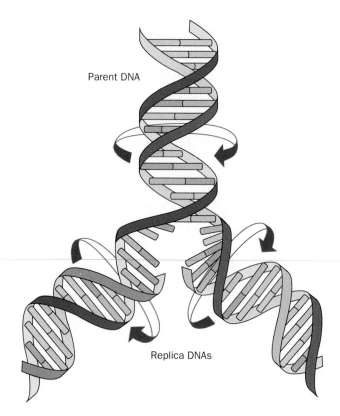

Figure 31-3
The replication of DNA.

parental strands accompanied by the synthesis of their complementary strands to yield two semiconservatively replicated duplex daughter strands (Fig. 31-3). DNA replication involving θ structures is known as *θ* **replication**.

A branch point in a replication eye at which DNA synthesis occurs is called a **replication fork**. A replication bubble may contain one or two replication forks (**unidirectional** or **bidirectional replication**). Autoradiographic studies have demonstrated that θ replication is almost always bidirectional (Fig. 31-4). Moreover, such experiments, together with genetic evidence, have established that prokaryotic and bacteriophage DNAs have but one **replication origin** (point where DNA synthesis is initiated).

B. Role of DNA Gyrase

The requirement that the parent DNA unwind at the replication fork (Fig. 31-3) presents a formidable topological obstacle. For instance, *E. coli* DNA is replicated at a rate of ~1000 nucleotides/s. If its 1300-μm long chromosome was linear, it would have to flail around within the confines of a 3-μm long *E. coli* cell at ~100 revolutions/s (recall that B-DNA has ~10 bp per turn). But since the *E. coli* chromosome is, in fact, circular, even this could not occur. Rather, the DNA molecule would accumulate +100 supercoils/s (see Section 28-5A for a

discussion of supercoiling) until it became too tightly coiled to permit further unwinding. Naturally occurring DNA's negative supercoiling promotes DNA unwinding but only to the extent of ~5% of its duplex turns (recall that naturally occurring DNAs are typically underwound by one supercoil per ~20 duplex turns; Section 28-5B). In prokaryotes, however, negative supercoils may be introduced into DNA through the action of a Type II topoisomerase (DNA gyrase; Section 28-5C) at the expense of ATP hydrolysis. This process is essential for prokarytic DNA replication as is demonstrated by the observation that DNA gyrase inhibitors, such as novobiocin and oxolinic acid, arrest DNA replication except in mutants whose DNA gyrase does not bind these antibiotics.

C. Semidiscontinuous Replication

The low resolution images provided by autoradiograms such as Figs. 31-2 and 31-4*b* suggest that duplex

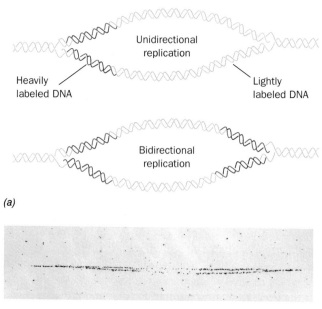

(a)

(b)

Figure 31-4
The autoradiographic differentiation of unidirectional and bidirectional θ replication of DNA. (a) An organism is grown for several generations in a medium that is lightly labeled with [³H]thymine so that all of its DNA will be visible in an autoradiogram. A large amount of [³H]thymine is then added to the medium for a few seconds before the DNA is isolated (**pulse labeling**) in order to label only those bases near the replication fork(s). Unidirectional DNA replication will exhibit only one heavily labeled branch point (*above*), whereas bidirectional DNA replication will exhibit two such branch points (*below*). (b) An autoradiogram of *E. coli* DNA so-treated demonstrating that it is bidirectionally replicated. [Courtesy of David M. Prescott, University of Colorado.]

DNA's two antiparallel strands are simultaneously replicated at an advancing replication fork. Yet, all known DNA polymerases can only extend DNA strands in the 5′ → 3′ direction. How, then, does DNA polymerase copy the parent strand that extends in the 5′ → 3′ direction past the replication fork? This question was answered in 1968 by Reiji Okazaki through the following experiments. If a growing *E. coli* culture is pulse labeled for 30 s with [³H]thymidine, much of the radioactive and hence newly synthesized DNA has a sedimentation coefficient in alkali of 7S to 11S. These so-called **Okazaki fragments** evidently consist of only 1000 to 2000 nucleotides. If, however, following the 30 s [³H]thymidine pulse, the *E. coli* are transferred to an unlabeled medium (a **pulse-chase** experiment), the resulting radioactively labeled DNA sediments at a rate that increases with the time that the cells had grown in the unlabeled medium. The Okazaki fragments must therefore become covalently incorporated into larger DNA molecules.

Okazaki interpreted his experimental results in terms of the **semidiscontinuous replication** model (Fig. 31-5). The two parent strands are replicated in different ways. *The newly synthesized DNA strand that extends 5′ → 3′ in the direction of replication fork movement, the so-called **leading strand**, is essentially continuously synthesized in its 5′ → 3′ direction as the replication fork advances. The other newly synthesized strand, the **lagging strand**, is also synthesized in its 5′ → 3′ direction but discontinuously as Okazaki fragments. The Okazaki fragments are only covalently joined together sometime after their synthesis in a reaction catalyzed by the enzyme **DNA ligase** (Section 31-2C).*

The semidiscontinuous model of DNA replication is corroborated by electron micrographs of replicating DNA showing single-stranded regions on one side of the replication fork (Fig. 31-6). In bidirectionally replicating DNA, moreover, the two single-stranded regions

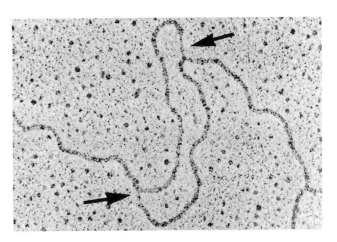

Figure 31-6
An electron micrograph of a replication eye in *Drosophila melanogaster* DNA. Note that the single-stranded regions (*arrows*) near the replication forks have the trans configuration consistent with the semidiscontinuous model of DNA replication. [From Kreigstein, H. J. and Hogness, D. S., *Proc. Natl. Acad. Sci.* **71**, 173 (1974).]

occur, as expected, on diagonally opposite sides of the replication bubble.

D. RNA Primers

DNA polymerases' all but universal requirement for a free 3′-OH group to extend a DNA chain poses a question that was emphasized by the establishment of the semidiscontinuous model of DNA replication: How is DNA synthesis initiated? Careful analysis of Okazaki fragments revealed that *their 5′ ends consist of RNA segments of 1 to 60 nucleotides (a length that is species dependent) that are complementary to the template DNA chain* (Fig. 31-7). *E. coli* has two enzymes that can catalyze the formation of these **RNA primers**: RNA polymerase, the enzyme that mediates transcription (Section 29-2), and the much smaller **primase** (60 kD), the monomeric product of the *dnaG* gene.

Primase is insensitive to the RNA polymerase inhibitor rifampicin (Section 29-2C). The observation that rifampicin inhibits only leading strand synthesis therefore indicates that *primase initiates the Okazaki fragment primers*. The initiation of leading strand synthesis in *E. coli*, a much rarer event than that of Okazaki fragments, can be mediated in *in vitro* by either RNA polymerase or primase alone but is greatly stimulated when both enzymes are present. It is therefore thought that these enzymes act synergistically *in vivo* to prime leading strand synthesis.

Mature DNA does not contain RNA. The RNA primers are removed during replication and the resulting single-strand gaps are filled in with DNA by a mechanism described in Section 31-2A.

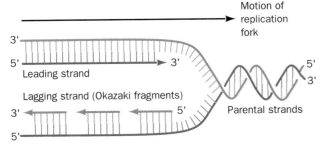

Figure 31-5
In DNA replication, both daughter strands (*red*) are synthesized in their 5′ → 3′ direction. The leading strand is synthesized continuously, whereas the lagging strand is synthesized discontinuously.

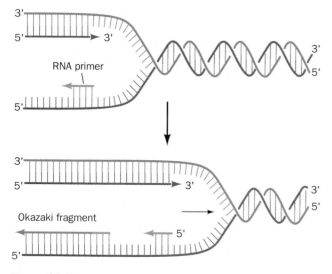

Figure 31-7
DNA synthesis is primed by short RNA segments.

2. ENZYMES OF REPLICATION

DNA replication is a complex process involving a great variety of enzymes. It requires, to list only its major actors in their order of appearance: (1) a DNA gyrase, (2) proteins to separate the DNA strands at the replication fork, (3) proteins to prevent them from reannealing before they are replicated, (4) enzymes to synthesize RNA primers, (5) a DNA polymerase, (6) an enzyme to remove the RNA primers, and (7) an enzyme to covalently link successive Okazaki fragments. In this section, we describe the properties and functions of many of these proteins.

A. DNA Polymerase I

In 1957, Arthur Kornberg reported that he had discovered an enzyme that catalyzes the synthesis of DNA in extracts of *E. coli* through its ability to incorporate the radioactive label from [^{14}C]thymidine triphosphate into DNA. This enzyme, which has since become known as **DNA polymerase I** or **Pol I,** consists of a single 928-residue polypeptide.

Pol I couples deoxynucleoside triphosphates on DNA templates (Fig. 31-1) in a reaction that occurs through the nucleophilic attack of the growing DNA chain's 3'-OH group on the α-phosphoryl of an incoming nucleoside triphosphate. The reaction is driven by the resulting elimination of PP$_i$ and its subsequent hydrolysis by inorganic pyrophosphatase. The overall reaction resembles that catalyzed by RNA polymerase (Section 29-2) but differs from it by the strict requirement that the incoming nucleotide be linked to a free 3'-OH group of a polynucleotide that is base paired to the template (recall that RNA polymerase initiates transcription by linking together two ribonucleotide triphosphates on a DNA template). The complementarity between the product DNA and the template was at first inferred through base composition and hybridization studies, and was eventually di-

rectly established by base sequence determinations. The error rate of Pol I in copying the template is quite low as was demonstrated by its *in vitro* replication of φX174 DNA to yield fully infective phage DNA.

The specificity of Pol I for an incoming base is thought to arise from the requirement that it form a Watson–Crick base pair with the template rather than direct recognition of the incoming base (recall that the four base pairs, A·T, T·A, G·C, and C·G, have nearly identical shapes; Section 28-2A). This accounts for the observation that Pol I may substitute 5-bromouracil only for thymine and hypoxanthine only for guanine. Pol I is said to be **processive** in that it catalyzes a series of successive polymerization steps, typically 20 or more, without releasing the template. Pol I can, of course, work in reverse by degrading DNA through pyrophosphorolysis. This reverse reaction, however, probably has no physiological significance because of the low *in vivo* concentration of PP$_i$ resulting from the action of inorganic pyrophosphatase.

Pol I Can Edit Its Mistakes
In addition to its polymerase activity, Pol I has two independent hydrolytic activities:

1. It can act as a $3' \rightarrow 5'$ exonuclease.

2. It can act as a $5' \rightarrow 3'$ exonuclease.

The $3' \rightarrow 5'$ exonuclease reaction differs chemically from the pyrophosphorolysis reaction only in that H$_2$O rather than PP$_i$ is the nucleotide acceptor. Kinetic and crystallographic studies (see below), however, indicate that these two catalytic activities occupy separate active sites. The $3' \rightarrow 5'$ exonuclease function is activated by an unpaired 3'-terminal nucleotide with a free OH group. If Pol I erroneously incorporates a wrong (unpaired) nucleotide at the end of a growing DNA chain, the polymerase activity is inhibited and the $3' \rightarrow 5'$ exonuclease excises the offending nucleotide (Fig. 31-8). The polymerase activity then resumes DNA replication.

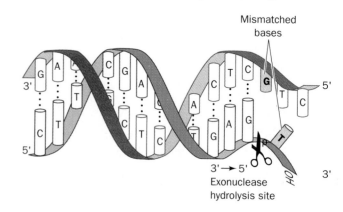

Figure 31-8
The $3' \rightarrow 5'$ exonuclease function of DNA polymerase I excises mispaired nucleotides from the 3' end of the growing DNA strand.

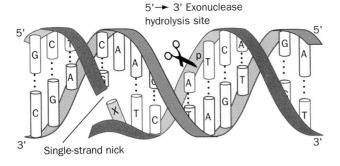

5′→ 3′ Exonuclease
hydrolysis site

Single-strand nick

Figure 31-9
The 5′ → 3′ exonuclease function of DNA polymerase I
excises up to 10 nucleotides from the 5′ end of a single-
strand nick. The nucleotide immediately past the nick (X)
may or may not be paired.

*Pol I therefore has the ability to proofread a DNA chain as it
is synthesized so as to correct its mistakes.* This explains
the great fidelity of DNA replication by Pol I despite the
relatively low free energy of a single base pairing interaction (the energetics of binding fidelity is discussed in
Section 30-2C). The price of this high fidelity is that
~ 10% of correctly incorporated nucleotides are also excised.

The Pol I 5′ → 3′ exonuclease binds to duplex DNA
at single-strand nicks with little regard to the character
of the 5′ nucleotide (5′-OH or phosphate group; base
paired or not). It cleaves the DNA in a base paired region
beyond the nick such that the DNA is excised as either

mononucleotides or oligonucleotides of up to 10 residues (Fig. 31-9). In contrast, the 3′ → 5′ exonuclease
removes only unpaired mononucleotides with 3′-OH
groups.

Pol I's Polymerase and Two Exonuclease Functions Each Occupy Separate Active Sites

The 5′ → 3′ exonuclease activity of Pol I is independent of both its 3′ → 5′ exonuclease and its polymerase
activities. In fact, as we saw in Section 28-6C, proteases
such as subtilisin or trypsin cleave Pol I into two fragments: The large or "Klenow" fragment (residues
324 – 928), which contains both the polymerase and the
3′ → 5′ exonuclease activities; and a smaller fragment
(residues 1 – 323), which contains the 5′ → 3′ exonuclease activity. Thus Pol I contains three active sites on a
single polypeptide chain.

The X-ray structure of the Klenow fragment in complex with dTMP, determined by Thomas Steitz, indicates that this protein consists of two domains (Fig.
31-10a). The smaller domain binds the dTMP whose 5′

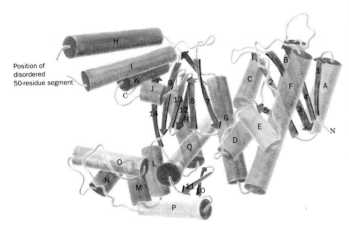

Position of disordered 50-residue segment

(a)

Figure 31-10
The X-ray structure of Pol I Klenow fragment. (*a*) A ribbon
drawing in which the break between helices H and I (*left*) indicates the position of a disordered 50-residue segment. The
division between the protein's small (*green*) and large (*purple*)
domains lies on the loop between helices F and G. [After
Ollis, L., Brick, P., Hamlin, R., Xuong, N. G., and Steitz, T. A.,
Nature **313,** 762 (1985).] (*b*) A space-filling drawing, based, in
part, on model building, of the Pol I Klenow fragment with
double-stranded B-DNA in its proposed DNA binding cleft (the
Pol I crystal does not contain DNA). The primer strand (*green*)
ends in the region of the polymerase's presumed active site
while the template strand (*yellow*) passes through the
enzyme. A dTMP molecule (*purple*) is shown bound at its
observed position in the 3′ → 5′ exonuclease active site.
[Courtesy of Thomas Steitz, Yale University.]

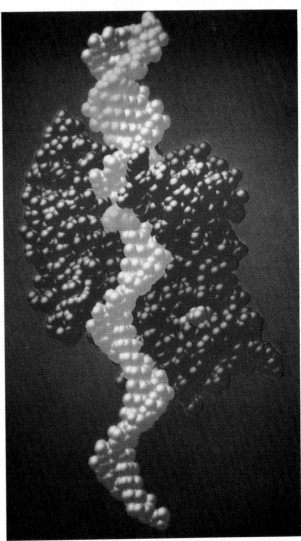

(b)

phosphate interacts with two bound divalent metal ions. The dTMP-binding site forms part of the $3' \rightarrow 5'$ exonuclease as was demonstrated by the absence of this function but not the polymerase activity in a genetically engineered Klenow fragment mutant that lacks metal ion chelating sites but which otherwise has a normal structure. The larger domain (helix G and beyond in Fig. 31-10*a*) contains a prominent cleft, the polymerase site, which model building indicates has the appropriate size and shape to bind a B-DNA molecule in a manner resembling a right hand grasping a rod (Fig. 31-10*b*). A 50-residue segment of the protein, which projects from the "thumb" of the "hand" (helices H and I in Fig. 31-10*a*), is not visible in the X-ray structure and is therefore thought to be conformationally disordered. Steitz has suggested that this segment is flexibly linked to the protein so as to close off the fourth side of the cleft when it contains DNA. This device would greatly reduce DNA's rate of dissociation from the polymerase and thereby account for Pol I's observed processivity. Model building studies, moreover, indicate that helices J and K extend into the DNA's major groove so as to fix the DNA's helical orientation in the Pol I active site much like the thread of a nut. If both structural features really exist, the DNA would have to screw to a new 3' terminus between polymerization steps.

In an effort to better characterize Pol I's polymerase function, Klenow fragment was crystallized with a DNA fragment designed to mimic the template–primer complex: An 8 bp DNA in which the "template" strand has a 3-nucleotide single-stranded overhang at its 5' end. The X-ray structure of this crystal revealed that the DNA, contrary to expectations, binds in the $3' \rightarrow 5'$ exonuclease site with only a 4-nucleotide segment visible (the remainder is presumably disordered) and extending towards the ~30-Å distant polymerase site. This suggests that the $3' \rightarrow 5'$ exonuclease site competes with the polymerase site for the newly synthesized DNA's 3' terminus (Fig. 31-11). Since the exonuclease site can only bind single-stranded DNA, a 3'-terminal mismatch, in promoting DNA melting, favors the binding of the newly synthesized strand to the exonuclease site. Mismatches are thereby preferentially excised.

Pol I Functions Physiologically to Repair DNA

For some 13 years after Pol I's discovery, it was generally assumed that this enzyme was *E. coli*'s DNA replicase because no other DNA polymerase activity had been detected in *E. coli*. This assumption was made untenable by Cairns and Paula DeLucia's isolation, in 1969, of a mutant *E. coli* whose extracts exhibit <1% of the normal Pol I activity (although it has nearly normal levels of the $5' \rightarrow 3'$ exonuclease activity) but which nevertheless reproduce at the normal rate. This mutant strain, however, is highly susceptible to the damaging effects of UV radiation and chemical mutagens. *Pol I*

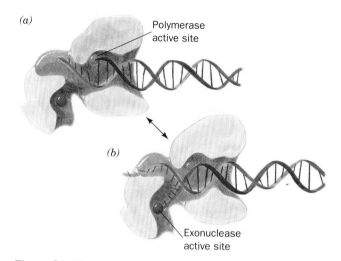

Figure 31-11

A schematic model of Klenow fragment with DNA binding to (*a*) the polymerase active site; and (*b*) the $3' \rightarrow 5'$ exonuclease active site. To slide from the polymerase to the exonuclease active sites, the 3' terminus of the daughter strand (*red*) must traverse the lengths of four nucleotides of double-stranded DNA and four of single-stranded DNA. The template strand is drawn in blue. [After Freemont, P. S., Friedman, J. M., Beese, L. S., Sanderson, M. R., and Steitz, T. A., *Proc. Natl. Acad. Sci.* **85**, 8927 (1988).]

evidently plays a central role in the repair of damaged (chemically altered) DNA.

Damaged DNA, as we discuss in Section 31-5, is detected by a variety of DNA repair systems. Many of them endonucleolytically cleave the damaged DNA on the 5' side of the lesion thereby activating Pol I's $5' \rightarrow 3'$ exonuclease. While excising this damaged DNA, Pol I simultaneously fills in the resulting single-strand gap through its polymerase activity. In fact, its $5' \rightarrow 3'$ exonuclease activity increases tenfold when the polymerase function is active. Perhaps the simultaneous excision and polymerization activities of Pol I protects DNA from the action of cellular nucleases that would further damage the otherwise gapped DNA.

Pol I Catalyzes Nick Translation

Pol I's combined $5' \rightarrow 3'$ exonuclease and polymerase activities can replace the nucleotides on the 5' side of a single-strand nick on otherwise undamaged DNA. These reactions, in effect, translate (move) the nick towards the DNA strand's 3' end without otherwise changing the molecule (Fig. 31-12). This **nick translation** process, in the presence of labeled deoxynucleoside triphosphates, is synthetically employed to prepare highly radioactive DNA (the required nicks may be generated by treating the DNA with a small amount of pancreatic DNase I).

Pol I's $5' \rightarrow 3'$ Exonuclease Functions Physiologically to Excise RNA Primers

The $5' \rightarrow 3'$ exonuclease also removes the RNA primers at the 5' ends of newly synthesized DNA and fills in the

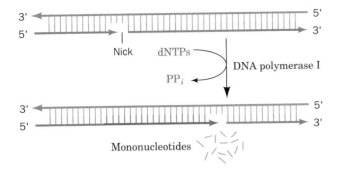

Figure 31-12
Nick translation as catalyzed by Pol I.

Table 31-2

Components of DNA Polymerase III Holoenzyme

Subunit	Mass (kD)	Structural Gene
α^a	130	*polC (dnaE)*
ε^a	27.5	*dnaQ*
θ^a	10	Unknown
τ	71	*dnaZX*[b]
γ	52	*dnaZ*[b]
δ	32	Unknown
β	40.6	*dnaN*

[a] Components of Pol III.
[b] The γ and τ subunits are encoded by the same gene sequence; the γ subunit comprises the N-terminal end of the τ subunit.

Source: McHenry, C. S., *Annu. Rev. Biochem.* **57**, 538 (1988).

resulting gaps. The importance of this function was demonstrated by the isolation of temperature sensitive *E. coli* mutants that are neither viable nor exhibit any $5' \rightarrow 3'$ exonuclease activity at the restrictive temperature of $\sim 43°C$ (the low level of polymerase activity in the Pol I mutant isolated by Cairns and DeLucia is apparently sufficient to carry out this essential gap-filling process during chromosome replication). Thus Pol I has an indispensable role in *E. coli* DNA replication although a different one than was first supposed.

B. DNA Polymerase III

The discovery of normally growing *E. coli* mutants that have very little Pol I activity stimulated the search for an additional DNA polymerizing activity. This effort was rewarded by the discovery of two more enzymes, designated, in the order they were discovered, **DNA polymerase II (Pol II)** and **DNA polymerase III (Pol III)**. The properties of these enzymes are compared with that of Pol I in Table 31-1. Pol II and Pol III had not previously been detected because their combined activities in the assays used are normally <5% that of Pol I.

Table 31-1

Properties of *E. coli* DNA Polymerases

Property	Pol I	Pol II	Pol III
Mass (kD)	109	120	140
Molecules/cell	400	?	10–20
Turnover number[a]	600	30	9000
Structural gene	*polA*	*polB*	*polC*
Conditionally lethal mutant	+	−	+
Polymerization: $3' \rightarrow 5'$	+	+	+
Exonuclease: $3' \rightarrow 5'$	+	+	+
Exonuclease: $5' \rightarrow 3'$	+	−	+

[a] Nucleotides polymerized min^{-1} molecule^{-1} at 37°C.

Source: Kornberg, A., *DNA Replication*, p. 169, Freeman (1980).

Mutants lacking Pol II activity have no known defects. Pol II's physiological function, if any, is therefore unknown.

Pol III Is *E. coli*'s DNA Replicase

The cessation of DNA replication in temperature sensitive *polC* mutants above the restrictive (high) temperature demonstrates that Pol III is *E. coli*'s *DNA replicase*. This enzyme has the subunit composition $\alpha\varepsilon\theta$ where α, the *polC* gene product (Table 31-2), contains the polymerase function. The catalytic properties of Pol III resemble those of Pol I (Table 31-1) except for Pol III's inability to replicate primed single-stranded or nicked duplex DNA. Rather, Pol III acts *in vitro* at single-strand gaps of <100 nucleotides, a situation that probably resembles the state of DNA at the replication fork. The Pol III $3' \rightarrow 5'$ exonuclease function, which resides on the enzyme's ε subunit, is DNA's primary editor during replication; it enhances the enzyme's replication fidelity by up to 200-fold. However, the Pol III $5' \rightarrow 3'$ exonuclease acts only on single-stranded DNA so that it cannot catalyze nick translation.

*Pol III functions in vivo as part of a complex and labile multisubunit enzyme, the **Pol III holoenzyme**, which consists of at least seven types of subunits (Table 31-2).* The latter four subunits in Table 31-2 act to modulate Pol III's activity. For example, the binding of Pol III holoenzyme to primed template requires ATP hydrolysis in a reaction involving the β subunit. This reaction clamps holoenzyme to the template to form a complex that has essentially unlimited processivity (>5000 residues). In contrast, Pol III alone, which does not hydrolyze ATP, has a processivity of 10 to 15 residues. It also seems likely that the holoenzyme's non-Pol III subunits provide some of the interaction sites with other proteins that participate in the DNA replication process (see Section 31-3).

C. Helicases, Binding Proteins, and DNA Ligases

Pol III holoenzyme, unlike Pol I, cannot unwind duplex DNA. Rather, *three proteins, **Rep protein, helicase II**, and **single-strand binding protein (SSB)** (Table 31-3), work in concert to unwind the DNA before an advancing replication fork (Fig. 31-13) in a process that is driven by ATP hydrolysis.* Rep protein, the *E. coli rep* gene product, separates the duplex DNA strands by moving along the leading strand template in the $3' \rightarrow 5'$ direction while consuming two ATPs per base pair separated. Helicase II similarly moves along the lagging strand template in the $5' \rightarrow 3'$ direction. The observation that *rep*⁻ mutations only slow DNA replication suggests that rep protein and helicase II act together at the replication fork to unwind DNA. The separated DNA strands behind the advancing helicase are prevented from reannealing by the binding of SSB. Numerous copies of this tetrameric protein cooperatively coat single-stranded DNA thereby maintaining it in an unpaired state. Note, however, that DNA must be stripped of SSB before it can be replicated by Pol III holoenzyme.

Table 31-3

E. coli **Unwinding and Binding Proteins**

Protein	Subunit Structure	Subunit Mass (kD)	Molecules/Cell
Rep protein	Monomer	65	50
Helicase II	Monomer	75	6000
SSB	Tetramer	19	800

Source: Kornberg, A., *DNA Replication, 1982 Supplement*, p. S83, Freeman (1982) and *DNA Replication*, p. 283, Freeman (1980).

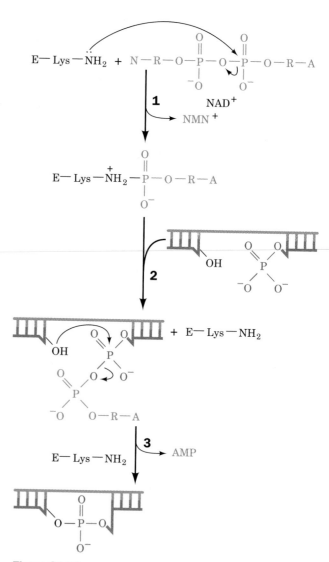

Figure 31-14
The reactions catalyzed by *E. coli* DNA ligase. In eukaryotic and T4 DNA ligases, NAD⁺ is replaced by ATP so that PP$_i$ rather than NMN⁺ is eliminated in the first reaction step.

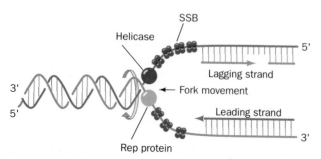

Figure 31-13
The unwinding of DNA by the combined action of Rep protein, helicase II, and SSB protein. The Rep protein moves along the leading strand template in the $3' \rightarrow 5'$ direction accompanied by helicase II on the lagging strand template moving in the $5' \rightarrow 3'$ direction. The separated DNA strands are prevented from reannealing by SSB binding.

DNA Ligase Seals Single-Strand Nicks

Pol I, as we saw in Section 31-1D, replaces the Okazaki fragments' RNA primers with DNA through nick translation. *The resulting single-strand nicks between adjacent Okazaki fragments, as well as that on circular DNA after leading strand synthesis, are sealed in a reaction catalyzed by **DNA ligase**.* The free energy required by this reaction is obtained, in a species-dependent manner, through the coupled hydrolysis of either NAD⁺ to NMN⁺ + AMP or ATP to PP$_i$ + AMP. The *E. coli* enzyme, a 77-kD monomer that utilizes NAD⁺, catalyzes a three-step reaction (Fig. 31-14):

1. The adenyl group of NAD⁺ is transferred to the ε-amino group of an enzyme Lys residue to form an

unusual phosphoamide adduct that is, nevertheless, readily isolated.

2. The adenyl group of this activated enzyme is transferred to the 5'-phosphoryl terminus of the nick to form an adenylated DNA. Here, AMP is linked to the 5'-nucleotide via a pyrophosphate rather than the usual phosphodiester bond.

3. DNA ligase catalyzes the formation of a phosphodiester bond by attack of the 3'-OH on the 5'-phosphoryl group, thereby sealing the nick and releasing AMP.

ATP-requiring DNA ligases, such as those of eukaryotes and T4 phage, release PP_i in the first step of the reaction rather than NMN^+. T4 ligase is also noteworthy in that, at high DNA concentrations, it can link together two duplex DNAs (**blunt end ligation**) in a reaction that is a boon to genetic engineering (Section 28-8B).

3. PROKARYOTIC REPLICATION MECHANISMS

Bacteriophages are among the simplest biological entities and their DNA replication mechanisms reflect this fact. Much of what we know about how DNA is replicated therefore stems from the study of this process in various phages. In this section we examine DNA replication in the coliphages **M13** and ϕX174 and then consider what is known about DNA replication in *E. coli* itself. Eukaryotic DNA replication is discussed in Section 31-4.

A. Bacteriophage M13

Bacteriophage M13 carries a 6408 nucleotide single-stranded circular DNA known as its **viral** or (**+**) **strand.** Upon infecting an *E. coli* cell, this strand directs the synthesis of its complementary or (**−**) **strand** to form the circular duplex **replicative form (RF),** which occurs either nicked (**RF II**) or supercoiled (**RF I**). This replication process (Fig. 31-15) may be taken as a paradigm for leading strand synthesis in duplex DNA.

As the M13 (+) strand enters the *E. coli* cell, it becomes coated with SSB except at a palindromic 57-nucleotide segment that forms a hairpin. RNA polymerase, in a process that requires a σ subunit for initiation-site recognition (Section 29-2B), commences primer synthesis six nucleotides before the start of the hairpin and extends the RNA 20 to 30 residues to form a segment of RNA–DNA hybrid duplex. The DNA that is displaced from the hairpin becomes coated with SSB so that when RNA polymerase reaches it, primer synthesis stops. Pol III holoenzyme then extends the RNA primer around the

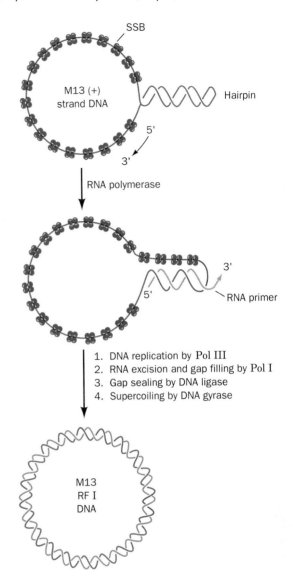

Figure 31-15
The synthesis of the M13 (−) strand DNA on a (+) strand template to form M13 RF I DNA.

circle to form the (−) strand. The primer is removed by Pol I-catalyzed nick translation, thereby forming RF II, which is converted to RF I by the sequential actions of DNA ligase and DNA gyrase.

B. Bacteriophage ϕX174

Bacteriophage ϕX174, as does M13, carries a small (5386 nucleotides) single-stranded circular DNA. Curiously, the *in vivo* conversion of the ϕX174 viral DNA to its replicative form is a much more complex process than that for M13 DNA in that ϕX174 replication requires the participation of a nearly 600-kD protein assembly known as a **primosome** (Table 31-4).

Table 31-4

Proteins of the Primosome[a]

Protein	Subunit Structure	Subunit Mass (kD)	Molecules/Cell
n	Dimer	14	80
n'	Monomer	76	70
n''	Monomer	17	
i	Trimer	22	50
DnaB	Hexamer	50	20
DnaC	Monomer	29	100
Primase	Monomer	60	50

[a] The complex of all primosome proteins but primase is known as the preprimosome.

Source: Kornberg, A., *DNA Replication, 1982 Supplement*, p. S123, Freeman (1982).

φX174 (−) Strand Replication Is a Paradigm for Lagging Strand Synthesis

φX174 (−) strand synthesis occurs in a six-step process (Fig. 31-16):

1. The reaction sequence begins in the same way as that for M13: The (+) strand is coated with SSB except for a 44-nucleotide hairpin near position 2300. A 55-nucleotide sequence containing this hairpin is then recognized and bound by the **n, n′,** and **n″** proteins.

2. The **i, DnaB,** and **DnaC** proteins add to this complex in an ATP-requiring process to form the **preprimosome.** The preprimosome, in turn, binds primase yielding the primosome.

3. The primosome is propelled in the 5′ → 3′ direction along the (+) strand by n′-catalyzed ATP hydrolysis. This motion, which displaces the SSB in its path, is opposite in direction to that of template reading during DNA chain propagation.

4. At randomly selected sites, the primosome reverses its migration while primase synthesizes an RNA primer. The initiation of primer synthesis requires the participation of DnaB protein which, through concomitant ATP hydrolysis, is thought to alter template DNA conformation in a manner required by primase.

5. Pol III holoenzyme extends the primers to form Okazaki fragments.

6. Pol I excises the primers and replaces them by DNA. The fragments are then joined by DNA ligase and supercoiled by DNA gyrase to form the φX174 RF I.

The primosome remains complexed with the DNA (Fig. 31-17) where it participates in (+) strand synthesis (see below).

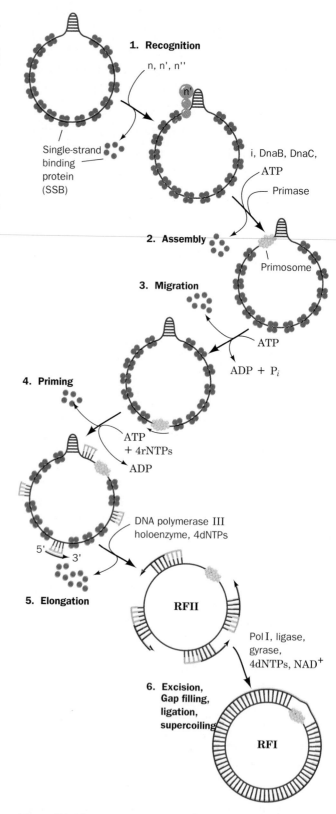

Figure 31-16

The synthesis of the φX174 (−) strand on a (+) strand template to form φX174 RF I DNA. [After Arai, K., Low, R., Kobori, J., Schlomai, J., and Kornberg, A., *J. Biol. Chem.* **256**, 5280 (1981).]

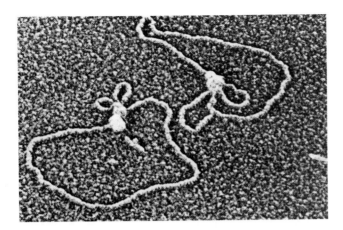

Figure 31-17
Electron micrograph of a primosome bound to a φX174 RF I
DNA. Such complexes always contain a single primosome
with one or two associated small DNA loops. [Courtesy of
Jack Griffith, Lineberger Cancer Research Center, University
of North Carolina.]

φX174 (+) Strand Replication Serves as a Model for Leading Strand Synthesis

One strand of a circular duplex DNA may be synthesized via the **rolling circle** or **σ replication** mode (so-
called because of the resemblance of the replicating
structure to the Greek letter sigma; Fig. 31-18). *The
φX174 (+) strand is synthesized on an RF I template by a
variation on this process, the **looped rolling circle mode**
(Fig. 31-19):*

1. (+) Strand synthesis begins with the primosome-
   aided binding of the phage-encoded enzyme **gene A
   protein** (60 kD) to its ~ 30 bp recognition site. There,
   gene A protein specifically cleaves the phosphodiester bond preceding (+) strand nucleotide 4306 by
   forming a covalent bond with its 5'-phosphoryl
   group, which conserves the cleaved bond's energy.

2. Rep protein (Section 31-2C) subsequently attaches to
   the (−) strand at the gene A protein and, with the aid
   of the primosome still associated with the (+)
   strand, commences unwinding the duplex DNA from the (+)
   strand's 5' end. The displaced (+) strand is coated
   with SSB, which prevents it from reannealing to the
   (−) strand. Rep protein is essential for φX174 DNA
   replication, in contrast to the situation for the *E. coli*
   chromosome, as is demonstrated by the inability of
   φX174 to multiply in *rep⁻ E. coli*. Pol III holoenzyme
   extends the (+) strand from its free 3'-OH group.

3. The extension process generates a **looped rolling
   circle** structure in which the 5' end of the old (+)
   strand remains linked to the gene A protein at the
   replication fork. It is thought that as the old (+) strand
   is peeled off the RF, the primosome synthesizes the

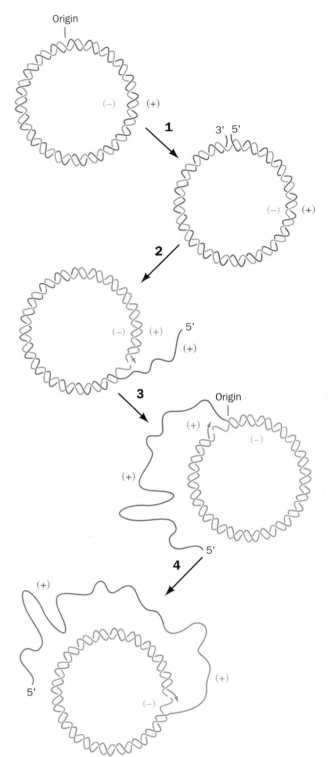

Figure 31-18
The rolling circle mode of DNA replication. The (+) strand
being synthesized is extended from a specific cut made at
the replication origin (1) so as to strip away the old (+)
strand (2 and 3). The continuous synthesis of the (+) strand
on a circular (−) strand template produces a series of
tandemly linked (+) strands (4), which may later be separated
by a specific exonuclease.

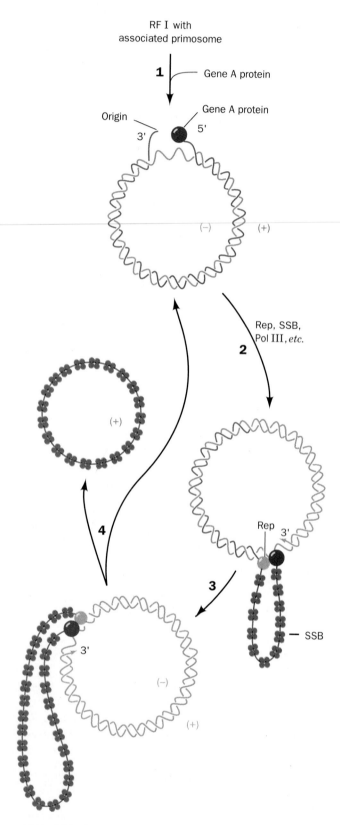

Figure 31-19
The synthesis of the φX174 (+) strand by the looped rolling circle mode.

primers required for the later generation of a new (−) strand.

4. Upon coming full circle around the (−) strand, the gene *A* protein again makes a specific cut at the replication origin so as to form a covalent linkage with the new (+) strand's 5′ end. Simultaneously, the newly formed 3′-terminal OH group of the old, looped-out (+) strand nucleophilically attacks its 5′-phosphoryl attachment to the gene *A* protein, thereby liberating a covalently closed (+) strand. Evidently, the gene *A* protein has two active sites that alternate in their attachment to the 5′ ends of successively synthesized (+) strands. The replication fork continues its progress about the duplex circle, producing new (+) strands in a manner reminiscent of linked sausages being pulled off a reel.

In the intermediate stages of a φX174 infection, each newly synthesized (+) strand directs the synthesis of the (−) strand to form RF I as described above. In the latter stages of infection, however, the newly formed (+) strands are packaged into phage particles.

C. *E. coli*

The E. coli chromosome replicates by the bidirectional θ mode from a single replication origin (Section 31-1A). The most plausible model for events at the *E. coli* replication fork (Fig. 31-20) is largely derived from studies on the simpler and more experimentally accessible DNA replication mechanisms of coliphages such as M13 and φX174. Duplex DNA is unwound by Rep protein on the leading strand template in concert with helicase II and the primosome on the lagging strand template. The separated single strands are immediately coated by SSB. Leading strand synthesis is catalyzed by Pol III holoenzyme as is that of the lagging strand after priming by primosome-associated primase. It is thought that both leading and lagging strand synthesis occur on a single multiprotein particle, the **replisome,** so that the lagging strand template must be looped around (Fig. 31-20). After completing the synthesis of an Okazaki fragment, the lagging strand holoenzyme relocates to a new primer near the replication fork, the primer heading the previously synthesized Okazaki fragment is excised by Pol I-catalyzed nick translation, and the nick is sealed by DNA ligase.

E. coli DNA Replication Is Initiated at oriC in a Process Mediated by DnaA Protein

The replication origin of the *E. coli* chromosome consists of a unique 245 bp segment known as the *oriC* locus. This sequence, segments of which are highly conserved among gram-negative bacteria, supports the bidirectional replication of the various plasmids into which it has been inserted. Based on experiments with such plasmids, Kornberg has proposed that replication

(a)

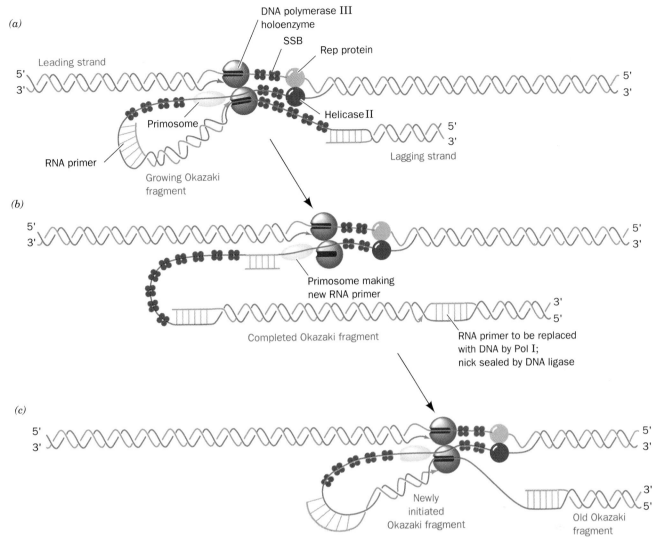

Figure 31-20
The replication of *E. coli* DNA. (a) The *E. coli* DNA replisome, which is thought to contain two DNA polymerase III holoenzyme complexes, synthesizes both the leading and lagging strands. The lagging strand template must loop around to permit holoenzyme to extend the primosome-primed lagging strand. (b) The holoenzyme releases the lagging strand template when it encounters the previously synthesized Okazaki fragment. This possibly signals the primosome to initiate the synthesis of lagging strand RNA primer. (c) Holoenzyme rebinds the lagging strand template and extends the RNA primer to form a new Okazaki fragment. Note that in this model, leading strand synthesis is always ahead of lagging strand synthesis.

initiation in *E. coli* occurs via the following multistep process (Fig. 31-21):

1. **DnaA protein** (52 kD) recognizes and binds up to four 9 bp repeats in *oriC* to form, as electron microscopy and DNase I protection experiments indicate, a complex of negatively supercoiled *oriC* DNA wrapped around a central core of 20 to 40 DnaA protein monomers. This process is facilitated by the histone-like **HU protein.**

2. The DnaA protein subunits then successively melt three tandemly repeated, 13 bp, AT-rich segments (consensus sequence 5'-GATCTnTTnTTTT-3' where n marks nonspecific positions) located near *oriC*'s

"left" boundary. The existence of the resulting ~45 bp open complex was established through its sensitivity to **P1 nuclease,** an endonuclease specific for single strands. The formation of the open complex requires the presence of DnaA protein and ATP (which DnaA protein tightly binds and hydrolyzes in a DNA-dependent manner) and only occurs above 22°C (at least *in vitro*). The AT-rich nature of the 13 bp repeats, no doubt, facilitates the melting process.

3. The DnaA protein then guides a complex of DnaB and DnaC proteins into the melted region to form a so-called **prepriming complex.**

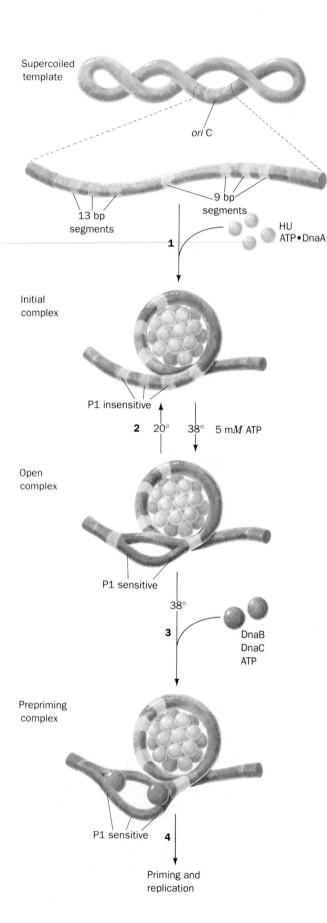

Figure 31-21
A model for DNA replication initiation at *oriC*. **(1)** DnaA proteins bind to the four 9-mers so as to wrap the suitably supercoiled *oriC* around a 20 to 40 subunit protein core. **(2)** The three AT-rich 13 bp repeats are then melted in an ATP-driven reaction to form an open complex to which **(3)** DnaB–DnaC complex binds. **(4)** The open complex is further unwound through the helicase action of DnaB protein thereby preparing the complex for priming and bidirectional replication. [After Bramhill, D. and Kornberg, A., *Cell* **52**, 752 (1988).]

4. In the presence of SSB and gyrase, DnaB, which has helicase activity, further unwinds the DNA in the prepriming complex in both directions so as to permit entry of primase and RNA polymerase. The participation of both these enzymes in leading strand primer synthesis (Section 31-1D) together with limitation of this process to the *oriC* site suggests that the RNA polymerase activates primase to synthesize the primer. This perhaps explains the similarity of *oriC*'s AT-rich 13-mers to RNA polymerase's transcriptional promoters (Section 29-2B).

The stage is thereby set for bidirectional DNA replication by Pol III holoenzyme as described above.

The Initiation of *E. coli* DNA Replication Is Strictly Regulated

Chromosome replication in E. coli occurs only once per cell division so that this process must be tightly controlled. The doubling time of *E. coli* at 37°C varies with growth conditions from <20 min to ~10 h. Yet the constant ~850 nucleotide/s rate of movement of each replication fork fixes the chromosome replication time, C at ~40 min (the *E. coli* chromosome consists of ~4×10^6 bp). Moreover, the segregation of cellular components and the formation of a septum between them, which must precede cell division, requires a constant time, $D = 20$ min, after the completion of the corresponding round of chromosome replication. *Cells with doubling times $<C + D = 60$ min must consequently initiate chromosome replication before the end of the preceding cell division cycle.* This results in the formation of **multiforked chromosomes** as is indicated in Fig. 31-22 for a cell division time of 35 min.

The above considerations indicate that there must be a signal that triggers each cycle of chromosome replication. The nature of this signal is unknown. However, the observations that (1) *oriC*'s 13 bp repeats each begin with the sequence GATC (see above), the sequence most commonly methylated in *E. coli* (Section 31-7), and (2) *E. coli* defective in the GATC methylation enzyme are very inefficiently transformed by *oriC*-containing plasmids, suggests that the DNA replication trigger in *E. coli* responds to the level of *oriC* methylation.

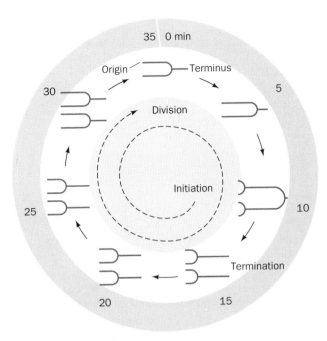

Figure 31-22
In cells that are dividing every 35 min, the fixed 60-min interval between the initiation of replication and cell division results in the production of multiforked chromosomes. [After Lewin, B., *Genes* (3rd ed.), p. 299, Wiley (1987).]

There is extensive morphological evidence, such as Fig. 31-23, that the *E. coli* chromosome is associated with the cell membrane. This association presumably permits the segregation of replicated chromosomes into different cells during cell division. There is, nevertheless, no direct evidence that any membrane component is required for DNA replication.

Replication Termination

Little is known about how DNA replication terminates in *E. coli*. Termination occurs in a specific region of

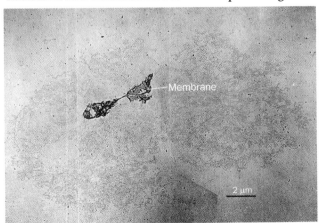

Figure 31-23
An electron micrograph of an intact and supercoiled *E. coli* chromosome attached to two fragments of the cell membrane. [From Delius H. and Worcel, A., *J. Mol. Biol.* **82,** 108 (1974).]

the *E. coli* chromosome, *terC,* about halfway around the chromosome from *oriC,* where the two oppositely moving replication forks meet. There is, however, scant indication as to the mechanism of the termination process or how the DNA products of the two colliding replication forks are untangled from each other.

D. Fidelity of Replication

Since a single polypeptide as small as the Pol I Klenow fragment can replicate DNA by itself, why does *E. coli* maintain a battery of >20 intricately coordinated proteins to replicate its chromosome? The answer apparently is *to ensure the nearly perfect fidelity of DNA replication required to preserve the genetic message's integrity from generation to generation.*

The rates of reversion of mutant *E. coli* or T4 phage to the wild-type indicates that only one mispairing occurs per 10^8 to 10^{10} base pairs replicated. This corresponds to ~1 error per 1000 bacteria per generation. Such high replication accuracy arises, in part, from the $3' \rightarrow 5'$ exonuclease functions of Pol I and Pol III that detect and eliminate the occasional errors made by their polymerase functions. In fact, mutations that increase a DNA polymerase's proofreading exonuclease activity decrease the rates of mutation of other genes.

The inability of a DNA polymerase to initiate chain elongation without a primer is thought to be a feature that increases DNA replication fidelity. The first few nucleotides of a chain to be coupled together are those most likely to be mispaired because of the cooperative nature of base pairing interactions (Section 28-3). The editing of a short duplex oligonucleotide is similarly an error-prone process. The use of RNA primers eliminates this source of error since the RNA is eventually replaced by DNA under conditions that permit accurate base pairing to be achieved.

One might wonder why cells have evolved the complex system of discontinuous lagging strand synthesis rather than a DNA polymerase that could simply extend DNA chains in their $3' \rightarrow 5'$ direction. Consideration of the chemistry of DNA chain extension also leads to the conclusion that this system promotes high fidelity replication. The linking of 5'-deoxynucleotide triphosphates in the $3' \rightarrow 5'$ direction would require the retention of the growing chain's 5'-terminal triphosphate group to drive the next coupling step (Fig. 31-24*a*). Upon editing a mispaired 5'-terminal nucleotide (Fig. 31-24*b*), this putative polymerase would, in analogy with Pol I, for example, excise the offending nucleotide leaving either a 5'-OH or a 5'-phosphate group. Neither of these terminal groups are capable of energizing further chain extension. A proofreading $3' \rightarrow 5'$ DNA polymerase would therefore have to be capable of reactivating its edited product. The inherent complexity of such a system has presumably selected against its evolution.

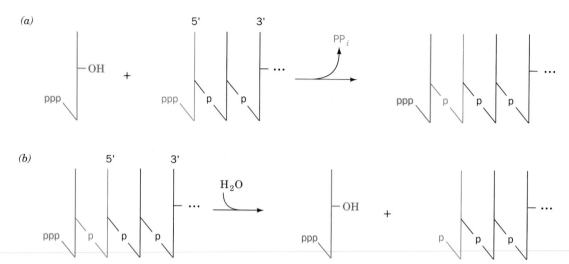

Figure 31-24
If a DNA polymerase could synthesize DNA in its 3' → 5' direction: (*a*) The coupling of each nucleoside triphosphate to the growing chain would be driven by the hydrolysis of the previously appended nucleoside triphosphate. (*b*) The editorial removal of an incorrect 5'-terminal nucleoside triphosphate would render the DNA chain incapable of further extension.

Any errors that remain in a duplex DNA after its replication or that subsequently arise through chemical and/or physical insults may be corrected through a variety of DNA repair processes (Section 31-5). These enable a cell to preserve its genetic heritage throughout its life cycle.

4. EUKARYOTIC DNA REPLICATION

It is becoming increasingly evident that *there is a remarkable degree of similarity between eukaryotic and prokaryotic DNA replication mechanisms.* There are, nevertheless, distinct differences between these two replication systems as a consequence of the vastly greater complexity of eukaryotes in comparison to prokaryotes. We consider these differences in this section.

A. The Cell Cycle

The **cell cycle,** the general sequence of events that occur during the lifetime of a eukaryotic cell, is divided into four distinct phases (Fig. 31-25):

1. Mitosis and cell division occur during the relatively brief **M phase** (for mitosis).

2. This is followed by the **G₁ phase** (for gap), which covers the longest part of the cell cycle.

3. G₁ gives way to the **S phase** (for synthesis) which, in contrast to events in prokaryotes, *is the only period in the cell cycle when DNA is synthesized.*

4. During the relatively short **G₂ phase**, the now tetra-ploid cell prepares for mitosis. It then enters M phase once again and thereby commences a new round of the cell cycle.

The cell cycle for cells in culture typically occupies a 16- to 24-h period. In contrast, cell cycle times for the different types of cells of a multicellular organism may vary from as little as 8 h to > 100 days. Most of this variation occurs in the G_1 phase. Moreover, many terminally differentiated cells, such as neurons or muscle cells, never divide; they assume as quiescent state known as the G_0 **phase.**

A cell's irreversible "decision" to proliferate is made

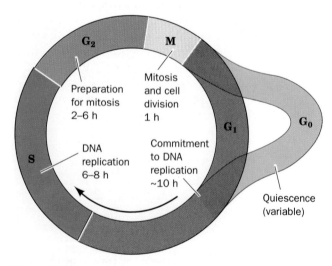

Figure 31-25
The eukaryotic cell cycle. Cells in G_1 may enter a quiescent phase (G_0) rather than continuing about the cycle.

Table 31-5
Properties of Animal DNA Polymerases

	α	β	γ	δ
Location	Nucleus	Nucleus	Mitochondrion	Nucleus
Mass (kD)	120–220	30–50	150–300	140–160
Inhibitors:				
Aphidicolin	Yes	No	No	Yes
Dideoxy NTPs	Weak	Strong	Strong	Weak
Arabinosyl NTPs	Strong	Weak	Weak	Strong
N-Ethylmaleimide (NEM)[a]	Strong	Weak	Strong	Strong

[a] A cysteine reagent (Table 6-3).

Source: Mostly Kornberg, A., *DNA Replication, p.* 204, Freeman (1980).

during G_1. Quiescence is maintained if, for example, nutrients are in short supply or the cell is in contact with other cells (**contact inhibition**). Conversely, DNA synthesis may be induced by various agents such as carcinogens or tumor viruses which trigger uncontrolled cell proliferation (cancer; Section 33-4C); by the surgical removal of a tissue, which results in its rapid regeneration; or by proteins known as **mitogens** that bind to cell surface receptors and somehow induce cell division. Growing cells contain cytoplasmic factors that stimulate DNA replication. For example, extracts of frog eggs induce DNA synthesis in frog spleen cells. The mode of action of these factors, however, is unknown (although see below).

B. Eukaryotic DNA Polymerases

Animal cells contain at least four distinct types of DNA polymerases, designated, in the order of their discovery, **DNA polymerases α, β, γ, and δ** (Table 31-5). Their functions were largely elucidated by their different responses to inhibitors (Table 31-5) because mutant forms of these enzymes have not, until recently, been available.

DNA polymerase α, which occurs only in the cell nucleus, participates in the replication of chromosomal DNA. This function was largely established through the use of its specific inhibitor **aphidicolin**

Aphidicolin

and by the observation that DNA polymerase α activity varies with the rate of cellular proliferation. This multisubunit protein (four types of subunit in *Drosophila;* five in rat liver), as do all DNA polymerases, replicates DNA by extending a primer in the $5' \rightarrow 3'$ direction under the direction of a single-stranded DNA template. DNA polymerase α has a tightly associated primase activity. However, it lacks exonuclease activity so that the DNA it replicates must be proofread by some other means.

DNA polymerase δ, a nuclear enzyme with inhibitor sensitivities similar to those of DNA polymerase α (Table 31-5), differs from the latter enzyme in that DNA polymerase δ lacks an associated primase but exhibits a proofreading $3' \rightarrow 5'$ exonuclease activity. Another difference is that DNA polymerase δ has apparently unlimited processivity (replicates the entire length of a template), whereas that of DNA polymerase α is only moderate (~ 100 nucleotides). It has therefore been suggested that DNA polymerase δ is the leading strand replicase (which requires high processivity but only occasional need of a primer), whereas DNA polymerase α is the lagging strand replicase (which requires frequent priming but a processivity of only 100–200 nucleotides).

DNA polymerase β is remarkable for its small size. The biological function of this nuclear enzyme is unknown although the observation that its level of activity does not vary with the rate of cell growth suggests that it participates in DNA repair processes. DNA polymerase γ occurs exclusively in the mitochondrion where it presumably replicates the mitochondrial DNA. Chloroplasts contain a similar enzyme.

Eukaryotic Chromosomes Consist of Numerous Replicons

Eukaryotic and prokaryotic DNA replication systems differ most obviously in that eukaryotic chromosomes have multiple replication origins in contrast to the single replication origin of prokaryotic chromosomes. DNA polymerase

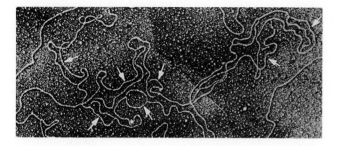

Figure 31-26
An electron micrograph indicating the multiple replication eyes (*arrows*) in a fragment of replicating *Drosophila* DNA. [From Kreigstein, H. J. and Hogness, D. S., *Proc. Natl. Acad. Sci.* **71**, 136 (1974).]

α synthesizes DNA at the rate of ~50 nucleotides/s (~20 times slower than prokaryotic DNA polymerases) as was determined by autoradiographically measuring the lengths of pulse-labeled sections of eukaryotic chromosomes. Since a eukaryotic chromosome typically contains 60 times more DNA than those of prokaryotes, its bidirectional replication from a single origin would require ~1 month to complete. Electron micrographs such as Fig. 31-26, however, show that eukaryotic chromosomes contain multiple origins, one every 3 to 300 **kb** (kilobase pairs) depending on both the species and the tissue, so that S phase usually occupies only a few hours.

Cytological observations indicate that the various chromosomal regions are not all replicated simultaneously but, rather, clusters of 20 to 80 adjacent **replicons** (replication units; DNA segments that are each served by a replication origin) are activated at once. New replicons are activated throughout the S phase until the entire chromosome has been replicated. During this process, replicons that have already been replicated are somehow distinguished from those that have not. Investigations of DNA synthesis in eukaryotic viruses suggests that replicon replication may be triggered by primer transcription under the control of replication origin-associated enhancers (enhancers are DNA sequences that regulate transcriptional initiation by binding RNA polymerase-stimulating proteins called transcription factors; Section 29-2F). Tissue-specific and cell cycle-specific transcription factors could then switch on particular replicons.

Mitochondrial DNA Is Replicated in D Loops

Mitochrondrial DNA is replicated by a process in which leading strand synthesis precedes lagging strand synthesis (Fig. 31-27). The leading strand therefore displaces the lagging strand template to form a **displacement** or **D loop**. The 15-kb circular mitochondrial chromosome of mammals normally contains a single 500 to

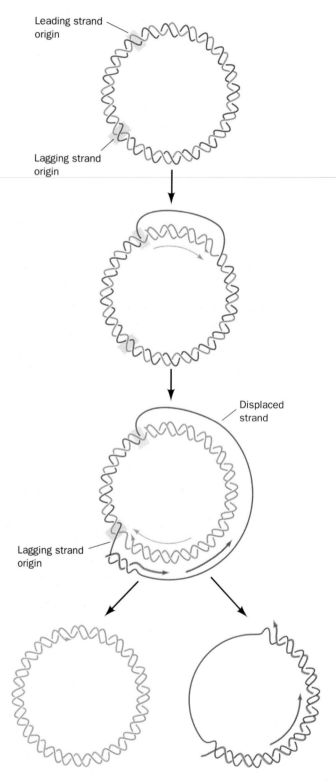

Figure 31-27
The D-loop mode of DNA replication.

600-nucleotide D loop that undergoes frequent cycles of degradation and resynthesis. During replication, the D loop is extended. When it has reached a point ~⅔ around the chromosome, the lagging strand origin is exposed and its synthesis proceeds in the opposite direction around the chromosome. Lagging strand synthesis is therefore only ~⅓ complete when leading strand synthesis terminates.

C. Reverse Transcriptase

The **retroviruses,** which are RNA-containing eukaryotic viruses such as certain tumor viruses and **human immunodeficiency virus (HIV,** the causitive agent of **AIDS),** contain an **RNA-directed DNA polymerase (reverse transcriptase).** This enzyme, which was independently discovered in 1970 by Howard Temin and David Baltimore, acts much like Pol I in that it synthesizes DNA in the $5' \rightarrow 3'$ direction from primed templates. In the case of reverse transcriptase, however, RNA is the template.

> The discovery of reverse transcriptase caused a mild sensation in the biochemical community because it was perceived by some as being heretical to the central dogma of molecular biology (Section 29-1). There is, however, no thermodynamic prohibition to the reverse transcriptase reaction; in fact, under certain conditions, Pol I can likewise copy RNA templates.

Reverse transcriptases from **avian myeloblastosis virus** and **Rous sarcoma virus** are $\alpha\beta$ dimers with subunit masses of 65 and 95 kD. Reverse transcriptase has two additional enzymatic activities:

1. It is an exoribonuclease that specifically degrades the RNA of an RNA–DNA hybrid (**RNase H** activity; H for hybrid).

2. It is a DNA-directed DNA polymerase.

During retrovirus infection, reverse transcriptase transcribes the viral RNA onto a complementary DNA strand to form an RNA–DNA hybrid. The enzyme then degrades the RNA and replicates the resulting single-stranded DNA to form the duplex DNA that directs the remainder of the viral infection.
    Reverse transcriptase has been a particularly useful tool in genetic engineering because of its ability to transcribe mRNAs to complementary strands of DNA **(cDNA).** In transcribing eukaryotic mRNAs, which have poly(A) tails (Section 29-4A), the primer can be oligo(dT). cDNAs have been used, for example, as probes in Southern transfer analysis (Section 28-4C) to identify the genes coding for their corresponding mRNAs. An RNA's base sequence can be easily determined by sequencing its cDNA (Section 28-6D).

5. REPAIR OF DNA

DNA is by no means the inert substance that might be supposed from naive consideration of the genome's stability. Rather, the reactive environment of the cell, the presence of a variety of toxic substances, and exposure to UV or ionizing radiation subjects it to numerous chemical insults that excise or modify bases and alter sugar–phosphate groups. Indeed, some of these reactions occur at surprisingly high rates. For example, under normal physiological conditions, the glycosidic bonds of some 10,000 purine nucleotides in the genome of each human cell hydrolyze spontaneously each day.

Any DNA damage must be repaired if the genetic message is to maintain its integrity. Such repair is possible because of duplex DNA's inherent information redundancy. The biological importance of **DNA repair** is indicated by the great variety of such pathways possessed by even relatively simple organisms such as *E. coli.* In fact, *the major DNA repair processes in E. coli and mammalian cells are chemically quite similar.* These processes are outlined in this section.

A. Direct Reversal of Damage

Pyrimidine Dimers Are Split by Photolyase

UV radiation (200–300 nm) promotes the formation of a cyclobutyl ring between adjacent thymine residues on the same DNA strand to form an intrastrand **thymine dimer** (Fig. 31-28). Similar cytosine and thymine–cytosine dimers are likewise formed but at lesser rates. Such **pyrimidine dimers** locally distort

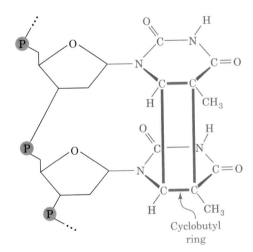

Figure 31-28
The cyclobutylthymine dimer that forms upon UV irradiation of two adjacent thymine residues on a DNA strand. The bonds joining the thymine rings (*red*) are much shorter than the normal 3.4-Å spacing between stacked rings in B-DNA thereby locally distorting the DNA.

DNA's base paired structure so that it can form neither a proper transcriptional nor replicational template.

Pyrimidine dimers may be restored to their monomeric forms through the action of a light-absorbing enzyme present in all life forms named **photoreactivating enzyme** or **photolyase**. The *E. coli* enzyme (54 kD) binds to pyrimidine dimers in a process that can occur in the dark and then, upon the absorption of 300 to 500-nm light by two noncovalently bound cofactors, a pterin and an $FADH_2$, splits the dimer.

Alkyltransferases Dealkylate Alkylated Bases

The exposure of DNA to alkylating agents such as **N-methyl-N'-nitro-N-nitrosoguanidine (MNNG)**

N-Methyl-*N'*-nitro-*N* -
nitrosoguanidine (MNNG)

O^6-Methylguanine residue

yields, among other products, O^6-**alkylguanine** residues. The formation of these derivatives is highly mutagenic because upon replication, they frequently cause the incorporation of thymine instead of cytosine.

O^6-Methylguanine and O^6-ethylguanine lesions of DNA in both *E. coli* and mammalian cells are repaired by O^6-**methylguanine–DNA methyltransferase,** which directly transfers the offending alkyl group to one of its own Cys residues. The reaction inactivates this protein, which therefore cannot be strictly classified as an enzyme. The alkyltransferase reaction has elicited considerable attention because carcinogenesis induced by methylating and ethylating agents is correlated with deficient repair of O^6-alkylguanine lesions.

B. Excision Repair

Pyrimidine dimers may also be mended by a process known as **excision repair.** *In such repair pathways, an oligonucleotide containing the lesion is excised from the DNA and the resulting single-strand gap filled in.* In *E. coli*, pyrimidine dimers are recognized by a multisubunit enzyme, the product of the *uvrA*, *uvrB*, and *uvrC* genes. This **UvrABC endonuclease,** in an ATP-dependent reaction, cleaves the dimer-containing DNA strand at the eighth and fourth phosphodiester bonds on the dimer's 5' and 3' sides, respectively (Fig. 31-29). The excised oligonucleotide is replaced through the action of a DNA polymerase, most probably Pol I, followed by that of DNA ligase.

UvrABC endonuclease excises other types of DNA lesions besides pyrimidine dimers. These lesions are characterized by the displacement of bases from their

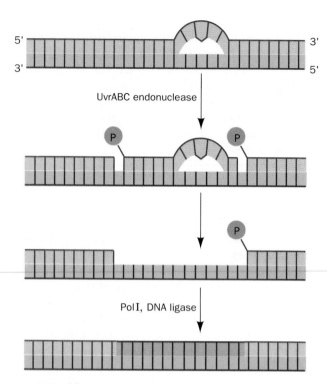

Figure 31-29
The mechanism of excision repair of pyrimidine photodimers. [After Haseltine, W. A., *Cell* **33**, 15 (1983).]

normal positions, as with pyrimidine dimers, or by the addition of a bulky substituent to a base. Evidently, UvrABC endonuclease is activated by a helix distortion rather than by the recognition of any particular group.

Xeroderma Pigmentosum Is Caused by Genetically Defective Excision Repair

In humans, the rare inherited disease *xeroderma pigmentosum* (Greek: *xeros*, dry + *derma*, skin) results from the inability of skin cells to repair UV-induced DNA lesions. Individuals suffering from this autosomal recessive condition are extremely sensitive to sunlight. During infancy they develop marked skin changes such as dryness, excessive freckling, and keratoses (a type of skin tumor), together with eye damage, such as opacification and ulceration of the cornea. Eventually they develop often fatal skin cancers.

Cultured skin fibroblasts from individuals with xeroderma pigmentosum are defective in the excision repair of pyrimidine dimers. Cell-fusion experiments with cultured cells taken from various patients have demonstrated that this disease results from a defect in any of nine complementation groups. Humans must therefore have at least nine gene products involved in their clearly important UV damage repair pathway.

Glycosylases Remove Altered Bases

DNA bases are modified by reactions that occur under normal physiological conditions as well as through the ac-

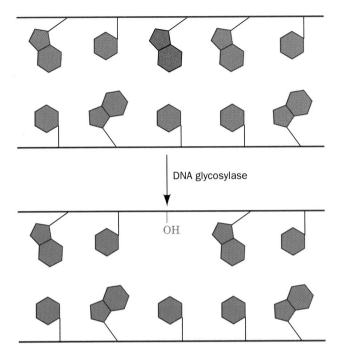

Figure 31-30
DNA glycosylases hydrolyze the glycosidic bonds of their corresponding altered base (*red*) to yield an AP site.

tion of environmental agents. For example, adenine and cytosine residues spontaneously deaminate at finite rates to yield hypoxanthine and uracil residues, respectively. *S*-Adenosylmethionine, a common metabolic methylating agent, occasionally nonenzymatically methylates a base to form derivatives such as 3-methyladenine and 7-methylguanine residues. Ionizing radiation can promote ring opening reactions in bases. Such changes modify or eliminate base pairing properties.

DNA containing a damaged base may be restored to its native state through a form of excision repair. Cells contain a variety of **DNA glycosylases** that each cleave the glycosidic bond of a corresponding specific type of altered nucleotide (Fig. 31-30) thereby leaving a deoxyribose residue in the backbone. Such **apurinic** or **apyrimidinic (AP)** sites are also generated under normal physiological conditions by the spontaneous hydrolysis of a glycosidic bond. The deoxyribose residue is then cleaved on one side by an **AP endonuclease,** the deoxyribose and several adjacent residues are removed by the action of DNA polymerase or some other cellular exonuclease, and the gap is filled in and sealed by DNA polymerase and DNA ligase.

Uracil in DNA Would Be Highly Mutagenic

For some time after the basic functions of nucleic acids had been elucidated there seemed no apparent reason why nature goes to the considerable metabolic effort of using thymine in DNA and uracil in RNA when these substances have virtually identical base pairing proper-

ties. This enigma was solved by the discovery of cytosine's penchant for conversion to uracil by deamination, either spontaneously or by reaction with nitrites (Section 30-1A). If U was a normal DNA base, the deamination of C would be highly mutagenic because there would be no indication of whether the resulting mismatched G · U base had initially been G · C or A · U. *Since T is DNA's normal base, however, any U in DNA is almost certainly a deaminated C.* U's that occur in DNA are efficiently excised by the DNA glycosylase **uracil *N*-glycosylase** and then replaced by C through excision repair.

Uracil *N*-glycosylase also has an important function in DNA replication. dUTP, an intermediate in dTTP synthesis, is present in all cells in small amounts (Section 26-4B). DNA polymerases do not discriminate well between dUTP and dTTP (recall that DNA polymerases select a base for incorporation into DNA according to its ability to base pair with the template) so that, despite the low dUTP level that cells maintain (Section 26-4B), newly synthesized DNA contains an occasional U. These U's are rapidly replaced by T through excision repair. However, since excision occurs more rapidly than repair, all newly synthesized DNA is fragmented. When Okazaki fragments were first discovered (Section 31-1C), it therefore seemed that all DNA was synthesized discontinuously. This ambiguity was resolved with the discovery of *E. coli* defective in uracil *N*-glycosylase. In these so-called *ung⁻* mutants, only about one half of the newly synthesized DNA is fragmented, strongly suggesting that DNA's leading strand is synthesized continuously.

C. Recombination Repair

Damaged DNA may undergo replication before the lesion can be eliminated by the previously described repair systems. The replication of DNA containing a pyrimidine dimer is interrupted by this template distortion and is only reinitiated at some point past the dimer site. The resulting daughter strand has a gap opposite the pyrimidine dimer (Fig. 31-31). This genetic lesion cannot be eliminated by excision repair, which requires an intact complementary strand. Yet, such an intact strand occurs in the sister duplex that was formed at the same replication fork. The lesion can therefore be corrected through a process that is alternatively known as **recombination** or **postreplication repair** (Fig. 31-31). *This pathway exchanges the corresponding segments of sister DNA strands thereby placing the gapped DNA segment in apposition to the undamaged strand where the gap can be filled in and sealed.* The pyrimidine dimer, which is likewise associated with its intact complementary strand, can then be eliminated by excision repair or photoreactivation. Recombination repair was directly detected by the observation that segments of isotopically labeled parental DNA are transferred into daughter strands.

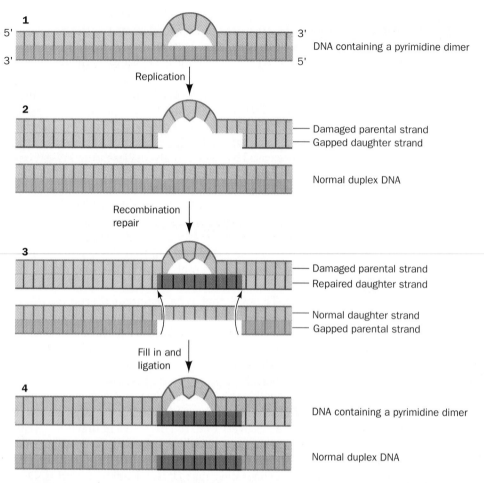

1 DNA containing a pyrimidine dimer

Replication

2
— Damaged parental strand
— Gapped daughter strand

Normal duplex DNA

Recombination repair

3
— Damaged parental strand
— Repaired daughter strand

— Normal daughter strand
— Gapped parental strand

Fill in and ligation

4
DNA containing a pyrimidine dimer

Normal duplex DNA

Figure 31-31
In recombination repair, the gap in a newly synthesized DNA strand opposite a damage site is filled by the corresponding segment from its sister duplex.

Recombination repair closely resembles genetic recombination. Indeed, both processes are mediated in *E. coli* by **RecA protein,** a 38-kD nuclease, which promotes sister strand exchange between homologous DNA segments. *E. coli* with a mutant *recA* gene are therefore deficient in both recombination repair (which makes them extremely sensitive to UV radiation) and genetic recombination. We consider the mechanism of recombination in Section 31-6A.

D. The SOS Response

Agents that damage DNA, such as UV radiation, alkylating agents, and cross-linking agents, induce a complex system of cellular changes in *E. coli* known as the **SOS response.** *E. coli* so treated cease dividing and increase their capacity to repair damaged DNA.

LexA Protein Represses the SOS Response

Clues as to the nature of the SOS response were provided by the observations that *E. coli* with mutant *recA* or *lexA* genes have their SOS response permanently switched on. Moreover, when wild-type *E. coli* are exposed to agents that damage DNA or inhibit DNA replication, their RecA specifically mediates the proteolytic cleavage of **LexA** protein (22 kD) at an Ala-Gly bond. RecA is so activated, at least *in vitro,* when it binds to single-stranded DNA (it was initially assumed that RecA directly proteolyzes LexA but subsequent experiments by John Little indicate that activated RecA stimulates LexA to cleave itself). Further genetic analysis indicated that LexA functions as a repressor of a number of operons including those of *recA* and *lexA.* Cynthia Kenyon identified these other LexA-controlled operons by inserting the *lacZ* gene (which codes for β-galactosidase; Section 29-1A) at random positions in the *E. coli* chromosome and examining the clones that had increased β-galactosidase activity in the presence of DNA-damaging agents. Altogether, 11 genes, including the excision repair genes *uvrA* and *uvrB,* lack normal function in these clones (the genes are inactivated by *lacZ* insertion). DNA sequence analyses of the LexA-repressible genes revealed that they are all preceded by a homologous 20 nucleotide sequence, the so-called **SOS**

Uninduced state

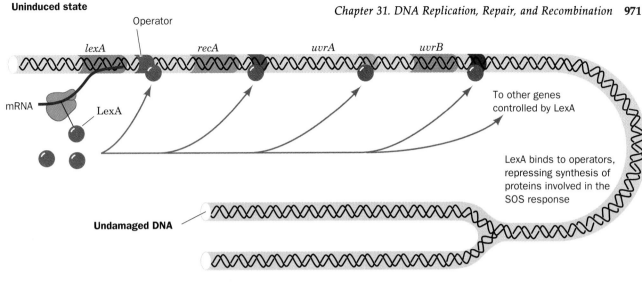

Induced state

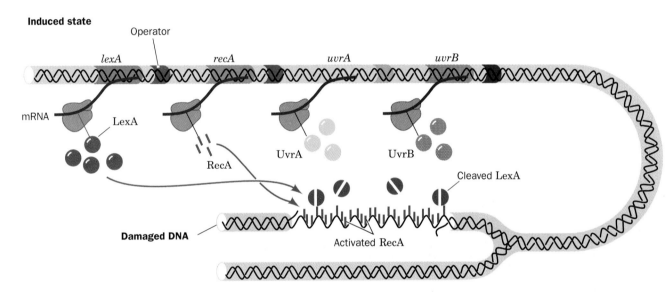

Figure 31-32
In a cell with undamaged DNA (*above*), LexA largely represses the synthesis of LexA, RecA, RecBCD, UvrABC, and other proteins involved in the SOS response. When there has been extensive DNA damage (*below*), RecA is activated, by binding to the resulting single-stranded DNA to stimulate LexA self-cleavage. The consequent synthesis of the SOS proteins results in the repair of the DNA damage.

box, that has the palindromic symmetry characteristic of operators (Section 29-3B). Indeed, LexA has been shown to directly bind the SOS boxes of *recA* and *lexA*.

The preceding information suggests a model for the regulation of the SOS response (Fig. 31-32). During normal growth, LexA largely represses SOS gene expression. When DNA damage has been sufficient to produce postreplication gaps, however, this single-stranded DNA binds to RecA so as to stimulate LexA cleavage. The LexA-repressible genes are consequently released from repression and direct the synthesis of SOS proteins including that of LexA (although this repressor continues to be cleaved through the influence of RecA). When the DNA lesions have been eliminated, RecA ceases stimulating LexA's autoproteolysis. The newly synthesized LexA can then function as a repressor, which permits the cell to return to normality.

SOS Repair Is Error Prone

SOS repair is an error prone and therefore mutagenic process. Yet, DNA damage that normally activates the SOS response is nonmutagenic in the *recA⁻ E. coli* that survive. This is because the intact SOS repair system will replace the bases at a DNA lesion even when there is no information as to which bases were originally present (via a poorly characterized process that involves the products of the SOS genes *umuC* and *umuD* together with RecA). The SOS repair system is therefore a testimonial to the proposition that survival with a chance of loss of function (and the possible gain of new ones) is advantageous, in the Darwinian sense, over death.

E. Identification of Carcinogens

Many forms of cancer are known to be caused by exposure to certain chemical agents that are therefore known as car-cinogens. It has been estimated that as much as 80% of human cancer arises in this fashion. There is consider-able evidence that the primary event in carcinogenesis is often damage to DNA (carcinogenesis is discussed in Section 33-4C). Carcinogens are consequently also likely to induce the SOS response in bacteria and thus act as indirect mutagenic agents. In fact, there is a high correlation between carcinogenesis and mutagenesis (recall, e.g., the progress of *xeroderma pigmentosum*; Section 31-5B).

There are presently some 60,000 man-made chemi-cals of commercial importance and ~1000 new ones are introduced each year. The standard animal tests for car-cinogenesis, exposing rats or mice to high levels of the suspected carcinogen and checking for cancer, are ex-pensive and require ~3 years to complete. Thus rela-tively few substances have been tested in this manner.

The Ames Test Assays for Probable Carcinogenicity

Bruce Ames devised a rapid and effective bacterial assay for carcinogenicity that is based on the high corre-lation between carcinogenesis and mutagenesis. He constructed special tester strains of *Salmonella typhi-murium* that are *his⁻* (cannot synthesize histidine so that they are unable to grow in its absence), have cell enve-lopes that lack the lipopolysaccharide coating which renders normal *Salmonella* impermeable to many sub-stances (Section 10-3B) and have an inactivated excision repair system. Mutagenesis in these tester strains is indi-cated by their reversion to the *his⁺* phenotype.

In the **Ames test,** ~10^9 tester strain bacteria are spread on a culture plate that lacks histidine. Usually a mixture of several *his⁻* strains is used so that both point and frameshift mutations can be detected. A mutagen placed in the culture medium causes some of these *his⁻* bacteria to revert to the *his⁺* phenotype, which is de-tected by their growth into visible colonies after 2 days at 37°C (Fig. 31-33). The mutagenicity of a substance is scored as the number of such colonies less the few spon-taneously revertant colonies that occur in the absence of the mutagen.

Many noncarcinogens are converted to carcinogens in the liver or in other tissues via a variety of detoxification reactions (Section 25-2D). A small amount of rat liver homogenate is therefore included in the Ames test me-dium in an effort to approximate the effects of mamma-lian metabolism.

Both Man-Made and Naturally Occurring Substances Can Be Carcinogenic

There is around an 80% correspondence between the

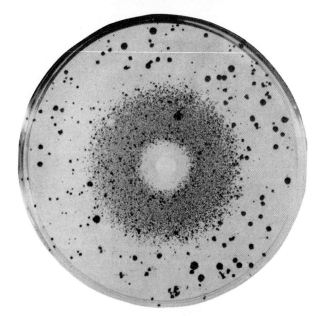

Figure 31-33
The Ames test for mutagenesis. A filter paper disk containing a mutagen, in this case the alkylating agent ethyl methanesulfonate, is centered on a culture plate containing *his⁻* tester strains of *Salmonella typhimurium* in a medium that lacks histidine. A dense halo of revertant bacterial colonies appears around the disk from which the mutagen diffused. The larger colonies distributed about the culture plate are spontaneous revertants. The bacteria near the disk have been killed by the toxic mutagen's high concentration. [From Devoret, R., *Sci. Am.* **241**(2): 46 (1979). Copyright © 1979 by Scientific American, Inc.]

compounds determined to be carcinogenic by animal tests and those found to be mutagenic by the Ames test. Dose-response curves, which are generated by testing a given compound at a number of concentrations, are al-most always linear indicating that *there is no threshold concentration for mutagenesis.* Several compounds to which humans have been extensively exposed that were found to be mutagenic by the Ames test were later found to be carcinogenic in animal tests. These include tris(2,3-dibromopropyl)phosphate, which was used as a flame retardant on children's sleepwear in the mid-1970s and can be absorbed through the skin; and furylfuramide, which was used in Japan in the 1960s and 1970s as an antibacterial additive in many prepared foods (and had passed two animal tests before it was found to be mutagenic). Carcinogens are not confined to only man-made compounds but also occur naturally. For example, carcinogens are contained in many plants that are common in the human diet, including alfalfa sprouts. **Aflatoxin B₁,**

Aflatoxin B₁

one of the most potent carcinogens known, is produced by molds that grow on peanuts and corn. Charred or browned food, such as occurs on broiled meats and toasted bread, contains a variety of DNA-damaging agents. Thus, with respect to carcinogenesis, as Ames has written, "Nature is not benign."

6. RECOMBINATION AND MOBILE GENETIC ELEMENTS

The chromosome is not just a simple repository of genetic information. If this were so, the unit of mutation would have to be an entire chromosome rather than a gene because there would be no means of separating a mutated gene from the other genes of the same chromosome. Chromosomes would therefore accumulate mostly deleterious mutations until they became nonviable.

It has, of course, been known from some of the earliest genetic studies that pairs of allelic genes may exchange chromosomal locations by a process known as **genetic recombination** (Section 27-1C). Mutated genes can thereby be individually tested, since their propagation is then not absolutely dependent on the propagation of the genes with which they had been previously associated. In this section, we consider the mechanisms by which genetic elements can move, both between chromosomes and within them.

A. General Recombination

General recombination is defined as the exchange of homologous segments between two DNA molecules. Both genetic and cytological studies have long indicated that such a crossing-over process occurs in higher organisms during meiosis (see Fig. 27-10). Bacteria, which are normally haploid, likewise have elaborate mechanisms for the interchange of genetic information. They can acquire foreign DNA through conjugation (Section 27-1D), transduction (Section 27-1E), and transformation (Section 27-2A). In all of these processes, the foreign DNA is installed in the recipient's chromosome or plasmid through general recombination (to be propagated, a DNA segment must be part of a replicon; that is, be associated with a replication origin such as occurs in a chromosome, a plasmid, or a virus).

Recombination Occurs via a Crossed-Over Intermediate

The prototypical model for general recombination (Fig. 31-34) was proposed by Robin Holliday in 1964 on the basis of genetic studies on fungi. The corresponding strands of two aligned homologous DNA duplexes are nicked, the nicked strands crossover to pair with the nearly complementary strands of the homologous duplex thereby forming a segment of **heteroduplex DNA,** after which the nicks are sealed (Fig. 31-34a–e). Model

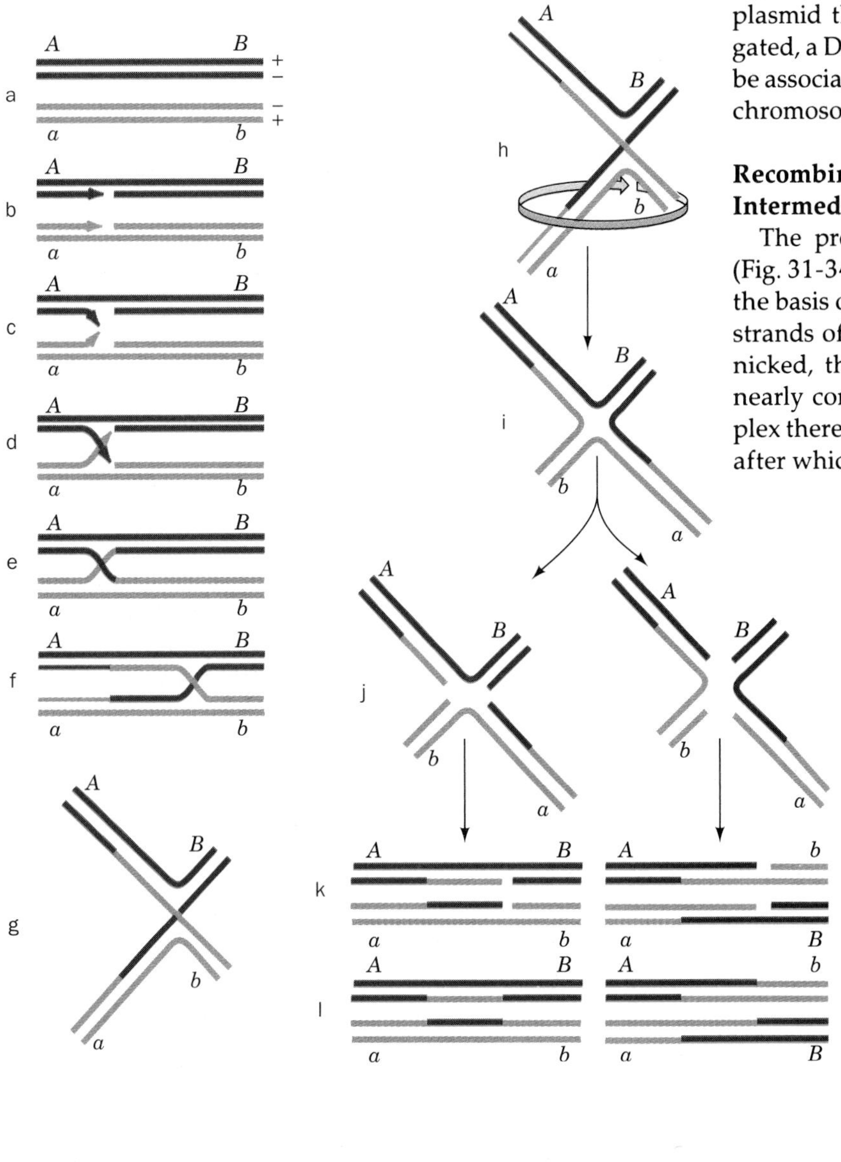

Figure 31-34
The Holliday model of general recombination between homologous DNA duplexes.

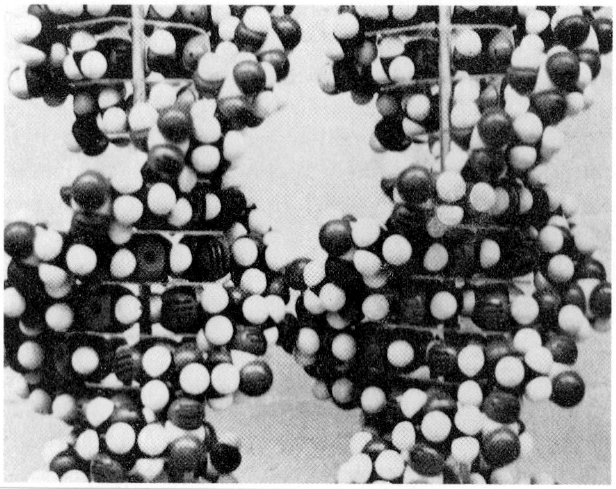

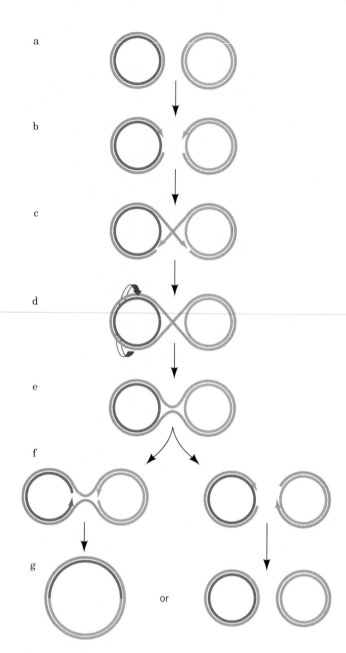

Figure 31-35
A molecular model indicating that the connections between two crossed-over duplex DNAs can be made without significant molecular distortions. [From Sigal, N. and Alberts, B., *J. Mol. Biol.* **71**, 792 (1972). Copyright © 1972 by Academic Press, Inc.]

building studies indicate, perhaps unexpectedly, that the bases of this four-stranded **Holliday structure** can all pair without steric strain so that its crossed-over structure is probably stable (Fig. 31-35). The cross-over point can move in either direction in a process known as **branch migration** (Fig. 31-34*e* and *f*).

The Holliday structure can be resolved into two duplex DNAs in two equally probable ways (Fig. 31-34*g–l*):

1. The cleavage of the strands that did not cross-over exchanges the ends of the original duplexes to form, after nick sealing, the traditional recombinant DNA molecule (right branch of Fig. 31-34*j–l*).

2. The cleavage of the strands that crossed-over exchanges a pair of homologous single-stranded segments (left branch of Fig. 34-34*j–l*).

The recombination of circular duplex DNAs results in the types of structures diagrammed in Fig. 31-36. Electron microscopic evidence for the existence of the postulated "figure-8" structures are shown in Fig. 31-37*a*. These figure-8 structures were shown not to be just twisted circles by cutting them with a restriction enzyme to yield **chi structures** (after their resemblance to the Greek letter χ) such as that pictured in Fig. 31-37*b*.

General Recombination in *E. coli* Is Catalyzed by RecA

The observation that *recA⁻ E. coli* have a 10^4-fold lower recombination rate than the wild-type indicates that *RecA protein has an important function in recombination.* Indeed, RecA greatly increases the rate at which complementary strands renature *in vitro*. This versatile protein (recall it also stimulates the autoproteolysis of

Figure 31-36
General recombination between two circular DNA duplexes. This process can result in the production of either two circles of the original sizes or in a single composite circle.

LexA to trigger the SOS response; Section 31-5D) coats single-stranded DNA to form a 120 Å in diameter right-handed helix in which each RecA monomer binds an estimated four nucleotides (Fig. 31-38). The formation of this complex stimulates the RecA to also bind duplex DNA, which it unidirectionally scans for a segment that is complementary to the single-stranded DNA. In doing so, RecA unwinds the duplex from the 10.5 bp/turn characteristic of B-DNA to 18.6 bp/turn. Upon encountering a complementary segment, RecA further unwinds the duplex and, in a reaction driven by RecA-catalyzed ATP hydrolysis, exchanges the single-stranded DNA with the corresponding strand on the duplex (Fig. 31-39). A model of how RecA might do so is diagrammed in Fig. 31-40.

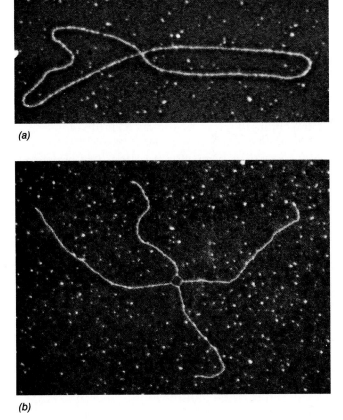

(a)

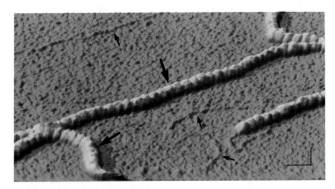

(b)

Figure 31-37
Electron micrographs of intermediates in the general
recombination of two plasmids. (*a*) A figure-8 structure. This
corresponds to Fig. 31-36*d*. (*b*) A chi structure that results
from the treatment of a figure-8 structure with a restriction
endonuclease. Note the single-stranded connections in the
cross-over region. [Courtesy of David Dressler and
Huntington Potter, Harvard Medical School.]

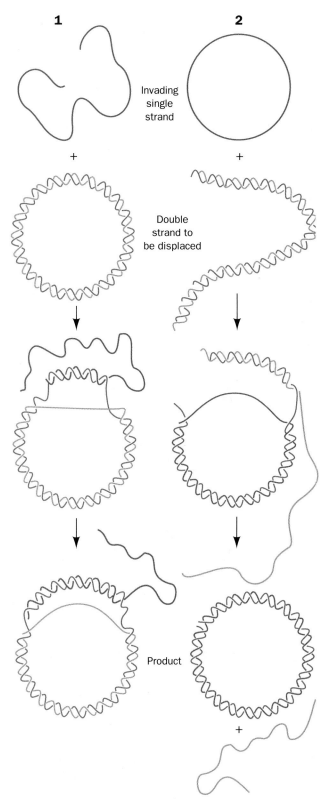

Figure 31-39
The assimilation of a single strand into a duplex DNA as
catalyzed by RecA. For this process to occur, one of the
strands must have a free end. (1) Linear strand + covalently
closed circular duplex. (2) Circular single strand + linear
duplex.

Figure 31-38
A scanning–tunneling microscopy image of RecA protein in
complex with DNA. The large arrows point to a RecA–DNA
complex; the small arrows point to free DNA. The bars
denote a distance of 20 nm in each direction. The striations
in the RecA–DNA complex result from the 10-nm pitch of
this helical structure. [From Amrein, R., Stasiak, A., Gross,
H. Stoll, E., and Travaglini, G., *Science* **240**, 515 (1988).
Copyright © 1988 by the American Society for the
Advancement of Science.]

(a)

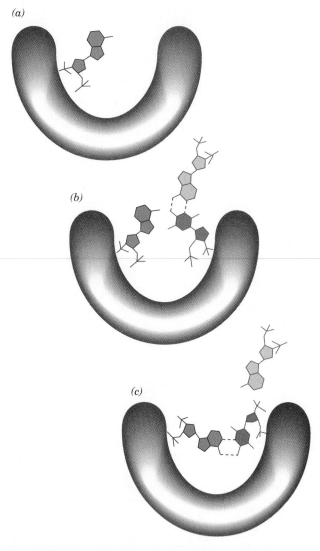

(b)

(c)

Figure 31-40

A hypothetical model for RecA-mediated pairing and strand exchange between single-stranded and duplex DNAs. (*a*) A single-stranded DNA binds to RecA to form an initiation complex. (*b*) Duplex DNA binds to the initiation complex so as to transiently form a triple helix that mediates the correct pairing of the homologous strands. (*c*) RecA rotates the bases of the aligned homologous strands to effect strand exchange. [After Howard-Flanders, P., West, S. C., and Stasiak, A., *Nature* **309,** 217 (1984).]

The RecA-mediated strand exchange process requires that one of the participating DNA strands have a free end and tolerates only a limited degree of mispairing. In the assimilation of a single-stranded circle by a linear duplex, the 3′ end of the duplex strand that pairs with the invading single strand must be complementary to this invading strand (Fig. 31-41). The invasion, moreover, cannot proceed past the 3′ end of a highly mismatched segment in the complementary strand. *The invasion of the single strand must therefore begin with its 5′ end.* Of course, two such strand exchange processes must simultaneously occur in a Holliday structure (Figs. 31-34 and

31-36). A model for the consequent branch migration process is diagrammed in Fig. 31-42.

RecBCD Initiates Recombination by Making Single-Strand Nicks

The single-strand nicks to which RecA binds are made by the **RecBCD** protein, the 330-kD product of the SOS genes *recB, recC,* and *recD* (Fig. 31-43). RecBCD does so by unwinding duplex DNA in an ATP-driven process and then rewinding it. However, RecBCD unwinds DNA faster than it rewinds it so that this protein is accompanied by two growing single-stranded loops as it unidirectionally advances down the helix. Upon encountering the sequence GCTGGTGG from its 3′ end (the so-called **Chi sequence,** which occurs about every 10 kb in the *E. coli* genome), RecBCD cleaves that strand 4 to 6 nucleotide past the 3′ side of Chi thereby yielding the single-stranded segment to which RecA binds. This explains the observation that Chi sequences occur in regions that have elevated rates of recombination.

RecBCD can only commence unwinding DNA at a free duplex end. Such ends are not normally present in *E. coli,* which has a circular genome, but become available during such recombinational processes as bacterial conjugation (Section 27-1D), viral transduction (Section 27-1E), and bacterial transformation (Section 27-2A).

Although the detailed mechanism of genetic recombination is still unclear, it is evident that other proteins besides RecA and RecBCD participate in the *E. coli* recombination process. Topoisomerase I (nicking–closing enzyme; Section 28-5C) is apparently necessary to relieve the supercoiling generated by the recombination process. SSB also participates in recombination: It helps maintain DNA in its single-stranded state and probably also modulates RecA function. The nuclease that cuts apart the two recombinant duplexes (Figs. 31-34*j* and 31-36*f*) has yet to be identified although it is suspected that it may be RecBCD.

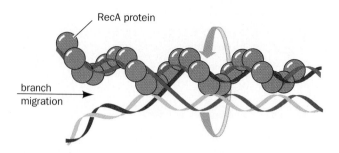

Figure 31-42

A model for RecA-mediated branch migration. RecA, which forms a nucleoprotein filament on one DNA duplex, promotes branch migration by rotating a second duplex, which is linked to the first via a crossover junction, around the outside of the filament. The reaction is unidirectionally driven by RecA-catalyzed ATP hydrolysis. [After Cox, M. M. and Lehman, I. R., *Annu. Rev. Biochem.* **56,** 252 (1987).]

Assimilation of 3' end of homologous DNA

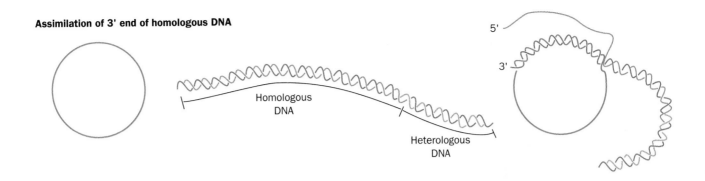

No assimilation of noncomplementary DNA

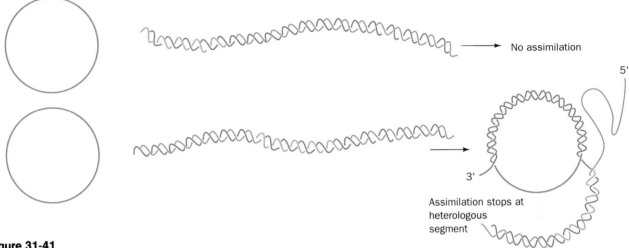

No assimilation

Assimilation stops at
heterologous
segment

Figure 31-41
The RecA-catalyzed assimilation of a single-stranded circle by a duplex DNA can occur
only if the duplex has a 3' end that can base pair with the circle. Strand assimilation
cannot proceed through a noncomplementary segment.

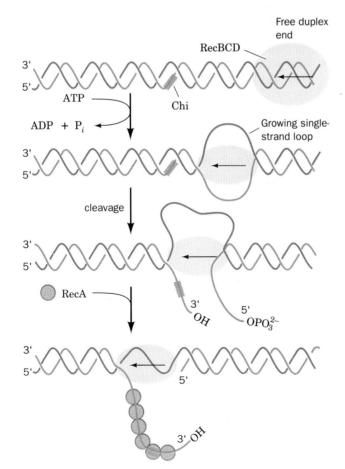

B. Transposition

In the early 1950s, on the basis of genetic analysis, Barbara McClintock reported that the variegated pigmentation pattern of **maize** (Indian corn) kernels results from the action of genetic elements that can move about the maize genome. This proposal was resoundingly ignored because it was contrary to the then held genetic orthodoxy that chromosomes consist of genes linked in fixed order. Another 20 years were to pass before evidence of mobile genetic elements was found in another organism, *E. coli*.

Transposons Move Genes between Unrelated Sites
*It is now known that **transposable elements** or **transposons** are common in both prokaryotes and eukaryotes*

Figure 31-43
A model for the generation of single-stranded DNA by RecBCD to initiate recombination. RecBCD binds to a free end of duplex DNA and, in an ATP-driven process, advances down the helix, unwinding the DNA and then rewinding it somewhat more slowly so as to form two single-stranded loops. RecBCD cleaves one of these loops 4 to 6 nucleotides on the 3' side of a properly oriented Chi sequence thereby yielding the potentially invasive single strand of DNA to which RecA initially binds.

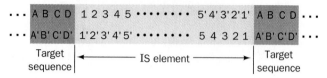

Figure 31-44
IS elements as well as other transposons have inverted terminal repeats (*numerals*) and are flanked by direct repeats of host DNA target sequences (*letters*).

Table 31-6

Properties of Some Insertion Elements

Insertion Element	Length (bp)	Inverted Terminal Repeat (bp)	Direct Repeat at Target (bp)
IS1	768	23	9
IS2	1327	41	5
IS4	1428	18	11 or 12
IS5	1195	16	4

Source: Lewin, B., *Genes* (3rd ed.), p. 591, Wiley (1987).

where they influence the variation of phenotypic expression over the short term and evolutionary development over the long term. Each transposon codes for the enzymes that specifically insert it into the recipient DNA. This process has been described as **illegitimate recombination** because it requires no homology between donor and recipient DNAs. In contrast to general recombination, however, transposition is a highly inefficient process: It occurs at a rate of only 10^{-7} to 10^{-4} events per generation.

Transposons with three levels of complexity have been characterized:

1. The simplest transposons, and the first to be characterized, are named **insertion sequences** or **IS elements.** They are designated by "IS" followed by an identifying number. IS elements are normal constituents of bacterial chromosomes and plasmids. For example, a common *E. coli* strain has eight copies of IS1 and five copies of IS2. IS elements generally consist of <2000 bp. These comprise a so-called **transposase** gene, and in some cases, a regulatory gene, flanked by short inverted (having opposite orientation) terminal repeats (Fig. 31-44 and Table 31-6). The inverted repeats are essential for transposition; their genetic alteration invariably prevents this process. An inserted IS element is flanked by a directly (having the same orientation) repeated segment of host DNA (Fig. 31-44). This suggests that an IS element is inserted in the host DNA at a staggered cut that is later filled in (Fig. 31-45). The length of this target sequence, most commonly 5 to 9 bp, but not its sequence, is characteristic of the IS element.

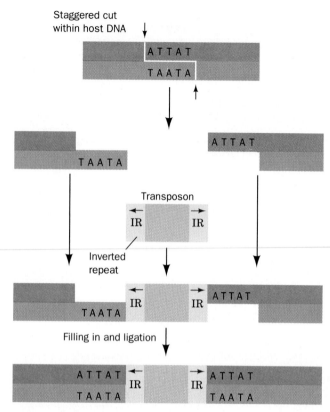

Figure 31-45
A model for the generation of direct repeats of the target sequence by transposon insertion.

2. *More complex transposons carry genes not involved in the transposition process, for example, antibiotic resistance genes.* Such transposons are designated "Tn" followed by an identifying number. For example, **Tn3** (Fig. 31-46) consists of 4957 bp and has inverted terminal repeats of 38 bp each. The central region of Tn3 codes for three proteins: (1) a 1015-residue transposase named **TnpA;** (2) a 185-residue protein known as **TnpR,** which functions as a repressor for the expression of both *tnpA* and *tnpR* as well as mediating the site-specific recombination reaction necessary for transposition (see below); and (3) a **β-lactamase** that inactivates ampicillin (Section 10-3B).

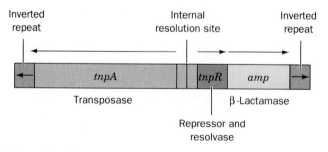

Figure 31-46
A map of transposon Tn3.

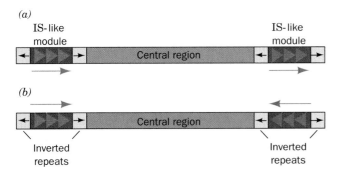

(a)

IS-like module ··· IS-like module

Central region

(b)

Central region

Inverted repeats ··· Inverted repeats

Figure 31-47
A composite transposon consists of two identical or nearly identical IS-like modules (*green*) flanking a central region carrying various genes. The IS-like modules may have either (*a*) direct or (*b*) inverted relative orientations.

The site-specific recombination occurs in an AT-rich region known as the **internal resolution site** that is located between *tnpA* and *tnpR*.

3. The so-called **composite transposons** (Fig. 31-47) consist of a gene-containing central region flanked by two identical or nearly identical IS-like modules that have either the same or an inverted relative orientation. It therefore seems that composite transposons arose by the association of two originally independent IS elements. Since the IS-like modules are themselves flanked by inverted repeats, the ends of either type of composite transposon must also be inverted repeats. Experiments demonstrate that composite transposons can transpose any sequence of DNA in their central region.

If a plasmid carrying a transposon is introduced into a bacterial cell carrying a plasmid that lacks the transposon, in some of the progeny cells both types of plasmid will contain the transposon (Fig. 31-48). Evidently,

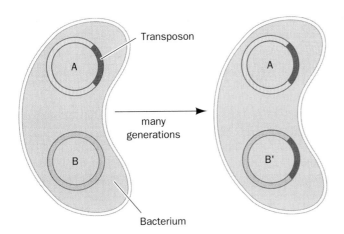

Figure 31-48
Transposition inserts a copy of the transposon at the target site while another copy remains at the donor site.

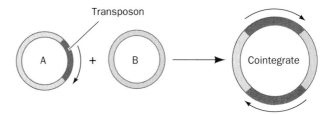

Transposon

A + B → Cointegrate

Figure 31-49
A cointegrate forms by the fusion of two plasmids, one carrying a transposon, such that both junctions of the original plasmid are spanned by transposons with the same orientation (*arrows*).

transposition involves the replication of the transposon into the recipient plasmid rather than, as early experiments suggested, its transfer from donor to recipient. In this case, the word "transposition" is a misnomer.

A Proposed Transposition Mechanism

Two plasmids, one containing a transposon, will occasionally fuse to form a so-called **cointegrate** containing like-oriented copies of the transposon at both junctions of the original plasmids (Fig. 31-49). Yet, some of the progeny of a cointegrate-containing cell lack the cointegrate but, instead, contain both original plasmids, each with one copy of the transposon (Fig. 31-48). The cointegrate must therefore be an intermediate in the transposition process.

The mechanism of transposition is unknown. However, a plausible model for this process (and there are several) that accounts for the foregoing observations consists of the following steps (Fig. 31-50):

1. A pair of staggered single-strand cuts, such as is diagrammed in Fig. 31-45, is made at the target sequence of the recipient plasmid. Similarly, single-strand cuts are made on opposite strands to either side of the transposon.

2. Each of the transposon's free ends is ligated to a protruding single strand at the insertion site. This forms a replication fork at each end of the transposon.

3. The transposon is replicated thereby yielding a cointegrate.

4. Through a site-specific crossover between the internal resolution sites of the two transposons, the cointegrate is resolved into the two original plasmids, each of which contains a transposon. This recombination process is catalyzed by a transposon-coded **resolvase** (TnpR in Tn3) rather than RecA; transposition proceeds normally in *recA⁻* cells (although RecA will resolve a cointegrate containing a transposon with either a mutant resolvase or an altered internal resolution site, albeit at a much reduced rate).

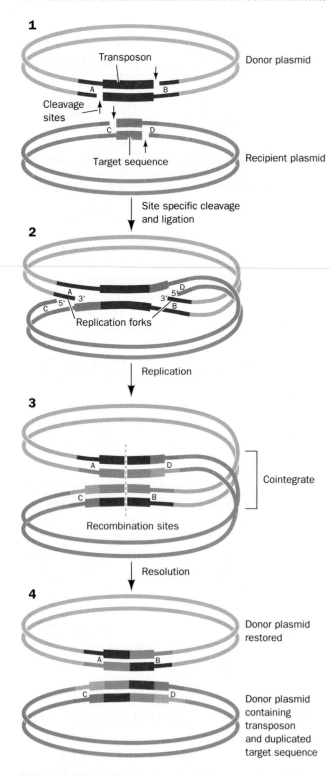

Figure 31-50
A model for transposition involving the intermediacy of a cointegrate. Here more lightly shaded bars represent newly synthesized DNA. [After Shapiro, J. A., *Proc. Natl. Acad. Sci.* **76**, 1934 (1979).]

Transposition Is Responsible for Much Genetic Remodeling

In addition to mediating their own insertion into DNA, *transposons promote inversions, deletions, and rearrangements of the host DNA.* Inversions can occur when the host DNA contains two copies of a transposon in inverted orientation. The recombination of these transposons inverts the region between them (Fig. 31-51*a*). If, instead, the two transposons have the same orientation, the resolution of this cointegratelike structure deletes the segment between the two transposons (Fig. 31-51*b*; if the deleted segment lacks a replication origin, it will not be propagated). The deletion of a chromosomal segment in this manner followed by its integration into the chromosome at a different site by a separate recombinational event results in chromosomal rearrangement.

Transposition appears to be important in chromosomal and plasmid evolution. Indeed, it has been suggested that transposons are nature's genetic engineering "tools." For example, the rapid evolution, since antibiotics came into common use, of plasmids that confer resistance to several antibiotics (Section 28-8A) resulted from the accumulation of the corresponding antibiotic-resistance transposons in these plasmids. Transposon-mediated rearrangements may well have been responsible for organizing originally distant genes into coordinately regulated operons as well as in forming new proteins by linking two formerly independent gene segments. Moreover, *the occurrence of identical transposons in unrelated bacteria indicates that the transposon-mediated transfer of genetic information between organisms is not limited to related species in contrast to genetic transfers mediated by general recombination.*

Phase Variation Is Mediated by Transposition

Phenotypic expression in bacteria can be regulated by a transposition mechanism. For example, certain strains of *Salmonella typhimurium* make two antigenically distinct versions of the protein **flagellin** (the major component of the whiplike flagella with which bacteria propel themselves; Section 34-3G) that are designated **H1** and **H2**. Only one of these proteins is expressed by any given cell but about once every 1000 cell divisions, in a process known as **phase variation,** a cell switches the type of flagellin it synthesizes. It is thought that phase variation enables *Salmonella* to evade its host's immunological defenses.

What is the mechanism of phase variation? The two flagellin genes reside on different parts of the bacterial chromosome. *H2* is linked to the *rh1* gene that encodes a repressor of H1 expression (Fig. 31-52). Hence, when the *H2-rh1* transcription unit is expressed, H1 synthesis is repressed; otherwise H1 is synthesized. Melvin Simon has shown that the expression of the *H2-rh1* unit is controlled by the orientation of a 995 bp segment that

(a)

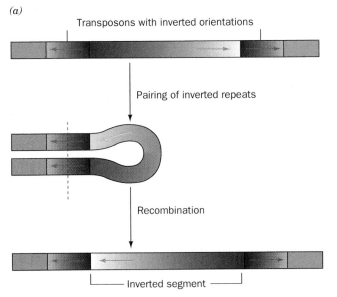

(b)

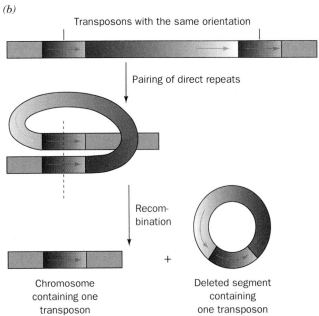

Figure 31-51
Recombinational models for: (*a*) The inversion of a DNA segment between two identical transposons with inverted orientations. (*b*) The deletion of a DNA segment between two identical transposons with the same orientation. This process parcels one transposon each to the resulting two DNA segments.

lies upstream of *H2* (Fig. 31-52). This segment, which is bounded by 14 bp inverted repeats, contains a promotor for *H2-rh1* expression. It also contains the **hin gene** that codes for a resolvase that mediates inversion of the DNA segment by a mechanism similar to that diagrammed in Fig. 31-51*a* (in fact, Hin protein is 33% homologous with TnpR, which indicates that these proteins have a common ancestor). In the Phase 2 orientation (Fig. 31-52*a*), the properly oriented promoter is just upstream of *H2* so that this gene and *rh1* are coordinately expressed thereby repressing H1 synthesis. In Phase 1 bacteria (Fig. 31-52*b*), however, this segment has the opposite orientation. Consequently, neither *H2* nor *rh1*, which then lacks a promoter, are expressed so that H1 is synthesized.

Figure 31-52
The mechanism of phase variation in *Salmonella*. (*a*) In Phase 2 bacteria, the *H2-rh1* promoter is oriented so that H2 flagellin and repressor are synthesized. Repressor binds to the *H1* gene thereby preventing its expression. (*b*) In Phase 1 bacteria, the segment preceding the *H2-rh1* transcription unit has been inverted relative to its orientation in Phase 2 bacteria. Hence this transcription unit cannot be expressed because it lacks a promoter. This releases *H1* from repression and results in the synthesis of H1 flagellin. The inversion of the segment preceding the *H2-rh1* transcription unit is mediated by the Hin protein which is expressed in either orientation by the *hin* gene.

(a) **Phase 2**

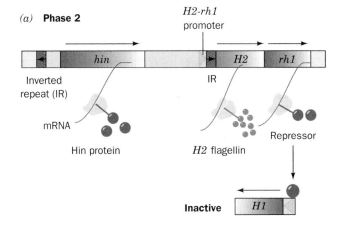

(b) **Phase 1**

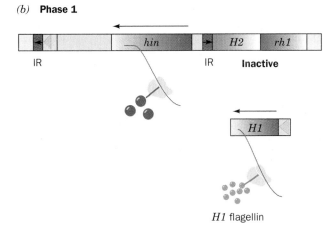

Transpositions in Eukaryotes Occur through RNA Intermediates

Transposons occur in such distantly related eukaryotes as yeast, maize, and fruit flies. In fact, ~3% of the *Drosphila* genome consists of transposons at various sites.

The similar base sequences of many eukaryotic transposons and retroviral genomes (and their dissimilarity to bacterial transposons) suggests that these transposons are degenerated retroviruses. They are therefore called **retrotransposons.** There is clear evidence that retrotransposons are transposed in a three-step process: (1) their transcription to RNA, (2) the reverse transcriptase-mediated copying of this RNA to DNA (Section 31-4C), and (3) the random insertion of this DNA into the host organism's genome [the retrovirus life cycle involves the integration (insertion) of the viral DNA into the host genome in a manner resembling that described in Section 32-3C for bacteriophage λ]. Thus, Gerald Fink remodeled **Ty,** the most common movable element in yeast (~35 copies of this 5.9-kb element are present in the yeast genome; Ty stands for *Transposon yeast*), to contain a yeast intron and to be preceded by a galactose-sensitive yeast promoter. When inserted into a yeast genome, the transposition rate of this remodeled Ty element varied with the galactose concentration in the medium and the transposed elements all lacked the intron thereby indicating that transposition occurs via an RNA intermediate. Moreover, of the two proteins that Ty encodes, one resembles a protein component of retrovirus particles and the other is homologous to retroviral reverse transcriptase. Indeed, these two proteins, together with Ty RNA and DNA, form viruslike particles in yeast cytoplasm. The Ty element, and presumably other retrotransposons, may therefore be considered an "internal virus" that can only move within a genome albeit at a very low rate compared to that of real viral infections.

7. DNA METHYLATION

The A and C residues of DNA may be methylated, in a species specific pattern, to form N^6-**methyladenine** and **5-methylcytosine** residues, respectively.

N^6-**Methyladenine residue**   **5-Methylcytosine residue**

These are the only types of modifications to which DNA is subjected in cellular organisms (although all the C residues of T-even phage DNAs are converted to **5-hydroxymethylcytosine** residues,

5-Hydroxymethylcytosine residue

which may, in turn, be glycosylated). These methyl groups project into B-DNA's major groove where they can interact with DNA-binding proteins. In most cells, only a few percent of the susceptible bases are methylated although this figure rises to >30% of the C residues in some plants.

Bacterial DNAs are methylated at their own particular restriction sites thereby preventing the corresponding restriction endonucleases from degrading the DNA (Section 28-6A). This phenomenon, however, accounts for only part of the methylation of bacterial DNAs. In *E. coli*, most DNA methylation is catalyzed by the products of the *dam* and *dcm* genes. The **Dam methylase** methylates the A residue in all GATC sequences, whereas the **Dcm methylase** methylates both C residues in CC^A_TGG. Note that both of these sequences are palindromic. The above enzymes, as do all known DNA methylases, use *S*-adenosylmethionine as their methyl donor.

Mismatch Repair

DNA methylation in prokaryotes functions most conspicuously in the repair of mismatched base pairs. Any replicational mispairing that has eluded the editing functions of Pol I and Pol III may still be corrected by a process known as **mismatch repair** (Pol I and Pol III have error rates of $10^{-6}-10^{-7}$ per base pair replicated but the observed mutational rates in *E. coli* are $10^{-8}-10^{-10}$ per base pair replicated). If this multienzyme system is to correct errors rather than perpetuate them, it must distinguish the parental DNA, which has the correct base, from the daughter strand, which has the incorrect although normal base. The observation that *dam⁻ E. coli* have a higher mutation rate than wild-type bacteria suggests how this distinction is made. A newly replicated daughter strand is undermethylated in comparison to the parental strand because DNA methylation lags behind DNA synthesis. Experiments with model DNAs have demonstrated that the mismatch repair system replaces an unmethylated strand between a mismatch and a nearby (up to ~1000 bp away) hemimethylated site.

DNA Methylation in Eukaryotes May Function in Gene Regulation

5-Methylcytosine (^{m}C) is the only methylated base in most eukaryotic DNAs including those of vertebrates. This modification occurs largely in the CG dinucleotide of various palindromic sequences (CG is present in the vertebrate genome at only about one fifth its randomly expected frequency). The degree of eukaryotic DNA methylation and its pattern is conveniently assessed by comparing the Southern blots of DNA cleaved by the restriction endonucleases *Hpa*II (which cleaves CCGG but not C^{m}CGG) and *Msp*I (which cleaves both). Such studies indicate that eukaryotic DNA methylation varies with the species, the tissue, and the position along a chromosome.

There is considerable circumstantial evidence that *DNA methylation switches off eukaryotic gene expression, particularly when it occurs in the control regions upstream of a genes' transcribed sequences.* For example, globin genes are less methylated in erythroid cells than they are in nonerythroid cells and, in fact, the specific methylation of the control region in a recombinant globin gene inhibits its transcription in transfected cells. In further support of the inhibitory effect of DNA methylation is the observation that **5-azacytosine (5-azaC),**

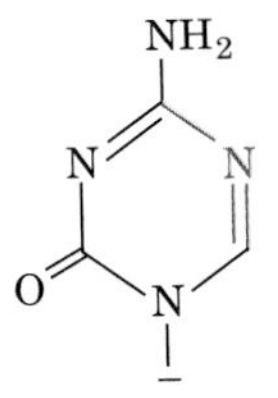

**5-Azacytosine
(5-azaC)**

a base analog that cannot be methylated at its N(5) position and that inhibits DNA methylases, stimulates the synthesis of several proteins and changes the cellular differentiation patterns of cultured eukaryotic cells. However, these effects may not be a consequence of gene demethylation: 5-azaC also induces unmethylated genes.

The way in which DNA methylation prevents gene expression is unknown. An intriguing possibility, however, has been raised by the observation that the methylation of synthetic poly(GC) stabilizes its Z-DNA conformation. Perhaps the formation of Z-DNA, which has been detected *in vivo* (Section 28-2B), acts as a conformational switch to turn off gene expression locally. Yet, since the DNAs of many highly differentiated invertebrates, including *Drosophila*, are not detectably methylated, DNA methylation cannot be the only means by which eukaryotic gene expression is regulated. Indeed, it has not been unequivocally demonstrated that gene expression is controlled by DNA methylation.

DNA Methylation in Eukaryotes Is Self-Perpetuating

The palindromic nature of DNA methylation sites has led to the hypothesis that the methylation pattern on a parental DNA strand in eukaryotes directs the generation of the same pattern in its daughter strand (Fig. 31-53). Such **maintenance methylation** would permit the stable "inheritance" of a methylation pattern in a cell line and hence cause these cells to all have the same differentiated phenotype. There is considerable experimental evidence in favor of this hypothesis including the observation that artificially methylated viral DNA, upon transfection into eukaryotic cells, maintains its methylation pattern for at least 30 cell generations. Little is known, however, about how a DNA methylation pattern is either established in germline cells or modified during differentiation.

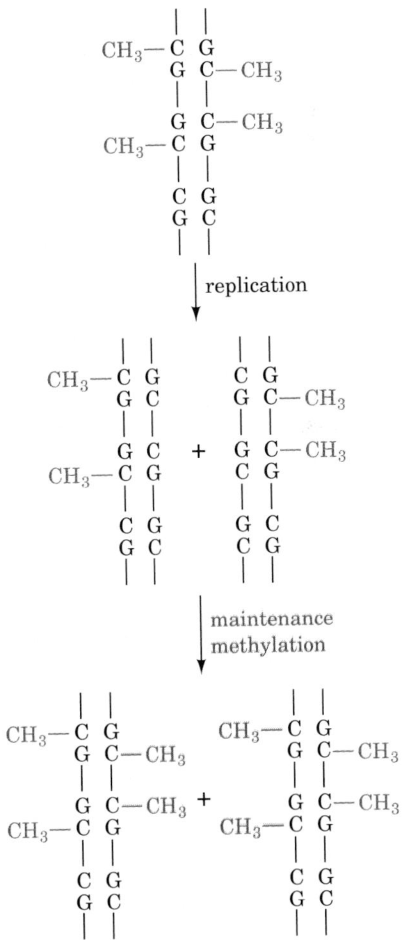

Figure 31-53
In maintenance methylation, the pattern of methylation on a parental DNA strand induces the corresponding methylation pattern in the complementary strand. In this way, a stable methylation pattern may be maintained in a cell line.

Chapter Summary

DNA is replicated in the 5′ → 3′ direction by the assembly of deoxynucleoside triphosphates on complementary DNA templates. Replication is initiated by the generation of short RNA primers, as mediated in *E. coli* by primase and RNA polymerase. The DNA is then extended from the 3′ ends of the primers through the action of DNA polymerase (Pol III in *E. coli*). The leading strand at a replication fork is synthesized essentially continuously, whereas the lagging strand is synthesized discontinuously by the formation of Okazaki fragments. RNA primers on newly synthesized DNA are excised and replaced by DNA through Pol I-catalyzed (in *E. coli*) nick translation. The single-strand nicks are then sealed by DNA ligase. Mispairing errors during DNA synthesis are corrected by the 3′ → 5′ exonuclease functions of both Pol I and Pol III. DNA synthesis in *E. coli* requires the participation of many auxilliary proteins including Rep, helicase II, SSB, and DNA gyrase.

DNA synthesis commences from specific sites known as replication origins. In the synthesis of the bacteriophage M13 (−) strand on the (+) strand template, the origin is recognized and primer synthesis is initiated by RNA polymerase. The analogous process in bacteriophage φX174, as well as in *E. coli*, is mediated by a complex primase-containing particle known as the primosome. φX174 (+) strands are synthesized according to the looped rolling circle mode of DNA replication on (−) strand templates of the replicative form in a process that is directed by the virus-specific gene *A* protein. The *E. coli* chromosome is bidirectionally replicated in the θ mode from a single origin, *oriC*. Leading strand synthesis is primed, probably by RNA polymerase and primase working together, whereas Okazaki fragments are primed by primase in the primosome. The great complexity of the DNA replication process apparently ensures the enormous fidelity necessary to maintain genome integrity.

In eukaryotes, DNA is synthesized during the S phase of the cell cycle. In animal cells, chromosomal DNA is bidirectionally replicated from multiple origins by DNA polymerases α and δ, which are thought to synthesize the lagging and leading strands, respectively. Mitochondrial DNA is replicated in the D-loop mode by DNA polymerase γ. Retroviruses produce DNA on RNA templates in a reaction sequence catalyzed by reverse transcriptase.

Cells have a great variety of DNA repair mechanisms. DNA damage may be directly reversed such as in the photoreactivation of UV-induced pyrimidine dimers or in the repair of O^6-methylguanine lesions. Pyrimidine dimers, as well as many other types of lesions, may also be removed by excision repair. DNA glycosylases specifically remove the corresponding chemically altered bases, including uracil, to form AP sites that are eliminated by excision repair. A lesion in a DNA strand resulting from its synthesis on a damaged template may be corrected through recombination repair. Large amounts of DNA damage induce the SOS response which involves an error prone DNA repair system. The high correlation between mutagenesis and carcinogenesis permits the detection of carcinogens by the Ames test.

Genetic information may be exchanged between homologous DNA sequences through general recombination. This process, which occurs according to the Holliday model, is mediated in *E. coli* by RecA. DNA may also be rearranged through the action of transposons. These DNA segments carry the genes coding for the proteins that mediate the transposition process as well as other genes. Transposition may be important in chromosomal and plasmid evolution and has been implicated in the control of phenotypic expression such as phase alternation in *Salmonella*. Transposons in eukaryotes appear to be degenerate retroviruses.

Prokaryotic DNA may be methylated at its A or C bases. This prevents the action of restriction endonucleases and permits the correct mismatch repair of newly replicated DNA. In eukaryotes, DNA methylation, which occurs through the formation of ^mC, has been implicated in the control of gene expression.

References

General

Adams, R. L. P., Knowler, J. T., and Leader, D. P., *The Biochemistry of the Nucleic Acids* (10th ed.), Chapters 6 and 7, Chapman & Hall (1986).

Boyer, P. D. (Ed.), *The Enzymes*, Vols. 14 and 15, Academic Press, (1981 *and* 1982). [Contains a series of authoritative reviews on many of the enzymes that mediate the replication, repair, and recombination of DNA.]

Freifelder, D., *Molecular Biology*, Chapters 8, 9, 18, and 19, Science Books International (1983).

Kornberg, A., *For Love of Enzymes: The Odyssey of a Biochemist*, Harvard University Press (1989). [A scientific autobiography.]

Kornberg, A., *DNA Replication* and *1982 Supplement to DNA Replication*, Freeman (1980 *and* 1982). [Highly informative source books on most aspects of DNA replication by the founder and one of the major contributors to the field.]

Lewin, B., *Genes* (3rd ed.), Chapters 13–15 and 28–30, Wiley (1987).

Watson, J. D., Hopkins, N. H., Roberts, J. W., Steitz, J. A., and Weiner, A. M., *Molecular Biology of the Gene* (4th ed.), Chapters 10–12, Benjamin/Cummings (1987).

DNA Replication

Bramhill, D. and Kornberg, A., A model for initiation at origins of DNA replication, *Cell* **54**, 915–918 (1988).

Campbell, J. L., Eukaryotic DNA replication, *Annu. Rev. Biochem.* **55**, 733–771 (1986).

Chase, J. W. and Williams K. R., Single-stranded DNA binding proteins required for DNA replication, *Annu. Rev. Biochem.* **55**, 103–136 (1986).

DePamphilis, M. L. and Wassarman, P. M., Replication of

chromosomal eukaryotic chromosomes: a close-up of the replication fork, *Annu. Rev. Biochem.* **49**, 627–666 (1980).

Freemont, P. S., Friedman, J. M., Beese, L. S., Sanderson, M. R., and Steitz, T. A., Cocrystal of an editing complex of Klenow fragment with DNA, *Proc. Natl. Acad. Sci.* **85**, 8924–8928 (1988).

Geider, K. and Hoffmann-Berling, H., Proteins controlling the helical structure of DNA, *Annu. Rev. Biochem.* **50**, 233–260 (1981).

Hübscher, U., DNA polymerase in prokaryotes and eukaryotes: mode of action and biological implications, *Experientia* **39**, 1–25 (1983).

Joyce, C. A. and Steitz, T. A., DNA polymerase I: from structure to function via genetics, *Trends Biochem. Sci.* **12**, 288–292 (1987).

Kornberg, A., DNA replication, *J. Biol. Chem.* **263**, 1–4 (1988).

Kornberg, A., DNA replication, *Trends Biochem. Sci.* **9**, 122–124 (1984).

Kunkel, T. A., Exonucleolytic proofreading, *Cell* **53**, 837–840 (1988).

Lehman, I. R. and Kagun, L. S. DNA polymerase α, *J. Biol. Chem.* **264**, 4265–4268 (1989).

Loeb, L. A., Liu, P. K., and Fry, M., DNA polymerase-α: enzymology, function, fidelity, and mutagenesis, *Prog. Nucleic Acid Res. Mol. Biol.* **33**, 57–110 (1986).

Loeb, L. A. and Kunkel, T. A., Fidelity of DNA synthesis, *Annu. Rev. Biochem.* **51**, 429–457 (1982).

Lohman, T. M., Bujalowski, W., and Overman, L. B., *E. coli* single strand binding protein, *Trends Biochem. Sci.* **13**, 250–255 (1988).

McHenry, C. S., DNA polymerase III holoenzyme of *Escherichia coli*, *Annu. Rev. Biochem.* **57**, 519–550 (1988).

Nagley, P., Termination of replication of bacterial chromosomes, *Trends Genet.* **2**, 221–222 (1986).

Nossal, N. G., Prokaryotic DNA replication systems, *Annu. Rev. Biochem.* **52**, 581–615 (1983).

Ogawa, T. and Okazaki, T., Discontinuous DNA replication, *Annu. Rev. Biochem.* **49**, 421–457 (1980).

Ollis, D. L., Brick, P., Hamlin, R., Xuong, N. G., and Steitz, T. A., Structure of large fragment of *Escherichia coli* DNA polymerase I complexed with dTMP, *Nature* **313**, 762–766 (1985).

So, A. G. and Downey, K. M., Mammalian DNA polymerases α and δ: current status in DNA replication, *Biochemistry* **27**, 4591–4595 (1988).

Varmus, H., Reverse transcription, *Sci. Am.* **257**(3): 56–64 (1987).

Watson, J. D. and Crick, F. H. C., Genetical implications of the structure of deoxyribonucleic acid, *Nature* **171**, 964–967 (1953). [The paper in which semiconservative DNA replication was first postulated.]

Repair of DNA

Ames, B. N., Identifying environmental chemicals causing mutations and cancer, *Science* **204**, 587–593 (1979).

Cairns, J., Robins, P., Sedgwick, B., and Talmud, P., The inducible repair of alkylated DNA, *Prog. Nucleic Acid. Res. Mol. Biol.* **26**, 237–244 (1981).

Cleaver, J. E., Xeroderma pigmentosum, *in* Stansbury, J. B., Wyngaarden, J. B., Fredrickson, D. S., Goldstein, J. L., and Brown, M. S. (Eds.), *The Metabolic Basis of Inherited Disease* (5th ed.), pp. 1227–1248, McGraw–Hill (1983).

Demple, B. and Karran, P., Death of an enzyme: suicide repair of DNA, *Trends Biochem. Sci.* **8**, 137–139 (1983).

Devoret, R., Bacterial tests for potential carcinogens, *Sci. Am.* **241**(2): 40–49 (1979).

Friedberg, E. C., *DNA Repair*, Freeman (1985). [An authoritative treatise.]

Haseltine, W. A., Ultraviolet light repair and mutagenesis revisited, *Cell* **33**, 13–17 (1983).

Howard-Flanders, P., Inducible repair of DNA, *Sci. Am.* **245**(5): 72–80 (1981).

Kenyon, C. J., The bacterial response to DNA damage, *Trends Biochem. Sci.* **8**, 84–87 (1983).

Lindahl, T., DNA repair enzymes, *Annu. Rev. Biochem.* **51**, 61–87 (1982).

Radman, M. and Wagner, R., The high fidelity of DNA replication, *Sci. Am.* **259**(2): 40–46 (1988).

Radman, M. and Wagner, R., Mismatch repair in *Escherichia coli*, *Annu. Rev. Genet.* **20**, 523–538 (1986).

Sancar, A. and Sancar, G. B., DNA repair enzymes, *Annu. Rev. Biochem.* **57**, 29–67 (1988).

Sedgwick, S. G. and Yarrangton, G. T., How cells in distress use SOS, *Nature* **296**, 606–607 (1982).

Walker, G. C., Inducible DNA repair systems, *Annu. Rev. Biochem.* **54**, 425–457 (1985).

Recombination and Mobile Genetic Elements

Cohen, S. N. and Shapiro, J. A., Transposable genetic elements, *Sci. Am.* **242**(2): 40–49 (1980).

Cox, M. M. and Lehman, I. R., Enzymes of general recombination, *Annu. Rev. Biochem.* **56**, 229–262 (1987). [A detailed review of RecA and RecBCD.]

Craig, N. L., The mechanism of conservative site-specific recombination, *Annu. Rev. Genet.* **22**, 77–105 (1988).

Dressler, D. and Potter, H., Molecular mechanisms in genetic recombination, *Annu. Rev. Biochem.* **51**, 727–761 (1982).

Grindley, N. D. F. and Reed, R. R., Transpositional recombination in prokaryotes, *Annu. Rev. Biochem.* **54**, 863–896 (1986).

Howard-Flanders, P., West, S. C., and Stasiak, A., Role of RecA protein in genetic recombination, *Nature* **309**, 215–220 (1984).

Kingsman, A. J. and Kingsman, S. M., Ty: a retroelement moving forward, *Cell* **53**, 333–335 (1988).

Kleckner, N., Transposable genetic elements in prokaryotes, *Annu. Rev. Genet.* **15**, 341–404 (1981).

Kowalczykowski, S. C., Mechanistic aspects of the DNA strand exchange activity of *E. coli* recA protein, *Trends Biochem. Sci.* **12**, 141–145 (1987).

Simon, M., Zieg, J., Silverman, M., Mandel, G., and Doolittle, R., Phase variation: evolution of a controlling element, *Science* **209**, 1370–1374 (1980).

Smith, G. R., Mechanism and control of homologous recombination in *Escherichia coli, Annu. Rev. Genet.* **21,** 179–201 (1987).

Stahl, F. W., Genetic recombination, *Sci. Am.* **256**(2): 90–101 (1987).

West, S. C., Protein-DNA interactions in genetic recombination, *Trends Genet.* **4,** 8–13 (1988).

DNA Methylation

Adams, R. L. P. and Burdon, R. H., *Molecular Biology of DNA Methylation,* Springer–Verlag (1985).

Bird, A. P., DNA methylation—how important in gene control, *Nature* **307**, 503–504 (1984).

Cedar, H., DNA methylation and gene activity, *Cell* **53,** 3–4 (1988).

Doerfler, W., DNA methylation and gene activity, *Annu. Rev. Biochem.* **52,** 93–124 (1983).

Holliday, R., A different kind of inheritance, *Sci. Am.* **260**(6): 60–73 (1989). [Discusses how DNA methylation may control gene-activity patterns and how these patterns might be passed from one cell generation to another.]

Marinus, M. G., DNA methylation in *Escherichia coli, Annu. Rev. Genet.* **21,** 113–131 (1987).

Messer, W. and Noyer-Weidner, W., Timing and targeting: the biological function of Dam methylation in E. coli, *Cell* **54,** 735–737 (1988).

Modrich, P., Methyl-directed DNA mismatch correction, *J. Biol. Chem.* **264,** 6597–6600 (1989).

Modrich, P., DNA mismatch correction, *Annu. Rev. Biochem.* **56,** 435–466 (1987).

Problems

1. Explain how certain mutant varieties of Pol I can be nearly devoid of DNA polymerase activity but retain almost normal levels of $5' \rightarrow 3'$ exonuclease activity.

2. Why haven't Pol I mutants been found that completely lack $5' \rightarrow 3'$ activity at low temperatures?

3. Why aren't Type I topoisomerases necessary in DNA replication?

*4. The $3' \rightarrow 5'$ exonuclease activity of Pol I excises only unpaired 3'-terminal nucleotides from DNA, whereas this enzyme's pyrophosphorolysis activity removes only properly paired 3'-terminal nucleotides. Discuss the mechanistic significance of this phenomenon in terms of the polymerase reaction.

5. You have isolated *E. coli* with temperature sensitive mutations in the following genes. What are their phenotypes above their restrictive temperatures? Be specific. (a) *dnaB*, (b) *dnaE*, (c) *dnaG*, (d) *lig*, (e) *polA*, (f) *rep*, (g) *ssb*, and (h) *recA*.

6. About how many Okazaki fragments are synthesized in the replication of an *E. coli* chromosome?

*7. What are the minimum and maximum number of replication forks that occur in a contiguous chromosome of an *E. coli* that is dividing every 25 min; every 80 min?

8. Why can't linear duplex DNAs, such as occurs in bacteriophage T7, be fully replicated by any of the mechanisms considered in this chapter?

*9. What is the half-life of a particular purine base in the human genome assuming that it is only subject to spontaneous depurination? What fraction of the purine bases in a human genome will have depurinated in the course of a single generation (assume 25 years)?

10. Why is the methylation of DNA to form O^6-methylguanine mutagenic?

11. There are certain sites in the *E. coli* chromosome known as **hot spots** that have unusually high rates of point mu-

tation. Many of these sites contain a 5-methylcytosine residue. Explain the existence of such hot spots.

12. Explain why the brief exposure of a cultured eukaryotic cell line to 5-azacytosine results in permanent phenotypic changes to the cells.

13. Explain why chi structures, such as that shown in Fig. 31-37b, have two pairs of equal length arms.

*14. Single-stranded circular DNAs containing a transposon have a characteristic stem-and-double-loop structure such as that shown in Fig. 31-54. What is the physical basis of this structure?

15. A composite transposon integrated in a circular plasmid occasionally transposes the DNA comprising the original plasmid rather than the transposon's central region. Explain how this is possible.

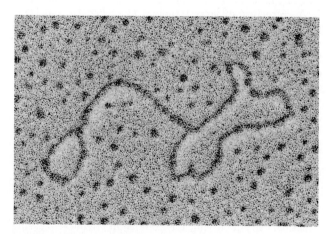

Figure 31-54
An electron micrograph of a single circular DNA containing a transposon. [Courtesy of Stanley Cohen, Stanford University School of Medicine.]

Chapter 32
VIRUSES: PARADIGMS FOR CELLULAR FUNCTIONS

Viruses are parasitic entities, consisting of nucleic acid molecules with protective coats which are replicated by the enzymatic machinery of suitable host cells. Since they lack metabolic apparatus, viruses are not considered to be alive (although this is a semantic rather than a scientific distinction). They range in complexity from **satellite tobacco necrosis virus (STNV),** whose genome has only one gene, to the **pox viruses,** which code for ~240 genes.

Viruses were originally characterized at the end of the nineteenth century as infectious agents that could pass through filters that held back bacteria. Yet viral diseases, varying in severity from small pox and rabies to the common cold, have no doubt plagued mankind since before the dawn of history. It is now known that viruses can infect plants and bacteria as well as animals. Each viral species has a very limited **host range;** that is, it can reproduce in only a small group of closely related species.

An intact virus particle, which is referred to as a **virion,** consists of a nucleic acid molecule encased by a protein **capsid.** In some of the more complex virions, the capsid is surrounded by a lipid bilayer and glycoprotein-containing **envelope,** which is derived from a host cell membrane. Since the small size of a viral nucleic acid severely limits the number of proteins that can be encoded by its genome, its capsid, as Francis Crick and James Watson pointed out in 1957, must be built-up of one or a few kinds of protein subunits that are arranged

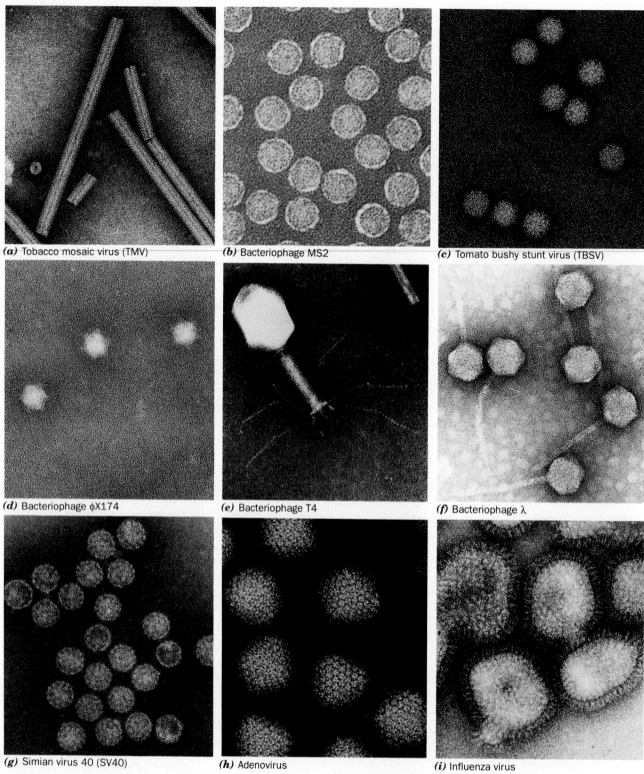

(a) Tobacco mosaic virus (TMV)   **(b)** Bacteriophage MS2   **(c)** Tomato bushy stunt virus (TBSV)

(d) Bacteriophage φX174   **(e)** Bacteriophage T4   **(f)** Bacteriophage λ

(g) Simian virus 40 (SV40)   **(h)** Adenovirus   **(i)** Influenza virus

Figure 32-1
Electron micrographs of a selection of viruses. TMV, MS2, TBSV, and influenza virus are single-stranded RNA viruses; φX174 is a single-stranded DNA virus, and λ, T4, SV40; and adenovirus are double-stranded DNA viruses. Bacteriophage M13, a filamentous, single-stranded DNA coliphage, is shown in Fig. 28-62. [Parts (a)–(d) and (f)–(i) courtesy of Robley Williams, University of California at Berkeley and Harold Fisher, University of Rhode Island; Part (e) courtesy of John Finch, Cambridge University.]

in a symmetrical or nearly symmetrical fashion. There are two ways that this can occur:

1. In the **helical viruses** (Section 32-1), the coat protein subunits associate to form helical tubes.

2. In the **spherical viruses** (Section 32-2), coat proteins aggregate as closed polyhedral shells.

In both cases, the viral nucleic acid occupies the capsid's central region. In many viruses, the coat protein subunits may be "decorated" by other proteins so that the capsid exhibits spikes and, in the larger bacteriophages, a complex tail. These assemblies are involved in recognizing the host cell and delivering the viral nucleic acid into its interior. Figure 32-1 is a "rogues gallery" of viruses of varying sizes and morphologies.

The great simplicity of viruses in comparison to cells makes them invaluable tools in the elucidation of gene structure and function, as well as our best-characterized models for the assembly of biological structures. Although all viruses use ribosomes and other host factors for the RNA-instructed synthesis of proteins, their modes of genome replication are far more varied than that of cellular life. In contrast to cells, in which the hereditary molecules are invariably double-stranded DNA, viruses contain either single- or double-stranded DNA or RNA. In RNA viruses, the viral RNA may be directly replicated or act as a template in the synthesis of DNA. The RNA of single-stranded RNA viruses may be the positive strand (the mRNA) or the negative strand (complementary to the mRNA). Viral DNA may replicate autonomously or be inserted in the host chromosome for replication with the host DNA. The DNA of eukaryotic viruses is either replicated and transcribed in the cell nucleus by cellular enzymes or in the cytoplasm by virally specified enzymes. In fact, in the case of negative strand RNA viruses, enzymes that mediate viral RNA transcription must be carried by the virion because most cells lack the ability to transcribe RNA.

This chapter is a discussion of the structures and biology of a variety of viruses. In it, we examine mainly **tobacco mosaic virus (TMV)**, a helical RNA virus; **tomato bushy stunt virus (TBSV)**, a spherical RNA virus; **bacteriophage λ**, a tailed DNA bacteriophage; and **influenza virus**, an enveloped RNA virus. These examples have been chosen to illustrate important aspects of viral structure, assembly, molecular genetics, and evolutionary strategy. *Much of this information is relevant to the understanding of the corresponding cellular phenomena.* The chapter ends with a discussion of **subviral pathogens**, recently discovered disease agents that are even simpler than viruses.

1. TOBACCO MOSAIC VIRUS

Tobacco mosaic virus causes leaf mottling and discol-

oration in tobacco and many other plants. It was the first virus to be discovered (by Dmitri Iwanowsky in 1892), the first virus to be isolated (by Wendell Stanley in 1935), and even now is the most extensively investigated and well-understood virus from the standpoint of structure and assembly. In this section, we discuss these aspects of TMV.

A. Structure

TMV is a rod-shaped particle (Fig. 32-1*a*) that is ~3000 Å long, 180 Å in diameter, and has a particle mass of 40 million D. Its ~2130 identical copies of coat protein subunits (158 amino acid residues each, 17.5 kD) are arranged in a hollow right-handed helix that has $16\frac{1}{3}$ subunits/turn, a pitch (rise per turn) of 23 Å, and a 40 Å in diameter central cavity (Fig. 32-2). TMV's single RNA strand (~6400 nucleotides; 2 million D) is coaxially wound within the turns of the coat protein helix such that three nucleotides are bound to each protein subunit (Fig. 32-2).

TMV Coat Protein Aggregates to Form Viruslike Helical Rods

The aggregation state of TMV coat protein is both pH and ionic strength dependent (Fig. 32-3). At slightly alkaline pH's and low ionic strengths, the coat protein forms complexes of only a few subunits. At higher ionic

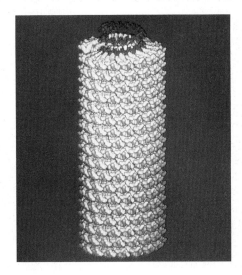

Figure 32-2
A model of TMV illustrating the helical arrangement of its coat protein subunits and RNA molecule. The RNA is represented by the red chain exposed at the top of the viral helix. Only 18 turns (415 Å) of the TMV helix are shown, which represent ~14% of the TMV rod. [Courtesy of Gerald Stubbs and Keiichi Namba, Vanderbilt University; and Donald Caspar, Brandeis University.]

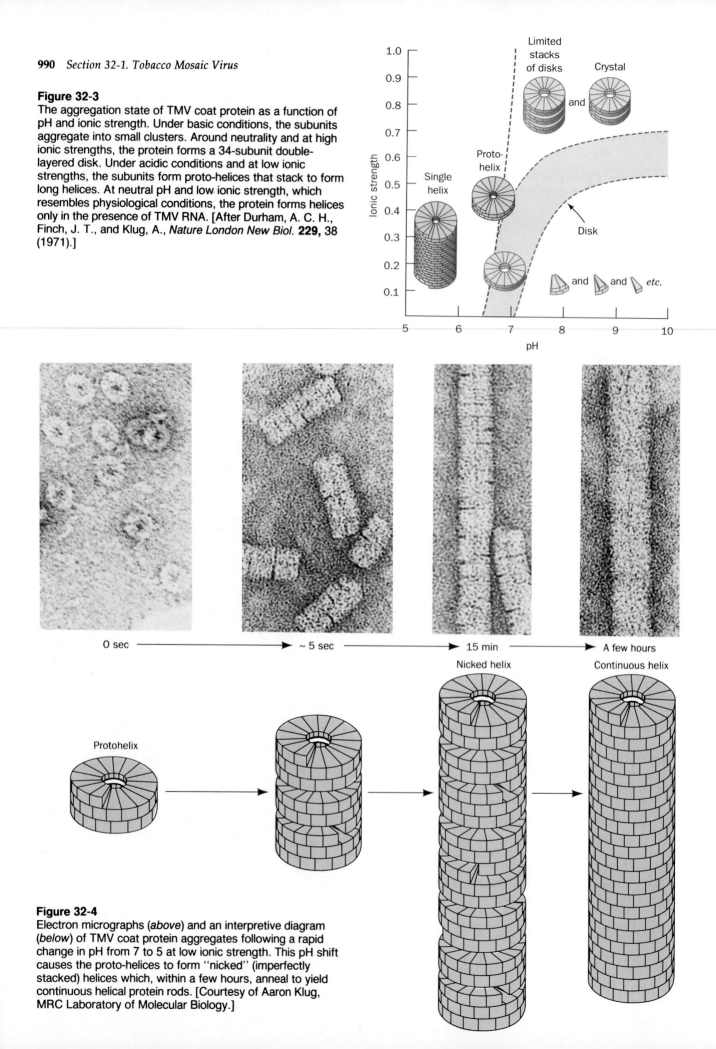

Figure 32-3
The aggregation state of TMV coat protein as a function of pH and ionic strength. Under basic conditions, the subunits aggregate into small clusters. Around neutrality and at high ionic strengths, the protein forms a 34-subunit double-layered disk. Under acidic conditions and at low ionic strengths, the subunits form proto-helices that stack to form long helices. At neutral pH and low ionic strength, which resembles physiological conditions, the protein forms helices only in the presence of TMV RNA. [After Durham, A. C. H., Finch, J. T., and Klug, A., *Nature London New Biol.* **229,** 38 (1971).]

0 sec ⟶ ~ 5 sec ⟶ 15 min ⟶ A few hours

Protohelix   Nicked helix   Continuous helix

Figure 32-4
Electron micrographs (*above*) and an interpretive diagram (*below*) of TMV coat protein aggregates following a rapid change in pH from 7 to 5 at low ionic strength. This pH shift causes the proto-helices to form "nicked" (imperfectly stacked) helices which, within a few hours, anneal to yield continuous helical protein rods. [Courtesy of Aaron Klug, MRC Laboratory of Molecular Biology.]

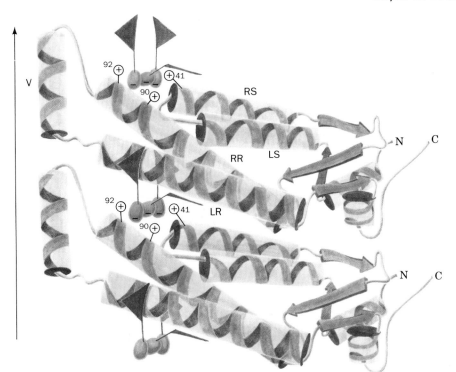

Figure 32-5
A ribbon diagram of two vertically stacked TMV subunits as viewed perpendicular to the virus helix axis (*vertical arrow on the left*). Each subunit has four more or less radially extending helices (LR, RR, LS, and RS), as well as a short vertical helix (V), which comprises part of the flexible loop in the disk structure (dashed lines in Fig. 32-7). Three successive turns of RNA are shown passing through their binding sites. Each subunit binds three nucleotides such that their three bases (*red triangles*) lie flat against the LR helix so as to grasp it in a clawlike manner. Arg residues 41, 90, and 92 neutralize the negative charges of the RNA's three sugar–phosphate groups (*ovals*). [After Casjens, S., *in* Casjens, S. (Ed.), *Virus Structure and Assembly*, p. 78, Jones and Bartlett (1985).]

strengths, however, the subunits associate to form a double-layered disk of 17 subunits/layer, a number that is nearly equal to the number of subunits per turn in the intact virion. At neutral pH and low ionic strengths, the subunits form short helices of slightly more than two turns (39 ± 2 subunits) termed "proto-helices" (also known as "lockwashers"). If the pH of these proto-helices is shifted to ~5, they stack in imperfect register and eventually anneal to form indefinitely long helical rods which, although they lack RNA, resemble intact virions (Fig. 32-4). These observations, as we shall see below, lead to the explanation of how TMV assembles.

TMV Coat Protein Interacts Flexibly with Viral RNA

X-ray studies of TMV have been pursued on two fronts. The virus itself does not crystallize but forms a highly oriented gel of parallel viral rods. The X-ray analysis of this gel by Kenneth Holmes and Gerald Stubbs has yielded a structure of sufficient resolution (2.9 Å) to reveal the folding of the protein and the RNA (Figs. 32-5 and 32-6). This study is complemented by Aaron Klug's X-ray crystal structure determination, at 2.8-Å resolution, of the 34 subunit coat protein disk (Fig. 32-7).

A major portion of each subunit consists of a bundle of four alternately parallel and antiparallel α helices that project more or less radially from the virus axis (Figs. 32-5–32-7). In the disk, one of the inner connections between these α helices, a 24-residue loop (residues 90–113; dotted line in Fig. 32-7), is not visible, apparently because it is highly mobile. This disordered loop is also present in the proto-helix as shown by NMR studies. In the virus, however, the loop adopts a definite

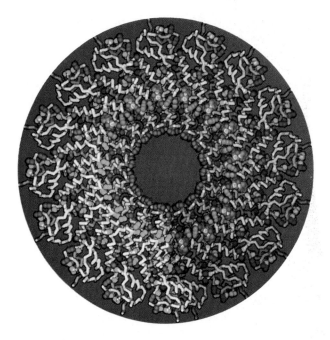

Figure 32-6
Top view of 17 TMV coat protein subunits comprising slightly more than one helical turn in complex with a 33-nucleotide RNA segment. The protein is represented by its C_α atoms, shown as connected 2.5 Å in diameter helical rods, together with its acidic side chains (Asp and Glu) in red and its basic side chains (Arg and Lys) in blue. The RNA's phosphate atoms are green and its bases are purple. Note that the protein's acidic side chains form a 25-Å radius helix that lines the virion's inner cavity while the basic side chains form a 40-Å radius helix that interacts with the RNA's anionic sugar–phosphate chain. [Courtesy of Gerald Stubbs and Keiichi Namba, Vanderbilt University; and Donald Caspar, Brandeis University.]

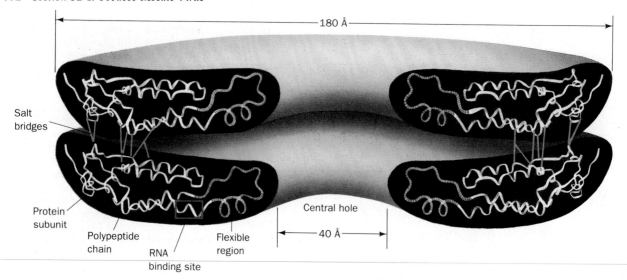

Figure 32-7
The structure of the TMV protein disk in cross-section showing its polypeptide chains as ribbon diagrams. The flexible regions are disordered loops of polypeptide chain that are therefore not visible in the disk X-ray structure. The stacked protein rings interact along their outer rims through a system of salt bridges (*red lines*). [After Butler, P. J. G. and Klug, A., *Sci. Am.* **239**(5): 67 (1978). Copyright © 1978 by Scientific American, Inc.]

conformation that includes a short α helix running parallel to the virus axis (V in Fig. 32-5). This conformational change, as we shall see, is an important aspect of virus assembly.

In the virus, the RNA is helically wrapped between the coat protein subunits at a radius of ∼40 Å. The triplet of bases binding to each subunit forms a clawlike structure around one of the radial helices (LR in Fig. 32-5) with each base occupying a hydrophobic pocket. Arg residues 90 and 92, which are invariant in the several known TMV strains and which are part of the disk and proto-helix's disordered loop, as well as Arg 41, form salt bridges to the RNA phosphate groups.

B. Assembly

How is the TMV virion assembled from its component RNA and coat protein subunits? *The assembly of any large molecular aggregate, such as a crystal or a virus, generally occurs in two stages: (1) **nucleation**, the largely random aggregation of subunits to form a quasi-stable nucleation complex, which is almost always the rate-determining step of the assembly process; followed by (2) **growth**, the cooperative addition of subunits to the nucleation complex in an orderly arrangement that usually proceeds relatively rapidly.* For TMV, it might reasonably be expected that the nucleation complex minimally consists of the viral RNA in association with the 17 or 18 subunits necessary to form a stable helical turn, which could then grow by the accumulation of subunits at one or both ends of the helix. The low probability for the formation of such a complicated nucleation complex from disaggregated subunits accounts for the observed 6-h time necessary to complete this *in vitro* assembly process. Yet, the *in vivo* assembly of TMV probably occurs much faster. A clue as to the nature of this *in vivo* process was provided by the observation that if proto-helices rather than disaggregated subunits are mixed with TMV RNA, complete virus particles are formed in 10 min. Other RNAs do not have this effect. Evidently, *the in vivo nucleation complex in TMV assembly is the association of a proto-helix with a specific segment of TMV RNA.* (It was previously assumed that the disk rather than the proto-helix formed the nucleating complex. It has since been shown, however, that the disk does not form under physiological conditions and that its rate of conversion to the proto-helix under these conditions is too slow to account for the rate of TMV assembly.)

TMV Assembly Proceeds by the Sequential Addition of Proto-Helices

The specific region of the TMV RNA responsible for initiating the virus particle's growth was isolated using the now classical nuclease protection technique. The RNA is mixed with a small amount of coat protein so as to form a nucleation complex that cannot grow because of the lack of coat protein. The RNA that is not protected by coat protein is then digested away by RNase leaving intact only the initiation sequence. This RNA fragment forms a hairpin loop whose 18-nucleotide apical sequence, AGAAGAAGUUGUUGAUGA, has a G at every third residue (recall that each coat protein subunit binds three nucleotides) but no C's (Fig. 32-8). Site-directed mutagenesis studies have confirmed that this initiation sequence is sufficient to direct TMV assembly and that the regularly spaced G's and lack of C's are important for its function. TMV's high binding affinity

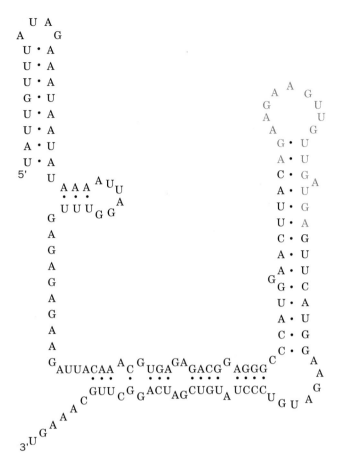

Figure 32-8
The initiation segment of TMV RNA. It probably forms a weakly base paired hairpin, as drawn, that is thought to begin TMV assembly by specifically binding to a coat protein proto-helix. Note that this RNA's loop region has an 18-nucleotide segment (*red*) with a G every third residue (each coat protein subunit binds three nucleotides) but no C's.

for this initiation sequence is explained, in part, by the observations that coat protein subunits bind every third nucleotide in the unusual syn conformation and that G assumes this conformation more easily than any other nucleotide (Section 28-3B). The lack of C's perhaps prevents the involvement of these G's in base pairing associations.

The above initiation complex is located some 1000 nucleotides from the 3' end of the TMV RNA. Hence, the simple model of viral assembly in which the RNA is sequentially coated by protein from one end to the other cannot be correct. Rather, the RNA initiation hairpin must insert itself between the proto-helix's protein layers from its central cavity (Fig. 32-9a). The RNA binding, for reasons explained below, induces the ordering of the disordered loop thereby trapping the RNA (Fig. 32-9b). Growth then proceeds by a repetition of this process at the "top" of the complex thereby incrementally pulling the RNA's 5' end up through the central cavity of the growing viral helix (Fig. 32-9c).

The above assembly model has been corroborated by several experimental observations:

1. Electron micrographs reveal that partially assembled rods (Fig. 32-10) have two RNA "tails" projecting from one end.

2. The length of the longer tail, presumably the 5' end, decreases linearly with the length of the rod, whereas the shorter tail maintains a more or less constant length.

3. Nuclease digestion experiments on partially assembled rods indicate that the RNA is protected in increments of ~100 nucleotides, as is expected for elonga-

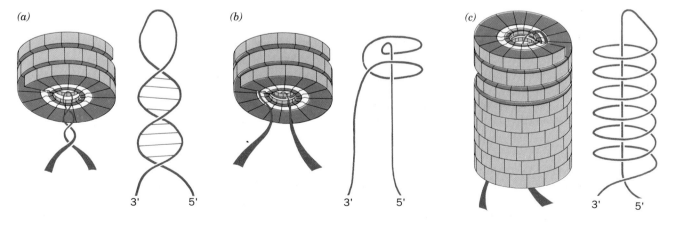

Figure 32-9
The assembly of TMV. (a) The process begins by the insertion of the hairpin loop formed by the initiation sequence of the viral RNA into the proto-helix's central cavity. (b) The RNA then intercalates between the layers of the proto-helix thereby inducing the ordering of the disordered loop and trapping the RNA. (c) Elongation proceeds by the step-wise addition of proto-helices to the "top" of the viral rod. The consequent binding of the RNA to each proto-helix, which converts them to the helical form, pulls the RNA's 5' end up through the virus' 40 Å in diameter central cavity to form a traveling loop at the viral rod's growing end. [After a drawing supplied by Aaron Klug, Cambridge University.]

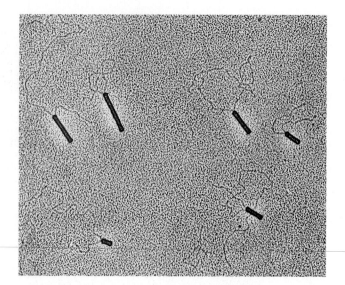

Figure 32-10
An electron mirograph of partially reconstituted TMV particles showing that their two RNA tails emerge from the same end of the growing viral rod. An analysis of these particles indicates that the length of one of the tails, probably the 3' end, is constant (720 ± 80 nucleotides), whereas that of the other tail is inversely proportional to the length of its incomplete rod. [Courtesy of K. E. Richards, CNRS.]

tion steps consisting of the addition of a proto-helix to the growing rod.

The coating of the 3' end of the RNA, which occurs by an unknown mechanism, is a late event in TMV assembly. It has been shown that the RNA, which acts as the viral mRNA, carries the gene specifying the coat protein near its 3' end. Perhaps this assembly mechanism allows coat protein synthesis during all but the final stages of assembly thereby permitting the completion of this process.

Electrostatic Repulsions and Steric Interactions Prevent Helix Formation in the Absence of RNA

What is the mechanism that prevents the formation of TMV coat protein helices in the absence of viral RNA but triggers virus assembly in its presence? Structural considerations suggest that the coat protein subunit's disordered loop sterically prevents the proto-helix from growing longer. Moreover, as we have seen (Fig. 32-3), the state of coat protein aggregation varies with pH. Titration studies show that each subunit has two ionizations with pK's near 7, which must each be attributed to anomalously basic carboxyl groups because coat protein has no His residues. The most plausible candidates for these anomalously basic carboxyls are two intersubunit pairs of carboxyl groups: Glu 95-Glu 106, disordered loop members that interact across a side-to-side subunit interface; and Glu 50-Asp 77, that interact across a top-to-bottom subunit interface. The electrostatic repulsions between these closely spaced negative charges pro-

motes the formation of the disordered loop and therefore favors the proto-helix conformation. The binding of the RNA initiation sequence to the protein apparently provides sufficient free energy to overcome these repulsions thereby triggering helix formation (a process that partially protonates the anomalously basic carboxyl groups; recall the similar conformationally induced pK changes in the Bohr effect of hemoglobin; Section 9-2E). Further growth of the viral rod can occur on RNA segments that lack this sequence as a consequence of the additional binding interactions between adjacent proto-helices. *The carboxyl groups evidently act as a negative switch to prevent the formation of a protein helix in the absence of RNA under physiological conditions.*

2. SPHERICAL VIRUSES

The simpler spherical viruses, being uniform molecular assemblies, crystallize in much the same way as proteins. The techniques of X-ray crystallography can therefore be brought to bear on determining virus structures. In this section we consider the results of such studies.

A. Virus Architecture

The very limited genomic resources of the simpler viruses in many cases limit them to having but one type of protein in their capsid. Since these coat protein subunits are chemically identical, they must all assume the same or nearly the same conformations and have similar interactions with their neighbors. What geometrical constraints does this limitation impose on viral architecture?

We have already seen that TMV solves this problem by assuming a helical geometry (Fig. 32-2). The coat protein subunits in such a long but finite helix, although geometrically distinguishable, have, with the exception of the subunits at the helix ends, virtually identical environments. Such subunits are said to be **quasi-equivalent** to indicate that they are not completely indistinguishable as they would be in an object whose elements are all related by exact symmetry.

Spherical Viruses Have Icosahedral Capsids

A second arrangement of equivalent subunits that can encapsulate a nucleic acid is that of a polyhedral shell. There are only three polyhedral symmetries in which all the elements are indistinguishable: those of a tetrahedron, a cube, and an icosahedron (Fig. 7-57c). Capsids with these symmetries would have 12, 24, or 60 subunits identically arranged on the surface of a sphere. For example, an icosahedron (Fig. 32-11a) has 20 triangular faces, each with threefold symmetry, for a total of $20 \times 3 = 60$ equivalent positions (each represented by a lobe in Fig. 32-11b). Of these polyhedra, the icosahe-

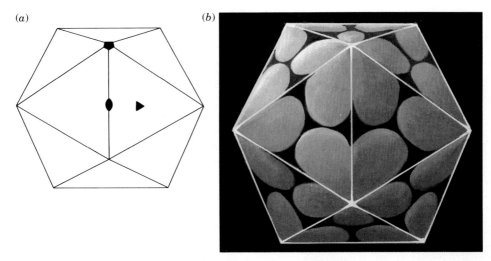

Figure 32-11

An icosahedron. (*a*) This regular polyhedron has 12 vertices, 20 equilateral triangular faces of identical size, and 30 edges. It has a fivefold axis of symmetry through each vertex, a threefold axis through the center of each face, and a twofold axis through the center of each edge. (*b*) A drawing of 60 identical subunits (*lobes*) arranged with icosahedral symmetry. [Drawing copyrighted © by Irving Geis.]

dron encloses the greatest volume per subunit. Indeed, electron microscopy of the so-called spherical viruses (such as Fig. 31-1*b–h*) has revealed that *all of them have icosahedral symmetry.*

Viral Capsids Resemble Geodesic Domes

A viral nucleic acid, if it is to be protected effectively against a hostile environment, must be completely covered by coat protein. Yet, many viral nucleic acids occupy so large a volume that their coat protein subunits would have to be prohibitively large if their capsids were limited to the 60 subunits required by exact icosahedral symmetry. In fact, nearly all viral capsids have considerably more than 60 chemically identical subunits. How is this possible?

Donald Caspar and Klug pointed out the solution to this dilemma. *The triangular faces of an icosahedron can be subdivided into integral numbers of equal sized equilateral triangles (e.g., Fig. 32-12a).* The resulting polyhedron, an **icosadeltahedron,** has "local" symmetry elements re-

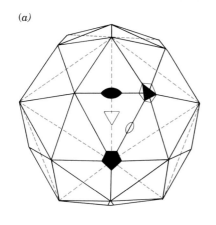

Figure 32-12

A $T = 3$ icosadeltahedron. (*a*) This polyhedron has the exact rotational symmetry of an icosahedron (*solid symbols*) together with local sixfold, threefold, and twofold rotational axes (*hollow symbols*). Note that the edges of the underlying icosahedron (*dashed red lines*), are not edges of this polyhedron and that its local sixfold axes are coincident with its exact threefold axes. (*b*) A drawing of a $T = 3$ icosadeltahedron showing its arrangement of 3 quasi-equivalent sets of 60 icosahedrally related subunits (*lobes*). The A lobes (*orange*) pack about the icosadeltahedron's exact fivefold axes, whereas the B and C lobes (*blue and green*) alternate about its local sixfold axes. TBSV's chemically identical coat protein subunits are arranged in this manner. [Drawing copyrighted © by Irving Geis.]

Figure 32-13
A geodesic dome built on the plan of a $T = 36$
icosadeltahedron. Two of its pentagonal vertices are visible
in this photograph. [Stanley Schoenberger/Grant Heilman.]

lating its subunits (lobes in Fig. 32-12*b*) in addition to its
exact icosahedral symmetry. By local symmetry, we
mean that the symmetry is only approximate so that, in
contrast to the case for exact symmetry, it breaks down
over larger distances. For instance, the subunits (lobes)
in Fig. 32-12*b* that are located at each newly generated
triangular vertex form clusters whose members are related by a local sixfold axis of symmetry. *Adjacent subunits in these clusters are not exactly equivalent; they are
quasi-equivalent.* In contrast, the subunits clustered
about the 12 fivefold axes of icosahedral symmetry are
exactly equivalent. The interactions between the subunits clustered about the local sixfold axes are therefore
essentially distorted versions of those about the exact
fivefold axes. Consequently, *the coat protein subunits of
any viral capsid with icosadeltahedral symmetry must
make alternative sets of intersubunit associations and/or
have sufficient conformational flexibility to accommodate
these distortions.*

Icosadeltahedra are actually familiar figures. The fa-
ceted surface of a soccer ball is an icosadeltahedron.
Likewise, **geodesic domes** (Fig. 32-13), which were
originally designed by Buckminster Fuller, are portions
of icosadeltahedra. It was, in fact, Fuller's designs that
inspired Caspar and Klug. *Geodesic domes are inherently
rigid shell-like structures that are constructed from a few
standard parts, make particularly efficient use of structural
materials, and can be rapidly and easily assembled.* Pre-

sumably the evolution of spherical virus capsids was guided
by these very principles.

The number of subunits in an icosadeltahedron is
$60T$, where T is called the **triangulation number** (it can
be shown that the permissible values of T are given by
$T = h^2 + hk + k^2$, where h and k are positive integers).
An icosahedron, the simplest icosadeltahedron, has
$T = 1$ ($h = 1, k = 0$) and therefore 60 subunits. The icosadeltahedron with the next level of complexity has a
triangulation number of $T = 3$ ($h = 1, k = 1$) and hence
180 subunits (Fig. 32-12). A capsid with this geometry
has three different sets of icosahedrally related subunits
that are quasi-equivalent to each other (lobes A, B, and
C in Fig. 12*b*). Viruses with capsids consisting of $T = 1$,
3, and 4 icosadeltahedra have been identified. Some of
the larger polyhedral viruses may form icosadeltahedra
with even greater triangulation numbers (although several of them have been shown to be based on somewhat
different assembly principles). The T value for any particular capsid, presumably, depends on its subunit's in-
nate curvature.

B. Tomato Bushy Stunt Virus

TBSV (Fig. 32-1*c*) is a $T = 3$ spherical virus that is
~175 Å in radius. It consists of 180 identical coat protein subunits, each of 386 residues (43 kD), encapsulat-

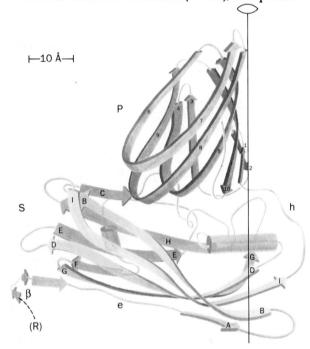

Figure 32-14
The TBSV coat protein subunit consists of three domains:
P, which projects from the virion's surface; S, which forms
the capsid; and R, which extends below the capsid surface
where it participates in binding the viral RNA. The P and S
domains are composed largely of antiparallel β sheets. The R
domain is not visible in the viral X-ray structure so that its
tertiary structure is unknown. [After Olsen, A. J., Bricogne,
G., and Harrison, S. C., *J. Mol. Biol.* **171,** 78 (1983).]

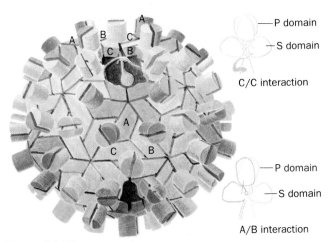

Figure 32-15
The *T* = 3 icosadeltahedral arrangement of TBSV's coat protein subunits. The subunits occur in three quasi-equivalent packing environments, A, B, and C. The A subunits pack around exact fivefold axes, whereas the B subunits alternate with the C subunits about the exact threefold axes (local sixfold axes). The C subunits are also disposed about the strict twofold axes, whereas the A and B subunits are related by local twofold axes. The subunits respond to the different conformational requirements of their three quasi-equivalent positions through flexion at the hinge region between their S and P domains (*right and in cutaways*). Compare this drawing to Fig. 32-12. [After Harrison, S. C., *Trends Biochem. Sci.* **9**, 348, 349 (1984).]

ing a single-stranded RNA molecule of ~4800 nucleotides (1500 kD; the positive or message strand) and a single copy of an ~85-kD protein. The X-ray crystal structure of TBSV, the first of a virus to be determined at high resolution, was reported in 1978 by Stephen Harrison. TBSV's coat protein subunits have three domains (Fig. 32-14): P, the C-terminal domain, which projects outwards from the virus; S, which forms the protein shell; and R, the protein's inwardly extending N-terminal domain, which is attached to the S domain via a connecting arm. The S and P domains are almost entirely composed of antiparallel β sheets.

TBSV's Identical Subunits Associate through Nonidentical Contacts

The chemically identical TBSV coat protein subunits occupy three symmetrically distinct environments denoted A, B, and C (Fig. 32-15). How does the protein accommodate the different contacts required by its several sets of analogous but nonidentical associations? TBSV's structure reveals that *analogous intersubunit contacts vary both through alternative sets of interactions and by conformational distortions of the same interactions.* Perhaps the most remarkable alternative interaction is the interdigitation of the arms connecting the R and S domains of the C subunits. These arms extend toward each icosahedral threefold axis (quasi-sixfold axis) in the clefts between the adjacent C and B subunits and then spiral downwards about this threefold axis to form a β sheetlike arrangement that resembles the overlapping flaps of a cardboard carton: chain 1 over chain 2 over chain 3 over chain 1 (Fig. 32-16*a*). This interaction, together with a strong association between neighboring C

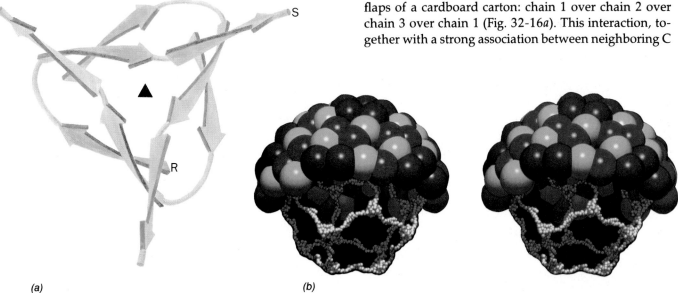

(a)   *(b)*

Figure 32-16
The architecture of the TBSV capsid. (*a*) The C subunit arms of TBSV protein pack about the capsid's exact threefold axes (*triangle*) and associate as β sheets. The view is from outside the capsid. [After a drawing by Jane Richardson, Duke University.] (*b*) A stereo cutaway drawing showing the capsid's internal scaffolding of C subunit arms. The chemically identical A (*dark blue*), B (*light blue*), and C

subunits (*red*) are represented by large spheres, whereas the residues comprising the C subunit arms are represented by small yellow spheres. The C subunit arms associate to form an icosahedral (*T* = 1) framework that apparently plays a major role in holding together the viral capsid. Directions for viewing stereo drawings are given in the appendix to Chapter 7. [Courtesy of Arthur Olson, Research Institute of Scripps Clinic.]

subunits across the icosahedral twofold axis (Fig. 32-15), organizes the 60 C subunits into a coherent network (Fig. 32-16b) that determines the triangulation number of the TBSV capsid: *The C subunits can be thought of as forming a T = 1 icosahedral shell whose gaps are filled in by the A and B subunits.* In response, the three sets of quasi-equivalent subunits assume somewhat different conformations: The 3- or 4-residue "hinge" connecting the S and P domains (h in Fig. 32-14) has an ~30° greater dihedral angle in the A and B subunits than in the C subunits (Fig. 32-15, *right*). This, in turn, permits the interactions between P domains to be identical in the AB and CC dimers (projecting knobs in Fig. 32-15). Evidently, interdomain associations between subunits are stronger in TBSV than those within subunits.

TBSV's RNA-Containing Core Is Disordered

The entire connecting arm between the R and S domains in the A and B subunits, as well as their first few residues in the C subunits, are not visible in TBSV's X-ray structure, thereby indicating that these polypeptide segments have no fixed conformations. The R domains are therefore flexibly tethered to the S domains so that they are also absent from the X-ray structure, even though these domains probably have a fixed conformation. Neutron scattering studies, nevertheless, suggest that protein, constituting perhaps one half of the R domains, forms a 50- to 80-Å radius inner shell. The remaining R domains are thought to project into the space between the inner and outer shells.

The viral RNA is absent from the X-ray structure, which indicates that it too is disordered. The above neutron scattering studies reveal that this RNA is sandwiched between the virus' inner and outer protein shells (Fig. 32-17). The volume constraints imposed by this arrangement require that the RNA be tightly packed. This packing is made possible because most of the negative charges of the RNA phosphate groups are neutralized by the numerous positively charged Arg and Lys residues of the R domains, the inner faces of the S domains, and their connecting arms.

Other RNA Plant Viruses Are Remarkably Similar to TBSV

The structures of several other RNA plant viruses have been elucidated, including those of **southern bean mosaic virus (SBMV)** by Michael Rossmann and **satellite tobacco necrosis virus (STNV)** by Bror Strandberg. SBMV is a T = 3 virus that closely resembles TBSV in its quaternary structure. Moreover, SBMV's 260-residue coat protein subunit, although it entirely lacks a P domain, has an S domain whose polypeptide backbone is nearly superimposable on that of TBSV (Fig. 32-18a). STNV's quaternary structure differs from those of TBSV or STNV; STNV is a T = 1 RNA virus whose 100-Å radius makes it among the smallest known virions [its

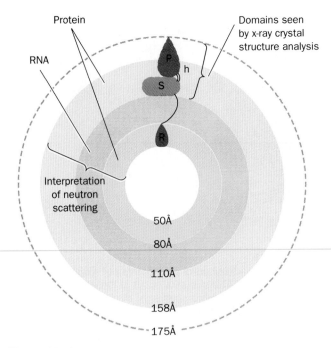

Figure 32-17
The radial organization of TBSV indicating the distribution of its protein and RNA components. The R domain positions are inferred from their known chain length. Only about one half of the R domains are contained in the inner protein shell. [After Harrison, S. C., *Biophys. J.* **32,** 140 (1980).]

1239-nucleotide RNA only encodes the gene for the viral coat protein; STNV can only multiply in cells that are coinfected with the more complex **tobacco necrosis virus (TNV)**]. Nevertheless, STNV's 195-residue coat protein, which also lacks a P domain, has an S domain that structurally resembles those of SBMV and TBSV. Evidently, these biochemically dissimilar viruses arose from a common ancestor. The RNA in all these viruses appears disordered.

C. Picornaviruses

The X-ray structures of two viral pathogens of humans were elucidated in 1985: that of **poliovirus,** the cause of **poliomyelitis,** by James Hogle; and that of **rhinovirus,** the cause of **infectious rhinitis** (the common cold), by Rossmann. Both pathogens are **picornaviruses,** a large family of animal viruses that also includes the agents causing human **hepatitis A** and **foot-and-mouth disease.** Picornaviruses (*pico*, small + *rna*) are among the smallest RNA-containing animal viruses: They have a particle mass of ~8.5 × 10⁶ D of which ~30% is a single-stranded RNA of ~7500 nucleotides. Their icosahedral protein shell, which is ~300 Å in diameter, contains 60 protomers, each consisting of 4 structural proteins, **VP1, VP2, VP3,** and **VP4.** These 4 proteins are synthesized by an infected cell as a single **polyprotein,** which is cleaved to the individual

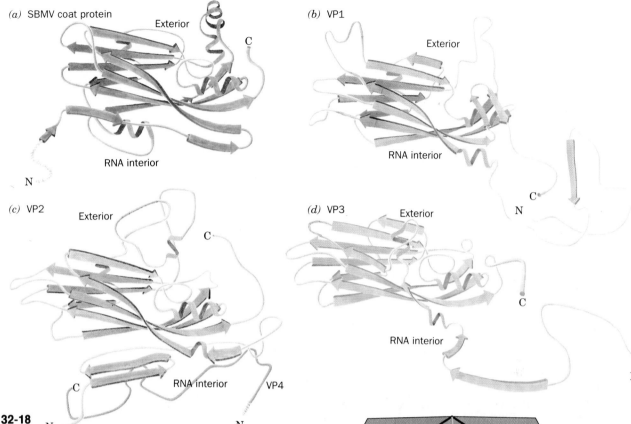

Figure 32-18
The structures of (a) SBMV coat protein, and the (b) VP1, (c) VP2 (together with VP4), and (d) VP3 proteins of human rhinovirus. Note the close structural similarities of their eight-stranded β-barrel cores and that of TBSV's S domain (Fig. 32-14). The VP1, VP2, and VP3 proteins of poliovirus are likewise similar. [After Rossmann, M. G., Arnold, E., Erickson, J. W., Frankenberger, E. A., Griffith, J. P., Hecht, H.-J., Johnson, J. E., Kamer, G., Luo, M., Mosser, A. G., Rueckert, R. R., Sherry, B., and Vriend, G., *Nature* **317,** 148 (1985).]

subunits during virion assembly. Picornaviruses can be highly specific as to the cells they infect; for example, poliovirus binds to receptors that occur only on certain types of primate cells.

The structures of poliovirus, rhinovirus, and **foot-and-mouth disease virus (FMDV;** whose structure was determined in 1988 by David Stuart) are remarkably alike, both to each other and to TBSV and SBMV. Although VP1, VP2, and VP3 of picornaviruses have no apparent sequence similarities with each other or with the coat proteins of TBSV and SBMV, these proteins all exhibit striking structural similarities (Figs. 32-14 and 32-18; VP4, which is much smaller than the other subunits, forms, in effect, an N-terminal extension of VP2). Indeed, the picornaviruses' chemically distinct VP1, VP2, and VP3 subunits are pseudosymmetrically related by pseudo-threefold axes passing through the center of each triangular face of the icosahedral ($T = 1$) virion (Fig. 32-19). The chemically identical but conformationally distinct A, B, and C subunits of the $T = 3$ plant viruses are likewise quasi-symmetrically related

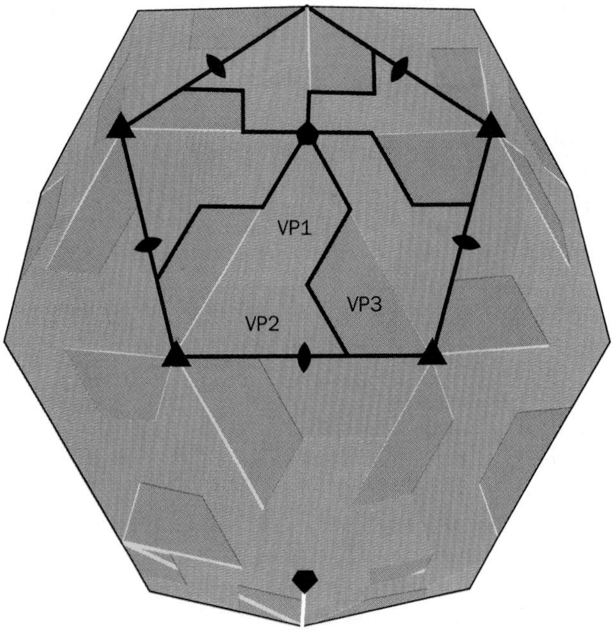

Figure 32-19
The arrangement of the 60 trimers (*triangles*) of pseudo-equivalent VP1, VP2, and VP3 subunits on human rhinovirus's icosahedral capsid. This arrangement resembles that of TBSV in which 180 chemically identical subunits are quasi-symmetrically related to form a $T = 3$ icosadeltahedron (Figs. 32-12 and 32-15). The positions of the icosahedron's exact fivefold, threefold, and twofold axes are marked. [After Rossmann, M. G., Arnold, E., Erickson, J. W., Frankenberger, E. A., Griffith, J. P., Hecht, H.-J., Johnson, J. E., Kamer, G., Luo, M., Mosser, A. G., Rueckert, R. R., Sherry, B., and Vriend, G., *Nature* **317,** 147 (1985).]

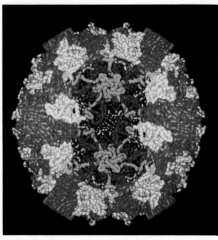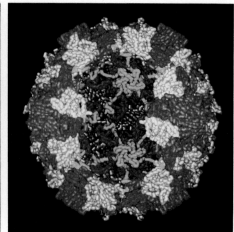

Figure 32-20
A stereo diagram of the poliovirus capsid in which the inner surface is revealed by the removal of two pentagonal faces. Here, the polypeptide chain is represented by a folded tube that approximates the volume of the protein and which is blue in VP1, yellow in VP2, red in VP3, and green in VP4. The VP4 subunits, which line the capsid's inner surface, associate about its fivefold axes of symmetry to form a framework similar to although geometrically distinct from that formed by the C subunit arms in TBSV (Fig. 32-16). [Courtesy of Arthur Olson, Research Institute of Scripps Clinic.]

by analogously located local threefold axes (Fig. 32-15). These structural similarities strongly suggest that the picornaviruses and the spherical plant viruses all diverged from a common ancestor.

The protein capsids of poliovirus, rhinovirus, and FMDV form a hollow shell enclosing a disordered core composed of the viral RNA and some protein, much as in the spherical plant viruses. This arrangement is vividly illustrated in Fig. 32-20, which shows both the inner and outer views of the poliovirus capsid. Note that VP4 largely lines the inside of the capsid. Also note the rugged topography of the capsid's outer surface. Some of its crevices form the receptor-binding site through which the virus is targeted to specific cells.

transport protein (the product of the *E. coli lamB* gene)

3. BACTERIOPHAGE λ

Bacteriophage λ (Figs. 32-1*f* and 32-21), a midsized (58 million D) coliphage, has a 55 nm in diameter icosahedral head and a flexible 15 × 135-nm tail that bears a single thin fiber at its end. The virion contains a 48,502 bp linear double-stranded B-DNA molecule of known sequence. Phage λ is, at present, the most extensively characterized complex virus with respect to its molecular biology. Indeed, as we shall see in this section, *its genetic regulatory mechanisms form our best paradigm for the control of development in higher organisms and its assembly is among our best characterized examples of the morphogenesis of biological structures.*

Bacteriophage λ adsorbs to *E. coli* through a specific interaction between the viral tail fiber and a maltose

that is a component of the bacterium's outer membrane. This interaction initiates a complex and poorly understood process in which the phage DNA is injected through the viral tail into the host cell. Soon after entering the host, the λ DNA, which has complementary single-stranded ends of 12 nucleotides (cohesive ends), circularizes and is covalently closed and supertwisted by the host DNA ligase and DNA gyrase (Fig. 32-22, Stages 1–4).

At this stage the virus has a "choice" of two alternative life styles (Fig. 32-22):

1. It can follow the familiar **lytic** mode in which the phage is replicated by the host such that, after 45 min at 37°C, the host lyses to release ~100 progeny phages.

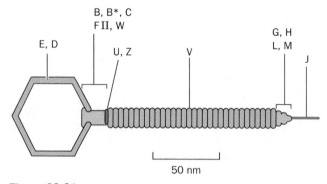

Figure 32-21
A sketch of bacteriophage λ indicating the locations of its protein components, The letters refer to specific proteins (gene products; see text). The bar represents 50 nm. [After Eiserling, F. A. *in* Fraenkel-Conrat, H. and Wagner, R. R. (Eds.), *Comparative Virlogy*, Vol. 13, *p.* 550, Plenum (1979).]

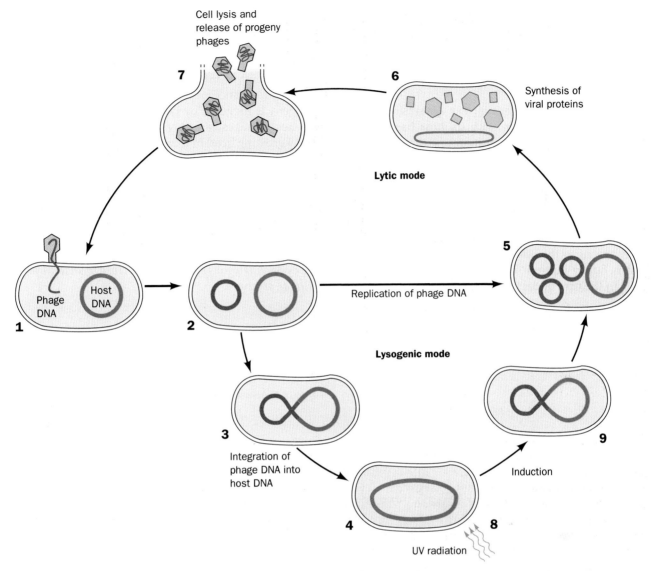

Figure 32-22
The λ phage life cycle. The infection of the bacterial host *E. coli* begins when the virus specifically adsorbs to the cell and injects its DNA (1). The linear DNA then circularizes (2) and commences directing the infection process. In the lysogenic mode, the phage DNA is stably integrated at a specific site in the host chromosome (3) and (4) so that it is passively replicated with the bacterial cell. Alternatively, the phage may take up the lytic mode in which the DNA directs its own replication (5), as well as the synthesis of viral proteins (6) so as to result in the lysis of the host cell with the release of ~100 progeny phages (7). DNA damage, as is caused, for example, by UV radiation (8), induces the excision of the prophage DNA from the lysogenic bacterial chromosome (9) and causes the phage to take up the lytic mode.

2. *The phage may take up the so-called* **lysogenic** *life cycle, in which its DNA is inserted at a specific site in the host chromosome such that the phage DNA passively replicates with the host DNA. Nevertheless, even after many bacterial generations, if conditions warrant, the phage DNA will be excised from the host DNA to initiate a lytic cycle in a process known as* **induction.**

How the phage chooses between the lytic and lysogenic modes is the subject of Section 32-3D.

Phage DNA that is following a lysogenic life cycle is described as a **prophage,** whereas its host is called a

lysogen. An intriguing property of lysogens is that they cannot be reinfected by phages of the type with which they are lysogenized: *They are* **immune to superinfection.** A bacteriophage that can follow either a lytic or a lysogenic life style is known as a **temperate phage,** whereas those that have only a lytic mode are described as **virulent.** Bacteriophages that are reproducing lytically are said to be engaged in **vegetative growth.**

Over 90% of the thousands of known types of phages are temperate and, conversely, most bacteria in nature are lysogens. Yet, the presence of prophages has frequently gone unnoticed because they have little apparent affect on their hosts. For example, the K12 strain of *E. coli* had been the subject of intensive investigations for > 20 years before 1951 when Ester Lederberg found it to be lysogenic for bacteriophage λ (which marks the discovery of this phage as well as the phenomenon of lysogeny).

The advantage of lysogeny is clear. A parasite that can form a stable association with its host has a better chance of long-term survival than one that invariably destroys its host. A virulent phage, on encountering a colony of its host bacteria, will multiply prodigiously. After the colony has been wiped out, however, it may be some time, if at all, before any of the progeny encounter another suitable host in a generally hostile world. In contrast, a prophage will multiply with its host indefinitely so long as the host remains viable. But what if the host is fatally injured? Does the parasite die with the host? In the case of bacteriophage λ, it is precisely such

traumatic conditions, exposure to agents that damage the host DNA or disrupt its replication, that induce the lytic phase. This has been described as the "lifeboat" response: The prophage escapes a doomed host through the formation of infectious viral particles that have at least some chance of further replication. Conversely, lysogeny is triggered by poor nutritional conditions for the host (phages can only lytically replicate in an actively growing host) or a large number of phages infecting each host cell (which signals that the phages are on the verge of eliminating the host).

This section describes the genetic system that controls the orderly formation of phage particles in the lytic mode, the mechanism through which these phage particles are assembled, and the regulatory mechanism through which bacteriophage λ selects and maintains its life cycle. *Analogous systems are believed to underlie many cellular processes.*

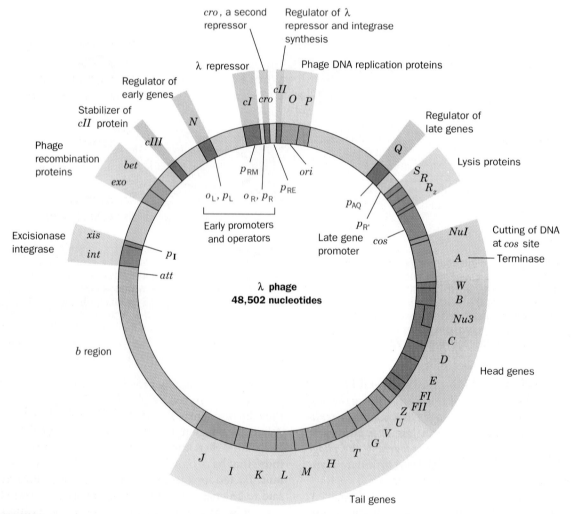

Figure 32-23
A genetic map of bacteriophage λ showing most of its structural genes (indicated outside the circle) and control sites (indicated inside the circle). The genes encoding regulatory proteins are shaded in red. Upon packaging into the virion, the circular chromosome is cut at the *cos* site yielding a linear DNA.

A. The Lytic Pathway

The bacteriophage λ genome, as its genetic map indicates (Fig. 32-23), encodes ~ 50 gene products and contains numerous control sites. Note the λ chromosome's organization. Its genes are clustered according to function. For example, the genes concerned with the synthesis of phage tail proteins are tandemly arranged on the bottom of Fig. 32-23. This organization, as we shall see, enables these genes to be transcribed together, that is, as an operon. The function of many of the λ genes and control sites, together with those of the host that are important in phage function, are tabulated in Table 32-1.

In the lytic replication of phage λ, as in love and war, proper timing is essential. This is because the DNA must be replicated in sufficient quantity before it is made unavailable by packaging into phage particles and because packaging must be completed before the host cell is enzymatically lysed. The transcription of the λ genome, which is carried out by host RNA polymerase, is controlled in both the lytic and the lysogenic programs by the regulatory genes that are shaded in red in Fig. 32-23.

The Lytic Mode Has Early, Delayed Early, and Late Phases

The lytic transcriptional program has three phases (Fig. 32-24):

1. Early transcription

Soon after phage infection or induction, E. coli RNA polymerase commences "leftward" transcription of the phage DNA starting at the promoter p_L and "rightward" transcription (and thus from the opposite DNA strand) from the promoters p_R and p'_R (Fig. 32-24a):

(i) The "leftward" transcript, L1, which terminates at termination site t_{L1}, encodes the N gene.

(ii) "Rightward" transcription from p_R terminates with $\sim 50\%$ efficiency at t_{R1}, to yield transcript R1, and otherwise at t_{R2} to yield transcript R2. R1 contains only the *cro* gene transcript, whereas R2 also contains the *cII*, *O*, and *P* gene transcripts.

(iii) "Rightward" transcription from p'_R terminating at t'_R yields a short transcript, R4, that specifies no protein.

L1, R1, and R2 are translated by host ribosomes to yield proteins whose functions are described below.

2. Delayed-early transcription

The second transcriptional phase commences as soon as a significant quantity of the protein **gp$N$** (gp for gene product) accumulates. *This protein, through a mechanism considered below, acts as a transcriptional antiterminator at termination sites t_{L1}, t_{R1}, and t_{R2} (Fig. 32-24b):*

Table 32-1

Important Genes and Genetic Sites for Bacteriophage λ

Phage genes

cI	λ Repressor; establishment and maintenance of lysogeny
cII, cIII	Establishment of lysogeny
cro	Repressor of *cI* and early genes
N, Q	Antiterminators for early and delayed early genes
O, P	Origin recognition in DNA replication
int	Prophage integration and excision
xis	Prophage excision
B, C, D, E, W, Nu3, FI, FII	Head assembly
G, H, I, J, K, L, M, U, V, Z	Tail assembly
A, Nu1	DNA packaging
R, S	Host lysis
b	Accessory gene region

Phage sites

*att*P	Attachment site for prophage integration
*att*L, *att*R	Prophage excision sites
cos	Cohesive end sites in linear duplex DNA
o_L, o_R	Operators
p_I, p_L, p_R, p_{RM}, p_{RE}, p'_R	Promoters
t_{L1}, t_{R1}, t_{R2}, t_{R3}, t'_R	Transcriptional termination sites
*nut*L, *nut*R	N utilization sites
qut	Q utilization sites
ori	DNA replication origin

Host genes

lamB	Host recognition protein
lig	DNA ligase
gyrA, gyrB	DNA gyrase
rpoA, rpoB, rpoC	RNA polymerase core enzyme
rho	Transcription termination factor
nusA, nusB, nusE	Necessary for gp$N$ function
groEL, groES	Head assembly
himA, himD	Integration host factor
hflA, hflB	Degrades gp*cII*
cap, cya	Catabolite repressor system
*att*B	Prophage integration site
recA	Induction of lytic growth.

(i) Leftward transcript L1 is extended to form L2, which additionally contains the transcripts of the *cIII*, *xis*, and *int* genes (which encode proteins involved in switching between the lytic and lysogenic modes; Sections 32-3C and D) together with the *b* region gene transcripts (which specify

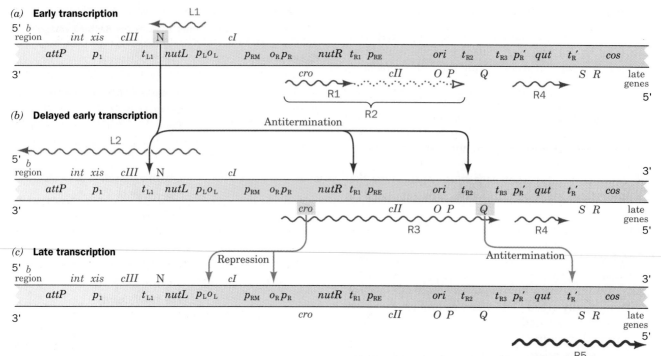

Figure 32-24

Gene expression in the lytic pathway of phage λ. Genes specifying proteins that are transcribed to the "left" and "right" are shown above and below the phage chromosome. Control sites are indicated between the DNA strands. The genetic map is not drawn to scale and not all of the genes or control sites are indicated. Transcripts are represented by wiggly arrows pointing in the direction of mRNA elongation; the actions of regulatory proteins are denoted by arrows pointing from each regulatory protein to the site(s) it controls. The lytic pathway has three transcriptional phases: (a) early transcription, (b) delayed early transcription, and (c) late transcription. Gene expression in each of the latter two phases is regulated by proteins synthesized in the preceding phase as is explained in the text. [After Arber, W., *in* Hendrix, R. W., Roberts, J. W.. Stahl, F. W., and Weisberg, R. A. (Eds.), *Lambda II, p.* 389, Cold Spring Harbor Laboratory (1983).]

the so-called **accessory proteins** which, although not essential for lytic growth, increase its efficiency).

(ii) Transcript R3, which includes R1 and R2, also encodes a second antiterminator, **gpQ**, whose function is discussed below. The continuing translation of R2 and later R3 to yield gpO and gpP, proteins that are both required for λ DNA replication, stimulates viral DNA production. Similarly, the translation of R1 and later R3 yields **Cro protein** (gpcro), a repressor of both the "rightward" and "leftward" genes (see below; *cro* stands for *c*ontrol of *r*epressor and *o*ther things).

At this stage, ~15-min post-infection, Cro protein has accumulated in sufficient quantity to bind to operators o_L *and* o_R, *thereby shutting off transcription from* p_L *and* p_R. This is more than just efficient use of resources; the overexpression of the early genes, as occurs in λcro⁻ phage, poisons the lytic cycle's late phase.

3. Late transcription

In the final transcriptional phase (Fig. 32-24c), *the*

antiterminator gpQ acts to extend the R4 transcript through t'_R *to form the R5 transcript.* The "gene dosage" effect of the ~30 copies of phage DNA that have accumulated by the beginning of this stage results in the rapid synthesis of the capsid-forming proteins (which are all encoded by late genes; their assembly to form mature phage particles is described in Section 32-3B), as well as **gpR** and **gpS**, which catalyze host cell lysis [gpR is an **endolysin**, an enzyme that hydrolyzes a peptide bond in the host cell wall peptidoglycan (Section 10-3B); gpS interacts with the cell membrane so as to induce pore formation]. The first phage particle is completed ~22-min post-infection.

Antitermination Requires the Action of Several Proteins

Transcriptional control in the λ lytic phase is exerted by gpN- and gpQ-mediated antitermination rather than by repressor binding at an operator site through which, for example, *lac* operon expression (Section 29-1B) is regulated. gpN (12 kD) acts at both rho-dependent and rho-independent termination sites (t_{L1} and t_{R1} are rho dependent, whereas t_{R2} is rho independent; transcriptional termination is discussed in Section 29-2E). Yet, gpN does not act at just any transcriptional termination site. Rather, genetic analysis of mutant phage defective for antitermination has established the existence of two so-called *nut* (for *N utilization*) sites that are required for antitermination: *nut*L, which is located between p_L and *N*, and *nut*R, which occurs between *cro* and t_{R1} (Fig. 32-24). These sites have closely similar 16- and 17-nucleotide sequences whose transcripts can form hydrogen bonded hairpin loops (Fig. 32-25*a* and *b*).

The λ promoters p_L and p_R play no role in gpN-mediated antitermination. This conclusion was reached through the construction of a plasmid containing the λ*nut*R and t_{R1} terminator site between a different promoter and a gene encoding tetracycline resistance. Antitermination would then confer tetracycline resistance on the host. When this plasmid was placed in an *E. coli* with an active chromosomal *N* gene (the bacterium was lysogenic for a mutant λ prophage defective for induction), the resulting strain was tetracycline resistant. Strains lacking either this *nut* site or an active *N* gene, however, cannot grow in the presence of tetracycline. *The nut site is therefore necessary and sufficient for gpN activity.*

The mechanism of antitermination is poorly understood. However, the observation that some *E. coli* defective in antitermination have mutations that map in the *rpoB* gene (which encodes the RNA polymerase β subunit), suggests that gpN acts at *nut* sites to render RNA polymerase resistant to termination. Indeed, gpN-modulated RNA polymerase will pass over many different terminators that it encounters either naturally or by experimental design. Genetic analyses have revealed that

antitermination requires several other host factors termed **Nus** (for *N utilization substance*) **proteins:** the gene products of *nusA* and *nusB*, both of which specifically bind gpN, and that of *nusE*, which probably also does so (and which, curiously, is ribosomal protein S10). Upon encountering a *nut* site, gpN forms a complex with the Nus proteins and RNA polymerase that travels with this enzyme during elongation and presumably prevents transcriptional termination. The observation that covering *nut* RNA with ribosomes prevents antitermination strongly suggests that gpN recognizes this site on RNA, not DNA.

gpQ, which overrides t'_R to permit late transcription, acts at a *qut* site (analogous to the *nut* sites) that is located some 20 bp downstream from p'_R and that can form an RNA hairpin similar to those of the *nut* sites (Fig. 32-25*c*). gpQ-mediated antitermination, like that of gpN, requires the participation of gp*nusA*.

gp*O* and gp*P* Participate in λ DNA Replication

The course of DNA replication in phage λ is diagrammed in Fig. 32-26. Electron microscopy indicates that in the early stages of lytic infection, λ DNA replication occurs both by the bidirectional θ mode (Section 31-1A) from a single replication origin (*ori*) that is located in the *O* gene, and by the rolling circle (σ) mode (Section 31-3B). By the late stage of the lytic program, however, DNA replication has completely switched, via an unknown mechanism, to the rolling circle mode (with the accompanying synthesis of the complementary strand). In the process of phage assembly (Section 32-3B), the resulting concatemeric (consisting of tandemly linked identical units) DNA is specifically cleaved

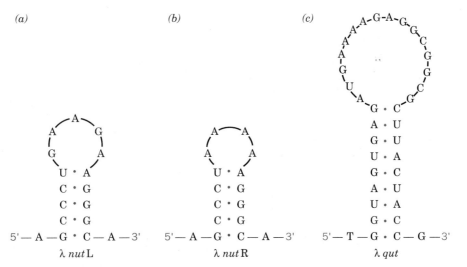

Figure 32-25
The RNA sequences of the homologous (*a*) *nut*L and (*b*) *nut*R sites and of the (*c*) *qut* site in phage λ. Each of these control sites is thought to form a base paired hairpin.

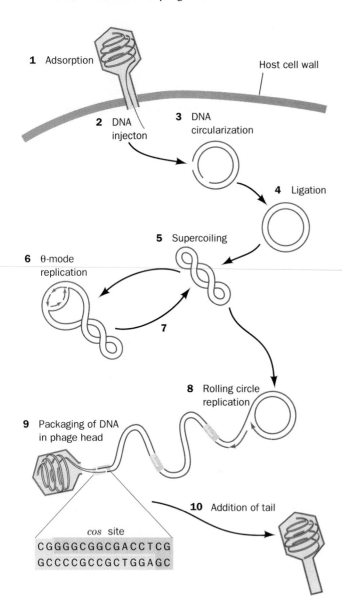

1 Adsorption

Host cell wall

2 DNA
injecton

3 DNA
circularization

4 Ligation

5 Supercoiling

6 θ-mode
replication

7

8 Rolling circle
replication

9 Packaging of DNA
in phage head

10 Addition of tail

cos site

CGGGGCGGCGACCTCG
GCCCCGCCGCTGGAGC

Figure 32-26
DNA replication in the lytic mode of bacteriophage λ. The
phage particle adsorbs to the host cell **(1)** and injects its
linear duplex DNA chromosome **(2)**. The DNA circularizes by
base pairing at its complementary single-stranded ends **(3)**,
and the resulting nicked circle is covalently closed **(4)** and
supercoiled **(5)** by the sequential actions of host DNA ligase
and host DNA gyrase. DNA replication commences
according to both the bidirectional θ mode **(6** and **7)** and the
rolling circle mode **(8)** but in the later stages of infection
occurs exclusively by the rolling circle mode. Here blue
arrows indicate the most recently synthesized DNA at the
replication forks and the arrowheads represent the 3′ ends
of the growing DNA chains. The concatemeric DNA
produced by the rolling circle mode is specifically cleaved at
its *cos* sites (*shaded boxes*) and is packaged into phage
heads **(9)**. The addition of tails **(10)** completes the assembly
of the mature phage particles, which are each capable of
initiating a new round of infection. [After Furth, M. E. and
Wickner, S. H., *in* Hendrix, R. W., Roberts, J. W., Stahl,
F. W., and Weisberg, R. A. (Eds.), *Lambda II, p.* 146, Cold
Spring Harbor Laboratory (1983).]

at its *cos* (for *co*hesive-end *s*ite) site to yield the linear
duplex DNA with complementary single-stranded ends
that are contained by mature phage particles. The staggered double-stranded scission is made by the so-called
terminase, which is a complex of the phage proteins
gp*A* and **gp***Nu1*.

Phage λ is replicated by the host DNA replication
machinery (Sections 31-1 and 2) with the participation
of only two phage proteins, **gp***O* and **gp***P*. gp*O* specifically binds to the phage DNA *ori* region, whereas gp*P*
interacts with both gp*O* and the DnaB protein of the
host primosome. gp*O* and gp*P*, it is thought, act analogously to host DnaA and DnaC proteins, which are required for the initiation of replication of *E. coli* DNA
(Section 31-3C) but not of λ DNA. Evidently, gp*O* and
gp*P* function to recognize the λ*ori* site.

B. Virus Assembly

The mature λ phage head contains two major proteins: **gp***E* (38 kD), which forms its polyhedral shell, and
gp*D* (12 kD), which "decorates" its surface. Electron
microscopy indicates that both these proteins are arranged in a *T* = 7 icosadeltahedron and are therefore
each present in 420 copies/phage. The λ head also con-
tain four major proteins, **gp***B*, **gp***C*, **gp***FII*, and **gp***W*,
in from 6 to 15 copies each, which form a cylindical
structure that attaches the tail to the head. This connector occurs at one of the head's fivefold vertices and
thereby breaks its icosahedral symmetry. The tail is a
tubular entity that consists of 32 stacked hexagonal
rings of **gp***V* (31 kD) for a total of 192 subunits. The tail
begins with a complex adsorption organelle composed
of 5 different proteins, **gp***G*, **gp***H*, **gp***L*, **gp***M*, and **gp***J*,
and ends with an assembly of **gp***U* and **gp***Z* (Fig. 32-21).

*The study of complex virus assembly has been motivated
by the conviction that it will provide a foundation for understanding the assembly of cellular organelles.* Phage assembly is studied through a procedure developed by
Robert Edgar and William Wood that combines genetics,
biochemistry, and electron microscopy. Conditionally
lethal mutations (either temperature sensitive mutants,
which appear normal at low temperatures but exhibit a
mutant phenotype at higher temperatures; or supres-
sor-sensitive *amber* mutants, Section 30-2E) are generated that, under nonpermissive conditions, block phage
assembly at various stages. This process results in the
accumulation of intermediate assemblies or side products that can be isolated and structurally characterized
through electron microscopy. The mutant protein can be
identified, through a process known as *in vitro* **complementation** (in analogy with *in vivo* genetic complementation; Section 27-1C), by mixing cell-free extracts
containing these structural intermediates with the corresponding normal protein to yield infectious phage
particles.

The assembly of bacteriophage λ occurs through a branched pathway in which the phage heads and tails are formed separately and then join to yield mature virions.

Phage Head Assembly

λ Phage head assembly occurs in five stages (Fig. 32-27, *right*):

1. Three phage proteins, gpB, gpC, and **gpNu3** (19 kD), together with two host proteins, **gpgroEL** (a 14-mer of identical 65-kD subunits) and **gpgroES** (a 6–8-mer of identical 15-kD subunits), interact to form an ill-characterized "initiator" that possibly consists of gpB, gpC, and gpNu3. This precursor of the mature phage head–tail connector apparently organizes the phage head's subsequent formation. *The host proteins facilitate the correct assembly of this connector precursor,* perhaps by preventing the formation of "improper" structures that might otherwise form as a consequence of the transient exposure of hydrophobic or charged surfaces during the assembly process. gpgroEL, which is one of *E. coli's* most abundant proteins, is nearly 50% homologous with a chloroplast protein that has been similarly implicated in the assembly of the oligomeric CO_2-fixing enzyme ribulose bisphosphate carboxylase of higher plants (Section 22-3A). Indeed, the presence of gpgroEL and gpgroES are both essential for the proper assembly of *E. coli* expressed ribulose bisphosphate carboxylase. gpgroEL-like proteins have also been identified in numerous bacteria and in the mitochondria of a wide variety of eukaryotes, thereby suggesting that these proteins are ubiquitous mediators of oligomeric protein assembly.

2. gpE and gpNu3 are thought to associate to form a structure called an immature **prohead.** If gpB, gpgroES, or gpgroEL are defective or absent, some gpE assembles into spiral or tubular structures, which suggests that the missing proteins guide the formation of a proper shell. The absence of gpNu3 results in the formation of but a few shells that contain only gpE. gpNu3 evidently facilitates proper shell construction and promotes the association of gpE with gpB and gpC.

Figure 32-27

The assembly of bacteriophage λ. The heads and tails are assembled in separate pathways before joining to form the mature phage particle. Within each pathway the order of the various reactions is obligatory for proper assembly to occur. gpE, gpNu3, gpD, and gpV are highlighted in red boxes to indicate that relatively large numbers of these proteins are required for phage assembly. [After Georgopoulos, C., Tilly, K., and Casjens, S.; *and* Katsura, I., *in* Hendrix, R. W., Roberts, J. W., Stahl, F. W., and Weisberg, R. A. (Eds.), *Lambda II, pp.* 288, 336 Cold Spring Harbor Laboratories (1983).]

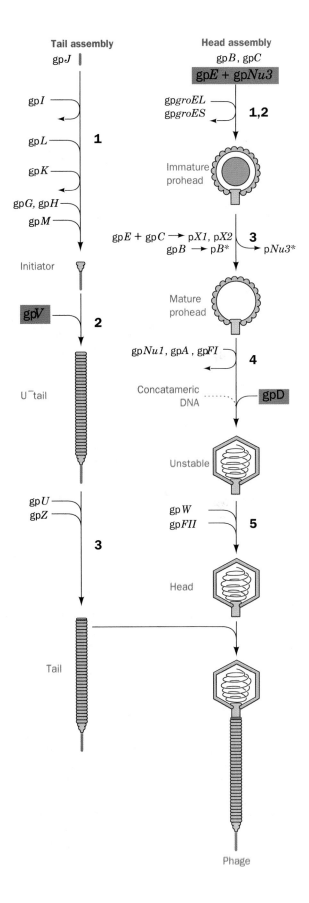

3. In the formation of the mature prohead, ~75% of the gp*B* (61 kD) is cleaved to form **gpB*** (56 kD); the gp*Nu3* is degraded and lost from the structure; and each gp*C* participates in a fusion–cleavage reaction with a gp*E* to form the hybrid proteins **pX1** and **pX2** (p for protein). This maturation process, which involves only phage gene products that are part of the immature prohead, requires that all of the prohead components be present and functional; that is, that the immature prohead be correctly assembled to start with. The enzyme(s) that catalyze this process have, nevertheless, not been identified.

4. The concatemeric viral DNA is packaged in the phage head and cleaved by mechanisms discussed below. During this process, the capsid proteins undergo a conformational change that results in an expansion of the phage head to twice its original volume (a process that occurs in $4M$ urea in the absence of DNA). gp*D* then adds to the capsid thereby partially stabilizing its expanded structure.

5. In the final stage of phage head assembly, gp*W* and gp*FII* add in that order to stabilize the head and form the tail-binding site.

These stages of phage head assembly, as well as some of their component reactions, must proceed in an obligatory order for proper assembly to occur. Of particular interest is that *the components of the mature phage head are not entirely self-assembling as are, for example, TMV (Section 32-1B) and ribosomes (Section 30-3A).* Rather, the *E. coli* proteins gp*groEL* and gp*groES*, as we saw, facilitate head–tail connector assembly. Moreover, gp*Nu3*, which occurs in ~200 copies inside the immature prohead but is absent from the mature prohead, evidently acts as a "scaffolding" protein that organizes gp*E* to form a properly assembled phage head. Finally, *since phage assembly involves several proteolytic reactions, it must also be considered to occur via enzyme-directed processes.*

DNA Is Tightly Packed in the Phage Head

An intriguing question of λ phage assembly is How does a 55 nm in diameter phage head package a 16,500 nm long and stiff DNA molecule? Two models have been proposed:

1. Electron microscopy of gently disrupted phages and X-ray scattering from phage solutions both suggest that the DNA is tightly wound in a spool-like structure (Fig. 32-28a). Since the DNA linearly enters the phage prohead through the head–tail connector (see below), it has been proposed that its stiffness would cause it to first coil against the inner wall of the rigid protein shell and then to wind concentrically inward, much like a spool of twine.

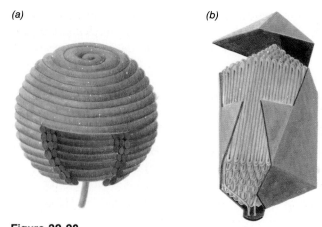

(a)   *(b)*

Figure 32-28
Models for the packing of double-stranded DNA inside a phage head: (*a*) The concentric shell model in which the DNA is wound inward like a spool of twine about the phage's long axis. [After Harrison, S. C., *J. Mol. Biol.* **171**, 579 (1983).] (*b*) The spiral-fold model in which the DNA strands run parallel to the phage's long axis with sharp 180° bends at the ends of the capsid. The folds themselves are radially arranged about the phage's long axis in spirally organized shells. [After Black, L. W., Newcomb, W. W., Boring, J. W., and Brown, J. C., *Proc. Natl. Acad. Sci.* **82**, 7963 (1985).]

2. Ion etching of phages (a process in which frozen phages are progressively worn away by bombardment with a beam of Ar$^+$ ions) in which only one end of the phage DNA is radioactively labeled indicates that the first DNA to enter the prohead is the most shielded from the ion beam. This observation suggests that the leading DNA segment is condensed in the center of the capsid thereby supporting the "spiral-fold" model of DNA packaging (Fig. 32-28b).

In both models, the DNA's detailed winding path varies randomly from particle to particle as is indicated by the observation that packaged DNA can be cross-linked to the capsid along its entire length. Both DNA packing models also predict that the injection of phage DNA into the host bacterium proceeds by a reversal of the packaging process.

DNA Is "Pumped" into the Phage Head by an ATP-Driven Process

The packaging of λ DNA begins by the recognition of a free cohesive end on the concatemeric DNA by the terminase (gp*A* + gp*Nu1*). The resulting complex then binds to the prohead so as to introduce the DNA into it through a 20 Å in diameter orifice in its head–tail connector. The "left" end of the DNA chromosome enters the prohead first as is indicated by the observation that only this end of the chromosome is packaged by an *in vitro* system when λ DNA restriction fragments are used.

The packing of DNA inside a phage head must be an enthalpically as well as entropically unfavorable process because of DNA's stiffness and its intramolecular charge repulsions. The observation that DNA packaging requires the presence of ATP therefore strongly suggests that DNA is actively "pumped" into the phage head by an ATP-driven process. In a particularly intriguing model of a DNA pump, for which there is no proof, ATP hydrolysis drives the rotation of the connector with respect to the prohead so as to screw the helically grooved DNA into the phage head much like a threaded rod in a nut. The injection of λ DNA into a host bacterium by a mature phage is presumably a spontaneous process that, once it has been triggered, is driven by the energy stored in the compacted DNA.

The final step in the DNA packaging process is the recognition and cleavage of the next *cos* site (Fig. 28-26) on the concatemeric DNA by terminase, possibly with the participation of **gpFI**. Phage λ therefore contains a unique segment of DNA (in contrast to some phages in which the amount of DNA packaged is limited by a "headful" mechanism that results in their containing somewhat more DNA than an entire chromosome). Indeed, the λ packaging system will efficiently package a DNA that is 75 to 105% the length of the wild-type λ DNA so long as it is flanked by *cos* sites (the central third of the phage DNA, which encodes the dispensable accessory genes, can be replaced by other sequences thereby making phage λ a useful cloning vector; Section 28-8A).

Tail Assembly

Tail assembly, which occurs independently of head assembly, proceeds, as a comparison of Figs. 32-21 and 32-27 indicates, from the tail fiber towards the head-binding end. This strictly ordered series of reactions can be considered to have three stages (Fig. 32-27, *left*):

1. The formation of the "initiator," which ultimately becomes the adsorption organelle, requires the sequential actions on gp*J* (the tail fiber protein) of the products of phage genes *I, L, K, G, H,* and *M,* respectively. Of these, only **gp*I*** and **gp*K*** are not components of the mature tail.

2. In this stage, the initiator forms the nucleus for the polymerization of gp*V,* the major tail protein, to form a stack of 32 hexameric rings. The length of this stack is thought to be regulated by gp*H* which, the available evidence suggests, becomes extended along the length of the growing tail and somehow limits its growth. λ Tail length is apparently specified in much the same way that the helical length of TMV is governed (Section 32-1B), although in TMV, the regulating template is an RNA molecule rather than a protein.

3. In the termination and maturation stage of tail assembly, gp*U* attaches to the growing tail thereby preventing its further elongation. The resultant immature tail has the same shape as the mature tail and can attach to the head. In order to form an infectious phage particle, however, the immature tail must be activated by the action of gp*Z* before joining the head.

The completed tail then spontaneously attaches to a mature phage head to form an infectious λ phage particle (Fig. 32-27, *bottom*).

The Assembly of Other Double-Stranded DNA Phages Resembles That of λ

The assembly of several other double-stranded DNA bacteriophages have been studied in detail, notably those of **coliphages T4, T7,** and the lambdoid (λ-like) phage **P22** (which grows on *Salmonella typhimurium*). All of them are formed in assembly processes that closely resemble that of phage λ. For example, their head assembly processes proceed in obligatory reaction sequences through an initiation stage, the scaffolded assembly of a prohead, an ATP-driven DNA packaging process in which the DNA assumes a tightly packed conformation and the prohead undergoes an expansion, and a final stabilization. The mature phages then form by the attachment of separately assembled tails to the completed and DNA-filled heads.

C. The Lysogenic Mode

Lysogeny is established by the integration of viral DNA into the host chromosome accompanied by the shutdown of all lytic gene expression. With phage λ, integration takes place through a **site-specific recombination** process that differs from general recombination (Section 31-6A) in that it only occurs between the chromosomal sites designated *att*P on the phage and *att*B on the bacterial host (Fig. 32-29). These two *att*achment sites have a 15 bp homology (Fig. 32-30) so that they can be represented as having the sequences POP' for *att*P and BOB' for *att*B where O denotes their common sequence. Phage integration occurs through a process that yields the inserted phage chromosome flanked by the sequence BOP' on the "left" (the *att*L site) and POB' on the "right" (the *att*R site; Fig. 32-29). The nature of the cross-over site was determined through the use of ^{32}P-labeled bacterial DNA and unlabeled phage DNA. The cross-over site occurs at a unique position on each strand that is displaced with respect to its complementary strand so as to form a staggered recombination joint (Fig. 32-30).

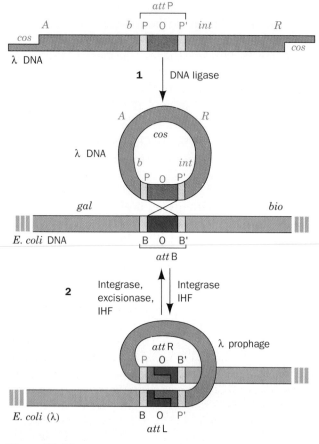

Figure 32-29
A schematic diagram showing: **(1)** the circularization of the linear phage λ DNA through base pairing between its complementary ends to form the *cos* site; and **(2)** the integration/excision of this DNA into/from the *E. coli* chromosome through site-specific recombination between the phage *att*P and host *att*B sites. The darker colored regions in the *att* sites represent the homologous 15 bp cross-over sequences (O), whereas the lighter colored regions symbolize the unique sequences of bacterial (B and B′) and phage (P and P′) origin. [After Landy, A. and Weisberg, R. A., *In* Hendrix, R. W., Roberts, J. W., Stahl, F. W., and Weisberg, R. A. (Eds.), *Lambda II, p.* 212, Cold Spring Harbor Laboratory (1983).]

Integrase Mediates λ DNA Integration whereas Excisionase Is Additionally Required for λ DNA Excision

Phage integration is mediated by a phage-specific **integrase**, the λ*int* gene product, acting in concert with a host protein named **integration host factor (IHF)**. Integrase, which specifically binds the region common to the cross-over sites (O), acts *in vitro* as a Type I topoisomerase (which nicks and reseals only one strand of double helical DNA; Section 28-5C) and has been shown to resolve synthetic Holliday structures (cross-over structures; Section 31-6A). It therefore seems likely that the breaking and resealing of DNA strands, which constitutes the cross-over event, is carried out by integrase. IHF has no demonstrable endonuclease or topoisomer-

ase activity but specifically binds to DNAs bearing various *att* sequences.

Since viral integration is not an energy-consuming process, why is phage integration not readily reversible? The answer is that the prophage excision requires the participation of an **excisionase**, the λ*xis* gene product, in concert with integrase and IHF. Apparently the λ recombination system has an inherent asymmetry that ensures the kinetic stability of the lysogenic integration product. The mechanism by which excisionase reverses the integration process in unknown although it has been shown that this protein specifically binds to POB′.

The Relative Levels of Cro Protein and cI Repressor Determine the λ Phage Life Cycle

The establishment of lysogeny in phage λ is triggered by high concentrations of **gpcII** *(see below).* This early gene product stimulates "leftward" transcription from two promoters, p_I (I for *i*ntegrase) and p_{RE}(RE for *r*epressor *e*stablishment; Fig. 32-31a):

1. Transcription initiated from p_I, which is located within the *xis* gene, results in the production of inte-

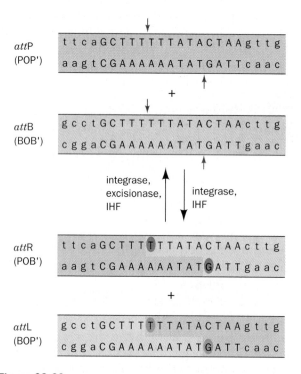

Figure 32-30
The site-specific recombination process that inserts/excises phage λ DNA into/from the chromosome of its *E. coli* host. Exchange occurs between the phage *att*P site (*blue*) and the bacterial *att*B site (*red*), and the prophage *att*L and *att*R sites. The strand breaks occur at the approximate positions indicated by the short arrows. The sources of the more darkly shaded bases in *att*R and *att*L are uncertain. The upper case letters represent bases in the O region common to the phage and bacterial DNA's, whereas lower case letters symbolize bases in the flanking B, B′, P, and P′ sites.

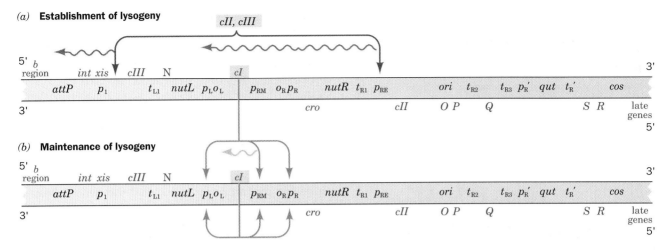

Figure 32-31
Gene expression in (*a*) the establishment and (*b*) the maintenance of lysogeny by bacteriophage λ. The symbols used are described in the legend of Fig. 32-24. [After Arber, W., *in* Hendrix, R. W., Roberts, J. W., Stahl, F. W., and Weisberg, R. A. (Eds.), *Lambda II, p.* 389, Cold Spring Harbor Laboratory (1983).]

grase but not excisionase. λ DNA is consequently integrated into the host chromosome to form the prophage.

2. The transcript initiated from p_{RE} encodes the *cI* gene whose product is called the λ or **cI repressor**. The λ repressor, as does Cro protein (Section 32-3A), binds to the o_L and o_R operators thereby blocking transcription from p_L and p_R, respectively (note that these operators are upstream from their corresponding promoters rather than downstream as in the *lac* operon; Fig. 29-3). *Both repressors therefore act to shut down the synthesis of early gene products, including Cro protein and gpcII.*

gpcII is metabolically unstable (see below) so that *cI* transcription from p_{RE} soon ceases. λ Repressor bound at o_R, but not Cro protein, however, stimulates "leftward" transcription of *cI* from p_{RM} (RM for repressor *maintenance*; Fig. 32-31*b*). In other words, *Cro protein represses all mRNA synthesis while λ repressor stimulates transcription of its own gene while repressing all other mRNA synthesis. This conceptually simple difference between the actions of λ repressor and Cro protein forms the basis of a genetic switch that stably maintains phage λ in either the lytic or the lysogenic state.* The molecular mechanism of this switch is described in Section 32-3D. In the following subsection we discuss how this switch is "thrown" from one state to another. You should recognize, however, that, *once the switch is thrown in favor of the lytic cycle; that is, when Cro protein occupies o_L and o_R, the phage is irrevocably committed to at least one generation of lytic growth.*

gpcII Is Inactivated When Phage Multiplicity Is High or Nutritional Conditions Are Poor

The reason why a high gpcII concentration is required to establish lysogeny is that this early gene product can stimulate transcription from p_I and p_{RE} only when it is in oligomeric form. This phenomenon accounts for the observation that lysogeny is induced when the **multiplicity of infection** (ratio of infecting phages to bacteria) is large (≥ 10) since this gene dosage effect results in gpcII being synthesized at a high rate.

gpcII is metabolically unstable because it is preferentially proteolyzed by host proteins, notably **gp*hflA*** and **gp*hflB***. However, gpcIII somehow protects gpcII from the action of gp*hflA*, which is why its presence enhances lysogenation (Fig. 32-31*a*). The activity of gp*hflA* is dependent on the host cAMP-activated catabolite repression system (Section 29-3C) as is indicated by the observation that *E. coli* mutants defective in this system lysogenize with less than normal frequency. Yet, if these mutant strains are also *hflA⁻*, they lysogenize with greater than normal frequency. Apparently the *E. coli* catabolite repression system, which is known to regulate the transcription of many bacterial genes, controls *hflA* activity, perhaps by directly repressing this protein's synthesis at high cAMP concentrations. *This explains why poor host nutrition, which results in elevated cAMP concentrations, stimulates lysogenation.*

Once a prophage has been integrated in the host chromosome, lysogeny is stably maintained from generation to generation by λ repressor. This is because λ repressor stimulates its own synthesis at a rate sufficient to maintain lysogeny in the progeny while repressing the transcription of all other phage genes. In fact, *λ repressor is synthesized in sufficient excess to also repress transcription from superinfecting λ phage thereby accounting for the phenomenon of immunity.* We shall see below how induction occurs.

D. Mechanism of the λ Switch

The lysogenic cycle is a highly stable mode of phage λ replication; under normal conditions lysogens spontaneously induce only about once per 10^5 cell divisions. Yet, transient exposure to inducing conditions triggers lytic growth in almost every cell of a lysogenic bacterial culture. In this section, we consider how this genetic switch, whose mechanism was largely elucidated by Mark Ptashne, can so tightly repress lytic growth and yet remain poised to efficiently turn it on.

o_R Consists of Three Homologous Palindromic Subsites

Both of the operators to which λ repressor and Cro protein bind, o_L and o_R, consist of three subsites (Fig. 32-32). These are designated o_{L1}, o_{L2}, and o_{L3} for o_L, and o_{R1}, o_{R2}, and o_{R3} for o_R. Each of these subsites consists of a homologous 17 bp segment that has approximate palindromic symmetry. Nevertheless, only the elements of o_R form components of the λ switch.

λ Repressor and Cro Protein Structurally Resemble Other Repressors

λ Repressor binds to DNA as a dimer so that its twofold symmetry matches those of the operator subsites to which it binds. The monomer's 236-residue polypeptide chain is folded into two roughly equal sized domains connected by an ~30-residue segment that is readily cleaved by proteolytic enzymes. The isolated N-terminal domains retain their ability to specifically bind to operators (although with only one half of the binding energy of the intact repressor), but cannot dimerize. The C-terminal domains can still dimerize but lack the capacity to bind DNA. Evidently, *repressor's N-terminal domain binds operator whereas its C-terminal domain provides the contacts for dimer formation.*

Although the λ repressor has not been crystallized, its N-terminal domain comprising residues 1 to 92, as excised by treatment with the papaya protease **papain,** does crystallize. The X-ray structure of this protein, both alone and in complex with a 20 bp DNA containing the o_{L1} sequence, has been determined by Carl Pabo. The N-terminal domain crystallizes as a symmetric dimer with each subunit containing an N-terminal arm and five α helices (Fig. 32-33). Two of these helices, α2 and α3, form a helix–turn–helix motif, much like those in other repressors of known structure (Sections 29-3C and E). The α3 helix, the "recognition" helix, protrudes from the protein surface such that the two α3 helices of the dimeric protein fit into successive major grooves of the operator DNA. Similar associations are observed in the X-ray structures of the closely related **bacteriophage 434 repressor** N-terminal fragment in complex with a 20 bp DNA containing its operator sequence (Fig. 29-21).

Cro protein also forms dimers. In contrast to λ or 434 repressor, however, this 66-residue polypeptide forms but one domain that contains both its operator recognition site and its dimerization contacts. The X-ray structure of Cro, determined by Brian Matthews, reveals that this dimer likewise contains a pair of helix–turn–helix units arranged such that they can bind to DNA (Fig. 32-34). This model is consistent with Ponzy Lu's NMR studies demonstrating that Cro only binds to one face of its operator and with the observation that Cro primarily protects operator subsites from chemical modification in two successive major grooves on one side of the DNA. The sequence-specific binding predicted by this model is further supported by Robert Sauer's genetic studies indicating that mutant varieties of Cro, in which the proposed DNA-contacting residues have been changed, are defective in operator binding. Finally, this model closely resembles the X-ray structure of the related **phage 434 Cro protein** in complex with a 20 bp DNA containing its operator sequence (Fig. 29-22).

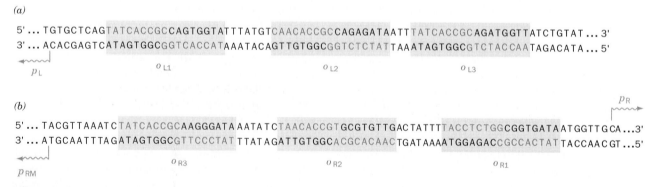

Figure 32-32
The base sequences of (*a*) the o_L and (*b*) the o_R regions of the phage λ chromosome. Each of these operators consist of three homologous 17 bp subsites separated by short AT-rich spacers. Each subsite has approximate palindromic symmetry as is demonstrated by the comparison of the two sets of red letters in each subsite. The wiggly arrows mark the transcriptional start sites and directions at the indicated promoters.

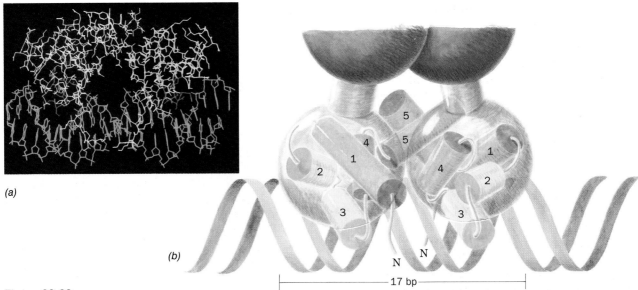

(a)

(b)

Figure 32-33

The X-ray structure of a dimer of λ repressor N-terminal domains in complex with B-DNA. (a) Computer graphics representation of the complex in which the DNA is blue, the two repressor N-terminal domains are yellow and violet, and their "recognition" helices are red. Note that the protein's N-terminal arms wrap around the DNA. This accounts for the observation that the G residues in the major groove on the repressor–operator complex's "back side" are protected from methylation only when these N-teminal arms are intact. [Courtesy of Carl Pabo, The Johns Hopkins University.] (b)

An interpretive drawing indicating how contacts between the repressor's C-terminal domains (not part of the X-ray structure) maintain the intact protein's dimeric character. The λ repressor binds to the 17 bp operator subsites of o_L and o_R as symmetric dimers with the N-terminal domain of each subunit specifically binding to a half-subsite. Note how the α3 "recognition" helices of the symmetry related α2–α3 helix–turn–helix units (*yellow*) fit into successive turns of the DNA's major groove. [After Ptashne, M., *A Genetic Switch, p.* 38, Cell Press (1986).]

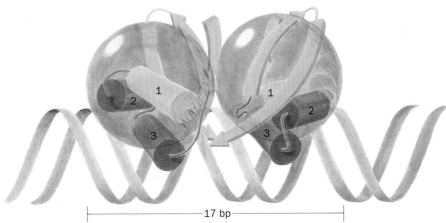

Figure 32-34

The X-ray structure of the Cro protein dimer shown in its presumed complex with B-DNA. Note that the λ repressor (Fig. 32-33), although otherwise dissimilar, contains helix–turn–helix units that also bind in successive turns of the DNA's major groove. [After Ptashne, M., *A Genetic Switch, p.* 40, Cell Press (1986).]

Repressor Stimulates Its Own Synthesis While Repressing All Other λ Genes

Chemical and nuclease protection experiments have indicated that λ repressor has the following order of intrinsic affinities for the subsites of o_R (Fig. 32-35):

$$o_{R1} > o_{R2} > o_{R3}$$

Despite this order, o_{R1} and o_{R2} are filled nearly together. This is because *λ repressor bound at o_{R1} cooperatively binds repressor at o_{R2} through associations between their C-terminal domains (Fig. 32-35c)*. o_{R1} and o_{R2} are therefore both occupied at low λ repressor concentrations,

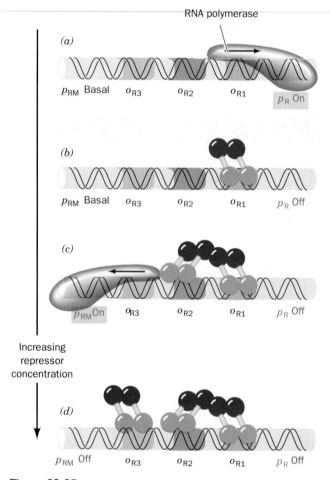

Figure 32-35
The binding of λ repressor to the three subsites of o_R. (*a*) In the absence of repressor, RNA polymerase initiates transcription at a high level from p_R (*right*) and at a basal level from p_{RM}. (*b*) Repressor has ~10 times higher affinity for o_{R1} than it does for o_{R2} or o_{R3}. Repressor dimer therefore first binds to o_{R1} so as to block transcription from p_R. (*c*) A second repressor dimer binds to o_{R2} at only slightly higher repressor concentrations due to specific binding between the C-terminal domains of neighboring repressors. In doing so, it stimulates RNA polymerase to initiate transcription from p_{RM} at a high level (*left*). (*d*) At high repressor concentrations, repressor binds to o_{R3} so as to block transcription from p_{RM}. [After Ptashne, M., *A Genetic Switch*, p. 23, Cell Press (1986).]

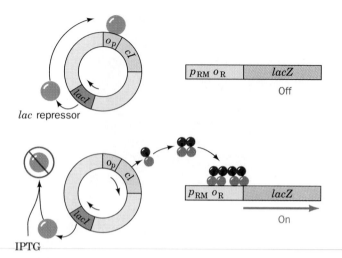

Figure 32-36
The genetic system used to study the effect of λ repressor on p_{RM}. The bacterium contains two hybrid operons. The first (*left*) is a plasmid bearing the *lac* operator–promoter (*Op*) fused to the *λcI* gene so as to provide a source of repressor. The *lacI* gene, which encodes *lac* repressor, is also incorporated in the plasmid so that the level of λ repressor in the bacterium may be controlled by the concentration of the *lac* inducer IPTG. The second operon (*right*) is carried on a prophage that contains the promoter p_{RM} fused to the *lacZ* gene. The level of β-galactosidase (gp*lacZ*) in these cells therefore reflects the activity of p_{RM}. In similar experiments, the *cro* gene was substituted for *λcI* and/or p_{RM} was replaced by p_R. [After Ptashne, M., *A Genetic Switch*, p. 89, Cell Press (1986).)]

whereas o_{R3} only becomes occupied at higher repressor concentrations.

The binding of λ repressor to o_R, as we previously mentioned, abolishes transcription from p_R and stimulates it from p_{RM} (Fig. 32-35c). At high concentrations of λ repressor, however, transcription from p_{RM} is also repressed (Fig. 32-35d). These phenomena were clearly demonstrated through the construction of a series of hybrid operons that permit the effect of λ repressor on a promoter to be studied in a controlled manner. The system has two elements (Fig. 32-36):

1. A plasmid bearing the *lacI* gene (which encodes *lac* repressor; Section 29-1A) and the *lac* operator–promoter sequence fused to the *cI* gene. This construct permits the amount of λ repressor produced to be directly controlled by varying the concentration of the *lac* inducer IPTG (Section 29-1A).

2. A prophage containing o_R and either p_{RM}, as Fig. 32-36 indicates, or p_R, fused to the *lacZ* gene. The amount of the *lacZ* gene product, β-galactosidase, produced, which can be readily assayed, reflects the activity of p_{RM} (or p_R).

The manipulation of these systems demonstrated that at intermediate λ repressor concentrations (when o_{R1} and o_{R2} are occupied), transcription from p_R is indeed re-

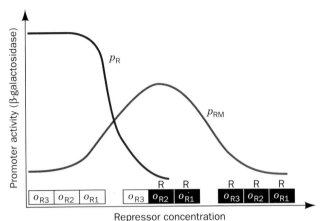

Figure 32-37
The response of p_{RM} and p_R to the λ repressor level. The p_{RM} curve was derived using the system diagrammed in Fig. 32-36, whereas the p_R curve was obtained using a similar system but with p_R rather than p_{RM} fused to *lacZ*. The amount of λ repressor that maximally stimulates p_{RM} is approximately that which occurs in a λ lysogen. At least fivefold more repressor is required to half-maximally repress p_{RM}. The boxes indicate the states of each o_R subsite at the various repressor concentrations; black represents repressor occupancy. [After Ptashne, M., *A Genetic Switch*, p. 90, Cell Press (1986).]

pressed, whereas that from p_{RM} is stimulated (Fig. 32-37). Transcription from p_{RM} only becomes repressed at high levels of λ repressor (when o_{R3} is also occupied). The stimulation of transcription from p_{RM} is abolished by mutations in o_{R2} that prevent repressor binding, whereas its repression at high repressor concentrations is relieved by mutations in o_{R3}. Thus, *occupancy of o_{R2} by*

λ repressor stimulates transcription from p_{RM}, whereas occupancy of o_{R3} prevents it (Fig. 32-35c and d). By the same token, occupancy of o_{R1} and/or o_{R2} prevents transcription from p_R. λ Repressor, in this way, prevents the synthesis of all phage gene products but itself. Yet, at high repressor concentrations, its synthesis is also repressed thereby maintaining the repressor concentration within reasonable limits.

What is the basis of λ repressor's remarkable property of inhibiting transcription from one promoter while stimulating it from another? Knowledge of the sizes and shapes of repressor and RNA polymerase, as well as their positions on the DNA as demonstrated by chemical protection experiments, indicate that repressor at o_{R2} and RNA polymerase at p_{RM} are in contact (Fig. 32-38). Evidently, *repressor stimulates RNA polymerase activity through their cooperative binding to DNA*. This model was corroborated by the analysis of repressor mutants that bind normally (or nearly so) to operators but fail to stimulate the binding of RNA polymerase: All of the mutated residues occur either in helix $\alpha2$ or in the link connecting it to helix $\alpha3$ and lie on the surface of the protein that is thought to face the RNA polymerase-binding site (Fig. 32-38).

Cro Protein Binding to o_R Represses All λ Genes

Cro protein binds to the subsites of o_R in an order opposite to that of λ repressor (Fig. 32-39):

$$o_{R3} > o_{R2} \approx o_{R1}$$

This binding is noncooperative. Through experiments similar to that diagrammed in Fig. 32-36, but with *cro* in place of *cI*, the binding of Cro protein to o_{R3} was shown

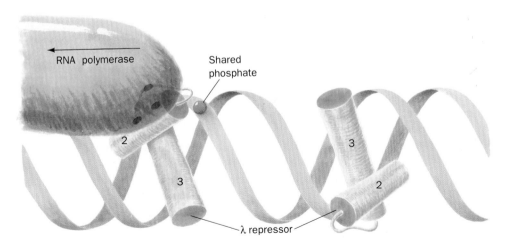

Figure 32-38
λ Repressor bound at o_{R2} is proposed to stimulate transcription at p_{RM} through a specific association with RNA polymerase that helps the polymerase bind to the promoter. This model is supported by the locations of the altered residues (*blue dots*) in three mutant repressors that bind

normaly to o_{R2} but fail to stimulate transcription at p_{RM}. The relative positions of repressor and RNA polymerase are established by the location of a phosphate group (*orange dot*) whose ethylation interferes with the binding of both proteins to the DNA. For the sake of clarity, only the $\alpha_2-\alpha_3$ helix–turn–helix units of the repressor dimer are shown.

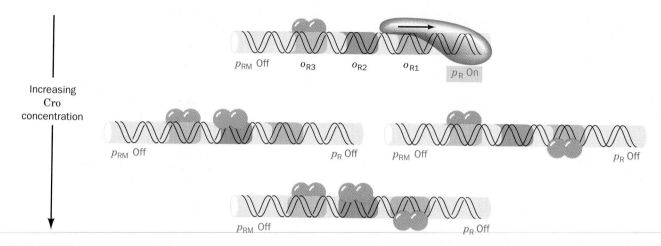

Figure 32-39
The binding of Cro protein to the three o_R subsites. o_{R3} binds Cro ~ 10 times more tightly than does o_{R1} or o_{R2}. Cro dimer therefore first binds to o_{R3}. A second dimer then binds to either o_{R1} or o_{R2} and in each case blocks transcription from p_R. At high Cro concentrations, all three operator subsites are occupied. Compare this binding sequence with that of λ repressor (Fig. 32-35). [After Ptashne, M., *A Genetic Switch*, p. 27, Cell Press (1986).]

to abolish transcription from p_{RM}. Additional Cro binding to o_{R2} and/or o_{R1} turns off transcription from p_R.

The SOS Response Induces the RecA-Mediated Cleavage of λ Repressor

A final piece of information allows us to understand the workings of the λ switch. *The lytic phase is induced by agents that damage host DNA or inhibit its replication.* These are just the conditions that induce *E. coli*'s SOS response: The resulting fragments of single-stranded DNA activate RecA protein to stimulate the self-cleavage of LexA protein, the SOS gene repressor, at an Ala-Gly bond (Section 31-5D). *Activated RecA protein likewise stimulates the autocatalytic cleavage of λ repressor monomer's Ala 111-Gly 112 bond (which occurs in the polypeptide segment linking the repressor's two domains).* Repressor's ability to cooperatively bind to o_{R2} is thereby abolished (Fig. 32-40a and b; the C-terminal domains can still dimerize but they no longer link the DNA-binding N-terminal domains). The consequent reduction in concentration of intact free monomers shifts the monomer–dimer equilibrium such that the operator-bound dimers dissociate to form monomers, which are then cleaved through the influence of activated RecA before they can rebind to their target DNA.

In the absence of repressor at o_R, the λ early genes, including *cro*, are transcribed (Fig. 32-40c). As Cro accumulates, it first binds to o_{R3} so as to block even basal levels of λ repressor synthesis (Fig. 32-40d). Thus, *there being no mechanism for selectively inactivating Cro, the phage irreversibly enters the lytic mode: The λ switch, once thrown, cannot be reset.* The prophage is subsequently excised from the host chromosome by the inte-

grase and excisionase that are produced in the delayed early phase.

The λ Switch's Responsiveness to Conditions Arises from Cooperative Interactions among Its Components

The complexity of the above switch mechanism endows it with a sensitivity that is not possible in simpler systems. The degree of repression at p_R is a steep function of repressor concentration (Fig. 32-41, *right*): The repression of p_R in a lysogen is normally 99.7% complete but drops to one half this level upon inactivation of 90% of the repressor. This steep sigmoid binding curve arises from the much greater operator affinity of repressor dimers compared to monomers. This situation, in turn, results from the cooperative linking of the monomer–dimer equilibrium, the binding of dimer to operator, and the association of dimers bound at o_{R1} and o_{R2}. In contrast, a 99.7% repressed promoter controlled by a stably oligomeric repressor binding to a single operator site, such as occurs in the *lac* system, requires 99% repressor inactivation for 50% expression (Fig. 32-41, *left*). *The cooperativity of λ repressor oligomerization and multiple operator site binding are therefore responsible for the remarkable responsiveness of the λ switch to the health of its host.*

4. INFLUENZA VIRUS

Influenza is one of the few common infectious diseases that is poorly controlled by modern medicine. Its annual epidemics, one of which was recorded by Hip-

(a) **Lysogenic mode**

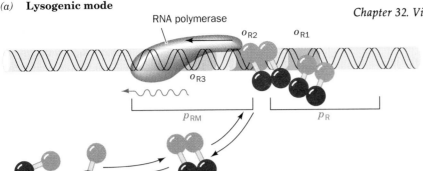

(b) **Induction (1)**

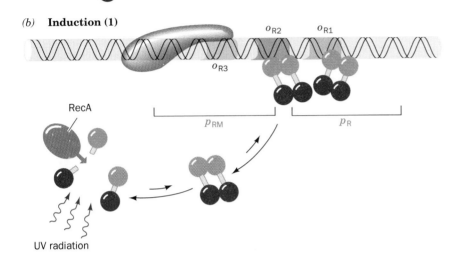

Figure 32-40

The λ switch. (a) In the lysogenic mode, two dimeric molecules of λ repressor cooperatively bind, through associations between their C-terminal domains, to o_{R1} and o_{R2}. This blocks host RNA polymerase from gaining access to p_R. However, the repressor bound to o_{R2} associates, through its N-terminal domain, with RNA polymerase at p_{RM} thereby inducing the transcription of *cl,* the λ repressor gene, from this promoter (*wiggly arrow*). (*b*) Damage to the host DNA, as caused, for example, by UV radiation, activates host RecA protein to stimulate the self-cleavage of λ repressor monomers at a specific Ala-Gly bond between their two domains. (*c*) The consequent degradation of repressor monomers shifts the monomer–dimer equilibrium so as to free the o_{R1} and o_{R2} subsites for binding by RNA polymerase. This results in the transcription of the early genes (Fig. 32-24), including *cro,* from p_R (*wiggly arrow*). (*d*) The Cro protein thus synthesized preferentially binds at o_{R3} so as to block further transcription of *cl* from p_{RM}. Lytic growth is thereby irreversibly induced.

(c) **Induction (2)**

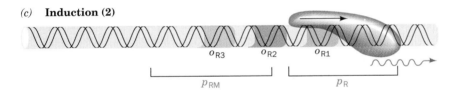

(d) **Early lytic growth**

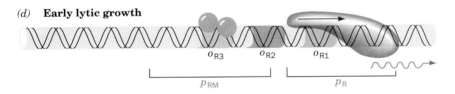

Figure 32-41

Theoretical repression curves for λp_R (*right*) and a simple repressor–operator system such as that of the *lac* operon (*left*). Note the greater sensitivity of the λ system to a change in repressor concentration. [After Johnson, A. D., Poteete, A. R., Lauer, G., Sauer, R. T., Ackers, G. K., and Ptashne, M., *Nature* **294**, 221 (1981).]

pocrates in 412 B.C., are occasionally punctuated by devastating pandemics. For example, the influenza pandemic of 1918, which killed over 20 million people and affected perhaps 100 times that number, was among the most lethal plagues ever recorded (in contrast, AIDS has killed < 100,000 people as of 1989). Since that time, there have been two other pandemics of lesser severity, the so-called Asian flu of 1957 and the Hong Kong flu of 1968. All of these pandemics were characterized by the appearance of a new strain of influenza virus to which the human population had little resistance and against which previously existing influenza virus vaccines were ineffective. Moreover, between pandemics, influenza virus undergoes a gradual antigenic variation that degrades the level of immunological resistance against renewed infection. What characteristics of the influenza virus permit it to evade man's immunological defenses in this manner? In this section we shall discuss this question and, in doing so, examine the structure and life cycle of the influenza virus.

A. Virus Structure and Life Cycle

Electron micrographs of influenza virus (Fig. 32-1*i*) reveal a nonuniform collection of spheroidal particles that are some 100 nm in diameter and whose surfaces are densely studded with radially projecting "spikes." The influenza virion, which grows by budding from the plasma membrane of an infected cell (Fig. 32-42), is an example of an enveloped virus. *Its outer envelope consists of a lipid bilayer of cellular origin that is pierced by*

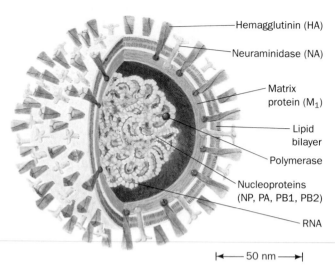

Figure 32-43
A cutaway diagram of the influenza virion. The HA and NA spikes are embedded in a lipid bilayer that forms the virion's outer envelope. Matrix protein, M_1, coats the underside of this membrane. The virion core contains the eight single-stranded RNA segments that comprise its genome in complex with the proteins NP, PA, PB1, and PB2 to form helical structures named nucleocapsids. [After Kaplan, M. M. and Webster, R. G., *Sci. Am.* **237**(6): 91 (1977). Copyright © 1977 by Scientific American, Inc.]

virally specified integral membrane glycoproteins, the "spikes." There are two types of these surface spikes (Fig. 32-43):

1. A rod-shaped spike composed of **hemagglutinin (HA)**, so-named because it causes erythrocytes to agglutinate (clump together). HA mediates influenza target cell recognition by specifically binding to cell surface receptors (glycophorin A molecules in erythrocytes; Section 11-3C) bearing terminal *N*-acetyl-neuraminic acid (sialic acid) residues. Each virion bears ~ 500 copies of HA.

2. A mushroom-shaped spike known as **neuraminidase (NA)**, which catalyzes the hydrolysis of the linkage joining a terminal sialic acid residue to a D-galactose or a D-galactosamine residue. NA probably facilitates the transport of the virus to and from the infection site by permitting its passage through mucin (mucus) and preventing viral self-aggregation. Each virion incorporates ~ 100 copies of NA.

Just beneath the viral membrane is a 6-nm thick protein shell composed of ~ 3000 copies of **matrix protein (M_1)**, the virion's most abundant protein.

The influenza virus genome is unusual in that it consists of 8 different-sized segments of single-stranded RNA. These RNA molecules are negative strands; that is, they are complementary to the viral mRNAs. In the viral core, these RNAs occur in complex with 4 different

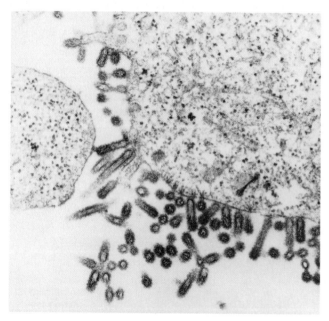

Figure 32-42
Electron micrograph of influenza viruses budding from infected chick embryo cells. [From Sanders, F. K., *The Growth of Viruses, p.* 15, Oxford University Press (1975).]

Table 32-2
The Influenza Virus Genome

RNA Segment	Length (nucleotides)	Polypeptides(s) Encoded
1	2341	PB2
2	2341	PB1
3	2233	PA
4	1778	HA
5	1565	NP
6	1413	NA
7	1027	M_1, M_2
8	890	NS_1, NS_2

Source: Lamb, R. A. and Choppin, P. W., *Annu. Rev. Biochem.* **52,** 473 (1983).

proteins: **nucleocapsid protein (NP),** which occurs in ~1000 copies, and 3 proteins, **PA, PB1,** and **PB2,** present in 30 to 60 copies each. The resulting **nucleocapsids** have the appearance of flexible rods.

The 8 viral RNAs, which vary in length from 890 to 2341 nucleotides, have all been sequenced. They code for the virus' 7 structural proteins (HA, NA, M_1, NP, PA, PB1, and PB2) and 3 nonstructural proteins that occur only in infected cells **(NS_1, NS_2, and M_2).** The sizes of the RNAs and the proteins they encode are listed in Table 32-2.

Virus Life Cycle

The influenza infection of a susceptible cell begins with the HA-mediated adsorption of the virus to specific cell-surface receptors. This is followed by uptake of the virus via an endocytotic mechanism (Section 11-4B) in which the viral and endosome membranes fuse through a process probably resembling that diagrammed in Fig. 11-44. The viral contents are thereby introduced into the cell. By some 20-min post-infection, the still intact nucleocapsids have been transported to the cell nucleus where they commence transcription of the viral RNAs **(vRNAs).** Cellular enzyme systems are incapable of mediating such RNA-directed RNA synthesis. Rather, it is carried out by a viral RNA transcriptase system that consists of the nucleocapsid proteins.

The transcription of the influenza virus genome is terminated if infected cells are treated with inhibitors of RNA polymerase II (which synthesizes cellular mRNA precursors; Section 29-2F) such as actinomycin D or α-amanitin. Yet, none of these agents affect the viral transcriptase's *in vitro* activity. The resolution of this seeming paradox is that *in vivo* viral mRNA synthesis is primed by newly synthesized cellular mRNA fragments consisting of a 7-methyl G cap (Section 29-4A) followed by a 10 to 13 nucleotide chain ending in A or G (Fig. 32-44, *top*). Cross-linking experiments, which indicate that PB2 and PB1, respectively, bind to the capped primer and the priming chain, suggest that PB2 functions to recognize the 5' cap whereas PB1 is the endonuclease that cleaves primers from the cellular mRNAs. The abundance of NP suggests that it has a structural role in the nucleocapsid. Thus, by a process of elimination, PA is thought to be the viral transcriptase component that catalyzes RNA chain elongation. Viral mRNAs, as do most mature cellular mRNAs, have poly(A) tails appended to their 3' ends (Section 30-4A).

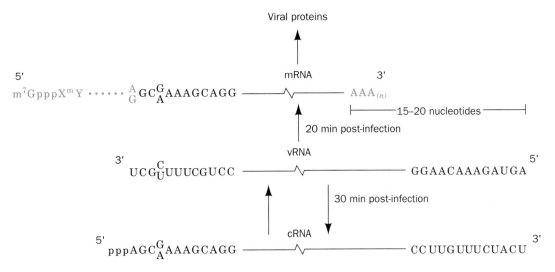

Figure 32-44
The biosynthesis of influenza vRNA, mRNA, and cRNA. The conserved nucleotides at the ends of the RNA segments are indicated. The viral mRNA's host-derived capped 5' head and 3' poly(A) tail are shown in color.
[After Lamb, R. A. and Choppin, P. W., *Annu. Rev. Biochem.* **52,** 490 (1983).]

The viral mRNAs lack 15 to 22 nucleotide segments that are complementary to the 5' ends of their parental vRNAs. They therefore cannot act as templates in vRNA replication. Rather, in an alternative transcription process that begins some 30-min post-infection, complete vRNA complements are synthesized. These so-called **cRNAs** (Fig. 32-44, *bottom*), whose synthesis does not require a primer, begin with pppA at their 5' ends and lack poly(A) tails. Hence cRNAs, unlike viral mRNAs, do not associate with polysomes in infected cells. The synthesis of cRNA, in contrast to that of viral mRNA, requires the continuing production of viral proteins. This observation suggests that the viral nonstructural proteins, NS_1, NS_2, and/or M_2, modify the viral transcriptase so as to render it primer independent thereby permitting it to fully transcribe the vRNAs. The cRNAs are the templates for vRNA synthesis in a process whose mechanism has not yet been elucidated.

The mechanism of influenza virus assembly is not well characterized. The viral spike glycoproteins, HA and NA, are ribosomally synthesized on the rough endoplasmic reticulum, further processed in the Golgi apparatus (Section 11-3F), and then transported, presumably in clathrin coated vesicles, to specific areas of the plasma membrane. There, they aggregate in sufficient numbers to exclude host proteins (Fig. 32-45a and b). The M_1 protein is thought to form a nucleocapsid-enclosing shell that binds to HA and NA on the inside of the plasma membrane (Fig. 32-45b). This binding process causes the entire assembly to bud from the cell surface thereby forming the mature virion (Fig. 32-45c). The complete infection cycle occupies ~ 8 to 12 h.

One of the mysteries of influenza virus assembly is how each virion acquires a complete set of the eight vRNAs. There is no evidence that the newly formed nucleocapsids are physically linked. On the contrary, in mixed infections with various influenza strains, the reassortment of their genomic segments occurs with high frequency. It has therefore been suggested that the nucleocapsids are randomly selected but that each virion contains sufficient numbers of vRNAs to ensure a reasonable probability that a given particle be infectious. This proposal is in agreement with the observation that aggregates of influenza virus have enhanced infectivity, a process that presumably occurs through the complementation of their vRNAs. Alternatively, the eight vRNAs may be selected by an ordered process, a hypothesis that is supported by the observation that mature viruses, but not infected cells, contain roughly equimolar amounts of the vRNAs.

B. Mechanism of Antigenic Variation

Influenza virus infects a wide variety of mammalian and avian species in addition to humans. Indeed, it is thought that migratory birds are the major vectors that

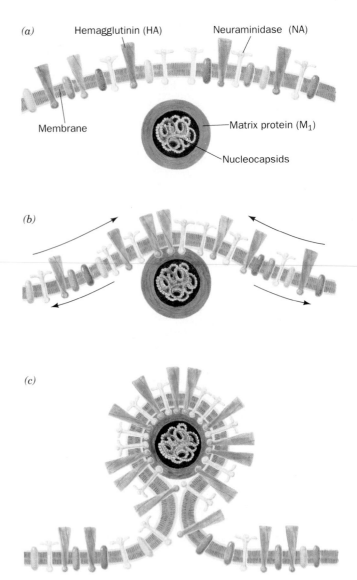

Figure 32-45
The budding of influenza virus from the host cell membrane. (a) The viral glycoproteins, HA and NA, are inserted into the plasma membrane of the host cell and the matrix protein, M_1, forms the nucleocapsid-containing shell. (b) The binding of the matrix protein to the cytoplasmic domains of HA and NA results in the aggregation of these glycoproteins so as to exclude host cell membrane proteins. (c) This binding process induces the membrane to envelop the matrix protein shell such that the mature virion buds from the host cell surface. [After Wiley, D. C., Wilson, I. A., and Skehel, J. J., *in* Jurnak, F. A. and McPherson, A. (Eds.), *Biological Macromolecules and Assemblies, Vol.* 1: *Virus Structures,* Wiley (1984).]

transport influenza viruses around the world. The species specificity of a particular viral strain presumably arises from the binding specificity of its HA for cell surface glycolipids. Influenza viruses are classified into three immunological types, A, B, and C, depending on the antigenic properties of their nucleoproteins and matrix proteins. The A virus has caused all of the major pandemics in humans and is the only influenza virus known to infect animals. It has therefore been more extensively investigated than the B and C viruses.

HA Residue Changes Are Responsible for Most of the Antigenic Variation in Influenza Viruses

HA, being the influenza virus' major surface protein, is largely responsible for stimulating the production of the antibodies that neutralize the virus. Consequently, the different influenza virus subtypes arise mainly through the variation of HA. Antigenic variation in NA, the virus' only other surface protein, also occurs but this has lesser immunological consequences.

Two-distinct mechanisms of antigenic variation have been observed in influenza-A viruses:

1. **Antigenic shift,** in which the gene encoding one HA species is replaced by an entirely new one. This change may or may not be accompanied by a replacement of NA. It is thought that these new viral strains arise from the reassortment of genes among animal and human flu viruses. *Antigenic shift is responsible for influenza pandemics because the human population's immunity against previously existing viral strains is ineffective against the newly generated strain.* Evidently, these viruses had retained the (unknown) genetic traits responsible for their virulence in humans.

2. **Antigenic drift,** which occurs through a succession of point mutations in the HA gene resulting in an accumulation of amino acid residue changes that attenuate the host's immunity. This process occurs in response to the selective pressure brought about by the buildup in the human population of immunity to the extant viral strains.

> In early 1976, at Fort Dix, New Jersey, there was an outbreak of an influenza strain that carried an HA type that occurs in swine flu virus. This viral subtype

is thought to have caused the great pandemic of 1918 (although influenza virus was not isolated until 1933, individuals who had contracted influenza during this pandemic have antibodies against swine flu virus in their serum). If this new strain was virulent, no one under the age of 50 at the time would have been immune to it. There was, consequently, grave concern that a deadly influenza pandemic would ensue. This situation led to a crash program in which well over a million people deemed to be at high risk (such as pregnant women and the elderly) were vaccinated against swine flu. Fortunately, the 1976 swine flu was not virulent; it did not spread beyond Fort Dix.

HA Is an Elongated Trimeric Trans-Membrane Glycoprotein

Influenza virus hemagglutinin plays a central role in both the viral infection process and in the immunological measures and countermeasures taken in the continuing biological contest between host and parasite. This has motivated considerable efforts to elucidate the structural basis of these properties. HA is a trimer of 550-residue identical subunits that is 19% carbohydrate by weight. The protein has three domains (Fig. 32-46):

1. A large hydrophilic, carbohydrate-containing domain that occupies the viral membrane's external surface and that contains its sialic acid-binding site.

2. A hydrophobic 24 to 28 residue membrane-spanning domain that is located near the polypeptide's C-terminus.

3. A hydrophilic domain that occurs on the membrane's inner side and that consists of the protein's 10 C-terminal residues.

HA is post-translationally cleaved by the excision of Arg 329 thereby yielding two chains, HA1 and HA2, that are linked by a disulfide bond. This cleavage, which does not affect HA's receptor-binding affinity, is required for

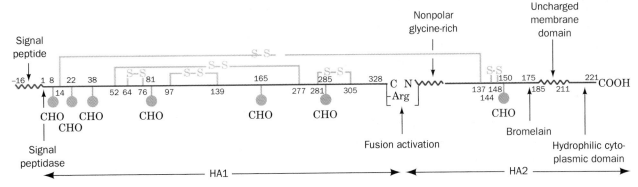

Figure 32-46
The 1968 Hong Kong influenza virus hemagglutinin amino acid sequence indicating its external domain (all of HA1 and HA2 through 185), its membrane anchoring domain (185–211 of HA2) and its cytoplasmic domain (212–221 of HA2).

The positions of the signal sequence directing the protein's insertion into the membrane, S—S bridges, carbohydrate (CHO) attachment sites, fusion-activation site, and bromelain cleavage site are given. [After Wilson, I. A., Skehel, J. J., and Wiley, D. C., *Nature* **289**, 367 (1981).]

fusion of the virus with the host cell and therefore activates viral infectivity.

HA can be removed from the virion by treatment with detergent but the resulting solubilized protein has not been made to crystallize. However, treatment of HA from a Hong Kong-type virus (influenza virus subtypes are named according to their site of discovery) with the pineapple protease **bromelain,** which cleaves the polypeptide just before the membrane-spanning segment, yields a water soluble protein that has been crystallized. X-ray analysis of these crystals by Don Wiley revealed an unusual structure (Fig. 32-47). The monomer consists

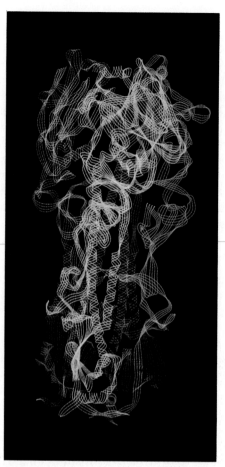

Figure 32-48
A ribbon diagram of the HA trimer. Each HA1 and HA2 chain is drawn in a different color. Compare this drawing with Fig. 32-47*a*. [Courtesy of Michael Carson, University of Alabama at Birmingham.]

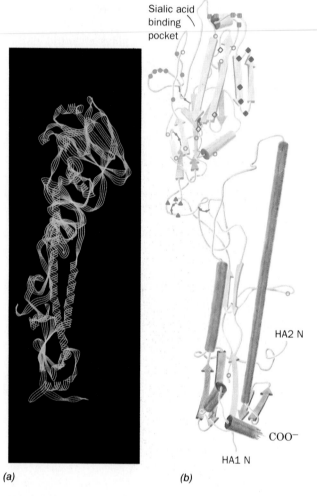

(a)　　　　　　　　(b)

Figure 32-47
The X-ray structure of the influenza hemagglutinin monomer. (*a*) The polypeptide backbone drawn as a ribbon. HA1 is green and HA2 is blue. [Courtesy of Michael Carson, University of Alabama at Birmingham.] (*b*) A cartoon diagram from a somewhat different point of view as part (*a*) in which HA2 is sketched in color. The pairs of small filled circles represent disulfide groups. The positions of the mutant residues at the four antigenic sites are indicated by filled circles, squares, triangles, and diamonds. Open symbols represent antigenically neutral residues. Note the position of the sialic acid-binding pocket. [After a drawing by Hidde Ploegh, *in* Wilson, I. A., Skehel, J. J., and Wiley, D. C., *Nature* **289,** 366 (1981).]

of a long fibrous stalk extending from the membrane surface upon which is perched a globular region. The fibrous stalk consists of residues from HA1 and HA2 and includes a remarkable 76-Å long (53 residues in 14 turns) α helix. The globular region, which is comprised of only HA1 residues, contains an eight-stranded antiparallel β sheet structure that forms the sialic acid-binding pocket.

The dominant interaction stabilizing HA's trimeric structure is a triple-stranded coiled coil consisting of the 76-Å α helices from each of its subunits (Fig. 32-48). The HA trimer is therefore an elongated molecule, some 135 Å in length, with a triangular cross-section that varies in radius from 15 to 40 Å. The carbohydrate chains, which are attached to the protein via N-glycosidic linkages at each of its seven Asn-X-Thr/Ser sequences (Section 10-3C), are located almost entirely along the trimer's lateral surfaces. The role of the carbohydrates is unclear despite the fact that they cover some 20% of the protein's surface. However, the observation that the mutational generation of a new oligosaccharide attachment site blocks antibody binding to HA suggests that carbohydrates modulate HA's antigenicity.

Antigenic Variation Results from Surface Residue Changes

HA's antigenic sites have been identified by mapping HA sequence changes on the protein's three-dimensional structure. The HA residues that mutated in an antigenically significant manner in Hong Kong-type viruses during the period 1968 to 1977 are indicated in Fig. 32-47*b*. *These residues all occur on the protein's surface, often in polypeptide loops, where their mutational variation affects the protein's surface character but probably not its overall structure or stability. The variable residues are clustered in four sites surrounding HA's receptor-binding pocket, which is formed from amino acid residues that are largely conserved in numerous influenza virus strains.* The strains responsible for the major flu epidemics between 1968 and 1975 had at least one mutation in each of these four antigenic sites. This degree of antigenic variation appears necessary to reinfect individuals previously infected with the same viral type. Evidently, *antibodies directed against even conserved residues in HA's receptor-binding pocket are dislodged by the antigenic variation that so readily occurs about its rim (we study antibody–antigen interactions in Section 34-2B).*

NA Is a Tetrameric Trans-Membrane Glycoprotein

Influenza virus neuraminidase is a tetrameric glycoprotein of identical 469-residue subunits. It has a box-shaped globular head attached to a slender stalk that is anchored in the viral membrane. Pronase digestion cleaves NA before residues 74 or 77, after the membrane attachment site, to yield an enzymatically active and crystallizable protein. The X-ray structure of this protein (Fig. 32-49) shows it to have fourfold symmetry. Each monomeric unit is composed of six topologically identical four-stranded antiparallel β sheets arranged like the blades of a propeller. Sugar residues are linked to the protein at four of its five potential Asn-X-Ser/Thr *N*-glycosylation sites.

NA's sialic acid-binding site is located in a large pocket on the top of each monomer (star on the upper right subunit in Fig. 32-49). It is surrounded by polar residues that are conserved in all known NA sequences. Sequence changes in antigenic variants of NA occur in seven chain segments that form a nearly continuous surface that encircles the catalytic site (squares on the lower right subunit in Fig. 32-49) in a manner similar to that of HA's receptor-binding site. Between 1968 and 1975, NA exhibited the same number of residue changes in its putative antigenic determinants as did HA. Antibodies against NA, nevertheless, do not neutralize infectivity. Rather, they restrict multiple cycles of viral replication and thus probably attenuate illness.

5. SUBVIRAL PATHOGENS

The development of concepts of pathogenesis has followed a common pattern in the evolution of human thought: A revolutionary idea is resisted by proponents of the established orthodoxy but once the idea has gained acceptance, it becomes the new orthodoxy. Thus, the hypothesis, developed in the 1870s and 1880s by Louis Pasteur and Robert Koch, that microorganisms can cause disease, was initially ridiculed by adherents of the theory of spontaneous generation. Yet, when the first viruses were discovered at the close of the nineteenth century, it was by no means easy to convince the proponents of the by then predominant microorganism theory of disease that these subcellular entities could also be pathogens. Presently, our ideas as to the nature of pathogenic agents are again evolving. It has recently been demonstrated that *subviral agents can also cause infectious diseases.* Two types of these substances have been discovered:

1. **Viroids,** which are small single-stranded RNA molecules.

2. **Prions,** which appear to be only protein molecules.

In this section we outline what is known about these surprising entities. In doing so, we shall see that some of these new discoveries appear to be challenging the basic principles of molecular biology and thereby leading to a deeper understanding of biological processes.

A. Viroids

The **potato spindle tuber disease,** which was first described in the 1920s, causes the production of gnarled elongated potatoes. It was soon established that the disease is contagious but that no microorganisms are associated with it. It was therefore assumed to be a viral disease. The fact that the putative virus could not be isolated was not doubt frustrating but not surprising; many viruses have been difficult to isolate.

Viroids Are Single-Stranded Circular RNAs

In the 1960s, it was discovered that the infectious agent of potato spindle tuber disease would grow in tomato plants so as to yield much more highly infectious extracts than were obtainable from infected potato plants. It was expected that the causative virus could be easily purified from these extracts by differential centrifugation. This was not the case; the infectious agent would not form a pellet even with a centrifugal acceleration of 100,000g for 4 h. The infectious agent must therefore be remarkably small.

Further investigations, largely by Theodor Diener, demonstrated that the potato spindle tuber agent is not only smaller than any known virus but is even smaller than any viral nucleic acid. The infectivity of these

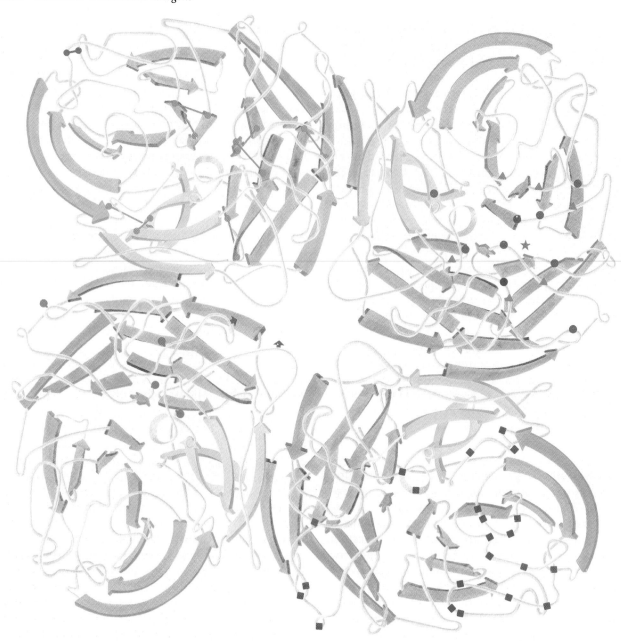

Figure 32-49

The influenza neuraminidase tetramer. In each monomer unit, each of the six topologically equivalent four-stranded antiparallel β sheets are differently shaded. The positions of the disulfide bonds are indicated in the upper left subunit. In the lower left subunit, the four carbohydrate attachment sites are indicated by filled circles and the Asp residues that ligand Ca^{2+} ions are represented by arrows. In the upper right monomer, the filled circles and triangles respectively represent the conserved acidic and basic residues surrounding the enzyme's sialic acid-binding site which is represented by a star. In the lower right monomer, the positions of the mutated residues in NA's antigenic variants are flagged by filled squares. [After Varghese, J. N., Laver, W. G., and Colman, P. M., *Nature* **303,** 35 (1983).]

agents is exquisitely sensitive to RNase but insensitive to DNase or proteases. *The infectious agent evidently consists of only RNA.* This observation, by itself, is not unprecedented; the naked nucleic acids of many viruses are infectious, albeit at very low efficiency. Yet, the RNA's very small size (~ 120 kD) together with the observation that it was not associated with protein, was indeed unique. Diener therefore named these infectious agents viroids to distinguish them from viruses.

Once **potato spindle tuber viroid (PSTV)** had been isolated, it became possible to characterize it. Electron micrographs (Fig. 32-50) indicate that it has an average length of 50 nm, much shorter than the nucleic acids of viruses, although its width is similar to that of the dou-

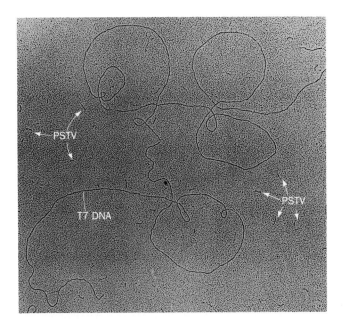

Figure 32-50
An electron micrograph of a mixture of PSTV (scattered short rods) and bacteriophage T7 DNA. The 359-nucleotide viroid is ~50 nm long, whereas the nearly 40,000 bp double-stranded T7 DNA, which is shown for comparison, is some 1400 nm long. [Courtesy of Jose Sogo, Research Institute of Scripps Clinic.]

Table 32-3
Some Viroids Causing Economically Important Diseases

Viroid	Number of Nucleotides
Avocado sun blotch viroid (ASBV)	247
Citrus exocortis viroid (CEV)	371
Chrysanthemum stunt viroid (CSV)	354
Coconut cadang–cadang viroid (CCCV)	246
Cucumber pale fruit viroid (CPFV)	303
Hop stunt viroid (HSV)	297
Potato spindle tuber viroid (PSTV)	359

Source: Riesner, D. and Gross, H. J., *Annu. Rev. Biochem.* **54,** 542–543 (1985).

ble-stranded DNAs in the same micrograph. Electron micrographs of denatured PSTV, however, exhibit circular structures so that *the viroid must be a heretofore unobserved species: single-stranded covalently closed RNA circles.* This observation was eventually confirmed by sequence analysis that established PSTV to be a highly self-complementary single-stranded circle of 359 nucleotides (Fig. 32-51). Its most probable secondary structure consists of short double-stranded regions alternating with shorter unpaired regions so as to form the apparently double-stranded linear structures observed in electron micrographs of the native viroid.

Viroids Are Replicated by Host Enzymes via the Intermediacy of cRNA

In the 1970s, shortly after the nature of PSTV had been established, several economically significant plant diseases previously assumed to be of viral origin were shown to be caused by viroids (Table 32-3). One of them, **coconut cadang–cadang viroid (CCCV),** has killed ~30 million coconut palms in the Philippine Islands and has therefore seriously affected an entire nation's economy. All these viroids are highly self-complementary, single-stranded, covalently closed RNA circles that are, in most cases, >50% homologous with PSTV. Most likely other plant diseases presently attributed to viruses will also be found to be caused by viroids.

How do viroids replicate? A reasonable hypothesis is that they are "defective" viruses that have lost the ability to specify coat protein. There is, however, no evidence that viroids can specify any proteins that do not occur in uninfected plants. Indeed, several viroids and their complements, including PSTV, are devoid of the

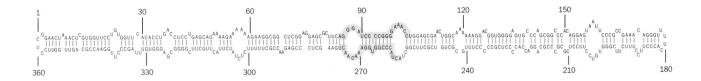

Figure 32-51
The base sequence of PSTV. Its most probable secondary structure is indicated above and the corresponding three-dimensional structure is sketched below. The shaded regions in the sequence are highly conserved in all known viroids, including CCCV, which is only 11% homologous to PSTV. [After Gross, H. J. and Riesner, D., *Angew. Chem. Int. Ed. Engl.* **19,** 237 (1980).]

translation initiation codon AUG, and none have any of mRNA's other characteristic control sequences.

An alternative hypothesis is that viroids somehow trigger the transcription of preexisting viroidal DNA sequences in susceptible plants. Yet, hybridization studies indicate that uninfected plants lack viroidal sequences. *Viroids must therefore be directly replicated by plant enzymes.* An obvious candidate for the viroidal replicase is the RNA-directed RNA polymerase that occurs in several higher plants (but not in animals; its normal function is unknown). Another possibility is RNA polymerase II since this enzyme, from certain plants, transcribes viroids *in vitro* with reasonable efficiency. The latter possibility is favored by the observations that viroid replication is inhibited by the RNA polymerase II inhibitors actinomycin D and α-amanitin which do not affect RNA-directed RNA polymerase. Whatever the enzyme responsible, the isolation, from viroid-infected plants, of RNAs complementary to these viroids indicates that they are replicated through the intermediate synthesis of cRNAs. *Viroids are therefore the only known autonomously replicating entities that do not specify at least one subunit of their replicating enzymes.*

Viroids May Be Escaped Introns

The mechanism of viroidal pathogenesis is by no means obvious. In fact, viroids that cause disease in certain plant species replicate harmlessly in others. It has been variously suggested that the replication of viroids in susceptible species somehow interferes with gene regulation, with the normal RNA maturation process, or perhaps with cellular differentiation. Recently, it has been demonstrated that a 68-kD plant protein exhibiting kinase activity is more heavily phosphorylated in PSTV-infected plants than in uninfected plants. This phosphoprotein resembles the protein kinase from interferon-treated cells that acts to inhibit ribosomal initiation in the presence of double-stranded RNA (Section 30-4B). Perhaps the viroid-affected plant protein adversely influences protein synthesis in susceptible plants. In any case, it is clear that viroids are pathogenic agents that are distinctly different from viruses.

The realization that all known viroidal diseases were only first detected after 1920, in contrast to most viral plant diseases, which were described in the nineteenth century or earlier, suggests that viroids are of recent origin. Indeed, the observation that only cultivated plants are known to be adversely affected by viroids additionally suggests that modern agricultural techniques are responsible for spreading these disease agents. How, then, did viroids originate? Sequence comparisons indicate that viroids are of similar size to, and share many features with, the class of introns that occur in mitochondrial and ribosomal RNA genes and which, in some cases, are self-splicing (Section 29-4B). In fact, these introns contain the shaded sequences in Fig. 32-51

that are common to all viroids suggesting that these two types of entities have similar secondary and tertiary structures (viroids *in vivo* are complexed with proteins so they are unlikely to assume the rodlike conformation they exhibit in the pure state). The above introns are normally sequestered in the nucleus or the mitochondrion so they probably never enter the cytoplasm. It therefore may be that *viroids are "escaped introns."*

B. Prions

Certain diseases that affect the mammalian central nervous systems were originally classified as being caused by "slow viruses" because they take months, years, or even decades to develop. Among them are **scrapie,** a neurological disorder of sheep and goats, so-named for the tendency of infected sheep to scrape off their wool; **Creutzfeldt–Jakob disease,** a rare, progressive, cerebellar disorder of humans; and **kuru,** a similar (possibly identical) degenerative brain disease occurring among the Fore people of Papua New Guinea and thought to be transmitted by ritual cannibalism. These diseases, all of which are ultimately fatal, have similar symptoms, which suggests that they are closely related. None of them exhibit any sign of an inflammatory process or fever which indicates that the immune system, which is not impaired by the disease, is not activated by it.

The classical technique for isolating an unknown disease agent involves the fractionation of diseased tissue as monitored by assays for the disease. The long incubation time for scrapie, the most extensively studied "slow virus" disease, has greatly hampered efforts to characterize its disease agent. Indeed, in the early work on scrapie, an entire herd of sheep and several years of observation were necessary to evaluate the results of a single fractionation. Assays for scrapie were greatly accelerated, however, by the discovery that hamsters develop the disease in a time, minimally 60 days, that decreases as the dose given is increased. Using a hamster assay, Stanley Prusiner has purified the scrapie agent to a high degree.

Scrapie Appears to Be Caused by Prion Protein

The scrapie agent apparently is a single species of protein. This astonishing conclusion was established by the observations that the scrapie agent is inactivated by substances that modify proteins, such as proteases, detergents, phenol, urea, and protein-specific reagents, whereas it is unaffected by agents that alter nucleic acids, such as nucleases, UV irradiation, and substances that specifically react with nucleic acids. For instance, scrapie agent is inactivated by treatment with diethylpyrocarbonate, which carboxyethylates the His residues of proteins (Table 6-3), but is unaltered by the cytosine-specific reagent hydroxylamine (Section 30-1A). In fact, the infectivity of diethylpyrocarbonate-inactivated scrapie agent is restored by treatment with hydroxylamine, presumably by the following known reaction:

$$CH_3CH_2OC-N\overset{\overset{\displaystyle CH_2}{|}}{}N \;+\; NH_2OH$$

Ethylcarboxamido-His   **Hydroxylamine**

$$CH_3CH_2OC-NHOH \;+\; \overset{\overset{\displaystyle CH_2}{|}}{}$$

His

The novel properties of the scrapie agent, which distinguish it from viruses, plasmids, and viroids, has resulted in its being termed a **prion** (for *pro*teinaceous *in*fectious particle). The scrapie protein, which is named **PrP** (for *Pr*ion *P*rotein), is an ~ 30-kD hydrophobic glycoprotein. This hydrophobicity apparently causes PrP to aggregate as clusters of rodlike particles (Fig. 32-52). There is a close resemblance between these clusters and the so-called **amyloid plaques** that are seen on microscopic examination of prion-infected brain tissue. In fact, brain tissue from victims of Creutzfeldt–Jakob disease contains protease-resistant protein that crossreacts with antibodies raised against scrapie PrP.

PrP Is the Normal Product of a Cellular Gene

The bizarre composition of prions immediately raises the question: How are they synthesized? Three possibilities have been suggested:

1. Despite all evidence to the contrary, prions might contain a nucleic acid genome that is somehow shielded from detection; that is, prions could be conventional viruses. The rapidly growing body of information concerning the nature of prions, however, makes this notion increasingly untenable.

2. Prions might somehow specify their own amino acid sequence by "reverse translation" to yield a nucleic acid that is normally translated by the cellular system. Such a process, of course, would directly contravene the "central dogma" of molecular biology (Section 29-1), which states that genetic information flows unidirectionally from nucleic acids to proteins. Alternatively, prions might directly catalyze their own synthesis. Such protein-directed protein synthesis is likewise unknown (although many small polypeptides are enzymatically rather than ribosomally synthesized; Section 30-7).

3. Susceptible cells carry a gene that codes for the corresponding PrP. Infection of such cells by prions activates this gene and/or alters its protein product in some autocatalytic way.

The latter hypothesis seems to be the most plausible mechanism of prion replication. Indeed, the use of oligonucleotide probes complementary to the PrP gene, as inferred from the amino acid sequence of PrP's N-terminus (Section 28-7), established that the brains of both scrapie-infected and normal mice and rats contain the PrP gene. The most surprising discovery, however, is that *the PrP gene is transcribed at similar levels in both normal and scrapie-infected brain tissue*. Moreover, the use of the above probes has revealed that PrP-like genes occur in many vertebrates, including humans, as well as in invertebrates such as *Drosophila* and possibly yeast. This evolutionary conservation suggests that PrP, whose deduced amino acid sequence indicates that it is probably a trans-membrane protein, has an important although presently unknown function.

Since PrP's from normal and infected tissues appear to have identical amino acid sequences, it seems likely that the difference between them is a post-translational modification, possibly a difference in glycosylation. This idea is supported by the observation that scrapie-derived PrP is less susceptible to protease degradation than is normal PrP. Moreover, pathological PrP, but not its normal homolog, aggregates into helically twisted rods that are similar, if not identical, to the scrapie-associated fibrils that appear in scrapie-infected brain tissue. How the difference between pathological and normal PrP is perpetuated, however, has yet to be determined.

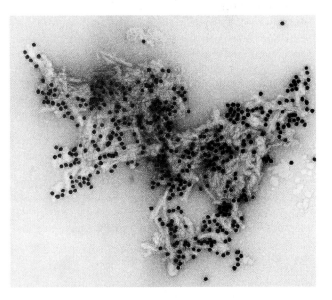

Figure 32-52
An electron micrograph of a cluster of prion rods. The black dots are colloidal gold beads that are coupled to anti-PrP antibodies adhering to the PrP. [Courtesy of Stanley Prusiner, University of California at San Francisco Medical Center.]

Perhaps pathological PrP induces the synthesis of enzymes that so-modify normal PrP.

There are numerous chronic degenerative diseases of humans whose causes are unknown. Pruisner has hypothesized that some of these diseases, such as the paralytic disease **amyotrophic lateral sclerosis,** may be a result of prion infections. Indeed, the discovery of a gene controlling the scrapie incubation time in mice (which is closely linked to that encoding PrP), combined with the lack of any indication that this disease is transmissable in the wild, suggests that scrapielike diseases are not normally infections but, rather, are degenerative conditions arising from genetic defects.

Chapter Summary

Viruses are complex molecular aggregates that exhibit many attributes of living systems. Their structural and genetic properties have therefore served as valuable paradigms for the analogous cellular functions. The TMV virion consists of a helix of identical and therefore largely quasi-equivalent coat protein subunits containing a coaxially wound single strand of RNA. X-ray studies of TMV gels reveal that this RNA is bound, with three nucleotides per subunit, between the subunits of the protein helix. In the absence of TMV RNA, the subunits aggregate at high ionic strengths to form two-layered disks, and at low ionic strengths to form proto-helices, which stack to form helical rods under acidic conditions. The virus' innermost polypeptide loop is disordered in both the disk and the proto-helix. Virus assembly is initiated when a proto-helix binds to the initiation sequence of TMV RNA, which is located ~1000 nucleotides from the RNA's 3' end. Interactions between the RNA and proto-helix trigger the ordering of the disordered loop thereby converting the proto-helix to the helical form. Elongation of the virus particle then proceeds by the sequential addition of proto-helices to the "top" of the assembly so as to pull the 5' end of the RNA up through the center of the growing viral helix.

Viral capsids are formed from one or a few types of coat protein subunits. These must be either helically arranged as in TMV, or quasi-equivalently arranged in a polyhedral shell so as to enclose the viral nucleic acid. The coat proteins of spherical viruses are arranged in icosadeltahedra consisting of $60T$ subunits. The coat protein of TBSV is arranged in a $T = 3$ icosadeltahedron so that TBSV subunits occupy three symmetrically distinct positions. The subunits must therefore associate through several sets of nonidentical intersubunit contacts. Some of the R domains form a structurally disordered inner protein shell. The viral RNA together with the remaining R domains are tightly packed in the space between the inner and outer protein shells. Other spherical plant viruses, SBMV and STNV, have tertiary and quaternary structures that are clearly related to those of TBSV. The structurally similar VP1, VP2, and VP3 coat proteins of poliovirus, rhinovirus, and FMDV are likewise icosahedrally arranged.

Lytic growth of bacteriophage λ in *E. coli* is controlled by the sequential syntheses of antiterminators. Thus gpN, which is synthesized in the early stage of growth, permits the synthesis of gpQ in the delayed early stage which, in turn, permits the synthesis of the capsid proteins in the late stage. Early gene transcription is repressed in the delayed early stage by Cro protein. DNA replication, which commences in the early stage, is mediated by the host DNA replication machinery with the aid of the phage proteins gpO and gpP. DNA synthesis initially occurs by both the θ and rolling circle modes but eventually switches entirely to the rolling circle mode.

The λ virion heads and tails are separately assembled. Head assembly is a complex process involving the participation of many phage gene products, not all of which are part of the mature virion. Phage heads are not self-assembling in that their formation is guided by a scaffolding protein and requires several enzymatically catalyzed protein modification reactions. The mature phage head is a $T = 7$ icosadeltahedron of gpE, which is decorated by an equal number of gpD subunits. Just before the final stage of its assembly, the phage head is filled with a linear double strand of DNA in a process that is driven by ATP hydrolysis. The packaged DNA is thought to be either wound in a spool or folded back and forth in a spirally arranged manner. Tail assembly occurs in a step-wise process from the tail fiber to the head-binding end. The body of the tail consists of a stack of hexameric rings of gpV. The completed heads and tails spontaneously join to form the mature virion.

Lysogeny is established by site-specific recombination between the phage *att*P site and the bacterial *att*B sites in a process mediated by phage integrase (gpint) and host IHF. Induction, in which this process is reversed, requires the additional action of phage excisionase (gpxis). Lysogeny is established by a high level of gpcII, which stimulates the transcription of *int* and the λ repressor gene, *cI*. Repressor, as does Cro, binds to the o_L and o_R operators to shut down early gene transcription, including that of *cro* and *cII*. Each of these dimeric proteins, like other repressors of known structure, contain two symmetrically related helix–turn–helix units that bind in successive turns of B-DNA's major groove. However, repressor, but not Cro, induces its own synthesis from the promoter p_{RM} by binding to o_{R2} so as to interact with RNA polymerase. The induction of repressor synthesis therefore throws the genetic switch which stably maintains the phage in the lysogenic state from generation to generation. Damage to host DNA, nevertheless, stimulates host RecA protein to mediate λ repressor cleavage so as to release repressor from o_L and o_R. This initiates the synthesis of early gene products, including gpint and gpxis, from p_L and p_R and thus triggers induction. If sufficient Cro protein is then synthesized to repress the synthesis of repressor, the phage becomes irrevocably committed to at least one generation of lytic growth. The tripartite character of o_R, the site of the λ switch, together with the cooperative nature of repressor binding to o_R, confers the λ switch with a remarkable sensitivity to the health of its host.

The influenza virion's enveloping membrane is studded with protein spikes consisting of hemagglutinin (HA), which mediates host recognition, and neuraminidase (NA), which facilitates the passage of the virus to and from the infection site. Inside the membrane is a shell of matrix protein that contains the virus' genome of eight single-stranded RNAs, each in a separate protein complex known as a nucleocapsid. These vRNAs are templates for the transcription of mRNAs as catalyzed by the nucleocapsid proteins. This process is primed by 7-methyl-G-capped host mRNA fragments. The viral mRNAs, which have poly(A) tails, lack the sequences complementary to the vRNA's 5' ends. The vRNAs, however, also act as templates for cRNA transcription which, in turn, are the templates for vRNA synthesis. The virus is assembled in and near the plasma membrane and forms by budding from the cell surface. Influenza viruses infect a variety of mammals besides humans as well as many birds. Variation in the antigenic character of HA has been mainly responsible for the different influenza subtypes. Antigenic variation in HA occurs by either antigenic shift, in which the HA gene from an animal virus replaces that from a human virus, and antigenic drift, which occurs by a succession of point mutations in the HA gene. NA may vary in a similar fashion. HA is an elongated trimeric glycoprotein. Its surface has four antigenic sites that surround its sialic acid-binding pocket and which, in the vi-

ruses that caused the major epidemics between 1968 and 1975, all exhibit at least one mutational change. NA is a mushroom-shaped tetrameric glycoprotein. Its antigenic variations occur on a surface that likewise encircles its active site.

Diseases may be caused by subviral pathogens. Viroids, which are only known to infect plants, are single-stranded but highly self-complementary RNA circles. They do not code for proteins. Rather, they are replicated, through an intermediate cRNA, by a plant enzyme, most probably either RNA polymerase II or an RNA-directed RNA polymerase. The homology between virions and certain types of introns suggests that virions are "escaped introns." Prions, which cause several degenerative neurological diseases in humans and other mammals, appears to consist of just a single species of protein named PrP. PrP is a hydrophobic protein that aggregates to form clusters of rodlike particles resembling the amyloid plaques that occur in prion-infected brain tissue. The mechanism of prion replication is presently unknown although the most plausible hypothesis is that prion infection of a susceptible cell somehow activates an enzyme system to post-translationally convert the product of the normal cellular PrP gene to a prion.

References

General

Casjens, S., *Virus Structure and Assembly*, Jones and Bartlett (1985).

Fraenkel-Conrat, H. and Kimball, P. C., *Virology*, Prentice–Hall (1982).

Freifelder, D., *Molecular Biology*, Chapters 15, 16, and 21, Science Books International (1983).

Jurnak, F. A. and McPherson, A. (Eds.), *Biological Macromolecules and Assemblies*, Vol. 1: *Virus Structures*, Wiley (1984); *and* Vol. 2: *Nucleic Acids and Interactive Proteins*, Wiley (1985).

Harrison, S. C., Virus structure: high resolution perspectives, *Adv. Virus Res.* **28**, 175–240 (1983).

Luria, S. E., Darnell, J. E., Jr., Baltimore, D., and Campbell, A., *General Virology* (3rd ed.), Wiley (1978).

Tobacco Mosaic Virus

Bloomer, A. C., Champness, J. N., Bricogne, G., Staden, R., and Klug, A., Protein disk of tobacco mosaic virus at 2.8 Å showing the interactions within and between subunits, *Nature* **276**, 362–368 (1978).

Butler, P. J. G., Tobacco mosaic virus, *J. Gen. Virol.* **65**, 253–279 (1984).

Butler, P. J. and Klug, A., The assembly of a virus, *Sci. Am.* **239**(5): 62–69 (1978).

Hirth, L. and Richards, K. E., Tobacco mosaic virus: model for structure and function of a simple virus, *Adv. Virus Res.* **26**, 145–199 (1981).

Holmes, K. C., Protein–RNA interactions during the assembly of tobacco mosaic virus, *Trends Biochem. Sci.* **5**, 4–7 (1980).

Lomonosoff, G. P. and Wilson, T. M. A., Structure and in vitro assembly of tobacco mosaic virus, *in* Davis, J. W. (Ed.), *Molecular Plant Virology*, Vol. I, *pp.* 43–83, CRC Press (1985).

Namba, K. and Stubbs, G., Structure of tobacco mosaic virus at 3.6 Å resolution: implications for assembly, *Science* **231**, 1401–1406 (1986).

Raghavendra, K., Kelly, J. A., Khairallah, L., and Schuster, T. M., Structure and function of disk aggregates of the coat protein of tobacco mosaic virus, *Biochemistry* **27**, 7583–7588 (1988). [Demonstrates that coat protein disks do not convert to the proto-helices that nucleate TMV assembly.]

Spherical Viruses

Abad-Zapetero, C., Abdel-Meguid, S. S., Johnson, J. E., Leslie, A. G. W., Rayment, I., Rossmann, M. G., Suck, D., and Tsukihara, T., Structure of southern bean mosaic virus at 2.8 Å resolution, *Nature* **286**, 33–39 (1980).

Acharya, R., Fry, E., Stuart, D., Fox, G., and Brown, F., The three-dimensional structure of foot-and-mouth disease virus at 2.9 Å resolution, *Nature* **337**, 709–716 (1989).

Caspar, D. L. D. and Klug, A., Physical principles in the construction of regular viruses, *Cold Spring Harbor Symp. Quant. Biol.* **27**, 1–24 (1962). [The classic paper formulating the geometric principles governing the construction of icosahedral viruses.]

Harrison, S. C., Multiple modes of subunit association in the structures of simple spherical viruses, *Trends Biochem. Sci.* **9**, 345–351 (1984).

Harrison, S. C., Olson, A. J., Schutt, C. E., Winkler, F. K., and Bricogne, G., Tomato bushy stunt virus at 2.9 Å resolution, *Nature* **276**, 368–373 (1978). [The first report of a high resolution virus structure.]

Hogle, J. M., Chow, M., and Filman, D. J., The structure of poliovirus, *Sci. Am.* **256**(3): 42–49 (1987).

Hogle, J. M., Chow, M., and Filman, D. J., Three-dimensional structure of poliovirus at 2.9 Å resolution, *Science* **229**, 1358–1365 (1985).

Hurst, C. J., Benton, W. H., and Enneking, J. M., Three dimensional model of human rhinovirus type 14, *Trends Biochem. Sci.* **12**, 460 (1987). [A "paper doll"-type cut-out with accompanying assembly directions for constructing an icosahedral model of human rhinovirus. This useful learning device may also be taken as a $T = 3$ icosadeltahedron.]

Liljas, L., The structure of spherical viruses, *Prog. Biophys. Mol. Biol.* **48**, 1–36 (1986). [A detailed review.]

Liljas, L., Unge, T., Jones, T. A., Fridborg, K., Lövgren, S., Skogland, U., and Strandberg, B., Structure of satellite tobacco necrosis virus at 3.0 Å resolution, *J. Mol. Biol.* **159**, 93–108 (1982).

Rossmann, M. G., Arnold, E., Erickson, J. W., Frankenberger, E. A., Griffith, J. P., Hecht, H.-J., Johnson, J. E., Kamer, G., Luo, M., Mosser, A. G., Rueckert, R. R., Sherry, B., and Vriend, G., Structure of a human common cold virus and relationship to other picornaviruses, *Nature* **317**, 145–153 (1985).

Rossmann, M. G., Arnold, E., Griffith, J. P., Kamer, G., Luo, M., Smith, T. J., Vriend, G., Rueckert, R. R., Sherry, B., McKinlay, M. A., Diana, G., and Otto, M., *Trends Biochem. Sci.* **12**, 313–318 (1987).

Rossmann, M. G. and Johnson, J. E., Icosahedral RNA virus structure, *Ann. Rev. Biochem.* **58**, 533–573 (1989).

Bacteriophage λ

Adhya, S. L., Garges, S., and Ward, D. F., Regulatory circuits of bacteriophage lambda, *Prog. Nucleic Acid Res. Mol. Biol.* **26**, 103–118 (1981).

Anderson, W. F., Ohlendorf, D. H., Takeda, Y., and Matthews, B. W., Structure of the Cro repressor from bacteriophage λ and its interaction with DNA, *Nature* **290**, 754–758 (1981).

Earnshaw, W. C. and Harrison, S. C., DNA arrangement in isometric phage heads, *Nature* **268**, 598–602 (1977).

Echols, H., Bacteriophage λ development: temporal switches and the choice of lysis or lysogeny, *Trends Genet.* **2**, 26–30 (1986).

Eiserling, F. A., Bacteriophage structure, *in* Fraenkel-Conrat, H. and Wagner, R. R. (Eds.), *Comprehensive Virology*, Vol. 13, pp. 543–580, Plenum Press (1979).

Harrison, S. C., Packaging of DNA into bacteriophage heads: a model, *J. Mol. Biol.* **171**, 577–580 (1983).

Hendrix, R. W., Roberts, J. W., Stahl, F. W., and Weisberg, R. A. (Eds.), *Lambda II*, Cold Spring Harbor Laboratory (1982). [An authoritative compendium of review articles on many aspects of bacteriophage λ.]

Herskowitz, I. and Hagen, D., The lysis-lysogeny decision of phage λ: explicit programming and responsiveness, *Annu. Rev. Genet.* **14**, 399–445 (1980).

Hohn, T. and Katsura, I., Structure and assembly of bacteriophage lambda, *Curr. Top. Microbiol. Immunol.* **78**, 69–110 (1977).

Johnson, A. D., Poteete, A. R., Lauer, G., Sauer, R. T., Ackers, G. K., and Ptashne, M., λ Repressor and Cro-components of an efficient molecular switch, *Nature* **294**, 217–223 (1981).

Jordan, S. R. and Pabo, C. O., Structure of the lambda complex at 2.5 Å resolution: details of the repressor–operator interactions, *Science* **242**, 893–899 (1988).

Landy, A., Dynamic, structural, and regulatory aspects of λ site-specific recombination, *Ann. Rev. Biochem.* **58**, 913–949 (1989).

Metzler, W. J. and Lu, P., λ *cro* Repressor complex with O_R3 operator DNA, *J. Mol. Biol.* **205**, 149–164 (1989).

Nash, H. A., Integration and excision of bacteriophage λ: the mechanism of conservative site specific recombination, *Annu. Rev. Genet.* **15**, 143–167 (1981).

Ohlendorf, D. H. and Matthews, B. W., Structural studies of protein–nucleic acid interactions, *Annu. Rev. Biophys. Bioeng.* **12**, 159–284 (1983).

Pabo, C. O. and Lewis, M., The operator-binding domain of λ repressor: structure and DNA recognition, *Nature* **298**, 443–447 (1982).

Ptashne, M., *A Genetic Switch*, Cell Press (1986). [An authoritative review of the λ switch.]

Ptashne, M., Johnson, A. D., and Pabo, C. O., A genetic switch in a bacterial virus, *Sci. Am.* **247**(5): 128–140 (1982).

Ptashne, M., Jeffrey, A., Johnson, A. D., Maurer, R., Meyer, B. J., Pabo, C. O., Roberts, T. M., and Sauer, R. T., How λ repressor and Cro work, *Cell* **19**, 1–19 (1980).

Roberts, J. W., Phage lambda and the regulation of transcription termination, *Cell* **52**, 5–6 (1988).

Ward, D. F. and Gottesman, M. E., Suppression of transcription termination by phage lambda, *Science* **216**, 946–951 (1982).

Influenza Virus

Air, G. M. and Laver, W. G., The molecular basis of antigenic variation in influenza virus, *Adv. Virus Res.* **31**, 53–102 (1986).

Kaplan, M. M. and Webster, R. G., The epidemiology of influenza, *Sci. Am.* **237**(6): 88–105 (1977).

Lamb, R. A. and Choppin, P. W., The gene structure and replication of influenza virus, *Annu. Rev. Biochem.* **52**, 467–506 (1983).

Skehel, J. J., Stevens, D. J., Daniels, R. S., Douglas, A. R., Knossow, M., Wilson, I. A., and Wiley, D. C., A carbohydrate side chain on hemagglutinins of Hong Kong influenza viruses inhibits recognition by a monoclonal antibody, *Proc. Natl. Acad. Sci.* **81**, 1779–1783 (1984).

Varghese, J. N., Laver, W. G., and Colman, P. M., Structure of the influenza virus glycoprotein antigen neuramimidase at 2.9 Å resolution, *Nature* **303**, 35–40 (1983); *and* Colman, P. M., Varghese, J. N., and Laver, W. G., Structure of the catalytic and antigenic sites in influenza virus neuraminidase, *Nature* **303**, 41–44 (1983).

Webster, R. G., Laver, W. G., Air, G. M., and Schild, G. C., Molecular mechanisms of variation in influenza viruses, *Nature* **296**, 115–121 (1982).

Weis, W., Brown, J. H., Cusack, S., Paulson, J. C., Skehel, J. J., and Wiley, D. C., Structure of the influenza virus haemagglutinin complexed with its receptor, sialic acid, *Nature* **333**, 426–431 (1988).

Wiley, D. C. and Skehel, J. J., The structure and function of the hemagglutinin membrane glycoprotein of influenza virus, *Annu. Rev. Biochem.* **56**, 365–394 (1987).

Wilson, I. A., Skehel, J. J., and Wiley, D. C., Structure of the haemagglutinin membrane glycoprotein of influenza virus at 3 Å resolution, *Nature* **289**, 366–373 (1981); *and* Wiley, D. C., Wilson, I. A., and Skehel, J. J., Structural identification of the antibody-binding sites of Hong Kong influenza haemaglutinin and their involvement in antigenic variation, *Nature* **289**, 373–378 (1981).

Subviral Pathogens

Brunori, M., Silvestrini, M. C., and Pocchiari, M., The scrapie agent and the prion hypothesis, *Trends Biochem. Sci.* **13**, 309–313 (1988).

Diener, T. O., (Ed.), *The Viroids,* Plenum Press (1987).

Diener, T. O., PrP and the nature of the scrapie agent, *Cell* **49**, 719–721 (1987).

Diener, T. O., Viroids, *Trends Biochem. Sci.* **9**, 133–136 (1984).

Diener, T. O., Viroids, *Sci. Am.* **244**(1): 66–73 (1981).

Diener, T. O., McKinley, M. P., and Prusiner, S. B., Viroids and prions, *Proc. Natl. Acad. Sci.* **79**, 5220–5224 (1982).

Dinter-Gottlieb, G., Viroids and virusoids are related to group I introns, *Proc. Natl. Acad. Sci.* **83**, 6250–6254 (1986).

Hiddinga, H. J., Crum, C. J., Hu, J., and Roth, D. A., Viroid-induced phosphorylation of a host protein related to a dsRNA-dependent protein kinase, *Science* **241**, 411–413 (1988).

Prusiner, S. B., Prions and neurodegenerative diseases, *New Engl. J. Med.* **317**, 1571–1581 (1987).

Prusiner, S. B., Prions causing degenerative neurological diseases, *Annu. Rev. Med.* **38**, 381–398 (1987).

Prusiner, S. B., Prions, *Sci. Am.* **251**(4): 50–59 (1984).

Reisner, D. and Gross, H. J., Viroids, *Annu. Rev. Biochem.* **54**, 531–564 (1985).

Semencik, J. S. (Ed.), *Viroids and Viroid-Like Pathogens,* CRC Press (1987).

Problems

1. Why does a pH shift from 7 to 5 at low ionic strengths cause TMV proto-helices to aggregate as helical rods?

*2. Explain why the number of vertices in an icosadeltahedron always ends in the numeral "2" (e.g., 12 for $T = 1$).

3. Sketch a $T = 9$ icosadeltahedron.

4. Why is it necessary to use conditionally lethal mutations in studying phage assembly rather than just lethal mutations?

5. Compare the volume contained by a λ phage head to that of λ DNA.

6. Virulent phages form clear plaques on a bacterial lawn whereas bacteriophage λ forms turbid (cloudy) plaques. Explain.

7. What is the mutual consensus sequence of the o_L and o_R half-subsites?

*8. λ Repressor binds cooperatively to o_{R1} and o_{R2} but independently to o_{R3}. However, if o_{R1} is mutationally altered so that it does not bind repressor, then repressor binds cooperatively to o_{R2} and o_{R3}. Explain.

9. Bacteriophage 434 is a lambdoid phage that has both a repressor and a Cro protein. A hybrid repressor has been constructed that consists of the 434 repressor with its $\alpha 3$ helix replaced by that from 434 Cro protein. Compare the pattern of contacts this hybrid protein makes with its operator, as indicated by chemical protection experiments, with those of the native 434 repressor and Cro proteins.

10. What is the probability that an influenza virion will have its proper complement of eight different RNAs if it has room for only eight nucleocapsids and binds them at random?

11. Single-stranded RNA normally does not act as a template for RNA polymerase II. What property(s) of viroids might enable them to do so?

Chapter 33

EUKARYOTIC GENE EXPRESSION

How does a fertilized ovum give rise to a highly differentiated multicellular organism? This question, of course, is just a sophisticated version of one that every child has asked: Where did I come from? Biologists began rational attempts to answer this question in the late nineteenth century and since that time have assembled an impressive body of knowledge concerning the general patterns of cellular differentiation and organismal development. Yet, we have only had the technical ability to study embryogenesis on the molecular level in the last 20 years or so.

In order to understand cellular differentiation we must first understand the workings of the eukaryotic cell. Eukaryotic cells are, for the most part, much larger and far more complex than prokaryotic cells (Section 1-2). However, *the basic difference between these two types of cells is that eukaryotes have a nuclear membrane that separates their chromosomes from their cytoplasm, thereby physically divorcing the eukaryotic transcriptional process from that of translation.* In contrast, the prokaryotic chromosome is embedded in the cytosol so that the initiation of protein synthesis often occurs on mRNAs that are still being transcribed. The transcriptional and translational control processes in eukaryotes are consequently fundamentally different from those of prokaryotes. This situation is reflected in both the packaging and genetic organization of eukaryotic DNA in comparison with that of prokaryotes. We therefore begin this chapter with a physical description of the eukaryotic chromosome. We then consider how the eukaryotic genome is organized and how it is expressed. Finally, we discuss cell differentiation and its aberration, cancer. In all these subjects, as we shall see, our knowledge is quite fragmentary. Eukaryotic molecular biology is under such intense scrutiny, however, that significant advances in its understanding are made almost daily. Thus, perhaps more so than for other subject matter considered in this text, it is important that the reader supplement the material in this chapter with that in the recent biochemical literature.

1. CHROMOSOME STRUCTURE

Eukaryotic chromosomes, which consist of a complex of DNA, RNA, and protein called **chromatin,** are dynamic entities whose appearance varies dramatically with the stage of the cell cycle. The individual chromosomes only assume their familiar condensed forms (Figs. 27-1 and 33-1) during cell division (M phase of the cell cycle; Section 31-4A). During interphase, the remainder of the cell cycle, when the chromosomal DNA is transcribed and replicated, the chromosomes of most cells become so highly dispersed that they cannot be individually distinguished (Fig. 33-2). Cytologists have long recognized that there are two types of this dispersed chromatin: a less densely packed variety named **euchromatin** and a more densely packed variety termed **heterochromatin** (Fig. 33-2). These two types of chromatin differ, as we shall see, in that euchromatin is genetically expressed whereas heterochromatin is not expressed.

The 46 chromosomes in a human cell each contain between 48 and 240 million bp so that their DNAs, which are most probably continuous (Section 28-2C), have contour lengths between 1.6 and 8.2 cm (3.4

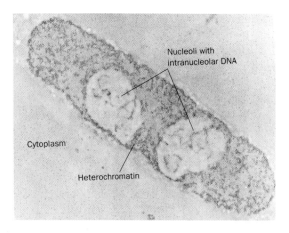

Figure 33-2
A thin section through a cell nucleus treated with **Feulgen reagent** (which reacts with DNA to form an intense red stain). Heterochromatin appears as dark-staining regions near the nucleolus and the nuclear membrane. The less darkly staining material is euchromatin. [Courtesy Edmund Puvion, CNRS.]

Å/bp). Yet, in metaphase, their most condensed state (Fig. 33-1), these chromosomes range in length from 1.3 to 10 μm. *Chromosomal DNA therefore has a **packing ratio** (ratio of its contour length to the length of its container) of > 8000.* How does the DNA in chromatin attain such a high degree of condensation? Structural studies have revealed that this results from three levels of folding. We discuss these levels below, starting with the lowest level. We begin, however, by studying the proteins responsible for much of this folding.

A. Histones

The protein component of chromatin, which comprises somewhat more than one half its mass, consists mostly of histones. There are five major classes of these proteins, **histones H1, H2A, H2B, H3,** and **H4,** all of which have a large proportion of positively charged residues (Arg and Lys; Table 33-1). These proteins therefore ionically bind

Table 33-1
Calf Thymus Histones

Histone	Number of Residues	Mass (kD)	% Arg	% Lys	UEP[a] (× 10⁻⁶ year)
H1	215	23.0	1	29	8
H2A	129	14.0	9	11	60
H2B	125	13.8	6	16	60
H3	135	15.3	13	10	330
H4	102	11.3	14	11	600

[a] Unit Evolutionary Period: The time for a protein's amino acid sequence to change by 1% after two species have diverged (Section 6-3B).

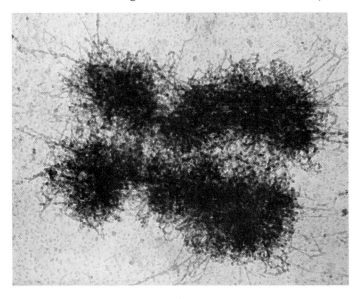

Figure 33-1
Electron micrograph of a human metaphase chromosome. [Courtesy of Gunther Bahr, Armed Forces Institute of Pathology.]

DNA's negatively charged phosphate groups. Indeed, histones may be extracted from chromatin by 0.5*M* NaCl, a salt solution of sufficient concentration to interfere with these electrostatic interactions.

Histones Are Evolutionarily Conserved

The amino acid sequences of histones H2A, H2B, H3, and H4 have remarkably high evolutionary stability (Table 33-1). For example, histones H4 from cows and peas, species that diverged 1.2 billion years ago, differ by only two conservative residue changes (Fig. 33-3) which makes histone H4, the most invariant histone, among the mostly evolutionarily conserved proteins known (Section 6-3B). *Such rigid evolutionary stability implies that the above four histones have critical functions to which their structures are so well tuned that they are all but intolerant to change.* The fifth histone, histone H1, is more variable than the other histones; we shall see below that its role differs from that of the other histones.

Histones May Be Modified

Histones are subject to post-translational modifications that include methylations, acetylations, and phosphorylations of specific Arg, His, Lys, Ser, and Thr residues. These modifications, many of which are reversible, all decrease the histones' positive charges thereby significantly altering histone–DNA interactions. Yet, despite the histones' great evolutionary stability, their degree of modification varies enormously with the species, tissue, and the stage of the cell cycle. A particularly intriguing modification is that 10% of the H2A's have an isopeptide bond between the ε-amino group of their Lys 119 and the terminal carboxyl group of the protein ubiquitin. Although such ubiquitination marks cytosolic proteins for degradation by cellular proteases (Section 30-6B), it is not known whether this is the case with H2A. It would be most surprising, however, if this ubiquitination, as well as the other histone modifications, do not somehow serve to modulate eukaryotic gene expression.

Many, if not all, eukaryotes, as Leonard Cohen, Kenneth Newrock, and Alfred Zweidler first discovered, have genetically distinct subtypes of histones H1, H2A, H2B, and H3 whose syntheses are switched on or off during specific stages of embryogenesis and in the development of certain cell types. The sequence variations of these subtypes are limited to only a few residues in H2A, H2B, and H3 but are much more extensive in H1. Indeed, the erythroid cells of chick embryos contain an H1 variant that differs so greatly from other H1's that it is named **histone H5**. Histone switching seems to be related to cell differentiation but the nature of this relationship is unknown.

B. Nucleosomes: The First Level of Chromatin Organization

The first level of chromatin organization was pointed out by Roger Kornberg in 1974 through the synthesis of several lines of evidence:

1. Chromatin contains roughly equal numbers of molecules of histones H2A, H2B, H3, and H4, and no more than one half that number of histone H1 molecules.

2. X-ray diffraction studies indicate that chromatin fibers have a regular structure that repeats about every 100 Å along the fiber direction. This same X-ray pattern is observed when purified DNA is mixed with equimolar amounts of all the histones except histone H1.

3. Electron micrographs of chromatin (Fig. 33-4) reveal that it consists of ~100 Å in diameter particles connected by thin strands of apparently naked DNA, rather like beads on a string. These particles are presumably responsible for the foregoing X-ray pattern.

Ac —	Ser —	Gly —	Arg —	Gly —	Lys —	Gly —	Gly —	Lys —	Gly — Leu —	10
	Gly —	Lys —	Gly —	Gly —	Ala —	Lys —	Arg —	His —	Arg — Lys —	20
	Val —	Leu —	Arg —	Asp —	Asn —	Ile —	Gln —	Gly —	Ile — Thr —	30
	Lys —	Pro —	Ala —	Ile —	Arg —	Arg —	Leu —	Ala —	Arg — Arg —	40
	Gly —	Gly —	Val —	Lys —	Arg —	Ile —	Ser —	Gly —	Leu — Ile —	50
	Tyr —	Glu —	Glu —	Thr —	Arg —	Gly —	Val —	Leu —	Lys — Val —	60
	Phe —	Leu —	Glu —	Asn —	Val —	Ile —	Arg —	Asp —	Ala — Val —	70
	Thr —	Tyr —	Thr —	Glu —	His —	Ala —	Lys —	Arg —	Lys — Thr —	80
	Val —	Thr —	Ala —	Met —	Asp —	Val —	Val —	Tyr —	Ala — Leu —	90
	Lys —	Arg —	Gln —	Gly —	Arg —	Thr —	Leu —	Tyr —	Gly — Phe —	100
	Gly —	Gly								102

Figure 33-3
The amino acid sequence of calf thymus histone H4. This 102-residue protein's 25 Arg and Lys residues are indicated in red. Pea seedling H4 differs from that of calf thymus by conservative changes at the two shaded residues: Val(60) → Ile and Lys(77) → Arg. The underlined residues are subject to post-translational modification: Ser(1) is invariably *N*-acetylated and may also be *O*-phosphorylated, Lys residues 5, 8, 12, and 16 may be *N*-acetylated, and Lys(20) may be mono- or di-*N*-methylated. [After DeLange, R. J., Fambrough, D. M., Smith, E. L., and Bonner, J., *J. Mol. Biol.* **244**, 330, 5678 (1969).]

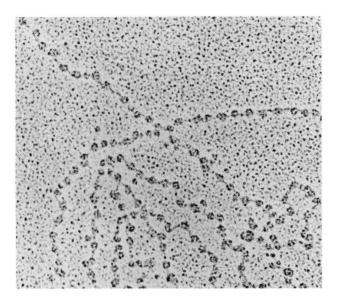

Figure 33-4
An electron micrograph of *D. melanogaster* chromatin
showing that its 100-Å fibers are strings of closely
spaced nucleosomes. [Courtesy of Oscar L. Miller, Jr.,
University of Virginia.]

4. Brief digestion of chromatin by **micrococcal nuclease** (which cleaves double-stranded DNA) cleaves the DNA between some of the above particles (Fig. 33-5*a*); apparently the particles protect the DNA closely associated with them from nuclease digestion. Gel electrophoresis indicates that each particle *n*-mer contains ~200*n* bp of DNA (Fig. 33-5*b*).

5. Chemical cross-linking experiments, such as are described in Section 7-5C, indicate that histones H3 and H4 associate to form the tetramer $(H3)_2(H4)_2$ (Fig. 33-6).

These observations led Kornberg to propose that *the chromatin particles, which are called **nucleosomes**, consist of the octamer $(H2A)_2(H2B)_2(H3)_2(H4)_2$ in association with ~200 bp of DNA.* The fifth histone, H1, was postulated to be associated in some manner with the outside of the nucleosome (see below).

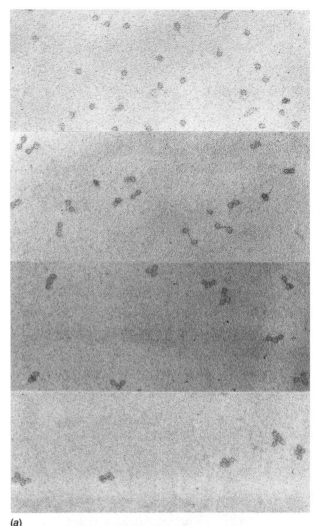

(*a*)

Figure 33-5
Defined lengths of calf thymus chromatin are obtained by
the sucrose density gradient ultracentrifugation of chromatin
that has been partially digested by micrococcal nuclease.
(*a*) Electron micrographs of sucrose density gradient
fractions containing, from top to bottom, nucleosome
monomers, dimers, trimers, and tetramers. (*b*) Gel
electrophoresis of DNA extracted from the nucleosome
multimers indicates that they are the corresponding multiples
of ~200 bp. The right most lane contains DNA from the
unfractionated nuclease digest. [Courtesy of Roger
Kornberg, MRC Laboratory of Molecular Biology.]

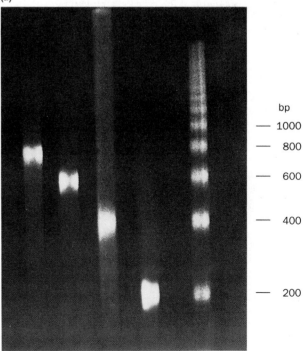

bp
— 1000
— 800
— 600
— 400
— 200

(*b*)

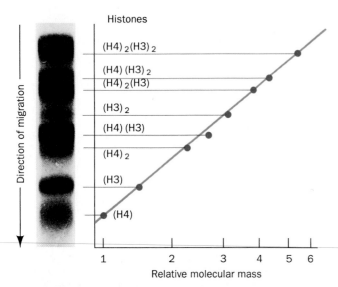

Figure 33-6
The SDS-gel electrophoresis of a mixture of calf thymus histones H3 and H4 that had been cross-linked by dimethylsuberimidate contains all the bands expected from an $(H3)_2(H4)_2$ tetramer. [Courtesy of Roger Kornberg, MRC Laboratory of Molecular Biology.]

Figure 33-7
The nucleosome core particle. (*a*) A schematic model viewed along the particle's twofold axis of symmetry and perpendicularly to the DNA supercoiling axis. (*b*) A stereo image of the 8-Å resolution electron density map viewed along the DNA supercoiling axis and showing slightly more than the upper half of the particle. The upper turn of the DNA (*orange-brown*) winds about the periphery of the particle, starting from position 0, such that the minor grooves face outwards at the positions marked by successive numbers. The histones that can be seen here comprise almost all of the upper H3 (*blue*) and H4 (*green*), small portions of the lower H3 and H4, and small portions of H2A (*violet*) and H2B (*dark brown*). Instructions for viewing stereo diagrams are given in the appendix to Chapter 7. (*c*) A portion of the electron density map as viewed along the core particle's twofold axis from 4 in Part (*b*). [Courtesy of Gerard Bunick and Edward Uberbacher, University of Tennesee and Oak Ridge National Laboratory.]

DNA Coils Around a Histone Octamer to Form the Nucleosome Core Particle

Micrococcal nuclease, as described above, initially degrades chromatin to single nucleosomes in complex with histone H1 (particles called **chromatosomes**). Upon further digestion, some of the chromatosomes' DNA is trimmed away in a process that releases histone H1. This yields the so-called **nucleosome core particle,** which consists of a 146 bp strand of DNA in association with the above histone octamer. The DNA removed by this digestion, which had previously joined neighboring nucleosomes, is named **linker DNA.** Its length has been found to vary between 8 and 114 bp from organism to organism and tissue to tissue although it is usually ~55 bp.

The X-ray structure of the nucleosome core particle was determined by Gerard Bunick and Edward Uberbacher (Fig. 33-7), and independently by Aaron Klug, John Finch, and Timothy Richmond to a resolution of ~8 Å. At this low resolution, gross molecular features such as α helices just begin to be discernible. The core particle is a twofold symmetric disk that is 70 Å thick

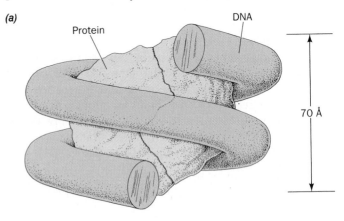

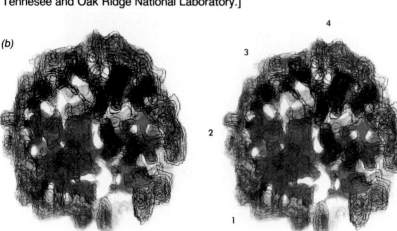

and 110 Å in diameter. The (H3)$_2$(H4)$_2$ tetramer constitutes the nucleus of the histone octamer with an H2A·H2B dimer forming the top and bottom of the disk. B-DNA is wrapped around the histone octamer in 1.8 turns of a flat left-handed superhelix of pitch 28 Å. Nevertheless, the DNA does not follow a smooth superhelical path but rather is bent fairly sharply at several locations such that there are large variations in the widths of its major and minor grooves. The protein–DNA interactions occur on the inside of the DNA superhelix; no protein appears to surround the DNA or protrude between the turns of the superhelix. Two symmetrically related pairs of rodlike features of H3, presumably α helices, contact the minor groove of the DNA in a manner that superficially resembles the proposed interaction between DNA and sequence-specific prokaryotic proteins such as Cro protein (Section 32-3D).

Histone H1 "Seals Off" the Nucleosome

In the micrococcal nuclease digestion of chromatosomes, the ~200 bp DNA is first degraded to 166 bp. Then there is a pause before histone H1 is released and the DNA is further shortened to 146 bp. The twofold symmetry of the core particle suggests that the reduction in length of the 166 bp DNA comes about by the removal of 10 bp from each of its two ends. Since the 146 bp DNA of the core particle makes 1.8 superhelical turns, the 166 bp intermediate should be able to make two full superhelical turns, which would bring its two ends as close together as possible. Klug has therefore proposed that histone H1 binds to nucleosomal DNA in a cavity formed by the central segment of its DNA and

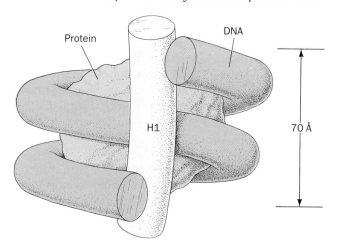

Figure 33-8
Histone H1 is thought to bind to the DNA of the 166 bp nucleosome. The DNA's two complete superhelical turns enable H1 to bind to the DNA's two ends and its middle. Here the histone octamer is represented by the central spheroid and the H1 molecule is represented by the cylinder.

the segments that enter and leave the core particle (Fig. 33-8). This model is supported by the observation that in chromatin filaments containing H1, the DNA enters and leaves the nucleosome on the same side (Fig. 33-9*a*), whereas in H1-depleted chromatin, the entry and exit points are more randomly distributed and tend to occur on opposite sides of the nucleosome (Fig. 33-9*b*). The model also suggests that the length of the linker DNA is controlled by the subspecies of histone H1 bound to it.

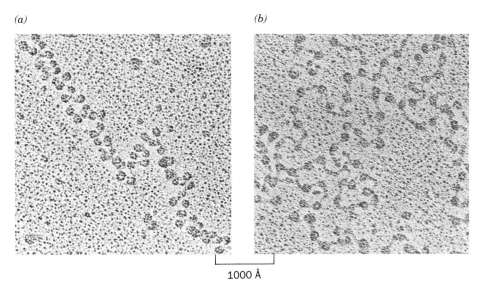

Figure 33-9
Electron micrographs of (*a*) H1-containing chromatin and (*b*) H1-depleted chromatin, both in 5 to 15 m*M* salt. [Courtesy of Fritz Thoma, Eidgenössische Technische Hochschule.]

Nucleosome Assembly Is Probably Factor Mediated

How are nucleosomes formed *in vivo? In vitro*, at high salt concentrations, nucleosomes self-assemble from the proper mixture of DNA and histones. In fact, when only H3, H4, and DNA are present, the mixture forms nucleosomelike particles that each contain an $(H3)_2(H4)_2$ tetramer. Presumably, nucleosome cores are formed by the addition of $H2A \cdot H2B$ dimers to these particles.

At physiological salt concentrations, *in vitro* nucleosome assembly occurs much more slowly than at high salt concentrations and, unless the histone concentrations are carefully controlled, is accompanied by considerable histone precipitation. However, in the presence of **nucleoplasmin,** an acidic protein that has been isolated from *Xenopus laevis* oocyte nuclei, and DNA topoisomerase I (nicking–closing enzyme; Section 28-5C), nucleosome assembly proceeds rapidly without histone precipitation. Nucleoplasmin binds to histones but neither to DNA nor to nucleosomes. This suggests that *nucleoplasmin participates in nucleosome assembly by acting as a "molecular chaperone" to bring histones and DNA together in a controlled fashion thereby preventing their nonspecific aggregation through their otherwise strong electrostatic interactions.* The nicking–closing enzyme, no doubt, acts to provide the nucleosome with its preferred level of supercoiling.

When eukaryotic DNA is replicated *in vivo*, both daughter strands are immediately incorporated into nucleosomes as is demonstrated by electron micrographs showing nucleosomes on both tines of active replication forks. When DNA is replicated in cells exposed to the ribosomal protein synthesis inhibitor cycloheximide (Section 30-3G), however, only one of the daughter strands has a beaded appearance while the other is just naked DNA (Fig. 33-10). This observation indicates that

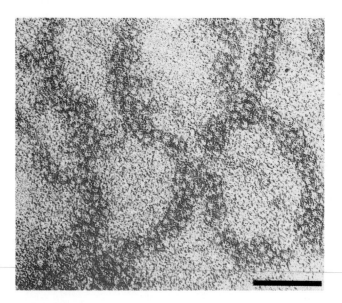

Figure 33-11
Electron micrograph of the 300-Å chromatin filaments. Note that the filaments are two to three nucleosomes across. The bar represents 1000 Å. [Courtesy of Jerome B. Rattner, University of Calgary.]

the parental strand transfers its nucleosomes to one daughter strand while the other acquires fresh nucleosomes. It has been argued that the old nucleosomes are passed to the daughter duplex containing the leading strand because it seems unlikely that nucleosomes could assemble on the single-stranded regions of DNA formed by lagging strand replication.

C. 300-Å Filaments: The Second Level of Chromatin Organization

The 166 bp nucleosomal DNA has a packing ratio of 10 (its 560-Å contour length is wound into a 56-Å high supercoil). Clearly, the 100-Å filament of nucleosome, which occurs at low ionic strengths, represents only the first level of chromosomal DNA compaction. Only at physiological ionic strengths does the next level of chromosomal organization become apparent.

As the salt concentration is raised, the H1-containing nucleosome filament initially folds to a zigzag conformation (Fig. 33-9a) whose appearance suggests that nucleosomes interact through contacts between their H1 molecules. Then, as the salt concentration approaches the physiological range, chromatin forms a 300-Å thick filament in which the nucleosomes are visible (Fig. 33-11). Klug proposed that the 300-Å filament is constructed by winding the 100-Å nucleosome filament into a solenoid with ~6 nucleosomes/turn and a pitch of 110 Å (the diameter of a nucleosome; Fig. 33-12). The solenoid is stabilized by H1 molecules which each consist of a conserved, nucleosome-binding globular core and relatively variable, extended N-terminal and C-ter-

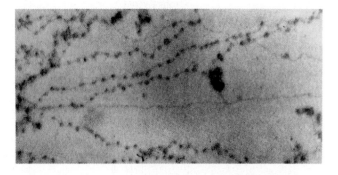

Figure 33-10
Electron micrograph of a DNA replication fork generated when the synthesis of new histones was blocked by the presence of the protein synthesis inhibitor cycloheximide. Just one of the two tines of the replication fork is assembled into nucleosomes thereby indicating that the nucleosomes associated with the parental DNA strand are transferred to only one of the daughter strands. [Courtesy of Harold Weintraub, Hutchinson Cancer Center.]

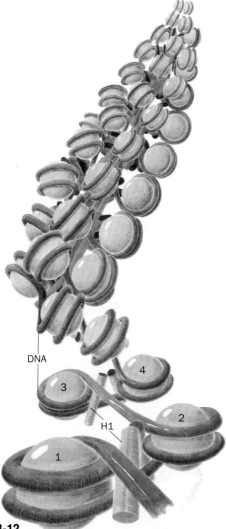

Figure 33-12
A proposed model of the 300-Å chromatin filament. The filament is represented (*bottom to top*) as it might form with increasing salt concentrations. The zigzag pattern of nucleosomes (*1, 2, 3, 4*) closes up to form a solenoid with ~6 nucleosomes/turn. The H1 molecules, which stabilize the structure, are thought to form a helical polymer running along the center of the solenoid.

minal arms, which are thought to contact adjacent nucleosomes. This model, which is consistent with the X-ray diffraction pattern of the 300-Å filaments, has a packing ratio of ~40 (6 nucleosomes, each with ~200 bp DNA, rising a total of 110 Å). Note, however, that several other plausible models for the 300-Å chromatin filament have also been formulated.

D. Radial Loops: The Third Level of Chromatin Organization

Histone-depleted metaphase chromosomes exhibit a central fibrous protein "scaffold" surrounded by an extensive halo of DNA (Fig. 33-13*a*). The strands of DNA that can be followed are observed to form loops that

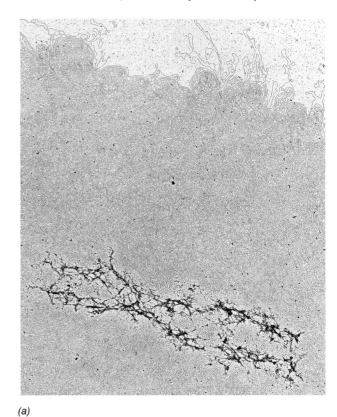

(a)

(b)

Figure 33-13
Electron micrographs of a histone-depleted metaphase human chromosome. (*a*) The central protein matrix (scaffold) serves to anchor the surrounding DNA.
(*b*) At higher magnification it can be seen that the DNA is attached to the scaffold in loops. [Courtesy of Ulrich Laemmli, University of Geneva.]

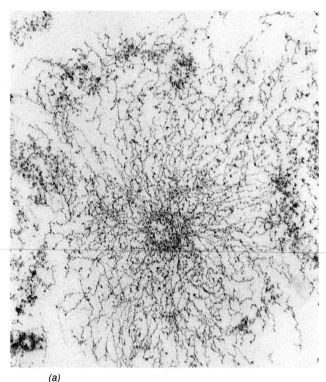

(a)

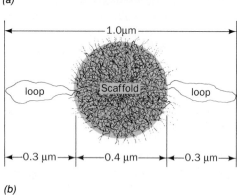

(b)

Figure 33-14
(a) Electron micrograph of a metaphase human chromosome in cross-section. Note the mass of chromatin fibers radially projecting from the central scaffold. [Courtesy of Ulrich Laemmli, University of Geneva.] (b) Interpretive diagram indicating how the 0.3-μm-long radial loops are thought to combine with the 0.4-μm wide scaffold to form the 1.0 μm in diameter metaphase chromosome.

back on itself to form a loop). Taking into account the 0.4 μm width of the scaffold, this model predicts the diameter of the metaphase chromosome to be 1.0 μm, in agreement with observation (Fig. 33-14b). A typical human chromosome, which contains ~140 million bp, would therefore have ~2000 of these ~70-kb radial loops. The 0.4-μm diameter scaffold of such a chromosome has sufficient surface area along its 6-μm length to bind this number of radial loops. The radial loop model therefore accounts for DNA's observed packing ratio in metaphase chromosomes.

Almost nothing is known about how the 300-Å filaments are organized to form radial loops nor about how metaphase chromosomes and the far more dispersed interphase chromosomes interconvert. Certainly, **nonhistone proteins,** whose hundreds or even thousands of varieties constitute some 10% of chromosomal proteins, must be involved in these processes. Moreover, there are intriguing indications that the radial loops are the chromosomal transcriptional units.

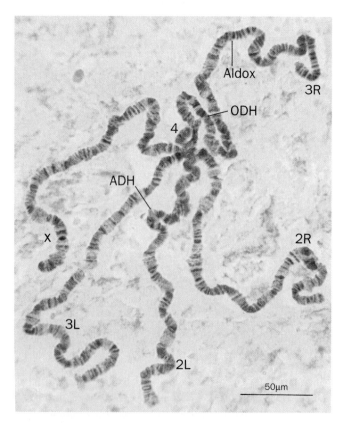

Figure 33-15
Photomicrograph of the stained polytene chromosomes from the *D. melanogaster* salivary gland. Such chromosomes consist of darkly staining bands interspersed with light staining interband regions. All four chromosomes in a single cell are held together by their centromeres. The chromosomal positions for the genes specifying alcohol dehydrogenase (ADH), aldehyde oxidase (Aldox), and octanol dehydrogenase (ODH) are indicated. [Courtesy of B. P. Kaufmann, University of Michigan.]

enter and exit the scaffold at nearly the same point (Fig. 33-13b). Most of these loops have lengths in the range 15 to 30 μm (which correspond to 45–90 kb), so that when condensed as 300-Å filaments they would be ~0.6 μm long. Electron micrographs of chromosomes in cross-section, such as Fig. 33-14a strongly suggest that the chromatin fibers of metaphase chromosomes are radially arranged. If the observed loops correspond to these radial fibers, they would each contribute 0.3 μm to the diameter of the chromosome (a fiber must double

E. Polytene Chromosomes

The diffuse structure of most interphase chromosomes (Fig. 33-2) makes it all but impossible to characterize them at the level of individual genes. Nature, however, has greatly ameliorated this predicament through the production of "giant" banded chromosomes in certain nondividing secretory cells of dipteran (two-winged) flies (Fig. 33-15). These chromosomes, of which those from the salivary glands of *Drosophila melanogaster* larvae are the most extensively studied, are produced by multiple replications of a synapsed (joined in parallel) diploid pair in which the replicas remain attached to one another and in register. Each diploid pair may replicate in this manner as many as nine times so that the final **polytene chromosome** contains up to to $2 \times 2^9 = 1024$ DNA strands.

Drosophila's 4 giant chromosomes have an aggregate length of ~2 mm so that its haploid genome of 1.65×10^8 bp has an average packing ratio in these chromosomes of almost 30. About 95% of this DNA is concentrated in chromosomal bands (Fig. 33-16). These bands (more properly, **chromomeres**), as microscopically visualized through staining, form a pattern that is characteristic of each *Drosophila* strain. Indeed, chromosomal rearrangements such as duplications, deletions, and inversions result in a corresponding change in the banding pattern. *A polytene chromosome's banding pattern therefore forms a cytological map that parallels its genetic map.*

Drosophila chromosomes exhibit ~5000 bands, more or less matching the estimated number of proteins that *Drosophila* produces. This correlation suggests that each chromosomal band corresponds to a single structural gene, a hypothesis corroborated by the application of *in situ* (on site) **hybridization**. In this technique, which Mary Lou Pardue and Joseph Gall developed, an immo-

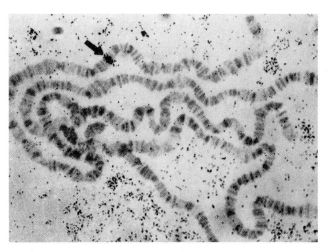

Figure 33-17
Autoradiograph of a *D. melanogaster* polytene chromosome that has been *in situ* hybridized with yolk protein cDNA. The dark grains (*arrow*) identify the chromosomal location of the yolk protein gene. [From Barnett, T., Pachl, C., Gergen, J. P., and Wensink, P. C., *Cell* **21**, 735 (1980). Copyright © 1980 by Cell Press.]

bilized chromosome preparation is treated with NaOH to denature its DNA, hybridized with a purified species of radioactively labeled mRNA (or its corresponding cDNA), and the chromosomal-binding site of the radioactive probe determined by autoradiography. A given mRNA hybridizes with one or, at most, a few chromosomal bands (Fig. 33-17). We shall see that these bands, which probably correspond to the radial loops of metaphase chromosomes, are the chromosome's transcriptional units (Section 33-3).

2. GENOMIC ORGANIZATION

Higher organisms contain a great variety of cells that differ not only in their appearances (e.g., Fig. 1-10) but in the proteins they synthesize. Pancreatic acinar cells, for example, synthesize copious amounts of digestive enzymes including trypsin and chymotrypsin but no insulin, whereas the neighboring pancreatic β cells produce large quantities of insulin but no digestive enzymes. Clearly, each of these different types of cells expresses different genes. Yet, most of a multicellular organism's somatic cells contain the same genetic information as the fertilized ovum from which they are descended (a phenomenon described as **totipotency**). This was demonstrated, for instance, by John Gurdon, who raised a normal adult frog from a fertilized frog egg whose nucleus he had replaced with the nucleus of a tadpole intestinal cell. In this section we describe the genetic organization of the eukaryotic chromosome, which permits its enormous expressional flexibility. How this genetic expression is controlled is the subject of Section 33-3.

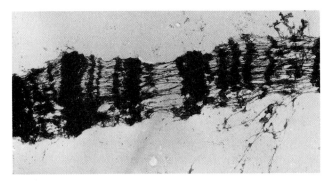

Figure 33-16
An electron micrograph of a segment of polytene chromosome from *D. melanogaster.* Note that its interband regions consist of chromatin fibers that are more or less parallel to the long axis of the chromosome, whereas its bands which contain ~95% of the chromosome's DNA, are much more highly condensed. [Courtesy of Gary Burkholder, University of Saskatechewan.]

A. The C-Value Paradox

One might reasonably expect the morphological complexity of an organism to be roughly correlated with its **C value,** the amount of DNA in its haploid genome. After all, the morphological complexity of an organism must reflect an underlying genetic complexity. Nevertheless, in what is known as the **C-value paradox,** many organisms have unexpectedly large C values (Fig. 33-18). For instance, the genomes of lungfish are 10 to 15 times larger than of those of mammals and those of some salamanders are yet larger. Moreover, the C-value paradox even applies to closely related species; for example, the C values for several species of *Drosophila* have a 2.5-fold spread. Does the "extra" DNA in the larger genomes have a function, and if not, why is it preserved from generation to generation?

The 4 million bp *E. coli* genome is thought to code for ~3000 gene products (about one half of which have been characterized). In contrast, the 2.9 billion bp haploid human genome, which is >700 times larger than that of *E. coli,* is estimated to code for 30 to 40 thousand proteins; that is, humans have only 10 to 13 times as many structural genes as do *E. coli.* Certainly the control of genetic expression in eukaryotes must be a far more elaborate process than it is in prokaryotes. Yet, does all the unexpressed DNA in the human genome, at least 98% of the total, function in the control of genetic expression?

At present, we are unable to satisfactorily answer the foregoing questions. As we shall see below, however, we are learning a considerable amount about the detailed genetic organization of the eukaryotic chromosome. This information, no doubt, will contribute heavily to finding the answers to these questions.

C_0t Curve Analysis Indicates DNA Complexity

The rate at which DNA renatures is indicative of the lengths of its unique sequences. If DNA is sheared into uniform fragments of 300 to 10,000 bp (Section 28-2C), denatured, and kept at a low concentration so that the effects of mechanical entanglement are small, the rate-determining step in renaturation is the collision of complementary sequences. Once the complementary se-

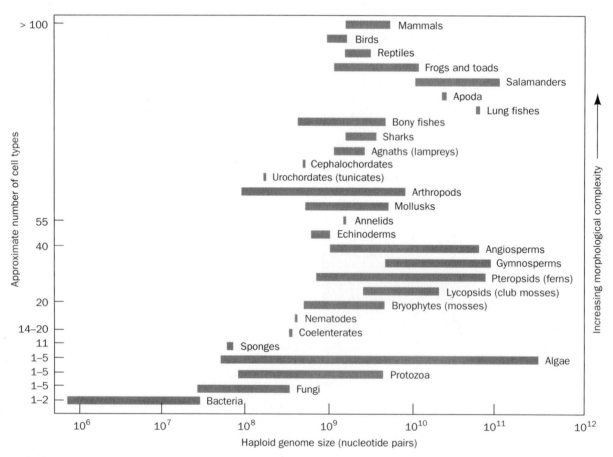

Figure 33-18
The range of haploid genome DNA contents in various categories of organisms indicating the C-value paradox. The morphological complexity of the organisms, as estimated according to their number of cell types, increases from bottom to top. [After Raff, R. A. and Kaufman, T. C., *Embryos, Genes, and Evolution,* p. 314, Macmillan (1983).]

quences have found each other through random diffusion, they rapidly zip up to form duplex molecules. The rate of renaturation of denatured DNA is therefore expressed

$$\frac{d[A]}{dt} = -k[A][B] \qquad [33.1]$$

where A and B represent complementary single-stranded sequences and k is a second-order rate constant (Section 13-1B). Since $[A] = [B]$ for duplex DNA, Eq. [33.1] integrates to

$$\frac{1}{[A]} = \frac{1}{[A]_0} + kt \qquad [33.2]$$

where $[A]_0$ is the initial concentration of A.

It is convenient to measure the fraction f of unpaired strands:

$$f = \frac{[A]}{[A]_0} \qquad [33.3]$$

Combining Eqs. [33.2] and [33.3] yields

$$f = \frac{1}{1 + [A]_0 kt} \qquad [33.4]$$

The concentration terms in these equations refer to unique sequences since the collision of noncomplementary sequences does not lead to renaturation. Hence, if C_0 is the initial concentration of base pairs in solution, then

$$[A]_0 = \frac{C_0}{x} \qquad [33.5]$$

where x is the number of base pairs in each unique sequence and is known as the DNA's **complexity**. For example, the repeating sequence $(AGCT)_n$, has a complexity of 4 whereas an *E. coli* chromosome, which consists of 4 million bp of unrepeated sequence, has a complexity of 4 million. Combining Eqs. [33.4] and [33.5] yields

$$f = \frac{1}{1 + C_0 kt/x} \qquad [33.6]$$

When one half of the molecules in the sample have renatured, $f = 0.5$ so that

$$C_0 t_{1/2} = \frac{x}{k} \qquad [33.7]$$

where $t_{1/2}$ is the time for this to occur. The rate constant k is characteristic of the rate at which single strands collide in solution under the conditions employed so that it is independent of the complexity of the DNA and, for reasonably short DNA fragments, the length of a strand. Consequently, *for a given set of conditions, the value of $C_0 t_{1/2}$ depends only on the complexity x of the DNA*. This situation is indicated in Fig. 33-19, which is a series of plots of f versus $C_0 t$ for various DNAs. Such plots are referred to as $C_0 t$ (pronounced "cot") curves. The complexities of the DNAs in Fig. 33-19 vary from 1 for the synthetic duplex poly(A)·poly(U) to ~3×10^9 for some fractions of mammalian DNAs. Their corresponding values of $C_0 t_{1/2}$ vary accordingly.

The speed and sensitivity of $C_0 t$ curve analysis is greatly enhanced through the hydroxyapatite fractionation of the renaturing DNA. Hydroxyapatite, it will be

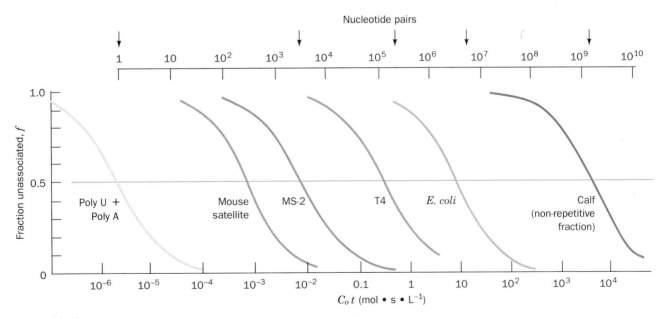

Figure 33-19

The reassociation ($C_0 t$) curves of duplex DNAs from the indicated sources. The DNA was dissolved in a solution containing 0.18M Na$^+$ and sheared to an average length of 400 bp. The upper scale indicates the genome sizes of some of the DNA's (**MS2** and **T4** are bacteriophages). [After Britten, R. J. and Kohne, D. E., *Science* **161**, 530 (1968).]

recalled (Section 28-4B), binds double-stranded DNA at a higher phosphate concentration than it binds single-stranded DNA. The single- and double-stranded DNAs in a solution of renaturing DNA may therefore be separated by hydroxyapatite chromatography and the amounts of each measured. The single-stranded DNA can then be further renatured and the process repeated. If the renaturing DNA is radioactively labeled, much smaller quantities of it can be detected than is possible by spectroscopic means. Thus, through the hydroxyapatite chromatography of radioactively labeled DNA, the C_0t curve analysis of a DNA of such a high complexity that its $t_{1/2}$ is days or weeks can be conveniently measured in a small fraction of that time.

B. Repetitive Sequences

Consider a sample of DNA that consists of sequences with varying degrees of complexity. Its C_0t curve, Fig. 33-20 for example, is the sum of the individual C_0t curves for each complexity class of DNA. C_0t *curve analysis has demonstrated that viral and prokaryotic DNAs have few, if any, repeated sequences (e.g., Fig. 33-19 for MS2, T4, and E. coli).* In contrast, *eukaryotic DNAs exhibit complicated C_0t curves (e.g., Fig. 33-21) that must arise from the presence of DNA segments of several different complexities.*

Kinetic analyses indicate that eukaryotic C_0t curves may be attributed to the presence of four somewhat arbitrarily defined classes of DNAs: (1) **unique sequences** (~1 copy/haploid genome), (2) **moderately repetitive sequences** (<10^6 copies/haploid genome), (3) **highly repetitive sequences** (>10^6 copies/haploid genome), and (4) **inverted repeats**. The sequences and

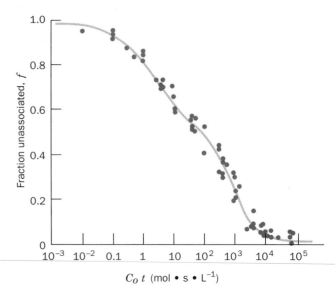

Figure 33-21
The (C_0t) curve of *Strongylocentrotus purpuratus* (a sea urchin) DNA. [After Galau, G. A., Britten, R. J., and Davidson, E. H., *Cell* **2**, 11 (1974).]

chromosomal distributions of these DNA segments vary with the species so that a unifying description of their arrangements cannot be made. Nevertheless, several broad generalizations are possible as we shall see below.

Inverted Repeats Form Foldback Structures

The most rapidly reassociating eukaryotic DNA, which represents as much as 10% of some genomes, renatures with first-order kinetics. Evidently, this DNA contains inverted (self-complimentary) sequences in close proximity, which can fold back on themselves to form hairpinlike **foldback structures** (Fig. 33-22*a*). Inverted sequences may be isolated by adsorbing the duplex DNA formed at very low C_0t values to hydroxyapatite and subsequently degrading its single-stranded loop and tails with **S1 nuclease** (an endonuclease from *Aspergillus oryzae* which preferentially cleaves single strands). The resulting inverted repeats range in length from 100 to 1000 bp, sizes much too large to have evolved at random. *In situ* hybridization studies on metaphase chromosomes using these inverted repeats as probes indicates that they are distributed at many chromosomal sites.

The function of inverted repeats, some 2 million of which occur in the human genome, is unknown. However, since the cruciform structures formed by paired foldback structures (Fig. 33-22*b*) are only slightly less stable than the corresponding normal duplex DNA, it has been suggested that the inverted repeats function in chromatin as some sort of molecular switch.

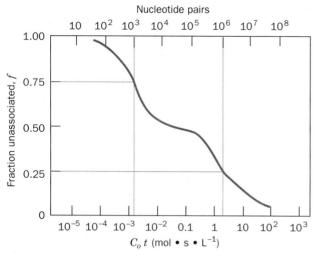

Figure 33-20
The (C_0t) curve of a hypothetical DNA molecule that, before fragmentation, was 2 million bp in length and consisted of a unique sequence of 1 million bp and 1000 copies of a 1000 bp sequence. Note the curve's biphasic nature.

Highly Repetitive DNA Is Clustered at Centromeres

Highly repetitive DNA consists of clusters of nearly

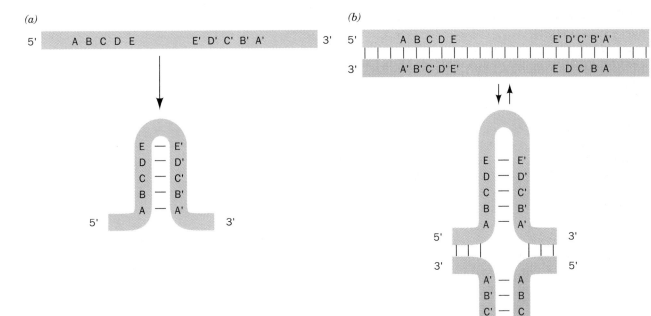

(a)

(b)

Figure 33-22
Foldback structures in DNA. (a) Single-stranded DNA containing an inverted repeat will, under renaturing conditions, form a base paired loop known as a foldback structure. Here A is complimentary to A', B is complimentary to B', *etc.* (b) An inverted repeat in duplex DNA could assume a cruciform conformation consisting of two opposing foldback structures. The stability of this structure would be less than that of the corresponding duplex but only by the loss of the base pairing energy in the unpaired loops.

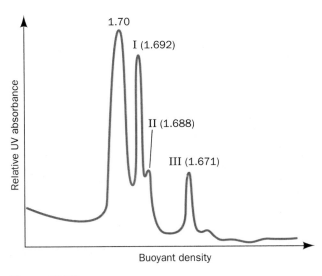

Figure 33-23
The buoyant density pattern of *Drosophila virilus* DNA centrifuged to equilibrium in neutral CsCl. Three prominent bands of satellite DNA ($\rho = 1.692$, 1.688, and 1.671) are present, in addition to the main DNA band ($\rho = 1.70$). [After Gall, J. G., Cohen, E. H., and Atherton, D. D., *Cold Spring Harbor Symp. Quant. Biol.* **38,** 417 (1973).]

identical sequences up to 10 bp long that are tandemly repeated thousands of times. Such **simple sequence DNAs** can often be separated from the bulk of the chromosomal DNA by shear degradation followed by density gradient ultracentrifugation in CsCl since their distinctive base compositions cause them to form "satellites" to the main DNA band (Fig. 33-23; recall that the buoyant density of DNA in CsCl increases with its G + C content; Section 28-4D). The sequences of these DNAs, which are also known as **satellite DNAs,** are species specific. For example, the crab *Cancer borealis* has a simple sequence DNA comprising 30% of its genome in which the repeating unit is the dinucleotide AT. The DNA of *Drosophila virilus* exhibits three satellite bands (Fig. 33-23), which each consist of a different although closely related repeating heptanucleotide sequence:

$$5'-A\overset{\cdots}{C}\overset{\cdots}{A}\overset{\cdots}{A}\overset{\cdots}{A}\overset{\cdots}{C}T-3'$$
$$3'-TGTTTGA-5'$$

Satellite I

$$5'-A\overset{\cdots}{T}\overset{\cdots}{A}\overset{\cdots}{A}\overset{\cdots}{A}\overset{\cdots}{C}T-3'$$
$$3'-TATTTGA-5'$$

Satellite II

$$5'-A\overset{\cdots}{C}\overset{\cdots}{A}\overset{\cdots}{A}\overset{\cdots}{A}\overset{\cdots}{T}T-3'$$
$$3'-TGTTTAA-5'$$

Satellite III

These comprise 25, 8, and 8% of the 4.4×10^7 bp *D. virilus* genome, so that these sequences are repeated 11, 3.6, and 3.6 million times, respectively.

The *in situ* hybridization of mouse chromosomes with ^{3}H-labeled RNA synthesized on mouse simple sequence DNA templates established that simple sequence DNA is concentrated in the heterochromatic region associated with the chromosomal centromere (Fig. 33-24). This observation suggests that simple sequence DNA, which is not transcribed *in vivo*, functions to align homologous chromosomes during meiosis and/or to facilitate their recombination. This hypothesis is supported by the observations that satellite DNAs are largely or entirely eliminated in the somatic cells of a variety of eukaryotes (which are consequently no longer totipotent) but not in their germ cells. The putative chromosomal proteins that specifically bind simple sequence DNAs have not been detected, however.

Moderately Repetitive DNAs Are Arranged in Dispersed Repeats

Moderately repetitive DNAs occur in segments of 100 to several thousand bp that are interspersed with larger blocks of unique DNA. Some of this repetitive DNA consists of tandemly repeated groups of genes that specify products that cells require in large quantities, such as ribosomal RNAs, tRNAs, and histones. The organization of these repeated genes is discussed in Section 33-2C. However, most moderately repetitive DNAs, although they may be transcribed, do not specify RNAs of known function. The best characterized such DNA is known as the **Alu family** because most of its ~300 bp segments contain a cleavage site for the restriction endonuclease *Alu*I (Table 28-5). The *Alu* family is the human genome's most abundant moderately repetitive DNA; the genome contains 300 to 500 thousand widely distributed *Alu* sequences that are, on average, 80 to 90% homologous with their consensus sequence. *Alu* DNA also occurs in monkeys and rodents, and *Alu*-like sequences occur in such distantly related organisms as slime molds, echinoderms, amphibians, and birds. Although the *Alu* family is the most prominent moderately repetitive DNA in many organisms, it is by no means the only one. Indeed, vertebrate genomes, as sequence analyses have shown, generally contain several different varieties of moderately repetitive DNAs.

Moderately Repetitive DNAs Have Unknown Functions

It seems likely, considering their ranges of segment lengths and copy numbers, that nonexpressed, moderately repetitive DNAs have several different functions. There is, however, little experimental evidence in support of any of the various proposals that have been put forward in this regard. The proposal that is usually given the most credence is that moderately repetitive DNAs function as control sequences that participate in coordinately activating nearby genes. Another possibil-

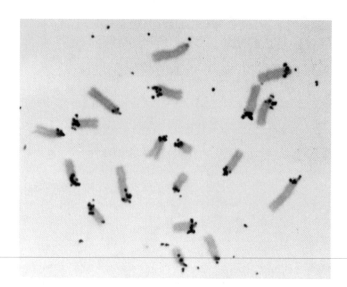

Figure 33-24
An autoradiograph of mouse chromosomes showing the location of their satellite DNA through *in situ* hybridization. [Courtesy of Joseph Gall, Carnegie Institution of Washington.]

ity, which is based on the observation that *Alu* DNA contains a segment that is homologous to the **papovavirus** replication origin, is that certain families of moderately repetitive DNAs act as DNA replication origins. A third class of proposed functions for moderately repetitive DNAs is that they increase the evolutionary versatility of eukaryotic genomes by facilitating chromosomal rearrangements and/or forming reservoirs from which new functional sequences can be recruited. Genetic evidence indicates that retrotransposons (Section 31-6B), which, for example, comprise ~3% of the *Drosophila* genome, indeed promote chromosomal rearrangements.

Considering both the enormous amount of repetitive DNA in most eukaryotic genomes and the dearth of confirmatory evidence for any of the above proposals, a possibility that must be seriously entertained is that much repetitive DNA serves no useful purpose whatever for its host. Rather, it is **selfish** or **junk DNA,** a molecular parasite that, over many generations, has disseminated itself throughout the genome through some sort of transpositional process. The theory of natural selection indicates that the increased metabolic burden imposed by the replication of an otherwise harmless selfish DNA would eventually lead to its elimination. Yet, for slowly growing eukaryotes, the relative disadvantage of replicating say an additional 1000 bp of selfish DNA in an ~1 billion bp genome would be so slight that its rate of elimination would be balanced by its rate of propagation. The C-value paradox may therefore simply indicate that a significant fraction, if not the great majority, of each eukaryotic genome is selfish DNA.

C. Tandem Gene Clusters

Most genes occur but once in an organism's haploid genome. This situation is feasible, even for genes specifying proteins required in large amounts, through the accumulation of their corresponding mRNAs. However, the great cellular demand for rRNAs (which comprise ~80% of a cell's RNA) and tRNAs, which are all transcription products, can only be satisfied through the expression of multiple copies of the genes specifying them. In the following subsections we discuss the organization of the genes coding for rRNAs and tRNAs. We shall also consider the organization of histone genes, the only protein-coding genes that occur in multiple identical copies.

rRNA Genes Are Organized into Repeating Sets

We have seen in Sections 29-4B and C that even the *E. coli* genome, which otherwise consists of unique sequences, contains multiple copies of rRNA and tRNA genes. In eukaryotes, the genes specifying the 18S, 5.8S, and 28S rRNAs are invariably arranged in this order, reading $5' \rightarrow 3'$ on the RNA strand, and separated by short transcribed spacers to form a single transcription unit of ~7500 bp (Fig. 33-25). (Recall that the primary transcript of this gene cluster is a 45S RNA from which the mature rRNAs are derived by post-transcriptional cleavage; Section 29-4B.) *Indeed, this rRNA gene arrangement is universal since the 5' end of prokaryotic 23S rRNA is homologous to eukaryotic 5.8S rRNA (Section 30-3A).*

Electron micrographs, such as Fig. 33-26, indicate that *the blocks of transcribed eukaryotic rRNA genes are arranged in tandem repeats that are separated by untranscribed spacers (Fig. 33-25).* These tandem repeats are typically ~12,000 bp in length although the untranscribed spacer varies in length between species, and to a lesser extent, from gene to gene. Quantitative measurements of the amounts of radioactively labeled rRNAs that can hybridize with the corresponding nuclear DNA (**rDNA**) indicate that these rRNA genes, which may be distributed among several chromosomes, vary in haploid number from less than 50 to over 10,000, depend-

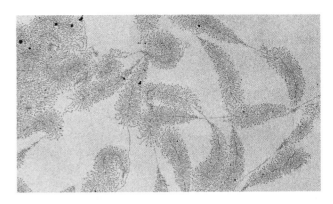

Figure 33-26
An electron micrograph of tandem arrays of actively transcribing 18S, 5.8S, and 28S rRNA genes from the nucleoli of the newt *Notophthalmus viridescens.* The axial fibers are DNA. The fibrillar "Christmas tree" matrices, which consist of newly synthesized RNA strands in complex with proteins, outline each transcriptional unit. Note that the longest ribonucleoprotein branches of each "Christmas tree" are only ~ 10% the length of their corresponding DNA stem. Apparently, the RNA strands are compacted through secondary structure interactions and/or protein associations. The matrix-free segments of DNA between the arrowheads are the untranscribed spacers. [Courtesy of Oscar L. Miller, Jr., University of Virginia.]

ing on the species. Humans, for example, have 50 to 200 blocks of rDNA spread over 5 chromosomes.

The Nucleolus Is the Site of rRNA Synthesis and Ribosome Assembly

In a typical interphase cell nucleus, the rDNA condenses to form a single nucleolus (Fig. 1-5). There, as Fig. 33-26 suggests, these genes are rapidly and continuously transcribed by RNA polymerase I (Section 29-2F). The nucleolus, as demonstrated by radioactive labeling experiments, is also the site where these rRNAs are post-transcriptionally processed and assembled with cytoplasmically synthesized ribosomal proteins into immature ribosomal subunits. Final assembly of the ribosomal subunits only occur as they are being transferred to the cytoplasm, which presumably prevents the premature translation of partially processed mRNAs (hnRNAs) in the nucleus.

5S rRNA Is Synthesized Separately from Other rRNAs

The genes coding for the 120-nucleotide 5S rRNAs, much like the other rRNA genes, are arranged in clusters that contain a total of several hundred to several hundred thousand tandem repeats distributed among one or more chromosomes. In *Xenopus laevis*, the organism whose 5S rRNA genes are best characterized, the repeating unit consists of the 5S rRNA gene, a nearby **pseudogene** (a 101 bp segment of the 5S rRNA

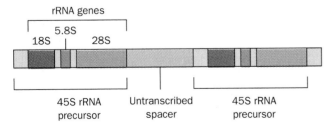

Figure 33-25
The 18S, 5.8S, and 28S rRNA genes are organized in tandem repeats in which sequences coding for the 45S rRNA precursor are interspersed by untranscribed spacers.

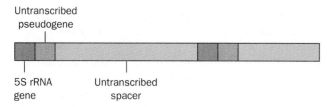

Figure 33-27
The organization of the 5S RNA genes in *Xenopus laevis*.
Each of the ~750 nucleotide-long tandemly repeated units
consists of a 5S rRNA gene trailed by an untranscribed
spacer in which a pseudogene closely follows the 5S gene.

gene which, curiously, is not transcribed), and an untranscribed spacer of variable length but averaging
~400 bp (Fig. 33-27). The 5S rRNA genes are transcribed outside of the nucleolus by RNA polymerase III
(Section 29-2F), an enzyme distinct from RNA polymerase I. 5S rRNA must therefore be transported into the
nucleolus for incorporation into the large ribosomal
subunit. The tRNA genes, which are likewise transcribed by RNA polymerase III, are also multiply reiterated and clustered, but the organization of these ~60
different gene types is largely unknown.

Histone Genes Are Reiterated

Histone mRNAs have relatively short cytoplasmic
lifetimes because of their lack of the poly(A) tails that
are appended to other eukaryotic mRNAs (Section
29-4A). Yet, histones must be synthesized in large
amounts during S phase of the cell cycle (when DNA is
synthesized). *This process is made possible through the
multiple reiteration of histone genes, which in most organisms are the only identically repeated genes that code for
proteins.* This organization, it is thought, permits the
sensitive control of histone synthesis through the coordinate transcription of sets of histone genes. Histone
genes also differ from most other eukaryotic genes in
that almost all histone sequences lack introns (noncoding intervening sequences; Section 29-4A). The significance of this observation is unknown.

There is little relationship between a genome's size
and its total number of histone genes. For example, birds
and mammals have 10 to 20 copies of each of the 5
histone genes, *Drosophila* has ~100, and sea urchins
have several hundred. This suggests that the efficiency
of histone gene expression varies with species. In many
organisms, as sequencing studies of cloned genes have
shown, the histone genes are organized into tandemly
repeated quintets consisting of a gene coding for each of
the 5 different histones interspersed by untranscribed
spacers (Fig. 33-28). The gene order and the direction of
transcription in these quintets is preserved over large
evolutionary distances. Corresponding spacer sequences vary widely among species and, to a limited
extent, among the repeating quintets within a genome.
In birds and mammals, this repetitive organization has

broken down; their histone genes occur in clusters but in
no particular order.

Reiterated Sequences May Be Generated and Maintained by Unequal Cross-Overs and/or Gene Conversion

A major question concerning reiterated genes is *How
do they maintain their identity?* The usual mechanism of
Darwinian selection would seem ineffective in doing so
since deleterious mutations in a few members of a multiply repeated set of identical genes would have little phenotypic effect. Indeed, many mutations do not affect the
function of a gene product and are therefore selectively
neutral. Reiterated gene sets must therefore maintain
their homogeneity through some additional mechanism. Two such mechanisms seem plausible:

1. In the **unequal cross-over** mechanism (Fig. 33-29*a*),
   recombination occurs between homologous segments of misaligned chromosomes thereby excising a
   segment from one of the chromosomes and adding it
   to the other. Computer simulations indicate that such
   repeated expansions and contractions of a chromosome will, by random processes, generate a cluster of
   reiterated sequences that have been derived from a
   much smaller ancestral cluster.

2. In the **gene conversion** mechanism (Fig. 33-29*b*),
   one member of a reiterated gene set "corrects" a
   nearby variant through a process resembling recombination repair (Section 31-5C).

Since point mutations are rare events compared to
cross-overs, either mechanism would eventually result
in a newly arisen variant copy of a repeated sequence
either being eliminated or taking over the entire cluster.
If a mutation that has been so-concentrated is deleterious, it will be eliminated by Darwinian selection. In
contrast, variant spacers, which are not as subject to

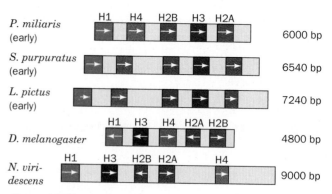

Figure 33-28
The organization and lengths of the histone gene cluster
repeating units in a variety of organisms (the top three
organisms are distantly related sea urchins). Coding regions
are indicated in color and spacers are gray. The arrows
denote the directions of transcription.

(a) Crossing-over

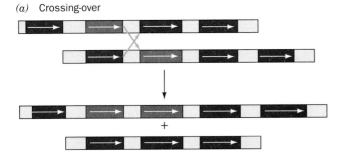

(b) Recombination repair

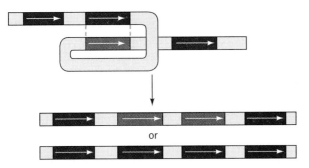

Figure 33-29
Two possible mechanisms for maintaining the homogeneity of a tandem multigene family. (a) Unequal crossing-over between mispaired but similar genes results in an unpaired DNA segment being deleted from one chromosome and added to the other. (b) Gene conversion "corrects" one member of a tandem array with respect to the other via a recombination repair mechanism. Repeated cycles of either process may either eliminate a variant gene or spread it throughout the entire tandem array.

selective pressure, would be eliminated at a slower rate. The existence of reiterated sets of identical genes separated by somewhat heterogeneous spacers may therefore be reasonably attributed to either homogenization model.

D. Gene Amplification

The selective replication of a particular set of genes, a process known as **gene amplification,** normally occurs only at specific stages of the life cycle of certain organisms. In the following subsections, we outline what is known about this phenomenon.

rRNA Genes Are Amplified During Oogenesis

The rate of protein synthesis during the early stages of embryonic growth is so great that in some species the normal genomic complement of rRNA genes cannot satisfy the demand for rRNA. In these species, notably certain insects, fish and amphibians, the rDNA is differentially replicated in developing oocytes (immature egg cells). In one of the most spectacular examples of this process, the rDNA in *Xenopus laevis* oocytes is amplified by ~1500 times its amount in somatic cells to yield some 2 million sets of rRNA genes comprising nearly 75% of

the total cellular DNA. The amplified rDNA occurs as extrachromosomal circles, each containing one or two transcription units, that are organized into hundreds of nucleoli (Fig. 33-30). Mature *Xenopus* oocytes therefore contain ~10^{12} ribosomes, 200,000 times the number in most larval cells. This is so many that mutant zygotes (fertilized ova), which lack nucleoli (and thus cannot synthesize new ribosomes; the oocyte's extra nucleoli are destroyed during its first meiotic division) survive to the swimming tadpole stage with only their maternally supplied ribosomes.

What is the mechanism of rDNA amplification? An important clue is that the untranscribed spacers from a given extrachromosomal nucleolus all have the same length, whereas we have seen that the corresponding chromosomal spacers exhibit marked length heterogeneities. This observation suggests that the rDNA circles in a single nucleolus are all descended from a single chromosomal gene. Gene amplification has been shown to occur in two stages: A low level of amplification in the first stage followed massive amplification in the second stage. It therefore seems likely that, in the first stage, no more than a few chromosomal rRNA genes are replicated by an unknown mechanism and the daughter strands released as extrachromosomal circles. Then, in the second stage, these circles are multiply replicated by the rolling circle mechanism (Section 31-3B). In support of this hypothesis are electron micrographs of amplified genes showing the "lariat" structures postulated to be rolling circle intermediates (Fig. 31-18).

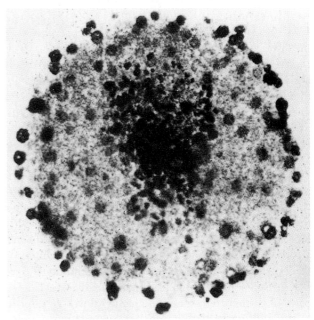

Figure 33-30
A photomicrograph of an isolated oocyte nucleus from *X. laevis.* Its several hundred nucleoli, which contain amplified rRNA genes, appear as darkly staining spots. [Courtesy of Donald Brown, Carnegie Institution of Washington.]

Figure 33-31
An electron micrograph of a chorion gene-containing chromatin strand from an oocyte follicle cell of *D. melanogaster*. The strand has undergone several rounds of partial replication (*arrows* at replication forks) to yield a multiforked structure containing several parallel copies of chorion genes. [Courtesy of Oscar L. Miller, Jr., University of Virginia.]

Chorion Genes Are Amplified

The only other known example of programmed gene amplification is that of the *Drosophila* ovarian follicle cell genes that code for **chorion** (egg shell) **proteins** (ovarian follicle cells surround and nourish the maturing egg). Prior to chorion synthesis, the entire haploid genome of each ovarian follicle cell is replicated 16-fold. This process is followed by an ~10-fold selective replication of only the chorion genes to form a multiply branched (partially polytene) structure in which the amplified chorion genes remain part of the chromosome (Fig. 33-31). Interestingly, chorion gene amplification does not occur in silk moth oocytes. Rather, this organism's genome has multiple copies of chorion genes.

Drug Resistance Can Result from Gene Amplification

In cancer chemotherapy, a common observation is that the continued administration of a cytotoxic drug causes an initially sensitive tumor to become increasingly drug resistant to the point that the drug loses its therapeutic efficacy. One mechanism by which a cell line can acquire such drug resistance is through the overproduction of the drug's target enzyme. Such a process can be observed, for example, by exposing cultured animal cells to the dihydrofolate analog methotrexate. This substance, it will be recalled, all but irreversibly binds to dihydrofolate reductase (DHFR) thereby inhibiting DNA synthesis (Section 26-4B). Slowly increasing the methotrexate dose yields surviving cells that ultimately contain up to 1000 copies of the DHFR gene and are thereby capable of tremendous overproduction of this enzyme — a clear laboratory demonstration of Darwinian selection. Members of some of these cell lines contain extrachromosomal elements known as **double minute chromosomes** that each bear one or more copies of the DHFR gene whereas in other cell lines, the additional DHFR genes are chromosomally integrated. The mechanism of gene amplification in either cell type is not well understood although it is worth noting that this phenomenon is only known to occur in cancer cells.

Both types of amplified genes are genetically unstable; further cell growth in the absence of methotrexate results in the gradual loss of the extra DHFR genes.

E. Clustered Gene Families: Hemoglobin Gene Organization

Few proteins in a given organism are really unique. Rather, like the digestive enzymes trypsin, chymotrypsin, and elastase (Section 14-3), or the various collagens (Section 7-2C), they are usually members of families of structurally and functionally related proteins. In many cases, the family of genes specifying such proteins are clustered together in a single chromosomal region. In the following subsections, we consider the organization of two of the best characterized clustered gene families, those coding for the two types of human hemoglobin subunits. The clustered gene families that encode immune system proteins are discussed in Section 34-2C.

Human Hemoglobin Genes Are Arranged in Two Developmentally Ordered Clusters

Human adult hemoglobin (HbA), consists of $\alpha_2\beta_2$ tetramers in which the α and β subunits are structurally related. The first hemoglobin made by the human embryo, however, is a $\zeta_2\varepsilon_2$ tetramer (**Hb Gower 1**) in which ζ and ε are α- and β-like subunits, respectively (Fig. 33-32). By around 8-weeks postconception, the embryonic subunits have been supplanted (in newly formed erythrocytes) by the α subunit and the β-like γ subunit to

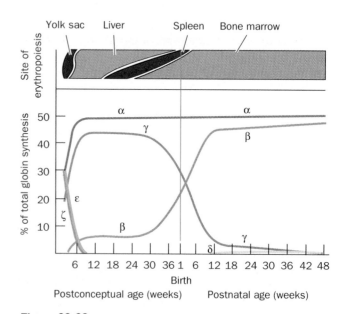

Figure 33-32
The progression of human globin chain synthesis with embryonic and fetal development. Note that any red blood cell contains only one type each of α- and β-like subunits. The sites of **erythropoiesis** (red cell formation) are indicated in the upper panel. [After Weatherall, D. J. and Clegg, J. B., *The Thalassaemia Syndromes* (3rd ed.), p. 64, Blackwell Scientific Publications (1981).]

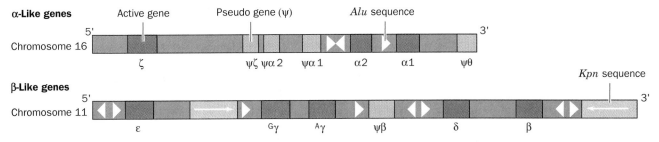

Figure 33-33
The organization of human globin genes on their respective coding strands. Red boxes represent active genes, green boxes represent pseudogenes, yellow boxes represent *Kpn* sequences with the arrows indicating their relative orientations, and triangles represent *Alu* sequences in their relative orientations. [After Karlsson, S. and Nienhuis, A. W., *Annu. Rev. Biochem.* **54,** 1074 (1985).]

form fetal hemoglobin (HbF), $\alpha_2\gamma_2$ (the hemoglobins present during the change-over period, $\alpha_2\varepsilon_2$, and $\zeta_2\gamma_2$, are named **Hb Gower 2** and **Hb Portland,** respectively). The γ subunit is gradually superseded by β starting a few weeks before birth. Adult blood normally contains ~97% HbA, 2% **HbA$_2$** ($\alpha_2\delta_2$ in which δ is a β variant), and 1% HbF.

In mammals, the genes specifying the α- and β-like hemoglobin subunits form two different gene clusters that occur on separate chromosomes. This distribution was largely determined through the sequence analysis of cloned hemoglobin genes, which were identified in genomic libraries (Section 28-8C) by Southern blotting (Section 28-4C). The probes used in this process were derived from hemoglobin mRNAs which, being the major mRNA products of reticulocytes, are readily isolated. In humans and many other mammals, the genes in each globin cluster are arranged, 5' → 3' on the coding strands, in the order of their developmental expression (Fig. 33-33). This ordering is common in mammals but not universal; in the mouse β gene cluster, for instance, the adult genes precede the embryonic genes.

The β-globin gene cluster (Fig. 33-33), which spans >60 kb, contains five functional genes: the embryonic ε gene, two fetal genes, $^G\gamma$ and $^A\gamma$ (duplicated genes that encode polypeptides differing only by having either Gly or Ala at their positions 136), and the two adult genes, δ

and β. The β-globin cluster also contains one **pseudogene,** $\psi\beta$ (an untranscribed relic of an ancient gene duplication that is ~75% homologous with the β gene), eight copies of the *Alu* family sequence, and two copies of the *Kpn* **family** (a 6.0-kb moderately repetitive DNA, so named because most of its ~10^4 members in the primate haploid genome have a cleavage site for the restriction endonuclease *Kpn*I).

The α-globin gene cluster (Fig. 33-33), which spans 28 kb, contains three functional genes, the embryonic ζ gene and two slightly different α genes, α1 and α2, which encode identical polypeptides. The α-cluster also contains four pseudogenes, $\psi\zeta$, $\psi\alpha$2, $\psi\alpha$1, and $\psi\theta$, and three *Alu* sequences.

Hemoglobin Genes All Have the Same Exon–Intron Structure

Protein-coding sequences represent <5% of either globin gene cluster. This situation is largely a consequence of the heterogeneous collection of untranscribed spacers separating the genes in each cluster. In addition, *all known vertebrate globin genes, including that of myoglobin and the hemoglobin pseudogenes, consist of three similarly placed coding sequences (exons) separated by two somewhat variable unexpressed intervening sequences (introns; Fig. 33-34).* This gene structure apparently arose quite early in vertebrate history, well over 500 million years ago.

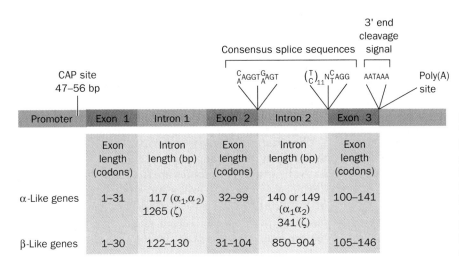

Figure 33-34
The structure of the prototypical hemoglobin gene indicating the conserved sequences at the exon–intron boundaries (splice sequences) and at the 3' end of the gene (polyadenylation site). The length of each exon (in codons) and each intron (in base pairs) is given. [After Karlsson, S. and Nienhuis, A. W., *Annu. Rev. Biochem.* **54,** 1079 (1985).]

	Exon length (codons)	Intron length (bp)	Exon length (codons)	Intron length (bp)	Exon length (codons)
α-Like genes	1–31	117 (α_1, α_2) 1265 (ζ)	32–99	140 or 149 ($\alpha_1\alpha_2$) 341 (ζ)	100–141
β-Like genes	1–30	122–130	31–104	850–904	105–146

DNA Polymorphisms Can Establish Genealogies

Unexpressed sequences, which are subject to little selective pressure, evolve so much faster than expressed sequences that they even accumulate significant numbers of sequence **polymorphisms** (variations) within a single species. Consequently, the evolutionary relationships among populations within a species can be established by determining how a series of polymorphic DNA sequences are distributed among them. For example, the genealogy of several diverse human populations has been inferred from the presence or absence of certain restriction sites [restriction-site length polymorphisms (RFLPs); Section 28-6A] in five segments of their β-globin gene clusters. This study has led to the construction of a "family tree" (Fig. 33-35), which indicates that non-African (Eurasian) populations are much more closely related to each other than they are to African populations. Fossil evidence indicates that anatomically modern man arose in Africa about 100,000 years ago and rapidly spread throughout that continent. This family tree therefore suggests that all Eurasian populations are descended from a surprisingly small "founder population" (perhaps only a few hundred individuals) that left Africa ~50,000 years ago. A similar analysis indicates that the sickle-cell variant of the β gene arose on at least three separate occasions in geographically distinct regions of Africa.

F. Significance of Introns

The rapidly growing body of known DNA sequences reveals that introns are rare in prokaryotic structural

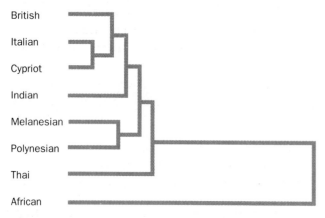

Figure 33-35
A family tree showing the lines of descent among eight human population groups as determined from the distribution of five restriction site polymorphisms in their β-globin gene clusters. The horizontal axis is indicative of the genetic distances between related populations and therefore of the times between their divergence. [After Wainscoat, J. S., Hill, A. V. S., Boyce, A. L., Flint, J., Hernandez, M., Thein, S. L., Old, J. M., Lynch, J. R., Falusi, A. G., Weatherall, D. J., and Clegg, J. B., *Nature* **319**, 493 (1986).]

genes, uncommon in lower eukaryotes such as yeast, and abundant in higher eukaryotes (the only known vertebrate structural genes lacking introns are those encoding histones and interferons). The exons in most interrupted genes have sizes in the range 100 to 250 bp. In contrast, intron sizes are broadly distributed from under 50 to over 20,000 bp. Moreover, the number of introns in a given gene can be surprisingly large; the most observed so far are the 51 introns in the 38-kb chicken $\alpha 2(I)$ collagen gene. Unexpressed sequences thereby constitute ~80% of a typical vertebrate structural gene.

What are the functions of introns? The argument that all introns are simply selfish DNA seems untenable since it would otherwise be difficult to rationalize why the evolution of splicing machinery offered any selective advantage over the simple elimination of the split genes. Yet, in most genes, introns have no obvious function (although introns may protect the integrity of the genes in gene families; see below, and act as regulatory elements in the transcription of certain genes; Section 33-3C). In fact, the number of introns in the gene encoding a given protein is not necessarily the same in all organisms or even within one organism. For instance, the rat has two functional insulin genes, one with two introns as do most rodent insulin genes, and the other in which one of these introns has been lost. It has therefore been proposed that introns had an essential function at an earlier stage of genetic evolution which is no longer important; that is, *introns may be genetic fossils.*

Introns Are the Products of Modular Gene Assembly

Walter Gilbert proposed that *protein-coding genes arose as collections of exons that were assembled by recombination between intron sequences. Modern introns are therefore the remnants of a process that facilitated protein evolution.* Considerable experimental evidence has accumulated that supports this hypothesis. For example, the triple helical region of chicken $\alpha 2(I)$ collagen, which consists of a 332-fold repeated triplet Gly-X-Y (Section 7-2C), is encoded by 42 of its gene's 52 exons. All of these exons are integral multiples of 9 bp (23 of them are 54 bp long with the rest consisting of either 45, 99, 108, or 162 bp) with each exon beginning with a Gly codon and ending with a Y codon. This distribution suggests that the gene segment specifying collagen's triple helical region evolved through multiple duplications of its repeating intron-flanked genetic element.

Pyruvate Kinase Exons Encode Discrete Structural Elements

The structural analysis of pyruvate kinase in terms of the chicken muscle gene's base sequence indicates that its exon–intron boundaries are functionally positioned. Each of this gene's 10 exons encodes a discrete element of protein secondary structure with most of the introns marking positions at which the polypeptide chain makes a reverse turn (Fig. 33-36). *The pyruvate kinase*

(a)

(b)

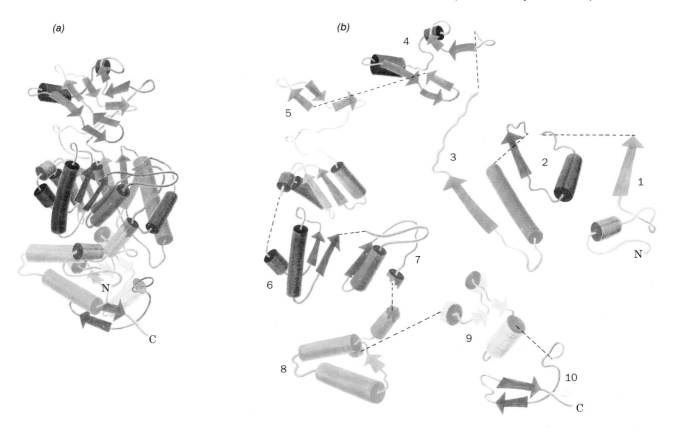

Figure 33-36
The structure of pyruvate kinase. (a) A single subunit of the tetrameric enzyme from cat muscle. (b) An exploded view of the cat muscle pyruvate kinase subunit in which the structural segments have been separated at the exon boundaries in the gene specifying the 88% homologous chicken muscle enzyme. The exons are numbered in the order they occur, 5′ → 3′, on the mRNA. [After Lonberg, N. and Gilbert, W., *Cell* **40**, 84 (1985).]

gene was apparently assembled by combining a series of smaller protein coding units and exploiting RNA splicing to express them as a single polypeptide.

The Exons of LDL Receptor Occur in Other Proteins

The gene sequence of the **LDL receptor** provides what is perhaps the most convincing evidence favoring Gilbert's hypothesis. This 839-residue plasma membrane protein functions to bind low density lipoprotein (LDL) to coated pits for transport into the cell via endocytosis (Section 11-4B). LDL receptor's 45-kb gene contains 18 exons, most of which encode specific functional domains of the protein. *The most intriguing aspect of this sequence, however, is that 13 of its exons specify polypeptide segments that are homologous to segments in other proteins:*

1. Five exons encode a sevenfold repeat of a 40-residue sequence that occurs once in **complement C9** (an immune system protein; Section 34-2F).

2. Three exons each encode a 40-residue repeat similar to that occurring four times in **epidermal growth factor (EGF) precursor** (EGF is a hormonally active polypeptide that stimulates cell proliferation) and

once each in three blood clotting system proteins: **factor IX, factor X,** and **protein C** (Section 34-1).

3. Five exons encode a 400-residue sequence that is 33% homologous with a polypeptide segment that is shared only with EGF precursor.

Evidently, the LDL receptor gene is modularly constructed from exons that also encode portions of other proteins. Numerous other eukaryotic proteins are similarly constituted.

Introns May Have Been Selectively Eliminated from the Genes of Lower Organisms

If introns are the remnants of a primordial gene shuffling process, why are they absent or nearly so in the "lower" forms of life from which the "higher" forms evolved (e.g., the gene for yeast pyruvate kinase, whose amino acid sequence is 45% homologous with that from chicken, has no introns)? The most plausible explanation of this observation, in light of the foregoing data, is that in lower forms of life, whose life styles place a premium upon efficiency (Section 1-2), introns have been selectively eliminated. In contrast, higher organisms, which are adapted to stable environments, have had much less selective pressure to do so (although the rat insulin genes discussed previously provide a clear example of intron elimination). Indeed, the large sizes of many vertebrate introns suggests that they have been invaded by selfish DNA.

Introns May Genetically Stabilize Gene Families

In addition to their role in facilitating the evolution of new proteins, *the introns in gene families may function to protect their neighboring exons from elimination via unequal crossing-over* (Fig. 33-29a). Duplicated genes are particularly susceptible to this form of degradation because their similar base sequences promote their mispairing. Their alteration with the much more variable and therefore less readily mispaired introns, however, inhibits this process. Prokaryotes and yeast, which have few gene families, have little need of such protection.

G. The Thalassemias: Genetic Disorders of Hemoglobin Synthesis

The study of mutant hemoglobins (Section 9-3) has provided invaluable insights into structure–function relationships in proteins. Likewise, the study of defects in hemoglobin expression has greatly facilitated our understanding of eukaryotic gene expression.

The most common class of inherited human disease results from the impaired synthesis of hemoglobin subunits. These anemias are named **thalassemias** (Greek: *thalassa*, sea) because they commonly occur in the region surrounding the Mediterranean Sea (although they are also prevalent in Central Africa, India, and the Far East). The observation that malaria is or was endemic in these same areas (Fig. 6-13) led to the realization that heterozygotes for thalassemic genes (who appear normal or are only mildly anemic; a condition known as **thalassemia minor**) are resistant to malaria. Thus, as we have seen in our study of sickle-cell anemia (Section 9-3B), mutations that are seriously debilitating or even lethal in homozygotes (who are said to suffer from **thalassemia major**) may offer sufficient selective advantage to heterozygotes to ensure the propagation of the mutant gene.

Thalassemia can arise from many different mutations, each of which causes a disease state of characteristic severity. In α^0- and β^0-thalassemias, the indicated globin chain is absent, whereas in α^+- and β^+-thalassemias, the normal globin subunit is synthesized in reduced amounts. In what follows, we shall consider thalassemias that are illustrative of several different types of genetic lesions.

α-Thalassemias

Most α-thalassemias are caused by the deletion of one or both of the α-globin genes in an α gene cluster (Fig. 33-33). A variety of such mutations have been cataloged. In the absence of equivalent numbers of α chains, the fetal γ chains and the adult β chains form homotetramers: **Hb Bart's** (γ_4) and **HbH** (β_4). Neither of these tetramers exhibit any cooperativity or Bohr effect (Sections 9-1C and D), which makes their oxygen affinities so high that they cannot release oxygen under physiological conditions. Consequently, α^0-thalassemia occurs with 4 degrees of severity depending on whether an individual has 1, 2, 3, or 4 missing α-globin genes:

1. **Silent-carrier state:** The loss of one α gene is an asymptomatic condition. The rate of expression of the remaining α genes largely compensates for the less-than-normal α gene dosage so that, at birth, the blood contains only ~1 to 2% Hb Bart's.

2. **α-Thalassemia trait:** With two missing α genes (either one each deleted from both α gene clusters or both deleted from one cluster), only minor anemic symptoms occur. The blood contains ~5% Hb Bart's at birth.

3. **Hemoglobin H disease:** Three missing α genes results in a mild to moderate anemia. Affected individuals can usually lead normal or nearly normal lives.

4. **Hydrops fetalis:** The lack of all four α genes is invariably lethal. Unfortunately, the synthesis of the embryonic ζ-chain continues well past the 8 weeks postconception when it normally ceases (Fig. 33-32) so that the fetus usually survives until around birth.

α-Thalassemias caused by nondeletion mutations are relatively uncommon. One of the best characterized such lesions changes the UAA termination codon of the $\alpha2$-globin gene to CAA (a Gln codon) so that protein synthesis continues for the 31 codons beyond this site to the next UAA. The resultant **Hb Constant Spring** is produced in only small amounts because, for unknown reasons, its mRNA is rapidly degraded in the cytosol. Another point mutation in the $\alpha2$ gene changes Leu H8(125)α to Pro, which no doubt disrupts the H helix. The consequent α^+-thalassemia results from the rapid degradation of this abnormal **Hb Quong Sze**.

β-Thalassemias

Heterozygotes of β-thalassemias are usually asymptomatic. Homozygotes become so severely anemic, once their HbF production has diminished, however, that many require frequent blood transfusions to sustain life and all require them to prevent the severe skeletal deformities caused by bone marrow expansion. The anemia results not only from the lack of β-chains but also from the surplus of α chains. The latter form insoluble membrane-damaging precipitates that cause premature red cell destruction (Section 9-3A). The coinheritance of α-thalassemia therefore tends to lessen the severity of β-thalassemia major.

In β-thalassemia, there may be an increased production of the δ- and γ-chains so that the consequent extra HbA₂ and HbF can compensate for some of the missing HbA. In **δβ-thalassemia,** the neighboring δ and β genes have both been deleted so that only increased production of the γ chain is possible. Yet many δβ-thalassemics, for reasons that are not understood, produce so much HbF as adults that they are asymptomatic. Such individuals are said to have **hereditary persistence of fetal hemoglobin (HPFH).**

β⁰-Thalassemias caused by deletions are rare compared to those causing α⁰-thalassemias. This is probably because the long repeated sequences in which the α-globin genes are embedded make them more prone to unequal crossing-over than the β-globin gene. Nevertheless, a β-thalassemic lesion causing the production of **Hb Lepore** is a particularly clear instance of this deletion mechanism. This lesion, the consequence of a deletion extending from within the δ gene to the corresponding position of its neighboring β gene, yields a δ/β hybrid subunit. Such deletions almost certainly arose through unequal crossovers between the β gene on one chromosome and the δ gene on another (Fig. 33-37). The second product of such crossovers, a chromosome containing a β/δ hybrid flanked by normal δ and β genes (Fig. 33-37) is known as **Hb anti-Lepore.** Homozygotes for Hb Lepore have symptoms similar to those of β-thalassemia major, whereas homozygotes for Hb anti-Lepore, which have the full complement of normal globin genes, are symptom-free and have only been detected through blood tests.

Most β-thalassemias are caused by a wide variety of point mutations that affect the production of β chains. These include:

1. Nonsense mutations that convert normal codons to the termination codon UAG.

2. Frameshift mutations that insert/delete one or more base pairs into/from an exon.

3. Point mutations in the β gene's promoter region, either in its TATA box or in its CACCC box (Section 29-2F). These attenuate transcriptional initiation.

4. Point mutations that alter the sequence at an exon–intron junction (Section 29-4A). These diminish/abolish splicing and/or activate a **cryptic splice site** (an exon–intron junctionlike sequence that normally is not spliced) to pair with the altered intron's unaltered end.

5. A point mutation that alters an intron's lariat-branch site (Section 29-4A). This activates a cryptic 3' splice upstream of the original site leading to the excision of a shorter-than-normal intron.

6. Point mutations that create new splice sites. These either compete with the neighboring normal splice site or pair with a nearby cryptic splice site.

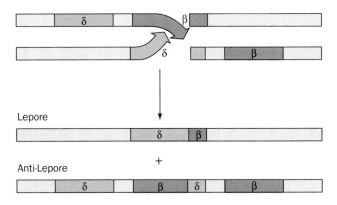

Figure 33-37
The formation of Hb Lepore and Hb anti-Lepore by unequal crossing-over between the β-globin gene on one chromosome and the δ-globin gene on its homolog.

7. A point mutation that alters the AAUAAA cleavage signal at the mRNAs 3' end (Section 29-4A).

Consideration of the effects of these mutations, particularly those involving gene splicing, has confirmed and extended our understanding of how eukaryotic genes are constructed and expressed.

3. CONTROL OF EXPRESSION

The elucidation of the mechanisms controlling gene expression in eukaryotes has lagged at least 20 years behind that of prokaryotes. This is largely because the types of genetic analyses that have been so useful in characterizing prokaryotic systems (which require the detection of very rare events) are precluded in multicellular organisms by their much slower reproductive rates. Compounding this problem are the difficulties in selecting for mutations in essential genes; the missing product of a defective enzyme in a multicellular organism usually cannot be replaced by simply adding that product to the diet as is often possible with, say, *E. coli.* This latter difficulty can be partially overcome by the rather laborious task of growing cells from multicellular organisms in **tissue culture.** Since somatic cells do not normally undergo genetic recombination, however, genetic manipulations cannot be carried out in tissue culture the way they can in a bacterial culture.

What has made genetic manipulations of eukaryotic systems feasible is the development, in the 1970s, of molecular cloning techniques (Section 28-8). The gene encoding a particular eukaryotic protein can be identified in genomic or cDNA libraries through Southern blotting (Section 28-4C) using an oligonucleotide probe whose sequence encodes a segment of the protein (a process termed "reverse genetics" because, in prokaryotes, genetics has been used to characterize proteins rather than *vice versa*). The gene may then be modified, for example, through site-directed mutagenesis (Section 28-7), and the effect of the modification analyzed in an expression vector such as *E. coli* or yeast, or alternatively, *in vitro*. The expression of cloned genes in multicellular organisms has been made possible through the development of a process in which DNA is microinjected into the nucleus of a fertilized ovum. Such DNA often integrates into the chromosome of the resulting zygote that then undergoes normal development to form a **transgenic** individual whose cells each contain the foreign genes (in *Xenopus*, this merely involves allowing the transfected egg to hatch whereas in mice, the fertilized ovum must be implanted in the uterus of a properly prepared foster mother; see Fig. 27-24 for a striking example of a transgenic mouse). This laborious process has been greatly simplified, at least in mice, by the recent discovery that sperm readily take up foreign DNA which they transfer to the ova they fertilize (Section 27-2A). The genome of already multicellular organisms may be altered, in a technique that holds the greatest promise for gene therapy, through the use of defective (unable to reproduce) retroviruses that contain the genes to be transferred. Thus, the eukaryotic genome can now be manipulated albeit with considerable clumsiness. We will, no doubt, become more adept at doing so as we gain further understanding of how eukaryotic chromosomes are organized and expressed.

In this section we consider the molecular basis of the enormous expressional variation that eukaryotic cells exhibit. In doing so, we shall first study the nature of transcriptionally active chromatin, then discuss how genetic expression in eukaryotes is mainly regulated through the control of transcriptional initiation, and finally consider the other means by which eukaryotes control genetic expression. In the next section, we take up what is known about the molecular basis of normal cell differentiation as well as its aberration, cancer.

A. Chromosomal Activation

Interphase chromatin, as is mentioned in Section 33-1, may be classified in two categories: the highly condensed and transcriptionally inactive heterochromatin, and the diffuse and transcriptionally active or activatable euchromatin (Fig. 33-2). Two types of heterochromatin have been distinguished:

1. **Constitutive heterochromatin,** which is permanently condensed in all cells, consists mostly of the highly repetitive sequences clustered near the chromosomal centromeres (Section 33-2B). Constitutive heterochromatin is therefore transcriptionally inert.

2. **Facultative heterochromatin,** which varies in a tissue-specific manner. Presumably the condensation of facultative heterochromatin functions to transcriptionally inactivate large chromosomal blocks.

Most Mammalian Cells Have Only One Active X Chromosome

Female mammalian cells contain two X chromosomes, whereas male cells have one X and one Y chromosome. *Female somatic cells, however, maintain only one of their X chromosomes in a transcriptionally active state.* The inactive X chromosome is visible during interphase as a heterochromatin structure known as a **Barr body** (Fig. 33-38). In marsupials (pouched mammals), the Barr body is always the paternally inherited X chromosome. In placental mammals, however, one randomly selected X chromosome in every somatic cell is inactivated when the embryo consists of only a few cells. The progeny of each of these cells maintain the same inactive X chromosomes. *Female placental mammals are therefore mosaics composed of clonal groups of cells in which the active X chromosome is either paternally or maternally inherited.* This situation is particularly evident in human females who are heterozygotes for the X-linked congenital sweat gland deficiency **anhidrotic ectodermal dysplasia.** The skin of these women consists of patches lacking sweat glands, in which only the X chromosome containing the mutant gene is active, alternating with normal patches in which only the other X chromosome is active. Similarly, calico cats, whose coats consist of patches of black fur and yellow fur, are almost always females whose two X chromosomes are allelic for black and yellow furs.

We do not know the mechanism of X chromosome inactivation, or how an X chromosome confers its state of activity on its progeny. One possibility, as we have discussed in Section 31-7, is that the DNA of the inactive

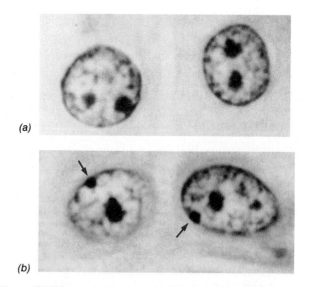

(a)

(b)

Figure 33-38
Photomicrographs of stained nuclei from human oral epithelial cells: (a) From a normal XY male showing no Barr body. (b) From a normal XX female showing a single Barr body (*arrows*). The presence of Barr bodies permits the rapid determination of an individual's chromosomal sex. [From Moore, K. L. and Barr, M. L., *Lancet* **2,** 57 (1955).]

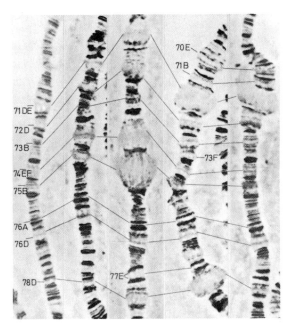

Figure 33-39
A series of photomicrographs showing the formation and regression of chromosome puffs (*lines*) in a *D. melanogaster* polytene chromosome over a 22-h period of larval development. Very large puffs are also known as **Balbiani rings.** [Courtesy of Michael Ashburner, Cambridge University.]

X chromosome is more heavily methylated than that of the active X chromosome and that the methylation state of each X chromosome is transmitted to its progeny via **maintenance methylation.** Indeed, there is an inverse correlation between the degree of methylation of a 5′ cluster of C's in the gene encoding hypoxanthine-guanine phosphoribosltransferase (HGPRT; Section 26-2B) and the degree of activity of the X chromosome that contains it.

Chromosome Puffs and Lampbrush Chromosomes Are Transcriptionally Active

The condensed state of facultative heterochromatin presumably renders it transcriptionally inactive by making its DNA inaccessible to the proteins mediating transcription. Conversely, *transcriptionally active chromatin must have a relatively open structure.* Such decondensed chromatin occurs in the **chromosome puffs** that emanate from single bands of giant polytene chromosomes (Fig. 33-39). These puffs reproducibly form and regress as part of the normal larval development program (Fig 33-39) and in response to physiological stimuli such as hormones and heat. Autoradiography studies with ³H-labeled uridine and immunofluorescence studies using antibodies against RNA polymerase II clearly demonstrate that *puffs are the major sites of RNA synthesis in polytene chromosomes.*

The analogous decondensation of nonpolytene chromosomes occurs most conspicuously in the so-called **lampbrush chromosomes** of amphibian oocytes (Fig. 33-40). During their prolonged meiotic prophase I (Fig. 27-3), these previously condensed chromosomes loop out segments of transcriptionally active DNA that electron micrographs such as Fig. 33-41 indicate are usually single transcription units.

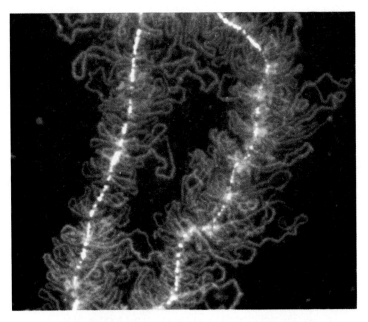

Figure 33-40
An immunofluorescence micrograph of lampbrush chromosome from an oocyte nucleus of the newt *Notophthalmus viridescens*. The chromosome's numerous transcriptionally active loops give rise to the name "lampbrush" (an obsolete implement for cleaning kerosene lamps). [From Roth, M. B. and Gall, J. G., *J. Cell Biol.* **105,** 1049 (1987). Copyright © 1987 by Rockefeller University Press.]

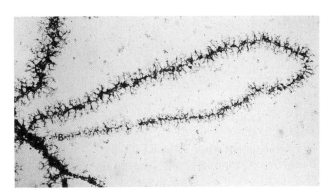

Figure 33-41
An electron micrograph of a single loop of a lampbrush chromosome. The ribonucleoprotein matrix coating the loop increases in thickness from one end of the loop (A) to the other (B), which indicates that the loop comprises a single transcriptional unit. [Courtesy of Oscar L. Miller, Jr., University of Virginia.]

B. Regulation of Transcriptional Initiation

The foregoing observations suggest that selective transcription is mainly responsible for the differential protein synthesis among the various types of cells in the same organism. It was not until 1981, however, that James Darnell actually demonstrated this to be the case as follows. Experimentally useful amounts of mouse liver genes were obtained by inserting the cDNAs of mouse liver mRNAs (some 95% of which are cytosolic) into plasmids and replicating them in *E. coli* (Section 28-8A). By hybridizing the resulting cloned cDNAs with radioactively labeled mRNAs from various mouse cell types, the *E. coli* colonies containing liver-specific genes were distinguished from colonies containing genes common to most mouse cells. After obtaining 12 liver-specific cDNA clones and three common cDNA clones, the question was asked, does a eukaryotic cell transcribe only the genes encoding the proteins it synthesizes or does it transcribe all of its genes but only properly process the transcripts it translates. This question was answered by hybridizing the cloned mouse genes with freshly synthesized and therefore unprocessed RNAs (hnRNAs) obtained from the nuclei of mouse liver, kidney, and brain cells (Fig. 33-42). Only the RNAs extracted from liver nuclei hybridized with the 12 liver-specific cDNAs that were probed. The RNAs from all three cell types, however, hybridized with the DNA from the three clones containing the common mouse genes. Evidently, *liver-specific genes are not transcribed by brain or kidney cells. This strongly suggests that the control of genetic expression in eukaryotes is primarily exerted at the level of transcription.*

Transcriptionally Active Chromatin Is Sensitive to Nuclease Digestion

The open structure of transcriptionally active chromatin presumably gives the transcriptional machinery access to the active genes. This hypothesis is corroborated by Harold Weintraub's demonstration that *transcriptionally active chromatin is more susceptible to digestion by pancreatic DNase I (a relatively nonspecific endonuclease) than is transcriptionally inactive chromatin.* For example, globin genes from chicken erythrocytes (avian red cells are nucleated) are more sensitive to DNase I digestion than are those from chicken oviduct (where eggs are made) as is indicated by the loss of the abilities of these genes to hybridize with a complementary DNA probe after DNase I treatment. Conversely, the gene coding ovalbumin (the major egg white protein) from oviduct is more sensitive to DNase I than is that from erythrocytes. Yet, nuclease sensitivity apparently reflects a gene's potential for transcription rather than transcription itself: The DNase I sensitivity of oviduct ovalbumin gene is independent of whether or not the oviduct has been hormonally stimulated to produce ovalbumin.

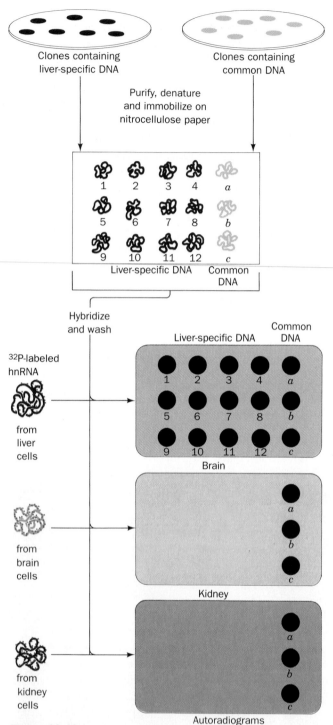

Figure 33-42

The primary role of selective transcription in the control of eukaryotic gene expression was established through gene cloning and hybridization techniques. Cloned cDNAs encoding 12 different mouse liver-specific proteins (*1 – 12*) and three different proteins common to most mouse cells (*a – c*) were purified, denatured, and spotted onto filter paper. The DNAs were hybridized with newly formed and therefore unprocessed radioactively labeled RNAs produced by either mouse liver, kidney, or brain nuclei.

Autoradiography showed that the liver RNAs hybridized with all 12 liver-specific cDNAs and all 3 common cDNAs but that the kidney and brain RNAs only hybridized with the common cDNAs.

Nonhistone Proteins Confer Nuclease Sensitivity

The variation of a given gene's transcriptional activity with the cell in which it is located indicates that chromosomal proteins participate in the gene activation process. Histones' chromosomal abundance and lack of variety, however, make it highly unlikely that they have the specificity required for this role. Among the most conspicuous nonhistone proteins are the members of the **high mobility group (HMG),** so-named because of their high electrophoretic mobilities in polyacrylamide gels. These highly conserved, low molecular mass (<30 kD) proteins have the unusual amino acid composition of ~25% basic side chains and 30% acidic side chains. Two of them, **HMG 14** and **HMG 17,** can be eluted from chick erythrocyte chromatin by 0.35M NaCl without gross structural changes to the nucleosomes. This treatment eliminates the preferential nuclease sensitivity of the erythrocyte globin genes. Their nuclease sensitivity can be restored, however, by adding HMGs 14 and 17, either individually or together, to the salt-extracted chromatin.

The HMGs are not tissue specific: HMGs 14 and 17 eluted from brain nuclei can also restore nuclease sensitivity to globin genes in HMG-depleted erythrocyte chromatin. Yet, the reverse process, adding HMGs 14 and 17 from erythrocytes to HMG-depleted chromatin from brain, neither induces nuclease sensitivity in the brain globin genes nor their selective transcription. These HMGs apparently do not recognize specific DNA sequences; rather, they must bind to tissue-specific chromatin components.

HMGs 14 and 17 bind directly to nucleosomes, possibly by displacing histone H1. Nucleosomal core particles in association with HMG 14 and/or 17 exhibit exactly the same nuclease sensitivity as does intact chromatin. *This observation indicates that genes need not be stripped of nucleosomes to be at least potentially transcriptionally active.*

Active Genes Have Nuclease Hypersensitive Control Sites

The very light digestion of transcriptionally active chromatin with DNase I and other nucleases has revealed the presence of **DNase I hypersensitive sites.** These specific DNA segments are mostly located in the 5'-flanking regions of transcriptionally active or activatable genes as well as in sequences involved in replication and recombination. Nuclease hypersensitive sites are apparently the "open windows" that allow proteins access to DNA control sequences. This is because *DNase I hypersensitive gene segments are free of nucleosomes.* For example, in SV40-infected cells, none of the ~24 nucleosomes that are complexed to the virus' 5.2-kb circular DNA (Fig. 33-43) incorporate the ~250 bp viral transcription initiation site thereby rendering that site nuclease hypersensitive.

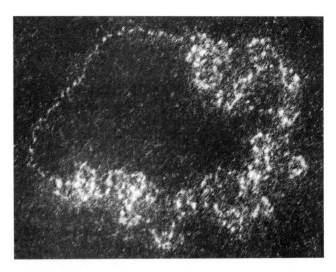

Figure 33-43
An electron micrograph of an SV40 minichromosome that has a nucleosome-free DNA segment. [Courtesy of Moshe Yaniv, Institut Pasteur.]

Gary Felsenfeld has similarly shown that the 5'-flanking region of the β^A-globin gene from chicken erythrocytes contains a 114 bp DNase I hypersensitive segment, that can be excised by the restriction endonuclease *Msp*I. The accessibility of such a long fragment indicates that it is not part of a nucleosome. Yet, since naked DNA is not nuclease hypersensitive, the special properties of nuclease hypersensitive chromatin must arise from the sequence-specific binding of proteins so as to exclude nucleosomes. In fact, two proteins, present in chicken erythrocytes but not oviducts, specifically bind to the β^A-globin gene so as to confer on it, when inserted in a plasmid and complexed with histones, the same nuclease hypersensitivity pattern it exhibits when isolated from erythrocytes. These proteins prevent the binding of histones to the hypersensitive site.

Transcriptional Initiation Is Mediated by Cell-Specific Factors Acting on Promoters and Enhancers

Differentiated eukaryotic cells possess a remarkable capacity for the selective expression of specific genes. The synthesis rates of a particular protein in two cells of the same organism may differ by as much as a factor of 10^9; that is, unexpressed eukaryotic genes are completely turned off. In contrast, simply repressible prokaryotic systems as the *E. coli lac* operon (Section 29-3B) exhibit no more than a thousandfold range in their transcriptional rates; they have significant basal levels of expression. Nevertheless, as we shall see below, *the basic mechanism of expressional control in eukaryotes resembles that in prokaryotes: the selective binding of proteins to specific genetic control sequences so as to modulate the rate of transcriptional initiation.*

The use of molecular cloning procedures has permit-

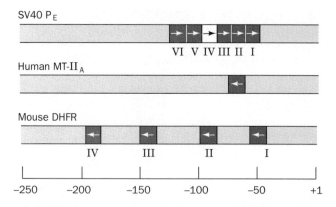

Figure 33-44
The arrangement and relative orientations of the GC boxes in the indicated promoters (each arrow represents the GC box sequence NGGGCGGNNN). The blue boxes represent Sp1-binding sites, whereas SV40 GC box IV is shown as a white box because Sp1 bound at GC box V prevents this transcription factor from efficiently binding to GC box IV. The transcription start site is designated by $+1$. DHFR = dihydrofolate reductase; MT = metallothionein. [After Kadonaga, J. T., Jones, K. A., and Tjian, R., *Trends Biochem. Sci.* **11**, 21 (1986).]

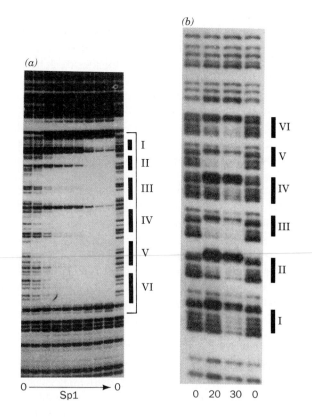

Figure 33-45
The identification of the Sp1-binding sites on the SV40 early promoter. (*a*) In a DNase I footprinting assay, a DNA segment that is ^{32}P end labeled on one strand is incubated with a binding protein and then lightly digested with DNase I such that, on average, the labeled DNA strand is nicked only once. The DNA is then denatured, the resulting labeled fragments separated according to size by electrophoresis on a sequencing gel, and detected by autoradiography. Unprotected DNA is cleaved more or less at random and therefore appears as a "ladder" of bands, each representing an additional nucleotide (as in a sequencing ladder; Sections 28-6B and C). In contrast, the DNA sequences that the protein protects from DNase I cleavage have no corresponding bands. In the above footprint, the lanes labeled "0" are the DNase I digestion pattern in the absence of Sp1 and in the other lanes the amount of Sp1 increases from left to right. The footprint boundary is delineated by the bracket and the positions of SV40 GC boxes I to VI are indicated. [From Kadonaga, J. T., Jones, K. A., and Tjian, R., *Trends Biochem. Sci.* **11**, 21 (1986). Copyright © 1986 by Elsevier Biomedical Press.] (*b*) In dimethyl sulfate (DMS) footprinting, a protein-complexed end-labeled DNA segment is treated with DMS, cleaved at its G residues, and the resulting fragments electrophoretically separated as in the chemical cleavage (Maxam–Gilbert) method of DNA sequence analysis (Section 28-6B). The DNA regions that the protein protects from methylation are not cleaved by this procedure and therefore are not represented in the resulting G residue "ladder". In the above autoradiogram, the number below each lane indicates the μL of an Sp1 fraction added to a given quantity of SV40 early promoter DNA. The positions of its GC boxes are indicated. [From Gidoni, D., Katonaga, J. T., Barrera-Saldaña, H., Takahashi, K., Chambon, P., and Tjian, R., *Science* **230**, 516 (1985). Copyright © 1985 by the American Society for the Advancement of Science.]

ted the demonstration that *eukaryotic promoter and enhancer elements mediate the expression of cell-specific genes* (recall that an enhancer is a gene sequence that is required for the full activity of its associated promoter but which may have a variable position and orientation with respect to that promoter; Section 29-2F). For example, William Rutter has linked the 5'-flanking sequences of either the insulin or the chymotrypsin gene to the sequence encoding **chloramphenicol acetyltransferase (CAT)**, an easily assayed enzyme not normally occurring in eukaryotic cells. A plasmid containing the insulin gene recombinant elicits expression of the CAT gene only when introduced into cultured cells that normally produce insulin. Likewise, the chymotrypsin recombinants are only active in chymotrypsin-producing cells. Dissection of the insulin control sequence indicates that the segment between its positions -103 and -333 contains an enhancer: In insulin producing cells only, it stimulates the transcription of the CAT gene with little regard to the enhancer's position and orientation relative to its promoter.

The foregoing indiates that cells contain specific factors that recognize the promoters and enhancers in the genes they transcribe. Numerous such **transcription factors** have recently been characterized. For instance, Robert Tjian has isolated a protein, **Sp1,** from cultured human cells that stimulates, by factors of 10 to 50, the transcription of cellular and viral genes containing at least one properly positioned GC box [GGGCGG (Section 29-2F); Fig. 33-44]. This protein binds, for example, to the 5'-flanking region of the SV40 virus early genes so as to protect its GC boxes from DNase I digestion (Fig.

33-45*a*; DNase I footprinting) and from methylation by dimethyl sulfate (Fig. 33-45*b*; DMS footprinting). Likewise, Sp1 specifically interacts with the four GC boxes in the upstream region of the mouse dihydrofolate reductase gene and with the single GC boxes in the human **metallothionein I$_A$** and **II$_A$** promoters (metallothioneins are metal ion-binding proteins that are implicated in heavy metal ion detoxification processes and whose synthesis is triggered by heavy metal ions).

In another well-characterized example of a transcription factor, the synthesis of the protective **heat-shock proteins** produced by eukaryotes in response to high temperatures is regulated by the 110-kD **heat-shock transcription factor (HSTF).** Footprinting studies have demonstrated that this protein specifically binds to a conserved partially palindromic segment (consensus sequence CNNGAANNTTCNNG) upstream of the many known heat-shock coding sequences. HSTF, like Sp1, when bound to its target sequence, stimulates the transcription of the associated gene. While protecting some regions of DNA from DNase I digestion, HSTF also renders portions of its bound DNA hypersensitive to this enzyme.

How do transcription factors work? *The enzyme responsible for eukaryotic mRNA synthesis, RNA polymerase II, unlike prokaryotic RNA polymerase holoenzyme (Section 29-2), lacks any inherent ability to bind to its promoters. It has therefore been postulated that the sequence-specific binding of several transcription factors in the vicinity of the transcription start site results in an interaction between these proteins and a putative **TATA-binding protein** that binds to the TATA box (Section 29-2F). These interactions stimulate RNA polymerase II to initiate transcription at that site* (Fig. 33-46). Transcription factors may bind cooperatively to each other and/or to RNA polymerase II in a manner resembling the binding of two λ repressor dimers and RNA polymerase to the o_R operator of bacteriophage λ (Section 32-3D). Indeed, molecular cloning experiments indicate that many enhancers consist of segments (modules) whose individual deletion reduces but does not eliminate enhancer activity. Such complex arrangements presumably permit transcriptional control systems to respond to a variety of stimuli in a graded manner.

The functional properties of eukaryotic transcription factors are surprisingly simple. They appear to have two domains:

1. A DNA-binding domain that binds to the protein's specified DNA sequence (and whose structural properties are discussed below).

2. A domain containing the transcription factor's activation function. Sequence analysis indicates that this activation domain has a conspicuously acidic surface region whose negative charge, if mutationally increased/decreased, raises/lowers the transcription factor's activity. Evidently, the association between a transcription factor and its putative TATA-binding target protein (which, in turn, is thought to stimulate transcriptional initiation by RNA polymerase II) is mediated by relatively nonspecific electrostatic interactions rather than by conformationally more demanding hydrogen bonds.

The DNA-binding and activation functions of eukaryotic transcription factors can be genetically separated (which is why they are thought to occur on different domains). Thus a genetically engineered hybrid protein containing the DNA-binding domain of one transcription factor and the activation domain of a second, activates the same genes as the first transcription factor. Indeed, it makes little functional difference as to whether the activation domain is placed on the N-terminal side of the DNA-binding domain or on its C-termi-

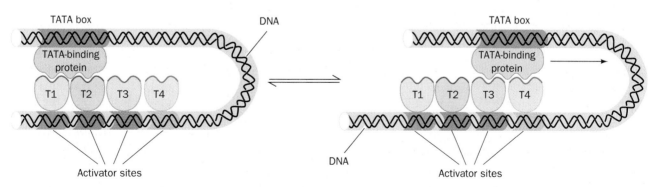

Figure 33-46
A model for the action of transcription factors. Here, four transcription factors, T1, T2, T3, and T4, are shown bound to their corresponding DNA sequences (activator sites) and simultaneously, in groups of two via relatively nonspecific electrostatic interactions (*red*), to their TATA-binding protein.

The transcription factor–TATA-binding protein association is probably rather weak but it is conjectured that continual sampling maintains this interaction so that RNA polymerase can attach to the TATA protein and initiate transcription. [After Ptashne, M., *Nature* **335,** 687 (1988).]

nal side. This geometric permissiveness in the binding between the activation domain and its target protein is also indicated by the observation that transcription factors are largely indifferent to the orientations and positions of their corresponding enhancers relative to the transcriptional start site (Section 29-2F). Of course, the DNA between an enhancer and its distant transcriptional start site must be looped out in the transcription factor–target protein complex (Fig. 33-46).

Steroid Receptors Are Transcription Factors

Eukaryotic cells express many cell-specific proteins in response to hormonal stimuli. Many of these hormones are **steroids,** cholesterol derivatives that mediate a wide variety of physiological and developmental responses (Section 34-4A). For example, the administration of **estrogens** (female sex hormones) such as **β-estradiol**

β-Estradiol

Ecdysone

causes chicken oviducts to increase their ovalbumin mRNA level from ~10 to ~50,000 molecules/cell while the amount of ovalbumin they produce rises from undetectable levels to a majority of their newly synthesized protein. Similarly, the insect steroid hormone **ecdysone** mediates several aspects of larval development (the temporal sequence of chromosome puffing shown in Fig. 33-39 can be induced by ecdysone administration).

Steroid hormones, which are nonpolar molecules, spontaneously pass through the plasma membranes of their target cells to the cytosol where they bind to their cognate receptors. The steroid–receptor complexes, in turn, enter the nucleus where they bind to specific chromosomal enhancers so as to induce, or in some cases repress, the transcription of their associated genes. For example, receptors for **glucocorticoids** (a class of steroids that affect carbohydrate metabolism; Section 34-4A) bind to specific 15 bp sequences in the upstream regions of many genes including those of metallothioneins. The actions of eukaryotic

steroid receptors therefore appear to resemble those of prokaryotic transcriptional regulators such as the *E. coli* CAP–cAMP complex (Section 29-3C) although eukaryotic systems are much more complex. For instance, different cell types may have the same receptor for a given steroid hormone and yet synthesize different proteins in response to the hormone. Apparently, only some of the genes inducible by a given steroid are made available for activation in each type of cell responsive to that steroid.

DNA-Binding Proteins Contain Zinc Fingers and Leucine Zippers

How do transcription factors recognize their target DNA sequences? Some of these proteins, as we shall see (Section 33-4B), contain a helix–turn–helix DNA-binding motif, a common component of prokaryotic gene regulators such as λ repressor (Section 32-3D). A second type of DNA-binding motif was discovered by Aaron Klug in *Xenopus* **transcription factor IIIA (TFIIIA),** a protein that binds to the internal control sequence of the 5S RNA gene (Section 29-2F). This complex then sequentially binds **TFIIIB, TFIIIC,** and RNA polymerase III which, in turn, initiates transcription of the 5S RNA gene. The 344-residue TFIIIA contains 9 similar, tandemly repeated, ~30-residue units, each of which contains two invariant Cys residues, two invariant His residues, and several conserved hydrophobic residues (Fig. 33-47). Each of these units binds a Zn^{2+} ion, which X-ray absorption measurements indicate is tetrahedrally liganded by the invariant Cys and His residues. This repeating motif is therefore dubbed the **zinc finger.** Sequence analysis has since revealed that zinc fingers occur 2 to 37 times each in a variety of eukaryotic transcription factors including Sp1, estrogen, and glucocorticoid receptors, several *Drosophila* developmental regulators (Section 33-4B), and the *Xenopus* **Xfin protein,** as well as in the *E. coli* UvrA protein (Section 31-5B) and certain retroviral nucleic acid binding proteins. In some of these proteins, the two Zn^{2+} liganding His residues are replaced by two additional Cys residues to form a second type of zinc finger motif.

How might a zinc finger protein bind to its target DNA? DNA protection experiments indicate that each zinc finger binds in DNA's major groove so as to interact with ~5 successive base pairs; that is, with about a half-turn of B-DNA. The most plausible model for the binding of a zinc finger protein to DNA is therefore one in which the protein binds along one face of the DNA with successive zinc fingers bound in the major groove on alternate sides of the double helix (Fig. 33-48). Evidently, zinc fingers, as do helix–turn–helix motifs, form structural "scaffolds" that match DNA's three-dimensional contour. Base sequence specificity is presumably provided by the particular sequence of each zinc finger's variable residues (Fig. 33-47).

(a)

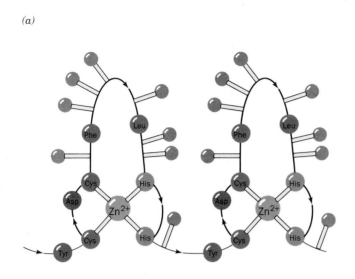

(b)

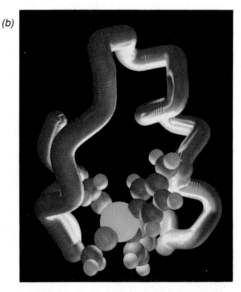

Figure 33-47
Zinc fingers. (*a*) A schematic diagram of tandemly repeated zinc finger motifs indicating their tetrahedrally liganded Zn^{2+} ions. Conserved amino acid residues are labeled. Gray balls represent the most probable DNA-binding side chains. [After Klug, A. and Rhodes, D., *Trends Biochem. Sci.* **12,** 465

(1988).] (*b*) A model of the zinc finger motif deduced from the NMR spectrum of a single zinc finger from the *Xenopus* protein Xfin. The Zn^{2+} ion together with the atoms of its His and Cys ligands are represented as spheres. [Courtesy of Michael Pique, Research Institute of Scripps Clinic. Based on an NMR structure by Peter E. Wright.]

Transcriptional activation requires, as we have seen, the cooperative association of several proteins that specifically bind to the DNA. Steven McKnight has suggested one way in which such associations might occur. He noticed that the rat liver protein named **C/EBP,** which specifically binds to the CCAAT box (Section 29-2F), has a Leu at every seventh position of a 28-residue segment in its DNA-binding domain. This sequence, when arranged in an α helix, forms a hydrophobic strip of Leu residues along one side and is rich in charged residues, particularly Arg, on its opposite face (Fig. 33-49*a*). Similar 7-residue repeats of Leu occur in a number of known DNA-binding proteins including the yeast gene regulatory protein **GCN4** and several DNA-binding proteins encoded by genes implicated in cancer

formation (Section 33-4C). McKnight therefore proposed that such proteins hydrophobically associate through the interdigitation of their Leu residue, much like the teeth of a zipper (Fig. 33-49*b*), an arrangement he consequently named the **leucine zipper.**

(a)   *(b)*

Figure 33-48
A model for the interaction of a zinc finger protein with its target DNA. The protein binds in the major groove such that successive zinc finger motifs (*cylinders*) bind to alternate sides of the DNA. The complex's structurally repeating unit is therefore two zinc fingers bound to a turn of DNA. [After Klug, A. and Rhodes, D., *Trends Biochem. Sci.* **12,** 468 (1987).]

Figure 33-49
The "leucine zipper." (*a*) A helical wheel projection of the 28-residue C-terminal segment of the DNA-binding protein C/EBP showing how 4 of its 5 Leu residues can form a hydrophobic strip along one side of the hypothetical helix (*top*). Helical wheels are discussed in Section 7-3B. (*b*) Schematic diagram indicating how the stacked Leu side chains of two α helices have been postulated to interdigitate to form a "leucine zipper". [After Landschultz, W. H., Johnson, P. F., and McKnight, S. L., *Science* **240,** 1759, 1763 (1988).]

Peter Kim's analysis of a synthetic polypeptide with the sequence of the Leu repeat-containing segment of GCN4 supports the "leucine zipper" model of protein association. NMR and spectroscopic observations indicate that this polypeptide assumes a stable α helix under physiological conditions and dimerizes with high affinity. Moreover, the dimer's two helices associate in a parallel rather than an antiparallel manner and therefore probably form a coiled-coil resembling that of the fibrous protein keratin (Section 7-2A). Thus, although there is considerable circumstantial evidence supporting the "leucine zipper" model, it seems unlikely that its Leu side chains actually interdigitate [since the side chains of an α helix extend downward as well as outward (Section 7-1B), the side chains of two parallel α helices cannot interdigitate as Fig. 33-49b suggests]. It is therefore thought that the Leu side chains hydrophobically associate with nonpolar side chains situated on the opposite helix in the coiled coil. Note that the leucine zipper is not directly implicated in DNA binding. Indeed, several DNA-binding proteins that contain zinc fingers also appear to contain leucine zippers.

C. Other Expressional Control Mechanisms

Most eukaryotic genes are specifically regulated only by the control of transcriptional initiation. Many viral genes and cellular genes, however, additionally respond to other types of control processes. The various mechanisms employed by these secondary systems are outlined below.

1. Selection of Alternative Initiation Sites

The expression of several eukaryotic genes is controlled, in part, through the selection of alternative transcriptional initiation sites. For example, identical molecules of α-amylase are produced by mouse liver and salivary gland but the corresponding mRNAs synthesized by these two organs differ at their 5' ends. Comparison of the sequences of these mRNAs with that of their corresponding genomic DNA indicates that the different mRNAs arise from separate initiation sites that are ~2.8-kb apart (Fig. 33-50). Thus, after being spliced, the liver and salivary gland α-amylase mRNAs have different untranslated 5' leaders but the same coding sequences. The two initiation sites, it is thought, support different rates of initiation. This hypothesis accounts for the observation that α-amylase mRNA comprises 2% of the polyadenylated mRNA in salivary gland but only 0.02% of that in liver.

2. Selection of Alternative Splice Sites

The expression of numerous cellular genes is modulated by the selection of alternative splice sites. Thus, certain exons in one type of cell may be introns in another. For example, a single rat gene encodes seven tissue-

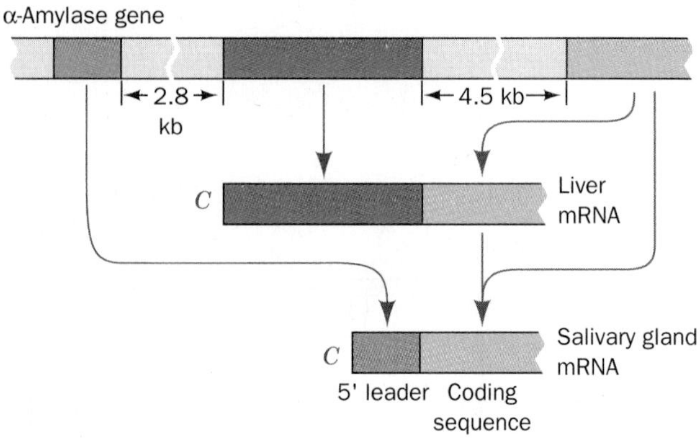

Figure 33-50
The transcription start site of the mouse α-amylase gene is subject to tissue-specific selection so as to yield mRNAs with different cap (C) and leader segments but the same coding sequences. [After Young, R. A., Hagenbüchle, O., and Schibler, U., *Cell* **23**, 454 (1981).]

specific variants of the muscle protein **α-tropomyosin** (Section 34-3B) through the selection of alternative splice sites (Fig. 33-51).

3. Translocational Control

The observation that only ~5% of nuclear RNA ever makes its way to the cytosol, probably less than can be accounted for by gene splicing, suggests that differential mRNA translocation to the cytosol may be an important expressional control mechanism in eukaryotes. Evidence is accumulating that this is, in fact, the case. Cellular RNA is never "naked" but rather is always in complex with a variety of conserved proteins. Intriguingly, nuclear and cytosolic mRNAs are associated with different sets of proteins indicating that there is protein exchange on translocating mRNA out of the nucleus.

4. Control of mRNA Degradation

The rates at which eukaryotic mRNAs are degraded in the cytosol vary widely. While most have half-lives of hours or days, some are degraded within 30 min of entering the cytosol. A given mRNA may also be subject to differential degradation. For example, the major egg yolk protein **vitellogenin** is synthesized in chicken liver in response to estrogens (in roosters as well as in hens) and transported via the bloodstream to the oviduct. Radioactive labeling experiments established that estrogen stimulation increases the rate of vitellogenin mRNA transcription by a factor of several hundred and that this mRNA has a cytosolic half-life of 480 h. When estrogen is withdrawn, the synthesis of vitellogenin mRNA returns to its basal rate and its cytosolic half-life falls to 16 h.

The poly(A) tails appended to nearly all eukaryotic mRNAs apparently help protect them from degradation (Section 29-4A). For example, histone mRNAs, which

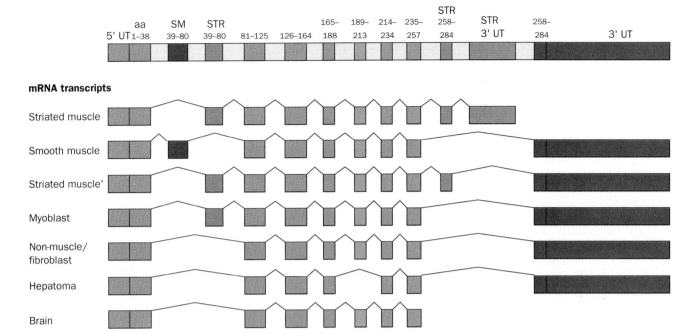

Figure 33-51

The organization of the rat α-tropomyosin gene and the seven alternative splicing pathways that give rise to cell-specific α-tropomyosin variants. The thin kinked lines indicate the positions occupied by the introns before they are spliced out to form the mature mRNAs. Tissue-specific exons are indicated together with the amino acid (aa) residues they encode: "constitutive" exons (those expressed in all tissues) are green; those expressed only in smooth muscle (SM) are red-brown; those expressed only in striated muscle (STR) are purple; and those variably expressed are yellow. Note that the smooth and striated muscle exons encoding amino acid residues 39 to 80 are mutually exclusive and, likewise, there are alternative 3'-untranslated (UT) exons. [After Breitbart, R. E., Andreadis, A., and Nadal-Ginard, B., *Annu. Rev. Biochem.* **56,** 481 (1987).]

lack poly(A) tails, have much shorter half-lives than most other mRNAs. Histones, in contrast to most other cellular proteins, are largely synthesized during the relatively short S phase of the cell cycle when they are required in massive amounts for chromatin replication (the small amounts of histones synthesized during the rest of the cell cycle are thought to be used for repair purposes). The short half-lives of histone mRNAs ensure that the rate of histone synthesis closely parallels the rate of histone gene transcription.

A structural feature that increases the rate at which mRNAs are degraded is the presence of certain AU-rich sequences in their untranslated 3' segments. These sequences, when grafted to mRNAs that lack them, decrease the mRNAs cytosolic lifetimes. By and large, however, the nature of the signals through which mRNAs are selected for degradation are poorly understood, in part, no doubt, because the nucleases that do so have not been identified.

5. Control of Translational Initiation Rates

The rates of translational initiation of eukaryotic mRNAs, as we have seen (Section 30-4), are responsive to the presence of certain substances including heme (in reticulocytes) and interferon, as well as to mRNA masking.

6. Selection of Alternative Post-Translational Processing Pathways

Polypeptides synthesized in both prokaryotes and eukaryotes are subject to proteolytic cleavage and covalent modification (Section 30-5). These post-translational processing steps are important regulators of enzyme activity (e.g., see Section 14-3E) and, in the case of glycosylations, are major determinants of a protein's final cellular destination (Sections 11-3F and 21-3B). The selective degradation of proteins (Section 30-6) is also a significant factor in eukaryotic gene expression.

In addition to the foregoing, most eukaryotic polypeptide hormones (whose functions are discussed in Section 34-4A) are synthesized as segments of large precursor polypeptides known as **polyproteins.** These are post-translationally cleaved to yield several, not necessarily different, polypeptide hormones. *The cleavage pattern of a particular polyprotein may vary among different tissues so that the same gene product can yield different sets of polypeptide hormones.* For example, the polyprotein **pro-opiomelanocortin (POMC),** which, in the rat, is synthesized in both the anterior and intermediate lobes of the pituitary gland, contains seven different polypeptide hormones (Fig. 33-52). In both of these

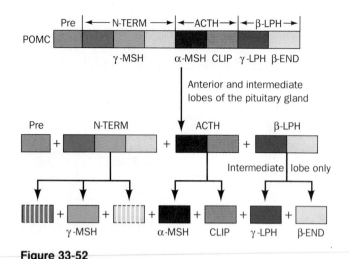

Figure 33-52
The tissue-specific post-transcriptional processing of POMC yields two different sets of polypeptide hormones. In both the anterior and intermediate lobes of the pituitary gland, POMC is proteolytically cleaved to yield its N-terminal fragment (N-TERM), **adrenocorticotropic hormone (ACTH)** and *β*-**lipotropin (*β*-LPH)**. In the intermediate lobe only, these polypeptide hormones are further cleaved to yield *γ*-**melanocyte stimulating hormone (*γ*-MSH)**, *α*-**MSH, corticotropin-like intermediate lobe peptide (CLIP, *γ*-LPH, and *β*-endorphin (*β*-END)**. [After Douglass, J., Civelli, O., and Herbert, E., *Annu. Rev. Biochem.* **53,** 698 (1984).]

lobes, which are functionally separate glands, post-translational processing of POMC yields an N-terminal fragment, ACTH and β-LPH. Processing in the anterior lobe ceases at this point. In the intermediate lobe, however, the N-terminal fragment is further cleaved to yield γ-MSH, ACTH is converted to α-MSH and CLIP, and β-LPH is split to γ-LPH and β-END (Fig. 33-52). These various hormones have different activities so that the products of the anterior and intermediate lobes of the pituitary are physiologically distinct.

Most of the cleavage sites in POMC and other polyproteins consist of pairs of basic amino acid residues, Lys-Arg, for example, which suggests that cleavage is mediated by enzymes with trypsin-like activity. Indeed, the enzymes that process POMC also activate other prohormones such as proinsulin. Moreover, the observation that a yeast protease that normally functions to activate a yeast prohormone, also properly processes POMC, suggests that prohormone processing enzymes are evolutionarily conserved.

4. CELL DIFFERENTIATION

Perhaps the most awe inspiring event in biology is the growth and development of a fertilized ovum to form an

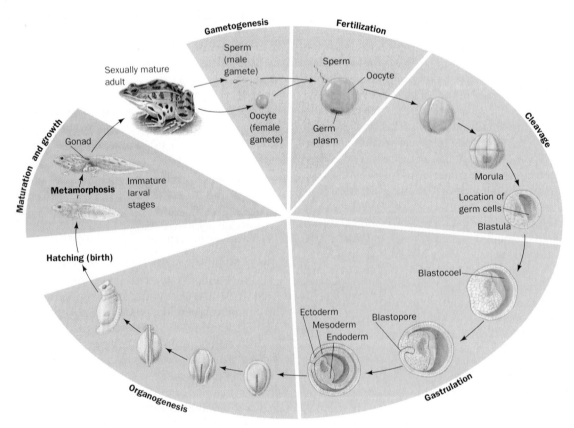

Figure 33-53
Embryogenesis in a representative animal, the frog.

extensively differentiated multicellular organism. No outside instruction is required to do so; *fertilized ova contain all the information necessary to form complex multicellular organisms such as human beings.* Since, contrary to the beliefs of the earliest microscopists, zygotes do not contain miniature adult structures, these structures must somehow be generated through genetic specification. In this section we discuss the little we presently know about this astounding process. We end by considering the genetic basis of cancer, a group of diseases caused by the proliferation of cells that have lost some of their developmental constraints.

A. Embryological Development

The formation of multicellular animals can be considered as occurring in four somewhat overlapping stages (Fig. 33-53):

1. **Cleavage,** in which the zygote undergoes a series of rapid mitotic divisions to yield many smaller cells arranged in a hollow ball known as a **blastula.**

2. **Gastrulation,** whereby the blastula, through a structural reorganization that includes the blastula's invagination, forms a triple-layered bilaterally symmetric structure called a **gastrula.** Cleavage and gastrulation together take from a few hours to several days depending on the organism.

3. **Organogenesis,** in which the body structures are formed in a process requiring various groups of proliferating cells to migrate from one part of the embryo to another in a complicated but reproducible choreography. Organogenesis occupies hours to weeks.

4. Maturation and growth, whereby the embryonic structures achieve their final sizes and functional capacities. This stage stretches into and sometimes throughout adulthood.

Cell Differentiation Is Mediated by Developmental Signals

As an embryo develops, its cells become progressively and irreversibly committed to specific lines of development. What this means is that these cells undergo sequences of self-perpetuating internal changes that distinguish them and their progeny from other cells. A cell and its descendents therefore "remember" their developmental changes even when placed in a new environment. For example, the dorsal (upper) ectoderm (outer layer) of an amphibian embryo (Fig. 35-54) is normally fated to give rise to brain tissue, whereas its ventral (lower) ectoderm becomes epidermis. If a block of an early gastrula's dorsal ectoderm is cut out and exchanged with a block of its ventral ectoderm, both blocks develop according to their new locations to yield a normal adult. If, however, this experiment is performed on the late gastrula, the transplanted tissues will differentiate as they

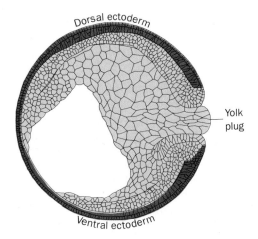

Figure 33-54
The dorsal and ventral ectoderm of an amphibian embryo.

had originally been fated, that is, as misplaced brain and epidermal tissues. Evidently, the dorsal and ventral ectoderms become committed to form brain and epidermal tissues sometime between the early and late gastrula stages.

How are developmental changes triggered; that is, What are the signals that induce two cells with identical genomes to follow different developmental pathways? To begin with, the zygote is not spherically symmetric. Rather, its yolk, as well as other substances, are concentrated towards one end. Consequently, the various cells in the early cleavage stages inherit different cytoplasmic determinants that apparently govern their further development. Even as early as an embryo's eight-cell stage, some of its cells are demonstrably different in their developmental potential from others. However, as the above transplantation experiments indicate, cells in later stages of development also obtain developmental cues from their embryonic positions.

Cells may obtain spatial information in two ways:

1. Through direct intercellular interactions.

2. From the gradients of diffusible substances called **morphogens** released by other cells.

For most developmental programs, the interacting tissues must be in direct contact, but this is not always the case. For example, mouse ectoderm fated to become eye lens will only do so in the presence of mesenchyme (embryonic tissue that gives rise to the muscle, skeleton, and connective tissue) but this process still occurs if the interacting tissues are separated by a porous filter. Lens development must therefore be mediated by diffusible substances.

Developmental signals may be recognized over great evolutionary distances. For instance, the epidermis from the back of a chick embryo, through interactions with the underlying dermis, forms feather buds that are arrayed in a characteristic hexagonal pattern. If embryonic chick epidermis is instead combined with dermis from the whiskered region of mouse embryo snout, the chick epidermis still forms feather buds but arranged in the pattern of mouse whiskers.

Even though mammals and birds diverged ~300 million years ago, mouse inducers can still activate the appropriate chicken genes although, of course, they cannot alter the products these genes specify. In an intriguing example of this phenomenon, combining epithelium from the jaw-forming region of a chick embryo with molar mesenchyme from mouse embryo, induces the chick tissue to grow teeth that are unlike those of mammals (Fig. 33-55). Apparently chickens, whose ancestors have been toothless for ~60 million years (the original bird, *Archaeopterix*, had teeth), retain the genetic potential to grow teeth even though they lack the developmental capacity to activate these genes. This observation corroborates the hypothesis that organismal evolution proceeds largely via mutations that alter developmental programs rather than the structural genes whose expression they control (Section 6-3B).

Developmental Signals Act in Combination

An additional developmental stimulus to a previously determined cell will modulate, but not reverse, its developmental state. Consider, for example, what happens in

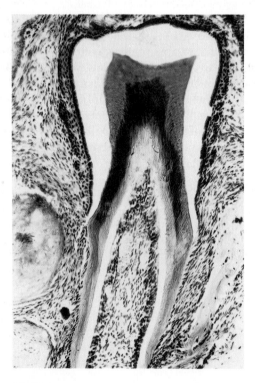

Figure 33-55
The proverbial ''hen's tooth'' forms in chick embryo jaw-forming epithelium under the influence of mouse embryo molar mesenchyme tissue. [Courtesy of Edward Kollar, University of Connecticut Health Center.]

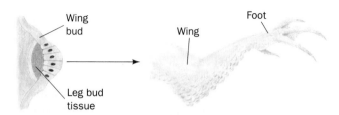

Figure 33-56
Presumptive thigh tissue from a chicken leg bud develops into a misplaced foot when implanted beneath the tip of a chicken wing bud.

a chicken embryo if undifferentiated tissue from the base of a leg bud, which normally gives rise to part of the thigh, is transplanted beneath the end of a wing bud which normally develops into the handlike wing tip. The transplant does not become a wing tip or even misplaced thigh tissue; instead it forms a foot (Fig. 33-56). Apparently the same stimulus that causes the end of a wing bud to form a wing tip causes tissue that is already committed to be part of a leg to form a leg's morphological equivalent to a wing tip, a foot. Evidently, the many different tissues of a higher organism do not each form in response to a tissue-specific developmental stimulus. Rather, *a given tissue results from the effects of a particular combination of relatively nonspecific developmental stimuli.* This situation, of course, greatly reduces the number of different developmental stimuli necessary to form a complex organism and therefore simplifies the regulation of the developmental process.

B. The Molecular Basis of Development

The study of the molecular basis of cell differentiation has only become possible in recent years with the advent of modern methods of molecular genetics. Much of what we know about this subject is based on studies of the fruit fly *Drosophila melanogaster*. We therefore begin this section with a synopsis of embryogenesis in this genetically best characterized multicellular organism.

Drosophila Development

Almost immediately after the *Drosophila* egg (Fig. 33-57a) is laid (which, rather than the earlier fertilization, triggers development), it commences a series of synchronized nuclear divisions, 1 every 6 to 10 min. The DNA must therefore be replicated at a furious rate, among the fastest known for eukaryotes. Most probably each of its replicons (Section 31-4B) are simultaneously active. The nuclear division process is unusual in that it is not accompanied by the formation of new cell membranes; the nuclei continue sharing their common cytoplasm to form a so-called **syncytium** (Fig. 33-57b). After the 8th round of nuclear division, the ~256 nuclei begin to migrate towards the cortex (outer layer) of the egg

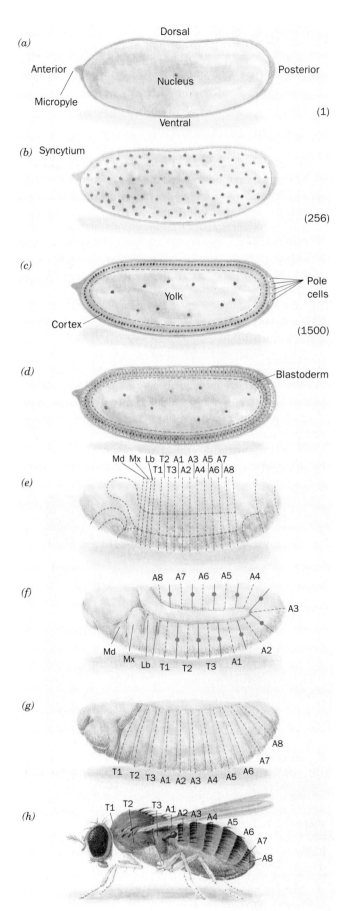

where, by around the 11th nuclear division, they have formed a single layer surrounding a yolk-rich core (Fig. 33-57c). At this stage, the mitotic cycle time begins to lengthen while the nuclear genes, which have heretofore been fully engaged in DNA replication, become transcriptionally active (a freshly laid egg contains an enormous store of mRNA that has been contributed by the developing oocyte's surrounding "nurse" cells). In the 14th nuclear division cycle, which lasts ~60 min, the egg's plasma membrane invaginates around each of the ~6000 nuclei to yield a cellular monolayer called a **blastoderm** (Fig. 33-57d). At this point, after ~2.5 h of development, genomic transcriptional activity reaches its maximum in the embryo, mitotic synchrony is lost, and gastrulation movements begin.

Until the blastoderm is formed, most of the embryo's nuclei maintain the ability to colonize any portion of the cortical cytoplasm and hence to form any part of the larva or adult except its germ cells [the germ cell progenitors, the **pole cells** (Fig. 33-57c), are set aside after the 9th nuclear division]. *Once the blastoderm has formed, however, its cells become progressively committed to ever narrower lines of development.* This has been demonstrated, for example, by tracing the developmental fates of small clumps of cells by excising them or ablating (destroying) them with a laser microbeam and characterizing the resultant deformity.

During the embryo's next few hours, it undergoes gastrulation and organogenesis. A striking aspect of this remarkable process, in *Drosophila* as well as in higher animals, is the division of the embryo into a series of segments corresponding to the adult organism's organization (Fig. 33-57e). The *Drosophila* embryo has at least three segments that eventually merge to form its head (Md, Mx, and Lb for mandibulary, maxillary, and labial), three thoracic segments (T1–T3), and eight abdominal segments (A1–A8). As development continues, the embryo elongates and several of its abdominal segments fold over its thoracic segments (Fig. 35-57f). At this stage, the segments become subdivided into anterior (forward) and posterior (rear) compartments. The embryo then shortens and unfolds to form a larva that hatches one day after beginning development (Fig. 33-57g). Over the next 5 days, the larva feeds, grows, molts twice, pupates, and commences metamorphosis to form an adult (**imago;** Fig. 33-57h). In this latter process, the larval epidermis is almost entirely replaced by the outgrowth of apparently undifferentiated patches of larval epithelium known as **imaginal**

Figure 33-57
Development in *Drosophila*. The various stages are explained in the text. Note that the embryos and newly hatched larva are all the same size, ~0.5 mm long. The adult is, of course, much larger.

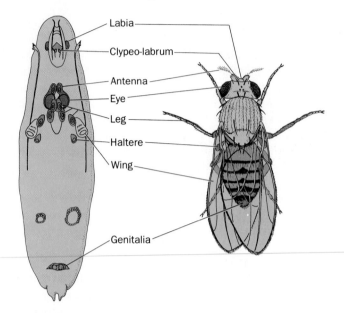

Figure 33-58
The locations and developmental fates, in *Drosophila*, of the imaginal disks (*left*), pouches of larval tissue that form the adult's outer structures. Only one of the three pairs of leg disks is shown. [After Fristrom, J. W., Raikow, R., Petri, W., and Stewart, D., *in* Hanly, E. W. (Ed.), *Problems in Biology: RNA in Development*, p. 382, University of Utah Press (1970).]

disks that are committed to their developmental fates as early as the blastoderm stage. These structures, which maintain the larva's segmental boundaries, form the adult's legs, wings, antennae, eyes, *etc.* (Fig. 33-58). About 10 days after commencing development, the adult emerges and, within a few hours, initiates a new reproductive cycle.

Developmental Patterns Are Genetically Mediated

What is the mechanism of embryonic pattern formation? Much of what we know about this process stems from genetic analyses of a series of bizarre mutations in three classes of *Drosophila* genes:

1. **Maternal-effect genes,** *which define the embryo's polarity, that is, its anteroposterior (head to tail) and dorsoventral (back to belly) axes.* Mutations of these genes globally alter the embryonic body pattern regardless of the paternal genotype. For instance, females homozygous for the *dicephalic* (two-headed) **mutation** lay eggs that develop into nonviable two-headed monsters. These are embryos with two anterior ends pointing in opposite directions and completely lacking posterior structures. Similarly, the *bicaudal* (two-tailed) and *snake* **mutations** give rise to mirror-symmetric embryos with two abdomens (Fig. 33-59*a*).

2. **Segmentation genes,** *which specify the correct number and polarity of embryonic body segments.* For example,

homozygous *fushi tarazu (ftz;* Japanese for not enough segments) **mutants** have only every second segment (Fig. 33-59*b*), whereas homozygous *engrailed (en;* indented with curved notches) **mutants** lack the posterior compartment of each segment. These embryos also die before hatching.

3. **Homeotic genes,** *which specify segmental identity;* their mutations transform one body part into another. For instance, *Antennapedia (Antp,* antennafoot) **mutants** have legs in place of antennae (Fig. 33-59*c* and *d*), whereas the mutations *bithorax (bx), anteriorbithorax (abx),* and *postbithorax (pbx)* each transform sections of halteres (vestigial wings that function as balancers), which normally occur only on segment T3, to the corresponding sections of wings, which normally occur only on segment T2 (Fig. 33-59*e*).

Maternal-Effect Gene Products Specify the Egg's Directionality

The properties of maternal-effect gene mutants suggest that maternal-effect genes specify substances whose distributions in the egg cytoplasm define the future embryo's spatial coordinate system. Indeed, immunofluorescence studies by Walter Gehring have demonstrated that **caudal protein** is distributed in a gradient that increases towards the posterior end of the normal embryo (Fig. 33-60*a*), whereas the *bicaudal* embryo lacks this protein

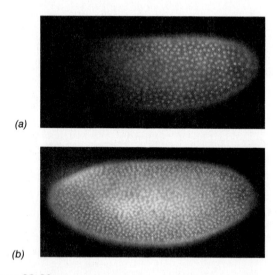

(a)

(b)

Figure 33-60
The distribution of *caudal* protein in *Drosophila* syncytial blastoderms as revealed by immunofluorescence. The embryos, whose nuclei are visible, were incubated with mouse anti-*caudal* protein antibodies and then secondarily stained with rhodamine-conjugated rabbit anti-mouse antibodies (rhodamine is a highly fluorescent compound). (a) In the wild-type embryo, the protein forms a concentration gradient that increases from the anterior (*left*) to the posterior end of the embryo. (b) This gradient is abolished in the *bicaudal* mutant. [Courtesy of Walter Gehring, University of Basel. From *Science*, **236**, 1249 (1987). Copyright © AAAS. Used by permission.]

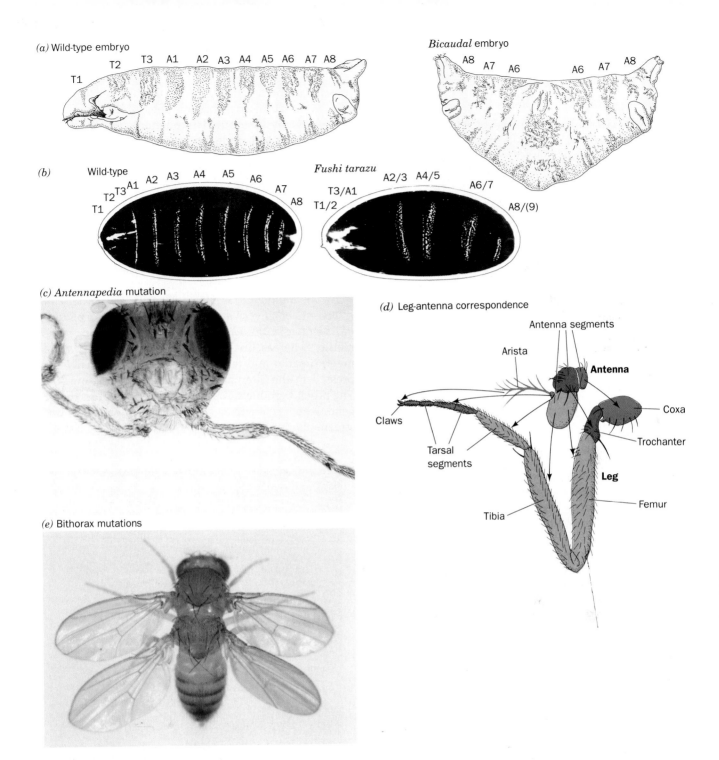

Figure 33-59

Developmental mutants of *Drosophila*. (*a*) The cuticle patterns of wild-type embryos (*left*) exhibit 11 body segments, T1 to T3 and A1 to A8 (the head segments have retracted into the body and hence are not visible here). In contrast, the nonviable "monsters" produced by homozygous *bicaudal* mutant females (*right*) develop only abdominal segments arranged with mirror symmetry. [After Gergen, P. J., Coulter, D., and Weischaus, E., *in* Gall, J. G., *Gametogenesis and the Early Embryo, p.* 200, Liss (1986).] (*b*) In the wild-type embryo (*left*), the anterior edge of each of the 11 abdominal and thoracic segments has a belt of tiny projections known as denticles (which help larvae crawl) that appear in these photomicrographs as white stripes. *Fushi tarazu* mutants (*right*) lack portions of alternate segments and the remaining segments are fused together (e.g., A2/3) yielding a nonviable embryo with only one half of the normal number of denticle belts. [Courtesy of Walter Gehring,

University of Basel. (*c*) Head of an adult fly that is homozygous for the homeotic *Antennapedia* mutation. Absence of the *Antp* gene product causes the imaginal disks that normally form antennae to develop as the legs that normally occur only on segment T2. [Courtesy of Walter Gehring, University of Basel.] (*d*) The correspondence (*arrows*) between antennae and the legs to which the *Antp* mutation transforms them. [After Postlethwait, J. H. and Schneiderman, H. A., *Dev. Biol.* **25,** 622 (1971).] (*e*) A four-winged *Drosophila* (it normally has two wings; Fig. 33-58) that results from the presence of three mutations in the bithorax complex, *abx, bx,* and *pbx*. These mutations cause the normally haltere-bearing segment T3 to develop as if it was the wing-bearing segment T2. This striking architectural change may reflect evolutionary history: *Drosophilia* evolved from more primitive insects that had four wings. [Courtesy of Edward B. Lewis, Caltech.]

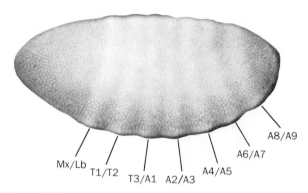

Figure 33-61
The pattern of *ftz* gene expression, as revealed by immunofluorescence, in a *Drosophila* embryo that is forming a cellular blastoderm (anterior end to the left). The resulting seven fluorescent bands, which are each around four cells wide, are centered on the boundaries between every second pair of primordial segments such that they span the future posterior compartment of one segment and the future anterior compartment of the following segment. [Courtesy of Sean Carroll and Matthew Scott, University of Colorado.]

pression of the *engrailed* gene. By the 13th nuclear division cycle, *en* transcripts become detectable but are more or less evenly distributed throughout the embryonic cortex. By the 14th cycle, however, they form a striking pattern of 14 stripes around blastoderm (one half the spacing of *ftz* expression). Continuing development reveals that these stripes are localized in the primordial posterior compartment of every segment (Fig. 33-62), just those compartments that are missing in homozygous *en*⁻ embryos. Thus, much like we saw for *ftz*, the *en* gene product somehow induces the posterior half of each segment to develop in a different fashion from its anterior half.

The distributions of segmentation gene products form in response to other gene products. For example, the pattern of *ftz* stripes in *bicaudal* mutant embryos is mirror-symmetric, whereas, in **hairy**⁻ embryos, the stripes are much broader and tend to fuse (*hairy* is normally expressed in the alternate segments not expressing *ftz*). The protein products of these latter genes repress the *ftz* gene, which sequencing studies indicate has three upstream controlling elements.

gradient (Fig. 33-60*b*). Remarkably, young embryos produced by females homozygous for maternal-effect mutations can often be "rescued" by the injection of cytoplasm, or in several cases just the mRNA, from early wild-type embryos. With some of these mutations, the polarity of the rescued embryo is determined by the site of the injection. The nature of the proteins specified by maternal-effect genes are largely unknown although the *snake* gene's base sequence suggests that it encodes a protein homologous to trypsin which is therefore likely to be a serine protease.

Segmentation Genes Are Expressed in the Regions They Affect

Gehring investigated the timing and distribution of *ftz* gene expression in embryos through *in situ* hybridization methods. *ftz* transcripts first appear in the nuclei lining the cortical cytoplasm during the embryo's 10th nuclear division cycle which makes *ftz* among the first of the embryonic genes to be expressed. The rate of *ftz* expression increases as the embryo develops until the 14th division cycle when the cellular blastoderm forms. At this stage, as immunofluorescence studies dramatically indicate, *ftz* is expressed in a pattern of seven belts around the blastoderm (Fig. 33-61), which correspond precisely to the missing regions in homozygous *ftz*⁻ embryos. Then, as the embryonic segments form, *ftz* expression subsides to undetectable levels (although it is later reactivated during the differentiation of specific nerve cells in which it is required to specify their correct "wiring" pattern). Evidently, the *ftz* gene must be expressed in alternate sections of the embryo for normal segmentation to occur.

Thomas Kornberg has similarly investigated the ex-

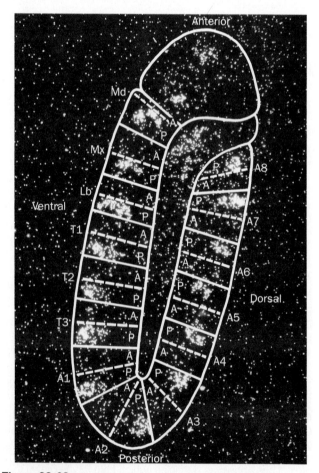

Figure 33-62
In situ hybridization demonstrates that the *Drosophila engrailed* gene is expressed in the posterior compartment of every embryonic segment. [Courtesy of Walter Gehring, University of Basel.]

Homeotic Genes Direct the Development of the Individual Body Segments

The structural components of developmentally analogous body parts, say *Drosophila* antennae and legs, are nearly identical; only their organizations differ (Fig. 33-59d). *Consequently, developmental genes must function to control the pattern of structural gene expression rather than simply turning these genes on or off.* We can therefore imagine that the expression of the structural genes charactistic of any given tissue is controlled by a complex network of regulatory genes. The homeotic genes, as we shall see, are the "master" genes in the control networks governing segmental differentiation.

Most homeotic mutations in *Drosophila* map into two large gene families: the **bithorax complex (BX-C)**, which controls the development of the thoracic and abdominal segments, and the **antennapedia complex (ANT-C)**, which primarily affects head and thoracic segments. *Recessive mutations in BX-C, when homozygous, cause one or more segments to develop as if they were more anterior segments.* Thus, the combined *bx*, *abx*, and *pbx* mutations cause segment T3 to develop as if it was segment T2 (Fig. 33-59e). Similarly, the entire deletion of BX-C causes all segments posterior to T2 to resemble T2; apparently T2 is the developmental "ground state" of these 10 segments. The evolution of such gene families, it is thought, permitted arthropods (the phylum containing insects) to arise from the more primitive annelids (segmented worms) in which all segments are nearly alike.

Detailed genetic analysis of BX-C led Edward B. Lewis to formulate a model for segmental differentiation (Fig. 33-63). BX-C, Lewis proposed, contains at least one gene for each segment from T3 to A8, which for simplicity are numbered 0 to 8 in Fig. 33-63. These genes, for reasons that are not understood, are arranged in the same order, from "left" to "right", as the segments whose development they influence. Starting with segment T3, progressively more posterior segments express successively more BX-C genes until, in segment A8, all of these genes are expressed. The developmental fate of a segment is thereby determined by its position in the embryo. This model, although probably oversimplified, largely explains the genetics of BX-C.

The sequential activation of the BX-C genes from the front to the rear of the embryo suggests that their differential expression is regulated by repressors whose concentration decreases from front to rear (Fig. 33-63). Several such putative repressors have, in fact, been identified; for instance, in embryos lacking the ***Polycomb (Pc)*** or the ***extra sex combs (esc)*** gene products, every segment develops A8-like character. The genes encoding these repressors are also homeotic since their mutations affect segmental differentiation.

Developmental Genes Have Common Sequences

In characterizing the *Antp* gene, Gehring and

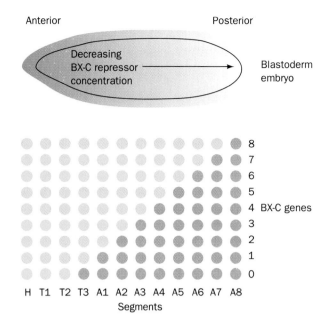

Figure 33-63

A model for the differentiation of embryonic segments in *Drosophila* as directed by the genes of the bithorax complex (BX-C). Segments T2, T3, and A1-8 in the embryo, as the lower drawing indicates, are each characterized by a unique combination of active (*purple circles*) and inactive (*yellow circles*) BX-C genes. These genes, here numbered 0 to 8, are thought to be sequentially activated from anterior to posterior in the embryo so that segment T2, the developmentally most primitive segment, has no active BX-C genes, while in segment A8, all of them are active. Such a pattern of gene expression may result from a gradient in the concentration of a BX-C repressor that decreases from the anterior to the posterior of the embryo (*upper drawing*). [After Ingham, P., *Trends Genet.* **1**, 113 (1985).]

Matthew Scott independently discovered that *Antp* cDNA hybridizes to both the *Antp* and the *ftz* gene and that, therefore, *these genes share a common base sequence.* This startling observation rapidly led to the discovery that *the Drosophila genome contains numerous such sequences, many of which occur in the homeotic gene complexes ANT-C and BX-C.* DNA sequencing studies of these genes revealed that each contains a 180 bp sequence, the so-called **homeobox,** which are 70 to 90% homologous to one another and which encode even more homologous 60-residue polypeptide segments (Fig. 33-64).

Further hybridization studies using homeobox probes led to the truly astonishing finding that *multiple copies of the homeobox are also present in the genomes of segmented animals ranging from annelids to vertebrates such as Xenopus, mice, and humans.* In some of these sequences the degree of homology is remarkably high; for example, the homeoboxes of the *Drosophila Antennapedia* gene and the *Xenopus MM3* **gene** encode polypeptides that have 59 of their 60 amino acids in common (Fig. 33-64).

Since vertebrates and invertebrates diverged over 600

	1																			20
Mouse *MO*-10	Ser	Lys	Arg	Gly	Arg	Thr	Ala	Tyr	Thr	Arg	Pro	Gln	Leu	Val	Glu	Leu	Glu	Lys	Glu	Phe
Frog *MM*3	Arg	Lys	Arg	Gly	Arg	Gln	Thr	Tyr	Thr	Arg	Tyr	Gln	Thr	Leu	Glu	Leu	Glu	Lys	Glu	Phe
Antennapedia	Arg	Lys	Arg	Gly	Arg	Gln	Thr	Tyr	Thr	Arg	Tyr	Gln	Thr	Leu	Glu	Leu	Glu	Lys	Glu	Phe
Fushi tarazu	Ser	Lys	Arg	Thr	Arg	Gln	Thr	Tyr	Thr	Arg	Tyr	Gln	Thr	Leu	Glu	Leu	Glu	Lys	Glu	Phe
Ultrabithorax	Arg	Arg	Arg	Gly	Arg	Gln	Thr	Tyr	Thr	Arg	Tyr	Gln	Thr	Leu	Glu	Leu	Glu	Lys	Glu	Phe

	21																			40
Mouse *MO*-10	His	Phe	Asn	Arg	Tyr	Leu	Met	Arg	Pro	Arg	Arg	Val	Glu	Met	Ala	Asn	Leu	Leu	Asn	Leu
Frog *MM*3	His	Phe	Asn	Arg	Tyr	Leu	Thr	Arg	Arg	Arg	Arg	Ile	Glu	Ile	Ala	His	Val	Leu	Cys	Leu
Antennapedia	His	Phe	Asn	Arg	Tyr	Leu	Thr	Arg	Arg	Arg	Arg	Ile	Glu	Ile	Ala	His	Ala	Leu	Cys	Leu
Fushi tarazu	His	Phe	Asn	Arg	Tyr	Ile	Thr	Arg	Arg	Arg	Arg	Ile	Asp	Ile	Ala	Asn	Ala	Leu	Ser	Leu
Ultrabithorax	His	Thr	Asn	His	Tyr	Leu	Thr	Arg	Arg	Arg	Arg	Ile	Glu	Met	Ala	Tyr	Ala	Leu	Cys	Leu

$\vdash$———— Helix 2 ————$\dashv$

	41																			60
Mouse *MO*-10	Thr	Glu	Arg	Gln	Ile	Lys	Ile	Trp	Phe	Gln	Asn	Arg	Arg	Met	Lys	Tyr	Lys	Lys	Asp	Gln
Frog *MM*3	Thr	Glu	Arg	Gln	Ile	Lys	Ile	Trp	Phe	Gln	Asn	Arg	Arg	Met	Lys	Tyr	Lys	Lys	Asp	Gln
Antennapedia	Thr	Glu	Arg	Gln	Ile	Lys	Ile	Trp	Phe	Gln	Asn	Arg	Arg	Met	Lys	Trp	Lys	Lys	Glu	Asn
Fushi tarazu	Ser	Glu	Arg	Gln	Ile	Lys	Ile	Trp	Phe	Gln	Asn	Arg	Arg	Met	Lys	Ser	Lys	Lys	Asp	Arg
Ultrabithorax	Thr	Glu	Arg	Gln	Ile	Lys	Ile	Trp	Phe	Gln	Asn	Arg	Arg	Met	Lys	Leu	Lys	Lys	Glu	Ile

$\vdash$———— Helix 3 ————$\dashv$

Figure 33-64

The amino acid sequences of the polypeptides encoded by the homeoboxes of 5 genes from mouse, *Xenopus*, and *Drosophila* (*Ultrabithorax* is a BX-C gene). Discrepancies between the polypeptide specified by the *Antp* homeobox and those of the other genes lack shading. Each polypeptide has a 19-residue segment (*red shading*), which is homologous to the DNA-binding helix–turn–helix fold of prokaryotic repressors. The positions of these proposed helices, numbered 2 and 3 as in the λ Cro protein, are indicated.

million years ago, this strongly suggests that the gene product of the homeobox has an essential function. What might this function be? The ~30% Arg + Lys content of homeobox polypeptides suggest that they bind DNA. Sequence comparisons and NMR studies further suggest that these polypeptide segments form helix–turn–helix motifs resembling those of prokaryotic gene regulators such as the *E. coli trp* repressor (Section 29-3C) and the λCro protein (Section 32-3D). Indeed, the polypeptide encoded by the homeobox of the *Drosophila engrailed* gene specifically binds to the DNA sequences just upstream from the transcription start sites of both the *en* and the *ftz* genes. Moreover, fusing the *ftz* gene's upstream sequence to other genes imposes *ftz*'s pattern of stripes (Fig. 33-61) on the expression of these genes in *Drosophila* embryos. *These observations suggest, in agreement with the idea that the products of developmental genes act to regulate the expression of other genes, that homeobox-containing genes encode transcription factors.* In fact, not all homeobox-encoded proteins are involved in regulating development. The homeobox is apparently a widespread genetic motif that specifies the DNA-binding segments of a variety of proteins.

It is tempting to speculate that vertebrate genes containing homeoboxes encode developmental regulators. As yet, however, the functions of these genes are unknown. That is not to say that vertebrates lack homeotic genes. For instance, the lethal **Rachiterata** mutation in mice, which transforms the seventh cervical (neck) vertebra into a first thoracic (chest) vertebra with a rib, may, in fact, be homeotic in character. Nevertheless, as the preceding discussion indicates, we are only just beginning to comprehend the molecular nature of the developmental process. Thus, although most of the maternal-effect, segmentation, and homeotic genes of *Drosophila* have probably been identified, few of the genes they control are yet known. Little, moreover, is known of the genes regulating later stages of embryonic development which probably form still more complex control networks than those governing the early stages. Clearly we have much to learn before we can even outline how the base sequence of DNA specifies the three-dimensional structure of an embryo and how this structure changes with time.

C. The Molecular Basis of Cancer: Oncogenes

The cells of the body normally remain under strict developmental control. Thus, during embryogenesis, cells must differentiate, proliferate, and even die in the correct spatial arrangement and temporal sequence to yield a normally functioning organism. In the adult, the cells of certain tissues, such as the intestinal epithelium and the blood-forming tissues of the bone marrow, continue to proliferate. Most adult body cells, however, remain quiescent, that is, in the G_0 phase of the cell cycle.

Cells occasionally lose their developmental controls and commence excessive proliferation. The resulting tumors can be of two types:

1. **Benign tumors,** such as warts and moles, grow by simple expansion and often remain encapsulated by a layer of connective tissue. Benign tumors are rarely life-threatening although if they occur in an enclosed space such as in the brain or secrete large amounts of certain hormones, they can prove fatal.

2. **Malignant tumors** or **cancers** grow in an invasive manner and shed cells that, in a process known as **metastasis,** colonize new sites in the body. Malignant tumors are almost invariably life-threatening; they are responsible for 20% of the mortalities in the United States.

Cancer, being one of the major human health problems, has received enormous biomedical attention over the past few decades. Around 100 different types of human cancers are recognized, methods of cancer detection and treatment are highly developed, and cancer epidemiology has been extensively characterized. We are, nevertheless, just beginning to understand the biochemical basis of this collection of diseases. In this section we outline what is known about this rapidly developing field of knowledge.

Cancer Cells Differ in Many Ways from Normal Cells

The most obvious and the medically most significant property of cancer cells is that they proliferate uncontrollably. For instance, when grown in a tissue culture dish, normal cells form a monocellular layer on the bottom of the dish and then, through a process termed **contact inhibition,** cease dividing (Fig. 33-65a). In contrast, the growth of malignant cells is unhampered by intercellular contacts; in culture they form multicellular layers (Fig. 33-65b). Moreover, even in the absence of contact inhibition, normal cells are far more limited in their capacity to reproduce than are cancer cells. Normal cells, depending on the species and age of the animal from which they were taken, will only divide in culture 20 to 60 times before they reach senescence and die (a phenomenon which, no doubt, is at the heart of the aging process). *Cancer cells, on the other hand, are immortal; there is no limit to the number of times they can divide.* In fact, some cancer cell lines have been maintained in culture through thousands of divisions spanning several decades. Immortal cells, however, are not necessarily malignant: *The hallmark of cancer is immortality combined with uncontrolled growth.*

The properties of cancer cells differ from those of the normal cells from which they are derived. The plasma membranes of malignant cells have a more fluid character than those of normal cells and have altered ratios of many of their cell surface components such as glycopro-

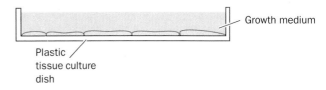

(a) **Normal cells**

Growth medium

Plastic tissue culture dish

(b) **Transformed cells**

Figure 33-65
The growth pattern of vertebrate cells in culture: (*a*) Normal cells stop growing through contact inhibition once they have formed a confluent monolayer. (*b*) In contrast, transformed cells lack contact inhibition; they pile up to form a multilayer.

teins and glycolipids. Internally, the cytoskeletons of cancer cells are less organized than those of normal cells. This, presumably, is why cancer cells have a more rounded appearance than the corresponding normal cells (Fig. 33-66). Metabolically, cancer cells have a high rate of glycolysis that results in a debilitating energy drain on the host. The conversion of a normal to a cancerous cell is therefore accompanied by a complex series of structural, biochemical and, as we shall see, genetic changes.

Cancer Is Caused by Carcinogens, Radiation, and Viruses

Most cancers are caused by agents that damage DNA or interfere with its replication or repair. These include a great variety of man made and naturally occurring substances known as chemical carcinogens (Section 31-5E), as well as radiation, both electromagnetic and particulate, with sufficient energy to break chemical bonds. In addition, *certain viruses induce the formation of malignant tumors in their hosts* (see below).

Almost all malignant tumors result from the **transformation** of a single cell (conversion to the cancerous state; this term should not be confused with the acquisition of genetic information from exogenously supplied DNA), which, being free of its normal developmental constraints, proliferates. Yet, considering, for example, that the human body consists of around 10^{14} cells, transformation must be a very rare event. One of the major reasons for this, as the age distribution of the cancer death rate indicates (Fig. 33-67), is that *transformation requires a cell or its ancestors to have undergone several independent and presumably improbable carcinogenic changes.* Consequently, exposure to a carcinogen may prime many cells for transformation but a malignant tumor may not form until decades later when one of these cells suffers a final transforming event.

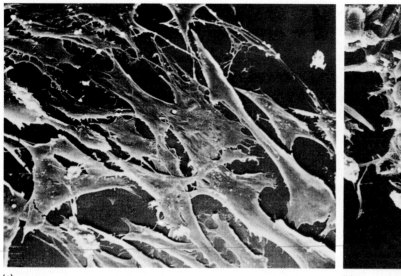
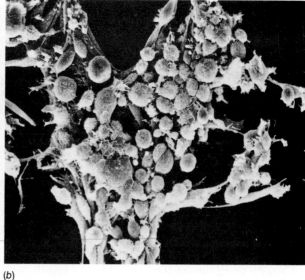

(a)　　　　　　　　　　　　　　　　　　　　　　　　*(b)*

Figure 33-66

The transformation of cultured chicken fibroblasts by Rous sarcoma virus: (*a*) Normal cells adhere to the surface of the culture dish where they assume a flat extended conformation. (*b*) Upon infection with RSV, these cells become rounded and cluster together in piles. [Courtesy of G. Steven Martin, University of California at Berkeley.]

Certain Retroviruses Carry Oncogenes

The viral induction of cancer was first observed in 1911 by Peyton Rous who demonstrated that cell-free filtrates from certain chicken **sarcomas** (malignant tumors arising from connective tissues) promote new sarcomas in chickens. Although decades were to pass before the significance of this work was appreciated (Rous was awarded the Nobel prize in 1966 at the age of 85), many other such **tumor viruses** have since been characterized.

The **Rous sarcoma virus (RSV),** as are all known RNA tumor viruses, is a **retrovirus** [an RNA virus that replicates its chromosome by copying it to DNA in a reaction mediated by virally specified reverse transcriptase (Section 31-4C), inserting the DNA into the host cell's genome, and then transcribing this DNA]. The base sequence of the RSV chromosome reveals that it contains four genes (Fig. 33-68), of which only three are essential for viral replication: *gag*, which encodes an inner capsid protein; *pol*, which specifies reverse transcriptase; and *env*, which codes for the outer envelope protein. The fourth gene, **v-*src*** ("v" for *v*iral, "*src*" for *sarcoma*), encodes a protein known as **pp60**$^{v\text{-}src}$ ("pp60" signifies that it is a 60-kD *p*hospoprotein), which medi-

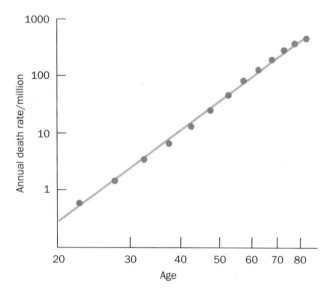

Figure 33-67

The variation of the cancer death rate in humans with age. The linearity of this log–log plot can be explained by the hypothesis that several randomly occurring mutations are required to generate a malignancy. The slope of the line suggests that, on the average, five such mutations are required for a malignant transformation.

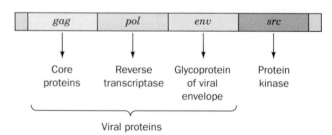

Figure 33-68

The genetic map of RSV. Its 9-kb genome encodes only four genes: *gag*, *pol*, and *env* are essential for viral replication, whereas v-*src* is an oncogene. The segments at the ends of the genome represent terminally redundant sequences.

ates host cell transformation. v-*src* has therefore been termed an **oncogene** (Greek: *onkos,* mass or tumor).

What is the origin of v-*src* and what is its viral function? Hybridization studies by Michael Bishop and Harold Varmus in 1976 led to the remarkable discovery that *uninfected chicken cells contain a gene,* **c-*src*** ("c" for **cel**lular), *that is homologous to* v-*src* (the two genes differ mainly in that c-*src* is interrupted by six introns, whereas v-*src* is uninterrupted). Moreover, c-*src* is highly conserved in a wide variety of eukaryotes that span the evolutionary scale from *Drosophila* to humans. This observation strongly suggests that c-*src,* which antibodies directed against pp60^{v-src} indicate is expressed in normal cells, is an essential cellular gene. In fact, as we shall see below, *both* pp60^{v-src} *and its normal cellular analog,* **pp60^{c-src},** *function to stimulate cell proliferation.* Apparently, v-*src* was originally acquired from a cellular source by an initially nontransforming ancestor of RSV. By maintaining the host cell in a proliferative state (cells are usually not killed by RSV infection), pp60^{v-src} presumably enhances the viral replication rate.

Viral Oncogene Products Mimic the Effects of Polypeptide Growth Factors and Hormones

In order to understand how oncogenes subvert the normal processes of cell division, we must first understand these processes. *Cell proliferation is stimulated by hormonelike polypeptide growth factors.* These **mitogens** (substances that induce mitosis), the best characterized of which are **epidermal growth factor (EGF)** and **platelet-derived growth factor (PDGF),** bind with high affinity to the extracellular domains of specific protein receptors that span the plasma membranes of certain types of cells (Fig. 33-69). In doing so, they activate the receptors' cytoplasmic domains to phosphorylate their target proteins which, in turn, are thought to act as intracellular messengers that stimulate cell division by as yet dimly perceived mechanisms. For example, **transforming growth factor α (TGF-α),** a protein synthesized by and required for the growth of epithelial cells, is produced in excessive amounts in the skin of individuals with **psoriasis,** a common skin disease characterized by epidermal hyperproliferation.

Many growth factor receptors are **tyrosine-specific** **protein kinases;** *that is, they phosphorylate specific Tyr OH groups in their target proteins* (Fig. 33-69). Most protein kinases, it should be noted, specifically phosphorylate Ser or Thr residues (recall, e.g., that phosphorylase kinase phosphorylates Ser 14 of glycogen phosphorylase; Section 17-3C); only about one phosphorylated amino acid residue in 2000 is Tyr. It is, nevertheless, becoming increasingly evident that tyrosine phosphorylation is of central importance in regulating a variety of basic cellular processes. Curiously, many activated tyrosine kinases phosphorylate themselves. This **autophosphorylation,** at least in several cases, further

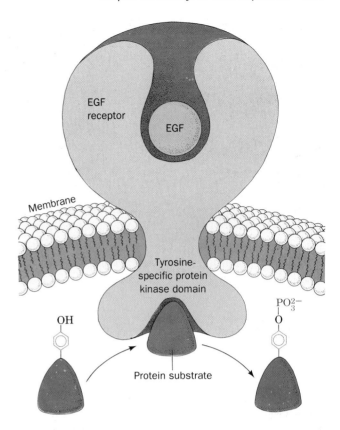

Figure 33-69
The binding of EGF to the external domain of EGF receptor activates its cytoplasmic tyrosine-specific protein kinase domain to phosphorylate specific Tyr residues of the receptor's protein substrates.

stimulates the tyrosine kinase activity of these activated receptors and therefore maintains the activated state after the growth factor has dissociated.

Hormones such as epinephrine and glucagon also profoundly affect the physiology of their target cells (Sections 17-3E – G and 34-4A and B). These hormones bind to specific receptors, thereby stimulating adenylate cyclase to catalyze the formation of cAMP, the second messenger that actually triggers the cellular response to the hormone. Hormone receptors, which face out from the plasma membrane, and adenylate cyclase, which is located on the membrane's cytoplasmic surface, are separate proteins that do not physically interact. Rather, they are functionally coupled by **G-protein** (Fig. 33-70), so-called because it specifically binds GTP and GDP. Adenylate cyclase is activated by G-protein but only when it is complexed with GTP. However, G-protein slowly hydrolyzes GTP to GDP + P$_i$ and thereby deactivates itself. G-protein is reactivated by a GDP–GTP exchange reaction that is catalyzed by hormone–receptor complex but not by unoccupied receptor. *G-protein therefore mediates the hormonal signal.* This process is discussed in greater detail in Section 34-4B.

The proteins encoded by many viral oncogenes are ana-

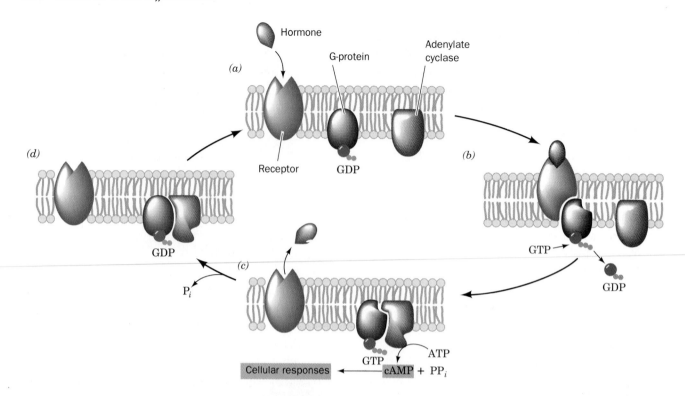

Figure 33-70

The activation/deactivation cycle for hormonally stimulated adenylate cyclase. (a) In the absence of hormone, G-protein binds GDP and adenylate cyclase is catalytically inactive. (b) The hormone–receptor complex stimulates G-protein to exchange its bound GDP for GTP. (c) The G-protein–GTP complex, in turn, binds to and thereby activates adenylate cyclase to produce cAMP. (d) The eventual G-protein catalyzed hydrolysis of its bound GTP to GDP causes G-protein to dissociate from and hence deactivate adenylate cyclase.

logs *of various growth factor and hormone system components.* For instance:

1. The **v-sis** oncogene of **simian sarcoma virus** encodes a protein secreted by infected cells that is nearly identical with one of the two subunits of PDGF. Hence, the uncontrolled growth of simian sarcoma virus-infected cells apparently results from the continuous and inappropriate presence of this PDGF homolog.

2. Nearly one half of the more than 20 known retrovirus oncogenes, including v-*src*, encode tyrosine-specific protein kinases. Indeed, the **v-erbB** oncogene specifies a truncated version of the **EGF receptor** that lacks the EGF-binding domain but retains its transmembrane segment and its protein kinase domain. *It seems likely, therefore, that oncogene-encoded protein kinases inappropriately phosphorylate the target proteins normally recognized by growth factor receptors thereby driving the afflicted cells to a state of unrestrained proliferation.*

3. The **v-ras** oncogene encodes a protein, **p21^{v-ras}**, that resembles G-protein; it is localized on the cytoplasmic side of the mammalian plasma membrane where it activates adenylate cyclase when binding GTP. Although p21^{v-ras} hydrolyzes GTP to GDP, it does so much more slowly than G-protein. The restraint to adenylate cyclase activation that GTP hydrolysis imposes on G-protein is therefore greatly reduced in p21^{v-ras}; the consequent continuous activation of adenylate cyclase apparently acts to transform the cell.

4. Even though several proteins are known to be phosphorylated by cellular and oncogene-encoded protein kinases, none has yet been shown to be a putative intracellular messenger that triggers cell division (although some of these proteins have been implicated in mediating other properties peculiar to transformed cells). However, several retroviral oncogenes, including **v-jun** and **v-fos**, encode nuclear proteins whose corresponding normal cellular analogs are synthesized in response to mitogenic signals. Many such proteins, including the v-*jun* and v-*fos* gene products, bind to DNA, strongly suggesting that they influence its transcription and/or replication. Indeed, v-*jun* is 80% homologous to the **proto-oncogene** (normal cellular analog of an oncogene) **c-jun**, which encodes a transcription factor named **AP-1**

(now also called **Jun**). Moreover, Jun/AP-1 forms a tight complex with the protein product of the proto-oncogene **c-*fos***, which sequence analysis and mutational alteration studies indicates is mediated by a leucine zipper (Section 33-3B). This association greatly increases the ability of Jun/AP-1 to stimulate transcription from Jun-responsive genes.

Oncogene products therefore appear to be functionally modified or inappropriately expressed components of elaborate control networks that regulate cell growth and differentiation. The complexity of these networks (cells generally respond to a variety of growth factors, hormones, and transcription factors in partially overlapping ways) is probably why malignant transformation requires several independent carcinogenic events.

Malignancies May Result from Specific Genetic Alterations

Although much of what we know concerning oncogenes stems from the study of retroviral oncogenes, few human cancers are caused by retroviruses. Nevertheless, *it seems likely that all cancers are caused by genetic alterations.* Robert Weinberg demonstrated this to be the case for mouse fibroblasts that had been transformed by a known carcinogen: Normal mouse fibroblasts in culture are transformed upon transfection with DNA from the transformed cells. Moreover, these newly transformed cells, when innoculated into mice, form tumors. Similar investigations indicate that DNAs from a wide variety of malignant tumors likewise have transforming activity.

What sorts of genetic changes can give rise to cancer? Several types of changes have been observed:

1. Altered Proteins

An oncogene, as we have seen, may give rise to a protein product with an anomalous activity relative to that of the corresponding proto-oncogene. This may even result from a simple point mutation. For example, Weinberg, Michael Wigler, and Mariano Barbacid showed that the *ras* oncogene isolated from a human bladder **carcinoma** (a malignant tumor arising from epithelial tissue) differs from its corresponding proto-oncogene (which is thought to encode a G-protein) by the mutation of the Gly 12 codon (GGC) to a Val codon (GTC). The resulting amino acid change attenuates the GTPase activity of **p21^{c-ras}**, evidently without affecting its ability to activate adenylate kinase, thereby prolonging the time this G-protein remains in the "on" state. Indeed, Sung-Hou Kim's X-ray structure determinations of normal human p21^{c-ras} (Fig. 33-71) and its oncogenic counterpart (Gly 12 → Val), both in complex with GDP, indicates that the mutation mainly alters the

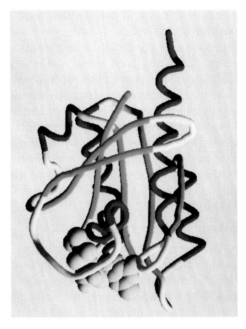

Figure 33-71
The X-ray structure of the human proto-oncogene protein, p21^{c-ras} in complex with GDP (*yellow*). The protein's helical and β sheet regions are drawn in red and green, respectively. The native protein's 18 C-terminal residues, which have no known biochemical function and which are thought to be flexible, were removed to facilitate the protein's crystallization. [Courtesy of Sung-Hou Kim, University of California at Berkeley.]

normal protein structure in the vicinity of its presumed GTPase function. Most other *ras* oncogene-activating mutations also change residues close to this site. Interestingly, *c-ras* is the most common proto-oncogene implicated in human cancers.

2. Altered Regulatory Sequences

Malignant transformation can result from the inappropriately high expression of a normal cellular protein. For example, the retroviral oncogene v-*fos* and its corresponding proto-oncogene c-*fos* encode similar proteins. These genes differ mainly in their regulatory sequences: v-*fos* has an efficient enhancer while c-*fos* has a 67-nucleotide AT-rich segment in its unexpressed 3'-terminal end that, when transcribed, promotes rapid mRNA degradation (Section 33-3C). Thus, c-*fos* can be converted to an oncogene by deleting its 3' end and adding the v-*fos* enhancer.

3. Chromosomal Rearrangements

An oncogene may be inappropriately transcribed when brought under the control of a foreign regulatory sequence through chromosomal rearrangement. For example, Carlo Croce found that the human cancer **Burkitt's lymphoma** (a lymphoma is an immune system cell malignancy) is characterized by an exchange of chromosomal segments in which the proto-oncogene **c-myc** is translocated from its normal position at one end of chromosome 8 to the end of chromosome 14 adjacent to certain immunoglobulin genes. The misplaced c-*myc* gene is thereby brought under the transcriptional control of the highly active (in immune system cells) immunoglobulin regulatory sequences. The consequent overproduction of the normal c-*myc* gene product (which encodes a DNA-binding protein whose transient increase is normally correlated with the onset of cell division) or alternatively, its production at the wrong time in the cell cycle, is apparently a major factor in cell transformation.

4. Gene Amplification

Oncogene overexpression can also occur when the oncogene is replicated multiple times, either as sequentially repeated chromosomal copies or as extra-chromosomal particles. The amplification of the c-*myc* gene, for example, has been observed in several types of human cancers. Gene amplification is usually an unstable genetic condition that can only be maintained under strong selective pressure such as that conferred by drugs (Section 33-2D). It is not known how oncogene amplification is stably maintained.

5. Viral Insertion into a Chromosome

Inappropriate oncogene expression may result from the insertion of a viral genome into a cellular chromosome such that the proto-oncogene is brought under the transcriptional control of a viral regulatory sequence. For instance, **avian leukosis virus,** a retrovirus that lacks an oncogene but which nevertheless induces lymphomas in chickens, has a chromosomal insertion site near c-*myc*. Some DNA tumor viruses also transform cells in this manner.

6. Loss or Inactivation of Anti-Oncogenes

The high incidence of particular cancers in certain families suggests that there are genetic predispositions towards these diseases. A clear-cut example of this phenomenon occurs in **retinoblastoma,** a cancer of the developing retina that therefore afflicts only infants and young children. The offspring of surviving retinoblastoma victims have a high probability of also developing this disease as well as several other types of malignancies. In fact, retinoblastoma is associated with the inheritance of a copy of chromosome 13 from which a particular segment has been deleted. Retinoblastoma develops, as Alfred Knudson first explained, through a somatic mutation in a **retinoblast** (a retinal precursor cell) that alters the same segment of the second, heretofore normal copy of chromosome 13. This is because *the affected chromosomal segment contains a gene, the Rb gene, which specifies a factor that somehow restrains uninhibited cell proliferation; that is, the Rb gene product acts as a tumor suppressor.* The *Rb* gene is therefore classified as an **antioncogene.** It encodes a 105-kD DNA-binding protein that is localized in the nucleus of normal retinal cells but does not occur in retinoblastoma cells. The specific function of this protein has yet to be determined although, presumably, it is involved in regulating gene expression. A strong indication that this is, in fact, the case is provided by cells that have been transformed by **adenovirus.** In these cells, the protein product of the viral oncogene **E1A,** which does not bind DNA, specifically complexes and thereby inactivates Rb protein. Thus, *an additional way that oncogenes can cause cancer is by inactivating the products of normal cellular anti-oncogenes.*

Mutations altering normal gene products, causing chromosomal rearrangements and deletions, and perhaps gene amplification, can all result from the actions of carcinogens on cellular DNA. Thus, normal cells bear the seeds of their own cancers. To date, ~40 viral and cellular oncogenes have been identified. It therefore seems likely that only a relatively small number of genes and perhaps an even smaller number of general molecular mechanisms are involved in oncogenesis. The characterization of these genes and the cellular processes they subvert should lead to rationally designed cancer therapies.

Chapter Summary

Eukaryotic chromatin consists of DNA, RNA, and proteins, the majority of which are the highly conserved histones. Chromatin is structurally organized in a hierarchial manner. In the first level of chromatin organization, ~200 bp of DNA are doubly wrapped around a histone octamer, $(H2A)_2(H2B)_2(H3)_2(H4)_2$, to form a nucleosome. Each nucleosome is associated with one molecule of histone H1. Nucleosomes probably form on the lagging strand of newly replicated DNA in a process that may be mediated by the protein nucleoplasmin. In the second level of chromatin organization, the nucleosome filaments coil into 300-Å thick filaments that probably contain 6 nucleosomes/turn. Then, in the third and final level of chromatin organization, the 300-Å thick filaments form 15 to 30-μm long radial loops that project from the axis of the metaphase chromosome. This accounts for DNA's packing ratio of > 8000 in the metaphase chromosome. The larvae of certain Dipteran flies, including *Drosophila*, contain banded polytene chromosomes, which consist of up to 1024 identical DNA strands in parallel register. The bands, as shown by *in situ* hybridization, correspond to the chromosomes' transcription units.

The complexity of a DNA sample can be determined from

its renaturation rate through C_0t curve analysis. Eukaryotic DNAs have complex C_0t curves that arise from the presence of unique, moderately repetitive and highly repetitive sequences, as well as from inverted repeats. The function of inverted repeats, which form foldback structures, is unknown. Highly repetitive sequences, which occur in the heterochromatic regions near the chromosomal centromeres, probably function to align homologous chromosomes during meiosis and/or to facilitate their recombination. Moderately repetitive DNAs, for the most part, have unknown functions; many of them may simply be selfish DNA. The genes specifying rRNAs and tRNAs are organized into tandemly repeated clusters. The rDNA condenses to form nucleoli, the sites of rRNA transcription by RNA polymerase I and of partial ribosomal assembly. The 5S RNA and tRNAs are transcribed outside the nucleoli by RNA polymerase III. The genes specifying histones, which are required in large quantities only during S phase of the cell cycle, are the only repeated protein-encoding genes. The identity of a series of repeated genes is probably maintained through unequal crossing-over and/or gene conversion. Many families of genes specifying related proteins are clustered into gene families. In mammals, the gene clusters encoding the α- and β-like hemoglobin subunits occur on separate chromosomes. Nevertheless, all vertebrate globin genes have the same exon–intron structure: three exons separated by two introns. The genes encoding modern proteins probably arose through the recombinational assembly of exons so that modern introns are the remnants of a process that facilitated protein evolution. This can be seen in the structure of chicken muscle pyruvate kinase whose 10 exons each encode discrete elements of protein secondary structure. Likewise, many of the polypeptide segments encoded by the 18 exons of the LDL receptor gene are homologous to segments in several other proteins. The scarcity of introns in the genes of lower organisms probably reflects the greater selective advantage that metabolic efficiency confers on them relative to higher organisms. In higher organisms, introns may function to protect the exons of gene families from elimination through unequal crossing-over. The thalassemias are inherited diseases caused by the genetic impairment of hemoglobin synthesis. Most α-thalassemias are caused by the deletion of one or more of the α-globin genes, whereas most β-thalassemias arise from point mutations that affect the transcription or the post-transcriptional processing of the β-globin mRNAs.

Heterochromatin may be subclassified as constitutive heterochromatin, which is never transcriptionally active, and facultative heterochromatin, whose activity varies in a tissue-specific manner. The Barr bodies in the cells of female mammals constitute a common form of facultative heterochromatin: One of each cell's two X chromosomes is permanently condensed and confers its state of inactivity on its progeny, possibly through maintenance methylation. Active chromatin, in contrast, has a relatively open structure that makes it available to the transcriptional machinery. Two well-characterized examples of transcriptionally active chromatin are the chromosome puffs that emanate from single bands in polytene chromosomes, and the lampbrush chromosomes of amphibian oocytes. The differential protein synthesis characteristic of the cells in a multicellular organism largely stems from the selective transcription of the expressed genes. Transcriptionally active chromatin is more susceptible to nuclease di-

gestion than is nontranscribing chromatin. This nuclease sensitivity appears to be promoted by the presence of nonhistone proteins such as HMGs 14 and 17, although these proteins are not tissue specific. Active genes also have nuclease hypersensitive sites that occur in nucleosome-free regions of DNA. Nuclease hypersensitivity is conferred on DNA by the binding of specific proteins that presumably make the genes accessible to the proteins mediating transcriptional initiation, replication, and recombination. The cell-specific expression of genes is mediated by the genes' promoter and enhancer elements. Consequently, cells must contain specific transcription factors that recognize these genetic elements. For example, Sp1 binds to the GC box that precedes many genes. Likewise, steroid hormones bind to their cognate receptors which, in turn, bind to specific enhancers so as to modulate their transcriptional activity. Transcription factors have two domains, a DNA-binding domain targeted to a specific sequence, and an activation domain which interacts with the transcription factor's target protein in a largely nonspecific manner via a negatively charged surface region. The cooperative binding of several transcriptional factors to DNA is thought to stimulate the binding of RNA polymerase II to the corresponding transcriptional initiation site. Such processes may be mediated by DNA-protein interactions involving zinc fingers and by protein–protein associations promoted by leucine zippers. Other forms of selective gene expression involve the use of alternative initiation sites in a single gene, the selection of alternative splice sites, the possible regulation of mRNA translocation across the nuclear membrane, the control of mRNA degradation, the control of translational initiation rates, and the selection of alternative post-translational processing pathways.

Embryogenesis occurs in four stages: cleavage, gastrulation, organogenesis, and maturation and growth. One of the most striking characteristics of embryological development is that cells become progressively and irreversibly committed to specific lines of development. The signals that trigger developmental changes, which are recognized over great evolutionary distances, may be transmitted through direct intercellular contacts or from the gradients of substances released by other embryonic cells. Developmental signals act combinatorially; that is, the developmental fate of a specific tissue is determined by several not necessarily unique developmental stimuli. In *Drosophila*, early embryonic development is governed by maternal-effect genes whose distribution imposes the embryo's spatial coordinate system. Segmentation genes, which are activated previous to blastoderm formation, specify the number and polarity of the larval and adult body segments. Homeotic genes, whose mutations transform one body part into another, then regulate the differentiation of the individual segments. Many of these regulatory genes, which are selectively expressed in the embryonic tissues whose development they control, have a common base sequence, the homeobox, which encodes a 60-residue polypeptide segment. The proteins encoded by these regulatory genes are thought to act as repressors and/or transcription factors in the selective expression of developmentally specified proteins. Homeoboxes have also been detected in vertebrate genes suggesting that these genes likewise have a developmental function. Cancer cells differ in a variety of structural, functional, and metabolic ways from the normal cells from which they are derived. Their medically most significant properties, their immortality and their ability to proliferate uncontrollably, endows them with the capacity to form invasive and metastatic tumors. Malignant tumors are caused by agents that alter DNA sequences: carcinogens, radiation, and viruses. However, several such changes must occur before a cell becomes transformed. Rous sarcoma virus, a retrovirus causing sarcomas in chickens, carries an oncogene, v-*src*, whose protein product, pp60^{v-src}, mediates the transformation of the host cell. Uninfected chicken cells contain a gene, c-*src*, that is homologous to v-*src*. Both genes encode tyrosine-specific protein kinases whose action is thought to stimulate cell division. The ~ 40 viral and cellular oncogenes that have been charac-

terized stimulate cell division by mimicking the effects of growth factors and certain hormones. The oncogene products include analogs of growth factors, growth factor receptors, nuclear proteins that stimulate transcription and/or cell division, and G-proteins. The types of genetic changes that distinguish oncogenes from their cellular homologs are point mutations, chromosomal rearrangements that bring oncogenes under the influence of inappropriate regulatory sequences, oncogene amplification, and the insertion of a viral genome in a position that places the cellular oncogene under the transcriptional control of viral regulatory sequences. The loss or inactivation of anti-oncogenes may also cause cancer.

References

General

Alberts, B., Bray, D., Lewis, J., Raff, M., Roberts, K., and Watson, J. D., *Molecular Biology of the Cell* (2nd ed.), Chapters 9, 10, 16, and 21, Garland Publishing (1989).

Bradbury, E. M., Maclean, N., and Matthews, H. R., *DNA, Chromatin and Chromosomes,* Wiley (1981).

Brown, D. D., Gene expression in eukaryotes, *Science* **211**, 667–674 (1981).

Darnell, J., Lodish, H., and Baltimore, D., *Molecular Cell Biology,* Chapters 9–12, 22, and 23, Scientific American Books (1986).

De Pomerai, D., *From Gene to Animal,* Cambridge University Press (1985).

Felsenfeld, G., DNA, *Sci. Am.* **253**(4): 58–67 (1985).

Lewin, B., *Genes* (3rd ed.), Chapters 17–19, 21, and 25–27, Wiley (1987).

Watson, J. D., Hopkins, N. H., Roberts, J. W., Steitz, J. A., and Weiner, A. M., *Molecular Biology of the Gene* (4th ed.), Chapters 20–22 and 25–27, Benjamin/Cummings (1987).

Chromosome Structure

Adolph, K. W., (Ed.), *Chromosomes and Chromatin,* Vols. I–III, CRC Press (1988).

Butler, P. J. G., The folding of chromatin, *CRC Crit. Rev. Biochem.* **15**, 57–91 (1983).

Cohen, L. H., Newrock, K. M., and Zweidler, A., Stage-specific switches in histone synthesis during embryogenesis, *Science* **190**, 994–997 (1975).

Eissenberg, J. C., Cartwright, I. L., Thomas, G. H., and Elgin, S. C. R., Selected topics in chromatin structure, *Annu. Rev. Genet.* **19**, 485–536 (1985). [Discusses DNase I hypersensitivity, nucleosome positioning and higher-order chromatin structure.]

Felsenfeld, G. and McGhee, J. D., Structure of the 30 nm chromatin fiber, *Cell* **44**, 375–377 (1986).

Gall, J. G., Chromosome structure and the C-value paradox, *J. Cell Biol.* **91**, 3s–14s (1981).

Igo-Kemenes, T., Hörz, W., and Zachau, H. G., Chromatin, *Annu. Rev. Biochem.* **51**, 89–121 (1982).

Kornberg, R. D., Chromatin structure: a repeating unit of histones and DNA, *Science* **184**, 868–871 (1974).

Kornberg, R. D. and Klug, A., The nucleosome, *Sci. Am.* **244**(2): 52–64 (1981).

Morse, R. H. and Simpson, R. T., DNA in the nucleosome, *Cell* **54**, 285–287 (1988).

Pedersen, D. S., Thoma, F., and Simpson, R. T., Core particle, fiber and transcriptionally active chromatin structure, *Annu. Rev. Cell Biol.* **2**, 117–147 (1986).

Richmond, T. J., Finch, J. T., Rushton, B., Rhodes, D., and Klug, A., Structure of the nucleosome core particle at 7 Å resolution, *Nature* **311**, 532–537 (1984).

Sperling, R. and Wachtel, E. J., The histones, *Adv. Protein Chem.* **34**, 1–60 (1980).

van Holde, K. E., *Chromatin,* Springer–Verlag (1989).

Uberbacher, E. C. and Bunick, G. J., Resolving the nucleosome structure controversy—a basis for comparison of nucleosome structures, *Comments Mol. Cell Biophys.* **4**, 339–348 (1988).

Widom, J., Toward a unified model of chromatin folding, *Ann. Rev. Biophys. Biophys. Chem.* **18**, 365–395 (1989).

Widom, J. and Klug, A., Structure of the 300 Å chromatin filament: X-ray diffraction from oriented samples, *Cell* **43**, 207–213 (1985).

Wu, R. S., Panusz, H. T., Hatch, C. L., and Bonner, W. M., Histones and their modifications, *CRC Crit. Rev. Biochem.* **20**, 201–263 (1986).

Genomic Organization

Britten, R. J. and Kohne, D., Repeated sequences in DNA, *Science,* **161**, 529–540 (1968).

Breathnach, R. and Chambon, P., Organization and expression of eucaryotic split genes coding for proteins, *Annu. Rev. Biochem.* **50**, 349–383 (1981).

Brutlag, D. L., Molecular arrangement and evolution of heterochromatic DNA, *Annu. Rev. Genet.* **14**, 121–144 (1980). [Deals with highly repetitive sequences.]

Collins, F. S. and Weissman, S. M., The molecular genetics of human hemoglobin, *Prog. Nucleic Acid Res. Mol. Biol.* **31**, 315–462 (1984).

Doolittle, R. F., The genealogy of some recently evolved vertebrate proteins, *Trends Biochem. Sci.* **10**, 233–237 (1985).

Doolittle, W. F. and Sapienza, C., Selfish genes, the phenotype paradigm and genome evolution, *Nature* **284**, 601–603 (1980).

Gilbert, G., Marchionni, M., and McKnight, G., On the antiquity of introns, *Cell* **46**, 151–154 (1986).

Hentschel, C. C. and Birnsteil, M. L., The organization and expression of histone gene families, *Cell* **25**, 301–313 (1981).

Jelinek, W. R. and Schmid, C. W., Repetitive sequences in eukaryotic DNA and their expression, *Annu. Rev. Biochem.* **51**, 813–844 (1982). [Concentrates on moderately repetitive sequences.]

Johnson, P. F. and McKnight, S. L., Eukaryotic transcriptional regulatory proteins, *Ann. Rev. Biochem.* **58**, 799–839 (1989).

Kafatos, F. C., Orr, W., and Delidakis, C., Developmentally regulated gene amplification, *Trends Genet.* **1**, 301–306 (1985).

Kan, Y. W., The thalassemias, *in* Stanbury, J. B., Wyngaarden, J. B., Fredrickson, D. S., Goldstein, J. L., and Brown, M. S. (Eds.), *The Metabolic Basis of Inherited Disease* (5th ed.), *pp.* 1711–1725, McGraw–Hill (1983).

Lonberg, N. and Gilbert, W., Intron/exon structure of the chicken pyruvate kinase gene, *Cell* **40**, 81–90 (1985).

Long, E. O. and Dawid, I. B., Repeated genes in eukaryotes, *Annu. Rev. Biochem.* **49**, 727–764 (1980). [Discusses structural genes that occur in multiple copies.]

Mandal, R. K., The organization and transcription of eukaryotic ribosomal RNA genes, *Prog. Nucleic Res. Mol. Biol.* **31**, 115–160 (1984).

Maxson, R., Cohn, R., and Kedes, L., Expression and organization of histone genes. *Annu. Rev. Genet.* **17**, 239–277 (1983).

Orgel, L. E. and Crick, F. H. C., Selfish DNA: the ultimate parasite, *Nature* **284**, 604–607 (1980).

Orkin, S. H. and Kazazian, H. H., Jr., The mutation and polymorphism of the human β-globin gene and its surrounding DNA, *Annu. Rev. Genet.* **18**, 131–171 (1984).

Schimke, R. T., Gene amplification in cultured cells, *J. Biol. Chem.* **263**, 5989–5982 (1988).

Schimke, R. T., Gene amplification in cultured animal cells, *Cell* **37**, 705–713 (1984).

Stamatoyannopoulos, G., Nienhuis, A. W., Leder, P., and Majerus, P. W. (Eds.), *The Molecular Basis of Blood Diseases*, Chapters 2–4, Saunders (1987). [Discusses hemoglobin genes and their normal and thalassemic expression.]

Stark, G. R. and Wahl, G. M., Gene amplification, *Annu. Rev. Biochem.* **53**, 447–491 (1984).

Südhof, T. C., Goldstein, J. L., Brown, M. S., and Russell, D. W., The LDL receptor gene: a mosaic of exons shared with different proteins, *Science* **228**, 815–828 (1985).

Ulla, E., The human *Alu* family of repeated DNA sequences, *Trends Biochem. Sci.* **7**, 216–219 (1982).

Wainscoat, J. S., Hill, A. V. S., Boyce, A. L., Flint, J., Hernandez, M., Thein, S. L., Old, J. M., Lynch, J. R., Falusi, A. G., Weatherall, D. J., and Clegg, J. B., Evolutionary relationships of human populations from an analysis of nuclear DNA polymorphisms, *Nature* **319**, 491–493 (1986).

Weatherall, D. J. and Clegg, J. B., *The Thalassemia Syndromes* (3rd ed.), Blackwell Scientific Publications (1981).

Control of Expression

Atchison, M. L., Enhancers: mechanisms of action and cell specificity, *Annu. Rev. Cell Biol.* **4**, 127–153 (1988).

Beato, M., Gene regulation by steroid hormones, *Cell* **56**, 335–344 (1989).

Breitbart, R. E., Andreadis, A., and Nadal-Ginard, B., Alternative splicing: a ubiquitous mechanism for the generation of multiple protein isoforms from single genes, *Annu. Rev. Biochem.* **56**, 467–495 (1987).

Darnell, J. E., Jr., Variety in the level of gene control in eukaryotic cells, *Nature* **297**, 365–371 (1982).

Douglass, J., Civelli, O., and Herbert, E., Polyprotein gene expression, *Annu. Rev. Biochem.* **53**, 665–715 (1984).

Dynan, W. S. and Tijan, R., Control of eukaryotic messenger RNA synthesis by sequence-specific DNA-binding proteins, *Nature* **316**, 774–778 (1985).

Elgin, S. C. R., The formation and function of DNase I hypersensitive sites in the process of gene activation. *J. Biol. Chem.* **263**, 19259–19262 (1988).

Evans, R. M., The steroid and thyroid hormone receptor superfamily, *Science* **240**, 889–895 (1988).

Evans, R. M. and Hollenberg, S. M., Zinc fingers: gilt by association, *Cell* **52**, 1–3 (1988).

Frohman, M. A. and Martin, G. R., Cut, paste, and save: new approaches to alternating specific genes in mice, *Cell* **56**, 145–147 (1989).

Gehring, U., Steroid hormone receptors: biochemistry, genetics and molecular biology, *Trends Biochem. Sci.* **12**, 399–402 (1987).

Green, S. and Chambon, P., Nuclear receptors enhance our understanding of transcription regulation, *Trends Genet.* **4**, 309–314 (1988).

Gross, D. S. and Garrard, W. T., Nuclease hypersensitive sites in chromatin, *Annu. Rev. Biochem.* **57**, 159–197 (1988).

Jeang, K.-T. and Khoury, G., The mechanistic role of enhancer elements in eukaryotic transcription, *BioEssays* **8**, 104–107 (1988).

Kadonaga, J. T., Jones, K. A., and Tijan, R., Promoter-specific activation of RNA polymerase II transcription by Sp1, *Trends Biochem. Res.* **11**, 20–23 (1986).

Klug, A. and Rhodes, D., 'Zinc fingers': a novel protein motif for nucleic acid recognition, *Trends Biochem. Sci.* **12**, 464–469 (1987).

Landschultz, W. H., Johnson, P. F., and McKnight, S. L., The leucine zipper: a hypothetical structure common to a new class of DNA binding proteins, *Science* **240**, 1759–1764 (1988).

Maniatis, T., Goodbourne, S., and Fischer, J. A., Regulation of inducible and tissue-specific gene expression, *Science* **236**, 1237–1245 (1987).

Martin, G. M., X-Chromosome inactivation in mammals, *Cell* **29**, 721–724 (1982).

O'Shea, E. K., Rutkowski, R., and Kim, P., Evidence that the leucine zipper is a coiled coil, *Science* **243**, 538–542 (1989).

Pelham, H., Activation of heat-shock genes in eukaryotes, *Trends Genet.* **1**, 31–35 (1985).

Ptashne, M., How gene activators work, *Sci. Am.* **260**(1): 41–47 (1989).

Ptashne, M., How eukaryotic transcriptional activators work. *Nature* **335**, 683–689 (1988).

Raghow, R., Regulation of messenger RNA turnover in eukaryotes, *Trends Biochem. Sci.* **12**, 358–360 (1987).

Sassone-Corsi, P. and Borrelli, E., Transcriptional regulation by *trans*-acting factors, *Trends Genet.* **2**, 215–219 (1986).

Schröder, H. C., Bachmann, M., Diehl-Siefert, B., and Müller, W. E. G., Transport of mRNA from nucleus to cytoplasm, *Prog. Nucleic Acid Res. Mol. Biol.* **34**, 89–142 (1987).

Schleif, R., DNA binding by proteins, *Science* **241**, 1182–1187 (1988).

Struhl, K., Molecular mechanisms of transcriptional regulation in yeast. *Ann. Rev. Biochem.* **58**, 1051–1077 (1989).

Struhl, K., Helix–turn–helix, zinc-finger, and leucine-zipper motifs for eukaryotic transcriptional regulatory proteins, *Trends Biochem. Sci.* **14**, 137–140 (1989).

Tullius, T. D., Physical studies of protein-DNA complexes by footprinting, *Ann. Rev. Biophys. Biophys. Chem.* **18**, 213–237 (1989).

Voss, S. D., Schlokat, U., and Gruss, P., The role of enhancers in the regulation of cell-type-specific transcriptional control, *Trends Biochem. Sci.* **11**, 287–289 (1986).

Weintraub, H., Tissue-specific gene expression and chromatin structure, *Harvey Lectures* **79**, 217–244 (1985).

Weisbrod, S., Active chromatin, *Nature* **297**, 289–295 (1982).

Yamamoto, K. R., Steroid receptor regulated transcription of specific genes and gene networks, *Annu. Rev. Genet.* **19**, 209–252 (1985).

Yaniv, M. and Cereghini, S., Structure of transcriptionally active chromatin, *CRC Crit. Rev. Biochem.* **21**, 1–26 (1986).

Young, R. A., Hagenbüchle, O., and Schibler, U., A single mouse α-amylase gene specifies two different tissue-specific mRNAs, *Cell* **23**, 451–458 (1981).

Cell Differentiation

Angier, N., *Natural Obsessions: The Search for the Oncogene,* Houghton Mifflin (1988). [A chronicle of how science is really done.]

Becker, W. M., *The World of the Cell,* Chapters 19 and 23, Benjamin/Cummings (1986).

Bishop, J. M., Viral oncogenes, *Cell* **42**, 23–38 (1985).

Croce, C. M. and Klein, G., Chromosome translocations and human cancer, *Sci. Am.* **252**,(3): 54–60 (1985).

Curran, T. and Franza, B. R., Jr., Fos and Jun: the AP-1 connection, *Cell* **55**, 395–397 (1988).

Davidson, E. H., *Gene Activity in Early Development* (4th ed.), Academic Press (1986).

Gehring, W., Homeo boxes in the study of development, *Science* **236**, 1245–1252 (1987).

Gehring, W. J., The molecular basis of development, *Sci. Am.* **253**(4): 152B–162 (1985).

Gilbert, S. F., *Developmental Biology* (2nd ed.), Chapters 7–18 and 20, Sinauer Associates (1988).

Haluska, F. G., Tsujimoto, Y., and Croce, C. M., Oncogene activation by chromosome translocation in human malignancy, *Annu. Rev. Genet.* **21**, 321–345 (1987).

Hunter, T., The proteins of oncogenes, *Sci. Am.* **251**(2): 70–79 (1984).

Hunter, T. and Cooper, J. A., Protein-tyrosine kinases, *Annu. Rev. Biochem.* **54**, 897–930 (1985).

Ingham, P., The regulation of the bithorax complex, *Trends Genet.* **1**, 112–116 (1985).

Jurnak F., The three-dimensional structure of c-H-ras p21., *Trends Biochem. Sci.* **13**, 195–198 (1988).

Levine, M. and Hoey, T., Homeobox proteins as sequence-specific transcription factors, *Cell* **55**, 537–540 (1988).

Linzer, D. I. H., The marriage of oncogenes and anti-oncogenes, *Trends Genet.* **4**, 245–247 (1988).

Mahowald, A. P. and Hardy, P. A., Genetics of Drosophila embryogenesis, *Annu. Rev. Genet.* **19**, 149–177 (1985).

Molecular Basis of Development, Cold Spring Harbor Symp. Quant. Biol. **50** (1985). [Contains numerous articles on pattern formation, homeotic mutations, and homeoboxes.]

Nüsslein-Volhard, C., Frohnhöfer, H. G., and Lehman, R., Determination of anteroposterior polarity in *Drosophila*, *Science* **238**, 1675–1681 (1987).

Reddy, E. P., Skulka, A. M., and Curran, T. (Eds.), *The Oncogene Handbook,* Elsevier (1988).

Ruddle, F. H., Hart, C. P., and McGinnis, W., Structural and functional aspects of the mammalian homeo-box sequences, *Trends Genet.* **1**, 48–51 (1985).

Scott, M. P. and Carroll, S. B., The segmentation and homeotic gene network in early Drosophila development, *Cell* **51**, 689–698 (1987).

Scott, M. P. and O'Farrell, P. H., Spatial programming of gene expression in early *Drosophila* embryogenesis, *Annu. Rev. Cell Biol.* **2**, 49–80 (1986).

Varmus, H., Cellular and viral oncogenes, *in* Stamatoyannopoulos, G., Nienhuis, A. W., Leder, P., and Majerus, P. W. (Eds.), *The Molecular Basis of Blood Diseases,* pp. 271–346, Saunders (1987).

Varmus, H. E., The molecular genetics of cellular oncogenes, *Annu. Rev. Genet.* **18**, 553–612 (1984).

de Vos, A. H., Tong, L., Milburn, M. V., Matius, P. M., Jancarik, J., Noguchi, S., Nishimura, S., Miura, K., Ohtsuka, E., and Kim, S.-H., Three-dimensional structure of an oncogene protein: catalytic domain of human c-H-*ras* p21, *Science* **239**, 888–893 (1988) *and* Tong, L., de Vos, A. M., Milburn, M. V., Jancarik, J., Noguchi, S., Nishimura, S., Miura, K., Ohtsuka, E., and Kim, S.-H., Structural differences between a *ras* oncogene protein and the normal protein, *Nature* **337**, 90–93 (1989).

Weinberg, R. A., Finding the anti-oncogene, *Sci. Am.* **259**(3): 44–51 (1988).

Weinberg, R. A., A molecular basis of cancer, *Sci. Am.* **249**(5): 126–142 (1983).

Yarden, Y. and Ullrich, A., Growth factor receptor tyrosine kinases, *Annu. Rev. Biochem.* **57**, 443–478 (1988).

Problems

1. What is the maximum possible packing ratio of a 10^6 bp segment of DNA; of a 10^9 bp segment of DNA? Assume the DNA is a 20 Å in diameter cylinder with a contour length of 3.4 Å/bp.

2. When an SV40 minichromosome (a closed circular duplex DNA in complex with nucleosomes) is relaxed so that it forms an untwisted circle and is then deproteinized, the consequent closed circular DNA has about -1 superhelical turn for each of the nucleosomes that it originally had. Explain the discrepancy between this observation and the fact that the DNA in each nucleosome is wrapped nearly twice about its histone octamer in a left-handed superhelix.

3. Explain why acidic polypeptides such as polyglutamate facilitate *in vitro* nucleosome assembly.

*4. Consider a 1 million bp DNA molecule that has 1500 tandem repeats of a 400 bp sequence with the remainder of the DNA consisting of unique sequences. Sketch the C_0t curve of this DNA when it is sheared into pieces averaging 1000 bp long; when they are 100 bp long.

5. Why do isolated foldback structures, when treated by an endonuclease that only cleaves single-stranded DNA and then denatured, yield complicated C_0t curves?

6. During its two month period of maturation, the *Xenopus* oocyte synthesizes $\sim 10^{12}$ ribosomes. The consequent tremendous rate of rRNA synthesis is only possible because the normal genomic complement of rDNA has been amplified by 1500-fold. (a) Why is it unnecessary to likewise amplify the genes encoding the ribosomal proteins? (b) Assuming that rRNA gene amplification occurs in a short time at the beginning of the maturation period, how long would oogenesis require if the rDNA was not amplified?

7. **Hb Kenya** is a β-thalassemia in which the β-globin cluster is deleted between a point in the $^A\gamma$-globin gene and the corresponding position in the β-globin gene. Describe the most probable mechanism for the generation of this mutation.

8. Figure 33-45a contains a single band just above the bracketed region that increases in density as the Sp1 concentration increases. What is origin of this band?

9. Why do the rare instances of male calico cats all have the abnormal XXY genotype?

10. In *Drosophila*, an *esc⁻* homozygote develops normally unless its mother is also an *esc⁻* homozygote. Explain.

11. Retroviruses bearing oncogenes will infect cells from their corresponding host animal but will usually not transform them. Yet, these retroviruses will readily transform immortalized cells derived from the same organism. Explain.

12. The fusion of cancer cells with normal cells often supresses the expression of the tumorigenic phenotype. Explain.

Chapter 34
MOLECULAR PHYSIOLOGY

Throughout this text we have been largely concerned with basic biochemical questions: What are the structures of biological molecules? How do enzymes work? How do cells extract energy from nutrients? How do they synthesize biomolecules? How are these various processes controlled? Yet, to describe biological processes beyond the level of the single cell we must also reach an understanding of the various biochemical tasks that higher organisms apportion among their specialized tissues and how they coordinate the activities of these tissues. The elucidation of these complex processes has traditionally been the realm of physiology (the study of biological function), cell biology (the study of cell structure), and anatomy (the study of organismal structure); their molecular descriptions seemed out of reach. The recent enormous gains in biomedical knowledge and technology, however, have greatly blurred the boundaries among these disciplines and biochemistry. In this final chapter we shall therefore be concerned with "molecular physiology," the study of the molecular basis of biological function. In particular, in what really are independent minichapters, we shall consider the workings of four of the better characterized complex biochemical systems, those mediating (1) blood clotting, (2) immunity, (3) biological motility, and (4) biochemical communications. Consider these sections to be this textbook's "dessert," an indication of modern biochemistry's rapidly advancing capacity to explain and influence the workings of multicellular organisms. There are, of course, numerous other physiological systems that we might equally well have chosen to describe.

1. BLOOD CLOTTING

The circulatory system must be self-sealing, otherwise continued blood loss from even the smallest injury would be life threatening. Normally, all but the most catastrophic bleeding is rapidly stopped, a process known as **hemostasis,** through several sequential processes. First, an injury stimulates **platelets** (unpigmented enucleated blood cells—really fragments of much larger progenitor cells named **megakarocytes**) to adhere to damaged blood vessels and then to each other so as to form a plug that can stop minor bleeding. This association is mediated by **von Willebrandt factor,** a large (up to ~10^4 kD) multimeric plasma glycoprotein of subunit mass 225 kD. This protein binds to both a specific receptor on the platelet membrane and to the collagen and possibly other components of the subendothelial membrane exposed by vascular injury. Then, as the platelets aggregate, they release several physiologically active substances including serotonin (5-hydroxytryptamine) and thromboxane A_2 (Section 23-7B) that stimulate vasoconstriction, thereby reducing the blood flow at the injury site. Finally, the aggregating platelets and the damaged tissue initiate **blood clotting** or **coagulation,** the body's major defense against blood loss. In this section, we discuss the chemical mechanism of blood clotting, how clotting is inhibited in the absence of injury, and how clots are dissolved as an injury heals.

*A blood clot (medically known as a **thrombus**) forms through the action of a bifurcated cascade of proteolytic reactions involving the participation of nearly 20 different substances, most of which are liver-synthesized plasma glycoproteins.* This cascade is diagrammed in Fig. 34-1 and many of its components are listed in Table 34-1. All but two of these factors are designated by both a roman numeral and a common name although, unfortunately, the order of the roman numerals has historical rather than mechanistic significance. Seven of the clotting factors are zymogens (inactive forms) of serine proteases that are proteolytically activated by serine proteases further up the cascade. Other clotting proteins, termed **accessory factors,** which are also activated by these serine proteases, enhance the rate of activation of some of the zymogens. In both cases, the active form of a factor is designated by the subscript *a*.

The blood clotting system, which occurs in recognizable form in all vertebrates, contains a number of homologous serine proteases and therefore appears to have arisen through a series of gene duplications. The C-terminal ~250 residues of these proteases, which comprise their catalytically active domains, are also homologous to the pancreatic serine proteases trypsin, chymotrypsin, and elastase (Section 14-3). Like these digestive enzymes, the blood clotting proteases are activated by proteolytic cleavages that precede their C-terminal segments (Section 14-3E). However, the clotting

Table 34-1
Human Blood Coagulation Factors

Factor Number	Common Name	Subunit Mass (kD)
I	Fibrinogen	340
II	Prothrombin	72
III	Tissue factor *or* thromboplastin	37
IV	Ca^{2+}	
V[a]	Proaccelerin	330
VII	Proconvertin	50
VIII	Antihemophilic factor	330
IX	Christmas factor	56
X	Stuart factor	56
XI	Plasma thromboplastin antecedent (PTA)	160
XII	Hageman factor	80
XIII	Fibrin-stabilizing factor (FSF)	320
	Prekallikrein	88
	High-molecular-weight kininogen (HMK)	150

[a] Factor V_a was once called factor VI; consequently there is no factor VI.

Source: Mostly Furie, B. and Furie, B. C., *Cell* **53,** 505 (1988).

proteases differ from the digestive enzymes in that the zymogen-to-protease conversion only takes place in the presence of Ca^{2+} and on an appropriate phospholipid membrane (see below); the resulting N-terminal fragments are quite large (150–582 residues); and, with the exception of prothrombin (Section 34-1B), are linked to their C-terminal segments via disulfide bonds so that these segments do not separate upon activation. These N-terminal segments are thought to be responsible, at least in part, for the exquisite specificities of the proteolytic blood clotting factors: *Their substrates are limited, as we shall see, to the few inactive factors they function to activate.*

In what follows, we describe the mechanism of blood clotting in humans from the bottom up, that is, starting with the formation of the clot itself and working backwards through the sequence of activation steps leading to this process.

A. Fibrinogen and Its Conversion to Fibrin

*Blood clots consist of arrays of cross-linked **fibrin** that form an insoluble fibrous network (Fig. 34-2). Fibrin is made from the soluble plasma protein **fibrinogen** (factor I) through a proteolytic reaction catalyzed by the serine protease **thrombin.*** Fibrinogen comprises 2 to 3% of plasma protein. A molecule of fibrinogen consists of three pairs of nonidentical but homologous polypeptide chains, Aα (610 residues), Bβ (461 residues) and γ (411 residues),

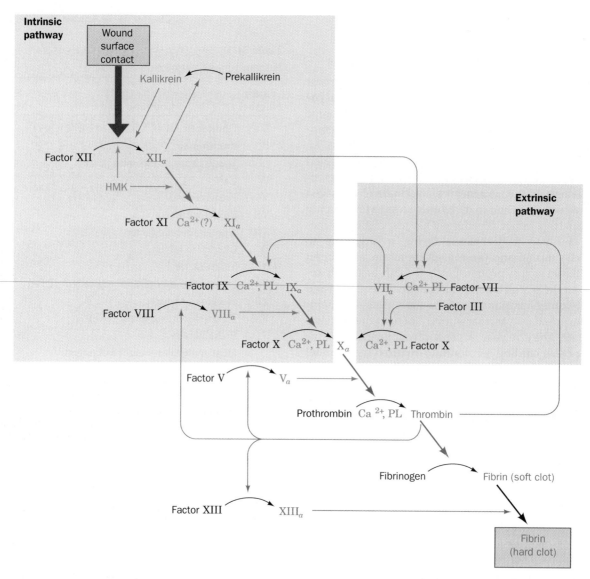

Figure 34-1
The blood clotting cascade in humans showing the division of its primary stages into the so-called intrinsic and extrinsic pathways. The clotting factors are named in Table 34-1. The active clotting factors, which with the exception of fibrin, are serine proteases, are indicated in red, whereas red arrows represent their proteolytic activation of other factors in the cascade. Similarly, active accessory factors, including Ca²⁺ and phospholipid membrane (PL), are indicated in green. Note the numerous feedback reactions that accelerate the clotting process and/or couple the intrinsic and extrinsic pathways.

and two pairs of *N*-linked oligosaccharides of ~ 2.5 kD each. Here A and B represent the 16- and 14-residue N-terminal **fibrinopeptides** that thrombin cleaves from fibrinogen so that a fibrin monomer is designated $\alpha_2\beta_2\gamma_2$. The reaction forming a blood clot from fibrinogen may therefore be represented

$$n(A\alpha)_2(B\beta)_2\gamma_2 \xrightarrow{\quad nA + nB \quad} n\alpha_2\beta_2\gamma_2 \longrightarrow (\alpha_2\beta_2\gamma_2)_n$$

Combined electron microscopic and low resolution X-ray crystallographic studies indicate that fibrinogen is a twofold symmetric elongated molecule, ~ 450 Å in length, that has two nodules at each end and one in the middle (Fig. 34-3). Its six polypeptide chains are joined by 17 disulfide bonds, 7 within each half of the dimer and 3 linking these two protomers (Fig. 34-4). Structural indications together with sequence-based conformation predictions suggest that the central region of each protomer consists mainly of a three-stranded coiled coil of α helices in which the α, β, and γ chains each contribute a strand (Fig. 34-5). The peripheral nodules diagrammed in Fig. 34-3 are formed by the C-terminal domains of the β and γ chains. The C-terminal segment of the α chain apparently lacks a definite conformation and is therefore not represented in Fig. 34-3.

How does fibrin polymerize to form a clot? Thrombin

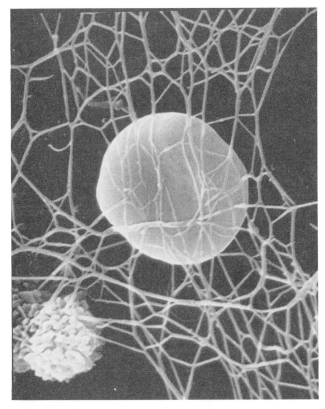

Figure 34-2
Scanning electron micrograph of a blood clot showing a red cell enmeshed in a fibrin network. [Manfred Kaga/Peter Arnold Inc.]

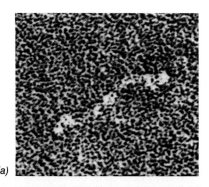

(a)

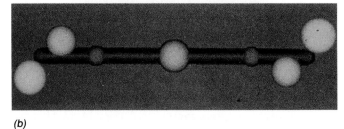

(b)

Figure 34-3
Fibrinogen. (*a*) An electron micrograph. (*b*) A 30-Å resolution model viewed along the molecule's twofold rotation axis. The light colored balls represent protuberances that are seen in combined electron microscopic and low resolution X-ray diffraction studies of fibrinogen. [Courtesy of John Weisel, University of Pennsylvania.]

Figure 34-4
A schematic diagram of the structure of fibrinogen, $(A\alpha)_2(B\beta)_2\gamma_2$, in which the interchain disulfide bonds are represented by yellow lines.

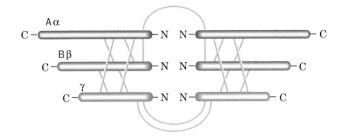

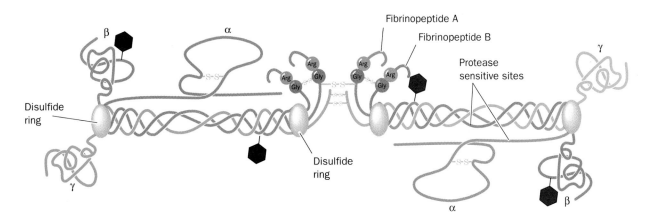

Figure 34-5
The proposed structure of fibrinogen based on low resolution structural studies, primary structure determinations, and chain-folding predictions. The so-called disulfide rings are regions containing three disulfide bonds cyclically linking homologous segments of the α, β, and γ chains. *N*-linked polysaccharides are represented by filled hexagons. The Arg-Gly bonds that are cleaved by thrombin in fibrin activation are indicated. [After the cover illustration, *Annu. N.Y. Acad. Sci.* **408** (1983).]

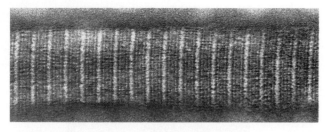

Figure 34-6
An electron micrograph of a fibrin fiber. The striations repeat every 225 Å, exactly one half of the length of a fibrin molecule. [Courtesy of John Weisel, University of Pennsylvania.]

specifically cleaves the Arg—X peptide bond (where X is Gly in most species) joining each fibrinopeptide to fibrin. Fibrin then spontaneously aggregates to form fibers that electron micrographs indicate have a banded structure that repeats every 225 Å (Fig. 34-6). This repeat distance is exactly one half the 450-Å length of a fibrin monomer, suggesting that fibrin monomers associate as a half-staggered array (Fig. 34-7). But why do fibrin monomers aggregate while fibrinogen, which has an all but identical structure, remains in solution? The main reason is that the loss of the fibrinopeptides exposes otherwise masked sites that mediate intermolecular association (Fig. 34-7). In addition, fibrinogen aggregation is inhibited by charge–charge repulsions: The fibrinopeptides are highly negatively charged; so much

so that fibrinogen's central region, where the fibrinopeptides reside, has a charge of -8, whereas that of fibrin is $+5$. Fibrinogen's end segments each have a similar charge of -4 but maintain that charge in fibrin. The repulsions between fibrinogen's like-charged segments helps prevent this protein from aggregating, whereas the attractions between fibrin's central and end segments promotes its specific association.

The diameters of fibrin fibers, which are fairly uniform (maximally ~50 nm), are important determinants of a clot's physical properties. What controls this fiber diameter? John Weisel and Lee Makowski have shown through electron microscopy studies that fibrin fibers are uniformly twisted. Consequently, the in-register molecules near the periphery of a twisted fiber must traverse a longer path than molecules near the fiber's center. The degree to which a molecule can stretch therefore limits the diameter of the fiber: Molecules add to the outside of a growing fiber until the energy to stretch an added molecule exceeds its energy of binding. A similar mechanism may limit the diameters of other biological fibers such as those of collagen (Section 7-2C).

Fibrin-Stabilizing Factor Cross-Links Fibrin Clots

The above "soft clot," as this name implies, is rather fragile. It is rapidly converted to a more stable "hard clot," however, by the covalent cross-linking of neighboring fibrin molecules in a reaction catalyzed by

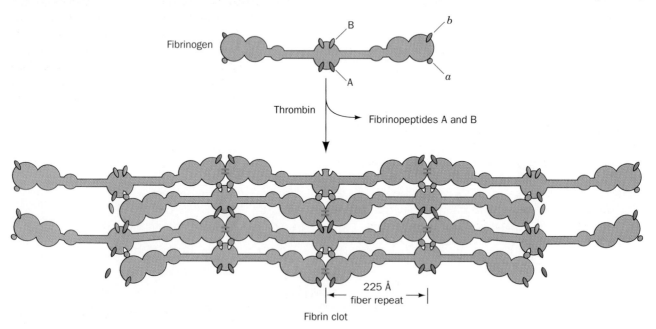

Figure 34-7
A model of the conversion of fibrinogen to a fibrin blood clot showing the half-staggered arrangement of its fibrin monomers. In "soft" clots, the fibrin monomers associate only by noncovalent interactions between knobs *a* and *b* that protrude from fibrin's γ chain and their complementary-binding sites that are exposed by the excision of fibrinopeptides A and B. In "hard" clots, however, fibrin monomers are also covalently cross-linked by isopeptide bonds between the C-terminal segments of the γ chains in neighboring molecules (*red lines*) as well as between the protruding segments of α chains (*not shown*). [After Weisel, J. W., *Biophys. J.* **50**, 1080 (1986).]

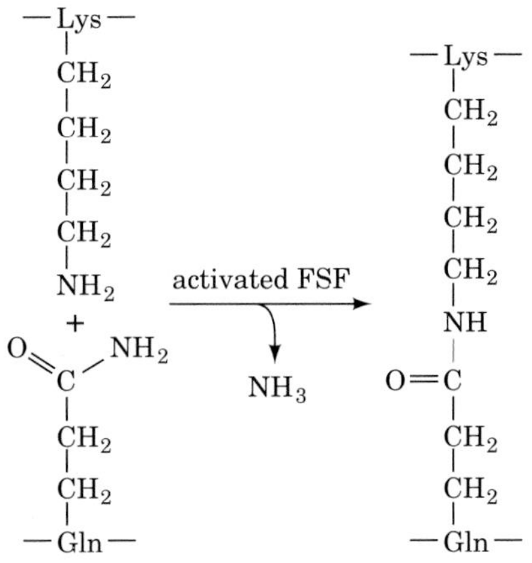

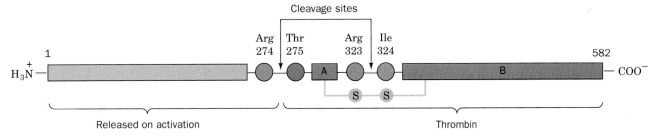

Figure 34-8
The transamidation reaction forming the isopeptide bonds cross-linking fibrin monomers in "hard" clots as catalyzed by activated fibrin-stabilizing factor (FSF, XIII$_a$).

fibrin-stabilizing factor (FSF or XIII$_a$). This transamidase initially joins the C-terminal segments of adjacent γ chains by forming isopeptide bonds between the side chains of a Gln residue on one γ chain and a Lys residue on another (Fig. 34-8). Two such symmetrically equivalent bonds are rapidly formed between each neighboring pair of γ chains (Fig. 34-7). The α chains are similarly cross-linked to one another but at a slower rate. The physiological importance of fibrin cross-linking is demonstrated by the observation that individuals deficient in FSF have a pronounced tendency to bleed.

FSF is present in both platelets and plasma. Platelet FSF consists of two 75-kD a chains, whereas plasma FSF additionally has two 88-kD b chains. Both species of FSF occur as zymogens that undergo thrombin-catalyzed cleavage of a specific Arg-Gly bond near the N-terminus of each a chain with the consequent release of a 37-residue propeptide. This treatment activates platelet FSF, designated a'_2, but plasma FSF, $a'_2 b_2$, remains inactive until its b chains dissociate, a process that is triggered by the binding of Ca^{2+} to the a' subunits. We shall see below that *Ca^{2+} is an essential factor in most stages of the blood clotting cascade.* The b subunits are thought to prolong the survival of plasma FSF in the circulation.

B. Thrombin Activation and the Function of Vitamin K

Thrombin is a serine protease that consists of two disulfide-linked polypeptide chains: a 49-residue A chain and a 259-residue B chain. The thrombin B chain is homologous to trypsin but is far more selective; it cleaves only particular Arg-Gly bonds on certain proteins.

*Thrombin is synthesized as a 582-residue zymogen, **prothrombin (II)**, which is activated by two proteolytic cleavages catalyzed by activated **Stuart factor (X$_a$)**, the product of the preceding step of the clotting cascade.* The cleavage of prothrombin's Arg 274-Thr 275 and Arg 323-Ile 324 bonds releases its N-terminal segment and separates the A and B chains (Fig. 34-9). The latter cleavage, it is thought, results in the formation of an ion pair much like that formed between Ile 16 and Asp 194 in the activation of chymotrypsinogen (Section 14-3E).

Vitamin K Is an Essential Cofactor in the Synthesis of γ-Carboxyglutamate

*Prothrombin, as well as the homologous factors VII, IX, and X, is synthesized in the liver in a process that requires an adequate dietary intake of **vitamin K** (Fig. 34-10; K for the Danish koagulation).* Lack of vitamin K or the presence of a competitive inhibitor such as **dicoumarol** (which was discovered in spoiled sweet clover because it causes fatal hemorrhaging in cattle) or **warfarin** (a rat poison) causes the production of an abnormal prothrombin that is activated by factor X$_a$ at only 1 to 2% of the normal rate. This observation was, at first, quite puzzling because normal and abnormal prothrombins seemed to have identical amino acid compositions. NMR studies eventually established, however, that normal prothrombin contains γ-**carboxyglutamate (Gla)** residues,

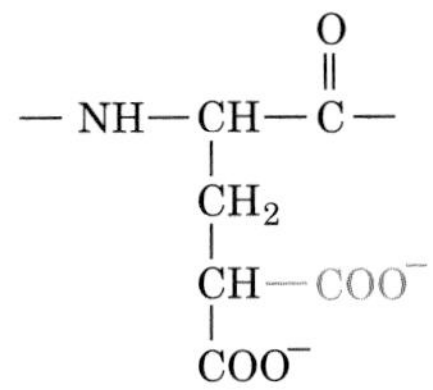

$$
\begin{array}{c}
\quad\quad\quad\quad O \\
\quad\quad\quad\quad \| \\
-\,NH\!-\!CH\!-\!C\!- \\
\quad\quad | \\
\quad\quad CH_2 \\
\quad\quad | \\
\quad\quad CH\!-\!COO^- \\
\quad\quad | \\
\quad\quad COO^-
\end{array}
$$

γ **-Carboxyglutamate (Gla)**

Figure 34-9
The structure of prothrombin showing the two peptide bonds that are cleaved by factor X$_a$ to form thrombin. The N-terminal peptide is released in this activation process while the A and B peptides of thrombin remain linked by a disulfide bond.

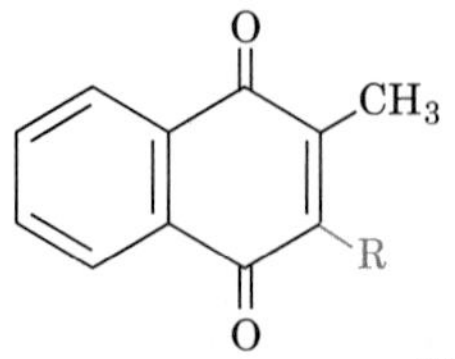

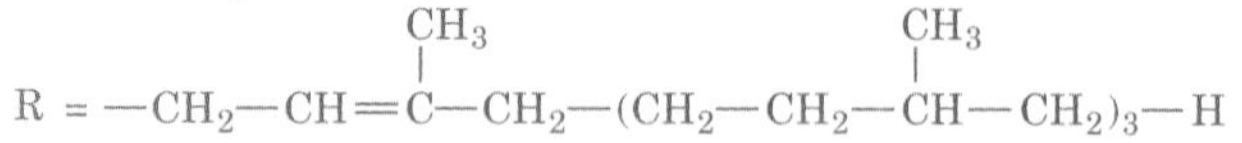

Vitamin K₁ (Phylloquinone)

$$R = -CH_2-CH=\overset{\overset{\displaystyle CH_3}{|}}{C}-CH_2-(CH_2-CH_2-\overset{\overset{\displaystyle CH_3}{|}}{CH}-CH_2)_3-H$$

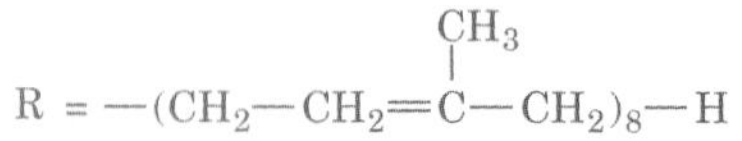

Vitamin K₂ (Menaquinone)

$$R = -(CH_2-CH_2=\overset{\overset{\displaystyle CH_3}{|}}{C}-CH_2)_8-H$$

Vitamin K₃ (Menadione)

$$R = -H$$

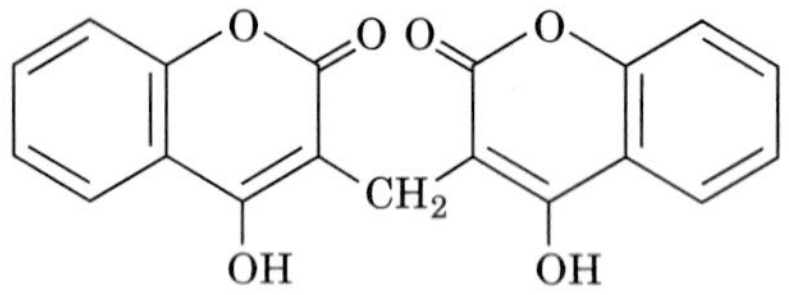

Dicoumarol

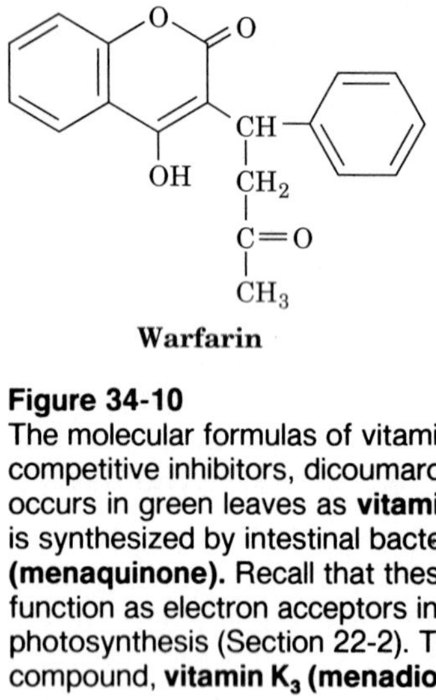

Warfarin

Figure 34-10
The molecular formulas of vitamin K and two of its competitive inhibitors, dicoumarol and warfarin. Vitamin K occurs in green leaves as **vitamin K₁ (phylloquinone)**, and is synthesized by intestinal bacteria as **vitamin K₂ (menaquinone)**. Recall that these forms of the vitamin function as electron acceptors in chloroplast and bacterial photosynthesis (Section 22-2). The body converts the parent compound, **vitamin K₃ (menadione)**, to a vitamin-active form.

10 of which occur between residues 6 and 33 in human prothrombin (Fig. 34-11). Abnormal prothrombin, in contrast, contains Glu in place of these Gla residues. *Vitamin K must therefore be a cofactor in the post-translational conversion of Glu to Gla.* The reason why prothrombin's Gla residues were not initially detected is because they decarboxylate to Glu under the conditions of acid hydrolysis normally used in amino acid composition determinations (Section 6-1D).

The liver reaction cycle that synthesizes Gla from Glu and then regenerates the vitamin K cofactor may be considered to occur in four reactions (Fig. 34-12):

1. Vitamin K, in its active hydroquinone form, abstracts a γ proton from Glu in an O_2-consuming reaction of unknown mechanism that yields the γ-carbanion of Glu and the 2,3-epoxide of vitamin K.

2. The Glu carbanion then reacts with CO_2 to yield Gla.

3. and 4. Vitamin K hydroquinone is regenerated in two sequential reactions, both apparently catalyzed by the same enzyme, that employ thiols such as lipoic acid. Dicoumarol and warfarin act by blocking both these steps. Reaction 4 may also be catalyzed by certain NADH- or NADPH-requiring reductases.

The discovery of Gla residues in clotting factors led to their discovery in other tissues, which must therefore also contain vitamin K-dependent carboxylases. We shall see below that the Gla residues of clotting factors function to bind Ca^{2+}. Presumably they have similar roles in other tissues.

Prothrombin Activation Is Accelerated in the Presence of Factor V_a, Ca^{2+}, and Phospholipid

Factor X_a, by itself, is an extremely sluggish prothrombin activator. Yet, in the presence of activated **proaccelerin (V_a)**, Ca^{2+}, and phospholipid membrane, its activity is enhanced 20,000-fold. The membrane surface in contact with the activation complex must contain negatively charged phospholipids such as phosphatidylserine in order to stimulate this rate enhancement. Such phospholipids occur almost exclusively on the cytoplasmic sides of cell membranes (Section 11-3B) which, of course, are normally not in contact with the blood plasma. Moreover, ~20% of the total factor V in blood is stored in the platelets and released only upon platelet activation. Consequently, *physiological prothrombin activation normally takes place at a significant rate only in the vicinity of an injury.*

Ca^{2+} *is required for either prothrombin or factor X_a to bind to phospholipid membranes;* these proteins are anchored to the membrane via Ca^{2+} bridges. Prothrombin and factor X_a from vitamin K-deficient animals have greatly reduced membrane-binding affinities compared to the corresponding normal proteins. Evidently, *the Gla side chains, which are much stronger Ca^{2+} chelators than*

$$\overset{+}{H_3N}-\overset{1}{Ala}-Asn-Lys-Gly-Phe-Leu-Gla-Gla-Val-\overset{10}{Arg}-$$
$$\overset{11}{Lys}-Gly-Asn-Leu-Gla-Arg-Gla-Cys-Leu-\overset{20}{Gla}-$$
$$\overset{21}{Gla}-Pro-Cys-Ser-Arg-Gla-Gla-Ala-Phe-\overset{30}{Gla}-$$
$$\overset{31}{Ala}-Leu-Gla-Ser-\overset{35}{Leu}-\cdots$$

Figure 34-11
The N-terminal sequence of prothrombin showing its 10 Gla residues.

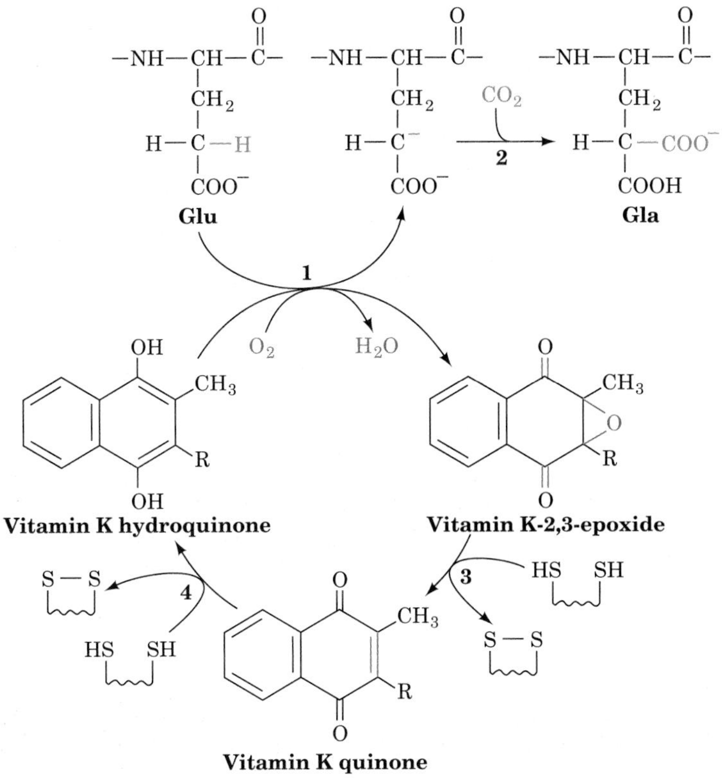

Figure 34-12
The vitamin K metabolism cycle in liver. Reactions 3 and 4, which are both inhibited by dicoumarol and warfarin, are thought to be catalyzed by the same enzyme. The R group is given in Fig. 34-10. [After Suttie, J. W., *Annu. Rev. Biochem.* **54,** 472 (1985).]

Glu, form the proteins' Ca^{2+}-binding sites. In fact, the 10 to 12 Gla residues that occur in each of the vitamin K-dependent clotting zymogens, prothrombin and factors VII, IX, and X, are contained in these proteins' highly homologous N-terminal segments (Fig. 34-11 for prothrombin). The excision of prothrombin's N-terminal segment by its activation releases the resulting thrombin from the phospholipid membrane so that it can activate fibrinogen in the plasma. Thrombin differs in this respect from the other vitamin K-dependent zymogens, which remain bound to the phospholipid membrane after their activation.

Activated proaccelerin (V_a), the accessory factor in prothrombin activation, is activated by a thrombin-catalyzed proteolytic cleavage. Prothrombin activation, in this indirect way, is thereby autocatalytic (thrombin, *in vitro*, can also directly activate prothrombin but this reaction has been shown to be physiologically insignificant). V_a, however, is subject to further thrombin-catalyzed proteolysis, which inactivates it. Moreover, thrombin can proteolytically inactivate other thrombin molecules. *Clot formation is therefore self-limiting, a safeguard that helps prevent blood clots from propagating away from the site of an injury.*

C. The Intrinsic Pathway

Factor X may be activated by two different proteases (Fig. 34-1):

1. By **factor IX_a**, the product of the **intrinsic pathway** (so-named because all of its protein components are contained in the blood).

2. By **factor VII_a**, the product of the **extrinsic pathway** (so-called because one of its important components occurs in the tissues).

We shall discuss these two pathways separately beginning with the intrinsic pathway.

Clotting Is Initiated by the Contact System

It has long been known that bringing blood into contact with negatively charged surfaces, such as those of glass or kaolin (a clay used to make porcelain), initiates clotting. *In vivo,* collagen and platelet membranes are thought to have the same effect. The nature of this so-called **contact system** has only been worked out in outline and its physiological significance is still under dispute.

The contact system consists of four glycoproteins: the serine protease zymogens named **Hageman factor (XII), prekallikrein, plasma thromboplastin antecedent (PTA or XI),** and **high-molecular-weight kininogen (HMK),** an accessory factor that is also a precursor of the nonapeptide hormone **bradykinin** (a potent vasodilator and diuretic factor). Adsorption to a suitable surface is thought to somehow activate Hageman factor which, in the presence of HMK, proteolyzes prekallikrein to form the active protease **kallikrein.** Kallikrein, in turn, proteolytically activates Hageman factor so that these two proteins reciprocally activate each other.

The nature of contact-activated Hageman factor is enigmatic; it is by no means certain that physical adsorption to a surface cleaves the same bond as does kallikrein or, for that matter, cleaves any bond at all. Much of the experimental difficulty in resolving this issue is a consequence of the contact-activation process's autocatalytic nature: Prekallikrein, contact-activated Hageman factor's substrate, is the zymogen of the protease that activates Hageman factor. Consequently, in any measurement of its activity, the nature of contact-activated Hageman factor is immediately obscured by large amounts of rapidly generated kallikrein-activated Hageman factor.

The final reaction mediated by the contact system is the proteolytic activation of factor XI by activated Hageman factor in a process that also uses HMK as an accessory factor. Although the contact system is clearly effective in initiating *in vitro* clot formation, its *in vivo* importance is in doubt because individuals deficient in Hageman factor, prekallikrein, or HMK do not suffer from bleeding problems.

The Last Two Steps of the Intrinsic Pathway Are Similar

The intrinsic pathway has two remaining steps leading to the activation of Stuart factor (X, Fig. 34-1). Factor XI_a catalyzes the proteolytic activation of **Christmas factor (IX),** a Gla-containing glycoprotein, in a Ca^{2+}-requiring reaction that takes place on a phospholipid membrane surface. No accessory factor is known for this reaction. Christmas factor may also be activated by activated **proconvertin (VII$_a$),** a product of the extrinsic pathway (Section 34-1D).

In the final step of the intrinsic pathway, factor X is proteolytically cleaved by activated Christmas factor (IX_a) on a phospholipid membrane surface in a reaction involving Ca^{2+} and the accessory factor activated **antihemophilic factor (VIII$_a$).** Antihemophilic factor, as is proaccelerin (V), is proteolytically activated by thrombin in a second autocatalytic process leading to prothrombin activation (Fig. 34-1). Not surprisingly, proaccelerin and antihemophilic factor are homologous proteins. Antihemophilic factor circulates in the plasma in complex with von Willebrandt factor; in fact, the activities of these two substances were initially attributed to a single protein.

Hemophilias Result from Clotting Factor Deficiencies

The discovery of antihemophilic factor came about through its deficiency in individuals with the most common clotting disorder, **hemophilia A,** a sex-linked inherited deficiency (~ 1 per 10,000 male births). Indeed, several of the clotting factors were discovered through the diagnosis of their deficiencies in various clotting disorders (the existence of Christmas factor was discovered through its absence in Stephen Christmas, a hemophiliac whose deficiency, **hemophilia B,** is the second most common form of hemophilia). Hemophiliacs may lose large amounts of blood from even the smallest injury and frequently hemorrhage without any apparent cause. However, the symptoms of their diseases may be alleviated by the intravenous administration of the deficient factor. In the past, this treatment was expensive and not without risk because large amounts of blood must be fractionated to obtain therapeutic doses of most clotting factors. Hemophiliacs were therefore inordinately subject to a variety of dangerous bloodborne viral diseases including hepatitis and AIDS. These difficulties have now been largely eliminated through the production of the required clotting factors by recombinant DNA techniques.

D. The Extrinsic Pathway

The extrinsic pathway (Fig. 34-1), the alternative arm of the clotting cascade, is initiated by the proteolysis of proconvertin (VII), a process that can be catalyzed by activated Hageman factor (XII_a) as well as by thrombin.

Activated proconvertin, in turn, mediates the activation of factor X in a process analogous to that of prothrombin activation (Section 34-1B) in that its rate is enhanced 16,000-fold by the presence of phospholipid membrane, Ca^{2+}, and an accessory factor named **tissue factor (III).** Intact proconvertin can also catalyze factor X activation but at only 2% the rate of activated proconvertin. Apparently this rate is so low that, in the absence of tissue factor, unactivated proconvertin cannot initiate *in vivo* clot formation.

Tissue factor is an integral membrane glycoprotein that occurs in many tissues and is particularly abundant in brain, lung, blood vessel walls, and placenta. Consequently, an injury that exposes blood to tissue rapidly initiates the extrinsic pathway. In fact, the addition of tissue factor to the extrinsic system causes clot formation in ~ 12 s, whereas the intrinsic system requires several minutes to do so. These observations suggest that the intrinsic pathway is normally of little significance. However, the severity of the hemophilias resulting from intrinsic pathway clotting factor deficiencies clearly establishes the importance of the intrinsic pathway in blood clotting. Of course the two pathways are not really independent since they are coupled through a number of reactions (Fig. 34-1).

E. Control of Clotting

The multilevel cascade of the blood clotting system permits enormous amplification of its triggering signals. Moving down the extrinsic pathway, for example, proconvertin (VII), Stuart factor (X), prothrombin, and fibrinogen are present in plasma in concentrations of $<1, 8, 150$, and up to $4000 \ \mu g \cdot mL^{-1}$, respectively. Yet, clotting must be very strictly regulated since even one inappropriate clot can have fatal consequences. Indeed, blood clots are the leading cause of strokes and heart attacks, the two major causes of human death in developed countries.

A Variety of Factors Limit Clot Growth

There are numerous physiological mechanisms that limit clot formation. We have seen that there are several interactions among the various clotting factors that inhibit blood coagulation (Fig. 34-1). The blood flow dilution of active clotting factors also does so as does their selective removal from the circulation by the liver. In addition, plasma contains several serine protease inhibitors whose presence prevents clots from spreading beyond the vicinity of an injury. For example, **antithrombin** (58 kD) inhibits all active proteases of the clotting system except VII_a by binding to them in tight $1:1$ complex (much like BPTI binds trypsin; Section 14-3D). The presence of **heparin,** a sulfated glycosaminoglycan (Section 10-2E), enhances the activity of antithrombin by several hundredfold. Heparin occurs almost exclusively in the intercellular granules of the mast cells that line certain blood vessels. Its release, presumably by injury, activates antithrombin thereby preventing runaway clot growth.

Protein C is another plasma protein that limits clotting. This Gla residue-containing 62-kD zymogen is activated by thrombin to proteolytically inactivate proaccelerin (V) and antihemophilic factor (VIII). Activated protein C attacks the activated forms of these accessory factors more readily than their nonactive forms. The importance of protein C is demonstrated by the observation that individuals who lack it often die in infancy of massive thrombotic complications.

Despite the foregoing, blood clotting *in vitro* is not self-limiting. This observation, which suggests the existence of additional *in vivo* clot-limiting factors, led to the discovery of **thrombomodulin**, a 74-kD glycoprotein that projects from the cell surface membranes of the vascular endothelium (inner lining). Thrombomodulin specifically binds thrombin so as to convert it to a form with decreased ability to catalyze clot formation but with a > 1000-fold increased capacity to activate protein C.

The control of clotting is a major medical concern. Heparin, the most frequently used anticoagulant, is administered before and after surgery to retard clot formation. For long-term control of hemostasis, dicoumarol is often employed. In the design of an artificial heart, the elimination of mechanically induced clots remains the major unsolved problem (the construction of an adequate pump is a relatively simple task). The prevention of clotting is also a concern of blood-sucking organisms. The leech *Hirudo medicinalis* solves this problem by secreting **hirudin**, a 65-residue protein, in its saliva. Hirudin, the most powerful naturally occurring anticoagulant known, specifically binds to thrombin so as to inactivate it. Consequently a leech bite, although a minor wound, bleeds quite freely.

F. Clot Lysis

Blood clots are only temporary patches; they must be eliminated as wound repair progresses. This is a particularly urgent need when a clot has inappropriately formed or has broken free into the general circulation. Fibrin is a molecule that is "designed" to be easily dismantled in a process termed **fibrinolysis**. The demolition agent is a plasma serine protease named **plasmin,** an enzyme that specifically cleaves fibrin's triple-stranded coiled coil segment and cuts away its covalently cross-linked α chain protuberances (Fig. 34-5). The rather open meshlike structure of a blood clot (Fig. 34-2) gives plasmin relatively free access to the polymerized fibrin molecules thereby facilitating clot lysis.

Plasmin is formed through the proteolytic cleavage of the 86-kD zymogen **plasminogen**, a protein that is homologous to the zymogens of the blood clotting cascade.

There are several serine proteases that activate plasminogen, most notably the 54-kD enzyme **urokinase,** which is synthesized by the kidney and occurs, as its name implies, in the urine, and the homologous 70-kD enzyme **tissue-type plasminogen activator (t-PA),** which occurs in vascular tissues. In addition, activated Hageman factor, in the presence of prekallikrein and HMK (the contact activation system) activates plasminogen although the physiological significance of the contact activation of the fibrinolytic system has not been determined. Nevertheless, the fibrinolysis system, as our experience might have led us to expect, is not so simple as just a zymogen and its activators. It also incorporates several inhibitors, principally the 70-kD glycoprotein α_2-**antiplasmin**, which forms an irreversible equimolar complex with plasmin that prevents it from binding to fibrin. The α_2-antiplasmin cross-links to fibrin α chains through the action of activated FSF (XIII$_a$, the enzyme that also cross-links fibrin), thereby making "hard" clots less susceptible to fibrinolysis than "soft" clots. The importance of this serine protease inhibitor (it also inhibits chymotrypsin) is indicated by the observation that homozygotes for a defective α_2-antiplasmin have a serious tendency to bleed.

Plasminogen activators are receiving considerable medical attention aimed at rapidly dissolving the blood clots responsible for heart attacks and strokes. **Streptokinase,** a 45-kD protein produced by certain streptococci, has shown considerable promise in this regard, particularly when administered together with aspirin (which inhibits platelet aggregation; Section 23-7B). Despite its name, streptokinase exhibits no enzymatic activity. Rather, it acts by forming a tight 1 : 1 complex with plasminogen that proteolytically activates other plasminogen molecules. The use of streptokinase to dissolve clots has the disadvantage that it activates plasmin to degrade fibrinogen as well as fibrin thereby increasing the risk of bleeding problems, particularly strokes. The therapeutic use of t-PA, which has been synthesized by recombinant DNA techniques, is thought to eliminate these problems because this enzyme activates plasminogen only in the presence of a blood clot (although the medical significance of these problems is under dispute).

2. IMMUNITY

All organisms are continually subject to attack by other organisms. In response to predators, animals have developed an enormous variety of defensive strategies. An even more insidious threat, however, is attack by disease-causing microorganisms and viruses (pathogens). In order to deal with them, animals have evolved an elaborate protective array known as the **immune system** (Latin: *immunis,* exempt). Pathogens that manage to breach the physical barrier presented by the skin and mucous membranes (a vital first line of defense) are identified as foreign invaders and destroyed. In this section we discuss how the immune system recognizes foreign invaders, how they are distinguished from normal components of self, and how they are destroyed. As we do so, keep in mind that the immune system exhibits many of the qualities that are characteristic of the nervous system such as the ability to detect and react to stimuli and to remember. Indeed, the size and complexity of the vertebrate immune system rivals that of the vertebrate nervous system.

A. The Immune Response

Immunity in vertebrates is conferred by certain types of white blood cells collectively known as **lymphocytes.** They arise, as do all blood cells, from common precursor cells **(stem cells)** in the bone marrow. Lymphocytes, however, in contrast to red blood cells, can leave the blood vessels and patrol the intercellular spaces for foreign intruders. They eventually return to the blood via the lymphatic vessels but not before interacting with specialized **lymphoid tissues** such as the thymus, the lymph nodes, and the spleen, the sites where much of the immune response occurs.

Two types of immunity have been distinguished:

1. **Cellular immunity,** which guards against virally infected cells, fungi, parasites, and foreign tissue, is mediated by *T* lymphocytes or *T* cells, so-called because their development occurs in the *t*hymus.

2. **Humoral immunity** (*humor* is an archaic term for fluid), which is most effective against bacterial infections and the extracellular phases of viral infections, is mediated by an enormously diverse collection of related proteins known as **antibodies** or **immunoglobulins.** Antibodies are produced by *B* lymphocytes or *B* cells which, in mammals, mature in the *b*one marrow.

We shall outline the operations and interactions of these systems as a prelude to discussing their biochemistry.

The Cellular Immune System

The immune response is triggered by the presence of foreign macromolecules, normally proteins, carbohydrates, and nucleic acids, known as **antigens.** This process occurs through a complex series of interactions among various types of *T* cells and *B* cells that specifically bind a particular antigen (Fig. 34-13). In the following paragraphs, italic numerals and letters refer to the corresponding drawing in Fig. 34-13.

The cellular immune response leads to the destruction of the offending cells. It begins when a **macrophage** (a type of white blood cell) engulfs *(1a, 1b)* and lysosomally partially digests *(2a, 2b)* a foreign antigen and then displays the resulting antigenic fragments on its surface *(3a,*

3b). There, it is thought, these fragments bind to one of two types of cell-surface proteins known as **major histocompatibility complex (MHC) proteins** (so-called because they are transcribed from a closely linked series of genes called the MHC; Section 34-2E). The MHC is remarkably polymorphic (has numerous alleles); so much so that any two unrelated individuals of the same species are highly unlikely to have an identical set of MHC proteins. *MHC proteins are therefore markers of individuality.*

Class I MHC proteins are displayed on the surfaces of nearly all nucleated vertebrate cells. Macrophages exhibiting Class I MHC proteins are recognized by proteins known as *T* **cell receptors,** which occur on the surfaces of immature **cytotoxic *T* cells.** *In order to bind to the antigen-displaying macrophage, however, these receptors must specifically complex the antigen together with the Class I MHC protein (4a);* neither molecule alone can do the job. In the same way, macrophages displaying antigenic fragments complexed to **Class II MHC proteins** are bound by immature **helper *T* (T_H) cells** bearing the cognate receptor *(4b).* This elaborate recognition system, as we shall see, focuses the attention of *T* cells on cell surfaces and thereby prevents the resources of the cellular immune system from being futilely squandered on noncellular targets.

T cells that bind to a macrophage-displayed antigen– MHC protein complex are induced to propagate, a process known as **clonal selection,** which was first recognized in the 1950s by Niels Kaj Jerne, Macfarlane Burnet, Joshua Lederberg, and David Talmadge. Consequently, *only those T cells that specifically recognize the intruding antigen are produced in quantity.* Clonal selection occurs because a macrophage bound to a *T* cell releases a polypeptide growth factor named **interleukin-1** *(5a, 5b),* which specifically stimulates *T* cells to proliferate and differentiate *(6a, 6b).* This process is enhanced by the *T* cells' autostimulatory secretion of **interleukin-2.** *T* cells only make **interleukin-2 receptor** so long as they remain bound to a macrophage *(7a, 7b)* thereby preventing unlimited *T* cell proliferation. Nevertheless, a large number of mature cytotoxic *T* cells *(8a),* which through their receptors are specifically targeted for host cells displaying both the foreign antigen and Class I MHC proteins, are generated starting a few days after the antigen is first encountered. The cytotoxic *T* cells, which are also known as **killer *T* cells,** live up to their name: *They bind to antigen-bearing host cells (9) and, at the point of contact, release a 70-kD protein,* **perforin,** *that lyses these target cells (10) by aggregating to form pores in their plasma membranes (Section 34-2F).*

The cellular immune system functions mainly to prevent the spread of a viral infection by killing virus-infected host cells (viral coat proteins are generally displayed on the surface of an animal cell during the latter stages of its viral infection; e.g., Section 32-4A). It is also effective

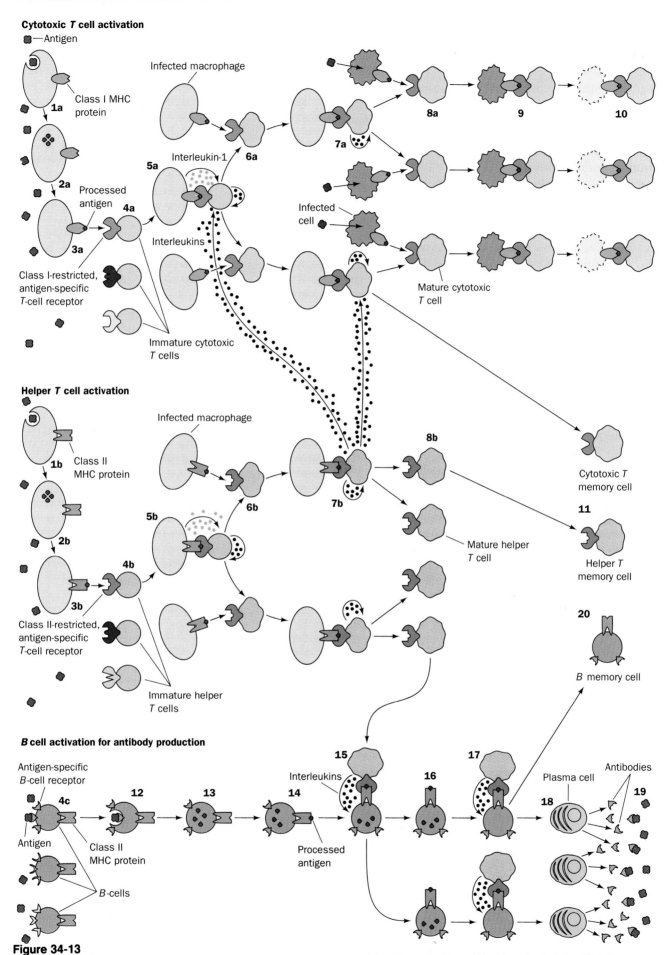

Figure 34-13

An outline of the immune response. See the text for an explanation. [After Marrack, P. and Kappler, J., *Sci. Am.* **254**(2): 38–39 (1986). Copyright © 1986 by Scientific American, Inc.]

against fungal infections, parasites, and certain types of cancers. Indeed, the cellular immune system's vital function has become painfully evident in recent years through the tragic spread of **acquired immune deficiency syndrome (AIDS),** whose causative agent, **human immunodeficiency virus (HIV),** acts by specifically attacking helper *T* cells. The cellular immune system is also responsible for various difficulties elicited by modern medicine that do not occur in nature such as the rejection of tissue and organ grafts from foreign donors. Such grafts, which are recognized as foreign because they almost always bear MHC proteins that differ from those of the host, have only been made possible by the development of drugs known as **immunosuppressants** that suppress the immune response (but not so much as to leave the body defenseless against pathogens).

The Humoral Immune System

B cells display both immunoglobulins and Class II MHC proteins *(4c)* on their surfaces. If a *B* cell encounters an antigen that binds to its particular immunoglobulin, it engulfs the complex *(12),* partially digests the antigen *(13),* and displays the fragments on its surface in complex with the Class II MHC protein *(14).* Mature helper *T* cells *(8b)* bearing receptors specific for this complex bind to the *B* cell *(15)* and, in response, release interleukins that stimulate the *B* cell to proliferate and differentiate *(16).* Cell division continues so long as the *B* cells are stimulated by the helper *T* cells *(17)* which, in turn, depends on the continuing presence of antigen *(1b – 8b). Most of the B cell progeny are **plasma cells** (18) that are specialized to secrete large amounts of the antigen-specific antibody. The antibodies bind to the available antigen (19), thereby marking it for destruction either by **phagocytosis** (ingestion by white cells known as **phagocytes**) or by activating the **complement system** (a series of interacting proteins that lyse cells and trigger local inflammatory reactions; Section 34-2F).*

Most *T* cells and *B* cells live only a few days unless stimulated by their corresponding antigen. Moreover, the proliferation of *B* cells is limited by their interactions with **suppressor *T* (T_S) cells,** an additional type of *T* lymphocyte progeny which have essentially the opposite function of helper *T* cells. Yet, one of the hallmarks of the immune system is that an animal is rarely infected twice by exactly the same type of pathogen; that is, *recovery from an infection by a pathogen renders an animal immune from that pathogen.* This so-called **secondary immune response** is mediated by long lived **memory *T* cells** *(11)* and **memory *B* cells** *(20)* which, upon reencountering their cognate antigen, perhaps decades after its previous appearance, proliferate much faster and more massively than do "virgin" *T* and *B* cells (those that have never encountered their corresponding antigen) as is indicated in Fig. 34-14. This characteristic of the immune system has been recognized since ancient

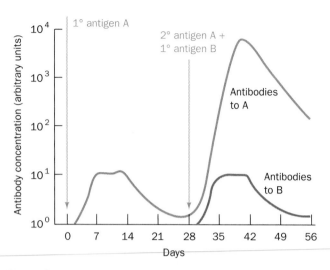

Figure 34-14
Primary and secondary immune responses: The rates of appearance of antibodies in the blood serum following primary (1°) immunization on day 0 with antigen A and secondary (2°) immunization on day 28 with antigens A and B. Antigen B is included in the secondary immunization to demonstrate the specificity of immunological memory for antigen A. Note that the secondary response to antigen A is both faster and greater than the primary response. *T* cell mediated responses exhibit similar immunological memory.

times: The Greek historian Thucydides noted over 2400 years ago that the sick could be treated by those who had recovered for a man was never attacked twice by the same disease.

The Immune System Is Self-Tolerant

Nearly all biological macromolecules are antigenic. To prevent self-destruction, an animal's immune system must therefore discriminate between self-antigens and foreign antigens. Such a process must be exquisitely selective. After all, a vertebrate, for example, has tens of thousands of different macromolecules, each with numerous distinctive antigenic sites.

What is the mechanism of immunological **self-tolerance?** The immune system in mammals becomes active around the time of birth. If a foreign antigen is implanted in an embryo before this time, the resulting animal is unable to mount an immune attack against that antigen. Apparently, the immune system eliminates the clones of *B* and *T* cells that recognize the antigens present during the critical period when the immune system becomes active. Yet, because new clones of lymphocytes, each with a nearly unique set of antigenic determinants, arise throughout an animal's lifetime (Section 34-2C; antibodies and *T* cell receptors are themselves antigenic), self-toleration must be an ongoing process. This process, although of dimly perceived mechanism, has been shown to occur in the thymus where, it appears, *only those virgin T cells displaying receptors that bind MHC proteins but which have no affin-*

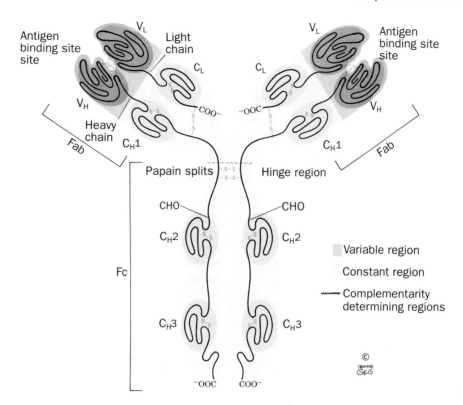

Figure 34-15
A diagram of the human IgG molecule. Each light (L) chain consists of two homologous units, V_L and C_L, where V and C indicate the polypeptide chain's variable and constant regions. Each heavy (H) chain is composed of four such units, V_H, C_H1, C_H2, and C_H3. Treatment of IgG by the proteolytic enzyme papain results in the cleavage of this immunoglobulin molecule in its "hinge" region yielding two Fab fragments and one Fc fragment. CHO represents carbohydrate chains. [Drawing copyrighted © by Irving Geis.]

ity for self-antigens are selected for further propagation. Indeed, only a small fraction of the lymphocytes that are processed by the thymus ever leave that organ.

Occasionally, the immune system loses tolerance to some of its self-antigens resulting in an **autoimmune disease.** For example, **myasthenia gravis,** an autoimmune disease in which individuals make antibodies against the **acetylcholine receptors** of their own skeletal muscles (acetylcholine is a neurotransmitter that triggers muscle contraction; Section 34-4C), results in a progressive and often fatal muscular weakness. Similarly, individuals with **systemic lupus erythematosis,** an often fatal inflammatory disease, produce antibodies against many of their own cellular components including certain ribonuclear proteins (Section 29-4A). Other common autoimmune diseases are **rheumatoid arthritis, insulin-dependent diabetes mellitus** (Section 25-3B), and **multiple sclerosis.**

B. Antibody Structures

The immunoglobulins form a related but yet enormously diverse group of proteins. In this section, we consider the structures of these essential molecules. How their diversity is generated is the subject of the following section.

There Are Five Classes of Immunoglobulins

Most immunoglobulins, and the basic building blocks of all of them, consist, as Gerald Edelman and Rodney Porter showed, of four subunits: two identical ~ 23-kD **light chains (L)** and two identical 53- to 75-kD **heavy chains (H).** These subunits associate via disulfide bonds as well as by noncovalent interactions to form, as electron micrographs indicate, a Y-shaped symmetric dimer, $(L - H)_2$ (Fig. 34-15). Immunoglobulins are glycoproteins; each heavy chain has an *N*-linked oligosaccharide.

Humans have five classes of secreted immunoglobulins, designated IgA (for immunoglobulin A), IgD, IgE, IgG, and IgM which differ in their corresponding types of heavy chains designated α, δ, ε, γ, and μ, respectively (Table 34-2). There are also two types of light chain, κ and λ, but these occur in immunoglobulins of all classes. IgD, IgE, and IgG exist only as $(L - H)_2$ dimers. IgM, however, consists of pentamers of its respective dimers and IgA occurs as monomers, dimers, and trimers of its corresponding dimers (Fig. 34-16). The dimeric units of these multimers are linked by disulfide bonds to each other and to an ~ 20-kD protein termed the **joining chain (J).** IgM also occurs in a *B* cell-displayed monomeric membrane-bound form. It is antigen binding by this latter form of IgM that triggers the humoral immune response.

Table 34-2
Classes of Human Immunoglobulins

Class	Heavy Chain	Light Chain	Subunit Structure	Subunit Mass (kD)
IgA	α	κ or λ	$(\alpha_2\kappa_2)_n J^a$ $(\alpha_2\lambda_2)_n J^a$	360–720
IgD	δ	κ or λ	$\delta_2\kappa_2$ $\delta_2\lambda_2$	160
IgE	ε	κ or λ	$\varepsilon_2\kappa_2$ $\varepsilon_2\lambda_2$	190
IgG[b]	γ	κ or λ	$\gamma_2\kappa_2$ $\gamma_2\lambda_2$	150
IgM	μ	κ or λ	$(\mu_2\kappa_2)_5 J$ $(\mu_2\lambda_2)_5 J$	950

[a] $n = 1$, 2, or 3.

[b] IgG has four subclasses, IgG1, IgG2, IgG3, and IgG4, which differ in their γ chains.

The various classes of secreted immunoglobulins have different physiological functions. IgM, which is largely confined to the blood, is most effective against invading microorganisms. It is the first immunoglobulin to be secreted in response to an antigen; its production begins 2 to 3 days after antigen is first encountered. IgG, the most common immunoglobulin, is equally distributed between the blood and the interstitial fluid. It is the only antibody that can cross the placenta (via receptor-mediated endocytosis) and thus provide the fetus with immunity. IgG production begins 2 to 3 days after IgM first appears. IgA occurs predominantly in the intestinal tract and in such secretions as saliva, sweat, and tears; it defends against invading pathogens by binding to their antigenic sites so as to block their attachment to epithelial (outer) surfaces. IgA is also the major antibody of milk and colostrum (the first milk secreted after pregnancy) and thereby protects nursing infants from gastrointestinal invasion by pathogens. IgE, which is normally present in the blood in minute concentrations, protects against parasites and has been implicated in allergic reactions. IgD, which is also present in blood in very small amounts, is of unknown function.

Immunoglobulin's Functional Segments May Be Proteolytically Separated

In 1959, Porter showed that IgG, the most common class of immunoglobulin, is cleaved, through limited proteolysis with papain, into three ~50-kD fragments: two identical **Fab fragments** and one **Fc fragment**. The Fab fragments, which form the arms of the Y-shaped IgG molecule and which each consist of an entire L chain and the N-terminal half of an H chain (Fig. 34-15), contain IgG's antigen-binding sites ("ab" stands for *an*tigen-*b*inding). Immunoglobulin's consequent divalent (or, for IgA and IgM, multivalent) antigen-binding char-

acter forms the basis of the **precipitin reaction,** a sensitive test that has long been used for determining the presence of antibody or antigen: A mixture of antibody and the antigen against which it is directed combine as an extended cross-linked lattice (most antigens have multiple antigenic determinants) that yields an easily detected precipitate (Fig. 34-17). *The formation of these cross-linked lattices enhances antibody–antigen binding through cooperative interactions and is required to trigger B cell proliferation.*

The Fc fragment (so-named because it is readily crystallized), derives from the stem of the Y and consists of the identical C-terminal segments of two H chains (Fig. 34-15). Fc fragments contain the effector sites that mediate the functions common to a particular class of immunoglobulins such as inducing phagocytosis, triggering the complement system, and directing the transport of immunoglobulins to their sites of action.

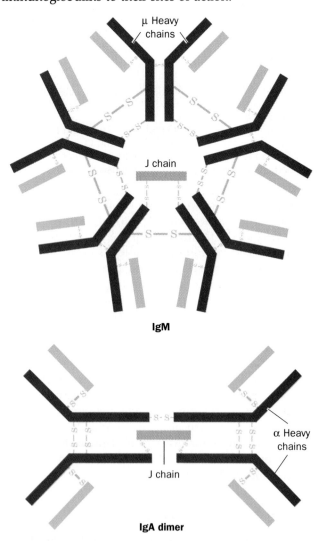

Figure 34-16
The five dimeric subunits of IgM (*top*) are held together by disulfide bonds. A single J chain joins two of the pentamer's μ heavy chains and is therefore thought to initiate assembly of this immunoglobulin. The J chain also participates in joining IgA chains to form dimers (*bottom*) and trimers.

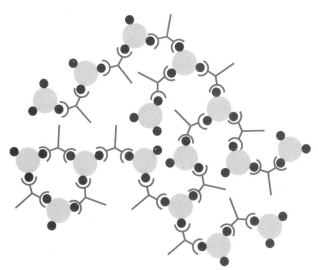

Figure 34-17
A divalent antibody (*green*) can cross-link its corresponding multivalent antigen (*red*) so as to form an extended lattice. Although the antigen is shown here as having three copies of only one type of antigenic determinant, most naturally occurring antigens, for example, a bacterium, have multiple copies of several different antigenic determinants. Such complex antigens are efficiently cross-linked by the mixtures of antibodies directed against their various antigenic sites.

IgG's Heavy and Light Chains Both Have Constant and Variable Regions

In order to characterize a molecule, it is necessary to obtain it in reasonably pure form. This requirement, at first, presented immunologists with a seemingly insurmountable obstacle. Exposing an animal to a particular antigen elicits the formation of numerous clones of plasma cells, each of which synthesizes a slightly different immunoglobulin molecule that binds the antigen. The resulting antibodies are therefore quite heteroge-

neous. This obstacle was largely removed by the demonstration, in the early 1960s, that an individual with **multiple myeloma**, a plasma cell cancer, synthesizes large amounts of a single species of immunoglobulin termed a **myeloma protein.** Some myelomas make excess light chains, which, when they are excreted in the urine, are known as **Bence Jones proteins** (after Henry Bence Jones who first described them in 1847).

The amino acid sequences of several different Bence Jones proteins, which each have 214 residues, revealed that *the sequence differences among light chains are largely confined to their N-terminal halves.* Light chains are therefore said to have a **variable region, V_L,** spanning residues 1 to 108, and a **constant region, C_L,** comprising residues 109 to 214 (Fig. 34-15). Similarly, comparisons of myeloma heavy chains, which have 446 residues, revealed that all the sequence differences among them occur between residues 1 to 125. *Thus heavy chains also have a variable region, V_H, and a constant region, C_H* (Fig. 34-15).

Additional sequence comparisons indicated that the C_H region consists of three ~110-residue segments, **C_H1, C_H2,** and **C_H3,** which are homologous to each other and to C_L. In fact, even the constant and variable sequences are related albeit not as closely as the members of each of these groups are related to each other. These homologies, together with the observation that each homology unit is cross-linked by a disulfide bond, correctly suggest (see below) that *an immunoglobulin molecule's 12 homology units each fold into an independent domain.* Apparently, modern light chain and heavy chain genes evolved through duplications of a primordial gene encoding an ~110-residue protein.

The V_L and V_H regions are not uniformly variable. Rather, most of their amino acid variations are concentrated into three short **hypervariable sequences** (Fig. 34-18). Elvin Kabat therefore predicted that *the hyper-*

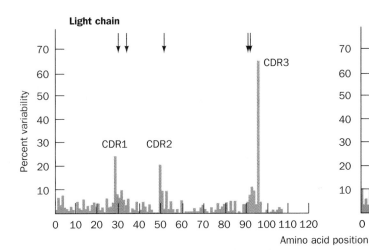

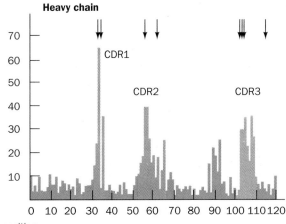

Figure 34-18
Sequence comparisons of a number of immunoglobulins indicates that their hypervariable segments (*orange bars*) are responsible for most of the sequence variation in the variable regions of both the light and heavy chains. The arrows mark the sites on anti-DNP antibodies that are derivatized by the affinity label *p*-nitrophenyldiazonium.

variable sequences line the immunoglobulin's antigen-binding site and that their amino acids determine its binding specificity.

Kabat's hypothesis was supported by affinity labeling experiments. Molecules of <5 kD are rarely antigenic. Yet, when small organic groups termed **haptens,** such as the **2,4-dinitrophenyl (DNP)** group, are covalently attached to a carrier protein such as bovine serum albumin (by reaction of fluorodinitrobenzene with its Lys residues)

$$\boxed{\text{Bovine serum albumin}}$$
$$|$$
$$(CH_2)_4$$
$$|$$
$$NH$$

DNP-hapten

and then injected into an animal, the animal produces antibodies that bind to the hapten in the absence of the carrier. If the DNP analog *p*-**nitrophenyldiazonium**

$$O_2N-\bigcirc-N_2^+$$

***p*-Nitrophenyldiazonium**

is combined with anti-DNP antibodies, the hapten's highly reactive diazonium group will form diazo bonds with the His, Lys, and Tyr side chains in the vicinity of the antibodies' DNP-binding sites (affinity labeling; Section 30-3A) . Most of the side chains so-derivatized

are, in fact, members of the antibodies' hypervariable sequences (Fig. 34-18) thereby indicating that *antigen-binding sites are lined with hypervariable residues.* Immunoglobulin's hypervariable segments are therefore also called **complementarity-determining regions (CDRs).**

Monoclonal Antibodies Are Indispensable Biomedical Tools

One might expect that homogeneous immunoglobulins could be obtained in quantity by simply cloning a single lymphocyte and harvesting the immunoglobulin the clone produced. Unfortunately, lymphocytes do not grow continuously in culture. In the late 1970s, however, Cesar Milstein and Georges Köhler developed a

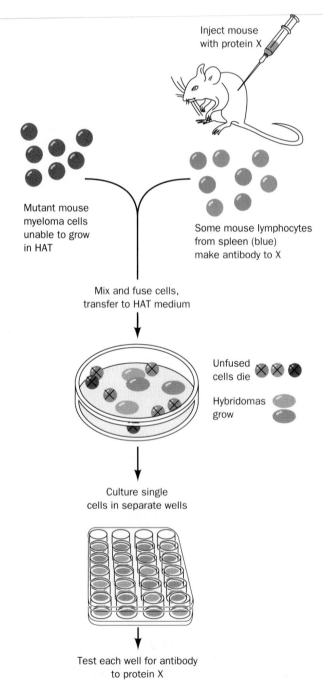

Figure 34-19
Procedure for producing monoclonal antibodies against an antigen, X. **HAT medium,** so-called because it contains *h*ypoxanthine, *a*methopterin (methotrexate, an antifolate; Section 26-4B), and *t*hymine, prevents the growth of mutant cell lines lacking hypoxanthine-guanine phosphoribosyl transferase (HGPRT; a purine salvage enzyme that catalyzes the formation of the AMP and GMP precursor IMP; Section 26-2D). The amethopterin blocks the *de novo* synthesis of purines, which *HGPRT⁻* cells cannot replace through salvage pathways, and thymine, which is available from the HAT medium. *HGPRT⁻* myeloma cells are fused with spleen-derived lymphocytes from a mouse immunized against X and the resultant preparation is transferred to a HAT medium. This treatment selects for fused cells (hybridomas): The *HGPRT⁻* myeloma cells cannot grow in HAT medium; lymphocytes, which make HGPRT, do not grow in culture; but the hybridoma cells, which have the lymphocytes' HGPRT and the myeloma cells' immortality, proliferate. Individual hybridoma cells are then cloned and screened for the production of anti-X antibody. A satisfactory clone can be grown in virtually unlimited quantities, either in culture or as a mouse tumor, so as to synthesize the desired amounts of monoclonal antibody.

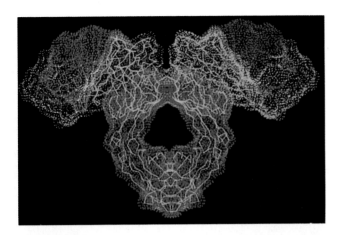

Figure 34-20

X-ray structure of an intact antibody. The antigen combining sites are located at the ends of the two horizontal Fab arms formed by the association of the light chains (whose surfaces are outlined with blue-green dots) and the heavy chains (outlined with blue dots). The C_α atoms of the variable and constant chains in the Fab arms are connected with red and green lines, respectively, while the heavy chain C_α atoms in the Fc segment are connected by white lines. The molecule's twofold axis of symmetry is vertical. Thus the light chain is in front of its associated heavy chain on the left and behind it on the right. Compare this figure with Fig. 34-15. [Courtesy of Arthur Olson, Research Institute of Scripps Institute. X-ray structure determined by David Davies.]

technique for immortalizing such clones (Fig. 34-19). *Monoclonal antibodies can now be obtained in virtually unlimited quantities and specific for almost any antigen by fusing myeloma cells with lymphocytes raised against that antigen (that is, isolated from an animal that has been immunized with the antigen). A clone of the resulting **hybridoma** (hybrid-myeloma) cell synthesizes the lymphocyte's immunoglobulin but has the myeloma cell's immortality. Monoclonal antibodies have become indis-*pensable biomedical tools; they can be used to assay for and to isolate extremely small amounts of nearly any specific biological substance.* For example, they have made possible the routine testing of blood for the presence of HIV (AIDS virus) thereby protecting the public blood supply.

Immunoglobulin Homology Units Are Similarly Folded

The proposed immunoglobulin structure was confirmed and extended by X-ray structure determinations of Fab and Fc fragments and entire myeloma proteins variously carried out by David Davies, Allan Edmondson, Robert Huber, Roberto Poljak, and many others. IgG is a molecule with twofold rotational symmetry whose homology units form separate domains (Fig. 34-20). Each of these domains is closely associated with a domain from another polypeptide chain so that the entire molecule may be considered to consist of six globular modules, two that form the stem of the Y (Fc region) and two that form each of its two arms (Fab regions; Fig. 34-15).

*The immunoglobulin homology units all have the same characteristic **immunoglobulin fold:** A barrel composed of a three- and a four-stranded antiparallel β sheet that are linked by a disulfide bond* (Fig. 34-21). The V domains differ from the C domains mainly by an additional polypeptide loop flanking each V domain's three-stranded β sheet.

Immunoglobulins, as both physicochemical studies and comparisons of their X-ray structures indicate, exhibit considerable intersegmental flexibility. This is particularly evident in the protein's so-called **"hinge" region,** the polypeptide segment joining each Fab region to the Fc region (Fig. 34-15; note that the IgG shown in Fig. 34-20 is better described as T shaped). Since the basic immunoglobulin structure must accommodate an enormous variety of antigens, its flexibility presumably

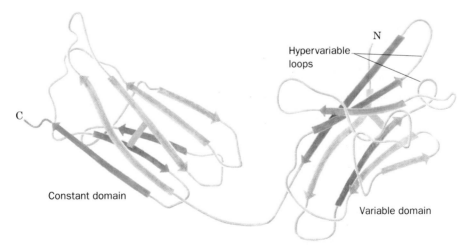

Figure 34-21

The chain folding of a myeloma protein light chain. Both its constant and variable domains assume the immunoglobin fold: a barrel-shaped sandwich of a four-stranded antiparallel β sheet (*blue arrows*) and a three-stranded antiparallel β sheet (*brown arrows*) that are linked by a disulfide bond (*yellow*). [After Schiffer, M., Girling, R. L., Ely, K. R., and Edmundson, A. B., *Biochemistry* **12,** 4628 (1973).]

facilitates antigen binding by permitting an optimal fit between the antigen and its combining site. The immunoglobulin's carbohydrate moiety is wedged between the C_H1 and C_H2 homology units and therefore also modulates the interactions between the Fab and Fc regions.

Antigen-Binding Sites Are Complementary to Their Corresponding Antigens

Myeloma proteins, upon which most of our structural knowledge of immunoglobulins is based, are produced by cancer cells that originally proliferated in response to unknown, if any, antigens. Nevertheless, haptens that bind to particular myeloma proteins have been identified by screening many different compounds.

The X-ray structures of several hapten–myeloma protein complexes indicate that an immunoglobulin's antigen-binding site is located at the tip of each Fab region in a crevice between its V_L and V_H units (Fig. 34-15). *The size and shape of this crevice depends on the amino acid sequences of the V_L and V_H units and its walls are formed, as predicted, by the six hypervariable segments (CDRs; Fig. 34-22).* Antibody–hapten complexes, not

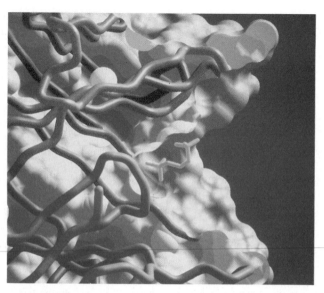

Figure 34-23
An X-ray structure showing the binding of a **phosphorylcholine** hapten (*green*) to the antigen-combining site of an Fab fragment. The C_α chains of the V_H and V_L subunits are represented by red and blue tubes. The surface of the protein is shaded pink when viewed from the outside and light blue when viewed from the inside. Note the precise complimentarity between the hapten and its binding site. [Courtesy of Arthur Olson, Research Institute of Scripps Clinic. X-ray structure determined by David Davies.]

surprisingly, resemble enzyme–substrate complexes (e.g., Fig. 34-23); both types of associations involve van der Waals interactions, hydrophobic forces, hydrogen bonding, and ionic interactions. Indeed, antibody–hapten complexes and enzyme–substrate complexes have similar ranges of dissociation constants, from 10^{-4} to $10^{-10}M$, which correspond to binding energies of 25 to 65 $kJ \cdot mol^{-1}$.

Antibody–hapten complexes are imperfect models of antibody–antigen complexes because a hapten only partially fills its corresponding antigen-binding site. Recently, however, the X-ray structures of three complexes of hen egg white (HEW) lysozyme with Fab's derived from different anti-HEW lysozyme monoclonal antibodies have been reported. Each of these Fab's binds to a largely independent, irregularly shaped, ~ 700-$Å^2$ surface patch on lysozyme such that one molecule's protruding side chains fit neatly into depressions on its mate (Fig. 34-24*a*). In each of these associations, all six Fab CDRs participate in lysozyme binding. These complexes, much like other known protein–protein associations, are cemented by highly complementary and thus solvent-excluding sets of van der Waals interactions, salt bridges, and hydrogen bonds. In several of these interactions, the lysozyme backbone and side chains maintain conformations identical to those in isolated lysozyme (Section 14-2A), but in others, there are significant local conformational variations (this comparison cannot be extended to the lysozyme-binding Fab's

Figure 34-22
The X-ray structures of seven different V_H units superimposed on their conserved framework residues. The C_α backbone is colored light gray for framework (non-CDR) residues, red for CDR1, yellow for CDR2, and green for CDR3. Note that the conformational variation among these structures resides almost entirely in their CDRs. The CDRs of the V_L units are similarly varied. [Courtesy of Elizabeth Getzoff, Victoria Roberts, Michael Pique, and John Tainer, Research Institute of Scripps Clinic.]

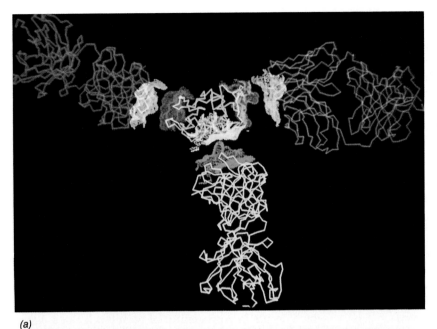

(a)

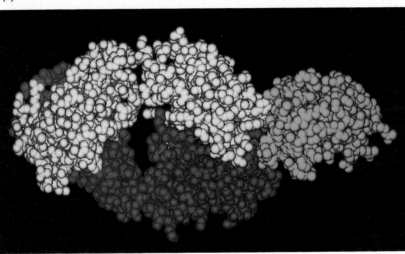

(b)

Figure 34-24
Hen egg white lysozyme in complex with the Fab fragments of monoclonal antibodies raised against it. (*a*) An exploded-view collage indicating how three different anti-lysozyme Fab's interact with lysozyme (*center*) in the X-ray structures of their respective complexes. The proteins are represented by their C_α chains and their interacting surfaces are outlined by juxtaposed dot surfaces. Note that the three crystal structures on which this diagram is based each contain only one Fab species; the Fab's do not crystallize together. [Courtesy of Steven Sheriff and David Davies, NIH.] (*b*) The X-ray structure of HEW lysozyme in complex with the anti-lysozyme Fab named D1.3 (Part *a, upper right*). In this space-filling representation, the Fab's L chain is yellow, its H chain is blue, the lysozyme molecule is green, and lysozyme Gln 121 is red. [From Amit, A. G., Mariuzza, R. A., Phillips, S. E. V., and Poljak, R. J., *Science* **233,** 749 (1986).]

because they have not crystallized by themselves).

The exquisite specificity of anti-lysozyme antibodies for their antigenic sites is demonstrated by the effect of a single amino acid change on the lysozyme contact surface. The dissociation constant of the anti-HEW lysozyme immunoglobulin named D1.3 with HEW lysozyme is $2.2 \times 10^{-8}M$. Yet, the dissociation constant of this monoclonal antibody with those of the nearly identical egg white lysozymes from partridge, California quail, and turkey are all $>10^{-5}M$. In all these latter lysozymes, Gln 121, which conspicuously protrudes from the HEW lysozyme surface into its Fab antigen-binding site (Fig. 34-24*b*), is replaced by His.

What are the special characteristics, if any, of the **epitopes** (antigenic sites) to which antibodies bind? All of the above lysozyme epitopes consist of 14 to 16 surface residues from two or more polypeptide segments. Some of these residues exhibit high mobility (Section 8-2) but

others do not. Thus, the observation that our sample of only three antibody–lysozyme complexes cover around one half of lysozyme's surface strongly suggests that *a protein's entire accessible surface is potentially antigenic.*

C. Generation of Antibody Diversity

The immune system has the capacity to generate antibodies against almost any antigen that it encounters; it can produce a virtually unlimited variety of antigen-binding sites. What is the origin of this enormous diversity? One might reasonably expect that immunoglobulin gene expression resembles that of other proteins in that every distinct H and L chain is encoded by a separate germline gene. If this were true, then to encode the billions of different antibodies each vertebrate appears capable of producing would require huge numbers of these genes. For example, it would require 10^3 H and L chain genes each to encode $10^3 \times 10^3 = 10^6$ different immunoglobulins. However, hybridization studies using radioactive cDNA probes transcribed from immunoglobulin mRNAs indicate that the mouse embryo genome, for example, contains far too few immunoglobulin genes to account for the mouse's observed level of antibody diversity. Consequently, this so-called **germ line hypothesis** must be rejected.

Two other models for the origin of antibody diversity have been seriously considered:

1. The **somatic recombination hypothesis,** which was originally formulated in 1965 by William Dreyer and Claude Bennett, proposes that *antibody diversity is generated by genetic recombination among a relatively few gene segments encoding the variable region of an immunoglobulin chain.* This process occurs via intrachromosomal recombination during B cell differentiation so that each B cell clone expresses an all but unique immunoglobulin.

2. The **somatic mutation hypothesis** proposes that *antibody diversity arises through an extraordinarily high rate of immunoglobulin gene mutation during B cell differentiation.*

We shall see below that both of these mechanisms contribute to antibody diversity.

κ Light Chain Genes Are Assembled from Three Sets of Gene Segments

DNA sequencing studies by Leroy Hood, Philip Leder, and Susumu Tonegawa have revealed that κ light chains are each encoded by four exons (Fig. 34-25):

1. A **leader** or L_κ **segment,** which encodes a 17 to 20-residue hydrophobic signal peptide. This polypeptide directs newly synthesized κ chains to the endoplasmic reticulum and is then excised (Section 11-3F).

2. A V_κ **segment,** which encodes the first 95 residues of the κ chain's 108-residue variable region.

3. A **joining** or J_κ **segment** (not to be confused with the J chain of IgA and IgM), which encodes the variable region's remaining 13 residues.

4. The C_κ **segment,** which encodes the κ chain's constant region.

The arrangement of these exons in human embryonic tissues (which do not make antibodies) differs strikingly from those in gene families we have previously encountered. The L_κ and V_κ segments are separated by an intron as occurs in other split genes. However, the κ chain gene family contains an array of ~150 of these ~400 bp $L_\kappa + V_\kappa$ units separated from each other by ~7-kb spacers. This sequence of exon pairs is followed, well downstream, by 5 J_κ segments at intervals of ~300 bp, a 2.4-kb spacer, and a single C_κ segment.

The assembly of a κ chain mRNA is a complex process involving both somatic recombination and selective gene splicing (Fig. 34-25). The first step of this process in mice, which occurs in a progenitor of each B cell clone, is an intrachromosomal recombination that joins an $L_\kappa - V_\kappa$ unit to a J_κ segment and deletes the intervening sequences. Then, in later cell generations, the entire modified gene is transcribed and selectively spliced so as to join the $L_\kappa - V_\kappa - J_\kappa$ unit to the C_κ segment. The L_κ and V_κ segments are also spliced together in this step, yielding an mRNA that encodes one of each of the elements of a κ chain gene.

Highly conserved sequences on the 3' side of each V_κ segment and on the 5' side of each J_κ segment suggest how the somatic recombination sites are selected. The V_κ sequence is immediately followed by the heptameric sequence CACAGTG, a 12 ± 1 nucleotide spacer, and an AT-rich nonamer. The J_κ chain is preceded by the complementary heptamer, a 23 ± 1 nucleotide spacer, and the complementary AT-rich nonamer. Evidently, these sequences can combine, under the influence of an as yet unidentified system of recombinatory enzymes, to form a stem-and-loop structure that acts as a recombination signal (Fig. 34-26).

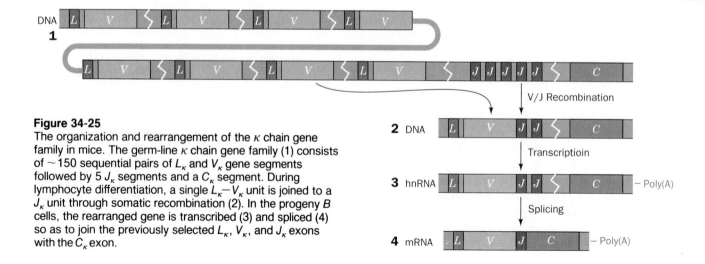

Figure 34-25
The organization and rearrangement of the κ chain gene family in mice. The germ-line κ chain gene family (1) consists of ~150 sequential pairs of L_κ and V_κ gene segments followed by 5 J_κ segments and a C_κ segment. During lymphocyte differentiation, a single L_κ–V_κ unit is joined to a J_κ unit through somatic recombination (2). In the progeny B cells, the rearranged gene is transcribed (3) and spliced (4) so as to join the previously selected L_κ, V_κ, and J_κ exons with the C_κ exon.

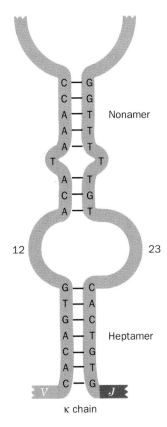

Figure 34-26
The germ-line κ gene family contains complementary heptamers and nonamers succeeding each V_κ segment and preceding each J_κ segment. These sequences are thought to mediate somatic recombination by forming the indicated stem-and-loop structure.

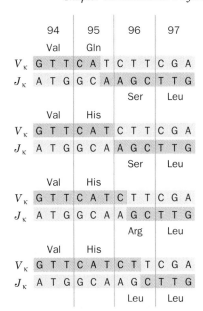

Figure 34-27
The cross-over point at which the V_κ and J_κ sequences somatically recombine varies by several nucleotides thereby giving rise to different nucleotide sequences (*brown bands*) in the active κ gene. For example, as is indicated here, amino acid 96, which occurs in the κ chain's third hypervariable region, can be Ser, Arg, or Leu.

Recombinational Flexibility Contributes to Antibody Diversity

The joining of 1 of 150 V_κ segments to 1 of 5 J_κ segments can generate only $150 \times 5 = 750$ different κ chains, far less than the number observed. However, studies of many joining events involving the same V_κ and J_κ segments revealed that *the V/J recombination site is not precisely defined; these two gene segments can join at different cross-over points (Fig. 34-27).* Consequently, the amino acids specified by the codons in the vicinity of the *V/J* recombination site depend on what part of the sequence is supplied by the germ-line V_κ segment and what part is supplied by the germ-line J_κ segment. Indeed, the amino acids specified by the codons surrounding the recombination junction form the light chain hypervariable region in the vicinity of residue 96 (CDR3; Fig. 34-18). Assuming that this recombinational flexibility increases the possible κ chain diversity by 10-fold, the expected number of possible different κ chains is increased to $150 \times 5 \times 10 = 7500$.

The imprecision of the *V/J* joining often results in the random loss of a few nucleotides from the ends of the V_κ and J_κ segments. Consequently, up to two thirds of the recombination products have an out-of-phase reading frame downstream from the recombination joint so that the resulting gene encodes a nonsense protein. Such proteins are not expressed. *A cell in which a nonproductive recombination event has occurred will attempt further κ gene rearrangements between its remaining $L_\kappa - V_\kappa$ and J_κ units and, if all of these fail, will rearrange its λ genes (see below).* This phenomenon accounts for the observation that κ expressing cells rarely have their λ genes rearranged, whereas λ expressing cells invariably have their κ genes rearranged. The mechanism by which the cell detects a productive recombination event is unknown.

λ Light Chains Derive from Multiple Constant Regions

The κ and λ chain gene families, which occur on different chromosomes, have different germ-line arrangements of their L, V, J, and C segments. Mice have only two $L_\lambda - V_\lambda$ segments, each followed by a pair of $J_\lambda - C_\lambda$ units (Fig. 34-28). Mice therefore have relatively little λ

Figure 34-28
The germ-line organization of the λ gene family in mice. $J_{\lambda 4}$ is a pseudogene segment.

Germ line heavy chain DNA

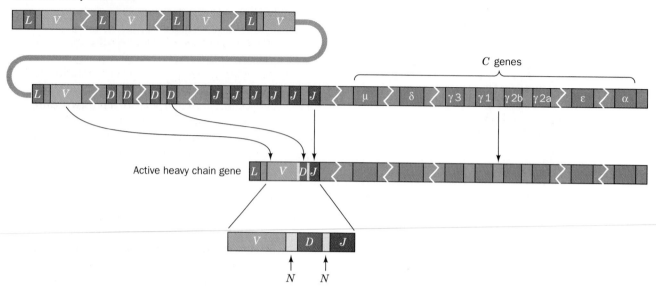

Figure 34-29

The organization and rearrangement of the heavy chain gene family in humans. This gene family consists of ~250 sequential pairs of L_H and V_H gene segments followed by ~10 D segments, 6 J_H segments and 8 C_H segments (one for each class or subclass of heavy chains). During lymphocyte differentiation, an $L_H - V_H$ unit is recombinationally joined to a D segment and a J_H segment. In this process, the D segment becomes flanked by short segments of random sequence called N regions. In the B cell and its progeny, transcription and splicing joins the $L_H - V_H - N - D - N - J_H$ unit to one of the 8 C_H gene segments.

chain diversity compared to that of their κ chains. This is probably why murine light chains are 95% κ and only 5% λ. Humans, in contrast, have many more $L_\lambda - V_\lambda$ and $J_\lambda - C_\lambda$ units than mice; human immunoglobulins contain approximately equal amounts of κ and λ chains.

Heavy Chain Genes Are Assembled from Four Sets of Gene Segments

Heavy chain genes are assembled in much the same way as are light chain genes but with the additional inclusion of an ~13 bp diversity or D segment between their V_H and J_H segments. The human heavy chain gene family, which occurs on a different chromosome from either of the light chain gene families, consists of clusters of ~250 different $L_H - V_H$ units, perhaps 10 D segments, 6 J_H segments, and 8 C_H segments, 1 for each of the 8 immunoglobulin classes and subclasses (Fig. 34-29). The D segments encode the core of the heavy chain's third hypervariable region (Fig. 34-18). Germ-line V_H, D, and J_H segments are flanked by heptamer–nonamer recombination signals similar to those that occur in light chain genes (Fig. 34-30). Moreover, heavy chain V/D and D/J joining sites are subject to the same recombinational flexibility as are light chain V/J sites. Assuming this recombinational flexibility contributes a factor of 100 towards heavy chain diversity, somatic recombination can generate some $250 \times 10 \times 6 \times 100 = 1.5 \times 10^6$ different heavy chains of a given class. Then, taking into account κ chain diversity (and neglecting that of λ chains), *there can be as many as $7500 \times 1.5 \times 10^6 = 11$*

billion different types of immunoglobulins of each class formed by somatic recombination among ~400 different gene segments.

Somatic Mutation Is a Further Source of Antibody Diversity

Despite the enormous antibody diversity generated by somatic recombination, *immunoglobulins are subject to even more variation through somatic mutations of two types:*

1. During V_H/D and D/J_H joining, a few nucleotides may be added or removed from the recombination joints. The added nucleotides, which form so-called **N regions,** yield *NDN* units of up to 30 bp that encode enormously variable heavy chain segments of 0 to 10 amino acid residues (Fig. 34-29). David Baltimore hypothesized that the N regions arise through the action of **terminal deoxynucleotidyl transferase,** a template-independent DNA polymerase present in the B cell progenitors that make the heavy chain joints but is probably absent in later cell generations when the light chain joints are formed.

2. The variable regions of both heavy and light chains are more diverse than is expected on the basis of comparisons of their amino acid sequences with their corresponding germ-line nucleotide sequences. Indeed, these regions mutate at rates of up to 10^{-3} base changes per nucleotide per cell generation, rates that are at least a millionfold higher than the rates of

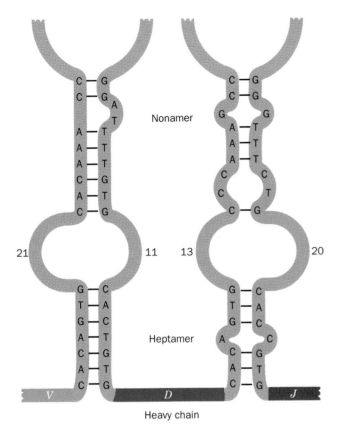

Figure 34-30
The stem-and-loop recombination sites in the germ-line heavy chain gene family that mediate somatic recombination between its V_H and D segments (*left*) and between its D and J_H segments (*right*). Compare them to the κ chain recombination signal (Fig. 34-26). The recombination system's requirement for both the 20/21 and the 11/13 bp spacers prevents it from inadvertently skipping the D segment by directly joining the V_H and J_H segments.

spontaneous mutation in other genes. *B* cells and/or their progenitors apparently possess enzymes that mediate this **somatic hypermutation** of immunoglobulin variable gene segments. Since the rate at which memory *B* cells are activated for proliferation increases with the affinity of their surface-displayed antibodies for antigen, *somatic hypermutation is thought to act, over many cell generations, to tailor antibodies to a particular antigen.*

These somatic mutation processes increase the possible number of different antibodies that humans can produce by many orders of magnitude beyond the 11 billion we estimated on the basis of somatic recombination alone. The final number is so large, probably $\sim 10^{10}$, that an individual synthesizes only a small fraction of its potential immunoglobulin repertoire. Somatic diversification arising from both recombination and mutation thereby permits an individual organism to cope, in a kind of Darwinian struggle, with the rapid mutational rates of pathogenic microorganisms.

Allelic Exclusion Ensures that Antibodies Are Monospecific

The immunoglobulins synthesized by a given *B* cell, as we have seen, consist of two identical heavy chains and two identical light chains. Such homogeneity is essential for the immune system's proper functioning because immunoglobulins consisting of two types of heavy and/or light chains would have two different antigen-combining sites and therefore could not form lattices of cross-linked antigens. Yet *B* cells, which like other somatic cells are diploid, contain two gene families specifying heavy chains (one maternal allele and one paternal allele) and four gene families encoding light chains (two κ's and two λ's). Apparently *B* cells are able to suppress the expression of all but one heavy chain allele and one light chain allele, a process known as **allelic exclusion,** by inhibiting further somatic recombination of heavy and light chain genes after a productive recombination has occurred. Allelic exclusion was experimentally demonstrated by microinjecting plasmids containing already recombined κ chain genes into fertilized mouse ova. The resulting transgenic mice suppress the somatic recombination of their endogenous κ chain genes. Analogous results were obtained for heavy chain genes. Although the mechanism of allelic exclusion is unknown, an intriguing possibility is that the protein products of a successful recombination event inhibit all further analogous recombinations.

The Switch from the Membrane Bound to the Secreted Form of an Antibody Involves a Change in Its Heavy Chain Transcript

The clonal selection model of antibody generation requires that the antibody displayed on the surface of a virgin *B* cell have the same specificity for antigen as the antibody secreted by its mature *B* cell progeny. Membrane-bound IgM (the antibody synthesized by virgin *B* cells) is anchored to the plasma membrane by a 41-residue hydrophobic polypeptide forming the C-terminus of its heavy chain (μ_m). In the secreted form of IgM (the first antibody secreted by mature *B* cells), the heavy chain (μ_s) has a different C-terminal segment but is otherwise identical. How does the *B* cell alter the synthesis of this heavy chain?

Somatically recombined heavy chain genes consist of eight exons (Fig. 34-31): an *L* segment that encodes a signal peptide leader; a *VDJ* unit that encodes the V_H domain; four exons that encode the C_H1 domain, the hinge region, the C_H2 domain, and the C_H3 domain (a further example of exons each specifying a structurally significant polypeptide unit); and two exons that collectively encode the trans-membrane tail of μ_m. In forming the mRNA specifying μ_m, the splicing system excludes the segment at the end of the C_H3 exon that specifies the μ_s tail and the entire transcript is terminated, as usual, by poly(A). In forming μ_s mRNA, however, the splicing

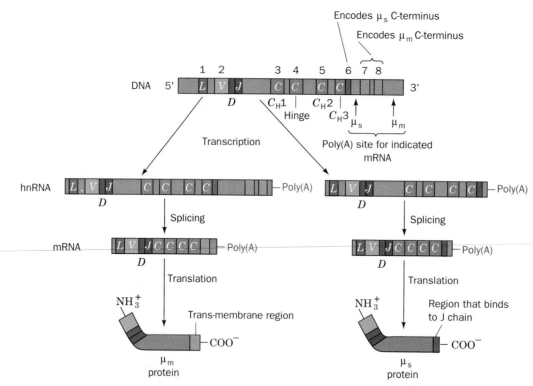

Figure 34-31
The C_μ gene specifies both the μ_m and the μ_s proteins through the selection of alternative splice and polyadenylation sites. In μ_m mRNA (*left*), the segment at the end of the C_H3 exon (6), which specifies the μ_s tail, has been spliced away and the transcript has been polyadenylated after the two exons specifying its trans-membrane segment (7 + 8). μ_s mRNA (*right*), on the other hand, is polyadenylated just past its retained μ_s tail segment.

system retains the μ_s segment and the transcript is polyadenylated after this point thereby eliminating the trans-membrane tail. How antigen-stimulated *B* cells switch between these alternative splice and polyadenylation sites is unknown.

B Cells Can Switch the Class of Immunoglobulin They Synthesize

Virgin *B* cells mostly synthesize membrane-bound IgM. Yet, the progeny of *B* cells that have been stimulated to proliferate may eventually synthesize immunoglobulins of different classes that have the same variable regions as the original IgM (recall that these different immunoglobulin classes have distinct physiological roles). The nucleic acid sequences specifying the variable region of the heavy chain must therefore become juxtaposed with the sequences specifying the constant regions of various types of heavy chains. What is the mechanism of this **class switching?**

The downstream regions of the human heavy chain gene family consist, as we have discussed, of eight segments encoding the constant regions for the various immunoglobulin classes and subclasses (Fig. 34-32). Class switching might occur either through RNA processing or through DNA processing. In fact, both mech-

anisms occur. In the RNA processing mechanism, it is uncertain whether the switching event is a change in transcriptional termination, polyadenylation, or splicing but, in any case, the result is the synthesis of heavy chain mRNAs with identical variable regions but different constant regions. The cell can therefore simultaneously synthesize two or more classes of immunoglobulins with identical antigen-binding sites.

The DNA processing mechanism of class switching occurs through somatic recombination between the *VDJ* unit and *C* region of choice. In doing so the intervening segment of DNA is deleted so that this mechanism is progressive and irreversible. For example, in recombinational switching from making IgM to making IgG1 (Fig. 34-32), a *B* cell loses its C_μ, C_δ, and $C_{\gamma3}$ segments so that its progeny cannot synthesize IgM, IgD, or IgG3. Yet, the progeny still have the potential to switch to IgG2, IgE, and IgA synthesis since the recombination does not disturb the $C_{\gamma2}$, C_ε, and C_α segments. Each of the C_H segments, with the exception of C_δ, is preceded by a **switch** or *S* **region** that consists of multiply repeated short complementary elements (C_δ is only expressed through RNA processing). These *S* regions are therefore thought to form the recombination signals used in class switching.

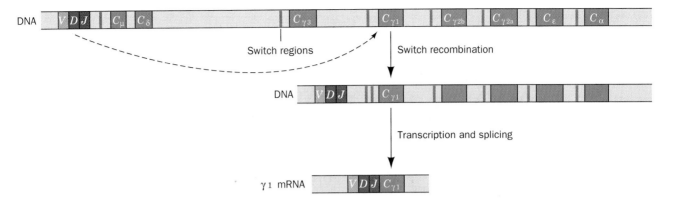

Figure 34-32
DNA-mediated class switching. An irreversible change in heavy chain synthesis from μ to a downstream constant region (shown here as $\gamma 1$) occurs through somatic recombination between the switch regions located upstream of all but one of the constant regions. Each constant region consists of multiple exons and encodes alternative secreted and membrane-binding C-termini (see Fig. 34-31).

D. *T* Cell Receptors

T cell receptors, as we have seen, are in many ways the cellular immunity system's analog of immunoglobulins. Like immunoglobulins, *T* cell receptors exhibit enormous specificity in binding antigens. Yet, *T* cell receptors have proven quite difficult to characterize because they only occur as cell surface proteins and are therefore present in small quantities. They were finally isolated in 1983 through the use of monoclonal antibodies directed against them and characterized through reverse genetics (Section 33-3).

The *T* cell receptor consists of two glycosylated polypeptide chains, α and β, which in humans have respective molecular masses of 50 and 39 kD. Each of these chains has a constant domain and a variable domain of approximately equal size as well as a C-terminal transmembrane segment (Fig. 34-33*a*). Not surprisingly, considering their similar functions, the *T* cell receptor subunits are sufficiently homologous to the immunoglobulin subunits (Fig. 34-33*b*) that their constant and variable domains are each predicted to assume the immunoglobulin fold. Moreover, the α and β subunit gene families are each organized into clusters of *V* and *J* regions, with the α family having an additional *D* region. The somatic recombination of these regions, as mediated by heptamer–nonamer sequences similar to those guiding the analogous process in immunoglobulin genes (Figs. 34-26 and 34-30), is a major source of *T* cell receptor diversity. The homology between the *T* cell receptor and immunoglobulin subunits indicates that these proteins all have a common ancestor. Nevertheless, the gene families encoding the α and β chains are distinct from the immunoglobulin gene families and, like them, reside on different chromosomes.

E. The Major Histocompatibility Complex

The membrane-bound proteins encoded by the major histocompatibility complex (MHC; *histo* refers to tissue), as we have seen, are the antigen-presenting markers used by the immune system to distinguish body cells from invading antigens (Class I MHC proteins) and immune system cells from other cells (Class II MHC proteins). In the following subsections, we outline the structures and genetic properties of these essential proteins.

Class I MHC Proteins

Tissues can be readily transplanted from one part of an individual's body to another or between genetically identical individuals. Yet, when tissues are transplanted between even closely related individuals, the graft is generally destroyed by the recipient's immune system (a phenomenon that is a major impediment to the transplantation of organs such as hearts and kidneys). Studies of such **graft rejection** led, nearly 50 years ago, to the discovery of the Class I MHC proteins, which are therefore also known as **transplantation antigens.**

The Class I MHC proteins are $\sim$44-kD trans-membrane glycoproteins that are displayed on the surfaces of nearly all nucleated vertebrate cells. These proteins' amino acid sequences suggest that they are folded into five domains that are, from C-terminus to N-terminus, an $\sim$30-residue cytoplasmic domain, a trans-membrane segment of $\sim$40 residues, and three external domains of $\sim$90 residues each designated α_3, α_2, and α_1 (Fig. 34-33*c*). The Class I MHC proteins are invariably noncovalently associated in a 1:1 ratio with β_2-**microglobulin** (β_2**m**; Fig. 34-33*c*), a 12-kD protein. The X-ray structure of a Class I MHC protein (Fig. 34-34*a*), which was elucidated by Don Wiley and Jack Strominger, indicates that its $\alpha 3$ domain as well as β_2-microglobulin, both of which are homologous with immunoglobulins, assume the immunoglobulin fold. *Evidently, all of these proteins, together with T cell receptors, are evolutionarily related and therefore form a **gene superfamily** (a set of evolutionarily related genes with divergent functions).*

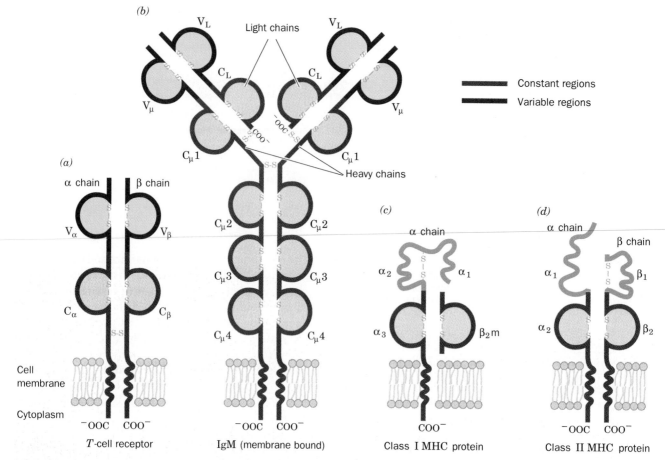

Figure 34-33
The members of the immunoglobulin gene superfamily, such as (*a*) the *T* cell receptor, (*b*) membrane-bound IgM, (*c*) Class I MHC protein, and (*d*) Class II MHC protein, all have similar domain structures. Each of these proteins contains multiple immunoglobulin homology units (*brown and purple regions*). Note that the IgM heavy chain has four constant regions compared with three in IgG (Fig. 34-15). Other members of the immunoglobulin gene superfamily include the **CD4** (alternatively, **T4) cell surface glycoprotein** of helper *T* cells, and the **CD8** (alternatively, **T8) cell surface glycoprotein** of cytotoxic and suppressor *T* cells. These proteins help their cells to recognize the Class II and Class I MHC proteins on other cells (Section 34-2A), presumably by binding to nonpolymorphic regions of these MHC proteins [CD4 is also the HIV (AIDS virus) receptor, which is why this virus specifically infects helper *T* cells]. Evidently, the members of the immunoglobulin gene superfamily are specialized for molecular recognition rather than limited to the immune system. Thus, the immunoglobulin fold-containing protein named **neuronal cell-adhesion molecule (N-CAM)** participates in cell–cell recognition but is not an immune system component.

The X-ray structure of the Class I MHC protein indicates that its homologous α_1 and α_2 domains form a relatively flat eight-stranded antiparallel β sheet that is parallel to the cell membrane and which is flanked by two α helices (Fig. 34-34*a* and *b*). *The resulting deep groove, which is of sufficient size to bind a 10 to 20 residue polypeptide, is a likely candidate for the binding site of a processed foreign antigen fragment that, together with the Class I MHC protein itself, is recognized by a T cell receptor* (Section 34-2A). Indeed, the X-ray structure contains an unknown "antigen" bound in the groove that must have copurified and cocrystallized with the Class I MHC protein.

Class II MHC Proteins

The discovery of Class II MHC proteins came about through Baruj Benacerraf's observation that certain *T* cell-dependent immune responses are mediated by gene products that are not antibodies. For example, when guinea pigs are inoculated with a simple antigen such as polylysine, some individuals mount a vigorous immunological response to the antigen, whereas others fail to respond. Immunological responsiveness to a given simple antigen is a dominant genetic trait. *A small number of so-called **immune response (Ir) genes** apparently govern how an individual responds to all simple antigens.* [A naturally occurring antigen, such as a protein, is complex,

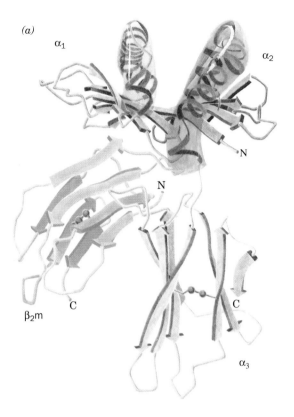

(a)

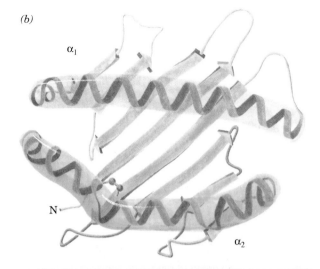

(b)

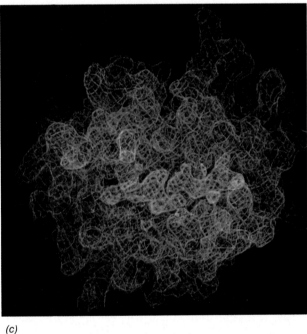

(c)

Figure 34-34

The X-ray structure of a human Class I MHC protein, HLA-A2. The protein was obtained by the papain digestion of plasma membranes from cultured human cells which cleaves the protein at residue 271, 13 residues past its trans-membrane segment. (a) A ribbon diagram of the protein indicating the relationships among its immunoglobulinlike domains, β_2m and α_3 (*bottom*), and its polymorphic α_1 and α_2 domains (*top*). The protein is oriented such that the plasma membrane from which it normally projects would be horizontal at the bottom of the drawing. The protein's antigen recognition site, which is thought to be located in the groove between the α_1 and α_2 helices, would therefore be readily accessible from outside the cell. Disulfide bonds are represented by connected yellow spheres. (b) Top view of the α_1 and α_2 domains showing the surface that presumably contacts the *T* cell receptor (viewed 90° from Part a). The domains, which each consist of four antiparallel β strands followed by a long helix, pair with pseudo-twofold symmetry to form a relatively flat eight-stranded antiparallel β sheet flanked by the two helices. The resulting groove presumably forms the binding site for the 10 to 20 residue antigen fragment that the Class I MHC protein presents to the *T* cell receptor. Most of the protein's polymorphic residues line the surface of this groove as do many residues critical for *T* cell receptor recognition. (c) The protein's van der Waals surface (*blue*), viewed as in Part b, showing the extra electron density (*red*) that is observed to occupy the protein's-binding site. This extra electron density is thought to represent an unidentified ''antigen'' that the protein had bound before its isolation. [Courtesy of Don Wiley, Harvard University.]

that is, it has numerous different epitopes (antigenic determinants; Section 34-2B). Consequently, an individual will almost always be able to mount an immune response against a naturally occurring antigen.]

The *Ir* genes map into the MHC so that they are now known as Class II MHC genes. They encode the two subunits of a heterodimeric trans-membrane glycoprotein composed of a 33-kD α chain and a 28-kD β chain, each of which consists of two domains (Fig. 34-33d). The amino acid sequences of these subunits indicate that the C-terminal α_2 and β_2 domains are members of the immunoglobulin gene superfamily. Moreover, their α_1 and β_1 domains can be convincingly aligned on the known structures of the Class I MHC protein's α_1 and α_2 domains, respectively. Evidently, Class I and Class II MHC proteins are structurally as well as functionally similar.

MHC Proteins Are Highly Polymorphic

The MHC has been extensively studied in both humans and mice. In humans, the Class I MHC proteins are encoded by three separate although homologous genetic loci, *HLA-A, HLA-B,* and *HLA-C* (Fig. 34-35; *HLA* stands for *human-leucocyte-associated* antigen since these proteins were first observed on leucocytes) so that each individual synthesizes up to six different Class I MHC proteins (see below). There are also three human Class II MHC proteins whose α and β chains are encoded by genes designated DP_α, DP_β, DQ_α, DQ_β, DR_α, and DR_β (Fig. 34-35). Mouse MHC genes, which occupy the *H-2* loci, are similarly arranged.

The most striking feature of the Class I and Class II MHC genes is their high level of polymorphism among individuals of the same species; *they are, in fact, the most polymorphic genes known in higher vertebrates.* For example, there are ~ 50 known alleles for each of the mouse Class I MHC genes; humans have a similar number. *Two unrelated individuals are therefore highly unlikely to have the same set of MHC genes.*

Most of the polymorphic residues are clustered in these proteins' antigen-building grooves (Fig. 34-34*b* for Class I MHC proteins and by inference for Class II MHC proteins) so that each polymorph binds a given antigenic fragment with characteristic affinity. The foregoing observations on the variation of immunological responses with Class II MHC (*Ir*) genes therefore suggest that certain Class II MHC protein polymorphs are less effective than others in associating with a given epitope. Indeed, *epidemiological studies indicate that certain polymorphs of MHC genes are associated with increased susceptibilities to particular infectious and/or autoimmune diseases.* For example, 95% of individuals with insulin-dependent diabetes mellitus (Section 25-3B) carry at least one *DR2* or *DR3* allele of the *DR* gene in comparison to 50% of normal individuals, whereas **celiac disease** (a violent intestinal upset resulting from eating wheat gluten) is 100% linked to the *DQw2* allele of the *DQ* gene.

What is the function of MHC protein polymorphism? It seems unlikely that it evolved only to prevent tissue grafts. Recall, however, that *T* cell receptors only recognize antigens when they are presented together with MHC proteins (Section 34-2A). If every member of a single vertebrate species had an identical set of MHC proteins, a pathogen whose epitopes interacted poorly with these MHC proteins would obliterate that species.

MHC gene polymorphism is thought to prevent pathogens from evolving the capacity to do so. Natural selection therefore tends to maintain a large variety of MHC proteins in a population.

F. The Complement System

Antibodies, for all their complications, only serve to identify foreign antigens. Other biological systems must then inactivate and dispose of the intruders. The **complement system,** a complex series of interacting plasma proteins, is one of these essential defensive systems. Indeed, it was named to indicate that it "complements" the function of antibodies in eliminating antigens. It does so in three ways:

1. It kills foreign cells by binding to and lysing their cell membranes, a process known as **complement fixation.**

2. It stimulates the phagocytosis of foreign particles, a process named **opsonization.**

3. It triggers a local acute inflammatory reaction that walls off the area and attracts phagocytotic cells.

In the following paragraphs, we describe the organization and function of the complement system.

The complement system consists of ~ 20 plasma proteins (Table 34-3) that interact in two related sets of reactions (Fig. 34-36): the antibody-dependent **classical pathway** and the antibody-independent **alternative pathway.** *Both pathways largely consist of the sequential activation of a series of serine proteases,* much like the blood clotting pathway (Section 34-1).

The complement system has its own peculiar nomenclature. Most complement protein names consist of the upper case letter "C" followed by a component number and, if the protein is either a subunit or fragment of a larger protein, a lower case letter. Active proteases are indicated by a bar over the component identifier. For example, $\overline{C4b}$ is a protease that has been activated by the proteolysis of C4.

The Classical Pathway Is Triggered by Antibody–Antigen Complexes

In the classical pathway, the complement proteins form three sequentially activated membrane-bound complexes (Fig. 34-36, *top*):

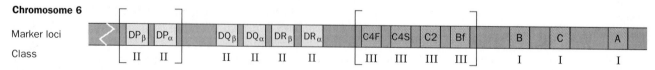

Figure 34-35
The genetic map of the MHC which, in humans, encodes the HLA proteins. The order of the loci enclosed in brackets is unknown. The Class III genes encode several complement system proteins (Section 34-2F).

Initiation **Amplification** **Membrane attack**

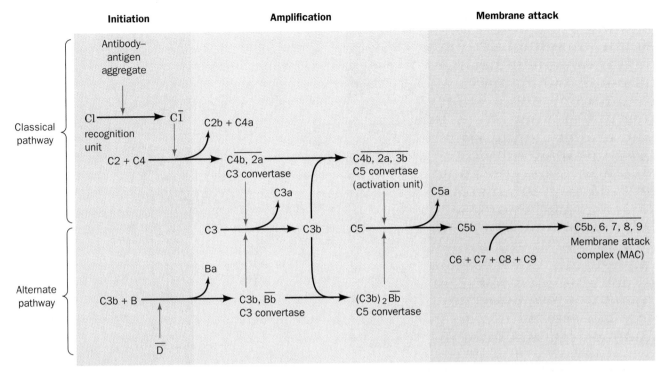

Figure 34-36
A schematic diagram of the complement system activation pathways. Colored arrows indicate proteolytic activations. Active proteases are indicated by a bar over the component number.

1. The **recognition unit,** which assembles on cell surface-bound antibody–antigen complexes.

2. The **activation unit,** which amplifies the recognition event through a proteolytic cascade.

3. The **membrane attack complex (MAC),** which punctures the antibody-marked cell's plasma membrane causing cell lysis and death.

The Recognition Unit

*The classical pathway is initiated when **C1**, the recognition unit, specifically binds to a cell-surface antigen–antibody aggregate.* C1 occurs in the plasma as a loosely bound complex of **C1q, C1r,** and **C1s.** C1q is a remarkable 18-polypeptide chain protein, $A_6B_6C_6$, in which the ~80 N-terminal residues of each chain has the repeating sequence Gly-X-Y characteristic of collagen where X is often Pro and Y is often 4-hydroxyproline or 5-hydroxylysine (Section 7-2C). C1q is therefore a bundle of six collagen-like triple helices that each end in a C-terminal globular domain so as to form an assembly that resembles a bunch of six tulips (Fig. 34-37). It is these globular domains that bind antigen-bound antibody through their recognition of the Fc regions of IgM and several subclasses of IgG (although how or even if the Fc region in an antibody–antigen complex conformationally differs from that in the free antibody is unknown). Moreover, C1 is only activated if at least two of its C1q

Table 34-3

Protein Components of the Complement System

Protein	Subunit Structure	Subunit Mass (kD)
Recognition unit (C1)		
C1q	$A_6B_6C_6$	460
C1r	α_2	166
C1s	α_2	166
Activation unit		
C2	Monomer	102
C3	$\alpha\beta$	185
C4	$\alpha\beta\gamma$	200
Membrane attack unit		
C5	$\alpha\beta$	191
C6	Monomer	120
C7	Monomer	110
C8	$\alpha\beta\gamma$	151
C9	Monomer	71
Alternative pathway		
Factor B	Monomer	92
Factor $\overline{D}$	Monomer	24
Properdin (P)	α_4	220
Regulatory proteins		
Factor H	Monomer	150
Factor I	$\alpha\beta$	88
C4b Binding protein	α_7	550
C$\overline{1}$ Inhibitor	Monomer	110
S Protein	Monomer	83

Source: Reid, K. B. M., *Essays Biochem.* **22,** 30 (1986).

(a)

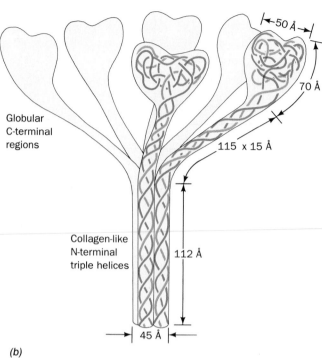

(b)

Figure 34-37
The structure of the complement protein C1q. (*a*) An electron micrograph of C1q seen in "side" view showing its 6 C-terminal domains attached to a central stalk. [Courtesy of Tibor Borso, USPHS–NIH.] (*b*) A schematic diagram of C1q. Its central stalk consists of a bundle of 6 collagen-like triple helices. The 80-residue triple helices are $80 \times 2.9 \approx 227$ Å in length. [After Porter, R. R. and Reid, K. B. M., *Nature* **275**, 701 (1978).]

heads are simultaneously bound to antibody, a process that requires the participation of at least two IgG's but only one IgM (recall that IgM is pentameric). IgM is therefore far more effective in activating the complement system than is IgG. A variety of foreign substances, including bacterial lipopolysaccharides and viral membranes, can also activate C1.

The remaining C1 components, C1r and C1s, are homologous serine protease zymogens which, like most of the blood clotting zymogens, are each activated by a single proteolytic cleavage that yields two disulfide-linked chains. The binding of antibody–antigen complex stimulates C1q to bind C1r and C1s more tightly which, in a Ca^{2+}-dependent process, results in the autoactivation of C1r. $\overline{C1r}$, in turn, specifically cleaves C1s, its only known substrate, to yield $\overline{C1s}$.

The Activation Unit

The activation unit consists of components derived from C2, C3, and C4. In the initial step forming the activation unit, $\overline{C1s}$ cleaves C4 and the larger of the resulting fragments, C4b, covalently binds to the cell membrane (as described below) in the vicinity of the recognition unit. Membrane-bound C4b, in association with $\overline{C1s}$, specifically cleaves C2. C2a, the resulting larger fragment, combines with C4b to yield $\overline{C4b,2a}$, a protease named **C3 convertase,** that cleaves C3 to C3a and C3b. Finally, C3b combines with C3 convertase to yield the activation unit, $\overline{C4b,2a,3b}$, also known as **C5 convertase,** which functions to proteolytically activate **C5.**

Both C4 and C3 have buried hyper-reactive thioester groups that, when exposed, can covalently link these proteins to the cell membrane. In C3, the thioester consists of a Cys thiol and a Glu γ-carboxyl group forming a macrocyclic ring of sequence Gly-Cys-Gly-Glu-Glu-Asn:

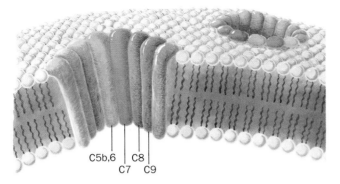

C5b,6 | C8
C7 | C9

Figure 34-38
The membrane attack complex (MAC) is a tubular structure that forms a trans-membrane pore in the target cell's plasma membrane. The perforin pores generated by cytotoxic *T* lymphocytes (Section 34-2A) have similar structures (perforin and C9 are homologous) but lack an analog of the C5b,6,7,8 complex.

Upon the cleavage of C3, the product C3b undergoes a conformational rearrangement that exposes its thioester group. The thioester then reacts with a nearby cell-surface amine or OH group to yield the corresponding amide or ester bond together with a free Cys sulfhydryl group. The function of this process is further discussed below. C4 is thought to behave in a similar manner on activation.

The activation of C3, C4, and C5 also triggers other immune system functions. C3b and, to a lesser extent, C4b are **opsonins,** substances that stimulate phagocytosis (opsonization), whereas C3a, C4a, and C5a (a product of the C5 convertase reaction) are **anaphylatoxins,** substances that trigger local acute inflammatory reactions and smooth muscle contraction.

The Membrane Attack Complex

C5b, the other product of the C5 convertase reaction, exhibits no proteolytic activity. Rather, it sequentially binds **C6** and **C7** to form a complex that spontaneously inserts into cell membranes. *This C5b,6,7 complex then binds one molecule of C8 followed by anywhere between 1 and 18 molecules of C9 to form the MAC. In this latter process, the C9 molecules polymerize to form a tubular membrane-embedded structure to which the C5b,6,7,8 complex is firmly attached (Figs. 34-38 and 34-39).* Cell lysis ensues both because the MAC forms a 30 to 100 Å in diameter aqueous channel that pierces the membrane (Fig. 34-38) and because it perturbs the surrounding membrane structure so as to increase its permeability. Both mechanisms permit the cell's small molecules, but not its macromolecules, to exchange with the surrounding medium. Water is therefore osmotically drawn into the cell causing it to swell and burst. MACs are efficient cell killers; very few, possibly only one, can lyse a cell.

The Alternative Pathway Is Antibody Independent

The alternative pathway of complement fixation (Fig. 34-36, bottom) uses many of the same components as the classical pathway and likewise causes formation of a C5 convertase that triggers MAC assembly. The two path-

ways differ in that the alternative pathway is antibody independent. It is therefore thought that the alternative pathway functions to defend against invading microorganisms before an immune response against them can be mounted (although the classical pathway may also do so). Once sufficient antibody has been synthesized, the alternative pathway assumes a secondary role relative to that of the classical pathway.

The alternative pathway is thought to always operate at a low level (see below) so as to continually produce small amounts of C3b, the same molecule produced by C3 convertase of the classical pathway. In the alternative pathway, however, C3b combines with the plasma protein **factor B** in a Mg^{2+}-dependent reaction. The resultant complex, C3b,B, is the only known substrate for the active plasma serine protease, **factor $\overline{D}$,** which cleaves the B subunit of C3b,B to yield C3b,$\overline{Bb}$. This latter complex is a C3 convertase that is equivalent to but distinct from that of the classical pathway. It cleaves C3 to C3b, which participates in the formation of more C3 convertase in a cyclically amplified process. The additional C3b also binds to C3 convertase to yield $(C3b)_2\overline{Bb}$, a C5 convertase distinct from that of the classical pathway but which likewise catalyzes the formation of the MAC.

What is the origin of the C3b that initiates the alternative pathway? It may, of course, be generated by the classical pathway in which case the alternative pathway acts as an amplification mechanism for antibody-induced complement activation. In the absence of this process, however, it is thought that the reactive but unexposed thioester bond of native C3 undergoes slow spontaneous hydrolysis to yield a C3b-like protein, **C3i,**

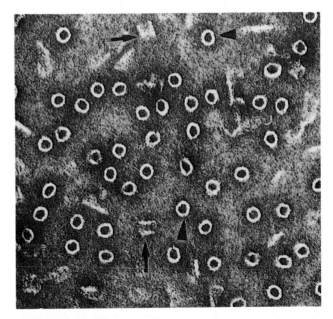

Figure 34-39
Electron micrograph of C9 ring complexes seen in side view (*arrows*) and top view (*arrow heads*). [Courtesy of Zanvil Cohn, Rockefeller University.]

in that it binds factor B and mediates its $\overline{\text{D}}$-catalyzed activation. The resulting C3 convertase, in turn, generates authentic C3b.

How does the alternative pathway target invading microorganisms? The C3b concentration in solution is limited by a plasma protein named **factor I** which, together with a second protein, **factor H,** forms a complex (I,H) that proteolytically degrades C3b in solution. When C3b is covalently bound to a surface, however, its degradation rate is greatly reduced. Moreover, the surface-bound C3 convertase complex is stabilized by the binding of the plasma protein **properdin (P),** which further protects C3b from I,H-mediated degradation as well as retards the dissociation of $\overline{\text{Bb}}$ from C3 convertase. Consequently, the faster C3b covalently attaches to a surface, the more slowly it is degraded. *Substances to which C3b efficiently attaches are therefore alternative pathway activators.* These include polymers of microbial origin such as the lipopolysaccharides of gram-negative bacteria known as **endotoxins** and cell wall teichoic acids from gram-positive bacteria (Section 10-3B), certain whole bacteria, fungi, and cells infected by certain viruses. The alternative pathway therefore provides an effective defense against invading microorganisms. Indeed, individuals who are genetically deficient in certain complement components are highly susceptible to various infections.

The Complement System Is Strictly Regulated

The inability of many complement components to discriminate between normal tissues and foreign substances requires that the complement system be maintained under tight control. Otherwise the complement system would destroy host cells. Indeed, the actual damage in many autoimmune diseases is caused by the complement system.

The complement system is regulated by the inactivation of its activated components. This occurs in three ways:

1. Complement components are inactivated through their spontaneous decay. For example, the hyperreactive thioesters of newly activated C3b and C4b react with water with half-lives of $\sim 60 \, \mu s$. These proteins are therefore lost to the classical pathway unless they attach to a membrane in the immediate vicinity of their activating recognition unit, that is, to the membranes of the invading microorganisms that triggered their activation (rather than to those of host cells). Similarly, classical pathway C3 convertase, C4b,2a, is but transiently active; its C2a component readily dissociates with the consequent loss of enzymatic activity.

2. Complement components are inactivated through their degradation by specific proteases. For instance, **C4b-binding protein** forms a complex with factor I that proteolytically inactivates C4b, much like, as we have seen, the I,H complex degrades C3b. Appar-

ently, C4b-binding protein and factor H act as cofactors that target factor I for C4b and C3b. From this point of view, C4b-binding protein limits the activities of classical pathway C3 convertase ($\overline{\text{C4b,2a}}$) and C5 convertase ($\overline{\text{C4b,2a,3b}}$) whereas, as we have seen, factor H does so for proteases containing C3b in both the classical and alternative pathways.

3. Complement components are inactivated through their association with specific binding proteins. For example, **C1 inhibitor** tightly binds $\overline{\text{C1r}}$ and $\overline{\text{C1s}}$ to form a complex that dissociates from and hence inactivates the recognition unit. Similarly, **S protein** attaches to MACs assembling in the plasma so as to prevent their later attachment to cell membranes. Such attachment is consequently limited to the site of complement activation.

The regulation of the complement system therefore functions to target foreign invaders while minimizing host cell damage.

3. MOTILITY: MUSCLES, CILIA, AND FLAGELLA

Perhaps the most striking characteristic of living things is their capacity for organized movement. Such phenomena occur at all structural levels and include such diverse vectorial processes as active transport through membranes, the translocation of DNA polymerase along DNA, the separation of replicated chromosomes during cell division, the beating of flagella and cilia, and, most obviously, the contraction of muscles. In this section we consider the structural and chemical basis of biological motility. In doing so we shall be mainly concerned with **striated muscle** since it is the most familiar and best understood motility system. We shall also briefly discuss smooth muscle and two entirely different types of biological motility systems: those of eukaryotic cilia and prokaryotic flagella.

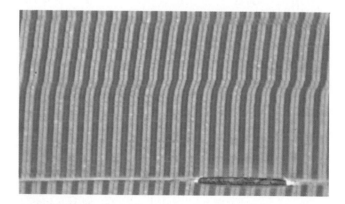

Figure 34-40
A photomicrograph of a muscle fiber in longitudinal section. Its transverse dark A bands and light I bands are clearly visible. [J. C. Révy/CNRI.]

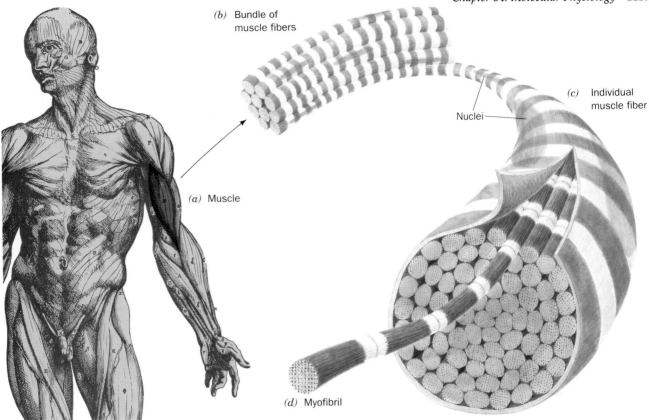

(b) Bundle of
muscle fibers

(c) Individual
muscle fiber

Nuclei

(a) Muscle

(d) Myofibril

Figure 34-41
Skeletal muscle organization. A muscle (a), consists of
bundles of muscle fibers (b), each of which is a long thin
multinucleated cell (c), that may run the length of the
muscle. Muscle fibers contain bundles of laterally aligned
myofibrils (d), which consist of bundles of alternating thick
and thin filaments.

A. Structure of Striated Muscle

The voluntary muscles, which include the skeletal
muscles, have a striated appearance when viewed under
the light microscope (Fig. 34-40). Such muscles consist
of long parallel bundles of 20 to 100 μm in diameter
muscle fibers (Fig. 34-41), which may span the entire
length of the muscle and which are actually giant multinucleated cells that arise during muscle development by
the end-to-end fusion of numerous precursor cells.
Muscle fibers are, in turn, composed of parallel bundles
of around one thousand 1 to 2 μm in diameter **myofibrils** (Greek: *myos,* muscle), which may extend the full
length of a fiber.

Electron micrographs show that muscle fiber striations arise from an underlying banded structure of multiple in-register myofibrils (Fig. 34-42). The bands are

Figure 34-42
An electron micrograph of skeletal muscle showing that the
myofibrils in muscle fiber are in register (the out of register
portion of the myofibril on the upper left is probably an
artifact of preparation). The ovoid object near the center is a
nucleus. Note its peripheral location in its fiber. [Courtesy of
Don Fawcett, Harvard University.]

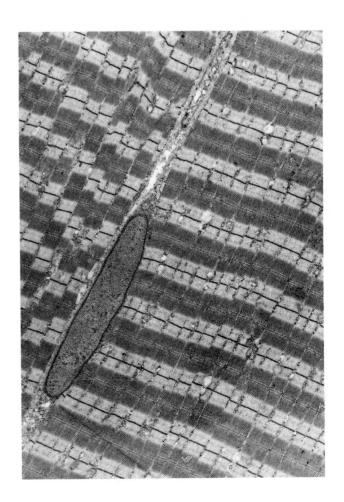

formed by alternating regions of greater and lesser electron density, respectively, named **A bands** and **I bands** (Fig. 34-43). The myofibril's repeating unit, the **sarcomere** (Greek: *sarkos*, flesh), which is 2.5 to 3.0 μm long in relaxed muscle but progressively shortens as the muscle contracts, is bounded by dark (electron dense) **Z disks** or **lines** at the center of each I band. The A band is centered on the lighter **H zone** which, in turn, is centered on the dark **M disk** or **line**.

Cross-sections through the sarcomere reveal the origin of its banded pattern. The H zones contain an array of parallel, hexagonally packed 150 Å in diameter **thick filaments**, whereas the lighter I bands consist of twice as many hexagonally arranged 70 Å in diameter **thin filaments** that are anchored to the Z disk. The darker areas at the ends of each A band mark the regions where the two sets of fibers interdigitate. The thick and thin filaments associate in this region by means of regularly spaced **cross-bridges** (Fig. 34-44). We shall see below that it is these associations that are responsible for the generation of muscular tension.

Thick Filaments Consist of Myosin

The major protein components of striated muscle are listed in Table 34-4. Vertebrate thick filaments are composed almost entirely of a single type of protein, **myosin,** which occurs in virtually every vertebrate cell. *Myosin molecules consist of six highly conserved polypeptide chains: two **heavy chains** and two pairs of different **light chains** designated LC1 and LC2.* Myosin is an unusual protein in that it has both fibrous and globular properties (Fig. 34-45). The N-terminal half of its heavy chain folds into an elongated globular head, around 55 $\times$ 200 Å, while its C-terminal half forms a long fibrous α helical tail. Two of these α helical tails associate to form a left-handed coiled coil yielding an ~1500-Å long rodlike segment with two globular heads. The amino acid sequence of myosin's α helical tail is characteristic of coiled coils: It has a seven-residue pseudorepeat, *a-b-c-d-e-f-g*, with nonpolar residues concentrated at positions *a* and *d*. Thus, much like in the coiled coils of the fibrous protein keratin (Section 7-2A), the myosin helix has a hydrophobic strip along one side that promotes its lengthwise association with another such helix. One of

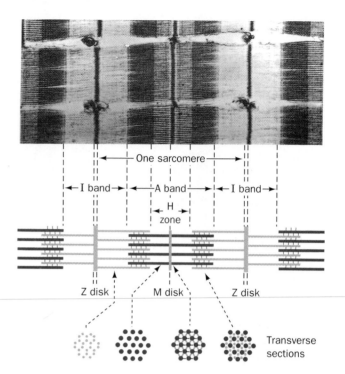

Figure 34-43

An electron micrograph of parts of three myofibrils in longitudinal section. The myofibrils are separated by horizontal gaps. A myofibril's major features, as indicated in the accompanying interpretive drawings, are the light I band, which contains only hexagonally arranged thin filaments; the A band, whose dark H zone contains only hexagonally packed thick filaments, and whose even darker outer segments contain overlapping thick and thin filaments; the Z disk, to which the thin filament is anchored; and the M disk, which arises from a bulge at the center of each thick filament. The myofibril's functional unit, the sarcomere, is the region between two successive Z disks. [Courtesy of Hugh Huxley, Brandeis University.]

each type of light chain are associated with each of the heavy chain dimer's globular heads.

Myosin only exists as single molecules at low ionic strengths. Under physiological conditions, these molecules form aggregates that resemble thick filaments. *Natural thick filaments consist of several hundred myosin molecules with their rodlike tails packed end-to-end in a*

Figure 34-44

An electron micrograph of deep-etched, freeze-fractured muscle showing its alternating thick and thin filaments. The knobs (cross-bridges) projecting from the thick filaments are helically arrayed. [Courtesy of John Heuser, Washington University School of Medicine.]

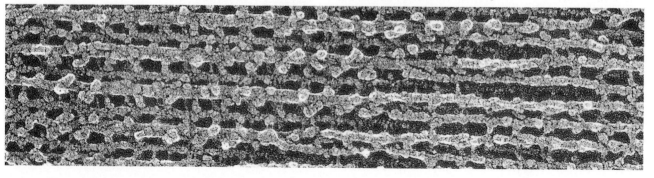

Table 34-4
Proteins of Striated Muscle

Protein	Subunit Mass (kD)
Myosin	540
Heavy chain	230
Light chain 1 (LC1)	~20
Light chain 2 (LC2)	~20
G-Actin	42
Tropomyosin	33
Troponin	72
TnC	18
TnI	23
TnT	31
α-Actinin	200
β-Actinin	34
Desmin	50
Vimentin	52
C-Protein	150
M-Protein	100

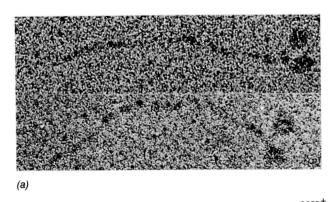

(a)

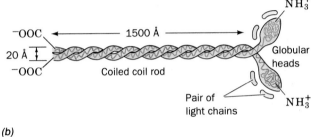

(b)

Figure 34-45
The myosin molecule. (a) An electron micrograph showing that the myosin molecule is a fibrous entity with two globular heads. [Courtesy of Henry Slayter, Harvard Medical School.] (b) Its rod-shaped tail is formed by the two extended α helices from each of its two identical heavy chains that wrap around each other to form a coiled coil. One of each type of myosin light chains, LC1 and LC2, is associated with each of myosin's identical globular heads although their exact positions are unknown.

regular staggered array (Fig. 34-46). The thick filament is therefore a bipolar entity in which the globular myosin heads project from either end leaving a bare central region. It is these myosin heads that form the cross-bridges that interact with the thin filaments in intact myofibrils.

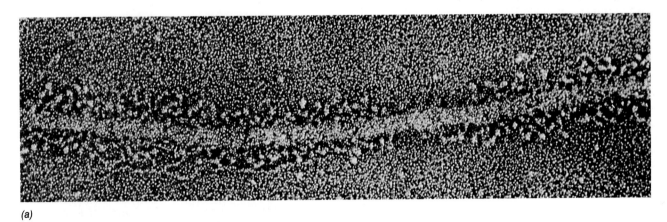

(a)

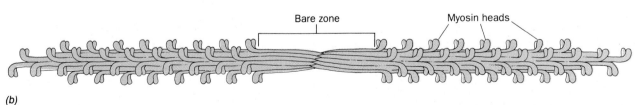

(b)

Figure 34-46
The thick filament of striated muscle. (a) An electron micrograph showing the myosin heads projecting from the thick filament's outer segments and its bare central zone. [From Trinick, J. and Elliott, A., *J. Mol. Biol.* **131**, 135 (1977).] (b) A thick filament typically contains several hundred myosin molecules organized in a repeating staggered array such that the myosin molecules are oriented with their globular heads pointing away from the filament's center.

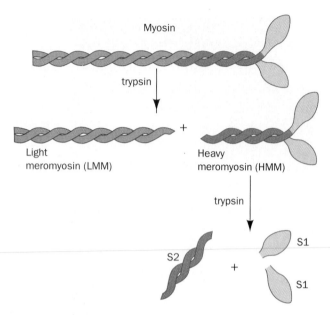

Figure 34-47
The enzymatic cleavage of myosin.

In addition to its structural function, the myosin heavy chain is an ATPase: It cleaves ATP to ADP and P_i in a reaction that powers muscle contraction. Muscle is therefore a device for transducing the chemical free energy of ATP hydrolysis to mechanical energy. The myosin light chains, through their level of phosphorylation, are thought to modulate the ATPase activity of their associated heavy chains.

In 1953, Andrew Szent-Györgi demonstrated that trypsin cleaves myosin into two fragments (Fig. 34-47):

1. **Light meromyosin (LMM),** an 850-Å long α helical rod that aggregates to form filaments but lacks both ATPase activity and the ability to associate with light chains.

2. **Heavy meromyosin (HMM),** which has a rodlike tail and two globular heads, does not aggregate but has ATPase activity and binds to light chains. HMM can be further split by treatment with papain to yield two identical globular subfragments (named **S1**) and a rod-shaped subfragment **(S2).** S1 contains myosin's ATPase activity and its thin filament-binding site.

Thin Filaments Consist of Actin, Tropomyosin, and Troponin

Actin, a ubiquitous and highly abundant eukaryotic protein, is the major constituent of thin filaments. At low ionic strengths, actin occurs as bilobal globular monomers called **G-actin** (G for globular). Under physiological conditions, however, *G-actin polymerizes to F-actin* (F for fibrous), *a right-handed double helical filament of pitch ~360 Å/turn that forms the thin filament's core (Fig. 34-48).* Each of F-actin's monomeric units is capable of binding a single myosin S1 head. Electron

micrographs of S1-decorated F-actin have the appearance of a series of head-to-tail arrowheads (Fig. 34-49a). F-actin must therefore be a polar entity, that is, all of its monomer units have the same orientation with respect to the fiber axis (Fig. 34-49b). The "arrowheads" in S1-decorated thin filaments that are still attached to their Z disk all point away from the Z disk indicating that *the thin filament bundles extending from the two sides of the Z disk have opposite orientations.*

Myosin and actin, the major components of muscle, account for 60 to 70% and 20 to 25% of total muscle protein, respectively. Of the remainder, two proteins that are associated with the thin filaments are particularly prominent (Fig. 34-50):

1. **Tropomyosin,** a heterodimer whose two homologous α helical subunits wrap around each other to form a coiled coil. These 400-Å long rod-shaped molecules are joined head-to-tail to form cables wound in the grooves of the F-actin helix such that each tropomyosin molecule contacts seven consecutive actin monomers in a quasi-equivalent manner.

2. **Troponin,** which consists of three subunits: **TnC,** a Ca^{2+}-binding protein (Fig. 34-51) that is 70% homologous to the Ca^{2+}-binding regulatory protein calmodulin (Section 18-3B); **TnI,** which binds to actin;

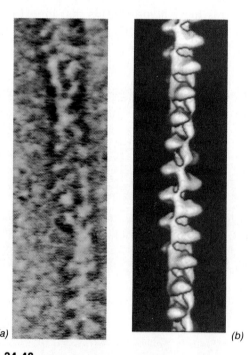

(a)   (b)

Figure 34-48
F-actin. (a) An electron micrograph of a thin filament from striated muscle showing its double-stranded character. [Courtesy of Hugh Huxley, Brandeis University.] (b) An actin fiber as visualized through image reconstruction from electron micrographs. Note the bilobal appearance of each monomeric (repeating) unit. [Courtesy of David J. DeRosier, Brandeis University.]

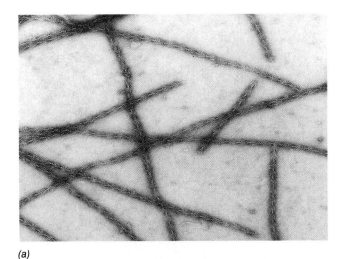

(a)

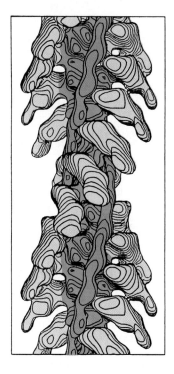

(b)

Figure 34-49
(a) An electron micrograph of a thin filament decorated with myosin S1 fragments. Note its resemblance to a series of arrowheads all pointing in the same direction along the filament. [Courtesy of Hugh Huxley, Brandeis University.] (b) Image reconstruction of S1-decorated actin filaments. The actin is colored green, the S1 fragments are pink, and the bound tropomyosin (see below) is orange. The helical filament has a pitch of 370 Å. [After a drawing provided by Ronald Milligan, Research Institute of Scripps Clinic, and Paula Flicker, University of California at San Francisco.]

and **TnT,** an elongated molecule, which binds to tropomyosin at its head-to-tail junction.

The tropomyosin–troponin complex, as we shall see, regulates muscle contraction by controlling the access of the myosin S1 cross-bridges to their actin-binding sites.

Figure 34-50
A model of the striated muscle thin filament based on the 15-Å resolution X-ray structure of tropomyosin and electron micrographic studies of F-actin. Tropomyosin, a coiled coil of two α helical subunits, wraps in the groove of the F-actin helix (*large blue spheres*) in a head-to-tail manner such that one tropomyosin molecule contacts seven consecutive bilobal actin monomer units (the red and blue regions of tropomyosin identify the seven homologous segments that are presumed to form its actin-binding sites). Each tropomyosin molecule binds a single troponin molecule at its head-to-tail joint (*left,* the small white spheres represent bound Ca^{2+} ions). The tropomyosin chain winding about the opposite side of the F-actin helix from the tropomyosin shown has been omitted for clarity. [Courtesy of George N. Phillips, Jr., Rice University.]

Minor Muscle Proteins Control Myofibril Assembly

The Z disk, which anchors two sets of oppositely oriented thin filaments (Fig. 34-43), is an amorphous entity that contains several fibrous proteins. For instance, **α-actinin,** which binds to the ends of F-actin filaments *in vitro,* is localized in the Z disk's interior (Fig. 34-52a). α-Actinin is therefore thought to attach thin filaments to the Z disk. In contrast, **β-actinin,** which also binds to F-actin, is believed to function as a length-controlling factor in thin filament assembly. Two other proteins,

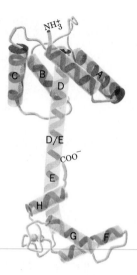

Figure 34-51
The X-ray structure of turkey skeletal muscle TnC. Its two
globular domains, which are connected by a nine-turn α
helix, can each bind two Ca²⁺ ions. Under the conditions of
the crystal structure determination, however, only the
Ca²⁺-binding sites of the C-terminal (*lower*) domain are
occupied (*spheres*). These latter sites remain occupied at
the lowest physiological Ca²⁺ concentrations in muscle so
that the regulatory effects exerted by TnC must arise from
the binding of Ca²⁺ to the N-terminal (*upper*) domain. Note
the resemblance of the TnC structure to that of the
homologous Ca²⁺-binding regulatory protein calmodulin (Fig.
18-21). [After Herzberg, O. and James, M. N. G. *Nature* **313**,
655 (1985).]

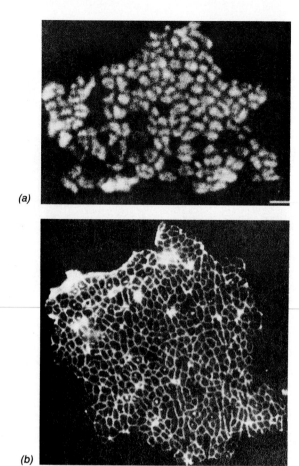

(a)

(b)

Figure 34-52
Indirect immunofluorescence micrographs of isolated
sheets of skeletal muscle Z disks: (*a*) Using antibodies to
α-actinin indicating that α-actinin occurs at the interior of the
Z disk. The bar represents 2.5 μm. (*b*) Using antibodies to
desmin showing that desmin is distributed about the Z-disk
periphery. Antibodies to vimentin exhibit the same
distribution. In the indirect immunofluorescence technique,
proteins are labeled with rabbit antibodies raised against
them. The bound rabbit antibodies are subsequently labeled
with goat anti-rabbit immunoglobulin antibodies to which
fluorescent molecules such as fluorescein are covalently
linked. The proteins are then observed under UV light. This
indirect approach of using two types of antibodies rather
than directly using fluorescently tagged rabbit antibodies
increases the sensitivity of the method because several
fluorescently labeled goat antibodies can bind to each rabbit
antibody. [From Lazarides, E., *Nature* **283**, 251 (1980).]

desmin and **vimentin,** largely occur at the Z disk periphery (Fig. 34-52*b*) where they apparently act to keep
adjacent myofibrils in lateral register.

The M disk (Fig. 34-43) arises from the local enlargement of in-register thick filaments. The two proteins
that are associated with this structure, **C-protein** and
M-protein, probably participate in thick filament assembly. Invertebrate thick filaments contain a core of
paramyosin which, in some muscles, is the dominant
component.

B. Mechanism of Muscle Contraction

So far we have simply described the components of
striated muscle. Now, like good engineers, we must ask
how do these components fit together and how do they
interact? In other words, how does muscle work?

Thick and Thin Filaments Slide Past Each Other During Muscle Contraction

Physiologists have long known that a contracted
muscle is as much as one third shorter than its fully
extended length. Electron micrographs have demonstrated that this shortening is a consequence of a decrease in the length of the sarcomere (Fig. 34-53). Yet,
during muscle contraction, the thick and the thin filaments maintain constant lengths as is indicated by the

observations that the width of the A band as well as the
distance between the Z disk and the edge of the adjacent
H zone do not change. Rather, sarcomere contraction is
accompanied by equal reductions in the widths of the I
band and the H zone. These observations were independently explained by Hugh Huxley and Jean Hanson
and by Andrew Huxley and R. Niedergerke who, in
1954, proposed the **sliding filament model:** *The force of
muscle contraction is generated by a process in which interdigitated sets of thick and thin filaments slide past each
other* (Fig. 34-53).

Actin Stimulates Myosin's ATPase Activity

The sliding filament model partially explains the me-

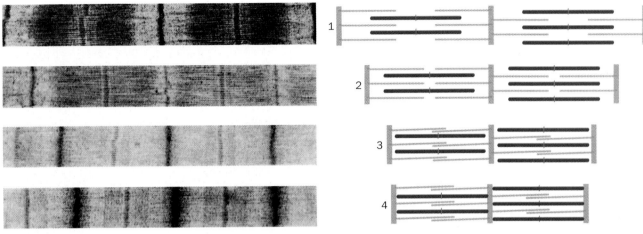

Figure 34-53
Electron micrographs with accompanying interpretive drawings of myofibrils in progressively more contracted states (1–4). Note that the widths of the I band and H zone decrease upon contraction, whereas the lengths of the thick and thin filaments remain constant. The interpenetrating sets of thick and thin filaments must therefore slide past each other, as drawn, upon myofibril contraction. [Courtesy of Hugh Huxley, Brandeis University.]

chanics of muscle contraction but not the origin of the contractile force. Albert Szent-Györgi's work in the 1940s pointed the way towards the elucidation of the contraction mechanism. The mixing of solutions of actin and myosin to form a complex known as **actomyosin** is accompanied by a large increase in the solution's viscosity. This viscosity increase is reversed, however, when ATP is added to the actomyosin solution. *Evidently, ATP reduces myosin's affinity for actin.*

Further insight into the role of ATP in muscle contraction was provided by kinetic studies. Isolated myosin's ATPase function has a turnover number of ~ 0.05 s^{-1}, far less than that in contracting muscle. Paradoxically, however, the presence of actin increases myosin's ATP hydrolysis rate to the physiologically more realistic turnover number of ~ 10 s^{-1}, a rate enhancement of ~ 200 (indeed, actin was so-named because it *act*ivates myosin). This is because isolated myosin rapidly hydrolyzes ATP

$$\text{ATP}^{4-} + \text{H}_2\text{O} \rightleftharpoons \text{ADP}^{3-} + \text{HPO}_4^{2-} + \text{H}^+$$

but only slowly releases the products ADP + P_i as is indicated by the observation that myosin-catalyzed ATP hydrolysis begins with a rapid burst of H$^+$, whereas free ADP and P_i appear much more slowly. Actin enhances myosin's ATPase activity by binding to the myosin–ADP–P_i complex and stimulating it to sequentially release P_i followed by ADP. The myosin–ADP–P_i complex cannot be formed by simply mixing myosin, ADP and P_i, which suggests that this complex is a "high-energy" intermediate in which the free energy of ATP hydrolysis has somehow been conserved.

The foregoing observations led Edwin Taylor to formulate a model for actomyosin-mediated ATP hydrolysis (Fig. 34-54):

Step 1 ATP binding to the myosin component of acto-myosin results in the dissociation of actin and myosin.

Step 2 The myosin-bound ATP is rapidly hydrolyzed to form a stable "high-energy" myosin–ADP–P_i complex.

Step 3 Actin binds to the myosin–ADP–P_i complex.

Step 4 In a process accompanied by a conformational relaxation to its resting state, the actin–myosin–ADP–P_i complex sequentially releases P_i followed by ADP yielding actomyosin that can undergo another round of ATP hydrolysis.

This ATP-driven alternate binding and release of actin by myosin provides, as we shall see, the vectorial force of muscle contraction.

Myosin Heads "Walk" Along Actin Filaments

In order to complete our description of muscle contraction we must determine how ATP hydrolysis is coupled to the sliding filament model. If the sliding filament

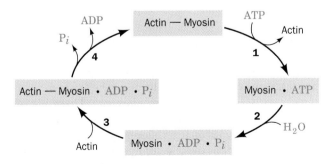

Figure 34-54
The reaction sequence in actomyosin-catalyzed ATP hydrolysis.

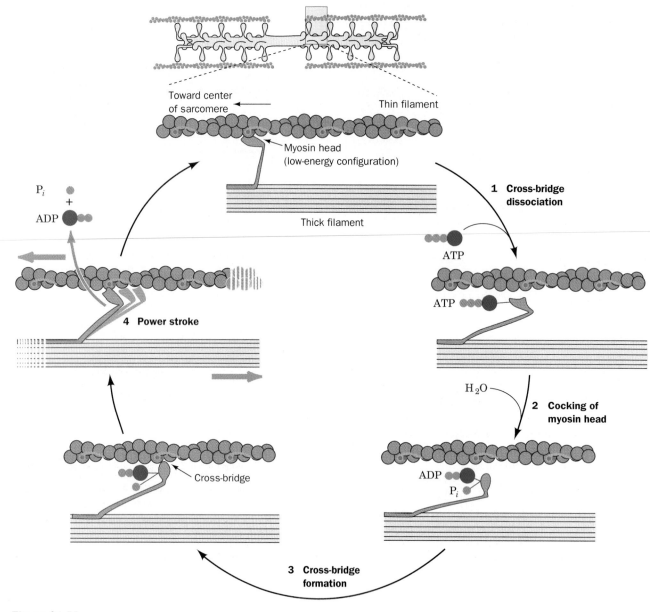

Figure 34-55

The mechanism of force generation in muscle. The myosin head "walks" up the actin thin filament through a cyclic vectorial process that is driven by ATP hydrolysis. Only one of myosin's two independent S1 heads is indicated. The nature of myosin's conformational changes are unknown although the proteolytic susceptability of the S1-S2 and S2-LMM joints suggests that these sites have the hingelike flexibility implied by the drawing.

model is correct then it would be impossible for a myosin cross-bridge to remain attached to the same point on a thin filament during muscle contraction. Rather, it must detach and then reattach itself at a new site further along the thin filament which, in turn, suggests that muscular tension is generated through the interaction of myosin cross-bridges with thin filaments. These ideas, in combination with the above ATP hydrolysis cycle, led to the development of the following widely accepted "rowboat" model of muscle contraction (Fig. 34-55):

1. ATP binding causes the S1 head on a thick filament to dissociate from an actin monomer unit on an adjacent thin filament.

2. The ATP hydrolyzes so as to shift myosin to a "high-energy" conformational state in which the long axis of the S1 head is approximately perpendicular to the thick filament. This process "cocks" the myosin molecule although this is the relaxed state of muscle.

3. Under stimulation to contract (see below), the S1 head–ADP–P_i complex attaches to an adjacent actin unit.

4. The actin–myosin interaction triggers the sequential release of P_i followed by ADP from the S1 myosin head with its concomitant conformational relaxation to the tilted orientation relative to the thick filament.

*This conformational shift, the so-called **power stroke** of muscle contraction, involves an ~45° tilt of the S1 head away from the neighboring Z disk so as to pull the thin filament ~100 Å in that direction relative to the thick filament.* The S1 head is thereby returned to its initial state ready to begin a new contraction cycle.

The ~500 S1 heads on every thick filament asynchronously cycle through this reaction sequence about five times per second each during a strong muscular contraction. The S1 heads thereby "walk" or "row" up adjacent thin filaments with the concomitant contraction of the muscle. The nature of the conformational states that enable myosin to carry out its function of transducing the chemical free energy of ATP hydrolysis to the mechanical energy of muscular motion is a subject of intense investigation.

C. Control of Muscle Contraction

Striated muscles are, for the most part, under voluntary control; that is, their contraction is triggered by motor nerve impulses. How do these nerve impulses trigger muscle contraction? To answer this question, let us begin at the level of the myofibril and work up.

Ca^{2+} Regulates Muscle Contraction in a Process Mediated by Troponin and Tropomyosin

It has been known since the 1940s that Ca^{2+} is somehow involved in controlling muscle contraction. It was not until the early 1960s, however, that Setsuro Ebashi demonstrated that the effect of Ca^{2+} is mediated by troponin and tropomyosin. He did so by showing that actomyosin extracted directly from muscle, and therefore bound to troponin and tropomyosin, contracts in the presence of ATP only when Ca^{2+} is also present, whereas actomyosin prepared from purified actin and myosin contracts in the presence of ATP regardless of the Ca^{2+} concentration. The addition of tropomyosin and troponin to the purified actomyosin system restored its sensitivity to Ca^{2+}. Indeed, it was through these experiments that troponin was discovered.

The TnC subunit of troponin (Fig. 34-51) is the only Ca^{2+}-binding component of the tropomyosin–troponin complex. Tropomyosin, as we saw, binds along the thin filament groove in relaxed muscle (Fig. 34-50) where it apparently blocks the attachment of S1 myosin heads to seven consecutive actin units. *X-ray diffraction studies indicate that when the $[Ca^{2+}]$ reaches a critical level, an allosteric interaction between Ca^{2+}-troponin and tropomyosin causes tropomyosin to move ~10-Å deeper into the thin filament groove (Fig. 34-56). This movement, it is thought, uncovers the actins' myosin-binding sites thereby switching on muscle contraction.* This switching mechanism may well be cooperative in that the binding of a myosin head to an actin subunit might push tropomyosin away from neighboring myosin-binding sites in a conformational change that could also increase the Ca^{2+}-binding affinity of its associated TnC subunit.

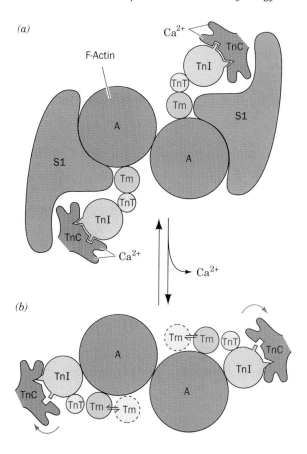

Figure 34-56
The control of skeletal muscle contraction by troponin and tropomyosin. (*a*) In contracting muscle, here diagrammed in cross-section, the myosin S1 heads freely interact with and thereby "walk" up the F-actin filaments (A). (*b*) Muscle relaxes when Ca^{2+} dissociates from troponin's TnC subunit thereby allosterically moving the tropomyosin (Tm) molecules to positions, which sterically block myosin–actin interactions. [After Zot, A. S. and Potter J. D., *Annu. Rev. Biophys. Biophys. Chem.* **16**, 555 (1987).]

Nerve Impulses Release Ca^{2+} from the Sarcoplasmic Reticulum

In order to understand how a nerve impulse affects the $[Ca^{2+}]$ in a myofibril we must further consider the anatomy of striated muscle fibers. A nerve impulse arriving at a **neuromuscular junction** is directly transmitted to each sarcomere by a system of **transverse** or **T tubules**, nervelike infoldings of the muscle fiber's plasma membrane that surround each myofibril at its Z disk (Fig. 34-57; nerve impulse transmission is the subject of Section 34-4C). All of a muscle's sarcomeres therefore receive the signal to contract within a few milliseconds of each other so that the muscle contracts as a unit. The electrical signal is transferred, in a poorly understood manner, to the **sarcoplasmic reticulum (SR)**, a system of flattened membranous vesicles derived from the endoplasmic reticulum that surround each myofibril rather like a net stocking. The SR membrane,

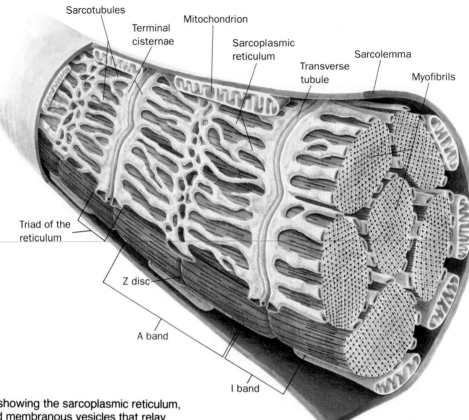

Figure 34-57
A diagram of myofibrils showing the sarcoplasmic reticulum, the system of connected membranous vesicles that relay contractile signals from nerve to myofibril.

which is normally impermeable to Ca^{2+}, contains a trans-membrane Ca^{2+}-ATPase (Section 18-3B) that pumps Ca^{2+} into the SR so as to maintain the cytosolic $[Ca^{2+}]$ of resting muscle below $10^{-7}M$ while that in the SR is over $10^{-3}M$. The SR's ability to store Ca^{2+} is enhanced by the presence of a highly acidic (37% Asp + Glu) 55-kD protein named **calsequestrin,** which has >40 Ca^{2+}-binding sites. *The arrival of a nerve impulse renders the SR permeable to Ca^{2+} which, in a few milliseconds, diffuses through specific Ca^{2+} channels into the myofibril so as to raise its internal $[Ca^{2+}]$ to $\sim 10^{-5}M$. This Ca^{2+} concentration is sufficient to trigger the conformational change in troponin–tropomyosin that permits muscle contraction.* Once nerve excitation has subsided, the SR membrane again becomes impermeable to Ca^{2+} so that the Ca^{2+} inside the myofibril is pumped back into the SR. Tropomyosin therefore resumes its resting conformation causing the muscle to relax.

D. Smooth Muscle

Vertebrates have two major types of muscle besides skeletal muscle, **cardiac muscle** and **smooth muscle.** Cardiac muscle, which is responsible for the heart's pumping action, is striated indicating the similarity of its organization to that of skeletal muscle. Cardiac and skeletal muscle differ mainly in their metabolism, with cardiac muscle, which must function continuously for a

lifetime, being much more dependent on aerobic metabolism than is skeletal muscle. Vertebrate heart muscle contraction is also spontaneously initiated by the heart muscle itself rather than through external nervous stimuli although the nervous system can influence this contractile response. Smooth muscle, which is responsible for the slow, long-lasting and involuntary contractions of such tissues as the intestinal walls, uterus, and large blood vessels, has a quite different organization from that of striated muscle. Smooth muscle consists of spindle-shaped, mononucleated cells whose thick and thin filaments are more or less aligned along the cells' long axes but which do not form myofibrils.

Smooth muscle myosin, a genetically distinct protein, is functionally distinct from striated muscle myosin in several ways:

1. Its maximum ATPase activity is only $\sim 10\%$ of that of striated muscle.

2. *It only interacts with actin when one of its light chains is phosphorylated at a specific Ser residue.*

3. It forms thick filaments whose cross-bridges lack the regular repeating pattern of striated muscle and are distributed along the thick filament's entire length.

Smooth Muscle Contraction Is Triggered by Ca^{2+}

The thin filaments of smooth muscle contain actin and tropomyosin but lack troponin. *Smooth muscle con-*

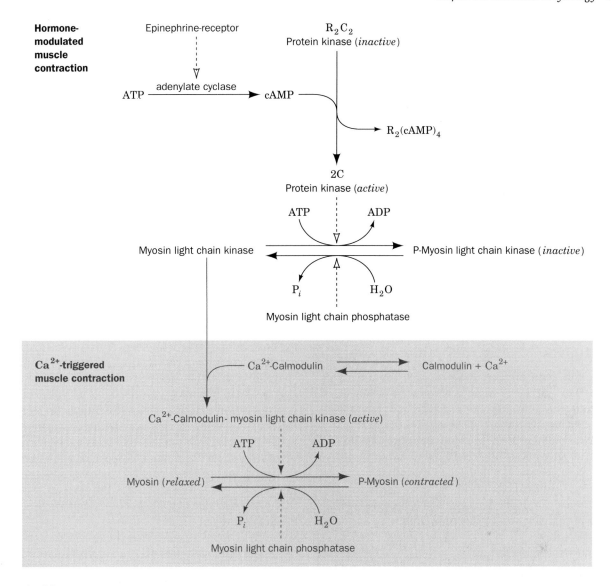

Figure 34-58
The control of smooth muscle contraction. Dashed arrows indicate stimulation or catalysis of a modification or demodification reaction. The lower part of the diagram (*shading*) indicates how Ca²⁺, whose intracellular concentration increases in response to nerve impulses, triggers muscle contraction. The upper part of the diagram indicates how hormones such as epinephrine modulate the contractile response.

traction is nevertheless triggered by Ca^{2+} because **myosin light chain kinase**, an enzyme that phosphorylates myosin light chains and thereby stimulates smooth muscle to contract, is enzymatically active only when it is associated with Ca^{2+}-calmodulin (Fig. 34-58, bottom; myosin light chain phosphorylation in skeletal muscle appears to modulate the degree of tension produced by contraction). The intracellular $[Ca^{2+}]$ varies with the permeability of the smooth muscle cell plasma membrane to Ca^{2+} which, in turn, is under the control of the autonomic (involuntary) nervous system. When the $[Ca^{2+}]$ rises to $\sim 10^{-5}M$, smooth muscle contraction is initiated as described. When the $[Ca^{2+}]$ falls to $\sim 10^{-7}M$ through the action of the plasma membrane's Ca^{2+}-ATPase, the myosin light chain kinase is deactivated, the myosin light chain is dephosphorylated by **myosin light chain phosphatase**,

and muscle relaxation ensues. *Thus Ca^{2+}, like cAMP, is a second messenger that transmits extracellular signals within the interior of a cell. In the many situations in which Ca^{2+} is a second messenger, calmodulin or a calmodulin-like protein is invariably the intracellular signal receiver.*

Smooth Muscle Activity Is Hormonally Modulated

Smooth muscles also respond to hormones such as epinephrine (Fig. 34-58, top). The binding of epinephrine to its plasma membrane-bound receptor activates adenylate cyclase. The cytosolic cAMP that is thereby generated binds to and causes the dissociation of the regulatory dimer, R_2, of an inactive protein kinase, R_2C_2, yielding active catalytic subunits, C, that phosphorylate myosin light chain kinase. This phosphorylated enzyme binds Ca^{2+}-calmodulin only weakly so that the extracellular presence of epinephrine causes smooth muscles to relax. Note the resemblance of this system to that controlling glycogen metabolism in skeletal muscle (Section 17-3). **Asthma,** a breathing disorder caused by the inappropriate contraction of bronchial smooth muscle, is often treated by the inhalation of an aerosol containing epinephrine, thereby relaxing the contracted bronchi.

The sequence of events culminating in smooth muscle contraction are inherently much slower than those leading to skeletal muscle contraction. Indeed, the structure and regulatory apparatus of smooth muscle suits it to its function: The maintenance of tension for prolonged periods while consuming ATP at a much lower rate than skeletal muscle performing the same task. The structural and functional resemblance of TnC to calmodulin therefore suggests that TnC is a calmodulin variant that evolved in skeletal muscle to provide a rapid response to the presence of Ca^{2+}.

E. Actin and Myosin in Nonmuscle Cells

Although actin and myosin are most prominent in muscle, they also occur in other tissues. In fact, actin is ubiquitous and is usually the most abundant cytoplasmic protein in eukaryotic cells, typically comprising 5 to 10% of their total protein. Myosin, in contrast, is usually present in only about one tenth the quantity of actin. This ratio reflects the fact that actin, in addition to its role in actomyosin-based contractile systems, participates in several myosin-independent motility systems as well as being a principal cytoskeleton component.

Actin Forms Microfilaments

Actin in muscles is entirely in the form of thin filaments. Nonmuscle actin, however, is about equally par-

titioned between soluble G-actin and F-actin fibers known as **microfilaments.** The actin content of microfilaments was established both through the immunofluorescence microscopy of living cells (Fig. 34-59) and because microfilaments can be decorated with S1 myosin heads to form arrowhead structures that are visually indistinguishable from those formed by muscle thin filaments (Fig. 34-49). Such decoration is possible because actin is highly conserved throughout the eukaryotic kingdom. For instance, actins from slime mold and rabbit muscle differ at only 17 of their 375 residues.

Actin *in vitro* is monomeric at low temperatures, low ionic strengths, and alkaline pH's. *Under physiological conditions, G-actin polymerizes in a process that is accelerated by the presence of ATP. In vivo,* microfilament assembly and disassembly is also influenced by numerous actin-binding proteins. For example, **profilin,** a 16-kD protein, binds G-actin in a 1 : 1 **profilactin** complex so as to prevent actin polymerization. A dramatic example of the effect of profilin occurs in many invertebrates upon the encounter of sperm and egg. A sea urchin sperm, for example, contains a reservoir of profilactin in its **acrosome,** a vesicle that lies just beneath the front of the sperm's head. Contact with the egg's jellylike coat triggers a reaction that dissociates the profilactin by raising the acrosomal pH. The newly liberated G-actin undergoes "explosive" polymerization so as to erect, in a matter of seconds, a 90-μm long bundle of F-actin filaments, the **acrosomal process,** that is projected outwards from the sperm head (Fig. 34-60). It is the acrosomal process that penetrates the egg's jelly coat to initiate the fusion of sperm and egg.

There is clear evidence that the assembly and disassembly of actin filaments plays an important role in such cellular motility processes as ameboid locomotion, phagocytosis, **cytokinesis** (the separation of daughter

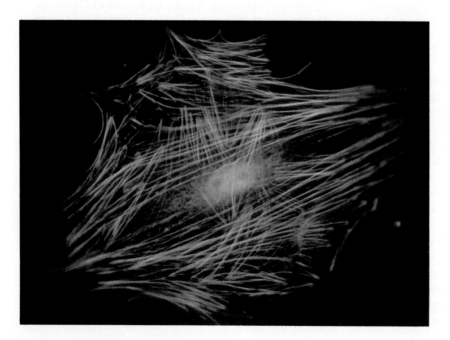

Figure 34-59
The microfilaments in a fibroblast resting on the surface of a culture dish as revealed by indirect immunofluorescence microscopy using anti-actin antibody. When the cell begins to move, the filaments disassemble to form a diffuse mesh thereby suggesting that actin plays a central role in cellular movement. [Courtesy of Elias Lazarides, California Institute of Technology.]

cells in the last stage of mitosis), and the extension and retraction of various cellular protuberances such as **microvilli** (fingerlike projections of cell surfaces) and neuronal axons. This evidence was obtained largely through the use of drugs that interfere with actin aggregation. For example, the fungal alkaloid **cytochalasin B**

Cytochalasin B

Phalloidin

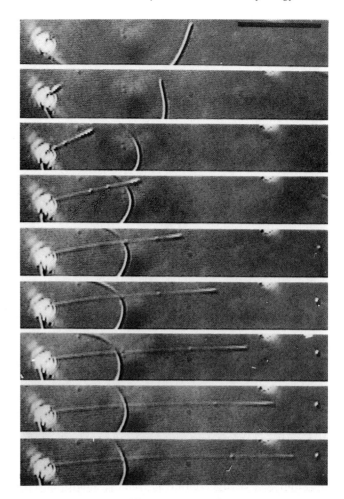

Figure 34-60
A series of light micrographs showing the elongation of the acrosomal process in a sea urchin sperm. The photographs were taken at 0.75-s intervals beginning 2 s after the sperm was artificially stimulated to begin the acrosomal reaction. The arc to the right of the sperm head is the sperm tail that has curved around outside the field of the micrograph. The acrosomal process' final length (*bottom frame*) is ~ 70 μm. [From Tilney, L. G. and Inoué, S., *J. Cell Biol.* **93**, 822 (1982).]

(which we have seen inhibits Na$^+$-independent glucose transport; Section 18-4A) blocks actin polymerization by specifically binding to the end of a growing F-actin filament (the "barbed" end of S1-decorated filaments) so as to inhibit actin polymerization from that end. In contrast, **phalloidin,** a bicyclic heptapeptide produced by the poisonous mushroom *Amanita phalloides* (which also synthesizes the chemically similar eukaryotic RNA polymerase inhibitor α-amanitin; Section 29-2F), blocks microfilament depolymerization by specifically binding to its actin units.

ATP accelerates actin polymerization by activating G-actin to add preferentially to a particular end of a growing actin filament. The ATP is eventually hydrolyzed but not until after polymerization has occurred. This vectorial process has the consequence that there is a certain "critical" G-actin concentration at which activated monomer units add predominantly (although not exclusively) to their preferred end of an actin filament at

the same rate that monomer units dissociate predominantly from the opposite end of the filament. Under these conditions, the actin filament neither grows nor shrinks; rather it assumes a steady state in which, through this **treadmilling** process, actin monomer units are continually translocated from one end of the actin filament to the other (Fig. 34-61).

Actomyosin Has Contractile Functions in Nonmuscle Cells

Myosin is not so well conserved a protein as is actin. Nevertheless, nonmuscle myosin forms thick filaments that participate in contractile processes with microfilaments. One of the best characterized such processes occurs during cytokinesis in animal cells. In the final stages of mitosis, a **cleavage furrow** forms around the

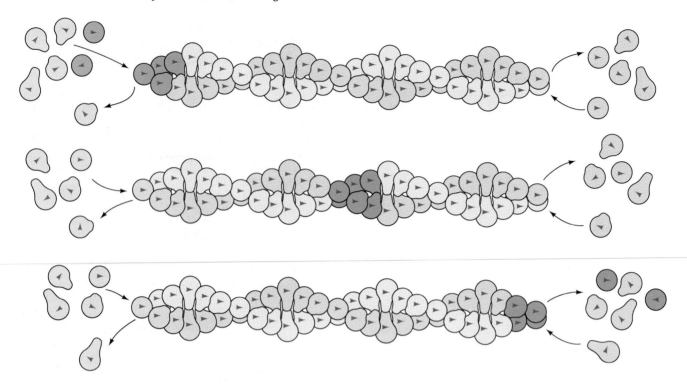

Figure 34-61
The ''treadmilling'' of actin monomer units along an actin filament. Actin monomers continually add to the left end of the filament, with eventual ATP hydrolysis, but dissociate at the same rate from the right end so that the filament maintains a constant length while its component monomer units translocate from left to right.

equator of the dividing cell in the plane perpendicular to the long axis of the mitotic spindle. Immunofluorescence microscopy demonstrates that the cleavage furrow is lined with an actomyosin belt (Fig. 34-62). Cell division is accomplished through the tightening of this so-called **contractile ring,** which disperses once cleavage has occurred. Blood platelets also contain actomyosin which, upon blood clot formation (Section 34-1), contracts so as to strengthen the clot. The contraction is initiated by the Ca^{2+}-calmodulin activation of myosin light chain kinase as occurs in smooth muscle.

F. Ciliary Motion and Vesicle Transport

Eukaryotes have two nearly ubiquitous but unrelated types of motility systems:

1. Microfilament-based systems, such as muscle, which contain actin.

2. **Microtubule**-based systems, such as cilia (see below), which contain the protein **tubulin.**

Microtubules (Fig. 34-63), as their name implies, are tubular structures, ~ 300 Å in diameter, which form a class of cytoskeletal components distinct from the ~ 70 Å in diameter microfilaments and the 100 to 150 Å in diameter **intermediate filaments** (the cytoskeleton's third major component, which apparently has only a structural role; Section 1-2A). *Microtubules comprise the*

major components of such cellular organelles as the mitotic spindle and cilia, and are thought to form the framework that organizes the cell.

Microtubules Are Composed of Tubulin

Microtubules are polymers of tubulin, a dimer of globular α- and β-tubulin subunits (55 kD each). Each of these types of subunits are highly homologous throughout the eukaryotic kingdom and, to a lesser extent, with each other. At low temperatures, and in the presence of Ca^{2+}*, tubulin assumes a soluble protomeric form (the αβ dimers dissociate only in the presence of denaturing*

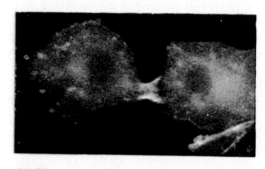

Figure 34-62
An indirect immunofluorescence micrograph, using anti-myosin antibodies, of a dividing chick embryo cell showing that its contractile ring contains myosin. [From Fujiwara, K., Porter, M. E., and Pollard, T. D., *J. Cell Biol.* **79,** 272 (1978).]

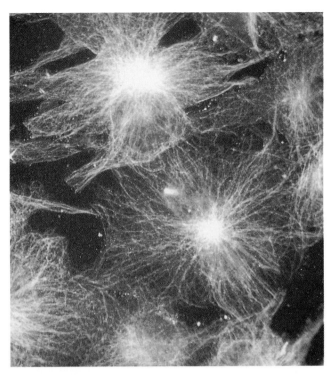

Figure 34-63
The networks of microtubules in fibroblasts as revealed by indirect immunofluorescence microscopy using anti-tubulin antibodies. [K. G. Marti/Visuals Unlimited.]

agents). Under physiological conditions, however, tubulin polymerizes to microtubules through a process in which each tubulin molecule binds 2 GTPs and hydrolyzes one of them to GDP + P_i during or shortly after the incorporation of the $\alpha\beta$ dimer into a microtubule. Electron microscopy and X-ray studies indicate that microtubules consist of 13 parallel but staggered protofilaments arranged about a hollow core (Fig. 34-64). The protofilaments, which consist of alternating head-to-tail-linked α- and β-tubulin subunits, all run in the same direction. Consequently, microtubules, like microfilaments, are polar entities: The end that grows most rapidly is called the plus end, whereas the other end is called the minus end. Microtubules are generally oriented in a cell with their minus ends towards a **centrosome** (the organizing centers from which they eminate; Fig. 34-63), and their plus ends towards the cell periphery.

Microtubules undergo continuous assembly and disassembly. Indeed, in a given population of microtubules, some may grow while others simultaneously shrink. This **dynamic instability** occurs because if the second GTP in a tubulin subunit at the microtubule's plus end becomes hydrolyzed to GDP before it is "capped" by another tubulin subunit, the resulting GDP-subunit rapidly dissociates from the microtubule. The balance between net microtubule growth or shrinkage in a cell therefore depends on the rate that tubulin hydrolyzes its second bound GTP together with the availability of GTP-tubulin subunits. Even when microtubules maintain a constant length, they are by no means static: They undergo GTP-driven treadmilling in which tubulin subunits add to the plus end at the same rate that they dissociate from the minus end. By regulating the rate of tubulin polymerization, cells presumably vary their shapes and induce the formation and dissolution of such cellular apparatus as the mitotic spindle.

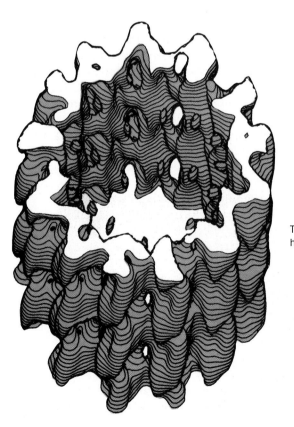

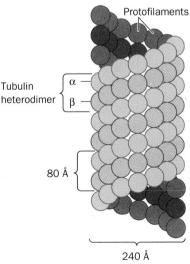

Figure 34-64
The 18-Å resolution X-ray structure of a microtubule together with an interpretive drawing. Microtubules may be considered to be composed of 13 parallel but staggered protofilaments which, in turn, consist of alternating α- and β-tubulin subunits linked head to tail. [Courtesy of Gerald Stubbs, Vanderbilt University; and Lorena Beese and Carolyn Cohen, Brandeis University.]

Antimitotic Drugs Inhibit Microtubule Formation

Colchicine,

Colchicine

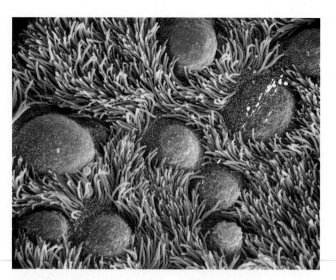

Figure 34-65
Scanning electron micrograph showing cilia lining the epithelial surface of a human bronchial tube as well as the rounded surfaces of a number of mucus-secreting **goblet cells.** Individuals with hereditary ciliary defects suffer from recurrent respiratory tract infections resulting from their reduced ability to clear away foreign particles. [From Kessel, R. G. and Kardon, R. H., *Tissues and Organs: A Text-Atlas of Scanning Electron Microscopy*, p. 210, Freeman (1979). Copyright ©1979 W. H. Freeman and Company. Reproduced by permission.]

an alkaloid produced by the meadow saffron, inhibits microtubule-dependent cellular processes by inhibiting the polymerization of tubulin protomers. For example, colchicine arrests mitosis in both plant and animal cells at metaphase (when the condensed and replicated chromosomes line up on the cell's equator; Section 27-1A) by preventing the formation of the mitotic spindle. It also inhibits cell motility.

> Colchicine has been used for centuries to treat acute attacks of gout (which result from elevated uric acid levels in body fluids; Section 26-5B). The lysosomes of the white cells that engulf urate microcrystals are ruptured by these needle-shaped crystals, causing cell lysis and triggering the local acute inflammatory reaction responsible for the exquisite pain characteristic of gout attacks. Colchicine, it is thought, slows the ameboid movements of white cells by inhibiting their microtubule-based systems.

The **vinca alkaloids, vinblastine** and **vincristine,**

Vinblastine: R = CH₃
Vincristine: R = CHO

products of the Madagascan periwinkle *Vinca rosea*, also inhibit microtubule polymerization by binding to tubulin. These substances are widely used in cancer chemotherapy since blocking mitosis preferentially kills fast growing cells. Curiously, colchicine is not selectively toxic to cancer cells.

Cilia and Eukaryotic Flagella Contain Organized Sheaves of Microtubules

Cilia are the hairlike organelles on the surfaces of many animal and lower plant cells that function to move

fluid over the cell's surface or to "row" single cells through a fluid. In humans, for example, epithelial cells lining the respiratory tract each bear ∼200 cilia that beat in synchrony to sweep mucus-entrained foreign particles towards the throat for elimination (Fig. 34-65; individuals with the inherited recessive disease **immotile-cilia syndrome** suffer from chronic respiratory disorders). Cilia are relatively short, operate with a whiplike motion, and occur in large numbers on a single cell. **Eukaryotic flagella** (as distinct from prokaryotic flagella; Section 34-3G), which occur on certain protozoa and comprise sperm tails, are much longer by comparison, carry out their propulsive function via undulatory motions, and occur in quantities of only one or a few per cell. Nevertheless, both types of organelles have the same basic architecture (males with immotile-cilia syndrome are usually sterile because their sperm are also immotile).

A cilium or flagellum consists of a plasma membrane-coated bundle of microtubules called an **axoneme.** Electron micrographs indicate that *an axoneme contains a ring of 9 double microtubules surrounding 2 single microtubules to form a common biological motif known as a 9 + 2 array* (Fig. 34-66). Each outer doublet consists of a ring of 13-protofilaments, **subfiber A,** fused to a C-shaped assembly of 10, or in some cases 11, protofilaments, **subfiber B** (Fig. 34-67). The 11 microtubules forming an axoneme

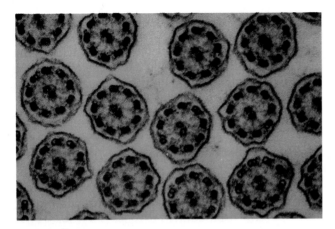

Figure 34-66
An electron micrograph of hamster oviduct cilia in cross-section. Two single microtubules are surrounded by 9 doublets to form a 9 + 2 array. [David M. Phillips/Visuals Unlimited.]

Figure 34-67
A schematic diagram showing the structure of a cilium in cross-section.

are held together by three types of connectors (Fig. 34-67):

1. Subfibers A are joined to the central microtubules by radial **spokes,** which each terminate in a knoblike feature termed a **spoke head.**

2. Adjacent outer doublets are joined by circumferential linkers that, in part, consist of a highly elastic protein named **nexin.**

3. The central microtubules are joined by a connecting bridge.

Each type of connector is repeated along the length of the axoneme with its own characteristic periodicity. Finally, every subfiber A bears two arms, an **inner dynein arm** and an **outer dynein arm** (Fig. 34-68), which both point clockwise when viewed from the base of the cilium (Fig. 34-67).

Ciliary Motion Results from the ATP-Powered "Walking" of Dynein Arms Along an Adjacent Subfiber B

An isolated flagellum (excised by a laser microbeam) whose plasma membrane has been removed by treat-

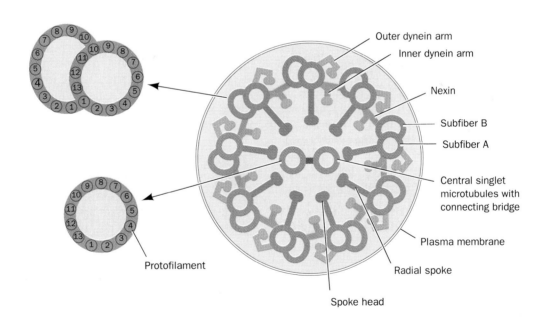

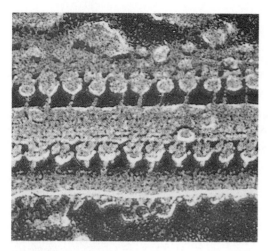

Figure 34-68
A freeze-etch electron micrograph of a flagellar microtubule from the unicellular algae *Chlamydomonas reinhardtii*, in transverse view, showing its dynein arms projecting like "lollipops." The outer dynein arms (*above*), which are spaced every 240 Å along the microtubule, have 100 Å in diameter heads attached to stalks that are <30 Å wide (an arrangement reminiscent of myosin's S1 heads). The spacing of the inner dynein arms (*below*) is less regular. [Courtesy of John Heuser and Ursala W. Goodenough, Washington University School of Medicine.]

ment with a nonionic detergent will continue to beat when supplied with ATP. Evidently, the eukaryotic flagellar "motor" is contained in the axoneme itself rather than at its base as occurs in bacterial flagella (Section 34-3G). What is the site of the eukaryotic flagellar motor? Several observations point to the dynein arms:

1. The dynein arms can be selectively extracted from naked axonemes by solutions containing high salt concentrations. This treatment immobilizes the axonemes while it solubilizes their ATPase activity (although it is much lower in solution than in intact axonemes). The addition of purified dynein to the salt-extracted axonemes restores their ability to beat.

2. In the absence of ATP, flagella become rigid. Electron micrographs indicate that the dynein arms in such ATP-deprived flagella are attached to their adjacent subfiber B.

3. Dynein resembles the S1 heads of myosin in both appearance and function. The outer dynein arms consist of either two-headed (~1200 kD) or three-headed (~1900 kD) entities, depending on the species, in which the globular heads are joined to a common base by flexible stems (Fig. 34-69; the inner dynein arms, which are not well characterized, probably have one or two globular heads). Dynein's ATPase functions are located in these heads.

4. Brief trypsin treatment, which selectively cleaves the radial spokes and nexin circumferential linkers, fol-

lowed by the addition of ATP, causes axonemes to elongate up to nine times their original length (Fig. 34-70). The elongation results from the telescoping of the axoneme's component microtubules out of the disrupted structure; the individual microtubules do not change in length.

These observations indicate that ciliary motion results from an ATP-driven process reminiscent of the sliding filament model of muscle contraction: *The dynein arms on one microtubule "walk" up the neighboring subfiber B so that these two microfilaments slide past each other.* However, the cross-links between microtubules in an intact cilium prevent neighboring microtubules from sliding past each another by more than a short distance. *These cross-links therefore convert the dynein-induced sliding motion to a bending motion of the entire axoneme.* This model is supported by electron microscopy studies showing that all the outer doublets in straight flagella have the same length and terminate at the same level but, in bent flagella, the doublets at the inside of the bend extend further than those on the outside of the bend (Fig. 34-71).

Dynein and Kinesin Motivate the Intracellular Transport of Vesicles and Organelles Along Microtubule Tracks

Eukaryotic cells, as we have seen (Sections 11-3F and 21-3B), transfer proteins and lipids between their various organelles via membranous vesicles. But how do these vesicles find their way to their proper destinations at a reasonable rate? The answer to this question was determined, in part, through the study of vesicle transport in the **axons** of neurons (cellular projections that extend from the cell body by up to 1 m; Fig. 1-10*d*). The use of **video-enhanced contrast microscopy** (subcellu-

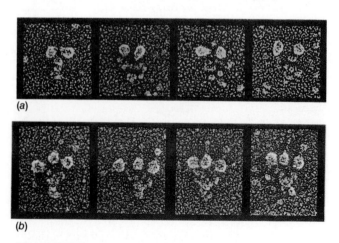

Figure 34-69
Electron micrographs of axonemal outer dynein arms. (*a*) Two-headed dyneins from sperm of the sea urchin *Strongylocentrotus purpuratus*. (*b*) Three-headed dyneins from *Chlamydomonus*. [Courtesy of Ursula Goodenough, Washington University School of Medicine.]

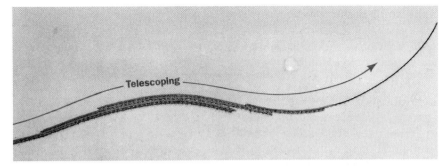

Figure 34-70
An electron micrograph of an isolated axoneme from a *Tetrahymena* cilium that has been briefly treated with trypsin to degrade its protein connectors and then exposed to ATP. The individual microtubule doublets telescope out from each other so that the axoneme elongates by up to a factor of 9. [From Warner, F. D. and Mitchell, D. R., *J. Cell Biol.* **89,** 36 (1981).]

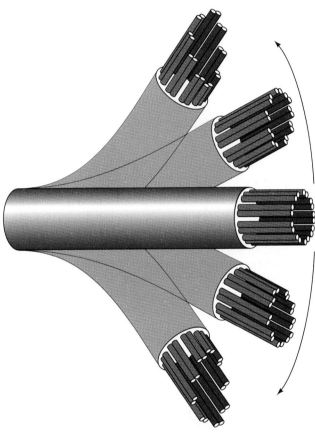

Figure 34-71
Diagram of sliding outer microtubules in a beating cilium. When the cilium is straight, all the outer doublets end at the same level (*center*). Cilium bending occurs when the doublets on the inner side of the bend slide beyond those on the outer side (*top* and *bottom*).

lar components are generally smaller than the resolution limit of light) revealed that vesicles and even entire organelles such as mitochondria are unidirectionally transported within axons at rates of 1 to 5 μm·s^{-1} so that they can traverse the length of even the longest axon in ~2 days. This apparently purposeful traffic, which simultaneously occurs in both directions (Fig. 34-72), moves along filamentous tracks that have been

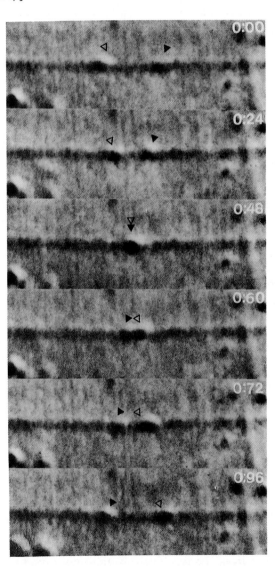

Figure 34-72
A series of successive video-enhanced contrast micrographs showing two organelles (*triangles*) moving in opposite directions along a microtubule and passing one another without colliding. The number in the upper right corner of each frame is the elapsed time in seconds from the top frame. [Courtesy of Bruce Schnapp, Boston Medical Center and Thomas S. Reese, NIH.]

identified as microtubules through their binding of specific antibodies.

What are the "motors" that drive vesicle and organelle transport? Two types have been identified:

1. Cytosolic dyneins, which resemble axonemal two-headed dyneins in appearance (Fig. 34-69a), transport vesicles and organelles from the plus to minus direction along microtubules (towards the cell center). This is corroborated by the observation that dynein which is immobilized by adsorption to a glass surface transports free microtubules in the direction of their plus end when supplied with ATP.

2. **Kinesin** transports vesicles and organelles from the minus to plus direction along microtubules (away from the cell center) and likewise, upon adsorption to a glass surface, transports free microtubules in the direction of their minus end. This ~600 kD, elongated (~1000 Å in length) protein resembles myosin and dynein in that it has twin globular ATPase-containing heads which power its motion.

Thus, eukaryotes have three classes of force-generating ATPases: myosins, dyneins, and kinesins.

G. Bacterial Flagella

The final aspect of biological motility that we shall consider is the nature of the bacterial flagellum. This remarkable propulsive organelle generates true rotary motion; *it is a propeller rather than a bending or a contractile device.*

Many species of bacteria, *E. coli,* for example (Fig. 1-3b), "swim" through solution via the action of a few flagella. Bacterial flagella, however, are entirely different from those of eukaryotes in both structure and chemistry. To begin with, bacterial flagella are only ~200 Å in diameter, less than the width of a single microtubule, and contain no tubulin. Electron micrographs (Fig. 34-73) indicate that bacterial flagella consist of three major segments (Fig. 34-74):

1. The **flagellar filament,** its most prominant portion, is a tightly coiled helix up to 10 μm long that consists only of subunits of the 55-kD protein **flagellin.**

2. The **flagellar hook,** which is assembled from subunits of the 42-kD **hook protein,** forms a short curved structure to which the flagellar filament is attached.

3. The **basal body,** which in gram-negative bacteria penetrates the outer membrane, the peptidoglycan cell wall and the inner (plasma) membrane, consists of a rod connecting several ringlike structures that anchor the flagellar hook to the bacterium.

Bacterial Flagella Rotate

Microscopic observations of swimming bacteria indi-

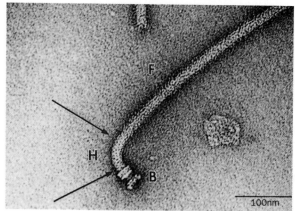

(a)

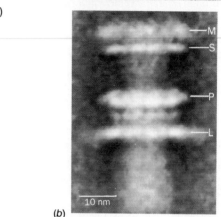

(b)

Figure 34-73
Electron micrographs of *Salmonella* flagella. (a) An intact flagellum. Its helical filament (F), which extends to the upper right, is connected via its hook (H) to its four-ringed basal body (B). [Courtesy of Robert McNab, Yale University.] (b) Close-up of the basal body as generated by computerized image enhancement. [Courtesy of David J. DeRosier, Brandeis University.]

cate that they are driven by what appear to be flagellar undulations. These undulations are an illusion; one cannot, by only watching through a light microscope, distinguish wave propagation along a helical flagellum from its rigid rotation. That bacterial flagella can freely rotate, a previously unknown and unexpected biological phenomenon, was established by the observation that when a bacterial flagellum is immobilized by "gluing" it to a microscope slide with anti-flagellin antibody, its attached bacterial cell slowly rotates. Similarly, if small latex beads are so-glued to a bacterial flagellum, they can be seen to rotate about the flagellum.

What is the nature of the flagellar "motor"? It cannot be located in either the flagellar filament or the flagellar hook since both flagellin and hook protein have no demonstrable enzymatic activity. The basal body must therefore form the rotary element. If this is so, then it must have the same mechanical elements as other rotary devices: a rotor (the rotating element) and a stator (the

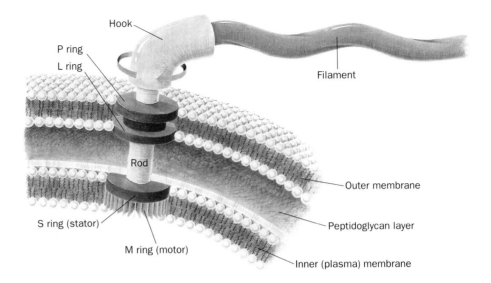

Figure 34-74

The structure of the gram-negative bacterial flagellum. The rotational "motor" consists of the basal body's M and S rings. The S ring is anchored to the peptidoglycan layer whereas the M ring, which is attached via the rod and hook to the flagellar filament, rotates freely in the plasma membrane. Torque is generated through the interaction of this "motor" with the electrochemical gradient across the plasma membrane. Gram-positive bacteria, which lack the outer membrane, also lack the "bushing" formed by the L and P rings.

stationary element). Indeed, as Figs. 34-73*b* and 34-74 indicate, the ringlike structures of the basal body form just such elements. The M ring is the rotor, the S ring, which is anchored to the peptidoglycan cell wall, is the stator, and the L and P rings form a bushing through which the rotating rod penetrates the bacterial outer membrane. The flagellar hook and filament are therefore chemically passive elements that mechanically convert rotary motion to linear thrust in the manner of a propeller.

A final question we shall ask is what makes the flagellar rotor rotate? One might guess, in analogy with muscles and cilia, that the M and S rings form an ATPase that acts as a mechanochemical transducer. However, the observation that bacterial swimming is unaffected by drastic reductions of the bacterial ATP pool requires that this hypothesis be abandoned. Rather, *it appears that the driving force behind flagellar rotation is the electrochemical proton gradient across the plasma membrane*; the same proton gradient that powers oxidative phosphorylation. This phenomenon was first demonstrated in *E. coli* mutants that lack an active F_0F_1 ATPase (the enzyme that mediates the proton gradient-driven synthesis of ATP; Section 20-3C) and are therefore unable to generate a proton gradient under anaerobic conditions (when an active F_0F_1 ATPase works in reverse). Such mutant bacteria can only swim under aerobic conditions, whereas normal bacteria can also swim under anaerobic conditions. The transduction of the proton gradient's free energy to mechanical work is thought to occur via the passage of protons across the M ring in a way that allows the protons to interact with suitably disposed charges fixed to the surface of the S ring. The elucidation of a detailed model of this process is a matter of active research.

4. BIOCHEMICAL COMMUNICATIONS: HORMONES AND NEUROTRANSMISSION

Living things coordinate their activities at every level of their organization through complex chemical signaling systems. Intracellular communications are maintained by the synthesis or alteration of a great variety of different substances that are often integral components of the processes they control; for example, metabolic pathways are regulated via the feedback control of allosteric enzymes by metabolites in those pathways or by the covalent modification of macromolecules. Intercellular signals occur through the mediation of chemical messengers known as **hormones** and, in higher animals, via neuronally transmitted electrochemical impulses. In this section we outline the operations of both these complex signaling systems.

A. The Endocrine System

Hormones are classified according to the distance over which they act (Fig. 34-75):

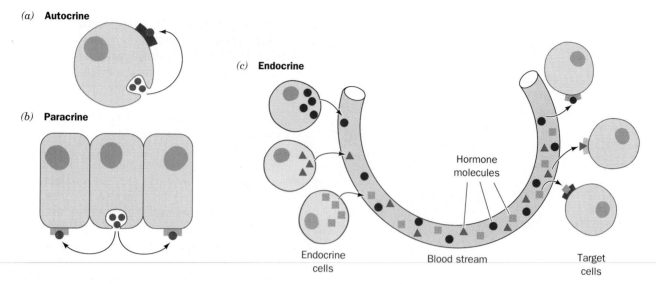

Figure 34-75

Hormonal communications are classified according to the distance over which the signal acts: (*a*) autocrine signals are directed at the cell that produced them, (*b*) paracrine signals are directed at nearby cells, and (*c*) endocrine signals are directed at distant cells through the intermediacy of the blood stream.

1. **Autocrine hormones** act on the same cell that released them. Interleukin-2, which stimulates *T* cell proliferation (Section 34-2A), is an autocrine hormone.

2. **Paracrine hormones** act only on cells close to the cell that released them. Prostaglandins (Section 23-7) and polypeptide growth factors (Section 33-4C) are examples of paracrine hormones.

3. **Endocrine hormones** act on cells distant from the site of their release. Endocrine hormones, for example, insulin and epinephrine, are synthesized and released in the bloodstream by specialized ductless **endocrine glands.**

We are already familiar with many aspects of hormonal control. For instance, we have considered how epinephrine, insulin, and glucagon regulate energy metabolism through the intermediacy of cAMP (Sections 17-3F and G), how steroid hormones activate transcription factors and thus influence protein synthesis (Section 33-3B), and how epinephrine relaxes smooth muscle through the intermediacy of Ca^{2+} (Section 34-3D). In this section we shall extend and systematize this information. Before we do so, it should be noted that biochemical communications are not limited to intracellular and intercellular signals. Many organisms release substances called **pheromones** that alter the behavior of other organisms of the same species in much the same way as hormones. Pheromones are commonly sexual attractants but some have other functions in species that have complex social interactions.

The human endocrine system (Fig. 34-76) secretes a wide variety of hormones (Table 34-5) that enable the body to:

1. Maintain homeostasis (e.g., insulin and glucagon maintain the blood glucose level within rigid limits during feast or famine).

2. Respond to a wide variety of external stimuli (such as the preparation for "fight or flight" engendered by epinephrine and norepinephrine).

3. Follow various cyclic and developmental programs (for instance, sex hormones regulate sexual differentiation, maturation, the menstrual cycle, and pregnancy).

Most hormones are either polypeptides, amino acid derivatives, or steroids although there are important exceptions to this generalization. In any case, only those cells with a specific receptor for a given hormone will respond to its presence even though nearly all cells in the body may be exposed to the hormone. Hormonal messages are therefore quite specifically addressed.

In what follows, we outline the hormonal functions of the various endocrine glands. Throughout this discussion keep in mind that these glands are not just a collection of independent secretory organs but form a complex and highly interdependent control system. Indeed, as we shall see, the secretion of many hormones is under feedback control through the secretion of other hormones to which the original hormone-secreting gland responds. Much of our understanding of hormonal function has come from careful measurements of hormone concentrations and the effects of changes of these concentrations on physiological functions. We begin, therefore, with a consideration of how physiological hormone concentrations are measured.

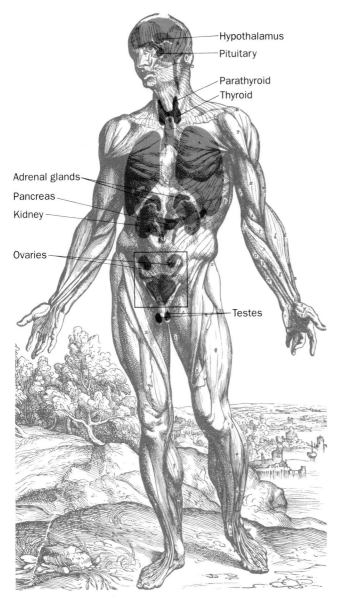

Figure 34-76
The major glands of the human endocrine system. Other tissues, the intestines, for example, also secrete endocrine hormones.

Hormone Concentrations Are Measured by Radioimmunoassays

The serum concentrations of hormones are extremely small, generally between 10^{-12} and $10^{-7}M$, so that they usually must be measured by indirect means. Biological assays have traditionally been employed for this purpose but they are generally slow, cumbersome, and imprecise. Such assays have therefore been largely supplanted by **radioimmunoassays.** In this technique, which was developed by Rosalyn Yalow, the unknown concentration of a hormone, H, is determined by measuring how much of a known amount of the radioactively labeled hormone, H*, binds to a fixed quantity of anti-H antibody in the presence of H. This competition reaction is easily calibrated by constructing a standard curve indicating how much H* binds to the antibody as a function of [H]. The high ligand affinity and specificity that antibodies possess gives radioimmunoassays the advantages of great sensitivity and specificity.

Pancreatic Islet Hormones Regulate the Storage and Release of Glucose and Fatty Acids

The pancreas is a large glandular organ, the bulk of which is an **exocrine gland** dedicated to producing digestive enzymes such as trypsin, RNase A, α-amylase, and phospholipase A_2 that it secretes via the pancreatic duct into the small intestine. However, ~1 to 2% of pancreatic tissue consists of scattered clumps of cells known as **islets of Langerhans** which comprise an endocrine gland that functions to maintain energy metabolite homeostasis. Pancreatic islets contain three types of cells, which each secrete a characteristic polypeptide hormone:

1. The α cells secrete glucagon (29 residues; Section 17-3E).

2. The β cells secrete insulin (51 residues; Fig. 8-4).

3. The δ cells secrete **somatostatin** (14 residues).

Insulin, which is secreted in response to high blood glucose levels, primarily functions, as we have seen (Section 25-2), to stimulate muscle, liver, and adipose cells to store glucose for later use by synthesizing glycogen, protein, and fat. Glucagon, which is secreted in response to low blood glucose, has essentially the opposite effects: It stimulates liver to release glucose through glycogenolysis and gluconeogenesis and it stimulates adipose tissue to release fatty acids through lipolysis. Somatostatin, which is also secreted by the hypothalamus (see below), inhibits the release of insulin and glucagon from their islet cells and is therefore thought to have a paracrine function in the pancreas.

Polypeptide hormones, as are other proteins destined for secretion, are ribosomally synthesized as preprohormones, processed in the rough endoplasmic reticulum and Golgi apparatus to form the mature hormone, and then packaged in secretory granules to await the signal for their release by exocytosis (Section 11-3F). The most potent physiological stimuli for the release of insulin and glucagon are, respectively, high and low blood glucose concentrations so that islet cells act as the body's primary glucose sensors. However, the release of these hormones is also influenced by the autonomic (involuntary) nervous system and by hormones secreted by the gastrointestinal tract (see below).

Gastrointestinal Hormones Regulate Digestion

The digestion and absorption of nutrients is a complicated process that is regulated by the autonomic nervous system in concert with a complex system of poly-

Table 34-5
Some Human Hormones

Hormone	Origin	Major Effects
Polypeptides		
Corticotropin-releasing factor (CRF)	Hypothalamus	Stimulates ACTH release
Gonadotropin-releasing factor (GnRF)	Hypothalamus	Stimulates FSH and LH release
Thyrotropin-releasing factor (TRF)	Hypothalamus	Stimulates TSH release
Growth hormone-releasing factor (GRF)	Hypothalamus	Stimulates growth hormone release
Somatostatin	Hypothalamus	Inhibits growth hormone release
Adrenocorticotropic hormone (ACTH)	Adenohypophysis	Stimulates the release of adrenocorticosteroids
Follicle-stimulating hormone (FSH)	Adenohypophysis	In ovaries, stimulates follicular development, ovulation, and estrogen synthesis; in testes, stimulates spermatogenesis
Lutinizing hormone (LH)	Adenohypophysis	In ovaries, stimulates oocyte maturation, and follicular synthesis of estrogens and progesterone; in testes, stimulates androgen synthesis
Chorionic gonadotropin (CG)	Placenta	Stimulates progesterone release from the *corpus luteum*
Thyrotropin (TSH)	Adenohypophysis	Stimulates T_3 and T_4 release
Somatotropin (growth hormone)	Adenohypophysis	Stimulates growth and synthesis of somatomedins
Met-enkephalin	Adenohypophysis	Opioid effects on central nervous system
Leu-enkephalin	Adenohypophysis	Opioid effects on central nervous system
β-Endorphin	Adenohypophysis	Opioid effects on central nervous system
Vasopressin	Neurohypophysis	Stimulates water resorption by kidney and increases blood pressure
Oxytocin	Neurohypophysis	Stimulates uterine contractions
Glucagon	Pancreas	Stimulates glucose release through glycogenolysis and stimulates lipolysis
Insulin	Pancreas	Stimulates glucose uptake through gluconeogenesis, protein synthesis, and lipogenesis
Gastrin	Stomach	Stimulates gastric acid and pepsinogen secretion
Secretin	Intestine	Stimulates pancreatic secretion of HCO_3^-
Cholecystokinin (CCK)	Intestine	Stimulates gallbladder emptying and pancreatic secretion of digestive enzymes and HCO_3^-
Gastric inhibitory peptide (GIP)	Intestine	Inhibits gastric acid secretion and gastric emptying; stimulates pancreatic insulin release
Parathyroid hormone	Parathyroid	Stimulates Ca^{2+} uptake from bone, kidney, and intestine
Calcitonin	Thyroid	Inhibits Ca^{2+} uptake from bone and kidney
Somatomedins	Liver	Stimulate cartilage growth; have insulin-like activity
Steroids		
Glucocorticoids	Adrenal cortex	Affect metabolism in diverse ways, decrease inflammation, increase resistance to stress
Mineralocorticoids	Adrenal cortex	Maintain salt and water balance
Estrogens	Gonads and adrenal cortex	Maturation and function of secondary sex organs, particularly in females
Androgens	Gonads and adrenal cortex	Maturation and function of secondary sex organs, particularly in males; male sexual differentiation
Progestins	Ovaries and placenta	Mediate the menstrual cycle and maintain pregnancy
Vitamin D	Diet and sun	Stimulates Ca^{2+} absorption from intestine, kidney, and bone
Amino acid derivatives		
Epinephrine	Adrenal medulla	Stimulates contraction of some smooth muscles and relaxes others, increases heart rate and blood pressure, stimulates glycogenolysis in liver and muscle, stimulates lipolysis in adipose tissue
Norepinephrine	Adrenal medulla	Stimulates arteriole contraction, decreases peripheral circulation, stimulates lipolysis in adipose tissue
Triiodothyronine (T_3)	Thyroid	General metabolic stimulation
Thyroxine (T_4)	Thyroid	General metabolic stimulation

peptide hormones. Indeed, gastrointestinal hormones are secreted into the bloodstream by a system of specialized cells lining the gastrointestinal tract whose aggregate mass is greater than that of the rest of the endocrine system. Four gastrointestinal hormones have been well characterized:

1. **Gastrin** (17 residues), which is produced by the gastric mucosa, stimulates the gastric secretion of HCl and **pepsinogen** (the zymogen of the digestive protease pepsin). Gastrin release is stimulated by amino acids and partially digested protein as well as by the vagus nerve (which innervates the stomach) in response to stomach distension. Gastrin release is inhibited by HCl and by other gastrointestinal hormones (see below).

2. **Secretin** (27 residues), which is produced by the upper small intestinal mucosa in response to acidification by gastric HCl, stimulates the pancreatic secretion of HCO_3^- so as to neutralize this acid.

3. **Cholecystokinin (CCK;** 33 residues), which is produced by the upper small intestine, stimulates gallbladder emptying, the pancreatic secretion of digestive enzymes and HCO_3^- (and thus enhances the effect of secretin), and inhibits gastric emptying. CCK is released in response to the products of lipid and protein digestion, that is, fatty acids, monoacylglycerols, amino acids, and peptides.

4. **Gastric inhibitory peptide (GIP;** 43 residues), which is produced by specialized cells lining the small intestine, is a potent inhibitor of gastric acid secretion, gastric mobility, and gastric emptying. However, GIP's major physiological function is to stimulate pancreatic insulin release. Indeed, the release of GIP is stimulated by the presence of glucose in the gut which accounts for the observation that, after a meal, the blood insulin level increases before the blood glucose level does.

These gastrointestinal hormones form families of related polypeptides: the C-terminal pentapeptides of gastrin and CCK are identical; secretin, GIP, and glucagon, are closely similar.

Several other polypeptides that affect gastrointestinal function have been isolated from the gut. However, the physiological roles of these so-called **candidate hormones** is unclear. Much of this difficulty stems from the diffuse distribution of gastrointestinal hormone secreting cells that precludes their excision, a procedure that is commonly used in controlled studies of the effects of other endocrine hormones.

Thyroid Hormones Are Metabolic Regulators

The thyroid gland produces two related hormones, triiodothyronine (T$_3$) and thyroxine (T$_4$),

X = H   **Triiodothyronine (T$_3$)**
X = I   **Thyroxine (T$_4$)**

that stimulate metabolism in most tissues (adult brain is a conspicuous exception). The production of these unusual iodinated amino acids begins with the synthesis of **thyroglobin,** a 660-kD globular protein. This protein is post-translationally modified in a series of biochemically unique reactions (Fig. 34-77):

1. Around 20% of thyroglobin's 140 Tyr residues are iodinated in an **iodoperoxidase**-catalyzed reaction forming **2,5-diiodo-Tyr** residues.

2. Two such residues are coupled to yield T$_3$ and T$_4$ residues. Mature thyroglobin itself is hormonally inactive; some five or six molecules of the active hormones, T$_3$ and T$_4$, are produced by the lysosomal proteolysis of thyroglobin upon hormonal stimulation of the thyroid (see below).

How do thyroid hormones work? T$_3$ and T$_4$, being nonpolar substances, are transported by the blood in complex with plasma carrier proteins, primarily **thyroxine-binding globin,** but also **prealbumin** and **albumin.** The hormones then pass through the cell membranes of their target cells into the cytosol where they bind to a specific protein. Since the resulting hormone–protein complex does not enter the nucleus, it is thought that this complex acts to maintain an intracellular reservoir of thyroid hormones. The true **thyroid hormone receptor** is a nonhistone chromosomal protein and therefore does not leave the nucleus. *The binding of T$_3$, and to a lesser extent T$_4$, activates this receptor as a transcription factor, resulting in increased rates of synthesis of numerous metabolic enzymes.* Indeed, thyroid hormone receptor is homologous to steroid hormone receptors (Section 33-3B). High affinity thyroid hormone-binding sites also occur on the inner mitochondrial membrane suggesting that these receptors may directly regulate O$_2$ consumption and ATP production.

Abnormal levels of thyroid hormones are common human afflictions. **Hypothyroidism** is characterized by lethargy, obesity, and cold dry skin, whereas **hyperthyroidism** has the opposite effects. The inhabitants of areas in which the soil has a low iodine content often develop hypothyroidism accompanied by an enlarged thyroid gland, a condition known as **goiter.** The small amount of NaI often added to commercially available table salt ("iodized" salt) easily prevents this iodine deficiency disease. Young mammals require thyroid hormone for normal growth and development: Hypothy-

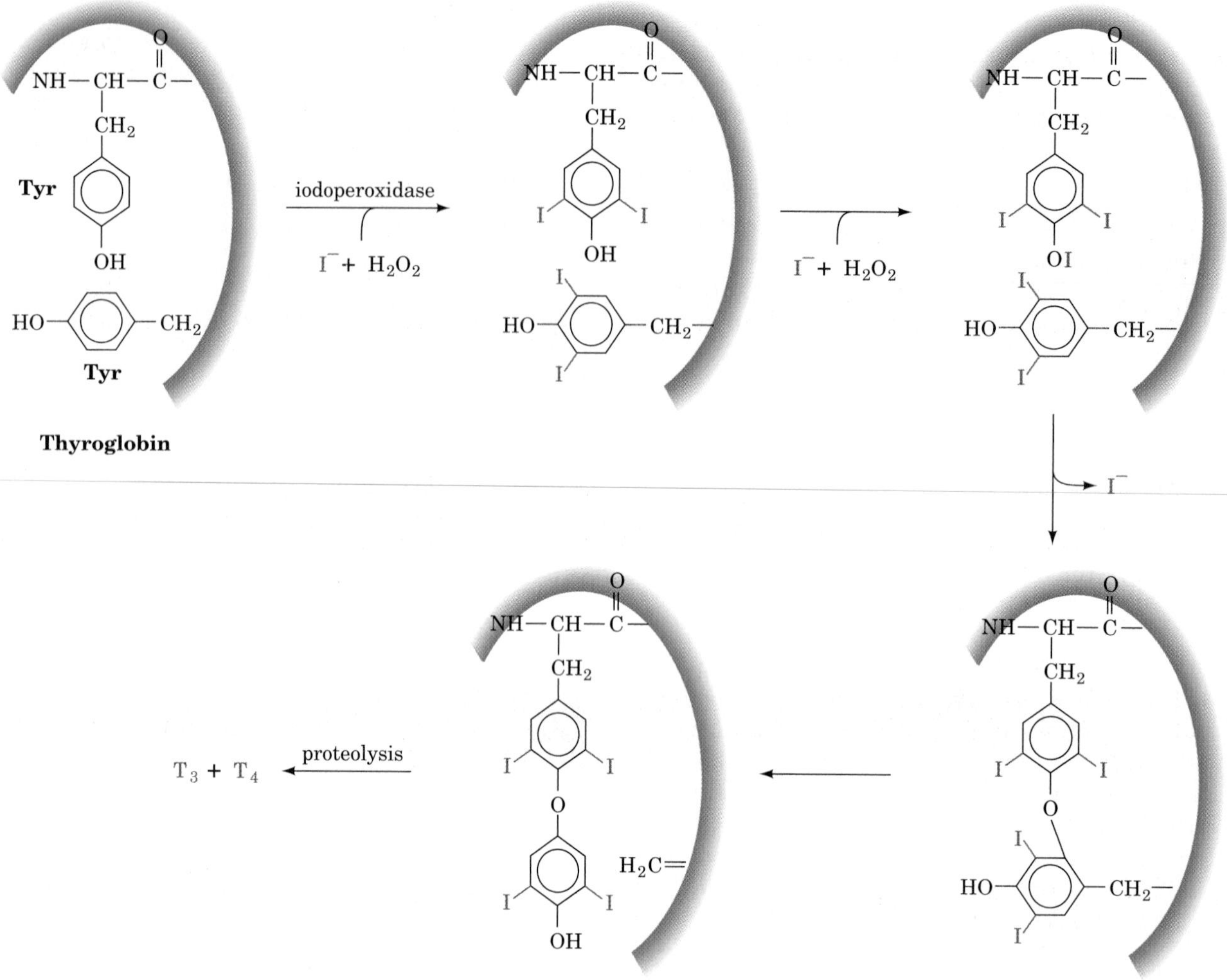

Figure 34-77
The biosynthesis of T_3 and T_4 in the thyroid gland through the iodination, rearrangement, and hydrolysis (proteolysis) of thyroglobin Tyr residues. The relatively scarce I^- is actively sequestered by the thyroid gland.

roidism during the fetal and immediate post-natal periods results in irreversible physical and mental retardation, a syndrome named **cretinism.**

Calcium Metabolism Is Regulated by Parathyroid Hormone, Vitamin D, and Calcitonin

Ca^{2+} forms **hydroxyapatite**, $Ca_5(PO_4)_3OH$, the major mineral constituent of bone, and is an essential element in many biological processes including the mediation of hormonal signals as a second messenger, the triggering of muscle contraction, the transmission of nerve impulses, and blood clotting. The extracellular $[Ca^{2+}]$ must therefore be closely regulated to keep it at its normal level of ~ 1.2 mM. Three hormones have been implicated in maintaining Ca^{2+} homeostasis (Fig. 34-78):

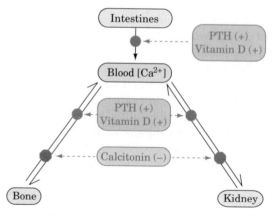

Figure 34-78
The roles of PTH, vitamin D, and calcitonin in controlling Ca^{2+} metabolism.

1. **Parathyroid hormone (PTH),** an 84-residue polypeptide secreted by the parathyroid gland, increases serum $[Ca^{2+}]$ by stimulating its resorption from bone and kidney and by increasing the dietary absorption of Ca^{2+} from the intestine.

2. **Vitamin D,** a group of steroid-like substances that act in a synergistic manner with PTH to increase serum $[Ca^{2+}]$.

3. **Calcitonin,** a 33-residue polypeptide synthesized by specialized thyroid gland cells, decreases serum $[Ca^{2+}]$ by inhibiting the resorption of Ca^{2+} from bone and kidney.

We shall briefly discuss the functions of these hormones.

The bones, the body's main Ca^{2+} reservoir, are by no means metabolically inert. They are continually "remodeled" through the action of two types of bone cells: **osteoblasts,** which synthesize the collagen fibrils that form the bulk of bone's organic matrix, the scaffolding upon which its $Ca_5(PO_4)_3OH$ mineral phase is laid down (Section 7-2C); and **osteoclasts,** which participate in bone resorption. *PTH inhibits collagen synthesis by osteoblasts and stimulates bone resorption by osteoclasts. The main effect of PTH, however, is to increase the rate that the kidneys excrete phosphate, the counterion of Ca^{2+} in bone.* The consequent decreased serum $[P_i]$ causes $Ca_5(PO_4)_3OH$ to leach out of bone through mass action and thus increase serum $[Ca^{2+}]$. Finally, PTH stimulates the production of the active form of vitamin D by kidney which, in turn, enhances the transfer of intestinal Ca^{2+} to the blood (see below).

Vitamin D is a group of dietary substances that prevent **rickets,** a disease of children characterized by stunted growth and deformed bones stemming from insufficient bone mineralization (vitamin D deficiency in adults is known as **osteomalacia,** a condition characterized by weakened, demineralized bones). Although rickets was first described in 1645, it was not until the early twentieth century that it was discovered that animal fats, particularly fish liver oils, are effective in preventing this deficiency disease. Moreover, rickets can also be prevented by exposing children to sunlight or just UV light in the wavelength range 230 to 313 nm, regardless of their diets.

The D vitamins, which we shall see are really hormones, are sterol derivatives in which the steroid B ring is disrupted at its 9,10 position. The natural form of the vitamin, **vitamin D₃ (cholecalciferol),** is nonenzymatically formed in the skin of animals through the photolytic action of UV light on **7-dehydrocholesterol.**

R = X **7-Dehydrocholesterol**
R = Y **Ergosterol**

UV radiation

spontaneous

R = X **Vitamin D₃ (cholecalciferol)**
R = Y **Vitamin D₂ (ergocalciferol)**

Vitamin D₂ (ergocalciferol), which differs from vitamin D₃ only by a side chain double bond and methyl group, is formed by the UV irradiation of the plant sterol **ergosterol.** Since vitamins D₂ and D₃ have essentially identical biological activities, vitamin D₂ is commonly used as a vitamin supplement, particularly in milk.

Vitamins D₂ and D₃ are hormonally inactive as such; they gain biological activity through further metabolic processing, first in the liver and then in the kidney (Fig. 34-79):

1. In the liver, vitamin D₃ is hydroxylated to form **25-hydroxycholecalciferol** in an O_2-requiring reaction catalyzed by **cholecalciferol-25-hydroxylase.**

2. The 25-hydroxycholecalciferol is transported to the kidney where it is further hydroxylated by a mitochondrial oxygenase, **25-hydroxycholecalciferol-1α-hydroxylase** to yield the active hormone **1α,25-dihydroxycholecalciferol [1,25(OH)₂D].** *The activity of 25-hydroxycholecalciferol-1α-hydroxylase is regu-*

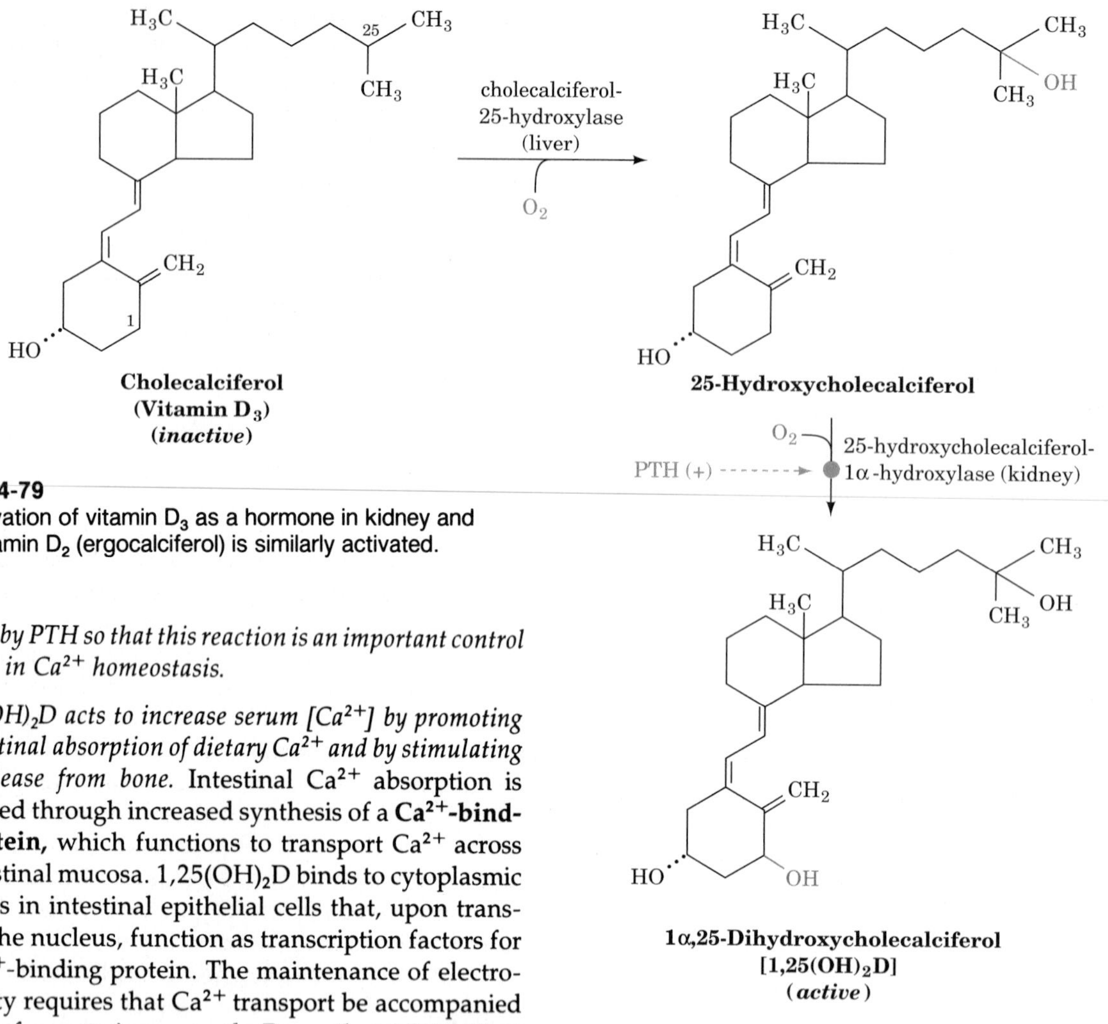

Figure 34-79
The activation of vitamin D_3 as a hormone in kidney and liver. Vitamin D_2 (ergocalciferol) is similarly activated.

lated by PTH so that this reaction is an important control point in Ca^{2+} homeostasis.

$1,25(OH)_2D$ acts to increase serum $[Ca^{2+}]$ by promoting the intestinal absorption of dietary Ca^{2+} and by stimulating Ca^{2+} release from bone. Intestinal Ca^{2+} absorption is stimulated through increased synthesis of a **Ca^{2+}-binding protein,** which functions to transport Ca^{2+} across the intestinal mucosa. $1,25(OH)_2D$ binds to cytoplasmic receptors in intestinal epithelial cells that, upon transport to the nucleus, function as transcription factors for the Ca^{2+}-binding protein. The maintenance of electroneutrality requires that Ca^{2+} transport be accompanied by that of counterions, mostly P_i, so that $1,25(OH)_2D$ also stimulates the intestinal absorption of P_i. The observation that $1,25(OH)_2D$ like PTH, stimulates the release of Ca^{2+} and P_i from bone seems paradoxical in view of the fact that low levels of $1,25(OH)_2D$ result in subnormal bone mineralization. Presumably the increased serum $[Ca^{2+}]$ resulting from $1,25(OH)_2D$-stimulated intestinal uptake of Ca^{2+} causes bone to take up more Ca^{2+} than it loses through direct hormonal stimulation.

Vitamin D, unlike the water-soluble vitamins, is retained by the body so that excessive intake of vitamin D over long periods causes **vitamin D intoxication.** The consequent high serum $[Ca^{2+}]$ results in aberrant calcification of a wide variety of soft tissues. The kidneys are particularly prone to calcification, a process that can lead to the formation of kidney stones and ultimately kidney failure. In addition, vitamin D intoxication promotes bone demineralization to the extent that bones are easily fractured. The observation that the level of skin pigmentation in indigenous human populations tends to increase with their proximity to the equator is explained by the hypothesis that skin pigmentation functions to prevent vitamin D intoxication by filtering out excessive solar radiation.

Calcitonin has essentially the opposite effect of PTH; it lowers serum $[Ca^{2+}]$. It does so primarily by inhibiting osteoclastic resorption of bone. Since PTH and calcitonin both stimulate the synthesis of cAMP in their target cells (see below), it is unclear how these hormones can oppositely affect osteoclasts. Calcitonin also inhibits kidney from resorbing Ca^{2+} but in this case the kidney cells that calcitonin influences differ from those that PTH stimulates to resorb Ca^{2+}.

The Adrenals Secrete Steroids and Catecholamines

The adrenal glands consist of two distinct types of tissue: the **medulla** (core), which is really an extension of the sympathetic nervous system (a part of the autonomic nervous system), and the more typically glandular **cortex** (outer layer). We shall first consider the hormones of the adrenal medulla and then those of the cortex.

The Adrenal Medulla Synthesizes Catecholamines

The adrenal medulla synthesizes two hormonally active catecholamines (amine-containing derivatives of catechol, 1,2-dihydroxybenzene), norepinephrine and its methyl derivative epinephrine.

HO—
HO—〈ring〉—CH—CH₂—⁺NH₂—R
|
OH

R = H   **Norepinephrine (adrenalin)**
R = CH₃   **Epiniphrine (noradrenalin)**

These hormones are synthesized from tyrosine as is described in Section 24-4B and stored in granules to await their exocytotic release under the control of the sympathetic nervous system.

The biological effects of catecholamines are mediated by two classes of plasma trans-membrane receptors, the **α- and the β-adrenoreceptors** (also known as **adrenergic receptors**). These glycoproteins were originally identified on the basis of their varying responses to certain **agonists** (substances that bind to a hormone receptor so as to evoke a hormonal response) and **antagonists** (substances that bind to a hormone receptor but fail to elicit a hormonal response thereby blocking agonist action). The β- but not the α-adrenoreceptors, for example, are stimulated by **isoproterenol** but blocked by **propranolol**, whereas α- but not β-adrenoreceptors are blocked by **phentolamine.**

Isoproterenol

Propranolol

Phentolamine

The α- and β-adrenoreceptors, which occur on separate tissues in mammals, generally respond differently and often oppositely to catecholamines. For instance, β-adrenoreceptors, which activate adenylate cyclase,

stimulate glycogenolysis and gluconeogenesis in liver and skeletal muscle (Sections 17-3E and G), lipolysis in adipose tissue, smooth muscle relaxation in the bronchi and the blood vessels supplying the skeletal muscles, and increased heart action. In contrast, α-adrenoreceptors, whose intracellular effects are mediated either by the inhibition of adenylate cyclase (α_2 receptors; Section 34-4B) or via the phosphoinositide cascade (α_1 receptors; Section 34-4B), stimulate smooth muscle contraction in blood vessels supplying peripheral organs such as skin and kidney, smooth muscle relaxation in the gastrointestinal tract, and blood platelet aggregation. *Most of these diverse effects are directed toward a common end: the mobilization of energy resources and their shunting to where they are most needed to prepare the body for sudden action.*

The varying responses and tissue distributions of the α- and β-adrenoreceptors and their subtypes to different agonists and antagonists has important therapeutic consequences. For example, propranolol is widely used for the treatment of high blood pressure and protects heart attack victims from further heart attacks, whereas epinephrine's bronchodilator effects make it clinically useful in asthma treatment (Section 34-3D).

β-Adrenoreceptors are trans-membrane glycoproteins which all contain 7 stretches of 20 to 28 hydrophobic amino acids that are each thought to form a membrane-spanning helix (Fig. 34-80). *This 7-helix bundle is a common receptor motif.* Sequencing studies indicate that it also occurs in α_2-adrenoreceptors, **rhodopsin** (the photoreceptor protein of retinal rod cells), and the muscarinic acetylcholine receptors of nerve synapses (Section 34-4C). We have previously encountered such a 7-helix bundle in our study of bacteriorhodopsin (Section 11-3A; Fig. 11-26).

The Adrenal Cortex Synthesizes a Variety of Steroids

The adrenal cortex produces at least 50 different adrenocortical steroids (whose synthesis is outlined in Section 23-6C). These have been classified according to the physiological responses they evoke:

1. The **glucocorticoids** affect carbohydrate, protein, and lipid metabolism in a manner nearly opposite to that of insulin, and influence a wide variety of other vital functions including inflammatory reactions and the capacity to cope with stress.

2. The **mineralocorticoids** largely function to regulate the excretion of salt and water by kidney.

3. The **androgens** and **estrogens** affect sexual development and function. They are made in larger quantities by the gonads and are therefore further discussed below.

Glucocorticoids, the most common of which are **cortisol** (also known as **hydrocortisone**) and **corticosterone**, and the mineralocorticoids, the most common of which is **aldosterone,** are all C_{21} compounds.

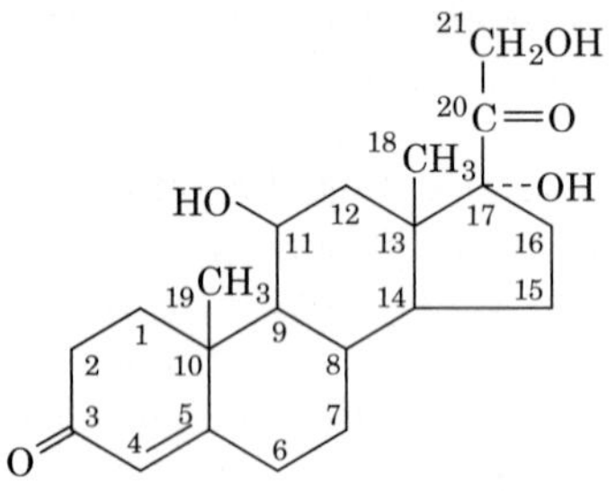

Cortisol (hydrocortisone)

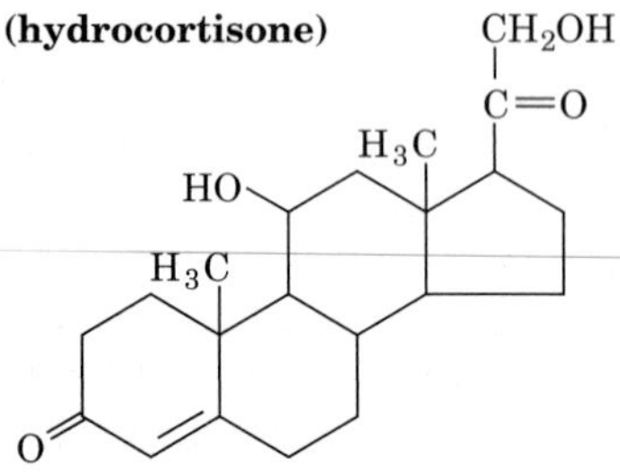

Corticosterone

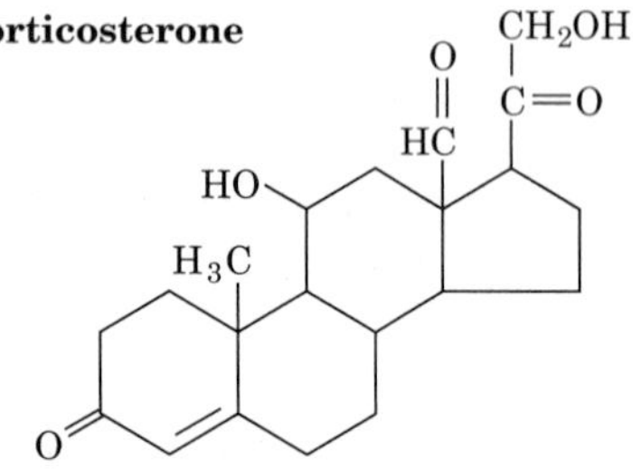

Aldosterone

Steroids, being water insoluble, are transported in the blood in complex with the glycoprotein **transcortin** and, to a lesser extent, by albumin. The steroids enter their target cells, apparently spontaneously, where they bind to their cytoplasmic receptors so as to activate them as transcription factors for specific proteins (Section 33-3B). Indeed, the glucocorticoids and the mineralocorticoids induce the synthesis of numerous metabolic enzymes in their respective target tissues.

Impaired adrenocortical function, either through disease or trauma, results in a condition known as **Addison's disease,** which is characterized by hypoglycemia, muscle weakness, Na^+ loss, K^+ retention, impaired cardiac function, loss of appetite, and a greatly increased susceptibility to stress. The victim, unless treated by the administration of glucocorticoids and mineralocorticoids, slowly languishes and dies without any particular pain or distress. The opposite problem, adrenocortical hyperfunction, which is usually caused by a tumor of the adrenal cortex or the pituitary gland (see below), results in **Cushing's syndrome,** which is characterized by fatigue, hyperglycemia, edema (water retention), and a redistribution of body fat to yield a characteristic "moon face." Long-term treatments of various diseases with synthetic glucocorticoids result in similar symptoms.

Gonadal Steroids Mediate Sexual Development and Function

*The **gonads** (testes in males, ovaries in females), in addition to producing sperm or ova, secrete steroid hormones (androgens and estrogens) that regulate sexual differentiation, the expression of secondary sex characteristics, and sexual behavior patterns.* Although testes and ovaries both synthesize androgens and estrogens, the testes predominantly secrete androgens, which are therefore known as male sex hormones, whereas ovaries produce mostly estrogens, which are consequently termed female sex hormones.

Androgens, of which **testosterone** is prototypic,

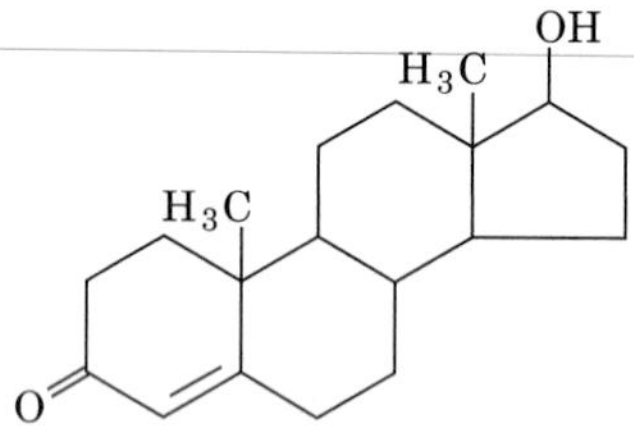

Testosterone

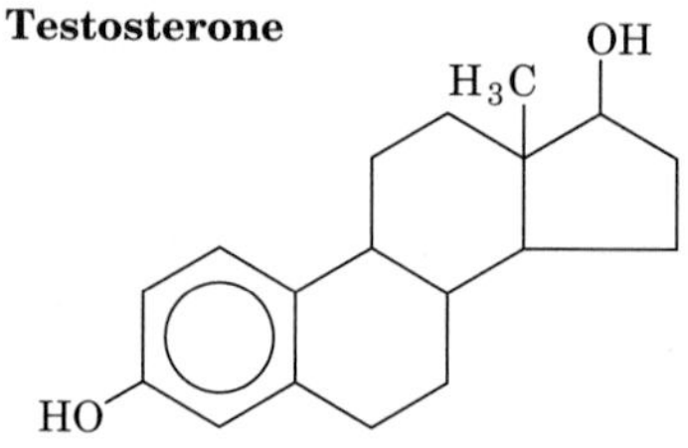

β-Estradiol

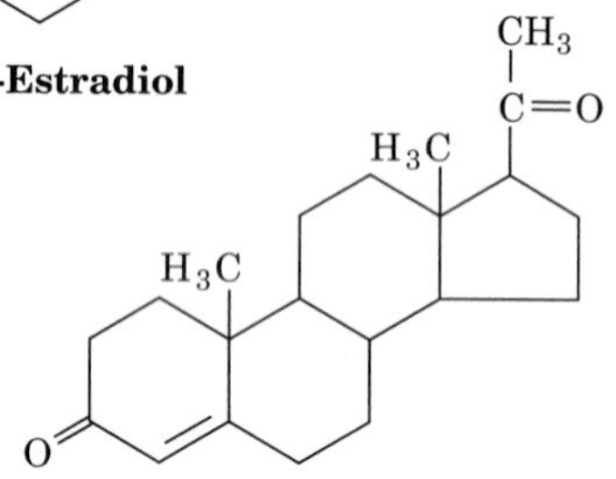

Progesterone

lack the C_2 substituent at C(17) occurring in glucocorticoids and are therefore C_{19} compounds. Estrogens, such as **β-estradiol,** resemble androgens but lack a C(10) methyl group because they have an aromatic A ring and are therefore C_{18} compounds. Interestingly, testosterone is an intermediate in estrogen biosynthesis (Fig. 23-51). A second class of ovarian steroids, C_{21} compounds called **progestins,** help mediate the menstrual cycle and pregnancy (see below). **Progesterone,** the most abundant progestin, is, in fact, a precursor of glucocorticoids, mineralocorticoids, and testosterone.

What factors control sexual differentiation? Normal individuals have the XY (male) or XX (female) genotypes; those with the abnormal genotypes XXY (**Klinefelter's syndrome**) and X0 (only one sex chromosome; **Turner's syndrome**) are, respectively, phenotypic males and phenotypic females although both are sterile.

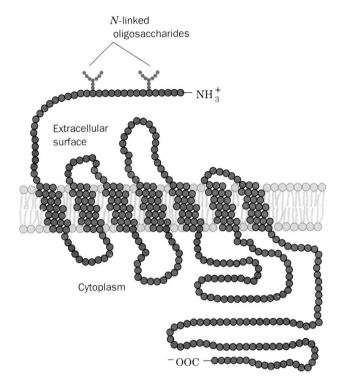

Figure 34-80
The amino acid sequence of human β-adrenoreceptor has 7 segments of ~24 hydrophobic residues each (*brown circles*), suggesting that this glycoprotein has 7 membrane-spanning helices as diagrammed. [After Dohlman, H. G., Caron, M. G., and Lefkowitz, R. J., *Biochemistry* **26,** 2660 (1987).]

Apparently, *the normal Y chromosome confers the male phenotype, whereas its absence results in the female phenotype.* There are, however, rare (1 in 20,000) XX males and XY females. DNA hybridization studies have revealed that these XX males (who are sterile and have therefore been identified through infertility clinics) have a small segment of a normal Y chromosome translocated onto one of their X chromosomes, whereas XY females are missing this segment. David Page showed that this segment contains a gene encoding a protein named **testis-determining factor (TDF)** which contains several zinc fingerlike sequences and is therefore thought to be a transcription factor. This strongly suggests that TDF directs the embryonic gonads to develop as testes rather than as ovaries (early male and female embryos—through the sixth week of development in humans—have identical undifferentiated genitalia) and hence is the first clear example of a gene that controls the development of an entire organ system in mammals. If, however, the gonads of an embryonic male animal are surgically removed, that individual will become a phenotypic female. Evidently, *embryonic mammals are programmed to develop as females unless subjected to the influence of testicular hormones.* Indeed, genetic males with absent or nonfunctional cytosolic androgen receptors are phenotypic females, a condition

named **testicular feminization.** Curiously, estrogens appear to play no part in embryonic female sexual development although they are essential for female sexual maturation and function.

The Hypothalamus Controls Pituitary Secretions that Regulate Other Endocrine Glands

The anterior lobe of the **pituitary gland** (the **adenohypophysis**) and the **hypothalamus,** a nearby portion of the brain, constitute a functional unit that hormonally controls much of the endocrine system. *The neurons of the hypothalamus synthesize a series of polypeptide hormones known as **releasing factors** and **release-inhibiting factors** which, upon delivery to the adenohypophysis via a direct circulatory connection, stimulate or inhibit the release of the corresponding **trophic hormones** into the blood stream. Trophic hormones, by definition, stimulate their target endocrine tissues to secrete the hormones they synthesize.* Four such hormonal systems are prominent in humans (Fig. 34-81; *left*):

1. **Corticotropin-releasing factor (CRF;** 41 residues) causes the adenohypophysis to release **adrenocorticotropic hormone (ACTH;** 39 residues), which stimulates the release of adrenocortical steroids.

$$
\underset{1}{Ser} - Gln - Glu - Pro - \underset{5}{Pro} - Ile - Ser - Leu - Asp - \underset{10}{Leu} -
$$
$$
\underset{11}{Thr} - Phe - His - Leu - \underset{15}{Leu} - Arg - Glu - Val - Leu - \underset{20}{Glu} -
$$
$$
\underset{21}{Met} - Thr - Lys - Ala - \underset{25}{Asp} - Gln - Leu - Ala - Gln - \underset{30}{Gln} -
$$
$$
\underset{31}{Ala} - His - Ser - Asn - \underset{35}{Arg} - Lys - Leu - Leu - Asp - \underset{40}{Ile} -
$$
$$
\underset{41}{Ala} - NH_2
$$

Sheep corticotropin-releasing factor (CRF)

$$
\underset{1}{Ser} - Tyr - Ser - Met - \underset{5}{Glu} - His - Phe - Arg - Trp - \underset{10}{Gly} -
$$
$$
\underset{11}{Lys} - Pro - Val - Gly - \underset{15}{Lys} - Lys - Arg - Arg - Pro - \underset{20}{Val} -
$$
$$
\underset{21}{Lys} - Val - Tyr - Pro - \underset{25}{Asn} - Gly - Ala - Glu - Asp - \underset{30}{Glu} -
$$
$$
\underset{31}{Ser} - Ala - Glu - Ala - \underset{35}{Phe} - Pro - Leu - Glu - \underset{39}{Phe}
$$

Human adrenocorticotropic hormone (ACTH)

The entire system is under feedback control: ACTH inhibits the release of CRF and the adrenocortical steroids inhibit the release of both CRF and ACTH. Moreover, the hypothalamus, being part of the brain, is also subject to neuronal control so that the hypothalamus forms the interface between the nervous system and the endocrine system. Recall that ACTH is synthesized together with several other hormones as a single polyprotein, pro-opiomelanocortin, that is post-translationally cleaved to yield the individual polypeptides (Section 33-3C).

2. **Thyrotropin-releasing factor (TRF),** a tripeptide with an N-terminal **pyroGlu** residue (a Glu derivative in which the side chain carboxyl group forms a

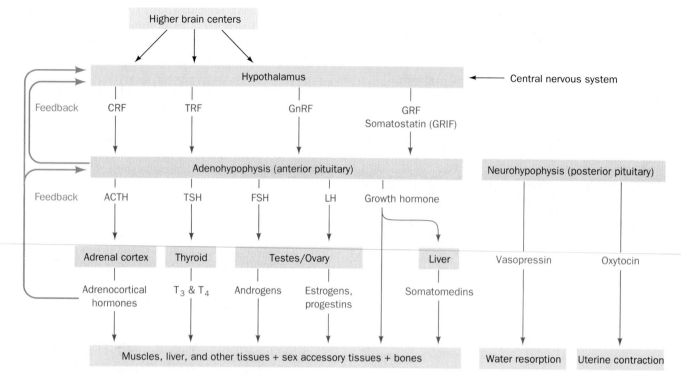

Figure 34-81
Hormonal control circuits indicating the relationships between the hypothalamus, the pituitary, and the target tissues. Releasing factors and release-inhibiting factors secreted by the hypothalamus signal the adenohypophysis to secrete or stop secreting the corresponding trophic hormones which, in turn, stimulate the corresponding endocrine gland(s) to secrete their respective target tissue hormones. The target tissue hormones, in addition to controlling the growth and metabolism of the corresponding target tissues, influence the secretion of releasing factors and trophic hormones through feedback inhibition. The levels of trophic hormones likewise influence the levels of their corresponding releasing factors.

peptide bond with its amino group),

$$\text{pyroGlu--His--Pro}^3\text{--NH}_2$$

Thyrotropin-releasing factor (TRF)

stimulates the adenohypophysis to release the trophic hormone **thyrotropin (thyroid-stimulating hormone; TSH)** which, in turn, stimulates the thyroid to synthesize and release T_3 and T_4. TRF, as are other releasing factors, is present in the hypothalamus in only vanishingly small quantities. It was independently characterized in 1969 by Roger Guillemin and Andrew Schally using extracts of the hypothalami from over 2 million sheep and 1 million pigs.

3. **Gonadotropin-releasing factor (GnRF;** 10 residues)

$$\overset{1}{\text{pyroGlu}}\text{-His-Trp-Ser-Tyr-Gly-Leu-Arg-Pro-Gly-}\overset{10}{\text{NH}_2}$$

Gonadotropin-releasing factor (GnRF)

stimulates the adenohypophysis to release **lutinizing hormone (LH)** and **follicle-stimulating hormone (FSH)**, which are collectively known as **gonadotropins.** In males, LH stimulates the testes to secrete androgens while FSH promotes spermatogenesis. In females, FSH stimulates the development of ovarian follicles (which contain the immature ova), whereas LH triggers ovulation.

4. **Growth hormone-releasing factor (GRF;** 44 residues) and somatostatin [14 residues; also known as **growth hormone release-inhibiting factor (GRIF)**], stimulate/inhibit the release of **growth hormone (GH)** from the adenohypophysis. GH (also called **somatotropin**), in turn, stimulates generalized growth (see Fig. 27-24 for a striking example of its effect). GH directly accelerates the growth of a variety of tissues (in contrast to TSH, LH, and FSH, which only act indirectly by activating endocrine glands) and induces the liver to synthesize a series of polypeptide growth factors termed **somatomedins** that stimulate cartilage growth and have insulin-like activities.

TSH, LH, and FSH are all $\alpha\beta$-glycoproteins which, in a given species, all have the same α subunit (92 residues) and a homologous β subunit (112, 112, and 118 resi-

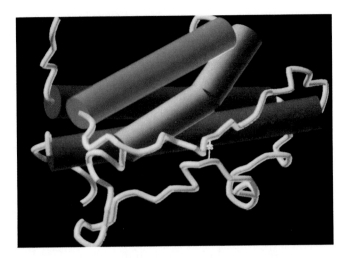

Figure 34-82
The X-ray structure of porcine growth hormone. The protein's four α helices are represented by cylindrical rods and its nonhelical segments are shown as thin tubes. The N-terminus is in the upper left corner and the C-terminus is in the lower left corner. Only one of the protein's two S—S bonds is visible in this view. [Courtesy of Sherin S. Abdel-Meguid, Monsanto Co.]

dues, respectively). GH consists of a single 191-residue polypeptide chain (Fig. 34-82), which is unrelated to TSH, LH, or FSH.

The menstrual cycle and pregnancy are particularly illustrative of the interactions among hormonal systems. The ~ 28 day human menstrual cycle (Fig. 34-83) begins during menstruation with a slight increase in the FSH level that initiates the development of a new ovarian follicle. As the follicle matures, it secretes estrogens that act to sensitize the adenohypophysis to GnRF. This process culminates in a surge of LH and FSH, which triggers ovulation. The ruptured ovarian follicle, the **corpus luteum,** secretes progesterone and estrogens, which inhibit further gonadotropin secretion by the adenohypophysis and stimulate the uterine lining to prepare for the implantation of a fertilized ovum. If fertilization does not occur, the *corpus luteum* regresses, progesterone and estrogen levels fall, and menstruation (the sloughing off of the uterine lining) ensues. The reduced steroid levels also permit a slight increase in the FSH level, which initiates a new menstrual cycle.

A fertilized ovum that has implanted into the hormonally prepared uterine lining soon commences synthesizing **chorionic gonadotropin (CG),** an αβ-glycoprotein hormone that has the same α subunit as LH and a β subunit that is 80% homologous. CG stimulates the *corpus luteum* to continue secreting progesterone rather than regressing and thus prevents menstruation. Present-day pregnancy tests utilize immunoassays that can detect CG in blood or urine within a few days after embryo implantation. Most female oral contraceptives

(birth control pills) contain progesterone derivatives whose ingestion induces a state of pseudopregnancy in that they inhibit the midcycle surge of FSH and LH so as to prevent ovulation.

Overproduction of GH, usually a consequence of a pituitary tumor, results in excessive growth. If this condition commences while the skeleton is still growing, that is, before its growth plates have ossified, then this excessive growth is of normal proportions over the entire body resulting in **gigantism.** Moreover, since excessive GH inhibits the testosterone production necessary for growth plate ossification, such "giants" continue growing throughout their abnormally short lives. If, however, the skeleton has already matured, GH stimulates only the growth of soft tissues resulting in enlarged hands and feet and thickened facial features, a condition named **acromegaly** (Fig. 34-84). The opposite problem, GH deficiency, which results in insufficient growth (**dwarfism**) can be treated before skeletal maturity by regular injections of human GH (animal GH is ineffective in humans). Since human GH was, at first, only available from the pituitaries of cadavers, it was in very short supply. Recently, however, human GH has been synthesized by recombinant DNA techniques and is therefore available in virtually unlimited amounts. Indeed, there is now concern that GH will be taken in an uncontrolled manner by individuals wishing to increase their athletic prowess.

Among the most intriguing hormones secreted by the adenohypophysis are polypeptides that have opiate-like effects on the central nervous system. These include the 31-residue **β-endorphin,** its N-terminal pentapep-

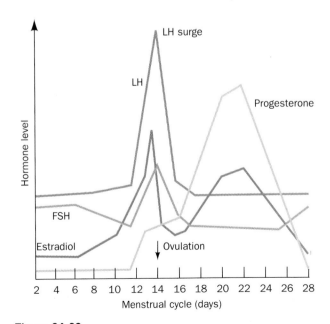

Figure 34-83
Patterns of hormone secretion during the menstrual cycle in the human female.

Figure 34-84
The characteristic enlarged features of Akhenaten, the Pharaoh who ruled Egypt in the years 1379–1262 B.C., strongly suggest that he suffered from acromegaly. [Agytisches Museum. Staadtliche Museen Preussicher Kulturbesitz, Berlin (West). Photo by Margarete Busing.]

tide, termed **methionine-enkephalin,** and the closely similar **leucine-enkephalin** (although the enkephalins are independently expressed).

$$\overset{1}{\text{Tyr}} - \text{Gly} - \text{Gly} - \text{Phe} - \overset{5}{\text{Met}} - \text{Thr} - \text{Ser} - \text{Glu} - \text{Lys} - \overset{10}{\text{Ser}}$$
$$\overset{11}{\text{Gln}} - \text{Thr} - \text{Pro} - \text{Leu} - \overset{15}{\text{Val}} - \text{Thr} - \text{Leu} - \text{Phe} - \text{Lys} - \overset{20}{\text{Asn}}$$
$$\overset{21}{\text{Ala}} - \text{Ile} - \text{Val} - \text{Lys} - \overset{25}{\text{Asn}} - \text{Ala} - \text{His} - \text{Lys} - \text{Lys} - \overset{30}{\text{Gly}}$$
$$\overset{31}{\text{Gln}} -$$

β-Endorphin

Tyr-Gly-Gly-Phe-Met

Methionine-enkephalin (Met-enkephalin)

Tyr-Gly-Gly-Phe-Leu

Leucine-enkephalin (Leu-enkephalin)

Morphine (an opiate)

These substances bind to **opiate receptors** in the brain and have been shown to be their physiological agonists. The role of these so-called **opioid peptides** has yet to be definitively established but it appears they are important in the control of pain and emotional states. Pain relief through the use of acupuncture and placebos as well as such phenomena as "runner's high" may be mediated by opioid peptides.

The posterior lobe of the pituitary, the **neurohypophysis,** which is anatomically distinct from the adenohypophysis, secretes two homologous nonapeptide hormones (Fig. 34-81, *right*); **vasopressin** [also known as **antidiuretic hormone (ADH)**], which increases blood pressure and stimulates the kidneys to retain water; and **oxytocin,** which causes contraction of uterine smooth muscle and therefore induces labor.

$$\overset{1}{\text{Cys}}-\text{Tyr}-\text{Phe}-\text{Gln}-\text{Asn}-\text{Cys}-\text{Pro}-\text{Arg}-\overset{9}{\text{Gly}}-\text{NH}_2$$
$$\underset{\text{S—S}}{\rule{0pt}{0pt}}$$

Human vasopressin

$$\overset{1}{\text{Cys}}-\text{Tyr}-\text{Ile}-\text{Gln}-\text{Asn}-\text{Cys}-\text{Pro}-\text{Leu}-\overset{9}{\text{Gly}}-\text{NH}_2$$
$$\underset{\text{S—S}}{\rule{0pt}{0pt}}$$

Human oxytocin

The rate of vasopressin release is largely controlled by osmoreceptors, which monitor the osmotic pressure of the blood.

B. Second Messengers

Since hormones are chemical signals, their actual chemical identities are themselves of little real significance. What is of importance is how the messages they carry are interpreted; that is, how hormone receptors receive hormonal messages and how they transmit these messages to the cellular machinery. Three classes of receptors are known:

1. Receptors epitomized by steroid receptors: Cytoplasmic and nuclear proteins that upon binding their corresponding hormones are activated as transcription factors for specific proteins (Section 33-3B).

2. Receptors represented by growth factor receptors such as insulin receptor: Membrane-spanning proteins whose cytoplasmic domains are activated as tyrosine-specific protein kinases by hormone binding to their external domains. The tyrosine-specific protein kinases, in turn, modulate the activities of specific cytoplasmic proteins (Section 33-4C).

3. Receptors exemplified by adrenoreceptors: Transmembrane proteins that upon hormone binding stimulate the synthesis and/or release of second messengers that mediate the hormonal signal within the cell.

In the following subsections we consider the formation and actions of second messengers in greater detail than

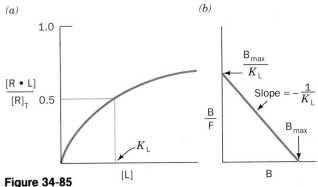

Figure 34-85
The binding of ligand to receptor: (*a*) A hyperbolic plot. (*b*) A Scatchard plot. Here, $B \equiv [R \cdot L]$, $F \equiv [L]$, and $B_{max} \equiv [R]_T$.

we have before. We begin, however, with a discussion of how receptor–ligand interactions are quantified.

Receptor Binding

Receptors, as do other proteins, bind their corresponding ligands (agonists and antagonists) according to the laws of mass action:

$$R + L \rightleftharpoons R \cdot L$$

Here R and L represent receptor and ligand, and the reaction's dissociation constant is expressed:

$$K_L = \frac{[R][L]}{[R \cdot L]} = \frac{([R]_T - [R \cdot L])[L]}{[R \cdot L]} \quad [34.1]$$

where the total receptor concentration, $[R]_T = [R] + [R \cdot L]$. Equation [34.1] may be rearranged to a form analogous to the Michaelis–Menten equation of enzyme kinetics (Section 13-2A):

$$Y = \frac{[R \cdot L]}{[R]_T} = \frac{[L]}{K_L + [L]} \quad [34.2]$$

where Y is the fractional occupation of the ligand-binding sites. Equation [34.2] represents a hyperbolic curve (Fig. 34-85*a*) in which K_L may be operationally defined as the ligand concentration at which the receptor is half-maximally occupied by ligand.

Although K_L and $[R]_T$ may, in principle, be determined from an analysis of a hyperbolic plot such as Fig. 34-85*a*, the analysis of a linear form of the equation is a more accurate procedure. Equation [34.1] may be rearranged to:

$$\frac{[R \cdot L]}{[L]} = \frac{([R]_T - [R \cdot L])}{K_L} \quad [34.3]$$

Now, in keeping with customary receptor-binding nomenclature, let us redefine $[R \cdot L]$ as B (for bound ligand), [L] as F (for free ligand), and $[R]_T$ as B_{max}. Then Eq. [34.3] becomes:

$$\frac{B}{F} = \frac{(B_{max} - B)}{K_L} \quad [34.4]$$

A plot of B/F versus B, which is known as a **Scatchard plot** (after George Scatchard, its originator), therefore

yields a straight line of slope $-1/K_L$ whose intercept on the B axis is B_{max} (Fig. 34-85*b*). Here, both B and F may be determined by filter-binding assays as follows. Most receptors are insoluble membrane-bound proteins and may therefore be separated from soluble free ligand by filtration (receptors that have been solubilized may be separated from free ligand by filtration, for example, through nitrocellulose; recall that proteins nonspecifically bind to nitrocellulose). Hence, through the use of radioactively labeled ligand, the values of B and F ($[R \cdot L]$ and [L]) may be determined, respectively, from the radioactivity on the filter and that remaining in solution. The rate of $R \cdot L$ dissociation is generally so slow (half-times of minutes to hours) as to cause insignificant errors when the filter is washed to remove residual free ligand.

Competitive-Binding Studies

Once the receptor-binding parameters for one ligand have been determined, the dissociation constant of other ligands for the same ligand-binding site may be determined through competitive-binding studies. The model describing this competitive binding is analogous to the competitive inhibition of a Michaelis–Menten enzyme (Section 13-3A):

$$\begin{array}{c} R + L \rightleftharpoons R \cdot L \\ + \\ I \\ \Updownarrow \\ R \cdot I + L \longrightarrow \text{No reaction} \end{array}$$

where I is the competing ligand whose dissociation constant with the receptor is expressed:

$$K_I = \frac{[R][I]}{[R \cdot I]} \quad [34.5]$$

Thus, in direct analogy with the derivation of the equation describing competitive inhibition:

$$[R \cdot L] = \frac{[R]_T [L]}{K_L \left(1 + \dfrac{[I]}{K_I}\right) + [L]} \quad [34.6]$$

The relative affinities of a ligand and an inhibitor may therefore be determined by dividing Eq. [34.6] in the presence of inhibitor with that in the absence of inhibitor:

$$\frac{[R \cdot L]_I}{[R \cdot L]_0} = \frac{K_L + [L]}{K_L \left(1 + \dfrac{[I]}{K_I}\right) + [L]} \quad [34.7]$$

When this ratio is 0.5 (50% inhibition), the competitor concentration is referred to as $[I_{50}]$. Thus, solving Eq. [34.7] for K_I at 50% inhibition:

$$K_I = \frac{[I_{50}]}{1 + \dfrac{[L]}{K_L}} \quad [34.8]$$

Through these procedures it has been shown, for example, that epinephrine tightly binds to β-adrenoreceptors ($K_L = 5 \times 10^{-6}M$), the agonist isoproterenol binds more tightly ($K_L = 0.4 \times 10^{-6}M$), whereas the antagonist propranolol binds yet more tightly ($K_L = 0.0034 \times 10^{-6}M$) even though it fails to activate adenylate cyclase (Section 34-4A). Further studies with a variety of epinephrine analogs indicate that epinephrine's amine function is essential for its tight binding to β-adrenoreceptor, whereas its catechol moiety is required for adenylate cyclase activation.

G Proteins Mediate Adenylate Cyclase Activation and Inhibition

Most adrenoreceptors and polypeptide hormone receptors mediate the formation of cAMP as a second (intracellular) messenger. cAMP, as we have seen for the case of glycogen metabolism (Section 17-3C), activates or inhibits various enzymes or cascades of enzymes by promoting their phosphorylation or dephosphorylation. In Section 33-4C, we saw that when the hormone receptor for such a system binds its cognate hormone, it activates

a membrane-bound G protein·GDP complex by stimulating it to exchange its GDP for GTP. The resulting G protein·GTP complex, in turn, activates adenylate cyclase but this activation is short lived because G protein hydrolyzes GTP to GDP + P_i at a rate of 2 to 3 min^{-1}, and, upon doing so, reverts to its inactive state. Nevertheless, *this system amplifies the hormonal signal because each hormone–receptor complex activates many G proteins before it becomes inactive and, during its lifetime, each G protein·GTP·adenylate cyclase complex catalyzes the formation of much cAMP.*

G protein, as shown by Alfred Gilman, is actually more complex than we previously indicated: It consists of three different subunits, α, β, and γ (45, 37, and 9 kD, respectively), of which it is G_α that binds GDP or GTP (Fig. 34-86). The binding of $G_\alpha \cdot GDP \cdot G_\beta G_\gamma$ to hormone –receptor complex induces G_α to exchange its bound GDP for GTP and, in so doing, dissociate from $G_\beta G_\gamma$. GTP binding also decreases G_α's affinity for activated hormone receptor while increasing its affinity for adenylate cyclase. Thus, *it is the binding of $G_\alpha \cdot GTP$ that activates adenylate cyclase.* Upon the eventual G_α-catalyzed hydrolysis of GTP, the resulting $G_\alpha \cdot GDP$ complex dis-

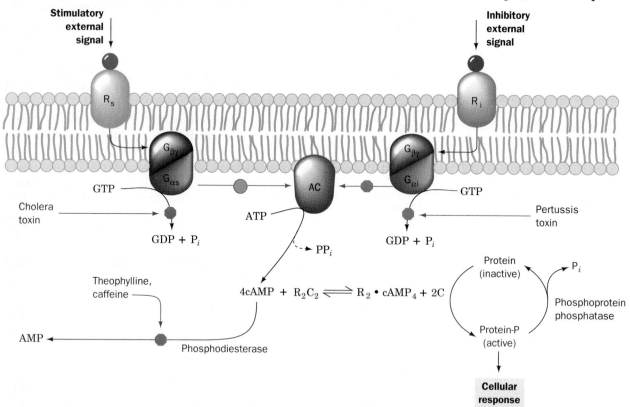

Figure 34-86
The mechanism of receptor-mediated activation/inhibition of adenylate cyclase (AC). The binding of hormone to a stimulatory receptor, R_s (*left*), induces it to bind G_s protein which, in turn, stimulates the $G_{s\alpha}$ subunit of this $G_{s\alpha}G_\beta G_\gamma$ trimer to exchange its bound GDP for GTP. The $G_{s\alpha} \cdot GTP$ complex then dissociates from $G_\beta G_\gamma$ and, until it catalyzes the hydrolysis of the bound GTP to GDP, stimulates

adenylate cyclase to convert ATP to cAMP. The binding of hormone to the inhibitory receptor, R_i (*right*), triggers an almost identical chain of events except that the presence of $G_{i\alpha} \cdot GTP$ complex inhibits AC from synthesizing cAMP. R_2C_2 represents cAMP-dependent protein kinase whose catalytic subunit, C, when activated by the dissociation of the regulatory dimer as $R_2 \cdot cAMP_4$, activates various cellular proteins by catalyzing their phosphorylation.

sociates from adenylate cyclase and reassociates with $G_\beta G_\gamma$ to reform inactive G protein.

Several types of hormone receptors may activate the same G protein. This occurs, for example, in liver cells in response to the binding of the corresponding hormone to glucagon receptors and to β-adrenoreceptors. In such cases, the amount of cAMP produced is the sum of that induced by the individual hormones. G proteins may also act in other ways than by activating adenylate cyclase: They are known, for example, to stimulate the opening of K^+ channels in heart cells and to participate in the phosphoinositide signaling system (see below).

Some hormone receptors inhibit rather than stimulate adenylate cyclase. These include the α_2-adrenoreceptor and receptors for somatostatin and opioids. The inhibitory effect is mediated by "inhibitory" G protein, G_i, which probably has the same β and γ subunits as does "stimulatory" G protein, G_s, but has a different α subunit, $G_{i\alpha}$ (41 kD). G_i acts analogously to G_s in that upon binding to its corresponding hormone–receptor complex, its $G_{i\alpha}$ subunit exchanges bound GDP for GTP and dissociates from $G_\beta G_\gamma$ (Fig. 34-86). $G_{i\alpha}$, however, inhibits rather than activates adenylate kinase through direct interactions and possibly because the liberated $G_\beta G_\gamma$ binds to and sequesters $G_{s\alpha}$. The latter mechanism is supported by the observation that liver cell membranes contain far more G_i than G_s. The activation of G_i in such cells would therefore release enough $G_\beta G_\gamma$ to bind the available $G_{s\alpha}$. Moreover, in some systems, $G_\beta G_\gamma$ itself may act as an intracellular mediator.

G_s and G_i are members of a large family of related signal-transducing G proteins. This family also includes:

1. G_p, which forms a link in the phosphoinositide signaling system (see below).

2. **Transducin (G_T),** which transduces visual stimuli by coupling the light-induced conformational change of the visual pigment rhodopsin to the activation of a phosphodiesterase that hydrolyzes **3′,5′-cyclic GMP (cGMP)** to GMP.

3. **G_0,** a G protein of unknown function that occurs in many tissues, notably brain and heart.

This heterogeneity in G proteins occurs in the β and γ subunits as well as in the α subunits. Indeed, a cell may contain several closely related G proteins of a given type that interact with varying specificities with receptors and effectors. This complex signaling system presumably permits cells to respond in a graded manner to a variety of stimuli.

Receptors Are Subject to Desensitization

One of the hallmarks of biological signaling systems is that they adapt to long-term stimuli by reducing their response to them, a process named **desensitization.**

These signaling systems therefore respond to changes in stimulation levels rather than to their absolute values. What is the mechanism of desensitization? In the case of β-adrenoreceptors, continuous exposure to epinephrine leads to the phosphorylation of one or more of the receptor's Ser residues. This phosphorylation, which is catalyzed by a specific kinase that acts on the hormone–receptor complex but not on the receptor alone, decreases the influence of hormone on G protein, at least in part, by reducing the receptor's epinephrine-binding affinity. The phosphorylated receptors, moreover, are endocytotically sequestered in specialized vesicles that are devoid of both G protein and adenylate cyclase thereby further attenuating the cell's response to epinephrine. If the epinephrine level is reduced, the receptor is slowly dephosphorylated by a phosphorylase and returned to the cell surface, thereby restoring the cell's epinephrine sensitivity.

Cholera Toxin Stimulates Adenylate Cyclase by Permanently Activating G_s Protein

The major symptom of **cholera,** an intestinal disorder caused by the bacterium *Vibrio cholerae,* is massive diarrhea that, if untreated, usually leads to death from dehydration. This dreaded disease is not an infection in the usual sense since the *vibrio* neither invades nor damages tissues but merely colonizes the intestine as does *E. coli.* Rather, the catastrophic fluid loss that cholera induces (over a liter per hour) occurs in response to a bacterial toxin. Indeed, merely replacing cholera victims' lost water and salts enables them to survive the few days necessary to immunologically eliminate the bacterial infestation.

Cholera toxin is an 87-kD protein of subunit composition AB_5 in which the B subunits form a pentagonal ring surrounding the A subunit (Fig. 34-87). Upon binding its cell surface receptor, ganglioside G_{M1} (Sections 11-1D and 23-8C), cholera toxin is taken into the cell via receptor-mediated endocytosis. This process triggers cholera toxin activation by the proteolytic cleavage and disulfide bond reduction of the A subunit to two fragments, A_1 (22 kD) and A_2 (5 kD), and A_1 is released into the cytosol (the B subunits do not penetrate the cell). There, A_1 catalyzes the transfer of the ADP-ribose unit from NAD^+ to an Arg side chain of $G_{s\alpha}$ (Fig. 34-88; recall that diphtheria toxin similarly ADP-ribosylates eukaryotic elongation factor eEF-2; Section 30-3G). *ADP-ribosylated $G_{s\alpha}$ can activate adenylate cyclase but is incapable of hydrolyzing GTP.* As a consequence, the adenylate cyclase remains "locked" in its active state. The epithelial cells of the small intestine normally secrete digestive fluid (an HCO_3^--rich salt solution) in response to small [cAMP] increases. The ~100-fold rise in intracellular [cAMP] induced by cholera toxin causes the symptoms of cholera by inducing these epithelial cells to pour out enormous quantities of digestive fluid. Cholera toxin

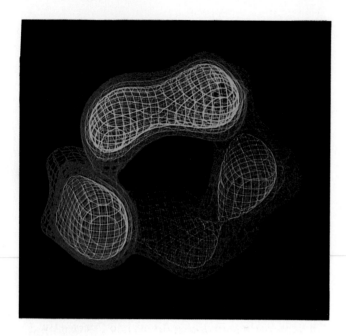

Figure 34-87
The three-dimensional structure of cholera toxin. The structure was determined by the electron micrographic analysis of two-dimensional crystals of cholera toxin that were formed by binding this AB_5 protein to its ganglioside G_{M1} receptor in a lipid membrane. The B subunits form a planar pentagonal ring (here viewed obliquely) surrounding the A subunit. [Courtesy of Hans Ribi, Stanford University School of Medicine.]

also affects other tissues *in vitro* but does not do so *in vivo* because cholera toxin is not absorbed from the gut into the bloodstream.

Bordetella pertussis, the bacterium that causes **pertussis** (whooping cough), produces a 76-kD heterodimeric protein, **pertussis toxin,** that ADP-ribosylates $G_{i\alpha}$. In doing so, it prevents $G_{i\alpha}$ from exchanging its bound GDP for GTP and therefore from inhibiting adenylate cyclase.

The Insulin Receptor Is a Tyrosine Kinase

Although the relationship between insulin and cellular metabolism has been under intense scrutiny since the 1920s (Section 25-3B), it is only with the recent characterization of the **insulin receptor** that we have come to have even a rudimentary understanding of this relationship on the molecular level. The insulin receptor is a trans-membrane glycoprotein that is present on the surfaces of nearly all mammalian cells in from 10^2 to 10^5 copies/cell depending on the tissue. The receptor's tight binding of insulin, a physiological necessity because of the low concentration of insulin in the blood, permitted Pedro Cuatrecasas to isolate it from a detergent-solubilized membrane preparation by affinity chromatography (Section 5-3D) on a column containing covalently bound insulin.

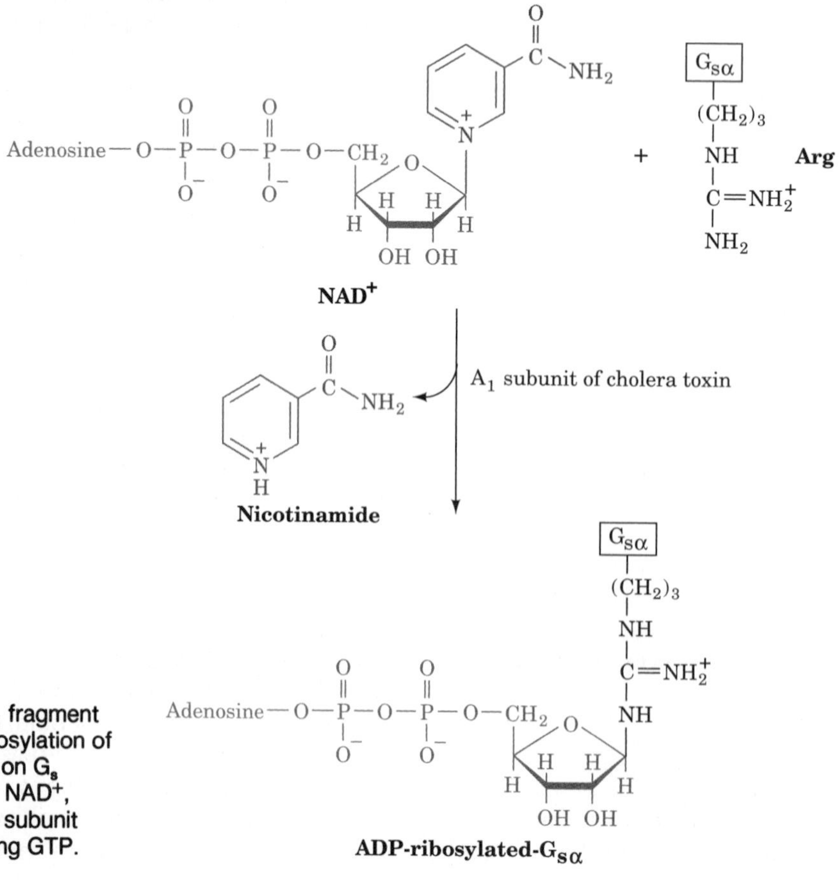

Figure 34-88
The cholera toxin's A_1 fragment catalyzes the ADP-ribosylation of a specific Arg residue on G_s protein's α subunit by NAD^+, thereby rendering this subunit incapable of hydrolyzing GTP.

NAD⁺

Nicotinamide

A_1 subunit of cholera toxin

ADP-ribosylated-$G_{s\alpha}$

The insulin receptor is an $\alpha_2\beta_2$ heterotetramer in which the extracellularly located α subunits (719 residues) and the trans-membrane β subunit (620 residues) are linked by disulfide bonds (Fig. 34-89). The $\alpha\beta$ unit, as deduced from the sequence of its cDNA clone, is synthesized as a single polypeptide chain in which a rapidly removed N-terminal signal sequence is followed by the α subunit sequence, a highly basic tetrapeptide of sequence Arg-Lys-Arg-Arg, which is eventually excised by a processing protease, and finally the β subunit sequence. Segments of this sequence are homologous to those of other growth factor receptors such as **epidermal growth factor (EGF) receptor.**

What are the molecular consequences of binding insulin to the insulin receptor? The first step in this reaction cascade is the autophosphorylation of two specific Tyr residues on each β subunit, which further activates the receptor's tyrosine kinase activity. This protein tyrosine kinase activity, which resembles that of other growth factor receptors, has been shown through mutational inactivation studies to be essential, although not necessarily sufficient, for insulin receptor signal transduction. The way in which the α subunit communicates the presence of bound insulin to the tyrosine kinase site on the β subunit is unknown. Perhaps it is a result of a protein conformational change that is propagated across the cell membrane. Alternatively, the aggregation of insulin receptors, which occurs in response to insulin binding, may somehow activate the tyrosine kinase (many cell-surface receptors are known to aggregate when they bind their corresponding ligands). Whatever the case, the mechanism by which ligand binding activates tyrosine kinase activity is common to other growth factor receptors. Thus, a chimeric protein genetically constructed by joining the insulin receptor's extracellular portion to the trans-membrane and cytoplasmic domains of EGF receptor exhibits tyrosine kinase activity in the presence of insulin.

The succeeding steps in the insulin signaling pathway are obscure. Many cellular enzymes, including glycogen phosphorylase (Section 17-3C), pyruvate dehydrogenase (Section 19-2B), and hormone-sensitive triacylglycerol lipase (Section 23-5) are dephosphorylated in response to insulin stimulation. Yet, insulin also promotes the phosphorylation of specific Ser and Thr residues on numerous other cellular proteins. One explanation of these observations is that the responsible protein kinases and phosphatases are activated by the insulin receptor-catalyzed phosphorylation of either their Tyr or their Ser and Thr residues. However, no protein has been unambiguously identified as a physiological target of the insulin receptor. It has therefore been suggested that the insulin receptor functions by activating protein kinase C of the phosphoinositide signaling system described below.

Ca^{2+}, Inositol Triphosphate, and Diacylglycerol Are Intracellular Second Messengers

Extracellular signals often cause a transient rise in the cytosolic $[Ca^{2+}]$ which, in turn, activates a great variety of enzymes through the intermediacy of calmodulin and its homologs. We have seen, for example, that an increase in cytosolic $[Ca^{2+}]$ triggers such diverse cellular processes as glycogenolysis and muscle contraction. What is the source of this Ca^{2+} and how does it enter the cytosol? In certain types of cells, neurons, for example (Section 34-4C), the Ca^{2+} originates in the extracellular fluid. However, the observation that the absence of extracellular Ca^{2+} does not inhibit certain Ca^{2+}-mediated processes led to the discovery that, in these cases, cytosolic Ca^{2+} is obtained from intracellular reservoirs, specifically specialized regions of the endoplasmic reticulum

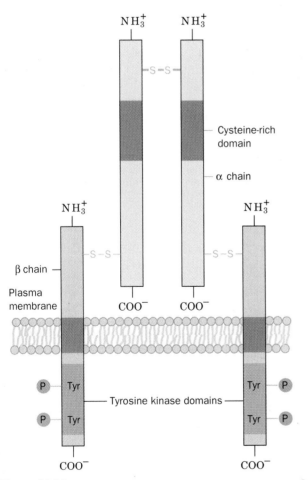

Figure 34-89
The insulin receptor is an $\alpha_2\beta_2$ heterotetramer whose extracellularly attached α subunits contain the insulin-binding sites and whose trans-membrane β subunits contain the tyrosine kinase function on their cytoplasmic sides. [After Ullrich, A., Bell, J. R., Chen, E. Y., Herrera, R., Petruzzelli, L. M., Dull, T. J., Gray, A., Coussens, L., Liao, Y.-C., Tsubokawa, M., Mason, A., Seeburg, P. H., Grunfeld, C., Rosen, O. M., and Ramachandran, J., *Nature* **313,** 760 (1985).]

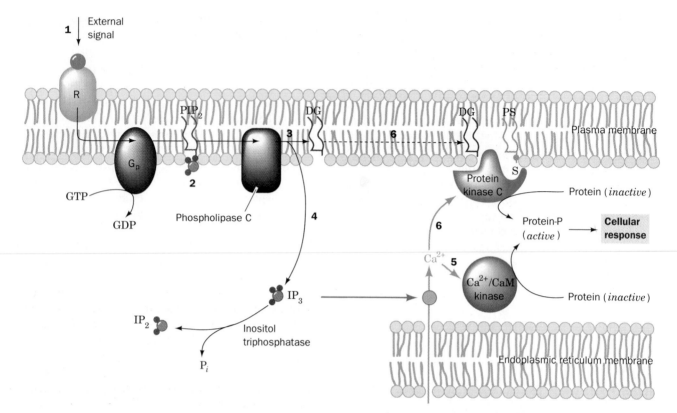

Figure 34-90

The role of PIP$_2$ in intracellular signaling. The (1) binding of agonists to a surface receptor, R, activates phospholipase C through the intermediacy of a G protein (2). Phospholipase C catalyzes the hydrolysis of PIP$_2$ to IP$_3$ and DG (3). The water soluble IP$_3$ stimulates the release of Ca^{2+} sequestered in the endoplasmic reticulum (4) which, in turn, activates numerous cellular processes through the intermediacy of calmodulin and its homologs (5). The nonpolar DG remains associated with the membrane where it activates protein kinase C to phosphorylate and thereby modulate the activities of a number of cellular proteins (6). This latter activation process also requires the presence of the membrane lipid phosphatidylserine (PS) and Ca^{2+}.

named **calcisomes** (the sarcoplasmic reticulum in muscle). Extracellular stimuli leading to Ca^{2+} release must therefore be mediated by an intracellular signal. The first clue as to the nature of this signal came from observations that the intracellular mobilization of Ca^{2+} and the turnover of **phosphatidylinositol-4,5-bisphosphate (PIP$_2$)**, a minor component of the plasma membrane's inner leaflet, are strongly correlated. This information led Robert Michell to propose, in 1975, that PIP$_2$ hydrolysis is somehow associated with Ca^{2+} release.

Investigations, notably by Mabel and Lowell Hokin, Michael Berridge, and Michell, have revealed that *PIP$_2$ is part of an important second messenger system, the phosphoinositide cascade, that mediates the transmission of numerous hormonal signals* including those of vasopressin, CRF, TRF, acetylcholine (a neurotransmitter; Section 34-4C), and epinephrine (with α_1-adrenoreceptors). Remarkably, this system yields up to three separate types of second messengers through the following sequence of events (Fig. 34-90):

1–3. Agonist–receptor interaction, via the intermediacy of the G protein, G$_p$, activates the integral membrane enzyme **phospholipase C** to hydro-

lyze PIP$_2$ to **inositol 1,4,5-triphosphate (IP$_3$)** and **sn-1,2-diacylglycerol (DG)**.

4. The water soluble IP$_3$, acting as a second messenger, diffuses through the cytoplasm to the ER from which it stimulates the release of Ca^{2+} into the cytoplasm, presumably through its interactions with an ER-bound Ca^{2+} transport system.

5. The Ca^{2+}, in turn, stimulates a variety of cellular processes through the intermediacy of calmodulin and its homologs.

6. The nonpolar DG is constrained to remain in the plane of the plasma membrane where it nevertheless also acts as a second messenger by activating **protein kinase C**. This membrane-bound enzyme (actually a large family of enzymes), in turn, phosphorylates and thereby modulates the activities of several different proteins including glycogen synthase (Section 17-3D) and smooth muscle myosin light chains (which, it will be recalled, are also phosphorylated by myosin light chain kinase; Section 34-3D). DG, which is predominantly **1-steroyl-2-arachidonoyl-sn-glycerol**, is further degraded in some cells to yield arachidonate, the obligatory substrate for the biosynthesis of prostaglandins, prostacyclins, thromboxanes, and leukotrienes. These paracrine hormones, as we have seen (Section 23-7), mediate or modulate a wide variety of physiological functions.

In order to be an effective second messenger, a substance must be rapidly eliminated once it has delivered its message. cAMP, for example, is hydrolyzed to AMP by a specific phosphodiesterase (Section 17-3C). Indeed, the methylated purine derivatives **caffeine,** an ingredient of coffee, and **theophylline,** which occurs in tea,

Caffeine
(1,3,7-trimethylxanthine)

Theophylline
(1,3-dimethylxanthine)

are stimulants, in part, because they inhibit this phosphodiesterase.

IP_3 and DG are rapidly recycled to form PIP_2 through the bicyclic metabolic pathway diagrammed in Fig. 34-91. It has been suggested that some of these inositol phosphates, as well as many not appearing in Fig. 34-91, may also act as signal molecules in certain cells, thereby increasing an organism's ability to respond to complex stimuli. It is of interest that the enzyme catalyzing the hydrolysis of **inositol-1-phosphate (IP_1), IP_1 phosphatase,** is inhibited by Li^+. The therapeutic efficacy of Li^+ in controlling the incapacitating mood swings of manic-depressive individuals therefore suggests that this mental illness is caused by an aberration in a phosphoinoside signaling system in the brain, possibly causing abnormal activation of Ca^{2+}-mobilizing receptors.

The activating effects of IP_3 and DG plausibly explain many cellular phenomena. In skeletal muscle, for example, IP_3 mobilizes Ca^{2+} from the sarcoplasmic reticulum. Since Ca^{2+} triggers muscle contraction (Section 34-3C), this observation suggests that nerve impulses mobilize Ca^{2+} by releasing neurotransmitters in the myofibril's T tubules, which then bind to phospholipase C-activating receptors. In a second example, several polypeptide growth factors, including platelet-derived growth factor (PDGF), act to mobilize IP_3 and DG which, in turn, stimulate cell proliferation. The v-*sis* oncogene, which it will be recalled specifies an analog of PDGF (Section 33-4C), may therefore act to permanently switch on PIP_2 degradation thereby forcing the cell into a state of continuous proliferation. Several other oncogene products, including the tyrosine-specific protein kinase specified by *src* (Section 33-4C), are thought to aberrantly activate the synthesis of PIP_2 from its precursors (Fig.

34-91). Similarly, **phorbol esters** such as **12-*O*-tetradecanoylphorbol-13-acetate,**

12-*O*-Tetradecanoylphorbol-13-acetate

which are potent activators of protein kinase C (they structurally resemble DG), are the most effective known **tumor promotors** (substances that are not in themselves carcinogenic but increase the potency of known carcinogens; phorbol esters induce the synthesis of transcription factor AP-1, the product of the *c-jun* proto-oncogene; Section 33-4C). Thus, although the functions of the receptor-mediated PIP_2 degradation system are just beginning to come to light, it has been clearly established that this system plays a central role in the control of cellular metabolism.

C. Neurotransmission

In higher animals, the most rapid and complex intercellular communications are mediated by nerve impulses. Neurons (nerve cells; e.g., Fig. 1-10*d*) electrically transmit these signals along their highly extended lengths (commonly over 1 m in larger animals) as traveling waves of ionic currents. Signal transmission between neurons as well as between neurons and muscles or glands, is usually chemically mediated by neurotransmitters. In the remainder of this section, we discuss both the electrical and chemical aspects of nerve impulse transmission.

Nerve Impulses Are Propagated by Action Potentials

Neurons, like other cells, generate ionic gradients across their plasma membranes through the actions of the corresponding ion-specific pumps. In particular, an (Na^+-K^+)-ATPase (Section 18-3A) pumps K^+ into and Na^+ out of the neuron to yield intracellular and extracellular concentrations of these ions similar to those listed in Table 34-6. The consequent membrane potential $\Delta\Psi$ across a cell membrane is described by the **Goldman equation,** an extension of Eq. [18.3] that explicitly takes into account the various ions' different membrane permeabilities:

$$\Delta\Psi = \frac{RT}{\mathscr{F}} \ln \frac{\Sigma P_c[C(out)] + \Sigma P_a[A(in)]}{\Sigma P_c[C(in)] + \Sigma P_a[A(out)]} \qquad [34.9]$$

Here, C and A represent cations and anions, respectively, and, for the sake of simplicity, we have made the

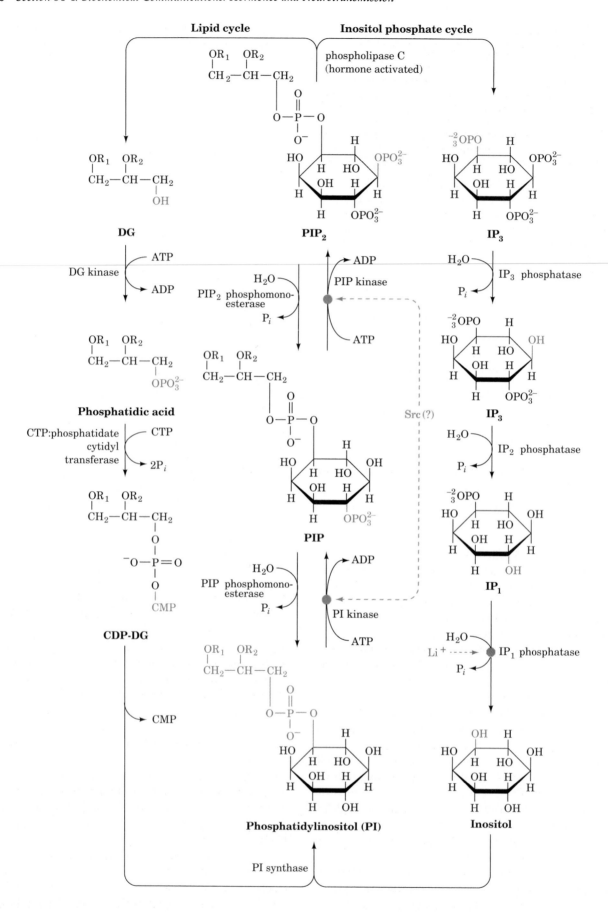

Figure 34-91 (*opposite*)
The metabolic cycles that dispose of IP$_3$ and DG and regenerate PIP$_2$. The IP$_3$ is converted to inositol through three consecutive phosphatase reactions (**inositol phosphate cycle,** *right*), the last of which is inhibited by Li$^+$. The DG is phosphorylated and then activated by cytidylation yielding CDP-DG (lipid cycle, *left*). The inositol and CDP-DG subsequently combine to yield phosphatidylinositol (PI) which, in turn, is phosphorylated in two sequential reactions to yield PIP$_2$ (*center*). These latter two reactions, which are thought to be inappropriately activated by the *src* oncogene product, may both be hydrolytically reversed through the action of the appropriate phosphomonoesterases thereby forming futile cycles that control the PIP$_2$ supply. The DG may also be hydrolyzed by a diacylglycerol lipase to yield monoacylglycerol and arachidonate (R$_2$ is most often an arachidonoyl group), the precursor of prostaglandins and related compounds. [After Berridge, M. J., *Nature* **312,** 316 (1984).]

Table 34-6

Ionic Concentrations and Membrane Permeability Coefficients in Mammals

Ion	Cell (mM)	Blood (mM)	Permeability Coefficient (cm · s^{-1})
K$^+$	139	4	5×10^{-7}
Na$^+$	12	145	5×10^{-9}
Cl$^-$	4	116	1×10^{-8}
X^{-a}	138	9	0

^{a} X$^-$ represents macromolecules that are negatively charged under physiological conditions.

Source: Darnell, J., Lodish, H., and Baltimore, D., *Molecular Cell Biology, pp.* 618 and 725, Scientific American Books (1986).

physiologically reasonable assumption that only monovalent ions have significant concentrations. The quantities P_c and P_a, the respective **permeability coefficients** for the various cations and anions, are indicative of how readily the corresponding ions traverse the membrane (each is equal to the corresponding ion's diffusion coefficient through the membrane divided by the membrane's thickness; Section 18-2A). Note that Eq. [34.9] reduces to Eq. [18.3] if the permeability coefficients of all mobile ions are assumed to be equal.

Applying Eq. [34.9] to the data in Table 34-6 and assuming a temperature of 25°C yields $\Delta\Psi = -83$ mV (negative inside), which is in good agreement with experimentally measured membrane potentials for mammalian cells. This value is somewhat greater than the K$^+$ equilibrium potential, the value of $\Delta\Psi = -91$ mV obtained assuming the membrane is permeable to only K$^+$ ions ($P_{Na^+} = P_{Cl^-} = 0$). The membrane potential is generated by a surprisingly small imbalance in the ionic distribution across the membrane; only ~1 ion pair/million is separated by the membrane with the anion going to the cytoplasmic side and the cation going to the external side. The resulting electric field is, nevertheless, enormous by macroscopic standards: Assuming

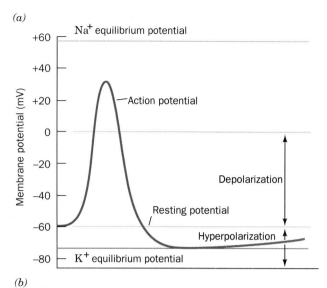

(*a*)

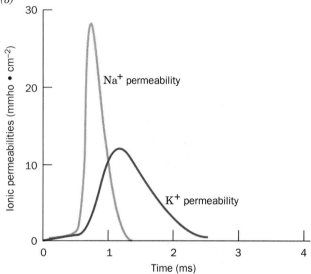

(*b*)

Figure 34-92
The time course of an action potential. (*a*) The axon membrane undergoes rapid depolarization followed by a nearly as rapid hyperpolarization and then a slow recovery to its resting potential. (*b*) The depolarization is caused by a transient increase in Na$^+$ permeability (conductance), whereas the hyperpolarization results from a more prolonged increase in K$^+$ permeability that begins a fraction of a millisecond later. [After Hodgkin, A. L. and Huxley, A. F., *J. Physiol.* **117,** 530 (1952).]

a typical membrane thickness of 50 Å, it is nearly 170,000 V · cm^{-1}.

A nerve impulse consists of a wave of transient membrane depolarization known as an **action potential** *that passes along a nerve cell.* A microelectrode implanted in an **axon** (the long process emanating from the nerve cell body) will record that during the first ~0.5 ms of an action potential, $\Delta\Psi$ increases from its resting potential of around -60 mV to about $+30$ mV (Fig. 34-92*a*). This depolarization is followed by a nearly as rapid repolarization past the resting potential to the K$^+$ equilibrium potential (hyperpolarization) and then a slower recov-

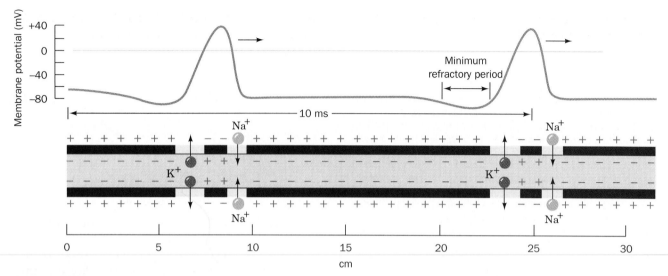

Figure 34-93

Action potential propagation along an axon. Membrane depolarization at the leading edge of an action potential triggers an action potential at the immediately downstream portion of the axon membrane by inducing the opening of its voltage-gated Na$^+$ channels. As the depolarization wave moves further downstream, the Na$^+$ channels close and the K$^+$ channels open to hyperpolarize the membrane. After a brief refractory period, during which the K$^+$ channels close and the hyperpolarized membrane recovers its resting potential, a second impulse can follow. The indicated impulse propagation speed is that measured in the giant axon of the squid which, because of its extraordinary width (~1 mm) is a favorite experimental subject of neurophysiologists. Note that the action potential in this figure appears backwards from that in Fig. 34-92 because this figure shows the distribution of the membrane potential along an axon at an instant in time, whereas Fig. 34-92 shows the membrane potential's variation with time at a fixed point on the axon.

ery to the resting potential. What is the origin of this complicated electrical behavior? In 1953, Alan Hodgkin and Andrew Huxley demonstrated that the action potential results from a transient increase in the membrane's permeability to Na$^+$ followed, within a fraction of a millisecond, by a transient increase in its K$^+$ permeability (Fig. 34-92b).

The ion-specific permeability changes that characterize an action potential result from the presence of trans-axonal membrane proteins that function as Na$^+$- and K$^+$-specific voltage-gated channels (also known as voltage sensitive channels). As a nerve impulse reaches a given patch of nerve cell membrane, the increased membrane potential induces the transient opening of the Na$^+$ channels so that Na$^+$ ions diffuse into the nerve cell at the rate of ~6000 ions · ms^{-1} per channel. This increase in P_{Na+} causes $\Delta\Psi$ to increase (Eq. [34.9]) which, in turn, induces more Na$^+$ channels to open, *etc.*, leading to an explosive entry of Na$^+$ into the cell. Yet, before this process can equilibrate at its Na$^+$ equilibrium potential of around +60 mV, the K$^+$ channels open (P_{K+} increases) while the Na$^+$ channels close (P_{Na+} returns to its resting value). $\Delta\Psi$ therefore reverses sign and overshoots its resting potential to approach its K$^+$ equilibrium value. Eventually the K$^+$ channels also close and the membrane patch regains its resting potential. The Na$^+$ channels, which remain open only 0.5 to 1.0 ms, will not reopen until the membrane has returned to its resting state thereby limiting the axon's firing rate.

An action potential is triggered by an ~20 mV rise in $\Delta\Psi$ to about −40 mV. Action potentials therefore propagate along an axon because the initially rising value of $\Delta\Psi$ in a given patch of axonal membrane triggers the action potential in an adjacent membrane patch that does so in an adjacent membrane patch, etc. (Fig. 34-93). The nerve impulse is thereby continuously amplified so that its signal amplitude remains constant along the length of the axon (in contrast, an electrical impulse traveling down a wire dissipates as a consequence of resistive and capacitive effects). Note, however, that since the relative ion imbalance responsible for the resting membrane potential is small, only a tiny fraction of a nerve cell's Na$^+$-K$^+$ gradient is discharged by a single nerve impulse. An axon can therefore transmit a nerve impulse every few milliseconds without letup. This capacity to fire rapidly is an essential feature of neuronal communications: Since nerve impulses all have the same amplitude, the magnitude of a stimulus is conveyed by the rate that a nerve fires.

The Voltage-Gated Na$^+$ Channel Is the Target of Numerous Neurotoxins

Neurotoxins have proven to be invaluable tools for dissecting the various mechanistic aspects of neurotransmission. Many neurotoxins, as we shall see, interfere with the action of neuronal voltage-gated Na$^+$

channels but, curiously, few are known that affect K^+ channels. **Tetrodotoxin,**

Tetrodotoxin

Saxitoxin

a paralytic poison of enormous potency, which occurs mainly in the skin, ovaries, liver, and intestines of the puffer fish (known in Japan as fugu where it is a delicacy that may only be prepared by chefs certified for their knowledge of puffer fish anatomy), acts by specifically blocking the Na^+ channel. The Na^+ channel is similarly blocked by **saxitoxin,** a product of marine dinoflagellates (a type of plankton known as the "red tide") that is concentrated by filter-feeding shell fish to such an extent that a small mussel can contain sufficient saxitoxin to kill 50 people. Both of these neurotoxins have a cationic guanidino group and both are effective only when applied to the external surface of a neuron (their injection into the cytoplasm elicits no response). It is therefore thought that these toxins specifically interact with an anionic carboxylate group located at the mouth of the Na^+ channel on its extracellular side.

The specific binding of radioactive tetrodotoxin to the detergent-solubilized Na^+ channel has greatly aided in the channel's purification. The Na^+ channel from mammalian brain is a heterotrimeric complex of α (270 kD), β_1 (36 kD), and β_2 (33 kD) subunits. This glycoprotein's large size no doubt reflects the complexity of the tasks it performs: It forms an ion-selective trans-membrane pore that contains two voltage-sensitive "gates," one to open the channel upon membrane depolarization and one to later close it. In fact, minute "gating currents" arising from the movements of these positively charged gates in opening and closing the Na^+ channel can be detected (electrical current is the movement of charge) if the much larger ionic currents through the membrane are first blocked by plugging the Na^+ and K^+ channels with tetrodotoxin and Cs^+ (the K^+ channel can be

blocked from its cytoplasmic side by high concentrations of Cs^+ or tetraethylammonium ion).

Batrachotoxin,

Batrachotoxin

a steroidal alkaloid secreted by the skin of a Columbian arrow-poison frog *Phyllobates aurotaenia*, is the most potent known venom (2-μg/kg body weight is 50% lethal in mice). This substance also specifically binds to the voltage-gated Na^+ channel but, in contrast to the actions of tetrodotoxin and saxitoxin, renders the axonal membrane highly permeably to Na^+. Indeed, batrachotoxin-induced axonal depolarization is reversed by tetrodotoxin. The observation that the repeated electrical stimulation of a neuron enhances the action of batrachotoxin indicates that this toxin binds to the Na^+ channel in its open state.

Venoms from American scorpions contain families of 60- to 70-residue protein neurotoxins that also act to depolarize neurons by binding to their Na^+ channels (Fig. 34-94; the different neurotoxins in the same venom appear to be specialized for binding to the Na^+ channels in the various species the scorpion is likely to encounter). Scorpion toxins and tetrodotoxin do not, however, compete with each other for binding to the Na^+ channel and therefore must bind at separate sites.

Nerve Impulse Velocity Is Increased by Myelination

The axons of the larger vertebrate neurons are sheathed with **myelin,** a biological "electrical insulating tape" that is wrapped about the axon (Fig. 34-95a) so as to electrically isolate it from the extracellular medium. Impulses in myelinated nerves propagate with velocities of up to 100 m $\cdot$ s^{-1}, whereas those in unmyelinated nerves are no faster than 10 m $\cdot$ s^{-1} (imagine the coordination difficulties that say a giraffe would have if it had to rely on only unmyelinated nerves). How does myelination increase the velocity of nerve impulses?

Myelin sheaths are interrupted every millimeter or so along the axon by narrow unmyelinated gaps known as **nodes of Ranvier** (Fig. 34-95b) where the axon contacts the extracellular medium. Binding studies using radioactive tetrodotoxin indicate that the voltage-gated Na^+ channels of unmyelinated axons have rather sparse although uniform distributions in the axonal membrane of $\sim$ 20 channels $\cdot$ μm^{-2}. In contrast, the Na^+ channels of myelinated axons only occur at the nodes of Ranvier where they are concentrated with a density of

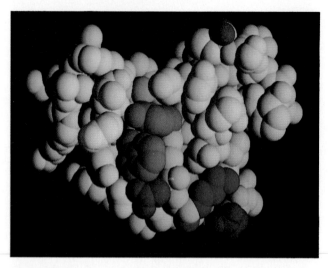

Figure 34-94
The X-ray structure of the 65-residue variant-3 toxin from the Southwestern American scorpion *Centruroides sculpturatus* Ewing. The residues shown in color are conserved among the various scorpion neurotoxins. These residues are clustered on the protein's near surface, which is therefore thought to form the site that binds to the voltage-gated Na⁺ channel. The color code used for the conserved residues only is: Pro and Gly side chains and the main chain are light blue; Asp and Glu side chains and the C-terminus are red; Arg and Lys side chains and the N-terminus are dark blue; uncharged polar side chains are purple; aliphatic side chains are green; and Cys side chains are yellow. [Courtesy of Michael Carson, University of Alabama at Birmingham; X-ray structure determined by Charles Bugg.]

$\sim 10^4 \cdot \mu m^{-2}$. The action potential of a myelinated axon evidently hops between these nodes, a process named **saltatory conduction** (Latin: *saltare*, to jump). Nerve impulse transmission between the nodes must therefore occur by the passive conduction of an ionic current, a mechanism that is inherently much faster than the continuous propagation of an action potential, but which is also dissipative. The nodes act as amplification stations to maintain the intensity of the electrical impulse as it travels down the axon. Without the myelin insulation, the electrical impulse would become too attenuated through trans-membrane ion leakage and capacitive effects to trigger an action potential at the next node. In fact, **multiple sclerosis,** an autoimmune disease that demyelinates nerve fibers in the brain and spinal cord, results in serious and often fatal neurological deficiencies.

Neurotransmitters Relay Nerve Impulses Across Synapses

The junctions at which neurons pass signals to other neurons, muscles, or glands are called **synapses.** In **electrical synapses,** which are specialized for rapid signal transmission, the cells are separated by a gap, the

synaptic cleft, of only 20 Å so that an action potential arriving at the presynaptic side of the cleft can sufficiently depolarize the postsynaptic membrane to directly trigger its action potential. However, the > 200-Å gap of most synapses is too great a distance for such direct electrical coupling. In these **chemical synapses,** the arriving action potential triggers the release from the presynaptic neuron of a specific substance known as a **neurotransmitter,** which diffuses across the cleft and binds to its corresponding receptors on the postsynaptic membrane. In **excitatory synapses,** neurotransmitter binding stimulates membrane depolarization thereby triggering an action potential on the postsynaptic membrane. Conversely, neurotransmitter binding in **inhibitory synapses** alters postsynaptic membrane permeability so as to inhibit an action potential and thus attenuate excitatory signals. What is the mechanism through which an arriving action potential stimulates the release of a neurotransmitter and by what means does its binding to a receptor alter the postsynaptic membrane's permeability? To answer these questions let us consider the workings of **cholinergic synapses;** that is, synapses that use **acetylcholine (ACh)** as a neurotransmitter.

$$CH_3-\overset{\overset{\displaystyle O}{\displaystyle \|}}{C}-CH_2-CH_2-\overset{+}{N}(CH_3)_3$$
Acetylcholine (ACh)

Muscarine

Nicotine

Two types of cholinergic synapses are known:

1. Those containing **nicotinic receptors** (receptors that respond to **nicotine**).

2. Those containing **muscarinic receptors** (receptors that respond to **muscarine,** an alkaloid produced by the poisonous mushroom *Amanita muscaria*).

In what follows, we shall focus on cholinergic synapses containing nicotinic receptors since this best characterized type of synapse occurs at all excitatory neuromuscular junctions in vertebrates and at numerous sites in the nervous system.

Electric Organs of Electric Fish Are Rich Sources of Cholinergic Synapses

The study of synaptic function has been greatly facilitated by the discovery that the homogenization of nerve

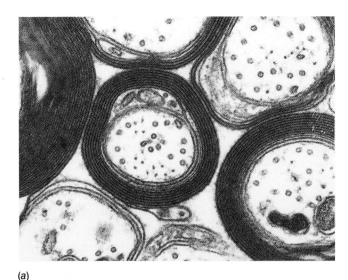

(a)

Figure 34-95

(a) An electron micrograph of myelinated nerve fibers in cross-section. The myelin sheath surrounding an axon is the plasma membrane of a **Schwann cell** which, as it spirally grows around an axon, extrudes its cytoplasm from between the layers. The resulting double bilayer, which makes between 10 and 150 turns about the axon, is a good electrical insulator because of its particularly high (79%) lipid content. [Courtesy of Cedric Raine, Albert Einstein College of Medicine of Yeshiva University.] (b) A schematic diagram of a myelinated axon in longitudinal cross-section indicating that in the nodes of Ranvier (the gaps between adjacent Schwann cells), the axonal membrane is in contact with the extracellular medium. A depolarization generated by an action potential at one node hops, via ionic conduction, down the myelinated axon (*red arrows*), to the neighboring node where it induces a new action potential. Nerve impulses in myelinated axons are therefore transmitted by saltatory conduction.

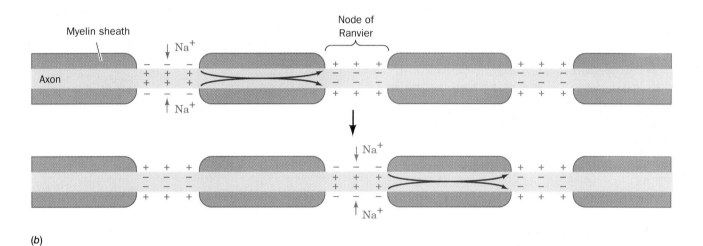

(b)

tissue causes its presynaptic endings to pinch off and reseal to form **synaptosomes.** The use of synaptosomes, which can be readily isolated by density gradient ultracentrifugation, has the advantage that they can be manipulated and analyzed without interference from other neuronal components.

The richest known source of cholinergic synapses is the electric organs of the freshwater electric eel *Electrophorus electricus* and saltwater electric fish of the genus *Torpedo*. Electric organs, which these organisms use to stun or kill their prey, consist of stacks of ~5000 thin flat cells called **electroplaques** that begin their development as muscle cells but ultimately lose their contractile apparatus. One side of an electroplaque is richly innervated and has high electrical resistance, whereas its opposite side lacks innervation and has low electrical resistance. Both sides maintain a resting membrane potential of around −90 mV. Upon neuronal stimulation, all the innervated membranes in a stack of electroplaques simultaneously depolarize to a membrane potential of around +40 mV yielding a potential difference across

each cell of 130 mV (Fig. 34-96). Since the 5000 electroplaques in a stack are "wired" in series like the batteries in a flashlight, the total potential difference across the stack is ~5000 × 0.130 V = 650 V, enough to kill a human being.

Acetylcholine Release Is Triggered by Ca^{2+}

ACh is synthesized near the presynaptic end of a neuron by the transfer of an acetyl group from acetyl-CoA to **choline** in a reaction catalyzed by **choline acetyltransferase.**

$$\underset{\textbf{Acetyl-CoA}}{H_3C-\overset{\overset{\textstyle O}{\|}}{C}-S-CoA} + \underset{\textbf{Choline}}{HO-CH_2-CH_2-\overset{+}{N}(CH_3)_3}$$

$$\Big\downarrow \substack{\text{choline} \\ \text{acetyltransferase}}$$

$$\underset{\textbf{Acetylcholine}}{H_3C-\overset{\overset{\textstyle O}{\|}}{C}-O-CH_2-CH_2-\overset{+}{N}(CH_3)_3} + HS-CoA$$

Much of this ACh is sequestered in ~ 400 Å in diameter

Motor
neuron

Electroplaques

0 mV $\left\{\begin{array}{l}-90 \text{ mV} \\ -90 \text{ mV}\end{array}\right.$

Resting state

$\left.\begin{array}{l}+40 \text{ mV} \\ -90 \text{ mV}\end{array}\right\}$ 130 mV

Depolarized

Figure 34-96
The simultaneous depolarization (*red*) of the innervated membranes in a stack of electroplaques "wired" in series results in a large voltage difference between the two ends of the stack. This is because the total voltage is the sum of the voltages generated by each electroplaque.

synaptic vesicles, which typically contain $\sim 10^4$ ACh molecules each.

*The arrival of an action potential at the presynaptic membrane triggers the opening of **voltage-gated Ca²⁺ channels.** The resulting influx of extracellular Ca²⁺ into the presynaptic terminal, in turn, stimulates the exocytosis of its synaptic vesicles so as to release their packets of ACh into the synaptic cleft (Fig. 34-97).* The black widow spider takes advantage of this system: Its highly neurotoxic venom protein, **α-latrotoxin** (130 kD), causes massive release of ACh at the neuromuscular junction, possibly by acting as a Ca²⁺ ionophore. In contrast, ACh release is inhibited by **botulinus toxin,** a mixture of eight 135 to 170-kD proteins produced by the anaerobic bacterium *Clostridium botulinum* and the agent responsible for the deadly food poisoning syndrome **botulism.**

The hypothesis that ACh is released by exocytosis of synaptic vesicles was proposed in 1952 by Bernard Katz on the basis of his observation that unstimulated neuromuscular junctions exhibit small (0.1–3.0 mV) randomly occurring depolarizing pulses known as **miniature end plate potentials** (the **end plate** is the neuromuscular junctions's postsynaptic membrane). The intensities of these pulses are consistent with the release of $\sim 10^4$ molecules of ACh each, which corresponds to the exocytosis of a single synaptic vesicle. A full end plate potential presumably results from the simultaneous release of as many as 400 such ACh packets. Although the mechanism of Ca²⁺-stimulated synaptic vesicle exocytosis is unknown, it is likely to involve **synapsin I,** a 75-kD synaptic vesicle membrane protein which, in brain, is a major substrate for Ca²⁺-calmodulin and cAMP-dependent protein kinases. Activated synapsin I presumably promotes fusion of the synaptic vesicle membrane with the plasma membrane.

Acetylcholine Receptor Is a Ligand-Gated Cation Channel

The **acetylcholine receptor** is an ~ 250-kD $\alpha_2\beta\gamma\delta$ trans-membrane glycoprotein whose four different sub-

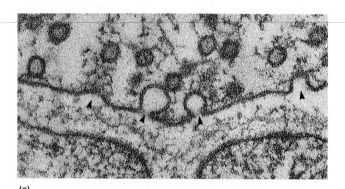

(a)

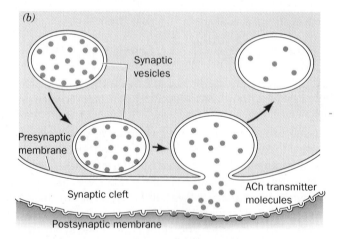

(b)

Synaptic
vesicles

Presynaptic
membrane

Synaptic cleft

ACh transmitter
molecules

Postsynaptic membrane

Figure 34-97
The transmission of nerve impulses across a synaptic cleft. (*a*) An electron micrograph of a frog neuromuscular junction in which synaptic vesicles are undergoing exocytosis (*arrows*) with the presynaptic membrane (*top*). [Courtesy of John Heuser, Washington University School of Medicine.] (*b*) This process discharges the ACh contents of the synaptic vesicles into the synaptic cleft where it diffuses in $<100 \mu s$ to the postsynaptic membrane and binds to its receptor. The synaptic vesicle membranes are later reclaimed by endocytosis and refilled with ACh.

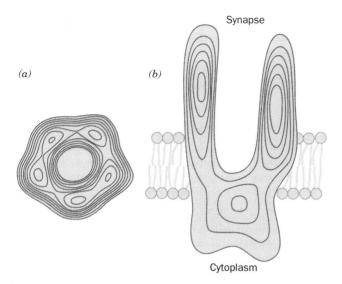

Figure 34-98
The structure of the acetylcholine receptor as determined through electron microscopic analysis of its two-dimensional crystals. *(a)* Section through the three-dimensional image parallel to the cell membrane and near its synaptic end. Note its nearly fivefold symmetry. *(b)* Section perpendicular to the cell membrane and through the center of the projection in Part *a*. The receptor's central channel is 25 to 30 Å in diameter. [After Brison, A. and Unwin, P. N., *Nature* **315**, 477, 476 (1985).]

units have similar sequences. X-ray diffraction and electron microscopy studies indicate that the ACh receptor is an 80 Å in diameter by 140-Å long molecule that protrudes ~55 Å into the synaptic space and ~15 Å into the cytoplasm (Fig. 34-98). Its five rodlike subunits are arranged with quasi-fivefold symmetry over much of their length. The ACh receptor's most striking structural feature is a 25 to 30 Å in diameter water-filled central channel that extends from the receptor's synaptic entrance to the level of the lipid bilayer where it becomes too narrow to be discerned. *The binding of two ACh molecules, one per α subunit, allosterically induces the opening of the channel through the bilayer to permit Na⁺ and K⁺ ions to diffuse, respectively, in and out of the cell at rates of ~20,000 of each type of ion per millisecond. The resulting depolarization of the postsynaptic membrane initiates a new action potential.* After 1 to 2 ms, the ACh spontaneously dissociates from the receptor and the channel closes.

The ACh receptor is the target of some of the most deadly known neurotoxins (death occurs through respiratory arrest) whose use has greatly aided in the elucidation of receptor function. **Histrionicatoxin,** an alkaloid secreted by the skin of the Columbian arrow-poison frog *Dendrobates histrionicus,* and *d*-**tubocurarine,** the active ingredient of the Amazonian arrow poison **curare** as well as a medically useful paralytic agent, are both ACh antagonists that prevent ACh receptor channel opening.

Histrionicatoxin

d-**Tubocurarine**

Similarly, a family of homologous 7 to 8-kD venom proteins from some of the worlds most poisonous snakes, including **α-bungarotoxin** from snakes of the genus *Bungarus,* **erabutoxin** from sea snakes, and **cobratoxin** from cobras (Fig. 34-99) prevent ACh receptor channel opening by binding specifically and all but irreversibly to its α subunits. Indeed, detergent-solubilized

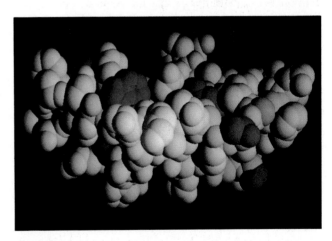

Figure 34-99
The X-ray structure of α-cobratoxin, a 71-residue neurotoxin from the venom of the cobra *Naja naja siamensis,* which specifically binds to the ACh receptor so as to inhibit its opening. The related snake venom neurotoxins α-bungarotoxin and erabutoxin have similar X-ray structures. The conserved residues thought to be essential for the toxicity of these neurotoxins are colored as is indicated in the legend to Fig. 34-94. [Courtesy of Michael Carson, University of Alabama at Birmingham; X-ray structure determined by Wolfram Saenger.]

ACh receptor has been purified by affinity chromatography on a column containing covalently attached cobratoxin.

Acetylcholine Is Rapidly Degraded by Acetylcholine Esterase

An ACh molecule that participates in the transmission of a given nerve impulse must be degraded in the few milliseconds before the potential arrival of the next nerve impulse. This essential task is accomplished by ***acetylcholinesterase****, a 75-kD fast acting enzyme that is bound to the surface of the postsynaptic membrane*

$$H_3C-\overset{\overset{\text{O}}{\|}}{C}-O-CH_2-CH_2-\overset{+}{N}(CH_3)_3 \ + \ H_2O$$
Acetylcholine

$$\downarrow \text{acetylcholinesterase}$$

$$H_3C-\overset{\overset{\text{O}}{\|}}{C}-O^- + \ HO-CH_2-CH_2-\overset{+}{N}(CH_3)_3 \ + \ H^+$$
Acetate   **Choline**

(the turnover number of acetylcholinesterase is $k_{cat} = 14{,}000 \ s^{-1}$; the enzyme's catalytic efficiency, $k_{cat}/K_M = 1.5 \times 10^8 M^{-1} \cdot s^{-1}$, is close to the diffusion-controlled limit so that it is a nearly perfect catalyst; Section 13-2B). The products, acetate and choline, are transported back into the presynaptic terminal for recycling to ACh.

Acetylcholinesterase is a serine esterase; that is, its catalytic mechanism resembles that of serine proteases such as trypsin. These enzymes, as we have seen in Section 14-3A, are irreversibly inhibited by alkylphosphofluoridates such as diisopropylphosphofluoridate (DIPF). Indeed, related compounds such as **tabun** and **sarin**

Tabun   **Sarin**

are military nerve gases because their efficient inactivation of human acetylcholinesterase causes paralysis stemming from cholinergic nerve impulse blockade and thus death by suffocation. **Succinylcholine,**

Succinylcholine

which is used as a muscle relaxant during surgery, is an ACh agonist that although rapidly released by ACh re-

ceptor, is but slowly hydrolyzed by acetylcholinesterase. Succinylcholine therefore produces persistent end plate depolarization. Its effects are short lived, however, because it is rapidly hydrolyzed by the relatively nonspecific liver and plasma enzyme **butyrylcholinesterase.**

Amino Acids and Their Derivatives Function as Neurotransmitters

The mammalian nervous system employs well over 30 substances as neurotransmitters. Some of these substances, such as glycine and glutamate, are amino acids; many others are amino acid decarboxylation products or their derivatives (often referred to as **biogenic amines**). Thus, as we saw in Section 24-4B, the catecholamines dopamine, norepinephrine, and epinephrine are sequentially synthesized from tyrosine, whereas γ-aminobutyric acid (GABA), histamine, and serotonin are derived from aspartate, histidine, and tryptophan, respectively (Fig. 34-100). You will recognize many of these compounds as hormonally active substances that are present in the blood stream. However, since the brain is largely isolated from the general circulation by a selective filtration system known as the **blood–brain barrier,** the presence of these substances in the blood has no direct effect on the brain. The use of the same compounds as hormones and neurotransmitters apparently has no physiological significance but, rather, is thought to reflect evolutionary opportunism in adapting already available systems to new roles.

The use of selective staining techniques has established that each of the different neurotransmitters is used in discrete and often highly localized regions of the nervous system. The various neurotransmitters are, nevertheless, not simply functional equivalents of acetylcholine. Rather, many of them have distinctive physiological roles. For example, both GABA and glycine are inhibitory rather than excitatory neurotransmitters. The receptors for these substances are ligand-gated channels that are selectively permeable to Cl^- so that their opening tends to hyperpolarize the membrane (make its membrane potential more negative) rather than depolarize it. A neuron inhibited in this manner must therefore be more intensely depolarized than otherwise to trigger an action potential (note that these neurons respond to more than one type of neurotransmitter). Ethanol, man's oldest and most widely used psychoactive drug, is thought to act by inducing GABA receptors in the brain to open their Cl^- channels.

The actual nature of a neuron's response to a neurotransmitter depends more on the characteristics of the corresponding receptor than on the neurotransmitter's identity. Thus, as we have seen, nicotinic ACh receptors, which trigger the rapid contraction of skeletal muscles, respond to ACh within a few milliseconds by depolarizing their postsynaptic membrane. In contrast, the

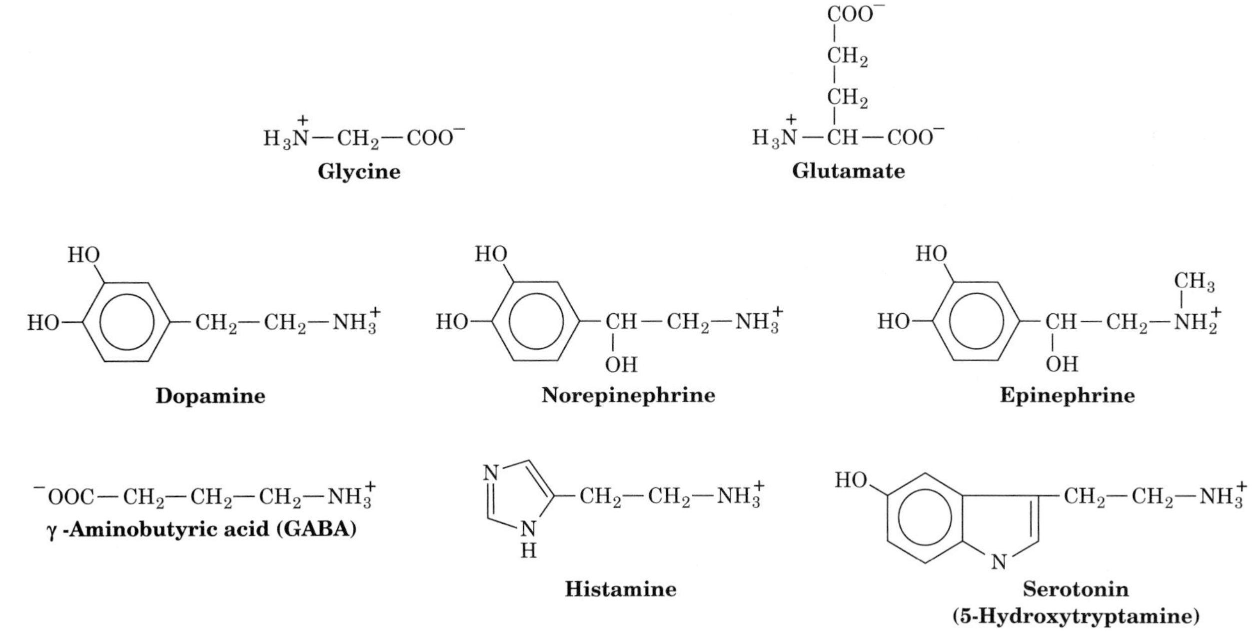

Figure 34-100
A selection of neurotransmitters.

binding of ACh to muscarinic ACh receptors in heart muscle inhibits muscle contraction over a period of several seconds (several heart beats). This is accomplished by hyperpolarizing the postsynaptic membrane through the closure of otherwise open K^+ channels. Slow-acting neurotransmitters may act by inducing the formation of a second messenger such as cAMP. In fact, the brain has the highest concentration of cAMP-dependent kinases in the body. The binding of catecholamines to their respective neuronal receptors, through the intermediacy of adenylate cyclase and cAMP, evidently activates protein kinases to phosphorylate ion channels so as to alter the neuron's electrical properties. The ultimate effect of this process can be either excitatory or inhibitory. Note that catecholamines, whether acting as hormones or neurotransmitters, have similar mechanisms of receptor activation.

Neuropeptides Are Neurotransmitters

A large and growing list of hormonally active polypeptides also act as neurotransmitters. Not surprisingly, perhaps, the opioid peptides β-endorphin, met-enkephalin, and leu-enkephalin, as well as the hypothalamic releasing factors TRF, GnRF, and somatostatin, are in this category. What is less expected is that several gastrointestinal polypeptides including the hormones gastrin, secretin, and cholecystokinin (CCK), may also act as neurotransmitters in discrete regions of the brain as do the pituitary hormones oxytocin, vasopressin, and possibly calcitonin-gene-related protein (CGRP; whose translation is described in Section 33-3C). such **neuropeptides** differ from the simpler neurotransmitters in that they seem to elicit complex behavior patterns. For example, intracranially injecting rats with a nanogram of vasopressin greatly enhances their ability to learn and remember new tasks. Similarly, injecting a male or a female rat with GnRF evokes the respective postures they require for copulation. Just how these neuropeptides operate is but one of the many enigmas of brain function and organization.

Chapter Summary

Blood Clotting

Blood coagulation occurs through a double cascade of proteolytic reactions, which result in the formation of cross-linked fibrin clots. Fibrin, which is formed from fibrinogen by the thrombin-catalyzed proteolytic excision of this soluble precursor's fibrinopeptides, self-associates to form a fibrous meshwork of half-staggered elongated molecules. The clot is later cross-linked by the formation of intermolecular Gln-Lys iso-

peptide bonds in a reaction catalyzed by the thrombin-activated fibrin-stabilizing factor (FSF or XIII_a).

Thrombin, which like all the proteolytic clotting factors is a serine protease, is synthesized as a zymogen that is proteolytically activated by Stuart factor (X_a). Thrombin, as do several other clotting factors, contains a number of γ-carboxylglutamate (Gla) residues whose post-translational synthesis from

Glu residues requires a vitamin K cofactor. The Gla residues chelate Ca^{2+} whose presence is required for prothrombin activation together with thrombin-activated proaccelerin (V_a) and a phosphatidylserine-containing membrane surface. The Gla residues apparently function to anchor X_a to the membrane via Ca^{2+} bridges.

Factor X may be activated either through the intrinsic pathway or the extrinsic pathway. In the intrinsic pathway, clotting is initiated by the contact system in which Hageman factor (XII), in the presence of high-molecular-weight kininogen (HMK), is activated by adsorption to a negatively charged surface such as glass to proteolyze prekallikrein to kallikrein. Kallikrein then reciprocally proteolyzes XII which, in the presence of HMK, proteolytically activates plasma thromboplastin antecedent (PTA or XI). In the final two steps of the intrinsic pathway, XI_a proteolytically activates Christmas factor (IX) which, in turn, proteolytically activates X in the presence of thrombin-activated antihemophilic factor (VIII_a). Both of these latter reactions must also take place on a phospholipid membrane surface in the presence of Ca^{2+}. The extrinsic pathway begins with the proteolytic activation of proconvertin (VII) by either XII_a or thrombin. VII_a, in turn, mediates the activation of X in the presence of Ca^{2+}, phospholipid membrane and tissue factor (III), a membrane glycoprotein that occurs in many tissues.

Clot formation is inhibited by numerous physiological mechanisms including the inactivation of all of the clotting system proteases but VII_a by the binding of antithrombin III. The presence of heparin, a component of many tissues, greatly stimulates this reaction. Similarly, the thrombin-activated protein C proteolytically inactivates V and VIII. Thrombin, in turn, is activated to activate protein C by binding to the cell surface protein thrombomodulin. Once wound healing is underway, clots are dismantled by the serine protease plasmin, a process termed fibrinolysis. Plasmin is formed by the proteolysis of plasminogen by several serine proteases notably urokinase and tissue-type plasminogen activator (t-PA). This process is limited through the inhibition of plasmin by α_2-antiplasmin.

Immunity

The immune response, which is conferred by lymphocytes in association with lymphoid tissues such as the thymus and the lymph nodes, results from cellular and humoral immunity. Cellular immunity, which guards against parasites, virally infected cells, cancers, and foreign tissue, is mediated by *T* cells, whereas humoral immunity, which is most effective against bacterial infections and extracellular viruses, is mediated by antibodies produced by *B* cells. A *T* cell is selected for proliferation if its *T* cell receptor simultaneously binds a foreign antigen and the host's Class I MHC protein. Some of the progeny, cytotoxic *T* cells, bind to antigen-bearing host cells and kill them by inserting pore-forming proteins in their plasma membranes. A *B* cell displaying an antibody that binds to a foreign antigen is similarly selected for proliferation and its plasma cell progeny secrete large amounts of that antibody. The immune system must be self-tolerant; failure to prevent the proliferation of *T* and *B* cells bearing antibodies against self-antigens results in autoimmune diseases.

Antibodies (immunoglobulins) are glycoproteins that consist of two identical light (L) chains and two identical heavy (H) chains which, in turn, each have a constant (C) region and a variable (V) region. The five classes of secreted immunoglobulins, IgA, IgD, IgE, IgG, and IgM, have different physiological functions and vary only in the identities of their H chains whose differences mainly affect the antibodies' Fc segments. The L chain in all of these immunoglobulins is either a κ or λ chain. An immunoglobulin's two identical Fab segments, which each consist of one C and one V domain from both the H and the L chains, contain the antibody's antigen-binding sites. The antigen-binding specificity of an immunoglobulin is largely dependent on the sequences of the hypervariable segments from both its H and L chains that line its antigen-binding sites. X-ray studies show that L and H chains are folded, respectively, into two and four domains that each have the characteristic immunoglobulin fold.

The immune system generates a virtually unlimited variety of antigen-binding sites through both somatic recombination and somatic mutation. The κ light chain is encoded by four exons known as the leader (L_κ), variable (V_κ), joining (J_κ), and constant (C_κ) segments. During *B* cell differentiation, one of the ~ 300 $L_\kappa + V_\kappa$ units contained in the embryonic human genome somatically recombines with one of the genome's five J_κ segments. Then, in later cell generations, the resulting $L_\kappa - V_\kappa - J_\kappa$ unit is transcribed and selectively spliced to the genome's single C_κ segment yielding κ chain mRNA. The V/J recombination joint in this process is not precisely defined leading to further variation in the κ chain. Heavy chain genes are similarly assembled but, in addition, have a *D* segment between their V_H and J_H segments that leads to even greater heavy chain diversity. Somatic mutation provides yet more diversity: Nucleotides may be added at random at the V_H/D and D/J_H joints through the action of terminal deoxynucleotidyl transferase and, furthermore, both heavy and light chain genes in *B* cell progenitors are subject to somatic hypermutation. *B* cells express only one heavy chain and one light chain allele, a phenomenon named as allelic exclusion, thereby ensuring that each cell expresses only one species of immunoglobulin. Differentiating *B* cells switch from the synthesis of membrane bound IgM, whose heavy chain has a hydrophobic trans-membrane tail, to a secreted IgM with the same antigen specificity but which lacks this C-terminal polypeptide. This switch occurs via the selection of alternative polyadenylation sites located before or after the exons specifying the trans-membrane tail. *B* cell progeny also progressively switch from the synthesis of IgM to other classes of immunoglobulins, a process known as class switching, through either selective splicing or somatic recombination leading to the expression of alternative C_H segments.

T cell receptors resemble immunoglobulins in that they consist of α and β chains that have constant and variable domains, each of which assumes the immunoglobulin fold. *T* cell receptor diversity is generated by the somatic selection of different V, J, and for α chains, D regions.

The major histocompatibility complex encodes a highly polymorphic group of membrane-bound proteins that act as individuality markers among members of the same species (Class I MHC proteins) and differentiate immune system cells from other body cells (Class II MHC proteins). MHC proteins have domains that structurally resemble those in immunoglobulins and *T* cell receptors and therefore the genes encoding all these proteins form a gene superfamily. A polypeptide fragment of the processed antigen is presented to *T* cell receptors in complex with the Class I MHC protein through its binding in the groove formed by the Class I MHC protein's α_1 and α_2 domains. The MHC gene's polymorphism is thought to prevent pathogens from evolving antigens that interact poorly with a particular MHC protein during the antigen recognition process.

The complement system defends against foreign invaders by killing foreign cells through complement fixation, inducing the phagocytosis of foreign particles (opsonization), and triggering local acute inflammatory reactions. The complement system consists of ~20 proteins that interact in the antibody-dependent classical pathway and in the antibody-independent alternative pathway. The classical pathway contains three sequentially activated protein complexes: The recognition unit, which assembles on cell surface-bound antibody–antigen complexes; the activation unit, which amplifies the recognition process through a proteolytic cascade involving a series of serine proteases; and the membrane attack complex (MAC), which punctures the antibody-marked cell's plasma membrane causing cell lysis and death. The alternative pathway, which is thought to defend against invading microorganisms before an effective immune response can be mounted, also leads to the assembly of the MAC but in a series of reactions that are activated by the presence of certain bacterially synthesized polymers, whole bacteria, and host cells that are infected by certain viruses. The complement system is tightly regulated by the structural instabilities of certain activated complement proteins, by the degradation of complement components through the actions of specific proteases such as C4b binding protein, and by the sequestering of complement components by their specific binding of proteins such as $\overline{\text{C1}}$ inhibitor and S protein.

Motility

Skeletal muscle fibers consist of banded myofibrils which are, in turn, comprised of interdigitated thick and thin filaments. The thick filaments are composed almost entirely of myosin, a dimeric protein with two globular heads and an elongated rodlike segment comprised of two α helices in a coiled coil. The myosin molecules aggregate end to end in a regular staggered array to form the bipolar thick filament. The myosin head is an ATPase. The thin filaments consist mainly of actin, a globular protein (G-actin) that polymerizes to form a right-handed double helical filament (F-actin) in which each monomer unit is capable of binding a single myosin head. The thin filament also contains two other proteins, tropomyosin and troponin. Tropomyosin is a heterodimeric protein that consists mainly of a coiled coil of two α helices that is wound in the grooves of the F-actin helix. Troponin consists of three subunits: TnC, a Ca^{2+}-binding calmodulin homolog; TnI, which binds actin; and TnT, which binds tropomyosin. The troponin–tropomysin complex regulates muscle contraction by varying the access of myosin heads to their actin-binding sites in response to the concentration of Ca^{2+}.

Structural studies indicate that the thick and thin filaments slide past each other during muscle contraction. Tension is generated through a four part reaction cycle: (1) ATP binding to a myosin head of a thick filament causes it to dissociate from an actin monomer unit of a neighboring thin filament; (2) the myosin-catalyzed hydrolysis of ATP to ADP + P_i conformationally "cocks" the myosin head by causing it to assume an orientation nearly perpendicular to the thick filament; (3) under stimulation to contract, the myosin head–ADP–P_i complex attaches to an adjacent actin unit; and (4) the actin–myosin interaction, in what is the cycle's power stroke, induces the sequential release of P_i and ADP with the concomitant relaxation of the myosin to its tilted conformation so as to pull the thick filament along the thin filament. Repeated such cycles cause the myosin heads to "walk" up the adjacent thin filaments resulting in muscle contraction.

Muscle contraction is triggered by an increase in $[Ca^{2+}]$. The Ca^{2+} binds to the TnC subunit of troponin with the resulting conformational change causing tropomyosin to move deeper into the thin filament groove, thereby exposing actin's myosin head-binding sites, and thus switching on muscle contraction. The Ca^{2+} is released from the sarcoplasmic reticulum (SR) in response to nerve impulses that render the SR membrane permeable to Ca^{2+}. The cytosolic $[Ca^{2+}]$ is otherwise maintained at a very low level through the action of SR membrane-bound Ca^{2+}-ATPases, which pump the Ca^{2+} into the SR thus terminating muscle contraction.

Smooth muscle, which is responsible for long-lasting and involuntary contractions, lacks the banded pattern of skeletal muscles. Its myosin heads only interact with actin when one of their light chains is phosphorylated at a specific Ser residue. Smooth muscle contraction is nevertheless triggered by Ca^{2+} because myosin light chain kinase, the enzyme that catalyzes the phosphorylation of myosin light chains, is only active when associated with Ca^{2+}-calmodulin. Myosin light chain phosphatase hydrolyzes myosin's activating phosphate group so that in the absence of active myosin light chain kinase, smooth muscle relaxation ensues. Nerve impulses increase the permeability of the smooth muscle cell plasma membrane to Ca^{2+}, which acts as an intracellular second messenger in stimulating smooth muscle contraction. Smooth muscles also respond to hormones such as epinephrine through the intermediacy of cAMP whose presence activates a protein kinase to phosphorylate myosin light chains.

Actin and myosin are also prominent in nonmuscle cells where they have both structural and functional roles. Nonmuscle actin is generally in a state of equilibrium between its monomeric G-actin form and polymeric F-actin microfilaments. The assembly and disassembly of microfilaments, as influenced by the presence of actin-binding proteins such as profilin, has an important role in cellular motility. Nonmuscle myosin also forms thick filaments which, in concert with microfilaments, participate in intracellular contractile processes such as the tightening of the contractile ring during cell division.

Ciliary motion is a microtubule-based phenomenon. Microtubules are formed from the protein tubulin, an $\alpha\beta$ dimer, that polymerizes with the concomitant hydrolysis of GTP. In cilia and eukaryotic flagella, the microtubules are arranged in a 9 + 2 array in which 9 double microtubules surround 2 single microtubules in an assembly that is cross-linked by three types of proteins. Subfibers A of the outer fibers each bear two dynein arms that "walk" up neighboring subfibers B in an ATP-powered process. However, the cross-links between neighboring fibers prevent these fibers from sliding past each other; rather, the cilia bend, which accounts for their oarlike motion. Cytosolic dyneins and kinesin oppositely motivate the transport of vesicles along microtubule tracks.

Bacterial flagella, which are responsible for bacterial propulsion, are entirely different from eukaryotic flagella. Bacterial flagella consist of a flagellin filament, a flagellar hook made of hook protein, and a complex basal body that is embedded in the bacterial plasma membrane. The basal body is a true rotary motor. Thus, the flagellar filament and hook are passive elements that, like a propeller, convert the rotary motion of the basal body to linear thrust. The basal body's rotary motion is directly powered by the discharge of the metabolically generated electrochemical proton gradient across the plasma membrane.

Biochemical Communications

Chemical messengers are classified as autocrine, paracrine, or endocrine hormones if they act on the same cell, cells that are nearby, or cells that are distant from the cell that secreted them. The body contains a complex endocrine system that controls many aspects of its metabolism. The pancreatic islet cells secrete insulin and glucagon, polypeptide hormones that induce liver and adipose tissue to store or release glucose and fat, respectively. Gastrointestinal polypeptide hormones coordinate various aspects of digestion. The thyroid hormones, T_3 and T_4, are iodinated amino acid derivatives that generally stimulate metabolism by activating cellular transcription factors. Ca^{2+} metabolism is regulated by the levels of PTH, vitamin D, and calcitonin. PTH and vitamin D induce an increase in blood $[Ca^{2+}]$ by stimulating Ca^{2+} release from bone and its adsorption from kidney and intestine, whereas calcitonin has the opposite effects. Vitamin D is a steroid derivative that must be obtained in the diet or by exposure to UV radiation. Vitamin D, after being sequentially processed in the liver and kidney to $1,25(OH)_2D$, stimulates the synthesis of a Ca^{2+}-binding protein in the intestinal epithelium. The adrenal medulla secretes the catecholamines epinephrine and norepinephrine, which bind to α- and β-adrenoreceptors on a great variety of cells so as to prepare the body for "fight or flight." The adrenal cortex secretes glucocorticoid and mineralocorticoid steroids. Glucocorticoids affect metabolism in a manner opposite to that of insulin as well as mediating a wide variety of other vital functions. Mineralocorticoids regulate the excretion of salt and water by kidney. The gonads secrete steroid sex hormones, the androgens (male hormones), and estrogens (female hormones), which regulate sexual differentiation, the development of secondary sex characteristics, and sexual behavior patterns. Ovaries, in addition, secrete progestins that help mediate the menstrual cycle and pregnancy.

The hypothalamus secretes a series of polypeptide releasing factors and release-inhibiting factors such as CRF, TRF, GnRF, and somatostatin that control the secretion of the corresponding trophic hormones from the adenohypophysis. Most of these trophic hormones, such a ACTH, TSH, LH, and FSH, stimulate their target endocrine glands to secrete the corresponding hormones. Growth hormone acts directly on tissues as well as stimulating liver to synthesize growth factors known as somatomedins. The menstrual cycle results from a complex interplay of hypothalamic, adenohypophyseal, and steroid sex hormones. The adenohypophysis also secretes opioid peptides that have opiate-like effects on the central nervous system. The neurohypophysis secretes the polypeptides vasopressin, which stimulates the kidneys to retain water, and oxytocin, which stimulates uterine contractions.

Receptors are membrane-bound proteins that bind their ligands according to the laws of mass action. The parameters describing the binding of a radiolabeled ligand to its receptor can be determined from Scatchard plots. The dissociation constants of additional ligands for the same receptor-binding site can then be determined through competitive binding studies. The intracellular effects of most polypeptide and catecholamine hormones are mediated by second messengers such as cAMP. Hormone binding to certain stimulatory receptors activates the $G_{s\alpha}$ subunit of a stimulatory G protein to replace its bound GDP with GTP, release its associated $G_\beta G_\gamma$ subunits, and activate adenylate cyclase to synthesize cAMP. Activation continues until $G_{s\alpha}$ spontaneously hydrolyzes its bound GTP to GDP and recombines with $G_\beta G_\gamma$. Several types of activated hormone receptors in a cell may stimulate the same G_s protein. There are also inhibitory G proteins, which have the same G_β and G_γ subunits as does G_s, but which have an inhibitory $G_{i\alpha}$ subunit that inactivates adenylate cyclase. Cholera toxin induces uncontrolled cAMP production by ADP-ribosylating $G_{s\alpha}$ so as to render it incapable of hydrolyzing GTP. PIP_2, a minor phospholipid component of the plasma membrane's inner leaflet, can yield up to three types of second messengers. Agonist–receptor interactions, through the intermediacy of a G protein, stimulate phospholipase C to hydrolyze PIP_2 to IP_3 and DG. The IP_3 stimulates the release of Ca^{2+} from the endoplasmic reticulum which, upon binding to calmodulin, activates a variety of cellular processes. The membrane-bound DG activates protein kinase C to phosphorylate numerous cellular proteins. DG may also be degraded to yield arachidonate, an obligate intermediate in the biosynthesis of prostaglandins and related compounds.

Nerve impulses are transmitted electrically along a neuron and, in most cases, chemically between neurons and from neurons to muscle or gland cells. Neurons, as do most cells, actively pump K^+ into and Na^+ out of the cell, so as to generate a membrane potential of around -60 mV. A nerve impulse consists of a wave of transient membrane depolarization known as an action potential that passes along an axon. An action potential begins with the opening of trans-membrane Na^+-specific voltage-gated channels followed, a fraction of a millisecond later, by the opening of K^+-specific voltage-gated channels and the closing of the Na^+ channels. As a consequence, the membrane potential increases from its resting potential to about $+30$ mV in a fraction of a millisecond and then, almost as rapidly, decreases past the resting potential to the membrane's K^+ equilibrium potential. As the K^+ channels close over the next few milliseconds, the resting potential is restored. The action potential is triggered by adjacent membrane depolarization so that a nerve impulse is a self-amplifying phenomenon that maintains its intensity along the length of an axon. The voltage-gated Na^+ channel is the target of many potent neurotoxins including tetrodotoxin, which blocks the channel, and batrachotoxin, which locks it open. Nerve impulses in myelinated neurons occurs by saltatory conduction in which the impulse is rapidly transmitted by ionic currents through the myelinated (electrically insulated) segments of the axon, and is renewed by the generation of an action potential at the unmyelinated nodes of Ranvier, which contain high concentrations of Na^+ channels.

Nerve impulses are chemically transmitted across most synapses by the release of neurotransmitters. Acetylcholine (ACh), the best characterized neurotransmitter, is packaged in synaptic vesicles that are exocytotically released into the synaptic cleft. This process is triggered by an increase in cytosolic $[Ca^{2+}]$ resulting from the arriving action potential opening voltage-gated Ca^{2+} channels. The ACh diffuses across the synaptic cleft where it binds to the ACh receptor, a transmembrane ion channel that opens in response to ACh binding. The resultant flow of Na^+ into and K^+ out of the postsynaptic cell depolarizes the post-synaptic membrane, which, if sufficient neurotransmitter had been released, triggers a post-synaptic action potential. The ACh receptor is the target of numerous deadly neurotoxins including histrionicatoxin, *d*-tubocurarine and cobratoxin, which all bind to the ACh receptor so as to prevent its opening. The ACh is rapidly degraded, before the possible arrival of the next nerve impulse, through the action of acetylcholinesterase. Nerve gases and succinylcholine inhibit acetylcholinesterase and therefore block nerve impulse transmission at cholinergic synapses. Many specific regions of the nervous system employ neurotransmitters other than ACh. Most of these neurotransmitters are amino acids such as glycine and glutamate, or their decarboxylation products and their derivatives including catecholamines, GABA, histamine, and serotonin. Many of these compounds are also hormonally active but they are excluded from the brain by the blood–brain barrier. Although many neurotransmitters, such as ACh, are excitatory, others are inhibitory. The latter act by hyperpolarizing the postsynaptic membrane so that it must be more highly depolarized than otherwise to trigger an outgoing action potential. There is also a growing list of polypeptide neurotransmitters, many of which are also polypeptide hormones. These polypeptides apparently elicit complex behavior patterns.

References

Blood Clotting

Colman, R. W., Hirsh, J., Marder, V. J., and Salzman, E. W. (Eds,), *Hemostasis and Thrombosis* (2nd ed.), Lipincott (1987).

Doolittle, R. F., Fibrinogen and fibrin, *Annu. Rev. Biochem.* **53**, 195–229 (1984).

Doolittle, R. F., Fibrinogen and fibrin, *Sci. Am.* **245**(6): 126–135 (1981).

Esmon, C. T., The regulation of natural anticoagulant pathways, *Science* **235**, 1348–1352 (1987).

Francis, C. W. and Marder, V. J., Concepts of clot lysis, *Annu. Rev. Med.* **37**, 187–204 (1986).

Furie, B. and Furie, B. C., The molecular basis of blood coagulation, *Cell* **53**, 505–518 (1988).

Lawn, R. M. and Vehar, G. A., The molecular genetics of hemophilia, *Sci. Am.* **254**(3): 48–54 (1986).

Mann, K. G., Jenny, R. J., and Krishnaswamy, S., Cofactor proteins in the assembly and expression of blood clotting enzyme complexes, *Annu. Rev. Biochem.* **57**, 915–956 (1988).

McKee, P. A., Hemostasis and disorders of blood coagulation, *in* Stanbury, J. W., Wyngaarden, J. B., Fredrickson, D. S., Goldstein, J. L., and Brown, M. S. (Eds.), *The Metabolic Basis of Inherited Disease* (5th ed.), pp. 1531–1560, McGraw–Hill (1983).

Stamatoyannopoulos, G., Nienhuis, A. W., Leder, P., and Majerus, P. W. (Eds.), *The Molecular Basis of Blood Diseases*, Chapters 15–18, Saunders (1987). [Authoritative discussions of hemostasis, hemophilia, fibrinogen, and fibrinolysis.]

Suttie, J. W., Vitamin K-dependent carboxylase, *Annu. Rev. Biochem.* **54**, 459–477 (1985).

Weisel, J. W., Nagaswami, C., and Makowski, L., Twisting of fibrin fibrils limits their radial growth, *Proc. Natl. Acad. Sci.* **84**, 8991–8995 (1987).

Weisel, J. W., Stauffacher, C. V., Bulitt, E., and Cohen, C., A model for fibrinogen: domains and sequences, *Science* **230**, 1388–1391 (1985).

Zwaal, R. F. A. and Hemker, H. C. (Eds.), *Blood Coagulation*, Elsevier (1986).

Immunity

Ada, G. L. and Nossal, G., The clonal-selection theory, *Sci. Am.* **257**(2): 62–69 (1987).

Alberts, B., Bray, D., Lewis, J., Raff, M., Roberts, K., and Watson, J. D., *Molecular Biology of the Cell* (2nd ed.), Chapter 18, Garland Publishing (1989).

Alt, F. W., Blackwell, T. K. and Yancopoulos, G. D., Development of the primary antibody repertoire, *Science* **238**, 1079–1087 (1987).

Amit, A. G., Mariuzza, R. A., Phillips, S. E. V., and Poljak, R. J., Three-dimensional structure of an antigen-antibody complex at 2.8 Å resolution, *Science* **233**, 747–753 (1986) *and* Mariuzza, R. A., Phillips, S. E. V., and Poljak, R. J., The structural basis of antigen-antibody recognition, *Annu. Rev. Biophys. Biophys. Chem.* **16**, 139–159 (1987). [The structure of a lysozyme-Fab complex.]

Amzel, L. M. and Poljak, R. J., Three-dimensional structure of immunoglobulins, *Annu. Rev. Biochem.* **48**, 961–997 (1979).

Benjamini, E. and Leskowitz, S., *Immunology. A Short Course*, Liss (1988).

Bjorkman, P. J., Saper, M. A., Samraoui, B., Bennett, W. S., Strominger, J. L., and Wiley, D. C., Structure of human class I histocompatibility antigen, HLA-A2. *and* The foreign antigen binding site and T cell recognition regions of class I histocompatibility antigens, *Nature* **329**, 506–512 *and* 512–518 (1987).

Blackwell, T. K. and Alt, F. W., Molecular characterization of the lymphoid V(D)J recombination activity, *J. Biol. Chem.* **264**, 10327–10330 (1989).

Brown, J. H., Jardetzky, T., Saper, M. A., Samraoui, B., Bjorkman, P. J., and Wiley, D. C., A hypothetical model of the foreign antigen binding site of Class II histocompatibility molecules, *Nature* **332**, 845–850 (1988).

Capra, J. D. and Edmundson, A. B., The antibody combining site, *Sci. Am.* **236**(1): 50–59 (1977).

Cohen, I. R., The self, the world and autoimmunity, *Sci. Am.* **258**(4): 52–60 (1988).

Colman, P. M., Structures of antibody-antigen complexes: implications for immune recognition, *Adv. Immunol.* **43**, 99–132 (1988).

Colman, P. M., Laver, W. G., Varghese, J. N., Baker, A. T., Tulloch, P. A., Air, G. M., and Webster, R. G., Three-dimensional structure of a complex of antibody with influenza virus neuraminidase, *Nature* **326**, 358–363 (1987).

Darnell, J., Lodish, H., and Baltimore, D., *Molecular Cell Biology*, Chapter 24, Scientific American Books (1986).

Davies, D. R., Sheriff, S., and Padlan, E. A., Antibody-antigen complexes, *J. Biol. Chem.* **263**, 10541–10544 (1988). [A comparison of the three lysozyme-Fab complexes.]

Davies, D. R. and Metzger, H., Structural basis of antibody function, *Annu. Rev. Immunol.* **1**, 87–117 (1983).

Davies, M. M. and Bjorkman, P. J., T-cell antigen receptor genes and T-cell recognition, *Nature* **334**, 395–402 (1988).

Flavell, R. A., Hamish, A., Burkly, L. C., Sherman, D. H., Waneck, G. L., and Widerea, G., Molecular biology of the H-2 histocompatibility complex, *Science* **22**, 437–443 (1986).

French, D. L., Laskov, R., and Scharff, M. D., The role of somatic hypermutation in the generation of antibody diversity, *Science* **244**, 1152–1157 (1989).

Getzoff, E. D., Tainer, J. A., Lerner, R. A., and Geysen, H. M., The chemistry and mechanism of antibody binding to protein antigens, *Adv. Immunol.* **43**, 1–98 (1988).

Honjo, T. and Habu, S., Origin of immune diversity: genetic variation and selection, *Annu. Rev. Biochem.* **54**, 803–830 (1985).

Hood, L. E., Weissman, I. L., Wood, W. B., and Wilson J. H., *Immunology* (2nd ed.), Benjamin/Cummings (1984).

Hunkapiller, T. and Hood, L., Diversity of the immunoglobulin gene superfamily, *Adv. Immunol.* **44**, 1–63 (1989).

Kappes, D. and Strominger, J. L., Human class II major histocompatibility genes and proteins, *Annu. Rev. Biochem.* **57**, 991–1028 (1988).

Kronenberg, M., Siu, G., Hood, L. E., and Shastri, N., The molecular genetics of the T-cell antigen receptor and T-cell antigen recognition, *Annu. Rev. Immunol.* **4**, 529–591 (1986).

Leder, P., The genetics of antibody diversity, *Sci. Am.* **246**(5): 102–115 (1982).

MacDonald, H. R. and Nabholz, M., T-cell activation, *Annu. Rev. Cell Biol.* **2**, 231–253 (1986).

Marrack, P. and Kappler, J., The T cell receptor, *Science* **238**, 1073–1079 (1987).

Marrack, P. and Kappler, J., The T cell and its receptor, *Sci. Am.* **254**(2): 36–45 (1986).

Müller-Eberhard, H. J., Molecular organization and function of the complement system, *Annu. Rev. Biochem.* **57**, 321–337 (1988).

Müller-Eberhard, H. J., The membrane attack complex of complement, *Annu. Rev. Immunol.* **4**, 503–528 (1986).

Nisonoff, A., *Introduction to Molecular Immunology* (2nd ed.), Sinauer Associates (1984).

Paul, W. E. (Ed.), *Fundamental Immunology*, Raven Press (1984).

Reid, K. B. M., Activation and control of the complement system, *Essays Biochem.* **22**, 27–68 (1986).

Ross, G. D. (Ed.), *Immunobiology of the Complement System*, Academic Press (1986).

Stamatoyannopoulos, G., Nienhuis, A. W., Leder, P., and Majerus, P. W. (Eds.), *The Molecular Basis of Blood Diseases*, Chapters 8 and 13, Saunders (1987). [Discussions of the immune and complement systems.]

Thomson, G., HLA disease associations, *Annu. Rev. Genet.* **22**, 31–50 (1988).

Todd, J. A., Acha-Orbea, H., Bell, J. I., Chao, N., Fronek, Z., Jacob, C. O., McDermott, M., Sinha, A. A., Timmerman, L., Steinman, L., and McDevitt, H. O., A molecular basis for MHC Class II-associated autoimmunity, *Science* **240**, 1003–1009 (1988).

Tonegawa, S., Somatic generation of immune diversity, *Angew. Chem. Int. Ed. Engl.* **27**, 1028–1039 (1988). [A Nobel lecture.]

Tonegawa, S., The molecules of the immune system, *Sci. Am.* **253**(4): 122–131 (1985).

Tonegawa, S., Somatic generation of antibody diversity, *Nature* **302**, 575–581 (1983).

Watson, J. D., Hopkins, N. H., Roberts, J. W., Steitz, J. A., and Weiner, A. M., *Molecular Biology of the Gene* (4th ed.), Chapter 25, Benjamin/Cummings (1987).

Yancopoulos, G. D. and Alt, F. W., Regulation of the assembly and expression of variable-region genes, *Annu. Rev. Immunol.* **4**, 339–368 (1986).

Young, J. D.-E. and Cohn, Z. A., How killer cells kill, *Sci. Am.* **258**(1): 38–44 (1988).

Motility

Adelstein, R. S. and Eisenberg, E., Regulation and kinetics of the actin–myosin–ATP interaction, *Annu. Rev. Biochem.* **49**, 921–956 (1980).

Alberts, B., Bray, D., Lewis, J., Raff, M., Roberts, K., and Watson, J. D., *Molecular Biology of the Cell* (2nd ed.), Chapter 11, Garland Publishing (1989).

Allen, R. D., The microtubule as an intracellular engine, *Sci. Am.* **256**(2): 42–49 (1987).

Amos, L. A., Structure of muscle filaments studied by electron microscopy, *Annu. Rev. Biophys. Biophys. Chem.* **14**, 291–313 (1985).

Berg, H. C., How bacteria swim, *Sci. Am.* **233**(2): 36–44 (1975).

Cohen, C., The protein switch of muscle contraction, *Sci. Am.* **233**(5): 36–45 (1975).

Cooke, R., The mechanism of muscle contraction, *CRC Crit. Rev. Biochem.* **21**, 53–118 (1986).

Darnell, J., Lodish, H., and Baltimore, D., *Molecular Cell Biology*, Chapters 18 and 19, Scientific American Books (1986).

Dustin, P., *Microtubules* (2nd ed.), Springer–Verlag (1984).

Dustin, P., Microtubules, *Sci. Am.* **243**(2): 67–76 (1980).

Franzini-Armstrong, C. and Peachey, L. D., Striated muscle—contractile and control mechanisms. *J. Cell Biol.* **91**, 166s–186s (1981).

Fleischer, S. and Inui, M., Biochemistry and biophysics of excitation-contraction coupling, *Ann. Rev. Biophys. Biophys. Chem.* **18**, 333–364 (1989).

Gibbons, I. R., Dynein ATPases as microtubule motors, *J. Biol. Chem.* **263**, 15837–15840 (1988).

Harrington, W. F. and Rodgers, M. E., Myosin, *Annu. Rev. Biochem.* **53**, 35–73 (1984).

Hibberd, M. G. and Trentham, D. R., Relationships between chemical and mechanical events during muscular contraction, *Annu. Rev. Biophys. Biophys. Chem.* **15**, 119–161 (1986).

Johnson, K. A., Pathway of the microtubule-dynein ATPase and the structure of dynein: a comparison with actomyosin, *Annu. Rev. Biophys. Biophys. Chem.* **14**, 161–188 (1985).

Korn, E. D. and Hammer, J. A., III, Myosins of nonmuscle cells, *Annu. Rev. Biophys. Biophys. Chem.* **17**, 23–45 (1988).

Lazarides, E. and Revel, J. P., The molecular basis of cell movement, *Sci. Am.* **240**(5): 100–113 (1979).

Macnab, R. M. and Aizawa, S.-I., Bacterial motility and the bacterial flagellar motor, *Annu. Rev. Biophys. Bioeng.* **13**, 51–83 (1984).

McIntosh, J. R. and Porter, M. E., Enzymes for microtubule-dependent motility, *J. Biol. Chem.* **264**, 6001–6004 (1989).

Ohtsuki, I., Maruyama, K., and Ebashi, S., Regulatory and cytoskeletal proteins of vertebrate skeletal muscle, *Adv. Protein Chem.* **38**, 1–67 (1986).

Phillips, G. N., Jr., Fillers, J. P., and Cohen, C., Tropomyosin crystal structure and muscle regulation, *J. Mol. Biol.* **192**, 111–131 (1986).

Pollard, T. D. and Cooper J. A., Actin and actin-binding proteins. A critical evaluation of mechanisms and functions, *Annu. Rev. Biochem.* **55**, 987–1035 (1986).

Sellers, J. R. and Adelstein, R. S., Regulation of contractile activity, *in* Boyer, P. D. and Krebs, E. G. (Eds.), *The Enzymes* (3rd ed.), Vol. 18, *pp.* 381–418, Academic Press (1987).

Squire, J., *The Structural Basis of Muscle Contraction*, Plenum Press (1981).

Timasheff, S. N. and Grisham, L. M., In vitro assembly of cytoplasmic microtubules, *Annu. Rev. Biochem.* **49**, 565–591 (1980).

Vale, R. D., Intracellular transport using microtubule-based motors, *Annu. Rev. Cell Biol.* **3**, 347–378 (1987).

Warrick, H. M. and Spudich, J. A., Myosin structure and function in cell motility, *Annu. Rev. Cell Biol.* **3**, 379–421 (1987).

Zot, A. S. and Potter, J. D., Structural aspects of troponin-tropomyosin regulation of skeletal muscle contraction, *Annu. Rev. Biophys. Biophys. Chem.* **16**, 535–559 (1987).

Communications

Alberts, B., Bray, D., Lewis, J., Raff, M., Roberts, K., and Watson, J. D., *Molecular Biology of the Cell* (2nd ed.) Chapters 12 and 19, Garland Publishing (1989).

Almassy, R. J., Fontecilla-Camps, J. C., Suddath, F. L., and Bugg, C. E., Structure of variant 3 scorpion neurotoxin from *Centruoides sculpturates* Ewing refined at 1.8 Å resolution, *J. Mol. Biol.* **170**, 497–527 (1983).

Bell, R. M., Protein kinase C activation by diacylglycerol second messengers, *Cell* **45**, 631–632 (1986).

Benovic, J. L., Bouvier, M., Caron, M. G., and Lefkowitz, R. J., Regulation of adenylyl cyclase-coupled β-adrenergic receptors, *Annu. Rev. Cell Biol.* **4**, 405–428 (1988).

Berridge, M. J., Inositol triphosphate and diacylglycerol: two interacting second messengers, *Annu. Rev. Biochem.* **56**, 159–193 (1987).

Berridge, M. J., The molecular basis of communication within the cell, *Sci. Am.* **253**(4): 142–152 (1985).

Berridge, M. J. and Irvine, R. F., Inositol triphosphate, a novel second messenger in cellular signal transduction, *Nature* **312**, 315–321 (1984).

Bradford, H. F., *Chemical Neurobiology*, Freeman (1986).

Casey, P. J. and Gilman, A. G., G protein involvement in receptor–effector coupling, *J. Biol. Chem.* **263**, 2577–2580 (1988).

Catt, K. J. and Balla, T., Phosphoinositide metabolism and hormone action, *Ann. Rev. Med.* **40**, 487–509 (1989).

Catterall, W. A., Structure and function of voltage-sensitive ion channels, *Science* **242**, 50–61 (1988).

Chard, T., An introduction to radioimmunoassay and related techniques, *in* Work, T. S. and Work, E. (Eds.), *Laboratory Techniques in Biochemistry and Molecular Biology*, Vol. 6, Part II, North–Holland (1978).

Cockcroft, S., Polyphosphoinositide phosphodiesterase: regulation by a novel guanine nucleotide binding protein, Gp, *Trends Biochem. Sci.* **12**, 75–78 (1980).

Crapo, L., *Hormones*, Freeman (1985). [A highly readable introductory work.]

Czech, M. P., Klarlund, J. K., Yagaloff, K. A., Bradford, A. P., and Lewis, R. E., Insulin receptor signaling, *J. Biol. Chem.* **263**, 11017–11020 (1988).

Darnell, J., Lodish, H., and Baltimore, D., *Molecular Cell Biology*, Chapters 16 and 17, Scientific American Books (1986).

DeLuca, H. F. and Schnoes, H. K., Vitamin D: recent advances, *Annu. Rev. Biochem.* **52**, 411–439 (1983).

Gillman, A. G., G proteins: transducers of receptor-generated signals, *Annu. Rev. Biochem.* **56**, 615–649 (1987).

Hadley, M., *Endocrinology* (2nd ed.), Prentice–Hall (1988).

Holmgren, J., Actions of cholera toxin and the prevention and treatment of cholera, *Nature* **292**, 413–417 (1981).

Jan, L. Y. and Jan, Y. N., Voltage-sensitive channels, *Cell* **56**, 13–25 (1989).

Kikkawa, U., Kishimoto, A., and Nishizuku, Y., The protein kinase C family: heterogeneity and its implications, *Ann. Rev. Biochem.* **58**, 31–44 (1989).

Kikkawa, U. and Nishizuka, Y., The role of protein kinase C in transmembrane signalling, *Annu. Rev. Cell Biol.* **2**, 149–178 (1986).

Lefkowitz, R. J. and Caron, M. G., Adrenergic receptors, *J. Biol. Chem.* **263**, 4993–4996 (1988).

Levitzki, A., From epinephrine to cyclic AMP, *Science* **241**, 800–806 (1988). [Discusses the mechanism of signal transduction via the β-adrenergic receptor.]

Lynch, D. R. and Snyder, S. H., Neuropeptides: multiple molecular forms, metabolic pathways and receptors, *Annu. Rev. Biochem.* **55**, 773–799 (1986).

Majerus, P. W., Connolly, T. M., Bansal, V. S., Inhorn, R. C., Ross, T. S., and Lips, D. L., Inositol phosphates: synthesis and degradation, *J. Biol. Chem.* **262**, 3051–3053 (1988).

Molecular Biology of Signal Transduction, Cold Spring Harbor Symp. Quant. Biol. **53** (1988). [A series of authoritative articles.]

Neer, E. J. and Clapham, D. E., Roles of G protein subunits in transmembrane signalling, *Nature* **333**, 129–134 (1988).

Norman, A. W. and Litwack, G., *Hormones,* Academic Press (1987).

Page, D. C., Mosher, R., Simpson, E. M., Fisher, E. M. C., Mardon, G., Pollack, J., McGillivray, B., de la Chapelle, A., and Brown, L. G., The sex-determining region of the human Y chromosome encodes a finger protein, *Cell* **51,** 1091–1104 (1987).

Pelech, S. L. and Vance, D. E., Signal transduction via phosphatidylcholine cycles, *Trends Biochem. Sci.* **14,** 28–30 (1989).

Pierce, J. G. and Parsons, T. F., Glycoprotein hormones: structure and function, *Annu. Rev. Biochem.* **50,** 465–495 (1981).

Poste, G. and Crooke, S. T. (Eds.), *Mechanisms of Receptor Regulation,* Plenum Press (1985).

Rasmussen, H., The cycling of calcium as an intracellular messenger, *Sci. Am.* **261**(4): 66–73(1989).

Reichardt, L. F. and Kelly, R. B., A molecular description of nerve terminal function, *Annu. Rev. Biochem.* **53,** 871–926 (1983).

Rhee, S. G., Suh, P.-G., Ryu, S.-H., and Lee, S. Y., Studies of inositol phospholipid-specific phospholipase C, *Science* **244,** 546–550 (1989).

Rosen, O., After insulin binds, *Science* **237,** 1452–1458 (1987).

Schally, A. V., Coy, D. H., and Meyers, C. A., Hypothalamic regulatory hormones, *Annu. Rev. Biochem.* **47,** 89–128 (1978).

Smith, E. L., Hill, R. L., Lehman, I. R., Lefkowitz, R. J., Handler, P., and White, A., *Principles of Biochemistry: Mammalian Biochemistry* (7th ed.), Chapters 7 and 10–20, McGraw–Hill (1983).

Snyder, S. H., The molecular basis of communication between cells, *Sci. Am.* **253**(4): 132–141 (1985).

Snyder, S. H., Drugs and neurotransmitter receptors in the brain, *Science* **224,** 22–31 (1984).

Stevens, C. F., The neuron, *Sci. Am.* **241**(3): 54–65 (1979).

Stroud, R. M. and Finer-Moore, J., Acetylcholine receptor structure, function, and evolution, *Annu. Rev. Cell Biol.* **1,** 317–351 (1985).

Stryer, L. and Bourne, H. R., G-Proteins: a family of signal transducers, *Annu. Rev. Cell Biol.* **2,** 391–419 (1986).

Taylor, P. and Weiner, N., Drugs acting at synaptic and neuroeffector junctional sites, *in* Gilman, G. G., Goodman, L. S., Rall, T. W., and Murad, F. (Eds.), *The Pharmacological Basis of Therapeutics* (7th ed.), Section II, *pp.* 66–235, Macmillan (1985).

Walkinshaw, M. D., Saenger, W., and Maelicke, A., Three-dimensional structure of the "long" neurotoxin from cobra venom, *Proc. Natl. Acad. Sci.* **77,** 2400–2404 (1980).

Wilson, J. D. and Griffin, J. E., Mutations that impair androgen action, *Trends Genet.* **1,** 335–339 (1985).

Problems

1. There is only one known symptom, in humans, of vitamin K deficiency. What is it?

2. Clotting is prevented in stored whole blood by mixing it with citrate. What is the function of the citrate?

3. Blood samples taken from patients with either hemophilia A or hemophilia B exhibit little tendency to clot. Yet, when these two types of blood are mixed, the resulting mixture has nearly normal clotting properties. Explain.

4. Explain why the following mixtures do not form a precipitate. (a) An Fab fragment with its corresponding antigen. (b) A hapten with the antibodies raised against it. (c) An antigen and its corresponding antibody when either component is in great excess.

5. Why do antibodies raised against a native protein rarely bind to the corresponding denatured protein?

6. Explain why: (a) Antibodies to antibodies raised against a particular enzyme occasionally specifically bind the enzyme's substrate. (b) Antibodies raised against a transition state analog of a particular reaction occasionally catalyze that reaction (such antibodies have been named **abzymes**).

*7. Although only ~0.1% of an individual's *T* cells will respond to a given antigen, as many as 10% of them respond to the MHC proteins of another individual. Suggest the reason(s) for this observation.

8. The injection of bacterial cell wall constituents into an animal can trigger many of the symptoms caused by an infection including fever and inflammation. Explain how these symptoms are elicited.

*9. When you hold a weight at arm's length, you are not doing any thermodynamic work but the muscles supporting the weight are nevertheless consuming energy. Describe, on the molecular level, how muscles might maintain such state of constant tension without contracting. Why does this state consume ATP?

10. When deprived of ATP, muscles assume a rigid and inextensible form known as the **rigor state** (after **rigor mortis,** the stiffening of the body after death). What is the molecular basis of rigor?

11. In nonmuscle cells, tropomyosin is associated with microfilaments that have a structural function but not with those that participate in contractile processes. Rationalize this observation.

12. Explain why the treatment of cells with colchicine results in the disappearance of their previously existing microtubules even though colchicine does not cause microtubule dissociation *in vitro.*

13. Explain the following observations: (a) Thyroidectomized rats, when deprived of food, survive for 20 days while normal rats starve to death within 7 days. (b) Cushing's syndrome, which results from excessive secretion of adrenocortical steroids, may be caused by a pituitary tumor. (c) **Diabetes insipidus,** which is characterized by unceasing urination and unquenchable thirst, results from an injury to the pituitary. (d) The growth of malignant tumors derived from sex organs may be slowed or even reversed by the surgical removal of the gonads and the adrenal glands.

14. How does the presence of the nonhydrolyzable GTP analog **guanylylimido diphosphate**

$$\text{GMP}-\text{O}-\overset{\displaystyle \overset{\text{O}}{\|}}{\underset{\displaystyle \underset{\text{O}^-}{|}}{\text{P}}}-\text{NH}-\overset{\displaystyle \overset{\text{O}}{\|}}{\underset{\displaystyle \underset{\text{O}^-}{|}}{\text{P}}}-\text{O}^-$$

Guanylylimido diphosphate

affect cAMP-dependent receptor systems?

15. What is the resting membrane potential across an axonic membrane at 25°C (a) in the presence of tetrodotoxin or (b) with a high concentration of Cs^+ inside the axon (use the data in Table 34-6)? How do these substances affect the axon's action potential?

16. Why don't nerve impulses propagate in the reverse direction?

17. **Decamethonium ion** is a synthetic muscle relaxant.

$$(\text{H}_3\text{C})_3\overset{+}{\text{N}}-(\text{CH}_2)_{10}-\overset{+}{\text{N}}(\text{CH}_3)_3$$

Decamethonium

What is its mechanism of action?

INDEX

Page numbers in **boldface** refer to a major discussion of the entry. F after a page number refers to a figure or a structural formula. T after a page number refers to a table. Positional and configurational designations in chemical names (e.g., 3-, α-, *N*-, *p*-, *trans*-, D-, *sn*-) are ignored in alphabetizing. Numbers and Greek letters are otherwise alphabetized as if they were spelled out.